MICROBIAL GLYCOBIOLOGY

MICROBIAL GLYCOBIOLOGY
STRUCTURES, RELEVANCE AND APPLICATIONS

Editor-in-Chief

ANTHONY P. MORAN

School of Natural Sciences,
National University of Ireland, Galway, Ireland;
Institute for Glycomics, Gold Coast Campus,
Griffith University, Queensland, Australia

Editors

OTTO HOLST

Division of Structural Biochemistry, Department of Immunochemistry
and Biochemical Microbiology, Research Center Borstel,
Leibniz-Center for Medicine and Biosciences, Borstel, Germany

PATRICK J. BRENNAN

Department of Microbiology, Immunology and Pathology, College of Veterinary and
Biomedical Sciences, Colorado State University, Fort Collins, Colorado, USA

MARK VON ITZSTEIN

Institute for Glycomics, Gold Coast Campus,
Griffith University, Queensland, Australia

AMSTERDAM • BOSTON • HEIDELBERG • LONDON • NEWYORK • OXFORD
PARIS • SAN DIEGO • SAN FRANCISCO • SINGAPORE • SYDNEY • TOKYO

Academic Press is an imprint of Elsevier

Academic Press is an imprint of Elsevier
32 Jamestown Road, London NW1 7BY, UK
30 Corporate Drive, Suite 400, Burlington, MA 01803, USA
525 B Street, Suite 1900, San Diego, CA 92101-4495, USA

First edition 2009

British Library Cataloguing in Publication Data
A catalogue record for this book is available from the British Library

Library of Congress Cataloging in Publication Data
A catalog record for this book is available from the Library of Congress

ISBN: 978-0-12-374546-0

For information on all Academic Press publications
visit our website at www.elsevierdirect.com

Typeset by Macmillan Publishing Solutions
www.macmillansolutions.com

Printed and bound in United States of America
09 10 11 12 13   10 9 8 7 6 5 4 3 2 1

To all those who contributed, knowingly or unknowingly, to this endeavour.

Contents

IV

BIOLOGICAL RELEVANCE OF MICROBIAL GLYCOSYLATED COMPONENTS

A. Environmental relevance

B. Medical relevance

V

BIOTECHNOLOGICAL AND MEDICAL APPLICATIONS

List of Contributors

Nawaf Al-Maharik College of Life Sciences, Division of Biological Chemistry and Drug Discovery, University of Dundee, Dundee, UK

Heidi Annuk School of Natural Sciences, National University of Ireland, Galway, Ireland

Michael A. Apicella Department of Microbiology, University of Iowa, Iowa City, Iowa, USA

Ben J. Appelmelk Department of Medical Microbiology and Infection Control, VU University Medical Center, Amsterdam, The Netherlands

Ivo G. Boneca Group Biology and Genetics of the Bacterial Cell Wall, Institut Pasteur, Paris, France

Petra Born Institute for Cell and Molecular Biosciences, University of Newcastle upon Tyne, Newcastle upon Tyne, UK

Klaus Brandenburg Division of Biophysics, Department of Immunochemistry and Biochemical Microbiology, Research Center Borstel, Leibniz-Center for Medicine and Biosciences, Borstel, Germany

Patrick J. Brennan Department of Microbiology, Immunology and Pathology, College of Veterinary and Biomedical Sciences, Colorado State University, Fort Collins, Colorado, USA

Volker Briken Department of Cell Biology and Molecular Genetics, University of Maryland, College Park, Maryland, USA

Eric D. Brown Department of Biochemistry and Biomedical Sciences and the Michael G. DeGroote Institute for Infectious Disease Research, McMaster University, Hamilton, Ontario, Canada

Stephen D. Carrington Veterinary Science Centre, University College Dublin, Belfield, Dublin, Ireland

Paola Cescutti Dipartimento di Scienze della Vita, Università di Trieste, Trieste, Italy

Delphi Chatterjee Department of Microbiology, Immunology and Pathology, College of Veterinary and Biomedical Sciences, Colorado State University, Fort Collins, Colorado, USA

Marguerite Clyne Health Science Centre, School of Medicine and Medical Science, University College Dublin, Belfield, Dublin, Ireland

Richard M. Cooper Department of Biology and Biochemistry, University of Bath, Bath, UK

Anthony P. Corfield Mucin Research Group, University of Bristol, Bristol, UK

Monica M. Cunneen School of Molecular and Microbial Biosciences, University of Sydney, New South Wales, Australia

Piet W. J. de Groot Swammerdam Institute of Life Sciences, University of Amsterdam, Amsterdam, The Netherlands

Gennaro De Libero Experimental Immunology, Department of Research, University Hospital, University of Basel, Basel, Switzerland

Clara G. de los Reyes-Gavilán Departamento de Microbiología y Bioquímica de Productos Lácteos, Instituto de Productos Lácteos de

Asturias, Consejo Superior de Investigaciones Científicas, Villaviciosa, Asturias, Spain

Tamara L. Doering Washington University School of Medicine, St. Louis, Missouri, USA

Max Dow BIOMERIT Research Centre, Department of Microbiology, University College Cork, Cork, Ireland

Nicole N. Driessen Department of Medical Microbiology and Infection Control, VU University Medical Center, Amsterdam, The Netherlands

Jeffrey C. Dyason Institute for Glycomics, Gold Coast Campus, Griffith University, Queensland, Australia

Eva Maria Egelseer Zentrum für NanoBiotechnologie, Universität für Bodenkultur Wien, Vienna, Austria

Alan D. Elbein Department of Biochemistry and Molecular Biology, University of Arkansas for Medical Sciences, Little Rock, Arkansas, USA

Mario F. Feldman Alberta Ingenuity Centre for Carbohydrate Science, Department of Biological Sciences, University of Alberta, Edmonton, Alberta, Canada

Eamonn Fitz Patrick Veterinary Science Centre, University College Dublin, Belfield, Dublin, Ireland

Yukari Fujimoto Department of Chemistry, Graduate School of Science, Osaka University, Osaka, Japan

Koichi Fukase Department of Chemistry, Graduate School of Science, Osaka University, Osaka, Japan

Patrick Garidel Institut für Physikalische Chemie, Martin-Luther-Universität Halle-Wittenberg, Halle/Saale, Germany

Jeroen Geurtsen Department of Medical Microbiology and Infection Control, VU University Medical Center, Amsterdam, The Netherlands

Marlyn Gonzalez Department of Biology, Brooklyn College of City University of New York, Brooklyn, New York, USA

I. Darren Grice Institute for Glycomics, Gold Coast Campus, Griffith University, Queensland, Australia

Patricia Guerry Enteric Diseases Department, Naval Medical Research Center, Silver Spring, Maryland, USA

Thomas Gutsmann Division of Biophysics, Department of Immunochemistry and Biochemical Microbiology, Research Center Borstel, Leibniz-Center for Medicine and Biosciences, Borstel, Germany

Holger Heine Division of Innate Immunity, Department of Immunology and Cell Biology, Borstel Research Center, Leibniz-Center for Medicine and Biosciences, Borstel, Germany

Joanne Heng Department of Biochemistry and Molecular Biology, University of Melbourne, Parkville, Victoria, Australia

Otto Holst Division of Structural Biochemistry, Department of Immunochemistry and Biochemical Microbiology, Borstel Research Center, Leibniz-Center for Medicine and Biosciences, Borstel, Germany

Harold J. Jennings Institute for Biological Sciences, National Research Council of Canada, Ottawa, Ontario, Canada

Michael P. Jennings School of Molecular and Microbial Sciences, University of Queensland, Brisbane, Queensland, Australia

Margaret I. Kanipes Department of Chemistry, North Carolina Agricultural and Technical State University, Greensboro, North Carolina, USA

Dennis L. Kasper Channing Laboratory, Department of Medicine, Brigham and Women's Hospital and Harvard Medical School, Boston, Massachusetts, USA

Wafa Khamri Department of Nephrology and Transplantation, King's College London School of Medicine, Guy's Hospital Campus, London, UK

Milton J. Kiefel Institute for Glycomics, Gold Coast Campus, Griffith University, Queensland, Australia

Dionne C. G. Klein Biophysics and Medical Technology, Department of Physics, The Norwegian University of Science and Technology, Trondheim; Department of Cancer Research and Molecular Medicine, The Norwegian University of Science and Technology, Trondheim, Norway

Frans M. Klis Swammerdam Institute of Life Sciences, University of Amsterdam, Amsterdam, The Netherlands

Yuriy A. Knirel N.D. Zelinsky Institute of Organic Chemistry, Russian Academy of Sciences, Moscow, Russia

Thomas Kohler Cellular and Molecular Microbiology Section, Medical Microbiology Department, University of Tübingen, Tübingen, Germany

Paul Kosma Department of Chemistry, University of Natural Resources and Applied Life Sciences Vienna, Vienna, Austria

Emir Kulauzovic Cellular and Molecular Microbiology Section, Medical Microbiology Department, University of Tübingen, Tübingen, Germany

Shoichi Kusumoto Suntory Institute for Bioorganic Research, Osaka, Japan

John R. Lawrence Aquatic Ecosystem Protection Research Division, Science and Technology Branch, Environment Canada, Saskatoon, Saskatchewan, Canada

Clifford A. Lingwood The Hospital for Sick Children, Molecular Structure & Function Programme, Toronto, Ontario, Canada

Peter N. Lipke Department of Biology, Brooklyn College of City University of New York, Brooklyn, New York, USA

Susan M. Logan Institute for Biological Sciences, National Research Council, Ottawa, Ontario, Canada

Radia Mahfoud The Hospital for Sick Children, Molecular Structure & Function Programme, Toronto, Ontario, Canada

Malcolm J. McConville Department of Biochemistry and Molecular Biology, University of Melbourne, Parkville, Victoria, Australia

Rachel M. McLoughlin Channing Laboratory, Department of Medicine, Brigham and Women's Hospital and Harvard Medical School, Boston, Massachusetts, USA

Paul Messner Zentrum für NanoBiotechnologie, Universität für Bodenkultur Wien, Vienna, Austria

Antonio Molinaro Dipartimento di Chimica Organica e Biochimica, Università di Napoli Federico II, Napoli, Italy

Anthony P. Moran School of Natural Sciences, National University of Ireland, Galway, Ireland; Institute for Glycomics, Gold Coast Campus, Griffith University, Queensland, Australia

Thomas Naderer Department of Biochemistry and Molecular Biology, University of Melbourne, Parkville, Victoria, Australia

Thomas R. Neu Department of River Ecology, Helmholtz Centre for Environmental Research – UFZ, Magdeburg, Germany

Mari-Anne Newman Department of Plant Biology, Faculty of Life Sciences, University of Copenhagen, Frederiksberg, Denmark

Andrei V. Nikolaev College of Life Sciences, Division of Biological Chemistry and Drug Discovery, University of Dundee, Dundee, UK

Christian M. Pedersen Fachbereich Chemie, Universität Konstanz, Konstanz, Germany

Mark P. Pereira Department of Biochemistry and Biomedical Sciences and the Michael G. DeGroote Institute for Infectious Disease Research, McMaster University, Hamilton, Ontario, Canada

Andreas Peschel Cellular and Molecular Microbiology Section, Medical Microbiology Department, University of Tübingen, Tübingen, Germany

Stefan Pöhlmann Institute of Virology, Hannover Medical School, Hannover, Germany

Robert A. Pon Institute for Biological Sciences, National Research Council of Canada, Ottawa, Ontario, Canada

Stuart A. Ralph Department of Biochemistry and Molecular Biology, University of Melbourne, Parkville, Victoria, Australia

Peter R. Reeves School of Molecular and Microbial Biosciences, University of Sydney, New South Wales, Australia

Anne N. Reid Microbiology Research Division, Bureau of Microbial Hazards, Food Directorate, Health Products and Food Branch, Health Canada, Ottawa, Ontario, Canada

Colm J. Reid Veterinary Science Centre, University College Dublin, Belfield, Dublin, Ireland

Morgann C. Reilly Washington University School of Medicine, St Louis, Missouri, USA

Sabine Riekenberg Division of Innate Immunity, Department of Immunology and Cell Biology, Research Center Borstel, Leibniz-Center for Medicine and Biosciences, Borstel, Germany

Patricia Ruas-Madiedo Departamento de Microbiología y Bioquímica de Productos Lácteos, Instituto de Productos Lácteos de Asturias, Consejo Superior de Investigaciones Científicas, Villaviciosa, Asturias, Spain

Nuria Salazar Departamento de Microbiología y Bioquímica de Productos Lácteos, Instituto de Productos Lácteos de Asturias, Consejo Superior de Investigaciones Científicas, Villaviciosa, Asturias, Spain

Christina Schäffer Zentrum für Nano-Biotechnologie, Universität für Bodenkultur Wien, Vienna, Austria

Richard R. Schmidt Fachbereich Chemie, Universität Konstanz, Konstanz, Germany

Ian C. Schoenhofen Institute for Biological Sciences, National Research Council, Ottawa, Ontario, Canada

Marit Sletmoen Biophysics and Medical Technology, Department of Physics, The Norwegian University of Science and Technology, Trondheim, Norway

Uwe B. Sleytr Zentrum für NanoBiotechnologie, Universität für Bodenkultur Wien, Vienna, Austria

Evelyn C. Soo Institute for Marine Biosciences, National Research Council, Halifax, Nova Scotia, Canada

Imke Steffen Institute of Virology, Hannover Medical School, Hannover, Germany

Daniel C. Stein Department of Cell Biology and Molecular Genetics, University of Maryland, College Park, Maryland, USA

Bjørn T. Stokke Biophysics and Medical Technology, Department of Physics, The Norwegian University of Science and Technology, Trondheim, Norway

Christine M. Szymanski Department of Biological Sciences, University of Alberta, Edmonton, Alberta, Canada

Joe Tiralongo Institute for Glycomics, Gold Coast Campus, Griffith University, Queensland, Australia

Jennifer A. Tee College of Life Sciences, Division of Biological Chemistry and Drug Discovery, University of Dundee, Dundee, UK

Mark R. Thursz Gastroenterology and Hepatology Research, Faculty of Medicine, Imperial College, St Mary's Campus, London, UK

M. Stephen Trent Section of Molecular Genetics and Microbiology, University of Texas, Austin, Texas, USA

Theodros S. Tsegaye Institute of Virology, Hannover Medical School, Hannover, Germany

Jean van Heijenoort Institut de Biochimie et Biophysique Moléculaire et Cellulaire, Université Paris-Sud, Orsay, France

Waldemar Vollmer Institute for Cell and Molecular Biosciences, University of Newcastle upon Tyne, Newcastle upon Tyne, UK

Mark von Itzstein Institute for Glycomics, Gold Coast Campus, Griffith University, Queensland, Australia

Jennifer C. Wilson Institute for Glycomics, Gold Coast Campus, Griffith University, Queensland, Australia

Guoqing Xia Cellular and Molecular Microbiology Section, Medical Microbiology Department, University of Tübingen, Tübingen, Germany

Preface

Glycobiology can be said to be sweet biology. The full appreciation of the role of sugars, glycomolecules and glycosylated structures and their biological functions has been a more recent one compared to that of nucleic acids and proteins, particularly in the specialization of microbiology and related fields. Understanding has grown that monosaccharides represent an alphabet of biological information similar to amino acids and nucleic acids, but with a greater, and potentially unsurpassed, coding capacity. Though it has been predicted that microorganisms can synthesize more sugar building blocks than their eukaryotic counterparts, e.g. for bacteria it is considered to be six-fold greater, this extensive coding capacity impacts the biological functioning of microbial molecules and also influences the interaction of microbes with their environment, including host structures.

The concept behind this book is to present, in an easy-to-read format, reviews of the important central aspects of microbial glycobiology, i.e. the study of carbohydrates as related to the biology of microorganisms. The importance of substitution of proteins by sugars (glycosylation) and the role played by glycosylated structures (glycoproteins, glycopeptides, glycolipids, lipoglycans, glycoconjugates, etc.) in disease development, immune recognition, and environmental processes have become well-established. Moreover, from the viewpoint of biotechnology industry, glycobiology, microbial glycobiology and microbial glycomics are important components. Microbial, especially bacterial, glycomes represent an excellent toolbox for glycobiologists to understand the fundamentals of glycosylation pathways, to develop new techniques for glycobiology, and to exploit glycosylation pathways for development of novel diagnostics and therapeutics. Despite the relevance of glycobiology in medical and environmental microbiology, and the potential to exploit this knowledge for industrial and medical processes, this field has only begun to be fully appreciated. In particular, glycomics – the applied biology and chemistry of the structures and functions of carbohydrates – and microbial glycomics – glycomics as related to microbial components – have become recognized as areas of emerging technological development. For instance, this area has been highlighted in the Massachusetts Institute of Technology Review as one of ten emerging technologies that will have a significant influence in the near future. Furthermore, it has been commented that the field of microbial glycobiology could fuel a revolution in biology and industry and aid biomedical development and drug discovery.

Indeed, the number of publications in this field has risen dramatically in recent years, making it extremely difficult for even the most diligent reader to stay abreast of progress. Additionally, in many areas of microbial glycobiology well-based and extensive reviews are lacking. Thus, we considered there was a major need to provide a book reviewing the range of topics relevant to microbial glycobiology since no such book has previously been available. It is our hope that this text distills the most important cutting-edge findings in the field to produce a timely and definitive overview, providing a useful introduction to the subject for new researchers, as well as an invaluable reference for experienced ones. Our goal has been to create a state-of-the-art compendium and to delineate the knowns and unknowns in the field.

Since microbial glycobiology represents a multidisciplinary and emerging area with implications for a range of basic and applied research fields, as well as having industrial, medical and biotechnological implications, care has been taken in the choice of topics to be covered. This volume cannot attempt to be completely comprehensive, but we believe, the central concepts and areas of intensive investigation have been covered as have aspects of the glycobiology of bacteria, viruses, fungi and protozoa. The approach has been to cover and link knowledge among microbiologists, synthetic and analytical chemists, biomedical and biopharmaceutical scientists and biotechnologists. The first section of the book introduces readers to the nature, structures and functions of glycomolecules and glycosylated components of microorganisms and infectious agents. This includes not only such components of bacteria but also those from viruses, fungi and protozoa. In the next section, the genetics and biosynthesis of these components, as well as the ability to chemically synthesize a number of these molecules are reviewed. The interaction with and recognition by the host of these molecules is considered subsequently, as are both the environmental and medical relevance of microbial glycosylated structures and glycomolecules. Finally, the biotechnological and medical applications of microbial glycosylation and glycosylated molecules is explored. Collectively, the chapters present basic science understanding of these molecules through to the applied science of exploitation and applications of microbial glycosylation, both industrially and biomedically.

We have been fortunate to have been joined by our colleagues, leaders in the field, who have contributed their ideas, experiences and insights in a free and open manner to yield what we believe are insightful, concise and stimulating chapters in a review format. The editors sincerely thank the numerous contributors of these chapters. Furthermore, special thanks is deserved by Mari Moran and Sharon Ackerman for their expert secretarial assistance during manuscript preparation at different stages of this project. Finally, and by no means least, we thank Lisa Tickner and Christine Minihane for their unwavering support in making this text a reality, Kristi Anderson for her editorial skills, and Claire Hutchins, Caroline Jones and their team for their efforts during the production phase.

On behalf of the editors,
Anthony P. Moran

PART I

MICROBIAL GLYCOLIPIDS, GLYCOPROTEINS AND GLYCOPOLYMERS

CHAPTER

1

Overview of the glycosylated components of the bacterial cell envelope

Otto Holst, Anthony P. Moran and Patrick J. Brennan

SUMMARY

Within this chapter, the various types of bacterial cell envelope and their carbohydrate-related molecules, as well as the glycoprotein S-layers that can be found in all bacteria except mycobacteria, are introduced. Whereas each of the Gram-positive, Gram-negative and mycobacterial cell envelopes possess a general and typical architecture, the archaeal cell envelope shows a broader variety of constructions and may be comprised of only a membrane. Apart from S-layers, carbohydrate-containing macromolecules like lipopolysaccharides, peptidoglycan, lipoteichoic acids, teichoic acids, capsule polysaccharides, lipoarabinomannan and others are briefly described. This chapter refers to other, more detailed, subsequent chapters that summarize the chemistry and biological function of the cell envelope macromolecules.

Keywords: Bacterial cell envelope; Lipopolysaccharide; Lipoteichoic acid; Teichoic acid; Lipoarabinomannan; Arabinogalactan; Peptidoglycan; Polysaccharide; Glycolipid; Glycoprotein

1. INTRODUCTION – THE BACTERIAL CELL ENVELOPE ENCOUNTERING ENVIRONMENTAL CHALLENGES

Bacteria of various species populate most of the environments encountered on Earth. As a group, these microorganisms cope with very low, even down to −4°C, temperatures (psycrophilic bacteria), medium range temperatures (mesophilic bacteria), warm to hot temperatures (thermophilic bactera) or very hot, up to 115°C, temperatures (hyperthermophilic bacteria). Also, they exhibit an ability to survive and withstand an impressive range of pH, between pH 0.7 and 9 (acido- and alkaliphilic bacteria), to survive high pressures (deep sea, barophilic bacteria), or they may need or have to survive high or higher salt concentrations (halotolerant and halophilic bacteria) in their environments. Moreover, they may need to live and proliferate in eukaryotic hosts (symbiotic or pathogenic bacteria). In addition, bacterial species have developed the means to overcome longer periods of dryness and/or starvation or exposure to strong UV radiation. In order to survive in a particular niche, bacteria have developed a large variety of intracellular physiological adaptations; also, most possess an outer cellular barrier, the cell envelope, by which they communicate with and protect themselves from the environment they inhabit (Seltmann and Holst, 2001).

In nearly all genera of the domain Bacteria and in several of those of the domain archaeal, this layer is present which, by definition, consists of the cytoplasmic membrane (CM), the cell wall and, if present, outer layers such as capsules or sheaths. The outer barrier of other genera of the domain archaeal is formed by a particular membrane structure which contains ether-linked lipids that span the whole membrane and are highly resistant to temperature and pH (Seltmann and Holst, 2001).

When considering the broad variety of environments inhabited and encountered by bacteria, it is quite astonishing that only a few general architectural forms of cell envelope have evolved to cope with these varied conditions. The common architectural principle of the CM and cell wall is present in all bacterial envelopes in the four great variations of the envelope, i.e. Gram-negative bacterial, Gram-positive bacterial, mycobacterial and archaeal, and which may possess further adaptation even at the species level. Importantly, all cell envelopes possess a significant proportion of carbohydrate-containing constituents. This chapter serves to introduce the various types of cell envelope and their carbohydrate-related molecules and the glycoprotein S-layers that can be found in all bacteria except mycobacteria. As subsequent chapters will present more detailed reviews of the cell envelope-related macromolecules, their synthesis and applications, extensive literature will not be cited here, merely the relevant chapters will be indicated.

2. THE GRAM-NEGATIVE CELL ENVELOPE

The Gram-negative bacterial world contains a broad variety of genera that may comprise only harmless species (e.g. phototrophic bacteria of genera like *Rhodobacter* or *Rhodomicrobium*) or both harmless and human- or animal-pathogenic species (e.g. *Escherichia coli* or *Acinetobacter* spp.) or plant-pathogenic species (e.g. *Erwinia* and *Xanthomonas* spp.). Independent of this, all Gram-negative bacteria possess a cell envelope of the same general architecture which is schematically depicted in Figure 1.1. The periplasmic "space" is present on top of the CM and contains peptidoglycan (PG), also known as murein, as a major constituent. This macromolecular sacculus is composed of sugar chains built from alternating 2-acetamido-2-deoxy-D-glucopyranose (Glc*p*NAc) and 2-acetamido-3-*O*-[(*R*)-1-carboxyethyl]-2-deoxy-D-glucopyranose, also termed *N*-acetylmuramic acid (MurNAc), residues which carry, and can be cross-linked by, smaller peptides – the amino acid composition of which varies in different species. The PG sacculus represents a rigid layer that determines the form of the bacterial cell and is important for osmotic stability and acts as a protective barrier. However, the PG sacculus is not a completely closed wall that inhibits transport of small molecules (e.g. nutrients) or their excretion. Although very stable, the PG matrix represents a mesh with holes large enough to guarantee the flow of molecules, including those that have to be transported to the outer membrane (OM). The three-dimensional architecture of PG has been an issue of discussion for a long time. The first model, which many scientists consider represents the correct one, is built from sugar chains that run in a parallel direction to the CM and which are interlinked by short peptide stems that are oriented perpendicular to this membrane. About ten years ago, an alternative, called the scaffold model, was proposed

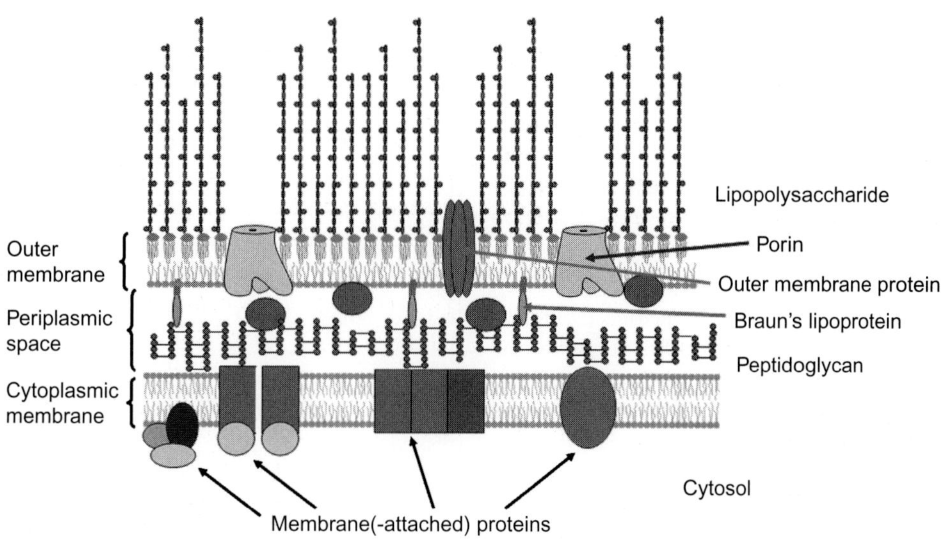

FIGURE 1.1 A schematic model of the Gram-negative cell envelope as present in *E. coli*. The cell envelope comprises the cytoplasmic membrane, which is a symmetric membrane consisting mainly of phospholipids in both leaflets, the periplasmic space and the outer membrane. Embedded in the cytoplasmic membrane, i.e. membrane-attached at the cytosolic leaflet or integrated into the membrane, are a number of proteins including those important for lipopolysaccharide or capsular polysaccharide biosynthesis and transport to the periplasmic space. The thin peptidoglycan layer represents the major constituent of the periplasmic space, but which also contains proteins including transport-proteins (indicated by the ovoid shapes in the periplasm). In *E. coli*, Braun's lipoprotein is present which is covalently bound to the peptidoglycan and anchored by its lipid moiety in the inner leaflet of the outer membrane. In contrast to the cytoplasmic membrane, the outer membrane represents an asymmetric membrane, i.e. comprising phospholipids in the inner and lipopolysaccharides in the outer leaflet. It contains a number of outer membrane proteins, including the porins that are water-filled channels important for the import of small molecules, like sugars or ions. Not shown are capsular polysaccharides and the enterobacterial common antigen which may also be anchored in the outer leaflet by a lipid structure. In addition, S-layer glycoproteins may be present in certain Gram-negative bacterial species (not shown).

with perpendicular sugar chains connected by peptides that run parallel to the CM. It should be noted that there is no clear proof for either model to date. Importantly, the Gram-negative PG is a rather thin construction making its examination difficult. A detailed review of PG is given in Chapter 2 and of its biosynthesis in Chapter 16.

Apart from PG, the periplasmic space contains various smaller molecules like mono- and oligosaccharides, amino acids and peptides, as well as the biosynthetic precursors and degradation products of PG. The concentration of all these substances is rather high, thus the periplasmic space represents a highly viscous solution that can be considered a gel-like matrix.

Outside the periplasmic space is located a second membrane, the OM, which had been considered unique to Gram-negative bacteria for a long time. Nevertheless, the presence of an analogous structure in mycobacteria has been postulated for quite some time and whose presence has finally been verified recently (see below) (Hoffmann *et al.*, 2008). Both leaflets of the OM represent lipid bilayers that are organized asymmetrically, i.e. the inner leaflet is composed of different molecules compared to the outer one. In the case of the OM of Gram-negative bacteria, the inner leaflet is comprised of phospholipids, whereas the outer leaflet is mainly constructed from lipopolysaccharides (LPSs)

(Holst and Müller-Loennies, 2007). In some cases, polysaccharide capsules are present that may be anchored by a lipid into the OM and this is also true for the enterobacterial common antigen (ECA) of enterobacteria. In addition, the OM contains various proteins (outer membrane proteins, OMPs), constituting up to 50% of the membrane and which often interact with LPS molecules, thereby yielding particular lipid–protein structural units. Several of the OMPs are channel-formers and are involved in diffusion of ions (e.g. diffusion of phosphate through PhoE channels in *E. coli*) and transport of mono- and small oligosaccharides (e.g. specific-channel forming proteins, like LamB that is involved in transport of maltose and maltodextrins) and, thus, they play important roles in the uptake of nutrients. Other OMPs represent structural proteins like OmpA and related proteins in various enterobacterial species and Braun's lipoprotein, also known as PG-associated lipoprotein, in *E. coli* and related species. The latter is covalently linked to PG and has its lipid moiety embedded in the inner leaflet of the OM, thus interconnecting PG and OM and providing structural stability.

Molecules of LPS are of high relevance not only for Gram-negative bacteria but also for an infected, eukaryotic host. In bacteria, these molecules are part of the protective barrier shielding the microbes from dangerous environmental compounds, like bile salts in the gut or antibiotics in *sensu latu*. In contrast, domains within LPS, specifically the core region and the O-specific polysaccharide, can also act as receptors for bacteriophages, thereby contributing indirectly to the destruction of the bacterial cell in such cases. In pathogenic bacteria that infect either humans or animals or plants, LPSs represent very important virulence factors. Moreover, this family of molecules is also called the endotoxins of Gram-negative bacteria; however, their toxicity is highly dependent on their structural properties, and it should be noted that not all LPSs are toxic molecules, not even those from pathogens. Importantly, in many chronically infecting bacterial species (e.g. *Helicobacter pylori*, *Porphyromonas gingivalis*, etc.), LPSs are of low toxic and immunological activity and this attribute is considered to aid the development of chronicity (Moran, 2007).

Chemically, LPSs, as exemplified by those of the *Enterobacteriaceae*, are lipoglycans. Three regions within LPS can be distinguished, according to their chemical structure, biosynthesis, genetics and function, i.e. the lipid A component (that anchors the whole molecule in the OM and which represents the endotoxically active moiety of toxic LPS), the core region which is covalently linked to lipid A and which may modulate lipid A toxicity and, finally, the O-specific polysaccharide which is covalently linked to the core region and represents a highly antigenic structure which is also called the O-antigen. Whereas the core region is comprised of an oligosaccharide of up to ≈15 sugars, the O-antigen is a polysaccharide mostly composed of repeating units containing 2–8 monosaccharide residues. In general, between one and about 40 repeating units are found, but this glycan may also be much longer. When all three regions are present in LPS, the resulting high-molecular-mass molecule is referred to as smooth- (S-) form LPS and is normally encountered in wild-type enterobacterial strains. A low-molecular-mass, rough- (R-) form LPS that lacks the O-antigen, thus consisting only of the lipid A and core regions, occurs in many laboratory-derived enterobacterial strains. Resembling this phenotype, but still a member of the LPS family of molecules, are the so-called lipo-oligosaccharide (LOS) molecules encountered in the wild-type strains of certain mucosal pathogens in particular (e.g. *Bordetella pertussis*, *Haemophilus influenzae* and *Campylobacter jejuni*). These molecules, composed of core and lipid A regions, exhibit greater structural variability in the core oligosaccharide than encountered in enterobacterial R-form LPS. Chemical and structural details of LPSs are discussed in Chapters 3 and 4, their biosyntheses in Chapters 17 and 18, chemical syntheses in Chapters 24 and 25 and functional aspects in Chapter 31.

As mentioned earlier, capsules composed of capsular polysaccharides (CPSs) may be anchored to the OM. Generally, it is considered that this is achieved by a glycerolipid structure, however, such a structure has only been identified in a few cases (e.g. *E. coli*, *C. jejuni*) (Schmidt and Jann, 1982; Corcoran *et al.*, 2006). In other cases, CPSs have been found linked to the core region of R-form LPS, yielding an "S-form-like LPS" structure which is referred to as K_{LPS}. Thus, probably all CPSs represent or are analogous to lipoglycans that, in many cases, contain repeating units like O-antigens. Consequently, both structures can only be distinguished by their genetics.

Rather than considering capsules of CPS as amorphous structures, since they are composed of 90% water they provide a gel-like hydrated mesh that protects the bacteria from dehydration. Polysaccharide capsules are present in many pathogenic bacteria (e.g. *E. coli*, *Klebsiella pneumoniae*, *C. jejuni*, etc.), in which CPSs function as important virluence factors, including aiding bacterial evasion of the host's defence system. The CPSs often represent antigenic structures called K-antigens and they are also involved in biofilm production and/or influence bacterial colonization. In plant–bacterial interactions, CPSs play also an important role, independent of whether the process is pathogenic or symbiotic.

A special case of CPS is that of ECA where the structure comprises the repeating unit → 3)-α-D-Fuc*p*NAc-(1→4)-β-D-Man*p*NAcA-(1→4)-α-D-Glc*p*NAc-(1→ (where D-Fuc*p*NAc, 2-acetamido-2-deoxy-D-fucopyranose; D-Man*p*NAcA, pyranosidic 2-acetamido-2-deoxy-D-mannuronic acid; D-Glc*p*NAc, 2-acetamido-2-deoxy-D-glucoyranose). Other polysaccharide structures belong to the so-called exocellular polymeric structures, also termed exopolysaccharides. Such molecules are often found in the surrounding milieu of bacteria (e.g. in the culture supernatant) and have importance in biofilm formation or in a variety of bacteria–host interactions. Mostly, exopolysaccharides are composed of repeating units of oligosaccharides. Chapter 6 provides an overview of CPS and these related molecules.

Importantly, all bacterial (lipo)glycans represent structures that are located in or on the outer bacterial surface and are synthesized in the cytosol and at the inner leaflet of the CM. Thus, bacteria have to deal with the great challenge of transport of hydro- or amphiphilic macromolecules through their lipid-like cell walls.

In the biosynthesis of LPS, which is best understood in *E. coli*, the syntheses of lipid A and the core oligosaccharide are genetically and biochemically connected. Beginning with uridine diphosphate- (UDP-) Glc*p*NAc, the tetra-acylated and bisphosphorylated lipid A precursor lipid IVa is synthesized in a series of six steps. Prior to the addition of the fifth and sixth fatty acid, two residues of 3-deoxy-D-*manno*-oct-2-ulopyranosonic acid (Kdo) are transferred by one Kdo-transferase and yield the disaccharide α-Kdo-(2→4)-α-Kdo which is linked to position C-6′ of β-D-Glc*p*NAcyl-3-*O*-Acyl-4-*P*-(1→6)-α-D-Glc*p*NAcyl-3-*O*-Acyl-1-*P*. Subsequently, the core region is completed by a series of glycosyltransferases that transfer activated monosaccharide precursors. This all occurs at the inner leaflet of the CM. Such completed R-form LPS is then transported (or "flipped") from the inner to the outer leaflet of the CM, a process which is achieved by the adenosine triphosphate- (ATP-) binding cassette (ABC)-transporter, MsbA. There, on the periplasmic side, the completed O-antigen is transferred to the core oligosaccharide and the completed LPS is then transferred to the OM. The molecular basis for this translocation process still remains an enigma (for more details see Chapter 17).

The biosynthesis of heteropolymeric O-antigen and of many such CPSs (e.g. *E. coli* capsule groups 1 and 4) follow the same general principle (see Chapter 18). The repeating units, built by the sequential action of glycosyltransferases, are linked to an undecaprenyl-phosphate carrier lipid. Lipid-linked repeats are then flipped across the CM and polymerized on the periplasmic side.

The O-antigen (and some CPS) are attached to lipid A-core molecules by a ligase.

With regard to the further transport of CPS to the surface, a number of participating components involved in the translocation of group 1 capsules and their interaction have been elucidated recently, confirming the earlier concept of zones of adhesion, or Bayer's fusion sites, published in the early 1980s (Bayer, 1981). These CPS molecules are transported to the surface via a complex of the two proteins Wzc and Wza, the latter of which is a channel-forming protein. Biosynthesis and export of *E. coli* group 2 and 3 capsules proceeds differently. Briefly, the polysaccharides are completely synthesized at the cytosolic surface of the CM, transferred to a diacylglycerophosphate anchor and the whole lipoglycan is translocated through the CM, by the ABC-transporter called KpsM, and through the periplasm and OM, by the protein complex of KpsE and KpsD. A review of the biosynthesis and assembly of CPSs is presented in Chapter 20.

3. THE GRAM-POSITIVE CELL ENVELOPE

As in Gram-negative bacteria, the Gram-positive group of bacteria comprise many harmless species and several important pathogenic ones, e.g. *Streptococcus pneumoniae*, *Staphylococcus aureus*, *Enterococcus faecalis* and *Clostridium tetani*. The Gram-positive cell envelope (Figure 1.2) is organized more simply than that of Gram-negative bacteria, but is as effective. The cell wall, layered on top of the CM, is mainly composed

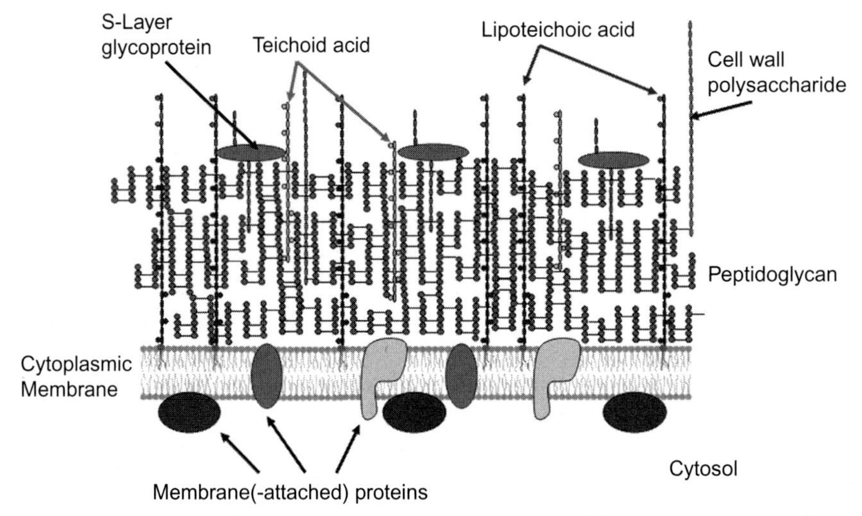

FIGURE 1.2 A schematic model of the Gram-positive cell envelope. This cell envelope comprises the cytoplasmic membrane, which is a symmetric membrane consisting mainly of phospholipids in both leaflets and the thick peptidoglycan layer. Embedded in the cytoplasmic membrane are a number of proteins, probably including those important for (lipo-)glycan and polysaccharide biosynthesis and transport to exterior regions. Peptidoglycan represents a major macromolecule outside the cytoplasmic membrane, however, also proteins, including transport-proteins and lipoproteins are present (not shown). Additionally, capsular polysaccharides (not depicted) and S-layer glycoproteins may be present in certain species. Various carbohydrate-based structures are present in the cell envelope, namely the lipoteichoic acids, which are anchored by a lipid in the cytoplasmic membrane, and the teichoic acids and transpose cell wall polysaccharides that are covalently linked to peptidoglycan. The latter polysaccharides may be neutral but often represent teichuronic acids which are mainly synthesized when phosphate is lacking in the environment and thus may replace teichoic acids.

of a thick layer of PG, generally 20–80 nm thick, compared to a ≈7 nm PG in Gram-negative bacteria. The general chemical structure of this PG is similar to that of Gram-negative bacteria. In addition, the overall conformation, i.e. 3-dimensional architecture of this PG, is discussed using the same bases of the two models mentioned above for Gram-negative PG.

The Gram-positive cell envelope does not contain a second membrane, i.e. the outer membrane as seen in Gram-negative bacteria. However, CPSs are often present in Gram-positive bacteria and represent, in many cases of pathogenic species, important virulence factors with similar functions as those mentioned for Gram-negative species. Moreover, the structures of CPSs of Gram-positive and -negative bacteria are generally similar. Again, in the former bacterial group, polysaccharide capsules are layered over the PG, but it remains unclear whether they are connected to the cell wall and, if so, how this is achieved. As far as has been investigated, Gram-positive CPSs are synthesized in a similar manner as *E. coli* group 1 capsules. Moreover, Gram-positive bacteria also produce exopolysaccharides which often are components of biofilms.

Three major carbohydrate structures are present in the Gram-positive cell envelope, namely the lipoteichoic acids (LTAs), wall teichoic acids (WTAs) and wall polysaccharides (WPSs). Several LTAs and WTAs are structurally very similar in their polymeric chain, which does not represent a classical polysaccharide but is, instead, produced from phosphodiester-bridged alditol units (of glycerol or ribitol) that are often substituted by sugars. Both macromolecules may contain D-alanine (D-Ala) which, in LTA, may substitute both the sugar(s) (as in *E. faecalis*) and the alditol residues. In LTA, this polymeric chain is linked to a short oligosaccharide which serves as a "linker" and binds the chain to a diacylglycerophosphate lipid anchor that is embedded in the CM. Thus, LTA molecules protrude through the PG matrix and reach the outer surface. The LTAs are considered immunomodulatory constituents of Gram-positive bacteria that react with cells of the innate immune system via Toll-like receptor (TLR-) 2 and induce cytokine expression. However, this view is heavily debated (Zähringer *et al.*, 2008).

In WTA, the polymeric chain is also bound to a "linker", which is a disaccharide composed of 2-acetamido-2-deoxy-mannopyranose (Man*p*NAc) and Glc*p*NAc that is covalently linked by a phosphodiester bridge to C-6 of MurNAc of PG. Both WTAs and LTAs play important roles in interactions with host receptors, maintaining cation homeostasis and providing physicochemical surface properties.

The biosyntheses of WTAs and LTAs occur in cyclic processes. In the case of WTA of *Bacillus subtilis*, the binding unit is synthesized first, linked to a lipid carrier. Subsequently, the glycerophosphate units are attached by a polymerase and part of this chain is subsequently substituted with glucopyranose (D-Glc*p*). The polymer is transported from the inner to the outer leaflet of the CM by means of a two-stage ABC-transporter. Then, other free hydroxyl-groups of the glycerol are substituted by D-Ala. Finally, the completed molecule is linked to MurNAc of PG. On the other hand, the biosynthesis of LTA begins with that of the lipid anchor, i.e. 1,2-diacylglycerol (e.g. in *S. aureus*), which is substituted with two residues of D-Glc*p*. To this linker, glycerophosphate units are attached up to the required chain length. Any further substituents can be incorporated during or after chain elongation. Chapters 5, 19 and 25 describe the structures and syntheses, both biological and chemical, of WTA and LTA in detail.

Another class of polysaccharides, WPS, namely the teichuronic acids, are usually synthesized in greater amounts under phosphate starvation. They are polysaccharides composed of at least one residue of uronic acid, the negative charge of which helps to overcome the lack of phosphate charges, e.g. in metal cation binding. Teichuronic acids are also linked by a phosphodiester bridge to MurNAc of PG via a linker structure. Their biosynthesis occurs in a cyclic process which involves a lipid carrier to which the linker is synthesized first, followed by transfer of the monosaccharide units and then polymerization.

4. THE MYCOBACTERIAL CELL ENVELOPE

Many mycobacteria live in soil and water and are harmless members of the genus, but three species are well known as human and animal pathogens, i.e. *Mycobacterium tuberculosis*, *Mycobacterium leprae* and the *Mycobacterium avium–Mycobacterium intracellulare* complex. In the host, these species live intracellularly, in particular in the phagosome of the most-feared defensive cell of the innate immune system, the macrophage. Such intracellular mycobacteria inhibit phagosome–lysosome fusion of macrophages and may survive for an extended period intracellularly. For this lifestyle, mycobacteria need a well-constructed protection barrier that renders them resistant to the various defence activities of eukaryotic hosts. Although mycobacteria yield slightly positive reactions in the Gram stain, the mycobacterial cell envelope (Figure 1.3) possesses more Gram-negative than -positive features, i.e. a less extensive PG matrix and an outer lipid bilayer which has been identified as like an outer membrane, the mycobacterial outer membrane (MOM) (Hoffmann *et al.*, 2008). Apart from such general similarities, the mycobacterial cell envelope represents a unique barrier that should be clearly distinguishable from other bacterial membrane barriers. The PG matrix is located on top of the CM and is organized similarly to those of Gram-positive and -negative bacteria.

As in other cell envelopes, the mycobacterial barrier contains several important glycan structures, namely; lipoarabinomannan (LAM)

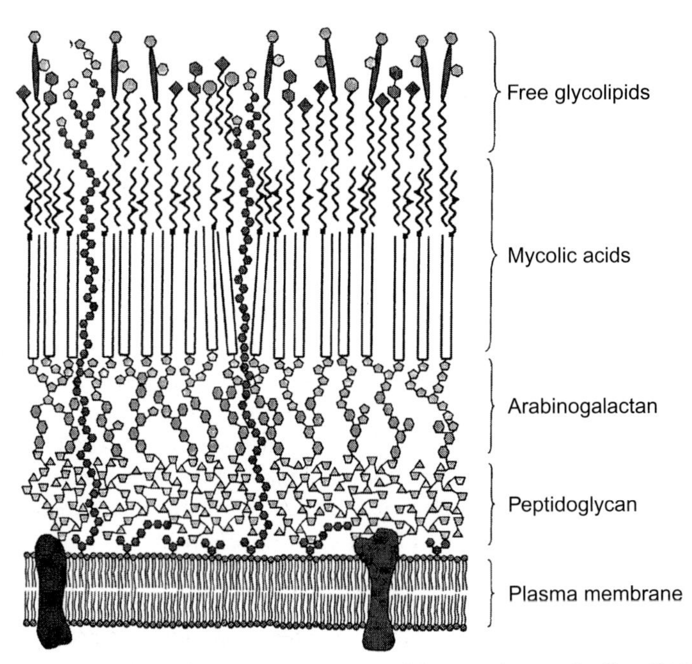

FIGURE 1.3 Schematic representation of the generic structure of the mycobacterial cell wall. Phosphatidylinositol mannosides and lipomannan are shown to be embedded and attached to the plasma membrane although their location is not known. The surface layer is covered with non-covalently associated free glycolipids.

and its biosynthetic precursors [the lipomannan (LM) and the phosphatidylinositol-mannosides (PIM)], the mycoloyl-arabinogalactan (MAG) complex and a capsular structure which comprises the polysaccharide arabinomannan (AM) (Daffé and Draper, 1998; Wittkowski *et al.*, 2007) and glucan (with proteins). Detailed overviews of the structures and functions of these components and their biosynthesis are presented in Chapters 9 and 21.

The LAM is a lipoglycan which is anchored in the CM by a phosphoinositol moiety that carries as characteristic fatty acids 9-hexadecenoic acid (16:1, palmitoleic acid) and 10-methyl-octadecanoic acid (tuberculostearic acid). In particular, bound to the inositol residue is an α-(1→6)-linked D-mannan main chain with α-(1→2)-linked mannopyranose (Man*p*) branches that are substituted by a longer branched arabinan composed of arabinofuranose (Ara*f*) residues at the second last Man*p* residue. The whole molecule is anchored in the CM by its lipid moiety and protrudes through the PG to the outside of the cell wall, similar to the LTA of Gram-positive bacteria. Here, at its non-reducing end, it may carry Man residues that substitute the terminal Ara sugars in monosaccharidic and disaccharidic substitutions. There is evidence also that part of the LAM is anchored in the MOM, probably in the outer leaflet, but proof to date is lacking. Mycobacterial LAM has been reported to possess important functions in the pathogenesis of mycobacteria, e.g. induction of phagocytosis of these microbes, phagosome–lysosome fusion or activation of innate immune responses. Nevertheless, some of the reports in these areas are conflicting.

The biosynthesis of LAM has not yet been completely elucidated. As a general pathway, it is believed that biosynthesis starts with PIM and that chain elongation to obtain the mannan chain proceeds from the Man*p* residue at O-6 of the inositol that builds up to yield LM. The molecule known as PIM_4, PIM containing Man*p* residues, is the likely precursor here. In the last step, the arabinan, a main chain of α-(1→5)-linked D-Ara*f* residues that may be branched at the non-reducing end, is produced and linked to LM.

The MAG is, as are the TA of Gram-positive bacteria, covalently linked via a disaccharide linker composed of L-rhamnopyranose (L-Rha*p*) and D-Glc*p*NAc and a phosphodiester bridge to C-6 of MurNAc of PG. The linker is substituted by a galactan chain, which is furnished exclusively from D-galactofuranose (D-Gal*f*) residues in alternating β-(1→5) and β-(1→6)-linkages, a polysaccharide feature that is rarely found in Nature. Quite early in this chain, branches of arabinan chains, i.e. α-(1→5)-linked D-Ara*f* residues, are present that protrude to the outer surface of the cell wall. At their reducing ends, a so-called hexa-arabinosyl motif is present as a branched structure, the terminal arabinose residues of which are substituted at C-5 by very long chain (up to 90 C-atoms) α-alkyl-β-hydroxy-fatty acids, the so-called mycolic acids that occur in a broad range of structural varieties (Barry *et al.*, 1998). These mycolic acids form the inner leaflet of the MOM, whereas the outer leaflet is believed to be composed of various lipids, including glycolipids, different types of which are present in different mycobacterial species (e.g. mycobacterial LOSs and glycopeptidolipids). The MAG complex, together with the MOM, represents cell envelope structures of mycobacteria that are highly important for cell wall architecture and rigidity, as well as protection. Thus, MAG (and LAM) biosynthesis represent potentially important drug targets.

The biosynthesis of the arabinogalactan proceeds stepwise, beginning with the linker and followed by the galactan chain which are linked to a decaprenylphosphoryl (C_{50}-*P*) lipid. The arabinan chains are attached by the sequential addition of Ara*f* residues that are transported as C_{50}-*P*-Ara*f* units.

The mycobacterial capsule has a different composition to those of Gram-negative and -positive bacteria; namely, it is not a purely polysaccharide capsule, but contains proteins, including glycoproteins, and mainly two polysaccharides, an arabinomannan and a glucan. The arabinomannan possesses a similar structure to the polysaccharide of LAM. So far, it remains unclear how this glycan is synthesized and transported to the outer surface. Although it has been reported several times to the contrary, a recent investigation shows that this molecule has no immunomodulatory properties. Instead, it has been shown to activate natural killer cells that then destroy tumour cells. Thus, the AM possesses cytotoxic properties. The glucan is structurally similar to glycogen, but no particular biological functions have been described so far.

5. THE ARCHAEAL CELL ENVELOPES AND S-LAYERS

As mentioned above, the archaeal cell envelope structures cannot be classified as a common architectural type as is the case with the other envelope types described thus far. Instead, the cell envelope shows a number of structural variants which are adapted to the particular or extreme living conditions.

Archaea can be subdivided into the kingdoms Euryarchaeota and Crenarchaeota, any further classification of which occurs according to the physiological properties of the organism in question. All members possess a membrane which, in several cases, consists of a tetra-ether lipid that spans the whole width of the membrane, i.e. no bilayer is present. Here, this membrane represents the whole "cell envelope". A so-called pseudomurein represents the cell-stabilizing component in Gram-positive Archaea. In other genera, a relatively thick layer of acidic polysaccharides constitute the cell wall. Finally, nearly all Archaea, and also many Gram-negative and -positive Eubacteria, contain cell surface layers (S-layers) which thus represent common structures of the prokaryotic cell envelope. S-layers appear in lattice-like arrangements that are formed by a self-assembly process and are attached to the bacterial cell wall by certain cell wall polymers. Also, the outer region of quite many S-layer proteins is often substituted by covalently linked glycan chains and, thus, glycoproteins are formed. The understanding of the mode of interaction between S-layer (glyco)proteins and secondary cell wall polymers is believed will pave the way towards the production of novel nanopatterned "biomaterials" for various applications in the fields of biomedicine and nanobiotechnology. Chapter 7 details the structure, function and applications of S-layer structures.

6. CONCLUSIONS

The cell envelope represents the outermost layer of the bacterial cell which has as general functions the protection of the cell, communication with the environment, maintenance of cellular shape, stability and rigidity of the cell, as well as allowing appropriate metabolism, growth and division of the cell. Important components of the cell envelope are carbohydrate-based and carbohydrate-containing macromolecules that contribute significantly to all of these functions, independent of which variation of the cell envelope occurs, i.e. Gram-positive, -negative, mycobacterial or archaeal. The structures, syntheses and functions of all these macromolecules are described in detail in subsequent chapters.

ACKNOWLEDGEMENTS

The authors thank Delphi Chatterjee for help with preparation of Figure 1.3.

References

Barry III, C.E., Lee, R.E., Mdluli, K., et al., 1998. Mycolic acids: structure, biosynthesis and physiological functions. Prog. Lipid Res. 37, 143–179.

Bayer, M.E., 1981. Structural and functional evidence of coopeartivity between membranes and cell wall in bacteria. Int. Rev. Cytol. Suppl. 12, 39–70.

Corcoran, A.T., Annuk, H., Moran, A.P., 2006. The structure of the lipid anchor of *Campylobacter jejuni* polysaccharide. FEMS Microbiol. Lett. 257, 228–235.

Daffé, M., Draper, P., 1998. The envelope layers of mycobacteria with reference to their pathogenicity. Adv. Microb. Physiol. 39, 131–203.

Hoffmann, C., Leis, A., Niederweis, M., Plitzko, J.M., Engelhardt, H., 2008. Disclosure of the mycobacterial outer membrane: cry-electron tomography and vitreous sections reveal the lipid bilayer structure. Proc. Natl. Acad. Sci. USA 105, 3963–3967.

Holst, O., Müller-Loennies, S., 2007. Microbial polysaccharides structures. In: Kamerling, J.P., Boons, G.-J., Lee, Y.C., Suzuki, A., Taniguchi, N., Voragen, A.G.J. (Eds.) Comprehensive Glycoscience. From Chemistry to Systems Biology, vol. 1. Elsevier Inc., Amsterdam, pp. 123–179.

Moran, A.P., 2007. Lipopolysaccharide in bacterial chronic infection: insights from *Helicobacter pylori* lipopolysaccharide and lipid A. Int. J. Med. Microbiol. 297, 307–319.

Schmidt, M.A., Jann, K., 1982. Phospholipid substitution of capsular (K) polysaccharide antigens from *Escherichia coli* causing extraintestinal infections. FEMS Microbiol. Lett. 14, 69–74.

Seltmann, G., Holst, O., 2001. The Bacterial Cell Wall. Springer, Heidelberg.

Wittkowski, M., Mittelstädt, J., Brandau, S., et al., 2007. Capsular arabinomannans from *Mycobacterium avium* with morphotype-specific structural differences but identical biological activity. J. Biol. Chem. 282, 19103–19112.

Zähringer, U., Lindner, B., Inamura, S., Heine, H., Alexander, C., 2008. TLR2 – promiscuous or specific? A critical re-evaluation of a receptor expressing apparent broad specificity. Immunobiology 213, 205–224.

CHAPTER

2

Bacterial cell envelope peptidoglycan

Waldemar Vollmer and Petra Born

SUMMARY

Peptidoglycan (murein) forms a bag-shaped sacculus in the cell envelope of most bacteria. It is essential for osmotic stability and determines the shape of a bacterial cell. Peptidoglycan is a heteropolymer consisting of glycan strands that carry short peptides. Peptides of neighbouring glycans may be connected by amide bonds resulting in a net-like architecture of the sacculus. Although there are some common structural features of peptidoglycan, there is great variation in the chemical composition of peptidoglycan isolated from different bacteria. The glycans are synthesized as oligo-(*N*-acetylglucosamine-*N*-acetylmuramic acid) strands of variable length and become modified shortly after their synthesis in many species. The peptides contain L- and D-amino acids and have a species-specific amino acid composition, sequence and type of cross-linkage. Maturation of peptidoglycan involves cleavage of amide bonds in the peptides resulting in a non-homogeneous structure of the polymer. Introduction of secondary modifications and maturation are the reasons for the complex composition of mature peptidoglycan often manifesting in a species-specific fine structure.

Keywords: Peptidoglycan; Murein; Sacculus; Bacterial cell wall; Peptide cross-links; De-*N*-acetylation; *O*-acetylation; *N*-acetylmuramic acid

1. INTRODUCTION

The bacterial peptidoglycan (murein) sacculus is a bag-shaped macromolecule which completely surrounds the cytoplasmic membrane and thereby protects the integrity of the cell against the turgor (Weidel and Pelzer, 1964; Vollmer *et al.*, 2008a). Peptidoglycan consists of glycan strands of alternating β-(1→4)-linked *N*-acetyl-D-glucosamine (GlcNAc) and *N*-acetylmuramic acid (MurNAc) residues cross-linked by peptide side chains, thus forming a rigid, net-like structure. The peptide subunit contains L- as well as D-amino acids and has the general structure L-Ala-D-iGlu-L-Lys (or *meso*-A_2pm)-D-Ala-D-Ala (where L- and D-Ala, L- and D-alanine; D-iGlu, D-*iso*-glutamic acid; L-Lys; L-lysine; *meso*-A_2pm, *meso*-2,6-diaminopimelic acid). However, the analysis of the peptidoglycan of different species has revealed a high degree of structural variation due to the presence of other amino acids, namely N^{γ}-acetyl-L-2,4-diaminobutyric acid, amidated *meso*-A_2pm, D-asparagine (D-Asn), D-aspartic acid (D-Asp), L- and D-2,4-diaminobutyric acid (L- and D-Dab); 2,6-diamino-3-hydroxypimelic acid; D-*iso*-glutamine (D-iGln); glycine (Gly); L-homoserine

(L-Hsr), *threo*-3-hydroxyglutamic acid (3-Hyg), *meso*-lanthionine, L-5-hydroxylysine, D-lysine (D-Lys); L- and D-ornithine (L- and D-Orn), L- and D-serine (L- and D-Ser); L-threonine (L-Thr) or D-lactic acid (D-Lac) and the frequent occurrence of modifications of the sugar residues. Besides its function as a stress-bearing layer, the sacculus serves as an anchor for other cell wall polymers such as surface proteins (Dramsi *et al.*, 2008), wall teichoic acids and capsular polysaccharides (Neuhaus and Baddiley, 2003; Vollmer, 2008).

Peptidoglycan is present in almost all eubacteria with the exception of *Chlamydia* spp., *Mycoplasma* spp. and *Orientia tsutsugamushi* (Moulder, 1993). Interestingly, although *Chlamydia* spp. lack detectable peptidoglycan, they possess a nearly complete set of peptidoglycan biosynthesis genes (Moulder, 1993; Chopra *et al.*, 1998; Bavoil *et al.*, 2000). Peptidoglycan is also present in the cyanelles of glaucophytes (Aitken and Stanier, 1979). Several peptidoglycan biosynthesis genes have been identified in *Arabidopsis thaliana* and the moss *Physcomitrella patens*. These genes are essential for chloroplast division, although peptidoglycan has not been detected in chloroplasts (Machida *et al.*, 2006).

This chapter provides an overview on the diversity and variability of the peptidoglycan structures found in different bacteria. We also summarize biophysical properties of peptidoglycan and discuss its architecture.

2. STRUCTURAL VARIATION IN BACTERIAL PEPTIDOGLYCAN

2.1. The glycan strands in peptidoglycan

Polymerization of the peptidoglycan precursor lipid II [GlcNAc-MurNAc(peptide)-pyrophosphoryl-undecaprenol] by peptidoglycan glycosyltransferases results in oligo-β-(1→4)-[GlcNAc-β-(1→4)-MurNAc(peptide)] glycan strands. Because MurNAc is a 3-D-lactyl-GlcNAc, the backbone of these glycan strands is an oligomer of β-(1→4)-linked GlcNAc residues and thus identical to the polysaccharide chains in chitin. Cleavage by muramidases or glucosaminidases in peptidoglycan may result in strands with an odd number of monosaccharides or with MurNAc and GlcNAc residing at the "wrong" end of the strand. In Gram-negative species, the glycan strands do not end with MurNAc but with a 1,6-anhydroMurNAc residue. In many species, these glycan strands become structurally modified shortly after their synthesis and they serve as sites for the attachment of other cell wall polymers such as teichoic acids and capsular polysaccharides (Vollmer, 2008).

2.1.1. Length distribution of the glycan strands

Depending on the species and growth conditions, the average length of the glycan strands in peptidoglycan varies between 10 and several hundred disaccharide units. Glycan strand length does not correlate with the thickness of the cell wall, as seen from the following examples from Gram-positive and Gram-negative species with thick and thin cell walls, respectively. Gram-positive *Bacillus* spp. (*Bacillus subtilis*, *Bacillus licheniformis* and *Bacillus cereus*) have an average chain length of 50–250 disaccharide units (Hughes, 1971; Warth and Strominger, 1971; Ward, 1973). In contrast, the glycan strands of *Staphylococcus aureus* are much shorter with an average chain length of about 18 disaccharide units (Tipper *et al.*, 1967; Ward, 1973). A major fraction of the glycan material from *S. aureus* has strands 3–10 disaccharide units in length. Longer glycan strands with more than 26 disaccharide units represent only 10–15% of the total glycan material (Boneca *et al.*, 2000). The L-ornithine-containing peptidoglycan of the Gram-positive *Deinococcus radiodurans* Sark has glycan strands with an average chain length of about 20 disaccharide units (Quintela *et al.*, 1999). In Gram-negative species, there is some variation in the average glycan strand length but the normal range lies between 20 and 40 disaccharide units (Glauner *et al.*, 1988; Tuomanen *et al.*, 1989; Harz *et al.*, 1990; Quintela *et al.*, 1995a,b). *Helicobacter pylori* has exceptionally short glycan strands with an average chain length of less than 10 disaccharide units (Costa *et al.*, 1999). These glycan strands become even shorter during transition from a spiral to coccoid cell shape in the stationary growth phase (Costa *et al.*, 1999; Chaput *et al.*, 2007).

2.1.2. Modifications in the glycan structure

Different types of glycan strand modifications have been found in a number of species (Figure 2.1). Many Gram-positive pathogens modify their glycan strands by *N*-glycolylation, de-*N*-acetylation and/or *O*-acetylation. These modifications, as well as the attachment of other cell wall polymers, contribute to higher levels of resistance to lysozyme, which is a glycan strand-cleaving enzyme and an important defence factor of the innate immune system.

The presence of a glycolyl residue (instead of acetate) at the 2-amino group of muramic acid is the hallmark of some closely related genera within the *Actinomycetales* (Uchida and Aida, 1979). In fact, glycolylated muramic acid- (MurNGlyc-) containing peptidoglycan is present in most genera with mycolic acids (the mycolata) including *Mycobacterium*, *Rhodococcus*, *Tsukamurella*, *Gordonia*, *Nocardia*, *Skermania* and *Dietzia* (Azuma *et al.*, 1970; Sutcliffe, 1998), but with the exception of the genus *Corynebacterium*, members of which contain "normal" acetylated muramic acid in their peptidoglycan (Azuma *et al.*, 1970; Uchida and Aida, 1979). In addition, there are several closely related genera within the *Actinomycetales* which contain MurNGlyc in their peptidoglycan but do not have mycolic acids (Evtushenko *et al.*, 2002; Matsumoto *et al.*, 2003; Li *et al.*, 2005 and references therein). In *Mycobacterium tuberculosis*, both the precursor pool and the peptidoglycan contained a mixture of MurNGlyc and MurNAc residues (Mahapatra *et al.*, 2005a,b).

The presence of de-*N*-acetylated amino sugars in the peptidoglycan has been detected, for example, in different *Bacillus* spp., *Streptococcus pneumoniae* and *Listeria monocytogenes* (Araki *et al.*, 1971; Vollmer and Tomasz, 2000; Boneca *et al.*, 2007). Either GlcNAc or MurNAc or both are de-*N*-acetylated to a variable extent by the activity of peptidoglycan deacetylases. The first gene encoding a peptidoglycan GlcNAc deacetylase, *pgdA*, was identified in *S. pneumoniae* (Vollmer and Tomasz, 2000). A *pgdA* mutant strain lacks de-*N*-acetylated amino sugars in its peptidoglycan. It shows no growth defect but higher sensitivity towards lysozyme and a significantly reduced virulence in the intraperitoneal mouse model of infection (Vollmer and Tomasz, 2002).

Spore peptidoglycan of *Bacillus* spp. and *Clostridium sporogenes* contains the spore-specific muramic acid δ-lactam structure, which is generated by intramolecular amide bond formation between the carboxyl group of the lactyl group at position 3 of de-*N*-acetylated MurN and the amino group at position 2. In *B. subtilis* about 50% of the MurNAc residues in the spore peptidoglycan are modified to the muramic acid δ-lactam and these residues appear to be distributed regularly at every second muramic acid position along the glycan strands (Atrih *et al.*, 1996; Popham *et al.*, 1996).

O-Acetylated peptidoglycan contains variable amounts of 2,6-*N*,*O*-diacetyl muramic acid which is MurNAc with an extra acetyl group linked to the O-6 position. This modification has been detected in a high number of species, for example in the Gram-positive *Micrococcus luteus* (Brumfitt *et al.*, 1958), *Enterococcus faecium* (Abrams, 1958), *S. pneumoniae* (Crisostomo *et al.*, 2006) and *S. aureus* (Ghuysen and Strominger, 1963) and in the Gram-negative *Neisseria meningitidis* (Antignac *et al.*, 2003), *H. pylori* (Weadge and Clarke, 2006) and *Proteus mirabilis* (Martin and Gmeiner, 1979). The extent of *O*-acetylation varies between less than 20% and 70% in different species and strains (Vollmer, 2008). The process of peptidoglycan *O*-acetylation is not well understood. Only more

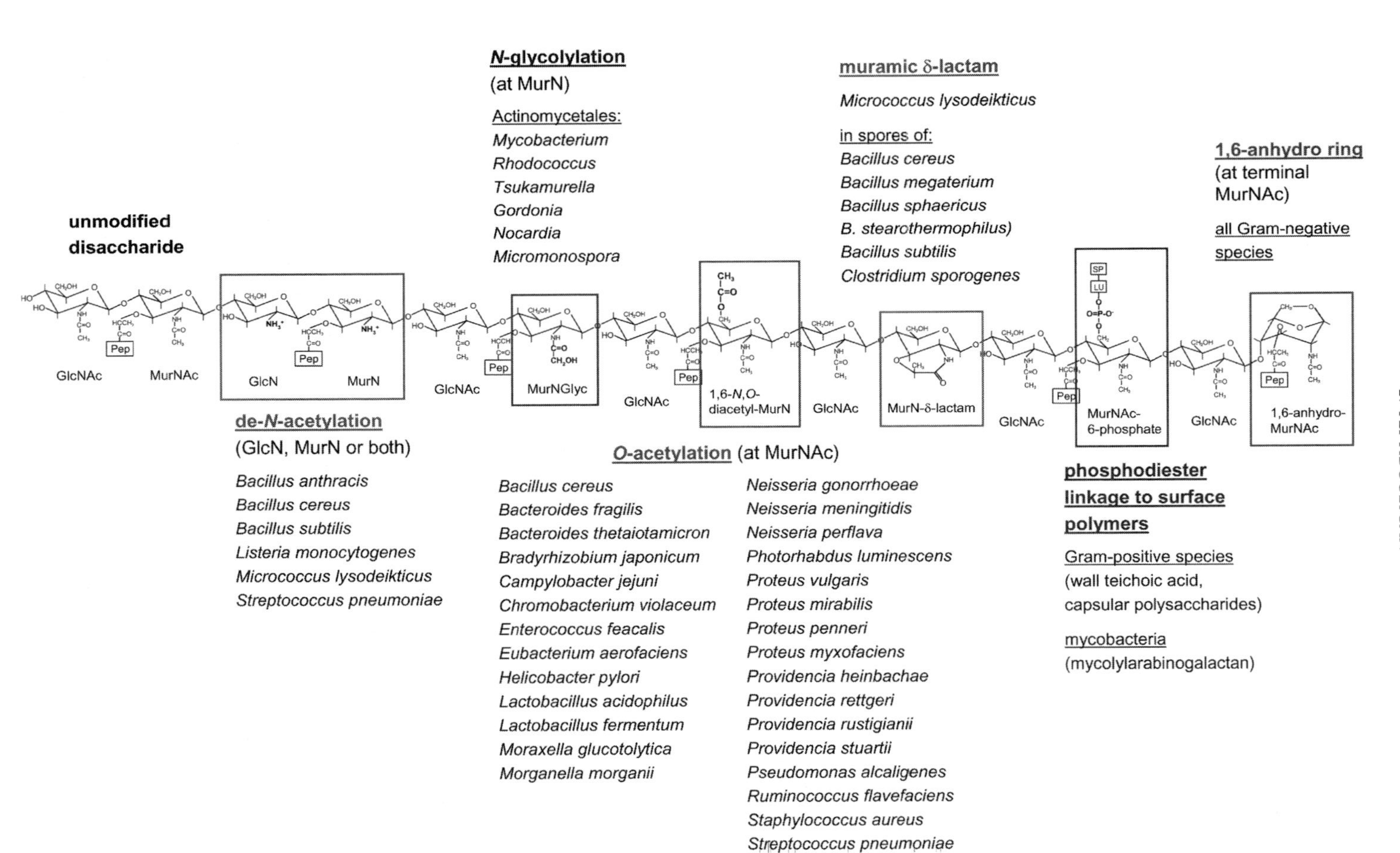

FIGURE 2.1 Modifications in the glycan strands. On the left side the unmodified disaccharide is shown. The names of species containing the modifications in their peptidoglycan are indicated. Pep, peptide at MurNAc; LU, linkage unit; SP, surface polymer.

recently, two genes, *oatA* and *adr*, encoding peptidoglycan *O*-acetyltransferases in *S. aureus* and *S. pneumoniae*, respectively, have been identified (Bera *et al.*, 2005; Crisostomo *et al.*, 2006).

The glycan strands in Gram-negative bacteria have a 1,6-anhydroMurNAc residue (instead of MurNAc) at their end. A low fraction of 1,6-anhydroMurNAc residues are also present in the Gram-positive *B. subtilis* (Atrih *et al.*, 1999), but are absent in *S. aureus* (Boneca *et al.*, 2000). The 1,6-anhydroMurNAc residues are generated by cleavage of glycan strands by lytic transglycosylases (Vollmer *et al.*, 2008b) and eventually also by synthetic glycosyltransferases when terminating the polymerization reaction (Höltje, 1998). All Gram-positive species covalently attach cell wall polymers (e.g. teichoic acid, teichuronic acid, capsular polysaccharides and/or arabinogalactan) to the glycan strands of peptidoglycan. Attachment occurs most often via a linkage unit and a phosphodiester bond to O-6 of MurNAc residues (Vollmer, 2008).

2.2. The peptides in peptidoglycan

2.2.1. Variation in the peptide structure

Structural variations in the stem peptide may be introduced at different steps of peptidoglycan biosynthesis, either on the level of the nucleotide activated precursors or on the level of the lipid intermediates (Barreteau *et al.*, 2008; Bouhss *et al.*, 2008) (Figure 2.2).

The amino acid residue in the first position of the peptide moiety is predominantly L-Ala and it is linked with its amino group to the lactyl group of the MurNAc residue. Only in a few species, as with glycine (Gly) in *Microbacterium lacticum*, *Mycobacterium leprae*, *Corynebacterium* or L-serine (L-Ser) in *Butyribacterium rettgeri* has another amino acid been found to occur instead of L-Ala (Schleifer and Kandler, 1972; Draper *et al.*, 1987).

A D-isoglutamic acid (D-iGlu) residue is linked in all species via its γ-carboxylic group to the amino group of the first amino acid. It often undergoes secondary modifications mainly on the lipid II level. For example, hydroxylation of D-iGlu results in the occurrence of *threo*-3-hydroxyglutamic acid in *M. lacticum*. Hydroxylation depends on the availability of oxygen during growth; cells grown in a microaerobic environment contain mainly non-hydroxylated D-iGlu (Schleifer *et al.*, 1967). The more abundant modifications affect the free α-carboxyl group of the D-iGlu residue; it can be amidated resulting in D-isoglutamine (D-iGln) that is common for Gram-positive species, or it can be substituted either by Gly (e.g. in *Thermus*

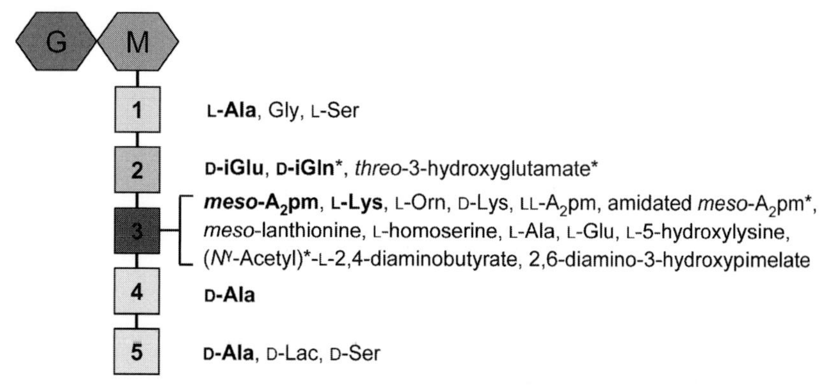

FIGURE 2.2 Variation in the structure of the stem peptide. The most common amino acids for each position are indicated in bold. Amino acids with a star (*) are formed by secondary modification during peptidoglycan biosynthesis. Abbreviations: G, GlcNAc; M, MurNAc; D-iGlu, D-*iso*-glutamic acid; D-iGln, D-isoglutamine; L-Orn, L-ornithine; *meso*-A_2pm, *meso*-2,6-diaminopimelic acid.

thermophilus and *Eubacterium nodatum*), glycine amide (in *Arthrobacter athrocyaneus*), D-alanine amide (in *Renibacterium salmoninarum*), or by polyamines like cadaverine (in *Selenomonas ruminantium*), putrescine (in *Veillonella alcalescens*), *N*-acetylputrescine (in the cyanelles of *Cyanophora paradoxa*) or spermidine (in *Anaerovibrio lipolytica*) (Schleifer and Kandler, 1972; Kamio *et al.*, 1981a,b; Fiedler and Draxl, 1986; Pittenauer *et al.*, 1993; Pfanzagl *et al.*, 1996a,b; Hirao *et al.*, 2000). Interestingly, in *T. thermophilus*, the Gly residue attached to D-iGlu can be modified itself by the addition of a phenylacetic acid residue (Quintela *et al.*, 1995b).

The greatest variation occurs at the third position of the peptide unit (see Figure 2.2). Usually, it is a diamino acid attached with its α-amino group to the γ-carboxyl group of the D-iGlu residue at position-2. The most widespread amino acids at this position are *meso*-A_2pm (in almost all Gram-negative species and species of *Bacillus*, *Lactobacilli* and *Mycobacterium*) or L-Lys (mainly in Gram-positive species). Other diamino acids (L-ornithine, D-Lys, *meso*-lanthionine, LL-A_2pm and L-2,4-diaminobutyric acid) or monoamino acids (L-homoserine, L-Ala, L-Glu) have been identified (Schleifer and Kandler, 1972). Like D-iGlu in position-2, *meso*-A_2pm and L-Lys can be further modified. Thus, the hydroxylated forms (2,6-diamino-3-hydroxypimelate and L-hydroxylysine) or amidated *meso*-A_2pm have been found in a few species (Shockman *et al.*, 1965; Perkins, 1969; Mahapatra *et al.*, 2005b).

The amino acids of position 4 and 5 are represented in most bacteria by D-Ala-D-Ala and are added during synthesis of the cytoplasmic precursor as a dipeptide. In contrast to the conserved D-Ala residue in position 4, the D-Ala in position 5 can be replaced in some species by either D-lactate (D-Lac) or D-Ser. The presence of these alternative residues correlates with a natural or acquired resistance to vancomycin of the related species, e.g. in *Enterococcus hirae* or *Lactobacillus plantarum* (Allen *et al.*, 1992; Arthur *et al.*, 1992; Ferain *et al.*, 1996). Under certain growth conditions, Gly can be incorporated in the peptide moiety instead of one of the terminal D-Ala residues (Glauner *et al.*, 1988). In *Escherichia coli*, the percentage of peptides with a Gly residue is relatively low (about 1%), but it can reach up to 19% in *Caulobacter crescentus* (Markiewicz *et al.*, 1983; Glauner *et al.*, 1988).

The peptide moiety functions as an anchor for covalent linkage of surface proteins (Dramsi *et al.*, 2008). In *E. coli* and related species, Braun's lipoprotein (Lpp) is linked by the ε-amino group of its C-terminal lysine residue to the α-amino group of the *meso*-A_2pm residue of the tripeptide unit in the peptidoglycan (Braun, 1975; Magnet *et al.*, 2007b). In Gram-positive bacteria, many proteins are found to be covalently attached to the peptidoglycan (e.g. protein A and adhesins). The protein precursor carries a C-terminal signal peptide containing the general sequence L-Leu-L-Pro-X-L-Thr-Gly (where L-Leu, L-leucine; L-Pro, L-proline; X, any L-amino acid). The membrane-bound protein sortase A cleaves between the Thr and Gly residues and links the protein with the α-carboxylic group of Thr to the amino group in the side chain of the diamino acid (position 3) of a lipid II molecule (Ton-That *et al.*, 2004).

2.2.2. *Structures of peptide cross-links*

Variations are found in the mode of cross-linkage and the composition of the interpeptide bridge (Figure 2.3). Two main types of peptide cross-links can be distinguished (Schleifer and Kandler, 1972). First, in 3–4 cross-links, the carboxyl group of the D-Ala on position-4 of one peptide (the acyl donor) is linked to the amino group of the side chain of the diamino acid in position-3 of another peptide (the acyl acceptor). This represents the main type of cross-linkage. The 3–4 cross-links are formed by DD-transpeptidases, the penicillin-binding proteins (PBPs), and can be either direct (mainly in Gram-negative species) or indirect via an interpeptide bridge (mainly in Gram-positive species). In some species, the

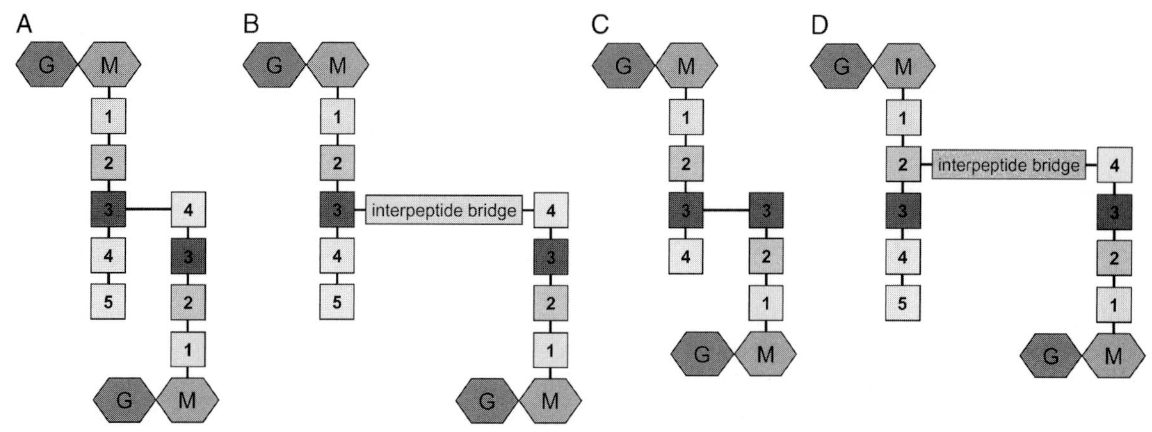

FIGURE 2.3 Types of peptide cross-linking. 3–4 cross-link between the diamino acid in position 3 and the D-Ala residue in position 4 can be either direct (A) or indirect via an interpeptide bridge (B). A direct 3–3 cross-bridge is shown in (C) and an indirect 2–4 cross-link is depicted in (D). G, GlcNAc; M, MurNAc.

peptidoglycan contains a mixture of different cross-links. For example, in *S. pneumoniae* there are direct 3–4 cross-links as well as indirect cross-links via L-Ala-L-Ala or L-Ala-L-Ser interpeptide bridges and the percentage of each type of cross-link varies with the strain and growth condition (Garcia-Bustos *et al.*, 1988). The second type of cross-link (termed 2–4) is only found in coryneform bacteria, especially in plant pathogenic species. Here the α-carboxyl group of the D-iGlu in position 2 of one peptide unit and the carboxyl group of D-Ala in position 4 of another unit are indirectly connected via an interpeptide bridge containing a diamino acid. The size of the interpeptide bridge varies between one to seven amino acids (Table 2.1). Only a few enzymes (so-called "branching enzymes") that are involved in the formation of the interpeptide bridge are characterized so far. The tRNA-dependent aminoacyl transferases, like the FmhB, FemA, FemB proteins of *S. aureus*, FemX of *Weissella viridescens*, the MurM and MurN proteins of *S. pneumoniae* or the BppA1 and BppA1 proteins of *Enterococcus faecalis* transfer the L-amino acids and Gly (activated as amino-acylated tRNAs) to the peptidoglycan precursors (Bouhss *et al.*, 2002; Fiser *et al.*, 2003; Biarrotte-Sorin *et al.*, 2004; Schneider *et al.*, 2004). Aspartate ligases belonging to the ATP-grasp superfamily activate a D-aspartate (D-Asp) residue as β-D-aspartyl-phosphate in an ATP-dependent reaction and, subsequently, link D-Asp to the ε-amino group of L-Lys at position-3 of a cytoplasmic precursor molecule. Some of the D-Asp residues are later amidated by an unknown enzyme (Bellais *et al.*, 2006).

Besides the two main types of cross-linking described above, also 3–3 cross-links can occur at a lesser frequency in species which normally exhibit 3–4 cross-links. The 3–3 cross-links were first described in mycobacteria (Wietzerbin *et al.*, 1974) and are formed by the more recently identified LD-transpeptidases (Mainardi *et al.*, 2005; Magnet *et al.*, 2007a). In *Mycobacterium smegmatis*, the 3–3 cross-links constitute about one-third of the total cross-links (Mainardi *et al.*, 2000). The 3–3 cross-links are present in small amounts in exponentially growing *E. coli* cells, but their abundance increases to about 16% of the total cross-links in stationary cells (Pisabarro *et al.*, 1985; Glauner *et al.*, 1988).

Interestingly, in *M. luteus* (formerly *Micrococcus lysodeikticus*), the stem peptide itself is utilized as the interpeptide bridge. The peptide unit L-Ala-D-iGlu(Gly)-L-Lys-D-Ala is removed from

TABLE 2.1 Examples of interpeptide bridges[a]

Cross-link	Interpeptide bridge(s)	Species
3–4	$(Gly)_5$	*Staphylococcus aureus*
	L-Ala-L-Ala	*Enterococcus faecalis*
	L-Ala-L-Ser; L-Ala-L-Ser-L-Ala	*Weissella viridescens*
	none; L-Ala-L-Ala; L-Ala-L-Ser	*Streptococcus pneumoniae*
	Gly-L-Thr	*Streptococcus salivarius*
	L-Ala-L-Thr-L-Ser	*Arthrobacter polychromogenes*
	D-Asp; D-Asn	*Enterococcus hirae, Lactococcus lactis*
	D-Glu(γ)-L-Ser	*Micrococcus cyaneus*
	D-Glu(γ)-D-Glu	*Micrococcus conglomeratus*
	$(\text{D-Ala-L-Lys-D-iGlu(Gly)-L-Ala})_n$	*Micrococcus luteus*
2–4	L-Orn	*Corynebacterium poinsettiae*
	D-Lys; D-Orn	*Butyribacterium rettgeri*
	D-diaminobutyrate	*Corynebacterium insidiosum*
	Gly-L-Lys	*Microbacterium lacticum*

[a]See text for references and Schleifer and Kandler (1972).

the MurNAc residue by an amidase activity and becomes incorporated between position 3 and 4 of two other peptide moieties. Up to four stem peptides can be linked in a "head-to-tail" arrangement to form a long interpeptide bridge (Ghuysen *et al.*, 1968).

2.3. Species- and strain-specific fine structure

Peptidoglycan is a dynamic structure which is enlarged during cell growth and division and which is subject to maturation and turnover caused by hydrolytic enzymes (Vollmer *et al.*, 2008b). This causes a high degree of structural diversity in the sacculus and is reflected by the great number of structurally different muropeptides released by a muramidase (like lysozme) from peptidoglycan of all species studied to date. For example, the peptidoglycan of *E. coli* is composed of more than 50 different muropeptide subunits that differ in the length of the peptide chain (i.e. di-, tri-, tetra- or pentapeptide), the presence of either D-Ala or Gly at positions-4 or -5, the state of oligomerization (i.e. monomer, dimer, trimer or tetramer), the type of cross-linkage (i.e. 3–4 or 3–3 links), the presence of 1,6-anhydroMurNAc residues (i.e. from the glycan strand termini) and the presence of a L-Lys-L-Arg dipeptide (i.e. linked to position-3) which remains after proteolytic digestion of Braun's lipoprotein (Glauner, 1988). The major muropeptides present in the murein of *E. coli* are the disaccharide tetrapeptide monomer (>30% of the total material) and the 3–4-cross-linked bis-disaccharide tetratetrapeptide dimer (>20% of the total material). The fine structure of the peptidoglycan varies with strain, growth phase and growth condition. New material incorporated into the sacculus has different muropeptide composition as compared to old peptidoglycan (Glauner *et al.*, 1988). In the Gram-positive *S. pneumoniae*, even greater structural diversity of the muropeptides exists due to the presence or absence of *O*-acetylated or de-*N*-acetylated MurNAc residues and the different possible types of cross-links (i.e. direct or indirect) mentioned above. Interestingly, the indirect type of cross-link is the hallmark of penicillin-resistant isolates of *S. pneumoniae*, whereas penicillin-sensitive strains contain mainly direct cross-links in their peptidoglycan (Garcia-Bustos and Tomasz, 1990). Thus, the muropeptide profile can dramatically vary within a species and is a useful parameter for the characterization of different strains and mutants.

3. BIOPHYSICAL PROPERTIES OF PEPTIDOGLYCAN

The bag-shaped peptidoglycan sacculi have unique biophysical properties which have been mainly studied for the Gram-negative model bacterium *E. coli* (Vollmer and Höltje, 2004). Sacculi from *E. coli* are extremely thin as compared to their length and diameter, but which are comparable to the dimensions of a bacterial cell. Hydrated sacculi from the Gram-negative *E. coli* are 2.5–7 nm thick, as determined by small-angle neutron scattering and atomic force microscopy (Labischinski *et al.*, 1991; Yao *et al.*, 1999). Yet, they have the strength to withstand the cell's turgor of several atmospheres. Sacculi are by no means rigid walls but are elastic and stretchable, allowing reversible expansion and shrinkage of cells upon osmolarity changes in the environment. Isolated *E. coli* sacculi are significantly more deformable in the direction of the long axis of the cell than in the direction perpendicular to the long axis (Yao *et al.*, 1999) which is consistent with the observation that volume changes of osmotically shocked *E. coli* cells are mainly due to changes in the cell length, whereas the cell diameter is virtually constant (van den Bogaart *et al.*, 2007). It has been suggested that the anisotropy in elasticity is the consequence of the predominant alignment of the (more flexible) peptides in the direction of the long axis of the cell and of the (more rigid) glycan strands perpendicular to the direction of the long axis. Such a network has been modelled and the theoretical calculations of the elastic moduli are in good agreement with the measured values (Boulbitch *et al.*, 2000).

The pores (or holes) present in peptidoglycan sacculi from *E. coli* or *B. subtilis* are similar in size and relatively uniform with a mean radius of about 2.1 nm (Demchick and Koch, 1996). These authors calculated that globular, uncharged proteins with molecular masses of 22 000–24 000 should be able to penetrate the isolated, relaxed peptidoglycan. Stretched peptidoglycan, as it would occur in the cell, might allow diffusion of globular proteins of a molecular mass of up to 50 000 or more. Thus, peptidoglycan from both Gram-positive and -negative bacteria has relatively wide pores enabling diffusion of large molecules such as proteins.

4. THE MOLECULAR ARCHITECTURE OF PEPTIDOGLYCAN

Peptidoglycan has a heterogeneous, flexible structure and is not crystalline, making it impossible to determine its three-dimensional structure at high resolution with the presently available techniques. Molecular modelling has suggested that the glycan strands are rather rigid structures, whereas the peptides are flexible and are able to adopt a bent or a more straight conformation (Barnickel *et al.*, 1979). Presumably, this flexibility of the peptides causes the observed elasticity of peptidoglycan. The anisotropy in elasticity observed in isolated sacculi could indicate that the flexible peptides run predominantly in the direction of the long axis of the rod-shaped *E. coli* cell, whereas the glycan strands run predominantly perpendicular to the long axis (Yao *et al.*, 1999). According to a model of peptidoglycan architecture, successive peptides protrude in a helical manner from the glycan strands, with about four disaccharide units being required for one complete turn (Barnickel *et al.*, 1979, 1983; Labischinski *et al.*, 1985). Monolayered, net-like structures can be modelled by cross-linking peptides from adjacent glycan strands. According to a model by Koch, the glycan strands are not straight but follow a zigzag line to form a layer with hexagonal "tessera" as the smallest pores (Koch, 1995, 1998). Recently, other "scaffold" models have been proposed in which the glycan strands have a vertical arrangement relative to the cytoplasmic membrane (Dmitriev *et al.*, 1999, 2003, 2004; Meroueh *et al.*, 2006). However, the majority of the glycan strands in the peptidoglycan are too long for a vertical arrangement in *E. coli* and other species (Vollmer and Höltje, 2004). To determine the molecular architecture of peptidoglycan will be a major task for future research.

5. CONCLUSIONS

Peptidoglycan is responsible for the osmotic stability by encasing the cytoplasmic membrane of most bacteria. It is a unique, net-like polymer made of glycan strands which are cross-linked by unusual peptides. There is large variability in the structure of bacterial peptidoglycan due to differences in the amino acid sequence, types of cross-links and the presence or absence of secondary modifications in both the glycan strands and peptides. High-resolution techniques have revealed high complexity in the fine structure of peptidoglycan, as well as compositional changes that occur when newly synthesized material matures. Sacculi from *E. coli* are highly elastic and have remarkably large pores allowing free diffusion of medium-sized proteins. The architecture of peptidoglycan cannot be determined by the available methods. Biophysical data support a model of a layered architecture in which the glycan strands are arranged in parallel to the cytoplasmic membrane. Further questions to be explored are detailed in the Research Focus Box.

ACKNOWLEDGEMENTS

This work was supported by the European Commission through the EUR-INTAFAR project (LSHM-CT-2004-512138) and the BBSRC (BB/F001231/1).

RESEARCH FOCUS BOX

The following major questions and tasks in peptidoglycan research remain to be solved:

- What is the molecular architecture of peptidoglycan? Is there a preferred direction of the glycan strands and peptides relative to the cell's axis? Is the architecture similar in Gram-positive and Gram-negative species?
- How are other cell surface polymers like teichoic acids interlaced in the peptidoglycan network in Gram-positive species?
- What is the function of modifications in the peptidoglycan structure? Are there other modifications in yet uncharacterized peptidoglycan structures? The fine structure of peptidoglycan is only known in a few model bacteria but is unknown in most species.
- Many enzymes responsible for peptidoglycan synthesis, hydrolysis and modification are poorly characterized.
- What is the molecular growth mechanism of the peptidoglycan sacculus?

References

Abrams, A., 1958. *O*-acetyl groups in the cell wall of *Streptococcus faecalis*. J. Biol. Chem. 230, 949–959.

Aitken, A., Stanier, R.Y., 1979. Characetrization of peptidoglycan from cyanelles of *Cyanophora paradoxa*. J. Gen. Microbiol. 112, 219–223.

Allen, N.E., Hobbs, J.N., Richardson, J.M., Riggin, R.M., 1992. Biosynthesis of modified peptidoglycan precursors by vancomycin-resistant *Enterococcus faecium*. FEMS Microbiol. Lett. 77, 109–115.

Antignac, A., Rousselle, J.C., Namane, A., Labigne, A., Taha, M.K., Boneca, I.G., 2003. Detailed structural analysis of the peptidoglycan of the human pathogen *Neisseria meningitidis*. J. Biol. Chem. 278, 31521–31528.

Araki, Y., Nakatani, T., Hayashi, H., Ito, E., 1971. Occurrence of non-*N*-substituted glucosamine residues in lysozyme-resistant peptidoglycan from *Bacillus cereus* cell walls. Biochem. Biophys. Res. Commun. 42, 691–697.

Arthur, M., Molinas, C., Bugg, T.D., Wright, G.D., Walsh, C.T., Courvalin, P., 1992. Evidence for *in vivo* incorporation of D-lactate into peptidoglycan precursors of vancomycin-resistant enterococci. Antimicrob. Agents Chemother. 36, 867–869.

Atrih, A., Zollner, P., Allmaier, G., Foster, S.J., 1996. Structural analysis of *Bacillus subtilis* 168 endospore peptidoglycan and its role during differentiation. J. Bacteriol. 178, 6173–6183.

Atrih, A., Bacher, G., Allmaier, G., Williamson, M.P., Foster, S.J., 1999. Analysis of peptidoglycan structure from vegetative cells of *Bacillus subtilis* 168 and role of PBP 5 in peptidoglycan maturation. J. Bacteriol. 181, 3956–3966.

Azuma, I., Thomas, D.W., Adam, A., et al., 1970. Occurrence of *N*-glycolylmuramic acid in bacterial cell walls. A preliminary survey. Biochim. Biophys. Acta 208, 444–451.

Barnickel, G., Labischinski, H., Bradaczek, H., Giesbrecht, P., 1979. Conformational energy calculation on the peptide part of murein. Eur. J. Biochem. 95, 157–165.

Barnickel, G., Naumann, D., Bradaczek, H., Labischinski, H., Giesbrecht, P., 1983. Computer aided molecular modelling of the three-dimensional structure of bacterial peptidoglycan. In: Hakenbeck, R., Höltje, J.-V., Labischinski, H. (Eds.), The Target of Penicillin: The Murein Sacculus of Bacterial Cell Walls. de Gruyter, Berlin/New York, pp. 61–66.

Barreteau, H., Kova , A., Boniface, A., Sova, M., Gobec, S., Blanot, D., 2008. Cytoplasmic steps of peptidoglycan biosynthesis. FEMS Microbiol. Rev. 32, 168–207.

Bavoil, P.M., Hsia, R., Ojcius, D.M., 2000. Closing in on *Chlamydia* and its intracellular bag of tricks. Microbiology 146, 2723–2731.

Bellais, S., Arthur, M., Dubost, L., et al., 2006. Asl_{fm}, the D-aspartate ligase responsible for the addition of D-aspartic acid onto the peptidoglycan precursor of *Enterococcus faecium*. J. Biol. Chem. 281, 11586–11594.

Bera, A., Herbert, S., Jakob, A., Vollmer, W., Götz, F., 2005. Why are pathogenic staphylococci so lysozyme resistant? The peptidoglycan *O*-acetyltransferase OatA is the major determinant for lysozyme resistance of *Staphylococcus aureus*. Mol. Microbiol. 55, 778–787.

Biarrotte-Sorin, S., Maillard, A.P., Delettre, J., Sougakoff, W., Arthur, M., Mayer, C., 2004. Crystal structures of *Weissella viridescens* FemX and its complex with UDP-MurNAc-pentapeptide: insights into FemABX family substrates recognition. Structure 12, 257–267.

Boneca, I.G., Huang, Z.H., Gage, D.A., Tomasz, A., 2000. Characterization of *Staphylococcus aureus* cell wall glycan strands, evidence for a new β-*N*-acetylglucosaminidase activity. J. Biol. Chem. 275, 9910–9918.

Boneca, I.G., Dussurget, O., Cabanes, D., et al., 2007. A critical role for peptidoglycan *N*-deacetylation in *Listeria* evasion from the host innate immune system. Proc. Natl. Acad. Sci. USA 104, 997–1002.

Bouhss, A., Josseaume, N., Severin, A., et al., 2002. Synthesis of the L-alanyl-L-alanine cross-bridge of *Enterococcus faecalis* peptidoglycan. J. Biol. Chem. 277, 45935–45941.

Bouhss, A., Trunkfield, A.E., Bugg, T.D.H., Mengin-Lecreulx, D., 2008. The biosynthesis of peptidoglycan lipid-linked intermediates. FEMS Microbiol. Rev. 32, 208–233.

Boulbitch, A., Quinn, B., Pink, D., 2000. Elasticity of the rod-shaped gram-negative eubacteria. Phys. Rev. Lett. 85, 5246–5249.

Braun, V., 1975. Covalent lipoprotein from the outer membrane of *Escherichia coli*. Biochim. Biophys. Acta 415, 335–377.

Brumfitt, W., Wardlaw, A.C., Park, J.T., 1958. Development of lysozyme-resistance in *Micrococcus lysodiekticus* and its association with an increased *O*-acetyl content of the cell wall. Nature 181, 1783–1784.

Chaput, C., Labigne, A., Boneca, I.G., 2007. Characterization of *Helicobacter pylori* lytic transglycosylases Slt and MltD. J. Bacteriol. 189, 422–429.

Chopra, I., Storey, C., Falla, T.J., Pearce, J.H., 1998. Antibiotics, peptidoglycan synthesis and genomics: the chlamydial anomaly revisited. Microbiology 144, 2673–2678.

Costa, K., Bacher, G., Allmaier, G., et al., 1999. The morphological transition of *Helicobacter pylori* cells from spiral to coccoid is preceded by a substantial modification of the cell wall. J. Bacteriol. 181, 3710–3715.

Crisostomo, M.I., Vollmer, W., Kharat, A.S., et al., 2006. Attenuation of penicillin resistance in a peptidoglycan *O*-acetyl transferase mutant of *Streptococcus pneumoniae*. Mol. Microbiol. 61, 1497–1509.

Demchick, P., Koch, A.L., 1996. The permeability of the wall fabric of *Escherichia coli* and *Bacillus subtilis*. J. Bacteriol. 178, 768–773.

Dmitriev, B.A., Ehlers, S., Rietschel, E.T., 1999. Layered murein revisited: a fundamentally new concept of bacterial cell wall structure, biogenesis and function. Med. Microbiol. Immunol. (Berlin) 187, 173–181.

Dmitriev, B.A., Toukach, F.V., Schaper, K.J., Holst, O., Rietschel, E.T., Ehlers, S., 2003. Tertiary structure of bacterial murein: the scaffold model. J. Bacteriol. 185, 3458–3468.

Dmitriev, B.A., Toukach, F.V., Holst, O., Rietschel, E.T., Ehlers, S., 2004. Tertiary structure of *Staphylococcus aureus* cell wall murein. J. Bacteriol. 186, 7141–7148.

Dramsi, S., Davison, S., Magnet, S., Arthur, M., 2008. Surface proteins covalently attached to peptidoglycan: examples from both Gram-positive and Gram-negative bacteria. FEMS Microbiol. Rev. 32, 307–320.

Draper, P., Kandler, O., Darbre, A., 1987. Peptidoglycan and arabinogalactan of *Mycobacterium leprae*. J. Gen. Microbiol. 133, 1187–1194.

Evtushenko, L.I., Dorofeeva, L.V., Krausova, V.I., Gavrish, E.Y., Yashina, S.G., Takeuchi, M., 2002. *Okibacterium fritillariae* gen. nov., sp. nov., a novel genus of the family *Microbacteriaceae*. Int. J. Syst. Evol. Microbiol. 52, 987–993.

Ferain, T., Hobbs, J.N., Richardson, J., et al., 1996. Knockout of the two *ldh* genes has a major impact on peptidoglycan precursor synthesis in *Lactobacillus plantarum*. J. Bacteriol. 178, 5431–5437.

Fiedler, F., Draxl, R., 1986. Biochemical and immunochemical properties of the cell surface of *Renibacterium salmoninarum*. J. Bacteriol. 168, 799–804.

Fiser, A., Filipe, S.R., Tomasz, A., 2003. Cell wall branches, penicillin resistance and the secrets of the MurM protein. Trends Microbiol 11, 547–553.

Garcia-Bustos, J., Tomasz, A., 1990. A biological price of antibiotic resistance: major changes in the peptidoglycan structure of penicillin-resistant pneumococci. Proc. Natl. Acad. Sci. USA 87, 5415–5419.

Garcia-Bustos, J.F., Chait, B.T., Tomasz, A., 1988. Altered peptidoglycan structure in a pneumococcal transformant resistant to penicillin. J. Bacteriol. 170, 2143–2147.

Ghuysen, J.M., Strominger, J.L., 1963. Structure of the cell wall of *Staphylococcus aureus*, strain Copenhagen. Ii. Separation and structure of disaccharides. Biochemistry 2, 1119–1125.

Ghuysen, J.M., Bricas, E., Lache, M., Leyh-Bouille, M., 1968. Structure of the cell walls of *Micrococcus lysodeikticus*. 3. Isolation of a new peptide dimer, *N*-alpha-[L-alanyl-gamma-(alpha-D-glutamylglycine)]-L-lysyl-D-alanyl-*N*-alpha-[L-alanyl-gamma-(alpha-D-glutamylglycine)]-L-lysyl-D-alanine. Bi. Biochemistry 7, 1450–1460.

Glauner, B., 1988. Separation and quantification of muropeptides with high-performance liquid chromatography. Anal. Biochem. 172, 451–464.

Glauner, B., Höltje, J.-V., Schwarz, U., 1988. The composition of the murein of *Escherichia coli*. J. Biol. Chem. 263, 10088–10095.

Harz, H., Burgdorf, K., Höltje, J.-V., 1990. Isolation and separation of the glycan strands from murein of *Escherichia coli* by reversed-phase high-performance liquid chromatography. Anal. Biochem. 190, 120–128.

Hirao, T., Sato, M., Shirahata, A., Kamio, Y., 2000. Covalent linkage of polyamines to peptidoglycan in *Anaerovibrio lipolytica*. J. Bacteriol. 182, 1154–1157.

Höltje, J.-V., 1998. Growth of the stress-bearing and shape-maintaining murein sacculus of *Escherichia coli*. Microbiol. Mol. Biol. Rev. 62, 181–203.

Hughes, R.C., 1971. Autolysis of *Bacillus cereus* cell walls and isolation of structural components. Biochem. J. 121, 791–802.

Kamio, Y., Itoh, Y., Terawaki, Y., 1981a. Chemical structure of peptidoglycan in *Selenomonas ruminantium*: cadaverine links covalently to the D-glutamic acid residue of peptidoglycan. J. Bacteriol. 146, 49–53.

Kamio, Y., Itoh, Y., Terawaki, Y., Kusano, T., 1981b. Cadaverine is covalently linked to peptidoglycan in *Selenomonas ruminantium*. J. Bacteriol. 145, 122–128.

Koch, A.L., 1995. Bacterial Growth and Form. Chapman & Hall, New York.

Koch, A.L., 1998. Orientation of the peptidoglycan chains in the sacculus of *Escherichia coli*. Res. Microbiol. 149, 689–701.

Labischinski, H., Barnickel, G., Naumann, D., Keller, P., 1985. Conformational and topological aspects of the three-dimensional architecture of bacterial peptidoglycan. Ann. Inst. Pasteur Microbiol. 136A, 45–50.

Labischinski, H., Goodell, E.W., Goodell, A., Hochberg, M.L., 1991. Direct proof of a 'more-than-single-layered' peptidoglycan architecture of *Escherichia coli* W7: a neutron small-angle scattering study. J. Bacteriol. 173, 751–756.

Li, W.J., Chen, H.H., Kim, C.J., et al., 2005. Microbacterium halotolerans sp. nov., isolated from a saline soil in the west of China. Int. J. Syst. Evol. Microbiol. 55, 67–70.

Machida, M., Takechi, K., Sato, H., et al., 2006. Genes for the peptidoglycan synthesis pathway are essential for chloroplast division in moss. Proc. Natl. Acad. Sci. USA 103, 6753–6758.

Magnet, S., Arbeloa, A., Mainardi, J.L., et al., 2007a. Specificity of L,D-transpeptidases from Gram-positive bacteria producing different peptidoglycan chemotypes. J. Biol. Chem. 282, 13151–13159.

Magnet, S., Bellais, S., Dubost, L., et al., 2007b. Identification of the L,D-transpeptidases responsible for attachment of the Braun lipoprotein to *Escherichia coli* peptidoglycan. J. Bacteriol. 189, 3927–3931.

Mahapatra, S., Scherman, H., Brennan, P.J., Crick, D.C., 2005a. *N*-Glycolylation of the nucleotide precursors of peptidoglycan biosynthesis of *Mycobacterium* spp. is altered by drug treatment. J. Bacteriol. 187, 2341–2347.

Mahapatra, S., Yagi, T., Belisle, J.T., et al., 2005b. Mycobacterial lipid II is composed of a complex mixture of modified muramyl and peptide moieties linked to decaprenyl phosphate. J. Bacteriol. 187, 2747–2757.

Mainardi, J.L., Legrand, R., Arthur, M., Schoot, B., van Heijenoort, J., Gutmann, L., 2000. Novel mechanism of beta-lactam resistance due to bypass of DD-transpeptidation in *Enterococcus faecium*. J. Biol. Chem. 275, 16490–16496.

Mainardi, J.L., Fourgeaud, M., Hugonnet, J.E., et al., 2005. A novel peptidoglycan cross-linking enzyme for a beta-lactam-resistant transpeptidation pathway. J. Biol. Chem. 280, 38146–38152.

Markiewicz, Z., Glauner, B., Schwarz, U., 1983. Murein structure and lack of DD- and LD-carboxypeptidase activities in *Caulobacter crescentus*. J. Bacteriol. 156, 649–655.

Martin, H.H., Gmeiner, J., 1979. Modification of peptidoglycan structure by penicillin action in cell walls of *Proteus mirabilis*. Eur. J. Biochem. 95, 487–495.

Matsumoto, A., Takahashi, Y., Shinose, M., Seino, A., Iwai, Y., Omura, S., 2003. *Longispora albida* gen. nov., sp. nov., a novel genus of the family *Micromonosporaceae*. Int. J. Syst. Evol. Microbiol. 53, 1553–1559.

Meroueh, S.O., Bencze, K.Z., Hesek, D., et al., 2006. Three-dimensional structure of the bacterial cell wall peptidoglycan. Proc. Natl. Acad. Sci. USA 103, 4404–4409.

Moulder, J.W., 1993. Why is *Chlamydia* sensitive to penicillin in the absence of peptidoglycan? Infect. Agents Dis. 2, 87–99.

Neuhaus, F.C., Baddiley, J., 2003. A continuum of anionic charge: structures and functions of D-alanyl-teichoic acids in Gram-positive bacteria. Microbiol. Mol. Biol. Rev. 67, 686–723.

Perkins, H.R., 1969. The configuration of 2,6-diamino-3-hydroxypimelic acid in microbial cell walls. Biochem. J. 115, 797–805.

Pfanzagl, B., Zenker, A., Pittenauer, E., et al., 1996a. Primary structure of cyanelle peptidoglycan of *Cyanophora paradoxa*: a prokaryotic cell wall as part of an organelle envelope. J. Bacteriol. 178, 332–339.

Pfanzagl, B., Allmaier, G., Schmid, E.R., de Pedro, M.A., Loffelhardt, W., 1996b. *N*-acetylputrescine as a characteristic constituent of cyanelle peptidoglycan in glaucocystophyte algae. J. Bacteriol. 178, 6994–6997.

Pisabarro, A.G., de Pedro, M.A., Vazquez, D., 1985. Structural modifications in the peptidoglycan of *Escherichia coli* associated with changes in the state of growth of the culture. J. Bacteriol. 161, 238–242.

Pittenauer, E., Schmid, E.R., Allmaier, G., et al., 1993. Structural characterization of the cyanelle peptidoglycan of *Cyanophora paradoxa* by 252Cf plasma desorption mass spectrometry and fast atom bombardment/tandem mass spectrometry. Biol. Mass Spectrom. 22, 524–536.

Popham, D.L., Helin, J., Costello, C.E., Setlow, P., 1996. Muramic lactam in peptidoglycan of *Bacillus subtilis* spores is required for spore outgrowth but not for spore dehydration or heat resistance. Proc. Natl. Acad. Sci. USA 93, 15405–15410.

Quintela, J.C., Caparros, M., de Pedro, M.A., 1995a. Variability of peptidoglycan structural parameters in Gram-negative bacteria. FEMS Microbiol. Lett. 125, 95–100.

Quintela, J.C., Pittenauer, E., Allmaier, G., Aran, V., de Pedro, M.A., 1995b. Structure of peptidoglycan from *Thermus thermophilus* HB8. J. Bacteriol 177, 4947–4962.

Quintela, J.C., Garcia-del Portillo, F., Pittenauer, E., Allmaier, G., de Pedro, M.A., 1999. Peptidoglycan fine structure of the radiotolerant bacterium *Deinococcus radiodurans* Sark. J. Bacteriol. 181, 334–337.

Schleifer, K.H., Kandler, O., 1972. Peptidoglycan types of bacterial cell walls and their taxonomic implications. Bacteriol. Rev. 36, 407–477.

Schleifer, K.H., Plapp, R., Kandler, O., 1967. Identification of threo-3-hydroxyglutamic acid in the cell wall of *Microbacterium lacticum*. Biochem. Biophys. Res. Commun. 28, 566–570.

Schneider, T., Senn, M.M., Berger-Bachi, B., Tossi, A., Sahl, H.G., Wiedemann, I., 2004. *In vitro* assembly of a complete, pentaglycine interpeptide bridge containing cell wall precursor (lipid II-Gly5) of *Staphylococcus aureus*. Mol. Microbiol. 53, 675–685.

Shockman, G.D., Thompson, J.S., Conover, M.J., 1965. Replacement of lysine by hydroxylysine and its effects on cell lysis in *Streptococcus faecalis*. J. Bacteriol. 90, 575–588.

Sutcliffe, I.C., 1998. Cell envelope composition and organisation in the genus *Rhodococcus*. Antonie van Leeuwenhoek 74, 49–58.

Tipper, D.J., Strominger, J.L., Ensign, J.C., 1967. Structure of the cell wall of *Staphylococcus aureus*, strain Copenhagen. VII. Mode of action of the bacteriolytic peptidase from *Myxobacter* and the isolation of intact cell wall polysaccharides. Biochemistry 6, 906–920.

Ton-That, H., Marraffini, L.A., Schneewind, O., 2004. Protein sorting to the cell wall envelope of Gram-positive bacteria. Biochim. Biophys. Acta 1694, 269–278.

Tuomanen, E., Schwartz, J., Sande, S., Light, K., Gage, D., 1989. Unusual composition of peptidoglycan in *Bordetella pertussis*. J. Biol. Chem. 264, 11093–11098.

Uchida, E., Aida, K., 1979. Taxonomic significance of cell-wall acyl type in *Corynebacterium-Mycobacterium-Nocardia* group by a glycolate test. J. Gen. Appl. Microbiol. 25, 169–183.

van den Bogaart, G., Hermans, N., Krasnikov, V., Poolman, B., 2007. Protein mobility and diffusive barriers in

Escherichia coli: consequences of osmotic stress. Mol. Microbiol. 64, 858–871.

Vollmer, W., 2008. Structural variation in the glycan strands of bacterial peptidoglycan. FEMS Microbiol. Rev. 32, 287–306.

Vollmer, W., Höltje, J.-V., 2004. The architecture of the murein (peptidoglycan) in Gram-negative bacteria: Vertical scaffold or horizontal layer(s)? J. Bacteriol. 186, 5978–5987.

Vollmer, W., Tomasz, A., 2000. The *pgdA* gene encodes for a peptidoglycan *N*-acetylglucosamine deacetylase in *Streptococcus pneumoniae*. J. Biol. Chem. 275, 20496–20501.

Vollmer, W., Tomasz, A., 2002. Peptidoglycan *N*-acetylglucosamine deacetylase, a putative virulence factor in *Streptococcus pneumoniae*. Infect. Immun. 70, 7176–7178.

Vollmer, W., Blanot, D., de Pedro, M.A., 2008a. Peptidoglycan structure and architecture. FEMS Microbiol. Rev. 32, 149–167.

Vollmer, W., Joris, B., Charlier, P., Foster, S., 2008b. Bacterial peptidoglycan (murein) hydrolases. FEMS Microbiol. Rev. 32, 259–286.

Ward, J.B., 1973. The chain length of the glycans in bacterial cell walls. Biochem. J. 133, 395–398.

Warth, A.D., Strominger, J.L., 1971. Structure of the peptidoglycan from vegetative cell walls of *Bacillus subtilis*. Biochemistry 10, 4349–4358.

Weadge, J.T., Clarke, A.J., 2006. Identification and characterization of *O*-acetylpeptidoglycan esterase: a novel enzyme discovered in *Neisseria gonorrhoeae*. Biochemistry 45, 839–851.

Weidel, W., Pelzer, H., 1964. Bagshaped macromolecules – a new outlook on bacterial cell walls. Adv. Enzymol. 26, 193–232.

Wietzerbin, J., Das, B.C., Petit, J.F., Lederer, E., Leyh-Bouille, M., Ghuysen, J.M., 1974. Occurrence of D-alanyl-(D)-*meso*-diaminopimelic acid and *meso*-diaminopimelyl-*meso*-diaminopimelic acid interpeptide linkages in the peptidoglycan of mycobacteria. Biochemistry 13, 3471–3476.

Yao, X., Jericho, M., Pink, D., Beveridge, T., 1999. Thickness and elasticity of Gram-negative murein sacculi measured by atomic force microscopy. J. Bacteriol. 181, 6865–6875.

CHAPTER

3

Core region and lipid A components of lipopolysaccharides

Otto Holst and Antonio Molinaro

SUMMARY

One of the major virulence factors of Gram-negative bacteria are lipopolysaccharides (LPSs), also known as endotoxins, intensive research on which has been performed for more than 50 years leading to an in-depth characterization of LPS structures and their functions, including immunological, pharmacological and pathophysiological effects displayed in eukaryotic hosts. Lipopolysaccharides represent amphiphilic molecules generally comprising three defined regions which are distinguished by their genetics, structures, function and biosynthesis. The first moiety, lipid A, is substituted by the second, the core region, which in turn carries a polysaccharide which may be the O-specific polysaccharide, a capsular polysaccharide, or the enterobacterial common antigen (only in *Enterobacteriaceae*). With regard to the structure–function relationships of LPSs, lipid A has been identified as the toxic principle of endotoxically active LPS. This toxicity depends mostly on lipid A chemical structure which defines the overall molecular conformation, but is also influenced by the core region. Thus, a complete structural analysis of lipid A and core region represent the prerequisite for the understanding of LPS functions. To date, quite a number of lipid A and core structures from LPSs of various Gram-negative bacteria have been published and summarized in several overviews. This chapter adds to this knowledge, however, due to space limitations, the collection of structures reviewed has to be incomplete.

Keywords: Lipopolysaccharide; Lipid A; Core region; Structural analysis; *Burkholderia cepacia,* Plant pathogenic bacteria; Pathogenesis; *Enterobacteriaceae*; *Pseudomonadaceae*

1. INTRODUCTION

Endotoxin was discovered and named by Richard Pfeiffer in the year 1892. Pfeiffer was working on cell lysates of *Vibrio cholerae* from which he identified this heat-resistant toxin that appeared to be attached to the bacteria unlike the exotoxins that were secreted. The term lipopolysaccharide (LPS) was introduced by Murray Shear in 1943 (Shear and Turner, 1943) characterizing a bacterial toxin which was constituted of polysaccharide and lipoidic material and used for the treatment of malignant tumours. Afterwards, this term was adopted by Otto Westphal and Otto Lüderitz, the two most important pioneer scientists of modern endotoxin research, to indicate

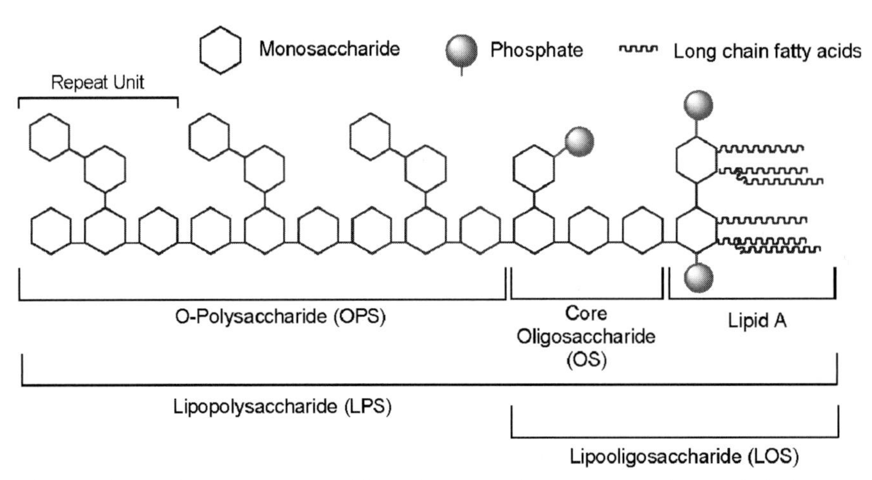

FIGURE 3.1 General chemical structure of an LPS from Gram-negative bacteria. All forms of LPS known to date consist of a lipid A domain and a covalently linked saccharide moiety composed of the core region (oligosaccharide) plus polysaccharide or oligosaccharide portion only (S- form LPS or R- form LPS/LOS, respectively). The polysaccharide region is termed the O-antigen, O-specific polysaccharide, O-side chain. According to preferential carbohydrate compositions in the core structure, an inner and an outer core region are commonly distinguished.

the pyrogenic fractions arising from enterobacteria, mainly *Escherichia coli* and *Salmonella enterica*, which they had isolated in sufficiently pure amounts for structural studies (Westphal *et al.*, 1952). They showed that the pyrogenic activity mainly resided in the lipopolysaccharide by testing all the bacterial components. From that era on the study of chemical and biological features of endotoxin has quickly developed and is nowadays in the molecular era in which the full chemical and biological characterization of endotoxin from various Gram-negative microorganisms can be achieved.

The interested reader will find detailed and fascinating historical perspectives in two other publications (Rietschel and Westphal, 1999; Beutler and Rietschel, 2003).

The lipopolysaccharides are amphiphilic components present in the cell wall of almost all Gram-negative bacteria, where in particular they are localized in the external leaflet of the outer membrane (OM) in which they represent approximately 75% of the material. The OM is responsible for the minor permeability of Gram-negative surface compared to that of Gram-positive bacteria which do not possess a second membrane in their cell wall. In general, the OM protects the bacteria against hydrophobic and higher molecular mass hydrophilic compounds owing to the very low fluidity of the highly ordered structure of the LPS monolayer. Furthermore, due to their external location, LPSs are involved in interactions with the external environment, in particular concerning host–bacterium interactions like recognition, adhesion and colonization. Lipopolysaccharides represent very important virulence factors of pathogenic Gram-negative bacteria, are major players in symbiosis and tolerance of commensal bacteria and, in the case of extremophile bacteria, are important for the survival under harsh conditions.

With regard to their structure, all LPSs possess the same general chemical architecture independent to the bacterial activity (pathogenic, symbiontic, commensal) or ecological niches: (human, animal, soil, plant, water) or growth conditions (Raetz and Whitfield, 2002; Holst and Müller-Loennies, 2007). Three different regions are usually identified in LPS (Figure 3.1), namely a polysaccharide, which may be the O-specific polysaccharide (OPS, known also as O-side chain or O-antigen), a capsular polysaccharide or, only in *Enterobacteriaceae*, the enterobacterial common antigen. The polysaccharide is covalently linked to an oligosaccharide named the core region, which in turn is linked to the lipid A. The lipid A anchors LPS in the outer leaflet of the OM, whereas the sugar moiety is oriented outwards (Raetz and Whitfield, 2002; Holst and Müller-Loennies, 2007). Complete LPS comprising all three regions is termed smooth- (S)-form LPS, whereas mutants lacking the OPS are called rough- (R)-form LPS (lipo-oligosaccharide, LOS).

This chapter will deal with the structural aspects of lipid A and the core region. The chemical features of OPS are illustrated in the following chapter.

2. GENERAL STRUCTURAL FEATURES OF THE LIPID A MOLECULE

The first lipid A structures, early investigated in the *Enterobacteriaceae* by Westphal and Lüderitz in the year 1954 (Westphal and Lüderitz, 1954), were finally established in 1983 (Takayama *et al.*, 1983; Imoto *et al.*, 1983) and consisted of a β-(1→6)-linked 2-amino-2-deoxy-D-glucopyranose (D-Glc*p*N, glucosamine) disaccharide that bore (*R*)-3-hydroxy fatty acid residues, two of which were ester-linked to the 3- and 3′-positions, and another two were amide-linked at the 2- and 2′-positions (primary fatty acid residues, e.g. 3-OH(14:0) in *E. coli*). The 3-hydroxy fatty acids were in turn further esterified by secondary fatty acids, e.g. tetradecanoic acid (14:0) and dodecanoic acid (12:0) in LPS of *E. coli*. The hydroxyl groups at position-4′ of the non-reducing GlcN II residue (distal unit) and that of the α-anomeric position of the reducing GlcN I residue (proximal unit) were both substituted by phosphate groups (Figure 3.2). Apart from such a general chemical architecture conserved in many bacterial LPSs, a number of subtle chemical differences have been identified in the past that are responsible for lipid A variation among bacterial LPSs. Furthermore, in both, mammalian and plant pathogen bacterial LPSs, chemical variations are responsible for the agonist/antagonist effect of lipid A in host innate immune response processes.

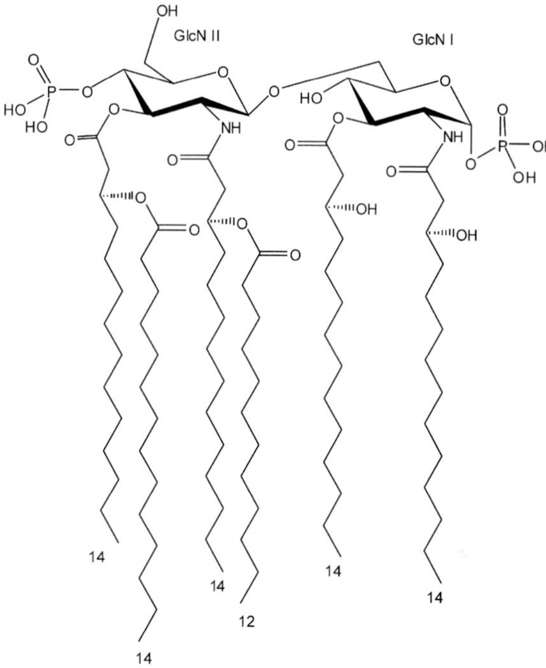

FIGURE 3.2 The chemical structure of lipid A from *Escherichia coli* LPS. The β-(1→6)-linked 2-amino-2-deoxy-D-glucopyranose (Glc*p*N) disaccharide bears primary (*R*)-3-OH(14:0) residues either ester- (at the 3- and 3′-positions) or amide-linked (at 2- and 2′-positions). Both fatty acid residues on the GlcN II are further esterified by a dodecanoic (12:0) and a tetradecanoic (14:0) acid residue. The hydroxyl at position-4′ of the non-reducing GlcN II residue (distal unit) and that of the α-anomeric position of the reducing GlcN I residue (proximal unit) are both phosphorylated.

The structures of lipid A residues shown in Table 3.1 are either those published after 2002, when the most recent review specifically addressing the chemical structure of lipid A appeared (Alexander and Zähringer, 2002), or those which had not been published in reviews of the years 2002–2008 (Munford and Varley, 2006; De Castro *et al.*, 2008; Munford, 2008).

TABLE 3.1 Substitution pattern on the two GlcN residues on different bacterial lipid A

Bacteria	GlcN II substitution			GlcN I substitution		
	O-4′	*O*-3′	*N*-2′	*O*-3	*N*-2	*O*-1
Acinetobacter radioresistens [a]	P	12:0(3-OH) 12:0	12:0(3-OH) 12:0	12:0(3-OH)	12:0/14:0 (3-OH) 12:0	P
Aeromonas salmonicida [b]	P	14:0(3-OH)[c] 16:1[c]	14:0 (3-OH) 12:0	14:0(3-OH)[c]	14:0(3-OH)	-
Agrobacterium tumefaciens [d]	P	14:0(3-OH)	16:0(3-OH) 28:0(27-OH) 14:0 (3-OH)[e]	14:0(3-OH)	16:0(3-OH)	P
Alteromonas macleodi [f]	P	12:0(3-OH)[c]	12:0(3-OH)[c] 12:0	12:0(3-OH)[c]	12:0(3-OH)[c]	P
Azospirillum lipoferum [g]	-	14:0(3-OH)	16:0(3-OH) 18:1/18:0	14:0(3-OH)	16:0(3-OH)	GalA
Bartonella henselae [h,j]	P	12:0(3-OH)	16:0(3-OH) 28:0(27-OH)	12:0(3-OH)	16:0(3-OH)	P
Bdellovibrio bacteriovorus [i,j]	Man	13:0(3-OH) 13:0(3-OH)	13:1(3-OH) 13:0(3-OH)	13:0(3-OH)	13:0(3,4-OH)	Man
Bordetella pertussis [k]	GalN-P	14:0(3-OH)	14:0(3-OH) 14:0	10:0(3-OH)	14:0(3-OH)	P-GalN
Burkholderia cepacia complex [l]	Ara4N-P[e]	14:0(3-OH)[e]	16:0 (3-O) 14:0	14:0(3-OH)[e]	16:0(3-OH)	Ara4N-P[e]
Francisella tularensis [m]	Man-P	-	18:0 (3-OH) 16:0	18:0(3-OH)	18:0(3-OH)	P-GalN
Halomonas magadiensis [n]	P	12:0(3-OH)[e] 18:1/16:0[e]	12:0(3-OH)[e] 14:0[e]	12:0(3-OH)	12:0(3-OH) 10:0	P[e]
Leptospira interrogans [o,j]	-	12:0(3-OH) 12:1/14:1	16:0(3-OH) 12:1/14:1	12:0(3-OH)	16:0(3-OH)	P-Methyl
Marinomonas vaga [p]	-	-	12:0(3-OH) 12:0(3-OH)	12:0(3-OH)	12:0(3-OH) 10:0/12:0	P

(Continues)

Photorabdus luminescens [q]	P	14:0/15:0 (3-OH) 14:0[a]	14:0/15:0(3-OH) 14:0[a]	14:0/15:0(3-OH)	14:0/15:0(3-OH)	P
Pseudoalteromonas nigrifaciens [r]	P	10:0 (3-OH)[b]	12:0(3-OH)	10:0(3-OH)[b]	12:0 (3-OH) 12:0	P
Shewanella pacifica [s]	P	13:0 (3-OH)[b] 13:0	12:0(3-OH)[b] 13:0	13:0(3-OH)[b]	13:0 (3-OH)[b]	P
Xanthomonas campestris [t]	P-P-EtN[a]	12:0 (3-OH)[b] 10:0[b]	12:0(3-OH)[b]	12:0(3-OH)[b] 10:0[b]	12:0 (3-OH)[b]	P-P-EtN[a]

Ara4N-P, 4-amino-4-deoxy-L-arabinopyranose-phosphate; GalA, galacturonic acid-phosphate; GalN, 2-amino-2-deoxy-D-galactose-phosphate; Man, mannose; P, phosphate; P-P-EtN, 2-aminoethanol diphosphate.

[a] Leone *et al.*, 2006.
[b] Wang *et al.*, 2006b.
[c] High structural heterogeneity in fatty acids.
[d] Silipo *et al.*, 2004a.
[e] Present in non-stoichiometric amount.
[f] Liparoti *et al.*, 2006.
[g] Choma and Komniecka, 2008.
[h] Zähringer *et al.*, 2004.
[i] Schwudke *et al.*, 2003.
[j] Both GlcN are replaced by GlcN3N.
[k] Marr *et al.*, 2008.
[l] Silipo *et al.*, 2005a.
[m] Kay *et al.*, 2006.
[n] Silipo *et al.*, 2004b.
[o] Raetz *et al.*, 2007.
[p] Krasikova *et al.*, 2004.
[q] Molinaro, personal communication.
[r] Krasikova *et al.*, 2001.
[s] Leone *et al.*, 2007.
[t] Silipo *et al.*, 2005d.

With regard to the lipid A carbohydrate backbone, a β-(1→6)-linked disaccharide has been always identified to date, in which a rather common variation is the replacement of one or both D-Glc*p*N residues by a residue of 2,3-diamino-2,3-dideoxy-D-glucopyranose (D-Glc*p*N3N) to which both primary fatty acids are linked as amide residues. In *Rhizobium etli* CE3, the proximal glucosamine unit is exceptionally replaced by its aldonic acid derivative, the 2-amino-2-deoxy-D-gluconate residue (Bhat *et al.*, 1994; Raetz *et al.*, 2007; De Castro *et al.*, 2008).

As for the polar heads of lipid A, these are largely represented by a phosphate group, to which in turn another phosphate group or 2-aminoethanol (EtN) can be further attached. A remarkable difference is observed when the phosphate group is substituted by a further carbohydrate moiety, i.e. 4-amino-4-deoxy-L-arabinopyranose (Ara*p*4N, arabinosamine) which has been found in a number of bacterial LPSs in recent years, in particular in all LPSs from the *Burkholderia cepacia* complex (see Table 3.1). Analogously, the 2-amino-2-deoxy-D-galactopyranose (Gal*p*N) has been found very recently to substitute phosphate in *Bordetella* LPS (Marr *et al.*, 2008). In a few cases, a phosphate group can be replaced by acidic monosaccharides like D-galactopyranosuronic acid (D-Gal*p*A) in the lipid A of *Aquifex pyrophylus* (Plötz *et al.*, 2000) and *Rhizobium* (Bhat *et al.*, 1994), or by neutral monosaccharide residues, e.g. D-mannopyranose (D-Man*p*) in the lipid A of *Bdellovibrio bacteriovorus* (Schwudke *et al.*, 2003).

Acyl residues may vary in number, type and distribution (see Table 3.1). The most frequent primary fatty acid residues found in lipid A are 3-OH(10:0), 3-OH(12:0), 3-OH(14:0), 3-OH(16:0) and 3-OH(18:0) and, as secondary fatty acids, the corresponding 3-deoxy acyl chains, some of which may be present as (*S*)-2-OH derivatives (Zähringer *et al.*, 1994; Raetz and Whitfield, 2002; Raetz *et al.*, 2007). Less frequently present are odd numbered, methyl branched and unsaturated fatty acids and long chain and (ω-1)-hydroxy fatty acids, the 27-OH(28:0) in *Rhizobiaceae* LPSs (Bhat *et al.*, 1994; Basu *et al.*, 2002; Vedam *et al.*, 2003; De Castro *et al.*, 2008), or in *Legionella* and *Brucella* LPSs (Zähringer *et al.*, 1994; Raetz and Whitfield, 2002; Raetz *et al.*, 2007). The grade of acylation and the fatty acid distribution between the Glc*p*N units determines the three-dimensional structure, i.e. the conical or cylindrical molecular shape, of lipid A (Seydel *et al.*, 1993, 2000; Brandenburg *et al.*, 1996) which is correlated to its biological activity, i.e. the binding to proteins of the innate immune system of both animals and plants (Medzhitov, 2001; Akira *et al.*, 2006).

3. LIPOPOLYSACCHARIDES OF MAMMALIAN PATHOGENIC BACTERIA: THE CASE OF *B. CEPACIA* COMPLEX

Among the lipid A from LPSs of pathogenic bacteria, that from *B. cepacia* complex (Bcc) deserves an own paragraph. Although a body of publications related to the biology and biochemistry of LPS/lipid A from Bcc has appeared in recent years, only scanty data are available on the molecular structure. The Bcc comprises ten closely related Gram-negative bacterial species, all of which appear to be capable of causing disease in humans. These organisms appear to be of particular relevance to patients with cystic fibrosis. Lipopolysaccharide is an important virulence determinant in these pathogens. An indisputable first step toward the understanding of the complex phenomenon of pathogenesis of Bcc is a complete characterization of the LPS structures of all ten species, followed by comprehensive studies of signal transduction and cytokine induction. The *in vivo* controls of Bcc lipid A acylation also require further investigations to define if mechanisms such as the magnesium responsive kinase system PhoP/Q as seen in enteric pathogens and *Pseudomonas* has a similar role in Bcc pathogenesis and affects TLR4 signalling.

Thus far, the established lipid A structures of Bcc LPSs possess a general architecture in which the following elements are always present: the D-GlcpN disaccharide carbohydrate backbone plus L-Arap4N; the phosphate groups, 3-OH(14:0) and 3-OH(16:0) as primary and 14:0 as secondary fatty acids. However, these elements are very often present in non-stoichiometric amounts giving rise to a wide array of lipid A primary structures that varies within the same and among the strains and depends on the environment from which the bacteria have been isolated. To date, the isolated structures of Bcc lipid A are penta- or tetra-acylated, but no hexa-acylated structures have been identified so far (De Soyza *et al.*, 2008).

The Bcc lipid A chemical structures possess strong similarities with other non-Bcc *Burkholderia* spp. lipid A, as for example the lipid A from the LPS of the plant pathogen *Burkholderia caryophylli* which possesses a nearly identical structure to that of *B. cepacia* LPS (Molinaro *et al.*, 2003a) (see Table 3.1). Collectively, these data suggest that the biosynthetic pathways for lipid A may not be unique to a particular genomovar but differential lipid A acylation and glycosylation by Ara*p*4N may depend on environmental pressures.

Both lipid A acylation and substitution by L-Ara*p*4N are pivotal for the survival of Gram-negative bacteria against host response. L-Ara*p*4N substitution in lipid A and the LPS inner core (see below) contributes to the overall net charge of the OM and plays a key role in pathogenesis. In fact, LPS from *Burkholderia* are frequently positively charged or in an isoelectric state. This is closely related to the abundance of L-Ara*p*4N residues present in the lipid A-inner core and confers resistance to antibiotic compounds and host cationic antimicrobial peptides. Under physiological conditions, L-Ara*p*4N moieties are positively charged and reduce the negative net charge on the OM, weakening the ionic attraction for respiratory tract antimicrobial defensins. The inherent resistance of these microorganisms to polymyxin B (PmB), a cyclic polycationic antibiotic peptide that possesses high affinity for negatively charged bacterial LPSs, is thought to depend on the L-Ara*p*4N content which prevents an increased permeability in the OM produced by PmB. Interestingly, other frequently occurring cationic substituents (e.g. EtN) have not been reported as substituents of lipid A from *B. cepacia*, suggesting a unique role for L-Ara*p*4N. Hence, the transferase which attaches L-Ara*p*4N residues to lipid A and earlier enzymatic steps involved in L-Ara*p*4N biosynthesis may be novel targets for future developments of drugs against *B. cepacia*. The gene cluster encoding L-Ara*p*4N biosynthesis enzymes has been mapped and shown to be essential for the viability of *Burkholderia cenocepacia*. Using a conditional mutation strategy with rhamnose supplementation in growth media, the group of M. Valvano demonstrated dramatic changes in bacterial cell morphology and ultrastructure and bacterial permeability and osmotic sensitivity, suggesting a fundamental cell membrane defect in the absence of L-Ara*p*4N (Ortega *et al.*, 2007).

4. PLANT PATHOGENIC *AGROBACTERIUM* AND *XANTHOMONAS* LPS AND THE ACTIVATION OF INNATE IMMUNE RESPONSE IN PLANTS

In animal and insect cells, innate immune defences are triggered by the perception of pathogen-associated molecular patterns (PAMPs), conserved and generally indispensable microbial structures including LPSs. The recognition of PAMPs by these cells is often mediated by LRR (leucine-rich repeat) proteins such as Toll in *Drosophila* and the Toll-like receptors (TLR) in mammals (see Chapter 31). Recognition of LPS occurs through the lipid A moiety which is responsible for most of the biological effects of LPSs in animals. Lipid A toxicity in animals strongly depends on its structure and is also influenced by the covalently linked core region, which possesses immunogenic properties. Lipopolysaccharides apparently play diverse roles in bacterial pathogenesis of plants. It can be recognized by plants to elicit or potentiate plant defence-related responses as part of a group of general elicitors that include flagellin and other pathogen surface effectors such as peptidoglycan.

In comparison with animal and human cells, little is known about the mechanisms of LPS perception by plants and cognate signal transduction pathway. The receptor for LPS in plant cells is not identified and cloning and characterization of LPS receptors in plants is currently a major goal in this area.

Despite the wealth of information on the structures of LPS and lipid A from *Rhizobiaceae* (Raetz *et al.*, 2007), no complete information was present until recently on *Agrobacterium* LPS and lipid A. The latter is composed of two main lipid A species (see Table 3.1), the first one of which is bearing two unsubstituted 3-OH(14:0) in ester linkage, two 3-OH(16:0) or one 3-OH(16:0) and one 3-OH(18:1) in amide linkage, one 27-OH(28:0) linked to 3-OH(16:0) on GlcN II, and one 3-OH(4:0) linked to 27-OH(28:0). The second species, present in minor amounts, is missing a 3-OH(14:0) on GlcN I. Other species are deriving from these two and lack phosphate or 3-OH(14:0).

Recently, the lipid A isolated from LPS of *Xanthomonas campestris* was identified. It consists of a *bis*-phosphorylated β-(1′→6)-linked D-Glc*p*N disaccharide backbone and, in the main hexa-acyl species, two unsubstituted 3-OH(12:0) and/or 3-OH(13:0) were linked to the disaccharide backbone in amide linkage and two primary fatty acids with different chain length [3-OH(10:0), 3-OH(11:0), 3-OH(12:0), or 3-OH(13:0)] were present in ester-linkage at C3 and C3′ of the Glc*p*N residues. These acyl chains were further substituted at their 3-hydroxy groups by 10:0 and 11:0 (see Table 3.1).

The importance of LPSs in plant pathogenesis has been recognized in past years, but the mechanisms involved are far from being fully understood, although there are interesting parallels with animals (Zipfel and Felix, 2005; Jones and Dangl, 2006; Newman *et al.*, 2007). Plants have evolved and maintained competence to recognize several general elicitors which are pathogen surface molecules and can thus be considered PAMPs. They bind to plant pattern recognition receptors and trigger the expression of immune response genes and the production of antimicrobial compounds. Lipopolysaccharides belong to such surface compounds that act as general elicitors of plant innate immunity (Dow *et al.*, 2000; Newman *et al.*, 2002, 2007; see also Chapter 40). The activation of plant innate immunity responses upon recognition of PAMPs resembles the activation mechanisms in mammals and insects (Medzhitov, 2001). The invasion of a bacterium activates a signal transduction pathway which may lead to the hypersensitive response (HR). The HR is a programmed cell death response, triggered by live bacteria and associated with plant host resistance. It is associated with a decline of the number of viable bacteria recovered in the tissue and follows a rapid necrosis of plant tissue representing the final stage of resistance, when stress signals induce strong defensive responses. One of the most widely studied effects of LPSs on plant cells is their ability to retard or completely block HR induced by avirulent bacteria. Pre-inoculation of leaves with heat-killed bacterial cells delays or prevents the disease symptoms expected with the subsequent inoculation with living bacteria (Dow *et al.*, 2000; Newman *et al.*, 2002).

Recent findings have shed light on the molecular aspect of these biological events. The lipid A moiety may be at least partially responsible for LPS perception by *Arabidopsis thaliana* leading to a rapid burst of nitric oxide, a hallmark of innate immunity in animals (Zeidler *et al.*, 2004). The minimal structural requirements for the elicitor activity can be different from a mammalian host. In the case of LOS, both lipid A and core oligosaccharide from a pathogenic bacterium can be recognized by plant receptors (Silipo *et al.*, 2005d) and are potent inducers of immune responses.

Consequently, a single pathogen-associated compound can cause multiple signals indicating the existence of multiple receptors of several general elicitors. Intact LOS, the lipid A and core oligosaccharide derived from *X. campestris* pv. *campestris* are all able to induce the defence-related genes *PR1* and *PR2* in *Arabidopsis* and to prevent HR caused by avirulent bacteria. Furthermore, these defensive cellular responses occur via a distinctive recognition event that occurs in different temporal phases (Silipo *et al.*, 2005d). As in the case of mammalian pathogenic bacteria, the structural variation in the lipid A region can strongly influence the ability of LPS to induce defence-related genes and to be recognized as PAMPs (Silipo *et al.*, 2008). The alteration of the acylation pattern and the modification of the polar heads seem to play a crucial role.

5. STRUCTURAL ELUCIDATION OF LIPID A

The complete structural elucidation of lipid A is of pivotal importance toward the full understanding of its biological properties, i.e. an (ant)agonistic action in the elicitation of the host innate immune response.

The lipid fraction obtained by precipitation after mild acid treatment of LPS (or LOS) is constituted of a mixture of intrinsically heterogeneous lipid A molecules. Any structural approach applies intensively mass spectrometric analyses by matrix-assisted laser desorption ionization time-of-flight mass spectrometry (MALDI-TOF-) and electrospray ionization (ESI-) mass spectrometry (MS) techniques on both the native and the selectively degraded lipid A fractions (see Chapter 13). The MALDI-TOF- and ESI-MS data give insight into the number of lipid A species present in the isolate, the presence of polar head groups and the distribution of acyl residues on each Glc*p*N units of the disaccharide backbone (Que *et al.*, 2000). This latter issue is of particular importance given that fatty acid composition and distribution strongly affect the endotoxic activity. Through the chemical and spectroscopical study of several lipids A and LPSs in our institutions, novel general approaches to gain structural insight in such molecules have been developed. In particular, combining MALDI-TOF- or ESI-MS and selective chemical degradation by ammonium hydroxide hydrolysis, we have developed a general and easy methodology to obtain the secondary fatty acid distribution, which is one of the most demanding issues in the structural determination of lipid A (Silipo *et al.*, 2002; Sforza *et al.*, 2004; Sturiale *et al.*, 2005). These approaches must be supported by classical chemical analysis of lipid A which implies the removal of ester-linked fatty acids by mild alkaline treatment or the complete removal of fatty acids by strong alkaline treatment (Holst, 2000) and gas-liquid chromatography-mass spectrometry (GLC-MS) analyses of fatty acids as methyl ester derivatives (Rietschel, 1976; Zähringer *et al.*, 1994)

NMR spectroscopy is still less applied but equally important (Ribeiro *et al.*, 1999; Brecker, 2003; Silipo *et al.*, 2004a). The amphiphilic nature of lipid A entails a very poor solubility of the sample in many solvent systems, however, when possible, it allows obtaining diagnostic information mainly about the saccharide backbone, i.e. the anomeric configuration of the carbohydrate residues and their sequence and the occurrence and position of the polar head groups.

6. GENERAL STRUCTURAL FEATURES OF THE CORE REGION

The structures of the core regions of LPSs have been reviewed in a number of overviews in the past 15 years (Holst and Brade, 1992; Knirel and Kochetkov, 1993; Holst, 1999, 2002, 2007; Bystrova *et al.*, 2006; Knirel *et al.*, 2006; De Soyza *et al.*, 2008). Owing to this and to space limitations, principles of core structures are given in this chapter, illustrated by structures published after the year 2002 and not yet included in the earlier reviews.

Chemical variation in core region structures occurs in a more limited range than in O-specific polysaccharides. Only one structural element is present in all core regions, which is the particular α-linked 3-deoxy-D-*manno*-oct-2-ulopyranosonic acid (Kdo) (Unger, 1983) residue that binds the core region to the lipid A. Several core regions possess in addition L-*glycero*-D-*manno*-heptopyranose (L,D-Hep) and the oligosaccharide L-α-D-Hep-(1→7)-L-α-D-Hep-(1→3)-L-α-D-Hep-(1→5)-[α-Kdo-(2→4)]-α-Kdo (Hep III, Hep II, Hep I, Kdo II and Kdo I, respectively), substitutions of which furnish structural variability. Other substituents may be sugars or phosphate residues and acetyl and amino acid residues may be present also. In addition to or instead of L,D-Hep, LPSs may contain D-*glycero*-D-*manno*-heptopyranose (D,D-Hep) which is the biosynthetic precursor of L,D-Hep. Several LPSs lack heptose residues at all. Both, Kdo I and II can be replaced by the stereochemical similar sugar D-*glycero*-D-*talo*-oct-2-ulopyranosonic acid (Ko) (Gass *et al.*, 1993), of which neither the biosynthesis nor the regulation of its exchange with Kdo have been elucidated to date. Other LPSs possess only Kdo I which may then be phosphorylated at O-4.

7. CORE STRUCTURES OF VARIOUS BACTERIA

7.1. Enterobacterial core structures

Presently, there are two types of enterobacterial core regions known, i.e. the *S. enterica* core regions and the core regions different to the *Salmonella* type. In the first, the common structural element L-α-D-Hep-(1→7)-L-α-D-Hep-(1→3)-L-α-D-Hep-(1→5)-α-Kdo is present which is substituted at O-3 of the second heptose by glucopyranose (Glc*p*). Heptose residues I and II are phosphorylated and O-4 of Hep I is not substituted by a saccharide. In the other core type, the common partial structure L-α-D-Hep-(1→7)-L-α-D-Hep-(1→3)-L-α-D-Hep-(1→5)-α-Kdo is present which is not substituted at O-3 of Hep II by Glc and in which heptose residues are not generally phosphorylated. Position O-4 of Hep I is substituted by a hexose residue or oligosaccharide. As examples for each type, the core regions of LPSs from *E. coli* R4 (in LPS from *E. coli*, five core regions of the *S. enterica* core type are known) and *Yersinia pestis* are shown in Figure 3.3. A detailed review on enterobacterial LPS core regions has been published recently (Holst, 2007).

7.2. Core structures of LPS from *Pasteurellaceae*

Of the family *Pasteurellaceae* (γ-proteobacteria), a good number of core structures have been published since the year 2002 from LPSs of the genera *Haemophilus*, *Histophilus* and *Pasteurella* and *Haemophilus somnus* (Cox *et al.*, 2002b, 2003a; Inzana *et al.*, 2002; Månsson *et al.*, 2002, 2003; Müller-Loennies *et al.*, 2002a; Bouchet *et al.*, 2003; Schweda *et al.*, 2003; Harper *et al.*, 2004; St Michael *et al.*, 2004a, 2005a,b,c,d, 2006; Tinnert *et al.*, 2005; Yildrim *et al.*, 2005; Lundström *et al.*, 2007).

In *Haemophilus influenzae*, the core region possesses as common partial structure the saccharide L-α-D-Hep*p*-(1→2)-[*P*Etn→6]-L-α-D-Hep*p*-(1→3)-[β-D-Glc*p*-(1→4)]-L-α-D-Hep*p*-(1→5)-[*PP*-Etn→4)]-α-Kdo (Figure 3.4). Generally, only this one Kdo residue is present that links the core region to lipid A. Further substitutions occur mainly at the β-D-Glc*p* and Hep III residues. In particular, core regions of non-typeable *H. influenzae* strains have been investigated

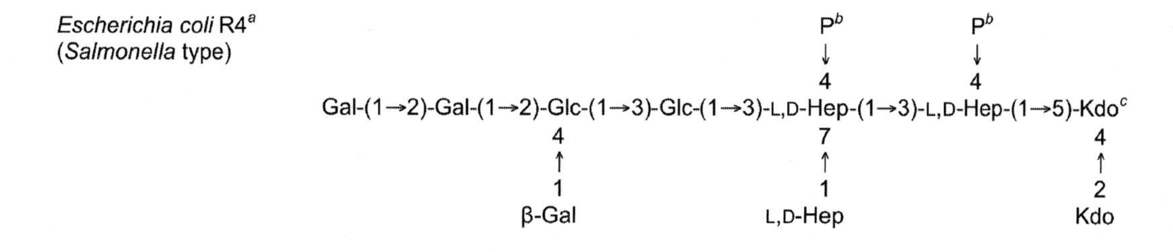

Yersinia pestis[d] (other than *Salmonella* type)

Various strains β-Gal or D,D-Hep-(1→7)-L,D-Hep

```
                    β-Gal or D,D-Hep-(1→7)-L,D-Hep
                                        1
                                        ↓
                                        7
  β-GlcNAc^e-(1→3)-L,D-Hep-(1→3)-L,D-Hep-(1→5)-Kdo^b
                                   4          4
                                   ↑          ↑
                                   1          2
                                 β-Glc    Kdo or Ko
```

FIGURE 3.3 Two examples of core region structures of enterobacterial LPS. Where not stated otherwise, sugars are α-D-pyranosides. [a] Müller-Loennies *et al.* (2002b). [b] Product obtained after de-*O*-acylation. Further substituents are not known. [c] This residue was identified to link the core region to GlcN II of lipid A. [d] Knirel *et al.* (2005). [e] Non-stoichiometric substitution.

in the past 5 years which are excellently summarized in a very recent review (Schweda *et al.*, 2008). Additionally, the functions of several biosynthetic genes (Griffin *et al.*, 2003; Fox *et al.*, 2005) and of the glycosyltransferase LpsA (Deadman *et al.*, 2006) have been elucidated.

The core region of *Histophilus somni* (*Haemophilus somnus*) strains possesses instead an α-Kdo-(2→4)-α-Kdo disaccharide and the common structure L-α-D-Hep*p*-(1→3)-[β-D-Glc*p*-(1→4)]-L-α-D-Hep*p*-(1→5)-α-Kdo. The β-D-Glc*p* is further substituted and Hep II carries either α-D-Glc*p*NAc or D-Gal*p* at O-2. Phosphocholine is not present, but Hep II may be substituted by one or two PEtn residues (see Figure 3.4). *N*-acetylneuraminic acid may be incorporated, leading to serum resistance and reduction of antibody binding (Inzana *et al.*, 2002).

The common partial structure of the core region of LPS from *Pasteurella multocida* may be described as L-α-D-Hep*p*-(1→2)-L-α-D-Hep*p*-(1→3)-[β-D-Glc*p*-(1→4)]-[α-D-Glc*p*-(1→6)]-L-α-D-Hep*p*-(1→5)-α-Kdo. Interestingly, an α-Kdo-(2→4)-α-Kdo disaccharide and Kdo-4→*PP*-Etn were present in the core region of strain VP161. Due to this variability, two acceptor-specific heptosyl I transferases are required in order to link Hep I either to the α-Kdo-(2→4)-α-Kdo-containing core or to the Kdo-4→*PP*-Etn one (Harper *et al.*, 2007). Core extensions which include a fourth L,D-Hep residue occur at the (1→4)-linked β-D-Glc*p* residue (Figure 3.5).

7.3. Core structures of LPSs from *Pseudomonadaceae*

7.3.1. Pseudomonas

The core regions of LPSs of *Pseudomonas aeruginosa* possess a Gal*p*N residue that is mostly amidated by alanine (in a few cases, it is present as Gal*p*NAc) and which substitutes (O-3 of) Hep II of the trisaccharide L,D-Hep-(1→3)-L,D-Hep-(1→5)-Kdo and two Kdo residues are present in the inner core region. The Hep II residue is further substituted at O-7 by a carbamoyl residue.

Most strikingly is the fact that the core regions are highly phosphorylated. Figure 3.6 presents some examples of these structures. Publications of the last six years (Bystrova *et al.*, 2003, 2004; Kooistra *et al.*, 2003; Choudhury *et al.*, 2005, 2008) and two recently published overviews (Bystrova *et al.*, 2006; Knirel *et al.*, 2006) expatiate on this field. Similar structures were identified in the core region of *Pseudomonas stutzeri* OX1 (Leone *et al.*, 2004a,b), however, no alanine was present. As a novel feature of a core region, two residue of pyruvic acid were identified (see Figure 3.6).

Another core structure which possesses features very similar to *Ps. aeruginosa* cores was identified in the LPS of *Pseudomonas syringae* pv. *phaseolicola* (Zdorovenko *et al.*, 2004). A completely unusual core structure lacking heptose and phosphate residues was identified in *Pseudomonas cichorii* which belongs to the RNA group I of *Pseudomonadaceae*, as do *Ps. aeruginosa*, *Ps. stutzeri* and *Ps. syringae* (De Castro *et al.*, 2004).

7.3.2. Burkholderia

The genus *Burkholderia* represents the RNA group II of *Pseudomonadaceae* and possesses other and significant structural features in its LPS core regions which are of chemotaxonomic value (Isshiki *et al.*, 2003; De Castro *et al.*, 2005). First of all, all core regions that have been identified so far possess in various amounts both the disaccharides α-Ko-(2→4)-α-Kdo (which has not yet been found in other genera of *Pseudomonadaceae*

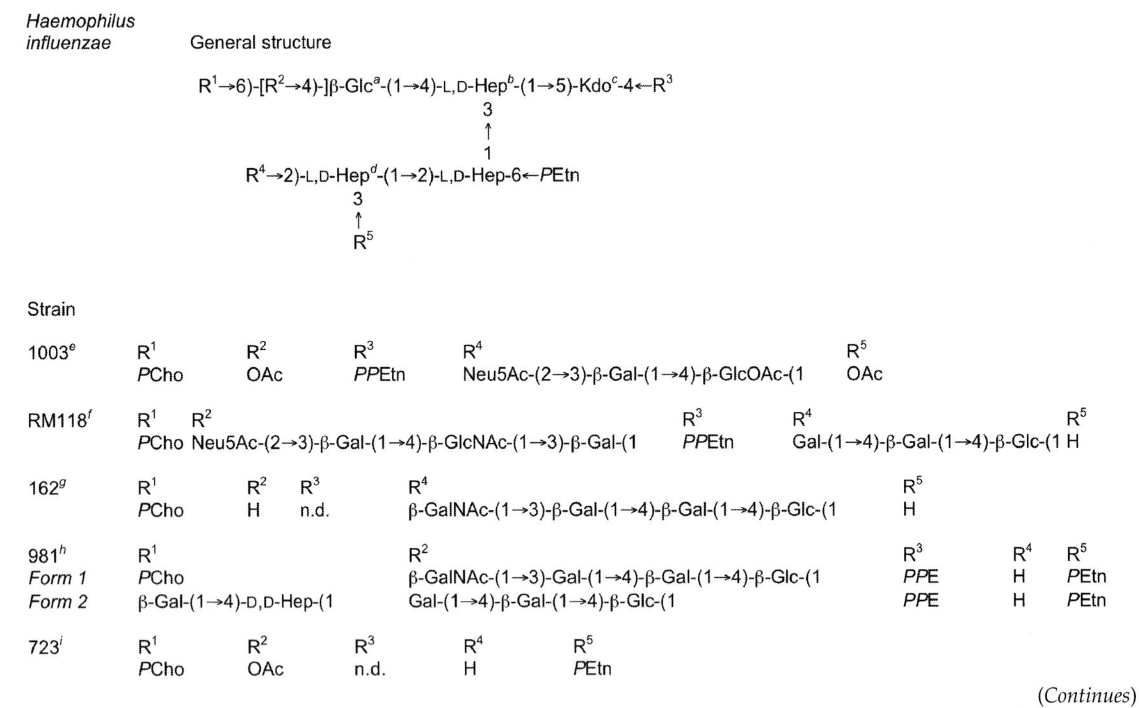

(*Continues*)

FIGURE 3.4 The structures of core regions of LPS from *Haemophilus*. Where not stated otherwise, sugars are α-D-pyranosides. Abbreviations: *P*Cho, 2-trimethylaminoethanol phosphate, *P*Etn, 2-aminoethanol phosphate, *PP*Etn, 2-aminoethanol diphosphate. [a] *O*-Acetylated at position 3 in strain 723. [b] *O*-Acetylated at position 2 in strain 723. [c] This residue was identified to link the core region to GlcN II of lipid A. [d] *O*-Acetylated at unknown position in strain 723. [e] Månsson *et al.* (2002a). [f] Cox *et al.* (2002b). [g] Schweda *et al.* (2003). [h] Tinnert *et al.* (2005). Månsson *et al.* (2002b). [i] Yildrim *et al.* (2005). [j] Lundsträm *et al.* (2007). [k] Cox *et al.* (2003a). [l] St Michael *et al.* (2004a). [m] St Michael *et al.* (2005b). [n] St Michael *et al.* (2006).

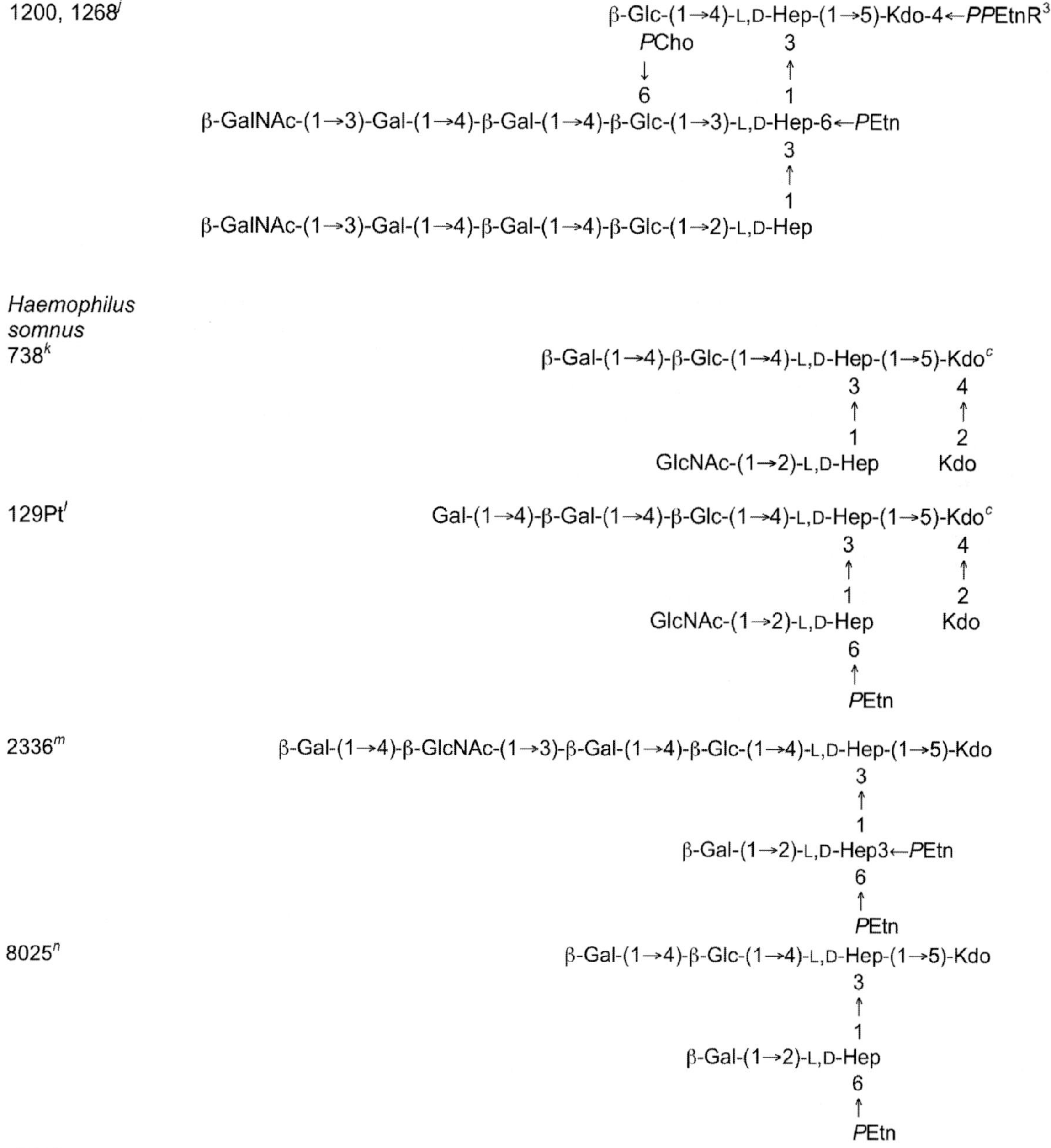

FIGURE 3.4 Continued

but in some enterobacteria, i.e. *Y. pestis* and *Serratia marcescens*) and α-Kdo-(2→4)-α-Kdo (see Figure 3.6). The common partial structure is L-α-D-Hep*p*-(1→7)-L-α-D-Hep*p*-(1→3)-[β-D-Glc*p*-(1→4)]-L-α-D-Hep*p*-(1→5)-α-Kdo. In *B. caryophylli*, O-3 of Hep II is substituted by a branched glycan to which one of the O-specific polysaccharides is linked (the caryan) (De Castro *et al.*, 2005).

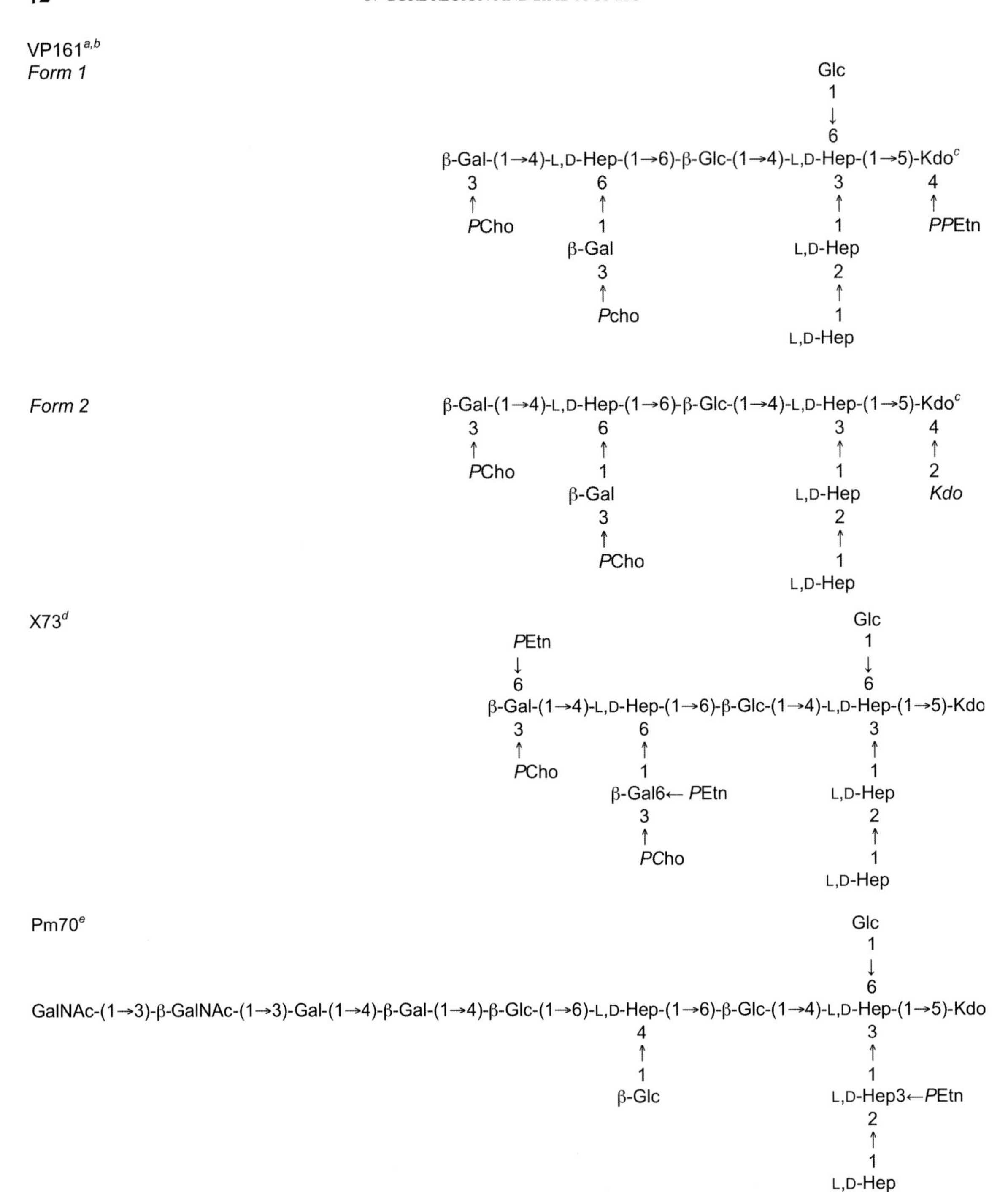

FIGURE 3.5 The structures of core regions of LPS from *Pasteurella multocida*. Where not stated otherwise, sugars are α-D-pyranosides. [a] St Michael *et al.* (2005a). [b] Harper *et al.* (2007). [c] This residue was identified to link the core region to GlcN II of lipid A. [d] St Michael *et al.* (2005c). [e] St Michael *et al.* (2005d).

This core region possessed as a novel feature two L-α-D-Hep*p*-(1→5)-α-Kdo moieties. The core region of the rough-type LPS of *Burkholderia pyrrocinia* possessed a complex oligosaccharide in which the Kdo unit was substituted at O-5 by the pentasaccharide α-Hep-(1→7)-[α-Rha-(1→2)]-α-Hep-(1→3)-[β-Glc-(1→4)]-α-Hep-(1→, a saccharide common in LPS of *B. cepacia* (Silipo *et al.*, 2006). The last heptose residue of this moiety was further substituted by a heptose trisaccharide, giving rise to a heptan pentasaccharide representing the remote region of the outer core. An excellent review on structural and biological features of LPS from *Burkholderia* has been published recently (De Soyza *et al.*, 2008).

7.3.3. Xanthomonas

It was shown that the oxidative burst in tobacco plants initiated by LOS from *X. campestris* pv. *campestris* was provoked by the inner core region (Braun *et al.*, 2005). The complete structure of the LOS was characterized (Silipo *et al.*, 2005d) (see Figure 3.6) and it was shown that plant cells recognize both core region and lipid A, however, these moieties induced defence-related gene transcription at different times. Later, even smaller core oligosaccharides from this LOS were shown to induce an oxidative burst in tobacco cells (Kaczynski *et al.*, 2007).

7.4. Core structures of LPSs from *Neisseria*

Additional information on the core structures from LPS of *Neisseria meningitidis* strain BZ157 galE (two *P*Etn residues at Hep I) and various immunotypes (the presence of glycine on O-7 of Hep II) have been published (Cox *et al.*, 2002a,c). A novel core structure was identified in strain 1000 (Cox *et al.*, 2003b) (Figure 3.7).

Pseudomonas aeruginosa[a]

PA103 *wbjE* mutant[b]

CONH2 ↓ 7 (on first L,D-Hep)

L-Rha-(1→6)-Glc-(1→4)-GalN[c]-(1→3)-L,D-Hep-(1→3)-L,D-Hep-(1→5)-Kdo

3 ↑ 1 (on GalN)

Glc-(1→6)-β-Glc

Pseudomonas syringae pv. *phaseolicola*[d]
Glycoform 1

CONH2 ↓ 7 (on first L,D-Hep); *P* ↓ 4, *P*Etn[e] ↓ 2 (on second L,D-Hep)

β-GlcNAc-(1→2)-Glc-(1→3)-GalNAla-(1→3)-L,D-Hep-(1→3)-L,D-Hep-(1→5)-Kdo[f]

4 ↑ 1 (on GalNAla): L-Rha-(1→6)-β-Glc; 6 ↑ *P* (on first L,D-Hep); 4 ↑ 2 (on Kdo): Kdo

(*Continues*)

FIGURE 3.6 The structures of core regions of LPS from *Pseudomonadaceae*. [a] Structure published after 2006. [b] Choudhury *et al.* (2008). [c] The amino function is either acetylated or alanylated. [d] Zdorovenko *et al.* (2004). [e] Non-stoichiometric substitution. [f] This residue was identified to link the core region to GlcN II of lipid A. [g] Leone *et al.* (2004a). [h] De Castro *et al.* (2004). [i] Silipo *et al.* (2006). [j] Isshiki *et al.* (2003). [k] Silipo *et al.* (2005d). [l] Kaczyski *et al.* (2007). [m] R, phosphoramide or galacturonic acid-phosphate-1 (GalA-1*P*).

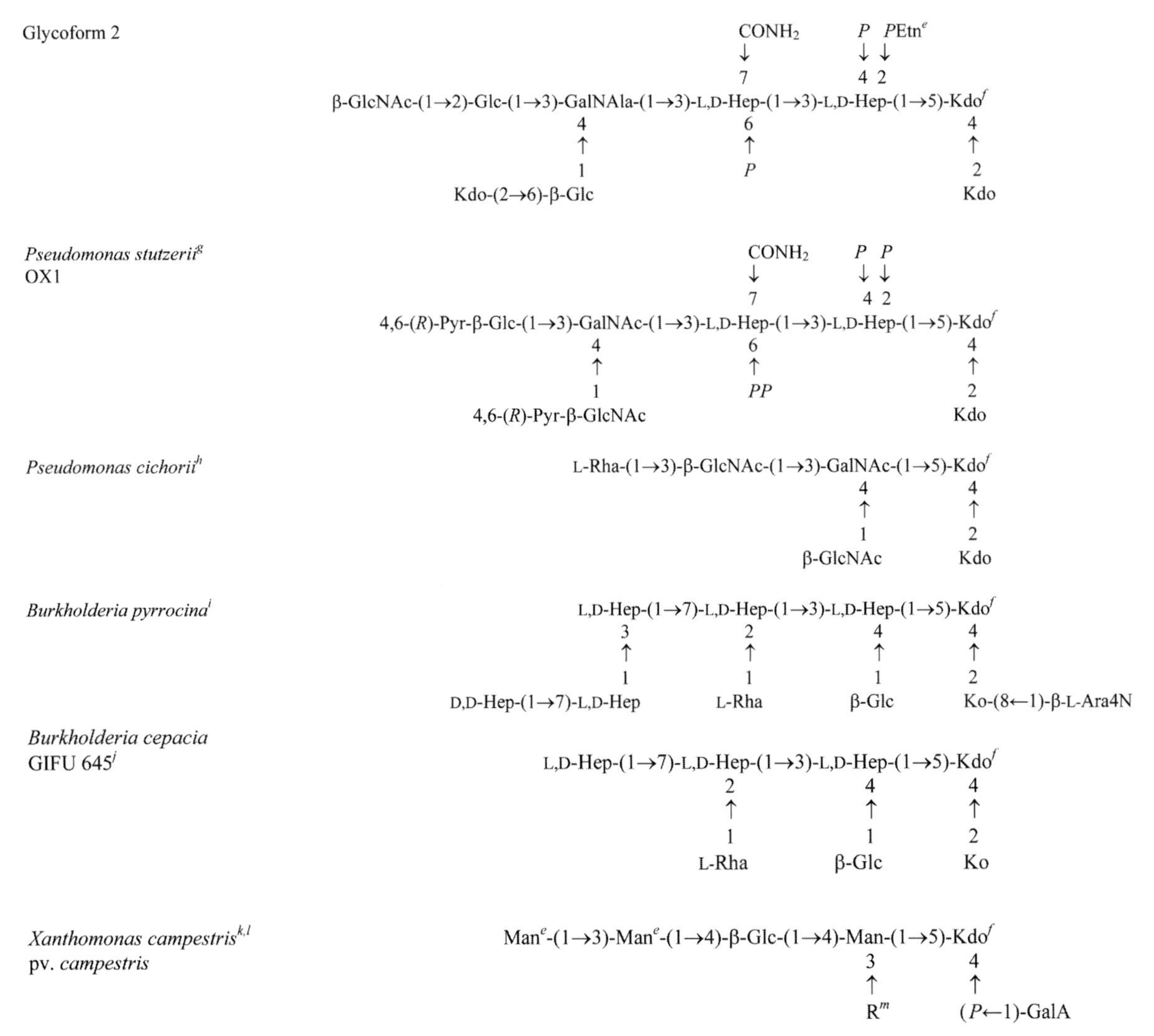

FIGURE 3.6 Continued

7.5. Core structures of LPSs from *Acinetobacter*

The core region of the LPS from *Acinetobacter radioresistens* S13 belonged to the Ko-lacking cores of *Acinetobacter* (Leone *et al.*, 2006) and possessed the branched trisaccharide α-Kdo-(2→4)-[α-Kdo-(2→5)-]α-Kdo, as found earlier in *Acinetobacter baumannii* strain ATCC 19606 (Vinogradov *et al.*, 2002a) and a Glc*p*-rich outer core region. In the core region of *Acinetobacter lwoffii* F78, the *Chlamydia*-specific epitope Kdo-(2→8)-Kdo was identified by chemical and serological methods, suggesting an explanation for the cross-reactivity between *Chlamydia* and *Acinetobacter* that had been described earlier (Brade and Brunner, 1979) (Figure 3.8).

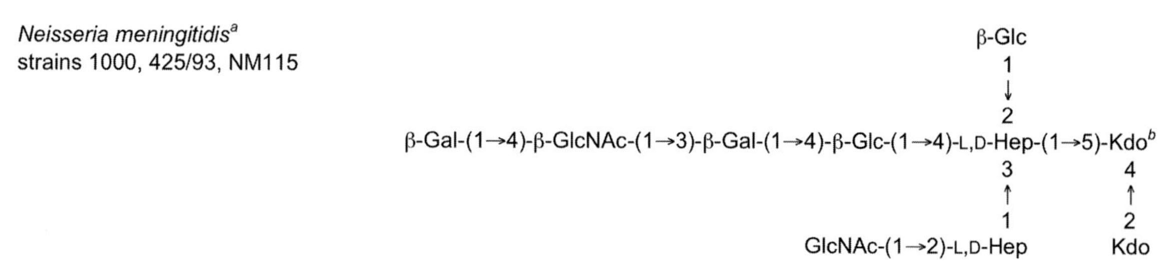

FIGURE 3.7 The structure of core regions of LPS from *Neisseriaceae*. [a] Cox *et al.* (2003b). [b] This residue was identified to link the core region to GlcN II of lipid A.

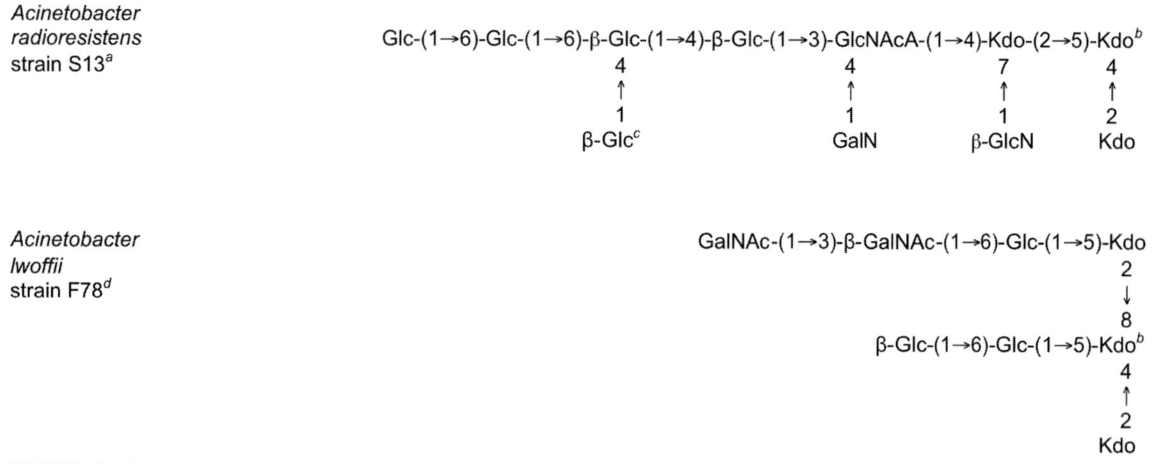

FIGURE 3.8 The structures of core regions of LPS from *Acinetobacter*. [a] Leone *et al.* (2006). [b] This residue was identified to link the core region to GlcN II of lipid A. [c] Non-stoichiometric substitution. [d] Hanuszkiewicz *et al.* (2008).

7.6. Core structures of LPSs from various families

7.6.1. Shewanella

This genus is of particular interest since it has the ability to take up various metals which are then excreted in an altered state. Thus, *Shewanella* bacteria are important for the cleaning of contaminated areas. Several core structures of LPSs from different *Shewanella* spp. have been published (Figure 3.9), two of which possessed the novel sugar 8-amino-Kdo4*P* which linked the core to the lipid A (Vinogradov *et al.*, 2003, 2004a), however, in *Shewanella putrefaciens* CN 32, an "ordinary" Kdo4*P* was identified (Vinogradov *et al.*, 2002b). The absolute configuration of 8-amino-Kdo has been identified recently (Leone *et al.*, 2007). To either residue a D,D-Hepp was linked at O-5. Further core elongations proceeded from various positions of this residue.

7.6.2. Alteromonadaceae

Two core structures from the LPSs of *Alteromonas macleodii* and *Pseudoalteromonas carrageenovora* have been elucidated (Silipo *et al.*, 2005b; Liparoti *et al.*, 2006). Both were of the deeper (3 and 4 sugar residues, respectively) rough type. The most prominent feature was the presence of a (terminal) β-linked Kdo residue which has been identified in LPS for the first time (see Figure 3.9).

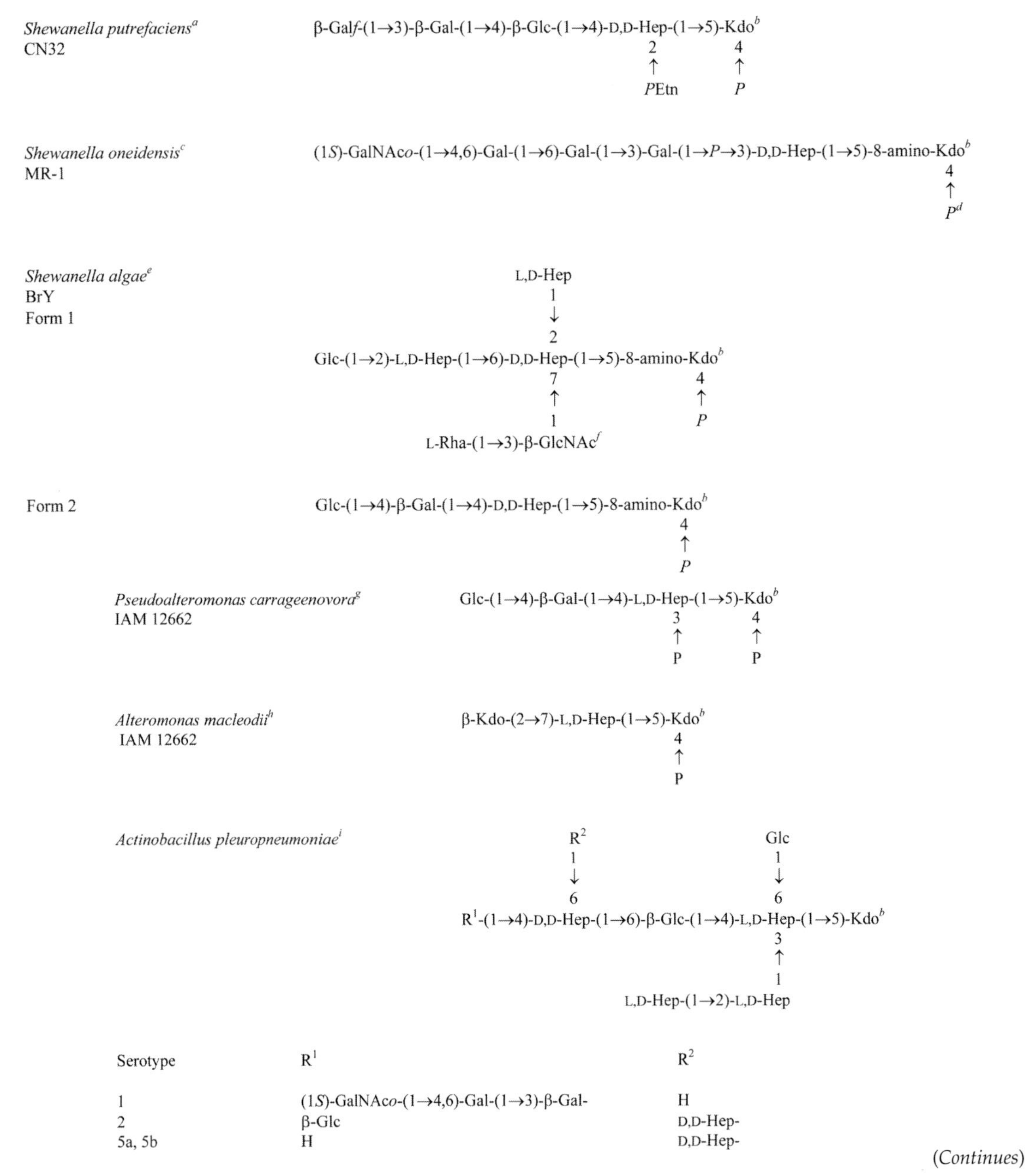

Serotype	R^1	R^2
1	(1S)-GalNAco-(1→4,6)-Gal-(1→3)-β-Gal-	H
2	β-Glc	D,D-Hep-
5a, 5b	H	D,D-Hep-

(Continues)

FIGURE 3.9 The structures of core regions of LPS from various species. [a] Vinogradov *et al.* (2002b). [b] This residue was identified to link the core region to GlcN II of lipid A. [c] Vinogradov *et al.* (2003). [d] Non-stoichiometric substitution by ethanolamine-phosphate. [e] Vinogradov *et al.* (2004a). [f] Non-stoichiometric substitution by Ac. [g] Silipo *et al.* (2005b). [h] Liparoti *et al.* (2006). [i] St Michael *et al.* (2004b). [k] Wang *et al.* (2006b). [l] Hashii *et al.* (2003). [m] 5-acetamido-7-(*N*-acetyl-D-alonyl) amino-3,5,7,9-tetradeoxy-D-galacto-non-zulosonic acid (Non5Ac7Ala), [n] Kay *et al.* (2006). [o] Molinaro *et al.* (2003b). [p] De Castro *et al.* (2006a). [q] De Castro *et al.* (2006a). [r] Zähringer *et al.* (2004). [s] Silipo *et al.* (2005c). [t] Non-stoichiometric substitution. [u] Pieretti *et al.* (2008). [v] Vinogradov *et al.* (2004b).

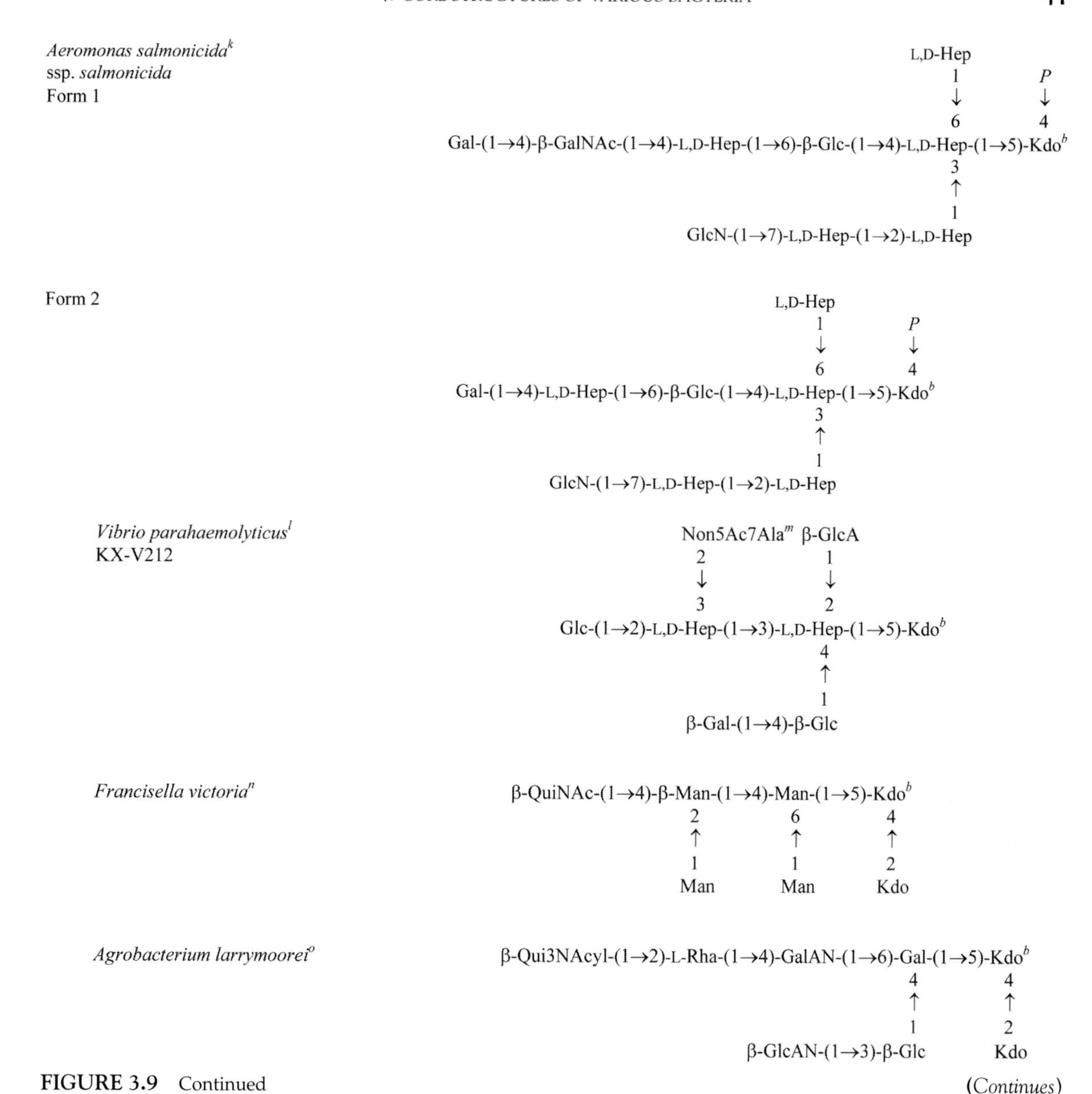

FIGURE 3.9 Continued

(Continues)

7.6.3. Actinobacillus

Four structures have been published in the past 6 years, namely those of the LPSs from *Actinobacillus pleuropneumoniae* serotypes 1, 2, 5a and 5b (St Michael *et al.*, 2004b). Their common feature is the oligosaccharide L-α-D-Hep*p*-(1→2)-L-α-D-Hep*p*-(1→3)-[D-α-D-Hep*p*-(1→6)-β-D-Glc*p*-(1→4)]-[α-D-Glc*p*-(1→6)]-L-α-D-Hep*p*-(1→5)-α-Kdo which is differently substituted at the D,D-Hep*p* residue (see Figure 3.9).

Agrobacterium tumefaciens[p]
A1

Man-(1→6)-Man-(1→5)-Kdo[b]
4
↑
2
β-Gal-(1→8)-Kdo

A. tumefaciens[q]
TT111

L-Rha[t]-(1→3)-L-Rha-(1→2)-L-Rha-(1→2)-L-Rha-(1→3)-L-Rha-(1→3)-Man-(1→5)-Kdo[b]
4
↑
2
β-GlcN[t]-(1→4)-β-Gal-(1→4)-Kdo-(8←1)-β-Gal

Bartonella henselae[r]
ATCC49882

Glc-(1→5)-Kdo[b]
4
↑
2
Kdo

Arenibacter certesii[s]
KMM3941

Rha[t]-(1→3)-Rha-(1→6)-Man-(1→6)-Man-(1→5)-Kdo[b]
4
↑
(*P*←1)-GalA

Halomonas pantellerensis[u]

P
↓
4
Glc-(1→4)-GlcNAc-(1→4)-β-Glc-(1→3)-L,D-Hep-(1→5)-Kdo[b]
4
↑
1
Glc-(1→7)-D,D-Hep-(1→2)-L,D-Hep-(6←1)-L,D-Hep[t]

Geobacter sulfurreducens[v]

L-Fuc-(1→6)-β-Man-(1→4)-L,D-Hep-(1→5)-Kdo[b]
3
↑
1
3-*O*-Me-L-QuiNAc[t]-(1→3)-GlcNAc-(1→2)-β-Man-(1→3)-β-ManNAc-(1→3)-L,D-Hep
7
↑
1
Glc-(1→7)-L,D-Hep

FIGURE 3.9 Continued

7.6.4. *Vibrionaceae*

The core region of *Aeromonas salmonicida* ssp. *salmonicida* (Wang *et al.*, 2006a) possesses a cluster of four L,D-Hep*p* residues, Hep*p* I of which is α-(1→5)-linked to Kdo4*P* and in turn carries the disaccharide L-α-D-Hep*p*-(1→6)-β-D-Glc*p* at O-4. The heptose residue of this unit is differently substituted at O-4, leading to heterogeneity (see Figure 3.9).

Another but incomplete structure was identified in the LPS from *Vibrio parahaemolyticus* isolated from a patient, containing alanylated nonulosonic acid (Hashii *et al.*, 2003) (see Figure 3.9).

7.7. Various genera

The core structures of the LPSs from the fish pathogen *Francisella victoria* (Kay *et al.*, 2006), the plant pathogens *Agrobacterium larrymoorei* (Molinaro *et al.*, 2003b) and *Agrobacterium tumefaciens* A1 and TT111 (De Castro *et al.*, 2006a,b), the zoonotic pathogen *Bartonella henselae* (Zähringer *et al.*, 2004), the marine bacterium *Arenibacter certesii* (Silipo *et al.*, 2005c), all of which possessed no heptose residues, and the heptose-containing core structures of the LPSs from the haloalkaliphilic bacterium *Halomonas pantellerensis* (Pieretti *et al.*, 2008) and the iron- and sulfur-reducing species *Geobacter sulfurreducens* (Vinogradov *et al.*, 2004b) are summarized in Figure 3.9.

8. CONCLUSIONS

From the chemical point of view, the lipid A structure still remains a rather conserved architectural principle within the LPSs of Gram-negative bacteria with only a low number of variants. Thus far, a β-(1→6)-linked amino sugar disaccharide has been found with no exception and the core oligosaccharide is linked to O-6 of the second (non-reducing) residue.

As for polar head groups, the principle of a charged lipid A is still valid, since phosphate residues are present in a good number of structures and may further be esterified (or replaced) by Gal*p*A. In some examples, these negative charges are counterbalanced through the esterification of the phosphate residue with positively charged groups as EtN, Ara*p*4N or Gal*p*N. An important but single exception of the dogma of a charged lipid A has been recently described in the LPS of *B. bacteriovorus* in which a totally neutral lipid A has been identified, containing α-D-mannose residues that stoichiometrically replace phosphate groups (Schwudke *et al.*, 2003).

With regard to the fatty acids, there is considerable variability either in length or in their number, to which is also correlated the biological activity of the lipid A. Both primary and secondary fatty acids between 10 and 16 carbon atoms represent the major species.

From the biological point of view, lipid A represents the endotoxic principle of the LPS and, thus, is responsible for LPS perception and biological activity in any system, i.e. vegetal or animal.

Since 2002, a good number of core structures from LPS of various bacterial species, including several which live under rather extreme environmental conditions, have been elucidated. The general principle of a negatively charged core region contributing to the rigidity of the Gram-negative cell wall through intermolecular cationic cross-links is still valid. Many core regions comprise as partial structure a tetrasaccharide L,D-Hep-(1→7)/(1→2)-L,D-Hep-(1→3)-L,D-Hep-(1→5)-Kdo and a common structural theme is often identified within a genus or strains of a species, varying by different substituents. Apart from the finding that a mutant of *E. coli* K-12 has been identified that is viable possessing only lipid IV_A as a minimal "LPS" structure, the common principle of LPS may still be seen as the core oligosaccharide and the lipid A, as identified in all other LPSs investigated so far. The expression of a (O-specific) polysaccharide in LPSs is not a prerequisite for bacterial survival, however, the finding that the polysaccharide portion in S-form LPS may be furnished either by the O-chain or a capsule or ECA indicates that such an LPS form is highly advantageous in many bacteria.

The binding of the core region to lipid A occurs always via a Kdo residue (except in *Acinetobacter*, where this Kdo may be replaced in non-stoichiometric amounts by Ko). A further general feature of LPS is that the core region is always negatively charged (provided by phosphoryl substituents and/or sugar acids like Kdo and uronic acids) which is believed to contribute to the rigidity of the Gram-negative cell wall through intermolecular cationic cross-links.

There are two types of core structures: those containing and those without heptoses. In the first type, L,D-Hep or D,D-Hep alone, or both may be present in a particular core structure. If present, D,D-Hep either decorates the inner core region (e.g. in *Y. enterocolitica*) or is attached to more remote parts of the carbohydrate chain. Still, the regulation of the distribution of L,D-Hep and D,D-Hep in the core region is not understood and it is not known whether L,D-Hep and D,D-Hep are transferred by different transferases.

Further areas for future investigation are listed in the Research Focus Box.

ACKNOWLEDGEMENTS

The research of O.H. is supported by the Deutsche Forschungsgemeinschaft (grants SFB-TR 22 A1 and A2) and the European Union (GABRIEL, GALTRAIN, GA^{2}LEN).

A.M. thanks M. Parrilli and R. Lanzetta for the useful discussions and help through these years.

References

Akira, S., Uematsu, S., Osamu, T., 2006. Pathogen recognition and innate immunity. Cell 124, 783–801.

Alexander, C., Zähringer, U., 2002. Chemical structure of lipid A – the primary immunomodulatory center of bacterial lipopolysaccharides. Trends Glycosci. Glycotech. 14, 69–86.

Basu, S.S., Karbarz, M.J., Raetz, C.H.R., 2002. Expression cloning and characterization of the C28 acyltransferase of lipid A biosynthesis in *Rhizobium leguminosarum*. J. Biol. Chem. 277, 28959–28971.

Beutler, B., Rietschel, E.Th., 2003. Innate immune sensing and its roots: the story of endotoxin. Nat. Rev. Immunol. 3, 169–176.

Bhat, U.R., Forsberg, L.S., Carlson, R.W., 1994. Structure of lipid A component of *Rhizobium leguminosarum* bv. *phaseoli* lipopolysaccharide. Unique nonphosphorylated lipid A containing 2-amino-2-deoxygluconate, galacturonate, and glucosamine. J. Biol. Chem. 269, 14402–14410.

Bouchet, V., Hood, D.W., Li, J., et al., 2003. Host-derived sialic acid is incorporated into *Haemophilus influenzae* lipopolysaccharide and is a major virulence factor in experimental otitis media. Proc. Natl. Acad. Sci. USA 100, 8898–8903.

RESEARCH FOCUS BOX

- Determination of the complete analysis of highly phosphorylated LPS.
- Identification of additional non-toxic and antagonistic lipid A structures would be advantageous.
- Investigative studies of the use of non-toxic/detoxified lipid A as a vaccine adjuvant.
- Genetic studies on LPS from non-enteric bacteria would aid the field.
- Additional molecular studies on the interaction of LPS with receptor molecules are required.
- Futher investigation of the interaction of LPS with the plant innate immune system.

Brade, H., Brunner, H., 1979. Serological cross-reactions between *Acinetobacter calcoaceticus* and chlamydiae. J. Clin. Microbiol. 10, 819–822.

Brandenburg, K., Seydel, U., Schromm, A.B., Loppnow, H., Koch, M.H.J., Rietschel, E.Th., 1996. Conformation of lipid A, the endotoxic center of bacterial lipopolysaccharides. J. Endotoxin Res. 3, 173–178.

Braun, S.G., Meyer, A., Holst, O., Pühler, A., Niehaus, K., 2005. Characterization of the *Xanthomonas campestris* pv. *campestris* lipopolysaccharide substructures essential for elicitation of an oxidative burst in tobacco cells. Mol. Plant Microbe Interact. 18, 674–681.

Brecker, L., 2003. Nuclear magnetic resonance of lipid A: the influence of solvents on spin relaxation and spectral quality. Chem. Phys. Lipids 125, 27–39.

Bystrova, O.V., Lindner, B., Moll, H., et al., 2003. Structure of the lipopolysaccharide of *Pseudomonas aeruginosa* O-12 with a randomly *O*-acetylated core region. Carbohydr. Res. 338, 1895–1905.

Bystrova, O.V., Lindner, B., Moll, H., et al., 2004. Full structure of the lipopolysaccahride of *Pseudomonas aeruginosa* immunotype 5. Biochemistry (Moscow) 69, 170–175.

Bystrova, O.V., Knirel, Y.A., Lindner, B., et al., 2006. Structures of the core oligosaccharides and O-units in the R- and SR-type lipopolysaccharides of reference strains of *Pseudomonas aeruginosa* O-serogroups. FEMS Immunol. Med. Microbiol. 46, 85–99.

Choma, A., Komaniecka, I., 2008. Characterization of a novel lipid A structure isolated from *Azospirillum lipoferum* lipopolysaccharide. Carbohydr. Res. 343, 799–804.

Choudhury, B., Carlson, R.W., Goldberg, J.B., 2005. The structure of the lipopolysaccharide from a *galU* mutant of *Pseudomonas aeruginosa* serogroup-O11. Carbohydr. Res. 343, 2761–2772.

Choudhury, B., Carlson, R.W., Goldberg, J.B., 2008. Characterization of the lipopolysaccharide from a *wbjE* mutant of the serogroup O11 *Pseudomonas aeruginosa* strain, PA103. Carbohydr. Res. 343, 238–248.

Cox, A.D., Li, J., Brisson, J.-R., Moxon, E.R., Richards, J.C., 2002a. Structural analysis of the lipopolysaccharide from *Neisseria meningitides* strain BZ157 *galE*: localization of two phosphoethanolamine residues in the inner core oligosaccharide. Carbohydr. Res. 337, 1435–1444.

Cox, A.D., Hodd, D.W., Martin, A., et al., 2002b. Identification and structural characterization of a sialylated lacto-*N*-neotetraose structure in the lipopolysaccharide of *Haemophilus influenzae*. Eur. J. Biochem. 269, 4009–4019.

Cox, A.D., Li, J., Richards, J.C., 2002c. Identification and localization of glycine in the inner core lipopolysaccharide of *Neisseria meningitidis*. Eur. J. Biochem. 269, 4169–4175.

Cox, A.D., Howard, M.D., Inzana, T.J., 2003a. Structural analysis of the lipooligosaccharide from the commensal *Haemophilus somnus* strain 1P. Carbohydr. Res. 338, 1223–1228.

Cox, A.D., Wright, J.C., Gidney, M.A.J., et al., 2003b. Identification of a novel inner-core oligosaccharide structure in *Neisseria meningitidis* lipopolysaccharide. Eur. J. Biochem. 270, 1759–1766.

Deadman, M.E., Lundström, S.L., Schweda, E.K.H., Moxon, E.R., Hood, D.W., 2006. Specific amino acids of the glycosyltransferase LpsA direct the addition of glucose or galactose to the terminal inner core heptose of *Haemophilus influenzae* lipopolysaccharide via alternative linkages. J. Biol. Chem. 281, 29455–29467.

De Castro, C., Molinaro, A., Nunziata, R., Lanzetta, R., Parrilli, M., Holst, O., 2004. A novel core region, lacking heptose and phosphate, of the lipopolysaccharide from the Gram-negative bacterium *Pseudomonas cichorii* (*Pseudomonadaceae* RNA group I). Eur. J. Org. Chem., 2427–2435.

De Castro, C., Molinaro, A., Lanzetta, R., Holst, O., Parrilli, M., 2005. The linkage between O-specific caryan and core region in the lipopolysaccharide of *Burkholderia caryophylli* is furnished by a primer monosaccharide. Carbohydr. Res. 340, 1802–1807.

De Castro, C., Carannante, A., Lanzetta, R., Liparoti, V., Molinaro, A., Parrilli, M., 2006a. Core oligosaccharide structure from the highly pathogenic *Agrobacterium tumefaciens* TT111 and conformational analysis of the putative rhamnan epitope. Glycobiology 16, 1272–1280.

De Castro, C., Carannante, A., Lanzetta, R., et al., 2006b. Structural characterisation of the core oligosaccharides isolated from the lipooligosaccharide fraction of *Agrobacterium tumefaciens* A1. Chem. Eur. J. 12, 4668–4674.

De Castro, C., Molinaro, A., Lanzetta, R., Silipo, A., Parrilli, M., 2008. Lipopolysaccharide structures from *Agrobacterium* and other *Rhizobiaceae* species. Carbohydr. Res. 343, 1924–1933.

De Soyza, A., Silipo, A., Lanzetta, R., Govan, J.R., Molinaro, A., 2008. The chemical and biological features of *Burkholderia cepacia* complex lipopolysaccharides. Innate Immun. 14, 127–144.

Dow, J.M., Newman, M.-A., von Roepenack, E., 2000. The induction and modulation of plant defence responses by bacterial lipopolysaccharides. Annu. Rev. Phytopathol. 38, 241–261.

Fox, K.L., Yildrim, H.H., Deadman, M.E., Schweda, E.K.H., Moxon, E.R., Hood, D.W., 2005. Novel lipopolysaccharide biosynthetic genes containing tetranucleotide repeats in *Haemophilus influenzae*, identification of a gene for adding *O*-acetyl groups. Mol. Microbiol. 58, 207–216.

Gass, J., Strobl, M., Loibner, A., Zähringer, U., 1993. Synthesis of allyl O-[sodium (α-D-*glycero*-D-*talo*-2-octulopyranosyl)onate]-(2→6)-2-acetamido-2-deoxy-β-D-glucopyranoside, a core constituent of the lipopolysaccharide from *Acinetobacter calcoaceticus* NCTC 10305. Carbohydr. Res. 244, 69–84.

Griffin, R., Cox, A.D., Makepeace, K., Richards, J.C., Moxon, E.R., Hood, D.W., 2003. The role of *lex2* in lipopolysaccharide biosynthesis in *Haemophilus influenzae* strains RM7004 and RM153. Microbiology 149, 3165–3175.

Hanuszkiewicz, A., Hübner, G., Vinogradov, E., et al., 2008. Structural and immunochemical analysis of the lipopolysaccharide from *Acinetobacter lwoffii* F78 located outside *Chlamydiaceae* with a *Chlamydia*-specific lipopolysaccharide epitope. Chem. Eur. J. 14, 10251–10258.

Harper, M., Cox, A.D., St Michael, F., Wilkie, I.W., Boyce, J.D., Adler, B., 2004. A heptosayltransferase mutant of *Pasteurella multocida* produces a truncated lipopolysaccharide structure and is attenuated in virulence. Infect. Immun. 72, 3436–3443.

Harper, M., Boyce, J.D., Cox, A.D., et al., 2007. *Pasteurella multocida* espresses two lipopolysaccharide glycoforms simultaneously, but only a single form is required for virulence: identification of two acceptor-specific heptosyl I transferases. Infect. Immun. 75, 3885–3893.

Hashii, N., Isshiki, Y., Iguchi, T., Kondo, S., 2003. Structural characterization of the carbohydrate backbone of the lipopolysaccharide of *Vibrio parahaemolyticus* O-untypeable strain KX-V212 isolated from a patient. Carbohydr. Res. 338, 2711–2719.

Holst, O., 1999. Chemical structure of the core region of lipopolysaccharides. In: Brade, H., Opal, S.M., Vogel, S.N., Morrison, D.C. (Eds.), Endotoxin in Health and Disease. Marcel Dekker, New York, pp. 115–154.

Holst, O., 2000. Deacylation of lipopolysaccharides and isolation of oligosaccharide phosphates. In: Holst, O. (Ed.), Methods in Molecular Biology: Bacterial Toxins. Humana Press, Totowa, pp. 345–353.

Holst, O., 2002. Chemical structure of the core region of lipopolysaccharides – an update. Trends. Glycosci. Glycotech. 14, 87–103.

Holst, O., 2007. The structures of core regions from enterobacterial lipopolysaccharides – an update. FEMS Microbiol. Lett. 271, 3–11.

Holst, O., Brade, H., 1992. Chemical structure of the core region of lipopolysaccharides. In: Morrison, D.C., Ryan, J.L. (Eds.), Bacterial Endotoxic Lipopolysaccharides, Vol. I. Boca Raton, CRC Press, pp. 135–170.

Holst, O., Müller-Loennies, S., 2007. Microbial polysaccharide structures. In: Kamerling, J.P., Boons, G.-J., Lee, Y.C., Suzuki, A., Taniguchi, N., Voragen, A.G.J. (Eds.), Comprehensive Glycoscience. From Chemistry to Systems Biology, Vol. 1. Elsevier Ltd, Oxford, pp. 123–179.

Imoto, M., Shiba, T., Naoki, H., et al., 1983. Chemical structure of *E. coli* lipid A: linkage site of acyl groups in the disaccharide backbone. Tetrahedron Lett. 24, 4017–4020.

Inzana, T.J., Glindemann, G., Cox, A.D., Wakarchuk, W., Howard, M.D., 2002. Incorporation of *N*-acetylneuraminic acid into *Haemophilus somnus* lipooligosaccharide (LOS): enhancement of resistance to serum and reduction of LOS antibody binding. Infect. Immun. 70, 4870–4879.

Isshiki, Y., Zähringer, U., Kawahara, K., 2003. Structure of the core oligosaccharide with a characteristic D-*glycero*-α-D-*talo*-oct-2-ulosonate-(2→4)-3-deoxy-D-*manno*-oct-2-ulosonate [α-Ko-(2→4)-Kdo] disaccharide in the lipopolysaccharide of *Burkholderia cepacia*. Carbohydr. Res. 338, 2659–2666.

Jones, J.D.G., Dangl, J.L., 2006. The plant immune system. Nature 444, 323–329.

Kaczyski, Z., Braun, S., Lindner, B., Niehaus, K., Holst, O., 2007. Investigation of the chemical structure and biological activity of oligosaccharides isolated from rough-type *Xanthomonas campestris* pv. *campestris* B100 lipopolysaccharides. J. Endotoxin Res. 13, 101–108.

Kay, W., Petersen, B.O., Duus, J.Ø., Perry, M.B., Vinogradov, E., 2006. Characterization of the lipopolysaccharide and β-glucan of the fish pathogen *Francisella victoria*. FEBS J. 273, 3002–3013.

Knirel, Y.A., Kochetkov, N.K., 1993. The structure of lipopolysaccharides of gram-negative bacteria. II. The structure of the core region: a review. Biokhimya 58, 182–201.

Knirel, Y.A., Lindner, B., Vinogradov, E., et al., 2005. Temperature-dependent variations and intraspecies diversity of the structure of the lipopolysaccharide of *Yersinia pestis*. Biochemistry 44, 1731–1743.

Knirel, Y.A., Bystrova, O.V., Kocharova, N.A., Zähringer, U., Pier, G.B., 2006. Conserved and variable structural features in the lipopolysaccharide of *Pseudomonas aeruginosa*. J. Endotoxin Res. 12, 324–336.

Kooistra, O., Bedoux, G., Brecker, L., et al., 2003. Structure of a highly phosphorylated core in the *ΔalgC* mutant derived from *Pseudomonas aeruginosa* wild-type strains PAO1 (serogroup O5) and PAC1R (serogroup O3). Carbohydr. Res. 338, 2667–2677.

Krasikova, I.N., Kapustina, N.V., Svetashev, V.I., et al., 2001. Chemical characterization of lipid A from some marine proteobacteria. Biochemistry (Moscow) 66, 1047–1054.

Krasikova, I.N., Kapustina, N.V., Isakov, V.V., et al., 2004. Detailed structure of lipid A isolated from lipopolysaccharide from the marine proteobacterium *Marinomonas vaga* ATCC 27119. Eur. J. Biochem. 271, 2895–2904.

Leone, S., Izzo, V., Silipo, A., et al., 2004a. A novel type of highly negatively charged lipooligosaccharide from *Pseudomonas stutzeri* OX1 possessing two 4,6-*O*-(1-carboxy)-ethylidene residues in the outer core region. Eur. J. Biochem. 271, 2691–2704.

Leone, S., Izzo, V., Sturiale, L., et al., 2004b. Structure of minor oligosaccharides from the lipopolysaccharide fraction of *Pseudomonas stutzeri* OX1. Carbohydr. Res. 339, 2657–2665.

Leone, S., Molinaro, A., Pessione, E., et al., 2006. Structural elucidation of the core-lipid A backbone from the lipopolysaccharide of *Acinetobacter radioresistens* S13, an organic solvent tolerating Gram-negative bacterium. Carbohydr. Res. 341, 582–590.

Leone, S., Molinaro, A., De Castro, C., et al., 2007. Absolute configuration of 8-amino-3,8-dideoxyoct-2-ulosonic acid, the chemical hallmark of lipopolysaccharides of the genus *Shewanella*. J. Nat. Prod. 70, 1624–1627.

Liparoti, V., Molinaro, A., Sturiale, L., et al., 2006. Structural analysis of the deep rough lipopolysaccharide from Gram negative bacterium *Alteromonas macleodii* ATCC 27126: the first finding of β-Kdo in the inner core of lipopolysaccharides. Eur. J. Org. Chem. 2006, 4710–4716.

Lundström, S.L., Twelkmeyer, B., Sagemark, M.K., et al., 2007. Novel globoside-like oligosaccharide expression patterns in non-typeable *Haemophilus influenzae* lipopolysaccharide. FEBS J. 274, 4886–4903.

Månsson, M., Hood, D.W., Li, J., Richards, J.C., Moxon, E.R., Schweda, E.K.H., 2002a. Structural analysis of the lipopolysaccharide from nontypeable *Haemophilus influenzae* strain 1003. Eur. J. Biochem. 269, 808–818.

Månsson, M., Hood, D.W., Moxon, E.R., Schweda, E.K.H., 2003. Structural characterization of a novel branching pattern in the lipopolysaccharide from nontypeable *Haemophilus influenzae*. Eur. J. Biochem. 270, 2979–2991.

Marr, N., Tirsoaga, A., Blanot, D., Fernandez, R., Caroff, M., 2008. Glucosamine found as a substituent of both phosphate groups in *Bordetellae* lipid A backbones: role of a BvgAS-activated ArnT ortholog. J. Bacteriol. 190, 4281–4290.

Medzhitov, R., 2001. Toll-like receptors and innate immunity. Nat. Rev. Immunol. 1, 135–145.

Molinaro, A., Lindner, B., De Castro, C., et al., 2003a. The structure of the lipid A of the lipopolysaccharide from *Burkholderia caryophylli* possessing a 4-amino-4-deoxy-L-arabinopyranose-1-phosphate residue exclusively in glycosidic linkage. Chem. Eur. J. 9, 1542–1548.

Molinaro, A., De Castro, C., Lanzetta, R., Parrilli, M., Raio, A., Zoina, A., 2003b. Structural elucidation of a novel core oligosaccharide backbone of the lipopolysaccharide from the new bacterial species *Agrobacterium larrymoorei*. Carbohydr. Res. 338, 2721–2730.

Müller-Loennies, S., Brade, L., Brade, H., 2002a. Chemical structure and immunoreactivity of the lipopolysaccharide of the deep rough mutant I-69 Rd^{-}/b^{+} of *Haemophilus influenzae*. Eur. J. Biochem. 269, 1237–1242.

Müller-Loennies, S., Lindner, B., Brade, H., 2002b. Structural analysis of deacylated lipopolysaccharide of *Escherichia coli* strains 2513 (R4 core-type) and F653 (R3 core-type). Eur. J. Biochem. 269, 5982–5991.

Munford, S.R., 2008. Sensing Gram-negative bacterial lipopolysaccharides: a human disease determinant? Infect. Immun. 76, 454–465.

Munford, S.R., Varley, A.W., 2006. Shield as signal: lipopolysaccharides and the evolution of immunity to Gram-negative bacteria. PLoS Pathog. 2, 467–471.

Newman, M.-A., von Roepenack-Lahaye, E., Parr, A., Daniels, M.J., Dow, J.M., 2002. Prior exposure to lipopolysaccharide potentiates expression of plant defences in response to bacteria. Plant J. 29, 485–497.

Newman, M.-A., Dow, J.M., Molinaro, A., Parrilli, M., 2007. Priming, induction and modulation of plant defence responses by bacterial lipopolysaccharides. J. Endotox. Res. 13, 69–84.

Ortega, X.P., Cardona, S.T., Brown, A.R., et al., 2007. A putative gene cluster for aminoarabinose biosynthesis is essential for *Burkholderia cenocepacia* viability. J. Bacteriol. 189, 3639–3644.

Pieretti, G., Corsaro, M.M., Lanzetta, R., et al., 2008. Structural characterization of the core region of the lipopolysaccharide from the haloalkaliphilic *Halomonas pantellerensis*: identification of the biological O-antigen repeating unit. Eur. J. Org. Chem. 2008, 721–728.

Plötz, B.M., Lindner, B., Stetter, K.O., Holst, O., 2000. Characterization of a novel lipid A containing D-galacturonic acid that replaces phosphate residues. The structure of the lipid A of the lipopolysaccharide from the hyperthermophilic bacterium *Aquifex pyrophilus*. J. Biol. Chem. 275, 11222–11228.

Que, N.L.S., Lin, S.H., Cotter, R.J., Raetz, C.H.R., 2000. Purification and mass spectrometry of six lipid a species from the bacterial endosymbiont *Rhizobium etli*. J. Biol. Chem. 275, 28006–28016.

Raetz, C.R.H., Whitfield, C., 2002. Lipopolysaccharide endotoxins. Annu. Rev. Biochem. 71, 635–700.

Raetz, C.R.H., Reynolds, C.M., Trent, M.S., Bishop, R.E., 2007. Lipid A modification systems in gram-negative bacteria. Annu. Rev. Biochem. 76, 295–329.

Ribeiro, A.A., Zhou, Z., Raetz, C.R.H., 1999. Multidimensional NMR structural analyses of purified Lipid X and Lipid A (endotoxin). Magn. Reson. Chem. 37, 620–630.

Rietschel, E.Th., 1976. Absolute configuration of 3-hydroxy fatty acids present in lipopolysaccharides from various bacterial groups. Eur. J. Biochem. 64, 423–428.

Rietschel, E.Th., Westphal, O., 1999. Endotoxins: historical perspectives. In: Brade, H., Morrison, D.C., Opal, S., Vogel, S. (Eds.), Endotoxin in Health and Disease. Marcel Dekker Inc, New York, pp. 1–30.

Schweda, E.K.H., Landerholm, M.K., Li, J., Moxon, E.R., Richards, J.C., 2003. Structural profiling of lipopolysaccharide glycoforms expressed by non-typeable *Haemophilus influenzae*: phenotypic similarities between NTHi strain 162 and the genome strain Rd. Carbohydr. Res. 338, 2731–2744.

Schweda, E.K.H., Twelkmeyer, B., Li, J., 2008. Profiling structural elements of short-chain lipopolysaccharide of non-typeable *Haemophilus influenzae*. Innate Immun. 14, 199–211.

Schwudke, D., Linscheid, M., Strauch, E., et al., 2003. The obligate predatory *Bdellovibrio bacteriovorus* possesses a neutral lipid A containing α-D-mannoses that replace phosphate residues: similarities and differences between the lipid As and the lipopolysaccharides of the wild type strain *B. bacteriovorus* HD100 and its host-independent derivative HI100. J. Biol. Chem. 278, 27502–27512.

Seydel, U., Labischinski, H., Kastowsky, M., Brandenburg, K., 1993. Phase behaviour, supramolecular structure, and molecular conformation of lipopolysaccharide. Immunobiology 187, 191–211.

Seydel, U., Oikawa, M., Fukase, K., Kusumoto, S., Brandenburg, K., 2000. Intrinsic conformation of lipid A is responsible for agonistic and antagonistic activity. Eur. J. Biochem. 267, 3032–3039.

Sforza, S., Silipo, A., Molinaro, A., Marchelli, R., Parrilli, M., Lanzetta, R., 2004. Determination of fatty acid positions in native lipid A by positive and negative electrospray ionization mass spectrometry. J. Mass Spectrom. 39, 378–383.

Shear, M.J., Turner, F.C., 1943. Chemical treatment of tumors. VII. Nature of hemorrhage-producing fraction from *Serrati marcescens* (*Bacillus prodigiosus*) culture filtrates. J. Natl. Cancer Inst. 4, 107–122.

Silipo, A., Lanzetta, R., Amoresano, A., Parrilli, M., Molinaro, A., 2002. Ammonium hydroxide hydrolysis: a valuable support in the MALDI-TOF mass spectrometry analysis of lipid A fatty acid distribution. J. Lipid Res. 43, 2188–2195.

Silipo, A., De Castro, C., Lanzetta, R., Molinaro, A., Parrilli, M., 2004a. Full structural characterization of the lipid A components from the *Agrobacterium tumefaciens* strain C58 lipopolysaccharide fraction. Glycobiology 14, 805–815.

Silipo, A., Sturiale, L., Garozzo, D., et al., 2004b. Structure elucidation of the highly heterogeneous lipid A from the lipopolysaccharide of the Gram-negative extremophile bacterium *Halomonas magadiensis* strain 21 M1. Eur. J. Org. Chem 2004, 2263–2271.

Silipo, A., Molinaro, A., Cescutti, P., et al., 2005a. Complete structural characterization of the lipid A fraction of a clinical strain of *Burkholderia cepacia* genomovar I lipopolysaccharide. Glycobiology 15, 561–570.

Silipo, A., Lanzetta, R., Parrilli, M., et al., 2005b. The complete structure of the core carbohydrate backbone from the LPS of marine halophilic bacterium *Pseudoalteromonas carrageenovora* type strain IAM 12662. Carbohydr. Res. 340, 1475–1482.

Silipo, A., Molinaro, A., Nazarenko, E.L., et al., 2005c. Structural characterization of the carbohydrate backbone of the lipooligosaccharide of the marine bacterium *Arenibacter certesii* strain KMM 3941. Carbohydr. Res. 340, 2540–2549.

Silipo, A., Molinaro, A., Sturiale, L., et al., 2005d. The elicitation of plant innate immunity by lipooligosaccharide of *Xanthomonas campestris*. J. Biol. Chem. 280, 33660–33668.

Silipo, A., Molinaro, A., Comegna, D., et al., 2006. Full structural characterization of the lipooligosaccharide of a *Burkholderia pyrrocina* clinical isolate. Eur. J. Org. Chem. 2006, 4874–4883.

Silipo, A., Sturiale, L., Garozzo, D., et al., 2008. The acylation and phosphorylation pattern of lipid A strongly influence its ability to trigger the innate immune response in *Arabidopsis*: *Xanthomonas campestris* pv. *campestris* mutant strain 8530 synthesizes a chemically different LPS that it is not recognized by innate immunity in plants. Chembiochem 9, 896–904.

St Michael, F., Howard, M.D., Li, J., Duncan, A.J., Inzana, T.J., Cox, A.D., 2004a. Structural analysis of the lipooligosaccharide from the commensal *Haemophilus somnus* genome strain 129Pt. Carbohydr. Res. 339, 529–535.

St Michael, F., Brisson, J.-R., Larocque, S., et al., 2004b. Structural analysis of the lipoopolysaccharide derived core oligosaccharides of *Actinobacillus pleuropneumoniae* serotypes 1, 2, 5a and the genome strain 5b. Carbohydr. Res. 339, 1973–1984.

St Michael, F., Li, J., Vinogradov, E., Larocque, S., Harper, M., Cox, A.D., 2005a. Structural analysis of the lipooligosaccharide of *Pasteurella multocida* strain VP161: identification of both Kdo-P and Kdo-Kdo species in the lipopolysaccharide. Carbohydr. Res. 340, 59–68.

St Michael, F., Li, J., Howard, M.D., Duncan, A.J., Inzana, T.J., Cox, A.D., 2005b. Structural analysis of the oligosaccharide of *Histophilus somni* (*Haemophilus somnus*) strain 2336 and identification of several lipooligosaccharide biosynthesis gene homologues. Carbohydr. Res. 340, 665–672.

St Michael, F., Li, J., Cox, A.D., 2005c. Structural analysis of the core oligosaccharide of *Pasteurella multocida* strain X73. Carbohydr. Res. 340, 1253–1257.

St Michael, F., Vinogradov, E., Li, J., Cox, A.D., 2005d. Structural analysis of the lipopolysaccharide of *Pasteurella multocida* strain Pm70 and identification of the putative lipopolysaccharide glycosyltransferases. Glycobiology 15, 323–333.

St Michael, F., Inzana, T.J., Cox, A.D., 2006. Structural analysis of the lipooligosaccharide-derived oligosaccharide of *Histophilus somni* (*Haemophilus somnus*) strain 8025. Carbohydr. Res. 341, 281–284.

Sturiale, L., Garozzo, D., Silipo, A., Lanzetta, R., Parrilli, M., Molinaro, A., 2005. MALDI mass spectrometry of native bacterial lipooligosaccharides. Rapid Commun. Mass Spectrom. 19, 1829–1834.

Takayama, K., Qureshi, N., Mascagni, P., 1983. Complete structures of lipid A obtained from the lipopolysaccharide of the heptoseless mutant of *Salmonella typhimurium*. J. Biol. Chem. 258, 12801–12803.

Tinnert, A.-S., Månsson, M., Yildrim, H.H., Hood, D.W., Schweda, E.K.H., 2005. Structural investigation of lipooligosaccharides from non-typeable *Haemophilus influenzae*:

investigation of inner-core phosphoethanolamine addition in NTHi strain 981. Carbohydr. Res. 340, 1900–1907.

Unger, F.M., 1983. The chemistry and biological significance of 3-deoxy-D-*manno*-2-octulosonic acid (Kdo). Adv. Carbohydr. Chem. Biochem. 348, 323–387.

Vedam, V., Kannenberg, E.L., Haynes, J.G., Sherrier, D., Datta., Carlson, R.W., 2003. A *Rhizobium leguminosarum* AcpXL mutant produces lipopolysaccharide lacking 27-hydroxyoctacosanoic acid. J. Bacteriol. 185, 1841–1850.

Vinogradov, E., Duus, J.Ø., Brade, H., Holst, O., 2002a. The structure of the carbohydrate backbone of the lipopolysaccharide from *Acinetobacter baumannii* strain ATCC 19606. Eur. J. Biochem 269, 422–430.

Vinogradov, E., Korenevsky, A., Beveridge, T.J., 2002b. The structure of the carbohydrate backbone of the LPS from *Shewanella putrefaciens* CN32. Carbohydr. Res. 337, 1285–1289.

Vinogradov, E., Korenevsky, A., Beveridge, T.J., 2003. The structure of the rough-type lipooligosaccharide from *Shewanella oneidensis* MR-1, containing 8-amino-8-deoxy-Kdo and an open-chain form of 2-acetamido-2-deoxy-D-galactose. Carbohydr. Res. 338, 1991–1997.

Vinogradov, E., Korenevsky, A., Beveridge, T.J., 2004a. The structure of the core region of the lipopolyosaccharide from *Shewanella algae* BrY, containing 8-amino-3,8-dideoxy-D-*manno*-oct-2-ulosonic acid. Carbohydr. Res. 339, 737–740.

Vinogradov, E., Korenevsky, A., Lovley, D.R., Beveridge, T.J., 2004b. The structure of the core region of the lipopolyosaccharide from *Geobacter sulfurreducens*. Carbohydr. Res. 339, 2901–2904.

Wang, Z., Li, J., Vinogradov, E., Altman, E., 2006a. Structural studies of the core region of *Aeromonas salmonicida* subsp. *salmonicida* lipopolysaccharide. Carbohydr. Res. 341, 109–117.

Wang, Z., Li, J., Altman, E., 2006b. Structural characterization of the lipid A region of *Aeromonas salmonicida* subsp. *salmonicida* lipopolysaccharide. Carbohydr. Res. 341, 2816–2825.

Westphal, O., Lüderitz, O., 1954. Chemische Erforschung von Lipopolysacchariden Gram-Negativer Bakterien. Angew. Chem. 66, 407–417.

Westphal, O., Lüderitz, O., Bister, F., 1952. Uber die Extraktion von Bakterien mit Phenol/Wasser. Z. Naturforsch. 7, 148–155.

Yildrim, H.H., Li, J., Richards, J.C., Hood, D.W., Moxon, E.R., Schweda, E.K.H., 2005. Complex *O*-acetylation in non-typeable *Haemophilus influenzae* lipopolysaccharide: evidence for a novel site of *O*-acetylation. Carbohydr. Res. 340, 2598–2611.

Zähringer, U., Lindner, B., Rietschel, E.Th., 1994. Molecular structure of lipid A, the endotoxic center of bacterial lipopolysaccharides. Adv. Carbohydr. Chem. Biochem. 50, 211–276.

Zähringer, U., Lindner, B., Knirel, Y.A., et al., 2004. Structure and biological activity of the short-chain lipopolysaccharide from *Bartonella henselae* ATCC 49882. J. Biol. Chem. 279, 21046–21054.

Zdorovenko, E.L., Vinogradov, E., Zdorovenko, G.M., et al., 2004. Structure of the core oligosaccharide of a rough-type lipopolysaccharide of *Pseudomonas syringae* pv. *phaseolica*. Eur. J. Biochem. 271, 4968–4977.

Zeidler, D., Zähringer, U., Gerber, I., et al., 2004. Innate immunity in *Arabidopsis thaliana*: lipopolysaccharides activate nitric oxide synthase (NOS) and induce defense genes. Proc. Natl. Acad. Sci. USA 101, 15811–15816.

Zipfel, C., Felix, G., 2005. Plants and animals: a different taste for microbes? Curr. Opin. Plant Pathol. 8, 353–360.

CHAPTER

4

O-Specific polysaccharides of Gram-negative bacteria

Yuriy A. Knirel

SUMMARY

The present chapter examines the composition and structure of the O-specific polysaccharides (O-antigens) of the lipopolysaccharides of Gram-negative bacteria. The occurrence of both monosaccharides and non-carbohydrate groups as O-specific polysaccharide components is surveyed. Various types of these structures are considered, including homopolymers of a monosaccharide, homo- and hetero-polysaccharides built up of oligosaccharide repeats (O-units) and block co-polymers. Structural features of regular O-specific polysaccharides and factors that mask their regularity are described. Emphasis is placed on the occurrence and the role of non-repetitive domains that are present at the reducing and non-reducing ends of some O-chains.

Key words: O-Specific polysaccharide; O-Antigen; O-Chain; O-Unit; Repeating unit; Lipopolysaccharide; Bacterial polysaccharide structure; Monosaccharide composition; Acyl group

1. INTRODUCTION

The O-specific polysaccharide (O-PS) or O-antigen represents the polymer chain of lipopolysaccharide (LPS), the major constituent of the outer membrane of Gram-negative bacteria (see Chapter 1). Composition and structure of O-PSs are highly diverse. Based on the O-antigens, serologically distinct strains of the same bacterial species are classified into O-serogroups or O-serotypes (e.g. Liu *et al.*, 1983; Nielsen, 1986; Ewing, 1986; Shimada *et al.*, 1994). The O-PS plays important roles in protection of the bacteria and their virulence in animals and humans (Iredell *et al.*, 1998; Erridge *et al.*, 2002; Skurnik and Bengoechea, 2003; Morona *et al.*, 2003; Lugo *et al.*, 2007; Kintz and Goldberg, 2008; Bravo *et al.*, 2008) as well as in microbe–plant interactions (Newman *et al.*, 2007). In recent decades the chemistry of O-PSs has made significant progress and the data obtained have helped a better understanding of the mechanisms of pathogenesis of infectious diseases and the development of improved vaccines (Cryz *et al.*, 1995; Favre *et al.*, 1996; Conlan *et al.*, 2002; Taylor *et al.*, 2004; Kubler-Kielb *et al.*, 2008; Döring and Pier, 2008) and diagnostic agents (Kawasaki *et al.*, 1995; Trautmann *et al.*, 1996; Jauho *et al.*, 2000; Thirumalapura *et al.*, 2005; Amano *et al.*, 2007; Blixt *et al.*, 2008).

Composition and structures of bacterial polysaccharides, including O-PSs, have been repeatedly surveyed (Kenne and Lindberg, 1983;

Jann and Jann, 1984; Lindberg, 1990, 1998; Knirel and Kochetkov, 1994; Wilkinson, 1996; Jansson, 1999; Corsaro *et al.*, 2001) and a number of reviews have been published that are devoted to O-PSs of particular bacterial species, including *Escherichia coli* (Stenutz *et al.*, 2006), *Shigella* spp. (Liu *et al.*, 2008), *Citrobacter* spp. (Knirel *et al.*, 2002a), *Proteus* spp. (Knirel *et al.*, 1999), *Hafnia alvei* (Romanowska, 2000), *Serratia marcescens* (Aucken *et al.*, 1998), *Yersinia* spp. (Ovodov and Gorshkova, 1988; Ovodov *et al.*, 1992; Bruneteau and Minka, 2003), *Pseudomonas aeruginosa* (Knirel, 1990; Knirel *et al.*, 2006), *Burkholderia cepacia* (Vinion-Dubiela and Goldberg, 2003); *Agrobacterium* and *Rhizobiaceae* spp. (De Castro *et al.*, 2008). Since 2005, an annually updated Bacterial Carbohydrate Structure Database is available via the Internet (http://www.glyco.ac.ru/bsdb). The present review updates data on the composition of O-PSs and considers selected topics related to their structural features. To avoid extensive citation, for topics already mentioned, the reviews cited above are referenced rather than original publications.

2. COMPOSITION OF O-PSs

2.1. Monosaccharides

D-Glucose and D-galactose are the most abundant neutral sugars in O-PSs and D-mannose is fairly common too. All three are present as pyranosides and D-galactofuranoside is not uncommon. Pentoses occur less often, D-ribose being the most common of them. D-Ribose and D-arabinose are known only as furanosides, D-xylose occurs in both forms and L-xylose always as a pyranoside. Other pentoses and hexoses have not been found in O-PSs. D-*glycero*-D-*manno*- and L-*glycero*-D-*manno*-Heptoses have been identified in various bacteria and D-*glycero*-D-*altro*-heptose has been reported in *Campylobacter jejuni* (Aspinall *et al.*, 1992). There are only a few examples of ketoses, including D-*threo*-pent-2-ulose (D-xylulose) (*Yersinia enterocolitica* O5,27 and *E. coli* O97), D-fructose (*Yersinia intermedia*) (Ovodov *et al.*, 1992; and Stenutz *et al.*, 2006).

A variety of 6-deoxy sugars occur in O-PSs, the most abundant being L-rhamnose and L-fucose; their D isomers as well as 6-deoxy-D- and -L-talose occur fairly often too. Less common are 6-deoxy-L-glucose (L-quinovose) (*Providencia stuartii* O44), 6-deoxy-L-altrose (*Y. enterocolitica, Pectinatus frisingensis*) and 6-deoxy-D-gulose (D-dantiorose) (*Y. enterocolitica*) (Ovodov *et al.*, 1992; Jansson, 1999; Kocharova *et al.*, 2005). D-Fucose and 6-deoxy-L-altrose are known as both pyranosides and furanosides, whereas the other 6-deoxyhexoses are always pyranosidic. From other deoxyhexoses, only 4-deoxy-D-*arabino*-hexopyranose has been reported as a component of several *Citrobacter* O-PSs (Knirel *et al.*, 2002a).

A group of 3,6-dideoxyhexoses has been identified, mainly in *Salmonella* (Jann and Jann, 1984) and *Yersinia pseudotuberculosis* (Ovodov and Gorshkova, 1988; Ovodov *et al.*, 1992). Most often occurring is the L-*xylo* isomer (colitose), which always exists as a pyranoside, just like the D- and L-*arabino,* D-*xylo* and D-*ribo* isomers (tyvelose, ascarylose, abequose and paratose, respectively). In *Y. pseudotuberculosis,* paratose has been found also as a furanoside.

There are several examples of the occurrence of 6-deoxy-D-*manno*-heptopyranose, e.g. in *Y. pseudotuberculosis* (Ovodov and Gorshkova, 1988) and 6-deoxy-D-*altro*-heptopyranose is present in *C. jejuni* (Aspinall *et al.*, 1992).

D-Glucosamine and D-galactosamine are widespread in O-PSs as the majority of Gram-negative bacteria employ either of them as the first monosaccharide of the O-unit (or repeating unit) whose transfer to a lipid carrier initiates biosynthesis of the O-antigen by the O-antigen polymerase (Wzy)-dependent pathway (Raetz and Whitfield, 2002) (see Chapter 18). Some 2-amino-2,6-dideoxy-D-hexoses(D-quinovosamine and D-fucosamine), as well as 2,4-diamino-2,4,6-trideoxy-D-glucose and -D-galactose, may play this role too. D-Mannosamine, 3-amino-3,6-dideoxy-D-glucose and -D-galactose as well as 4-amino-4,6-dideoxy-D-hexoses (the *gluco*, *manno* and *galacto* isomers) occur fairly often.

L-Quinovosamine and L-fucosamine are components of numerous O-PSs, whereas some other L isomers have been identified only in a few bacteria, namely L-glucosamine in *B. cepacia* O1 (Vinion-Dubiel and Goldberg, 2003), L-rhamnosamine in *Proteus vulgaris* O55 (Kondakova *et al.*, 2003) and *E. coli* O3 (Stenutz *et al.*, 2006), 3-amino-3,6-dideoxy-L-glucose in *Vibrio anguillarum* (Knirel and Kochetkov, 1994) and 4-amino-4,6-dideoxy-L-mannose in *Vibrio cholerae* O76 and O144 (Jansson, 1999). The sugar 2,3-diamino-2,3,6-trideoxy-L-mannose has been found in *Proteus penneri* O66 (Knirel *et al.*, 1999) and its D enantiomer in *E. coli* O119 (Anderson *et al.*, 1992), but the latter finding requires additional confirmation.

A biosynthetic precursor of 2-amino-2,6-dideoxy-D-hexoses, 2-amino-2,6-dideoxy-D-*xylo*-hex-4-ulose, has been identified in several O-PSs, e.g. in *Flavobacterium columnare* (MacLean *et al.*, 2003) and *Pseudoalteromonas rubra* (Kilcoyne *et al.*, 2005). All amino sugars occurring in O-chains are known exclusively as pyranosides, except for D-galactosamine, which is known also as a furanoside, e.g. in *P. penneri* O63 (Knirel *et al.*, 1999).

D-Glucuronic and D-galacturonic acids are common and their C-5 epimers have been reported, e.g. L-iduronic acid in *Pseudoalteromonas haloplanktis* (Hanniffy *et al.*, 1998) and *E. coli* (Perepelov *et al.*, 2008) and L-altruronic acid in *Proteus mirabilis* O10 (Knirel *et al.*, 1999). From 2-amino-2-deoxyhexuronic acids, the D-*galacto*, D-*manno* and L-*gulo* isomers occur rather often in O-chains, whereas the L-*galacto* and L-*altro* isomers have been found in *P. aeruginosa* (Knirel, 1990) and *Shigella sonnei* (Kenne *et al.*, 1980), respectively. In particular, 2-amino-2-deoxy-D-mannuronic acid is also a constituent of the enterobacterial common antigen (Kuhn *et al.*, 1988).

A group of 2,3-diamino-2,3-dideoxyhexuronic acids has been identified, three of them, the D-*gluco*, D-*manno* and L-*gulo* isomers, as components of various *P. aeruginosa* O-PSs (Knirel, 1990). However, 2,3-diamino-2,3-dideoxy-D-galacturonic acid has only been found in the genus *Bordetella* (Vinogradov, 2002; Preston *et al.*, 2006) and the L-*galacto* isomer originally reported in *Vibrio ordalii* and *V. anguillarum* has been revised later to the L-*gulo* isomer (Kilcoyne *et al.*, 2005). A 2,4-diamino-2,4-dideoxyglucuronic acid has been found in a *Thiobacillus* spp. (Jansson., 1999), though its absolute configuration remains unknown to date. All glycuronic acids that have been reported in O-PSs are pyranosidic.

Additionally, all glyculosonic acids found in O-PSs are 3-deoxy sugars; 3-deoxy-D-*manno*-oct-2-ulosonic acid (Kdo) is a component of the O-PS of *Providencia alcalifaciens* O36 (Kocharova *et al.*, 2007) and neuraminic acid has been identified, e.g. in some other enteric bacteria (Gamian and Kenne, 1993). The 5,7-diamino-3,5,7,9-tetradeoxy-non-2-ulosonic acids are unique to bacteria (Knirel *et al.*, 2003) and four stereoisomers thereof are known. These are the L-*glycero*-L-*manno* (pseudaminic acid), D-*glycero*-D-*galacto* (legionaminic acid), L-*glycero*-D-*galacto* (8-epilegionaminic acid) and D-*glycero*-D-*talo* (4-epilegionaminic acid) isomers, the originally reported configurations of the last three being subsequently revised (Knirel *et al.*, 2003). Also, a 5,7,8-triamino-3,5,7,8,9-pentadeoxy-L- or -D-*glycero*-L-*manno*-non-2-ulosonic acid has been found in *Flexibacter maritimus* (Vinogradov *et al.*, 2003d), but the configuration at C-8 remains to be determined.

Monosaccharides with a branched skeleton occur rarely. The simplest of these is 3-*C*-methyl-D-mannose found in *Helicobacter pylori* (Kocharova *et al.*, 2000). Structures of the other representative sugars are shown in Figure 4.1;

their names are derived from the bacteria in which they were found. Two branched isomeric monosaccharides, yersiniose A **1** and yersiniose B **2** occur mainly in *Yersinia* spp. (Ovodov and Gorshkova, 1988; Ovodov *et al.*, 1992); their structures have been confirmed by chemical synthesis of the authentic sugars (Zubkov *et al.*, 1992). The highest branched sugars, 10-carbon erwiniose **3** and 12-carbon caryophyllose **4**, have been found in *Erwinia carotovora* (Senchenkova *et al.*, 2005) and *Burkholderia caryophylli* (Adinolfi *et al.*, 1996b), respectively. A branched 9-carbon ketoamino glycuronamide **5**, called shewanellose, has been identified in two distantly related bacteria *Shewanella putrefaciens* (Shashkov *et al.*, 2002) and *Morganella morganii* (Kilcoyne *et al.*, 2002). In the former, shewanellose exists as a pyranoside, whereas in the latter, shewanellose occurs as both pyranoside and furanoside. The most unusual of the known O-PS components appears to be a carbocyclic sugar caryose **6**, which as caryophyllose **4**, is present in *B. caryophylli* (Adinolfi *et al.*, 1996a).

2.2. Non-carbohydrate components

O-PSs include a great variety of non-sugar constituents, which can be linked to sugars in various manners: by the ether linkage (as in partially methylated monosaccharides and ethers with lactic acids), acetal linkage (acetals with pyruvic acid), ester linkage (*O*-acylated and phosphorylated monosaccharides) or amidic linkage (*N*-acylated amino sugars and amides of uronic acids).

O-Acetyl groups are very common and an *O*-linked propanoyl group has been found in *V. anguillarum* (Knirel and Kochetkov, 1994). The occurrence of various *O*-methylated

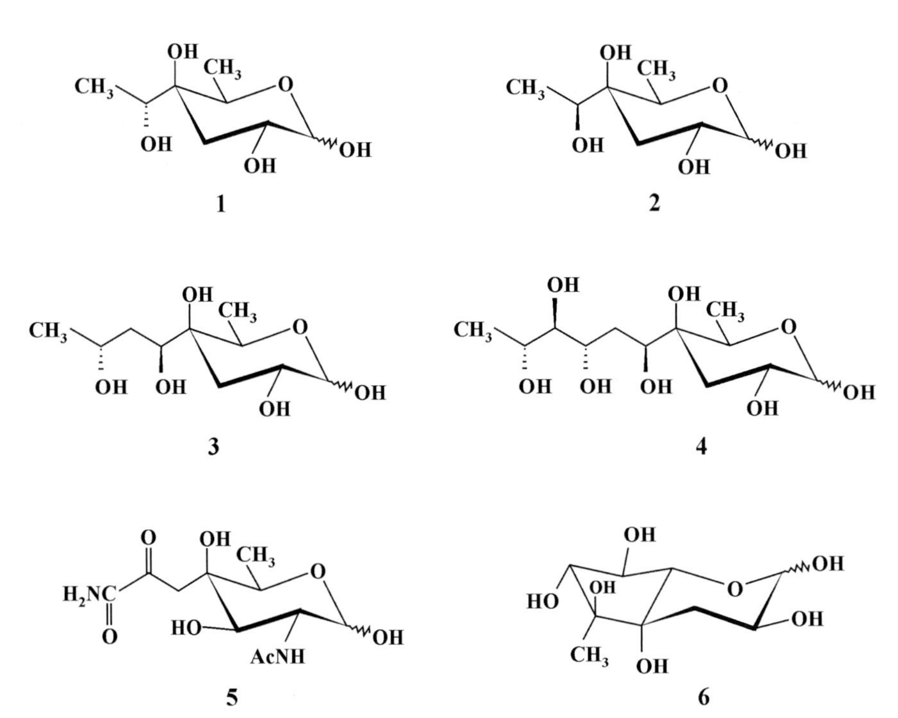

FIGURE 4.1 Structures of branched and carbocyclic monosaccharide components of various O-PSs: yersiniose A **1**; yersiniose B **2**; erwiniose **3**; caryophyllose **4**; shewanellose **5**; and caryose **6**.

monosaccharides in LPSs has been deduced (Lindberg, 1990), but it is not always clear if the methylated sugar definitely occurs in the O-PS. However, in several cases this has been unambiguously established (Wilkinson, 1996).

Ethers with (*R*)- and (*S*)-lactic acids called glycolactilic acids are components of several O-PSs. The parent sugars are hexoses, 6-deoxyhexoses or hexosamines. For example, 4-*O*-[(*R*)-1-carboxyethyl]-D-glucose and 3-*O*-[(*R*)-1-carboxyethyl]-L-rhamnose occur in *Shigella dysenteriae* (Liu *et al.*, 2008) and 2-amino-3-*O*-[(*R*)-1-carboxyethyl]-2-deoxy-D-glucose (muramic acid), its (*S*)-stereoisomer (isomuramic acid) and their regioisomers with an ether group at position-4 have been identified in *Proteus* (Knirel *et al.*, 1999; Perepelov *et al.*, 2002) and *Providencia* (Kocharova *et al.*, 2002).

Cyclic (*R*)- or (*S*)-acetals of pyruvic acid are common in exopolysaccharides (Kenne and Lindberg, 1983), but occur less often in O-PSs as the only non-sugar carbonyl components. Usually, pyruvic acid is linked to O-4 and O-6 of various monosaccharides but, in a few cases, it has been found on O-3 and O-4 of D-galactose, e.g. in *P. mirabilis* O24 (Shashkov *et al.*, 2000) and *S. dysenteriae* type 9 (Zhao *et al.*, 2007).

Amino sugars with a free amino group are uncommon; among the few exceptions are 4-amino-4,6-dideoxy-D-glucose and 2-amino-2-deoxy-L-guluronic acid in *Idiomarina zobellii* (Kilcoyne *et al.*, 2004). Moreover, 2,4-diamino-2,4,6-trideoxy-D-galactose occurs always with the free amino group at position-4, e.g. in *S. sonnei* where it was found for the first time (Kenne *et al.*, 1980).

The 2-amino group of amino sugars is usually acetylated but formyl and acetimidoyl (Knirel, 1990), (*R*)-3-hydroxybutanoyl (Romanowska, 2000) and L-alanyl (Knirel *et al.*, 1999) groups have been found in *P. aeruginosa*, *H. alvei* 1187 and *P. penneri* 25 (O69), respectively. Amino groups in other positions of 6-deoxyamino and diamino sugars are acylated with various functionalized carboxylic acids much more often.

Of the hydroxy derivatives, (*R*)- and (*S*)-3-hydroxybutanoic acids are the most abundant, but the others are: (*R*)- and (*S*)-2-hydroxypropanoic in *V. cholerae* O76 and O144 (Jansson, 1999); 3-hydroxypropanoic in *V. cholerae* 1875, 4-hydroxybutanoic in *Yersinia ruckeri* O1, L-glyceric in *Citrobacter* O32, (*S*)-2,4-dihydroxybutanoic in *V. cholerae* O1 (Kenne and Lindberg, 1983) and (*S,S*)-3,5-dihydroxyhexanoic in *Flavobacterium psychrophilum* (MacLean *et al.*, 2001). There are also examples of *N*-linked dicarboxylic acids occurring, such as malonic and succinic in *P. mirabilis* (Kondakova *et al.*, 2004), L-malic in *Shewanella algae* (Vinogradov *et al.*, 2003a) and (*S*)-2-hydroxyglutaric in *F. maritimus* (Vinogradov *et al.*, 2003d).

A number of *N*-linked amino acids have been identified and, among them, glycine in *S. dysenteriae* O7 (Liu *et al.*, 2008), D- and L-alanine in *Proteus* spp. (Knirel *et al.*, 1999), L-serine in *E. coli* O114 (Stenutz *et al.*, 2006), D- and L-aspartic acids in *Providencia* and *Proteus* spp. (Kocharova *et al.*, 2004) and a group of 5-oxoproline derivatives: (*R,R*)-3-hydroxy-3-methyl in *V. cholerae* O5), 3-hydroxy-2,3-dimethyl in *Pseudomonas fluorescens*), 2,4-dihydroxy-3,3,4-trimethyl in *V. anguillarum* (Knirel and Kochetkov, 1994; Wilkinson, 1996) or 4-amino-3-hydroxy-3-methyl in *Francisella victoria* (Kay *et al.*, 2006).

A hydroxyl group of *N*-linked hydroxy acids may be *O*-acetylated (Kilcoyne *et al.*, 2005), *O*-methylated (Preston *et al.*, 2006) or glycosylated with a neighbouring monosaccharide (Knirel, 1990; Vinogradov *et al.*, 2003d), in the latter case, monomers in the polysaccharide are linked by both glycosidic and amidic linkages. Amino groups of *N*-linked amino acids are usually *N*-acetylated or substituted with another acyl group, such as a formyl (Preston *et al.*, 2006), (*R*)-3-hydroxybutanoyl (Perepelov *et al.*, 2001) or 3-aminobutanoyl (Kay *et al.*, 2006) group.

Uronic acids may occur as amides with 2-aminopropane-1,3-diol, e.g. in *Shigella boydii* type 8 (Lindberg, 1990) or with various amino acids, including glycine in *E. coli* O91 (Stenutz *et al.*, 2006), L-alanine, L-serine, L-threonine and L-lysine all in *Proteus* spp. (Knirel *et al.*, 1999) and D-allothreonine in *H. alvei* 1206 (Romanowska, 2000). Derivatives of L-lysine belonging to the class of opines, N^{ε}-[(*R*)- and (*S*)-1-carboxyethyl]-L-lysine, have been identified in *Proteus* and *Providencia* spp. (Knirel *et al.*, 1999; Kocharova *et al.*, 2003). Phosphoric esters occur fairly often. Phosphate may interlink O-units in the polysaccharide chain. Alternatively, it may attach as a substituent or incorporate an alcohol or amino alcohol into the main chain. Most common phosphate-linked non-sugar constituents are glycerol, ribitol and 2-aminoethanol, i.e. ethanolamine (Lindberg, 1990), but there are other examples, such as arabinitol in *H. alvei* 1191 (Romanowska, 2000), 2-[(*R*)-1-carboxyethy lamino]ethanol (Knirel *et al.*, 1999) and choline (Fudala *et al.*, 2003) in *P. mirabilis* O14 and O18, respectively, or the more complex compound, 2-amino-3,4,5-trihydroxy-2-methylpentanoic acid in *Fusobacterium necrophorum* (Knirel and Kochetkov, 1994). Unsubstituted phosphate is uncommon in O-PSs but a cyclic phosphate has been found in *V. cholerae* O139 (Jansson, 1999).

3. REPETITIVE O-PS STRUCTURES

3.1. Homopolysaccharides

Long known homopolysaccharides are linear mannans having tri- to penta-saccharide O-units that are characteristic for some enterobacteria. The O-PS of *E. coli* O9, *Klebsiella pneumoniae* O3 and *H. alvei* PCM 1223 is D-mannan **7** with alternating α-(1→2)- and α-(1→3)-linkages in the pentasaccharide repeating unit (Katzenellenbogen *et al.*, 2001; Stenutz *et al.*, 2006) as shown in Figure 4.2. Mannan **8** with a truncated tetrasaccharide O-unit has been found in *E. coli* O9a and mannan **9** having a trisaccharide repeating unit with both α- and β-linked mannose residues is shared by *E. coli* O8 and *K. pneumoniae* O5 (Stenutz *et al.*, 2006).

The LPSs of the phytopathogenic bacterium *Pseudomonas syringae* possess various linear D- and L-rhamnans **10–13**, including a pair of enantiomeric homopolysaccharides **11** and **12** (Ovod *et al.*, 1999, 2004) (see Figure 4.2). The L-rhamnans **12** and **13** differ in the number of α-(1→2)- and α-(1→3)-linkages in the tetrasaccharide O-units that are present together in the same strain (Ovod *et al.*, 2004). In addition,

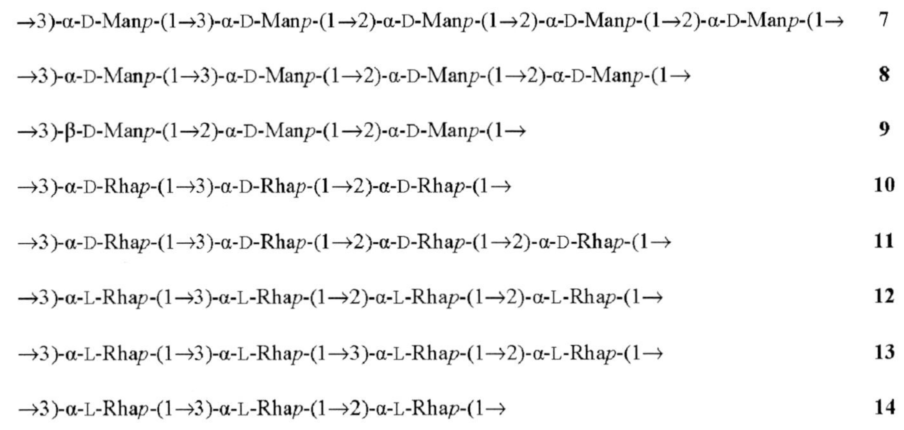

FIGURE 4.2 Structures of D-mannan- **7–9**, D-rhamnan- **10**, **11** and L-rhamnan- **12–14** containing O-PSs of various bacteria.

D-rhamnan **10** represents the common polysaccharide antigen of *P. aeruginosa* which, as the O-antigen chain, is linked to the core-lipid A moiety (Rocchetta *et al.*, 1998). Finally, the L-rhamnan **14** that is enantiomeric to D-rhamnan **10** has been reported as the O-PS of *Rhizobium* sp. NGR234 (Reuhs *et al.*, 2005).

In some bacterial species, all strains produce only one or several related homopolysaccharides and, accordingly, are characterized by a limited sero-variation. For instance, strains of *Legionella pneumophila* possess homopolymers of α-(1→4)-interlinked 5-acetamidino-7-acetamido-3,5,7,9-tetradeoxy-non-2-ulosonic acid residues, either 8-*O*-acetylated or not, which have the D-*glycero*-D-*galacto* configuration in serogroup 1 or the D-*glycero*-D-*talo* configuration in all other serogroups (Knirel *et al.*, 2003).

Also in *Brucella,* two homopolymers have been found, one an α-(1→2)-interlinked poly(4,6-dideoxy-4-formamido-D-mannopyranose), whereas the other includes both α-(1→2)- and α-(1→3)-linkages. They are characteristic for antigenically different A-dominant and M-dominant *Brucella* strains, respectively and may coexist in one strain (Meikle *et al.*, 1989). The former polysaccharide is characteristic also for *Y. enterocolitica* Hy 128 (O9) (Wilkinson, 1996) that accounts for false-positive serological reactions in the serodiagnostics of these diseases.

Homopolymers of 4-amino-4,6-dideoxy-D-mannopyranose (D-Rha*p*4N) with other *N*-acyl groups are known too, e.g. (*S*)-2,4-dihydroxybutanoyl in *V. cholerae* O1 (Kenne and Lindberg, 1983) and acetyl in *V. cholerae* 1875 (Wilkinson, 1996). The O-PSs of *V. cholerae* O76 and O144 possess α-(1→2)-linked homopolymers of the enantiomer sugar L-Rha*p*4N, which bears an *N*-linked (*S*)- or (*R*)-2-hydroxypropanoyl group, respectively (Jansson, 1999).

More examples of homopolysaccharides with a monosaccharide O-unit are presented in the review by Wilkinson (1996).

3.2. Heteropolysaccharides

In many cases, e.g. in *S. dysenteriae* and *S. boydii* (Liu *et al.*, 2008), the O-PSs of serologically distinct strains have quite different structures but, in some bacterial species, they are related to a various extent. A classical example is *Salmonella enterica* serogroups A–D (Jann and Jann, 1984). Their O-PSs share the main chain, which is either unsubstituted (serogroup E) or decorated with paratose, abequose and tyvelose in serogroups A, B and D, respectively. Being exposed on the surface, the 3,6-dideoxyhexoses define the serospecificity of these *Salmonella* strains.

The O-PSs of *P. aeruginosa* O1-O13 (Figure 4.3) also share some peculiar features (Knirel, 1990; Knirel *et al.*, 2006). In all O-serogroups, the first monosaccharide of the O-unit is a 6-deoxy-D-hexosamine, 2-acetamido-2,6-dideoxy-D-glucose (D-QuiNAc), 2-acetamido-2,6-dideoxy-D-galactose (D-FucNAc) or 2,4-diacetamido-2,4,6-trideoxy-D-glucose (D-QuiNAc4Ac), which are synthesized in a combined pathway. The O-units are linear tri- or tetrasaccharides that are enriched in L-rhamnose, deoxyhexosamines, amino- and diamino-hexuronic and diamino-nonulosonic acids.

Simple O-serogroups include strains having the same O-PS structure, whereas complex O-serogroups are divided into subgroups with different, but more closely related, O-PSs than between O-serogroups. For example, the O-PSs of all O6 subgroups have the same O-unit but differ in the mode of their connection to each other, i.e. by α-(1→2)-, α-(1→3)- or β-(1→3)-linkage. In serogroup O2, each of the two 2,3-diamino-2,3-dideoxyhexuronic acids may have either the D-*manno* or L-*gulo* configuration. The repeating units in O4 serogroup differ in the first monosaccharide, which is either D-QuiNAc or D-FucNAc. The O7 polysaccharides have the same carbohydrate structure but different 5-*N*-acyl groups on pseudaminic acid, either acetyl or (*R*)-3-hydroxybutanoyl groups. Further

O1	→4)-α-D-GalNAc-(1→4)-β-D-GlcNAc3NAcA-(1→3)-α-D-FucNAc-(1→3)-α-D-QuiNAc-(1→
O2a,d	→4)-β-D-ManNAc3NAmA-(1→4)-β-D-ManNAc3NAcA-(1→3)-α-D-FucNAc-(1→
O2a,d,e	→4)-β-D-ManNAc3NAmA-(1→4)-α-L-GulNAc3NAcA-(1→3)-α-D-FucNAc-(1→
O2a...	→4)-α-L-GulNAc3NAmA-(1→4)-β-D-ManNAc3NAcA-(1→3)-α-D-FucNAc-(1→
O3a,d	→3)-α-L-Rha-(1→6)-α-D-GlcNAc-(1→4)-α-L-GalNAcA-(1→3)-β-D-QuiNAc4NAc-(1→
O4a,b	→2)-α-L-Rha-(1→3)-α-L-FucNAc-(1→3)-α-L-FucNAc-(1→3)-α-D-QuiNAc-(1→
O4a,c	→2)-α-L-Rha-(1→3)-α-L-FucNAc-(1→3)-α-L-FucNAc-(1→3)-α-D-FucNAc-(1→
O6a,b	→2)-α-L-Rha-(1→4)-α-D-GalNAcA-(1→4)-α-D-GalNFoA-(1→3)-α-D-QuiNAc-(1→
O6a,c	→3)-α-L-Rha-(1→4)-α-D-GalNAcA-(1→4)-α-D-GalNFoA-(1→3)-α-D-QuiNAc-(1→
O6a,d	→3)-α-L-Rha-(1→4)-α-D-GalNAcA-(1→4)-α-D-GalNFoA-(1→3)-β-D-QuiNAc-(1→
O7a,b,c	→4)-α-Pse5*R*Hb7Fo-(2→4)-β-D-Xyl-(1→3)-α-D-FucNAc-(1→
O7a,b,d	→4)-α-Pse5Ac7Fo-(2→4)-β-D-Xyl-(1→3)-α-D-FucNAc-(1→
O9a,b,d	→3)-*R*Hb-(1→7)-β-Pse5Ac-(2→4)-α-D-FucNAc-(1→3)-β-D-QuiNAc-(1→
O10a,c	→3)-α-L-Rha-(1→4)-α-L-GalNAcA-(1→3)-α-D-QuiNAc-(1→
O11	→2)-β-D-Glc-(1→3)-α-L-FucNAc-(1→3)-β-D-FucNAc-(1→
O12	→8)-α-8eLeg5Ac7Ac-(2→3)-α-L-FucNAm-(1→3)-α-D-QuiNAc-(1→
O13a,b	→2)-α-L-Rha-(1→3)-α-L-Rha-(1→4)-α-D-GalNAcA-(1→3)-β-D-QuiNAc-(1→

FIGURE 4.3 Structures of the O-polysaccharides of selected *P. aeruginosa* O-serogroups and subgroups. All monosaccharides are in the pyranose form. Amidation of hexuronic acids and *O*-acetylation are not shown. Abbreviations: GalNA, 2-amino-2-deoxygalacturonic acid; GlcN3NA, GulN3NA, ManN3NA, 2,3-diamino-2,3-dideoxygluc-, gul- and mann-uronic acids, respectively; Pse, 5,7-diamino-3,5,7,9-tetradeoxy-L-*glycero*-L-*manno*-non-2-ulosonic (pseudaminic) acid; 8eLeg, 5,7-diamino-3,5,7,9-tetradeoxy-L-*glycero*-D-*galacto*-non-2-ulosonic (8-epilegionaminic) acid; Am, acetimidoyl; Fo, formyl; RHb, (*R*)-3-hydroxybutanoyl.

differences are added by *O*-acetylation and amidation (see next section).

As with homopolysaccharides, some heteropolysaccharides are shared by different bacteria. Thus, more than half of the 34 *Shigella* O-PSs have been found in various *E. coli* clones which is explained by the recent origin of the former bacterium from the latter (Liu *et al.*, 2008). Close relatedness of the O-PSs between various other enterobacterial genera are known as well and could be due to the origin of these strains from a common ancestor.

Nevertheless, sometimes similar or even identical O-chains are found in taxonomically distant bacteria, even in those belonging to different families (Knirel and Kochetkov, 1994; Wilkinson, 1996). For instance, the O-PSs of *Francisella tularensis*, *V. anguillarum*, *S. dysenteriae* type 7 and *P. aeruginosa* O6 possess a common trisaccharide fragment of linear tetrasaccharide units, either α-D-GalNA-(1→4)-α-D-GalNA-(1→3)-D-QuiN, β-D-QuiN-(1→4)-α-D-GalNA-(1→4)-α-D-GalNA or both (Knirel and Kochetkov, 1994). A likely reason for this phenomenon is a horizontal transfer of polysaccharide gene clusters or their parts.

An O-PS may have the same structure as a capsular or extracellular polysaccharide produced by the same bacterium (Wilkinson, 1996) or, sometimes, by different bacteria. For example, as shown in Figure 4.4, an identical structure **15** has been established for the pyruvic acid (*R*)-acetal-containing O-PS of *S. dysenteriae* type 9 and the capsular polysaccharide of *E. coli* K47 (Zhao *et al.*, 2007).

In addition, in members of the *Enterobacteriaceae*, a specific O-PS may be replaced or coexist with the enterobacterial common antigen, a heteropolysaccharide having a trisaccharide repeating unit, which is known in several forms, including a core-lipid A-linked form (Kuhn *et al.*, 1988).

O-PSs may be synthesized by a combined pathway with other glycoconjugates. For instance, in *P. aeruginosa* O7, a single pseudaminic acid-containing trisaccharide O-unit is either polymerized to give the O7a,b,c polysaccharides (see Figure 4.3) or *O*-linked to a serine residue of pilin, a glycoprotein of the somatic pili (Castric *et al.*, 2001).

3.3. Block copolysaccharides

O-PSs may consist of blocks of O-units having different structures. In *K. pneumoniae* O1, two homopolysaccharide domains called galactan I and II are connected to afford a copolymer **16** (Vinogradov *et al.*, 2002) (Figure 4.5). Galactan II is necessary for serum resistance and O1 strains are most prevalent in clinical isolates. The O-PS of *K. pneumoniae* O2a,c **17** includes a galactan I domain associated with a heteropolysaccharide domain.

The O-PS of *S. marcescens* O19 **18** (see Figure 4.5) consists of two blocks built of similar disaccharide repeating units that differ in the

```
 Rpyr
 3/\4
→2)-β-D-Galp-(1→4)-β-D-Manp-(1→4)-α-D-Galp-(1→3)-β-D-GlcpNAc-(1→     15
```

FIGURE 4.4 The identical structure of the (*R*)-acetal-pyruvic acid (*R*pyr) containing O-PS of *S. dysenteriae* type 9 and the capsular polysaccharide of *E. coli* K47 **15**.

```
–[→3)-β-D-Galp-(1→3)-α-D-Galp-(1–]n–[→3)-β-D-Galf-(1→3)-α-D-Galp-(1–]m→     16
          galactan II                         galactan I

–[→5)-β-D-Galf-(1→3)-β-D-GlcpNAc-(1–]n–[→3)-β-D-Galf-(1→3)-α-D-Galp-(1–]m→     17
                                               galactan I

–[→3)-α-L-Rhap-(1→3)-β-D-GlcpNAc-(1–]n–[→4)-α-L-Rhap-(1→3)-β-D-GlcpNAc-(1–]m→     18
```

FIGURE 4.5 Structures of the block copolymer-type O-PSs of *K. pneumoniae* **16**, **17** and *S. marcescens* **18**.

position of glycosylation of the rhamnose residue (Vinogradov *et al.*, 2003c). It was suggested that domains of the same repeating units occur also in serotypes O1 and O17 and serological differences between strains of these O-serotypes may be due to variations in the lengths and position of the blocks.

4. NON-REPETITIVE MOTIFS

4.1. Irregular carbohydrate structures

The O-PS of *Xanthomonas campestris* bv. *vitians* shows a rare example of the lack of regularity in the main chain, which is due to a random distribution of α- and β-linked L-rhamnopyranose residues (Molinaro *et al.*, 2002). Two O-PSs have been isolated, one linear **19** and the other branched **20** (Figure 4.6), in which some α-L-rhamnose residues are glycosylated with residues of 3-acetamido-3,6-dideoxy-α-D-galactose (α-D-Fuc3NAc). In both O-PSs, the majority of α-L-rhamnose blocks include two sugar residues, but can include just a single residue and blocks of three residues occur as well. The irregular structure of the main chain may result from a relaxed specificity of rhamnosyltransferases involved.

Similarly, caryophyllan, the major O-PS of *B. caryophylli,* consists of α-(1→7)- and β-(1→7)-linked caryopyllose **4** residues in a 8:1 ratio, respectively (De Castro *et al.*, 1998), but no data of their sequence in the O-chain have been obtained to date.

4.2. Non-stoichiometric O-unit modifications

A common reason for masking the repetitive O-PS structure is a non-stoichiometric modification of the O-unit by glycosylation, acetylation, methylation, amidation or phosphorylation that is added in the course of or after polymerization.

Most often glycosylation is D-glucosylation but substitution by other monosaccharides, such as L-xylose, D-galactose, D- and L-fucose, D- and L-rhamnose, are known to occur as well. A repeating unit may possess more than one site of post-polymerization modification. For instance, both the main chain and the side chain of the O-PS are non-stoichiometrically D-glucosylated in *H. alvei* PCM 1189 (Katzenellenbogen *et al.*, 2005). In addition, three sites of L-xylosylation have been reported in the main chain of the O-unit of *X. campestrus* pv. *pruni* (Molinaro *et al.*, 2003).

Multiple *O*-acetylation of repeating units occurs often. This is usually site-specific but may be random within one monosaccharide residue, e.g. within 6-deoxy-L-talose in *Aeromonas hydrophila* O34 (Knirel *et al.*, 2002b) or caryophyllose in *B. caryophylli* (De Castro *et al.*, 2001). *O*-Methylation and phosphorylation are site-specific in all known cases. For example, only one of four α-D-rhamnose residues is replaced with 3-*O*-methyl-α-D-rhamnose (D-acofriose) in ≈30% of repeating units of *P. syringae* pv. *phaseolicola* (Zdorovenko *et al.*, 2001). Multiple phosphorylation has been reported with up to three phosphate groups linking

$$-[\rightarrow 3)\text{-}\alpha\text{-L-Rha}p\text{-}(1-]_n\rightarrow 3)\text{-}\beta\text{-L-Rha}p\text{-}(1\rightarrow$$

19

$$\begin{array}{c}\alpha\text{-D-Fuc}p\text{3NAc}\\ 1\\ \downarrow\\ 2\\ -[\rightarrow 3)\text{-}\alpha\text{-L-Rha}p\text{-}(1-]_n\rightarrow 3)\text{-}\beta\text{-L-Rha}p\text{-}(1\rightarrow\end{array}$$

20

FIGURE 4.6 Irregular structure motifs of the O-PS of *Xanthomonas campestris* bv. *vitians* **19**, **20**. Abbreviations: Fuc3NAc, 3-acetamido-3,6-dideoxygalactose.

one ribitol and two ethanolamine residues in the tetrasaccharide O-unit of *P. mirabilis* O41 (Senchenkova *et al.*, 2004).

Amidation of the carboxyl group occurs usually when several amino- and diamino- hexuronic acids are present in the repeating unit (Knirel, 1990; Vinogradov, 2002) and is believed to help to adjust the optimal charge of the cell surface.

Another reason of structural heterogeneity in O-PSs is a non-stoichiomteric epimerization at C-5 of hexuronic acids, which is well known to occur in bacterial alginate. An example is provided by *P. aeruginosa* O(2a),2d,2f that possess comparable amounts of trisaccharide O-units of two types which differ in the occurrence of either 2,3-diamino-2,3-dideoxy-D-mannuronic acid or 2,3-diamino-2,3-dideoxy-L-guluronic acid (Knirel, 1990).

Finally, incorporation to the O-PS of amino sugars with different *N*-acyl substituents has been reported, e.g. in *Acinetobacter baumannii* strain 24 the amino group at C-4 of 2,4-diamino-2,4,6-trideoxy-D-glucose is acetylated in about two-thirds of repeating units of the O-PS and bears an (*S*)-3-hydroxybutanoyl group in the rest (Vinogradov *et al.*, 2003b).

Non-stoichiometric modifications may not be distributed evenly within the polysaccharide chain. Mainly blockwise *O*-acetylation has been reported in caryan, a minor O-PS of *B. caryophylli* (Molinaro *et al.*, 2000). The O-units of *P. syringae* pv. *phaseolicola* that contain 3-*O*-methyl-D-rhamnose are linked to each other nearby the non-reducing end of the O-PS (Zdorovenko *et al.*, 2001). The occurrence of two populations of D-glucosylated and non-glucosylated O-PSs has been reported in *Salmonella enterica* sv. Thompson (Lindberg *et al.*, 1988).

4.3. Non-repetitive terminal domains

In homopolysaccharides, the non-reducing terminus is often occupied by an *O*-methylated monosaccharide, which is usually one of the O-unit components. Examples are D-acofriose and L-acofriose in the corresponding rhamnans or 3-*O*-methyl-D-mannose in D-mannans (Wilkinson, 1996). In *V. cholerae* O1, a loss of the *O*-methyl group from the terminal 2-*O*-methylated 4-amino-4,6-dideoxy-D-mannopyranose residue results in seroconversion from Inaba to Ogawa (Stroeher *et al.*, 1992). Terminal *O*-methylation may occur also in heteropolysaccharides, e.g. in the O-PS of *Bordetella hinzii* that consists of derivatives of 2,3-diamino-2,3-dideoxy-D-glucuronic and -galacturonic acids (Vinogradov, 2002).

In some other cases, the O-PSs are terminated with a monosaccharide that is different from the repeating unit components. Thus, various *N*-acyl derivatives of 2,3,4-triamino-2,3,4-trideoxy-D-galacturonamide occupy the non-reducing end of the O-PS of *Bordetella bronchiseptica* and *Bordetella parapertussis* which is a homopolymer of 2,3-diacetamido-2,3-dideoxy-D-galacturonamide (Vinogradov *et al.*, 2000; Preston *et al.*, 2006). The O-PSs of *S. marcescens* O19 (Vinogradov *et al.*, 2002) and *K. pneumoniae* O12 (Vinogradov *et al.*, 2002) are terminated with a β-linked Kdo residue and that of *K. pneumoniae* O4 (Vinogradov *et al.*, 2002) has a terminal α-linked Kdo; all these O-PSs have disaccharide repeating units.

The non-reducing end of the O-PS may be the subject of phase variation, e.g. in the variable O-PS of the gastric bacterium *H. pylori*, which is composed of occasionally glycosylated *N*-acetyl-β-lactosamine units (Moran, 2008). The terminal non-reducing O-unit often carries one or two α-L-fucose residues giving rise to Lewis x and Lewis y antigen determinants. Although less often, sialyl-Lewis x is observed too. Moreover, the β-(1→4)-linkage of the terminal non-reducing *N*-acetyllactosamine unit may be replaced with an isomeric disaccharide, in which D-galactose and D-*N*-acetylglucosamine are β-(1→3)-interlinked and whose α-L-fucosylation affords Lewis a, Lewis b or H type I blood group antigens (Moran, 2008).

→4)-β-D-ManNAc3NAcAN-(1→4)-β-D-GlcNAc3NAcAN-(1→4)-α-D-GalNAc-(1→

4)-β-D-ManNAc3NAcA-(1→3)-β-D-FucNAc4N-(1→6)-α-D-GlcN-(1→ **21**

FIGURE 4.7 Structure of the pentasaccharide domain between the O-PS and the core in the LPS of *B. parapertussis* and *B. hinzii* **21**. Abbreviations: GlcNAc3NAcAN, ManNAc3NAcAN, 2,3-diacetamido-2,3-dideoxygluc-, mann-uronamides, respectively; FucNAc4N, 2-acetamido-4-amino-2,4,6-trideoxygalactose.

A different monosaccharide, a so-called primer, occurs at the reducing end of homopolysaccharides. The most common primer is β-D-Glc*p*NAc, which is present, e.g. in all *K. pneumoniae* O-serotypes (Vinogradov *et al.*, 2002) and the common polysaccharide antigen of *P. aeruginosa* (Rocchetta *et al.*, 1998). Other sugars can be involved as a primer, e.g. 2-acetamido-2,6-dideoxy-β-D-glucose in caryan of *B. caryophylli* (De Castro *et al.*, 2005).

In some bacteria, a more complex non-repetitive domain occurs between the O-PS and the core. In *K. pneumoniae* O5 and O3, the O-unit is not linked directly to the primer but, instead, the O-PS includes a bridging (→3)-α-D-Man*p*-(1→3)-α-D-Man*p*-(1→3)- disaccharide called an adaptor, which is located between the primer and the first O-unit (Vinogradov *et al.*, 2002). The O-PSs of *B. parapertussis* and *B. hinzii* are linked to the core through a specific pentasaccharide **21** (Preston *et al.*, 2006; Vinogradov, 2007) (Figure 4.7). This domain is not present in natural LPS variants without a repetitive O-PS.

5. CONCLUSIONS

O-PSs are extraordinarily diverse in composition and structure giving rise to a high degree of antigenic heterogeneity both between and within Gram-negative bacterial species. Numerous monosaccharides and non-sugar constituents have been identified as their components; some of them are widespread in Nature, others are less common and some occur in O-PSs only. The structures of most unusual monosaccharides have been established unambiguously using modern analytical methods but the stereochemistry of some requires confirmation, especially when they cannot be isolated in the free state as, e.g. as in the case of higher sugars. Many O-PS structures established by older methods should be confirmed, which can result in significant revision, e.g. as has happened when *S. dysenteriae* and *S. boydii* O-antigen structures have been reinvestigated (Liu *et al.*, 2008).

Heteropolysaccharides are more widespread in bacteria than homopolysaccharides and their topology is more diverse. About half of the known heteropolysaccharides are branched, the whole O-repeating unit varying from di- to octa-saccharide, the number of side chains varying from 1 to 3 and the side chain length from a mono- to a tetra-saccharide. Homopolysaccharides are usually linear and either represent a polymerized monosaccharide or have an oligosaccharide (up to a pentasaccharide) repeating unit.

Many, if not all, homopolysaccharides and linear heteropolysaccharides with disaccharide O-units are synthesized by an ATP-binding cassette (ABC) transporter-dependent pathway including a sequential transfer of single monosaccharides to the growing chain (see Chapter 18). Other heteropolysaccharides are products of polymerization of pre-assembled O-units by the O-antigen polymerase (Wzy)-dependent pathway (Raetz and Whitfield, 2002) (see Chapter 18). An intriguing question is how bacteria regulate the O-PS chain length, which is modal and appears to be fine-tuned giving bacteria advantages in particular econiches.

Several O-PSs consisting of two repetitive homo- or hetero-polysaccharide domains have been discovered. One can expect that more structures of these exist but, in most cases, when more than one O-PS is found, their connection to each other is neither proved nor disproved. Alternatively, one of the multiple isolated polysaccharides may not be related to the LPS as has happened with one of the "O-PSs" of *Burkholderia pseudomallei* (Isshiki *et al.*, 2001).

A common theme in the field is the occurrence of the same or closely related O-PSs in different bacteria. In some particular cases, this phenomenon may complicate reliable serodiagnosis of infectious disease.

As a result of long coevolution with the host, some bacteria, e.g. *H. pylori* (Moran, 2008), have acquired an ability to express on the O-PS various domains that are similar to host carbohydrates. Such molecular mimicry helps the bacterium to escape adaptive immune response and may be a reason of autoimmune diseases (see Chapters 42 and 43).

Although many O-PSs are regular polymers, non-repetitive motifs are rather common, which result either from post-polymerization modifications or, in the case of ABC transporter-dependent synthesis pathway, from different initiation and termination steps of biosynthesis. For instance, an addition of an *O*-methylated sugar or a different monosaccharide to the non-reducing end appears to be a signal for the cessation of the O-PS chain, which allows termination of the O-chain at a specific sugar residue rather than at any residue.

Another non-repetitive domain that is derived from the O-antigen biosynthesis pathway occurs between the O-PS and the core. It usually includes a primer (a 2-*N*-acetylamino sugar) whose transfer to a lipid carrier initiates the O-antigen synthesis. In O-PSs that are synthesized by the Wzy-dependent pathway, the first sugar of the O-unit plays the role of the primer. More complex reducing-end domains (so-called adaptors) have been found to be present in a few O-PSs but may be much more common. A biological role for this domain remains to be determined. These findings point to a higher complexity of LPSs than had been expected based on the impressive progress in structure elucidation of their isolated components at the end of the 20th century.

References

Adinolfi, M., Corsaro, M.M., De Castro, C., et al., 1996a. Caryose: a carbocyclic monosaccharide from *Pseudomonas caryophylli*. Carbohydr. Res. 284, 111–118.

Adinolfi, M., Corsaro, M.M., De Castro, C., et al., 1996b. Analysis of the polysaccharide components of the lipopolysaccharide fraction of *Pseudomonas caryophylli*. Carbohydr. Res. 284, 119–133.

Amano, K., Yatsuyanagi, J., Saito, S., 2007. Development of O-serogroup serodiagnosis for patients with hemolytic uremic syndrome by Ec-LPS array. Kansenshogaku Zasshi. 81, 26–32.

Anderson, A.N., Richards, J.C., Perry, M.B., 1992. Structure of the O-antigen of *Escherichia coli* O119 lipopolysaccharide. Carbohydr. Res. 237, 249–262.

Aspinall, G.O., McDonald, A.G., Pang, H., 1992. Structures of the O chains from lipopolysaccharides of *Campylobacter jejuni* serotypes O:23 and O:36. Carbohydr. Res. 231, 13–30.

Aucken, H.M., Wilkinson, S.G., Pitt, T.L., 1998. Re-evaluation of the serotypes of *Serratia marcescens* and separation into two schemes based on lipopolysaccharide (O) and capsular polysaccharide (K) antigens. Microbiology 144, 639–653.

Blixt, O., Hoffmann, J., Svenson, S., Norberg, T., 2008. Pathogen specific carbohydrate antigen microarrays: a chip for detection of *Salmonella* O-antigen specific antibodies. Glycoconj. J. 25, 27–36.

Bravo, D., Silva, C., Carter, J.A., et al., 2008. Growth-phase regulation of lipopolysaccharide O-antigen chain length influences serum resistance in serovars of *Salmonella*. J. Med. Microbiol. 57, 938–946.

Bruneteau, M., Minka, S., 2003. Lipopolysaccharides of bacterial pathogens from the genus *Yersinia*: a mini-review. Biochimie 85, 145–152.

Castric, P., Cassels, F.J., Carlson, R.W., 2001. Structural characterization of the *Pseudomonas aeruginosa* 1244 pilin glycan. J. Biol. Chem. 276, 26479–26485.

Conlan, J.W., Shen, H., Webb, A., Perry, M.B., 2002. Mice vaccinated with the O-antigen of *Francisella tularensis* LVS lipopolysaccharide conjugated to bovine serum albumin develop varying degrees of protective immunity against systemic or aerosol challenge with virulent type A and type B strains of the pathogen. Vaccine 20, 3465–3471.

Corsaro, M.M., De Castro, C., Molinaro, A., Parrilli, M., 2001. Structure of lipopolysaccharides from phytopathogenic Gram-negative bacteria. Recent Res. Devel. Phytochem. 5, 119–138.

Cryz Jr, S.J., Que, J.O., Cross, A.S., Fürer, E., 1995. Synthesis and characterization of a polyvalent *Escherichia coli* O-polysaccharide-toxin A conjugate vaccine. Vaccine 13, 449–453.

De Castro, C., Evidente, A., Lanzetta, R., et al., 1998. Presence of β-glycosyl linkages in caryophyllan: the main polysaccharide from the *Pseudomonas caryophylli* LPS fraction. Carbohydr. Res. 307, 167–172.

De Castro, C., Lanzetta, R., Molinaro, A., Parrilli, M., Piscopo, V., 2001. Acetyl substitution of the O-specific polysaccharide caryophyllan from the phenol phase of *Pseudomonas (Burkholderia) caryophylli*. Carbohydr. Res. 335, 205–211.

De Castro, C., Molinaro, A., Lanzetta, R., Holst, O., Parrilli, M., 2005. The linkage between O-specific caryan and core region in the lipopolysaccharide of *Burkholderia caryophylli* is furnished by a primer monosaccharide. Carbohydr. Res. 340, 1802–1807.

De Castro, C., Molinaro, A., Lanzetta, R., Silipo, A., Parrilli, M., 2008. Lipopolysaccharide structures from *Agrobacterium* and *Rhizobiaceae* species. Carbohydr. Res. 343, 1924–1933.

Döring, G., Pier, G.B., 2008. Vaccines and immunotherapy against *Pseudomonas aeruginosa*. Vaccine 26, 1011–1024.

Erridge, C., Bennett-Guerrero, E., Poxton, I.R., 2002. Structure and function of lipopolysaccharides. Microbes Infect. 4, 837–851.

Ewing, W.H., 1986. Edwards and Ewing's Identification of *Enterobacteriaceae*. Elsevier, New York.

Favre Jr, D., Cryz, S.J., Viret, J.F., 1996. Development of *Shigella sonnei* live oral vaccines based on defined rfb_{Inaba} deletion mutants of *Vibro cholerae* expressing the *Shigella* serotype D O polysaccharide. Infect. Immun. 64, 576–584.

Fudala, R., Kondakova, A.N., Bednarska, K., et al., 2003. Structure and serological characterization of the O-antigen of *Proteus mirabilis* O18 with a phosphocholine-containing oligosaccharide phosphate repeating unit. Carbohydr. Res. 338, 1835–1842.

Gamian, A., Kenne, L., 1993. Analysis of 7-substituted sialic acid in some enterobacterial lipopolysaccharides. J. Bacteriol. 175, 1508–1513.

Hanniffy, O.M., Shashkov, A.S., Senchenkova, S.N., et al., 1998. Structure of a highly acidic O-specific polysaccharide from *Pseudoalteromonas haloplanktis* KMM 223 (44-1) containing L-iduronic acid. Carbohydr. Res 307, 291–298.

Iredell, J.R., Stroeher, U.H., Ward, H.M., Manning, P.A., 1998. Lipopolysaccharide O-antigen expression and the effect of its absence on virulence in *rfb* mutants of *Vibrio cholerae* O1. FEMS Immunol. Med. Microbiol. 20, 45–54.

Isshiki, Y., Matsuura, M., Dejsirilert, S., Ezaki, T., Kawahara, K., 2001. Separation of 6-deoxy-heptane from a smooth-type lipopolysaccharide preparation of *Burkholderia pseudomallei*. FEMS Microbiol. Rev. 199, 21–25.

Jann, K., Jann, B., 1984. Structure and biosynthesis of O-antigens. In: Rietschel, E.T. (Ed.), Handbook of Endotoxin, Vol. 1. Chemistry of Endotoxin. Elsevier, Amsterdam, pp. 138–186.

Jansson, P.-E., 1999. The chemistry of O-polysaccharide chains in bacterial lipopolysaccharides. In: Brade, H., Opal, S.M., Vogel, S.N., Morrison, D.C. (Eds.). Endotoxin in Health and Disease. Marcel Dekker, New York, pp. 155–178.

Jauho, E.S., Boas, U., Wiuff, C., et al., 2000. New technology for regiospecific covalent coupling of polysaccharide antigens in ELISA for serological detection. J. Immunol. Methods. 242, 133–143.

Katzenellenbogen, E., Kocharova, N.A., Zatonsky, G.V., et al., 2001. Structural and serological studies on *Hafnia alvei* O-specific polysaccharide of α-D-mannan type isolated from the lipopolysaccharide of strain PCM 1223. FEMS Immunol. Med. Microbiol. 30, 223–227.

Katzenellenbogen, E., Kocharova, N.A., Zatonsky, G.V., et al., 2005. Structure of the O-polysaccharide of *Hafnia alvei* strain PCM 1189 that has hexa- to octa-saccharide repeating units owing to incomplete glucosylation. Carbohydr. Res. 340, 263–270.

Kawasaki, M., Takamatsu, N., Ansai, T., Yamashita, Y., Takehara, T., Koga, T., 1995. An enzyme-linked immunosorbent assay for measuring antibodies to serotype-specific polysaccharide antigens of *Actinobacillus actinomycetemcomitans*. J. Microbiol. Methods 21, 181–192.

Kay, W., Petersen, B.O., Duus, J., Perry, M.B., Vinogradov, E., 2006. Characterization of the lipopolysaccharide and β-glucan of the fish pathogen *Francisella victoria*. FEBS J. 273, 3002–3013.

Kenne, L., Lindberg, B., 1983. Bactertial polysaccharides. In: Aspinall, G.O. (Ed.), The Polysaccharides, Vol. 2. Academic Press, New York, pp. 287–363.

Kenne, L., Lindberg, B., Petersson, K., Katzenellenbogen, E., Romanowska, E., 1980. Structural studies of the O-specific side-chains of the *Shigella sonnei* phase I lipopolysaccharide. Carbohydr. Res. 78, 119–126.

Kilcoyne, M., Shashkov, A.S., Senchenkova, S.N., et al., 2002. Structural investigation of the O-specific polysaccharides of *Morganella morganii* consisting of two higher sugars. Carbohydr. Res. 337, 1697–1702.

Kilcoyne, M., Perepelov, A.V., Tomshich, S.V., et al., 2004. Structure of the O-polysaccharide of *Idiomarina zobellii* KMM 231^T containing two unusual amino sugars with the free amino group, 4-amino-4,6-dideoxy-D-glucose and 2-amino-2-deoxy-L-guluronic acid. Carbohydr. Res. 339, 477–482.

Kilcoyne, M., Shashkov, A.S., Knirel, Y.A., et al., 2005. The structure of the O-polysaccharide of the *Pseudoalteromonas rubra* ATCC 29570^T lipopolysaccharide containing a keto sugar. Carbohydr. Res. 340, 2369–2375.

Kintz, E., Goldberg, J.B., 2008. Regulation of lipopolysaccharide O antigen expression in *Pseudomonas aeruginosa*. Future Microbiol. 3, 191–203.

Knirel, Y.A., 1990. Polysaccharide antigens of *Pseudomonas aeruginosa*. CRC Crit. Rev. Microbiol. 17, 273–304.

Knirel, Y.A., Kochetkov, N.K., 1994. The structure of lipopolysaccharides of Gram-negative bacteria. III. The structure of O-antigens. Biochemistry (Moscow) 59, 1325–1383.

Knirel, Y.A., Kaca, W., Rozalski, A., Sidorczyk, Z., 1999. Structure of the O-antigenic polysaccharides of *Proteus* bacteria. Polish J. Chem. 73, 895–907.

Knirel, Y.A., Kocharova, N.A., Bystrova, O.V., Katzenellenbogen, E., Gamian, A., 2002a. Structures and serology of the O-specific polysaccharides of bacteria of the genus *Citrobacter*. Arch. Immunol. Ther. Exp. 50, 379–391.

Knirel, Y.A., Shashkov, A.S., Senchenkova, S.N., Merino, S., Tomas, J.M., 2002b. Structure of the O-polysaccharide of *Aeromonas hydrophila* O:34; a case of random *O*-acetylation of 6 deoxy-L-talose. Carbohydr. Res. 337, 1381–1386.

Knirel, Y.A., Shashkov, A.S., Tsvetkov, Y.E., Jansson, P.-E., Zähringer, U., 2003. 5,7 Diamino-3,5,7,9-tetradeoxynon-2-ulosonic acids in bacterial glycopolymers: chemistry and biochemistry. Adv. Carbohydr. Chem. Biochem. 58, 371–417.

Knirel, Y.A., Bystrova, O.V., Kocharova, N.A., Zähringer, U., Pier, G.B., 2006. Conserved and variable structural features of the *Pseudomonas aeruginosa* lipopolysaccharide. J. Endotoxin Res. 12, 324–336.

Kocharova, N.A., Knirel, Y.A., Widmalm, G., Jansson, P.-E., Moran, A.P., 2000. Structure of an atypical O-antigen polysaccharide of *Helicobacter pylori* containing a novel monosaccharide 3-*C*-methyl-D-mannose. Biochemistry 39, 4755–4760.

Kocharova, N.A., Zatonsky, G.V., Bystrova, O.V., et al., 2002. Structure of the O-specific polysaccharide of *Providencia alcalifaciens* O16 containing *N*-acetylmuramic acid. Carbohydr. Res. 337, 1667–1671.

Kocharova, N.A., Zatonsky, G.V., Torzewska, A., et al., 2003. Structure of the O-specific polysaccharide of *Providencia rustigianii* O14 containing N^{ε}-[(S)-1-carboxyethyl]-N^{α}-(D-galacturonoyl)-L-lysine. Carbohydr. Res. 338, 1009–1016.

Kocharova, N.A., Senchenkova, S.N., Kondakova, A.N., et al., 2004. D- and L-Aspartic acids: new non-sugar components of bacterial polysaccharides. Biochemistry (Moscow) 69, 103–107.

Kocharova, N.A., Bushmarinov, I.S., Ovchinnikova, O.G., et al., 2005. The structure of the O-polysaccharide from the lipopolysaccharide of *Providencia stuartii* O44 containing L-quinovose, a 6-deoxy sugar rarely occurring in bacterial polysaccharides. Carbohydr. Res. 340, 1419–1423.

Kocharova, N.A., Ovchinnikova, O.G., Torzewska, A., Shashkov, A.S., Knirel, Y.A., Rozalski, A., 2007. The structure of the O-polysaccharide from the lipopolysaccharide of *Providencia alcalifaciens* O36 containing 3-deoxy-D-*manno*-oct-2-ulosonic acid. Carbohydr. Res. 342, 665–670.

Kondakova, A.N., Zych, K., Senchenkova, S.N., Sidorczyk, Z., Shashkov, A.S., Knirel, Y.A., 2003. Structure of the *N*-acetyl-L-rhamnosamine-containing O-polysaccharide of *Proteus vulgaris* TG 155 from a new *Proteus* serogroup, O55. Carbohydr. Res. 338, 1999–2004.

Kondakova, A.N., Lindner, B., Fudala, R., et al., 2004. New stuctures of the O-specific polysaccharides of *Proteus*. Part 4. Polysaccharides containing unusual acidic *N*-acyl derivatives of 4-amino-4,6-dideoxy-D-glucose. Biochemistry (Moscow) 69, 1034–1043.

Kubler-Kielb, J., Vinogradov, E., Ben Menachem, G., Pozsgay, V., Robbins, J.B., Schneerson, R., 2008. Saccharide/protein conjugate vaccines for *Bordetella* species: Preparation of saccharide, development of new conjugation procedures, and physico-chemical and immunological characterization of the conjugates. Vaccine 26, 3587–3593.

Kuhn, H.-M., Meier-Dieter, U., Mayer, H., 1988. ECA, the enterobacterial common antigen. FEMS Microbiol. Lett. 54, 195–222.

Lindberg, B., 1990. Components of bacterial polysaccharides. Adv. Carbohydr. Chem. Biochem. 48, 279–318.

Lindberg, B., 1998. Bacterial polysaccharides: components. In: Dumitriu, S. (Ed.), Polysaccharides: Structural Diversity and Functional Versatility. Marcel Dekker, New York, pp. 237–273.

Lindberg, B., Leontein, K., Lindquist, U., et al., 1988. Structural studies of the O-antigen polysaccharide of *Salmonella thompson*, serogroup C_1 (6,7). Carbohydr. Res. 174, 313–322.

Liu, B., Knirel, Y.A., Feng, L., et al., 2008. Structure and genetics of *Shigella* O antigens. FEMS Microbiol. Rev. 32, 627–653.

Liu, P.V., Matsumoto, H., Kusama, H., Bergan, T., 1983. Survey of heat-stable major somatic antigens of *Pseudomonas aeruginosa*. Int. J. Syst. Bacteriol. 33, 256–264.

Lugo, J.Z., Price, S., Miller, J.E., et al., 2007. Lipopolysaccharide O-antigen promotes persistent murine bacteremia. Shock 27, 186–191.

MacLean, L.L., Vinogradov, E., Crump, E.M., Perry, M.B., Kay, W.W., 2001. The structure of the lipopolysaccharide O-antigen produced by *Flavobacterium psychrophilum* (259-93). Eur. J. Biochem. 268, 2710–2716.

MacLean, L.L., Perry, M.B., Crump, E.M., Kay, W.W., 2003. Structural characterization of the lipopolysaccharide O-polysaccharide antigen produced by *Flavobacterium columnare* ATCC 43622. Eur. J. Biochem. 270, 3440–3446.

Meikle, P.J., Perry, M.B., Cherwonogrodzky, J.W., Bundle, D.R., 1989. Fine structure of A and M antigens from *Brucella* biovars. Infect. Immun. 57, 2820–2828.

Molinaro, A., De Castro, C., Petersen, B.O., Duus, J.O., Parrilli, M., Holst, O., 2000. Acetyl substitution of the O-specific caryan from the lipopolysaccharide of

Pseudomonas (Burkholderia) caryophylli leads to a block pattern. Angew. Chem. Int. Ed. Engl. 39, 156–160.

Molinaro, A., De Castro, C., Lanzetta, R., et al., 2002. NMR and MS evidences for a random assembled O-specific chain structure in the LPS of the bacterium *Xanthomonas campestris* pv. *vitians*. Eur. J. Biochem. 269, 4185–4193.

Molinaro, A., Evidente, A., Lo Cantore, P., et al., 2003. Structural determination of a novel O-chain polysaccharide of the lipopolysaccharide from the bacterium *Xanthomonas campestris* pv. *Pruni*. Eur. J. Org. Chem. 2254–2259.

Moran, A.P., 2008. Relevance of fucosylation and Lewis antigen expression in the bacterial gastroduodenal pathogen *Helicobacter pylori*. Carbohydr. Res. 343, 1952–1965.

Morona, R., Daniels, C., Van den Bosch, L., 2003. Genetic modulation of *Shigella flexneri* 2a lipopolysaccharide O antigen modal chain length reveals that it has been optimized for virulence. Microbiology 149, 925–939.

Newman, M.A., Dow, J.M., Molinaro, A., Parrilli, M., 2007. Priming, induction and modulation of plant defence responses by bacterial lipopolysaccharides. J. Endotoxin Res. 13, 69–84.

Nielsen, R., 1986. Serological characterization of *Actinobacillus pleuropneumoniae* strains and proposal of a new serotype: serotype 12. Acta Vet. Scand. 27, 453–455.

Ovod, V., Knirel, Y.A., Samson, R., Krohn, K., 1999. Immunochemical characterization and taxonomic evaluation of the O polysaccharides of the lipopolysaccharides of *Pseudomonas syringae* serogroup O1 strains. J. Bacteriol. 181, 6937–6947.

Ovod, V., Zdorovenko, E.L., Shashkov, A.S., Kocharova, N.A., Knirel, Y.A., 2004. Structural diversity of O-polysaccharides and serological classification of *Pseudomonas syringae* pv. *garcae* and other strains of genomospecies 4. Mikrobiologiya 73, 666–677.

Ovodov, Y.S., Gorshkova, R.P., 1988. Lipopolysaccharides of *Yersinia pseudotuberculosis*. Khim. Prirod. Soed., 163–171.

Ovodov, Y.S., Gorshkova, R.P., Tomshich, S.V., et al., 1992. Chemical and immunochemical studies on lipopolysaccharides of some *Yersinia* species. A review of some recent investigations. J. Carbohydr. Chem. 11, 21–35.

Perepelov, A.V., Babicka, D., Senchenkova, S.N., et al., 2001. Structure of the O-specific polysaccharide of *Proteus vulgaris* O4 containing a new component of bacterial polysaccharides, 4,6-dideoxy-4-{*N*-[(*R*)-3-hydroxybutyryl]-L-alanyl}amino-D-glucose. Carbohydr. Res. 331, 195–202.

Perepelov, A.V., Senchenkova, S.N., Shashkov, A.S., Rozalski, A., Knirel, Y.A., 2002. Structure of the O-specific polysaccharide of the bacterium *Proteus vulgaris* O15 containing a novel regioisomer of *N*-acetylmuramic acid. Carbohydr. Res. 337, 2463–2468.

Perepelov, A.V., Liu, B., Senchenkova, S.N., et al., 2008. Structure of the O-polysaccharide of *Escherichia coli* O112ab containing L-iduronic acid. Carbohydr. Res. 343, 571–575.

Preston, A., Petersen, B.O., Duus, J.Ø., et al., 2006. Complete structure of *Bordetella bronchiseptica* and *Bordetella parapertussis* lipopolysaccharides. J. Biol. Chem. 281, 18135–18144.

Raetz, C.R.H., Whitfield, C., 2002. Lipopolysaccharide endotoxins. Annu. Rev. Biochem. 71, 635–700.

Reuhs, B.L., Relic, B., Forsberg, L.S., et al., 2005. Structural characterization of a flavonoid-inducible *Pseudomonas aeruginosa* A-band-like O antigen of *Rhizobium* sp. strain NGR234, required for the formation of nitrogen-fixing nodules. J. Bacteriol. 187, 6479–6487.

Rocchetta, H.L., Burrows, L.L., Pacan, J.C., Lam, J.S., 1998. Three rhamnosyltransferases responsible for assembly of the A-band D-rhamnan polysaccharide in *Pseudomonas aeruginosa*: a fourth transferase, WbpL, is required for the initiation of both A-band and B-band lipopolysaccharide synthesis. Mol. Microbiol. 28, 1103–1119.

Romanowska, E., 2000. Immunochemical aspects of *Hafnia alvei* O antigens. FEMS Immunol. Med. Microbiol. 27, 219–225.

Senchenkova, S.N., Perepelov, A.V., Cedzynski, M., et al., 2004. Structure of a highly phosphorylated O-polysaccharide of *Proteus mirabilis* O41. Carbohydr. Res. 339, 1347–1352.

Senchenkova, S.N., Shashkov, A.S., Knirel, Y.A., Ahmed, M., Mavridis, A., Rudolph, K., 2005. Structure of the O-polysaccharide of *Erwinia carotovora* ssp. *atroseptica* GSPB 9205, containing a new higher branched monosaccharide. Russ. Chem. Bull. 1276–1281.

Shashkov, A.S., Senchenkova, S.N., Vinogradov, E.V., et al., 2000. Full structure of the O-specific polysaccharide of *Proteus mirabilis* O24 containing 3,4-*O*-[(*S*)-1-carboxyethylidene]-D-galactose. Carbohydr. Res. 329, 453–457.

Shashkov, A.S., Torgov, V.I., Nazarenko, E.L., et al., 2002. Structure of the phenol-soluble polysaccharide from *Shewanella putrefaciens* strain A6. Carbohydr. Res. 337, 1119–1127.

Shimada, T., Arakawa, E., Itoh, K., et al., 1994. Expanded serotyping scheme for *Vibrio cholerae*. Curr. Microbiol. 28, 175–178.

Skurnik, M., Bengoechea, J.A., 2003. The biosynthesis and biological role of lipopolysaccharide O-antigens of pathogenic *Yersiniae*. Carbohydr. Res. 338, 2521–2529.

Stenutz, R., Weintraub, A., Widmalm, G., 2006. The structures of *Escherichia coli* O-polysaccharide antigens. FEMS Microbiol. Rev. 30, 382–403.

Stroeher, U.H., Karageorgos, L.E., Morona, R., Manning, P.A., 1992. Serotype conversion in *Vibrio cholerae*. Proc. Natl. Acad. Sci. USA 89, 2566–2570.

Taylor, R.K., Kirn, T.J., Bose, N., et al., 2004. Progress towards development of a cholera subunit vaccine. Chem. Biodivers. 1, 1036–1057.

Thirumalapura, N.R., Morton, R.J., Ramachandran, A., Malayer, J.R., 2005. Lipopolysaccharide microarrays for the detection of antibodies. J. Immunol. Methods 298, 73–81.

Trautmann, M., Cross, A.S., Reich, G., Held, H., Podschun, R., Marre, R., 1996. Evaluation of a competitive ELISA method for the determination of *Klebsiella* O antigens. J. Med. Microbiol. 44, 44–51.

Vinion-Dubiel, A.D., Goldberg, J.B., 2003. Lipopolysaccharide of *Burkholderia cepacia* complex. J. Endotoxin Res. 9, 201–213.

Vinogradov, E., 2002. Structure of the O-specific polysaccharide chain of the lipopolysaccharide of *Bordetella hinzii*. Carbohydr. Res. 337, 961–963.

Vinogradov, E., 2007. The structure of the core-O-chain linkage region of the lipopolysaccharide from *Bordetella hinzii*. Carbohydr. Res. 342, 638–642.

Vinogradov, E., Peppler, M.S., Perry, M.B., 2000. The structure of the nonreducing terminal groups in the O-specific polysaccharides from two strains of *Bordetella bronchiseptica*. Eur. J. Biochem. 267, 7230–7236.

Vinogradov, E., Frirdich, E., MacLean, L.L., et al., 2002. Structures of lipopolysaccharides from *Klebsiella pneumoniae*. Eluicidation of the structure of the linkage region between core and polysaccharide O chain and identification of the residues at the non-reducing termini of the O chains. J. Biol. Chem. 277, 25070–25081.

Vinogradov, E., Korenevsky, A., Beveridge, T.J., 2003a. The structure of the O-specific polysaccharide chain of the *Shewanella algae* BrY lipopolysaccharide. Carbohydr. Res. 338, 385–388.

Vinogradov, E.V., Brade, L., Brade, H., Holst, O., 2003b. Structural and serological characterisation of the O-antigenic polysaccharide of the lipopolysaccharide from *Acinetobacter baumannii* strain 24. Carbohydr. Res. 338, 2751–2756.

Vinogradov, E., Petersen, B.O., Duus, J.Ø., Radziejewska-Lebrecht, J., 2003c. The structure of the polysaccharide part of the LPS from *Serratia marcescens* serotype O19, including linkage region to the core and the residue at the non-reducing end. Carbohydr. Res. 338, 2757–2761.

Vinogradov, E., MacLean, L.L., Crump, E.M., Perry, M.B., Kay, W.W., 2003d. Structure of the polysaccharide chain of the lipopolysaccharide from *Flexibacter maritimus*. Eur. J. Biochem. 270, 1810–1815.

Wilkinson, S.G., 1996. Bacterial lipopolysaccharides. Themes and variations. Prog. Lipid Res. 35, 283–343.

Zdorovenko, E.L., Ovod, V., Zatonsky, G.V., Shashkov, A.S., Kocharova, N.A., Knirel, Y.A., 2001. Location of the *O*-methyl groups in the O polysaccharide of *Pseudomonas syringae* pv. *phaseolicola*. Carbohydr. Res. 330, 505–510.

Zhao, G., Perepelov, A.V., Senchenkova, S.N., et al., 2007. Structural relation of the antigenic polysaccharides of *Escherichia coli* O40, Shigella dysenteriae type 9, and E. coli K47. Carbohydr. Res. 342, 1275–1279.

Zubkov, V.A., Gorshkova, R.P., Ovodov, Y.S., Sviridov, A.F., Shashkov, A.S., 1992. Synthesis of 3,6-dideoxy-4-*C*-(4^1-hydroxyethyl)hexopyranoses (yersinioses) from 1,6-anhydro-β-D-glycopyranose. Carbohydr. Res. 225, 189–207.

CHAPTER

5

Teichoic acids, lipoteichoic acids and related cell wall glycopolymers of Gram-positive bacteria

Thomas Kohler, Guoqing Xia, Emir Kulauzovic and Andreas Peschel

SUMMARY

Most Gram-positive bacteria contain characteristic membrane- or peptidoglycan-attached carbohydrate-based polymers in their cell wall whose structure, function and biosynthesis are only superficially understood. The composition of these cell wall glycopolymers (CWGs) is highly variable and often species and strain specific. Recent studies have yielded a growing picture of the biosynthetic pathways for the cell wall-anchored wall teichoic acid (WTA) and the membrane-anchored lipoteichoic acid (LTA) in *Bacillus subtilis* and *Staphylococcus aureus*. Besides zwitterionic LTA and WTA, further CWG types such as anionic teichuronic acids or uncharged lipoglycans occur in certain Gram-positive bacteria. The CWGs play important roles in bacterial physiology and various potential functions, such as control of autolytic enzymes, regulation of divalent cations, attachment of surface proteins or protection against antibacterial molecules, have been described. In those bacteria who colonize or infect animal hosts, CWGs have been implicated in adherence to host cells and activation of immune responses, e.g. via the Toll-like receptors. In addition, CWGs are very important targets for antimicrobials, vaccines and diagnostics. Accordingly, CWGs represent a very important topic for research.

Keywords: Teichoic acid; Cell wall glycopolymers; Lipoteichoic acid; Teichuronic acid; Lipoarabinomannan; Pathogen-associated molecular pattern; Glycobiology; *Staphylococcus aureus*; *Bacillus subtilis*; Innate immunity; Vaccines; Antibiotics; Diagnostics

1. INTRODUCTION

Most Gram-positive bacterial cell walls are composed of two types of structural components, the peptidoglycan (PG) with highly conserved composition and additional cell wall glycopolymers (CWGs), which vary extensively between species and even between individual strains (Figure 5.1). While the functions of PG in maintaining the mechanical stability of the bacterial cells are well studied, the roles of CWGs are still far from clear. Most Gram-positive bacteria have two types of CWG, one a PG-anchored polymer (P-CWG) and one a membrane-anchored polymer (M-CWG) (Fischer, 1988; Neuhaus and Baddiley, 2003; Weidenmaier and Peschel, 2008). Some bacteria even have three or four types of CWGs. The CWG polymers are composed of repeating units containing one or more sugar building blocks plus non-sugar residues such as phosphate, alanine, succinate, pyruvate, choline or mycolic acids to name a few. Those CWGs containing phosphate in the polymer backbone are traditionally referred to as teichoic acids (TAs) according to James Baddiley's nomenclature from the 1950s (Armstrong *et al.*, 1959). Most TAs have zwitterionic properties since, in addition to the negatively charged phosphate groups, the repeating units contain D-alanine residues with free, positively charged amino groups (Fischer, 1988; Neuhaus and Baddiley, 2003). Many Gram-positive bacteria produce CWGs without phosphate groups in repeating units. Such polymers are either uncharged, e.g. in many actinobacteria and certain bacilli, or are anionic because of the presence of uronic acid, pyruvyl or succinyl groups (Powell *et al.*, 1975; Ward, 1981; Greenberg *et al.*, 1996; Delmas *et al.*, 1997; Soldo *et al.*, 1999; Schäffer and Messner, 2005). This review will mainly focus on TAs, their structures, biosynthetic pathways and functions while other types of CWGs will be mentioned only briefly.

2. TEICHOIC ACID STRUCTURES

2.1. Wall teichoic acids and other P-CWGs

Most P-CWGs are attached to PG via linkage units with more or less conserved structures (Figure 5.2). These units often start with *N*-acetylglucosamine (GlcNAc) residues, which are linked to the *N*-acetylmuramic acid of PG via a phosphodiester bond and are connected to further sugars of the linkage unit or directly connected to the CWG polymer (Araki and Ito, 1989; Naumova and Shashkov, 1997). The TAs of the P-CWG type are referred to as wall teichoic acids (WTAs), which can account for up to 50% of the entire cell wall mass. The linkage unit of *Staphylococcus aureus* WTA, which is particularly well studied, is composed of one GlcNAc residue, one *N*-acetylmannosamine and two glycerolphosphate (Gro-P) residues. The WTA repeating units contain sugars of various sizes ranging from trioses to hexoses, which are frequently reduced to corresponding polyols such as glycerol or ribitol. The WTA structures are particularly diverse and differ profoundly between species or even strains (Endl *et al.*, 1983; Potekhina *et al.*, 1993; Fischer *et al.*, 1993; Naumova and Shashkov, 1997). The WTA polymer of most *S. aureus* strains is formed by up to 40 repeating units of ribitolphosphate (Endl *et al.*, 1983). Simpler WTA structures occur, e.g. in *Staphylococcus epidermidis*, where Gro-P units are not only found in the linkage unit but form the entire polymer. More complex structures occur, e.g. in *Staphylococcus hyicus*, whose repeating units are composed of glycerolphosphateglycosylphosphate (Endl *et al.*, 1983). The most complicated WTA-like molecule among the known molecules is probably found in *Streptococcus agalactiae*. This polymer is extensively branched and is composed of different types of repeating units forming individual branches of the molecule (Sutcliffe *et al.*, 2008). Of note, this polymer represents the species-specific antigen used for serological differentiation of streptococci according to Rebecca Lancefield (Lancefield and Freimer, 1966). In most but not all WTA molecules, the repeating units are further substituted with a diverse set of additional sugars and with D-alanine (Neuhaus and Baddiley, 2003), other amino acids such as glutamate or lysine (in certain actinobacteria) (Potekhina *et al.*, 1993; Shashkov *et al.*, 2006) or

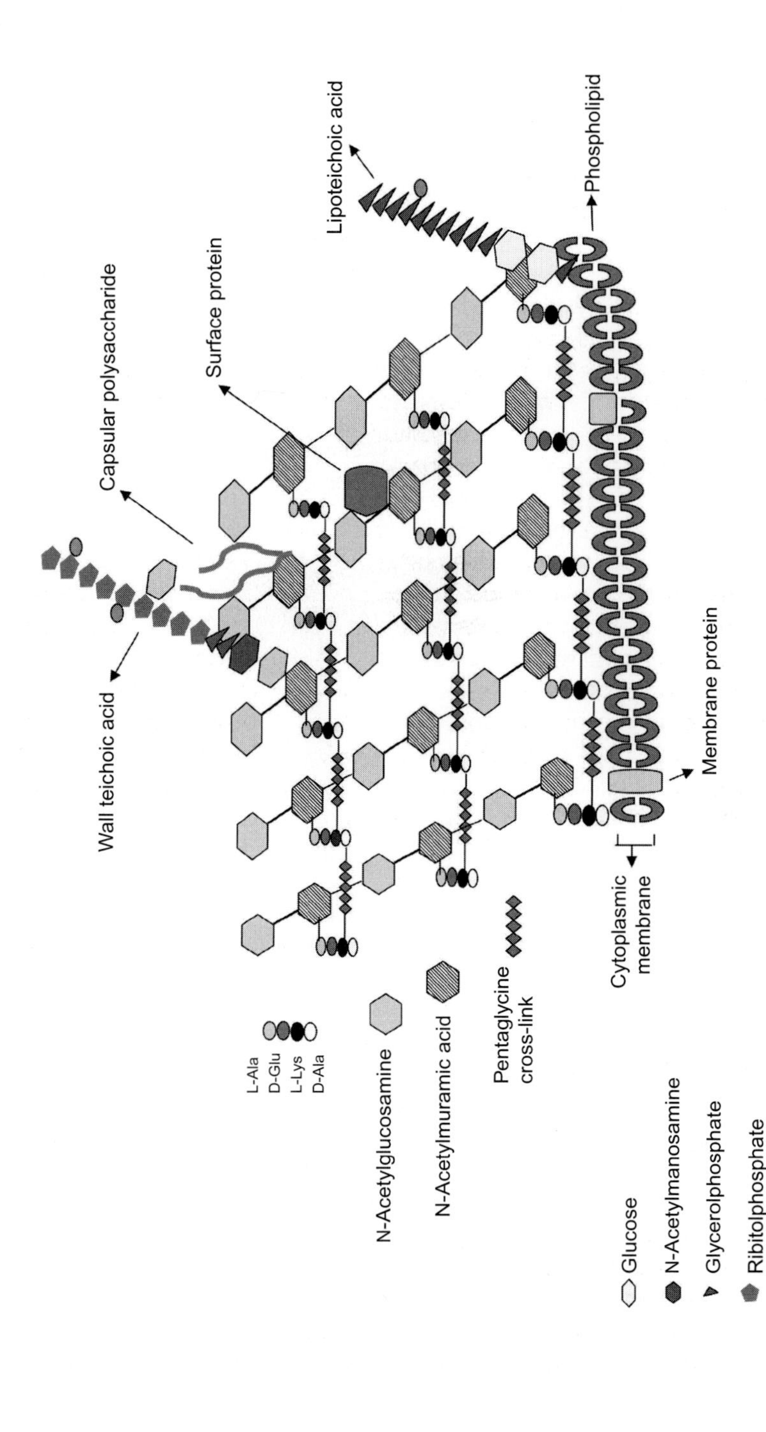

FIGURE 5.1 Molecular structure of the cell wall of *S. aureus*, composed of peptidoglycan, teichoic acid, proteins and capsular polysaccharide.

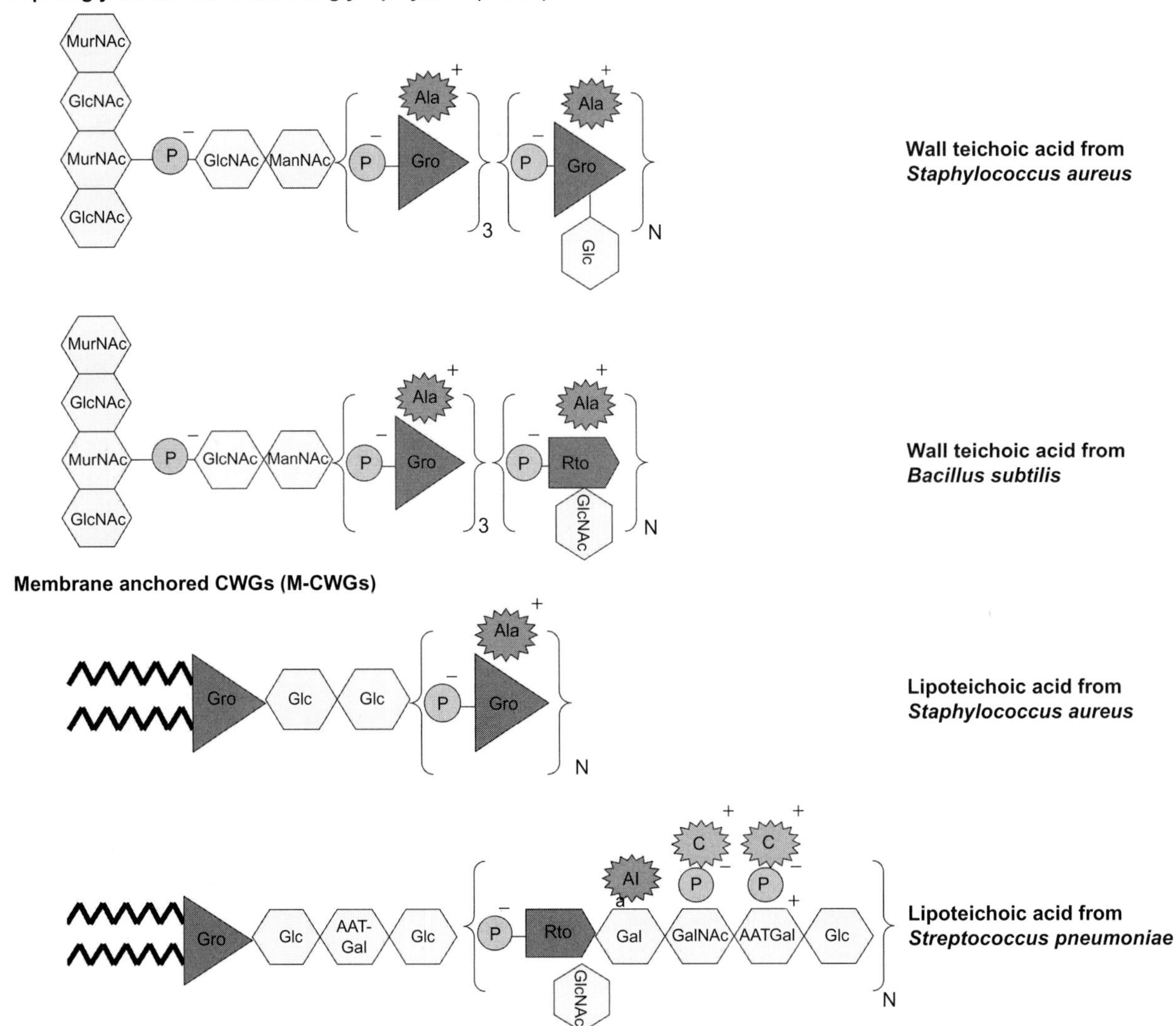

FIGURE 5.2 Selected WTA and LTA structures. Trioses, pentoses and hexoses are shown as triangles, pentagons and hexagons, respectively. Fatty acids are shown as zigzag lines. Non-glycosyl residues: Ala, D-alanine; C, choline; P, phosphate. Glycosyl-residues: AATGal, 2-acetamido-4-amino-2,4,6-trideoxy-D-galactose; Gal, galactose; GalNAc, *N*-acetylgalactosamine; Glc, glucose; GlcNAc, *N*-acetylglucosamine; Gro, glycerol; ManNAc, *N*-acetylmannosamine; Rto, ribitol.

phosphocholine (*S. pneumoniae*) (Fischer, 2000). The incorporation of D-alanine leads to a reduction or compensation of the negative charge derived from the phosphate groups (Neuhaus and Baddiley, 2003).

Anionic P-CWG lacking phosphate represent a less well-studied group of so-called "secondary cell wall polymers". They include the teichuronic acids (TUA), which are distinguished by the presence of uronic acid residues in repeating units (e.g. in many bacilli and micrococci) (Ward, 1981; Soldo *et al.*, 1999) and pyruvylated polymers (e.g. in *Bacillus anthracis* and *Bacillus cereus*) (Schäffer and Messner, 2005; Choudhury *et al.*, 2006; Leoff *et al.*, 2007). *Bacillus subtilis* expresses TUA when grown under phosphate limitation instead of WTA. The *B. subtilis* TUA has a similar linkage unit as WTAs and its linkage units consist of GlcNAc and glucuronic acid (Soldo *et al.*, 1999). The pyruvylated *B. anthracis* P-CWG has only recently been characterized and found to consist of repeating units formed by several hexoses (Choudhury *et al.*, 2006). Uncharged, often branched P-CWGs are produced by many actinobacteria, particularly in those with an outer membrane-like mycolic acid layer. *Mycobacterium tuberculosis*, for instance, has a branched arabinogalactan polymer, which is covalently bound to mycolic acid thereby connecting PG and the mycolic acid membrane (Brennan, 2003; Takayama *et al.*, 2005).

2.2. Lipoteichoic acid (LTA) and other M-CWGs

The M-CWGs, such as LTA, are attached to cytoplasmic membranes by linking the polymer to glycolipids. In *S. aureus*, the lipid anchor consists of a diglycosylated diacylglycerol (Fischer, 1988). The backbone is formed by Gro-P repeating units in *S. aureus* and many other Gram-positive bacteria. In contrast to WTAs, LTA is usually less diverse, which is probably due to the particular biosynthetic pathway (see below) (Fischer, 1994) (see Figure 5.2). However, a very complicated LTA structure is found, e.g. in *Streptococcus pneumoniae*, the repeating units of which are identical to those of pneumococcal WTAs and contain phosphocholine (Fischer *et al.*, 1993). As in WTAs, most LTA polymers are substituted with D-alanine at the 2′-hydroxy group of the glycerol and with additional sugars.

Of note, M-CWGs without phosphate in the repeating units are produced by many actinobacteria. Such polymers are often anionic because of substitution with succinyl groups (Powell *et al.*, 1975; Greenberg *et al.*, 1996; Delmas *et al.*, 1997). Uncharged, branched lipoarabinomannans (LAM) are found in many mycolic acid-producing bacteria (Briken *et al.*, 2004; Sutcliffe, 2005).

3. BIOSYNTHESIS OF WTAs AND LTA

Despite their structural similarity, WTAs and LTA usually use profoundly different precursor molecules and biosynthetic pathways. Because of the enormous structural diversity of P-CWGs, which implies that most bacterial species need different sets of genes, it has been difficult to identify biosynthetic genes and pathways by comparative genomics. The rather simple Gro-P WTA of *B. subtilis 168* requires only ca. 13 genes (Qian *et al.*, 2006) while synthesis of the extremely complex WTA-like polymer of *S. agalactiae* is proposed to depend on more than 120 genes (Sutcliffe *et al.*, 2008). Anyway, certain principles are widely conserved among P-CWG-producing bacteria, such as the enzymes mediating the first steps of linkage unit biosynthesis, the use of a C_{55} lipid carrier during the assembly process and the allocation of biosynthetic genes in clusters. Most knowledge on genetics and biosynthesis has been gathered in *B. subtilis* and *S. aureus* as outlined in the chapters below (Bhavsar and Brown, 2006; Xia and Peschel, 2008).

3.1. WTA biosynthesis in *B. subtilis* and *S. aureus*

Pioneering *in vitro* reconstitution studies of certain WTA-biosynthetic steps demonstrated the involvement of nucleotide-activated precursor molecules such as cytidine diphosphate-(CDP-)

glycerol, CDP-ribitol and uridine diphosphate-(UDP)-GlcNAc in WTA biosynthesis (Nathenson *et al.*, 1966; Brooks *et al.*, 1971; Bracha *et al.*, 1978; Baddiley, 1989). They also demonstrated that biosynthesis occurs initially at the inside of the cytoplasmic membrane on the universal undecaprenyl pyrophosphate lipid carrier (C_{55}), which is also used for peptidoglycan or capsular polysaccharide biosynthesis (Anderson *et al.*, 1972). The first gene cluster involved in biosynthesis of WTA, *tagABDEFGH,* has been identified in *B. subtilis 168* by analysis of temperature-sensitive mutants by the laboratory of Dimitri Karamata (Pooley and Karamata, 1994). Further genes such as *tagO* and the more complex WTA gene clusters of *B. subtils W23* and *S. aureus*, both of which produce ribitolphosphate (Rbo-P) WTA, were identified by large-scale sequencing projects and comparative genomics in the last couple of years (Qian *et al.*, 2006). Biochemical studies with crude enzymatic preparations derived from temperature-sensitive mutant strains, or with recombinant enzymes produced in *Escherichia coli*, led to functional predictions for most of the WTA-biosynthetic gene products as outlined below. The CWG-biosynthetic steps are shown in Figure 5.3. They can be divided into four groups:

(i) The first group of WTA-biosynthetic enzymes encompasses those that mediate synthesis of linkage units that usually have conserved structures and connect the polymers with peptidoglycan or membrane (Araki and Ito, 1989). Accordingly, involved genes such as *tagO* and *tagA* are well conserved among Gram-positive bacteria and allow an easy identification of CWG gene clusters (Weidenmaier *et al.*, 2004; Ginsberg *et al.*, 2006). The protein TagO transfers GlcNAc phosphate to C_{55} phosphate (Soldo *et al.*, 2002a) and TagA adds a ManNAc unit using UDP-GlcNAc and UDP-ManNAc precursors, respectively (Ginsberg *et al.*, 2006).

(ii) The second group of enzymes includes enzymes involved in generation of special nucleotide-activated precursor molecules such as UDP-ManNAc (MnaA) (Soldo *et al.*, 2002b), CDP-glycerol (TagD) (Park *et al.*, 1993) or CDP-ribitol (TarI, TarJ) (Pereira and Brown, 2004). Many of these genes share conserved domains because of interaction with the nucleotides.

(iii) The products of the third group of genes incorporate the preformed repeating units into CWG and encompass both priming and polymerizing enzymes. The Gro-P WTA polymer of *B. subtilis* 168 depends on the primase TagB, which adds the first repeating unit to the C_{55}-bound linkage unit and the polymerase TagF adding the further Gro-P units (Ginsberg *et al.*, 2006). The situation is more complicated in *S. aureus* Rbo-P WTA biosynthesis. Here, the TagB reaction is followed by addition of only a second Gro-P unit by the TarF enzyme (Brown *et al.*, 2008). The RboP polymer is subsequently synthesized by the TarL polymerase (Brown *et al.*, 2008; Pereira *et al.*, 2008). While no Rbo-P primase seems to be involved in *S. aureus*, such an enzyme has been implicated in Rbo-P WTA biosynthesis in *B. subtilis W23* (TarK) (Bhavsar and Brown, 2006). Three genes involved in Rbo-P generation and incorporation seem to be duplicated in *S. aureus* (Qian *et al.*, 2006), which has been a major obstacle in studying enzyme functions. It appears now that the two TarL enzymes mediate the same types of reaction albeit leading to WTA of different chain length and electrophoretic migration. One of the *tarL* genes (also been named *tarK*) is regulated by the agr quorum sensing system indicating that *S. aureus* can control WTA structure according to bacterial density and environmental changes (Brown *et al.*, 2008).

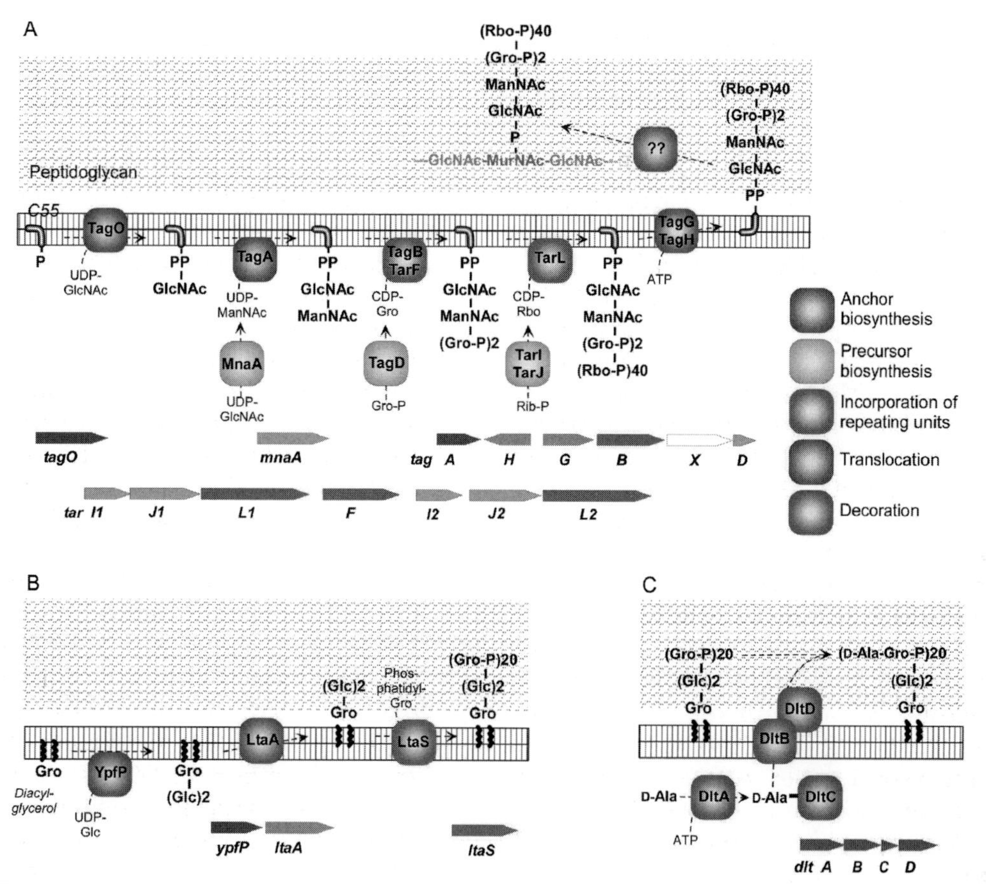

FIGURE 5.3 Pathways of *S. aureus* wall teichoic acid (WTA) biosynthesis (A), lipoteichoic acid (LTA) biosynthesis (B) and D-alanine incorporation into LTA and WTA (C). CDP-Gro, cytidyldiphosphate-glycerol; CDP-Rbo, cytidyldiphosphate-ribitol; Glc, glucose; GlcNAc, *N*-acetylglucosamine; Gro, glycerol; Gro-P, glycerolphosphate; ManNAc, *N*-acetylmannosamine; MurNAc, *N*-acetylmuramic acid; Rbo-P, ribitol-phosphate; Rib-P, ribulose-5-phosphate; UDP-Glc, undecaprenylphoshate-glucose; UDP-GlcNAc, uridine-5′-diphosphate-glucose; UDP-GlcNAc, uridine-5′-*N*-acetylglucosamine; UDP-ManNAc, uridine-5′-*N*-acetylmannosamine. (See colour plates section.)

(iv) Proteins mediating the transfer of the WTA polymers to the outer membrane leaflet (TagG, TagH, forming an ABC transporter) (Lazarevic and Karamata, 1995) and from C_{55} to the acceptor molecules (responsible proteins still unknown) are allocated into group iv.

Further proteins involved in decoration of WTA/LTA repeating units with sugars, D-alanine, cholin, pyruvate or other residues constitute a further group (v). Different biosynthetic pathways only sharing the *tagO* and (in many cases) *tagA* genes with the above described ones have been proposed for WTA with more complex, hexose-containing repeating units such as the minor WTA of *B. subtilis* (Freymond *et al.*, 2006) or the branched WTA-like polymer of *S. agalactiae* (Sutcliffe *et al.*, 2008).

3.2. LTA biosynthesis in *S. aureus*

The Gro-P repeating units of LTA are not derived from a nucleotide-activated precursor but from the phospholipid phosphatidylglycerol, a major constituent of bacterial membranes (Glaser and Lindsay, 1974). A second difference between LTA and WTA biosynthesis is the fact that LTA is not polymerized on C55 but directly on the glycolipid that serves as the LTA membrane anchor (Koch *et al.*, 1984; Fischer, 1988). As the glycolipids differ between species, different genes have been implicated in glycolipid biosynthesis of *S. aureus* and *S. agalactiae* (Kiriukhin *et al.*, 2001; Doran *et al.*, 2005). *S. aureus* LTA is linked to diglucosyldiacylglycerol. The YpfP enzyme generates this lipid by adding two glucose residues from UDP-glucose to diacylglycerol (Jorasch *et al.*, 1998, 2000; Kiriukhin *et al.*, 2001). Efficient LTA biosynthesis depends on the *ltaA* gene which encodes a membrane protein that is thought to be a flippase that translocates the glycolipid from the inner leaflet of the membrane to the outer leaflet (Grundling and Schneewind, 2007b). The LTA polymerase LtaS, recently discovered by Angelika Grundling and Olaf Schneewind, utilizes Gro-P units from phosphatidylglycerol to synthesize the LTA polymer at the outer surface of the cytoplasmic membrane (Grundling and Schneewind, 2007a). One or several *ltaS*-related genes are found in most bacteria known to produce LTA indicating that LTA biosynthesis is a rather conserved process. Amazingly, deletion of *ypfP* does not lead to a halt of LTA biosynthesis but to synthesis of LTA whose polymer is attached to diacylglygcerol (Kiriukhin *et al.*, 2001; Fedtke *et al.*, 2007). Another unexpected finding is the fact that, depending on the *S. aureus* strain background, *ypfP* mutants produce strongly reduced or unaltered amounts of the altered LTA, compared to wild-type strains for unknown reasons (Fedtke *et al.*, 2007). In conclusion, LTA biosynthesis requires far less genes and is less complex compared to WTA biosynthesis. The limited diversity of LTA molecules reflects the use of the universal lipid phosphatidylglycerol as repeating unit donor. In any case, bacteria such as *S. pneumoniae* produce highly complex LTA polymers (Fischer *et al.*, 1993; Draing *et al.*, 2006), which are most probably synthesized in a C_{55}-dependent fashion.

3.3. Incorporation of D-alanine into WTAs and LTA

While most other constituents of WTAs and LTA are variable, the modification with D-alanine is a very constant trait of most TA polymers. Only those TA molecules without polyol constituents in repeating units, such as the minor WTA of *B. subtilis*, seem to lack these substituents (Freymond *et al.*, 2006). In accord with the high prevalence of D-alanylation, the *dltABCD* genes responsible for D-alanine activation and incorporation into WTA and LTA are highly conserved and seem always to form an operon (Neuhaus *et al.*, 1996; Neuhaus and Baddiley, 2003). D-Alanine has a profound impact on TA net charge with crucial consequences for resistance to antimicrobial peptides, adhesion to host cell receptors and biofilm formation (Weidenmaier and Peschel, 2008). Interestingly, the *dlt* operons of *S. aureus* and *S. epidermidis* are controlled in response to antimicrobial peptide challenge via the ApsXRS (also named GraXRS) regulatory system and in response to cell wall stress (Li *et al.*, 2007; Herbert *et al.*, 2007; Kraus *et al.*, 2008). D-Alanine is incorporated after biosynthesis of the TA polymers is completed and, since the D-alanine esters are rather labile and get easily lost, D-alanine can be repeatedly incorporated into a given molecule (Koch *et al.*, 1985). The transfer of D-alanine into teichoic acids requires four proteins (DltA, -B, -C, -D) forming a pathway that involves activation of D-alanine in the cytoplasm, linkage to a D-alanine carrier protein (DltC), translocation and incorporation of D-alanine into teichoic acids (Neuhaus and Baddiley, 2003). The first step requires hydrolysis of ATP. It is catalysed by the DltA protein, which is homologous to the activating domains of peptide synthetases (Heaton and Neuhaus, 1992; Neuhaus *et al.*, 1996). In fact, D-alanine activation and transfer to a carrier protein resembles the biosynthesis of non-ribosomally synthesized peptides or fatty acids. All these pathways involve intermediates linked to the phosphopantetheine prosthetic groups of dedicated carrier proteins by energy-rich thioester bonds. The last steps of D-alanine transfer into teichoic acids are less well understood. They seem to require DltC, DltB, an integral membrane protein, and DltD, a membrane-tethered hydrophilic protein (Debabov *et al.*, 2000). The Dlt proteins represent interesting targets for inhibitory compounds that block D-alanylation. Accordingly, specific inhibitors of DltA have recently been described to render bacteria more susceptible to cationic antimicrobial molecules (May *et al.*, 2005) and to be very efficient in clearing bacterial infections *in vivo* (Escaich *et al.*, 2007).

4. ROLES OF WTAs AND LTA IN BACTERIAL PHYSIOLOGY

The universal presence of CWGs in Gram-positive bacteria and the fact that the bacterial cells commit considerable amounts of energy and genetic information into their biosynthesis indicates very important roles of WTAs and LTA-like polymers for bacterial integrity and fitness. Only recently, defined mutants lacking CWGs (Weidenmaier *et al.*, 2004; D'Elia *et al.*, 2006) or with altered CWG structures (Peschel *et al.*, 1999; Doran *et al.*, 2005; Kristian *et al.*, 2005) have become available that allow a thorough characterization of CWG functions. Recently, WTA has been shown to be dispensable for viability of *S. aureus* and *B. subtilis* (Weidenmaier *et al.*, 2004; D'Elia *et al.*, 2006). These mutants seemed to have only minor phenotypic defects when grown in the laboratory. In contrast, LTA is indispensable in *S. aureus* (Grundling and Schneewind, 2007a) but the LTA content in the cell envelope can be strongly reduced without affecting the *in vitro* growth behaviour (Fedtke *et al.*, 2007). Accordingly, many of the functions assigned to CWGs seem to be non-essential and may be critical only in certain instances, such as exposure to environmental stresses or to host defence factors (Weidenmaier and Peschel, 2008). In fact, many reports point toward a function of WTAs and LTA in protecting the cell envelope from penetration of harmful molecules such as host defence molecules, bacteriocins and antibiotics. This protective function may be direct by clogging pores and cavities between peptidoglycan layers or indirect by attachment of outer protection layers such as S-layer proteins (e.g. *B. anthracis*) (Mesnage *et al.*, 2000) or mycolic acids (e.g. *M. tuberculosis*) (Brennan, 2003). Accordingly, *S. aureus* WTA contributes to lysozyme resistance (Bera *et al.*, 2007) and mutants lacking D-alanine in WTA and LTA are more susceptible to cationic antimicrobial peptides (Peschel *et al.*, 1999, 2000; Peschel and Sahl, 2006). In certain cases, however, WTA increases the susceptibility for harmful molecules such as phages that use WTA as a receptor (Park *et al.*, 1974; Lopez *et al.*, 1982; Wendlinger *et al.*, 1996) or human antimicrobial defensin hBD3 and secretory group IIA phospholipase A2 (Koprivnjak *et al.*, 2008).

Certain bacterial proteins are non-covalently anchored to the cell wall by CWGs. *S. pneumoniae* anchors many of its virulence factors to the phosphocholine residues of its CWGs (Bergmann and Hammerschmidt, 2006). *B. anthracis* and relatives use pyruvylated CWGs to attach S-layer proteins to the bacterial surface (Mesnage *et al.*, 2000). The autolysins of *S. aureus* and other bacteria have high affinities for WTA and LTA (Giudicelli and Tomasz, 1984; Bierbaum and Sahl, 1987) but *S. aureus* mutants with reduced LTA had no reduced amounts of autolysins (Fedtke *et al.*, 2007) suggesting that autolysins bind to the cell wall independently of CWGs. However, WTA and LTA play a profound role in controlling autolysin activity by interactions that are only partially understood and may involve CWG-bound bivalent cations. Depleting *S. aureus* from LTA leads to bacterial cells with distorted shapes and division sites (Grundling and Schneewind, 2007a) suggesting that LTA interacts with components of the membrane-bound cell division machinery and contributes to its proper placement or regulation.

The zwitterionic and anionic CWG types have ion-exchanger-like properties that have been proposed to have important roles in shaping the ionic milieu in the cell wall. The TAs have particularly high affinities for magnesium ions and are regarded as magnesium ion storage molecules (Heptinstall *et al.*, 1970). The hydrophilic nature of surface-exposed CWGs has a strong impact on the physicochemical properties of bacterial cell surfaces. Accordingly, *S. aureus* and *Enterococcus faecalis* mutants with altered CWGs are strongly affected in biofilm formation on biomaterials and virulence attenuated in animal models of foreign body infections (Fabretti *et al.*, 2006; Fedtke *et al.*, 2007; Gross *et al.*, 2001; Kristian *et al.*, 2003). The CWGs can play a second role in biofilm formation when they are released to form part of the biofilm matrix that protects and glues bacterial cells together (Sadovskaya *et al.*, 2004; Vinogradov *et al.*, 2006). How CWGs are shed from their cell wall or membrane anchors has remained unknown.

5. TEICHOIC ACIDS AND HOST CELL RECEPTOR INTERACTION

The CWGs seem to play crucial roles in microbe/host interaction in bacteria colonizing or infecting animal hosts (Figure 5.4). Such roles have been studied thoroughly in *S. aureus* whose WTAs and LTA appear to shape the entire infection process from initial colonization to activation of innate immunity and to recognition by the adaptive immune system (Weidenmaier and Peschel, 2008). *S. aureus* mutants lacking WTA are abrogated in binding to epithelial and endothelial cells and lose the ability to colonize the nose in animal models (Weidenmaier *et al.*, 2004, 2008) or to leave the bloodstream and infect subendothelial tissues in endovascular infections (Weidenmaier *et al.*, 2005a). Altering TA structure by disrupting the D-alanylation pathway in *S. aureus* (Weidenmaier *et al.*, 2004, 2005b), *Streptococcus pyogenes* (Kristian *et al.*, 2005), or *Listeria monocytogenes* (Abachin *et al.*, 2002) or altering LTA membrane anchoring in *S. agalactiae* (Doran *et al.*, 2005) has similar impacts on bacterial host cell binding. Several lines of evidence point toward a direct binding of WTA to yet-to-be identified receptors on epithelial and endothelial cells. Of note, these interactions seem to contribute to *S. aureus* host cell attachment to a similar extent as the staphylococcal adhesion proteins (Weidenmaier *et al.*, 2008), which bind, for example, keratin, fibronectin or fibrinogen (Foster and Hook, 1998; Navarre and Schneewind, 1999; Mongodin *et al.*, 2002). The WTA-binding host receptors are still unknown but the fact that WTA-mediated binding can be inhibited by polyinosinic acid, an established inhibitor of scavenger receptors (SR) indicates that SR-like receptors play a major role in this type of interaction (Weidenmaier *et al.*, 2008). Several members of the SR family have been identified on mammalian cells and some of them have been shown to bind purified CWGs and intact *S. aureus* cells. The SCARA5 and LOX1 are expressed by airway epithelial and endothelial cells, respectively, and represent candidate receptors for WTA-mediated binding (Shimaoka *et al.*, 2001; Jiang *et al.*, 2006).

While SRs appear to mediate bacterial attachment, other host receptors have been shown to stimulate inflammatory processes upon CWG binding. In fact, LTA, mycobacterial LAM and other M-CWGs have been identified as

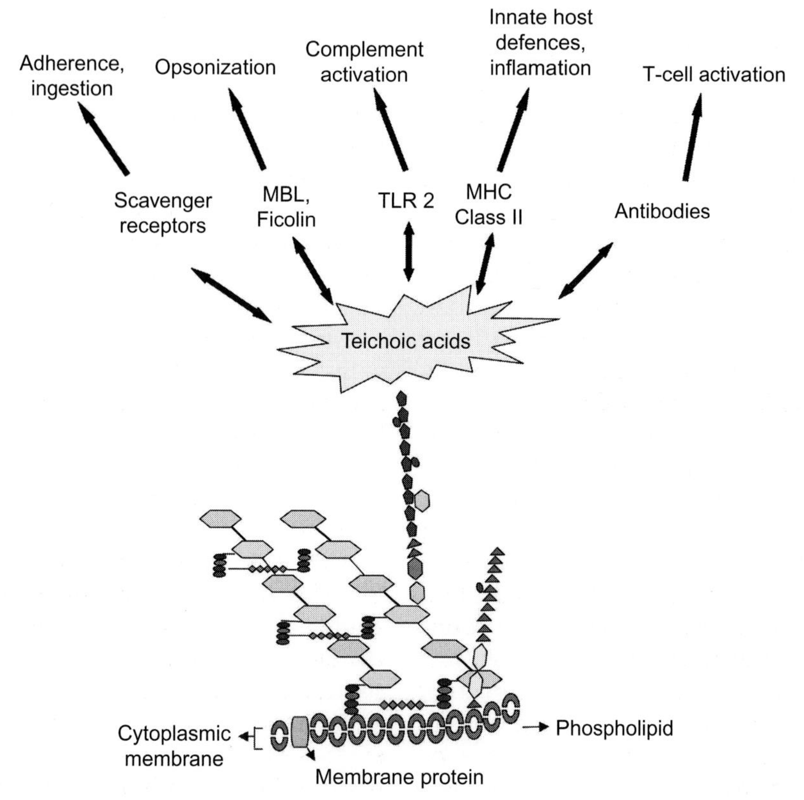

FIGURE 5.4 Interaction of LTA and WTA with host molecules. Scavenger receptors, mannose-binding lectin and ficolins interact with different CWGs, this leads to binding and internalization by host cells or complement activation. Some M-CWGs seem to elicit proinflammatory responses through TLR-2. MBL, Mannose-binding lectin; TLR2, Toll-like receptor 2; MHC, Major histocompatibility complex.

pathogen-associated molecular pattern (PAMP) molecules that activate the innate immune system via the Toll-like receptor-2 (TLR-2) (Hermann *et al.*, 2002; Sugawara *et al.*, 2003; Tapping and Tobias, 2003; Hoebe *et al.*, 2005). The co-receptors TLR-6, CD14, LBP and CD36 seem to contribute to LTA-mediated TLR-2 activation (Chavakis *et al.*, 2002; Han *et al.*, 2003; Hoebe *et al.*, 2005; Henneke *et al.*, 2005). The proinflammatory potency of M-CWGs is still a matter of debate since many of the commonly used prepations have been shown to be contaminated with lipopeptides, which account for a large percentage of the activity (Hashimoto *et al.*, 2006, 2007). Nevertheless, synthetic LTA analogues have been shown to stimulate TLR-2 (Deininger *et al.*, 2003). Soluble C-type lectins such as the mannose-binding lectin (MBL), L-ficolin and the lung surfactant proteins A and D also bind LTA or LAM (Polotsky *et al.*, 1996, 1997; Ferguson *et al.*, 1999; Sidobre *et al.*, 2000; van de Wetering *et al.*, 2001; Lynch *et al.*, 2004). MBL and L-ficolin activate the lectin-initiated complement pathway upon binding to CWGs on the bacterial surface leading to bacterial opsonization and release of chemotactic complement split products (Endo *et al.*, 2007; Takahashi *et al.*, 2007).

The CWGs, such as *S. aureus* LTA, have been known to be a major target for antibodies for many decades (Verbrugh *et al.*, 1981; Verhoef *et al.*, 1983; Kumar *et al.*, 2005). Accordingly, CWGs have been proposed as vaccination targets and promising results have been obtained with active vaccination directed against enterococcal LTA (Theilacker *et al.*, 2006) or passive vaccination with a humanized monoclonal antibody specific for staphylococcal LTA (Weisman, 2007). While glycopolymers have traditionally been regarded only as T-cell-independent antigens, there is increasing evidence that zwitterionic glycopolymers can activate T-cells upon processing and presentation via MHC class II by antigen-presenting cells (Kalka-Moll *et al.*, 2002; Mazmanian and Kasper, 2006). Most detailed studies concerning this new pathway are available for *Bacteroides fragilis* capsular polysaccharides, but recent studies with *S. aureus* WTA, which is equally zwitterionic, suggest that WTA may serve as a T-cell-independent antigen (Tzianabos *et al.*, 2001; McLoughlin *et al.*, 2006). Such studies are still in their infancy but, if the concept is confirmed, CWGs should be regarded increasingly as vaccine candidates with the capacity to elicit immunological memory. Another way how CWGs might stimulate specific T-cells comes from mycobacterial LAM which has been shown to stimulate T-cells restricted to the MHC-like molecule CD1, which is known to present certain lipid antigens (Prigozy *et al.*, 1997). Thus, it is tempting to propose that M-CWGs could prove to be the general subjects of CD1 presentation than had been previously thought.

6. CONCLUSIONS AND PERSPECTIVES

The TAs and other CWGs have represented a field of active research following their discovery in the 1960s, but only a small group of scientists has studied the chemical, biochemical and genetic basis of CWG biosynthesis later on. Only in recent years, after certain CWGs turned out to be of pivotal importance in microbe–host interaction, did these polymers receive increasing attention in the scientific community. Moreover, the desperate need for new antimicrobial target structures has put CWG chemistry and biology on the list of major scientific challenges. In fact, several promising studies have recently demonstrated the suitability of CWGs as targets for new vaccines or antibiotics (Mikusova *et al.*, 1995; Theilacker *et al.*, 2004; May *et al.*, 2005). With improved glycochemical methodology it should be possible to obtain a broader view on the diversity and variability of CWG structures. The availability of genomic and metagenomic databases represents a valuable basis for predicting CWG biosynthetic pathways by bioinformatic methods. Moreover, recent progress in the *in vitro* reconstitution of biosynthetic steps will help to confirm predicted enzyme functions and to use the reactions in high-throughput screening programmes in the search for new antibiotics. The WTA biosynthetic enzymes seem to form a membrane-associated complex in *B. subtilis* indicating that CWG biosynthesis is a highly organized process (Formstone *et al.*, 2008). It can be assumed that the CWGs and PG biosynthetic machineries are coordinated in sophisticated ways, maybe in cooperation with cytoskeletal elements and with the cell division apparatus.

With the increasing availability of defined mutants it should be possible to correlate structural features of CWGs with certain functions in cell wall physiology or host interaction. Major open questions needing to be answered concern the proinflammatory capacity of M-CWGs and the potential of CWGs to activate specific T-cells upon processing and presentation in MHC class II or CD1 molecules. Many host receptors recognizing and binding CWGs remain to be identified. It is tempting to speculate that the vast diversity of CWG structures play a role in bacterial cell and host tropism according to cell- and species-specific differences in expression of CWG-binding molecules. Thus, a number of areas for further investigation remain (Research Focus Box).

RESEARCH FOCUS BOX

- Complete clarification of the biosynthetic pathways of the different types of CWGs is required.
- *In vitro* reconstitution of biosynthetic steps to confirm predicted enzyme functions could prove useful.
- CWGs could act as a target for new vaccines or antibiotics and are worthy of examination.
- The role of M-CWG and P-CWG as proinflammatory stimuli of the immune system requires further intense investigation.
- A deeper understanding of the role of CWGs in bacterial physiology and host interaction could prove beneficial.

ACKNOWLEDGEMENTS

Our research is supported by grants from the German Research Foundation (TR34, FOR449, GRK685, SFB685, SFB766), the European Union (LSHM-CT-2004-512093), the German Ministry of Education and Research (NGFN2, SkinStaph) and the IZKF programme of the Medical Faculty, University of Tübingen, to AP.

References

Abachin, E., Poyart, C., Pellegrini, E., et al., 2002. Formation of D-alanyl-lipoteichoic acid is required for adhesion and virulence of *Listeria monocytogenes*. Mol. Microbiol. 43, 1–14.

Anderson, R.G., Hussey, H., Baddiley, J., 1972. The mechanism of wall synthesis in bacteria. The organization of enzymes and isoprenoid phosphates in the membrane. Biochem. J. 127, 11–25.

Araki, Y., Ito, E., 1989. Linkage units in cell walls of Gram-positive bacteria. Crit. Rev. Microbiol. 17, 121–135.

Armstrong, J.J., Baddiley, J., Buchanan, J.G., Davision, A.L., Kelemen, M.V., Neuhaus, F.C., 1959. Composition of teichoic acids from a number of bacterial walls. Nature 184, 247–248.

Baddiley, J., 1989. Bacterial cell walls and membranes. Discovery of the teichoic acids. Bioessays 10, 207–210.

Bera, A., Biswas, R., Herbert, S., et al., 2007. Influence of wall teichoic acid on lysozyme resistance in *Staphylococcus aureus*. J. Bacteriol. 189, 280–283.

Bergmann, S., Hammerschmidt, S., 2006. Versatility of pneumococcal surface proteins. Microbiology 152, 295–303.

Bhavsar, A.P., Brown, E.D., 2006. Cell wall assembly in *Bacillus subtilis*: how spirals and spaces challenge paradigms. Mol. Micrbiol. 60, 1077–1090.

Bierbaum, G., Sahl, H.G., 1987. Autolytic system of *Staphylococcus simulans* 22: influence of cationic peptides on activity of *N*-acetylmuramoyl-L-alanine amidase. J. Bacteriol. 169, 5452–5458.

Bracha, R., Chang, M., Fiedler, F., Glaser, L., 1978. Biosynthesis of teichoic acids. Methods Enzymol. 50, 387–402.

Brennan, P.J., 2003. Structure, function, and biogenesis of the cell wall of *Mycobacterium tuberculosis*. Tuberculosis (Edinburgh) 83, 91–97.

Briken, V., Porcelli, S.A., Besra, G.S., Kremer, L., 2004. Mycobacterial lipoarabinomannan and related lipoglycans: from biogenesis to modulation of the immune response. Mol. Microbiol. 53, 391–403.

Brooks, D., Mays, L.L., Hatefi, Y., Young, F.E., 1971. Glucosylation of teichoic acid: solubilization and partial characterization of the uridine diphosphoglucose: polyglycerolteichoic acid glucosyl transferase from membranes of *Bacillus subtilis*. J. Bacteriol. 107, 223–229.

Brown, S., Zhang, Y.H., Walker, S., 2008. A revised pathway proposed for *Staphylococcus aureus* wall teichoic acid biosynthesis based on *in vitro* reconstitution of the intracellular steps. Chem. Biol. 15, 12–21.

Chavakis, T., Hussain, M., Kanse, S.M., et al., 2002. *Staphylococcus aureus* extracellular adherence protein serves as anti-inflammatory factor by inhibiting the recruitment of host leukocytes. Nat. Med. 8, 687–693.

Choudhury, B., Leoff, C., Saile, E., et al., 2006. The structure of the major cell wall polysaccharide of *Bacillus anthracis* is species-specific. J. Biol. Chem. 281, 27932–27941.

D'Elia, M.A., Millar, K.E., Beveridge, T.J., Brown, E.D., 2006. Wall teichoic acid polymers are dispensable for cell viability in *Bacillus subtilis*. J. Bacteriol. 188, 8313–8316.

Debabov, D.V., Kiriukhin, M.Y., Neuhaus, F.C., 2000. Biosynthesis of lipoteichoic acid in *Lactobacillus rhamnosus*: role of DltD in D-alanylation. J. Bacteriol. 182, 2855–2864.

Deininger, S., Stadelmaier, A., von Aulock, S., Morath, S., Schmidt, R.R., Hartung, T., 2003. Definition of structural prerequisites for lipoteichoic acid-inducible cytokine induction by synthetic derivatives. J. Immunol. 170, 4134–4138.

Delmas, C., Gilleron, M., Brando, T., et al., 1997. Comparative structural study of the mannosylated-lipoarabinomannans from *Mycobacterium bovis* BCG vaccine strains: characterization and localization of succinates. Glycobiology 7, 811–817.

Doran, K.S., Engelson, E.J., Khosravi, A.M.H., et al., 2005. Group B *Streptococcus* blood-brain barrier invasion depends upon proper cell surface anchoring of lipoteichoic acid. J. Clin. Invest. 115, 2499–2507.

Draing, C., Pfitzenmaier, M., Zummo, S., et al., 2006. Comparison of lipoteichoic acid from different serotypes of *Streptococcus pneumoniae*. J. Biol. Chem. 281, 33849–33859.

Endl, J., Seidl, H.P., Fiedler, F., Schleifer, K.H., 1983. Chemical composition and structure of the cell wall teichoic acids of staphylococci. Arch. Microbiol. 135, 215–223.

Endo, Y., Matsushita, M., Fujita, T., 2007. Role of ficolin in innate immunity and its molecular basis. Immunobiology 212, 371–379.

Escaich, S., Moreau, F., Vongsouthi, V.S.C., et al., 2007. Discovery of new Gram-positive antivirulence drugs: the first antivirulence molecule active *in vivo*. ICAAC Conference, abstract F2-958.

Fabretti, F., Theilacker, C., Baldassarri, L., et al., 2006. Alanine esters of enterococcal lipoteichoic acid play a role in biofilm formation and resistance to antimicrobial peptides. Infect. Immun. 74, 4164–4171.

Fedtke, I., Mader, D., Kohler, T., et al., 2007. A *Staphylococcus aureus ypfP* mutant with strongly reduced lipoteichoic acid (LTA) content: LTA governs bacterial surface properties and autolysin activity. Mol. Microbiol. 65, 1078–1091.

Ferguson, J.S., Voelker, D.R., McCormack, F.X., Schlesinger, L.S., 1999. Surfactant protein D binds to *Mycobacterium tuberculosis* bacilli and lipoarabinomannan via carbohydrate-lectin interactions resulting in reduced phagocytosis of the bacteria by macrophages. J. Immunol. 163, 312–321.

Fischer, W., 1988. Physiology of lipoteichoic acids in bacteria. Adv. Microbiol. Physiol. 29, 233–302.

Fischer, W., 1994. Lipoteichoic acids and lipoglycans. In: Ghuysen, J.-M., Hakenbeck, R. (Eds.), Bacterial Cell Wall. Elsevier Science B.V., Amsterdam, pp. 199–215.

Fischer, W., 2000. Phosphocholine of pneumococcal teichoic acids: role in bacterial physiology and pneumococcal infection. Res. Microbiol. 151, 421–427.

Fischer, W., Behr, T., Hartmann, R., Peter-Katalinic, J., Egge, H., 1993. Teichoic acid and lipoteichoic acid of *Streptococcus pneumoniae* possess identical chain structures. A reinvestigation of teichoic acid (C polysaccharide). Eur. J. Biochem. 215, 851–857.

Formstone, A., Carballido-Lopez, R., Noirot, P., Errington, J., Scheffers, D.J., 2008. Localization and interactions of teichoic acid synthetic enzymes in *Bacillus subtilis*. J. Bacteriol. 190, 1812–1821.

Foster, T.J., Hook, M., 1998. Surface protein adhesions of *Staphylococcus aureus*. Trends Microbiol. 6, 484–488.

Freymond, P.P., Lazarevic, V., Soldo, B., Karamata, D., 2006. Poly(glucosyl-*N*-acetylgalactosamine 1-phosphate), a wall teichoic acid of *Bacillus subtilis* 168: its biosynthetic pathway and mode of attachment to peptidoglycan. Microbiology 152, 1709–1718.

Ginsberg, C., Zhang, Y.H., Yuan, Y., Walker, S., 2006. *In vitro* reconstitution of two essential steps in wall teichoic acid biosynthesis. ACS Chem. Biol. 1, 25–28.

Giudicelli, S., Tomasz, A., 1984. Attachment of pneumococcal autolysin to wall teichoic acids, an essential step in enzymatic wall degradation. J. Bacteriol. 158, 1188–1190.

Glaser, L., Lindsay, B., 1974. The synthesis of lipoteichoic acid carrier. Biochem. Biophys. Res. Commun. 59, 1131–1136.

Greenberg, J.W., Fischer, W., Joiner, K.A., 1996. Influence of lipoteichoic acid structure on recognition by the macrophage scavenger receptor. Infect. Immun. 64, 3318–3325.

Gross, M., Cramton, S., Götz, F., Peschel, A., 2001. Key role of teichoic acid net charge in *Staphylococcus aureus* colonization of artificial surfaces. Infect. Immun. 69, 3423–3426.

Grundling, A., Schneewind, O., 2007a. Synthesis of glycerol phosphate lipoteichoic acid in *Staphylococcus aureus*. Proc. Natl. Acad. Sci. USA. 104, 8478–8483.

Grundling, A., Schneewind, O., 2007b. Genes required for glycolipid synthesis and lipoteichoic acid anchoring in *Staphylococcus aureus*. J. Bacteriol. 189, 2521–2530.

Han, S.H., Kim, J.H., Martin, M., Michalek, S.M., Nahm, M.H., 2003. Pneumococcal lipoteichoic acid (LTA) is not as potent as staphylococcal LTA in stimulating Toll-like receptor 2. Infect. Immun. 71, 5541–5548.

Hashimoto, M., Tawaratsumida, K., Kariya, H., et al., 2006. Not lipoteichoic acid but lipoproteins appear to be the dominant immunobiologically active compounds in *Staphylococcus aureus*. J. Immunol. 177, 3162–3169.

Hashimoto, M., Furuyashiki, M., Kaseya, R., et al., 2007. Evidence of immunostimulating lipoprotein existing in the natural lipoteichoic acid fraction. Infect. Immun. 75, 1926–1932.

Heaton, M.P., Neuhaus, F.C., 1992. Biosynthesis of D-alanyl-lipoteichoc acid: cloning, nucleotide sequence, and expression of the *Lactobacillus casei* gene for the D-alanine-activating enzyme. J. Bacteriol. 174, 4707–4717.

Henneke, P., Morath, S., Uematsu, S., et al., 2005. Role of lipoteichoic acid in the phagocyte response to group B *streptococcus*. J. Immunol. 174, 6449–6455.

Heptinstall, S., Archibald, A.R., Baddiley, J., 1970. Teichoic acids and membrane function in bacteria. Nature 225, 519–521.

Herbert, S., Bera, A., Nerz, C., et al., 2007. Molecular basis of resistance to muramidase and cationic antimicrobial peptide activity of lysozyme in staphylococci. PLoS Pathog. 3, 981–994.

Hermann, C., Spreitzer, I., Schroder, N.W., et al., 2002. Cytokine induction by purified lipoteichoic acids from various bacterial species – role of LBP, sCD14, CD14 and failure to induce IL-12 and subsequent IFN-gamma release. Eur. J. Immunol. 32, 541–551.

Hoebe, K., Georgel, P., Rutschmann, S., et al., 2005. CD36 is a sensor of diacylglycerides. Nature 433, 523–527.

Jiang, Y., Oliver, P., Davies, K.E., Platt, N., 2006. Identification and characterization of murine SCARA5, a novel class A scavenger receptor that is expressed by populations of epithelial cells. J. Biol. Chem. 281, 11834–11845.

Jorasch, P., Warnecke, D.C., Lindner, B., Zähringer, U., Heinz, E., 2000. Novel processive and nonprocessive glycosyltransferases from *Staphylococcus aureus* and *Arabidopsis thaliana* synthesize glycoglycerolipids, glycophospholipids, glycosphingolipids and glycosylsterols. Eur. J. Biochem. 267, 3770–3783.

Kalka-Moll, W.M., Tzianabos, A.O., Bryant, P.W., Niemeyer, M., Ploegh, H.L., Kasper, D.L., 2002. Zwitterionic polysaccharides stimulate T cells by MHC class II-dependent interactions. J. Immunol. 169, 6149–6153.

Kiriukhin, M.Y., Debabov, D.V., Shinabarger, D.L., Neuhaus, F.C., 2001. Biosynthesis of the glycolipid anchor in lipoteichoic acid of *Staphylococcus aureus* RN4220: role of YpfP, the diglucosyldiacylglycerol synthase. J. Bacteriol. 183, 3506–3514.

Koch, H.U., Haas, R., Fischer, W., 1984. The role of lipoteichoic acid biosynthesis in membrane lipid metabolism of growing *Staphylococcus aureus*. Eur. J. Biochem. 138, 357–363.

Koch, H.U., Doker, R., Fischer, W., 1985. Maintenance of D-alanine ester substitution of lipoteichoic acid by reesterification in *Staphylococcus aureus*. J. Bacteriol. 164, 1211–1217.

Koprivnjak, T., Weidenmaier, C., Peschel, A., Weiss, J.P., 2008. Wall teichoic acid deficiency in *Staphylococcus aureus* confers selective resistance to mammalian group IIA phospholipase A(2) and human beta-defensin 3. Infect. Immun. 76, 2169–2176.

Kraus, D., Herbert, S., Kristian, S.A., et al., 2008. The GraRS regulatory system controls Staphylococcus aureus susceptibility to antimicrobial host defenses. BMC Microbiol. 8, 85.

Kristian, S.A., Lauth, X., Nizet, V., et al., 2003. Alanylation of teichoic acids protects *Staphylococcus aureus* against Toll-like receptor 2-dependent host defense in a mouse tissue cage infection model. J. Infect. Dis. 188, 414–423.

Kristian, S.A., Datta, V., Weidenmaier, C., et al., 2005. D-alanylation of teichoic acid promotes group A *Streptococcus* antimicrobial peptide resistance, neutrophil survival, and epithelial cell invasion. J. Bacteriol. 187, 6719–6725.

Kumar, A., Ray, P., Kanwar, M., Sharma, M., Varma, S., 2005. A comparative analysis of antibody repertoire against *Staphylococcus aureus* antigens in patients with deep-seated versus superficial staphylococcal infections. Int. J. Med. Sci. 2, 129–136.

Lancefield, R.C., Freimer, E.H., 1966. Type-specific polysaccharide antigens of group B streptococci. J. Hyg. (London) 64, 191–203.

Lazarevic, V., Karamata, D., 1995. The *tagGH* operon of *Bacillus subtilis* 168 encodes a two-component ABC transporter involved in the metabolism of two wall teichoic acids. Mol. Microbiol. 16, 345–355.

Leoff, C., Saile, E., Sue, D., et al., 2007. Cell wall carbohydrate compositions of strains from the *B. cereus* group of species correlate with phylogenetic relatedness. J. Bacteriol. 190, 112–121.

Li, M., Lai, Y., Villaruz, A.E., Cha, D.J., Sturdevant, D.E., Otto, M., 2007. Gram-positive three-component antimicrobial peptide-sensing system. Proc. Natl. Acad. Sci. USA 104, 9469–9474.

Lopez, R., Garcia, E., Garcia, P., Ronda, C., Tomasz, A., 1982. Choline-containing bacteriophage receptors in *Streptococcus pneumoniae*. J. Bacteriol. 151, 1581–1590.

Lynch, N.J., Roscher, S., Hartung, T., et al., 2004. L-ficolin specifically binds to lipoteichoic acid, a cell wall constituent of Gram-positive bacteria, and activates the lectin pathway of complement. J. Immunol. 172, 1198–1202.

May, J.J., Finking, R., Wiegeshoff, F., et al., 2005. Inhibition of the D-alanine:D-alanyl carrier protein ligase from *Bacillus subtilis* increases the bacterium's susceptibility to antibiotics that target the cell wall. FEBS J. 272, 2993–3003.

Mazmanian, S.K., Kasper, D.L., 2006. The love-hate relationship between bacterial polysaccharides and the host immune system. Nat. Rev. Immunol. 6, 849–858.

McLoughlin, R.M., Solinga, R.M., Rich, J., et al., 2006. CD4+ T cells and CXC chemokines modulate the pathogenesis of *Staphylococcus aureus* wound infections. Proc. Natl. Acad. Sci. USA. 103, 10408–10413.

Mesnage, S., Fontaine, T., Mignot, T., Delepierre, M., Mock, M., Fouet, A., 2000. Bacterial SLH domain proteins are non-covalently anchored to the cell surface via a conserved mechanism involving wall polysaccharide pyruvylation. EMBO J. 19, 4473–4484.

Mikusova, K., Slayden, R.A., Besra, G.S., Brennan, P.J., 1995. Biogenesis of the mycobacterial cell wall and the site of action of ethambutol. Antimicrob. Agents Chemother. 39, 2484–2489.

Mongodin, E., Bajolet, O., Cutrona, J., et al., 2002. Fibronectin-binding proteins of *Staphylococcus aureus* are involved in adherence to human airway epithelium. Infect. Immun. 70, 620–630.

Nathenson, S.G., Ishimoto, N., Anderson, J.S., Strominger, J.L., 1966. Enzymatic synthesis and immunochemistry of α- and β-*N*-acetylglucosaminylribitol linkages in teichoic acids from several strains of *Staphylococcus aureus*. J. Biol. Chem. 241, 651–658.

Naumova, I.B., Shashkov, A.S., 1997. Anionic polymers in cell walls of Gram-positive bacteria. Biochemistry (Moscow) 62, 809–840.

Navarre, W.W., Schneewind, O., 1999. Surface proteins of Gram-positive bacteria and mechanisms of their targeting to the cell wall envelope. Microbiol. Mol. Biol. Rev. 63, 174–229.

Neuhaus, F.C., Baddiley, J., 2003. A continuum of anionic charge: structures and functions of D-alanyl-teichoic acids in Gram-positive bacteria. Microbiol. Mol. Biol Rev. 67, 686–723.

Neuhaus, F.C., Heaton, M.P., Debabov, D.V., Zhang, Q., 1996. The *dlt* operon in the biosynthesis of D-alanyllipoteichoic acid in *Lactobacillus casei*. Microb. Drug Resist. 2, 77–84.

Park, J.T., Shaw, D.R., Chatterjee, A.N., Mirelman, D., Wu, T., 1974. Mutants of staphylococci with altered cell walls. Ann. NY Acad. Sci. 236, 54–62.

Park, Y.S., Sweitzer, T.D., Dixon, J.E., Kent, C., 1993. Expression, purification, and characterization of CTP: glycerol-3-phosphate cytidylyltransferase from *Bacillus subtilis*. J. Biol. Chem. 268, 16648–16654.

Pereira, M.P., Brown, E.D., 2004. Bifunctional catalysis by CDP-ribitol synthase: convergent recruitment of reductase and cytidylyltransferase activities in *Haemophilus influenzae* and *Staphylococcus aureus*. Biochemistry 43, 11802–11812.

Pereira, M.P., Delia, M.A., Troczynska, J., Brown, E.D., 2008. Duplication of teichoic acid biosynthetic genes in *Staphylococcus aureus* leads to functionally redundant poly(ribitol phosphate) polymerases. J. Bacteriol. 190 (16), 5642–5649.

Peschel, A., Sahl, H.G., 2006. The co-evolution of host cationic antimicrobial peptides and microbial resistance. Nat. Rev. Microbiol. 4, 529–536.

Peschel, A., Otto, M., Jack, R.W., Kalbacher, H., Jung, G., Götz, F., 1999. Inactivation of the *dlt* operon in *Staphylococcus aureus* confers sensitivity to defensins, protegrins and other antimicrobial peptides. J. Biol. Chem. 274, 8405–8410.

Peschel, A., Vuong, C., Otto, M., Götz, F., 2000. The D-alanine residues of *Staphylococcus aureus* teichoic acids alter the susceptibility to vancomycin and the activity of autolysins. Antimicrob. Agents Chemother. 44, 2845–2847.

Polotsky, V.Y., Fischer, W., Ezekowitz, A.B., Joiner, K.A., 1996. Interactions of human mannose-binding protein with lipoteichoic acids. Infect. Immun. 64, 380–383.

Polotsky, V.Y., Belisle, J.T., Mikusova, K., Ezekowitz, R.A., Joiner, K.A., 1997. Interaction of human mannose-binding protein with *Mycobacterium avium*. J. Infect. Dis. 175, 1159–1168.

Pooley, H.M., Karamata, D., 1994. Teichoic acid synthesis in *Bacillus subtilis*: genetic organization and biological roles. In: Ghuysen, J.-M., Hakenbeck, R. (Eds.) Bacterial Cell Wall. Elsevier Science B.V., Amsterdam, pp. 187–197.

Potekhina, N.V., Tul'skaya, E.M., Naumova, I.B., Shashkov, A.S., Evtushenko, L.I., 1993. Erythritolteichoic acid in the cell wall of *Glycomyces tenuis* VKM Ac-1250. Eur. J. Biochem. 218, 371–375.

Powell, D.A., Duckworth, M., Baddiley, J., 1975. A membrane-associated lipomannan in micrococci. Biochem. J. 151, 387–397.

Prigozy, T.I., Sieling, P.A., Clemens, D., et al., 1997. The mannose receptor delivers lipoglycan antigens to endosomes for presentation to T cells by CD1b molecules. Immunity 6, 187–197.

Qian, Z., Yin, Y., Zhang, Y., Lu, L., Li, Y., Jiang, Y., 2006. Genomic characterization of ribitol teichoic acid synthesis in *Staphylococcus aureus*: genes, genomic organization and gene duplication. BMC Genomics 7, 74.

Sadovskaya, I., Vinogradov, E., Li, J., Jabbouri, S., 2004. Structural elucidation of the extracellular and cell-wall teichoic acids of *Staphylococcus epidermidis* RP62A, a reference biofilm-positive strain. Carbohydr. Res. 339, 1467–1473.

Schäffer, C., Messner, P., 2005. The structure of secondary cell wall polymers: how Gram-positive bacteria stick their cell walls together. Microbiology 151, 643–651.

Shashkov, A.S., Streshinskaya, G.M., Senchenkova, S.N., et al., 2006. Cell wall teichoic acids of streptomycetes of the phenetic cluster '*Streptomyces fulvissimus*'. Carbohydr. Res. 341, 796–802.

Shimaoka, T., Kume, N., Minami, M., et al., 2001. LOX-1 supports adhesion of Gram-positive and Gram-negative bacteria. J. Immunol. 166, 5108–5114.

Sidobre, S., Nigou, J., Puzo, G., Riviere, M., 2000. Lipoglycans are putative ligands for the human pulmonary surfactant protein A attachment to mycobacteria. Critical role of the lipids for lectin-carbohydrate recognition. J. Biol. Chem. 275, 2415–2422.

Soldo, B., Lazarevic, V., Pagni, M., Karamata, D., 1999. Teichuronic acid operon of *Bacillus subtilis* 168. Mol. Microbiol. 31, 795–805.

Soldo, B., Lazarevic, V., Karamata, D., 2002a. *tagO* is involved in the synthesis of all anionic cell-wall polymers in *Bacillus subtilis* 168. Microbiology 148, 2079–2087.

Soldo, B., Lazarevic, V., Pooley, H.M., Karamata, D., 2002b. Characterization of a *Bacillus subtilis* thermosensitive teichoic acid-deficient mutant: gene *mnaA* (*yvyH*) encodes the UDP-*N*-acetylglucosamine 2-epimerase. J. Bacteriol. 184, 4316–4320.

Sugawara, I., Yamada, H., Li, C., Mizuno, S., Takeuchi, O., Akira, S., 2003. Mycobacterial infection in TLR2 and TLR6 knockout mice. Microbiol. Immunol. 47, 327–336.

Sutcliffe, I., 2005. Lipoarabinomannans--structurally diverse and functionally enigmatic macroamphiphiles of mycobacteria and related actinomycetes. Tuberculosis (Edinburgh) 85, 205–206.

Sutcliffe, I.C., Black, G.W., Harrington, D.J., 2008. Bioinformatic insights into the biosynthesis of the Group B carbohydrate in *Streptococcus agalactiae*. Microbiology 154, 1354–1363.

Takahashi, M., Mori, S., Shigeta, S., Fujita, T., 2007. Role of MBL-associated serine protease (MASP) on activation of the lectin complement pathway. Adv. Exp. Med. Biol. 598, 93–104.

Takayama, K., Wang, C., Besra, G.S., 2005. Pathway to synthesis and processing of mycolic acids in *Mycobacterium tuberculosis*. Clin. Microbiol. Rev. 18, 81–101.

Tapping, R.I., Tobias, P.S., 2003. Mycobacterial lipoarabinomannan mediates physical interactions between TLR1 and TLR2 to induce signaling. J. Endotoxin. Res. 9, 264–268.

Theilacker, C., Krueger, W.A., Kropec, A., Huebner, J., 2004. Rationale for the development of immunotherapy regimens against enterococcal infections. Vaccine 22 (Suppl. 1), S31–S38.

Theilacker, C., Kaczynski, Z., Kropec, A., et al., 2006. Opsonic antibodies to *Enterococcus faecalis* strain 12030 are directed against lipoteichoic acid. Infect. Immun. 74, 5703–5712.

Tzianabos, A.O., Wang, J.Y., Lee, J.C., 2001. Structural rationale for the modulation of abscess formation by *Staphylococcus aureus* capsular polysaccharides. Proc. Nat. Acad. Sci. USA 98, 9365–9370.

van de Wetering, J.K., van, E.M., van Golde, L.M., Hartung, T., Van Strijp, J.A., Batenburg, J.J., 2001. Characteristics of surfactant protein A and D binding to lipoteichoic acid and peptidoglycan, 2 major cell wall components of Gram-positive bacteria. J. Infect. Dis. 184, 1143–1151.

Verbrugh, H.A., Peters, R., Rozenberg-Arska, M., Peterson, P.K., Verhoef, J., 1981. Antibodies to cell wall peptidoglycan of *Staphylococcus aureus* in patients with serious staphylococcal infections. J. Infect. Dis. 144, 1–9.

Verhoef, J., Musher, D.M., Spika, J.S., Verbrugh, H.A., Jaspers, F.C., 1983. The effect of staphylococcal peptidoglycan on polymorphonuclear leukocytes *in vitro* and *in vivo*. Scand. J. Infect. Dis. (Suppl. 41), 79–86.

Vinogradov, E., Sadovskaya, I., Li, J., Jabbouri, S., 2006. Structural elucidation of the extracellular and cell-wall teichoic acids of *Staphylococcus aureus* MN8 m, a biofilm forming strain. Carbohydr. Res. 341, 738–743.

Ward, J.B., 1981. Teichoic and teichuronic acids: biosynthesis, assembly and location. Microbiol. Rev. 45, 211–243.

Weidenmaier, C., Peschel, A., 2008. Teichoic acids and related cell-wall glycopolymers in Gram-positive physiology and host interactions. Nat. Rev. Microbiol. 6, 276–287.

Weidenmaier, C., Kokai-Kun, J.F., Kristian, S.A., et al., 2004. Role of teichoic acids in *Staphylococcus aureus* nasal colonization, a major risk factor in nosokomial infections. Nat. Med. 10, 243–245.

Weidenmaier, C., Peschel, A., Xiong, Y.Q., et al., 2005a. Lack of wall teichoic acids in *Staphylococcus aureus* leads to reduced interactions with endothelial cells and to attenuated virulence in a rabbit model of endocarditis. J. Infect. Dis 191, 1771–1777.

Weidenmaier, C., Peschel, A., Kempf, V.A., Lucindo, N., Yeaman, M.R., Bayer, A.S., 2005b. DltABCD- and MprF-mediated cell envelope modifications of *Staphylococcus aureus* confer resistance to platelet microbicidal proteins and contribute to virulence in a rabbit endocarditis model. Infect. Immun. 73, 8033–8038.

Weidenmaier, C., Kokai-Kun, J.F., Kulauzovic, E., et al., 2008. Differential roles of sortase-anchored surface proteins and wall teichoic acid in *Staphylococcus aureus* nasal colonization. Int. J. Med. Microbiol. 298, 505–513.

Weisman, L.E., 2007. Antibody for the prevention of neonatal nosocomial staphylococcal infection: a review of the literature. Arch. Pediatr. 14 (Suppl. 1), S31–S34.

Wendlinger, G., Loessner, M.J., Scherer, S., 1996. Bacteriophage receptors on *Listeria monocytogenes* cells are the *N*-acetylglucosamine and rhamnose substituents of teichoic acids or the peptidoglycan itself. Microbiology 142, 985–992.

Xia, G., Peschel, A., 2008. Toward the pathway of *S. aureus* WTA biosynthesis. Chem. Biol. 15, 95–96.

CHAPTER

6

Bacterial capsular polysaccharides and exopolysaccharides

Paola Cescutti

SUMMARY

Bacteria often produce an external layer of polysaccharides, characterized by a definite primary structure, which in turn is responsible for sometimes remarkable physicochemical properties. Although the number of monosaccharides which constitute the polymers is rather low, the great number of different polysaccharides defined up to now shows the capacity of the microbes to exploit the possible isomers and linkage types of the building blocks. Furthermore, variability is often introduced by the presence of non-carbohydrate groups linked to hydroxyl, carboxyl or amine functions. In this chapter, examples of polysaccharides produced by Gram-negative and Gram-positive pathogenic bacteria are given, together with a description of those polymers that are interesting for industrial and biotechnological purposes. Some discussion is also devoted to the general features of shapes that polysaccharides may adopt in solution. The biological functions of these biomolecules are discussed particularly in relation to their role in human infection processes. The structures of the polysaccharides produced by species of the *Burkholderia cepacia* complex is reported as an example of a current investigation devoted to the understanding of the role of these biopolymers in lung infections.

Keywords: Capsular polysaccharides; Exopolysaccharides; Primary structure; Biological properties; *Burkholderia cepacia* complex

1. INTRODUCTION

Bacterial cells are often surrounded by a polysaccharidic layer which constitutes the interface with the environment. Sometimes this structure is referred to as the "glycocalyx". Bacterial polysaccharides are classified into two different types: capsular polysaccharides (K-antigens) (CPS) and exopolysaccharides (EPS). Thus, CPS are defined as polymers linked to the cell surface via covalent bond to phospholipid or lipid A molecules, while EPS appear to be released on the cell surface with no attachment to the cell and they are often sloughed off to form slime. However, there is no clear cut definition for CPS and EPS, because the former can also be released into the growth medium and the latter may be closely associated with the cell surface (Taylor and Roberts, 2005). Despite this ambiguity, these are the most common definitions found in the literature and they will be used as such in this context. It is worth mentioning that Costerton *et al.* (1981) gave a more detailed definition of capsule, distinguishing four different types of CPS:

(i) rigid, which excludes particles, like India ink;
(ii) flexible, sufficiently deformable that it does not exclude particles;
(iii) integral, intimately associated with the cell; and
(iv) peripheral, may remain associated with the cell and may be shed into the medium.

In this view, EPS probably would be defined as capsule type (iv).

Structurally, CPS and EPS can be homopolysaccharides, like dextran and cellulose, or heteropolysaccharides, and can have a linear or branched structure. They are constituted of oligomers, ranging in size from 2 to 10 sugar residues, which repeat regularly a given number of times to yield a polymer of generally high molecular mass (10^5–10^6). These macromolecules may be neutral or charged and the polyanionic ones prevail. Cyanobacteria are an exception in this context because they usually have very complicated repeating units which may not have a strictly repeating regular structure (De Philippis and Vincenzini, 1998).

In general, CPS and EPS are rather widespread and are synthesized by plant, animal and human pathogen bacteria, by soil, fresh water and seawater microbes and by cyanobacteria. Being the interface with the "outer" world for microbes, their biological role must be related to the interaction with the environment. Furthermore, it was demonstrated that most bacteria, except those found in deep groundwater and abyssal oceans, live associated with surfaces in the form of multicellular aggregates commonly referred to as biofilm and that CPS and EPS are involved in the biofilm architecture (see Chapters 37 and 39).

Sometimes bacterial polysaccharides exhibit extremely interesting solution and gelling properties, characteristics strictly related to their primary structure and conformation. Therefore, some of them are used in several industrial applications or are exploited as potential new biomaterials (Kumar *et al.*, 2007). Their commercial use is linked to their origin; in general those already on the market are environmental species, while CPS and EPS from human pathogenic bacterial are used to prepare conjugated vaccine (see Chapter 48).

The availability of powerful techniques, like nuclear magnetic resonance (NMR) spectroscopy and mass spectrometry, has led to an enormous increase in the amount of available polysaccharide structures in the last 20 years. Therefore, a choice on the content of this chapter was inevitable and it was concluded that a description of only some CPS and EPS related to human disease and to industrial exploitation was possible. A number of references, especially reviews, are provided and the reader is encouraged to refer to them for more detailed data.

2. CARBOHYDRATE COMPONENTS OF CAPSULAR AND EXO-POLYSACCHARIDES

The most common components of CPS and EPS are:

(i) pentoses: ribose and arabinose, the latter found especially in *Mycobacterium* species;
(ii) hexoses: glucose, mannose, galactose and fructose;
(iii) deoxysugars: fucose and rhamnose;
(iv) uronic acids: glucuronic and galacturonic acids; and
(v) aminosugar: glucosamine and galactosamine, often *N*-acetylated.

Occasionally, other sugars are detected: 2-acetamido-2,6-dideoxygalactose, 3-deoxy-D-*manno*-oct-2-ulosonic acid (Kdo), *N*-acetyl neuraminic acid, 2-acetamido-2-deoxy-guluronic acid, 2-acetamido-2-deoxy-mannuronic acid, guluronic acid and mannuronic acid. *O*-Methylated sugars are also common. Apart from ribose, fructose and galactose, which sometimes adopt the furanosidic ring structure, the other monosaccharides are usually in the pyranosidic form. Although the number of monosaccharides used by microbes to synthesize polysaccharides is relatively small, the great variety of CPS and EPS structures is achieved through the versatility of the glycosidic linkages, where variation occurs in terms of both configuration at the anomeric carbon atom (α or β) and choice of the carbon atom engaged in the linkage. CPS and EPS usually are negatively charged due to the presence of uronic acids or to ionic non-carbohydrate substituents (see below). In the past years, excellent reviews (Kenne and Lindberg 1983; Lindberg, 1990; Griffiths and Davies 1991a,b) have listed with great detail the carbohydrate components of bacterial polysaccharides.

3. NON-CARBOHYDRATE SUBSTITUENTS OF CAPSULAR AND EXOPOLYSACCHARIDES

The most common non-carbohydrate groups are: *N*- and *O*-acetyl, pyruvic acid ketal and phosphoric acid linked with diester linkages. Less common are succinic acid, amino acids, glycerol, ribitol, lactic acid, to name a few (Kenne and Lindberg, 1983). *N*-acetyl groups are linked through an amide bond and are thus much more resistant to hydrolysis than the *O*-acetyl substituents. Substitution with *N*- or *O*-acetyl groups changes the hydrophilic character of sugars which normally is dominated by hydroxyl groups. Moreover, *O*-acetyl groups on a sugar may migrate, unless they are impeded by other substituents. Therefore, if *O*-acetyl groups are found in two positions in a sugar residue, *O*-acetylation may have occurred in a single position during biosynthesis. Pyruvic acid is most often linked to the 4- and 6-position of a hexopyranose ring and its absolute configuration can be determined by ^{13}C-NMR spectroscopy (Garegg *et al.*, 1980). In addition, pyruvic acid can also be linked to vicinal hydroxyl functions, for example 3- and 4-position in galactopyranosyl residues (Lindberg, 1990). Some non-carbohydrate substituents are present in an even ratio with the repeating unit (e.g. pyruvate), while others may exist in stoichiometric as well as non-stoichiometric amounts (e.g. *O*-acetyl). When charged, the substituents turn a neutral polymer into a polyelectrolyte, with important consequences on its macromolecular properties; alternatively, they might increase the charge density on an already anionic polysaccharide.

4. STRUCTURE OVERVIEW OF BACTERIAL POLYSACCHARIDES

4.1. Capsular polysaccharides of Gram-negative bacteria

The cell envelope of Gram-negative bacteria is made of an inner membrane, separated from the peptidoglycan by a periplasmic space and, finally, an outer membrane containing the LPS. Some bacteria also possess an extracellular envelope called a capsule, or K-antigen (Kapselantigene) which, being the most external layer, imparts its dominant immunological properties to the whole cell. Some pathogenic bacteria are capsulated and their virulence is closely related with the structure of the polysaccharide. The knowledge of the polymer primary structure is important in the definition of its immunological activity, a first step towards the preparation of efficacious vaccines.

The *Enterobacteriaceae* is a family of Gram-negative bacteria that normally live in the intestines of humans and other animals, while others are found in water or soil, or are parasites on a variety of different animals and plants. Some species can cause infections of the digestive tract or of other organs of the body. In this family, *Escherichia coli* and *Klebsiella pneumoniae* are remarkable for the copious number of different CPS they produce. *E. coli* is an opportunistic human pathogen because certain strains can cause intestinal or urinary tract infections and neonatal septicaemia and meningitis. This species is divided into more than 80 serotypes, according to the number of different capsular antigens synthesized. These CPS were characterized and classified in group I and group II according to their structure (Jann and Jann, 1990). Group I is rather homogeneous and is further subdivided according to the presence or absence of amino sugars. Group II contains polymers that differ widely in composition and in general structural features and this group can be subdivided depending on the acidic component. The biosynthesis and regulation of the CPS in *E. coli* has been extensively investigated (Whitfield, 2006).

K. pneumoniae is the most important clinical species of its genus, because it is an opportunistic human pathogen responsible of nosocomial infection outbreaks, like pneumonia, bacteraemia and infections of the urinary tract, wound, soft tissue and respiratory tract. This species also synthesizes more than 80 different K-antigens which have been fully characterized (Kenne and Lindberg, 1983; Dutton, 1984; Griffiths and Davies, 1991a; Ovodov, 2006). The structure of the repeating units of *K. pneumoniae* CPS shows great variability. They contain from three to seven sugars, arranged in linear or branched oligomers, with negative charges carried by uronic acids and/or pyruvate groups, and may be acetylated. Some examples are shown in Figure 6.1.

Another Gram-negative bacterium interesting as a human pathogen is *Haemophilus influenzae*. It is present in the nasopharynx of about 75% of healthy children and adults in the form of encapsulated strains. This species is divided into six types (a–f) which express six different K-antigens (Griffiths and Davies, 1991b; Ovodov, 2006). The repeating units of these CPS are constituted of two or three sugar residues, some of which are amino sugars, and may contain phosphate esters and *N*- and *O*-acetyl groups. The importance of this species resides in the dangerousness of *H. influenzae* type b (Hib) infection, which can result in meningitis and other severe infections (e.g. pneumonia, bacteraemia, cellulitis, septic arthritis and epiglottitis) primarily among infants and children <5 years of age. In the early 1990s, the introduction of carbohydrate-based Hib conjugate vaccines was followed by a dramatic decrease of meningitis in children (Weintraub, 2003). The success of vaccination was also ascribed to a reduction of Hib carriage, which probably decreased the transmission rates to children devoid of protective antibodies.

Neisseria meningitidis are Gram-negative aerobic bacteria, commonly known as meningococci and are exclusively human pathogens causing meningitis. The major virulence factor is the K-antigen: on the basis of its structure, twelve different K-types were identified (A, B, C, H, I, K, L, X, Y, Z, 29E and W135) (Griffiths and Davies, 1991b). The different repeating units contain from one to three sugar residues, often amino sugars and neuraminic acid; acetyl groups and phosphate esters are also present as non-carbohydrate substituents. More than 90% of meningitis cases are caused by serotypes A, B, C, W135 and Y. As with Hib vaccine, A and C glycoconjugate vaccines showed to be efficacious and are available for disease prevention.

4.2. Capsular polysaccharides of Gram-positive bacteria

Gram-positive bacteria are characterized by an inner membrane surrounded by a thick layer

```
[3)-α-D-Galp-(1→2)-[3,4-O-(1-carboxyethylidene)]-α-L-Rhap-(1→

                                   →3)-β-L-Rhap-(1→4)-α-L-Rhap-(1→]n
```

K32[a]

```
[3)-α-L-Rhap-(1→2)-α-L-Rhap-(1→4)-α-D-GlcpA-(1→

             →2)-α-D-Manp-(1→2)-α-D-Manp-(1→3)-α-D-Galp-(1→]n
```

K40[b]

```
[3)-β-D-Glcp-(1→3)-β-D-GlcpA-(1→)-α-D-Galp-(1→3)-α-D-Manp-(1→]n
                4                   2                  2
                ↑                   ↑                  ↑
                1                   1                  1
            α-D-Glcp            β-D-Glcp           β-D-Glcp
```

K60[c]

```
[3)-α-L-Rhap-(1→3)-β-D-Glcp-(1→2)-α-L-Rhap-(1→2)-α-L-Rhap-(1→]n
                                        3
                                        ↑
                                        1
                     β-D-GlcAp-(1→3)-α-L-Rhap
                                        2
                                        ↑
                                        1
                                    β-D-Glcp
```

K71[d]

FIGURE 6.1 Examples of *Klebsiella pneumoniae* CPS repeating units. *O*-Acetyl substituents are not considered in the structures: [a] Bebault *et al.* (1978); [b] Cescutti *et al.* (1993); [c] Dutton and Di Fabio (1980); [d] Jackson *et al.* (1990).

of peptidoglycan, which in turn may be covered with a polysaccharidic capsule. Also, these microorganisms present a large variability of the CPS structure. Among the *Streptococcus* genus there are several species which are human opportunistic pathogens. *Streptococcus pneumoniae* often causes severe infections, especially in infants, elderly persons and patients with debilitating conditions. The most common types of infections include middle ear infections, pneumonia, bacteraemia, sinus infections and meningitis. The pneumococcal capsule is indispensable for virulence (Llull *et al.*, 2001). More than 90 different CPS are known (Weintraub, 2003), but only 23, those considered the most virulent, are used for the formulation of a polysaccharide vaccine (van Dam *et al.*, 1990). The structure of their CPS is very diverse ranging from a simple disaccharidic repeating unit (type 3) to more complex ones containing higher number of monomers in the repeating unit and non-carbohydrate substituents, like phosphate, pyruvate, glycerol, ribitol and acetyl (van Dam *et al.*, 1990; Ovodov, 2006). Contrary to the Gram-negative bacteria, the CPS of streptococci, and most probably of other Gram-positive bacteria, is covalently attached to the bacterial cell wall and detailed data are available for serotype III of group B streptococci (Llull *et al.*, 2001).

Staphylococcus aureus is another capsulated opportunistic pathogen of humans and animals. It plays an important role in nosocomial and community acquired infections. Control of *S. aureus* infections is becoming more difficult due to the prevalence of antibiotic resistant strains and to the emergence of vancomycin resistant clinical isolates. CPS is listed among the virulence factors of this species. At least 18 different CPS were described and all contain hexosaminuronic acids, but only four were completely characterized. Besides the acidic component, they contain *N*-acetylated amino sugars and sometimes *O*-acetyl or alanyl groups (O'Riordan and Lee, 2004).

Mycobacterium tuberculosis is the aetiologic agent of tuberculosis, an old disease that is still a threat for humans. The envelope which surrounds bacterial cells is really unique; it is constituted of long chain fatty acids (mycolic acids), glycolipids, lipoglycans and polysaccharides. Such an envelope contributes to bacterial survival in the infected host; the lipoglycan was shown to modulate the host immune response and, in particular, to have a powerful anti-inflammatory activity. Lipoglycans are known as lipoarabinomannans (LAM) and they are composed of three domains: a polysaccharide backbone, the mannosyl-phosphatidyl-*myo*-inositol (MPI) anchor and the capping motifs. These are ubiquitous structural features which are then subjected to strain dependent structural variation. The polysaccharide part is constituted of two homopolysaccharides, a D-mannan and a D-arabinan; the MPI is linked to the mannan polymer and is acylated; the capping motifs are essentially two: a manno-oligosaccharide and a phosphoinositol. According to the cap structure, LAM are classified into ManLAM and PILAM, the former being present in the slow growing microbes (e.g. *M. tuberculosis*) and the latter in the fast growing ones (Nigou *et al.*, 2003; Ovodov, 2006).

4.3. Exopolysaccharides

Exopolysaccharides are produced by a great variety of microorganisms, opportunistic human pathogens as well as environmental microbes.

Among the pathogens, the case of *Pseudomonas aeruginosa,* a Gram-negative free-living bacterium, commonly found in soil and water, is very well known, because it constitutes a life threat for cystic fibrosis patients, causing infections characterized by high morbidity and mortality. In fact, it colonizes the respiratory tract of cystic fibrosis (CF) patients and switches from a non-mucoid to a mucoid phenotype, producing copious amounts of alginate, as extracellular polymer. Alginate is a linear copolymer of β-D-mannuronic acid (ManA) and its epimer, α-L-guluronic acid (GulA), linked via (1→4)-glycosidic linkages. An important characteristic of alginate is that it does not possess a regular repeating unit, it is highly negatively charged and it is *O*-acetylated on the mannuronic acid residues (Gacesa and Russell, 1990). In general, alginate from the *Pseudomonas* genus is characterized by the lack of contiguous sequences of the two uronic acids, and by a mannuronic and guluronic acid (ManA:GulA) ratio between 2 and 5. *Azotobacter vinelandi*, a Gram-negative soil microbe, is another species which produces alginate: in this case the polymer is more similar to the algal product because it contains contiguous sequences of both mannuronic and guluronic acids as well as mixed random sequences (Sutherland, 1994; Remminghorst and Rehm, 2006).

Exopolysaccharides also surround bacterial cells which live in seawater; they constitute part of the reduced carbon reservoir in the ocean and directly protect marine bacteria by creating a controlled physicochemical environment around the cells. Moreover, EPS are abundant in harsh environments, like the Antarctic Ocean, where they may contribute to bacterial survival in extreme temperatures, salinity and scarce-nutrient niches. The EPS from marine bacteria are usually heteropolysaccharides containing hexoses, pentoses and uronic acids and are formed by repeating units of 10 or less residues: they may be linear or branched and also comprise non-sugar substituents, like acetyl, phosphate, sulfate and pyruvate (Mancuso Nichols *et al.*, 2005).

For their physicochemical and gelling properties, some EPS from environmental species are used for technological applications, including the biotechnology industry. Curdlan is a linear β-(1→3)-glucan, produced by several different Gram-negative bacteria, including strains of *Agrobacterium* spp. and *Rhizobium* spp. It is used as a gelling agent to improve the textural quality, water-holding capacity and thermal stability of various foods (Sutherland, 1999). Xanthan is synthesized by *Xanthomonas campestris* and also by other *Xanthomonas* species. This polysaccharide is composed of a pentasaccharide repeating unit (Figure 6.2) and, depending on the bacterial strain, it may be *O*-acetylated and pyruvylated (Sutherland, 1994). It produces high viscosity water solutions characterized by high pseudoplasticity, properties that render it amenable for many industrial applications. Xanthan is used in crude-oil recovery, in paints, pesticide and detergent formulations, cosmetics, pharmaceuticals and in food often in combination with guar gum. Gellan is produced by strains of *Sphingomonas paucimobilis* and has a linear tetrasaccharide repeating unit (see Figure 6.2). Removal of L-glyceryl and acetyl groups gives a strong and clear gel, comparable to that formed by agarose and, therefore, modified gellan is used in bacterial culture media. Dextran is a homopolymer of α-(1→6)-linked glucose with a varying degree of branching at C-3. The main applications for dextran are in the pharmaceutical and medical areas, as a blood plasma substitute and as a substitute for heparin, after being modified through the introduction of sulfate groups. Cross-linked dextrans are widely used in chromatographic media, while the use of dextrans in the field of bakery is less exploited (Kumar *et al.*, 2007; Lacaze *et al.*, 2007). Hyaluronan is constituted of a disaccharidic repeating unit (see Figure 6.2) and is widely used in the cosmetic and pharmaceutical industries (e.g. in skin moisturizers, for osteoarthritis treatment, in ophthalmic surgery, in adhesion prevention after abdominal surgery and wound healing). It is obtained commercially from rooster combs and some attenuated strains of group C *Streptococcus*. However, the polysaccharide from rooster combs can cause side effects in persons allergic to avian products, while that from *Streptococcus* spp. can contain potentially dangerous toxins. Recently, to overcome these undesirable effects, hyaluronan was produced by expressing the *hasA* gene, which encodes the enzyme hyaluronan synthase in *Streptococcus equisimilis*, in *Bacillus subtilis*, as many of its products were awarded a GRAS (generall recognized as safe) definition (Widner *et al.*, 2005).

Cellulose is a linear homopolysaccharide constituted of β-(1→4)-glucopyranosyl residues. *Acetobacter xylinum* is the most prolific cellulose-producing bacterial species. Microbial cellulose is currently used in food and wound care products, paper and paper products (patent JP 63295793). Even audio speaker diaphragms were made using microbial cellulose (US Patent 4,742,164).

The exopolysaccharide named succinoglycan (see Figure 6.2) was first isolated from *Alcaligenes faecalis* var. *myxogenes*. It was then discovered that succinoglycans are also synthesized by strains of *Agrobacterium tumefaciens*, *Agrobacterium radiobacter* and *Rhizobium meliloti* and by many strains of *Pseudomonas* spp.

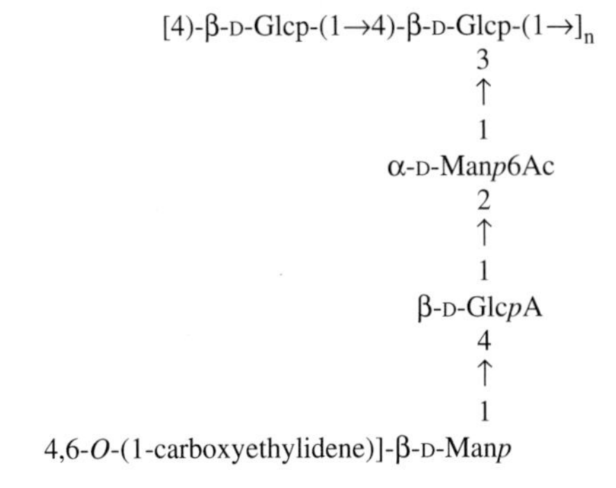

xanthan[a]

[4)-α-L-Rhap-(1→3)-β-D-Glcp-(1→4)-β-D-GlcAp-(1→4)-β-D-Glcp-(1→]$_n$

gellan[b]

[3)-β-D-GlcNAcp-(1→4)-β-D-GlcAp-(1→]$_n$

hyaluronic acid

[4)-β-D-Glcp-(1→4)-β-D-Glcp-(1→3)-β-D-Galp-(1→4)-β-D-Glcp-(1→]$_n$
6
↑
1
[4,6-O-(1-carboxyethylidene)]-β-D-Glcp-(1→3)-β-D-Glcp-(1→3)-β-D-Glcp-(1→6)-β-D-Glcp

succinoglycan[c]

FIGURE 6.2 Structures of the repeating units of EPS (mentioned in this chapter) and used in industrial applications. [a] The EPS produced by *Xanthomonas campestris*: usually the internal α-mannosyl residue is fully acetylated, while the pyruvil group, in the *S* absolute configuration, is present on about 30% of the terminal β-mannosyl residues. [b] The 3-linked glucose residue is substituted on O-2 with L-glyceryl and on O-6 with acetyl groups. [c] Succinoglycan contains also acetyl and succinyl substituents depending on the bacterial strain (Zevenhuizen, 1997); pyruvyl groups are in the *S* absolute configuration.

isolated from activated sludge and from other sources. Succinoglycans produced by diverse species differ for the content of pyruvate, succinate and acetyl groups (Zevenhuizen, 1997). Succinoglycan is used in agrochemical formulations, oil and gas field chemicals and products for personal and home care.

5. POLYSACCHARIDE SHAPES

The biological function and the properties of polysaccharide are closely related to their primary structure and conformation, which in turn depends stringently on the former. The overall chain conformation of the carbohydrate polymers depends on the conformation of their building blocks. In fact, the size (penta-, hexa-, heptasaccharide), the type of ring (pyranose or furanose), the anomeric configuration (α or β) greatly influence the conformation of the monosaccharide (boat, skew boat or chair) and, consequently, that of the polysaccharide. Furthermore, the multiplicity of possible glycosidic linkages between sugar residues introduces further variability in the conformation of the final macromolecules. Glycosidic linkages involve C-1 of one residue and one of the hydroxyls on the other residues, originating several possibilities: (1→2)-, (1→3)-, (1→4)-, and (1→6)-linkages. The glycosidic bond can also have two possible orientations, axial or equatorial, thus originating four possibilities for each linkage type. For example, in the case of a pyranose ring in the $^4C_1$(D) conformation, the four possibilities for a (1→4)-linkage are: 1e→4e, 1a→4e, 1e→4a, 1a→4a, where 1a is the α anomer and 1e is the β one. Each of these possibilities has consequences on the conformation of the whole macromolecule. For example, there are eleven known disaccharides constituted only of glucose residues, but differing in the anomeric configuration and in the carbon atoms involved in the glycosidic bond. In contrast, from one type of amino acid only one dipeptide can be formed. It is straightforward that conformational complexity and variability increases by increasing the number of residues linked together. Without entering into details about the overall shapes of carbohydrate molecules, it can be generally stated that the rather high rigidity of the glycosidic ring, its steric hindrance and the limitation of the rotation degree of freedom around the glycosidic bonds [except for (1→6)-linkages which exhibit three rotational degrees of freedom, one of them having low rotational energy barriers] confer to these macromolecules rather elongated shapes. Therefore, their hydrodynamic volume is definitely larger than that of a globular protein with the same molar mass. Generally, polysaccharides do not exhibit a tertiary structure, as globular protein do, but their overall conformation dynamically changes according to the shape of the conformational-energy surface of local conformers. The concept of secondary structure for saccharides is also rather far from that applied to proteins. Although some structure regularity may exist, it usually dynamically involves stretches of the polymer and it is not restricted to specific sequences. Often, the presence of secondary structures is related to particular chemical constraints, like junction zones between different chains in gel phase (Morris *et al.*, 1978), or interaction with specific agents (e.g. amylose-I_2 complexes or specific cations interactions with polyelectrolytes (Chandrasekaran and Thailambal, 1990)).

Several experimental methodologies are used in structure analysis: fibre X-ray diffraction, NMR spectroscopy and computational methods (Rao *et al.*, 1998). In recent years, atomic force microscopy has given a notable contribution to the investigation of polysaccharide shapes (see Chapter 14).

6. BIOLOGICAL FUNCTIONS OF CAPSULAR AND EXOPOLYSACCHARIDES

The functions of the exopolymers can be distinguished in general and specific ways. With general functions, those related to the physicochemical characteristic of the polysaccharide molecules are intended and therefore they may apply to all bacteria with an exosaccharide layer. Polysaccharides are highly hydrated molecules and, therefore, they can provide an anti-dessication milieu. Thus, CPS and EPS, having quite a high molecular mass and a large number of possible hydrogen bond-forming groups, easily give rise to a very viscous aqueous solution, eventually leading to the formation of an interconnected polymeric matrix.

These structures constitute a physical barrier towards potential enemies (e.g. macromolecules, bacteriophages and cell components of the host immune system), preventing the recognition and attachment to epitopes exposed on the underneath layer (outer membrane or peptidoglycan). When polyanionic, a further charge effect adds to the simple barrier, capturing positively charged species and repulsing negatively charged molecules, like serum components and cells with similar surface charge. For example, encapsulated bacteria pathogens show resistance to non-specific host immunity, like complement-mediated killing. In the absence of specific antibodies, the CPS barrier masks the underlying cell surface components that are usually potent activators of the complement alternative pathway (Taylor and Roberts, 2005). Another example is related to cationic antimicrobial peptides which are components of the humoral innate host defences and are produced by neutrophils; they can interact with bacterial membranes causing ultimately cell lysis. It was shown that alginate formed complexes with some antimicrobial peptides, keeping them distant from the bacterial membrane and thus protecting *P. aeruginosa* from lysis (Herasimenka *et al.*, 2005).

In addition, CPS and EPS can promote the adherence of bacteria to surfaces as well as to each other, facilitating the formation of a biofilm and the colonization of different ecological niches (fresh and seawater surfaces, dental surfaces, medical devices, industrial pipes and vessels) (Donlan and Costerton, 2002; Taylor and Roberts, 2005).

Specific functions are those related to particular chemical and stereochemical motifs of the CPS and EPS. Hence, CPS that contain *N*-acetylneuraminic acid (Neu5Ac) are poor activators of the alternative pathway of the complement cascade because Neu5Ac binds a complement factor, resulting in breaking the amplification cascade and failure of the membrane attack complex formation (Michalek *et al.*, 1988). Moreover, CPS that have a primary structure identical to components of the host are not recognized as non-self and therefore they elicit no or a poor immune response. *E. coli* K1 and *N. meningitidis* serogroup B synthesize a homopolymer of α-(2→8) sialic acid. Its low immunogenicity is attributable to immunologic tolerance induced by fetal exposure to polysialylated glycoproteins (Weintraub, 2003). Other examples are *E. coli* K5 CPS [4)-β-D-GlcA*p*-(1→4)-α-D-GlcNAc*p*-$(1\rightarrow]_n$, which has a structure identical to an intermediate in the biosynthesis of heparin and *Streptococcus pyogenes*, one of the most pathogenic streptococci, with a capsule of hyaluronan, identical to the polysaccharide produced in higher animals.

Alginate produced by *P. aeruginosa* has been extensively investigated to elucidate its biological function in the context of CF. It has been reported that alginate scavenges reactive oxygen species produced by macrophages as well as by cell free systems (Learn *et al.*, 1987; Simpson *et al.*, 1989). Most importantly, it was found that the conversion of non-mucoid to mucoid phenotype was the response of the bacteria to the release of H_2O_2 by activated polymorphonuclear leukocytes, as a defence mechanism towards this toxic chemical (Mathee *et al.*, 1999). A role for the *O*-acetyl groups was also postulated; their presence was associated with dampening of the complement activation, with a consequent resistance to antibody-independent phagocytic killing (Pier *et al.*, 2001). A further interesting example of a specific effect exerted by a CPS is that related to the *Salmonella* genus. *Salmonella* strains are responsible for typhoid fever (*S. enterica* sv. Typhi) and gastroenteritis (*S. enterica* serovars Typhimurium and Enteritidis) worldwide. It was demonstrated that neutrophil infiltration, observed in typhoid fever and not in gastroenteritis, was associated with the Vi capsular antigen, which was expressed by *S. enterica* sv. Typhi and not by serovar Typhimurium. (Raffatellu *et al.*, 2006).

7. EXOPOLYSACCHARIDES OF THE *BURKHOLDERIA CEPACIA* COMPLEX: A CASE STUDY

B. cepacia was first described as *Pseudomonas cepacia* by Walter Burkholder in 1950 (Burkholder, 1950) and recognized as the phytopathogen responsible for the bacterial rot of onions. Before the early 1980s, *B. cepacia* was reported to cause infections only in hospitalized patients but, by the beginning of the 1980s, a rising incidence of pulmonary infection in CF patients was observed. Since then, *B. cepacia* was recognized as one of the bacterial species which can cause serious pulmonary infections in CF patients reducing their survival and sometimes leading to the "cepacia syndrome", a rapidly fatal necrotizing pneumonia (Govan and Deretic, 1996). Detailed typing of microorganisms classified as *B. cepacia* revealed enough diversity to divide them in 16 different species *B. cepacia, B. multivorans, B. cenocepacia, B. stabilis, B.vietnamensis, B. dolosa, B. ambifaria, B. anthina, B. pyrrocinia, B. latens, B. diffusa, B. arboris, B. seminalis, B. metallica B. contaminans* and *B. lata*, that altogether constitute the *B. cepacia* complex (BCC), which includes environmental bacteria, as well as opportunistic pathogens.

Most of the BCC species are mucoid due to the presence of exopolysaccharides when cultivated on mannitol-yeast extract solid medium (Sage *et al.*, 1990), while this character is not always evident on other media. The presence of EPS as an indispensable requirement to express virulence is under debate. An investigation carried out on all species of the BCC showed that they express the mucoid phenotype, although among clinical isolates, *B. cenocepacia* was most frequently non-mucoid. Interestingly, isolates from patients with chronic infections were shown to convert from mucoid to non-mucoid (Zlosnik *et al.*, 2008). Structural studies of several species revealed that members of the BCC can produce at least five different EPS (Figure 6.3) and that sometimes the type produced depends on the medium used. Moreover, very often these microorganisms produce mixtures of up to three of the EPS indicated. Most of the strains examined produced the EPS named cepacian (see Figure 6.3), which has a rather unusual structure. The heptasaccharidic repeating unit has three residues in the main chain, two single and one double unit side chains, the rhamnose is in the D absolute configuration and the glucuronic acid is completely substituted on its hydroxyl functions (Cérantola *et al.*, 1999). A three-dimensional sketch of three repeating units of cepacian is shown in Figure 6.4 as an example of polymeric chain complexity of EPS. Cepacian bears also from 1 to 3 acetyl substituents per repeating unit, depending on the strain examined. Regarding the remaining four known EPS, the one named PS-I (Cérantola *et al.*, 1996) was reported to be produced only by two strains, while the one constituted of galactose and Kdo (Cescutti *et al.*, 2003) is identical to the EPS produced by *B. pseudomallei*, the causative agent of mieloidosis. Levan was reported to be synthesized only by one strain of the BCC when mannitol was used instead of glycerol as a carbon source (Cescutti *et al.*, 2003); interestingly, the bacteria did not need sucrose to produce it, as reported in the literature for other species. Dextran was demonstrated to be produced together with cepacian and PS-I only by *B. cenocepacia* strain C9343 (Conway *et al.*, 2004).

Molecular modelling calculations indicated a large intramolecular network of hydrogen bonds, especially between the main chain and the side chains. Overall, the chain showed a rather elongated shape, with bends characterized by a large bending radius (Sampaio Nogueira *et al.*, 2005). In water, cepacian is also capable of forming double stranded stretches, which can be disentangled by addition of dimethylsulphoxide, as evidenced by viscometry and atomic force microscopy (AFM) studies (Figure 6.5). These double-chain segments do not give perfect matches between the two different chains, thus leading a single chain to interact

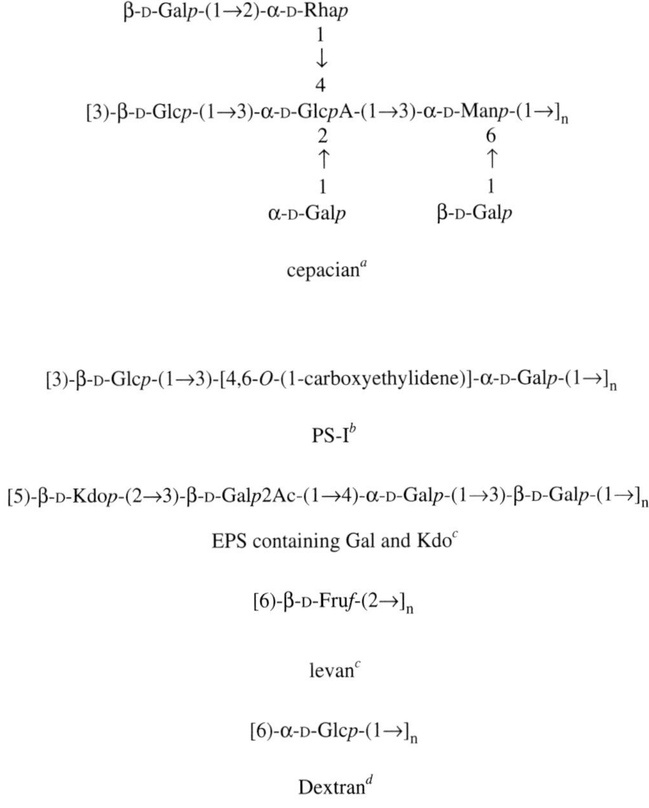

FIGURE 6.3 Structures of the EPS repeating units produced by *Burkholderia cepacia* complex bacteria. Cepacian contains from 1 to 3 *O*-acetyl groups per repeating unit, depending on the strain: [a] Cérantola *et al.* (1999); [b] Cérantola *et al.* (1996); [c] Cescutti *et al.* (2003); [d] Conway *et al.* (2004).

with more than one chain, with the consequent formation of a polymer network (Herasimenka *et al.*, 2008). In particular, the chain rigidity and the tendency to aggregate makes cepacian a good candidate in establishing biofilm architecture. Although the primary structure of cepacian is very different from that of alginate, these two EPS share some similarities in their macromolecular properties: they give rise to a rather viscous aqueous solution and can form polymeric network, as visualized by AFM.

The biological functions of EPS of BCC have just begun to be explored. Bylund *et al.* (2006) demonstrated that a mixture of cepacian, PS-I and dextran scavenged reactive oxygen species, with a potency similar to alginate, and inhibited neutrophil chemotaxis. Recent experiments also indicated a protective effect of cepacian towards hypochlorite, another reactive oxygen species produced by neutrophils to kill microbes. In fact it was shown that hypochlorite caused an abrupt reduction of the molecular mass, a reaction which occurs mainly in the first two hours of incubation, by cleaving mainly the main chain of the polymer; alginate behaves in a very similar manner (P. Cescutti, unpublished results). Further definition of the biological properties of the EPS produced by strains of the BCC will lead to a better understanding of the interaction between bacterial pathogens and

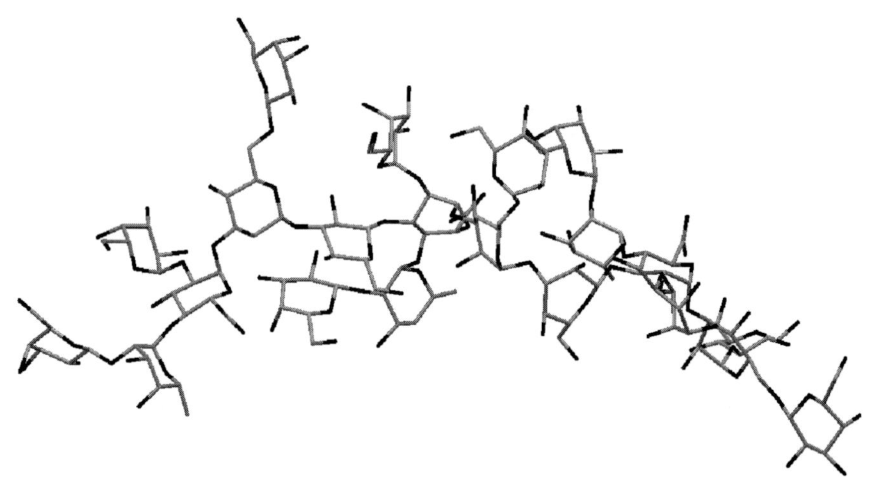

FIGURE 6.4 Snapshot of three repeating units of cepacian taken during molecular dynamic simulation. Reprinted with permission, from Sampaio Nogueira *et al.* (2005).

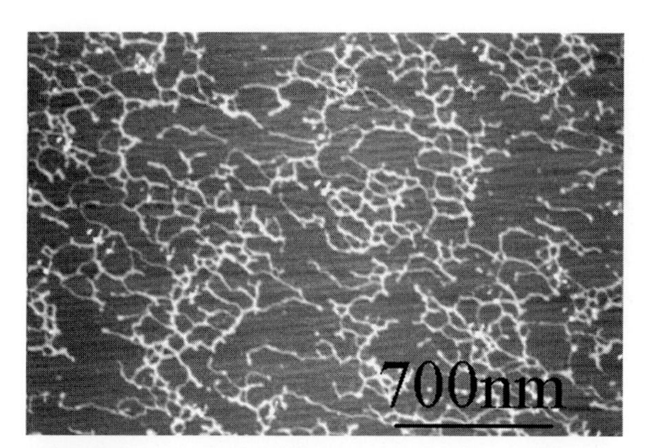

FIGURE 6.5 AFM image of native cepacian recorded in non-contact mode after spray-drying onto a mica surface. The polymer concentration was 30 μg/ml in water. The network of intertwined polysaccharidic chains is visible as white strings on a dark background.

host, a fundamental knowledge to fight and defeat microbial infections.

8. CONCLUSION

In the last thirty years, there has been an enormous increase in the amount of polysaccharides structures determined, mainly due to the advancement in the field of NMR and mass spectrometry. These studies on one hand evidenced the striking variability of the synthesized polymers, making one think that their number might be very large. On the other hand, the availability of these two very powerful techniques changed the way the primary structure of polysaccharides is determined. In fact, even if it is far from being automated, as it is for proteins, the full determination of the primary structure nowadays is achieved with small amounts of samples, because little chemical derivatization is necessary, and more NMR spectroscopy and mass spectrometry experiments, compared to some decades ago. It is worth emphasizing that the knowledge of the primary structure is fundamental to understand the biological functions of polysaccharides and their physicochemical properties.

Future work should focus on understanding what are the general characteristics of EPS and CPS that render bacteria more resistant within their environmental niches and thus allowing their survival. The association of mucoid phenotype with pathogenicity has been proven for several bacteria, e.g. *P. aeruginosa*, *S. aureus* and

K. pneumoniae. However, in the case of bacteria belonging to the BCC, this association is still controversial.

An important issue is the biological role of non-carbohydrate substituents. Certainly their presence is reflected on the shape the polysaccharide assumes, thus influencing the interaction with other molecules. A possible consequence of acetate binding to hydroxyl functions on the polysaccharide is the subtraction of linkage sites to complement opsonins (Pier *et al.*, 2001). Another example is associated with enzymatic activity: it was reported that a lyase specific for cepacian was much more active on the deacetylated polysaccharide than on the native one. Considering that both bacteria, those producing cepacian and those synthesising the enzyme, were environmental, one could suppose that the *O*-acetyl groups conferred some kind of protection to the former, impeding degradation of the EPS by the latter which, in turn, metabolized the enzymatic products (Cescutti *et al.*, 2006).

A characteristic that recently emerged and needs scientific investigation is the ability of some bacterial strains to synthesize more than one polysaccharide simultaneously; one example is given by some strains of the BCC. This capacity may confer a biological advantage to the microbes, like a wider range of responses, leading to the hypothesis that polysaccharide biosynthesis can be modulated depending on the type of molecular interactions.

Further open questions for investigations on CPS and EPS are indicated in the Research Focus Box.

RESEARCH FOCUS BOX

- Is primary structure the key to understanding the biological properties?
- Are there yet unknown general properties of CPS and EPS which are beneficial to microbes?
- What is the importance of bacterial polysaccharides in biofilm structure?
- What are the roles of non-carbohydrate substituents?
- What advantage is conferred to the microbe by the production of a polysaccharide mixture?

ACKNOWLEDGEMENTS

I wish to thank Professor R. Rizzo for useful discussion throughout this work. The research on BCC polysaccharides was supported by the Italian Cystic Fibrosis Foundation (grant FFC # 11/2006) and by a grant of the Friuli Venezia Giulia Regional Government (LR11/2003, project n. 200502027001).

References

Bebault, G.M., Dutton, G.G.S., Funnell, N.A., Mackie, K.L., 1978. Structural investigation of *Klebsiella* serotype K32 polysaccharide. Carbohydr. Res. 63, 183–192.

Burkholder, W.H., 1950. Sour skin, a bacterial rot of onion bulbs. Phytopathology 40, 115–117.

Bylund, J., Burgess, L.A., Cescutti, P., Ernst, R.K., Speert, D.P., 2006. Exopolysaccharides from *Burkholderia cenocepacia* inhibit neutrophil chemotaxis and scavenge reactive oxygen species. J. Biol. Chem. 281, 2526–2532.

Cérantola, S., Marty, N., Montrozier, H., 1996. Structural studies of the acidic exopolysaccharide produced by a mucoid strain of *Burkholderia cepacia*, isolated from cystic fibrosis. Carbohydr. Res. 285, 59–67.

Cérantola, S., Lemassu-Jacquier, A., Montrozier, H., 1999. Structural elucidation of a novel exopolysaccharide produced by a mucoid clinical isolate of *Burkholderia cepacia*. Characterization of a trisubstituted glucuronic acid residue in a heptasaccharide repeating unit. Eur. J. Biochem. 260, 373–383.

Cescutti, P., Toffanin, R., Kvam, B.J, Paoletti, S., Dutton, G.G.S., 1993. Structural determination of the capsular polysaccharide produced by *Klebsiella pneumoniae* serotype K40. NMR studies of the oligosaccharide obtained upon depolymerisation of the polysaccharide with a bacteriophage-associated endoglycanase. Eur. J. Biochem. 213, 445–453.

Cescutti, P., Impallomeni, G., Garozzo, D., et al., 2003. Novel exopolysaccharides produced by a clinical strain of *Burkholderia cepacia* isolated from a cystic fibrosis patient. Carbohydr. Res. 338, 2687–2695.

Cescutti, P., Scussolin, S., Herasimenka, Y., Impallomeni, G., Bicego, M., Rizzo, R., 2006. First report of a lyase for cepacian, the polysaccharide produced by *Burkholderia cepacia* complex bacteria. Biochem. Biophys. Res. Commun. 339, 821–826.

Chandrasekaran, R., Thailambal, V.G., 1990. The influence of calcium ions, acetate and L-glycerate groups on the gellan double-helix. Carbohydr. Polym. 12, 431–442.

Conway, B.D., Chu, K.K., Bylund, J., Altman, E., Speert, D.P., 2004. Production of exopolysaccharide by *Burkholderia cenocepacia* results in altered cell-surface interactions and altered bacterial clearance in mice. J. Infect. Dis. 190, 957–966.

Costerton, J.W., Irwin, R.T., Cheng, K.-J., 1981. The bacterial glycocalix in nature and disease. Annu. Rev. Microbiol. 35, 299–324.

De Philippis, R., Vincenzini, M., 1998. Exocellular polysaccharides from cyanobacteria and their possible applications. FEMS Microbiol. Rev. 22, 151–175.

Donlan, R.M., Costerton, J.W., 2002. Biofilms: survival mechanisms of clinically relevant microorganisms. Clin. Microbiol. Rev. 15, 167–193.

Dutton, G.G.S., 1984. Capsular polysaccharides of Gram negative bacteria. In: Crescenzi, V., Dea, I.C.M., Stivala, S.S. (Eds.), New Developments in Industrial Polysaccharides. Gordon and Breach Science Publishers, Amsterdam, pp. 7–26.

Dutton, G.G.S., Di Fabio, J., 1980. The capsular polysaccharide of *Klebsiella* serotype K60; a novel structural pattern. Carbohydr. Res. 87, 129–139.

Gacesa, P., Russell, N.J., 1990. *Pseudomonas* Infection and Alginates: Biochemistry, Genetics and Pathology. Chapman and Hall, London.

Garegg, P.J., Jansson, P.-E., Lindberg, B., et al., 1980. Configuration of the acetal carbon atom of pyruvic acid acetals in some bacterial polysaccharides. Carbohydr. Res. 78, 127–132.

Govan, J.R.W., Deretic, V., 1996. Microbial pathogenesis in cystic fibrosis: mucoid *Pseudomonas aeruginosa* and *Burkholderia cepacia*. Microbiol. Rev. 60, 539–574.

Griffiths, A.J., Davies, D.B., 1991a. Type specific carbohydrate antigens of pathogenic bacteria. Part 1: Enterobacteriaceae. Carbohydr. Polym. 14, 241–279.

Griffiths, A.J., Davies, D.B., 1991b. Type specific carbohydrate antigens of pathogenic bacteria. Part 2. Carbohydr. Polym. 14, 339–365.

Herasimenka, Y., Benincasa, M., Mattiuzzo, M., Cescutti, P., Gennaro, R., Rizzo, R., 2005. Interaction of antimicrobial peptides with bacterial polysaccharides from lung pathogens. Peptides 26, 1127–1132.

Herasimenka, Y., Cescutti, P., Sampaio Noguera, C.E., et al., 2008. Macromolecular properties of cepacian in water and in dimethylsulfoxide. Carbohydr. Res. 343, 81–89.

Jackson, G.E., Ravenscroft, N., Stephen, A.M., 1990. The use of bacteriophage-mediated depolymerisation in investigations of the structure of the capsular polysaccharide from *Klebsiella* serotype K71. Carbohydr. Res. 200, 409–428.

Jann, B., Jann, K., 1990. Structure and biosynthesis of the capsular antigens of *Escherichia coli*. Curr. Top. Microbiol. 150, 19–42.

Kenne, L., Lindberg, B., 1983. Bacterial polysaccharides. In: Aspinall, G. (Ed.), The Polysaccharides, Vol. 2. Academic Press Inc, New York, pp. 287–363.

Kumar, A.S., Mody, K., Jha, B., 2007. Bacterial exopolysaccharides – a perception. J. Basic Microb. 47, 103–117.

Lacaze, G., Wick, M., Cappelle, S., 2007. Emerging fermentation technologies: development of novel sourdoughs. Food Microbiol. 24, 155–160.

Learn, D.B., Brestel, E.P., Seetharama, S., 1987. Hypochlorite scavenging by *Pseudomonas aeruginosa* alginate. Infect. Immun. 55, 1813–1818.

Lindberg, B., 1990. Components of bacterial polysaccharides. Adv. Carbohydr. Chem. Biochem. 48, 279–318.

Llull, D., López, R., García, E., 2001. Genetic bases and medical relevance of capsular polysaccharide biosynthesis in pathogenic streptococci. Curr. Mol. Med. 1, 475–491.

Mancuso Nichols, C.A., Guezennec, J., Bowman, J.P., 2005. Bacterial exopolysaccharides from extreme marine environments with special consideration on the Southern ocean, sea ice, and deep-sea hydrothermal vents: a review. Mar. Biotechnol. 7, 253–271.

Mathee, K., Ciofu, O., Sternberg, C., et al., 1999. Mucoid conversion of *Pseudomonas aeruginosa* by hydrogen peroxide: a mechanism for virulence activation in the cystic fibrosis lung. Microbiology 145, 1349–1357.

Michalek, M.T., Mold, C., Bremer, E.G., 1988. Inhibition of the alternative pathway of human complement by structural analogues of sialic acid. J. Immunol. 140, 1588–1594.

Morris, E.R., Rees, D.A., Thom, D., Boyd, J., 1978. Chiroptical and stoichiometric evidence of a specific, primary dimerisation process in alginate gelation. Carbohydr. Res. 66, 145–154.

Nigou, J., Gilleron, M., Puzo, G., 2003. Lipoarabinomannans: from structure to biosynthesis. Biochimie 85, 153–166.

O'Riordan, K., Lee, J.C., 2004. *Staphylococcus aureus* capsular polysaccharides. Clin. Microbiol. Rev. 17, 218–234.

Ovodov, Y.S., 2006. Bacterial capsular antigens. Structural patterns of capsular antigens. Biochemistry (Moscow) 71, 937–954.

Pier, G.B., Coleman, F., Grout, M., Franklin, M., Ohman, D.E., 2001. Role of alginate O acetylation in resistance of mucoid *Pseudomonas aeruginosa* to opsonic phagocytosis. Infect. Immun. 69, 1895–1901.

Raffatellu, M., Chessa, D., Wilson, R.P., Tükel, Ç., Akçelik, M., Bäumler, A.J., 2006. Capsule-mediated immune evasion: a new hypothesis explaining aspects of typhoid fever pathogenesis. Infect. Immun. 74, 19–27.

Rao, V.S.R., Qasba, P.K., Balaji, P.V., Chandrasekaran, R., 1998. Conformation of Carbohydrates. Harwood Academic Publishers, Amsterdam.

Remminghorst, U., Rehm, B.H.A., 2006. Bacterial alginates: from biosynthesis to applications. Biotechnol. Lett. 28, 1701–1712.

Sage, A., Linker, A., Evans, L.R., Lessie, T.G., 1990. Hexose phosphate metabolism and exopolysaccharide formation in *Pseudomonas cepacia*. Curr. Microbiol. 20, 191–198.

Sampaio Nogueira, C.E., Ruggiero, J.R., Sist, P., Cescutti, P., Urbani, R., Rizzo, R., 2005. Conformational features of cepacian: the exopolysaccharides produced by clinical strains of *Burkholderia cepacia*. Carbohydr. Res. 340, 1025–1037.

Simpson, J.A., Smith, S.E., Dean, R.T., 1989. Scavenging by alginate of free radicals released by macrophages. Free Radical Bio. Med. 6, 347–353.

Sutherland, I.W., 1994. Structure-function relationships in microbial exopolysaccharides. Biotechnol. Adv. 12, 393–448.

Sutherland, I.W., 1999. Microbial polysaccharide products. Biotechnol. Genet. Eng. 16, 217–229.

Taylor, C.M., Roberts, I.S., 2005. Capsular polysaccharides and their role in virulence. Contrib. Microbiol. 12, 55–66.

van Dam, J.E.G., Fleer, A., Snippe, H., 1990. Immunogenicity and immunochemistry of *Streptococcus pneumoniae* capsular polysaccharides. Antonie van Leeuwenhoek 58, 1–47.

Weintraub, A., 2003. Immunology of bacterial polysaccharide antigens. Carbohydr. Res. 338, 2539–2547.

Whitfield, C., 2006. Biosynthesis and assembly of capsular polysaccharides in *Escherichia coli*. Annu. Rev. Biochem. 75, 39–68.

Widner, B., Behr, R., Von Dollen, S., et al., 2005. Hyaluronic acid production in *Bacillus subtilis*. Appl. Environ. Microb. 71, 3747–3752.

Zevenhuizen, L.P.T.M., 1997. Succinoglycan and galactoglucan. Carbohydr. Polym. 33, 139–144.

Zlosnik, J.E., Hird, T.J., Fraenkel, M.C., Moreira, L.M., Henry, D.A., Speert, D.P., 2008. Differential mucoid exopolysaccharide production by members of the *Burkholderia cepacia* complex. J. Clin. Microbiol. 46, 1470–1473.

CHAPTER

7

Bacterial surface layer glycoproteins and "non-classical" secondary cell wall polymers

Paul Messner, Eva Maria Egelseer, Uwe B. Sleytr and Christina Schäffer

SUMMARY

Cell surface layers, S-layers, are common structures of the prokaryotic cell envelope with a lattice-like appearance. They are formed by a self-assembly process and are attached to the bacterial cell via a special class of secondary cell wall polymers. Frequently, the constituting S-layer proteins are modified with covalently linked glycan chains facing the extracellular environment. S-layer glycoproteins by far exceed the display found with eukaryal or viral glycoproteins, making them ideal targets for studying the molecular basis of bacterial protein glycosylation. While in the past, research was focused on the elucidation of S-layer glycan structures and on their biochemistry, now, a molecular picture of the S-layer glycosylation process is evolving. This is the basis for the envisaged combination of S-layers with heterologous glycosylation systems to produce functional S-layer neoglycoproteins that, according to their intrinsic property to self-assemble into crystalline arrays, allow controlled and oriented display of glycan motifs with nanometer-scale periodicity. Together with the developing understanding of the mode of interaction between S-layer (glyco)proteins and secondary cell wall polymers this will pave the way towards the production of novel nanopatterned "biomaterials" for various applications in the fields of biomedicine and nanobiotechnology, using carbohydrate recognition.

Keywords: Cell wall component; Crystalline array; Glycoprotein; Molecular basis; Nanobiotechnology; Nanopatterning; Prokaryotic glycosylation; Secondary cell wall polymer; S-layer

1. INTRODUCTION

A major issue for the survival of bacteria in a competitive habitat was the development of highly specific cell envelope structures as the primary site of interaction with the environment. These were even more diversified through incorporation and attachment of saccharide structures. This chapter focuses on unique, (glyco)protein self-assembly structures, forming two-dimensional crystalline arrays on bacterial cell surfaces (S-layers) (Sleytr and Beveridge, 1999; Sleytr *et al.*, 2007) and on their cell wall-anchoring polysaccharides (Mesnage *et al.*, 2000; Sára, 2001; Schäffer and Messner, 2005).

2. BACTERIAL AND ARCHAEAL S-LAYERS

S-layers are recognized as one of the most common cell surface structures in bacteria and represent an almost universal feature of Archaea. In most Archaea, the S-layer is either integrated into or penetrating the plasma membrane (Baumeister and Lembke, 1992). In Gram-positive bacteria and Archaea, the S-layer lattice assembles on the surface of the rigid peptidoglycan or the pseudomurein, respectively (Sleytr *et al.*, 1999). In Gram-negative bacteria, a specific fraction of "smooth" lipopolysaccharide (LPS) is required for attachment of the S-layer to the outer membrane (Awram and Smit, 2001). S-layer synthesis and assembly resemble a very efficient system since, at a generation time of ≈20 minutes, at least 500 copies of a single (glyco)protein species have to be synthesized per second, translocated to the cell surface and incorporated into the S-layer lattice (Messner and Sleytr, 1992; Sára and Sleytr, 2000; Sleytr *et al.*, 2007). This corresponds to up to 20% of the total protein synthesis effort of an organism being devoted to S-layer (glyco)protein production.

With a few exceptions, S-layers are composed of a single (glyco)protein species endowed with the ability to assemble into two-dimensional arrays on the cell surface during all stages of the cellular life cycle. High resolution electron microscopy studies revealed that S-layer lattices can exhibit oblique (p1, p2), square (p4) or hexagonal (p3, p6) symmetry with centre-to-centre spacing of the morphological units of ≈3 to 35 nm. Depending on the lattice type, the morphological units consist of one, two, three, four and six identical (glyco)protein subunits with molecular masses ranging from 40 000 to 170 000 (Sleytr and Beveridge, 1999; Sleytr *et al.*, 2002).

S-layer proteins can undergo diverse co- or post-translational modifications. Examples of covalent modifications include glycosylation, lipid attachment, phosphorylation and methylation (Claus *et al.*, 2002; Schäffer and Messner, 2004; Eichler and Adams, 2005). About 30 years ago, first evidence was provided that S-layer proteins of Archaea and bacteria may contain covalently linked carbohydrate chains (Mescher and Strominger, 1976; Sleytr and Thorne, 1976). Indeed, glycosylation is the most frequent modification of S-layer proteins, with the glycosylation degree varying between 1 and 10% (w/w) (Sumper and Wieland, 1995; Messner and Schäffer, 2003; Eichler and Adams, 2005; Messner *et al.*, 2008). Glycosylation is energetically costly as co- or post-translational modification adds significantly to the potential functional spectrum of S-layer proteins (Sumper and Wieland, 1995; Claus *et al.*, 2005; Eichler and Adams, 2005; Messner *et al.*, 2008). Besides Gram-positive bacteria and Archaea, for which S-layer glycosylation has been confirmed by various research groups in the past, only recently there were first reports on Gram-negative microorganisms possessing an S-layer glycoprotein layer; these are the pathogenic species *Tannerella forsythia* (Lee *et al.*, 2006) and *Bacteroides distasonis* (Fletcher *et al.*, 2007). The relation of S-layer protein glycosylation to pathogenicity is shedding new light in a potential *in vivo* function of S-layer glycans.

3. GENERAL FEATURES OF GLYCOSYLATED S-LAYER PROTEINS

3.1. Bacterial S-layer glycoproteins

Presently, about 15 different S-layer glycoprotein glycan structures have been fully or at least partially elucidated and according to biochemical evidence, there are currently ≈30 further indications for S-layer glycoproteins (Figure 7.1). It is evident that prokaryotic S-layer glycoproteins by far exceed the display found with eukaryotic and viral glycoproteins, regarding composition, structure and linkage regions to the polypeptide. Usually, bacterial S-layer glycan chains are long linear or branched homo- or heterosaccharides with 50 to 150 glycoses that constitute about 15 to 50 repeating units. The repeating unit-wise concept of their biosynthesis leads to S-layer glycan heterogeneity, referred to as nanoheterogeneity (Schäffer and Messner, 2004). The monosaccharide constituents of bacterial S-layer glycan chains include a wide range of neutral hexoses, 6-deoxyhexoses and amino sugars. In addition, rare sugars, such as 3-*N*-acetylquinovosamine (Qui*p*3NAc), 3-*N*-acetylfucosamine (Fuc*p*3NAc), D-rhamnopyranose (D-Rha*p*), D-fucopyranose (D-Fuc*p*) or D-*glycero*-D-*manno*-heptose, which are otherwise typical constituents of LPS O-antigens (Raetz and Whitfield, 2002), are found. The typical linkages of bacterial S-layer glycans to the protein portion are *O*-glycosidic linkages to serine, threonine and tyrosine; so far *N*-glycans have been found only in Archaea. Analysis of the corresponding S-layer protein sequences using PSIPREDView in combination with biochemical data has revealed that S-layer glycans are bound to surface-accessible loops of the protein, with two to four identified glycosylation sites per protein (see Figure 7.1).

Bacterial S-layer glycoproteins possess a tripartite structure (Schäffer and Messner, 2004), which compares to LPS O-antigens (Raetz and Whitfield, 2002). In some S-layer glycoproteins, capping of the terminal sugar residue at the non-reducing end of the glycan chain with non-carbohydrate constituents, such as *O*-methyl-groups (2-*O*-Me, 3-*O*-Me) or (*R*)-*N*-acetylmuramic acid occurs (Schäffer *et al.*, 1999b, 2002; Kählig *et al.*, 2005). It may be speculated that these modifications are involved in glycan chain length termination during S-layer glycan biosynthesis (Raetz and Whitfield, 2002; Clarke *et al.*, 2004).

3.2. Archaeal S-layer glycoproteins

S-layers of archaeal organisms are directly exposed to their often extreme natural environments. As a consequence, intrinsic resistances against environmental stresses, such as high salt, acidity and temperature, are characteristic features of archaeal S-layers. The molecular mechanisms for these phenomena are still poorly understood. Archaeal S-layer (glyco)proteins are generally characterized by a predominance of non-polar amino acid residues, but some species contain an increased portion of charged residues (Claus *et al.*, 2002). The S-layer glycoproteins of the hyperthermophiles *Methanothermus fervidus* and *Mt. sociablis* seem, however, to be exceptions to the rule, containing more asparagine than aspartate residues (Bröckl *et al.*, 1991). Additional charges can be introduced into these S-layer proteins upon glycosylation (Claus *et al.*, 2002). It has been proposed that an increase of charged residues can confer thermal stability to S-layer (glyco)proteins through enhanced electrostatic interactions (Cambilliau and Claverie, 2000).

While most archaeal S-layer proteins seem to be glycosylated, complete S-layer glycan structures were only determined for the S-layer glycoproteins of *Halobacterium salinarum* (Sumper and Wieland, 1995), *Haloferax volcanii* (Sumper *et al.*, 1990) and *Methanothermus fervidus* (Kärcher *et al.*, 1993) (see Figure 7.1). Like eukaryal glycoproteins, archaeal S-layer glycoproteins can undergo

Bacterial S-Layer Glycoprotein Glycans

Geobacillus stearothermophilus NRS 2004/3a (formerly *Bacillus stearothermophilus*)

2-*O*Me-α-L-Rha*p*-(1→3)-β-L-Rha*p*-(1→2)-α-L-Rha*p*-(1→[2)-α-L-Rha*p*-(1→3)-β-L-Rha*p*-(1→2)-α-L-Rha*p*-(1→]$_{n\sim15}$ 2)-
α-L-Rha*p*-(1→3)-α-L-Rha*p*-(1→3)-β-D-Gal*p*-(1→*O*)-Thr$_{590,620}$ and/or →*O*)-Ser$_{794}$

Geobacillus tepidamans GS5-97[T]

α-D-MurNAc-(1→3)-α-L-Rha*p*-(1→2)-α-D-Fuc*p*-(1→[3)-α-L-Rha*p*-(1→2)-α-D-Fuc*p*-(1→]$_{n\sim10}$ 3)-
2
↑
1
β-D-GlcNAc
α-L-Rha*p*-(1→3)-α-L-Rha*p*-(1→3)-α-L-Rha*p*-(1→3)-β-D-Gal*p*-(1→*O*)-Ser and/or D-Gal→*O*-Thr

Aneurinibacillus thermoaerophilus L420-91[T] (DSM 10154) and GS4-97 (formerly *Bacillus thermoaerophilus*)

3-*O*Me-α-D-Rha*p*-(1→3)-α-D-Rha*p*-(1→2)-α-D-Rha*p*-(1→2)-α-D-Rha*p*-(1→[3)-α-D-Rha*p*-(1→3)-α-D-Rha*p*-(1→2)-α-D-Rha*p*-
(1→2)-α-D-Rha*p*-(1→]$_{n\sim15}$ [3)-α-D-Rha*p*-(1→]$_{n=1-3}$ 3)-β-D-Gal*p*NAc-(1→*O*)-Thr$_{67}$

Aneurinibacillus thermoaerophilus (DSM 10155) (formerly *Bacillus thermoaerophilus*)

[→4)-α-L-Rha*p*-(1→3)-β-D-*glycero*-D-*manno*-Hep*p*-(1→]$_{n\sim18}$ [3)-α-L-Rha*p*-(1→]$_{n=0-2}$ 3)-β-D-Gal*p*NAc-(1→*O*)-Ser$_{69}$ and/or →*O*)-Thr$_{471}$

Paenibacillus alvei (CCM 2051) (formerly *Bacillus alvei*)

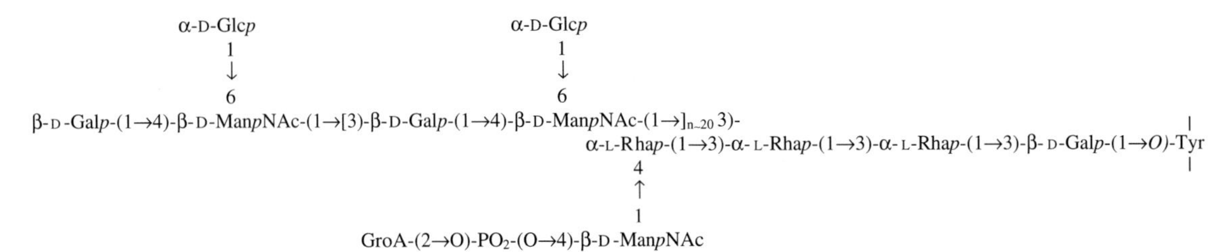

Thermoanaerobacter thermohydrosulfuricus L111-69 and L110-69 (DSM 568) (formerly *Clostridium thermohydrosulfuricum*)

3-*O*Me-α-L-Rha*p*-(1→4)-α-D-Man*p*-(1→[3)-α-L-Rha*p*-(1→4)-α-D-Man*p*-(1→]$_{n\sim27}$ 3)-
α-L-Rha*p*-(1→3)-α-L-Rha*p*-(1→3)-α-L-Rha*p*-(1→3)-β-D-Gal*p*-(1→*O*)-Tyr

Thermoanaerobacter thermohydrosulfuricus S102-70 (formerly *Clostridium thermohydrosulfuricum*)

β-D-Gal*f*-(1→3)-α-D-Gal*p*-(1→2)-α-L-Rha*p*-(1→3)-α-L-Man*p*-(1→3)-α-L-Rha*p*-(1→3)-β-L-Glc*p*-(1→*O*)-Tyr

(*Continues*)

FIGURE 7.1 Glycan structures of bacterial and archaeal S-layer proteins. Abbreviations: Glc*p*, glucopyranose; Gal*f*, galactofuranose; Man, mannose; Rha, rhamnose; GlcNAc, *N*-acetylglucosamine; GalNAc, *N*-acetylgalactosamine; ManNAc, *N*-acetylmannosamine; Fuc3NAc, 3-*N*-acetylfucosamine (3-acetamido-3,6-dideoxygalactose); Qui3NAc, 3-*N*-acetylquinovosamine (3-acetamido-3,6-dideoxyglucose), D-*glycero*-D-*manno*-Hepp, D-*glycero*-D-*manno*-heptose; MurNAc, *N*-acetylmuramic acid; GlcA, glucuronic acid; GalA, galacturonic acid; ManA, mannuronic acids; IdA, iduronic acid; 3-OMe-GalA, 3-*O*-methylgalacturonic acid; *O*Me, O-methyl; PO_4, phosphate; SO_4^{2-}, sulfate; Asn, asparagine; Thr, threonine; Ser, serine; Tyr, tyrosine; Ala, alanine; *X*, interchangeable amino acid; n, number of repeats.

Thermoanaerobacter thermohydrosulfuricus L77-66 (DSM 569) and L92-71 (formerly *Clostridium thermohydrosulfuricum*)

[→3)-α-D-Gal*p*NAc-(1→3)-α-D-Gal*p*NAc-(1→]$_{n\sim25}$ — *O*-glycosidic bond via Tyr

4
↑
1
α-D-Glc*p*NAc-(1→2)-β-D-Man*p*

Thermoanaerobacterium thermosaccharolyticum D120-70 (formerly *Clostridium thermosaccharolyticum*)

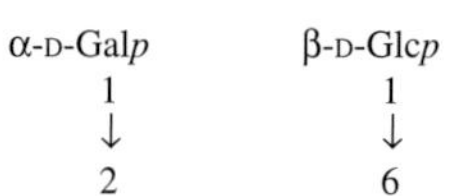

α-D-Gal*p* (1→2), β-D-Glc*p* (1→6); α-D-Gal*p* (1→2), β-D-Glc*p* (1→6)

α-L-Rha*p*-(1→3)-α-D-Man*p*-(1→4)-β-L-Rha*p*-(1→3)-α-D-Glc*p*-(1→[4)-α-L-Rha*p*-(1→3)-α-D-Man*p*-(1→4)-β-L-Rha*p*-(1→3)-α-D-Glc*p*-
(1→]$_{n=3\text{-}10}$ 4)-α-L-Rha*p*-(1→3)-α-D-Man*p*-(1→4)-β-D-Rha*p*-(1→3)-β-D-Glc*p*-(1→*O*)-Tyr

2 ↑ 1 α-D-Gal*p*; 6 ↑ 1 β-L-Glc*p*

Thermoanaerobacterium thermosaccharolyticum E207-71 (*Clostridium thermosaccharolyticum*)

[→4)-β-D-Gal*p*-(1→4)-β-D-Glc*p*-(1→4)-β-D-Man*p*-(1→]$_{n\sim17}$ — *O*-glycosidic bond via Tyr

3
↑
1
β-D-Qui*p*3NAc-(1→6)-β-D-Gal*f*-(1→4)-α-D-Rha*p*

Lactobacillus buchneri 41021/251

α-D-Glc*p*-(1→6)-[α-D-Glc*p*-(1→6)-]$_{n=4\text{-}6}$ α-D-Glc*p*-(1→O)-Ser

Archaeal S-Layer Glycoprotein Glycans

Halobacterium halobium R_1M_1

OSO_3^- (overline), OSO_3^-, Ala-NH$_2$

[→4)-GlcNAc-(1→4)-GalA-(1→3)-GalNAc-(1]$_{n=10\text{-}15}$ →*N*)-Asn

6 ↑ 1 3-*O*Me-GalA; 3 ↑ 1 Gal*f*; Asn | Ala | Ser

GlcA-(1→4)-GlcA-(1→4)-GlcA-(1→4)-β-D-Glc-(1→*N*)-Asn — 1/3 of GlcA residues can be replaced by IdA

3 ↑ OSO_3^-; 3 ↑ OSO_3^-; 3 ↑ OSO_3^-; Asn | X | Thr/Ser

α-D-Glc-(1→2)-Gal-(1→*O*)-Thr

(*Continues*)

FIGURE 7.1 Continued

Haloferax volcanii DS2

β-D-Glc-(1→[4)-β-D-Glc-(1→]$_{n=8}$ 4)-β-D-Glc-(1→*N*)-Asn

Glc-(1→2)-Gal-(1→O)-Thr

Methanothermus fervidus V24S

3-*O*Me-α-D-Man*p*-(1→6)-3-*O*Me-α-D-Man*p*-(1→[2)-α-D-Man*p*-(1→]$_{n=3}$ 4)-D-GalNAc-(1→*N*)-Asn

Methanosaeta soehngenii FE (formerly *Methanothrix soehngenii*)

Oligosaccharide-Rha-(1→*N*)-Asn

FIGURE 7.1 Continued

both *N*- and *O*-glycosylation (see Figure 7.1). Often, the sugar residues are modified by methylation or sulfation. Summarizing, differences between archaeal and bacterial S-layer glycoproteins are:

(i) short versus long glycan chains;
(ii) multiple versus few glycosylation sites;
(iii) different *versus* single type of S-layer glycoprotein glycans; and
(iv) predominantly *N*-glycans versus exclusively *O*-glycans (so far).

In addition to S-layer protein glycosylation, glycosylation has also been reported for archaeal flagellins. The *Hb. salinarum* flagellins contain sulfated *N*-glycans similar to the glycan moieties found on the S-layer glycoprotein (Wieland *et al.*, 1985). More recently, *Methanococcus voltae* flagellins have been shown to contain a novel *N*-linked trisaccharide -[β-D-Man*p*NAcA6Thr-(1→4)-β-D-Glc*p*NAc3NAcA-(1→3)-β-D-Glc*p*NAc]-, with an unusual amide-linked amino acid modification of Man*p*NAcA (Voisin *et al.*, 2005). Interestingly, the *Mc. voltae* S-layer glycoprotein also shows this novel trisaccharide structure, suggesting a common glycosylation process for the two *Mc. voltae* surface proteins. This observation indicates that, in *Hb. salinarum* and *Mc. voltae* and presumably also in other, not yet identified, prokaryotes, common pools of glycan precursors are shared for biosynthesis of cell surface glycoconjugates.

4. GENETICS

4.1. Bacterial S-layer glycoproteins

S-layer protein glycosylation of the investigated Gram-positive bacteria was found to be encoded by S-layer glycosylation (*slg*) gene clusters. Currently, most data are available from the organisms *G. stearothermophilus* NRS 2004/3a (GenBank AF328862), *G. tepidamans* GS5-97^T (AY883421), *A. thermoaerophilus* strains L420-91^T (AY442352) and DSM 10155/G$^+$ (AF324836). In addition, a partial *slg* gene cluster sequence is available from *T. thermosaccharolyticum* E207-71 (AY422724) (Messner *et al.*, 2008). Depending on the complexity of the encoded S-layer glycan, the clusters are ≈16 to 25 kb in size, corresponding to 16–21 open reading frames, and are transcribed as polycistronic units (Novotny *et al.*, 2004). They code for nucleotide sugar pathway genes that are arranged consecutively, glycosyl transferase genes, glycan processing genes and transporter genes. From the assigned genes, it is evident that S-layer protein glycosylation additionally requires the participation of housekeeping genes that map outside the cluster.

The comparison of the *slg* gene clusters of *G. stearothermophilus* NRS 2004/3a and *G. tepidamans* GS5-97^T, whose S-layer glycans both possess an extended, tripartite structure, revealed that the clusters are organized in a similar way (Zayni *et al.*, 2007). It seems that the central, variable part is responsible for the biosynthesis of individual repeating units and terminating elements of the S-layer glycan, whereas the region with higher homology encodes proteins involved in the assembly of the core region, the transport of the glycan to the cell surface and its ligation to distinct amino acids on the S-layer protein. This resembles the organization of O-antigen gene clusters in Gram-negative organisms, where the variability of O-antigens is considered as a result of recombination events in the central region of the O-antigen gene clusters (Kaplan *et al.*, 2001). A complete comparison of all known *slg* gene clusters, including the assignment of individual open reading frames and their designation according to the bacterial polysaccharide gene nomenclature, has been published recently (Messner *et al.*, 2008).

4.2. Archaeal S-layer glycoproteins

With the availability of complete genome sequences, research now focuses on the functional characterization of genes involved in the glycosylation processes of archaeal S-layer proteins. Most of the molecular understanding of S-layer protein glycosylation in archaeal organisms was obtained for *Hf. volcanii* during recent years (Abu-Qarn and Eichler, 2006; Abu-Qarn *et al.*, 2007, 2008). In the course of deciphering the *N*-glycosylation pathway, a number of genes homologous to human and mouse genes have been identified. The encoded proteins are, for instance, possibly responsible for loading activated glucose and mannose residues onto dolichol carriers in the endoplasmic reticulum (ER) membrane (Burda and Aebi, 1999). Among the annotated proteins are also homologues of *Campylobacter jejuni* glycosyltransferases involved in protein *N*-glycosylation (Nothaft *et al.*, 2008). In *C. jejuni*, the transfer of the completed oligosaccharide structure to selected Asn residues is mediated by PglB (Feldman *et al.*, 2005), the Stt3 homologue of the eukaryal oligosaccharyl transferase (OST) complex (Yan and Lennarz, 2005). In *Hf. volcani*, Stt3, now renamed AglB (Chaban *et al.*, 2006), is a predicted multi-spanning membrane protein, containing the canonical WWDYG motif (Abu-Qarn and Eichler, 2006). Involvement of AglB in *N*-glycosylation was verified by transcription of mRNA and differential expression of the protein in RT-PCR experiments. The *aglD* gene encodes a glycosyltransferase that is involved in the addition of the terminal hexose residue to a pentasaccharide decorating the *Hf. volcanii* S-layer protein, at least at sequon Asn13 and Asn83 (Abu-Qarn *et al.*, 2007). The topology of AglD is reminiscent of eukaryal glycosyltransferases responsible for the transfer of mannose or glucose residues from charged dolichol carriers (Burda and Aebi, 1999). Another glycosyltransferase, AglE, was demonstrated to add a so far uncharacterized 190-Da sugar residue to the *Hf. volcanii* S-layer protein. The topological analysis assigned AglE as an integral membrane protein with the N-terminus and the putative active site located at the cytoplasmic face of the plasma membrane (Abu-Qarn *et al.*, 2008). Together, these findings corroborate the developing picture about the *N*-glycosylation of the *Hf. volcanii* S-layer protein in particular, and about *N*-glycosylation of archaeal proteins in general.

5. BIOSYNTHESIS

5.1. Bacterial S-layer glycoproteins

Unravelling molecular details about S-layer protein glycosylation is a prerequisite for S-layer protein based carbohydrate engineering, a strategy in which S-layer protein glycosylation and exogenous, functional glycosylation pathways shall converge (Steiner *et al.*, 2008a). S-layer glycoprotein biosynthesis is a very complex process in which the glycosylation event has to be coordinated with the synthesis of the S-layer protein and of the secondary cell wall polymer (SCWP) anchor, its translocation through the cell wall and the incorporation into the existing S-layer lattice.

Based on the common principle that sugars are incorporated into growing glycan chains from the respective nucleotide-activated precursors, genes coding for nucleotide sugar pathway enzymes were identified in the different *slg* gene clusters (Messner *et al.*, 2008). From these data, it is obvious that S-layer protein glycosylation pathways provide a spectrum of enzymes that may be used for carbohydrate engineering.

The set-up of functional assays for the initiation enzyme WsaP and for four assigned transferases, named WsaC through WsaF, encoded by the *slg* gene cluster of *G. stearothermophilus* NRS 2004/3a, has allowed drawing a first, full picture of how the S-layer glycan is assembled (Figure 7.2) (Steiner *et al.*, 2007, 2008a). Our current model implicates WsaP in the first step of biosynthesis, whereby galactose is transferred from its nucleotide-activated form (UDP-Gal) to a membrane-associated lipid carrier at the cytoplasmic face of the plasma membrane (Steiner *et al.*, 2007). Chain extension continues by consecutive addition of rhamnose residues from dTDP-β-L-rhamnose to the non-reducing terminus of the lipid-linked glycan chain by the action of the enzymes WsaC through WsaF, catalysing the formation of the α-(1→2)-, α-(1→3)- and β-(1→2)-linkages as present in the S-layer glycan, leading to the core saccharide and repeating unit-like structures (see Figure 7.1). Chain growth is predicted to be terminated by 2-*O*-methylation of the distal, terminal repeating unit, catalysed by a distinct domain of the multifunctional

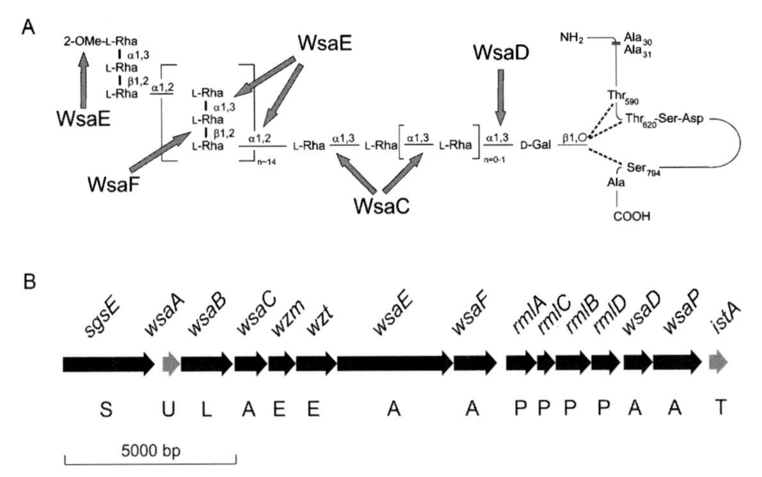

FIGURE 7.2 Schematic representation of the S-layer glycoprotein glycan of *G. stearothermophilus* NRS 2004/3a (A) and organization of the corresponding S-layer glycosylation gene cluster (B). Gene nomenclature: *sgsE*, the S-layer structural gene; *wsaA*, a gene of unknown function; *wsaB*, oligosaccharyl:protein transferase; *wsaC*, rhamnosyltransferase; *wzm*, ABC transporter integral membrane protein; *wzt*, ATP-binding cassette (ABC) transporter nucleotide-binding protein; *wsaE*, trifunctional methyl-rhamnosyltransferase; *wsaF*, rhamnosyltransferase; *rmlA*; glucose-1-phosphate thymidylyltransferase; *rmlC*, dTDP-dehydrorhamnose 3,5-epimerase; *rmlB*, dTDP-D-glucose 4,6-dehydratase; *rmlD*, dTDP-dehydrorhamnose reductase; *wsaD*, rhamnosyltransferase, *wsaP*, UDP-galactose-lipid carrier transferase; *istA*, transposase. Other abbreviations: S, S-layer protein; U, unkown function; L, ligation; E, export; A, assembly; P, sugar precursor biosynthesis, T, transposase. The site of action of the individual enzyme proteins for S-layer glycan biosynthesis is indicated in (A). Adapted, with permission, from Steiner *et al.* (2008a).

methylrhamnosyltransferase WsaE. The complete glycan chain would then be transported across the membrane by a process involving an ABC transporter and eventually transferred to the S-layer protein by the oligosaccharyl:protein transferase WsaB. With the identification of the initiation enzyme as a UDP-Gal:phosphoryl-polyprenol Gal-1-phosphate transferase and the presence of a dedicated ABC transporter, it is evident that S-layer protein glycosylation utilizes modules from either LPS O-antigen biosynthesis pathways (Raetz and Whitfield, 2002). As several other *slg* gene clusters contain an ABC-transporter and numerous glycan chains are modified at the non-reducing end, the model described here for S-layer protein glycosylation might be widely valid.

5.2. Archaeal S-layer glycoproteins

Genetic investigations (see Section 3.2) clearly revealed that the AglB, AglD and AglE proteins

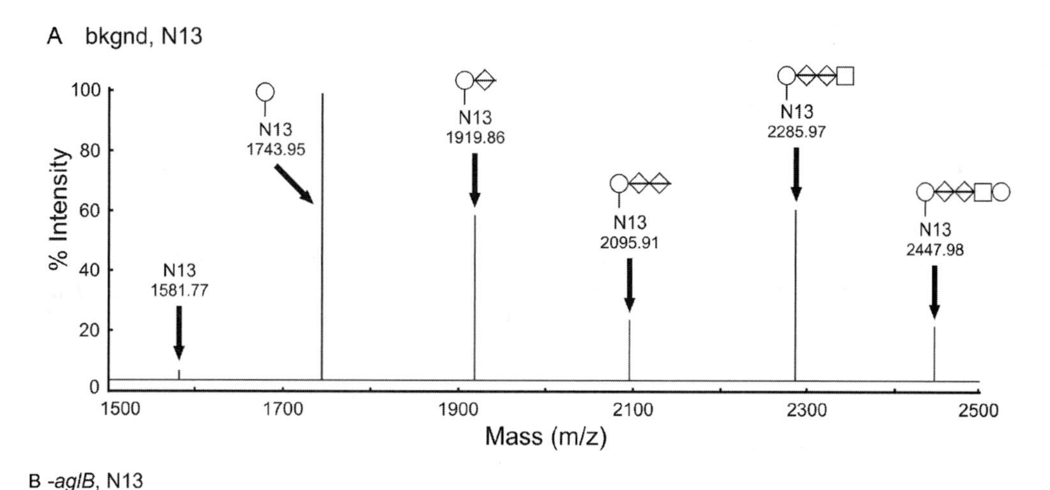

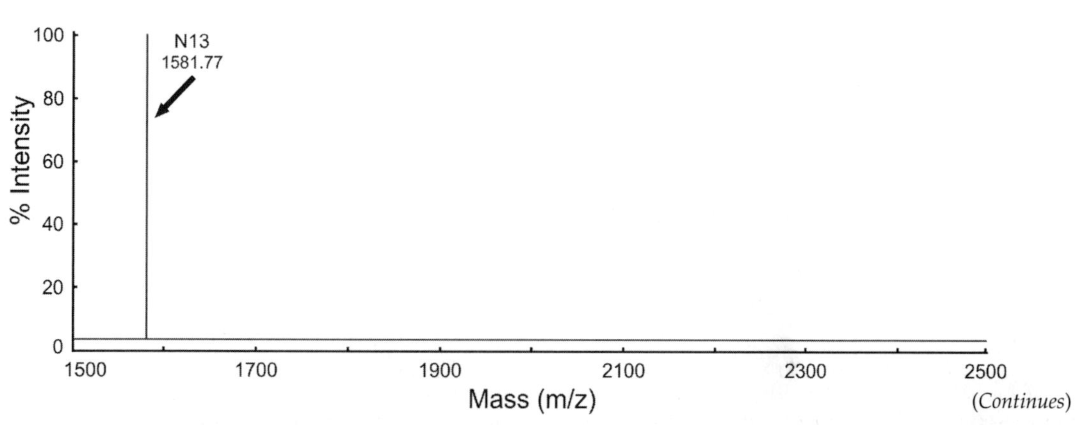

(*Continues*)

FIGURE 7.3 Schematic presentation of the matrix-assisted laser desorption ionization time-of-flight (MALDI-TOF) analysis of the Asn13-containing, *Haloferax volcanii* S-layer glycoprotein-derived glycopeptide. The MALDI-TOF spectra of (A) the background, (B) the *aglB*-deleted, (C) the *aglD*-deleted and (D) the *aglE*-deleted strains are shown. The components of the glycopeptide-associated sugar residues are: ○, 162 m/z; ⬖, 176 m/z; □, 190 m/z. The glycan moieties decorating the peptide peaks are marked on the MALDI-TOF spectra, accordingly. Adapted, with permission, from Abu-Qarn *et al.* (2007, 2008).

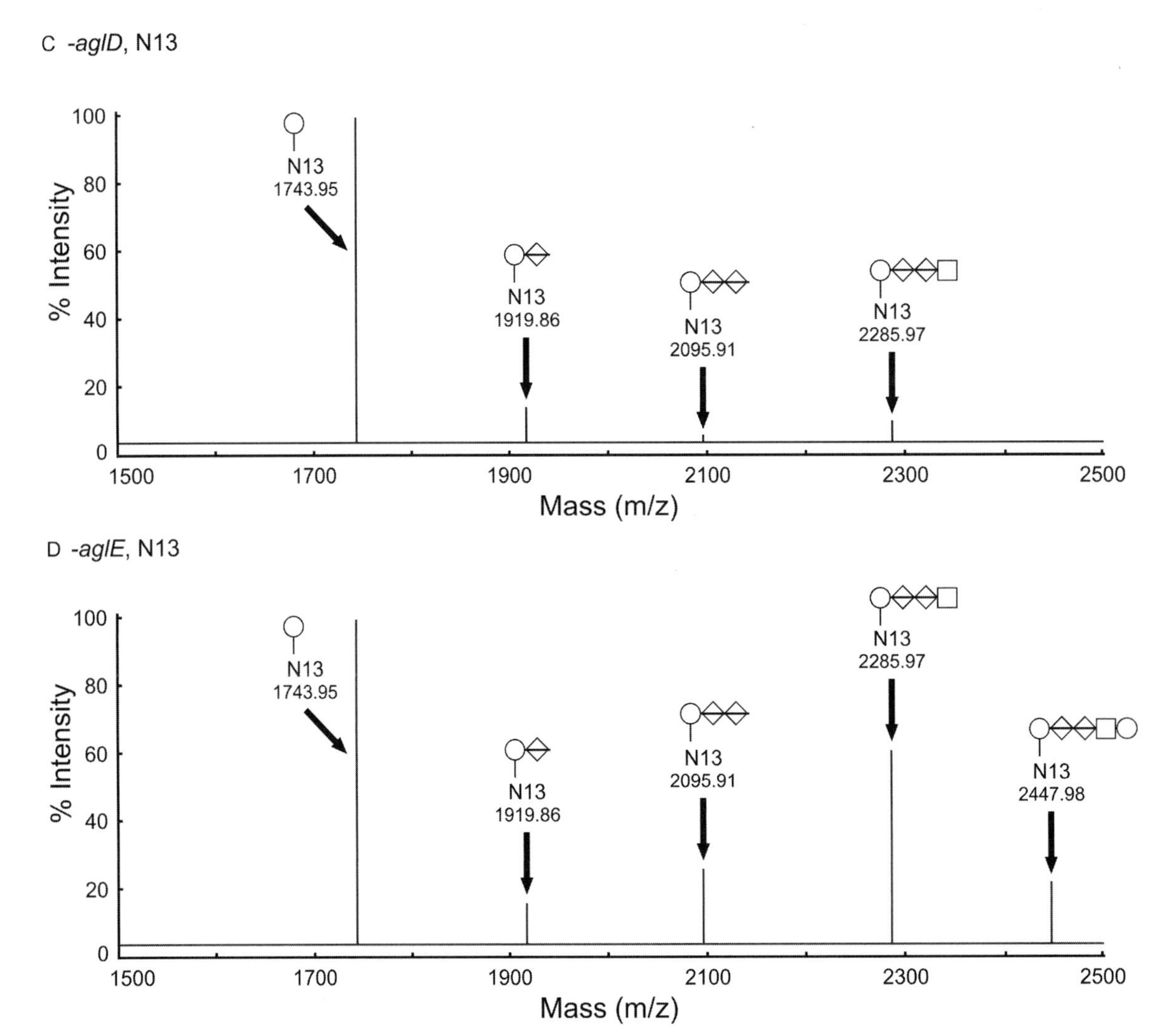

FIGURE 7.3 Continued

of *Hf. volcanii* (Chaban *et al.*, 2006) are involved in S-layer protein *N*-glycosylation (Abu-Qarn *et al.*, 2007, 2008). The glycosylation profile of this S-layer protein comprises both *N*- and *O*-linked glycan structures (Sumper *et al.*, 1990). Seven proposed *N*-linked glycans are thought to contain either glucose alone or a mixture of glucose, galactose and idose. Recently, mass spectrometry was employed to provide insight into the composition of the S-layer glycan moieties decorating amino acids Asn13 and Asn83 of the S-layer in *Hf. volcanii* deletion mutants (Abu-Qarn *et al.*, 2007, 2008) (Figure 7.3). Deletion of *aglB* completely eliminated glycosylation at Asn13 (see Figure 7.3B) and Asn83, which confirms that AglB is solely responsible for oligosaccharyl transferase activity and that *N*-glycosylation is not essential for the organism. The glycan moieties naturally linked to Asn13 and Asn83 contain pentasaccharides consisting of two hexoses, two hexuronic acids and a 190 m/z subunit, possibly corresponding to a methylated hexuronic acid or a di-methylated hexose (Abu-Qarn *et al.*, 2007, 2008) (see Figure 7.3A).

The AglD protein is involved in the addition of the final hexose to the pentasaccharides decorating at least Asn13 (see Figure 7.3C) and Asn83. The S-layer glycoprotein of an *aglD*-deletion mutant indicated reduced glycosylation and displayed altered protease sensitivity and an increased degree of structural disorder (Abu-Qarn *et al.*, 2007).

The observation that AglB can also transfer truncated glycan structures to the S-layer protein suggests relaxed substrate specificity of the enzyme, as reported for eukaryal OSTs (Munoz *et al.*, 1994), *C. jejuni* PglB (Glover *et al.*, 2005) and *Mc. voltae* AglB (Chaban *et al.*, 2006).

Although the precise role of AglE in *N*-glycosylation is still unknown, it is involved in the addition of a 190-Da sugar residue. This modification has been found at position 4 of the pentasaccharide decorating at least two S-layer protein sequons (Abu-Qarn *et al.*, 2007, 2008) (see Figure 7.3D). The AglE protein is a homologue of eukaryal Dpm1, which is responsible for the addition of GDP-mannose to dolichol pyrophosphate to yield phosphodolichol mannose (Burda and Aebi, 1999). It is conceivable that AglE is responsible for loading the 190-Da sugar residue from the *Hf. volcanii* S-layer glycoprotein pentasaccharide onto a lipid carrier, from where it is then transferred onto the 176-Da sugar residue of the three-membered pentasaccharide precursor via the actions of a distinct transferase (Abu-Qarn *et al.*, 2008), indicating also exocytosolic glycosylation reactions as proposed in *Hb. salinarum*. Different to *aglB* and *aglD* mutants, in *aglE*-deficient cells, no difference in cell growth, in S-layer glycoprotein release or in proteolytic susceptibility of the S-layer was observed under test conditions. The most pronounced shift in the apparent molecular mass of the *Hf. volcanii* S-layer glycoprotein and the biggest differences in S-layer architecture and protease susceptibility were observed with the *aglD* deletion strain, where only the final 162-Da sugar residue of the *N*-linked pentasaccharide was missing. It can be speculated that this sugar residue plays a critical role in stabilizing the S-layer glycan structure.

As in eukarya and bacteria, the *N*-glycosylation process of proteins in *Hf. volcanii*, and likely other Archaea, can obviously be divided into two parts:

(i) the assembly of a lipid-linked oligosaccharide starting on the cytoplasmic side of the membrane; and
(ii) the transfer of the glycan moiety to its target on the external side of the membrane (Sumper and Wieland, 1995).

The exact mechanism, however, remains to be elucidated. Aberrant and absent *N*-glycosylation compromise the ability of *Hf. volcanii* cells to grow at high salt concentrations. Thus, *N*-glycosylation can be considered a post-translational modification designed to allow the organism to remain intact in a hypersaline environment (Sumper and Wieland, 1995; Abu-Qarn *et al.*, 2007, 2008).

6. THE "NON-CLASSICAL" GROUP OF SECONDARY CELL WALL POLYMERS

6.1. Structure

The presence of polysaccharide compounds in the cell wall of bacteria, where they are linked to the peptidoglycan scaffold, has been known since the early days of cell wall research. These mostly anionic polymers have been classified into (i) teichoic acids and (ii) teichuronic acids (Araki and Ito, 1989); both classes together constituting the group of "classical" secondary cell wall polymers, which possess diverse "secondary roles" in cell wall function (Archibald *et al.*, 1993; Navarre and Schneewind, 1999). A third, rather new class of SCWPs comprises compounds that have another, discrete function, which is mediation of the non-covalent attachment of S-layer (glyco)proteins to the bacterial cell wall (Mesnage *et al.*, 2000; Sára, 2001; Cava *et al.*, 2004). This finding, together with their structural features (see below), let us classify these SCWPs as "non-classical" SCWPs (Schäffer and Messner, 2005) that, per definition, are neither teichoic nor teichuronic acids (Araki and Ito, 1989).

While initial investigations on "non-classical" SCWPs were focused on the elucidation of polysaccharide structures and on the identification of the linkage to the peptidoglycan layer (Schäffer and Messner, 2005), recent research mainly concerns refinement of the data on the interaction between SCWPs and S-layer (glyco)proteins.

In the investigated bacteria of the *Bacillaceae* family, the "non-classical" SCWPs were found to account for a substantial amount of 7–15% (by weight) of the peptidoglycan layer, which may be taken as an indication for their pivotal role in normal cell function. Interestingly, per microorganism, one distinct type of "non-classical" SCWP is present, making it a strain-specific feature. The linkage of these SCWPs to the peptidoglycan layer has been investigated by nuclear magnetic resonance (NMR) spectroscopy using intact SCWP-peptidoglycan complexes and provided evidence that they are linked to the C-6 atom of *N*-acetylmuramic acid residues via phosphodiester or pyrophosphate bridges, with about 20–25% of muramyl residues being substituted with SCWP (Schäffer *et al.*, 1999a; Steindl *et al.*, 2002, 2005). The direct association with peptidoglycan corroborates the suitability of "non-classical" SCWPs for their proposed *in vivo* anchoring function of S-layer (glyco)proteins (Sára, 2001).

Comparison of the overall composition of all "non-classical" SCWPs analysed so far revealed some common features: SCWP glycans:

(i) are heteropolysaccharides, with constituting glycoses including Glc*p*NA, Man*p*NAc, Gal*p*NAc, Man*p*NAc3NAcA, Man*p*NAc3NAcNH_2, Glc*p*, and Rib*f*;
(ii) may be anionic or neutral;
(iii) possess possible non-carbohydrate substituents including pyruvate, phosphate or acetate;
(iv) are, in most cases, composed of linear or branched repeating units of two to five glycoses; and
(v) possess an average molecular mass of 4000–6000 (Table 7.1).

Based on their structural motifs, these SCWPs are placed into three groups (Schäffer and Messner, 2005). Group I comprises a [→3)-β-D-Man*p*NAc-(1→4)-β-D-Glc*p*NAc-(1→] disaccharide backbone, which may additionally contain pyruvic acid substituents at the Man*p*NAc residues as shown for *Paenibacillus alvei* CCM 2051 (Schäffer *et al.*, 2000) and *Lysinibacillus sphaericus* CCM 2177 (Ilk *et al.*, 1999) or may contain substoichiometric amounts of ribofuranosyl side chains (Altman *et al.*, 1996). Other *Geobacillus* strains harbouring a tetrasaccharide backbone with the structure [→4)-β-D-Man*p*NAc3NAcA-(1→6)-α-D-Glc*p*-(1→4)-β-D-Man*p*NAc3NAcA-(1→3)-α-D-Glc*p*NAc(1→], in which the 2,3-diacetamido-2,3-dideoxymannuronic acid may also be modified by amidation, as in the case of *G. tepidamans* GS5-97^T (Steindl *et al.*, 2005), are compiled in group II. One example of an SCWP isolated from *Aneurinibacillus thermoaerophilus* DSM 10155 was reported with an unusual biantennary structure of defined chain length and has been placed into group III (Steindl *et al.*, 2002). According to this classification scheme, the recently elucidated SCWP of *G. stearothermophilus* PV72/p2, possessing a [→3)-β-D-Man*p*NAcA-(1→4)-β-D-Glc*p*N(Ac)$_{0.3}$-(1→6)-α-D-Glc*p*NAc-(1→] backbone resembles to some extent group II (Petersen *et al.*, 2008). In this case, however, it has to be considered, that the material used for structure elucidation has been liberated from the cell wall by hydrofluoric acid (HF) treatment, which can induce polysaccharide degradation. Coming back to the *in vivo* situation of the "non-classical" SCWPs, it is conceivable that glycan chain length variation due to the

TABLE 7.1 SCWPs as anchors for S-layer proteins to the bacterial cell wall[a]

SCWP	S-layer protein			
Organism structure	**Charge**	**Accession no.**	**Site of interaction with SCWP**	**References**
Geobacillus stearothermophilus **PV72/p2** (4[DGNβ3] $_{\sim0.3}$ DMNAβ3DGN(Ac) $_{\sim0.3}$ β6[(*S*)Pyr-4, 6-DMN4α]DGNα) (partial structure)	Negative	SbsB (X98095)	N-terminus 3-SLH motifs	Ries *et al.*, 1997 Mader *et al.*, 2004 Rünzler *et al.*, 2004 Petersen *et al.*, 2008
Lysinibacillus sphaericus **CCM 2177** (3[Pyr-4,6] $_{\sim0.5}$ DMNβ4DGNβ) $_{n=8-9}$	Negative	SbpA (AF211170)	N-terminus 3-SLH motifs	Ilk *et al.*, 1999
Paenibacillus alvei **CCM 2051** (3[Pyr-4,6]DMNβ4DGNβ) $_{n\sim11}$ 3[Pyr-4,6]DMNβ4DGNα*OP*	Negative	SpaA (submitted to GenBank)	Unknown	Schäffer *et al.*, 2000
Bacillus anthracis *Structure unknown; presence of pyruvyl residues*	Negative	Sap (P49051) EA1 (P94217)	N-terminus (proposed) 3-SLH motifs	Mesnage *et al.*, 2000
Geobacillus stearothermophilus **NRS 2004/3a** (4DM2,3diNAβ6DGα4DM2,3diNAβ3DGNα) $_{n\sim6}$ *OP*	Negative	SgsE (AF328864)	N-terminus no SLH-motifs	Schäffer *et al.*, 1999a, 2002
Geobacillus stearothermophilus **PV72/p6** Same structure proposed as for *G. stearothermophilus* NRS 2004/3a	Negative	SbsA (X71092)	N-terminus (proposed) no SLH-motifs	
Geobacillus stearothermophilus **ATCC 12980** Same structure proposed as for *G. stearothermophilus* NRS 2004/3a	Negative	SbsC (AF055578)	N-terminus no SLH-motifs	Jarosch *et al.*, 2001 Ferner-Ortner *et al.*, 2007 Pavkov *et al.*, 2008
Geobacillus tepidamans **GS5-97**^{T} (4DM2,3diNARβ6DGα4DM2,3diNARβ3DGNα) $_{n\sim6}$ *OP*	Neutral	SgtA (AY883421)	N-terminus (proposed) no SLH-motifs	Steindl *et al.*, 2005
Aneurinibacillus thermoaerophilus **DSM 10155/G**$^{+}$ [DGNα3DMNβ4DGalNβ3DGNα3DMNβ4DGalNβ3DGNα3] DGNα3DMNβ4DGalNβ3DGNα3DMNβ4DGalNβ3DGNα4 DMNβ3DGNα3DMNβ3DGNα3DGNα*OP*	Neutral	SatB (AY395579)	C-terminus (proposed) no SLH-motifs	Steindl *et al.*, 2002

(*Continues*)

TABLE 7.1 (*Continued*)

SCWP	S-layer protein			
Organism structure	Charge	Accession no.	Site of interaction with SCWP	References
***Thermoanaerobacterium thermosaccharolyticum* D120-70**	Neutral	Unknown	Unknown	Schäffer *et al.*, unpublished
(3[Rib*f* $_{0.5}$]DMNβ4DGNβ) $_n$				
***Thermoanaerobacterium thermosaccharolyticum* E207-71**	Neutral	Unknown	Unknown	Altman *et al.*, 1996
(3[Rib*f* $_{0.5}$]DMNβ4DGNβ) $_n$				Schäffer *et al.*, unpublished

[a]G, glucose; GN, *N*-acetylglucosamine; GalN, *N*-acetylgalactosamine; MN, *N*-acetylmannosamine; MNA, *N*-acetylmannosaminuronic acid; M2,3diNA, 2,3-acetamido-2,3-dideoxy-mannuronic acid; M2,3diNAR, 2,3-acetamido-2,3-dideoxy-mannuronic acid with modified carboxyl group; all sugars are in the pyranose form; P, phosphate; Pyr, pyruvic acid; Rib*f*, ribofuranose; n, number of repeats.

repeating unit-wise composition of the glycans may be beneficial for anchoring of the S-layer (glyco)protein by providing more spatial flexibility to the protein during the initial phase of 2D crystallization on the cell surface. In the case of the chemically homogeneous biantennary SCWP of *A. thermoaerophilus* DSM 10155, doubling of binding motifs may be required to obtain a physiologically relevant binding strength for anchoring the "water-soluble" S-layer protein SatB to the cell wall (Steindl *et al.*, 2002).

6.2. Functions

Detailed investigations of the binding mechanism between S-layer (glyco)proteins and SCWPs revealed that, in general, S-layer proteins have two functional regions: a cell-wall-targeting domain which, in most of the organisms investigated thus far, is located at the N-terminus, and a C-terminal self-assembly domain. The existence of an N-terminal cell wall-targeting domain was substantiated by the identification of so called S-layer homology (SLH) motifs, consisting of 50 to 60 amino acids each, which are mostly found in triplicate at the N-terminus of S-layer proteins (Lupas *et al.*, 1994). If present, SLH motifs interact with group I SCWPs carrying pyruvic acid residues (Ries *et al.*, 1997; Lemaire *et al.*, 1998; Ilk *et al.*, 1999; Mesnage *et al.*, 2000; Cava *et al.*, 2004; Mader *et al.*, 2004; Rünzler *et al.*, 2004). Recent data indicated that a highly conserved TRAE motif plays a key role in the binding function of SLH domains (May *et al.*, 2006). For SLH-mediated binding, the construction of knock-out mutants in *Bacillus anthracis* and *Thermus thermophilus* demonstrated that the addition of pyruvic acid residues to the SCWP was a necessary modification (Mesnage *et al.*, 2000; Cava *et al.*, 2004). This was also confirmed by surface plasmon resonance (SPR) spectroscopy measurements using the S-layer protein SbsB of *Geobacillus stearothermophilus* PV72/p2 and the corresponding native and depyruvylated SCWP for interaction studies (Mader *et al.*, 2004). SPR data analysis revealed the presence of at least two binding sites on a single SCWP molecule with an overall K_d of 7.7×10^{-7} M (Mader *et al.*, 2004). Contrary to SbsB, in SbpA, the S-layer protein of *L. sphaericus* CCM 2177, the functional SCWP-binding domain is formed by three SLH-motifs and an additional 58-amino acid SLH-like motif (Huber *et al.*, 2005).

A further type of binding mechanism between S-layer (glyco)proteins and SCWPs has been described for the *G. stearothermophilus* wild-type strains PV72/p6, NRS 2004/3a and ATCC 12980 (Egelseer *et al.*, 1998; Schäffer *et al.*, 1999a; Jarosch *et al.*, 2001). There, non-pyruvylated group II SCWP interacts with a highly conserved N-terminal region of the S-layer protein devoid of SLH motifs (Egelseer *et al.*, 1998; Jarosch *et al.*, 2001). First affinity studies using different N- or C-terminally truncated forms of the S-layer protein SbsC from *G. stearothermophilus* ATCC 12980 indicated that the N-terminal part comprising aa 31 to 257 is exclusively responsible for cell wall binding (Jarosch *et al.*, 2001). This result is corroborated by the fact that the N-terminus contains a surplus of arginine, lysine and tyrosine residues, which are known to interact with carbohydrates via direct electrostatic interactions and hydrogen bonds. In a very recent study, the first high-resolution structure (2.4 Å resolution) of a bacterial S-layer protein could be elucidated from crystals of the water-soluble, C-terminally truncated form $rSbsC_{31-844}$ (Pavkov *et al.*, 2008). The crystal structure revealed a novel fold, consisting of six separate domains, which are connected by short flexible linkers. The N-terminal SCWP binding domain, comprising amino acids (aa) 32 to 260, consists of seven α-helices organized in three left-handed anti-parallel triple-helical bundles connected by short loops. According to a BLAST search, the N-terminus of SbsC has high similarity not only to S-layer proteins of *G. stearothermophilus* strains but also to those of *Geobacillus kaustophilus* and *Geobacillus tepidamans* GS5-97^T (SgtA). The SCWP of *G. tepidamans* GS5-97^T recognized by SgtA differs from that bound by SbsC therein that the carboxyl groups of the Man*p*NAc3NAcA residues carry modifications that turn the anionic character of this glycan into a neutral one, while that of *G. stearothermophilus* ATCC 12980 is negatively charged (Schäffer and Messner, 2005). Recently, the basic mechanism for anchoring an S-layer protein devoid of SLH motifs to the rigid cell wall layer was systematically investigated by SPR biosensor technology using the S-layer protein SbsC and the corresponding SCWP of *G. stearothermophilus* ATCC 12980 as a model system (Ferner-Ortner *et al.*, 2007). The SPR data confirmed that the interaction is highly specific and that the N-terminal region comprising aa 31 to 270 is exclusively responsible for SCWP binding. Analysis of data from the setup, in which SCWP was immobilized on a sensor chip and either $rSbsC_{31-270}$ or $rSbsC_{31-443}$ represented the soluble analytes, indicated a binding behaviour with low ($K_d = 9.32 \times 10^{-5}$ M and 2.95×10^{-6} M), medium ($K_d = 4.8 \times 10^{-9}$ M and 1.22×10^{-8} M) and high ($K_d = 1.94 \times 10^{-12}$ M and 2.05×10^{-12}M) affinity (Ferner-Ortner *et al.*, 2007). Furthermore, isothermal titration calorimetry (ITC) suggested a strong binding of two S-layer protein molecules to one SCWP molecule with overall binding constants in the range of 10^6–10^7 and CD measurements showed that binding of the SCWP dramatically stabilizes the proteins (Pavkov *et al.*, 2008).

A cell-wall targeting domain is not necessarily located in the N-terminal region of S-layer proteins. In the S-layer proteins SlpA of *Lactobacillus acidophilus* ATCC 4356 and CbsA of *Lactobacillus crispatus* JCM 5810, a cell wall-binding domain has been identified in the C-terminal one third of these S-layer proteins and sequence alignment studies revealed a putative carbohydrate-binding repeat comprising approximately the last 130 C-terminal amino acids, which were suggested to be involved in cell wall binding (Smit *et al.*, 2001). Cell wall anchoring via the C-terminal region seems to be present also in the S-layer proteins of *A. thermoaerophilus* strains DSM 10155 and L420-91^T according to sequence alignments.

To conclude, the binding strength and specificity between S-layer (glyco)proteins and SCWPs is of high relevance for generating and maintaining a dynamic protein crystal on a bacterial cell surface during all stages of cell growth and division since it guarantees a defined orientation for incorporated S-layer (glyco)proteins while allowing enough flexibility for recrystallization of S-layer subunits to continuously assume a low free energy arrangement (Sleytr *et al.*, 2005).

7. OUTLOOK

Considering that glycans are ubiquitous biomolecules which, in many cases, are key to protein functions, it is evident that they are useful means for addressing various questions in the fields of basic and applied research, relating to the areas of biomimetics, drug targeting, vaccine design or diagnostics. In this context, high-density, periodic and controllable display of glycans, which through random chemical coupling reactions of glycans to various supports cannot be fully accomplished, plays a pivotal role. Exactly for this reason, the S-layer protein self-assembly system offers an attractive solution (Sleytr *et al.*, 2007; Messner *et al.*, 2008). Future research will likely focus on deepening the understanding of the molecular basis of the S-layer protein glycosylation process and of the mode of interaction between S-layer (glyco)proteins and their native anchoring structures to the cell wall, i.e. the "non-classical" SCWPs (Research Focus Box). Optimal strategies will be set up to produce modular S-layer neoglycoproteins by a combined protein/carbohydrate engineering approach that are equipped with functional, tailor-made glycan motifs (Steiner *et al.*, 2008b). Nanometre-controlled glycan display can be envisaged *in vivo* on selected bacterial species or, by applying a biomimetic approach including "non-classical" SCWPs, on diverse supports. Engineering of such novel, nanopatterned composites by bottom-up strategies using an S-layer based biomolecular construction kit will decisively open up new possibilities in influencing and controlling complex biological systems.

RESEARCH FOCUS BOX

- Deepening the understanding the molecular basis of the S-layer glycosylation process.
- Better understanding the mode of interaction between S-layer (glyco)proteins and the anchoring structures to the cell wall, i.e. "non-classical" secondary cell wall polymers.
- Development of production of modular S-layer (neo)glycoproteins with functional tailor-made glycan motifs.
- Development of nanometre-controlled glycan display *in vivo* and *in vitro*.

ACKNOWLEDGEMENTS

Financial support came from the Austrian Science Fund, projects P18013-B10 and P20745-B11 (to P.M.), projects P19047-B12 and P20605-B12 (to C.S.), project P18510-B12 (to E.-M.E.), the Hochschuljubiläumsstiftung der Stadt Wien, project H-1809/2006 (to C.S.) and the US Air Force Office of Scientific Research, project F49620-03-1-0222 (to U.B.S.).

References

Abu-Qarn, M., Eichler, J., 2006. Protein N-glycosylation in Archaea: defining *Haloferax volcanii* genes involved in S-layer glycoprotein glycosylation. Mol. Microbiol. 61, 511–525.

Abu-Qarn, M., Yurist-Doutsch, S., Giordano, A., et al., 2007. *Haloferax volcanii* AglB and AglD are involved in *N*-glycosylation of the S-layer glycoprotein and proper assembly of the surface layer. J. Mol. Biol. 374, 1224–1236.

Abu-Qarn, M., Giordano, A., Battaglia, F., et al., 2008. Identification of AglE, a second glycosyltransferase involved in *N*-glycosylation of the *Haloferax volcanii* S-layer glycoprotein. J. Bacteriol. 190, 3140–3146.

Altman, E., Schäffer, C., Brisson, J.-R., Messner, P., 1996. Isolation and characterization of an amino sugar-rich glycopeptide from the surface layer glycoprotein of *Thermoanaerobacterium thermosaccharolyticum* E207-71. Carbohydr. Res. 295, 245–253.

Araki, Y., Ito, E., 1989. Linkage units in cell walls of Gram-positive bacteria. CRC Crit. Rev. Microbiol. 17, 121–135.

Archibald, A.R., Hancock, I.C., Harwood, C.R., 1993. Cell wall structure, synthesis and turnover. In: Sonenshein, A., Hoch, J.A., Losick, R. (Eds.), *Bacillus subtilis* and Other Gram-Positive Bacteria. ASM Press, Washington, DC, pp. 381–410.

Awram, P., Smit, J., 2001. Identification of lipopolysaccharide O antigen synthesis genes required for attachment of the S-layer of *Caulobacter crescentus*. Microbiology 147, 1451–1460.

Baumeister, W., Lembcke, G., 1992. Structural features of archaebacterial cell envelopes. J. Bioenerg. Biomembr. 24, 567–575.

Bröckl, G., Behr, M., Fabry, S., et al., 1991. Analysis and nucleotide sequence of the genes encoding the surface-layer glycoproteins of the hyperthermophilic methanogens *Methanothermus fervidus* and *Methanothermus sociabilis*. Eur. J. Biochem. 199, 147–152.

Burda, P., Aebi, M., 1999. The dolichol pathway of N-linked glycosylation. Biochim. Biophys. Acta 1426, 239–257.

Cambillau, C., Claverie, J.M., 2000. Structural and genomic correlates of hyperthermostability. J. Biol. Chem. 275, 32383–32386.

Cava, F., de Pedro, M.A., Schwarz, H., Henne, A., Berenguer, J., 2004. Binding to pyruvylated compounds as an ancestral mechanism to anchor the outer envelope in primitive bacteria. Mol. Microbiol. 52, 677–690.

Chaban, B., Voisin, S., Kelly, J., Logan, S.M., Jarrell, K.F., 2006. Identification of genes involved in the biosynthesis and attachment of *Methanococcus voltae* N-linked glycans: insight into *N*-linked glycosylation pathways in Archaea. Mol. Microbiol. 61, 259–268.

Clarke, B.R., Cuthbertson, L., Whitfield, C., 2004. Nonreducing terminal modifications determine the chain length of polymannose O-antigens of *Escherichia coli* and couple chain termination to polymer export via an ATP-binding cassette transporter. J. Biol. Chem. 279, 35709–35718.

Claus, H., Akça, E., Debaerdemaeker, T., Evrard, C., Declercq, J.-P., König, H., 2002. Primary structure of selected archaeal mesophilic and extremely thermophilic outer surface layer proteins. Syst. Appl. Microbiol. 25, 3–12.

Claus, H., Akça, E., Debaerdemaeker, T., et al., 2005. Molecular organization of selected prokaryotic S-layer proteins. Can. J. Microbiol. 51, 731–743.

Egelseer, E.M., Leitner, K., Jarosch, M., et al., 1998. The S-layer proteins of two *Bacillus stearothermophilus* wild-type strains are bound via their *N*-terminal region to a secondary cell wall polymer of identical chemical composition. J. Bacteriol. 180, 1488–1495.

Eichler, J., Adams, M.W.W., 2005. Posttranslational protein modification in Archaea. Microbiol. Mol. Biol. Rev. 69, 393–425.

Feldman, M.F., Wacker, M., Hernandez, M., et al., 2005. Engineering *N*-linked protein glycosylation with diverse O-antigen lipopolysaccharide structures in *Escherichia coli*. Proc. Natl. Acad. Sci. USA 102, 3016–3021.

Ferner-Ortner, J., Mader, C., Ilk, N., Sleytr, U.B., Egelseer, E.M., 2007. High-affinity interaction between the S-layer protein SbsC and the secondary cell wall polymer of *Geobacillus stearothermophilus* ATCC 12980 determined by surface plasmon resonance technology. J. Bacteriol. 189, 7154–7158.

Fletcher, C.M., Coyne, M.J., Bentley, D.L., Villa, O.F., Comstock, L.E., 2007. Phase-variable expression of a family of glycoproteins imparts a dynamic surface to a symbiont in its human intestinal ecosystem. Proc. Natl. Acad. Sci. USA 104, 2413–2418.

Glover, K.J., Weerapana, E., Numao, S., Imperiali, B., 2005. Chemoenzymatic synthesis of glycopeptides with PglB, a bacterial oligosaccharyl transferase from *Campylobacter jejuni*. Chem. Biol. 12, 1311–1315.

Huber, C., Ilk, N., Rünzler, D., et al., 2005. The three S-layer-like homology motifs of the S-layer protein SbpA of *Bacillus sphaericus* CCM 2177 are not sufficient for binding to the pyruvylated secondary cell wall polymer. Mol. Microbiol. 55, 197–205.

Ilk, N., Kosma, P., Puchberger, M., et al., 1999. Structural and functional analyses of the secondary cell wall polymer of *Bacillus sphaericus* CCM 2177 that serves as an S-layer-specific anchor. J. Bacteriol. 181, 7643–7646.

Jarosch, M., Egelseer, E.M., Huber, C., et al., 2001. Analysis of the structure-function relationship of the S-layer protein SbsC of *Bacillus stearothermophilus* ATCC 12980 by producing truncated forms. Microbiology 147, 1353–1363.

Kählig, H., Kolarich, D., Zayni, S., et al., 2005. *N*-acetylmuramic acid as capping element of α-D-fucose-containing S-layer glycoprotein glycans from *Geobacillus tepidamans* GS5-97^{T}. J. Biol. Chem. 280, 20292–20299.

Kaplan, J.B., Perry, M.B., MacLean, L.L., Furgang, D., Wilson, M.E., Fine, D.H., 2001. Structural and genetic analyses of O polysaccharide from *Actinobacillus actinomycetemcomitans* serotype f. Infect. Immun. 69, 5375–5384.

Kärcher, U., Schröder, H., Haslinger, E., et al., 1993. Primary structure of the heterosaccharide of the surface glycoprotein of *Methanothermus fervidus*. J. Biol. Chem. 268, 26821–26826.

Lee, S.W., Sabet, M., Um, H.S., Yang, J., Kim, H.C., Zhu, W., 2006. Identification and characterization of the genes encoding a unique surface (S-) layer of *Tannerella forsythia*. Gene 371, 102–111.

Lemaire, M., Miras, I., Gounon, P., Béguin, P., 1998. Identification of a region responsible for binding to the cell wall within the S-layer protein of *Clostridium thermocellum*. Microbiology 144, 211–217.

Lupas, A., Engelhardt, H., Peters, J., Santarius, U., Volker, S., Baumeister, W., 1994. Domain structure of the *Acetogenium kivui* surface layer revealed by electron crystallography and sequence analysis. J. Bacteriol. 176, 1224–1233.

Mader, C., Huber, C., Moll, D., Sleytr, U.B., Sára, M., 2004. Interaction of the crystalline bacterial cell surface layer protein SbsB and the secondary cell wall polymer of *Geobacillus stearothermophilus* PV72 assessed by real-time surface plasmon resonance biosensor technology. J. Bacteriol. 186, 1758–1768.

May, A., Pusztahelyi, T., Hoffmann, N., Fischer, R.J., Bahl, H., 2006. Mutagenesis of conserved charged amino acids in SLH domains of *Thermoanaerobacterium thermosulfurigenes* EM1 affects attachment to cell wall sacculi. Arch. Microbiol. 185, 263–269.

Mescher, M.F, Strominger, J.L., 1976. Purification and characterization of a prokaryotic glucoprotein from the cell envelope of *Halobacterium salinarium*. J. Biol. Chem. 251, 2005–2014.

Mesnage, S., Fontaine, T., Mignot, T., Delepierre, M., Mock, M., Fouet, A., 2000. Bacterial SLH domain proteins are non-covalently anchored to the cell surface via a conserved mechanism involving wall polysaccharide pyruvylation. EMBO J. 19, 4473–4484.

Messner, P., Schäffer, C., 2003. Prokaryotic glycoproteins. In: Herz, W., Falk, H., Kirby, G.W. (Eds.), Progress in the Chemistry of Organic Natural Products, Vol. 85. Springer-Verlag, Wien, pp. 51–124.

Messner, P., Sleytr, U.B., 1992. Crystalline bacterial cell-surface layers. Adv. Microb. Physiol. 33, 213–275.

Messner, P., Steiner, K., Zarschler, K., Schäffer, C., 2008. S-layer nanoglycobiology of bacteria. Carbohydr. Res. 343, 1934–1951.

Munoz, M.D., Hernandez, L.M., Basco, R., Andaluz, E., Larriba, G., 1994. Glycosylation of yeast exoglucanase sequons in alg mutants deficient in the glucosylation steps of the lipid-linked oligosaccharide. Presence of glucotriose unit in Dol-PP-GlcNAc2-Man9Glc3 influences both glycosylation efficiency and selection of *N*-linked sites. Biochim. Biophys. Acta 1201, 361–366.

Navarre, W.W., Schneewind, O., 1999. Surface proteins of gram-positive bacteria and mechanisms of their targeting to the cell wall envelope. Microbiol. Mol. Biol. Rev. 63, 174–229.

Nothaft, H., Amber, S., Aebi, M., Szymanski, C.M., 2008. *N*-linked protein glycosylation in *Campylobacter*. In: Nachamkin, I., Szymanski, C.M., Blaser, M.J. (Eds.) *Campylobacter*, 3[ed] edn. ASM Press, Washington, DC, pp. 447–469.

Novotny, R., Pföstl, A., Messner, P., Schäffer, C., 2004. Genetic organization of chromosomal S-layer glycan biosynthesis loci of *Bacillaceae*. Glycoconj. J. 20, 435–447.

Pavkov, T., Egelseer, E.M., Tesarz, M., Svergun, D.I., Sleytr, U.B., Keller, W., 2008. The structure and binding behavior of the bacterial cell surface layer protein SbsC. Structure 16, 1226–1237.

Petersen, B.O., Sára, M., Mader, C., et al., 2008. Structural characterization of the acid-degraded secondary cell wall polymer of *Geobacillus stearothermophilus* PV72/p2. Carbohydr. Res. 343, 1346–1358.

Raetz, C.R.H., Whitfield, C., 2002. Lipopolysaccharide endotoxins. Annu. Rev. Biochem. 71, 635–700.

Ries, W., Hotzy, C., Schocher, I., Sleytr, U.B., Sára, M., 1997. Evidence that the N-terminal part of the S-layer protein from *Bacillus stearothermophilus* PV72/p2 recognizes a secondary cell wall polymer. J. Bacteriol. 179, 3892–3898.

Rünzler, D., Huber, C., Moll, D., Köhler, G., Sára, M., 2004. Biophysical characterization of the entire bacterial surface layer protein SbsB and its two distinct functional domains. J. Biol. Chem. 279, 5207–5215.

Sára, M., 2001. Conserved anchoring mechanisms between crystalline cell surface S-layer proteins and secondary cell wall polymers in Gram-positive bacteria. Trends Microbiol. 9, 47–49.

Sára, M., Sleytr, U.B., 2000. S-layer proteins. J. Bacteriol. 182, 859–868.

Schäffer, C., Messner, P., 2004. Surface-layer glycoproteins: an example for the diversity of bacterial glycosylation with promising impacts on nanobiotechnology. Glycobiology 14, 31R–42R.

Schäffer, C., Messner, P., 2005. The structure of secondary cell wall polymers: how Gram-positive bacteria stick their cell walls together. Microbiology 151, 643–651.

Schäffer, C., Kählig, H., Christian, R., Schulz, G., Zayni, S., Messner, P., 1999a. The diacetamidodideoxyuronic-acid-containing glycan chain of *Bacillus stearothermophilus* NRS 2004/3a represents the secondary cell-wall polymer of wild-type *B. stearothermophilus* strains. Microbiology 145, 1575–1583.

Schäffer, C., Müller, N., Christian, R., Graninger, M., Wugeditsch, T., Scheberl, A., Messner, P., 1999b. Complete glycan structure of the S-layer glycoprotein of *Aneurinibacillus thermoaerophilus* GS4-97. Glycobiology 9, 407–414.

Schäffer, C., Müller, N., Mandal, P.K., Christian, R., Zayni, S., Messner, P., 2000. A pyrophosphate bridge links the pyruvate-containing secondary cell wall polymer of *Paenibacillus alvei* CCM 2051 to muramic acid. Glycoconj. J. 17, 681–690.

Schäffer, C., Wugeditsch, T., Kählig, H., Scheberl, A., Zayni, S., Messner, P., 2002. The surface layer (S-layer) glycoprotein of *Geobacillus stearothermophilus* NRS 2004/3a. Analysis of its glycosylation. J. Biol. Chem. 277, 6230–6239.

Sleytr, U.B., Beveridge, T.J., 1999. Bacterial S-layers. Trends Microbiol. 7, 253–260.

Sleytr, U.B., Thorne, K.J.I., 1976. Chemical characterization of the regularly arranged surface layers of *Clostridium thermosaccharolyticum* and *Clostridium thermohydrosulfuricum*. J. Bacteriol. 126, 377–383.

Sleytr, U.B., Messner, P., Pum, D., Sára, M., 1999. Crystalline bacterial cell surface layers (S-layers): from supramolecular cell structure to biomimetics and nanotechnology. Angew. Chem. Int. Ed. Engl. 38, 1034–1054.

Sleytr, U.B., Sára, M., Pum, D., Schuster, B., Messner, P., Schäffer, C., 2002. Self-assembly protein systems: microbial S-layers. In: Steinbüchel, A., Fahnestock, S.R. (Eds.), Biopolymers, Vol. 7: Polyamides and Complex Proteinaceous Matrices I. Wiley-VCH Verlag, Weinheim, pp. 285–338.

Sleytr, U.B., Sára, M., Pum, D., Schuster, B., 2005. Crystalline bacterial cell surface layers (S-layers): a versatile self-assembly system. In: Ciferri, A. (Ed.), Supramolecular Polymers, 2[ed] edn. Taylor & Francis Inc, Boca Raton, pp. 583–612.

Sleytr, U.B., Egelseer, E.M., Ilk, N., Pum, D., Schuster, B., 2007. S-Layers as a basic building block in a molecular construction kit. FEBS J. 274, 323–334.

Smit, E., Oling, F., Demel, R., Martinez, B., Pouwels, P.H., 2001. The S-layer protein of *Lactobacillus acidophilus* ATCC 4356: identification and characterisation of domains responsible for S-protein assembly and cell wall binding. J. Mol. Biol. 305, 245–257.

Steindl, C., Schäffer, C., Wugeditsch, T., et al., 2002. The first biantennary bacterial secondary cell wall polymer from bacteria and its influence on S-layer glycoprotein assembly. Biochem. J. 368, 483–494.

Steindl, C., Schäffer, C., Smrecki, V., Messner, P., Müller, N., 2005. The secondary cell wall polymer of *Geobacillus tepidamans* GS5-97^T: structure of different glycoforms. Carbohydr. Res. 340, 2290–2296.

Steiner, K., Novotny, R., Patel, K., et al., 2007. Functional characterization of the initiation enzyme of S-layer glycoprotein glycan biosynthesis in *Geobacillus stearothermophilus* NRS 2004/3a. J. Bacteriol. 189, 2590–2598.

Steiner, K., Novotny, R., Werz, D.B., et al., 2008a. Molecular basis of S-layer glycoprotein glycan biosynthesis in Geobacillus stearothermophilus. J. Biol. Chem. 283, 21120–21133.

Steiner, K., Hanreich, A., Kainz, B., et al., 2008b. Recombinant glycans on an S-layer self-assembly protein: a new dimension for nanopatterned biomaterials. Small 4, 1728–1740.

Sumper, M., Wieland, F.T., 1995. Bacterial glycoproteins. In: Montreuil, J., Vliegenthart, J.F.G., Schachter, H. (Eds.), Glycoproteins. Elsevier Science B.V., Amsterdam, pp. 455–473.

Sumper, M., Berg, E., Mengele, R., Strobl, I., 1990. Primary structure and glycosylation of the S-layer protein of *Haloferax volcanii*. J. Bacteriol. 172, 7111–7118.

Voisin, S., Houliston, R.S., Kelly, J., et al., 2005. Identification and characterization of the unique N-linked glycan common to the flagellins and S-layer glycoprotein of *Methanococcus voltae*. J. Biol. Chem. 280, 16586–16593.

Wieland, F., Paul, G., Sumper, M., 1985. Halobacterial flagellins are sulfated glycoproteins. J. Biol. Chem. 260, 15180–15185.

Yan, A., Lennarz, W.J., 2005. Unraveling the mechanism of protein N-glycosylation. J. Biol. Chem. 280, 3121–3124.

Zayni, S., Steiner, K., Pföstl, A., et al., 2007. The dTDP-4-dehydro-6-deoxyglucose reductase encoding *fcd* gene is part of the surface layer glycoprotein glycosylation gene cluster of *Geobacillus tepidamans* GS5-97^{T}. Glycobiology 17, 433–443.

CHAPTER

8

Glycosylation of bacterial and archaeal flagellins

Susan M. Logan, Ian C. Schoenhofen and Evelyn C. Soo

SUMMARY

The biosynthesis, assembly and regulation of the flagellar organelle has been extensively described over many decades and has focused primarily on the peritrichous flagella of *Escherichia coli* and *Salmonella enterica*. More recently, the characterization of flagellar systems from other bacterial and archaeal species has revealed distinct differences in flagellar composition and mode of assembly. Glycosylation of the flagellin structural protein has been identified as an important feature of numerous systems and has been shown to play an integral role in flagellar assembly or in virulence of a number of pathogenic species. This chapter focuses on the structural diversity of flagellar glycans, methods for characterization of flagellin glycoproteins and novel glycan biosynthetic pathways. The relevance of the glycosylation process to assembly as well as other novel biological roles is discussed.

Keywords: Bacterial flagellin; Archaeal flagellin; Protein glycosylation; *O*-Linked glycan, *N*-Linked glycan

1. INTRODUCTION

Flagellar motility is one of the most extensively studied processes of prokaryotic microbiology (for review see Macnab, 2004). The rich body of work directed towards this nanomachine encompasses many fields of biology including genetics, physiology, bioenergetics and structural biology, and has provided exquisite detail on the structure and function of this unique organelle. In more recent years, the body of work has been expanded to include the field of glycobiology that describes the production of glycosylated flagellins from a diverse range of bacterial and archaeal species. The flagellar glycans produced by these microorganisms are novel and the role of this glycosylation process in both flagellar assembly, as well as in biological interactions, is becoming clearly established. In contrast, the mechanistic basis of the process remains poorly defined.

1.1. Flagellar structures

Although the extensively studied peritrichous flagella of *Escherichia coli* and *Salmonella enterica* sv. Typhimurium do not appear to produce glycosylated flagellins, glycosylation of flagellins in organisms producing polar flagella appears more commonplace (reviewed in Logan, 2006). Polar flagella are made by many bacteria and can be either simple (i.e. composed of only a single flagellin protein monomer) or complex (i.e. composed of more than one flagellin protein monomer) and are assembled either as a single filament (monotrichous) or in a bundle (multitrichous) at one or both ends of the cell. In some cases, these flagella may be covered with a sheath. Lateral flagella (LAF) are visible as a number of filaments extending at random points on the cell surface, but not from either pole, and are usually comprised of more than one structural protein. The periplasmic flagellum of spirochaetes is attached subterminally to each end of the cell cylinder and the flagella rotate within the periplasmic space conferring a unique form of motility to spirochaete cells (Charon and Goldstein, 2002).

Flagellar-based motility is also common to the Archaea, although the flagellar structure produced by these microorganisms is quite distinct to that produced by bacteria. Archaeal flagella are more closely related to bacterial type IV pili in structure and mode of assembly (Thomas *et al.*, 2001; Bardy *et al.*, 2004). The flagellar filament is generally complex in nature and the archaeal flagellin monomers are produced with a leader peptide which is cleaved by a signal peptidase to produce mature flagellin. This protein is then incorporated into the base of the filament which lacks a central channel (Trachtenberg and Cohen-Krausz, 2006). This process is in distinct contrast to the eubacterial system where flagellin is incorporated at the distal end of the growing filament following export through the central lumen (Iino, 1969; Emerson *et al.*, 1970).

1.2. O- versus N-linked glycans

Glycans are commonly attached to proteins through either *N*- or *O*-linkage. Examination of eubacterial glycosylated flagellins has revealed that, in all examples studied to date, the glycans are attached to the protein via *O*-linkage to either serine or threonine residues. As is the case for *O*-glycosylation in eukaryotes, no consensus sequence has been identified for bacterial *O*-linked glycosylation. However, the sites of glycosylation of bacterial flagellins appear to be restricted to the central variable region of the flagellin protein monomer rather than at the more conserved N- and C-terminals. Elegant X-ray crystallography and electron cryomicroscopy studies have provided a three-dimensional structure of *S. enterica* sv. Typhimurium flagellin and identified four structural domains, D0, D1, D2 and D3 (Samatey *et al.*, 2000; Yonekura *et al.*, 2003, 2005) (Figure 8.1). Both the N- and C-termini which comprise the D0 and D1 domains are rich in hydrophobic residues and these regions are known to interact hydrophobically with other flagellin subunits in the assembled filament (Yonekura *et al.*, 2005). The D0 and D1 domains are highly conserved among bacteria and are critical for subunit interactions, whereas the central variable regions which comprise the D2 and D3 domains do not appear to be critical to subunit interactions. These domains are surface-exposed in the folded, assembled filament and only limited subunit interaction has been described for the D2 domain of *S. enterica* sv. Typhimurium. The D2 and D3 domains vary in sequence and length among eubacterial flagellins and are where the sites of glycosylation have been shown to reside (see Figure 8.1).

In contrast, the two archaeal flagella studied to date produce flagellins where the glycan is *N*-glycosidically linked (Wieland *et al.*, 1985; Voisin *et al.*, 2005). Extensive mass spectrometry (MS) characterization of the four flagellin structural proteins from *Methanococcus voltae* revealed that the glycan was attached through the asparagine (Asn) residue of a typical eukaryotic *N*-linked sequon (Asn-Xaa-Ser/Thr, where X does not equal proline) rather than via the more recently characterized bacterial *N*-linked sequon (Asp/Glu-Yaa-Asn-Xaa-Ser/Thr, where X and Y do not equal proline) (Kowarik *et al.*, 2006). In summary, archaeal flagellins show little homology to eubacterial flagellins.

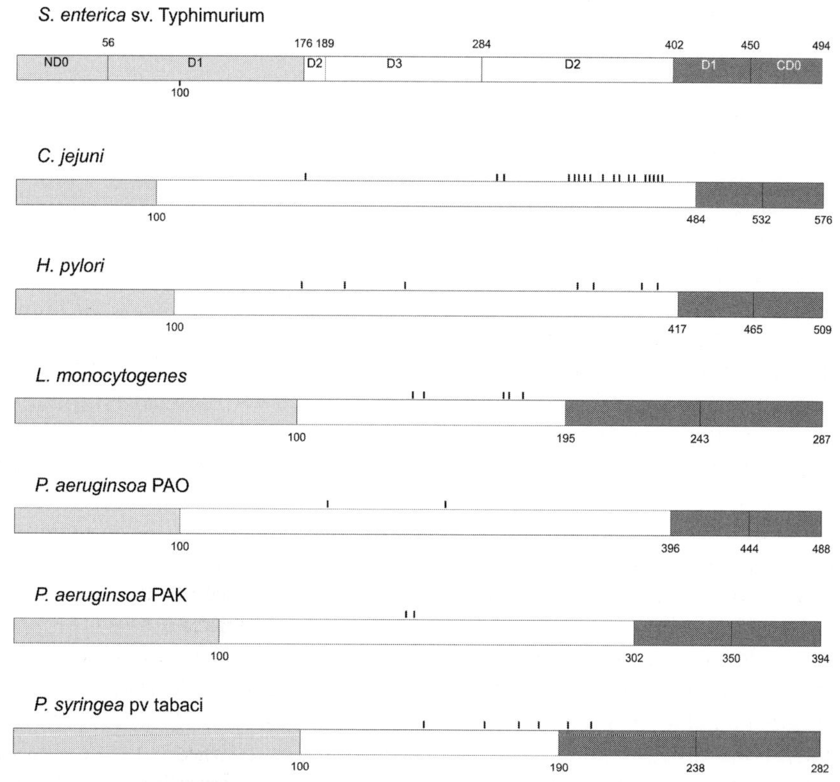

FIGURE 8.1 *O*-Linked sites of flagellar glycan attachment to flagellin monomers. Position and region of various structural domains in *S. enterica* sv. Typhimurium FliC with alignment of N- (■) and C-terminal regions (■/■) of glycosylated flagellins from *C. jejuni* 81–176, *H. pylori* 1061, *L. monocytogenes*, *P. aeruginosa* PAO, *P. aeruginosa* PAK, and *P. syringae* pv. *tabaci*. Precise sites of glycosylation on each flagellin monomer are indicated by vertical lines.

2. FLAGELLAR GLYCAN STRUCTURES

2.1. Archaea

The first structural characterization of a flagellar glycan was completed for the flagellin of *Halobacterium halobium* (Wieland *et al.*, 1985). This flagellin protein was shown to contain *N*-glycosidically linked oligosaccharides of sulfated (1→4)-linked glucuronic acid. More recently, the complete structural characterization of each of the four flagellin proteins FlaA, FlaB1, FlaB2 and FlaB3 from *M. voltae* revealed a novel trisaccharide of mass 779 attached in *N*-linkage to a total of 14 sites. The structure of this novel glycan was determined by nuclear magnetic resonance (NMR) spectroscopy and shown to be composed of 2-acetamido-2-deoxy-6-*N*-threonyl-mannopyranosiduronic acid-β-(1→4)-2,3-dideoxy-2,3-diacetamido-glucopyranosiduronicacid-β-(1→3)-2-deoxy-2-acetamido-glucopyranosyl (β-Man*p*NAcA6Thr-(1→4)-β-Glc*p*NAc3NAcA-(1→3)-βGlc*p*NAc) linked to Asn (Voisin *et al.*, 2005). In both of these studies, the same glycan was found to decorate the respective S-layer protein implicating a common *N*-linked glycosylation pathway for these two major surface structures in Archaea.

Genetic studies in *M. voltae* have identified genes involved in the biosynthesis and transfer of the flagellar glycan to the flagellin protein. The *aglA* gene encodes a glycosyltransferase responsible for attachment of the terminal sugar to the glycan while *aglB*, which encodes an STT3 oligosaccharyltransferase homologue, is involved in the transfer of the complete glycan to the flagellins and S-layer proteins from a dolichol lipid carrier (Chaban *et al.*, 2006). A third gene from *M. voltae*, *aglH*, encodes a protein that displays significant homology to Alg7 from yeast which catalyses the first step in the yeast *N*-linked glycosylation pathway by performing the enzymatic attachment of *N*-acetyl-D-glucosamine (GlcNAc) to a dolichol lipid carrier. Complementation of the *alg7* conditional lethal mutation in yeast with the *aglH* gene provided the first evidence that the function of the *M. voltae*-encoded protein is an *N*-acetyl-D-glucosamine-1-phosphate transferase (Shams-Eldin *et al.*, 2008). Mutational analysis of glycan biosynthetic genes revealed that the glycans played a significant role in the flagellar assembly process. In *M. voltae*, a mutant strain which was unable to transfer glycan to flagellin protein from the lipid linked carrier was non-motile and non-flagellated, whereas a second mutant strain, in which attachment of the terminal sugar of the glycan was disrupted, was poorly flagellated (Chaban *et al.*, 2006).

2.2. Gram-positive bacteria

2.2.1. Listeria monocytogenes

L. monocytogenes produces 4–6 simple, peritrichous flagella when grown at <30°C. Structural characterization revealed that the FlaA structural protein monomer is covalently modified with *O*-linked β-GlcNAc at 3–6 sites per subunit (Schirm *et al.*, 2004b). Modification with GlcNAc by the eukaryotic *O*-GlcNAc transferase (OGT) is commonly found on many nuclear and cytoplasmic eukaryotic proteins and is critical for a number of regulatory processes within the cell. In *L. monocytogenes*, GmaR (*lmo0688*), which lies close to the flagellin structural gene, is the first flagellin glycosyltransferase and first prokaryotic OGT for which enzymatic activity has been directly demonstrated (Shen *et al.*, 2006). Interestingly, GmaR also has been shown to function as an anti-repressor for MogR (a DNA binding transcriptional repressor) and plays a critical role in transcriptional regulation of flagellar biosynthesis. *L. monocytogenes* is a facultative intracellular pathogen that downregulates flagellar gene expression upon encountering physiological temperatures (37°C) and, consequently, flagella and related motility are not currently recognized as virulence factors in human infection (Peel *et al.*, 1988). Flagellar motility is known to be critical for biofilm formation by *L. monocytogenes* and in attachment to environmental surfaces, although glycosylation does not appear to play a role in this process (Lemon *et al.*, 2007). Glycosylation of flagellin is not required for flagellar formation by this bacterium and it remains to be determined what role glycosylation of the flagellin protein plays.

2.2.2. Clostridia *spp.*

Motility is a recognized feature of a number of members of the genus *Clostridium* including the important human pathogens *Clostridium botulinum* and *Clostridium difficile*. Bioinformatic analyses of completed genome sequences of *C. botulinum*, *C. difficile* and *Clostridium tetani* have identified a number of glycan biosynthetic genes within the respective flagellar loci indicating glycosylation is likely to be an integral part of the clostridial motility system. Preliminary evidence for glycosylation of the flagellin protein from *Clostridium tyrobutyricum* has been presented (Arnold *et al.*, 1998; Bedouet *et al.*, 1998). In the case of *Clostridium acetobutylicum* flagellin, the glycan was sensitive to neuraminidase treatment indicating a "sialic acid-like" sugar as a glycan component (Lyristis *et al.*, 2000).

2.3. Gram-negative bacteria

2.3.1. Pseudomonas *spp.*

Pseudomonads are ubiquitous environmental organisms and, in some cases, opportunistic pathogens of both humans and plants. Motility has been recognized as critical to the pathogenic process for the human pathogen *Pseudomas aeruginosa* as well as the plant pathogen *Pseudomonas syringae,* and both of these organisms produce glycosylated flagella. *P. aeruginosa* produces a simple polar flagellum and its flagellins are classified as type a or type b based on amino acid sequence, antigenicity and molecular mass. Structural analysis of both type a and b flagellins has shown that novel glycans are attached at two sites in the central surface exposed region of each flagellin monomer (Schirm *et al.*, 2004a; Verma *et al.*, 2006). Flagellin from strain PAK (flagellin type a) is modified with a heterogeneous glycan of up to 11 monosaccharide units with rhamnose (Rha) as the linking sugar. In contrast, the glycan from PAO flagellin (type b) is simpler and appears also to have a deoxyhexose linking sugar. Distal to this monosaccharide is a unique modification of 209 Da, the structure of which has yet to be determined (Verma *et al.*, 2006). A flagellar glycosylation island (GI) has been identified in both strains lying between *flgL* and *fliC*. The GI of the PAK strain consists of 14 open-reading frames (ORFs), many of which encode products with similarities to enzymes involved in carbohydrate biosynthesis. Polymorphism in genetic content of this island has been shown among other *P. aeruginosa* strains producing type a flagellins (Arora *et al.*, 2001, 2004). In contrast, the GI of the PAO strain contains only four ORFs (i.e. PA1088, PA1089, PA1090 and PA1091), which is likely reflective of the simplicity of the glycan found on strains producing type b flagellins (Verma *et al.*, 2006). Both GIs contain a gene encoding a glycosyltransferase (PA1091, *orfN*), which has been shown to be critical for addition of glycan to the protein (Schirm *et al.*, 2004a; Verma *et al.*, 2006). Mutagenesis has demonstrated that glycosylation is not required for filament assembly or subsequent motility, but has been shown to be important in virulence. In the burned mouse model of infection, genetically defined mutants which produced non-glycosylated yet functional flagella were significantly attenuated (Arora *et al.*, 2005). Additionally, glycosylation of the flagellin protein appears to play a role in the proinflammatory action of *P. aeruginosa* flagellin (Verma *et al.*, 2005).

The flagellins of some plant pathogens including *Pseudomonas syringae* pv. *tabaci*, *Pseudomonas syringae* pv. *glycinea* and *Pseudomonas syringae* pv. *tomato* have also been found to be glycosylated (Taguchi *et al.*, 2006a,b; Takeuchi *et al.*, 2007). The pathovar classification of *P. syringae* is due to virulence towards different host plant species and flagellin has been shown to be an elicitor of a hypersensitive reaction (HR) towards non-host plants. Glycosylation of flagellin has been shown to induce this HR (Taguchi *et al.*, 2003). A tuft of polar flagella is produced by these organisms and these flagella are composed of a single FliC protein monomer. Six sites of glycosylation were identified in *P. syringae* pv. *tabaci* (Taguchi *et al.*, 2006b) and mutation of these sites demonstrated that glycosylation was essential for adhesion, swarming and virulence on host tobacco leaves but was not required for flagellar assembly (Taguchi *et al.*, 2003, 2006b, 2008). Structural analysis of the flagellin from *P. syringae* pv. *tabaci* 6605 revealed that the flagellin was glycosylated with a novel trisaccharide consisting of two rhamnosyl (Rha) residues and one modified 4-amino-4,6-dideoxyglucosyl (Qui4N) residue, 4-amino-4,6-dideoxy-3-(hydroxyl-1-oxobutyl)-2-Me-β-D-glucopyranosyl-(β-D-Qui*p*4N(3-hydroxy-1-oxobutyl)2Me-(1→3)-α-L-Rha*p*-(1→2)-α-L-Rha*p*). In contrast, although the flagellin proteins from the two strains are identical, the trisaccharide isolated from *P. syringae* pv. *glycinea* race 4 contained both L-Rha and D-Rha (Takeuchi *et al.*, 2007). It remains to be determined if this change in chirality of the Rha residues contributes to the unique elicitor activity of each flagellin with respect to non-host plants.

A *P. syringae* GI lying between *flgL* and *fliC* contains three ORFs; *orf1* and *orf2* which display significant homology to *orfN* from the *P. aeruginosa* GI and *orf3* which shares homology to a putative 3-oxoacyl- (acyl carrier protein) synthase of *Pseudomonas putida*. Deletion of these genes reduced both virulence for respective host plants and HR-inducing activity in both *P. syringae* pv. *tabaci* 6605 and *P. Syringae* pv. *glycinea* strains. Only *orf1* and *orf2* had a direct effect on flagellar glycosylation (Takeuchi *et al.*, 2003; Ishiga *et al.*, 2005; Taguchi *et al.*, 2006b), while *orf3* was shown to be required for the production of acyl-homoserine lactones, molecules that are utilized for quorum sensing (Taguchi *et al.*, 2006a).

2.3.2. Campylobacter jejuni/coli

Campylobacter spp. produce bipolar, complex flagella composed of two structural proteins, FlaA and FlaB. Motility of campylobacters is essential for colonization of the viscous intestinal mucus in the human host and the flagellin protein is the immunodominant antigen on the cell surface (Black *et al.*, 1988). Extensive genetic and structural studies have characterized the process of flagellar glycosylation in these microorganisms and have shown that it is essential for flagellar assembly, subsequent motility and virulence (Guerry *et al.*, 1996; Linton *et al.*, 2000; Thibault *et al.*, 2001; Logan *et al.*, 2002; Goon *et al.*, 2003). The flagellin structural proteins have been shown to be one of the most heavily glycosylated prokaryotic proteins (see Figure 8.1) with up to 19 sites per monomers. A number of derivatives of the novel "sialic acid-like" nonulosonate sugars, such as pseudaminic acid, i.e. 5,7-diacetamido-3,5,7,9-tetradeoxy-L-*glycero*-α-L-*manno*-nonulosonic acid (Pse5Ac7Ac) and legionaminic acid, i.e. 5,7-diacetamido-3,5,7,9-tetradeoxy-D-*glycero*-D-*galacto*-nonulosonic acid (Leg5Ac7Ac), have been shown to comprise the glycan modifications. These include the unique acetamidino derivatives 5-acetamidino-7-acetamido-3,5,7,9-tetradeoxy-D-*glycero*-D-*galacto*-nonulosonic acid (Leg5Am7Ac), 5-*E*/*Z*-*N*-(*N*-methylacetimidoyl)-7-acetamido-3,5,7,9-tetradeoxy-D-*glycero*-D-*galacto*-nonulosonic acid (Leg5AmNMe7Ac) and 5-acetamido-7-acetamidino-3,5,7,9-tetradeoxy-L-*glycero*-L-*manno*-nonulosonic acid (Pse5Ac7Am) (Thibault *et al.*, 2001; Logan *et al.*, 2002; McNally *et al.*, 2006, 2007). Genomic comparisons have revealed that the *Campylobacter* GI of approximately 50 genes is one of the most variable loci. Within this island are a number of genes that encode proteins with homology to carbohydrate biosynthetic genes, including those similar to sialic acid or *N*-acetylneuraminic acid (Neu5Ac) biosynthetic enzymes NeuB, NeuC and NeuA. Mutational analysis has revealed a role for many of these genes in motility and additional studies utilizing novel metabolomics approaches and functional studies on recombinant enzymes have defined the precise function of each in nonulosonate sugar biosynthetic pathways (see below). In addition, a unique feature of this GI is the presence of multiple copies of hypothetical genes encoding proteins that belong to the motility accessory family (Maf) of flagellin associated proteins (Karlyshev *et al.*, 2002). The role of these proteins in glycosylation is currently unknown and homologues are found in only a limited number of other bacterial species including *Wolinella, Helicobacter* and *Clostridium*. Comparative phylogenomics of *C. jejuni* isolates has revealed a cluster of six genes within this locus which are a specific marker for a "livestock" clade, although the corresponding structure of the unique glycoform made by these genes remains to be determined (Champion *et al.*, 2005). The flagellar glycans have also been shown to mediate autoagglutination of campylobacter cells (Guerry *et al.*, 2006). This process is recognized to be the preliminary step for many organisms in formation of microcolonies and biofilms and is associated with a virulence phenotype of *C. jejuni* (Misawa and Blaser, 2000; Golden and Acheson, 2002).

2.3.3. Helicobacter pylori

This bacterium produces a polar tuft of sheathed, complex flagella at one pole of the cell. Motility is essential for colonization of the mucosal surface of the human stomach prior to development of gastritis and duodenal ulcers (Eaton *et al.*, 1996). The flagellar structural proteins FlaA and FlaB are glycosylated in a similar manner to *C. jejuni* with Pse5Ac7Ac found at seven sites on FlaA and ten sites on FlaB. As with *C. jejuni,* glycosylation is essential for filament formation and subsequent motility and so the process offers potential as a therapeutic target. In contrast to *C. jejuni/coli*, no heterogeneity in glycan composition has been found on *H. pylori* flagellin which may be due to the presence of a flagellar sheath covering the filament. This observation correlates with the simpler genetic content in *H. pylori* with respect to flagellar glycan biosynthetic genes (Creuzenet *et al.*, 2000; Josenhans *et al.*, 2002; Schirm *et al.*, 2003; Merkx-Jacques *et al.*, 2004; Schoenhofen *et al.*, 2006a). *Helicobacter felis*, which is naturally found in the gastric mucosa of cats and dogs, produces bipolar bundles of sheathed flagellar filaments and was the first member of this genus shown to produce a glycosylated flagellin (Josenhans *et al.*, 1999).

2.3.4. Aeromonas *spp.*

These organisms are ubiquitous aquatic bacteria which have been shown to cause disease in fish, amphibians and reptiles; the mesophilic members of this species are significant human pathogens, causing predominantly gastrointestinal infections as well as wound infections and septicaemia. A polar, unsheathed flagellum is expressed in broth cultures (Rabaan *et al.*, 2001) and, when cells are grown on solid media, lateral flagella are produced (Kirov *et al.*, 2002). Polar flagellar glycosylation has been demonstrated in the human pathogens *Aeromonas caviae* and *Aeromonas hydrophila* (Rabaan *et al.*, 2001; Schirm *et al.*, 2005) with glycosylation being essential for filament assembly (Gryllos *et al.*, 2001). Characterization by mass spectrometry of the flagellin protein has revealed that six sites on FlaA and seven sites on FlaB carry a novel derivative of a nonulosonic acid sugar of mass 373 Da (Schirm *et al.*, 2005). The lateral flagella of *Aeromonas* spp. also appear to be glycosylated, although the structure of the glycan on these flagellar filaments is not known (Gavin *et al.*, 2002).

2.3.5. Caulobacter crescentus

C. crescentus produces a unipolar, sheathed, complex flagellum in its swarmer cell progeny, as part of a unique developmental life cycle. While the identification of a flagellar glycosylation locus of this organism has been made and shown to contain genes that affect flagellar assembly (Johnson *et al.*, 1983; Leclerc *et al.*, 1998), no structural analysis of the flagellin proteins has been described to date. However, some of the genes identified in the *C. crescentus* flagellar GI (FlmA, FlmB, FlmC, FlmD and FlmH) have orthologues in GIs of a number of bacterial species which do produce glycosylated flagellins (e.g. *C. jejuni, H. pylori* and *A. caviae*).

2.3.6. *Spirochaete periplasmic flagella*

While no detailed structural analysis of any spirochaete flagellar glycan has been completed, a number of studies have provided preliminary evidence for glycosylation of the flagellin structural proteins from *Spirochaeta aurantia* (Brahamsha and Greenberg, 1988), *Serpulina hyodysenteriae* (Li *et al.*, 1993), *Treponema pallidum* (Wyss, 1998) and *Borrelia burgdorferi* (Ge *et al.*, 1998). While reactivity of *B. burgdorferi* FlaA protein with the lectins SNA (*Sambucus nigra* agglutinin) and GNA (*Galanthus nivalis* agglutinin) indicated a novel complex glycan and sensitivity to *N*-glycosidase F suggested the glycan was attached in *N*-linkage, an extensive structural analysis of both FlaA and FlaB proteins from *B. burgdorferi* B31 has demonstrated that neither protein carries an *N*-linked glycan (Sterba *et al.*, 2008).

3. STRUCTURAL ANALYSIS OF FLAGELLAR GLYCANS

The use of mass spectrometry for characterization of flagellin glycan structures has been substantial. Typically, structural analysis begins with accurate mass measurements of the purified intact glycoprotein to determine if the measured mass of the flagellin is consistent with the predicted protein mass and to deduce the mass of the carbohydrate residues. This is then followed by determination of the nature of the carbohydrate modifications and the sites at which the modifications occur. Two approaches have been used to facilitate this work.

The first, a "bottom-up" proteomics approach, involves proteolytic digestion of the flagellin glycoprotein followed by separation of peptides by liquid chromatography and collection of peptide fractions at regular intervals for analysis by liquid chromatography-mass spectrometry (LC-MS). Capillary liquid chromatography coupled to nanoelectrospray mass spectrometry (cLC-ESMS) is generally used to obtain the sensitivity that is required for the analysis of the suspected glycopeptides. By performing tandem mass spectrometry (MS-MS) experiments on the multiply-charged ions of the glycopeptides, cleavage of the glycosidic bonds takes place and allows the subsequent detection of characteristic oxonium ions relating to the carbohydrate modifications. The putative structures of the carbohydrate modifications are then deduced through further MS-MS experiments on the oxonium ions.

A second complementary approach, relying upon optimization of collisional activation conditions for specific cleavage of the carbohydrate moieties from the intact glycoprotein, is known as "top-down" mass spectrometry (Fridriksson *et al.*, 2000). Schrim and co-workers have employed this novel approach to characterize flagellar glycans (Schrim *et al.*, 2005; McNally *et al.*, 2007). Typically, low collision energies do not cause significant collision-induced dissociation of the intact multiply-charged protein ions. However, because of the labile nature of the glycosidic bond, it is possible to gradually increase the collision energy and determine the optimal collision voltage at which the carbohydrate residues can be cleaved from the intact glycoprotein, without cleaving the protein backbone itself. The production of second-generation product ions of the intact flagellin multiply-charged ions can be readily achieved to provide structural information on the carbohydrate residues.

To identify the location of the carbohydrate modifications on the bacterial flagellin, it is necessary to perform a proteolytic digestion of the flagellin protein followed by base-catalysed hydrolysis of the tryptic peptides to yield modified amino acids of neutral mass which can then be readily identified by a corresponding mass shift in the product ion spectrum of the β-eliminated products. For example, in the case of *O*-linked flagellin glycosylation where the carbohydrate modifications occur at Ser and Thr residues, alkaline hydrolysis would yield modified amino acids of neutral mass 86 and 100, respectively.

The top-down approach offers a rapid and cost-effective alternative to the bottom-up approach for the structural analysis of flagellin glycans, as well as an opportunity for high-throughput application. Nevertheless, to determine the precise structural configuration of the carbohydrate moieties, it is necessary to use NMR spectroscopy. Early investigations of the novel glycans observed on the flagellin of

C. jejuni 81–176 (Thibault *et al.*, 2001) revealed that it was difficult to obtain sufficient quantities of purified glycopeptide for precise structural characterization of the unique flagellin glycans by NMR analysis. This is in part due to the labile nature of the glycan modifications but also the limitations in current NMR technology. As such, Soo *et al.* (2004) explored the use of metabolomics as a novel means to further the understanding of Pse5Ac7Ac biosynthesis in *C. jejuni* 81–176. They observed that intracellular pools of cytidine monophosphate- (CMP-) linked precursors had accumulated in wild-type cells of *C. jejuni* 81–176. The identification of these novel biosynthetic intermediates within cell lysates of *C. jejuni* presented an exciting opportunity to use the metabolome as a source for biosynthetic precursors of relevance to the flagellin glycosylation process and to achieve precise structural information on novel flagellin glycans by NMR spectroscopy through their corresponding CMP-linked intermediates. In order to isolate the CMP-linked intermediates from the metabolome, a method employing hydrophilic interaction liquid chromatography and mass spectrometry (HILIC-MS) with precursor ion scanning for fragment ions related to CMP (m/z 322) was developed.

These metabolomics studies (McNally *et al.*, 2006, 2007) in combination with NMR spectroscopy have yielded invaluable information on the identity of the biosynthetic CMP-activated precursors. Additionally, they have helped to determine the roles of genes involved in the biosynthesis of sialic-acid like sugars, such as those required for CMP-Pse5Ac7Am biosynthesis in *C. jejuni* 81–176. Furthermore, they offer an alternate method to the proteomics based bottom-up and top-down approaches for the structural determination of novel flagellar glycans. In contrast to purification of the glycopeptides for NMR analysis, the large-scale purification (1 μg per litre of culture) of the CMP-linked intermediates is relatively easy to achieve. The use of this approach has permitted the purification of the CMP-linked precursors, CMP-Pse5Ac7Ac, CMP-Pse5Ac7Am, CMP-Leg5Am7Ac and CMP-Leg5AmNMe7Ac from either *C. jejuni* 81–176 or *C. coli* VC167. The structures of these metabolites are presented in Figure 8.2.

4. FLAGELLAR GLYCAN BIOSYNTHETIC PATHWAYS

The most extensively studied flagellin glycosylation systems are those from the gastrointestinal pathogens, *H. pylori* and *C. jejuni/coli*, exemplified by the wealth of detailed structural information summarized in Table 8.1. Since flagellins of *Helicobacter* and *Campylobacter* spp. are glycosylated via *O*-linkage with monosaccharides resembling sialic acid, this section will focus on the biosynthesis of these sialic acid-like molecules.

Sialic acids are a family of α-keto sugars with a common nine carbon backbone and, by definition, are *N*- and *O*-acyl derivatives of 5-amino-3,5-dideoxy-D-*glycero*-β-D-*galacto*-nonulosonic acid or neuraminic acid (Neu). These sugars are typically found as the outermost moiety of oligosaccharides on vertebrate glycolipids and glycoproteins and have been shown to mediate a multitude of cell–cell and cell–molecule interactions (Vimr *et al.*, 2004). Importantly, the biosynthesis of sialic acid in prokaryotes requires two enzymes: a UDP-GlcNAc-hydrolysing 2-epimerase, NeuC, which produces *N*-acetyl-D-mannosamine, and NeuB which condenses this *N*-acetylhexoamine with the three-carbon molecule pyruvate, forming the nine-carbon sialic acid, Neu5Ac. Nucleotide activation of Neu5Ac is performed by NeuA resulting in CMP-Neu5Ac, which is the required substrate for sialyltransferases that complete the assembly of sialoglycoconjugates.

It is only more recently that the biosynthetic steps leading to bacterial sialic acid-like sugars, or 5,7-diacetamido-3,5,7,9-tetradeoxy-nonulosonate derivatives, have been described.

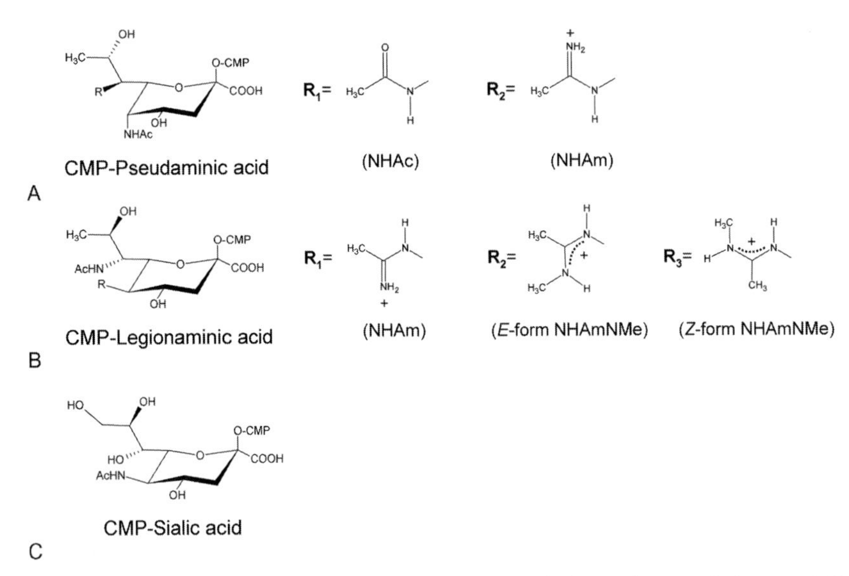

FIGURE 8.2 The structures of CMP-sialic acid-like sugars identified in the metabolome of *Campylobacter* and characterized by NMR spectroscopy. (A) CMP-pseudaminic acid or CMP-5,7-diacetamido-3,5,7,9-tetradeoxy-L-*glycero*-L-*manno*-nonulosonic acid and derivatives. (B) CMP-legionaminic acid or CMP-5,7-diacetamido-3,5,7,9-tetradeoxy-D-*glycero*-D-*galacto*-nonulosonic acid and derivatives. (C) CMP-sialic acid or CMP-5-acetamido-3,5-dideoxy-D-*glycero*-D-*galacto*-nonulosonic acid (CMP-Neu5Ac) is shown for comparison purposes only. Of note, sialic acid and legionaminic acid exhibit the same D-*glycero*-D-*galacto* absolute configuration. Known R-groups are shown for both (A) and (B), illustrating the diversity of functional groups observed for each. Confirmation of the α- or β-linkage of (A) and (B) to flagellin has yet to be determined, although preliminary evidence suggests both are α-glycosidically linked.

These sugars appear to be unique to microorganisms and may exhibit configurational differences compared with Neu5Ac (Knirel *et al.*, 2003). Pseudaminic acid, Pse5Ac7Ac, was the first bacterial sialic acid-like sugar identified and was found as a component of *P. aeruginosa* lipopolysaccharides (Knirel *et al.*, 1984). Leading up to the complete elucidation of the CMP-pseudaminic acid pathway in *Helicobacter* (Schoenhofen *et al.*, 2006a), a number of studies provided critical initial biosynthetic characterization and are summarized below. First, assuming that the biosynthesis of the bacterial sialic acid-like sugars follows a similar route to that for bacterial Neu5Ac, Pse5Ac7Ac biosynthesis would be expected to involve condensation of a HexNAc derivative with pyruvate. In fact, Chou *et al.* (2005) demonstrated this was the case using a *C. jejuni* NeuB homologue, now designated PseI, a synthetically derived form of 2,4-diacetamido-2,4,6-trideoxy-L-altropyranose, and pyruvate, to chemi-enzymatically synthesize Pse5Ac7Ac. However, it is the HexNAc derivative that largely confers the precise stereochemical and functional nature of the respective nonulosonate the biosynthesis of which was previously ill-defined.

Second, characterization of the initial two enzymes was important as there was considerable uncertainty surrounding their functions (Creuzenet *et al.*, 2000; Goon *et al.*, 2003; Creuzenet, 2004; Merkx-Jacques *et al.*, 2004; Obhi and Creuzenet, 2005). Because of its lability and ring flexibility, the product of the initial pathway enzyme, PseB, was overlooked. By accounting for product lability and flexibility,

TABLE 8.1 Flagellin glycan structures associated with prokaryotic groups

Organism	Flagellin glycan structure	Flagella type and glycan linkage	Possible relevance
Gastrointestinal pathogens			
C. coli	Pse5Ac7Ac[a] Leg5Am7Ac[b] Leg5AmNMe7Ac[c]	Complex *O*-linkage (Ser[i]/Thr[j])	Mimicry with host cell-surface sialic acids Host-pathogen interactions, including adherence and invasion Immune evasion
C. jejuni	Pse5Ac7Ac Pse5Ac7Am[d]		Filament assembly
H. pylori	Pse5Ac7Ac		
Intracellular pathogens			
L. monocytogenes	GlcNAc[e]	Simple *O*-linkage (Ser/Thr)	Mimicry with host intracellular *O*-GlcNAcylation Immune evasion Modulation of host cell physiology
Plant pathogens			
P. syringae pv. *tabaci* pv. *glycinea*	Qui4N(3-hydroxy-1-oxobutyl)2Me-Rha-Rha[f]	Simple *O*-linkage (Ser/Thr)	Host/pathogen interactions influencing host specificity Filament stabilization
Extremophiles			
H. halobium	Glc-GlcASO_4-GlcASO_4-GlcASO_4[g]	Type IV pilin-like	Protection from harsh environments, in addition to a similarly glycosylated S-layer
M. voltae	ManNAcA6Thr-GlcNAc3NAcA-GlcNAc[h]	*N*-linkage (Asn[k])	Filament assembly

[a]5,7-Diacetamido-3,5,7,9-tetradeoxy-L-*glycero*-L-*manno*-nonulosonic acid;
[b]5-Acetamidino-7-acetamido-3,5,7,9-tetradeoxy-D-*glycero*-D-*galacto*-nonulosonic acid;
[c]5-*E/Z*-*N*-(*N*-methylacetimidoyl)-7-acetamido-3,5,7,9-tetradeoxy-D-*glycero*-D-*galacto*-nonulosonic acid;
[d]5-Acetamido-7-acetamidino-3,5,7,9-tetradeoxy-L-*glycero*-L-*manno*-nonulosonic acid;
[e]2-acetamido-2-deoxy-glucopyranosyl;
[f]4-amino-4,6-dideoxy-3-(hydroxyl-1-oxobutyl)-2-Me-β-D-glucopyranosyl-α-(1→3)-L-rhamnopyranosyl-α-(1→2)-L-rhamnopyranosyl;
[g]Glucopyranosyl-(1→4)-2-*O*-sulfo-glucopyranosiduronic acid-(1→4)-2-*O*-sulfo-glucopyranosiduronic acid;
[h]2-Acetamido-2-deoxy-6-*N*-threonyl-mannopyranosiduronic acid-β-(1→4)-2,3-dideoxy-2,3-diacetamido-glucopyranosiduronic acid-β-(1→3)-2-deoxy-2-acetamido-glucopyranosyl;
[i]Serine;
[j]Threonine;
[k]Asparagine.

the PseB and PseC reaction products were then unequivocally determined (Schoenhofen *et al.*, 2006b), providing a solid foundation for defining the complete CMP-Pse5Ac7Ac pathway. Finally, as with any biosynthetic pathway elucidation, identification of the "catalytic players" or enzymes was critical. This was accomplished by a holistic approach including genomic, metabolomic and functional analyses.

Utilizing the knowledge gained from these studies, the elucidation of the CMP-Pse5Ac7Ac pathways of both *C. jejuni* and *H. pylori* was completed. This was accomplished by monitoring sequential enzymatic reactions, including their expected co-factors and co-substrates, directly with NMR spectroscopy (Schoenhofen *et al.*, 2006a). Here, the entire CMP-Pse5Ac7Ac pathway, starting from UDP-GlcNAc, was determined and proceeds as indicated in Figure 8.3. In summary, this pathway requires the actions of six biosynthetic enzymes: a C5-epimerase/C4,6-dehydratase, aminotransferase, *N*-acetyltransferase, UDP-sugar hydrolase, Pse5Ac7Ac synthase or aldolase and CMP-Pse5Ac7Ac synthetase, designated PseB, PseC, PseH, PseG, PseI and PseF, respectively. Further studies also support these findings, including a detailed comprehensive characterization of PseG function (Liu and Tanner, 2006) as well as a metabolomics study implicating all of these enzymes as necessary for the production of CMP-Pse5Ac7Ac (McNally *et al.*, 2006). Interestingly, the CMP-Pse5Ac7Ac pathway exhibits feedback regulation involving the initial pathway enzyme PseB and the final pathway product CMP-Pse5Ac7Ac (McNally *et al.*, 2008); CMP-Pse5Ac7Ac was found to be a potent competitive inhibitor of PseB [Ki(app) = 8.7 μM]. Surprisingly, the regulation of CMP-Pse5Ac7Ac biosynthesis in *Campylobacter* and *Helicobacter* more closely resembles that for CMP-Neu5Ac production in eukaryotes, which displays similar feedback regulation (Kornfeld *et al.*, 1964). This is in contrast to CMP-Neu5Ac production in *E. coli*, which is controlled by an aldolase (Vimr and Troy, 1985).

The biosynthetic pathway of the related nonulosonate, legionaminic acid (Leg5Ac7Ac) would be expected to be similar to that for Neu5Ac since they share the same absolute configuration (see Figure 8.2). As such, the Leg5Ac7Ac NeuC-like enzyme would catalyse the conversion of UDP-2,4-diacetamido-2,4,6-trideoxy-α-D-glucose (UDP-6-deoxy-GlcNAc4NAc → 6-deoxy-ManNAc4NAc) similar to the conversion performed by the hydrolysing 2-epimerase, NeuC, involved in Neu5Ac biosynthesis (UDP-GlcNAc → ManNAc) (see Figure 8.3). In addition to *C. coli*, *Legionella pneumophila* also possesses legionaminic acid and, in fact, this sugar was first identified as a component of *L. pneumophila* lipopolysaccharides (Knirel *et al.*, 2003). In efforts to reconstitute the CMP-Leg5Ac7Ac pathway from *L. pneumophila*, Glaze *et al.* (2008) described its biosynthesis using NeuC-, NeuB- and NeuA-like enzymes from *L. pneumophila* and UDP-2,4-diacetamido-2,4,6-trideoxy-α-D-glucose produced from *Campylobacter* general protein glycosylation enzymes (PglF, PglE and PglD). Although they reported modest synthesis via this route, the non-requirement for PglE in the production of CMP-Leg5Ac7Ac derivatives from *Campylobacter* (McNally *et al.*, 2007) casts doubt on the pathway proposed. Further studies will be required to address this incongruence.

5. CONCLUSIONS AND FUTURE PERSPECTIVES

A striking observation is the finding that eubacterial species possessing complex flagella (i.e. more than one flagellin subunit) glycosylate their flagellins with sialic acid-like sugars (see Table 8.1). In addition, it is only in these species that glycosylation has been found to be required for flagellar filament assembly. In contrast, eubacterial species possessing simple flagella (i.e. only one flagellin subunit) do not appear to produce sialic acid-like flagellin

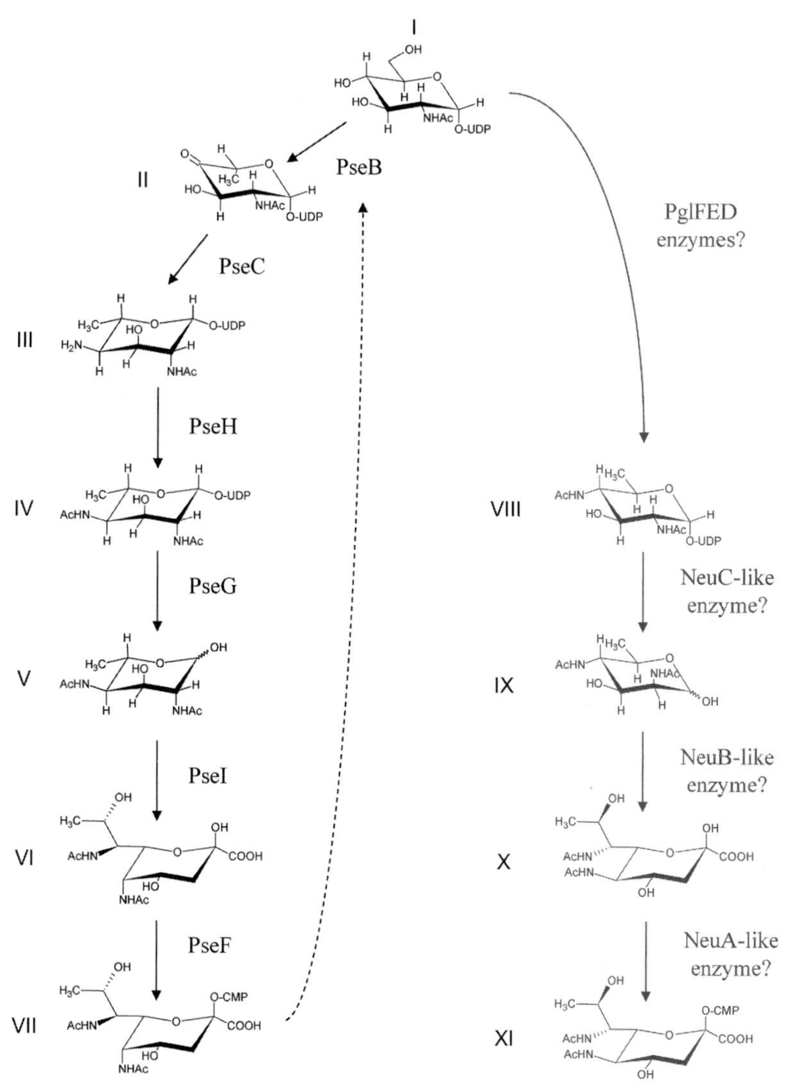

FIGURE 8.3 Biosynthesis of sialic acid-like sugars. The CMP-pseudaminic acid biosynthetic pathway in *H. pylori* and *C. jejuni* (left) versus that proposed for CMP-legionaminic acid in *C. coli* (right). The enzymes of the CMP-pseudaminic acid pathway in order are: PseB, NADP-dependent dehydratase/epimerase; PseC, pyridoxal phosphate (PLP)-dependent aminotransferase; PseH, *N*-acetyltransferase; PseG, nucleotide diphosphate (NDP)-sugar hydrolase; PseI, pseudaminic acid synthase; PseF, CMP-pseudaminic acid synthetase. The biosynthetic intermediates are (I) Uridine diphosphate (UDP)-GlcNAc; (II) UDP-2-acetamido-2,6-dideoxy-β-L-*arabino*-hexos-4-ulose; (III) UDP-4-amino-4,6-dideoxy-β-L-AltNAc; (IV) UDP-2,4-diacetamido-2,4,6-trideoxy-β-L-altropyranose; (V) 2,4-diacetamido-2,4,6-trideoxy-L-altropyranose; (VI) pseudaminic acid; (VII) CMP-pseudaminic acid. Pyranose rings are shown as their predominant chair conformation in solution determined from nuclear Overhauser effects (NOEs) and $J_{H,H}$ coupling constants. In addition, this pathway is feedback regulated (dashed line). The CMP-legionaminic acid biosynthetic pathway in *C. coli* is proposed to include a UDP-2,4-diacetamido-2,4,6-trideoxy-α-D-glucopyranose intermediate; (VIII), possibly produced via the general protein glycosylation enzymes PglF, PglE and PglD. The final putative steps would include NeuC-, NeuB- and NeuA-like enzymes catalysing the formation of 2,4-diacetamido-2,4,6-trideoxy-D-mannopyranose; (IX), legionaminic acid; (X) and CMP-legionaminic acid; (XI), respectively.

modifications and the alternative glycosylation utilized by these organisms is not required for flagellar filament assembly. Structural comparison of *C. jejuni* complex filaments, obtained by electron cryomicroscopy reconstruction, with the atomic model for *S. enterica* sv.Typhimurium simple filaments has indicated that these two filament classes exhibit distinctive topographical features (Galkin *et al.*, 2008). By cross-sectional examination, it was shown that *C. jejuni* flagellar filaments had seven protofilaments surrounding the central lumen, as opposed to eleven protofilaments in the filament from *S. enterica* sv. Typhimurium. It was suggested that the distinctive morphology of the filament types results from alternative packing of the D1 domains (see Figure 8.1). The D1 domains of *C. jejuni* and related organisms have diverged from those of *S. enterica* sv. Typhimurium, such that they no longer contain a Toll-like receptor 5 (TLR5) signalling motif (see Chapter 31) and these changes may cause alterations in the folding of the D1 domain and overall ultrastructure. Despite the differences in subunit packing, the lumen size is comparable between the two types. It is tempting to suggest that the loose packing observed for *C. jejuni* complex filaments is facilitated by the sialic acid-like sugar modifications, whereby incorporation of these glycosylations has now become obligatory for assembly. Although non-glycosylated simple flagellin of *P. syringae* still assemble into filaments, their ability to promote motility in highly viscous media is signicantly reduced (Taguchi *et al.*, 2008). Here, glycosylation is thought to stabilize the filament structure since lack of glycosylation results in excess bundling that could impede proper rotation of the filaments. Finally, although the flagellar filaments of Archaea more closely resemble type IV pili in structure and mode of assembly, the stability and assembly of these filaments are also dependent on glycosylation status (Chaban *et al.*, 2006).

Aside from its requirement for filament assembly and function, flagellin glycosylation has also been suggested to play important roles in adhesion, invasion, immune activation and evasion. For example, autoagglutination is a recognized marker of virulence for bacterial pathogens and is associated with flagellar glycosylation in *C. jejuni* (Misawa and Blaser, 2000; Golden and Acheson, 2002). Importantly, the composition of glycans present on the filament is critical to this process as evidenced by an autoagglutination-negative phenotype for filaments containing only Pse5Ac7Ac and lacking Pse5Ac7Am (Guerry *et al.*, 2006). This may result directly from carbohydrate interactions, or indirectly from their ability to introduce subtle filament ultrastructural changes as suggested above. In addition, flagellar glycans of *Pseudomonas* spp. are important virulence determinants, decreasing host lethality (LD_{50} values) in certain cases 10 000-fold and have been shown to stimulate inflammation as well as determine host specificity (Takeuchi *et al.*, 2003, 2007; Arora *et al.*, 2005; Verma *et al.*, 2005).

Furthermore, the similarity of flagellar glycans from gastrointestinal species to host sialic acids may confer upon these pathogens the ability to interact with specific eukaryotic receptors or lectins, facilitating host–pathogen interactions. Alternatively, these modifications may contribute to immune evasion by mimicking host cell surface composition, thus masking highly immunogenic epitopes. This correlates with the extensive repertoire of nonulosonate derivatives produced by *Campylobacter* spp., such as Leg5Am7Ac, Leg5AmNMe7Ac, Pse5Ac7Ac and Pse5Ac7Am, possibly enabling continued immune avoidance, whereas *H. pylori*, which covers its filament with a sheath, only produces Pse5Ac7Ac. Finally, it remains to be established the biological role of the *O*-GlcNAc modifications found on the flagellins of the intracellular pathogen *L. monocytogenes* (see Table 8.1), but it cannot be overlooked that the same modification is present on eukaryote nuclear and cytoplasmic proteins where they are vital to proper cellular physiology.

As a number of significant human pathogens produce glycosylated flagella essential for virulence, the development of inhibitors targeting flagellin glycosylation has great potential for providing novel antibacterials, especially towards gastrointestinal pathogens where motility is essential for colonization. In closing, it is clear that glycosylation is important for the proper functioning of flagella in the microorganisms listed in Table 8.1. With the continued addition of new examples of organisms possessing flagellar glycans and the structural determination of these glycans, a pattern may emerge to further our understanding of their role in biological- or virulence-associated processes, as well as in the filament asembly process (Research Focus Box).

RESEARCH FOCUS BOX

- When are eubacterial glycans attached to flagellin monomers and how?
- What is the role of glycans in flagella filament assembly, especially for complex flagella?
- For complex flagella do "sialic acid-like" sugars influence filament shape and stability?
- Do the "sialic acid-like" sugars found on flagellins of gastrointestinal pathogens play a role in specific host cell interactions?
- What is the role of the GlcNAc modification found of the flagellins of the intracellular pathogen *L. monocytogenes*?

ACKNOWLEDGEMENTS

We are grateful to T. Devecseri for help with figure preparation.

References

Arnold, F., Bedouet, L., Batina, P., et al., 1998. Biochemical and immunological analyses of the flagellin of *Clostridium tyrobutyricum* ATCC 25755. Microbiol. Immunol. 42, 23–31.

Arora, S.K., Bangera, M., Lory, S., Ramphal, R., 2001. A genomic island in *Pseudomonas aeruginosa* carries the determinants of flagellin glycosylation. Proc. Natl. Acad. Sci. USA 98, 9342–9347.

Arora, S.K., Wolfgang, M.C., Lory, S., Ramphal, R., 2004. Sequence polymorphism in the glycosylation island and flagellins of *Pseudomonas aeruginosa*. J. Bacteriol. 186, 2115–2122.

Arora, S.K., Neely, A.N., Blair, B., Lory, S., Ramphal, R., 2005. Role of motility and flagellin glycosylation in the pathogenesis of *Pseudomonas aeruginosa* burn wound infections. Infect. Immun. 73, 4395–4398.

Bardy, S.L., Ng, S.Y., Jarrell, K.F., 2004. Recent advances in the structure and assembly of the archaeal flagellum. J. Mol. Microbiol. Biotechnol. 7, 41–51.

Bedouet, L., Arnold, F., Robreau, G., Batina, P., Talbot, F., Binet, A., 1998. Evidence for an heterogeneous glycosylation of the *Clostridium tyrobutyricum* ATCC 25755 flagellin. Microbios 94, 183–192.

Black, R.E., Levine, M.M., Clements, M.L., Hughes, T.P., Blaser, M.J., 1988. Experimental *Campylobacter jejuni* infection in humans. J. Infect. Dis. 157, 472–479.

Brahamsha, B., Greenberg, E.P., 1988. Biochemical and cytological analysis of the complex periplasmic flagella from *Spirochaeta aurantia*. J. Bacteriol. 170, 4023–4032.

Chaban, B., Voisin, S., Kelly, J., Logan, S.M., Jarrell, K.F., 2006. Identification of genes involved in the biosynthesis and attachment of *Methanococcus voltae N*-linked glycans: insight into *N*-linked glycosylation pathways in Archaea. Mol. Microbiol. 61, 259–268.

Champion, O.L., Gaunt, M.W., Gundogdu, O., et al., 2005. Comparative phylogenomics of the food-borne pathogen *Campylobacter jejuni* reveals genetic markers predictive of infection source. Proc. Natl. Acad. Sci. USA 102, 16043–16048.

Charon, N.W., Goldstein, S.F., 2002. Genetics of motility and chemotaxis of a fascinating group of bacteria: the spirochetes. Annu. Rev. Genet. 36, 47–73.

Chou, W.K., Dick, S., Wakarchuk, W.W., Tanner, M.E., 2005. Identification and characterization of NeuB3 from *Campylobacter jejuni* as a pseudaminic acid synthase. J. Biol. Chem. 43, 35922–35928.

Creuzenet, C., 2004. Characterization of CJ1293, a new UDP-GlcNAc C6 dehydratase from *Campylobacter jejuni*. FEBS Lett. 559, 136–140.

Creuzenet, C., Schur, M.J., Li, J., Wakarchuk, W.W., Lam, J.S., 2000. FlaA1, a new bifunctional UDP-GlcNAc C6 dehydratase/C4 reductase from *Helicobacter pylori*. J. Biol. Chem. 275, 34873–34880.

Eaton, K.A., Suerbaum, S., Josenhans, C., Krakowka, S., 1996. Colonization of gnotobiotic piglets by *Helicobacter pylori* deficient in two flagellin genes. Infect. Immun. 64, 2445–2448.

Emerson, S.U., Tokuyasu, K., Simon, M.I., 1970. Bacterial flagella: polarity of elongation. Science 169, 190–192.

Fridriksson, E.K., Beavil, A., Holowka, D., Gould, H.J., Baird, B., McLafferty, F.W., 2000. Heterogeneous glycosylation of immunoglobulin E constructs characterized by top-down high-resolution 2-D mass spectrometry. Biochemistry 39, 3369–3376.

Galkin, V.E., Yu, X., Bielnicki, J., et al., 2008. Divergence of quaternary structures among bacterial flagellar filaments. Science 320, 382–385.

Gavin, R., Rabaan, A.A., Merino, S., Tomas, J.M., Gryllos, I., Shaw, J.G., 2002. Lateral flagella of *Aeromonas* species are essential for epithelial cell adherence and biofilm formation. Mol. Microbiol. 43, 383–397.

Ge, Y., Li, C., Corum, L., Slaughter, C.A., Charon, N.W., 1998. Structure and expression of the FlaA periplasmic flagellar protein of *Borrelia burgdorferi*. J. Bacteriol. 180, 2418–2425.

Glaze, P.A., Watson, D.C., Young, N.M., Tanner, M.E., 2008. Biosynthesis of CMP-*N,N*′-diacetyllegionaminic acid from UDP-*N,N*′-diacetylbacillosamine in *Legionella pneumophila*. Biochemistry 47, 3272–3282.

Golden, N.J., Acheson, D.W., 2002. Identification of motility and autoagglutination *Campylobacter jejuni* mutants by random transposon mutagenesis. Infect. Immun. 70, 1761–1771.

Goon, S., Kelly, J.F., Logan, S.M., Ewing, C.P., Guerry, P., 2003. Pseudaminic acid, the major modification on *Campylobacter* flagellin, is synthesized via the *Cj1293* gene. Mol. Microbiol. 50, 659–671.

Gryllos, I., Shaw, J.G., Gavin, R., Merino, S., Tomas, J.M., 2001. Role of *flm* locus in mesophilic *Aeromonas* species adherence. Infect. Immun. 69, 65–74.

Guerry, P., Doig, P., Alm, R.A., Burr, D.H., Kinsella, N., Trust, T.J., 1996. Identification and characterization of genes required for post-translational modification of *Campylobacter coli* VC167 flagellin. Mol. Microbiol. 19, 369–378.

Guerry, P., Ewing, C.P., Schirm, M., et al., 2006. Changes in flagellin glycosylation affect *Campylobacter* autoagglutination and virulence. Mol. Microbiol. 60, 299–311.

Iino, T., 1969. Polarity of flagellar growth in salmonella. J. Gen. Microbiol. 56, 227–239.

Ishiga, Y., Takeuchi, K., Taguchi, F., et al., 2005. Defense responses of *Arabidopsis thaliana* inoculated with *Pseudomonas syringae* pv. *tabaci* wild type and defective mutants for flagellin (*fliC*) and flagellin glycosylation (*orf1*). J. Gen. Plant Pathol. 71, 302–307.

Johnson, R.C., Ferber, D.M., Ely, B., 1983. Synthesis and assembly of flagellar components by *Caulobacter crescentus* motility mutants. J. Bacteriol. 154, 1137–1144.

Josenhans, C., Ferrero, R.L., Labigne, A., Suerbaum, S., 1999. Cloning and allelic exchange mutagenesis of two flagellin genes of *Helicobacter felis*. Mol. Microbiol. 33, 350–362.

Josenhans, C., Vossebein, L., Friedrich, S., Suerbaum, S., 2002. The neuA/flmD gene cluster of *Helicobacter pylori* is involved in flagellar biosynthesis and flagellin glycosylation. FEMS Microbiol. Lett. 210, 165–172.

Karlyshev, A.V., Linton, D., Gregson, N.A., Wren, B.W., 2002. A novel paralogous gene family involved in phase-variable flagella-mediated motility in *Campylobacter jejuni*. Microbiology 148, 473–480.

Kirov, S.M., Tassell, B.C., Semmler, A.B., O'Donovan, L.A., Rabaan, A.A., Shaw, J.G., 2002. Lateral flagella and swarming motility in *Aeromonas* species. J. Bacteriol. 184, 547–555.

Knirel, Y.A., Vinogradov, E.V., L'vov, V.L., et al., 1984. Sialic acids of a new type from the lipopolysaccharides of *Pseudomonas aeruginosa* and *Shigella boydii*. Carbohydr. Res. 133, C5–C8.

Knirel, Y.A., Shashkov, A.S., Tsvetkov, Y.E., Jansson, P.E., Zahringer, U., 2003. 5,7-diamino-3,5,7,9-tetradeoxynon-2-ulosonic acids in bacterial glycopolymers: chemistry and biochemistry. Adv. Carbohydr. Chem. Biochem. 58, 371–417.

Kornfeld, S., Kornfeld, R., Neufeld, E.F., O'Brien, P.J., 1964. The feedback control of sugar nucleotide biosynthesis in liver. Proc. Nat. Acad. Sci. USA 52, 371–379.

Kowarik, M., Young, N.M., Numao, S., et al., 2006. Definition of the bacterial *N*-glycosylation site consensus sequence. EMBO J. 25, 1957–1966.

Leclerc, G., Wang, S.P., Ely, B., 1998. A new class of *Caulobacter crescentus* flagellar genes. J. Bacteriol. 180, 5010–5019.

Lemon, K.P., Higgins, D.E., Kolter, R., 2007. Flagellar motility is critical for *Listeria monocytogenes* biofilm formation. J. Bacteriol. 189, 4418–4424.

Li, Z., Dumas, F., Dubreuil, D., Jacques, M., 1993. A species-specific periplasmic flagellar protein of *Serpulina (Treponema) hyodysenteriae*. J. Bacteriol. 175, 8000–8007.

Linton, D., Karlyshev, A.V., Hitchen, P.G., et al., 2000. Multiple *N*-acetyl neuraminic acid synthetase (*neuB*) genes in *Campylobacter jejuni*: identification and characterization of the gene involved in sialylation of lipo-oligosaccharide. Mol. Microbiol. 35, 1120–1134.

Liu, F., Tanner, M.E., 2006. PseG of pseudaminic acid biosynthesis: a UDP-sugar hydrolase as a masked glycosyltransferase. J. Biol. Chem. 281, 20902–20909.

Logan, S.M., 2006. Flagellar glycosylation – a new component of the motility repertoire? Microbiology 152, 1249–1262.

Logan, S.M., Kelly, J.F., Thibault, P., Ewing, C.P., Guerry, P., 2002. Structural heterogeneity of carbohydrate modifications affects serospecificity of *Campylobacter* flagellins. Mol. Microbiol. 46, 587–597.

Lyristis, M., Boynton, Z.L., Petersen, D., Kan, Z., Bennett, G.N., Rudolph, F.B., 2000. Cloning, sequencing and characterisation of the gene encoding *flaC* and the post-translational modification of flagellin from *Clostridium acetobutylicum* ATCC824. Anaerobe 6, 69–79.

Macnab, R.M., 2004. Type III flagellar protein export and flagellar assembly. Biochim. Biophys. Acta 1694, 207–217.

McNally, D.J., Hui, J.P., Aubry, A.J., et al., 2006. Functional characterization of the flagellar glycosylation locus in *Campylobacter jejuni* 81–176 using a focused metabolomics approach. J. Biol. Chem. 281, 18489–18498.

McNally, D.J., Aubry, A.J., Hui, J.P., et al., 2007. Targeted metabolomics analysis of *Campylobacter coli* VC167 reveals legionaminic acid derivatives as novel flagellar glycans. J. Biol. Chem. 282, 14463–14475.

McNally, D.J., Schoenhofen, I.C., Houliston, R.S., et al., 2008. CMP-pseudaminic acid is a natural potent inhibitor of PseB, the first enzyme of the pseudaminic acid pathway in *Campylobacter jejuni* and *Helicobacter pylori*. Chem. Med. Chem. 3, 55–59.

Merkx-Jacques, A., Obhi, R.K., Bethune, G., Creuzenet, C., 2004. The *Helicobacter pylori flaA1* and *wbpB* genes control lipopolysaccharide and flagellum synthesis and function. J. Bacteriol. 186, 2253–2265.

Misawa, N., Blaser, M.J., 2000. Detection and characterization of autoagglutination activity by *Campylobacter jejuni*. Infect. Immun. 68, 6168–6175.

Obhi, R.K., Creuzenet, C., 2005. Biochemical characterization of the *Campylobacter jejuni* Cj1294, a novel UDP-4-keto-6-deoxy-GlcNAc aminotransferase that generates UDP-4-amino-4,6-dideoxy-GalNAc. J. Biol. Chem. 280, 20902–20908.

Peel, M., Donachie, W., Shaw, A., 1988. Temperature-dependent expression of flagella of *Listeria monocytogenes* studied by electron microscopy, SDS-PAGE and western blotting. J. Gen. Microbiol. 134, 2171–2178.

Rabaan, A.A., Gryllos, I., Tomas, J.M., Shaw, J.G., 2001. Motility and the polar flagellum are required for *Aeromonas caviae* adherence to HEp-2 cells. Infect. Immun. 69, 4257–4267.

Samatey, F.A., Imada, K., Vonderviszt, F., Shirakihara, Y., Namba, K., 2000. Crystallization of the F41 fragment of flagellin and data collection from extremely thin crystals. J. Struct. Biol. 132, 106–111.

Schirm, M., Soo, E.C., Aubry, A.J., Austin, J., Thibault, P., Logan, S.M., 2003. Structural, genetic and functional characterization of the flagellin glycosylation process in *Helicobacter pylori*. Mol. Microbiol. 48, 1579–1592.

Schirm, M., Arora, S.K., Verma, A., et al., 2004a. Structural and genetic characterization of glycosylation of type a flagellin in *Pseudomonas aeruginosa*. J. Bacteriol. 186, 2523–2531.

Schirm, M., Kalmokoff, M., Aubry, A., Thibault, P., Sandoz, M., Logan, S.M., 2004b. Flagellin from *Listeria monocytogenes* is glycosylated with beta-*O*-linked *N*-acetylglucosamine. J. Bacteriol. 186, 6721–6727.

Schirm, M., Schoenhofen, I.C., Logan, S.M., Waldron, K.C., Thibault, P., 2005. Identification of unusual bacterial glycosylation by tandem mass spectrometry analyses of intact proteins. Anal. Chem. 77, 7774–7782.

Schoenhofen, I.C., McNally, D.J., Brisson, J.R., Logan, S.M., 2006a. Elucidation of the CMP-pseudaminic acid pathway in *Helicobacter pylori*: synthesis from UDP-*N*-acetylglucosamine by a single enzymatic reaction. Glycobiology 16, 8C–14C.

Schoenhofen, I.C., McNally, D.J., Vinogradov, E., et al., 2006b. Functional characterisation of dehydratase/aminotransferase pairs from *Helicobacter* and *Campylobacter*: enzymes distinguishing the pseudaminic acid and bacillosamine biosynthetic pathways. J. Biol. Chem. 281, 723–732.

Shams-Eldin, H., Chaban, B., Niehus, S., Schwarz, R.T., Jarrell, K.F., 2008. Identification of the archaeal alg7 gene homolog (encoding *N*-acetylglucosamine-1-phosphate transferase) of the *N*-linked glycosylation system by cross-domain complementation in *Saccharomyces cerevisiae*. J. Bacteriol. 190, 2217–2220.

Shen, A., Kamp, H.D., Grundling, A., Higgins, D.E., 2006. A bifunctional *O*-GlcNAc transferase governs flagellar motility through anti-repression. Genes Dev. 20, 3283–3295.

Soo, E.C., Aubry, A.J., Logan, S.M., et al., 2004. Selective detection and identification of sugar nucleotides by CE-electrospray-MS and its application to bacterial metabolomics. Anal. Chem. 76, 619–626.

Sterba, J., Vancova, M., Rudenko, N., et al., 2008. Flagellin and outer surface proteins from *Borrelia burgdorferi* are not glycosylated. J. Bacteriol. 190, 2619–2623.

Taguchi, F., Shimizu, R., Nakajima, R., Toyoda, K., Shiraishi, T., Ichinose, Y., 2003. Differential effects of flagellins from *Pseudomonas syringea* pvs. *tabaci*, *tomato* and *glycinea* on plant defence response. Plant Physiol. Biochem. 41, 165–174.

Taguchi, F., Ogawa, Y., Takeuchi, K., et al., 2006a. A homologue of the 3-oxoacyl-(acyl carrier protein) synthase III gene located in the glycosylation island of *Pseudomonas syringae* pv. *tabaci* regulates virulence factors via *N*-acyl homoserine lactone and fatty acid synthesis. J. Bacteriol. 188, 8376–8384.

Taguchi, F., Takeuchi, K., Katoh, E., et al., 2006b. Identification of glycosylation genes and glycosylated amino acids of flagellin in *Pseudomonas syringae* pv. *tabaci*. Cell Microbiol. 8, 923–938.

Taguchi, F., Shibata, S., Suzuki, T., et al., 2008. Effects of glycosylation on swimming ability and flagellar polymorphic transformation in *Pseudomonas syringae* pv. *tabaci* 6605. J. Bacteriol. 190, 764–768.

Takeuchi, K., Taguchi, F., Inagaki, Y., Toyoda, K., Shiraishi, T., Ichinose, Y., 2003. Flagellin glycosylation island in *Pseudomonas syringea* pv. *glycinea* and its role in host specificity. J. Bacteriol. 185, 6658–6665.

Takeuchi, K., Ono, H., Yoshida, M., et al., 2007. Flagellin glycans from two pathovars of *Pseudomonas syringae* contain rhamnose in D and L configurations in different ratios and modified 4-amino-4,6-dideoxyglucose. J. Bacteriol. 189, 6945–6956.

Thibault, P., Logan, S.M., Kelly, J.F., et al., 2001. Identification of the carbohydrate moieties and glycosylation motifs in *Campylobacter jejuni* flagellin. J. Biol. Chem. 276, 34862–34870.

Thomas, N.A., Bardy, S.L., Jarrell, K.F., 2001. The archaeal flagellum: a different kind of prokaryotic motility structure. FEMS Microbiol. Rev. 25, 147–174.

Trachtenberg, S., Cohen-Krausz, S., 2006. The archaeabacterial flagellar filament: a bacterial propeller with a pilus-like structure. J. Mol. Microbiol. Biotechnol. 11, 208–220.

Verma, A., Arora, S.K., Kuravi, S.K., Ramphal, R., 2005. Roles of specific amino acids in the N terminus of *Pseudomonas aeruginosa* flagellin and of flagellin glycosylation in the innate immune response. Infect. Immun. 73, 8237–8246.

Verma, A., Schirm, M., Arora, S.K., Thibault, P., Logan, S. M., Ramphal, R., 2006. Glycosylation of b-Type flagellin of *Pseudomonas aeruginosa*: structural and genetic basis. J. Bacteriol. 188, 4395–4403.

Vimr, E.R., Troy, F.A., 1985. Regulation of sialic acid metabolism in *Escherichia coli*: role of *N*-acylneuraminate pyruvate-lyase. J. Bacteriol. 164, 854–860.

Vimr, E.R., Kalivoda, K.A., Deszo, E.L., Steenbergen, S.M., 2004. Diversity of microbial sialic acid metabolism. Microbiol. Mol. Biol. Rev. 68, 132–153.

Voisin, S., Houliston, R.S., Kelly, J., et al., 2005. Identification and characterization of the unique *N*-linked glycan common to the flagellins and S-layer glycoprotein of *Methanococcus voltae*. J. Biol. Chem. 280, 16586–16593.

Wieland, F., Paul, G., Sumper, M., 1985. Halobacterial flagellins are sulfated glycoproteins. J. Biol. Chem. 260, 15180–15185.

Wyss, C., 1998. Flagellins, but not endoflagellar sheath proteins of *Treponema pallidum* and of pathogen-related oral spirochetes are glycosylated. Infect. Immun. 66, 5751–5754.

Yonekura, K., Maki-Yonekura, S., Namba, K., 2003. Complete atomic model of the bacterial flagellar filament by electron cryomicroscopy. Nature 424, 643–650.

Yonekura, K., Maki-Yonekura, S., Namba, K., 2005. Building the atomic model for the bacterial flagellar filament by electron cryomicroscopy and image analysis. Structure 13, 407–412.

CHAPTER

9

Glycosylated components of the mycobacterial cell wall: structure and function

Delphi Chatterjee and Patrick J. Brennan

SUMMARY

Pathogenic mycobacteria, such as *Mycobacterium tuberculosis* and *Mycobacterium leprae*, the causative agents of tuberculosis and leprosy, are responsible for considerable morbidity and mortality worldwide. The distinctive cell wall (lipo)glycans are centrepiece to the underlying pathology. In addition, a hallmark of chronic mycobacterial infections is the tendency of these pathogens to maintain long-term, asymptomatic infections. The capability of mycobacteria to survive for extended periods in the macrophage has been attributed to components of the cell wall. This structure provides a permeability barrier, shielding the bacterium from degradative forces as well as modulating host immune responses to its own needs and presenting special problems for disease chemotherapy. The cell envelope possesses two major structural components: the plasma membrane and the wall proper. The plasma membrane is characterized by the presence of the phosphatidylinositol mannosides, lipomannan and lipoarabinomannan but, in other respects, appears typical of bacterial membranes. The cell wall however, is highly distinctive characterized by a "core" consisting of covalently linked mycolic acids, the unusual hetero-polysaccharide D-arabinogalactan and peptidoglycan, apparently common to all members of the genus *Mycobacterium* and related genera, in one form or another. The physical arrangement of the mycolic acids in conjunction with a variety of soluble trehalose-, phthiocerol- and peptide-containing glycolipids, varying among species and subspecies, generates a distinctive outer membrane, responsible for many of the physiological and pathogenic features of *Mycobacterium*.

Keywords: Mycobacterium spp.; Cell wall; Glycolipids; Glycans; Structure; Functions; Lipoarabinomannan; Lipomannan (LM); Phosphatidylinositol-mannosides; *N*-glycolyl-muramic acid (MurNGlyc); Arabinogalactan; Trehalose

1. CHARACTERISTIC FEATURES OF *MYCOBACTERIUM* SPP.

The mycobacteria are a diverse prokaryotic genus containing over 100 members. The majority of these species are non-pathogenic environmental bacteria related to the soil bacteria *Streptomyces* and *Actinomycetes*. Among these few species are extremely successful pathogens. These include *Mycobacterium tuberculosis*, *Mycobacterium leprae* and *Mycobacterium ulcerans*, the respective aetiological agents of tuberculosis, leprosy and Buruli (Bairnsdale) ulcer. Their key to success perhaps lies with their ability to establish residence and proliferate inside the host macrophages despite the anti-mycobacterial properties of these cells. *M. tuberculosis* is a major human pathogen causing significant human suffering worldwide. As the leading cause of death from a single bacterial infection, it is imperative to understand the biology, physiology and pathogenicity of the organism in order to improve our chance of eradicating the disease. Emergence of new forms of tuberculosis, such as multidrug resistant tuberculosis and extensively-drug resistant tuberculosis, has led to consolidated research on the genome and chemical phenotype of *M. tuberculosis* towards new drug targets, vaccines and the molecular basis of pathogenesis.

M. tuberculosis contains features which most likely contribute to its pathogenicity and potential to establish latent infections. Among these, the lipid and carbohydrate-dominated cell wall plays a critical part in interaction with the host, being the first line of defence against immune attack. In addition, cell wall impermeability contributes to the intrinsic resistance of mycobacteria to many antibiotics.

M. leprae, a non-culturable organism is the causative agent of leprosy (Hansen's disease), a highly variable disease with a spectrum of clinical manifestations that ranges between two polar forms. Leprosy primarily affects superficial tissues, the skin and peripheral nerves, responsible for a peripheral neuropathy with the potential for visible disabilities and the stigma long-associated with leprosy.

There are many different classes of non-tuberculous mycobacteria widespread in the environment, generally not transmitted from person to person, but responsible for serious opportunistic infections. The usual sources of infection are water, soil and aerosols developed from these sources. The most prominent of these opportunistic pathogens are members of the *Mycobacterium avium* complex (MAC), including *M. avium*, *Mycobacterium intracellulare* and *Mycobacterium scrofulaceum* responsible for the majority of bacterial infections in advanced human immunodeficiency virus- (HIV-) acquired immunodeficiency syndrome (AIDS).

Mycobacteria also causes devastating diseases in livestock, with *Mycobacterium bovis* and *Mycobacterium paratuberculosis* causing bovine tuberculosis in cattle and Johne's disease in sheep, cattle, goats and deer respectively. Alongside, several of the mycobacterium members of the genus are non-pathogenic, soil dwelling saprophytes, such as *Mycobacterium smegmatis* and *Mycobacterium chelonae*.

The purpose of this chapter is to familiarize readers with various glycan-containing entities that dominate the mycobacterial cell wall and are implicated in disease processes.

2. THE MYCOBACTERIAL ENVELOPE

The mycobacterial cell envelope differs significantly from that of the Gram-positive and Gram-negative bacteria for its low permeability and hence resistance to common antibiotics and chemotherapeutic agents. The envelope acts as a permeability barrier to toxic molecules (Brennan and Nikaido, 1995), allows specific interaction with the host during infection and is responsible for the evasion of the host immune responses and pathogenicity (Chatterjee and Khoo, 1998; Briken *et al.*, 2004). Although ostensibly classified as Gram-positive bacteria, the mycobacterial cell wall shares similarities with Gram-negative bacteria. At the electron microscopy level, the mycobacterial cell wall appears to have, beyond the plasma membrane, at least three distinct layers of peptidoglycan (PG), arabinogalactan and an outer, asymmetric lipid bilayer composed of mycolic acids and multiple classes of phospholipids and glycolipids (Brennan and Nikaido, 1995) (see Chapter 1). Two recent publications (Hoffmann *et al.*, 2008; Zuber *et al.*, 2008) have provided a new perspective on the mycobacterial "outer membrane". Cryo-electron microscopy of vitreous sections has shown a distinct lipid bilayer structure analogous to that of Gram-negative bacteria, but dominated by the presence of mycolic acids.

The PG in covalent linkage to the arabinogalactan, in turn covalently attached to the mycolic acids that dominate the outer membrane, represent the mycobacterial cell wall skeleton (Brennan and Nikaido, 1995; Daffe and Draper, 1998; Brennan and Crick, 2007). Associated with the cell envelope, either the outer or inner membranes, are other important polymers including the lipoarabinomannan (LAM), lipomannan (LM), the phosphatidylinositol-mannosides (PIMs), the trehalose-, and phthiocerol-containing glycolipids (Daffe and Draper, 1998; Brennan and Crick, 2007) and species-specific surface glycolipids, e.g. the trehalose-containing lipo-oligosaccharides and the glycopeptidolipids (Aspinall *et al.*, 1995; Chatterjee and Khoo, 2001). In non-tuberculosis mycobacteria, such as members of the MAC, these latter glycolipids are the major mycobacterial antigens providing the basis for serovariation.

3. THE MYCOBACTERIAL CELL WALL SKELETON: THE MYCOLYLARABINOGALACTAN-PG COMPLEX

Experimentally, the mycolylarabinogalactan-PG (mAGP) complex is the cell wall residue remaining after all other components have been extracted. It is a large polymer consisting of covalently attached PG, the arabinogalactan hetero-polysaccharide and mycolic acids (Figure 9.1) and is regarded as crucial for the viability of the organism and provides many essential enzyme targets for drug development (Brennan and Crick, 2007).

3.1. Peptidoglycan

Mycobacteria synthesize a layer of PG that is deposited immediately above the plasma membrane. The mycobacterial PG consists of chains of alternating β-(1→4)-linked *N*-acetyl-D-glucosamine (GlcNAc) and *N*-glycolyl-muramic acid (MurNGlyc) residues which are cross-linked with a tetrapeptide moiety of L-alanyl-D-isoglutamyl-*meso*-diaminopimelyl-D-alanine (L-Ala-D-Glu-A_2pM-D-Ala). This structure is different to PG found in Gram-negative bacteria as the muramic acid is *N*-glycolylated, rather than acylated and unusual cross-links occur between two chains of mycobacterial PG (Adam *et al.*, 1969). The overall degree of cross-linking in *Mycobacterium* spp. is 70–80% (Matsuhashi, 1966) compared to 30–50% in *E. coli* (Vollmer and Höltje, 2004). About two-thirds of the cross-links are between the carboxyl group of a terminal D-Ala and the amino group of the D-centre of the A_2pM residue of another peptide. Approximately one-third of the peptide cross-bridges occur between the carboxyl group of the L-centre of one *meso*-diaminopimelic acid (DAP) residue and the amino group of the D-centre of another DAP to form an L-D cross-link. In *M. leprae*, the L-Ala is replaced by glycine (Draper *et al.*, 1987) and the muramic acid residues are exclusively *N*-acetylated (Mahapatra *et al.*, 2008). The non-cross-linked peptide side chains of *M. leprae* consist of tetra- and tripeptides, some of which contain additional glycine residues. The carbon C-6 of some of the muramic acid residues form phosphodiester bonds to C-1 of α-D-GlcNAc which in turn is (1→3)-linked to α-L-rhamnopyranose (Rha*p*) providing the "linker unit" between the galactan of arabinogalactan (AG) and PG; theoretically, this is a perfect drug target, in that inhibition of any aspect of its synthesis should disrupt mAGP integrity with untoward consequences

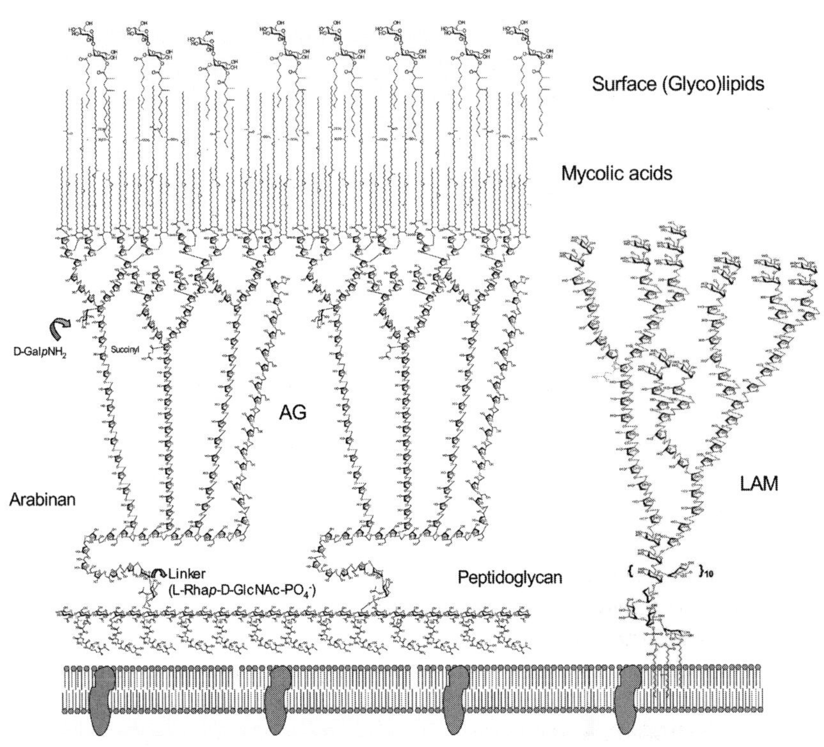

FIGURE 9.1 Structure of the mycolylarabinogalactan-peptidoglycan (mAGP) complex of the mycobacterial cell wall. The arabinogalactan (AG) contains 32–35 Gal residues in an unbranched, alternating α-(1→5) and α-(1→6) linkage. Three arabinan chains, each with 22 residues, are linked to the galactan backbone. Compare to the lipoarabinomannan (LAM). Abbreviations: L-Rha*p*, L-rhamnopyranose; D-GlcNAc, *N*-acetyl-D-glucosamine; D-GalpNH$_2$, 2-amino-2-deoxy-D-galactopyranose.

of the effects of osmotic pressure on the plasma membrane.

3.2. Arabinogalactan

The mAGP complex is dominated by the AG hetero-polysaccharide, constituting about 35% of the cell wall mass. This complex was initially isolated by base solubilization of insoluble cell wall fraction (Misaki *et al.*, 1974) and shown to contain D-arabinofuranosyl (Ara*f*) and D-galactofuranosyl (Gal*f*) residues. The present view of the structure of mycobacterial AG is illustrated in Figure 9.1. The polysaccharide is composed of three arabinan chains attached to the homogalactan core (consisting of 32 D-Gal*f* units) of linear alternating 5- and 6-linked β-D-Gal*f* residues. The homoarabinan chains are composed of linear α-D-Ara*f* residues with branching produced by 3,5-linked α-D-Ara*f* residues units substituted at both positions by α-D-Ara*f* residues. Characterization of larger per-*O*-alkylated oligosaccharides by fast-atom bombardment mass spectrometry (Besra *et al.*, 1995) demonstrated that the arabinan chains are attached to the homogalactan core in a region near the reducing end of the molecule. Galactosamine (2-amino-2-deoxygalactose; GalN) has been reported as a component of mycobacterial envelopes and, interestingly, was found only in the walls of slow-growing mycobacteria, but not in the rapid-growing species such as that of *M. smegmatis*. The GalN substituent is attached to an Ara*f* unit of AG and occurs as the free amine rather than the usual *N*-acetyl compound (Draper *et al.*, 1997). In recent studies (Lee *et al.*, 2006; Bhamidi *et al.*, 2008), an endogenous arabinanase secreted by *M. smegmatis* was utilized to release large oligoarafuranosides to solubilize the arabinan region of the AG. Using matrix-assisted laser desorption ionization time-of-flight (MALDI-TOF) mass spectrometry (MS) with time-of-flight MS/MS and nuclear magnetic resonance (NMR) spectroscopy, it was demonstrated that succinyl esters are present on 2-position of the inner-branched 1,3,5-α-D-Ara*f* residues (see Figure 9.1). In addition, an inner arabinan segment of 14 linear α-(1→5)-linked Ara*f* residues has been identified. These and earlier results now allow the presentation of a model of the entire primary structure of the mycobacterial mycolyl arabinogalactan highlighted by three arabinan chains of 31 residues each (Lee *et al.*, 2006; Bhamidi *et al.*, 2008). The mycolic acids are located in clusters of four on the terminal [β-D-Ara*f*-(1→2)-α-D-Ara*f*]$_2$-3,5-α-D-Ara*f*-(1→5)-α-D-Ara*f* hexa-arabinofuranosides, however, only about two-thirds of these are mycolylated (McNeil *et al.*, 1991).

4. THE SOLUBLE CROSS-SPECIES GLYCOCONJUGATES OF THE MYCOBACTERIAL CELL WALL

Whereas the mAGP complex is regarded as essential for bacterial survival, and hence various aspects of its synthesis are prime targets for drugs (Brennan and Crick, 2007), the soluble entities of the cell wall are regarded as the effectors of the immune response and disease pathogenesis in such as tuberculosis.

4.1. Phosphatidylinositol (PI)-based LAM, LM and PIMs

4.1.1. Lipoarabinomannan

Indications of the presence of a polysaccharide predominantly containing D-mannopyranose (Man*p*) and D-Ara*f* residues existing in both acylated and non-acylated forms were provided almost 70 years ago (Heidelberger, 1939). These polysaccharides were later shown to be ubiquitous in all *Mycobacterium* spp. and contained an α-(1→6)-D-Man*p* backbone to which was attached an immunodominant α-(1→5)-linked D-Ara*f*-containing arabinan (Azuma *et al.*, 1968; Misaki *et al.*, 1977). Our current understanding of the cell wall-associated lipoglycans, LAM and LM stems from Hunter, Brennan and Chatterjee (Hunter *et al.*, 1986; Hunter and Brennan, 1990; Chatterjee *et al.*, 1992a) who have described the isolation of native acylated forms of LAM/LM devoid of proteins and neutral polysaccharides and first reported on the presence of phosphatidylinositol in the lipoglycans, suggesting that LAM/LM are multiglycosylated extensions of the PIMs (Figure 9.2). Using octyl-Sepharose, LAM/LM have been shown to bind to the hydrophobic matrix through their fatty acyl chains in low concentrations of propanol and salt. Additionally, LAM/LM have been shown to bind to the hydrophobic matrix of octyl-Sepharose through their fatty acyl chains in low concentrations of propanol and salt (Leopold and Fischer, 1993), a property exploited to separate LAM/LM from non-acylated forms or contaminating polysaccharides (Khoo *et al.*, 1996), but allowing separation based on size differences in the glycan moieties and by the number of fatty acids (Leopold and Fischer, 1993). Moreover, arabinan chains of LAMs from all *M. tuberculosis* strains that were examined, as well as the attenuated *M. bovis* BCG vaccine strain and *M. leprae*, the aetiological agent of leprosy, are Man*p*-capped to varying degree (40–70%). In contrast, the majority of the arabinan termini from the rapidly growing, non-infective *M. smegmatis*

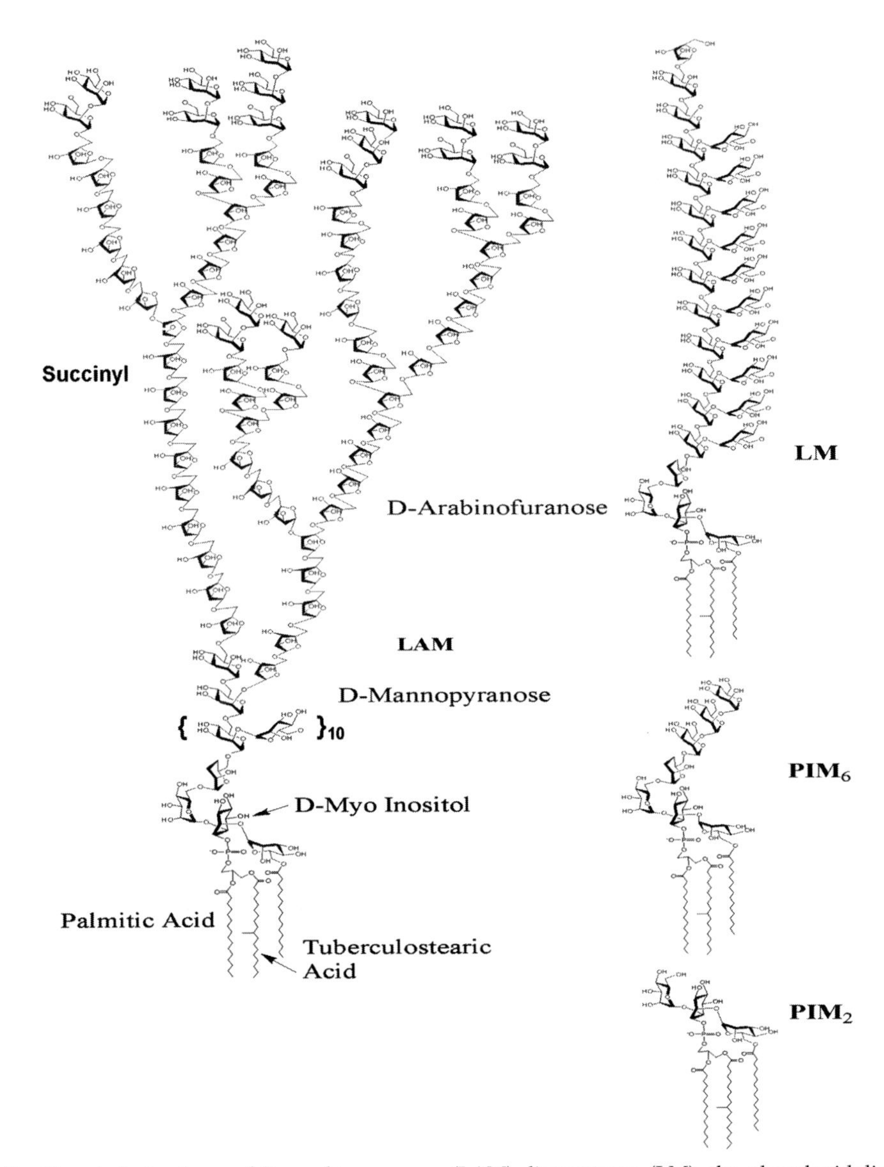

FIGURE 9.2 Chemical structures of lipoarabinomannan (LAM), lipomannan (LM), the phosphatidylinositol-mannosides (PIMs) of the cell wall/membrane of *M. tuberculosis*. The non-reducing termini of LAM and arabinogalactan (AG) are Ara_4 and Ara_6 as released by *Cellulomonas arabinanse*. In D-mannopyranose-capped LAM (ManLAM), there are mannose caps, as depicted, attached to the termini of the arabinan. A hallmark of LAM arabinan, as distinct from AG arabinan, is the presence of the linear Ara_4.

are uncapped, while a minor portion terminates in inositol-phosphate (Chatterjee *et al.*, 1992b; Prinzis *et al.*, 1993; Khoo *et al.*, 1995b; Gilleron *et al.*, 1997).

Irrespective of the capping functions, the underlying α-(1→5)-linked arabinan chains terminate in either one of the two well-defined motifs, namely a branched Ara_6 or a linear Ara_4 (see Figure 9.2), the relative proportion of which reflects the amount of α-(1→3)-linkages branching off the α-(1→5)-linked arabinan backbone, prior to termination by β-(1→2)-arabinosylation (Chatterjee *et al.*, 1991; Khoo *et al.*, 1995a). Constituent Ara*f* residues in the setting of α-(1→5)-elongation, α-(1→3)-linkages branching off a 3,5-disubstituted-α-Ara*f* residue and β-(1→2)-chain termination is essentially conserved in LAMs from all species. Conventional analytical methods have also led to identification of other structural indices, such as the succinylation (whose exact location remains unresolved) (Delmas *et al.*, 1997; Guerardel *et al.*, 2003) or the presence of α-methylthioxylofurnose (Treumann *et al.*, 2002; Turnbull *et al.*, 2004) along with the common presence of varying degree of Man*p* capping. Two-dimensional gel electrophoresis experiments indicate that isoelectric resolution of LAM into isoforms could be based on arabinan architecture, rather than capping or fatty acyl functions, suggesting that the arabinan exerts significant effect on the isoelectric point of a heterogeneous molecule such as LAM (Torrelles *et al.*, 2004).

Biologically, LAM has been included in a general class of bacterial virulence factors named modulins, which operate by inducing host cytokines (Henderson *et al.*, 1996). For example, LAM stimulates macrophages to produce tumour necrosis factor-α (TNF-α) and the granulocyte-macrophage-colony stimulating factor, interleukin- (IL-) 1α, IL-1β, IL-6 and IL-10. Moreover, LAM induces several "early genes" involved in the activation of macrophages and stimulates production of nitric oxide, a microbicidal substance, synergistically with interferon-γ (see Chatterjee and Khoo, 1998; Strohmeier and Fenton, 1999; Briken *et al.*, 2004 and references cited therein). In some of these systems there were intriguing indications that the precise structure of LAM, which differs somewhat between mycobacteria, was important in determining the biological effect. The Man*p*-capped LAM (ManLAM) produced by *M. tuberculosis* is generally less effective at stimulating macrophages than the non-capped LAM (AraLAM), which may owe its potency entirely to inositol-phosphate capping (Khoo *et al.*, 1995a; Gilleron *et al.*, 1997). Nevertheless, the activity of bare LAM, i.e. devoid of any capping function on its arabinan, has not been tested against ManLAM, in systems where the latter demonstrates significant activities, to resolve the precise roles of Man*p* capping. On the other hand, the presence of Man*p* cap on LAM has been implicated to be essential in mediating the binding to human macrophage Man*p* receptor (Schlesinger, 1993; Schlesinger *et al.*, 1994, 1996; Taylor *et al.*, 2005), thus presenting an alternative, and probably favourable, mode of phagocytosis of whole bacteria. Interestingly, uptake via the Man*p* receptors was shown to mediate delivery of internalized LAM to late endosomes for eventual presentation to T-cells by CD1b molecules (Prigozy *et al.*, 1997). The successful isolation of CD1 restricted cytotoxic T-cells from the peripheral blood of patients with tuberculosis (Stenger *et al.*, 1997) that recognize lipoglycan antigens of *M. tuberculosis*, argues that, as in the case of cutaneous leprosy lesion (Sieling *et al.*, 1995), LAM can be an effective T-cell antigen in host defence and may be targeted as novel vaccine (Jullien *et al.*, 1997).

Evidence is emerging that surface mannosylation of *M. tuberculosis* bacilli is an important host adaptive mechanism for directing phagocytic and post-phagocytic processes within the environment of lung alveoli and mannosylation of the *M. tuberculosis* surface aids in host cell recognition. The ligand is ManLAM which has been shown to intercalate into the membrane (Ilangumaran *et al.*, 1995) and to traffic within the endomembranes of cells infected with mycobacteria (Beatty *et al.*, 2000).

The ManLAM Man*p* caps also bind to the dendritic cell (DC)-specific intracellular adhesion molecule-3-grabbing non-integrin (DC-SIGN) on DCs (Maeda *et al.*, 2003; Tailleux *et al.*, 2003). Studies show that the Man*p* receptor and DC-SIGN, both C-type lectins that recognize *M. tuberculosis* ManLAM, regulate phagosome trafficking differently. Thus, terminal components of ManLAM are very important in host cell recognition and response.

The body of evidence implicating LAM in modulating immune response, such as that manifested by macrophage cytokine induction, is compelling. However, despite many of the structural studies emerging from our laboratory and many others, there had been a general deficit in two areas, namely the structure to function relationship of LAM and the identification of the biosynthetic steps in its formation. The reason is the inherent extreme heterogeneity of LAM and a general lack of functional mutants. With respect to mutants in LAM, major achievements in dissecting biosynthetic steps are discussed in detail in Chapter 21.

4.1.2. Lipomannan

Lipomannan is another phosphorylated polysaccharide-associated with the cell envelope and is considered to be the multimannosylated form of PIM which is primarily located in the plasma membrane. In electrophoresis (SDS-PAGE), LM migrates as a broad diffuse band around a mass of 16000 (Hunter *et al.*, 1986). Structurally, LM is composed of two segments, a PI anchor to which is attached an α-D-mannan domain (see Figure 9.2); both play a key role in inducing cytokine production by phagocytic cells. The mannan core consists of a linear α-(1→6)-linked mannan backbone extending from the C-6 of the *myo*-inositol; the mannan chain is further substituted by α-(1→2)-Man*p* side branches. There are about 20–30 Man*p* residues in all in *M. tuberculosis* LM. The LM-like molecules are proinflammatory and often their activities are associated with the degree of glycosylation and acylation on the PI anchor (Gilleron *et al.*, 2006) and the latter is abrogated after deacylation of LM by alkaline treatment (Gibson *et al.*, 2005).

4.1.3. Phosphatidylinositol mannosides

Among the phospholipids found in *Mycobacterium* (cardiolipin/diphosphatidylglycerol; phosphatidylethanolamine) (Pangborn and McKinney, 1966), the most unusual and highly characteristic members are PI and its mannosylated forms (Brennan, 1989; Chen *et al.*, 1991). Anderson first isolated a mixture of phospholipids from tubercle bacilli which, upon hydrolysis, yielded glycerophosphoric acid, inositol and Man (Anderson, 1939). The structures of this set of molecules were firmly established almost 25 years later by Lee and Ballou (1964). The PIMs are thought to be localized in both the plasma membrane and the overlying cell envelope (Gaylord and Brennan, 1987; Ortalo-Magné *et al.*, 1996).

The PIMs with conserved fatty acid composition substituted with up to six Man residues. *M. leprae* produces PIM_2 as the predominant PIM species (Khoo *et al.*, 1995b), whereas in *M. tuberculosis, Mycobacterium phlei, M. bovis* both PIM_2 and PIM_5 (Ballou and Lee, 1964; Lee and Ballou, 1965) are plentiful. *M. smegmatis* produces PIM_2 and PIM_6 as most abundant molecules (Chatterjee *et al.*, 1992a; Severn *et al.*, 1998).

Structurally, PIM_1 has a Man*p* substituted on the 2-position of the *myo*-inositol and PIM_2 contains a second mannose on C-6 of *myo*-inositol (Lee and Ballou, 1965). The PIM_4-PIM_6 contain some α-(1→2)-linked Man*p* and, therefore, differ in structure from the other PIMs (Besra *et al.*, 1997; Khoo *et al.*, 1995b). The extent of acylation of PIM_1 and PIM_2 varies from the diacyl form consisting of palmitate and tuberculostearic acids to the triacyl and tetracal forms that contain additional palmitate moieties. The first two acyl functions are a part of PI, while the third acyl group is added to O-6 of the first Man*p* substituted on PI (Brennan and Ballou, 1967, 1968; Khoo *et al.*, 1995b). In the case of the tetra-acylated forms of PIM_2 (Ac_4PIM_2), it has been suggested that there may be acylation of the inositol ring since, logically, the addition of a fatty acyl group to the second Man*p* of PIM_2 would result in termination of the glycosidic chain, thus preventing synthesis of PIM_3-PIM_6 (Khoo *et al.*, 1995b). Studies have clearly established that the glycerol-phosphate moiety was attached to the L-1 position of the *myo*-inositol ring and the Man*p* residues were glycosidically linked to the 2- and the 6-positions of the *myo*-inositol.

Purified PIM species have shown various immunomodulatory activities. They activate macrophages to induce TNF-α secretion through Toll-like receptor-2 (TLR-2) (Gilleron *et al.*, 2003) and can induce granuloma formation by recruiting natural killer T-cells (Fischer *et al.*, 2004). The PIMs have been identified as natural antigens of CD1d-restricted or CD1b-restricted T-cells (Fischer *et al.*, 2004; de la Salle *et al.*, 2005). Interestingly, PIMs have been shown to stimulate fusion between phagosomes and early endosomes when presented from the luminal side (*in vivo*) or the cytosolic side (*in vitro*). Two models have been proposed as to how PIM could exert this kind of activity, either directly on its target after translocation to the cytosolic side of the membrane, or indirectly by modifying the biophysical properties of the membrane bilayer (Vergne *et al.*, 2004).

4.2. Trehalose-containing lipids

4.2.1. *Trehalose mono- and di-mycolates*

Middlebrook *et al.* (1959) recognized that virulent tubercle bacilli differed from avirulent and attenuated strains in the unusual growth (serpentine cords) pattern. A search for a marker substance implicated in cord formation and virulence led Bloch (1950) to the trehalose-based lipids, primarily "cord factor", i.e. trehalose 6,6′-dimycolate (Figure 9.3). Accurate structures were later developed by Kato and Asselineau (1971). Although there is little evidence that cord factors or sulfolipids contribute to this cord formation, these families of lipids have been implicated in several disease-related events (Goren, 1972).

Trehalose [α-D-Glc (1↔1)-α-D-Glc] (see Chapter 11) is ubiquitous among all mycobacterial species and these disaccharides are acylated with long chain fatty acids, e.g. mycolic acids to form trehalose monomycolates (TMM) and trehalose dimycolates (TDM). The structures of cord factors from *M. tuberculosis* were established through a series of classic studies conducted by Noll *et al.* (1956) and Goren and Brennan (1979). The heterogeneity of cord factors is associated with mycolic acid composition; symmetrical α-α, methoxy-methoxy and keto-keto and the asymmetrical combination of α-methoxy, methoxy-keto and α-keto were all found. *M. bovis* cord factor lacks the methoxy substituted class. Various biological functions have been attributed to the cord factors (Goren and Brennan, 1979). Most of them seemingly related to the ability of cord factors to induce cytokine-mediated events such as systemic toxicity, granulomagenic activity and macrophage released chemotactic factors (Goren and Brennan, 1979; Minnikin, 1982).

4.2.2. *Sulfolipids*

The sulfolipids/sulfatides are also members of the acylated trehalose family. Sulfolipids were originally isolated from *M. tuberculosis* by Middlebrook *et al.* (1959) and chemically defined by Goren and Brennan (1979). All of these are tri- or tetra-acylated trehalose-2-sulfate. The acyl functions of the sulfatides may be either palmitate, stearate, phthioceranate or hydroxyphthioceranate (Brennan, 1988) (Figure 9.3). These lipids were originally believed to inhibit phagosome–lysosome fusion (Goren *et al.*, 1976), however, later evidence suggested that sulfatides inhibited phagosome activation (Pabst *et al.*, 1988).

4.3. Phthiocerol-containing glycolipids – the phenolic glycolipids (PGLs)

This class of glycolipids was originally described by Smith *et al.* (1954) in *M. bovis* and *Mycobacterium kansasii*, as "mycosides that lack

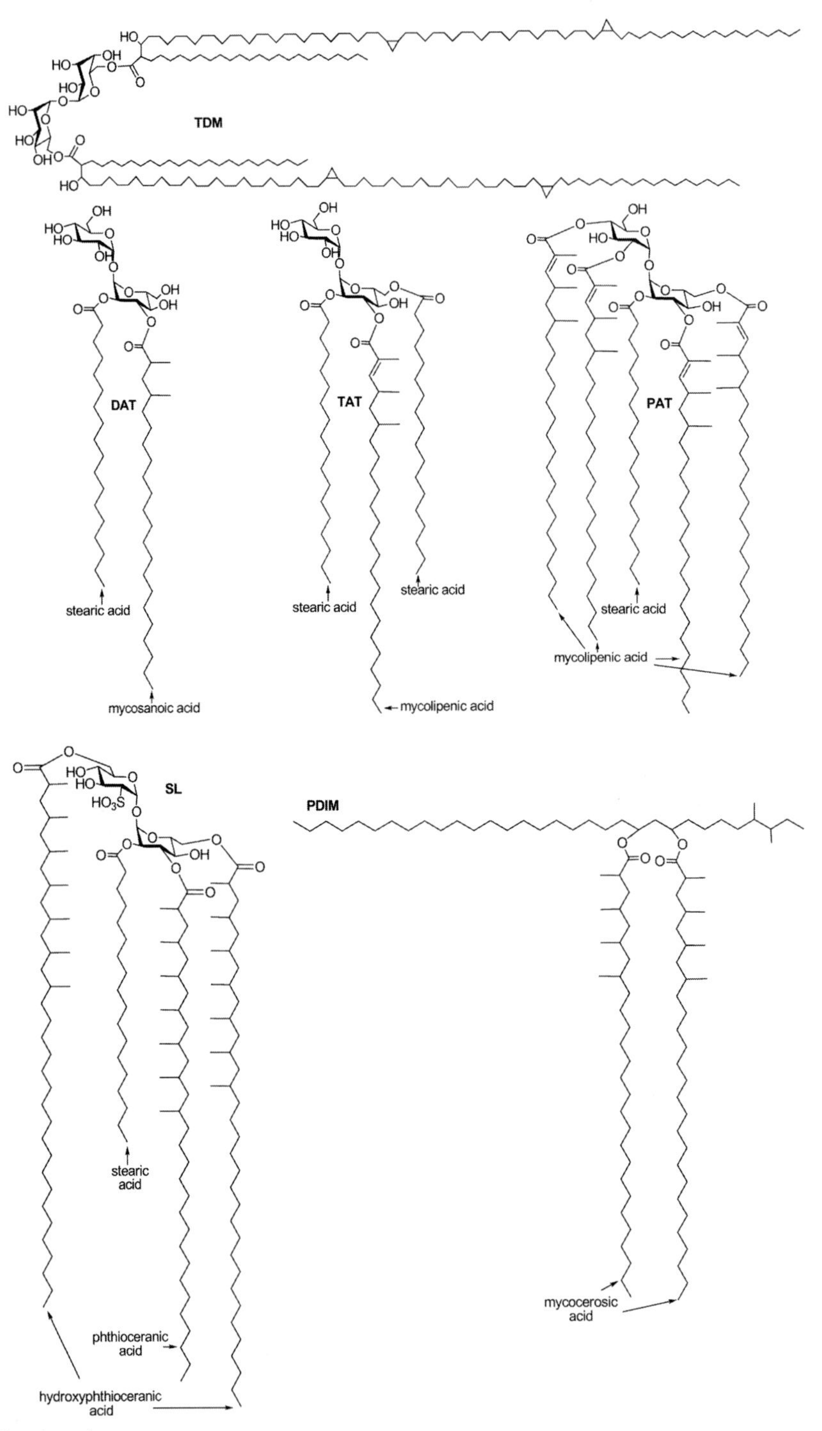

FIGURE 9.3 Continued

(Continues)

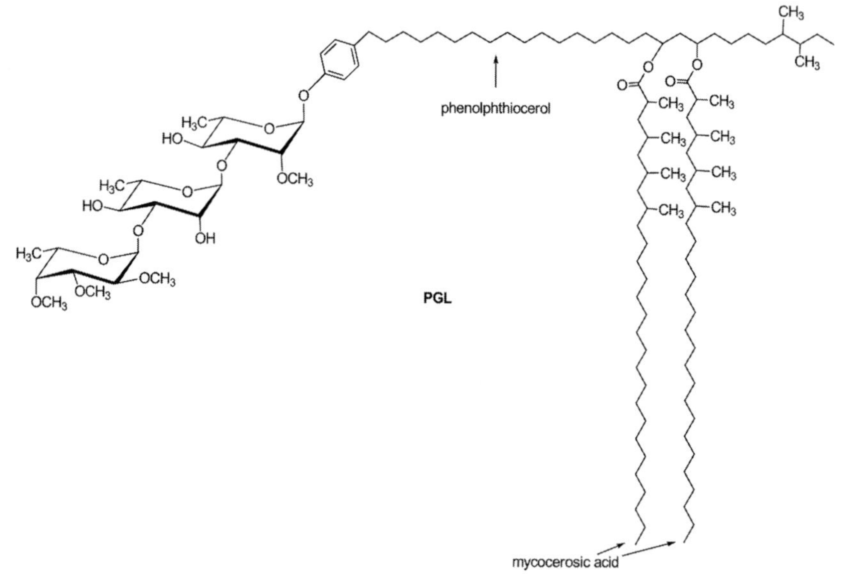

FIGURE 9.3 Structures of some representative glycolipids of the *M. tuberculosis* cell wall. Trehalose dimycolate (TDM); diacyltrehalose (DAT); triacyltrehalose (TAT); polyacyltrehalose (PAT); trehalose-containing sulfolipid I (SL); phthiocerold-imycocerosate (PDIM); phenolic glycolipid (PGL).

amino acids" and were established as glycosides of phenolphthiocerol dimycocerosate (see Figure 9.3). The lipid core is composed of a family of long-chain β-diols (C_{33}–C_{41}), esterified by polymethyl-branched (C_{27}–C_{34}) fatty acids (mycocerosic and phthioceranic acids). Later on, these were renamed as phenolic glycolipids after the discovery by Brennan (1988) of the most significant phenolic glycolipid I (PGL-I), a species-specific glycolipid antigen of *M. leprae*. Until that time, although structurally considered unusual, dominated by the highly unusual secondary alcohol phthiocerol, they had no known immunological and physiological role. The PGLs are present in significant amounts in *M. leprae*, *M. kansasii*, a few strains of *M. tuberculosis*, *M. bovis* and a few other slow-growing mycobacteria (*M. ulcerans*, *Mycobacterium marinum*, *Mycobacterium gastri*, *Mycobacterium microti* and *Mycobacterium haemophillum*). The major chemical differences and immunological distinctiveness of individual PGLs from various species lies in their variable glycosyl residues (Brennan, 1988) (see Figure 9.3 and Table 9.1). The unique triglycosyl unit of PGL-I of *M. leprae* contains methylated glucopyranose (Glc*p*) and Rha*p* residues in 3,6-di-*O*-methyl-β-D-Glc*p*-(1→4)-2,3-di-*O*-methyl-α-L-Rha*p*-(1→4)-3-*O*-methyl-α-L-Rha*p*. In *M. kansasii*, the phenolic glycolipid comprises a fucopyranose- (Fuc*p*-) containing triglycosyl component 2,6-dideoxy-4-*O*-methyl-Man*p*-(1→3)-4-*O*-acetyl-2-*O*-methyl-Fuc*p*-(1→3)-2,4-di-*O*-methyl-Rha*p*. Mycoside B is characteristic of certain bovine strains of *M. tuberculosis* and *M. bovis* (Smith *et al.*, 1954; 1960a,b; Noll and Jackim, 1958; Maclennan *et al.*, 1961) and contains but one sugar component, a 2-*O*-methyl-α-L-Rha*p* (Chatterjee *et al.*, 1989).

A methoxylated phenolphthiocerol, the so-called "attenuation indicator lipid", was isolated from some attenuated tubercle bacilli (Goren and Brennan, 1979). In addition, aliphatic (and non-glycosylated) derivatives, diesters of phthiocerols, are produced by several mycobacterial species;

TABLE 9.1 The phthiocerol-containing phenolphthiocerolglycolipids of some *Mycobacterium* spp.

Mycobacterial species	Structure of carbohydrate domain[a]	References
M. leprae	3,6-di-*O*-methyl-β-D-Glc*p*-(1→4)-2,3-di-*O*-methyl-α-L-Rha*p*-(1→4)-3-*O*-methyl-α-L-Rha*p*	Hunter and Brennan, 1981; Hunter *et al.*, 1983
M. bovis	2-*O*-methyl-Rha*p*	Chatterjee *et al.*, 1989
M. tuberculosis	2,3,4 tri-*O*-methyl-α-L-Fuc*p*-(1→3)-α-L-Rha*p*-(1→3)-2-*O*-methyl-α-L-Rha*p*	Daffe *et al.*, 1991
M. kansasii	2,6-dideoxy-4-*O*-methyl-Man*p*-(1→3)-4-*O*-acetyl-2-*O*-methyl-Fucp-(1→3)-2,4-di-*O*-methyl-Rha*p*	
M. marinum	3-*O*-methyl-Rha*p*	Dobson *et al.*, 1990

[a]Fuc*p*, fucopyranose; Glc*p*, glucopyranose; Man*p*, mannopyranose; Rha*p*, rhamnopyranose.

originally characterized from *M. tuberculosis,* these compounds are present in all the PGL-containing strains examined so far (Daffe and Laneelle, 1989). Among many strains of *M. tuberculosis*, the Canetti strains contain a major triglycosyl phenolphthiocerol, the glycosyl portion of which is 2,3,4-tri-*O*-methyl-α-L-Fuc*p*-(1→3)-α-L-Rha*p*-(1→3)-2-*O*-Me-α-L-Rha*p*.

The phenolic glycolipids from two strains of *M. marinum* have been isolated and characterized. The glycolipids from *M. marinum* MNC 170 were principally glycosides of diacyl C_{37}, C_{39} and C_{41} phenolphthiocerols A but, in *M. marinum* MNC 842, these lipids were accompanied by glycosides of diacyl phenolphthiodiolones A and novel phthiotriols A with the same overall chain lengths. The main acyl components of the phenolic glycolipids from *M. marinum* MNC 170 were C_{26} dimethyl and C_{27} and C_{29} trimethyl-branched fatty acids but, in the lipids of *M. marinum* MNC 842, the C_{27} trimethyl acid was the only principal component. The sugar composition of all these glycolipids had been previously shown to correspond to 3-*O*-methyl-Rha*p*.

All PGLs are serologically active and they have several biological activities which may be significant in mycobacterial diseases. For instance, *M. leprae* PGL-I has been shown to play a role in the disease process of leprosy by its ability to scavenge OH^- radicals and superoxide anions, thus protecting the bacilli from phagocytic killing. Also, PGL-I has been shown to have a role in the involvement in the invasion of Schwann cells through basal lamina in a laminin-2-dependent pathway (Ng *et al.*, 2000).

5. THE SPECIES AND SUB-SPECIES SPECIFIC SOLUBLE GLYCOCONJUGATES OF MYCOBACTERIAL CELL WALLS

5.1. Glycopeptidolipids (GPLs)

The most intricate class of glycolipids and the most widely studied of all mycobacterial glycolipids are the GPLs or C mycosides (Brennan, 1988; Aspinall *et al.*, 1995). Smith *et al.* (1954, 1960a,b) were the first to identify GPLs by infrared spectroscopy of chromatographically fractionated organic solvent extracts from mycobacteria and named them as J substances and then as C-mycosides. In the later part of the 1960s, W.B. Schaefer identified antigens which could be used to serotype isolates of the MAC (Schaefer, 1965) resulting in the recognition of the existence in the environment and in immune compromised patients of at least 31 antigenically-distinct serotypes/serovariants in the MAC.

Independently, Marks and co-workers described a number of lipids which gave distinct patterns on thin-layer chromatography and gave a basis for grouping isolates of the MAC (Marks *et al.*, 1970, 1971; Jenkins *et al.*, 1971). Structurally, the GPLs of MAC can be divided into two general classes: the non-polar classes expansively synthesized by all serovariants of MAC, and the polar GPLs, which differ between serovars. The basic structures of the non-polar C-mycosides or GPLs from *M. avium*, *M. intracellulare*, *M. scrofulaceum* and other mycobacteria had been first described by the French workers in the 1960s (Chaput *et al.*, 1961, 1962, 1963; Gastambide-Odier *et al.*, 1965; Vilkas and Lederer, 1968). The three amino acids were established as of the D series, linked by the N-terminal group to 3-hydroxy-C_{28} fatty acid and by the C-terminal group to an amino alcohol, alaninol. Also, a diacetylated 6-deoxy-L-talose (6-dTal) was glycosidically linked to the hydroxyl group of the D-*allo* threonine and 2,3,4-tri-*O*-methyl-L-Rha*p* to that of alaninol. In the rhamnosyl residue, a 3-*O*-methyl or 3,4-di-*O*-methyl group can replace the 2,3,4-tri-*O*-methyl group. However, it was not until Brennan and Goren (1979) conclusively demonstrated that the Schaefer antigens and Marks-Jenkins lipids were structurally alike and formally renamed these first as peptidolipids and later, more accurately as GPLs. The molecular basis of the serological differences between MAC strains, observed by Schaefer (Schaefer, 1965), was shown due to the unique variable oligosaccharide sequences (Brennan *et al.*, 1981) elaborated on the 6-dTal. Only these polar variable GPLs were able to confer sero-specificity.

All GPLs have in common an *N*-acylated lipopeptide core that bears a rhamnosylated alaninyl C terminus. The chemical difference between the non-polar (nsGPL) and the polar GPLs (ssGPL) lies in the structure of the oligosaccharide attached to the *allo*-threonine residue (Table 9.2). In non-polar GPLs, this carbohydrate is a 6-deoxytalose residue, whereas in polar GPLs, the 6-deoxytalose residue is further glycosylated with oligosaccharide. Currently, polar GPLs define 28 serovars of *Mycobacterium simiae*, *Mycobacterium chelonei* and *Mycobacterium fortuitum* (Brennan, 1988). The polar C-mycoside GPLs provide perfect phenotypic markers of individual serovars and, as polar surface immunogens, can be used in such as enzyme-linked immunosorbent assay to trace infections due to *M. avium* and *M. intracellulare*. The modification of MAC GPLs beyond the core non-polar structure led to identification of 31 serotypes using antibodies specific for the different GPLs (for review see Chatterjee and Khoo, 2001). The GPLs of MAC have been associated with the capsular-like matrix surrounding the bacilli, thus forming an electron transparent zone in phaogocytic vesicles (Brennan, 1988). The GPLs have been clearly demonstrated to survive within the intraphagosomal environment where they are resistant to degradation by lysosomal enzymes and inhibit phagosome–lysosome fusion (Frehel and Rastogi, 1989). The function that GPLs play in *M. avium* pathogenesis is currently unknown, although they are major stimulators of innate immunity through interaction with TLR-2 (Sweet *et al.*, 2008). Certain serotypes have been isolated with high frequency from AIDS patients, i.e. serotypes 1, 4 and 8, but this is probably a reflection of their dominance in the environment, rather than increased virulence. These glycolipids are present in abundant amounts on the cell surface and several lines of evidence indicate that GPLs modulate the surface chemistry of mycobacterial cells. The extent of GPL expression affects the morphology of mycobacterial colonies (Patterson *et al.*, 2000) and mediates the ability of mycobacterial colonies to slide across agar (Martinez *et al.*, 1999; Recht *et al.*, 2000). Both the virulent SmT and avirulent SmO colony forms express GPLs, whereas the rough variants, which have been variously described as virulent (Schaefer *et al.*, 1970) or avirulent (Moehring and Solotorovsky, 1965), are completely devoid of GPLs (Barrow and Brennan, 1982). Furthermore, the SmT variants have been consistently shown to have an increased resistance to antibiotics. Ambiguous results have been obtained using rough morphotypes of *M. avium*.

TABLE 9.2 Grouping of haptens of the glycopeptidolipids antigens of M. *avium* complex, based on structural similarities

Serovar number	Structure of oligosaccharide[a]
Group 1	
2	4-*O*-Ac-2,3-di-*O*-Me-α-L-Fuc*p*-(1→3)-α-L-Rha*p*-(1→2)-6-deoxy-L-Tal
4	4-*O*-Me-α-L-Rha*p*-(1→4)-*O*-Me-α-L-Fuc*p*-(1→3)-α-L-Rha*p*(1→2)-6-deoxy-L-Tal
9	4-*O*-Ac-2,3-di-*O*-Me-α-L-Fuc*p*-(1→4)-β-D-Glc*p*A-(1→4)-2,3-di-*O*-Me-α-L-Fuc*p*-(1→3)- α-L-Rha*p*(1→2)-6-deoxy-L-Tal
14	4-formamido-4,6-dideoxy-2-*O*-Me-3-*C*-Me-α-Man*p*-(1→3)-2-*O*-Me-α-D-Rha*p*-(1→3)-2-*O*-Me-α-L-Fuc*p*-(1→3)-α-L-Rha*p*(1→2)-6-deoxy-L-Tal
20	2-*O*-Me-α-D-Rha*p*-(1→3)-2-*O*-Me-α-L-Fuc*p*-(1→3)-α-L-Rha*p*(1→2)-6-deoxy-L-Tal
25	2-*O*-Me-α-D-Fuc*p*4*N*Ac-(1→4)-β-D-Glc*p*A-(1→4)-2-*O*-Me-α-L-Fuc*p*-(1→3)-α-L-Rhap(1→2)-6-deoxy-L-Tal
26	2,4-di-*O*-Me-α-L-Fuc*p*-(1→4)-β-D-Glc*p*A-(1→4)-2-*O*-Me-α-L-Fuc*p*-(1→3)-α-L-Rhap(1→2)-6-deoxy-L-Tal
Group 2	
12	4-(2-hydroxy)propanamido-4,6-dideoxy-3-*O*-Me-β-D-Glc*p*-(1→3)-4-*O*-Me-α-L-Rha*p*-(1→3)-α-L-Rha*p*-(1→2)-6-deoxy-L-Tal
17	3-(3-hydroxy-2-methyl)butanamido-3,6-dideoxy-β-D-Glc*p*-(1→3)-4-*O*-Me-α-L-Rha*p*-(1→3)-α-L-Rha*p*-(1→2)-6-deoxy-L-Tal
19	3,4-di-*O*-Me-β-D-Glc*p*A-(1→3)-3-*C*-Me-2,4-di-*O*-Me-α-L-Rha*p*-(1→3)-α-L-Rha*p*-(1→2)-6-deoxy-L-Tal
Group 3	
8	4,6-*O*-(1-carboxyethylidene)-3-*O*-Me-β-D-Glc*p*-(1→3)- α-L-Rha*p*(1→2)-6-deoxy-L-Tal
21	4,6-*O*-(1-carboxyethylidene)- β-D-Glc*p*-(1→3)-α-L-Rha*p*-(1→2)-6-deoxy-L-Tal

[a]Ac, acetyl group; Fuc*p*, fucopyranose; Glc*p*, glucopyranose; Man*p*, mannopyranose; Me, methyl group; Rha*p*, rhamnopyranose; Tal, talose.

Some rough morphotypes of *M. avium* have been shown to be avirulent compared to their isogenic SmT isolate, whereas some are of similar virulence. The differences in GPL composition between these Rg and SmT morphotypes have not been defined. Increased permeability of GPL-deficient *M. smegmatis* to chenodeoxycholate, a negatively charged hydrophobic molecule, often used to study mycobacteria cell wall fluidity (Etienne *et al.*, 2002), has been noted.

5.2. Lipooligosaccharides (LOS)

Recognition of this trehalose-based class of glycolipids, LOSs, originated from the works of Saadat and Ballou (1983) who named these as pyruvylated glycolipids from *M. smegmatis*. Independently, Brennan *et al.* (1982) identified serologically active lipid extracts of *M. kansasii* and *M. szulgai* and several other mycobacteria that were alkali-labile unlike the lipid extracts from *M. avium* serocomplex. The basic structure of LOS is a multiply-acylated oligosaccharide comprised of residues of xylose, 3-*O*-Me-Rha*p*, Fuc*p* and, in the case of *M. kansasii*, *N*-acyl-kansosamine (4-acylamido-4,6-dideoxy-3-*C*-methyl-2-*O*-L-Rhamnose) linked to a common core of tetra-Glc*p* (Table 9.3) (Hunter *et al.*, 1984). Trehalose-containing LOS was also identified in *M. malmoense*, an organism that has been implicated in pulmonary infection

TABLE 9.3 Structure of major trehalose-containing lipo-oligosaccharides of *Mycobacterium* spp.

Species	Trivial name	Structure of oligosaccharide[a]	Positions of acyl residues	Reference
M. smegmatis	(Acidic oligosaccharide A)	4,6-*O*-(1-carboxyethylidene)-3-*O*-Me-β-D-Glc*p*-(1→3)-4,6-*O*-(1-carboxyethylidene)-β-D-Glc*p*-(1→4)-β-D-Glc*p*-(1→6)-α-D-Glc*p* (1↔1)-α-D-Glc*p*	4 and 6-hydroxyls of terminal trehalose	Hunter *et al.*, 1983
M. smegmatis	(Acidic oligosaccharide B_1)	4,6-*O*-(1-carboxyethylidene)-β-D-Glc*p*-(1→4)-β-D-Glc*p*-(1→6)-α-D-Glc*p*(1↔1)-α-D-Glc*p*	ND[a]	
M. smegmatis	(Acidic oligosaccharide B_2)	4,6-*O*-(1-carboxyethylidene)-3-*O*-Me-β-D-Glc*p*-(1→3)-β-D-Glc*p*-(1→4)-β-D-Glc*p*-(1→6)-α-D-Glc*p*(1↔1)-α-D-Glc*p*	ND	
M. kansasii	LOS I′	3-*O*-Me-α-L-Rha*p*-(1→3)-β-D-Glc*p*-(1→3)-β-D-Glc*p*-(1→4)-α-D-Glc*p*(1↔1)-α-D-Glc*p*	ND	Hunter *et al.*, 1983, 1984
M. kansasii	LOS I	β−D-*Xylp*-(1→4)-3-*O*-Me-α-L-Rha*p*-(1→3)-β-D-Glc*p*-(1→3)-β-D-Glc*p*-(1→4)-α-D-Glc*p*(1↔1)-α-D-Glc*p*	ND	
M. kansasii	LOS II, III	$(\beta-$D-*Xylp*$)_2$-(1→4)-3-*O*-Me-α-L-Rha*p*-(1→3)-β-D-Glc*p*-(1→3)-β-D-Glc*p*-(1→4)-α-D-Glc*p*(1↔1)- α-D-Glc*p*	ND	
M. kansasii	LOS IV, V, VI	Kan*N*acyl -(1→3)-Fuc-(1→4)-(β−D-Xyl*p*)$_4$-(1→4)-3-*O*-Me-α-L-Rha*p*-(1→3)-β-D-Glc*p*-(1→3)-β-D-Glc*p*-(1→4)-α-D-Glc*p* (1↔1)-α-D-Glc*p*[b]	ND	
M. kansasii	LOS VII, VIII	Kan*N*acyl -(1→3)-Fuc*p*-(1→4)-[(β−D-Xyl*p*)$_4$-(1→4)$_6$-3-*O*-Me-α-L-Rha*p*-(1→3)]-β-D-Glc*p*-(1→3)-β-D-Glc*p*-(1→4)-α-D-Glc*p* (1↔1)-α-D-Glc*p*[b]	3-, 4-, and 6-hydroxyls of terminal Glc*p* unit of terminal trehalose	
M. malmoense	LOS II	α-D-Man*p*-(1→3)-α-D-Man*p*-(1→2)-α-L-Rha*p*-(1→2)-[3-*O*-Me-α − L-Rha*p*)-(1→2)]$_2$-α-L-Rha*p*-(1→3)-α-D-Glc*p*-(1↔1)-α-D-Glc*p*	3-, 4-, and 6-hydroxyls of terminal Glc*p* unit of terminal trehalose	McNeil *et al.*, 1987

[a]Ac, acetyl group; Fuc*p*, fucopyranose; Glc*p*, glucopyranose; Kan*N*acyl, N-acyl-kansosamine; Man*p*, mannopyranose; Me, methyl group; Rha*p*, rhamnopyranose; Tal, talose; Xyl*f*, Xylopyronose.

(McNeil *et al.*, 1987). Variations of LOS include differences in acylation, pyruvylation of some residues and glycosylation patterns.

The polar characteristic of these glycolipids, combined with evidence that they are not membrane-bound, suggests that the LOSs may be a major component of the lipid barrier, however, their sporadic presence in only a few strains seems to argue against this. *M. leprae*, *M. tuberculosis* and *M. avium* are all devoid of LOS (Minnikin, 1982).

5.3. Lipid-linked sugar donors

Cytoplasmic membranes of all actinomycetes and related bacteria contain small amounts of glycosyl-phosphopolyisoprenols involved in the synthesis of cell wall glycans. Polyprenol mannosyl donor (DPM) was first identified by Takayama *et al.* (1973) and was determined by Yokoyama and Ballou (1989) to be responsible for α-(1→6)-linked manno-oligosaccharide synthesis in mycobacteria. Synthesis of DPM was shown to be sensitive to amphomycin, which inhibits formation of DPM by preventing the transfer of Man*p* from guanosine diphosphate- (GDP-) Man*p* to decaprenylphosphate (Besra *et al.*, 1997). The arabinan portion of LAM and AG is mediated by decaprenylphosphoryl-Ara*f* (DPA) (Wolucka *et al.*, 1994). The latter is the only known precursor of polymerized Ara*f* residues in mycobacteria (Schultz and Elbein, 1974). Polyprenylphosphoryl sugars are typically synthesized through the transfer of a sugar from a sugar nucleotide precursor, as is for decaprenylphosphoryl-D-Man*p* and decaprenylphosphoryl-D-Glc*p* (Takayama and Goldman, 1969, 1970). Interestingly, DPA is synthesized from 5-phosphoribose diphosphate (pRpp) via the pentose-phosphate shunt (Scherman *et al.*, 1995, 1996; Klutts *et al.*, 2002).

An unusual mycolic acid-containing phospholipid, the 6-*O*-mycolyl-β-D-mannopyranosyl-1-phosphoheptaprenol has been shown to be present in *M. smegmatis* and *M. tuberculosis* (Besra *et al.*, 1994). This component was initially identified without mycolic acid substitution and thought to be involved in the transfer of Man*p* residues to phosphatidyl-mannosides (Lee and Ballou, 1965).

6. CONCLUSIONS

Mycobacterium spp. are characterized by a cell wall that is extraordinary among prokaryotes in the array of highly complex and unique glycolipids and glycopolymers expressed. The mAGP complex provides the cell wall framework, equivalent in function to the Gram-positive teichoic acid-PG assemblage (see Chapter 1). This is the site of action of existing first-line anti-tuberculosis drugs and of new potential ones. Most distinctive of all features is the peculiar outer membrane, the so-called lipid barrier, constructed in part by the mycolic acids and in part by such compounds as the trehalose-, phthiocerol-, phosphatidylinositol-containing lipids/glycoconjugates and, depending on the species, glycopeptidolipids and mycobacterial-specific LOSs. Importantly, the characteristic outer membrane of *Mycobacterium* determines much of the physiology, pathogenicity, immunoreactivity and chemotherapeutic profile of some of these major bacterial pathogens of our times.

ACKNOWLEDGEMENTS

The authors wish to acknowledge funding from grants and contracts from the National Institute of Allergy and Infectious Diseases of the National Institutes of Health.

References

Adam, A., Petit, J.F., Wietzerbin-Falszpan, J., Sinay, P., Thomas, D.W., Lederer, E., 1969. L'acide *N*-glycolyl-muramique, constituant des parois de *Mycobacterium smegmatis*: identification par spectrometrie de masse. FEBS Lett. 4, 87–92.

Anderson, R.J., 1939. The chemistry of the lipids of the tubercle bacillus and certain other microorganisms. Prog. Chem. Org. Nat. Prod. 3, 145–202.

Aspinall, G.O., Chatterjee, D., Brennan, P.J., 1995. The variable surface glycolipids of mycobacteria: structures, synthesis of epitopes, and biological properties. In: Horton, D. (Ed.), Advances in Carbohydrate Chemistry and Biochemistry, vol. 51. Academic Press, Washington DC, pp. 169–242.

Azuma, I., Kimura, H., Ninaka, T., Aoki, I., Yamamura, Y., 1968. Chemical and immunological studies on mycobacterial polysaccharides. I. Purification and properties of polysaccharides from human tubercle bacilli. J. Bacteriol. 95, 263–271.

Ballou, C.E., Lee, Y.C., 1964. The structure of a *myo*-inositol mannoside from *Mycobacterium tuberculosis* glycolipid. Biochemistry 3, 682–685.

Barrow, W.W., Brennan, P.J., 1982. Immunogenicity of type-specific C-mycoside glycopeptidolipids of mycobacteria. Infect. Immun. 36, 678–684.

Beatty, W.L., Rhoades, E.R., Ullrich, H.J., Chatterjee, D., Heuser, J.E., Russell, D.G., 2000. Trafficking and release of mycobacterial lipids from infected macrophages. Traffic 1, 235–247.

Besra, G.S., Sievert, T., Lee, R.E., Slayden, R.A., Brennan, P.J., Takayama, K., 1994. Identification of the apparent carrier in mycolic acid synthesis. Proc. Natl. Acad. Sci. USA 91, 12735–12739.

Besra, G.S., Khoo, K.-H., McNeil, M.R., Dell, A., Morris, H.R., Brennan, P.J., 1995. A new interpretation of the structure of the mycolyl-arabinogalactan complex of *Mycobacterium tuberculosis* as revealed through characterization of oligoglycosylalditol fragments by fast-atom bombardment mass spectrometry and ^{1}H nuclear magnetic resonance spectroscopy. Biochemistry 34, 4257–4266.

Besra, G.S., Morehouse, C.B., Rittner, C.M., Waechter, C.J., Brennan, P.J., 1997. Biosynthesis of mycobacterial lipoarabinomannan. J. Biol. Chem. 272, 18460–18466.

Bhamidi, S., Scherman, M.S., Rithner, C.D., et al., 2008. The identification and location of succinyl residues and the characterization of the interior arabinan region allow for a model of the complete primary structure of *Mycobacterium tuberculosis* mycolyl arabinogalactan. J. Biol. Chem. 283, 12992–13000.

Bloch, H., 1950. Studies on the virulence of tubercle bacilli; isolation and biological properties of a constituent of virulent organisms. J. Exp. Med. 91, 197–218.

Brennan, P.J., 1988. *Mycobacterium* and other actinomycetes. In: Wilkinson, R.C., Wilkinson, S.G. (Eds.) Mycobacterial Lipids, Vol. I. Academic Press, London, pp. 203–298.

Brennan, P.J., 1989. Structure of mycobacteria: recent developments in defining cell wall carbohydrates and proteins. Rev. Infect. Dis. 11, S420–S430.

Brennan, P.J., Ballou, C.E., 1967. Biosynthesis of mannophosphoinositides by *Mycobacterium phlei.* The family of dimannosylphosphoinositides. J. Biol. Chem. 242, 3046–3056.

Brennan, P.J., Ballou, C.E., 1968. Biosynthesis of mannophosphoinositides by *Mycobacterium phlei.* Enzymatic acylation of the dimannophosphoinositides. J. Biol. Chem. 243, 2975–2984.

Brennan, P.J., Crick, D.C., 2007. The cell-wall core of *Mycobacterium tuberculosis* in the context of drug discovery. Curr. Top. Med. Chem. 7, 475–488.

Brennan, P.J., Goren, M.B., 1979. Structural studies on the type-specific antigens and lipids of the *Mycobacterium avium-Mycobacterium intracellulare Mycobacterium scrofulaceum* complex serocomplex. J. Biol. Chem. 254, 4205–4211.

Brennan, P.J., Nikaido, H., 1995. The envelope of mycobacteria. Annu. Rev. Biochem. 64, 29–63.

Brennan, P.J., Heifets, M., Ullom, B.P., 1982. Thin-layer chromatography of lipid antigens as a means of identifying non-tuberculous mycobacteria. J. Clin. Microbiol. 15, 447–455.

Brennan, P.J., Mayer, H., Aspinall, G.O., Nam Shin, J.E., 1981. Structures of the glycopeptidolipid antigens from serovars in the *Mycobacterium avium/Mycobacterium intracellulare/ Mycobacterium scrofulaceum* serocomplex. Eur. J. Biochem. 115, 7–15.

Briken, V., Porcelli, S.A., Besra, G.S., Kremer, L., 2004. Mycobacterial lipoarabinomannan and related lipoglycans: from biogenesis to modulation of the immune response. Mol. Microbiol. 53, 391–403.

Chaput, M., Michel, G., Lederer, E., 1961. On the mycoside Cm, a new peptido-glycolipid isolated from *Mycobacterium marianum*. Experientia 17, 107–108.

Chaput, M., Michel, G., Lederer, E., 1962. [Structure of mycoside Cm, peptidoglycolipid of *Mycobacterium marianum*]. Biochim. Biophys. Acta 63, 310–326.

Chaput, M., Michel, G., Lederer, E., 1963. [Structure of mycoside C2 of *Mycobacterium Avium*]. Biochim. Biophys. Acta 78, 329–341.

Chatterjee, D., Khoo, K.H., 1998. Mycobacterial lipoarabinomannan: an extraordinary lipoheteroglycan with profound physiological effects. Glycobiology 8, 113–120.

Chatterjee, D., Khoo, K.-H., 2001. The surface glycopeptidolipids of mycobacteria:structures and biological properties. Cell. Mol. Life Sci. 58, 2018–2042.

Chatterjee, D., Bozic, C.M., Knisley, C., Cho, S.N., Brennan, P.J., 1989. Phenolic glycolipids of *Mycobacterium bovis*: new structures and synthesis of a corresponding seroreactive neoglycoprotein. Infect. Immun. 57, 322–330.

Chatterjee, D., Bozic, C.M., McNeil, M., Brennan, P.J., 1991. Structural features of the arabinan component of the lipoarabinomannan of *Mycobacterium tuberculosis*. J. Biol. Chem. 266, 9652–9660.

Chatterjee, D., Hunter, S.W., McNeil, M., Brennan, P.J., 1992a. Lipoarabinomannan. Multiglycosylated form of the mycobacterial mannosylphophatidylinositols. J. Biol. Chem. 267, 6228–6233.

Chatterjee, D., Roberts, A.D., Lowell, K., Brennan, P.J., Orme, I.M., 1992b. Structural basis of capacity of lipoarabinomannan to induce secretion of tumor necrosis factor. Infect. Immun. 60, 1249–1253.

Chen, Q., Brglez, I., Boss, W.F., 1991. Inositol phospholipids as plant second messengers. Symp. Soc. Exp. Biol. 45, 159–175.

Daffe, M., Draper, P., 1998. The envelope layers of mycobacteria with reference to their pathogenicity. Adv. Microb. Physiol. 39, 131–203.

Daffe, M., Laneelle, M.A., 1989. Diglycosyl phenol phthiocerol diester of *Mycobacterium leprae*. Biochim. Biophys. Acta 1002, 333–337.

Daffe, M., Cho, S.N., Chatterjee, D., Brennan, P.J., 1991. Chemical synthesis and seroreactivity of a neoantigen containing the oligosaccharide hapten of the Mycobacterium tuberculosis-specific phenolic glycolipid. J. Infect. Dis. 163, 161–168.

de la Salle, H., Mariotti, S., Angenieux, C., et al., 2005. Assistance of microbial glycolipid antigen processing by CD1e. Science 310, 1321–1324.

Delmas, C., Gilleron, M., Brando, T., et al., 1997. Comparative structural study of the mannosylated-lipoarabinomannans from *Mycobacterium bovis* BCG vaccine strains: characterization and localization of succinates. Glycobiology 7, 811–817.

Dobson, G., Minnikin, D.E., Besra, G.S., Mallet, A.I., Magnusson, M., 1990. Characterisation of phenolic glycolipids from *Mycobacterium marinum*. Biochim. Biophys. Acta 1042, 176–181.

Draper, P., Kandler, O., Darbre, A., 1987. Peptidoglycan and arabinogalactan of *Mycobacterium leprae*. J. Gen. Microbiol. 133, 1187–1194.

Draper, P., Khoo, K.H., Chatterjee, D., Dell, A., Morris, H.R., 1997. Galactosamine in walls of slow-growing mycobacteria. Biochem. J. 327, 519–525.

Etienne, G., Villeneuve, C., Billman-Jacobe, H., Astarie-Dequeker, C., Dupont, M.A., Daffe, M., 2002. The impact of the absence of glycopeptidolipids on the ultrastructure, cell surface and cell wall properties, and phagocytosis of *Mycobacterium smegmatis*. Microbiology 148, 3089–3100.

Fischer, K., Scotet, E., Niemeyer, M., et al., 2004. Mycobacterial phosphatidylinositol mannoside is a natural antigen for CD1d-restricted T cells. Proc. Natl. Acad. Sci. USA 101, 10685–10690.

Frehel, C., Rastogi, N., 1989. Phagosome-lysosome fusions in macrophages infected by *Mycobacterium avium*: role of mycosides-C and other cells surface components. Acta Leprol. 7 (Suppl. 1), 173–174.

Gastambide-Odier, M., Lederer, E., Sarda, P., 1965. [Structure of the aglycones of mycosides A and B]. Tetrahedron. Lett. 35, 3135–3143.

Gaylord, H., Brennan, P.J., 1987. Leprosy and the leprosy bacillus: recent developments in characterization of antigens and immunology of the disease. Annu. Rev. Microbiol. 41, 645–675.

Gibson, K.J., Gilleron, M., Constant, P., et al., 2005. A lipomannan variant with strong TLR-2-dependent proinflammatory activity in Saccharothrix aerocolonigenes. J. Biol. Chem. 280, 28347–28356.

Gilleron, M., Himoudi, N., Adam, O., et al., 1997. *Mycobacterium smegmatis* phosphatidylinositols-glyceroarabinomannans. J. Biol. Chem. 272, 117–124.

Gilleron, M., Quesniaux, V.F., Puzo, G., 2003. Acylation state of the phosphatidylinositol hexamannosides from *Mycobacterium bovis* bacillus Calmette Guerin and *Mycobacterium tuberculosis* H37Rv and its implication in Toll-like receptor response. J. Biol. Chem. 278, 29880–29889.

Gilleron, M., Nigou, J., Nicolle, D., Quesniaux, V., Puzo, G., 2006. The acylation state of mycobacterial lipomannans modulates innate immunity response through Toll-like receptor 2. Chem. Biol. 13, 39–47.

Goren, M.B., 1972. Mycobacterial lipids: selected topics. Bacteriol. Rev. 36, 33–64.

Goren, M.B., Brennan, P.J., 1979. Mycobacterial lipids: chemistry and biological activities. In: Youmans, G.P. (Ed.), Tuberculosis. W.B. Saunders, Philadelphia, pp. 69–193.

Goren, M.B., D'Arcy Hart, P., Young, M.R., Armstrong, J.A., 1976. Prevention of phagosome-lysosome fusion in cultured macrophages by sulfatides of *Mycobacterium tuberculosis*. Proc. Natl. Acad. Sci. USA 73, 2510–2514.

Guerardel, Y., Maes, E., Briken, V., et al., 2003. Lipomannan and lipoarabinomannan from a clinical isolate of *Mycobacterium kansasii*: novel structural features and apoptosis-inducing properties. J. Biol. Chem. 278, 36637–36651.

Heidelberger, M., 1939. Quantitative absolute methods in the study of antigen-antibody reactions. Bacteriol. Rev. 3, 49–95.

Henderson, B., Poole, S., Wilson, M., 1996. Bacterial modulins: a novel class of virulence factors which cause host tissue pathology by inducing cytokine synthesis. Microbiol. Rev. 60, 316–341.

Hoffmann, C., Leis, A., Niederweis, M., Plitzko, J.M., Engelhardt, H., 2008. Disclosure of the mycobacterial outer membrane: cryo-electron tomography and vitreous sections reveal the lipid bilayer structure. Proc. Natl. Acad. Sci. USA 105, 3963–3967.

Hunter, S.W., Brennan, P.J., 1981. A novel phenolic glycolipid from *Mycobacterium leprae* possibly involved in immunogenicity and pathogenicity. J. Bacteriol. 147, 728–735.

Hunter, S.W., Brennan, P.J., 1990. Evidence for the presence of a phosphatidylinositol anchor on the lipoarabinomannan and lipomannan of *Mycobacterium tuberculosis*. J. Biol. Chem. 265, 9272–9279.

Hunter, S.W., Murphy, R.C., Clay, K., Goren, M.B., Brennan, P.J., 1983. Trehalose-containing lipooligosaccharides. A new class of species-specific antigens from Mycobacterium. J. Biol. Chem. 258, 10481–10487.

Hunter, S.W., Fujiwara, T., Murphy, R.C., Brennan, P.J., 1984. *N*-acylkansosamine. A novel *N*-acylamino sugar from the trehalose-containing lipooligosaccharide antigens of *Mycobacterium kansasii*. J. Biol. Chem. 259, 9729–9734.

Hunter, S.W., Gaylord, H., Brennan, P.J., 1986. Structure and antigenicity of the phosphorylated lipopolysaccharide antigens from the leprosy and tubercle bacilli. J. Biol. Chem. 261, 12345–12351.

Ilangumaran, S., Arni, S., Poincelet, M., et al., 1995. Integration of mycobacterial lipoarabinomannans into glycosylphosphatidylinositol-rich domains of lymphomonocytic cell plasma membranes. J. Immunol. 155, 1334–1342.

Jenkins, P.A., Marks, J., Schaefer, W.B., 1971. Lipid chromatography and seroagglutination in the classification of rapidly growing mycobacteria. Am. Rev. Respir. Dis. 103, 179–187.

Jullien, D., Stenger, S., Ernst, W.A., Modlin, R.L., 1997. CD1 presentation of microbial nonpeptide antigens to T cells. J. Clin. Invest. 99, 2071–2074.

Kato, M., Asselineau, J., 1971. Chemical structure and biochemical activity of cord factor analogs. 6,6′-Dimycoloyl sucrose and methyl 6-mycoloyl-D-glucoside. Eur. J. Biochem. 22, 364–370.

Khoo, K.-H., Dell, A., Morris, H.R., Brennan, P.J., Chatterjee, D., 1995a. Inositol phosphate capping of the nonreducing termini of lipoarabinomannan from rapidly growing strains of *Mycobacterium*. J. Biol. Chem. 270, 12380–12389.

Khoo, K.-H., Dell, A., Morris, H.R., Brennan, P.J., Chatterjee, D., 1995b. Structural definition of acylated phosphatidylinositol mannosides from *Mycobacterium tuberculosis*: definition of a common anchor for lipomannan and lipoarabinomannan. Glycobiology 5, 117–127.

Khoo, K.H., Douglas, E., Azadi, P., et al., 1996. Truncated structural variants of lipoarabinomannan in ethambutol drug-resistant strains of *Mycobacterium smegmatis* – inhibition of arabinan biosynthesis by ethambutol. J. Biol. Chem. 271, 28682–28690.

Klutts, J.S., Hatanaka, K., Pan, Y.T., Elbein, A.D., 2002. Biosynthesis of D-arabinose in *Mycobacterium smegmatis*: specific labeling from D-glucose. Arch. Biochem. Biophys. 398, 229–239.

Lee, A., Wu, S.W., Scherman, M.S., et al., 2006. Sequencing of oligoarabinosyl units released from mycobacterial arabinogalactan by endogenous arabinanase: identification of distinctive and novel structural motifs. Biochemistry 45, 15817–15828.

Lee, Y.C., Ballou, C.E., 1964. Structural studies on the *myo*-inositol mannosides from the glycolipids of *Mycobacterium tuberculosis* and *Mycobacterium phlei*. J. Biol. Chem. 239, 1316–1327.

Lee, Y.C., Ballou, C.E., 1965. Complete structures of the glycophospholipids of mycobacteria. Biochemistry 4, 1395–1404.

Leopold, K., Fischer, W., 1993. Molecular analysis of the lipoglycans of *Mycobacterium tuberculosis*. Anal. Biochem. 208, 57–64.

Maclennan, A.P., Randall, H.M., Smith, D.W., 1961. The occurence of methyl ethers of rhamose and fucose in specific glycolipids of certain mycobacteria. Biochem. J. 80, 309–318.

Maeda, N., Nigou, J., Herrmann, J.L., et al., 2003. The cell surface receptor DC-SIGN discriminates between *Mycobacterium* species through selective recognition of the mannose caps on lipoarabinomannan. J. Biol. Chem. 278, 5513–5516.

Mahapatra, S., Crick, D.C., McNeil, M.R., Brennan, P.J., 2008. Unique structural features of the peptidoglycan of *Mycobacterium leprae*. J. Bacteriol. 190, 655–661.

Marks, J., Jenkins, P.A., Schaefer, W.B., 1970. Identification and incidence of a third type of *Mycobacterium avium*. Am. Rev. Respir. Dis. 102, 499–506.

Marks, J., Jenkins, P.A., Schaefer, W.B., 1971. Thin-layer chromatography of mycobacterial lipids as an aid to classification: technical improvements: *Mycobacterium avium, M. intracellulare* (Battey bacilli). Tubercle 52, 219–225.

Martinez, A., Torello, S., Kolter, R., 1999. Sliding motility in mycobacteria. J. Bacteriol. 181, 7331–7338.

Matsuhashi, M., 1966. [Biosynthesis in the bacterial cell wall]. Tanpakushitsu Kakusan Koso 11, 875–886.

McNeil, M., Tsang, A.Y., McClatchy, J.K., Stewart, C., Jardine, I., Brennan, P.J., 1987. Definition of the surface antigens of *Mycobacterium malmoense* and use in studying the etiology of a form of mycobacteriosis. J. Bacteriol. 169, 3312–3320.

McNeil, M., Daffe, M., Brennan, P.J., 1991. Location of the mycolyl ester substituents in the cell walls of mycobacteria. J. Biol. Chem. 266, 13217–13223.

Middlebrook, G., Colemann, C., Schaeffer, W.B., 1959. Sulfolipids from virulent tubercle bacilli. Proc. Natl. Acad. Sci. USA 45, 1801–1804.

Minnikin, D.E., 1982. Lipids: complex lipids, their chemistry, biosynthesis and roles. In: Ratledge, C., Stanford, J. (Eds.) The Biology of Mycobacteria, Vol. 2. Academic Press, London, pp. 95–184.

Misaki, A., Seto, N., Azuma, I., 1974. Structure and immunological properties if D-arabino-D-galactan isolated from cell walls of *Mycobacterium* species. J. Biochem. 76, 15–27.

Misaki, A., Azuma, I., Yamamura, Y., 1977. Structural and immunochemical studies on D-arabino-D-mannans and D-mannans of *Mycobacterium tuberculosis* and other *Mycobacterium* species. J. Biochem. 82, 1759–1770.

Moehring, J.M., Solotorovsky, M.R., 1965. Relationship of colonial morphology to virulence for chickens of *Mycobacterium avium* and the nonphotochromogens. Am. Rev. Respir. Dis. 92, 704–713.

Ng, V., Zanazzi, G., Timpl, R., et al., 2000. Role of the cell wall phenolic glycolipid-1 in the peripheral nerve predilection of *Mycobacterium leprae*. Cell 103, 511–524.

Noll, H., Jackim, E., 1958. The chemistry of the native constituents of the acetone-soluble fat of *Mycobacterium tuberculosis* (Brevannes). I. Glycerides and phosphoglycolipides. J. Biol. Chem. 232, 903–917.

Noll, H., Bloch, H., Asselineau, J., Lederer, E., 1956. The chemical structure of the cord factor of *Mycobacterium tuberculosis*. Biochim. Biophys. Acta 20, 299–309.

Ortalo-Magné, A., Lemassu, A., Lanéelle, M.A., et al., 1996. Identification of the surface-exposed lipids on the cell envelopes of *Mycobacterium tuberculosis* and other mycobacterial species. J. Bacteriol. 178, 456–461.

Pabst, M.J., Gross, J.M., Brozna, J.P., Goren, M.B., 1988. Inhibition of macrophage priming by sulfatide from *Mycobacterium tuberculosis*. J. Immunol. 140, 634–640.

Pangborn, M.C., McKinney, J.A., 1966. Purification of serologically active phosphoinositides of *Mycobacterium tuberculosis*. J. Lipid Res. 7, 627–633.

Patterson, J.H., McConville, M.J., Haites, R.E., Coppel, R.L., Billman-Jacobe, H., 2000. Identification of a methyltransferase from *Mycobacterium smegmatis* involved in glycopeptidolipid synthesis. J. Biol. Chem. 275, 24900–24906.

Prigozy, T.I., Sieling, P.A., Clemens, D., et al., 1997. The mannose receptor delivers lipoglycan antigens to endosomes for presentation to T cells by CD1b molecules. Immunity 6, 187–197.

Prinzis, S., Chatterjee, D., Brennan, P.J., 1993. Structure and antigenicity of lipoarabinomannan from *Mycobacterium bovis* BCG. J. Gen. Microbiol. 139, 2649–2658.

Recht, J., Martinez, A., Torello, S., Kolter, R., 2000. Genetic analysis of sliding motility in *Mycobacterium smegmatis*. J. Bacteriol. 182, 4348–4351.

Saadat, S., Ballou, C.E., 1983. Pyruvylated glycolipids from *Mycobacterium smegmatis*. Structures of two oligosaccharide components. J. Biol. Chem. 258, 1813–1818.

Schaefer, W.B., 1965. Serologic identification and classification of the atypical mycobacteria by their agglutination. Am. Rev. Respir. Dis. 92, 85–93.

Schaefer, W.B., Davis, C.L., Cohn, M.L., 1970. Pathogenicity of transparent, opaque, and rough variants of *Mycobacterium avium* in chickens and mice. Am. Rev. Respir. Dis. 102, 499–506.

Scherman, M., Weston, A., Duncan, K., et al., 1995. The biosynthetic origin of the mycobacterial cell wall arabinosyl residues. J. Bacteriol. 177, 7125–7130.

Scherman, M.S., Kalbe-Bournonville, L., Bush, D., Xin, Y., Deng, L., McNeil, M., 1996. Polyprenylphosphate-pentoses in mycobacteria are synthesized from 5-phosphoribose pyrophosphate. J. Biol. Chem. 271, 29652–29658.

Schlesinger, L.S., 1993. Macrophage phagocytosis of virulent but not attenuated strains of *Mycobacterium tuberculosis* is mediated by mannose receptors in addition to complement receptors. J. Immunol. 150, 2920–2925.

Schlesinger, L.S., Hull, S.R., Kaufman, T.M., 1994. Binding of the terminal mannosyl units of lipoarabinomannan from a virulent strain of *Mycobacterium tuberculosis* to human macrophages. J. Immunol. 152, 4070–4079.

Schlesinger, L.S., Kaufman, T.M., Iyer, S., Hull, S.R., Marchiando, L.K., 1996. Differences in mannose receptor-mediated uptake of lipoarabinomannan from virulent and attenuated strains of *Mycobacterium tuberculosis* by human macrophages. J. Immunol. 157, 4568–4575.

Schultz, J., Elbein, A.D., 1974. Biosynthesis of mannosyl- and glucosyl-phosphoryl polyprenols in *Mycobacterium smegmatis*. Evidence for oligosaccharide-phosphoryl-polyprenols. Arch. Biochem. Biophys. 160, 311–322.

Severn, W.B., Furneaux, R.H., Falshaw, R., Atkinson, P.H., 1998. Chemical and spectroscopic characterisation of the phosphatidylinositol manno-oligosaccharides from *Mycobacterium bovis* AN5 and WAg201 and *Mycobacterium smegmatis* mc2 155. Carbohydr. Res. 308, 397–408.

Sieling, P.A., Chatterjee, D., Porcelli, S.A., et al., 1995. CD1-restricted T cell recognition of microbial lipoglycan antigens. Science 269, 227–230.

Smith, D.W., Harrell, W.K., Randall, H.M., 1954. Correlation of biologic properties of strains of *Mycobacterium* with infrared spectrum III, differentiation of bovine and human varieties of *M. tuberculosis* by means of infrared spectrum. Am. Rev. Tuberc. 69, 505–510.

Smith, D.W., Randall, H.M., MacLennan, A.P., Putney, R.K., Rao, S.V., 1960a. Detection of specific lipids in mycobacteria by infrared spectroscopy. J. Bacteriol. 79, 217–229.

Smith, D.W., Randall, H.M., MaclLennan, A.P., Lederer, E., 1960b. Mycosides: a new class of type specific glycolipids of mycobacteria. Nature (London) 186, 887–888.

Stenger, S., Mazzaccaro, R.J., Uyemura, K., et al., 1997. Differential effects of cytolytic T cell subsets on intracellular infection. Science 276, 1684–1687.

Strohmeier, G.R., Fenton, M.J., 1999. Roles of lipoarabinomannan in the pathogenesis of tuberculosis. Microb. Infect. 1, 709–717.

Sweet, L., Zhang, W., Torres-Fewell, H., Serianni, A., Boggess, W., Schorey, J., 2008. *Mycobacterium avium* glycopeptidolipids require specific acetylation and methylation patterns for signaling through TLR2. J. Biol. Chem. 283, 33221–33231.

Tailleux, L., Schwartz, O., Herrmann, J.L., et al., 2003. DC-SIGN is the major *Mycobacterium tuberculosis* receptor on human dendritic cells. J. Exp. Med. 197, 121–127.

Takayama, K., Goldman, D.S., 1969. Pathway for the synthesis of mannophospholipids in *Mycobacterium tuberculosis*. Biochim. Biophys. Acta 176, 196–198.

Takayama, K., Goldman, D.S., 1970. Enzymatic synthesis of mannosyl-1-phosphoryl-decaprenol by a cell-free system of *Mycobacterium tuberculosis*. J. Biol. Chem. 245, 6251–6257.

Takayama, K., Schnoes, H.K., Semmler, E.J., 1973. Characterization of the alkali-stable mannophospholipids of *Mycobacterium smegmatis*. Biochim. Biophys. Acta 316, 212–221.

Taylor, P.R., Gordon, S., Martinez-Pomares, L., 2005. The mannose receptor: linking homeostasis and immunity through sugar recognition. Trends Immunol. 26, 104–110.

Torrelles, J.B., Khoo, K.H., Sieling, P.A., et al., 2004. Truncated structural variants of lipoarabinomannan in *Mycobacterium leprae* and an ethambutol-resistant strain of *Mycobacterium tuberculosis*. J. Biol. Chem. 279, 41227–41239.

Treumann, A., Xidong, F., McDonnell, L., et al., 2002. 5-Methylthiopentose: a new substituent on lipoarabinomannan in *Mycobacterium tuberculosis*. J. Mol. Biol. 316, 89–100.

Turnbull, W.B., Shimizu, K.H., Chatterjee, D., Homans, S.W., Treumann, A., 2004. Identification of the 5-methylthiopentosyl substituent in *Mycobacterium tuberculosis* lipoarabinomannan. Angew. Chem. Int. Ed. Engl. 43, 3918–3922.

Vergne, I., Fratti, R.A., Hill, P.J., Chua, J., Belisle, J., Deretic, V., 2004. *Mycobacterium tuberculosis* phagosome maturation arrest: mycobacterial phosphatidylinositol analog phosphatidylinositol mannoside stimulates early endosomal fusion. Mol. Biol. Cell 15, 751–760.

Vilkas, E., Lederer, E., 1968. *N*-methylation of peptides by the method of Hakomori. Structure of mycoside Cbl. Tetrahedron Lett. 26, 3089–3092.

Vollmer, W., Höltje, J.V., 2004. The architecture of the murein (peptidoglycan) in gram-negative bacteria: vertical scaffold or horizontal layer(s)? J. Bacteriol. 186, 5978–5987.

Wolucka, B.A., McNeil, M.R., de Hoffmann, E., Chojnacki, T., Brennan, P.J., 1994. Recognition of the lipid intermediate for arabinogalactan/arabinomannan biosynthesis and its relation to the mode of action of ethambutol on mycobacteria. J. Biol. Chem. 269, 23328–23335.

Yokoyama, K., Ballou, C.E., 1989. Synthesis of α-1-6 mannooligosaccharide in *M. smegmatis*. J. Biol. Chem. 264, 21621–21628.

Zuber, B., Chami, M., Houssin, C., Dubochet, J., Griffiths, G., Daffé, M., 2008. Direct visualization of the outer membrane of mycobacteria and corynebacteria in their native state. J. Bacteriol. 190, 5672–5680.

CHAPTER

10

Glycoconjugate structure and function in fungal cell walls

Marlyn Gonzalez, Piet W. J. de Groot, Frans M. Klis and Peter N. Lipke

SUMMARY

Complex cell walls distinguish fungi from animals, and the roles of these walls include osmotic support, selective permeability and interaction with the environment. Fungal walls consist of covalently cross-linked complexes of polysaccharides (usually β-glucans and chitin) with glycoproteins. Several different polysaccharides are covalently cross-linked through glycosidic bonds. They are also cross-linked to two types of glycoproteins (CWPs). The Pir-CWPs are multiply cross-linked through ester bonds between deamidated glutamine residues and hydroxyl groups in the major β-(1→3)-glucan. The GPI_t-CWPs are defined by their linkage through transglycosylation of the truncated glycosylphosphatidylinositol (GPI) glycan to β-(1→6)-glucan. These GPI_t-CWPs are highly *N*- and *O*-mannosylated and may also have ester-glucan cross-links like the Pir-CWPs. GPI_t-CWPs have a common modular structure, with an N-terminal secretion signal, then a well-folded functional enzymatic or ligand binding domain, a middle region of tandem repeats, a C-terminal serine- and threonine-rich highly glycosylated "stalk" and a C-terminal GPI addition signal. Their functional regions include enzymes active in cell wall biogenesis and maintenance and nutrient acquisition; transport functions; and cell adhesion proteins that mediate interactions between fungal cells, pathogen and host and in biofilms. Therefore, fungal walls have functions including those assumed by membrane, cytoskeleton and extracellular matrix in other organisms.

Keywords: β-Glucan; Chitin; Glycoprotein; Glycosylphosphatidylinositol anchor; *N*-Glycosylation; *O*-Glycosylation; Glucan-protein ester

1. INTRODUCTION

Among the defining characteristics of fungi are cell walls with complex architecture. Fungal walls are substantially thicker than bacterial walls and normally make up 10–30% of the biomass. They are freely permeable to small molecules and so solute transport systems and signalling receptors remain in the membrane, and are homologous to those in other eukaryotes. Walls have obvious functions in osmotic and mechanical support of the cells, which have less extensive cytoskeletons than do motile eukaryotes. Walls are also the site of cell contact with the environment, so the outer surface of the wall is adapted to participate in the cell–cell interactions characteristic of colony and biofilm formation, as well as host–fungal adherence in symbiotic, commensal and pathogenic states.

A variety of activities reside in glycoproteins embedded within the walls and linked with fungal wall polysaccharides. Many of these proteins are adhesins; while others cross-link cell wall polysaccharides or are carbohydrate-active enzymes. Such enzymes have functions such as wall biogenesis and turnover, modulation of wall structure in response to stress and to fungal morphogenesis and, conceivably, also in the formation of the extracellular matrix of biofilms. Finally, some cell wall glycoproteins are involved in iron and sterol acquisition or in coping with oxidative stress (Yin *et al.*, 2008). In addition, many, if not all, of the major wall components can elicit immune responses or other host responses as antigens and modulators. Thus, the walls of fungi are complex organelles with extensive cellular, metabolic and environmental functions.

This review will briefly discuss the major polysaccharides in fungal walls and, in particular, focus on wall glycoproteins.

2. OVERALL STRUCTURE

Our knowledge is based primarily on studies in baker's yeast, *Saccharomyces cerevisiae*, and the commensal/opportunistic pathogenic yeast, *Candida albicans*. Although these yeasts apparently diverged many hundreds of millions of years ago, they both belong to the phylum Ascomycota and their wall structures are similar. Gene products in these two yeasts are referenced by their genus and species initials, followed by the locus name, e.g. CaAls1 is the product of the *ALS1* locus in *C. albicans*. Occasionally, we will also reference studies in other fungal groups. The general observation is that wall glycan structures differ somewhat between fungal organisms, but that the same basic designs are widespread. Indeed, a recent bioinformatics study showed that the sequences of fungal wall glycoproteins differed greatly between Ascomycetes, but that many cell wall biosynthetic enzymes and cellular processes of biosynthesis were highly conserved and ancestral to all of the fungi (Coronado *et al.*, 2007). Similarly, the details of wall polysaccharides and of *N*- and *O*-linked protein side-chains differ between ascomycetous fungi, but the general arrangement is likely to be conserved.

Breaking fungal cells with glass beads releases a covalently bonded cell wall "ghost" whose shape is maintained and which retains most of its constituents after extraction at 100°C in the presence of sodium dodecyl sulfate (SDS) and reducing agents (Pitarch *et al.*, 2008). Figure 10.1 is a cartoon of the fungal wall structure based mostly on what is known from the ascomycetous yeasts. There is a fibrous and flexible polysaccharide network (structures 1–3), decorated and cross-linked by glycoproteins (structures 4–7). On average, the polysaccharide layer is proximal to the membrane and most glycoproteins are at the exterior surface of the wall. There are minor components, not illustrated or discussed here, including glycolipids and metal ions.

3. WALL POLYSACCHARIDES

3.1. Fibrous polysaccharides

Ninety percent or more of the wall mass is accounted for by polysaccharides and by carbohydrate side-chains of cell wall glycoproteins. In baker's yeast, about half of this mass is accounted for by the structural polysaccharide β-(1→3)-glucan (see Figure 10.1, structure 2).

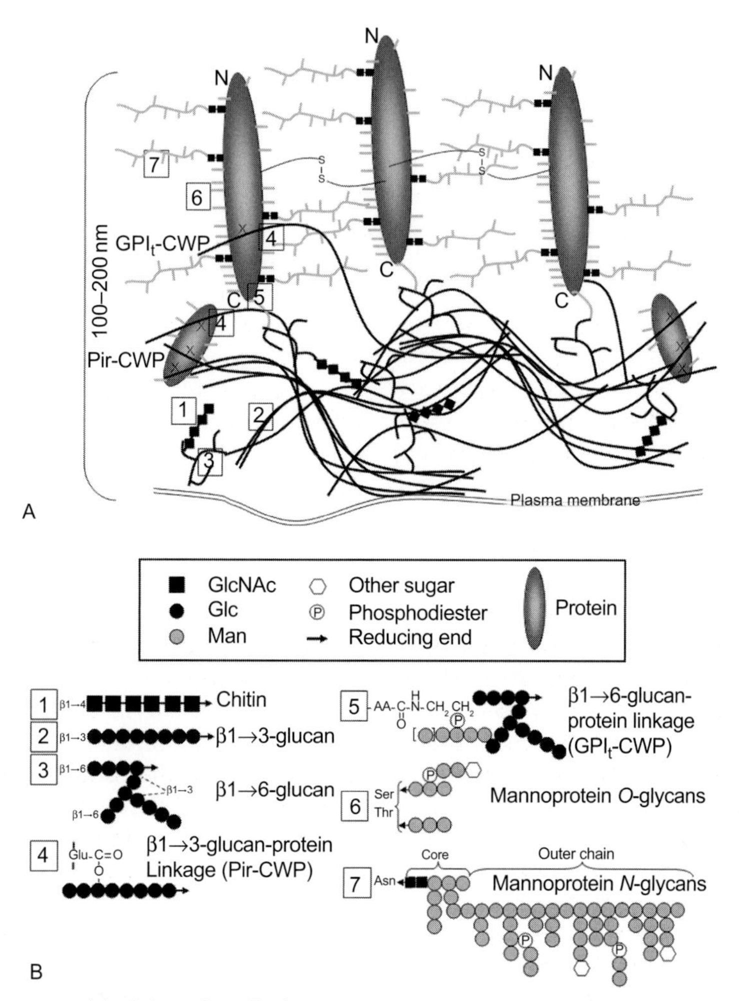

FIGURE 10.1 Cartoon model of the cell wall of an ascomycetous yeast. (A) Shows the general arrangement, including polysaccharides and glycoproteins. Disulfide cross-links are shown as vertically paired "S"; X designates ester bonds between deamidated Gln residues on CWPs and hydroxyl groups in β-(1→3) glucans. (B) A key and sketches of glycan structures. Many linkage details are not shown. Mannosyl linkages are in the α-configuration except the left-most mannose in the core of the *N*-glycan and in some species-specific groups discussed in the text. For details, readers are referred to the text.

β-(1→3)-Glucan forms single and triple-stranded helices through regular H bonding between sugars. Because its average degree of polymerization is ≥1000, and there are a few branches, there develops a network of interacting fibres in the wall. Probably, this network is reinforced by β-glucan cross-linking proteins such as the Pir proteins (see Figure 10.1, structure 4), which will be discussed later. Chitin, an unbranched polymer of β-(1→4)-linked *N*-acetylglucosamine (see Figure 10.1, structure 1), is present in the walls of almost all fungi, at least in some stages of the life cycle and, indeed, chitin is a conserved character in walled cysts of many eukaryote kingdoms (Cavalier-Smith and Chao, 2003). The ratio of chitin to other polysaccharides varies widely between species. Whereas chitin is absent from walls of vegetative cells of *Schizosaccharomyces pombe* and usually present only in low amounts in the walls of *C. albicans* and *S. cerevisiae*, it is, in some fungi, the major cell wall polysaccharide. Chitin forms insoluble ribbons of stacked polysaccharide chains with great tensile strength. In other organisms, cellulose [β-(1→4)-glucan], chitosan [β-(1→4)-polyglucosamine] or α-(1→3)-glucan are the major fibrous wall polysaccharides. Each of these can form ribbons or helical structures. Thus, the fibrous polysaccharides constitute a cross-linked framework that gives the wall its insolubility and mechanical strength.

A β-(1→6)-glucan with β-(1→3) branches is a minor but essential component of yeast cell walls (see Figure 10.1, structure 3). This highly branched glucan, which is water-soluble and consists of a few hundred monosaccharide units per molecule, probably does not have a regular secondary structure, but serves as a flexible linking molecule, which is covalently bonded to β-(1→3)-glucan chains, chitin chains and glycosylphosphatidylinositol- (GPI-) linked glycoproteins through transglycosylation reactions. These cross-linking reactions apparently take place within the growing wall, so they are exterior to the plasma membrane. Plasma membrane-bound GPI proteins and cell wall resident proteins, such as the family of Gas transglycosylases, ScBgl2, ScCrh1, Scw1 and ScDfg5, have been implicated in these activities, but details are not yet known.

The wall polysaccharides elicit responses in host organisms, presumably because they are markers of fungal infections. Thus, chitin and chitosan oligosaccharides are elicitors of plant host defence pathways. Similarly, β-(1→3)-glucan binds to the immunomodulatory C-type lectin dectin, leading to proinflammatory reactions aimed at clearing the infection (Dennehy and Brown, 2007; Sun and Zhao, 2007). Recent work implies that glucan becomes more exposed on cell surfaces as the fungi are damaged, thus contributing to increased inflammation as the infection is contained (Wheeler and Fink, 2006).

3.2. Capsules

Capsules are characteristic of the basidiomycete *Cryptococcus neoformans* and a few other species. *C. neoformans* capsules consist of long linear chains of α-(1→3)-linked mannosyl residues, decorated with highly antigenic 6-*O*-acetyl groups, as well as β-linked glucuronic acid and β-linked xylopyranose units. The capsules form a fibrous network that can double or triple the apparent cell diameter and, like bacterial capsules, they render the cells resistant to immune-system surveillance and killing. *C. neoformans* capsules are released from the cell following treatment with dimethyl sulfoxide. This result implies that they are non-covalently bound to the walls, rather than being part of the cross-linked structure (Cherniak *et al.*, 1998; Frases *et al.*, 2008; Rodrigues *et al.*, 2008; Yoneda and Doering, 2008). A recent report implicates chitinous structures in the wall as anchor sites for the capsule (Rodrigues *et al.*, 2008).

4. CELL WALL GLYCOPROTEINS OF ASCOMYCETOUS FUNGI

The major bioactive components of fungal cell walls are the glycoproteins, which form an exterior layer in the wall and, thus, are the major subject of this chapter. For a recent review of cell wall proteins in Basidiomycetes see Yin *et al.* (2008). The cell walls of ascomycetous fungi often display an external, fibrillar layer that consists of glycoproteins emanating into the environment (De Groot *et al.*, 2005; Yin *et al.*, 2008). This fibrillar layer is covalently linked to the underlying skeletal polysaccharides that form the internal layer of the wall. The external glycoprotein layer seems to consist largely of individual proteins. However, in some fungi, so-called fimbriae have been described, which are multi-subunit protein complexes that can be released from the wall by shearing forces (Yu *et al.*, 1994). For example, very long appendages can be seen to emerge from *C. albicans* walls, apparently mediating contact with buccal epithelial cells. Partial purification of these fibrillar structures revealed a 66-kDa glycoprotein as its major component.

It has been estimated that a haploid cell of baker's yeast in the exponential phase of growth possesses about two million covalently linked cell wall proteins (CWPs). There are several classes of CWP molecules (Yin *et al.*, 2008):

(i) The GPI-CWPs (see Figure 10.1, structures 5–7) possess a trimmed form of a GPI anchor (GPI_t) through which they are covalently bound to the skeletal framework, thereby forming protein–polysaccharide complexes such as CWP-GPI_t $\rightarrow$ β-(1$\rightarrow$6)-glucan $\rightarrow$ β-(1$\rightarrow$3)-glucan or CWP-GPI_t $\rightarrow$ β-(1$\rightarrow$6)-glucan $\leftarrow$ chitin (see Figure 10.1, structure 5, where the arrows in the complexes indicate the direction of the glycosidic linkages interconnecting the macromolecules of the complex). This seems to be the main class of covalently linked CWPs. They can be released from the wall by using phosphodiesterases or hydrofluoric acid-pyridine treatment, which cleave the phosphodiester bond in the GPI_t structure, or by cell wall hydrolases, e.g. β-(1$\rightarrow$6)-glucanase, β-(1$\rightarrow$3)-glucanase or chitinase. The mature protein is shorter than the predicted amino acid sequence as it lacks both the N-terminal signal peptide and the C-terminal GPI anchor signal sequence (see Figure 10.1). Both cleavage sites can be reliably predicted using suitable algorithms (De Groot *et al.*, 2005). When by molecular genetic means a C-terminally truncated form of a GPI protein is created without the GPI anchor signal sequence, the resulting protein is generally secreted into the growth medium, thus facilitating its purification.

(ii) The Pir-CWPs are secreted proteins that possess multiple internal tandem repeats. The repeats are believed to be responsible for cross-linking β-(1$\rightarrow$3)-glucan chains through an (alkali-sensitive) ester linkage between a specific de-amidated glutamine residue in the repeats and hydroxyl groups of the glucosyl residues (Ecker *et al.*, 2006) (marked by "X" in Figure 10.1, structure 4). Interestingly, CaPir1 is essential for cell viability, emphasizing the role of Pir-CWPs in cell wall integrity. Some GPI-CWPs in *S. cerevisiae* also possess a Pir-repeat, thus allowing them to act as a cross-linking cell wall protein as well (in Figure 10.1, the left-most GPI-CWP contains both linkages, structures 4 and 5) (Klis *et al.*, 2006).

(iii) A third group of secretory CWPs, e.g. CaMP65, CaTos1 and ScScw4, ScScw10 and ScTos1, can also be released from hot detergent-extracted cell walls by using mild alkali but they lack the repeats that are characteristic for Pir-CWPs (De Groot *et al.*, 2005). The precise nature of their linkage to the wall is so far unknown.

(iv) Disulfide-linked secreted CWPs ScAga2, the subunit of the sexual agglutinin of mating type α-cells, is involved in recognition of such cells. This subunit is linked through two disulfide bridges to the GPI-CWP ScAga1 (Cappellaro *et al.*, 1994). Conceivably, the fimbriae observed in *C. albicans* and other fungi are also kept together by disulfide bonds.

(v) Secretory CWPs such as ScBgl2 that are non-covalently bound to cell wall polysaccharides, possibly, by the presence of a glycan-binding domain.

(vi) Some enzymes that hydrolyse nutrients, including phosphatases and glycosidases like invertase are physically entrapped in the periplasmic region and perhaps within the wall matrix. These glycoproteins can be solubilized from walls by heating in SDS sample buffer and are in the soluble fraction after breakage of the walls (Kapteyn *et al.*, 1997).

There are also numerous reports in the literature describing the presence of non-covalently bound, non-glycosylated, cytosolic proteins in the fungal cell wall. Unfortunately, these observations are rife with artefacts and possible export mechanisms for such proteins remain unclear (Yin *et al.*, 2008); these observations fall therefore out of the scope of this chapter.

5. CHARACTERISTICS OF FUNGAL CELL WALL GLYCOPROTEINS

Some important characteristics of the cell wall proteomes of ascomycetous fungi are the following:

(i) Many fungi seem to possess a wide arsenal of covalently linked CWPs, as indicated by mass spectrometric and biochemical analysis of the cell wall proteome. For example, the cell walls of *S. cerevisiae* and *C. albicans* contain at any time more than 20 different covalently linked CWPs in their walls (Yin *et al.*, 2008).

(ii) Some CWPs are among the most abundantly expressed proteins of the cell.

(iii) The CWPs often display a modular organization (see Figures 10.1 and 10.2). In GPI-CWPs, the effector domain is generally found in the N-terminal part of the protein, which extends into the medium. This is followed by a serine- (Ser-) and threonine- (Thr-) rich region; the C-terminal residue of the mature protein is linked through a GPI-remnant to the skeletal framework of the wall. The Ser- and Thr-rich region often contains an internal tandem series of repeats (see Figure 10.2). The multiple short *O*-linked carbohydrate side-chains in the Ser- and Thr-rich regions seem to rigidify these protein regions, allowing them to act as a spacer domain (Dranginis *et al.* 2007). In CWPs that lack a GPI anchor, the effector domain is not necessarily found in the N-terminal domain (Linder and Gustafsson, 2008).

(iv) As discussed in the previous section, some CWPs, such as the Pir-CWPs and the GPI-protein ScCwp1, are involved in cross-linking of cell wall polysaccharides. Thus, these CWPs serve covalently to cross-link walls into a single covalent complex that remains intact following boiling in SDS (see Figures 10.1 and 10.3). In agreement with this, both ScCWP1 and ScPIR genes are upregulated in response to cell wall stress.

(v) The CWPs also mediate homotypic and heterotypic cell aggregation, endocytosis, (mixed) biofilm formation, coping with oxidative stress and iron and sterol acquisition (De Groot *et al.*, 2005; Dranginis *et al.*, 2007; Yin *et al.*, 2008).

(vi) The CWPs as an external layer in the wall also have collective functions (see Figure 10.1). For example, the external glycoprotein layer limits the permeability of the cell wall to macromolecules and, thus, hampers recognition of the internal β-glucan layer of the wall by dectins, which are membrane-bound lectin-like receptor proteins belonging to the innate immune system of the host. The extended *N*-linked protein side-chains found in many CWPs seem to be largely responsible for limiting the permeability of the walls (De Groot *et al.*, 2005). In addition, disulfide links between CWPs may also be involved in limiting the permeability of walls to macromolecules (De Nobel *et al.*, 1990). Both *N*- and *O*-linked side-chains of CWPs may contain phosphodiester linkages (see Figure 10.1). At most physiological pHs, these phosphodiester groups are negatively

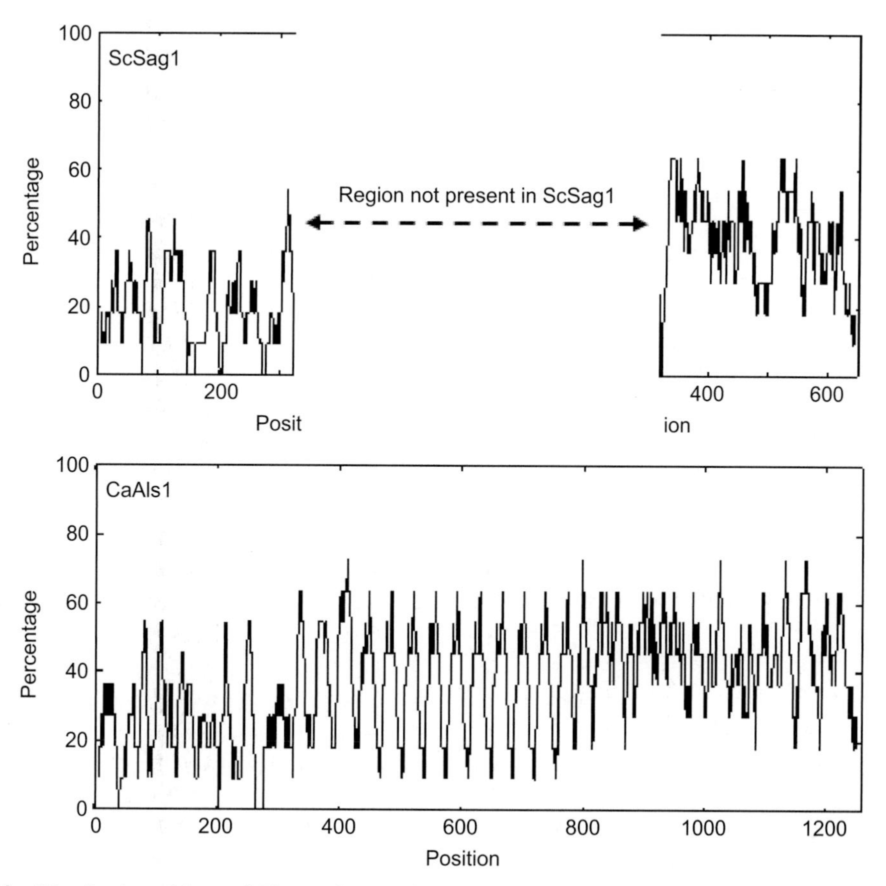

FIGURE 10.2 Distribution of Ser and Thr residues in the Ig-like GPI_t-CWP adhesins ScSag1 and CaAls1. Note that the Ser and Thr content of the first 300 amino acids, which corresponds to the effector domain, is relatively low. The repeats in the middle part of the CaAls1 protein are clearly recognizable. Upper panel: ScSag1; predicted sequence length, 650 amino acids; predicted secretory signal peptide, 1–19; predicted GPI signal sequence, 628–650; predicted S/T-content, 29.8%. Lower panel: CaAls1; predicted sequence length, 1260 amino acids; predicted secretory signal peptide, 1–17; GPI signal sequence, 1239–1260; predicted S/T-content, 36.8%. There are tandem repeats in the middle of CaAls1 that are not present in ScSag1, which has been split to align the similar regions. A sliding window of 11 amino acids was used.

charged and are thus responsible for the numerous negative charges at the fungal cell surface. Therefore, fungal cell walls act as cation exchangers, thus allowing them to bind positively charged proteins, e.g. proteins released from the cytosol. The change also explains the stainability of many fungi with the positively charged dye Alcian blue (De Groot *et al.*, 2005).

(vii) The CWPs are stable proteins that show only a limited release into the medium (Klis *et al.*, 2006).

(viii) The incorporation of specific CWPs in newly formed walls is often temporally and spatially controlled and also strongly depends on environmental conditions as further discussed below (Klis *et al.*, 2006).

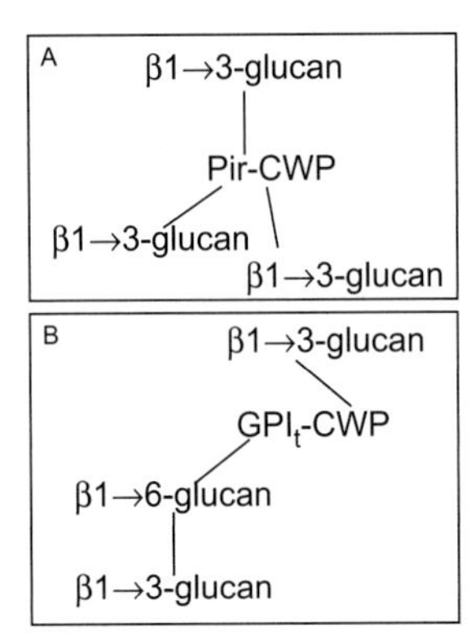

FIGURE 10.3 Potential cross-linking functions of Pir-CWPs and hybrid GPI-CWPs. Pir-CWPs possess a number of repeats, allowing them to cross-link multiple β-(1→3)-glucan chains. This is consistent with their location in the inner part of the cell wall. The ScCwp1 and other GPI-CWPs possess a single Pir repeat, allowing them to cross-link a β-(1→6)-glucan to a β-(1→3)-glucan. (A) Pir-CWP cross-links to multiple β-(1→3) glucans through tandem repeats. (B) GPI_t-CWP cross-linking to β-(1→3) glucan through an ester bond and to β-(1→6) glucan through the truncated and transglycosylated GPI.

6. THE MANNOPROTEIN GLYCANS

In many mannoproteins, the glycan part outweighs the peptide part by a 10:1 ratio. There are three types of glycosylation known in fungal cell wall proteins: *O*-glycans (see Figure 10.1, structure 6), *N*-glycans (structure 7) and GPI anchors (structure 5). Each type of glycosylation shares characteristics of structure and biosynthesis with mammals and other eukaryotes, with some characteristic fungal differences. In most cases, the *O*-linked glycosyl side-chains consist of manno-oligosaccharides that are occasionally decorated with other sugars, whereas the *N*-linked side-chains are hypermannosylated, but contain a core structure similar to mammalian *N*-chains. The core structure is often extended with a very large branched α-mannan with an α-(1→6)-linked backbone that may be decorated with various other sugars as well. GPI anchors are characteristic of a large number of wall proteins and, in fungi, they are covalently bonded to the wall polysaccharides.

6.1. O-Glycans

See Lehle *et al.* (2006) and Goto (2007) for excellent reviews on this subject. Invariably, proteins that traverse the secretory pathway to the cell surface acquire some or all of the glycan modifications just described. Protein glycosylation is therefore tightly linked to the cell's secretory pathway. In ascomycetous fungi, secretory proteins that become part of the cell membrane and the cell wall receive their first *O*-linked mannosyl residue at the endoplasmic reticulum. This first mannose is co- or post-translationally added to the protein, with dolichylphosphate-mannose as the sugar donor and protein *O*-mannosyltransferases (PMTs) as the enzymes that catalyse the covalent attachment of the mannose residue to target amino acids. In two independent studies, *O*-glycosylations in GPI-CWPs were mapped and shown to be absent from the N-terminal domains of the proteins (Chen *et al.*, 1995; Rauceo *et al.*, 2006).

Following addition of the first mannosyl residue at the endoplasmic reticulum (ER), glycoproteins are transported to the Golgi apparatus where *O*-mannosyltranferases of the *MNT* family catalyse addition of the second, third and subsequent mannosyl groups in Mn^{2+} dependent reactions that use GDP-Man as the sugar donor (Lehle *et al.*, 2006). These enzymes involved in the elongation of *O*-linked chains reside in the Golgi apparatus with their large catalytic carboxy-terminal domains luminally oriented.

Several *MNT* mannosyltranferases have been identified that show specificity for the type of linkage they catalyse and for the target protein. For example, three proteins, Ktr1, Ktr3 and Kre2/Mnt1 catalyse the transfer of the second, α-(1→2)-linked mannosyl residue and Mnt1 is also responsible for the attachment of the third, α-(1→2)-linked mannose. Finally in *S. cerevisiae*, the fourth and fifth mannose, both α-(1→3)-linked, are thought to be transferred by Mnn1, a homolog of Mnt1 that has been reported to participate in both *N*- and *O*-glycosylation (Gemmill and Trimble, 1999). That *O*-glycan extension proceeds in the Golgi apparatus was first demonstrated by the use of several conditional *sec* mutants in which vesicular transport from the ER to the Golgi apparatus is blocked. For example, in the temperature-sensitive *sec18* mutant, 90% of newly formed *O*-linked saccharides exist as single mannosyl residues at the non-permissive temperature (Ferro-Novick *et al.*, 1984).

Protein *O*-glycosylation in fungi is diverse with respect to sugar components and to the mode of linkage connecting the sugars. In yeasts, such as *S. cerevisiae*, *C. albicans* and *Pichia pastoris*, *O*-linked glycans consist of mannose residues that are linked through both α-(1→2), α-(1→3) and β-(1→2) linkages. In contrast to *N*-linked glycans, which can reach lengths of up to 200 mannose units and be highly branched, *O*-linked glycans are short (3–7 mannoses) and mostly linear. An exception occurs in *S. pombe*, in which galactopyranose branches may be added to the mannose chains and contribute to the formation of a mildly branched structure. Similarly branched structures form when phosphorylated mannoses are added to the first or second mannosyl residues and serve as branching points for the addition of new mannoses (see Figure 10.1, structure 6). This type of chain modification may occur in all ascomycetous fungi and impart a negative charge to the cell wall.

In filamentous fungi, such as *Aspergillus* species and *Trichoderma reseei*, protein *O*-linked glycans are highly variable in their sugar components and in the linkages connecting them. In addition to mannose, filamentous fungi may also incorporate glucose, galactofuranose and galactopyranose to their *O*-linked chains and form more highly branched structures than those observed in yeasts. In addition to the α-(1→2), α-(1→3) and β-(1→2) linkages common in yeast, sugar residues of *O*-glycans in filamentous fungi may be linked through α-(1→6), β-(1→5) (for galactofuranoses) and β-(1→6) linkages (Goto, 2007).

Such variability in chain structure and sugar components has been proposed to serve as an identifying feature for glycoprotein sorting and specific cellular function. Protein *O*-linked chains may modulate glycoprotein function by promoting protein stability since these can protect proteins from protease degradation and prevent the formation of protein aggregates.

Protein *O*-mannosylation is vital; complete abolition of this process results in cell death, whereas deficiencies lead to defects in important physiological processes such as cell budding, protein secretion and to activation of the cell wall integrity pathway. The cell wall integrity pathway protects the cell from wall damage caused by environmental stress or genetic defects. In pathogenic fungi, such as the dimorphic *C. albicans*, defects in *O*-linked glycosylation lead to reduced growth rates and to the formation of cellular aggregates. Furthermore, *O*-linked glycosylation mutants of *C. albicans* exhibit defective hyphal morphogenesis, reduced adherence to epithelial cells and defective biofilm formation. These mutants also show hypersensitivity to commonly used antifungal agents (Willer *et al.*, 2003).

Unlike the tripeptide Asn-Xaa-Ser/Thr signal (where Asn, asparagine; Ser, serine; Thr, threonine) for *N*-linked glycosylation, no consensus has been described for *O*-linked sugars. Nevertheless, proline (Pro) at position − 1 or + 3 from a target Ser or Thr appears to favour *O*-linked glycosylation. Also, in experiments with synthetic acceptor peptides, it was shown that at least a tripeptide is necessary for *O*-linked glycosylation to proceed and that the process is favoured by increasing the peptide length.

6.2. N-Glycans

N-glycans can occur in the N-terminal functional domains, the mid-piece tandem repeat regions and the C-terminal Ser- and Thr-rich stalk regions of GPI-CWPs and other classes of CWPs (Dranginis *et al.*, 2007). In structured regions, *N*-glycans tend to occur in the loop regions in compact protein domains (Grigorescu *et al.*, 2000).

Yeast *N*-glycans are high-mannose structures based on a core structure common with mammalian *N*-glycans. The core structures are added to the nascent protein in the rough ER. The core Man_9Glc_3 is trimmed in the ER, then re-elongated in the Golgi apparatus. The mature *N*-glycan structure is presented in Figure 10.1, structure 7. The outer branch structures are similar to the *O*-glycans (for reviews, see Dean, 1999; Gemmill and Trimble, 1999). The non-reducing termini of the branches often have species-specific terminal sugars. Some examples include β-(1→2)-linked mannose chains in *C. albicans* or α-linked *N*-acetylglucosamine in *Kluyveromyces lactis*. Phosphodiesters of mannose are also common in the branches.

The *N*-glycans of wall mannoprotein carry out many functions mentioned above and below. The *N*-glycans contain many of the phosphodiester groups that give the cell surface its charge and their great size occupies much of the volume of the wall matrix thus restricting wall permeability. Both *N*- and *O*-glycosylations contribute to protease resistance of the glycosylated regions (Chen *et al.*, 1995; Klis *et al.*, 2006).

6.3. GPI anchors

Mannoproteins with a carboxy-terminal signal peptide for GPI-anchoring receive a GPI anchor in the lumen of the ER (Orlean and Menon, 2007). This step is mediated by a transamidase, which cleaves the protein at about 20–30 amino acids from the C-terminus and adds a pre-assembled glycolipid anchor to the new carboxyterminal residue. In the trans-amidase reaction, an amide linkage is formed between the protein and the ethanolamine phosphate of the third mannose residue. The GPI-anchoring ties a protein with its C-terminus to the membrane, allowing it to travel to the cell surface within the lumen of cellular vesicles. The GPI-anchoring of mannoproteins is an essential function in ascomycetous yeasts but not, e.g. in the clinically important, filamentous fungus *Aspergillus fumigatus* (Li *et al.*, 2007). The preassembled glycolipid anchor consists of a phosphatidylinositol group, a glucosamine residue (GlcN, obtained by deacetylation of GlcNAc) and four or five mannose residues (Figure 10.4). The first three mannose residues each receive an ethanolamine phosphate group, added by three different enzymes of the Mcd4 family. Modification of the GPI anchor has been reported to occur in the Golgi apparatus during the translocation of GPI-proteins to the cell surface. In baker's yeast, about 50% of the GPI-protein molecules become covalently linked to the cell wall, the remainder being retained at the plasma membrane. Some GPI-proteins are known as typical cell wall resident proteins (e.g. ScCwp1, ScSag1, CaAls1, CaHwp1) and others as abundant plasma membrane proteins (ScEcm33, ScGas1), however, intermediate distributions also occur. The distribution between wall and plasma membrane depends on characteristics of the sequence immediately upstream of the GPI addition site. Consecutive basic residues in this region seem to stimulate plasma membrane retention, however, this may be overridden when they are preceded by stretches of Ser and/or Thr residues. Cleavage of GPI anchors of proteins destined for wall incorporation occurs between the inositol and the first mannose residue (see Figure 10.4). In the same or in a subsequent reaction, the active sugar end of the trimmed anchor is cross-linked to a non-reducing end of β-(1→6)-glucan. This positions the protein in the outermost layer of the cell while covalently linking it to the cell wall framework. Which protein or protein complex

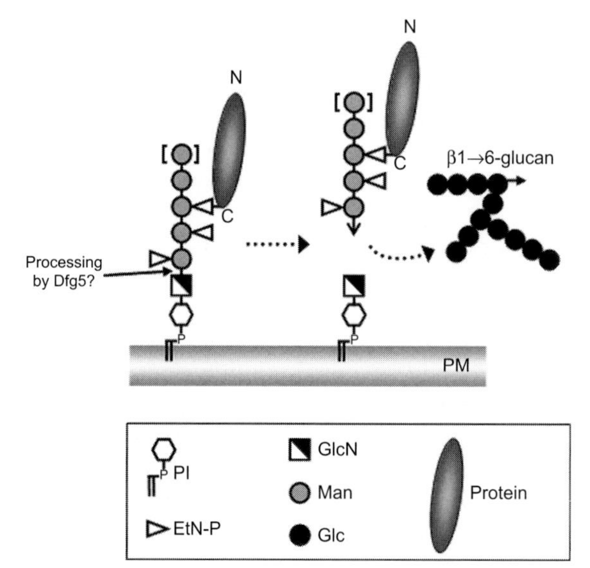

FIGURE 10.4 GPI truncation and transglycosylation of a GPI_t-CWP. The carbohydrate active enzyme Dfg5 is believed to be involved in at least the first step. Key: PM, plasma membrane; PI, phosphatidyl inositol; EtNP, ethanolamine phosphate; GlcN, glucosamine; Man, mannose; Glc, glucose. The N- and C-termini of the polypeptide are marked. The presence or absence of the non-reducing-terminal mannose residue has not been established, so it is shown in brackets.

is responsible for anchor cleavage is still uncertain. Interesting candidates are the paralogues Dfg5 and Dcw1, belonging to subfamily 76 according to the carbohydrate-active enzyme classification. Both are plasma membrane-localized GPI proteins and homologues of bacterial endomannosidases. Their coupled deletion leads to synthetic lethality in both *S. cerevisiae* and *C. albicans* (Kitagaki *et al.*, 2002).

7. FUNCTIONS OF WALL GLYCOPROTEINS

7.1. Fungal adhesins

An important and large group of GPI-CWPs is involved in adhesion, both homotypic and heterotypic. Two major classes can be distinguished, the immunoglobulin- (Ig-) like adhesins and the lectin-like adhesins and there are also adhesins such as CaHwp1 and CaEap1 that do not fall into either category. In *S. cerevisiae*, *C. albicans* and *C. glabrata*, many adhesins share a common overall design with an N-terminal, well-folded ligand-binding domain, a Thr-rich midpiece composed of tandem repeats and a Ser- and Thr-rich tail that is probably in an extended conformation with little regular structure (Dranginis *et al.*, 2007; De Groot and Klis, 2008).

7.1.1. Ig-like adhesins

The Ig-like adhesins are found in *S. cerevisiae* as mating adhesins (e.g. ScSag1) and in *C. albicans* the Als (Agglutinin-like sequence) family, which consists of eight members (Dranginis *et al.*, 2007; Hoyer *et al.*, 2008). Figure 10.2A shows a plot of the Ser and Thr content of ScSag1. The mating adhesin ScSag1 has a two-domain structure: the N-terminal half of about 300 amino acids is relatively poor in these two residues and is homologous to the immunoglobulin (Ig) superfamily (Dranginis *et al.*, 2007), whereas the C-terminal half is enriched in Ser and Thr. As illustrated in Figure 10.2B, CaAls1 has a three-domain structure with an N-terminal domain which is also homologous to the immunoglobulin family, a middle part which is enriched in Thr residues and consists of a series of repeats and a third part which is enriched in serine and threonine residues but lacks a regular structure (Hoyer *et al.*, 2008). The number of repeats in the middle part may vary, resulting in considerable allelic variation (Hoyer *et al.*, 2008). Recent modelling work has predicted that the Thr-rich tandem repeats in Als proteins form a string of compact subdomains that have non-specific hydrophobic interactions with a variety of fungal and host protein structures (A. Frank *et al.*, unpublished results). In contrast, the C-terminal Ser- and

Thr-rich sequences probably form an extended, unstructured stalk that elevates the interaction domains away from the cell surface. It is tempting to extrapolate these observations to other GPI-CWPs with a similar domain organization and, in particular, to ScMuc1. The number of repeats in the middle part of ScMuc1/Flo11 is correlated with the hydrophobicity of the protein and thus the capacity of the cells to adhere to hydrophobic surfaces, and with homotypic aggregation of the cells, indicating that the middle part of the protein not only contributes to allelic variability but has other important physiological functions as well (Dranginis *et al.*, 2007).

7.1.2. Lectin-like adhesins

Lectin-like adhesins are found in *S. cerevisiae*, where they are involved in flocculation, the Flo (flocculin) family, and in *C. glabrata*, where they are involved in adhesion to host cells, the Epa (Epithelial adhesion) family. The Ser and Thr distribution plots show that the members of both families consist of an N-terminal effector domain followed by a Ser- and Thr-rich middle part consisting of a variable number of tandem repeats and a Ser- and Thr-rich C-terminal part without a regular structure (De Groot and Klis, 2008).

Amazingly, the Epa lectin family in *C. glabrata* includes 23 members (Zupancic *et al.*, 2008). To study the function of individual adhesins in the absence of their family members, they (or their N-terminal domains) are often heterologously expressed as GPI-CWPs in a non-adhesive strain of *S. cerevisiae*, where they are incorporated in the cell wall in a similar way as in the donor organism (Rauceo *et al.*, 2006; Dranginis *et al.*, 2007; Hoyer *et al.*, 2007; Zupancic *et al.*, 2008). Domain deletion and domain exchange experiments in this so-called *S. cerevisiae* surface display model have shown that the N-terminal region of fungal adhesins is largely responsible for substrate binding. In a seminal paper, Zupancic *et al.* (2008) combined the *S. cerevisiae* surface display model with the use of glycan microarrays for determining the glycan specificity of the effector domains of three Epa adhesins of *C. glabrata*. This allowed them to establish that the PA14 (anthrax Protective Antigen) domain, a short conserved domain within the N-terminal effector region and, in particular, a pentapeptide sequence in a predicted surface loop of the PA14 domain, governs glycan specificity and cell adhesion. All Epa proteins possess a PA14 domain and, except for ScMuc1/Flo11, a similar PA14 domain is present in the N-terminal effector domain of the flocculin family of *S. cerevisiae*, but is missing from the Ig-like and other adhesins (de Groot and Klis, 2008).

8. DYNAMICS OF THE FUNGAL WALL PROTEOME

The availability of genomic sequences in combination with microarray studies and mass spectrometric approaches has rapidly advanced our knowledge about the dynamics of the cell wall proteomes of ascomycetous fungi. Genomic transcription studies clearly show that the transcript levels of CWP-encoding genes are tightly controlled. This raises the question how well transcript levels and protein levels are correlated. Immunological analysis indicates that transcript levels of CWP-encoding genes tend to be correlated with the corresponding protein levels. As mentioned below, this has also been confirmed by a quantitative mass spectrometric analysis of the cell wall proteome of a wall mutant of *S. cerevisiae*, suggesting that transcript levels of CWP-encoding genes have predictive value for the corresponding protein levels in the wall (Yin *et al.*, 2008). However, the available evidence is still limited.

Both internal and external cues affect the composition of the fungal wall glycoproteome, both quantitatively and qualitatively. Some examples of these are:

(i) Phase of the cell cycle. Many CWP-encoding genes are only expressed at specific phases of the cell cycle and the corresponding proteins are predominantly found in specific areas of the wall. This has been studied in most detail in the yeast *S. cerevisiae* (Klis *et al.*, 2006).

(ii) Mating pheromones. Not surprisingly, the expression of the mating adhesins is strongly induced in response to the mating pheromone of the opposite mating partner (Dranginis *et al.*, 2007).

(iii) Cell wall stress. It is known for *S. cerevisiae* that, in response to cell wall stress, a specific set of CWP-encoding genes is strongly upregulated and that this response is mediated by the cell wall integrity signalling pathway (Klis *et al.*, 2006). As mentioned above, this group of genes includes the *PIR* family genes and *CWP1*, which both encode β-glucan-crosslinking CWPs. It has further been established that, in yeast cells deleted for *GAS1*, a gene that encodes a plasma membrane-bound β-(1→3)-transglucosylase, the cell wall integrity pathway is constitutively activated. Quantitative mass spectrometric analysis of the cell wall proteome of such cells indeed revealed considerable changes as predicted by earlier transcript profiling studies (Yin *et al.*, 2008).

(iv) Hypoxic conditions. Both in *C. albicans* and in *S. cerevisiae* considerable changes in the cell wall proteome occur in response to low levels of oxygen (Klis *et al.*, 2006; Sosinska *et al.*, 2008; Yin *et al.*, 2008).

(v) Iron limitation. Iron limitation results in the increased incorporation of iron-acquisition proteins in the wall of *C. albicans* and *S. cerevisiae* (Klis *et al.*, 2006; Sosinska *et al.*, 2008).

(vi) Neutral versus acidic pH. Transcript profiling indicates that, both in *C. albicans* and in *S. cerevisiae*, the external pH strongly affects the cell wall proteome and this is supported by the available mass spectrometric and immunological evidence (Klis *et al.*, 2006; Sosinska *et al.*, 2008; Yin *et al.*, 2008).

9. CONCLUSIONS

We have briefly illustrated the structures and the roles for glycoconjugates in fungal cell walls. The overall model is of a covalently cross-linked complex of polysaccharides and glycoproteins. The polysaccharides comprise a fibrous matrix of long H-bonded fibres. Some of the fibres are cross-linked to wall glycoproteins through modified GPI anchors, carboxyl esters and some unknown linkages. Individual glycoprotein molecules can be cross-linked to several polysaccharide chains and they are also cross-linked to each other through disulfide bonds. The matrix also physically entraps some proteins. The wall glycoproteins themselves are highly *N*- and *O*-glycosylated and they carry out many cellular activities including adhesion, proteolysis, extracellular metabolism and wall assembly, maintenance and remodelling. Thus, the overall result is that fungal cell walls are covalently cross-linked cell surface complexes of glycoconjugates with a wide variety of functions.

This chapter points to several areas where additional research can add to our basic understanding in glycobiology. The areas and questions that are still unclear and potentially will be the focus for future research are listed in the Research Focus Box.

RESEARCH FOCUS BOX

- What are the mechanisms of synthesis and assembly of multi-molecular polysaccharide fibrils in fungal cell walls, including β-glucans and chitin?
- What are the cellular sites and mechanisms of synthesis of the β-(1→6)-glucan that cross-links wall components to each other?
- What are the genes encoding the proteins and mechanisms that cross-link cell wall glycoproteins to the polysaccharides?
- Why is *O*-glycosylation essential in yeast?
- How do the differing ligand specificities of fungal adhesins and lectins affect host-pathogen interactions?
- How and why do fungal cells regulate the incorporation of different glycoproteins into their cell walls?

ACKNOWLEDGEMENTS

Work at Brooklyn College was supported by NIH SCORE grant S06 GM075158 and work at the Swammerdam Institute for Life Sciences was supported by the EU programme FP7-2214004, FINSysB.

References

Cappellaro, C., Baldermann, C., Rachel, R., Tanner, W., 1994. Mating type-specific cell-cell recognition of *Saccharomyces cerevisiae*: cell wall attachment and active sites of **a**- and alpha-agglutinin. EMBO J. 13, 4737–4744.

Cavalier-Smith, T., Chao, E.E., 2003. Phylogeny of choanozoa, apusozoa, and other protozoa and early eukaryote megaevolution. J. Mol. Evol. 56, 540–563.

Chen, M.H., Shen, Z.M., Bobin, S., Kahn, P.C., Lipke, P.N., 1995. Structure of *Saccharomyces cerevisiae* alpha-agglutinin. Evidence for a yeast cell wall protein with multiple immunoglobulin-like domains with atypical disulfides. J. Biol. Chem. 270, 26168–26177.

Cherniak, R., Valafar, H., Morris, L.C., Valafar, F., 1998. *Cryptococcus neoformans* chemotyping by quantitative analysis of 1H nuclear magnetic resonance spectra of glucuronoxylomannans with a computer-simulated artificial neural network. Clin. Diagn. Lab. Immunol. 5, 146–159.

Coronado, J.E., Mneimneh, S., Epstein, S.L., Qiu, W.G., Lipke, P.N., 2007. Conserved processes and lineage-specific proteins in fungal cell wall evolution. Eukaryot. Cell 6, 2269–2277.

De Groot, P.W., Klis, F.M., 2008. The conserved PA14 domain of cell wall-associated fungal adhesins governs their glycan-binding specificity. Mol. Microbiol. 68, 535–537.

De Groot, P.W., Ram, A.F., Klis, F.M., 2005. Features and functions of covalently linked proteins in fungal cell walls. Fungal Genet. Biol. 42, 657–675.

De Nobel, J.G., Klis, F.M., Munnik, T., Priem, J., van den Ende, H., 1990. An assay of relative cell wall porosity in *Saccharomyces cerevisiae, Kluyveromyces lactis* and *Schizosaccharomyces pombe*. Yeast 6, 483–490.

Dean, N., 1999. Asparagine-linked glycosylation in the yeast Golgi. Biochim. Biophys. Acta 1426, 309–322.

Dennehy, K.M., Brown, G.D., 2007. The role of the β-glucan receptor Dectin-1 in control of fungal infection. J. Leukoc. Biol. 82, 253–258.

Dranginis, A.M., Rauceo, J.M., Coronado, J.E., Lipke, P.N., 2007. A biochemical guide to yeast adhesins: glycoproteins for social and antisocial occasions. Microbiol. Mol. Biol. Rev. 71, 282–294.

Ecker, M., Deutzmann, R., Lehle, L., Mrsa, V., Tanner, W., 2006. Pir proteins of *Saccharomyces cerevisiae* are attached to β-1,3-glucan by a new protein-carbohydrate linkage. J. Biol. Chem. 281, 11523–11529.

Ferro-Novick, S., Novick, P., Field, C., Schekman, R., 1984. Yeast secretory mutants that block the formation of active cell surface enzymes. J. Cell Biol. 98, 35–43.

Frases, S., Nimrichter, L., Viana, N.B., Nakouzi, A., Casadevall, A., 2008. Cryptococcus neoformans capsular polysaccharide and exopolysaccharide fractions manifest physical, chemical, and antigenic differences. Eukaryot. Cell 7, 319–327.

Gemmill, T.R., Trimble, R.B., 1999. Overview of *N*- and *O*-linked oligosaccharide structures found in various yeast species. Biochim. Biophys. Acta 1426, 227–237.

Goto, M., 2007. Protein *O*-glycosylation in fungi: diverse structures and multiple functions. Biosci. Biotechnol. Biochem. 71, 1415–1427.

Grigorescu, A., Chen, M.H., Zhao, H., Kahn, P.C., Lipke, P.N., 2000. A CD2-based model of yeast alpha-agglutinin elucidates solution properties and binding characteristics. IUBMB Life 50, 105–113.

Hoyer, L.L., Green, C.B., Oh, S.H., Zhao, X., 2008. Discovering the secrets of the *Candida albicans* agglutinin-like sequence (ALS) gene family – a sticky pursuit. Med. Mycol. 46, 1–15.

Kapteyn, J.C., Ram, A.F., Groos, E.M., et al., 1997. Altered extent of cross-linking of β1,6-glucosylated mannoproteins to chitin in *Saccharomyces cerevisiae* mutants with reduced cell wall beta1,3-glucan content. J. Bacteriol. 179, 6279–6284.

Kitagaki, H., Wu, H., Shimoi, H., Ito, K., 2002. Two homologous genes, *DCW1* (YKL046c) and *DFG5*, are essential for cell growth and encode glycosylphosphatidylinositol (GPI)-anchored membrane proteins required for cell wall biogenesis in *Saccharomyces cerevisiae*. Mol. Microbiol. 46, 1011–1022.

Klis, F.M., Boorsma, A., De Groot, P.W., 2006. Cell wall construction in *Saccharomyces cerevisiae*. Yeast 23, 185–202.

Lehle, L., Strahl, S., Tanner, W., 2006. Protein glycosylation, conserved from yeast to man: a model organism helps elucidate congenital human diseases. Angew. Chem. Int. Ed. Engl. 45, 6802–6818.

Li, H., Zhou, H., Luo, Y., Ouyang, H., Hu, H., Jin, C., 2007. Glycosylphosphatidylinositol (GPI) anchor is required in *Aspergillus fumigatus* for morphogenesis and virulence. Mol. Microbiol. 64, 1014–1027.

Linder, T., Gustafsson, C.M., 2008. Molecular phylogenetics of ascomycotal adhesions – a novel family of putative cell-surface adhesive proteins in fission yeasts. Fungal Genet. Biol. 45, 485–497.

Orlean, P., Menon, A.K., 2007. Thematic review series: lipid posttranslational modifications. GPI anchoring of protein in yeast and mammalian cells, or: how we learned to stop worrying and love glycophospholipids. J. Lipid Res. 48, 993–1011.

Pitarch, A., Nombela, C., Gil, C., 2008. Cell wall fractionation for yeast and fungal proteomics. Methods Mol. Biol. 425, 217–239.

Rauceo, J.M., De Armond, R., Otoo, H., et al., 2006. Threonine-rich repeats increase fibronectin binding in the *Candida albicans* adhesin Als5p. Eukaryot. Cell 5, 1664–1673.

Rodrigues, M.L., Alvarez, M., Fonseca, F.L., Casadevall, A., 2008. Binding of the wheat germ lectin to *Cryptococcus neoformans* suggests an association of chitinlike structures with yeast budding and capsular glucuronoxylomannan. Eukaryot. Cell 7, 602–609.

Sosinska, G.J., de Groot, P.W., Teixeira de Mattos, M.J., et al., 2008. Hypoxic conditions and iron restriction affect the cell-wall proteome of *Candida albicans* grown under vagina-simulative conditions. Microbiology 154, 510–520.

Sun, L., Zhao, Y., 2007. The biological role of dectin-1 in immune response. Int. Rev. Immunol. 26, 349–364.

Wheeler, R.T., Fink, G.R., 2006. A drug-sensitive genetic network masks fungi from the immune system. PLoS Pathog. 2, e35.

Willer, T., Valero, M.C., Tanner, W., Cruces, J., Strahl, S., 2003. *O*-mannosyl glycans: from yeast to novel associations with human disease. Curr. Opin. Struct. Biol. 13, 621–630.

Yin, Q.Y., de Groot, P.W., de Koster, C.G., Klis, F.M., 2008. Mass spectrometry-based proteomics of fungal wall glycoproteins. Trends Microbiol. 16, 20–26.

Yoneda, A., Doering, T.L., 2008. Regulation of *Cryptococcus neoformans* capsule size is mediated at the polymer level. Eukaryot. Cell 7, 546–549.

Yu, L., Lee, K.K., Ens, K., et al., 1994. Partial characterization of a *Candida albicans* fimbrial adhesin. Infect. Immun. 62, 2742–2834.

Zupancic, M.L., Frieman, M., Smith, D., Alvarez, R.A., Cummings, R.D., Cormack, B.P., 2008. Glycan microarray analysis of *Candida glabrata* adhesin ligand specificity. Mol. Microbiol. 68, 547–559.

CHAPTER

11

Cytoplasmic carbohydrate molecules: trehalose and glycogen

Alan D. Elbein

SUMMARY

Trehalose and glycogen are two important glucose derivatives long considered to function as storehouses of glucose. However, trehalose is a diverse molecule having a number of other roles including: (i) protecting proteins and membranes against stress; (ii) acting as an allosteric inhibitor of carbohydrate metabolism and a transcriptional regulator; and (iii) serving as an essential component of mycobacterial cell walls and a donor of mycolic acid. Mycobacteria have three different pathways to produce trehalose and mutants missing all these pathways cannot grow without exogenous trehalose. Nevertheless, glycogen is not essential for growth but it does aid microbial survival when carbon and energy sources are absent. Bacterial glycogen is synthesized from adenosine diphosphate- (ADP-) glucose rather than uridine diphosphate- (UDP-) glucose, which is the precursor for animal glycogen synthesis. Furthermore, the bacterial enzyme that synthesizes ADP-glucose is a regulatory enzyme that controls glycogen formation. Activity of this enzyme is increased by glycolytic intermediates, such as pyruvate and fructose-phosphates, and inhibited by adenosine monophosphate (AMP). Thus, when glycolytic intermediates are plentiful, energy is available to synthesize glycogen and ADP-glucose is synthesized. However, when AMP levels are high, there is insufficient energy to produce glycogen and ADP-glucose production is decreased. In mycobacteria, glycogen can be converted to trehalose, probably to increase trehalose levels when they become dangerously low in the cytoplasm. On the other hand, mycobacteria can also convert trehalose to glycogen and this may occur when trehalose levels become dangerously high, since high trehalose levels may be toxic.

Keywords: Trehalose; Maltose; Trehalose synthase; Glycogen; Adenosine diphosphate- (ADP-) glucose; Regulation of metabolism; Glycogen-trehalose interconversion; Amylase

1. INTRODUCTION

The major intracellular sugars found in most, if not all, bacteria are glucose (Glc), trehalose and glycogen. Trehalose [α-D-glucopyranosyl α-D-glucopyranoside or α-D-Glc*p*-(1→1)-α-D-Glc*p*] and glycogen are higher homologues of Glc, containing two (trehalose), or many (glycogen), Glc residues glycosidically-linked to each other. Until more recent times, both of these compounds were considered to be produced strictly as storage forms of Glc that the cell could call upon as an immediate source of carbon and/or energy

(Elbein, 1974). For example, glycogen follows the three guidelines suggested by Wilkinson (1959) to demonstrate that a compound serves an energy-storage function. First, the given compound should accumulate intracellularly during times when the cell has excess energy. Second, the reserve material should be utilized as a source of carbon and energy when the cell no longer has exogenous sources of carbon or energy. Third, the organism should be able to use the reserve material for survival when it is in a non-supportive environment. In many bacteria, glycogen fulfils these requirements, whereas trehalose only fulfils some of these in some organisms and at certain times, but it is more suspect as a reserve than is glycogen. Trehalose can have a number of other important functions in living cells, which will be discussed below, in addition to serving as a reservoir of Glc. In this chapter, the occurrence and distribution, structure, biosynthesis, degradation and biological function(s) of these two very important Glc oligomers will be discussed. For certain microorganisms, such as mycobacteria, some of the trehalose metabolic pathways are apparently closely related and interconnected with the glycogen metabolic pathways; those inter-relations are discussed subsequently.

2. OCCURRENCE, DISTRIBUTION AND FUNCTION OF GLYCOGEN

Glycogen is found in many bacteria and usually accumulates either during the stationary phase or during periods of limited growth, when there is a readily available and sufficient source of energy and carbon to use for producing glycogen. However, some bacteria do synthesize glycogen efficiently, even when undergoing exponential growth (Govons *et al.*, 1973). Glycogen is not an essential component for bacterial growth or survival; deletion mutants in the biosynthetic genes for glycogen synthetic enzymes and, therefore, not able to produce this polymer, grow as well as the normal, glycogen-producing, parental strains (Dietzler *et al.*, 1974). However, these glycogen-deficient mutants have a shorter survival time when placed in media with no exogenous carbon source when compared with glycogen-containing strains. Furthermore, wild-type *Escherichia coli*, that synthesizes and contains glycogen, does not degrade its RNA or protein when subjected to conditions where a usable carbon source is unavailable due to its glycogen acting as an endogenous energy source (Preiss, 1989). Conversely, a glycogen-deficient mutant releases ammonia under these conditions indicating that protein degradation is occurring and that the bacterium is using its proteins for energy (Preiss, 1989). These data suggest that, under starvation conditions, glycogen serves as the principal energy source when it is available. Another suggested function for glycogen is to provide the energy and carbon required for spore formation (Mackey and Morris, 1971). Various *Clostridium* spp. accumulate large amounts of glycogen prior to or during the initiation of spore formation and this glycogen is rapidly degraded during spore formation (Mackey and Morris, 1971). On the other hand, glycogen-deficient strains are poor spore formers (Russell, 1992). Glycogen or other similar α-(1$\rightarrow$4)-glucans have been identified in at least 40 different bacterial species and this polymer is found in Gram-positive and Gram-negative bacteria, as well as in the Archae (Preiss and Romeo, 1989).

A number of studies have demonstrated that the series of events leading to glycogen accumulation during the stationary phase begins as a consequence of the depletion of an essential nutrient such as nitrogen or perhaps sulfur and this nutrient deficiency then results in the cessation of growth (Russell, 1992). Since many of these cultures will still have sufficient amounts of energy available, but are unable to grow in the absence of nitrogen, these bacteria store this energy and carbon in the form of glycogen. The results of studies with many different bacteria indicate that the greatest accumulation of glycogen occurs under conditions of limiting amounts of nitrogen (Lowry *et al.*, 1971).

3. STRUCTURE OF GLYCOGEN

Bacterial glycogen is similar in terms of its polysaccharide structure to that of mammalian glycogen (Manners, 1971). Thus, glycogen from either source is composed entirely of Glc molecules, most of which are linked in α-(1→4)-glycosidic bonds. However, this polymer is a highly-branched glucan, with each branch being initiated by a Glc attached in an α-(1→6)-glucosidic-bond to one of the 4-linked Glc residues (Manners, 1991). Each of these branches then contains a variable number of α-(1→4)-linked Glc residues. In bacteria, ≈8–16% of the total glycosidic linkages are α-(1→6)-bonds and most of the branches are ≈6–10 Glc residues in length (Yang *et al.*, 1996) (Figure 11.1).

One apparent difference between mammalian glycogen and bacterial glycogen is that, in animal cells, the glycogen molecule is attached to a protein called glycogenin (Krisman and Barengo, 1975). In this glycoprotein, the Glc at the reducing end of the glycogen chain is linked in an *O*-glycosidic linkage to the hydroxyl group of a tyrosine (Tyr) in the glycogenin. In contrast to animal glycogen, the bacterial glycogen appears to be present as a free polysaccharide without attachment to protein (see Figure 11.1) and glycogenin has not been found naturally occurring in bacteria (Yang *et al.*, 1996).

4. BIOSYNTHESIS OF GLYCOGEN

4.1. Biosynthesis of the glycogen precursor, adenosine diphosphate (ADP)-Glc

The precursor for the biosynthesis of glycogen in animal cells is uridine diphosphate-D-Glc (UDP-Glc), whereas the glucosyl donor for formation of bacterial glycogen is ADP-Glc (Preiss and Romeo, 1994). The mammalian glycogen synthase cannot use ADP-Glc as a substitute for UDP-Glc, nor can the bacterial glycogen synthase utilize UDP-Glc.

The enzyme that synthesizes ADP-Glc, i.e. ADP-Glc pyrophosphorylase or adenosine triphosphate (ATP):α-D-Glc-1-phosphate adenyltransferase, is a key enzyme in glycogen synthesis, since it represents the major regulatory step in the biosynthesis of glycogen (Preiss and Romeo, 1994). This enzyme catalyses the following reaction:

ATP + α-D-Glc-1-phosphate
↔ ADP-Glc + PPi (inorganic pyrophosphate)

Bacterial ADP-Glc pyrophosphorylases are all allosteric enzymes and regulation of their activity is a major controlling factor in the synthesis and/or accumulation of bacterial glycogen (Preiss, 1984). This regulation in most bacterial species involves the activation or stimulation of the pyrophosphorylase by various glycolytic intermediates and its inhibition by adenosine monophosphate (AMP), ADP or inorganic phosphate. Table 11.1 lists some bacterial ADP-Glc pyrophosphorylases, the activating compounds and the major bacterial pathway of carbon metabolism related to the specific activator of each pyrophosphorylase. In most cases, the synthesis of ADP-Glc is increased by glycolytic intermediates and inhibited by ADP, AMP and inorganic phosphate. High levels of various glycolytic intermediates in the cell are an indication of carbon and/or energy excess (Preiss, 2000). Thus, when cell growth is inhibited as a result of limiting nitrogen in the medium but excess carbon is available, the accumulation of glycolytic intermediates serves as signals for activation of the synthesis of ADP-Glc which serves mainly as a precursor for production of glycogen (Swedes *et al.*, 1975). The activation of

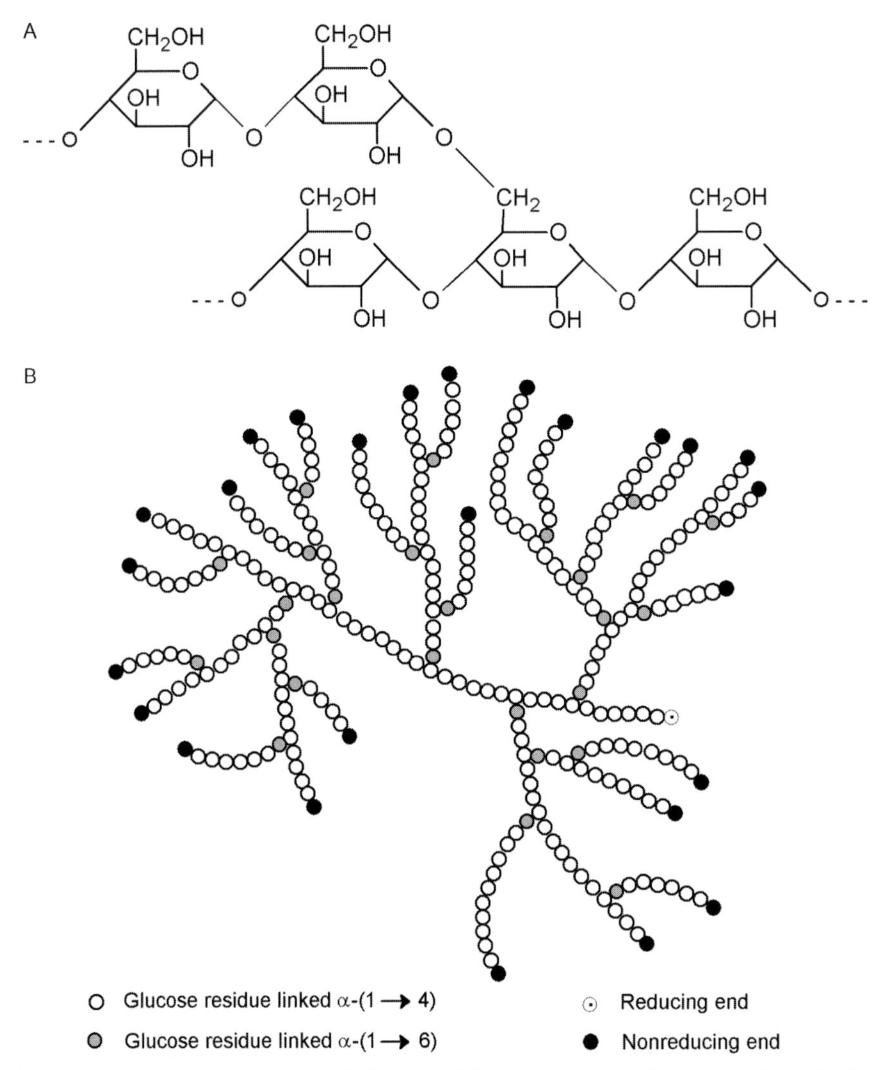

FIGURE 11.1 Generalized structure of bacterial glycogen. (A) A portion of the glycogen molecule is shown to highlight the nature of the glycosidic linkages between individual glucose (Glc) residues. The most common sugar linkage is α-(1→4), although each branch is connected by an α-(1→6)-glycosidic linkage. (B) The overall structure of bacterial glycogen is depicted by representing each Glc residue by a circle. Open circles, Glc residues in α-(1→4)-linkages; closed circles, Glc residues in α-(1→6)-linkages at branch points.

ADP-Glc pyrophosphorylases is likely to be due to increases in the affinity of the enzyme for the substrates ATP and Glc-1-phosphate. As shown in Table 11.1, variation in the specificity of the activator for each of the bacterial ADP-Glc pyrophosphorylases correlates with the type of carbon assimilation pathway occurring in that bacterium. For example, *E. coli* and *Salmonella enterica* sv. Typhimurium obtain their energy mostly through glycolysis and the primary activator for their ADP-Glc pyrophosphorylases is fructose-1,6-bisphosphate, with AMP being

TABLE 11.1 Activator specificity of various bacterial ADP-Glc pyrophosphorylases

Bacterial species	Activators	Carbon assimilation pathway
Rhodospirillum spp.	Pyruvate	Reductive pyruvate cycle
Rhodocyclus purpureus		
Cyanobacteria	3-Phosphoglycerate	Photosynthesis (Calvin cycle)
Agrobacterium spp.		Entner-Doudoroff cycle
Arthrobacter spp.	Pyruvate, fructose-6-phosphate	Reductive pyruvate cycle
Rhodopseudomonas spp.	Pyruvate, fructose-6-phosphate	Entner-Doudoroff, glycolysis, reductive pyruvate cycle
Mycobacterium smegmatis	Fructose-6-phosphate Fructose-1,6-bisphosphate	Glycolysis

the major inhibitor of the enzyme (Okita *et al.*, 1981). On the other hand, in photosynthetic bacteria, e.g. cyanobacteria, the major activator is 3-phosphoglyceric acid and the major pathway is the Calvin cycle (Levi and Preiss, 1976).

A number of ADP-Glc pyrophosphorylases have been purified from various bacteria and, in most cases, the enzyme was found to be a homotetramer composed of a 56 kDa subunit (Preiss *et al.*, 1976). Interestingly, there is a strong and direct correlation between the affinity of ADP-Glc pyrophosphorylase for its positive effector and the accumulation of glycogen in the cell. Thus, a number of *E. coli* mutants have been isolated with variant ADP-Glc pyrophosphorylases having different affinities for the activator fructose-1,6-bisphosphate. Studies with these mutants have shown that the higher the affinity for the activator, the more glycogen accumulated in that mutant (Govons *et al.*, 1973). Chemical modifications with azido-labelled substrates, as well as site-directed mutagenesis studies, have been used to locate the activator binding site, the inhibitor binding site and the substrate binding site in this enzyme. In addition, various mutated ADP-Glc pyrophosphorylase genes (*glgC*) have been cloned and expressed in various mutants of *E. coli* and *S. enterica* sv. Typhimurium. Analysis of the allosteric properties of these mutant enzymes have provided important and extensive information on structure–function relationships of this enzyme (Preiss and Romeo, 1994).

4.2. Biosynthesis of α-(1→4)-linked glucan chain (glycogen precursor)

The bacterial glycogen synthases are enzymes that add Glc from ADP-Glc to existing glycogen chains and link these Glc residues in α-(1→4)-glycosidic bonds (Shen *et al.*, 1964). Thus, this enzyme, by itself, produces a long, linear α-(1→4)-linked glucan that is unbranched. The bacterial synthases are specific for the glucosyl donor, ADP-Glc (Sigal *et al.*, 1964). In those cases where the enzyme has been examined in more detail, the subunit size of the protein is about 50 kDa and the active enzyme is either a dimer or a tetramer (Fox *et al.*, 1976). The structural gene, *glgA*, for glycogen synthase, has been cloned from a number of bacteria (i.e. *Bacillus stearothermophilus, E. coli, S. enterica* sv. Typhimurium, *Agrobacterium tumefaciens*) and their nucleotide sequences have been determined (Kumar *et al.*, 1986). The *E. coli* sequence comprises 1431 base pairs (bp) that specify a protein of 477 amino acids with a molecular mass of 52412. Having this information has allowed site-directed mutagenesis to be undertaken to determine the essential amino acids in the substrate binding sites as well as other functional aspects of the various synthases. For example, studies on substrate binding using a substrate affinity analogue, ADP-pyridoxal, identified Lys15 in the sequence Lys-X-Gly-Gly (where Lys = lysine and Gly = glycine) as an essential amino acid in binding the substrate ADP-Glc (Tagaya *et al.*, 1985). Site-directed mutagenesis indicated that this Lys is probably involved in binding to the phosphate residue that is adjacent to the glycosidic linkage holding the Glc portion of ADP-Glc, but this Lys is not directly involved in the catalytic event. Hence, changing this Lys for a glutamic acid raised the Km for the ADP-Glc substrate ≈30- to 50-fold (Furukawa *et al.*, 1994).

The bacterial glycogen synthase differs from the mammalian enzyme in at least two major aspects. First, the bacterial synthase is not a regulatory protein and does not exist in an active and inactive form, as do the mammalian synthases. Thus, there is no evidence for either phosphorylation or dephosphorylation of the bacterial glycogen synthases, as occurs with the mammalian glycogen synthases (Chock *et al.*, 1980), nor is there any indication of other enzyme catalysed modifications of bacterial glycogen synthases. Second, as mentioned earlier, the bacterial enzyme uses ADP-Glc as the glycosyl donor (Preiss and Greenberg, 1965), whereas the mammalian enzymes all use UDP-Glc (Leloir *et al.*, 1961; Krisman and Barengo, 1975). In one report, the *E. coli* glycogen synthase showed <1% activity with either UDP-Glc, cytidine diphosphate- (CDP-) Glc or guanosine diphosphate- (GDP-) Glc, as compared to ADP-Glc (Holmes and Preiss, 1979). Of note, various α-glucans can serve as effective primers for the bacterial glycogen synthase, including glycogen from either animal or bacterial sources, as well as plant amylose or amylopectin (Preiss and Romeo, 1989).

4.3. Branching of the α-glucan chain

Glycogen synthases are α-(1→4)-glucan synthases and, as such, they catalyse the transfer of Glc residues from nucleoside diphosphate glucoses to glucan chains with the formation of α-(1→4) glycosidic bonds (Cori and Cori, 1939). However, glycogen is a highly branched polymer and each branch is attached by an α-(1→6)-glycosidic linkage. Therefore, a branching enzyme is necessary to convert this linear glucan into a highly branched glucan. This branching enzyme cleaves an α-(1→4)-glucosidic linkage 6–10 glucose from the non-reducing end of the newly synthesized α-(1→4)-glucan and then transfers this oligosaccharide to the main glucan chain and attaches it to the –OH group at position-6 of one of the Glc residues somewhere in the main chain to produce a (1→6)-linked branch. The genes for the branching enzymes of a number of different bacteria have been cloned (Baecker *et al.*, 1986). The *E. coli glgB* gene contains 2181 bp, specifying a protein of 727 amino acids, with a molecular mass of 84231. The amino acid sequence of the *E. coli* branching enzyme has been compared to the amino acid sequences of amylolytic enzymes, e.g. α-amylase, pullulanase, glucosyltransferase and cyclodextrin glucanotransferase (Romeo *et al.*, 1988). A marked conservation was noted in the amino acid sequences of the catalytic sites of the four amylolytic enzymes with a sequence in the bacterial branching enzyme, as well as in the branching enzymes from plant and animal sources. It is not surprising that the bacterial branching enzymes show such strong similarity to amylases, since they also catalyse the cleavage of α-(1→4)-glycosidic bonds (Robyt and French, 1970).

5. DEGRADATION OF GLYCOGEN

A number of enzymes in bacteria exist that can degrade glycogen (MacGregor *et al.*, 2001). For endogenous glycogen to be utilized as an energy source, many bacterial species have a glycogen phosphorylase that can catalyse the following reaction:

$$\text{Glycogen} + H_3PO_4 \leftrightarrow \text{Glc-1-phosphate} + \text{Glycogen (-1 Glc residue)}$$

This enzyme removes one Glc residue from the non-reducing end of any branch of glycogen, yielding Glc-1-phosphate, in a phosphorolysis reaction using inorganic phosphate rather than ATP as the phosphate donor (Chen and Segal, 1968). Thus, this enzymatic reaction saves the cell one ATP for every Glc released from glycogen. Mammalian glycogen phosphorylases are regulatory enzymes that are activated by phosphorylation, i.e. catalysed by a protein kinase, but the bacterial phosphorylases are not known to be regulatory enzymes (Dietzler and Strominger, 1973).

Bacteria also possess various types of amylases, i.e. enzymes that cleave internal α-(1→4)-glucosidic bonds of glycogen or maltodextrins to produce oligosaccharides (OSs) of various sizes (MacGregor *et al.*, 2001). Exoamylases, or β-amylases, catalyse a successive hydrolysis of alternate α-(1→4)-linkages with the stepwise release of maltose units whose reducing ends have the β-configuration (Henrissat, 1991). Degradation by β-amylase starts at the non-reducing end of a glucan chain and ceases when the amylase gets close to the branch, i.e. the α-(1→6)-linkage point. On the other hand, endoamylases, or α-amylases, cleave internal α-(1→4)-bonds of linear or branched glucans into smaller OSs and maltose products that possess a new reducing group in the α-configuration (Qian *et al.*, 1994). Some amylases are secreted from bacteria to degrade α-glucans, e.g. starch and glycogen in the environment, for use by the bacteria as energy sources. Certain amylases are cytoplasmic enzymes that participate in the turnover and utilization by the bacterial cell of its own glycogen for energy requirements (Kiel *et al.*, 1994). Other potential degradative enzymes that can act on glycogen are amyloglucosidases and various α-glucosidases that are present in the cytoplasm of various bacterial species (Vikinen and Mantsala, 1989).

6. OCCURRENCE AND DISTRIBUTION OF TREHALOSE

Trehalose is a non-reducing disaccharide in which two Glc residues are linked in α-(1→1)-linkage. Although three different anomers are possible, and all three have been synthesized chemically, only the αα-anomer (Figure 11.2) has been isolated from, and synthesized in, living cells (Elbein, 1974). This naturally occurring disaccharide is widespread throughout the biological community, being found in the cytoplasm of many bacteria, as well as in yeast and fungal species, where it occurs in spores, fruiting bodies and vegetative cells. Spores and macrocysts of *Dictyostelium mucoroides* have been reported to contain as much as 7% trehalose on a dry-weight basis (Cleeg and Filosa, 1961), while the ascospores of *Neurospora tetrasperma* have as much as 10% trehalose (Sussman and Lingappa, 1959). This disaccharide also occurs in lichens and algae and is quite prevalent in mushrooms (Lindberg, 1955). Trehalose has also been identified in many lower animals including protozoa, insects, shrimp, nematodes, roundworms, etc. (Becker

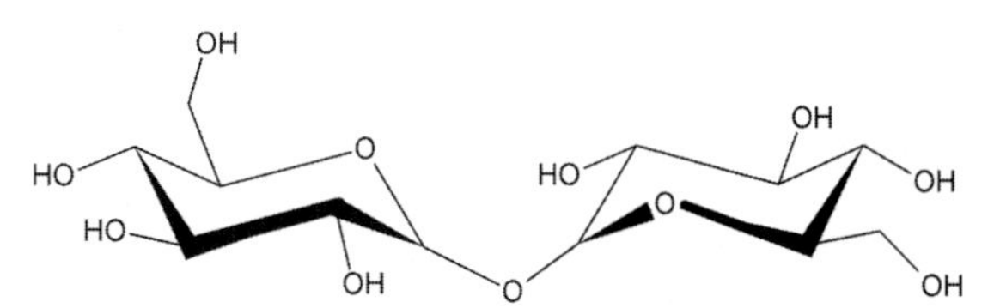

FIGURE 11.2 Structure of α,α-trehalose. Although three different isomers of trehalose have been chemically synthesized (α,α; α,β; β,β), only the α,α-isomer has been found in biological systems.

et al., 1996). Finally, this sugar also occurs in many different plants (Avonce *et al.*, 2005). In contrast, trehalose is not synthesized or present in mammals, although many mammals, including man, do contain a trehalase that cleaves trehalose into two Glc residues (Ruf *et al.*, 1990). This enzyme is located in the brush border membranes of intestinal epithelial cells and its function is probably to degrade the trehalose that is present in many consumed foods, such as mushrooms.

7. BIOSYNTHESIS OF TREHALOSE

There are at least three different pathways (DeSmet *et al.*, 2000) that can be utilized for the biosynthesis of trehalose in microorganisms (Figure 11.3). In addition, there are several degradative pathways of trehalose metabolism that involve readily reversible reactions that could be utilized, at least in principle, for the biosynthesis of trehalose. These various reactions are discussed below.

7.1. Enzymatic synthesis of trehalose-6-phosphate

The most widely reported and best studied pathway for the synthesis of trehalose involves two enzymes, named trehalose-phosphate synthase (TPS), also referred to as OtsA in *E. coli* (Kaasen *et al.*, 1992), and trehalose-phosphate phosphatase (TPP), also referred to as OtsB in *E. coli* (Kaasen *et al.*, 1994). These two enzymes catalyse the reactions summarized below (and also shown in Figure 11.3):

(i) UDP-Glc + Glc-6-phosphate $\xrightarrow{\text{TPS}}$ Trehalose-6-phosphate + UDP

(ii) Trehalose-6-phosphate $\xrightarrow{\text{TPP}}$ Trehalose + Phosphate (inorganic)

The enzyme TPS was first isolated from the yeast *Saccharomyces cerevesiae* by Cabib and Leloir (1958) and its reaction was demonstrated using cell-free extracts of this yeast. That reaction has subsequently been shown in cell-free extracts prepared from numerous insects and other lower animals, as well as with extracts from *Dictyostelium discoideum*, plants and a number of different bacterial species (Elbein *et al.*, 2003). On the other hand, purified trehalose-phosphate synthase purified from *Streptomyces hygroscopicus* and several other streptomycetes catalyses a reaction similar to *reaction (i)* above but, in these cases, the glucosyl donor is GDP-Glc, rather than UDP-Glc (Elbein, 1968). Importantly, in the case of yeast, slime moulds, insects or certain bacterial species, UDP-Glc appears to be the only Glc donor that these TPSs utilize, whereas in streptomycetes, the only Glc donor that has been found active is GDP-Glc (Elbein, 1968).

Mycobacterium smegmatis, *Mycobacterium tuberculosis* and other mycobacteria also contain significant amounts of trehalose which is utilized for a number of different functions (Elbein and Mitchell, 1973; Woodruff *et al.*, 2004). In these microorganisms, TPS can utilize any of the Glc sugar nucleotides, i.e. ADP-Glc, CDP-Glc, GDP-Glc, thymidine diphospate- (TDP-) Glc, UDP-Glc as glucosyl donors for the production of trehalose-phosphate (Lapp *et al.*, 1971). The 58 kDa TPS has been purified to apparent homogeneity from *M. smegmatis* and the properties of the enzyme were determined, as well as the effects of polyanions on the activity. The protein was also subjected to amino acid sequencing and several peptides were sequenced (Pan *et al.*, 1996). Based on that information, the gene was identified in the *M. tuberculosis* genome and was cloned and expressed in *E. coli* as an active enzyme (Pan *et al.*, 2002). This recombinant TPS showed the same broad specificity of glucosyl donor, as substrates for the production of trehalose-phosphate, as the enzyme from *M. smegmatis*. Hence, the trehalose-phosphate

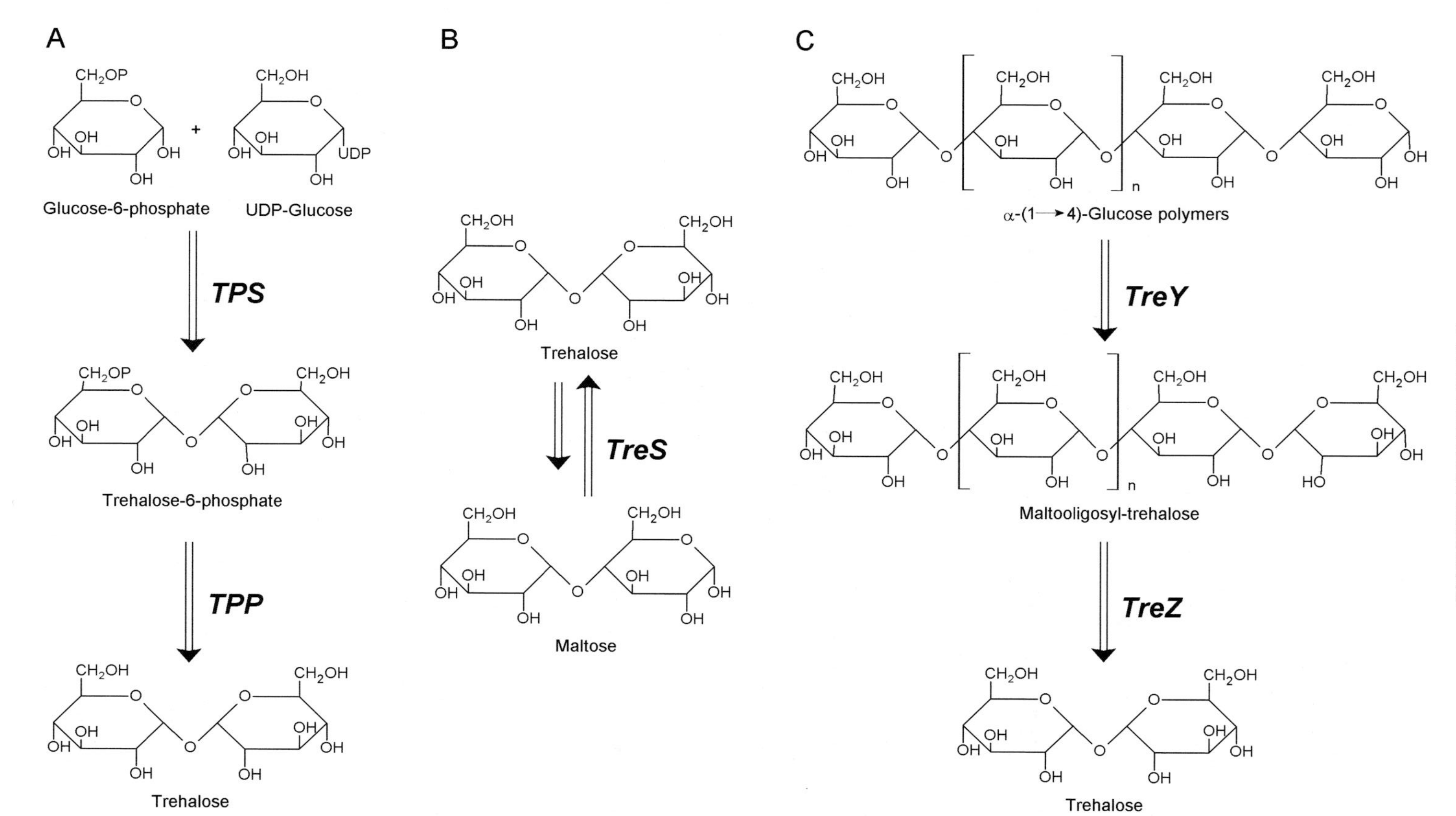

FIGURE 11.3 Pathways of biosynthesis of trehalose in bacteria: (A) TPS/TPP or OtsA/OtsB pathway; (B) TreS pathway; and (C) the TreY/TreZ pathway.

synthases of mycobacteria are similar to the plant sucrose synthase since both transferases have broad specificity for nucleoside diphosphate Glc donors that they utilize as substrates for formation of their disaccharide product. Sucrose synthase can utilize all of the Glc sugar nucleotides for the transfer of Glc to the acceptor fructose to produce sucrose; UDP-Glc, ADP-Glc and TDP-Glc are the preferred substrates (Delmer and Albersheim, 1970). This contrasts with most other glycosyltransferases which are quite specific for the sugar being transferred and the purine/pyrimidine base of the nucleotide donor (Coutinho *et al.*, 2003). It can be hypothesized that the broad substrate specificities of bacterial TPS and plant sucrose synthase may reflect the importance to the organism of these disaccharides as storage compounds, stabilizers and structural components.

The second enzyme in this biosynthetic pathway, TPP, is a specific phosphatase and organisms that contain TPS also appear to have a TPP that converts the trehalose-phosphate to free trehalose, *reaction (ii)* above (Matula *et al.*, 1971). Although many non-specific acid and alkaline phosphatases can cleave the phosphate from trehalose-phosphate, the TPPs that have been isolated are highly specific enzymes that will not cleave other sugar phosphates, e.g. Glc-1-phosphate or Glc-6-phosphate or even other phosphorylated substrates, i.e. the common phosphatase substrate, *p*-nitrophenyl-phosphate (Klutts *et al.*, 2003). The gene-encoding TPP is in the same operon as that encoding TPS in *E. coli* (Horlacher and Boos, 1997), while in *S. cerevesiae*, TPP is one of the subunits of a multi-enzyme complex that synthesizes trehalose (Bell *et al.*, 1998). The TPP-encoding gene of *M. smegmatis* has been cloned and the recombinant protein expressed in *E. coli* as an active enzyme, a 27 kDa protein, which shows an almost absolute requirement for Mg^{2+} ions, but is somewhat active in the presence of Mn^{2+} ions. The enzyme is specific for trehalose-phosphate as substrate and has no activity on any other sugar-phosphates or any other phosphorylated compounds (Klutts *et al.*, 2003). The TPP of *M. tuberculosis* has very similar enzymatic properties, i.e. pH optimum, requirement for divalent cations and substrate specificity. However, the *M. tuberculosis* enzyme is a 45 kDa protein and has less than 25% identity at the amino acid level to the TPP from *M. smegmatis* (Edavana *et al.*, 2004).

In yeast, trehalose-phosphate can apparently serve as a regulator of carbohydrate metabolism by inhibiting the action of hexokinase (Blazquez *et al.*, 1993; Hohmann *et al.*, 1996). It is also a regulator of carbohydrate utilization and growth in *Arabidopsis thaliana* (Schluepmann *et al.*, 2003). This suggests that the enzyme, TPP, could play an important role in the regulation of Glc metabolism by controlling the level of trehalose-phosphate in these cells (Elbein *et al.*, 2003).

7.2. Other pathways for the synthesis of trehalose

The TPS/TPP pathway is probably the major route for the production of trehalose in mycobacteria, as well as in most other organisms. However, mycobacteria, corynebacteria, streptomycetes and a number of other microbes have several other alternate pathways that also can produce trehalose (DeSmet *et al.*, 2000; Elbein *et al.*, 2003). In mycobacteria and perhaps other microorganisms, the synthesis and control of the amounts of trehalose and glycogen are closely linked and interrelated and may be coordinately controlled (Pan *et al.*, 2008). This interaction may allow these microorganisms to convert excess trehalose to glycogen when cytoplasmic trehalose levels get "dangerously" high or to use glycogen as a source of trehalose when trehalose levels get "dangerously" low. Studies on a mutant of *M. smegmatis* that is missing all three synthetic pathways outlined in Figure 11.3 have shown that this mutant cannot grow in any media without exogenous trehalose (Woodruff *et al.*, 2004). Thus, trehalose is essential for the growth of mycobacteria, but high concentrations of trehalose also appear to be toxic to cells.

The second trehalose biosynthetic pathway produces this molecule from glycogen or malto-oligosaccharides. This pathway involves the action of two enzymes, referred to as TreY and TreZ (see Figure 11.3). The TreY enzyme, malto-oligosyltrehalose synthase, catalyses removal of the α-(1→4)-Glc at the reducing end of the bacterial glycogen chain and then the re-addition of this Glc to form the "α-Glc-(1→1)-α-Glc" glycosidic bond of trehalose (Maruta *et al.*, 1996a). Thus, TreY basically changes the maltose at the reducing end of the glycogen chain to a trehalose. Subsequently, the second enzyme TreZ, malto-oligosyltrehalose trehalohydrolase, cleaves off the trehalose by hydrolysing the next α-(1→4)-bond, thereby yielding one free trehalose and leaving a glycogen missing two Glc residues from its reducing end (Maruta *et al.*, 1996b). This pathway was first described in archaebacteria belonging to the *Sulfolobus* genus (Maruta *et al.*, 1996b), but has since been found in a number of mycobacteria, as well as in corynebacteria, streptomyces and other bacteria (DeSmet *et al.*, 2000; Elbein *et al.*, 2003). The pathway presumably enables these bacteria to maintain and monitor the level of trehalose in their cytoplasm by degrading glycogen. Nevertheless, nothing is known about the regulation of this pathway, or whether low trehalose levels increase the synthesis or activity of either TreY or TreZ.

The third pathway for producing trehalose is from maltose and involves the enzyme, trehalose synthase (TreS). This enzyme is a 68 kDa protein that catalyses the interconversion of maltose and trehalose (Nishimoto *et al.*, 1995). The exact equilibrium of this reaction has been difficult to determine because the Km for each substrate is so high, i.e. 10 mM for maltose, but 90 mM for trehalose (Pan *et al.*, 2004). Despite this, TreS has a greater affinity for maltose as substrate, suggesting that the cellular role of the enzyme is to supply trehalose. The difficulty is that *M. smegmatis* grows very poorly on maltose and thus this sugar is unlikely to be an exogenous source of trehalose. However, the purified or recombinant TreS protein has been shown to have amylase activity in addition to its maltose-trehalose interconverting activity (Pan *et al.*, 2008). The amylase activity is much lower than the maltose to trehalose activity but, nevertheless, is sufficient to allow the enzyme to cleave malto-oligosaccharides or glycogen to maltose and then convert the maltose to trehalose.

In addition, TreS also appears to be involved in the interconversion of trehalose to glycogen. This is based upon studies with a large number of *M. smegmatis* trehalose-deficient mutants that have been prepared (J.D. Carroll *et al.*, unpublished results): some are single-mutants missing one of the enzymes from the three pathways shown in Figure 11.3; others are double-mutants, missing one enzyme from each of two pathways; and, finally, there are several triple-mutants which cannot synthesize trehalose because they have deletions in all three pathways and are missing either the TPS or TPP of that pathway and the TreY or TreZ of the glycogen pathway.

When wild-type *M. smegmatis* is grown in a mineral salts medium in the presence of high concentrations of trehalose (2% or higher), this organism accumulates very large amounts of glycogen, i.e. $\approx$10 to 30-fold increases over the amount of glycogen produced in the absence of trehalose or in other media such as Trypticase Soy broth. Thus, this increase in glycogen does not occur in the presence of high concentrations of other sugars, e.g. Glc or sucrose. Furthermore, any of the above trehalose mutants that still contain TreS, but are missing any other enzymes of the other pathways, still accumulate large amounts of glycogen when grown in the presence of 2% trehalose. In contrast, mutants missing TreS do not accumulate glycogen under these conditions (Pan *et al.*, 2008). These results show that TreS, as well as trehalose, is necessary to convert trehalose to glycogen, but the steps involved in, or the mechanism of this conversion is not known at this time.

7.3. Other possibilities for producing trehalose

There are several other pathways in bacteria that are probably degradative in nature, but the reactions are reversible and could, under the right conditions, catalyse the formation of trehalose or a trehalose-derivative. One such pathway is a phosphorolysis pathway catalysed by the enzyme trehalose phosphorylase (Wannet *et al.*, 1998). This enzyme catalyses the following reaction:

$$\text{Trehalose} + H_3PO_4 \leftrightarrow \text{Glc-1-phosphate} + \text{Glc}$$

This reaction is analogous to other phosphorylases, e.g. glycogen phosphorylase or maltose phosphorylase and, in the presence of high levels of Glc-1-phosphate and Glc, it could produce free trehalose.

Another reaction that may give rise to trehalose involves the phosphotransferase system. For example, *E. coli* has two systems for metabolizing trehalose (Boos *et al.*, 1990). Under conditions of high osmolarity, trehalose is hydrolysed to Glc by a periplasmic trehalase, encoded by *treA*, and the Glc is then taken up by the phosphotransferase system (PTS) as Glc-6-phosphate and used in glycolysis (Boos *et al.*, 1990). This periplasmic trehalase, TreA, is induced by 250 mM NaCl, but not by trehalose. On the other hand, at low osmolarity, trehalose is transported via a trehalose-specific enzyme II of the PTS system, encoded by *treB*. This trehalose-6-phosphate produced within the cell is then hydrolysed to Glc and Glc-6-phosphate by trehalose-6-phosphate hydrolase, encoded by *treC* (Rimmele and Boos, 1994). Besides the TPS/TPP pathway for synthesis of trehalose-6-phosphate and then trehalose, the above PTS system is the only other known mechanism for producing trehalose-6-phosphate. Since trehalose-6-phosphate has been implicated as a regulator of carbohydrate metabolism (Blazquez *et al.*, 1993; Muller *et al.*, 1999), the PTS pathway may also be involved in control of Glc metabolism.

8. FUNCTIONS OF TREHALOSE

8.1. Source of energy and carbon

As discussed earlier, trehalose is widespread in nature and trehalose concentrations vary greatly in cells depending on the stage of growth, the nutritional state of the organism and the environmental conditions prevailing at the time of measurement. In a number of instances, it clearly has been implicated as functioning as an energy source. For example, in insects, trehalose is a major sugar in the haemolymph and thorax muscles and is consumed during flight (Becker *et al.*, 1996). In fungal spores, trehalose hydrolysis is a major event during early germination and trehalose is considered a major source of carbon for synthesis of macromolecules as well as providing energy for growth and synthesis (Thevelein, 1984).

8.2. Protectant of proteins and membranes during stress

Trehalose has the capacity to protect proteins from denaturation whether occurring as a result of heat shock, dehydration, oxidative stress or freezing and thawing (Hottinger *et al.*, 1987; Strom and Kaasen, 1993). The ability of trehalose to stabilize proteins and membranes is due to its structure and stereochemistry (Leopold, 1986). Thus, X-ray crystallization studies have shown that trehalose fits well between the polar head groups of membrane phospholipids and that this interaction results in stabilization (Crowe and Crowe, 1988). Also, as in the case of membranes, trehalose interacts well with proteins by forming hydrogen bonds between its properly oriented hydroxyl groups and the polar residues of the protein (Crowe *et al.*, 1992). Nevertheless, it is still unclear why or how this interaction stabilizes proteins. However, it is interesting to note that as a non-reducing disaccharide, trehalose does not cause modification of proteins, as occurs with reducing sugars. That is to say, since trehalose does not have a free aldehyde group, it cannot form a Schiff base with the free amino groups of Lys residues or the N-terminal amino groups of proteins and therefore is not involved in non-enzymic glycation reactions.

8.3. Sensing compound and/or growth regulator

In yeast, trehalose-phosphate has been shown to be an allosteric inhibitor of the enzyme hexokinase and is postulated to be a regulator of carbohydrate metabolism (Blazquez *et al.*, 1993). Somewhat similar effects of trehalose or its metabolites have been observed in plants (Muller *et al.*, 1999). For example, insertion of the genes for trehalose synthetic enzymes (TPS and/or TPP) into tobacco plants results in severe growth defects, e.g. dwarfism and aberrant root development (Pilon-Smits *et al.*, 1998). Moreover, in *Bacillus* spp., there is evidence to indicate that trehalose functions as a transcriptional regulator (Burklen *et al.*, 1998).

8.4. Structural component of the bacterial cell wall

In mycobacteria and corynebacteria, trehalose is the basic component of a number of cell wall glycolipids (Brennan and Nikaido, 1995). The best known, and most widely studied, of these trehalose lipids is cord factor, a cell wall lipid of *M. tuberculosis* (see Chapter 9) which contains the unusual long-chain and branched fatty acids called mycolic acids that are esterified to the –OH groups at position-6 of each Glc (Lederer, 1976). This lipid, called trehalose dimycolate, is considered to be one of the major toxic components of the cell wall and may be the primary reason that trehalose is essential for these microorganisms (Brennan and Nikaido, 1995). Trehalose mono- and dimycolates are also largely responsible for the low permeability of mycobacterial cell walls which confers considerable drug resistance to these organisms (Liu and Nikaido, 1999).

9. CONCLUSIONS

Trehalose and glycogen are both derivatives of glucose and play important roles in living cells. Glycogen is mainly produced and stored as a glucose reserve when the cell has excess energy and carbon and this glycogen is then converted back to glucose when the cell needs a source of glucose and/or energy. However, glycogen also serves an essential role in providing energy and carbon for spore germination and other energy-requiring activities. Bacterial glycogen differs from animal glycogen in several respects. First, bacterial glycogen is not a glycoprotein, i.e. there is no protein attached to the glycogen. Second, the glucose donor for bacterial glycogen is ADP-glucose, rather than UDP-glucose. Third, bacterial glycogen synthesis is regulated by glycolytic intermediates or other metabolites that affect the activity of the allosteric enzyme that produces ADP-glucose, i.e. ADP-glucose pyrophosphorylase. On the other hand, animal glycogen synthesis and degradation is regulated by hormonal action which results in activation of protein kinases that cause the phosphorylation of glycogen synthase and glycogen phosphorylase resulting in increased degradation and decreased synthesis of glycogen.

Structurally, trehalose is a much simpler molecule then glycogen but, interestingly, it is physiologically more diverse. It does function as a reservoir of glucose for some organisms and is utilized as an energy source for insect flight muscle and for spore germination in fungi. It is also a very important protectant for cells during times of stress, since it protects proteins and membranes from denaturation and other damage. In many different organisms, synthesis of trehalose is greatly increased when these organisms are subjected to desiccation, heat shock or oxidative stress. Another role of trehalose is as a regulatory molecule. Thus, in yeast and some plants, trehalose-phosphate has been shown to function as a regulator of carbohydrate metabolism and trehalose also appears to function as a transcriptional regulator in *Bacillus subtilis*.

In mycobacteria and related organisms, trehalose is essential for growth and survival and mutants that are unable to synthesize trehalose cannot grow unless exogenous trehalose is added to the media. In these organisms, trehalose may function in some or all of the above capacities, but it also has another very important role as a structural component of the cell wall. In this case, trehalose is acylated with various fatty acids such as mycolic acid to form very hydrophobic glycolipids such as trehalose-dimycolate. These lipids protect the cell from various toxins, such as antibiotics and other noxious chemicals, by preventing them from penetrating the cell wall. Most organisms that contain trehalose synthesize it from UDP-glucose and glucose-6-phosphate with the intermediate formation of trehalose-6-phosphate. But some organisms, such as mycobacteria and corynebacteria, have other pathways that allow them to convert glycogen or glycogen degradation products to trehalose. How these various pathways interact and/or communicate with each other to control trehalose and glycogen levels is not known, but it does represent a very interesting metabolic question. The answer to that question may also disclose new and novel targets for chemoptherapy against important pathogens, such as *M. tuberculosis*.

In conclusion, studies on trehalose and glycogen have provided a wealth of information on synthesis and function of these two important glucose derivatives. Moreover, a number of intriguing and significant questions that need research have been revealed (Research Focus Box).

RESEARCH FOCUS BOX

- How are the levels of trehalose and glycogen controlled in mycobacteria with three different pathways of synthesis? Are they coordinately controlled? Do they communicate with each other?
- What is the mechanism involved in the conversion of trehalose to glycogen in mycobacteria? Is trehalose synthase a regulatory enzyme and, if so, how is it regulated?
- Is trehalose synthase a sensor of trehalose levels in the cell?
- Is trehalose metabolism a reasonable target site for chemotherapy against mycobacteria or other microorganisms? If so, how can it be exploited?
- Can too much trehalose, or too much glycogen, be a bad thing?
- If you could produce the ideal organism, would it synthesize and utilize trehalose?
- How? What about glycogen? Would it have the capacity to interconvert glycogen and trehalose? Why?

References

Avonce, N., Leyman, B., Thevelein, J.M., Iturriaga, G., 2005. Trehalose metabolism and glucose sensing in plants. Biochem. Soc. Trans. 33, 276–279.

Baecker, P.A., Greenberg, E., Preiss, J., 1986. Biosynthesis of bacterial glycogen. Primary structure of *Escherichia coli* 1,4-α-D-glucan:1,4-α-D-glucan 6-α-D-(1,4-α-D-glucano)-transferase as deduced from the nucleotide sequence of the *glgB* gene. J. Biol. Chem. 268, 8738–8743.

Becker, A., Schloeder, P., Steele, J.E., Wegener, G., 1996. The regulation of trehalose metabolism in insects. Experientia 52, 433–439.

Bell, W., Sun, W., Hohmann, S., et al., 1998. Composition and functional analysis of the *Saccharomyces cerevisiae* trehalose synthase complex. J. Biol. Chem. 272, 33311–33319.

Blazquez, M.A., Lagunas, R., Gancedo, C., Gancedo, J.M., 1993. Trehalose-6-phosphate, a new regulator of yeast glycolysis that inhibits hexokinase. FEBS Lett. 329, 51–54.

Boos, W., Ehmann, U., Forki, W., Klein, M., Rimmele, M., Postma, P., 1990. Trehalose transport and metabolism in *Escherichia coli*. J. Bacteriol. 172, 3450–3461.

Brennan, P.J., Nikaido, H., 1995. The envelope of mycobacteria. Annu. Rev. Biochem. 64, 29–63.

Burklen, L., Schock, F., Dahl, M.K., 1998. Molecular analysis of the interaction between the *Bacillus subtilis* trehalose repressor TreR and the *tre* operator. Molec. Gen. Genet. 260, 48–55.

Cabib, E., Leloir, L.F., 1958. The biosynthesis of trehalose-phosphate. J. Biol. Chem. 231, 259–275.

Chen, G.S., Segal, H., 1968. *Escherichia coli* polyglucose phosphorylases. Arch. Biochem. Biophys. 127, 164–174.

Chock, P.B., Rhee, S.G., Stadtman, E.R., 1980. Interconvertible enzyme cascades in cellular regulation. Annu. Rev. Biochem. 490, 813–843.

Cleeg, J.S., Filosa, M.F., 1961. Trehalose in the cellular slime mold, *Dictyostelium mucoroides*. Nature 192, 1077–1078.

Cori, C.F., Cori, G.R., 1939. The activating effect of glycogen on the enzymatic synthesis of glycogen from glucose-1-phosphate. J. Biol. Chem. 131, 397–398.

Coutinho, P.M., Deleury, E., Davies, G.J., Henrissat, B., 2003. An evolving hierarchical family classification of glycosyltransferases. J. Mol. Biol. 328, 307–317.

Crowe, L.M., Crowe, J.H., 1988. Trehalose and dry dipalmitoyl phosphatidylcholine revisited. Biochim. Biophys. Acta 946, 193–201.

Crowe, J.H., Hoekstra, F.A., Crowe, L.M., 1992. Anhydrobiosis. Annu. Rev. Physiol. 54, 579–599.

Delmer, D.P., Albersheim, P., 1970. The biosynthesis of sucrose and nucleoside diphosphate glucoses in *Phaseolus aureus*. Plant Physiol. 45, 782–786.

DeSmet, K.A.L., Weston, A., Brown, I.N., Young, D.B., Robertson, B.D., 2000. Three pathways for trehalose biosynthesis in mycobacteria. Microbiology 146, 199–208.

Dietzler, D.N., Strominger, J.L., 1973. Purification and properties of the adenosine diphosphoglucose:glycogen transglucosylase of *Pasteurella pseudotuberculosis*. J. Bacteriol. 113, 946–952.

Dietzler, D.N., Leckie, M.P., Lais, C.J., Magnani, J.L., 1974. Evidence for the regulation of bacterial glycogen biosynthesis *in vivo*. Arch. Biochem. Biophys. 162, 602–606.

Edavana, V.K., Pastuszak, I., Carroll, J.D., Thampi, P., Abraham, E.C., Elbein, A.D., 2004. Cloning and expression of the trehalose-phosphate phosphatase of *Mycobacterium tuberculosis*: comparison to the enzyme from *Mycobacterium smegmatis*. Arch. Biochem. Biophys. 426, 250–257.

Elbein, A.D., 1968. Trehalose-phosphate synthesis in *Streptomyces hygroscopicus*. Purification of GDP-glucose: glucose-6-phosphate 1-glucosyl transferase. J. Bacteriol. 96, 1623–1631.

Elbein, A.D., 1974. The metabolism of α,α-trehalose. Adv. Carbohydr. Chem. Biochem. 30, 227–256.

Elbein, A.D., Mitchell, M., 1973. Levels of glycogen and trehalose in *Mycobacterium smegmatis*, and the purification and properties of the glycogen synthase. J. Bacteriol. 113, 863–873.

Elbein, A.D., Pan, Y.T., Pastuszak, I., Carroll, D., 2003. New insights on trehalose: a multifunctional molecule. Glycobiology 13, 17R–27R.

Fox, J., Kawaguchi, K., Greenberg, E., Preiss, J., 1976. Biosynthesis of bacterial glycogen. 13. Purification and properties of the *Escherichia coli* B ADP glucose: 1,4-α-D-glucan 4- α-glucosyltransferase. Biochemistry 15, 849–857.

Furukawa, K., Tagaya, M., Tanaziwa, K., Fukui, T., 1994. Identification of Lys277 at the active site of *Escherichia coli* glycogen synthase. Application of affinity labeling combined with stie-directed mutagenesis. J. Biol. Chem. 269, 868–871.

Govons, S., Gentner, N., Greenberg, E., Preiss, J., 1973. Biosynthesis of bacterial glycogen. XI. Kinetic characterization of an altered ADP-glucose synthase from a "glycogen-excess" mutant of *Escherichia coli*. J. Biol. Chem. 248, 1731–1740.

Henrissat, A., 1991. A classification of glycosyl hydrolases based on amino acid sequence similarities. Biochem. J. 280, 309–316.

Hohmann, S., Bell, S., Neves, M.J., Valckx, D., Thevelein, J.M., 1996. Evidence for trehalose-6-phosphate-dependent and -independent mechanisms in control of sugar influx into yeast glycolysis. Mol. Microbiol. 20, 981–991.

Holmes, E., Preiss, J., 1979. Characterization of *Escherichia coli* B glycogen synthase enzymatic reactions and products. Arch. Biochem. Biophys. 196, 436–448.

Horlacher, R., Boos, W., 1997. Characterization of TreR, the major regulator of the *Escherichia coli* trehalose system. J. Biol. Chem. 272, 13026–13032.

Hottinger, T., Boller, T., Wiemken, A., 1987. Rapid changes of heat and dessication tolerance correlated with changes of trehalose content in *Saccharomyces cerevesiae*. FEBS Lett. 220, 113–115.

Kaasen, I., Falkenberg, P., Sryrvold, O.B., Strom, A.R., 1992. Molecular cloning and physical mapping of the *otsBA* genes upon exposure, which encode the osmoregulatory trehalose pathway of *Escherichia coli*: evidence that transcription is activated by *katF* (AppR). J. Bacteriol. 174, 889–898.

Kaasen, I., McDougall, J., Strom, A.R., 1994. Analysis of the *otsBA* operon for osmoregulatory trehalose synthesis in *Escherichia coli* and homology of the OtsA and OtsB proteins to the yeast trehalose-6-phosphate synthase/phosphatase complex. Gene 145, 9–15.

Kiel, J.A.K., Boels, J.M., Beldman, G., Venema, G., 1994. Glycogen in *Bacillus subtilis*: molecular characterization of an operon encoding enzymes involved in glycogen biosynthesis and degradation. Mol. Microbiol. 11, 203–218.

Klutts, S., Pastuszak, I., Korath-Edavana, V., et al., 2003. Purification, cloning, expression and properties of the mycobacterial trehalose-phosphate phosphatase. J. Biol. Chem. 278, 2093–2100.

Krisman, C.R., Barengo, R., 1975. A precursor of glycogen biosynthesis: α-1,4-glucan-protein. Eur. J. Biochem. 52, 117–123.

Kumar, A., Larsen, E., Preiss, J., 1986. Biosynthesis of bacterial glycogen. Primary structure of Escherichia coli ADP-glucose: α-1,4-glucan,4-glucosyltransferase as deduced from the nucleotide sequence of the glgA gene. J. Biol. Chem. 261, 16256–16259.

Lapp, D., Patterson, B.W., Elbein, A.D., 1971. Propeties of a trehalose-phosphate synthetase from *Mycobacterium smegmatis*. J. Biol. Chem. 246, 4567–4579.

Lederer, E., 1976. Cord factor and related trehalose esters. Chem. Phys. Lipids 16, 91–106.

Leloir, L.F., Rongine deFerrente, M.A., Cardini, C.E., 1961. Starch and oligosaccharide synthesis from uridine diphosphate glucose. J. Biol. Chem. 236, 636–641.

Leopold, A.C., 1986. Membranes, Metabolism and Dry Organisms. Cornell University Press, Ithaca.

Levi, C., Preiss, J., 1976. Regulatory properties of the ADP-glucose pyrophosphorylase of the blue-green bacterium *Synechococcus* 6301. Plant Physiol. 58, 753–756.

Lindberg, B., 1955. Studies on the chemistry of lichens. Investigation of a *Dermatocarpon* and some *Roccella* species. Acta Chem. Scand. 9, 917–919.

Liu, J., Nikaido, H., 1999. A mutant of *Mycobacterium smegmatis* defective in the biosynthesis of mycolic acid accumulates meromycolate. Proc. Natl. Acad. Sci. USA 96, 4011–4016.

Lowry, O.H., Carter, J., Ward, J.B., Glaser, L., 1971. The effect of carbon and nitrogen sources on the level of metabolic intermediates in *Escherichia coli*. J. Biol. Chem. 246, 6511–6521.

MacGregor, E.A., Janecek, S., Svensson, B., 2001. Relationship of sequence to specificity in the α-amylase family of enzymes. Biochim. Biophys. Acta 1546, 1–20.

Mackey, B.M., Morris, J.G., 1971. Ultrastructural changes during sporulation in *Clostridium pasteurianum*. J. Gen. Microbiol. 63, 13.

Manners, D.J. (1971). The structure of glycogen. In: Rose, A.H., Harrison, J.S. (Eds.), The Yeasts, Vol. 2. Academic Press, NewYork, pp. 419–440.

Manners, D.J., 1991. Recent developments in our understanding of glycogen structure. Carbohydr. Polym. 16, 37–82.

Maruta, K., Hattori, K., Nakada, T., Kubota, M., Sugimoto, T., Kurimoto, M., 1996a. Cloning and sequencing of trehalose biosynthesis genes from *Rhizobium* sp. M-11. Biosci. Biotechnol. Biochem. 60, 717–720.

Maruta, K., Mitsuzumi, H., Nakada, T., et al., 1996b. Cloning and sequencing of a cluster of genes encoding novel enzymes of trehalose biosynthesis from thermophilic archaebacterium *Sulfolobus acidocaldarius*. Biochim. Biophys. Acta 1291, 177–181.

Matula, M., Mitchell, M., Elbein, A.D., 1971. Trehalose-phosphate from *Mycobacterium smegmatis*. J. Bacteriol. 107, 217–223.

Muller, J., Wiemken, A., Aeschbacher, R., 1999. Trehalose metabolism in sugar sensing and plant development. Plant Sci. 147, 37–47.

Nishimoto, T., Nakano, M., Bnakada, T., et al., 1995. Purification and properties of a novel enzyme, trehalose synthase, from *Pimelobacter* sp. R48. Biosci. Biotechnol. Biochem. 60, 640–644.

Okita, T.W., Rodriguez, R.L., Preiss, J., 1981. Biosynthesis of bacterial glycogen. Cloning of the glycogen biosynthetic enzyme structural genes of *Escherichia coli*. J. Biol. Chem. 256, 6944–6952.

Pan, Y.T., Drake, R.R., Elbein, A.D., 1996. Trehalose-P synthase of *Mycobacteria*. Its substrate specificity is affected by polyanions. Glycobiology 6, 453–461.

Pan, Y.T., Carroll, J.D., Elbein, A.D., 2002. Trehalose-phosphate synthase of *Mycobacterium tuberculosis*. Cloning, expression and properties of the recombinant enzyme. Eur. J. Biochem. 269, 6091–6100.

Pan, Y.T., Edavana, V.K., Jourdian, W.J., et al., 2004. Trehalose synthase of *Mycobacterium smegmatis*. Purification, cloning, expression and properties of the enzyme. FEBS J. 271, 4259–4269.

Pan, Y.T., Carroll, J.D., Asano, N., Pastuszak, I., Korath-Edavana, V., Elbein, A.D., 2008. Trehalose synthase converts glycogen to trehalose. FEBS J. 275, 3408–3420.

Pilon-Smits, E.A.H., Terry, N., Sears, T., et al., 1998. Trehalose-producing transgenic tobacco plants show improved growth performance under drought stress. J. Plant Physiol. 152, 525–532.

Preiss, J., 1984. Bacterial glycogen synthesis and its regulation. Annu. Rev. Microbiol. 38, 419–458.

Preiss, J., 1989. Glycogen biosynthesis In: Poindexter, J.S., Ledbetter, E. (Eds.), Bacteria in Nature, Vol. 3. Plenum Press, New York, pp. 189–256.

Preiss, J., 2000. Glycogen biosynthesis. In: Lederberg, J., Alexander, M., Bloom, B.R. et al. (Eds.) Encyclopedia of Microbiology, Vol. 2. Academic Press, New York, pp. 541–556.

Preiss, J., Greenberg, E., 1965. Biosynthesis of bacterial glycogen. 3. The adenosine diphosphate-glucose: α-4-glucosyl transferase of *Escherichia coli* B. Biochemistry 4, 2328–2334.

Preiss, J., Romeo, T., 1989. Physiology, biochemistry and genetics of bacterial glycogen synthesis. Adv. Bacter. Physiol. 30, 184–238.

Preiss, J., Romeo, T., 1994. Molecular biology and regulatory aspects of glycogen biosynthesis in bacteria. Prog. Nucleic Acid Res. Mol. Biol. 47, 300–329.

Preiss, J., Lammel, C., Greenberg, E., 1976. Biosynthesis of bacterial glycogen. Kinetic studies of a glucose-1-P adenylyltrasferase (EC 2.7.7.27) from a glycogen-excess mutant of *Escherichia coli* B. Arch. Biochem. Biophys. 174, 105–119.

Qian, M., Haser, R., Buisson, G., Duee, E., Payan, F., 1994. The active center of a mammalian α-amylase. Biochemistry 33, 6284–6294.

Rimmele, M., Boos, W., 1994. Trehalose-6- hydrolase of *Escherichia coli*. J. Bacteriol. 176, 5654–5664.

Robyt, J.F., French, D., 1970. Multiple attack and polarity of action of porcine pancreatic α-amylase. Arch. Biochem. Biophys. 13, 622–670.

Romeo, T., Kumar, A., Preiss, J., 1988. Analysis of the *Esherichia coli* glycogen gene cluster suggests that catabolic enzymes are encoded among the biosynthetic genes. Gene 70, 363–376.

Ruf, J., Wacker, H., James, P., et al., 1990. Rabbit small intestine trehalase. Purification, cDNA cloning, expression, and verification of GPI-anchoring. J. Biol. Chem. 265, 15034–15040.

Russell, J.B., 1992. Glucose toxicity and inability of *Bacteroides ruminicola* to regulate glucose transport and utilization. Appl. Environ. Microbiol. 58, 2040–2045.

Schluepmann, H., Pellny, T., van Dijken, A., Smeekens, S., Paul, M., 2003. Trehalose 6-phosphate is indispensable for carbohydrate utilization and growth in *Arabidopsis thaliana*. Proc. Natl. Acad. Sci. USA, 100, 6849–6854.

Shen, L., Ghosh, H.P., Greenberg, E., Preiss, J., 1964. Adenosine diphosphate glucose-glycogen transglucosylase in *Arthrobacter* sp. NRRL B. Biochim. Biophys. Acta 89, 370–372.

Sigal, N., Cattaneo, J., Segal, I.H., 1964. Glycogen accumulation by wild-type and uridine diphosphate glucose pyrophosphorylase-negative strains of *Escherichia coli*. Arch. Biochem. Biophys. 108, 440–451.

Strom, A.R., Kaasen, I., 1993. Trehalose metabolism in *Escherichia coli*: stress protection and stress regulation of gene expression. Mol. Microbiol. 8, 205–210.

Sussman, A.S., Lingappa, B.T., 1959. Role of trehalose in ascospores of *Neurospora tetrasperma*. Science 130, 1343–1344.

Swedes, J.S., Sedo, R.J., Atkinson, D.E., 1975. Relation of growth and protein synthesis to the adenylate energy charge in an adenine-requiring mutant of *Escherichia coli*. J. Biol. Chem. 250, 6930–6938.

Tagaya, M., Nakano, K., Fukui, T., 1985. A new affinity labeling reagent for the active site of glycogen synthase. Uridine diphosphopyridoxal. J. Biol. Chem. 260, 6670–6676.

Thevelein, J.M., 1984. Regulation of trehalose metabolism in fungi. Microbiol. Rev. 48, 42–59.

Vikinen, M., Mantsala, P., 1989. Microbial amylolytic enzymes. Crit. Rev. Biochem. Mol. Biol. 24, 329–418.

Wannet, W.J.B., Op den Camp, H.J.M., Wisselink, H.W., van der Drift, C., Van Griensven, L.J., Vogels, G.D., 1998. Purification and characterization of trehalose phosphorylase from the commercial mushroom *Agaricus bisporus*. Biochim. Biophys. Acta 1425, 177–188.

Wilkinson, J.F., 1959. The problem of energy-storage compounds in bacteria. Exptl. Cell Res. (Suppl. 7), 111–130.

Woodruff, P.J., Carlson, B.L., Siridechadilok, B., et al., 2004. Trehalose is required for growth of *Mycobacterium smegmatis*. J. Biol. Chem. 279, 28835–28843.

Yang, H., Liu, M.Y., Romeo, T., 1996. Coordinate genetic regulation of glycogen catabolism and biosynthesis in *Escherichia coli* via the CsrA gene product. J. Bacteriol. 178, 1012–1017.

CHAPTER

12

Glycosylated compounds of parasitic protozoa

Joanne Heng, Thomas Naderer, Stuart A. Ralph and Malcolm J. McConville

SUMMARY

Parasitic protists belong to a range of deeply diverging eukaryotic taxa and are the cause of many important diseases in humans. These organisms are capable of surviving in multiple vertebrate and arthropod host environments and, in some cases, as free-living organisms. All parasitic protists express a range of glycoconjugates that form protective protein-rich or carbohydrate-rich surface coats. Protein-rich coats are typically found on developmental stages that inhabit non-hydrolytic niches, such as the bloodstream and non-acidified intracellular vacuoles. These coats are commonly dominated by a limited repertoire of antigenically diverse proteins that are commonly, but not always, glycosylphosphatidylinositol- (GPI-) anchored and modified with *N*- or *O*-glycans. Carbohydrate-rich coats are commonly found on developmental stages that dwell within hydrolytic environments, such as vertebrate and arthropod digestive tracts and lysosomal vacuoles. These coats are dominated by GPI-anchored glycoproteins that are heavily modified with *N*-glycans, *O*-glycans or phosphoglycans. Free GPI glycolipids (not attached to protein) can also be abundant or dominant components of these coats. Some parasitic protists can also form highly resistant cyst stages encased within polysaccharide-rich cell walls. Considerable progress has been made in defining the structures of the surface and intracellular glycans of the parasitic protists, their biosynthesis and the role that individual components play in parasite infectivity.

Keywords: Protozoan parasites; *N*-Glycosylation; *O*-Glycosylation; Glycosylphosphatidylinositol; Phosphoglycosylation

1. INTRODUCTION

Parasitic protists comprise a highly diverse group of single-celled eukaryotes that are responsible for a range of important human and veterinary diseases (Table 12.1). The major parasite groups belong to the Apicomplexa (*Plasmodium, Theileria, Babesia, Toxoplasma* and *Cryptosporidium* spp.), the Kinetoplastida (*Trypanosoma* and *Leishmania* spp.), the Amoebozoa (*Entamoeba histolytica* and *Acanthamoeba* spp.) and the Metamonada (*Giardia* and *Trichomonas* spp.). While organisms

TABLE 12.1 Glycan classes synthesized by different parasitic protists

Parasite	*N*-glycan[a]	GPI[b]	*O*-glycan[c]		P-glycans[d]	CHO reserve[e]	Oocyst wall[f]
			SP	cyto			
Kinetoplastidae							
Typanosoma brucei	Yes	Yes	No	No	Yes	No	No
Trypanosoma cruzi	Yes	Yes	Yes	No	Yes	?	No
Leishmania spp.	Yes	Yes	No	Yes	Yes	Man	No
Apicomplexa							
Plasmodium spp.	Yes	Yes	No	No	No	?	No
Babesia bovis	Yes	Yes	?	?	No	?	No
Theileria spp	No	Yes	?	?	No	?	No
Toxoplasma gondii	Yes	Yes	No	No	No	G/AP	Yes
Cryptosporidium spp.	Yes	Yes	No	No	No	G/AP	Yes
Metamonada							
Trichomonas vaginalis	Yes	No	?	?	Yes	G/AP	Yes
Giardia lamblia	Yes	Yes	Yes	No	No	G/AP	Yes
Archamoeba							
Entamoeba histolytica	Yes	Yes	Yes	No	Yes	G/AP	Yes

[a]*N*-linked glycans.
[b]Glycosylphosphatidylinositols.
[c]*O*-Glycosylation of proteins in the secretory pathway (SP) or cytosol (cyto).
[d]Phosphoglycosylation on proteins or a lipid anchor.
[e]Carbohydrate reserves comprising glycogen/amylopectin-like glucans (G/AP) or mannogen (Man).
[f]Oocyt/cyst stages containing protein or carbohydrate-rich walls.

within each clade have shared characteristics, they can also differ markedly from each other with regard to host specificities, tissue tropisms and developmental programmes. For example, the Apicomplexan parasites, *Plasmodium*, *Theileria* and *Babesia* spp. infect blood cells and are transmitted to vertebrates by a hematophagus arthropod definitive host. These parasites target several tissues and differentiate through multiple developmental stages. In contrast, *Cryptosporidium parvum* has a relatively simple life cycle that involves a single host and invasion of a single cell type, primarily intestinal epithelial cells. Despite their diverse biology, all parasitic protozoa produce complex surface coats that are dominated by glycolipids, glycoproteins and/or proteo(phospho)glycans. These surface and secreted molecules provide protection from environmental stresses and microbicidal processes and are responsible for mediating essential host–parasite recognition, adhesion and invasion steps. Parasite glycosylation pathways, such as *N*-glycosylation and glycosylphosphatidylinositol (GPI) biosynthesis are similar to those found in yeast, animals and plants, although significant differences can occur in the substrate specificities and regulatory properties of parasite and mammalian glycosylation enzymes that might be exploitable in developing new anti-parasite therapies. Other parasite glycosylation pathways are entirely novel. For example, the *trans*-sialytion of *O*-linked glycans, the phosphoglycosylation

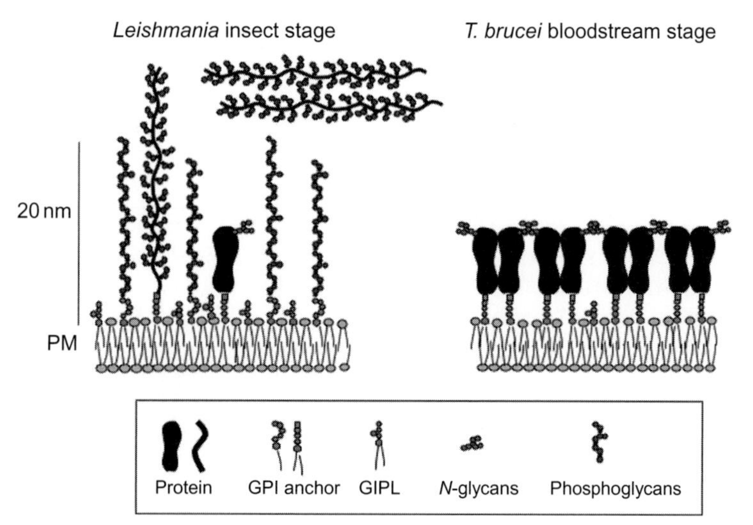

FIGURE 12.1 Protein- and carbohydrate-rich coats of parasitic protists. The VSG coat of *T. brucei* bloodstream forms is the paradigm for a protein-rich surface coat. In contrast, the insect (i.e. promastigote) stages of *Leishmania* express a carbohydrate-rich surface coat. Both coats are dominated by GPI-anchored macromolecules.

of proteins and glycolipids and the assembly of novel cyst wall polysaccharides and intracellular reserve glycans. The diversity of parasite glycosylation reflects the evolutionary diversity of this group of eukaryotes, the opportunity for lateral gene transfer from prokaryotes and various arthropod and vertebrate hosts and the need to generate compositionally-diverse surface coats at specific stages in their life cycles. In this chapter, we describe the range of glycan structures and pathways that are found in different parasitic protozoa. The reader is also referred to other reviews on this topic for more details (see McConville *et al.*, 2002; Mendonca-Previato *et al.*, 2005; von Itzstein *et al.*, 2008).

2. THE SURFACE COATS OF PARASITIC PROTOZOA – AN OVERVIEW

The surface coats of parasitic protozoa are highly diverse, but can be broadly classified as being protein-rich or carbohydrate-rich (McConville and Ferguson, 1993) (Figure 12.1). The variant surface glycoprotein (VSG) coat of *Trypanosoma brucei* bloodstream trypomastigotes (the causative agent of African sleeping sickness) is the prototypic protein-rich surface coat (Mehlert *et al.*, 1998). The bloodstream stages are coated with a monolayer of $\approx 10^6$ VSG dimers that effectively shield other surface proteins (transporters, nutrient receptors etc) and membrane components from antibody- and complement-opsonization (see Figure 12.1). The successive expression of structurally-related, but antigenically-distinct VSGs allow these parasites to avoid being cleared by the host antibody responses. All VSGs are GPI-anchored, *N*-glycosylated glycoproteins (Mehlert *et al.*, 1998). Both modifications are important for maintenance of the VSG surface coat. The use of GPI anchors is likely to facilitate the high-density packing of VSG, as well as the constitutive internalization of immunoglobulin-bound VSG as a mechanism for evading the mammalian immune system (Engstler *et al.*, 2007). The VSG *N*-glycans may facilitate the folding of diverse VSGs and have important space-filling roles in the surface coat (Mehlert *et al.*, 2002). Similar protein-rich coats are assembled by intracellular stages of the Apicomplexan parasites, *Plasmodium* and *Toxoplasma* spp. (Boothroyd *et al.*, 1998; Sanders *et al.*, 2005) and the intestinal parasite *Giarida lamblia* (Nash, 2002). Interestingly, all of these developmental stages occupy relatively non-hydrolytic environments (i.e. bloodstream or non-hydrolytic vacuoles) and/or niches that are directly exposed to the adaptive immune system. A major advantage of the protein-rich coats is that it is possible to generate an almost infinite antigenic diversity, a key strategy used by some of these parasites (i.e. *T. brucei*, *Giardia* and *Plasmodium falciparum*) to avoid being cleared by the immune system (Nash, 2002).

In contrast, parasite stages that reside within proteolytically active niches typically express carbohydrate-rich surface coats. The insect (promastigote) stage of *Leishmania* parasites that proliferate within the digestive tract of the sandfly vector produce the prototypic carbohydrate-rich surface coat. The promastigote surface coat is dominated by lipophosphoglycan (LPG, a GPI-anchored phosphoglycan), free GPI glycolipids and a limited repertoire of GPI-anchored glycoproteins (McConville *et al.*, 2002) (see Figure 12.1). These lipoglycoconjugates contain unusual terminal and internal glycosidic linkages that are likely to contribute to the intrinsic resistance of these stages to host hydrolyases. Carbohydrate-rich surface coats are also expressed by the extracellular insect stages of *T. brucei* and *Trypansoma cruzi* (Buscaglia *et al.*, 2006), the mammalian stage of *Leishmania* and *T. cruzi* (Buscaglia *et al.*, 2006) and the intestinal parasites, *E. histolytica* and *Trichomonas vaginalis* (Moody-Haupt *et al.*, 2000; Bastida-Corcuera *et al.*, 2005), all of which reside in highly lytic environments (insect or mammalian digestive tracts, (phago)lysosome of mammalian cells). It is important to note that the composition of these surface coats varies enormously. For example, intracellular amastigote stages of *Leishmania*, which reside in the mature phagolysosomes of mammalian macrophages, lack major surface macromolecules, but retain a densely packed layer of endogenously synthesized GPI glycolipids and host-derived glycosphingolipids (McConville and Blackwell, 1991; Schneider *et al.*, 1993; Winter *et al.*, 1994). The uptake and surface expression of host glycolipids may constitute a form of host mimicry and allow this stage to avoid detection by the host immune system. It is also important to note that the same species can express protein-rich and carbohydrate-rich surface coats on different developmental stages and that the remodelling of these coats can be very rapid as parasites transition from one host environment to another. The following sections describe the assembly of the glycan components of these different surface coats in more detail.

3. PROTEIN-LINKED AND FREE GPI GLYCOLIPIDS

The GPI (or related) glycolipids are a dominant class of glycoconjugates in all parasitic protists. Functionally, GPI glycolipids typically anchor the major surface proteins of these organisms to the plasma membrane and can also be the dominant class of free glycolipids. Most parasitic protists have a canonical GPI pathway that leads to the assembly of protein anchor precursors with the structure EtN-P-Man_3GlcN-inositol-phospholipid (where EtN-P, ethanolamine-phosphate; Man, mannose; GlcN, glucosamine), although considerable variation can occur in the nature of the lipid moiety and side chain modifications that are added before or after attachment to protein (Figure 12.2). While *Trichomonas* spp. lack this pathway (Carlton *et al.*, 2007), pathogenic stages of these parasites synthesize lipoglycoconjugates with a related inositol-lipid anchor.

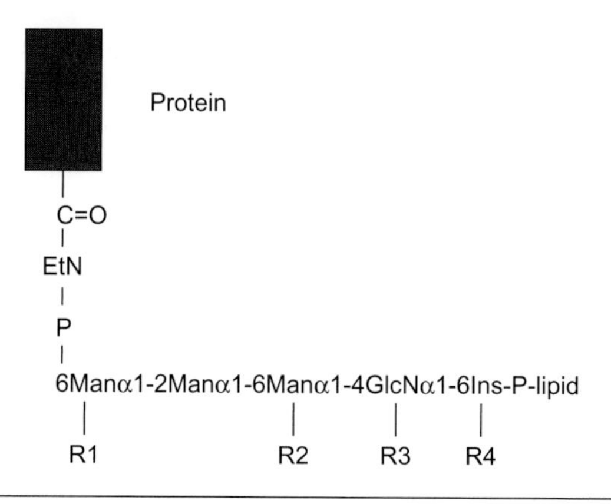

Species	R1	R2	R3	R4	Lipid
T. brucei					
VSG	-	$[\alpha\text{-Gal}]_{1.5}$		-	$\text{diacyl}_{14:0,14:0}\text{-Grol}$
procyclin	-	[2]pNAL		$\text{acyl}_{18:0}$	*lyso*-$\text{acyl}_{18:0}\text{-Grol}$
T. cruzi	α-(1-2)-Man	-	[2]+/-AEP	-	$\text{alkyl}_{16:0}\text{acyl}_{16:0}\text{-Grol}$/ *lyso*-alkyl-Grol/ceramide
Leishmania	-	-		-	$\text{alkyl}_{24:0}\text{acyl}_{14:0}\text{-Grol}$
P. falciparum	α-(1-2)-Man	-		$\text{acyl}_{18:0,14:0}$	diacyl-Grol
T. gondii		$\alpha\text{-GalNAc}_{1\text{-}4}$		-	$\text{diacyl}_{16:0,18:1}\text{-Grol}$
[1]*E. histolytica*		$[\alpha\text{-Gal}]_n$		+/-$\text{acyl}_{16:0}$	*lyso*-$\text{acyl}_{28:0,30:1}$

[1]The GPI anchor of the *E. histolytica* PPG contains a divergent Gal_1Man_2 GlcN-backbone. The site of side chain attachment to this backbone is unknown.
[2]AEP, aminoethylphosphonate; pNAL, poly-*N*-acetyllactosamine

FIGURE 12.2 Structures of the GPI protein anchors of parasitic protists.

3.1. The GPI protein anchors of kinetoplastid parasites

The GPI glycolipids anchor both the dominant surface glycoproteins and a number of other less abundant glycoproteins to the plasma membranes of *T. brucei*, *T. cruzi* and *Leishmania* spp. (McConville and Ferguson, 1993; McConville *et al.*, 2002). Many aspects of GPI biosynthesis were first delineated in *T. brucei* bloodstream forms because of the high flux through this pathway and the ease of obtaining material for biochemical studies. The biosynthesis of GPI protein anchors is essential for viability of *T. brucei* bloodstream stages, although remarkably not for the procyclic insect stages, and compounds that selectively inhibit the *T. brucei* enzymes have recently been identified (Nagamune *et al.*, 2000; Smith *et al.*, 2004; Stokes *et al.*, 2008). The biosynthesis of the protein anchor precursors in *T. brucei* bloodstream stages involves the following steps:

(i) The addition of *N*-acetylglucosamine (GlcNAc), transferred from uridine diphosphate- (UDP-) GlcNAc, to an endoplasmic reticulum (ER) pool of phosphatidylinositol (PI) that is biosynthetically separate from bulk membrane PI (Martin and Smith, 2006).

(ii) The de-*N*-acetylation of GlcNAc-PI to form GlcN-PI.

(iii) The transfer of GlcN-PI from the cytosolic to the luminal face of the ER and addition of three mannose residues donated from Dol-P-Man.

(iv) The acylation of the inositol ring and addition of EtN-P.

(v) The sequential remodelling of *sn*-1 and *sn*-2 fatty acids with myristate.

(vi) The reversible acylation of the inositol group in mature GPI intermediates, to generate steady state pools of both inositol acylated (glycolipid C) and non-inositol acylated (glycolipid A) GPIs.

However, only glycolipid A is added to VSG and RNAi knock-down of the enzyme that deacylates the inositol headgroup results in reduced expression of VSG (Hong *et al.*, 2006). The transfer of glycolipid A to VSG and other proteins is mediated by the GPI:protein transamidase complex, comprising GPI8 (catalytic subunit), Gaa1p, GPI16 and the kinetoplastid-specific subunits, TTA1 and TTA2 (Nagamune *et al.*, 2003). Finally, the glycan backbone of the VSG anchor can be variably modified with 0–10 α-galactose (α-Gal) residues in the Golgi apparatus, while the diacylglycerol lipid moiety can be further remodelled with myristate at the plasma membrane (Buxbaum *et al.*, 1996; Roper *et al.*, 2002) (see Figure 12.2).

T. brucei procyclic stages have a similar GPI pathway, although they lack the myristoyltransferases that complete the fatty acid remodelling steps and accumulate mature GPI anchor precursors with an inositol-acylated *lyso*-PI lipid moiety (Mehlert *et al.*, 1998) (see Figure 12.2). These inositol acylated GPIs are attached to the procyclins, a family of surface glycoproteins that contain a negatively-charged polypeptide backbone. While the procyclins are minimally modified with a single, *N*-linked glycan, the glycan backbone of the procyclin GPI anchor can be extensively modified with branched poly-*N*-acetyllactosamine (poly-NAL) side chains (Mehlert *et al.*, 1998). These side chains are added in the Golgi apparatus and can be further elaborated with sialic acid residues when the GPI-anchored procyclins reach the cell surface. The sialylation of poly-NAL chains is catalysed by a surface trans-sialidase enzyme that transfers sialic acid from host sialylglycoconjugates (Nagamune *et al.*, 2004; Montagna *et al.*, 2006). The polypeptide backbone of the procyclins likely forms a relatively diffuse protein coat, while the sialylated poly-NAL side chains may form a dense glycocalyx over the plasma membrane. Remarkably, *T. brucei* mutants lacking GPI-anchored procyclins are still able to survive and infect the tsetse fly vector (Nagamune *et al.*, 2000). These mutants appear to compensate for the loss of the procyclin coat by upregulating the expression of free GPI glycolipids that carry the poly-NAL chains (Lillico *et al.*, 2003). Significantly, *T. brucei* mutants lacking all GPIs (protein-linked and free) are severely impaired in their ability to colonize the tsetse fly (Guther *et al.*, 2006) indicating that the GPI glycocalyx is essential for survival in this host. Unexpectedly, the GPI-negative mutants retain a surface glycocalyx, although the composition of this GPI-free coat remains to be determined (Guther *et al.*, 2006).

The cell surfaces of the major *T. cruzi* developmental stages are dominated by two large superfamilies of GPI-anchored glycoproteins, the trans-sialidase/gp85 glycoproteins and the mucin-like glycoproteins (Buscaglia *et al.*, 2006). Other surface glycoproteins (e.g. IG7, Ssp4) are also GPI anchored. The glycan backbones of these anchors are modified in a stage- and protein-specific manner with an α-(1→2)-linked mannose side chain or the charged group, aminoethylphosphonate (see Figure 12.2). The lipid moieties of these anchors can also vary. While GPI anchor precurors generally have alkylacyl- or diacylglycerol lipid moieties, alkylacylglycerol, lyso-alkylglycerol or ceramide lipid moieties occur in mature protein-linked GPIs (Lederkremer and Bertello, 2001). These observations suggest ceramide is incorporated into mature GPI precursors or GPI-anchored proteins by lipid remodelling reactions as is proposed to occur in yeast (Heise *et al.*, 1996).

The GPI-anchored proteins (e.g. gp63, PSA2) are also abundant on the surface of the promastigote (insect) stages of *Leishmania*, although the expression of these and other macromolecules is downregulated in the intracellular amastigote stage. *Leishmania* GPI anchors lack side chain modifications, but may undergo lipid remodelling reactions similar to those that occur in *T. brucei* (Ralton and McConville, 1998). The GPI-anchored proteins are not essential for promastigote growth in culture or infectivity in the mammalian host (Hilley *et al.*, 2000).

3.2. Non-protein-linked GPIs in kinetoplastid parasites

All trypanosomatid parasites synthesize more GPI glycolipids than are required for protein anchoring and free GPIs are the major class of glycolipids in these parasites. In *T. brucei* bloodstream and procyclic stages, excess GPI glycolipids are transported to the cell surface without additional modifications (Figure 12.3). In *T. cruzi* and *Leishmania*, specific pools of GPI precursors appear to be channelled into the synthesis of free GPIs or glycoinsitol-phospholipids (GIPLs) which can be >100-fold more abundant than GPI-anchored proteins (Heise *et al.*, 1996; Ralton and McConville, 1998; McConville *et al.*, 2002; Previato *et al.*, 2004). The major *T. cruzi* GIPLs retain the Man_3GlcN_1-inositol-lipid backbone of the protein anchors, but are additionally modified with α-(1$\rightarrow$2)-linked mannose (Man), galactopyranose or galactofuranose residues (Previato *et al.*, 2004) (see Figure 12.3). In *Leishmania*, distinct classes of GIPLs are synthesized in a species and stage-specific manner. Type-1 and type-2 GIPLs are structurally related to protein and LPG anchors, respectively, while hybrid type GIPLs have branched glycan stuctures with both protein and LPG anchor motifs (Zawadzki *et al.*, 1998; Mullin *et al.*, 2001) (see Figure 12.3). Despite being the major glycoconjugates on the surface of intracellular amastigote stages, *Leishmania* mutants lacking GIPLs are viable in culture and remain virulent in susceptible animal models (Zufferey *et al.*, 2003). *Leishmania* amastigotes are able to scavenge glycosphingolipids from the mammalian host (McConville and Blackwell, 1991; Schneider *et al.*, 1993; Winter *et al.*, 1994) and it is possible that these glycolipids may form an alternative glycocalyx and compensate for the loss of GIPLs *in vivo*.

The insect-dwelling promastigote stages of *Leishmania* also synthesize a hyperglycosylated GPI, termed lipophosphoglycan (LPG) (McConville *et al.*, 2002). The GPI anchor of LPG contains a divergent glycan core that is subsequently modified with linear chains of β-Gal-(1$\rightarrow$4)-α-Man-1-PO_4 repeating units (see Figure 12.3). These units are assembled by the sequential transfer of Man-1-PO_4 from guanosine diphosphate- (GDP-) Man and Gal from UDP-Gal in the Golgi apparatus (Capul *et al.*, 2007b) and can be further elaborated with monosaccharide (Gal, Man or oligosaccharide side chains) in a species- and growth-dependent manner (McConville *et al.*, 1992). The regulation of LPG elongation and side chain modification appears to be functionally important for parasite development in the sandfly vector. *Leishmania major* mutants lacking the capacity to synthesize LPG are able to differentiate in the sandfly midgut, but are unable to bind to midgut epithelial cells and are cleared from the gut when the sandfly excretes the remains of the bloodmeal (Sacks *et al.*, 2000). The β-Gal side chains on *L. major* LPG are bound by a lectin on the surface of epithelial cells, promoting parasite binding to the midgut wall (Pimenta *et al.*, 1992; Kamhawi *et al.*, 2004). As promastigotes develop

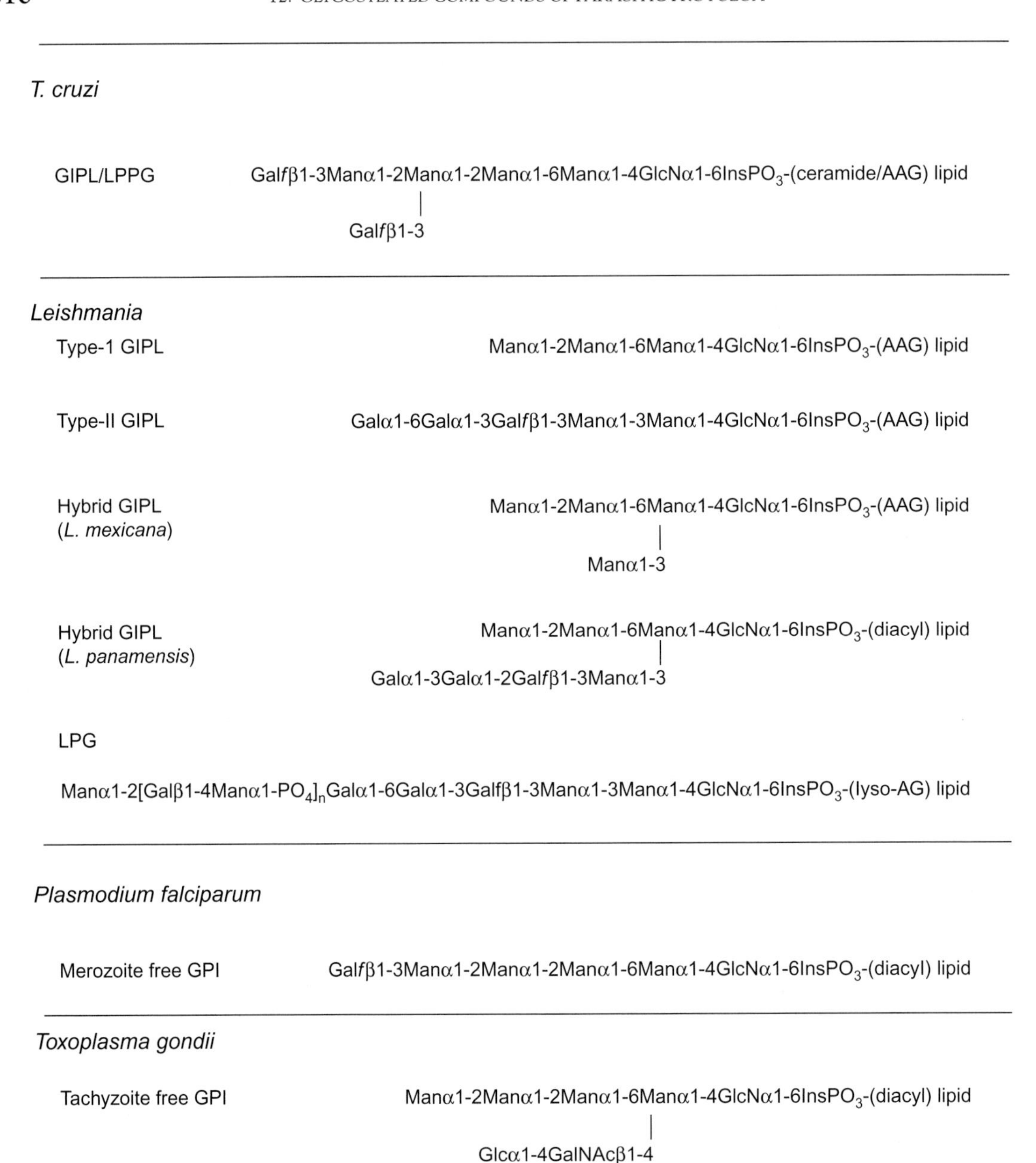

FIGURE 12.3 Structures of the free GPI glycolipids of parasitic protists.

in the sandfly midgut, structurally-distinct LPG chains are incorporated into the glycocalyx that have side chains capped with arabinopyranose residues (McConville *et al.*, 1992) facilitating promastigote detachment from the epithelial layer and anterior migration along the digestive tract. The longer LPG chains of these stages may also confer resistance to complement lysis.

Finally, LPG may play a role in establishing promastigote infection in the mammalian host (Spath *et al.*, 2003a). *L. major* mutants lacking LPG are slow to establish lesions in susceptible animal models, reflecting a high rate of clearance during the early stages of infection and appear to be more susceptible to oxidative stress prior to their differentiation to the obligate intracellular amastigote stage (Spath *et al.*, 2003a; Naderer and McConville, 2008). However, LPG biosynthesis is dramatically decreased in amastigotes stages and *L. major* mutant strains lacking LPG establish lesions in susceptible mice with the same kinetics as wild-type parasites if the infection is initated with amastigotes. Some other species of *Leishmania* do not require LPG for either promastigote or amastigote infectivity, possibly reflecting differences in sensitivity to oxidative stress (Ilg, 2000a; Naderer and McConville, 2008).

3.3. The GPI glycolipids of Apicomplexan parasites

Biochemical and genomic studies suggest that all studied members of the Apicomplexa have a well-developed GPI biosynthetic pathway (Delorenzi *et al.*, 2002; Templeton *et al.*, 2008). The major mammalian infective-stages of *Plasmodium* spp. (sporozoites and asexual red blood stages) reside within non-hydrolytic vacuoles and only the transiently-extracellular infective merozoites are coated by GPI-anchored proteins (Gowda *et al.*, 1997; Sanders *et al.*, 2005). At least twelve GPI-anchored proteins, including the dominant merozoite surface protein-1 (MSP-1), are expressed by asexual red blood stages of *P. falciparum* (Sanders *et al.*, 2005). Most of these surface proteins lack other types of glycosylation and the biosynthesis of GPI anchor precursors is the major pathway of glycosylation in these parasite stages (Naik *et al.*, 2000). As in the trypanosomatids, *P. falciparum* merozoites synthesize a two- to five-fold excess of GPI precursors over that needed for protein anchoring and these excess intermediates are thought to accumulate in the plasma membrane (see Figure 12.3). It has been proposed that GPI-anchored proteins and/or these free GPI glycolipids have potent immunoregulatory properties and are responsible for inducing many of the symptoms associated with malaria (Naik *et al.*, 2000; Vijaykumar *et al.*, 2001; von Itzstein *et al.*, 2008).

The rapidly dividing tachyzoite stage of *Toxoplasma gondii* also reside within non-hydrolytic vacuoles, although they infect a much wider range of mammalian cells than *Plasmodium* spp. Like the infective merozoites of *Plasmodium* spp., *T. gondii* tachyzoites are coated with GPI-anchored proteins, termed surface antigens (SAG1-4), SAC-related sequences (SRS) and SAG-unrelated proteins (SUSA) (Boothroyd *et al.*, 1998; Pollard *et al.*, 2008). The GPI anchors of these proteins contain single *N*-acetylgalactosamine (GalNAc) side chain modification (Striepen *et al.*, 1997), while the free GPIs that accumulate on the surface of tachyzoites are modified with α-Glc-(1→4)-GalNAc (where Glc, glucose; GalNAc, *N*-acetylgalactosamine) (Striepen *et al.*, 1997) (see Figures 12.2 and 12.3). The glucosylated side chain of the free GPIs is highly immunogenic in humans and appears to be widely distributed across different strains of *T. gondii* (Striepen *et al.*, 1997). It was originally proposed that the GPI anchors of SAG-1 and other surface proteins were required for surface transport and/or to allow the shedding of antibody-opsonized proteins from the cell surface. However, GPI-anchored proteins have been identified in apical intracellular organelles and a modified SAG-1 containing a polypeptide transmembrane domain was still transported to the cell surface and shed with similar kinetics to the GPI-anchored SAG-1 (Seeber *et al.*, 1998). On the other hand, replacement of the GPI anchor with a transmembrane polypeptide reduced SAG-1 dimerization and the incorporation of this protein into large surface complexes, while also increasing the turnover and degradation of this chimeric protein. The utilization of a GPI-anchor may minimize steric constraints interfering with protein–protein interactions and/or increase the lateral mobility of SAG-1 in the parasite plasma membrane (Seeber *et al.*, 1998).

The pathways of GPI biosynthesis in *P. falciparum* and *T. gondii* have been delineated and share many common characteristics (Gerold *et al.*, 1999; Schmidt *et al.*, 1998; Smith *et al.*, 2007). In particular, acylation of the early-intermediate GlcN-PI appears to be essential for subsequent mannosylation and ethanolamine-phosphate additions. In *P. falciparum*, the inositol acyl chain is predominantly myristate (Gerold *et al.*, 1999). In both *Plasmodium* and *Toxoplasma*, the side chain additions (Man, GalNAc, respectively) are added prior to the attachment of these precursors to protein (Schmidt *et al.*, 1998; Kimmel *et al.*, 2006; Smith *et al.*, 2007). In *T. gondii*, but not in *Plasmodium*, the inositol acyl chain is removed prior to GPI attachment to protein (Smith *et al.*, 2007).

In contrast to the other Apicomplexan parasites, *C. parvum* sporozoites have a complex surface coat of mucin-like glycoproteins and abundant GIPLs (Priest *et al.*, 2003, 2006). The latter have a conserved tri-mannose backbone and a heterogeneous diacylglycerol PI lipid moiety (Riggs *et al.*, 1999; Priest *et al.*, 2003). The *C. parvum* GIPLs are recognized by antibodies in patient's serum and a monoclonal antibody against these glycolipids neutralizes sporozoite infectivity and reduces *C. parvum* infection in oocyst-challenged mice (Riggs *et al.*, 1999). While the genome of *C. parvum* contains most of the expected genes for GPI biosynthesis, it lacks a canonical GlcNAc deacetylase that catalyses the second step in GPI biosynthesis. An unrelated bacterial-type sugar deacetylase is suggested to have replaced this enzyme (Templeton *et al.*, 2008).

3.4. The GPI-anchored proteins in *E. histolytica*

E. histolytica trophozoites are coated with a heavily glycosylated GPI-anchored proteophosphoglycan (PPG, previously referred to as LPG). Trophozoites also express an abundant surface Gal/GalNAc-binding lectin comprising a type-1 membrane protein (heavy chain) and a GPI-anchored light chain (McCoy *et al.*, 1993; Moody-Haupt *et al.*, 2000; Santi-Rocca *et al.*, 2008). The GPI-anchor of the PPG is highly unusual in containing the glycan backbone sequence α-Gal(1→?)-α-Man-(1→6)-α-Man-(1→4)-GlcN and long chains of α-Gal (see Figure 12.2). The precise point of attachment of these side chains, the linkage between the protein and the GPI and the nature of the lipid moiety have yet to be determined. The nature of the lipid moiety has been established recently as 1-*O*-[(28:0/30:1)-*lyso-glycero*-3-phosphatidyl-]2-*O*-(16:0)-Ins (Lotter *et al.*, 2009). Knock-down of genes involved in the de-*N*-acetylation of GlcNAc-PI and addition of the first mannose residue to GlcN-PI result in downregulation of PPG expression, increased sensitivity to complement lysis and decreased capacity for amoebic abscess formation (Vats *et al.*, 2005; Weber *et al.*, 2008), highlighting the importance of these molecules in pathogenesis.

3.5. Synthesis of GPI-related glycoconjugates in *Trichomonas*

T. vaginalis lacks homologues of GPI biosynthetic enzymes and the capacity to synthesize the essential donor, dolichol-P-Man (Samuelson *et al.*, 2005; Carlton *et al.*, 2007) (see Figure 12.3). However, flagellated stages of *T. vaginalis* produce an abundant LPG that may form a carbohydrate-rich surface coat. The *T. vaginalis* LPG contains an inositol-phosphoceramide lipid anchor and a complex array of sugars: Glc, Gal, Man, arabinose (Ara), GlcNAc, GalNAc and rhamnose (Rha) (Singh, 1993; Bastida-Corcuera *et al.*, 2005). *T. vaginalis* mutants expressing truncated LPGs have been isolated after chemical mutagenesis. These mutants are no longer agglutinated by Gal- or GlcNAc-binding lectins and are less adherent and cytotoxic to human vaginal ectocervical cells than wild-type parasites (Bastida-Corcuera *et al.*, 2005).

4. N-LINKED GLYCANS

N-glycans are commonly added to eukaryotic proteins that are destined for the cell surface, secretion or other compartments in the secretory pathway. *N*-glycan precursors are built up in the ER on a dolichol pyrophosphate (Dol-PP) lipid carrier and then transferred *en-bloc* to asparagine (Asn) residues within the sequon, Asn-X-Ser/Thr (where Ser, serine; Thr, threonine), in newly synthesized proteins (Figure 12.4). The *en-bloc* transfer occurs during or after protein translocation into the ER lumen and is mediated by the oligosaccharyltransferase (OST) which, in fungi and metazoans, comprises a multiprotein complex. Following transfer to protein, *N*-glycans can be extensively processed by a range of glycosidases and glycosyltransferases in the ER and Golgi apparatus to generate more complex structures. Bioinformatic analysis of recently sequenced parasite genomes suggests that most parasitic protists have the capacity to synthesize dolichol-linked oligosaccharides, although these are commonly truncated due to extensive secondary loss of asparagine-linked glycosylation (ALG) genes (Samuelson *et al.*, 2005) (see Figure 12.4). With the exception of the trypanosomes, high mannose or pauci-mannose *N*-glycans are generally not elaborated to more complex structures.

4.1. N-Glycosylation in trypanosomatid parasites

The trypanosomatid parasites retain the most complete *N*-glycosylation pathway, lacking only the Dol-P-Glc-dependent glucosyltransferases (Alg6, 8, 10), Dol-P-Glc synthase and, in the case of *Leishmania*, one of the Dol-P-Man-dependent mannosyltransferases (Alg 12) (Parodi, 1993; Samuelson *et al.*, 2005). While both *T. brucei* and *T. cruzi* have the capacity to synthesize $Man_9GlcNAc_2$-PP-Dol, these parasites also transfer shorter $Man_{5-7}GlcNAc_2$ oligosaccharides to protein (Doyle *et al.*, 1986; Manthri *et al.*, 2008). $Man_9GlcNAc_2$ and $Man_5GlcNAc_2$ oligosaccharides are added to Asn residues in VSG in a site-specific manner, possibly by distinct OST complexes (Manthri *et al.*, 2008). Following their addition to protein, the high mannose and pauci-mannose *N*-glycans of trypanosomatid parasites are variably modified by the unfolded glycoprotein glucosyltransferase (UGGT), a soluble ER protein that recognizes unfolded glycoproteins (Jones *et al.*, 2005). The UGGT-mediated glucosylation of *N*-glycans leads to the recruitment of calreticulin, a protein chaperone with a lectin domain that facilitates protein folding. Subsequent removal of the terminal Glc by the ER glucosidase-II can result in the disassembly of this complex and ER export of the folded protein or reglucosylation by UGGT (Jones *et al.*, 2005). The proteins required for this cycle (UGGT, calreticulin and glucosidase II) are present in all the trypanosomatid genomes, suggesting that *N*-glycosylation in these protists has been retained to engage this protein-folding machinery. In support of this notion, calreticulin has been shown to bind to major lysosomal glycoproteins, while genetic deletion of UGGT or chemical inhibition of glucosidase-II interfere with the lysosomal transport of cysteine proteases in *T. cruzi* (Conte *et al.*, 2003). Moreover, *T. brucei*, *T. cruzi* and *L. mexicana* mutants lacking components of the calreticulin machinery remain viable in culture, possibly reflecting redundancy in the ER chaperone network, but display a variable capacity to establish infections in the mammalian host (Conte *et al.*, 2003; Jones *et al.*, 2005). In particular, *L. mexicana* mutants lacking the capacity to synthesize mannosylated *N*-glycans are viable in culture, but are highly sensitive to elevated temperatures encountered in the mammalian host and are avirulent in animal models (Garami and Ilg, 2001; Ralton *et al.*, 2003). Interestingly, inhibition of core *N*-glycosylation with tunicamycin (a specific inhibitor of Alg7) often has a more profound effect on the growth and viability of typanosomatid parasites than the mutants mentioned above, suggesting that *N*-glycosylation

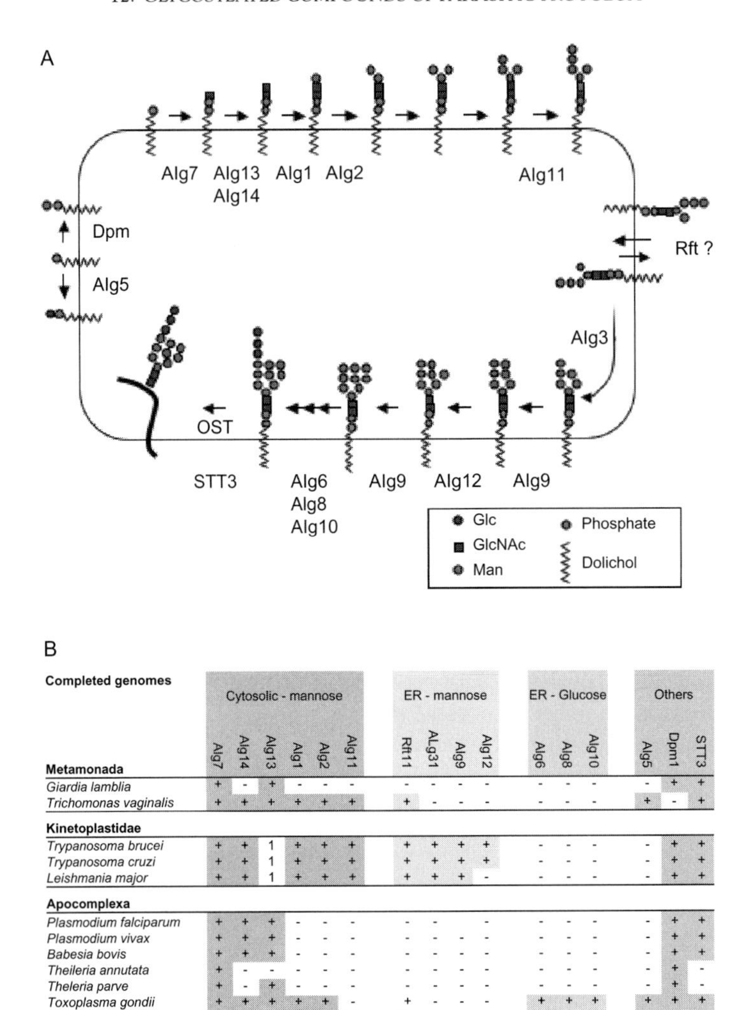

Completed genomes	Cytosolic - mannose						ER - mannose				ER - Glucose			Others		
	Alg7	Alg14	Alg13	Alg1	Alg2	Alg11	Rft11	ALg31	Alg9	Alg12	Alg6	Alg8	Alg10	Alg5	Dpm1	STT3
Metamonada																
Giardia lamblia	+	-	+	-	-	-	-	-	-	-	-	-	-	-	+	+
Trichomonas vaginalis	+	+	+	+	+	+	+	-	-	-	-	-	-	+	-	+
Kinetoplastidae																
Trypanosoma brucei	+	+	1	+	+	+	+	+	+	+	-	-	-	-	+	+
Trypanosoma cruzi	+	+	1	+	+	+	+	+	+	+	-	-	-	-	+	+
Leishmania major	+	+	1	+	+	+	+	+	+	-	-	-	-	-	+	+
Apocomplexa																
Plasmodium falciparum	+	+	+	-	-	-	-	-		-	-	-	-	-	+	+
Plasmodium vivax	+	+	+	-	-	-	-	-	-	-	-	-	-	-	+	+
Babesia bovis	+	+	+	-	-	-	-	-	-	-	-	-	-	-	+	+
Theileria annutata	+	-	-	-	-	-	-	-	-	-	-	-	-	-	+	-
Theleria parve	+	-	+	-	-	-	-	-	-	-	-	-	-	-	+	-
Toxoplasma gondii	+	+	+	+	+	-	+	-	-	-	+	+	+	+	+	+
Cryptosporidium hominus	+	+	-	+	+	+	+	-	-	-	+	+	-	+	+	+
Cryptosporidium parvum	+	+	-	+	+	+	+	-	-	-	+	+	-	+	+	+
Archamoebae																
Entamoeba histolytica	+	+	1	+	+	+	+	-	-	-	-	-	-	-	+	+

FIGURE 12.4 Dolichol-linked oligosaccharide biosynthesis in parasitic protists. (A) The pathway of oligosaccharide-PP-Dol synthesis in yeast and metazoan organisms. (B) Presence of ALG genes in the genomes of parasitic protozoa. Adapted from Samuelson *et al.* (2005) but with additional genome annotations. The analyses provide strong evidence for extensive secondary loss of ALG genes in different parasitic protozoa. Symbols: +, Presence; -, absence. 1, Kinetoplastidae and *E. histolytica* encode *alg13* and *alg14* fused as a single polypeptide.

may be required for protein folding or function independent of calreticulin binding (Nolan and Farrell, 1985; Kink and Chang, 1987). Long-term treatment of *Leishmania* parasites with tunicamycin results in loss of viability under culture conditions, as well as the amplification of the Alg7 gene encoding the GlcNAc-1-P transferase (Kink and Chang, 1987). Similarly, disruption of hexosamine metabolism and *N*-glycosylation in *L. major* was lethal even when parasites were cultivated at 27°C (Naderer *et al.*, 2008). These observations suggest that the addition of the minimal

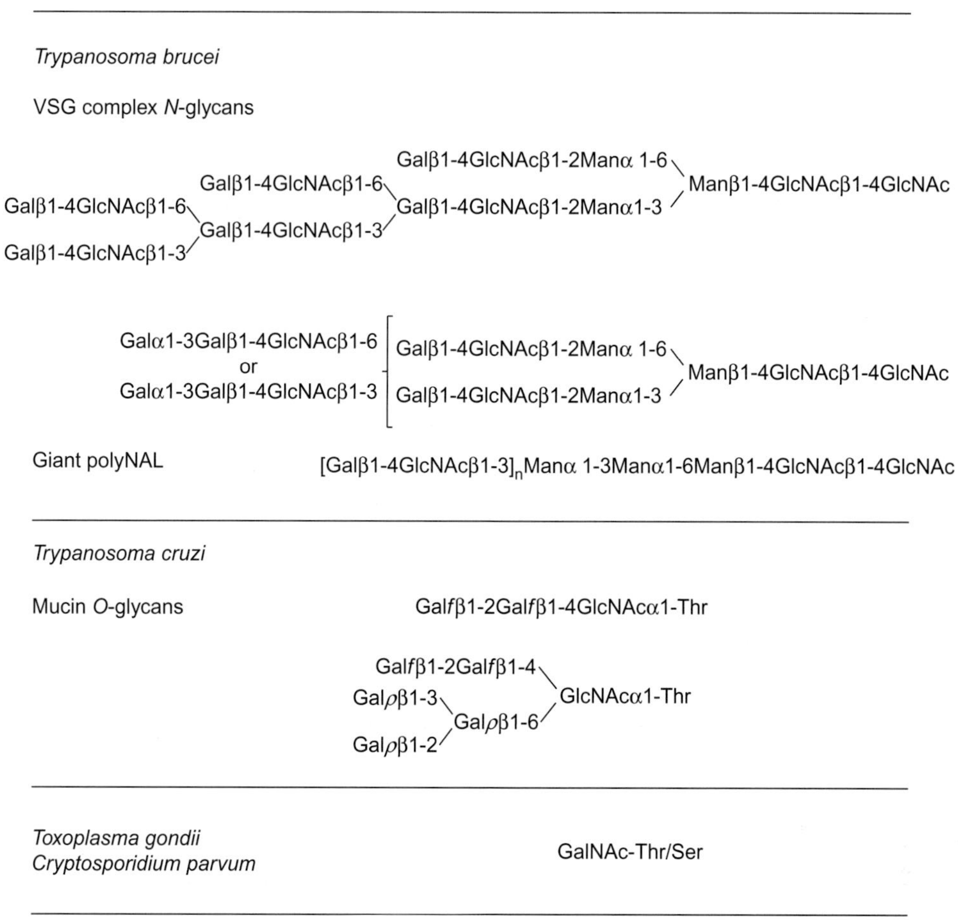

Trypanosoma brucei

VSG complex *N*-glycans

Galβ1-4GlcNAcβ1-6 \ Galβ1-4GlcNAcβ1-3 / Galβ1-4GlcNAcβ1-6 \ Galβ1-4GlcNAcβ1-3 / Galβ1-4GlcNAcβ1-2Manα1-3 \ Galβ1-4GlcNAcβ1-2Manα 1-6 / Manβ1-4GlcNAcβ1-4GlcNAc

[Galα1-3Galβ1-4GlcNAcβ1-6 or Galα1-3Galβ1-4GlcNAcβ1-3] [Galβ1-4GlcNAcβ1-2Manα 1-6 \ Galβ1-4GlcNAcβ1-2Manα1-3 /] Manβ1-4GlcNAcβ1-4GlcNAc

Giant polyNAL [Galβ1-4GlcNAcβ1-3]$_n$Manα 1-3Manα1-6Manβ1-4GlcNAcβ1-4GlcNAc

Trypanosoma cruzi

Mucin *O*-glycans Gal*f*β1-2Gal*f*β1-4GlcNAcα1-Thr

Gal*p*β1-3 \ Gal*p*β1-2 / Gal*p*β1-6 \ Gal*f*β1-2Gal*f*β1-4 / GlcNAcα1-Thr

Toxoplasma gondii
Cryptosporidium parvum GalNAc-Thr/Ser

FIGURE 12.5 *N*- and *O*-glycosylation in parasitic protists. The trypanosomes (*T. brucei* and *T. cruzi*) are the only parasitic protists to synthesize complex *N*-glycans. The structures of the complex polyNAL *N*-glycans on *T. brucei* VSG and the giant polyNAL chains on a flagellar protein are shown. *T. cruzi* is the only parasitic protist in which complex *O*-linked glycans have been characterized. Representative *O*-glycan structures of the major surface mucins are shown. The *O*-GalNAc has been identified in some Apicomplexan parasites.

N-glycan GlcNAc$_2$-core is essential for *Leishmania* viability at 27°C, while addition of mannosylated *N*-glycans are required for viability at elevated temperatures.

The high mannose and pauci-mannose *N*-glycans of all trypanosomatid parasites can be further processed by Golgi α-mannosidases. However, only in *T. brucei*, and to a lesser extent in *T. cruzi*, are *N*-glycans converted to more complex structures. In *T. brucei* bloodstream stages, Man$_9$GlcNAc$_2$-glycans on VSG are processed to triantennary Man$_5$GlcNAc$_2$, while conventional biantennary Man$_5$GlcNAc$_2$- are first processed to Man$_3$GlcNAc$_3$- before being elaborated with short or long chains of pNAL (Manthri *et al.*, 2008) (Figure 12.5). The pNAL chains on VSG glycans contain two to eight β-Gal-(1→4)-GlcNAc and β-Gal-(1→3)-GlcNAc repeating units and, in some cases, an α-Gal-(1→3)-Gal capping unit (see Figure 12.5). However, other polypeptides can be modified with larger *N*-glycan chains. A ≈50-kDa protein in *T. brucei* bloodstream forms was found to contain giant pNAL chains (≈54 repeating units) with a large number of β-GlcNAc-(1→6)-Gal inter-repeat linkages (Atrih *et al.*, 2005) (see Figure 12.5). The giant pNAL-containing glycoproteins were localized to the flagellar pocket and intracellular endocytic vacuoles and may form a gel-like matrix in these compartments. Inhibition of Gal metabolism and pNAL synthesis in *T. brucei* results in disruption of membrane transport and eventual loss of viability, suggesting that Golgi glycosylation is essential for the viability of *T. brucei* bloodstream forms (Urbaniak *et al.*, 2006). Apart from a potential role in maintaining the permeability barrier properties of the VSG coat, pNAL-*N*-glycans attached to a number of nutrient receptors in the flagellar pocket are thought to be required for their endocytosis and function (Nolan *et al.*, 1999).

The dominant surface glycoproteins of the insect-dwelling *T. brucei* procyclic stages are exclusively modified with high mannose (tri-antennary $Man_{5-9}GlcNAc_2$-) *N*-linked glycans, indicating that the *N*-glycosylation machinery of these parasites is regulated in a stage-specific manner. However, disruption of the ALG3 gene in procyclic stages (Manthri *et al.*, 2008), or selection for ConA resistance (Hwa and Khoo, 2000), results in the assembly of complex *N*-glycans on the procyclins indicating that this developmental stage retains the Golgi enzymes needed to make pNAL chains. A similar situation may occur in some other trypanosomatids. For example, while most cell surface and lysosomal proteins in *T. cruzi* are modified with high mannose *N*-glycans (Atwood *et al.*, 2006), biantennary complex and sulfated *N*-glycans have been identified on the major lysosomal enzyme, cruzipain (Parodi *et al.*, 1995; Barboza *et al.*, 2005). In contrast, there is no evidence that the high mannose *N*-glycans of *Leishmania* spp. are elaborated to complex structures in any stage (Ilg *et al.*, 1994; Funk *et al.*, 1997).

4.2. N-Glycosylation in the Apicomplexa

The secondary loss of ALG genes appears to have been common in the Apicomplexa, as most apicomplexan taxa are predicted to synthesize highly truncated dolichol-linked oligosaccharides (see Figure 12.4). Only *T. gondii* and *Cryptosporidium* are predicted to have the capacity to make mannosylated dolichol-linked oligosaccharides, while all apicomplexan taxa lack obvious mannosyltransferases for the assembly of $Man_9GlcNAc_2$-PP-Dol in the ER lumen (see Figure 12.4). Intriguingly, biochemical analysis of total protein extracts of *T. gondii* tachyzoite and bradyzoites indicate that many proteins in these stages are modified with both high mannose ($Man_{5-9}GlcNAc_2$) and pauci-mannose *N*-glycans (Fauquenoy *et al.*, 2008; Luk *et al.*, 2008). As the *T. gondii* genome lacks obvious homologues for ALG3, 9 and 12 (see Figure 12.4), other glycosyltransferases may catalyse these reactions. Alternatively, the contamination of parasite extracts with host glycoproteins may account for the presence of high mannose *N*-glycans. Interestingly, tunicamycin has a delayed effect on *T. gondii* tachyzoite growth in host cells. Tachyzoites develop normally during the first cycle of intracellular infection following tunicamycin treatment, but are unable to develop normally during second and third rounds of infections in the absence of drugs. A number of other drug treatments induce a similar delayed death phenotype, but the basis of this phenomenon is unknown. Tunicamycin has a similar effect on the intracellular growth of the distantly related apicomplexan parasite, *P. falciparum* (Naik *et al.*, 2001). Given that the *N*-glycans of these two parasites differ markedly (see below), a question remains as to whether the effect of tunicamycin on parasite infectivity is due to inhibition of *N*-glycosylation or other metabolic processes.

Plasmodium, *Babesia* and *Theileria* are predicted to have lost most of the genes in the *N*-glycosylation pathway, including all of the ALG genes involved in the assembly of mannosylated dolichol-linked oligosaccharides (see Figure 12.4). While *Plasmodium*, *Babesia* and *Theileria parva* have retained the two enzymes involved in the synthesis of $GlcNAc_2$-PP-Dol, *Theileria annulata* lacks homologues for Alg13, the catalytic subunit of the second *N*-acetylglucosaminyltransferase (see Figure 12.3). Intriguingly, *Plasmodium* and *Babesia* have the catalytic STT3 component of the OST, suggesting that these parasites transfer $GlcNAc_2$- to protein. The *Plasmodium* OST may have a refined specificity for these small oligosaccharides, as early studies suggested that conventional dolichol-linked oligosaccharides are not transferred to synthetic peptides in *in vitro* assays (Dieckmann-Schuppert *et al.*, 1992, 1994). In contrast, no candidates for STT3 could be identified in either of the recently sequenced *Theileria* genomes, raising the possibility that these parasites may be completely deficient in protein *N*-glycosylation (see Figure 12.4).

4.3. N-Glycosylation in the Metamonada and Archamoebae

T. vaginalis and *E. histolytica* are both predicted to transfer simple $Man_5GlcNAc_2$- glycans to proteins (Samuelson *et al.*, 2005) (see Figure 12.4). In *E. histolytica*, these glycans can be modified with a single terminal Glc residue by the ER UGGT, processed to $Man_3GlcNAc$ by a swainsonine-sensitive α-mannosidase and further elaborated with short chains of α-Gal-(1→2)- and α-Glc-(1→6)-α-Gal-(1→2)- (Magnelli *et al.*, 2008). In contrast, the intestinal parasite, *G. lamblia* lacks all ALG mannosyltransferases, but retains the enzymes needed to make GlcNAc-PP-Dol and GlcNAc2-PP-Dol. The synthesis of both these lipid precursors and their transfer to synthetic peptide acceptors has been confirmed using *G. lamblia* cell extracts (Samuelson *et al.*, 2005). The precise function of these highly truncated *N*-glycans remains to be determined.

5. O-LINKED GLYCANS

A number of parasite stages are coated by mucin-like glycoproteins that are heavily modified with *O*-linked glycans. *O*-Glycosylation in these organisms is initiated by the transfer of α-GalNAc (mammalian-like) or α-GlcNAc to threonine or serine residues in the polypeptide backbone. There are no examples of *O*-mannosylation, as occurs in many pathogenic fungi, or of the reversible addition of β-GlcNAc to cytoplasmic proteins (*O*-GlcNAcation), as occurs in metazoans.

5.1. O-Glycosylation in the Kinetoplastida

The major developmental stages of *T. cruzi* are coated with a family of mucin-like glycoproteins that are extensively modified with *O*-linked glycans (Previato *et al.*, 1994; Buscaglia *et al.*, 2004, 2006; Alves and Colli, 2007) (see Figure 12.5). *O*-Glycosylation is initiated with the transfer of GlcNAc to Ser or Thr residues in the polypeptide backbone. The core *O*-GlcNAc is further extended with 1–5 Gal residues to form short, linear and branched, glycan chains (see Figure 12.5). The terminal β-Gal residues in these glycans are the major acceptors for the cell surface trans-sialidases that transfer sialic acid (1–2 residues per chain) from host glycoconjugates (Schenkman *et al.*, 1993). Other terminal modifications can vary in different strains. For example, all the *O*-linked Gal residues are in the pyranose configuration in the *T. cruzi* Y-strain, while in the G-strain some of these residues are in the furanose configuration (Almeida *et al.*, 1994). Moreover, some of the *O*-glycans on the high molecular mass mucins of *T. cruzi* amastigote stages are modified with *O*-glycans that terminate in α-Gal-(1→3)-β-Gal-(1→4)-GlcNAc (Almeida *et al.*, 1994). These α-Gal epitopes are highly immunogenic to humans and are recognized by lytic antibodies in the serum of acute and chronic Chagasic patients (Almeida *et al.*, 1994).

The GlcNAc-transferase that initiates *O*-glycosylation in *T. cruzi* is associated with purified

Golgi membranes, utilizes UDP-GlcNAc as the sugar donor and adds GlcNAc residues to Thr residues in a synthetic dodecapeptide containing the consensus sequence of the *T. cruzi* MUC gene (Previato *et al.*, 1998). Partial disruption of Gal metabolism in *T. cruzi* insect-stage epimastigotes by single allele deletion of the gene encoding UDP-Glc-4′-epimerase, that is responsible for converting UDP-Glc to UDP-Gal, causes a severe reduction in mucin *O*-glycosylation and profound defects in cell morphology (MacRae *et al.*, 2006). It is likely that mucin *O*-glycosylation is important for the integrity and structure of the surface glycocalyx (Di Noia *et al.*, 1996). The sialylated *O*-glycans may also protect the mammalian stages of *T. cruzi* from complement lysis and the parasite's own haemolysin, prior to invasion of new host cells and escape from host cell lysosomes to the cytosol, respectively (Pereira-Chioccola *et al.*, 2000).

Conventional *O*-glycosylation has not been detected in *T. brucei* or *Leishmania*. However, a cytoplasmic protein in *L. major* has been reported to have *O*-glycans containing terminal GlcNAc (Handman *et al.*, 1993). The *Leishmania* genomes lack candidate genes for reversible cytoplasmic *O*-GlcNAcation or modification of hydroxyproline (see below) and nothing is known about the enzymology of this reaction.

5.2. O-Glycosylation in the Apicomplexa

Genomic and biochemical evidence has been obtained for canonical "mammalian-like" *O*-glycosylation in *T. gondii* and *C. parvum* (Winter *et al.*, 2000; Templeton *et al.*, 2008). The abundant GPI-anchored surface glycoproteins of *C. parvum* sporozoites are modified with single *O*-linked GalNAc residues linked to Ser and Thr residues (Winter *et al.*, 2000) (see Figure 12.5). Similar modification may occur in *T. gondii* based on:

(i) the presence of UDP-GalNAc: polypeptide *N*-acetylgalactosaminyltransferases homologues in the *T. gondii* genome;
(ii) the expression of these enzymes in tachyzoite and bradyzoite developmental stages; and
(iii) the demonstration that these enzymes are functionally active when expressed in *Drosophila* (Stwora-Wojczyk *et al.*, 2004a,b).

However, the endogenous polypeptide substrates for these enzymes have not yet been identified. Early reports suggested that *P. falciparum* may also *O*-glycosylate some proteins *in vivo* (Dayal-Drager *et al.*, 1991). However, analysis of the *Plasmodium* genomes has failed to reveal candidate glycosyltransferases and subsequent biochemical analyses of *in vivo*-labelled glycans indicate that *O*-glycosylation is negligible in intra-erythrocytic stages of *P. falciparum* (Gowda *et al.*, 1997).

A novel pathway of cytoplasmic glycosylation has recently been proposed to occur in *T. gondii* (West *et al.*, 2004). This pathway was originally identified in the slime mould, *Dictyosteleum discoideum*, and involves the addition of α-GlcNAc to hydroxyproline residues in the cytoplasmic protein, Skp1. Homologues of the Skp1 protein, the prolyl-hydroxylase and the UDP-GlcNAc: hydroxyproline *N*-acetylglucosaminyltransferase have been identified in the *T. gondii* genome (West *et al.*, 2006). Moreover, *D. discoideum* Skp1 is modified with hydroxyproline-linked GlcNAc when incubated with *T. gondii* lysates (West *et al.*, 2004, 2006). However, the function of *T. gondii* Skp-1 and this cytoplasmic glycosylation has not been defined.

5.3. O-Glycosylation in Metamonada and Archamoebae

G. lamblia trophozites produce a protein-rich surface coat dominated by acylated (non-GPI anchored) variant surface proteins (VSPs) and a GPI-anchored invariant surface protein (Das *et al.*, 1991; Nash, 2002). Antigenically distinct VSPs are expressed each time trophozoites re-emerge from the cyst stage, providing a mechanism for generating new antigenic coats in the same or different hosts (Nash, 2002). The modification of VSP4A1 with short *O*-linked oligosaccharides containing terminal GlcNAc was demonstrated by compositional analysis and labelling with UDP-[^{3}H]Gal and galactosyltransferase (Papanastasiou *et al.*, 1997). *O*-Glycosylation of the VSP proteins may be common and contribute to the antigenic diversity of these surface coats (Hiltpold *et al.*, 2000). Interestingly, *E. histolytica* trophozoites also express an abundant type-1 surface membrane protein, termed SREHP (serine rich E. histolytica protein) that is modified with *O*-GlcNAc (Stanley *et al.*, 1995). Recent studies suggest that SREHP is recognized by host cell receptors and mediates host cell adhesion, independent of the interactions mediated by the abundant surface Gal/GalNAc lectin (Teixeira and Huston, 2008).

6. PHOSPHOGLYCOSYLATION

The surface proteins of many parasitic protists are modified with glycans that are linked to Ser/Thr residues in the polypeptide backbone via phosphodiester linkages. This novel type of modification, termed phosphoglycosylation, was first detected in *D. discoideum*, where Ser residues in the lysosomal proteinase-I are modified with α-GlcNAc-1-PO_4^- (Gustafson and Gander, 1984). In parasitic protists, distinct forms of phosphoglycosylation can constitute the major type of protein post-translational modification in the secretory pathway (Haynes, 1998; McConville *et al.*, 2002) (Figure 12.6).

6.1. Phosphoglycosylation in the Kinetoplastida

Protein phosphoglycosylation is the most abundant form of protein glycosylation in *Leishmania* spp. (Ilg, 2000b; McConville *et al.*, 2002). Early immunochemical and biochemical analyses indicated that the secreted acid phosphatases (SAPs) of different *Leishmania* species were modified with similar phosphoglycans to those found on the LPG (Ilg *et al.*, 1991). Subsequent studies showed that serine residues in the protein backbone of the secreted acid phosphatase (SAP) were modified with Man-1-PO_4 which was extended with short linear chains of α-(1→2)- linked mannose or longer chains of repeating β-Gal-(1→4)-α-Man-1-PO_4^- phosphodisaccharides (Ilg *et al.*, 1994; Lippert *et al.*, 1999) (see Figure 12.6). The SAPs belong to a heterogeneous family of proteophosphoglycans (PPGs), which include the promastigote filamentous PPG (fPPG), the GPI-anchored cell surface PPG (mPPG) and the non-filamentous amastigote PPG (aPPG) (Ilg, 2000b). The distinct polypeptide backbones of these PPGs contain Ser/Thr-rich or Ser/Proline- (Pro-) rich domains that are extensively modified with phosphoglycan chains (Ilg *et al.*, 1996, 1998). The length and composition of the phosphoglycan chains vary enormously in different PPGs. While the 100-kDa form of *L. mexicana* SAP is primarily modified with short (1–6 residues long) mannose oligosaccharides, the secreted PPGs of the amastigote stage are elaborated with exceedingly complex branched phosphoglycans (Ilg *et al.*, 1994, 1998) (see Figure 12.6). The assembly of phosphoglycan chains may depend on sequence motifs within the protein carrier, the rate at which these proteins are transported through the secretory pathway and the complement of glycosyltransferases expressed in each species.

Protein phosphoglycosylation is essential for *Leishmania* survival in the sandfly vector. *L. major* mutants lacking the Golgi GDP-Man transporter (encoded by the LPG2 gene) are deficient in all phosphoglycans and are rapidly cleared from the sandfly midgut with the remains of the blood-meal (Sacks *et al.*, 2000). The clearance of the phosphoglycan-deficent parasites is not due to loss of LPG, as *L. major* mutants lacking surface LPG are able to survive the first few days in the midgut, although they

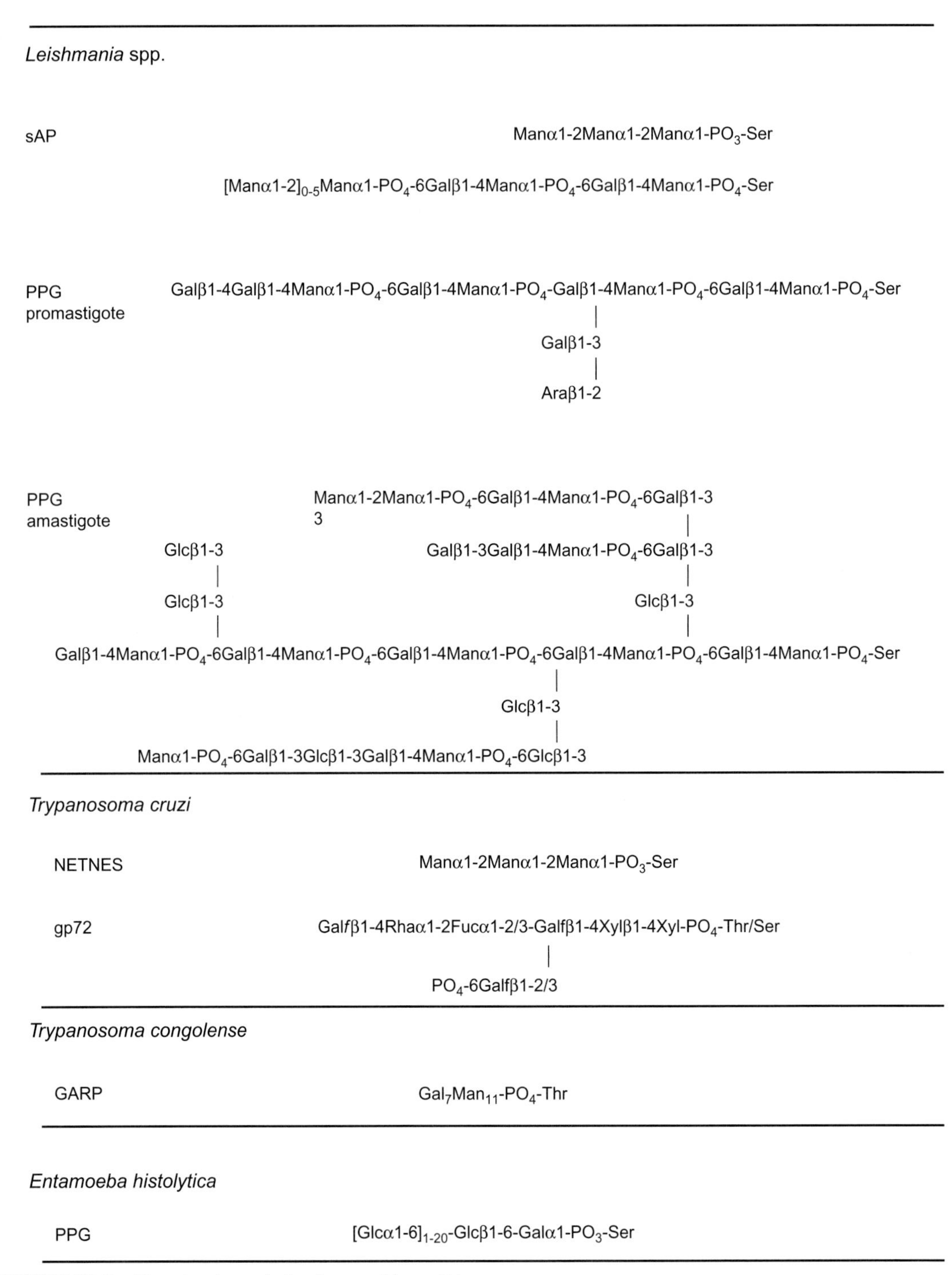

FIGURE 12.6 Phosphoglycosylation in parasitic protists.

are eventually cleared due to their inability to bind to midgut epithelial cells (Sacks *et al.*, 2000). These analyses suggest that the secreted proteophosphoglycans protect the newly-transformed promastigotes from complement lysis or other microbicidal processes. Secreted proteophosphoglycans may also promote aggregation of infective metacyclic promastigotes of *Leishmania* in the fore-gut of the sandfly and facilitate transmission of promastigotes to the mammalian host (Rogers *et al.*, 2004). Conflicting results have been obtained as to whether protein phosphoglycosylation is essential for infectivity in the mammalian host. Initial studies with the *L. major ΔLPG2* mutants indicated that phosphoglycosylation was required for initial promastigote survival in macrophages, as well as for growth of the intracellular amastigote stage that perpetuates disease in the mammalian host (Spath *et al.*, 2003b). However, other *L. major* mutants have since been generated that lack two Golgi UDP-Gal transporter isoforms (LPG4 and LPG5) and are also deficient in protein (and LPG) phosphoglycosylation (Capul *et al.*, 2007b). Unlike the *L. major Δlpg2* mutants, the *L. major Δlpg4/5* mutants are as virulent as wild-type parasites in highly susceptible animal models following their transformation to the amastigote (Capul *et al.*, 2007a). These mutants exhibit a characteristic delay in lesion development because of their deficiency in LPG biosynthesis that is required for promastigote infectivity. Similarly, *L. mexicana* mutants with deficiencies in Golgi phosphoglycans are as virulent as wild-type parasites (Ilg *et al.*, 2001). Collectively, these analyses suggest phosphoglycosylation is essential for promastigote infectivity in the sandfly vector and important for promastigote transmission to the mammalian host. However, despite the fact that intracellular amastigotes secrete abundant amounts of PPGs, this modification does not appear to be essential for parasite survival in the mammalian host.

A number of proteins in the *T. cruzi* insect (epimastigote and metacyclic trypomastigote) stages have been shown to be phosphoglycosylated. The abundant surface glycoprotein, gp72, is heavily modified with glycans (50% w/w) containing Rha, Xyl, Fuc, Gal_p, Gal_f and GlcNAc. These glycans are released by mild acid hydrolysis and are linked to phosphothreonine and some phosphoserine residues in the polypeptide backbone (Haynes, 1998; Ferguson, personal communication). Gp72 homologues in both *T. cruzi* and *T. brucei* appear to be required for attachment of the flagellum to the cell body. Interestingly, the *T. brucei* homologue, Fla-1 is modified with both *N*-glycans and acid-labile oligosaccharides suggesting that it is phosphoglycosylated (Nozaki *et al.*, 1996). RNAi knockdown of a key enzyme in fucose metabolism in both *T. brucei* procyclic and bloodstream stages is lethal and, in the case of the former, results in disruption of flagellum adhesion to the cell body (Turnock *et al.*, 2007). These data raise the possibility that fucosylation of Gp72/Fla-1 proteins in both *T. cruzi* and *T. brucei* is required for adhesion of the emergent flagellum with the cell body (possibly via a carbohydrate-lectin interaction) and infectivity or survival, respectively. Recent studies have revealed further diversity in the phosphoglycosylation machinery of *T. cruzi*. The GPI-anchored surface glycoprotein, NETNES, is modified with short chains of α-(1→2)-linked Man that are bound via phosphodiester linkage to Ser residues (Macrae *et al.*, 2005) (see Figure 12.6). These structures are identical to those found in *Leishmania,* suggesting that the GDP-Man:α-Man-1-PO_4-transferase is likely to be conserved in these trypanosomatids.

6.2. Phosphoglycosylation in *E. histolytica*

Protein phosphoglycosylation has not yet been reported in the Apicomplexa or in the Metamonada. However, the infective trophozoite stage of *E. histolytica* is coated with an abundant GPI-anchored PPG that is heavily modified with phosphoglycans. The *E. histolytica* phosphoglycans contain an αGal-1-PO_4-Ser linker region that is variably modified with linear chains of α-(1→6)-linked Glc (Moody-Haupt *et al.*, 2000). Trophozoites of the pathogenic HM-1:IMSS strain synthesize two distinct classes of PPG which have polydisperse molecular masses of 50 000–180 000 (PPG-1) and 35 000–60 000 (PPG-2) and are modified with glucan side-chains of different average lengths. In contrast, the non-pathogenic Rahman strain synthesizes one class of PPG which is only elaborated with the short disaccharide side-chain β-Glc-(1→6)-Gal (Moody-Haupt *et al.*, 2000). Differences in phosphoglycosylation may thus impact on the density of the surface glycocalyx and/or exposure of other surface molecules and trophozoite virulence. While the transferases responsible for initiating and elongating these phosphoglycans have not been identified yet, a number of UDP-Glc and UDP-Gal transporters have been localized to a putative Golgi fraction in *E. histolytica* trophozoites, suggesting that these modifications occur following export of the GPI-anchored PPG from the ER (Bredeston *et al.*, 2005).

7. PARASITE CYST WALL POLYSACCHARIDES

The intestinal parasites *G. lamblia* and *Entamoeba* spp. survive adverse environmental conditions by forming cysts that are surrounded by carbohydrate-rich walls (Eichinger, 2001). Encystment is induced in the host intestine by a variety of factors, including nutritional (Glc, cholesterol) shortages, exposure to bile salts and the presence of specific carbohydrate ligands (Eichinger, 2001). The resulting cyst stages are excreted in the faeces and are the primary stage responsible for transmission and persistence in the environment. The fibrillar wall of *Giardia* cysts contains both protein and polysaccharide (1:2 w/w). The polysaccharide comprises linear polymers of β-(1→3)-linked GalNAc that are predicted to form ordered helical structures (Gerwig *et al.*, 2002). Encystation requires the induction of enzymes involved in *de novo* hexosamine synthesis, the formation of UDP-GalNAc from UDP-GlcNAc and a novel cyst wall synthase (Jarroll *et al.*, 2001; Karr and Jarroll, 2004). Cyst wall synthesis is also associated with the upregulation of secretory transport and the formation of specialized encystation-specific secretory vesicles (ESVs) that are derived from the Golgi apparatus and contain preformed GalNAc polymers (Lujan *et al.*, 1998). In contrast, the cyst walls of *E. histolytica* and *Entamoeba invadens*, contain a mixture of chitin, $[\beta\text{-GlcNAc-}(1\rightarrow4)\text{-GlcNAc}]_n$, and chitosan, $[\beta\text{-GlcN-}(1\rightarrow4)\text{-GlcN}]_n$, as a major fibrillar component (Das *et al.*, 2006). Cyst wall synthesis requires the induction of two chitin synthases (Campos-Gongora *et al.*, 2004), a family of chitinases (that are presumably involved in remodelling nascent chitin chains) and chitin de-*N*-acetylases (Campos-Gongora *et al.*, 2004). Chitin polymers are pre-assembled in the Golgi apparatus and deposited at multiple sites over the plasma membrane. Samuelson and colleagues have shown that the expression of an abundant glycoprotein, termed Jacob, is also induced during encystation (Frisardi *et al.*, 2000). They have proposed that Jacob is bound to the plasma membrane of encysting trophozoites by the surface Gal/GalNAc-lectin and directs the deposition of chitin by virtue of containing Cys-rich chitin-binding domains (Frisardi *et al.*, 2000). The cysts of *Acanthamoeba*, another pathogenic amoeba that is more related to slime moulds than *Entamoeba*, contain cellulose, i.e. polymers of β-(1→4)-linked Glc, rather than chitin (Lorenzo-Morales *et al.*, 2008).

Some of the Apicomplexan parasites can also form resistant cyst stages that allow these parasites to persist outside the host. The wall of coccidian oocysts are generally protein-rich and composed of specialized oocyst wall proteins (COWP). Members of the multigene COWP family of *Cryptosporidium* and *T. gondii* contain cysteine-rich motifs that may facilitate intra- and inter-molecular cross-linking (Smith *et al.*, 2005). Similar sequences are absent from the genomes of apicomplexan parasites that lack an environmentally-resistant cyst stage (i.e. *Plasmodium, Babesia, Theileria*). The oocyst walls of *C. parvum* may also contain a carbohydrate component, comprising GlcNAc, GalNAc, Man and Gal (Smith *et al.*, 2005). In *Eimeria*, another coccidian parasite, the predominant cyst wall proteins contain tyrosine-rich domains that are cross-linked by dityrosine bridges to form highly rigid structures (Belli *et al.*, 2003).

8. INTRACELLULAR RESERVE GLYCANS

Bioinformatic analyses suggest that many members of the Apicomplexa (*T. gondii, C. parvum* and *Eimeria* spp.), the Archamoebae (*E. histolytica* and *Acanthamoeba* spp.) and the Metamonada (*Trichomonas* and *Giardia* spp.) synthesize intracellular α(1→4)-/α-(1→6)-glucans that are structurally related to glycogen and/or amylopectin (Wang *et al.*, 1975; Karkhanis *et al.*, 1993; Wu and Muller, 2003; Harris *et al.*, 2004). For example, homologues of glycogen phosphorylases, amylopectin and starch-binding proteins, amylases and debranching enzymes, and an R1 α-glucan water dikinase have been identified in *C. parvum* (Templeton *et al.*, 2008). Glycogen phosphorylase genes have also been identified in *Trichomonas, Entamoeba, Acanthamoeba* and *Giardia*. Biochemical analyses have confirmed the presence of amylopectin-like glucans in *Eimeria* cysts and all stages of *T. gondii* (Guerardel *et al.*, 2005). The high molecular mass amylopectin fraction of *T. gondii* forms crystalline inclusions in the cytoplasm that are prominent in slow growing bradyzoites and sporozoite stages. However, amylopectin can also accumulate in intracellular trophozoite stages, during invasion of less permissive host cells, suggesting a possible relationship between amylopectin accumulation and the host nutrient environment (Guerardel *et al.*, 2005). It is likely that these glucans can be mobilized for glycolysis and/or the synthesis of carbohydrate-rich polymers. Silencing of the *Acanthamoeba* gene encoding a glycogen phosphorylase prevented the formation of mature cysts, suggesting that the mobilization of these intracellular glucans is needed to sustain high levels of cellulose biosynthesis during cyst formation (Lorenzo-Morales *et al.*, 2008).

Some of the trypanosomatid parasites lack obvious reserve polysaccharides. On the other hand, others, including *Leishmania, Crithidia* and *Herpetomonas*, synthesize a novel β-(1→2)-mannan, which we have recently termed mannogen (Gorin *et al.*, 1979; Mendonca-Previato *et al.*, 1979; Ralton *et al.*, 2003). *Leishmania* mannogen is assembled on a novel α-Man-(1→4)-cyclic phosphate primer by the sequential transfer of mannose residues from GDP-Man (Sernee *et al.*, 2006). Mature mannogen lacks phosphate indicating that the phosphate at the reducing end is removed during synthesis (Sernee *et al.*, 2006). The synthesis and degradation of mannogen is highly dynamic and responsive to external carbon source availability indicating a primary reserve function (Ralton *et al.*, 2003). *Leishmania* mutants lacking mannogen due to global defects in mannose metabolism are very sensitive to elevated temperature and are avirulent in animal models (Ralton *et al.*, 2003), suggesting that mannogen is required for intracellular growth in the sugar-poor environment of mammalian macrophages (Naderer *et al.*, 2006).

9. CONCLUSIONS

Parasitic protists synthesize a range of glycans and glycoconjugates that are constituents of cell surface coats, cyst walls and intracellular carbohydrate reserves. The glycoconjugates that make up the cell surface coats are highly diverse and, in many cases, have been shown to play important roles in parasite infectivity.

The surface coats of parasitic protists have the following properties:

(i) They can be classified as protein-rich or carbohydrate-rich depending on the nature of the major surface macromolecules and/or surface glycolipids. The development of compositionally different coats is likely to reflect the distinct needs of each parasite stage for antigenic diversity, host cell recognition and protection against host hydrolases (proteases, glycosidases and lipases) within the specific niches that they occupy.

(ii) Surface coats with similar physical properties can contain structurally distinct glycoconjugates. For example, the dominant surface glycoconjugates of *Leishmania* promastigotes (LPG), *T. brucei* procyclics (procyclins) and *E. histolytica* (PPG), all have different backbone structures but contribute to the formation of highly polyanionic surface glycocalyxes.

(iii) Nearly all the protist surface coats are dominated by GPI-anchored macromolecules or free GPI glycolipids. The use of GPI anchors may facilitate the surface packing of coat components, while also insulating the cytoplasm from external insult. GPI anchors may promote the oligomerization of surface proteins and regulate their secretory or endocytic transport. It is likely that the high demand for GPI precursors led to the evolution of free GPIs as the major glycolipids of these organisms. *Trichomonas*, the only parasitic protist to lack conventional GPIs, synthesizes a highly abundant surface lipoglycoconjugate that is anchored to the plasma membrane via a related inositol-lipid moiety.

(iv) The carbohydrate-rich surface coats often contain novel glycan modifications. Examples include the sialylation of *O*-linked glycans and GPI anchors in *T. cruzi* and *T. brucei*, respectively, and the phosphoglycosylation of polypeptides in trypanosomatids and *E. histolytica*. These modifications contribute to the strong polyanionic characteristics of these coats and their intrinsic resistance to host hydrolyases.

(v) With the exception of the trypanosomes, parasitic protists generally have relatively simple *N*- and *O*-glycosylation pathways. In the case of *N*-glycosylation, this reflects extensive secondary loss of ALG genes rather than a primitive state. The high/pauci-mannose *N*-glycans may facilitate protein folding via the UGGT-calreticulin chaperone pathway and/or chaperone-independent pathways.

(vi) Finally, the analysis of parasite glycosylation mutants has revealed marked species- and stage-specific differences in the requirement for individual coat components. For example, the synthesis of GPI protein anchor precursors is essential for the viability and virulence of *T. brucei* bloodstream stage and the asexual red blood cell stages of *Plasmodium*, but not for the *T. brucei* procyclic stage or any of the *Leishmania* developmental stages. These studies highlight potential redundancy in the function of some classes of cell surface glycoconjugates and/or the possibility that some of these molecules are only essential under field conditions that are difficult to replicate in the laboratory. They also suggest that inhibitors of both eukaryotic-general and parasite-specific glycosylation pathways may have considerable potential as new therapeutics, but that the requirement for individual glycosylation steps should be validated for all relevant (infective) parasite stages.

Areas and questions that will potentially be the topics for future research are listed in the Research Focus Box.

RESEARCH FOCUS BOX

- Developments in mass spectrometry and glycan profiling can now be used to investigate the stage-specific regulation of individual glycosylation pathways and measure levels of expression of different glycoconjugates in the respective *in vivo* host systems.
- There is a need to develop and refine bioinformatic and biochemical searches for detecting and functionally characterizing glycosylation enzymes in the protozoa.
- Targeted gene deletion or knock-down approaches of enzymes involved in glycan synthesis (and catabolism) are needed to establish the physiological roles of glycoconjugates and glycan modifications in parasite biology.
- Functional redundancy has complicated studies on the function of parasite glycoconjugates. The generation of parasite mutants that exhibit loss of highly specific, as well as multiple, glycan classes is needed to define levels of redundancy and potential compensatory changes in glycosylation.
- Notwithstanding the focused efforts on some parasite glycosylation pathways (i.e. GPI biosynthesis in trypanosomatids), comparatively little is known about the enzymology and regulation of glycosylation pathways and sugar nucleotide biosynthesis in parasitic protozoa.

ACKNOWLEDGEMENTS

The work from the authors' laboratory described in this review was supported by the Australian National Health and Medical Research Council Principal Research Fellow. JH is supported by a Melbourne University Research Scholarship. SR is a University of Melbourne CR Roper Research Fellow. MJM is an NH&MRC Principal Research Fellow.

References

Almeida, I.C., Ferguson, M.A., Schenkman, S., Travassos, L.R., 1994. Lytic anti-α-galactosyl antibodies from patients with chronic Chagas' disease recognize novel *O*-linked oligosaccharides on mucin-like glycosyl-phosphatidylinositol-anchored glycoproteins of *Trypanosoma cruzi*. Biochem. J. 304, 793–802.

Alves, M.J., Colli, W., 2007. *Trypanosoma cruzi*: adhesion to the host cell and intracellular survival. IUBMB Life 59, 274–279.

Atrih, A., Richardson, J.M., Prescott, A.R., Ferguson, M.A., 2005. *Trypanosoma brucei* glycoproteins contain novel giant poly-*N*-acetyllactosamine carbohydrate chains. J. Biol. Chem. 280, 865–871.

Atwood 3rd, J.A., Minning, T., Ludolf, F., et al., 2006. Glycoproteomics of *Trypanosoma cruzi* trypomastigotes using subcellular fractionation, lectin affinity, and stable isotope labeling. J. Proteome Res. 5, 3376–3384.

Barboza, M., Duschak, V.G., Fukuyama, Y., et al., 2005. Structural analysis of the *N*-glycans of the major cysteine proteinase of *Trypanosoma cruzi*. Identification of sulfated high-mannose type oligosaccharides. FEBS J. 272, 3803–3815.

Bastida-Corcuera, F.D., Okumura, C.Y., Colocoussi, A., Johnson, P.J., 2005. *Trichomonas vaginalis* lipophosphoglycan mutants have reduced adherence and cytotoxicity to human ectocervical cells. Eukaryot. Cell 4, 1951–1958.

Belli, S.I., Wallach, M.G., Luxford, C., Davies, M.J., Smith, N.C., 2003. Roles of tyrosine-rich precursor glycoproteins and dityrosine- and 3,4-dihydroxyphenylalanine-mediated protein cross-linking in development of the oocyst wall in the coccidian parasite *Eimeria maxima*. Eukaryot. Cell 2, 456–464.

Boothroyd, J.C., Hehl, A., Knoll, L.J., Manger, I.D., 1998. The surface of *Toxoplasma*: more and less. Intl. J. Parasitol. 28, 3–9.

Bredeston, L.M., Caffaro, C.E., Samuelson, J., Hirschberg, C.B., 2005. Golgi and endoplasmic reticulum functions take place in different subcellular compartments of *Entamoeba histolytica*. J. Biol. Chem. 280, 32168–32176.

Buscaglia, C.A., Campo, V.A., Di Noia, J.M., et al., 2004. The surface coat of the mammal-dwelling infective trypomastigote stage of *Trypanosoma cruzi* is formed by highly diverse immunogenic mucins. J. Biol. Chem. 279, 15860–15869.

Buscaglia, C.A., Campo, V.A., Frasch, A.C., Di Noia, J.M., 2006. *Trypanosoma cruzi* surface mucins: host-dependent coat diversity. Nat. Rev. Microbiol. 4, 229–236.

Buxbaum, L.U., Milne, K.G., Werbovetz, K.A., Englund, P.T, 1996. Myristate exchange on the *Trypanosoma brucei* variant surface glycoprotein. Proc. Natl. Acad. Sci. USA 93, 1178–1183.

Campos-Gongora, E., Ebert, F., Willhoeft, U., Said-Fernandez, S., Tannich, E., 2004. Characterization of chitin synthases from *Entamoeba*. Protist 155, 323–330.

Capul, A.A., Barron, T., Dobson, D.E., Turco, S.J., Beverley, S.M., 2007a. Two functionally divergent UDP-Gal nucleotide sugar transporters participate in phosphoglycan synthesis in *Leishmania major*. J. Biol. Chem. 282, 14006–14017.

Capul, A.A., Hickerson, S., Barron, T., Turco, S.J., Beverley, S.M., 2007b. Comparisons of mutants lacking the Golgi UDP-galactose or GDP-mannose transporters establish that phosphoglycans are important for promastigote but not amastigote virulence in *Leishmania major*. Infect. Immun. 75, 4629–4637.

Carlton, J.M., Hirt, R.P., Silva, J.C., et al., 2007. Draft genome sequence of the sexually transmitted pathogen *Trichomonas vaginalis*. Science 315, 207–212.

Conte, I., Labriola, C., Cazzulo, J.J., Docampo, R., Parodi, A.J., 2003. The interplay between folding-facilitating mechanisms in *Trypanosoma cruzi* endoplasmic reticulum. Mol. Biol. Cell 14, 3529–3540.

Das, S., Traynor-Kaplan, A., Reiner, D.S., Meng, T.C., Gillin, F.D., 1991. A surface antigen of *Giardia lamblia* with a glycosylphosphatidylinositol anchor. J. Biol. Chem. 266, 21318–21325.

Das, S., Van Dellen, K., Bulik, D., et al., 2006. The cyst wall of *Entamoeba invadens* contains chitosan (deacetylated chitin). Mol. Biochem. Parasitol. 148, 86–92.

Dayal-Drager, R., Hoessli, D.C., Decrind, C., et al., 1991. Presence of *O*-glycosylated glycoproteins in the *Plasmodium falciparum* parasite. Carbohydr. Res. 209, C5–C8.

Delorenzi, M., Sexton, A., Shams-Eldin, H., Schwarz, R.T., Speed, T., Schofield, L., 2002. Genes for glycosylphosphatidylinositol toxin biosynthesis in *Plasmodium falciparum*. Infect. Immun. 70, 4510–4522.

Di Noia, J.M., Pollevick, G.D., Xavier, M.T., et al., 1996. High diversity in mucin genes and mucin molecules in *Trypanosoma cruzi*. J. Biol. Chem. 271, 32078–32083.

Dieckmann-Schuppert, A., Bender, S., Odenthal-Schnittler, M., Bause, E., Schwarz, R.T., 1992. Apparent lack of *N*-glycosylation in the asexual intraerythrocytic stage of *Plasmodium falciparum*. Eur. J. Biochem. 205, 815–825.

Dieckmann-Schuppert, A., Bause, E., Schwarz, R.T., 1994. Glycosylation reactions in *Plasmodium falciparum*, *Toxoplasma gondii*, and *Trypanosoma brucei* probed by the use of synthetic peptides. Biochim. Biophys. Acta 1199, 37–44.

Doyle, P., de la Canal, L., Engel, J.C., Parodi, A.J., 1986. Characterization of the mechanism of protein glycosylation and the structure of glycoconjugates in tissue culture trypomastigotes and intracellular amastigotes of *Trypanosoma cruzi*. Mol. Biochem. Parasitol. 21, 93–101.

Eichinger, D., 2001. Encystation in parasitic protozoa. Curr. Opin. Microbiol. 4, 421–426.

Engstler, M., Pfohl, T., Herminghaus, S., et al., 2007. Hydrodynamic flow-mediated protein sorting on the cell surface of trypanosomes. Cell 131, 505–515.

Fauquenoy, S., Morelle, W., Hovasse, A., et al., 2008. Proteomics and glycomics analyses of *N*-glycosylated structures involved in *Toxoplasma gondii*–host cell interactions. Mol. Cell. Proteomics 7, 891–910.

Frisardi, M., Ghosh, S.K., Field, J., et al., 2000. The most abundant glycoprotein of amebic cyst walls (Jacob) is a lectin with five Cys-rich, chitin-binding domains. Infect. Immun. 68, 4217–4224.

Funk, V.A., Thomas-Oates, J.E., Kielland, S.L., Bates, P.A., Olafson, R.W., 1997. A unique, terminally glucosylated oligosaccharide is a common feature on *Leishmania* cell surfaces. Mol. Biochem. Parasitol. 84, 33–48.

Garami, A., Ilg, T., 2001. Disruption of mannose activation in *Leishmania mexicana*: GDP-mannose pyrophosphorylase is required for virulence, but not for viability. EMBO J. 20, 3657–3666.

Gerold, P., Jung, N., Azzouz, N., Freiberg, N., Kobe, S., Schwarz, R.T., 1999. Biosynthesis of glycosylphosphatidylinositols of *Plasmodium falciparum* in a cell-free incubation system: inositol acylation is needed for mannosylation of glycosylphosphatidylinositols. Biochem. J. 344, 731–738.

Gerwig, G.J., van Kuik, J.A., Leeflang, B.R., et al., 2002. The *Giardia intestinalis* filamentous cyst wall contains a novel β(1-3)-*N*-acetyl-D-galactosamine polymer: a structural and conformational study. Glycobiology 12, 499–505.

Gorin, P.A., Previato, J.O., Mendonca-Previato, L., Travassos, L.R., 1979. Structure of the D-mannan and D-arabino-D-galactan in *Crithidia fasciculata*: changes in proportion with age of culture. J. Protozool 26, 473–478.

Gowda, D.C., Gupta, P., Davidson, E.A., 1997. Glycosylphosphatidylinositol anchors represent the major carbohydrate modification in proteins of intraerythrocytic stage *Plasmodium falciparum*. J. Biol. Chem. 272, 6428–6439.

Guerardel, Y., Leleu, D., Coppin, A., et al., 2005. Amylopectin biogenesis and characterization in the protozoan parasite *Toxoplasma gondii*, the intracellular development of which is restricted in the HepG2 cell line. Microbes Infect. 7, 41–48.

Gustafson, G.L., Gander, J.E., 1984. O β-(*N*-acetyl-α-glucosamine-1-phosphoryl)serine in proteinase I from *Dictyostelium discoideum*. Methods Enzymol. 107, 172–183.

Guther, M.L., Lee, S., Tetley, L., Acosta-Serrano, A., Ferguson, M.A., 2006. GPI-anchored proteins and free GPI glycolipids of procyclic form *Trypanosoma brucei* are nonessential for growth, are required for colonization of the tsetse fly, and are not the only components of the surface coat. Mol. Biol. Cell 17, 5265–5274.

Handman, E., Barnett, L.D., Osborn, A.H., Goding, J.W., Murray, P.J., 1993. Identification, characterisation and genomic cloning of an *O*-linked *N*-acetylglucosamine-containing cytoplasmic *Leishmania* glycoprotein. Mol. Biochem. Parasitol. 62, 61–72.

Harris, J.R., Adrian, M., Petry, F., 2004. Amylopectin: a major component of the residual body in *Cryptosporidium parvum* oocysts. Parasitology 128, 269–282.

Haynes, P.A., 1998. Phosphoglycosylation: a new structural class of glycosylation? Glycobiology 8, 1–5.

Heise, N., Raper, J., Buxbaum, L.U., Peranovich, T.M., de Almeida, M.L., 1996. Identification of complete precursors for the glycosylphosphatidylinositol protein anchors of *Trypanosoma cruzi*. J. Biol. Chem. 271, 16877–16887.

Hilley, J.D., Zawadzki, J.L., McConville, M.J., Coombs, G.H., Mottram, J.C., 2000. *Leishmania mexicana* mutants lacking glycosylphosphatidylinositol (GPI):protein transamidase provide insights into the biosynthesis and functions of GPI-anchored proteins. Mol. Biol. Cell. 11, 1183–1195.

Hiltpold, A., Frey, M., Hulsmeier, A., Kohler, P., 2000. Glycosylation and palmitoylation are common modifications of *Giardia* variant surface proteins. Mol. Biochem. Parasitol. 109, 61–65.

Hong, Y., Nagamune, K., Morita, Y.S., et al., 2006. Removal or maintenance of inositol-linked acyl chain in glycosylphosphatidylinositol is critical in trypanosome life cycle. J. Biol. Chem. 281, 11595–11602.

Hwa, K.Y., Khoo, K.H., 2000. Structural analysis of the asparagine-linked glycans from the procyclic *Trypanosoma brucei* and its glycosylation mutants resistant to concanavalin A killing. Mol. Biochem. Parasitol. 111, 173–184.

Ilg, T., 2000a. Lipophosphoglycan is not required for infection of macrophages or mice by *Leishmania mexicana*. EMBO J. 19, 1953–1962.

Ilg, T., 2000b. Proteophosphoglycans of *Leishmania*. Parasitol. Today 16, 489–497.

Ilg, T., Stierhof, Y.D., Etges, R., Adrian, M., Harbecke, D., Overath, P., 1991. Secreted acid phosphatase of *Leishmania mexicana*: a filamentous phosphoglycoprotein polymer. Proc. Natl. Acad. Sci. USA 88, 8774–8778.

Ilg, T., Overath, P., Ferguson, M.A., Rutherford, T., Campbell, D.G., McConville, M.J., 1994. *O*- and *N*-glycosylation of the *Leishmania mexicana*-secreted acid phosphatase. Characterization of a new class of phosphoserine-linked glycans. J. Biol. Chem. 269, 24073–24081.

Ilg, T., Stierhof, Y.D., Craik, D., Simpson, R., Handman, E., Bacic, A., 1996. Purification and structural characterization of a filamentous, mucin-like proteophosphoglycan secreted by *Leishmania* parasites. J. Biol. Chem. 271, 21583–21596.

Ilg, T., Craik, D., Currie, G., Multhaup, G., Bacic, A., 1998. Stage-specific proteophosphoglycan from *Leishmania mexicana* amastigotes. Structural characterization of novel mono-, di-, and triphosphorylated phosphodiester-linked oligosaccharides. J. Biol. Chem. 273, 13509–13523.

Ilg, T., Demar, M., Harbecke, D., 2001. Phosphoglycan repeat-deficient *Leishmania mexicana* parasites remain infectious to macrophages and mice. J. Biol. Chem. 276, 4988–4997.

Jarroll, E.L., Macechko, P.T., Steimle, P.A., et al., 2001. Regulation of carbohydrate metabolism during *Giardia* encystment. J. Eukaryot. Microbiol. 48, 22–26.

Jones, D.C., Mehlert, A., Guther, M.L., Ferguson, M.A., 2005. Deletion of the glucosidase II gene in *Trypanosoma brucei* reveals novel *N*-glycosylation mechanisms in the biosynthesis of variant surface glycoprotein. J. Biol. Chem. 280, 35929–35942.

Kamhawi, S., Ramalho-Ortigao, M., Pham, V.M., et al., 2004. A role for insect galectins in parasite survival. Cell 119, 329–341.

Karkhanis, Y.D., Allocco, J.J., Schmatz, D.M., 1993. Amylopectin synthase of *Eimeria tenella*: identification and kinetic characterization. J. Eukaryot. Microbiol. 40, 594–598.

Karr, C.D., Jarroll, E.L., 2004. Cyst wall synthase: *N*-acetylgalactosaminyltransferase activity is induced to form the novel *N*-acetylgalactosamine polysaccharide in the *Giardia* cyst wall. Microbiology 150, 1237–1243.

Kimmel, J., Smith, T.K., Azzouz, N., et al., 2006. Membrane topology and transient acylation of *Toxoplasma gondii* glycosylphosphatidylinositols. Eukaryot. Cell 5, 1420–1429.

Kink, J.A., Chang, K.P., 1987. Tunicamycin-resistant *Leishmania mexicana amazonensis*: expression of virulence associated with an increased activity of *N*-acetylglucosaminyltransferase and amplification of its presumptive gene. Proc. Natl. Acad. Sci. USA 84, 1253–1257.

Lederkremer, R.M., Bertello, L.E., 2001. Glycoinositolphospholipids, free and as anchors of proteins, in *Trypanosoma cruzi*. Curr. Pharm. Des. 7, 1165–1179.

Lillico, S., Field, M.C., Blundell, P., Coombs, G.H., Mottram, J.C., 2003. Essential roles for GPI-anchored proteins in

African trypanosomes revealed using mutants deficient in GPI8. Mol. Biol. Cell 14, 1182–1194.

Lippert, D.N., Dwyer, D.W., Li, F., Olafson, R.W., 1999. Phosphoglycosylation of a secreted acid phosphatase from *Leishmania donovani*. Glycobiology 9, 627–636.

Lorenzo-Morales, J., Kliescikova, J., Martinez-Carretero, E., et al., 2008. Glycogen phosphorylase in *Acanthamoeba* spp.: determining the role of the enzyme during the encystment process using RNA interference. Eukaryot. Cell 7, 509–517.

Lotter, H., González-Roldán, N., Lindner, B., Winau, F., Isibasi, A., Moreno-Lafont, M., Ulmer, A.J., Holst, O., Tannich, E., Jacobs, T., 2009. Natural killer T cells activated by a lipopeptidophosphoglycan from *Entamoeba histolytica* are creitically important to control amebic liver abscess. PLoS Pathog. 5, e1000434.

Lujan, H.D., Mowatt, M.R., Nash, T.E., 1998. The molecular mechanisms of *Giardia* encystation. Parasitol. Today 14, 446–450.

Luk, F.C., Johnson, T.M., Beckers, C.J., 2008. *N*-linked glycosylation of proteins in the protozoan parasite *Toxoplasma gondii*. Mol. Biochem. Parasitol. 157, 169–178.

Macrae, J.I., Acosta-Serrano, A., Morrice, N.A., Mehlert, A., Ferguson, M.A., 2005. Structural characterization of NETNES, a novel glycoconjugate in *Trypanosoma cruzi* epimastigotes. J. Biol. Chem. 280, 12201–12211.

MacRae, J.I., Obado, S.O., Turnock, D.C., et al., 2006. The suppression of galactose metabolism in *Trypanosoma cruzi* epimastigotes causes changes in cell surface molecular architecture and cell morphology. Mol. Biochem. Parasitol. 147, 126–136.

Magnelli, P., Cipollo, J.F., Ratner, D.M., et al., 2008. Unique ASN-linked oligosaccharides of the human pathogen *Entamoeba histolytica*. J. Biol. Chem. 283, 18355–18364.

Manthri, S., Guther, M.L., Izquierdo, L., Acosta-Serrano, A., Ferguson, M.A., 2008. Deletion of the TbALG3 gene demonstrates site-specific *N*-glycosylation and *N*-glycan processing in *Trypanosoma brucei*. Glycobiology 18, 367–383.

Martin, K.L., Smith, T.K., 2006. The glycosylphosphatidylinositol (GPI) biosynthetic pathway of bloodstream-form *Trypanosoma brucei* is dependent on the *de novo* synthesis of inositol. Mol. Microbiol. 61, 89–105.

McConville, M.J., Blackwell, J.M., 1991. Developmental changes in the glycosylated phosphatidylinositols of *Leishmania donovani*. Characterization of the promastigote and amastigote glycolipids. J. Biol. Chem. 266, 15170–15179.

McConville, M.J., Ferguson, M.A., 1993. The structure, biosynthesis and function of glycosylated phosphatidylinositols in the parasitic protozoa and higher eukaryotes. Biochem. J. 294, 305–324.

McConville, M.J., Turco, S.J., Ferguson, M.A., Sacks, D.L., 1992. Developmental modification of lipophosphoglycan during the differentiation of *Leishmania major* promastigotes to an infectious stage. EMBO J. 11, 3593–3600.

McConville, M.J., Mullin, K.A., Ilgoutz, S.C., Teasdale, R.D., 2002. Secretory pathway of trypanosomatid parasites. Microbiol. Mol. Biol. Rev. 66, 122–154.

McCoy, J.J., Mann, B.J., Vedvick, T.S., Pak, Y., Heimark, D.B., Petri, W.A., 1993. Structural analysis of the light subunit of the *Entamoeba histolytica* galactose-specific adherence lectin. J. Biol. Chem. 268, 24223–24231.

Mehlert, A., Zitzmann, N., Richardson, J.M., Treumann, A., Ferguson, M.A., 1998. The glycosylation of the variant surface glycoproteins and procyclic acidic repetitive proteins of *Trypanosoma brucei*. Mol. Biochem. Parasitol. 91, 145–152.

Mehlert, A., Bond, C.S., Ferguson, M.A., 2002. The glycoforms of a *Trypanosoma brucei* variant surface glycoprotein and molecular modeling of a glycosylated surface coat. Glycobiology 12, 607–612.

Mendonca-Previato, L., Gorin, P.A., Previato, J.O., 1979. Investigations on polysaccharide components of cells of *Herpetomonas samuelpessoai* grown on various media. Biochemistry 18, 149–154.

Mendonca-Previato, L., Todeschini, A.R., Heise, N., Previato, J.O., 2005. Protozoan parasite-specific carbohydrate structures. Curr. Opin. Struct. Biol. 15, 499–505.

Montagna, G.N., Donelson, J.E., Frasch, A.C., 2006. Procyclic *Trypanosoma brucei* expresses separate sialidase and trans-sialidase enzymes on its surface membrane. J. Biol. Chem. 281, 33949–33958.

Moody-Haupt, S., Patterson, J.H., Mirelman, D., McConville, M.J., 2000. The major surface antigens of *Entamoeba histolytica* trophozoites are GPI-anchored proteophosphoglycans. J. Mol.Biol. 297, 409–420.

Mullin, K.A., Foth, B.J., Ilgoutz, S.C., et al., 2001. Regulated degradation of an endoplasmic reticulum membrane protein in a tubular lysosome in *Leishmania mexicana*. Mol. Biol. Cell 12, 2364–2377.

Naderer, T., McConville, M.J., 2008. The *Leishmania*-macrophage interaction: a metabolic perspective. Cell Microbiol. 10, 301–308.

Naderer, T., Ellis, M.A., Sernee, M.F., et al., 2006. Virulence of *Leishmania major* in macrophages and mice requires the gluconeogenic enzyme fructose-1,6-bisphosphatase. Proc. Natl. Acad. Sci. USA 103, 5502–5507.

Naderer, T., Wee, E., McConville, M.J., 2008. Role of hexosamine biosynthesis in *Leishmania* growth and virulence. Mol. Microbiol. 69, 858–869.

Nagamune, K., Nozaki, T., Maeda, Y., et al., 2000. Critical roles of glycosylphosphatidylinositol for *Trypanosoma brucei*. Proc. Natl. Acad. Sci. USA 97, 10336–10341.

Nagamune, K., Ohishi, K., Ashida, H., et al., 2003. GPI transamidase of *Trypanosoma brucei* has two previously uncharacterized (trypanosomatid transamidase 1 and 2) and three common subunits. Proc. Natl. Acad. Sci. USA 100, 10682–10687.

Nagamune, K., Acosta-Serrano, A., Uemura, H., et al., 2004. Surface sialic acids taken from the host allow trypanosome survival in tsetse fly vectors. J. Exp. Med. 199, 1445–1450.

Naik, R.S., Branch, O.H., Woods, A.S., et al., 2000. Glycosy lphosphatidylinositol anchors of *Plasmodium falciparum*: molecular characterization and naturally elicited antibody response that may provide immunity to malaria pathogenesis. J. Exp. Med. 192, 1563–1576.

Naik, R.S., Venkatesan, M., Gowda, D.C., 2001. *Plasmodium falciparum*: the lethal effects of tunicamycin and mevastatin on the parasite are not mediated by the inhibition of *N*-linked oligosaccharide biosynthesis. Exp. Parasitol. 98, 110–114.

Nash, T.E., 2002. Surface antigenic variation in *Giardia lamblia*. Mol. Microbiol. 45, 585–590.

Nolan, T.J., Farrell, J.P., 1985. Inhibition of *in vivo* and in vitro infectivity of *Leishmania donovani* by tunicamycin. Mol. Biochem. Parasitol. 16, 127–135.

Nolan, D.P., Geuskens, M., Pays, E., 1999. *N*-linked glycans containing linear poly-*N*-acetyllactosamine as sorting signals in endocytosis in *Trypanosoma brucei*. Curr. Biol. 9, 1169–1172.

Nozaki, T., Haynes, P.A., Cross, G.A., 1996. Characterization of the *Trypanosoma brucei* homologue of a Trypanosoma cruzi flagellum-adhesion glycoprotein. Mol. Biochem. Parasitol. 82, 245–255.

Papanastasiou, P., McConville, M.J., Ralton, J., Kohler, P., 1997. The variant-specific surface protein of *Giardia*, VSP4A1, is a glycosylated and palmitoylated protein. Biochem. J. 322, 49–56.

Parodi, A.J., 1993. *N*-glycosylation in trypanosomatid protozoa. Glycobiology 3, 193–199.

Parodi, A.J., Labriola, C., Cazzulo, J.J., 1995. The presence of complex-type oligosaccharides at the C-terminal domain glycosylation site of some molecules of cruzipain. Mol. Biochem. Parasitol. 69, 247–255.

Pereira-Chioccola, V.L., Acosta-Serrano, A., Correia de Almeida, I., et al., 2000. Mucin-like molecules form a negatively charged coat that protects *Trypanosoma cruzi* trypomastigotes from killing by human anti-α-galactosyl antibodies. J. Cell Sci. 113, 1299–1307.

Pimenta, P.F., Turco, S.J., McConville, M.J., Lawyer, P.G., Perkins, P.V., Sacks, D.L., 1992. Stage-specific adhesion of *Leishmania* promastigotes to the sandfly midgut. Science 256, 1812–1815.

Pollard, A.M., Onatolu, K.N., Hiller, L., Haldar, K., Knoll, L.J., 2008. Highly polymorphic family of glycosylphosphatidylinositol-anchored surface antigens with evidence of developmental regulation in *Toxoplasma gondii*. Infect. Immun. 76, 103–110.

Previato, J.O., Jones, C., Goncalves, L.P., Wait, R., Travassos, L.R., Mendonca-Previato, L., 1994. *O*-glycosidically linked *N*-acetylglucosamine-bound oligosaccharides from glycoproteins of *Trypanosoma cruzi*. Biochem. J. 301, 151–159.

Previato, J.O., Sola-Penna, M., Agrellos, O.A., et al., 1998. Biosynthesis of *O*-*N*-acetylglucosamine-linked glycans in *Trypanosoma cruzi*. Characterization of the novel uridine diphospho-*N*-acetylglucosamine:polypeptide *N*-acetylglucosaminyltransferase-catalyzing formation of *N*-acetylglucosamine $\alpha 1 \rightarrow O$-threonine. J. Biol. Chem. 273, 14982–14988.

Previato, J.O., Wait, R., Jones, C., et al., 2004. Glycoinositol phospholipid from *Trypanosoma cruzi*: structure, biosynthesis and immunobiology. Adv. Parasitol. 56, 1–41.

Priest, J.W., Mehlert, A., Arrowood, M.J., Riggs, M.W., Ferguson, M.A., 2003. Characterization of a low molecular weight glycolipid antigen from *Cryptosporidium parvum*. J. Biol. Chem. 278, 52212–52222.

Priest, J.W., Mehlert, A., Moss, D.M., Arrowood, M.J., Ferguson, M.A., 2006. Characterization of the glycosylphosphatidylinositol anchor of the immunodominant *Cryptosporidium parvum* 17-kDa antigen. Mol. Biochem. Parasitol. 149, 108–112.

Ralton, J.E., McConville, M.J., 1998. Delineation of three pathways of glycosylphosphatidylinositol biosynthesis in *Leishmania mexicana*. Precursors from different pathways are assembled on distinct pools of phosphatidylinositol and undergo fatty acid remodeling. J. Biol. Chem. 273, 4245–4257.

Ralton, J.E., Naderer, T., Piraino, H.L., Bashtannyk, T.A., Callaghan, J.M., McConville, M.J., 2003. Evidence that intracellular β1-2 mannan is a virulence factor in Leishmania parasites. J. Biol. Chem. 278, 40757–40763.

Riggs, M.W., McNeil, M.R., Perryman, L.E., Stone, A.L., Scherman, M.S., O'Connor, R.M., 1999. *Cryptosporidium parvum* sporozoite pellicle antigen recognized by a neutralizing monoclonal antibody is a β-mannosylated glycolipid. Infect. Immun. 67, 1317–1322.

Rogers, M.E., Ilg, T., Nikolaev, A.V., Ferguson, M.A., Bates, P.A., 2004. Transmission of cutaneous leishmaniasis by sand flies is enhanced by regurgitation of fPPG. Nature 430, 463–467.

Roper, J.R., Guther, M.L., Milne, K.G., Ferguson, M.A., 2002. Galactose metabolism is essential for the African sleeping sickness parasite *Trypanosoma brucei*. Proc. Natl. Acad. Sci. USA 99, 5884–5889.

Sacks, D.L., Modi, G., Rowton, E., et al., 2000. The role of phosphoglycans in *Leishmania*-sand fly interactions. Proc. Natl. Acad. Sci. USA 97, 406–411.

Samuelson, J., Banerjee, S., Magnelli, P., et al., 2005. The diversity of dolichol-linked precursors to Asn-linked glycans likely results from secondary loss of sets of glycosyltransferases. Proc. Natl. Acad. Sci. USA 102, 1548–1553.

Sanders, P.R., Gilson, P.R., Cantin, G.T., et al., 2005. Distinct protein classes including novel merozoite surface antigens in raft-like membranes of *Plasmodium falciparum*. J. Biol. Chem. 280, 40169–40176.

Santi-Rocca, J., Weber, C., Guigon, G., Sismeiro, O., Coppee, J.Y., Guillen, N., 2008. The lysine- and glutamic acid-rich protein KERP1 plays a role in *Entamoeba histolytica* liver abscess pathogenesis. Cell Microbiol. 10, 202–217.

Schenkman, S., Ferguson, M.A., Heise, N., de Almeida, M.L., Mortara, R.A., Yoshida, N., 1993. Mucin-like glycoproteins linked to the membrane by glycosylphosphatidylinositol anchor are the major acceptors of sialic acid in a reaction catalyzed by trans-sialidase in metacyclic forms of *Trypanosoma cruzi*. Mol. Biochem. Parasitol. 59, 293–303.

Schmidt, A., Schwarz, R.T., Gerold, P., 1998. *Plasmodium falciparum*: asexual erythrocytic stages synthesize two structurally distinct free and protein-bound glycosylphosphatidylinositols in a maturation-dependent manner. Exp. Parasitol. 88, 95–102.

Schneider, P., Rosat, J.P., Ransijn, A., Ferguson, M.A., McConville, M.J., 1993. Characterization of glycoinositol phospholipids in the amastigote stage of the protozoan parasite *Leishmania major*. Biochem. J. 295, 555–564.

Seeber, F., Dubremetz, J.F., Boothroyd, J.C., 1998. Analysis of *Toxoplasma gondii* stably transfected with a transmembrane variant of its major surface protein, SAG1. J. Cell Sci. 111, 23–29.

Sernee, M.F., Ralton, J.E., Dinev, Z., et al., 2006. *Leishmania* β-1,2-mannan is assembled on a mannose-cyclic phosphate primer. Proc. Natl. Acad. Sci. USA 103, 9458–9463.

Singh, B.N., 1993. Lipophosphoglycan-like glycoconjugate of *Tritrichomonas foetus* and *Trichomonas vaginalis*. Mol. Biochem. Parasitol. 57, 281–294.

Smith, H.V., Nichols, R.A., Grimason, A.M., 2005. *Cryptosporidium* excystation and invasion: getting to the guts of the matter. Trends Parasitol. 21, 133–142.

Smith, T.K., Crossman, A., Brimacombe, J.S., Ferguson, M.A., 2004. Chemical validation of GPI biosynthesis as a drug target against African sleeping sickness. EMBO J. 23, 4701–4708.

Smith, T.K., Kimmel, J., Azzouz, N., Shams-Eldin, H., Schwarz, R.T., 2007. The role of inositol acylation and inositol deacylation in the *Toxoplasma gondii* glycosylphosphatidylinositol biosynthetic pathway. J. Biol. Chem. 282, 32032–32042.

Spath, G.F., Garraway, L.A., Turco, S.J., Beverley, S.M., 2003a. The role(s) of lipophosphoglycan (LPG) in the establishment of *Leishmania major* infections in mammalian hosts. Proc. Natl. Acad. Sci. USA 100, 9536–9541.

Spath, G.F., Lye, L.F., Segawa, H., Sacks, D.L., Turco, S.J., Beverley, SM., 2003b. Persistence without pathology in phosphoglycan-deficient *Leishmania major*. Science 301, 1241–1243.

Stanley Jr., S.L., Tian, K., Koester, J.P., Li, E., 1995. The serine-rich *Entamoeba histolytica* protein is a phosphorylated membrane protein containing *O*-linked terminal *N*-acetylglucosamine residues. J. Biol. Chem. 270, 4121–4126.

Stokes, M., Guther, M., Turnock, D., et al., 2008. The synthesis of UDP-*N*-acetylglucosamine is essential for bloodstream form *Tryposomma brucei in vitro* and *in vivo* and UDP-*N*-acetylglucosamine starvation reveals a hierachy in parasite protein glycosylation. J. Biol. Chem. 283, 16147–16161.

Striepen, B., Zinecker, C.F., Damm, J.B., et al., 1997. Molecular structure of the "low molecular weight antigen" of *Toxoplasma gondii*: a glucose α1-4 N-acetylgalactosamine makes free glycosyl-phosphatidylinositols highly immunogenic. J. Mol. Biol. 266, 797–813.

Stwora-Wojczyk, M.M., Dzierszinski, F., Roos, D.S., Spitalnik, S.L., Wojczyk, B.S., 2004a. Functional characterization of a novel *Toxoplasma gondii* glycosyltransferase: UDP-*N*-acetyl-D-galactosamine:polypeptide *N*-acetylgalactosaminyltransferase-T3. Arch. Biochem. Biophys. 426, 231–240.

Stwora-Wojczyk, M.M., Kissinger, J.C., Spitalnik, S.L., Wojczyk, B.S., 2004b. *O*-glycosylation in *Toxoplasma gondii*: identification and analysis of a family of UDP-GalNAc: polypeptide *N*-acetylgalactosaminyltransferases. Int. J. Parasitol. 34, 309–322.

Teixeira, J.E., Huston, C.D., 2008. Participation of the serine-rich *Entamoeba histolytica* protein in amebic phagocytosis of apoptotic host cells. Infect. Immun. 76, 959–966.

Templeton, T.J., Iyer, L.M., Anantharaman, V., et al., 2008. Comparative analysis of Apicomplexa and genomic diversity in eukaryotes. Genome Res. 14, 1686–1695.

Turnock, D.C., Izquierdo, L., Ferguson, M.A., 2007. The de novo synthesis of GDP-fucose is essential for flagellar adhesion and cell growth in Trypanosoma brucei. J. Biol. Chem. 282, 28853–28863.

Urbaniak, M.D., Turnock, D.C., Ferguson, M.A., 2006. Galactose starvation in a bloodstream form *Trypanosoma brucei* UDP-glucose 4′-epimerase conditional null mutant. Eukaryot. Cell 5, 1906–1913.

Vats, D., Vishwakarma, R.A., Bhattacharya, S., Bhattacharya, A., 2005. Reduction of cell surface glycosylphosphatidylinositol conjugates in *Entamoeba histolytica* by antisense blocking of *E. histolytica* GlcNAc-phosphatidylinositol deacetylase expression: effect on cell proliferation, endocytosis, and adhesion to target cells. Infect. Immun. 73, 8381–8392.

Vijaykumar, M., Naik, R.S., Gowda, D.C., 2001. *Plasmodium falciparum* glycosylphosphatidylinositol-induced TNF-α secretion by macrophages is mediated without membrane insertion or endocytosis. J. Biol. Chem. 276, 6909–6912.

von Itzstein, M., Plebanski, M., Cooke, B.M., Coppel, R.L., 2008. Hot, sweet and sticky: the glycobiology of *Plasmodium falciparum*. Trends Parasitol. 24, 210–218.

Wang, C.C., Weppelman, R.M., Lopez-Ramos, B., 1975. Isolation of amylopectin granules and identification of amylopectin phosphorylase in the oocysts of *Eimeria tenella*. J. Protozool. 22, 560–564.

Weber, C., Blazquez, S., Marion, S., et al., 2008. Bioinformatics and functional analysis of an *Entamoeba histolytica* mannosyltransferase necessary for parasite complement resistance and hepatical Infection. PLoS Negl. Trop. Dis. 2, e165.

West, C.M., Van Der Wel, H., Sassi, S., Gaucher, E.A., 2004. Cytoplasmic glycosylation of protein-hydroxyproline and its relationship to other glycosylation pathways. Biochim. Biophys. Acta 1673, 29–44.

West, C.M., van der Wel, H., Blader, I.J., 2006. Detection of cytoplasmic glycosylation associated with hydroxyproline. Methods Enzymol. 417, 389–404.

Winter, G., Fuchs, M., McConville, M.J., Stierhof, Y.D., Overath, P., 1994. Surface antigens of *Leishmania mexicana* amastigotes: characterization of glycoinositol phospholipids and a macrophage-derived glycosphingolipid. J. Cell Sci. 107, 2471–2482.

Winter, G., Gooley, A.A., Williams, K.L., Slade, M.B., 2000. Characterization of a major sporozoite surface glycoprotein of *Cryptosporidum parvum*. Funct. Integr. Genomics. 1, 207–217.

Wu, G., Muller, M., 2003. Glycogen phosphorylase sequences from the amitochondriate protests, *Trichomonas vaginalis, Mastigamoeba balamuthi, Entamoeba histolytica* and Giardia intestinalis. J. Eukaryot. Microbiol. 50, 366–372.

Zawadzki, J., Scholz, C., Currie, G., Coombs, G.H., McConville, M.J., 1998. The glycoinositolphospholipids from *Leishmania panamensis* contain unusual glycan and lipid moieties. J. Mol. Biol. 282, 287–299.

Zufferey, R., Allen, S., Barron, T., et al., 2003. Ether phospholipids and glycosylinositolphospholipids are not required for amastigote virulence or for inhibition of macrophage activation by *Leishmania major*. J. Biol. Chem. 278, 44708–44718.

CHAPTER

13

Analytical approaches towards the structural characterization of microbial wall glycopolymers

I. Darren Grice and Jennifer C. Wilson

SUMMARY

Outlined in this chapter are strategies for the use of the analytical techniques of nuclear magnetic resonance (NMR) spectroscopy and mass spectrometry (MS) towards the elucidation of the structures of microbial wall glycans. Specifically, the determination of the structure of capsular polysaccharide, core oligosaccharide or O-antigen utilizing 1D, 1D selective and 2D homonuclear and heteronuclear NMR experiments and database resources is presented. Described is the application of high resolution magic angle spinning techniques for studying glycopolymer structures. MS sample derivatization strategies along with associated gas chromatography-mass spectrometry, sugar, linkage, and absolute configuration analyses are outlined. The use of fast atom bombardment, matrix-assisted laser desorption ionization and electrospray-mass spectrometry ionization techniques along with their associated mass analysers are presented for characterizing oligosaccharide, lipid A, lipo-oligosaccharide, lipopolysaccharide and capsular polysaccharide glycan structures.

Keywords: Nuclear magnetic resonance (NMR) spectroscopy; Mass spectrometry; Carbohydrates; Glycans; Lipo-oligosaccharide; Lipopolysaccharide; O-polysaccharide; Lipid A; Capsular polysaccharide; Oligosaccharide

1. INTRODUCTION

The surface and cell wall components of microbial species are covered in glycomolecules (polysaccharides, oligosaccharides or the carbohydrate-moiety of a glycoconjugate). These structures play diverse and important roles in host interactions, colonization and virulence (Comstock and Kasper, 2006). Carbohydrates and, in particular, monosaccharides, possess features that result in a high level of structural complexity. Over 100 different monosaccharides have been identified in bacterial polysaccharides (Lindberg, 1998). Monosaccharides also display structural isomerism (e.g. stereoisomers, cyclic, anomeric) and there is a multiplicity of possible linkages between monomeric carbohydrate units and, furthermore, polysaccharides are often branched. Moreover, substitution of the sugar rings with up to 50 various non-sugar functionalities (e.g. acetate, phosphate, etc.)

provides virtually limitless structural variation (Lindberg, 1998). All these factors make the task of assigning a unique structure to a particular microbial isolate challenging.

Although surface glycans are found universally, the glycan cell wall components of Gram-negative bacteria will be discussed predominantly herein as examples illustrating the analytical approaches employed in the characterization of glycosylated microbial wall components. The most ubiquitous Gram-negative bacterial cell wall components are lipopolysaccharide (LPS) or lipo-oligosaccharide (LOS) (Comstock and Kasper, 2006). The LPS molecule contains three distinct regions: lipid A, core oligosaccharide (OS) and O-specific polysaccharide (O-PS) each with integral carbohydrate structures (see Chapters 3 and 4), whereas LOS contains lipid A and core OS (Raetz and Whitfield, 2002). Additionally, capsular polysaccharide (CPS) is a cell-surface glycan (see Chapter 6), both of Gram-positive and -negative bacteria. Although other glycan structures exist on the surface of microbial cell walls, including glycoproteins such as those of the glycosylated pili, e.g. in *Pseudomonas aeruginosa* (Voisin *et al.*, 2007), *Neisseria gonorrhoeae* (Banerjee and Ghosh, 2003; Aas *et al.*, 2007) and *Neisseria meningitidis* (Banerjee and Ghosh, 2003; Power *et al.*, 2006), or flagella (Logan, 2006), e.g. in *Helicobacter pylori* (Schirm *et al.*, 2003), *Campylobacter jejuni* (McNally *et al.*, 2006a; Logan *et al.*, 2008) and *P. aeruginosa* (Dasgupta *et al.*, 2004), this chapter will focus on strategies for investigating LPS/LOS and CPS structures. Nuclear magnetic resonance (NMR) spectroscopy and mass spectrometry (MS) have played a pivotal role in elucidating their primary structural features and will be discussed.

2. ISOLATION AND PURIFICATION OF BACTERIAL GLYCAN STRUCTURES

Characterization of glycans is dependent on having effective sample preparation prior to analysis (Haslam *et al.*, 2003). Information on the class of structure will determine the strategy employed to isolate the glycan of interest. Figure 13.1 provides some general indications of the pathway often used for isolating glycan

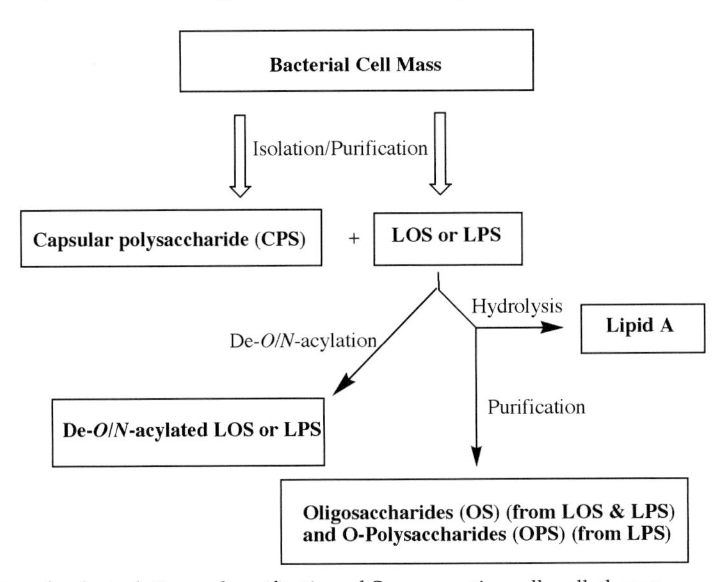

FIGURE 13.1 Strategy for the isolation and purification of Gram-negative cell wall glycans.

structures. Selected detailed procedures to isolate LPS, LOS, OS, O-PS, CPS and lipid A from a diverse range of bacteria are outlined in Table 13.1. For instance, an aqueous-based isolation of *Staphylococcus haemolyticus* bacterial cell mass followed by dialysis provided access to CPS (Flahaut *et al.*, 2008). A phenol-chloroform-petroleum ether extraction of *Yersinia pestis* bacterial cell mass yielded LOS (Hitchen *et al.*, 2002).

Ultimately, sample preparation requires a purification step to afford high quality NMR and MS data to make characterization of the samples more straightforward.

3. NMR TECHNIQUES EMPLOYED FOR STRUCTURAL CHARACTERIZATION OF GLYCANS

3.1. Introduction

NMR analysis of carbohydrate-containing structures can be challenging, primarily due to spectral overlap; carbohydrate ring protons generally resonate over a relatively small chemical shift range ($\delta = 3.0$–$4.0\,\text{ppm}$). This complication is in addition to the other factors that add complexity to carbohydrate structure already discussed. Microbes add to this inherent complexity by introducing further structural diversity with mechanisms such as phase variation (Comstock and Kasper, 2006). Developments in NMR technology, such as ultra-high magnetic field strengths (Blundell *et al.*, 2006) and cryogenically cooled probeheads (Feng *et al.*, 2002; Martin, 2002; Yamamoto, 2002; Chapdelaine and Cleon, 2003; Voehler *et al.*, 2006) have led to sensitivity increases. The introduction of pulsed field gradients has also led to sensitivity gains because gradient-enhanced spectra are less prone to artefacts, allow better solvent suppression and require shorter phase cycles (Keeler *et al.*, 1994). These advances have all contributed to making NMR analysis of often heterogeneous, small-scale glycan samples feasible.

More recently, there has been a number of excellent reviews of the process of NMR determination of polysaccharide structures (Duus *et al.*, 2000; Uhrin and Brisson, 2000; Bubb, 2003; Schweda and Richards, 2003; Hricovini, 2004; Vliegenthart, 2006; Stenutz *et al.*, 2006). For an in-depth coverage of carbohydrate structure determination these reviews and the literature cited therein should be consulted.

3.2. Sample preparation

The quantity of sample for NMR analysis is often dictated by practical factors, such as bacterial growth rate, heterogeneity or sample loss in the hydrolysis or purification procedures and, therefore, sometimes, suboptimal amounts of sample are available for analysis. In these cases, it is beneficial to analyse the sample with a cryogenically cooled probe that increases sensitivity by a factor of 3–4. Strategies also exist for examining small volumes of sample, for instance using microprobes (2.5 or 3 mm) or shigemi tubes that allow small volumes of $\approx 200\,\mu\text{l}$ to be examined.

3.2.1. *Lipid A*

Lipid A is a membrane-associated molecule with a tendency to aggregate due to its amphipathic nature, contributing to its limited solubility in organic solvents and poor solubility in aqueous media (Raetz and Whitfield, 2002). Purified lipid A is unstable in solution being prone to hydrolysis at the anomeric phosphate under acidic conditions. Chemical modification by methylation, dephosphorylation or deacylation may improve stability (Ribeiro *et al.*, 1999). Solvents previously used for lipid A NMR analysis include: deuterated dimethyl sulfoxide (d_6DMSO), deuterated chloroform ($CDCl_3$) or 2:3:1 (v/v) deuterated chloroform-deuterated methanol-deuterium oxide ($CDCl_3$-CD_3OD-D_2O).

TABLE 13.1 Selected representative structural characterization strategies

Bacterium investigated	*Yersinia pestis*[a]	*Campylobacter jejuni*[b]	*Pseudomonas aeruginosa*[c]	*Moraxella bovis*[d]	*Haemophilus influenzae*[e]	*Helicobacter pylori*[f]	*Plesiomonas shigelloides O74:H5*[g]	*Agrobacterium tumefaciens A1*[h]	*Moraxella catarrhalis*[i]	*Pseudomonas syringae pv.*[j]	*Campylobacter jejuni (GBS, MFS assoc.)*[k]	*Azospirillum brasilense S17*[l]	*Staphylococcus haemolyticus*[m]
Structure(s) characterized	LOS	LOS CPS	OS O-PS	CPS	Lipid A	OS O-PS Lipid A	LPS	LOS	OS	LPS	LOS	O-PS	CPS
Isolation													
Hot phenol/H_2O			⊕	⊙	⊕	⊕	⊕					⊕	
PCP	⊗				⊕			⊗	⊗				
Dialysis					⊕		⊕						⊙
Other		⊗ ⊙	⊕					⊗		⊕	⊗		
SEC			⊘ ○	⊙		⊕	⊘ ○ ⊘*		⊘	⊘ ○		○	⊙
Chemical manipulations													
De-*O*-acylation		⊗					⊕	⊗			⊗		
De-*N*-acylation							⊕	⊗					
Acid hydrolysis	⊗		⊕	⊙	⊕	⊕	⊕		⊗	⊕	⊕	⊕	
Permethylation	⊘			⊙		⊘ ○	⊘ ○	⊘*		⊘ ○			
Compositional analysis													
GC-MS: Sugar	⊘ ¤			⊙		⊘ ○ ¤	⊘ ○	⊘*		⊘ ○		○	⊙
GC-MS: Linkage	⊘			⊙		⊘ ○		⊘*		⊘ ○		○	
GC-MS: Absolute configuration	⊘ ¤			⊙		⊘ ○	⊘ ○ ¤	⊘*		⊘ ○		○	⊙
GC-MS: Fatty Acid	¤				¤	¤	¤			¤			
MALDI-MS													
TOF				⊙			⊘ ○ ¤						

(*Continues*)

FAB-MS	⊘ ¤												
ESI-MS						⊘							
CE/CE-MS/MS		⊗/⊗									⊗/⊗		
IT/IT-MS/MS					– / ¤		– / ¤		⊘/ –				
HRFT								⊗					
FT-ICR			⊘ ○										
Q-TOF (MS/MS)	⊘												
LD-MS						¤							
NMR		●	⊘ ○	●		○ ¤	⊘ ○⊘*	⊘*	⊘	⊘ ○		○	●

Symbols: ●, Capsular polysaccharide (CPS); ¤, lipid A; ⊗, lipo-oligosaccharide (LOS); ⊕, lipo-polysaccharide (LPS); ⊘, oligosaccharide (OS); ○, O-specific polysaccharide (O-PS); *, cases where the OS is the full sugar backbone with lipid A fatty acyl moieties stripped from GlcN sugars.

Abbreviations: CE, capillary electrophoresis; GBS, Guillain-Barré syndrome; FT-ICR – Fourier transform-ion cyclotron resonance; HRFT, high resolution Fourier transform; IT, ion trap; LD-MS, laser desorption-mass spectrometry; MFS, Miller Fisher syndrome; PCP, phenol-chloroform-petroleum ether; SEC, size exclusion chromatography.

[a]Hitchen *et al.* (2002); [b]Szymanski *et al.* (2003); [c]Bystrova *et al.* (2004); [d]Wilson *et al.* (2005); [e]Mikhail *et al.* (2005); [f]Khamri *et al.* (2005); [g]Lukasiewicz *et al.* (2006); [h]De Castro *et al.* (2006); [i]Wilson *et al.* (2006); [j]Zdorovenko *et al.* (2007); [k]Godschalk *et al.* (2007); [l]Fedonenko *et al.* (2008); [m]Flahaut *et al.* (2008).

Moreover, it has been suggested that addition of deuterated ethylenediaminetetraacetic acid (EDTA) and sodium dodecyl sulfate (SDS) to a sample improves signal resolution markedly. Increasing sample temperature to 60–80°C can also lead to sharper signals and improve resolution (Schweda and Richards, 2003).

3.2.2. *OS/CPS/O-PS*

Most often spectra are acquired in D_2O. The sample should be dissolved in 100% D_2O under a nitrogen atmosphere, then lyophilized and re-dissolved in 100% D_2O (twice). This ensures, that all residual water is removed from the sample so the intensity of the residual $^1HO^2H$ signal does not obscure any carbohydrate signals. If it is suspected that a carbohydrate resonance is located beneath the $^1HO^2H$ signal, either increasing or decreasing the temperature at which the spectra are acquired will reveal any hidden anomeric signals (Schweda and Richards, 2003).

3.3. One-dimensional (1D) spectra and the structural reporter region

Although 1D ^1H spectra are easy to acquire, they reveal a surprising amount of information. Visual inspection of a glycan spectrum can immediately provide information regarding the purity, concentration, number of sugar units and the heterogeneity of the sample. For lipid A samples, aggregation state is immediately obvious. In addition, for OS/CPS/O-PS samples, the anomeric configuration of the sugars and whether the sample contains additional non-carbohydrate signals, such as acetate or methyl substituents, can be determined readily from 1D spectra. Samples containing sialic acid or 3-deoxy-D-*manno*-oct-2-ulosonic acid (Kdo) will be immediately obvious as well. This is largely due to the fact that there are two distinct sets of signals that are generally found in glycan samples. As mentioned previously, the bulk of non-anomeric protons in a carbohydrate-containing sample resonate over a relatively narrow frequency range and constitute one of these groups of signals. Conveniently, however, anomeric proton signals are found outside this region between ≈4.5 and 5.5 ppm and provide an estimate of the number of sugar units contained within the sample. The integration of the anomeric signals will allow estimation of relative proportions of components in heterogeneous samples. Other signals, such as those arising from the effects of substitution by phosphates, sulfates or acyl groups are also found outside the 3–4 ppm spectral window. In fact, this second group of signals, distinct from the bulk region of the carbohydrate ring protons, gives rise to the so-called "structural reporter-group signals", a term introduced by Vliegenthart (2006). In the structural reporter approach, the chemical shift and coupling patterns of the structural reporter groups can be exploited to provide structural information by comparison to patterns of chemical shift and coupling constant tabulated for reference compounds, as long as pH, temperature and solvent are carefully calibrated.

The chiral carbon next to the ring oxygen that is formed when an aldohexose or ketohexose cyclizes to form a hemiacetal or hemiketal is called the anomeric centre (C-1 in Figure 13.2). The relative orientation of the hydroxyl group that this carbon bears defines which anomer is formed; the α- and β-anomers of glucose are shown in Figure 13.2. Since anomers are diastereomers, they can be readily distinguished from their NMR spectra both with regard to chemical shift ($\delta\alpha > \delta\beta$) and coupling constants. The vicinal coupling constant between H-1 and H-2 if they have an axial-equatorial orientation is $^3J_{1,2}\ \alpha \approx 3\,Hz$ (α-anomer) and if they have an axial-axial orientation $^3J_{1,2}\ \beta \approx 7\,Hz$ (β-anomer). The exceptions are mannose and rhamnose that have an equatorial proton at H-2. Carbon chemical shifts and the ^{13}C-H coupling constants

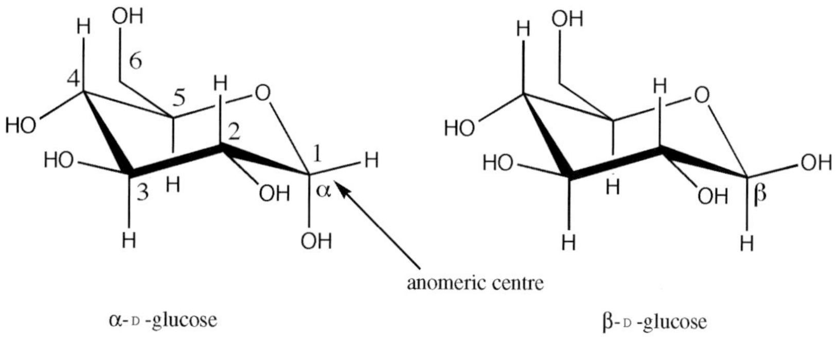

anomeric centre

α-D-glucose

β-D-glucose

FIGURE 13.2 The α- and β-anomers of glucose.

are also indicative of anomeric configuration; $^{1}J_{C\text{-}1H\text{-}1} \approx 170\,Hz$ = α-anomer and $^{1}J_{C-1H-1} \approx 160\,Hz$ = β-anomer. For sialic acid, which does not have an anomeric proton at C-1, the anomeric configuration can be determined by the H-3_{eq}/H-3_{ax} chemical shift (Wilson *et al.*, 1995).

3.4. 2D homonuclear techniques

As previously mentioned, the majority of ring protons from the sugar residues resonate in a relatively narrow spectral region (3.2–4.0 ppm) and, therefore, this region of the 1D spectrum is often severely overlapped for all components of LPS samples, thereby making structural assignment challenging. For this reason, it is necessary to use the 2D homonuclear techniques, such as correlated spectroscopy (COSY) or total correlated spectroscopy (TOCSY) to reduce the overlap by resolving the proton signals over two dimensions. In these experiments, the well-separated anomeric resonances provide a convenient starting point for the assignment process.

A 2D COSY spectrum shows correlations only between directly coupled protons and, therefore, it becomes immediately obvious from inspection of a 2D COSY that the proton attached to the anomeric carbon must correlate with protons in the C-2 position. Problems arise for strongly coupled spins such as H-6 and H-6′ from the hydroxymethyl moiety that resonate close to the spectrum diagonal and for galactopyranoses where the small coupling constant between H-4 and H-5 results in H-4/H-5 correlation being absent. Hence, double quantum filtered COSY (DQF-COSY) or gradient-enhanced COSY (gCOSY) spectra are typically acquired.

The other typical "workhorse" homonuclear technique is the 2D TOCSY experiment. In these experiments, magnetization can be transferred through the coupled network of spins that typifies monosaccharides from the anomeric position H-1 around through the ring protons and even beyond in heptoses or octoses via TOCSY transfer. Typically, long mixing times experiments (120–200 ms) should be acquired to ensure maximal TOCSY transfer. Again, the anomeric protons provide a convenient reference point for assignment and examination of 1D slices through a 2D TOCSY spectrum at each of the anomeric chemical shifts yielding data on the individual monosaccharide coupled spin systems.

The 2D nuclear Overhauser enhancement spectroscopy (NOESY) spectra provide through-space information and, in most cases, the glycosidic linkages can be confirmed by observation of a nuclear Overhauser enhancement (NOE) between the anomeric proton of one sugar unit and the aglycone proton of the next sugar in the sequence. Ultimately, this information can be used to establish the order of monosaccharide units in a polysaccharide sequence. Patterns of

NOE data also provide information regarding anomeric configuration, with particular sets of data being observed for different anomeric configurations (Uhrin and Brisson, 2000).

3.5. 2D heteronuclear techniques

^{1}H detected ^{1}H-^{13}C correlated NMR methods are valuable in structure determination of polysaccharides because they take advantage of the chemical shift dispersion of the ^{13}C nucleus and are more sensitive than ^{13}C direct detection techniques. Spectra can often be acquired on small quantities of sample (≈100 μg) using gradient-selected sensitivity-enhanced versions of these spectra. The heteronuclear single quantum coherence (HSQC) or hetereonuclear multiple quantum correlation (HMQC) spectra correlates 1-bond ^{1}H to ^{13}C nuclei and is used extensively for LPS and capsular polysaccharide primary structure determination (Duus *et al.*, 2000; Uhrin and Brisson, 2000; Bubb, 2003; Schweda and Richards, 2003; Hricovini, 2004; Vliegenthart, 2006; Stenutz *et al.*, 2006). This experiment is particularly useful for establishing or confirming glycosidic linkage since glycosylation leads to frequency shifts of between 4 and 10 ppm for carbons at the anomeric and linked positions (Bubb, 2003). Two- and three-bond correlations can be obtained from heteronuclear multiple bond correlation (HMBC) spectra but have the added advantage of revealing through-bond correlations, such as that through the glycosidic linkage, although they are less sensitive than HSQC spectra and require longer acquisition times. Of particular interest and widespread use is the 2D ^{1}H-^{13}C HSQC-TOCSY spectrum that combines the heteronuclear correlation experiment with the TOCSY experiment providing optimal chemical shift dispersion. Furthermore, edited versions of ^{1}H-^{13}C HSQC and ^{1}H-^{13}C HSQC-TOCSY experiments (Willker *et al.*, 1993) are useful for discriminating methylene resonances from methyl or methine resonances and, therefore, aid the investigation of overlap involving hydroxymethyl groups of monosaccharide residues. This is especially important when ^{13}C glycosylation effects shift the hydroxymethyl resonances (C-6 galactose/glucose) into the 66–77 ppm region of the spectrum where the bulk of methine resonances are found.

A new heteronuclear experiment (Petersen *et al.*, 2006 and references therein), entitled the heteronuclear 2-bond correlation (H2BC), was introduced that identifies two bond correlations, particularly C-2 to H-1 correlations (compared with HSQC/HMQC 1-bond correlations and HMBC 2- and 3-bond correlations). The utility of this method was illustrated with the full structure determination of the 30 sugar-residue LOS from *Francisella victoria*. The advantage of this experiment over the conventional HMBC experiment is that it allows ready assignment of 2-bond correlations without complicating 3-bond correlations. Moreover, due to small C-2 to H-1 couplings some of these correlations would be missing in an HMBC spectrum. While this information can also be obtained from an HSQC-TOCSY experiment, the advantage of the H2BC is that it does not contain all the intra-ring correlations that increase the complexity of HSQC-TOCSY spectra, thus preventing unequivocal assignment of C-2 carbons from this type of spectrum.

3.6. Selective experiments

One of the advantages of the anomeric resonances resonating outside the bulk of carbohydrate signals is that their separation can be conveniently exploited by the use of selective NMR experiments. These experiments involve selective excitation over a small frequency range typically using a 90° shaped pulse. There are many shapes of excitation pulses (e.g. Gaussian, q-sneeze) available and the most suitable should produce clean selection of the signals of interest. Significant contributions to the arsenal of selective experiments have been provided by Uhrin and co-workers (Uhrinova *et al.*, 1991; Uhrin *et al.*, 1993, 1994; Uhrin, 1997; Uhrin and Brisson, 2000). Typically, a 90° selective-shaped pulse can be applied to the selective region in which each of the anomeric protons resonate and, in turn, in combination with either COSY, TOCSY and NOESY sequences produce 1D spectra that are quickly acquired and interpreted. For example, St Michael *et al.* (2004), in elucidating the core OS of *Actinobacillus pleuropneumoniae* serotypes, used selective 1D TOCSY experiments with mixing times of 150 ms for heptose residues and 90 ms for hexose residues at anomeric resonances to aid the assignment process.

3.7. Application of high resolution magic angle spinning (HR-MAS) to determination of glycopolymer structures

The technique of HR-MAS has been used to examine modifications of bacterial CPS on intact *C. jejuni* bacterial cells (St Michael *et al.*, 2002; McNally *et al.*, 2005, 2006b, 2007; Karlyshev *et al.*, 2005). In these studies, the prevalence of an unusual *O*-methyl-phosphoramidate [$CH_3OP(O)(NH_2)(OR)$ or MeOPN] group was investigated. The NMR signals arising from the OCH_3 protons of MeOPN, have a distinctive chemical shift and scalar coupling to ^{31}P compared to other CPS signals, making them ideal to monitor the presence or absence of this CPS modification. Spectra of ^{1}H and 1D ^{1}H-^{31}P HSQC HR-MAS NMR were used rapidly to detect MeOPN in microlitre samples of bacterial cells (St Michael *et al.*, 2002; McNally *et al.*, 2005, 2006b, 2007; Karlyshev *et al.*, 2005). A rapid, high throughput HR-MAS NMR screening method was developed to examine animal and human isolates from a range of environmental and clinical settings (McNally *et al.*, 2007). These studies were aimed at determining the abundance of this CPS modification and its potential as a diagnostic marker. Thereby, it was determined that 68% of *C. jejuni* strains expressed MeOPN and that MeOPN expression was highly prevalent in isolates associated with enteritis (82%) and in isolates from patients with *C. jejuni*-associated Guillain-Barré (80%) or Miller-Fisher (100%) neurological syndromes.

3.8. Determination of nuclei other than ^{1}H

The location of substituents is determined primarily by perturbations to chemical shifts of carbohydrate ring protons, as compared to literature values. Phosphorylation is the most common substitution of polysaccharides and 2D long-range ^{1}H-^{31}P HMQC or HSQC can reveal the position of phosphorylation (Kenne *et al.*, 1988). An example concerns studies on the LOS of *Plesiomonas shigelloides* (Lukasiewicz *et al.*, 2006), a bacterium associated with waterborne infections, whereby a decasaccharide was isolated by de-*N,O*-acylation of *P. shigelloides* O74 LPS (achieved by mild hydrazinolysis followed by KOH treatment). The decasaccharide consisted of the inner core oligosaccharide linked via Kdo to the de-*N,O*-acylated lipid A (→6)-β-D-Glc*p*N-(1→6)-α-D-Glc*p*N-1-*P*), and showed two signals in its ^{31}P spectrum (δ = 1.58 and 1.76 ppm) typical of phosphate monoesters. The location of these phosphates was determined from a ^{1}H-^{31}P HMBC spectrum.

Another illustrative example concerns studies on the structure of mild acid-hydrolysed and de-lipidated *P. aeruginosa* immunotype 5 LPS (Mikhail *et al.*, 2005). Mild hydrolysis conditions were used to prevent hydrolysis of diphosphate or acyl groups, however, acid hydrolysis of susceptible glycosidic linkages, such as those involving Kdo residues, can still occur. Analysis using MS and NMR of the obtained OSs established that the core region is highly phosphorylated, with the major species containing two monophosphate groups and one ethanolamine diphosphate (*PP*EtN) group. In a $^1H^{31}P$ HMQC spectrum, the diphosphate signals correlated with signals for the CH_2O of the ethanolamine and H-2 of Hep1. A ^1H-^{31}P HMQC-TOCSY experiment confirmed *PP*EtN attachment at position 2 of the first heptose residue (HepI) of the core. The signals of the monophosphate groups showed correlations to the H-4 of HepI and H-6 of the second heptose (HepII).

A third example, concerns the 1D ^1H NMR spectrum of OS from *Haemophilus influenzae* strain Rd that shows a strong methyl signal at ≈3.24 ppm that is indicative of phosphocholine substitution, the location of which has been determined using ^1H-^{31}P HSQC spectra (Schweda *et al.*, 2000).

3.9. Databases of chemical shift information

Tabulated chemical shift data for monosaccharides and polysaccharides are an invaluable resource for aiding structural assignment of polysaccharides. Publicly accessible databases have been previously reviewed (Duus *et al.*, 2000). CASPER has been published (Jansson *et al.*, 2006; Loss *et al.*, 2006) with a web interface (http://www.casper.organ.su.se/casper). This database has been tested extensively and used to simulate ^{13}C spectra of greater than 200 structures with excellent reliability. Other databases of chemical shift data, include SUGABASE that combines CarbBank and Complex Carbohydrate Structure Data with proton and carbon chemical shifts in a search routine (http://www.boc.chem.uu.nl/sugabase/databases.html), Bacterial Carbohydrate Structure DataBase (http://www.glyco.ac.ru/bcsdb/start.shtml). Sweetdb, is an initiative of glycosciences.de containing NMR and MS data are also available publicaly (http://www.glycosciences.de/sweetdb/Sweetdatabase).

4. MASS SPECTROMETRY OF GLYCANS

4.1. Introduction

Electron-impact ionization MS (EI-MS) was first used in the 1960s to perform structural characterization studies on sugars: permethylated sugars (Kochetkov and Chizhov, 1965); peracetylated sugars (Biemann *et al.*, 1963); and glycosides (de Jongh and Biemann, 1963). Derivatization (e.g. methylation of the labile hydrogens on −OH groups) resulted in an increased volatility of the sample, which allowed the sample to ionize rather than undergo pyrolysis (i.e. decompose due to heat) at the higher temperatures (200–300°C) required for volatization. Trimethylsilane (TMS) has become a common derivatization reagent for analysis by EI-MS and gas chromatography-MS (GC-MS) (de Jongh *et al.*, 1967). Much of this derivatization methodology has remained in current use, with only minor improvements being incorporated (Settineri and Burlingname, 1995).

Major advancements in carbohydrate analysis came with development of the soft ionization techniques, fast-atom bombardment mass spectrometry (FAB-MS), liquid-secondary ion mass spectrometry (LSI-MS) and field desorption mass spectrometry (FD-MS). Often FAB-MS gives molecular ions in addition to fragmentation that makes possible the determination of sequence information of glycoconjugates (Reinhold and Carr, 1983). Systematic nomenclature for the fragmentation of glycoconjugates by FAB-MS and collision-induced dissociation (CID) MS/MS was subsequently proposed and reported on by Dell (1987) and Domon and Costello (1988) and are currently in use. In the late 1980s to early 1990s, electrospray ionization mass spectrometry (ES-MS) (Fenn *et al.*, 1989; Smith *et al.*, 1990) and matrix-assisted laser desorption ionization time-of-flight mass spectrometry (MALDI-TOF MS) (Karas *et al.*, 1987; Hillenkamp *et al.*, 1991) were developed and facilitated analysis at pmol, at times fmol, levels (Settineri and Burlingame, 1995). Modern MS-based structural characterization strategies, namely, ES-MS, MALDI-TOF MS, collision-activated dissociation/collision-induced dissociation (CAD/CID) MS and FAB-MS with accompanying derivatization techniques offer speed, high-throughput, precision and high sensitivity (Haslam *et al.*, 2003). Extensive reviews of MS techniques relating to glycans can be found in the literature (Settineri and Burlingame, 1995; Harvey, 2008; Morelle and Michalski, 2005).

4.2. MS structural characterization

There is no singular MS strategy that can be applied to characterize the structure of all glycans. In principle, if a researcher has access to suitable MS instrumentation and associated technical and interpretational skills, structural characterization of glycans can be achieved by MS.

The fundamental basis of the MS strategy involves the formation of gas-phase molecular and fragment ions from the sample matrix followed by accurate mass measurement. However, the specific sample preparation required to obtain these gas-phase ions is almost exclusively dictated by the ionization method itself (Haslam *et al.*, 2003). Therefore, the choice of ionization method is an integral part of the strategy. Glycans often require chemical manipulations prior to analysis. We briefly outline here some of these manipulations plus some of the accompanying GC-MS analyses. The technique of GC-MS is used routinely for the determination of constituent monosaccharides, linkage analysis and absolute configuration of the sugars.

4.2.1. *Permethylation derivatization*

This increases the lipophilicity of the glycan to aid ionization and facilitates a more consistent fragmentation process via known pathways, which is important particularly for FAB-MS and ES-MS/MS analysis (where MS/MS refers to successive stages of fragmentation of an isolated precursor). Permethylation replaces the free hydroxy groups (-OH) with methoxyl (-OCH_3) groups (Dell, 1987; Dell *et al.*, 1994). Deuteromethyl derivatization can also be carried out, whereby -OCD_3 groups replace hydroxyl groups. This gives a characteristic three mass unit shift relative to non-deuterated methyl groups further aiding structural characterization.

4.2.2. *Sugar analysis*

Such analysis provides fundamental information on the constituent monosaccharides and their respective ratios in a glycan sample. The sample is initially chemically hydrolysed into constituent monosaccharides and then commonly converted to volatile alditol acetate or TMS derivatives, that can readily be analysed in high resolution, highly sensitive GC- and EI-MS experiments (Albersheim *et al.*, 1967; de Jongh *et al.*, 1969). The monosaccharides are identified by retention times on the GC column and by comparison to MS spectra of monosaccharide standards.

4.2.3. *Linkage analysis*

This analysis provides information on the glycosidic linkages. The sample is first derivatized to form acid-stable methyl ethers (permethylation), then hydrolysed and readuced to produce monosaccharides with hydroxyl groups at the linkage positions that are then peracetylated (Lindberg and Lönngren, 1978). Derivatized monosaccharides are identified by retention times on the GC column and by comparison to MS spectra of standards (Albersheim *et al.*, 1967).

4.2.4. *D and L absolute configuration analysis*

For this analysis, prior to samples being TMS derivatized, samples are subjected to hydrolysis with a chiral reagent (e.g. (*S*)-(+)-2-butanolic HCl) and are then re-*N*-acetylated. Analysis is carried out with the absolute configuration determined by retention time on a GC column and by comparison to MS spectra of standards (Albersheim *et al.*, 1967). Another widely applied protocol has been described by Gerwig *et al.* (1979).

4.2.5. *Chemical hydrolysis*

This treatment is used to cleave certain glycosidic linkages that are particularly susceptible to hydrolysis with specific reagents. For example, incubation of LPS with sodium acetate buffer (or acetic acid) will cleave the acid-labile ketosidic linkage between the core OS and lipid A (Moran *et al.*, 2002; Mikhail *et al.*, 2005). Incubation of native OS sample with aqueous hydrofluoric acid (HF) will cleave fucose residues or phosphate moieties (Haslam *et al.*, 2000). Treatment with HCl hydrolyses fucose residues from permethylated samples leaving a hydroxyl group. Mild periodate treatment cleaves carbon-carbon bonds of sugars if they possess vicinal hydroxyl groups to enable assignment of the linkage position of the reducing end sugar (Haslam *et al.*, 2003).

4.2.6. *Exo-glycosidase digestion*

Such enzymatic digestion is used to assign the anomeric glycan linkage configurations based on the specificity of the utilized enzymes. Some enzymes are even specific for varying linkages in the sugars (i.e. 3-linked as opposed to a 4-linked sugar). The digestion products are then examined by linkage and MS analysis (Settineri and Burlingame, 1994).

4.3. MS techniques

The present focus in MS of glycans is on ES and MALDI ionization techniques (Settineri and Burlingame, 1995; Klein, 2005; Didraga *et al.*, 2006; Harvey, 2008), although it is worth noting that there also exist at least 20 different commercial mass spectrometers. All, however, have three basic components:

(i) an ionization source where sample molecules are given an electrical charge;
(ii) a mass analyser where ions are separated on the basis of mass-to-charge ratio; and
(iii) a detector where ions are observed and counted.

EI-MS and FAB-MS have specific features that make them important in the analysis of glycans. EI is most useful for analysis of highly volatile small molecules, with the EI-MS instrument usually coupled to a GC, e.g. for sugar analysis (de Jongh *et al.*, 1969) and linkage analysis (Bjorndal *et al.*, 1970).

Our discussion will subsequently focus on the principal techniques presently used to characterize glycans, namely FAB-MS, MALDI-MS and ES-MS, and the most important analysers associated with these techniques.

4.3.1. *FAB-MS*

FAB-MS (Barber *et al.*, 1981; Dell *et al.*, 1983) is the only soft ionization technique that is capable of providing molecular ions as well as adequate fragmentation ions without employing MS/MS. The FAB experiment itself involves an accelerated beam of atoms (usually xenon) or ions (usually caesium) being fired towards a small metal target from an atom or ion gun. The metal target is mounted on a probe and loaded with a viscous liquid matrix (usually thioglycerol or *m*-nitrobenzyl alcohol) in which the sample to be analysed is dissolved. When the atom or ion beam collides with the matrix many surface molecules of the sample are ionized and sputtered into the high vacuum and then moved to the detector (Dell, 1987; Dell *et al.*, 2000).

Advantages of the technique are that a significant fragmentation is produced along with the molecular ion; positive (usually protonated) and negative (usually deprotonated or with addition of Cl^-) ions can be produced. The method is very effective and sensitive for profiling glycan mixtures, almost as effective as MALDI-TOF, and interpretation of spectra to obtain molecular mass information is uncomplicated, due to singly charged molecular ions that are usually formed.

Limitations are those of inferior sensitivity at higher mass ranges compared to MALDI-MS or ES-MS or when non-derivatized or native glycan samples are being analysed. In practical terms, relatively hydrophilic carbohydrates require derivatization to enhance surface activity and spectra are characterized by a high level of chemical noise.

4.3.2. *MALDI-MS*

For MALDI (Karas and Hillenkamp, 1988), the analyte is required to be embedded in a low molecular mass UV-absorbing matrix, normally 2,5-dihydroxybenzoic acid, in vast excess over the glycan sample (Harvey *et al.*, 1996). A laser pulse is fired at the matrix, which absorbs enough energy to transfer some of this to the sample, essentially volatilizing individual isolated monomeric ions. The ions produced are typically analysed by TOF detectors, where ions separate according to their mass/charge ratio, heavier ions drift or move more slowly and lighter ions drift more rapidly towards the detector. In a linear TOF detector, little fragmentation is seen without resorting to analysis of post-source decay ions or additional CAD in the MS/MS mode. Further advances in sensitivity (i.e. order of magnitude) can be achieved by permethylation, usually eliminating the tendency to lose negatively-charged residues when analysis is carried out in the positive ion mode (Haslam *et al.*, 2003).

Advantages are that the method is more sensitive than FAB-MS, especially at higher mass range, and there is an uncomplicated interpretation due to singly charged molecular ions usually being formed in the linear mode but, in the CID mode, parent ion selection can be achieved at high-resolution and sensitivity (Harvey *et al.*, 1996). Moreover, the TOF analyser has almost unlimited mass range.

Limitations of the technique are that it is not as useful as FAB-MS for profiling small oligosaccharides below *m/z* 600; the MALDI spectrum is overwhelmed by matrix peaks in this region. Also, there is a lack of sequence-informative fragment ions in the low-mass region without the use of post-source decay or MS/MS inclusion (Haslam *et al.*, 2003) and the difficulty of obtaining protonated ions in MALDI-MS.

4.3.3. *ESI-MS*

In an ES experiment, a solution containing the analyte is introduced into the ES apparatus, by means of loop injection-liquid chromatography (-LC), direct infusion or syringe pump, whereby upon spraying the solution into a high electric field, microdroplets are desolvated to produce gas-phase ions. To cater for small sample quantities, the trend is to use nanoES for higher sensitivity (Wuhrer *et al.*, 2005; Robbe *et al.*, 2006).

Numerous separation techniques have been coupled to ES-MS for in-line separation and then characterization of glycans. Li and Richards (2007) reported on the use of capillary electrophoresis ES-MS for the structural characterization of bacterial LPS. Extensive research has been conducted on glycans employing LC in conjunction with ES-MS (Black and Fox, 1996). The ES-MS technique is an ideal method for interfacing to capillary electrophoresis or LC as it is a flow technique that operates at atmospheric pressure.

The technique of ESI-MS is favoured over FAB and MALDI for its ease of implementing automated LC-MS/MS and does not require derivatization of samples. Further advantages are sensitivity to pmol, or even fmol, quantities of material (Harvey *et al.*, 1996; Haslam *et al.*, 2003) and ionization is gentle. The limitations are that raw data can be complex, due to characteristic multiple charging in the ES spectra, with very few fragment ions and little or no sequence information unless MS/MS employed.

Although FAB-MS, MALDI-MS and ES-MS do not prerequisitely demand derivatization, more reliable fragmentation can be obtained by functionalizing hydroxyl groups of glycans (e.g. methyl or acetyl derivatives) rather than analysing the native sample. Furthermore, by choosing an appropriate experimental strategy, most structural problems can successfully be addressed using these ionization methods (Dell *et al.*, 2000).

4.3.4. *MS/MS and mass analysers*

A demand for higher sensitivity and more effective MS/MS implementation has led to the development of numerous new mass analysers. Two analysers of principal importance are the ion-trap (IT) and the quadrupole-orthogonal acceleration time-of-flight (Q-TOF) instruments. Both have found application in coupling to ES or nanoES (Sheeley and Reinhold 1998; Weiskopf *et al.*, 1998; Wuhrer *et al.*, 2002; Matamoros, 2007) and MALDI sources (Zehl *et al.*, 2007). Due to the nature of FAB instrumentation, coupling to IT or Q-TOF analysers has not been possible to date (Haslam *et al.*, 2003).

The IT instrumentation allows MS/MS due to its unique strength as an ion storage device. The Q-TOF has both quadrupole and TOF analysers arranged orthogonally with a collision cell separating the two. With this configuration, the Q-TOF-MS instrumentation is more powerful than ES-MS and CID MS/MS. These mass analysers produce extensive fragmentation enabling comprehensive characterization abilities. Signal to noise ratios in the MS/MS mode allow sample analysis in the low fmol to amol range (Dell *et al.*, 2000).

Fourier-transform-ion cyclotron resonance (FT-ICR) analysers are being used to extend further the capabilities of IT-MS in structural profiling. These instruments offer ultra-high sensitivity and resolution for structural characterization of glycans coupled to ES and MALDI ionization sources (Park and Lebrilla, 2005; Bereman *et al.*, 2007).

4.4. Structural characterization procedures of selected bacterial glycans classes

No singular approach exists for the structural characterization of microbal glycans, however, a generalized approach that offers useful insights is shown in Figure 13.3. Also, an example from the literature of the MS structural characterization of LOS from *Yersinia pestis* (*phoP* mutant) will be examined (Figure 13.4). Hitchen *et al.* (2002) employed the strategy outlined below to characterize the structure of an oligosaccharide and lipid A from *Y. pestis*. A final structure was deduced by collating partial information from the varying experiments performed.

Interpretation of the FAB-MS spectrum of the permethylated *phoP* mutant OS showed signals at *m/z* 2108 and 1802 of compositions $HexNAcHexHep_4KoKdo$-ol and $HexNAcHexHep_4Kdo$-ol (where HexNAc, *N*-acetylhexosamine; Hex, hexose; Hep, heptose; Ko, D-*glycero*-D-*talo*-oct-2-ulosonic acid), respectively. An A-type ion at *m/z* 1004 was attributed to $HexNAcHep_3^+$ with a signal at *m/z* 767 attributed to the double cleavage product Hep_3 (see Figure 13.4, FAB-MS (OS)). Interpretation of the FAB-MS of lipid A spectrum showed a sodiated molecular ion at *m/z* 1267, which was attributed to a glucosamine disaccharide backbone substituted with four 3-hydroxytetradecanoic acid, 14:O(3-OH), residues. Signals at *m/z* 1041 and 815 were assigned to moieties lacking one or two 14:O(3-OH) acyl residues smaller than the *m/z* 1267. The signal at *m/z* 1685 arose from the presence of two phosphate groups plus an additional 16:1 acyl residue on the *m/z* 1267 species [see Figure 13.4, FAB-MS (Lipid A)].

The ESI-CAD MS/MS spectrum of the core OS was carried out on a Q-TOF MS to give an unambiguous sequence and branching pattern

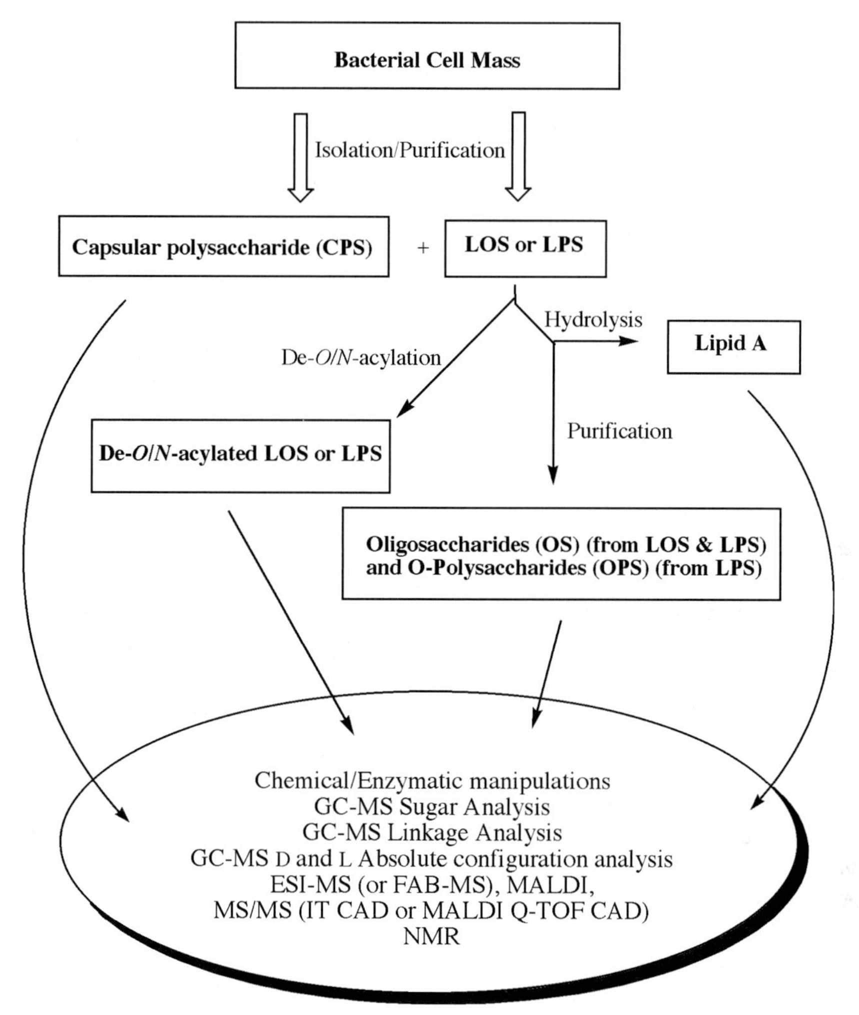

FIGURE 13.3 Overview of characterization strategies for glycans of Gram-negative bacteria.

information for the *phoP* mutant OS. The principal assignments were an ion at *m/z* 1065 from the permethylated sample (see Figure 13.4, ESI-CAD MS/MS (OS-A)) being assigned to $HexNAcHexHep_4KoKdo\text{-}ol$ $[M + 2Na]^{2+}$ and *m/z* 1165 from the deuteromethylated sample (see Figure 13.4, ESI-CAD MS/MS (OS-B)) from $HexNAcHexHep_4KoKdo\text{-}ol$ $[M + 2Na]^{2+}$. Comparison of the permethylated and deuteromethylated samples revealed the number of hydroxyl groups in the native sample plus additional more subtle information. Both the $[M + 2Na]^{2+}$ (i.e. permethylated and deuteromethylated) ions were selected for CAD MS/MS analysis. Noteworthy product ions were at *m/z* 361 and 375 (see Figure 13.4, spectrum OS-A) and *m/z* 379 and 396 (see Figure 13.4, spectrum OS-B) and correspond to Ko product ions.

Table 13.1 provides some selected, representative glycan structural characterization strategies. Comprehensive details of procedures can be found in the literature cited therein.

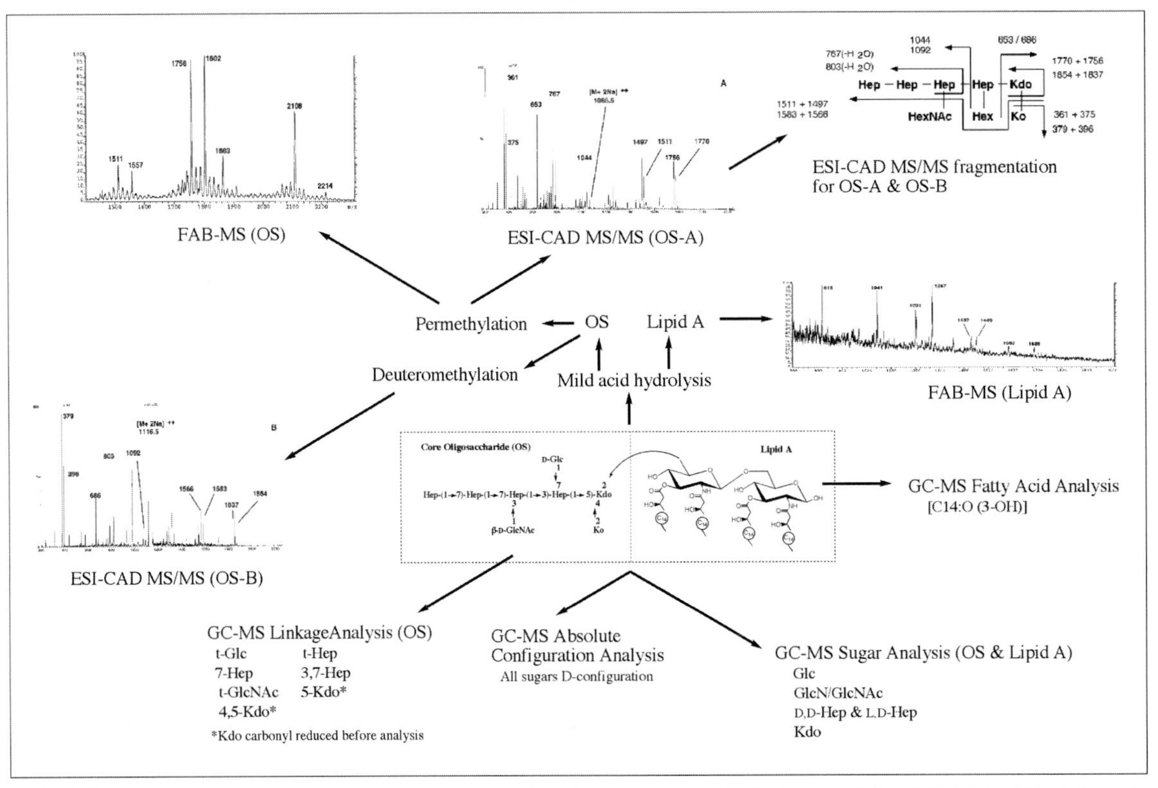

FIGURE 13.4 Overview of MS strategies used to define the structures of both the core OS and lipid A of *Yersinia pestis* LOS. Abbreviations: Glc, glucose; GlcN, glucosamine; GlcNAc, *N*-acetylglucosamine; Hep, heptose; Kdo, 3-deoxy-D-*manno*-oct-2-ulosonic acid. Adapted, with permission, from Hitchen *et al.*, 2002.

4.5. Databases of mass spectral information

A critical key to the structural characterization of complex glycans are the informatics methods for the interpretation of MS and MS fragmentation data. The use and development of tools and databases for the interpretation of these data have increased greatly in recent years (von der Lieth *et al.*, 2006). However, in relation to proteomics and genomics, the field is still regarded as being in its infancy (Packer *et al.*, 2007). Currently, extensive efforts are underway to formulate centralized carbohydrate structure databases with the ultimate goal of a single central database, but obstacles still exist to achieving that worthwhile goal (Toukach *et al.*, 2007). Nevertheless, there is a range of programs available on the Internet that can be used to interpret MS data originating from glycans. Reviews by von der Lieth *et al.* (2006) and others (Matthiesen, 2007; Maass *et al.*, 2007; Toukach *et al.*, 2007) provide excellent resources for accessing programs to analyse and interpret MS spectral information.

5. CONCLUSIONS

Glycans are structurally complex molecules and, as such, their structural characterization is a challenging task. The precise strategy to be applied is dictated by the level of information required and the expertise of the research team involved, however, the most comprehensive and informative studies employ aspects of enzymatic, chemical, chromatographic, MS and NMR methods to complete the characterization process. New advances in mass spectrometry and NMR spectroscopy in terms of hardware, software and databases will further accelerate the discovery of novel glycan structures, leading to a greater understanding of the glycopolymers of bacteria.

References

Aas, F., Vik, A., Vedde, J., Koomey, M., Egge-Jacobsen, W., 2007. *Neisseria gonorrhoeae O*-linked pilin glycosylation: functional analyses define both the biosynthetic pathway and glycan structure. Mol. Microbiol. 65, 607–624.

Albersheim, P., Nevins, D.J., English, P.D., Karr, A., 1967. Analysis of sugars in plant cell-wall polysaccharides by gas-liquid chromatography. Carbohydr. Res. 5, 340–345.

Banerjee, A., Ghosh, S.K., 2003. The role of pilin glycan in Neisserial pathogenesis. Mol. Cell. Biochem. 253, 179–190.

Barber, M., Bordoli, R.S., Sedgwick, R.D., Tyler, A.N., 1981. Fast atom bombardment of solids (F.A.B.): a new ion source for mass spectrometry. J. Chem. Soc. Chem. Commun. 7, 325–327.

Bereman, M.S., Williams, T.I., Muddiman, D.C., 2007. Carbohydrate analysis by desorption electrospray ionization fourier transform ion cyclotron resonance mass spectrometry. Anal. Chem. 79, 8812–8815.

Biemann, K., DeJongh, D.C., Schnoes, H.K., 1963. Application of mass spectrometry to structure problems. XIII. Acetates of pentoses and hexoses. J. Am. Chem. Soc. 85, 1763–1770.

Bjorndal, H., Hellerqvist, C.G., Lindberg, B., Svensson, S., 1970. Gas-liquid chromatography and mass spectrometry in methylation analysis of polysaccharides. Angew. Chem. Int. Ed. Engl. 9, 610–619.

Black, G.E., Fox, A., 1996. Liquid chromatography with electrospray ionization tandem mass spectrometry. Profiling carbohydrates in whole bacterial cell hydrolyzates. ACS Symp. Ser. 619, 81–105.

Blundell, C.D., Reed, M.A.C., Overduin, M., Almond, A., 2006. NMR spectra of oligosaccharides at ultra-high field (900MHz) have better resolution than expected due to favorable molecular tumbling. Carbohydr. Res. 341, 1985–1991.

Bubb, W.A., 2003. NMR spectroscopy in the study of carbohydrates: characterizing the structural complexity. Concepts Magn. Reson. 19A, 1–19.

Bystrova, O.V., Lindner, B., Moll, H., et al., 2004. Full structure of the lipopolysaccharide of *Pseudomonas aeruginosa* immunotype 5. Biochemistry (Moscow) 69, 170–175.

Chapdelaine, G., Cleon, P., 2003. NMR cryogenic probe: a new step to sensitivity. Spectra Anal. 32, 35–38.

Comstock, L.E., Kasper, D.L., 2006. Bacterial glycans: key mediators of diverse host immune responses. Cell 126, 847–850.

Dasgupta, N., Arora, S.K., Rampal, R., 2004. The flagellar system of *Pseudomonas aeruginosa*. Pseudomonas 1, 675–698.

De Castro, C., Carannante, A., Lanzetta, R., et al., 2006. Structural characterisation of the core oligosaccharides isolated from the lipooligosaccharide fraction of *Agrobacterium tumefaciens* A1. Chem.-Eur. J. 12, 4668–4674.

de Jongh, D.C., Biemann, K., 1963. Application of mass spectrometry to structure problems. XIV. Acetates of partially methylated pentoses and hexoses. J. Am. Chem. Soc. 85, 2289–2294.

de Jongh, D.C., Hribar, J.D., Hanessian, S., Woo, P.W.K., 1967. Mass spectrometric studies on aminocyclitol antibiotics. J. Am. Chem. Soc. 89, 3364–3365.

de Jongh, D.C., Radford, T., Hribar, J.D., et al., 1969. Analysis of trimethylsilyl derivatives of carbohydrates by gas chromatography and mass spectrometry. J. Am. Chem. Soc. 91, 1728–1740.

Dell, A., 1987. F.A.B.-mass spectrometry of carbohydrates. Adv. Carbohydr. Chem. Biochem. 45, 19–72.

Dell, A., Morris, H.R., Egge, H., von Nicolai, H., Strecker, G., 1983. Fast-atom-bombardment mass spectrometry for carbohydrate-structure determination. Carbohydr. Res. 115, 41–52.

Dell, A., Reason, A.J., Khoo, K.H., Panico, M., McDowell, R. A., Morris, H.R., 1994. Mass spectrometry of carbohydrate-containing biopolymers. Methods Enzymol. 230, 108–132.

Dell, A., Morris, H.R., Easton, R., et al., 2000. Structural analysis of oligosaccharides: FAB-MS, ES-MS and MALDI-MS. Carbohydr. Chem. Biol. 2, 915–945.

Didraga, M., Barroso, B., Bischoff, R., 2006. Recent developments in proteoglycan purification and analysis. Curr. Pharm. Anal. 2, 323–337.

Domon, B., Costello, C.E., 1988. A systematic nomenclature for carbohydrate fragmentations in FAB-MS/MS spectra of glycoconjugates. Glycoconj. J. 5, 397–409.

Duus, J., Gotfredsen, C.H., Bock, K., 2000. Carbohydrate structural determination by NMR spectroscopy: modern methods and limitations. Chem. Rev. 100, 4589–4614.

Fedonenko, Y.P., Konnova, O.N., Zdorovenko, E.L., et al., 2008. Structural analysis of the O-polysaccharide from the lipopolysaccharide of *Azospirillum brasilense* S17. Carbohydr. Res. 343, 810–816.

Feng, R., Xie, H., Ren, D.-Z., 2002. Application of cryogenic NMR probes. Bopuxue Zazhi 19, 447–454.

Fenn, J.B., Mann, M., Meng, C.K., Wong, S.F., Whitehouse, C.M., 1989. Electrospray ionization for mass spectrometry of large biomolecules. Science 246, 64–71.

Flahaut, S., Vinogradov, E., Kelley, K.A., Brennan, S., Hiramatsu, K., Lee, J.C., 2008. Structural and biological characterization of a capsular polysaccharide produced by *Staphylococcus haemolyticus*. J. Bacteriol. 190, 1649–1657.

Gerwig, G.J., Kamerling, J.P., Vliegenthart, J.F., 1979. Determination of the absolute configuration of monosaccharides in complex carbohydrates by capillary G.L.C. Carbohydr. Res. 88, 10–17.

Godschalk, P.C.R., Kuijf, M.L., Li, J., et al., 2007. Structural characterization of *Campylobacter jejuni* lipooligosaccharide outer cores associated with Guillain-Barre and Miller Fisher syndromes. Infect. Immun. 75, 1245–1254.

Harvey, D.J., 2008. Analysis of carbohydrates and glycoconjugates by matrix-assisted laser desorption/ionization mass spectrometry: an update covering the period 2001–2002. Mass Spectrom. Rev. 27, 125–201.

Harvey, D.J., Naven, T.J.P., Kuster, B., 1996. Identification of oligosaccharides by matrix-assisted laser desorption ionization and electrospray MS. Biochem. Soc. Trans. 24, 905–912.

Haslam, S.M., Coles, G.C., Morris, H.R., Dell, A., 2000. Structural characterization of the *N*-glycans of *Dictyocaulus viviparus*: discovery of the Lewis^x structure in a nematode. Glycobiology 10, 223–229.

Haslam, S.M., Khoo, K.-H., Dell, A., 2003. Sequencing of oligosaccharides and glycoproteins. Carbohydr.-Based Drug Discov. 2, 461–482.

Hillenkamp, F., Karas, M., Beavis, R.C., Chait, B.T., 1991. Matrix-assisted laser desorption/ionization mass spectrometry of biopolymers. Anal. Chem. 63, 1193A–1203A.

Hitchen, P.G., Prior, J.L., Oyston, P.C.F., et al., 2002. Structural characterization of lipooligosaccharide (LOS) from *Yersinia pestis*: regulation of LOS structure by the PhoPQ system. Mol. Microbiol. 44, 1637–1650.

Hricovini, M., 2004. Structural aspects of carbohydrates and the relation with their biological properties. Curr. Med. Chem. 11, 2565–2583.

Jansson, P.-E., Stenutz, R., Widmalm, G., 2006. Sequence determination of oligosaccharides and regular polysaccharides using NMR spectroscopy and a novel web-based version of the computer program CASPER. Carbohydr. Res. 341, 1003–1010.

Karas, M., Hillenkamp, F., 1988. Laser desorption ionization of proteins with molecular masses exceeding 10000 daltons. Anal. Chem. 60, 2299–2301.

Karas, M., Bachmann, D., Bahr, U., Hillenkamp, F., 1987. Matrix-assisted ultraviolet laser desorption of non-volatile compounds. Int. J. Mass Spectrom. Ion Proc. 78, 53–68.

Karlyshev, A.V., Champion, O.L., Churcher, C., et al., 2005. Analysis of *Campylobacter jejuni* capsular loci reveals multiple mechanisms for the generation of structural diversity and the ability to form complex heptoses. Mol. Microbiol. 55, 90–103.

Keeler, J., Clowes, R.T., Davis, A.L., Laue, E.D., 1994. Pulsed-field gradients: theory and practice. Methods Enzymol. 239, 145–207.

Kenne, L., Lindberg, B., Schweda, E., Gustafsson, B., Holme, T., 1988. Structural studies of the O-antigen from *Vibrio cholerae* O:2. Carbohydr. Res. 180, 285–294.

Khamri, W., Moran, A.P., Worku, M.L., et al., 2005. Variations in *Helicobacter pylori* lipopolysaccharide to evade the innate immune component surfactant protein D. Infect. Immun. 73, 7677–7686.

Klein, M.L., 2005. HPLC and mass spectrometry approaches to protein and carbohydrate characterization. Am. Pharm. Rev. 8, 108, 110–114.

Kochetkov, N.K., Chizhov, O.S., 1965. Mass spectrometry of methylated methyl glycosides. Principles and analytical application. Tetrahedron 21, 2029–2047.

Li, J., Richards, J.C., 2007. Application of capillary electrophoresis mass spectrometry to the characterization of bacterial lipopolysaccharides. Mass Spectrom. Rev. 26, 35–50.

Lindberg, B., 1998. Bacterial polysaccahrides: components. In: Dimitriu, S. (Ed.), Polysaccahrides – Structural Diversity and Functional Versatility. Marcel Dekker Inc., New York, pp. 237–274.

Lindberg, B., Lönngren, J., 1978. Methylation analysis of complex carbohydrates: general procedure and application for sequence analysis. Adv. Carbohydr. Chem. Biochem. 33, 295–322.

Logan, S., 2006. Flagellar glycosylation: a new component of the motility repertoire? Microbiology 152, 1249–1262.

Logan, S., Schoenhofen, I.C., Guerry, P., 2008. *O*-linked flagellar glycosylation in Campylobacter. In: Nachamkin, I., Szymanski, C.M., Blaser, M.J. (Eds.), *Campylobacter*, 3rd edn. American Society for Microbiology, Washington DC, pp. 471–481.

Loss, A., Stenutz, R., Schwarzer, E., von der Lieth, C.W., 2006. GlyNest and CASPER: two independent approaches to estimate ^{1}H and ^{13}C NMR shifts of glycans available through a common web-interface. Nucl. Acids Res. 34, W733–W737.

Lukasiewicz, J., Dzieciatkowska, M., Niedziela, T., et al., 2006. Complete lipopolysaccharide of *Plesiomonas shigelloides* O74:H5 (strain CNCTC 144/92). 2. Lipid A, its structural variability, the linkage to the core oligosaccharide, and the biological activity of the lipopolysaccharide. Biochemistry 45, 10434–10447.

Maass, K., Ranzinger, R., Geyer, H., von der Lieth, C.-W., Geyer, R., 2007. "Glyco-Peakfinder" – de novo composition analysis of glycoconjugates. Proteomics 7, 4435–4444.

Martin, G.E., 2002. Cryogenic NMR probes: applications. Encycl. Nucl. Magn. Reson. 9, 33–35.

Matamoros, F.L.E., 2007. Introduction to ion trap mass spectrometry: application to the structural characterization of plant oligosaccharides. Carbohydr. Polym. 68, 797–807.

Matthiesen, R., 2007. Useful mass spectrometry programs freely available on the Internet. Methods Mol. Biol. 367, 303–312.

McNally, D.J., Jarrell, H.C., Li, J., et al., 2005. The HS:1 serostrain of *Campylobacter jejuni* has a complex teichoic acid-like capsular polysaccharide with nonstoichiometric fructofuranose branches and O-methyl phosphoramidate groups. FEBS J. 272, 4407–4422.

McNally, D.J., Hui, J.P., Aubry, A.J., et al., 2006a. Functional characterization of the flagellar glycosylation locus in *Campylobacter jejuni* 81–176 using a focused metabolomics approach. J. Biol. Chem. 281, 18489–18498.

McNally, D.J., Jarrell, H.C., Khieu, N.H., et al., 2006b. The HS:19 serostrain of *Campylobacter jejuni* has a hyaluronic acid-type capsular polysaccharide with a nonstoichiometric sorbose branch and O-methyl phosphoramidate group. FEBS J. 273, 3975–3989.

McNally, D.J., Lamoureux, M.P., Karlyshev, A.V., et al., 2007. Commonality and biosynthesis of the O-methyl phosphoramidate capsule modification in *Campylobacter jejuni*. J. Biol. Chem. 282, 28566–28576.

Mikhail, I., Yildirim, H.H., Lindahl, E.C.H., Schweda, E.K.H., 2005. Structural characterization of lipid A from nontypeable and type f *Haemophilus influenzae*: variability of fatty acid substitution. Anal. Biochem. 340, 303–316.

Moran, A.P., Knirel, Y.A., Senchenkova, S.Y.N., Widmalm, G., Hynes, S.O., Jansson, P.-E., 2002. Phenotypic variation in molecular mimicry between *Helicobacter pylori* lipopolysaccharides and human gastric epithelial cell surface glycoforms. Acid-induced phase variation in LewisX and LewisY expression by *H. pylori* lipopolysaccharides. J. Biol. Chem. 277, 5785–5795.

Morelle, W., Michalski, J.-C., 2005. The mass spectrometric analysis of glycoproteins and their glycan structures. Curr. Anal. Chem. 1, 29–57.

Niedziela, T., Dag, S., Lukasiewicz, J., et al., 2006. Complete lipopolysaccharide of *Plesiomonas shigelloides* O74: H5 (strain CNCTC 144/92). 1. Structural analysis of the highly hydrophobic lipopolysaccharide, including the O-antigen, its biological repeating unit, the core oligosaccharide, and the linkage between them. Biochemistry 45, 10422–10433.

Packer, N.H., von der Lieth, C.-W., Aoki-Kinoshita, K.F., et al., 2007. Frontiers in glycomics: bioinformatics and biomarkers in disease: an NIH White Paper prepared from discussions by the focus groups at a workshop on the NIH campus, Bethesda MD (September 11–13, 2006). Proteomics 8, 8–20.

Park, Y., Lebrilla, C.B., 2005. Application of Fourier transform ion cyclotron resonance mass spectrometry to oligosaccharides. Mass Spectrom. Rev. 24, 232–264.

Petersen, B.O., Vinogradov, E., Kay, W., et al., 2006. H2BC: a new technique for NMR analysis of complex carbohydrates. Carbohydr. Res. 341, 550–556.

Power, P.M., Seib, K.L., Jennings, M.P., 2006. Pilin glycosylation in *Neisseria meningitidis* occurs by a similar pathway to wzy-dependent O-antigen biosynthesis in *Escherichia coli*. Biochem. Biophys. Res. Comm. 347, 904–908.

Raetz, C.R.H., Whitfield, C., 2002. Lipopolysaccharide endotoxins. Annu. Rev. Biochem. 71, 635–700.

Reinhold, V.N., Carr, S.A., 1983. New mass-spectral approaches to complex carbohydrate structure. Mass Spectrom. Rev. 2, 153–221.

Ribeiro, A.A., Zhou, Z.M., Raetz, C.R.H., 1999. Multi-dimensional NMR structural analyses of purified Lipid X and Lipid A (endotoxin). Magn. Reson. Chem. 37, 620–630.

Robbe, C., Michalski, J.-C., Capon, C., 2006. Structural determination of O-glycans by tandem mass spectrometry. Methods Mol. Biol. 347, 109–123.

Schirm, M., Soo, E.C., Aubry, A.J., Austin, J., Thibault, P., Logan, S.M., 2003. Structural, genetic and functional characterization of the flagellin glycosylation process in *Helicobacter pylori*. Mol. Microbiol. 48, 1579–1592.

Schweda, E.K.H., Richards, J.C., 2003. Structural profiling of short-chain lipopolysaccharides from *Haemophilus influenzae*. Methods Mol. Med. 71, 161–183.

Schweda, E.K.H., Brisson, J.R., Alvelius, G., et al., 2000. Characterization of the phosphocholine-substituted oligosaccharide in lipopolysaccharides of type b *Haemophilus influenzae*. Eur. J. Biochem. 267, 3902–3913.

Settineri, C.A., Burlingame, A.L., 1994. Strategies for the characterisation of carbohydrates from glycoproteins by mass spectrometry. In: Crabb, J.W. (Ed.), Techniques in Protein Chemsitry V. Academic Press, Inc., San Diego, pp. 97–104.

Settineri, C.A., Burlingame, A.L., 1995. Mass spectrometry of carbohydrates and glycoconjugates. J. Chromatogr. Libr. 58, 447–514.

Sheeley, D.M., Reinhold, V.N., 1998. Structural characterization of carbohydrate sequence, linkage, and branching in a quadrupole ion trap mass spectrometer: neutral oligosaccharides and *N*-linked glycans. Anal. Chem. 70, 3053–3059.

Smith, R.D., Loo, J.A., Edmonds, C.G., Barinaga, C.J., Udseth, H.R., 1990. New developments in biochemical mass spectrometry: electrospray ionization. Anal. Chem. 62, 882–899.

St Michael, F., Szymanski, C.M., Li, J., et al., 2002. The structures of the lipooligosaccharide and capsule polysaccharide of *Campylobacter jejuni* genome sequenced strain NCTC 11168. Eur. J. Biochem. 269, 5119–5136.

St Michael, F., Brisson, J.-R., Larocque, S., et al., 2004. Structural analysis of the lipopolysaccharide derived core oligosaccharides of *Actinobacillus pleuropneumoniae* serotypes 1, 2, 5a and the genome strain 5b. Carbohydr. Res. 339, 1973–1984.

Stenutz, R., Weintraub, A., Widmalm, G., 2006. The structures of *Escherichia coli* O-polysaccharide antigens. FEMS Microbiol. Rev. 30, 382–403.

Szymanski, C.M., St Michael, F., Jarrell, H.C., et al., 2003. Detection of conserved *N*-linked glycans and phase-variable lipooligosaccharides and capsules from *Campylobacter* cells by mass spectrometry and high resolution magic angle spinning NMR spectroscopy. J. Biol. Chem. 278, 24509–24520.

Toukach, P., Joshi, H.J., Ranzinger, R., Knirel, Y., von der Lieth, C.-W., 2007. Sharing of worldwide distributed carbohydrate-related digital resources: online connection of the Bacterial Carbohydrate Structure DataBase and GLYCOSCIENCES.de. Nucl. Acids Res. 35, D280–D286.

Uhrin, D., 1997. Concatenation of polarization transfer steps in 1D homonuclear chemical shift correlated experiments. Application to oligo- and polysaccharides. Anal. Spectrosc. Libr. 8, 51, 53–89.

Uhrin, D., Brisson, J.-R., 2000. Structure determination of microbial polysaccharides by high resolution NMR spectroscopy. NMR Microbiol., 165–190.

Uhrin, D., Brisson, J.R., Bundle, D.R., 1993. Pseudo-3D NMR spectroscopy: application to oligo- and polysaccharides. J. Biomol. NMR 3, 367–373.

Uhrin, D., Brisson, J.-R., Kogan, G., Jennings, H.J., 1994. 1D analogs of 3D NOESY-TOCSY and 4D TOCSY-NOESY-TOCSY. Application to polysaccharides. J. Magn. Reson. Ser. B 104, 289–293.

Uhrinova, S., Uhrin, D., Liptaj, T., Batta, G., 1991. Detection of long-range couplings in oligomers and evaluation of coupling constants in polymers. Application of the 1D COSY technique. Magn. Reson. Chem. 29, 22–28.

Vliegenthart, J.F.G., 2006. Introduction to NMR spectroscopy of carbohydrates. ACS Symp. Ser. 930, 1–19.

Voehler, M.W., Collier, G., Young, J.K., Stone, M.P., Germann, M.W., 2006. Performance of cryogenic probes as a function of ionic strength and sample tube geometry. J. Magn. Reson. 183, 102–109.

Voisin, S., Kus, J.V., Houliston, S., St-Michael, F., Watson, D., Cvitkovitch, D.G., Kelly, J., Brisson, J.R., Burrows, L.L., 2007. Glycosylation of *Pseudomonas aeruginosa* strain Pa5196 type IV pilins with mycobacterium-like α-1,5-linked D-Ara*f* oligosaccharides. J. Bacteriol. 189, 151–159.

von der Lieth, C.-W., Luetteke, T., Frank, M., 2006. The role of informatics in glycobiology research with special emphasis on automatic interpretation of MS spectra. Biochim. Biophys. Acta 1760, 568–577.

Weiskopf, A.S., Vouros, P., Harvey, D.J., 1998. Electrospray ionization-ion trap mass spectrometry for structural analysis of complex N-linked glycoprotein oligosaccharides. Anal. Chem. 70, 4441–4447.

Willker, W., Leibfritz, D., Kerssebaum, R., Bermel, W., 1993. Gradient selection in inverse heteronuclear correlation spectroscopy. Magn. Reson. Chem. 31, 287–292.

Wilson, J.C., Angus, D.I., von Itzstein, M., 1995. ^{1}H NMR evidence that *Salmonella typhimurium* sialidase hydrolyzes sialosides with overall retention of configuration. J. Am. Chem. Soc. 117, 4214–4217.

Wilson, J.C., Hitchen, P.G., Frank, M., et al., 2005. Identification of a capsular polysaccharide from *Moraxella bovis*. Carbohydr. Res. 340, 765–769.

Wilson, J.C., Collins, P.M., Klipic, Z., Grice, I.D., Peak, I. R., 2006. Identification of a novel glycosyltransferase involved in LOS biosynthesis of *Moraxella catarrhalis*. Carbohydr. Res. 341, 2600–2606.

Wuhrer, M., Kantelhardt, S.R., Dennis, R.D., Doenhoff, M.J., Lochnit, G., Geyer, R., 2002. Characterization of glycosphingolipids from *Schistosoma mansoni* eggs carrying Fuc(α 1-3)GalNAc-, GalNAc(β 1-4)[Fuc(α 1-3)]GlcNAc- and Gal(β 1-4)[Fuc(α 1-3)]GlcNAc- (Lewis X) terminal structures. Eur. J. Biochem. 269, 481–493.

Wuhrer, M., Deelder, A.M., Hokke, C.H., 2005. Protein glycosylation analysis by liquid chromatography-mass spectrometry. J. Chromatogr., B: Anal. Technol. Biomed. Life Sci. 825, 124–133.

Yamamoto, A., 2002. Nuclear magnetic resonance spectroscopy: the recent developments of hardware: the current status of cryogenic probe. Bunseki 12, 320–321.

Zdorovenko, G.M., Zdorovenko, E.L., Varbanets, L.D., 2007. Composition, structure, and biological properties of lipopolysaccharides from different strains of *Pseudomonas syringae* pv. *atrofaciens*. Microbiology 76, 683–697.

Zehl, M., Pittenauer, E., Jirovetz, L., et al., 2007. Multistage and tandem mass spectrometry of glycosylated triterpenoid saponins isolated from *Bacopa monnieri*. Comparison of the information content provided by different techniques. Anal. Chem. 79, 8214–8221.

CHAPTER

14

Single-molecule characterization of microbial polysaccharides

Marit Sletmoen[1]*, Dionne C.G. Klein*[1] *and Bjørn T. Stokke*

SUMMARY

Single-molecule techniques have paved the way for a more detailed molecular understanding of various biological functions of single molecules. The impact of single-molecule approaches has been particularly large for studies of cellular functions, e.g. related to replication and the function of motor proteins, and there is emerging evidence that this strategy also will provide important information on microbial polysaccharides. Single-molecule characterization of the mechanical responses of microbial polysaccharides has revealed reduced entropy associated with low-force stretching to "fingerprints" reflecting the deformation of the constituting carbohydrates and their linkage pattern. Molecular interactions and structure rearrangements are readily determined within the sub-nanonewton and nanometre range. The molecular understanding of the role of polysaccharides in various biological processes, e.g. recognition and cell adhesion, is expected to gain momentum by further application of single-molecule techniques.

Keywords: Single-molecule; Atomic force microscopy; Fluorescence spectroscopy; Force spectroscopy; Molecular recognition

[1]Footnote: These authors contributed equally to this work.

1. INTRODUCTION

Single-molecule techniques that have emerged over the past 20 years have provided new and important molecular insight into the function of biological macromolecules. Important examples include macromolecular behaviour involved in e.g. cellular transport (Svoboda *et al.*, 1993), generation of adenosine triphosphate (ATP) (Yasuda *et al.*, 2001) and DNA transcription, recombination and replication (Bustamante *et al.*, 2003; van Oijen, 2007). This approach allows insight into the distribution of molecular properties and is complementary to most classical characterization techniques that reveal average properties of the molecular ensembles studied. The additional information obtained from single-molecule studies opens up a new level of understanding of complex systems since distributions of a given property, derived from the collection of individual observations using a single-molecule approach, may reveal deviating subpopulations and their properties.

The importance of building knowledge at the single-molecule level within molecular biology is nowadays well accepted. Two important examples are single-enzyme kinetics that permit a statistical description of binding trajectories and transition states (Schenter *et al.*, 1999) and understanding the mechanisms influencing the target localization speed of many gene-regulating proteins which exceed that set by the bimolecular collision limit imposed by the diffusion of the compounds (Stanford *et al.*, 2000; Gowers *et al.*, 2005). In general, although single-molecule studies are not as widely applied to microbial polysaccharides (PSs) as within other areas of molecular biology, there is an increasing literature documenting their potential for use on microbial PSs and which clearly shows that the novelty of the available information is important also for these materials. Additionally, the more abundant literature on single-molecule studies of other biological macromolecules may serve as an inspiration for the application of the techniques to microbial PSs.

Single-molecule properties are accessible using experimental set-ups such as atomic force microscopy, optical or magnetic tweezers, fluorescence resonance energy transfer and fluorescence correlation spectroscopy (Figure 14.1) or others. Though the studies require properly extracted and further prepared samples of molecules, they support the determination of microbial PS chain trajectories, force-elongation profiles, interaction strengths with their specific receptors, co-localization or proximity and co-migration with other molecular components. Examples of the application of these techniques to the field of microbial PSs are discussed in this chapter.

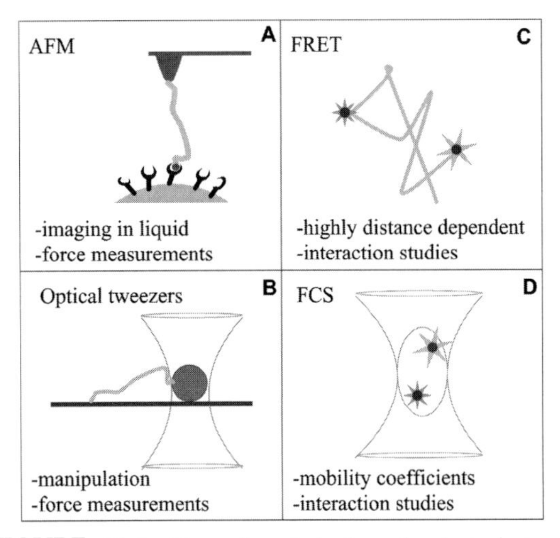

FIGURE 14.1 Examples of single-molecule techniques applicable for studying microbial PSs: (A) atomic force microscopy (AFM); (B) optical and magnetic tweezers; (C) fluorescence resonance energy transfer (FRET); (D) fluorescence correlation spectroscopy (FCS).

2. ATOMIC FORCE MICROSCOPY AND ITS APPLICATION TO MICROBIAL PSs

In an atomic force microscope (AFM) (Binnig *et al.*, 1986), a sharp tip at the end of a miniature cantilever is scanned over a surface using a piezoelectrical scanner, while a laser records the deflection of the cantilever. In the contact mode, the deflection of the cantilever is kept constant. In the oscillating mode, either with the tip intermittently touching the surface for a fraction of the oscillation cycle or oscillating within the attractive force field of the surface, the cantilever is driven close to its resonance frequency and the amplitude of the response of the cantilever is kept constant while scanning the surface. Tapping mode is preferred for imaging of biological samples, as the applied lateral forces are minimized. An AFM enables studies of microbial cell surfaces and their PSs under physiological conditions with nanometre resolution, combined with elasticity and molecular interaction studies (Gaboriaud and Dufrêne, 2007). The samples need to be attached to a surface for the AFM imaging. The resolution depends on the radius of curvature of the scanning tip and the imaging conditions; these may be as good as a lateral resolution of 1 nm and a vertical resolution of 0.1 nm under optimal conditions. Because an AFM can be used in liquid medium, it offers the possibility of studying the function (e.g. adhesion and specific interaction) of biomolecules together with their structure under near physiological conditions.

The imaging capabilities of an AFM are to some extent similar to that achievable using electron microscopy, but the strength within the AFM sets it apart because it can be operated under conditions closer to physiological ones. Figure 14.2 shows a tapping mode AFM image depicting the biological precursor PSs of alginate as they are being converted to alginate by the action of enzymes (Sletmoen *et al.*, 2004). These enzymes are identified as the higher globular domains in the image. This image illustrates the resolution that can be obtained when viewing single PS strands using an AFM.

In (dynamic) force spectroscopy, the determination of the deflection of the cantilever, while performing approach–retract cycles at the same lateral position, yields information about the interaction force between the tip and surface with pico-newton sensitivity. In this way, elasticity, adhesion and molecular interactions can be measured at the single-molecule level. Thus, atomic force microscopy has been applied to study the topography and elastic properties as well as molecular interactions of microbial cell surfaces (Gaboriaud and Dufrêne, 2007).

FIGURE 14.2 Tapping mode AFM height topograph of mannuronan, the biological precursor PSs of alginate, as such molecules are being converted to alginate by the action of the C-5-mannuronan epimerase, AlgE4. The enzymes are identified as the higher globular domains in the image. (See colour plates section.)

3. MEASUREMENTS OF MECHANICAL PROPERTIES OF SINGLE PSs

Generally, PSs possess both mechanical and biochemical functions and PSs excreted by microorganisms influence the viscosity of the microenvironment surrounding the cell. The macroscopic properties of such PSs depend on the primary chemical composition, structure and conformation of the molecules, as well as their interactions at the single-molecule level. Additionally, PSs play key roles in protein sorting, cell–cell communication, cell adhesion and molecular recognition in the immune system. Researchers have more recently become aware of a coupling between the mechanical forces exerted on biopolymers and their biochemical function (Bustamante *et al.*, 2004). This has led biochemists to apply external forces to these processes in an attempt to alter the extent, or even the fate, of these reactions and, thus, reveal their underlying molecular mechanisms.

Measurements of the mechanical properties of single biopolymers can be performed using ultrasensitive force probes, e.g. an AFM or optical or magnetic tweezers. When using these techniques, the applied forces and molecular extensions typically range from tens to hundreds of piconewtons (pN) and a few to tens of nanometres, respectively. While most nano-mechanical probing techniques support the determination of force-elongation profiles, magnetic tweezers can also be used to probe torsional properties of extended biopolymers.

The force–extension relationships of PSs are most often obtained by stretching single molecules that have been adsorbed between a substrate and the cantilever tip (Rief *et al.*, 1997, 1998; Li *et al.*, 1998, 1999; Marszalek *et al.*, 1998, 1999). The gradual extension of the molecules is obtained by continuously increasing the tip-surface separation controlled by the z-piezo of the AFM-scanner and carried out in the aqueous or other relevant solution.

Knowledge obtained of the viscoelastic properties of PS at the single-molecule level are considered essential in order to understand their properties and functions in living systems (Kawakami *et al.*, 2004). The elastic response of pyranose rings of PSs that are involved in load-bearing structures, or that experience mechanical forces in other contexts, is one important aspect of this. Single-molecule techniques have afforded new information concerning how sugar rings respond to these stresses (Marszalek *et al.*, 2001). Most of the studies presented in the following text include investigations of the properties of microbial PSs, but data obtained on other PSs are also included because they serve as model compounds.

Polysaccharide force–extension relations are characterized by an elastic retracting force at low forces that arises from the reduction of chain conformational entropy as the molecules approach their contour length (Marszalek *et al.*, 2001). At a certain degree of extension, an enthalpic contribution to the molecular elasticity becomes dominant, resulting in a more abrupt increase in force with further increase in tip-surface separation distance. Generic models of polymer elasticity, e.g. the freely jointed or the worm-like chain model, work to a certain point for several PSs (Marszalek *et al.*, 2001). Force-induced mechanical deformation, and even transitions of the monosaccharides, leaves characteristic fingerprints in the force–extension curves when the forces are increased. The transitions may involve a change of the pyranose ring from the chair to the boat-like or inverted chair conformation. Such a conformational change increases the displacement between consecutive glycosidic oxygen atoms (Rief *et al.*, 1997; Marszalek *et al.*, 1998). The force drives these new conformations by reducing the activation energy of conformations that are not populated at 20–22°C (Bell, 1978; Evans and Ritchie, 1997; Rief *et al.*, 1998). By controlling the applied force, the probability for a certain transition to occur is increased or decreased. In the force–extension curve this is observed as a plateau region (Figure 14.3). Once all the monomer units have gone through the conformational transition, a deformation of the pyranose rings and bending of the covalent bonds in the linkages between the pyranose rings can give rise to a further increase in the length of the molecule.

Examination of a family of force–extension curves can reveal contour lengths of PS segments stretched by the AFM that vary over a wide range, because the AFM tip picks molecules randomly with respect to their ends. Nevertheless, general features related to the deformation of the constituting monosaccharides and their linkage geometry is clearly discernible. Cellulose, polymeric β-(1→4)-D-glucose, displays a force–extension behaviour conforming to a purely entropic polymer model (Marszalek *et al.*, 2001). The force–elongation relationships for other PSs display pronounced deviations from the entropic elasticity that is visible as sudden changes in the curvature of their force–extension curves (see Figure 14.3). Amylose, α-(1→4)-D-glucan and dextran, α-(1→4)-D-glucan, show a single transition that occurs at ≈280 and ≈850 pN, respectively. This transition involves a transition of the pyranose ring from a chair-to-boat conformation, a conformational change which provides a 20% increase in chain length due to an increase in distance between glycosidic oxygen atoms (Marszalek *et al.*, 1998; Rief *et al.*, 1998). The spectrum obtained for pullulan is a combination of the spectra of amylose and dextran, in agreement with its structural composition. The two transitions in pectin, α-(1→4)-D-galacturonan, at ≈300 and 900 pN has been explained by a force-induced two-step chair inversion transition in the α-D-galactopyranuronic acid ring, which increases the separation of the O-1 to O-4 oxygen atoms by ≈20% (Marszalek *et al.*, 1998).

In the experiments described above, single PSs are stretched at a constant rate, while the resulting force changes over wide ranges. Since monosaccharide rings undergo force-dependent

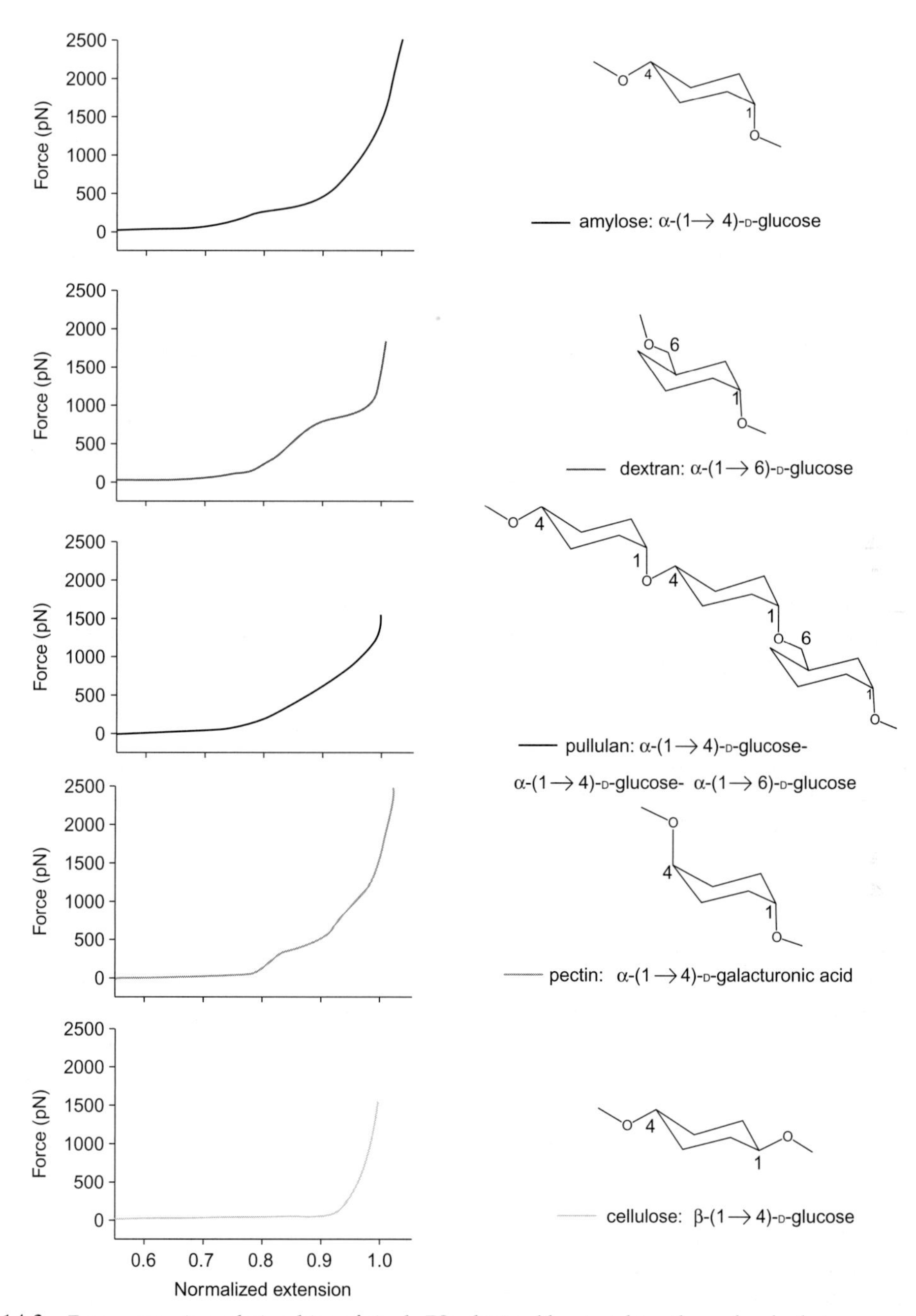

FIGURE 14.3 Force–extension relationships of single PSs obtained by stretching the molecules between a substrate and an AFM cantilever. The monomer and the type of glycosidic linkages in the PS for which the force–extension curves were obtained are displayed on the left. All force–extension curves, except that of cellulose, display marked deviations from the purely entropic elasticity. Reproduced, with permission, from Marszalek *et al.* (2001).

transitions, this approach (where the force is a variable) introduces an undesirable level of complexity in the results (Marszalek *et al.*, 2002). No analytical model allowing the analysis of the observed chair-to-boat transition has been developed for a controlled extension due to the associated difficulties. An alternative approach of controlling the force applied to the molecule, therefore, has been suggested, potentially yielding results set apart from that obtained using the constant extension approach (e.g. Oberhauser *et al.*, 2001). Force-ramp AFM captures the ring transitions under conditions where the entropic elasticity of the molecule is separated from its conformational transitions. This approach enables a quantitative analysis of the data with a two-state model that directly provides the physicochemical characteristics of the ring transitions, e.g. the width of the energy barrier, the relative energy of the conformers and their enthalpic elasticity (Marszalek *et al.*, 2002).

The variance in length of a single PS can be used to establish a thermodynamic equilibrium and to detect conformational changes and underlying dynamics not directly observable with other methods. In constant-velocity experiments, the length variance is calculated from the fluctuations in the measured force in the subsequent stretching of a molecule. This force variance is converted into the variance in the length of the molecule using the spring constant of the cantilever. A further analysis of the variance in the transition region between the chair and the boat conformation has revealed the existence of an intermediate state approximately half-way between the chair and the boat conformation (Lee *et al.*, 2004). These results are interpreted by the aid of single-molecule dynamics (SMD) simulations. The additional degree of freedom associated with the C-5–C-6 bond of (1→6)-linked PSs has been shown to increase the complexity of the mechanical behaviour of these polymers (Lee *et al.*, 2004). Subsequent simulations revealed that the plateau in the force–extension curve observed in experiments can be explained by a transition of the glucopyranose rings in the dextran monomers from the chair ($^4C_1$) to the inverted chair ($^1C_4$) conformation (Neelov *et al.*, 2006).

Such experiments provide the free energy difference and distance between states. However, unlike the observation of hopping between folded and unfolded states of an RNA hairpin (Liphardt *et al.*, 2001), the dynamics of the pyranose conformational transition cannot at present be probed, since each monomeric hopping processes is too small (<1 Å) and too fast for the stretching experiments to probe.

In many of the above-mentioned studies, a constant loading rate has been applied to the molecules, supporting the determination of the elastic response of the molecules. However, most biomolecular interactions exhibit elastic and viscous forces (Kawakami *et al.*, 2004), which should be analysed independently to allow a deeper understanding of their contribution to dynamic biological processes. The viscoelastic responses of single biomolecules can be determined by force spectroscopy in an AFM by sinusoidally varying the cantilever position with the molecule extended between the tip and the surface. Monitoring the amplitude and phase response of the cantilever allows conservative (elastic) and dissipative (viscous) interactions to be distinguished. For example, Kawakami *et al.* (2004) determined the viscoelastic behaviour of single dextran molecules from the frequency power spectral density (PSD) of the Brownian oscillations when the molecule was clamped at a given tension between the cantilever and the substrate (Kawakami *et al.*, 2004). Importantly, the system under investigation is close to equilibrium throughout the measurement and measurement relies on the thermally induced molecular dynamics that are inherent to biological functions.

Later studies showed that single-molecule viscoelastic data can also be directly extracted from the thermally driven cantilever motion during conventional force–extension measurements (Bippes *et al.*, 2006). The results obtained using this approach showed good agreement with those extracted from the PSD and were acquired with a time resolution of one order of magnitude faster than the approaches based on the analysis of the PSD of the cantilever motion.

The dynamics of the transition from chair-to-boat conformation cannot be probed due to the small size of each hopping process and its high speed. It has, however, been argued that the determination of local dissipation would give access to finer-scale conformational dynamics (Khatri *et al.*, 2007). This is in analogy to the macroscopic rheology of complex fluids where dissipative mechanical spectra reflect dynamics of various structural and topological transitions.

The observations presented in some of the papers cited above (Kawakami *et al.*, 2004; Bippes *et al.*, 2006; Khatri *et al.*, 2007) have revealed that the measured frictions are, in the dissipative part of the spectra, many orders of magnitude larger than solvent friction. These observations suggest an internal source of dissipation. In the case of dextran, the effective friction to elongation exhibits a minimum at a force that coincides with the plateau in the force–extension trace (Humphris *et al.*, 2000, 2002; Kawakami *et al.*, 2004). These observations indicate that the internal friction arises through a process related to the local internal conformational transitions in the chain (Khatri *et al.*, 2007).

4. ATOMIC FORCE MICROSCOPY AS A TOOL TO INVESTIGATE FUNCTION OF MICROBIAL PSs

Molecular recognition is important in cell adhesion and aggregation. Also, in pathogenesis, molecular recognition plays a key function, as the first steps in the infectious process are caused by interaction of microbial molecules with receptors on the host cell surface (Gaboriaud and Dufrêne, 2007). The technique of atomic force microscopy offers the possibility to study single molecules, molecular complexes and the surfaces of whole cells under physiological conditions at nanoscale resolution (Müller and Engel, 1999). In addition, specific interactions between molecules can be measured directly, which makes it possible to study the function of certain biomolecules (e.g. adhesion) and relate this to topographical features (Dupres *et al.*, 2005). Examples of both AFM imaging and interaction studies on various microbial species and their PSs are outlined below.

In force spectroscopy measurement in an AFM, a ligand tethered to the AFM tip is allowed to interact with receptors on a surface allowing determination of ligand–receptor interactions at the single-molecule level. Building a molecular understanding of such interactions can be accomplished by varying the ligand or introducing a mutation of the molecules used in the force-spectroscopic approach (Hinterdorfer and Dufrêne, 2006). Control measurements verifying the specificity can be achieved, e.g. by adding free ligand to the AFM fluid cell, which strongly reduces the frequency of specific interaction events (Sletmoen *et al.*, 2004). Also, molecular linkers are used to couple ligands to AFM tips and/or receptors to samples (Hinterdorfer and Dufrêne, 2006). These linkers reduce non-specific interaction and make it possible for the ligand to orient itself such that it can access the receptor-binding pocket. Measured interactions range from 20 to 400 pN depending on the type of interaction (e.g. Butt *et al.*, 2005; Samori, 2006). In force-volume imaging, spatially-resolved interaction maps are determined recording force–distance curves at multiple lateral positions. Molecular recognition imaging employs molecular recognition during imaging and is a fast technique with high-resolution (Raab *et al.*, 1999).

The application of atomic force microscopy for structural imaging, chemical imaging, e.g. as applied to measure hydrophobic forces, and molecular recognition imaging as part of nanoscale microbe analysis has been reviewed by

Dufrêne (2008). Various microbes can be imaged under physiological conditions using this microscopy with emphasis on the effects of drugs on cell surface ultrastructure. For example, major ultrastructural changes on mycobacterial cells upon treatment with four different drugs have been reported and these changes reflect inhibition of synthesis of major cell wall constituents, i.e. arabinans (Dufrêne, 2008).

Analyses with an atomic force microscope have a high impact on our understanding of the structure–function relationship of microbial surfaces and new applications can be developed for atomic force microscopy in medicine and biotechnology (Dufrêne, 2004). Images, obtained using AFM analysis, of bacteria mutants possessing different, or even lacking, extracellular PSs show clear differences in topology (Figure 14.4). The strength of adhesion caused by fibril material has been determined by force spectroscopy and the role of extracellular PSs in cellular cohesion and the social behaviour of *Myxococcus xanthus* has been demonstrated (Pelling *et al.*, 2005). With respect to fungi, Dague *et al.* (2008) followed the structural changes of *Aspergillus fumigatus* that occur during germination and recorded a change from nanofibrillar structures on the cells into an amorphous layer (Figure 14.5). Adhesion force measurements revealed a transition from hydrophobic to hydrophilic surface properties corresponding to the structural changes.

In vivo cell–cell interactions have been investigated at the single-molecule level, e.g. cells of *Dictyostelium discoideum* are engaged in development of a multicellular organism, in which the glycoprotein csA plays a central role in cell–cell adhesion (Benoit *et al.*, 2000). For investigation, one *D. discoideum* cell was picked up with a tipless AFM cantilever that was coated with a lectin binding the cell, contact between the cell on the cantilever and one of the cells in the Petri dish was established and then the cantilever retracted

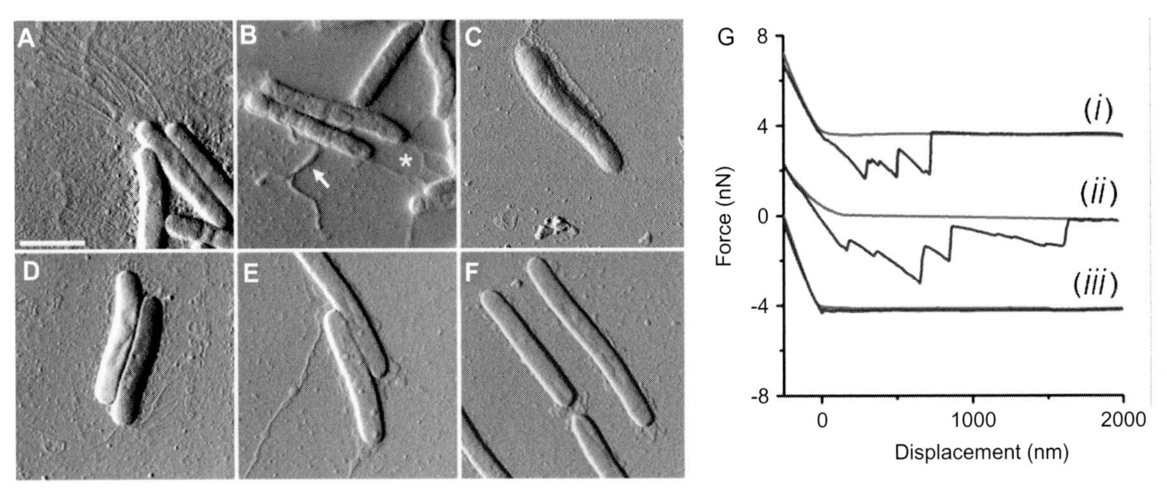

FIGURE 14.4 Atomic force microscopy deflection-mode images of large social groups and individual *M. xanthus* cells in air (A–F) and force–displacement curves (G). Individual cells of *M. xanthus* mutants (Scale bar, 2 μm): (A) wild-type showing polar pili; (B) wild-type displaying slime-like substances (*) and extruding blebs (arrow); (C) a pilA mutant showing the absence of pili at the cell pole; (D) a *dif* mutant showing the presence of long pili that bend toward the cell body; (E) a *stk* mutant displaying an excess of extracellular substances in the form of filaments with diameters from 15 to 65 nm; (F) an LPS O-antigen mutant. (G) Force curves measured on living wild-type (curve i), *stk* (curve ii) and glutaraldehyde-fixed wild-type *M. xanthus* (curve iii) cells (the curves are shifted 4 nN for clarity). Reproduced, with permission, from Pelling *et al.* (2005).

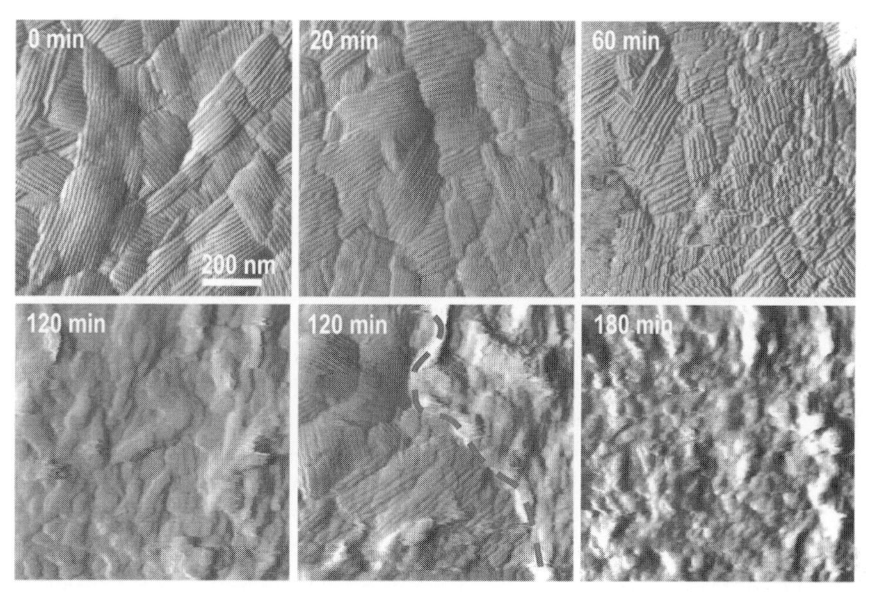

FIGURE 14.5 Real-time imaging of cell surface dynamics. A series of high-resolution AFM deflection images recorded on a single *A. fumigatus* cell during germination. Within less than 3 h, the crystalline rodlet layer changed to a layer of amorphous material, presumably reflecting inner cell wall PSs. Reproduced, with permission, from Dague *et al.* (2008).

while monitoring the adhesion forces. The forces that are required to break the adhesion between the two cells were 23 pN and multiples of this value (Benoit *et al.*, 2000). This indicates that a single molecular species is involved in most of the adhesive interaction between *D. discoideum* cells. Control measurements were performed on genetically modified cells, in which the csA gene was selectively inactivated and dissociation was highly reduced for these cells. The value of 23 pN is lower than for most antibody–antigen or lectin–sugar interactions; the force measured for the adhesive interaction between a single mannose molecule and the lectin concanavalin A was 47±9 pN (Ratto *et al.*, 2004). This can explain why motile cells of *D. discoideum* can still glide against each other as they become part of the multicellular structure.

Bacterial lipopolysaccharides (LPSs) (see Chapters 1 and 3) can cause septic shock, a life-threatening complication (Waage *et al.*, 1989). The LPS-binding protein (LBP) binds LPS transferring it to CD14 and LBP and CD14, together with TLR-4 and MD-2, are involved in the immune response related to septic shock (see Chapter 31). The His-tagged LBP that had been attached to AFM tips and exposed to LPS from *Salmonella enterica* sv. Minnesota, with varying saccharide lengths, was absorbed onto the LBP-functionalized tips (Kim *et al.*, 2007). The interaction between the LBP-LPS-modified AFM tips and model lipid bilayers containing CD14 was studied. Longer LPS saccharide regions resulted in higher interaction forces. Above a certain concentration, the polycationic, antimicrobial peptide polymyxin B inhibited the specific binding interaction between LBP-LPS and CD 14 molecules (Kim *et al.*, 2007).

The interaction between AFM tips coated with organic macromolecules and strains of the Gram-negative bacterium *Pseudomonas aeruginosa* have been measured (Abu-Lail *et al.*, 2007). Two strains, deviating in their ability to express PSs, were covalently bound to different glass slides. Separate AFM cantilevers with 1 μm silica spheres were coated with either poly(methacrylic acid) or one of two different humic acids. Different adhesion forces between the various organic macromolecular probes and bacterial strains were observed and suggested to be caused by electrostatic interactions, cell softness, LPS heterogeneity and hydrophobicity (Abu-Lail *et al.*, 2007).

Nanoscale investigations of the adhesive properties of the terrestrial algae *Prasiola linearis* using an AFM have revealed that extracellular polymeric substances contain both PSs and a small amount of protein (Mostaert *et al.*, 2006). Using atomic force microscopy pulling techniques, it was concluded that amyloid protein structures are important for the strength of this EPS.

The use of AFM in studies of interactions of bacterial biofilms with different surfaces has been overviewed by Beech *et al.* (2002). Both the bacterial cells and the exopolymers, extracellular polymeric substances (EPSs) (see Chapter 37) of the biofilms have been clearly resolved and applied in the study of adhesion and cellular properties of biofilms.

The EPSs produced by diatoms, which are an important component of marine biofilms, consist of PSs, proteins and glycoproteins (Arce *et al.*, 2004). Although it is not known which of these molecules are crucial in the attachment of diatoms to surfaces and to each other, atomic force microscopy studies have contributed to understanding the interaction forces between EPSs and surfaces and the distribution of different types of EPS on the diatom surface. Arce *et al.* (2004) attached a single diatom cell to the end of a tipless cantilever and measured the interaction with two different surfaces, intersleek and mica. They found that the interaction forces depend more strongly on the cell used than on the growth stage of the cell, were similar for the two surfaces examined, and concluded that macromolecular specificity of diatom EPSs play a pivotal role in adhesion. Secretion of mucilage from diatom pores was vizualized for the first time by Higgins *et al.* (2002) (Figure 14.6). In order to make imaging of the pores possible,

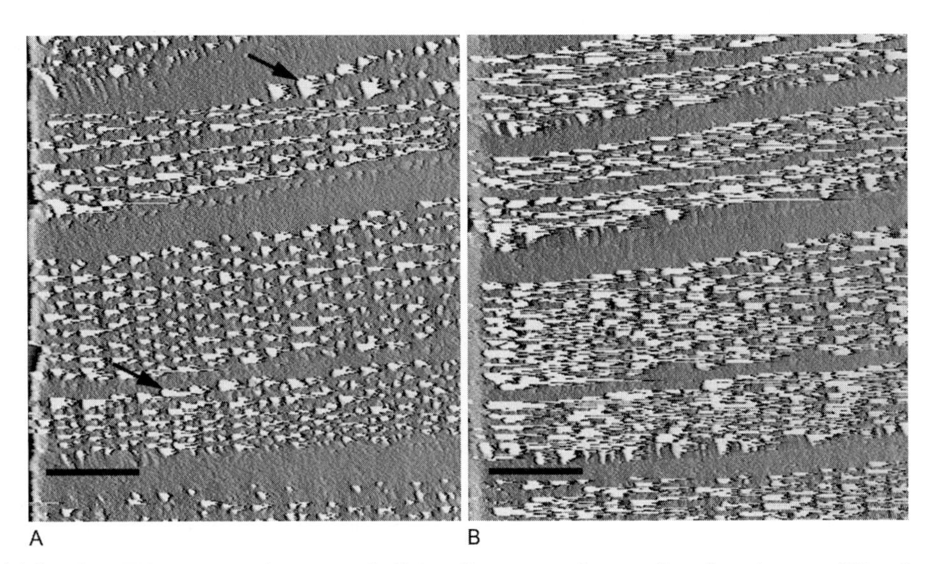

FIGURE 14.6 The AFM contact mode images of a living diatom secreting mucilage from its pores (A) and a subsequent scan showing an increase in mucilage secretion (B). Scale bars, 1 μm. Reproduced, with permission, from Higgins *et al.* (2002).

the EPS layer was gently removed by the scanning AFM tip. Two types of mucilage with distinct adhesive properties were distinguished, with measured adhesive forces of ≈4 nN (girdle region) and ≈2 nN (valve region). This study illustrates the possibility of using atomic force microscopy to study topography and related function of microbial structures.

Single adhesins can be mapped on living bacteria (Dupres *et al.*, 2005), as shown in Figure 14.7. Living *Mycobacterium bovis* BCG cells were immobilized on a membrane and an AFM tip functionalized with heparin was used to record force–distance curves in a grid pattern across the cell surface. The lateral resolution was 20 nm. The recorded forces were in the range of single-molecule interactions. Interestingly, the distribution of the adhesins was not homogeneous, but localized to nano-domains that may have very specific biological functions.

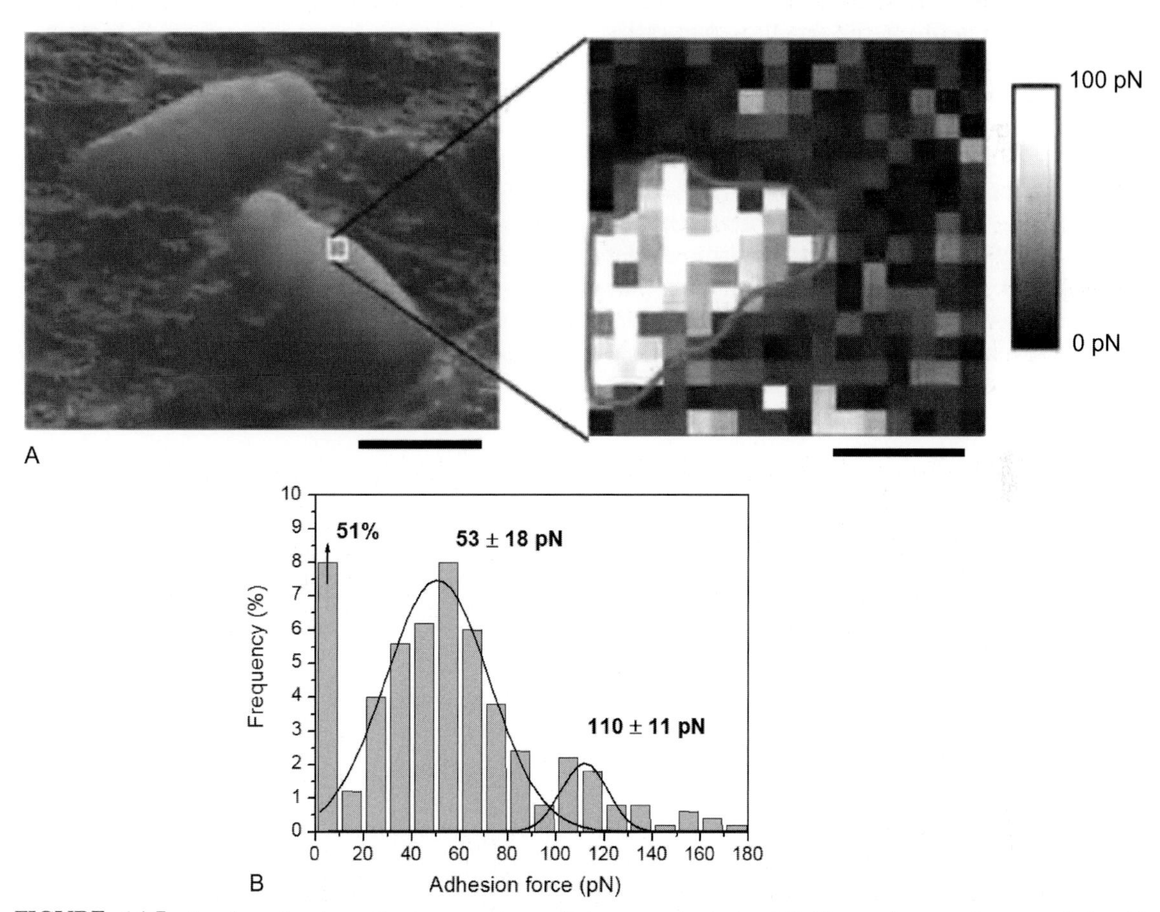

FIGURE 14.7 Single-molecular adhesion events on living mycobacteria. (A) Nanoscale mapping recorded with a heparin AFM tip. Notice that the adhesion is concentrated in a nanodomain. (B) Corresponding histogram from 1024 force curves of adhesion forces. The first peak (53 pN) reflects the detection of single monomers, whereas the second peak (110 pN) may correspond to single dimers or two monomers. Reproduced, with permission, from Hinterdorfer and Dufrêne (2006) and Dupres *et al.* (2005).

5. SINGLE-MOLECULE STUDIES OF MICROBIAL PSs USING OPTICAL TECHNIQUES

In fluorescence resonance energy transfer (FRET), the molecules of interest are labelled with a donor and an acceptor chromophore. Energy transfer from the donor to the acceptor results in a FRET signal. This signal is highly distance-dependent and the FRET technique can be applied as a "molecular ruler". Moreover, FRET can be implemented to provide either ensemble average properties or single-pair FRET (spFRET) (Weiss, 2000). The ensemble FRET approach is implemented in highly sensitive and specific assays and sensors. Single-pair FRET is realized by intramolecular labelling, thus the intermolecular dynamics of single molecules are accessible, e.g. enzyme structural changes during catalysis. Alternatively, intermolecular labelling makes it possible to follow enzyme–substrate association/dissociation at the single-molecule level. As an example, see Ditzler *et al.* (2007) for a review on single-molecule RNA enzymology.

Proteolytic enzymes can be investigated in continous assays by using internally quenched peptide substrates, i.e. fluorescently labelled substrates that mimic the target of the enzyme (Warfield *et al.*, 2006). In addition to this donor group, a second fluorophore attached to these substrates makes up the acceptor. When an enzyme cleaves the peptide chain that holds together donor and acceptor, the distance is increased and the FRET effect is very much reduced with an associated decrease in FRET and an increase in donor fluorescence. This is a sensitive assay for enzyme activity and a low concentration of substrate is sufficient. For exmple, Warfield *et al.* (2006) have developed internally quenched peptide substrates for FRET assays of lysostaphin, an antibiotic for *Staphylococcus aureus*; the substrates are based on the target for lysostaphin, the bacterial penta-glycine cross-bridge.

Sub-micromolar detection of LPS has been accomplished within a FRET-based sensor (Voss *et al.*, 2007). A peptide corresponding to the amino acids in the LPS-binding domain of CD14 was used which was labelled with a fluorescent donor and acceptor. The change in FRET efficiency upon interaction with LPS was used as the detection signal. The sensor was claimed to yield an increase in sensitivity of three orders of magnitude compared to previous LPS sensors and able to detect LPS in the presence of up to 50% fetal calf serum, which is important for application of the sensor to clinical samples. See also Lager *et al.* (2006) and Ha *et al.* (2007) for other applications.

Jones *et al.* (1999) developed a model that describes an amphiphatic molecule with its lipophilic region integrated into a detergent micelle and its hydrophilic region, containing the donor, extending outward. Experimental verification was achieved with LPS from different bacterial strains, which resulted in differences in measured donor to micelle separation caused by differing lengths of various *O*-specific PS chains. Thus, this system offers the possibility of studying the molecular conformation of reconstituted lipophilic molecules such as LPS.

Fluorescence correlation spectroscopy (FCS) is a highly sensitive optical technique to determine diffusion constants, sizes of molecules and complexes, molecular interactions and fast internal dynamics. In addition to studies of fluorescently-labelled molecules in solution and unlabelled or differently labelled interacting molecules, trafficking of biomolecules in living cells can also be determined (Schwille and Haustein, 2002). A confocal laser scanning microscope equipped with a highly sensitive photodetector is used and measurements are performed in the confocal volume, <1 femtolitre (smaller than an *Escherichia coli* cell). Concentrations of fluorescent molecules used are in the nM range, such that on average one (or very few) single-molecule is present in the detection volume. Fluctuations in fluorescence intensity are recorded and, from the auto-correlation or cross-correlation function, the local concentration of fluorescently-labelled molecules and mobility coefficient can be derived. Very short data acquisition times are involved in this technique (Schwille and Haustein, 2002).

Examples of the applications of FCS in microbial PS include reports on transport properties of schizophyllan and succinoglycan. The dynamic properties of schizophyllan in a dilute environment as determined by FCS were used as a basis to derive molecular parameters, e.g. as chain length and stiffness (Leng *et al.*, 2001). The diffusion coefficient of succinoglycan, as measured by FCS, was found to decrease with decreasing pH, at constant ionic strength (Duval *et al.*, 2006). The effective hydrodynamic radius increased from 10.3nm at high pH to 14.5nm at low pH. This increase is likely due to aggregation following a decrease in molecular charge. Masuda *et al.* (2005) developed a special FCS technique, where the size of the effective confocal volume was varied, that enables monitoring of the spatio-temporal dependence of the diffusion coefficient. This so-called sampling-volume-controlled (SVC)-FCS makes it possible directly to observe anomalous subdiffusion (ASD) in non-homogeneous media, in this case hyaluronan aqueous solutions. It was shown that moderate ASD may occur even in the presence of a small amount of hyaluronan in extracellular matrices.

The diffusion of bacteriophages in the *Stenotrophonas maltophilia* biofilm and the influence of EPS on this diffusion and reactivity were studied using two-photon excitation- (TPE-) FCS (Lacroix-Gueu *et al.*, 2005). With TPE, photobleaching was reduced and, by using infra-red wavelengths, deeper specimen penetration was achieved. Moreover, the bacteriophages were found to penetrate the EPS matrix of the biofilm and, in this way, use the protective effect of the biofilm toward antimicrobials.

6. CONCLUSIONS

Single-molecule techniques applied to microbial PSs have proven capable of providing novel information necessary for the understanding of structure–function relationships of this class of molecule. In particular, the possibility of determining these properties under mechanical tension provides important insights into molecular behaviour that is not readily accessible by combined classical approaches.

Nevertheless, single-molecule studies of microbial PSs is probably awaiting a richer future than seen in past years due to the potential migration of single-molecule techniques, which have already been established in other fields, to applications with microbial PSs (Research Focus Box). This is the case for techniques such as optical tweezers or optical traps (Ashkin, 1992, 1997) that provide a unique means of manipulating and controlling biological objects and for magnetic tweezers that are capable of inducing mechanical torsion on single molecules. When using optical tweezers, force is sensed by the displacement of a microsphere trapped in a narrowly focused beam of laser light. While the AFM allows measurement of forces in the range of 5–10^3pN, optical tweezers allow reliable determination of forces ranging from 0.1 to 100pN (Neuman *et al.*, 2007). Interesting studies have been conducted in which adhesive forces between individual bacterial surfaces and protein-coated surfaces have been reported (Simpson *et al.*, 2003; Castelain *et al.*, 2007) but, to our knowledge, optical or magnetic tweezers have not yet been used to investigate single PS molecules.

The present lack of examples of published optical tweezers-based single-molecule studies investigating PSs may reflect practical challenges faced when developing this field. The use of optical tweezers for the study of PSs requires further development of suitable attachment methods for the PSs onto the colloidal probes,

RESEARCH FOCUS BOX

- Sample preparation standarization is required from extraction of microbial samples to preparation for single-molecule studies, including, e.g immobilization strategies.
- Microbial PS standards for single-molecule studies are needed.
- Extension of the accessible force range and measurement capabilities with the application of additional single-molecule techniques to microbial polysaccharides.
- Automation and high-throughput data analysis would be advantageous.
- Developments towards single-molecule studies with additional molecules present, i.e. steps that would aid *in vivo* applications.

including methods that allow specific attachment of the end of the molecule. The use of molecular linkers, as developed for AFM single-molecule stretching and interaction studies, may find use also in optical tweezer studies. Despite the practical challenges faced, the potential of this technique in revealing insights into the involvement of PSs in intra- and intermolecular interactions should inspire the future use of this technique in PS research.

ACKNOWLEDGEMENTS

D.C.G.K. acknowledges support from the Norwegian Research Council, Grant number 170521/V40.

References

Abu-Lail, L.I., Liu, Y., Atabek, A., Camesano, T.A., 2007. Quantifying the adhesion and interaction forces between *Pseudomonas aeruginosa* and natural organic matter. Environ. Sci. Tech. 41, 8031–8037.

Arce, F.T., Avci, R., Beech, I.B., Cooksey, K.E., Wigglesworth-Cooksey, B., 2004. A live bioprobe for studying diatom-surface interactions. Biophys. J. 87, 4284–4297.

Ashkin, A., 1992. Forces of a single-beam gradient laser trap on a dielectric sphere in the ray optics regime. Biophys. J. 61, 569–582.

Ashkin, A., 1997. Optical trapping and manipulation of neutral particles using lasers. Proc. Natl. Acad. Sci. USA 94, 4853–4860.

Beech, I.B., Smith, J.R., Steele, A.A., Penegar, I., Campbell, S.A., 2002. The use of atomic force microscopy for studying interactions of bacterial biofilms with surfaces. Colloids Surf. B, Biointerf. 23, 231–247.

Bell, G.I., 1978. Models for the specific adhesion of cells to cells. A theoretical framework for adhesion mediated by reversible bonds between cell surface molecules. Science 200, 618–627.

Benoit, M., Gabriel, D., Gerisch, G., Gaub, H.E., 2000. Discrete interactions in cell adhesion measured by single-molecule force spectroscopy. Nat. Cell Biol. 2, 313–317.

Binnig, G., Quate, C.F., Gerber, C., 1986. Atomic force microscope. Phys. Rev. Lett. 56, 930–933.

Bippes, C.A., Humphris, A.D.L., Stark, M., Müller, D.J., Janovjak, H., 2006. Direct measurement of single-molecule visco-elasticity in atomic force microscope force-extension experiments. Eur. Biophys. J. 35, 287–292.

Bustamante, C., Bryant, Z., Smith, S.B., 2003. Ten years of tension: single-molecule DNA mechanics. Nature 421, 423–427.

Bustamante, C., Chemla, Y.R., Forde, N.R., Izhaky, D., 2004. Mechanical processes in biochemistry. Annu. Rev. Biochem. 73, 705–748.

Butt, H.-J., Cappella, B., Kappl, M., 2005. Force measurements with the atomic force microscope: Technique, interpretation and applications. Surf. Sci. Rep. 59, 1–152.

Castelain, M., Pignon, F., Piau, J.M., Magnin, A., Mercier-Bonin, M., Schmitz, P., 2007. Removal forces and adhesion properties of *Saccharomyces cerevisiae* on glass substrates probed by optical tweezer. J. Chem. Phys. 127, 135104–135114.

Dague, E., Alsteens, D., Latgé, J-P., Dufrêne, Y.F., 2008. High-resolution cell surface dynamics of germinating *Aspergillus fumigatus* conidia. Biophys. J. 94, 656–660.

Ditzler, M.A., Aleman, E.A., Rueda, D., Walter, N.G., 2007. Single molecule RNA enzymology. Biopolymers 87, 302–316.

Dufrêne, Y.F., 2004. Using nanotechniques to explore microbial surfaces. Nat. Rev. Microbiol. 2, 451–460.

Dufrêne, Y.F., 2008. AFM for nanoscale microbe analysis. Analyst 133, 297–301.

Dupres, V., Menozzi, F.D., Locht, C., et al., 2005. Nanoscale mapping and functional analysis of individual adhesins on living bacteria. Nat. Methods 2, 515–520.

Duval, J.F.L., Slaveykova, V.I., Hosse, M., Buffle, J., Wilkinson, K.J., 2006. Electrohydrodynamic properties of succinoglycan as probed by fluorescence correlation spectroscopy, potentiometric titration and capillary electrophoresis. Biomacromolecules 7, 2818–2826.

Evans, E., Ritchie, K., 1997. Dynamic strength of molecular adhesion bonds. Biophys. J. 72, 1541–1555.

Gaboriaud, F., Dufrêne, Y.F., 2007. Atomic force microscopy of microbial cells: application to nanomechanical properties, surface forces and molecular recognition forces. Colloids Surf. B Biointerfaces 54, 10–19.

Gowers, D.M., Wilson, G.G., Halford, S.E., 2005. Measurement of the contributions of 1D and 3D pathways to the translocation of a protein along DNA. Proc. Natl. Acad. Sci. USA 102, 15883–15888.

Ha, J-S., Song, J.J., Lee, Y-M., et al., 2007. Design and application of highly responsive fluorescence resonance energy transfer biosensors for detection of sugar in living *Saccharomyces cerevisiae* cells. Appl. Environ. Microbiol. 73, 7408–7414.

Higgins, M.J., Crawford, S.A., Mulvaney, P., Wetherbee, R., 2002. Characterization of the adesive mucilages secreted by live diatom cells using atomic force microscopy. Protist 153, 25–38.

Hinterdorfer, P., Dufrêne, Y.F., 2006. Detection and localization of single molecular recognition events using atomic force microscopy. Nat. Methods 3, 347–355.

Humphris, A.D.L., Tamayo, J., Miles, M.J., 2000. Active quality factor control in liquids for force spectroscopy. Langmuir 16, 7891–7894.

Humphris, A.D.L., Antognozzi, M., McMaster, T.J., Miles, M.J., 2002. Transverse dynamic force spectroscopy: a novel approach to determining the complex stiffness of a single molecule. Langmuir 18, 1729–1733.

Jones, G.M., Wofsy, C., Aurell, C., Sklar, L.A., 1999. Analysis of vertical fluorescence resonance energy transfer from the surface of a small-diameter sphere. Biophys. J. 76, 517–527.

Kawakami, M., Byrne, K., Khatri, B.S., McLeish, T.C.B., Radford, S.E., Smith, D.A., 2004. Viscoelastic properties of single polysaccharide molecules determined by analysis of thermally driven oscillations of an atomic force microscope cantilever. Langmuir 20, 9299–9303.

Khatri, B.S., Kawakami, M., Byrne, K., Smith, D.A., McLeish, T.C.B., 2007. Entropy and barrier-controlled fluctuations determine conformational viscoelasticity of single biomolecules. Biophys. J. 92, 1825–1835.

Kim, J.S., Jang, S., Kim, U., Cho, K., 2007. AFM studies of inhibition effect in binding of antimicrobial peptide and immune proteins. Langmuir 23, 10438–10440.

Lacroix-Gueu, P., Briandet, R., Lévêque-Fort, S., Bellon-Fontaine, M-N., Fontaine-Aupart, M-P., 2005. In situ measurements of viral particles diffusion inside mucoid biofilms. C. R. Biol. 328, 1065–1072.

Lager, I., Looger, L.L., Hilpert, M., Lalonde, S., Frommer, W.B., 2006. Conversion of a putative *Agrobacterium* sugar-binding protein into a FRET sensor with high selectivity for sucrose. J. Biol. Chem. 13, 30875–30883.

Lee, G., Nowak, W., Jaroniec, J., Zhang, Q., Marszalek, P.E., 2004. Molecular dynamics simulations of forced conformational transitions in 1,6-linked polysaccharides. Biophys. J. 87, 1456–1465.

Leng, X., Starchev, K., Buffle, J., 2001. Applications of fluorescence correlation spectroscopy: measurement of size-mass relationship of native and denatured schizophyllan. Biopolymers 59, 290–299.

Li, H., Rief, M., Oesterhelt, F., Gaub, H.E., 1998. Single-molecule force spectroscopy on xanthan by AFM. Adv. Mater. 3, 316–319.

Li, H., Rief, M., Oesterhelt, F., Gaub, H.E., Zhang, X., Shen, J., 1999. Single-molecule force spectroscopy on polysaccharides by AFM-nanomechanical fingerprint of α̃-(1,4)-linked-polysaccharides. Chem. Phys. Lett. 305, 197–201.

Liphardt, J., Onoa, B., Smith, S.B., Tinoco, I., Bustamante, C., 2001. Reversible unfolding of single RNA molecules by mechanical force. Science 292, 733–737.

Marszalek, P.E., Oberhauser, A.F., Pang, Y-P., Fernandez, J.M., 1998. Polysaccharide elasticity governed by chair-boat transitions of the glucopyranose ring. Nature 396, 661–664.

Marszalek, P.E., Pang, Y-P., Li, H., Yazal, J.E., Oberhauser, A.F., Fernandez, J.M., 1999. Atomic levers control pyranose ring conformations. Proc. Natl. Acad. Sci. USA 96, 7894–7898.

Marszalek, P.E., Li, H., Fernandez, J.M., 2001. Fingerprinting polysaccharides with single-molecule atomic force microscopy. Nat. Biotechnol. 19, 258–262.

Marszalek, P.E., Li, H., Oberhauser, A.F., Fernandez, J.M., 2002. Chair-boat transitions in single polysaccharide molecules observed with force-ramp AFM. Proc. Natl. Acad. Sci. USA 99, 4278–4283.

Masuda, A., Ushida, K., Okamoto, T., 2005. New fluorescence correlation spectroscopy enabling direct observation of spatiotemporal dependence of diffusion constants as an evidence of anomalous transport in extracellular matrices. Biophys. J. 88, 3584–3591.

Mostaert, A., Higgins, M.J., Fukuma, T., Rindi, F., Jarvis, S.P., 2006. Nanoscale mechanical characterisation of amyloid fibrils discovered in a natural adhesive. J. Biol. Physics. 32, 393–401.

Müller, D.J., Engel, A., 1999. pH and voltage induced structural changes of porin OmpF explains channel closure. J. Mol. Biol. 285, 1347–1351.

Neelov, I.M., Adolf, D.B., McLeish, T.C.B, Paci, E., 2006. Molecular dynamics simulation of dextran extension by constant force in single molecule AFM. Biophys. J. 91, 3579–3588.

Neuman, K.C., Lionnet, T., Allemand, J-F., 2007. Single-molecule micromanipulation techniques. Annu. Rev. Mat. Sci. 37, 33–67.

Oberhauser, A.F., Hansma, P.K., Carrion-Vazquez, M., Fernandez, J.M., 2001. Stepwise unfolding of titin under force-clamp atomic force microscopy. Proc. Natl. Acad. Sci. USA 98, 468–472.

Pelling, A.E., Li, Y., Shi, W., Gimzewski, J.K., 2005. Nanoscale visualization and characterization of *Myxococcus xanthus* cells with atomic force microscopy. Proc. Natl. Acad. Sci. USA 102, 6484–6489.

Raab, A., Han, W., Badt, D., et al., 1999. Antibody recognition imaging by force microscopy. Nat. Biotechn. 17, 902–905.

Ratto, T.V., Langry, K.C., Rudd, R.E., Balhorn, R.L., Allen, M.J., McElfresh, M.W., 2004. Force spectroscopy of the double-tethered concanavalin-A mannose bond. Biophys. J. 86, 2430–2437.

Rief, M., Oesterhelt, F., Heymann, B., Gaub, H.E., 1997. Single molecule force spectroscopy on polysaccharides by atomic force microscopy. Science 275, 1295–1297.

Rief, M., Fernandez, J.M., Gaub, H.E., 1998. Elastically coupled two-level systems as a model for biopolymer extensibility. Phys. Rev. Lett. 81, 4764–4767.

Samori, P. (Ed.), 2006. Scanning Probe Microscopies Beyond Imaging. Manipulation of Molecules and Nanostructures. Wiley-VCH Verlag, Weinheim.

Schenter, G.K., Lu, H.P., Xie, X.S., 1999. Statistical analyses and theoretical models of single-molecule enzymatic dynamics. J. Phys. Chem. A. 103, 10477–10488.

Schwille, P., Haustein, E., 2002. Fluorescence correlation spectroscopy: a tutorial for the biophysics textbook online. <http://www.biophysics.org/education/resources.htm>.

Simpson, K.H., Bowden, G., Höök, M., Anvari, B., 2003. Measurement of adhesive forces between individual *Staphylococcus aureus* MSCRAMMs and protein-coated surfaces by use of optical tweezers. J. Bacteriol. 185, 2031–2035.

Sletmoen, M., Skjåk-Bræk, G., Stokke, B.T., 2004. Single-molecular pair unbinding studies of mannuronan C-5 epimerase AlgE4 and its polymer substrate. Biomacromolecules 5, 1288–1295.

Stanford, N.P., Szczelkun, M.D., Marko, J.F., Halford, S.E., 2000. One- and three-dimensional pathways for proteins to reach specific DNA sites. EMBO J. 19, 6546–6557.

Svoboda, K., Schmidt, C.F., Schnapp, B.J., Block, S.M., 1993. Direct observation of kinesin stepping by optical trapping interferometry. Nature 365, 721–727.

van Oijen, A.M., 2007. Single-molecule studies of complex systems: the replisome. Mol. Biosyst. 3, 117–125.

Voss, S., Fisher, R., Jung, G., Wiesmüller, K-H., Brock, R., 2007. A fluorescence-based synthetic LPS sensor. J. Am. Chem. Soc. 129, 554–561.

Waage, A., Brandtzaeg, P., Halstensen, A., Kierulf, P., Espevik, T., 1989. The complex pattern of cytokines in serum from patients with meningococcal septic shock. Association between interleukin 6, interleukin 1, and fatal outcome. J. Exp. Med. 169, 333–338.

Warfield, R., Bardelang, P., Saunders, H., et al., 2006. Internally quenched peptides for the study of lysostaphin: An antimicrobial protease that kills *Staphylococcus aureus*. Org. Biomol. Chem. 4, 3626–3638.

Weiss, S., 2000. Measuring conformational dynamics of biomolecules by single molecule fluorescence spectroscopy. Nat. Struct. Biol. 7, 724–729.

Yasuda, R., Noji, H., Yoshida, M., Kinosita, K., Itoh, H., 2001. Resolution of distinct rotational substeps by submillisecond kinetic analysis of F-1-ATPase. Nature 410, 898–904.

CHAPTER

15

Viral surface glycoproteins in carbohydrate recognition: structure and modelling

Jeffrey C. Dyason and Mark von Itzstein

SUMMARY

The importance of carbohydrate-recognizing viral surface glycoproteins in a range of clinically important viral infections has been identified and a number of these proteins have been investigated as possible drug discovery targets. As a part of these investigations several of these proteins have had their three-dimensional structures determined by either nuclear magnetic resonance spectroscopic or X-ray crystallographic methods. This structural information has provided an excellent basis for structure-assisted inhibitor design using computational chemistry methods. This chapter describes some of the most significant developments in the field of structure-based investigations of viral surface-resident carbohydrate-recognizing proteins. Specifically, an overview of these carbohydrate-recognizing proteins from four important human viruses, including influenza, dengue, rotavirus and parainfluenza, and associated structural investigations will be presented.

Keywords: Virus; Carbohydrate; Molecular modelling; Influenza; Dengue virus; Rotavirus; Parainfluenza

1. INTRODUCTION

For viruses to be able to replicate, they need to be able to adhere, then bind and invade and, finally, to escape from a susceptible host cell. Animal host cell surfaces are highly decorated by a range of glycan structures that are made up of both charged and neutral carbohydrates (Gagneux and Varki, 1999). These carbohydrates are involved in a range of different functions, including cell protection, a host of recognition events and cell–cell communication (Gagneux and Varki, 1999). As a consequence of this elaborate host cell surface carbohydrate display, viruses have had to develop a variety of strategies that can deal with this protective barrier to enable successful viral propagation. In fact, many viruses actively use these carbohydrates in the recognition and invasion process of their life cycle. Understanding the structure of these viral surface proteins is vital not only to help understand their function but also to enable the design of new drugs capable of mitigating the effects caused during infection.

2. INFLUENZA VIRUS

Influenza virus has been an affliction of mankind for centuries, with several major pandemics being reported, including the Spanish Flu which, at the end of World War I (1918–1919), killed an estimated 40 million people (Taubenberger *et al.*, 2001). More recently, the emergence of the highly pathogenic avian influenza virus strain, H5N1, although not easily transmissible between humans, has a reported mortality rate in man of over 50% (WHO, 2008).

Influenza virus contains a single-stranded segmented RNA genome and is a member of the orthomyxoviridae family that is further subdivided into three distinct types, influenza virus A, B and C (van Regenmortel *et al.*, 2000). Of most concern to the human population are types A and B, with type A the most likely to cause pandemics (Luscher-Mattli, 2000). Both types A and B have two surface-based carbohydrate recognizing proteins involved in the infectious cycle of the virus, namely haemagglutinin (HA), which has a lectin and fusion function and sialidase (also known as neuraminidase, NA) which has a glycohydrolase function. Type C has only one major carbohydrate recognizing protein and that is haemagglutinin-esterase-fusion protein (HEF), which combines the roles of the HA and NA of influenza A and B (Herrler *et al.*, 1988).

Currently, the major way of combating influenza virus A and B infection is via an annual vaccine injection, however, this is not foolproof and generally provides no protection against new strains of influenza virus, like that which may arise from the avian H5N1 strains. Also available are several drugs, the most successful of which are the NA inhibitors (von Itzstein, 2007), Relenza® and Tamiflu® (see below). However, mutant influenza strains resistant to Tamiflu® have been recently detected that potentially limit the available armament and clearly necessitates the ongoing search for new treatments (von Itzstein, 2007). While influenza virus C is not common, it can cause respiratory tract infections generally in young children, however, in some cases, this may lead to more severe infections such as bronchiolitis and pneumonia (Wagaman *et al.*, 1989; Crescenzo-Chaigne and van der Werf, 2007). There is currently no vaccine or specific drug treatment for influenza C.

2.1. Haemagglutinin as a drug discovery target

Influenza virus HA is a homo-trimeric glycosylated protein located on the surface of the virion (Figure 15.1) and is responsible for the initial

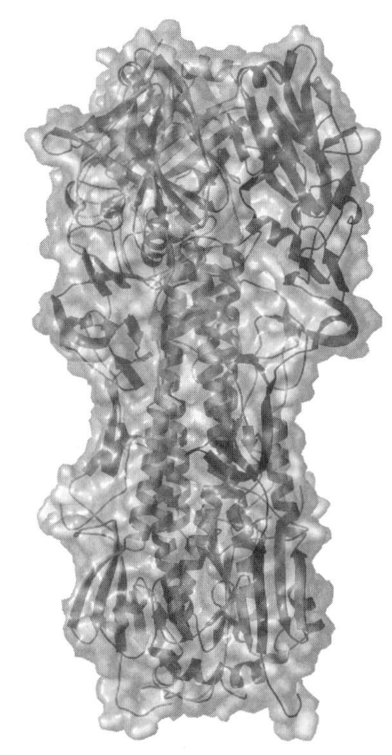

FIGURE 15.1 The secondary structure of an influenza virus haemagglutinin trimer is shown with a translucent surface. The surface is coloured by electrostatic potential with red denoting areas of negative charge while blue denotes areas of positive charge. (See colour plates section.)

attachment of the virion to the host cell surface, by recognition of terminal sialic acid moieties. A change in conformation of the protein is then responsible for the initiation of fusion with the cell membrane. The first structure of influenza virus A HA was published in 1981 (Wilson *et al.*, 1981). Since that time, over 70 structures of various influenza virus HAs have been deposited in the Research Collabatory for Structural Bioinformatics (RCSB) Protein Data Bank (PDB) (Berman *et al.*, 2000). During the 1980s and early 1990s, there were a number of attempts at using the available crystal structures of influenza virus HA to design small molecule HA binders that would hopefully inhibit the replication of influenza virus. A number of excellent reviews have appeared that detail these attempts (Luscher-Mattli, 2000; Matrosovich and Klenk, 2003).

However, it was realized that the low affinity of the sialic acid binding site would not produce the tight binding inhibitor required for a drug, leading to the design and synthesis of multivalent and polymeric inhibitors. These inhibitors have, however, not produced a suitable drug candidate, although work is still being undertaken (Ogata *et al.*, 2007; Marra *et al.*, 2008). A recent study by Nandi (2008) described an *in silico* search of a small subset in the ZINC database (Irwin and Shoichet, 2005) and led to the suggestion of some possible small molecule inhibitors of HA binding, although biological evaluation of these compounds is not yet reported.

Avian influenza virus HA preferentially recognizes terminal α-(2→3)-linked sialic acid containing glycoconjugates, whereas human-adapted influenza virus HA preferentially recognizes terminal α-(2→6)-linked sialic acid glycoconjugates. More recently, the crystal structures of HAs from the virus responsible for the 1918–1919 pandemic (Gamblin *et al.*, 2004) and from the highly pathogenic avian influenza H5N1 strain (Stevens *et al.*, 2006) have been elucidated. Knowledge of these structures has led to a greater understanding of the sialic acid specificity of these strains and it appears as if only a few amino acid mutations are needed to enable the avian influenza virus HA to switch its specificity and recognize human receptors (Stevens *et al.*, 2006; Yamada *et al.*, 2006). Recent reports that influenza virus may recognize receptors other than the accepted α-(2→6)-linked sialic acids observed in the human lower respiratory tract (Rapoport *et al.*, 2006; Nicholls *et al.*, 2007), has inspired studies that may more completely characterize novel asialoglycan recognition by HA. For example, the von Itzstein group described a nuclear magnetic resonance (NMR) spectroscopy-based study that enables the interrogation of H5 (from H5N1), as a surface protein of virus-like particles, with novel inhibitors and potential novel glycan structures (Haselhorst *et al.*, 2008).

2.2. Sialidase as a drug discovery target

The NA of influenza virus A and B has been the major focus in the development of drugs to combat influenza infection with two drugs already available on the market, namely the structure-based designed zanamivir (Relenza®, GlaxoSmithKline) and oseltamivir (Tamiflu®, Roche). The development of these drugs has been adequately reviewed elsewhere (Laver, 2006; von Itzstein, 2007, 2008). Influenza virus NA is a receptor-destroying enzyme which cleaves the HA receptor, sialic acid, from a range of sialyloligosaccharides allowing newly formed viral progeny to escape the surface of infected cells and go on to infect more cells.

Influenza virus NA is a glycosylated homotetramer which is tethered to the virus surface by a long protein stalk at its C terminal (Figure 15.2). The first structure of influenza virus NA was published in 1983 (Colman *et al.*, 1983; Varghese *et al.*, 1983) and there exists nearly 100 structures in the current PDB, including a number of mutant proteins as well as a number of inhibitor-enzyme complexes.

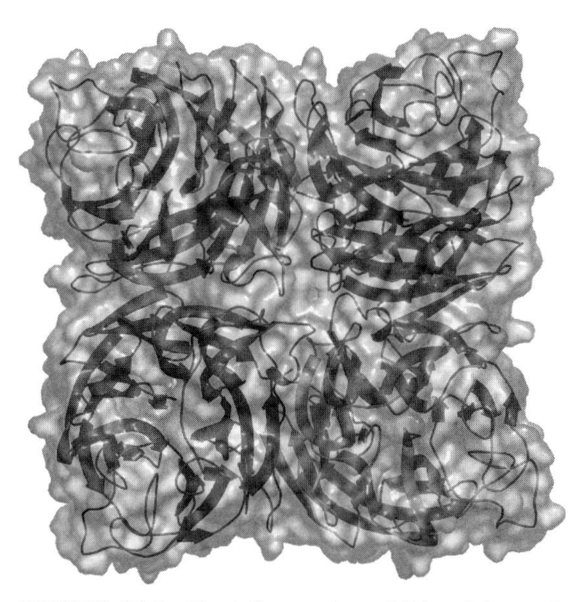

FIGURE 15.2 The influenza virus sialidase tetramer is shown with a translucent surface, coloured by electrostatic potential overlaying the monomer secondary structure. Each monomer has six four-stranded anti-parallel β-sheets arranged as if on the blades of a propeller. (See colour plates section.)

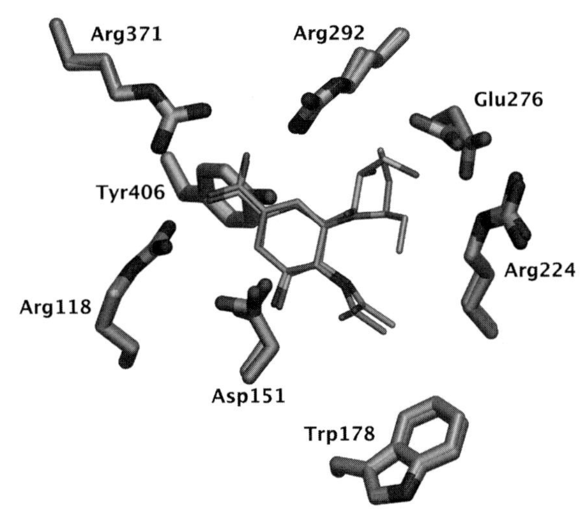

FIGURE 15.3 Superimposition of two complexed N9 sialidase structures of influenza virus showing the induced fit of Glu276 when oseltamivir carboxylate (orange carbon atoms) is bound, compared with Neu5Ac2en (green carbon atoms). (See colour plates section.)

The inhibitor-enzyme complexes in particular allow further study to understand the interactions involved and the design of new inhibitors to combat the existence of mutated influenza viruses that are less susceptible to the existing drugs, particularly Tamiflu®. This wealth of structural information has facilitated the use of structure-based drug design techniques, such as docking, molecular mechanics/Poisson-Boltzmann surface area (MM/PBSA) simulations, comparative molecular field analysis (CoMFA), etc., in providing not only an explanation of how inhibitors bind to the influenza virus NAs but also in the design of new inhibitors. A selection of publications gives an overview of the use of these techniques (Stoll *et al.*, 2003; Masukawa *et al.*, 2003; Armstrong *et al.*, 2006; Platis *et al.*, 2006; Zhang *et al.*, 2006; Abu Hammad *et al.*, 2007).

One aspect that is not handled consistently in many of these studies is the ability of the protein to undergo induced-fit when the glycerol side chain of the sialic acid-based inhibitor zanamivir is replaced with a hydrophobic group, e.g. the active form of oseltamivir, oseltamivir carboxylate (Figure 15.3). In this example, to accommodate the hydrophobic group of oseltamivir carboxylate, glutamic acid residue 276 (Glu276) flips to form an internal interaction with arginine 224 (Arg224) that results in a hydrophobic pocket where, previously, there was a negatively-charged hydrogen bonding group. Such induced-fit phenomena are not easily predicted or indeed adequately handled by theoretical models. Relying on this induced fit, as oseltamivir carboxylate does, also makes it a "hot-spot" for the virus to generate mutants that render the drug less able to inhibit the receptor-destroying enzyme function of the NA. This outcome has already been observed in a histidine-tyrosine (His274Tyr) N1 mutant where oseltamivir carboxylate is 265-fold less potent

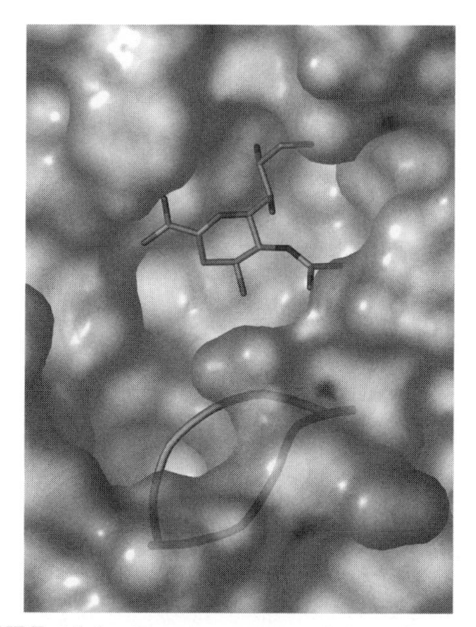

FIGURE 15.4 Crystal structures of the influenza virus N8 sialidase. The surface and the magenta coloured 150 loop are from a structure of the uncomplexed protein while the Neu5Ac2en and green-coloured 150 loop are from a structure where Neu5Ac2en was soaked into the protein crystals. The surface is coloured by electrostatic potential. (See colour plates section.)

compared to zanamivir which is only a factor of 2 less potent (Collins *et al.*, 2008).

One of the most interesting advances to come from recent NA structure determination and analysis is the existence of two distinct groups of influenza virus NAs, group 1 and 2 (Russell *et al.*, 2006). The main difference between the groups appears to be a very flexible active site loop that includes one of the main catalytic residues, aspartic acid 151 (Asp151). The group 1 NAs consisting of N1, N4, N5 and N8 subtypes have the very flexible 150-loop which opens up a large new pocket in the active site in their uncomplexed form, whereas the group 2 NAs consisting of N2, N3, N6, N7 and N9 show no movement of this 150-loop (Figure 15.4). While the current commercially available NA-targeting drugs have all been designed against the group 2 NAs, they generally appear to be equally effective against the group 1 NAs (Govorkova *et al.*, 2001; Mishin *et al.*, 2005), although some strain variability has been observed. A complication with the current avian H5N1 strain (an example of a group 1 NA) induced disease is that in humans it manifests differently to the normally observed human influenza disease resulting in a higher mortality. In fact, the H5N1 strain appears to have more in common with the Spanish flu strain of 1918–1919 (H1N1) where healthier individuals appear to be more susceptible than the usual influenza target population of the old, young and immunocompromised. It has been proposed that this is a direct consequence of these pandemic or pandemic-like viruses' ability to cause an immune event referred to as a cytokine storm (Peiris *et al.*, 2007).

2.3. Haemagglutinin esterase fusion protein

The HEF of influenza virus C is structurally very similar to HA found on the surface of influenza virus A and B, in that it is a homotrimeric glycosylated surface protein. The major differences are in:

(i) the receptor that is recognized, sialic acid-containing glycoconjugates are recognized by HA while 9-*O*-acetylated sialic acid-glycoconjugates are recognized by HEF; and
(ii) the fact that HEF contains both a lectin function as well as a receptor-destroying enzyme, in contrast to influenza virus A and B that have two distinct proteins for these functions.

The receptor-destroying enzyme of HEF is an esterase and cleaves an *O*-acetate group from the 9′-position of sialic acids. The crystal structure of HEF was published in 1998 (Rosenthal *et al.*, 1998) with the receptor-destroying enzyme active site being very similar in nature to a serine esterase and being completely separate to the receptor-binding site. Modelling experiments using a forced glycosidic angle torsion search have helped explain NMR results obtained recently within the von Itzstein group (Mayr *et al.*, 2008) for both HEF and bovine coronavirus esterase. This led to the understanding of the essential pharmacophoric groups required for binding to the receptor-destroying enzyme active site and the knowledge that the aglycon group does not appear to be involved in substrate binding to this site. Similar haemagglutinin esterases (≈30% homology) have also been identified in corona and toroviruses (de Groot, 2006) and, although these viruses are not closely related to influenza virus C, an understanding of the structure and function of the haemagglutinin esterase will provide more information on the mechanism of cell recognition and entry.

2.4. Future directions

Influenza virus is very adaptable, with resistant mutants already appearing for at least one of the commercially available NA inhibitors. The ongoing structural and computational studies on the surface-resident carbohydrate recognizing proteins will help not only to identify new and potent drugs but also to understand more fully the mechanism of action of these proteins. This is especially of interest to the group 1 NAs including the current avian influenza (H5N1), where a flexible loop in the active site provides anti-influenza drug designers with another target to improve the potency of existing drugs and the chance to develop new group 1-specific drugs.

3. PARAINFLUENZA

The paramyxoviridae family includes as members the human parainfluenza viruses, Newcastle disease virus and the measles and mumps viruses (Lamb *et al.*, 1996). Of particular interest are human parainfluenza viruses and Newcastle disease virus, both of which have a surface-resident carbohydrate recognizing protein known as haemagglutinin-neuramindase (HN) which is involved in both cellular recognition and as a receptor-destroying enzyme (Lamb *et al.*, 1996). There are four human parainfluenza viruses (hPIV) strains identified, hPIV-1, hPIV-2, hPIV3 and hPIV-4, all of which cause respiratory tract symptoms, e.g. coughing and wheezing, predominantly in infants. The human parainfluenza viruses are considered the second most important cause of lower respiratory tract disease in young children, behind respiratory syncytial virus (Henrickson, 2003). Strains hPIV-1 and hPIV-2 are commonly associated with the respiratory tract disease croup, while hPIV-3 can cause bronchiolitis and pneumonia and hPIV-4 is rarely detected (Henrickson, 2003). Newcastle disease virus is also known as avian paramyxovirus 1 and has a devastating effect on the poultry industry worldwide (Seal *et al.*, 2000).

3.1. Haemagglutinin-neuraminidase as a drug discovery target

Similar to the HEF of influenza virus C, HN includes both a recognition element for terminal sialic acids as well as a receptor-destroying (sialidase/neuraminidase) function to facilitate virion budding. The first crystal structures of a HN from Newcastle disease virus were published in 2000 (Crennell *et al.*, 2000), showing it to be very similar to the influenza virus NA in its overall shape and active site configuration, however, the active site appeared to be more flexible, probably allowing it to function as both a sialic acid-recognizing lectin and a NA. A more recent Newcastle disease virus HN structure (Zaitsev *et al.*, 2004) has revealed the presence of a second sialic acid binding site which is proposed to be involved in the fusion process but not in the initial recognition of sialic acids. At that time, these Newcastle disease virus HN structures were the only available structures for structure-based drug design of inhibitors against human parainfluenza viruses and led to two reports of designed inhibitors for this enzyme system

(Alymova *et al.*, 2004; Tindal *et al.*, 2007). These studies made use of the 2-deoxy-2,3-didehydro-*N*-acetylneuraminic acid (Neu5Ac2en) template which was also the precursor of the influenza virus NA inhibitor Relenza® and both also targeted a large cavity around the O-4 binding region of the Neu5Ac2en template. The best of these inhibitors led to micromolar inhibition of the NA function of the most serious of the human parainfluenza viruses (i.e. hPIV-3). A crystal structure has been reported for one of these designed inhibitors (Ryan *et al.*, 2006) in complex with Newcastle disease virus HN and demonstrates that the ligand is bound into the active site in the predicted manner (Figure 15.5).

More recently, several crystal structures of the HN from hPIV-3 have been published (Lawrence *et al.*, 2004), although, to the best of our knowledge, no structure-based drug design studies have been reported. Furthermore, crystal structures of the HN from PIV5, formerly known as simian virus 5, have been published (Yuan *et al.*, 2005) and reveal an overall architecture that appears to be similar to those of Newcasle disease virus and hPIV-3.

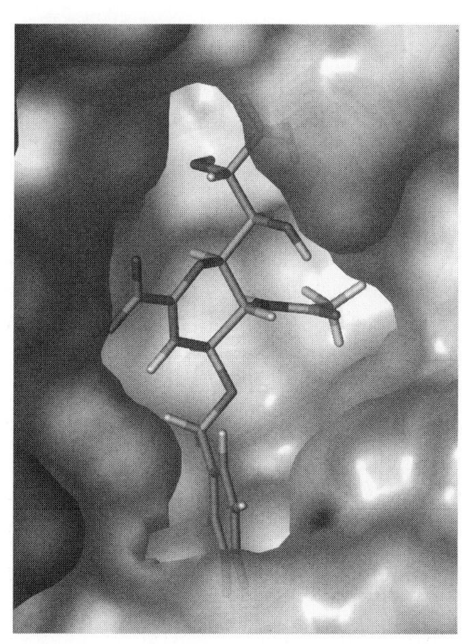

FIGURE 15.5 Crystal structure of 4-*O*-benzyl-Neu5Ac2en soaked into the haemagglutinin-neuraminidase of Newcastle disease virus, showing the large pocket targeted by the benzyl group of the designed inhibitor. (See colour plates section.)

3.2. Future directions

In addition to providing drug design opportunities, these HN structures, along with structures of the fusion (F) protein, provide more information on the mechanism of cell entry by this family of viruses. This will lead to a greater understanding of how these viruses propagate, making it easier to combat their spread. The original structure-based drug design studies for an anti-human parainfluenza virus drug were based on the available Newcastle disease virus HN structures, more detailed design work can now be undertaken using the hPIV-3 structures. Also, there are a number of non-sialic acid based templates that have been used in the development of influenza virus NA (e.g. cyclohexene, cyclopentane and pyrrolidine), these alternative templates could be used in the structure-based drug design process to extend the range of compounds as potential HN inhibitors.

4. DENGUE VIRUS

The flaviviruses are members of the *Flaviviridae* family and are transmitted to humans by either mosquito or tick bites (Solomon and Mallewa, 2001). Flavivrus infection can lead to a number of severe and debilitating symptoms, e.g. fever-arthralgia-rash, viral haemorrhagic fever and neurological disease (Solomon and Mallewa, 2001). The mosquito-borne viruses include Dengue virus (DENV), of which there are four strains (DENV1, DENV2, DENV3 and DENV4), yellow fever virus, West Nile virus and Japanese encephalitis virus, while the tick-borne viruses include tick-borne encephalitis virus, Langat virus and Omsk haemorrhagic fever virus.

4.1. E-Glycoprotein and E-glycoprotein-domain 3

The surface of flavivirus is predominantly covered by E-glycoprotein (EGP) which forms dimers that are tethered to the surface by a stalk region. The EGP-associated glycan of the virus has been shown to be important for DENV infection through its interaction with dendritic cell-specific ICAM-3-grabbing non-integrin, DC-SIGN (Navarro-Sanchez *et al.*, 2003; Tassaneetrithep *et al.*, 2003) (see Chapter 34). This interaction and its involvement in the life cycle have been recently reviewed (Perera *et al.*, 2008). Of note, EGP is the major site of host cell receptor binding and of host-mediated antibody neutralization. Both the mosquito and tick-borne virus EGPs share significant sequence and structural homology. The crystal structures of EGP from several sources show that it is comprised of three distinct domains (Figure 15.6); namely DENV2 (Modis *et al.*, 2003; Zhang *et al.*, 2004), DENV3 (Modis *et al.*, 2005) and tick-borne encephalitis virus (Rey *et al.*, 1995). After binding and during the fusion process, the EGP dimers dissociate and reform as trimers; the structures of this trimeric fusion state are also available, i.e. DENV2 (Modis *et al.*, 2004), tick-borne encephalitis virus (Bressanelli *et al.*, 2004) and West Nile virus (Nybakken *et al.*, 2006).

Of these three domains, it has been found that Domain 3 (D3) is mostly responsible for the virus binding to target host cells (Crill and Roehring, 2001). Also, D3 by itself has been shown to block virus binding. In addition to the crystal structures mentioned above, there have been structures of the D3 sub-unit from DENV2 (Huang *et al.*, 2008), DENV4 (Volk *et al.*, 2007), West Nile virus (Volk *et al.*, 2004), Japanese encephalitis virus (Wu *et al.*, 2003), Langat virus (Wu *et al.*, 2003) and haemorrhagic fever virus (Volk *et al.*, 2006) determined by NMR spectroscopic techniques.

While the receptor for EGP has not been definitively established, previous studies have identified the glycosaminoglycan, heparan sulfate and a range of protein species as proposed mammalian cell surface receptors (Clyde *et al.*, 2006). A recent work (Aoki *et al.*, 2006) has shown that there is an association between DENV and the mammalian cell surface glycolipid, paragloboside. This suggests that EGP and, in particular D3, are carbohydrate-recognizing proteins with small variations in sequence between the members of this family leading to differing cell receptors.

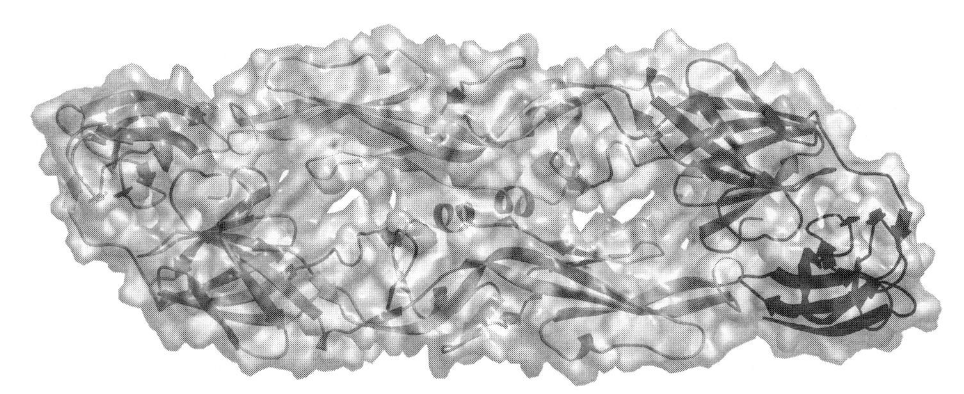

FIGURE 15.6 Crystal structure of the E-glycoprotein dimer of dengue virus. The red-coloured ribbon is Domain 1, the yellow-coloured ribbon is Domain 2, while the blue-coloured ribbon is Domain 3. The surface is coloured by electrostatic potential. (See colour plates section.)

The available D3 structures from different flaviviruses also allows us to come to a greater understanding of how various neutralizing antibodies bind to D3 and EGP (Wu *et al.*, 2003; Nybakken *et al.*, 2005; Kaufmann *et al.*, 2006; Gromowski and Barrett, 2007) and facilitate the design of possible therapeutics capable of halting the spread of these arthropod-borne viruses (Figure 15.7).

4.2. Future directions

Access to the available structures and molecular modelling techniques, for example blind autodocking (Hetenyi and van der Spoel, 2002) and saturation transfer difference (STD-) NMR spectroscopic techniques (Mayer and Meyer, 2001; Haselhorst *et al.*, 2007a,b) in combination with glycan array studies (Day *et al.*, 2007), will provide exciting opportunities for the identification of not only the host cell based glycan receptor but also characterization of the binding site on D3. This information combined with structure-based drug design techniques can then be applied to search for a possible inhibitor of the initial recognition and binding events involved in cellular invasion by these arthropod-borne viruses. As presented in the influenza virus HA example above, small molecule-based inhibitors may not bind with sufficient affinity to produce an entity capable of preventing virus replication, however, multivalent inhibitors may provide valuable alternatives.

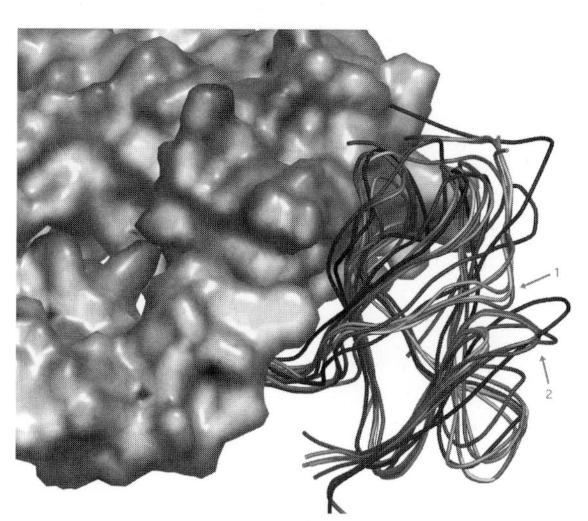

FIGURE 15.7 Superimposition of the available flavivirus Domain 3 structures. Green, DENV2; red, DENV3; blue, Japenese encephalitis virus; cyan, Langat virus; magenta, Omsk haemorrhagic fever virus; yellow, tick-borne encephalitis virus; and brown, West Nile virus. The surface represents the remaining portion of the EGP dimer and is coloured by electrostatic potential. A major difference can be seen in loop 1 for tick-borne encephalitis virus, Omsk haemorrhagic fever virus and Langat virus which bulges out because there is a 4 residue deletion in the loop 2 portions of these proteins. (See colour plates section.)

5. ROTAVIRUS

Rotavirus is a member of the *Reoviridae* family and is the most frequent causative agent of severe diarrhoea in young children accounting for 25–55% of all hospital admissions for diarrhoea and 600 000 deaths every year (Glass *et al.*, 2006). While advances in the development of vaccines have been made (Heyse *et al.*, 2008) for the human population, questions about safety and efficacy still remain (Franco and Greenberg, 2001). Furthermore, rotavirus also impacts on productivity in the livestock industry (Yu *et al.*, 2008) and therefore further work is needed to better understand the rotavirus lifecycle and carbohydrate involvement for drug and new vaccine development opportunities.

5.1. VP8*

The rotavirus virion surface spike protein VP4 is responsible for receptor binding and cell penetration (Ludert *et al.*, 1996; Zarate *et al.*, 2000; Arias *et al.*, 2002). The protein VP4 is cleaved by trypsin into the N-terminal VP8* fragment which has been shown to be responsible for haemagglutination and the VP5* fragment which is believed to be involved in cellular fusion (Fiore *et al.*, 1991; Ciarlet *et al.*, 2002; Graham *et al.*, 2003). There has been some controversy over the absolute requirement of sialic acid in the initial binding of VP8* of the virus to the cell surface (Delorme *et al.*, 2001; Ciarlet *et al.*, 2002; Isa *et al.*, 2006). Currently, rotavirus strains are classified as either sialidase-sensitive, i.e. they no longer infect sialidase-treated cells inferring a sialic acid dependence of infection, or sialidase-insensitive as they infect cells regardless of sialidase treatment, thereby suggesting sialic acid independence of infection.

To examine carbohydrate recognition further, the first structures (NMR and X-ray) of a VP8* from a Rhesus rotavirus strain (sialidase-sensitive) was published in 2002 (Dormitzer *et al.*, 2002). Subsequently, the crystal structures of a number of native and site-directed mutant Rhesus rotavirus VP8* proteins have been reported (Kraschnefski *et al.*, 2009). The only other VP8* structure from a sialidase-sensitive strain that has been published is from the porcine CRW-8 strain (Blanchard *et al.*, 2007). All of the crystal structures of the above VP8* proteins have included the ligand Neu5Acα2Me (Figure 15.8). The binding site shows that the sugar ring is positioned between two tyrosine (Tyr) residues, while the carboxylic acid moiety of the sialic acid forms hydrogen bonds with a serine (Ser) residue and finally an arginine (Arg) residue forms hydrogen bonds to the glycerol side chain of the sialic acid. This sialic acid binding motif is different to that seen in influenza virus haemagglutinin and sialidase, but it has been reported in several sialyl-Lewis X recognizing bacterial toxins (Baker *et al.*, 2007).

Further to these structural investigations, and more relevant to human rotavirus infection, two structures of VP8* proteins from the sialidase-insensitive human strains Wa (Kraschnefski *et al.*, 2005; Blanchard *et al.*, 2007) and DS-1 (Monnier *et al.*, 2006) have been determined. Finally, a recent publication (Yu *et al.*, 2008) has described the crystallization and preliminary X-ray diffraction analysis of VP8* protein from the sialidase-sensitive bovine rotavirus strain NCDV.

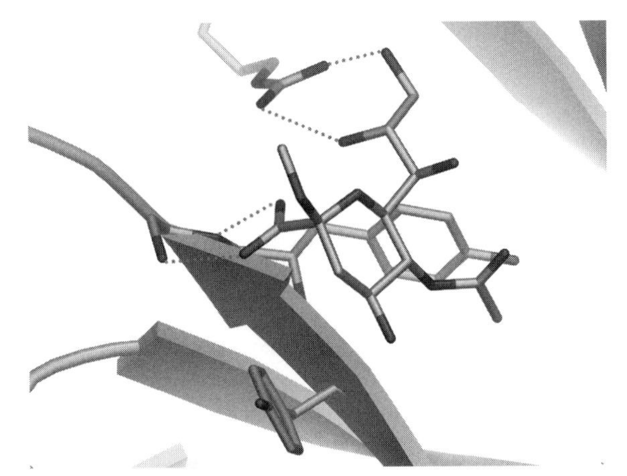

FIGURE 15.8 Crystal structure of Neu5Acα2Me (cyan-coloured carbons) bound to the VP8* protein of rotavirus strain CRW-8. Hydrogen bonds to the serine (Ser) and arginine (Arg) are shown with magenta-coloured dotted lines while the two tyrosine (Tyr) residues which help position the sugar ring are also shown. (See colour plates section.)

Molecular modelling and STD NMR experiments within our group (Haselhorst *et al.*, 2007a) have shown a very good correlation between the binding of a number of sialic acid derivatives to Rhesus rotavirus VP8* and the degree to which these compounds can inhibit rotavirus infection of cells. Also, information obtained on the binding epitope of sialic acid disaccharides showed that the penultimate sugar residue does not interact with the protein, but remains predominantly solvated. This is in very good agreement with molecular dynamics calculations that showed that this sugar (galactose) was unable to adopt a conformation where it is in close proximity to the surface (Haselhorst *et al.*, 2007a).

RESEARCH FOCUS BOX

- Identification of next generation influenza virus sialidase inhibitors is required to replace existing drugs as resistance develops.
- Understanding the carbohydrate recognition processes involved in dengue virus and rotavirus infection is necessary to enable the development of useful chemotherapeutic agents.
- A systematic investigation of host–virus carbohydrate recognition phenomena would greatly enhance the field.

5.2. Future directions

The knowledge of the binding site and ligand for the VP8* protein from human rotavirus strains provides exciting opportunities for the design of inhibitors to prevent virus entry and infection. For example, the use of computational chemistry methods in exploring the available structural data in combination with other techniques, such as NMR spectroscopy, will be invaluable in advancing the role of carbohydrates in the rotavirus lifecycle.

6. CONCLUSIONS

A large number of clinically important viruses utilize carbohydrate viral surface protein interactions to propagate infection and disease. The identification and structural characterization of the proteins and glycans essential in these recognition phenomena opens up new avenues for both drug discovery and vaccine development. While only a limited number of viral surface protein structures have been determined, already structure-based drug design techniques have afforded new classes of drugs. Although significant advances have been made in targeting virus surface glycoproteins for the discovery of drugs, there remains an urgent need for further investigations (Research Focus Box). The science of glycovirology is still in its infancy but shows great promise for the delivery of new weapons against pathogens that have very few drug treatments available.

References

Abu Hammad, A.M., Afifi, F.U., Taha, M.O., 2007. Combining docking, scoring and molecular field analyses to probe influenza neuraminidase-ligand interactions. J. Mol. Graph. Model. 26, 443–456.

Alymova, I.V., Taylor, G., Takimoto, T., et al., 2004. Efficacy of novel hemagglutinin-neuraminidase inhibitors BCX 2798 and BCX 2855 against human parainfluenza viruses in vitro and in vivo. Antimicrob. Agents Chemother. 48, 1495–1502.

Aoki, C., Hidari, K.I., Itonori, S., et al., 2006. Identification and characterization of carbohydrate molecules in mammalian cells recognized by dengue virus type 2. J. Biochem. 139, 607–614.

Arias, C.F., Isa, P., Guerrero, C.A., et al., 2002. Molecular biology of rotavirus cell entry. Arch. Med. Res. 33, 356–361.

Armstrong, K.A., Tidor, B., Cheng, A.C., 2006. Optimal charges in lead progression: a structure-based neuraminidase case study. J. Med. Chem. 49, 2470–2477.

Baker, H.M., Basu, I., Chung, M.C., Caradoc-Davies, T., Fraser, J.D., Baker, E.N., 2007. Crystal structures of the staphylococcal toxin SSL5 in complex with sialyl Lewis X reveal a conserved binding site that shares common features with viral and bacterial sialic acid binding proteins. J. Mol. Biol. 374, 1298–1308.

Berman, H.M., Westbrook, J., Feng, Z., et al., 2000. The Protein Data Bank. Nucleic Acids Res. 28, 235–242.

Blanchard, H., Yu, X., Coulson, B.S., von Itzstein, M., 2007. Insight into host cell carbohydrate-recognition by human and porcine rotavirus from crystal structures of the virion spike associated carbohydrate-binding domain (VP8*). J. Mol. Biol. 367, 1215–1226.

Bressanelli, S., Stiasny, K., Allison, S.L., et al., 2004. Structure of a flavivirus envelope glycoprotein in its low-pH-induced membrane fusion conformation. EMBO J. 23, 728–738.

Ciarlet, M., Ludert, J.E., Iturriza-Gomara, M., et al., 2002. Initial interaction of rotavirus strains with *N*-acetylneuraminic (sialic) acid residues on the cell surface correlates with VP4 genotype, not species of origin. J. Virol. 76, 4087–4095.

Clyde, K., Kyle, J.L., Harris, E., 2006. Recent advances in deciphering viral and host determinants of dengue virus replication and pathogenesis. J. Virol. 80, 11418–11431.

Collins, P.J., Haire, L.F., Lin, Y.P., et al., 2008. Crystal structures of oseltamivir-resistant influenza virus neuraminidase mutants. Nature 453, 1258–1261.

Colman, P.M., Varghese, J.N., Laver, W.G., 1983. Structure of the catalytic and antigenic sites in influenza virus neuraminidase. Nature 303, 41–44.

Crennell, S., Takimoto, T., Portner, A., Taylor, G., 2000. Crystal structure of the multifunctional paramyxovirus hemagglutinin-neuraminidase. Nat. Struct. Biol. 7, 1068–1074.

Crescenzo-Chaigne, B., van der Werf, S., 2007. Rescue of influenza C virus from recombinant DNA. J. Virol. 81, 11282–11289.

Crill, W.D., Roehrig, J.T., 2001. Monoclonal antibodies that bind to domain III of dengue virus E glycoprotein are the most efficient blockers of virus adsorption to Vero cells. J. Virol. 75, 7769–7773.

Day, C., Tiralongo, J., Hartnell, R.D., von Itzstein, M., Wilson, J.C., Korolik, V., 2007. Optimisation of glycan and small molecule arrays for analysis of *Campylobacter* chemotaxis and adherence. Zoonoses Pub. Health 54 (Suppl. 1), 99.

de Groot, R.J., 2006. Structure, function and evolution of the hemagglutinin-esterase proteins of corona- and toroviruses. Glycoconj. J. 23, 59–72.

Delorme, C., Brussow, H., Sidoti, J., et al., 2001. Glycosphingolipid binding specificities of rotavirus: identification of a sialic acid-binding epitope. J. Virol. 75, 2276–2287.

Dormitzer, P.R., Sun, Z.Y., Wagner, G., Harrison, S.C., 2002. The rhesus rotavirus VP4 sialic acid binding domain has a galectin fold with a novel carbohydrate binding site. EMBO J. 21, 885–897.

Fiore, L., Greenberg, H.B., Mackow, E.R., 1991. The VP8 fragment of VP4 is the rhesus rotavirus hemagglutinin. Virology 181, 553–563.

Franco, M.A., Greenberg, H.B., 2001. Challenges for rotavirus vaccines. Virology 281, 153–155.

Gagneux, P., Varki, A., 1999. Evolutionary considerations in relating oligosaccharide diversity to biological function. Glycobiology 9, 747–755.

Gamblin, S.J., Haire, L.F., Russell, R.J., et al., 2004. The structure and receptor binding properties of the 1918 influenza hemagglutinin. Science 303, 1838–1842.

Glass, R.I., Parashar, U.D., Bresee, J.S., et al., 2006. Rotavirus vaccines: current prospects and future challenges. Lancet 368, 323–332.

Govorkova, E.A., Leneva, I.A., Goloubeva, O.G., Bush, K., Webster, R.G., 2001. Comparison of efficacies of RWJ-270201, zanamivir, and oseltamivir against H5N1, H9N2, and other avian influenza viruses. Antimicrob. Agents Chemother. 45, 2723–2732.

Graham, K.L., Halasz, P., Tan, Y., et al., 2003. Integrin-using rotaviruses bind $\alpha2\beta1$ integrin $\alpha2$ I domain via VP4 DGE sequence and recognize $\alpha X\beta2$ and $\alpha V\beta3$ by using VP7 during cell entry. J. Virol. 77, 9969–9978.

Gromowski, G.D., Barrett, A.D., 2007. Characterization of an antigenic site that contains a dominant, type-specific neutralization determinant on the envelope protein domain III (ED3) of dengue 2 virus. Virology 366, 349–360.

Haselhorst, T., Blanchard, H., Frank, M., et al., 2007a. STD NMR spectroscopy and molecular modeling investigation of the binding of *N*-acetylneuraminic acid derivatives to rhesus rotavirus VP8* core. Glycobiology 17, 68–81.

Haselhorst, T., Munster-Kuhnel, A.K., Oschlies, M., Tiralongo, J., Gerardy-Schahn, R., von Itzstein, M., 2007b. Direct detection of ligand binding to Sepharose-immobilised protein using saturation transfer double difference (STDD) NMR spectroscopy. Biochem. Biophys. Res. Commun. 359, 866–870.

Haselhorst, T., Garcia, J.M., Islam, T., et al., 2008. Avian influenza H5-containing virus-like particles (VLPs): host-cell receptor specificity by STD NMR spectroscopy. Angew. Chem. Int. Ed. Engl. 47, 1910–1912.

Henrickson, K.J., 2003. Parainfluenza viruses. Clin. Microbiol. Rev. 16, 242–264.

Herrler, G., Durkop, I., Becht, H., Klenk, H.D., 1988. The glycoprotein of influenza C virus is the haemagglutinin, esterase and fusion factor. J. Gen. Virol. 69, 839–846.

Hetenyi, C., van der Spoel, D., 2002. Efficient docking of peptides to proteins without prior knowledge of the binding site. Protein Sci. 11, 1729–1737.

Heyse, J.F., Kuter, B.J., Dallas, M.J., Heaton, P., 2008. Evaluating the safety of a rotavirus vaccine: the REST of the story. Clin. Trials 5, 131–139.

Huang, K.C., Lee, M.C., Wu, C.W., Huang, K.J., Lei, H.Y., Cheng, J.W., 2008. Solution structure and neutralizing antibody binding studies of domain III of the dengue-2 virus envelope protein. Proteins 70, 1116–1119.

Irwin, J.J., Shoichet, B.K., 2005. ZINC – a free database of commercially available compounds for virtual screening. J. Chem. Infect. Model. 45, 177–182.

Isa, P., Arias, C.F., Lopez, S., 2006. Role of sialic acids in rotavirus infection. Glycoconj. J. 23, 27–37.

Kaufmann, B., Nybakken, G.E., Chipman, P.R., et al., 2006. West Nile virus in complex with the Fab fragment of a neutralizing monoclonal antibody. Proc. Natl. Acad. Sci. USA 103, 12400–12404.

Kraschnefski, M.J., Scott, S.A., Holloway, G., Coulson, B.S., von Itzstein, M., Blanchard, H., 2005. Cloning, expression, purification, crystallization and preliminary X-ray diffraction analysis of the VP8* carbohydrate-binding protein of the human rotavirus strain Wa. Acta Crystallogr. Sect. F Struct. Biol. Cryst. Commun. 61, 989–993.

Kraschnefski, M.J., Bugarcic, A., Fleming, F., et al., 2009. Effects on sialic acid recognition of amino acid mutations in the carbohydrate-binding cleft of the rotavirus spike protein. Glycobiology 19, 194–200.

Lamb, R.A., Kolakofsky, D., 1996. Paramyxoviridae: the viruses and their replication. In: Fields, B.N., Knipe, D.M., Howley, P.M. (Eds.), Fundamental Virology. Lippincott-Raven, Philadelphia, pp. 577–604.

Laver, W., 2006. Antiviral drugs for influenza: Tamiflu® past, present and future. Future Virol. 1, 577–586.

Lawrence, M.C., Borg, N.A., Streltsov, V.A., et al., 2004. Structure of the haemagglutinin-neuraminidase from human parainfluenza virus type III. J. Mol. Biol 335, 1343–1357.

Ludert, J.E., Feng, N., Yu, J.H., Broome, R.L., Hoshino, Y., Greenberg, H.B., 1996. Genetic mapping indicates that VP4 is the rotavirus cell attachment protein in vitro and in vivo. J. Virol. 70, 487–493.

Luscher-Mattli, M., 2000. Influenza chemotherapy: a review of the present state of art and of new drugs in development. Arch. Virol. 145, 2233–2248.

Marra, A., Moni, L., Pazzi, D., Corallini, A., Bridi, D., Dondoni, A., 2008. Synthesis of sialoclusters appended to calix[4]arene platforms via multiple azide-alkyne cycloaddition. New inhibitors of hemagglutination and cytopathic effect mediated by BK and influenza A viruses. Org. Biomol. Chem. 6, 1396–1409.

Masukawa, K.M., Kollman, P.A., Kuntz, I.D., 2003. Investigation of neuraminidase-substrate recognition using molecular dynamics and free energy calculations. J. Med. Chem. 46, 5628–5637.

Matrosovich, M., Klenk, H.D., 2003. Natural and synthetic sialic acid-containing inhibitors of influenza virus receptor binding. Rev. Med. Virol. 13, 85–97.

Mayer, M., Meyer, B., 2001. Group epitope mapping by saturation transfer difference NMR to identify segments of a ligand in direct contact with a protein receptor. J. Am. Chem. Soc. 123, 6108–6117.

Mayr, J., Haselhorst, T., Langereis, M.A., et al., 2008. Influenza C virus and bovine coronavirus esterase reveal a similar catalytic mechanism: new insights for drug discovery. Glycoconj. J. 25, 393–399.

Mishin, V.P., Hayden, F.G., Gubareva, L.V., 2005. Susceptibilities of antiviral-resistant influenza viruses to novel neuraminidase inhibitors. Antimicrob. Agents Chemother. 49, 4515–4520.

Modis, Y., Ogata, S., Clements, D., Harrison, S.C., 2003. A ligand-binding pocket in the dengue virus envelope glycoprotein. Proc. Natl. Acad. Sci. USA 100, 6986–6991.

Modis, Y., Ogata, S., Clements, D., Harrison, S.C., 2004. Structure of the dengue virus envelope protein after membrane fusion. Nature 427, 313–319.

Modis, Y., Ogata, S., Clements, D., Harrison, S.C., 2005. Variable surface epitopes in the crystal structure of dengue virus type 3 envelope glycoprotein. J. Virol. 79, 1223–1231.

Monnier, N., Higo-Moriguchi, K., Sun, Z.Y., Prasad, B.V., Taniguchi, K., Dormitzer, P.R., 2006. High-resolution molecular and antigen structure of the VP8* core of a sialic acid-independent human rotavirus strain. J. Virol. 80, 1513–1523.

Nandi, T., 2008. Proposed lead molecules against hemagglutinin of avian influenza virus (H5N1). Bioinformation 2, 240–244.

Navarro-Sanchez, E., Altmeyer, R., Amara, A., et al., 2003. Dendritic-cell-specific ICAM3-grabbing non-integrin is essential for the productive infection of human dendritic cells by mosquito-cell-derived dengue viruses. EMBO Rep. 4, 723–728.

Nicholls, J.M., Chan, M.C., Chan, W.Y., et al., 2007. Tropism of avian influenza A (H5N1) in the upper and lower respiratory tract. Nat. Med. 13, 147–149.

Nybakken, G.E., Oliphant, T., Johnson, S., Burke, S., Diamond, M.S., Fremont, D.H., 2005. Structural basis of West Nile virus neutralization by a therapeutic antibody. Nature 437, 764–769.

Nybakken, G.E., Nelson, C.A., Chen, B.R., Diamond, M.S., Fremont, D.H., 2006. Crystal structure of the West Nile virus envelope glycoprotein. J. Virol. 80, 11467–11474.

Ogata, M., Murata, T., Murakami, K., et al., 2007. Chemoenzymatic synthesis of artificial glycopolypeptides containing multivalent sialyloligosaccharides with a gamma-polyglutamic acid backbone and their effect on inhibition of infection by influenza viruses. Bioorg. Med. Chem. 15, 1383–1393.

Peiris, J.S., de Jong, M.D., Guan, Y., 2007. Avian influenza virus (H5N1): a threat to human health. Clin. Microbiol. Rev. 20, 243–267.

Perera, R., Khaliq, M., Kuhn, R.J., 2008. Closing the door on flaviviruses: entry as a target for antiviral drug design. Antiviral Res. 80, 11–22.

Platis, D., Smith, B.J., Huyton, T., Labrou, N.E., 2006. Structure-guided design of a novel class of benzyl-sulfonate inhibitors for influenza virus neuraminidase. Biochem. J. 399, 215–223.

Rapoport, E.M., Mochalova, L.V., Gabius, H.J., Romanova, J., Bovin, N.V., 2006. Search for additional influenza virus to cell interactions. Glycoconj. J. 23, 115–125.

Rey, F.A., Heinz, F.X., Mandl, C., Kunz, C., Harrison, S.C., 1995. The envelope glycoprotein from tick-borne encephalitis virus at 2 A resolution. Nature 375, 291–298.

Rosenthal, P.B., Zhang, X., Formanowski, F., et al., 1998. Structure of the haemagglutinin-esterase-fusion glycoprotein of influenza C virus. Nature 396, 92–96.

Russell, R.J., Haire, L.F., Stevens, D.J., et al., 2006. The structure of H5N1 avian influenza neuraminidase suggests new opportunities for drug design. Nature 443, 45–49.

Ryan, C., Zaitsev, V., Tindal, D.J., et al., 2006. Structural analysis of a designed inhibitor complexed with the hemagglutinin-neuraminidase of Newcastle disease virus. Glycoconj. J. 23, 135–141.

Seal, B.S., King, D.J., Sellers, H.S., 2000. The avian response to Newcastle disease virus. Dev. Comp. Immunol. 24, 257–268.

Solomon, T., Mallewa, M., 2001. Dengue and other emerging flaviviruses. J. Infect 42, 104–115.

Stevens, J., Blixt, O., Tumpey, T.M., Taubenberger, J.K., Paulson, J.C., Wilson, I.A., 2006. Structure and receptor specificity of the hemagglutinin from an H5N1 influenza virus. Science 312, 404–410.

Stoll, V., Stewart, K.D., Maring, C.J., et al., 2003. Influenza neuraminidase inhibitors: structure-based design of a novel inhibitor series. Biochemistry 42, 718–727.

Tassaneetrithep, B., Burgess, T.H., Granelli-Piperno, A., et al., 2003. DC-SIGN (CD209) mediates dengue virus infection of human dendritic cells. J. Exp. Med. 197, 823–829.

Taubenberger, J.K., Reid, A.H., Janczewski, T.A., Fanning, T.G., 2001. Integrating historical, clinical and molecular genetic data in order to explain the origin and virulence of the 1918 Spanish influenza virus. Philos. Trans. Roy. Soc. Lond., Sect. B, Biol. Sci. 356, 1829–1839.

Tindal, D.J., Dyason, J.C., Thomson, R.J., et al., 2007. Synthesis and evaluation of 4-*O*-alkylated 2-deoxy-2,3-didehydro-*N*-acetylneuraminic acid derivatives as inhibitors of human parainfluenza virus type-3 sialidase activity. Bioorg. Med. Chem. Lett. 17, 1655–1658.

van Regenmortel, M.H.V., Fauquet, C.M., Bishop, D.H.L. et al., (Eds.), 2000. Virus Taxonomy: Seventh Report of the International Committee on Taxonomy of Viruses. Academic Press, San Diego.

Varghese, J.N., Laver, W.G., Colman, P.M., 1983. Structure of the influenza virus glycoprotein antigen neuraminidase at 2.9 Å resolution. Nature 303, 35–40.

Volk, D.E., Beasley, D.W., Kallick, D.A., Holbrook, M.R., Barrett, A.D., Gorenstein, D.G., 2004. Solution structure and antibody binding studies of the envelope protein domain III from the New York strain of West Nile virus. J. Biol. Chem. 279, 38755–38761.

Volk, D.E., Chavez, L., Beasley, D.W., Barrett, A.D., Holbrook, M.R., Gorenstein, D.G., 2006. Structure of the envelope protein domain III of Omsk hemorrhagic fever virus. Virology 351, 188–195.

Volk, D.E., Lee, Y.C., Li, X., et al., 2007. Solution structure of the envelope protein domain III of dengue-4 virus. Virology 364, 147–154.

von Itzstein, M., 2007. The war against influenza: discovery and development of sialidase inhibitors. Nat. Rev. Drug Discov. 6, 967–974.

von Itzstein, M., 2008. Avian influenza virus, a very sticky situation. Curr. Opin. Chem. Biol. 12, 102–108.

Wagaman, P.C., Spence, H.A., O'Callaghan, R.J., 1989. Detection of influenza C virus by using an in situ esterase assay. J. Clin. Microbiol. 27, 832–836.

WHO. (2008). World Health Organization – Avian Influenza. http://www.who.int/csr/disease/avian_influenza/en/index.html.

Wilson, I.A., Skehel, J.J., Wiley, D.C., 1981. Structure of the haemagglutinin membrane glycoprotein of influenza virus at 3 Å resolution. Nature 289, 366–373.

Wu, K.P., Wu, C.W., Tsao, Y.P., et al., 2003. Structural basis of a flavivirus recognized by its neutralizing antibody: solution structure of the domain III of the Japanese encephalitis virus envelope protein. J. Biol. Chem. 278, 46007–46013.

Yamada, S., Suzuki, Y., Suzuki, T., et al., 2006. Haemagglutinin mutations responsible for the binding of H5N1 influenza A viruses to human-type receptors. Nature 444, 378–382.

Yu, X., Guillon, A., Szyczew, A.J., et al., 2008. Crystallization and preliminary X-ray diffraction analysis of the carbohydrate-recognizing domain (VP8*) of bovine rotavirus straon NCDV. Acta Crystallogr. Sect. F Struct. Biol. Cryst. Commun. F64, 509–511.

Yuan, P., Thompson, T.B., Wurzburg, B.A., Paterson, R.G., Lamb, R.A., Jardetzky, T.S., 2005. Structural studies of the parainfluenza virus 5 hemagglutinin-neuraminidase tetramer in complex with its receptor, sialyllactose. Structure 13, 803–815.

Zaitsev, V., von Itzstein, M., Groves, D., et al., 2004. Second sialic acid binding site in Newcastle disease virus hemagglutinin-neuraminidase: implications for fusion. J. Virol. 78, 3733–3741.

Zarate, S., Espinosa, R., Romero, P., Mendez, E., Arias, C.F., Lopez, S., 2000. The VP5 domain of VP4 can mediate attachment of rotaviruses to cells. J. Virol. 74, 593–599.

Zhang, J., Yu, K., Zhu, W., Jiang, H., 2006. Neuraminidase pharmacophore model derived from diverse classes of inhibitors. Bioorg. Med. Chem. Lett. 16, 3009–3014.

Zhang, Y., Zhang, W., Ogata, S., et al., 2004. Conformational changes of the flavivirus E glycoprotein. Structure 12, 1607–1618.

PART II

SYNTHESIS OF MICROBIAL GLYCOSYLATED COMPONENTS

A. Biosynthesis and biosynthetic processes

CHAPTER

16

Biosynthesis of bacterial peptidoglycan

Jean van Heijenoort

SUMMARY

Bacterial peptidoglycan has been extensively investigated owing to its importance as an essential cell wall structural component. Its biosynthesis is a two-stage process, which is the target of well-known antibiotics and which is closely correlated with cellular morphogenesis. First, the monomer unit is assembled by a sequence of reactions in the cytoplasm and at the inner side of the cytoplasmic membrane leading to the lipid II intermediate. The second stage involves extracytoplasmic polymerization reactions catalysed by glycosyltransferases and transpeptidases using lipid II as substrate after its translocation through the membrane. The different steps of the pathway are described, as well as their variations, their *in vivo* functioning and their inhibition by antibiotics. The areas calling for more thorough investigation concern the peptidoglycan-synthesizing complexes functioning in coordination with cell growth and cell division, the role played by peptidoglycan hydrolases and the examination of yet unexplored structural modifications of peptidoglycan. The need to understand better the mechanisms of resistance to antibiotics and to use specific peptidoglycan targets for the search of novel antibacterials also fully warrants the continued study of peptidoglycan biosyntheis.

Keywords: Uridine diphosphate- (UDP-) *N*-acetylglucosamine; UDP-*N*-acetylmuramic acid; UDP-*N*-acetylmuramic acid-peptides; Lipid intermediates; Mur synthetases; MraY transferase; MurG transferase; Penicillin-binding proteins

1. INTRODUCTION

Over the past fifty years, bacterial peptidoglycan has been extensively investigated owing to its importance as an essential structural cell wall component (see Chapter 2), to its involvement in cellular morphogenesis (Scheffers and Pinho, 2005; Vicente *et al.*, 2006; den Blaauwen *et al.*, 2008; Zapun *et al.*, 2008a) and to the fact that steps of its biosynthesis are specifically inhibited by well-known antibiotics and are potential targets for the search for novel antibacterials (Gale *et al.*, 1981; Green, 2002; Kotnik *et al.*, 2007; Barreteau *et al.*, 2008). The various steps of the biosynthesis were studied in different species and a two-stage pathway valid for most eubacteria emerged (Rogers *et al.*, 1980; Ward, 1984). First, the peptidoglycan monomer unit is assembled in the final form of a lipid intermediate by enzymes located in the cytoplasm or at the inner side of the cytoplasmic membrane (Bugg and Walsh, 1992; van Heijenoort, 2001b, 2007; Barreteau *et al.*, 2008; Bouhss *et al.*, 2008). The second stage involves extracytoplasmic polymerization reactions using as substrate the lipid intermediate after its translocation through the membrane. Two major types of membrane-bound activities are involved in polymerization: glycosyltransferases which catalyse the formation of the linear glycan chains (Goffin and Ghuysen, 1998; van Heijenoort, 2001a, 2007) and transpeptidases which catalyse the cross-linking between the peptide subunits of the new chains and the binding to the pre-existing cell wall (Goffin and Ghuysen, 2002; Macheboeuf *et al.*, 2006; Sauvage *et al.*, 2008; Zapun *et al.*, 2008b).

2. ASSEMBLY OF THE MONOMER UNIT

The monomer unit is assembled by a linear sequence of reactions extending from D-fructose-6-phosphate to the final lipid intermediate (Figure 16.1). The pathway was established by characterizing its precursors and developing a specific enzymatic assay for each step. The genes involved in the process were identified, mapped in a few regions of the bacterial chromosome, cloned, sequenced and most shown to be essential. Their products were over-produced, characterized and many crystallized. This in turn enabled structural and mechanistic studies. The literature on the pathway has been reviewed in detail (Rogers *et al.*, 1980; Ward, 1984; Bugg and Walsh, 1992; van Heijenoort, 2001b; Barreteau *et al.*, 2008; Bouhss *et al.*, 2008). For convenience, the successive formations (see Figure 16.1) of uridine diphosphate- (UDP-) *N*-acetylglucosamine (GlcNAc), UDP-*N*-acetylmuramic acid (UDP-MurNAc), UDP-MurNAc-peptides and lipids intermediates will be considered.

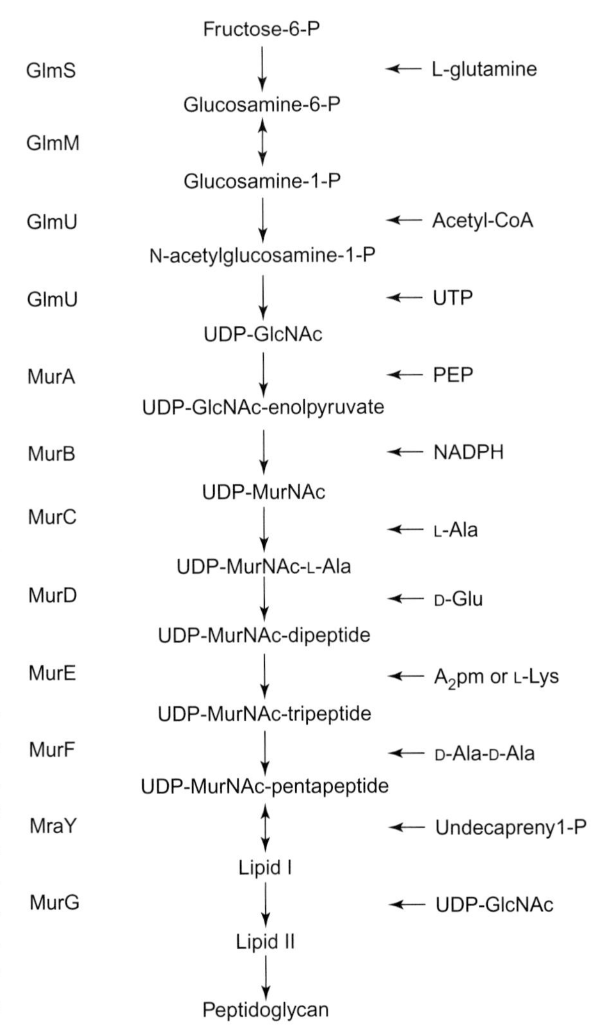

FIGURE 16.1 Stepwise assembly of the peptidoglycan monomer unit. Abbreviations: A_2pm, diaminopimelic acid; acetyl-CoA, acetyl co-enzyme A; D-, L-Ala, D-, L-alanine; D-Glu, D-glutamic acid; L-Lys, L-lysine; NADPH, nicotinamide adenine dinucleotide phosphate; PEP, phosphoenolpyruvate; undecaprenyl-P, undecaprenyl-phosphate; UDP, uridine diphosphate; UDP-GlcNAc, UDP-*N*-acetylglucosamine; UDP-MurNAc, UDP-*N*-acetylmuramic acid; UTP, uridine diphosphate. Reproduced, with permission, from van Heijenoort (2007).

2.1. Formation of UDP-GlcNAc

In eubacteria, UDP-GlcNAc is used not only for peptidoglycan synthesis, but also for that of other cell wall polymers (see Chapters 17 to 22). Its synthesis from D-fructose-6-phosphate (Figure 16.2) requires four steps (Dobrogosz, 1968;

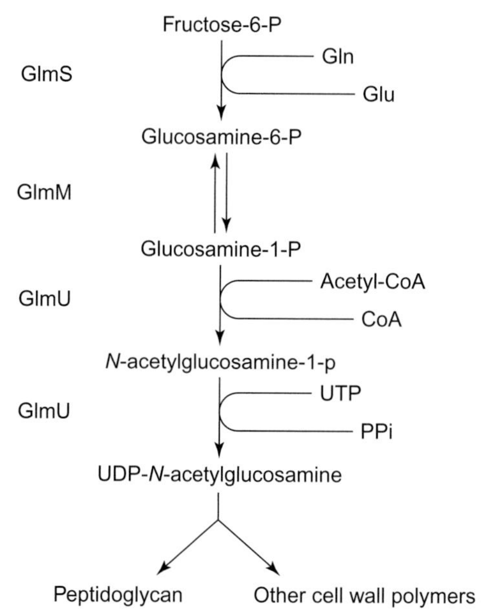

FIGURE 16.2 Biosynthesis of UDP-*N*-acetylglucosamine in bacteria. Abbreviations: acetyl-CoA, acetyl co-enzyme A; CoA, co-enzyme A; gln, glutamine; glu, glutamic acid; PPi, inorganic pyrophosphate; UDP, uridine diphosphate; UTP, uridine diphosphate.

White, 1968). First, the *glmS* gene product converts D-fructose-6-phosphate to D-glucosamine-6-phosphate. The N-terminal domain of GlmS catalyses hydrolysis of L-glutamine to L-glutamate and ammonia, whereas the C-terminal domain utilizes released ammonia to form D-glucosamine-6-phosphate. The 3D-structures of the two domains were determined and catalytic mechanisms proposed (Durand *et al.*, 2008 and references therein). In the second step (see Figure 16.2), the interconversion of D-glucosamine-6-phosphate and D-glucosamine-1-phosphate is catalysed by phosphoglucosamine mutase GlmM (Mengin-Lecreulx and van Heijenoort, 1996). The mutase was shown to be active only after phosphorylation of a specific serine and catalysis proceeds according to a ping-pong bi-bi mechanism involving D-glucosamine-1,6-diphosphate as intermediate (Jolly *et al.*, 1999).

The last two steps concern the acetylation and uridylation reactions (see Figure 16.2) carried out by the bifunctional *glmU* gene product (Mengin-Lecreulx and van Heijenoort, 1994). The C-terminal domain of GlmU catalyses acetylation of D-glucosamine-1-phosphate, whereas the N-terminal domain catalyses uridylation of the formed *N*-acetylglucosamine-1-phosphate to yield UDP-GlcNAc (Mengin-Lecreulx and van Heijenoort, 1994; Gehring *et al.*, 1996). The two domains are functionally independent and each one is essential. Crystal structure determinations revealed that the two domains are connected by a long α-helical arm and that three GlmU molecules form a homotrimeric arrangement (Mochalkin *et al.*, 2007 and references therein).

2.2. Formation of UDP-MurNAc

The formation of UDP-MurNAc (see Figure 16.1) is a two-step process. First, the addition of enolpyruvate from phosphoenolpyruvate to position-3 of *N*-acetylglucosamine in UDP-GlcNAc is catalysed by transferase MurA to yield UDP-GlcNAc-enolpyruvate. In the second step, the reduction of the enolpyruvate moiety to D-lactoyl is catalysed by reductase MurB to yield UDP-MurNAc. In general, Gram-negative bacteria have only one *murA* gene, whereas low-G + C Gram-positive bacteria have two. The MurA and MurB enzymes were purified from various organisms and their 3D structures determined (van Heijenoort, 2001b; Barreteau *et al.*, 2008 and references therein).

The reaction pathway of MurA was studied using rapid kinetics, phosphoenolpyruvate (PEP) analogues and site-directed mutagenesis (Skarzynski *et al.*, 1998 and references therein). An addition–elimination mechanism was proposed and proceeds through a tetrahedral intermediate (Figure 16.3). The ketal adduct formed between the two substrates is non-covalently bound to the enzyme and the stereochemical course of the enzymatic enolpyruvyl transfer was

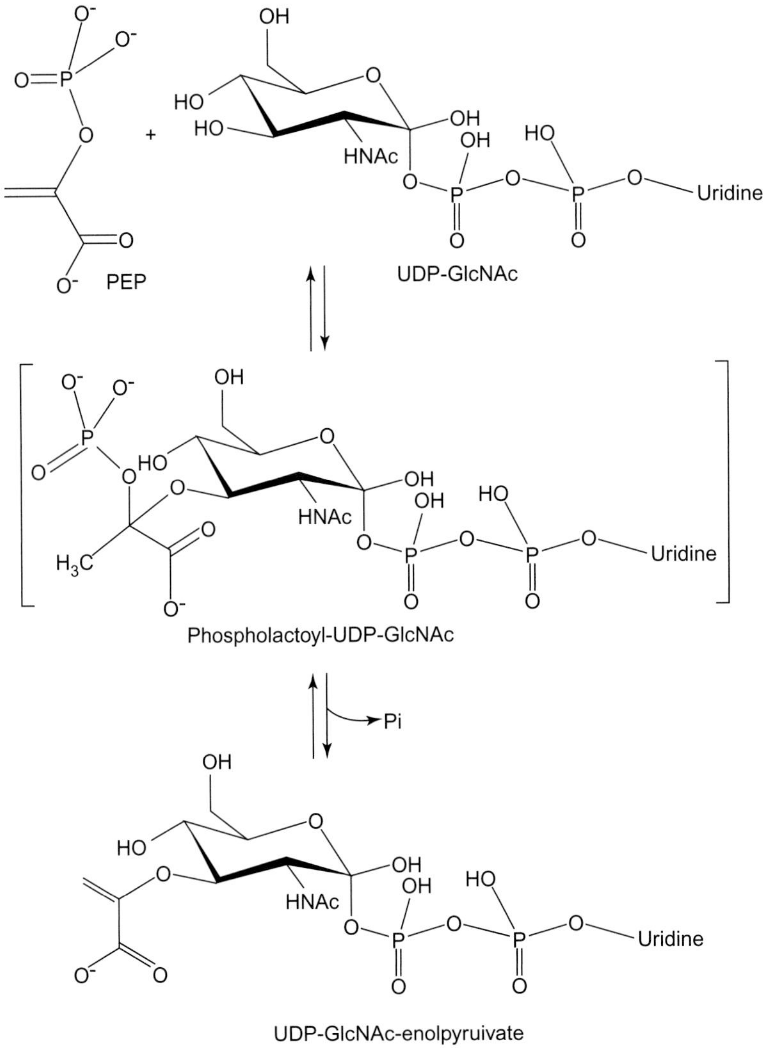

FIGURE 16.3 MurA-catalysed formation of UDP-*N*-acetylglucosamine-enolpyruvate. Abbreviations: PEP, phosphoenolpyruvate; P_i, inorganic phosphate; UDP, uridine diphosphate; UDP-GlcNAc, UDP-*N*-acetylglucosamine.

determined. Structural studies with *Escherichia coli* MurA revealed that it can assume three conformational states corresponding to different stages of the catalytic process.

Reductase MurB from *E. coli* is a flavoprotein and its mechanism (Figure 16.4) involves two half-reactions (Benson *et al.*, 1997 and references therein). First, tightly bound flavine adenine dinucleotide (FAD) is reduced by the two-electron transfer from nicotinamide adenine dinucleotide phosphate (NADPH) (see Figure 16.4A). Second (see Figure 16.4B), the two-electron transfer from E-$FADH_2$ to the C-3 of the enol ether of UDP-GlcNAc-enolpyruvate is followed by the delivery at C-2 of a proton from the active-site serine of MurB.

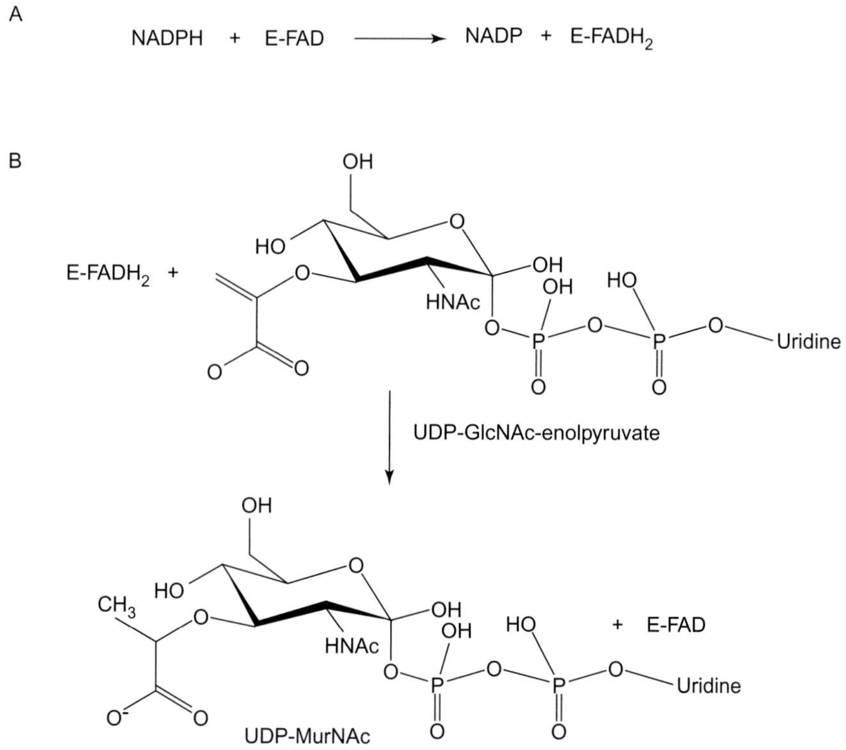

A

$$NADPH + E\text{-}FAD \longrightarrow NADP + E\text{-}FADH_2$$

B

$E\text{-}FADH_2$ + UDP-GlcNAc-enolpyruvate → UDP-MurNAc + E-FAD

FIGURE 16.4 MurB-catalysed reduction of UDP-*N*-acetylglucosamine-enolpyruvate to UDP-*N*-acetylmuramic acid. Abbreviations: FAD, flavine adenine dinucleotide (oxidized form); $FADH_2$, flavine adenine dinucleotide (reduced form); NAD, nicotinamide adenine dinucleotide; NADPH, nicotinamide adenine dinucleotide phosphate (reduced form); UDP, uridine diphosphate; UDP-GlcNAc, UDP-*N*-acetylglucosamine; UDP-MurNAc, UDP-*N*-acetylmuramic acid.

2.3. Formation of the UDP-MurNAc-peptide precursors

The formation of the UDP-MurNAc-peptide precursors proceeds (see Figure 16.1) by the stepwise addition of L-alanine (less frequently glycine), D-glutamic acid, a diamino acid (usually L-lysine or diaminopimelic acid, less frequently another diamino acid or homoserine) and D-alanyl-D-alanine (less frequently D-alanyl-D-lactate or D-alanyl-D-serine) onto the D-lactoyl group of UDP-MurNAc (van Heijenoort, 2001b; Smith, 2006; Barreteau *et al.*, 2008 and references therein). At each step a specific cytoplasmic enzyme, designated as Mur synthetase, catalyses the formation of an amide or peptide bond with concomitant cleavage of adenosine triphosphate (ATP) into adenosine diphosphate (ADP) and inorganic phosphate (Figure 16.5).

The isolation of conditional-lethal mutants, characterized by a cell-lysis phenotype, was developed in different organisms. In *E. coli*, the Mur synthetase genes (*murC, murD, murE,* and *murF*) are unique, essential and are located in the *dcw* cluster containing both peptidoglycan synthesis and cell division genes. Cognate *mur* gene sequences from a wide variety of bacterial genera are now known, many of them also belonging to a division-cell wall cluster. Although the Mur synthetases have limited overall sequence identity, several specific motifs with high identity reveal the existence of common invariants.

MurC UDP-MurNAc + L-Ala + ATP ⟶ UDP-MurNAc-L-Ala + ADP + Pi

MurD UDP-MurNAc-L-Ala + D-Glu + ATP ⟶ UDP-MurNAc-L-Ala-D-Glu + ADP + Pi

MurE UDP-MurNAc-L-Ala-D-Glu + L-Lys (or A_2pm) + ATP ⟶ UDP-MurNAc-tripeptide + ADP + Pi

MurF UDP-MurNAc-tripeptide + D-Ala-D-Ala + ATP ⟶ UDP-MurNAc-pentapeptide + ADP + Pi

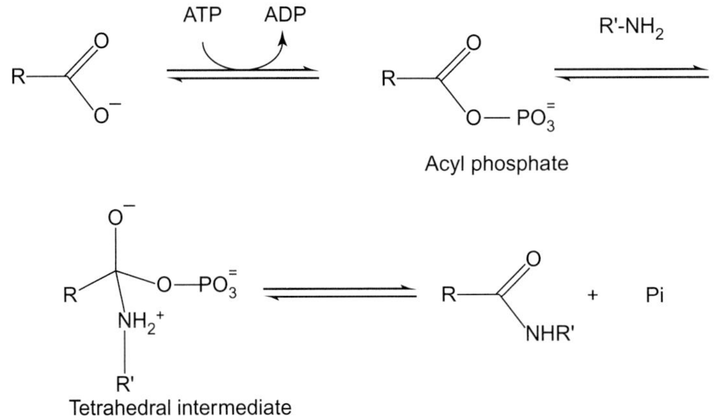

ATP ADP

R'-NH_2

Acyl phosphate

+ Pi

Tetrahedral intermediate

FIGURE 16.5 Stepwise assembly of the peptide subunit and reaction mechanism of the Mur synthetases, where R = UDP-MurNAc or UDP-MurNAc-peptide and R′ = amino acid or dipeptide. Abbreviations: A_2pm, diaminopimelic acid; ADP, adenosine diphosphate; ATP, adenosine triphospahte; L-Ala, L-alanine; D-Glu, D-glutamic acid; L-lys, L-lysine; Pi, inorganic phosphate; UDP-MurNAc, UDP-*N*-acetylmuramic acid.

The structures of the four Mur synthetases were found to share the same three-domain topology with an N-terminal domain responsible for binding the UDP-MurNAc-peptide, a central ATP-binding domain and a C-terminal domain associated with binding the incoming amino acid or dipeptide. Domains 2 and 3 have conserved topology whereas that of the N-terminal domain differs from one synthetase to another. The Mur synthetases have also a similar active site architecture lying at the junction of the three structural domains and comprising specific binding pockets for the three substrates distributed around the catalytic centre.

The Mur synthetases all operate by a similar catalytic mechanism (see Figure 16.5) entailing the carboxy activation of a C-terminal amino acid residue of the nucleotide substrate to an acyl phosphate intermediate followed by the nucleophilic attack by the amino group of the condensing amino acid or dipeptide, with the elimination of phosphate and subsequent amide or peptide bond formation. The reversible formation of the acyl phosphate intermediate was substantiated and the possible presence of a tetrahedral transition state following the acyl phosphate was suggested. Being closely structurally and functionally related proteins, the Mur synthetases thus appear as a well-defined subfamily of the ATP-dependent ligases (Smith, 2006).

2.4. Lipid intermediates in the biosynthesis of peptidoglycan

The study of cell-free peptidoglycan-synthesizing systems using UDP-GlcNAc, radiolabelled UDP-MurNAc-pentapeptide and particulate preparations from *Staphylococcus aureus* and *Micrococcus luteus* led to the discovery of lipid intermediates I and II and to the determination of their respective role in the pathway (Chatterjee and Park, 1964; Meadow *et al.*, 1964; Anderson *et al.*, 1965). The first membrane step involves the transfer of the phospho-MurNAc-pentapeptide moiety of UDP-MurNAc-pentapeptide to undecaprenyl phosphate to yield lipid I and uridine monophosphate (UMP) (Figure 16.6). Thereafter, *N*-acetylglucosamine is added onto lipid I to yield lipid II. The structures of lipids I and II

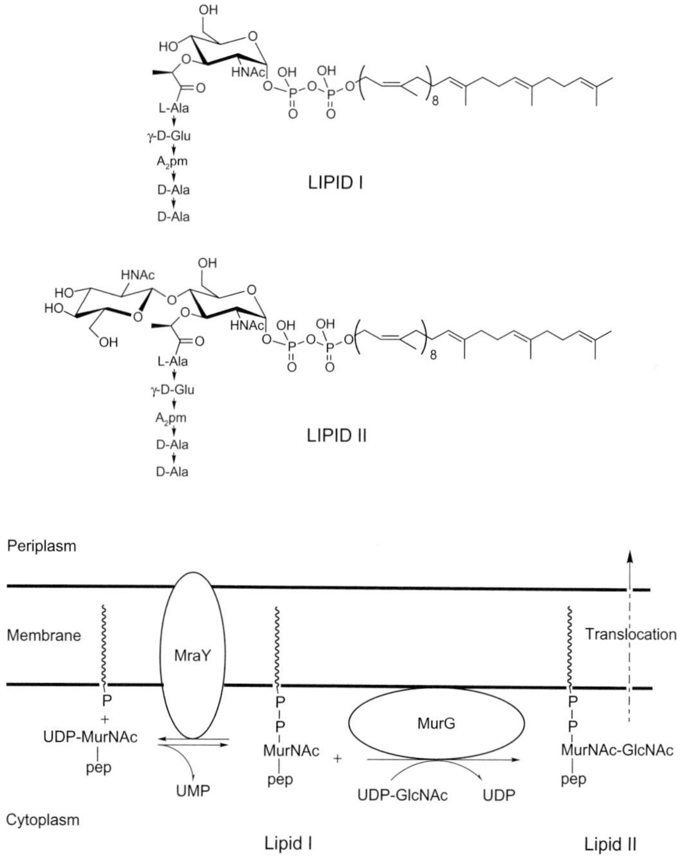

FIGURE 16.6 Structure of lipids I and II, and their synthesis by transferases MraY and MurG. Pep = pentapeptide. Abbreviations: UDP, uridine diphosphate; UDP-GlcNAc, UDP-*N*-acetylglucosamine; UDP-MurNAc, UDP-*N*-acetylmuramic acid; UMP, uridine monophosphate. Reproduced, with permission, from van Heijenoort (2007).

(see Figure 16.6) were established after isolation from cell-free systems (Higashi *et al.*, 1967). Subsequently, they were characterized from various organisms. Their low pool levels, their limited accumulation in cell-free systems and the tedious work of their isolation restricted for decades their availability for the study of the membrane steps of the pathway. These difficulties have now been overcome by their chemical or enzymatic synthesis (Welzel, 2005; van Heijenoort, 2007; Bouhss *et al.*, 2008 and references therein).

2.5. Biosynthesis of lipid I by transferase MraY

The formation of lipid I is catalysed by integral membrane protein MraY which is essential and unique (Ikeda *et al.*, 1991; Boyle and Donachie, 1998). A common 2D membrane topology model was established for the *E. coli* and *S. aureus* MraYs (Bouhss *et al.*, 1999), which has ten transmembrane segments, five cytoplasmic domains and six periplasmic domains including the N- and C-terminal ends. The cytoplasmic domains are involved in substrate recognition and catalysis, thereby indicating that lipid I is formed at the inside surface of the membrane (Bouhss *et al.*, 1999; Price and Momany, 2005).

The MraY enzyme belongs to the UDP-D-*N*-acetylhexosamine: polyprenol phosphate D-*N*-acetylhexosamine-1-P transferase family (Price and Momany, 2005). A multistep catalytic mechanism (Figure 16.7) was proposed (Neuhaus, 1971; Ikeda *et al.*, 1991; Lloyd *et al.*, 2004; Price and Momany, 2005). The MraY reaction is fully reversible and takes place with conservation of the α-anomeric configuration of the MurNAc residue. In most eubacteria, undecaprenyl phosphate is assumed to be the substrate of MraY. However, the recent structural analysis of lipids I and II from *Mycobacterium smegmatis* demonstrated the predominant presence of decaprenol (Mahapatra *et al.*, 2005). Purification of MraY from *Bacillus subtilis* (Bouhss *et al.*, 2004) has not yet led to its crystallization.

2.6. Biosynthesis of lipid II by transferase MurG

The transfer of *N*-acetylglucosamine from UDP-GlcNAc onto the C4 hydroxyl of the MurNAc unit of lipid I is catalysed by transferase MurG (Mengin-Lecreulx *et al.*, 1991). The formation of the β-(1→4)-linkage is accompanied by inversion of the anomeric configuration of *N*-acetylglucosamine (Figure 16.8). The MurG protein belongs to the GT-B glycosyltransferase (GT) superfamily (Ünligil and Rini, 2000). It is unique and essential. Mainly MurG from *E. coli* has been studied (van Heijenoort, 2001a, 2007; Bouhss *et al.*, 2008 and references therein) and shown to be associated with the inner side of the cytoplasmic membrane, thereby establishing that the entire peptidoglycan monomer unit is assembled inside the cell prior to translocation across the membrane. Thus, MurG is a key enzyme at the junction between the two stages of peptidoglycan synthesis.

The crystal structure of MurG consists of two domains separated by a deep cleft (Ha *et al.*, 2000). The binding site for UDP-GlcNAc was proposed to be in the C-terminal domain and that for lipid I in the N-terminal domain in which a hydrophobic patch surrounded by basic residues could be the site of interaction with the negatively charged membrane. Sequence comparison of various orthologues confirmed the extrinsic and cationic characters of MurG and its functioning as a moderately hydrophobic peripheral protein. Detailed enzymatic studies were carried out with lipid I analogues. The partial reversibility of the reaction was established, the specificities of the acceptor and donor sites were investigated and an ordered bi-bi mechanism was demonstrated.

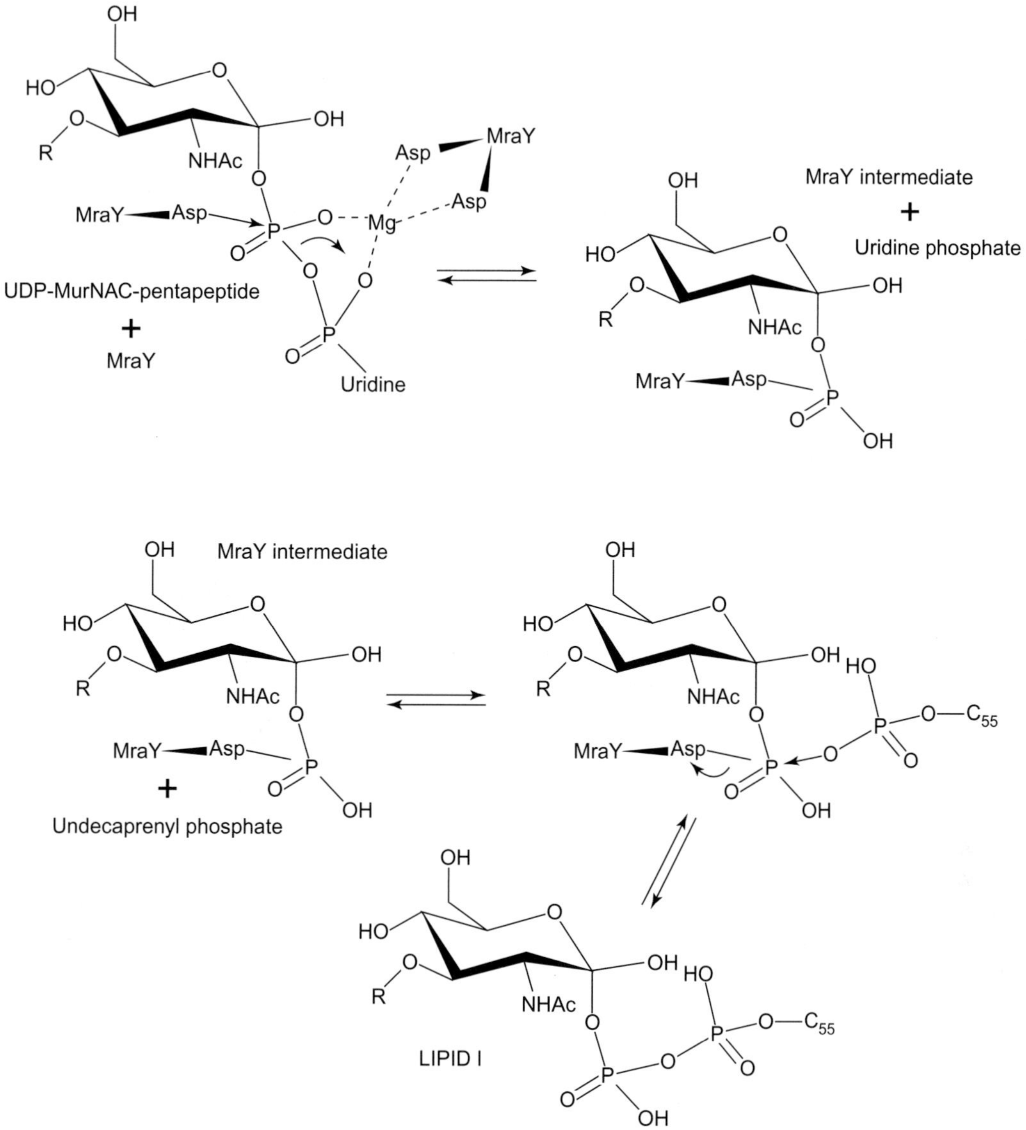

FIGURE 16.7 Multistep MraY-catalysed formation of lipid I. Abbreviations: Asp, aspartic acid; UDP, uridine diphosphate; UDP-MurNAc, UDP-*N*-acetylmuramic acid. Reproduced, with permission, from van Heijenoort (2007).

2.7. Side pathways

Besides the UDP-MurNAc-peptide and lipid intermediate precursors, the functioning of the pathway is dependent on a number of other substrates (see Figure 16.1). Some, like acetyl co-enzyme A (acetyl-CoA), uridine triphosphate (UTP), PEP, ATP, L-alanine, L-lysine and diaminopimelic acid, are involved in various metabolic reactions. Other ones, like D-alanine, D-glutamic acid, dipeptide D-alanyl-X, undecaprenyl phosphate, are more

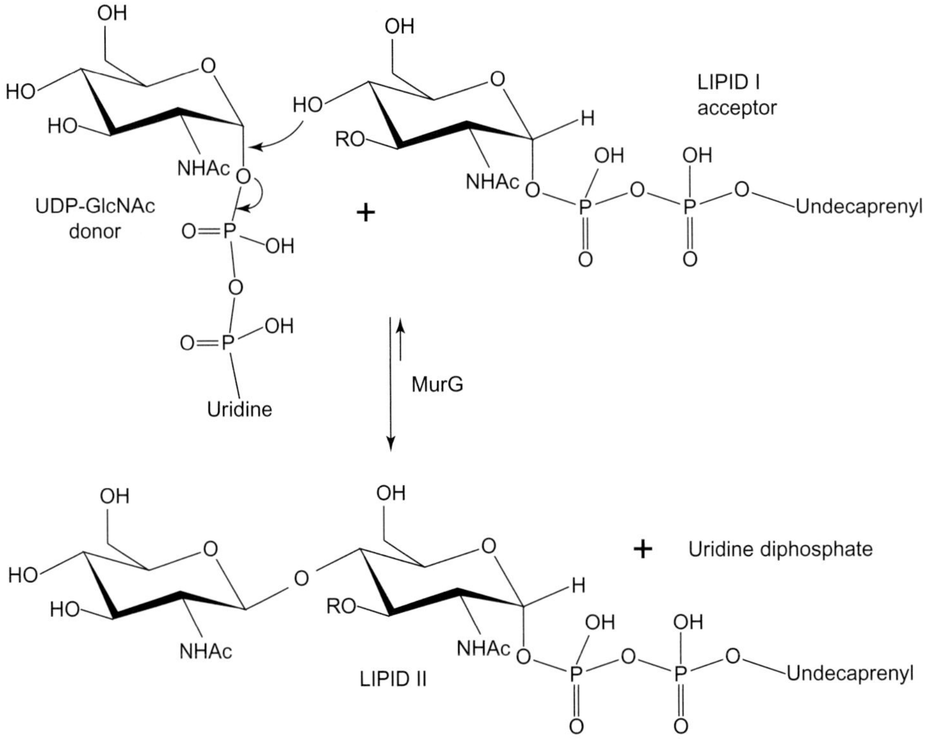

FIGURE 16.8 MurG-catalysed formation of lipid II. Abbreviations: UDP, uridine diphosphate; UDP-GlcNAc, UDP-*N*-acetyl-glucosamine. Reproduced, with permission, from van Heijenoort (2007).

specific of the biosynthesis of peptidoglycan or other cell wall polymers and their formation has been reviewed in detail (van Heijenoort, 2001b; Barreteau *et al.*, 2008; Bouhss *et al.*, 2008).

3. TRANSLOCATION OF THE MONOMER UNIT

Prior to its use as substrate in extracytoplasmic polymerization, lipid II is translocated through the hydrophobic environment of the cytoplasmic membrane (van Heijenoort, 2001a, 2007; Bouhss *et al.*, 2008 and references therein). Considering its low pool level, its translocation must be a fast and unidirectional process in order to sustain a steady peptidoglycan synthesis in growing cells. More than four decades after its discovery, the mechanism of its translocation remains unknown. Fluorescence spectrometry experiments carried out in lipid vesicles with a fluorescent lipid II analogue showed that there was no spontaneous move of the lipid intermediate across the bilayer (van Dam *et al.*, 2007). Nevertheless, translocation was observed when *E. coli* inner membrane vesicles were used, revealing that an intact translocation machinery was present and likely composed of membrane proteins. Additional experiments indicated that lipid II translocation was perhaps coupled to ongoing transglycosylation. Earlier results had suggested that translocation of lipid II was dependent on phospholipid synthesis and was likely to be more than just a flip-flop mechanism (Ehlert and Höltje, 1996). Interestingly, protein-mediated mechanisms were reported for the transfer of enterobacterial common antigen and O-antigen units across the membrane (van Heijenoort, 2007; Bouhss *et al.*, 2008 and references therein).

4. POLYMERIZATION OF THE MONOMER UNIT

4.1. Glycan chain formation by transglycosylation

Peptidoglycan glycan chains are assembled by polymerization of the GlcNAc-β-(1→4)-MurNAc-disaccharide-peptide unit of lipid II with formation of β-(1→4)-linkages (Figure 16.9). The transglycosylation reaction is accompanied by inversion of the α-anomeric configuration of the MurNAc residue and thus leads to linear chains containing exclusively β-(1→4)-linkages (van Heijenoort, 2001a and references therein). In several cases, it was established that the reaction takes place at the reducing end of the growing chain which is transferred as the donor substrate to the C4 hydroxyl of the GlcNAc unit of lipid II acting as the acceptor substrate (van Heijenoort, 2001a; Perlstein *et al.*, 2007 and references therein).

The GTs responsible for the polymerization reaction come in two forms (Goffin and Ghuysen, 1998; van Heijenoort, 2001a, 2007 and references therein), namely as an N-terminal module in bifunctional class A penicillin-binding proteins (PBPs), which also contain a C-terminal transpeptidase (TP) module, and as monofunctional glycosyltransferases. They

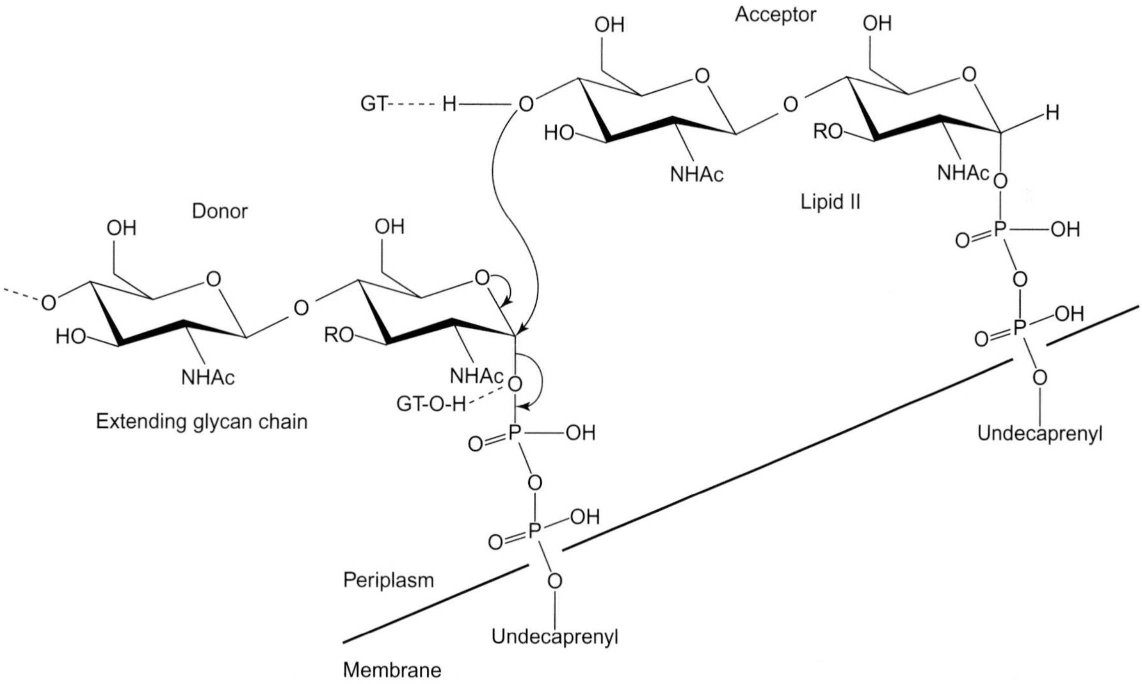

FIGURE 16.9 Mechanism of transglycosylation with chain elongation at the reducing end. R = D-lactoyl-peptide. Abbreviation: GT, glycosyltransferase. Reproduced, with permission, from van Heijenoort (2007).

show high sequence similarity, belong to the GT51 family in the sequence-based classification of GTs and possess five conserved motifs (Goffin and Ghuysen, 1998; Coutinho *et al.*, 2003). Presently, more than 15 GTs have been purified and most assayed for their activity with lipid II or analogues as substrate (van Heijenoort, 2007; Sauvage *et al.*, 2008 and references therein).

Analysis of the reaction products formed by purified PBP1b from *E. coli* revealed an average glycan chain length of over 25 disaccharide units and almost 50% of cross-linked peptides (Bertsche *et al.*, 2005). The recent resolution of the crystal structures of PBP2 from *S. aureus* (Lovering *et al.*, 2007) and of the GT domain of *Aquifex aeolicus* (Yuan *et al.*, 2007) opens the way to more systematic studies of the GT catalytic site and reaction mechanism. PBP2a from *S. aureus* has a bi-lobal fold with the GT and TP domains separated by a short β-rich linker region (Lovering *et al.*, 2007).

4.2. Cross-bridge formation by transpeptidation

The cross-linking between the peptide subunits of the glycan chains is essential for the formation of the 3D structure of peptidoglycan (see Chapter 2). It involves two types of transpeptidation reactions (Goffin and Ghuysen, 2002). Type (4→3) or DD-transpeptidation (Figure 16.10) is the main one and entails the formation of a

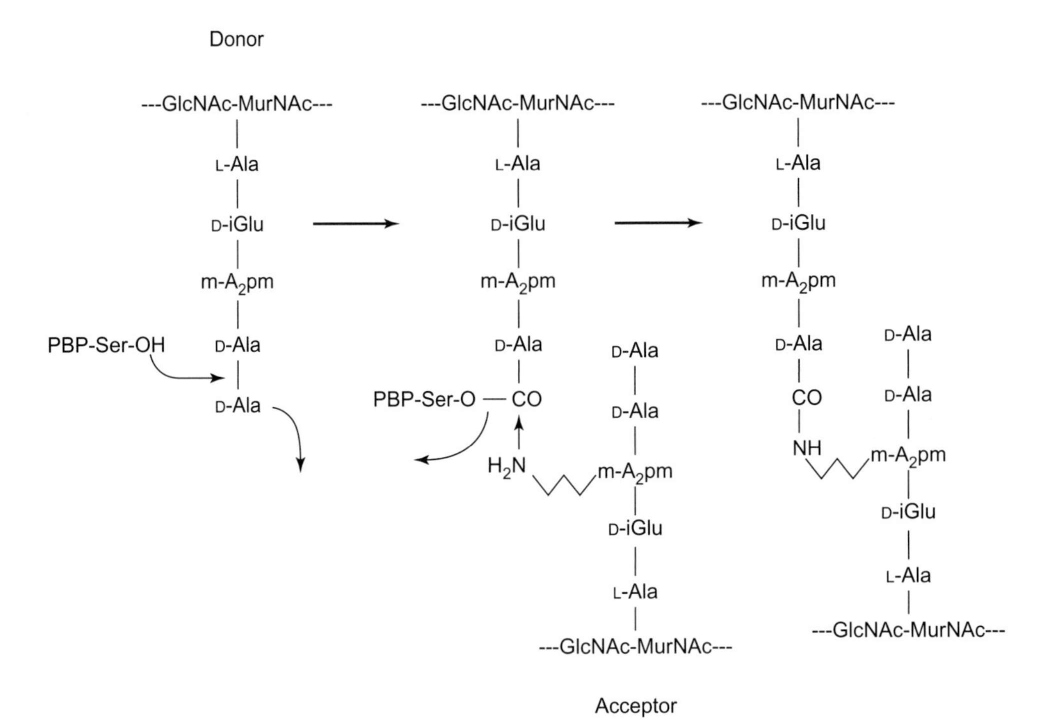

FIGURE 16.10 Cross-linking by DD-transpeptidation between donor and acceptor peptide subunits. Abbreviations: A_2pm, diaminopimelic acid; D-, L-Ala, D-, L-alanine; D-Glu, D-glutamic acid; D-iGlu, D-isoglutamic acid; GlcNAc, *N*-acetylglucosamine; MurNAc, *N*-acetylmuramic acid; PBP, penicillin-binding protein; ser, serine. Reproduced, with permission, from Zapun *et al.* 2008b. Penicillin-binding proteins and β-lactam resistance. FEMS Microbiol. Rev. 32, 361–385.

peptide bond between the carboxyl group of the D-alanine in position-4 of a donor peptide subunit and the N^{ε} amino group of L-lysine or diaminopimelic acid in position-3 of an acceptor peptide subunit, concomitantly with the release of the donor C-terminal D-alanine. The DD-transpeptidases responsible for this reaction belong to the PBP family. The PBPs are acyltransferases specifically inhibited by the covalent binding of β-lactam antibiotics to their active site (Goffin and Ghuysen, 2002; Macheboeuf *et al.*, 2006; Sauvage *et al.*, 2008; Zapun *et al.*, 2008b). They come in two forms: as C-terminal modules in class A PBPs in association with an N-terminal GT module or without it as in class B PBPs. Furthermore, they are bound to the cytoplasmic membrane with the bulk of the polypeptide chain exposed to the outside. Like all PBPs, DD-transpeptidases have the three conserved motifs of the SXXK serine enzyme superfamily. In the known 3D structures of TP domains, the three motifs are close to each other near the boundary of the catalytic centre. The active site serine of the SXXK motif is the catalytic nucleophile giving rise to an acyl-enzyme intermediate with the donor peptide subunit prior to the formation of the cross-linking with the acceptor subunit (see Figure 16.10).

Type (3→3) or LD-transpeptidation functions in parallel with DD-transpeptidation in a number of organisms (Goffin and Ghuysen, 2002). This entails the formation of a peptide bond between the carboxyl of L-lysine or diaminopimelic acid in position-3 of a donor tetrapeptide subunit and the side chain amino group of the diamino acid in position-3 of the acceptor peptide subunit, concomitantly with the release of the donor C-terminal D-alanine. With the exception of carbapenems, LD-transpeptidases are insensitive to β-lactams. Present in many organisms, they play a critical role in the mechanisms of resistance to antibiotics (Mainardi *et al.*, 2008) and in the covalent binding of proteins to peptidoglycan (Dramsi *et al.*, 2008).

5. VARIATIONS IN PEPTIDOGLYCAN BIOSYNTHESIS

Modifications of the basic structure of peptidoglycan are encountered in most eubacteria (see Chapter 2; Vollmer, 2008; Vollmer *et al.*, 2008). Moreover, variations in a given bacterium are observed under many circumstances (growth phase, growth conditions, antibiotic treatments, mutations, etc.). The physiological significance of many structural modifications is yet poorly understood. They are introduced either on the precursors of the monomer unit or during and after the polymerization reactions. They imply a certain variability of the specificities of the enzymes of the pathway or the presence of additional activities. Since the enzymatic specificities are not quite the same from one step to another, any change at a given step must be accepted by the following ones. Therefore, the cumulative effect of the different specificities along the pathway determines the extent of flexibility of peptidoglycan biosynthesis and that of its structural variability. It should be noted that the reactions responsible of modifications are not necessarily complete. Consequently, the presence of homologous precursors can lead to the functioning in parallel of more than one pathway (van Heijenoort, 2001b, 2007; Barreteau *et al.*, 2008 and references therein). It should also be stressed that certain mechanisms of resistance to β-lactam and glycopeptide antibiotics imply modifications in the biosynthesis of peptidoglycan (Mainardi, 2008; Zapun *et al.*, 2008b).

6. *IN VIVO* FUNCTIONING OF THE MONOMER UNIT ASSEMBLY

The mechanisms regulating the expression of the genes of the pathway have been studied to a limited extent. The expression of the *murC, murD, murE, murF, murG* and *mraY* genes of the *E. coli dcw* cluster was shown to be mainly dependent on one promoter, namely P_{mra} (Mengin-Lecreulx *et al.*, 1998). More recently, different riboregulatory mechanisms were described in *B. subtilis* and *E. coli* for the control of the expression of the *glmS* gene responsible for the synthesis of D-glucosamine-6-phosphate (Winkler *et al.*, 2004; Urban *et al.*, 2007).

By comparing precursor pool levels, specific enzyme activities and rates of peptidoglycan synthesis, an overall view of the *in vivo* functioning of the *E. coli* pathway was established (reviewed in van Heijenoort, 2001b). Many enzymes were found to be more or less constitutive, their specific activity varying little with growth rate and only to a certain extent with growth phase. The specific activities are in excess or at least adjusted to the requirements of fast-growing cells. Possible control mechanisms by feed-back inhibition involving specific cell effectors were proposed for different steps. Furthermore, the functioning of the pathway from UDP-GlcNAc to UDP-MurNAc-pentapeptide (see Figure 16.1) was shown to depend on the rate of synthesis of UDP-GlcNAc. The formation of lipid I is limited by the pool of undecaprenyl phosphate and the reversibility of the MraY reaction (see Figure 16.6). During growth of many organisms, a turnover of peptidoglycan with release of soluble fragments into the medium is observed. However, in some Gram-negative bacteria, the turnover is accompanied by a recycling process whereby released fragments are incorporated back into the cell and reused in the assembly of the monomer unit (Park and Ueharan, 2008 and references therein).

7. *IN VIVO* FUNCTIONING OF THE POLYMERIZATION PROCESS

The correlation between the transglycosylation and transpeptidation reactions was investigated with numerous organisms and cell-free systems (Rogers *et al.*, 1980; Ward, 1984; van Heijenoort, 2001a and references therein). In growing cells, they are tightly coupled in agreement with the bifunctionality of the class A PBPs. However, transglycosylation can proceed independently from transpeptidation as exemplified for instance by the formation of soluble material after inhibition of transpeptidation by β-lactams. This clearly suggests that transpeptidation is necessary for the attachment of new material to pre-existing peptidoglycan. No transpeptidation was observed when transglycosylation was specifically inhibited by moenomycin (van Heijenoort, 2001a and references therein) or when GT-inactivated PBP1a from *E. coli* was used in a polymerization assay with lipid II (Born *et al.*, 2006).

Peptidoglycan being an essential continuous structural component of the cell envelope, the incorporation of new material is necessarily closely correlated with the complex machineries of cell growth and cell division and thus involves both spatial and temporal constraints. The localization of incorporation was analysed by various labelling and microscopy approaches (den Blaauwen *et al.*, 2008; Zapun *et al.*, 2008a and references therein). The results are consistent with the presence of different peptidoglycan-synthesizing systems. In rod-shaped organisms, one is associated with cell elongation whereas another one is involved in cell septation (den Blaauwen *et al.*, 2008). In cocci, most material is incorporated during septation, but additional systems were proposed to account for preseptal, peripheral and thickening processes (Zapun *et al.*, 2008a). A major difficulty with the study of the peptidoglycan-synthesizing complexes is the presence in many organisms of a multiplicity of peptidoglycan GT and TP activities (Goffin and Ghuysen, 1998, 2002). Efforts are made to specify with which system each activity is associated and several models of their *in vivo* functioning have been proposed (Scheffers and Pinho, 2005; Sauvage *et al.*, 2008; Zapun *et al.*, 2008a and references therein).

8. INHIBITION OF PEPTIDOGLYCAN BIOSYNTHESIS

An historically important aspect of peptidoglycan biosynthesis is its specific inhibition by antibiotics, some of which are well-known drugs in clinical use (β-lactams, glycopeptides, fosfomycin, bacitracin, etc.) or in use as animal growth promoters (moenomycin, etc.). The elucidation of the mechanisms of action of these drugs and that of numerous other synthesized inhibitors often paralleled the study of individual steps of the pathway. Inhibitors for practically every step have now been described and they act by targeting an enzyme or a precursor.

Non-covalent binding with an enzyme was observed with many steps of the assembly of the monomer unit (Dini, 2005; Barreteau *et al.*, 2008; Bouhss *et al.*, 2008 and references therein) and also with inhibition of GT activities by moenomycin (Welzel, 2005 and references therein). Covalent binding to an enzyme with formation of stable intermediates is exemplified with fosfomycin and D-cycloserine (Gale *et al.*, 1981; Barreteau *et al.*, 2008) and, most of all, with β-lactams acting as suicide substrates of the PBPs (Goffin and Ghuysen, 2002; Macheboeuf *et al.*, 2006; Sauvage *et al.*, 2008; Zapun *et al.*, 2008b). The formation of non-covalent complexes between certain antibiotics (glycopeptides, lantibiotics, ramoplanin and others) and lipid II or nascent peptidoglycan interferes with the polymerization reactions (Kahne *et al.*, 2005; Walker *et al.*, 2005; Chatterjee *et al.*, 2005; Mainardi *et al.*, 2008 and references therein).

The cellular effect of a peptidoglycan inhibitor is dependent on target accessibility. By their size and/or polar structure, many potent inhibitors of isolated enzymes cannot penetrate the cytoplasmic membrane. The efficiency of fosfomycin and D-cycloserine is due to the use of endogeneous transport systems (Gale *et al.*, 1981). The extracytoplasmic location of the GT and TP activities, as well as that of lipid II or nascent peptidoglycan, explains in part their efficient use as targets. In growing cells, the inhibition of a specific step of the pathway will bring about an accumulation of the upstream precursors and a decrease of the downstream precursor pools with arrest of the polymeriztion reactions. This in turn leads to bacteriostasis or to a bactericidal effect often accompanied with morphological changes and cell lysis.

9. CONCLUDING REMARKS

Although many aspects of peptidoglycan biosynthesis have been studied, important areas call for more thorough investigations. Understanding the organization of the peptidoglycan-synthesizing complexes and their functioning in coordination with cell growth and cell division is undoubtedly the most challenging problem and will need many years of sustained efforts. Closely related with this aspect is the presence in most organisms of numerous specific peptidoglycan hydrolases. They have only been studied in some cases and more systematic work is required for determining the roles they play. Furthermore, the genetics, biochemistry and physiological meaning of many structural modifications of peptidoglycan (amidation, methylation, phosphorylation, etc.) remain unexplored. Finally, various mechanisms of resistance to antibiotics involve modifications in peptidoglycan synthesis, several of which are not yet clearly understood. The need to overcome resistance mechanisms and to use specific targets for the search of novel antibacterials also fully warrants the continued study of peptidoglycan biosynthesis.

ACKNOWLEDGEMENTS

This work was supported by grant UMR8619 from the Centre National de la Recherche Scientifique.

References

Anderson, J.S., Matsuhashi, M., Haskin, M.A., Strominger, J.L., 1965. Lipid-phosphoacetylmuramyl-pentapeptide and lipid-phosphodisaccharide-pentapeptide: presumed membrane transport intermediates in cell wall synthesis. Proc. Natl. Acad. Sci. USA 53, 881–889.

Barreteau, H., Kovač, A., Boniface, A., Sova, M., Gobec, S., Blanot, D., 2008. Cytoplasmic steps of peptidoglycan biosynthesis. FEMS Microbiol. Rev. 32, 168–207.

Benson, T.E., Walsh, C.T., Massey, V., 1997. Kinetic characterization of wild-type and S229A mutant MurB: evidence for the role of Ser 229 as general acid. Biochemistry 36, 796–805.

Bertsche, U., Breukink, E., Kast, T., Vollmer, W., 2005. *In vitro* murein (peptidoglycan) synthesis by dimers of the bifunctional transglycosylase-transpeptidase PBP1B from *Escherichia coli*. J. Biol. Chem. 280, 38096–38101.

Born, P., Breukink, E., Vollmer, W., 2006. *In vitro* synthesis of cross-linked murein and its attachment to sacculi by PBP1A from *Escherichia coli*. J. Biol. Chem. 281, 26985–26993.

Bouhss, A., Mengin-Lecreulx, D., Le Beller, D., van Heijenoort, J., 1999. Topological analysis of the MraY protein catalysing the first membrane step of peptidoglycan synthesis. Mol. Microbiol. 34, 576–585.

Bouhss, A., Crouvoisier, M., Blanot, D., Mengin-Lecreulx, D., 2004. Purification and characterization of the bacterial MraY translocase catalyzing the first membrane step of peptidoglycan synthesis. J. Biol. Chem. 279, 29974–29980.

Bouhss, A., Trunkfield, A.E., Bugg, T.D.H., Mengin-Lecreulx, D., 2008. The biosynthesis of peptidoglycan lipid-linked intermediates. FEMS Micriob. Rev. 32, 208–233.

Boyle, D.S., Donachie, W.D., 1998. *mraY* is an essential gene for cell growth in *Escherichia coli*. J. Bacteriol. 180, 6429–6432.

Bugg, T.D.H., Walsh, C.T., 1992. Intracellular steps of bacterial cell wall peptidoglycan synthesis: enzymology, antibiotics and antibiotic resistance. Nat. Prod. Rep. 9, 199–215.

Chatterjee, A.N., Park, J.T., 1964. Biosynthesis of cell wall mucopeptide by a particulate fraction from *Staphylococcus aureus*. Proc. Natl. Acad. Sci. USA 51, 9–16.

Chatterjee, C., Paul, M., Xie, L., van der Donk, W.A., 2005. Biosynthesis and mode of action of lantibiotics. Chem. Rev. 105, 633–683.

Coutinho, P.M., Deleury, E., Davies, G.J., Henrissat, B., 2003. An evolving hierarchical family classification of glycosyltransferases. J. Mol. Biol. 328, 307–317.

den Blaauwen, T., de Pedro, M.A, Nguyen-Distèche, M., Ayala, J.A., 2008. Morphogenesis of rod-shaped sacculi. FEMS Microbiol. Rev. 32, 321–344.

Dini, C., 2005. MraY inhibitors as novel antibacterial agents. Curr. Top. Med. Chem. 5, 1221–1236.

Dobrogosz, W.J., 1968. Effect of amino sugars on catabolite repression in *Escherichia coli*. J. Bacteriol. 95, 578–584.

Dramsi, S., Magnet, S., Davison, S., Arthur, M., 2008. Covalent attachment of proteins to peptidoglycan. FEMS Microbiol. Rev. 32, 307–320.

Durand, P., Golinelli-Pimpaneau, B., Mouilleron, S., Badet, B., Badet-Denisot, M.-A., 2008. Highlights of glucosamine-6P synthase catalysis. Arch. Biochem. Biophys. 474, 302–317.

Ehlert, K., Höltje, J.-V., 1996. Role of precursor translocation in coordination of murein and phospholipid synthesis in *Escherichia coli*. J. Bacteriol. 178, 6766–6771.

Gale, E.F., Cundiffe, E., Reynolds, P.E., Richmond, M.H., Warning, M.J., 1981. The Molecular Basis of Antibiotic Action. John Wiley and Sons, London.

Gehring, A.M., Lees, W.J., Mindiola, D.J., Walsh, C.T., Brown, E.D., 1996. Acetyltransfer precedes uridylyltransfer in the formation of UDP-*N*-acetylglucosamine in separable active sites of the bifunctional GlmU protein of *Escherichia coli*. Biochemistry 35, 579–585.

Goffin, C., Ghuysen, J.-M., 1998. Multimodular penicillin-binding proteins: an enigmatic family of orthologs and paralogs. Microbiol. Mol. Biol. Rev. 62, 1079–1093.

Goffin, C., Ghuysen, J.-M., 2002. Biochemistry and comparative genomics of SxxK superfamily acyltransferases offer a clue to the mycobacterial paradox: presence of penicillin-susceptible target proteins versus lack of efficiency of penicillin as therapeutic agent. Microbiol. Mol. Biol. Rev. 66, 702–738.

Green, D.W., 2002. The bacterial cell wall as a source of antibacterial targets. Expert Opin. Ther. Targets 6, 1–19.

Ha, S., Walker, D., Shi, Y., Walker, S., 2000. The 1.9 Å crystal structure of *Escherichia coli* MurG, a membrane-associated glycosyltransferase involved in peptidoglycan biosynthesis. Protein Sci. 9, 1045–1052.

Higashi, Y., Strominger, J.L., Sweeley, C.C., 1967. Structure of a lipid intermediate in cell wall peptidoglycan synthesis: a derivative of a C_{55} isoprenoid alcohol. Proc. Natl. Acad. Sci. USA 57, 1878–1884.

Ikeda, M., Wachi, M., Jung, H.K., Ishino, F., Matsuhashi, M., 1991. The *Escherichia coli mraY* gene encoding UDP-*N*-acetylmuramoyl-pentapeptide:undecaprenyl-phosphate phospho-N-acetylmuramoyl-pentapeptide transferase. J. Bacteriol. 173, 1021–1026.

Jolly, L., Ferrari, P., Blanot, D., van Heijenoort, J., Fassy, F., Mengin-Lecreulx, D., 1999. Reaction mechanism of phosphoglucosamine mutase from *Escherichia coli*. Eur. J. Biochem. 262, 202–210.

Kahne, D., Leimkuhler, C., Lu, W., Walsh, C., 2005. Glycopeptide and lipoglycopeptide antibiotics. Chem. Rev. 105, 425–448.

Kotnik, M., Anderluh, P.S., Prezelj, A., 2007. Development of novel inhibitors targeting intracellular steps of peptidoglycan biosynthesis. Curr. Pharma. Design 13, 2283–2309.

Lloyd, A.J., Brandish, P.E., Gilbey, A.M., Bugg, T.D.H., 2004. Phospho-*N*-acetyl-muramyl-pentapeptide translocase from *Escherichia coli*: catalytic role of conserved aspartic acid residues. J. Bacteriol. 186, 1747–1757.

Lovering, A.L., de Castro, L.H., Lim, D., Strynadka, N.C.J., 2007. Structural insight into the transglycosylation step of bacterial cell-wall biosynthesis. Science 315, 1402–1405.

Macheboeuf, P., Contreras-Martel, C., Job, V., Dideberg, O., Dessen, A., 2006. Penicillin binding proteins: key players in bacterial cell cycle and drug resistance processes. FEMS Microbiol. Rev. 30, 673–691.

Mahapatra, S., Yagi, T., Belisle, J.T., et al., 2005. Mycobacterial lipid II is composed of a complex mixture of modified muramyl and peptide moieties linked to decaprenyl phosphate. J. Bacteriol. 187, 2747–2757.

Mainardi, J.-L., Villet, R., Bugg, T.D., Mayer, C., Arthur, M., 2008. Evolution of peptidoglycan biosynthesis under the selective pressure of antibiotics in Gram-positive bacteria. FEMS Microbiol. Rev. 32, 386–408.

Meadow, P.M., Anderson, J.S., Strominger, J.L., 1964. Enzymatic polymerization of UDP-acetylmuramyl.L-ala.D-glu.L-lys.D-ala.D-ala and UDP-acetylglucosamine by a particulate enzyme from *Staphylococcus aureus* and its inhibition by antibiotics. Biochem. Biophys. Res. Commun. 14, 382–387.

Mengin-Lecreulx, D., van Heijenoort, J., 1994. Copurification of glucosamine-1-phosphate acetyltransferase and *N*-acetylglucosamine-1-phosphate uridyltransferase activities of *Escherichia coli*: characterization of the *glmU* gene product as a bifunctional enzyme catalyzing two subsequent steps in the pathway for UDP-*N*-acetylglucosamine synthesis. J. Bacteriol. 176, 5788–5795.

Mengin-Lecreulx, D., van Heijenoort, J., 1996. Characterization of the essential gene *glm*M encoding phosphoglucosamine mutase in *Escherichia coli*. J. Biol. Chem. 271, 32–39.

Mengin-Lecreulx, D., Texier, L., Rouseau, M., van Heijenoort, J., 1991. The *murG* gene of *Escherichia coli* codes for the UDP-*N*-acetylglucosamine: *N*-acetylmuramyl-(pentapeptide) pyrophosphoryl-undecaprenol *N*-acetylglucosamine transferase involved in the membrane steps of peptidoglycan synthesis. J. Bacteriol. 173, 4625–4636.

Mengin-Lecreulx, D., Ayala, J., Bouhss, A., van Heijenoort, J., Parquet, C., Hara, H., 1998. Contribution of the P_{mra} promoter to expression of genes in the *Escherichia coli mra* cluster of cell envelope biosynthesis and cell division genes. J. Bacteriol. 180, 4406–4412.

Mochalkin, I., Lightle, S., Zhu, Y., et al., 2007. Characterization of substrate binding and catalysis in the potential antibacterial target *N*-acetylglucosamine-1-phosphate uridyltransferase (GlmU). Protein Sci. 16, 2657–2666.

Neuhaus, F.C., 1971. Initial translocation reaction in the biosynthesis of peptidoglycan by bacterial membranes. Acc. Chem. Res. 4, 297–303.

Park, J.T., Uehara, T., 2008. How bacteria consume their own exoskeletons (turnover and recycling of cell wall peptidoglycan). Microbial. Mol. Biol. Rev. 72, 211–227.

Perlstein, D.L., Zhang, Y., Wang, T.-S., Kahne, D.E., Walker, S., 2007. The direction of glycan chain elongation by peptidoglycan glycosyltransferases. J. Amer. Chem. Soc. 129, 12674–12675.

Price, N.P., Momany, F.A., 2005. Modeling bacterial UDP-HexNAc:polyprenol-P HexNAc-1-P transferases. Glycobiology 15, 29R–42R.

Rogers, H.J., Perkins, H.R., Ward, J.B., 1980. Microbial Cell Walls and Membranes. Chapman & Hall Ltd., London.

Sauvage, E., Kerff, F., Terrak, M., Ayala, J.A., Charlier, P., 2008. The penicillin-binding proteins: structure and role in peptidoglycan biosynthesis. FEMS Microbiol. Rev. 32, 234–258.

Scheffers, D.-J., Pinho, M.G., 2005. Bacterial cell wall synthesis: new insights from localization studies. Microbiol. Mol. Biol. Rev. 69, 585–607.

Skarzynski, T., Kim, D.H., Lees, W.J., Walsh, C.T., Duncan, K., 1998. Stereochemical course of enzymatic enolpyruvyl transfer and catalytic conformation of the active site revealed by the crystal structure of the fluorinated analogue of the reaction tetrahedral intermediate bound to the active site of the C115A mutant of MurA. Biochemistry 37, 2572–2577.

Smith, C.A., 2006. Structure, function and dynamics in the *mur* family of bacterial cell wall ligases. J. Mol. Biol. 362, 640–655.

Ünligil, U.M., Rini, J.M., 2000. Glycosyltransferase structure and mechanism. Curr. Opin. Struct. Biol. 10, 510–517.

Urban, J.H., Papenfort, K., Thomsen, J., Schmitz, R.A., Vogel, J., 2007. A conserved small RNA promotes discoordinate expression of the *glmUS* operon mRNA to activate GlmS synthesis. J. Mol. Biol. 373, 521–528.

van Dam, V., Sijbrandi, R., Kol, M., Swiezewska, E., de Kruijff, B., Breukink, E., 2007. Transmembrane transport of peptidoglycan precursors across model and bacterial membranes. Mol. Microbiol. 64, 1105–1114.

van Heijenoort, J., 2001a. Formation of the glycan chains in the synthesis of bacterial peptidoglycan. Glycobiology 11, 25R–36R.

van Heijenoort, J., 2001b. Recent advances in the formation of the bacterial peptidoglycan monomer unit. Nat. Prod. Rep. 18, 503–519.

van Heijenoort, J., 2007. Lipid intermediates in the biosynthesis of bacterial peptidoglycan. Microbiol. Mol. Biol. Rev. 71, 620–635.

Vicente, M., Rico, A.L., Martinez-Arteaga, R., Mingorance, J., 2006. Septum enlightenment: assembly of bacterial division proteins. J. Bacteriol. 188, 19–27.

Vollmer, W., 2008. Structural variation in the glycan strands of bacterial peptidoglycan. FEMS Microbiol. Rev. 32, 287–306.

Vollmer, W., Blanot, D., de Pedro, M.A., 2008. Peptidoglycan structure and architecture. FEMS Microbiol. Rev. 32, 149–167.

Walker, S., Chen, L., Hu, Y., Rew, Y., Shin, D., Boger, D.L., 2005. Chemistry and biology of ramoplanin: a lipoglycodepsipeptide with potent antibiotic activity. Chem. Rev. 105, 449–475.

Ward, J.B., 1984. Biosynthesis of peptidoglycan: points of attack by wall inhibitors. Pharmac. Ther. 25, 327–369.

Welzel, P., 2005. Syntheses around the transglycosylation step in peptidoglycan biosynthesis. Chem. Rev. 105, 4610–4660.

Winkler, W.C., Nahvi, A., Roth, A., Collins, J.A., Breaker, R.R., 2004. Control of gene expression by a natural metabolite-responsive ribozyme. Nature 428, 281–286.

White, R.J., 1968. Control of amino sugar metabolism in *Escherichia coli* and isolation of mutants unable to degrade amino sugars. Biochem. J. 106, 847–858.

Yuan, Y., Barrett, D., Zhang, Y., Kahne, D., Sliz, P., Walker, S., 2007. Crystal structure of a peptidoglycan glycosyltransferase suggests a model for processive glycan chain synthesis. Proc. Natl. Acad. Sci. USA 104, 5348–5353.

Zapun, A., Vernet, T., Pinho, M.G., 2008a. The different shapes of cocci. FEMS Microbiol. Rev. 32, 345–360.

Zapun, A., Contreras-Martel, C., Vernet, T., 2008b. Penicillin-binding proteins and β-lactam resistance. FEMS Microbiol. Rev. 32, 361–385.

CHAPTER

17

Biosynthesis and membrane assembly of lipid A

M. Stephen Trent

SUMMARY

Lipopolysaccharide is the major surface molecule of most Gram-negative bacteria and is a potent stimulator of the innate immune response. The lipid anchor of lipopolysaccharide, known as lipid A, is a unique glucosamine-based saccharolipid that forms the outer monolayer of the outer membrane. Lipid A is synthesized on the cytoplasmic side of the inner membrane by nine constitutive enzymes that are highly conserved among diverse Gram-negative bacterial species. Once the lipid anchor is assembled, it is substituted with an oligosaccharide core and transported across the inner membrane by MsbA, a conserved ATP-binding cassette (ABC) transporter. The oligosaccharide can then be extended further by the addition of the O-specific polysaccharide. Recently, several proteins involved in the transport of nascent lipopolysaccharide to the outer membrane were identified. Although the synthesis of lipid A is a highly conserved process, Gram-negative bacteria have evolved mechanisms to modify the structure of lipid A during transit to the bacterial surface. Lipid A modifications are variable from organism to organism, are often regulated and play an important role in pathogenesis.

Keywords: Lipopolysaccharide; Lipid A; Outer membrane; Lpx; MsbA

1. INTRODUCTION

Gram-negative bacteria are further guarded from their environment by an asymmetric outer membrane that encapsulates their peptidoglycan layer. The outer membrane consists of phospholipids, lipopolysaccharides (LPS) and proteins (Figure 17.1). The inner leaflet of the outer membrane is composed of phospholipids, whereas the outer leaflet is composed primarily of LPS (Kamio and Nikaido, 1976), a conserved microbial component that is a potent stimulator of the innate immune response (Galanos *et al.*, 1985; Kotani *et al.*, 1985; Raetz and Whitfield, 2002). In *Enterobacteriaceae*, LPS is anchored within the outer membrane by a glucosamine-based saccharolipid termed lipid A. The oligosaccharide (OS) core and O-specific polysaccharide (O-PS) of LPS are linked to the lipid moiety via an eight-carbon sugar called 3-deoxy-D-*manno*-oct-2-ulosonic acid (Kdo). Notably, it is the lipid A domain of LPS that is detected by the Toll-like receptor- (TLR-) 4/MD2 receptor of the mammalian innate immune system (Galanos *et al.*, 1984; Hoshino *et al.*, 1999; Aderem and Ulevitch, 2000) (see Chapter 31). Herein, we will discuss the biosynthesis of the Kdo_2-lipid A region of LPS

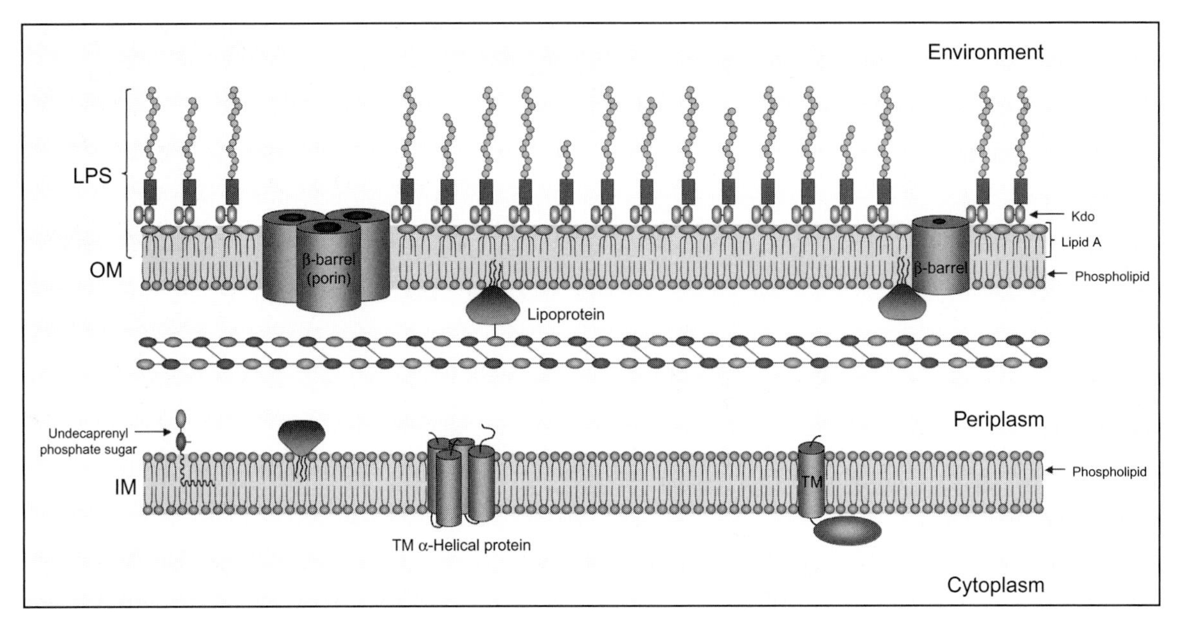

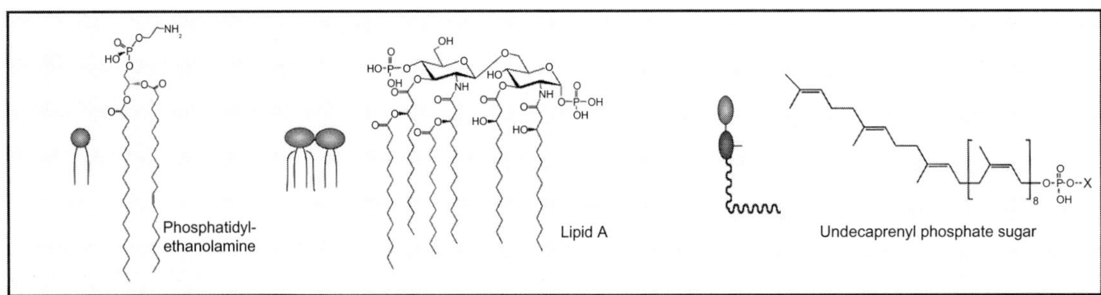

FIGURE 17.1 Schematic of the Gram-negative cell envelope typified by *E. coli*. Unlike Gram-positive organisms, Gram-negative bacteria are further guarded from their environment by an asymmetric outer membrane (OM), interspersed with proteins (top panel). The inner leaflet is composed of glycerophospholipids (PL) whereas the outer leaflet of the outer membrane is composed almost exclusively of lipopolysaccharide (LPS). Lipid A serves as the hydrophobic anchor of LPS and the Kdo (3-deoxy-D-*manno*-oct-2-ulosonic acid) residues serve as a bridge to the remaining saccharide moiety. The chemical structures of lipid A, phosphatidylethanolamine and undecaprenyl-phosphate are shown (bottom panel). Undecaprenyl phosphate is an essential carrier lipid required for the synthesis of various bacterial polymers.

and also describe how the molecule is modified and assembled into the outer membrane.

2. THE CONSTITUTIVE LIPID A BIOSYNTHETIC PATHWAY

In *Escherichia coli*, nine enzymatic steps are required to produce Kdo_2-lipid A, a *bis*-phosphorylated dissacharide of glucosmaine that is hexa-acylated and glycoslyated with two Kdo residues (Figure 17.2). Most Gram-negative bacteria have the ability to synthesize a lipid similar to that of *E. coli* Kdo_2-lipid A via a highly conserved pathway referred to as the Raetz pathway (named after Christian R. H. Raetz). However, examination of the lipid A structures from various organisms shows an impressive amount of diversity (Trent *et al.*, 2006; Raetz *et al.*, 2007). These differences arise from the action of

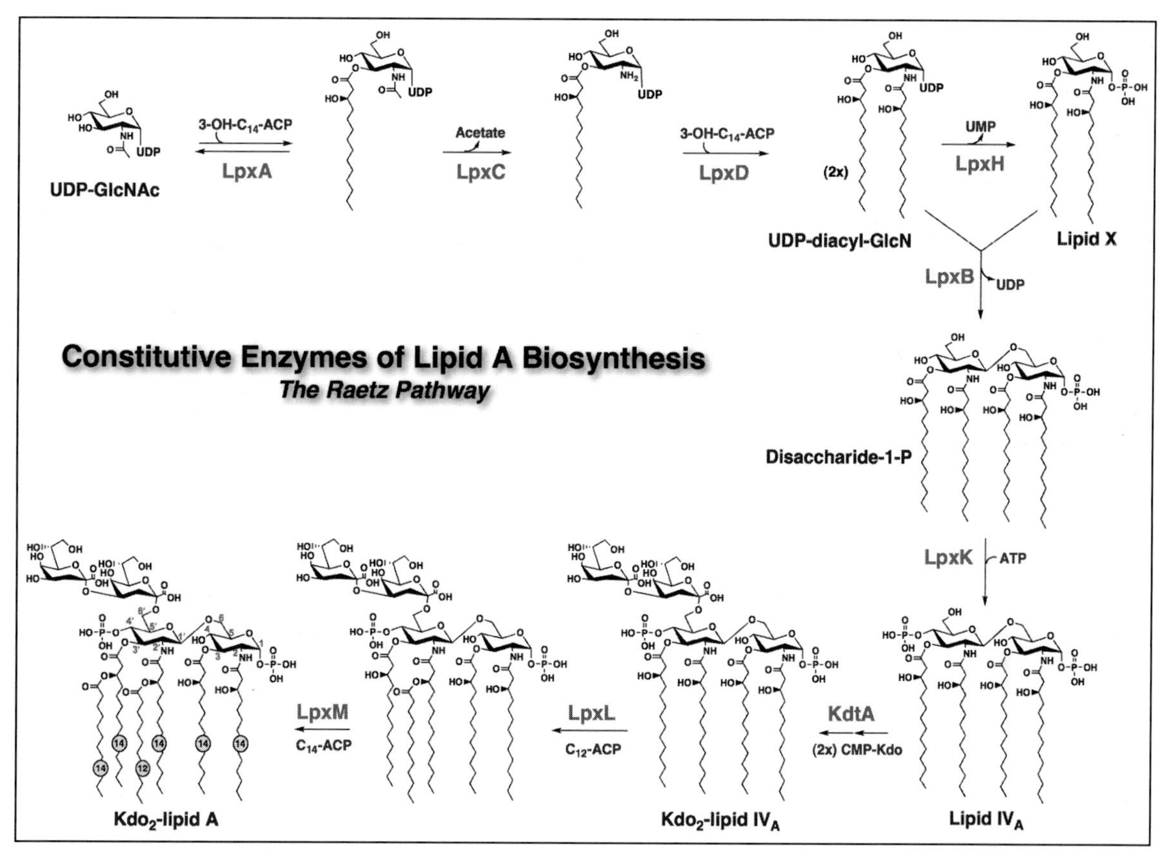

FIGURE 17.2 The constitutive biosynthetic pathway of the Kdo$_2$-lipid A domain of LPS. The structures shown are for the biosynthesis of Kdo$_2$-lipid A in *E. coli* K-12 and *S. enterica* sv. Typhimurium. The names of the major intermediates are indicated and the nine enzymes are listed below reaction arrows. Acyl-acyl carrier protein serves as the obligate acyl donor for the various acyltransferases. The pathway is also referred to as the Raetz pathway, named after its discoverer Christian R.H. Raetz.

latent modification enzymes that together make up the variable component of lipid A biosynthesis and which is discussed later in the chapter.

E. coli LpxA catalyses the first step of the constitutive pathway transferring a 3-hydroxytetradecanoic acid (β-hydroxymyristate) [14:0(3-OH)] chain to uridine diphosphate- (UDP-) *N*-acetylglucosamine (UDP-GlcNAc) (Galloway and Raetz, 1990), a precursor of both peptidoglycan and lipid A. The second and first committed step of the pathway is catalysed by the zinc metalloenzyme LpxC (Young *et al.*, 1995; Jackman *et al.*, 1999). The enzyme LpxC catalyses the irreversible hydrolysis of the amide linkage at position-2 of the backbone glucosamine by removing an acetyl group. Also, LpxC is highly conserved in all Gram-negative bacteria and, since the enzyme has no homology to mammalian deacetylases or amidases, it has been an attractive target for the development of novel antibacterial compounds (Kline *et al.*, 2002; McClerren *et al.*, 2005).

Once deacetylated, a second 14:0(3-OH) chain is added by LpxD to form UDP-2,3-diacylglucosamine (Kelly *et al.*, 1993). Both LpxA and LpxD utilize acyl-acyl carrier protein (ACP) as the obligate acyl donor (Galloway and Raetz, 1990; Kelly *et al.*, 1993). Unlike acyltransferases involved in phospholipid biosynthesis, these enzymes usually have a high degree of specificity for their acyl donor with regards to fatty acyl chain length and extent of saturation. Detailed biochemical investigations of LpxA, including a crystal structure with bound lipid product (Williams and Raetz, 2007), support a precise "hydrocarbon ruler" contributing to the extraordinary chain length selectivity of these enzymes.

The fourth step of Kdo_2-lipid A biosynthesis is cleavage of the pyrophosphate linkage of UDP-2,3-diacylglucosamine by LpxH to form uridine monophosphate (UMP) and 2,3-diacylglucosamine 1-phosphate (lipid X) (Babinski *et al.*, 2002) (see Figure 17.2). Next, LpxB condenses one molecule of UDP-2,3-diacylglucoasmine with one molecule of lipid X and releases UDP resulting in the formation of the characteristic β-(1′→6)-glycosidic linkage present in lipid A molecules (Crowell *et al.*, 1986). The early steps of the pathway occur in the cytoplasm, however, both LpxB and LpxH are peripherally associated with the inner membrane. The final steps of the constitutive pathway are catalysed by four integral membrane proteins: LpxK, KdtA (WaaA), LpxL (HtrB) and LpxM (MsbB). Since these reactions all require cytolosic factors (see Figure 17.2), the active sites of these enzymes are presumably in the cytoplasm.

A specific kinase, LpxK, catalyses the phosphorylation of the 4′-OH group of the tetraacyldisaccharide-1-phosphate intermediate thereby producing lipid IV_A, a key lipid A precursor (Garrett *et al.*, 1997). In *E. coli*, this is followed by the transfer of two Kdo residues onto position-6′ of the dissacharide backbone. The reaction is catalysed by the bifunctional glycosyltransferase KdtA (WaaA) and utilizes CMP-Kdo as the nucleotide sugar donor (Clementz and Raetz, 1991). Addition of Kdo to the lipid anchor represents the first step of core biosynthesis as Kdo is present in nearly all Gram-negative bacteria. Interestingly, the number of Kdo residues transferred by KdtA differs in various bacterial species. For example, the Kdo transferase of *Haemophilus influenzae* (White *et al.*, 1997) transfers a single Kdo residue, whereas KdtA of *Chlamydia* and *Chlamydophila* spp. can transfer three or even four residues to the dissacharide backbone (Brabetz *et al.*, 2000). Remarkably, comparison of the secondary structure of the Kdo transferases (Raetz and Whitfield, 2002) gives no clues as to the number of glycosidic linkages that can be generated by these enzymes.

The final steps of the *E. coli* biosynthetic pathway involve the addition of two secondary acyl chains by LpxL and LpxM. The protein LpxL (HtrB) transfers a dodecanoic acid (laurate) (12:0) to the 2′-position (Clementz *et al.*, 1996) and LpxM (MsbB) a tetradecanoic acid (myristate) (14:0) to the 3′-position (Clementz *et al.*, 1997). These enzymes utilize acyl-ACPs as their substrate and function only after the addition of the Kdo residues. Like the early acyltransferases (LpxA and LpxD), LpxL and LpxM are very selective for a specific acyl chain length and degree of saturation (Clementz *et al.*, 1996, 1997; Vorachek-Warren *et al.*, 2002). In some bacteria, a third acyl transferase, termed LpxP, is expressed during cold shock and replaces the function of LpxL in the constitutive pathway (Carty *et al.*, 1999). The LpxP protein incorporates a 9-hexadecenoic acid (palmitoleate) (16:1) chain which is thought to aid in adjusting the fluidity of the outer membrane when shifted to lower growth temperatures. Finally, it is important to note that not all Gram-negative bacteria contain two late acyltransferases. For example, *Chlamydia trachomatis* contains a single LpxL orthologue and produces a lipid A with a single secondary acyl chain at the 2′-position (Rund *et al.*, 1999). Interestingly, some Gram-negative organisms, like *Campylobacter jejuni*, synthesize a lipid A bearing two secondary acyl chains (Moran *et al.*, 1991), but contain a single late acyltransferase orthologue. The differences in the acyl chain specificity and number of late acyltransferases contributes greatly to the diversity of lipid A structures found in Gram-negative bacteria. Since the secondary acyl chains of lipid A are critical for the endotoxicity of the molecule (Golenbock *et al.*, 1991; Somerville *et al.*, 1996; Nichols *et al.*, 1997) these differences are also of significant biological importance.

3. TRANSPORT

Both the lipid A domain and core OS, including the Kdo residues, are assembled on the cytoplasmic side of the inner membrane. Once translocated across the inner membrane, the O-specific polysaccharide (O-PS) is ligated to the core-lipid A moiety completing LPS assembly (Figure 17.3). Nascent LPS must then be shuttled across the periplasm and eventually translocated to the outer leaflet of the outer membrane where it resides as a major surface molecule. Although progress has been made towards identifying the molecular machinery required for transport of LPS to the bacterial cell surface (Bos *et al.*, 2007), this process is not fully understood.

Transport of the core-lipid A across the inner membrane is mediated by a highly conserved ATP-binding cassette (ABC) transporter, called MsbA (Doerrler *et al.*, 2001) (see Figure 17.3). In *E. coli*, MsbA is essential for growth and inactivation of the transporter in temperature-sensitive mutants results in the accumulation of lipid A and glycerophospholipids in the inner membrane (Doerrler *et al.*, 2001). Further supporting the role of MsbA as a LPS flippase is the observation that loss of MsbA function results in the loss of lipid A modifications known to occur on the periplasmic face of the inner membrane (Doerrler *et al.*, 2004). Unlike *E. coli*, *Neisseria meningitidis* is not dependent upon LPS synthesis for growth (Steeghs *et al.*, 1998). Loss of MsbA-dependent transport in this bacterial species does not result in cell death or the loss of outer membrane biogenesis (Tefsen *et al.*, 2005). For this reason, it remains unclear if MsbA plays a direct role in the flip-flop of glycerophospholipids.

Recent work has identified several proteins involved in the periplasmic transport of LPS, including the periplasmic protein LptA, and a complex of proteins composed of LptBFG that function together as an ABC protein complex. The LptBFG complex is thought to extract LPS from the inner membrane en route to the outer membrane. The bitopic inner membrane protein LptC is also required for removal of LPS from the inner membrane (Sperandeo *et al.*, 2007, 2008; Tran *et al.*, 2008) (see Figure 17.3). Since LptA is a soluble protein, it may function as a chaperone shuttling LPS across the periplasm where it is transferred to the LptDE complex (Figure 17.3). Indeed, *E. coli* LptA has been shown directly to bind the lipid A portion of LPS (Tran *et al.*, 2008). The outer membrane protein LptD (formerly termed Imp) and the lipoprotein LptE (formerly termed RlpB) are thought to function together to flip nascent LPS within the outer membrane to the bacterial cell surface (Braun and Silhavy, 2002; Bos *et al.*, 2004; Wu *et al.*, 2006). Depletion of any single LPS transport (Lpt) protein in *E. coli* results in the accumulation of abnormal membrane structures resulting in cell death.

4. MODIFICATION OF THE Kdo-LIPID A DOMAIN OF LPS

Following the conserved pathway, the Kdo-lipid A domain of LPS can be modified by latent enzymes that make up the variable component of lipid A biosynthesis (Trent *et al.*, 2006). In most cases, these modifications occur at the periplasmic face of the inner membrane or in the outer membrane, separating the constitutive and variable pathways. How these modifications are regulated and the number and type of modifications vary for each bacterial species contributing towards the diversity of microbial cell surfaces (Trent *et al.*, 2006; Raetz *et al.*, 2007). For this section, we will focus primarily on the lipid A modifications known to occur in *E. coli* and *Salmonella enterica* sv. Typhimurium.

Modification to the disaccharide backbone of lipid A centres on the removal or decoration of the lipid A phosphate groups. The lipid A disaccharide of *E. coli* K-12 typically contains a monophosphate group at positions 1 and 4′, however,

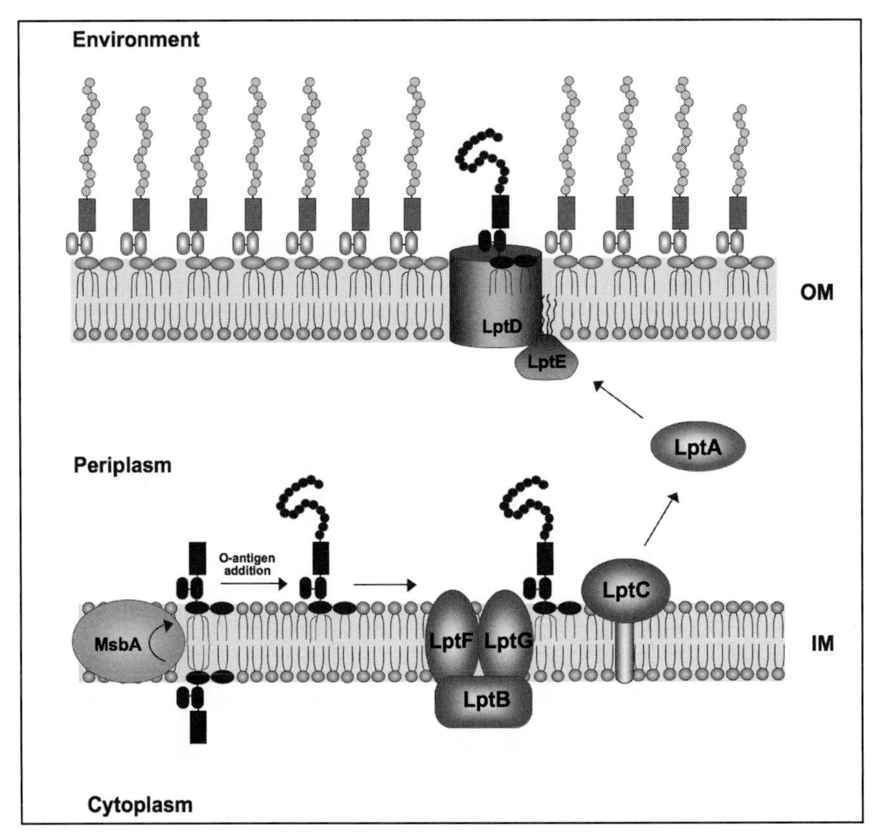

FIGURE 17.3 Possible model for the transport of LPS through the bacterial cell envelope. Core-lipid A is synthesized on the cytoplasmic side of the inner membrane (IM) and transported over the inner membrane by the ABC transporter, MsbA (Doerrler *et al.*, 2004). The O-PS is assembled separately (not shown) from core-lipid A and is transferred to the core-lipid A domain within the periplasm (Raetz and Whitfield, 2002). How LPS is transported to the cell surface remains unclear, but several essential proteins involved in this process have been identified. The ABC transporter comprised of LptBFG along with LptC and the periplasmic protein LptA participate in transport of LPS from the periplasmic side of the inner membrane to the inner leaflet of the outer membrane (Sperandeo *et al.*, 2007, 2008; Ruiz *et al.*, 2008; Tran *et al.*, 2008). The proteins LptDE are required to transfer LPS to the cell surface (Bos *et al.*, 2004; Wu *et al.*, 2005), perhaps by acting as a flippase complex. It is very likely that additional unidentified protein components are involved in this process.

approximately 20–30% of the lipid A contains an unsubstituted diphosphate group at the 1-position (Raetz and Whitfield, 2002). *E. coli* LpxT phosphorylates the 1-position of Kdo_2-lipid A (Figure 17.4) forming a 1-diphosphate lipid A species (Touze *et al.*, 2008). The reaction occurs on the periplasmic side of the inner membrane and undecaprenyl pyrophosphate (C_{55}-PP) serves as the phosphate donor releasing undecaprenyl-P (C_{55}-P). In this way, LpxT also contributes to the total cellular pool of C_{55}-P. The latter is an essential carrier lipid (see Figure 17.1) required for the synthesis of various bacterial polymers (Bouhss *et al.*, 2008), such as peptidoglycan and the O-antigen domain of LPS.

Two additional enzymes, ArnT and EptA, modify the phosphate groups of *E. coli* and *S. enterica* lipid A catalysing the addition of amine-containing residues. The protein ArnT transfers

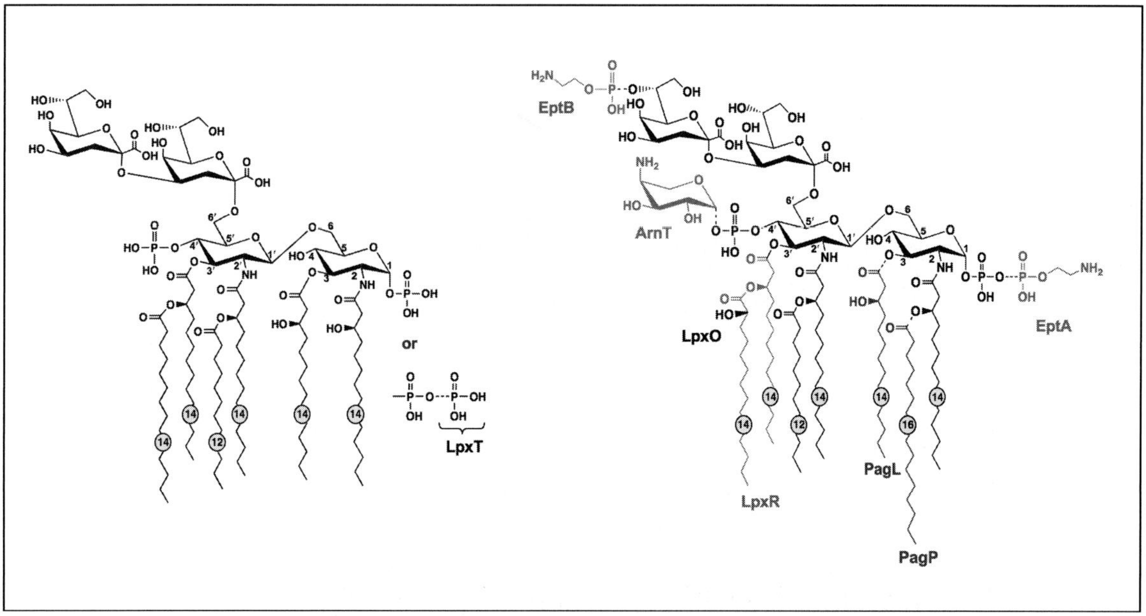

FIGURE 17.4 Comparison of the modified and unmodified forms of Kdo$_2$-lipid A. The structure on the left shows the unmodified Kdo$_2$-lipid A of *E. coli* K-12 or *S. typhimurium* that is hexa-acylated containing monophosphate groups at positions-1 and -4′. An additional phosphate group may also be present during growth in nutrient rich broth forming a diphosphate at the 1-position (Touze *et al.*, 2008). The PagL and PagP modifications are regulated by the PhoP/PhoQ system (Guo *et al.*, 1998; Trent *et al.*, 2001a) whereas modifications requiring the enzymes ArnT (Trent *et al.*, 2001b) and EptA (Trent and Raetz, 2002) are regulated by the transcription factor PmrA. Removal of the 3′-acyloxyacyl linked fatty acyl chains by LpxR and addition of phosphoethanolamine by EptB to the outer Kdo sugar require the presence of Ca^{2+} in the growth medium (Reynolds *et al.*, 2005, 2006). The proteins LpxO, LpxR and PagL are not present in *E. coli* K-12 (Gibbons *et al.*, 2000; Trent *et al.*, 2001a; Reynolds *et al.*, 2006). However, a homologue of LpxR can be found in some pathogenic strains of *E. coli.*

the sugar, 4-amino-4-deoxy-L-arabinopyranose (L-Ara*p*4N) (Trent *et al.*, 2001b) and EptA serves as a phosphoethanolamine transferase (Trent and Raetz, 2002; Lee *et al.*, 2004). Expression of ArnT and EptA is under the control of the transcription factor PmrA that is induced during growth under conditions of low pH and high Fe^{3+} concentration (Groisman *et al.*, 1997; Wosten *et al.*, 2000). Attachment of these positively charged residues to lipid A can provide resistance to some cationic antimicrobial peptides including polymyxin B (Helander *et al.*, 1994; Nummila *et al.*, 1995; Gunn *et al.*, 1998; Cox *et al.*, 2003; Tzeng *et al.*, 2005). It is important to note that formation of the 1-diphosphate species by the action of LpxT does not occur under growth conditions that promote the addition of L-Ara*p*4N and phosphoethanolamine (Guo *et al.*, 1997; Zhou *et al.*, 1999, 2001).

Like LpxT, both ArnT and EptA contain multiple transmembrane domains and modify lipid A on the outer surface of the inner membrane (Doerrler *et al.*, 2004). ArnT utilizes the sugar donor undecaprenyl-phosphate-α-L-Ara*p*4N (Trent *et al.*, 2001c) that is first synthesized within the cytoplasm and transported across the inner membrane (Raetz *et al.*, 2007). The L-Ara*p*4N residue is transferred primarily to the 4′-phosphate group of *E. coli* and *S. enterica* sv. Typhimurium lipid A (Trent *et al.*, 2001b). In pathogens lacking the phosphoethanolamine modification, e.g. *Yersinia pestis* and *Pseudomonas aeruginosa*, L-Ara*p*4N can be found attached to both phosphate groups

(Trent *et al.*, 2006) (Figure 17.5). Interestingly, most Gram-negative bacteria produce either phosphoethanolamine-modified or L-Ara*p*4N-modified lipid A, whereas *E. coli* and *S. enterica* sv. Typhimurium contain both residues (see Figures 17.4 and 17.5). In different organisms, L-Ara*p*4N addition to lipid A has been correlated with increased resistance to cationic antimicrobial peptides and L-Ara*p*4N-deficient mutants of *S. enterica* sv. Typhimurium show reduced virulence (Gunn *et al.*, 1998; McCoy *et al.*, 2001; Moskowitz *et al.*, 2004; Winfield *et al.*, 2005).

The addition of phosphoethanolamine groups to lipid A occurs in a large number of pathogenic bacteria. Phosphatidylethanolamine (see Figure 17.1) serves as the phosphoethanolamine donor substrate (Trent *et al.*, 2001c; Trent and Raetz, 2002). In *E. coli* and *S. enterica*, EptA adds phosphoethanolamine predominantly to the 1-phosphate group (Zhou *et al.*, 2000). *S. enterica eptA* mutants show only slight increases in susceptiblity to polymyxin B suggesting that phosphoethanolamine addition is not critical for resistance to cationic antimicrobial peptides (Lee *et al.*, 2004). On the other hand, loss of phosphoethanolamine addition in *N. meningitidis* (Tzeng *et al.*, 2005) or *C. jejuni* (T.W. Cullen and M.S. Trent, unpublished results) leads to significant polymxyin sensitivity. Both *N. meningitidis* and *C. jejuni* lack the enzymatic machinery to produce L-Ara*p*4N-modified lipid A and these bacteria modify both lipid A phosphate groups with

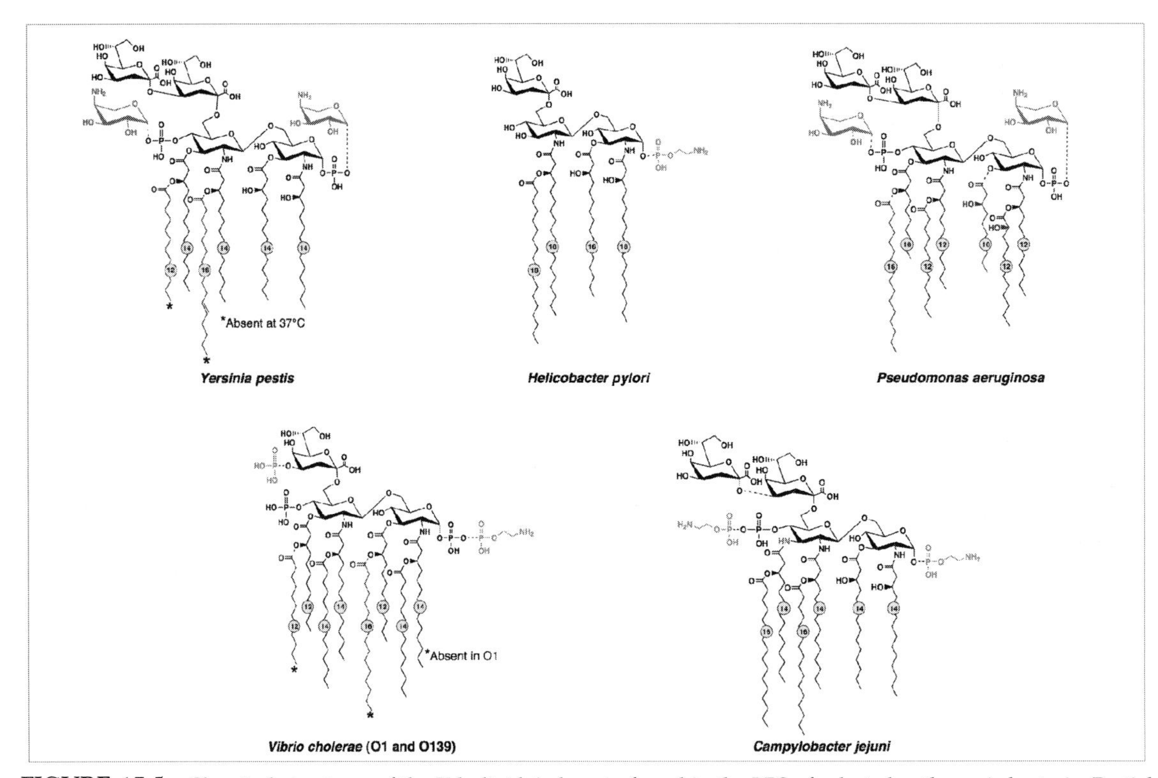

FIGURE 17.5 Chemical structures of the Kdo-lipid A domain found in the LPS of selected pathogenic bacteria. Partial covalent modifications are indicated with dashed bonds and the length of each fatty acyl chain is indicated by the enclosed circles. Note that the dissacharide of *C. jejuni* lipid A contains a diamino-sugar resulting in three acyl chains in an amide linkage (Moran *et al.*, 1991). The biochemical origin of the diamino-sugar precedes the constitutive lipid A pathway with the conversion of UDP-GlcNAc to its diamino-anolog, that can be acylated by LpxA (Sweet *et al.*, 2004a,b).

phosphoethanolamine. Further investigation will be required to determine the roles of phosphoethanolamine modification in cationic antimicrobial peptide resistance and pathogenesis.

Several enzymes that alter the acylation pattern of the lipid A moiety of LPS have also been described. A small outer membrane enzyme, PagP, transfers a hexadecanoic acid (palmitate) (16:0) chain to the 2-position of *E. coli* or *S. enterica* lipid A (Guo *et al.*, 1998; Bishop *et al.*, 2000) (see Figure 17.4). Addition of a palmitate to lipid A increases resistance to certain cationic antimicrobial peptides, presumably by increasing the packing of the fatty acyl chains in the outer membrane. The active site of PagP faces the exterior and the enzyme utilizes phospholipids (e.g. phosphatidylethanolamine) as palmitoyl donors (Hwang *et al.*, 2002). Since phospholipids are typically not present on the outer surface, PagP activity is thought to be regulated by the migration of phospholipids into the outer leaflet. Expression of *pagP* is under the control of the PhoP/PhoQ two-component regulatory system which is activated during exposure of the bacterium to cationic antimicrobial peptides or during Mg^{2+} limitation (Guo *et al.*, 1997; Bader *et al.*, 2005). In the case of *S. enterica*, the PhoP/PhoQ system is activated following phagocytosis by macrophages alerting the bacterium to its presence inside the phagolysosome (Alpuche Aranda *et al.*, 1992). Homologues of PagP can be found in several pathogenic bacteria including *Shigella flexneri*, *Legionella pneumophila* and species of *Bordetella* and *Yersinia*.

Two additional outer membrane enzymes, PagL and LpxR, also modify the lipid moiety of LPS by removing specific fatty acyl chains (see Figure 17.4). Though PagL is a PhoP-activated lipase and removes the 14:0(3-OH) chain at position-3 (Trent *et al.*, 2001a), LpxR removes the intact 3′-acyloxyacyl moiety of the lipid A domain (Reynolds *et al.*, 2006). Moreover, LpxR is not dependent on PhoP activation, but does require the divalent cation Ca^{2+} for enzymatic activity. Interestingly, both PagL and LpxR remain inactive within the *S. enterica* outer membrane, unless over-expressed, suggesting the presence of endogenous inhibitors. Although the function of these enzymes is unknown, reduction in the number of acyl chains reduces the endotoxicity of LPS. Therefore, activation of PagL and/or LpxR within the outer membrane may function to aid in evading the innate immune response. Finally, in some bacterial species, the inner membrane enzyme LpxO may hydroxylate the secondary linked acyl chains of lipid A (Gibbons *et al.*, 2000, 2008). In *S. enterica*, LpxO hydroxylates the 3′-secondary tetradecanoic acid (myristate) (14:0) in the presence of oxygen, using Fe^{2+} and α-ketoglutarate as co-factors. The reaction occurs on the cytoplasmic side of the inner membrane and is not under the control of the PhoP/PhoQ system. Orthologues of LpxO, LpxR, and PagL are present in several Gram-negative bacteria; however, these enzymes are not found in *E. coli* K-12.

The core region of LPS, including the Kdo residues, is also subject to structural modification. For example, *E. coli* and *S. enterica* grown in the presence of 5 mM Ca^{2+} modify the outer Kdo residue of their LPS with a phosphoethanolamine residue (Kanipes *et al.*, 2001) (see Figure 17.4). The protein EptB, a homologue of EptA, catalyses the transfer of phosphoethanolamine to Kdo from phosphatidylethanolamine releasing diacylglycerol. Loss of EptB function in deep-rough mutants of *E. coli* results in hypersensitivity to Ca^{2+} suggesting that EptB plays a role in maintaining outer membrane stability under certain conditions (Reynolds *et al.*, 2005). Of note, EptB is not regulated by PmrA, but is under the control of the σ^E transcription factor (Figueroa-Bossi *et al.*, 2006).

As described earlier, some organisms express a mono-functional Kdo transferase (KdtA). Typically, bacteria having a single Kdo residue in their inner core, such as *H. influenzae* (Helander *et al.*, 1988; Phillips *et al.*, 1992), *Vibrio cholerae* (Chatterjee and Chaudhuri, 2003) and *Bordetella pertussis* (Caroff *et al.*, 2001), phosphorylate their Kdo at the same position where the outer Kdo residue is normally present. Phosphorylation of Kdo takes place in the cytoplasm, does not appear to be regulated and is catalysed by a specific ATP-dependent kinase, called KdkA (White *et al.*, 1999). The inner core region of *Helicobacter pylori* also has a single Kdo residue, however, the Kdo transferase of this bacterium is bifunctional. Interestingly, following transport of the core-lipid A across the *H. pylori* inner membrane, a Kdo hydrolase removes the outer Kdo residue (Stead *et al.*, 2005). A putative Kdo-trimming enzyme has also been reported in membrane of *Francisella novicida* (Wang *et al.*, 2004). The level of structural diversity found in the Kdo-lipid A domain of LPS is quite remarkable, allowing Gram-negative bacteria to express numerous chemical subtypes of lipid A on their surface. Although not discussed here, other modification enzymes include phosphatases that remove the 1 and 4′ phosphate groups (Tran *et al.*, 2004; Wang *et al.*, 2004, 2006) and methyl transferases that use *S*-adenosylmethionine to methylate the 1-phosphate group (Boon Hinckley *et al.*, 2005). As the lipid A structure of more Gram-negative organisms is investigated, it is likely that additional lipid A modification systems will be discovered.

5. CONCLUSIONS

In 1979, the laboratory of Christian R. H. Raetz reported the discovery of lipid X, a key precursor in the lipid A biosynthetic pathway (Nishijima *et al.*, 1981). From this initial finding evolved the elucidation of the constitutive biosynthetic pathway of Kdo_2-lipid A (see Figure 17.2) representing a major achievement in understanding the biogenesis of the bacterial cell envelope. More recently, progress has also been made towards understanding the transport of LPS and the major proteins involved in this process have likely been identified. One of the major challenges remaining in this area is the development of reconstituted *in vitro* systems allowing the biochemical characterization of these proteins. Finally, many of the molecular mechanisms involved in the modification of the Kdo-lipid A domain of LPS have been elucidated. However, since these modification systems are highly variable from organism to organism, more investigation is required to determine the specific role these modifications play in the maintenance of the outer membrane and in pathogenesis (Research Focus Box).

RESEARCH FOCUS BOX

- What regulatory mechanisms are involved in maintaining the perfect balance that leads to the asymmetrical nature of the Gram-negative outer membrane? Little is known about the regulation of lipid A biosynthesis within the bacterial cell and whether its synthesis is coordinated with that of glycerophospholipids.
- *N. meningitidis* is not dependent upon LPS synthesis for growth. Is this true for other Gram-negative organisms?
- Can lipid A modification systems be exploited for our benefit? Lipid A modification systems have the potential to be critical in the development of new adjuvants, LPS antagonists for treatment of sepsis and for engineering whole cell vaccines.

ACKNOWLEDGEMENTS

M.S. Trent is supported by National Institutes of Health Grants AI064184 and AI076322.

References

Aderem, A., Ulevitch, R.J., 2000. Toll-like receptors in the induction of the innate immune response. Nature 406, 782–787.

Alpuche Aranda, C.M., Swanson, J.A., Loomis, W.P., Miller, S.I., 1992. *Salmonella typhimurium* activates virulence gene transcription within acidified macrophage phagosomes. Proc. Natl. Acad. Sci. USA 89, 10079–10083.

Babinski, K.J., Ribeiro, A.A., Raetz, C.R., 2002. The *Escherichia coli* gene encoding the UDP-2,3-diacylglucosamine pyrophosphatase of lipid A biosynthesis. J. Biol. Chem. 277, 25937–25946.

Bader, M.W., Sanowar, S., Daley, M.E., et al., 2005. Recognition of antimicrobial peptides by a bacterial sensor kinase. Cell 122, 461–472.

Bishop, R.E., Gibbons, H.S., Guina, T., Trent, M.S., Miller, S.I., Raetz, C.R., 2000. Transfer of palmitate from phospholipids to lipid A in outer membranes of gram-negative bacteria. EMBO J. 19, 5071–5080.

Boon Hinckley, M., Reynolds, C.M., Ribeiro, A.A., et al., 2005. A *Leptospira interrogans* enzyme with similarity to yeast Ste14p that methylates the 1-phosphate group of lipid A. J. Biol. Chem. 280, 30214–30224.

Bos, M.P., Tefsen, B., Geurtsen, J., Tommassen, J., 2004. Identification of an outer membrane protein required for the transport of lipopolysaccharide to the bacterial cell surface. Proc. Natl. Acad. Sci. USA 101, 9417–9422.

Bos, M.P., Robert, V., Tommassen, J., 2007. Biogenesis of the gram-negative bacterial outer membrane. Annu. Rev. Microbiol. 61, 191–214.

Bouhss, A., Trunkfield, A.E., Bugg, T.D., Mengin-Lecreulx, D., 2008. The biosynthesis of peptidoglycan lipid-linked intermediates. FEMS Microbiol. Rev. 32, 208–233.

Brabetz, W., Lindner, B., Brade, H., 2000. Comparative analyses of secondary gene products of 3-deoxy-D-*manno*-oct-2-ulosonic acid transferases from *Chlamydiaceae* in *Escherichia coli* K-12. Eur. J. Biochem. 267, 5458–5465.

Braun, M., Silhavy, T.J., 2002. Imp/OstA is required for cell envelope biogenesis in *Escherichia coli*. Mol. Microbiol. 45, 1289–1302.

Caroff, M., Aussel, L., Zarrouk, H., et al., 2001. Structural variability and originality of the *Bordetella* endotoxins. J. Endotoxin Res. 7, 63–68.

Carty, S.M., Sreekumar, K.R., Raetz, C.R., 1999. Effect of cold shock on lipid A biosynthesis in *Escherichia coli*. Induction At 12 degrees C of an acyltransferase specific for palmitoleoyl-acyl carrier protein. J. Biol. Chem. 274, 9677–9685.

Chatterjee, S.N., Chaudhuri, K., 2003. Lipopolysaccharides of *Vibrio cholerae*. I. Physical and chemical characterization. Biochim. Biophys. Acta 1639, 65–79.

Clementz, T., Raetz, C.R., 1991. A gene coding for 3-deoxy-D-*manno*-octulosonic-acid transferase in *Escherichia coli*. Identification, mapping, cloning, and sequencing. J. Biol. Chem. 266, 9687–9696.

Clementz, T., Bednarski, J.J., Raetz, C.R., 1996. Function of the *htrB* high temperature requirement gene of *Escherichia coli* in the acylation of lipid A: HtrB catalyzed incorporation of laurate. J. Biol. Chem. 271, 12095–12102.

Clementz, T., Zhou, Z., Raetz, C.R., 1997. Function of the *Escherichia coli msbB* gene, a multicopy suppressor of *htrB* knockouts, in the acylation of lipid A. Acylation by MsbB follows laurate incorporation by HtrB. J. Biol. Chem. 272, 10353–10360.

Cox, A.D., Wright, J.C., Li, J., Hood, D.W., Moxon, E.R., Richards, J.C., 2003. Phosphorylation of the lipid A region of meningococcal lipopolysaccharide: identification of a family of transferases that add phosphoethanolamine to lipopolysaccharide. J. Bacteriol. 185, 3270–3277.

Crowell, D.N., Anderson, M.S., Raetz, C.R., 1986. Molecular cloning of the genes for lipid A disaccharide synthase and UDP-*N*-acetylglucosamine acyltransferase in *Escherichia coli*. J. Bacteriol. 168, 152–159.

Doerrler, W.T., Reedy, M.C., Raetz, C.R., 2001. An *Escherichia coli* mutant defective in lipid export. J. Biol. Chem. 276, 11461–11464.

Doerrler, W.T., Gibbons, H.S., Raetz, C.R., 2004. MsbA-dependent translocation of lipids across the inner membrane of *Escherichia coli*. J. Biol. Chem. 279, 45102–45109.

Figueroa-Bossi, N., Lemire, S., Maloriol, D., Balbontin, R., Casadesus, J., Bossi, L., 2006. Loss of Hfq activates the σ^E-dependent envelope stress response in *Salmonella enterica*. Mol. Microbiol. 62, 838–852.

Galanos, C., Lehmann, V., Lüderitz, O., et al., 1984. Endotoxic properties of chemically synthesized lipid A part structures. Comparison of synthetic lipid A precursor and synthetic analogues with biosynthetic lipid A precursor and free lipid A. Eur. J. Biochem. 140, 221–227.

Galanos, C., Lüderitz, O., Rietschel, E.T., et al., 1985. Synthetic and natural *Escherichia coli* free lipid A express identical endotoxic activities. Eur. J. Biochem. 148, 1–5.

Galloway, S.M., Raetz, C.R., 1990. A mutant of *Escherichia coli* defective in the first step of endotoxin biosynthesis. J. Biol. Chem. 265, 6394–6402.

Garrett, T.A., Kadrmas, J.L., Raetz, C.R., 1997. Identification of the gene encoding the *Escherichia coli* lipid A 4′-kinase. Facile phosphorylation of endotoxin analogs with recombinant LpxK. J. Biol. Chem. 272, 21855–21864.

Gibbons, H.S., Lin, S., Cotter, R.J., Raetz, C.R., 2000. Oxygen requirement for the biosynthesis of the *S*-2-hydroxymyristate moiety in *Salmonella typhimurium* lipid A. Function of LpxO, A new Fe^{2+}/α-ketoglutarate-dependent dioxygenase homologue. J. Biol. Chem. 275, 32940–32949.

Gibbons, H.S., Reynolds, C.M., Guan, Z., Raetz, C.R., 2008. An inner membrane dioxygenase that generates the 2-hydroxymyristate moiety of *Salmonella* lipid A. Biochemistry 47, 2814–2825.

Golenbock, D.T., Hampton, R.Y., Qureshi, N., Takayama, K., Raetz, C.R., 1991. Lipid A-like molecules that antagonize the effects of endotoxins on human monocytes. J. Biol. Chem. 266, 19490–19498.

Groisman, E.A., Kayser, J., Soncini, F.C., 1997. Regulation of polymyxin resistance and adaptation to low-Mg^{2+} environments. J. Bacteriol. 179, 7040–7045.

Gunn, J.S., Lim, K.B., Krueger, J., et al., 1998. PmrA-PmrB-regulated genes necessary for 4-aminoarabinose lipid A modification and polymyxin resistance. Mol. Microbiol. 27, 1171–1182.

Guo, L., Lim, K.B., Gunn, J.S., et al., 1997. Regulation of lipid A modifications by *Salmonella typhimurium* virulence genes phoP-phoQ. Science 276, 250–253.

Guo, L., Lim, K.B., Poduje, C.M., et al., 1998. Lipid A acylation and bacterial resistance against vertebrate antimicrobial peptides. Cell 95, 189–198.

Helander, I.M., Lindner, B., Brade, H., et al., 1988. Chemical structure of the lipopolysaccharide of *Haemophilus influenzae* strain I-69 Rd^{-}/b^{+}. Description of a novel deep-rough chemotype. Eur. J. Biochem. 177, 483–492.

Helander, I.M., Kilpelainen, I., Vaara, M., 1994. Increased substitution of phosphate groups in lipopolysaccharides and lipid A of the polymyxin-resistant pmrA mutants of *Salmonella typhimurium*: a ^{31}P-NMR study. Mol. Microbiol. 11, 481–487.

Hoshino, K., Takeuchi, O., Kawai, T., et al., 1999. Cutting edge: Toll-like receptor 4 (TLR4)-deficient mice are hyporesponsive to lipopolysaccharide: evidence for TLR4 as the Lps gene product. J. Immunol. 162, 3749–3752.

Hwang, P.M., Choy, W.Y., Lo, E.I., et al., 2002. Solution structure and dynamics of the outer membrane enzyme PagP by NMR. Proc. Natl. Acad. Sci. USA 99, 13560–13565.

Jackman, J.E., Raetz, C.R., Fierke, C.A., 1999. UDP-3-*O*-(*R*-3-hydroxymyristoyl)-*N*-acetylglucosamine deacetylase of *Escherichia coli* is a zinc metalloenzyme. Biochemistry 38, 1902–1911.

Kamio, Y., Nikaido, H., 1976. Outer membrane of *Salmonella typhimurium*: accessibility of phospholipid head groups to phospholipase c and cyanogen bromide activated dextran in the external medium. Biochemistry 15, 2561–2570.

Kanipes, M.I., Lin, S., Cotter, R.J., Raetz, C.R., 2001. Ca^{2+}-induced phosphoethanolamine transfer to the outer 3-deoxy-D-*manno*-octulosonic acid moiety of *Escherichia coli* lipopolysaccharide. A novel membrane enzyme dependent upon phosphatidylethanolamine. J. Biol. Chem. 276, 1156–1163.

Kelly, T.M., Stachula, S.A., Raetz, C.R., Anderson, M.S., 1993. The *firA* gene of *Escherichia coli* encodes UDP-3-*O*-(*R*-3-hydroxymyristoyl)-glucosamine *N*-acyltransferase. The third step of endotoxin biosynthesis. J. Biol. Chem. 268, 19866–19874.

Kline, T., Andersen, N.H., Harwood, E.A., et al., 2002. Potent, novel *in vitro* inhibitors of the *Pseudomonas aeruginosa* deacetylase LpxC. J. Med. Chem. 45, 3112–3129.

Kotani, S., Takada, H., Tsujimoto, M., et al., 1985. Synthetic lipid A with endotoxic and related biological activities comparable to those of a natural lipid A from an *Escherichia coli* Re-mutant. Infect. Immun. 49, 225–237.

Lee, H., Hsu, F.F., Turk, J., Groisman, E.A., 2004. The PmrA-regulated *pmrC* gene mediates phosphoethanolamine modification of lipid A and polymyxin resistance in *Salmonella enterica*. J. Bacteriol. 186, 4124–4133.

McClerren, A.L., Endsley, S., Bowman, J.L., et al., 2005. A slow, tight-binding inhibitor of the zinc-dependent deacetylase LpxC of lipid A biosynthesis with antibiotic activity comparable to ciprofloxacin. Biochemistry 44, 16574–16583.

McCoy, A.J., Liu, H., Falla, T.J., Gunn, J.S., 2001. Identification of *Proteus mirabilis* mutants with increased sensitivity to antimicrobial peptides. Antimicrob. Agents Chemother. 45, 2030–2037.

Moran, A.P., Zahringer, U., Seydel, U., Scholz, D., Stutz, P., Rietschel, E.T., 1991. Structural analysis of the lipid A component of *Campylobacter jejuni* CCUG 10936 (serotype O:2) lipopolysaccharide. Description of a lipid A containing a hybrid backbone of 2-amino-2-deoxy-D-glucose and 2,3-diamino-2,3-dideoxy-D-glucose. Eur. J. Biochem. 198, 459–469.

Moskowitz, S.M., Ernst, R.K., Miller, S.I., 2004. PmrAB, a two-component regulatory system of *Pseudomonas aeruginosa* that modulates resistance to cationic antimicrobial peptides and addition of aminoarabinose to lipid A. J. Bacteriol. 186, 575–579.

Nichols, W.A., Raetz, C.R.H., Clementz, T., et al., 1997. *htrB* of *Haemophilus influenzae*: determination of biochemical activity and effects on virulence and lipooligosaccharide toxicity. J. Endotoxin Res. 4, 163–172.

Nishijima, M., Bulawa, C.E., Raetz, C.R., 1981. Two interacting mutations causing temperature-sensitive phosphatidylglycerol synthesis in *Escherichia coli* membranes. J. Bacteriol. 145, 113–121.

Nummila, K., Kilpeläinen, I., Zähringer, U., Vaara, M., Helander, I.M., 1995. Lipopolysaccharides of polymyxin B-resistant mutants of *Escherichia coli* are extensively substituted by 2-aminoethyl pyrophosphate and contain aminoarabinose in lipid A. Mol. Microbiol. 16, 271–278.

Phillips, N.J., Apicella, M.A., Griffiss, J.M., Gibson, B.W., 1992. Structural characterization of the cell surface lipooligosaccharides from a nontypable strain of *Haemophilus influenzae*. Biochemistry 31, 4515–4526.

Raetz, C.R., Whitfield, C., 2002. Lipopolysaccharide endotoxins. Annu. Rev. Biochem. 71, 635–700.

Raetz, C.R., Reynolds, C.M., Trent, M.S., Bishop, R.E., 2007. Lipid A modification systems in Gram-negative bacteria. Annu. Rev. Biochem. 76, 295–329.

Reynolds, C.M., Kalb, S.R., Cotter, R.J., Raetz, C.R., 2005. A phosphoethanolamine transferase specific for the outer 3-deoxy-D-*manno*-octulosonic acid residue of *Escherichia coli* lipopolysaccharide. Identification of the *eptB* gene and Ca^{2+} hypersensitivity of an *eptB* deletion mutant. J. Biol. Chem. 280, 21202–21211.

Reynolds, C.M., Ribeiro, A.A., McGrath, S.C., Cotter, R.J., Raetz, C.R., Trent, M.S., 2006. An outer membrane enzyme encoded by *Salmonella typhimurium lpxR* that removes the 3′-acyloxyacyl moiety of lipid A. J. Biol. Chem. 281, 21974–21987.

Ruiz, N., Gronenberg, L.S., Kahne, D., Silhavy, T.J., 2008. Identification of two inner-membrane proteins required for the transport of lipopolysaccharide to the outer membrane of *Escherichia coli*. Proc. Natl. Acad. Sci. USA 105, 5537–5542.

Rund, S., Lindner, B., Brade, H., Holst, O., 1999. Structural analysis of the lipopolysaccharide from *Chlamydia trachomatis* serotype L2. J. Biol. Chem. 274, 16819–16824.

Somerville Jr, J.E., Cassiano, L., Bainbridge, B., Cunningham, M.D., Darveau, R.P., 1996. A novel *Escherichia coli* lipid A mutant that produces an antiinflammatory lipopolysaccharide. J. Clin. Invest. 97, 359–365.

Sperandeo, P., Cescutti, R., Villa, R., et al., 2007. Characterization of *lptA* and *lptB*, two essential genes implicated in lipopolysaccharide transport to the outer membrane of *Escherichia coli*. J. Bacteriol. 189, 244–253.

Sperandeo, P., Lau, F.K., Carpentieri, A., et al., 2008. Functional analysis of the protein machinery required for the transport of lipopolysaccharide to the outer membrane of *Escherichia coli*. J. Bacteriol. 190, 4460–4469.

Stead Jr, C., Tran, A., Ferguson, D., McGrath, S., Cotter, R., Trent, S., 2005. A novel 3-deoxy-D-*manno*-octulosonic acid (Kdo) hydrolase that removes the outer Kdo sugar of *Helicobacter pylori* lipopolysaccharide. J. Bacteriol. 187, 3374–3383.

Steeghs, L., den Hartog, R., den Boer, A., Zomer, B., Roholl, P., van der Ley, P., 1998. Meningitis bacterium is viable without endotoxin. Nature 392, 449–450.

Sweet, C.R., Ribeiro, A.A., Raetz, C.R., 2004a. Oxidation and transamination of the 3″-position of UDP-*N*-acetylglucosamine by enzymes from *Acidithiobacillus ferrooxidans*. Role in the formation of lipid A molecules with four amide-linked acyl chains. J. Biol. Chem. 279, 25400–25410.

Sweet, C.R., Williams, A.H., Karbarz, M.J., et al., 2004b. Enzymatic synthesis of lipid A molecules with four amide-linked acyl chains. LpxA acyltransferases selective for an analog of UDP-*N*-acetylglucosamine in which an amine replaces the 3″-hydroxyl group. J. Biol. Chem. 279, 25411–25419.

Tefsen, B., Bos, M.P., Beckers, F., Tommassen, J., de Cock, H., 2005. MsbA is not required for phospholipid transport in *Neisseria meningitidis*. J. Biol. Chem. 280, 35961–35966.

Touze, T., Tran, A.X., Hankins, J.V., Mengin-Lecreulx, D., Trent, M.S., 2008. Periplasmic phosphorylation of lipid A is linked to the synthesis of undecaprenyl phosphate. Mol. Microbiol. 67, 264–277.

Tran, A.X., Karbarz, M.J., Wang, X., et al., 2004. Periplasmic cleavage and modification of the 1-phosphate group of *Helicobacter pylori* lipid A. J. Biol. Chem. 279, 55780–55791.

Tran, A.X., Trent, M.S., Whitfield, C., 2008. The LptA protein of *Escherichia coli* is a periplasmic lipid A-binding protein involved in the lipopolysaccharide export pathway. J. Biol. Chem. 283, 20342–20349.

Trent, M.S., Raetz, C.R.H., 2002. Cloning of EptA, the lipid A phosphoethanolamine transferase associated with polymyxin resistance. J. Endotoxin Res. 8, 159.

Trent, M.S., Pabich, W., Raetz, C.R., Miller, S.I., 2001a. A PhoP/PhoQ-induced lipase (PagL) that catalyzes 3-*O*-deacylation of lipid A precursors in membranes of *Salmonella typhimurium*. J. Biol. Chem. 276, 9083–9092.

Trent, M.S., Ribeiro, A.A., Lin, S., Cotter, R.J., Raetz, C.R., 2001b. An inner membrane enzyme in *Salmonella* and *Escherichia coli* that transfers 4-amino-4-deoxy-L-arabinose to lipid A: induction on polymyxin-resistant mutants and role of a novel lipid-linked donor. J. Biol. Chem. 276, 43122–43131.

Trent, M.S., Ribeiro, A.A., Doerrler, W.T., Lin, S., Cotter, R. J., Raetz, C.R., 2001c. Accumulation of a polyisoprene-linked amino sugar in polymyxin-resistant *Salmonella typhimurium* and *Escherichia coli*: structural characterization and transfer to lipid A in the periplasm. J. Biol. Chem. 276, 43132–43144.

Trent, M.S., Stead, C.M., Tran, A.X., Hankins, J.V., 2006. Diversity of endotoxin and its impact on pathogenesis. J. Endotoxin Res. 12, 205–223.

Tzeng, Y.L., Ambrose, K.D., Zughaier, S., et al., 2005. Cationic antimicrobial peptide resistance in *Neisseria meningitidis*. J. Bacteriol. 187, 5387–5396.

Vorachek-Warren, M.K., Ramirez, S., Cotter, R.J., Raetz, C.R., 2002. A triple mutant of *Escherichia coli* lacking secondary acyl chains on lipid A. J. Biol. Chem. 277, 14194–14205.

Wang, X., Karbarz, M.J., McGrath, S.C., Cotter, R.J., Raetz, C.R., 2004. MsbA transporter-dependent lipid A 1-dephosphorylation on the periplasmic surface of the inner membrane: topography of *Francisella Novicida* LpxE expressed in *Escherichia coli*. J. Biol. Chem. 279, 49470–49478.

Wang, X., McGrath, S.C., Cotter, R.J., Raetz, C.R., 2006. Expression cloning and periplasmic orientation of the *Francisella novicida* lipid A 4′-phosphatase LpxF. J. Biol. Chem. 281, 9321–9330.

White, K.A., Kaltashov, I.A., Cotter, R.J., Raetz, C.R., 1997. A mono-functional 3-deoxy-D-*manno*-octulosonic acid (Kdo) transferase and a Kdo kinase in extracts of *Haemophilus influenzae*. J. Biol. Chem. 272, 16555–16563.

White, K.A., Lin, S., Cotter, R.J., Raetz, C.R., 1999. A *Haemophilus influenzae* gene that encodes a membrane bound 3-deoxy-D-*manno*-octulosonic acid (Kdo) kinase. Possible involvement of kdo phosphorylation in bacterial virulence. J. Biol. Chem. 274, 31391–31400.

Williams, A.H., Raetz, C.R., 2007. Structural basis for the acyl chain selectivity and mechanism of UDP-*N*-acetylglucosamine acyltransferase. Proc. Natl. Acad. Sci. USA 104, 13543–13550.

Winfield, M.D., Latifi, T., Groisman, E.A., 2005. Transcriptional regulation of the 4-amino-4-deoxy-L-arabinose biosynthetic genes in *Yersinia pestis*. J. Biol. Chem. 280, 14765–14772.

Wosten, M.M., Kox, L.F., Chamnongpol, S., Soncini, F.C., Groisman, E.A., 2000. A signal transduction system that responds to extracellular iron. Cell 103, 113–125.

Wu, T., Malinverni, J., Ruiz, N., Kim, S., Silhavy, T.J., Kahne, D., 2005. Identification of a multicomponent complex required for outer membrane biogenesis in *Escherichia coli*. Cell 121, 235–245.

Wu, T., McCandlish, A.C., Gronenberg, L.S., Chng, S.S., Silhavy, T.J., Kahne, D., 2006. Identification of a protein complex that assembles lipopolysaccharide in the outer membrane of *Escherichia coli*. Proc. Natl. Acad. Sci. USA 103, 11754–11759.

Young, K., Silver, L.L., Bramhill, D., et al., 1995. The envA permeability/cell division gene of *Escherichia coli* encodes the second enzyme of lipid A biosynthesis. UDP-3-*O*-(*R*-3-hydroxymyristoyl)-*N*-acetylglucosamine deacetylase. J. Biol. Chem. 270, 30384–30391.

Zhou, Z., Lin, S., Cotter, R.J., Raetz, C.R., 1999. Lipid A modifications characteristic of Salmonella typhimurium are induced by NH_4VO_3 in *Escherichia coli* K12. Detection of 4-amino-4-deoxy-L-arabinose, phosphoethanolamine and palmitate. J. Biol. Chem. 274, 18503–18514.

Zhou, Z., Ribeiro, A.A., Raetz, C.R., 2000. High-resolution NMR spectroscopy of lipid A molecules containing 4-amino-4-deoxy-L-arabinose and phosphoethanolamine substituents. Different attachment sites on lipid A molecules from NH_4VO_3-treated *Escherichia coli* versus *kdsA* mutants of *Salmonella typhimurium*. J. Biol. Chem. 275, 13542–13551.

Zhou, Z., Ribeiro, A.A., Lin, S., Cotter, R.J., Miller, S.I., Raetz, C.R., 2001. Lipid A modifications in polymyxin-resistant *Salmonella typhimurium*: PMRA-dependent 4-amino-4-deoxy-L-arabinose, and phosphoethanolamine incorporation. J. Biol. Chem. 276, 43111–43121.

CHAPTER

18

Biosynthesis of O-antigen chains and assembly

Peter R. Reeves and Monica M. Cunneen

SUMMARY

The O-antigen (O-specific polysaccharide) chain is a major component of the surface lipopolysaccharide of Gram-negative bacteria and is highly variable in structure. Contributing to this diversity are the sugar constituents and the glycosyl linkages present within and between the identical repeating units of this polysaccharide, which is the antigenic basis for the many recognized serogroups now known for bacterial species like *Escherichia coli*. This diversity reflects the polymorphisms present in the genes responsible for the biosynthesis of these structures. The genes are generally in a single cluster known as the O-antigen gene cluster, with the chromosomal locus usually conserved within a species. There are three main biosynthesis pathways by which O-antigens can be synthesized, namely Wzx/Wzy, adenosine triphosphate- (ATP-) binding cassette-transporter and Synthase, which share the initiating steps of synthesis but diverge in the processing steps. The genes and pathways of these three systems are discussed herein using well-studied examples from *E. coli*, *Salmonella enterica* and *Pseudomonas aeruginosa*, and the current knowledge of these systems is reviewed to highlight potential areas for future research.

Keywords: O-antigen; Repeating unit; O-Unit; Wzx/Wzy; Adenosine-triphosphate-(ATP-)binding cassette (ABC); ABC-transporter; Synthase; *Escherichia coli*, *Salmonella enterica*; *Pseudomonas aeruginosa*

1. INTRODUCTION

O-Antigens (also known as O-specific polysaccharides or O-side chains) are repeating unit polysaccharide extensions of the lipopolysaccharide (LPS) of Gram-negative bacteria (see Chapter 4). They vary in the sugars and linkages present, even within a species, and many of the genes involved are in an O-antigen gene cluster. Within a species they commonly map to the same site on the chromosome and hence constitute a major polymorphism.

The genes for synthesis of the sugars comprise a major class that are usually easily recognized. Those for sugars specific to the O-antigen are in the gene cluster, but genes for sugars also found in other structures, or otherwise used in the metabolism of that species, are usually synthesized by genes at other loci. O-Antigens are synthesized by one of three different pathways: the Wzx/Wzy pathway, the adenosine triphosphate- (ATP-) binding cassette (ABC) transporter pathway, or Synthase pathway. Each pathway requires proteins for assembly and for processing, most of which have considerable specificity for the O-antigen, and are generally in the gene cluster. The third class of genes is the glycosyl transferases (GTs) that link the component sugars together. These linkages tend to be specific to the O-antigen and again the genes involved are usually in the gene cluster. These gene clusters thus contain the genes that distinguish one O-antigen from another.

The *Escherichia coli* O16 O-antigen and then some *Salmonella enterica* O-antigens will be examined to illustrate the patterns of biosynthesis observed in many such groups, followed by an overview of O-antigen biosynthesis step by step, starting with the initial (sugar) transferases (ITs) common to all O-antigen biosynthesis pathways, followed by the steps found in the Wzx/Wzy, ABC transporter and Synthase pathways, and conclude with a discussion of the biochemical and genetic basis of the enormous diversity of O-antigens.

O-Antigen synthesis has been reviewed many times often within LPS reviews. Further information on these can be found in Valvano (2003), Samuel and Reeves (2003) and Raetz and Whitfield (2002). In this review, we will treat *Shigella* strains as part of *E. coli* when giving statistics, as that reflects their natural status and many of the O-antigens are common to both.

2. THE *E. COLI* K-12 (O16) O-ANTIGEN

The *E. coli* O16 O-antigen is an example of a biosynthesis system that is fairly well understood (but it should be noted that for O16 synthesis to occur in the laboratory strain K-12, complementation of the *wbbL* mutation is required (Liu and Reeves, 1994)). There are four sugars in the basic O16 structure (Figure 18.1). Precursors uridine diphosphate- (UDP-) *N*-acetylglucosamine (GlcNAc) and UDP-D-glucose (UDP-Glc) are found in all *E. coli* and the synthesis genes for these reside at other loci, but those for deoxythymidine-diphosphate-L-rhamnose(dTDP-Rha)and UDP-D-galactofuranose (UDP-Gal*f*) are within the O-antigen gene cluster, as are those for the GTs, other than for GlcNAc, and processing steps (Stevenson *et al.*, 1994).

O-Unit synthesis is initiated by WecA transferring GlcNAc-Phosphate (GlcNAc-P) from UDP-GlcNAc to undecaprenol phosphate (UndP), to give UndPP-GlcNAc. The WecA protein is present in all *E. coli*, and, indeed generally in the *Enterobacteriaceae*, and the gene is in the gene cluster for the enterobacterial common antigen (ECA), which also has GlcNAc as the first sugar of its repeating unit (Meier-Dieter *et al.*, 1992). The other sugars are added sequentially to give the linear oligosaccharide. The WbbL enzyme is the second transferase, transferring Rha from dTDP-Rha onto the UndPP-GlcNAc moiety (Stevenson *et al.*, 1994) and WbbI is known to transfer Gal*f* (Wing *et al.*, 2006), the fourth sugar of the O-unit, so by elimination, the third [glucose(Glc)] transferase, must be WbbK, although we have no direct evidence for this to date.

The K-12 O16 O-antigen has two side-branch moieties, a Glc residue on the GlcNAc residue and an acetyl group on the Rha residue. The Glc residue is assumed to be added after translocation, by WbeR (YfdI), which is encoded outside of the gene cluster as part of a three-gene *gtr* set (see below), within cryptic prophage CPS-53 (http://www.ecogene.org/geneInfo.php?eg_id=EG14133). The protein WbbJ is a putative *O*-acetyl transferase and there is strong support for it being the gene involved (Yao and Valvano, 1994). It is not uncommon for Glc and *O*-acetyl groups to be added to O-antigens by genes outside of the main gene cluster, so the additional evidence for *wbbJ* is important. The *wzx* and *wzy* genes for translocation and polymerization (see below) are also present in the gene cluster, which is at the usual locus between *galF* and *gnd* for *E. coli, S. enterica* and related species, while the *wzz* gene for determination of chain length is about 3 kb downstream in these species.

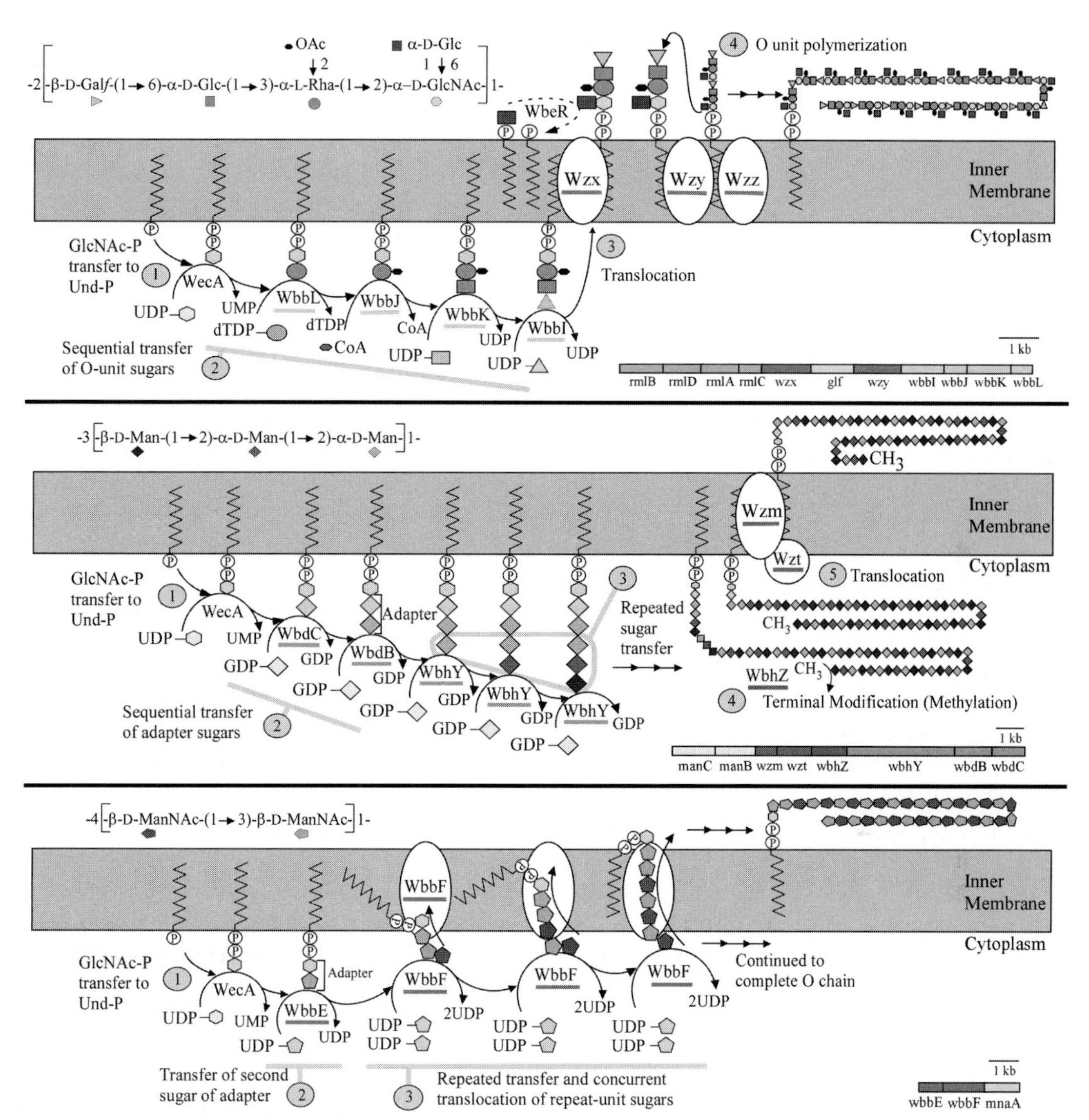

FIGURE 18.1 O-Antigen biosynthesis pathways. Top panel: *E. coli* O16 (Wzx/Wzy), centre panel: *E. coli* O8 (ABC-transporter), and bottom panel: *S. enterica* O54 (Synthase), with each panel also indicating the gene cluster involved and the final repeating unit structure. Synthesis in all three pathways is initiated by transfer of GlcNAc-P onto Und-P. Pathway steps are numbered sequentially. Each sugar reisdue is represented by a different symbol. Symbols for each nucleotide diphosphate-linked sugar, and the respective genes, have the same colouring. Transferases and processing proteins, and respective genes, are also colour coded. Residue colour varies in the repeating unit if more than one residue of that sugar is present, to indicate the different linkage formed. Each pathway has an independent colour scheme, except for UDP-GlcNAc, that is common to all three. O-Unit modification in *E. coli* O16 occurs after repeat unit synthesis and is indicated by a dashed line. The ligation of the O-antigen from Und-PP to the lipid-A-core by WaaL and the final export of the LPS molecule to the outer membrane, both common to all three pathways, are not shown (see Chapter 17). Not drawn to scale. (See colour plates section.)

3. *S. ENTERICA* LT2 AND A FAMILY OF O-ANTIGENS

Some O-antigens form groups with related structures and, where available, the sequences are also generally related. In this section, three *S. enterica* O-antigens, B, C2 and D1, are examined from one such group consisting of the A, B, C2/3, D1, D2, D3 and E O-antigens. The type strain for *S. enterica* genetic studies was LT2 which, unlike K-12, produced an O-antigen and, for this reason, the genetics of group B O-antigen synthesis was first unravelled for LT2. The O-antigen gene cluster was then sequenced (Wyk and Reeves, 1989; Jiang *et al.*, 1991; Liu *et al.*, 1993, 1995) and found to include 16 genes, comprising 11 for three nucleotide sugar pathways, three GT genes (Liu *et al.*, 1993, 1995), one IT gene (Wang *et al.*, 1996) and what is now known as the *wzx* gene. The *wzy* gene had been shown in early studies to be at a separate locus, originally known as *rfc*, but a *wzy* remnant was later found in the group B gene cluster (Wang *et al.*, 2002). It appears that the original *wzy* gene was functionally replaced at some time by an imported gene, making the original one redundant and now much reduced in size. The structure and gene cluster are shown in Figure 18.2, with the relevant IT or GT for each linkage indicated on the structure.

The first thing to note is that it is not WecA that initiates O-unit synthesis, but WbaP, encoded in the O-antigen gene cluster, which transfers galactose (Gal). Groups B and D1 have the same structural backbone but differ in the presence of abequose (Abe) or tyvelose (Tyv) side-branches, respectively, that reflects the presence of an *abe* gene in group B and *prt* and *tyv* genes in group D1 (see Figure 18.2). Major sequence differences are present in the *wbaV* genes, encoding dideoxyhexose transfer, and also in the *wzx* genes, but clones of only the *abe* gene, or *prt* plus *tyv* genes, allow the additional sugar to be incorporated (Wyk and Reeves, 1989; Verma and Reeves, 1989), so the *wbaV* and *wzx* genes still complement. In contrast, group B and C2 O-antigens have the same sugars, but with differing sequence order and linkages. Again, these differences are reflected in the genes, with the same sugar pathway genes, but with GTs *wbaN*, *wbaU* and *wbaV* of group B replaced by *wbaZ*, *wbaW*, *wbaQ* and *wbaR* in group C2 (Brown *et al.*, 1992; Liu *et al.*, 1993, 1995). There is also an *O*-acetyl transferase gene (*wbaL*) and *wzy* genes in the group C2 gene cluster (Liu *et al.*, 1995) and a very different *wzx* gene compared to that of group B.

What is striking is that for both comparisons, the genes that determine the structural differences group together, with separate synthesis modules for the dTDP-Rha, guanosine diphosphate- (GDP-) D-mannose (Man) and cytidine diphosphate- (CDP-) dideoxyhexoses, e.g. CDP-Tyv and CDP-Abe, a set of genes for assembly of the O-unit, including *wzx* and *wzy* (or remnant) and all GT genes between the *ddh* and *man* genes. Also remarkable is that the GT genes, and the IT gene, map inversely to function order. These gene cluster traits are quite common, but not universal, phenomena. Indeed, the distribution of GT and processing genes is quite variable and there seems to be no general trends for these genes.

4. INITIAL TRANSFERASES THAT INITIATE O-ANTIGEN SYNTHESIS

O-Antigen synthesis is initiated by transfer of a sugar phosphate from a Nucleotide diphosphate (NDP-)sugar to Und-P. The ITs involved fall into two families, the PNPT (*p*olyisoprenyl-phosphate *N*-acetyl-hexosamine-1-*p*hosphate *t*ransferase) family, including WecA and the PHPT (*p*olyisoprenyl-phosphate *h*exose-1-*p*hosphate *t*ransferase) family, including WbaP (Price and Momany, 2005).

There are relatively few ITs. For *E. coli*, most O-antigens have the Wzx/Wzy pathway and all but three appear to have GlcNAc or *N*-acetyl-galactosamine (GalNAc) as first sugar, with WecA

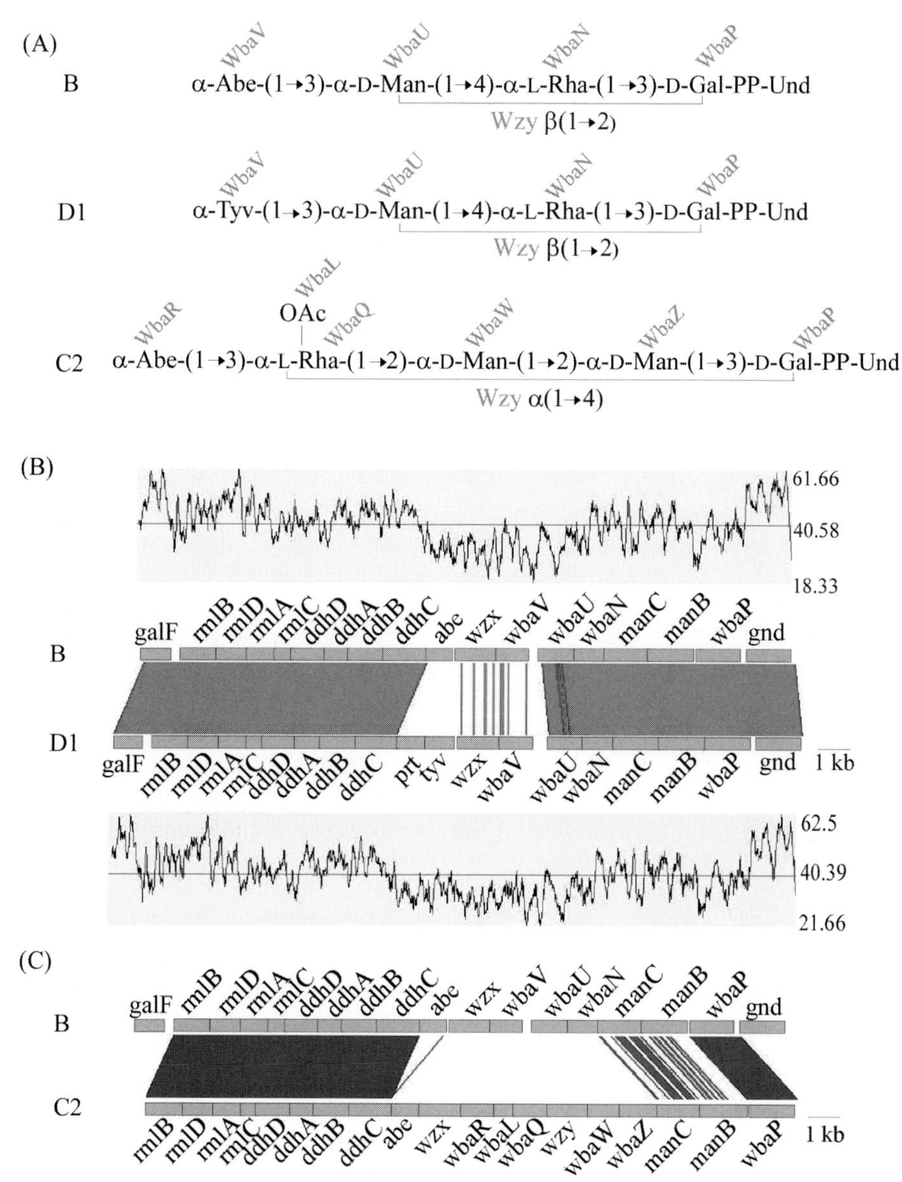

FIGURE 18.2 *S. enterica* B, D1 and C2 O antigens and gene clusters. (A) Repeating units of B, D1 and C2 O antigens attached to UndPP, with the protein responsible for each linkage shown in red and the polymerase linkage indicated below each structure. Only sugars and linkages in the basic structure encoded in the O-antigen gene clusters are shown, but most published structures include one or more additional Glc or *O*-acetyl side-branch residues. (B) Comparison of B and D1 gene clusters and flanking genes, generated by ARTEMIS, with G + C% distribution for each cluster. (C) Comparison of B and C2 gene clusters. (See colour plates section.)

as the IT. There are gene cluster sequences for two of the three with known O-antigen structures that lack GlcNAc or GalNAc. The *Shigella sonnei* O-unit is a dimer of 2-acetamido-4-amino-2, 4-dideoxy-D-fucose (FucNAc4N) and 2-acetamido-2-deoxy-L-altruronic acid (AltNAcA) (Kenne *et al.*, 1980), with WbgY, encoded in the gene cluster, inferred to be the IT for transfer of D-FucNAc4N, the first sugar (Xu *et al.*, 2002). The gene cluster is on a plasmid and thought to have been transferred from *Plesiomonas shigelloides*. The second, *E. coli* O45, is discussed below with *Pseudomonas aeruginosa*. To our knowledge, these are the only *E. coli* gene clusters to have an IT gene in the gene cluster, confirming the role of *wecA* in the others. Most *S. enterica* and *Yersinia* spp. O-antigens also use WecA, which is present in most *Enterobacteriaceae* genomes, so this pattern may apply throughout this family.

P. aeruginosa also has a single IT gene, *wbpL*, but WbpL can accept a variety of *N*-acetyl sugars as donor (see below). There are homologues of *wbpL* in the gene clusters for the *Y. enterocolitica* O3 outer core (*wbcO*) and *E. coli* O45 O-antigen (*wbhQ*), with both structures having 2-acetamido-2-deoxy-D-fucose (FucNAc) as the first sugar (Skurnik, 2003; DebRoy *et al.*, 2005).

All PNPTs have similar hydrophobicity plots and WecA has been shown to have 11 transmembrane (TM) segments (Lehrer *et al.*, 2007) (Figure 18.3). There is also a proposed reaction mechanism (Price and Momany, 2005). In addition, PHPTs are integral membrane proteins with conserved hydrophobicity plots, but no sequence similarity to PNPT proteins. The WbaP protein has five TM segments (see Figure 18.3), but mutants expressing only the C-terminal domain, including the fifth TM segment, can carry out the Gal-phosphate (Gal-P) transferase function (Wang *et al.*, 1996). The mutants were also affected in further processing of the O-unit and it was proposed that WbaP is a bifunctional protein, with the N-terminal domain involved in further processing of the UndPP-O-units. The role of the C-terminal region was confirmed by Salidas *et al.* (2008), who also recognized the periplasmic loop between TM segments 4 and 5 as an additional domain that was proposed to have a role in chain length determination. There were differences in the observations and proposals for the role of the N-terminal domain, but it is clear that WbaP is not simply a Gal-P transferase and more work is needed to define its other roles, that probably involve interaction with proteins such as Wzx, Wzy and Wzz.

5. OVERVIEW OF THE Wzx/Wzy PATHWAY

Most of the ≈180 *E. coli* O-antigens are thought to be synthesized by the Wzx/Wzy pathway and this also appears to be the case for most genera, as very few of the known O-antigen structures are of the type normally synthesized by the ABC transporter pathway. In this pathway, O-unit sugars and other constituents are transferred sequentially to the first sugar by linkage specific enzymes to form an O-unit that commonly consists of 4–6 sugars, which is then translocated by Wzx across the inner membrane to put the O-unit on the periplasmic face, where it is polymerized by Wzy to generate the polymer (see Figure 18.1) with the number of O-units in the final O-antigen being regulated by Wzz. The complete O-antigen chain is then transferred by the ligase, WaaL, to lipid A-core to make a complete LPS molecule. The *E. coli* O16 and *S. enterica* O-antigens discussed above are examples of this. We now look at GTs and then O-unit processing by Wzx, Wzy and Wzz.

5.1. Completing the O-unit in the Wzx/Wzy pathway: glycosyl and other transferases

In the Wzx/Wzy pathway, sugars are added sequentially by GTs to complete the basic O-unit.

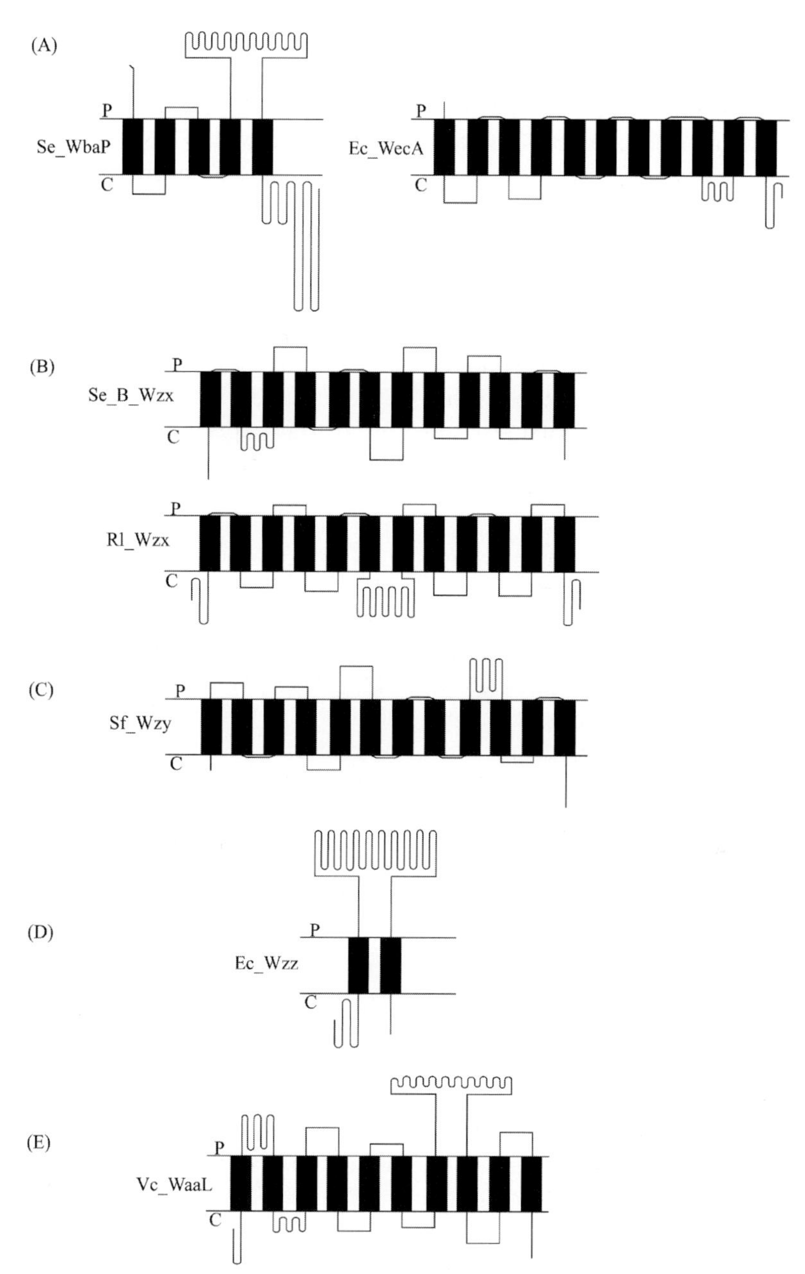

FIGURE 18.3 The TM segment topology models of O-antigen biosynthesis proteins. (A) Initiating transferases: *S. enterica* WbaP (Saldias *et al.*, 2008) and *E. coli* WecA (Lehrer *et al.*, 2007), (B) Flippase proteins: *S. enterica* B Wzx (Cunneen and Reeves, 2008) with that of *R. leguminosarum* bv. trifolii (also called PssL), for exopolysaccharide synthesis, included for comparison (Mazur *et al.*, 2005), (C) O-antigen polymerase: *S. flexneri* 2a Wzy (Daniels *et al.*, 1998), (D) Chain-length determination: *E. coli* O86 Wzz (Tang *et al.*, 2007), (E) Ligation: *V. cholerae* O1 WaaL (Schild *et al.*, 2005). Topology diagrams were drawn using as a guide TOPO2 (http://www.sacs.ucsf.edu/TOPO-run/wtopo.pl) outputs, based on the published topology data for each protein; TOPO2 generates the loops to scale.

In doing this, each GT determines the sugar to be added, the acceptor sugar, the target carbon and the anomerism of the linkage. Each of the NDP-sugar precursors has either an α- or β- linkage and GTs either retain or invert that anomerism during transfer. The GTs may also play a role in the choice of the first sugar. The WecA protein can transfer GlcNAc-P or GalNAc-P and it seems that WecA itself does not differentiate between these as clones with both expressed in K-12. This reaction is reversible and presumably a pool of each is made when the *gne* gene is present for UDP-GalNAc synthesis and it is the next GT that selects either GlcNAc or GalNAc as acceptor.

Glycosyltransferases have been divided into 90 families in the CAZy classification (Lairson *et al.*, 2008), with all members of a given family either retaining or inverting. Most GTs have either the GT-A or GT-B fold, but the two properties are not correlated. Glycosyltransferases are recognized by sequence as being in a given CAZy family and so either retaining or inverting. In some cases, GTs can be associated with a specific linkage as discussed in the examples above but, in general, this has not been done. A few GTs have been studied in detail and the biochemistry is reviewed in Davies *et al.* (2005), Schuman *et al.* (2007) and Lairson *et al.* (2008). Most GTs are responsible for a single linkage, but some such as WbdA/WbhY (discussed later) add two or more residues. In addition to GTs, other transferases can be present for other O-unit additions, such as acetyl or pyruvyl moieties and also sugar phosphates. Finally, as for the Glc residue in *E. coli* O16 described above, Glc, *O*-acetyl and perhaps other moieties can be added after completion of the basic O-unit.

5.2. Translocation (flipping) of the O-unit across the membrane by Wzx

The O-unit is synthesized on the cytoplasmic face of the inner membrane, but polymerized on the periplasmic face (see below). The protein Wzx carries out the necessary translocation, although the evidence for Wzx involvement is largely indirect (Liu *et al.*, 1996; Rick *et al.*, 2003).

The role of Wzx was first inferred (Reeves, 1994) from the observation that gene clusters with *wzy* always included another gene encoding a very hydrophobic protein with about 12 predicted TM segments, that was named *wzx*. For the *S. enterica* LT2 O-antigen Wzx, the 12 TM segment prediction has been confirmed by PhoA and LacZ fusions (Cunneen and Reeves, 2008) and the larger *Rhizobium leguminosarum* bv. trifolii exopolysaccharide Wzx, also called PssL, similarly has 12 TM segments, but with a long cytoplasmic loop between TM segments 6 and 7 (Mazur *et al.*, 2005).

The translocation mechanism is unknown. However, Rick *et al.* (2003) showed that Wzx_{ECA} facilitated the movement across membranes of everted vesicles of GlcNAc-PP-Nerol, a water-soluble synthetic analogue of GlcNAc-PP-Und with only two isoprene units. This did not appear to require a proton-motive force and Wzx proteins do not have an ATP-binding site, suggesting no requirement for energy, and uptake and efflux took about the same time to equilibrium in agreement with a facilitated diffusion model. These experiments provide a model for UndPP-O-unit translocation. The substrate GlcNAc-PP-Nerol is water-soluble while the UndPP-O-unit is firmly embedded in the membrane but, if we assume that the movement of GlcNAc-PP-Nerol involves molecules that had inserted into the membrane, allowing translocation of the hydrophilic PP-O-unit through Wzx, while the hydrophobic Nerol component enters and remains within the hydrophobic inner membrane environment, then we have a good model for UndPP-O-unit translocation, with the direction in Nature being driven by the subsequent polymerization of the O-unit and/or its incorporation into LPS.

The *wzx* genes are very diverse in sequence and that was attributed to the diversity of O-unit substrates. However, in experimental studies, the specificity is largely for the first sugar, as shown in two studies from the Valvano laboratory. Feldman *et al.* (1999) showed that after inactivation of a GT gene then LPS with a single incomplete O-unit was observed. This was *wzx*-dependent establishing that the K-12 Wzx protein could translocate incomplete O-units. In the second study (Marolda *et al.*, 2004), it was shown that *wzx* genes from the *E. coli* O7, O111 and O157 gene clusters could replace the *E. coli* O16 *wzx* gene. The O-antigens all have GlcNAc or GalNAc as first sugar, but *wzx* from gene clusters for the *P. aeruginosa* O5 and *S. enterica* Group B O-antigens, which have FucNAc and Gal respectively, could complement only partially and *wzx* from the gene cluster for *E. coli* colanic acid, with Glc as first sugar, did not complement at all.

Nonetheless, while demonstrated specificity is for the first sugar, the extreme sequence diversity needs explanation. In Nature, it is essentially only the complete O-unit that is ligated and polymerized. Polymerization is by Wzy and the *in vitro* data suggest that Wzx facilitates diffusion rather than drives translocation, so translocation specificity may not be critical. However, it may be selectively advantageous if translocation is more efficient for complete than incomplete O-units, driving selection for preference for the complete O-unit. Particularly interesting is the situation with the Wzx protein from the *E. coli* ECA gene cluster which could complement a wzx_{Ec_O16} deletion, although not fully, but *only* if the ECA gene cluster was deleted (Marolda *et al.*, 2006). The ECA *wzy* and *wzz* genes were responsible for this effect, either of which alone could reduce the ability to complement the wzx_{Ec_O16} deletion. This suggests that Wzx interacts with Wzy and Wzz and that they probably act as a protein complex. Similar data were obtained by complementation with *E. coli* O7 *wzx* suggesting that this is a general phenomenon. The diversity of Wzx may reflect the diversity of Wzy, with which it could interact, and perhaps that interaction also includes the IT, gene with the complex conferring more specificity than observed in the experiments in which there is no competition between incomplete and complete O-units, thereby increasing the efficiency of the whole process.

5.3. Polymerization of O-unit to O-antigen by Wzy

The Wzy protein is responsible for the polymerization of O-units on UndPP on the periplasmic face of the inner membrane (see Figure 18.1), although evidence for location is still indirect (Mulford and Osborn, 1983; McGrath and Osborn, 1991). The polymerization occurs at the reducing end (Bray and Robbins, 1967) and, in effect, the growing polymer must be transferred to the non-reducing end of a single O-unit on UndPP. The role of Wzy was first shown, and has been confirmed in many strains and species, by the effect of *wzy* knockouts on the LPS molecules produced which contain only one O-unit.

The Wzy protein confers the specificity of the polymerization linkage, involving the first sugar as the donor sugar, determination of the acceptor sugar which does not have to be the terminal sugar of the O-unit, determination of the target carbon in the acceptor sugar and the anomerism of the linkage. If, as for example in the *S. enterica* O-antigens B, D1 and C2, the linkage is not with the terminal sugar, then the terminal residue or residues become a side-branch.

Although Wzy is very hydrophobic, hydropathy and TM segment analysis often indicate a lower predicted number of TM segments present than for Wzx and there is always a large loop present in Wzy proteins (Bastin *et al.*, 1993; Morona *et al.*, 1994). The only available topology model for an O-antigen Wzy is for *Shigella flexneri* (Daniels *et al.*, 1998), which indicates that 12 TM segments are present and confirms the presence of a large loop on the periplasmic face. However, there are no Wzy structures or information on reaction mechanisms available for this protein.

5.4. Ligation of O-antigen to lipid A-core

The addition of O-antigen to a specific site on the LPS core is carried out on the periplasmic face of the membrane by WaaL, encoded by a gene in the *waa* LPS core gene cluster in *E. coli* and *S. enterica* (Whitfield, 1995). Thus, *waaL* mutants have the expected phenotype in lacking O-antigen on the LPS.

The WaaL proteins, although diverse, have a characteristic hydropathy plot and the predicted topology was confirmed for a *V. cholerae* protein (Schild *et al.*, 2005) (see Figure 18.3). The *P. aeruginosa* WaaL protein has been purified and shown to ligate the lipid A-core and UndPP-associated O-antigen molecules that accumulate in a *waaL* mutant, the reaction being ATP-dependent (Abeyrathne and Lam, 2007). In *E. coli*, which has several forms of outer core, *waaL* is thought to be specific to the core structure and maps in the LPS core gene cluster as it does in other genera (Amor *et al.*, 2000).

5.5. Determination of O-antigen chain length

In general, LPS seems invariably to be a mixture of molecules with and without O-antigen. For the former, the distribution of chain lengths is readily observed on sodium dodecyl sulfate polyacrylamide gel electrophoresis (SDS-PAGE) gels. A common pattern is of decreasing band intensity with increasing number of O-units, followed by increased intensity for a cluster of bands, with maximal intensity near the middle of this cluster (Figure 18.4). This is generally referred to as a modal distribution. In some cases, there are no visible bands beyond this but, in other cases, there are one or more additional clusters of bands to give a bimodal or trimodal distribution.

This distinctive pattern is conferred by Wzz, discovered by the effect of its absence which changes the distribution from modal to random (see Figure 18.4). For *E. coli* O111, this distribution is as expected if, after addition of each O-unit, the probability of ligation to lipid A-core is 0.065 and 0.935 for extending the chain length by one, before the same choice arises again (Bastin *et al.*, 1993). The effect of Wzz is to impose a non-random distribution of O-antigen chain length on LPS, by varying, in a chain length-related manner, the probabilities of ligation or a further round of polymerization. Moreover, Wzz mutants have been shown to have the same effect on chain length distribution in other species, e.g. in *P. aeruginosa* B-band O-antigen (Burrows *et al.*, 1997) and, while Wzz is the only protein known to have a major effect on chain length in Nature, it can be affected in more subtle ways by other proteins including RfaH (Carter *et al.*, 2007).

In *Yersinia*, *wzz* is within the O-antigen gene cluster, usually as the last gene, and in *P. aeruginosa* as the first gene but, in *E. coli* and *S. enterica*, it is found a few kb downstream of it. In *E. coli*, the modal chain length can vary and, in one study (Franco *et al.*, 1998), cloned *wzz* genes were shown to confer a distribution of O-antigen chain lengths that corresponded to those of the donor strain. Further, minor changes in the *wzz* gene could change the distribution in the direction predicted from the sequences of the two Wzz proteins. It is clear that Wzz not only confers a modal chain length but also can affect the modal value that is conferred. This may account for *wzz* being a short distance from the O-antigen gene cluster as it can be transferred by homologous recombination either independently or in association with the gene cluster, making O-antigen structure and chain lengths potentially independent variables in recombination. An unusual variation is found in *Shigella flexneri* which has an additional plasmid encoded *wzz* gene that confers a very long chain length component to the LPS (Stevenson *et al.*, 1995) and a second example was found on the chromosome in *S. enterica* (Murray *et al.*, 2003).

The Wzz protein includes two TM segments and a large periplasmic loop (Morona *et al.*,

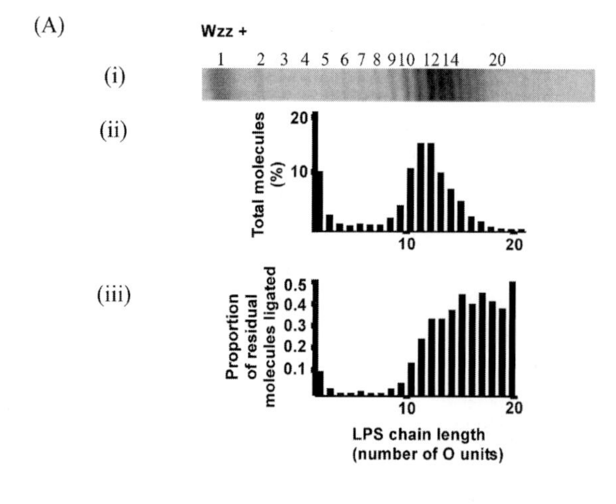

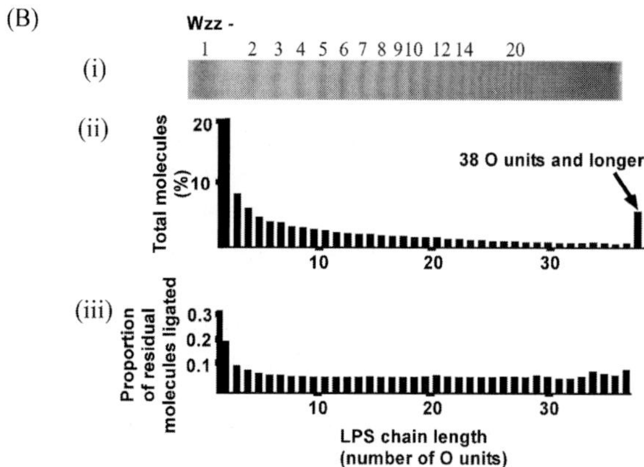

FIGURE 18.4 Role of Wzz in O-antigen synthesis. The LPS profiles from (A) wild-type Wzz and (B) mutant Wzz strains are shown. (i) PhosphoImager image of an SDS-PAGE gel of LPS labelled with C^{14}-Gal with the number of O-units in each LPS band (up until 20) given above the PhosphorImager image; (ii) percentage of total O-antigen molecules by chain length; and (iii) the corresponding proportion of residual O-antigen molecules that were ligated. Modified, with permission, from Bastin *et al.* (1993).

1995) (see Figure 18.3) and we now have structures for the periplasmic loops of Wzz_{FepE_Ec} (chain length >80), $Wzz_{Oag\_Se\_LT2}$ (16–35) and $Wzz_{ECA\_Ec\_O157}$ (5–7) (Tocilj *et al.*, 2008), based on X-ray diffraction data. The three structures are similar with a membrane-proximal domain and a domain extending into the periplasm with a helical hairpin containing an ≈100-Å-long helix. Elongated particles were observed by electron microscopy and models for bell-shaped structures were developed with oligomers of nine, five and eight monomers, respectively. A later study using electron microscopy on full length $Wzz_{Oag\_Se\_LT2}$, $Wzz_{FepE\_Se\_LT2}$, and Wzz_{K40_Ec}, in reconstituted lipid bilayers, found only hexameric structures (Larue *et al.*, 2009). There is clearly variation under different experimental conditions in the association of Wzz to form complexes, and the size of the functional complex is yet to be determined.

There are three models for Wzz action, all of which are speculative. It was proposed, on the basis of the structures (Tocilj *et al.*, 2008), that Wzz associates with a number of Wzy molecules, with polymer intermediates shuttled successively between them, with the chain length determined by the number of associated Wzy molecules, which would in turn correlate with the size of the PCP oligomer. The current uncertainty over the size of the functional oligomer complicates evaluation of this model. An earlier model (Bastin *et al.*, 1993), also without experimental support, is based on the assumption that the growing chain and the single O-unit, both on UndPP, bind to different active sites on the one Wzy protein and, in the manner of a ribosome, Wzy moves the newly extended molecule to the acceptor site for another round on the same Wzy molecule. The Wzz protein was proposed to influence the number of cycles that are undertaken by reducing the probability of ligation early in polymerization but then increasing it dramatically over the default level, leading to ligation of most molecules at a similar stage in chain extension. The option proposed was a timed period for extension. The two models are sufficiently different in the way in which Wzx and Wzy interact to make it possible to test them. The third model involves Wzz controlling the interaction of Wzy and WaaL, but this seems to be ruled out by two reports that the modal chain length is observed in the absence of WaaL (Daniels *et al.*, 2002) so is established before ligation. For the other models this would require the extended chain to be stable after extension was terminated.

6. THE ABC TRANSPORTER PATHWAY

The ABC transporter pathway differs from the Wzx/Wzy pathway in that the complete O-antigen chain is synthesized on the cytoplasmic face of the inner membrane and translocation of the UndPP-O-antigen (not a UndPP-O-unit) is carried out by ABC transporter proteins Wzm and Wzt. The polysaccharides are usually homopolymers or have only two constituent sugars, with the repeating unit often determined by a pattern of linkages. This pathway is rare among O-antigens, but common among capsules. *Klebsiella* O-antigens are an exception as the four known gene clusters all have the ABC transporter pathway and the known structures all have only one or two sugar species, consistent with all having that pathway. *E. coli* O8, O9a and O9 have been studied in some detail (Kido *et al.*, 1995; Clarke *et al.*, 2004). All are polymannans, with repeating units of three, four or five Man residues, respectively, distinguished by specific patterns of α-(1$\rightarrow$4), β-(1$\rightarrow$2) and/or α-(1$\rightarrow$3)-linkages. The gene cluster is at a locus between *gnd* and the *his* operon, close to that for the Wzx/Wzy pathway gene clusters. Structures identical to *E. coli* O8 and O9 are found in *K. pneumoniae* (O5 and O3, respectively), with the same gene clusters at the same locus (Sugiyama *et al.*, 1997) and the *E. coli* O8 and O9 antigens are thought to be derived from *Klebsiella.*

The first step in synthesis is the addition of GlcNAc-P to UndP by WecA, followed by addition of two α-(1$\rightarrow$3)-linked Man residues by WbdC and WbdB to give a three-sugar adapter present in all three gene clusters (Clarke *et al.*, 2004). The genes have very similar sequences in all three. The WbdA (O9 and 9a) and WbhY (O8) proteins are clearly involved in synthesis of the multiple repeating units and, although both encoding genes were originally named *wbdA*, the sequences are quite different, reflecting the differences in structure. The WbdB protein is also reported to be involved in O9a repeating unit synthesis (Kido *et al.*, 1995), but interpretation of those data is complicated by later findings that the strain involved was O9a and not O9 as thought at the time. From clarification of the adaptor structure, however, it seems clear that the enzymes involved in extension add more than one residue of the repeating unit.

The *E. coli* O8 O-antigen has a terminal methyl group on the last Man residue, apparently added by WbhZ (first named WbdD), that terminates addition of further repeating units, thereby determining chain length. This addition is also required for export (Clarke *et al.*, 2004). A similar situation occurs in O9 and O9a, but the terminal residues are thought to comprise phosphate and methyl groups (Clarke *et al.*, 2004). The proteins WbhZ of O8 and WbdD of O9/O9a have low sequence similarity and are now given different names. They share a conserved methyltransferase motif in the N-terminus, with the latter also having a kinase motif.

The membrane component, Wzm, that presumably carries out the export process, is interchangeable between O8 and O9a, although its activity is compromised (Cuthbertson *et al.*, 2007), and presumably O9. Wzt is the nucleotide-binding component with two domains. The N-terminal domain is typical of other ABC proteins, whereas the C-terminal has a unique domain that binds the polysaccharide and, being specific for the cognate O-antigen, is critical for specificity (Cuthbertson *et al.*, 2007). For O9a, at least, Wzt binding requires methylation and phosphorylation of the non-reducing end by WbdD. Cognate domains can function when expressed separately, but do not cross complement, so that the N-terminal domain may also be specific for the substrate, or it may be that the docking does not work properly for non-cognate partners.

7. THE SYNTHASE PATHWAY

The Synthase pathway is the simplest, with a single integral membrane protein responsible both for sequential addition of the sugar(s) and the concurrent extrusion of the nascent polymer across the cell membrane. *S. enterica* O54 (Keenleyside *et al.*, 1994) is the only example that we are aware of for O-antigens, but there are two well-studied cases in *S. pneumoniae* capsules and some major eucaryote structural polysaccharides, with that for hyaluronic acid particularly well studied.

The repeating units in Synthase pathways are generally homopolymers or have two sugars, i.e. a disaccharide repeating unit, in which the two sugars may alternate or one may form the main chain and the other a side-branch on each main chain residue. The O54 repeating unit is a dimer, →4)-β-D-Man*p*NAc-(1→3)-β-D-Man*p*NAc-(1→.

The O54 antigen gene cluster is on a small mobilizable plasmid and O54-positive isolates sometimes express another O-antigen. In the O54 type strain, this is not the case due to inactivation of the chromosomal O-antigen locus. The O54 gene cluster includes *mnaA* for synthesis of precursor UDP-*N*-acetylmannosamine (UDP-Man*p*NAc), *wbbE* for transfer of the first Man*p*NAc to UndPP-GlcNAc to complete an adapter and *wbbF* as the synthase gene for repeat-unit synthesis and export (see Figure 18.1).

8. CONCLUSIONS

We have looked at and reviewed the biosynthesis of O-antigens using examples from representative species. Most are synthesized by the Wzx/Wzy pathway, a number by the ATP transporter pathway and one known example by the Synthase pathway. Many of the nucleotide-sugar pathways are now known but not as well studied as those for eukaryotes. However, each sugar, or other constituent, is usually synthesized by the same pathway wherever found and genes from one source will generally complement other genes with the same function. These genes can generally be identified from sequence once the pathway is known and are generally given pathway related names with the fourth letters ideally in function order, e.g. *rmlA*, *rmlB*, *rmlC* and *rmlD* are genes for synthesis of dTDP-L-Rha, though the map order is usually different (see Samuel and Reeves (2003) and the Bacterial Polysaccharide Gene Database (BPGD) web site for more details: http://www.mmb.usyd.edu.au/BPGD/).

The membrane-based O-antigen processing genes *wzx*, *wzy*, *wzz*, *wzm*, and *wzt* and also *waaL* are generally identifiable from sequences alone, and their functions are all understood except for Wzz. However, only three Wzz structures are known and their mechanisms of interaction are unknown. There is much to be done, especially as protein–protein interactions seem to be involved for all of them, making the determination of structures a requirement to properly understand any in the two major pathways. These are all classes of genes with similar function, but are specific to one, or a group, of O-antigens. They can be extremely diverse in sequence, but blast searches usually hit related genes of the same class. Nevertheless, they can vary enormously in specificity, in the case of Wzy generating a wide range of glycosidic linkages, and different forms can be distinguished by addition of suffixes.

The GTs are responsible for most of the very high level of diversity and it is rarely possible to assign specific linkages to all of the GTs in a gene cluster from sequence alone. It has been difficult to assign function experimentally as only the more common precursors are available and the natural substrates are UndPP-associated incomplete O-units. A recent paper offers a solution to this problem by looking at the intermediates accumulated after mutation of each GT gene, enabling at least provisional allocation of function to each, provided the O-antigen structure is known (Stevenson *et al.*, 2008).

Most species have only one O-antigen locus but *P. aeruginosa* is unusual in having two O-antigen forms, A-band and B-band, that map to different loci. The A-band O-antigen, synthesized by the ABC transporter pathway, is present in most isolates and always has the same structure (Rocchetta *et al.*, 1998), while B-band O-antigen, synthesized by the Wzx/Wzy pathway, varies in the manner of the single O-antigen in most other species. The only IT gene is *wbpL* in the B-band gene cluster. Of the 13 B-band serogroups (Lanyi-Bergan), 11 of the O-units include an *N*-acetylhexosamine as the first sugar: three contain FucNAc, one has 2-acetamido-4-[(*S*)-3-hydroxybutyrylamino]-2,4,5-trideoxyglucose (QuiNAc4NSHb) and seven contain 2-acetamido-2,6-dideoxy-D-glucose (QuiNAc), with the remaining two found not to produce an O-unit (Bystrova *et al.*, 2006), while the A-band O-unit has GlcNAc as the adapter (see below). The WbpL protein can recognize a range of substrates and usually only one is present in the B-band. In two cases where both QuiNAc and FucNAc are present, the former is used as the first sugar. *E. coli* and *Yersinia* spp. also have two loci, one for Wzx/Wzy pathway gene clusters and the other ABC transporter pathway gene clusters. Nonetheless, usually only one is expressed and, in *Y. enterocolitica*, the major O-antigen is encoded at the ABC transporter locus upstream of *galF/gnd*, but there is also an "outer core" which is, in effect, a single repeating unit encoded at the Wzx/Wzy locus between *hemH* and *gsk* and added to the basic core at a different site (Skurnik, 2003).

There are many other contributions to diversity but we will conclude with side-branch Glc and *O*-acetyl residues. Side-branch Glc residues are often encoded outside of the gene cluster as part of a three-gene *gtr* set. The GrtB protein forms UndP-Glc and GtrA translocates it to place the Glc residue on the periplasmic face and a locus specific gene encodes the GT for transfer to a specific site, which occurs in the periplasm after O-unit translocation. The genes are often on phages or other genetic elements and not always present, as in the case of the side-branch Glc in the O16 O-antigen of K-12, which is not present in the type O16 strain. Allison and Verma (2000) give details of one such group in *Shigella flexneri*, in which most of the serotypes are based on O-antigen variation due to side-branch Glc or *O*-acetyl moieties encoded on prophage genomes. A study of the other *Shigella* shows that of 33 O-antigens, 15 have *O*-acetyl residues, but only four have a putative transferase in the gene cluster (Liu *et al.*, 2008). Transferases for *O*-acetyl side-branches are also often encoded by genes outside of the gene cluster, again often on phages or other genetic elements.

Areas for further research that will aid development in this field are detailed in the Research Focus Box.

RESEARCH FOCUS BOX

- Determination of the roles, interactions, structures and reaction mechanisms of O-antigen biosynthesis proteins, particularly of initiating transferases and processing proteins.
- Determination of additional structures of O-antigen chains, particularly those of non-enterobacterial species.

References

Abeyrathne, P.D., Lam, J.S. 2007. WaaL of *Pseudomonas aeruginosa* utilizes ATP *in vitro* ligation of O antigen onto lipid A-core. Mol. Microbiol. 65, 1345–1359.

Allison, G.E., Verma, N.K. 2000. Serotype-converting bacteriophages and O-antigen modification in *Shigella flexneri*. Trends Microbiol. 8, 17–23.

Amor, K., Heinrichs, D.E., Frirdich, E., Ziebell, K., Johnson, R., Whitfield, C. 2000. Distribution of core oligosaccharide types in lipopolysaccharides from *Escherichia coli*. Infect. Immun. 68, 1116–1124.

Bastin, D.A., Brown, P.K., Haase, A., Stevenson, G., Reeves, P.R. 1993. Repeat unit polysaccharides of bacteria: a model for polymerisation resembling that of ribosomes and fatty acid synthetase, with a novel mechanism for determining chain length. Mol. Microbiol. 7, 725–734.

Bray, D., Robbins, P.W. 1967. The direction of chain growth in *Salmonella anatum* O-antigen biosynthesis. Biochem. Biophys. Res. Commun. 28, 334–339.

Brown, P.K., Romana, L.K., Reeves, P.R. 1992. Molecular analysis of the *rfb* gene cluster of *Salmonella* serovar Muenchen (strain M67): genetic basis of the polymorphism between groups C2 and B. Mol. Microbiol. 6, 1385–1394.

Burrows, L.L., Chow, D., Lam, J.S. 1997. *Pseudomonas aeruginosa* B-band O-antigen chain length is modulated by Wzz (Rol). J. Bacteriol. 179, 1482–1489.

Bystrova, O.V., Knirel, Y.A., Lindner, B. et al., 2006. Structures of the core oligosaccharide and O-units in the R- and SR-type lipopolysaccharides of reference strains of *Pseudomonas aeruginosa* O-serogroups. FEMS Immunol. Med. Microbiol. 46, 85–99.

Carter, J.A., Blondel, C.J., Zaldivar, M. et al., 2007. O-Antigen modal chain length in Shigella flexneri 2a is growth-regulated through RfaH-mediated transcriptional control of the wzy gene. Microbiology 153, 3499–3507.

Clarke, B.R., Cuthbertson, L., Whitfield, C. 2004. Nonreducing terminal modifications determine the chain length of polymannose O antigens of *Escherichia coli* and couple chain termination to polymer export via an ATP-binding cassette transporter. J. Biol. Chem. 279, 35709–35718.

Cunneen, M.M., Reeves, P.R. 2008. Membrane topology of the *Salmonella enterica* serovar Typhimurium Group B O-antigen translocase Wzx. FEMS Microbiol. Lett. 287, 76–84.

Cuthbertson, L., Kimber, M.S., Whitfield, C. 2007. Substrate binding by a bacterial ABC transporter involved in polysaccharide export. Proc. Natl. Acad. Sci. USA 104, 19529–19534.

Daniels, C., Vindurampulle, C., Morona, R. 1998. Overexpression and topology of the *Shigella flexneri* O-antigen polymerase (Rfc/Wzy). Mol. Microbiol. 28, 1211–1222.

Daniels, C., Griffiths, C., Cowles, B., Lam, J.S. 2002. *Pseudomonas aeruginosa* O-antigen chain length is determined before ligation to lipid A core. Environ. Microbiol. 12, 883–897.

Davies, G.J., Gloster, T.M., Henrissat, B., 2005. Recent structural insights into the expanding world of carbohydrate-active enzymes. Curr. Opin. Struct. Biol. 15, 637–645.

DebRoy, C., Fratamico, P.M., Roberts, E., Davis, M.A., Liu, Y. 2005. Development of PCR assays targeting genes in O-antigen gene clusters for detection and identification of *Escherichia coli* O45 and O55 serogroups. Appl. Environ. Microbiol. 71, 4919–4924.

Feldman, M.F., Marolda, C.L., Monteiro, M.A., Perry, M.B., Parodi, A.J., Valvano, M.A. 1999. The activity of a putative polyisoprenol-linked sugar translocase (Wzx) involved in *Escherichia coli* O antigen assembly is independent of the chemical structure of the O repeat. J. Biol. Chem. 274, 35129–35138.

Franco, V.A., Liu, D., Reeves, P.R. 1998. The Wzz (Cld) protein in *Escherichia coli*: amino acid sequence variation determines O antigen chain length specificity. J. Bacteriol. 180, 2670–2675.

Jiang, X.M., Neal, B., Santiago, F., Lee, S.J., Romana, L.K., Reeves, P.R. 1991. Structure and sequence of the *rfb* (O antigen) gene cluster of *Salmonella* serovar typhimurium (strain LT2). Mol. Microbiol 5, 695–713.

Keenleyside, W.J., Perry, M.B., MacLean, L.L., Poppe, C., Whitfield, C. 1994. A plasmid-encoded *rfb*O:54 gene cluster is required for biosynthesis of the O:54 antigen in *Salmonella enterica* serovar Borreze. Mol. Microbiol. 11, 437–448.

Kenne, L., Lindberg, B., Petersson, K., Katzenellenbogen, E., Romanowska, E. 1980. Structural studies of the O-specific side-chains of *Shigella sonnei* phase I lipopolysaccharide. Carbohydr. Res. 78, 119–126.

Kido, N., Torgov, V.I., Sugiyama, T. et al. 1995. Expression of the O9 polysaccharide of *Escherichia coli*: sequencing of the *E. coli* O9 *rfb* gene cluster, characterization of mannosyl transferases, and evidence for an ATP-binding cassette transport system. J. Bacteriol. 177, 2178–2187.

Lairson, L.L., Henrissat, B., Davies, G.J., Withers, S.G. 2008. Glycosyltransferases: structures, functions, and mechanisms. Annu. Rev. Biochem. 77, 521–555.

Larue, K., Kimber, M.S., Ford, R., Whitfield, C. 2009. Biochemical and structural analysis of bacterial O-antigen chain length regulator proteins reveals a conserved quaternary structure. J. Biol. Chem. 284, 7395–7403.

Lehrer, J., Vigeant, K.A., Tatar, L.D., Valvano, M.A. 2007. Functional characterization and membrane topology of *Escherichia coli* WecA, a sugar-phosphate transferase initiating the biosynthesis of enterobacterial common antigen and O-antigen lipopolysaccharide. J. Bacteriol. 189, 2618–2628.

Liu, B., Knirel, Y.A., Feng, L. et al. 2008. Structure and genetics of Shigella O antigens. FEMS. Microbiol. Rev 32, 627–653.

Liu, D., Reeves, P.R. 1994. *Escherichia coli* regains its O antigen. Microbiology 140, 49–57.

Liu, D., Haase, A.M., Lindqvist, L., Lindberg, A.A., Reeves, P.R. 1993. Glycosyl transferases of O-antigen biosynthesis in *Salmonella enterica*: identification and characterization of transferase genes of groups B, C2, and E1. J. Bacteriol. 175, 3408–3413.

Liu, D., Lindquist, L., Reeves, P.R. 1995. Transferases of O-antigen biosynthesis in *Salmonella enterica*: dideoxhexosyl transferases of groups B and C2 and acetyltransferase of group C2. J. Bacteriol. 177, 4084–4088.

Liu, D., Cole, R., Reeves, P.R. 1996. An O-antigen processing function for Wzx(RfbX): a promising candidate for O-unit flippase. J. Bacteriol. 178, 2102–2107.

Marolda, C.L., Vicarioli, J., Valvano, M.A. 2004. Wzx proteins involved in biosynthesis of O antigen function in association with the first sugar of the O-specific lipopolysaccharide subunit. Microbiology 150, 4095–4105.

Marolda, C.L., Tatar, L.D., Alaimo, C., Aebi, M., Valvano, M.A. 2006. Interplay of the Wzx translocase and the corresponding polymerase and chain length regulator proteins in the translocation and periplasmic assembly of lipopolysaccharide o antigen. J. Bacteriol. 188, 5124–5135.

Mazur, A., Marczak, M., Krol, J.E., Skorupska, A. 2005. Topological and transcriptional analysis of *pssL* gene product: a putative Wzx-like exopolysaccharide translocase in *Rhizobium leguminosarum* bv. trifolii TA1. Arch. Microbiol. 184, 1–10.

McGrath, B.C., Osborn, M.J. 1991. Localisation of the terminal steps of O-antigen synthesis in *Salmonella typhimurium*. J. Bacteriol. 173, 649–654.

Meier-Dieter, U., Barr, K., Starman, R., Hatch, L., Rick, P.D. 1992. Nucleotide sequence of the *Escherichia coli rfe* gene involved in the synthesis of enterobacterial common antigen. J. Biol. Chem. 267, 746–753.

Morona, R., Mavris, M., Fallarino, A., Manning, P.A. 1994. Characterisation of the *rfc* region of *Shigella flexneri*. J. Bacteriol. 176, 733–747.

Morona, R., Van Den Bosch, L., Manning, P. 1995. Molecular, genetic, and topological characterization of O antigen chain regulation in *Shigella flexneri*. J. Bacteriol. 177, 1059–1068.

Mulford, C.A., Osborn, M.J. 1983. A intermediate step in translocation of lipopolysaccharide to outer membrane of *Salmonella typhimurium*. Proc. Natl. Acad. Sci. USA 80, 1159–1163.

Murray, G.L., Attridge, S.R., Morona, R. 2003. Regulation of *Salmonella typhimurium* lipopolysaccharide O antigen chain length is required for virulence; identification of FepE as a second Wzz. Mol. Microbiol. 47, 1395–1406.

Price, N.P., Momany, F.A. 2005. Modeling bacterial UDP-HexNAc: polyprenol-P HexNAc-1-P transferases. Glycobiology 15, 29R–42R.

Raetz, C.R.H., Whitfield, C. 2002. Lipopolysaccharide endotoxins. Annu. Rev. Biochem. 71, 635–700.

Reeves, P.R. 1994. Biosynthesis and assembly of lipopolysaccharide. In: Neuberger, A., van Deenen, L.L.M. (Eds.), Bacterial Cell Wall, Vol. 27. Elsevier Science Publishers, Amsterdam, pp. 281–314.

Rick, P.D., Barr, K., Sankaran, K., Kajimura, J., Rush, J.S. Waechter, C.J. 2003. Evidence that the *wzxE* Gene of *Escherichia coli* K-12 encodes a protein involved in the transbilayer movement of a trisaccharide-lipid intermediate in the assembly of Enterobacterial common antigen. J. Biol. Chem. 278, 16534.

Rocchetta, H.L., Burrows, L.L., Pacan, J.C., Lam, J.S. 1998. Three rhamnosyltransferases responsible for assembly of the A-band D-rhamnan polysaccharide in *Pseudomonas aeruginosa*: a fourth transferase, WbpL, is required for the initiation of both A-band and B-band lipopolysaccharide synthesis. Mol. Microbiol. 28, 1103–1119.

Saldias, M.S., Patel, K., Marolda, C.L., Bittner, M., Contreras, I., Valvano, M.A. 2008. Distinct functional domains of the *Salmonella enterica* WbaP transferase that is involved in the initiation reaction for synthesis of the O antigen subunit. Microbiology 154, 440–453.

Samuel, G., Reeves, P.R. 2003. Biosynthesis of O-antigens: genes and pathways involved in nucleotide sugar precursor synthesis and O-antigen assembly. Carbohydr. Res. 338, 2503–2519.

Schild, S., Lamprecht, A., Reidl, J. 2005. Molecular and functional characterization of O antigen transfer in Vibrio cholerae. J. Biol. Chem. 280, 25936–25947.

Schuman, B., Alfaro, J.A., Evans, S.V. 2007. Glycosyltransferase structure and function. Top. Curr. Chem. 272, 217–257.

Skurnik, M., 2003. Molecular genetics, biochemistry and biological role of *Yersinia* lipopolysaccharide. In: Skurnik, M., Bengoechea, J.A., Granfors, K. (Eds.), The Genus *Yersinia*: Entering the Functional Genomics Era. Kluwer Plenum, New York, pp. 187–197.

Stevenson, G., Neal, B., Liu, D. et al. 1994. Structure of the O-antigen of *E. coli* K-12 and the sequence of its *rfb* gene cluster. J. Bacteriol. 176, 4144–4156.

Stevenson, G., Diekelmann, M., Reeves, P.R. 2008. Determination of glycosyltransferase specificities for the *Escherichia coli* O111 O antigen by a generic approach. Appl. Environ. Microbiol. 74, 1294–1298.

Stevenson, G.S., Kessler, A., Reeves, P.R. 1995. A plasmid-borne O-antigen chain length determinant and its relationship to other chain length determinants. FEMS Microbiol. Lett. 125, 23–30.

Sugiyama, T., Kido, N., Kato, Y., Koide, N., Yoshida, T., Yokochi, T. 1997. Evolutionary relationship among rfb gene clusters synthesizing mannose homopolymer as O-specific polysaccharides in *Escherichia coli* and *Klebsiella*. Gene 198, 111–113.

Tang, K.H., Guo, H., Yi, W., Tsai, M.D., Wang, P.G. 2007. Investigation of the conformational states of Wzz and the Wzz O-antigen complex under near-physiological conditions. Biochemistry (Moscow) 46, 11744–11752.

Tocilj, A., Munger, C., Proteau, A. et al. 2008. Bacterial polysaccharide co-polymerases share a common framework for control of polymer length. Nat. Struct. Mol. Biol. 15, 121–123.

Valvano, M.A. 2003. Export of O-specific lipopolysaccharide. Front. Biosci. 8, S452–S471.

Verma, V., Reeves, P.R. 1989. Identification and sequence of *rfbS* and *rfbE*, which determine antigenic specificity of group A and group D *Salmonella*. J. Bacteriol. 171, 5694–5701.

Wang, L., Liu, D., Reeves, P.R. 1996. C-terminal half of *Salmonella enterica* WbaP (RfbP) is the galactosyl-1-phosphate transferase domain catalysing the first step of O antigen synthesis. J. Bacteriol. 178, 2598–2604.

Wang, L., Andrianopoulos, K., Liu, D., Popoff, M.Y., Reeves, P.R. 2002. Extensive variation in the O-antigen gene cluster within one Salmonella serogroup reveals an unexpected complex history. J. Bacteriol. 184, 1669–1677.

Whitfield, C., 1995. Biosynthesis of lipopolysaccharide O antigens. Trends Microbiol. 3, 178–185.

Wing, C., Errey, J.C., Mukhopadhyay, B., Blanchard, J.S., Field, R.A. 2006. Expression and initial characterization of WbbI, a putative D-Galf:α-D-Glc β-1,6-galactofuranosyltransferase from Escherichia coli K-12. Org. Biomol. Chem. 4, 3945–3950.

Wyk, P., Reeves, P.R. 1989. Identification and sequence of the gene for abequose synthase which confers antigenic specificity on group B salmonellae: homology with galactose epimerase. J. Bacteriol. 171, 5687–5693.

Xu, D.Q., Cisar, J.O., Ambulos, N., Burr, D.H., Kopecko, D.J. 2002. Molecular cloning and characterization of genes for *Shigella sonnei* form I O polysaccharide: proposed biosynthetic pathway and stable expression in a live salmonella vaccine vector. Infect. Immun. 70, 4414–4423.

Yao, Z., Valvano, M.A. 1994. Genetic analysis of the O-specific lipopolysaccharide biosysnthesis region (*rfb*) of *Escherichia coli* K-12 W3110: identification of genes that confer group 6-specificity to *Shigella flexneri* serotypes Y and 4a. J. Bacteriol. 176, 4133–4143.

CHAPTER

19

Biosynthesis of cell wall teichoic acid polymers

Mark P. Pereira and Eric D. Brown

SUMMARY

Wall teichoic acids are phosphate-rich polymers that are part of the complex meshwork that makes up the Gram-positive cell wall. These polymers are essential to the proper rod-shaped morphology of *Bacillus subtilis* and have been shown to be an important virulence determinant in the opportunistic pathogen *Staphylococcus aureus*. Although the biosynthesis of the teichoic acid polymer has been the subject of decades of investigation, only through recent advances in the development of chemical tools necessary for functional assays has the model of teichoic acid biosynthesis in *B. subtilis* and *S. aureus* been solidified. Here, we discuss the use of these tools to characterize the enzymes involved in the membrane-localized steps of biosynthesis through both *in vitro* reconstitution of entire biosynthetic pathways, as well as careful mechanistic studies of each reaction. Mechanistic and structural studies aimed at understanding the biosynthesis of soluble precursors necessary for each step are also outlined. In addition, we discuss recent work on understanding the enzymatic polymerization of glycerol-phosphate and ribitol-phosphate in terms of catalytic mechanism and processivity. Finally, the current challenges and unknowns in the biosynthesis and insertion of teichoic acid in the bacterial cell wall are summarized.

Keywords: Teichoic acid; Bacterial cell wall; Anionic polymer; Gram-positive bacteria; Poly(glycerol-phosphate)

1. INTRODUCTION

In the model Gram-positive organism *Bacillus subtilis*, the vegetative cell wall is primarily made up of a thick peptidoglycan layer, proteins and covalently associated anionic polymers. The major anionic polymer is a phosphate-rich molecule termed wall teichoic acid (WTA) that can account for more than 50% of the cell wall dry weight (Hancock, 1997). These molecules are typically phosphodiester-linked polyol-phosphate repeats that are coupled to peptidoglycan via a conserved linkage unit. The structure of WTA can vary depending on organism or particular strain. *B. subtilis* 168 has a glycerol-phosphate polymer, while both *Staphylococcus aureus* and *B. subtilis* strain W23 have a ribitol-phosphate polymer. A comparison of the chemical structures of these molecules is depicted in Figure 19.1. (See Chapter 5 for an introductory review to these molecules.)

Although peptidoglycan is known to function in cell shape determination and to provide rigidity and stability to resist osmotic pressure, to date, the function of teichoic acid cannot be linked to any one specific process (Bhavsar and Brown, 2006). It has long been suggested that WTA plays a role in providing bacteriophage attachment sites (Young, 1967; Yasbin *et al.*, 1976), regulating autolysins in cell wall turnover and remodelling (Herbold and Glaser, 1975) or simply as a reserve for excess phosphate (Grant, 1979). Recently, WTA has also been linked to cell shape by suggestions that it adds considerable rigidity to the cell wall (Matias and Beveridge, 2005). This idea is supported by a loss of rod shape in *B. subtilis* mutants that lack WTA (D'Elia *et al.*, 2006b). A possible clinical relevance of WTA has also begun to emerge with the discovery that this polymer is an important factor in biofilm formation (Gross *et al.*, 2001) and virulence of *S. aureus* (Weidenmaier *et al.*, 2004).

In this chapter, we focus on recent advances in the study of the biosynthesis of the most common and studied teichoic acid polymers, the major WTAs from *B. subtilis* and *S. aureus*. This in-depth review details synthesis of the required nucleotide-activated precursors as well as our emerging understanding of priming and polymerization of teichoic acid polymers. We discuss how the availability of active recombinant proteins and soluble substrate analogues has facilitated some exciting advances in our knowledge of teichoic acid biosynthesis. We also discuss the outstanding questions and hurdles that currently preclude a complete understanding of WTA biosynthesis.

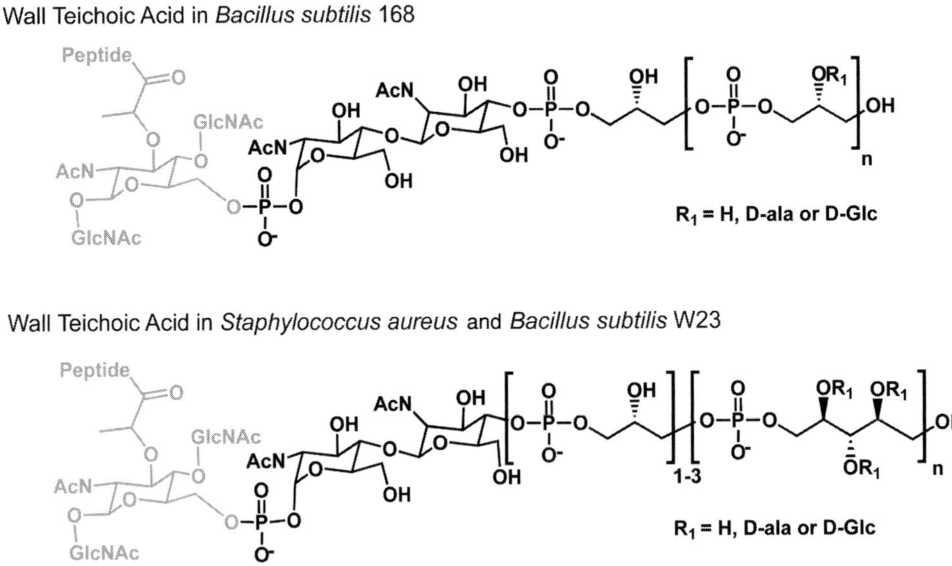

FIGURE 19.1 Chemical structure of common wall teichoic acid polymers. Wall teichoic acid from *B. subtilis* 168 has a glycerol-phosphate repeating unit, while both *S. aureus* and *B. subtilis* W23 have a repeating unit of ribitol-phosphate. Each of the polyol-phosphate polymers are linked to the C-6 of muramic acid in cell wall peptidoglycan via a conserved disaccharide linkage unit. Abbreviations: D-Ala, D-alanine; D-Glc, D-glucose; GlcNAc, *N*-acetyl-D-glucosamine.

2. MODEL OF TEICHOIC ACID BIOSYNTHESIS

Until recently, much of our comprehension of teichoic biosynthesis has come from vintage biochemical studies done decades ago using impure cellular lysates and exhaustive chemical analysis of the Gram-positive cell walls. With the discovery and identification of the genes involved in WTA biosynthesis, *tagABDEFGHO* in *B. subtilis* 168 and *tarABDFGHIJKLO* in *B. subtilis* W23 and *S. aureus* (Mauël *et al.*, 1991; Lazarevic *et al.*, 2002; Qian *et al.*, 2006), new and better informed models of teichoic acid biosynthesis began to emerge. Indeed, the availability of temperature-sensitive mutations in some of the *tag* genes led to putative functional assignments (Pooley *et al.*, 1991, 1992). Bioinformatic studies analysing sequence homology between the *tag* genes of *B. subtilis* 168 and the ribitol-phosphate *tar* genes of *B. subtilis* W23 and *S. aureus* aided in the functional predictions for genes linked to ribitol-phosphate teichoic acid biosynthesis (Lazarevic *et al.*, 2002; Qian *et al.*, 2006). Figure 19.2 outlines the current model for poly(glycerol-phosphate) WTA biosynthesis in the organism *B. subtilis* 168.

In addition to the polymer shown in Figure 19.1, two additional anionic polymers can be present in the *B. subtilis* cell wall: a minor teichoic acid polymer [poly(glucopyranosyl-*N*-acetylgalactosamine-1-phosphate)] and the phosphate starvation induced polymer teichuronic acid [poly(glucuronyl-*N*-acetylgalactosamine)]. Comparatively little is known about the assembly of these polymers. It is believed that enzymes encoded in the *gga* locus catalyse the synthesis of minor teichoic acid using nucleotide-activated precursors in a similar manner to teichoic acid (Estrela *et al.*, 1991). Likewise, the products *tua* gene cluster have been implicated in the synthesis of teichuronic acid, an anionic polymer devoid of phosphate in *B. subtilis* in phosphate limiting conditions (Ellwood and Tempest, 1972). Dispensability studies suggest although these polymers may be present in the cell wall, they cannot functionally replace teichoic acid in either phosphate limiting (Bhavsar *et al.*, 2004) or phosphate replete conditions (Bhavsar *et al.*, 2001).

3. BIOSYNTHESIS OF WTA PRECURSORS

The nucleotide precursors involved in polymer biosynthesis were discovered before the polymer itself was known. In the mid 20th century, Baddiley and co-workers described the presence of the cytidine-containing compounds cytidine diphosphate- (CDP-) glycerol and CDP-ribitol in the cell (Baddiley *et al.*, 1956). It was further discovered that the formation of teichoic acid is dependent on the concurrent synthesis of these and other nucleotide-activated precursors as substrates for glycosyl transfer. Due to the soluble nature of enzymes involved in precursor synthesis and the availability of functional assays, the study of activated precursor biosynthesis is by far the most advanced area of teichoic acid research.

3.1. Linkage unit precursors

According to the biosynthetic model of teichoic acid, synthesis begins with the sequential addition of *N*-acetylglucosamine-1-phosphate (GlcNAc-1-P) and *N*-acetylmannosamine (ManNAc) to the lipid carrier undecaprenyl-phosphate, forming Lipidα and Lipidβ, respectively. Each of these additions is dependent on the activated precursor uridine diphosphate- (UDP-) *N*-acetylglucosamine (UDP-GlcNAc) formed in two successive reactions by the bifunctional enzyme, GlmU. Careful enzymological studies have revealed that the substrate glucosamine-1-phoshate is acetylated with the co-factor acetyl-co-enzyme A prior to activation by uridine triphosphate (UTP) without channelling of intermediates between the active

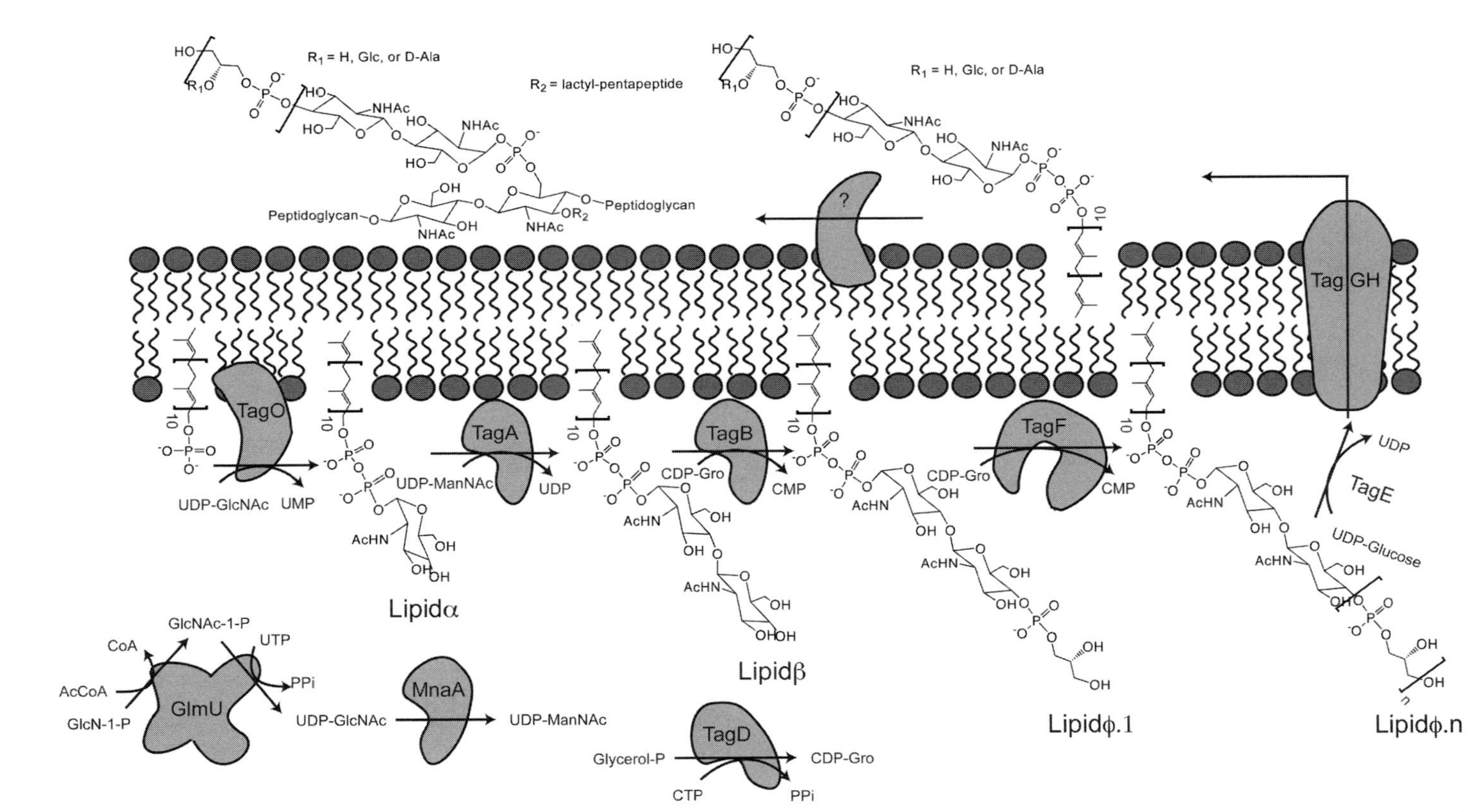

FIGURE 19.2 Model of teichoic acid biosynthesis in *B. subtilis* 168. Assembly of teichoic acid is accomplished on a membrane-bound undecaprenyl-phosphate. A linkage unit comprised of *N*-acetyl-D-glucosamine-1-phosphate (GlcNAc-1-P) and *N*-acetyl-D-mannosamine (ManNAc) is added to the intracellular face by the concerted action of TagO and TagA. The enzymes TagB and TagF have primase and polymerase activities, respectively. The TagB enzyme adds the first glycerol-phosphate and subsequently TagF polymerizes poly(glycerol-phosphate) to form the membrane-bound product Lipidϕ.n. Glucosylation of the polymer is thought to be catalysed by the TagE enzyme and export by TagGH is believed to occur before attachment to peptidoglycan by an unknown transferase. It is also believed that D-alanylation of the polymer occurs extracytoplasmically; for an excellent review on alanylation of these polymers, see Neuhaus and Baddiley (2003). Abbreviations: AcCoA, acetyl-co-enzyme A; CoA, co-enzyme A; D-Ala, D-alanine; D-Glc, D-glucose; GlcN-1-P, D-glucosamine-1-phospahte; Glycerol-P, glycerol phosphate; CDP-Gro, cytidine diphosphate-glycerol; CTP, cytidine triphosphate; UMP, uridine monophosphate; UDP, uridine diphosphate; UDP-Glc, UDP-glucose; UDP-GlcNAc, UDP-*N*-acetyl-D-glucosamine; UTP, uridine triphosphate; PPi, inorganic pyrophosphate.

sites (Gehring *et al.*, 1996). Although little work has been done on the *B. subtilis* orthologue, considerable structural and mechanistic data are available for the *Escherichia coli* and *Streptococcus pneumoniae* enzymes (Sulzenbacher *et al.*, 2001; Olsen *et al.*, 2007). As UDP-GlcNAc is a precursor for a variety of extracellular carbohydrates, including teichoic acid and peptidoglycan, GlmU is arguably a reasonable target for antibacterial drug discovery. Indeed, the left-handed β-helix fold of the acetyltransferase domain is unique to prokaryotic UDP-GlcNAc synthases (Olsen and Roderick, 2001).

The activated precursor for the addition of the second sugar, UDP-*N*-acetylmannosamine (UDP-ManNAc), is an epimer at the 2 position of UDP-GlcNAc. Through a thermosensitive mutation in the gene *mnaA*, Soldo *et al.* (2002b) assigned this locus to the UDP-GlcNAc 2-epimerase activity. The mechanism of epimerization is believed to progress through *anti*-elimination of UDP and the subsequent *syn*-addition of UDP and 2-acetamidoglucal intermediates to form UDP-ManNAc (Tanner, 2002). A recent structure of the *Bacillus anthracis* MnaA orthologue reveals an allosteric interaction of UDP-GlcNAc that controls solvent accessibility to the active site (Velloso *et al.*, 2008). The interaction was shown to be necessary in maintaining the enzyme in an active conformation (Velloso *et al.*, 2008), thus potentially modulating the flux of this important precursor between peptidoglycan and teichoic acid biosynthesis. This notion is supported by the maintenance of a 12:1 ratio of UDP-GlcNAc to UDP-ManNAc at equilibrium by MnaA accounting for a higher need of UDP-GlcNAc for murein synthesis (Soldo *et al.*, 2002b).

3.2. Teichoic acid polymer precursors

The activation of glycerol-3-phosphate with cytidine triphosphate (CTP), to form CDP-glycerol and pyrophosphate, is necessary for WTA synthesis in both *B. subtilis* and *S. aureus* strains. The *tagD* locus was originally linked to CDP-glycerol synthesis when it was determined that a mutation in this gene diminished glycerol incorporation into both teichoic acid and cellular CDP-glycerol levels when cells were grown at a non-permissive temperature. Mutations at other loci (*tagB* and *tagF*) also decreased glycerol incorporation into teichoic acids, however, CDP-glycerol concentrations were increased. From this the authors suggested that *tagD* encoded the glycerol-3-phosphate cytidylyltransferase (Pooley *et al.*, 1991).

Both the TagD enzyme of *B. subtilis* 168 and the *S. aureus* orthologue TarD have been the subject of extensive structural and biochemical study. Although each of these enzymes catalyses the synthesis of CDP-glycerol from the same substrates, the kinetic characterization of the purified enzymes show significant mechanistic differences. Steady state kinetic analyses of these enzymes indicated that both follow a sequential reaction mechanism that requires the formation of a ternary complex of enzyme and substrates for catalysis. Nevertheless, these enzymes differed in other respects. Using product inhibition studies, Park *et al.* (1993) described a mechanism where a particular order of substrate binding or product release was not a prerequisite for TagD catalysis while, curiously, Badurina *et al.* (2003) showed that the *S. aureus* orthologue TarD catalyses cytidylyl-transfer by an ordered mechanism. The kinetic constants for these reactions were also markedly different. The Michaelis constants for the TagD enzyme were also reported to be 100-fold higher than that of the TarD enzyme (Park *et al.*, 1993; Badurina *et al.*, 2003). This discrepancy may be due to the negative cooperativity in substrate binding observed only for the TagD enzyme (Park *et al.*, 1993; Sanker *et al.*, 2001).

The X-ray crystallographic structural analyses of TagD bound with the substrate CTP or the product CDP-glycerol have described the amino acids potentially involved in substrate binding (Weber *et al.*, 1999; Pattridge *et al.*, 2003). These structural analyses along with site-directed mutagenesis indicated two motifs, HXGH and RTXGISTT, as having a role in catalysis. The motif HXGH is believed to bind CTP (Weber *et al.*, 1999) and stabilize a putative pentacoordinate transition state (Park *et al.*, 1997); the motif also places TagD as a member of a nucleotidyl-transferase superfamily including class I aminoacyl-tRNA synthetases, pantothenate synthases and CTP:choline-phosphate cytidylyltransferases (Weber *et al.*, 1999). As the TarD and TagD enzymes share these motifs and a high level of amino acid sequence identity (69%), the observed mechanistic and kinetic differences remain a curiosity. Nevertheless, Badurina *et al.* (2003) have demonstrated that *tarD* can functionally replace *tagD in vivo* by complementing a *tagD* null.

Comparatively less is known about the synthesis of CDP-ribitol than the analogous CDP-glycerol synthase reaction. The proteins TarI and TarJ were predicted to be involved in this reaction based on their similarities to enzymes involved in Gram-negative capsule synthesis (Lazarevic *et al.*, 2002). Akin to the bifunctional *Haemophilis influenzae* enzyme Bcs1, CDP-ribitol synthesis in *S. aureus* and *B. subtilis* W23 occurs in a two-step reaction where D-ribulose-5-phosphate is reduced using nicotinamide adenine dinucleotide phosphate (NADPH) prior to cytidylyl-transfer from CTP (Zolli *et al.*, 2001; Pereira and Brown, 2004). Interestingly, co-expression of *tarI* and *tarJ* open reading frames resulted in association of the gene products allowing for co-purification of the enzymes (Pereira and Brown, 2004). It was further determined that TarI and TarJ formed a stable hetero-tetramer, i.e. TarIJ, to catalyse efficiently the two reactions. Steady state kinetic analysis and product inhibition studies indicated ordered mechanisms for both the reductase activity and the cytidylyltransferase activity. In each case, the co-factor bound first and exited last from the active site (Pereira and Brown, 2004). Unlike TagD and TarD, currently no structural information is available for the enzyme complex.

4. STUDYING MEMBRANE ACTIVITIES

In studying teichoic acid biosynthesis, our understanding of biochemical activities that are membrane-localized has lagged significantly behind that of soluble precursor synthesis. Historically, membrane-localized activities had to be assayed using crude or partially purified fractions as a source of enzymes and insoluble membrane fractions as a source of substrates. An *in vitro* assay, using recombinant teichoic acid biosynthetic proteins, partially solved this problem by defining the enzyme variable (Schertzer and Brown, 2003), however, by using membranes as a source of acceptors, the chemical composition of substrates had not been clearly defined. Soluble undecaprenyl analogues have emerged in the last five years as a very important tool in the study of enzymes that act upon membrane-embedded substrates. Such analogues have been created to study the substrate specificity of peptidoglycan glycosyltransferase MurG (Cudic *et al.*, 2001; Liu *et al.*, 2003) and transglycosylase PBP1A (Ye *et al.*, 2001; Barrett *et al.*, 2007), as well as lipopolysaccharide synthesizing enzymes (Montoya-Peleaz *et al.*, 2005). Recently, the Walker group has addressed the lack of available tools for teichoic acid research through the synthesis of soluble analogues of membrane-bound teichoic acid intermediates. In the TagO product Lipidα, the C-55 undecaprenyl group has been replaced with that of either a C-13 saturated hydrocarbon moiety (Ginsberg *et al.*, 2006) or a C-15 farnesyl moiety (Zhang *et al.*, 2006; Brown *et al.*, 2008). Steady state kinetic analysis of the farnesyl substitution with the TagA enzyme (Zhang *et al.*, 2006) and the C-13 substitution with the TagF enzyme (Pereira *et al.*, 2008a) allowed for the determination of specificity constants that support the suitability of these analogues for enzymatic study. Interestingly, Pereira *et al.* (2008a) reported a 100-fold increase in TagF turnover when a soluble analogue was used in place of inverted membrane vesicles, highlighting the success of these molecules as tools for teichoic acid study. The seminal research utilizing these analogues for *in vitro* pathway reconstitution and mechanistic study is outlined in the text. The use of soluble acceptor molecules will perhaps revolutionize the study of teichoic biosynthetic enzymes as the mechanistic details of these enzymes are only beginning to appear in the literature. Note that we have proposed here a new naming scheme to differentiate the teichoic acid undecaprenyl-linked intermediates from those involved in peptidoglycan and Gram-negative outer membrane biogenesis. To simplify, the designation of each intermediate corresponds with the enzyme to which it interacts: Lipidα (TagA), Lipidβ (TagB) and Lipidϕ (TagF) (see Figure 19.2).

5. LINKAGE UNIT GLYCOSYLTRANSFERASES

5.1. TagO-UDP-GlcNAc–undecaprenyl-phosphate GlcNAc-1-P transferase

The first step in teichoic acid synthesis is the addition of GlcNAc-1-P from UDP-GlcNAc to the membrane-localized substrate undecaprenyl-phosphate. To date, this activity has been poorly characterized biochemically. Much evidence on the role of this reaction in teichoic acid biosynthesis comes from observations that radiolabelled UDP-GlcNAc is incorporated into a substrate for further catalysis by the teichoic acid polymerases (Yokoyama *et al.*, 1989). Homology to the MraY glycosyltransferase involved in peptidoglycan biosynthesis led to the assertion by Soldo *et al.* (2002a) that the transmembrane enzyme TagO is involved in this first step. It was through gene deletion studies that the role of TagO was solidified. Soldo *et al.* (2002a) first observed an impact on phosphate and glycerol-phosphate incorporation into the *B. subtilis* cell wall in cells depleted of TagO while peptidoglycan synthesis was unaffected, supporting a role in teichoic acid biosynthesis. It was subsequently discovered that an uncomplemented *tagO* null could be made abolishing teichoic acid synthesis (Weidenmaier *et al.*, 2004; D'Elia *et al.*, 2006a) and resulting in a loss of rod-shaped morphology (D'Elia *et al.*, 2006b). From a comprehensive deletion study of teichoic acid biosynthetic genes in *S. aureus*, D'Elia *et al.*, (2006a) discovered that genes linked to late stage synthesis became non-essential in a *tagO* null. That a deletion in *tagO* would have such a profound impact on the dispensability of the late genes was remarkable and further supported a role of TagO in the first step of teichoic acid biosynthesis.

Much of what is known for the mechanism of catalysis of TagO is from bioinformatic comparison to other members in its family of GlcNAc-1-phosphate transferases (MraY, WecA, and WbpL). Each of these enzymes differs in their carbohydrate specificity which is determined by C-terminal carbohydrate recognition domains on a cytoplasmic loop (Anderson *et al.*, 2000). It is believed that these prenyl-phosphate glycosyltransferases operate via a double displacement mechanism where an Asp residue acts as an active site nucleophile cleaving the pyrophosphate bond in UDP-GlcNAc to form an acyl-enzyme intermediate before transfer to undecaprenyl-phosphate. For an in-depth review see Price and Momany (2005).

5.2. TagA-UDP-ManNAc–Lipidα ManNAc transferase

The synthesis of a soluble analogue of the TagO product (Lipidα) through the substitution of insoluble 55-carbon undecaprenyl moiety with a soluble saturated 13-carbon moiety allowed for the characterization of the mechanism and products of the TagA reaction. Ginsberg *et al.* (2006) showed that the TagA enzyme was able to catalyse the transfer of ManNAc to the soluble Lipidα analogue to form Lipidβ. This product was further confirmed to be suitable for catalysis by the enzyme TagB. Zhang *et al.* (2006) continued this work in a study investigating the lipid preference and steady-state kinetic mechanism of the enzyme. While it was established that TagA was sensitive to lipid length and hydrophobicity, it was found that the enzyme was not sensitive to lipid structure. The level of branching, bond saturation or double bond geometry did not greatly affect the substrate specificity constant. As membranes were not present in the reaction, these studies also established that, although the TagA protein and the undecaprenyl-linked substrate are membrane-localized (Leaver *et al.*, 1981), biological membranes are not required for the enzymatic activity.

The use of the soluble Lipidα and Lipidβ analogues for steady-state kinetic analysis allowed for the determination of the kinetic mechanism of TagA. In an ordered mechanism where UDP-ManNAc binds the active site first and UDP is the last product to exit, the product Lipidβ was found to form dead-end complexes with both the apo and the UDP-ManNAc-bound form of the enzyme (Zhang *et al.*, 2006). The TagA enzyme belongs to a family of glycosyltransferases that are for the most part uncharacterized. The purification and mechanistic characterization of this enzyme will surely aid in a structural biological effort to study this class of enzymes.

6. TEICHOIC ACID POLYMER BIOSYNTHESIS

Much like the study of the teichoic acid linkage unit glycosyltransferases, biochemical study of the enzymes involved in the polyol-phosphate polymer formation has been hampered due to a lack of chemically defined substrates. Only until very recently, poly(glycerol-phosphate) and poly(ribitol-phosphate) activities were studied using crude preparations of insoluble materials (Burger and Glaser, 1964; Mauck and Glaser, 1972a). However, with the genetic association of the *tagF* allele to poly(glycerol-phosphate) polymerase activity and the *tagB* allele to glycerol-phosphate primase activity (Pooley *et al.*, 1991, 1992), biochemical characterization of these activities became a possibility. Through the use of an *in vitro* membrane-pelleting assay, developed by Schertzer and co-workers, it was demonstrated that pure recombinant TagB and TagF could catalyse the addition of glycerol-phosphate to a membrane-linked acceptor molecule using the activated precursor CDP-glycerol (Schertzer and Brown, 2003; Bhavsar *et al.*, 2005). Analytical characterization of the reaction products indicated that the TagF enzyme did in fact catalyse the formation of a polymer of glycerol-phosphate while TagB added only a single glycerol-phosphate (Schertzer and Brown, 2003; Bhavsar *et al.*, 2005).

6.1. Primase-polymerase model of teichoic acid biosynthesis

While the membrane pelleting assays supported the model that *tagB* would encode a primase and *tagF* a polymerase, these activities were unambiguously defined with the use of soluble substrate analogues. It was confirmed that the TagB enzyme catalyses the addition of one glycerol-phosphate to the product of the TagA reaction (Ginsberg *et al.*, 2006). Using the products of both the TagA (Lipidβ) and TagB (Lipidϕ.1) reaction, it was determined that TagF would only catalyse glycerol-phosphate transfer to a lipid-linked substrate that had already been "primed" with a single polyol-phosphate unit (Pereira *et al.*, 2008a). Chemical logic supports the notion that a separate enzyme activity would be necessary for the addition of the first glycerol-phosphate due to the differences in active site architecture necessary to accommodate the ManNAc for catalysis. The necessity of a priming reaction for TagF polymerization is consistent with the genetic indispensability of the *tagB* gene (Bhavsar *et al.*, 2004). Interestingly, the poly(ribitol-phosphate) polymerase of *S. aureus*, TarL, does not require a separate enzyme for the addition of the first ribitol-phosphate residue and subsequent polymerization (Brown *et al.*, 2008). This result is not entirely surprising since the ribitol-phosphate polymer is built on an oligomer of glycerol-phosphate (see Figure 19.1) that potentially could be accommodated in the TarL active site in place of ribitol-phosphate. Perhaps most interesting is that dispensability studies suggest the *B. subtilis* W23 orthologue of TarL cannot use an "unprimed" substrate (Meredith *et al.*, 2008; Pereira *et al.*, 2008b). As it is possible for *S. aureus* TarL to catalyse addition to the terminal hydroxyl of both glycerol-phosphate and ribitol-phosphate, careful steady-state analysis to determine specificity constants for lipid-linked substrates that are "primed" and "unprimed" with ribitol-phosphate are needed to understand how the relaxed specificity necessary for accepting two chemically distinct substrates for ribitol-phosphate transfer affects enzyme catalysis.

6.2. Mechanism of catalysis

Although the demonstration of *in vitro* biochemical activity for these enzymes using pure recombinant proteins and chemically defined substrates has proven to be a landmark stepping stone in WTA biosynthesis, much still remains to be learned about these enzymes. A lack of homology to characterized enzymes and an absence of structural information for these macromolecules remain an obstacle to deeper understanding of their structure and function. Site-directed mutagenesis studies of the TagF and TagB enzymes aimed at understanding the mechanism of catalysis has implicated two conserved histidine residues that were shown to be essential for *in vivo* and *in vitro* function (Bhavsar *et al.*, 2005; Schertzer *et al.*, 2005). High levels of sequence similarity in the C-terminus among polyol-phosphate transferases and identification of conserved residues that affected catalysis in both TagB and TagF is highly suggestive that these enzymes operate via a conserved mechanism (Figure 19.3A). From pH dependence of TagF activity, Schertzer *et al.* (2005) suggested that a histidine may be acting as an active site base, however, further study is needed to confirm this model. Histidine could also reasonably function as a nucleophile in catalysis (Figure 19.3B). In the absence of structural data for any member of this enzymatic class, further kinetic analysis into the steady-state kinetic mechanism is required for the understanding of how this enzyme catalyses the formation of phosphodiester linkages.

The construction of a defined length polymer by the TagF enzyme is particularly curious. *In vivo* isolated teichoic acid and polymers synthesized on membrane vesicles have both been estimated at lengths between 30 and 60 units (Pollack and Neuhaus, 1994; Schertzer and Brown, 2003). In contrast, *in vitro* polymer synthesis using a soluble analogue of Lipid ϕ.1 has produced mixed results. In a study by Brown *et al.* (2008) of the poly(ribitol-phosphate) polymerase of *S. aureus*, TarL, the polymeric product formed was approximately 30 units. Varying the ratios of CDP-ribitol to soluble acceptor could not modulate this length, suggestive of a processive assembly of defined length polymer. In contrast, Pereira *et al.* (2008a) showed that *B. subtilis* 168 TagF poly(glycerol-phosphate) polymerase formed a polymer with lengths determined by the ratio of the provided substrates. This study suggested that TagF's substrate K_ms (2.6 μM for Lipid ϕ.1 and 152 μM for CDP-glycerol) might dictate polymer length through the availability of substrates and further indicated that TagF may have a distributive polymerase activity. Figure 19.3C describes processive and distributive polymerization. It is

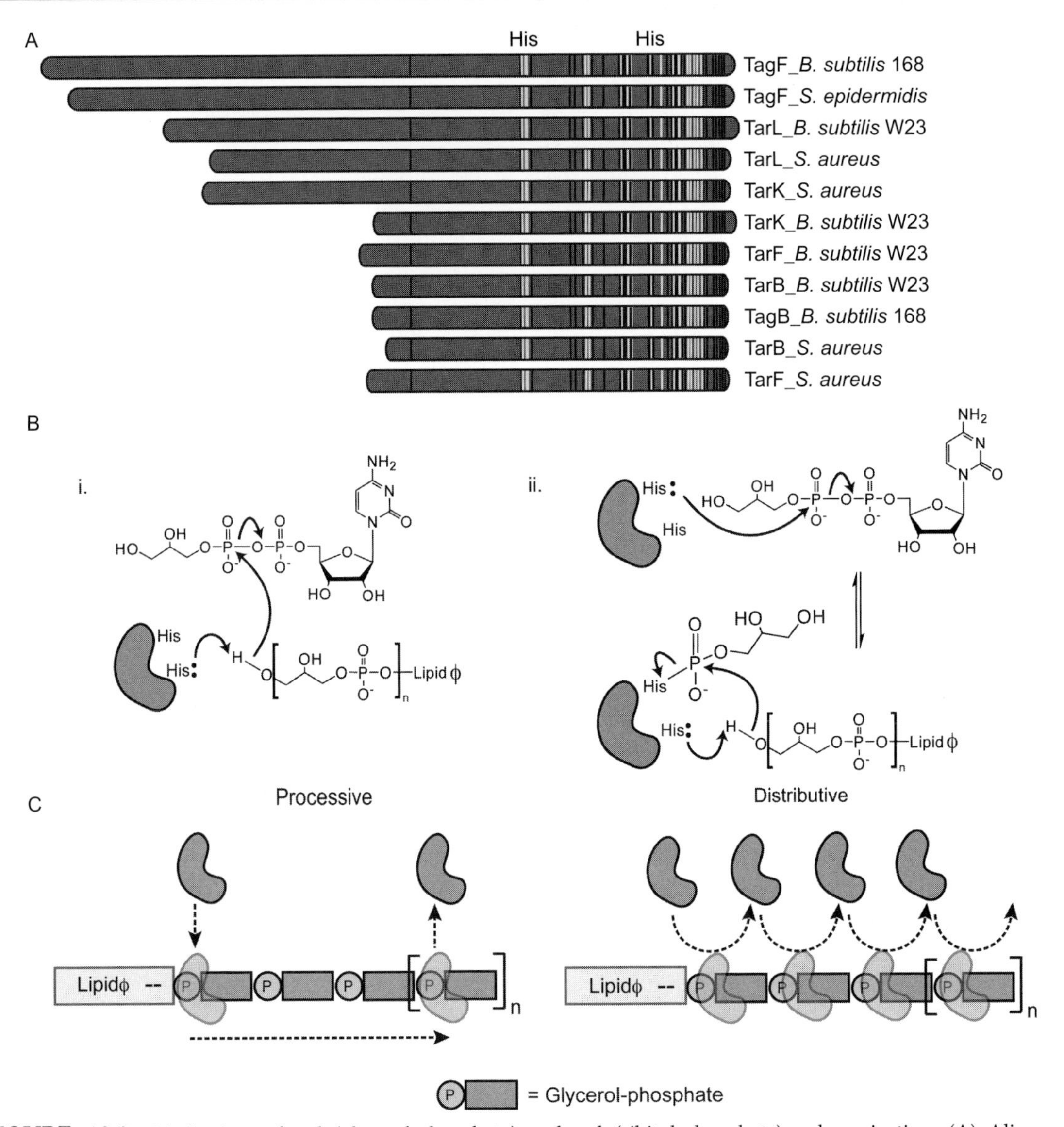

FIGURE 19.3 Mechanism of poly(glycerol-phosphate) and poly(ribitol-phosphate) polymerization. (A) Alignment schematic of proteins involved in teichoic acid polyol-phosphate transfer. Highest levels of identity (yellow) and similarity (black) among the amino acid sequences are present in the C-terminal region. Highlighted are the conserved histidines Schertzer *et al.* (2005) implicated in TagF and TagB activity. One point of curiosity is the function of the extensive *N*-terminal region of TagF that is not homologous to any other sequenced protein. (B) Two potential catalytic mechanisms for glycerol-phosphate transfer are shown schematically. It is hypothesized that a histidine residue may function to deprotonate the terminal hydroxyl of the growing polymer for nucleophilic attack on the β-phosphate of cytidine diphosphate- (CDP-) glycerol. It is also conceivable that the histidine directly acts as a nucleophile forming a phosphor-histidine intermediate with glycerol-phosphate from CDP-glycerol to promote catalysis through a ping-pong mechanism. (C) The factors regulating polymer length are currently unknown. It is conceivable that the level of processivity of a polymerase may determine the length of polymer that is formed. A processive polymerase will remain associated with the growing polymer through the addition of many units in a single binding event, while a distributive polymerase dissociates and re-associates to the polymer repeatedly adding only as few as one unit in each binding event. Preliminary data studying the processivity of two teichoic acid polymerases provided differing models of elongation. The TarL enzyme from *S. aureus* was suggested to be processive (Brown *et al.*, 2008), while the TagF enzyme from *B. subtilis* did not appear to maintain association with the growing polymer during catalysis (Pereira *et al.*, 2008a). (See colour plates section.)

conceivable that the TarL and TagF enzymes work through different mechanisms, however, sequence conservation of this enzyme class would appear to predict otherwise. Further investigations into the mechanism of length regulation and processivity are clearly necessary.

7. OUTSTANDING ISSUES

7.1. Export and assembly of WTA

To date, biochemical studies have focused on intracellular steps for teichoic acid biosynthesis. Relatively little is known about polymer export, even less is known about the attachment of teichoic acid to the C-6 of muramic acid in peptidoglycan. In the two final steps following polymerization, the entire prenyl-linked polymer is hypothesized to be flipped to the outer face of the cell membrane by the TagGH enzymes. These enzymes are proposed to carry out the polysaccharide transport role because of their homology to other ATP-binding cassette (ABC) polysaccharide transporters (Lazarevic and Karamata, 1995). Indeed, depletion of TagGH in *B. subtilis* 168 yielded a phenotype that was consistent with depletion of other late acting *B. subtilis* Tag proteins (Lazarevic and Karamata, 1995). It has also been demonstrated that a deletion in *tarG* in *S. aureus*, while deleterious in a wild-type background, was dispensable in a *tarO* null, consistent with characteristic genetic interactions that have been observed between late acting genes and the first step in teichoic acid synthesis. The identity of the enzyme catalysing the covalent attachment of teichoic acid to peptidoglycan has remained elusive. This activity is thought to occur on the extracytoplasmic face of the cell membrane in chorus with peptidoglycan strand synthesis (Mauck and Glaser, 1972b). Recent advances in the synthesis of lipid-linked teichoic acid *in vitro* will surely aid in the molecular characterization of the putative transporter TagGH and in the discovery of the enzyme(s) involved in transfer to peptidoglycan.

7.2. Localization of enzymes and formation of biosynthetic complex

The localization of WTA biosynthetic enzymes may provide clues to the molecular details of teichoic acid incorporation into the Gram-positive cell wall. Similar to the proposed notion of a murein biosynthetic complex, the formation of a complex of WTA biosynthetic enzymes is a possibility. This idea is not a new one as Leaver *et al.* (1981) described the co-fractionation of detergent solubilized enzymes involved in linkage unit synthesis during sucrose density centrifugation. Interestingly, these enzymes also fractionated with prenyl-phosphate containing molecules that could subsequently be used for poly(glycerol) polymerization. Recently, it was discovered that disruption of the amphipathic helix in the membrane targeting domain of TagB, while abolishing function, did not affect the membrane localization of the enzyme (Bhavsar *et al.*, 2007). Cellular localization of green fluorescent protein-tagged enzymes revealed that the teichoic acid machinery showed a similar helical localization pattern (Formstone *et al.*, 2008) to that of nascent peptidoglycan (Daniel and Errington, 2003). Indeed, two hybrid studies revealed a complex network of interactions between WTA machinery and shape determinants MreC and MreD. As it has been reported that teichoic acid incorporation into the cell wall occurs with non-cross-linked peptidoglycan (Mauck and Glaser, 1972b), it may be surmised that synthesis and export of WTA is mediated within a large multienzyme complex associated with the peptidoglycan synthesis.

8. CONCLUSIONS

Our understanding of WTA synthesis has developed considerably in recent years. With the *in vitro* reconstitution of teichoic acid biosynthetic pathways in both *B. subtilis* and *S. aureus*, functional roles for many of the gene products are now emerging. Additionally, the development of assays using chemically defined substrate analogues provides new opportunities for careful kinetic and mechanistic study of teichoic acid biosynthetic enzymes. Early work on mechanism has revealed some surprising and fascinating results, particularly around priming and polymerization steps, that will surely be subject of further and on-going study. Further areas that will potentially be the topics for future research are listed in the Research Focus Box.

RESEARCH FOCUS BOX

- The determination of the additional protein factors necessary for export and cell wall attachment is essential to understanding the complete pathway of teichoic acid biosynthesis.
- Further study of polyol-phosphate transferases is required to allow understanding of both the chemical mechanism of this family of enzymes and the factors governing polymer length.
- Biochemical study is needed to determine the possible coordination of teichoic acid and peptidoglycan biosynthesis in a biosynthetic complex.

ACKNOWLEDGEMENTS

Work in our laboratory has been supported by the Canadian Institutes of Health Research Grant MOP-15496. E.D. Brown holds a Canada Research Chair and M.P. Pereira holds a Canadian Institutes of Health Research Canada Graduate Scholarship.

References

Anderson, M.S., Eveland, S.S., Price, N.P., 2000. Conserved cytoplasmic motifs that distinguish sub-groups of the polyprenol phosphate: *N*-acetylhexosamine-1-phosphate transferase family. FEMS Microbiol. Lett. 191, 169–175.

Baddiley, J., Buchanan, J.G., Carss, B., Mathias, A.P., 1956. Cytidine diphosphate ribitol from *Lactobacillus arabinosus*. J. Chem. Soc. 64, 4583–4588.

Badurina, D.S., Zolli-Juran, M., Brown, E.D., 2003. CTP: glycerol 3-phosphate cytidylyltransferase (TarD) from *Staphylococcus aureus* catalyzes the cytidylyl transfer via an ordered Bi-Bi reaction mechanism with micromolar Km values. Biochim. Biophys. Acta 1646, 196–206.

Barrett, D., Wang, T.S., Yuan, Y., Zhang, Y., Kahne, D., Walker, S., 2007. Analysis of glycan polymers produced by peptidoglycan glycosyltransferases. J. Biol. Chem. 282, 31964–31971.

Bhavsar, A.P., Brown, E.D., 2006. Cell wall assembly in *Bacillus subtilis*: how spirals and spaces challenge paradigms. Mol. Microbiol. 60, 1077–1090.

Bhavsar, A.P., Beveridge, T.J., Brown, E.D., 2001. Precise deletion of tagD and controlled depletion of its product, glycerol 3-phosphate cytidylyltransferase, leads to irregular morphology and lysis of *Bacillus subtilis* grown at physiological temperature. J. Bacteriol. 183, 6688–6693.

Bhavsar, A.P., Erdman, L.K., Schertzer, J.W., Brown, E.D., 2004. Teichoic acid is an essential polymer in *Bacillus subtilis* that is functionally distinct from teichuronic acid. J. Bacteriol. 186, 7865–7873.

Bhavsar, A., Truant, R., Brown, E., 2005. The TagB protein in Bacillus subtilis 168 is an intracellular peripheral membrane protein that can incorporate glycerol phosphate onto a membrane-bound acceptor *in vitro*. J. Biol. Chem. 280, 36691–36700.

Bhavsar, A.P., D'Elia, M.A., Sahakian, T.D., Brown, E.D., 2007. The amino terminus of *Bacillus subtilis* TagB possesses separable localization and functional properties. J. Bacteriol. 189, 6816–6823.

Brown, S., Zhang, Y.H., Walker, S., 2008. A revised pathway proposed for *Staphylococcus aureus* wall teichoic acid biosynthesis based on in vitro reconstitution of the intracellular steps. Chem. Biol. 15, 12–21.

Burger, M., Glaser, L., 1964. The synthesis of teichoic acids. I. Polyglycerolphosphate. J. Biol. Chem. 239, 3168–3177.

Cudic, P., Behenna, D.C., Yu, M.K., Kruger, R.G., Szewczuk, L.M., McCafferty, D.G., 2001. Synthesis of P(1)-Citronellyl-P(2)- α-D-pyranosyl pyrophosphates as potential substrates for the E. coli undecaprenyl-pyrophosphoryl-*N*-acetylglucoseaminyl transferase MurG. Bioorg. Med. Chem. Lett. 11, 3107–3110.

D'Elia, M.A., Pereira, M.P., Chung, Y.S., et al., 2006a. Lesions in teichoic acid biosynthesis in *Staphylococcus aureus* lead to a lethal gain of function in the otherwise dispensable pathway. J. Bacteriol. 188, 4183–4189.

D'Elia, M.A., Millar, K.E., Bevridge, T.J., Brown, E.D., 2006b. Wall teichoic acid polymers are dispensable for cell viability in *Bacillus subtilis*. J. Bacteriol. 188, 8313–8316.

Daniel, R.A., Errington, J., 2003. Control of cell morphogenesis in bacteria: two distinct ways to make a rod-shaped cell. Cell 113, 767–776.

Ellwood, D.C., Tempest, D.W., 1972. Influence of culture pH on the content and composition of teichoic acids in the walls of *Bacillus subtilis*. J. Gen. Microbiol. 73, 395–402.

Estrela, A.I., Pooley, H.M., de Lencastre, H., Karamata, D., 1991. Genetic and biochemical characterization of *Bacillus subtilis* 168 mutants specifically blocked in the synthesis of the teichoic acid poly(3-*O*-β-D-glucopyranosyl-*N*-acetylgalactosamine 1-phosphate): *gneA*, a new locus, is associated with UDP-*N*-acetylglucosamine 4-epimerase activity. J. Gen. Microbiol. 137, 943–950.

Formstone, A., Carballido-Lopez, R., Noirot, P., Errington, J., Scheffers, D.J., 2008. Localization and interactions of teichoic acid synthetic enzymes in *Bacillus subtilis*. J. Bacteriol. 190, 1812–1821.

Gehring, A.M., Lees, W.J., Mindiola, D.J., Walsh, C.T., Brown, E.D., 1996. Acetyltransfer precedes uridylyltransfer in the formation of UDP-*N*-acetylglucosamine in separable active sites of the bifunctional GlmU protein of *Escherichia coli*. Biochemistry 35, 579–585.

Ginsberg, C., Zhang, Y.H., Yuan, Y., Walker, S., 2006. *In vitro* reconstitution of two essential steps in wall teichoic acid biosynthesis. ACS Chem. Biol. 1, 25–28.

Grant, W.D., 1979. Cell wall teichoic acid as a reserve phosphate source in *Bacillus subtilis*. J. Bacteriol. 137, 35–43.

Gross, M., Cramton, S.E., Gotz, F., Peschel, A., 2001. Key role of teichoic acid net charge in *Staphylococcus aureus* colonization of artificial surfaces. Infect. Immun. 69, 3423–3426.

Hancock, I.C., 1997. Bacterial cell surface carbohydrates: structure and assembly. Biochem. Soc. Trans. 25, 183–187.

Herbold, D.R., Glaser, L., 1975. Interaction of *N*-acetylmuramic acid L-alanine amidase with cell wall polymers. J. Biol. Chem. 250, 7231–7238.

Lazarevic, V., Karamata, D., 1995. The tagGH operon of *Bacillus subtilis* 168 encodes a two-component ABC transporter involved in the metabolism of two wall teichoic acids. Mol. Microbiol. 16, 345–355.

Lazarevic, V., Abellan, F., Möller, S., Karamata, D., Mauël, C., 2002. Comparison of ribitol and glycerol teichoic acid genes in *Bacillus subtilis* W23 and 168: identical function, similar divergent organization, but different regulation. Microbiology 148, 815–824.

Leaver, J., Hancock, I.C., Baddiley, J., 1981. Fractionation studies of the enzyme complex involved in teichoic acid synthesis. J. Bacteriol. 146, 847–852.

Liu, H., Ritter, T.K., Sadamoto, R., Sears, P.S., Wu, M., Wong, C.H., 2003. Acceptor specificity and inhibition of the bacterial cell-wall glycosyltransferase MurG. Chem. Bio. Chem. 4, 603–609.

Matias, V.R., Beveridge, T.J., 2005. Cryo-electron microscopy reveals native polymeric cell wall structure in *Bacillus subtilis* 168 and the existence of a periplasmic space. Mol. Microbiol. 56, 240–251.

Mauck, J., Glaser, L., 1972a. An acceptor-dependent polyglycerolphosphate polymerase. Proc. Natl. Acad. Sci. USA 69, 2386–2390.

Mauck, J., Glaser, L., 1972b. On the mode of *in vivo* assembly of the cell wall of *Bacillus subtilis*. J. Biol. Chem. 247, 1180–1187.

Mauël, C., Young, M., Karamata, D., 1991. Genes concerned with synthesis of poly(glycerol phosphate), the essential teichoic acid in *Bacillus subtilis* strain 168, are organized in two divergent transcription units. J. Gen. Microbiol. 137, 929–941.

Meredith, T.C., Swoboda, J.G., Walker, S., 2008. Late-stage polyribitol phosphate wall teichoic acid biosynthesis in *Staphylococcus aureus*. J. Bacteriol. 190, 3046–3056.

Montoya-Peleaz, P.J., Riley, J.G., Szarek, W.A., Valvano, M.A., Schutzbach, J.S., Brockhausen, I., 2005. Identification of a UDP-Gal: GlcNAc-R galactosyltransferase activity in *Escherichia coli* VW187. Bioorg. Med. Chem. Lett. 15, 1205–1211.

Neuhaus, F.C., Baddiley, J., 2003. A continuum of anionic charge: structures and functions of D-alanyl-teichoic acids in Gram-positive bacteria. Microbiol. Mol. Biol. Rev. 67, 686–723.

Olsen, L.R., Roderick, S.L., 2001. Structure of the *Escherichia coli* GlmU pyrophosphorylase and acetyltransferase active sites. Biochemistry 40, 1913–1921.

Olsen, L.R., Vetting, M.W., Roderick, S.L., 2007. Structure of the E. coli bifunctional GlmU acetyltransferase active site with substrates and products. Protein Sci. 16, 1230–1235.

Park, Y.S., Sweitzer, T.D., Dixon, J.E., Kent, C., 1993. Expression, purification, and characterization of CTP:

glycerol-3-phosphate cytidylyltransferase from *Bacillus subtilis*. J. Biol. Chem. 268, 16648–16654.

Park, Y.S., Gee, P., Sanker, S., Schurter, E.J., Zuiderweg, E.R., Kent, C., 1997. Identification of functional conserved residues of CTP: glycerol-3-phosphate cytidylyltransferase. Role of histidines in the conserved HXGH in catalysis. J. Biol. Chem. 272, 15161–15166.

Pattridge, K.A., Weber, C.H., Friesen, J.A., Sanker, S., Kent, C., Ludwig, M.L., 2003. Glycerol-3-phosphate cytidylyltransferase. Structural changes induced by binding of CDP-glycerol and the role of lysine residues in catalysis. J. Biol. Chem. 278, 51863–51871.

Pereira, M.P., Brown, E.D., 2004. Bifunctional catalysis by CDP-ribitol synthase: convergent recruitment of reductase and cytidylyltransferase activities in *Haemophilus influenzae* and *Staphylococcus aureus*. Biochemistry 43, 11802–11812.

Pereira, M.P., Schertzer, J.W., D'Elia, M.A., et al., 2008a. The wall teichoic acid polymerase TagF efficiently synthesizes poly(glycerol phosphate) on the TagB product lipid III. Chem. Bio. Chem. 9, 1385–1390.

Pereira, M.P., D'Elia, M.A., Troczynska, J., Brown, E.D., 2008b. Duplication of teichoic acid biosynthetic genes in *Staphylococcus aureus* leads to functionally redundant poly(Ribitol Phosphate) polymerases. J. Bacteriol. 190, 5642–5649.

Pollack, J., Neuhaus, F., 1994. Changes in wall teichoic acid during the rod-sphere transition of *Bacillus subtilis* 168. J. Bacteriol. 176, 7252–7259.

Pooley, H.M., Abellan, F.X., Karamata, D., 1991. A conditional-lethal mutant of *Bacillus subtilis* 168 with a thermosensitive glycerol-3-phosphate cytidylyltransferase, an enzyme specific for the synthesis of the major wall teichoic acid. J. Gen. Microbiol. 137, 921–928.

Pooley, H.M., Abellan, F.X., Karamata, D., 1992. CDP-glycerol: poly(glycerophosphate) glycerophosphotransferase, which is involved in the synthesis of the major wall teichoic acid in *Bacillus subtilis* 168, is encoded by tagF (rodC). J. Bacteriol. 174, 646–649.

Price, NP., Momany, F.A., 2005. Modeling bacterial UDP-HexNAc: polyprenol-P HexNAc-1-P transferases. Glycobiology 15, 29R–42R.

Qian, Z., Yin, Y., Zhang, Y., Lu, L., Li, Y., Jiang, Y., 2006. Genomic characterization of ribitol teichoic acid synthesis in *Staphylococcus aureus*: genes, genomic organization and gene duplication. BMC Genomics 7, 74.

Sanker, S., Campbell, H.A., Kent, C., 2001. Negative cooperativity of substrate binding but not enzyme activity in wild-type and mutant forms of CTP: glycerol-3-phosphate cytidylyltransferase. J. Biol. Chem. 276, 37922–37928.

Schertzer, J.W., Brown, E.D., 2003. Purified, recombinant TagF protein from *Bacillus subtilis* 168 catalyzes the polymerization of glycerol phosphate onto a membrane acceptor *in vitro*. J. Biol. Chem. 278, 18002–18007.

Schertzer, J.W., Bhavsar, A.P., Brown, E.D., 2005. Two conserved histidine residues are critical to the function of the TagF-like family of enzymes. J. Biol. Chem. 280, 36683–36690.

Soldo, B., Lazarevic, V., Karamata, D., 2002a. tagO is involved in the synthesis of all anionic cell-wall polymers in *Bacillus subtilis* 168. Microbiology 148, 2079–2087.

Soldo, B., Lazarevic, V., Pooley, H.M., Karamata, D., 2002b. Characterization of a *Bacillus subtilis* thermosensitive teichoic acid-deficient mutant: gene *mnaA* (*yvyH*) encodes the UDP-*N*-acetylglucosamine 2-epimerase. J. Bacteriol. 184, 4316–4320.

Sulzenbacher, G., Gal, L., Peneff, C., Fassy, F., Bourne, Y., 2001. Crystal structure of *Streptococcus pneumoniae N*-acetylglucosamine-1-phosphate uridyltransferase bound to acetyl-coenzyme A reveals a novel active site architecture. J. Biol. Chem. 276, 11844–11851.

Tanner, M.E., 2002. Understanding nature's strategies for enzyme-catalyzed racemization and epimerization. Acc. Chem. Res. 35, 237–246.

Velloso, L.M., Bhaskaran, S.S., Schuch, R., Fischetti, V.A., Stebbins, C.E., 2008. A structural basis for the allosteric regulation of non-hydrolysing UDP-GlcNAc 2-epimerases. EMBO Rep. 9, 199–205.

Weber, C.H., Park, Y.S., Sanker, S., Kent, C., Ludwig, M.L., 1999. A prototypical cytidylyltransferase: CTP: glycerol-3-phosphate cytidylyltransferase from *Bacillus subtilis*. Structure 7, 1113–1124.

Weidenmaier, C., Kokai-Kun, J.F., Kristian, S.A., et al., 2004. Role of teichoic acids in *Staphylococcus aureus* nasal colonization, a major risk factor in nosocomial infections. Nat. Med. 10, 243–245.

Yasbin, R.E., Maino, V.C., Young, F.E., 1976. Bacteriophage resistance in *Bacillus subtilis* 168, W23, and interstrain transformants. J. Bacteriol. 125, 1120–1126.

Ye, X.Y., Lo, M.C., Brunner, L., Walker, D., Kahne, D., Walker, S., 2001. Better substrates for bacterial transglycosylases. J. Am. Chem. Soc. 123, 3155–3156.

Yokoyama, K., Mizuguchi, H., Araki, Y., Kaya, S., Ito, E., 1989. Biosynthesis of linkage units for teichoic acids in Gram-positive bacteria: distribution of related enzymes and their specificities for UDP-sugars and lipid-linked intermediates. J. Bacteriol. 171, 940–946.

Young, F.E., 1967. Requirement of glucosylated teichoic acid for adsorption of phage in *Bacillus subtilis* 168. Proc. Natl. Acad. Sci. USA 58, 2377–2384.

Zhang, Y.H., Ginsberg, C., Yuan, Y., Walker, S., 2006. Acceptor substrate selectivity and kinetic mechanism of *Bacillus subtilis* TagA. Biochemistry 45, 10895–10904.

Zolli, M., Kobric, D.J., Brown, E.D., 2001. Reduction precedes cytidylyl transfer without substrate channeling in distinct active sites of the bifunctional CDP-ribitol synthase from *Haemophilus influenzae*. Biochemistry 40, 5041–5048.

CHAPTER

20

Biosynthesis and assembly of capsular polysaccharides

Anne N. Reid and Christine M. Szymanski

SUMMARY

Capsular polysaccharides are important surface structures necessary for bacterial survival in the host and in the environment. This chapter describes the biosynthesis, export and surface attachment of bacterial capsules. Its primary focus is the expression of *Escherichia coli* capsules, though related polysaccharides of Gram-positive and Gram-negative bacteria are also discussed. Two distinct modes of assembly are presented. In the Wzy-dependent system, repeating units are assembled at the cytoplasmic face of the inner membrane. The Wzx protein is believed to mediate the flipping of repeating units across the inner membrane, while Wzy appears to effect their polymerization, though mechanistic data for these processes are lacking. The precise role of reversible phosphorylation of Wzc in capsule assembly remains elusive, despite evidence implicating it in regulating polymer chain-length and in modulating the structure of the outer membrane channel for capsule export. In adenosine triphosphate-(ATP-) binding cassette (ABC) transporter-dependent systems, the polysaccharide is synthesized to its full length in the cytoplasm and exported to the periplasm via an ABC transporter. Early stages of the biosynthesis of these polymers remain poorly understood. Export of polysaccharides across the outer membrane of Gram-negative bacteria is discussed in light of recent structural data for the Wza complex. Genetic, subcellular localization, chemical cross-linking and bacterial two-hybrid studies suggest the existence of a multienzyme complex which couples polymer biosynthesis to its export.

Keywords: Capsular polysaccharides; Exopolysaccharides; Wzy protein; adenosine triphosphate-(ATP-) cassette binding cassette (ABC) transport; Glycoconjugate biosynthesis, Polysaccharide export; Glycosyltransferase; Wzc protein; Chain-length regulation; Wza protein

1. INTRODUCTION

Capsular polysaccharides (CPSs) form carbohydrate layers surrounding many bacteria. Individual polysaccharide chains can be linear or branched homo- or hetero-polymers and can include non-carbohydrate moieties. The CPSs are firmly attached to the bacterial cell surface, which distinguishes them from the loosely-associated exopolysaccharides (EPSs) produced by some bacteria (e.g. colanic acid produced by many strains of *Escherichia coli*).

Capsular polysaccharides and EPSs are generally acidic and form a hydrated layer around bacterial cells which protects against desiccation and other environmental stresses. The importance of CPSs for bacterial pathogenesis and commensalism has long been appreciated. Mutants unable to express CPSs are impaired for adherence to and colonization of host tissues, as well as invasion of host cells and evasion of predatory bacteriophages. In addition, the capsule protects bacteria from early stages of the immune response by impairing phagocytosis and/or complement-mediated killing (see Comstock and Kasper 2006, for a minireview on this topic). Some bacteria express CPSs whose structures are indistinguishable from host polysaccharides (e.g. sialic acid-containing capsules produced by *E. coli* K1 and group B *Streptococcus*), allowing the bacteria to evade the immune system altogether. In the case of group B streptococci, removal of the sialic acid from the glycoconjugate vaccine allowed switching from an IgM to IgG antibody response which was protective in Rhesus monkey challenge studies (Guttormsen *et al.*, 2008).

Capsular polysaccharide structures have been exploited for bacterial classification and form the basis of numerous serotyping schemes. For instance, *E. coli* assembles greater than 80 distinct capsular (K) antigens (Orskov and Orskov, 1992), *Campylobacter jejuni* expresses greater than 60 different Penner serotypes, of which the CPS is a major antigen (Karlyshev *et al.*, 2000, 2005) and *Streptococcus pneumoniae* produces 91 different CPSs that define the serotype (Bentley *et al.*, 2006). While bacteria assemble an impressive array of complex and distinct CPS structures, the mechanisms underlying their biosynthesis and export are remarkably few and are shared with the lipopolysaccharide (LPS) O-antigen (see Chapter 18) and EPS assembly systems. The methods used by bacteria to assemble and export CPSs are an important component of the Whitfield and Roberts classification scheme, which recognizes four groups of *E. coli* CPSs (groups 1 to 4) (Whitfield and Roberts, 1999). While this scheme was devised for *E. coli* CPSs, it is increasingly evident that the CPSs and EPSs of many other bacteria can be assigned to one of these groups, suggesting a limited number of ways in which these polymers can be made and expressed on the cell surface.

2. BIOSYNTHESIS AND TRANSPORT OF CPSs ACROSS THE INNER MEMBRANE

Two distinct modes of CPS biosynthesis are recognized, depending on whether the polysaccharide is destined for export via a Wzy-dependent system or an adenosine triphosphate- (ATP-) binding cassette (ABC) transporter-dependent system. Capsular polysaccharides assembled by a Wzy-dependent system are built as individual repeating units, which are often branched structures containing a variety of sugars and linkages. In contrast, CPSs exported by an ABC transporter are synthesized to their full length in the cytoplasm and tend to be simple unbranched structures (Figure 20.1). However, this trend does not apply for all bacteria. For example, *C. jejuni* assembles complex, branched CPSs that are exported by an ABC transporter (Figure 20.2). The assembly and export of *E. coli* capsules has been studied extensively and, as such, will be the primary focus of this chapter.

2.1. Wzy-dependent polymerization system

The Wzy-dependent polymerization system is exemplified by group 1 and 4 CPSs of *E. coli* (see Figure 20.1), but is also used to assemble LPS O-antigens and EPSs from a number of bacteria. This pathway, from initiation of repeating unit biosynthesis to export across the outer membrane, is depicted in Figure 20.3.

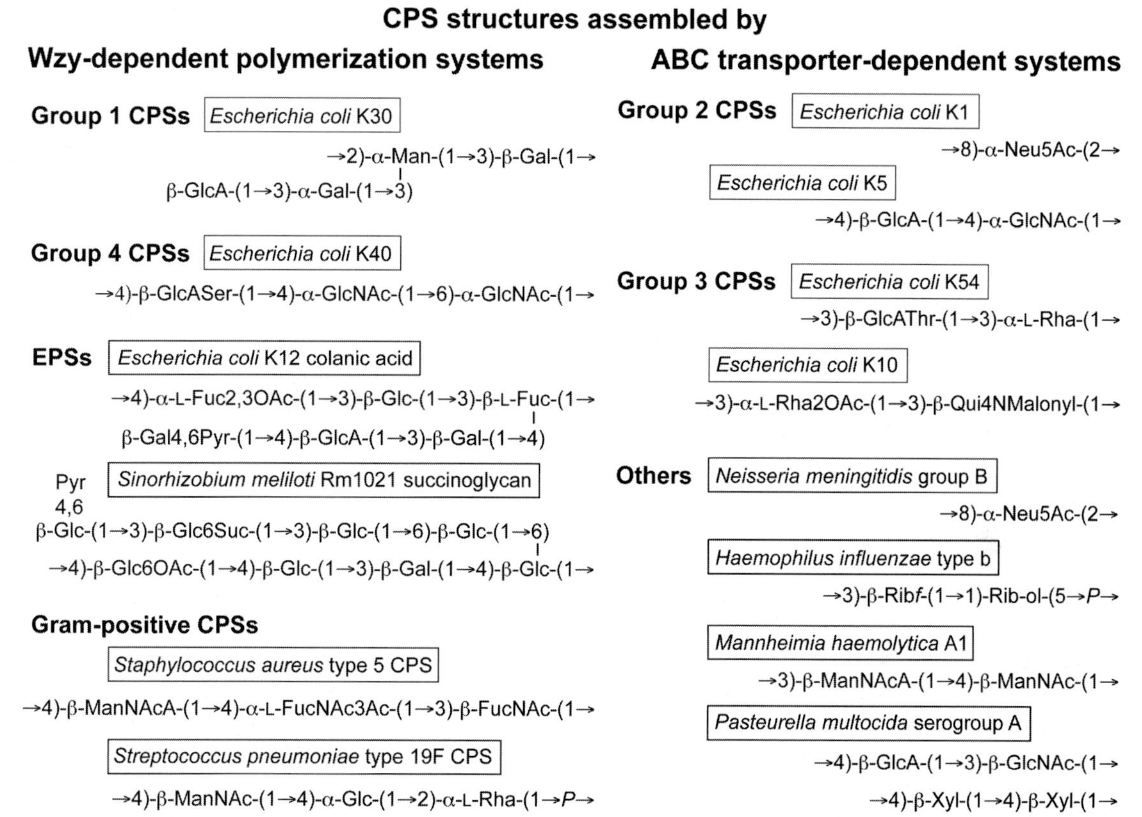

CPS structures assembled by

Wzy-dependent polymerization systems

Group 1 CPSs *Escherichia coli* K30

→2)-α-Man-(1→3)-β-Gal-(1→
β-GlcA-(1→3)-α-Gal-(1→3)

Group 4 CPSs *Escherichia coli* K40

→4)-β-GlcASer-(1→4)-α-GlcNAc-(1→6)-α-GlcNAc-(1→

EPSs *Escherichia coli* K12 colanic acid

→4)-α-L-Fuc2,3OAc-(1→3)-β-Glc-(1→3)-β-L-Fuc-(1→
β-Gal4,6Pyr-(1→4)-β-GlcA-(1→3)-β-Gal-(1→4)

Sinorhizobium meliloti Rm1021 succinoglycan

Pyr 4,6
β-Glc-(1→3)-β-Glc6Suc-(1→3)-β-Glc-(1→6)-β-Glc-(1→6)
→4)-β-Glc6OAc-(1→4)-β-Glc-(1→3)-β-Gal-(1→4)-β-Glc-(1→

Gram-positive CPSs

Staphylococcus aureus type 5 CPS

→4)-β-ManNAcA-(1→4)-α-L-FucNAc3Ac-(1→3)-β-FucNAc-(1→

Streptococcus pneumoniae type 19F CPS

→4)-β-ManNAc-(1→4)-α-Glc-(1→2)-α-L-Rha-(1→*P*→

ABC transporter-dependent systems

Group 2 CPSs *Escherichia coli* K1

→8)-α-Neu5Ac-(2→

Escherichia coli K5

→4)-β-GlcA-(1→4)-α-GlcNAc-(1→

Group 3 CPSs *Escherichia coli* K54

→3)-β-GlcAThr-(1→3)-α-L-Rha-(1→

Escherichia coli K10

→3)-α-L-Rha2OAc-(1→3)-β-Qui4NMalonyl-(1→

Others *Neisseria meningitidis* group B

→8)-α-Neu5Ac-(2→

Haemophilus influenzae type b

→3)-β-Rib*f*-(1→1)-Rib-ol-(5→*P*→

Mannheimia haemolytica A1

→3)-β-ManNAcA-(1→4)-β-ManNAc-(1→

Pasteurella multocida serogroup A

→4)-β-GlcA-(1→3)-β-GlcNAc-(1→
→4)-β-Xyl-(1→4)-β-Xyl-(1→

FIGURE 20.1 Representative CPS and EPS structures assembled by Wzy- and ABC transporter-dependent systems. Structures were obtained from the following references: *E. coli* K30 (Chakraborty *et al.*, 1980); *E. coli* K40 (Dengler *et al.*, 1986); colanic acid (Garegg *et al.*, 1971); succinoglycan (Reinhold *et al.*, 1994; Chouly *et al.*, 1995); *S. aureus* type 5 (Jones, 2005b); *S. pneumoniae* type 19F (Abeygunawardana *et al.*, 2000); *E. coli* K1 (McGuire and Binkley, 1964); *E. coli* K5 (Vann *et al.*, 1981); *E. coli* K54 (Hofmann *et al.*, 1985); *E. coli* K10 (Sieberth *et al.*, 1993); *N. meningitidis* B (Liu *et al.*, 1971; Jennings *et al.*, 1987); *H. influenzae* b (Jones, 2005a); *Mannheimia haemolytica* A1 (Beesley *et al.*, 1984); *Pasteurella multocida* A (Rosner *et al.*, 1992). All sugars are D-pyranosides, unless otherwise indicated. Where present, amino acids are amide-linked to the carboxyl group of glucuronic acid (GlcA). Abbreviations: L-Fuc, L-fucose; FucNAc, fucosamine; Gal, galactose; Glc, glucose; GlcNAc, *N*-acetylglucosamine; Man, mannose; ManNAcA, *N*-acetylmannuronic acid; Neu5Ac, *N*-acetylneuraminic acid; OAc, *O*-acetyl group; Pyr = pyruvyl group; Qui4NMalonyl, 4,6-dideoxy-4-malonylamino-D-glucose; L-Rha, L-rhamnose; Rib, ribose; Rib-ol, ribitol; Ser, serine; Suc, succinyl group; Thr, threonine; Xyl, xylose.

Much of the information concerning early stages of Wzy-dependent CPS biosynthesis has been inferred from O-antigen systems. Synthesis of CPS chains originates in the cytoplasm from pools of activated sugar-nucleotide precursors. These can be shared between CPS, lipopolysaccharide/lipo-oligosaccharide (LPS/LOS) and/or peptidoglycan (PG) assembly pathways. To begin CPS synthesis, an initiating glycosyltransferase (GT) transfers the first sugar, as a sugar-1-phosphate (-1-P), to undecaprenyl-phosphate (Und-P), the lipid carrier on which the CPS is built. Two distinct families of initiating transferases have been identified. Polyisoprenyl-P hexose-1-P transferases are unique to bacteria and transfer a hexose-1-P to Und-P. The

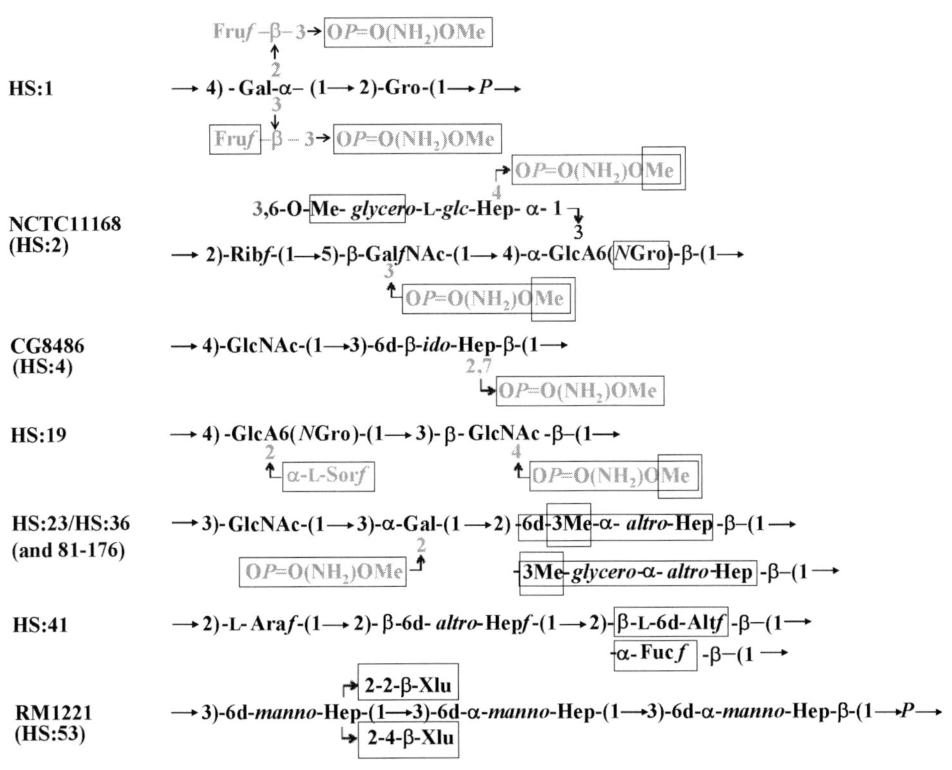

FIGURE 20.2 Examples of CPS structures produced by *Campylobacter jejuni*. In contrast to the structures shown in Figure 20.1, the CPSs are very complex in this ABC transporter-dependent organism. Structural information was obtained from the following references: *C. jejuni* HS:1 serostrain (Moran *et al.*, 2000; McNally *et al.*, 2005), NCTC 11168 (St Michael *et al.*, 2002; McNally *et al.*, 2007); CG8486 (Chen *et al.*, 2008); HS:19 serostrain (Aspinall *et al.*, 1994; McNally *et al.*, 2006); HS:23/HS:36 serostrains and 81–176 (Karlyshev *et al.*, 2005; Kanipes *et al.*, 2008); HS:41 serostrain (Moran *et al.*, 2000; Karlyshev *et al.*, 2005); and RM1221 (Gilbert *et al.*, 2007). Residues in black were described in the original reference while residues in grey were only apparent when less harsh purification conditions were used. These alternate purification conditions were considered when differences were observed in magic-angle spinning NMR spectra compared to the original publications (Karlyshev *et al.*, 2005). This figure also illustrates the remarkable level of heterogeneity in *C. jejuni* CPS structures due to phase variable (i.e. non-stoichiometric) expression of carbohydrate modifications (indicated with boxes). For example, strain NCTC 11168 has eight variable modifications resulting in >40 000 possible variations in CPS structure. All sugars are D-pyranosides, unless otherwise indicated. Abbreviations: Alt*f*, altrofuranose; Ara, arabinose; Fru*f*, fructofuranose; 6d-, 6-deoxy-; Gro, glycerol; Hep, heptose; OPO(NH_2)OMe, *O*-methyl phosphoramidate; NGro = aminoglycerol (which can be substituted with ethanolamine in strain NCTC 11168); Rib*f*, ribofuranose; Sor*f*, sorbofuranose; Xlu, xylulose. Also, see Figure 20.1 for further abbreviations.

prototype enzyme for this family is WbaP, whose homologues transfer galactose- (Gal-) 1-P or glucose- (Glc-) 1-P to the lipid carrier (Wang *et al.*, 1996; Drummelsmith and Whitfield, 1999; Cartee *et al.*, 2005). Polyisoprenyl-P *N*-acetylhexosamine- (HexNAc-) 1-P transferases are found in both prokaryotes and eukaryotes and catalyse the transfer of HexNAc-1-P to a lipid acceptor. The WecA protein and homologues belonging to this family transfer *N*-acetylglucosamine- (GlcNAc-) 1-P or *N*-acetylgalactosamine- (GalNAc-) 1-P to Und-P

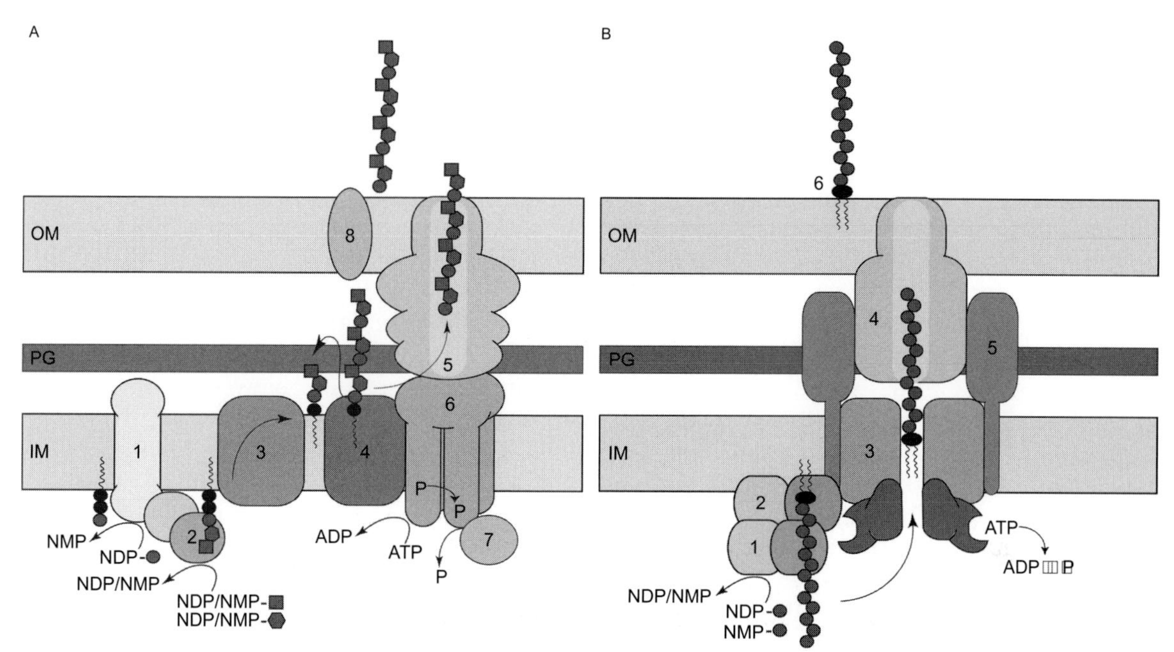

FIGURE 20.3 Models for Wzy- and ABC transporter-dependent CPS/EPS biosynthesis and surface expression in Gram-negative bacteria. IM, Inner (cytoplasmic) membrane; PG, peptidoglycan layer; OM, Outer membrane. (A) Wzy-dependent CPS/EPS biosynthesis is initiated by transfer of a sugar-1-P to undecaprenyl-P on the cytoplasmic face of the inner membrane, by either a WecA or WbaP homologue (1). Other glycosyltransferases (GTs) complete the repeating unit by transferring additional sugars from nucleotide-activated precursors (2). Then Wzx "flips" the repeating units to the periplasmic face of the inner membrane, using an as yet undefined mechanism (3). The Wzy enzyme catalyses polymerization of the repeating units by transferring the growing chain to a new repeating unit; growth occurs at the reducing end of the polymer (4). The completed polymer is translocated across the outer membrane via a channel formed by Wza (5). The tyrosine autokinase, Wzc, is capable of auto-phosphorylation and *trans*-phosphorylation reactions (6). The phosphatase, Wzb, dephosphorylates Wzc (7). Thus, CPS/EPS expression appears to require both phosphorylation and dephosphorylation of Wzc. The Wzc protein interacts with the Wza multimer and may mediate channel opening. In the case of group 1 CPSs, the Wzi protein in the outer membrane is required for normal surface-association of the polysaccharide (8). See text for additional details. (B) Initiation of ABC transporter-dependent CPSs is poorly understood, but may involve assembly on 1,2-diacylglycerol phosphate (shown) or another acceptor molecule and generation of a primer (not shown). One or more GTs assemble the CPS molecule by addition of sugars, from nucleotide-activated pools, to the non-reducing end of the chain (1). The roles of KpsC and KpsS are poorly understood, but both are membrane-associated and KpsC interactions with biosynthetic enzymes and KpsE are believed to couple polymer synthesis and export (2). The completed polymer is then transported across the inner membrane via an ABC transporter comprised of KpsM and KpsT homologues (3). In some systems, the polymer is known to be linked to diacylglycerol prior to transport. Transport across the outer membrane is likely via a channel formed by KpsD (shown) or Wza (4), although the existence of a KpsD channel has not been shown to date. The membrane fusion protein, KpsE, may serve as a bridge connecting processes occurring at the inner membrane with those occurring at the outer membrane (5). At least for some systems, surface attachment of the CPS is via linkage to diacylglycerol phosphate (6). See Dong *et al.* (2006) and Whitfield and Roberts (1999) and text for additional details. Abbreviations: ADP, adenosine diphosphate; ATP, adenosine triphosphate; NDP, nucleotide diphosphate; NMP, nucleotide monophosphate; P, inorganic phosphate.

(Meier-Dieter *et al.*, 1990, 1992; Alexander and Valvano, 1994; Rick *et al.*, 1994; Amor and Whitfield, 1997; Rush *et al.*, 1997; Zhang *et al.*, 1997; Wang and Reeves, 1998; Lehrer *et al.*, 2007). Initiating transferases are large enzymes with multiple transmembrane domains. The WbaP proteins have five transmembrane domains, a large periplasmic loop and a cytoplasmic C-terminal domain. The C-terminal region (including the fifth transmembrane domain) is sufficient for glycosyltranferase activity (Wang *et al.*, 1996; Saldias *et al.*, 2008), while the N-terminal domain is required for membrane localization (Saldias *et al.*, 2008) and may play a role in releasing the Und-P-P-Gal product (Wang *et al.*, 1996). The WecA enzymes are predicted to contain 11 transmembrane domains flanking five periplasmic and five cytoplasmic loops (Amer and Valvano, 2000, 2001; Lehrer *et al.*, 2007). Truncation of the N-terminal domain (Amer and Valvano, 2000) or replacement of aspartate residues in the second and third cytoplasmic loops (Amer and Valvano, 2002; Lehrer *et al.*, 2007) leads to reduced transferase activity. Basic domains in the C-terminal cytoplasmic loop of the bacterial polyisoprenyl-P HexNAc-1-P transferases have been identified whose sequences are conserved among proteins with similar substrate specificities (Anderson *et al.*, 2000) and who appear to be involved in substrate binding (Amer and Valvano, 2001). After initiation of polymer synthesis, other GTs complete the repeating unit by sequential addition of sugars to the non-reducing end. Unlike the integral initiating transferases, these enzymes are either soluble or membrane-associated.

Following their synthesis, lipid-linked repeating units are transferred to the periplasmic face of the inner membrane by the putative flippase, Wzx. Consistent with this, a *wzx* deletion mutant in an O-antigen system accumulates individual lipid-linked repeating units in the cytoplasm (Liu *et al.*, 1996). Further evidence implicating Wzx as a transmembrane flippase is the finding that Wzx is required for the transport of a water-soluble Und-P-P-GlcNAc analogue into everted cytoplasmic membrane vesicles (Rick *et al.*, 2003). Cross-complementation studies revealed that Wzx may recognize the first sugar of the repeating unit, thus differentiating between polymers initiated by a WbaP or WecA homologue (Feldman *et al.*, 1999; Marolda *et al.*, 2004). However, the presence of the flippase's cognate Wzy (polymerase) or Wzz (chain length regulator) in these trans-complementation studies impaired Wzx function, suggesting that interactions between these proteins might sequester Wzx (Marolda *et al.*, 2006). The mechanism used by Wzx to effect the transmembrane "flipping" of lipid-linked repeating units remains unknown. One possibility is that Wzx may facilitate diffusion of these molecules across the inner membrane (Rick *et al.*, 2003). An alternate hypothesis is based on the finding that polyisoprenoid lipids can alter membrane structure, which could conceivably lead to the formation of a channel via which these CPS intermediates could pass (Zhou and Troy, 2005).

Following transfer of repeating units to the periplasm, polymerization occurs to generate high-molecular mass (HMM) polysaccharides (Mulford and Osborn, 1983; McGrath and Osborn, 1991). The putative polymerase, Wzy, is implicated in this process and is thought to transfer the growing CPS chain to a new repeating unit with chain elongation occurring at the reducing end of the polymer. The Wzy proteins have nine or more predicted transmembrane domains (Daniels *et al.*, 1998) and share limited sequence homology. Mutation of *wzy* in O-antigen systems leads to the expression of LPS molecules containing single O-antigen repeating units (Daniels *et al.*, 1998; Bengoechea *et al.*, 2002; Tao *et al.*, 2004; Cheng *et al.*, 2007; Nakhamchik *et al.*, 2007). Consistent with this, a *wzy* mutation in *E. coli* K30 results in the loss of HMM K30 CPS and the accumulation of single K30 repeating units ligated to lipid A-core (Drummelsmith and Whitfield, 1999). Ligation of K antigen to lipid A-core has only been observed in *E. coli* and its biological significance is unclear. Of note, the mechanism used by Wzy to effect polymerization is not currently known, owing in part to the difficulties inherent with over-expression, purification and manipulation of integral membrane proteins with multiple trans-membrane domains. However, sequence comparisons between Wzy and WaaL (O-antigen ligase) suggest these proteins may employ similar mechanisms to catalyse polymer transfer (Schild *et al.*, 2005).

It is not clear what dictates the extent of polymerization of the growing CPS chain or what signal terminates polymerization and initiates export. The Wzy-dependent O-antigen systems display a preference for a given length of O-chains, referred to as modal chain length. Chain length is influenced by Wzz, a member of the *p*olysaccharide *c*o-*p*olymerase (PCP-1) family (Whitfield *et al.*, 1997; Morona *et al.*, 2000b). A mutation in *wzz* abolishes O-antigen modality and favours the production of short O-chains, in amounts inversely proportional to the length of the chain (Bastin *et al.*, 1993; Bengoechea *et al.*, 2002; Murray *et al.*, 2003). It has been proposed that Wzz may serve as a molecular timer, allowing polymerization to occur for a preset amount of time, before ligation to lipid A-core (Bastin *et al.*, 1993) or, alternatively, that Wzz acts as a molecular chaperone mediating the interaction between the polymerase and ligase, with the stoichiometry of Wzy/WaaL dictating the preferred chain length (Morona *et al.*, 1995; Daniels *et al.*, 1998). Consistent with the latter model, levels of Wzz and Wzy in *Shigella flexneri* 2a affect O-antigen chain length distribution (Carter *et al.*, 2007). More recently, it has been proposed that Wzz interacts with both Wzy and Wzx (Marolda *et al.*, 2006) to control the extent of O-antigen polymerization. Oligomers of Wzz have been observed *in vivo* in several bacteria (Daniels and Morona, 1999; Daniels *et al.*, 2002; Stenberg *et al.*, 2005). Nuclear magnetic resonance (NMR) spectroscopy-established structures of the periplasmic domains of three different Wzz proteins also revealed oligomers whose multimeric state and helix packing differed for each Wzz protein (Tocilj *et al.*, 2008). These authors proposed that Wzz oligomers act as a focal point to gather Wzy molecules and that chain elongation occurs by transfer of the growing polymer from one Wzy molecule to another. The size of the Wzz oligomer would dictate the number of associated Wzy proteins, which would in turn dictate polysaccharide chain length.

Unlike their O-antigen counterparts, Wzy-dependent CPS expression systems do not include a Wzz protein. However, a member of the PCP-2 family (Wzc) is found in these systems (Morona *et al.*, 2000b), though its role is clearly not limited to CPS chain length determination. Both Wzc and Wzz share topological features, including two transmembrane domains and a periplasmic loop with propensity for coiled-coil formation (Morona *et al.*, 1995, 2000b; Wugeditsch *et al.*, 2001; Doublet *et al.*, 2002). Unlike Wzz, Wzc has an additional cytoplasmic tyrosine autokinase domain, which harbours C-terminal tyrosine residues that are accessible to phosphorylation (Morona *et al.*, 2000a, 2003; Niemeyer and Becker, 2001; Wugeditsch *et al.*, 2001; Doublet *et al.*, 2002; Grangeasse *et al.*, 2002; Bender *et al.*, 2003; Nakar and Gutnick, 2003). Deletion of *wzc* abolishes or reduces HMM CPS and EPS expression (Drummelsmith and Whitfield, 1999; Morona *et al.*, 2000a; Vincent *et al.*, 2000; Cieslewicz *et al.*, 2001; Wugeditsch *et al.*, 2001; Bender *et al.*, 2003; Nakar and Gutnick, 2003; Minic *et al.*, 2007). Loss or reduction of CPS expression also arises from mutation of the kinase catalytic site (Morona *et al.*, 2000a; Niemeyer and Becker, 2001; Wugeditsch *et al.*, 2001; Obadia *et al.*, 2007) and from removal or replacement of the C-terminal tyrosine residues (Becker *et al.*, 1995; Morona *et al.*, 2000a, 2003; Niemeyer and Becker, 2001; Wugeditsch *et al.*, 2001; Paiment *et al.*, 2002; Nakar and Gutnick, 2003; Obadia *et al.*, 2007), suggesting that phosphorylation of Wzc is required for its activity in CPS expression. Site-directed mutagenesis studies in *E. coli* K30 (Paiment *et al.*, 2002) and *S. pneumoniae* (Morona *et al.*, 2000a, 2003) suggested that the overall level of Wzc phosphorylation, rather than the modification of specific residues, is important for CPS expression. Interactions between Wzc molecules were first suggested by the finding that phosphorylated Wzc could be obtained by co-expressing a catalytically-inactive Wzc derivative with a Wzc derivative that lacked the C-terminal tyrosine residues (Wugeditsch *et al.*, 2001). This early evidence of transphosphorylation reactions was supported by subsequent *in vitro* studies (Grangeasse *et al.*, 2002). The 3D-structure of the Wzc oligomer was solved by single particle analysis of cryo-negatively stained samples and revealed a tetramer with the appearance of an extracted molar tooth (Collins *et al.*, 2006). The periplasmic domains form the "crown" region, which appears to contain the site(s) of intermonomer interaction, while the "roots" contain the tyrosine autokinase domains. The Wzc molecules in these oligomers are heterogeneously phosphorylated, though phosphorylation is not required for oligomerization. This supports earlier *in vivo* cross-linking data which showed that complex formation by Wzc occurred regardless of the phosphorylation state of the protein (Doublet *et al.*, 2002; Paiment *et al.*, 2002).

While phosphorylation of Wzc is essential for HMM CPS expression in Wzy-dependent systems, dephosphorylation by Wzb appears to be equally important. Importantly, Wzb is a phosphotyrosine protein phosphatase, capable of dephosphorylating Wzc (Grangeasse *et al.*, 1998; Vincent *et al.*, 1999, 2000; Wugeditsch *et al.*, 2001; Preneta *et al.*, 2002; Klein *et al.*, 2003; Nakar and Gutnick, 2003). This phosphatase belongs to the *p*hospho*t*yrosine protein *p*hosphatase (PTP) family whose members are widely distributed in Nature and share high levels of homology. Of note, Wzb homologues are required for expression of wild-type levels of CPSs and EPSs in many systems (Bugert and Geider, 1995; Vincent *et al.*, 2000; Wugeditsch *et al.*, 2001; Nakar and Gutnick, 2003). The requirement for both a phosphorylation-competent kinase and a functional phosphatase led to the hypothesis that Wzc cycling between phosphorylated and dephosphorylated states is required for CPS expression (Wugeditsch *et al.*, 2001; Paiment *et al.*, 2002). However, others have proposed that phosphorylated Wzc is a negative regulator of CPS expression (Morona *et al.*, 2000a; Vincent *et al.*, 2000). While the majority of the data regarding Wzc phosphorylation and CPS/EPS expression are in agreement, important differences have been noted. These do not likely reflect inherent differences between the kinases and phosphatases themselves, as Wzb and Wza plus Wzc from the colanic acid system (denoted with subscript CA from this point onward) could complement *wzb* and *wzc* mutations in *E. coli* K30, respectively (Reid and Whitfield, 2005). The use of different genetic backgrounds, which have been shown in some instances to yield conflicting results for the same CPS system (Bender *et al.*, 2003) and the acquisition of compensatory mutations in some mutants (Morona *et al.*, 2006; Xayarath and Yother, 2007), complicate data interpretation and comparisons between systems.

Gram-positive Wzc proteins are encoded by two distinct polypeptides, one an integral membrane protein (e.g. *Streptococcus* spp. CpsC) and the other a soluble cytoplasmic protein containing the tyrosine autokinase domain (e.g. *Streptococcus* spp. CpsD) (Guidolin *et al.*, 1994; Sau *et al.*, 1997). This two-protein organization does not likely reflect differences in protein function, as some Gram-negative bacteria are predicted to encode a 2-part Wzc and Wzc from *E. coli* K30 (Wzc_{K30}) is active when expressed as two separate polypeptides (Wugeditsch *et al.*, 2001), though the same is not true for Wzc_{CA} (Obadia *et al.*, 2007). The periplasmic loop is also shorter in Gram-positive Wzc proteins and the C-terminal domain contains fewer tyrosine residues. The CPS-associated phosphatases (e.g. *Streptococcus* spp. CpsB) in the Gram-positive systems are unrelated to Wzb and belong to the *poly**h**istidinol **p**hosphatase*- (PHP-) family (Bender and Yother, 2001; Morona *et al.*, 2002). While these phosphatases dephosphorylate CpsD (Bender and Yother, 2001; Morona *et al.*, 2002), they may also modulate its kinase activity (Bender and Yother, 2001). Deletion of *cpsB* led to reduced CPS production in *Streptococcus agalactiae* (Cieslewicz *et al.*, 2001) and *S. pneumoniae* Rx1-19F (type 19F CPS) (Morona *et al.*, 2000a) and Rx1-2 (type 2 CPS) (Bender *et al.*, 2003). While a *cpsB* mutation did not affect the amount of cell wall-associated CPS in *S. pneumoniae* D39 (Bender *et al.*, 2003; Morona *et al.*, 2006), this strain actually produced only 60% of the total (cell-associated plus cell-free) CPS produced by the wild-type (Morona *et al.*, 2006). Reversible phosphorylation of Wzc is important for virulence of *S. pneumoniae*, as mutations in *wzc* (*cpsC* or *cpsD*) (Morona *et al.*, 2004) or *cpsB* (Bender *et al.*, 2003; Morona *et al.*, 2004, 2006) caused reduced virulence in animal models. It is important to note that Wzb homologues have been identified in some Gram-positive organisms, e.g. *Staphylococcus aureus* (Soulat *et al.*, 2002), but they have not to date been ascribed a role in CPS/EPS production.

While Wzc clearly shares topological features with Wzz, the evidence implicating Wzc in CPS chain length regulation is limited. The difficulty in assigning such a role to Wzc is due in part to the acapsular phenotype of many *wzc* deletion mutants. In *S. agalactiae*, mutation of the *wzc* homologue resulted in an ≈50% decrease in the molecular mass of the CPS produced (Cieslewicz *et al.*, 2001). Altered EPS size distribution was also observed upon expression of various mutant *wzc* homologues in *E. coli* K-12 (Obadia *et al.*, 2007). Finally, in several systems, loss of HMM polysaccharide caused by deletion of *wzc* is accompanied by increased levels of low-molecular mass (LMM) polymers (Drummelsmith and Whitfield, 1999; Niemeyer and Becker, 2001; Wugeditsch *et al.*, 2001; Bender *et al.*, 2003). Both Wzz and Wzc share a proline-glycine-rich motif just upstream of and within the second transmembrane domain. Mutations within this domain in Wzz led to the production of O chains of altered length (Daniels and Morona, 1999). In Wzc homologues involved in succinoglycan, emulsan and colanic acid production, amino acid replacements in the Pro-Gly-rich motif altered the molecular mass of the associated EPS, leading to larger polymers in *Acinetobacter venetianus* (Dams-Kozlowska and Kaplan, 2007) and smaller material in *Sinorhizobium meliloti* and *E. coli* K-12 (Becker and Puhler, 1998; Obadia *et al.*, 2007). Emulsan of increased molecular mass was also produced by replacement of *A. venetianus wza*, *wzb* and *wzc* genes with *E. coli* K-12 homologues (Dams-Kozlowska *et al.*, 2008). If indeed Wzc plays a role in Wzy-dependent CPS chain length determination, it may not be the sole component involved, as data suggest that the periplasmic domain of the initiating GT, WbaP, influences O-antigen chain length (Saldias *et al.*, 2008).

The Wzc protein can act on additional (non-Wzc) substrates which further complicates elucidation of its role in CPS and EPS expression. The Wzc-mediated phosphorylation of uridine diphosphate- (UDP-) glucose dehydrogenase (Ugd) has been detected *in vitro* in *E. coli* K-12 and enhances the enzyme's *in vitro* activity (Grangeasse *et al.*, 2003). The Ugd enzyme converts UDP-Glc to UDP-glucuronic acid (GlcA), which is a building block for colanic acid biosynthesis, though it is not clear whether these *in vitro* observations are relevant to *in vivo* EPS expression. Similarly, UDP-sugar dehydrogenases in *Bacillus subtilis* (Mijakovic *et al.*, 2003) and *S. aureus* (Soulat *et al.*, 2006) can also be phosphorylated by Wzc homologues. In *Streptococcus thermophilus*, initiating transferase (EpsE) activity is decreased in a *wzc* (*epsC* or *epsD*) mutant and in a strain expressing EpsE lacking a tyrosine residue (Minic *et al.*, 2007). While tyrosine phosphorylation of EpsE was not conclusively shown, the authors suggest

EpsE phosphorylation by Wzc may be required for full transferase activity (Minic *et al.*, 2007).

2.2. The ABC transporter-dependent system

The ABC transporter-dependent system moves full-length CPSs from the cytoplasmic to the periplasmic face of the inner membrane. In *E. coli*, ABC transporters export CPSs belonging to groups 2 and 3, which include the polysialic acid capsule of *E. coli* K1 and the *N*-acetylheparosan capsule of *E. coli* K5 (see Figure 20.1). A schematic diagram depicting ABC transporter-dependent CPS assembly is presented in Figure 20.3.

Initiation of CPS biosynthesis in ABC transporter-dependent systems remains poorly understood. Neither the acceptor upon which the CPS chains are built nor the enzyme(s) responsible for initiation of biosynthesis have been identified. Evidence suggests that ABC transporter-dependent CPSs of *E. coli* and *Neisseria meningitidis* are covalently anchored to the outer membrane via a 1,2-diacylglycerol moiety (Gotschlich *et al.*, 1981; Schmidt and Jann, 1982; Tzeng *et al.*, 2005). Accumulation of intracellular CPSs bearing this lipid anchor (Tzeng *et al.*, 2005) suggests CPS chains are either synthesized directly on diacylglycerophosphate or that they are transferred from a different acceptor (e.g. Und-P) to this lipid prior to export. The KpsC and KpsS proteins (LipA and LipB in *N. meningitidis*) were initially implicated in lipid modification of group 2 capsules (Bronner *et al.*, 1993b; Frosch and Muller, 1993; Roberts, 1996; Rigg *et al.*, 1998) as expression of an *N. meningitidis cps* cluster deleted for *lipA* and/or *lipB* in *E. coli* K-12 resulted in the intracellular accumulation of non-lipidated polysialic acid (Frosch and Muller, 1993). However, equivalent mutations in *E. coli* K-12 harbouring the K1 *kps* cluster and in *N. meningitidis* strain NMB, resulted in the accumulation of lipidated CPS (Cieslewicz and Vimr, 1996; Tzeng *et al.*, 2005). In light of these findings, the identity of the enzymes responsible for lipid modification of group 2 capsules is not known. In *E. coli* K1, the minimum requirement for *de novo* CPS biosynthesis includes cytidine monophosphate- (CMP-) *N*-acetylneuraminic acid, the polysialyltransferase (NeuS), an accessory membrane protein (NeuE) and KpsC (Andreishcheva and Vann, 2006). The additional presence of KpsS greatly enhanced the amount of polymer synthesized (Andreishcheva and Vann, 2006). While the role(s) of KpsS and KpsC remain unknown, the authors raise the possibility that they are involved in the formation of a primer which could then be extended by the polysialyltransferase.

There is also limited evidence to suggest that 3-deoxy-D-*manno*-oct-2-ulosonic acid (Kdo) serves as a linker between the lipid and the CPS in some group 2 systems (Schmidt and Jann, 1982; Finke *et al.*, 1991; Bronner *et al.*, 1993a). This is supported by the presence of genes whose products are involved in CMP-Kdo synthesis (*kpsF* and *kpsU*) in group 2 CPS loci (Finke *et al.*, 1989; Pazzani *et al.*, 1993; Rosenow *et al.*, 1995). However, KpsF and KpsU are not essential for group 2 CPS expression (Cieslewicz and Vimr, 1996, 1997) and loci for group 3 CPSs in *E. coli* or group 2/3 CPSs in *C. jejuni* lack one or both of these genes. But, it has not been ruled out that homologues present in LPS biosynthetic loci compensate for loss or absence of these genes from the respective CPS loci (Whitfield, 2006).

Following polysaccharide initiation, membrane-associated GTs extend the polymer by sequential addition of sugars to the non-reducing end of the chain (Bliss and Silver, 1996). The activities of a number of these GTs have been studied using purified protein and exogenous acceptors. For instance, the α-(2$\rightarrow$8)-polysialyltransferases (PSTs) from *E. coli* K1 and *N. meningitidis* are able to extend exogenous acceptors *in vitro* (Steenbergen and Vimr, 1990; Steenbergen *et al.*, 1992; Cho and Troy, 1994; Freiberger *et al.*, 2007; Willis *et al.*, 2008). Efficient elongation requires the presence of a sialic acid primer on the acceptor (Steenbergen and Vimr, 1990; Ferrero *et al.*, 1991; Chao *et al.*, 1999; Freiberger *et al.*, 2007; Willis *et al.*, 2008), which suggests that these enzymes are not able to initiate polymer biosynthesis and is consistent with the *in vivo* minimum requirement for K1 CPS biosynthesis (Andreishcheva and Vann, 2006). Polymer synthesis by polysialyltransferases was proposed to occur via a processive mechanism, as reaction intermediates could not be detected *in vivo* (Steenbergen and Vimr, 2003). While reaction intermediates were detected from *in vitro* reactions using *N. meningitidis* and *E. coli* K92 PST (Freiberger *et al.*, 2007; Vionnet and Vann, 2007), it remains possible that other cellular factors increase processivity of these enzymes (Freiberger *et al.*, 2007).

The ABC transporter for CPS export is a member of the ABC-2 family and is comprised of a transmembrane domain (TMD) homo-dimer (KpsM) and a nucleotide-binding domain (NBD) homo-dimer (KpsT) (Pigeon and Silver, 1994; Bliss and Silver, 1996). Deletion of either *kpsM* or *kpsT* leads to the intracytoplasmic accumulation of polysaccharide (Kroncke *et al.*, 1990; Pavelka *et al.*, 1994; Pigeon and Silver, 1994). The transporters for CPS export do not appear to recognize the polysaccharide, as the individual components (KpsM or KpsT) and the complete transporter (KpsM and KpsT) are functionally interchangeable between the K1 and K5 systems (Roberts *et al.*, 1986, 1988; Pearce and Roberts, 1995; Cuthbertson *et al.*, 2005), as well as between bacterial species (Ward and Inzana, 1997; Bacon *et al.*, 2001; Lo *et al.*, 2001). This feature distinguishes the CPS transporters from their O-antigen counterparts, whose NBD components (designated Wzt) cannot be exchanged (Cuthbertson *et al.*, 2005).

As for the Wzy-dependent system, it is not clear what signals the end of chain elongation and initiates polysaccharide transport. Early suggestions that lipid modification of the CPS is sufficient for export do not hold in light of mutants that accumulate intracellular, lipidated CPS. In the ABC transporter-dependent O-antigen systems of *E. coli* O8 and O9a, chain elongation is terminated by transfer of phosphate and/or methyl groups to the non-reducing end of the polymer (Clarke *et al.*, 2004). These terminal modifications are recognized by an extended domain in the NBD component of the ABC transporter (Cuthbertson *et al.*, 2007) and are required for polysaccharide transport (Clarke *et al.*, 2004). However, a comparable system for CPS biosynthesis has not been described.

Another unanswered question in ABC transporter-dependent CPS export concerns the mechanism used to transport a large, charged polymer linked to a hydrophobic membrane anchor. Studies of the ABC transporter for lipid A-core export (MsbA) have suggested that the hydrophilic domains of the transported material are moved through the lumen of the transporter while the hydrophobic portion is dragged through the surrounding lipid bilayer (Reyes and Chang, 2005). Such an arrangement could conceivably accommodate transport of lipidated CPSs (Whitfield, 2006).

3. CAPSULAR POLYSACCHARIDE EXPORT ACROSS THE OUTER MEMBRANE

In Gram-negative bacteria, the CPSs need to traverse the outer membrane before reaching the cell surface. This area of research has seen exciting developments over the last few years, with the determination of the structures for the outer membrane channel formed by Wza (Beis *et al.*, 2004; Dong *et al.*, 2006) and the complex formed by the Wza/Wzc pair (Collins *et al.*, 2007).

Given the large size of CPS molecules, their transport across the outer membrane must require a dedicated transporter. Capsular polysaccharide operons typically encode either a Wza- or a KpsD-type outer membrane protein. Despite sharing limited sequence homology, these proteins have been grouped together within the *o*uter *m*embrane *a*uxiliary (OMA) protein family (Paulsen *et al.*, 1997) whose members share a polysaccharide biosynthesis/export motif (PES; Pfam02563) (Dong *et al.*, 2006). While it was initially proposed that Wza proteins transport CPSs assembled by Wzy-dependent systems and that KpsD is the outer membrane transporter for ABC-transporter-dependent systems, Wza homologues have been found in ABC-transporter-dependent systems in several bacteria, e.g. *Haemophilus influenzae* (Kroll *et al.*, 1990) and *N. meningitidis* (Frosch *et al.*, 1992).

Furthermore, Wza is an outer membrane lipoprotein processed by signal peptidase II (Paulsen *et al.*, 1997; Drummelsmith and Whitfield, 1999). Deletion of *wza* in *E. coli* K30 leads to an acapsular phenotype, but does not lead to intracellular polymer accumulation (Drummelsmith and Whitfield, 2000). This suggests that, in the absence of a functional export apparatus, a feedback mechanism prevents extensive CPS polymerization. Expression of a non-acylated Wza derivative did cause accumulation of periplasmic HMM K30 CPS (Nesper *et al.*, 2003), thus uncoupling polymer synthesis and translocation.

Multimer formation by Wza was first noted in sodium dodecyl sulfate polyacrylamide gel electrophoresis (SDS-PAGE) profiles which showed that purified Wza formed heat-labile but SDS-resistant aggregates (Drummelsmith and Whitfield, 2000). When visualized by electron microscopy, these aggregates formed ring-like structures with a central pore (Drummelsmith and Whitfield, 2000; Nesper *et al.*, 2003). A single particle analysis study of the Wza multimer revealed that the Wza complex is an octamer, organized as a tetramer of dimers (Beis *et al.*, 2004). The crystal structure of the mature, acylated Wza octamer has significantly advanced our understanding of CPS transport across the outer membrane (Dong *et al.*, 2006). When viewed from the side, the bulk of the complex appears as three distinct rings in the periplasm, with a narrow neck located in the outer membrane. The neck region forms an α-helical barrel, rather than the β-barrel structure typical of integral outer membrane proteins (Dong *et al.*, 2006). The structure is sealed at the periplasmic end by eight loop regions, suggesting the complex exists predominantly in a closed state. An interior channel runs the length of the structure and is lined with hydrophilic residues. Rather than Wza binding the CPS chains, the authors propose that the Wza multimer forms a water-filled channel for CPS export. This proposed lack of polymer recognition is consistent with the finding that Wza from the colanic acid locus in *E. coli* K-12 is a functional homologue of Wza from *E. coli* K30, despite dramatic differences in the structures of their cognate polysaccharides (Reid and Whitfield, 2005) (see Figure 20.1).

Compared with the wealth of information now available for Wza, little is known about the function of KpsD in CPS export. Nevertheless, KpsD homologues have been found exclusively in ABC-transporter-dependent CPS systems to date, including *E. coli* K1 (Silver *et al.*, 1987; Wunder *et al.*, 1994), K5 (Pazzani *et al.*, 1993) and *C. jejuni* (Karlyshev *et al.*, 2000). Initial studies suggested KpsD was a periplasmic protein (Silver *et al.*, 1987; Wunder *et al.*, 1994), which delayed recognition of this protein as a putative outer membrane channel for CPS export. Localization of KpsD within the cell is affected by the presence/absence of CPS biosynthesis and export components (Arrecubieta *et al.*, 2001), though surface expression of KpsD was later confirmed in whole cells of *E. coli* K5 (McNulty *et al.*, 2006). The homology between Wza and KpsD is largely limited to the N-terminal region (McNulty *et al.*, 2006), which includes the PES domain (part of the innermost ring in the Wza complex structure). Contrary to Wza,

KpsD contains a signal peptidase I consensus sequence and is not acylated (Pazzani *et al.*, 1993; Wunder *et al.*, 1994). Despite this difference, heat-labile and SDS-stable KpsD interactions have also been detected in immunoblots (McNulty *et al.*, 2006).

4. BRIDGING THE GAP BETWEEN THE INNER AND OUTER MEMBRANES

4.1. Coordinated biosynthesis and export

Early visualization of CPS export by electron microscopy led to the coining of the term "Bayer's bridges", i.e. sites where the inner and outer membrane were in close apposition. Using a conditional *E. coli* mutant, sites of nascent group 1 CPS export were identified using CPS-specific phage. Examination of thin sections by electron microscopy revealed that sites of nascent polymer export often coincided with Bayer's bridges (Bayer and Thurow, 1977). Export of group 2 CPSs also appeared to coincide with these sites (Kroncke *et al.*, 1990). The existence of these zones of adhesion has been controversial (Kellenberger, 1990), but a number of bacterial systems have been shown to employ multienzyme complexes that span both bacterial membranes (e.g. type III and IV protein secretion systems) (Christie, 2004; He *et al.*, 2004).

Other lines of evidence support coordinated biosynthetic and export activities. In *E. coli* K30, inactivation of genes involved in late stages of CPS export (e.g. *wza, wzb, wzc*) does not lead to accumulation of HMM CPS, as might be expected, but instead appears to feed back to prevent extensive repeating unit polymerization (Drummelsmith and Whitfield, 1999; Wugeditsch *et al.*, 2001; Nesper *et al.*, 2003). In addition, deletion mutants in *wza, wzy* or *wzc* in *E. coli* K30 have reduced galactosyltransferase activity in isolated membrane preparations, suggesting that repeating unit synthesis is impaired when polymerization and/or export of the final product are prevented (A. Reid and C. Whitfield, unpublished observations).

4.2. Evidence for a multienzyme complex in ABC transporter-dependent CPS systems

Clusters for synthesis of ABC transporter-dependent CPSs encode a *m*embrane *p*eriplasmic *a*uxiliary protein (MPA-2), KpsE, also referred to as a *m*embrane *f*usion *p*rotein (MFP). This predominantly periplasmic protein is anchored in the inner membrane (Arrecubieta *et al.*, 2001; Phoenix *et al.*, 2001) and is believed to exist as a dimer (Arrecubieta *et al.*, 2001). Proteins belonging to this family are found in systems such as bacterial efflux pumps, where they are proposed to span the periplasm, linking inner and outer membrane proteins.

Early evidence from subcellular localization studies suggested that components of the CPS biosynthetic machinery in ABC-transporter-dependent systems interact with the export machinery. Several biosynthetic enzymes (KfiA, KfiC and KfiD of *E. coli* K5) were found associated with the ABC transporter components, as well as with KpsS and KpsC (Rigg *et al.*, 1998). Similarly, subcellular localization of KpsD was altered in *kpsE* and *kfiC* mutants (Arrecubieta *et al.*, 2001). Additionally, membranes isolated from *kpsT* mutants lose the ability to incorporate sialic acid onto endogenous acceptors *in vitro*, suggesting communication between biosynthesis and transport (Vimr *et al.*, 1989; Bronner *et al.*, 1993a). These data were extended by *in vivo* studies in *E. coli* K5, where chemical cross-linking was coupled to protein identification by matrix-assisted laser desorption ionization time-of-flight (MALDI-TOF) mass spectrometry (McNulty *et al.*, 2006). In these studies, histidine- (His-) tagged KpsS was used to pull down protein complexes which included KpsD, KpsE, KpsM, KpsT, as well as KfiA. Also, RhsA, a predicted outer membrane protein with a putative carbohydrate-binding motif, was identified in the complex. An *rhsA* mutant makes less K5 CPS, exhibits reduced UDP-GlcA transferase activity *in vitro* and shows loss of polar localization of KpsE and KpsD, but not KpsS (McNulty *et al.*, 2006).

Bacterial two-hybrid studies have also recently been used to probe interactions between the components involved in *E. coli* K1 CPS expression (Steenbergen and Vimr, 2008). A strong interaction between the polysialyltranferase (NeuS) and KpsC was detected. Contrary to the *E. coli* K5 system, where KpsC is required for membrane localization of the polysaccharide synthase (Rigg *et al.*, 1998), membrane localization of the K1 synthase does not require any Neu or Kps proteins (Steenbergen *et al.*, 1992). It is, nevertheless, possible that efficient targeting of NeuS to the CPS export apparatus may rely on this interaction. Homo-dimeric interactions between KpsC proteins were also observed using the two-hybrid system, as were interactions between KpsC and the membrane fusion protein KpsE (Steenbergen and Vimr, 2008). These support the hypothesis that KpsC serves as a bridge to couple NeuS biosynthetic activity to polysaccharide transport (Steenbergen and Vimr, 2008). The use of various truncated KpsC derivatives as bait in the two-hybrid system revealed that a domain near the N-terminus appears to be responsible for KpsC–KpsC interactions, while the C-terminus of the protein is required for KpsC–KpsE interactions (Steenbergen and Vimr, 2008).

In wild-type *E. coli* K1, nascent polysaccharide is protected from degradation by plasmid-encoded endosialidase (Steenbergen and Vimr, 2008). However, polymer accumulating in the cytoplasm of an *E. coli* K1 *kpsC* deletion mutant appears to be sensitive to endosialidase digestion (Steenbergen and Vimr, 2008). This supports the existence of a multienzyme complex coordinating biosynthesis and export, where polymer would either be entirely sequestered or only transiently exposed to other cytoplasmic constituents.

4.3. The Wza-Wzc interactions in Wzy-dependent CPS systems

In vivo chemical cross-linking studies in *E. coli* K30 initially revealed that Wza and Wzc were in close proximity within the cell (Nesper *et al.*, 2003). Genetic support for this interaction was later provided using trans-complementation experiments. In these experiments, Wzc from the colanic acid biosynthetic operon in *E. coli* K-12 (Wzc_{CA}) was unable to complement a *wzc* deletion mutant in *E. coli* K30, unless it was co-expressed with its cognate Wza (Wza_{CA}) (Reid and Whitfield, 2005). A chimeric Wzc protein expressing the N-terminal domain (periplasmic domain plus two transmembrane domains) of Wzc_{CA} and the cytoplasmic domain of Wzc_{K30} also required co-expression with Wza_{CA} to restore K30 CPS surface expression (Reid and Whitfield, 2005). These data provided further support to the idea that Wzc in the inner membrane interacts with Wza in the outer membrane, perhaps lending functional significance to the extended periplasmic loop found in Gram-negative Wzc proteins.

A snapshot of this interaction was later obtained by single particle analysis of cryo-negatively stained Wza–Wzc complexes (Collins *et al.*, 2007). In this complex, which resembles an elongated molar tooth, a heterogeneously-phosphorylated Wzc tetramer is interacting with an acylated Wza octamer. The Wza octamer within this complex has a 22 Å opening caused by a large conformational change in the PES domain. The authors proposed that interactions between Wza and Wzc may allow the periplasmic polysaccharide to enter the Wza channel. The dimensions of the complex further suggest that the periplasm is compressed at sites of complex formation, which supports the existence of Bayer's bridges or fusion sites.

4.4. Crossing the PG layer

One seldom-discussed physical barrier to CPS export is the stress-bearing PG layer. Associated with some transenvelope protein complexes, such as those for type II, III and IV secretion, is a lytic transglycosylase enzyme, believed selectively to cleave the PG layer to allow insertion of the protein complex (reviewed in Koraimann, 2003). While PG-cleaving enzymes have not been found to date in loci for CPS expression, this does not preclude the involvement of lytic transglycosylases encoded elsewhere on the chromosome.

Using immunofluorescence microscopy, McNulty and coworkers found that KpsD, KpsE and KpsS from *E. coli* K5 localize to the cell poles (McNulty *et al.*, 2006). Polar localization of CPS export in this strain was further confirmed by following the emergence of nascent CPS chains using immunofluorescence microscopy and an anti-K5 mAb (McNulty *et al.*, 2006). The poles represent former sites of cell division and, as such, may have unique properties. For instance, it is accepted that the PG at the poles is metabolically inert (Burman *et al.*, 1983; de Pedro *et al.*, 1997, 2001). This feature may make the cell poles more tolerant to insertion of envelope-spanning structures. In fact, it is becoming apparent that a growing number of proteins and cellular processes are targeted to the cell poles (see Janakiraman and Goldberg, 2004 for a review).

5. CELL-SURFACE ATTACHMENT OF THE CPS

One of the main distinctions between CPSs and EPSs lies in their degree of surface attachment. Capsular polysaccharides are firmly anchored to the cell surface and remain in association with the cells through manipulations such as centrifugation and resuspension of cell pellets, while EPSs are generally sloughed off into the culture supernatant or easily washed from the cell surface. The likely anchors for the capsule are covalent bonds to either lipids or proteins embedded in the outer membrane. As previously discussed, the attachment of some group 2 CPSs appears to be via 1,2-diacylglycerol phosphate (Gotschlich *et al.*, 1981; Schmidt and Jann, 1982; Tzeng *et al.*, 2005).

A terminal lipid moiety has not to date been detected for group 1 CPSs. The group 1 CPS cluster of *E. coli* K30 encodes an outer membrane lipoprotein designated Wzi (Rahn *et al.*, 2003). Deletion of *wzi* leads to increased production of cell-free CPSs at the expense of cell-associated CPSs and prevents assembly of a coherent capsular structure on the cell surface (Rahn *et al.*, 2003). While these studies suggest that Wzi may mediate surface attachment of K30 CPSs, this does not appear to be a widely-used mechanism as the only known homologues of this gene are present in *Klebsiella pneumoniae* isolates that assemble group 1 capsules (Rahn *et al.*, 2003).

In contrast, CPSs from Gram-positive bacteria are either covalently linked to the PG layer (De Cueninck *et al.*, 1982; Yeung and Mattingly, 1983a,b; Sorensen *et al.*, 1990; Deng *et al.*, 2000) or are anchored to the membrane via a phosphatidylglycerol moiety (Bender *et al.*, 2003; Cartee *et al.*, 2005).

6. CONCLUSIONS

Significant inroads made in the last decade have improved our understanding of how CPSs and EPSs are synthesized and exported from the cell. The coupling of structural, genetic and biochemical data has provided a more complete picture of how Wzy-dependent polymers are

RESEARCH FOCUS BOX

- What are the mechanisms used by Wzx and Wzy to flip and polymerize repeating units, respectively? What are the flippases specifically recognizing?
- What is/are the precise role(s) of Wzc? Do all roles require reversible phosphorylation of Wzc?
- How is the extent of polymerization determined? What is the termination signal?
- How are Wzy-dependent CPSs anchored to the cell surface in Gram-negative bacteria?
- Are all ABC-dependent CPSs anchored to the membrane through diacylglycerol phosphate?
- How is polymer synthesis initiated in ABC transporter-dependent systems? What is the nature of the primer? What is the initial acceptor? Does Kdo serve as a linker between the lipid and the polymer?
- What is the structure of the ABC-transporter complex?
- What are the roles of KpsC and KpsS?
- Does KpsD form an outer membrane channel for CPS export? If so, how does its structure differ from that of Wza?
- Do KpsD and KpsE interact? Does the PES domain serve as the site of interaction between these proteins?
- How do Wza-Wzc and KpsD-KpsE complexes form across the peptidoglycan layer?
- Why do bacteria express such a diverse assortment of CPS structures and phase variable modifications?

exported. Given the presence of Wza in many ABC transporter-dependent systems, this information may have broader relevance. Despite a wealth of data concerning Wzc and Wzb/CpsB in Wzy-dependent systems, the precise role of Wzc phosphorylation remains poorly understood. Important questions remain concerning the initiation of biosynthesis of ABC transporter-dependent CPSs and the role of various proteins encoded in their loci (see Research Focus Box). Structural data for KpsD may be required to determine whether this protein indeed forms a channel in the outer membrane and whether its structure differs from that of Wza. Future work on these and related glycoconjugate pathways will be crucial to further our understanding of the role of these systems in bacterial survival and exploitation for novel antimicrobial targets.

ACKNOWLEDGEMENTS

The authors would like to thank Evgeny Vinogradov for helpful comments.

References

Abeygunawardana, C., Williams, T.C., Sumner, J.S., Hennessey, J.P., 2000. Development and validation of an NMR-based identity assay for bacterial polysaccharides. Anal. Biochem. 279, 226–240.

Alexander, D.C., Valvano, M.A., 1994. Role of the *rfe* gene in the biosynthesis of the *Escherichia coli* O7-specific lipopolysaccharide and other O-specific polysaccharides containing *N*-acetylglucosamine. J. Bacteriol. 176, 7079–7084.

Amer, A.O., Valvano, M.A., 2000. The N-terminal region of the *Escherichia coli* WecA (Rfe) protein, containing three predicted transmembrane helices, is required for function but not for membrane insertion. J. Bacteriol. 182, 498–503.

Amer, A.O., Valvano, M.A., 2001. Conserved amino acid residues found in a predicted cytosolic domain of the lipopolysaccharide biosynthetic protein WecA are implicated in the recognition of UDP-*N*-acetylglucosamine. Microbiology 147, 3015–3025.

Amer, A.O., Valvano, M.A., 2002. Conserved aspartic acids are essential for the enzymic activity of the WecA protein initiating the biosynthesis of O-specific lipopolysaccharide and enterobacterial common antigen in *Escherichia coli*. Microbiology 148, 571–582.

Amor, P.A., Whitfield, C., 1997. Molecular and functional analysis of genes required for expression of group IB K antigens in *Escherichia coli*: characterization of the *his*-region containing gene clusters for multiple cell-surface polysaccharides. Mol. Microbiol. 26, 145–161.

Anderson, M.S., Eveland, S.S., Price, N.P., 2000. Conserved cytoplasmic motifs that distinguish sub-groups of the polyprenol phosphate: *N*-acetylhexosamine-1-phosphate transferase family. FEMS Microbiol. Lett. 191, 169–175.

Andreishcheva, E.N., Vann, W.F., 2006. Gene products required for *de novo* synthesis of polysialic acid in *Escherichia coli* K1. J. Bacteriol. 188, 1786–1797.

Arrecubieta, C., Hammarton, T.C., Barrett, B., et al., 2001. The transport of group 2 capsular polysaccharides across the periplasmic space in *Escherichia coli*. Roles for the KpsE and KpsD proteins. J. Biol. Chem. 276, 4245–4250.

Aspinall, G.O., McDonald, A.G., Pang, H., 1994. Lipopolysaccharides of *Campylobacter jejuni* serotype O:19: structures of O antigen chains from the serostrain and two bacterial isolates from patients with the Guillain-Barre syndrome. Biochemistry 33, 250–255.

Bacon, D.J., Szymanski, C.M., Burr, D.H., Silver, R.P., Alm, R.A., Guerry, P., 2001. A phase-variable capsule is involved in virulence of *Campylobacter jejuni* 81–176. Mol. Microbiol. 40, 769–777.

Bastin, D.A., Stevenson, G., Brown, P.K., Haase, A., Reeves, P.R., 1993. Repeat unit polysaccharides of bacteria: a model for polymerization resembling that of ribosomes and fatty acid synthetase, with a novel mechanism for determining chain length. Mol. Microbiol. 7, 725–734.

Bayer, M.E., Thurow, H., 1977. Polysaccharide capsule of *Escherichia coli*: microscope study of its size, structure, and sites of synthesis. J. Bacteriol. 130, 911–936.

Becker, A., Puhler, A., 1998. Specific amino acid substitutions in the proline-rich motif of the *Rhizobium meliloti* ExoP protein result in enhanced production of low-molecular-weight succinoglycan at the expense of high-molecular-weight succinoglycan. J. Bacteriol. 180, 395–399.

Becker, A., Niehaus, K., Puhler, A., 1995. Low-molecular-weight succinoglycan is predominantly produced by *Rhizobium meliloti* strains carrying a mutated ExoP protein characterized by a periplasmic N-terminal domain and a missing C-terminal domain. Mol. Microbiol. 16, 191–203.

Beesley, J.E., Orpin, A., Adlam, C., 1984. An evaluation of the conditions necessary for optimal protein A-gold labelling of capsular antigen in ultrathin methacrylate sections of the bacterium *Pasteurella haemolytica*. Histochem. J. 16, 151–163.

Beis, K., Collins, R.F., Ford, R.C., Kamis, A.B., Whitfield, C., Naismith, J.H., 2004. Three-dimensional structure of Wza, the protein required for translocation of group 1 capsular polysaccharide across the outer membrane of *Escherichia coli*. J. Biol. Chem. 279, 28227–28232.

Bender, M.H., Yother, J., 2001. CpsB is a modulator of capsule-associated tyrosine kinase activity in *Streptococcus pneumoniae*. J. Biol. Chem. 276, 47966–47974.

Bender, M.H., Cartee, R.T., Yother, J., 2003. Positive correlation between tyrosine phosphorylation of CpsD and capsular polysaccharide production in *Streptococcus pneumoniae*. J. Bacteriol. 185, 6057–6066.

Bengoechea, J.A., Pinta, E., Salminen, T., et al., 2002. Functional characterization of Gne (UDP-*N*-acetylglucosamine-4-epimerase), Wzz (chain length determinant), and Wzy (O-antigen polymerase) of *Yersinia enterocolitica* serotype O:8. J. Bacteriol. 184, 4277–4287.

Bentley, S.D., Aanensen, D.M., Mavroidi, A., et al., 2006. Genetic analysis of the capsular biosynthetic locus from all 90 pneumococcal serotypes. PLoS Genet. 2, e31.

Bliss, J.M., Silver, R.P., 1996. Coating the surface: a model for expression of capsular polysialic acid in *Escherichia coli* K1. Mol. Microbiol. 21, 221–231.

Bronner, D., Sieberth, V., Pazzani, C., et al., 1993a. Expression of the capsular K5 polysaccharide of *Escherichia coli*: biochemical and electron microscopic analyses of mutants with defects in region 1 of the K5 gene cluster. J. Bacteriol. 175, 5984–5992.

Bronner, D., Sieberth, V., Pazzani, C., et al., 1993b. Synthesis of the K5 (group II) capsular polysaccharide in transport-deficient recombinant *Escherichia coli*. FEMS Microbiol. Lett. 113, 279–284.

Bugert, P., Geider, K., 1995. Molecular analysis of the *ams* operon required for exopolysaccharide synthesis of *Erwinia amylovora*. Mol. Microbiol. 15, 917–933.

Burman, L.G., Raichler, J., Park, J.T., 1983. Evidence for diffuse growth of the cylindrical portion of the *Escherichia coli* murein sacculus. J. Bacteriol. 155, 983–988.

Cartee, R.T., Forsee, W.T., Bender, M.H., Ambrose, K.D., Yother, J., 2005. CpsE from type 2 *Streptococcus pneumoniae* catalyzes the reversible addition of glucose-1-phosphate to a polyprenyl phosphate acceptor, initiating type 2 capsule repeat unit formation. J. Bacteriol. 187, 7425–7433.

Carter, J.A., Blondel, C.J., Zaldivar, M., et al., 2007. O-antigen modal chain length in *Shigella flexneri* 2a is growth-regulated through RfaH-mediated transcriptional control of the *wzy* gene. Microbiology 153, 3499–3507.

Chakraborty, A.K., Friebolin, H., Stirm, S., 1980. Primary structure of the *Escherichia coli* serotype K30 capsular polysaccharide. J. Bacteriol. 141, 971–972.

Chao, C.F., Chuang, H.C., Chiou, S.T., Liu, T.Y., 1999. On the biosynthesis of alternating alpha-2,9/alpha-2,8 heteropolymer of sialic acid catalyzed by the sialyltransferase of *Escherichia coli* Bos-12. J. Biol. Chem. 274, 18206–18212.

Chen, Y.H., Poly, F., Pakulski, Z., Guerry, P., Monteiro, M.A., 2008. The chemical structure and genetic locus of *Campylobacter jejuni* CG8486 (serotype HS:4) capsular polysaccharide: the identification of 6-deoxy-D-*ido*-heptopyranose. Carbohydr. Res. 343, 1034–1040.

Cheng, J., Liu, B., Bastin, D.A., Han, W., Wang, L., Feng, L., 2007. Genetic characterization of the *Escherichia coli* O66 antigen and functional identification of its *wzy* gene. J. Microbiol. 45, 69–74.

Cho, J.W., Troy, F.A., 1994. Polysialic acid engineering: synthesis of polysialylated neoglycosphingolipids by using the polysialyltransferase from neuroinvasive *Escherichia coli* K1. Proc. Natl. Acad. Sci. USA 91, 11427–11431.

Chouly, C., Colquhoun, I.J., Jodelet, A., York, G., Walker, G.C., 1995. NMR studies of succinoglycan repeating-unit octasaccharides from *Rhizobium meliloti* and *Agrobacterium radiobacter*. Int. J. Biol. Macromol. 17, 357–363.

Christie, P.J., 2004. Type IV secretion: the *Agrobacterium* VirB/D4 and related conjugation systems. Biochim. Biophys. Acta 1694, 219–234.

Cieslewicz, M., Vimr, E., 1996. Thermoregulation of *kpsF*, the first region 1 gene in the *kps* locus for polysialic acid biosynthesis in *Escherichia coli* K1. J. Bacteriol. 178, 3212–3220.

Cieslewicz, M., Vimr, E., 1997. Reduced polysialic acid capsule expression in *Escherichia coli* K1 mutants with chromosomal defects in *kpsF*. Mol. Microbiol. 26, 237–249.

Cieslewicz, M.J., Kasper, D.L., Wang, Y., Wessels, M.R., 2001. Functional analysis in type Ia group B *Streptococcus* of a cluster of genes involved in extracellular polysaccharide production by diverse species of streptococci. J. Biol. Chem. 276, 139–146.

Clarke, B.R., Cuthbertson, L., Whitfield, C., 2004. Nonreducing terminal modifications determine the chain length of polymannose O antigens of *Escherichia coli* and couple chain termination to polymer export via an ATP-binding cassette transporter. J. Biol. Chem. 279, 35709–35718.

Collins, R.F., Beis, K., Clarke, B.R., et al., 2006. Periplasmic protein-protein contacts in the inner membrane protein Wzc form a tetrameric complex required for the assembly of *Escherichia coli* group 1 capsules. J. Biol. Chem. 281, 2144–2150.

Collins, R.F., Beis, K., Dong, C., et al., 2007. The 3D structure of a periplasm-spanning platform required for assembly of group 1 capsular polysaccharides in *Escherichia coli*. Proc. Natl. Acad. Sci. USA 104, 2390–2395.

Comstock, L.E., Kasper, D.L., 2006. Bacterial glycans: key mediators of diverse host immune responses. Cell 126, 847–850.

Cuthbertson, L., Powers, J., Whitfield, C., 2005. The C-terminal domain of the nucleotide-binding domain protein Wzt determines substrate specificity in the ATP-binding cassette transporter for the lipopolysaccharide O-antigens in *Escherichia coli* serotypes O8 and O9a. J. Biol. Chem. 280, 30310–30319.

Cuthbertson, L., Kimber, M.S., Whitfield, C., 2007. Substrate binding by a bacterial ABC transporter involved in polysaccharide export. Proc. Natl. Acad. Sci. USA 104, 19529–19534.

Dams-Kozlowska, H., Kaplan, D.L., 2007. Protein engineering of Wzc to generate new emulsan analogs. Appl. Environ. Microbiol. 73, 4020–4028.

Dams-Kozlowska, H., Sainath, N., Kaplan, D.L., 2008. Construction of a chimeric gene cluster for the biosynthesis of apoemulsan with altered molecular weight. Appl. Microbiol. Biotechnol. 78, 677–683.

Daniels, C., Morona, R., 1999. Analysis of *Shigella flexneri* *wzz* (Rol) function by mutagenesis and cross-linking: Wzz is able to oligomerize. Mol. Microbiol. 34, 181–194.

Daniels, C., Vindurampulle, C., Morona, R., 1998. Overexpression and topology of the *Shigella flexneri* O-antigen polymerase (Rfc/Wzy). Mol. Microbiol. 28, 1211–1222.

Daniels, C., Griffiths, C., Cowles, B., Lam, J.S., 2002. *Pseudomonas aeruginosa* O-antigen chain length is determined before ligation to lipid A core. Env. Microbiol. 4, 883–897.

De Cueninck, B.J., Shockman, G.D., Swenson, R.M., 1982. Group B, type III streptococcal cell wall: composition and structural aspects revealed through endo-*N*-acetylmuramidase-catalyzed hydrolysis. Infect. Immun. 35, 572–581.

de Pedro, M.A., Quintela, J.C., Holtje, J.V., Schwarz, H., 1997. Murein segregation in *Escherichia coli*. J. Bacteriol. 179, 2823–2834.

de Pedro, M.A., Donachie, W.D., Holtje, J.V., Schwarz, H., 2001. Constitutive septal murein synthesis in *Escherichia coli* with impaired activity of the morphogenetic proteins RodA and penicillin-binding protein 2. J. Bacteriol. 183, 4115–4126.

Deng, L., Kasper, D.L., Krick, T.P., Wessels, M.R., 2000. Characterization of the linkage between the type III capsular polysaccharide and the bacterial cell wall of group B *Streptococcus*. J. Biol. Chem. 275, 7497–7504.

Dengler, T., Jann, B., Jann, K., 1986. Structure of the serine-containing capsular polysaccharide K40 antigen from *Escherichia coli* O8:K40:H9. Carbohydr. Res. 150, 233–240.

Dong, C., Beis, K., Nesper, J., et al., 2006. Wza the translocon for *E. coli* capsular polysaccharides defines a new class of membrane protein. Nature 444, 226–229.

Doublet, P., Grangeasse, C., Obadia, B., Vaganay, E., Cozzone, A.J., 2002. Structural organization of the protein-tyrosine autokinase Wzc within *Escherichia coli* cells. J. Biol. Chem. 277, 37339–37348.

Drummelsmith, J., Whitfield, C., 1999. Gene products required for surface expression of the capsular form of the group 1 K antigen in *Escherichia coli* (O9a:K30). Mol. Microbiol. 31, 1321–1332.

Drummelsmith, J., Whitfield, C., 2000. Translocation of group 1 capsular polysaccharide to the surface of *Escherichia coli* requires a multimeric complex in the outer membrane. EMBO J. 19, 57–66.

Feldman, M.F., Marolda, C.L., Monteiro, M.A., Perry, M.B., Parodi, A.J., Valvano, M.A., 1999. The activity of a putative polyisoprenol-linked sugar translocase (Wzx) involved in *Escherichia coli* O antigen assembly is independent of the chemical structure of the O repeat. J. Biol. Chem. 274, 35129–35138.

Ferrero, M.A., Luengo, J.M., Reglero, A., 1991. H.p.l.c. of oligo(sialic acids). Application to the determination of the minimal chain length serving as exogenous acceptor in the enzymic synthesis of colominic acid. Biochem. J. 280, 575–579.

Finke, A., Roberts, I., Boulnois, G., Pzzani, C., Jann, K., 1989. Activity of CMP-2-keto-3-deoxyoctulosonic acid synthetase in *Escherichia coli* strains expressing the capsular K5 polysaccharide implication for K5 polysaccharide biosynthesis. J. Bacteriol. 171, 3074–3079.

Finke, A., Bronner, D., Nikolaev, A.V., Jann, B., Jann, K., 1991. Biosynthesis of the *Escherichia coli* K5 polysaccharide, a representative of group II capsular polysaccharides: polymerization *in vitro* and characterization of the product. J. Bacteriol. 173, 4088–4094.

Freiberger, F., Claus, H., Gunzel, A., et al., 2007. Biochemical characterization of a *Neisseria meningitidis* polysialyltransferase reveals novel functional motifs in bacterial sialyltransferases. Mol. Microbiol. 65, 1258–1275.

Frosch, M., Muller, A., 1993. Phospholipid substitution of capsular polysaccharides and mechanisms of capsule formation in *Neisseria meningitidis*. Mol. Microbiol. 8, 483–493.

Frosch, M., Muller, D., Bousset, K., Muller, A., 1992. Conserved outer membrane protein of *Neisseria meningitidis* involved in capsule expression. Infect. Immun. 60, 798–803.

Garegg, P.J., Lindberg, B., Onn, T., Sutherland, I.W., 1971. Comparative structural studies on the M-antigen from *Salmonella typhimurium*, *Escherichia* coli and *Aerobacter* cloacae. Acta Chem. Scand. 25, 2103–2108.

Gilbert, M., Mandrell, R.E., Parker, C.T., Li, J., Vinogradov, E., 2007. Structural analysis of the capsular polysaccharide from *Campylobacter jejuni* RM1221. Chembiochem. 8, 625–631.

Gotschlich, E.C., Fraser, B.A., Nishimura, O., Robbins, J.B., Liu, T.Y., 1981. Lipid on capsular polysaccharides of gram-negative bacteria. J. Biol. Chem. 256, 8915–8921.

Grangeasse, C., Doublet, P., Vincent, C., et al., 1998. Functional characterization of the low-molecular-mass phosphotyrosine-protein phosphatase of *Acinetobacter johnsonii*. J. Mol. Biol. 278, 339–347.

Grangeasse, C., Doublet, P., Cozzone, A.J., 2002. Tyrosine phosphorylation of protein kinase Wzc from *Escherichia coli* K12 occurs through a two-step process. J. Biol. Chem. 277, 7127–7135.

Grangeasse, C., Obadia, B., Mijakovic, I., Deutscher, J., Cozzone, A.J., Doublet, P., 2003. Autophosphorylation of the *Escherichia coli* protein kinase Wzc regulates tyrosine phosphorylation of Ugd, a UDP-glucose dehydrogenase. J. Biol. Chem. 278, 39323–39329.

Guidolin, A., Morona, J.K., Morona, R., Hansman, D., Paton, J.C., 1994. Nucleotide sequence analysis of genes essential for capsular polysaccharide biosynthesis in *Streptococcus pneumoniae* type 19F. Infect. Immun. 62, 5384–5396.

Guttormsen, H.K., Paoletti, L.C., Mansfield, K.G., Jachymek, W., Jennings, H.J., Kasper, D.L., 2008. Rational chemical design of the carbohydrate in a glycoconjugate vaccine enhances IgM-to-IgG switching. Proc. Natl. Acad. Sci. USA 105, 5903–5908.

He, S.Y., Nomura, K., Whittam, T.S., 2004. Type III protein secretion mechanism in mammalian and plant pathogens. Biochim. Biophys. Acta 1694, 181–206.

Hofmann, P., Jann, B., Jann, K., 1985. Structure of the amino acid-containing capsular polysaccharide (K54 antigen) from *Escherichia coli* O6:K54:H10. Carbohydr. Res. 139, 261–271.

Janakiraman, A., Goldberg, M.B., 2004. Recent advances on the development of bacterial poles. Trends Microbiol. 12, 518–525.

Jennings, H.J., Beurret, M., Gamian, A., Michon, F., 1987. Structure and immunochemistry of meningococcal lipopolysaccharides. Antonie van Leeuwenhoek 53, 519–522.

Jones, C., 2005a. NMR assays for carbohydrate-based vaccines. J. Pharm. Biomed. Anal. 38, 840–850.

Jones, C., 2005b. Revised structures for the capsular polysaccharides from *Staphylococcus aureus* Types 5 and 8, components of novel glycoconjugate vaccines. Carbohydr. Res. 340, 1097–1106.

Kanipes, M.I., Tan, X., Akelaitis, A., et al., 2008. Genetic analysis of lipooligosaccharide core biosynthesis in *Campylobacter jejuni* 81–176. J. Bacteriol. 190, 1568–1574.

Karlyshev, A.V., Linton, D., Gregson, N.A., Lastovica, A.J., Wren, B.W., 2000. Genetic and biochemical evidence of a *Campylobacter jejuni* capsular polysaccharide that accounts for Penner serotype specificity. Mol. Microbiol. 35, 529–541.

Karlyshev, A.V., Champion, O.L., Churcher, C., et al., 2005. Analysis of *Campylobacter jejuni* capsular loci reveals multiple mechanisms for the generation of structural diversity and the ability to form complex heptoses. Mol. Microbiol. 55, 90–103.

Kellenberger, E., 1990. The 'Bayer bridges' confronted with results from improved electron microscopy methods. Mol. Microbiol. 4, 697–705.

Klein, G., Dartigalongue, C., Raina, S., 2003. Phosphorylation-mediated regulation of heat shock response in *Escherichia coli*. Mol. Microbiol. 48, 269–285.

Koraimann, G., 2003. Lytic transglycosylases in macromolecular transport systems of Gram-negative bacteria. Cell. Mol. Life Sci. 60, 2371–2388.

Kroll, J.S., Loynds, B., Brophy, L.N., Moxon, E.R., 1990. The *bex* locus in encapsulated *Haemophilus influenzae*: a chromosomal region involved in capsule polysaccharide export. Mol. Microbiol. 4, 1853–1862.

Kroncke, K.D., Boulnois, G., Roberts, I., et al., 1990. Expression of the *Escherichia coli* K5 capsular antigen: immunoelectron microscopic and biochemical studies with recombinant *E. coli*. J. Bacteriol. 172, 1085–1091.

Lehrer, J., Vigeant, K.A., Tatar, L.D., Valvano, M.A., 2007. Functional characterization and membrane topology of *Escherichia coli* WecA, a sugar-phosphate transferase initiating the biosynthesis of enterobacterial common antigen and O-antigen lipopolysaccharide. J. Bacteriol. 189, 2618–2628.

Liu, D., Cole, R.A., Reeves, P.R., 1996. An O-antigen processing function for Wzx (RfbX): a promising candidate for O-unit flippase. J. Bacteriol. 178, 2102–2107.

Liu, T.Y., Gotschlich, E.C., Dunne, F.T., Jonssen, E.K., 1971. Studies on the meningococcal polysaccharides. II. Composition and chemical properties of the group B and group C polysaccharide. J. Biol. Chem. 246, 4703–4712.

Lo, R.Y., McKerral, L.J., Hills, T.L., Kostrzynska, M., 2001. Analysis of the capsule biosynthetic locus of *Mannheimia (Pasteurella) haemolytica* A1 and proposal of a nomenclature system. Infect. Immun. 69, 4458–4464.

Marolda, C.L., Vicarioli, J., Valvano, M.A., 2004. Wzx proteins involved in biosynthesis of O antigen function in association with the first sugar of the O-specific lipopolysaccharide subunit. Microbiology 150, 4095–4105.

Marolda, C.L., Tatar, L.D., Alaimo, C., Aebi, M., Valvano, M.A., 2006. Interplay of the Wzx translocase and the corresponding polymerase and chain length regulator proteins in the translocation and periplasmic assembly of lipopolysaccharide o antigen. J. Bacteriol. 188, 5124–5135.

McGrath, B.C., Osborn, M.J., 1991. Localization of the terminal steps of O-antigen synthesis in *Salmonella typhimurium*. J. Bacteriol. 173, 649–654.

McGuire, E.J., Binkley, S.B., 1964. The structure and chemistry of colominic acid. Biochemistry 3, 247–251.

McNally, D.J., Jarrell, H.C., Li, J., et al., 2005. The HS:1 serostrain of *Campylobacter jejuni* has a complex teichoic acid-like capsular polysaccharide with nonstoichiometric fructofuranose branches and *O*-methyl phosphoramidate groups. FEBS J. 272, 4407–4422.

McNally, D.J., Jarrell, H.C., Khieu, N.H., et al., 2006. The HS:19 serostrain of *Campylobacter jejuni* has a hyaluronic acid-type capsular polysaccharide with a nonstoichiometric sorbose branch and *O*-methyl phosphoramidate group. FEBS J. 273, 3975–3989.

McNally, D.J., Lamoureux, M.P., Karlyshev, A.V., et al., 2007. Commonality and biosynthesis of the *O*-methyl phosphoramidate capsule modification in *Campylobacter jejuni*. J. Biol. Chem. 282, 28566–28576.

McNulty, C., Thompson, J., Barrett, B., Lord, L., Andersen, C., Roberts, I.S., 2006. The cell surface expression of group 2 capsular polysaccharides in *Escherichia coli*: the role of KpsD, RhsA and a multi-protein complex at the pole of the cell. Mol. Microbiol. 59, 907–922.

Meier-Dieter, U., Starman, R., Barr, K., Mayer, H., Rick, P.D., 1990. Biosynthesis of enterobacterial common antigen in *Escherichia coli*. Biochemical characterization of Tn10 insertion mutants defective in enterobacterial common antigen synthesis. J. Biol. Chem. 265, 13490–13497.

Meier-Dieter, U., Barr, K., Starman, R., Hatch, L., Rick, P.D., 1992. Nucleotide sequence of the *Escherichia coli rfe* gene involved in the synthesis of enterobacterial common antigen. Molecular cloning of the *rfe-rff* gene cluster. J. Biol. Chem. 267, 746–753.

Mijakovic, I., Poncet, S., Boel, G., et al., 2003. Transmembrane modulator-dependent bacterial tyrosine kinase activates UDP-glucose dehydrogenases. EMBO J. 22, 4709–4718.

Minic, Z., Marie, C., Delorme, C., et al., 2007. Control of EpsE, the phosphoglycosyltransferase initiating exopolysaccharide synthesis in *Streptococcus thermophilus*, by EpsD tyrosine kinase. J. Bacteriol. 189, 1351–1357.

Moran, A.P., Penner, J.L., Aspinall, G.O., 2000. *Campylobacter* lipopolysaccharides. In: Nachamkin, I., Blaser, M.J. (Eds.,) *Campylobacter*, 2nd edn. American Society for Microbiology, Washington, DC, pp. 241–257.

Morona, R., van den Bosch, L., Manning, P.A., 1995. Molecular, genetic, and topological characterization of O-antigen chain length regulation in *Shigella flexneri*. J. Bacteriol. 177, 1059–1068.

Morona, J.K., Paton, J.C., Miller, D.C., Morona, R., 2000a. Tyrosine phosphorylation of CpsD negatively regulates capsular polysaccharide biosynthesis in *Streptococcus pneumoniae*. Mol. Microbiol. 35, 1431–1442.

Morona, R., Van Den Bosch, L., Daniels, C., 2000b. Evaluation of Wzz/MPA1/MPA2 proteins based on the presence of coiled-coil regions. Microbiology 146, 1–4.

Morona, J.K., Morona, R., Miller, D.C., Paton, J.C., 2002. *Streptococcus pneumoniae* capsule biosynthesis protein CpsB is a novel manganese-dependent phosphotyrosine-protein phosphatase. J. Bacteriol. 184, 577–583.

Morona, J.K., Morona, R., Miller, D.C., Paton, J.C., 2003. Mutational analysis of the carboxy-terminal (YGX)4 repeat domain of CpsD, an autophosphorylating tyrosine kinase required for capsule biosynthesis in *Streptococcus pneumoniae*. J. Bacteriol. 185, 3009–3019.

Morona, J.K., Miller, D.C., Morona, R., Paton, J.C., 2004. The effect that mutations in the conserved capsular polysaccharide biosynthesis genes *cpsA*, *cpsB*, and *cpsD* have on virulence of *Streptococcus pneumoniae*. J. Infect. Dis. 189, 1905–1913.

Morona, J.K., Morona, R., Paton, J.C., 2006. Attachment of capsular polysaccharide to the cell wall of *Streptococcus pneumoniae* type 2 is required for invasive disease. Proc. Natl. Acad. Sci. USA 103, 8505–8510.

Mulford, C.A., Osborn, M.J., 1983. An intermediate step in translocation of lipopolysaccharide to the outer membrane of *Salmonella typhimurium*. Proc. Natl. Acad. Sci. USA 80, 1159–1163.

Murray, G.L., Attridge, S.R., Morona, R., 2003. Regulation of *Salmonella typhimurium* lipopolysaccharide O antigen chain length is required for virulence; identification of FepE as a second Wzz. Mol. Microbiol. 47, 1395–1406.

Nakar, D., Gutnick, D.L., 2003. Involvement of a protein tyrosine kinase in production of the polymeric bioemulsifier emulsan from the oil-degrading strain *Acinetobacter lwoffii* RAG-1. J. Bacteriol. 185, 1001–1009.

Nakhamchik, A., Wilde, C., Rowe-Magnus, D.A., 2007. Identification of a Wzy polymerase required for group IV capsular polysaccharide and lipopolysaccharide biosynthesis in *Vibrio vulnificus*. Infect. Immun. 75, 5550–5558.

Nesper, J., Hill, C.M., Paiment, A., et al., 2003. Translocation of group 1 capsular polysaccharide in *Escherichia coli* serotype K30. Structural and functional analysis of the outer membrane lipoprotein Wza. J. Biol. Chem. 278, 49763–49772.

Niemeyer, D., Becker, A., 2001. The molecular weight distribution of succinoglycan produced by *Sinorhizobium meliloti* is influenced by specific tyrosine phosphorylation and ATPase activity of the cytoplasmic domain of the ExoP protein. J. Bacteriol. 183, 5163–5170.

Obadia, B., Lacour, S., Doublet, P., Baubichon-Cortay, H., Cozzone, A.J., Grangeasse, C., 2007. Influence of tyrosine-kinase Wzc activity on colanic acid production in *Escherichia coli* K12 cells. J. Mol. Biol. 367, 42–53.

Orskov, F., Orskov, I., 1992. *Escherichia coli* serotyping and disease in man and animals. Can. J. Microbiol. 38, 699–704.

Paiment, A., Hocking, J., Whitfield, C., 2002. Impact of phosphorylation of specific residues in the tyrosine autokinase, Wzc, on its activity in assembly of group 1 capsules in *Escherichia coli*. J. Bacteriol. 184, 6437–6447.

Paulsen, I.T., Beness, A.M., Saier, M.H., 1997. Computer-based analyses of the protein constituents of transport systems catalysing export of complex carbohydrates in bacteria. Microbiology 143, 2685–2699.

Pavelka, M.S., Hayes, S.F., Silver, R.P., 1994. Characterization of KpsT, the ATP-binding component of the ABC-transporter involved with the export of capsular polysialic acid in *Escherichia coli* K1. J. Biol. Chem. 269, 20149–20158.

Pazzani, C., Rosenow, C., Boulnois, G.J., Bronner, D., Jann, K., Roberts, I.S., 1993. Molecular analysis of region 1 of the *Escherichia coli* K5 antigen gene cluster: a region encoding proteins involved in cell surface expression of capsular polysaccharide. J. Bacteriol. 175, 5978–5983.

Pearce, R., Roberts, I.S., 1995. Cloning and analysis of gene clusters for production of the *Escherichia coli* K10 and K54 antigens: identification of a new group of *serA*-linked capsule gene clusters. J. Bacteriol. 177, 3992–3997.

Phoenix, D.A., Brandenburg, K., Harris, F., Seydel, U., Hammerton, T., Roberts, I.S., 2001. An investigation into the membrane-interactive potential of the *Escherichia coli* KpsE C-terminus. Biochem. Biophys. Res. Comm. 285, 976–980.

Pigeon, R.P., Silver, R.P., 1994. Topological and mutational analysis of KpsM, the hydrophobic component of the ABC-transporter involved in the export of polysialic acid in *Escherichia coli* K1. Mol. Microbiol. 14, 871–881.

Preneta, R., Jarraud, S., Vincent, C., et al., 2002. Isolation and characterization of a protein-tyrosine kinase and a phosphotyrosine-protein phosphatase from *Klebsiella pneumoniae*. Comp. Biochem. Physiol. B, Biochem. Mol. Biol. 131, 103–112.

Rahn, A., Beis, K., Naismith, J.H., Whitfield, C., 2003. A novel outer membrane protein, Wzi, is involved in surface assembly of the *Escherichia coli* K30 group 1 capsule. J. Bacteriol. 185, 5882–5890.

Reid, A.N., Whitfield, C., 2005. Functional analysis of conserved gene products involved in assembly of *Escherichia coli* capsules and exopolysaccharides: evidence for molecular recognition between Wza and Wzc for colanic acid biosynthesis. J. Bacteriol. 187, 5470–5481.

Reinhold, B.B., Chan, S.Y., Reuber, T.L., Marra, A., Walker, G.C., Reinhold, V.N., 1994. Detailed structural characterization of succinoglycan, the major exopolysaccharide of *Rhizobium meliloti* Rm1021. J. Bacteriol. 176, 1997–2002.

Reyes, C.L., Chang, G., 2005. Structure of the ABC transporter MsbA in complex with ADP.vanadate and lipopolysaccharide. Science 308, 1028–1031.

Rick, P.D., Hubbard, G.L., Barr, K., 1994. Role of the *rfe* gene in the synthesis of the O8 antigen in *Escherichia coli* K-12. J. Bacteriol. 176, 2877–2884.

Rick, P.D., Barr, K., Sankaran, K., Kajimura, J., Rush, J.S., Waechter, C.J., 2003. Evidence that the *wzxE* gene of *Escherichia coli* K-12 encodes a protein involved in the transbilayer movement of a trisaccharide-lipid intermediate in the assembly of enterobacterial common antigen. J. Biol. Chem. 278, 16534–16542.

Rigg, G.P., Barrett, B., Roberts, I.S., 1998. The localization of KpsC, S and T, and KfiA, C and D proteins involved in the biosynthesis of the *Escherichia coli* K5 capsular polysaccharide: evidence for a membrane-bound complex. Microbiology 144, 2905–2914.

Roberts, I., Mountford, R., High, N., et al., 1986. Molecular cloning and analysis of genes for production of K5, K7, K12, and K92 capsular polysaccharides in *Escherichia coli*. J. Bacteriol. 168, 1228–1233.

Roberts, I.S., 1996. The biochemistry and genetics of capsular polysaccharide production in bacteria. Ann. Rev. Microbiol. 50, 285–315.

Roberts, I.S., Mountford, R., Hodge, R., Jann, K.B., Boulnois, G.J., 1988. Common organization of gene clusters for production of different capsular polysaccharides (K antigens) in *Escherichia coli*. J. Bacteriol. 170, 1305–1310.

Rosenow, C., Roberts, I.S., Jann, K., 1995. Isolation from recombinant *Escherichia coli* and characterization of CMP-Kdo synthetase, involved in the expression of the capsular K5 polysaccharide (K-CKS). FEMS Microbiol. Lett. 125, 159–164.

Rosner, H., Grimmecke, H.D., Knirel, Y.A., Shashkov, A.S., 1992. Hyaluronic acid and a (1→4)-beta-D-xylan, extracellular polysaccharides of *Pasteurella multocida* (Carter type A) strain 880. Carbohydr. Res. 223, 329–333.

Rush, J.S., Rick, P.D., Waechter, C.J., 1997. Polyisoprenyl phosphate specificity of UDP-GlcNAc:undecaprenyl phosphate *N*-acetylglucosaminyl 1-P transferase from *E. coli*. Glycobiology 7, 315–322.

Saldias, M.S., Patel, K., Marolda, C.L., Bittner, M., Contreras, I., Valvano, M.A., 2008. Distinct functional domains of the *Salmonella enterica* WbaP transferase that is involved in the initiation reaction for synthesis of the O antigen subunit. Microbiology 154, 440–453.

Sau, S., Bhasin, N., Wann, E.R., Lee, J.C., Foster, T.J., Lee, C.Y., 1997. The *Staphylococcus aureus* allelic genetic loci for serotype 5 and 8 capsule expression contain the type-specific genes flanked by common genes. Microbiology 143, 2395–2405.

Schild, S., Lamprecht, A.K., Reidl, J., 2005. Molecular and functional characterization of O antigen transfer in *Vibrio cholerae*. J. Biol. Chem. 280, 25936–25947.

Schmidt, M.A., Jann, K., 1982. Phospholipid substitution of capsular (K) polysaccharide antigens from *Escherichia coli* causing extraintestinal infections. FEMS Microbiol. Lett. 14, 69–74.

Sieberth, V., Jann, B., Jann, K., 1993. Structure of the K10 capsular antigen from *Escherichia coli* O11:K10:H10, a polysaccharide containing 4,6-dideoxy-4-malonylamino-D-glucose. Carbohydr. Res. 246, 219–228.

Silver, R.P., Aaronson, W., Vann, W.F., 1987. Translocation of capsular polysaccharides in pathogenic strains of *Escherichia coli* requires a 60-kilodalton periplasmic protein. J. Bacteriol. 169, 5489–5495.

Sorensen, U.B., Henrichsen, J., Chen, H.C., Szu, S.C., 1990. Covalent linkage between the capsular polysaccharide and the cell wall peptidoglycan of *Streptococcus pneumoniae* revealed by immunochemical methods. Microb. Pathog. 8, 325–334.

Soulat, D., Vaganay, E., Duclos, B., Genestier, A.L., Etienne, J., Cozzone, A.J., 2002. *Staphylococcus aureus* contains two low-molecular-mass phosphotyrosine protein phosphatases. J. Bacteriol. 184, 5194–5199.

Soulat, D., Jault, J.M., Duclos, B., Geourjon, C., Cozzone, A. J., Grangeasse, C., 2006. *Staphylococcus aureus* operates protein-tyrosine phosphorylation through a specific mechanism. J. Biol. Chem. 281, 14048–14056.

St Michael, F., Szymanski, C.M., Li, J., et al., 2002. The structures of the lipooligosaccharide and capsule polysaccharide of *Campylobacter jejuni* genome sequenced strain NCTC 11168. Eur. J. Biochem. 269, 5119–5136.

Steenbergen, S.M., Vimr, E.R., 1990. Mechanism of polysialic acid chain elongation in *Escherichia coli* K1. Mol. Microbiol. 4, 603–611.

Steenbergen, S.M., Vimr, E.R., 2003. Functional relationships of the sialyltransferases involved in expression of the polysialic acid capsules of *Escherichia coli* K1 and K92 and *Neisseria meningitidis* groups B or C. J. Biol. Chem. 278, 15349–15359.

Steenbergen, S.M., Vimr, E.R., 2008. Biosynthesis of the *Escherichia coli* K1 group 2 polysialic acid capsule occurs within a protected cytoplasmic compartment. Mol. Microbiol. 68, 1252–1267.

Steenbergen, S.M., Wrona, T.J., Vimr, E.R., 1992. Functional analysis of the sialyltransferase complexes in *Escherichia coli* K1 and K92. J. Bacteriol. 174, 1099–1108.

Stenberg, F., Chovanec, P., Maslen, S.L., et al., 2005. Protein complexes of the *Escherichia coli* cell envelope. J. Biol. Chem. 280, 34409–34419.

Tao, J., Feng, L., Guo, H., Li, Y., Wang, L., 2004. The O-antigen gene cluster of *Shigella boydii* O11 and functional identification of its *wzy* gene. FEMS Microbiol. Lett. 234, 125–132.

Tocilj, A., Munger, C., Proteau, A., et al., 2008. Bacterial polysaccharide co-polymerases share a common framework for control of polymer length. Nat. Struct. Mol. Biol. 15, 130–138.

Tzeng, Y.L., Datta, A.K., Strole, C.A., Lobritz, M.A., Carlson, R.W., Stephens, D.S., 2005. Translocation and surface expression of lipidated serogroup B capsular polysaccharide in *Neisseria meningitidis*. Infect. Immun. 73, 1491–1505.

Vann, W.F., Schmidt, M.A., Jann, B., Jann, K., 1981. The structure of the capsular polysaccharide (K5 antigen) of urinary-tract-infective *Escherichia coli* 010:K5:H4. A polymer similar to desulfo-heparin. Eur. J. Biochem. 116, 359–364.

Vimr, E.R., Aaronson, W., Silver, R.P., 1989. Genetic analysis of chromosomal mutations in the polysialic acid gene cluster of *Escherichia coli* K1. J. Bacteriol. 171, 1106–1117.

Vincent, C., Doublet, P., Grangeasse, C., Vaganay, E., Cozzone, A.J., Duclos, B., 1999. Cells of *Escherichia coli* contain a protein-tyrosine kinase, Wzc, and a phosphotyrosine-protein phosphatase, Wzb. J. Bacteriol. 181, 3472–3477.

Vincent, C., Duclos, B., Grangeasse, C., et al., 2000. Relationship between exopolysaccharide production and protein-tyrosine phosphorylation in gram-negative bacteria. J. Mol. Biol. 304, 311–321.

Vionnet, J., Vann, W.F., 2007. Successive glycosyltransfer of sialic acid by *Escherichia coli* K92 polysialyltransferase in elongation of oligosialic acceptors. Glycobiology 17, 735–743.

Wang, L., Reeves, P.R., 1998. Organization of *Escherichia coli* O157 O antigen gene cluster and identification of its specific genes. Infect. Immun. 66, 3545–3551.

Wang, L., Liu, D., Reeves, P.R., 1996. C-terminal half of *Salmonella enterica* WbaP (RfbP) is the galactosyl-1-phosphate transferase domain catalyzing the first step of O-antigen synthesis. J. Bacteriol. 178, 2598–2604.

Ward, C.K., Inzana, T.J., 1997. Identification and characterization of a DNA region involved in the export of capsular polysaccharide by *Actinobacillus pleuropneumoniae* serotype 5a. Infect. Immun. 65, 2491–2496.

Whitfield, C., 2006. Biosynthesis and assembly of capsular polysaccharides in *Escherichia coli*. Annu. Rev. Biochem. 75, 39–68.

Whitfield, C., Roberts, I.S., 1999. Structure, assembly and regulation of expression of capsules in *Escherichia coli*. Mol. Microbiol. 31, 1307–1319.

Whitfield, C., Amor, P.A., Koplin, R., 1997. Modulation of the surface architecture of gram-negative bacteria by the action of surface polymer:lipid A-core ligase and by determinants of polymer chain length. Mol. Microbiol. 23, 629–638.

Willis, L.M., Gilbert, M., Karwaski, M.F., Blanchard, M.C., Wakarchuk, W.W., 2008. Characterization of the alpha-2,8-polysialyltransferase from *Neisseria meningitidis* with synthetic acceptors, and the development of a self-priming polysialyltransferase fusion enzyme. Glycobiology 18, 177–186.

Wugeditsch, T., Paiment, A., Hocking, J., Drummelsmith, J., Forrester, C., Whitfield, C., 2001. Phosphorylation of Wzc, a tyrosine autokinase, is essential for assembly of group 1 capsular polysaccharides in *Escherichia coli*. J. Biol. Chem. 276, 2361–2371.

Wunder, D.E., Aaronson, W., Hayes, S.F., Bliss, J.M., Silver, R.P., 1994. Nucleotide sequence and mutational analysis of the gene encoding KpsD, a periplasmic protein involved in transport of polysialic acid in *Escherichia coli* K1. J. Bacteriol. 176, 4025–4033.

Xayarath, B., Yother, J., 2007. Mutations blocking side chain assembly, polymerization, or transport of a Wzy-dependent *Streptococcus pneumoniae* capsule are lethal in the absence of suppressor mutations and can affect polymer transfer to the cell wall. J. Bacteriol. 189, 3369–3381.

Yeung, M.K., Mattingly, S.J., 1983a. Biosynthesis of cell wall peptidoglycan and polysaccharide antigens by protoplasts of type III group B *Streptococcus*. J. Bacteriol. 154, 211–220.

Yeung, M.K., Mattingly, S.J., 1983b. Isolation and characterization of type III group B streptococcal mutants defective in biosynthesis of the type-specific antigen. Infect. Immun. 42, 141–151.

Zhang, L., Radziejewska-Lebrecht, J., Krajewska-Pietrasik, D., Toivanen, P., Skurnik, M., 1997. Molecular and chemical characterization of the lipopolysaccharide O-antigen and its role in the virulence of *Yersinia enterocolitica* serotype O:8. Mol. Microbiol. 23, 63–76.

Zhou, G.P., Troy, F.A., 2005. NMR studies on how the binding complex of polyisoprenol recognition sequence peptides and polyisoprenols can modulate membrane structure. Curr. Protein Pept. Sci. 6, 399–411.

CHAPTER

21

Biosynthesis of the mycobacterial cell envelope components

Delphi Chatterjee and Patrick J. Brennan

SUMMARY

The cell wall of *Mycobacterium* spp. is based upon a skeleton composed of covalently attached mycolic acids, arabinogalactan and peptidoglycan. Arranged on and in this complex are numerous genus- and species-specific glycolipids and lipoglycans, such as the phenolic glycolipids, glycopeptidolipids, lipo-oligosaccharides, acylated and glycosylated trehaloses, phosphatidylinositol-mannosides, lipomannan and lipoarabinomannan. Definition of the biosynthesis of these constituents and their underlying genetics has benefited not only from the availability of the genomic sequences of several strains of *Mycobacterium* and related genera, but also from the creation of tools for genetic exchange, deletion and inactivation, that are particularly invaluable for generation of mutant strains. Mycobacteria, as for all prokaryotes, adhere to certain common biochemical principles in the synthesis of this range of glycopolymers; for instance, the nucleotide versions of D-mannopyranose (Man*p*), D-galactofuranose, D-glucopyranose and the many deoxyhexoses are the instigators of polymerization. Yet, some of the Man*p* and all of the D-arabinofuranose (Ara*f*) residues originate in polyprenyl-phosphate-linked forms, the Ara*f* itself arising in the pentose phosphate pathway. Likewise, decaprenyl-phosphate is invariably the lipid/membrane carrier, not the prokaryote-wide undecaprenyl-phosphate. In addition, a distinctive feature of *Mycobacterium* spp. is the major role played by glycosyltransferases (GTs) of the multiple transmembrane domain polyprenyl-phosphate-sugar-requiring GT-C class. This chapter discusses major advances in defining the biosynthesis of the individual glycopolymers of the mycobacterial cell wall. However, to date, we have little comprehension of the assembly of the constituents into the whole complex structure of the cell wall.

Keywords: Mycobacterium; Mycobacterium tuberculosis; Cell wall biosynthesis; Glycolipids; Lipoglycans; Glycans; Glycosyltransferases

1. MYCOBACTERIAL GLYCOSYLTRANSFERASES

The majority of the glycosyltransferases (GTs) involved in the synthesis of the common and rare sugars of the cell envelope of *Mycobacterium* spp. are essential for viability and, conceivably, are suitable targets for new drug development against multidrug and extensively-drug resistant tuberculosis. In fact, ethambutol, a first-line anti-tuberculosis drug, targets cell wall D-arabinan synthesis (Mikusova *et al.*, 1995). The GTs are generally

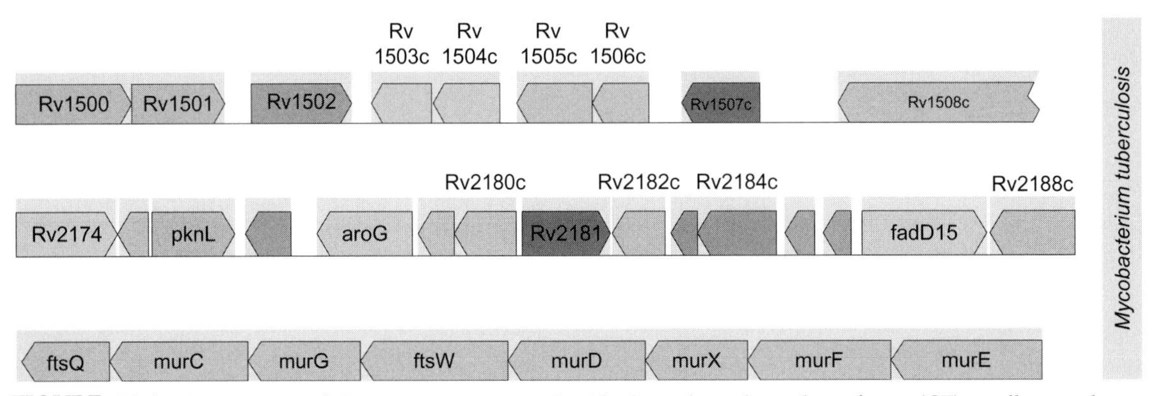

FIGURE 21.1 Organization of the putative sugar nucleotide-dependent glycosyltransferase (GT) small gene cluster. Only one gene, *Rv1500*, has been clearly annotated as a GT and is involved in mycobacterial LOS biosynthesis (Burguiere *et al.*, 2005).

classified as either retaining or inverting, depending on the stereochemical outcome at the anomeric centre relative to that of the donor sugar. They are also classified into sequence similarity based families, which are kept updated on the Carbohydrate-Active Enzymes Database (CAZy) (see http://afmb.cnrs-mrs.fr/CAZY). While the mechanism of inverting GTs is generally accepted, that of retaining enzymes remains a topic of debate. According to CAZy, among the approximately 3900 open-reading frames (ORFs) found in the *Mycobacterium tuberculosis* H37Rv genome (Cole *et al.*, 1998) about 41, i.e. approximately 1%, encode putative GTs (Berg *et al.*, 2007). The majority of these, within the GT-A and -B families, have a requirement for nucleotide-sugar donors. In addition, based on fold recognition analysis, an additional 15 proteins have been identified that have a predicted similarity to the structural folds of GT-A and GT-B, however, these have not yet been shown to be *bona fide* GTs (Wimmerova *et al.*, 2003). In contrast, polyprenyl-dependent GTs of the GT-C classification are more confined to *Actinomycetales* and, mostly, form their own distinct GT families.

Most of the characterized and uncharacterized GT genes are evenly distributed on the *M. tuberculosis* H37Rv chromosome. Nevertheless, there are at least two obvious GT-containing gene clusters, each holding nine proposed GT genes; one is located in the region of *Rv1500* to *Rv1508c* (Figure 21.1) and the other spans from *Rv3779* to *Rv3809c* (Figure 21.2). The former contains mostly GTs proposed to utilize nucleotide diphosphate-sugars, but their functions are still not known, except for LosA (encoded by *Rv1500*) which was recently annotated as a GT involved in mycobacterial lipo-oligosaccharide (LOS) biosynthesis (Burguiere *et al.*, 2005). The other is contained within the "the cell wall biosynthetic cluster" (Cole *et al.*, 1998; Belanger and Inamine, 2000) and includes several now well characterized proteins implicated in arabinogalactan (AG), lipoarabinomannan (LAM) and mycolic acid biosyntheses (Mikusova *et al.*, 2000; Dover *et al.*, 2004; Takayama *et al.*, 2005). A third smaller cluster with three GT genes, *Rv2174*, *Rv2181*, and *Rv2188c* (see Figure 21.1), is located downstream of the *fts*/*mur*-gene cluster (*Rv2150c*-*Rv2158c*), which carries a large number of genes involved in peptidoglycan (PG) biosynthesis and cell division. In *M. tuberculosis*, *Rv2174* is an essential gene that encodes a mannosyltransferase (ManT) responsible for lipomannan (LM) elongation (Kaur *et al.*, 2007). The *Rv2182c* gene, a non-essential *M. tuberculosis* gene, encodes a putative

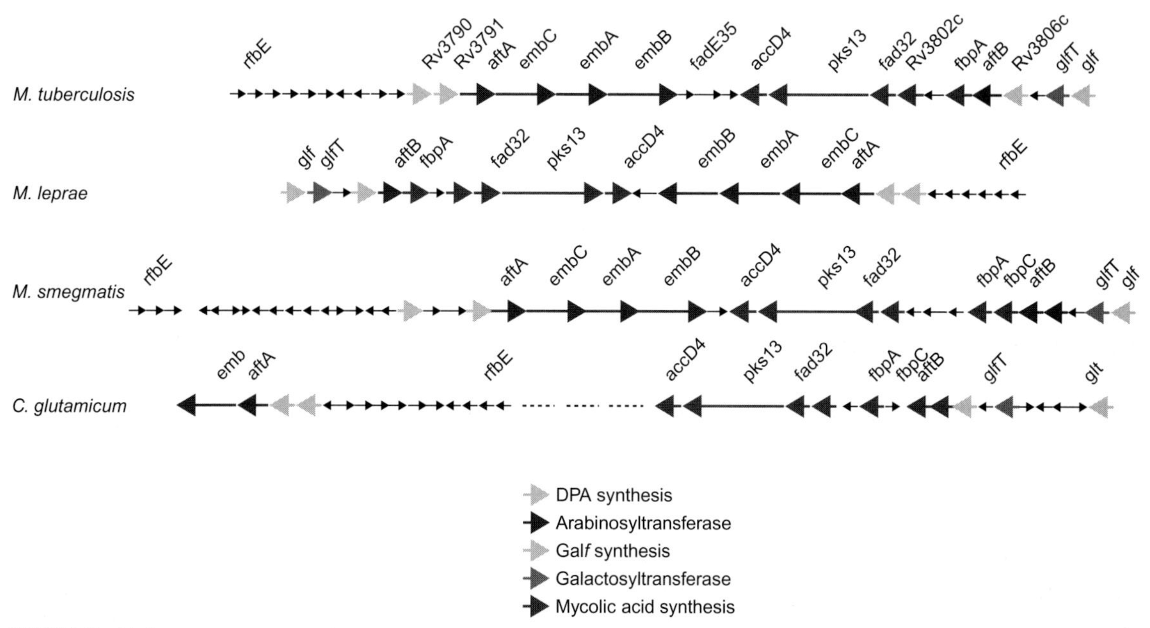

FIGURE 21.2 Organization of the cell wall synthetic gene cluster of *M. tuberculosis*, *M. leprae*, *M. smegmatis* and *C. glutamicum* containing genes with confirmed or suspected roles in cell wall synthesis. See text for details of the functions of some of the annotated genes. (See colour plates section.)

acyltransferase and is also part of this cluster. Gene *Rv2181* has been shown to be a ManT involved in LM biosynthesis (Kaur *et al.*, 2006).

2. BIOCHEMISTRY AND GENETICS OF PEPTIDOGLYCAN (PG) SYNTHESIS

Peptidoglycan biosynthesis in *Escherichia coli* has been widely investigated and the majority of the genes involved in its synthesis are now well known (Ghuysen, 1977) (see Chapter 16). In mycobacteria, however, PG carries some distinctive modifications (Mahapatra *et al.*, 2005). For instance, in *M. tuberculosis* and *Mycobacterium smegmatis*, the muramic acid residues contain a mixture of the *N*-acetyl and *N*-glycolyl derivatives, a modification suggested to take place after the synthesis of uridine diphosphate-(UDP-) *N*-acetyl-muramic acid (MurNAc), but before the formation of the UDP-*N*-glycolyl-muramic acid (MurNGlyc)/Ac-pentapeptides, probably at the earliest stage, as a partial conversion of UDP-MurNAc to UDP-MurNGlyc. The *namH* gene in *M. tuberculosis* that is responsible for the conversion of the -NAc to the -NGlyc derivatives has been identified (Raymond *et al.*, 2005); incidentally, *namH* is represented by a pseudogene in *Mycobacterium leprae* and, hence, its mature PG contains only MurNAc units. Although not investigated in detail, it is generally assumed that the genes involved in mycobacterial PG synthesis are similar to those in *E. coli* and, therefore, the topic will not be discussed in detail here. The *M. tuberculosis*

MurA, a potential drug target, has been overexpressed and partially characterized (De Smet *et al.*, 1999). Also, in an effort to explain the presence of a glycine (Gly) residue in place of L-alanine (L-Ala) on the tetrapeptide side chain of *M. leprae,* the MurC protein, a UDP-MurNGlyc/Ac-Ala ligase from *M. tuberculosis* and *M. leprae*, which is responsible for the normal addition of the L-Ala to the lactyl moiety of UDP-MurNGlyc/Ac, has been overexpressed and partially characterized (Mahapatra *et al.*, 2000). The MurD protein from *M. tuberculosis*, a UDP-MurNAc-L-Ala:D-glutamic acid (Glu) ligase, which is responsible for the addition of D-iso-glutamic acid (D-isoGlu) to the carboxyl terminus of the L-Ala residue of UDP-MurNGlyc/Ac-L-Ala, is similar to the enzyme from *E. coli* (Bertrand *et al.*, 2000). Among the other enzymes involved in UDP-MurNGly/Ac-pentapeptide synthesis in mycobacteria (Figure 21.3), the D-Ala racemase and D-Ala:D-Ala ligase from *M. smegmatis* have been studied (Feng and Barletta, 2003). The *Rv0050* and *Rv3682* genes are believed to encode the two penicillin-binding bifunctional proteins, carrying transglycosylase and transpeptidase activities, which are responsible for the polymerization stages of PG synthesis (Bhakta and Basu, 2002).

3. BIOCHEMISTRY AND GENETICS OF AG SYNTHESIS

As discussed in Chapter 9, mycobacterial AG, constituting about 35% of the cell wall mass, can be conveniently divided into three segments: the linker region consisting of a disaccharide of L-rhamnose (L-Rha) and *N*-acetyl-D-glucosamine (D-GlcNAc), namely→4)-L-Rha-(1→3)-D-GlcNAc-(1→PO_3^- (McNeil *et al.*, 1990); a D-galactofuran attached to the linker region consisting of alternating β-(1→5)- and β-(1→6)-linked D-galactofuranose (D-Gal*f*) residues (Daffe *et al.*, 1990); and, finally, the complex D-arabinofuran covalently linked to the galactan. Years ago, it was predicted that a cluster comprised of 31 genes is perhaps involved in AG synthesis (Belanger *et al.*, 1996). The *Rv3792*, *Rv 3805c*, *embCAB*, the *glf* and the mycolyltransferases (*fbp*) genes are all within this large gene cluster. Interspersed throughout this cluster are genes encoding proteins with similarity to other polysaccharide biosynthetic proteins. Several genes with unknown function are arranged in potential operons and could very well be involved in undefined aspects of AG synthesis.

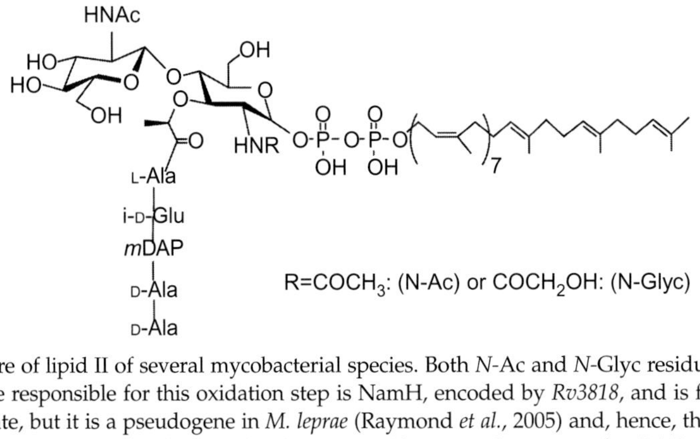

HNAc
HO
HO
OH
O
O
OH
O
O
=O
HNR
O-P-O-P-O
OH OH
7
L-Ala
i-D-Glu
*m*DAP
D-Ala
D-Ala
R=$COCH_3$: (N-Ac) or $COCH_2OH$: (N-Glyc)

FIGURE 21.3 Structure of lipid II of several mycobacterial species. Both *N*-Ac and *N*-Glyc residues are found in mycobacterial lipid II. The enzyme responsible for this oxidation step is NamH, encoded by *Rv3818,* and is found in all mycobacterial genomes sequenced to date, but it is a pseudogene in *M. leprae* (Raymond *et al.*, 2005) and, hence, the PG of *M. leprae* contains Ac residues only. Abbreviations: D- and L-Ala, D- and L-alanine; i-D-Glu; iso-D-glutamic acid; *m*DAP, *meso*-diaminopimelic acid.

Nevertheless, many genes involved in the AG biosynthetic pathway have been identified. In *E. coli*, thymidine diphosphate- (TDP-) Rha is synthesized from α-D-glucose-1-phosphate (α-D-Glc-1-P) and deoxythymidine triphosphate (dTTP) encoded by the gene products of *rmlABCD* (Stevenson *et al.*, 1994). The deoxythymidine diphosphate- (dTDP-) β-L-rhamnopyranose (Rha*p*) is synthesized by this well characterized Rml-pathway (Ma *et al.*, 1997). The biosynthesis of AG itself (Bhamidi *et al.*, 2008) (Figure 21.4) is initiated with the formation of the linker unit, through the transfer of GlcNAc-1-P and Rha from their respective sugar nucleotides, to form decaprenyl-P-P-GlcNAc (an activity attributable to *Rv1302*) and decaprenyl-P-P-GlcNAc-Rha (Mikusova *et al.*, 1996); the rhamnosyltransferase/WbbL activity is attributable to *Rv3265c*. Decaprenyl-P-P-GlcNAc-Rha then serves as an acceptor for sequential addition of Gal*f* residues donated by UDP-α-D-Gal*f* ultimately to form decaprenyl-P-P-GlcNAc-Rha-$Gal_{\sim 30}$. The UDP-α-D-Gal*f* is derived directly from UDP-α-D-galactopyranose (Gal*p*) catalysed by the enzyme Glf, a UDP-α-D-Gal*p* mutase (*Rv3809c*) (Weston *et al.*, 1997); the UDP-Gal*p* is synthesized from UDP-Glc*p* by UDP-Gal*p* epimerase, likely encoded by *Rv3634*. The galactosyltransferase-A (Gal*f*T-A), encoded by *Rv3782*, catalyses the first stages of galactan synthesis transferring the first two Gal*f* residues onto the linker unit (Mikusova *et al.*, 2006), whereas *Rv3808c* (Gal*f*T-B) is responsible for the subsequent polymerization (Mikusova *et al.*, 2000; Kremer *et al.*, 2001; Rose *et al.*, 2006; Belanova *et al.*, 2008). The Rv3808c protein is a bifunctional Gal*f*T that catalyses the synthesis of the alternating 5- and 6-linked linear galactofuran in a processive manner. It is thought to be responsible for the polymerization of the bulk of the galactofuran. Some of the intermediate of decaprenyl-P-P-GlcNAc-Rha-$Gal_{\sim 30}$ apparently serves as the substrate(s) for addition of D-Ara*f* residues from β-D-arabinofuranosyl-1-monophosphoryl-decaprenol (C_{50}-P-Ara*f*) in the formation of the D-arabinan segment of AG. The key steps leading to the biosynthesis of C_{50}-P-Ara*f*, from 5-phospho-ribofuranosyl-pyrophosphate (pRpp) and decaprenol-P have been characterized (Mikusova *et al.*, 2005; Huang *et al.*, 2005, 2008). The known polyprenol-P-linked sugars in *Mycobacterium* spp. are β-D-glucopyranosyl-1-monophosphoryl decaprenol (C_{50}-P-Glc*p*) (Schultz and Elbein, 1974), β-D-mannopyranosyl-1-monophosphoryl-decaprenol (C_{50}-P-Man*p*) (Yokoyama and Ballou, 1989; Besra *et al.*, 1994) and C_{50}-P-Ara*f* (Wolucka *et al.*, 1994); small amounts of the corresponding C_{35} and C_{55} derivatives are also present. The C_{50}-P-Ara*f* is the only proven donor of the Ara*f* units of mycobacterial LAM and AG, whereas both C_{50}-P-Man*p* and guanosine diphosphate- (GDP-) mannose (Man) are involved in the synthesis of the mannan segments of such molecules as LM and LAM.

Information on the enzymology and the events involved in the elongation of the D-arabinan structures of AG and LAM is limited. Since C_{50}-P-Ara*f* is the only known Ara*f* donor, it is expected that the arabinosylation of AG and LAM is catalysed by membrane-associated polyprenyl-dependent GTs on the periplasmic side of the plasma membrane (Berg *et al.*, 2007). Arabinofuranosyltransferases (Ara*f*Ts) characterized to date include AftA (Rv3792), apparently responsible for the transfer of the first Ara*f* residue to the galactan domain of AG (Alderwick *et al.*, 2006; Shi *et al.*, 2008), the terminal β-(1→2)-capping Ara*f*T, AftB (Rv3805c) (Seidel *et al.*, 2007), the Rv2673 protein (AftC) involved in the internal α-(1→3)-branching of the arabinan domain of AG (Birch *et al.*, 2008) and the EmbA and EmbB proteins involved in the formation of the $[\beta$-D-Ara*f*-(1→2)-α-D-Ara*f*$]_2$-3,5-α-D-Ara*f*-(1→5)-α-D-Ara*f*-Ara_6 terminii of arabinan (Escuyer *et al.*, 2001; Khasnobis *et al.*, 2006). Although EmbA and EmbB, acting alone or as heterodimers, have also been proposed to participate in the α-(1→5)-elongation of the linear portion of arabinan, experimental evidence for this assumption is still lacking (Bhamidi *et al.*, 2008). We and others

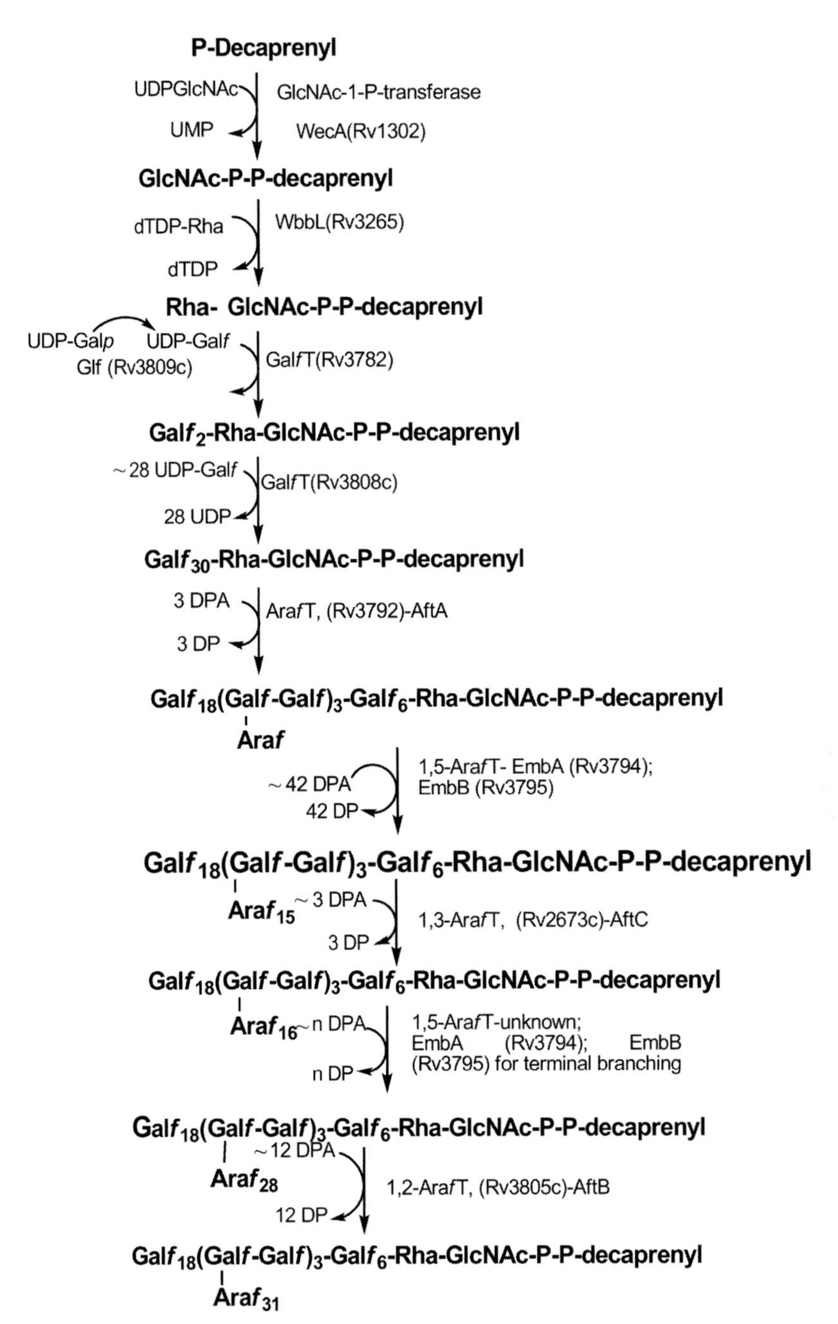

FIGURE 21.4 Biosynthesis of the arabinogalactan (AG) segment of the mycolylarabinogalactan-peptidoglycan (mAGP) complex of *M. tuberculosis*. Most of the earlier steps are described in the text. See also the text for definition of the various abbreviations. The steps in mature arabinan synthesis following the addition of the first three Ara*f* residues to the galactan backbone are speculative. Three chains, each of 31 residues of Ara*f* units are present per galactan chain (Bhamidi *et al.*, 2008). Only one chain of arabinan is shown in this Figure. It is also unclear at what stage mycolic acids are transferred to the terminal Ara*f* motifs, or when and how the AG complex is ligated to peptidoglycan (PG).

have recently reported on the existence of 17 potential GT-C membranous, polyprenyl-P-sugar-requiring enzymes in *M. tuberculosis* H37Rv, 13 of which are known to be involved in the glycosylation of various proteins, glycolipids or polysaccharides (Berg *et al.*, 2007; Seidel *et al.*, 2007; Kaur *et al.*, 2007, 2008; Mishra *et al.*, 2007, 2008; Birch *et al.*, 2008). Likewise, the enzymes and donors responsible for the transfer of the galactosamine and succinyl units onto some of the interior α-(1→5)-Ara*f* residues of arabinan are not yet known. At some point, and at an undetermined subcellular location, apparently the AG-lipid intermediate, whether fully or only partially glycosylated is not clear, is transglycosylated to PG and mycolylated (Yagi *et al.*, 2003). Further results and discussion on the enzymology, topology and genetics of AG synthesis have been reviewed recently (Kaur *et al.*, 2009).

4. BIOSYNTHESIS AND GENETICS OF THE PHOSPHATIDYLINOSITOL- (PI-) CONTAINING PHOSPHATIDYLINOSITOL-MANNOSIDES (PIMS), LMs AND LAMs

The pathway leading to the synthesis of PI in mycobacteria is very similar to that of eukaryotic cells (Salman *et al.*, 1999). Specifically, the *de novo* synthesis of inositol involves the cyclization of glucose-6-P, catalysed by inositol-1-P synthase (encoded by *INO1*, *Rv0046c*) (Bachhawat and Mande, 1999), dephosphorylation through the action of inositol monophosphate phosphatase (IMP) (Parish *et al.*, 1997) and final condensation of inositol with cytidine diphosphate- (CDP-) diacylglycerol (Paulus and Kennedy, 1960). The *M. tuberculosis* gene *pgsA* (*Rv2612c*) appears to encode PI synthase (PIS), as overexpression of this gene in *M. smegmatis* resulted in elevated levels of PI synthesis (Jackson *et al.*, 2000); PIS is apparently an essential enzyme, since the *pgsA* homologue in *M. smegmatis* could only be deleted in the presence of an episomal copy of *pis*. Collectively, these studies indicate that the synthesis of PI in mycobacteria occurs through a pathway identical to that described in eukaryotes.

Early studies implicated direct mannosylation of PI by GDP-Man in the subsequent synthesis of the PIMs through simple incubation of cell free extracts of *Mycobacterium phlei* and *M. tuberculosis*, however, and importantly, a number of PIM_2 variants differing in their degree of acylation were found (Brennan and Ballou, 1967; Takayama and Goldman, 1969). A GDP-Man-dependent mannosyltransferase (encoded by *pimA*, *Rv2610c*) in *M. tuberculosis* is responsible for the addition of the first Man*p* residue to the 2-OH of the inositol of PI to form PIM_1, thereby catalysing the first committed step in PIM biosynthesis (Kordulakova *et al.*, 2002) (Figure 21.5). The PimA protein is essential for growth of *M. smegmatis*, as inactivation of *pimA* results in cell death. The *pimA* and *pgsA* genes are located close by on the chromosome, suggesting that PI and PIM biosynthesis may be co-regulated. The acyltransferase that modifies PIM_1 to Ac_1-PIM_1 by attaching an acyl residue to the 6-OH of the Man has also been described (Rv2611c) (Kordulakova *et al.*, 2002); the *Rv2611c* gene is located between *pgsA* and *pimA* on the chromosome. Recently, the crystal structure of PimA from *M. smegmatis* has been solved (Guerin *et al.*, 2007). A second ManT (Rv0557) with significant homology to PimA was thought to be responsible for the formation of Ac_1-PIM_2 (Schaeffer *et al.*, 1999). This step was not inhibited by amphomycin, indicating that GDP-Man is involved as a donor rather than polyprenyl-P-Man. Since then, and only recently, *Rv2188c* in *M. tuberculosis* has been shown to encode a ManT involved in the conversion of $AcPIM_1$ to $AcPIM_2$; this step is essential for the synthesis of LAM (Lea-Smith *et al.*, 2008). Based on experiments with the *C. glutanicum* homologue it now seems likely that the Rv0557 protein is responsible for the transfer of Man to diacylglycerol-containing glycolipids and LMs, rather than PI-based PIMs and LM (Tatituri *et al.*,

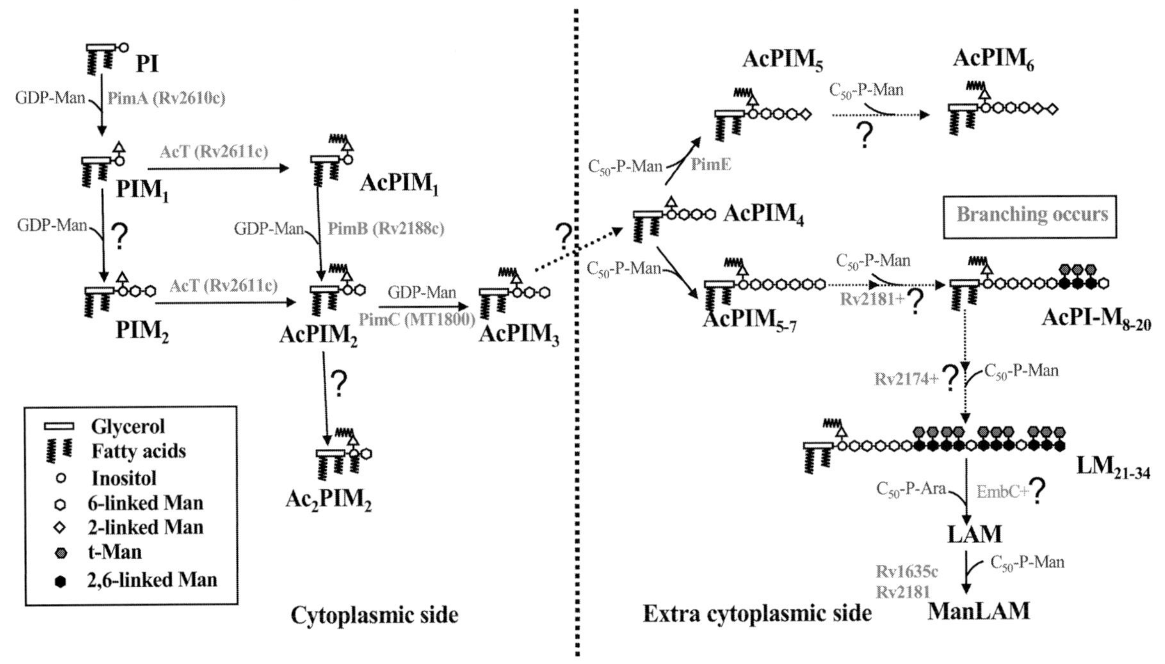

FIGURE 21.5 Biosynthesis of phosphatidylinositol-mannosides (PIMs), lipomannan (LM) and lipoarabinomannan (LAM). Most of the mannosyltransferases (ManTs) involved in PIM and LM synthesis are now known; these are described in this review. For details of the steps in the synthesis and associated abbreviations please refer to text. The EmbC protein is the only arabinosyltransferase that is thought to be selectively involved in the synthesis of the arabinan of LAM (Zhang *et al.*, 2003).

2007; Lea-Smith *et al.*, 2008). A third ManT, PimC, has been identified in *M. tuberculosis* CDC 1551. Overexpression of *pimC* in *M. smegmatis* led to the formation of Ac_1-PIM_3 in an *in vitro* assay and, since this ManT reaction was resistant to amphomycin, GDP-Man rather than C_{50}-P-Man was implicated as the direct Man donor. However, the H37Rv genome lacks a homologue of PimC suggesting that synthesis of Ac-PIM_3 may be carried out by a second protein (Kremer *et al.*, 2002a). The ManTs that catalyse the transfer of a Man*p* residue onto PIM_3 to produce PIM_4 remain to be identified. None of these known ManTs contains predicted transmembrane domains, despite the fact that the products/substrates are likely intercalated into the plasma membrane. Interestingly, many of the GTs involved in the O-antigen and core assembly of *E. coli* lipopolysaccharide (LPS) also lack transmembrane domains but are detected in the membrane fractions (Raetz and Whitfield, 2002), in contrast to findings in eukaryotic systems, where most of the GTs involved in glycolipid synthesis are integral membrane proteins (Colley, 1997; Berger, 2002).

Steps leading to the formation of higher PIMs (PIM_4–PIM_6), LM and LAM are only now being unravelled (Zhang *et al.*, 2003; Kaur *et al.*, 2006, 2007). Early studies suggested that the synthesis of the mannan chain of LAM involves polyprenyl-P-Man donor species (Schultz and Takayama, 1975; Yokoyama and Ballou, 1989). However, studies with amphomycin showed inhibition of the synthesis of PIM_4, PIM_5 and PIM_6, suggesting that these enzymatic steps utilize C_{50}-P-Man as a donor substrate (Morita *et al.*, 2004). The PimE (Rv1159) protein was recently identified as a probable polyprenyl-P-Man-dependent ManT responsible for the synthesis of PIM_5 from PIM_4 (Morita *et al.*, 2006). Also, recent bioinformatic approaches identified putative integral membrane proteins, encoded by *Rv2181* in *M. tuberculosis* and *MSMEG4250* in *M. smegmatis*, with predicted 10 transmembrane domains and a GT motif, features that are common to eukaryotic ManTs of the GT-C superfamily that rely on polyprenyl-linked, rather than nucleotide-linked, sugar donors. Inactivation of *MSMEG4250* led to an LM-depleted strain and the accumulation of a truncated LAM characterized by an unbranched mannan core. These data and detailed structural analysis of the products implicated Rv2181 and MSMEG4250 as responsible for addition of α-(1→2)-Man*p* branches to the mannan core of LM/LAM (Kaur *et al.*, 2006). The proximity of *Rv2181* and *Rv2188c* on the chromosome points to a LAM biosynthesis cluster mainly comprised of ManTs (see Figure 21.1). Another essential *M. tuberculosis* gene *Rv2174*, encoding a ManT responsible for LM elongation (Kaur *et al.*, 2007) and *Rv2182c*, a non-essential *M. tuberculosis* putative acyltransferase, may also be a part of this cluster.

Recently, structural analyses of the LM and LAM variants produced by a *M. tuberculosis Rv2181* knockout mutant revealed complete absence of α-(1→2)-linked Man*p* branching on the mannan backbones of LM and LAM. In addition, the non-reducing end of LAM was affected in that only a single Man*p* residue on the non-reducing arabinan termini of LAM was found. Co-expression of *Rv2181* and *Rv1635c* [*Rv1635c* encodes the protein responsible for the addition of the first Man*p* residue on Man*p*-capped LAM (ManLAM) of *M. tuberculosis*] in *M. smegmatis* resulted in synthesis of ManLAM with termini capped with α-(1→2)-Man*p*-linked di-and tri-Man, confirming that the protein Rv2181 plays a dual role in Man-capping and mannan-core branching (Dinadayala *et al.*, 2006) (see Figure 21.5).

The topographical location of the various steps in the biosynthesis of the higher PIMs, LM and LAM is worthy of consideration even if only on the basis of the requirements for polyprenyl-based sugar donors. The initial steps involving PI and PIM_2-PIM_3 synthesis are likely to occur on the cytoplasmic face of the plasma membrane, as they utilize GDP-Man and ManTs that are apparently cytosolic. The polyprenol phosphate-(Pol-P-) donor species could also be synthesized on the cytosolic face of the membrane, given the membrane localization of *ppm*1, the gene encoding polyprenyl-P-Man*p* of *M. tuberculosis* (Gurcha *et al.*, 2002) (Figure 21.6). In the case of LAM/LM synthesis, it is unclear whether these precursors are actively transported through the mycobacterial cell membrane/wall or whether they diffuse passively through pores to the surface. The model proposed in Figure 21.6 is based partially on knowledge of the final phases of LPS synthesis of Gram-negative bacteria (Raetz and Whitfield, 2002).

5. BIOSYNTHESIS AND GENETICS OF THE GLYCOPEPTIDOLIPIDS

Glycopeptidolipid synthesis is initiated by the actions of a mycobacterial peptide synthetase (MPS) which mediates the non-ribosomal assembly of the tri-peptide amino alcohol core (Billman-Jacobe *et al.*, 1999). This large multisubunit enzyme contains four distinct modules, each containing domains for co-factor binding, amino acid recognition and adenylation. Three of the modules contain racemase domains, capable of converting L-phenylalanine (L-Phe), L-threonine (L-Thr) and L-Ala to their respective D-isomers. It is not known whether the peptide is assembled on the amide-linked lipid or the lipid is transferred to the peptide after synthesis. A methylated Rha*p* and an acylated 6-deoxy-L-talopyranose (6-dTal*p*) residue are then attached through the action of unknown transferases (Figure 21.7). Both deoxy sugar residues can be synthesized from an identical precursor, dTDP-6-deoxy-L-*lyxo*-4-hexulose and, hence, their synthesis may

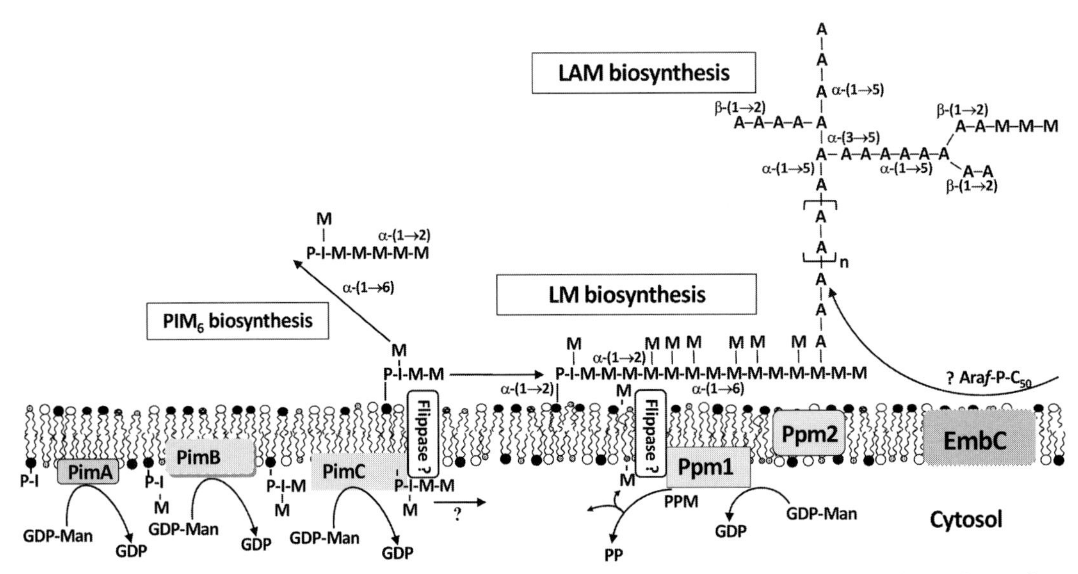

FIGURE 21.6 Postulated topography of the main glycosyltransferase- (GT-) mediated events in the synthesis of some of the major glycan components of the cell wall of *M. tuberculosis*. Please refer to the text for details of the GTs.

be co-regulated (Liu and Thorson, 1994; Ma *et al.*, 1997; Giraud *et al.*, 1999; Graninger *et al.*, 1999; Nakano *et al.*, 2000). Glycopeptidolipids from *Mycobacterium avium* can be further modified with serovar-specific oligosaccharides that are attached to the 2-position of the 6-dTal residue. In the case of *M. avium* serovar 2, all of the enzymes required for the modification of the core 6-dTal*p* and Rha*p* residues have also been described (Jeevarajah *et al.*, 2002), suggesting a shared genetic theme for glycopeptidolipid core and haptenic side chain synthesis.

6. BIOSYNTHESIS AND GENETICS OF THE PHTHIOCEROL-CONTAINING LIPIDS

Phenolic glycolipids (PGLs) are polyketide-derived virulence factors produced by many pathogenic mycobacteria, notably *M. tuberculosis*, *Mycobacterium bovis* and *M. leprae*. Biosynthesis of the phenolthiocerol moiety of the PGLs involves the Pks15/1-PpsABCDE type I polyketide synthase system (Kolattukudy *et al.*, 1997; Azad *et al.*, 1997; Kremer *et al.*, 2002b; Trivedi *et al.*, 2004, 2005), a trans-acting enoyl reductase (Simeone *et al.*, 2007b) and two enzymes that convert phenolthiodiolones to phenolthiocerols (Perez *et al.*, 2004a; Onwueme *et al.*, 2005; Simeone *et al.*, 2007a).

Three GTs and a methyltransferase have been shown to be involved in the formation of the glycosyl moiety of both PGLs (Figure 21.8) and the *p*-hydroxybenzoic acid precursor derivative (p-HBAD I) (Perez *et al.*, 2004a,b). The genes encoding these enzymes are *Rv2962c*, *Rv2957*, *Rv2958c*, and *Rv2959c*. These genes have been individually disrupted in *M. tuberculosis* H37Rv and the corresponding mutants analysed (Perez *et al.*, 2004a,b). These mutants accumulated several biosynthetic intermediates and were unable to synthesize full length PGL. Analysis of these

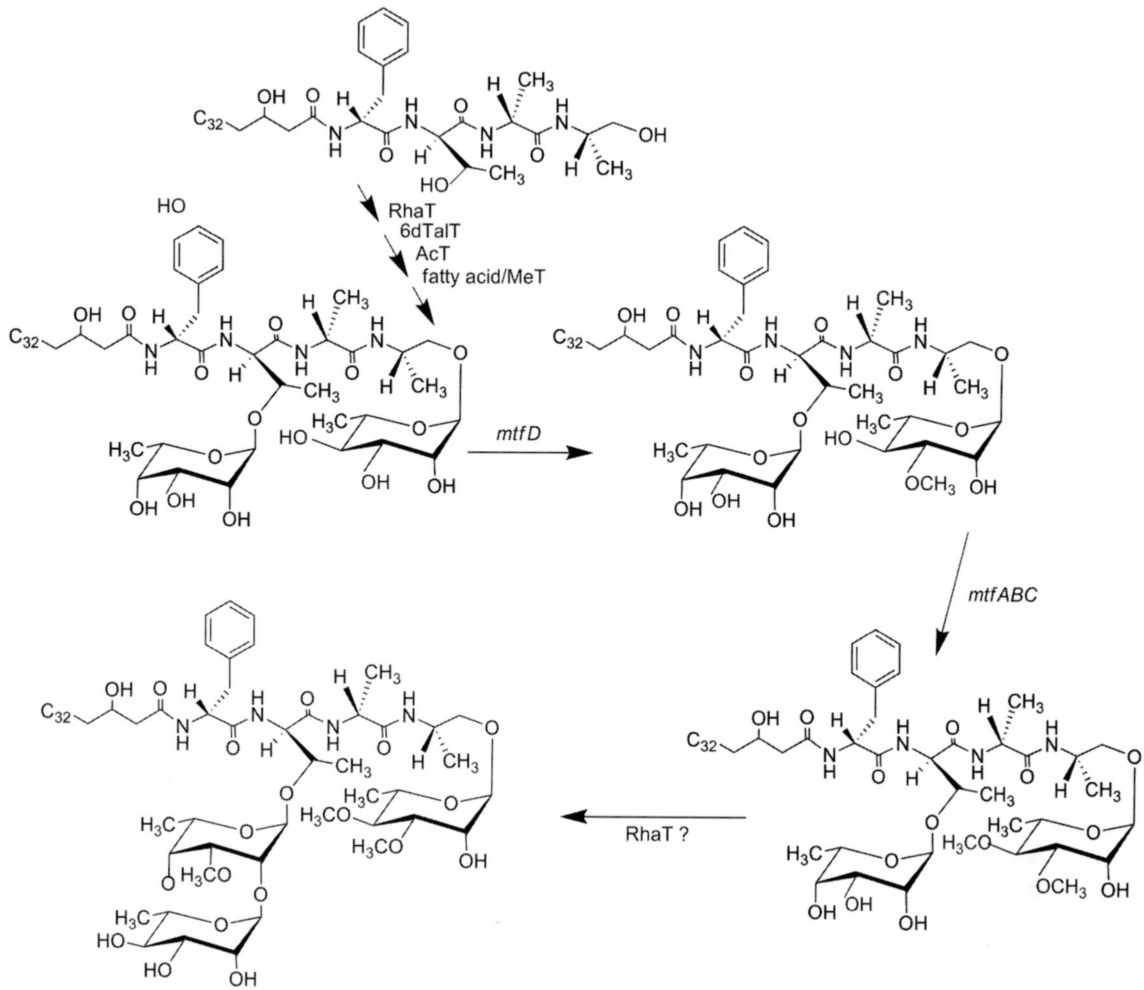

FIGURE 21.7 Biosynthesis of the glycopetidolipids (GPLs) of some non-tuberculous mycobacteria. The GPL synthesis is initiated by the actions of a mycobacterial peptide synthetase (MPS); see the text for further details. It is not known whether the peptide is assembled on the amide-linked lipid or if the lipid is transferred to the peptide after synthesis. Methylated rhamnopyranose and an acylated 6-deoxy-L-talopyranose residue are then attached through the action of unknown glycosyltransferases. The 6-deoxy-L-talopyranose is sometimes methylated at the 3-position (as drawn), as in serovar 1 of the *M. avium* complex. Abbreviations: AcT, acyltransferase; MeT, methyltransferase; RhaT, rhamnosyltransferase; 6dTalT, 6-deoxy-L-talose transferase.

mutants allowed assignment of functions to these genes and the Rv2962c protein was determined to be involved in transferring the first Rha on the *p*-hydroxyphenolphthiocerol dimycocerosates and *p*-methoxyphenolphthiocerol dimycocerosates. Knockout mutants of *Rv2957* and *Rv2958c* synthesized mono-glycosylated glycolipids, implicating these two GTs in the synthesis of the terminal sugar residues. An unmarked mutation in *Rv2959c* was constructed in *M. tuberculosis* in which the *pks15/1* mutation was complemented by a functional *pks15/1* gene

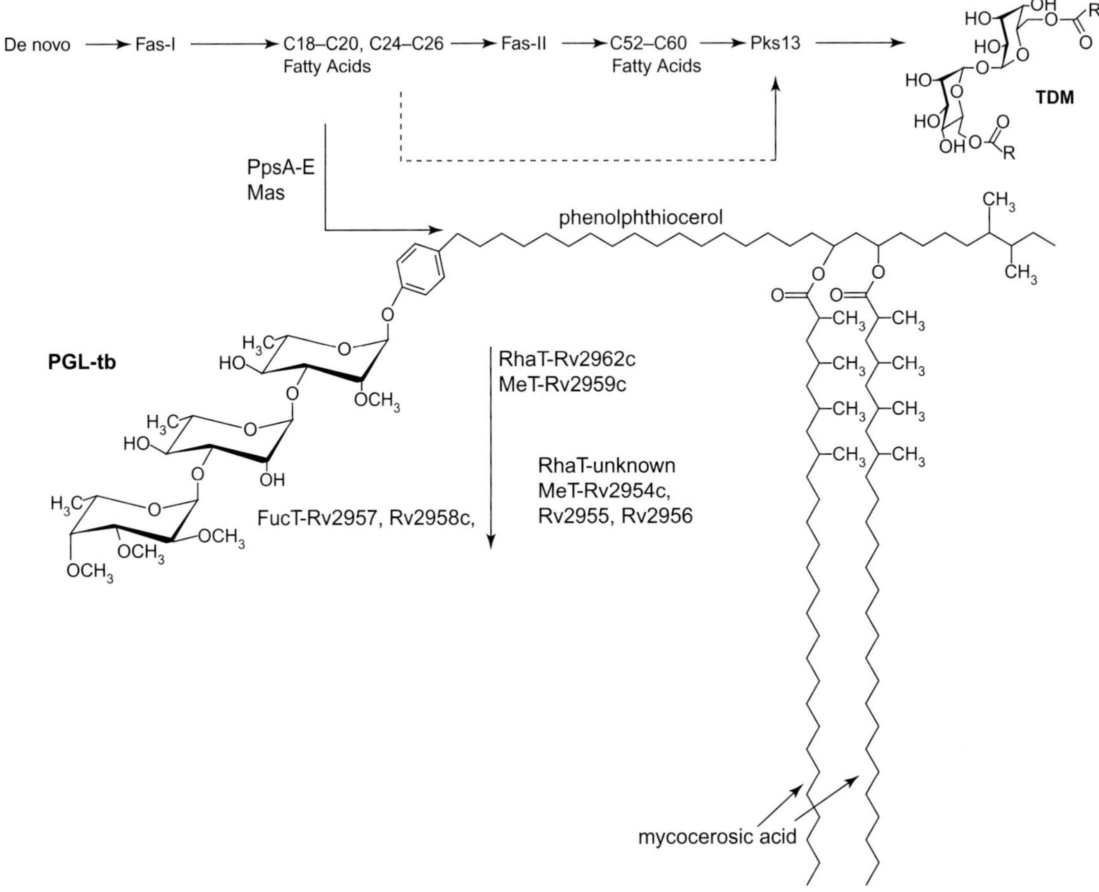

FIGURE 21.8 Biosynthesis of the phthiocerol-containing lipids of *M. tuberculosis*. A cascade of enzymes is required to complete the synthesis of the triglycosylated phenolic glycolipid synthesized by *M. tuberculosis* (PGL-tb) and the intermediate phenol-phthiocerol dimycocerosates, as discussed in this review. Only a few are shown. Abbreviation: TDM, trehalose dimycolates.

(Perez *et al.*, 2004b). This mutant produced only unmethylated Rha-phenolphthiocerol dimycocerosates and a tri-*O*-Me-Fuc-di-Rha-phenolphthiocerol dimycocerosate suggesting that *Rv2959c* encodes a methyltransferase (MeT) for alkylation of the 2-position of the first Rha residue of the PGL synthesized by *M. tuberculosis* (PGL-tb) and p-HBAD. Other MeTs are required to complete PGL-tb synthesis; *Rv29549, reRv2955c* and *Rv2956* are all in the relevant chromosomal locus and have been implicated as encoding these MeTs.

Monoglycosylated PGL formed by *M. bovis* differs from PGL-tb, the triglycosylated PGL synthesized by *M. tuberculosis*, because of two genetic defects: a frameshift mutation within the gene *Rv2958c* and a deletion of a region that encompasses two genes that encode a GDP-D-mannose 4,6-dehydratase and a GDP-4-keto-6-deoxy-D-mannose-3,5-epimerase/reductase, required for the formation of activated L-fucose. Expression of these three genes in *M. bovis* strain BCG allowed synthesis of PGL-tb in this recombinant strain. Additionally, all *M. bovis*,

Mycobacterium microti, *Mycobacterium pinnipedii* and some *Mycobacterium africanum* strains harbour the same frameshift mutation in their *Rv2958c* orthologues. Consistently, the structure of PGLs purified from *M. africanum* that harbour the *Rv2958c* mutation and *M. pinnipedii* strains have revealed that these compounds are monoglycosylated PGLs. These findings explain the specificity of PGL-tb production by some strains of the *M. tuberculosis* complex and have important implications for our understanding of the evolution of this complex (Malaga *et al.*, 2008).

7. BIOSYNTHESIS OF THE TREHALOSE-CONTAINING GLYCOLIPIDS

The acyltrehaloses found in the cell envelope of *M. tuberculosis* include sulfatides (SL), diacyltrehaloses (DAT), triacyltrehaloses (TAT), polyacyltrehaloses (PAT), trehalose monomycolate (TMM) and trehalose dimycolates (TDM). In addition, many of the non-tuberculous mycobacteria and a few select strains of *M. tuberculosis*, generally with atypical smooth morphology, produce highly polar species-specific LOSs based on acylated trehalose (Brennan *et al.*, 1970; Daffe *et al.*, 1991).

Mycobacteria are unusual among microorganisms in possessing three pathways for trehalose synthesis (De Smet *et al.*, 2000) (see also Chapter 11). One route involves condensation of glucose-6-phosphate with UDP-glucose to form trehalose-6-phosphate followed by dephosphorylation to release the free disaccharide. These reactions are catalysed by trehalose-6-phosphate synthase (OtsA, Rv3490) (Pan *et al.*, 2002) and trehalose-6-phosphate phosphatase (OtsB2, Rv3372). The second pathway generates trehalose from glycogen in a two-step mechanism involving the malto-oligosyltrehalose synthase (TreY, Rv1653c) and the malto-oligosyltrehalose trehalohydrolase (TreZ, Rv1562c). The third pathway consists of the conversion of maltose to trehalose by the trehalose synthase (TreS, Rv0126). While the three pathways are functionally redundant in *M. smegmatis* (Woodruff *et al.*, 2004), the OtsAB pathway was found to be predominant in *M. tuberculosis* (Murphy *et al.*, 2005). Importantly, OtsB2 was demonstrated to be strictly essential for growth in *M. tuberculosis* (Murphy *et al.*, 2005). The synthesis of TDM from two TMM molecules and the transfer of mycolates to the non-reducing ends of AG have been shown to involve the mycolyltransferases synonymous with antigens 85A, 85B and 85C (Belisle *et al.*, 1997; Jackson *et al.*, 1999). However, whether this is the true physiological reaction is not clear; moreover, the origin of TMM is not certain. Neither have the reactions leading to the acylation of trehalose with short-chain fatty acyl substituents and long-chain multi-methyl-branched fatty acids to form SL, DAT, TAT and PAT been defined. The only two acyltransferases characterized to date are PapA2, involved in the transfer of a palmitoyl group to the 2′-position of trehalose-2-sulfate in SL biosynthesis, and PapA1, apparently responsible for the transfer of the first (hydroxy)phthioceranoyl group onto the product of PapA2 (Kumar *et al.*, 2007). Both acyltransferases are essential for the synthesis of SL-1 (Bhatt *et al.*, 2007; Kumar *et al.*, 2007).

The Stf0 (Rv0295c) protein was characterized as the sulfotransferase responsible for the formation of the trehalose-2-sulfate moiety of SL (Mougous *et al.*, 2004). Little is known of the biosynthesis of the specific LOS antigens of non-tuberculous mycobacteria and some strains of *M. tuberculosis* (Mougous *et al.*, 2004; Kaur *et al.*, 2008).

8. CONCLUSIONS

Mycobacteria synthesize a range of usual and unusual glycosyl residues themselves often as part of unusual glycolipids and lipoglycans, notably AG as part of the mycolylarabinogalactan-PG complex, D-arabinofuranoside-D-mannopyranoside as part of LAM, the phthiocerol-containing, glycopeptide-containing and trehalose-containing glycolipids. Stimulated by knowledge of the genomes of several species and strains of *Mycobacterium* and major advances in genetic manipulations and analytical chemistry, we now have a most impressive understanding not only of complete structure, but also of underlying genetics and synthesis, down to the role of individual GTs. As a consequence of these developments, particularly the creation of isogenic mutants, comprehension of the roles of the individual wall constituents in the immunopathogenesis of tuberculosis is rapidly emerging. Many of the essential enzymes are being targeted in the search for new antidotes against rampant multidrug resistant tuberculosis. However, we know little of the assembly of the individual entities into the whole complex cell wall which, itself, is the major phenotypic determinant of the disease induction process and requires further intensive investigation (see Research Focus Box). This is the great challenge for the next generation of mycobacterial molecular biologists.

ACKNOWLEDGEMENTS

Research conducted in the authors' laboratories was supported by grants from the National Institute of Allergy and Infectious Diseases, National Institutes of Health (AI018357 and AI037139).

RESEARCH FOCUS BOX

- Continued creation of isogenic mutants to aid the comprehension of the roles of the individual cell wall components in the immunopathogenesis of tuberculosis.
- A concerted investigation of the assembly of the individual constituents into the mycobacterial complex cell wall is needed since this is the major phenotypic determinant of the disease induction process.
- Exploitation of the knowledge gained by targeting the many essential enzymes in mycobacterial cell wall biosynthesis and assembly as objects for anti-mycobacterial therapies, particularly against multidrug resistant tuberculosis.

References

Alderwick, L.J., Seidel, M., Sahm, H., Besra, G.S., Eggeling, L., 2006. Identification of a novel arabinofuranosyltransferase (AftA) involved in cell wall arabinan biosynthesis in *Mycobacterium tuberculosis*. J. Biol. Chem. 281, 15653–15661.

Azad, A.K., Sirakova, T.D., Fernandes, N.D., Kolattukudy, P.E., 1997. Gene knockout reveals a novel gene cluster for the synthesis of a class of cell wall lipids unique to pathogenic mycobacteria. J. Biol. Chem. 272, 16741–16745.

Bachhawat, N., Mande, S.C., 1999. Identification of the INO1 gene of *Mycobacterium tuberculosis* H37Rv reveals a novel class of inositol-1-phosphate synthase enzyme. J. Mol. Biol. 291, 531–536.

Belanger, A.E., Inamine, J.I., 2000. Genetics of cell wall biosynthesis. In: Hatfull, G.F., Jacobs, W.R. (Eds.), Molecular Genetics of Mycobacteria. ASM Press, Washington, DC, pp. 191–202.

Belanger, A.E., Besra, G.S., Ford, M.E., et al., 1996. The *embAB* genes of *Mycobacterium avium* encode an arabinosyl transferase involved in cell wall arabinan biosynthesis that is the target for the antimycobacterial drug ethambutol. Proc. Natl. Acad. Sci. USA 93, 11919–11924.

Belanova, M., Dianiskova, P., Brennan, P.J., et al., 2008. Galactosyl transferases in mycobacterial cell wall synthesis. J. Bacteriol. 190, 1141–1145.

Belisle, J.T., Vissa, V.D., Sievert, T., Takayama, K., Brennan, P.J., Besra, G.S., 1997. Role of the major antigen of *Mycobacterium tuberculosis* in cell wall biogenesis. Science 276, 1420–1422.

Berg, S., Kaur, D., Jackson, M., Brennan, P.J., 2007. The glycosyltransferases of *Mycobacterium tuberculosis*; roles in the synthesis of arabinogalactan, lipoarabinomannan, and other glycoconjugates. Glycobiology 17, 35R–56R.

Berger, E.C., 2002. Ectopic localization of golgi glycosyltransferase. Glycobiology 12, 29R–36R.

Bertrand, J.A., Fanchon, E., Martin, L., et al., 2000. "Open" structures of MurD: domain movements and structural similarities with folylpolyglutamate synthetase. J. Mol. Biol. 301, 1257–1266.

Besra, G.S., Gurcha, S.S., Khoo, K.H., et al., 1994. Characterization of the specific antigenicity of representatives of *M. senegalense* and related bacteria. Zentralbl. Bakteriol. 281, 415–432.

Bhakta, S., Basu, J., 2002. Overexpression, purification and biochemical characterization of a class A high-molecular-mass penicillin-binding protein (PBP), PBP1* and its soluble derivative from *Mycobacterium tuberculosis*. Biochem. J. 361, 635–639.

Bhamidi, S., Scherman, M.S., Rithner, C.D., et al., 2008. The identification and location of succinyl residues and the characterization of the interior arabinan region allow for a model of the complete primary structure of *Mycobacterium tuberculosis* mycolyl arabinogalactan. J. Biol. Chem. 283, 12992–13000.

Bhatt, K., Gurcha, S.S., Bhatt, A., Besra, G.S., Jacobs, W.R., 2007. Two polyketide-synthase-associated acyltransferases are required for sulfolipid biosynthesis in *Mycobacterium tuberculosis*. Microbiology 153, 513–520.

Billman-Jacobe, H., McConville, M.J., Haites, R.E., Kovacevic, S., Coppel, R.L., 1999. Identification of a peptide synthetase involved in the biosynthesis of glycopeptidolipids of *Mycobacterium smegmatis*. Mol. Microbiol. 33, 1244–1253.

Birch, H.L., Alderwick, L.J., Bhatt, A., et al., 2008. Biosynthesis of mycobacterial arabinogalactan: identification of a novel $\alpha(1\rightarrow3)$ arabinofuranosyltransferase. Mol. Microbiol. 69, 1191–1206.

Brennan, P., Ballou, C.E., 1967. Biosynthesis of mannophosphoinositides by *Mycobacterium phlei*. The family of dimannophosphoinositides. J. Biol. Chem. 242, 3046–3056.

Brennan, P.J., Rooney, S.A., Winder, F.G., 1970. The lipids of *Mycobacterium tuberculosis* BCG: fractionation, composition, turnover and the effects of isoniazid. Ir. J. Med. Sci. 3, 371–390.

Burguiere, A., Hitchen, P.G., Dover, L.G., et al., 2005. LosA, a key glycosyltransferase involved in the biosynthesis of a novel family of glycosylated acyltrehalose lipooligosaccharides from *Mycobacterium marinum*. J. Biol. Chem. 280, 42124–42133.

Cole, S.T., Brosch, R., Parkhill, J., 1998. Deciphering the biology of *Mycobacterium tuberculosis* from the complete genome sequence. Nature 393, 537–544.

Colley, K.J., 1997. Golgi localization of glycosyltransferases: more questions than answers. Glycobiology 7, 1–13.

Daffe, M., Brennan, P.J., McNeil, M., 1990. Predominant structural features of the cell wall arabinogalactan of *Mycobacterium tuberculosis* as revealed through characterization of oligoglycosyl alditol fragments by gas chromatography/mass spectrometry and by ^{1}H and ^{13}C NMR analyses. J. Biol. Chem. 265, 6734–6743.

Daffe, M., McNeil, M., Brennan, P.J., 1991. Novel type-specific lipooligosaccharides from *Mycobacterium tuberculosis*. Biochemistry 30, 378–388.

De Smet, K.A., Kempsell, K.E., Gallagher, A., Duncan, K., Young, D.B., 1999. Alteration of a single amino acid residue reverses fosfomycin resistance of recombinant MurA from *Mycobacterium tuberculosis*. Microbiology 145, 3177–3184.

De Smet, K.A., Weston, A., Brown, I.N., Young, D.B., Roberts, B.D., 2000. Three pathways for trehalose biosynthesis in mycobacteria. Microbiology 146, 199–208.

Dinadayala, P., Kaur, D., Berg, S., et al., 2006. Genetic basis for the synthesis of the immunomodulatory mannose caps of lipoarabinomannan in *Mycobacterium tuberculosis*. J. Biol. Chem. 281, 20027–20035.

Dover, L.G., Cerdeno-Tarraga, A.M., Pallen, M.J., Parkhill, J., Besra, G.S., 2004. Comparative cell wall core biosynthesis in the mycolated pathogens, *Mycobacterium tuberculosis* and *Corynebacterium diphtheriae*. FEMS Microbiol. Rev. 28, 225–250.

Escuyer, V.E., Lety, M.A., Torrelles, J.B., et al., 2001. The role of the embA and embB gene products in the biosynthesis of the terminal hexaarabinofuranosyl motif of *Mycobacterium smegmatis* arabinogalactan. J. Biol. Chem. 276, 48854–48862.

Feng, Z., Barletta, R.G., 2003. Roles of *Mycobacterium smegmatis* D-alanine: D-alanine ligase and D-alanine racemase in the mechanisms of action of and resistance to the peptidoglycan inhibitor D-cycloserine. Antimicrob. Agents Chemother. 47, 283–291.

Ghuysen, J.M., 1977. Biosynthesis and assembly of bacterial cell walls. In: Poste, G., Nicholson, G.L. (Eds.), Cell Surface Reviews. North-Holland Publishing Co., Amsterdam, pp. 463–595.

Giraud, M.F., McMiken, H.J., Leonard, G.A., Messner, P., Whitfield, C., Naismith, J.H., 1999. Overexpression, purification, crystallization and preliminary structural study of dTDP-6-deoxy-L-lyxo-4-hexulose reductase (RmlD), the fourth enzyme of the dTDP-L-rhamnose synthesis pathway, from *Salmonella enterica* serovar Typhimurium. Acta Crystallogr. D Biol. Crystallogr. 55, 2043–2046.

Graninger, M., Nidetzky, B., Heinrichs, D.E., Whitfield, C., Messner, P., 1999. Characterization of dTDP-4-dehydrorhamnose 3,5-epimerase and dTDP-4-dehydrorhamnose reductase, required for dTDP-L-rhamnose biosynthesis in *Salmonella enterica* serovar Typhimurium LT2. J. Biol. Chem. 274, 25069–25077.

Guerin, M.E., Kordulakova, J., Schaeffer, F., et al., 2007. Molecular recognition and interfacial catalysis by the essential phosphatidylinositol mannosyltransferase PimA from mycobacteria. J. Biol. Chem. 282, 20705–20714.

Gurcha, S.S., Baulard, A.R., Kremer, L., et al., 2002. Ppm1, a novel polyprenol monophosphomannose synthase from *Mycobacterium tuberculosis*. Biochem. J. 365, 441–450.

Huang, H., Scherman, M.S., D'Haeze, W., et al., 2005. Identification and active expression of the *Mycobacterium tuberculosis* gene encoding 5-phospho-α-D-ribose-1-diphosphate: decaprenyl-phosphate 5-phosphoribosyltransferase, the first enzyme committed to decaprenylphosphoryl-D-arabinose synthesis. J. Biol. Chem. 280, 24539–24543.

Huang, H., Berg, S., Spencer, J.S., et al., 2008. Identification of amino acids and domains required for catalytic activity of DPPR synthase, a cell wall biosynthetic enzyme of *Mycobacterium tuberculosis*. Microbiology 154, 736–743.

Jackson, M., Raynaud, C., Laneelle, M.A., et al., 1999. Inactivation of the antigen 85C gene profoundly affects the mycolate content and alters the permeability of the *Mycobacterium tuberculosis* cell envelope. Mol. Microbiol. 31, 1573–1587.

Jackson, M., Crick, D.C., Brennan, P.J., 2000. Phosphatidylinositol is an essential phospholipid of mycobacteria. J. Biol. Chem. 275, 30092–30099.

Jeevarajah, D., Patterson, J.H., McConville, M.J., Billman-Jacobe, H., 2002. Modification of glycopeptidolipids by an *O*-methyltransferase of *Mycobacterium smegmatis*. Microbiology 148, 3079–3087.

Kaur, D., Berg, S., Dinadayala, P., et al., 2006. Biosynthesis of mycobacterial lipoarabinomannan: role of a branching mannosyltransferase. Proc. Natl. Acad. Sci. USA 103, 13664–13669.

Kaur, D., McNeil, M.R., Khoo, K.H., et al., 2007. New insights into the biosynthesis of mycobacterial lipomannan arising from deletion of a conserved gene. J. Biol. Chem. 282, 27133–27140.

Kaur, D., Obregon-Henao, A., Pham, H., Chatterjee, D., Brennan, P.J., Jackson, M., 2008. Lipoarabinomannan of *Mycobacterium*: mannose capping by a multifunctional terminal mannosyltransferase. Proc. Natl. Acad. Sci. USA 105, 17973–17977.

Kaur, D., Guerin, M.E., Skovierova, H., Brennan, P.J., Jackson, M., 2009. Biogenesis of glycoconjugates in *Mycobacterium tuberculosis*. Adv. Appl. Microbiol. (in press)

Khasnobis, S., Zhang, J., Angala, S.K., et al., 2006. Characterization of a specific arabinosyltransferase activity involved in mycobacterial arabinan biosynthesis. Chem. Biol. 13, 787–795.

Kolattukudy, P.E., Fernandes, N.D., Azad, A.K., Fitzmaurice, A.M., Sirakova, T.D., 1997. Biochemistry and molecular genetics of cell-wall lipid biosynthesis in mycobacteria. Mol. Microbiol. 24, 263–270.

Kordulakova, J., Gilleron, M., Mikusova, K., et al., 2002. Definition of the first mannosylation step in phosphatidylinositol mannoside synthesis. PimA is essential for growth of mycobacteria. J. Biol. Chem. 277, 31335–31344.

Kremer, L., Dover, L.G., Morehouse, C., et al., 2001. Galactan biosynthesis in *Mycobacterium tuberculosis*: Identification of a bifunctional UDP-Galactofuranosyltransferase. J. Biol. Chem. 276, 26430–26440.

Kremer, L., Gurcha, S.S., Bifani, P., et al., 2002a. Characterization of a putative α-mannosyltransferase involved in phosphatidylinositol trimannoside biosynthesis in *Mycobacterium tuberculosis*. Biochem. J. 363, 437–447.

Kremer, L., Dover, L.G., Carrere, S., et al., 2002b. Mycolic acid biosynthesis and enzymic characterization of the β-ketoacyl-ACP synthase A-condensing enzyme from *Mycobacterium tuberculosis*. Biochem. J. 364, 423–430.

Kumar, P., Schelle, M.W., Jain, M., et al., 2007. PapA1 and PapA2 are acyltransferases essential for the biosynthesis of the *Mycobacterium tuberculosis* virulence factor sulfolipid-1. Proc. Natl. Acad. Sci. USA 104, 11221–11226.

Lea-Smith, D.J., Martin, K.L., Pyke, J.S., et al., 2008. Analysis of a new mannosyltransferase required for the synthesis of phosphatidylinositol mannosides and lipoarbinomannan reveals two lipomannan pools in corynebacterineae. J. Biol. Chem. 283, 6773–6782.

Liu, H., Thorson, J.S., 1994. Pathways and mechanisms in the biogenesis of novel deoxysugars by bacteria. Annu. Rev. Microbiol. 48, 223–256.

Ma, Y.F., Mills, J.A., Belisle, J.T., et al., 1997. Determination of the pathway for rhamnose biosynthesis in mycobacteria: Cloning, sequencing and expression of the *Mycobacterium tuberculosis* gene encoding α-D-glucose-1-phosphate thymidylyltransferase. Microbiology 143, 937–945.

Mahapatra, S., Crick, D.C., Brennan, P.J., 2000. Comparison of the UDP-*N*-acetylmuramate: L-alanine ligase enzymes from *Mycobacterium tuberculosis* and *Mycobacterium leprae*. J. Bacteriol. 182, 6827–6830.

Mahapatra, S., Yagi, T., Belisle, J.T., et al., 2005. Mycobacterial lipid II is composed of a complex mixture of modified muramyl and peptide moieties linked to decaprenyl phosphate. J. Bacteriol. 187, 2747–2757.

Malaga, W., Constant, P., Euphrasie, D., et al., 2008. Deciphering the genetic bases of the structural diversity of phenolic glycolipids in strains of the *Mycobacterium tuberculosis* complex. J. Biol. Chem. 283, 15177–15184.

McNeil, M., Daffe, M., Brennan, P.J., 1990. Evidence for the nature of the link between the arabinogalactan and peptidoglycan of mycobacterial cell walls. J. Biol. Chem. 265, 18200–18206.

Mikusova, K., Slayden, R.A., Besra, G.S., Brennan, P.J., 1995. Biogenesis of the mycobacterial cell wall and the site of action of ethambutol. Antimicrob. Agents Chemother. 39, 2484–2489.

Mikusova, K., Mikus, M., Besra, G.S., Hancock, I., Brennan, P.J., 1996. Biosynthesis of the linkage region of the mycobacterial cell wall. J. Biol. Chem. 271, 7820–7828.

Mikusova, K., Yagi, T., Stern, R., et al., 2000. Biosynthesis of the galactan component of the mycobacterial cell wall. J. Biol. Chem. 275, 33890–33897.

Mikusova, K., Huang, H., Yagi, T., et al., 2005. Decaprenylphosphoryl arabinofuranose, the donor of the D-arabinofuranosyl residues of mycobacterial arabinan, is formed via a two-step epimerization of decaprenylphosphoryl ribose. J. Bacteriol. 187, 8020–8025.

Mikusova, K., Belanova, M., Kordulakova, J., et al., 2006. Identification of a novel galactosyl transferase involved in biosynthesis of the mycobacterial cell wall. J. Bacteriol. 188, 6592–6598.

Mishra, A.K., Alderwick, L.J., Rittmann, D., et al., 2007. Identification of an $\alpha(1\rightarrow6)$ mannopyranosyltransferase (MptA), involved in Corynebacterium glutamicum lipomanann biosynthesis, and identification of its orthologue in *Mycobacterium tuberculosis*. Mol. Microbiol. 65, 1503–1517.

Mishra, A.K., Alderwick, L.J., Rittmann, D., et al., 2008. Identification of a novel $\alpha(1\rightarrow6)$ mannopyranosyltransferase MptB from *Corynebacterium glutamicum* by deletion of a conserved gene, NCgl1505, affords a lipomannan- and lipoarabinomannan-deficient mutant. Mol. Microbiol. 68, 1595–1613.

Morita, Y.S., Patterson, J.H., Billman-Jacobe, H., McConville, M.J., 2004. Biosynthesis of mycobacterial phosphatidylinositol mannosides. Biochem J. 378, 589–597.

Morita, Y.S., Sena, C.B., Waller, R.F., et al., 2006. PimE is a polyprenol-phosphate-mannose-dependent mannosyltransferase that transfers the fifth mannose of phosphatidylinositol mannoside in *Mycobacteria*. J. Biol. Chem. 281, 25143–25155.

Mougous, J.D., Petzold, C.J., Senaratne, R.H., et al., 2004. Identification, function and structure of the mycobacterial sulfotransferase that initiates sulfolipid-1 biosynthesis. Nat. Struct. Mol. Biol. 11, 721–729.

Murphy, H.N., Stewart, G.R., Mischenko, V.V., et al., 2005. The OtsAB pathway is essential for trehalose biosynthesis in *Mycobacterium tuberculosis*. J. Biol. Chem. 280, 14524–14529.

Nakano, Y., Yoshida, Y., Suzuki, N., Yamashita, Y., Koga, T., 2000. A gene cluster for the synthesis of serotype d-specific polysaccharide antigen in *Actinobacillus actinomycetemcomitans*. Biochim. Biophys. Acta 1493, 259–263.

Onwueme, K.C., Vos, C.J., Zurita, J., Soll, C.E., Quadri, L.E., 2005. Identification of phthiodiolone ketoreductase, an enzyme required for production of mycobacterial diacyl phthiocerol virulence factors. J. Bacteriol. 187, 4760–4766.

Pan, Y.T., Carroll, J.D., Elbein, A.D., 2002. Trehalose-phosphate synthase of *Mycobacterium tuberculosis*. Cloning, expression and properties of the recombinant enzyme. Eur. J. Biochem. 269, 6091–6100.

Parish, T., Liu, J., Nikaido, H., Stoker, N.G., 1997. A *Mycobacterium smegmatis* mutant with a defective inositol monophosphate phosphatase gene homolog has altered cell envelope permeability. J. Bacteriol. 179, 7827–7833.

Paulus, H., Kennedy, E.P., 1960. The enzymatic synthesis of inositol monophosphatide. J. Biol. Chem. 235, 1303–1311.

Perez, E., Constant, P., Lemassu, A., Laval, F., Daffe, M., Guilhot, C., 2004a. Characterization of three glycosyltransferases involved in the biosynthesis of the phenolic glycolipid antigens from the *Mycobacterium tuberculosis* complex. J. Biol. Chem. 279, 42574–42583.

Perez, E., Constant, P., Laval, F., et al., 2004b. Molecular dissection of the role of two methyltransferases in the biosynthesis of phenolglycolipids and phthiocerol dimycoserosate in the *Mycobacterium tuberculosis* complex. J. Biol. Chem. 279, 42584–42592.

Raetz, C.R.H., Whitfield, C., 2002. Lipopolysaccharide endotoxins. Ann. Rev. Biochem. 71, 635–700.

Raymond, J.B., Mahapatra, S., Crick, D.C., Pavelka, M.S., 2005. Identification of the namH gene, encoding the hydroxylase responsible for the *N*-glycolylation of the mycobacterial peptidoglycan. J. Biol. Chem. 280, 326–333.

Rose, N.L., Completo, G.C., Lin, S.J., McNeil, M., Palcic, M.M., Lowary, T.L., 2006. Expression, purification, and characterization of a galactofuranosyltransferase involved in *Mycobacterium tuberculosis* arabinogalactan biosynthesis. J. Am Chem Soc. 128, 6721–6729.

Salman, M., Lonsdale, J.T., Besra, G.S., Brennan, P.J., 1999. Phosphatidylinositol synthesis in mycobacteria. Biochim. Biophys. Acta 1436, 437–450.

Schaeffer, M.L., Khoo, K.H., Besra, G.S., et al., 1999. The pimB gene of *Mycobacterium tuberculosis* encodes a mannosyltransferase involved in lipoarabinomannan biosynthesis. J. Biol. Chem. 274, 31625–31631.

Schultz, J., Elbein, A.D., 1974. Biosynthesis of mannosyl- and glucosyl-phosphoryl polyprenols in *Mycobacterium smegmatis*. Evidence for oligosaccharide-phosphorylpolyprenols. Arch. Biochem. Biophys. 160, 311–322.

Schultz, J.C., Takayama, K., 1975. The role of mannosylphosphorylpolyisoprenol in glycoprotein biosynthesis in *Mycobacterium smegmatis*. Biochim. Biophys. Acta. 381, 175–184.

Seidel, M., Alderwick, L.J., Birch, H.L., Sahm, H., Eggeling, L., Besra, G.S., 2007. Identification of a novel arabinofuranosyltransferase AftB involved in a terminal step of cell wall arabinan biosynthesis in *Corynebacterianeae*, such as *Corynebacterium glutamicum* and *Mycobacterium tuberculosis*. J. Biol. Chem. 282, 14729–14740.

Shi, L., Zhou, R., Liu, Z., Lowary, T.L., et al., 2008. Transfer of the first arabinofuranose residue to galactan is essential for *Mycobacterium smegmatis* viability. J. Bacteriol. 190, 5248–5255.

Simeone, R., Constant, P., Malaga, W., Guilhot, C., Daffe, M., Chalut, C., 2007a. Molecular dissection of the biosynthetic relationship between phthiocerol and phthiodiolone dimycocerosates and their critical role in the virulence and permeability of *Mycobacterium tuberculosis*. FEBS J. 274, 1957–1969.

Simeone, R., Constant, P., Guilhot, C., Daffe, M., Chalut, C., 2007b. Identification of the missing trans-acting enoyl reductase required for phthiocerol dimycocerosates and phenolglycolipids biosynthesis in *M. tuberculosis*. J. Bacteriol. 189, 4597–4602.

Stevenson, G., Neal, B., Liu, D., et al., 1994. Structure of the O antigen of *Escherichia coli* K-12 and the sequence of its *rfb* gene cluster. J. Bacteriol. 176, 4144–4156.

Takayama, K., Goldman, D.S., 1969. Pathway for the synthesis of mannophospholipids in *Mycobacterium tuberculosis*. Biochim. Biophys. Acta 176, 196–198.

Takayama, K., Wang, C., Besra, G.S., 2005. Pathway to synthesis and processing of mycolic acids in *Mycobacterium tuberculosis*. Clin. Microbiol. Rev. 18, 81–101.

Tatituri, R.V., Illarionov, P.A., Dover, L.G., et al., 2007. Inactivation of *Corynebacterium glutamicum* NCgl0452 and the role of MgtA in the biosynthesis of a novel mannosylated glycolipid involved in lipomannan biosynthesis. J. Biol. Chem. 282, 4561–4572.

Trivedi, O.A., Arora, P., Sridharan, V., Tickoo, R., Mohanty, D., Gokhale, R.S., 2004. Enzymic activation and transfer of fatty acids as acyl-adenylates in mycobacteria. Nature 428, 441–445.

Trivedi, O.A., Arora, P., Vats, A., et al., 2005. Dissecting the mechanism and assembly of a complex virulence mycobacterial lipid. Mol. Cell 17, 631–643.

Weston, A., Stern, R.J., Lee, R.E., et al., 1997. Biosynthetic origin of mycobacterial cell wall galactofuranosyl residues. Tuber. Lung. Dis. 78, 123–131.

Wimmerova, M., Engelsen, S.B., Bettler, E., Breton, C., Imberty, A., 2003. Combining fold recognition and exploratory data analysis for searching for glycosyltransferases in the genome of *Mycobacterium tuberculosis*. Biochimie 85, 691–700.

Wolucka, B.A., McNeil, M.R., de Hoffmann, E., Chojnacki, T., Brennan, P.J., 1994. Recognition of the lipid intermediate for arabinogalactan/arabinomannan biosynthesis and its relation to the mode of action of ethambutol on mycobacteria. J. Biol. Chem. 269, 23328–23335.

Woodruff, P.J., Carlson, B.L., Siridechadilok, B., et al., 2004. Trehalose is required for growth of *Mycobacterium smegmatis*. J. Biol. Chem. 279, 28835–28843.

Yagi, T., Mahapatra, S., Mikusova, K., Crick, D.C., Brennan, P.J., 2003. Polymerization of mycobacterial arabinogalactan and ligation to peptidoglycan. J. Biol. Chem. 278, 26497–26504.

Yokoyama, K., Ballou, C.E., 1989. Synthesis of α-1–6 mannooligosaccharide in *M. smegmatis*. JBC 264, 21621–21628.

Zhang, N., Torrelles, J.B., McNeil, M.R., et al., 2003. The Emb proteins of mycobacteria direct arabinosylation of lipoarabinomannan and arabinogalactan via an N-terminal recognition region and a C-terminal synthetic region. Mol. Microbiol. 50, 69–76.

CHAPTER

22

Biosynthesis of fungal and yeast glycans

Morgann C. Reilly and Tamara L. Doering

SUMMARY

The glycobiology of eukaryotes presents a tremendous diversity of structures and functions, which is well-exemplified within the fungal kingdom. Glycans of fungi include a broad variety of protein-linked structures, extensive polymers that comprise cell walls, molecules for energy storage and elaborate surface structures that help pathogens evade the host immune response. Many of these molecules and the processes they mediate are essential to the survival of the organism. This chapter addresses the biosynthesis of these fascinating compounds. The most detailed studies of glycan and glycoconjugate biosynthesis in fungi have been performed in the model organism *Saccharomyces cerevisiae*; many of the discoveries made in this yeast have provided insight into related mechanisms in other fungi as well as eukaryotes in general. These studies serve as the core of this chapter, with attention drawn to known variations on the biosynthetic processes as they occur in other fungal species. Significant differences in glycan biosynthesis between fungi and mammals are also highlighted, as these provide clues to potential drug targets in the case of fungal pathogens.

Keywords: Fungi; Yeast; *N*-Linked glycosylation; *O*-Linked glycosylation; Glycophosphatidylinositol; Glycosylinositol phosphorylceramide; Glycosylceramide; Glucan; Chitin; Glycogen; Trehalose; Capsule

1. INTRODUCTION

Fungi produce a remarkable range of glycans and glycoconjugates. The pathways responsible for the synthesis of these structures differ greatly from those of prokaryotic organisms because of the distinct structures formed and to accommodate the topology of eukaryotic organelles. Fungal pathways leading to the formation of some glycans, such as those linked to protein, present variations on general eukaryotic themes, while pathways forming other structures, such as the cell wall, are unique to these organisms.

In this chapter, an overview of fungal glycan biosynthesis is presented, beginning with the precursor molecules that contribute to biosynthetic pathways and proceeding to individual classes of glycoconjugates. Detailed studies of the model yeast *Saccharomyces cerevisiae* form the core of our knowledge of these processes; this genetically manipulable system has been extensively investigated with regard to its carbohydrate structures. Beyond this single species, however, lies a diverse world of fungi that has been largely unexplored. As space permits, differences in glycosylation and biosynthetic machinery between *S. cerevisiae* and other fungal species are described and variations between fungi and higher eukaryotes are mentioned. The latter is particularly important in the case of pathogenic fungi, as unique aspects of biosynthesis may present opportunities for the development of selective anti-fungal chemotherapies.

2. PRECURSORS FOR GLYCAN SYNTHESIS

In both eukaryotes and prokaryotes, glycan synthesis requires sugar donor molecules. These active precursors are usually either nucleotide triphosphate sugars or dolichol-phosphate sugars, depending on the synthetic context. For the purpose of this discussion, the focus will be on mannose (Man), one of the more common monosaccharides in fungal glycans.

2.1. Nucleotide-Linked sugar synthesis and localization

The activated nucleotide sugar donor of mannose (Man), guanosine diphosphate- (GDP-) Man, is synthesized from guanosine triphosphate (GTP) and Man-1-phosphate (Man-1-P) in a reaction catalysed by GDP-Man pyrophosphorylase. The Man-1-P is derived either from the phosphorylation of Man obtained from the environment or originates within the cell when generated from other sugar phosphates or via other metabolic processes.

GDP-Mannose, like many other nucleotide sugars, is made in the cytosol. This presents a topological problem for cells that produce most of their glycoconjugates in the organelles of the secretory pathway. The cellular solution to this dilemma is to produce specific nucleotide sugar transporters (NSTs) to move these charged donor molecules across the membrane of the endoplasmic reticulum (ER) or Golgi apparatus. The NSTs are multimembrane spanning proteins that act as antiporters, importing nucleotide sugars into an organelle in exchange for the corresponding nucleotide monophosphate. For example, the Golgi-localized GDP-Man transporter, Vrg4p in *S. cerevisiae* (Poster and Dean, 1996), imports GDP-Man in exchange for guanosine monophospahte (GMP); the latter is derived from the cleavage of the GDP produced when Man is transferred from the nucleotide sugar to a growing glycoconjugate.

The *S. cerevisiae* protein Vrg4p is essential, as are its homologues in the pathogens *Candida albicans* and *Candida glabrata* (Nishikawa *et al.*, 2002a,b). In contrast, *Cryptococcus neoformans*, a basidiomycetous fungal pathogen of humans with two functional GDP-Man transporters, is able to survive, albeit with poor growth, even in the absence of both proteins (Cottrell *et al.*, 2007). Importantly, higher eukaryotes have no such proteins as NSTs: mammalian Golgi mannosylation uses lipid-linked Man as the donor for comparable secretory protein modification (see below). These transporters in pathogenic fungi thus represent potential drug targets. The occurrence and specificity of NSTs for other nucleotide sugars are discussed in several excellent reviews (Hirschberg *et al.*, 1998; Berninsone and Hirschberg, 2000; Gerardy-Schahn *et al.*, 2001).

2.2. Dolichol-Linked sugar synthesis

Dolichol-linked sugars are critical donors of monosaccharide residues in the synthesis of eukaryotic glycan structures (reviewed in Maeda and Kinoshita, 2008). Again using Man as an example, dolichol phosphate mannose (Dol-P-Man) is formed on the cytosolic leaflet of the ER membrane by the enzyme dolichol phosphate mannose synthase (Dpm1p), which catalyses the transfer of Man from cytosolic GDP-Man to membrane-associated dolichol monophosphate (Dol-P). *S. cerevisiae* and the plant pathogen *Ustilago maydis* encode a Dpm1p with a hydrophobic region that localizes the protein to the ER. In contrast, the model fission yeast *Schizosaccharomyces pombe* encodes a Dpm1p without a transmembrane domain. To mediate the same reaction, the *S. pombe* Dpm1p therefore requires the presence of two additional proteins, Dpm2p and Dpm3p, which are thought to stabilize and localize the catalytic subunit (Maeda and Kinoshita, 2008). The lipid moiety of Dol-P-Man allows an unidentified flippase to translocate the active sugar donor to the lumenal leaflet of the ER membrane where it is appropriately situated to participate in reactions of glycan synthesis. Dolichol may also be used as a platform for the assembly of larger biosynthetic intermediates; these compounds are discussed below in the context of protein glycosylation.

3. FUNGAL PROTEIN GLYCOSYLATION

The carbohydrate modifications of a glycoprotein can contribute to the final conformation, stability, function and localization of the polypeptide. Glycans are typically associated with fungal proteins in one of three ways:

(i) *N*-glycosylation, where the glycan is linked to an asparagine (Asn) residue;
(ii) *O*-glycosylation, where the glycan is linked to the hydroxy group of serine (Ser) or threonine (Thr); and
(iii) glycosylphosphatidylinositol (GPI) anchors, where the glycolipid is linked to a C-terminal amino acid.

The transfer of these carbohydrate structures is initiated during or soon after the translocation of nascent polypeptide chains into the ER lumen. The core glycans are further elaborated by the actions of glycosidases and glycosyltransferases in the ER and Golgi as the protein traverses the secretory pathway, yielding a diverse array of structures.

3.1. N-Linked glycan synthesis

N-Glycosylation begins with the assembly of a dolichol-linked oligosaccharide precursor at the cytoplasmic leaflet of the ER membrane (recently reviewed in Weerapana and Imperiali, 2006) (Figure 22.1). In *S. cerevisiae*, the glycosylphosphotransferase Alg7p first transfers *N*-acetylglucosamine (GlcNAc)-phosphate from UDP-GlcNAc to membrane-bound Dol-P to form Dol-PP-GlcNAc. A second GlcNAc residue is added by the dimer Alg13p/Alg14p (Bickel *et al.*, 2005; Gao *et al.*, 2005) and the structure is further modified by a series of mannosyltransferases (Alg1p, Alg2p and Alg11p) which add five mannose residues derived from GDP-Man (O'Reilly *et al.*, 2006). The resulting Dol-PP-$GlcNAc_2Man_5$ molecule is then translocated by the flippase Rft1p to the lumenal leaflet of the ER (Helenius *et al.*, 2002).

Synthesis of the *N*-glycan precursor continues with the transfer of four Man residues from Dol-P-Man to the oligosaccharide core by the mannosyltransferases Alg3p, Alg12p and Alg9p. Finally, the Dol-PP-$GlcNAc_2Man_9$ structure is capped with three glucose (Glc) residues (from Dol-P-Glc) via the actions of the glucosyltransferases Alg6p, Alg8p and Alg10p. The resulting Dol-PP-$GlcNAc_2Man_9Glc_3$ structure contains the complete core glycan that can be transferred from its dolichol anchor to a protein. This polypeptide modification occurs at the Asn residue of the sequence motif Asn-X-Ser/Thr (where X is any amino acid except for proline) and is catalysed by the oligosaccharyltransferase (OST) protein complex. In *S. cerevisiae*, the OST is comprised of at least eight proteins; the highly conserved Stt3p harbours the transferase activity of the complex and the other proteins mediate association of the diverse reaction substrates (Lennarz, 2007).

Following transfer of the $GlcNAc_2Man_9Glc_3$ core to the nascent polypeptide, two lumenal glucosidases act upon the core glycan (Figure 22.2, top): α-glucosidase I (encoded by

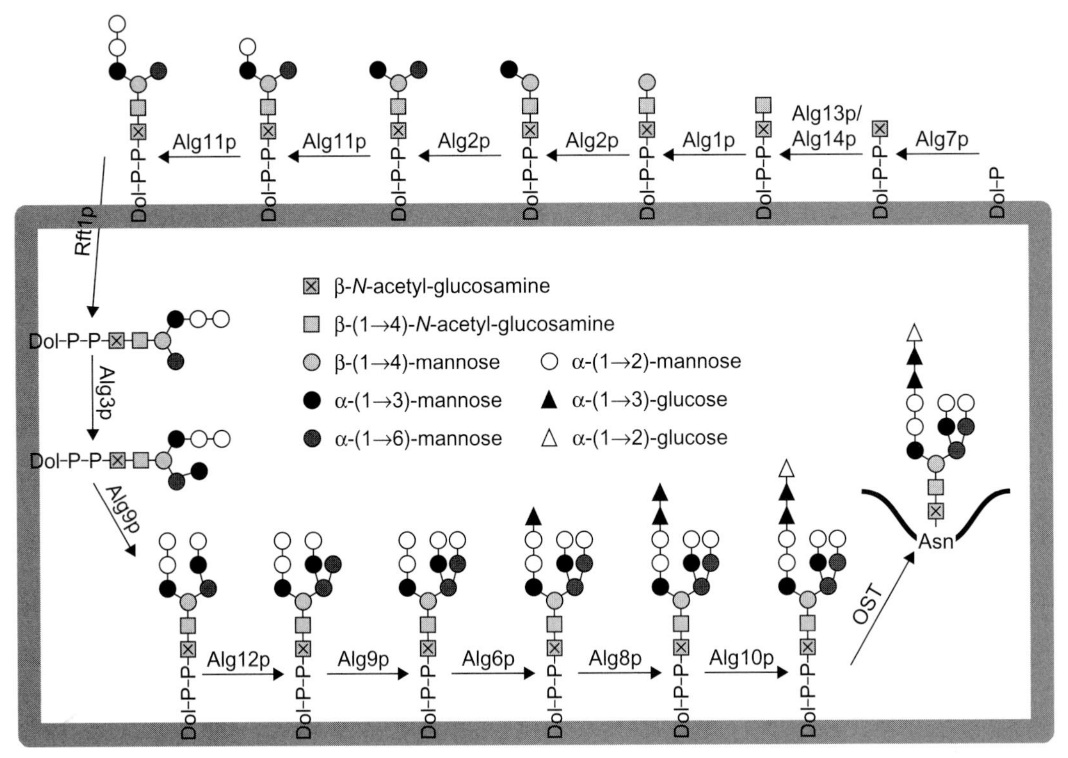

FIGURE 22.1 *N*-Linked glycosylation in *S. cerevisiae*: core synthesis and transfer. The grey rectangle indicates the ER membrane; Dol, dolichol; P, phosphate; Asn, the acceptor amino acid (asparagine) of a schematically indicated nascent polypeptide. Other symbols are indicated within the figure.

CWH41) removes the terminal α-(1→2)-Glc, while α-glucosidase II (encoded by *ROT2*) removes the distal α-(1→3)-Glc to yield $GlcNAc_2Man_9Glc$ (reviewed in Herscovics, 1999). In this mono-glucosylated state, the unfolded polypeptide is recognized by molecular chaperones that assist in glycoprotein folding. (The role of *N*-linked glycans in protein quality control in the ER is an extensive subject that is beyond the scope of this chapter; the topic is thoroughly addressed in a review by Moreman and Molinari, 2006). Once the glycoprotein has achieved its proper conformation, it is released by the molecular chaperones and Rot2p removes the final Glc residue from the *N*-linked oligosaccharide. In *S. cerevisiae*, the resulting structure is further trimmed by the mannosidase Mns1p, which removes a single α-(1→2)-Linked Man residue. The folded protein, with its $GlcNAc_2Man_8$ modification, is then transferred from the ER to the Golgi.

Fungi do not generate the "complex-type" *N*-glycans typical of mammalian systems (Varki *et al.*, 1999). Instead, the core oligosaccharide structures of *N*-linked glycans are either minimally modified ("core-type") or receive extensive modifications ("highly mannosylated"). In *S. cerevisiae*, protein modifications involve Man addition exclusively. When a $GlcNAc_2Man_8$ modified polypeptide arrives in the Golgi apparatus, Och1p adds a single α-(1→6)-Man to the structure. A glycoprotein that will retain a core-type *N*-linked glycan structure is further modified at this new Man by

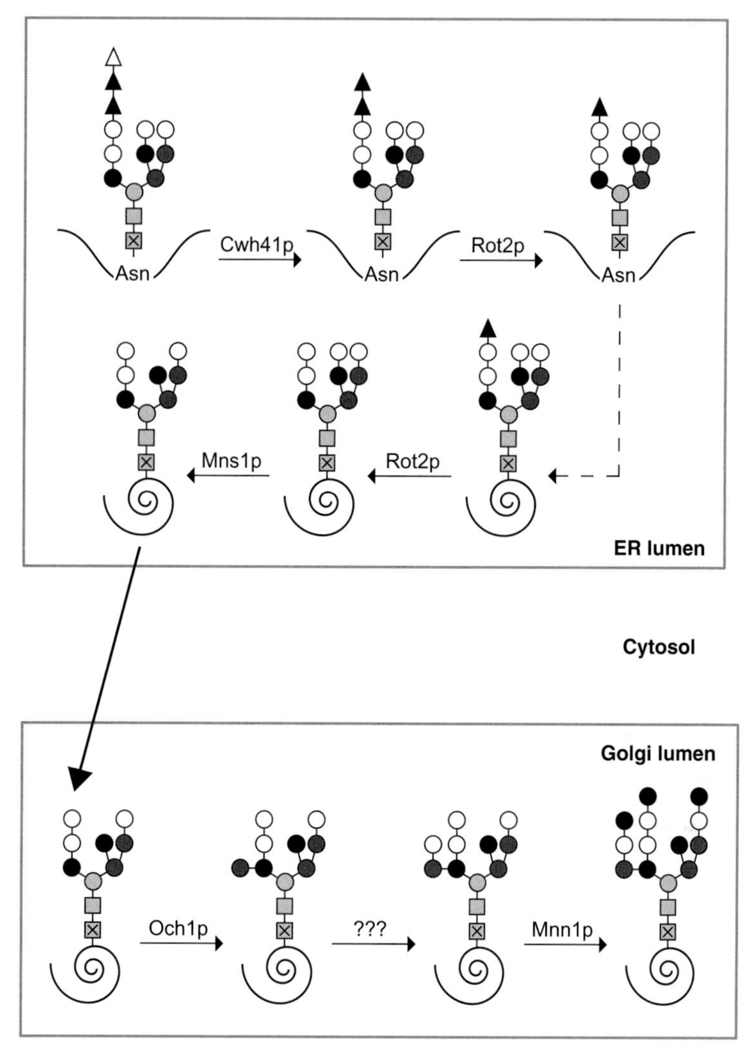

FIGURE 22.2 *N*-Linked glycosylation in *S. cerevisiae*: ER core processing (top) and Golgi processing of core-type structures (bottom). The spiral indicates that the polypeptide has been properly folded; the dashed arrow indicates protein folding; and the large arrow indicates progression to the Golgi. All other symbols are as in Figure 22.1.

an α-(1→2)-mannosyltransferase (the protein responsible has not yet been identified) that adds a single Man and subsequent capping of secondary branches by the α-(1→3)-mannosyltransferase, Mnn1p (see Figure 22.2, bottom). In contrast to these modest alterations, highly mannosylated *N*-linked glycans receive extensive modifications beyond the actions of Och1p (Figure 22.3). First, the enzyme complex mannan polymerase I (M-Pol I) modifies the new Man with a linear branch of α-(1→6)-linked Man that is upwards of ten residues in length; a second mannan polymerase complex (M-Pol II) extends this branch with up to fifty more α-(1→6)-Man residues. These α-(1→6)-Man are elaborated by the addition of α-(1→2)-Man by Mnn2p or Mnn5p and Man-P by the oligomer Mnn4p/Mnn6p. Finally, the secondary branches of these extensive Man chains are capped with an α-(1→3)-linked Man residue by Mnn1p. Proteins that receive extensive Man chains, so-called mannans, frequently localize to the cell wall where covalent linkages form between mannan and other cell wall glycan polymers (see below).

While most *N*-glycan synthetic events are conserved among fungi, there is some variation between species. With regard to the core oligosaccharide, for example, the genome of *C. neoformans* lacks any homologues of the Alg6p, Alg8p and Alg10p glucosyltransferases and the organism accordingly generates a truncated core glycan (Samuelson *et al.*, 2005). *S. pombe* and *Kluyveromyces lactis*, a yeast used in industry, do not appear to have a functional homologue of the mannosidase Mns1p, such that proteins are transferred from the ER to the Golgi with intact $GlcNAc_2Man_9$ structures. The greatest diversity with regard to the number and linkages of Man residues occurs within the Golgi apparatus, where elaboration of the linear α-(1→6)-Man branch can create a structure of up to two hundred Man residues. Here, the model mould *Neurospora crassa* and several members of the *Aspergillus* genus are unusual in their incorporation of galactofuranose (Gal*f*) residues into

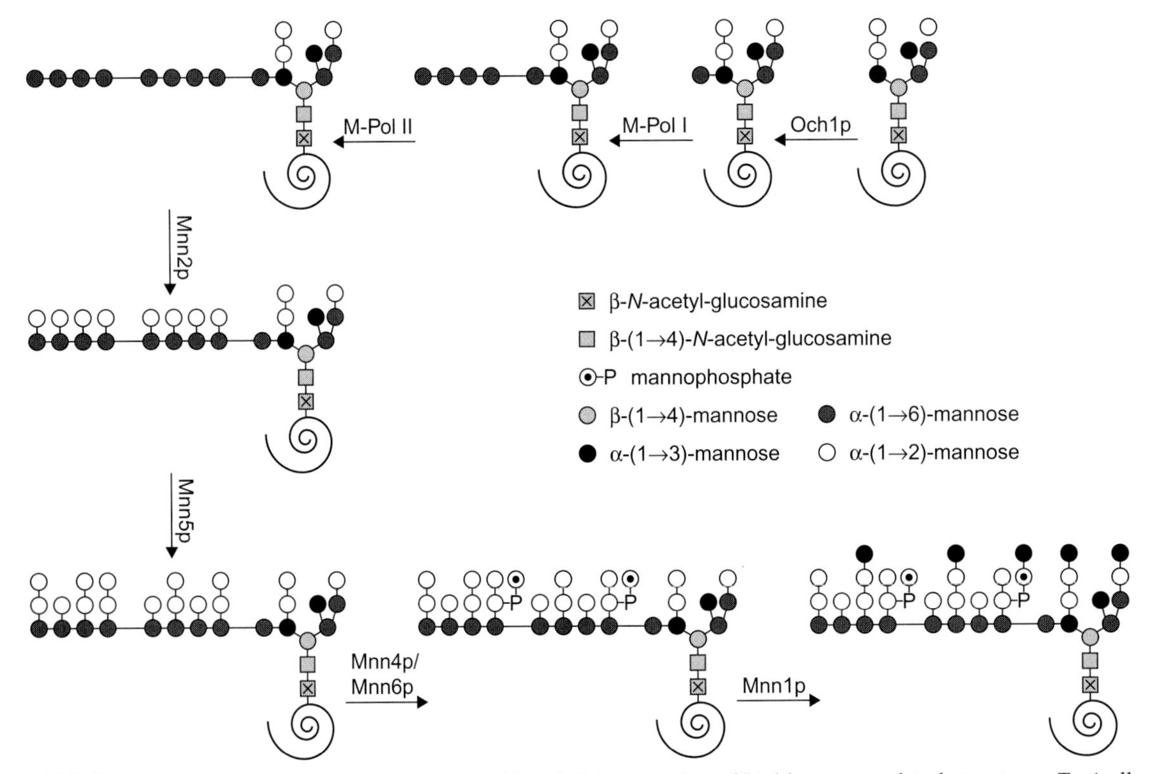

FIGURE 22.3 *N*-Linked glycosylation in *S. cerevisiae*: Golgi processing of highly mannosylated structures. Typically, M-Pol I adds upwards of ten Man residues and M-Pol II adds upwards of fifty Man residues; for the purposes of this illustration, only four Man are shown for each. Modifications by Mnn5p, Mnn6p and Mnn1p are variable; a sample structure is shown. Symbols are indicated on the diagram.

some *N*-linked glycan structures (Nakajima *et al.*, 1984; Wallis *et al.*, 2001; Morelle *et al.*, 2005).

3.2. O-Linked glycan synthesis

Fungal *O*-glycosylation is initiated in the ER lumen with the transfer of a Man residue from Dol-P-Man to a Ser or Thr residue by protein *O*-mannosyltransferases (PMTs) (Figure 22.4, top). Unlike *N*-glycosylation, no consensus sequence dictating which residues in a polypeptide will be *O*-glycosylated has been elucidated. Seven PMTs have been identified in *S. cerevisiae*. These integral membrane proteins exhibit 50–80% homology and are classified into three major subfamilies based on similarities in hydropathy profiles: the PMT1 family (Pmt1p and Pmt5); the PMT2 family (Pmt2p, Pmt3p and Pmt6p); and the PMT4 family (Pmt4p) (detailed in Willer *et al.*, 2003; Goto, 2007). Members of the PMT1 and PMT2 families dimerize with one another while Pmt4p forms homomeric complexes; the resulting Pmt complexes exhibit varying substrate specificities. In *S. cerevisiae*, PMT activity is essential, although redundancy in the proteins allows viability of some single and double mutants (discussed in Goto, 2007). Attesting to the importance of *O*-glycosylation in fungi, strains with PMT defects exhibit alterations in growth, cell wall integrity, morphology, development and virulence. These effects may be a direct consequence of glycan loss or may be due to the role of these modifications in mediating protein stability, localization or function.

Echoing the progress of *N*-glycosylation, the core Man of *O*-glycans that is added to protein in the ER is elaborated in the Golgi apparatus, where one to six Man residues derived from GDP-Man may be added to extend the linear chain (see Figure 22.4, bottom). Mannose may be added in α-(1→2)-linkages through the actions of the KTR family (Ktr1p, Ktr3p and Kre2p) or in α-(1→3)-linkages by the MNN1 family (Mnn1p, Mnt2p and Mnt3p). *S. cerevisiae* also links Man-P to the second Man in some *O*-glycan structures (Jigami and Odani, 1999).

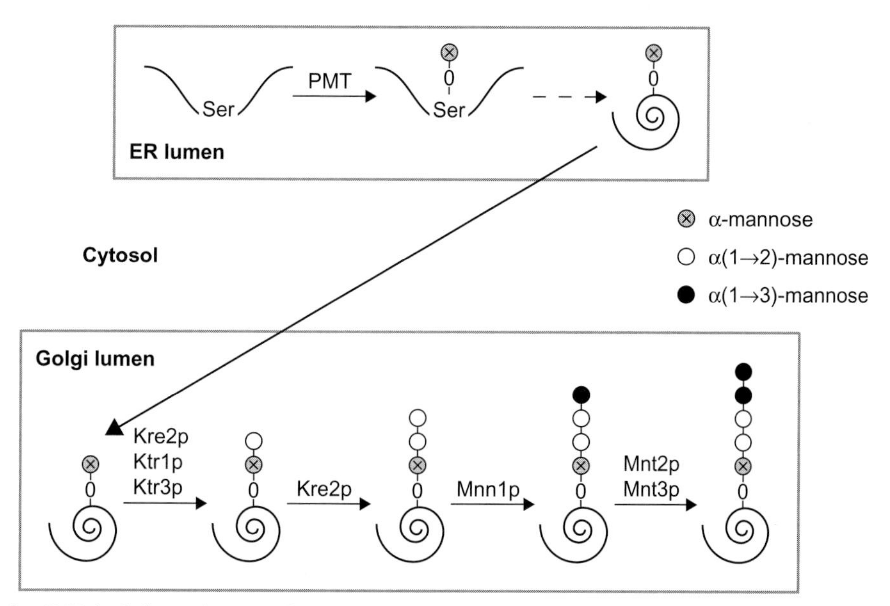

FIGURE 22.4 *O*-Linked glycosylation in *S. cerevisiae*: ER (top) and Golgi (bottom) steps in synthesis of a Ser-linked glycan are shown. Symbols are as in Figure 22.2 or indicated on the diagram.

The process of *O*-glycosylation described above for *S. cerevisiae* is thought to proceed similarly in other fungal species, although some variations have been identified. For example, while all fungi have members of each of the three PMT subfamilies identified in *S. cerevisiae*, many do not exhibit the same extent of redundancy. In contrast to *S. cerevisiae*, *C. albicans* has only five PMTs; other species, including *S. pombe* and the filamentous fungus *Aspergillus nidulans*, encode just one enzyme in each subfamily (Prill *et al.*, 2005).

Variation in *O*-glycan structures suggests that *O*-glycan synthesis in many fungi involves enzymes beyond those identified in *S. cerevisiae*, which tends to utilize a simple repertoire of sugars. For example, *S. pombe* adds galactose (Gal) residues to the non-reducing end of short α-(1→2)-Man chains (Gemmill and Trimble, 1999). Studies of the environmental yeast *Cryptococcus laurentii* have defined three *O*-linked glycans: α-Gal-(1→6))$_{10}$-β-Gal-Man; α-Man-(1→2)-α-Man-(1→2)-Man; and α-Man-(1→2)-α-Man(1→6)-α-Man-(1→3)[β-Xyl-(1→2)]-Man (where Xyl is xylose). Biochemical studies have allowed detection of enzyme activities potentially involved in most of the synthetic steps required to generate these structures (reviewed in Schutzbach *et al.*, 2007). In contrast, simple *O*-glycans containing only α-(1→2)-Man residues are found in *C. albicans* and the methylotropic yeast *Pichia pastoris* (Gemmill and Trimble, 1999).

3.3. Synthesis of the GPI anchor

A third major form of protein glycosylation that occurs in fungi is the addition of GPI anchors (for a detailed review, see Orlean and Menon 2007). The basic structure of GPI anchors is conserved across all eukaryotes; in fungi, these glycoconjugates undergo several unique processing events, including remodelling of the GPI lipid to ceramide (Cer) and transfer of anchored polypeptides from the GPI moiety to covalent linkage with cell wall glycans (see Chapter 10).

Synthesis of the GPI anchor begins in the cytoplasmic leaflet of the ER membrane, where GlcNAc is transferred from UDP-GlcNAc to phosphatidylinositol (PI) (Figure 22.5). This process is catalysed in *S. cerevisiae* by the transmembrane protein Gpi3p in association with five other polypeptides that form the GPI-GlcNAc transferase complex. The resulting GlcNAc-PI is then de-*N*-acetylated by Gpi12p to yield GlcN-PI and transferred to the lumenal leaflet of the ER membrane by an unidentified flippase. Once in the lumen, the inositol moiety is palmitoylated by Gwt1p. This is followed by the addition of up to four Man residues (from Dol-P-Man) and up to three phosphoethanolamine (EtnP) moieties. The presence of EtnP on the third Man is absolutely required for the association of the GPI anchor with a protein. Therefore, both Gpi13p (which adds this EtnP) and Smp3p (which adds the fourth Man, whose presence is required for the actions of Gpi13p) are essential. This is not the case in mammalian cells, which require only three Man residues for EtnP addition.

There is no specific amino acid sequence that directs GPI anchorage of a protein, but the residues at and near the addition site (termed the "ω" site) have been analysed in detail. The C-terminus of a protein that will receive a GPI anchor consists of a sequence of ten polar amino acids preceding ω; a glycine (Gly), alanine (Ala), Ser, Asn, aspartic acid (Asp) or Cys at ω; a Gly, Ala or Ser at ω + 1; six or more moderately polar amino acids; and a final stretch of hydrophobic residues that form a transmembrane region. This pattern is recognized by GPI transamidase, a complex of five membrane proteins. The catalytic protein of this complex, Gpi8p, displaces the GPI signal sequence from the target protein (which is initially anchored to the ER membrane by its C-terminal hydrophobic sequence) and transfers the protein molecule to the GPI structure.

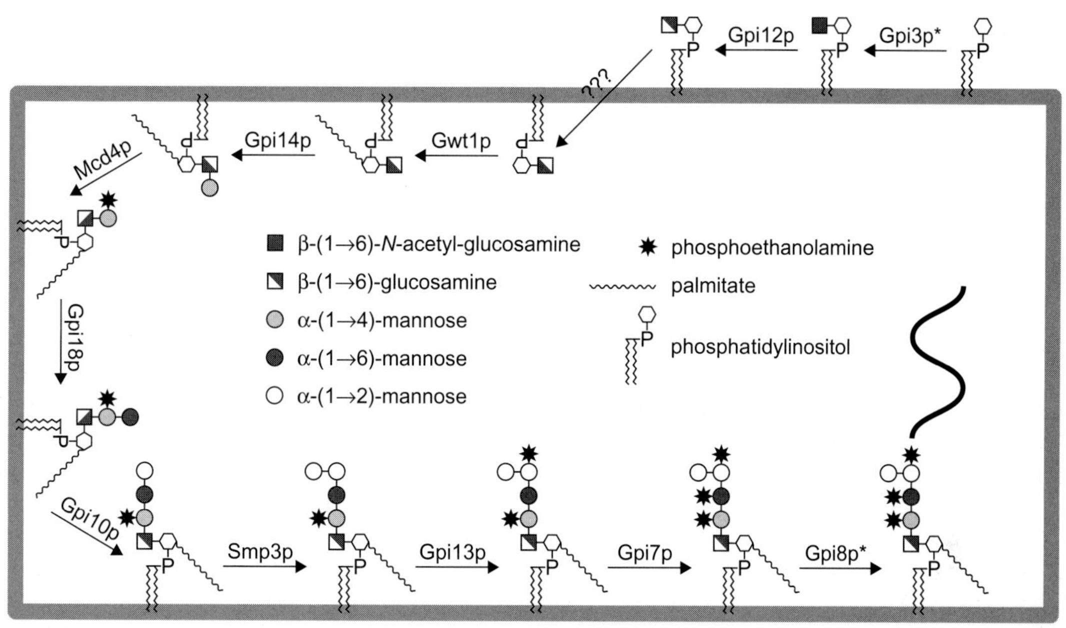

FIGURE 22.5 *S. cerevisiae* GPI-anchor synthesis and addition to protein. The grey rectangle symbolizes the ER membrane and the dark line in the last structure indicates the protein being anchored. Asterisk (*) indicates the catalytic protein of a larger complex described in the text and other symbols are indicated within the figure. Variations in the precise order of lumenal reactions are not shown; for a review see Orlean and Menon (2007).

Both the lipid and glycan components of *S. cerevisiae* GPI anchors are subject to modification. Lipid remodelling promotes the efficient transport and membrane localization of anchored proteins. These steps occur in the ER, beginning with the deacylation of inositol by Bst1p. One fatty acid of the diacylglycerol moiety is then removed by Per1p and replaced with a longer fatty acid by Gup1p; a similar replacement may occur at the other fatty acid chain of the diacylglycerol. In a fungal-specific process, most GPI diacylglycerol moieties are replaced by phytoceramide, though the enzymes responsible have not yet been identified. Further modifications occur following the transport of the GPI-associated protein from the ER to the Golgi apparatus, such as the replacement of the phytoceramide with other Cer species (Reggiori *et al.*, 1997). The glycan portion of GPIs are also sometimes modified by the addition of another α-(1→2)- or α-(1→3)-Man to the fourth Man of the anchor by an unknown mannosyltransferase.

Glypiated proteins in fungi frequently undergo one final transformation, whereby the polypeptide and most of the GPI glycan are transferred from the anchor to a covalent linkage with cell wall glucans. This process has been demonstrated in multiple fungi, including *C. albicans*, *S. pombe*, *Aspergillus niger*, *C. glabrata* and *C. neoformans* as well as in *S. cerevisiae*, and serves to localize proteins to the cell wall. This transglycosylation process does not occur outside of the fungal kingdom and has yet to be defined in terms of enzymology and regulation.

The structure and biosynthesis of GPIs in fungi other than *S. cerevisiae* appear for the most part to be well conserved. Still, work by Franzot and Doering (1999) determined that a range of fatty acids beyond palmitate can modify the inositol group in both *C. neoformans* and *S. cerevisiae*. Fontaine *et al.* (2003) have also reported that the glycan portion of *Aspergillus fumigatus* GPI anchors consistently contains a fifth Man residue (Man_5GlcN), one more than typically found in *S. cerevisiae*.

4. FUNGAL GLYCOLIPIDS

The major glycolipids in fungi are glycosphingolipids. These glycoconjugates function as essential components of the yeast cell membrane, contributing to its fluidity and permeability. Fungal glycosphingolipids consist of one or more monosaccharide residues joined to Cer, a fatty acid-linked to a sphingosine (a long-chain aliphatic amino alcohol). Glycosphingolipids can be divided into two classes: one in which the glycosyl moiety is linked to Cer via inositol phosphate, i.e. glycosylinositol phosphorylceramide (GIPC); and another in which the glycosyl moiety is linked directly to Cer, i.e. glycosylceramide. The former occur as free membrane lipids and as membrane anchors of covalently bound proteins (see discussion of GPI anchors above), while glycosylceramides are associated with the fungal cell wall and are thought to play a role in cell cycle and differentiation (Barreto-Bergter *et al.*, 2004). Importantly, there are significant differences between the structures, and thus the biosynthetic pathways, of fungal glycosphingolipids and those of mammals. This suggests that glycosphingolipid synthesis has potential as a target for antifungal compounds.

4.1. The synthesis of GIPC

Synthesis of *S. cerevisiae* glycosphingolipids (Figure 22.6) begins in the ER with the actions of the serine palmitoyltransferase (SPT) complex. This complex catalyses the condensation of palmitoyl coenzyme A and Ser to generate 3-ketosphinganine. The ketone group of 3-ketosphinganine is reduced in a nicotinamide adenine dinucleotide phosphate- (NADPH-) dependent reaction by Tsc10p to produce sphinganine and this molecule can then be hydroxylated by Sur2p to form phytosphingosine. The Cer synthase complex (Lip1p in combination with either Lac1p or Lag1p) next adds a very long chain fatty acid (C_{24}–C_{26}) to phytosphingosine, generating phytoceramide. Alternatively, sphinganine may first be acylated by the ceramide synthase complex to form dihydroceramide and subsequently hydroxylated by Sur2p to form phytoceramide.

Phytoceramide made in the ER is transported to the Golgi where, in a synthetic step unique to fungi, inositol phosphate is transferred by IPC synthase (Aur1p) from phosphatidylinositol to the Cer. The resulting inositol phosphorylceramide (IPC) is then acted upon by one of two mannosylation complexes (Sur1p and Csg2 or Csh1p and Csg2) that catalyses the transfer of Man from GDP-Man to form mannosylinositol phosphorylceramide (MIPC). Finally, a second inositol phosphate can be transferred from phosphatidylinositol to form mannosyl di-inositol phosphorylceramide [$M(IP)_2C$] in a reaction that requires the products of the *IPT1* and *SKN1* genes. From the Golgi, GIPCs are transported to the outer leaflet of the plasma membrane. The reactions of GIPC synthesis are reviewed elsewhere (Dickson *et al.*, 2006; Cowart and Obeid, 2007).

Some fungal species have evolved variations on the GIPC synthesis pathway outlined above. For example, while the proteins Lcb1p, Lcb2p and Tsc3p are necessary for optimal activity of the SPT complex in *S. cerevisiae*, other fungal species do not appear to require Tsc3p. With regard to variations in structure, the pathogen *Sporothrix shenckii* synthesizes glucosaminyl IPC in addition to MIPC (Toledo *et al.*, 2001) and several fungi, including *C. neoformans* and

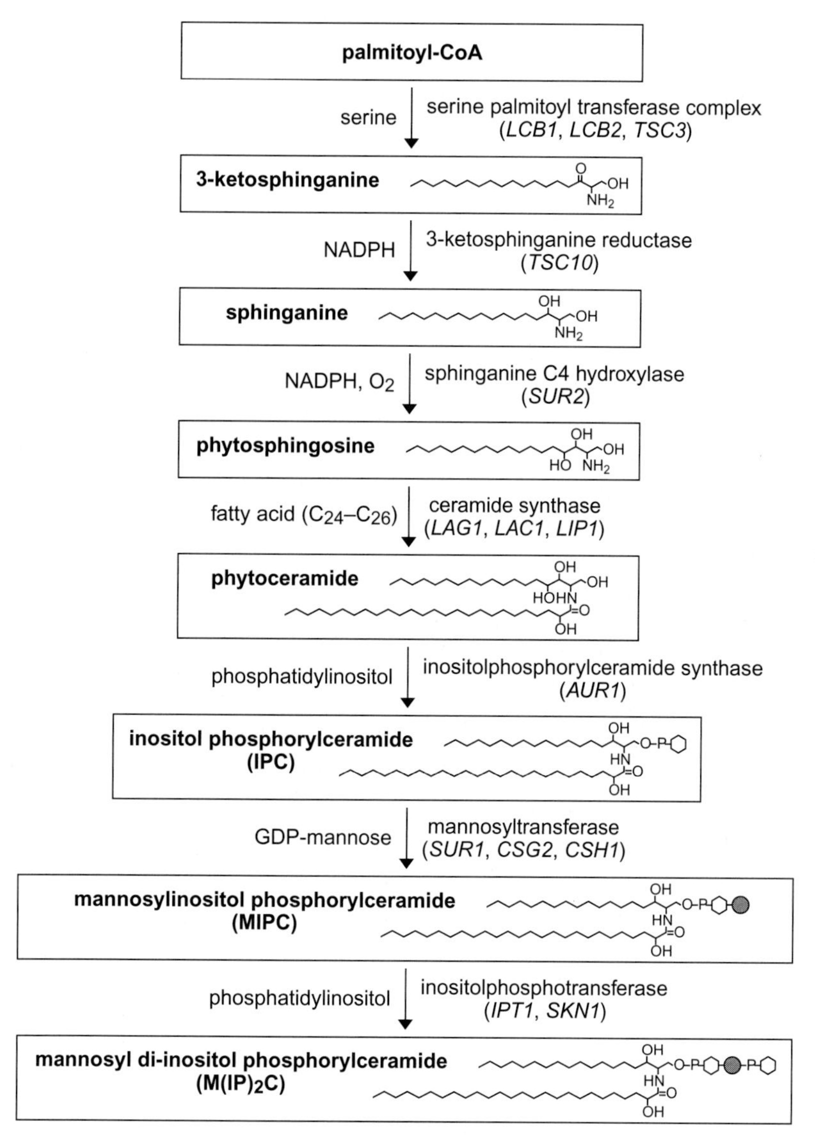

FIGURE 22.6 GIPC synthesis in *S. cerevisiae*. Biosynthetic intermediates are boxed and gene names are indicated in parentheses. Open hexagon indicates inositol; shaded circle indicates mannose. Abbreviations: GDP, guanosine diphosphate; NADPH, nicotinamide adenine dinucleotide phosphate.

A. fumigatus, generate derivatives of MIPC with additional sugar residues such as Man, Xyl, Gal*f* and glucosamine (Heise *et al.*, 2002; Toledo *et al.*, 2007). *C. albicans* incorporates additional phosphate residues into GIPCs: Man-P is added to the Man of MIPC and afterwards modified with a linear chain of β-linked Man residues, generating a surface molecule referred to as phospholipomannan (PLM) (Trinel *et al.*, 2002). Significantly, *C. albicans* yeast strains that are deficient in this pathway are unable to avoid macrophage lysis *in vitro* and exhibit reduced pathogenicity in animal models of candidiasis (Mille *et al.*, 2004).

4.2. Glycosylceramide synthesis

The generation of glycosylceramides proceeds as described above for GIPCs through the generation of phytosphingosine (see Figure 22.6, third step). The pathways then diverge upon the addition of a shorter fatty acid (C_{16}–C_{18}) by the Cer synthase complex (Figure 22.7). The resulting phytoceramide undergoes additional lipid modifications of desaturation and methylation and ultimately is glycosylated. Helpful reviews of fungal glycosphingolipid synthesis include those of Warnecke and Heinz (2003) and Rhome *et al.* (2007).

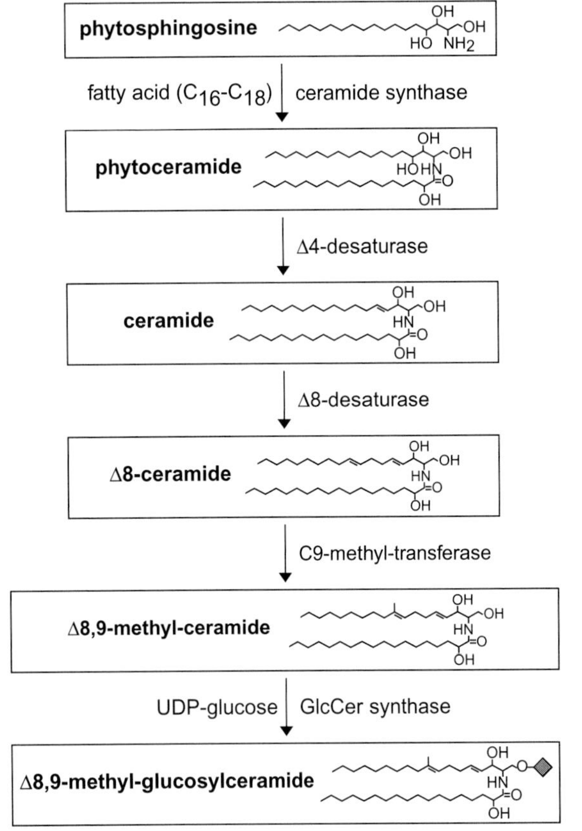

FIGURE 22.7 Glycosylceramide synthesis in fungi. Biosynthetic intermediates are boxed. Shaded diamond indicates glucose. Abbreviations: GlcCer, glucosylceramide; UDP, uridine diphosphate.

There are numerous variations on the synthesis of fungal glycosylceramides. Most fungi incorporate a Glc residue in the last step of synthesis through the actions of Gsc1p, a glucosylceramide synthase (see Figure 22.7). However, a subset of species, including *S. cerevisiae*, instead utilizes an unidentified Cer galactosyltransferase to attach a Gal residue. The monosaccharide headgroups can also be elongated, as when unidentified glucosylceramide galactosyltransferases form GalGlcCer in the plant pathogen *Magnaporthe griseae* (Maciel *et al.*, 2002) or Gal_3GlcCer in *N. crassa* (Lester *et al.*, 1974). In addition, the lipid portions of fungal glycosylceramides may be varied, as by the actions of a fatty acid Δ3-desaturase (Warnecke and Heizne, 2003). Creative studies by Leipelt *et al.* (2001), in which the endogenous *GCS1* of *P. pastoris* was replaced with homologues from other fungi, demonstrated the ability of these enzymes to glycosylate sphingolipids with longer chain fatty acids (C_{24}–C_{26}); similar structures containing elongated fatty acids were seen when *GCS1* homologues were expressed in *S. cerevisiae* (which cannot synthesize GlcCer *de novo*). The utilization of these compounds, which in *S. cerevisiae* are generally directed to GIPC synthesis, further increases glycosylceramide diversity. Broader investigation of fungal glycolipids will undoubtedly expose additional variations on this synthetic theme.

5. FUNGAL CELL WALL POLYMERS

The fungal cell wall is a complex and dynamic structure of glycans and glycoproteins. It is primarily composed of polymers of Glc, Man and GlcNAc (glucans, mannans and chitin, respectively), with extensive cross-linking between these elements (see Chapter 10). Both the degree of interconnection and the distribution of the wall components depend on the fungal species, developmental stage and growth conditions. Here, we present an overview of the synthesis of the major glycan components of fungal cell walls (reviewed in Klis *et al.*, 2006; Lesage and Bussey, 2006).

5.1. β-(1→3)-Glucan synthesis

The dominant fungal cell wall component is β-(1→3)-glucan, a polymer of ≈1500 Glc residues that is branched via β-(1→6)-linkages (Lesage and Bussey, 2006). This polymer is notably absent in mammals yet required for viability in yeast, a combination that has led to the development of effective antifungal drugs targeting its synthesis (Kartsonis *et al.*, 2003). Despite its relevance, there is much that is not understood about the synthesis of β-(1→3)-glucan, including whether this process requires a primer molecule and how polymer length is regulated. The Glc donor for synthesis is likely to be cytoplasmic UDP-Glc, which is utilized by plasma membrane-bound enzymes under the regulatory control of GTP-binding proteins. The catalytic subunits of the activity in *S. cerevisiae* are believed to be Fks1p and Gsc2p. These integral membrane proteins localize to the plasma membrane at sites of polar growth and enable nascent glucan chains to be transported across the membrane for incorporation into the cell wall. In *S. cerevisiae*, Fks1p is the dominant β-(1→3)-glucan synthase and is expressed during mitotic growth, while Gsc2p is active under conditions of nutritional or environmental stress. A third homologue, Fks3p, may be involved in cell wall construction during developmental processes such as mating and spore formation.

Genome analysis has identified sequences encoding multiple FKS homologues in fungi including *C. albicans*, *S. pombe*, the cotton pathogen *Ashbya gossypii* and members of the *Saccharomyces* genus. In contrast, only a single, essential *FKS1* has been identified in *Yarrowia lipolytica*, *C. neoformans*, the dimorphic pathogen *Coccidioides posadasii* and the acquired immunodeficiency syndrome (AIDS)-defining pathogen *Pneumocystis carinii*. Excellent reviews of β-(1→3)-glucan synthesis have been published by Douglas (2001) and Lesage and Bussey (2006). An unusual variation related to β-(1→3)-glucan is the linear β-(1→3)/(1→4)-glucan found in *A. fumigatus* (Latge *et al.*, 2005). Although not previously described in fungi, this glycan represents 10% of total β-glucan in that organism.

5.2. β-(1→6)-Glucan synthesis

β-(1→6)-Glucan interconnects other cell wall components in *S. cerevisiae*, thus playing a central role in cell wall structure. The degree of polymerization of this glucan is generally lower than that of β-(1→3)-glucan, with ≈350 Glc residues per chain. The structure is highly branched by the introduction of β-(1→3)-linkages; the frequency of these branches ranges from 7% of the backbone residues in *C. albicans* to ten-fold that frequency in *S. pombe*, suggesting species-dependent variation in branching or cross-linking activities (Lesage and Bussey, 2006). The synthesis of β-(1→6)-glucans (reviewed in Shahinian and Bussey, 2000) presents numerous questions, starting with its localization. Genetic manipulation of *S. cerevisiae* and analysis of the resulting levels of β-(1→6)-glucan has implicated a broad array of genes in this pathway, including genes encoding proteins of the secretory pathway and at the cell surface. Supporting the resulting hypothesis that β-(1→6)-glucan is made intracellularly, immuno-electron microscopy studies have detected this polymer in the Golgi and secretory vesicles of *S. pombe* (Humbel *et al.*, 2001), but the degree of polymerization and form of the glucan represented by this localization is not clear. Cell-free synthesis of β-(1→6)-glucan has been achieved in crude membrane preparations from *S. cerevisiae* (Vink *et al.*, 2004); continued effort in this biochemical direction should help clarify this intriguing research area.

5.3. Synthesis of other cell wall glucans

Beyond the β-glucans described above, many fungal cell walls also include α-glucans. These compounds do not occur in *S. cerevisiae* or *C. albicans*, but are highly abundant (up to 95% of the glucans) in the cell wall of other yeasts, such as the pathogens *Paracoccidioides brasiliensis* and *Blastomyces dermatitidis*. Synthesis of α-(1→3)-glucan has been studied in *S. pombe*, where it exists as a linear polymer of ≈200 Glc residues per chain (Hochstenbach *et al.*, 1998). This compound is synthesized by the product of a single essential gene, *AGS1* (alpha-glucan synthase 1), which may further join polymer chains by linkers of α-(1→4)-glucan (Garcia *et al.*, 2006). Three homologues of Ags1p are required for *S. pombe* spore wall maturation.

Similar to *S. pombe*, an *AGS* gene family is also found in *A. fumigatus*, with data suggesting the homologues have distinct roles in growth, development and virulence (Beauvais *et al.*, 2005). Simpler synthetic machinery is present in *H. capsulatum* and *C. neoformans*, where a single *AGS1* gene is required for normal virulence (Rappleye *et al.*, 2004; Reese *et al.*, 2007). The α-(1→3)-glucan plays a special role in the latter pathogen, where it is required for association of the fungal capsule with the cell wall (Reese and Doering, 2003).

5.4. Chitin synthesis

Chitin is a polymer of β-(1→4)-linked GlcNAc, typically composed of more than 1000 residues, which self-associates to form microfibrils. This relatively minor but critical component of the cell wall is deposited at the bud neck of yeast and at fungal septa in a highly regulated manner and can be deacetylated to form another cell wall polymer, chitosan. Chitin is generated from UDP-GlcNAc by synthases that translocate the polymeric product through the plasma membrane (Bowman and Free, 2006). In accordance with their function, these enzymes are integral proteins of the cytoplasmic leaflet of the plasma membrane. Because chitin synthesis primarily occurs at sites of active growth and cell wall remodelling, it is both temporally and spatially regulated. In *S. cerevisiae*, three chitin synthases (Chs1p, Chs2p and Chs3p) have been described, which play specific roles in cell growth: Chs1p repairs the site of daughter cell separation from the parent; Chs2p forms chitin in the septum during cell division; and Chs3p makes chitin at the bud neck, the lateral cell wall and spore cell walls. Interestingly, the activity of Chs3p is regulated by subcellular localization: it is stored in small microsomal vesicles (chitosomes) that deliver it to the plasma membrane as needed. Multiple gene products have been implicated in the regulation and localization of chitin synthesis in *S. cerevisiae*, but describing them is beyond the scope of this chapter; useful reviews of chitin synthesis include those of Munro and Gow (2001) and Roncero (2002).

Most fungi have multiple chitin synthase genes which appear to have distinct, although occasionally overlapping, functions (reviewed in Roncero, 2002). *S. cerevisiae* offers a relatively simple case, compared to the eight Chs proteins in *C. neoformans* (Banks *et al.*, 2005) or the seven in *A. fumigatus* (Mellado *et al.*, 2003). These enzymes have been classified based on sequence motifs and homology; notably, some families are restricted to filamentous fungi. Deletion of the genes encoding many of these enzymes yields striking phenotypic changes, including altered morphology, stress resistance or virulence in animal and plant pathogens (Banks *et al.*, 2005). Some fungi secrete deacetylases that modify chitin to chitosan, a more soluble cationic polymer. The genes encoding two such enzymes in *S. cerevisiae* (*CDA1* and *CDA2*) are expressed only during sporulation and contribute to spore wall formation. In contrast, chitosan is a normal cell wall component of *C. neoformans* and the three cryptococcal chitin deacetylases are required for normal cell integrity and bud separation (Baker *et al.*, 2007).

6. INTRACELLULAR GLYCANS

Glycogen and trehalose are the main stores of glucose in *S. cerevisiae*. Glycogen functions as a reserve carbohydrate that is synthesized and stored under nutrient-rich conditions and then degraded during periods of nutrient deprivation. Trehalose has been postulated to serve as a chemical chaperone that protects proteins and membranes from stress-induced denaturation (reviewed in Elbein *et al.*, 2003). Synthesis of both glycogen and trehalose is reviewed extensively elsewhere (François and Parrou, 2001; see also Chapter 11).

6.1. Glycogen synthesis

Glycogen is a branched polymer of up to 100 000 Glc residues that is distributed throughout the fungal cytosol. Synthesis (reviewed in François and Parrou, 2001; Lomako *et al.*, 2004) occurs in the cytosol, beginning with the actions of the protein glycogenin (either Glg1p or Glg2p in *S. cerevisiae*). Interestingly, glycogenin itself serves as the primer for glycogen polymerization: Glg1p and Glg2p are each capable of utilizing UDP-Glc as a donor for auto-glucosylation at one of several tyrosine (Tyr) residues, generating a short α-(1$\rightarrow$4)-Glc chain. Although Glg1p and Glg2p are capable of autoglucosylating, the proteins exist as a dimer *in vivo* and the reaction is believed to occur intermolecularly. The protein-linked chain of up to 10 Glc residues generated by glycogenin is elongated by the action of Gsy1p or the more dominant Gsy2p (glycogen synthase isoforms 1 and 2, respectively), which adds Glc in an α-(1$\rightarrow$4)-linkage to the non-reducing end of the initial oligomer. Following elongation, the linear α-(1$\rightarrow$4)-glucans are ramified by Glc3p, a branching enzyme that adds seven α-(1$\rightarrow$4)-Glc residues in an α-(1$\rightarrow$6)-linkage. In a departure from the glycogen synthesis pathway of *S. cerevisiae*, the glycogenin and glycogen synthase enzymes of *N. crassa* are each encoded by just one protein (de Paula *et al.*, 2005).

6.2. Trehalose synthesis

Trehalose is an unusual disaccharide of glucose linked "head to head" in an α-(1$\rightarrow$1)-linkage (see Chapter 11) and notably is absent from mammalian cells. Trehalose synthesis in *S. cerevisiae* proceeds by two sequential reactions: first, trehalose-phosphate synthase (Tps1p) catalyses the transfer of Glc from UDP-Glc to Glc-6-phosphate, forming trehalose-6-phosphate. In a subsequent reaction, trehalose-6-phosphate phosphatase (Tps2p) acts on this product to generate free trehalose. Together these enzymes form a complex with two other proteins, Tps3p and Tsl1p, which are thought to have regulatory functions. *C. neoformans* strains that lack Tps1p and Tps2p are temperature sensitive and a *tps1* mutant is avirulent (Petzold *et al.*, 2006).

7. EXOPOLYSACCHARIDES

The processes described earlier in this chapter broadly apply to all fungi. Some fungi, however, are unique in their generation of additional extracellular glycan structures (exopolysaccharides). One example is the extensive polysaccharide capsule of *C. neoformans*, the organism's main virulence factor (reviewed in Chayakulkeeree and Perfect, 2006). This structure consists primarily of two polysaccharides: galactoxylomannan (GalXM; polymer size of $\approx 1 \times 10^5$ Da) and glucuronoxylomannan (GXM; polymer size of $\approx 1\text{–}7 \times 10^6$ Da). Both polymers are composed of repeating subunits (Figure 22.8). The GalXM polysaccharide has a linear backbone of α-(1$\rightarrow$6)-linked Gal residues with oligomeric side chains of Gal, Man and Xyl (James and Cherniak, 1992); similar polymers have been reported in other cryptococcal species. The GXM polysaccharide has a linear backbone of α-(1→3)-Man residues that are 6-*O*-acetylated and substituted with residues of Xyl, linked β-(1→2) and β-(1→4), and glucuronic acid (GlcA) linked β-(1→2) (Cherniak *et al.*, 1998). The pattern of GXM modifications varies among different serotypes of *C. neoformans* (see Figure 22.8) and between related species. For example, studies of *Cryptococcus flavescens* revealed a similar high-molecular mass polymer that is substituted more frequently with GlcA (Ikeda and Maeda, 2004). This polymer is also substituted at the 6-position of Man with chains consisting of β-Man-(1→4)-Xyl or β-Man-(1→4)-β-Xyl-(1→4)-Xyl.

The cryptococcal polysaccharides are of interest because of their unique structures and the role of the capsule in disease. Studies of mutant *C. neoformans* that are deficient in secretion suggest that GXM biosynthesis begins within the cell, with products exported via the secretory pathway, but the nature of these products remains to be established (Yoneda and Doering, 2006). Little is known about specific enzymes involved in synthesis of capsule polymers. A mannosyltransferase activity capable of modifying Xylα-CH_3 has been described in membrane preparations of *C. laurentii* (Schutzbach and Ankel, 1971); this could potentially be involved in the formation of GXM structures like those of *C. flavascens* described above, although the linkage formed has not been determined. Recently, a β-(1→2)-xylosyltransferase that participates in synthesis of both GXM and GalXM has been purified and cloned from *C. neoformans* (Klutts *et al.*, 2007; Klutts and Doering, 2008). Intriguingly,this enzyme is also required for cryptococcal GIPC synthesis (Castle *et al.*, 2008), raising the possibility that the two biosynthetic pathways are linked. The enzymes required for other capsule biosynthetic steps are not yet known. Once capsule polymers are made and exported they become associated with the cell surface in a process that is dependent on cell wall α-(1→3)-glucan (Reese and Doering, 2003; Reese *et al.*, 2007), but the mechanism of this association has not been defined. Future investigations are clearly required to determine the biosynthetic pathways of cryptococcal capsule polysaccharides.

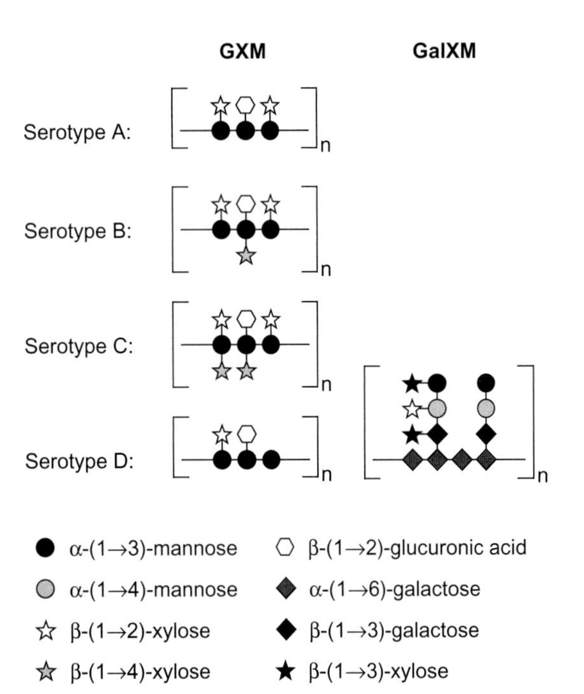

FIGURE 22.8 Repeating units of GXM and GalXM capsular polysaccharides of *C. neoformans*. Symbols are indicated on the diagram.

8. CONCLUSIONS

This survey of fungal glycan biosynthesis emphasizes themes common to all fungi as well as the myriad of ways in which these themes are embodied in members of this diverse kingdom. The genetically tractable model yeast *S. cerevisiae* has enabled ground-breaking studies of glycan biosynthetic pathways with impact far beyond fungi. The greater world of fungal organisms further offers fascinating biology and potential application with broad relevance to public health and agriculture (see Research Focus Box).

RESEARCH FOCUS BOX

- Flippases comprise an intriguing class of enzymes that are relevant to the synthesis of glycans and glycoconjugates. Further identification and characterization of elusive flippases such as those that translocate Dol-P-Man or GlcN-PI will advance the understanding of biosynthetic pathways.
- The fungal kingdom is tremendously diverse. Broadening research efforts beyond the traditional *S. cerevisiae* model to include a greater variety of fungal organisms will both help track common biosynthetic mechanisms across members of this kingdom and identify novel biology.
- Pathogenic fungi exert detrimental effects on the economy and society by causing disease in plants, animals and humans. Elucidation of glycan synthetic processes that differ between fungi and their plant or mammalian hosts, e.g. synthesis of β-(1→6)-glucan or GPI lipid remodelling, will help in the development of needed antifungal agents.

ACKNOWLEDGEMENTS

The authors thank Aki Yoneda for helpful suggestions on the manuscript and Maurizio del Poeta for comments on the glycolipid section.

References

Baker, L.G., Specht, C.A., Donlin, M.J., Lodge, J.K., 2007. Chitosan, the deacetylated form of chitin, is necessary for cell wall integrity in *Cryptococcus neoformans*. Eukaryot. Cell 6, 855–867.

Banks, I.R., Specht, C.A., Donlin, M.J., Gerik, K.J., Levitz, S.M., Lodge, J.K., 2005. A chitin synthase and its regulator protein are critical for chitosan production and growth of the fungal pathogen *Cryptococcus neoformans*. Eukaryot. Cell 4, 1902–1912.

Barreto-Bergter, E., Pinto, M.R., Rodrigues, M.L., 2004. Structure and biological functions of fungal cerebrosides. An Acad. Bras. Cienc. 76, 67–84.

Beauvais, A., Maubon, D., Park, S., et al., 2005. Two α(1-3) glucan synthases with different functions in *Aspergillus fumigatus*. Appl. Environ. Microbiol. 71, 1531–1538.

Berninsone, P.M., Hirschberg, C.B., 2000. Nucleotide sugar transporters of the Golgi apparatus. Curr. Opin. Struct. Biol. 10, 542–547.

Bickel, T., Lehle, L., Schwarz, M., Aebi, M., Jakob, C.A., 2005. Biosynthesis of lipid-linked oligosaccharides in *Saccharomyces cerevisiae*: Alg13p and Alg14p form a complex required for the formation of GlcNAc(2)-PP-dolichol. J. Biol. Chem. 280, 34500–34506.

Bowman, S.M., Free, S.J., 2006. The structure and synthesis of the fungal cell wall. Bioessays 28, 799–808.

Castle, S.A., Owuor, E.A., Thompson, S.H., et al., 2008. β1,2-xylosyltransferase Cxt1p is solely responsible for xylose incorporation into *Cryptococcus neoformans* glycosphingolipids. Eukaryot. Cell 7, 1611–1615.

Chayakulkeeree, M., Perfect, J.R., 2006. Cryptococcosis. Infect. Dis. Clin. North Am. 20, 507–544.

Cherniak, R., Valafar, H., Morris, L.C., Valafar, F., 1998. *Cryptococcus neoformans* chemotyping by quantitative analysis of 1H nuclear magnetic resonance spectra of glucuronoxylomannans with a computer-simulated artificial neural network. Clin. Diagn. Lab. Immunol. 5, 146–159.

Cottrell, T.R., Griffith, C.L., Liu, H., Nenninger, A.A., Doering, T.L., 2007. The pathogenic fungus *Cryptococcus neoformans* expresses two functional GDP-mannose transporters with distinct expression patterns and roles in capsule synthesis. Eukaryot. Cell 6, 776–785.

Cowart, L.A., Obeid, L.M., 2007. Yeast sphingolipids: recent developments in understanding biosynthesis, regulation, and function. Biochim. Biophys. Acta 1771, 421–431.

de Paula, R.M., Wilson, W.A., Terenzi, H.F., Roach, P.J., Bertolini, M.C., 2005. GNN is a self-glucosylating protein involved in the initiation step of glycogen biosynthesis in *Neurospora crassa*. Arch. Biochem. Biophys. 435, 112–124.

Dickson, R.C., Sumanasekera, C., Lester, R.L., 2006. Functions and metabolism of sphingolipids in *Saccharomyces cerevisiae*. Prog. Lipid Res. 45, 447–465.

Douglas, C.M., 2001. Fungal β(1,3)-D-glucan synthesis. Med. Mycol. 39 (Suppl. 1), 55–66.

Elbein, A.D., Pan, Y.T., Pastuszak, I., Carroll, D., 2003. New insights on trehalose: a multifunctional molecule. Glycobiology 13, 17R–27R.

Fontaine, T., Magnin, T., Melhert, A., Lamont, D., Latge, J.P., Ferguson, M.A., 2003. Structures of the glycosylphosphatidylinositol membrane anchors from *Aspergillus fumigatus* membrane proteins. Glycobiology 13, 169–177.

François, J., Parrou, J.L., 2001. Reserve carbohydrates metabolism in the yeast *Saccharomyces cerevisiae*. FEMS Microbiol. Rev. 25, 125–145.

Franzot, S.P., Doering, T.L., 1999. Inositol acylation of glycosylphosphatidylinositols in the pathogenic fungus *Cryptococcus neoformans* and the model yeast *Saccharomyces cerevisiae*. Biochem. J. 340, 25–32.

Gao, X.D., Tachikawa, H., Sato, T., Jigami, Y., Dean, N., 2005. Alg14 recruits Alg13 to the cytoplasmic face of the endoplasmic reticulum to form a novel bipartite UDP-*N*-acetylglucosamine transferase required for the second step of *N*-linked glycosylation. J. Biol. Chem. 280, 36254–36262.

Garcia, I., Tajadura, V., Martin, V., Toda, T., Sanchez, Y., 2006. Synthesis of α-glucans in fission yeast spores is carried out by three α-glucan synthase paralogues, Mok12p, Mok13p and Mok14p. Mol. Microbiol. 59, 836–853.

Gemmill, T.R., Trimble, R.B., 1999. Overview of *N*- and *O*-linked oligosaccharide structures found in various yeast species. Biochim. Biophys. Acta 1426, 227–237.

Gerardy-Schahn, R., Oelmann, S., Bakker, H., 2001. Nucleotide sugar transporters: biological and functional aspects. Biochimie 83, 775–782.

Goto, M., 2007. Protein *O*-glycosylation in fungi: diverse structures and multiple functions. Biosci. Biotechnol. Biochem. 71, 1415–1427.

Heise, N., Gutierrez, A.L., Mattos, K.A., et al., 2002. Molecular analysis of a novel family of complex glycoinositolphosphoryl ceramides from *Cryptococcus neoformans*: structural differences between encapsulated and acapsular yeast forms. Glycobiology 12, 409–420.

Helenius, J., Ng, D.T., Marolda, C.L., Walter, P., Valvano, M.A., Aebi, M., 2002. Translocation of lipid-linked oligosaccharides across the ER membrane requires Rft1 protein. Nature 415, 447–450.

Herscovics, A., 1999. Processing glycosidases of *Saccharomyces cerevisiae*. Biochim. Biophys. Acta 1426, 275–285.

Hirschberg, C.B., Robbins, P.W., Abeijon, C., 1998. Transporters of nucleotide sugars. ATP, and nucleotide sulfate in the endoplasmic reticulum and Golgi apparatus. Annu. Rev. Biochem. 67, 49–69.

Hochstenbach, F., Klis, F.M., van den Ende, H., van Donselaar, E., Peters, P.J., Klausner, R.D., 1998. Identification of a putative α-glucan synthase essential for cell wall construction and morphogenesis in fission yeast. Proc. Natl. Acad. Sci. USA 95, 9161–9166.

Humbel, B.M., Konomi, M., Takagi, T., Kamasawa, N., Ishijima, S.A., Osumi, M., 2001. In situ localization of β-glucans in the cell wall of *Schizosaccharomyces pombe*. Yeast 18, 433–444.

Ikeda, R., Maeda, T., 2004. Structural studies of the capsular polysaccharide of a non-*neoformans Cryptococcus* species identified as *C. laurentii,* which was reclassified as *Cryptococcus* flavescens, from a patient with AIDS. Carbohydr. Res. 339, 503–509.

James, P.G., Cherniak, R., 1992. Galactoxylomannans of *Cryptococcus neoformans*. Infect. Immun. 60, 1084–1088.

Jigami, Y., Odani, T., 1999. Mannosylphosphate transfer to yeast mannan. Biochim. Biophys. Acta 1426, 335–345.

Kartsonis, N.A., Nielsen, J., Douglas, C.M., 2003. Caspofungin: the first in a new class of antifungal agents. Drug Resist. Update 6, 197–218.

Klis, F.M., Boorsma, A., De Groot, P.W., 2006. Cell wall construction in *Saccharomyces cerevisiae*. Yeast 23, 185–202.

Klutts, J.S., Doering, T.L., 2008. Cryptococcal xylosyltransferase 1 (Cxt1p) from *Cryptococcus neoformans* plays a direct role in the synthesis of capsule polysaccharides. J. Biol. Chem. 283, 14327–14334.

Klutts, J.S., Levery, S.B., Doering, T.L., 2007. A β-1,2-xylosyltransferase from *Cryptococcus neoformans* defines a new family of glycosyltransferases. J. Biol. Chem. 282, 17890–17899.

Latge, J.P., Mouyna, I., Tekaia, F., Beauvais, A., Debeaupuis, J.P., Nierman, W., 2005. Specific molecular features in the organization and biosynthesis of the cell wall of *Aspergillus fumigatus*. Med. Mycol. 43 (Suppl. 1), S15–S22.

Leipelt, M., Warnecke, D., Zahringer, U., et al., 2001. Glucosylceramide synthases, a gene family responsible for the biosynthesis of glucosphingolipids in animals, plants, and fungi. J. Biol. Chem. 276, 33621–33629.

Lennarz, W.J., 2007. Studies on oligosaccharyl transferase in yeast. Acta Biochim. Pol. 54, 673–677.

Lesage, G., Bussey, H., 2006. Cell wall assembly in *Saccharomyces cerevisiae*. Microbiol. Mol. Biol. Rev. 70, 317–343.

Lester, R.L., Smith, S.W., Wells, G.B., Rees, D.C., Angus, W.W., 1974. The isolation and partial characterization of two novel sphingolipids from *Neurospora crassa*: di(inositolphosphoryl)ceramide and ((gal)3glu)ceramide. J. Biol. Chem. 249, 3388–3394.

Lomako, J., Lomako, W.M., Whelan, W.J., 2004. Glycogenin: the primer for mammalian and yeast glycogen synthesis. Biochim. Biophys. Acta 1673, 45–55.

Maciel, D.M., Rodrigues, M.L., Wait, R., Villas Boas, M.H., Tischer, C.A., Barreto-Bergter, E., 2002. Glycosphingolipids from *Magnaporthe grisea* cells: expression of a ceramide dihexoside presenting phytosphingosine as the long-chain base. Arch. Biochem. Biophys. 405, 205–213.

Maeda, Y., Kinoshita, T., 2008. Dolichol-phosphate mannose synthase: structure, function and regulation. Biochim. Biophys. Acta 1780, 861–868.

Mellado, E., Dubreucq, G., Mol, P., et al., 2003. Cell wall biogenesis in a double chitin synthase mutant (*chsG⁻/chsE⁻*) of *Aspergillus fumigatus*. Fungal Genet. Biol. 38, 98–109.

Mille, C., Janbon, G., Delplace, F., et al., 2004. Inactivation of CaMIT1 inhibits *Candida albicans* phospholipomannan β-mannosylation, reduces virulence, and alters cell wall protein β-mannosylation. J. Biol. Chem. 279, 47952–47960.

Morelle, W., Bernard, M., Debeaupuis, J.P., Buitrago, M., Tabouret, M., Latge, J.P., 2005. Galactomannoproteins of *Aspergillus fumigatus*. Eukaryot. Cell 4, 1308–1316.

Moremen, K.W., Molinari, M., 2006. *N*-linked glycan recognition and processing: the molecular basis of endoplasmic reticulum quality control. Curr. Opin. Struct. Biol. 16, 592–599.

Munro, C.A., Gow, N.A., 2001. Chitin synthesis in human pathogenic fungi. Med. Mycol. 39 (Suppl. 1), 41–53.

Nakajima, T., Yoshida, M., Hiura, N., Matsuda, K., 1984. Structure of the cell wall proteogalactomannan from *Neurospora crassa*. I. Purification of the proteoheteroglycan and characterization of alkali-labile oligosaccharides. J. Biochem. 96, 1005–1011.

Nishikawa, A., Poster, J.B., Jigami, Y., Dean, N., 2002a. Molecular and phenotypic analysis of *CaVRG4*, encoding an essential Golgi apparatus GDP-mannose transporter. J. Bacteriol. 184, 29–42.

Nishikawa, A., Mendez, B., Jigami, Y., Dean, N., 2002b. Identification of a *Candida glabrata* homologue of the *S. cerevisiae VRG4* gene, encoding the Golgi GDP-mannose transporter. Yeast 19, 691–698.

O'Reilly, M.K., Zhang, G., Imperiali, B., 2006. In vitro evidence for the dual function of Alg2 and Alg11: essential mannosyltransferases in *N*-linked glycoprotein biosynthesis. Biochemistry 45, 9593–9603.

Orlean, P., Menon, A.K., 2007. Thematic review series: lipid posttranslational modifications. GPI anchoring of protein in yeast and mammalian cells, or: how we learned to stop worrying and love glycophospholipids. J. Lipid Res. 48, 993–1011.

Petzold, E.W., Himmelreich, U., Mylonakis, E., et al., 2006. Characterization and regulation of the trehalose synthesis pathway and its importance in the pathogenicity of *Cryptococcus neoformans*. Infect. Immun. 74, 5877–5887.

Poster, J.B., Dean, N., 1996. The yeast *VRG4* gene is required for normal Golgi functions and defines a new family of related genes. J. Biol. Chem. 271, 3837–3845.

Prill, S.K., Klinkert, B., Timpel, C., Gale, C.A., Schroppel, K., Ernst, J.F., 2005. PMT family of *Candida albicans*: five protein mannosyltransferase isoforms affect growth, morphogenesis and antifungal resistance. Mol. Microbiol. 55, 546–560.

Rappleye, C.A., Engle, J.T., Goldman, W.E., 2004. RNA interference in *Histoplasma capsulatum* demonstrates a role for α-(1,3)-glucan in virulence. Mol. Microbiol. 53, 153–165.

Reese, A.J., Doering, T.L., 2003. Cell wall α-1,3-glucan is required to anchor the *Cryptococcus neoformans* capsule. Mol. Microbiol. 50, 1401–1409.

Reese, A.J., Yoneda, A., Breger, J.A., et al., 2007. Loss of cell wall α(1-3) glucan affects *Cryptococcus neoformans* from ultrastructure to virulence. Mol. Microbiol. 63, 1385–1398.

Reggiori, F., Canivenc-Gansel, E., Conzelmann, A., 1997. Lipid remodeling leads to the introduction and exchange of defined ceramides on GPI proteins in the ER and Golgi of *Saccharomyces cerevisiae*. EMBO J. 16, 3506–3518.

Rhome, R., McQuiston, T., Kechichian, T., et al., 2007. Biosynthesis and immunogenicity of glucosylceramide in *Cryptococcus neoformans* and other human pathogens. Eukaryot. Cell 6, 1715–1726.

Roncero, C., 2002. The genetic complexity of chitin synthesis in fungi. Curr. Genet. 41, 367–378.

Samuelson, J., Banerjee, S., Magnelli, P., et al., 2005. The diversity of dolichol-linked precursors to Asn-linked glycans likely results from secondary loss of sets of glycosyltransferases. Proc. Natl. Acad. Sci. USA 102, 1548–1553.

Schutzbach, J.S., Ankel, H., 1971. Multiple mannosyl transferases in *Cryptococcus laurentii*. J. Biol. Chem. 246, 2187–2194.

Schutzbach, J., Ankel, H., Brockhausen, I., 2007. Synthesis of cell envelope glycoproteins of *Cryptococcus laurentii*. Carbohydr. Res. 342, 881–893.

Shahinian, S., Bussey, H., 2000. β-1,6-Glucan synthesis in *Saccharomyces cerevisiae*. Mol. Microbiol. 35, 477–489.

Toledo, M.S., Levery, S.B., Straus, A.H., Takahashi, H.K., 2001. Sphingolipids of the mycopathogen *Sporothrix schenckii*: identification of a glycosylinositol phosphorylceramide with novel core GlcNH2α1→2Ins motif. FEBS Lett. 493, 50–56.

Toledo, M.S., Levery, S.B., Bennion, B., et al., 2007. Analysis of glycosylinositol phosphorylceramides expressed by the opportunistic mycopathogen *Aspergillus fumigatus*. J. Lipid Res. 48, 1801–1824.

Trinel, P.A., Maes, E., Zanetta, J.P., et al., 2002. *Candida albicans* phospholipomannan, a new member of the fungal mannose inositol phosphoceramide family. J. Biol. Chem. 277, 37260–37271.

Varki, A., Cummings, R., Esko, J., Hart, G., Marth, J. (Eds.), 1999. Essentials of Glycobiology. Cold Spring Harbor Laboratory Press, Cold Spring Harbor, New York.

Vink, E., Rodriguez-Suarez, R.J., Gerard-Vincent, M., et al., 2004. An *in vitro* assay for (1→6)-β-D-glucan synthesis in *Saccharomyces cerevisiae*. Yeast 21, 1121–1131.

Wallis, G.L., Easton, R.L., Jolly, K., Hemming, F.W., Peberdy, J.F., 2001. Galactofuranoic-oligomannose *N*-linked glycans of α-galactosidase A from *Aspergillus niger*. Eur. J. Biochem. 268, 4134–4143.

Warnecke, D., Heinz, E., 2003. Recently discovered functions of glucosylceramides in plants and fungi. Cell. Mol. Life Sci. 60, 919–941.

Weerapana, E., Imperiali, B., 2006. Asparagine-linked protein glycosylation: from eukaryotic to prokaryotic systems. Glycobiology 16, 91R–101R.

Willer, T., Valero, M.C., Tanner, W., Cruces, J., Strahl, S., 2003. *O*-mannosyl glycans: from yeast to novel associations with human disease. Curr. Opin. Struct. Biol. 13, 621–630.

Yoneda, A., Doering, T.L., 2006. A eukaryotic capsular polysaccharide is synthesized intracellularly and secreted via exocytosis. Mol. Biol. Cell 17, 5131–5140.

PART II

SYNTHESIS OF MICROBIAL GLYCOSYLATED COMPONENTS

B. Chemical synthesis

CHAPTER

23

Chemical synthesis of bacterial lipid A

Shoichi Kusumoto, Koichi Fukase and Yukari Fujimoto

SUMMARY

Chemical synthesis has played essential and important roles in the history of endotoxin research. The recent status of chemical syntheses of lipid A analogues is described after a brief introduction of the historically important early synthetic works which proved that lipid A is the endotoxic principle of LPS. The targets of the syntheses mentioned in this chapter include both natural-type lipid A analogues as obtained from bacterial cells and several artificial structures related to the former. Owing to many methodological improvements after the first synthesis, chemical synthesis of this group of complex amphiphilic glycoconjugates is now quite powerful and efficient. The overview presented herein allows the deduction that various lipid A analogues are now synthetically accessible; homogeneous preparations of any desired structures can become available when the synthetic route to the target is carefully designed. Synthesis of ^{3}H- or ^{13}C-labelled derivatives can also be achieved and used for biological investigations of interactions with the receptor complex of hosts or for conformational analysis. The use of various synthetic compounds with both natural and unnatural structures will continue to contribute to our deeper understanding of the mechanism and meanings of biological events induced by endotoxin.

Keywords: Lipid A; Lipopolysaccharide Glycolipid; Endotoxin; Endotoxic principle; Antagonist; Fatty acid; Phosphate; Protecting groups; Synthesis; Amphiphile

1. INTRODUCTION

The term "lipid A" is a name given to the lipophilic partial structure of lipopolysaccharide (LPS) by Westphal and Lüderitz, who first described its preparation (Rietschel and Westphal, 1999). Lipid A never exists in its free form in Nature: it is only liberated by mild acid hydrolysis of LPS (Westphal and Lüderitz, 1954); the ketosidic linkage between lipid A and the 3-deoxy-D-*manno*-oct-2-ulosonic acid (Kdo) residue, which represents the proximal terminal of the hydrophilic polysaccharide part of LPS, is susceptible to acid and thus can be selectively cleaved. Simultaneously, lipid A was shown to be the chemical entity responsible for the endotoxic activity of LPS (see Chapter 3).

The basic structure of enterobacterial lipid A was reported to be *N,O*-polyacylated derivatives of the hydrophilic backbone composed of 1,4′-bisphosphate of β-(1→6) disaccharide of 2-amino-2-deoxy-D-glucose (D-glucosamine) (Gmeiner *et al.*, 1969; Hase and Rietschel, 1976). Even though the exact number, composition and location of acyl groups linked to the backbone were not readily determined, lipid A attracted interests of many researchers including chemists because of its potent and important biological activities and complex molecular architecture. Thus, a few research groups had started their synthetic approaches (Inage *et al.*, 1981; Kusumoto *et al.*, 1983; Kiso and Hasegawa, 1983) toward its plausible structures aiming at production of homogeneous molecular species possessing the endotoxic activity. These preliminary synthetic efforts were not successful, no endotoxic compounds being obtained. However, as the consequence of these works, further intensive structural studies were undertaken and led to independent determination of the complete structure of *Escherichia coli* and *Salmonella enterica* sv. Typhimurium lipid A (**1** shown in Figure 23.1; Takayama *et al.*, 1983; Imoto *et al.*, 1983, 1984b).

2. EARLY CHEMICAL SYNTHESES OF BACTERIAL LIPID A

The first natural-type lipid A analogue synthesized was a disaccharide biosynthetic precursor **2** to LPS (see Figure 23.1), designated precursor Ia or recently, more frequently, lipid IVa (Hansen-Hagge *et al.*, 1985; Raetz and Whitfield, 2002). This compound contains four (*R*)-3-hydroxytetradecanoic acids (D-β-hydroxymyristic acids) linked to the backbone. Because of the symmetrical distribution of these acyl groups, synthesis was simpler than that for **1** (Imoto *et al.*, 1984a, 1987a): a suitably protected disaccharide 4′-phosphate was first prepared and each two ester- and amide-linked acyl groups were introduced. Final hydrogenolytic deprotection after glycosyl phosphorylation gave the product **2**. The important constituent, optically pure (*R*)-3-hydroxytetradecanoic acid, was prepared by enantioselective reduction of the corresponding ketoester (Tai *et al.*, 1980) and its hydroxy group protected by benzylation before coupling.

The first synthesis of mature lipid A of *E. coli* **1** was then actualized. The following strategy basically similar to the above work was employed, except for the steps of acylations (Imoto *et al.*, 1984b, 1987b):

(i) Only those persistent protecting groups that can be removed by hydrogenolysis at the final step were employed. This was expected to facilitate the purification of the final amphiphilic product.
(ii) All the acyl groups and the protected 4′-phosphate were introduced before the disaccharide formation in order to reduce the number of protecting groups required.
(iii) Only the *N*-acyl group of the distal glucosamine residue was introduced

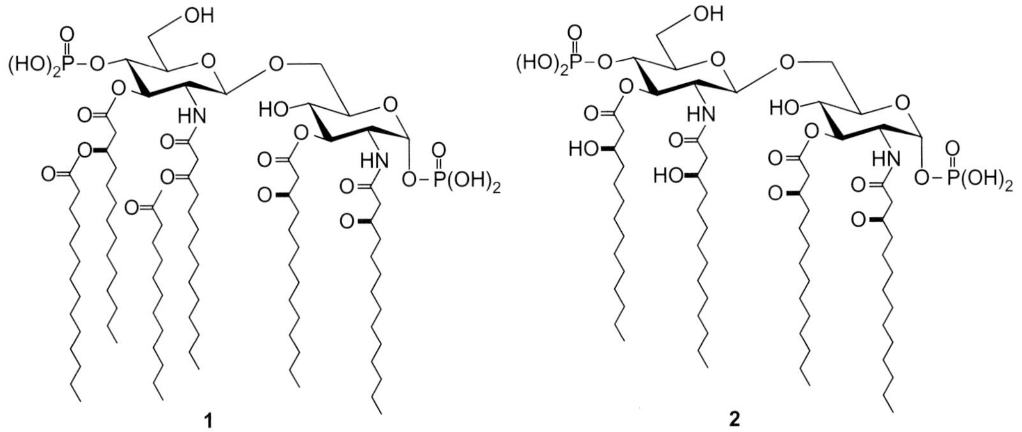

FIGURE 23.1 Structures of *Escherichia coli* lipid A **1** and its biosynthetic precursor **2**.

after the formation of the disaccharide in order to avoid β-elimination of the *N*-3-acyloxyacyl group during the glycosylation reaction. The particular 2-amino group was protected during the glycosylation step by 2,2,2-trichloroethoxycarbonyl (Troc) group. This group assures the formation of the desired β-glycoside and can be selectively removed later.

(iv) The chemically labile glycosyl phosphate moiety was introduced at the final synthetic stage just before the final deprotection.

The synthesis of **1** was achieved as illustrated in Scheme 23.1. Coupling of a 2-*N*-Troc-glycosyl bromide **3** with an acceptor **4** proceeded with a selective reaction at the primary 6-hydroxy group of the latter to give the β-(1→6)-backbone disaccharide. Cleavage of the *N*- and *O*-Troc groups followed by *N*-acylation gave the fully acylated disaccharide **5**. The 6′-hydroxy group was then protected again and the 1-*O*-allyl group cleaved. Selective lithiation and subsequent reaction with dibenzyl phosphorochloridate gave the desired 1-α-phosphate (Inage *et al.*, 1982). Hydrogenolytic removal of all the

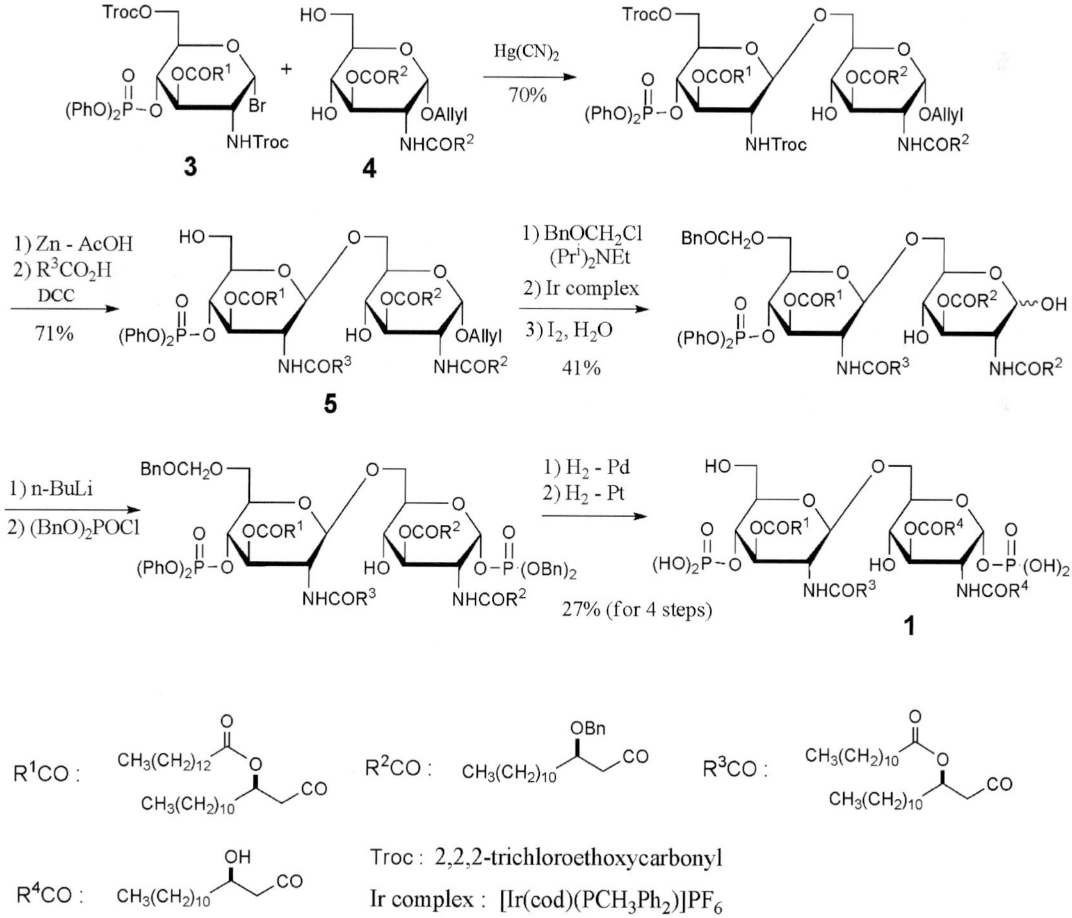

SCHEME 23.1 The first synthesis of *E. coli* lipid A **1**.

benzyl-type protecting groups with a palladium catalyst was followed by hydrogenolysis with a platinum catalyst to remove the phenyl esters of the 4′-phosphate to give the desired synthetic *E. coli* lipid A **1**.

Biological test proved that the synthetic **1** exhibited all the endotoxic activities, including both beneficial and detrimental ones, described for bacterial LPS and lipid A (Galanos *et al.*, 1985; Kotani *et al.*, 1985).

Although the efficiency of the above syntheses of **1** and **2** was not high as compared to the later improved procedures described in the following section, these were historically important works which provided unequivocal evidence clearly supporting the basic concept that lipid A is the endotoxic principle (Galanos *et al.*, 1984, 1985; Kotani *et al.*, 1984, 1985). Even many years after the first report of Westphal and Lüderitz, this concept had remained obscure until the above synthesis: owing to the amphiphilic nature and intrinsic heterogeneity of lipid A, possible influence of contaminants from bacterial origin was not completely excluded. The fact that synthetic preparations showed the expected endotoxic activities decisively excluded this concern. Thus, the first and final direct proof was obtained that lipid A is the real active entity of bacterial endotoxin.

3. IMPROVED SYNTHESIS OF LIPID A ANALOGUES

The above successful syntheses of lipid A not only proved the chemical structure of the endotoxic principle but also showed that similar structures can also be synthetically accessible. In the meantime, methods for purification and structural study and, in particular, nuclear magnetic resonace (NMR) spectroscopy and mass spectrometry (MS) analyses, have made great advances. Lipid A analogues from various bacterial species have been isolated and their structure elucidated (see Chapter 3). Discussions on their biological activities in relation to their detailed chemical structures have become possible. In such a situation, chemical synthesis reserves its important role in preparing various definite structural analogues of lipid A for precise study of the structure–activity relationship of this complex glycoconjugate family. Thus, many analogous structures, including both natural and unnatural, were prepared. Of these, a series or sets of compounds that differed from each other only in the acyl chain lengths or the positions of acylation on the backbone were prepared. The other series contained structures which either lack one of the phosphates or have polar substituents on the phosphates. Some examples from such syntheses will be described in this section.

During the decades after the first synthesis of lipid A, synthetic procedures have been improved in many aspects, such as conversion yields and reduction of reaction steps, so that the total efficiency to reach the final products became much higher. There have appeared several review articles on the synthesis of lipid A and its analogues (Kusumoto *et al.*, 1999, 2007, 2008). Yet, the basic strategy in the above early synthesis was employed in most of the new syntheses. The major points retained were:

(i) the use of benzyl-type groups for persistent protections which enable the final hydrogenolytic deprotection;
(ii) the use of *N*-Troc glycosyl donors; and
(iii) introduction of the glycosyl phosphate at the latest synthetic stage before the final deprotection.

As one of the typical examples, new synthesis of *E. coli* lipid A **1** is illustrated in Scheme 23.2 (Liu *et al.*, 1999). The following are the major improved points, which made the synthesis efficient:

(i) Direct preparation of a 6-*O*-benzylated glucosamine derivative via regioselective reductive opening of a 4,6-*O*-benzylidene

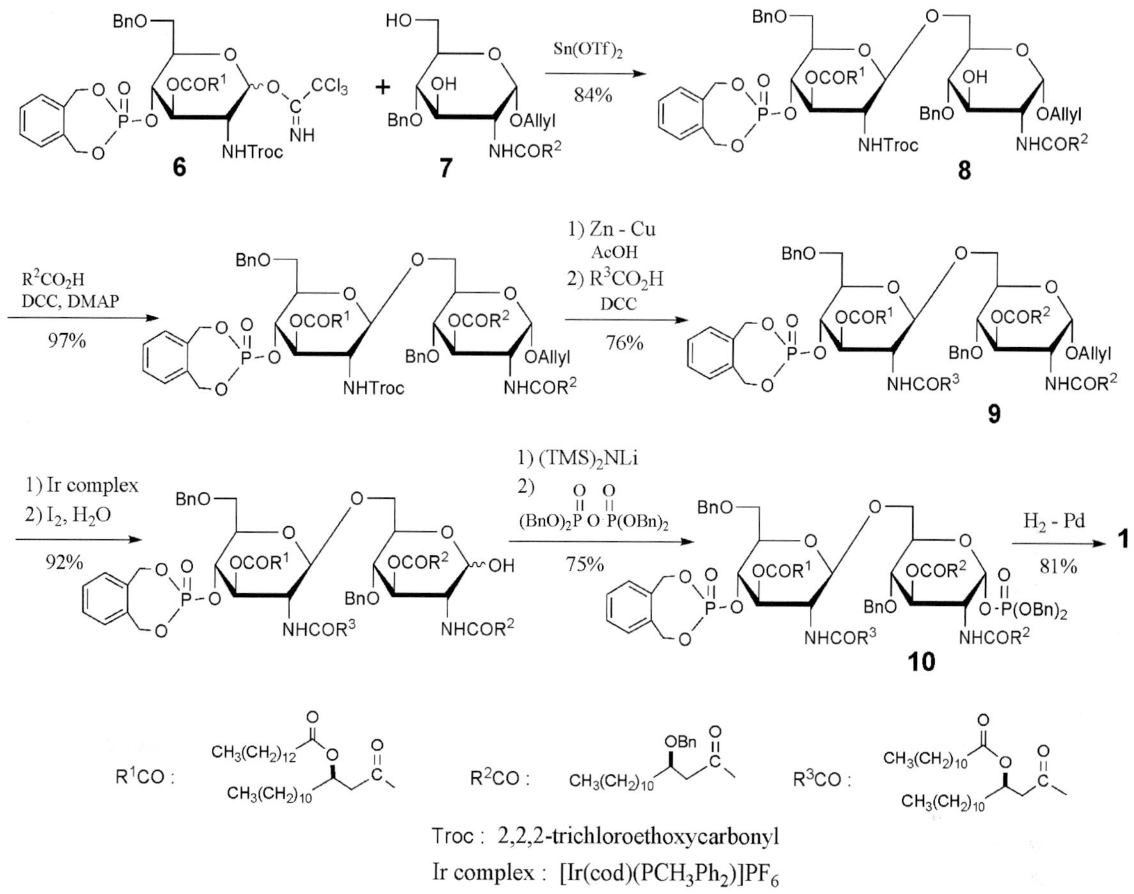

Troc : 2,2,2-trichloroethoxycarbonyl
Ir complex : $[Ir(cod)(PCH_3Ph_2)]PF_6$

SCHEME 23.2 Improved synthesis of *E. coli* lipid A **1**.

ring in one step. The glycosyl donor **6** was prepared through this procedure.

(ii) a cyclic benzyl-type diester (Watanabe *et al.*, 1990) was used for the protection of the 4′-phosphate.

(iii) an *N*-Troc-glucosamine imidate was employed as a glycosyl donor.

The cyclic benzyl-type xylidene ester of a phosphate is stable enough to survive the long steps of total synthesis as compared to the corresponding simple dibenzyl ester which is not suitable for the same purpose because its partial cleavage cannot be avoided during the multistep synthesis. In addition, the use of the xylidene protection for the 4′-phosphate enabled the final hydrogenolytic deprotection in one step with palladium catalyst alone.

Coupling of the imidate donor **6** with a new acceptor **7** afforded the desired β-(1→6)-disaccharide **8**. After stepwise introduction of the acyl groups at 3-*O*- and 2′-*N*-positons of the disaccharide, the fully protected product **9** was then converted to 1-*O*-phosphate **10**, which was subjected to hydrogenolytic deprotection in one step to give **1**.

Several new procedures were also reported for preparation of optically pure 3-hydroxy fatty acids of both configurations (Noyori *et al.*, 1987; Oikawa and Kusumoto, 1995; Liu *et al.*, 1997). By the use of (*S*)-3-hydroxytetradecanoic acid, an artificial analogue of *E. coli*-type lipid A **11** (Figure 23.2) was thus obtained which contains unnatural (*S*)-hydroxy acids in place of the (*R*)-acids present in natural lipid A **1** (Liu *et al.*, 1997). Needless to say, such unnatural structural analogues are available only by chemical synthesis.

Liquid–liquid partition chromatography was applied to practical purification of the deprotected lipid A after hydrogenolysis. Separation by the principle of partition proved to be particularly effective for amphiphilic molecules like lipid A which have a strong tendency to aggregate in solutions and are often strongly retained on both polar and lipophilic surfaces for column chromatographic separations. For example, an artificial *E. coli*-type lipid A analogue **11** with (*S*)-hydroxytetradecanoic acids was purified by the use of centrifugal partition chromatography (CPC) with a two-phase solvent system of butanol:THF:water:triethylamine = 45:35:100:22(Liu *et al.*, 1997). More convenient and effective purification by the same principle of partition can be achieved soon after that with a column of Sephadex LH-20 using similar solvent systems.

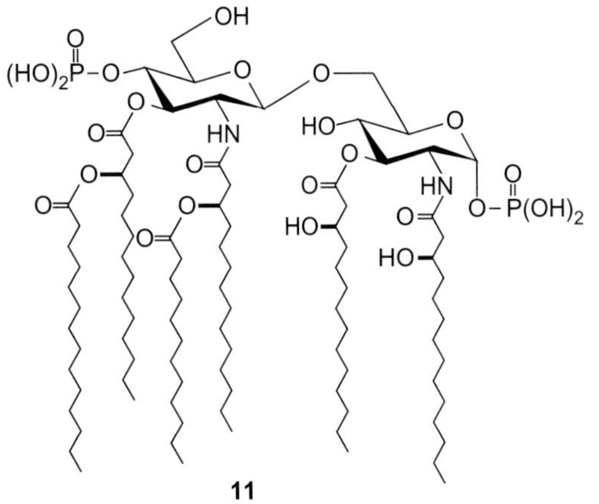

FIGURE 23.2 Structure of *E. coli*-type lipid A **11** containing (*S*)-hydroxy acids.

The biosynthetic precursor **2** prepared by the new efficient route was purified successfully by partition chromatography with a solvent system of chloroform:methanol:2-propanol: water:triethylamine = 40:40:5:45:0.01 (Oikawa *et al.*, 1997).

Chemical structures of lipid A from various bacteria have been elucidated (see Chapter 3). Most of them share the same bis-phosphorylated disaccharide as their hydrophilic backbone with the typical enterobacterial lipid A, but some lack one or both of the phosphates. The fatty acyl groups in lipid A also vary depending on the bacterial species and sometimes also growing conditions. Various natural and unnatural lipid A analogues have been synthesized in order to confirm their structures and biological activities. Some of such synthetic works are described below. Biological activities of structural analogues synthesized before 1992 have been summarized elsewhere (Takada and Kotani, 1992).

There are some natural lipid As which lack the 4′-phosphate. Lipid A of *Helicobacter pylori* produces lipid A of this class but isolated preparations usually consist of a mixture of 1-monophosphate with or without an ethanolamine substituent **24a** and **24b** (Figure 23.3). Biological activity of these lipid A analogues attracted much interest in relation to the pathogenicity of this bacterium which causes gastric diseases. These structures were separately synthesized and their functions investigated (Sakai *et al.*, 2000).

Another example of a unique lipid A analogue recently synthesized is that from *Rhizobium* sin-1. This lipid A **25** (see Figure 23.3) contains a characteristic β-(1→6)-disaccharide composed of glucosamine and 2-aminogluconolactone and lacks both phosphates. The location of total five acyl groups is also unique containing one very long C_{28} (ω-1)-hydroxy acid. This lipid A is reported to be not endotoxic but to inhibit the action of *E. coli* lipid A, thus representing another example of antagonists similar to lipid A of *Rhodobacter sphaeroides* and *Rh. capsulatus* (see below). A divergent synthetic route to **25** was reported and

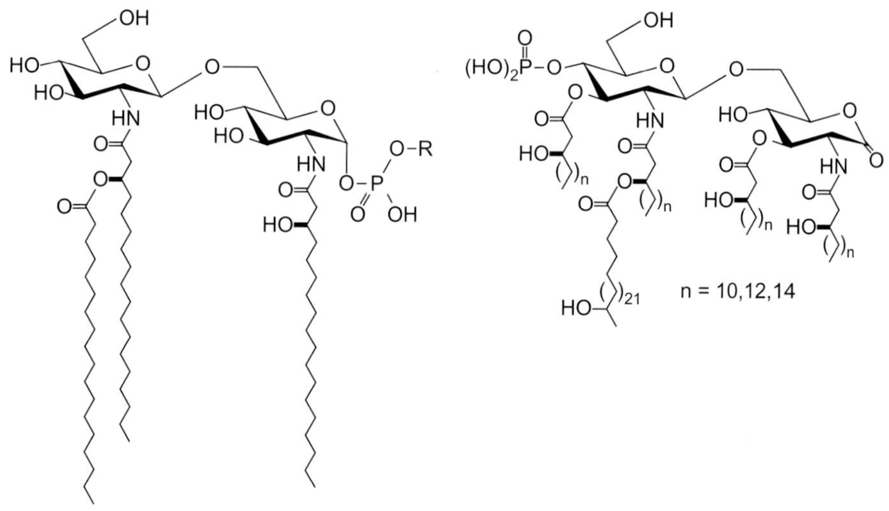

24

a : R = $OCH_2CH_2NH_2$

b : R = H

25

n = 10,12,14

FIGURE 23.3 Structures of *Helicobacter pylori* lipid A **24** and a synthetic counterpart **25** of *Rhizobiumm* sin-1 lipid A.

the antagonistic activity to suppress the effect of *E. coli* lipid A was confirmed for a synthetic product (Demchenko *et al.*, 2003).

Synthetic substitutions of the glycosyl phosphate with other acidic functionalities were attempted for the purpose of finding more readily accessible compounds which may retain the beneficial potencies of lipid A, such as antitumour activity. The presence of the chemically labile glycosyl phosphate makes the synthesis and purification of natural type lipid A derivatives quite difficult. This is particularly the case when dealing with a small amount of this amphiphilic substance which tends to aggregate. A novel analogue of lipid A created is a phosphono-oxyethyl (PE) analogue **26** (Kusama *et al.*, 1990) (Figure 23.4), which exhibits potent endotoxic activity indistinguishable from that of natural counterpart **1**. The PE derivative **27** with four hydroxyacyl groups corresponding to the biosynthetic precursor **2** was also synthesized (Kusama *et al.*, 1991). As expected from the activity of the latter, which is now known to act as an antagonist to suppress the endotoxic action of LPS and lipid A in human cells (Loppnow *et al.*, 1986), **27** turned out to be an antagonist also. Because the phosphono-oxyethyl group is stable enough as compared to the glycosyl phosphate, the synthetic routes to **26** and **27** can be quite flexible. Though their first syntheses were achieved by a similar procedure to the early synthesis of natural type lipid A described in the previous section above, these PE analogues can then be prepared via more efficient improved routes. The purification of the final deprotected products is also much easier than the natural type counterparts (Liu *et al.*, 1999).

This situation led to the first successful synthesis of radiolabelled endotoxic lipid A analogues **26** and **27** (Fukase *et al.*, 2001). Radiolabelled lipid A has been desired for the elucidation of the action mechanism of lipid A. Because of its strong endotoxicity, a labelled derivative of a very high specific radioactivity is inevitably required: the labelled lipid A has to be detected at a concentration below the level where host cells are not killed by its toxicity. Such a high radioactivity can never be attained by a biosynthetic procedure. In addition, labelling of the fatty acyl groups and phosphate should be avoided in order to exclude false information because these functionalities can be cleaved off in living systems when used as a tracer. Scheme 23.3 illustrates the synthesis of a phosphono-oxyethyl analogue **26a** tritium-labelled at the ethylene glycol part.

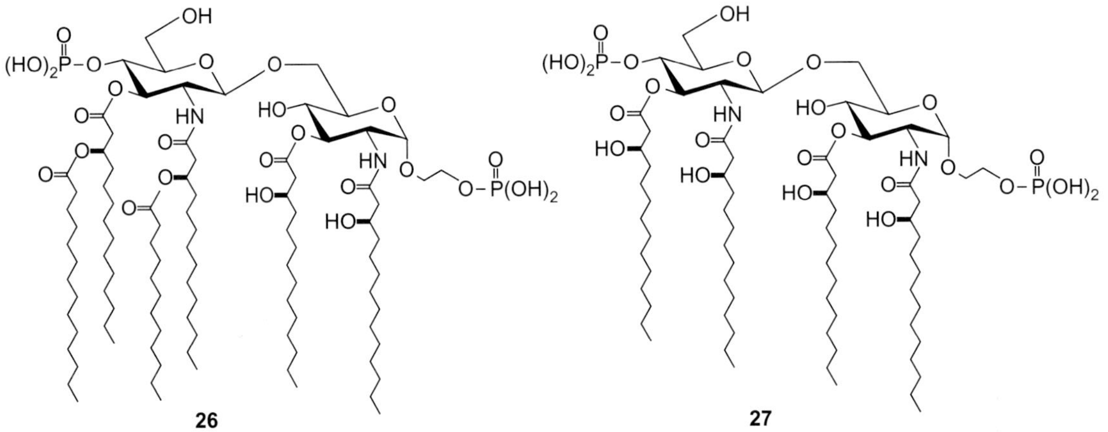

FIGURE 23.4 Structures of phosphono-oxyethyl (PE) analogues of *E. coli*-type lipid A **26** and biosynthetic precursor **27**.

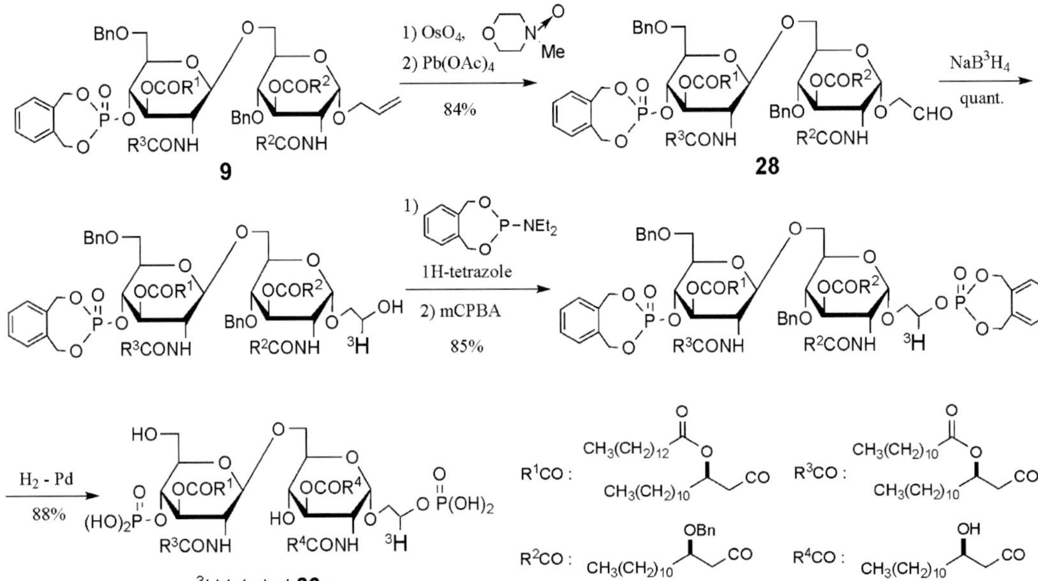

SCHEME 23.3 Synthesis of ^{3}H-labelled phosphono-oxyethyl analogue **26a** of *E. coli* lipid A.

The same fully acylated disaccharide 4′-phosphate **9** used as a synthetic intermediate to *E. coli* lipid A (see Scheme 23.2) served as the starting material to the tritium-labelled PE-type lipid A. Contrary to the usual procedure for synthesis of lipid A, where the allyl group is cleaved and the resulting 1-hydroxy group phosphorylated to give the lipid A structure, the allyl group of the disaccharide **9** was oxidized to an aldehyde **28** in this synthesis. Reduction of the aldehyde with a tritium-labelled borohydride reagent gave the α-glycosidically linked radiolabelled hydroxyethyl function, which was then phosphorylated by the phosphoroamidite procedure followed by oxidation. Hydrogenolysis after intensive purification of the protected precursor gave the product **26a** with high specific radioactivity in a highly pure state. The corresponding biosynthetic precursor-type labelled PE derivative was also obtained similarly (Fukase *et al.*, 2001). The synthetic radiolabelled compounds were utilized to investigate the mode of interaction of lipid A with its receptor proteins (Kobayashi *et al.*, 2006).

A carboxymethyl group proved to be another acidic substituent of the glycosyl phosphate by synthesis of the carboxymetyl analogue **29** (Figure 23.5) of *E. coli* lipid A (Liu *et al.*, 1999).

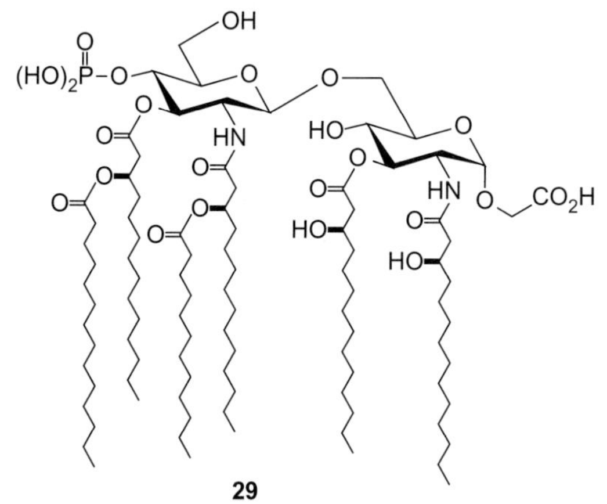

29

FIGURE 23.5 Structure of carboxymethyl analogue of *E. coli*-type lipid A **29**.

A series of carboxymethyl analogues with various distribution patterns of acyl groups were synthesized and their biological activities discussed in relation to their molecular conformations (Fukase *et al.*, 2008).

4. SYNTHESIS OF LIPID A CONTAINING AN UNSATURATED FATTY ACYL GROUP

Catalytic hydrogenolysis was employed for the final deprotection in all the syntheses of the lipid A analogues described above. This general strategy has been satisfactory for two reasons. The first reason is that most lipid A isolated from natural bacterial sources contains only saturated fatty acids as its components. Another reason is that hydrogenolysis of benzyl, xylidene and, in earlier syntheses, phenyl groups, form only volatile byproducts which are readily removable leaving the final free lipid A in pure states without damaging other functionalities including the unstable glycosidic phosphate.

Such a synthetic strategy based on the benzyl-type protection is not applicable to lipid A analogues which contain unsaturated acyl groups. A typical example of this is the *Rhodobacter sphaeroides* lipid A **30** (Figure 23.6), which is reported to share the same hydrophilic backbone consisting of the 1,4′-bisphosphorylated β-(1→6)-glucosamine disaccharide as in many other bacteria but contains unusual fatty acids: a 3-keto acid on the 2-amino group and an unsaturated acid in the 3-acyloxyacyl group linked to the 2′-amino group of the backbone (Salimath *et al.*, 1983; Takayama *et al.*, 1989; Kirkland *et al.*, 1991).

Because of its potent antagonistic activity to suppress the endotoxic function of LPS, *R. sphaeroides* lipid A attracts much attention: it may have a possible clinical application to therapy against sepsis and shock syndrome caused by LPS in the case of Gram-negative infections.

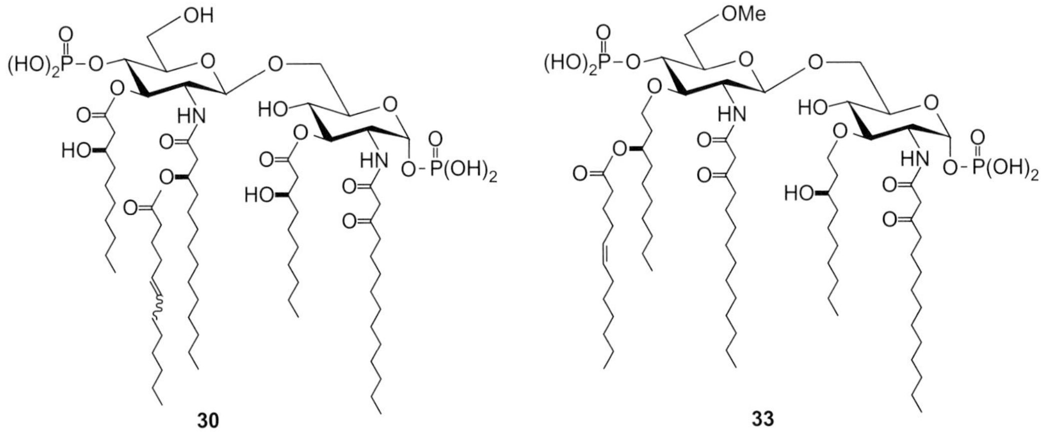

FIGURE 23.6 Proposed structures of *Rhodobacter sphaeroides* lipid A **30** and its synthesized stable artificial analogue **33**.

A new synthetic route was elaborated to such unsaturated lipid A based on allyl-type protections (Christ *et al.*, 1994). These protecting groups are known to be removable by transition metal-catalysed reactions thereby leaving the isolated double bond in the molecule intact.

In their synthesis, the allyloxycarbonyl (Alloc) group and diallyl ester were employed for persistent protection of hydroxy groups and the 4′-phosphate, respectively, as shown in Scheme 23.4. Thus, the glycosyl trichloroacetimidate **31** of a 2-azide sugar was coupled with an acceptor **32**, whose glycosidic position was temporarily protected as the *t*-butyldimethylsily (TBS) glycoside. The 2-azide function of the disaccharide obtained was then reduced and converted to the acylamido goup. The TBS group was then selectively cleaved and the glycosyl phosphate introduced as its diallyl ester by means of the phosphoramidite method. The all allyl-type protecting groups were removed cleanly in one step with a palladium(0) catalyst to give the first synthetic lipid A analogues **30a** and **30b** each having a double bond.

The synthetic products **30a** and **30b** proved to have potent activities to suppress the toxic effect of LPS in a human monocyte system, though neither of them was identical with the natural lipid A obtained from *R. sphaeroides* cells. The same authors then designed a novel artificial derivative **33** which is more resistant than **30** against biological degradation and expected to be of therapeutic value in future (Christ *et al.*, 1995).

5. CONCLUDING REMARKS

In this chapter are described typical examples of recent chemical syntheses of lipid A analogues including both natural types corresponding to those obtained from bacterial cells and several artificial structural analogues related to the former. In the first instance, an overview of some key earlier published works has been provided, followed by a review of subsequent methodological improvements which made the synthesis of this group of biologically interesting glycoconjugate molecules practical and efficient. One may agree that the complex amphiphilic architectures of structural analogues are now synthetically accessible. Homogeneous preparations of any desired structure become available

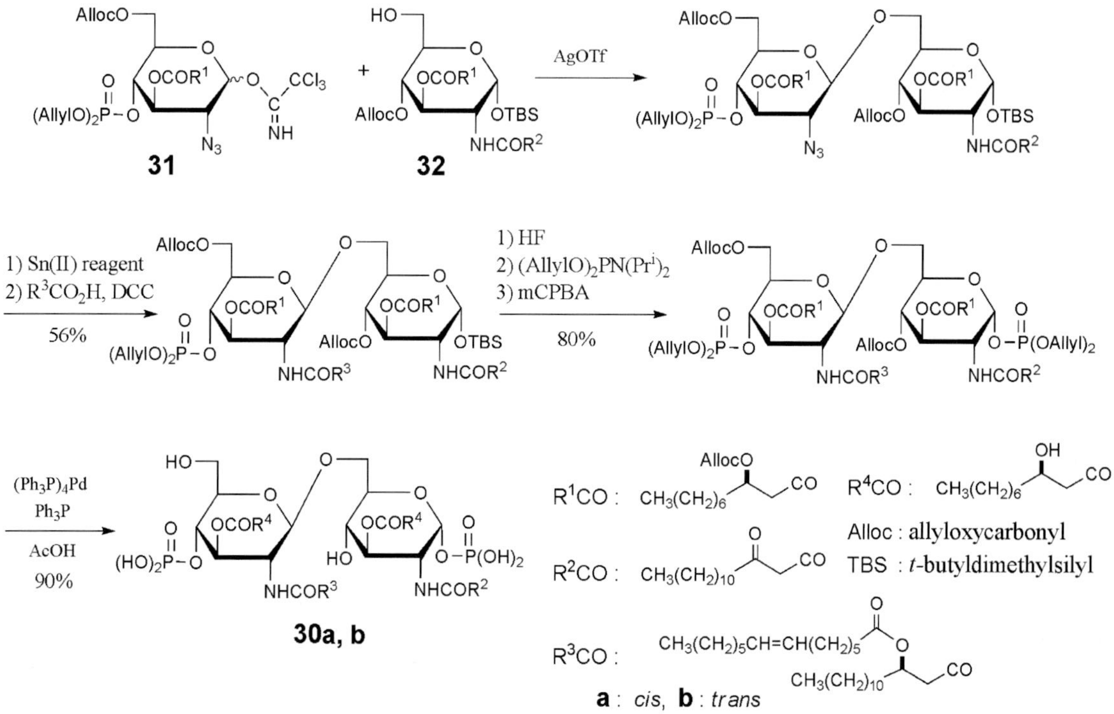

SCHEME 23.4 Synthesis of a proposed structure of *Rhodobacter sphaeroides* lipid A **30**.

when the synthetic route to the target is carefully designed. As mentioned above, both ^{3}H-labelled endotoxic and antagonistic lipid A analogues were prepared and utilized for the study of their interactions with the receptor complex that induces innate immune responses of hosts on contact with LPS and lipid A. Regiospecifically ^{13}C-labelled lipid A analogues were also prepared and used for detailed conformational analysis by NMR spectroscopy (Fukase *et al.*, 1999; Oikawa *et al.*, 1999). Synthesis was also extended from lipid A to more complex structures such as *E. coli* Re-type LPS which contains two Kdo residues sequentially linked to lipid A (Yoshizaki *et al.*, 2001). Synthesis and biological activity of similar Kdo-containing derivatives of *H. pylori* lipid A have been recently reported as well (Fujimoto *et al.*, 2007).

By the aid of modern methods of purification, lipid A analogues from various bacterial species have been isolated in high homogeneity and their structures readily elucidated by new powerful spectroscopic methods. Even in such a situation, the important role of chemical synthesis still remains in that preparations completely free from any contaminants of bacterial origin can only be supplied by synthesis. Beside such standard materials, labelled or modified derivatives to be used as tracers or probes for biological investigations or spectroscopic conformational analyses have also to be synthetically prepared. To fulfill such requirements, methodologies for chemical synthesis should be continuously improved with regards to time requirement and efficiency of transformations as well as from ecological viewpoints.

References

Christ, W.J., McGuiness, P.D., Asano, O., et al., 1994. Total synthesis of the proposed structure of *Rhodobacter sphaeroides* lipid A resulting in the synthesis of new potent lipopolysaccharide antagonists. J. Am. Chem. Soc. 116, 3637–3638.

Christ, W.J., Asano, O., Robidoux, A.L., et al., 1995. E5531, a pure endotoxin antagonist of high potency. Science 268, 80–83.

Demchenko, A.V., Wolfelt, M.A., Santhanam, B., Moore, J.N., Boons, G.-J., 2003. Synthesis and biological evaluation of *Rhizobium* sin-1 lipid A derivatives. J. Am. Chem. Soc. 125, 6103–6112.

Fujimoto, Y., Iwata, M., Imakita, N., et al., 2007. Synthesis of immunoregulatory *Helicobacter pylori* lipopolysaccharide partial structures. Tetrahedron Lett. 48, 6577–6581.

Fukase, K., Oikawa, M., Suda, Y., et al., 1999. New synthesis and conformational analysis of lipid A: biological activity and supramolecular assembly. J. Endotoxin Res. 5, 46–51.

Fukase, K., Kirikae, T., Kirikae, F., et al., 2001. Synthesis of [^{3}H]-labeled bioactive lipid A analogs and their use for detection of lipid A-binding proteins on murine macrophages. Bull. Chem. Soc. Jpn. 74, 2189–2197.

Fukase, Y., Fujimoto, Y., Adachi, Y., Suda, Y., Kusumoto, S., Fukase, K., 2008. Synthesis of *Rubrivivax gelatinosus* lipid A and analogues for investigation of the structural basis for immunostimulating and inhibitory activities. Bull. Chem. Soc. Jpn. 81, 796–819.

Galanos, C., Lehman, V., Lüderitz, O., et al., 1984. Endotoxic properties of chemically synthesized lipid A part structures – comparison of synthetic lipid A precursor and synthetic analogues with biosynthetic lipid A precursor and free lipid A. Eur. J. Biochem. 140, 221–227.

Galanos, C., Lüderitz, O., Rietschel, E.Th., et al., 1985. Synthetic and natural *Escherichia coli* free lipid A express identical endotoxic activities. Eur. J. Biochem. 148, 1–5.

Gmeiner, L., Lüderitz, O., Westphal, O., 1969. Biochemical studies on lipopolysaccharide of *Salmonella* R mutant 6. Investigation on the structure of the lipid A component. Eur. J. Biochem. 7, 270–379.

Hansen-Hagge, T., Lehman, V., Seydel, U., Lindner, B., Zähringer, U., 1985. Isolation and characterization of eight lipid A precursors from a 3-deoxy-D-*manno*-octulosonic acid-deficient mutant of *Salmonella typhimurium*. J. Biol. Chem. 260, 16080–16088.

Hase, S., Rietschel, E.Th., 1976. Isolation and analysis of the lipid A backbone. Lipid A structure of lipopolysaccharide from various bacterial groups. Eur. J. Biochem. 63, 101–107.

Imoto, M., Kusumoto, S., Shiba, T., et al., 1983. Chemical structure of *E. coli* lipid A: linkage site of acyl groups in the disaccharide backbone. Tetrahedron Lett. 24, 4017–4020.

Imoto, M., Yoshimura, H., Yamamoto, M., Shimamoto, T., Kusumoto, S., Shiba, T., 1984a. Chemical synthesis of phosphorylated tetraacyl disaccharide corresponding to a biosynthetic precursor of lipid A. Tetrahedron Lett. 25, 2667–2670.

Imoto, M., Yoshimura, H., Kusumoto, S., Shiba, T., 1984b. Total synthesis of lipid A, active principle of bacterial endotoxin. Proc. Jpn. Acad., Ser. B, Phys. Biol. Sci. 60, 285–288.

Imoto, M., Yoshimura, H., Yamamoto, M., Shimamoto, T., Kusumoto, S., Shiba, T., 1987a. Chemical synthesis of a biosynthetic precursor of lipid A with a phosphorylated tetraacyl disaccharide structure. Bull. Chem. Soc. Jpn. 60, 2197–2204.

Imoto, M., Yoshimura, H., Shimamoto, T., Sakaguchi, N., Kusumoto, S., Shiba, T., 1987b. Total synthesis of *Escherichia coli* lipid A, the endotoxically active principle of cell-surface lipopolysaccharide. T. Bull. Chem. Soc. Jpn. 60, 2205–2214.

Inage, M., Chaki, H., Kusumoto, S., Shiba, T., 1981. Chemical synthesis of phosphorylated fundamental structure of lipid A. Tetrahedron Lett. 22, 2281–2284.

Inage, M., Chaki, H., Kusumoto, S., Shiba, T., 1982. A convenient preparative method of carbohydrate phosphates with butyllithium and phosphorochloridate. Chem. Lett., 1281–1284.

Kirkland, T.N., Qureshi, N., Takayama, K., 1991. Diphosphoryl lipid A derivative from lipopolysaccharide (LPS) of *Rhodopseudomonas sphaeroides* inhibits activation of 70Z/3 cells by LPS. Infect. Immun. 59, 131–136.

Kiso, M., Hasegawa, M., 1983. Synthetic studies on the lipid A component of bacterial lipopolysaccharide. In: Anderson, L., Unger, F.M. (Eds.) Bacterial Lipopolysaccharide. ACS Symp. Ser. 231. American Chemical Society, Washington DC, pp. 277–300.

Kobayashi, M., Saitoh, S., Tanimura, N., et al., 2006. Lipid A antagonist, lipid IVa, is distinct from lipid A in interaction with Toll-like receptor 4 (TLR4)-MD-2 and ligand-induced TLR4 oligomerization. J. Immunol. 176, 6211–6218.

Kotani, S., Takada, H., Tsujimoto, M., et al., 1984. Immunobiologically active lipid A analogs synthesized according to a revised structural model of natural lipid A. Infect. Immun. 45, 293–296.

Kotani, S., Takada, H., Tsujimoto, M., et al., 1985. Synthetic lipid A with endotoxic and related biological activities comparable to those of a natural lipid A from an *Escherichia coli* Re-mutant. Infect. Immun. 49, 225–237.

Kusama, T., Soga, T., Shioya, E., et al., 1990. Synthesis and antitumor activity of lipid A analogs having a phosphonooxyethyl group with α- or β-configuration at position 1. Chem. Pharm. Bull. 38, 3366–3372.

Kusama, T., Soga, T., Ono, Y., et al., 1991. Synthesis and biological activity of a lipid A biosynthetic precursor: 1-*O*-phosphonooxyethyl-4′-*O*-phosphono-disaccharices with (*R*)-3-hydroxytetradecanoyl or tetradecanoyl groups at positions 2, 3, 2′, and 3′. Chem. Pharm. Bull. 39, 1994–1999.

Kusumoto, S., Inage, M., Chaki, H., Shimamoto, T., Shiba, T., 1983. Chemical synthesis of lipid A for the elucidation of structure-activity relationships. In: Anderson, L., Unger, F.M. (Eds.), Bacterial Lipopolysaccharide. ACS Symp. Ser. 231. American Chemical Society, Washington DC, pp. 237–254.

Kusumoto, S., Fukase, K., Oikawa, M., 1999. The chemical synthesis of lipid A. In: Brade, H., Opal, S.M., Vogel, S.N., Morrison, D.C. (Eds.), Endotoxin in Health and Disease. Marcel Dekker, New York, pp. 243–256.

Kusumoto, S., Fukase, K., Fujimoto, Y., 2007. Synthesis of lipopolysaccharide, peptidoglycan, and lipoteichoic acid fragments. In: Kamerling, J.P. (Ed.), Comprehensive Glycosicience, Vol. 1. Elsevier, Amsterdam, pp. 685–711.

Kusumoto, S., Hashimoto, M. Kawahara, K., 2008. Structure and synthesis of lipid A. In: Jeannin J.-F. (Ed.), Lipid A in Cancer Therapy. Landes Bioscience, Austin. epub ahead of print (http://www.eurekah.com/chapter/3709).

Liu, W.-C., Oikawa, M., Fukase, K., et al., 1997. Enzymatic preparation of (*S*)-3-hydroxytetradecanoic acid and synthesis of unnatural analogues of lipid A containing the (*S*)-acid. Bull. Chem. Soc. Jpn. 70, 1441–1450.

Liu, W.-C., Oikawa, M., Fukase, K., Suda, Y., Kusumoto, S., 1999. A divergent synthesis of lipid A and its chemically stable unnatural analogues. Bull. Chem. Soc. Jpn. 72, 1377–1385.

Loppnow, H., Brade, L., Brade, H., et al., 1986. Induction of human interleukin 1 by bacterial and synthetic lipid A. Eur. J. Immunol. 16, 1263–1267.

Noyori, R., Ohkuma, T., Kitamura, M., et al., 1987. Asymmetric hydrogenation of β-keto carboxylic acid esters. A practical, purely chemical access to β-hydroxy esters in high enantiomeric purity. J. Am. Chem. Soc. 109, 5856–5858.

Oikawa, M., Kusumoto, S., 1995. On a practical synthesis of β-hydroxy fatty acid derivatives. Tetrahedron Asymm. 6, 961–966.

Oikawa, M., Wada, A., Yoshizaki, H., Fukase, K., Kusumoto, S., 1997. New efficient synthesis of a biosynthetic precursor of lipid A. Bull. Chem. Soc. Jpn. 70, 1435–1440.

Oikawa, M., Shintaku, T., Sekljic, H., Fukase, K., Kusumoto, S., 1999. Synthesis of ^{13}C-labeled biosynthetic precursor of lipid A and its analogue with shorter acyl chains. Bull. Chem. Soc. Jpn. 72, 1857–1867.

Salimath, P.V., Weckesser, J., Strittmatter, W., Mayer, H., 1983. Structural studies on nontoxic lipid A from Rhodobacter sphaeroides ATCC 17023. Eur. J. Biochem. 136, 195–200.

Raetz, C.R., Whitfield, C., 2002. Lipopolysaccharide endotoxins. Annu. Rev. Biochem. 71, 635–700.

Rietschel, E.Th., Westphal, O., 1999. Endotoxin: historical perspectives. In: Brade, H., Opal, S.M., Vogel, S.N., Morrison, D.C. (Eds.), Endotoxin in Health and Disease. Marcel Dekker, New York, pp. 1–31.

Sakai, Y., Oikawa, M., Yoshizaki, H., et al., 2000. Synthesis of *Helicobacter pylori* lipid A and its analogue using *p*-(trifluoromethyl)benzyl protecting group. Tetrahedron Lett. 41, 6843–6847.

Tai, A., Nakahata, M., Harada, H., et al., 1980. A facile method for preparation of the optically pure 3-hydroxytetradecanoic acid by an application of asymmetrically modified nickel catalyst. Chem. Lett. 1125–1126.

Takada, H., Kotani, S., 1992. Structure-function relationships of lipid A. In: Morrison, D.C., Ryan, J.L. (Eds.), Bacterial Endotoxic Lipopolysaccharides Vol. I: Molecular Biochemistry and Cellular Biology. CRC Press, Boca Raton, pp. 43–65.

Takayama, K., Qureshi, N., Mascagni, P., 1983. Complete structure of lipid A obtained from *Salmonella typhimurium*. J. Biol. Chem. 258, 12801–12803.

Takayama, K., Qureshi, N., Beutler, B., Kirkland, T.H., 1989. Diphosphoryl lipid A from *Rhodopseudomonas sphaeroides* ATCC 1702 blocks induction of cachectin in macrophage by lipopolysaccharide. Infect. Immun. 57, 1336–1338.

Watanabe, Y., Kodama, Y., Ebysuya, K., Ozaki, S., 1990. An efficient phosphorylation method using a new phophitylating agent, 2-diethylamino-1,3,2-benzodioxaphosphepane. Tetrahedron Lett. 31, 255–256.

Westphal, O., Lüderitz, O., 1954. Chemische Erforschung von Lipopolysacchariden Gram-negativer Bacterien. Angew. Chem. 66, 401–407.

Yoshizaki, H., Fukuda, N., Sato, K., et al., 2001. First total synthesis of the Re-type lipopolysaccharide. Angew. Chem. Int. Ed. Engl. 40, 1475–1480.

CHAPTER

24

Chemical synthesis of the core oligosaccharide of bacterial lipopolysaccharide

Paul Kosma

SUMMARY

The chapter reviews major strategies towards the chemical synthesis of the core region of Gram-negative bacteria illustrated by representative examples. The focus is on the synthesis of structural units and neoglycoconjugates of the heptose- and 3-deoxy-D-*manno*-oct-2-ulosonic acid-containing inner core region listed according to the corresponding bacterial species. Additional subtopics comprise synthesis of phosphate-substituted determinants as well as selected outer-core units.

Keywords: Lipopolysaccharide; Lipo-oligosaccharide; Heptose; 3-deoxy-D-*manno*-oct-2-ulosonic acid (Kdo); D-*glycero*-D-*talo-oct-2-ulosonic acid* (Ko); Core region; Synthesis; Neoglycoconjugate

1. INTRODUCTION

Several in-depth reviews on the synthesis of higher-carbon bacterial sugars and of oligosaccharide structures of the core region of bacterial lipopolysaccharides (LPSs) have been published previously covering important aspects of chemical synthesis, protecting group strategies and preparation of neoglycoconjugates (Kosma, 1999; Hansson and Oscarson, 2000; Pozsgay, 2003). The preparation of the major structural units of the enterobacterial core architecture had been accomplished mostly in the period of ≈1985–1995. Meanwhile, the powerful tools of modern genomics provide access to a broad spectrum of complex bacterial oligosaccharides. Thus, synthetic targets have been re-focused towards specific ligands needed for diagnostics or therapeutic applications, as well as to the preparation of non-natural analogues. The synthesis of minor structural modifications in core oligosaccharides, however, has received much less attention by carbohydrate chemists. The present chapter is focused on an update of previous comprehensive summaries and will mainly address the relevant literature of the past decade. The reader is recommended to consult the earlier work for a full overview of the field.

2. SYNTHESIS OF 3-DEOXY-D-MANNO-OCT-2-ULOSONIC ACID (Kdo)- AND D-*GLYCERO*-D-*TALO*-OCT-2-ULOSONIC ACID (Ko)-CONTAINING CORE STRUCTURES

2.1. Synthesis of Kdo, Kdo glycosyl donors and Kdo oligosaccharides

Since the seminal review on the chemical and biological significance of Kdo by Unger (1981), an impressive collection of various chemical syntheses of this important bacterial sugar has accumulated in the chemical literature and has only partly been covered in a recent summary (Li and Wu, 2003). Although the number of papers dealing with the synthesis of Kdo – mostly utilizing two- or three-carbon chain elongation reactions from appropriately protected precursors – is impressive, practical syntheses of multigram amounts mostly rely on the improved Cornforth condensation of D-arabinose with oxaloacetic acid and on the reaction of α-keto carboxylic acid synthons to suitably protected mannose precursors, respectively. Whereas the former approach makes further installation of protecting groups necessary, the latter one allows for direct access to various glycosyl donors of Kdo. Commonly used donors of Kdo comprise the peracetylated Kdo bromide **1**, the isopropylidene-protected Kdo benzyl ester fluoride **2** and, less frequently, Kdo-thioglycosides **3** (Mannerstedt *et al.*, 2007). Major problems encountered in the synthesis of Kdo-glycosides reside in the acid sensitivity of the ketosidic bond, the low reactivity of the anomeric centre due to the deactivating effect exerted by the carboxylic group, the lack of stereocontrol due to the absence of anchimeric assistance from the 3-position and the propensity of Kdo donors to form glycal esters such as **4** as by-products (Figure 24.1).

Whereas a substantial number of auxiliary groups at position 3 as well as activating groups at the anomeric centre of Kdo have been studied, there is still room for improvement with respect to yield, anomeric selectivity in glycosylation reactions and the study of other types of anomeric leaving groups. The early chemical syntheses of the Kdo region and of α-Kdo-(2→4)-α-Kdo units linked to the glucosamine disaccharide backbone of lipid A were achieved by Paulsen and Unger (Paulsen *et al.*, 1987). The main features of these approaches comprised the regioselective protection of the 7,8-diol side chain of Kdo by introduction of ester, acetal or silyl ether groups followed by preferred reaction of the equatorial 4-OH group with Kdo glycosyl donors under carefully optimized glycosylation conditions. Development of Kdo-donors such as **2** (or as 4,5-di-*O*-*tert*-butyldimethylsilyl- protected derivatives) allowing for final deprotection under non-alkaline conditions provided the basis for the ensuing full assembly of Kdo units linked to acylated and phosphorylated lipid A as synthetic highlights in natural product

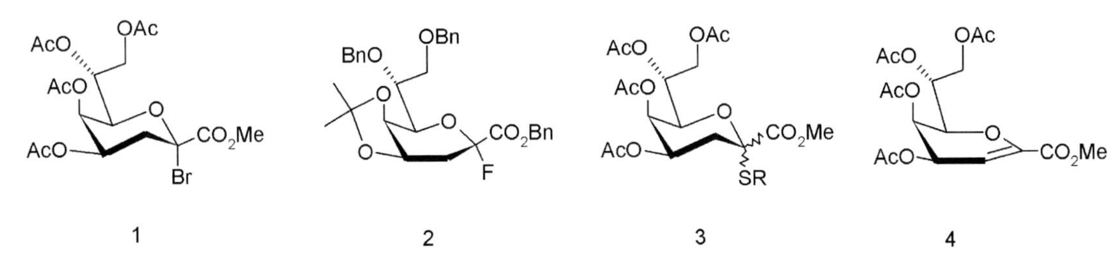

FIGURE 24.1 Commonly used glycosyl donors of Kdo.

synthesis (see Chapter 23, Yoshizaki *et al.*, 2001; Fujimoto *et al.*, 2007; Zhang *et al.*, 2008).

2.2. Synthesis of core structures of *Chlamydiaceae*

All *Chlamydiaceae* share a common carbohydrate epitope **5**, formerly called the genus-specific epitope, attached to lipid A and which resembles the deep rough mutant LPS structures of *Enterobacteriaceae*. For *Chlamydia*, however, the Kdo region constitutes an immunodominant epitope of the sequence α-Kdo-(2→8)-α-Kdo-(2→4)-α-Kdo-(2→6)-lipid A, wherein the (2→8)-linkage confers *Chlamydia*-specificity (Brade *et al.*, 1987). Additionally, species-specific Kdo oligosaccharides are present in the LPS of *Chlamydophila psittaci* such as the α-(2→4)-interlinked Kdo trisaccharide **6** and the branched Kdo tetrasaccharide α-Kdo-(2→4)-[α-Kdo-(2→8)]-α-Kdo-(2→4)-α-Kdo **7** which have been isolated and characterized from recombinant strains expressing the respective CMP-Kdo transferase (Holst *et al.*, 1995; Rund *et al.*, 2000) (Figure 24.2).

For the generation of Kdo-specific monoclonal antibodies and characterization of their epitope specificities, a series of allyl glycosides and bovine serum albumin-(BSA)-conjugates corresponding to the Kdo tri- and tetrasaccharide as well as the corresponding subunits has been prepared (Kosma *et al.*, 1999; Brade *et al.*, 2000). As an example, the synthesis of the branched Kdo tri- and tetrasaccharide structures (Kosma *et al.*, 2000) is outlined in Scheme 24.1. In order to simplify the assembly of the Kdo units and avoid multistep protection, regioselective reactions have been employed. Thus, coupling of the 7,8-diol β-benzyl ketoside acceptor **8** with the Kdo bromide **1** under Helferich conditions proceeded in remarkable regioselectivity to give the α-(2→8)-linked trisaccharide derivative **9** in 59% isolated yield. The branched trisaccharide was then converted into the bromide donor **10** and condensed with the 7,8-carbonate derivative **11**

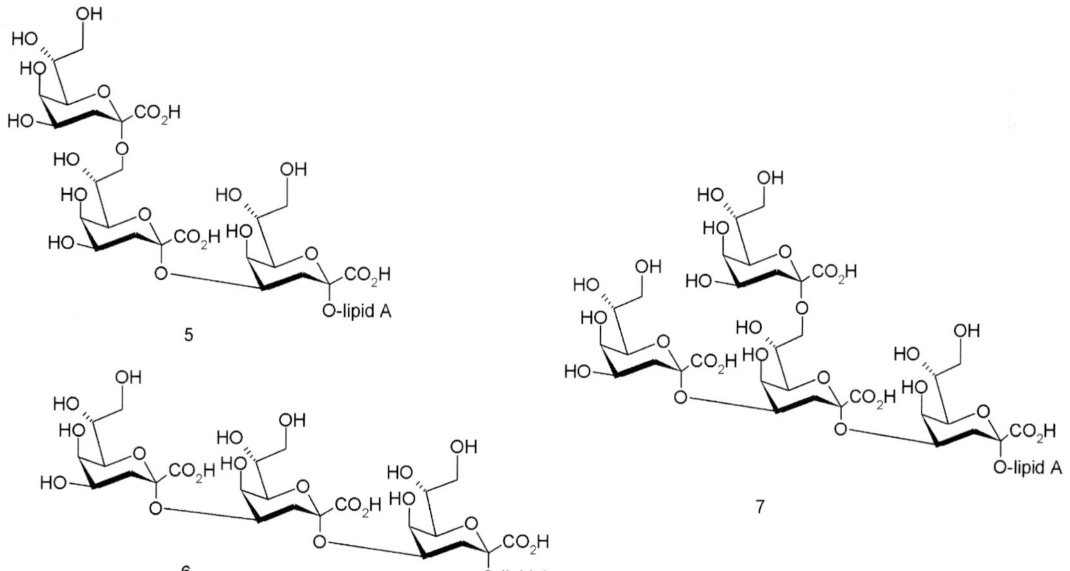

FIGURE 24.2 Chemical structure of the Kdo region from *Chlamydiaceae*. Reprinted, with permission, from Kosma *et al.* (2008) Lipopolysaccharide antigens of *Chlamydia*. In: Roy, R. (ed.), Carbohydrate Vaccines, ACS Symposium Series 989. American Chemical Society, Washington DC, pp. 211–224. American Chemical Society © 2008.

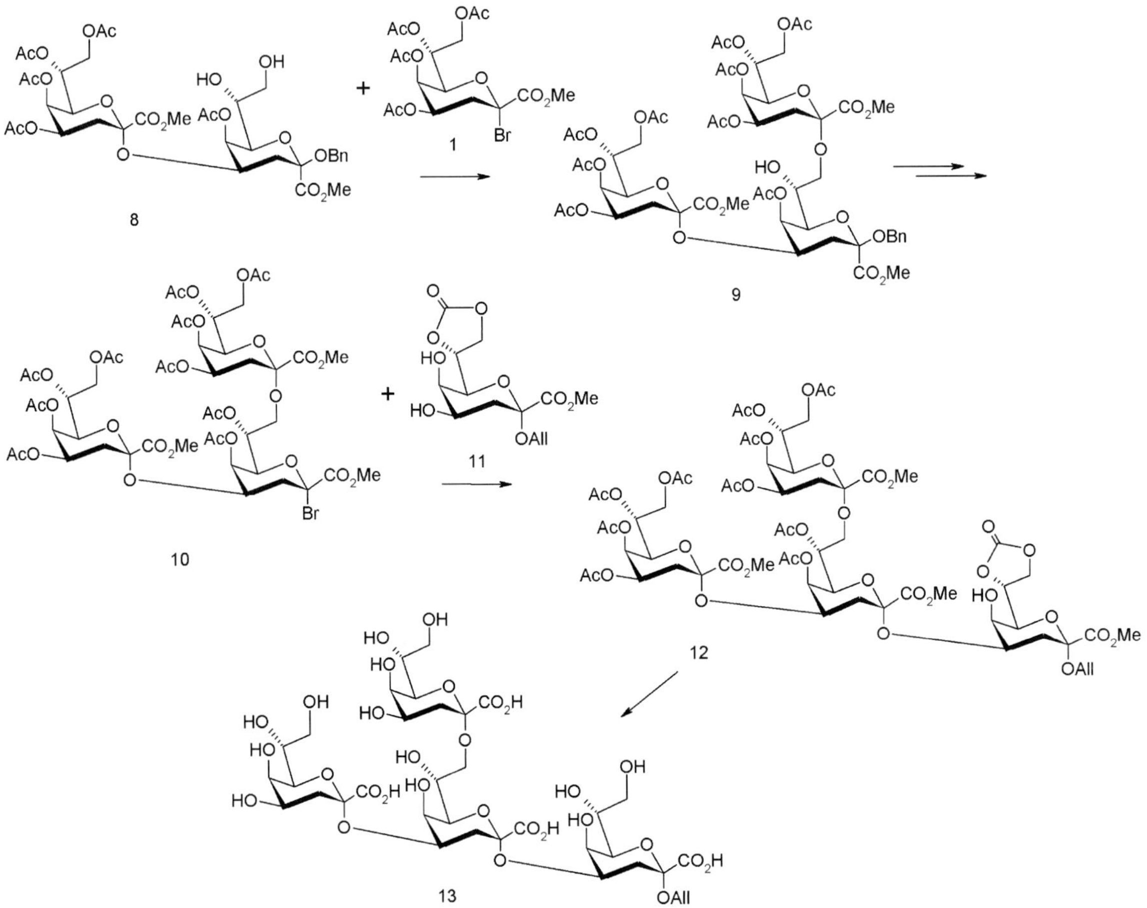

SCHEME 24.1 Synthesis of the branched Kdo tetrasaccharide corresponding of *Chl. psittaci*. Reprinted, with permission, from Kosma *et al.* (2008) Lipopolysaccharide antigens of *Chlamydia*. In: Roy, R. (Ed.), Carbohydrate Vaccines, ACS Symposium Series 989. American Chemical Society, Washington DC, pp. 211–224. American Chemical Society © 2008.

to give the branched tetrasaccharide derivative **12** in 32% yield. Separation from the by-products and deprotection of **12** afforded the Kdo tetrasaccharide allyl glycoside **13**.

Conversely, the branched trisaccharide **19** was synthesized by first assembling the α-(2→8)-disaccharide part, again exploiting a regioselective glycosylation of the 7,8-diol **15**, which gave the α-(2→8)-linked product **16** in 25% yield (Scheme 24.2). Acidic cleavage of the 4,5-*O*-isopropylidene group furnished the glycosyl acceptor **17** which was condensed with the third Kdo unit to give the branched trisaccharide **18**. Subsequent deprotection gave the trisaccharide allyl glycoside **19**. Both oligosaccharides **13** and **19** were reacted with cysteamine, then activated with thiophosgene and conjugated to BSA. A murine monoclonal antibody raised against the neoglycoconjugate derived from **19** displayed high specificity for *Chl. psittaci* LPS and did not cross-react with the linear α-(2→4)-linked Kdo trisaccharide **6** (Müller-Loennies *et al.*, 2006).

Oligosaccharides isolated from recombinant strains expressing chlamydial CMP Kdo

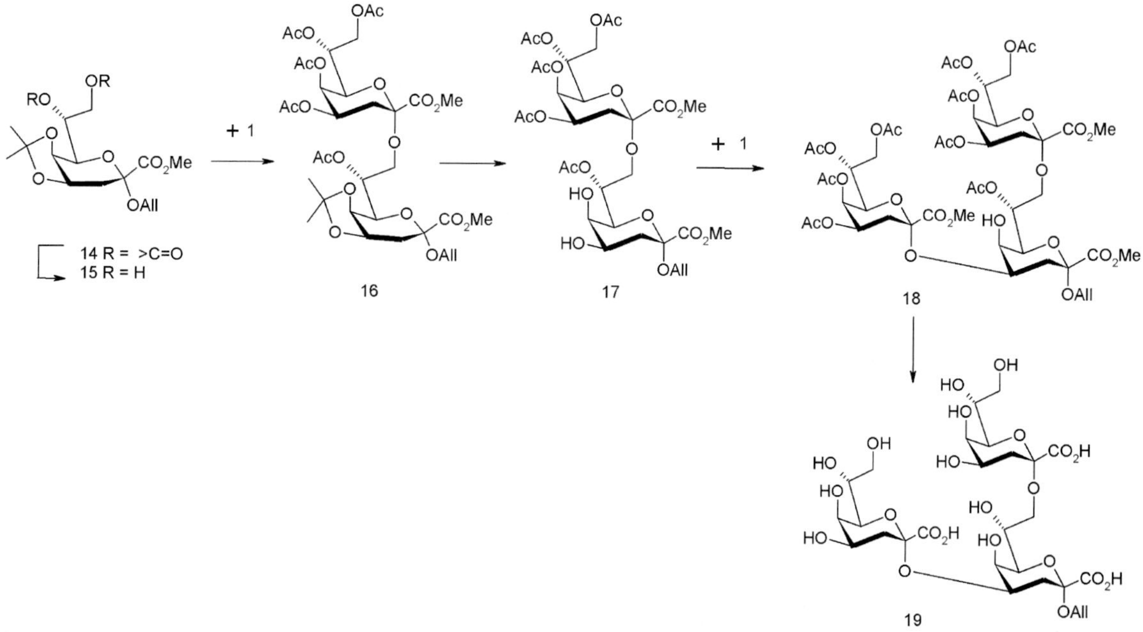

SCHEME 24.2 Synthesis of the branched Kdo-trisaccharide epitope of *Chl. psittaci*.

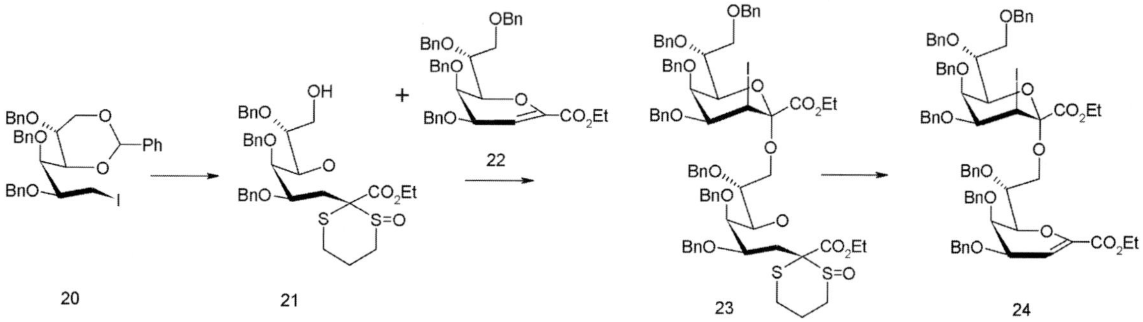

SCHEME 24.3 Iterative synthesis of α-(2→8)-linked Kdo oligosaccharides. Adapted, with permission, from Tanaka *et al.* (2006) Stereoselective synthesis of oligo-α-(2,8)-3-deoxy-D-*manno*-2-octulosonic acid derivatives. Angew. Chem. Int. Ed. Engl. 45, 770–773. © Wiley-VCH Verlag GmbH & Co. KGaA.

transferases as well as synthetic Kdo ligands and Kdo analogues have been utilized as ligands for binding and co-crystallization studies with the Fab-fragments of *Chlamydia*-specific antibodies (Ngyuen *et al.*, 2003). The major interactions seen in ligated complexes of monoclonal antibodies S25-2 and S45-18 reside in the distal Kdo unit with the carboxylic group, OH-4 and OH-5 providing major contributions. Relaxed binding interactions of the antibody were observed for other positions such as C-7 and C-3 in liganded crystal structures with Kdo analogues (Brooks *et al.*, 2008).

As an alternative approach for the construction of α-(2→8)-linked Kdo oligomers, iodoalkylation of open-chain Kdo derivatives has been elaborated (Scheme 24.3). The 5,8-dihydroxy-2-thio-2-sulfoxide Kdo derivative **21** obtained from the 1-iodo-mannitol precursor **20**, was linked to the glycal ester **22** followed by ring closure and elimination to afford **24**. The method was extended to prepare homo-oligomers of α-(2→8)-linked Kdo units via subsequent reduction of the 3-iodo group (Tanaka *et al.*, 2006).

2.3. Synthesis of core structures of *Moraxella catarrhalis*

The inner core region of *Moraxella catarrhalis* contains a truncated and highly branched LPS, with glucose residues being attached to O-5 of Kdo. Oscarson and co-workers achieved the synthesis of a glucosyl tetrasaccharide in 81% yield by one-step elongation of the 3-, 4- and 6-positions of a 2-*O*-benzyl protected glucopyranoside acceptor derivative with benzobromoglucose (Ekelöf and Oscarson, 1995a). Alternatively, stepwise addition was achieved from the β-(1→3)-linked disaccharide **25** (Ekelöf and Oscarson, 1996). Regioselective reductive opening of the benzylidene ring followed by silylation of the 6-OH group was used to install the β-(1→4)-linked glucose unit which was further elongated at the 2-position with the 2-azido-2-deoxy thioglycoside donor **27** to afford **28**. Desilylation of the 6-position was followed by glycosylation of **29** with a 2-*O*-chloroacetyl-protected donor **30**. Selective cleavage of the 2-*O*-chloroacetate from **31** gave the pentasaccharide alcohol **32**, which was further condensed with perbenzylated mono-, di- and trisaccharide donors to give the corresponding protected oligosaccharides **33–35**. Full deprotection of **35** yielded octasaccharide **36** suitable for conversion into immunogenic neoglycoconjugates (Scheme 24.4). Along similar lines, a spacer-equipped Kdo glycosyl acceptor was derived from the glycal ester **4** and was coupled with a suitably protected ethyl-1-thio glucopyranoside donor to give an α-Glc*p*-(1→5)-Kdo disaccharide in 84% yield, again demonstrating the versatile use of thioglycoside donors (Ekelöf and Oscarson, 1995a). After removal of acetyl protecting groups at the glucose unit, simultaneous glucosylation at O-3, O-4 and O-6 was accomplished in a single step to afford the pentasaccharide in 38% isolated yield. Removal of the protecting groups finally furnished the spacer-containing pentasaccharide core determinant **37**.

2.4. Synthesis of glycosides of Ko occurring in *Acinetobacter*, *Yersinia*, *Serratia* and *Burkholderia* LPS

A 3-hydroxy-derivative of Kdo of the D-*glycero*-D-*talo*-configuration, termed Ko, has been detected in several bacteria. This sugar may replace Kdo as the linking sugar to lipid A, but may also occur as lateral substituent in the core and is remarkably stable toward acid hydrolysis. The biosynthesis of this sugar has not been elucidated thus far and needs to be studied in full detail.

The sugar Ko was prepared from the bis-isopropylidene protected mannose **38** by condensation of the reducing end with the lithium salt of methyl glyoxalate diethyl mercaptal **39** followed by ring closure to furnish the D-*glycero*-D-*galacto*-octulosonic acid **40** (Scheme 24.5). Following allylation at the anomeric position with sodium hydride/allyl bromide in high yield, conversion into the 3-epimeric glycoside **41** had to rely on an oxidation/reduction sequence (Reiner and Schmidt, 2000). A second approach towards Ko has been developed from the glycal ester **4**, readily available from the peracetylated Kdo methyl ester by elimination using TMSO-triflate (Scheme 24.6). Epoxidation of **4** followed by ring opening of **42** gave a moderate yield of the D-*glycero*-D-*talo*-pyranose derivative which had to be separated from the concomitantly formed furanose isomers. Nitrobenzoylation of the 2,3-diol, however, resulted in the formation of 2,4- and 2,3-di-*O*-*p*-nitrobenzoyl derivatives **43**

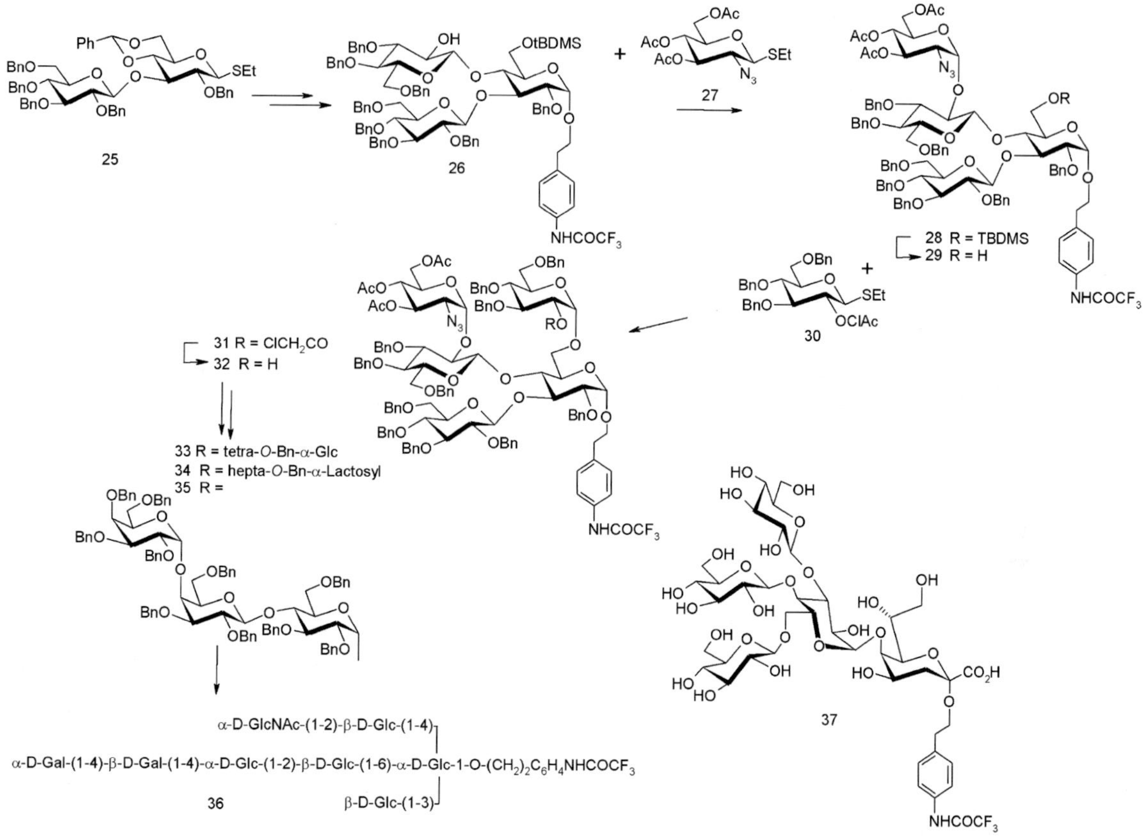

SCHEME 24.4 Synthesis of oligosaccharides related to the core region of *Moraxella catarrhalis*. Reprinted, with permission, from Ekelöf and Oscarson (1996) Synthesis of oligosaccharide structures from the lipopolysaccharide of *Moraxella catarrhalis*. J. Org. Chem. 61, 7711–7718.

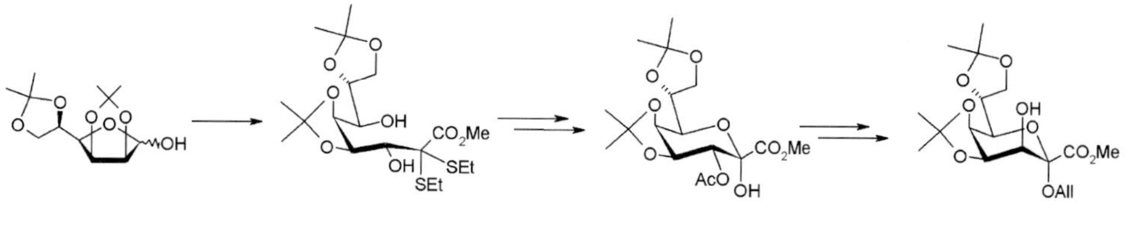

SCHEME 24.5 Synthesis of Ko-glycosides. Adapted, with permission, from Reiner and Schmidt (2000) Synthesis of 3-deoxy-D-*manno*-2-octulosonic acid (Kdo) and D-*glycero*-D-*talo*-2-octulosonic acid (Ko) and their α-glycosides. Tetrahedron Asymm. 11, 319–335. © Elsevier 2000.

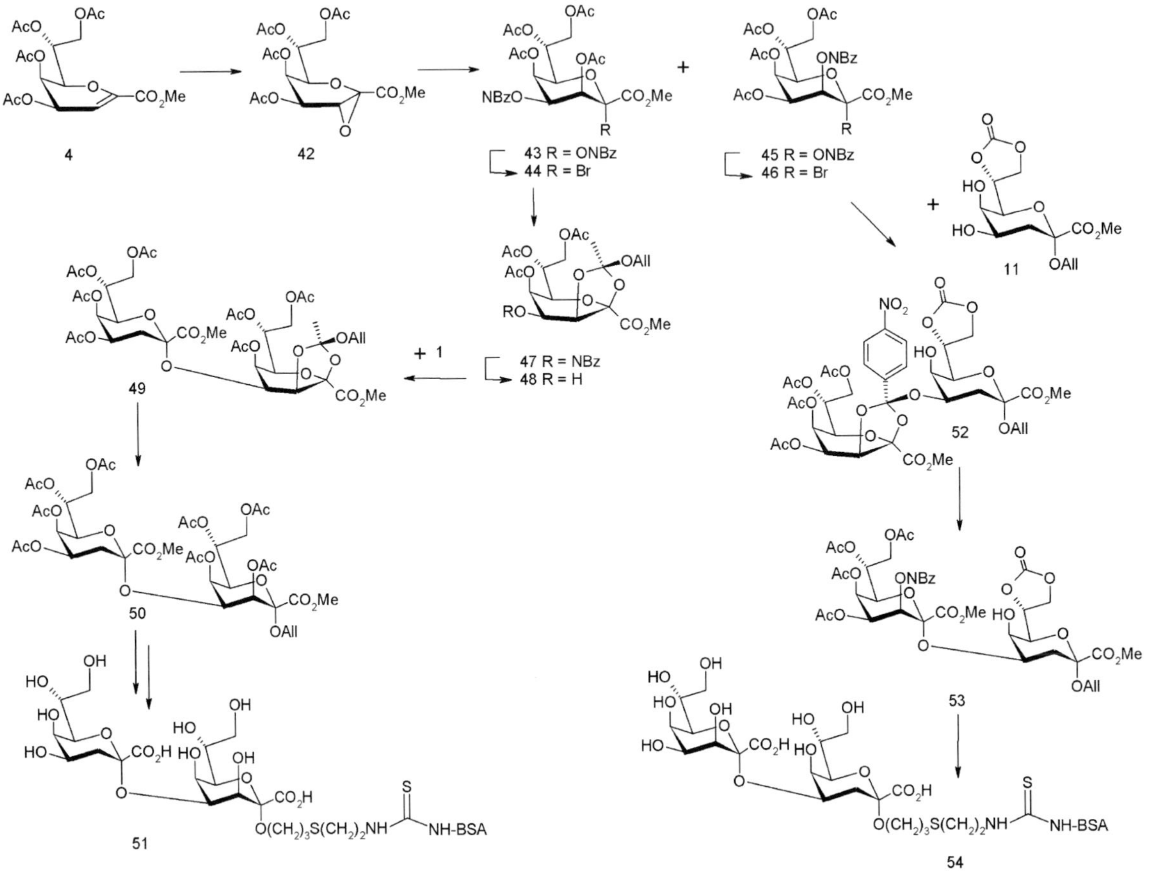

SCHEME 24.6 Synthesis of Ko-containing disaccharides from the core region of *Burkholderia* and *Acinetobacter*. Adapted, with permission, from Wimmer *et al.* (2000) Synthesis of neoglycoproteins containing D-*glycero*-D-*talo*-oct-2-ulopyranosylonic acid (Ko) ligands corresponding to core units from *Burkholderia* and *Acinetobacter* lipopolysaccharide. Carbohydr. Res. 329, 549–560. © Elsevier 2000.

and **45**, respectively, which were separated and converted into the bromide donors **44** and **46**, respectively. Derivative **44** was then transformed into the allyl orthoester derivative **47** which was selectively deprotected at O-4 under mild conditions. Coupling of the resulting acceptor **48** with the Kdo bromide **1** afforded the α-(2→4)-linked disaccharide **49** in 87% yield and in a 5:1 α/β-ratio (Wimmer *et al.*, 2000). The orthoester **49** was subsequently rearranged into the allyl glycoside **50** in 80% yield and eventually converted into the neoglyconconjugate **51**. Although additional steps were needed, this approach gave a better overall yield than the previously used direct coupling of the Kdo bromide **1** to a 7,8-carbonate derivative of allyl Ko (Kosma *et al.*, 1999). Nevertheless, more efficient approaches for the synthesis of Ko glycosides are definitely needed.

For the synthesis of the inverted sequence Ko-(2→4)-Kdo, the 7,8-*O*-carbonate protected Kdo acceptor **11** was reacted with the Ko-bromide **1** to give the *exo*-nitrobenzylidene orthoester derivative **52** in 85% yield, which was rearranged into the Ko-glycoside **53** using TMSO-triflate.

Disaccharide **53** was fully deprotected and converted – via a cysteamine spacer – into BSA-neoglycoconjugate **54** (see Scheme 24.6). A highly specific antibody, S67-9 could be raised against the Ko-(2→4)-Kdo BSA antigen **54**, which did not cross-react with the related Kdo-(2→4)-Kdo epitope (Brade *et al.*, 2002). The antibody, however, did not react with the wild-type LPS-core of *Burkholderia*, since this epitope is further masked by 4-amino-4-deoxy-L-arabinose at O-8 of Ko. Interestingly, co-crystallization experiments of Ko ligands with Kdo-specific antibodies showed a binding mode similar to Kdo in the binding site with an extra hydrogen bond extending from OH-3 (Brooks *et al.*, 2008).

2.5. Synthesis of glycosides containing 4-amino-4-deoxy-L-arabinose residues

4-Amino-4-deoxy-L-arabinose (Ara4N) is a frequently detected sugar unit in the lipid A part of various bacterial LPSs. Substitution by Ara4N, however, has also been reported for the inner core region of LPS, where Ara4N either is linked to the 8-position of an inner Kdo unit (in *Proteus mirabilis* R45) or to O-8 of a distal Ko (in *B. cepacia*). Recently, a reliable method to generate multigram amounts of Ara4N has been elaborated (B. Müller, M. Blaukopf and P. Kosma, unpublished results), which is suited for the preparation of spacer glycosides and of Ara4N glycosyl donors. Starting from commercially available methyl β-xylopyranoside **55**, regioselective tosylation or nosylation at O-4 was accomplished via an intermediate stannylidene acetal followed by benzoylation and introduction of the 4-azido moiety by S_N-type displacement to afford **57** (Scheme 24.7). Direct transglycosylation reactions on the methyl glycoside **58** have been developed to introduce allyl as well as bromohexyl spacer groups. Reduction of the 4-azido-group of **59** furnished the allyl arabinopyranoside **60**. Alternatively, the bromohexyl glycoside was converted into the spacer derivative **61** containing a terminal

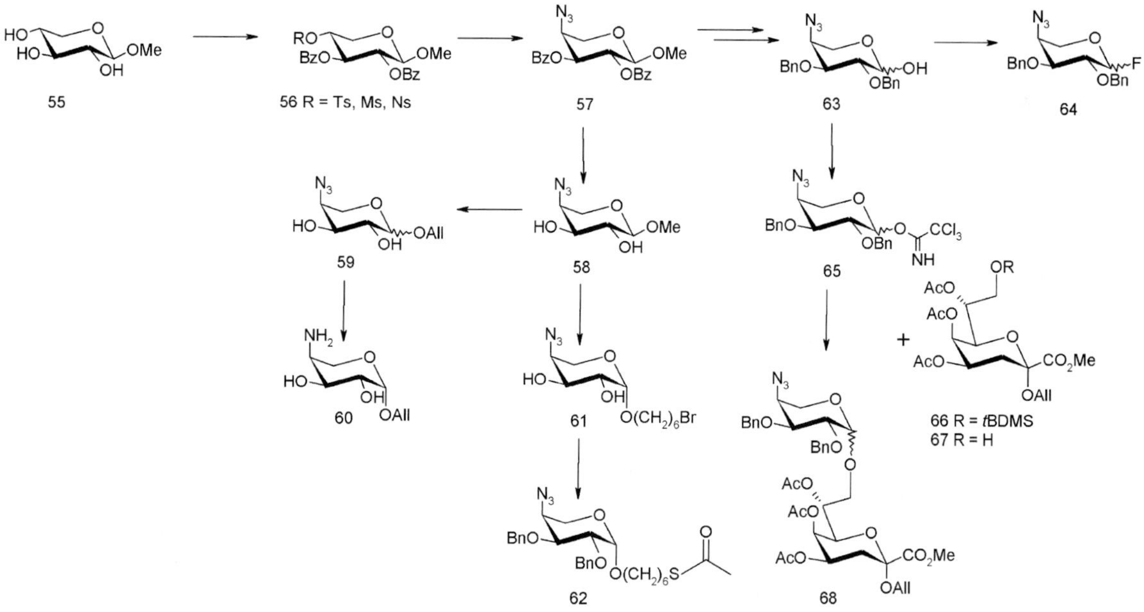

SCHEME 24.7 Synthesis of glycosides of 4-amino-4-deoxy-L-arabinose.

thioester function for subsequent preparation of neoglycoconjugates. Compound **57** was hydrolysed under acidic conditions to give the reducing derivative **63** as precursor of the fluoride and trichloroacetimidate donor **64** and **65**, respectively. Preliminary experiments to connect the aminoarabinose donors with the 8-*O*-silyl or 8-OH groups of allyl glycosides **66** and **67**, respectively, proceeded in fair yields but with modest anomeric selectivities. The deprotected compounds correspond to inner core units of *Proteus* and *Burkholderia* LPS, respectively.

3. SYNTHESIS OF HEPTOSE-CONTAINING CORE STRUCTURES

3.1. Synthesis of heptose and heptose glycosyl donors

Next to Kdo, heptoses are important higher carbon sugars in the inner core of LPS. Numerous synthetic methods have been developed for the preparation of the 6-epimeric *manno*-heptoses with a major focus on the synthesis of the L-*glycero*-D-*manno*-heptose as the most common heptose constituent in the inner core of LPS. The majority of these approaches is based on chain elongation, either from the anomeric position or terminal aldehyde group of the corresponding lower carbon chain precursors using Grignard reagents as well as Wittig-type homologation reactions. Since the heptosidic linkages in many bacterial LPS units are in the α-anomeric configuration, the glycosylations usually proceed with good stereocontrol in the presence of a neighbouring acyl protecting group at position 2. The most commonly used heptosyl donors are the anomeric thioglycoside as well as trichloroacetimidate derivatives. Conceptually novel protocols have been elaborated for the selective formation of β-heptoside and 6-deoxy-heptoside derivatives as well (Crich and Banerjee, 2006). These methods, and the synthesis of important core structures from *Salmonella*, *Haemophilus* and *Neisseria* have been reviewed previously, and will not be covered comprehensively in this chapter (Oscarson, 1997; Hansson and Oscarson, 2000; Pozsgay, 2003; Kosma, 2008). Although heptoses have generally been regarded as biologically inocuous carbohydrates with low immunogenic potential, evidence has recently been obtained for lectin-type binding of both 6-epimeric forms of *glycero*-D-*manno*-heptoses to proteins such as lung surfactant protein D and concanavalin A, respectively (Wang *et al.*, 2008; Jaipuri *et al.*, 2008).

3.2. Synthesis of core structures of *Haemophilus* LPS

A large variety of synthetic heptosyl oligosaccharides containing functional spacer groups for covalent conjugation to proteins has been prepared by the group of Oscarson.

For the synthesis of the 3,4-dibranched heptose unit occurring in the inner core of *Haemophilus influenzae*, Oscarson elaborated a 1,6-anhydro-intermediate in order to avoid steric crowding for the subsequent glycosylation steps at O-3 and O-4. Thus, the 6-*O*-trimethylsilyl protected thioglycoside **69** was reacted with NIS/triflic acid to give the 1,6-anhydro compound **70** in 85% yield (Scheme 24.8). Protecting group manipulation then produced the acceptor **71** which was connected to a perbenzoylated β-glucosyl unit at position 4, followed by cleavage of the 2,3-*O*-isopropylidene group and selective installation of a 2-*O*-benzyl group to afford the 3-OH derivative **72**. Subsequent glycosylation with the perbenzoylated heptosyl thioglycoside donor **73** provided the trisaccharide **74** in 75% yield. Acetolysis of **74** gave **75**, which was subsequently converted into an ethyl-1-thio glycosyl donor **76**. The thioglycoside **76** was exploited for the introduction of a 2-(*p*-trifluoroacetamidophenyl)ethyl aglycon followed

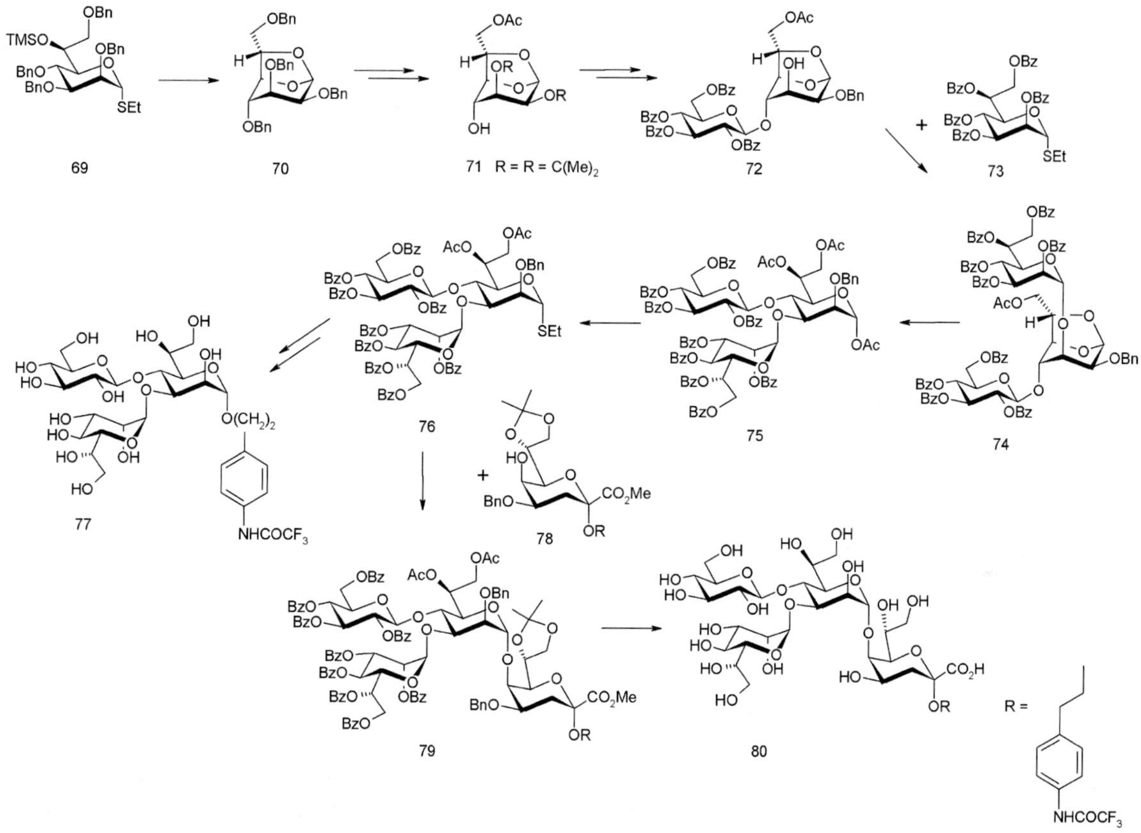

SCHEME 24.8 Synthesis of core units related to *Haemophilus influenzae* LPS. Reprinted, with permission, from Oscarson (2001) Synthesis of *Haemophilus influenzae* carbohydrate surface antigens. Carbohydr. Polym. 44, 305–311. © Elsevier 2000.

by deprotection to furnish the spacer-derivative **77**. In addition, the trisaccharide donor **76** was successfully coupled to the Kdo-acceptor **78** to afford **79** followed by full deprotection to furnish the tetrasaccharide **80** (Oscarson, 2001). Recently, a 2-aminoethyl glycoside analogue of **77** has been extended with an α-(1→2)-linked heptose moiety followed by introduction of a phosphoethanolamine substituent at O-6 of the internal heptose residue, providing a common tetrasaccharide unit of *Haemophilus influenzae* LPS (Mannerstedt *et al.*, 2008).

As an alternative promoter for activation of thioglycosides, dimethyl(methylthio)sulfonium triflate (DMTST) had been utilized in the synthesis of another *Haemophilus* tetrasaccharide structure containing galactose (Bernlind and Oscarson, 1998). First, the heptosyl acceptor **82** was reacted with the 2-*O*-*p*-methoxybenzyl heptosyl donor **81** to give the α-(1→3)-linked disaccharide **83** in 89% yield (Scheme 24.9). The *p*-methoxybenzyl group was then oxidatively removed to allow for a stepwise assembly of the α-(1→2)-linkages. This way the third heptosyl moiety was attached to position 2 of the terminal heptose in disaccharide **84** in good yield to give **85** which was fully deprotected giving the spacer derivative **87**. Coupling of the heptotrioside acceptor **86** with the perbenzoylated galactosyl bromide **88** promoted by silver triflate furnished the

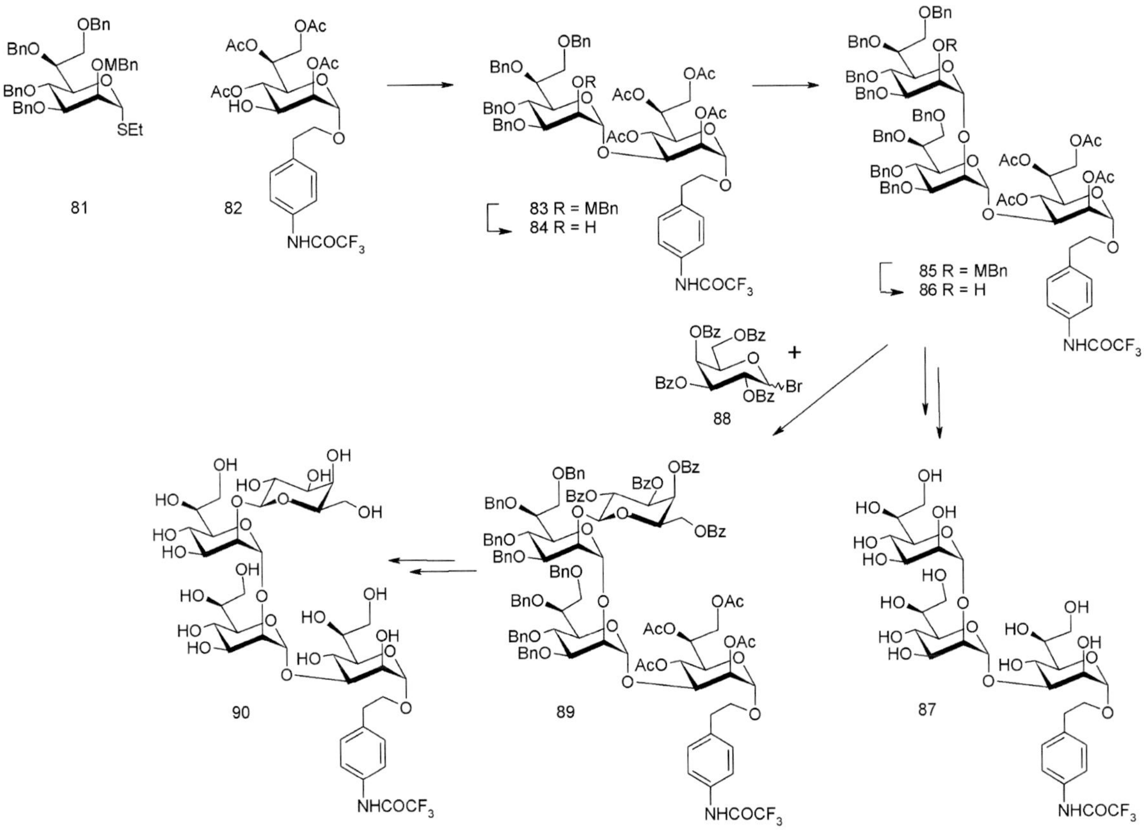

SCHEME 24.9 Synthesis of core units related to *Haemophilus influenzae*. Reproduced, with permission, from Bernlind and Oscarson (1998) Synthesis of a branched heptose- and Kdo-containing common tetrasaccharide core structure of *Haemophilus influenzae* lipopolysaccharides via a 1,6-anhydro-L-*glycero*-β-D-*manno*-heptopyranose intermediate. J. Org. Chem. 63, 7780–7788. American Chemical Society © 1998.

β-(1→2″)-linked tetrasaccharide **89** in 74% yield and eventually the deprotected compound **90**. An alternative, convergent pathway using a blockwise approach failed, since reaction of the bromide **88** with a 2-OH heptoside monosaccharide acceptor exclusively furnished an α-(1→2)-linked disaccharide donor despite the presence of benzoyl protecting groups.

In addition, trisaccharide ligands containing β-(1→4)-linked cellobiosyl and lactosyl heptoside derivatives have been synthesized (Bernlind and Oscarson, 1997).

The synthesis of a linear hexasaccharide from *H. ducreyi* LOS containing both 6-epimeric *manno*-heptoses has been achieved in 3 + 3 blockwise assembly by Oscarson, demonstrating the versatility of thioglycoside donors (Bernlind *et al.*, 2000). First, the ethyl-1-thio mannoside **91** was elaborated into both L-*glycero*-and D-*glycero*-D-*manno*-heptoside derivatives **92** and **93** allowing the further transformation into the *N*-trifluoroacetamidophenylethyl spacer derivative **94** as well as the D-*glycero*-D-*manno*-heptosyl donor **98** (Scheme 24.10). The 6-*O*-chloroacetyl

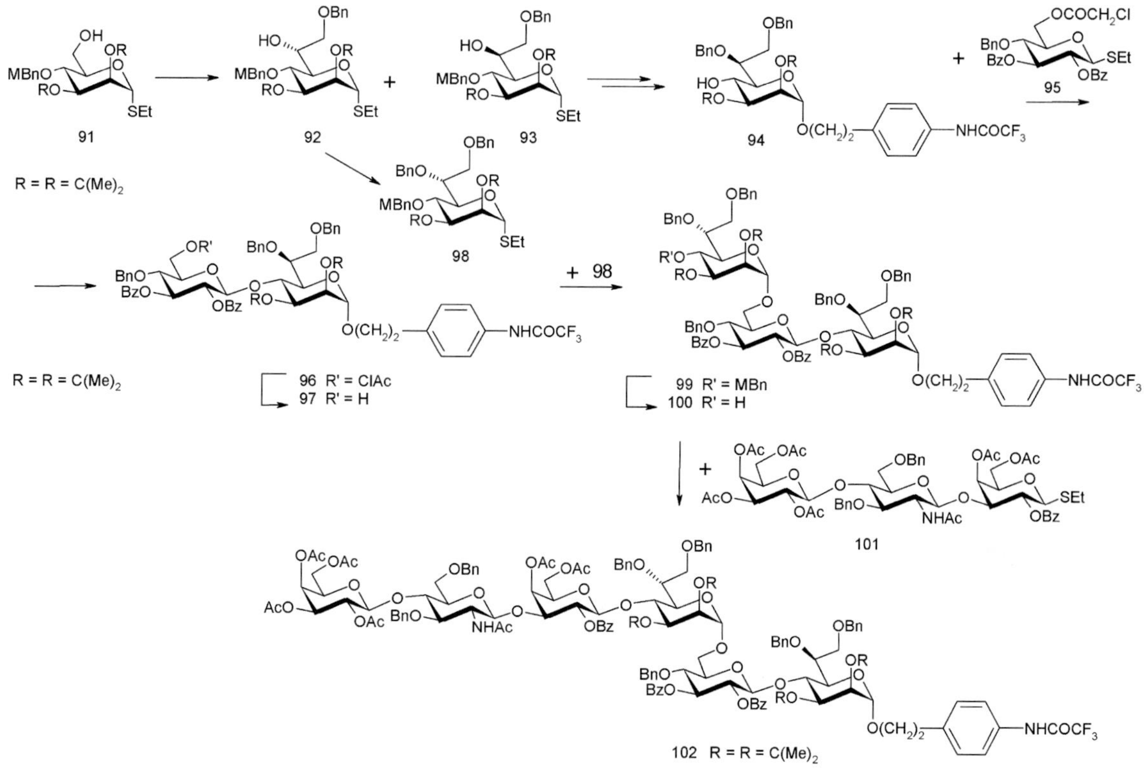

SCHEME 24.10 Synthesis of a lipo-oligosaccharide from *Haemophilus ducreyi*. Reprinted, with permission, from Bernlind *et al.* (2000) Synthesis of a D,D- and L,D-heptose-containing hexasaccharide corresponding to a structure from *Haemophilus ducreyi* lipopolysaccharides. Tetrahedron Asymm. 11, 481–492. © Elsevier 2000.

glucosyl donor **95** was coupled to the heptoside acceptor **94** in excellent yield followed by selective deprotection of the 6-*O*-chloroacetate and subsequent condensation of **97** with the thioheptoside **98** to give **99**. In a separate series, *N*-acetyllactosamine was linked to O-3 of a trimethylsilylethyl galactoside and further transformed into the corresponding trisaccharide ethyl-1-thio donor **101**. Subsequent coupling with the trisaccharide acceptor **100** in the presence of *N*-iodosuccinimide/triflic acid produced the protected hexasaccharide **102** in 45% yield. By using an *N,N*-diacetylamino derivative of donor **101**, an improved yield (75%) of the corresponding hexasaccharide was achieved.

3.3. Synthesis of core structures of *Neisseriaceae*

Numerous core units from *Neisseria meningitidis* have been prepared in the group of van Boom and used for the investigation of the immune response against meningococcal immunotype L1, L2 and L3, 7, 9 LPSs (Boons *et al.*, 1992; Verheul *et al.*, 1991). A concise synthetic strategy was developed for the assembly of structurally related branched heptose units in *N. gonorrhoeae* 15253 (Kubo *et al.*, 2004). For the synthesis of the peracetylated tetrasaccharide **110**, the benzylidene-protected methyl mannopyranoside **103** was regioselectivey glycosylated at O-3 with the heptosyl donor **104** to give the α-(1→3)-linked disaccharide **105** (Scheme 24.11). Following regioselective reductive opening of the benzylidene group and ensuing protection of the 6-OH group, the lactosyl trichloroacetimidate donor **107** was reacted with O-4 of the mannose unit of **106** to afford the β-(1→4)-linked tetrasaccharide **108** in 59% yield. Eventually, the mannose unit in **108** was elaborated into the heptose moiety via exchange of acetates with benzyl groups, selective cleavage of the silyl ether group followed by a reaction sequence comprising Swern oxidation, Grignard-type elongation of the intermediate aldehyde to afford **109** with ensuing osmylation and periodate oxidation/reduction to provide **110**. A related approach had been employed to produce the methyl α-lactosyl-(1→3)-heptoside **111** (Ishii *et al.*, 2002). Previously, a different strategy based on transformations of an intermediate 1,6-anhydro-heptopyranose intermediate was used by Oscarson, since glycosylation of neither OH-3 nor OH-4 heptose acceptor derivatives could be accomplished for the direct formation of the 3,4-branched heptoside structures (Bernlind and Oscarson, 1998).

The synthesis of the second, 2,3-dibranched heptose moiety in *N. gonorrhoeae* 15253 was based

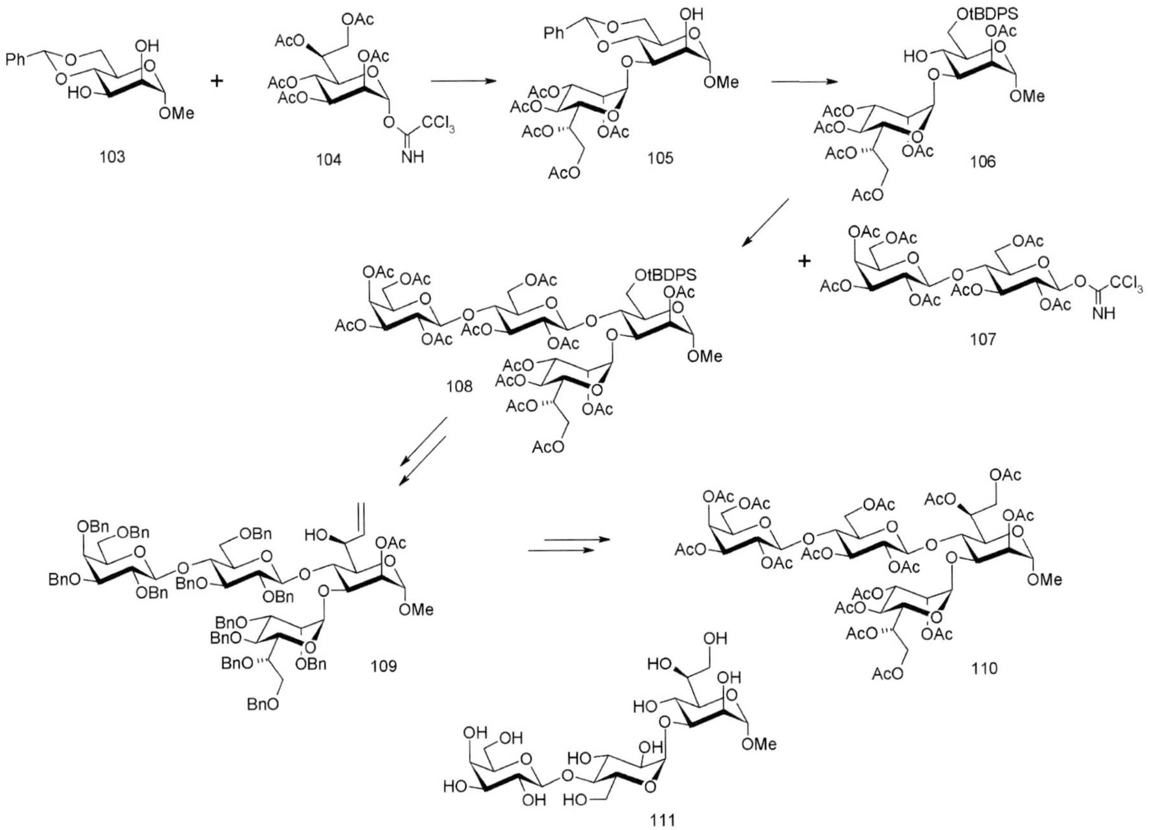

SCHEME 24.11 Synthesis of oligosaccharides related to the core region of *Neisseria gonorrhoeae* 15253. Reproduced, with permission, from Kubo *et al.* (2004) Synthesis of a 3,4-di-*O*-substituted heptose structure: a partial oligosaccharide expressed in Neisserial lipopolysaccharide. Eur. J. Org. Chem. 1202–1213. © Wiley-VCH Verlag GmbH & Co. KGaA.

on the selective protection of the 3,4-diequatorial hydroxy groups of **112** as dimethoxybutane-acetal followed by chain elongation of **113** via **114** (Yamasaki *et al.*, 2001). Hydrogenolytic removal of the 2-*O*-benzyl group furnished the acceptor **115**, which was coupled with the 2-azido-2-deoxy-glucose trichloroacetimidate donor **116** to give an optimum yield of 80% of the disaccharide **117** (α/β ratio 2.5:1). Selective liberation of the 3-OH group of the heptose unit was accomplished via selective silylation at O-3, acetylation and desilylation to give **118** (Scheme 24.12). Similar to the synthesis of the methyl α-lactosyl-(1→3)-heptoside **111**, reaction with the perbenzylated lactosyl β-trichloroacetimidate **119** furnished the α-linked product **120** in good yield, which was finally converted into the peracetate **121** (Ishii *et al.*, 2004).

3.4. Synthesis of core structures of *Pseudomonas aeruginosa*

P. aeruginosa is an opportunistic pathogen of clinical relevance in immunocompromised patients and in patients affected by cystic fibrosis. A characteristic feature of several *P. aeruginosa* strains is the presence of various phosphate, diphosphate and triphosphate substituents in the heptose region (Kooistra *et al.*, 2003). In addition, a 7-*O*-carbamoyl residue at

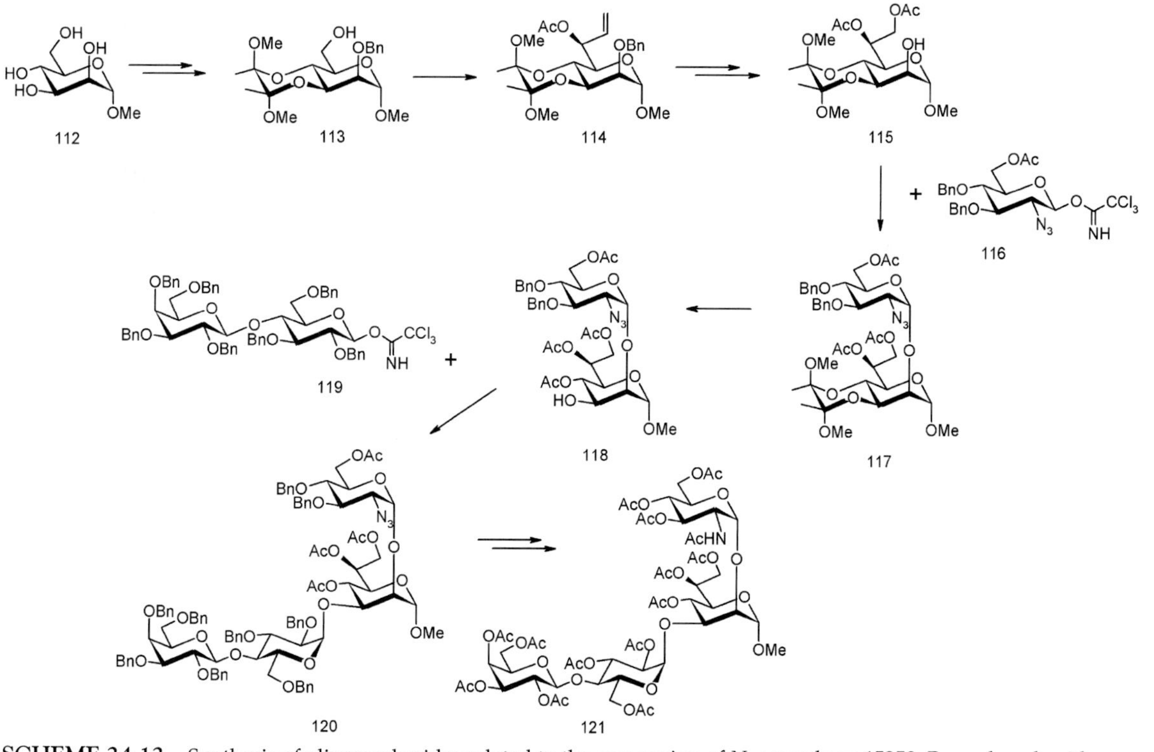

SCHEME 24.12 Synthesis of oligosaccharides related to the core region of *N. gonorrhoeae* 15253. Reproduced, with permission, from Ishii *et al.* (2004) Synthesis of α-lactosyl-(1→3)-L-*glycero*-α-D-*manno*-heptopyranoside, a partial oligosaccharide structure expressed within the lipo-oligosaccharide produced by *Neisseria gonorrhoeae* strain 15253. Carbohydr. Res. 337, 11–20. © Elsevier 2004.

the distal heptose residue has been identified (Beckmann *et al.*, 1995).

The introduction of the 7-*O*-carbamoyl group was based on a regioselective ring opening of a 6,7-*O*-carbonate group introduced in good yield by reaction of **122** with trichloromethyl chloroformate (Scheme 24.13). Subsequent conversion of **123** furnished glycosyl donor **125** and the allyl disaccharide **126** in good yields. Reaction of **126** with NH_4HCO_3/aq. NH_3 led to formation of the 7-carbamate and simultaneous removal of the acetate protecting groups giving **127**. Further transformation of the allyl group into an amino-spacer group allowed the preparation of neoglycoconjugate **128** and the characterization of epitope specificities of murine monoclonal antibodies binding to the carbamoyl-heptose (Reiter *et al.*, 1999). The GalNAc-epitope **133** was synthesized along similar lines from donor **129** and heptosyl carbonate **130** (Reiter *et al.*, 2001). The synthesis of the highly phosphorylated carbamoyl-heptose region remains as a major synthetic challenge.

3.5. Synthesis of core structures of *Campylobacter jejuni*

The distal regions of core oligosaccharides of several *Campylobacter jejuni* strains contain epitopes which are involved in neurological disorders as sequelae of *C. jejuni* infections due to their similarity with human gangliosides GM_1, GM_2, GD_{1a} and GQ_{1b}. Similar to the assembly of core units of *Neisseriaceae*, a central 3,4-dibranched heptose unit had to be synthesized with subsequent attachment of a trisaccharide containing neuraminic acid. Starting from a suitably protected 2-trimethylsilylethyl (SE) heptoside **134**, elongation at O-4 with benzoyl-protected glucopyranose was accomplished to afford the disaccharide **136** in 89% yield (Hori *et al.*, 2003). The heptoside **134** was transformed into the corresponding methyl-1-thioglycoside donor **138** and coupled in the presence of NIS/TMSO-triflate to O-3 of the disaccharide acceptor **137** to give trisaccharide **139** in 54% yield. This sequence constitutes an efficient route

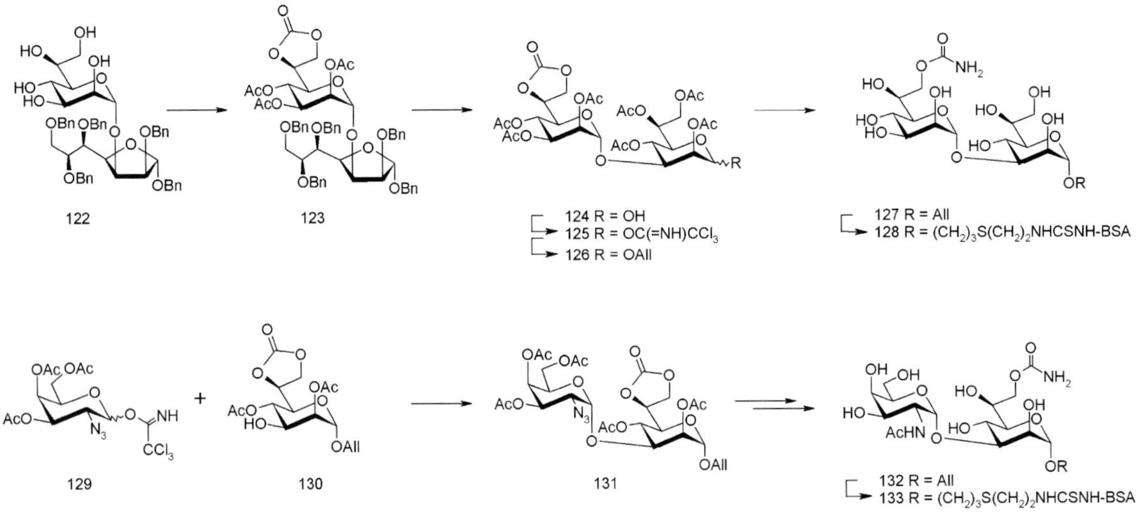

122 123 124 R = OH; 125 R = OC(=NH)CCl_3; 126 R = OAll; 127 R = All; 128 R = $(CH_2)_3S(CH_2)_2NHCSNH$-BSA

129 130 131 132 R = All; 133 R = $(CH_2)_3S(CH_2)_2NHCSNH$-BSA

SCHEME 24.13 Synthesis of carbamoyl-substituted heptosides from *Pseudomonas aeruginosa*. Adapted, with permission, from Reiter *et al.* (1999) Synthesis of *Pseudomonas aeruginosa* core antigens containing 7-*O*-carbamoyl-L-*glycero*-α-D-*manno*-heptopyranosyl residues. Carbohydr. Res. 317, 39–52. © Elsevier 1999.

towards the branched central unit and allowed further glycosylation at position 3 after selective deprotection of a 3-*O*-levulinoyl protecting group. Thus, the disialylgalactose donor **141** was coupled to the trisaccharide acceptor **140** to produce the protected hexasaccharide lactone **142** in 48% yield (Scheme 24.14).

3.6. Synthesis of core structures of *Klebsiella*

The LPS of the rough mutant *K. pneumoniae* ssp. R20 contains a novel α-(1→2)-interlinked heptoglycan of D-*glycero*-D-*manno*-heptose units. The distal disaccharide end of this glycan chain was synthesized starting from methyl α-D-mannopyranoside **112** which was converted via protecting group manipulation into compound **143** containing a 2-*O*-*p*-methoxybenzyl group and a free terminal hydroxy group (Gurjar and Talukdar, 2004). Elongation to the heptoside was accomplished via oxidation followed by olefination of the intermediate 6-aldehyde giving **144** and subsequent asymmetric dihydroxylation according to the Sharpless methodology. The resulting 6,7-diol **145** was further transformed into the heptosyl acceptor **146** and donor derivative **147**, respectively (Scheme 24.15).

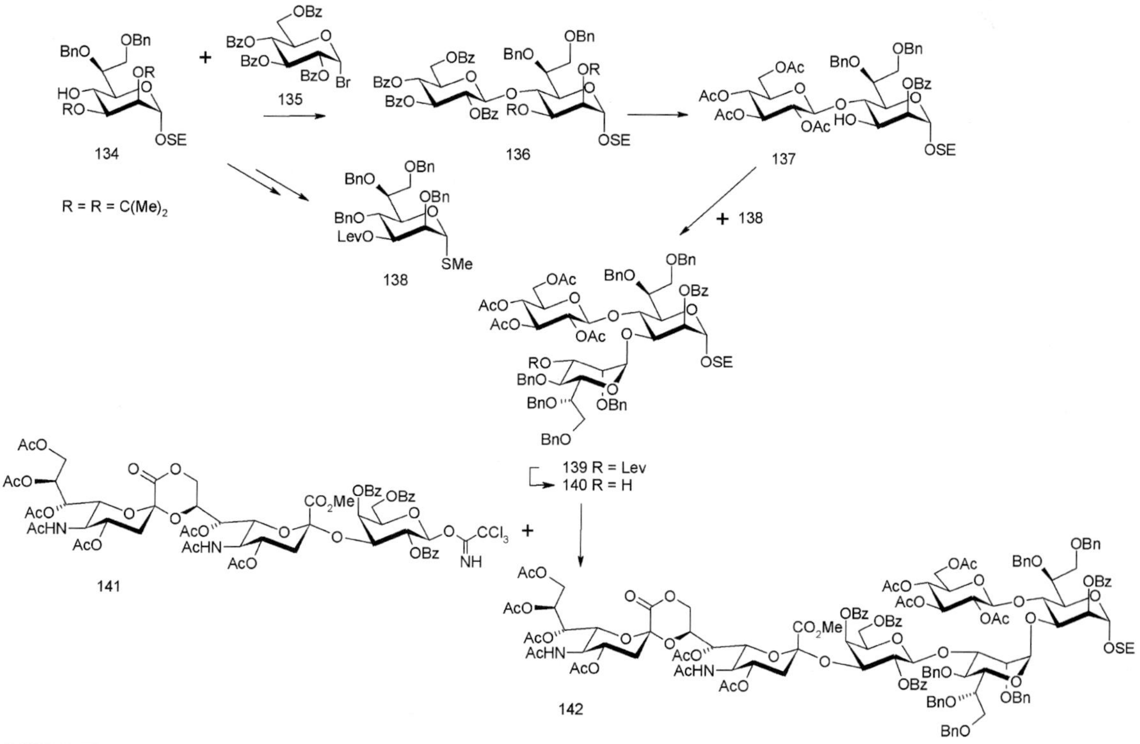

SCHEME 24.14 Synthesis of a GD$_3$-related hexasaccharide related to lipo-oligosaccharides of *Campylobacter jejuni*. Adapted, with permission, from Hori *et al.* (2003) Synthetic study on *Campylobacter jejuni* lipopolysaccharides: an improved synthesis of a branched, heptose-containing trisaccharide core structure and its conversion into ganglioside GD3 related hexasaccharide. Eur. J. Org. Chem. 3752–3760. © Wiley-VCH Verlag GmbH & Co. KGaA.

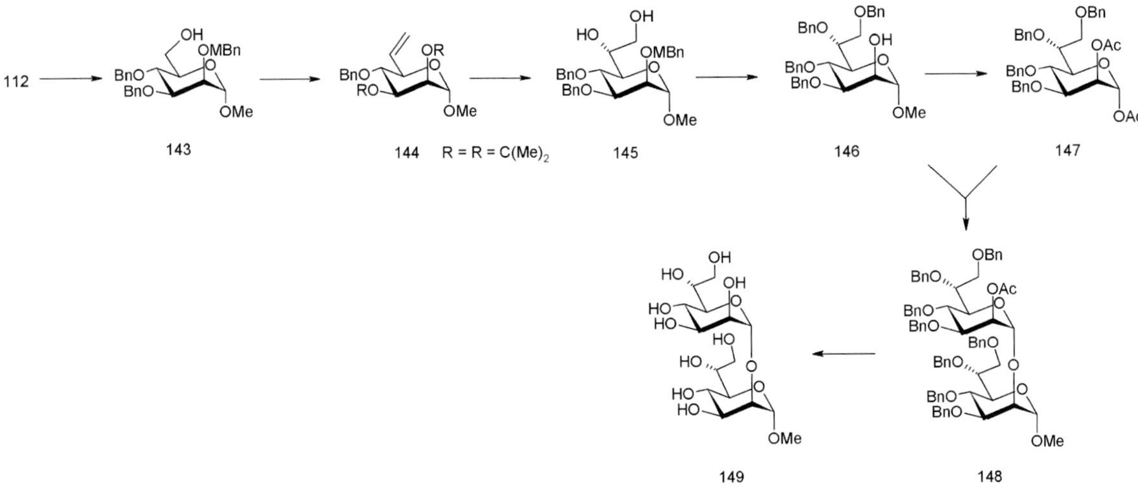

SCHEME 24.15 Synthesis of a terminal disaccharide corresponding to the heptan of *Klebsiella pneumoniae* ssp. R20. Reproduced, with permission, from Gurjar and Talukdar (2004) Synthesis of terminal disaccharide unit of *Klebsiella pneumoniae* ssp. R20. Tetrahedron 60, 3267–3271. © Elsevier 2004.

Coupling of the alcohol **146** with the anomeric heptosyl acetate **147** under Lewis acid catalysis gave a low yield of the disaccharide **148**, which was fully deprotected yielding the α-(1→2)-linked disaccharide **149**. The strategy enables further chain elongation upon removal of the 2′-*O*-acetyl group of **148**.

3.7. Synthesis of core structures of *Vibrio parahaemolyticus*

A trisaccharide fragment corresponding to a part structure of *Vibrio parahaemolyticus* serotype O2 containing both 6-epimeric *manno*-heptoses was prepared by the group of van Boom (van Straten *et al*., 1997). First, as described for the *Klebsiella* heptosides, a side chain double bond was introduced at C-6 which was then oxidatively cleaved to give preferentially the D-*glycero*-D-*manno*-heptoside derivative, which was further processed into the 3-OH acceptor **150**. Conversely, a "masked" precursor of the L-*glycero*-D-*manno*-heptoside was generated in excellent diastereoselectivity from a Grignard reaction of a *manno*-hexo-1,6-dialdoside with (phenyldimethylsilyl)methyl magnesium chloride and was subsequently elaborated into the glycosyl donor **151**. Coupling of the donor **151** to the benzyl-protected D-*glycero*-D-*manno*-heptoside acceptor **150** proceeded in 76% yield. Oxidative cleavage of the silyl group of **152**, protection of the resulting alcohol with a benzyloxymethyl group and cleavage of the 2′-*O*-benzoate gave the acceptor **153**. The final introduction of a glucuronyl residue at O-2′ was accomplished via glucosylation employing thioglycoside donor **154** affording **155**, followed by selective oxidation at C-6 of the glucosyl unit to furnish the trisaccharide α-GlcA-(1→2)-α-L,D-Hep-(1→3)-α-D,D-Hep-(1→OMe) **157** after hydrogenolytic removal of all benzyl protecting groups of trisaccharide **156** (Scheme 24.16).

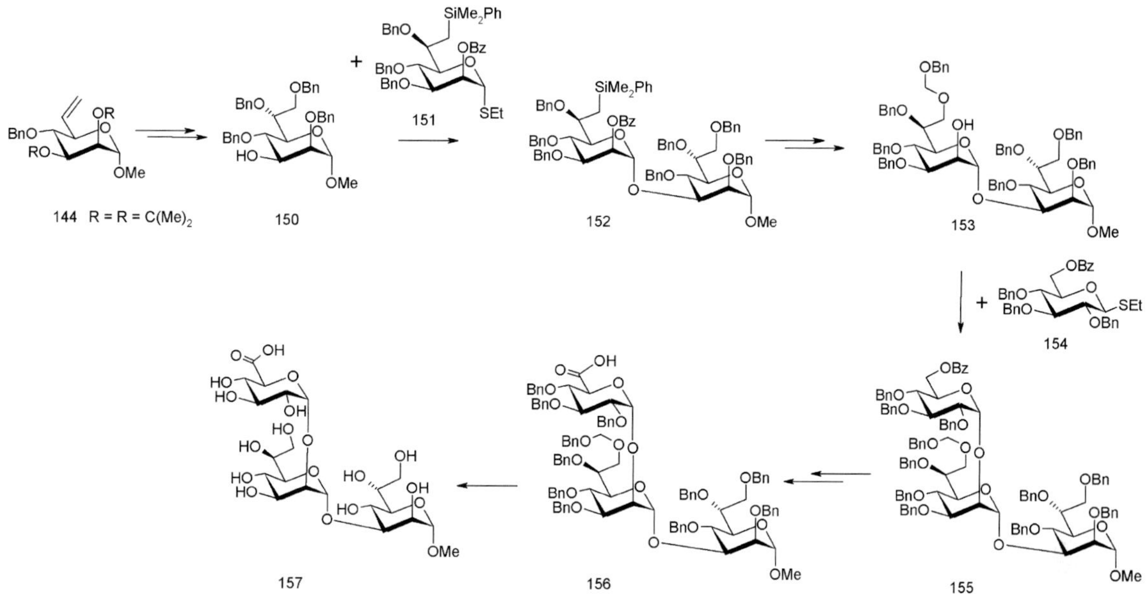

SCHEME 24.16 Synthesis of trisaccharide units from *Vibrio parahaemolyticus*.

4. SYNTHESIS OF PHOSPHORYLATED CORE UNITS

4.1. Synthesis of Kdo phosphates

The minimal core structure occurring in deep rough mutants has been found for *Haemophilus influenzae* I-69 Rd$^-$/b$^+$, which contains a phosphate residue linked to either O-4 or O-5 of a single Kdo residue. Starting from an α-(2→6)-interlinked Kdo-glucosamine disaccharide **158**, the Kdo moiety was selectivey protected by a 7,8-*O*-isopropylidene group to furnish **159**, followed by selective acetylation at O-4 and phosphotriester formation at O-5 using the phosphoramidite protocol. Deprotection of **160** gave the allyl glycoside which was further processed into the BSA-glycoconjugate **161**. For the synthesis of the 4-*O*-phosphate, diol **159** was subjected to regioselective methoxybenzylation leading to the 4-*O*-*p*-methoxybenzyl ether **162** and the (1→5′)-lactone **163** (Scheme 24.17). Oxidative removal of the 4-*O*-*p*-methoxybenzyl ether was followed by phosphorylation to provide **164**, which was deprotected and converted into the neoglycoconjugate **165** (Sekljic *et al.*, 1996).

Introduction of the heptosyl residue to O-5 of **162** using the trichloroacetimidate heptosyl donor **104**, however, failed due to glycosyl imidate formation at the *N*-acetyl group of **162**. Hence, the β-benzyl Kdo ketoside **166** was first reacted with donor **104** to give a good yield of the disaccharide **167**, which was then elaborated into the disaccharide bromide donor **168**. Coupling of the donor with the *N*-acetylglucosamine acceptor **169** furnished the α-(2→6)-linked trisaccharide **170** (Scheme 24.18). Subsequent selective removal of the 4-*O*-chloroacetate protecting group followed by phosphotriester formation and further processing gave the neoglycoconjugate **172** containing trisaccharide α-L,D-Hep-(1→5)-α-Kdo-4-phosphate-(2→6)-β-GlcNAc ligands (Sekljic *et al.*, 1997). Antibodies raised against the

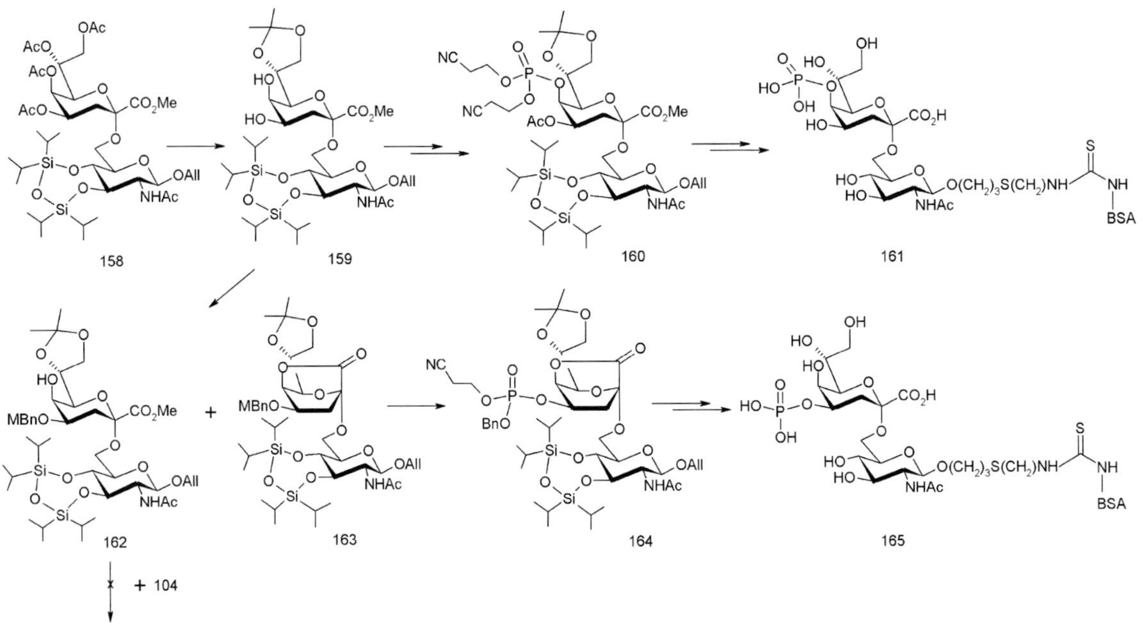

SCHEME 24.17 Synthesis of Kdo-4- and 5-phosphates corresponding to a deep-rough mutant of *Haemophilus influenzae*. Reprinted, with permission, from Sekljic *et al.* (1996) Synthesis of neoglycoproteins containing 5-*O*-phosphorylated Kdo monosaccharide, 4-*O*- and 5-*O*-phosphorylated α-Kdo-(2→6)-2-acetamido-2-deoxy-β-D-glucopyranosyl disaccharide residues. J. Endotoxin Res. 3, 151–164.

LPS of this *Haemophilus* mutant were specific for Kdo-4-phosphate, Kdo-5-phosphate and Kdo-phosphates linked to the phosphorylated lipid A backbone (Rozalski *et al.*, 1997).

4.2. Synthesis of phosphorylated heptose core structures

The heptose region of the inner core in *Enterobacteriaceae* is frequently substituted with phosphate, phosphoethanolamine and other phosphate-containing moieties. The major phosphorylated units have been chemically synthesized by Oscarson and co-workers (Ekelöf and Oscarson, 1995b). Methyl and (trifluoroacetamidophenyl)ethyl spacer heptosides **174a** and **174b** were used as common intermediates for the assembly of the disaccharide phosphates. First, the glycosyl acceptors were condensed with the thioglycoside donor **173** to produce the α-(1→3)-linked heptobiosides **175a** and **175b** in excellent yields. A 4-*O*-chloroacetyl protecting group had been installed to differentiate between the two heptose units. Selective removal of the chloroacetate provided the alcohols **176a** and **176b**, whereas debenzylation and isopropylidenation at the distal heptose residue of the disaccharides afforded the corresponding 4′-OH derivatives **177a** and **177b**, and **178a** and **178b**. Phosphorylation of the alcohols with phosphorus trichloride and imidazole followed by benzylation gave the corresponding phosphotriester derivatives which were fully deprotected to give the heptobioside 3-phosphates **179a** and **179b**, the 4′-monophosphates **181a** and **181b** and the 4,4′-bisphosphate derivatives **180a** and **180b**, respectively (Scheme 24.19).

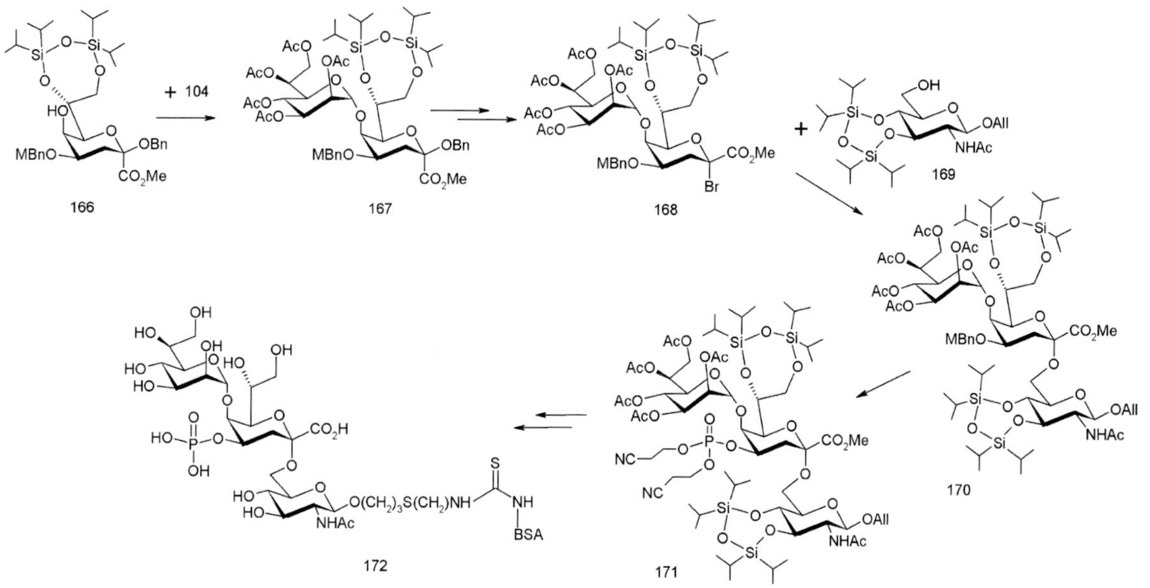

SCHEME 24.18 Synthesis of phosphorylated inner core units related to LPS structures of *Haemophilus, Bordetella pertussis, Bacteroides* and *Vibrionaceae*. Reproduced, with permission, from Sekljic *et al.* (1997) Synthesis of neoglycoproteins containing L-*glycero*-α-D-*manno*-heptopyranosyl-(1→4)- and (1→5)-linked 3-deoxy-α-D-*manno*-2-octulopyranosylonic acid (Kdo) phosphate determinants. J. Chem. Soc. Perkin Trans. I, 1973–1982.

Further synthetic work targeting larger LPS core fragments could be helpful in elucidating the physiological and immunochemical role of phosphate substituents in the heptose region.

5. SYNTHESIS OF OUTER CORE UNITS

The chemical synthesis of oligosaccharides corresponding to the Ra core of *Salmonella* was accomplished by the group of Oscarson and Garegg in the early 1990s (Oscarson and Ritzén, 1994). A trisaccharide corresponding to the outer core of *E. coli* K-12 was synthesized as allyl glycoside and converted into neoglycoconjugates (Antonov *et al.*, 1998). One approach comprised the assembly of the α-(1→6)-linked disaccharide **183**, which was transformed – via deacetylation/regioselective silylation at position 7 and desilylation of the heptose residue – into the alcohol **184**. Subsequent coupling with the thioglycoside donor **185** produced the trisaccharide **186** in 70% yield. Alternatively, the donor **185** was linked to a benzyl-protected heptoside acceptor **187**, to give the β-(1→7)-linked disaccharide **188**. Next, compound **188** was converted into a trichloroacetimidate donor **189** and coupled to the allyl glucoside acceptor in 78% yield (Scheme 24.20). Using neoglycoconjugates derived from the cysteamine adducts of the fully deprotected allyl trisaccharide **190**, it could be shown that rabbit polyclonal sera contain antibodies binding to the terminal glucosamine unit.

Recently, a common trisaccharide fragment of the outer core from *P. aeruginosa* containing L-alanine substituted galactosamine was

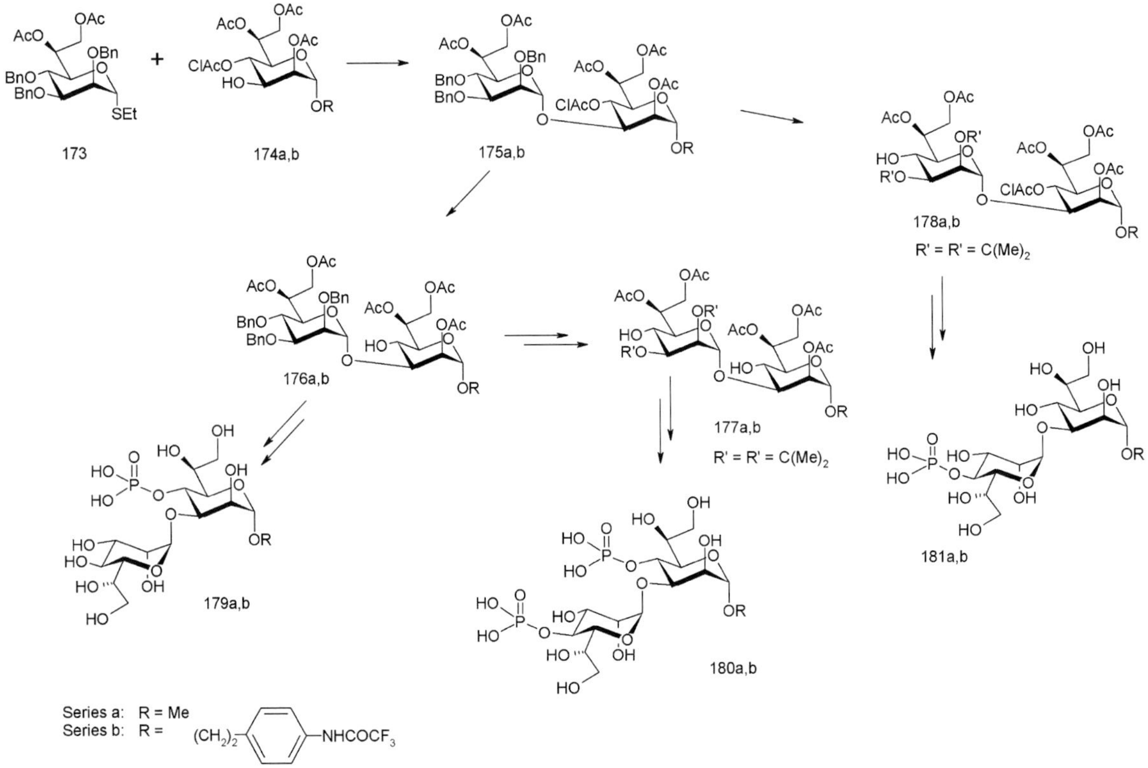

Series a: R = Me
Series b: R = $(CH_2)_2$–C_6H_4–$NHCOCF_3$

SCHEME 24.19 Synthesis of phosphorylated heptobiosides.

synthesized. Coupling of the protected disaccharide intermediate **191** at position 4 with the glucosyl trifluoroacetimidate donor **192** furnished the branched trisaccharide **193** in 62% yield (Komarova *et al.*, 2006). Deprotection of **193** with concomitant reduction of the azide group gave the amine **194** which was *N*-alanylated to give the trisaccharide methyl glycoside **195** (Scheme 24.21).

6. CONCLUDING REMARKS

The synthetic repertoire of carbohydrate chemists has expanded rapidly over the past ten years and an impressive series of oligosaccharides and glycoconjugates has been prepared, albeit only by a limited number of groups. Although improvements in methodology should still be pursued, the toolbox of glycochemistry nowadays allows tackling the assembly of larger fragments of the inner core and outer core regions of LPS, in particular, those which are not available from native sources (Research Focus Box). Thus, detailed studies to unravel the complex interaction of these carbohydrate ligands towards their protein targets will certainly aid in future developments for improved diagnostic tools and therapeutic measures against Gram-negative bacterial infections.

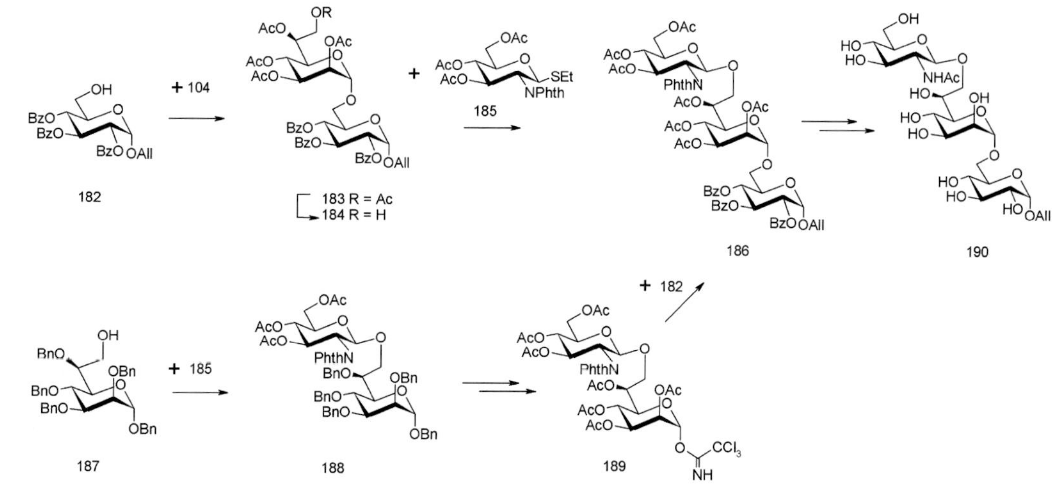

SCHEME 24.20 Synthesis of a trisaccharide fragment from the outer core of *Escherichia coli* K-12 LPS. Reprinted, with permission, from Antonov *et al.* (1998) Synthesis and serological characterization of L-*glycero*-α-D-*manno*-heptopyranose-containing di- and trisaccharides of the non-reducing terminus of the *Escherichia coli* K-12 LPS core oligosaccharide. Carbohydr. Res. 314, 85–93. © Elsevier 1998.

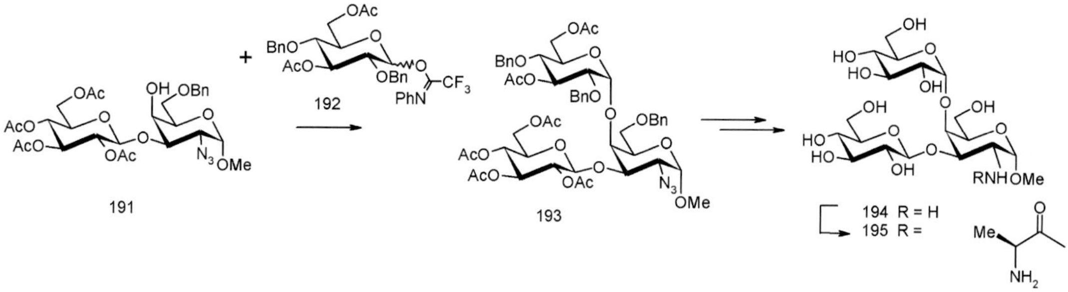

SCHEME 24.21 Synthesis of a trisaccharide fragment from the outer core of *Pseudomonas aeruginosa*. Reproduced, with permission, from Komarova *et al.* (2006) Synthesis of a common trisaccharide fragment of glycoforms of the outer core region of the *Pseudomonas aeruginosa* lipopolysaccharide. Tetrahedron Lett. 47, 3583–3587. © Elsevier 2006.

RESEARCH FOCUS BOX

- While efficient methods for the stereoselective synthesis of α- and β-heptosides have been developed, Kdo glycoside syntheses still need to be improved with respect to yields and anomeric selectivity.
- Published chemical syntheses of Ko and Ko glycosides are lengthy. More versatile and practical approaches are definitely needed.
- Worthwhile synthetic targets reside within the manifold additional structural modifications

RESEARCH FOCUS BOX (*cont'd*)

of the core oligosaccharides by acyl, carbamoyl, phosphoryl, phosphoethanolamine or 4-amino-4-deoxy-arabinose substituents.

- In the context of structural biology, analogues are of increasing importance to unravel deatils of protein–carbohydrate interactions of the core region with components of the adaptive and innate immune system and are worthy of further development.
- Novel and orthogonal spacer groups need to be developed which are required for reliable and defined conjugation of multifunctional core saccharides to protein carriers.

ACKNOWLEDGEMENTS

Financial support by FWF (grants P17407 and P19295) for the work performed in the author's laboratory is gratefully acknowledged.

References

Antonov, K., Backinowsky, L.V., Grzeszczyk, B., Brade, L., Holst, O., Zamojski, A., 1998. Synthesis and serological characterization of L-*glycero*-α-D-*manno*-heptopyranose-containing di- and trisaccharides of the non-reducing terminus of the *Escherichia coli* K-12 LPS core oligosaccharide. Carbohydr. Res. 314, 85–93.

Beckmann, H., Moll, H., Jäger, K.E., Zähringer, U., 1995. 7-O-Carbamoyl-L-*glycero*-α-D-*manno*-heptose: a new core constituent in the lipopolysaccharide of *Pseudomonas aeruginosa*. Carbohydr. Res. 267, C3–C5.

Bernlind, C., Oscarson, S., 1997. Synthesis of L-*glycero*-D-*manno*-heptopyranose-containing oligosaccharide structures found in lipopolysaccharides from *Haemophilus influenzae*. Carbohydr. Res. 297, 251–260.

Bernlind, C., Oscarson, S., 1998. Synthesis of a branched heptose- and Kdo-containing common tetrasaccharide core structure of *Haemophilus influenzae* lipopolysaccharides via a 1,6-anhydro-L-*glycero*-β-D-*manno*-heptopyranose intermediate. J. Org. Chem. 63, 7780–7788.

Bernlind, C., Bennett, S., Oscarson, S., 2000. Synthesis of a D,D- and L,D-heptose-containing hexasaccharide corresponding to a structure from *Haemophilus ducreyi* lipopolysaccharides. Tetrahedron Asymm. 11, 481–492.

Boons, G.J.P.H., van Delft, F.L., van der Klein, P.A.M., van der Marel, G.A., van Boom, J.H., 1992. Synthesis of LD-Hep*p* and KDO containing di- and trisaccharide derivatives of *Neisseria meningitidis* inner-core region via iodonium ion promoted glycosidations. Tetrahedron 48, 885–904.

Brade, H., Brade, L., Nano, F., 1987. Chemical and serological investigations on the genus-specific epitope of *Chlamydia*. Proc. Natl. Acad. Sci. USA 84, 2508–2512.

Brade, L., Rozalski, A., Kosma, P., Brade, H., 2000. A monoclonal antibody recognizing the 3-deoxy-D-*manno*-oct-2-ulosonic acid (Kdo) trisaccharide Kdo(2→4) Kdo(2→4) Kdo of *Chlamydia psittaci* 6BC lipopolysaccharide. J. Endotoxin Res. 6, 361–368.

Brade, L., Gronow, S., Wimmer, N., Kosma, P., Brade, H., 2002. Monoclonal antibodies against 3-deoxy-α-D-*manno*-oct-2-ulosonic acid (Kdo) and D-*glycero*-D-*talo*-oct-2-ulosonic acid (Ko). J. Endotoxin Res. 8, 357–364.

Brooks, C., Müller-Loennies, S., Brade, L., et al., 2008. Exploration of specificity of a germline monoclonal antibody in the recognition of natural and synthetic bacterial epitopes. J. Mol. Biol. 377, 450–468.

Crich, D., Banerjee, A., 2006. Stereocontrolled synthesis of the D- and L-*glycero*-β-D-*manno*-heptopyranosides and their 6-deoxy analogues. Synthesis of methyl α-L-*rhamno*-pyranosyl-(1→3)-D-*glycero*-β-D-*manno*-heptopyranosyl-(1→3)-6-deoxy-*glycero*-β-D-*manno*-heptopyranosyl-(1→4)-α-L-rhamno-pyranoside, a tetrasaccharide subunit of the lipopolysaccharide from *Plesomonas shigelloides*. J. Am. Chem. Soc. 128, 8078–8086.

Ekelöf, K., Oscarson, S., 1995a. Synthesis of 2-(4-aminophenyl)ethyl 3-deoxy-5-*O*-(3,4,6-tri-*O*-β-D-glucopyranosyl-α-D-glucopyranosyl)-α-D-*manno*-oct-2-ulopyranosiduronic acid, a highly branched pentasaccharide corresponding to structures found in lipopolysaccharides from *Moraxella catarrhalis*. Carbohydr. Res. 278, 289–300.

Ekelöf, K., Oscarson, S., 1995b. Synthesis of 4- and/or 4'-phosphate derivatives of methyl 3-*O*-L-*glycero*-α-D-*manno*-heptopyranosyl-L-*glycero*-α-D-*manno*-heptopyranoside and their 2-(4-trifluoroacetamidophenyl)ethyl glycoside analogues. J. Carbohydr. Chem. 14, 299–315.

Ekelöf, K., Oscarson, S., 1996. Synthesis of oligosaccharide structures from the lipopolysaccharide of *Moraxella catarrhalis*. J. Org. Chem. 61, 7711–7718.

Fujimoto, Y., Iwata, M., Imakita, N., et al., 2007. Synthesis of immunoregulatory *Helicobacter pylori* lipopolysaccharide partial structures. Tetrahedron Lett. 48, 6577–6581.

Gurjar, M.K., Talukdar, A., 2004. Synthesis of terminal disaccharide unit of *Klebsiella pneumoniae* ssp. R20. Tetrahedron 60, 3267–3271.

Hansson, J., Oscarson, S., 2000. Complex bacterial carbohydrate surface antigen structures: syntheses of Kdo- and heptose-containing lipopolysaccharide core structures and anomerically linked phosphodiester-linked oligosaccharide structures. Curr. Org. Chem. 4, 535–564.

Holst, O., Bock, K., Brade, L., Brade, H., 1995. The structures of oligosaccharide bisphosphates isolated from the lipopolysaccharide of a recombinant *Escherichia coli* strain expressing the gene *gseA* [3-deoxy-D-*manno*-octulopyranosonic acid (Kdo) transferase] of *Chlamydia psittaci* 6BC. Eur. J. Biochem. 229, 194–200.

Hori, K., Sawada, N., Ando, H., Ishida, H., Kiso, M., 2003. Synthetic study on *Campylobacter jejuni* lipopolysaccharides: an improved synthesis of a branched, heptose-containing trisaccharide core structure and its conversion into ganglioside GD3 related hexasaccharide. Eur. J. Org. Chem., 3752–3760.

Ishii, K., Kubo, H., Yamasaki, R., 2002. Synthesis of α-lactosyl-(1→3)-L-*glycero*-α-D-*manno*-heptopyranoside, a partial oligosaccharide structure expressed within the lipooligosaccharide produced by *Neisseria gonorrhoeae* strain 15253. Carbohydr. Res. 337, 11–20.

Ishii, K., Esumi, Y., Iwasaki, Y., Yamasaki, R., 2004. Synthesis of a 2,3-di-*O*-substituted heptose structure by regioselective 3-*O*-silylation of a 2-*O*-substituted heptose derivative. Eur. J. Org. Chem., 1214–1227.

Jaipuri, F.A., Collet, B.Y.M., Pohl, N.L., 2008. Synthesis and quantitative evaluation of *glycero*-D-*manno*-heptose binding to concanavalin A by fluorous-tag assistance. Angew. Chem. Int. Ed. Engl. 47, 1707–1710.

Komarova, B.S., Tsvetkov, Y.E., Knirel, Y.A., Zähringer, U., Pier, G.B., Nifantiev, N.E., 2006. Synthesis of a common trisaccharide fragment of glycoforms of the outer core region of the *Pseudomonas aeruginosa* lipopolysaccharide. Tetrahedron Lett. 47, 3583–3587.

Kooistra, O., Bedoux, G., Brecker, L., et al., 2003. Structure of a highly phosphorylated lipopolysaccharide core in the ΔalgC mutants derived from *Pseudomonas aeruginosa* wild-type strains PAO1 (serogroup O5) and PAC1R (serogroup O3). Carbohydr. Res. 338, 2667–2677.

Kosma, P., 1999. Chemical synthesis of core structures. In: Brade, H., Opal, S.M., Vogel, S.N., Morrison, D.C. (Eds.) Endotoxin in Health and Disease. Marcel Dekker Inc, New York, pp. 257–281.

Kosma, P., 2008. Occurrence, synthesis and biosynthesis of bacterial heptoses. Curr. Org. Chem. 12, 1021–1039.

Kosma, P., Reiter, A., Zamyatina, A., Wimmer, N., Glück, A., Brade, H., 1999. Synthesis of inner core antigens related to *Chlamydia, Pseudomonas* and *Acinetobacter* LPS. J. Endotoxin Res. 5, 157–163.

Kosma, P., Reiter, A., Hofinger, A., Brade, L., Brade, H., 2000. Synthesis of neoglycoproteins containing Kdo epitopes specific for *Chlamydophila psittaci* lipopolysaccharides. J. Endotoxin Res. 6, 57–69.

Kosma, P., Brade, H., Evans, S.V., 2008. Lipopolysaccharide antigens of Chlamydia. In: Roy, R. (Ed.), Carbohydrate Vaccines, ACS Symposium Series 989. American Chemical Society, Washington DC, pp. 211–224.

Kubo, H., Ishii, K., Koshino, H., Toubetto, K., Naruchi, K., Yamasaki, R., 2004. Synthesis of a 3,4-di-*O*-substituted heptose structure: a partial oligosaccharide expressed in Neisserial lipopolysaccharide. Eur. J. Org. Chem., 1202–1213.

Li, L.S., Wu, Y.L., 2003. Recent progress in the syntheses of higher 3-deoxy-octulosonic acids and their derivatives. Curr. Org. Chem. 7, 447–475.

Mannerstedt, K., Ekelöf, K., Oscarson, S., 2007. Evaluation of Kdo as glycosyl donors. Carbohydr. Res. 342, 631–637.

Mannerstedt, K., Segerstedt, E., Olsson, J., Oscarson, S., 2008. Synthesis of a common tetrasaccharide motif of *Haemophilus influenzae* LPS inner core structures. Org. Biomol. Chem. 6, 1087–1091.

Müller-Loennies, S., Gronow, S., Brade, L., MacKenzie, R., Kosma, P., Brade, H., 2006. A monoclonal antibody that recognizes an epitope present in the lipopolysaccharide of *Chlamydiales* differentiates *Chlamydophila psittaci* 6BC from *Chlamydophila pneumoniae* and *Chlamydia trachomatis*. Glycobiology 16, 184–196.

Ngyuen, H.P., Seto, N.O.L., MacKenzie, R., et al., 2003. Murine germline antibodies recognize multiple carbohydrate epitopes by flexible utilization of binding site residues. Nat. Struct. Biol. 10, 1019–1025.

Oscarson, S., 1997. Synthesis of oligosaccharides of bacterial origin containing heptoses, uronic acids and fructofuranoses as synthetic challenges. Topics Curr. Chem. 186, 171–202.

Oscarson, S., 2001. Synthesis of *Haemophilus influenzae* carbohydrate surface antigens. Carbohydr. Polym. 44, 305–311.

Oscarson, S., Ritzén, H., 1994. Synthesis of a hexasaccharide corresponding to part of the heptose-hexose region of the *Salmonella* Ra core, and a penta- and a tetra-saccharide that compose parts of this structure. Carbohydr. Res. 254, 81–90.

Paulsen, H., Stiem, M., Unger, F.M., 1987. Synthesis of the sequence α-KDO-(2→4)-α-KDO-(2→6)-D-GlcN of the

"Inner Core" structure of lipopolysaccharides. Liebigs Ann. Chem., 273–281.

Pozsgay, V., 2003. Chemical synthesis of bacterial carbohydrates. In: Wong, S.Y.C., Arsequell, G. (Eds.), Immunobiology of Carbohydrates. Kluwer Academic/ Plenum Publishers, New York, pp. 192–273.

Reiner, M., Schmidt, R.R., 2000. Synthesis of 3-deoxy-D-*manno*-2-octulosonic acid (Kdo) and D-*glycero*-D-*talo*-2-octulosonic acid (Ko) and their α-glycosides. Tetrahedron Asymm. 11, 319–335.

Reiter, A., Zamyatina, A., Schindl, H., Hofinger, A., Kosma, P., 1999. Synthesis of *Pseudomonas aeruginosa* core antigens containing 7-*O*-carbamoyl-L-*glycero*-α-D-*manno*-heptopyranosyl residues. Carbohydr. Res. 317, 39–52.

Reiter, A., Brade, L., Sanchez-Carballo, P., Brade, H., Kosma, P., 2001. Synthesis and immunochemical characterization of neoglycoproteins containing epitopes of the inner core region of *Pseudomonas aeruginosa* RNA group 1 lipopolysaccharide. J. Endotoxin Res. 7, 125–131.

Rozalski, A., Brade, L., Kosma, P., Moxon, R., Kusumoto, S., Brade, H., 1997. Characterization of monoclonal antibodies recognizing three distinct, phosphorylated carbohydrate epitopes in the lipopolysaccharide of the deep rough mutant I-69 Rd^-/b^+ of *Haemophilus influenzae*. Mol. Microbiol. 23, 569–577.

Rund, S., Lindner, B., Brade, H., Holst, O., 2000. Structural analysis of the lipopolysaccharide from *Chlamydophila psittaci* strain 6BC. Eur. J. Biochem. 267, 5717–5726.

Sekljic, H., Kosma, P., Bartek, J., Fukase, K., Kusumoto, S., Brade, H., 1996. Synthesis of neoglycoproteins containing 5-O-phosphorylated Kdo monosaccharide, 4-O- and 5-O-phosphorylated α-Kdo-(2→6)-2-acetamido-2-deoxy-β-D-glucopyranosyl disaccharide residues. J. Endotoxin Res. 3, 151–164.

Sekljic, H., Wimmer, N., Hofinger, A., Brade, H., Kosma, P., 1997. Synthesis of neoglycoproteins containing L-glycero-α-D-manno-heptopyranosyl-(1→4)- and (1→5)-linked 3-deoxy-α-D-manno-2-octulopyranosylonic acid (Kdo) phosphate determinants. J. Chem. Soc. Perkin Trans. I, 1973–1982.

Tanaka, H., Takahashi, D., Takahashi, T., 2006. Stereoselective synthesis of oligo-α-(2,8)-3-deoxy-D-*manno*-2-octulosonic acid derivatives. Angew. Chem. Int. Ed. Engl. 45, 770–773.

Unger, F.M., 1981. The chemistry and biological significance of 3-deoxy-D-*manno*-2-octulosonic acid (KDO). Adv. Carbohydr. Chem. Biochem. 38, 323–388.

van Straten, N.C.R., Kriek, N.M.A.J., Timmers, C.M., Wigchert, S.C.M., van der Marel, G.A., van Boom, J.H., 1997. Synthesis of a trisaccharide fragment corresponding to the lipopolysaccharide region of *Vibrio parahaemolyticus*. J. Carbohydr. Chem. 16, 947–966.

Verheul, A.F.M., Boons, G.J.P.H., van der Marel, G.A., et al., 1991. Minimal oligosaccharide structures required for induction of immune responses against meningococcal immunotype L1, L2, and L3,7,9 lipopolysaccharides determined by using synthetic oligosaccharide-protein conjugates. Infect. Immun. 59, 3566–3573.

Wang, H., Head, J., Kosma, P., et al., 2008. Recognition of heptoses and the inner core of bacterial lipopolysaccharides by surfactant protein D. Biochemistry 47, 710–720.

Wimmer, N., Brade, H., Kosma, P., 2000. Synthesis of neoglycoproteins containing D-*glycero*-D-*talo*-oct-2-ulopyranosylonic acid (Ko) ligands corresponding to core units from *Burkholderia* and *Acinetobacter* lipopolysaccharide. Carbohydr. Res. 329, 549–560.

Yamasaki, R., Takajyo, A., Kubo, H., Matsui, T., Ishii, K., Yoshida, M., 2001. Convenient synthesis of methyl L-*glycero*-D-*manno*-heptopyranoside. J. Carbohydr. Chem. 20, 171–180.

Yoshizaki, H., Fukuda, N., Sato, K., et al., 2001. First total synthesis of the Re-type lipopolysaccharide. Angew. Chem. Int. Ed. Engl. 40, 1475–1480.

Zhang, Y., Gaekwad, J., Wolfert, M.A., Boons, G.J., 2008. Innate immune responses of synthetic lipid A derivatives of *Neisseria meningitidis*. Chem. Eur. J. 14, 558–569.

CHAPTER

25

Chemical synthesis of lipoteichoic acid and derivatives

Christian Marcus Pedersen and Richard R. Schmidt

SUMMARY

The roles of lipoteichoic acid (LTA) in Gram-positive bacterial cell walls has led to some debate in the last decades. Biological studies have shown several activities of LTA but, due to contaminations in the isolated LTA, the results of early studies are doubtful. Synthetic derivatives have played a major role in the discussions and led to differing conclusions during the successive years. In the first syntheses, the main interest was the chemistry and not the biological role. Later work was more focused on the investigation of the biological activity. The early approaches to lipoteichoic acid synthesis were lacking some fundamental structural functions and, therefore, the target molecules were not biologically active. Biological studies showed that D-alanine residues play a crucial role in the activity of type I LTA and these observations were confirmed and underlined by fully active synthetic derivatives. By systematic work in the field, the minimal structural requirement for biological activity has been revealed. In this chapter, the focus will be on chemical synthesis with the description and discussion of the different synthetic approaches. The conclusions from the various biological studies will be summarized and the synthesis of structurally, closely related derivatives will be described briefly.

Keywords: Lipoteichoic acid; Synthesis; Gram-positive bacteria; Biologial activity; Immunity; Modifications

1. INTRODUCTION

Throughout their lives, higher organisms fight against pathogen invasion from bacteria, viruses, fungi etc. Bacteria are the most common pathogens that cause infections in mammalian organisms. Diseases such as tuberculosis, pneumonia, anthrax etc. are well known and infections caused by bacteria such as *Escherichia coli, Salmonella enterica, Streptococcus pneumoniae and Staphylococcus aureus* are common concerns in daily life.

Bacteria are divided into two main groups by the Gram stain method: Gram-negative and Gram-positive bacteria. The main difference between Gram-negative and Gram-positive bacteria is the construction of their cell wall (see Chapter 1). Where Gram-negative bacteria have a relatively complex and thin cell wall, the opposite is the case with Gram-positive bacteria. Upon infection, the innate immune system is activated by receptor recognition followed by release of a variety of cytokines. For Gram-negative bacteria, the biological pathway has been well understood for more than 50 years (Westphal *et al.*, 1952) and the most important actors in the cytokine release are lipopolysaccharides (LPSs). The innate immune system is triggered by binding of these LPSs to the toll-like receptor-4 (TLR-4) (see Chapter 31).

The mechanism of infection is, however, not well understood in Gram-positive bacteria, even though the structural counterpart of LPS, called lipoteichoic acid (LTA), has been known for some time (see Chapter 5); representative structures of LTA of types I–IV (Fischer, 1990) are shown in Figure 25.1. It has not been possible, until recently, to observe a clear connection between innate immune stimulatory response and LTA. One of the main reasons for the lack of knowledge about Gram-positive bacterial influence on the innate immune system and wrong conclusions originate from inappropriate preparation methods of LTA resulting in impure or partly decomposed material.

In 2001, Hartung and co-workers (Morath *et al.*, 2001) showed that with appropriate purification of LTA from *S. aureus*, the biological activity could be preserved. Furthermore, it was recognized that the D-alanine (D-Ala) substituents of LTA play a crucial role in the activity (Fischer, 1981) and, due to their base lability, these D-Ala residues are often lost through standard preparation methods, since cleavage is observed already at pH 8.5.

Due to the great interest in exploring the immune stimulation effects of LTA together with its biological role, several attempts to synthesize these very complex compounds have been undertaken. Until recently, none of the synthesized analogues showed significant biological activity. Van Boom and co-workers (Oltvoort *et al.*, 1982, 1984) were the first to synthesize a fragment of LTA from *S. aureus* containing a disaccharide glycolipid unit coupled to a glycerol phosphate trimer linked by interglyceridic (1→3)-phosphoric diester bonds. The next contribution to the synthesis of an LTA derivative came from Kusumoto's group (Fukase *et al.*, 1992, 1994), but the synthetic LTA was found to have no immune stimulatory activity. Together with the before mentioned difficulties in isolating active LTA from Gram-positive bacteria, these observations have put into doubt the general role of LTA as an innate immune stimulator. The misleading conclusions in the field of LTA were put to an end in 2002 when Schmidt and co-workers (Morath *et al.*, 2002; Stadelmaier *et al.*, 2003) finished the synthesis of LTA of *S. aureus* – the first synthetic fully active LTA. Together with very pure natural LTA isolated by Hartung's improved method (Morath *et al.*, 2001), the immunostimulatory effects of LTA could not be put into question anymore.

Since this fruitful synthesis, the Schmidt group (Figueroa-Perez *et al.*, 2005, 2006; Stadelmaier *et al.*, 2006) has prepared structural variants of LTA of *S. aureus* and the importance of the D-Ala residues has been underlined by this work. In this chapter, the above-mentioned syntheses will be discussed and an overview on the status of LTA synthesis will be given.

2. VAN BOOM'S SYNTHESIS OF *S. AUREUS* LTA TYPE I

The first synthesis of an LTA derivative, by van Boom and colleagues, appeared in the literature in the early 1980s (Oltvoort *et al.*, 1982, 1984). The target was an LTA type I from *S. aureus* **1** (see Figure 25.1) (Greenberg *et al.*, 1996), which is the most common cause of staphylococcal infections and, hence, an interesting target to investigate. The structure of the target had been resolved by Baddiley (1979) and later further investigated by Fischer *et al.* (Fischer *et al.*, 1980; Fischer, 1994a,b). The LTA from *S. aureus* is a group of ampliphilic molecules consisting of a lipophilic end and a glycerol-phosphate unit, which is negatively charged. The lipophilic part contains a β-(1→6)-linked diglucoside (gentiobioside) with a β-linked

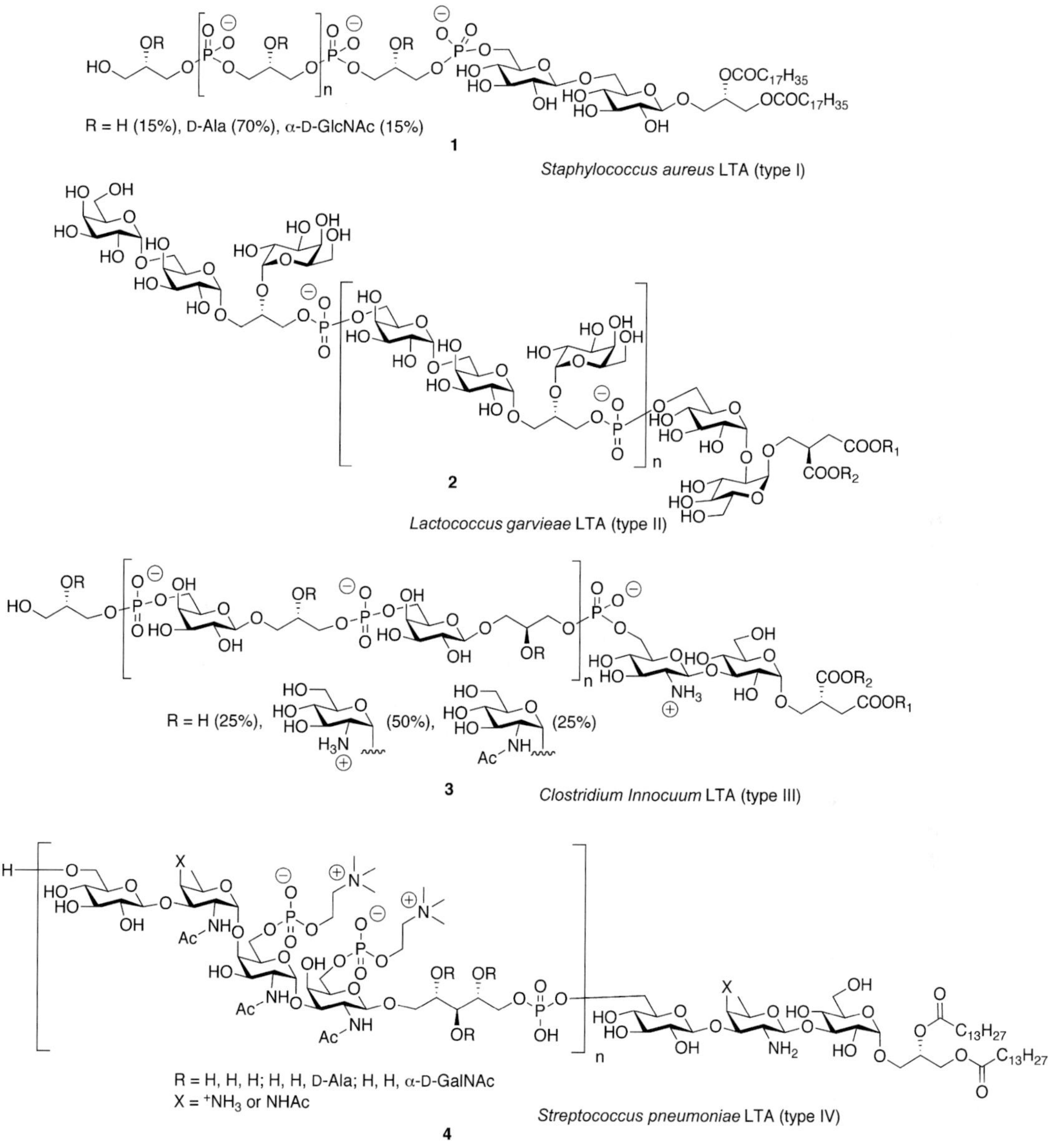

FIGURE 25.1 The different types of LTA represented by structures isolated from the respective bacteria.

diglyceride. The acyl chains vary between saturated and unsaturated acid residues.

In the first work, the fragment **5** containing major functionalities of the LTA was synthesized (Scheme 25.1). The fragment contains the diglucoside with the saturated fatty diacyl glyceride and a single glycerol unit connected to the disaccharide via a phosphodiester.

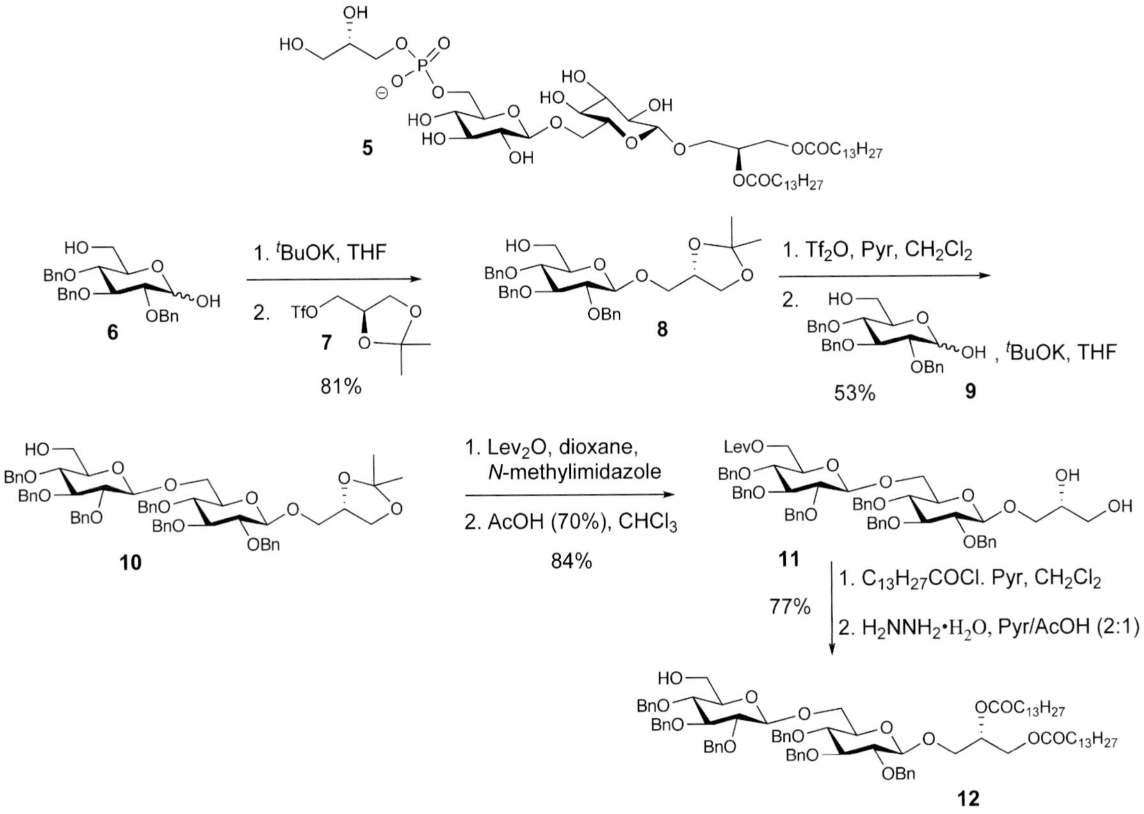

SCHEME 25.1 Synthesis of protected disaccharide building block using the anomeric *O*-alkylation method.

The strategy to synthesize the target molecule is divided into three distinct parts. The first deals with the synthesis of the glycolipid part derivative having a free hydroxy group at the 6″ position. The second part is the synthesis of the phosphorylated *sn*-glycerol from a chiral precursor having the proper protecting groups. The final part deals with the assembly of the two building blocks to give the fully protected LTA derivative which, after stepwise deprotection, gives the desired product.

2.1. Preparation of the glycolipid part

The two β-glucosidic bonds in the gentiobiose-diglyceride **10** moiety were obtained by using "Schmidt's anomeric *O*-alkylation method" (Schmidt *et al.*, 1980), where only one glucose building block had to be prepared (see Scheme 25.1). According to the method, the β-glucosidic bond can be obtained by reaction of the glucosyl oxide anion with the acceptor triflate, due to the higher reactivity of the β-oxide. The 2,3,4-tri-*O*-benzyl-D-glucose **6** was treated with potassium *tert*-butoxide in tetrahydrofuron (THF) followed by the addition of 1,2-*O*-isopropylidene-*sn*-glycerol triflate **7**, which gave **8** in 81% yield. Then **8** could be transformed into the acceptor triflate **9** and added to a solution of the donor **6** resulting in a very good yield of the gentiobiose derivative **10**. The 6″-OH was then protected with levulinoyl (Lev) followed by acidic cleavage of the isopropylidene group and introduction of the fatty acids from the

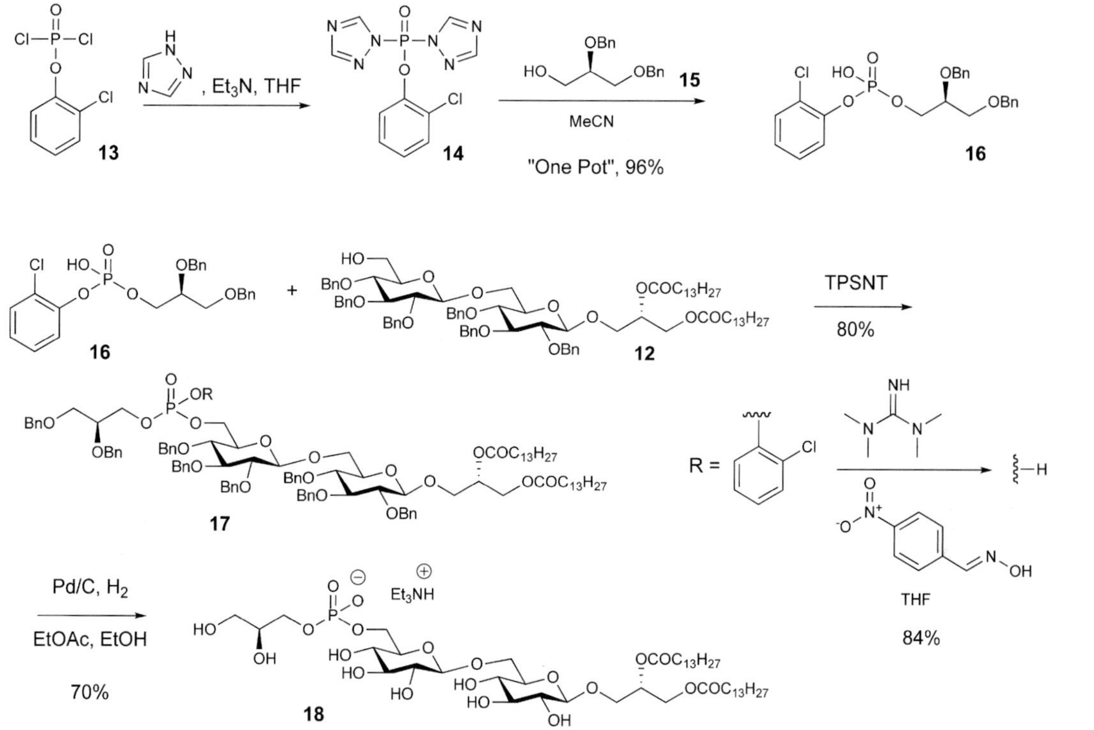

SCHEME 25.2 Synthesis of the phosphorylation reagent and assembling of the building blocks to give the desired *S. aureus* LTA type I fragment.

corresponding acid chloride to give the protected glyco-lipid part in high yield. The Lev group could then selectively be removed with hydrazine in pyridine/acetic acid giving **12** in 80% yield.

2.2. Preparation of the glycerol-phosphate unit

The 1,2-di-*O*-benzyl-*sn*-glycerol **15** was phosphorylated with an excess of (2-chlorophenyl)-phosphoro-di-(1,2,4-triazolide) **14**, which was readily obtained from the dichloride **13** and 1,2,4-triazole (Scheme 25.2).

The condensation of the two building blocks **12** and **16** was then carried out using 1-(2,4,6-triisopropylbenzenesulfonyl)-3-nitro-1,2,4-triazole (TPSNT) to activate the glycerol-phosphate unit giving the fully protected LTA derivative **17**.

Deprotections were carried out in two steps. The 2-chlorophenyl group was removed by treating **17** with N^1,N^1,N^2,N^2-tetramethylguanidine and *syn*-4-nitrobenzaldoxime in THF giving the intermediate phosphate (Reese *et al.*, 1978; Reese and Tau, 1978), after which the benzyl groups were removed by hydrogenolysis using Pd/C as the catalyst giving the desired LTA fragment **18** in 59% yield (for the deprotection steps).

The synthesized fragment **18** contains some of the fundamental functionalities of the native LTA obtained from *S. aureus* but it was, however, still quite far from the average structure found in Nature. First, the teichoic part is not a polymeric chain with a glycerol backbone, but consists only of one glycerol unit connected to the glycolipid via a phosphodiester. Second, there are no substituents in the glycerol part. The native

LTA is to a high degree substituted with D-Ala or D-glucosamine. Another difference is in the fatty acid chains, which are partly unsaturated in the LTA isolated from Nature.

As a logical continuation of this work, van Boom and co-workers continued with the synthesis of other derivatives having a polymeric teichoic acid unit, as well as other naturally occurring derivatives, of the fundamental structure (see below).

2.3. Expanding the teichoic acid part

As a natural development in the synthesis of LTA fragments, the next step was to synthesize derivatives which better resemble the native compounds. The most striking lack in the first synthetic LTA is the teichoic chain, which has a length of up to fifty in the natural LTA and only one in the synthesized analogue. van Boom and co-workers decided to synthesize *S. aureus* LTA having three phosphodiester linkages, but still without substituents on the glycerol backbone.

In order to synthesize the glycerol phosphate trimer **36**, three properly protected glycerol building blocks had to be prepared **30** (Scheme 25.3). One terminal unit **19** having a free 3-OH group and benzyl protection on the other hydroxy groups and two non-terminal units **26** and **29** having the 2-OH group benzylated and orthogonal protecting groups on the remaining hydroxy groups were required.

The terminal unit was prepared from 3,4-*O*-isopropylidene-D-mannitol, whereas L-serine (L-Ser) **21** (Lok *et al.*, 1976) was used as the chiral source for non-terminal units. The L-Ser **21** was transformed into the 2,3-*O*-isopropylidene-*sn*-glycerol **22** in four steps (see Scheme 25.3) followed by allylation of the 1-OH group. After acidic removal of the isopropylidene group, the 3-OH group could be temporally protected with trityl followed by benzylation of the 2-OH group and detritylation to give the key intermediate **24**. To obtain the last building block, the allyl protecting group was isomerized to 1-propenyl **26** (Oltvoort *et al.*, 1981) and treated by 4-oxovaleric anhydride (Levulinic anhydride) to give **27**, followed by removal of the 1-*O*-propenyl group effected by mercuric oxide to give **29**.

The two non-terminal units were firstly coupled using the phosphorylation agent **28** together with 1-methylimidiazole to afford the dimer **30** in high yield. The 1-*O*-propenyl group was then removed by mercuric oxide (Nashed and Anderson, 1982) and the dimer **31** could be coupled to the terminal unit using the phosphorylation agent **32** and 1-methylimidiazole. After hydrazinolysis of the 4-oxovaleryl group, the teichoic acid fragment **34** was coupled to the glycolipid part **12** using **14** as the phosphorylating agent (Scheme 25.4).

With the fully protected LTA **35** in hand, all that was left was a three-step deprotection sequence. The tribromoethyl group was deprotected using Zn in the presence of triisopropylbenzenesulfonic acid, followed by removal of the 2-chlorophenyl groups mediated by (*E*)-pyridine-2-carbaldehyde oxime and N^1,N^1,N^3,N^3-tetramethylguanidine. Finally, hydrogenolysis of the benzyl groups in the presence of Pd/C afforded the LTA fragment **36**.

The biological role of the synthetic LTA fragment was unfortunately neither discussed nor studied. But, due to the knowledge we have now, the effect of these compounds on the immune response would have been minor since important functionalities for the activity are missing in the structures, i.e. D-Ala as well as *N*-acetylglucosamine (GlcNAc) in the glycerol backbone.

2.4. Related structures

van Boom and co-workers (van Boeckel and van Boom, 1985; Veeneman *et al.*, 1989; Smid *et al.*, 1993) have synthesized a few other structures related to LTA (Figure 25.2). In connection with LTA, the most interesting modification

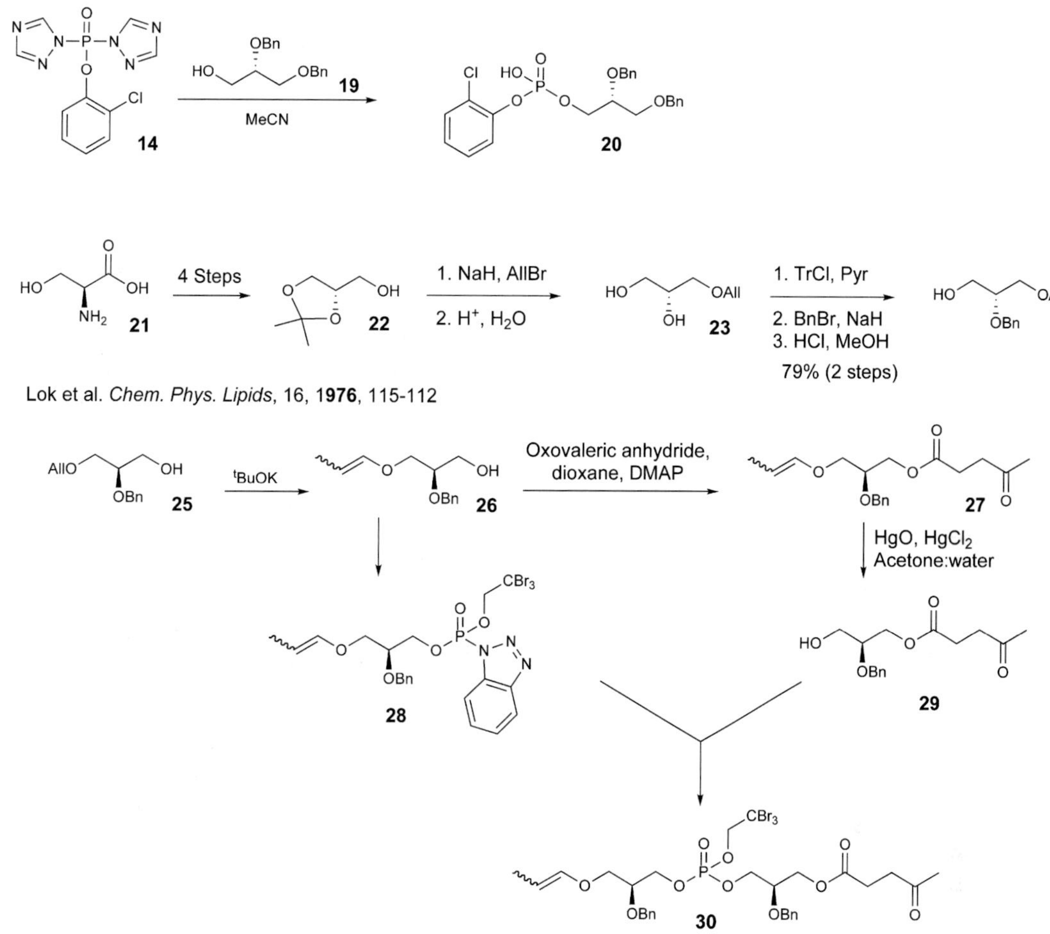

Lok et al. *Chem. Phys. Lipids*, 16, **1976**, 115-112

SCHEME 25.3 Synthesis of trimeric glycerophosphate.

was performed in the synthesis of streptococci phosphatidyl-α-diglucosyldiglyceride **38** containing a phospholipid part (see Figure 25.2). The synthesis is an early and elegant example of the use of 4′,6′-tetraisopropyl-disiloxane-1,3-diyl (TIPS) as a dynamic protecting group. Another example is the synthesis of teichoic acid fragments **37** which have similarities to the structures found in LTA, i.e. a fragment from *S. pneumoniae* type 1 (see Figure 25.2) and which is closely related to LTA (type IV) **4** (see Figure 25.1) from the same species.

3. KUSUMOTO'S SYNTHESIS OF LTA FRAGMENTS FROM *ENTEROCOCCUS HIRAE* AND *STREPTOCOCCUS PYOGENES*

The structure of *S. pyogenes* and *E. hirae* LTA (Figure 25.3, **39** and **40**) was proposed by Fischer and Koch (1981) and, as shown in Figure 25.3, there are many similarities with the LTA of *S. aureus* **1** (see Figure 25.1) described above. The glycolipid part consists of α-(1→2)-D-glucose

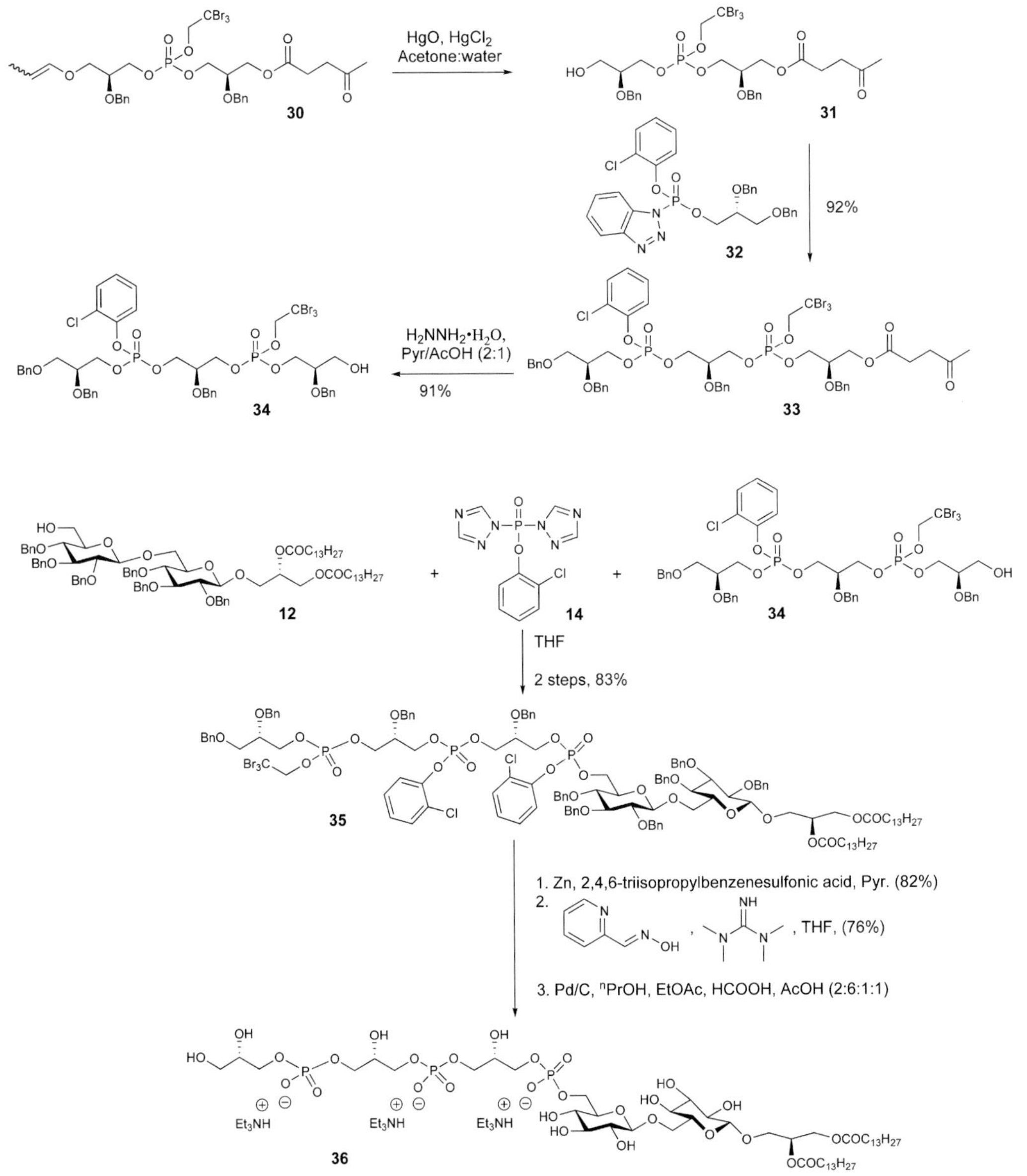

SCHEME 25.4 Assembly of the building blocks to give the LTA type I derivative having trimeric teichoic acid part.

disaccharide (kojibiose) which is linked via an α-glycosidic bond to 1,2-di-*O*-acyl-*sn*-glycerol. The acyl moiety can either be palmitic or oleic acids. The hydrophilic teichoic acid part consists of a glycerol backbone (1→3)-linked via phosphodiester bonds. The 2-OH groups are partially esterified with D-Ala groups or glycosidically linked to D-glucose. The length is proposed to be up to fifty repeating units.

The native LTA showed potent anti-tumour activity, which was attributed to induction of tumour necrosis factor (TNF) (Tsutsui *et al.*, 1991).

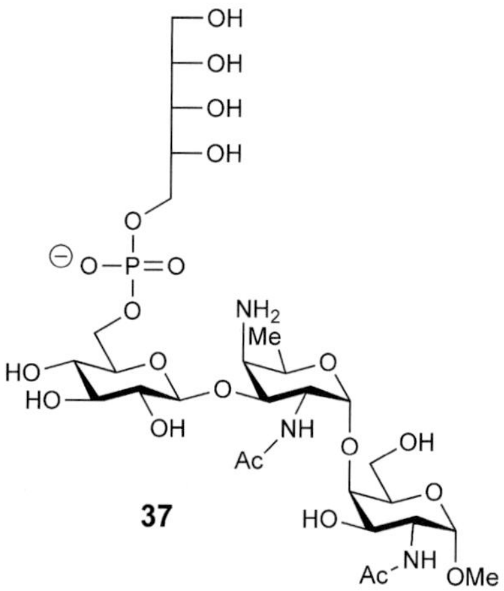

FIGURE 25.2 Structures closely related to LTA synthesized by van Boom and co-workers (van Boeckel and van Boom, 1985; Veeneman *et al.*, 1989; Smid *et al.*, 1993).

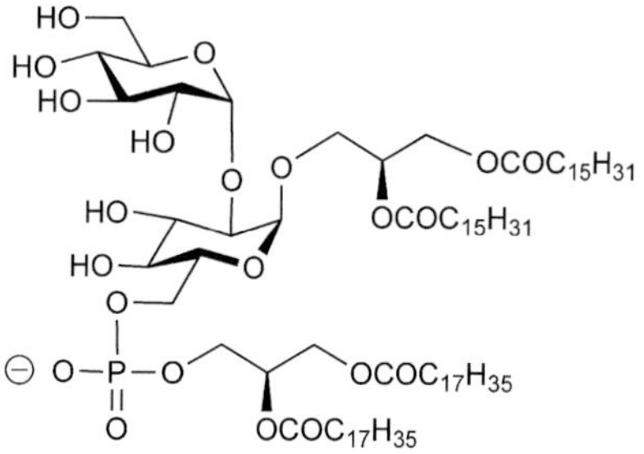

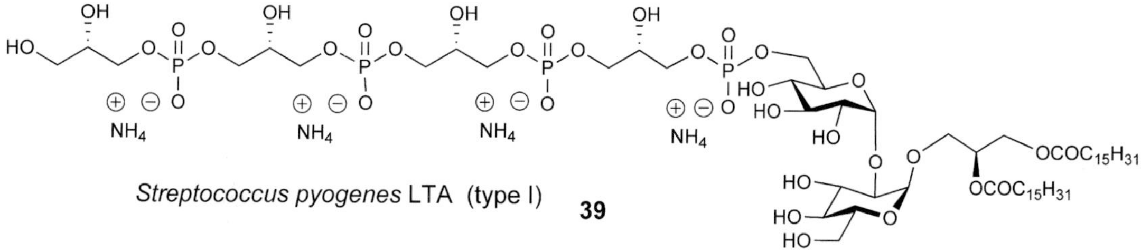

FIGURE 25.3 LTA derivatives synthesized by Kusumoto (Fukase *et al.*, 1992, 1994).

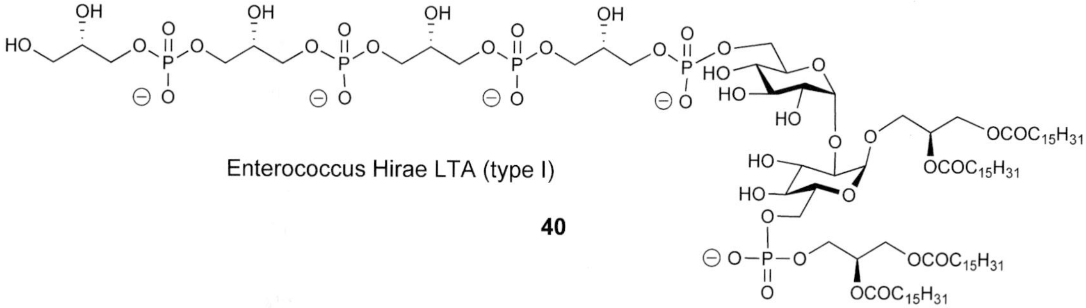

When the isolated LTA was partially hydrolysed under acidic conditions, the activity remained, whereas basic hydrolysis resulted in loss of activity, hence the acyl groups are important for the TNF induction. The biological studies showed that the whole molecule is not necessary for activity, and hence, a chemical study in order to find the minimal structure required for TNF induction was of great interest.

From the biological results it was decided to synthesize a fragment containing the fundamental structural moieties, but without substituents in the glycerol backbone and a significantly shorter glycerol phosphate chain consisting of only four units (Fukase *et al.*, 1992).

Due to the protecting strategy involving benzyl groups (Fukase *et al.*, 1990), the LTA fragment containing the unsaturated palmitoyl residues was prepared.

3.1. Synthesis of the poly(glycerol phosphate) part

As was the case in van Boom's synthesis, benzyl protection was used as the persistent protecting group of the 2-OH group and for the phosphoric acid as well (Scheme 25.5). The two non-terminal glycerol building blocks were synthesized from chiral 1,2-*O*-isopropylidene-*sn*-glycerol **22** (see Scheme 25.5) which, after introduction of *para*-methoxybenzyl (PMB) protection at *O*-3 and treatment with acetic acid (90%), gave the diol **41** followed by introduction of benzylidene protection. The benzylidene group was regioselectively opened to give the desired 2-*O*-benzyl derivative **42** as the major product (11:1 as an inseparable mixture) that was protected with *tert*-butyldimethylsilyl at the 1-OH group. Selective deprotection of the PMB group mediated by dichloro-dicyanobenzoquinone (DDQ) oxidation gave the building block **45**, whereas treatment with tetrabutylammonium fluoride (TBAF) afforded pure **42**. Alternatively, the building blocks could be prepared from the mannitol derivative **46**.

With the building blocks in hand, the assembly of the phosphate tetramer could be carried out using the phosphoramidite method with an excess of reagent **47** and 1 *H*-tetrazole catalysis to transfer **42** into the phosphoramidite **48** which, when condensed with **45**, followed by oxidation, gave the dimer **49** (Scheme 25.6). By selective deprotection with TBAF, **50** could be obtained and, by using DDQ oxidation, **51** was provided in good yields. Repeating the phosphoramidite procedure and deprotection of the *tert*-butyldimethylsilyl (TBDMS) group gave **53** ready to be coupled to the glycolipid part **62**.

3.2. Preparation of the glycolipid part

Two glucosyl donors were needed in the synthesis of the glycolipid part. One having a temporary 2-*O* protecting group and one with a 6-*O* protecting group to be removed later for the coupling with the phosphate tetramer **53**. For the 2-*O*-protection, nitrobenzyl (NPM) was chosen, since it does not participate in the glycosylation

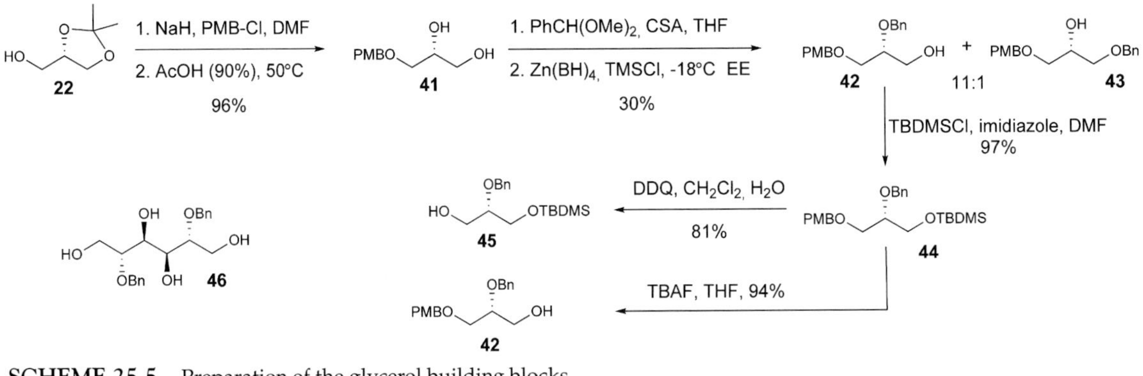

1. NaH, PMB-Cl, DMF
2. AcOH (90%), 50°C
96%
1. PhCH(OMe)$_2$, CSA, THF
2. Zn(BH)$_4$, TMSCl, -18°C EE
30%
11:1
TBDMSCl, imidiazole, DMF 97%
DDQ, CH_2Cl_2, H_2O 81%
TBAF, THF, 94%
22 41 42 43 44 45 46 42

SCHEME 25.5 Preparation of the glycerol building blocks.

and can be readily removed by a two-step sequence involving reduction to the amine, followed by oxidative cleavage using DDQ or an electrochemical procedure (Fukase *et al.*, 1990, 1992).

As a common starting material for the donors, diacetone-D-glucose **54** was used, which was transformed into the two building blocks **55** and **60** using standard protecting group modifications (Scheme 25.7). By using Noyori's glycosylation procedure (Hashimato *et al.*, 1984) having 1,2-di-*O*-allyl-3-*O*-trimethylsilyl-*sn*-glycerol **56** as the acceptor and trimethylsilyl trifluoromethanesulfonate (TMSOTf) as the promoter, the α-glucoside **57** was obtained as the major product (3.4:1). After deallylation using the before mentioned method, the palmitic ester residues could be introduced with the corresponding acid chlorides to give the glyceroglycolipid **58**. The NPM group was then removed by a two-step procedure; first reduction of the nitro functionality by zinc–copper couple and acetylacetone to give the aniline derivative, which was oxidized by anodic oxidation to give the 2-OH acceptor **59**.

The glycosylation reaction between the glycolipid acceptor **59** and the glycosyl fluoride donor **60** was investigated using Mukaiyama's procedure (Mukaiyama *et al.*, 1981) ($SnCl_2$ and $AgClO_4$ in ether) and Suzuki's procedure (Matsumoto *et al.*, 1988) (Cp_2ZrCl_2 and $AgClO_4$ in toluene), with the latter giving the best α-selectivity (3:1) compared with the former (2.3:1). The Troc group of **61** was removed using zinc reduction to give the protected 6-OH glycolipid part **62**, that was ready for the subsequent coupling to the polyglycerol phosphate part **53** (Scheme 25.8), which was carried out by the phosphoramidite method described in the synthesis of the polyglycerol phosphate part (see Scheme 25.6). The fully protected fundamental structure **63** of *S. pyogenes* LTA was globally deprotected by palladium catalysed hydrogenolysis to give the target compound **64**.

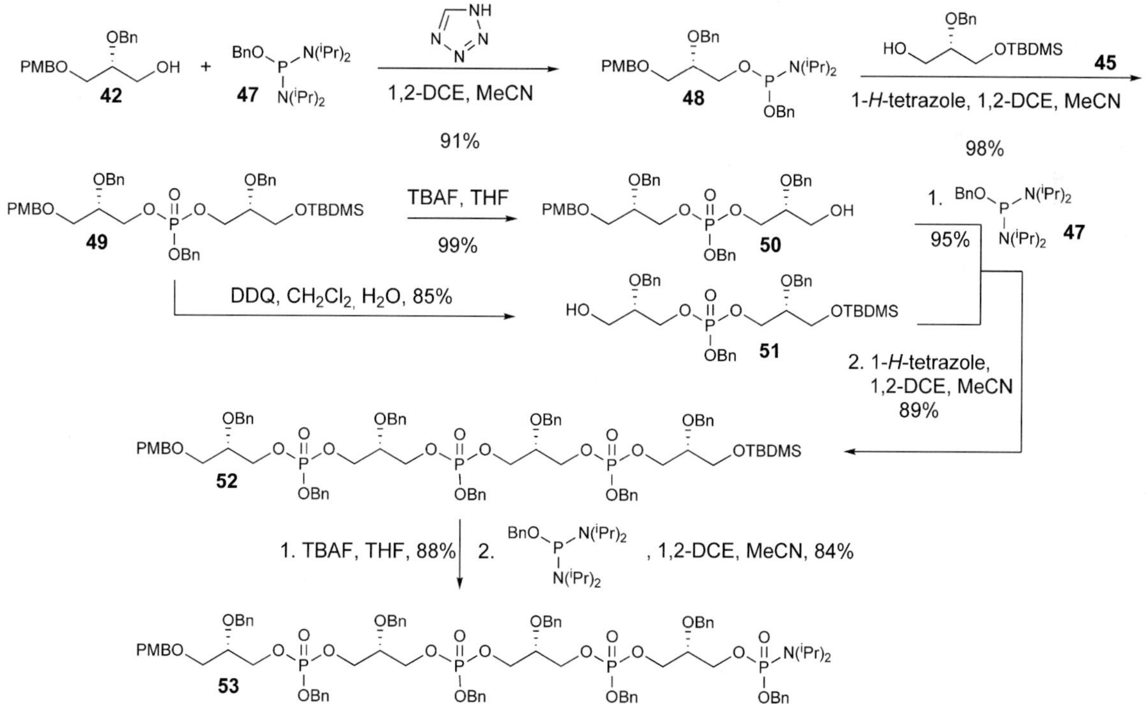

1,2-DCE, MeCN

91%

1-*H*-tetrazole, 1,2-DCE, MeCN

98%

TBAF, THF

99%

DDQ, CH_2Cl_2, H_2O, 85%

1. 47, 95%

2. 1-*H*-tetrazole, 1,2-DCE, MeCN 89%

1. TBAF, THF, 88% 2. $BnOP(N(^iPr)_2)_2$, 1,2-DCE, MeCN, 84%

SCHEME 25.6 Assembly of the glycerol building blocks to give a tetrameric backbone.

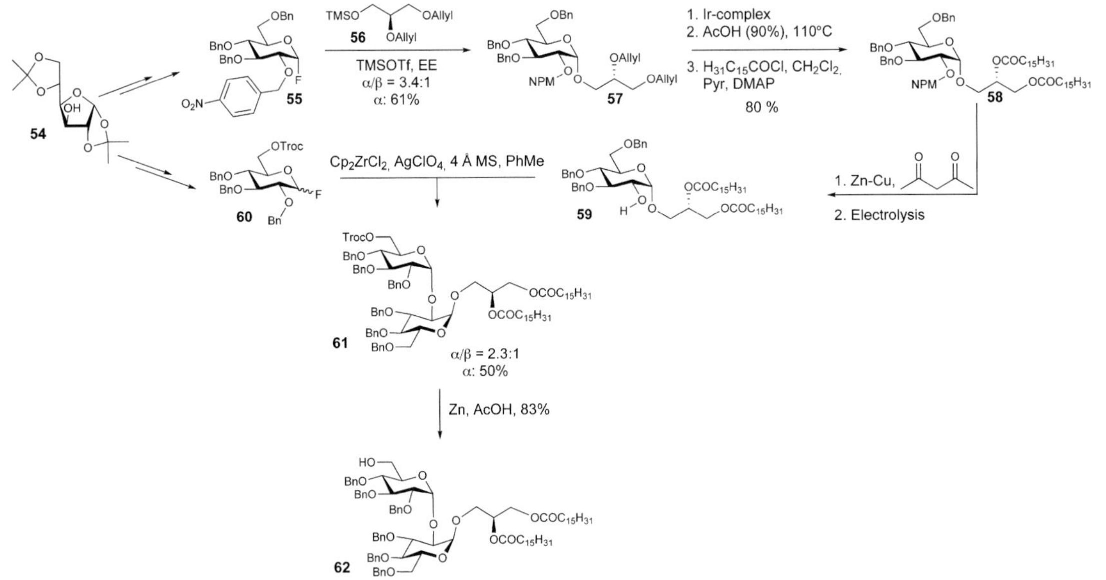

SCHEME 25.7 Synthesis of the diacyl-disaccharide building block.

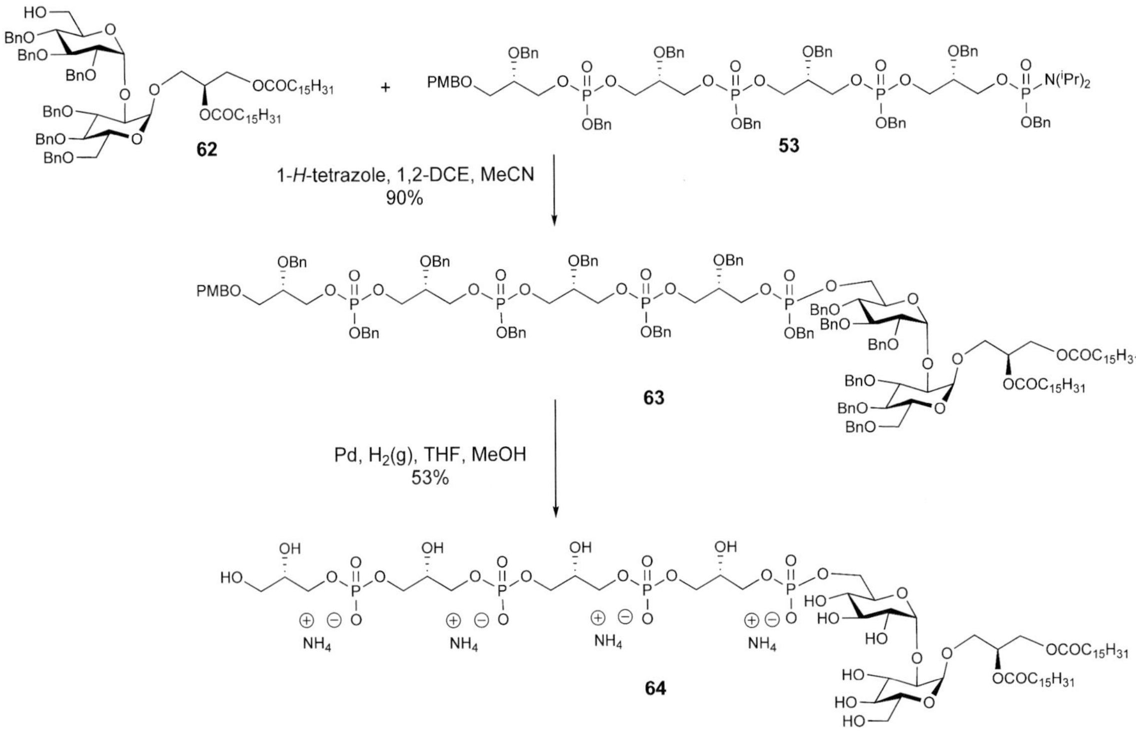

SCHEME 25.8 Assembly of the building blocks followed by deprotection to give the final product of LTA type I from *S. pyogenes*.

The product was mainly analysed by mass spectroscopy and elemental analysis. The glycolipid part without the glycerolphosphate polymer was studied by nuclear magnetic resonance (NMR) spectroscopy and, compared with native glycolipid isolated from *S. pyogenes*, showed good correlation except for the saturated acyl moieties, which are partly unsaturated in the native LTA.

3.3. Synthesis of the fundamental structure of *Enterococcus hirae* LTA

The next target in the Kusumoto group was the LTA isolated from *E. hirae* (Fukase *et al.*, 1994) which, in biological studies, showed biological activity and anti-tumour potency *in vivo* (Tsutsui *et al.*, 1991); the structure of this LTA was proposed by Fischer *et al.* (1980). The LTAs found in *E. hirae* **40** are closely related to the ones found in *S. pyogenes* **39** but there is, however, one additional LTA (LTA 2) in *E. hirae* (see Figure 25.3) which, besides the kojibiose part α-linked to 1,2-diacyl-*sn*-glycerol, has an additional diacylglycerol unit linked to the 6-*O* of glucose in the glycolipid part via a phosphodiester. As is the case in *S. pyogenes*, the length of the polyglycerol phosphate part is deduced to be approximately twenty repeating units and contains D-Ala and glycosyl substituents on the secondary glycerol hydroxy group. Since biological studies revealed that the biological activity was retained in D-Ala-free LTA and the undefined substituent pattern, it was decided to synthesize the fragment without a functionalized glycerol backbone.

Due to the structural similarities with LTA from *S. pyogenes*, the synthesis of *E. hirae* has many similarities. Some of the weak points in the previous synthesis, e.g. low selectivity in the two glycosylations, however, were solved and the installation of the temporary 6-*O* protecting group is different in the later synthesis.

Glycosylation of the new donor **65** with acceptor **66** using Noyori's method gave the desired α-anomer (α:β = 11.5:1) in 81% yield (Scheme 25.9), whereas glycosylation with the donor having 6-*O*-benzyl protection **55** gave low stereo-selectivity (α:β = 3.4:1). The increase in selectivity was attributed to the Troc protection. After removal of the Troc protection using NaOMe, *p*-pivaloylaminobenzyl (PAB) was introduced as a temporary protecting group to give **67**. The allyl groups in **67** were isomerized and hydrolysed to give the diol which, upon treatment with palmitoyl chloride, gave the fully protected di-palmitoyl derivative. The NPM was removed by the above mentioned two-step procedure to give the acceptor **68**, which was glycosylated with the donor **60** using Suzuki's procedure to afford **70**. Again, a directing effect of the Troc group was observed resulting in good α-selectivity (α:β = 7.3:1) and a yield of 72 %, compared with the corresponding 6-*O*-benzyl donor, which gave lower yield (56 %) and selectivity (α:β = 2.2:1) (see Scheme 25.9).

Selective deprotection of the PAB group by oxidation with DDQ afforded the 6′-OH glycolipid, which was treated with the phosphoramidite reagent **70**, prepared from 1,2-di-*O*-palmitoyl-*sn*-glycerol and benzyloxybis(diisopropylamino) phosphane, followed by *m*CPBA oxidation to give glycolipid **71**. The structural similarities with native LTA isolated from *E. hirae* were studied by NMR spectroscopy. In order to improve the resolution in the NMR spectra, the benzyl groups were substituted by methyl groups. The obtained derivative was identical to the methylated natural LTA except from the saturation in the acyl groups.

Introduction of the glycerol phosphate tetramer was mediated using the phosphoramidite method as described in Scheme 25.8.

Finally, global deprotection using palladium-catalysed hydrogenolysis gave the LTA fragment, which could be confirmed by fast atom bombardment (FAB) mass spectra.

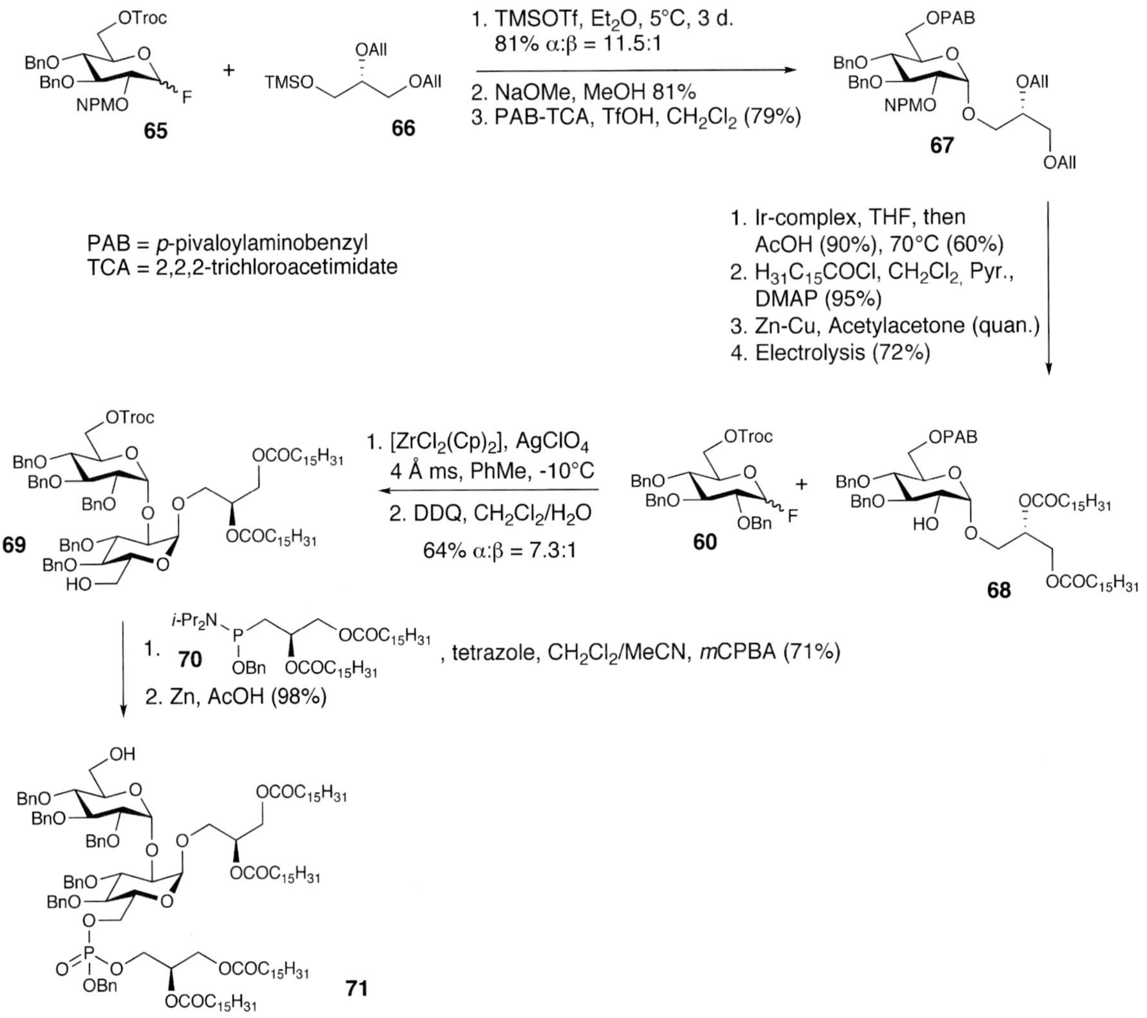

PAB = *p*-pivaloylaminobenzyl
TCA = 2,2,2-trichloroacetimidate

SCHEME 25.9 Synthesis of glycolipid part from *E. hirae*.

3.4. Biological test of the synthetic LTA fragments

Biological tests (Suda *et al.*, 1995; Takada *et al.*, 1996) of the synthetic LTA derivatives and their glycolipid parts revealed, however, that neither anti-tumour activity nor cytokine induction were observed. These observations led to doubts about the biological role of LTA and it was suggested that the biological activity observed from native LTA is mainly due to contaminants. The lack of activity was also explained by the structural differences between the synthetic and natural LTA; either by the missing substitution in the glycerol backbone, the saturation of the fatty acids or the significantly shorter polyglycerol phosphate chain in the synthetic derivative (four versus up to fifty units in the natural LTA).

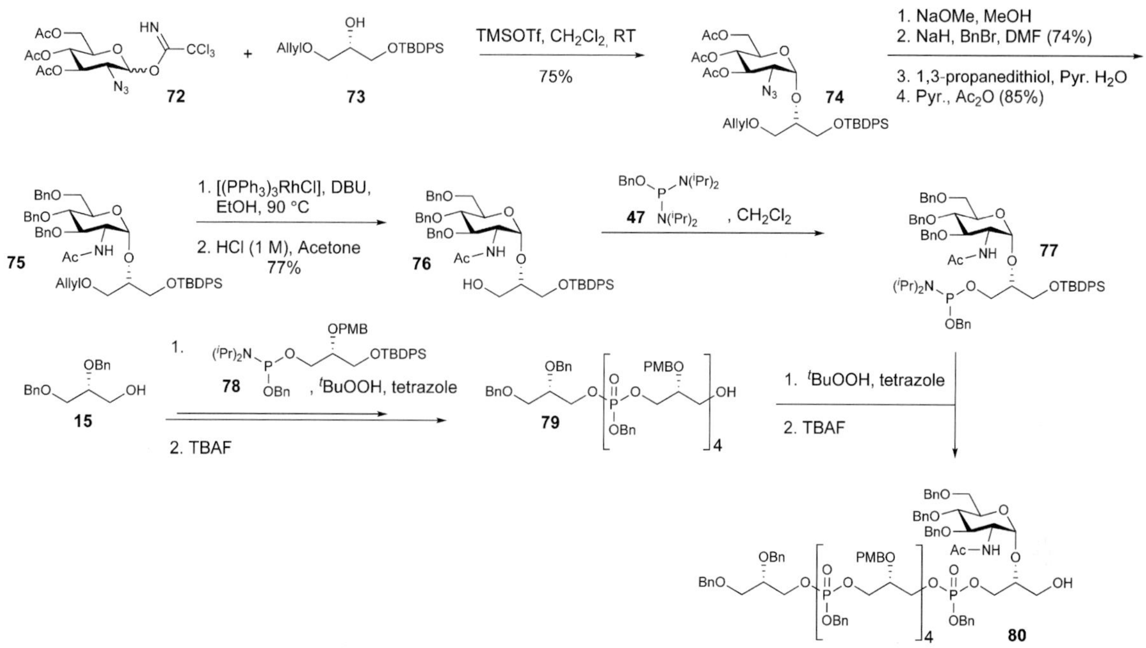

SCHEME 25.10 Preparation of the substituted glycerophosphate backbone building block.

4. SCHMIDT'S SYNTHESIS OF LTA TYPE I FROM *S. AUREUS*

From the structural analysis by biologists (Fischer and Koch, 1981; Morath *et al.*, 2001, 2002; Deininger *et al.*, 2007), the crucial role of the D-Ala residues at the polyglycerophosphate backbone was revealed and it was clear that, in order to study the biological activity of LTA from *S. aureus*, it was necessary to synthesize an LTA fragment containing the D-Ala residues.

In contrast to earlier syntheses of LTA derivatives, the introduction of side chains in the glycerophosphate backbone makes the synthesis particularly demanding, because the D-Ala residues are readily cleaved at pH 8.5 and thus have to be introduced very late in the synthesis.

The target compound **86** (see below and Scheme 25.12) was based on the knowledge about average substitution in the glycerophosphate backbone, where the D-Ala content is about 70%, α-D-GlcNAc 15% and hydrogen 15% (Morath *et al.*, 2001). This led to the compound **86** having a hexameric glycerophosphate backbone with four D-Ala units, one α-D-GlcNAc and one hydrogen. In a retrosynthetic perspective, it was obvious to disconnect between the glycolipid part and the teichoic acid part, as was the case in the previously discussed syntheses. The glycerophosphate backbone could then be disconnected into three building blocks; one terminal with a 2-*O*-benzyl protection (hydrogen in the final product), a *glycero* derivative having a 2-*O*-PMB group, which can be selectively deprotected and, finally, a unit having the α-D-GlcNAc substituent installed.

4.1. Synthesis

The versatile building block **73** (Scheme 25.10) was obtained from commercial 1,2-*O*-isopropylidene-*sn*-glycerol which, after 3-*O*-allylation, deprotection of the isopropylidene group and

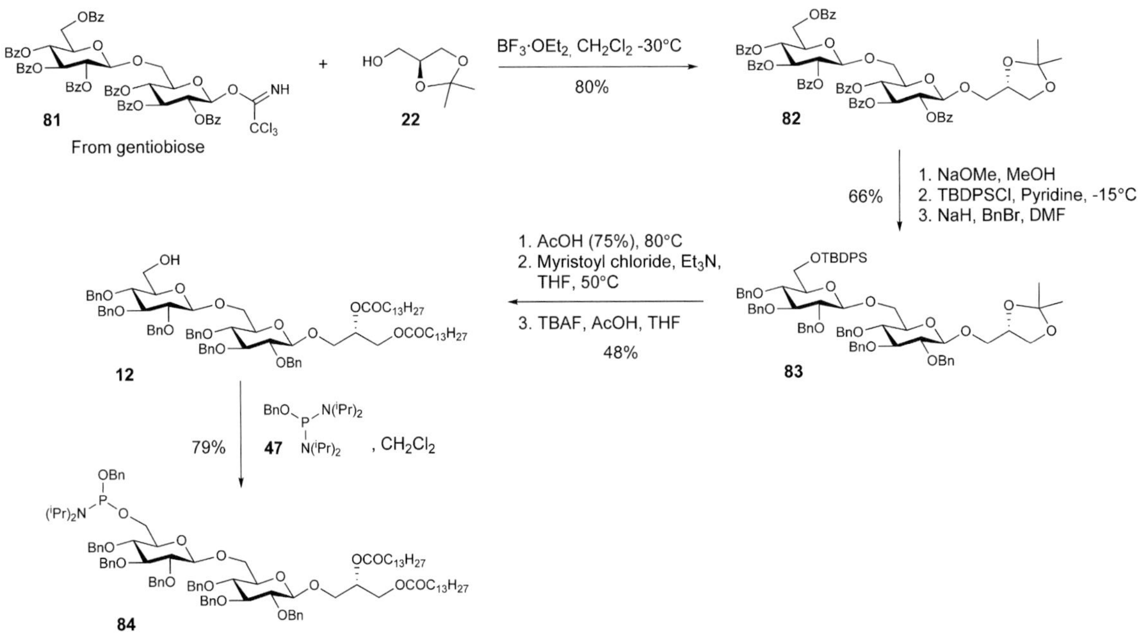

From gentiobiose

$BF_3 \cdot OEt_2$, CH_2Cl_2 -30°C, 80%

1. NaOMe, MeOH
2. TBDPSCl, Pyridine, -15°C
3. NaH, BnBr, DMF
66%

1. AcOH (75%), 80°C
2. Myristoyl chloride, Et_3N, THF, 50°C
3. TBAF, AcOH, THF
48%

47, CH_2Cl_2, 79%

SCHEME 25.11 Synthesis of the glycolipid part.

1-*O*-*tert*-butyldiphenylsilyl (TBDPS) protection, gave the intermediate **73**. This was either 2-*O*-PMB protected, deallylated and phosphitylated to give **78**, or glycosylated with the azidoglucosyl trichloroacetimidate donor **72** (Grundler and Schmidt, 1984) under TMSOTf catalysis, which led to the α-connected glucoside **74** in 75% yield, followed by deacetylation and introduction of benzyl groups. Azide reduction and *N*-acetylation gave the fully protected glycerol derivative **75** in 85% yield. Deallylation and then phosphitylation provided compound **77**. With the three building blocks in hand, the construction of **80** could begin. Thus, **15** and **78** were coupled under tetrazole catalysis and the addition of *tert*-butyl hydroperoxide afforded the phosphorus triester derivative which, after removal of the TBDPS protecting group mediated by TBAF in THF, and reaction with **78**, repeatedly led to the pentameric compound **79**. After desilylation, this intermediate was reacted with **77** to give the fully protected teichoic acid derivative, which after another TBAF-mediated desilylation, gave **80** in an overall very good yield.

The synthesis of the glycolipid part (Scheme 25.11) was effectively carried out by starting from gentiobiose, which was per-benzoylated, the anomeric oxygen de-*O*-benzoylated and the trichloroacetimidate group introduced to afford the glycosyl donor **81** which, upon reaction with 1,2-*O*-isopropylidene-*sn*-glycerol **22**, gave the fully protected glycoside **82** in overall 59% yield. The benzoyl protecting groups were removed under Zemplén conditions followed by selective protection of the 6′-OH group with TBDPS and benzylation of the remaining hydroxy groups to give **83**. The isopropylidene group was removed under acidic conditions followed by treating the diol with myristoyl chloride to give the protected glycolipid. The TBAF-mediated desilylation and phosphitylation of the 6′-OH group gave the desired compound **84**.

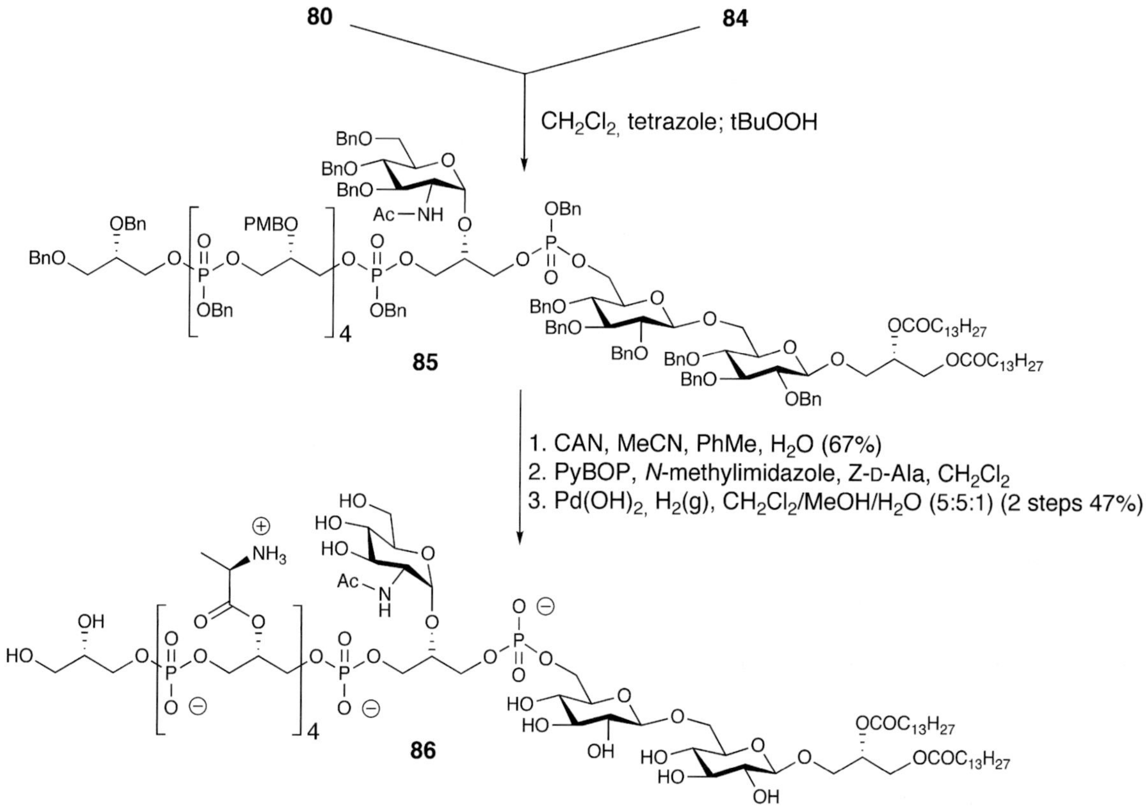

80 84
CH_2Cl_2, tetrazole; tBuOOH
85
1. CAN, MeCN, PhMe, H_2O (67%)
2. PyBOP, N-methylimidazole, Z-D-Ala, CH_2Cl_2
3. $Pd(OH)_2$, $H_2(g)$, $CH_2Cl_2/MeOH/H_2O$ (5:5:1) (2 steps 47%)
86

SCHEME 25.12 Synthesis of the LTA type I from *S. aureus* having a substituted glycerophosphate backbone.

The ligation of the two fragments **80** and **84** was performed with tetrazole as catalyst followed by oxidation with *tert*-butyl hydroperoxide to give the fully protected LTA fragment **85** (Scheme 25.12). Selective deprotection of the PMB-protecting groups by cerium ammonium nitrate (CAN) oxidation liberated four of the glycerol hydroxy groups, to which the Z-protected D-alanyl residues were attached using PyBOP/*N*-methylmorpholine as the coupling reagents. Hydrogenolysis using Pearlman's catalyst and purification using hydrophobic interaction (HI) chromatography afforded the final product **86** in 47% yield. Similarly the L-alanyl derivative was prepared starting from **83**.

4.2. Biological analysis of the synthetic LTA derivative

The evaluation of the biological activity of the LTA derivative **86** revealed that the D-alanyl residues have practically the same activity in terms of induction of cytokine release in human whole blood cells as the native LTA from *S. aureus*, whereas the L-alanyl derivative was about 10- to 100-fold less potent (Morath *et al.*, 2002; Deininger *et al.*, 2003). The glycolipid part alone has practically no activity. Taken together, the biological studies showed that LTA represents a Gram-positive endotoxin similar to LPS found in Gram-negative bacteria.

In order to find the minimum structural requirements, it was necessary to synthesize more derivatives containing the crucial D-alanyl substituents as well as the glycophospholipids anchor.

4.3. Synthesis of LTA derivatives – investigation of the structural requirements for biological activity

For elucidation of the structural requirements for biological activity, several derivatives of LTA of *S. aureus* were synthesized (Figure 25.4, **92–96**) lacking different parts of the fundamental LTA structure such as the phosphorylgentiobiose part (Figueroa-Perez *et al.*, 2005) or the α-D-GlcNAc substituent in the glycerolphosphate chain. In order to improve the stability of the D-alanyl residues, the amide-linked derivative **94** was prepared (Figueroa-Perez *et al.*, 2005). Due to the importance of the D-alanyl residues in the recognition of the LTA, it was hypothesized that having two diacylglycerol units **96**, one at each end of the molecule, would increase the biological activity by displaying the D-alanyl units better and hence a compound with these features was designed (Stadelmaier *et al.*, 2006). Furthermore, it was investigated how many D-alanyl groups are needed to induce enhanced biological activity; hence derivatives having different polyglycerolphosphate length were synthesized and analysed.

Due to the convergent synthesis strategy of LTA from *S. aureus*, some of the above mentioned derivatives could be obtained with a few modifications in the procedure. Variations in the polyglycerophosphate chain length were easily obtained by the number of repetitions in the stepwise generation of the chain. In this way, LTA with two, three and four D-alanyl units were synthesized.

The derivative **92**, lacking the gentiobiose part, was synthesized from the previously prepared hexameric polyglycerophosphate **80**, followed by reaction with the known phosphite derivative (Wickberg, 1958; Baeschlin *et al.*, 1998).

The amide derivative **94** demanded the preparation of a new building block as a substitute for the orthogonally protected glycerol derivative, starting from the D-Ser methyl ester **87**, which was condensed with Z-protected D-Ala **88**, mediated by HOBt, DCC and Hünig's base to give the dipeptide (Scheme 25.13). The primary hydroxy group was then protected with monomethoxytrityl (MMT) to give **89**, followed by reduction of the methyl ester and introduction of a TBDPS group to give **90** after acidic removal of the MMT protection. Phosphitylation and then sequential prolongation of the polymer could be carried out starting from the terminal di-*O*-benzylglycerol **15**. After tetrazole-mediated condensation and oxidation by *tert*-butyl hydroperoxide, the silyl protecting group was removed with TBAF and the next coupling cycle could take place. This was repeated five times to give the hexameric derivative which was finally coupled to the diacylglycerol unit to give the fully protected heptamer having a terminal acyl anchor. Global deprotection was carried out using Pearlman's catalyst and, after HI chromatography, **94** was obtained in 35% yield.

4.4. Biological activity of the derivatives

The biological studies (Morath *et al.*, 2005; Deininger *et al.*, 2007) of these derivatives gave some interesting insight into the minimal structural requirements for biological activity. As before, the induction of cytokine release by human blood leukocytes was measured.

The glycolipid part alone was not sufficient to induce cytokine release, however, when three or more unsubstituted glycerophosphate units were added the activity was observed. When D-alanyl substituents (minimum two, **93**) were present, the TLR-2 dependent activation was amplified by a factor of 10, but this was not the case, however, with L-alanyl substituents. Removal of the α-D-GlcNAc substituent did not influence the activity significantly (**92** versus **93**).

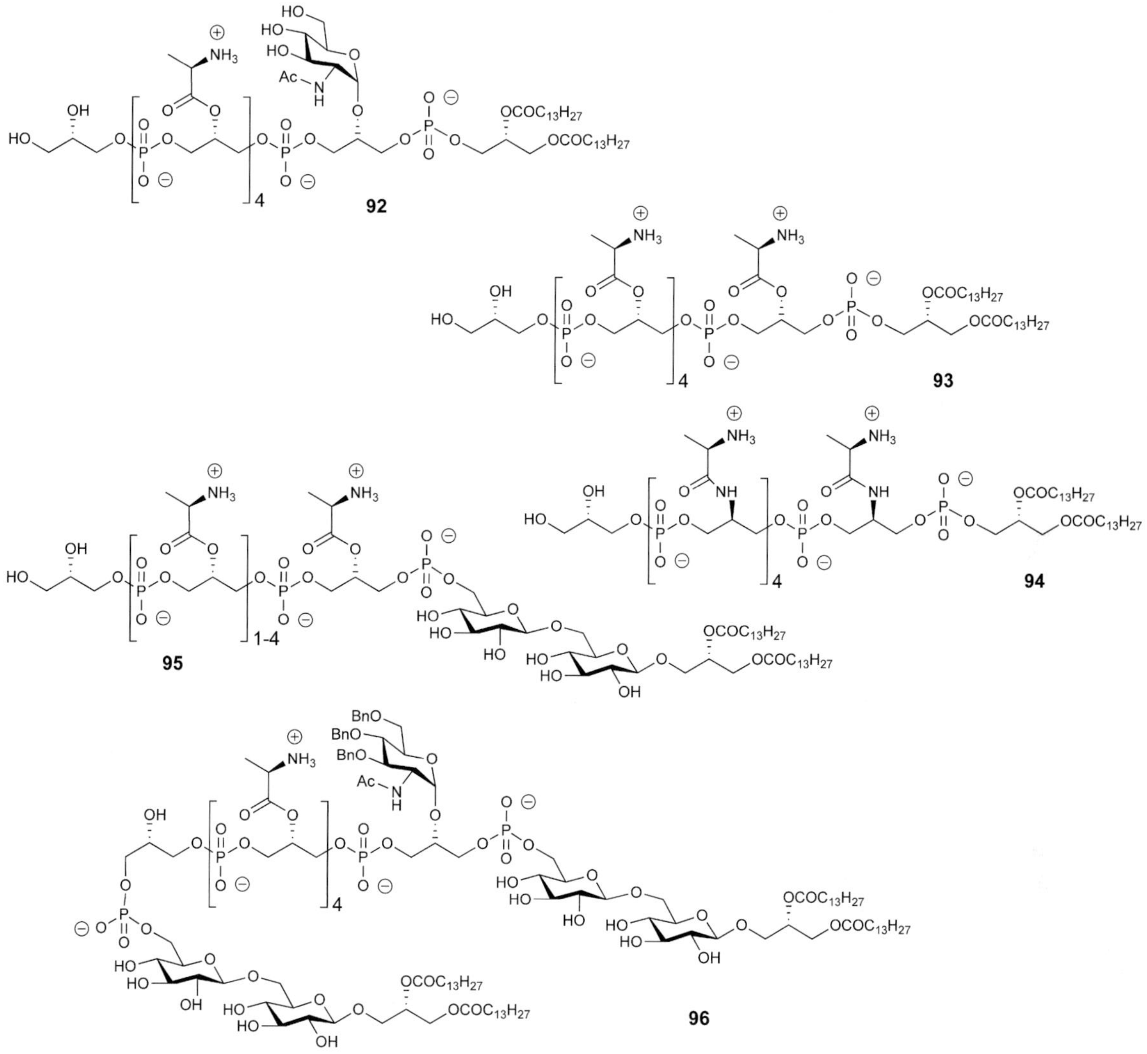

FIGURE 25.4 Derivatives of LTA from *S. aureus* prepared in order to determine the minimal structural requirements.

The derivative missing the gentiobiose part, but having the lipid anchor **93**, showed practically unchanged activity when the D-alanyl was connected to the backbone, having opposite stereochemistry, and with amide bonds **94** the activity remained – hence the activity is mainly dependent on the D-alanyl residues and the lipid anchor. This was further demonstrated by the bisamphiphilic LTA derivative **96** that displayed a 10-fold increase in cytokine release when compared with its monoamphiphilic counterpart **86**.

5. CONCLUSION

In this chapter, developments in the synthesis of LTA have been described from the most fundamental structures to more complicated

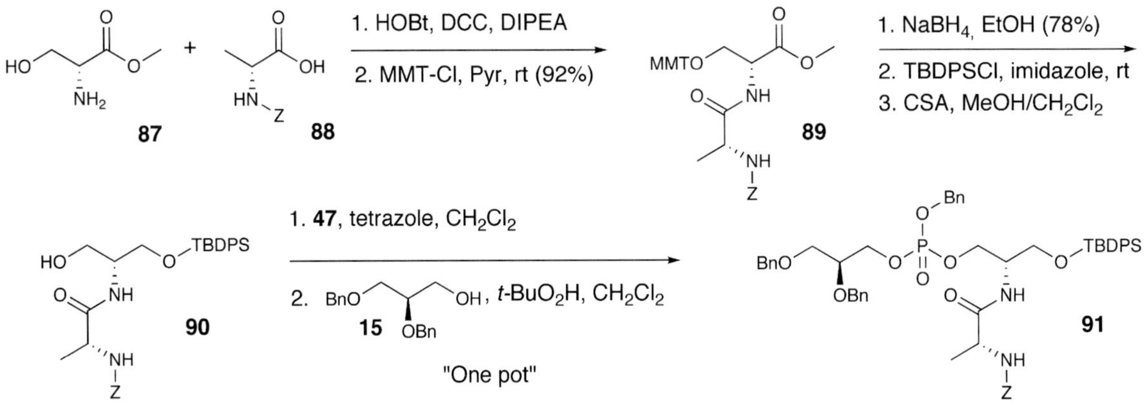

SCHEME 25.13 Synthesis of amide-linked D-alanine residues in the glycerolphosphate backbone.

derivatives. The early synthetic studies did not result in any biological activity but, with the increasing complexity and, hence, compounds closer to native LTA, it was demonstrated that synthetic LTA indeed has biological activity. The crucial role of the D-alanyl residues in *S. aureus* LTA was confirmed by the synthetic derivatives when compared with the pure natural product. Until now, the main focus has been on the synthesis of type I LTA, since it is the least complex form. It is of major interest to synthesize derivatives of the other three types of LTA, since the structural diversity between them is large and their biology only poorly understood. Studies towards the synthesis of *S. pneumoniae* LTA (Figueroa-Perez, 2007) as well as *Lactococcus garvieae* LTA are in progress by the Schmidt group. Further topics for studies on synthesis and related investigations are outlined in the Research Focus Box.

RESEARCH FOCUS BOX

- Improved preparation methods have provided biologically active type I LTA from *S. aureus* and D-alanine residues seem to play a crucial role in this biological activity. This needs confirmation using synthetic LTA.
- Structural modifications of type I LTA have enabled structure–activity relationship studies and should be explored further.
- Four different LTA types (types I–IV) have been assigned quite different structures. Very little is known of their biological properties, hence further synthetic work on these is required as it is difficult to isolate pure material from natural sources.
- The physiological role of LTA is still a matter of debate. With both structurally defined and designed LTA made available by synthesis, the role of LTA could be elucidated.

References

Baddiley, J., 1979. Polymers containing phosphorus in bacterial cell walls. In: Stec, W.J. (Ed.), Lectures Presented at the International Symposium on Phosphorus Chemistry Directed Towards Biology, Burzenin, Poland, 25–28 September 1979. Pergamon, Oxford, pp. 1–7.

Baeschlin, D.K., Chaperon, A.K., Charbonneau, V., et al., 1998. Rapid assembly of oligosaccharides: total synthesis of a glycosylphosphatidylinositol anchor of *Trypanosoma brucei*. Angew. Chem. Int. Ed. Engl. 37, 3423–3428.

Deininger, S., Stadelmaier, A., von Aulock, S., Morath, S., Schmidt, R.R., Hartung, T., 2003. Definition of structural prerequisites for lipoteichoic acid-inducible cytokine induction by synthetic derivatives. J. Immunol. 170, 4134–4138.

Deininger, S., Figueroa-Perez, I., Sigel, S., et al., 2007. Definition of the cytokine inducing minimal structure of lipoteichoic acid using synthetic derivatives. Clin. Vaccine Immunol. 14, 1629–1633.

Figueroa-Perez, I., 2007. Synthesis of fragments and one analogue of *S. aureus* LTA: investigations toward the synthesis of *S. pneumoniae* LTA. PhD thesis, University of Konstanz, Germany.

Figueroa-Perez, I., Stadelmaier, A., Morath, S., Hartung, T., Schmidt, R.R., 2005. Synthesis of structural variants of *Staphylococcus aureus* lipoteichoic acid (LTA). Tetrahedron Assym. 16, 493–506.

Figueroa-Perez, I., Stadelmaier, A., Deininger, S., von Aulock, S., Hartung, T., Schmidt, R.R., 2006. Synthesis of *Staphylococcus aureus* lipoteichoic acid derivatives for determining the minimal structural requirements for cytokine induction. Carbohydr. Res. 341, 2901–2911.

Fischer, W., 1981. Glycerophosphoglycolipids presumptive biosynthetic precursors of lipoteichoic acids. In: Shockman, G.D., Wicken, A.J. (Eds.), Chemistry and Biological Activities of Bacterial Surface Amphiphiles. Academic Press, New York, pp. 209–228.

Fischer, W., 1990. Bacterial phosphoglycolipids and lipoteichoic acids. In: Kates, M. (Ed.), Handbook of Lipid Research, Vol. 6. Plenum Press, New York, pp. 123–234.

Fischer, W., 1994a. Lipoteichoic acid and lipids in the membrane of *Staphylococcus aureus*. Med. Microbiol. Immunol. 183, 61–76.

Fischer, W., 1994b. Lipoteichoic acids and lipoglycans. In: Ghuysen, J.-M. (Ed.), Bacterial Cell Wall. Elsevier, Amsterdam, pp. 199–215.

Fischer, W., Koch, U., 1981. Alanine ester substitution and its effect on the biological properties of lipoteichoic acids. In: Shockman, G.D., Wicken, A.J. (Eds.), Chemistry and Biological Activities of Bacterial Surface Amphiphiles. Academic Press, New York, pp. 181–194.

Fischer, W., Koch, H.U., Rösel, P., Fiedler, F., Schmuk, L., 1980. Structural requirements of lipoteichoic acid carrier for recognition by the poly(ribitol phosphate) polymerase from *Staphylococcus aureus* H. A study of various lipoteichoic acids, derivatives, and related compounds. J. Biol. Chem. 225, 4457–4562.

Fukase, K., Tanaka, H., Torii, S., Kusumoto, S., 1990. 4-Nitrobenzyl group for protection of hydroxyl functions. Tetrahedron Lett. 31, 389–392.

Fukase, K., Matsumoto, T., Ito, N., Yoshimura, T., Kotani, S., Kusumoto, S., 1992. Synthetic study on lipoteichoic acid of gram positive bacteria. I. Synthesis of proposed fundamental structure of *Streptococcus pyogenes* lipoteichoic acid. Biological role of synthetic LTA. Bull. Chem. Soc. Jpn. 65, 2643–2654.

Fukase, K., Yoshimura, T., Kotani, S., Kusumoto, S., 1994. Synthetic study of lipoteichoic acid of gram positive bacteria. II. Synthesis of the proposed fundamental structure of *Enterococcus hirae* lipoteichoic acid. Bull. Chem. Soc. Jpn. 67, 473–482.

Greenberg, J.W., Fischer, W., Joiner, K.A., 1996. Influence of lipoteichoic acid structure on recognition by macrophage scavenger receptor. Infect. Immunol. 64, 3318–3325.

Grundler, G., Schmidt, R.R., 1984. Anwendung des Trichloracetimidat-Verfahrens auf 2-Azidoglucose- und 2-Azidogalactose-Derivate. Liebigs Ann. Chem., 1826–1847.

Hashimato, S., Hayashi, M., Noyori, R., 1984. Glycosylation using glucopyranosyl fluorides and silicon-based catalysts. Solvent dependency of the stereoselection. Tetrahedron Lett. 25, 1379–1382.

Lok, C.M., Ward, J-P., van Dorp, D.A., 1976. The synthesis of chiral glycerides starting from D- and L-serine. Chem. Phys. Lipids 16, 115–122.

Matsumoto, T., Maeta, H., Suzuki, K., 1988. New glycosidation reaction 1: combinational use of Cp_2ZrCl_2-$AgClO_4$ for activation of glycosyl fluorides and application to highly β-selective gylcosidation of D-mycinose. Tetrahedron Lett. 29, 3567–3570.

Morath, S., Geyer, A., Hartung, T., 2001. Structure-function relationship of cytokine induction by lipoteichoic acid from *Staphylococcus aureus*. J. Exp. Med. 193, 393–397.

Morath, S., Stadelmaier, A., Geyer, A., Schmidt, R.R., Hartung, T., 2002. Synthetic lipoteichoic acid from *Staphylococcus aureus* is a potent stimulus of cytokine release. J. Exp. Med. 195, 1635–1640.

Morath, S., von Aulock, S., Hartung, T., 2005. Structure/function relationships of lipoteichoic acids. J. Endotoxin Res. 11, 348–356.

Mukaiyama, T., Murai, Y., Shoda, S., 1981. An efficient method for glucosylation of hydroxy compounds using glucopyranosyl fluoride. Chem. Lett., 431–432.

Nashed, M.A., Anderson, L., 1982. Iodine as a reagent for the ready hydrolysis of prop-1-enyl glycosides, or their conversion into oxazolines. J. Chem. Soc. Chem. Commun. 18, 1274–1276, and references.

Oltvoort, J.J., van Boeckel, C.A.A., de Koning, J.H., van Boom, J.H., 1981. Use of the cationic iridium complex 1,5-cyclooctadiene-bis[methyldiphenylphosphine]-iridium hexafluorophosphate in carbohydrate chemistry: smooth isomerization of allyl ethers to 1-propenyl ethers. Synthesis, 305–308.

Oltvoort, J.J., van Boeckel, C.A.A., de Koning, J.H., van Boom, J.H., 1982. A simple approach to the synthesis of membrane teichoic acid fragment of *Staphylococcus aureus*. Recl. Trav. Chim. Pays-Bas 101, 87–91.

Oltvoort, J.J., Kloosterman, M., van Boeckel, C.A.A., van Boom, J.H., 1984. Synthesis of a lipoteichoic acid-carrier fragment of *Staphylococcus aureus*. Carbohydr. Res. 130, 147–163.

Reese, C.B., Tau, L., 1978. Reaction between 4-nitrobenzaldoximate ion and phosphotriesters. Tetrahedron Lett. 19, 4443–4446.

Reese, C.B., Titmas, R.C., Yau, L., 1978. Oximate ion promoted unblocking of oligonucleotide phosphotriester intermediates. Tetrahedron Lett. 19, 2727–2730.

Schmidt, R.R., Moering, U., Reichrath, M., 1980. Synthesis of α- and β-glycopyranosides via 1-*O*-alkylation. Tetrahedron Lett. 21, 3565–3568.

Smid, P., de Zwart, M., Jörning, W.P.A., van der Marel, G.A., van Boom, J.H., 1993. Stereoselective synthesis of a tetrameric fragment of *Streptococcus pneumoniae* type 1 containing an α-linked 2-acetamido-4-amino-2,4,6-trideoxy-d-galactopyranose (SUG*p*) unit. J. Carbohydr. Chem. 12, 1073–1090.

Stadelmaier, A., Morath, S., Hartung, T., Schmidt, R.R., 2003. Synthesis of the first fully active lipoteichoic acid. Angew. Chem. Int. Ed. Engl. 42, 916–920.

Stadelmaier, A., Figueroa-Perez, I., Deininger, S., von Aulock, S., Hartung, T., Schmidt, R.R., 2006. A *Staphy-lococcus aureus* lipoteichoic acid (LTA) derived structural variant with two diacylglycerol residues. Bioorg. Med. Chem. 14, 6239–6254.

Suda, Y., Tochio, H., Kawano, K., et al., 1995. Cytokine-inducing glycolipids in the lipoteichoic acid fraction from *Enterococcus Lirae* ATCC 9790. FEMS Immunol. Microbiol 12, 97–112.

Takada, H., Kawabata, Y., Arakaki, R., et al., 1996. Molecular and structural requirements of a lipoteichoic acid from *Enterococcus hirae* ATCC 9790 for cytokine-inducing, antitumor, and antigenic activities. Infect. Immun. 63, 57–65.

Tsutsui, O., Kokeguchib, S., Matsumura, T., Kato, K., 1991. Relationship of the chemical structure and immunobiological activities of lipoteichoic acid from *Streptococcus faecalis (Enterococcus hirae)* ATCC 9790. FEMS Microbiol. Lett. 76, 211–218.

van Boeckel, C.A.A., van Boom, J.H., 1985. Synthesis of *Steptococci* phosphatidyl-α-diglycosyldiglyceride and related glycolipids. Application of the tetraisopropyldisiloxane-1,3-diyl (TIPS) protecting group in sugar chemistry. Part V. Tetrahedron 41, 4567–4575.

Veeneman, G.H., Gomes, L.J.F., van Boom, J.H., 1989. Synthesis of fragments of a *Streptococcus pneumoniae* type-specific capsular polysaccharide. Tetrahedron 45, 7433–7448.

Westphal, O., Lüderitz, O., Keiderling, W., 1952. Effects of bacterial toxins; biochemical analysis of inflammation. Zentralbl. Bakteriol. Parasitenkd. Infectionskr. Hyg. 158, 152–160.

Wickberg, B., 1958. Synthesis of l-glyceritol-D-galactopyranosides. Acta Chem. Scand. 12, 1187–1201.

CHAPTER

26

Chemical synthesis of parasitic glycoconjugates and phosphoglycans

Nawaf Al-Maharik, Jennifer A. Tee and Andrei V. Nikolaev

SUMMARY

Glycosylinositolphospholipids and glycosylphosphatidylinositol-anchored glycoproteins (mucins) and phosphoglycans are abundant and form a dense protective layer, i.e. a glycocalyx, on the surface of various protozoan parasites. They are crucial for the life cycle of the parasites and are involved in the formation of the parasites resistance barrier against hostile forces. The diversity within the parasitic glycoconjugate structures and the scope of their functions, from host cell invasion to the deception of the immune system of the host, is astonishing. This comprehensive review highlights the progress and recent achievements in the chemical synthesis of the glycoconjugate structures like glycosylphosphatidylinositol anchors and phosphoglycans of the protozoan parasites *Trypanosoma brucei, Trypanosoma cruzi, Toxoplasma gondii, Plasmodium falciparum* and *Leishmania*.

Keywords: Protozoan parasites; Carbohydrates; Glycoconjugates; Glycolipids; Phosphoglycans; Synthesis; *Trypanosoma; Toxoplasma; Plasmodium; Leishmania*

1. INTRODUCTION

The surface of many protozoan parasites is covered with glycoconjugates, glycosylinositolphospholipids (GIPLs) and glycosylphosphatidylinositol- (GPI-) anchored glycoproteins (mucins), as well as GPI-anchored lipophosphoglycans (LPGs) (McConville and Ferguson, 1993). The tremendous diversity of glycoconjugates has implicated their critical importance in the life cycle of these organisms. The survival strategies of many protozoan parasites (*Leishmania, Trypanosoma*) involve the formation of an elaborate and dense cell-surface glycocalyx composed of diverse stage-specific glycoconjugates that form a protective barrier (Ferguson, 1997, 1999). The GIPLs form a dense layer on the surface of the parasite with the mucins and LPGs projecting out and upwards from this layer (Almeida and Gazzinelli, 2001).

Historically, the first structural identification of a GPI anchor was the one of *Trypanosoma brucei* variant surface glycoproteins (VSGs) (Ferguson *et al.*, 1988) followed by the structure of a GPI anchor of rat brain Thy-1 glycoprotein (Homans *et al.*, 1988). Since then, about 50 GPI anchors have been structurally characterized, all have a common core structure of Man-α-($1\rightarrow4$)-$GlcNH_2$-α-($1\rightarrow6$)-*myo*-Ino-1-OPO_3-lipid ($GlcNH_2$, D-glucosamine; *myo*-Ino, *myo*-inositol; Man, D-mannose) (Ferguson and Williams, 1988). Two classes of GPI anchors have been identified (Figure 26.1). Type I anchors have the above core structure and are usually found in protozoan parasites. These GPIs attach non-protein-bound glycoconjugates, such as GIPLs and LPGs, onto parasite cell membranes. Type II anchors contain the expanded core structure, $H_2NC_2H_4OPO_3$-6Man-α-(1→2)-Man-α-(1→6)-Man-α-(1→4)-GlcNH$_2$-α-(1→6)-*myo*-Ino-1-OPO_3-lipid, and are found in mammals and lower eukaryotes including protozoan parasites and yeasts. The type II GPIs attach proteins or glycoproteins through their C-termini to the ethanolamine-phosphate group at the non-reducing end of the glycan core.

Many of type II GPIs diverge from their basic core structure and contain one or more species-specific side chains (McConville and Ralton, 1997) linked to specific positions of the core shown in Figure 26.1. The side chains can be comprised of mono- or oligosaccharides, an extra ethanolamine-phosphate residues (specific for higher eukaryotes), 2-aminoethylphosphonate (specific for *Trypanosoma cruzi* GPIs) or an additional fatty acid residue. Another modification site is the lipid moiety where various structures, e.g. diacylglycerol, acylalkylglycerol, *lyso*-alkylglycerol and ceramide, can be found.

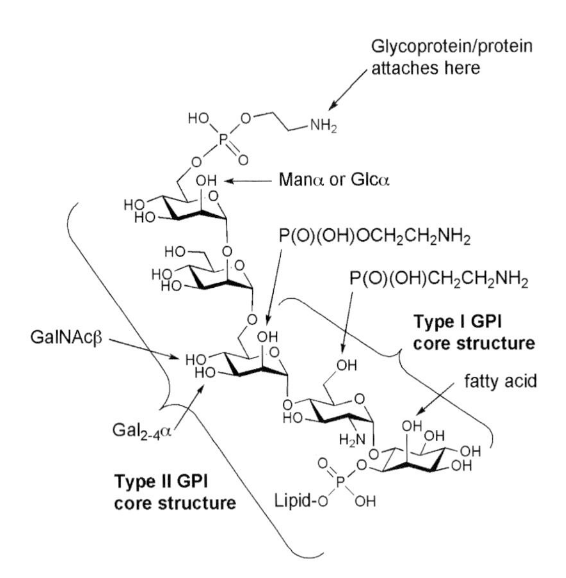

FIGURE 26.1 Structure of glycosylphosphatidylinositol (GPI) anchors. Abbreviations: Galα, α-linked galactose; GalNAcβ, β-linked *N*-acetylgalactosamine; Glcα, α-linked glucose; Manα, α-linked mannose.

The function of GPI anchors (in addition to the clear one of linking the above biopolymers to membranes) has been extensively discussed (Varma and Mayor, 1998; Ferguson, 1999; McConville and Menon, 2000) and there is also evidence that GPIs and/or metabolites of them can act as secondary messengers, modulating biological events including insulin production, insulin-mediated signal transduction, cellular proliferation and cell–cell recognition. The discovered role as mediators of regulatory processes makes the chemical preparation of the compounds and their analogues of great interest.

Among the most complex of the GPI-related structures is the LPG from *Leishmania* parasites, which is a predominant cell-surface glycoconjugate of *Leishmania* promastigotes. It consists of -6)-Gal-β-(1→4)-Man-α-(1→OPO_3 repeating units (Gal, D-galactose) (Figure 26.2), where the chain length and the nature of the X and Y branching substituents varies in a species-, strain- and stage-specific manner (McConville and Ferguson, 1993). All the LPG molecules are capped with a mannobiosyl phosphate unit at the non-reducing terminus and contain a glycan core linked to a type I GPI anchor at the reducing end of the chain. In addition to LPG molecules, *Leishmania* promastigotes produce a number of secreted glycoproteins (like filamentous proteophosphoglycans forming a viscous gel surrounding the cells and *Leishmania donovani* specific secreted acid phosphatase), where a peptide core is glycosylated with LPG-type phosphoglycans at serine/threonine domains, as well as hydrophilic phosphoglycans, which are essentially the corresponding LPG molecules without the glycan core-GPI region. *Leishmania mexicana* amastigotes synthesize large amounts of a macromolecular amastigote specific proteophosphoglycan and secrete it into the phagolysosome of the mammalian macrophage. There are several recent reviews about the structure and biological functions of *Leishmania* phosphoglycans α-D-Man*p*-(1→2)-α-D-Man*p*-(1-PO_3H-[-6]-β-D-Gal*p*-(1→4)-α-D-Man*p*-(1-PO_3H-]$_n$

(X attached at position 3 of Gal*p*; Y attached at position 2 of Man*p*)

L. donovani promastigote,	X = Y = H (100%)
L. mexicana promastigote,	X = β-D-Glc*p*-1- (25%), Y = H (100%)
L. major procyclic promastigote,	X = β-D-Gal*p*-1- (52%),
β-D-Gal*p*-(1→3)-β-D-Gal*p*-1- (25%)	β-D-Ara*p*-(1→2)-β-D-Gal*p*-1- (9%); Y = H (100%)
L.major metacyclic promastigote,	X = β-D-Gal*p*-1- (31%),
β-D-Gal*p*-(1→3)-β-D-Gal*p*-1- (6%)	β-D-Ara*p*-(1→2)-β-D-Gal*p*-1- (45%); Y = H (100%)
L. aethiopica promastigote,	X = β-D-Gal*p*-(1→3)-β-D-Gal*p*-1-or β-D-Gal*p*-1-;
	Y = α-D-Man*p*-1- (35%)

FIGURE 26.2 Structure of the phosphoglycan region of lipo- and proteo-phosphoglycans of *Leishmania*. Abbreviations: Ara*p*, arabinopyranose; Gal*p*, galactopyranose; Glc*p*, glucopyranose; Man*p*, mannopyranose.

(Ilg, 2000; Guha-Niyogi *et al.*, 2001; Turco *et al.*, 2001; Späth *et al.*, 2003).

In this chapter, we discuss various methodologies and specific features for the chemical preparation of parasitic GPI compounds and parasitic phosphoglycans. To date, two reviews have been published describing the chemical synthesis of GPI-related derivatives (Gigg and Gigg, 1997; Guo and Bishop, 2004) and two review papers have dealt with the synthesis of phosphoglycan structures (Hansson and Oscarson, 2000; Nikolaev *et al.*, 2007).

2. CHEMICAL SYNTHESIS OF PARASITIC GLYCOCONJUGATES (GPI ANCHORS)

GPIs are among the most complex classes of natural products as they combine lipids, carbohydrates, *myo*-inositol and phosphate groups. Their structural complexity and recently discovered biological importance have inspired widespread chemical interest and a number of synthetic approaches towards GPIs (*T. brucei*, *T. cruzi*, *Plasmodium falciparum* and *Leishmania*) have been reported. Several problems face GPI synthesis, such as synthesis of protected and optically pure *myo*-inositol derivatives, the stereoselective construction of the glycan and the regioselective introduction of the side chains and phosphate moieties.

Positioning of side groups complicates the protecting group strategy. Positions on the glycan core have to be differentiated using orthogonal protecting groups that enable regioselective inclusion of the side groups (such as phosphoethanolamine and phospholipid) at later stages in the synthesis. Branching sugar chains also complicate the protecting group strategy of the main glycan backbone assembly and add the further problem of their stereoselective introduction. This can be affected by the nature of protecting groups on both the glycosyl donor and glycosyl acceptor and by the steric hindrance caused by other sugar residues or side groups in the oligosaccharide. If the right glycosidic linkage (α or β) is not obtained, this can result in a complete change to the synthetic strategy. In general, the stereochemistry of the glycosylation can be resolved by ingenious use of protecting groups as well as by varying glycosylation methods and conditions (e.g. experimentation with different promoters and solvents).

GPI structures are generally synthesized by linear or convergent means. The linear approach can be used to build the oligosaccharide from individual monosaccharides in a stepwise manner. This strategy relies on the selective activation of a monoglycosyl donor over the growing oligosaccharide chain. The convergent (or blockwise) approach constructs the oligosaccharide from smaller building blocks, which results in a fewer number of protecting group manipulations within the oligosaccharide chain.

Synthesis of the glycan core usually proceeds with the synthesis of an optically pure differentially protected *myo*-inositol derivative followed by the addition of appropriately protected α-linked $GlcNH_2$. The next stage involves the formation of the remaining glycan core either by linear or convergent means. The glycan core is then decorated with the ethanolamine-phosphate and phospholipid moieties before the global deprotection to provide the final GPI structure.

2.1. Syntheses of a GPI anchor of *T. brucei*

2.1.1. *The RIKEN group approach*

The first synthesis of a glycosylphosphatidylinositol was achieved by the RIKEN group (Murakata and Ogawa, 1991, 1992), who published the synthess of a GPI anchor of *T. brucei* VSG. As indicated in Scheme 26.1, the synthetic plan was to assemble the glycan core **2** and then introduce phosphoethanolamine first followed by diacylglycerol phosphate through the H-phosphonate precursors **4** and **3**, correspondingly. Glycan core **2** was further disconnected into four smaller building blocks **5–8**. This synthesis of a GPI anchor did not plan the synthesis of the pseudo-disaccharide (i.e. azidoglucose-inositol) block separately. Instead the *myo*-inositol acceptor **12** was glycosylated with the disaccharide glycosyl fluoride **16** to form a pseudo-trisaccharide **5** (Scheme 26.2).

The *myo*-inositol acceptor **12** (see Scheme 26.2) was constructed starting from the previously synthesized racemic dicyclohexylidene derivative **9** (Garegg *et al.*, 1984). The latter was subjected to dibutyltin oxide-assisted regioselective 6-*O*-methoxybenzylation, 1-*O*-allylation, followed by acidic removal of cyclohexylidene protecting groups and per-*O*-benzylation of the hydroxyl groups to afford racemate **10**. This was de-*O*-allylated and treated with (1*S*)-(-)-camphanic acid chloride to afford the corresponding diastereomeric mixture of (1*S*)-(-)-camphanyl-D,L-*myo*-inositols.

Thus, enantiomeric resolution of *myo*-inositol was achieved by addition of a chiral side group (1*S*)-(-)-camphanate (Murakata and Ogawa, 1990) and then separation of the diastereomeric mixture by silica gel chromatography. Conversion of the chiral *myo*-inositol derivative **11** to the required glycosyl acceptor **12** was readily achieved in 83% yield over five steps (de-*O*-methoxybenzylation, *O*-vinylation, hydrolysis of the camphanate, introduction of a *p*-methoxybenzyl (PMB) group at the 1-OH group and then cleavage of the vinyl ether with acetic acid).

The glycobiosyl fluoride **16** was constructed from the methyl thiomannoside donor **15** (Ogawa and Sasajima, 1981) and azidoglucose acceptor **14** as described in Scheme 26.2. Azidoglucose acceptor **14** was obtained in 61% overall yield via sequential 4,6-*O*-benzylidenation of **13**, (Kinzy and Schmidt, 1985) 3-*O*-benzylation and finally regioselective cleavage of the benzylidene group. Copper(II) bromide-Bu_4NBr-AgOTf-promoted glycosylation of **14** with methyl thiomannoside donor **15**, followed by removal of the anomeric silyl group and conversion of the hemiacetal intermediate to glycosyl fluoride gave the glycobiosyl fluoride

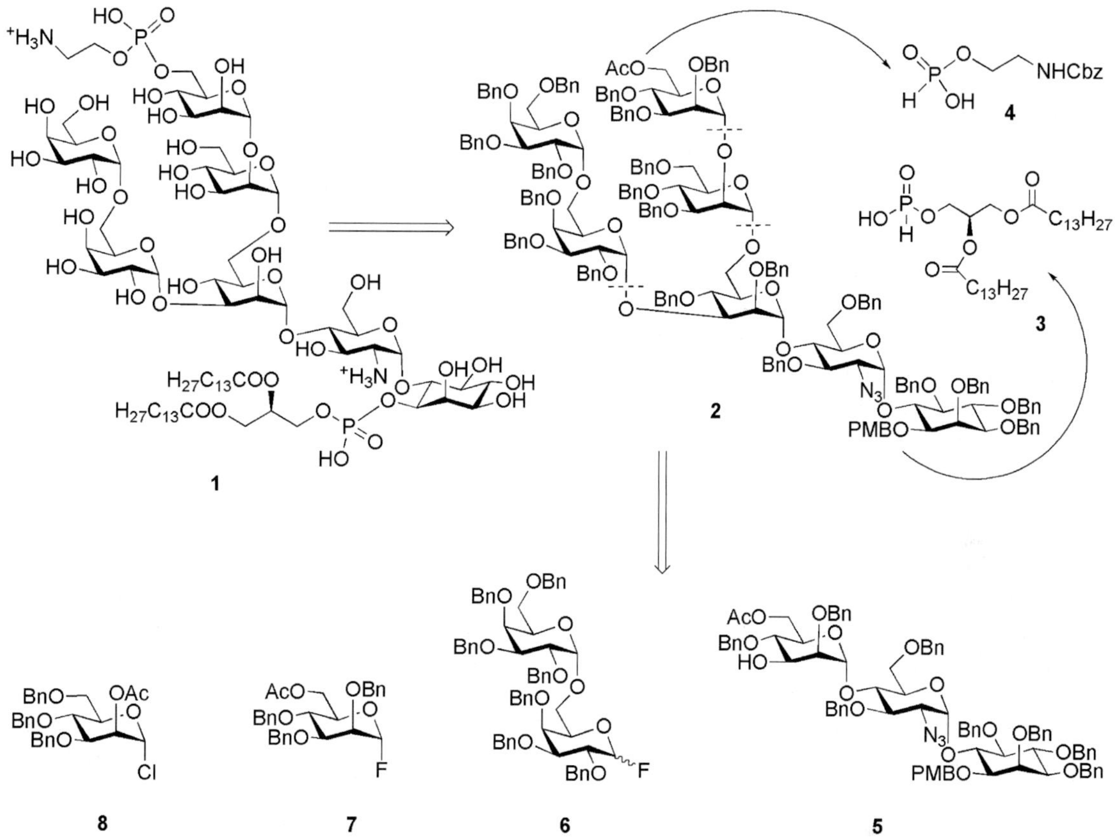

SCHEME 26.1 Retrosynthesis of a GPI anchor of *T. brucei* according to the strategy of the RIKEN group.

donor **16** (89%; α,β ratio 2:3). Cp_2ZrCl_2-$AgClO_4$ affected coupling of the glycosyl fluoride **16** with the *myo*-inositol acceptor **12** furnished the corresponding α-linked pseudo-trisaccharide in 73% yield (plus 20% of the β-anomer), in which the two acetyl groups were removed and the 6-hydroxyl of Man was selectively acetylated to afford pseudo-trisaccharide **5** in 95% yield.

The galactobiose synthon **6** was constructed starting from peracetylated Gal **17** as depicted in Scheme 26.3. Glycosylation of *p*-methoxyphenol under the influence of trimethylsilyl trifluoromethanesulfonate (TMSOTf) with the donor **17**, followed by sequential de-*O*-acetylation, selective 6-*O*-tritylation with 4,4′-dimethoxytrityl chloride (DMTCl), followed by *O*-benzylation and finally removal of the DMT group afforded the desired acceptor **18** in a 51% overall yield. This was glycosylated with the methyl thiogalactoside **19** (Koike *et al.*, 1987) in presence of copper(II) bromide-Bu_4NBr to afford the corresponding α-(1→6)-linked galactobiose, which was further subjected to ammonium cerium(IV) nitrate- (CAN-) mediated cleavage of *p*-methoxyphenyl group followed by conversion of the hemiacetal intermediate to the glycosyl fluoride donor **6** in 62% yield (α,β ratio 2:3).

The tricky coupling between the galactobiose **6** and the pseudo-trisaccharide **5** was achieved under the influence of Cp_2ZrCl_2-$AgClO_4$ in 76%

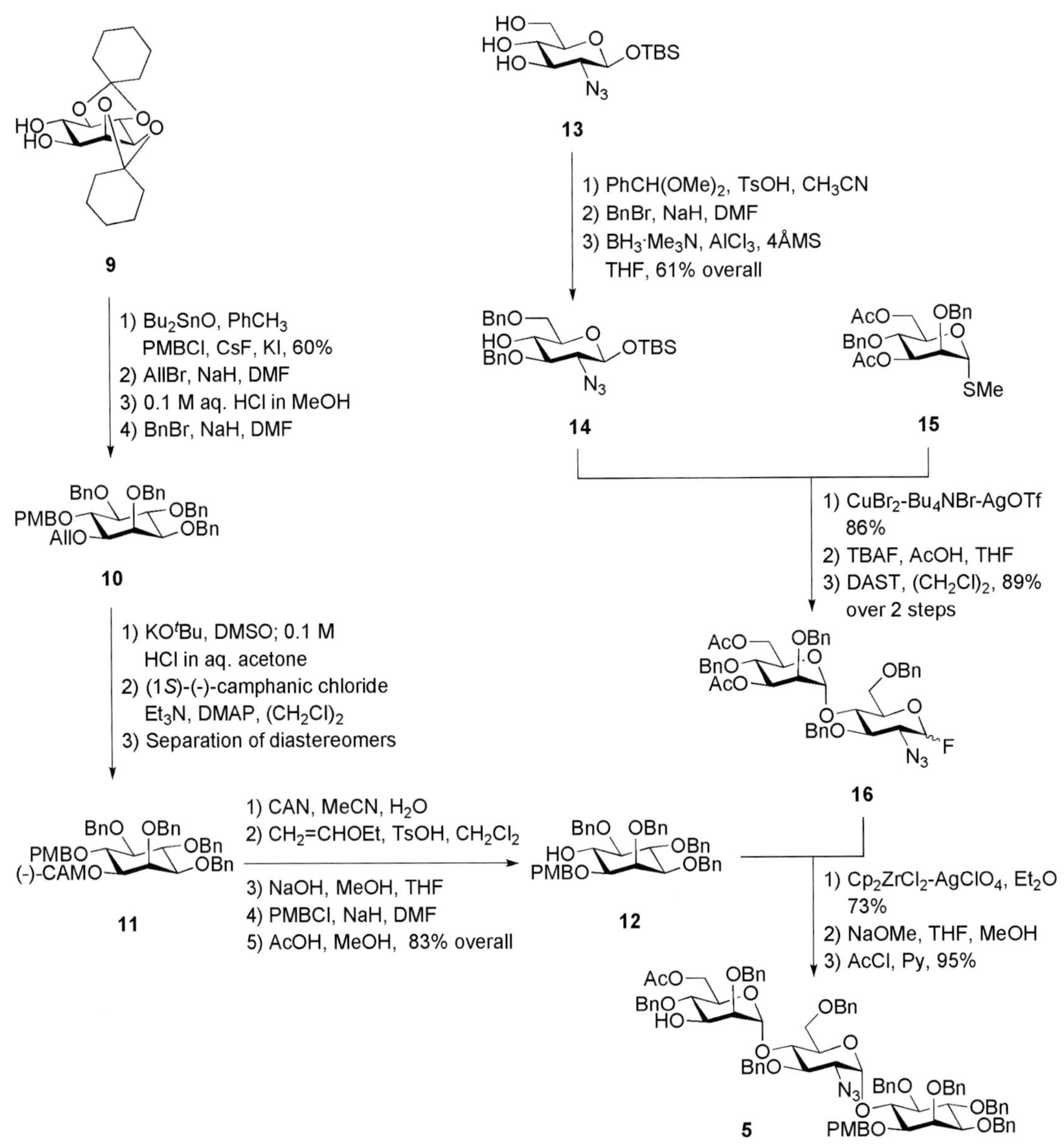

SCHEME 26.2 *myo*-Inositol enantiomeric resolution and the synthesis of a pseudo-trisaccharide according to the procedure of the RIKEN group.

yield forming preferentially the α-linked product (α:β = 9:1). After the removal of the acetyl group, the required α-linked pseudo-pentasaccharide **20** was isolated. The latter was glycosylated with the mannosyl chloride **8** in the presence of a mixture of mercury(II) bromide and mercury(II) cyanide as promoter to offer the corresponding branched pseudo-hexasaccharide in 89%

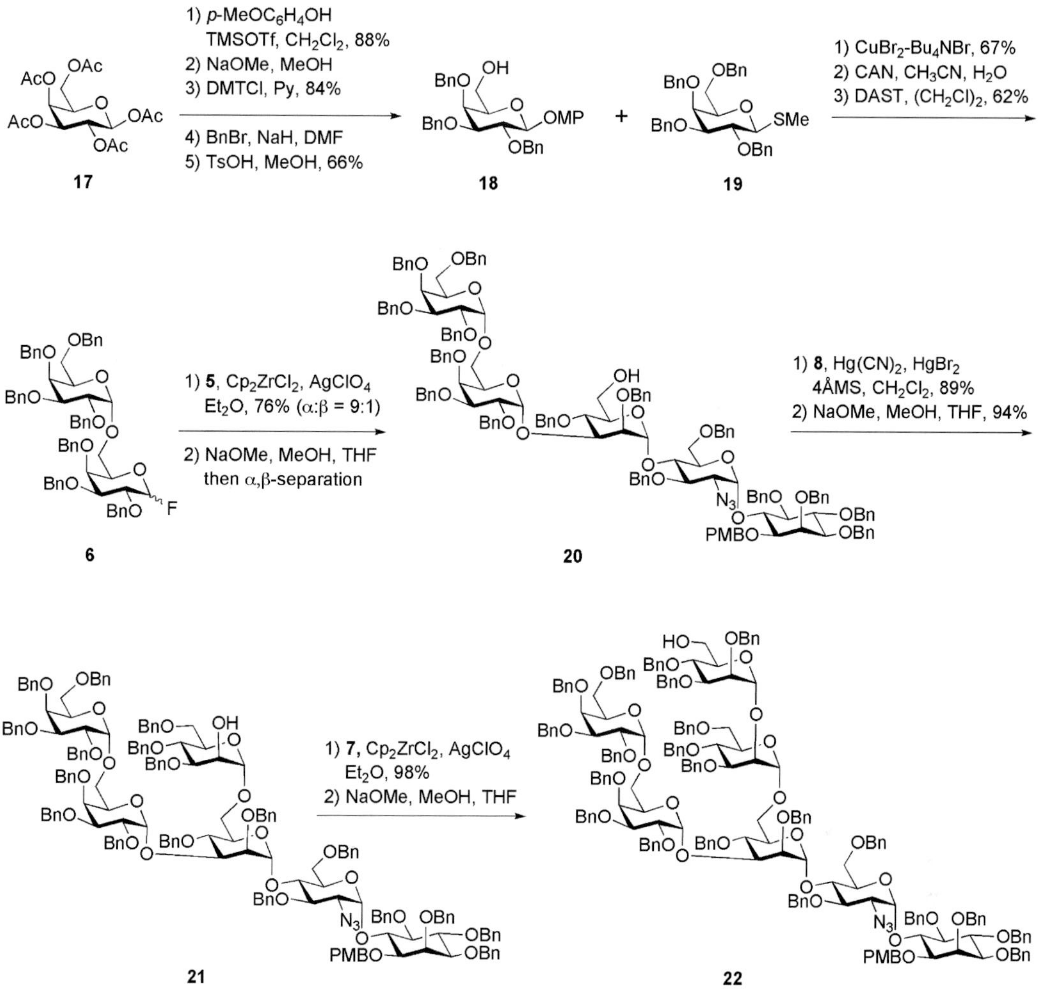

SCHEME 26.3 *T. brucei* GPI branched glycan core assembly after the strategy of the RIKEN group.

yield, in which the acetyl group was removed to give the acceptor **21**. The branched pseudo-heptasaccharide **22** was then obtained by glycosylation of **21** with mannosyl fluoride **7** using the Cp_2ZrCl_2-$AgClO_4$ mixture as promoter followed by removal of the acetyl group.

Due to the problems associated with the CAN-assisted removal of the PMB group from C-1 of the *myo*-inositol unit after the introduction of the ethanolamine-phosphate at C-6 in Man-3, the synthesis was completed (Scheme 26.4) initially by introduction of the phospholipid and then addition of the ethanolamine-phosphate moiety. 6-*O*-Chloroacetylation in Man-3, then TMSOTf-mediated removal of the methoxybenzyl group at C-1 of *myo*-inositol followed by pivaloyl chloride-assisted condensation with the H-phosphonate **3** afforded the H-phosphonate **23**. The chloroacetyl group of the latter compound was cleaved with thiouria to yield **24**. This was then phosphitylated with *N*-Cbz-protected aminoethyl H-phosphonate **4** to give the corresponding pseudo-heptasaccharide (40%; as a mixture of four diastereomers at P atoms), in which the two H-phosphonate moieties were then oxidized with iodine in aqueous (aq.) pyridine to give the fully protected GPI **25** in a 68% yield. Finally, global deprotection by hydrogenolysis in the presence of $Pd(OH)_2/C$ in $CHCl_3$-MeOH-H_2O solution gave the first synthetic GPI anchor in 23% yield.

It was proved later (Cottaz *et al.*, 1993; Ogawa, 1994; Gigg and Gigg, 1997) that the above synthetic study was based on the wrong stereoisomer of *myo*-inositol derivative **11** (see Scheme 26.2), which was chosen by the RIKEN group erroneously and thus provided the wrong stereoisomer of the glycosyl acceptor **12**. Because of this misassignment of the absolute configuration of the *myo*-inositol derivatives used, the final deprotected synthetic GPI anchor appeared to be a stereoisomer (in

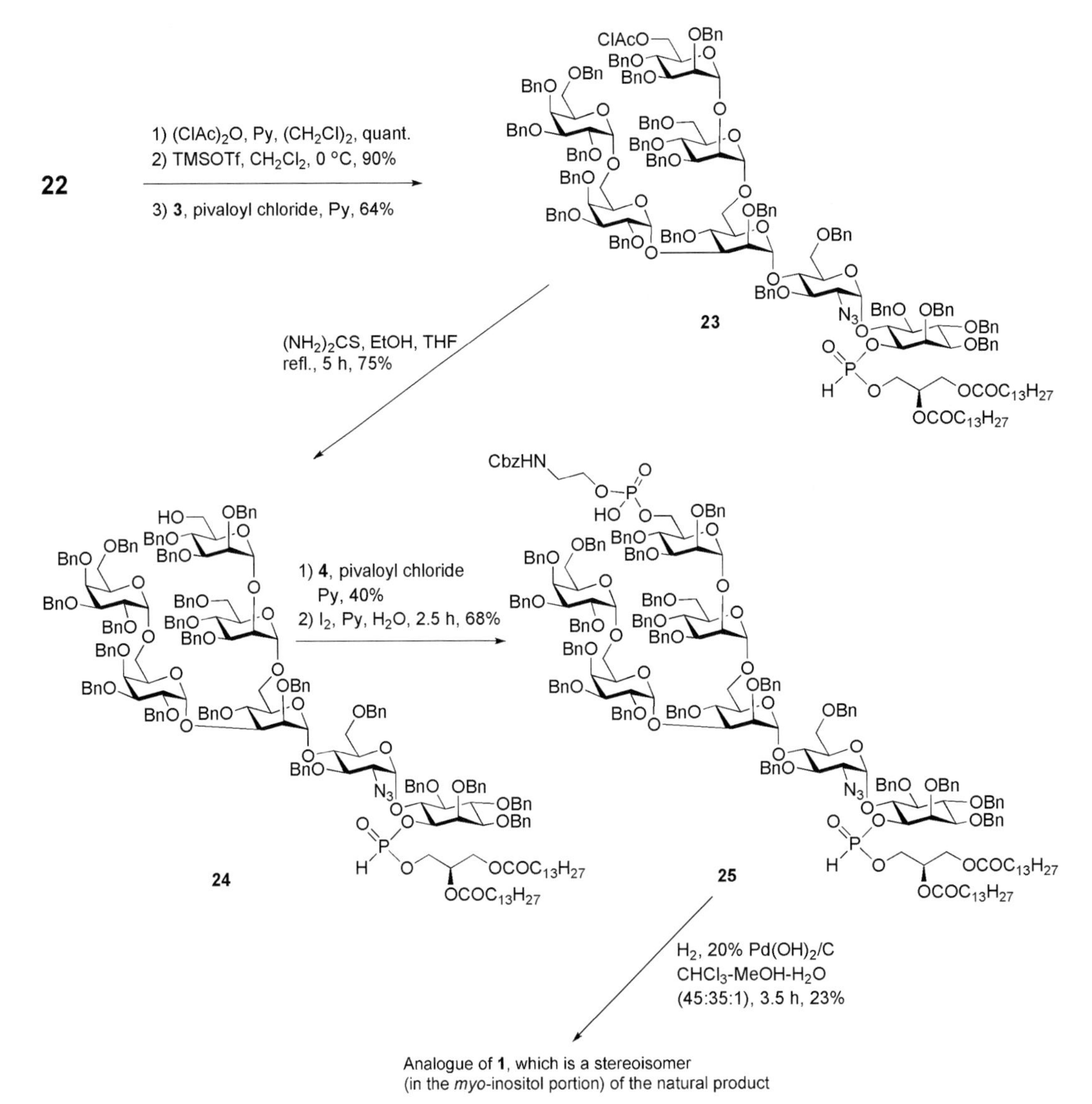

SCHEME 26.4 Successive phosphorylation steps and global deprotection according to the strategy of the RIKEN group.

the *myo*-inositol portion) of the natural *T. brucei* GPI anchor **1**. In summary, the first total synthesis of a GPI anchor was achieved employing glycosyl fluoride, glycosyl chloride and thioglycoside derivatives as glycosyl donors and benzyl groups as permanent protecting group, as well as the use of the H-phosphonate chemistry to introduce two different phosphodiester linkages.

2.1.2. The Cambridge group approach

The Cambridge group's (Ley and co-workers) synthesis of the GPI anchor **1** (Baeschlin *et al.*, 1998, 2000a) appeared to be the first chemical preparation of a real GPI anchor of *T. brucei* VSG. The synthetic strategy relied on the use of:

(i) butanediacetal (BDA) groups, chloroacetic esters and benzyl ethers as permanent protecting groups;
(ii) selenoglycosides and thioglycosides as glycosyl donors that allowed as few manipulations on the mounting oligosaccharide core as possible; and
(iii) a chiral bis(dihydropyran) for enantiomeric resolution of *myo*-inositol.

Ley brought together the 1,2-di-*O*-acetal protecting group studies from his laboratory to try to simplify the puzzle of putting together an oligosaccharide chain. As indicated in Scheme 26.5, the two phosphate substituents in the GPI anchor **1** were retrosynthetically disconnected as phosphoramidites **27** and **28**, leaving corresponding protected glycan core **26** for further simplification. The group followed a convergent route for assembling the pseudo-heptasaccharide **26**. Protecting groups may exhibit distinctive influence on the reactivity of glycosyl donors. The Cambridge group successfully used this effect for rapid assembly of the glycal core **26** with just one protecting group manipulation needed. The core was assembled in just six steps from the six building blocks **29–34**, taking advantage from the reactivity tuning effects caused by protecting groups and the use of appropriate anomeric leaving groups.

The anomeric PhSe group in compounds **31–34** bears higher reactivity towards electrophilic activation than their EtS counterpart in **30** due to the greater polarizability of the selenium atom. The selenoglycosides **33** and **34** are more reactive than the selenoglycosides **31** and **32** due to the nature of "arming–disarming" effects of the protecting groups. The *O*-benzyl "arming" groups in **33** and **34** enhance the reactivity of the donors, whereas the butanediacetal (BDA) and *O*-chloroacetyl in acceptors **31** and **32** exhibit deactivating ("disarming") effects on their anomeric leaving group.

De-symmetrization or chiral enantiomeric resolution of *myo*-inositol molecule remains a problem despite significant efforts invested towards establishing methods to obtain enantiomerically pure *myo*-inositol derivatives. The Cambridge group reported a route to both chiral D-*myo*-inositols and L-*myo*-inositols employing a chiral bis(dihydropyran). This method was illustrated by the synthesis of D-*myo*-inositol derivative **43** (Scheme 26.6). Regioselective transformation of *myo*-inositols **35** to the bis (dispoke) acetal **36** was easily achieved on treatment with butane-2,3-dione (Riley *et al.*, 1998). The side product **37** was removed by simple re-crystallization. It is important to note that butane-2,3-dione forms 1,2-*trans*-diacetal derivatives (i.e. involving di-equatorial 1,2-diols), contrasting with the isopropylidene and benzylidene acetal formation, when 1,2-*cis*-acetals (i.e. involving axial/equatorial 1,2-diols) are favoured products. Butane 2,3-dione initially adds to the most reactive hydroxyl group in **35** (i.e. the equatorial hydroxyl groups vicinal to the axial hydroxyl-2) and then cyclizes with a vicinal equatorial OH to form a six-membered ring. *O*-Silylation of the 1,6:3,4-bis(diacetal) **36** with *tert*-butyldiphenylsilyl (TBDPS) chloride, followed by removal of the BDA groups with aq. trifluoroacetic acid (TFA), gave tetraol **38**, which in turn was desymmetrized with chiral bis(dihydropyran) **39** yielding **40** as a single diastereoisomer in 81% yield. This product was subsequently *O*-desilylated, per-*O*-benzylated

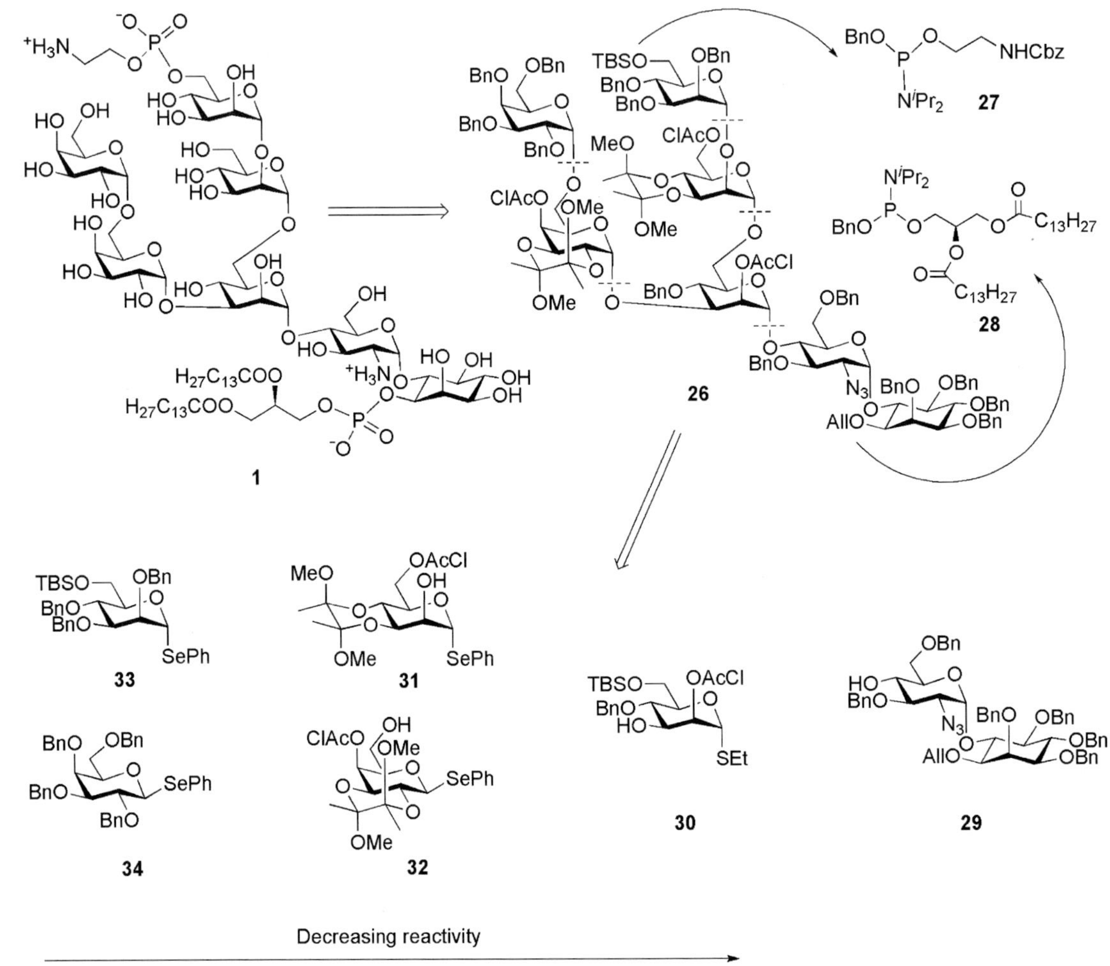

SCHEME 26.5 Retrosynthesis of a GPI anchor of *T. brucei* according to the procedure of the Cambridge group.

and oxidized to form bis(phenylsulfone) **41**. Exposure of **41** to lithium hexamethyldisilazide removed the dispiroacetal furnishing the chiral diol **42**, which then was selectively allylated on the C-1 hydroxyl group via the dibutyltin acetal formation to offer the desired *myo*-inositol acceptor **43**.

Building block **47** (Scheme 26.7) required for the construction of pseudo-disacchride **29** was synthesized from the known phthalimide **44** (Nicolaou *et al.*, 1990). *O*-Benzylation, reductive benzylidene acetal ring opening, followed by hydrazine-assisted cleavage of the phthalimido moiety transform **44** into the amine **45**. Further transformation into the 2-azide **46** was accomplished (98%) employing triflic azide (TfN_3) in presence of 4-(dimethylamine)-pyridine (DMAP). Initial *O*-silylation of the alcohol **46** with *tert*-butyldimethylsilyl chloride (TBSCl), followed by bromination gave glycosyl donor **47**, which was coupled with *myo*-inositol acceptor **43** using Lemieux's halide inversion protocol (Lemieux *et al.*, 1975). This afforded only the desired α-linked pseudo-disaccharide

SCHEME 26.6 *myo*-inositol chiral de-symmetrization after the strategy of the Cambridge group.

(65%) which, on de-*O*-silylation, furnished the pseudo-disaccharide acceptor **29**.

The readily available phenyl 1-selenogalactoside **48** (Zuurmond *et al.*, 1993; Mallet *et al.*, 1994) was converted into two required building blocks **32** and **34** employing standard chemistry (Scheme 26.8). Consecutive 2,3-*O*-protection with butane-2,3-dione (→**49**), 6-*O*-silylation with TBSCl, *O*-chloroacylation (→**50**) and de silylation gave the acceptor **32**. Conventional *O*-benzylation of the same tetraol **48** afforded the galactoside donor **34**. The Man building blocks **31** and **33** were synthesized from phenyl 1-selenomannosides **51** (Grice *et al.*, 1997). The BDA protection of the 3,4-diol, followed by 6-*O*-chloroacetylation via the 6-tributyltin ether formation produced the acceptor **31**. Compound **33** was prepared by selective 6-*O*-silylation of **51**, followed by per-*O*-benzylation.

Synthesis of the central thiomannoside **30** (Scheme 26.9) required chloroacetate group at 2-position to allow further anchimeric assistance and differentiation of 3- and 6-OH groups for regioselective glycosylation. The tetraol **52** was silylated at 6-OH with TBDPSCl, the 2,3-diol was isopropylidenated and the 4-OH group was benzylated. Successive desilylation, acidic removal of the acetonide and silylation at 6-position with TBSCl gave the 2,3-diol **53** in 65% overall yield. Selective silylation of the equatorial 3-hydroxyl

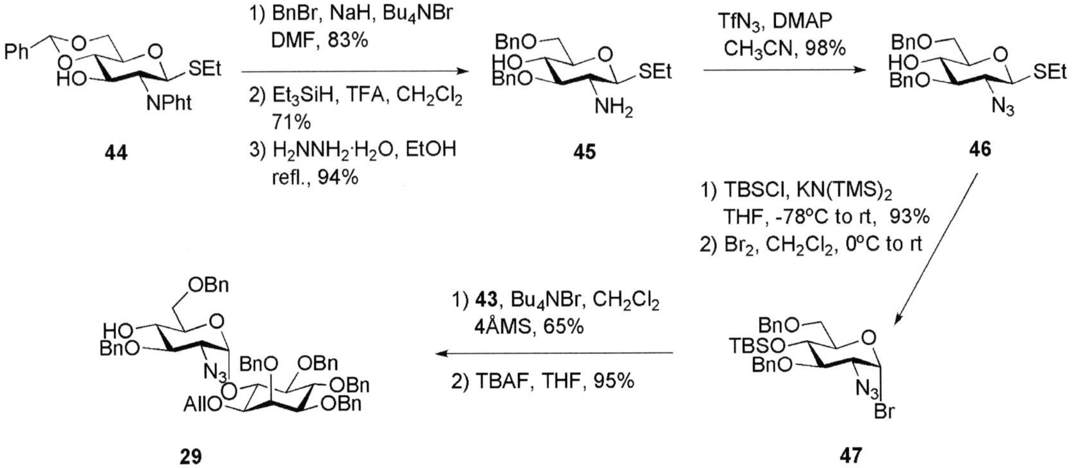

SCHEME 26.7 The synthesis of a pseudo-disaccharide according to the strategy of the Cambridge group.

group with trimethylsilyl chloride (TMSCl) followed by standard introduction of the chloroacetyl group at 2-position and de-3-*O*-silylation furnished the desired middle building block **30**.

With all the building blocks in hand, the protected glycan core **26** was then assembled (Scheme 26.10). The assembly began with the stereoselective coupling of the fully benzylated donor **34** with the acceptor **32** under the influence of *N*-iodosuccinimide (NIS) and TMSOTf to afford the targeted α-linked disaccharide fragment **54** (71%). The combined deactivating effects of the BDA and the chloroacetate groups in **32** are crucial in avoiding any homo-coupling. The thiomannoside acceptor **30** was then 3-*O*-glycosylated with the galactobioside donor **54** under the influence of methyl triflate (MeOTf) (76%) followed by de-6-*O*-silylation to furnish the requisite trisaccharide alcohol **55**. Glycosylation of the acceptor **55** with excessive mannobioside selenophenyl donor **56** (5 eq.), prepared by coupling (87%) of **33** and **31** in a similar fashion to the disaccharide **54** formation, under the influence of MeOTf, furnished the branched pentasaccharide **57** (75%). Finally, coupling of the pseudo-disaccharide acceptor **20** and the pentasaccharide donor **57** was promoted with NIS/TfOH to offer the branched pseudo-heptasaccharide core **26** (50%).

The glycan core **26** was further elaborated to the protected GPI anchor **59** (Scheme 26.11) by making use of phosphoramidite chemistry, which has been successfully applied earlier in syntheses of GPIs of yeast (Mayer *et al.*, 1994; Mayer and Schmidt, 1999) and rat brain Thy-1 (Campbell and Fraser-Reid, 1995). Removal of TBS group at 6-position in the Man-3 moiety in **26** with aq. HF enabled the introduction of the ethanolamine-phosphate moiety via 1*H*-tetrazole-assisted phosphitylation with phosphoramidite **27** and subsequent oxidation of the formed phosphite triester with *m*-chloroperbenzoic acid (*m*-CPBA), afforded the phosphodester **58** (89%). This was subjected to de-*O*-allylation with $PdCl_2$ and 1*H*-tetrazole-assisted phospholipidation with diacylglyceryl phosphoramidite **28**, followed by oxidation with *m*-CPBA to furnish the fully protected GPI anchor **59** (81%) as a mixture of four diastereoisomers at the P atoms. The global deprotection sequence involving, first, palladium-catalysed hydrogenolysis to remove *O*-benzyl and the *N*-Cbz groups and to reduce the azide to an amino group, followed by de-*O*-chloroacetylation and

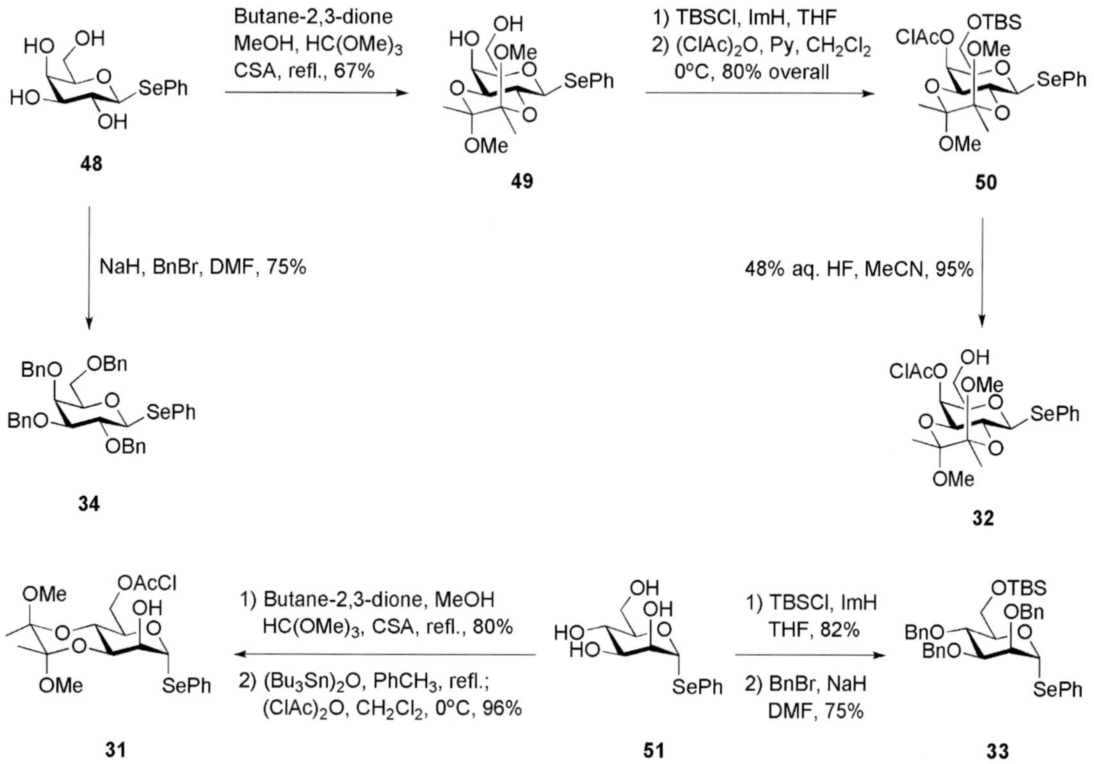

SCHEME 26.8 The synthesis of D-galactose and D-mannose building blocks after the strategy of the Cambridge group.

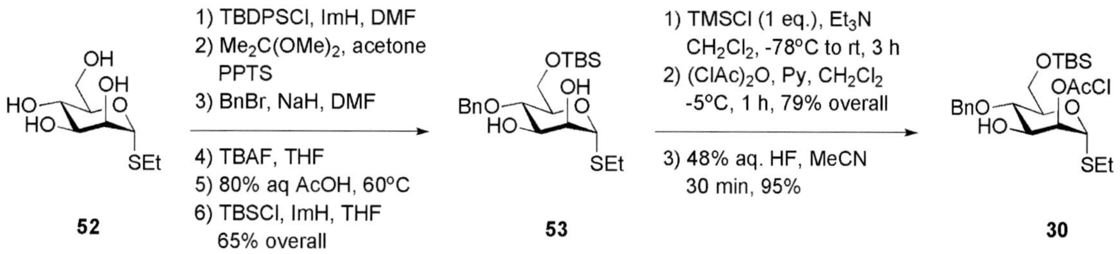

SCHEME 26.9 The synthesis of a central D-mannose building block according to the procedure of the Cambridge group.

TFA-assisted deacetalization, gave the required GPI anchor **1** in 90% overall yield.

To summarize, the Cambridge group developed an efficient and convergent synthetic strategy of GPI anchor **1**, which is adaptable to the assembly of other GPI anchors. The strategy is based on (3 + 2 + 2) building blocks for construction of glycan core, use of BDA and chloroacetate protecting groups to tune the reactivity of leaving groups in the glycosyl donors and the use of bis(dihydropyran) to desymmetrize *myo*-inositol.

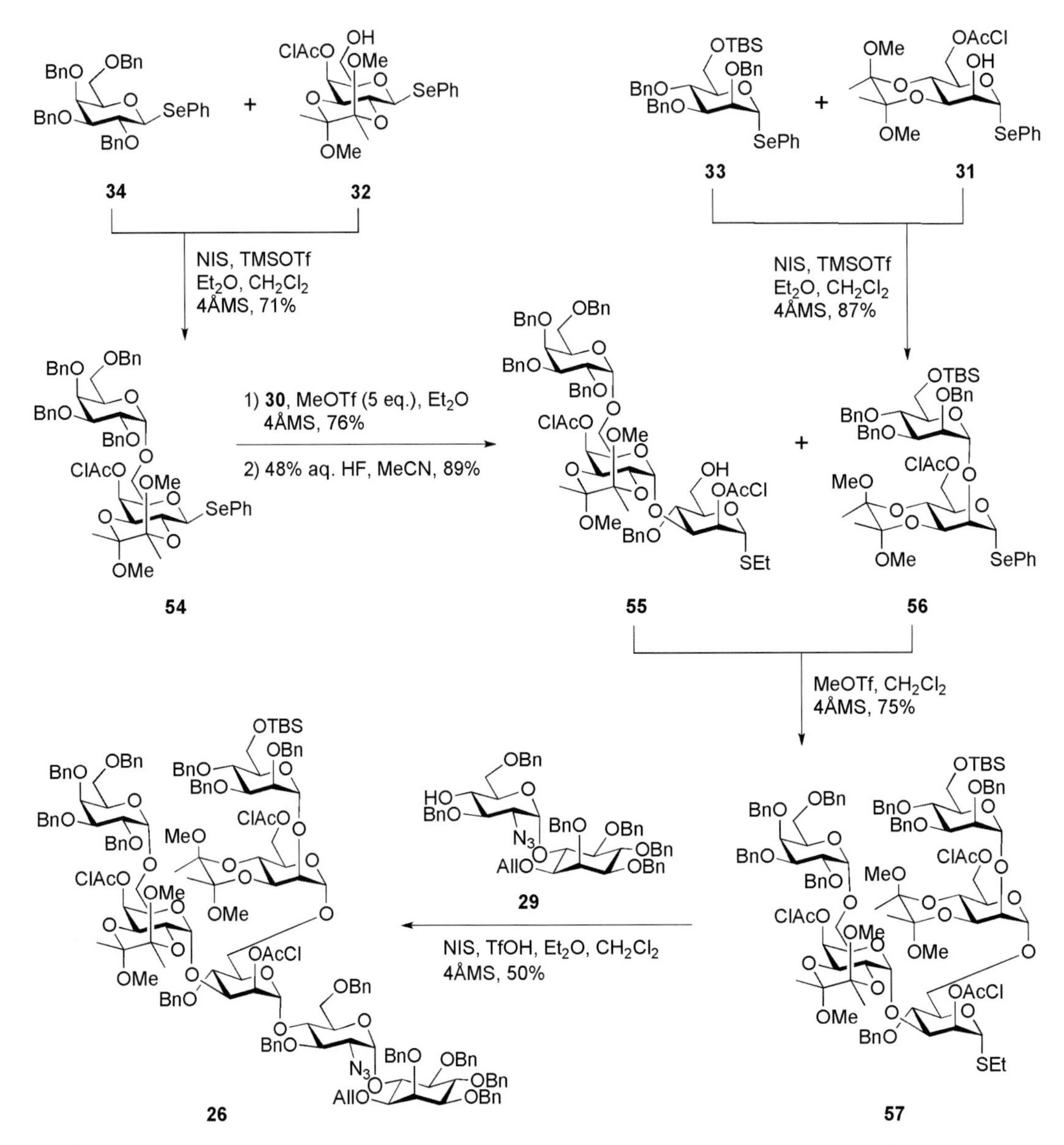

SCHEME 26.10 *T. brucei* GPI branched glycan core assembly according to the strategy of the Cambridge group.

2.2. Synthesis of a glycoconjugate of *Toxoplasma gondii*

T. gondii is a ubiquitous protozoan parasite causing congenital infection and severe and often lethal encephalitis in the course of the acquired immunodeficiency syndrome (AIDS). An oligosaccharide-containing small antigen has been illustrated to express immunological characteristics suitable for serological diagnosis of acute toxoplasmosis (Striepen *et al.*, 1997). The structure of this antigen was identified to be a family

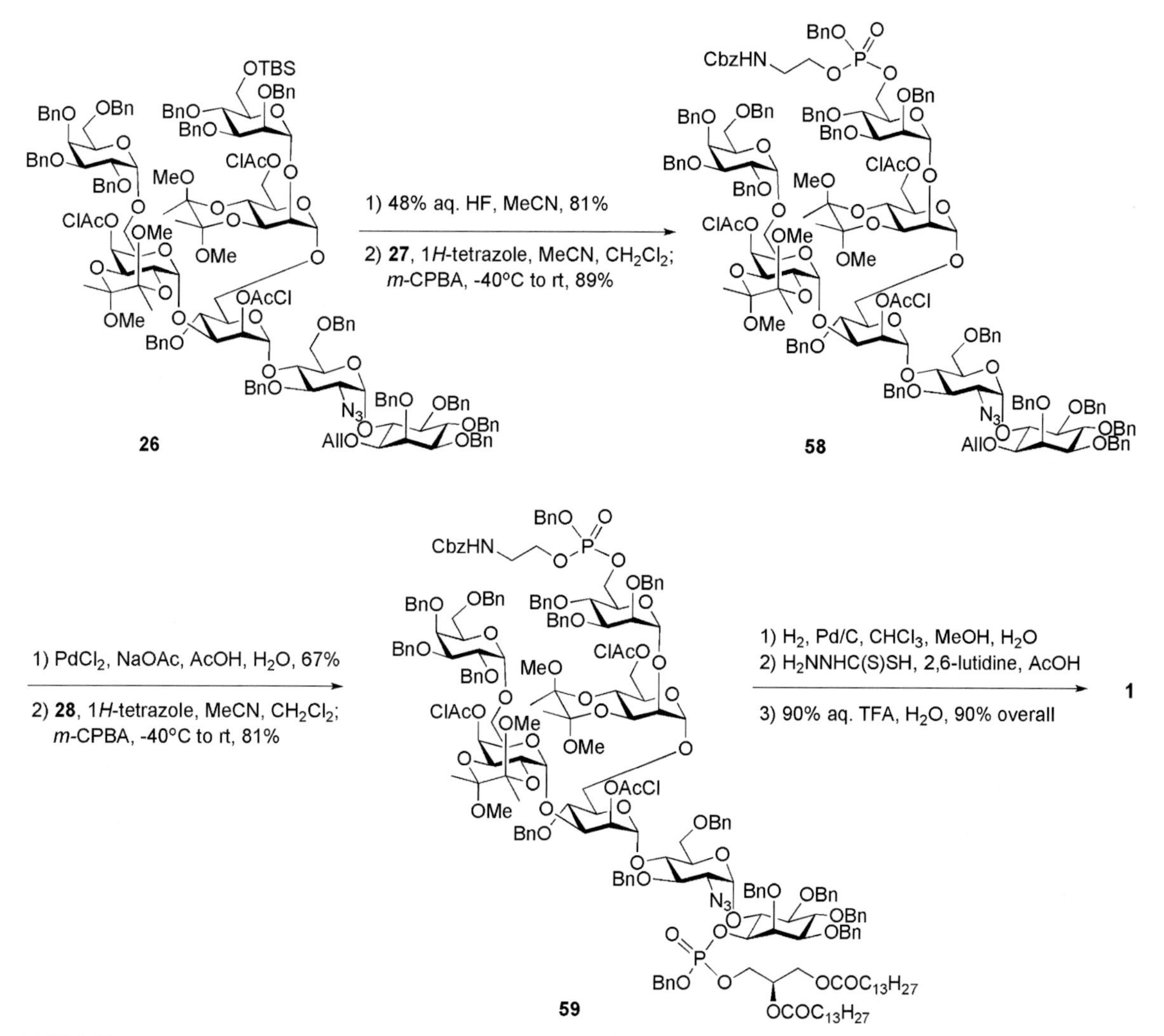

SCHEME 26.11 Successive phosphorylation steps and global deprotection after the strategy of the Cambridge group.

of protein-free GPI glycolipids from which the structures of two core glycans (A and B, not shown here) were elucidated. Immunological studies revealed that only type B GPIs containing Glc-α-(1→4)-GalNAc side chain (Glc, D-glucose; GalNAc, *N*-acetyl-D-galactosamine) linked to the first Man moiety were recognized by sera from infected humans, suggesting that the unique Glc modification is essential for immunogenicity (Striepen *et al.*, 1999). In order to understand this phenomenon, synthesis of the type A GPI glycan of *T. gondii* (containing a GalNAc side chain instead; compound **60**, Scheme 26.12) was required.

A convergent and versatile strategy for the synthesis of phosphorylated GPI anchor pseudohexasaccharide **60** was developed in R. Schmidt's laboratory (Pekari *et al.*, 2001). A strategy that

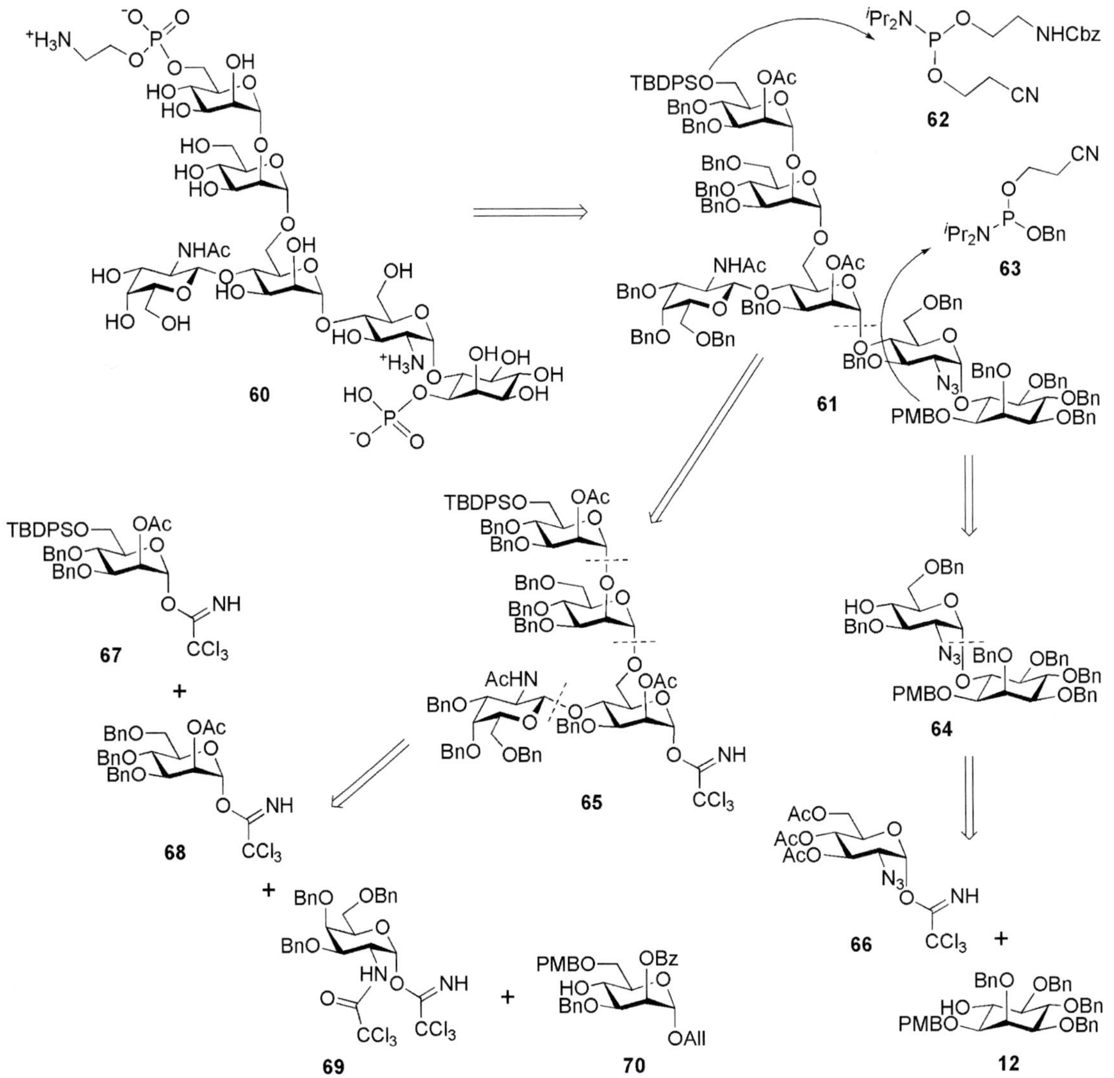

SCHEME 26.12 Retrosynthesis of a GPI-related glycoconjugate of *T. gondii*.

relies on late stage phosphorylation, after the assembly of the glycan core **61**, with *N,N*-diisopropylphosphoramidites **62** and **63** and the use of appropriate protecting groups is shown in Scheme 26.12. The Schmidt group's syntheses of GPI anchors in general (Mayer *et al.*, 1994; Mayer and Schmidt, 1999; Pekari *et al.*, 2001; Pekari and Schmidt, 2003) rely on trichloroacetimidate chemistry together with participation of benzoyl and/or acetyl at C-2 (of Man) to confer α-selectivity in every glycosylation.

In retrosynthesis of the branched glycan core **61**, the α-glycosidic linkage between central Man and $GlcNH_2$ residues is disconnected affording the building blocks **64** and **65**. The tetrasaccharide **65** could be easily prepared from building blocks **67–70**, while the synthesis of pseudo-disaccharide **64** is challenging.

Synthesis of the pseudo-disaccharide **64** is based upon the conversion of the chiral *myo*-inositol derivative **71** into the required product (Scheme 26.13). *O*-Cyclohexylidenation (Garegg *et al.*, 1984; Massy and Wyss, 1990) of *myo*-inositol **35**, followed by enantiomeric resolution via addition of a chiral (-)-menthylformate group and then separation of diastereomers by crystallization gave the desired D-*myo*-inositol derivative **71** (Mayer *et al.*, 1994; Mayer and Schmidt, 1997). Subsequent consecutive 6-*O*-allylation, replacement of the (-)-methyloxycarbonyl group with a *p*-methoxybenzyl group, and finally acid-catalysed cleavage of the cyclohexylidene groups offered the required tetraol. Perbenzylation of the OH groups in **72**, followed by selective de-6-*O*-allylation using Wilkinson's catalyst and acidic treatment offered the 6-OH *myo*-inositol acceptor **12**.

The latter was glycosylated with trichloroacetimidate donor **66** (Grundler and Schmidt, 1984) in the presence of TMSOTf to furnish the desired α-(1→6)-linked pseudo-disaccharide **73** in 70% yield. The 2-azido group in **66** served as a non-participating and a protected amino group that assists in the formation of the α-glycoside linkage. Compound **73** was subjected to sequential de-*O*-acetylation, 4,6-*O*-benzylidenation, 3-*O*-benzylation and, finally, reductive ring opening of the benzylidene acetal with $NaBH_4$ in the presence of HCl to furnish the pseudo-disaccharide acceptor **64**.

The branched trisaccharide **81** (Scheme 26.14) was assembled from three monosaccharide synthons **68**, **69** and **70**. Mannosyl donor **68** (Mayer and Schmidt, 1999) derived from 1,2-orthoester **79**, available from Man (Kaur and Hindsgaul, 1992) via the 1,2-di-*O*-acetate **80** (Ponpipom, 1977). Compound **70** was prepared from Man **76** in five steps (Goebel *et al.*, 1997), while the D-galactosamine ($GalNH_2$) donor **69** was made from the 2-azido derivative **74** (Grundler and Schmidt, 1984). Anomeric *O*-allylation of **76** and subsequent dibutyltin oxide-mediated selective 3-*O*-benzylation afforded the desired triol **77** (61% overall yield). This was treated with *p*-methoxybenzaldehyde dimethylacetal to form 4,6-*O*-arylidene intermediate, which was 2-*O*-benzoylated with benzoyl cyanide and then the 4,6-acetal ring was reductively opened using $NaBH_3CN$-TFA, thereby furnishing the 4-OH Man derivative **70**. The nature of protecting groups in the acceptor **70** offers the desired regio- and stereoselective glycosylation sequence and further transformation into a glycosyl donor permits the subsequent α-selective glycosylation of the pseudo-disaccharide **64** (see Scheme 26.13).

The $GalNH_2$ donor **69** was derived from the azide **74** (see Scheme 26.14) via silylation of the 1-OH group with thexyldimethylsilyl chloride (TDSCl) (Tailler *et al.*, 1999), reduction of the 2-azido group and trichloracetylation of the amine to furnish **75** which, on successive de-1-*O*-silylation and anomeric *O*-trichloroacetimidation, yielded **69**. The *N*-trichloroacetyl protecting group was introduced in order to ensure high glycosyl donor properties and anchimeric assistance for β-glycoside bond formation. The trichloroacetimidate **69** and the acceptor **70** were coupled under the influence of $BF_3 \cdot Et_2O$ to afford stereoselectively the corresponding disaccharide (85%), in which the *p*-methoxybenzyl group was removed (→**78**), enabling TMSOTf-promoted coupling with the mannosyl donor **68** to yield a branched trisacchaide (95%) that was de-*O*-acetylated to give the trisaccharide acceptor **81**.

The third Man building block **67** (Scheme 26.15) was derived (Mayer and Schmidt, 1999) from the orthoester **79**. 6-*O*-silylation, *O*-benzylation, subsequent acid catalysed orthoester opening and then *O*-acetylation afforded **82**. Selective anomeric deacetylation with hydrazinium acetate gave the hemiacetal, which was treated with trichloroacetonitrile in the presence of DBU to give the trichloroacetimidate donor **67**.

The TMSOTf-promoted stereoselective glycosylation of the branched acceptor **81** with the donor **67** gave the corresponding tetrasaccharide

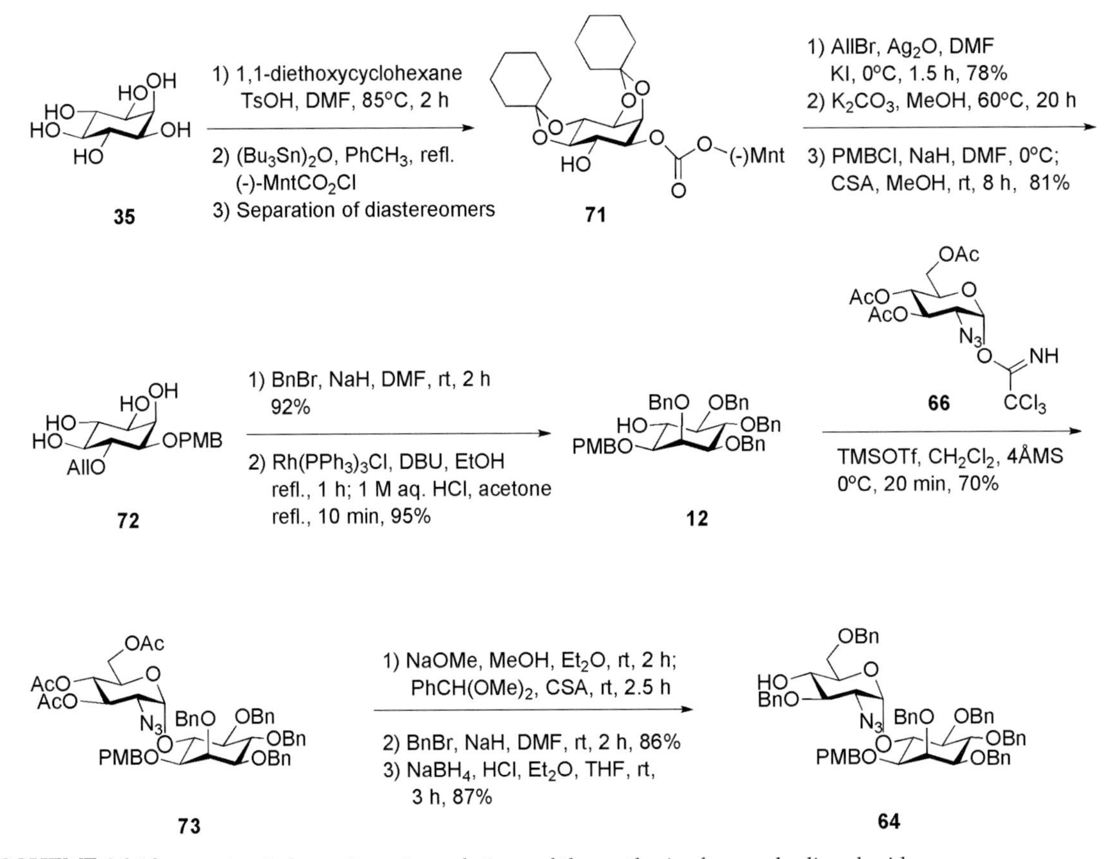

SCHEME 26.13 *myo*-inositol enantiomeric resolution and the synthesis of a pseudo-disaccharide.

(92%), in which the benzoyl protecting group was replaced with an acetyl, via a standard de-*O*-acylation/*O*-acetylation procedure (→**83**) (see Scheme 26.15). Transformation of the *N*-trichloroacetyl group into the *N*-acetyl group by the homolytic reduction, removal of the anomeric allyl group under the influence of Wilkinson's catalyst, followed by the reaction of the hemiacetal with trichloroacetonitrile afforded trichloroacetimidate donor **65**. The pseudo-disaccharide **64** was then coupled onto the growing chain (→**61**, 74%) to furnish, after de-*O*-silylation, the pseudo-hexasaccharide **84**.

Introduction of phosphate residues onto **84** was achieved via 1*H*-tetrazole-assisted phosphitylation reactions. First, treatment with the phosphoramidite **62** furnished the phosphite intermediate, which was oxidized into the phosphotriester with *m*-CPBA followed by treatment with Et_3N to remove the cyanoethyl P-protecting group. The following CAN-assisted removal of the *p*-methoxybenzyl group afforded the *N*-Cbz-amino-ethyl phosphate **85**. Phosphorylation of **85** with benzyl cyanoethyl-*N*,*N*-diisopropylaminophosphoramidite **63** in a similar fashion, followed by cyanoethyl cleavage with Me_2NH and de-*O*-acetylation under Zemplén conditions furnished the diphosphorylated compound **86** which, by hydrogenolytic de-*O*-benzylation, afforded the targeted compound **60** in good yield and purity.

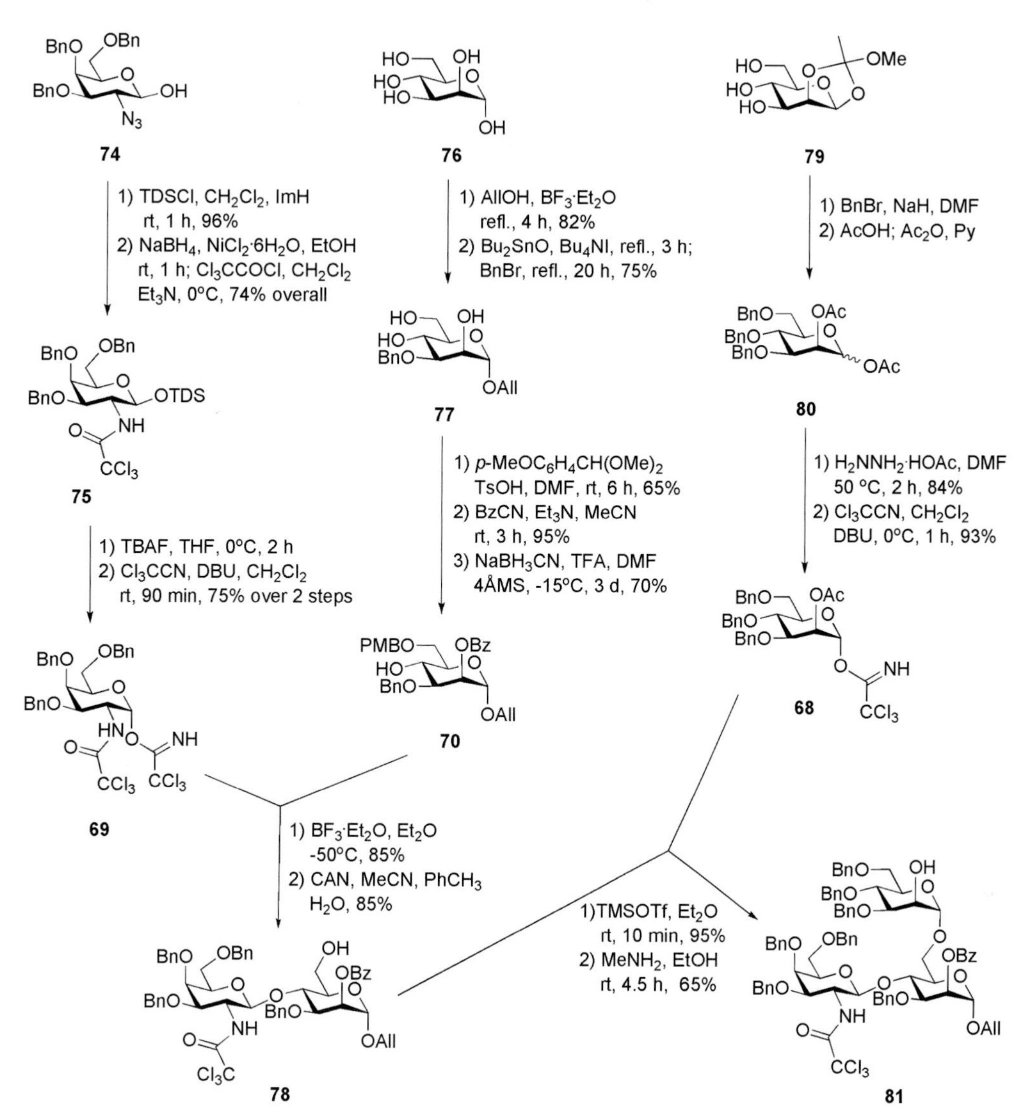

SCHEME 26.14 The synthesis of monosaccharide building blocks and assembly of a branched trisaccharide.

2.3. Synthesis of a glycoconjugate of *Leishmania*

Stockholm University and Linköping University laboratories (Ruda *et al.*, 2000a) reported the synthesis of *Leishmania* LPG core phosphorylated heptasaccharyl *myo*-inositol **87** (Scheme 26.16) found in the surface structures of *Leishmania* parasites. The synthesis was accomplished using a convergent (3 + 2 + 2 + 1) synthetic strategy, in which the introduction of an anomeric phosphodiester, due to its instablilty towards acidic conditions, was accomplished at a late stage, just before the global deprotection. The preparation of the trisaccharide **88** and the disaccharide **89** building blocks is shown in Scheme 26.17.

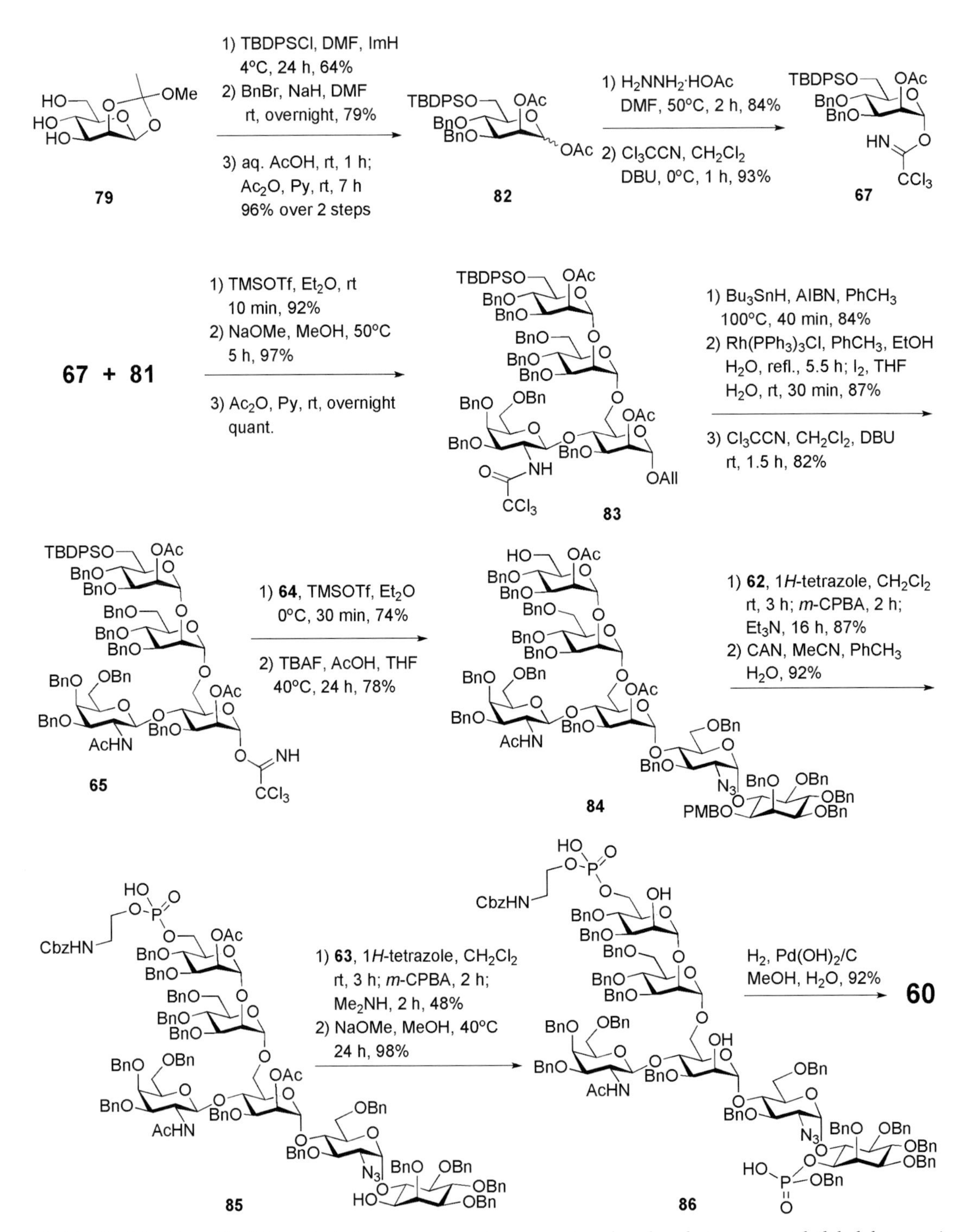

SCHEME 26.15 *T. gondii* GPI branched glycan core assembly, successive phosphorylation steps and global deprotection.

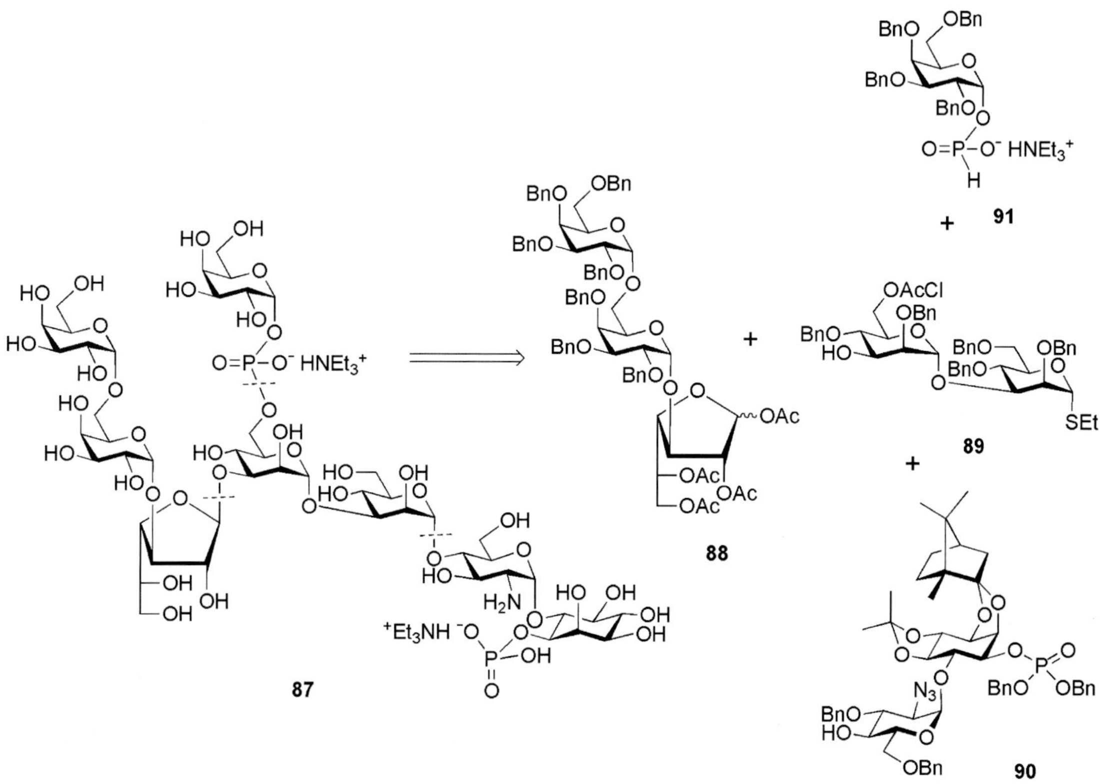

SCHEME 26.16 Retrosynthesis of a GPI-related glycoconjugate of *Leishmania*.

Ethyl 1-thio-β-D-galactopyranoside **92** (see Scheme 26.17) was 6-*O*-tritylated and then per-*O*-benzylated, followed by acidic *O*-detritylation to yield derivative **94**. Per-*O*-benzylation of **92** and replacement of the ethanethiol with bromine gave the galactosyl bromide **93** that was coupled with **94** in Lemieux's conditions (Lemieux *et al.*, 1975) to give the galactobiose disaccharide **95**. The galactofuranose acceptor **98** was prepared from the 5,6-diol **96** (Morris and Kiely, 1987). *O*-acetylation, to preserve the furanose ring, TFA-assisted cleavage of isopropylidene acetal and *O*-acetylation gave **97**. After hydrogenolysis, the acceptor **98** was glycosylated with the thioglycoside donor **95** under the influence of dimethyl(methylthio)sulfonium trifluoromethanesulfonate (DMTST) to yield the trisaccharide **88** (67%). The mannobiose disaccharide **89** was synthesized (Ruda *et al.*, 2000b) from Man derivative **99** (Garegg *et al.*, 1993). *O*-benzylation followed by reductive benzylidene ring cleavage gave **100**. This was 6-*O*-benzylated and de-3-*O*-silylated to furnish the acceptor **102**. 6-*O*-chloroacetylation of **100**, followed by 1-bromination gave the mannosyl bromide **101** that was coupled with **102** in the presence of AgOTf to give the corresponding disaccharide (79%), in which the TBS group was removed to furnish **89**. This was coupled with the trisaccharide donor **88** under the influence of TMSOTf to yield the pentasaccharide **103** (85%).

The *myo*-inositol containing building block **90** (Scheme 26.18) was synthesized via AgOTf-dicyclopentadienylzirconium dichloride (Cp_2ZrCl_2)-mediated coupling (48%) of glycosyl fluoride **106**, acquired in five steps from ethyl 2-azido-2-deoxy-1-thio-β-D-glucopyranoside **104** (Buskas *et al.*, 1994), to chiral *myo*-inositol derivative

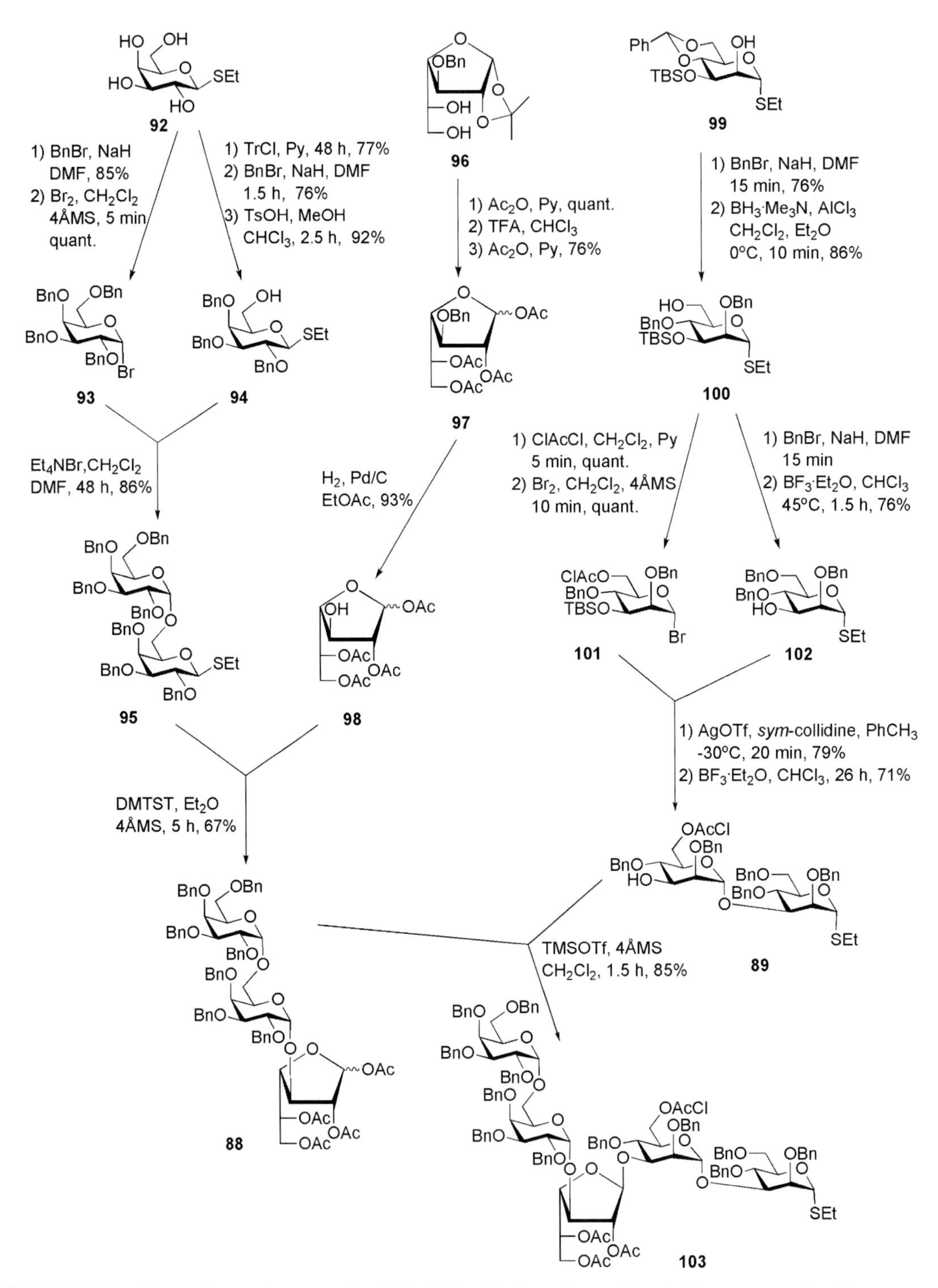

SCHEME 26.17 The synthesis of monosaccharide building blocks and assembly of a linear pentsaccharide.

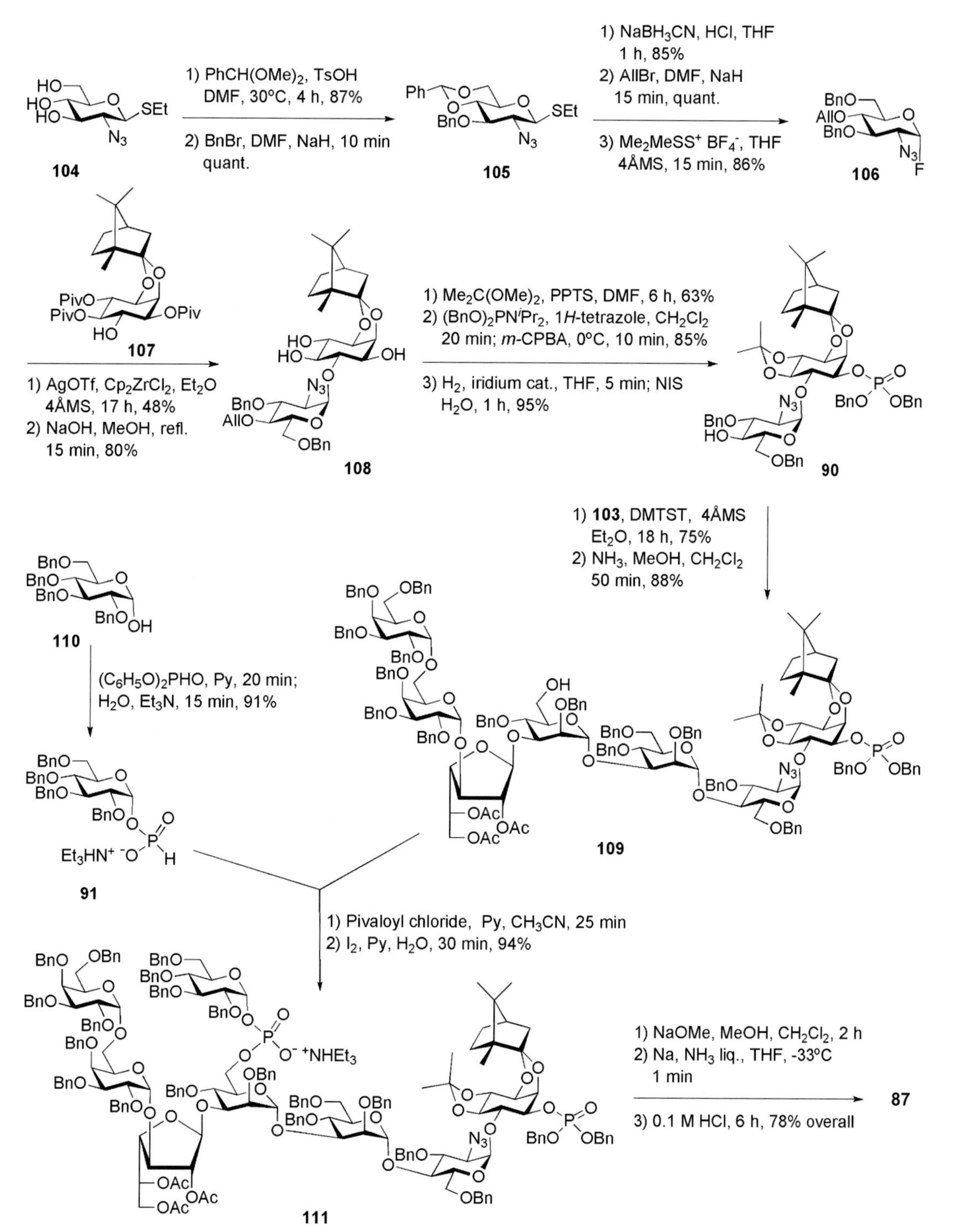

SCHEME 26.18 *Leishmania* GPI glycan core assembly (including synthesis of a pseudo-disaccharide and phosphorylation steps) and global deprotection.

107 (Bruzik and Tsai, 1992; Garegg *et al.*, 1997), followed by subsequent removal of pivaloyl groups (→**108**, 80%), 4,5-*O*-isopropylidene acetal introduction, phosphorylation of 1-OH and, finally, removal of the allyl group. The pentasaccharide thioglycoside donor **103** was activated with DMTST in ether and coupled to the pseudo-disaccharide **90** to give the corresponding pseudo-heptasaccharide (75%), in which the chloroacetyl group was selectively removed to give **109**. The latter compound was subjected to pivaloyl chloride-mediated coupling to α-D-glucopyranosyl H-phosphonate **91**, prepared by the reaction of the hemiacetal **110** with diphenyl phosphite (Jankowska *et al.*, 1994) and subsequent oxidation to produce the protected compound **111** (94%). Conventional deacetylation, followed by de-*O*-benzylation by Birch reduction and subsequent acid hydrolysis of the acetal groups gave the phosphorylated glycan-*myo*-inositol **87** in 78% yield.

2.4. Syntheses of GPI anchors and of a glycoconjugate of *P. falciparum*

2.4.1. *The Durham group approach*

A step-by-step synthetic route for preparation of fully lipidated GPI anchor **112** present on the cell surface of the malaria pathogen *Plasmodium falciparum* was developed by the Durham group (Fraser-Reid and co-workers; Lu *et al.*, 2004a,b). The synthetic strategy relies on the use of *n*-pentenyl orthoesters as glycosyl donors and on an appropriately protected carbohydrate core to allow site-specific deprotection and phosphorylation and acylation at an appropriate stage. The strategy exploited trityl ether and cyclohexylidene acetal as orthogonal blocking groups, while benzyl group was chosen for permanent *O*-protection. In the retrosynthesis of **112** (Scheme 26.19), the glycan core **113** was assembled from Man 1,2-orthoesters **115–118** and the 6-*O*-unprotected *myo*-inositol acceptor **119**. Late phosphorylation was achieved by making use of the phosphoramidites **27** and **114**.

The group developed (Jia *et al.*, 1998) a novel methodology for preparation of optically pure and differentially protected *myo*-inositol acceptor **119** starting from methyl glucoside **120** (Scheme 26.20). The procedure relied on the discovery (Bender and Budhu, 1991; Estevez and Prestwich, 1991) that enol esters undergo the Ferrier reaction with excellent stereocontrol. The tetraol **120** was converted to 6-*O*-unprotected compound **121** and the primary hydroxyl transformed into an aldehyde using Moffat oxidation conditions, followed by enol acetylation to afford the enol acetate **122**. Lewis acid-promoted re-arrangement employing Ferrier's conditions (Ferrier and Middleton, 1993) furnished the corresponding cyclohexan-5-one **123**, which was further subjected to reduction, de-*O*-acetylation and *O*-cyclohexylidenation to furnish the desired optically pure 6-hydroxy-*myo*-inositol derivative **119**.

Building blocks **115–118** were synthesized from the readily available *n*-pentenyl 1,2-orthoester **124** (Mach *et al.*, 2002) (Scheme 26.21). Tin-mediated selective double benzylation of the C-6 and C-3 OH-groups in the triol **124**, followed by *p*-methoxybenzylation of the remaining 4-OH afforded the donor **118**. Conventional 6-*O*-tritylation of **124**, followed by acetylation gave **115**, while consecutive 6-*O*-silylation and benzylation provided **117**.

The glycan core assembly (Scheme 26.22) started from the glycosylation of **119** with the orthoester **118**, in the presence of ytterbium(III) triflate and NIS, to give the corresponding pseudo-disaccharide, in which the 2-benzoate group was replaced with the 2-triflate group, to furnish the pseudo-disaccharide **125**. Conversion of the Man-inositol **125** into the azidoglucose-inositol **126** was achieved by azide displacement (Soli *et al.*, 1999) of the triflate, followed by $BF_3 \cdot Et_2O$-assisted removal of the *p*-methoxybenzyl group. Glycosylation of the acceptor **126** with the orthoester donor **117**, under the influence of NIS and $BF_3 \cdot Et_2O$ (79%), followed by replacement of the 2″-benzoate with a 2″-*O*-benzyl group and de-6″-*O*-silylation furnished the pseudo-trisaccharide **127**. This was then coupled with the orthoester **116** to give the corresponding oligomer (99%) which, on conventional debenzoylation, afforded the requisite pseudo-tetrasaccharide **128**. The latter was coupled with the donor **115** (75%), followed by subsequent de-*O*-acylation and *O*-benzylation to furnish the desired pseudo-pentasaccharide **113**.

Mild acidic removal of the trityl group in **113** (Scheme 26.23) enabled *O*-phosphitylation with phosphoramidite **27** and 1*H*-tetrazole followed by oxidation to provide the *N*-Cbz-aminoethyl phosphotriester **129**. Removal of cyclohexylidene protection in **129**, followed by myristoylation of the diol with trimethylorthomyristate (Presova and Smrt, 1989) gave the corresponding 1,2-cyclic orthoester, which was opened efficiently and with good regioselectivity employing $Yb(OTf)_3$ to provide a 4.7:1 ratio of the axial **131** (71%) and the equatorial **130** (15%) myristic esters. The axial ester **131** was phospholipidated by making use of the diacylglyceryl phosphoamidite **114** (Campbell and Fraser-Reid, 1994) to afford the fully protected GPI **132** (72%) which, on global deprotection by hydrogenolysis, first, in organic solvent, then followed by addition of water, gave the desired GPI anchor **112**.

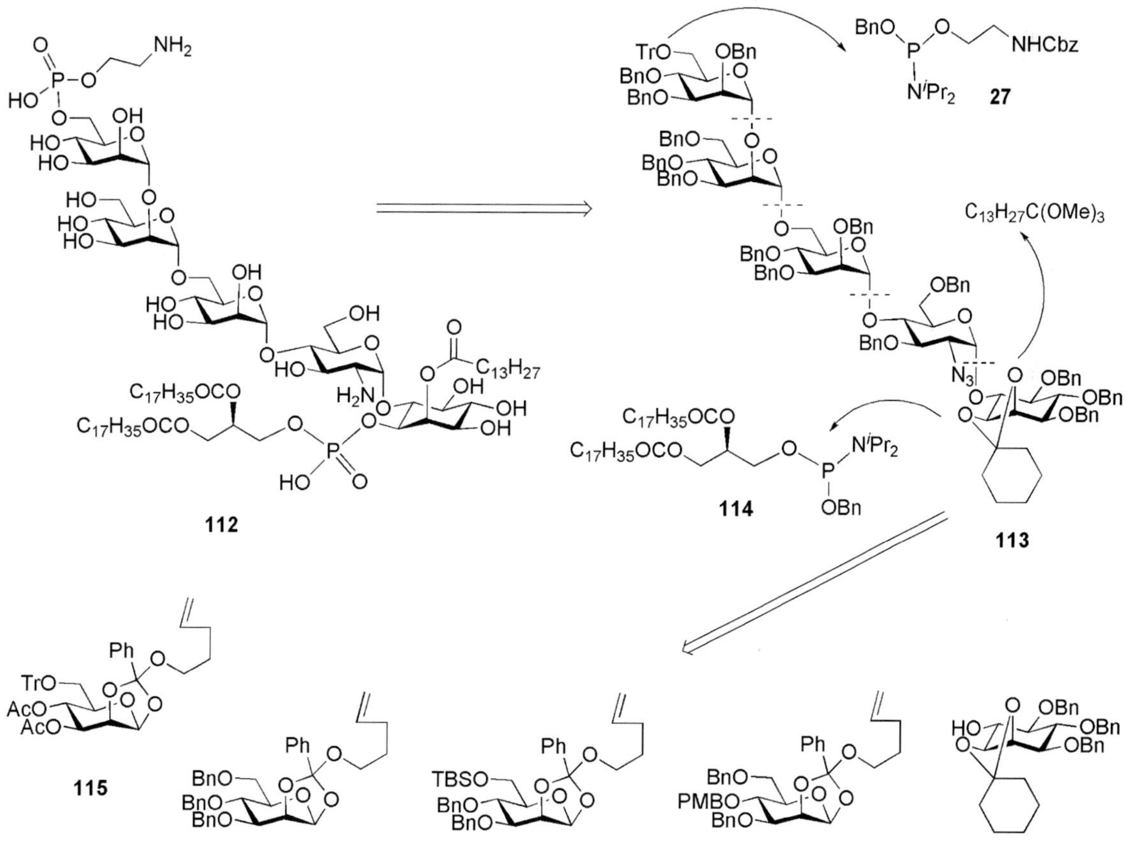

SCHEME 26.19 Retrosynthesis of a GPI anchor of *P. falciparum* according to the strategy of the Durham group.

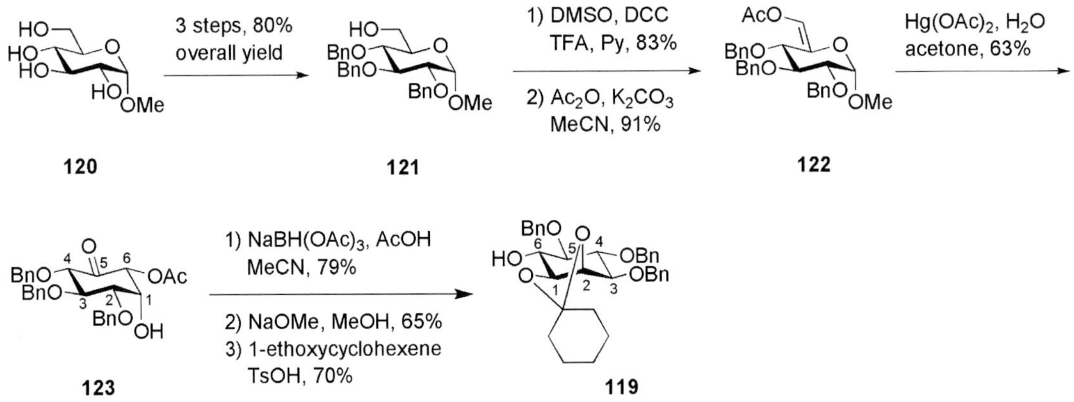

3 steps, 80% overall yield

1) DMSO, DCC TFA, Py, 83%
2) Ac_2O, K_2CO_3 MeCN, 91%

$Hg(OAc)_2$, H_2O acetone, 63%

120 121 122

1) $NaBH(OAc)_3$, AcOH MeCN, 79%
2) NaOMe, MeOH, 65%
3) 1-ethoxycyclohexene TsOH, 70%

123 119

SCHEME 26.20 *De-novo* synthesis of an optically pure chiral *myo*-inositol derivative according to the strategy of the Durham group.

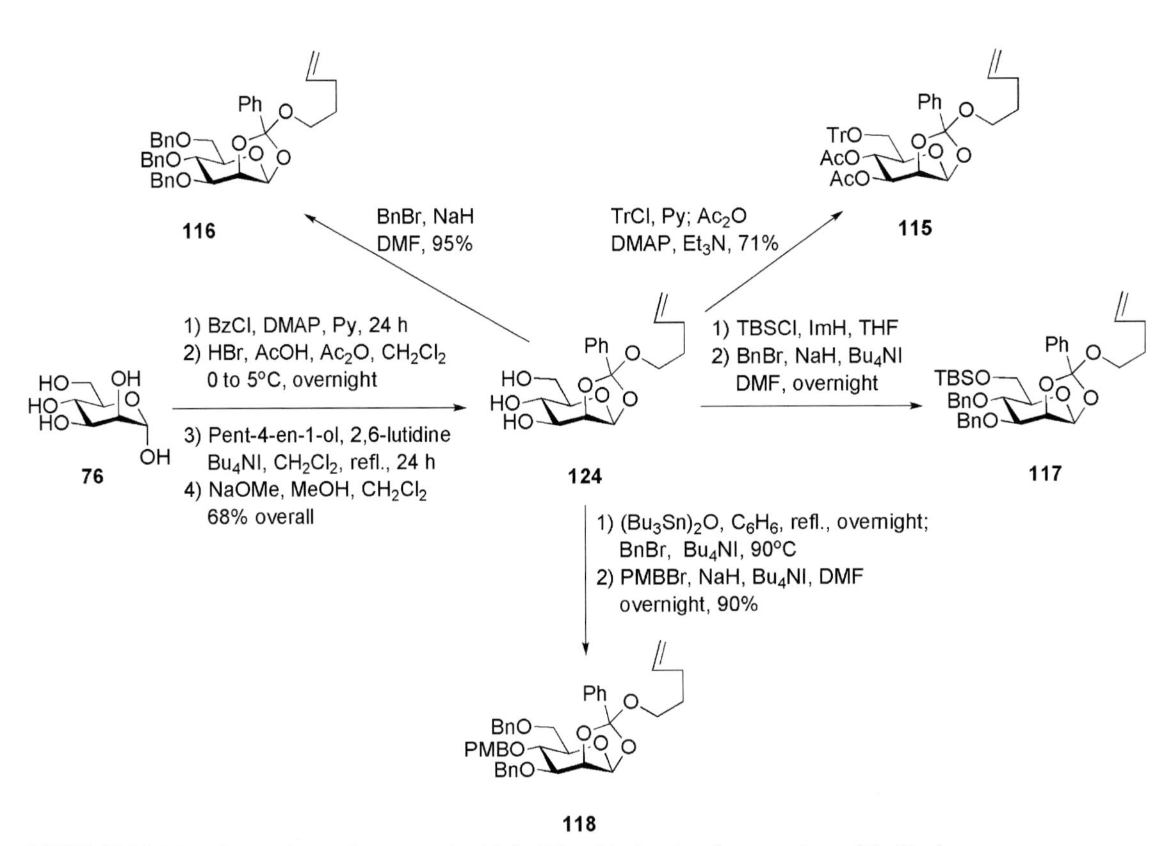

SCHEME 26.21 The synthesis of monosaccharide building blocks after the procedure of the Durham group.

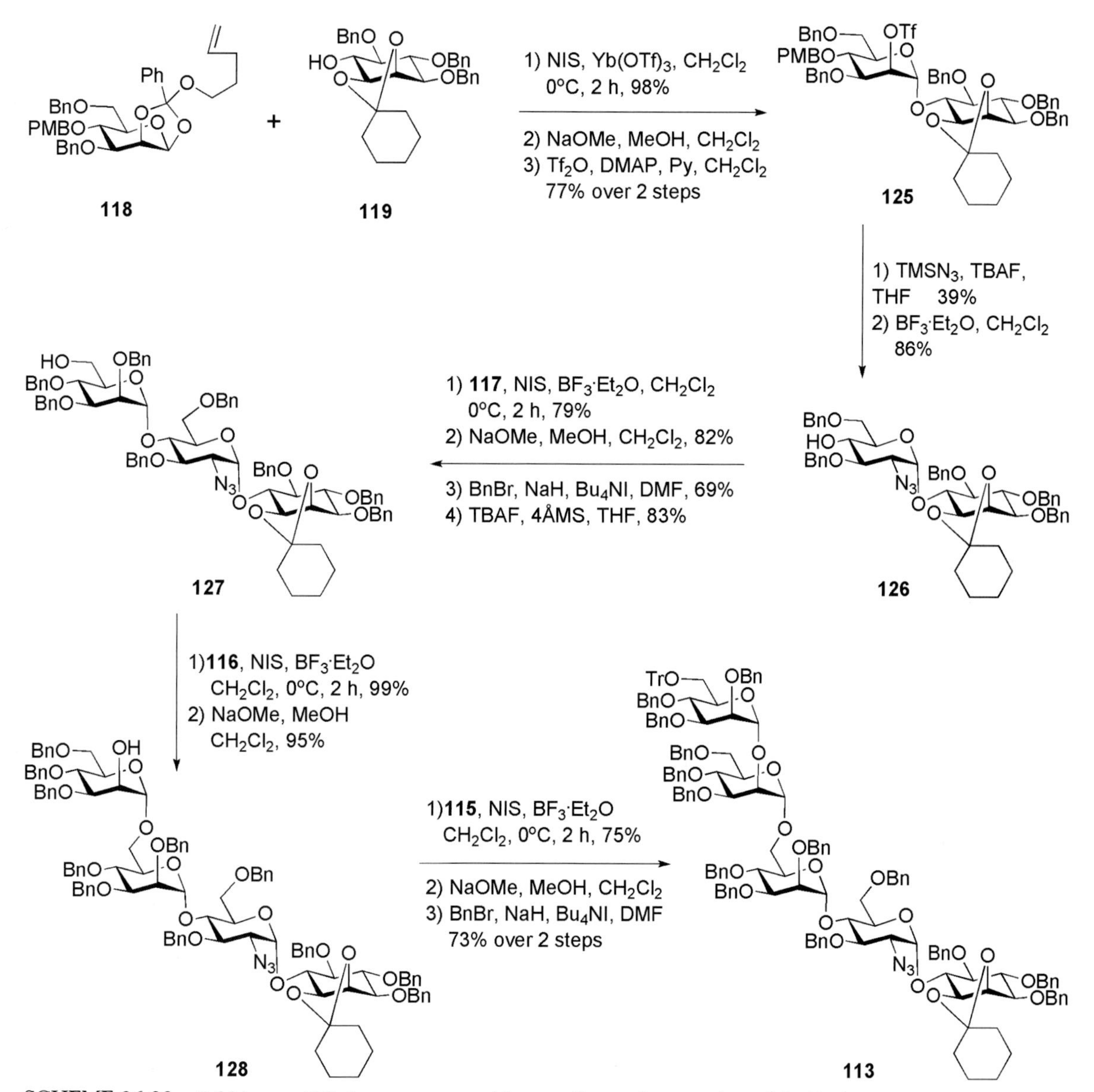

SCHEME 26.22 *P. falciparum* GPI glycan core assembly according to the procedure of the Durham group.

2.4.2. The Cambridge (Massachusetts)-Zürich group approach

The Cambridge (Massachusetts)-Zürich group (Seeberger and co-workers) synthesized another GPI anchor **133** (Scheme 26.24) of *P. falciparum* (Liu *et al.*, 2005) with a similar structure to **112**, but with an extra Man residue α-(1→2)-linked to Man-3 and differing in the *myo*-inositol acyl moiety. The glycan core **134** was assembled using a convergent (4 + 2) synthetic strategy via the coupling of the tetramannosyl trichloroacetimidate **137** and the pseudo-disaccharide **136**.

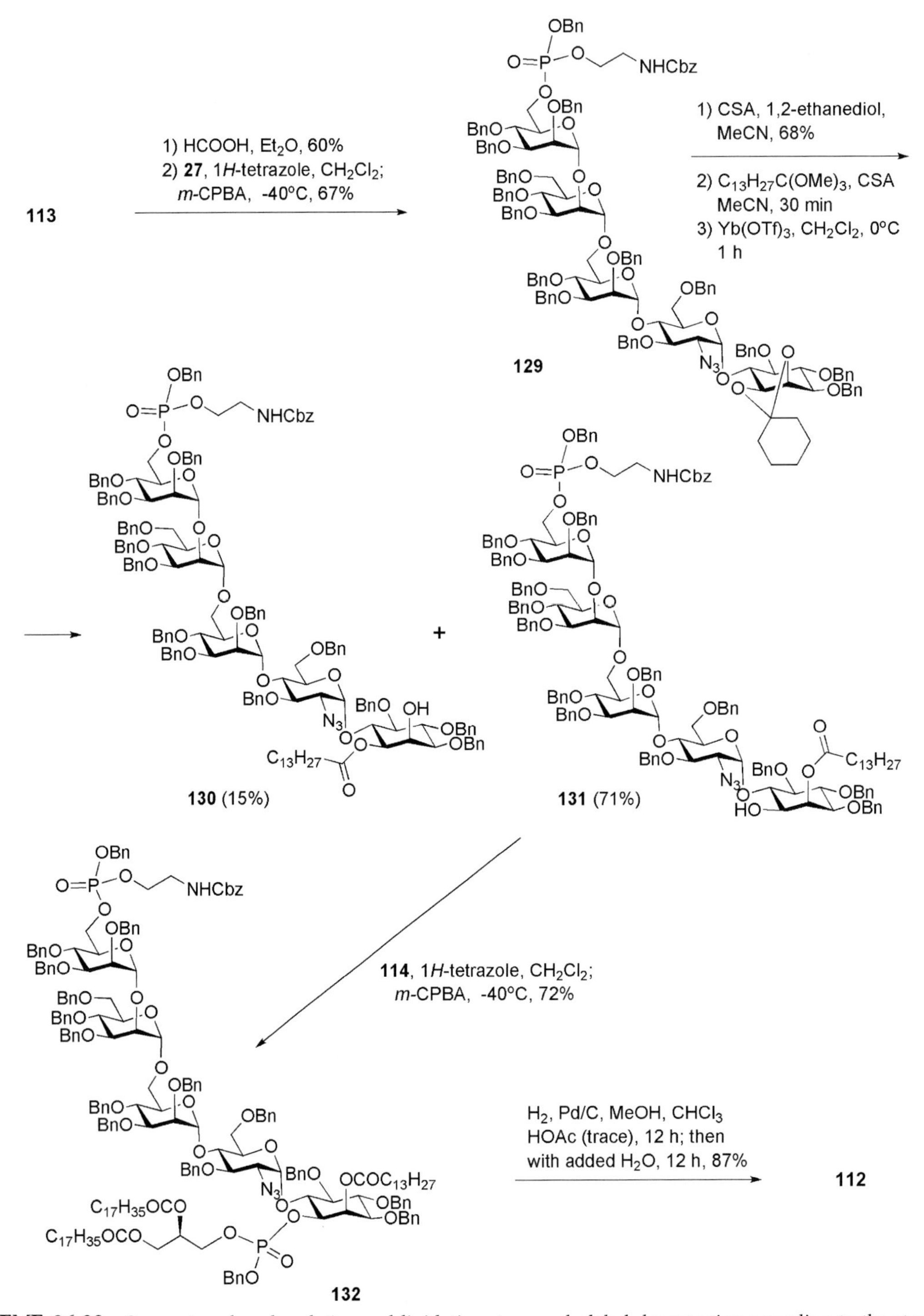

SCHEME 26.23 Successive phosphorylation and lipidation steps and global deprotection according to the strategy of the Durham group.

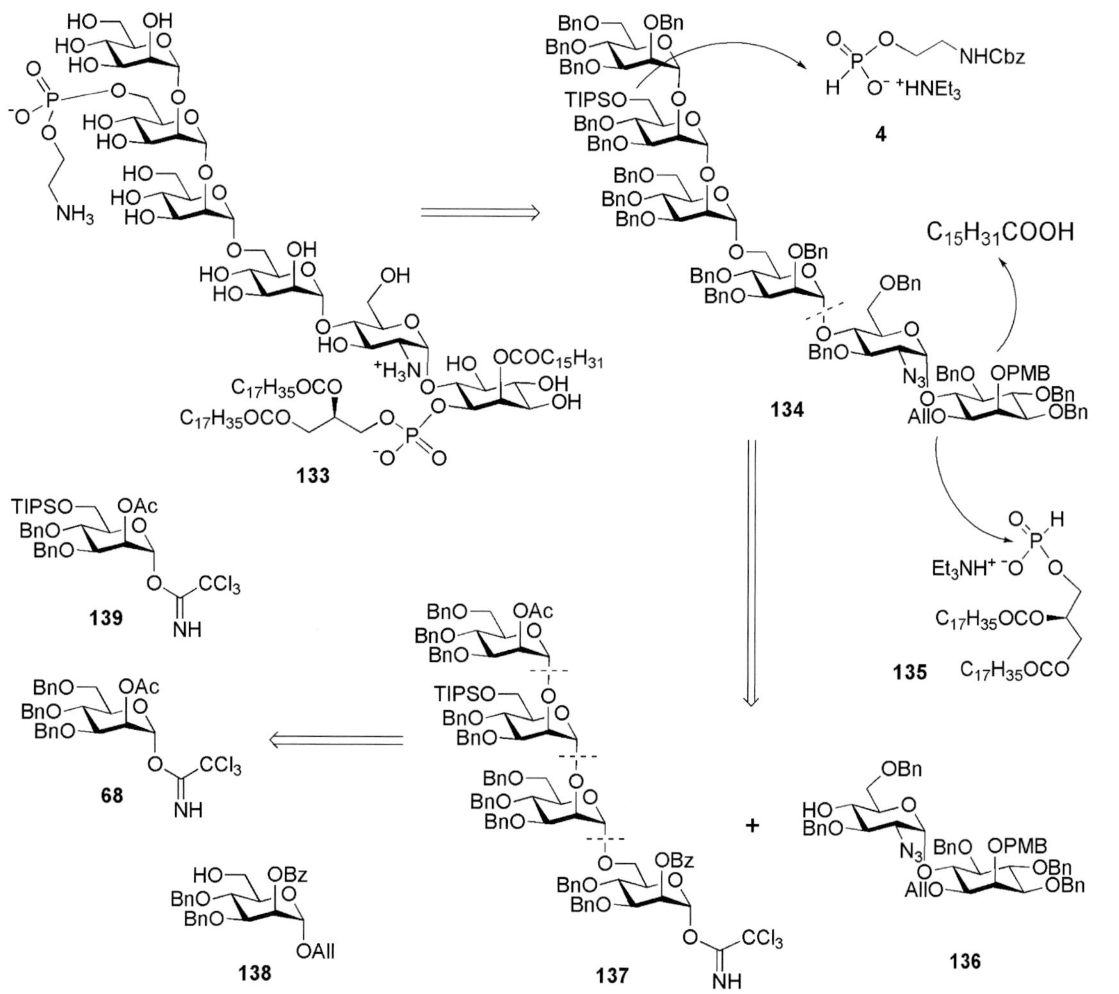

SCHEME 26.24 Retrosynthesis of a GPI anchor of *P. falciparum* according to the procedure of the Cambridge (Massachusetts)-Zürich group.

The strategy exploited orthogonal protecting groups, such as PMB ether for the C-2 *myo*-inositol acylation site, TIPS ether for ethanolamine-phosphate in Man-3 and allyl ether for phospholipids at C-1 of *myo*-inositol. Late phosphorylation was achieved by making use of the H-phosphonate derivatives **4** and **135**.

The pseudo-disaccharide **136** (Scheme 26.25) was constructed starting from compound **140** (Jia *et al.*, 1998; Liu *et al.*, 2006), used earlier by the Durham group. De-*O*-acetylation of **140** and subsequent Bu_2SnO-assisted regioselective 1-*O*-allylation and 2-*O*-methoxybenzylation gave the *myo*-inositol acceptor **141**. This was glycosylated with the known trichloroacetimidate **66** (Grundler and Schmidt, 1984; Orgueira *et al.*, 2003) to give the corresponding pseudo-disaccharide as inseparable anomeric mixture (α:β = 4:1), which was de-*O*-acetylated to provide triol **142** (89%). Installation of 4,6-benzylidene acetal followed by 3-*O*-benzylation and reductive opening of the 4,6-*O*-benzylidene protection with

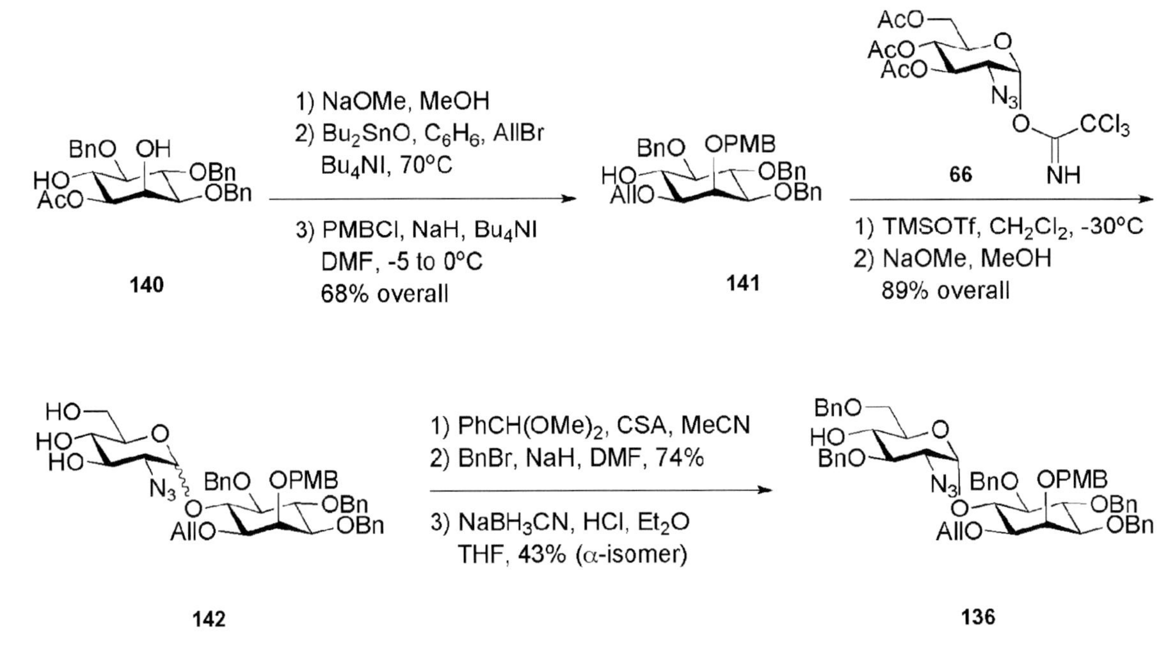

SCHEME 26.25 The synthesis of a pseudo-disaccharide after the strategy of the Cambridge (Massachusetts)-Zürich group.

$NABH_3CN$-HCl gave the pseudo-disaccharide, while the separation of the α- (**136**) and β-anomers became possible.

The tetrasaccharide block **137** (Seeberger *et al.*, 2004; Kwon *et al.*, 2005) was formed (Scheme 26.26) by making use of trichloroacetimidate donors, employing a similar approach to Schmidt (see Section 2.2) towards the construction of α-(1→2)-linked Man residues. The Man building blocks **68**, **138** and **139** (Schofield *et al.*, 2002; Hewitt *et al.*, 2002; Liu *et al.*, 2006) were derived from the orthoesters **143**, **144** and **79**, respectively. Construction of tetramannoside **137** started with the TMSOTf-catalysed coupling of acceptor **138** with donor **68** (91%), followed by removal of the acetic ester in the presence of benzoate by acidic methanolysis (Byramova *et al.*, 1983), to furnish the disaccharide **145**. The latter was further elongated to the tetrasaccharide **147** employing iterative glycosylations with donors **139** (93%, followed by deacetylation to give **146**) and **68** (80%). Removal of the anomeric allyl group and introduction of the trichloroacetimidate gave the tetrasaccharide glycosyl donor **137**.

Stereoselective TMSOTf-catalysed coupling of **137** and the pseudo-disaccharide acceptor **136** (Scheme 26.27) gave the corresponding pseudo-hexasaccharide (94%) (the 2-*O*-benzoyl group in the donor assisted the α-selectivity) in which the acetyl and benzoyl groups were replaced with benzyl groups to yield the protected glycan core **134**.

Oxidative cleavage of the PMB protecting group enabled the introduction of a palmitoyl moiety at O-2 of *myo*-inositol using the DCC-assisted acylation. Subsquent removal of the allyl group and introduction of the phospholipid by pivaloyl chloride-assisted phosphitylation with the H-phosphonate **135**, followed by oxidation, gave the phosphodiester **148** (72%). Finally, after removal of the silyl ether in Man-3 with the help of scandium(III) triflate, the

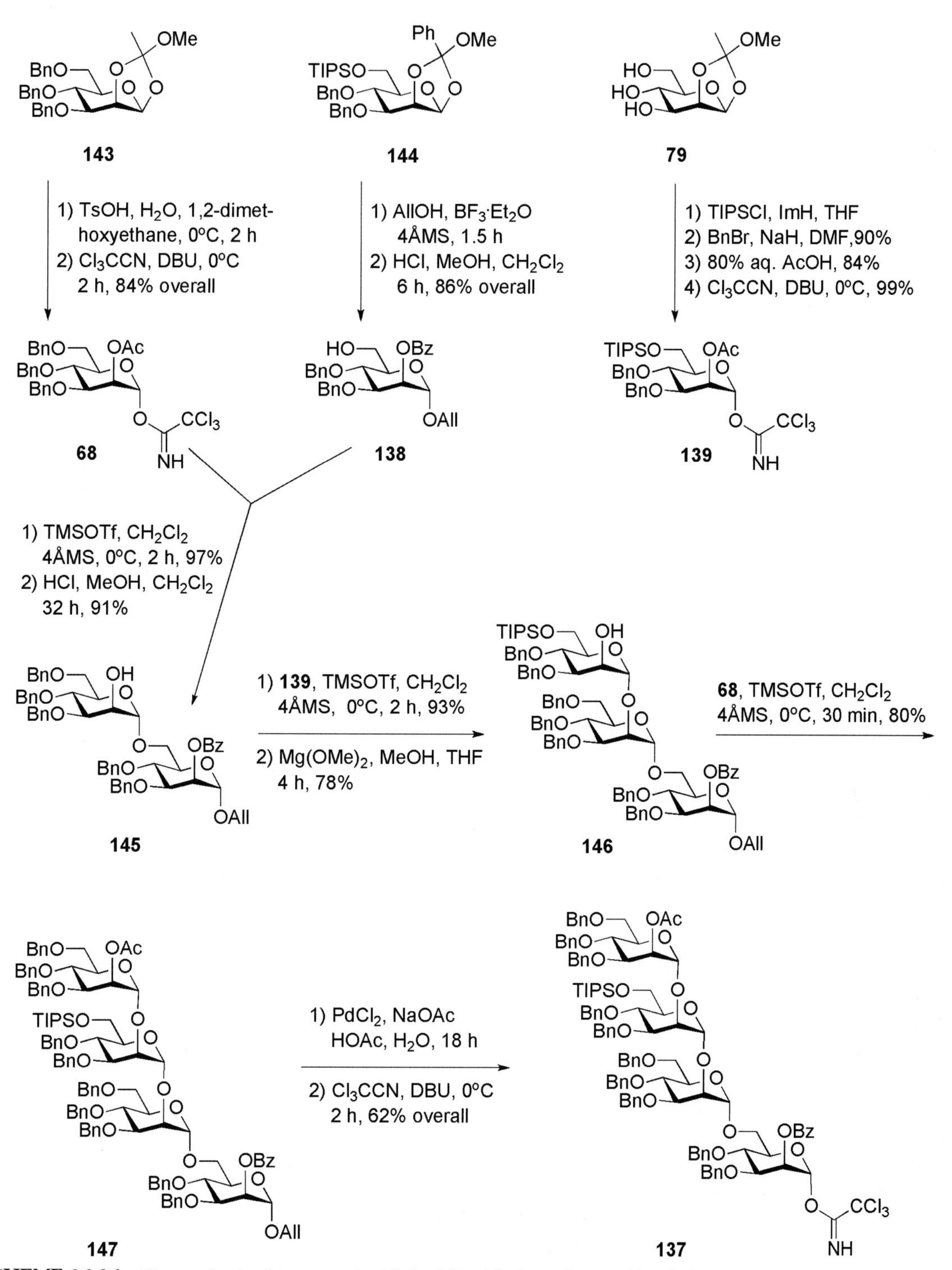

SCHEME 26.26 The synthesis of monosaccharide building blocks and assembly of a linear tetramannoside according to the strategy of the Cambridge (Massachusetts)-Zürich group.

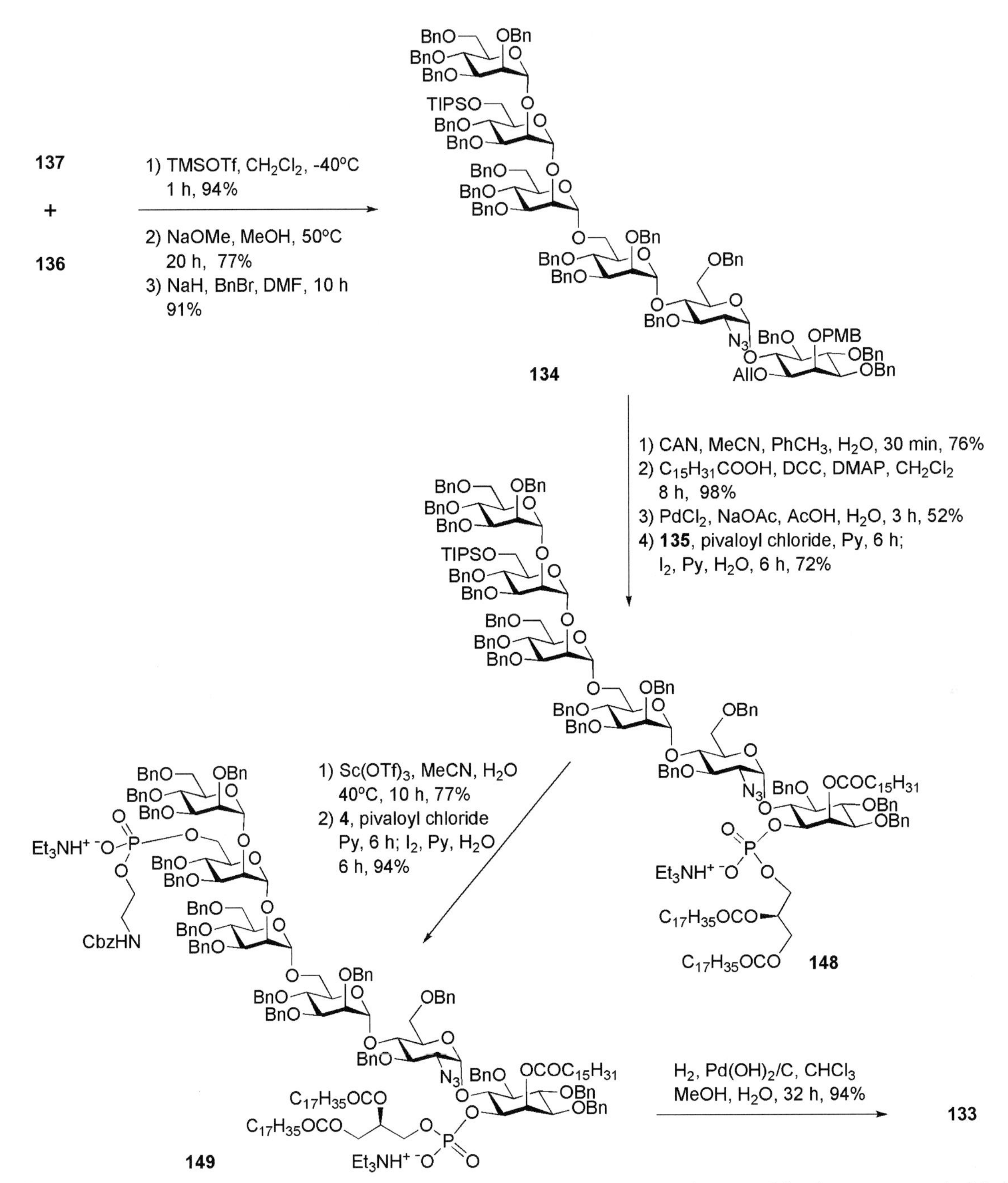

SCHEME 26.27 *P. falciparum* GPI glycan core assembly, successive phosphorylation and lipidation steps and global deprotection after the procedure of the Cambridge (Massachusetts)-Zürich group.

ethanolamine-phosphate was installed in a similar manner using the H-phosphonate **4** to provide the fully protected GPI **149** (94%). The global deprotection was then achieved by hydrogenolysis over the Pearlmann's catalyst to give the GPI anchor **133** in a 94% yield.

2.4.3. Solid-phase synthesis of pseudo-hexasaccharide malarial toxin

The Cambridge (Massachusetts)-Zürich group has also synthesized phosphorylated GPI anchor pseudo-hexasaccharide **150** (Scheme 26.28) of *P. falciparum* (Hewitt *et al.*, 2002) employing automated solid-phase synthesis for the preparation of the tetramannnoside **152**. The glycan core **151** was assembled utilizing (4 + 2) strategy via the coupling of a tetramannose trichloroacetimidate made from **152** and the pseudo-disaccharide **153**. Late phosphate group introduction was achieved using methyl dichlorophosphate and the phosphoramidite **62**.

The tetramannoside **152** was assembled on solid phase using octenediol-functionalized Merrifield resin **156** (Scheme 26.29) and trichloroacetimidate mannosyl donors **68**, **139**, **154** and **155**. Each chain elongation cycle relied on double glycosylations (using catalytic TMSOTf) to secure high coupling efficiencies followed by removal of the acetic ester with NaOMe. Three consecutive elongations using donors **154**, **68** and **139** provided polymer-bound acceptor **157** which, on further chain extension with the donor **155** followed by cleavage of the octendiol linker with Grubbs' catalyst (Schwab *et al.*, 1996) in an atmosphere of ethylene, afforded *n*-pentenyl tetrasaccharide **152**. The aglycon then was hydrolysed with NBS and the hemiacetal was converted to the corresponding trichloroacetimidate prior to the coupling with the pseudo-disaccharide **153**, thus forming the glycan core **151** (32%).

Further protecting group remodelling released the 1,2-diol in *myo*-inositol moiety that was cyclophosphorylated with $MeOP(O)Cl_2$, followed by de-*O*-methylation with aq. HCl and standard de-*O*-silylation to provide the glycan cyclophosphate **158** (70%). Final phosphorylation was achieved via 1*H*-tetrazole catalysed coupling with **62** followed by oxidation with ^{t}BuOOH to give the protected compound **159** (84%). Cleavage of the cyanoethyl ester with DBU followed by Birch reduction for global deprotection gave the required phosphoglycan **150**. The chemically made compound **150** was then bound to a protein carrier and the conjugate was tested as malarial toxin glycovaccine in mice, which appeared to be substantially protected by the vaccination from death caused by malaria parasites (Schofield *et al.*, 2002).

2.5. Syntheses of GPI anchors and of a glycoconjugate of *Trypanosoma cruzi*

2.5.1. The Dundee group approaches

The protozoan parasite *T. cruzi* is a causative agent of Chagas' disease, which is widespread in South and Central America. Throughout the life cycle, *T. cruzi* produce both common and stage-specific GPI-anchored cell-surface macromolecules (Almeida and Gazzinelli, 2001). Local release of GPI-anchored mucins by the bloodstream trypomastigote stage of the parasite is believed to be responsible for development of parasite-elicited inflammation causing cardiac and other pathologies associated with acute and chronic phases of Chagas' disease. It has been discovered (Almeida *et al.*, 2000) that purified GPI fraction of *T. cruzi* trypomastigote mucins (trypomastigote GPI or tGPI) revealed extraordinary proinflammatory activities, comparable to those of bacterial lipopolysaccharide. The extreme biological activity was allegedly associated with the presence of unsaturated fatty acids in the *sn*-2 position of alkylacylglycerophosphate moiety (structures **160a** and **160b** in Scheme 26.30). The content of fatty acid components in the biologically active tGPI anchor fraction was found to be: oleic acid (C18:1, 31%), linoleic acid (C18:2, 21%) and palmitic acid (C16:0, 37%).

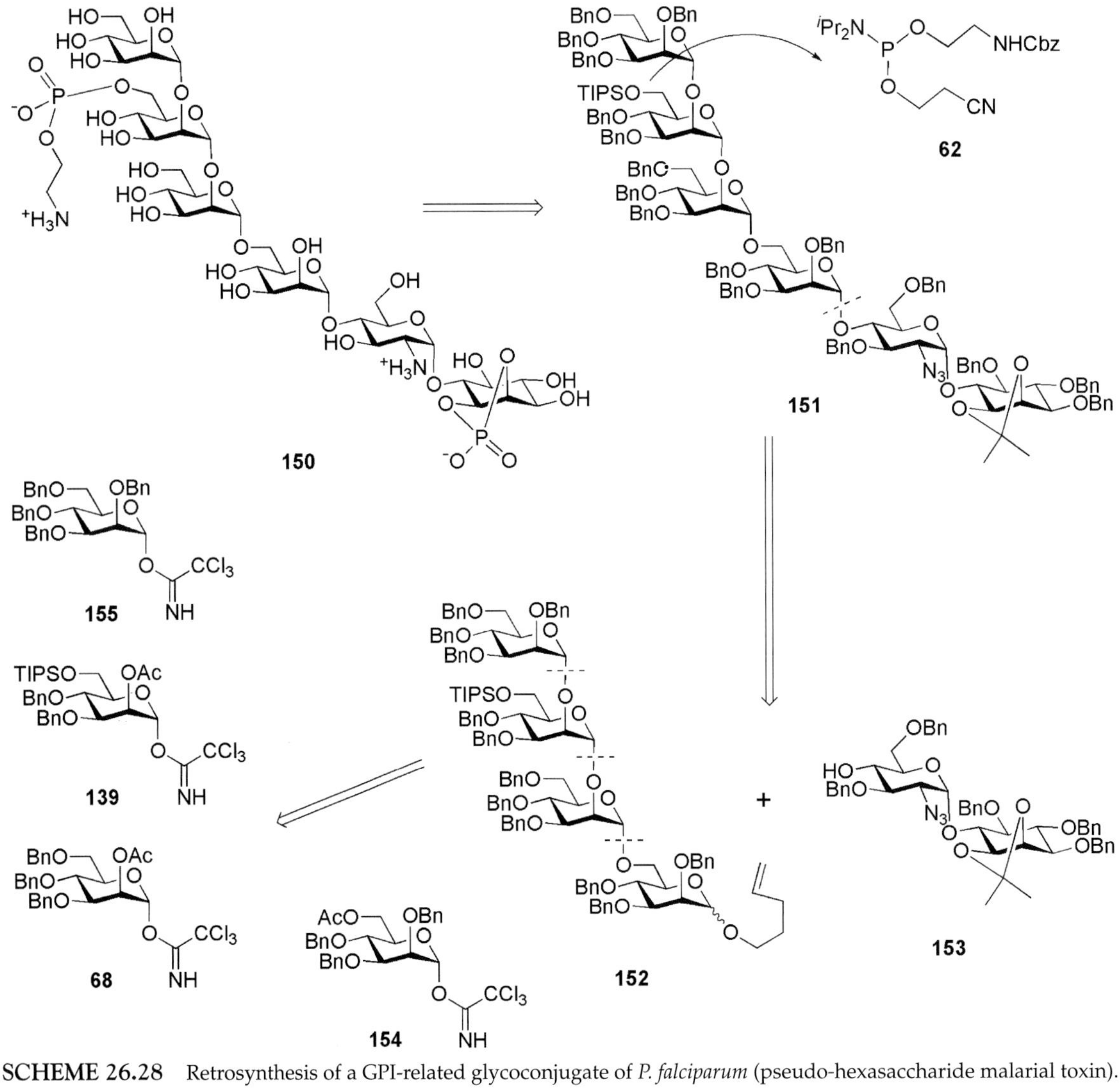

SCHEME 26.28 Retrosynthesis of a GPI-related glycoconjugate of *P. falciparum* (pseudo-hexasaccharide malarial toxin).

The Dundee University group (Nikolaev and co-workers) reported recently chemical syntheses of tGPIs from *T. cruzi* bearing oleic **160a** and linoleic **160b** acid moieties.

There are two major structural features by which compounds **160a** and **160b** differ from the GPIs synthesized previously:

(i) the presence of unsaturated fatty acids in the lipid moiety instead of saturated ones; and

(ii) the presence of 2-aminoethylphosphonate at C-6 of $GlcNH_2$ moiety, which is a parasite specific substituent for *T. cruzi* only.

Since the presence of double bonds was not compatible with the use of benzyl ethers (widely used before) as permanent *O*-protecting groups, novel strategies were developed. *Strategy A* was designed to imply mild base labile (esters) and acid labile (acetals and *N*-Boc) permanent protecting

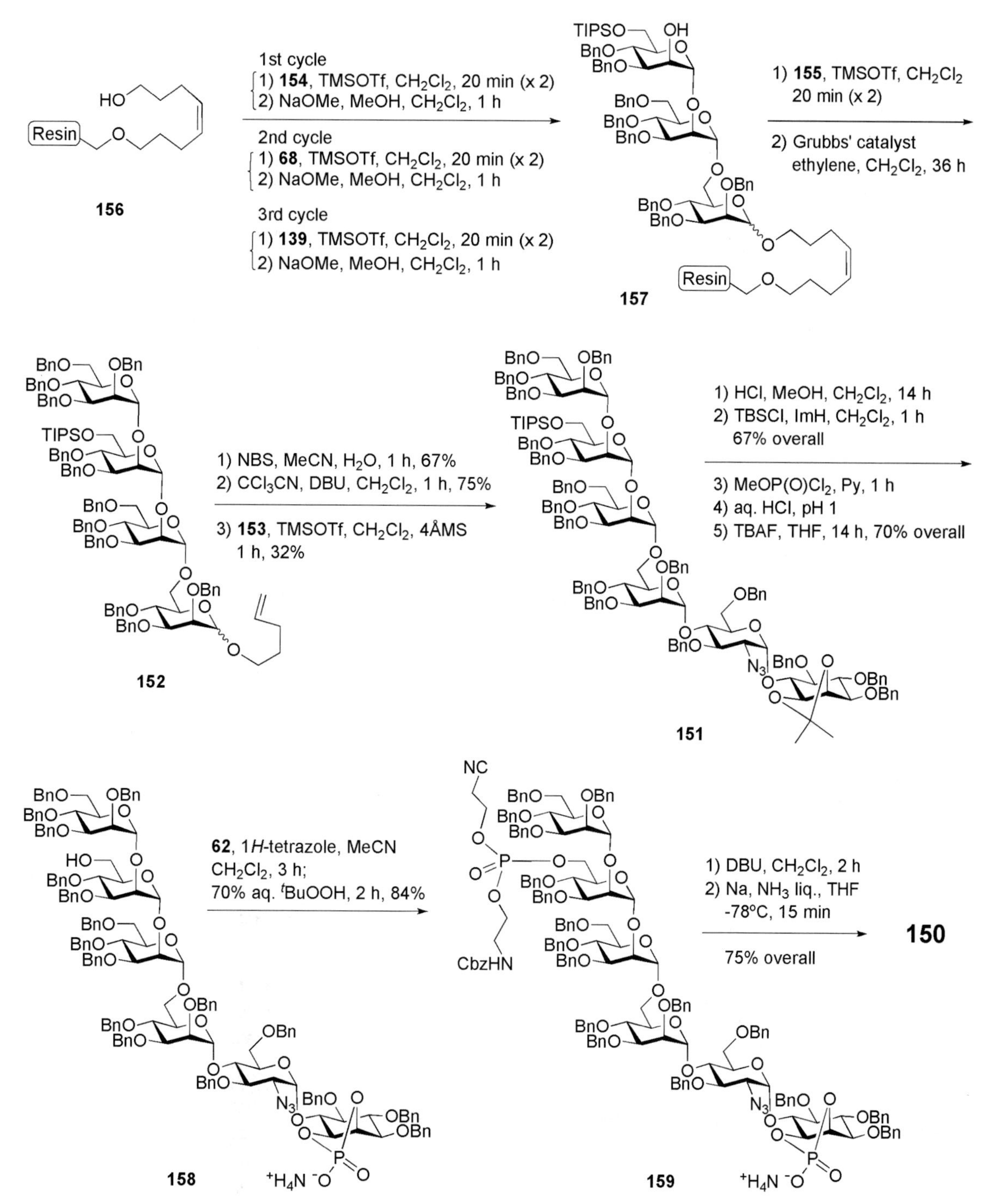

SCHEME 26.29 Solid-phase synthesis of a pseudo-hexasaccharide malarial toxin.

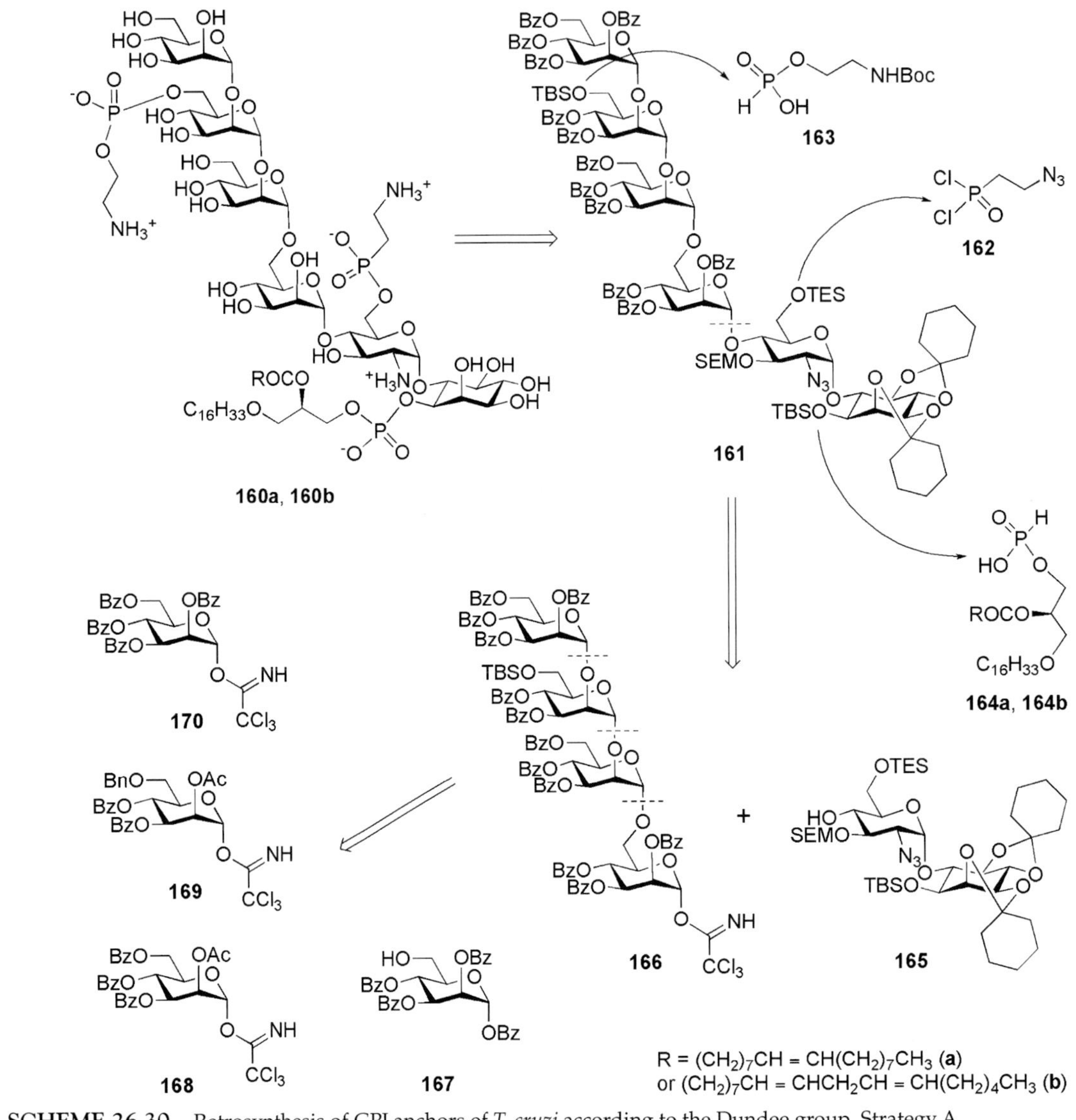

R = $(CH_2)_7CH = CH(CH_2)_7CH_3$ (**a**)
or $(CH_2)_7CH = CHCH_2CH = CH(CH_2)_4CH_3$ (**b**)

SCHEME 26.30 Retrosynthesis of GPI anchors of *T. cruzi* according to the Dundee group, Strategy A.

groups, whereas *Strategy B* suggested the use of acid labile (diacetals, acetals and *N*-Boc) and fluoride ion labile (primary and secondary TBS ethers) groups for this purpose. *Strategy C* implied exploration of acid labile (diacetals, acetals, PMB ethers and *N*-Boc) permanent protecting groups only and was used for the preparation of tGPI from the *T. cruzi* bearing palmitic acid moiety **160c**.

Synthetic strategy A (Yashunsky et al., 2006)

The C-phosphonate and phosphate linkers in the GPI anchors **160a** and **160b** (see Scheme 26.30) were retrosynthetically disconnected sequentially as phosphonodichloridate **162**, *N*-Boc-ethanolamine H-phosphonate **163** and acylakylglycerol H-phosphonates **164a** and **164b**, leaving the corresponding glycan core **161** for further simplification. The core was assembled in a (4 + 2) manner from benzoylated mannotetraose building block **166** and acetal-protected pseudo-disaccharide **165**. Various silyl ethers were employed as orthogonal blocking groups for C-6 of $GlcNH_2$ (TES), C-6 of Man-3 (primary TBS) and C-1 of *myo*-inositol (secondary TBS) to ensure further introduction of the P-containing esters. The tetramannoside **166** was assembled stepwise from four building blocks **167–170** (see Scheme 26.30) in an upstream manner.

The acceptor **167** and the donor **170** were prepared from Man in three common steps each. The trichloroacetimidate donors **168** and **169** were derived (Scheme 26.31) also from Man via the 1,2-orthoester **171** (Dais *et al.*, 1983). Successive de-*O*-acetylation, 6-*O*-silylation, *O*-benzoylation and TFA-catalysed opening of the orthoester cycle gave the hemiacetal **172**, which was *O*-acetylated at the anomeric position, de-*O*-silylated and then 6-*O*-benzylated with benzyl trichloroacetimidate to give **173**. Subsequent anomeric de-*O*-acetylation of the latter compound and treatment with Cl_3CCN in presence of Cs_2CO_3 furnished the trichloroacetimidate donor **169**.

Replacement of the acetyl groups with benzoates in the orthoester **171**, followed by opening of the orthoester ring and reaction with Cl_3CCN in presence of Cs_2CO_3 offered the α-trichloroacetimidate donor **168**. Coupling of **168** with the acceptor **167** under the influence of TMSOTf, followed by selective removal of the lone acetyl group in the presence of benzoates with HCl in methanol-dichloromethane (Byramova *et al.*, 1983) gave the disaccharide acceptor **174** which, on glycosylation with the trichloroacetimidate donor **169** and de-*O*-acetylation, produced the trisaccharide **175**. One more glycosylation of **175** with **170**, followed by replacement of the lone benzyl ether with TBS protecting group gave the tetrasaccharide **176**, which was subjected to anomeric debenzoylation with ethylenediamine and trichloroacetimidation to furnish the tetrasaccharide building block **166**.

Synthesis of the pseudo-disaccharide acceptor **165** (Scheme 26.32) followed on from the Schmidt group's synthetic work. Glycosylation of the diastereomeric mixture (Mayer and Schmidt, 1997) of *myo*-inositol derivatives **177** and **71** with glycosyl donor **66** gave the required pseudo-disaccharide **179** (D-D) in 64% yield, which was separable from the diastereoisomer **178** (D-L) (30%) via flash column chromatography. Manipulation of the protecting group pattern in **179** gave the desired pseudo-disaccharide acceptor **165**. This was achieved via consecutive mild selective de-*O*-acetylation in presence of (-)-menthylcarbonate, 4′,6′-orthoesterification and introduction of the acid labile trimethylsilylethoxymethyl (SEM) group at 3′-OH to give fully protected azidoglucose-inositol **180**. Replacement of the (-)-menthylcarbonate with the TBS group, ring opening of the orthoester (thus forming a mixture of 4′- and 6′-benzoates), followed by standard de-*O*-benzoylation and selective 6-*O*-silylation with Et_3SiCl furnished **165**.

With both building blocks in hand, the assembly of the GPI glycan backbone was only one step away (Scheme 26.33). Glycosylation of the pseudo-disaccharide acceptor **165** with the tetrasaccharide trichloroacetimidate donor **166**, under the influence of TMSOTf catalyst, gave the pseudo-hexasaccharide **161** (71%). The corresponding 6′-OH derivative was produced (85%) after selective cleavage of the TES ether with TBAF. 1*H*-Tetrazole-assisted esterification of the 6′-OH group with phosphonodichloridate **162**, followed by methanolysis gave the corresponding phosphonic acid diester (79%), which was subjected to successive reduction of the two azido groups employing Staudinger conditions and *N*-protection with Boc anhydride to afford **181**. Selective removal of the primary TBS group, phosphitylation with the H-phosphonate **163** and subsequent oxidation with iodine gave phosphate–phosphonate block **182** (88%). Cleavage of the secondary TBS group in the *myo*-inositol moiety enabled the final addition of the phospholipid by coupling

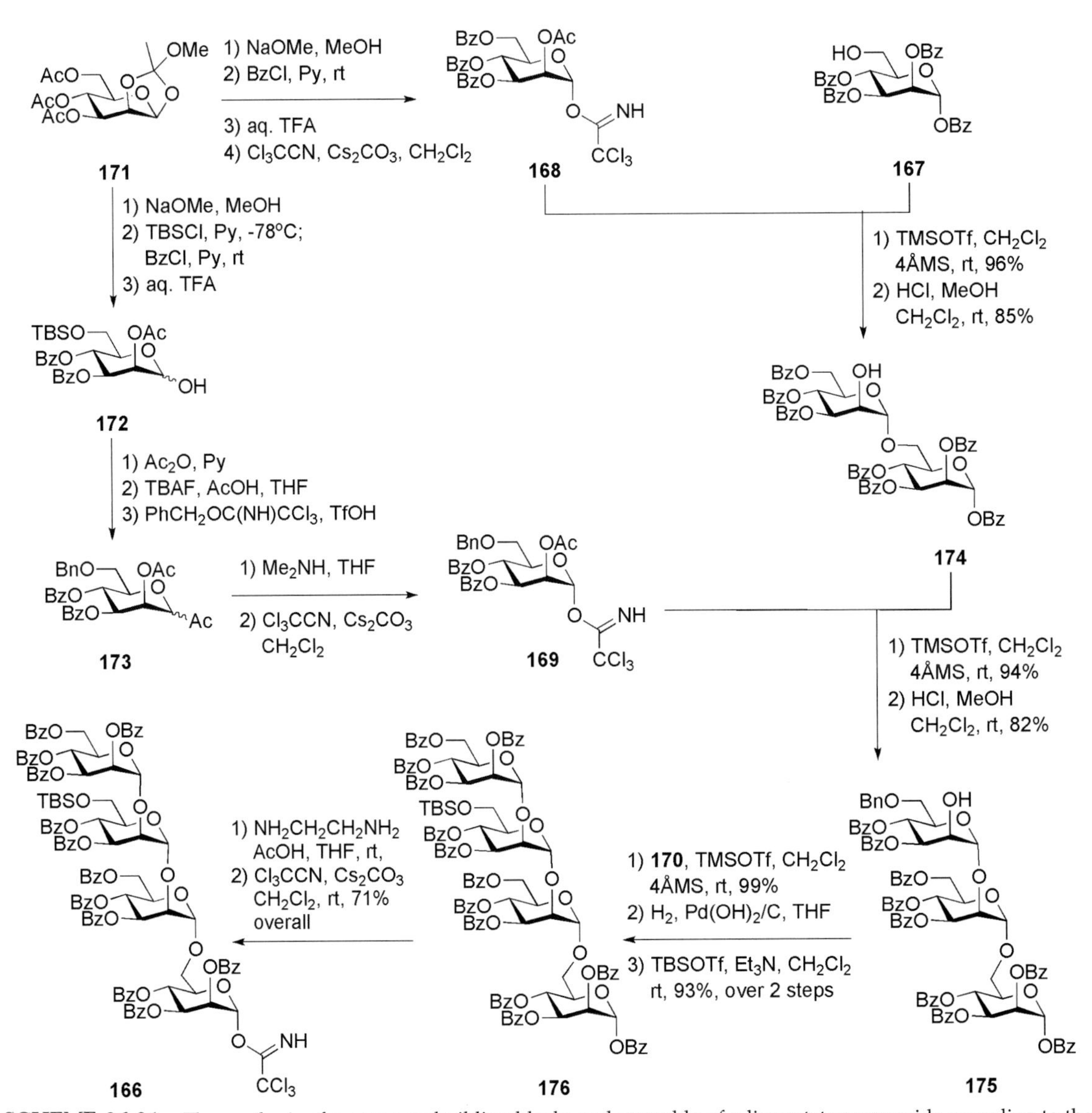

SCHEME 26.31 The synthesis of D-mannose building blocks and assembly of a linear tetramannoside according to the Dundee group, Strategy A.

with the corresponding acylalkylglycerol H-phosphonate (**164a** or **164b**) to afford fully protected GPIs, which were demethylated at the aminoethylphosphonate moiety to furnish **183a** and **183b**. Global deprotection involved first controlled de-*O*-benzoylation, which provided corresponding debenzoylated GPIs (38–40%) isolated by flash column chromatography. Mild basic treatment in polar solvent cleaved benzoic esters preferentially and left the fatty ester of the lipid mostly intact, probably, because of the micelle formation. Final removal of the acid labile acetal and *N*-Boc protections with aq. TFA gave the required products **160a** and **160b**.

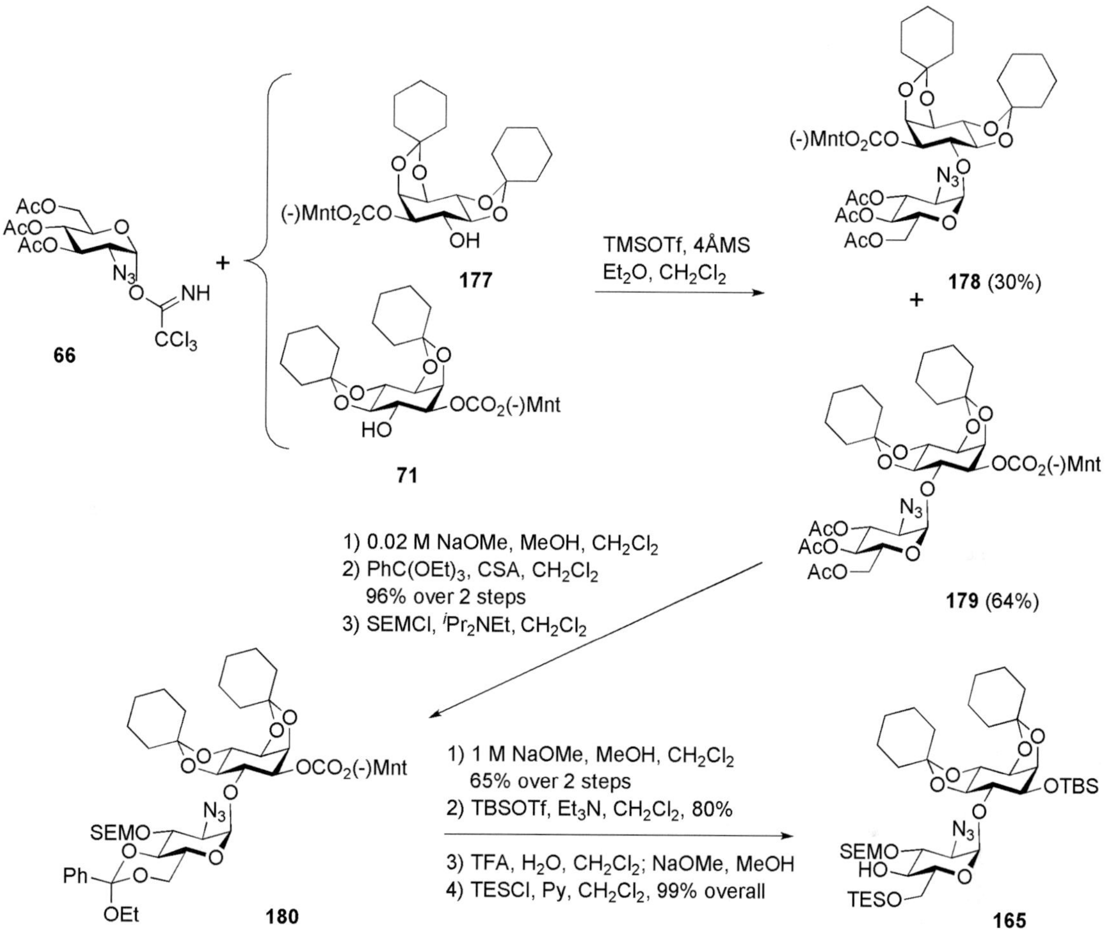

SCHEME 26.32 *myo*-inositol enantiomeric resolution and the synthesis of a pseudo-disaccharide after the Dundee group, Strategy A.

Synthetic strategy B (Yashunsky et al., 2007)

In the second approach, the pseudo-hexasaccharide core **184** (Scheme 26.34) was assembled by (3 + 3) coupling of derivatives **185** and **186**. These building blocks were designed utilizing acid and fluoride ion labile permanent protecting groups while also using orthogonal protecting groups (TES ether at C-6 of $GlcNH_2$, phenoxyacetate at C-6 of Man-3 and benzoic ester at C-1 of *myo*-inositol) required to achieve further "P-decoration" at late stages. The pseudo-trisaccharide **186** was formed from the pseudo-disaccharide **190** and the thiomannoside donor **191**. The trimannoside **185** was made up of three mannosyl derivatives **187–189**.

The trimannoside **185** was constructed (Scheme 26.35) starting from coupling the trichloroacetimidate **188** and the acceptor **189** (83%), followed by de-*O*-benzyolation and 6-*O*-phenoxyacetylation to furnish the disaccharide acceptor **192** which, on glycosylation with the thioglycoside **187** in the presence of MeOTf, gave the trimannoside **185** (65%). The pseudo-trisaccharide block **186** was assembled from the thiomannoside donor **191** and azidoglucose-inositol derivative **190**. The latter was made from

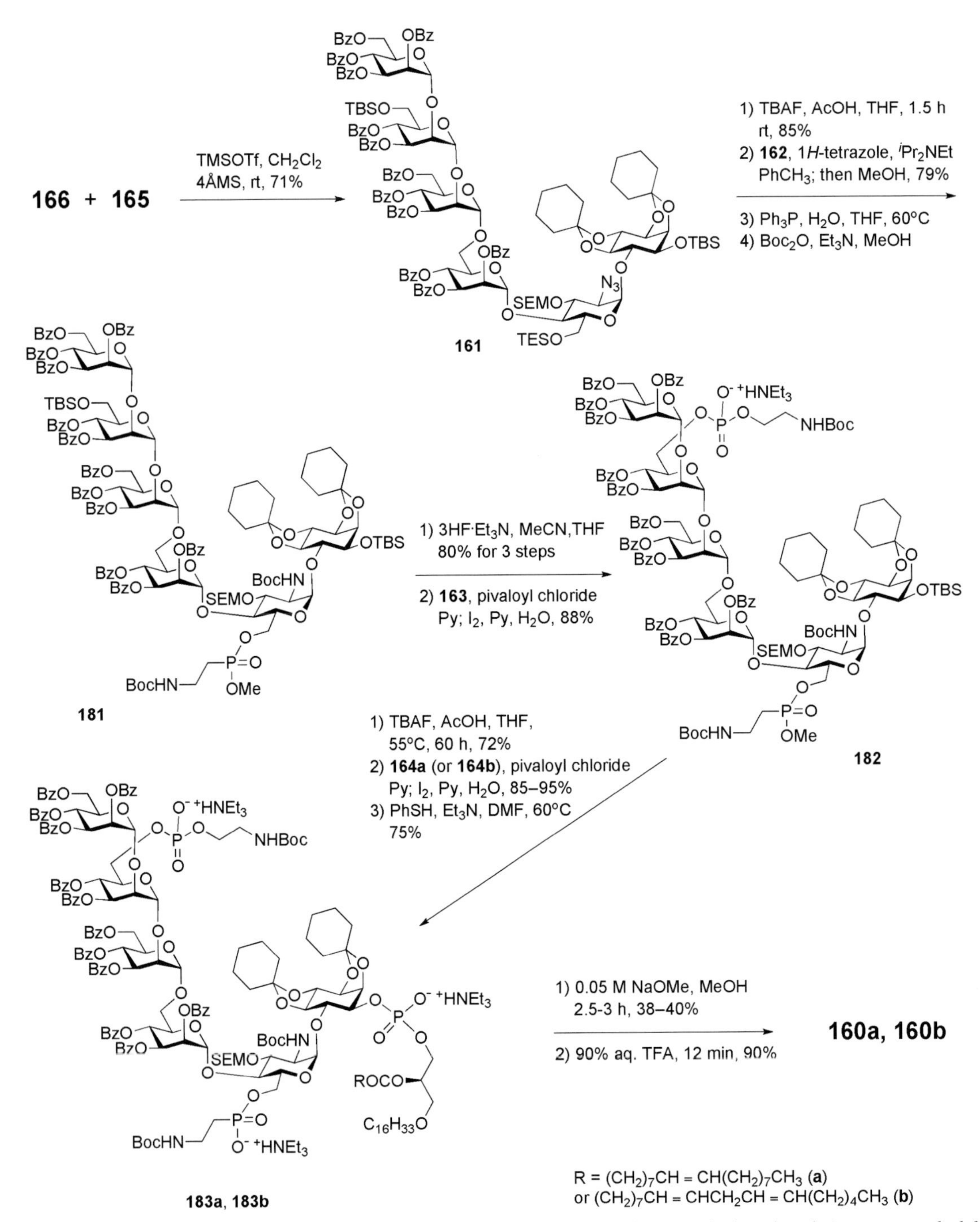

SCHEME 26.33 *T. cruzi* GPI glycan core assembly, successive phosphonylation and phosphorylation steps and global deprotection according to the Dundee group, Strategy A.

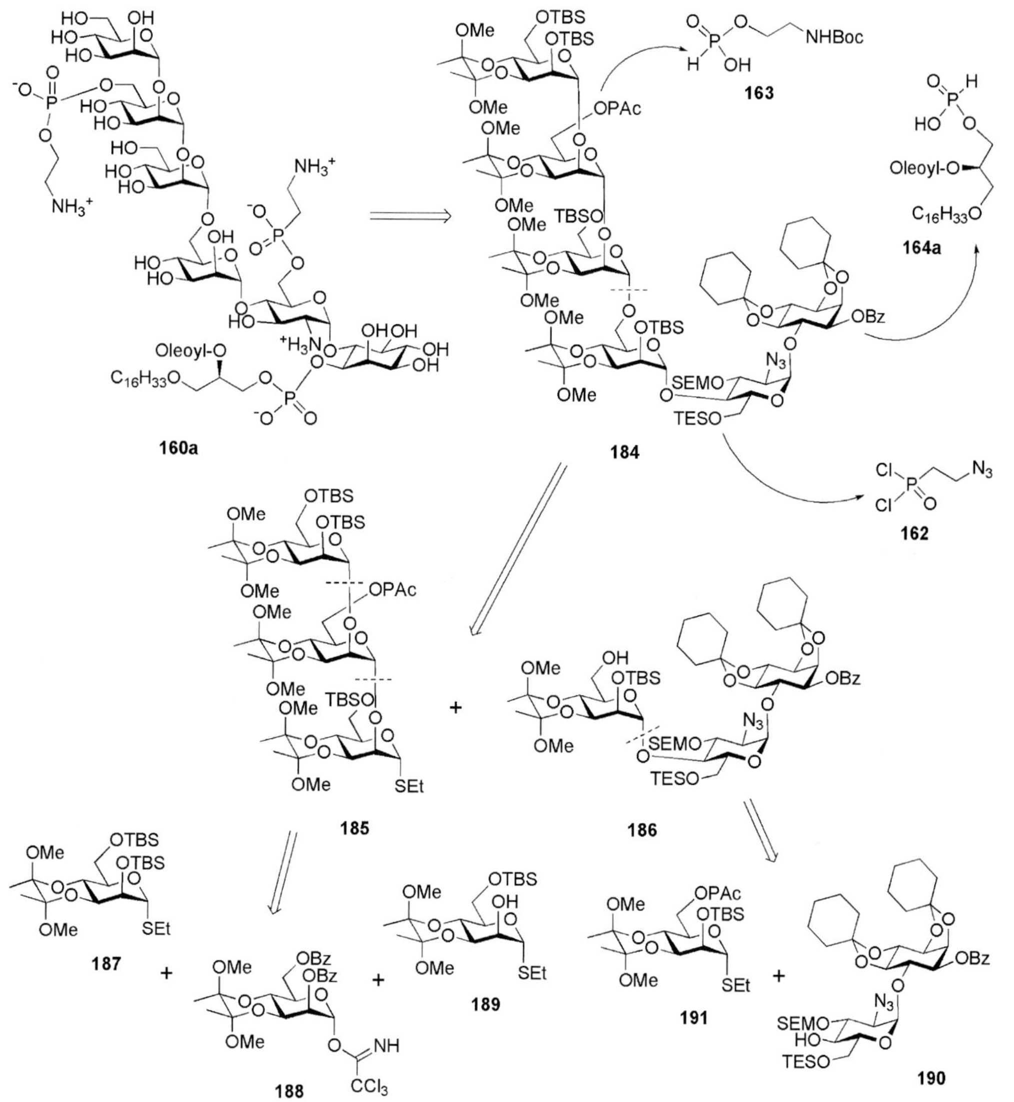

SCHEME 26.34 Retrosynthesis of a GPI anchor of *T. cruzi* after the Dundee group, Strategy B.

the pseudo-disaccharide **179** via manipulation of the protecting group pattern as described in Scheme 26.35. It was also discovered that more efficient separation of the diastereomers **178** and **179** (see Scheme 26.32) could be achieved after addition of the tetraisopropyldisiloxane (TIPDS) 4′,6′-*O*-protection. Common replacement of the (-)-menthylcarbonate with benzoate in the TIPDS-protected derivative **193**, followed by TIPDS cleavage (with TBAF) and selective 6′-*O*-silylation with TESCl furnished **190**. Methyl triflate-promoted glycosylation of the pseudo-disaccharide acceptor **190** with the thioglycoside **191**, in the presence of 2,6-di-*tert*-butyl-4-methylpyridine (DTBMP) to avoid cleavage of the acid-labile groups, followed by removal of the

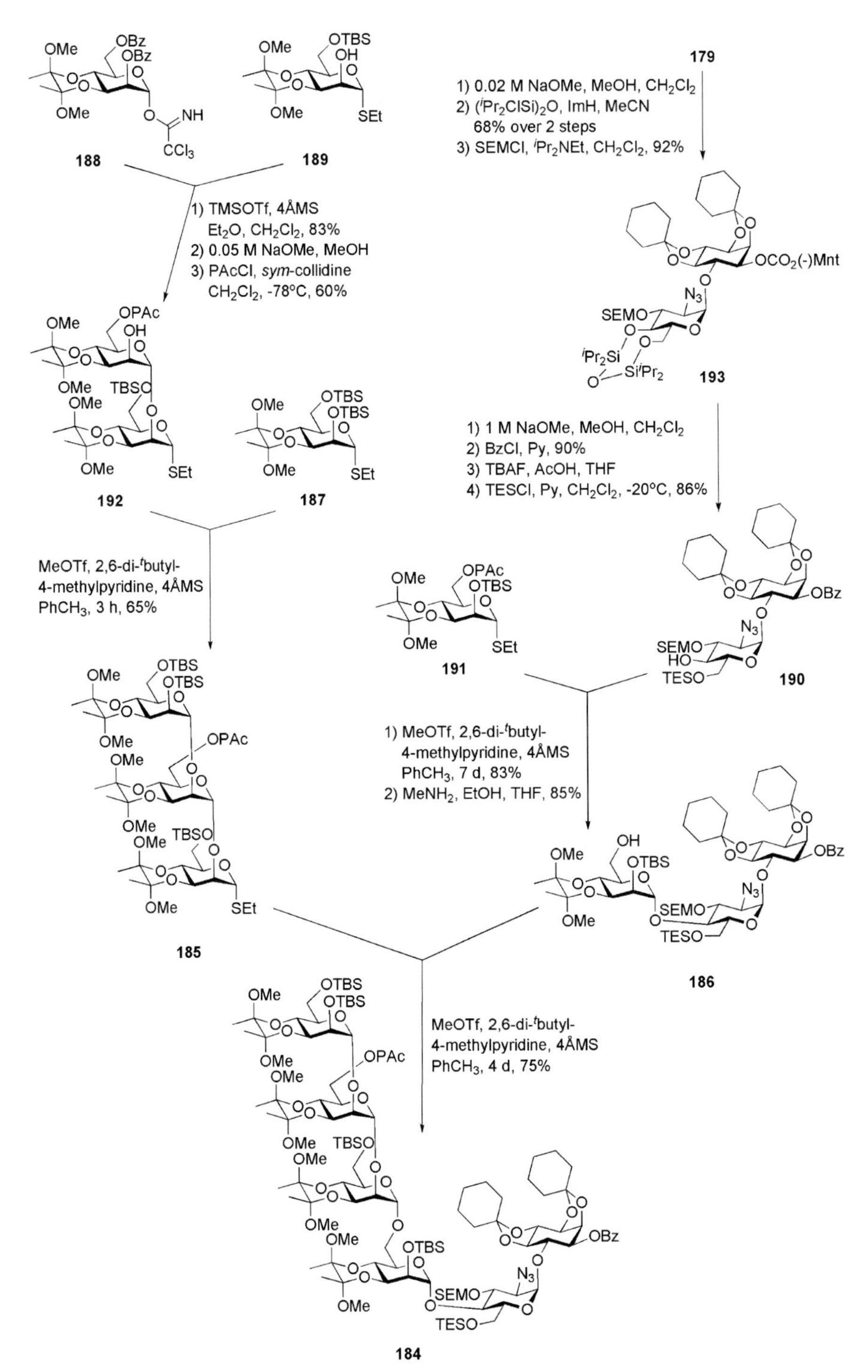

SCHEME 26.35 The synthesis of a linear trimannoside and pseudo-trisaccharide; *T. cruzi* GPI glycan core assembly according to the Dundee group, Strategy B.

phenoxyacetate ($MeNH_2$), gave the pseudo-trisaccharide **186** (71% yield over two steps). The latter was coupled to the trimannoside **185** under the conditions described above to furnish the pseudo-hexasaccharide **184** (75%).

Selective removal of the TES ether in the azidoglucose moiety in **184** (Scheme 26.36), followed by introduction of the azidoethyl-phosphonate with **162** (77%) and conversion of the azido groups into Boc-protected amines, as performed in Strategy A, gave the phosphonic acid ester **194**. Selective cleavage of the phenoxyacetate ($MeNH_2$), followed by introduction of the *N*-Boc-protected ethanolamine-phosphate by the use of the H-phosphonate **163** gave the phosphate–phosphonate block **195** (80%). Removal of the benzoate in the *myo*-inositol moiety (NaOMe, MeOH, 95%) enabled the final addition of the phospholipid (using the H-phosphonate **164a**, 65%) followed by demethylation (with thiophenol) to give the protected GPI **196**. Global deprotection involved removal of silyl ethers with $3HF.Et_3N$ and the acid labile acetals and *N*-Boc groups with aq. TFA to give the required product **160a**.

Synthetic strategy C (N. Al-Maharik and A.V. Nikolaev, unpublished results)

In the third approach, the glycan core **197** (Scheme 26.37) was retrosynthetically broken down into three major building blocks: the mannobiosides **198** and **199** and the pseudo-disaccharide **190**. The disaccharides **198** and **199** were further disconnected to the mannosyl derivatives **200–203**. These building blocks were designed in such a way that only acid labile protecting groups, such as cyclohexylidene and SEM acetals, butanediacetal (BDA), *p*-methoxy-benzyl ethers and primary TBS ether, which are known to be stable enough towards miscellaneous reaction conditions, were used for permanent OH-protection. Their removal at once in the final deprotection step was anticipated. The orthogonal protecting groups required to execute further "P-decoration" of the glycan core **197** and to prepare the GPI anchor **160c** were identical to those used in Strategy B (see Scheme 26.34).

The derivatives **200–203** derived from BDA-protected thiomannoside **204** (Scheme 26.38) prepared, in turn, by selective incorporation of the BDA motif at the 3,4-vicinal diol (Baeschlin *et al.*, 2000b). Selective 6-*O*-silylation of **204** and subsequent benzylation with *p*-methoxybenzyl chloride and desilylation gave **203**. 2,6-Di-*O*-methoxybenzylation of **204** followed by subsequent NIS-TfOH-mediated hydrolysis of the ethanethiol group and treatment with Cl_3CCN in presence of DBU furnished the trichloroacetimidate **200**. Coupling of the donor **200** with acceptor **201** (prepared by selective 6-*O*-phenoxyacetylation of **200**) under the influence of TMSOTf gave the corresponding disaccharide (88%) which, upon hydrolysis of the ethanethiol and installation of the trichloroacetimidate, gave the disaccharide glycosyl donor **198**. The dimannoside building block **199** was assembled by TMSOTf activated glycosylation of **203** with the trichloroacetimidate **202**, prepared from **204** by successive 6-*O*-silylation, 2-*O*-phenoxyacetylation, anomeric deprotection and trichloroacetimidation.

The protected mannobiose **199** intermediate is ideally suited to further glycan chain extension either upstream or downstream. For upstream extension, a portion of the compound was transformed into the trichloroacetimidate donor **205** (see Scheme 26.38). This was used for TMSOTf-promoted glycosylation of the pseudo-disaccharide **190** (Scheme 26.39) to give the corresponding pseudo-tetrasaccharide (87%) in which the phenoxyacetyl group was removed to yield the acceptor **208**. This was further glycosylated with the donor **198** in the presence of TMSOTf to give the glycan core **197** (88%), in which the TES group was removed with TBAF to yield the pseudo-hexasaccharide **209**. A second portion of **199** was converted to the acceptor **206** (required for downstream chain assembly) by removal of the phenoxyacetyl group. The glycosylation of **206** with the donor **198** (93%),

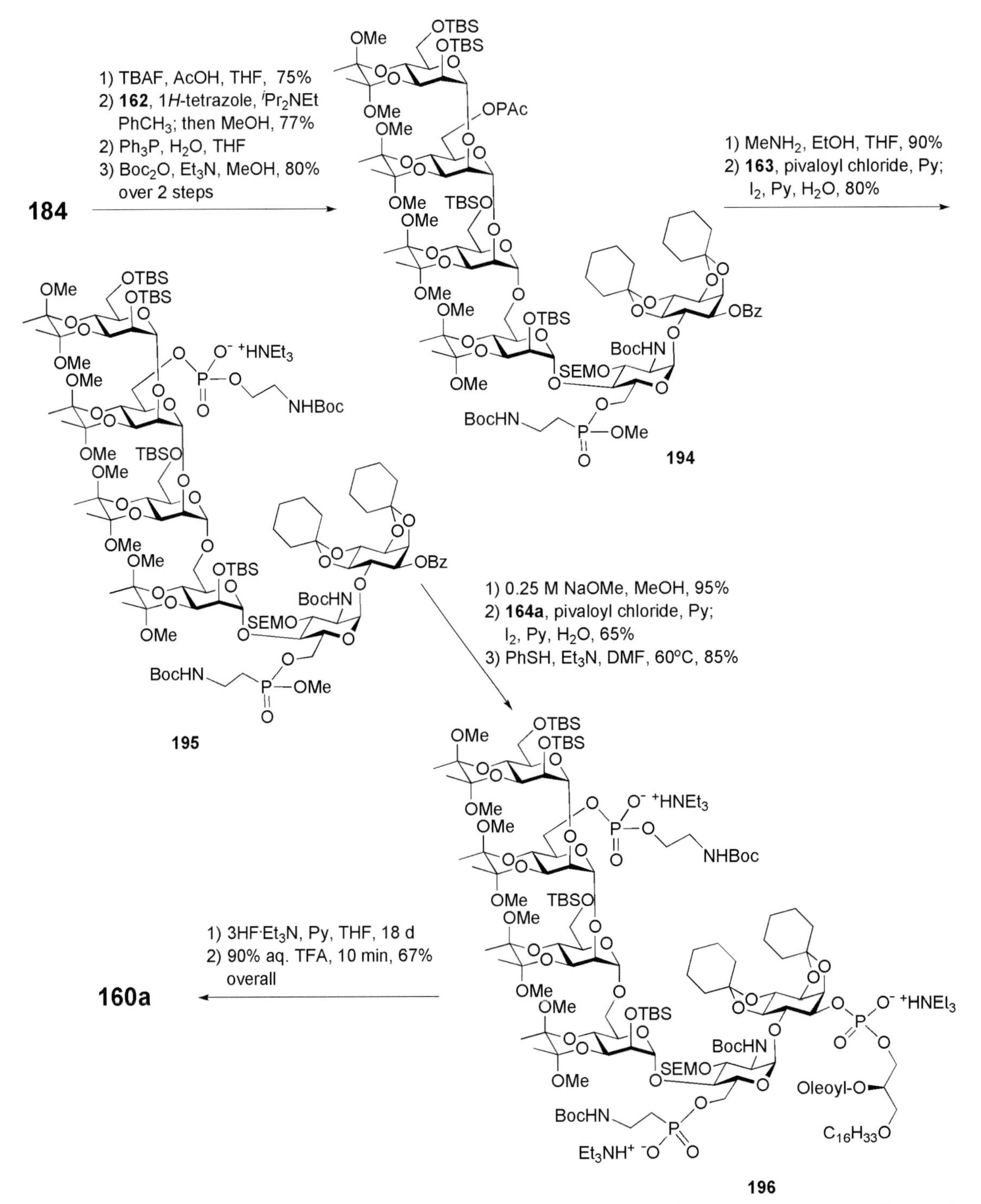

SCHEME 26.36 Successive phosphonylation and phosphorylation steps and global deprotection according to the Dundee group, Strategy B.

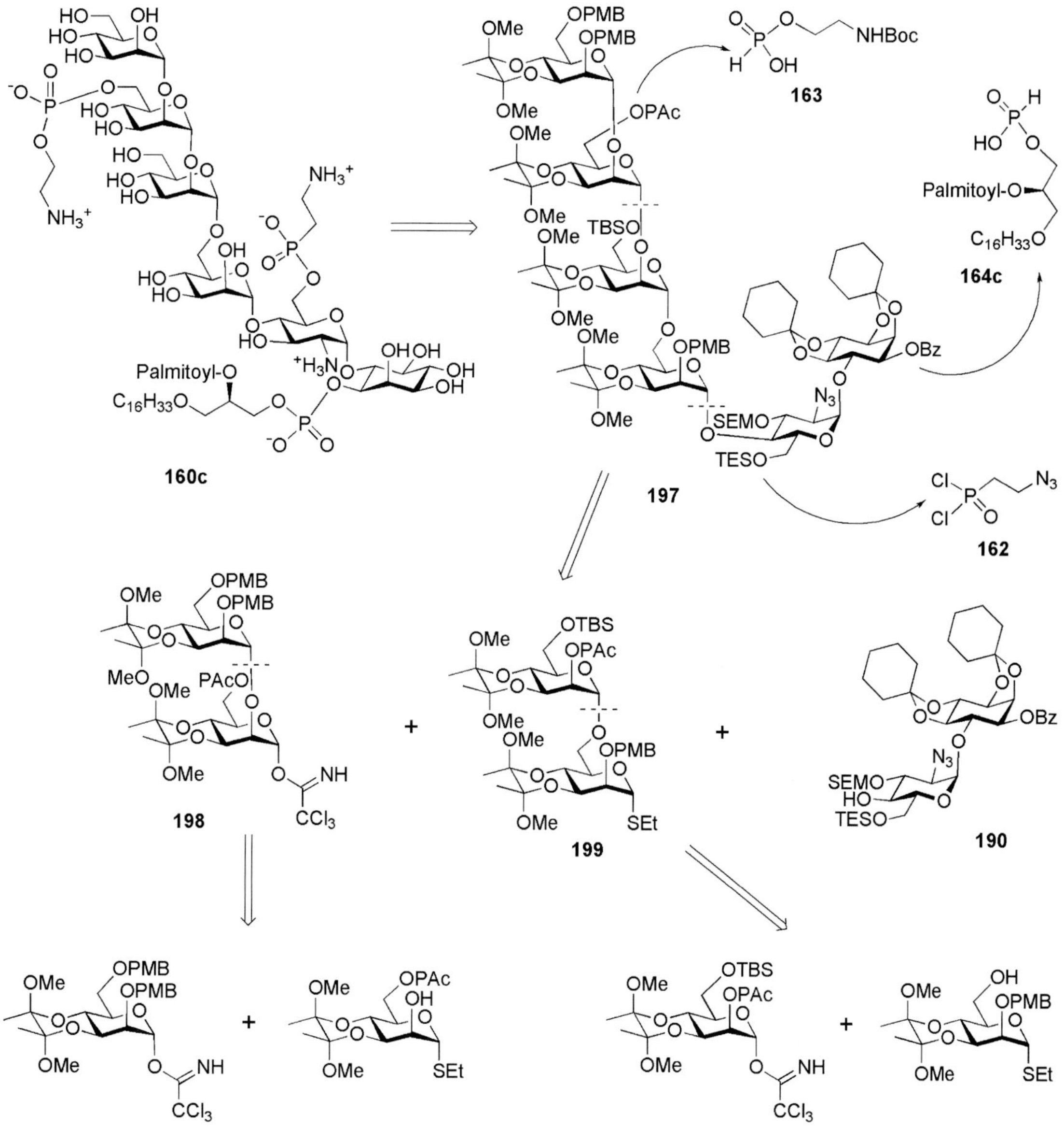

SCHEME 26.37 Retrosynthesis of a GPI anchor of *T. cruzi* after the Dundee group, Strategy C.

followed by anomeric deprotection and trichloroacetimidation of the hemiacetal afforded the tetrasaccharide trichloroacetimidate **207**. The latter was coupled (TMSOTf) with the acceptor **190** (→**197**, 92%), followed by selective TES removal to afford the pseudo-hexasaccharide **209**.

Succesive "P-decoration" of the glycan core (Scheme 26.40) was achieved as described above in Strategy B. First, 1*H*-tetrazole-assisted phosphonylation of the alcohol function in **209** with phosphonodichloridate **162** followed by methanolysis afforded the phosphonic acid

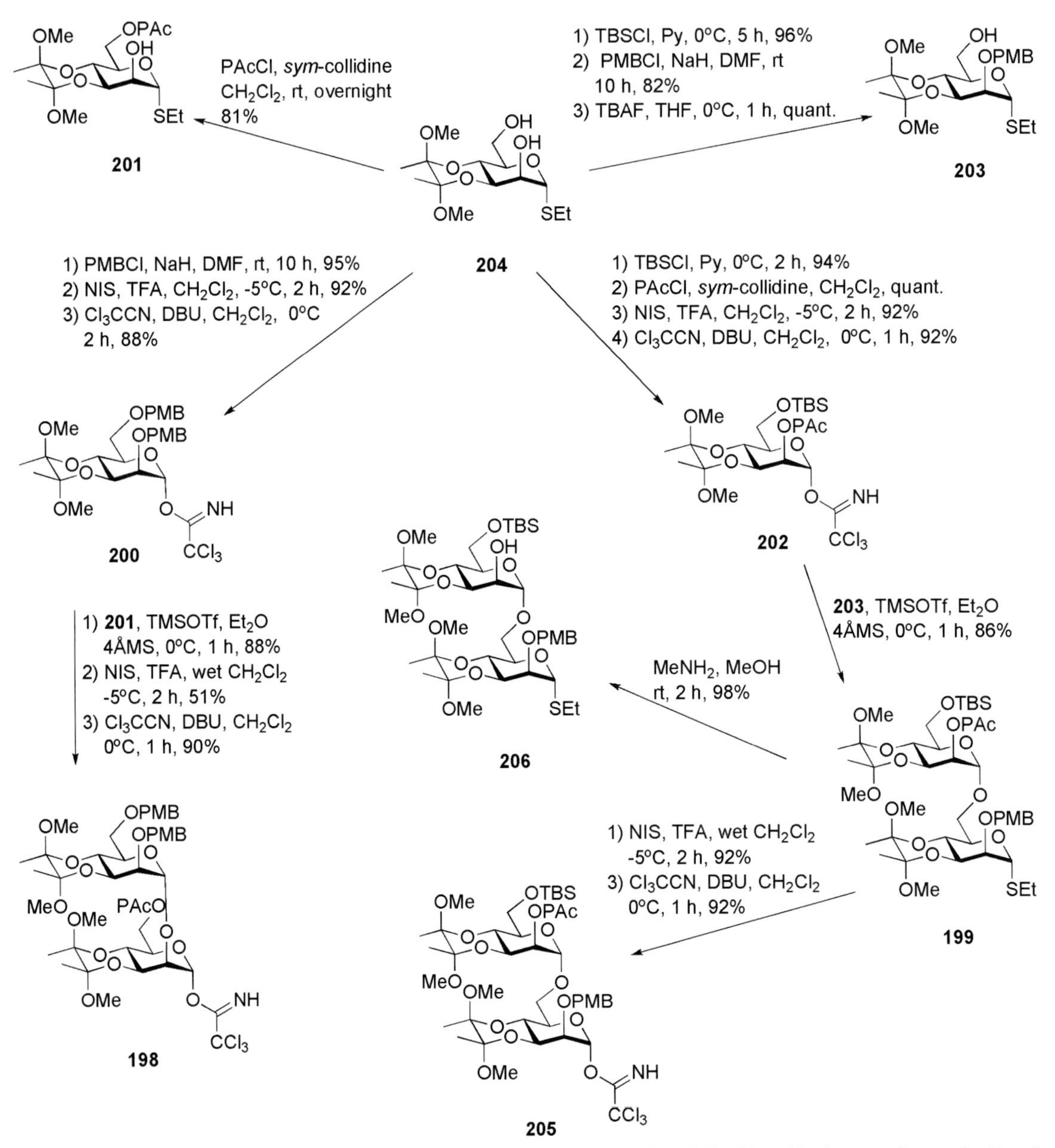

SCHEME 26.38 The synthesis of D-mannose monosaccharide and disaccharide building blocks according to the Dundee group, Strategy C.

diester (60%) in which the two azido groups were reduced with PPh_3 and *N*-protected with Boc-anhydride to afford the phosphonylated block **210**. Removal of the PAc group in **210** and subsequent pivaloyl chloride-assisted phosphitylation with the H-phosphonate **163**, followed by oxidation of the intermediate with iodine gave the phosphate-phosphonate block **211** (79%).

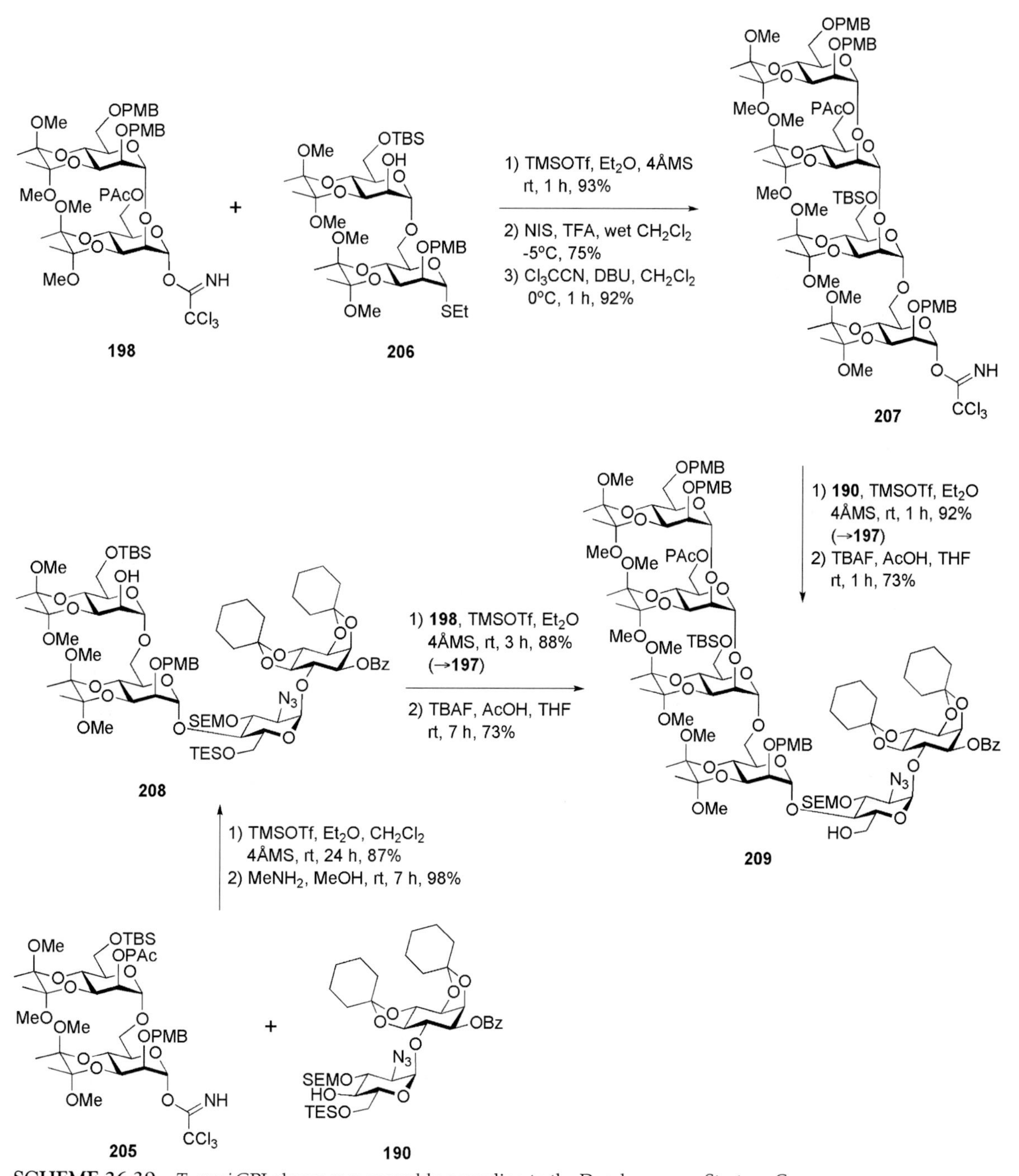

SCHEME 26.39 *T. cruzi* GPI glycan core assembly according to the Dundee group, Strategy C.

Cleavage of the benzoic ester in the *myo*-inositol enabled the successful phospholipidation (86%) using the palmitoylhexadecylglycerol H-phosphonate **164c**. The *O*-methyl group in the aminoethylphosphonate moiety was removed with thiophenol to afford **212** (85%). Global deprotection of **212** was successfully carried out in a one-step procedure with 90% aq. TFA to provide **160c** in almost quantitative yield.

2.5.2. The New Delhi group approach

The New Delhi group (Vishwakarma and co-workers) reported (Ali *et al.*, 2005) a downstream (2 + 2+ 2) approach to synthesize the GPI anchor **213** of the *T. cruzi* IG7 antigen (Scheme 26.41) via the efficient construction of the glycan core **214** from the azidoglucose-inositol **64** and mannotetraose **216** intermediates. The tetramannoside **216** was retrosynthetically broken down into two disaccharide building blocks **217** and **218**, which were further disconnected to Man monosaccharide derivatives.

The assembly of the tetramannoside **216** (Scheme 26.42) began from coupling of acceptor **219** with trichloroacetimidate **155** to afford a disaccharide intermediate (81%); this was subjected to the anomeric *O*-allyl and the 4,6-*O*-benzylidene removals followed by *O*-acetylation to afford the disaccharide **220**. Anomeric deacetylation followed by Schmidt activation (CCl_3CN, DBU) provided the required glycosyl donor **217**. Glycosylation of the acceptor **221** with the orthoester donor **116** in the presence of TESOTf-NIS (72%), followed by removal of the *O*-benzoyl group from the position-2 gave mannobioside **218**. The latter was coupled with the donor **217** (69%), followed by anomeric deallylation and trichloroacetimidation to give the tetramannose donor **216**. The TMSOTf-activated glycosylation of the pseudo-disaccharide **64** (Pekari *et al.*, 2001) with **216** afforded the pseudo-hexasaccharide intermediate (60%), in which the *O*-acetyl groups were removed and the primary 6″″-OH was first selectively silylated prior to benzylation of 4″″-OH to provide the glycan core **214**. Successive de-6″″-O-silylation, 1*H*-tetrazole-assisted phosphitylation with *N*-Cbz-aminoethyl phosphoramidite **27**, followed by *m*-CPBA oxidation gave the phosphotriester **222** (60%). Removal of the PMB group from the position-1 of the *myo*-inositol moiety, followed by phospholipidation with 1-*O*-stearyl-2-*O*-stearoylglycerol H-phosphonate **215**, provided **223** (55%) which, on global deprotection by hydrogenolysis in the presence of $Pd(OH)_2$/C, gave the targeted GPI anchor **213**.

2.5.3. The Linköping group approach

The Linköping University group (Hederos and Konradsson, 2005, 2006) developed a convergent methodology for preparation of the phosphorylated heptasaccharyl *myo*-inositol **224** (Scheme 26.43) found in the *T. cruzi* lipopeptidophosphoglycan (de Lederkremer *et al.*, 1991). This glycan is attached to a ceramide lipid anchor and constitutes a major cell membrane component in the proliferative epimastigote stage of *T. cruzi* that is found in the insect vector. Structurally, the phosphoglycan **224** consists of the basic GPI sequence Man_4GlcNH_2Ino with extra 2-aminoethylphosphonic ester at C-6 of $GlcNH_2$ and two β-D-galactofuranose branches linked to Man-3 and Man-4 residues. The synthetic plan relied on (4 + 2 + 2) downstream assembly strategy and the use of the tetrasaccharide **225** and the disaccharide **226** building blocks, whereas the phosphonate–phosphate decorated pseudo-disaccharide **227** was introduced just before the global deprotection steps. Tetrasaccharide block **225** was further disconnected into two smaller Gal*f*-Man building blocks **228** and **229**.

Construction of the branched hexasaccharide **236** (Scheme 26.44) began with the regioselective glycosylation of the diol **233** (Sarbajna *et al.*, 1998) with galactofuranose donor **232** (Lerner, 1996). The 2-*O*-acetyl participating group in the donor assists the reaction toward the formation

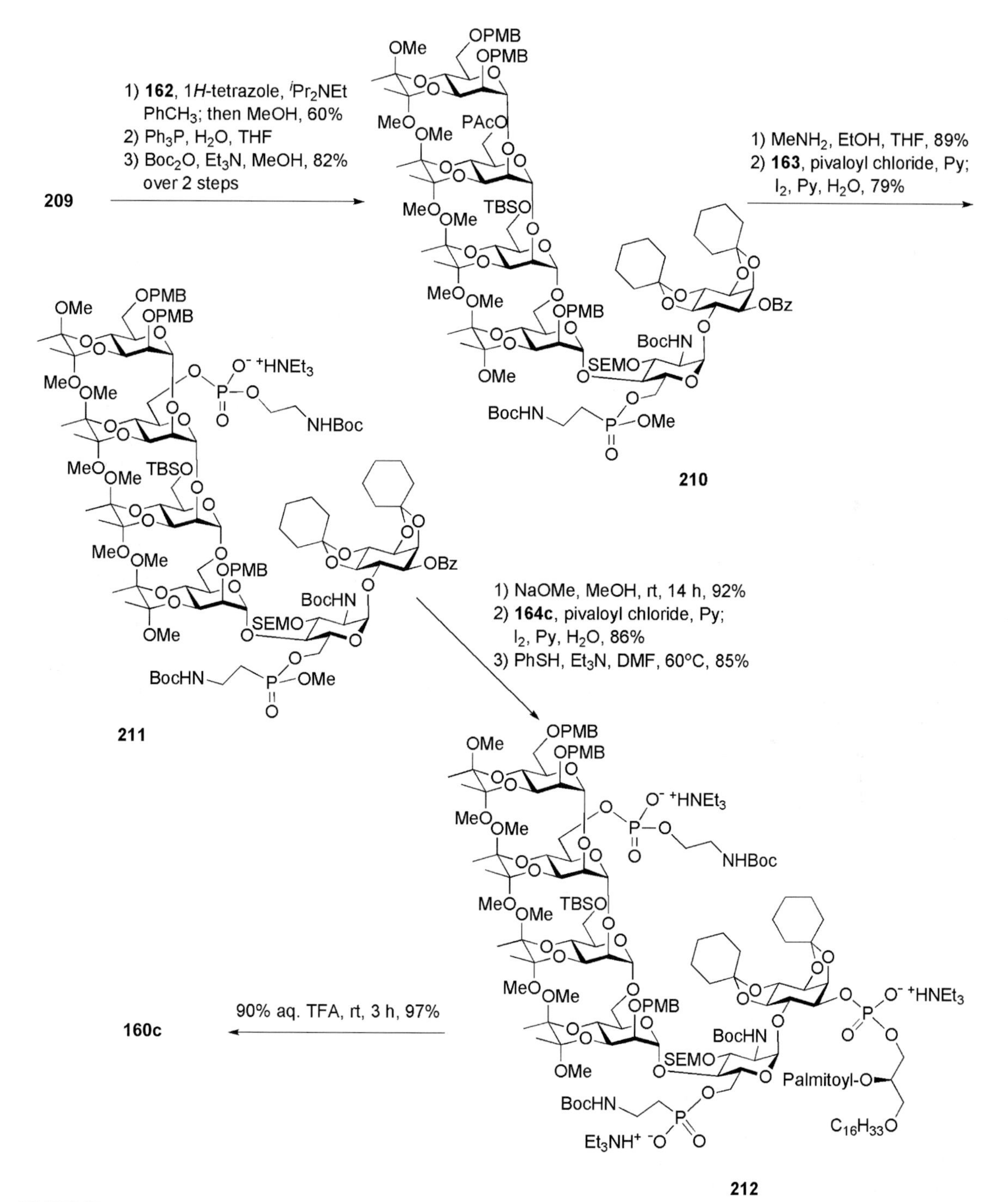

SCHEME 26.40 Successive phosphonylation and phosphorylation steps and global deprotection after the Dundee group, Strategy C.

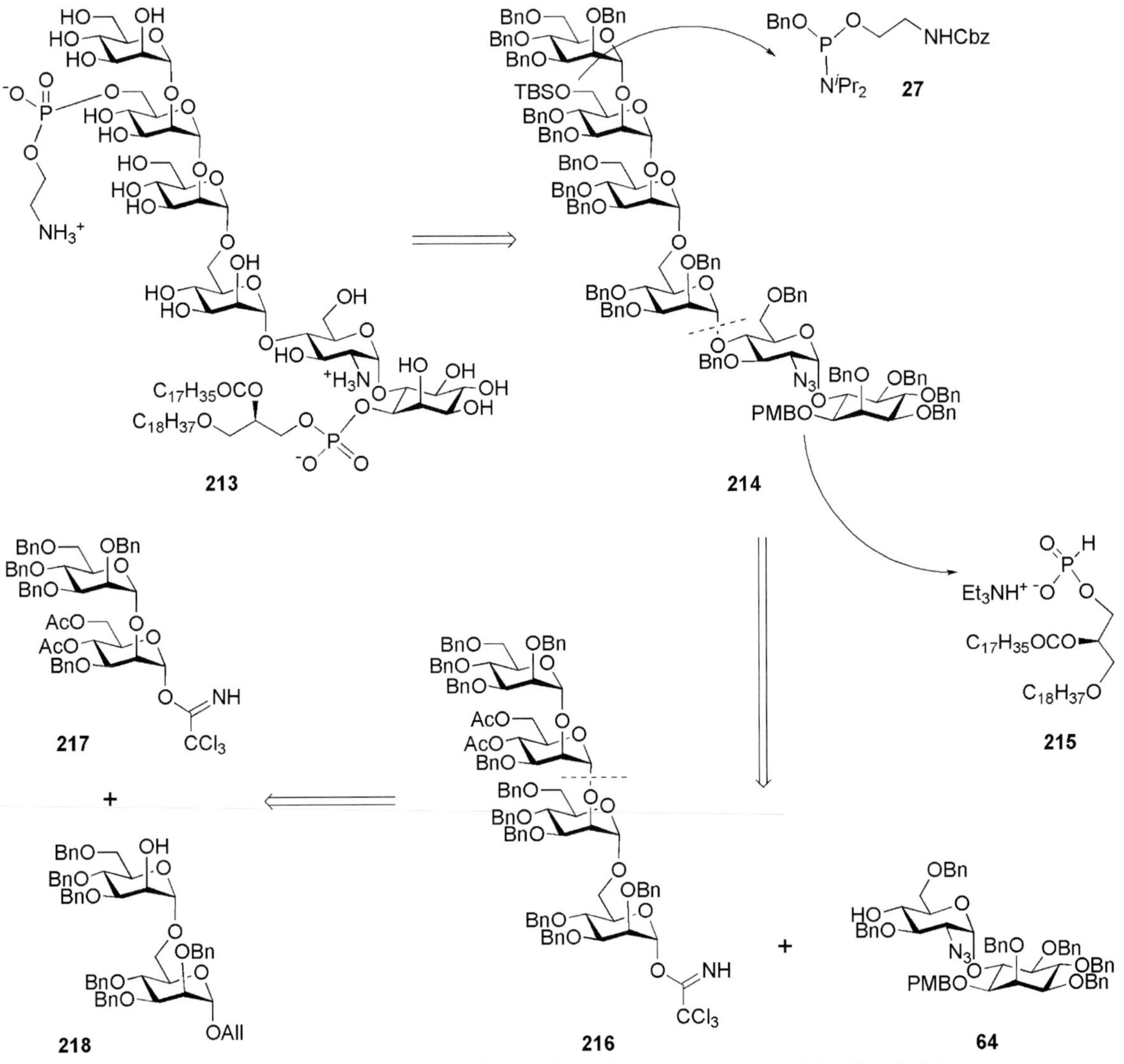

SCHEME 26.41 Retrosynthesis of a GPI anchor of *T. cruzi* according to the strategy of the New Delhi group.

of the targeted β-(1→3)-linked disaccharide **229**. The acetyl groups in **229** were then replaced with benzyl ethers followed by conversion into diglycosyl bromide donor **228** by treatment with bromine. Compound **228** was coupled with acceptor **229** using AgOTf as promoter to give the corresponding tetrasaccharide (60%), in which the anomeric ethanethiol group was replaced with the trichloroacetimidate group to furnish the tetrasaccharide donor **225**. This was used in glycosylation of the disaccharide **226** using TMSOTf as a catalyst to give the hexasaccharide **236** (89%). Compound **226** was, in turn, prepared by condensation of glycosyl bromide **234** (Elie *et al.*, 1990) with Man acceptor **235** (Ottosson, 1990) followed by de-2′-*O*-benzoylation.

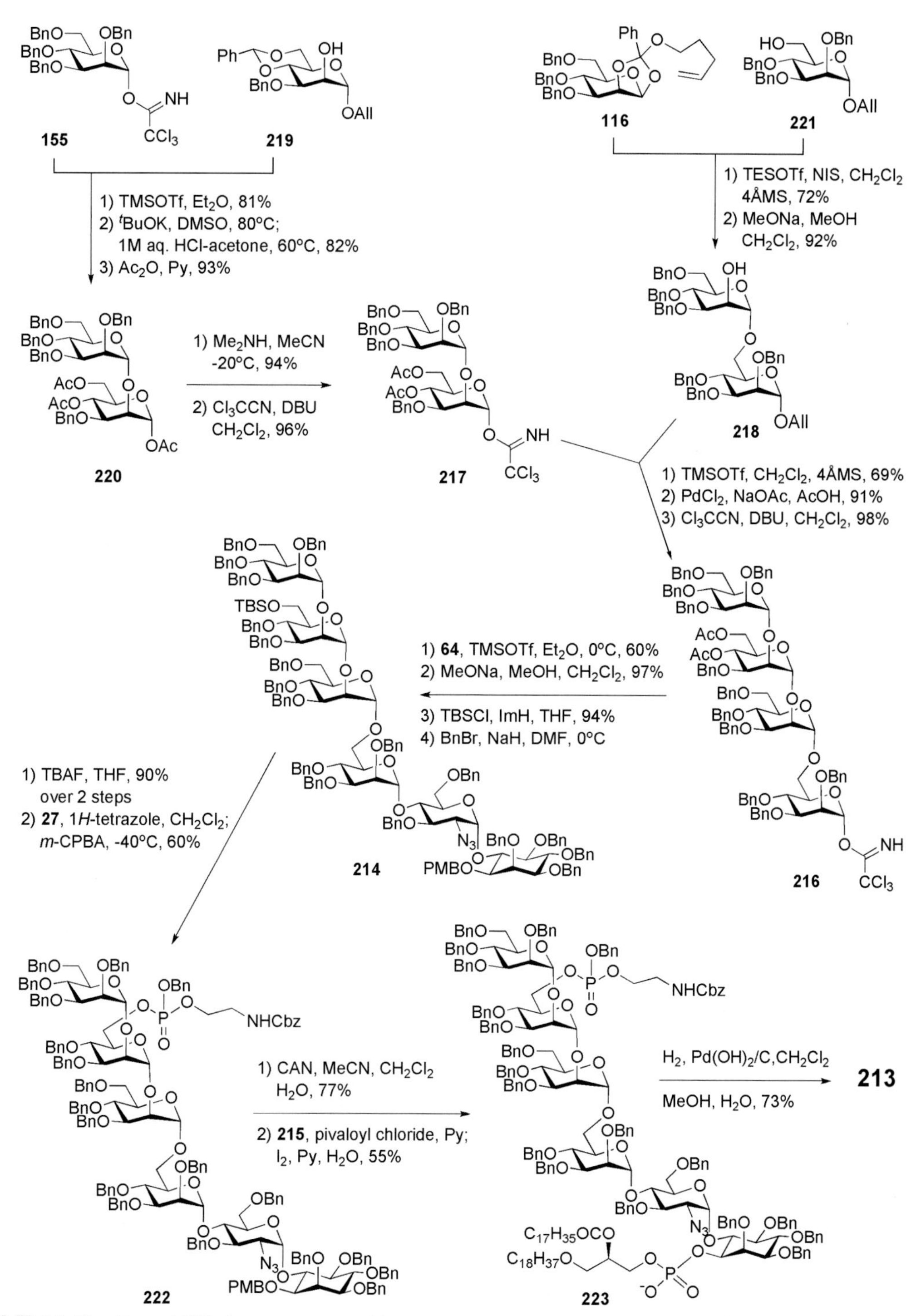

SCHEME 26.42 *T. cruzi* GPI glycan core assembly, successive phosphorylation steps and global deprotection according to the procedure of the New Delhi group.

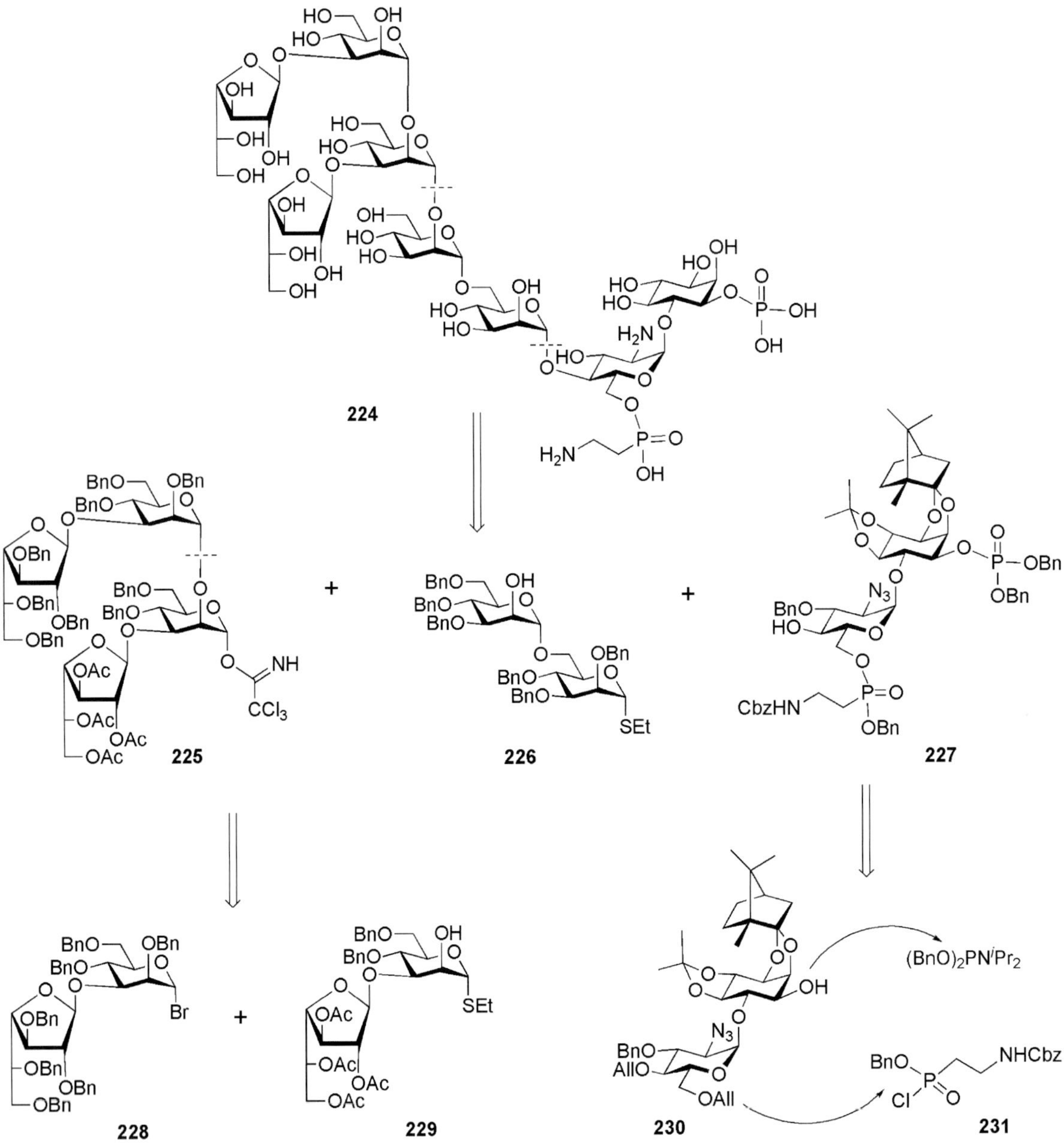

SCHEME 26.43 Retrosynthesis of a GPI-related glycoconjugate of *T. cruzi* after the strategy of the Linköping group.

Glycosylation of the chiral D-camphor acetal-protected *myo*-inositol acceptor **107** (see Section 2.3) with glycosyl sulfoxide donor **237** (Scheme 26.45) using triflic anhydride (Tf_2O) as promoter gave the pseudo-disaccharide **238**. Protecting group manipulation by the removal of pivalic esters and introduction of the 4,5-*O*-isopropylidene acetal gave the required pseudo-disaccharide synthon **230** that was further 1-*O*-phosphorylated (→**239**, 85%) using phosphoramidite

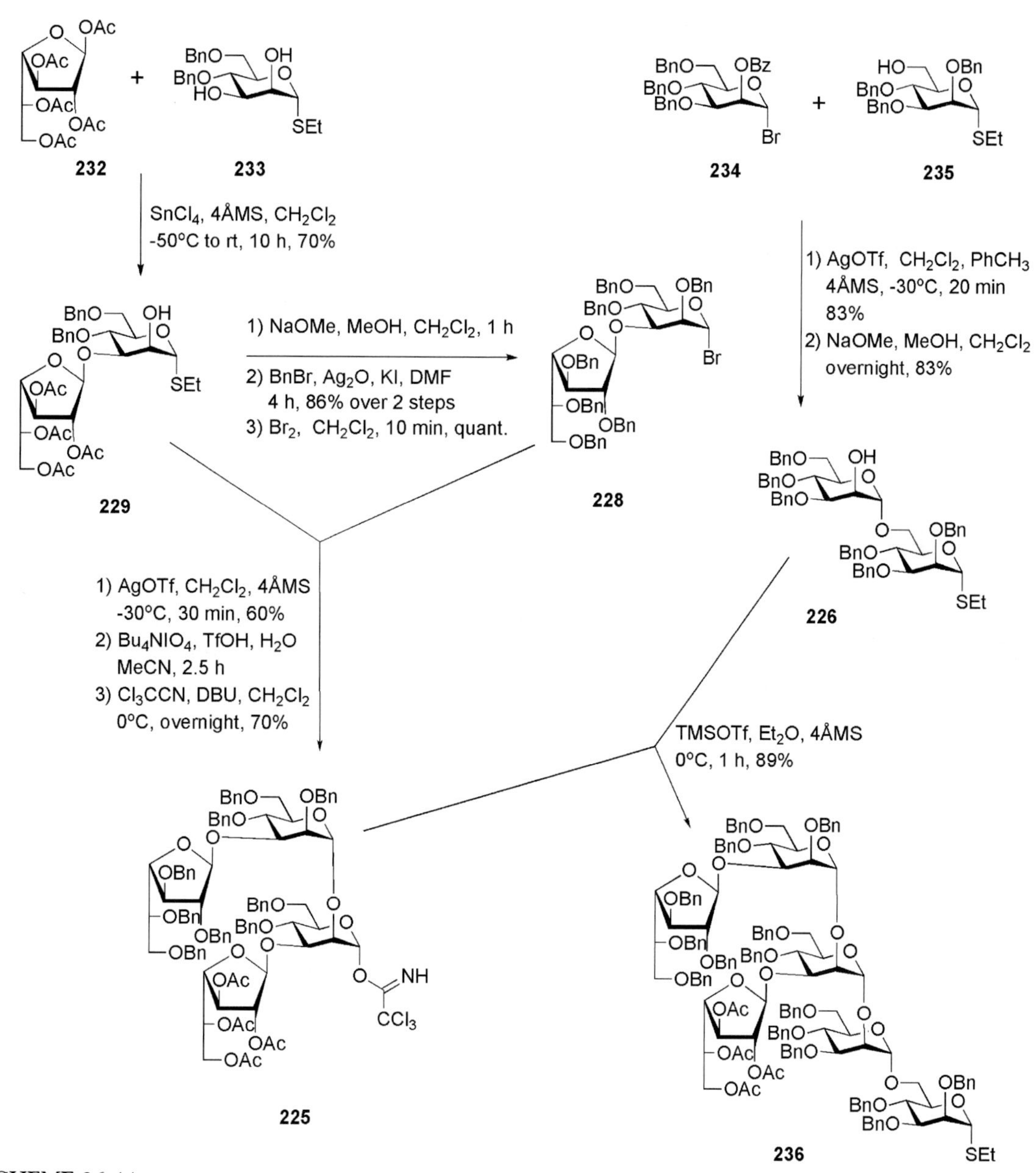

SCHEME 26.44 The synthesis of monosaccharide disaccharide building blocks and assembly of a branched hexasaccharide according to the procedure of the Linköping group.

chemistry. The prepared **239** was de-*O*-allylated using iridium(I) catalyst and then regioselectively 6′-*O*-phosphonylated with phosphonochloridate **231** to furnish "P-decorated" glycosyl acceptor **227** (82%). The latter was glycosylated with branched hexasaccharide donor **236** utilizing dimethyl(methylthio)sulfonium trifluoromethanesulfonate (DMTST) as a promoter to afford the fully protected pseudo-octasaccharide **240** (71%).

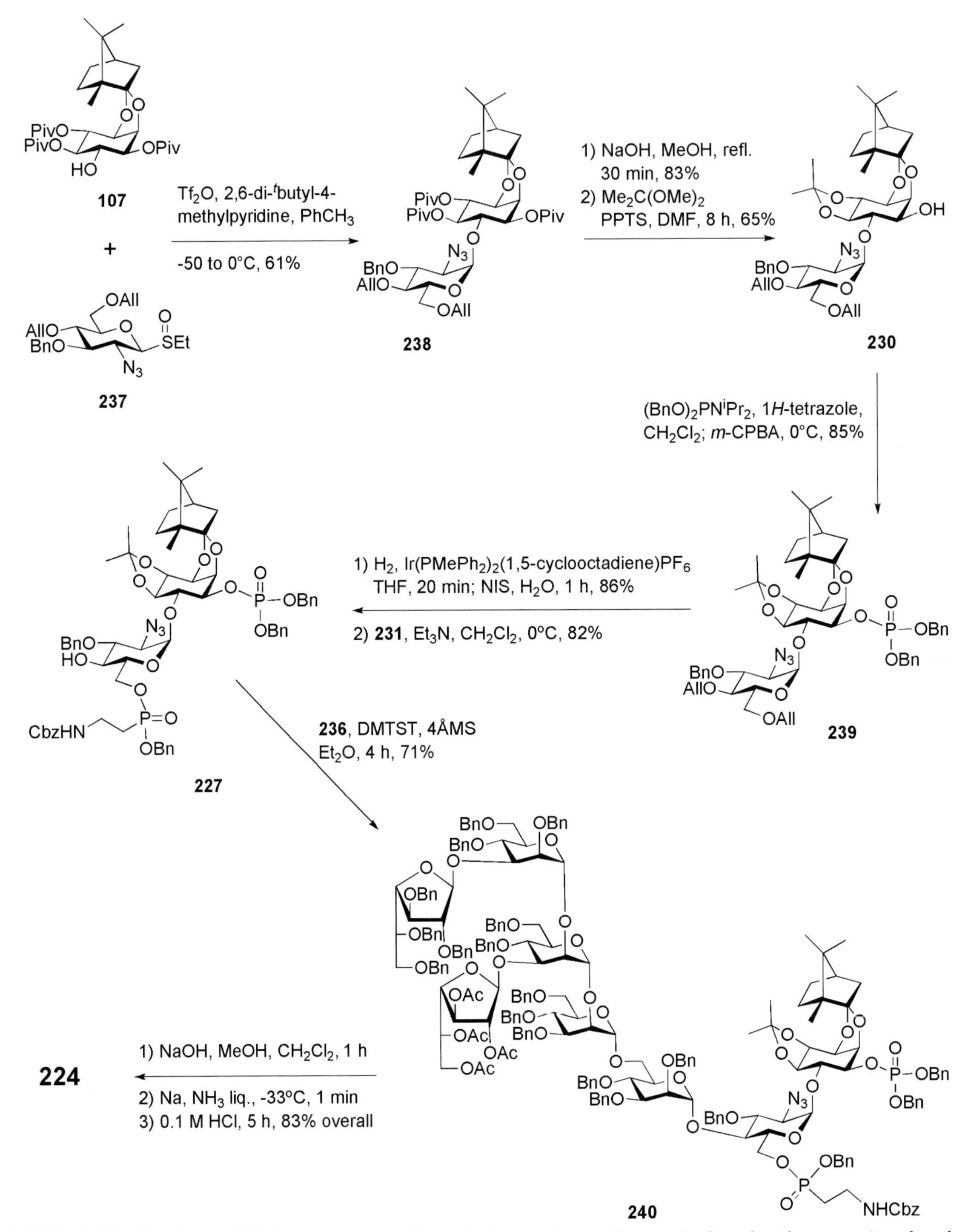

SCHEME 26.45 *T. cruzi* GPI glycan core assembly, including synthesis of a pseudo-disaccharide, successive phosphonylation and phosphorylation steps and global deprotection according to the strategy of the Linköping group.

Global deprotection started with *O*-deacetylation using Zemplen conditions followed by de-*O*-benzylation with Na in liquid NH_3 and, finally, removal of the acetals by acid hydrolysis to provide the desired phosphoglycan **224** in 83% yield.

3. CHEMICAL SYNTHESIS OF PARASITIC PHOSPHOGLYCANS OF *LEISHMANIA*

The most distinctive part of *Leishmania* LPGs is their variable phosphoglycan domain, made of phosphodisaccharide repeats of the β-Gal-(1→4)-α-Man-(1→OPO_3) structure linked to each other through a phosphodiester group between the anomeric OH of the Man of one repeat and the 6-OH of the Gal of the adjoining repeat (see the Introduction section). The dynamic structure of the phosphoglycans and their role in host–parasite interaction led to significant biological interest. This upholds the need to develop routes for the chemical preparation of these biopolymers. The development of synthetic approaches to phosphoglycans is challenging, because the phosphoglycans are hydrolytically labile due to anomeric phosphodiester linkage present between the repeating units. Synthesis of anomeric phosphodiesters is complicated as both the correct stereochemistry at C-1 and the lability of anomeric phosphodiester linkages must be taken into consideration. For these reasons, only few syntheses of anomerically-linked phosphoglycans have been reported (Nikolaev *et al.*, 2007).

3.1. Syntheses of phoshoglycans using stepwise chain elongation

The Dundee group reported the first successful syntheses of *Leishmania* phosphoglycans making use of the H-phosphonate chemistry (Nikolaev *et al.*, 1994, 1995b). The glycosyl H-phosphonate route appeared to be the method of choice for the efficient and reliable assembly of various phosphodiester linkages in natural phosphoglycans (Nikolaev *et al.*, 2007). The *L. donovani* phosphoglycans **241** and **242** were constructed using the disaccharide building blocks **244**, **245** and **250** (Scheme 26.46). Adamantane-1-carbonyl chloride-assisted condensation of the disaccharide H-phosphonate **244** (prepared by phosphitylation of the hemiacetal **243** with tri-imidazolylphosphine and subsequent mild hydrolysis) with the alcohol **245**, followed by *in situ* oxidation with iodine in aqueous pyridine and subsequent mild acid hydrolysis of the DMT group afforded the phosphotetrasaccharide **246** (81%). The disaccharides **243** and **245** were prepared from the corresponding monosaccharide synthons followed by protecting group re-modelling. Coupling of **246** with the same H-phosphonate **244**, employing the prescribed route of the oxidation and de-*O*-tritylation, provided the phosphohexasaccharide **247** (75%) which, on coupling with the mannobiosyl H-phosphonate **250** (made up from compounds **248** and **249**) followed by oxidation and global deprotection using MeONa in MeOH, furnished the phospho-octasaccharide **242**. Compound **247** was also globally deprotected to provide the phosphoglycan **241**. The final phosphoglycans contain a dec-9-enyl aglycon moiety to enable the preparation of neoglycoconjugates via successive ozonolysis of the double bond and coupling to protein carrier by reductive amination (Routier *et al.*, 2000).

The branched heptaglycosyl triphosphate fragment **251** (Scheme 26.47) of the LPG from *L. mexicana* was assembled via stepwise chain elongation from the H-phosphonate derivatives **244** and **257** (Higson *et al.*, 1998). The trisaccharide **257** was prepared starting from the glucosyl bromide **252** and the disaccharide acceptor **253** via their coupling and acid hydrolysis (→**254**), followed by the orthogonal 6′-*O*-DMT protection and anomeric debenzoylation (→**255**). The hemiacetal **255** was converted to the H-phosphonate **257** by phosphitylation with tri-imidazolylphosphine and subsequent hydrolysis. Condensation of the disaccharide H-phosphonate **244** and dec-9-en-1-ol in the presence of a condensing reagent

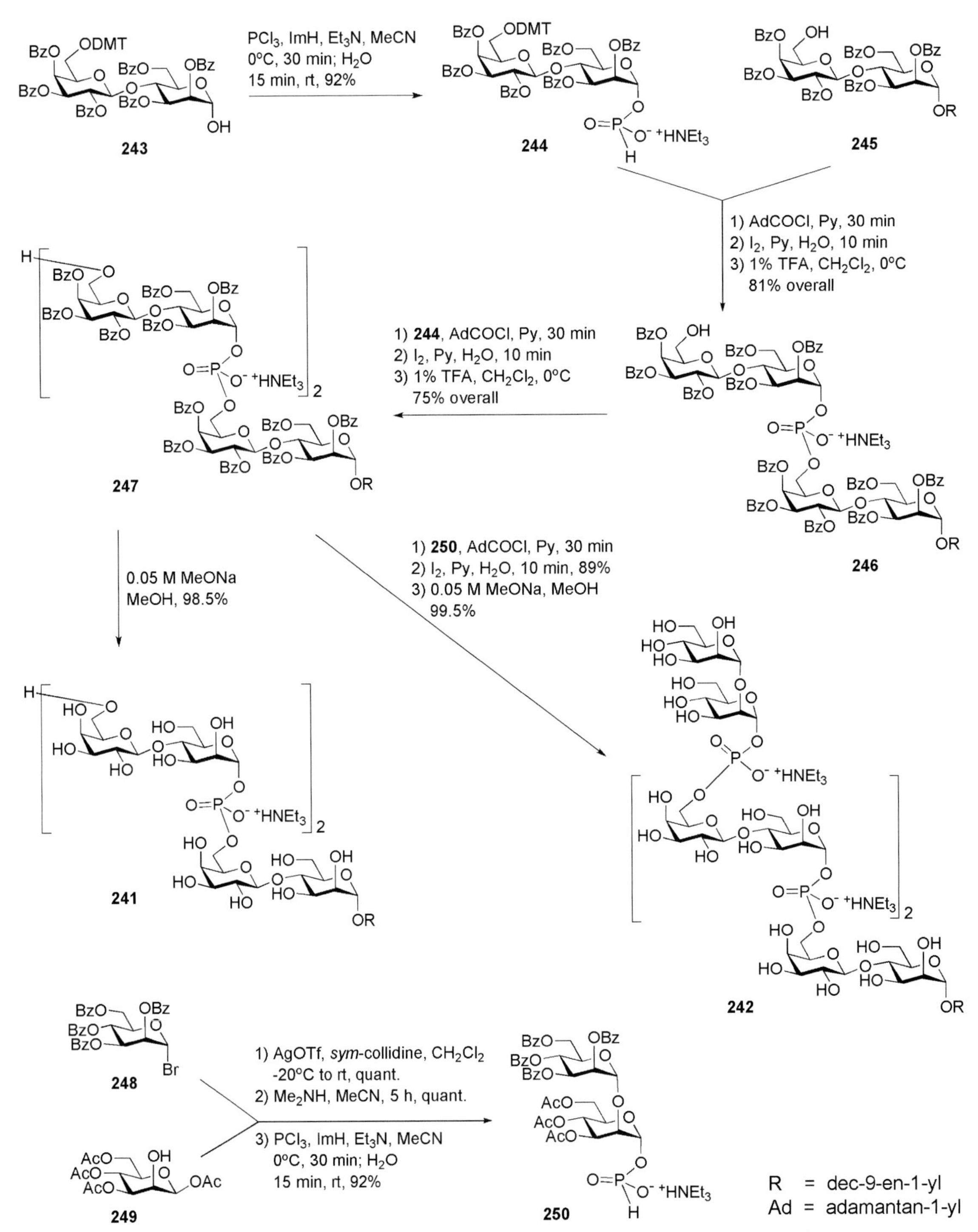

SCHEME 26.46 Stepwise synthesis of phosphoglycans of *L. donovani* after the strategy of the Dundee group.

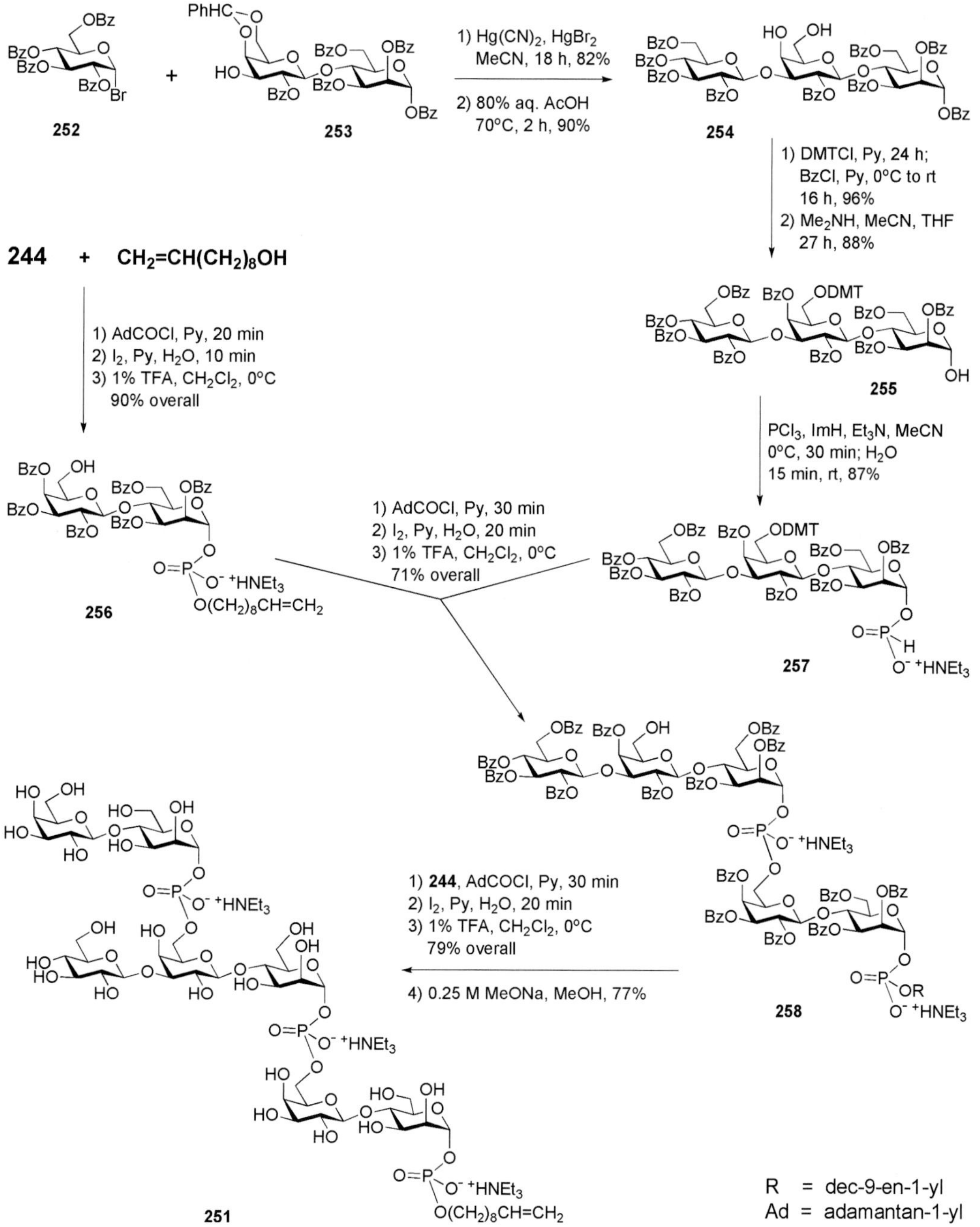

252 + 253

1) $Hg(CN)_2$, $HgBr_2$ MeCN, 18 h, 82%
2) 80% aq. AcOH 70°C, 2 h, 90%

254

1) DMTCl, Py, 24 h; BzCl, Py, 0°C to rt 16 h, 96%
2) Me_2NH, MeCN, THF 27 h, 88%

255

PCl_3, ImH, Et_3N, MeCN 0°C, 30 min; H_2O 15 min, rt, 87%

257

244 + $CH_2{=}CH(CH_2)_8OH$

1) AdCOCl, Py, 20 min
2) I_2, Py, H_2O, 10 min
3) 1% TFA, CH_2Cl_2, 0°C
90% overall

256

1) AdCOCl, Py, 30 min
2) I_2, Py, H_2O, 20 min
3) 1% TFA, CH_2Cl_2, 0°C
71% overall

258

1) **244**, AdCOCl, Py, 30 min
2) I_2, Py, H_2O, 20 min
3) 1% TFA, CH_2Cl_2, 0°C
79% overall
4) 0.25 M MeONa, MeOH, 77%

251

R = dec-9-en-1-yl
Ad = adamantan-1-yl

SCHEME 26.47 Stepwise synthesis of a branched phosphoglycan of *L. mexicana*.

(adamantane-1-carbonyl chloride) followed by oxidation (I_2) and mild acidic detritylation gave the phosphodiester 6′-OH block **256** (90%). Coupling of the disaccharide phosphate **256** with the trisaccharide H-phosphonate **257**, followed by oxidation and de-O-tritylation furnished **258** (71%) which, on further elongation with the H-phosphonate **244** (79%), afforded after global deprotection, the desired branched phosphoglycan **251**. This was then coupled to a protein carrier (recombinant tetanus toxin fragment C) employing the ozonolysis/reductive amination technique (Routier *et al.*, 2000). The *L. mexicana*–TetC glycoconjugate (a novel synthetic glycovaccine) was shown to be protective in mice against the bite of *Leishmania*-infected sand flies (Rogers *et al.*, 2006).

A similar strategy, but adamantane-1-carbonyl chloride being replaced with pivaloyl chloride as a condensing reagent, was employed in the assembly of the branched phosphoglycans **261–263** (Higson *et al.*, 1999, 2005) and **264–266** (Yashunsky *et al.*, 2001, 2003; D.V. Yashunsky and A.V. Nikolaev, unpublished results) from *Leishmania major* (Scheme 26.48). The phosphoglycan molecules were synthesized from the disaccharide **244**, trisaccharide **259** and tetrasaccharide **260** H-phosphonate building blocks and the monohydroxyl phosphodiester derivative **256**.

The synthesis of the decaglycosyl triphosphate **266**, which is the largest molecule of the set, is shown in Scheme 26.49. Compound **256** was first elongated with the tetrasaccharide H-phosphonate **260** to provide the monohydroxyl phosphoglycan derivative **274** (89%). One more chain extension using **260** (83%), followed by global deprotection, gave the targeted phosphoglycan **266**. The H-phosphonate **260** was prepared by making use of the thiogalactoside **267**, galactosyl trichloroacetimidate **270** and arabinosyl chloride **272** (Lemieux *et al.*, 1975) as glycosylation donors in a 16-step linear synthesis starting from the Gal donor **267** and the Man acceptor **268** (Yashunsky *et al.*, 2001). Their coupling in the presence of MeOTf (80%) and subsequent de-*O*-chloroacetylation gave the disaccharide **269** which, in turn, was glycosylated with the trichloroacetimidate **270** (69%), followed by protecting group re-modelling to form the trisaccharide acceptor **271**. This was coupled with the arabinosyl donor **272** (76%), followed by partial deprotection and introduction of the 6′-TBS ether (→**273**). Tetrasaccharide **273** was benzoylated, followed by replacement of the TBS protection with a 6′-*O*-DMT group and standard introduction of the H-phosphonate moiety at C-1 to produce the required **260**.

The New Delhi group (Ruhela and Vishwakarma, 2001, 2003) reported an iterative synthesis of the *L. donovani* phosphoglycans by solution and solid-phase (see Section 3.3) approaches starting from lactal **275** (Scheme 26.50) which, in turn, was easily prepared from lactose in four steps (Upreti *et al.*, 2000).

Dibutyltin oxide-mediated selective 6-*O*-silylation of the Gal moiety in **275**, followed by *m*-CPBA-assisted D-glucal to Man transformation and conventional acetylation led to the desired disaccharide **276**, which served as a single material for the disaccharide H-phosphonate **277** as well as the alcohol **278** preparation for iterative assembly of the phosphoglycan repeats. Selective de-*O*-acetylation at the anomeric position with Me_2NH, followed by phosphitylation with tri-imidazolylphosphine provided the H-phosphonate **280** which, in pivaloyl chloride-assisted coupling with the alcohol **278** and subsequent oxidation with iodine, led to the fully protected phosphotetrasaccharide **279** (75%). The latter compound was subjected to further chain elongation either upstream (at the non-reducing 6‴-end) or downstream (at the reducing 1-OH end). After removal of the 6‴-*O*-TBS group the resulting tetrasaccharide alcohol was coupled with the H-phosphonate **277** to provide the phosphohexasaccharide **280** (63%). On the other hand, selective deacetylation at the reducing-end anomeric position of **279**, followed by direct conversion to the corresponding glycosyl H-phosphonate (86%) and

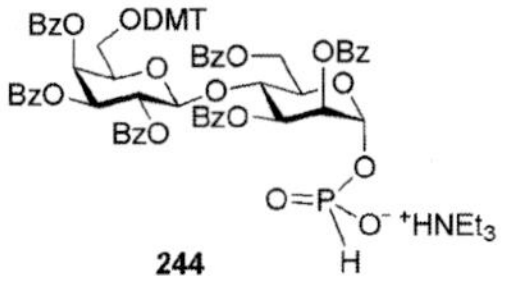

244 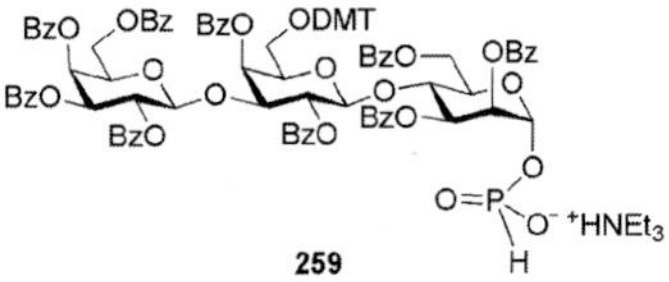 **259** 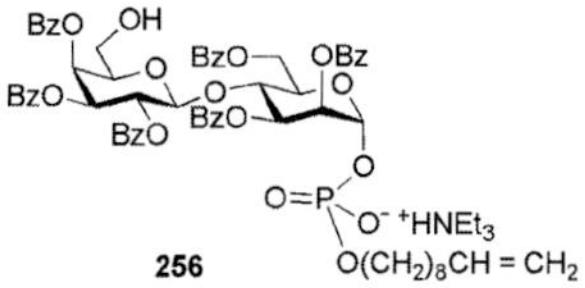 **256** 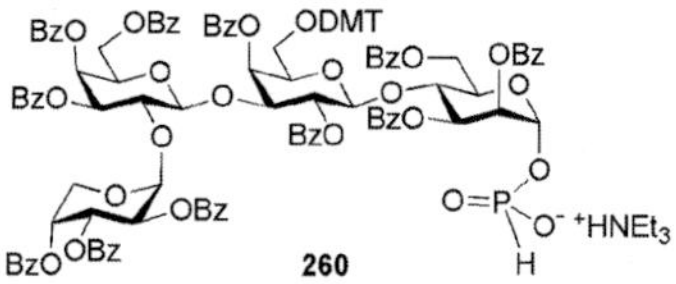 **260**

β-D-Gal*p*-(1→4)-α-D-Man*p*-(1-PO_3H-6)-β-D-Gal*p*-(1→4)-α-D-Man*p*-(1-PO_3H-6)-β-D-Gal*p*-(1→4)-α-D-Man*p*-1-PO_3H-O$[CH_2]_8CH = CH_2$
(second β-D-Gal*p*: ↑3 ← 1 β-D-Gal*p*)

261

β-D-Gal*p*-(1→4)-α-D-Man*p*-(1-PO_3H-6)-β-D-Gal*p*-(1→4)-α-D-Man*p*-(1-PO_3H-6)-β-D-Gal*p*-(1→4)-α-D-Man*p*-1-PO_3H-O$[CH_2]_8CH = CH_2$
(first β-D-Gal*p*: ↑3 ← 1 β-D-Gal*p*)

262

β-D-Gal*p*-(1→4)-α-D-Man*p*-(1-PO_3H-6)-β-D-Gal*p*-(1→4)-α-D-Man*p*-(1-PO_3H-6)-β-D-Gal*p*-(1→4)-α-D-Man*p*-1-PO_3H-O$[CH_2]_8CH = CH_2$
(first β-D-Gal*p*: ↑3 ← 1 β-D-Gal*p*; second β-D-Gal*p*: ↑3 ← 1 β-D-Gal*p*)

263

β-D-Gal*p*-(1→4)-α-D-Man*p*-(1-PO_3H-6)-β-D-Gal*p*-(1→4)-α-D-Man*p*-(1-PO_3H-6)-β-D-Gal*p*-(1→4)-α-D-Man*p*-1-PO_3H-O$[CH_2]_8CH = CH_2$
(first β-D-Gal*p*: ↑3 ← 1 β-D-Gal*p*; second β-D-Gal*p*: ↑3 ← 1 β-D-Gal*p* ↑2 ← 1 β-D-Ara*p*)

264

β-D-Gal*p*-(1→4)-α-D-Man*p*-(1-PO_3H-6)-β-D-Gal*p*-(1→4)-α-D-Man*p*-(1-PO_3H-6)-β-D-Gal*p*-(1→4)-α-D-Man*p*-1-PO_3H-O$[CH_2]_8CH = CH_2$
(first β-D-Gal*p*: ↑3 ← 1 β-D-Gal*p* ↑2 ← 1 β-D-Ara*p*; second β-D-Gal*p*: ↑3 ← 1 β-D-Gal*p*)

265

β-D-Gal*p*-(1→4)-α-D-Man*p*-(1-PO_3H-6)-β-D-Gal*p*-(1→4)-α-D-Man*p*-(1-PO_3H-6)-β-D-Gal*p*-(1→4)-α-D-Man*p*-1-PO_3H-O$[CH_2]_8CH = CH_2$
(first β-D-Gal*p*: ↑3 ← 1 β-D-Gal*p* ↑2 ← 1 β-D-Ara*p*; second β-D-Gal*p*: ↑3 ← 1 β-D-Gal*p* ↑2 ← 1 β-D-Ara*p*)

266

SCHEME 26.48 Key building blocks and *L. major* phosphoglycan structures prepared by stepwise chain elongation. Abbreviations: Ara*p*, arabinopyranose; Gal*p*, galactopyranose; Man*p*, mannopyranose.

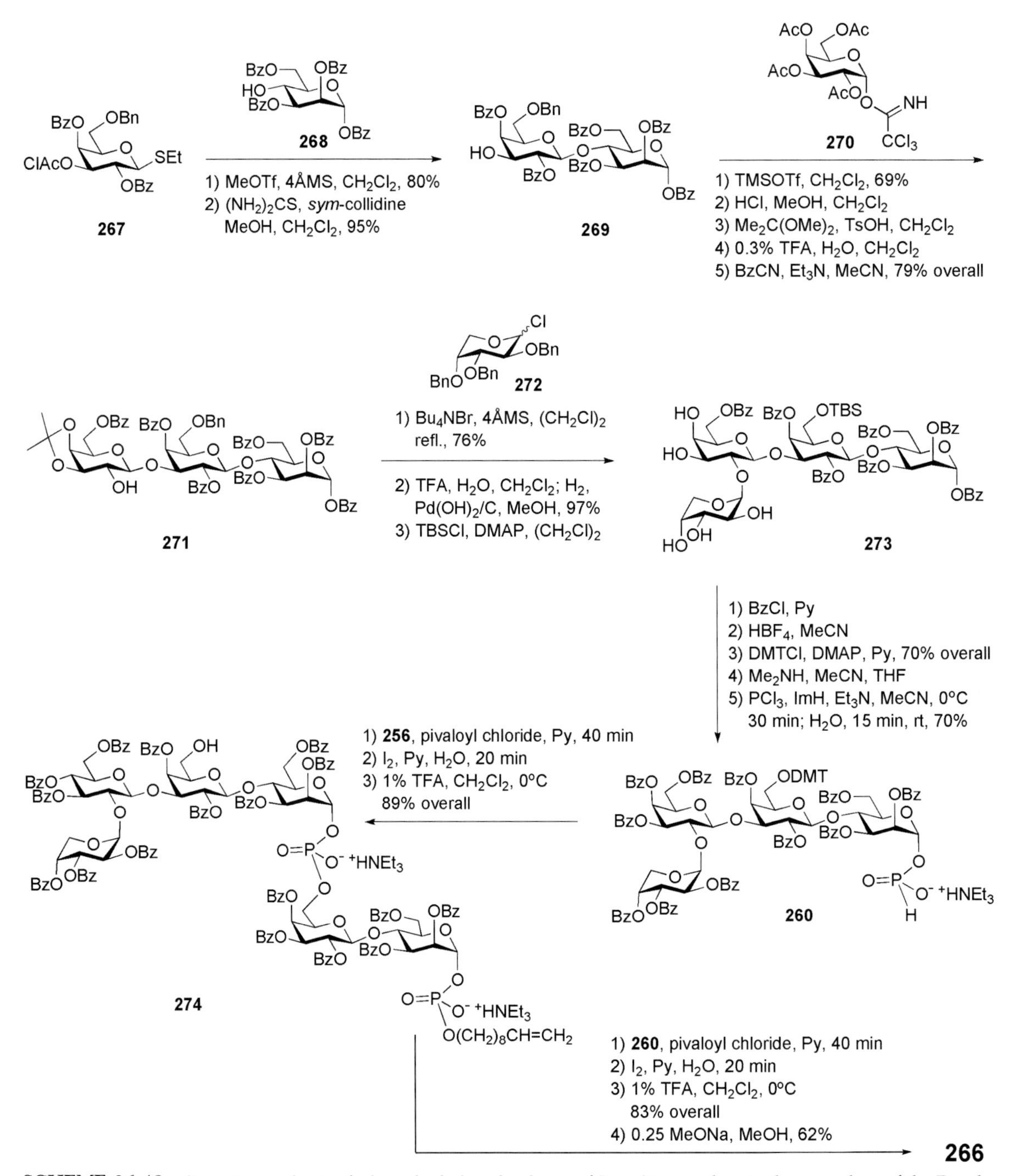

SCHEME 26.49 Stepwise synthesis of a branched phosphoglycan of *L. major* according to the procedure of the Dundee group.

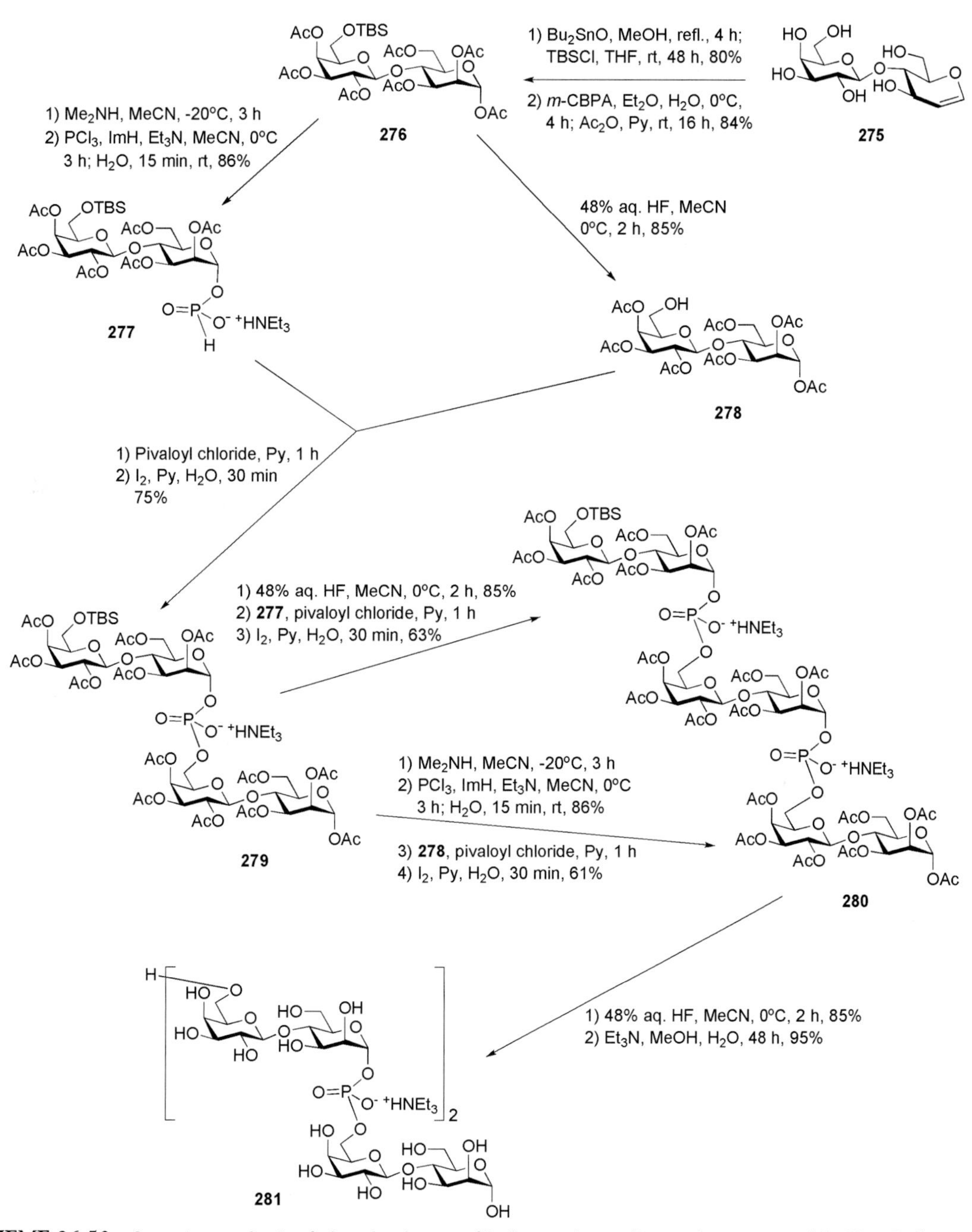

SCHEME 26.50 Stepwise synthesis of phosphoglycans of *L. donovani* according to the strategy of the New Delhi group.

its coupling with the alcohol **278** provided the same protected hexasaccharide diphosphate **280** (61%) which, on successive desilylation with aq. HF in MeCN and deacetylation with Et_3N in MeOH-water, provided the targeted phosphoglycan **281**.

The same starting material lactal **275** and similar synthetic route were employed in the synthesis of branched phosphoglycan of *L. major* **287** (Scheme 26.51), with flexibility to extend further the phosphoglycan repeats at either end (Ruhela and Vishwakarma, 2004). Dibutyltin oxide-assisted selective 3′-*O*-methoxybenzylation of **275** and then 6′-*O*-silylation (→**282**), followed by *m*-CPBA catalysed D-glucal to Man transformation and standard acetylation furnished the corresponding fully protected disaccharide. The PMB group was removed and the resulting alcohol acceptor **283** was glycosylated with the D-glactosyl donor **270** (see Scheme 26.49) to furnish the trisaccharide **284** (74%) that served as a single synthon to access both the trisaccharide H-phosphonate **285** and the alcohol acceptor **286**. Condensation of **285** and **286**, followed by oxidation and global deprotection led to the phosphohexasaccharide **287** (75% overall).

3.2. Blockwise chain elongation

In addition to the stepwise chain elongation approach using protected oligosyl H-phosphonates for the successive introduction of the sugar phosphate residues, the Dundee group (Nikolaev *et al.*, 1996) has also developed a blockwise approach that involves condensation of two phosphodiester blocks (Scheme 26.52). This attractive approach was successfully adapted in the synthesis of the hexaglycosyl triphosphate **291**, a terminal fragment of the phosphoglycan potion of *L. donovani* LPG, using adamantane-1-carbonyl chloride-mediated condensation of the tetrasaccharide H-phosphonate phosphodiester **290** and the monohydroxyl phosphodiester block **256**, followed by oxidation with iodine and global deacylation (59% overall). Compound **290**, in turn, was prepared by coupling of the H-phosphonate **250** and disaccharide **288**, followed by anomeric deprotection (→**289**, 78%) and conversion of 1-OH to the anomeric H-phosphonate.

3.3. Polymer-supported syntheses

Synthesis of the *L. donovani* phosphohexasaccharide **296** (Scheme 26.53) was successfully developed by the Dundee group using a polymer-supported methodology (Ross *et al.*, 2000) that employed the glycosyl H-phosphonates **244** and **295** for consecutive chain elongations, DMT as the orthogonal protecting group and monomethyl polyethylene glycol (MPEG), with mass of 5000, as a polymer support.

The advantage of MPEG is that it allows for both solid-phase and solution-phase techniques: the MPEG-bound products are soluble in most organic solvents and water, however, they can be precipitated (quantitatively) from ether or cold ethanol greatly simplifying their purification. The succinic ester linker was used for binding the galactoside **292** to the polymer in the presence of 1-(2-mesitylenesulfonyl)-3-nitro-1,2,4-triazole (MS-NT) in 91% yield. Successive acetylation of any residual OH-groups and acidic detritylation afforded the MPEG-bound hydroxyl acceptor **293**. Standard chain-elongation cycle (i.e. condensation, oxidation and detritylation) engaging the disaccharide H-phosphonate **244** (see Scheme 26.46) was applied twice to provide the polymer **294**. The mannosyl H-phosphonate **295** was then used to terminate the chain extension. After total deprotection with methanolic NaOMe, which also cleaved the product from MPEG, the hexaglycosyl triphosphate **296** (57% from **292**, corresponding to an average 95% per step) was isolated by anion-exchange chromatography.

Alternatively, a solid-phase synthesis of the *L. donovani* phosphohexasaccharide **299** (Scheme 26.54) was reported by the New Delhi group (Ruhela and Vishwakarma, 2003), which used modified Merrifield resin and the disaccharide

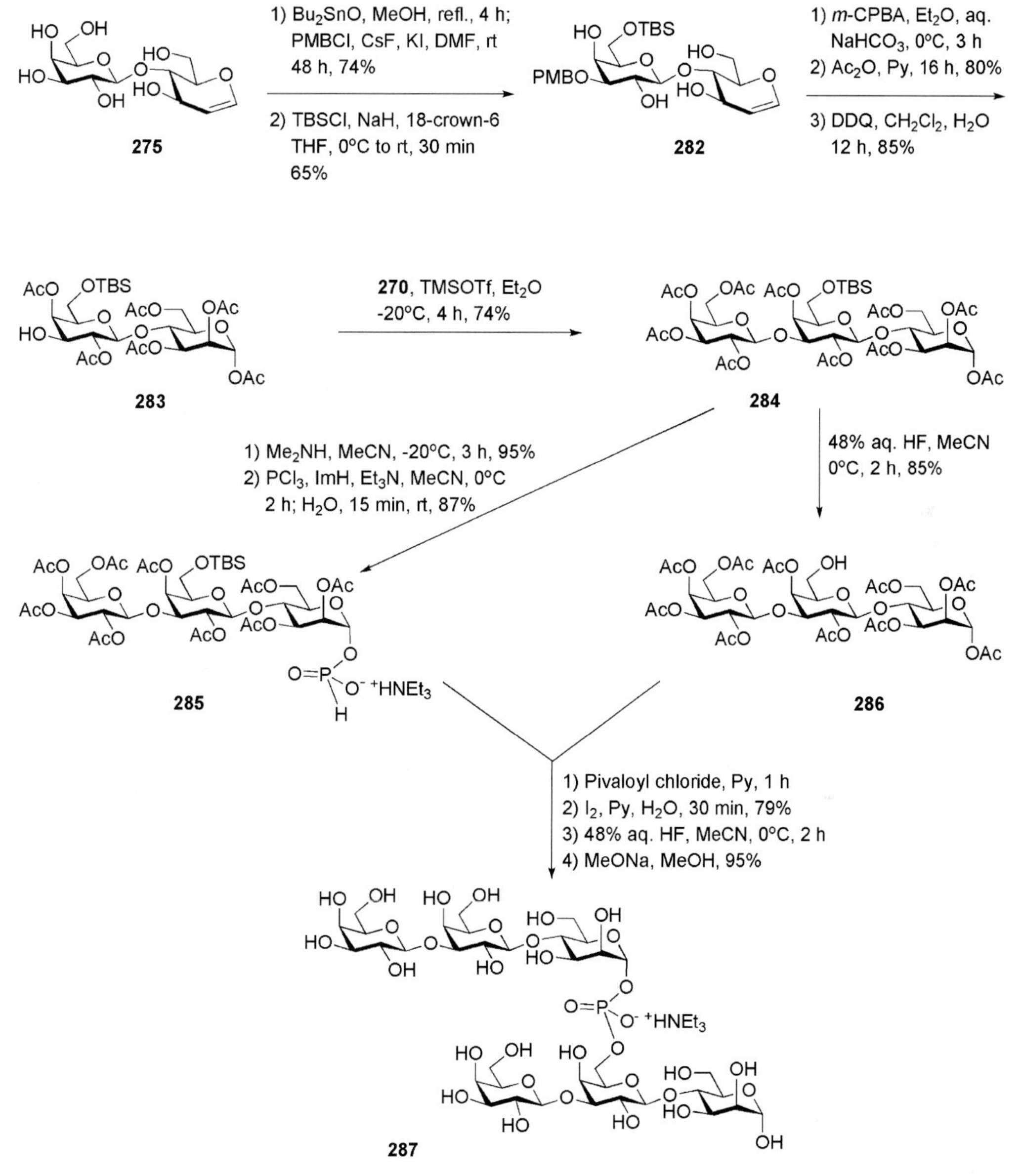

SCHEME 26.51 Synthesis of a branched phosphoglycan of *L. major* according to the procedure of the New Delhi group.

H-phosphonate **277**. Their methodology was based on the application of a *cis*-but-2-en-1,4-diyl-1-*O*-phosphoryl linker that enabled the selective cleavage of the first anomeric phosphodiester linkage from the resin with Wilkinson's catalyst without affecting the glycosyl phosphate bond. The pivaloyl chloride-mediated coupling efficiency of each of the three iterative cycles (→ **297**; →**298**) was more than 90%. Cleavage of the phosphoglycan from the resin, followed by

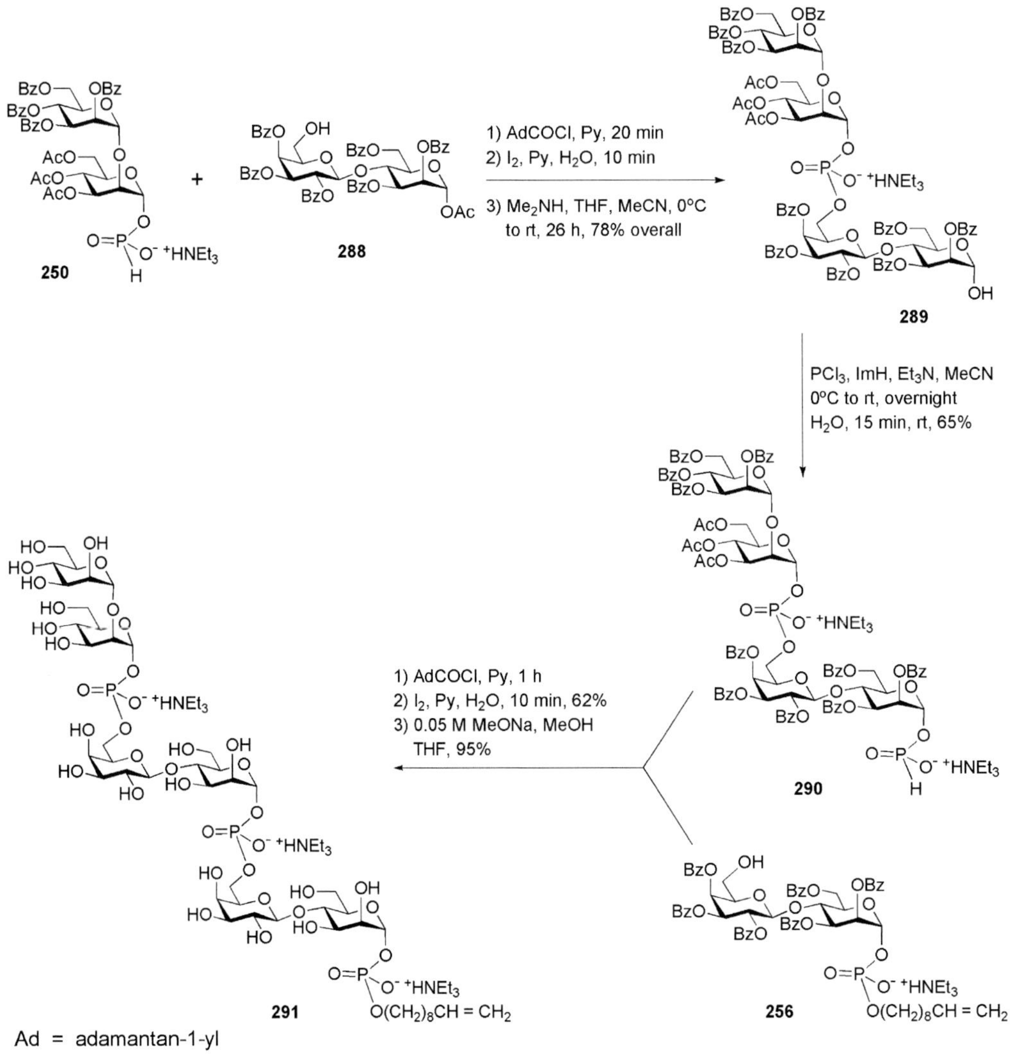

Ad = adamantan-1-yl

SCHEME 26.52 Blockwise synthesis of a phosphoglycan of *L. donovani*.

desilylation (aq. HF in CH_3CN) and deacetylation, provided the targeted hexasaccharide triphosphate **299** in 70% yield.

3.4. Phoshoglycan syntheses by polycondensation

Both groups, Dundee and New Delhi, described the construction of larger phosphoglycans **302** having 10 repeats (Nikolaev *et al.*, 1995a) and **303** having 19–22 repeats (Ruhela and Vishwakarma, 2003) in a one-pot synthesis by polycondensation of the monohydroxyl H-phosphonate derivatives **300** and **301**, respectively, which served as bifunctional monomer building blocks (Scheme 26.55). Polycondensation of the monomer **301**, containing *O*-acetyl protecting groups (rather than *O*-benzoyl in **300**) seems to

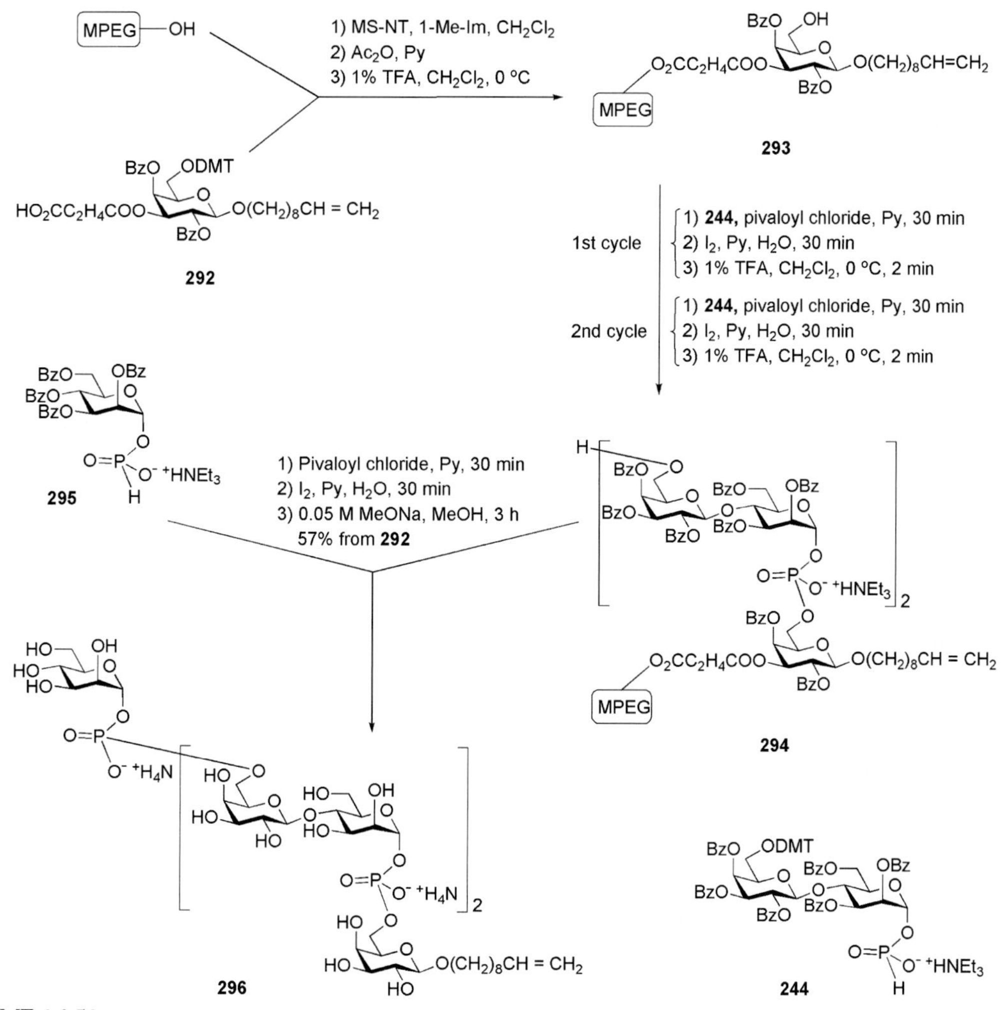

SCHEME 26.53 Polymer-supported solution-phase synthesis of a phosphoglycan of *L. donovani*.

favour the formation of longer polymer chains. After deacetylation, the polymers **302** and **303** were isolated by anion-exchange chromatography in 85 and 58% yield, respectively.

4. CONCLUSIONS AND FUTURE PERSPECTIVES

The GPIs are a class of natural glycosylphospholipids that anchor proteins, glycoproteins and LPGs to the membrane of eukaryotic cells. These anchors are widely present in parasitic protozoa where GPI-anchored mucins and phosphoglycans are abundant and form a dense protective layer, i.e. a glycocalyx, on the surface of the parasites. This type of anchor appears to be present in the parasites with a much higher frequency than in higher eukaryotes (McConville and Ferguson, 1993).

Synthetic strategies towards GPI anchors and GPI-related glycoconjugates have matured

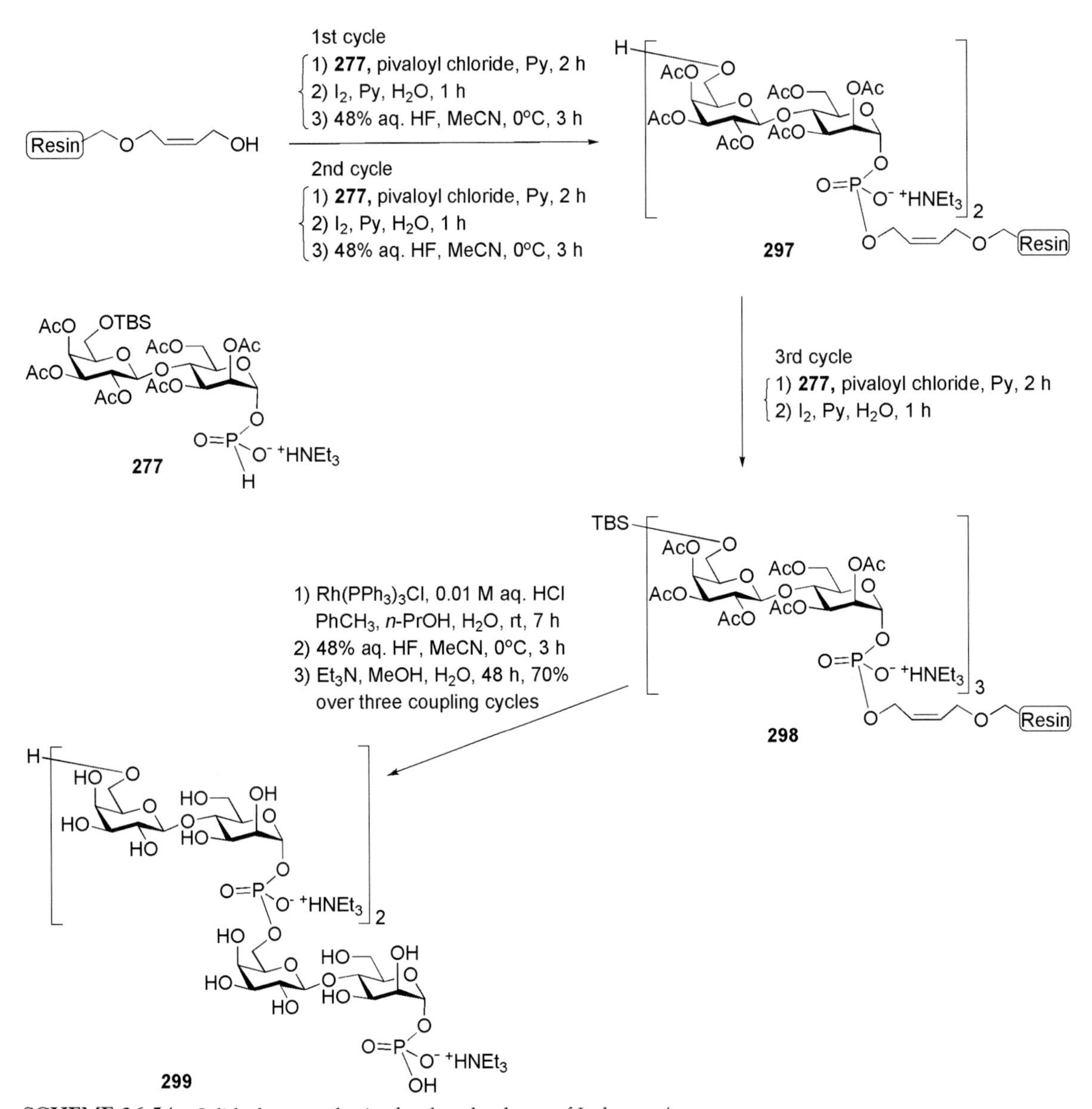

SCHEME 26.54 Solid-phase synthesis of a phosphoglycan of *L. donovani*.

since the first synthesis published by Murakata and Ogawa in the early 1990s (Murakata and Ogawa, 1990, 1991, 1992). Several attractive synthetic routes were used for the preparation of the twelve parasitic structures discussed in this chapter. In addition, a GPI anchor found in the yeast *Saccharomyces cerevisiae* (Mayer *et al.*, 1994; Mayer and Schmidt, 1999) and a few structures from the higher eukaryotes have been synthesized as well. The latter reports include a GPI anchor of rat brain Thy-1 glycoprotein prepared independently in the laboratories of B. Fraser-Reid (Campbell and Fraser-Reid, 1995) and R. Schmidt (Tailler *et al.*, 1999; Pekari and Schmidt,

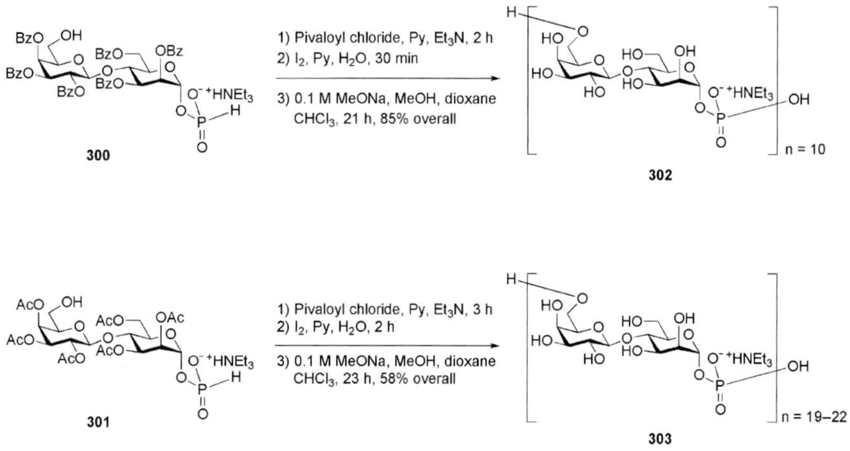

SCHEME 26.55 Syntheses of *L. donovani* phosphoglycans by polycondensation.

2003) and GPI anchors and related glycopeptide conjugates of sperm CD52 glycopeptide antigen (see Guo and Bishop, 2004 and references cited therein; Wu and Guo, 2007; Wu *et al.*, 2008).

A common feature in most of the strategies used (with the exception of CD52 GPIs) is the initial construction of the GPI glycan core having the appropriate orthogonal protecting groups, followed by subsequent introduction of the phosphate moieties. Almost all strategies used benzyl ethers as permanent *O*-protecting groups, except the Cambridge group (where butanediacetals and chloroacetic esters in addition to benzyl groups were used) and, recently, the Dundee group, who successfully mastered the use of arrays of the easily removable permanent *O*-protecting group (i.e. benzoic esters plus acid labile groups or, alternatively, a combination of acid-sensitive butanediacetals, cyclohexylidene and SEM acetals and *p*-methoxybenzyl ethers) while retaining the carbon–carbon double bonds in the lipid chain. For the addition of the phospholipid and the ethanolamine-phosphate, both the H-phosphonate and phosphoramidite methods were used. With the appreciable increase in the number of characterized GPIs and with mounting knowledge of their biological function, it is expected that, in the near future, there will be a substantial growth in research exploring structural and biological studies as well as the chemical synthesis of GPIs.

An anomeric phosphodiester linkage formed by a glycosyl phosphate unit and a hydroxyl group of another monosaccharide was found in many glycopolymers of the cell envelope of bacteria, i.e. both in capsular and lipopolysaccharides, parasitic protozoa and yeasts. The importance of the phosphoglycans composed of glycosyl phosphate (or oligosyl phosphate) repeat units as defensive and immunologically active components of the cell wall and/or capsule of these microorganisms upholds the need to develop routes for their chemical preparation. Synthetic approaches are challenging due to the extreme lability of the phosphoglycans caused by the anomeric phosphodiester linkages between the repeat units.

Historically, the first syntheses of glycosyl phosphosaccharides, i.e. phosphodiesters where phosphorus is linked to one monosaccharide through the hemiacetal hydroxyl group at C-1 and to another one through an alcoholic OH-group, were attempted in 1970–1980s using the phosphodiester and the phosphite triester methods. The glycosyl hydrogenphosphonate method, developed in the late 1980s, appeared to be the most efficient for constructing phosphodiester linkages of this type (Nikolaev *et al.*, 2007 and references cited therein). Furthermore, the method was applied for the first syntheses of natural phosphoglycans of bacterial and parasitic origin using both stepwise and blockwise chain elongation. Polycondensation tactics and polymer-supported syntheses were also developed for the preparation of *Leishmania* phosphoglycans. The synthetic phosphoglycans are of great interest for biological experiments designed to probe the function, immunology, biosynthesis, conformation and mode of action of these complex cell-surface polymeric structures.

Despite the appreciable progress in the chemical synthesis of GPI anchors and phosphoglycans composed of glycosyl phosphate repeating units, there are still several unresolved issues that prevent both from being routinely synthesized and, therefore, require special attention and experience (see Research Focus Box).

RESEARCH FOCUS BOX

- Additional side chains in GPIs, or positional alteration in side chains, results in total reform of the intact plan, particularly in the protection strategy.
- All the employed GPI synthetic plans are time-consuming; therefore reduction of the number of steps is a necessity. This is achievable by adopting and developing one-pot strategies that enable the construction of the glycan core by means of sequential multiple glycosylations under catalytic activation.
- Installation and removal of protecting groups are the most time-consuming aspects in GPI glycan core synthesis. Minimizing their use by using the donor/acceptor match concept is a valuable goal, since there is no need to protect a hydroxyl that does not match the specific donor used.
- A combined chemo-enzymatic approach to synthetic GPIs would be an interesting challenge. It may require chemical preparation of a pseudo-disaccharide or a pseudo-trisaccharide, followed by further chain elongation with the help of recombinant enzymes, e.g. glycosyltransferases, phospholipases, phosphoryltransferases and acyltransferases.
- In spite of effective use of H-phosphonate chemistry in phosphoglycan synthesis, novel methods towards anomeric phosphodiesters are needed. This may also include a search for novel activating (for the H-phosphonate coupling) and oxidizing reagents. The oxidation step seems to be a problem for the preparation of sterially hindered glycosyl phosphosaccharides as the intermediate H-phosphonic acid diester, instead of its mild transformation to the phosphodiester, may undergo hydrolysis of the glycosidic linkage as a concurrent reaction.
- The overwhelming majority of glycosyl phosphate fragments from the known parasitic and bacterial structures have α-D- or α-L-hexopyranose configuration, which is favoured by the anomeric effect thanks to the axial position of the anomeric phosphate. This requires the preparation of a pure α-glycosyl H-phosphonate, which can be tricky if started from the α,β-hemiacetal derivative. Thus, novel methods for stereoselective synthesis of anomeric α-(H-phosphonates) would be appreciated. These may include (i) a search for novel α-selective H-phosphonylating reagents, (ii) for mild electrophilic reagents capable of transforming the anomeric α,β-(H-phosphonate) mixtures to the required α-compounds, as well as (iii) the use of glycosylation reactions.

References

Ali, A., Gowda, D.C., Vishwakarma, R.A., 2005. A new approach to construct full-length glycosylphosphatidylinositols of parasitic protozoa and [4-deoxy-Man-III]-GPI analogues. Chem. Commun., 519–521.

Almeida, I.C., Gazzinelli, R.T., 2001. Proinflammatory activity of glycosylphosphatidylinositol anchors derived from *Trypanosoma cruzi*: structural and functional analysis. J. Leukoc. Biol. 70, 467–477.

Almeida, I.C., Camargo, M.M., Procopio, D.O., et al., 2000. Highly purified glycosylphosphatidylinositols from *Trypanosoma cruzi* are potent proinflammatory agents. EMBO J. 19, 1476–1485.

Baeschlin, D.K., Chaperon, A.R., Charbonneau, V. et al., 1998. Rapid assembly of oligosaccharides: total synthesis of a glycosylphosphatidylinositol anchor of *Trypanosoma brucei*. Angew. Chem. Int. Ed. Eng. 37, 3423–3428.

Baeschlin, D.K., Chaperon, A.R., Green, L.G., Hahn, M.G., Ince, S.J., Ley, S.V., 2000a. 1,2-Diacetals in synthesis: total synthesis of a glycosylphosphatidylinositol anchor of *Trypanosoma brucei*. Chem. Eur. J. 6, 172–186.

Baeschlin, D.K., Green, L.G., Hahn, M.G., Hinzen, B., Ince, S.J., Ley, S.V., 2000b. Rapid assembly of oligosaccharides: 1,2-diacetal-mediated reactivity tuning in the coupling of glycosyl fluorides. Tetrahedron: Asymm. 11, 173–197.

Bender, S.L., Budhu, R.J., 1991. Biomimetic synthesis of enantiomerically pure D-*myo*-inositol derivatives. J. Am. Chem. Soc. 113, 9883–9885.

Bruzik, K.S., Tsai, M.D., 1992. Efficient and systematic syntheses of enantiomerically pure and regiospecifically protected *myo*-inositols. J. Am. Chem. Soc. 114, 6361–6374.

Buskas, T., Garegg, P.J., Konradsson, P., Maloisel, J.-L., 1994. Facile preparation of glycosyl donors for oligosaccharide synthesis: 2-azido-2-deoxyhexopyranosyl building blocks. Tetrahedron Asymm. 5, 2187–2194.

Byramova, N.E., Ovchinnikov, M.V., Backinowsky, L.V., Kochetkov, N.K., 1983. Selective removal of *O*-acetyl groups in the presence of *O*-benzoyl groups by acid-catalysed methanolysis. Carbohydr. Res. 124, c8–c11.

Campbell, A.S., Fraser-Reid, B., 1994. Support studies for installing the phosphodiester residues of the Thy-1 glycoprotein membrane anchor. Bioorg. Med. Chem. 2, 1209–1219.

Campbell, A.S., Fraser-Reid, B., 1995. First synthesis of a fully phosphorylated GPI membrane anchor: rat brain Thy-1. J. Am. Chem. Soc. 117, 10387–10388.

Cottaz, S., Brimacombe, J.S., Ferguson, M.A.J., 1993. Parasite glycoconjugates, Part 1. The synthesis of some early and related intermediates in the biosynthetic pathway of glycosylphosphatidylinositol anchors. J. Chem. Soc., Perkin Trans I, 2945–2951.

Dais, P., Shing, T.K.M., Perlin, A.S., 1983. Proton spin-lattice relaxation rates and nuclear Overhauser enhancement, in relation to the stereochemistry of β-D-mannopyranose 1,2-orthoacetates. Carbohydr. Res. 122, 305–313.

de Lederkremer, R.M., Lima, C., Ramirez, M.I., Ferguson, M.A.J., Homans, S.W., Thomas-Oates, J., 1991. Complete structure of the glycan of lipopeptidophosphoglycan from *Trypanosoma cruzi* epimastigotes. J. Biol. Chem. 266, 23670–23675.

Elie, C.J.J., Verduyn, R., Dreef, C.E., Brounts, D.M., van der Marel, G.A., van Boom, J.H., 1990. Synthesis of 6-*O*-(α-D-mannopyranosyl)-D-*myo*-inositol: a fragment from *Mycobacteria* phospholipids. Tetrahedron 46, 8243–8254.

Estevez, V., Prestwich, G.D., 1991. Synthesis of enantiomerically pure, P-1-tethered inositol tetrakis(phosphate) affinity labels via a Ferrier rearrangement. J. Am. Chem. Soc. 113, 9885–9887.

Ferguson, M.A.J., 1997. The surface glycoconjugates of trypanosomatid parasites. Philos. Trans. R. Soc. Lond., Sect. B, Biol. Sci. 352, 1295–1302.

Ferguson, M.A.J., 1999. The structure, biosynthesis and functions of glycosylphosphatidylinositol anchors, and the contributions of trypanosome research. J. Cell Sci. 112, 2799–2809.

Ferguson, M.A.J., Williams, A.F., 1988. Cell-surface anchoring of proteins via glycosyl-phosphatidylinositol structures. Annu. Rev. Biochem. 57, 285–320.

Ferguson, M.A.J., Homans, S.W., Dwek, R.A., Rademacher, T.W., 1988. Glycosyl-phosphatidylinositol moiety that anchors *Trypanosoma brucei* variant surface glycoprotein to the membrane. Science 239, 753–759.

Ferrier, R.J., Middleton, S., 1993. The conversion of carbohydrate derivatives into functionalized cyclohexanes and cyclopentanes. Chem. Rev. 93, 2779–2831.

Garegg, P.J., Iversen, T., Johansson, R., Lindberg, B., 1984. Synthesis of some mono-*O*-benzyl- and penta-*O*-methyl-*myo*-inositols. Carbohydr. Res. 130, 322–326.

Garegg, P.J., Olsson, L., Oscarson, S.J., 1993. Synthesis of oligosaccharides corresponding to structures found in capsular polysaccharides of *Cryptococcus neoformans*. Part 1. J. Carbohydr. Chem. 12, 955–967.

Garegg, P.J., Konradsson, P., Oscarson, S., Ruda, K., 1997. Synthesis of part of a proposed insulin second messenger glycosylinositol phosphate and the inner core of glycosylphosphatidylinositol anchors. Tetrahedron 53, 17727–17734.

Gigg, R., Gigg, J., 1997. Synthesis of glycosylphosphatidylinositol anchors. In: Large, D.G., Warren, C.D. (Eds.), Glycopeptides and Related Compounds. Marcel Dekker, Inc., New York, pp. 327–392.

Goebel, M., Nothofer, H.-G., Ross, G., Ugi, I., 1997. A facile synthesis of per-*O*-alkylated glycono-δ-lactones from per-*O*-alkylated glycopyranosides and a novel ring contraction for pyranoses. Tetrahedron 53, 3123–3134.

Grice, P., Ley, S.V., Pietruszka, J., Osborn, H.M.I., Priepke, H. W.M., Warriner, S.L., 1997. A new strategy for oligosaccharide assembly exploiting cyclohexane-1,2-diacetal methodology: an efficient synthesis of a high mannose type nonasaccharide. Chem. Eur. J. 3, 431–440.

Grundler, G., Schmidt, R.R., 1984. Glycosylimidate, 13. Anwendung des trichloracetimidat-verfahrens auf 2-azidoglucose- und 2-azidogalactose-derivative. Liebigs Ann. Chem., 1826–1847.

Guha-Niyogi, A., Sullivan, D.R., Turco, S.J., 2001. Glycoconjugate structures of parasitic protozoa. Glycobiology 11, 45R–59R.

Guo, Z., Bishop, L., 2004. Chemical synthesis of GPIs and GPI-anchored glycopeptides. Eur. J. Org. Chem., 3585–3596.

Hansson, J., Oscarson, S., 2000. Complex bacterial carbohydrate surface antigen structures: syntheses of Kdo- and heptose-containing lipopolysaccharide core structures and anomerically phosphodiester-linked oligosaccharide structures. Curr. Org. Chem. 4, 535–564.

Hederos, M., Konradsson, P., 2005. Synthesis of the core tetrasaccharide of *Trypanosoma cruzi* glycoinositolphospholipids. J. Org. Chem. 70, 7196–7207.

Hederos, M., Konradsson, P., 2006. Synthesis of the *Trypanosoma cruzi* LPPG heptasaccharyl *myo*-inositol. J. Am. Chem. Soc. 128, 3414–3419.

Hewitt, M.C., Snyder, D.A., Seeberger, P.H., 2002. Rapid synthesis of a glycosylphosphatidylinositol-based malaria vaccine using automated solid-phase oligosaccharide synthesis. J. Am. Chem. Soc. 124, 13434–13436.

Higson, A.P., Tsvetkov, Y.E., Ferguson, M.A.J., Nikolaev, A.V., 1998. Parasite glycoconjugates. Part 8. Chemical synthesis of a heptaglycosyl triphosphate fragment of *Leishmania mexicana* lipo- and proteo-phosphoglycan and of a phosphorylated trisaccharide fragment of *Leishmania donovani* surface lipophosphoglycan. J. Chem. Soc., Perkin Trans. I, 2587–2595.

Higson, A.P., Tsvetkov, Y.E., Ferguson, M.A.J., Nikolaev, A.V., 1999. The synthesis of *Leishmania major* phosphoglycan fragments. Tetrahedron Lett. 40, 9281–9284.

Higson, A.P., Ross, A.J., Tsvetkov, Y.E., et al., 2005. Synthetic fragments of antigenic lipophosphoglycans from *Leishmania major* and *Leishmania mexicana* and their use for characterisation of the *Leishmania elongating* α-D-mannopyranosylphosphate transferase. Chem. Eur. J. 11, 2019–2030.

Homans, S.W., Ferguson, M.A.J., Dwek, R.A., Rademacher, T.W., Anand, R., Williams, A.F., 1988. Complete structure of the glycosyl phosphatidylinositol membrane anchor of rat brain Thy-1 glycoprotein. Nature 333, 269–272.

Ilg, T., 2000. Proteophosphoglycans of *Leishmania*. Parasitol. Today 16, 489–501.

Jankowska, J., Sobkowski, M., Stawinski, J., Kraszewski, A., 1994. Studies on aryl H-phosphonates. I. An efficient method for the preparation of deoxyribo- and ribonucleoside 3′-H-phosphonate monoesters by transesterification of diphenyl H-phosphonate. Tetrahedron Lett. 35, 3355–3358.

Jia, Z.J., Olsson, L., Fraser-Reid, B., 1998. Ready routes to key *myo*-inositol component of GPIs employing microbial arene oxidation or Ferrier reaction. J. Chem. Soc., Perkin Trans I, 631–632.

Kaur, K.J., Hindsgaul, O., 1992. Combined chemical-enzymic synthesis of a dideoxypentasaccharide for use in a study of the specificity of *N*-acetylglucosaminyltransferase-III. Carbohydr. Res. 226, 219–231.

Kinzy, K., Schmidt, R.R., 1985. Glycosylimidate, 16. Synthese des Trisaccharids aus der "repeating unit" des Kapselpolysaccharids von *Neisseria meningitidis* (Serogruppe L). Liebigs Ann. Chem., 1537–1545.

Koike, K., Sugimoto, M., Sato, S., Ito, Y., Nakabara, Y., Ogawa, T., 1987. Total synthesis of globotriaosyl-E and Z-ceramides and isoglobotriaosyl-E-ceramide. Carbohydr. Res. 163, 189–208.

Kwon, Y.-U., Soucy, R.L., Snyder, D.A., Seeberger, P.H., 2005. Assembly of a series of malarial glycosylphosphatidylinositol anchor oligosaccharides. Chem. Eur. J. 11, 2493–2504.

Lemieux, R.U., Hendriks, K.B., Stick, R.V., James, K., 1975. Halide ion catalyzed glycosidation reactions. Syntheses of α-linked disaccharides. J. Am. Chem. Soc. 97, 4056–4062.

Lerner, L.M., 1996. The acetolysis of D-galactose diethyl dithioacetal. Carbohydr. Res. 282, 189–192.

Liu, X., Kwon, Y.-U., Seeberger, P.H., 2005. Convergent synthesis of a fully lipidated glycosylphosphatidylinositol anchor of *Plasmodium falciparum*. J. Am. Chem. Soc. 127, 5004–5005.

Liu, X., Stocker, B.L., Seeberger, P.H., 2006. Total synthesis of phosphatidylinositol mannosides of *Mycobacterium tuberculosis*. J. Am. Chem. Soc. 128, 3638–3648.

Lu, J., Jayaprakash, K.N., Fraser-Reid, B., 2004a. First synthesis of a malarial prototype: a fully lipidated and phosphorylated GPI membrane anchor. Tetrahedron Lett. 45, 879–882.

Lu, J., Jayaprakash, K.N., Schlueter, U., Fraser-Reid, B., 2004b. Synthesis of a malaria candidate GPI structure: a strategy for fully inositol acylated and phosphorylated GPIs. J. Am. Chem. Soc. 126, 7540–7547.

Mach, M., Schlueter, U., Mathew, F., Fraser-Reid, B., Hazen, K.C., 2002. Comparing *n*-pentenyl orthoesters and *n*-pentenyl glycosides as alternative glycosyl donors. Tetrahedron 58, 7345–7354.

Mallet, A., Mallet, J.M., Sinay, P., 1994. The use of selenophenyl galactopyranosides for the synthesis of α and β-(1→4)-C-disaccharides. Tetrahedron Asymm. 5, 2593–2608.

Massy, D.J.R., Wyss, P., 1990. The synthesis of DL-1-(hexadecanoyloxy)methyl- and 1-*O*-hexadecanoylinositols as potential inhibitors of phospholipase C. Helv. Chim. Acta 73, 1037–1057.

Mayer, T.G., Schmidt, R.R., 1997. An efficient synthesis of galactinol and isogalactinol. Liebigs Ann./Recueil., 859–863.

Mayer, T.G., Schmidt, R.R., 1999. Glycosylphosphatidylin ositol (GPI) anchor synthesis based on versatile building blocks. Total synthesis of a GPI anchor of yeast. Eur. J. Org. Chem., 1153–1165.

Mayer, T.G., Kratzer, B., Schmidt, R.R., 1994. Synthesis of a GPI anchor of the yeast *Saccharomyces cerevisiae*. Angew. Chem. Int. Ed. Engl. 33, 2177–2181.

McConville, M.J., Ferguson, M.A.J., 1993. The structure, biosynthesis and function of glycosylated phosphatidylinositols in the parasitic protozoa and higher eukaryotes. Biochem. J. 294, 305–324.

McConville, M.J., Ralton, J.E., 1997. Analysis of GPI protein anchors and related glycopeptides. In: Large, D.G., Warren, C.D. (Eds.), Glycopeptides and Related Compounds. Marcel Dekker, Inc., New York, pp. 393–425.

McConville, M.J., Menon, A.K., 2000. Recent developments in the cell biology and biochemistry of glycosylphosphatidylinositol lipids (review). Mol. Membr. Biol. 17, 1–16.

Morris, P.E., Kiely, D.E., 1987. Ruthenium tetraoxide phase-transfer-promoted oxidation of secondary alcohols to ketones. J. Org. Chem. 52, 1149–1152.

Murakata, C., Ogawa, T., 1990. Synthetic study on glycophosphatidyl inositol (GPI) anchor of *Trypanosoma brucei*: glycoheptaosyl core. Tetrahedron Lett. 31, 2439–2442.

Murakata, C., Ogawa, T., 1991. Stereoselective total synthesis of glycophosphatidyl inositol (GPI) anchor of *Trypanosoma brucei*. Tetrahedron Lett. 32, 671–674.

Murakata, C., Ogawa, T., 1992. Stereoselective total synthesis of glycosyl phosphatidylinositol (GPI) anchor of *Trypanosoma brucei*. Carbohydr. Res. 235, 95–114.

Nicolaou, K.C., Caulfield, T.J., Kataoka, H., Stylianides, N.A., 1990. Total synthesis of the tumor-associated Le^x family of glycosphingolipids. J. Am. Chem. Soc. 112, 3693–3695.

Nikolaev, A.V., Rutherford, T.J., Ferguson, M.A.J., Brimacombe, J.S., 1994. The chemical synthesis of *Leishmania donovani* phosphoglycan fragments. Bioorg. Med. Chem. Lett. 4, 785–788.

Nikolaev, A.V., Chudek, J.A., Ferguson, M.A.J., 1995a. The chemical synthesis of *Leishmania donovani* phosphoglycan via polycondensation of a glycobiosyl hydrogenphosphonate monomer. Carbohydr. Res. 272, 179–189.

Nikolaev, A.V., Rutherford, T.J., Ferguson, M.A.J., Brimacombe, J.S., 1995b. Parasite glycoconjugates. Part 4. Chemical synthesis of disaccharide and phosphorylated oligosaccharide fragments of *Leishmania donovani* antigenic lipophosphoglycan. J. Chem. Soc., Perkin Trans. I, 1977–1987.

Nikolaev, A.V., Rutherford, T.J., Ferguson, M.A.J., Birmacombe, J.S., 1996. Parasite glycoconjugates. Part 5. Blockwise approach to oligo(glycosyl phosphates): chemical synthesis of a terminal tris(glycobiosyl phosphate) fragment of *Leishmania donovani* antigenic lipophosphoglycan. J. Chem. Soc., Perkin Trans. I, 1559–1566.

Nikolaev, A.V., Botvinko, I.V., Ross, A.J., 2007. Natural phosphoglycans containing glycosyl phosphate units: structural diversity and chemical synthesis. Carbohydr. Res. 342, 297–344.

Ogawa, T., 1994. Haworth Memorial Lecture. Experiments directed towards glycoconjugate synthesis. Chem. Soc. Rev. 23, 397–407.

Ogawa, T., Sasajima, K., 1981. Synthetic studies on cell surface glycans. Part 4. Reconstruction of glycan chains of glycoprotein. Branching mannopentaoside and mannohexaoside. Tetrahedron 37, 2787–2792.

Orgueira, H.A., Bartolozzi, A., Schell, P., Litjens, R.E.J.N., Palmacci, E.R., Seeberger, P.H., 2003. Modular synthesis of heparin oligosaccharides. Chem. Eur. J. 9, 140–169.

Ottosson, H., 1990. Synthesis of *p*-trifluoroacetamidophenyl 3,6-di-*O*-{2-*O*-[α-D-mannopyranosyl 6-(disodium phosphate)]-α-D-mannopyranosyl}-α-D-mannopyranoside. Carbohydr. Res. 197, 101–107.

Pekari, K., Schmidt, R.R., 2003. A variable concept for the preparation of branched glycosyl phosphatidyl inositol anchors. J. Org. Chem. 68, 1295–1308.

Pekari, K., Tailler, D., Weingart, R., Schmidt, R.R., 2001. Synthesis of the fully phosphorylated GPI anchor pseudohexasaccharide of *Toxoplasma gondii*. J. Org. Chem. 66, 7432–7442.

Ponpipom, M.M., 1977. Synthesis of 3-*O*-substituted D-mannoses. Carbohydr. Res. 59, 311–317.

Presova, M., Smrt, J., 1989. Oligonucleotidic compounds. Part LXXII. Synthesis of 2′-end lipophilized derivatives of 2′-5′-triadenylates. Collect. Czech. Chem. Commun. 54, 487–497.

Riley, A.M., Jenkins, D.J., Potter, B.V.L., 1998. A concise synthesis of *neo*-inositol. Carbohydr. Res. 314, 277–281.

Rogers, M.E., Sizova, O.V., Ferguson, M.A.J., Nikolaev, A.V., Bates, P.A., 2006. Synthetic glycovaccine protects against the bite of *Leishmaia*-infected sand flies. J. Infect. Dis. 194, 512–518.

Ross, A.J., Ivanova, I.A., Higson, A.P., Nikolaev, A.V., 2000. Application of MPEG soluble polymer support in the synthesis of oligo-phosphosaccharide fragments from the *Leishmania* lipophosphoglycan. Tetrahedron Lett. 41, 2449–2452.

Routier, F.H., Nikolaev, A.V., Ferguson, M.A.J., 2000. The preparation of neoglycoconjugates containing intersaccharide phosphodiester linkages as potential anti-*Leishmania* vaccines. Glycoconjugate J. 16, 773–780.

Ruda, K., Lindberg, J., Garegg, P.J., Oscarson, S., Konradsson, P., 2000a. Synthesis of the *Leishmania* LPG core heptasaccharyl *myo*-inositol. J. Am. Chem. Soc. 122, 11072–11076.

Ruda, K., Lindberg, J., Garegg, P.J., Oscarson, S., Konradsson, P., 2000b. Synthesis of an inositol phosphoglycan fragment *Leishmania* parasites. Tetrahedron 56, 3969–3975.

Ruhela, D., Vishwakarma, R.A., 2001. Efficient synthesis of the antigenic phosphoglycans of the *Leishmania* parasite. Chem. Commun. 2001, 2024–2025.

Ruhela, D., Vishwakarma, R.A., 2003. Iterative synthesis of *Leishmania* phosphoglycans by solution, solid-phase, and polycondensation approaches without involving any glycosylation. J. Org. Chem., 4446–4456.

Ruhela, D., Vishwakarma, R.A., 2004. A facile and novel route to the antigenic branched phosphoglycan of the protozoan *Leishmania major* parasite. Tetrahedron Lett. 45, 2589–2592.

Sarbajna, S., Misra, A.K., Roy, N., 1998. Synthesis of a di- and a trisaccharide related to the antigen from *Klebsiella* type 43. Synth. Commun. 28, 2559–2570.

Schofield, L., Hewit, M.C., Evans, K., Siomos, M.-A., Seeberger, P.H., 2002. Synthetic GPI as a candidate antitoxic vaccine in a model of malaria. Nature 418, 785–789.

Schwab, P., Grubbs, R.H., Ziller, J.W., 1996. Synthesis and applications of $RuCl_2(=CHR')(PR_3)_2$: the influence of the alkylidene moiety on metathesis activity. J. Am. Chem. Soc. 118, 100–110.

Seeberger, P.H., Soucy, R.L., Kwon, Y.-U., Snyder, D.A., Kanemitsu, T., 2004. Convergent, versatile route to two synthetic conjugate anti-toxin malaria vaccine. Chem. Commun. 1706–1707.

Soli, E.D., Manoso, A.E, Patterson, M.C., et al., 1999. Azide and cyanide displacements via hypervalent silicate intermediates. J. Org. Chem. 64, 3171–3177.

Späth, G.F., Lye, L.-F., Segawa, H., Sacks, D., Turco, S.J., Beverley, S.M., 2003. Persistence without pathology in phosphoglycan-deficient *Leishmaina major*. Science 301, 1241–1243.

Striepen, B., Zinecker, C.F., Damm, J.B.L., et al., 1997. Molecular structure of the "low molecular weight antigen" of *Toxoplasma gondii*: a glucose α 1-4 *N*-acetylgalactosamine makes free glycosylphosphatidylinositols highly immunogenic. J. Mol. Biol. 266, 797–813.

Striepen, B., Dubremetz, J.-F., Schwarz, R.T., 1999. Glucosylation of glycosylphosphatidylinositol membrane anchors: identification of uridine diphosphate-glucose as the direct donor for side chain modification in *Toxoplasma gondii* using carbohydrate analogs. Biochemistry 38, 1478–1487.

Tailler, D., Ferrieres, V., Pekari, K., Schmidt, R.R., 1999. Synthesis of the glycosyl phosphatidyl inositol anchor of rat brain Thy-1. Tetrahedron Lett. 40, 679–682.

Turco, S.J., Späth, G.F., Beverley, S.M., 2001. Is lipophosphoglycan a virulence factor? A surprising diversity between *Leishmaina* species. Trends Parasit. 17, 223–226.

Upreti, M., Ruhela, D., Vishwakarma, R.A., 2000. Synthesis of the tetrasaccharide cap domain of the antigenic lipophosphoglycan of *Leishmania donovani* parasite. Tetrahedron 56, 6577–6584.

Varma, R., Mayor, S., 1998. GPI-anchored proteins are organized in submicron domains at the cell surface. Nature 394, 798–801.

Wu, X., Guo, Z., 2007. Convergent synthesis of a fully phosphorylated GPI anchor of the CD52 antigen. Org. Lett. 9, 4311–4313.

Wu, X., Shen, Z., Zeng, X., Lang, S., Palmer, M., Guo, Z., 2008. Synthesis and biological evaluation of sperm CD52 GPI anchor and related derivatives as binding acceptors of pore-forming CAMP factor. Carbohydr. Res. 343, 1718–1729.

Yashunsky, D.V., Higson, A.P., Ross, A.J., Nikolaev, A. V., 2001. An efficient and stereoselective synthesis of β-D-Ara*p*-(1-2)-β-D-Gal*p*-(1-3)-β-D-Gal*p*-(1-4)-α-D-Man*p*, a tetrasaccharide fragment of *Leishmania major* lipophosphoglycan. Carbohydr. Res. 336, 243–248.

Yashunsky, D.V., Higson, A.P., Sizova, O.V., Ferguson, M.A.J., Nikolaev, A.V., 2003. Synthetic fragments of *Leishmania major* lipophosphoglycan containing β-D-Ara*p* side chains and the preparation of neoglycoproteins therefrom. In: Perez, S. (Ed.), Book of Abstracts of the 12th European Carbohydrate Symposium, Grenoble, France, July 2003. Imprimerie des Ecureuils, Gieres, p. 116.

Yashunsky, D.V., Borodkin, V.S., Ferguson, M.A.J., Nikolaev, A.V., 2006. The chemical synthesis of bioactive glycosylphosphatidylinositols from *Trypanosoma cruzi* containing an unsaturated fatty acid in the lipid. Angew. Chem. Int. Ed. Engl. 45, 468–474.

Yashunsky, D.V., Borodkin, V.S., McGivern, P.G., Ferguson, M.A.J., Nikolaev, A.V., 2007. The chemical synthesis of glycosylphosphatidylinositol anchors from *Trypanosoma cruzi* trypomastigote mucins: exploration of ester and acetal type permanent protecting groups. In: Demchenko, A.V. (Ed.), Frontiers in Modern Carbohydrate Chemistry. American Chemical Society, Washington DC, pp. 285–306.

Zuurmond, H.M., van der Meer, P.H., van der Klein, P.A.M., van der Marel, G.A., van Boom, J.H., 1993. Iodonium-promoted glycosylations with phenyl selenoglycosides. J. Carbohydr. Chem. 12, 1091–1103.

PART III

MICROBE–HOST GLYCOSYLATED INTERACTIONS

CHAPTER

27

Bacterial lectin-like interactions in cell recognition and adhesion

Joe Tiralongo and Anthony P. Moran

SUMMARY

To colonize host animals, many pathogenic bacteria express proteinaceous surface molecules called adhesins that facilitate binding to host cells and tissue. Host cell adhesion by pathogenic bacteria plays a critical role in the establishment of disease, allowing bacteria to overcome the natural cleansing systems of the host organism that include urine flow in the urinary tract and the flow of mucosal secretions in the gastrointestinal tract. Lectins, which bind carbohydrate structures on the surface of host tissue, represent the most common adhesin used by bacteria for adhesion. The carbohydrate specificity of diverse lectins to a large extent dictates bacteria tissue tropism by mediating specific attachment to unique host sites expressing the corresponding carbohydrate receptor. This chapter describes the best characterized examples of bacterial lectins, particularly the structural basis for specific carbohydrate recognition, and the role of lectins in uropathogenic *Escherichia coli* tissue tropism. The utilization of glycan array technology towards the further identification and characterization of novel bacterial lectins, and the development of anti-adhesion agents targeting bacteria-carbohydrate interactions, are also described.

Keywords: Adhesins; Anti-adhesion agents; Cell recognition; Lectins; Tissue tropism; Glycan array; Uropathogenic *Escherichia coli*; Pathogenesis; Virulence

1. INTRODUCTION

Bacterial surface components that mediate adherence are collectively called adhesins. Often, these adhesins are associated with fimbriae or pili and have been correlated with pathogenicity (Sharon, 2006). Several bacterial species utilize specific adhesins, or proteinaceous lectins, that bind carbohydrate structures on the surface of host tissues to facilitate attachment. A growing list of bacteria have been reported to mediate host cell adherence through sialic acid-containing glycans (discussed in Chapter 29) which have been the subject of extensive reviews (Angata and Varki, 2002; Lehmann *et al.*, 2006) and, therefore, will not be further discussed here. Table 27.1 presents a summary of the specificity, localization and tropism of other bacterial lectins.

TABLE 27.1 Bacterial lectins and their specificity, localization and tissue tropism

Carbohydrate specificity	Species	Lectin, localization	Tropism	Reference
Man				
α-Man-(1→3)-α-Man-(1→6)-Man	*Escherichia coli*	FimH, Type 1 fimbriae	Commensal	Sokurenko *et al.*, 1997
Man	*E. coli*	FimH, Type 1 fimbriae	Urinary	Sokurenko *et al.*, 1997
α-Man-(1→2/3/6)-Man	*Burkholderia cenocepacia*	BclA, Intracellular	Lung	Lameignere *et al.*, 2008
$Man_9(GlcNAc)_2$	*Enterobacter cloacae*	FimH, Type 1 pili	Gastrointestinal	Pan *et al.*, 1997
Man	*Vibrio cholerae*	MSHA, Type 4 pili	Gastrointestinal	Jonson *et al.*, 1994
Man	*Salmonella enterica* sv. Typhimurium	FimH, Type 1 fimbriae	Gastrointestinal	Kukkonen *et al.*, 1993
Man	*Klebsiella pneumoniae*	FimH, Type 1 fimbriae	Lung	Sharon, 1996
Man	*Bacillus thuringiensis*	HA-A[a], Parasporal inclusions	Insect	Wasano *et al.*, 2003
Man	*Shigella flexneri*	CA-A[b], Type 1 fimbriae	Eye	Gbarah *et al.*, 1993
Fuc				
β-Gal-(1→3)-[α-Fuc-(1→4)-]GlcNAc	*Pseudomonas aeruginosa*	PA-IIL, Intracellular	Lung	Imberty *et al.*, 2004
β-Gal-(1→4)-[α-Fuc-(1→3)-]GlcNAc	*S. enterica* sv. Typhimurium	PefA subunit, Type 1 fimbriae	Gastrointestinal	Chessa *et al.*, 2008
α-Fuc-(1→2)-β-Gal-(1→4)-GlcNAc	*Campylobacter jejuni*	CA-A, Cell surface	Gastrointestinal	Ruiz-Palacios *et al.*, 2003
α-Fuc-(1→2)-β-Gal-(1→3)-[α-Fuc-(1→4)-] Gal	*Helicobacter pylori*	BabA, Cell surface	Gastrointestinal	Falk *et al.*, 1993
Fuc	*Chromobacterium violaceum*	CV-IIL, Intracellular	Lung	Zinger-Yosovich *et al.*, 2006
α-Fuc-(1→2)-Gal and β-GalNAc-(1→3)-Gal	*Ralstonia solanacearum*	RSL, Intracellular	Plant	Kostlanova *et al.*, 2005
Fuc = Man	*Ralstonia solanacearum*	RS-IIL, Intracellular	Plant	Sudakevitz *et al.*, 2004
Fuc	*Porphyromonas gingivalis*	HA-A, Fimbriae	Oral	Sojar *et al.*, 2004
Fuc	*Azospirillum brasilense*	HA-A, Cell surface	Plant	Castellanos *et al.*, 1998
	Azospirillum lipoferum			
Fuc		HA-A, Cell surface	Plant	Castellanos *et al.*, 1998
Gal				
β-Gal-(1→3/4/6)-Gal	*P. aeruginosa*	PA-IL, Intracellular	Gastrointestinal	Imberty *et al.*, 2004
β-Gal-(1→4)-Glc	*Haemophilus ducreyi*	DltA, Cell surface	Genital	Leduc *et al.*, 2004
β-Gal-(1→4)-GlcNAc	*E. coli*	Bundlin, Bundle-forming pili	Gastrointestinal	Hyland *et al.*, 2008

(*Continues*)

TABLE 27.1 (*Continued*)

Carbohydrate specificity	Species	Lectin, localization	Tropism	Reference
α-Gal-(1→4)-Gal	*E. coli*	PapG, P fimbriae	Kidney	Dodson *et al.*, 2001
α-Gal-(1→4)-Gal	*Streptococcus suis*	PN & PO adhesin, Cell surface	Lung	Haataja *et al.*, 1996
GlcNAc				
GlcNAc	*E. coli*	F17-G, F17 fimbriae	Gastrointestinal	Buts *et al.*, 2003
β-GlcNAc-(1→4)-GlcNAc	*E. coli*	CA-A, K1 fimbriae	Endothelial	Sharon, 1996
GalNAc				
β-GalNAc-(1→4)-Gal	*E. coli*	SPB-A[c], F1C fimbriae	Urinary	Khan *et al.*, 2000
β-GalNAc-(1→4)-Gal	*P. aeruginosa*	SPB-A, Type IV pili	Lung	Sheth *et al.*, 1994
GalNAc	*Eikenella corrodens*	HA-A, Type IV fimbriae	Oral	Azakami *et al.*, 2006
GalNAc	*Actinomyces naeslundi*	HA-A, Type 2 fimbriae	Oral	Sharon, 1996
GalNAc	*B. thuringiensis*	HA-A, Parasporal inclusions	Insect	Akao *et al.*, 1999

[a]HA-A, Haemagglutinin activity.
[b]CA-A, Cell adherence activity.
[c]SPB-A, Solid phase binding activity.
Abbreviations: Fuc, L-fucose; Gal, D-galactose; GalNAc, *N*-acetyl-D-galactosamine; GlcNAc, *N*-acetyl-D-glucosamine; Man, D-mannose.

The cellular localization of bacterial lectins is quite diverse. Some are secreted as soluble proteins such as toxins (see Chapter 30) or are located intracellularly, e.g. *Pseudomonas aeruginosa* PA-IL and PA-IIL (Imberty *et al.*, 2004) and *Ralstonia solanacearum* RSL and RS-IIL (Sudakevitz *et al.*, 2004; Kostlanova *et al.*, 2005). However, the vast majority of known bacterial lectins are associated with pili or fimbriae, with the lectin domain located at the tip of these structures (Figure 27.1). In general, bacterial lectins have a low affinity for their carbohydrate receptor. Nevertheless, bacteria compensate for this low affinity by using fimbriae to form multimeric lectin displays or by taking advantage of receptor structures that cluster in the membrane of the host cell. Both examples provide multivalency that increases the overall binding strength by engaging many individual carbohydrate receptors simultaneously. An analogy frequently used to illustrate this concept is the interaction of the two faces of a Velcro™ strip (Esko, 1999; Mulvey *et al.*, 2001) (Figure 27.2A).

As shown in Table 27.1, lectins of varying specificity are associated with fimbrial structures. These include:

(i) the mannose- (Man-) specific lectins associated with the type 1 fimbriae of *Escherichia coli, Enterobacter cloacae, Salmonella enterica* sv. Typhimurium, *Klebsiella pneumoniae* and *Shigella flexneri*;
(ii) the galabiose-, α-Gal-(1→4)-Gal-, specific lectin associated with the *E. coli* P-fimbriae;

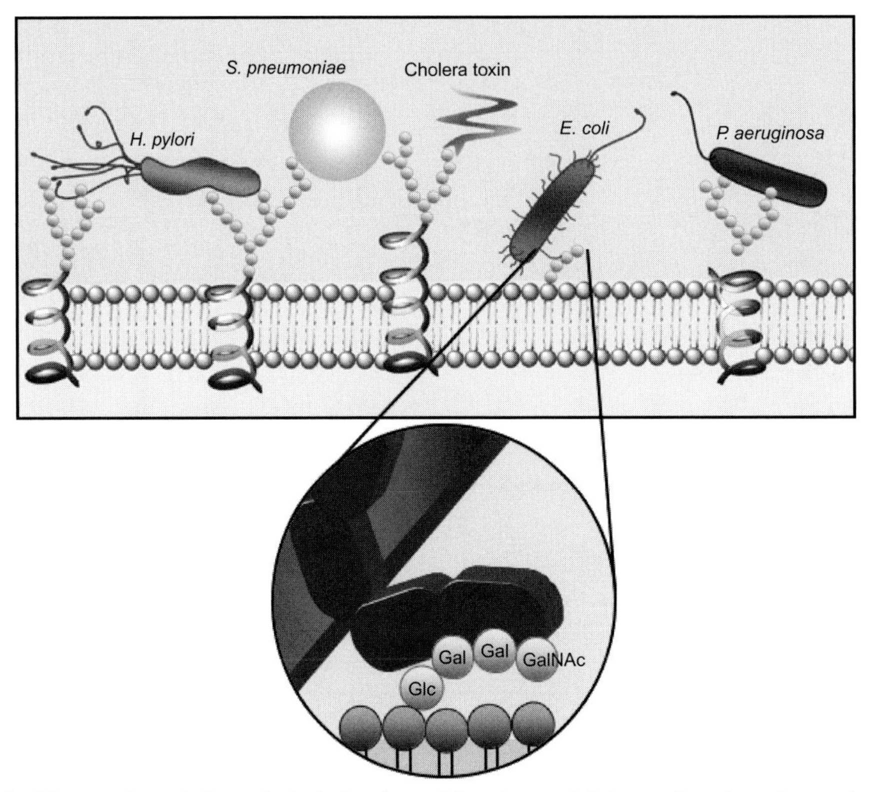

FIGURE 27.1 Many pathogenic bacteria, including bacterial toxins, exploit host cell-surface glycoconjugates as receptors for attachment, colonization and invasion. Although the cellular localization of bacterial lectins is quite diverse, the vast majority of known bacterial lectins are associated with pili or fimbriae. The insert illustrates the proposed "side-on" binding of the PapG adhesin located at the tip of P fimbriae to the galabiose, β-Gal-(1→4)-Gal, core of Gb_4.

(iii) the *N*-acetyl-D-galactosamine- (GalNAc-) specific lectins associated with the *E. coli* F1C-fimbriae and *Eikenella corrodens* type IV fimbriae; and
(iv) the *N*-acetyl- D-glucosamine- (GlcNAc-) specific lectin associated with the F-17 *E. coli*.

For the vast majority, bacterial lectins and their corresponding carbohydrate specificities have been inferred from inhibition studies, whereby specificity has been revealed by the ability of carbohydrates of defined structure to inhibit agglutination of erythrocytes or adherence to animal cells. It is noteworthy that only a small number of bacterial lectins are well characterized at the primary sequence and biochemical level and crystal structure data are available for an even smaller number.

This chapter summarizes the best characterized examples of bacterial lectins specific for Man, fucose (Fuc), GlcNAc, galactose (Gal) and GalNAc and describes the role of three of these lectins in uropathogenic *E. coli* (UPEC) tissue tropism. The utilization of glycan array technology towards the identification and characterization of novel bacterial lectins, and the development of anti-adhesion agents targeting bacteria–carbohydrate interactions, are also discussed.

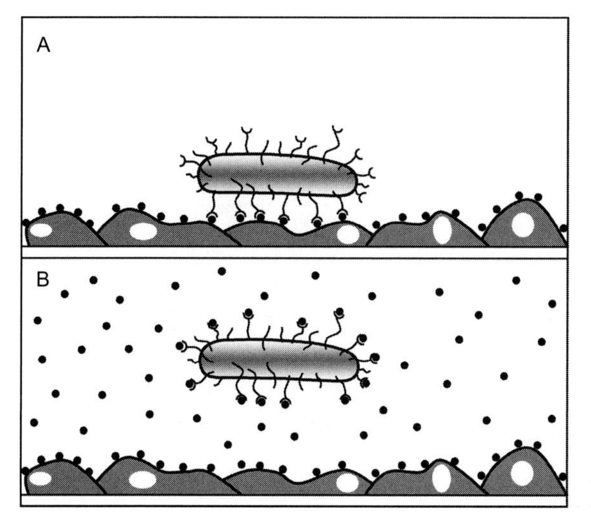

FIGURE 27.2 Binding of bacteria to host cells and the concept of anti-adhesion therapy. (A) The binding of bacterial to host cells is analogous to that between the two faces of a Velcro™ strip, where the resulting multivalency increases the overall binding strength by engaging many individual carbohydrate receptors simultaneously. (B) The concept of anti-adhesion therapy is based on the ability of carbohydrate-mediated adhesion to host cells being prevented through the interaction of natural saccharides or synthetic analogues, present in saturating amounts, with the target bacterial lectin.

2. MANNOSE-SPECIFIC BACTERIAL LECTINS

Mannose-specific lectins have been described for a number of bacterial species (see Table 27.1), but the *E. coli* FimH, a 30-kDa protein found at the distal end of type I fimbriae, represents the best studied example (Mulvey, 2002; Westerlund-Wikström and Korhonen, 2005). The FimH adhesin is composed of two domains, a C-terminal pilin domain that incorporates FimH into the fimbrial tip and an N-terminal receptor-binding domain (Choudhury *et al.*, 1999). The receptor-binding domain is an 11-stranded elongated β-barrel with an overall jellyroll topology that is connected to the pilin domain through an extended linker (Choudhury *et al.*, 1999). The latter apparently provides FimH with the required flexibility to position correctly the Man-binding site for attachment to the host receptor.

The available crystal structure for FimH complexed with D-mannopyranoside shows that Man is buried deep in a negatively charged pocket located at the tip of the binding domain, thereby making multiple interactions with the lectin (Hung *et al.*, 2002). The numerous hydrogen bonding and hydrophobic contacts that are made account for the unusually high affinity of FimH for Man compared to other lectins. A ridge of hydrophobic residues surrounding the binding site further increases affinity by helping to direct Man into the pocket. The mutation of residues within the Man-binding pocket of FimH leads, in all but one case, to the complete loss of FimH binding to human bladder cells (Hung *et al.*, 2002). The finding that residues within the binding pocket are invariant in over 200 uropathogenic isolates of *E. coli* (Hung *et al.*, 2002) highlights the importance of Man-dependent FimH binding in mediating *E. coli* adherence to human bladder cells. This will be discussed further below.

3. FUCOSE-SPECIFIC BACTERIAL LECTINS

The PA-IIL adhesin, the best described Fuc-binding bacterial lectin, expressed by the opportunistic human pathogen *P. aeruginosa*, is involved in host cell adherence and biofilm formation through specific interactions with Lewis^a-, β-Gal-1→3)-[Fucα1→4)-]GlcNAc-, related structures (Imberty *et al.*, 2004; Tielker *et al.*, 2005; Wu *et al.*, 2006). Infections by *P. aeruginosa* are particularly prominent in immunocompromised individuals and cystic fibrosis sufferers. The interaction of PA-IIL with Fuc is also of unusually high affinity, i.e. low μM range, and is dependent on the presence of divalent cations, with the crystal structure showing that two Ca^{2+} ions within the binding site are involved in Fuc coordination (Imberty *et al.*, 2004). The involvement of PA-IIL in host cell attachment and biofilm formation makes it an attractive target for anti-*Pseudomonas* therapy. In particular, oligosaccharides and glycoconjugates from human milk have attracted special attention due to their ability to block PA-IIL binding activity (Lesman-Movshovich *et al.*, 2003; Imberty *et al.*, 2004). Interestingly, significant differences between cow and human milk glycan PA-IIL blocking activity have been noted; specifically only human milk glycans containing Lewis structures inhibited PA-IIL haemagglutination activity (Lesman-Movshovich *et al.*, 2003). On the other hand, further development of anti-*Pseudomonas* therapies has been hampered by the lack of a suitable animal model for studying experimental *Pseudomonas* bronchopulmonary infections. The recent description and characterization of mink as an animal model potentially overcomes this deficiency, thereby providing a suitable model to investigate *P. aeruginosa* adherence *in vivo* (Kirkeby *et al.*, 2007).

4. GALACTOSE AND GalNAc-SPECIFIC BACTERIAL LECTINS

P. aeruginosa also expresses a second soluble lectin involved in host cell adherence, PA-IL. This lectin binds Gal with moderate affinity in the high μM range (Imberty *et al.*, 2004) or binds disaccharides possessing terminal Gal residues, with β-Gal-(1→3/4/6)-Gal displaying the highest affinities (Chen *et al.*, 1998). Whereas the involvement of *P. aeruginosa* in lung infections is relatively well studied, its effect on the intestinal tract-related disease is poorly understood (Wu *et al.*, 2003). Nevertheless, what has been established is that PA-IL facilitates the adhesion of *P. aeruginosa* to the intestinal epithelium leading to a defect in the intestinal epithelial barrier that makes it permeable to the cytotoxin, exotoxin A, resulting in gut-derived sepsis (Laughlin *et al.*, 2000; Alverdy *et al.*, 2000; Wu *et al.*, 2003).

Unlike bacterial lectins associated with type 1 fimbriae that bind carbohydrate structures on glycoproteins, PapG, which is present on the flexible tip of the *E. coli* P fimbriae, specifically binds glycolipid receptors consisting of a galabiose, α-Gal-(1→4)-Gal, core (Dodson *et al.*, 2001). The PapG adhesin comprises a C-terminal pilin domain and an N-terminal carbohydrate-binding domain. Dodson *et al.* (2001) reported that the α-Gal-(1→4)-Gal-binding site is relatively large, shallow and extended, and is located on the side of the PapG molecule (Dodson *et al.*, 2001). Moreover, these investigators postulated that the binding site on the side of PapG, together with the location of PapG on a flexible tip, is required to present the binding site surface "side-on" to the host membrane (see Figure 27.1). Structural models of glycoshingolipids which indicate that the α-Gal-(1→4)-Gal moiety of the receptor is displayed at a right angle to the ceramide (Cer) group (Pascher *et al.*, 1992) support this model.

5. N-ACETYLGLUCOSAMINE-SPECIFIC BACTERIAL LECTINS

The GlcNAc-specific F17-G adhesin associated with F17 fimbriae plays a central role in the attachment of enterotoxigenic *E. coli* to intestinal microvilli (reviewed in Sharon, 2006 and references therein). Based on the available fimbrial lectin crystal structures, the carbohydrate-binding site of F17-G is distinct from FimH, but similar to PapG, in so far as the binding sites of both are located laterally on the lectin domain allowing sideways interaction with their receptors (Buts *et al.*, 2003). The relative position of the lectin-binding site of F17-G, PapG and FimH is believed to reflect the diverse tissue tropism displayed by the bacterial strains harbouring these lectins. That is, the long and flexible F17 fimbriae of enterotoxigenic *E. coli* infiltrate between the epithelial microvilli, with the shallow binding site of the F17-G lectin domain interacting laterally with GlcNAc-containing receptors, thus resulting in F17 fimbrial attachment (Buts *et al.*, 2003). This is in contrast to the deep Man-binding site of FimH, located at the tip of the lectin domain, which allows high affinity binding of UPEC to terminal Man on glycoproteins in the urinary tract (Hung *et al.*, 2002). Of note, F17-G, PapG and FimH all have a remarkably similar overall shape and size, despite sharing only low sequence identity (Sharon, 2006).

6. TISSUE TROPISM OF UPEC

As an example of the important role played by bacterial lectins in determining tissue tropism, infection of the urinary tract by UPEC strains will be considered. Importantly, UPEC strains account for over 80% of urinary tract infections. These UPEC isolates can vary significantly in their ability to adhere, colonize and persist within different niches of the urinary tract. The carbohydrate specificity of diverse bacterial lectins, to a large extent, dictates this tissue tropism by mediating specific attachment of UPEC to unique sites along the urinary tract that express the corresponding carbohydrate receptor (Wright and Hultgren, 2006; Wiles *et al.*, 2008) (Figure 27.3). This section describes the contribution of three bacterial lectins in UPEC tissue tropism.

6.1. FimH

The binding of *E. coli* FimH to tri-Man structures has been observed for all FimH adhesins, but the level of binding to mono-Man varies considerably among phenotypic variants. Significantly, high affinity mono-Man-binding FimH has been shown to be associated with UPEC variants, whereas isolates from the intestine of healthy individuals express low affinity mono-Man-binding FimH (Sokurenko *et al.*, 1997, 1998). The latter phenotype is believed to provide an advantage in colonizing the gastrointestinal tract, whereas high affinity mono-Man-binding is advantageous in colonizing the urinary tract (Sokurenko *et al.*, 1998). The conversion of FimH from low to high affinity mono-Man-binding frequently involves mutations within the adhesin domain that stabilizes the protein loops carrying residues required for Man-binding (Sokurenko *et al.*, 1998; Schembri *et al.*, 2000).

A key receptor for FimH on the apical surface of the bladder is uroplakin 1a, an integral membrane glycoprotein (Zhou *et al.*, 2001), however, FimH can also bind many other host proteins (Westerlund-Wikström and Korhonen, 2005). One recent example is the identification of Man-dependent binding of $\alpha_3\beta_1$ integrin subunits by FimH, a process that appears to mediate the uptake of type 1 piliated *E. coli* (Eto *et al.*, 2007). The FimH interaction with $\alpha_3\beta_1$ integrin subunits, which are expressed by many host cell types including urothelial cells, triggers signalling cascades and adaptor proteins that stimulate actin rearrangement, a process implicated in pathogen entry into host cells (Wiles *et al.*, 2008). Both uroplakin 1a and $\alpha_3\beta_1$ integrin subunits are commonly found clustered on the lumenal surface of the bladder as uroplakin plaques of 0.3–0.5 μm in diameter, or within cholesterol-rich lipid rafts, respectively (Mulvey, 2002; Wiles *et al.*, 2008). The FimH interaction with Man is of quite high affinity compared to most lectins; on the other hand, the clustering of receptors in the plane of the membrane can facilitate even stronger FimH interactions with host cells. Interestingly, the binding of FimH to mono-Man is even further enhanced under shear stress conditions, leading to an increase in bacterial adhesion strength (Thomas *et al.*, 2004). This phenomenon, which is analogous to the shear-dependent rolling of leukocytes on the endothelial cell surface, is dependent on the structural properties of FimH (Thomas *et al.*, 2002). The differential level and strength of *E. coli* adhesion

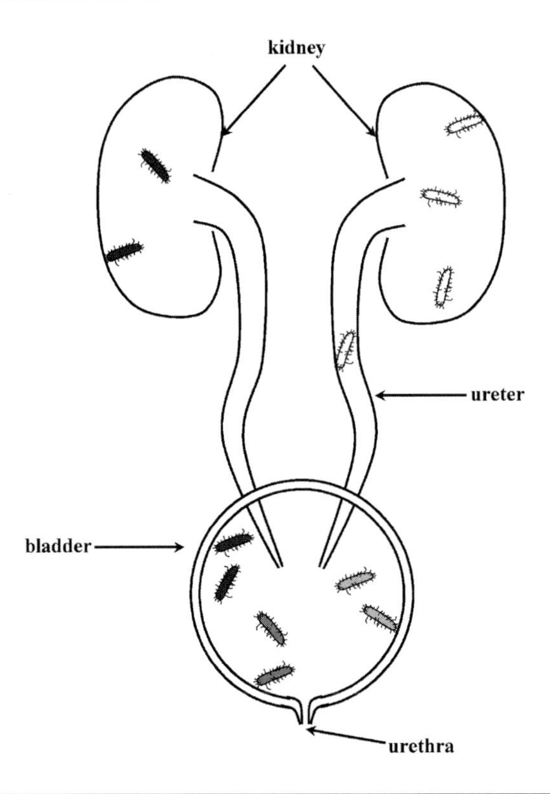

UPEC	Lectin	Receptor	Specificity	Tropism
	PapG II	Gb_4	α-Gal-(1→4)-Gal	Kidney
	PapG III	Gb_5	α-Gal-(1→4)-Gal	Bladder
	FimH	Uroplakin Ia/α3β1 Integrin	Man	Bladder
	SfaS	Uroplakin III/Sialosyl Gb_4	Sia	Bladder/Kidney

FIGURE 27.3 The tissue tropism of four UPEC isolates expressing PapG II, PapG III, FimH or SfaS, their known receptors, carbohydrate specificities and distribution along the urinary tract are illustrated. Depending on the tropism of the UPEC isolates either human pyelonephritis, i.e. a urinary tract infection that has reached the kidney, or human cystitis, i.e. inflammation of the bladder, results. Abbreviations: Gal, D-galactose; Man, D-mannose; Sia, silaic acid.

under shear stress to diverse Man-containing receptors present in different niches may also dictate host cell tropism of UPEC.

6.2. PapG

The PapG adhesins recognize glycolipid receptors, specifically globotriaosylceramide (Gb_3) isotypes, that are expressed on the surface of erythrocytes and target cells present in the kidney and urinary tract. Three distinct PapG adhesins have been identified (PapG I, II and III), with each recognizing a unique receptor; however, the core galabiose structure, α-Gal-(1→4)-Gal, that is linked by a β-glucose (Glc) to Cer, which anchors the receptor to the membrane, is common throughout (Roberts *et al.*, 1994; Dodson *et al.*, 2001). The PapG II adhesion which binds globotetraosylceramide (Gb_4), β-GalNAc-(1→3)-α-Gal-(1→4)-β-Gal-(1→4)-β-Glc-(1→1)-Cer, is present on urothelial cells, including kidney cells, and is associated with the development of human pyelonephritis, i.e. an ascending urinary tract infection that has reached the kidney (Dodson *et al.*, 2001). The PapG III variant, however, correlates with human cystitis, i.e. an inflammation of the urinary bladder, with the receptor being Gb_5, α-GalNAc-(1→3)-β-GalNAc-(1→3)-α-Gal-(1→4)-β-Gal-(1→4)-β-Glc-(1→1)-Cer (Stromberg *et al.*, 1990). The distribution of the Gb_3 receptor isotypes in different host tissues, and the binding specificity of the various PapG variants, suggest that these interactions may be involved in host tissue tropism of UPEC.

6.3. SfaS

The adhesin SFaS, a minor component of *E. coli* S-fimbriae, can mediate interactions with sialic acid residues on receptors expressed by kidney epithelial and urothelial cells (Lehmann *et al.*, 2006). A detailed analysis of the specificity of S-fimbriae towards glycolipids has demonstrated a preferential binding to gangliosides carrying α-Neu5Gc-(2→3)-Gal- and α-Neu5Gc-(2→8)-Neu5Ac-structures (where Neu5Gc, *N*-glycolylneuraminic acid; Neu5Ac, *N*-acetylneuraminic acid) (Hanisch *et al.*, 1993). Interestingly, both α-(2→3)- and α-(2→6)-linked sialic acid residues have been found on uroplakin III which, together with uroplakin 1a, 1b and II, assemble into uroplakin plaques and galactosyl-Gb_4, β-Gal-(1→3)-β-GalNAc-(1→3)-α-Gal-(1→4)-β-Gal-(1→4)-β-Glc-(1→1)-Cer, which are expressed on the mammalian bladder and urinary tract, respectively (Stapleton *et al.*, 1998; Malagolini *et al.*, 2000).

As stated earlier, the formation of uroplakin plaques on the lumenal surface of the bladder facilitates strong Man-specific FimH binding to uroplakin 1a. The presence of sialic acid residues on uroplakin III suggests that the *E. coli* S-fimbriae may also play a role in the development of urinary bladder infection, i.e. cystitis. Furthermore, Gb_4 acts as the receptor of PapG II, a binding event that correlates with human cystitis. Stapleton *et al.* (1998) showed that sialosyl and disialosyl galactosyl-Gb_4 can act as receptors for a number of UPEC isolates expressing each of the three PapG adhesins. Indeed, binding of sialosyl and disialosyl galactosyl-Gb_4 by UPEC was of higher avidity than that observed for Gb_4 and Gb_3 (Stapleton *et al.*, 1998), suggesting that the presence of a charged sugar may increase or modulate binding strength.

7. UTILIZING GLYCAN ARRAY TECHNOLOGY TO IDENTIFY AND CHARACTERIZE NOVEL BACTERIA–CARBOHYDRATE INTERACTIONS

The interaction and specificity of lectins to their ligand can be effectively interpreted by probing arrays, consisting of glycans and glycoproteins of known structure covalently immobilized onto glass-slides through a 6- to 12-carbon linker. Typically, glycans are printed using robotic printing technology onto an appropriately functionalized glass-slide allowing covalent attachment to the glass surface. Specific lectin–sugar interactions can then be visualized following interrogation of the glycan array, e.g. with a fluorescent-labelled lectin. The advantages of glycan array technology are that it provides a sensitive, high-throughput, easy-to-use platform that is able to provide novel information concerning lectin–sugar interactions. The identification and characterization of lectins has been greatly enhanced through the development of glycan array technology (Blixt *et al.*, 2004). Glycan arrays are fast becoming the technique of choice for identifying and elucidating the specificity of lectins (Blixt *et al.*, 2004; Angeloni *et al.*, 2005; Stevens *et al.*, 2006) and the technology has been successfully used to identify and characterize the interactions of bacteria with their glycan receptors (Disney and Seeberger, 2004; Walz *et al.*, 2005; Chessa *et al.*, 2008; Day *et al.*, 2009). Two recent reports serve as examples of the use of glycan array technology to probe bacteria–carbohydrate interactions.

Lewis structures, specifically Lewis^x, β-Gal-(1→4)-[α-Fuc-(1→3)-]GlcNAc, were identified as the ligand for the plasmid-encoded PefA fimbrial subunit of *S. enterica* sv. Typhimurium using glycan array technology (Chessa *et al.*, 2008). The expression of the *S. enterica* sv. Typhimurium *pef* operon in *E. coli* was used to generate flexible fibrillae composed of the major fimbrial subunit, PefA. Glycan array analysis was then used to probe the binding specificity of purified plasmid-encoded PefA fimbriae. This general methodological approach should prove valuable for the future identification and characterization of other fimbriae-associated lectins.

The culture collection strain NCTC 11168 was the first strain of microaerophilic *Campylobacter jejuni* to be sequenced (Parkhill *et al.*, 2000) and has been widely used for studying *C. jejuni* pathogenesis. However, the continuous passaging of this strain has been shown to affect dramatically and adversely its colonization potential (Ringoir and Korolik, 2003; Gaynor *et al.*, 2004). Recently, glycan array analysis on fluorescently labelled *C. jejuni* NCTC 11168 has been performed using the frequently passaged, non-colonizing, genome-sequenced strain (11168-GS) and, for comparison, the infrequently passaged, original, virulent strain (11168-O) grown or maintained under various conditions (Day *et al.*, 2009). Glycan structures recognized and bound by *C. jejuni* included terminal Man, sialic acid, Gal and Fuc residues. Significantly, it was found that only when challenged with normal oxygen levels at room temperature did 11168-O consistently bind to sialic acid or terminal Man structures, whereas 11168-GS bound these structures regardless of growth or maintenance conditions. These binding differences were confirmed by the ability of specific lectins to inhibit competitively the adherence of *C. jejuni* to a Caco-2 intestinal cell line. These data suggest that the binding of Man and/or sialic acid may provide the initial interactions important for colonization following environmental exposure (Day *et al.*, 2009). Therefore, this example illustrates the power of glycan array technology to explore variations in the carbohydrate binding ability of diverse bacterial strains and isolates.

8. INHIBITORS OF LECTIN-MEDIATED ADHESION OF BACTERIA TO HOST CELLS

The well documented increase in antibiotic resistant, clinically relevant bacteria has focused attention on alternative therapies to combat bacterial pathogenesis. One such approach is anti-adhesion therapy using carbohydrates that block binding of pathogenic bacteria to host cells (Sharon, 2006). This general concept is illustrated in Figure 27.2B. The basis of the concept is that carbohydrate-mediated adhesion to host cells is prevented through the interaction of natural saccharides or synthetic analogues, present in saturating amounts, with the target bacterial lectin. Given that many saccharides shown to inhibit bacterial adhesion are naturally occurring, as discussed earlier for human milk oligosaccharide inhibition of *P. aeruginosa* adhesion, and are not bactericidal, they are seen as safe and less likely to lead to the generation of therapy-resistant bacteria (Newburg *et al.*, 2005; Sharon, 2006).

There are a number of well described examples of carbohydrates or their derivatives acting as anti-adhesion agents in experimental models:

(i) the Man and Man-derivatives that inhibit *E. coli* type 1 adhesion to intestinal cells and prevent urinary tract and gastrointestinal tract infections in mice (reviewed in Sharon, 2006 and references therein);

(ii) 3′-sialyllactose that inhibits *H. pylori* adhesion to intestinal cells and reduces *H. pylori* colonization in monkeys (reviewed in Sharon and Ofek, 2000 and references therein);
(iii) 2′-fucosylated oligosaccharides protecting human milk-fed infants from *Campylobacter*-induced diarrhoea (Newburg *et al.*, 2005); and
(iv) α-Gal-(1→4)-Gal-based monovalent and multivalent inhibitors of PapG (Ohlsson *et al.*, 2002; Salminen *et al.*, 2007) and PN and PO adhesins (Ohlsson *et al.*, 2005; Salminen *et al.*, 2007).

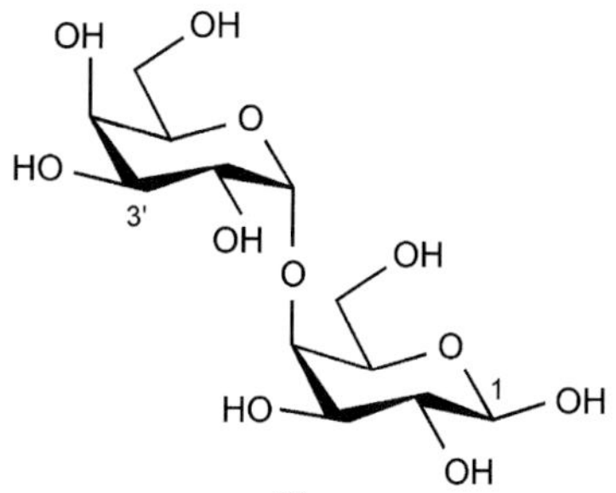

The multivalent interaction between bacteria and host cells, akin to the interaction between two faces of a Velcro™ strip, has the effect of significantly increasing the avidity of the overall interaction, making it extremely difficult for monovalent inhibitors to disrupt bacterial adhesion. To overcome this problem monovalent inhibitors of significantly high affinity, i.e. in the sub-μM range, or inhibitors that mimic Nature's multivalent display, are necessary. The development of monovalent and multivalent PapG I inhibitors represents an excellent case in point.

A commonly used approach towards developing carbohydrate-based anti-adhesives is the systematic replacement of hydroxyl groups involved in hydrogen bonding interactions within the lectin-binding pocket with hydrophobic or other non-hydrogen bonding moieties. The idea behind this approach is that the resulting carbohydrate analogue(s) would increase the relative contributions of non-hydrogen bonding interactions leading to an enhancement in the affinity of the analogues for the bacterial lectin (Mulvey *et al.*, 2001). This strategy has been employed to develop the most potent, monovalent galabiose, α-Gal-(1→4)-Gal)-based **1**, inhibitor of PapG I (Figure 27.4). Through the replacement of the hydroxyl groups at positions C-1 and C-3′ with aromatic substituents, a PapG I inhibitor has been generated that is 20–30 times more potent than the natural receptor Gb_4 (Ohlsson *et al.*, 2002). However, with an IC_{50} of 4.1 μM (IC_{50} is defined as the amount of compound required to inhibit a given biological process by 50%), this falls well short of the potency required for a therapeutically useful anti-adhesion agent. In the search for more potent PapG I inhibitors, Salminen *et al.* (2007) screened a number of multivalent unsubstituted galabiose derivatives to investigate the effect of multivalency on PapG inhibition. An octavalent inhibitor was found to be the most effective with an IC_{50} of 2 μM, i.e. 40–50 times more potent than the corresponding monovalent compound, and which compared well with the monovalent C-1- and C-3′-substituted galabiose analogue described by Ohlsson *et al.* (2002). It will be interesting to see whether more potent PapG I inhibitors can be generated by constructing multivalent displays of galabiose with aromatic substituents at C-1 and C-3′.

(**1**)

FIGURE 27.4 Development of the monovalent galabiose-, α-Gal-(1→4)-Gal)-, based inhibitor of the adhesin PapG I. Through the replacement of the hydroxyl groups at position C-1 and C-3′ (as indicated) with aromatic substituents, a PapG I inhibitor has been generated that is 20–30 times more potent than the natural receptor Gb_4.

Despite the exciting potential of anti-adhesion therapy based on carbohydrates suggested from *in vitro* and *in vivo* studies, their clinical efficacy has yet to be demonstrated. Two separate clinical trials investigating the effect of 3′-sialyllacto-*N*-neotetraose on acute otitis media (Ukkonen *et al.*, 2000) and 3′-sialyllactose on *H. pylori* infection (Parente *et al.*, 2003), however showed no beneficial effect of anti-adhesion oligosaccharides.

9. CONCLUSIONS

Adhesion, mediated by carbohydrate-specific lectins, plays a central role in bacterial colonization and tissue tropism. Although a variety of such bacterial lectins have been described, only a small number are well characterized at the primary sequence and biochemical level, and crystal structure data are available for a limited number only. Nevertheless, as a model of tissue tropism, the expression of adhesins by UPEC strains acts as a useful paradigm. Based on the specificity of lectin–carbohydrate interaction, anti-adhesion agents have been developed but, for those that have undergone clinical trials, no beneficial effects were observed. Thus, for future development of this field, the points listed in the Research Focus Box could prove worthy of consideration.

ACKNOWLEDGEMENTS

The authors gratefully acknowledge the financial support of Griffith University, Science Foundation Ireland and the EU Marie Curie Programme (grant no. MTKD-CT-2005-029774) for their studies.

RESEARCH FOCUS BOX

- The role of bacteria–carbohydrate interactions in dictating tissue tropism, particularly in UPEC, should be more intensively studied. This would be aided by the identification of additional carbohydrate–receptors used by UPEC isolates for adherence, colonization and invasion.
- Utilization of glycan array technology for the identification and characterization of novel bacterial lectins should be more widely applied.
- Development of potent carbohydrate-based multivalent inhibitors of bacterial adhesion could prove therapeutically worthwhile. This would be aided by the structural elucidation of bacterial lectins and establishment of crystal structures.

References

Akao, T., Mizuki, E., Yamashita, S., Saitoh, H., Ohba, M., 1999. Lectin activity of *Bacillus thuringiensis* parasporal inclusion proteins. FEMS Microbiol. Lett. 179, 415–421.

Alverdy, J., Holbrook, C., Rocha, F., et al., 2000. Gut-derived sepsis occurs when the right pathogen with the right virulence genes meets the right host: evidence for *in vivo* virulence expression in *Pseudomonas aeruginosa*. Ann. Surg. 232, 480–489.

Angata, T., Varki, A., 2002. Chemical diversity in the sialic acids and related α-keto acids: an evolutionary perspective. Chem. Rev. 102, 439–469.

Angeloni, S., Ridet, J.L., Kusy, N., et al., 2005. Glycoprofiling with micro-arrays of glycoconjugates and lectins. Glycobiology 15, 31–41.

Azakami, H., Akimichi, H., Noiri, Y., Ebisu, S., Kato, A., 2006. Plasmid-mediated genomic recombination at the pilin gene locus enhances the *N*-acetyl-D-galactosamine-specific haemagglutination activity and the growth rate of *Eikenella corrodens*. Microbiology 152, 815–821.

Blixt, O., Head, S., Mondala, T., et al., 2004. Printed covalent glycan array for ligand profiling of diverse glycan binding proteins. Proc. Natl. Acad. Sci. USA 101, 17033–17038.

Buts, L., Bouckaert, J., De Genst, E., et al., 2003. The fimbrial adhesin F17-G of enterotoxigenic *Escherichia coli* has an immunoglobulin-like lectin domain that binds *N*-acetylglucosamine. Mol. Microbiol. 49, 705–715.

Castellanos, T., Ascencio, F., Bashan, Y., 1998. Cell-surface lectins of *Azospirillum* spp. Curr. Microbiol. 36, 241–244.

Chen, C.P., Song, S.C., Gilboa-Garber, N., Chang, K.S., Wu, A. M., 1998. Studies on the binding site of the galactose-specific agglutinin PA-IL from *Pseudomonas aeruginosa*. Glycobiology 8, 7–16.

Chessa, D., Dorsey, C.W., Winter, M., Baumler, A.J., 2008. Binding specificity of *Salmonella* plasmid-encoded fimbriae assessed by glycomics. J. Biol. Chem. 283, 8118–8124.

Choudhury, D., Thompson, A., Stojanoff, V., et al., 1999. X-ray structure of the FimC-FimH chaperone-adhesin complex from uropathogenic *Escherichia coli*. Science 285, 1061–1066.

Day, C.J., Tiralongo, J., Hartnell, R.D., et al., 2009. Differential carbohydrate recognition by *Campylobacter jejuni* strain 11168: influences of temperature and growth conditions. PloS One, 4, e 4927.

Disney, M.D., Seeberger, P.H., 2004. The use of carbohydrate microarrays to study carbohydrate–cell interactions and to detect pathogens. Chem. Biol. 11, 1701–1707.

Dodson, K.W., Pinkner, J.S., Rose, T., Magnusson, G., Hultgren, S.J., Waksman, G., 2001. Structural basis of the interaction of the pyelonephritic *E. coli* adhesin to its human kidney receptor. Cell 105, 733–743.

Esko, J.D, 1999. Microbial carbohydrate-binding proteins. In: Varki, A., Cummings, H.J., Esko, J.D., Freeze, H., Hart, G.W., Marth, J.D. (Eds.), Essentials of Glycobiology. Cold Spring Harbor Laboratory Press, Cold Spring Harbor, New York, pp. 429–440.

Eto, D.S., Jones, T.A., Sundsbak, J.L., Mulvey, M.A., 2007. Integrin-mediated host cell invasion by type 1-piliated uropathogenic *Escherichia coli*. PLoS Pathog. 3, e100.

Falk, P., Roth, K.A., Borén, T., Westblom, T.U., Gordon, J.I., Normark, S., 1993. An *in vitro* adherence assay reveals that *Helicobacter pylori* exhibits cell lineage-specific tropism in the human gastric epithelium. Proc. Natl. Acad. Sci. USA 90, 2035–2039.

Gaynor, E.C., Cawthraw, S., Manning, G., MacKichan, J.K., Falkow, S., Newell, D.G., 2004. The genome-sequenced variant of *Campylobacter jejuni* NCTC 11168 and the original clonal clinical isolate differ markedly in colonization, gene expression, and virulence-associated phenotypes. J. Bacteriol. 186, 503–517.

Gbarah, A., Mirelman, D., Sansonetti, P.J., Verdon, R., Bernhard, W., Sharon, N., 1993. *Shigella flexneri* transformants expressing type 1 (mannose-specific) fimbriae bind to, activate, and are killed by phagocytic cells. Infect. Immun. 61, 1687–1693.

Haataja, S., Tikkanen, K., Hytonen, J., Finne, J., 1996. The Galα1-4 Gal-binding adhesin of *Streptococcus suis*, a gram-positive meningitis-associated bacterium. Adv. Exp. Med. Biol. 408, 25–34.

Hanisch, F.G., Hacker, J., Schroten, H., 1993. Specificity of S fimbriae on recombinant *Escherichia coli*: preferential binding to gangliosides expressing NeuGcα(2-3)Gal and NeuAcα(2-8)NeuAc. Infect. Immun. 61, 2108–2115.

Hung, C.S., Bouckaert, J., Hung, D., et al., 2002. Structural basis of tropism of *Escherichia coli* to the bladder during urinary tract infection. Mol. Microbiol. 44, 903–915.

Hyland, R.M., Sun, J., Griener, T.P., et al., 2008. The bundlin pilin protein of enteropathogenic *Escherichia coli* is an *N*-acetyllactosamine-specific lectin. Cell. Microbiol. 10, 177–187.

Imberty, A., Wimmerova, M., Mitchell, E.P., Gilboa-Garber, N., 2004. Structures of the lectins from *Pseudomonas aeruginosa*: insight into the molecular basis for host glycan recognition. Microbes Infect. 6, 221–228.

Jonson, G., Lebens, M., Holmgren, J., 1994. Cloning and sequencing of *Vibrio cholerae* mannose-sensitive haemagglutinin pilin gene: localization of mshA within a cluster of type 4 pilin genes. Mol. Microbiol. 13, 109–118.

Khan, A.S., Kniep, B., Oelschlaeger, T.A., Van Die, I., Korhonen, T., Hacker, J., 2000. Receptor structure for F1C fimbriae of uropathogenic *Escherichia coli*. Infect. Immun. 68, 3541–3547.

Kirkeby, S., Wimmerova, M., Moe, D., Hansen, A.K., 2007. The mink as an animal model for *Pseudomonas aeruginosa* adhesion: binding of the bacterial lectins (PA-IL and PA-IIL) to neoglycoproteins and to sections of pancreas and lung tissues from healthy mink. Microbes Infect. 9, 566–573.

Kostlanova, N., Mitchell, E.P., Lortat-Jacob, H., et al., 2005. The fucose-binding lectin from *Ralstonia solanacearum*. A new type of β-propeller architecture formed by oligomerization and interacting with fucoside, fucosyllactose, and plant xyloglucan. J. Biol. Chem. 280, 27839–27849.

Kukkonen, M., Raunio, T., Virkola, R., et al., 1993. Basement membrane carbohydrate as a target for bacterial adhesion: binding of type I fimbriae of *Salmonella enterica* and *Escherichia coli* to laminin. Mol. Microbiol. 7, 229–237.

Lameignere, E., Malinovska, L., Slavikova, M., et al., 2008. Structural basis for mannose recognition by alectin from opportunistic bacteria *Burkholderia cenocepacia*. Biochem. J. 411, 307–318.

Laughlin, R.S., Musch, M.W., Hollbrook, C.J., Rocha, F.M., Chang, E.B., Alverdy, J.C., 2000. The key role of *Pseudomonas aeruginosa* PA-I lectin on experimental gut-derived sepsis. Ann. Surg. 232, 133–142.

Leduc, I., Richards, P., Davis, C., Schilling, B., Elkins, C., 2004. A novel lectin, DltA, is required for expression of a full serum resistance phenotype in *Haemophilus ducreyi*. Infect. Immun. 72, 3418–3428.

Lehmann, F., Tiralongo, E., Tiralongo, J., 2006. Sialic acid-specific lectins: occurrence, specificity and function. Cell. Mol. Life Sci. 63, 1331–1354.

Lesman-Movshovich, E., Lerrer, B., Gilboa-Garber, N., 2003. Blocking of *Pseudomonas aeruginosa* lectins by human milk glycans. Can. J. Microbiol. 49, 230–235.

Malagolini, N., Cavallone, D., Wu, X.R., Serafinicessi, F., 2000. Terminal glycosylation of bovine uroplakin III, one of the major integral-membrane glycoproteins of mammalian bladder. Biochim. Biophys. Acta 1475, 231–237.

Mulvey, G., Kitov, P.I., Marcato, P., Bundle, D.R., Armstrong, G.D., 2001. Glycan mimicry as a basis for novel anti-infective drugs. Biochimie 83, 841–847.

Mulvey, M.A., 2002. Adhesion and entry of uropathogenic *Escherichia coli*. Cell. Microbiol. 4, 257–271.

Newburg, D.S., Ruiz-Palacios, G.M., Morrow, A.L., 2005. Human milk glycans protect infants against enteric pathogens. Annu. Rev. Nutr. 25, 37–58.

Ohlsson, J., Jass, J., Uhlin, B.E., Kihlberg, J., Nilsson, U.J., 2002. Discovery of potent inhibitors of PapG adhesins from uropathogenic *Escherichia coli* through synthesis and evaluation of galabiose derivatives. ChemBioChem 3, 772–779.

Ohlsson, J., Larsson, A., Haataja, S., et al., 2005. Structure–activity relationships of galabioside derivatives as inhibitors of *E. coli* and *S. suis* adhesins: nanomolar inhibitors of *S. suis* adhesins. Org. Biomol. Chem. 3, 886–900.

Pan, Y.T., Xu, B., Rice, K., Smith, S., Jackson, R., Elbein, A.D., 1997. Specificity of the high-mannose recognition site between *Enterobacter cloacae* pili adhesin and HT-29 cell membranes. Infect. Immun. 65, 4199–4206.

Parente, F., Cucino, C., Anderloni, A., Grandinetti, G., Bianchi Porro, G., 2003. Treatment of *Helicobacter pylori* infection using a novel antiadhesion compound (3'sialyllactose sodium salt). A double blind, placebo-controlled clinical study. Helicobacter 8, 252–256.

Parkhill, J., Wren, B.W., Mungall, K., et al., 2000. The genome sequence of the food-borne pathogen Campylobacter jejuni reveals hypervariable sequences. Nature 403, 665–668.

Pascher, I., Lundmark, M., Nyholm, P.G., Sundell, S., 1992. Crystal structures of membrane lipids. Biochim. Biophys. Acta 1113, 339–373.

Ringoir, D.D., Korolik, V., 2003. Colonisation phenotype and colonisation potential differences in *Campylobacter jejuni* strains in chickens before and after passage in vivo. Vet. Microbiol. 92, 225–235.

Roberts, J.A., Marklund, B.I., Ilver, D., et al., 1994. The Gal(alpha 1-4)Gal-specific tip adhesin of *Escherichia coli* P-fimbriae is needed for pyelonephritis to occur in the normal urinary tract. Proc. Natl. Acad. Sci. USA 91, 11889–11893.

Ruiz-Palacios, G.M., Cervantes, L.E., Ramos, P., Chavez-Munguia, B., Newburg, D.S., 2003. *Campylobacter jejuni* binds intestinal H(O) antigen (Fucα1,2Galβ1,4GlcNAc), and fucosyloligosaccharides of human milk inhibit its binding and infection. J. Biol. Chem. 278, 14112–14120.

Salminen, A., Loimaranta, V., Joosten, J.A., et al., 2007. Inhibition of P-fimbriated *Escherichia coli* adhesion by multivalent galabiose derivatives studied by a live-bacteria application of surface plasmon resonance. J. Antimicrob. Chemother. 60, 495–501.

Schembri, M.A., Sokurenko, E.V., Klemm, P., 2000. Functional flexibility of the FimH adhesin: insights from a random mutant library. Infect. Immun. 68, 2638–2646.

Sharon, N., 1996. Carbohydrate–lectin interactions in infectious disease. Adv. Exp. Med. Biol. 408, 1–8.

Sharon, N., 2006. Carbohydrates as future anti-adhesion drugs for infectious diseases. Biochim. Biophys. Acta 1760, 527–537.

Sharon, N., Ofek, I., 2000. Safe as mother's milk: carbohydrates as future anti-adhesion drugs for bacterial diseases. Glycoconjugate J. 17, 659–664.

Sheth, H.B., Lee, K.K., Wong, W.Y., et al., 1994. The pili of *Pseudomonas aeruginosa* strains PAK and PAO bind specifically to the carbohydrate sequence βGalNAc(1-4)βGal found in glycosphingolipids asialo-GM1 and asialo-GM2. Mol. Microbiol. 11, 715–723.

Sojar, H.T., Sharma, A., Genco, R.J., 2004. *Porphyromonas gingivalis* fimbriae binds to neoglycoproteins: evidence for a lectin-like interaction. Biochimie 86, 245–249.

Sokurenko, E.V., Chesnokova, V., Doyle, R.J., Hasty, D.L., 1997. Diversity of the *Escherichia coli* type 1 fimbrial lectin. Differential binding to mannosides and uroepithelial cells. J. Biol. Chem. 272, 17880–17886.

Sokurenko, E.V., Chesnokova, V., Dykhuizen, D.E., et al., 1998. Pathogenic adaptation of *Escherichia coli* by natural variation of the FimH adhesin. Proc. Natl. Acad. Sci. USA 95, 8922–8926.

Stapleton, A.E., Stroud, M.R., Hakomori, S.I., Stamm, W.E., 1998. The globoseries glycosphingolipid sialosyl galactosyl globoside is found in urinary tract tissues and is a preferred binding receptor in vitro for uropathogenic *Escherichia coli* expressing pap-encoded adhesins. Infect. Immun. 66, 3856–3861.

Stevens, J., Blixt, O., Paulson, J.C., Wilson, I.A., 2006. Glycan microarray technologies: tools to survey host specificity of influenza viruses. Nat. Rev. Microbiol. 4, 857–864.

Stromberg, N., Marklund, B.I., Lund, B., et al., 1990. Host-specificity of uropathogenic *Escherichia coli* depends on differences in binding specificity to Galα1-4Gal-containing isoreceptors. Embo J. 9, 2001–2010.

Sudakevitz, D., Kostlanova, N., Blatman-Jan, G., et al., 2004. A new *Ralstonia solanacearum* high-affinity mannose-binding lectin RS-IIL structurally resembling the *Pseudomonas aeruginosa* fucose-specific lectin PA-IIL. Mol. Microbiol. 52, 691–700.

Thomas, W.E., Trintchina, E., Forero, M., Vogel, V., Sokurenko, E.V., 2002. Bacterial adhesion to target cells enhanced by shear force. Cell 109, 913–923.

Thomas, W.E., Nilsson, L.M., Forero, M., Sokurenko, E.V., Vogel, V., 2004. Shear-dependent 'stick-and-roll' adhesion of type 1 fimbriated *Escherichia coli*. Mol. Microbiol. 53, 1545–1557.

Tielker, D., Hacker, S., Loris, R., et al., 2005. *Pseudomonas aeruginosa* lectin LecB is located in the outer membrane and is involved in biofilm formation. Microbiology 151, 1313–1323.

Ukkonen, P., Varis, K., Jernfors, M., et al., 2000. Treatment of acute otitis media with an antiadhesive oligosaccharide: a randomised, double-blind, placebo-controlled trial. Lancet 356, 1398–1402.

Walz, A., Odenbreit, S., Mahdavi, J., Boren, T., Ruhl, S., 2005. Identification and characterization of binding properties of *Helicobacter pylori* by glycoconjugate arrays. Glycobiology 15, 700–708.

Wasano, N., Ohgushi, A., Ohba, M., 2003. Mannose-specific lectin activity of parasporal proteins from a lepidoptera-specific *Bacillus thuringiensis* strain. Curr. Microbiol. 46, 43–46.

Westerlund-Wikström, B., Korhonen, T.K., 2005. Molecular structure of adhesin domains in *Escherichia coli* fimbriae. Int. J. Med. Microbiol. 295, 479–486.

Wiles, T.J., Kulesus, R.R., Mulvey, M.A., 2008. Origins and virulence mechanisms of uropathogenic *Escherichia coli*. Exp. Mol. Pathol. 85, 11–19.

Wright, K.J., Hultgren, S.J., 2006. Sticky fibers and uropathogenesis: bacterial adhesins in the urinary tract. Future Microbiol. 1, 75–87.

Wu, A.M., Wu, J.H., Singh, T., Liu, J.H., Tsai, M.S., Gilboa-Garber, N., 2006. Interactions of the fucose-specific *Pseudomonas aeruginosa* lectin, PA-IIL, with mammalian glycoconjugates bearing polyvalent Lewis^a and ABH blood group glycotopes. Biochimie 88, 1479–1492.

Wu, L., Holbrook, C., Zaborina, O., et al., 2003. *Pseudomonas aeruginosa* expresses a lethal virulence determinant, the PA-I lectin/adhesin, in the intestinal tract of a stressed host: the role of epithelia cell contact and molecules of the Quorum Sensing Signaling System. Ann. Surg. 238, 754–764.

Zhou, G., Mo, W.J., Sebbel, P., et al., 2001. Uroplakin Ia is the urothelial receptor for uropathogenic *Escherichia coli*: evidence from in vitro FimH binding. J. Cell Sci. 114, 4095–4103.

Zinger-Yosovich, K., Sudakevitz, D., Imberty, A., Garber, N.C., Gilboa-Garber, N., 2006. Production and properties of the native *Chromobacterium violaceum* fucose-binding lectin (CV-IIL) compared to homologous lectins of *Pseudomonas aeruginosa* (PA-IIL) and *Ralstonia solanacearum* (RS-IIL). Microbiology 152, 457–463.

CHAPTER

28

Lectin-like interactions in virus–cell recognition: human immunodeficiency virus and C-type lectin interactions

Imke Steffen, Theodros S. Tsegaye and Stefan Pöhlmann

SUMMARY

Recognition of pathogen-specific glycostructures by lectins on immune cells is an important means of host immune defence but may also be exploited by some pathogens to promote their spread. The calcium-dependent lectin dendritic cell-specific ICAM-grabbing non-integrin (DC-SIGN) is involved in human immunodeficiency virus (HIV) interactions with dendritic cells. Attachment of HIV to dendritic cells can potentiate viral infectivity for adjacent T-cells and it has been postulated that this process contributes to the dissemination of sexually transmitted virus. However, more recent research has revealed that the consequences of lectin-dependent HIV interactions with dendritic cells are diverse and can include uptake for major histocompatibility complex presentation, productive infection and transfer of virus to T-cells. In this chapter DC-SIGN and other cellular lectins known to recognize HIV are introduced, and how lectin binding might impact viral dissemination is discussed.

Keywords: Human immunodeficiency virus (HIV); Dendritic cell-specific intercellular adhesion molecule 3-grabbing non-integrin (DC-SIGN); C-Type lectin; Dendritic cell; Langerin; Platelet; Dissemination; Pathogenesis

1. INTRODUCTION

Many enveloped viruses hijack the host cell glycosylation machinery to ensure appropriate carbohydrate modification of their surface proteins. The efficiency and type of glycans added to viral membrane proteins can determine recognition of viruses by cellular lectins and the humoral immune response which, in turn, can have profound consequences for viral spread and pathogenicity (Vigerust and Shepherd, 2007). Conversely, several viral glycoproteins function as lectins and employ cellular glycans for infectious entry into target cells. Most prominently, the influenza haemagglutinin binds to sialic acid present on surface structures of target cells

(Vigerust and Shepherd, 2007) and the nature of the sialic acid linkage determines if cells are susceptible to infection by human viruses (which bind to α-(2→6)-linked sialic acid) or avian viruses (which recognize α-(2→3)-linked sialic acid) (see Chapter 15).

The interactions of viruses with calcium-dependent (C-type) lectins have received particular attention. This lectin family comprises membrane-bound and soluble members which can promote cell adhesion and/or sense pathogens (van Kooyk and Geijtenbeek, 2003; Ji *et al.*, 2006; van Kooyk and Rabinovich, 2008) (see Chapter 34). One would expect that virion capture by C-type lectins invariably promotes establishment of an effective immune response. However, several lines of evidence suggest that certain viruses and non-viral pathogens specifically target C-type lectins to slip detection by the immune system (van Kooyk and Geijtenbeek, 2003). The most prominent example might be the interaction of human immunodeficiency virus (HIV) with dendritic cell-specific intercellular adhesion molecule 3-grabbing non-integrin (DC-SIGN, CD209), a C-type lectin expressed at high levels on dendritic cells (DCs). Binding of HIV to DC-SIGN can potentiate viral infectivity or induce viral uptake and degradation. The molecular mechanisms underlying these processes and their consequences for HIV dissemination in and between individuals are the topic of the remainder of this review. As stated above, targeting DCs and other immune cells via DC-SIGN or related lectins is not a particular trait of HIV (Table 28.1). Thus, for example, DC-SIGN promotes DC infection, e.g. by dengue virus and measles virus, and the

TABLE 28.1 Viruses that are recognized by DC-SIGN and DC-SIGNR

Virus	DC-SIGN	DC-SIGNR
Human immunodeficiency virus	Curtis *et al.*, 1992; Geijtenbeek *et al.*, 2000b	Pöhlmann *et al.*, 2001a; Bashirova *et al.*, 2001; Mummidi *et al.*, 2001
Human T-cell leukemia virus	Ceccaldi *et al.*, 2006	?
Human herpes virus 8	Rappocciolo *et al.*, 2006a	?
Human cytomegalovirus	Halary *et al.*, 2002	Halary *et al.*, 2002
Herpes simplex virus	de Jong *et al.*, 2008	?
Measles virus	de Witte *et al.*, 2006	?
Influenza virus	Wang *et al.*, 2008	?
Dengue virus	Navarro-Sanchez *et al.*, 2003; Tassaneetrithep *et al.*, 2003	Tassaneetrithep *et al.*, 2003
West Nile virus	Davis *et al.*, 2006a,b	Davis *et al.*, 2006a,b
Hepatitis C virus	Gardner *et al.*, 2003; Pöhlmann *et al.*, 2003; Lozach *et al.*, 2003	Gardner *et al.*, 2003; Pöhlmann *et al.*, 2003; Lozach *et al.*, 2003
Sindbis virus	Klimstra *et al.*, 2003	Klimstra *et al.*, 2003
Ebola virus	Alvarez *et al.*, 2002; Simmons *et al.*, 2003	Alvarez *et al.*, 2002; Simmons *et al.*, 2003
Marburg virus	Marzi *et al.*, 2004	Marzi *et al.*, 2004
SARS-coronavirus	Marzi *et al.*, 2004; Yang *et al.*, 2004	Marzi *et al.*, 2004; Yang *et al.*, 2004
Human coronavirus NL63	Hofmann *et al.*, 2006	Hofmann *et al.*, 2006

DC-SIGN-related protein DC-SIGNR (also termed L-SIGN, CD209L) may concentrate hepatitis C virus in liver sinusoidal endothelial cells, thereby potentially promoting infection of adjacent hepatocytes (see Table 28.1). Moreover, Ebola- and Marburg virus, which induce a lethal haemorrhagic fever in humans, employ several C-type lectins for augmentation of infectivity and lectin engagement may determine the discrete cell and organ tropism observed at various stages of filovirus infection. In summary, an intricate interplay between viruses and C-type lectins impacts the balance between viral attack and host defence, as specified below for HIV, and elucidation of the underlying mechanisms can provide important insights into the pathogenesis of viral infections and may uncover attractive targets for therapy and prevention.

2. MAKING IT STICK: Env MEDIATES HIV ATTACHMENT AND ENTRY INTO HOST CELLS

Human immunodeficiency virus is the causative agent of acquired immunodeficiency syndrome (AIDS). In 2007, HIV and AIDS afflicted 33.2 million people with devastating socioeconomic consequences (Cohen *et al.*, 2008). Interindividual spread of HIV mainly occurs via the sexual route. It is believed that capture of sexually transmitted virus by mucosal DCs (see below) is important for subsequent dissemination to lymphoid tissue (Wu and KewalRamani, 2006; Piguet and Steinman, 2007). The gut-associated lymphoid tissue is the first and the principal target of HIV infection (Veazey and Lackner, 2004). During a phase of clinical latency, virally destroyed T-cells are constantly replaced by fresh cells. However, after several years (in the absence of therapy), the capacity of the host to replenish T-cells gradually decreases and the decline in T-cell numbers is paralleled by an increasing susceptibility to opportunistic infections, which are ultimately fatal.

The HIV envelope protein (Env) allows the virus to recognize and access the host cell (Pöhlmann and Reeves, 2006). The Env protein is synthesized in the secretory pathway of infected cells. An N-terminal signal sequence earmarks nascent Env for import into the endoplasmatic reticulum, where the protein is extensively modified with *N*-linked mannose-rich glycans (Scanlan *et al.*, 2007). Upon transport of Env in the Golgi apparatus, these glycans are further processed to a complex and hybrid type in a host cell-dependent fashion. However, less than half of the oligosaccharides are completely processed due to their recessed location and/or dense packaging (Scanlan *et al.*, 2007). It has been shown that Env is also *O*-glycosylated (Bernstein *et al.*, 1994), but target sites and biological relevance are largely unclear. Extensive glycosylation of surface exposed regions shields underlying epitopes from recognition by antibodies and critically contributes to immune evasion (Scanlan *et al.*, 2007). Moreover, Env glycosylation is essential for interaction with cellular lectins, which can promote or inhibit viral spread, as discussed below.

Infectious cellular entry of HIV is initiated by Env interactions with the CD4 receptor, which is expressed on T-cells, macrophages, monocytes and DCs, all of which are susceptible to HIV (Pöhlmann and Reeves, 2006) (Figure 28.1). Binding to CD4 triggers conformational changes in the surface unit gp120, which lead to the formation and/or exposure of a co-receptor binding site. Engagement of a chemokine co-receptor, usually CCR5 or CXCR4, activates the membrane fusion machinery located in the transmembrane unit gp41 (Pöhlmann and Reeves, 2006), which undergoes a series of conformational changes resulting in the fusion of the viral and the host cell membrane (see Figure 28.1).

The interactions between Env on virions and CD4 and co-receptor on target cells are cell type-independent and indispensable for infectious entry. Consequently, they are attractive targets for therapeutic intervention (Este and Telenti, 2007). However, a constantly accumulating body

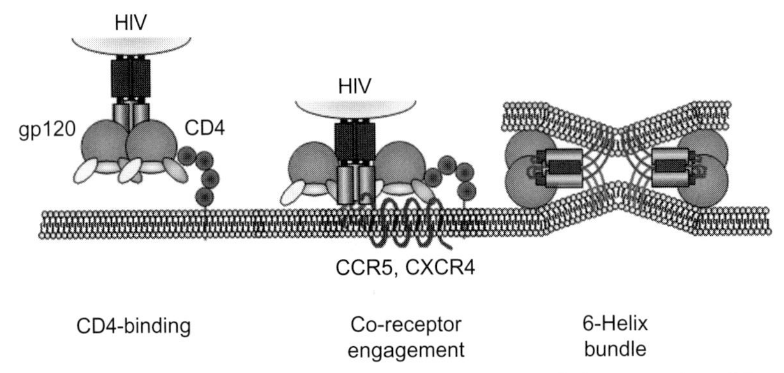

FIGURE 28.1 Cellular entry of HIV. Infectious entry of HIV commences with the interaction of the surface unit gp120 of the viral envelope protein (Env) with the primary receptor CD4. Binding to CD4 triggers conformational changes in gp120 which allow engagement of a chemokine co-receptor, usually CCR5 and/or CXCR4. Sexually transmitted viruses almost invariably use CCR5 for entry. Viruses that use CXCR4 evolve in about 40% of infected individuals and their emergence is associated with progression towards AIDS. The interaction of gp120 with a co-receptor induces conformational changes in the transmembrane unit, gp41, namely the formation of a six-helix bundle structure, which is intimately associated with membrane fusion. Adapted, with permission, from Reeves *et al.* (2006) Evaluation of current strategies to inhibit HIV entry, integration and maturation. In: Bogner, E., Holzenburg, A. (Eds), New Concepts of Antiviral Therapy. Springer, Dordrecht, pp. 213–254. © Springer 2006.

of evidence suggests that CD4- and co-receptor-independent interactions of HIV with target cells, albeit being ultimately dispensable for infectious entry, can profoundly augment infection efficiency. Thus, fragments of prostatic acidic phosphatase, which form amyloid fibrils in human semen, boost HIV infectivity by concentrating virions onto target cells (Münch *et al.*, 2007) and attachment of HIV to DCs potentiates infectivity for adjacent T-cells (Wu and KewalRamani, 2006; Piguet and Steinman, 2007). The latter process may particularly promote dissemination of sexually transmitted HIV, because DCs might not only promote mucosal spread of HIV by facilitating infection of adjacent susceptible cells but may also ferry the virus into lymph nodes where it has ample access to target T-cells, as discussed below. It has been proposed that calcium-dependent (C-type) lectins on DCs are intimately involved in HIV capture and transfer to T-cells (Wu and KewalRamani, 2006; Piguet and Steinman, 2007). This review will introduce lectins participating in HIV capture and will discuss how lectin binding may modulate HIV spread.

3. PROMOTION OF HIV CAPTURE, *TRANS*-INFECTION AND DISSEMINATION BY DC-SIGN – THE PARADIGM REVISITED

Dendritic cells are divided into different subsets, which can be of myeloid and lymphoid origin, and are intimately involved in innate responses, tolerance induction and adaptive immunity (Liu, 2001). Langerhans cells and dermal DCs line the major surfaces of the human body and are uniquely equipped to recognize, take up and process antigen. Upon acquisition of antigen, both dendritic cell types migrate into lymphoid tissue and undergo a process termed maturation, during which expression of the antigen capture machinery is downregulated while production of factors required for antigen presentation is upregulated. In lymphoid tissues, mature DCs present antigen to T-cells and, due to their unique capability to stimulate naïve T-cells, are intimately involved in the induction of adaptive responses (Banchereau and Steinman, 1998).

Cell culture studies undertaken in the early 1990s indicated that DCs, despite their key role in the immune system, might promote HIV spread. Thus, it was demonstrated that HIV-exposed DCs could catalyse efficient infection of T-cells, apparently without being productively infected (Cameron *et al.*, 1992; Pope *et al.*, 1994). A possible interpretation of these findings was the existence of a so far unidentified factor on DCs which captures HIV and facilitates transmission of the virus to adjacent susceptible cells, a process termed "infection in *trans*". Two reports by Geijtenbeek and colleagues, which showed that the C-type lectin DC-SIGN is expressed on DCs and promotes HIV *trans*-infection, supported this concept (Geijtenbeek *et al.*, 2000a,b) (Table 28.2). Much effort has subsequently been devoted to the definition of the DC-SIGN/HIV interface and to the analysis of the molecular mechanisms underlying DC-SIGN-facilitated HIV *trans*-infection.

Sequence comparison and functional analysis of DC-SIGN defined the following domain structure: an N-terminal cytoplasmic domain, a transmembrane region, a repeat (also termed neck) region consisting of 7.5 repeats of a 23 amino acid comprising sequence and a C-type lectin domain. The C-type lectin domain, whose atomic structure has been determined (Feinberg *et al.*, 2001; Guo *et al.*, 2004), recognizes mannose- and fucose-containing glycans and is responsible for DC-SIGN binding to appropriately glycosylated ligands like the HIV Env protein, as discussed below. While monomers of the carbohydrate recognition domain (CRD) bind to ligands, DC-SIGN tetramerization, which is mainly driven by the repeat region, is required for high avidity binding (Mitchell *et al.*, 2001; Feinberg *et al.*, 2005; Snyder *et al.*, 2005). Recognition of ligands by DC-SIGN is solely carbohydrate-dependent (Lin *et al.*, 2003; Snyder *et al.*, 2005), albeit evidence to the contrary has been reported

TABLE 28.2 Cellular C-type lectins and lectin-like receptors discussed in this review[a]

Lectin	Tissue	Cell type	Gp120 binding	HIV *trans*-infection	HIV-degradation
DC-SIGN (CD209)	Mucosa, dermis, lymph node, spleen, bone marrow, placenta, lung	Dendritic cell, megakaryocyte, macrophage	Yes	Yes	Yes
DC-SIGNR (CD209L, L-SIGN)	Liver, lymph node, lung, intestine, placenta, bone marrow	Sinusoidal and capillary endothelial cell, alveolar cell	Yes	Yes	?
Langerin (CD207)	Mucosa, dermis	Langerhans cell	Yes	No	Yes
LSECtin	Liver, lymph node, bone marrow	Sinusoidal endothelial cells	Yes	No	?
CLEC-2	Bone marrow, liver	Megakaryocyte, platelet, sinusoidal endothelial cell	No	Yes	?

[a]See text for references.

(Geijtenbeek *et al.*, 2002) and this has two important consequences. First, ligands which do not exhibit appreciable amino acid sequence homology but display appropriate glycans in an adequate spatial configuration can be recognized by DC-SIGN. Indeed, a wide spectrum of viral and non-viral pathogens with glycosylated surface structures has been found to interact with DC-SIGN and, for many, it has been suggested that targeting DCs via DC-SIGN might promote their spread (Khoo *et al.*, 2008). Second, due to the cell type-dependent nature of glycosylation, the cellular background used for generation of HIV and other pathogens will profoundly impact the interaction with DC-SIGN. For example, the Env protein of HIV produced in T-cells is efficiently modified with high-mannose glycans and the respective viruses are robustly transmitted to target cells in a DC-SIGN-dependent manner (Lin *et al.*, 2003). In contrast, incorporation of mannose-rich glycans and *trans*-infection driven by DC-SIGN is inefficient if viruses are generated in macrophages (Lin *et al.*, 2003). The cell type used for pathogen amplification thus needs to be taken into account when evaluating the potential *in vivo* relevance of DC-SIGN interactions with pathogens observed *in vitro*.

How does binding of HIV Env to DC-SIGN facilitate infection of adjacent target cells? A straightforward explanation could be that DC-SIGN tethers virions on the cell surface and thereby increases the chance of virus transfer to susceptible cells, once virus-loaded cells and target cells make random contacts. However, several studies indicated that DC-SIGN-mediated *trans*-infection may be more complex. For one, analyses of mutant lectins revealed that HIV binding and transmission are dissociable functions, indicating that mere concentration of virions on target cells is insufficient for DC-SIGN-dependent *trans*-infection (Baribaud *et al.*, 2001; Pöhlmann *et al.*, 2001c). In addition, Geijtenbeek and colleagues reported that binding of HIV to a DC-SIGN-expressing cell line conserves viral infectivity over several days (Geijtenbeek *et al.*, 2000b), a remarkable finding, considering that infectivity of cell free virus is lost within hours. Finally, Kwon and co-workers provided evidence that DC-SIGN-dependent *trans*-infection involves DC-SIGN-driven uptake of virions into low pH compartments, where infectivity is preserved and from which virus is regurgitated upon contact of virus-containing cells with T-cells (Kwon *et al.*, 2002). Cumulatively, these observations fitted with a model suggesting that DCs in the submucosa, which express high levels of DC-SIGN, may take up sexually transmitted HIV in a DC-SIGN-dependent fashion and might subsequently transport HIV into lymph nodes, where the virus could be transferred to T-cells (Geijtenbeek *et al.*, 2000b). In such a scenario, DC-SIGN-expressing DCs would act as Trojan horses which shield the virus from the immune system by conserving particles in intracellular vesicles and which promote HIV dissemination due to their natural capability to migrate into lymphoid tissues, the major target sites of HIV infection. However, several key aspects of the "Trojan horse model" have subsequently been challenged, as discussed below.

3.1. Several receptors contribute to HIV capture by dendritic cells

Geijtenbeek and colleagues demonstrated that anti-DC-SIGN antibodies and the mannose-polymer mannan profoundly inhibited HIV interactions with DC-SIGN-positive cell lines (Geijtenbeek *et al.*, 2000b). Reduction of HIV *trans*-infection by monocyte-derived DCs (MDDCs) was also observed but inhibition was less robust compared to experiments with cell lines (Geijtenbeek *et al.*, 2000b). These studies have subsequently been repeated by several groups and a wide spectrum of effects was observed, ranging from DC-SIGN being responsible for the vast majority of dendritic cell-mediated HIV *trans*-infection to DC-SIGN not being involved in this process at all (Baribaud *et al.*, 2002; Wu *et al.*, 2002a,b; Trumpfheller *et al.*, 2003; Gummuluru *et al.*, 2003; Arrighi *et al.*, 2004b; Granelli-Piperno *et al.*, 2005; van Montfort *et al.*, 2007; Boggiano *et al.*, 2007). The reasons for these discrepancies are unclear but may involve usage of different viruses and virus producer cells as well as different DCs. In fact, Turville and colleagues showed that different types of DCs bind HIV Env via different receptors, with both CD4 and C-type lectins contributing to Env capture (Turville *et al.*, 2002). A recent study complemented these findings by demonstrating that CD4 expression negatively regulates DC-SIGN-mediated *trans*-infection, with viruses exposed to cells co-expressing CD4 and DC-SIGN being mainly sorted into late endosomal compartments (Wang *et al.*, 2007). Finally, variable *trans*-infection results may have been due to the ability of some DCs to support HIV infection, as discussed below.

3.2. DC-SIGN – not specific for dendritic cells?

Most studies on DC-SIGN function used a THP cell line engineered to express DC-SIGN. The THP cells are of monocytic origin and can be differentiated into macrophages upon phorbol myristate acetate treatment. Consequently, these cells may mirror some aspects of MDDCs. However, Wu and colleagues discovered that the THP cells widely used for DC-SIGN expression were indeed of B-cell origin and are most likely identical to Raji B-cells (the cell line is now termed B-THP) (Wu *et al.*, 2004a). In fact, analysis of true THP-DC-SIGN cells revealed that these cells are not able to mediate HIV *trans*-infection with appreciable efficiency (Wu *et al.*, 2004a). The reason for the cell type-dependence of DC-SIGN-driven HIV *trans*-infection is at present unclear. It has been noted that *trans*-infection requires cell-to-cell contact and can be diminished by contact of transmitting cells with certain cell types (Wu *et al.*, 2004b). Yet, the factors governing *trans*-infection efficiency remain to be elucidated on a molecular level. Of note, misidentification of DC-SIGN-expressing cells might also have occurred upon analysis of human and macaque tissue sections. Thus, initial studies indicated that DC-SIGN is a marker for DCs and that DC-SIGN-positive DCs are found in lymph nodes (Wu and KewalRamani, 2006; Piguet and Steinman, 2007). In contrast to this view, Granelli-Piperno and colleagues provided evidence that DC-SIGN-positive cells in normal lymph nodes are almost exclusively of macrophage origin (Granelli-Piperno *et al.*, 2005). Similarly, DC-SIGN-positive macrophages were detected in rheumatoid arthritis synovium (van Lent *et al.*, 2003), in the lung (Soilleux *et al.*, 2002) and in lesions of leprosy patients (Krutzik *et al.*, 2005). A potential misidentification of DC-SIGN-positive macrophages as DCs in lymph nodes and maybe in other tissues would have important implications for the contribution of DC-SIGN to HIV transmission, considering that macrophages but not DCs are readily susceptible to infection by CCR5-tropic viruses. A more detailed characterization of the nature of the DC-SIGN-positive cells in tissues, particularly in the anogenital mucosa, may therefore be required.

3.3. Enhancement of viral infectivity by DC-SIGN – *trans*-infection versus productive infection of transmitting cells

Analyses of DC-SIGN-mediated HIV *trans*-infection were based on the assumption that neither the commonly used B-THP DC-SIGN cell line nor MDDCs were susceptible to productive HIV infection. As it turned out, both assumptions were wrong. Nobile and colleagues demonstrate that B-THP cells are susceptible to infection by CXCR4-tropic HIV (Nobile *et al.*, 2005), most likely due to expression of CXCR4 and low levels of CD4. Thus, the reported DC-SIGN-mediated preservation of HIV infectivity by B-THP cells (Geijtenbeek *et al.*, 2000b; Kwon *et al.*, 2002) may have been due to release of infectious progeny viruses from transmitting cells and not to transfer of captured HIV. Notably, analogous observations were made with MDDCs. Thus, it is now established that both immature and mature DCs are susceptible to productive infection with CCR5-using viruses (albeit with different efficiencies) and that, apart from receptor expression, restriction by APOBEC3G mainly regulates susceptibility of DCs to HIV infection (Wu and KewalRamani, 2006; Piguet and Steinman, 2007). In the light of these observations, the consequences of HIV interactions with DCs were re-analysed. These studies revealed that input virus is transmitted only during a short-time window (hours) after HIV exposure of DCs, while all subsequent transmission events (days) are due to release of progeny viruses (Figure 28.2), suggesting that DCs might not be capable of storing infectious HIV over prolonged time periods (Turville *et al.*, 2004; Nobile *et al.*, 2005; Burleigh *et al.*, 2006).

It was proposed that DC-SIGN-driven uptake of HIV into acidic intracellular vesicles is a prerequisite to efficient *trans*-infection and the LL motif in the cytoplasmic domain of DC-SIGN has been shown to facilitate DC-SIGN internalization upon ligand uptake (Kwon *et al.*, 2002; Engering *et al.*, 2002). However, subsequent studies could not confirm a role for intracellular acidic pH or DC-SIGN internalization in *trans*-infection (Nobile *et al.*, 2005; Burleigh *et al.*, 2006). In fact, the vast majority of virus internalized by DCs was found to be processed for major histocompatibility complex (MHC) presentation (Moris *et al.*, 2004, 2006), while mainly particles located at the surface of DCs were transferred to T-cells (Cavrois *et al.*, 2007), albeit the latter finding is controversial (Piguet and Steinman, 2007) (see Figure 28.2). Internalization and conservation of infectious HIV particles by different types of DCs therefore warrants further assessment.

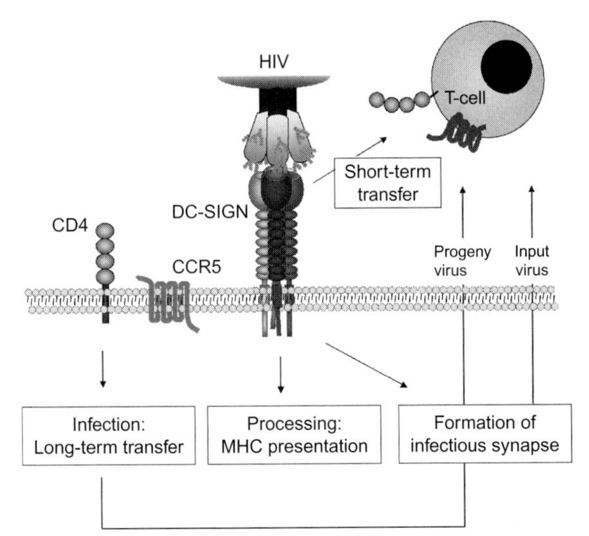

FIGURE 28.2 Attachment of HIV to DC-SIGN on dendritic cells (DCs). High levels of DC-SIGN are expressed on monocyte-derived DCs (MDDC) and cells in the anogenital mucosa and DC-SIGN contributes to HIV capture by these cells. Binding of HIV to DC-SIGN may have several consequences. Virus bound to the cell surface can be transmitted to adjacent target cells during a short-time window (hours, short-term transfer). Infectious entry via CD4 and CCR5 may also be promoted by DC-SIGN and release of progeny virions from infected cells is responsible for long-term HIV transfer to T-cells (days). The majority of DC-SIGN-bound HIV seems to be processed for MHC presentation. However, a fraction of the internalized virus evades degradation and can be transmitted to T-cells at the infectious synapse. Reproduced, with permission, from Pöhlmann and Tremblay (2007) Attachment of human immunodeficiency virus to cells and its inhibition. In: Reeves, J.D., Derdeyn, C.A. (Eds), Milestones in Drug Therapy: Entry Inhibitors in HIV Therapy. Birkhäuser, Basel, pp. 31–48.

3.4. Contribution of DC-SIGN to formation of infectious synapses

Transmission of HIV from DCs to T-cells occurs most efficiently at sites of intimate cell-to-cell contact. Thus, DCs were shown to accumulate internalized HIV particles at the site of contact to T-cells which, in turn, concentrate CD4 and co-receptor at the cell–cell interface (McDonald *et al.*, 2003; Garcia *et al.*, 2005). As a consequence, a microenvironment, termed infectious synapse, is established that provides optimal conditions for HIV *trans*-infection. DC-SIGN seems to be involved in the formation of infectious synapses for transmission of CXCR4-tropic viruses (Arrighi *et al.*, 2004a) (see Figure 28.2). However, potential interaction partners on the T-cell surface required for DC-SIGN-promoted synapse formation are unclear. Expression of ICAM-3 on target cells is dispensable for efficient *trans*-infection (Wu *et al.*, 2002b) and engineered expression of components of the immunological synapse on DC-SIGN positive, *trans*-infection negative cell lines does not rescue the defect in transmission (Wu *et al.*, 2004b), indicating a role for so far unappreciated factors.

4. BINDING OF HIV TO DC-SIGN ON B-CELLS AND PLATELETS – MODULATION OF IMMUNE RESPONSES AND *TRANS*-INFECTION OF T-CELLS

It was initially postulated that DC-SIGN is a DC-specific marker. However, subsequent analyses demonstrated a considerably broader expression pattern, with DC-SIGN-protein being detectable on megakaryocytes, platelets, B-cells, certain tissue macrophages and endothelial cells (Soilleux *et al.*, 2001, 2002; Rappocciolo *et al.*, 2006b; Lai *et al.*, 2006; Boukour *et al.*, 2006; Chaipan *et al.*, 2006; He *et al.*, 2006). Several of these cell types could impact HIV dissemination in a DC-SIGN-dependent fashion. Consistent with this speculation, polymorphisms in the DC-SIGN promoter were found to impact the risk of acquiring HIV via the parenteral but not the sexual route (Martin *et al.*, 2004), suggesting that DC-SIGN-positive cells might impact viral spread once the virus entered the blood stream. Here, we will describe potential consequences of DC-SIGN expression on B-cells and platelets for HIV spread.

At first sight there is no obvious connection between HIV and platelets, except that thrombocytopenia is frequently observed in HIV/AIDS patients. A closer look, however, reveals several links. For one, HIV productively infects megakaryocytes and this process is potentially promoted by DC-SIGN (Scaradavou, 2002). Moreover, an association between platelet counts and viral load/disease progression has been reported (Rieg *et al.*, 2007) and, most importantly, a substantial fraction of HIV in the blood of infected individuals has been found to be associated with platelets (Lee *et al.*, 1998). The molecular mechanism behind HIV binding to platelets has recently been uncovered, when two groups independently demonstrated that platelets, or at least a substantial fraction of these cell fragments, express DC-SIGN and capture HIV via this receptor (Boukour *et al.*, 2006; Chaipan *et al.*, 2006). Platelets exposed to HIV were found to promote *trans*-infection of T-cells in a DC-SIGN-dependent fashion (Chaipan *et al.*, 2006). However, a fraction of the bound virus may also be degraded since both intact and inactivated particles were detected in platelets and were localized to anatomically distinct compartments (Boukour *et al.*, 2006). Therefore, the consequences of HIV capture by platelets for HIV infectivity require further assessment. Besides DC-SIGN, the C-type lectin-like receptor 2 (CLEC-2) contributed to HIV capture by platelets, by interacting with one or more cellular factors incorporated into the viral envelope upon release of progeny particles from infected cells (Chaipan *et al.*, 2006). Identification of the responsible factor(s) may yield further insights into the interplay between HIV and platelets. In summary, platelets express HIV attachment factors and, simply because of their high concentration in human blood, are likely to modulate viral spread. A quantitative analysis of HIV binding to permissive and non-permissive blood cells, including platelets, the investigation of lectin expression levels on platelets of healthy and HIV-infected individuals and the correlation of the data with viral load and disease progression might help to clarify the role of platelets in HIV spread.

Infection of B-cells is normally inefficient in HIV/AIDS patients but these cells may be able to transfer the virus to T-cells (De Milito, 2004). This capability has been linked to DC-SIGN expression (Rappocciolo *et al.*, 2006b), which was detected on a subset of B-cells from blood and tonsils (Rappocciolo *et al.*, 2006b; He *et al.*, 2006), albeit these data are not undisputed (Geijtenbeek *et al.*, 2000a). Expression of DC-SIGN was enhanced by treatment of cells with IL-4 and CD40L and was found to be responsible for *trans*-infection of T-cells with CXCR4- and CCR5-tropic viruses (Rappocciolo *et al.*, 2006b). Notably, DC-SIGN on B-cells may not only promote HIV infection but compromise the humoral immune response of the infected host. Thus, it has been shown that binding of HIV Env to C-type lectins, particularly DC-SIGN, on a subset of B-cells induces class switch DNA recombination in these cells, which is further enhanced by IL-4 and IL-10 (He *et al.*, 2006). This phenomenon may explain why hyperactivation of B-cells is frequently seen in HIV/AIDS patients and leads to production of non-protective antibodies with specificity for HIV Env or irrelevant specificity.

5. IMPACT OF DC-SIGN POLYMORPHISMS ON THE SUSCEPTIBILITY TO HIV INFECTION

The neck region of the DC-SIGN-related protein DC-SIGNR (also termed L-SIGN, CD209L) is highly polymorphic and polymorphisms may affect the risk of HIV infection, as discussed below. In contrast, the DC-SIGN neck region was found to be rarely polymorphic when patients in US-based cohorts were analysed (Liu *et al.*, 2004a). Nevertheless, polymorphisms were detected more often in multiple exposed seronegative individuals compared to HIV-infected patients (Liu *et al.*, 2004a). These findings were extended by a subsequent study showing that polymorphisms in the DC-SIGN neck region are more frequent in the Chinese population compared to the US or the worldwide population and are associated with reduced risk of HIV infection (Zhang *et al.*, 2008). Albeit independent analyses reached different conclusions (Wichukchinda *et al.*, 2007; Rathore *et al.*, 2008b), the above discussed studies indicate that variations in the neck region impact DC-SIGN interactions with HIV, but the molecular basis for this finding is not entirely clear. Engineered alterations in the number and configuration of repeat units in the neck region, as well as *N*-glycosylation of the N-terminal repeat unit, can impact DC-SIGN multimerization and carbohydrate binding (Feinberg *et al.*, 2005; Serrano-Gomez *et al.*, 2008). Polymorphic DC-SIGN variants analysed in one study retained the ability to form homo-oligomers but did not multimerize appreciably with wild-type DC-SIGN (Serrano-Gomez *et al.*, 2008), suggesting that cells from heterozygous individuals might express less (or less stable) DC-SIGN homo-oligomers on the cell surface. However, it remains to be proven that such a reduction in wild-type DC-SIGN homo-oligomers indeed impacts the interaction with pathogens. The observation that at least MDDCs express DC-SIGN copy numbers in excess of these required for highly efficient HIV *trans*-infection by cells lines (Pöhlmann *et al.*, 2001b; Baribaud *et al.*, 2002) suggests that this may not be the case. In fact, transient co-expression of wild-type DC-SIGNR with DC-SIGNR neck region variants did not reduce HIV *trans*-infection compared to cells expressing wild-type DC-SIGNR alone (Gramberg *et al.*, 2006).

The polymorphism 336G in the DC-SIGN promoter reduces promoter activity by altering a binding site for the transcription factor SP1 and impacts the risk of acquiring dengue fever but not dengue haemorrhagic fever (Sakuntabhai *et al.*, 2005). Notably, individuals carrying the polymorphism −336C were found to be more susceptible to HIV infection by the parenteral route while susceptibility to sexually transmitted HIV was not affected (Martin *et al.*, 2004). This finding suggests that DC-SIGN-positive cells impact HIV spread if the virus directly enters the blood stream. In this scenario, high DC-SIGN expression levels seem to be beneficial for the host, provided that the −336G and −336C polymorphisms are identical.

6. LANGERIN ON LANGERHANS CELLS – BARRIER AGAINST HIV TRANSMISSION?

Langerhans cells in the top layer of the mucosal epithelium are among the very first cell types to be exposed to sexually transmitted HIV. It is, therefore, conceivable that HIV interactions with these cells might impact transmission efficiency. Indeed, Langerhans cells, which constitutively express CD4 but not DC-SIGN (Soilleux and Coleman, 2001), were shown to be permissive in *in vitro* and *ex vivo* systems and were found to be infected in HIV-positive individuals (Kawamura *et al.*, 2005). Also, Langerhans cells were among the first cell types infected after intravaginal challenge of macaques (Hu *et al.*, 2000), albeit independent studies reached different conclusions (Spira *et al.*, 1996; Zhang *et al.*, 1999). However, experimental infection of Langerhans cells is typically inefficient and mostly limited to CCR5-tropic viruses, probably due to absence or low expression of CXCR4 on immature Langerhans cells (Kawamura *et al.*, 2005). Hence, it has been speculated that low susceptibility of Langerhans cells to HIV infection might account for the infrequent transmission of HIV upon sexual encounters (Kawamura *et al.*, 2005).

Recently, a molecular mechanism has been identified by which susceptibility of Langerhans cells to HIV infection may be regulated. De Witte and colleagues demonstrated that langerin, a Langerhans cell-specific C-type lectin previously identified as a gp120 binding partner (Turville *et al.*, 2002), promotes HIV uptake by cell lines expressing exogenous langerin and by Langerhans cells (de Witte *et al.*, 2007) (see Table 28.2). Virus captured by langerin is transported to Birbeck granules, a Langerhans cell-specific intracellular compartment, where virions are degraded (de Witte *et al.*, 2007). Since HIV Env may preferentially bind to langerin compared to CD4 (Turville *et al.*, 2002), one can envision that langerin constitutes a powerful barrier against acquisition of HIV infection by the mucosal route. On the other hand, it has been reported that langerin-expressing 293 cells bind to soluble gp120 but not to HIV particles (Gramberg *et al.*, 2008) and it can be speculated that particularly high langerin expression levels might be required for virus capture while reduced levels may be sufficient for gp120 binding. More importantly, however, it has recently been demonstrated that Langerhans cells were responsible for the vast majority of HIV dissemination driven by emigrants of infected human skin explants (Kawamura *et al.*, 2008). Dissemination was dependent on availability of CCR5 but not C-type lectins and, out of three dendritic cell subsets analysed, only Langerhans cells were found to be infected (Kawamura *et al.*, 2008). The potency of langerin as a potential barrier against sexually transmitted HIV therefore requires further assessment.

7. DC-SIGNR AND LSECtin – CONSEQUENCES OF HIV CAPTURE BY VASCULAR ENDOTHELIAL CELLS

Shortly after the discovery of DC-SIGN as an HIV binding factor on DCs, a related molecule, termed DC-SIGNR, has been discovered (Bashirova *et al.*, 2001; Pöhlmann *et al.*, 2001a;

Mummidi *et al.*, 2001). The domain organization and carbohydrate specificity of DC-SIGNR is similar to that of DC-SIGN, albeit DC-SIGNR seems to exhibit exclusive specificity for high-mannose carbohydrates (Guo *et al.*, 2004) and both lectins interact with much the same ligands. However, DC-SIGNR but not DC-SIGN augments infectious entry of West Nile virus with high efficiency and binding of virions to DC-SIGNR seems to depend on recognition of complex carbohydrates (Davis *et al.*, 2006a,b). Despite similarities in structure and ligand specificity, DC-SIGN and DC-SIGNR differ in their expression patterns, with DC-SIGNR being expressed by liver and lymph node sinusoidal endothelial cells, placental macrophages and alveolar type II cells (Bashirova *et al.*, 2001; Pöhlmann *et al.*, 2001a; Jeffers *et al.*, 2004; Gramberg *et al.*, 2008) (see Table 28.2). On liver and lymph node sinusoidal endothelial cells, DC-SIGNR is co-expressed with the related C-type lectin LSECtin, which binds to soluble gp120 (Liu *et al.*, 2004b; Gramberg *et al.*, 2008). In addition, liver sinusoidal endothelial cells (LSECs) also express the HIV attachment factor CLEC-2 (Chaipan *et al.*, 2006). Despite the expression of various attachment factors, a role for these endothelial cells in HIV infection is not obvious. However, evidence has been reported that LSECs are permissive to HIV infection *in vitro* (Steffan *et al.*, 1992) and possibly *in vivo*. In addition, DC-SIGNR-mediated binding of soluble gp120 and Ebola glycoprotein to LSECs (Dakappagari *et al.*, 2006) and facilitated hepatitis C virus transmission by LSECs (Lai *et al.*, 2006), suggesting that DC-SIGNR on these cells may promote *cis*- and *trans*-infection of HIV and other pathogens. *Cis*-infection might result in constant virus release into the blood stream by infected LSECs, while LSEC-dependent *trans*-infection might promote HIV spread to T-cells in the blood or to susceptible Kupffer cells in the liver. It is at present unclear to what degree LSECtin contributes to HIV interactions with LSECs, since expression of this lectin on cell lines promotes binding of soluble gp120 but does not facilitate HIV capture and *trans*-infection (Gramberg *et al.*, 2008). The reasons for this defect remain to be identified.

In contrast to DC-SIGN, the repeat region of DC-SIGNR is highly polymorphic and several studies assessed whether polymorphisms impact the risk of HIV infection (Lichterfeld *et al.*, 2003; Liu *et al.*, 2006; Wichukchinda *et al.*, 2007; Rathore *et al.*, 2008a). Overall, the results suggest that heterozygosity for DC-SIGNR may be associated with reduced risk of HIV infection, while homozygosity for the wild-type variant may increase the risk of acquiring the virus (Liu *et al.*, 2006; Wichukchinda *et al.*, 2007). Notably, one study observed these associations only in females (Wichukchinda *et al.*, 2007), indicating that DC-SIGNR differentially impacts HIV susceptibility of females and males, an observation that deserves further investigation. How can a lectin expressed in liver and lymph node sinusoids impact the risk of acquiring HIV? Importantly, HIV present in low amounts in the blood may be concentrated in lymph nodes in a DC-SIGNR-dependent fashion. Moreover, DC-SIGNR transcripts were detected at sites of mucosal transmission (Liu *et al.*, 2005), suggesting the DC-SIGNR may be involved in the very early events in HIV transmission.

8. CONCLUSIONS

Cellular lectins can modulate HIV infection in cell culture. DC-SIGN on DCs may impact dissemination of sexually transmitted HIV. However, the consequences of HIV interactions with DC-SIGN on DCs are more diverse than were initially appreciated. The majority of virus seems to be degraded for MHC presentation while a minor portion can be transmitted to T-cells. Transmission does not seem to require transport of virions into acidic intracellular compartments and may not involve preservation of viral infectivity by the transmitting cells. Whether the short-time window during which bound virus can be transmitted to T-cells in a DC-SIGN-dependent fashion is sufficient to impact spread of sexually transmitted HIV is at present unclear. The observation that virus-loaded DCs are detectable in lymph nodes of vaginally challenged macaques as early as 30 minutes post challenge argues that *trans*-infection could indeed impact dissemination (Kawamura *et al.*, 2005). The contribution of DC-SIGN to HIV transmission by DCs is still a matter of debate. It is becoming clear, however, that a contribution of DC-SIGN to dendritic cell-mediated HIV *trans*-infection might be at least in part due to the lectins' involvement in the formation of infectious synapses (Hodges *et al.*, 2007), specialized microenvironments that serve as conduits for HIV transfer to T-cells. Finally, langerin-dependent HIV degradation and interference of DC-SIGN with antibody-mediated HIV and simian immunodeficiency virus (SIV) neutralization are novel and so far little explored functions, which highlight the multiple consequences of HIV binding to cellular lectins. Future research areas and open questions for investigation are listed in the Research Focus Box.

RESEARCH FOCUS BOX

- Determination of whether HIV internalization into DCs is required for viral transmission to T-cells via the infectious synapse.
- Detailed analyses of the nature of the DC-SIGN-positive cells found *in vivo*: DC versus macrophage phenotype.
- Does DC-SIGN on platelets and B-cells promote viral spread *in vivo* or is bound virus mainly degraded?
- Does DC-SIGN promote SIV dissemination in the SIV-macaque model?
- Does DC-derived langerin block mucosal transmission of SIV in the macaque model?
- Role of the vascular endothelium in HIV dissemination?

ACKNOWLEDGEMENTS

We apologize to all colleagues whose contributions could not be cited because of space limitations. We thank Thomas F. Schulz for support and the Center of Infection Biology and the MD/PhD program of molecular medicine at Hannover Medical School, as well as SFB587 for funding.

References

Alvarez, C.P., Lasala, F., Carrillo, J., Muniz, O., Corbi, A.L., Delgado, R., 2002. C-type lectins DC-SIGN and L-SIGN mediate cellular entry by Ebola virus in *cis* and in *trans*. J. Virol. 76, 6841–6844.

Arrighi, J.F., Pion, M., Garcia, E., et al., 2004a. DC-SIGN-mediated infectious synapse formation enhances X4 HIV-1 transmission from dendritic cells to T cells. J. Exp. Med. 200, 1279–1288.

Arrighi, J.F., Pion, M., Wiznerowicz, M., et al., 2004b. Lentivirus-mediated RNA interference of DC-SIGN expression inhibits human immunodeficiency virus transmission from dendritic cells to T cells. J. Virol. 78, 10848–10855.

Banchereau, J., Steinman, R.M., 1998. Dendritic cells and the control of immunity. Nature 392, 245–252.

Baribaud, F., Pöhlmann, S., Sparwasser, T., et al., 2001. Functional and antigenic characterization of human, rhesus macaque, pigtailed macaque, and murine DC-SIGN. J. Virol. 75, 10281–10289.

Baribaud, F., Pöhlmann, S., Leslie, G., Mortari, F., Doms, R.W., 2002. Quantitative expression and virus transmission analysis of DC-SIGN on monocyte-derived dendritic cells. J. Virol. 76, 9135–9142.

Bashirova, A.A., Geijtenbeek, T.B., van Duijnhoven, G.C., et al., 2001. A dendritic cell-specific intercellular adhesion molecule 3-grabbing nonintegrin (DC-SIGN)-related protein is highly expressed on human liver sinusoidal endothelial cells and promotes HIV-1 infection. J. Exp. Med. 193, 671–678.

Bernstein, H.B., Tucker, S.P., Hunter, E., Schutzbach, J.S., Compans, R.W., 1994. Human immunodeficiency virus type 1 envelope glycoprotein is modified by *O*-linked oligosaccharides. J. Virol. 68, 463–468.

Boggiano, C., Manel, N., Littman, D.R., 2007. Dendritic cell-mediated trans-enhancement of human immunodeficiency virus type 1 infectivity is independent of DC-SIGN. J. Virol. 81, 2519–2523.

Boukour, S., Masse, J.M., Benit, L., Dubart-Kupperschmitt, A., Cramer, E.M., 2006. Lentivirus degradation and DC-SIGN expression by human platelets and megakaryocytes. J. Thromb. Haemost. 4, 426–435.

Burleigh, L., Lozach, P.Y., Schiffer, C., et al., 2006. Infection of dendritic cells (DCs), not DC-SIGN-mediated internalization of human immunodeficiency virus, is required for long-term transfer of virus to T cells. J. Virol. 80, 2949–2957.

Cameron, P.U., Freudenthal, P.S., Barker, J.M., Gezelter, S., Inaba, K., Steinman, R.M., 1992. Dendritic cells exposed to human immunodeficiency virus type-1 transmit a vigorous cytopathic infection to CD4+ T cells. Science 257, 383–387.

Cavrois, M., Neidleman, J., Kreisberg, J.F., Greene, W.C., 2007. In vitro derived dendritic cells trans-infect CD4 T cells primarily with surface-bound HIV-1 virions. PLoS Pathog. 3, e4.

Ceccaldi, P.E., Delebecque, F., Prevost, M.C., et al., 2006. DC-SIGN facilitates fusion of dendritic cells with human T-cell leukemia virus type 1 infected cells. J. Virol. 80, 4771–4780.

Chaipan, C., Soilleux, E.J., Simpson, P., et al., 2006. DC-SIGN and CLEC-2 mediate human immunodeficiency virus type 1 capture by platelets. J. Virol. 80, 8951–8960.

Cohen, M.S., Hellmann, N., Levy, J.A., DeCock, K., Lange, J., 2008. The spread, treatment, and prevention of HIV-1: evolution of a global pandemic. J. Clin. Invest. 118, 1244–1254.

Curtis, B.M., Scharnowske, S., Watson, A.J., 1992. Sequence and expression of a membrane-associated C-type lectin that exhibits CD4-independent binding of human immunodeficiency virus envelope glycoprotein gp120. Proc. Natl. Acad. Sci. USA 89, 8356–8360.

Dakappagari, N., Maruyama, T., Renshaw, M., et al., 2006. Internalizing antibodies to the C-type lectins, L-SIGN and DC-SIGN, inhibit viral glycoprotein binding and deliver antigen to human dendritic cells for the induction of T cell responses. J. Immunol. 176, 426–440.

Davis, C.W., Nguyen, H.Y., Hanna, S.L., Sanchez, M.D., Doms, R.W., Pierson, T.C., 2006a. West Nile virus discriminates between DC-SIGN and DC-SIGNR for cellular attachment and infection. J. Virol. 80, 1290–1301.

Davis, C.W., Mattei, L.M., Nguyen, H.Y., Ansarah-Sobrinho, C., Doms, R.W., Pierson, T.C., 2006b. The location of asparagine-linked glycans on West Nile virions controls their interactions with CD209 (dendritic cell-specific ICAM-3 grabbing nonintegrin). J. Biol. Chem. 281, 37183–37194.

de Jong, M.A., de Witte, L., Bolmstedt, A., van Kooyk, Y., Geijtenbeek, T.B., 2008. Dendritic cells mediate herpes simplex virus infection and transmission through the C-type lectin DC-SIGN. J. Gen. Virol. 89, 2398–2409.

De Milito, A., 2004. B lymphocyte dysfunctions in HIV infection. Curr. HIV Res. 2, 11–21.

de Witte, L., Abt, M., Schneider-Schaulies, S., van Kooyk, Y., Geijtenbeek, T.B., 2006. Measles virus targets DC-SIGN to enhance dendritic cell infection. J. Virol. 80, 3477–3486.

de Witte, L., Nabatov, A., Pion, M., et al., 2007. Langerin is a natural barrier to HIV-1 transmission by Langerhans cells. Nat. Med. 13, 367–371.

Engering, A., Geijtenbeek, T.B., Van Vliet, S.J., et al., 2002. The dendritic cell-specific adhesion receptor DC-SIGN internalizes antigen for presentation to T cells. J. Immunol. 168, 2118–2126.

Este, J.A., Telenti, A., 2007. HIV entry inhibitors. Lancet 370, 81–88.

Feinberg, H., Mitchell, D.A., Drickamer, K., Weis, W.I., 2001. Structural basis for selective recognition of oligosaccharides by DC-SIGN and DC-SIGNR. Science 294, 2163–2166.

Feinberg, H., Guo, Y., Mitchell, D.A., Drickamer, K., Weis, W. I., 2005. Extended neck regions stabilize tetramers of the receptors DC-SIGN and DC-SIGNR. J. Biol. Chem. 280, 1327–1335.

Garcia, E., Pion, M., Pelchen-Matthews, A., et al., 2005. HIV-1 trafficking to the dendritic cell-T-cell infectious synapse uses a pathway of tetraspanin sorting to the immunological synapse. Traffic 6, 488–501.

Gardner, J.P., Durso, R.J., Arrigale, R.R., et al., 2003. L-SIGN (CD 209L) is a liver-specific capture receptor for hepatitis C virus. Proc. Natl. Acad. Sci. USA 100, 4498–4503.

Geijtenbeek, T.B., Torensma, R., Van Vliet, S.J., et al., 2000a. Identification of DC-SIGN, a novel dendritic cell-specific ICAM-3 receptor that supports primary immune responses. Cell 100, 575–585.

Geijtenbeek, T.B., Kwon, D.S., Torensma, R., et al., 2000b. DC-SIGN, a dendritic cell-specific HIV-1-binding protein that enhances trans-infection of T cells. Cell 100, 587–597.

Geijtenbeek, T.B., van Duijnhoven, G.C., Van Vliet, S.J., et al., 2002. Identification of different binding sites in the dendritic cell-specific receptor DC-SIGN for intercellular adhesion molecule 3 and HIV-1. J. Biol. Chem. 277, 11314–11320.

Gramberg, T., Zhu, T., Chaipan, C., et al., 2006. Impact of polymorphisms in the DC-SIGNR neck domain on the interaction with pathogens. Virology 347, 354–363.

Gramberg, T., Soilleux, E., Fisch, T., et al., 2008. Interactions of LSECtin and DC-SIGN/DC-SIGNR with viral ligands: differential pH dependence, internalization and virion binding. Virology 373, 189–201.

Granelli-Piperno, A., Pritsker, A., Pack, M., et al., 2005. Dendritic cell-specific intercellular adhesion molecule 3-grabbing nonintegrin/CD209 is abundant on macrophages in the normal human lymph node and is not required for dendritic cell stimulation of the mixed leukocyte reaction. J. Immunol. 175, 4265–4273.

Gummuluru, S., Rogel, M., Stamatatos, L., Emerman, M., 2003. Binding of human immunodeficiency virus type 1 to immature dendritic cells can occur independently of DC-SIGN and mannose binding C-type lectin receptors via a cholesterol-dependent pathway. J. Virol. 77, 12865–12874.

Guo, Y., Feinberg, H., Conroy, E., et al., 2004. Structural basis for distinct ligand-binding and targeting properties of the receptors DC-SIGN and DC-SIGNR. Nat. Struct. Mol. Biol. 11, 591–598.

Halary, F., Amara, A., Lortat-Jacob, H., et al., 2002. Human cytomegalovirus binding to DC-SIGN is required for dendritic cell infection and target cell trans-infection. Immunity 17, 653–664.

He, B., Qiao, X., Klasse, P.J., et al., 2006. HIV-1 envelope triggers polyclonal Ig class switch recombination through a CD40-independent mechanism involving BAFF and C-type lectin receptors. J. Immunol. 176, 3931–3941.

Hodges, A., Sharrocks, K., Edelmann, M., et al., 2007. Activation of the lectin DC-SIGN induces an immature dendritic cell phenotype triggering Rho-GTPase activity required for HIV-1 replication. Nat. Immunol. 8, 569–577.

Hofmann, H., Simmons, G., Rennekamp, A.J., et al., 2006. Highly conserved regions within the spike proteins of human coronaviruses 229E and NL63 determine recognition of their respective cellular receptors. J. Virol. 80, 8639–8652.

Hu, J., Gardner, M.B., Miller, C.J., 2000. Simian immunodeficiency virus rapidly penetrates the cervicovaginal mucosa after intravaginal inoculation and infects intraepithelial dendritic cells. J. Virol. 74, 6087–6095.

Jeffers, S.A., Tusell, S.M., Gillim-Ross, L., et al., 2004. CD209L (L-SIGN) is a receptor for severe acute respiratory syndrome coronavirus. Proc. Natl. Acad. Sci. USA 101, 15748–15753.

Ji, X., Chen, Y., Faro, J., Gewurz, H., Bremer, J., Spear, G. T., 2006. Interaction of human immunodeficiency virus (HIV) glycans with lectins of the human immune system. Curr. Protein Pept. Sci. 7, 317–324.

Kawamura, T., Kurtz, S.E., Blauvelt, A., Shimada, S., 2005. The role of Langerhans cells in the sexual transmission of HIV. J. Dermatol. Sci. 40, 147–155.

Kawamura, T., Koyanagi, Y., Nakamura, Y., et al., 2008. Significant virus replication in Langerhans cells following application of HIV to abraded skin: relevance to occupational transmission of HIV. J. Immunol. 180, 3297–3304.

Khoo, U.S., Chan, K.Y., Chan, V.S., Lin, C.L., 2008. DC-SIGN and L-SIGN: the SIGNs for infection. J. Mol. Med. 86, 861–874.

Klimstra, W.B., Nangle, E.M., Smith, M.S., Yurochko, A.D., Ryman, K.D., 2003. DC-SIGN and L-SIGN can act as attachment receptors for alphaviruses and distinguish between mosquito cell- and mammalian cell-derived viruses. J. Virol. 77, 12022–12032.

Krutzik, S.R., Tan, B., Li, H., et al., 2005. TLR activation triggers the rapid differentiation of monocytes into macrophages and dendritic cells. Nat. Med. 11, 653–660.

Kwon, D.S., Gregorio, G., Bitton, N., Hendrickson, W.A., Littman, D.R., 2002. DC-SIGN-mediated internalization of HIV is required for trans-enhancement of T cell infection. Immunity 16, 135–144.

Lai, W.K., Sun, P.J., Zhang, J., et al., 2006. Expression of DC-SIGN and DC-SIGNR on human sinusoidal endothelium: a role for capturing hepatitis C virus particles. Am. J. Pathol. 169, 200–208.

Lee, T.H., Stromberg, R.R., Heitman, J.W., Sawyer, L., Hanson, C.V., Busch, M.P., 1998. Distribution of HIV type 1 (HIV-1) in blood components: detection and significance of high levels of HIV-1 associated with platelets. Transfusion 38, 580–588.

Lichterfeld, M., Nischalke, H.D., van Lunzen, J., et al., 2003. The tandem-repeat polymorphism of the DC-SIGNR gene does not affect the susceptibility to HIV infection and the progression to AIDS. Clin. Immunol. 107, 55–59.

Lin, G., Simmons, G., Pöhlmann, S., et al., 2003. Differential *N*-linked glycosylation of human immunodeficiency virus and Ebola virus envelope glycoproteins modulates interactions with DC-SIGN and DC-SIGNR. J. Virol. 77, 1337–1346.

Liu, H., Hwangbo, Y., Holte, S., et al., 2004a. Analysis of genetic polymorphisms in CCR5, CCR2, stromal cell-derived factor-1, RANTES, and dendritic cell-specific intercellular adhesion molecule-3-grabbing nonintegrin in seronegative individuals repeatedly exposed to HIV-1. J. Infect. Dis. 190, 1055–1058.

Liu, H., Hladik, F., Andrus, T., et al., 2005. Most DC-SIGNR transcripts at mucosal HIV transmission sites are alternatively spliced isoforms. Eur. J. Hum. Genet. 13, 707–715.

Liu, H., Carrington, M., Wang, C., et al., 2006. Repeat-region polymorphisms in the gene for the dendritic cell-specific intercellular adhesion molecule-3-grabbing nonintegrin-related molecule: effects on HIV-1 susceptibility. J. Infect. Dis. 193, 698–702.

Liu, W., Tang, L., Zhang, G., et al., 2004b. Characterization of a novel C-type lectin-like gene, LSECtin: demonstration of carbohydrate binding and expression in sinusoidal endothelial cells of liver and lymph node. J. Biol. Chem. 279, 18748–18758.

Liu, Y.J., 2001. Dendritic cell subsets and lineages, and their functions in innate and adaptive immunity. Cell 106, 259–262.

Lozach, P.Y., Lortat-Jacob, H., de Lacroix, d.L., et al., 2003. DC-SIGN and L-SIGN are high affinity binding receptors for hepatitis C virus glycoprotein E2. J. Biol. Chem. 278, 20358–20366.

Martin, M.P., Lederman, M.M., Hutcheson, H.B., et al., 2004. Association of DC-SIGN promoter polymorphism with increased risk for parenteral, but not mucosal, acquisition of human immunodeficiency virus type 1 infection. J. Virol. 78, 14053–14056.

Marzi, A., Gramberg, T., Simmons, G., et al., 2004. DC-SIGN and DC-SIGNR interact with the glycoprotein of Marburg virus and the S protein of severe acute respiratory syndrome coronavirus. J. Virol. 78, 12090–12095.

McDonald, D., Wu, L., Bohks, S.M., KewalRamani, V.N., Unutmaz, D., Hope, T.J., 2003. Recruitment of HIV and its receptors to dendritic cell-T cell junctions. Science 300, 1295–1297.

Mitchell, D.A., Fadden, A.J., Drickamer, K., 2001. A novel mechanism of carbohydrate recognition by the C-type lectins DC-SIGN and DC-SIGNR. Subunit organization and binding to multivalent ligands. J. Biol. Chem. 276, 28939–28945.

Moris, A., Nobile, C., Buseyne, F., Porrot, F., Abastado, J.P., Schwartz, O., 2004. DC-SIGN promotes exogenous MHC-I-restricted HIV-1 antigen presentation. Blood 103, 2648–2654.

Moris, A., Pajot, A., Blanchet, F., Guivel-Benhassine, F., Salcedo, M., Schwartz, O., 2006. Dendritic cells and HIV-specific CD4+ T cells: HIV antigen presentation, T-cell activation, and viral transfer. Blood 108, 1643–1651.

Mummidi, S., Catano, G., Lam, L., et al., 2001. Extensive repertoire of membrane-bound and soluble dendritic cell-specific ICAM-3-grabbing nonintegrin 1 (DC-SIGN1) and DC-SIGN2 isoforms. Inter-individual variation in expression of DC-SIGN transcripts. J. Biol. Chem. 276, 33196–33212.

Münch, J., Rucker, E., Standker, L., et al., 2007. Semen-derived amyloid fibrils drastically enhance HIV infection. Cell 131, 1059–1071.

Navarro-Sanchez, E., Altmeyer, R., Amara, A., et al., 2003. Dendritic-cell-specific ICAM3-grabbing non-integrin is essential for the productive infection of human dendritic cells by mosquito-cell-derived dengue viruses. EMBO Rep. 4, 723–728.

Nobile, C., Petit, C., Moris, A., et al., 2005. Covert human immunodeficiency virus replication in dendritic cells and in DC-SIGN-expressing cells promotes long-term transmission to lymphocytes. J. Virol. 79, 5386–5399.

Piguet, V., Steinman, R.M., 2007. The interaction of HIV with dendritic cells: outcomes and pathways. Trends Immunol. 28, 503–510.

Pöhlmann, S., Reeves, J.D., 2006. Cellular entry of HIV: evaluation of therapeutic targets. Curr. Pharm. Des. 12, 1963–1973.

Pöhlmann, S., Tremblay, M.J., 2007. Attachment of human immunodeficiency virus to cells and its inhibition. In: Reeves, J.D., Derdeyn, C.A. (Eds.), Milestones in Drug Therapy: Entry Inhibitors in HIV Therapy. Birkhäuser, Basel, pp. 31–48.

Pöhlmann, S., Soilleux, E.J., Baribaud, F., et al., 2001a. DC-SIGNR, a DC-SIGN homologue expressed in endothelial cells, binds to human and simian immunodeficiency viruses and activates infection in trans. Proc. Natl. Acad. Sci. USA 98, 2670–2675.

Pöhlmann, S., Baribaud, F., Lee, B., et al., 2001b. DC-SIGN interactions with human immunodeficiency virus type 1 and 2 and simian immunodeficiency virus. J. Virol. 75, 4664–4672.

Pöhlmann, S., Leslie, G.J., Edwards, T.G., et al., 2001c. DC-SIGN interactions with human immunodeficiency virus: virus binding and transfer are dissociable functions. J. Virol. 75, 10523–10526.

Pöhlmann, S., Zhang, J., Baribaud, F., et al., 2003. Hepatitis C virus glycoproteins interact with DC-SIGN and DC-SIGNR. J. Virol. 77, 4070–4080.

Pope, M., Betjes, M.G., Romani, N., et al., 1994. Conjugates of dendritic cells and memory T lymphocytes from skin facilitate productive infection with HIV-1. Cell 78, 389–398.

Rappocciolo, G., Jenkins, F.J., Hensler, H.R., et al., 2006a. DC-SIGN is a receptor for human herpesvirus 8 on dendritic cells and macrophages. J. Immunol. 176, 1741–1749.

Rappocciolo, G., Piazza, P., Fuller, C.L., et al., 2006b. DC-SIGN on B lymphocytes is required for transmission of HIV-1 to T lymphocytes. PLoS Pathog. 2, e70.

Rathore, A., Chatterjee, A., Sivarama, P., Yamamoto, N., Dhole, T.N., 2008a. Role of homozygous DC-SIGNR 5/5 tandem repeat polymorphism in HIV-1 exposed seronegative North Indian individuals. J. Clin. Immunol. 28, 50–57.

Rathore, A., Chatterjee, A., Sood, V., et al., 2008b. Risk for HIV-1 infection is not associated with repeat-region polymorphism in the DC-SIGN neck domain and novel genetic DC-SIGN variants among North Indians. Clin. Chim. Acta 391, 1–5.

Reeves, J.D., Barr, S.D., Pöhlmann, S., 2006. Evaluation of current strategies to inhibit HIV entry, integration and maturation. In: Bogner, E., Holzenburg, A. (Eds.), New Concepts of Antiviral Therapy. Springer, Dordrecht, pp. 213–254.

Rieg, G., Yeaman, M., Lail, A.E., Donfield, S.M., Gomperts, E.D., Daar, E.S., 2007. Platelet count is associated with plasma HIV type 1 RNA and disease progression. AIDS Res. Hum. Retroviruses 23, 1257–1261.

Sakuntabhai, A., Turbpaiboon, C., Casademont, I., et al., 2005. A variant in the CD209 promoter is associated with severity of dengue disease. Nat. Genet. 37, 507–513.

Scanlan, C.N., Offer, J., Zitzmann, N., Dwek, R.A., 2007. Exploiting the defensive sugars of HIV-1 for drug and vaccine design. Nature 446, 1038–1045.

Scaradavou, A., 2002. HIV-related thrombocytopenia. Blood Rev. 16, 73–76.

Serrano-Gomez, D., Sierra-Filardi, E., Martinez-Nunez, R.T., et al., 2008. Structural requirements for multimerization of the pathogen receptor dendritic cell-specific ICAM3-grabbing non-integrin (CD209) on the cell surface. J. Biol. Chem. 283, 3889–3903.

Simmons, G., Reeves, J.D., Grogan, C.C., et al., 2003. DC-SIGN and DC-SIGNR bind Ebola glycoproteins and enhance infection of macrophages and endothelial cells. Virology 305, 115–123.

Snyder, G.A., Ford, J., Torabi-Parizi, P., et al., 2005. Characterization of DC-SIGN/R interaction with human immunodeficiency virus type 1 gp120 and ICAM molecules favors the receptor's role as an antigen-capturing rather than an adhesion receptor. J. Virol. 79, 4589–4598.

Soilleux, E.J., Coleman, N., 2001. Langerhans cells and the cells of Langerhans cell histiocytosis do not express DC-SIGN. Blood 98, 1987–1988.

Soilleux, E.J., Morris, L.S., Lee, B., et al., 2001. Placental expression of DC-SIGN may mediate intrauterine vertical transmission of HIV. J. Pathol. 195, 586–592.

Soilleux, E.J., Morris, L.S., Leslie, G., et al., 2002. Constitutive and induced expression of DC-SIGN on dendritic cell and macrophage subpopulations in situ and in vitro. J. Leukoc. Biol. 71, 445–457.

Spira, A.I., Marx, P.A., Patterson, B.K., et al., 1996. Cellular targets of infection and route of viral dissemination after an intravaginal inoculation of simian immunodeficiency virus into rhesus macaques. J. Exp. Med. 183, 215–225.

Steffan, A.M., Lafon, M.E., Gendrault, J.L., et al., 1992. Primary cultures of endothelial cells from the human liver sinusoid are permissive for human immunodeficiency virus type 1. Proc. Natl. Acad. Sci. USA 89, 1582–1586.

Tassaneetrithep, B., Burgess, T.H., Granelli-Piperno, A., et al., 2003. DC-SIGN (CD209) mediates dengue virus infection of human dendritic cells. J. Exp. Med. 197, 823–829.

Trumpfheller, C., Park, C.G., Finke, J., Steinman, R.M., Granelli-Piperno, A., 2003. Cell type-dependent retention and transmission of HIV-1 by DC-SIGN. Int. Immunol. 15, 289–298.

Turville, S.G., Cameron, P.U., Handley, A., et al., 2002. Diversity of receptors binding HIV on dendritic cell subsets. Nat. Immunol. 3, 975–983.

Turville, S.G., Santos, J.J., Frank, I., et al., 2004. Immuno-deficiency virus uptake, turnover, and 2-phase transfer in human dendritic cells. Blood 103, 2170–2179.

van Kooyk, Y., Geijtenbeek, T.B., 2003. DC-SIGN: escape mechanism for pathogens. Nat. Rev. Immunol. 3, 697–709.

van Kooyk, Y., Rabinovich, G.A., 2008. Protein-glycan interactions in the control of innate and adaptive immune responses. Nat. Immunol. 9, 593–601.

van Lent, P.L., Figdor, C.G., Barrera, P., et al., 2003. Expression of the dendritic cell-associated C-type lectin DC-SIGN by inflammatory matrix metalloproteinase-producing macrophages in rheumatoid arthritis synovium and interaction with intercellular adhesion molecule 3-positive T cells. Arthritis Rheum. 48, 360–369.

van Montfort, T., Nabatov, A.A., Geijtenbeek, T.B., Pollakis, G., Paxton, W.A., 2007. Efficient capture of antibody neutralized HIV-1 by cells expressing DC-SIGN and transfer to CD4+ T lymphocytes. J. Immunol. 178, 3177–3185.

Veazey, R.S., Lackner, A.A., 2004. Getting to the guts of HIV pathogenesis. J. Exp. Med. 200, 697–700.

Vigerust, D.J., Shepherd, V.L., 2007. Virus glycosylation: role in virulence and immune interactions. Trends Microbiol. 15, 211–218.

Wang, J.H., Janas, A.M., Olson, W.J., KewalRamani, V.N., Wu, L., 2007. CD4 coexpression regulates DC-SIGN-mediated transmission of human immunodeficiency virus type 1. J. Virol. 81, 2497–2507.

Wang, S.F., Huang, J.C., Lee, Y.M., et al., 2008. DC-SIGN mediates avian H5N1 influenza virus infection in cis and in trans. Biochem. Biophys. Res. Commun. 373, 561–566.

Wichukchinda, N., Kitamura, Y., Rojanawiwat, A., et al., 2007. The polymorphisms in DC-SIGNR affect susceptibility to HIV type 1 infection. AIDS Res. Hum. Retroviruses 23, 686–692.

Wu, L., KewalRamani, V.N., 2006. Dendritic-cell interactions with HIV: infection and viral dissemination. Nat. Rev. Immunol. 6, 859–868.

Wu, L., Bashirova, A.A., Martin, T.D., et al., 2002a. Rhesus macaque dendritic cells efficiently transmit primate lentiviruses independently of DC-SIGN. Proc. Natl. Acad. Sci. USA 99, 1568–1573.

Wu, L., Martin, T.D., Vazeux, R., Unutmaz, D., KewalRamani, V.N., 2002b. Functional evaluation of DC-SIGN monoclonal antibodies reveals DC-SIGN interactions with ICAM-3 do not promote human immunodeficiency virus type 1 transmission. J. Virol. 76, 5905–5914.

Wu, L., Martin, T.D., Carrington, M., KewalRamani, V.N., 2004a. Raji B cells, misidentified as THP-1 cells, stimulate DC-SIGN-mediated HIV transmission. Virology 318, 17–23.

Wu, L., Martin, T.D., Han, Y.C., Breun, S.K., KewalRamani, V.N., 2004b. Trans-dominant cellular inhibition of DC-SIGN-mediated HIV-1 transmission. Retrovirology 1, 14.

Yang, Z.Y., Huang, Y., Ganesh, L., et al., 2004. pH-dependent entry of severe acute respiratory syndrome coronavirus is mediated by the spike glycoprotein and enhanced by dendritic cell transfer through DC-SIGN. J. Virol. 78, 5642–5650.

Zhang, J., Zhang, X., Fu, J., et al., 2008. Protective role of DC-SIGN (CD209) neck-region alleles with <5 repeat units in HIV-1 transmission. J. Infect. Dis. 198, 68–71.

Zhang, Z., Schuler, T., Zupancic, M., et al., 1999. Sexual transmission and propagation of SIV and HIV in resting and activated CD4+ T cells. Science 286, 1353–1357.

CHAPTER

29

Sialic acid-specific microbial lectins

Joe Tiralongo

SUMMARY

Sialic acids are a family of structurally diverse acidic nine-carbon sugars that are typically located at the terminal positions of a variety of glycoconjugates. Due to their unique physiochemical properties, diversity of structure and exposed position, sialic acids have been implicated in numerous essential biological processes, including microbial pathogenesis. Sialoglycoconjugates expressed on the surface of host cells serve as receptors or ligands for microbial sialic acid-specific lectins. This interaction permits attachment of pathogenic microbes to host tissue and, therefore, represents a critical initial step in pathogenesis. This chapter summarizes the role of sialic acid-specific lectins in mediating host cell attachment by viruses, bacteria, protozoa and micro-fungi and provides insights into the common binding site structural elements required for sialic acid recognition. Not only do microbes express sialic acid-specific lectins to aid in host cell attachment, but they also decorate their cell surfaces with sialic acid. Gram-negative bacteria utilize transport systems to acquire sialic acid from the environment that is then channelled into the bacterial sialylation pathway. To acquire sialic acid from the environment bacteria have evolved diverse transport systems, however, they all specifically recognize and bind sialic acid. Therefore, this chapter also describes the nature of the sialic acid-binding domains associated with bacterial sialic acid transport systems.

Keywords: Sialic acids; Lectin; Adhesin; Haemagglutinin; Sialic acid-binding site; Pathogenesis; Infectious disease

1. INTRODUCTION

Sialic acids are a family of 9-carbon α-keto acids found predominantly at the non-reducing end of oligosaccharide chains on glycoproteins and glycolipids. Sialic acids can occur free in Nature, but are generally found glycosidically linked to either galactose (Gal), *N*-acetylglucosamine (GlcNAc), *N*-acetylgalactosamine (GalNAc) or as α-(2→8)-linked homopolymers known as polysialic acid (Schauer and Kamerling 1997; Angata and Varki, 2002) (Figure 29.1A).

Sialic acids show remarkable structural diversity, with the family currently comprising over 50 naturally occurring members. The largest structural variations of naturally occurring sialic

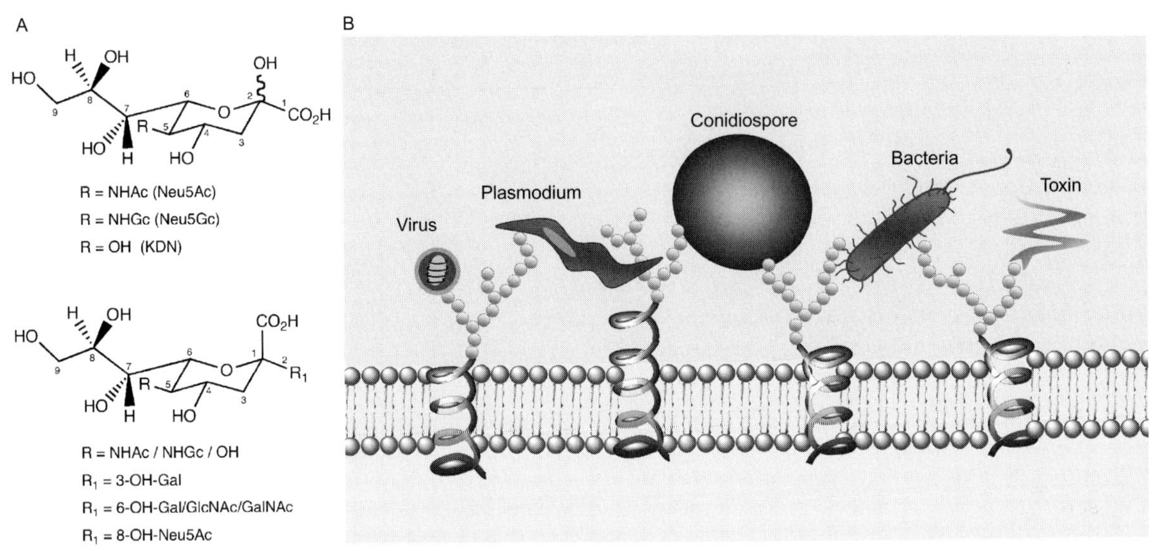

FIGURE 29.1 (A) The structural diversity of sialic acids is generated by a combination of variations at C-5 with modifications of any of the hydroxyl groups at C-4, C-7, C-8 and C-9. Sialic acids are predominantly found glycosidically linked via α-(2→3)-, α-(2→6)-, or α-(2→8)-linkages to underlying sugars as shown. (B) Sialic acids, which frequently occupy the terminal position of glycan chains on glycoproteins or glycolipids (the individual sugars are represented by spheres), participate in numerous host–pathogen recognition events through sialic acid-specific lectins.

acids are at carbon 5, which can be substituted with either an acetamido, hydroxyacetamido or hydroxyl moiety to form 5-*N*-acetylneuraminic acid (Neu5Ac), 5-*N*-glycolylneuraminic acid (Neu5Gc) or deaminoneuraminic acid (KDN), respectively (see Figure 29.1(A). Further structural diversity is generated primarily by a combination of the above-mentioned variations at C-5 with modifications of any of the hydroxyl groups located at C-4, C-7, C-8 and C-9 (Schauer and Kamerling 1997; Angata and Varki, 2002).

Due to their unique physiochemical properties, diversity of structure and exposed position, sialic acids have been implicated in numerous essential biological processes, such as neural cell growth and embryogenesis, stem cell biology, immune system regulation, human evolution, cancer progression and microbial pathogenesis (Schauer and Kamerling, 1997; Lehmann *et al.*, 2006; Lanctot *et al.*, 2007; Varki and Varki, 2007). In the vast majority of these processes, sialic acid acts as a ligand for specific sialic acid-binding lectins, with some of the best characterized lectins being those associated with pathogenic microbes (Figure 29.1B).

2. ROLE OF SIALIC ACID-SPECIFIC MICROBIAL LECTINS IN HOST CELL ATTACHMENT

2.1. Viruses

Members of at least eight different virus families – *Orthomyxoviridae*, *Paramyxoviridae*, *Coronaviridae*, *Reoviridae*, *Picornaviridae*, *Parvoviridae*, *Papovaviridae* and *Adenoviridae* – including enveloped and non-enveloped viruses and RNA and DNA viruses, exploit sialoglycoconjugates for attachment. Some viruses bind preferentially to sialic acids attached via a particular glycosidic linkage and this specificity may contribute to virus host range, tissue tropism and pathogenesis (comprehensively reviewed in Angata and Varki, 2002; Lehmann *et al.*, 2006).

Influenza virus A is the most well-known and best-studied example in the field. The sialic acid-specific lectin (or haemagglutinin, HA) of the human influenza virus A predominantly binds α-Neu5Ac-(2→6)-Gal structures present on non-ciliated cells of the human trachea. The H5N1 avian influenza A virus, however, exclusively binds α-Neu5Ac-(2→3)-Gal, thus limiting the host range to those species possessing α-Neu5Ac-(2→3)-Gal receptor structures on tracheal cells (e.g. birds, horses and pigs) (Skehel and Wiley, 2000; Suzuki, 2005). Due to the serious consequences associated with the emergence of a human adapted H5N1 virus, understanding and rapidly assessing the virus receptor specificity, particularly using glycan array technology, has become a major focus (Stevens *et al.*, 2006).

Rotaviruses, the leading cause of gastroenteritis in humans, possess an outermost layer composed of two proteins, VP4 and VP7. Treatment of the virus with trypsin results in the specific cleavage of VP4 into the polypeptides denoted as VP8* and VP5*. It is generally accepted that Neu5Ac is required by several animal rotavirus strains for attachment to the cell surface. The infectivity of some of these strains is greatly diminished by the treatment of cells with sialidase; consequently, these strains are termed sialidase-sensitive. By contrast, many animal-derived strains and most strains isolated from humans are sialidase-resistant (Ciarlet and Estes, 1999). This is believed to be due to the ability of these strains to bind gangliosides that possess internal sialic acid residues that are resistant to sialidase treatment (Delorme *et al.*, 2001), but this viewpoint is controversial.

Recent crystal structure data (Table 29.1 provides a list of microbial sialic acid-specific lectins of known structure) have done little to resolve the exact involvement, if any, of sialic acids in sialidase-insensitive rotavirus recognition of host cells. Neither human sialidase-insensitive strains DS-1 or Wa VP8* have been crystallized with sialic acid or sialic acid-derivatives bound to the proposed sialic acid-binding site that was identified from the crystal structure of sialidase-sensitive VP8* (Monnier *et al.*, 2006; Blanchard *et al.*, 2007). However, both studies identified a novel ligand-binding site located within a cleft or groove that is wider in the human sialidase-insensitive VP8* compared with animal sialidase-sensitive VP8* (Blanchard *et al.*, 2007). Monnier *et al.* (2006) observed that this groove bound the N-terminal residues of a neighbouring molecule, suggesting that it may act as a peptide-binding site. More recently, the same groove was found in the porcine rotavirus CRW-8 VP8* to be compatible with a sialic acid-binding site, however, the equivalent groove in the Wa VP8* was unable to accommodate sialic acid. Due to the comparable general nature of the groove region between CRW-8 VP8* and Wa VP8*, the authors suggested that the Wa VP8* groove may still have a carbohydrate-binding role (Blanchard *et al.*, 2007).

Based on the available evidence – at least for sialidase-sensitive rotavirus strains – gangliosides such as GM_1 and GM_3 and the Gal component of glycoprotein receptors, as well as integrins $\alpha_2\beta_1$ and $\alpha_4\beta_1$ all play a role in attachment and entry of rotaviruses into host cells, indicating that the rotavirus functional receptor is a complex of several cell components (Arias *et al.*, 2002). Lopez and Arias (2004) suggested a model where the initial contact of the virus with the cell surface is through the binding of the VP8* domain of VP4 to a ganglioside receptor which induces a conformational change in VP4, thus allowing the virus to interact with integrin $\alpha_2\beta_1$ through VP5*. Following this second interaction, one to three additional interactions take place involving VP5* and VP7, integrins $\alpha v\beta_3$ and $\alpha x\beta_2$ and probably other proteins (Lopez and Arias, 2004).

TABLE 29.1 Microbial sialic acid-specific lectins of known structure

Species	Lectin	Specificity	Reference
Viruses[a]			
Human influenza virus A	HA	α-Neu5Ac-(2→3)-Gal	Wilson *et al.*, 1981
Avian influenza virus A	HA	α-Neu5Ac-(2→3)-Gal	Ha *et al.*, 2001
Murine polyoma virus	VP1	α-Neu5Ac-(2→3)-Gal	Stehle and Harrison, 1997
Adenovirus type 37	Fiber knob	Neu5Ac	Burmeister *et al.*, 2004
Theiler's murine encephalomyelitis virus BeAn	VP2	Neu5Ac	Zhou *et al.*, 2000
Rhesus rotavirus	VP8*	Neu5Ac	Dormitzer *et al.*, 2002
Porcine rotavirus	VP8*	Neu5Ac	Blanchard *et al.*, 2007
Human rotavirus	VP8*	Neu5Ac (?)	Monnier *et al.*, 2006; Blanchard *et al.*, 2007
Gram-negative bacteria			
Neisseria meningitidis	OpcA	Neu5Ac	Prince *et al.*, 2002; Cherezov *et al.*, 2008
Vibrio cholerae	VCNA	Neu5Ac	Moustafa *et al.*, 2004
Bacterial Toxins			
Vibrio cholerae	Cholera toxin	GM_1	Merritt *et al.*, 1998
Clostridium botulinum	Neurotoxin A-F	1b gangliosides	Swaminathan and Eswaramoorthy, 2000
Clostridium tetani	Tetanus toxin	GT_{1b}, GQ_{1b}	Fotinou *et al.*, 2001
Bordetella pertussis	Pertussis toxin	GD_{1a}	Stein *et al.*, 1994
Staphylococcus aureus	SSL5	Sialyl-Lewis^x	Baker *et al.*, 2007
Staphylococcus aureus	SSL11	Sialyl-Lewis^x	Chung *et al.*, 2007
Protozoa			
Plasmodium falciparum	EBA-175	α-Neu5Ac-(2→3)-Gal	Tolia *et al.*, 2005

[a]Only a representative list of viral sialic-specific lectin structures are provided, for a comprehensive listing see Lehmann *et al.* (2006).

Abbreviations: Gal, D-galactose; Neu5Ac, small *N*-acetylneuramnic acid.

2.2. Bacteria and bacterial toxins

As is the case with viral infections, adhesion of bacteria to host tissues represents an initial and essential step in pathogenesis. Bacterial surface components that mediate adherence are collectively called adhesins. Often, these adhesins are associated with fimbriae or pili, which have been correlated with pathogenicity (Sharon, 2006). Several bacteria have been reported to use sialic acid-containing glycans as ligands, although the identity of the specific adhesin remains uncertain in many cases (Angata and Varki, 2002; Lehmann *et al.*, 2006).

Pathogenic *Escherichia coli* express several classes of fimbria-associated adhesins (Sharon, 2006), strains shown to use sialoglycoconjugates as attachment sites express the S fimbrial adhesins Sfa I or II, K99-fimbriae, the F41-fimbriae or one of the colonization factor antigens CFA I and CFA II (Hacker *et al.*, 1993; Lehmann *et al.*, 2006). The S-fimbriae preferentially bind gangliosides carrying α-Neu5Gc-(2→3)-Gal and α-Neu5Ac-(2→8)-Neu5Ac-structures with the C-8 and C-9 hydroxyl groups on the sialic acids necessary for recognition (Hanisch *et al.*, 1993). Mutagenesis studies performed on SFaS, a minor component of the multisubunit S-fimbriae, suggest that amino acids lysine (Lys) 116 and arginine (Arg) 118 influence SfaS binding to sialic acids. Notably, these amino acids are part of a stretch of conserved amino acids that are also found in other bacterial sialic acid-binding lectins, such as CFAI and K99 adhesins of *E. coli* and the *Vibrio cholerae* toxin B subunit (Morschhauser *et al.*, 1990).

In contrast to S-fimbriae, where the adhesin SfaS is only a minor component, in K99-fimbriae, the sialic acid-binding site is found in the major subunit. The presence of a hydrophobic region close to the binding site seems to enhance sialic acid-binding affinity (Lindahl *et al.*, 1988; Ono *et al.*, 1989), which favours Neu5Gc over Neu5Ac. The specific recognition of Neu5GcLacCer by K99-fimbriated *E. coli* might contribute to host specificity, since humans and animals that lack Neu5Gc cannot be infected (Teneberg *et al.*, 1993). Very little is known about the receptors or binding structures for the different colonization factor antigens (CFA), however, CFAI has been shown to bind to free sialic acid (Evans *et al.*, 1979), sialoglycoproteins (Wenneras *et al.*, 1990) and GM_2 (Lindahl *et al.*, 1982). Furthermore, purified CS2 antigen belonging to CFAII has been shown to be a sialic acid-dependent lectin inhibited specifically by sialyllactose (Sjoberg *et al.*, 1988). Whereas CFAI is a single fimbrial antigen, CFAII is composed of antigenically distinct structures called coli surface antigens (Evans *et al.*, 1975; Levine *et al.*, 1984).

Helicobacter pylori exhibits an unusual complexity in carbohydrate-binding specificity with interactions through sialylated and non-sialylated glycoconjugates. Among other *H. pylori* adhesins, three have been shown to interact in a sialic acid-dependent manner, SabA, HP-NAP and HP0721 (Lehmann *et al.*, 2006). All bind terminal α-(2→3)-linked sialic acid residues, with α-Neu5Ac-(2→3)-Gal representing the minimal binding epitope for SabA (Aspholm *et al.*, 2006) and HP0721 (Bennett and Roberts, 2005), while HP-NAP (*H. pylori* neutrophil-activating protein) binds α-Neu5Ac-(2→3)-β-Gal-(1→4)-β-GlcNAc-(1→3)-β-Gal-(1→4)-GlcNAc structures exclusively (Teneberg *et al.*, 1997).

Models for the colonization of mucosal cells by *Neisseria* propose a two-stage process with primary adherence occurring through pili followed by a second, closer range contact between outer membrane proteins and host cell surface receptors (Nassif *et al.*, 1999). OpcA, an integral outer membrane protein from *Neisseria meningitidis*, is believed to play an important role in this process through a high affinity interaction with sialic acid-containing oligosaccharides (Moore *et al.*, 2005). Recent crystal structure data show that OpcA adopts a 10-stranded β-barrel structure with 5 extensive loop regions forming an open channel on the extracellular side of the membrane. Docking simulations suggest a sialic acid-binding site at the base of the loop region adjacent to a tyrosine (Tyr) residue that directly interacts, together with an Arg residue, with the C-2 carboxyl of sialic acid (Cherezov *et al.*, 2008).

In addition to adhesins, some bacterial pathogens express soluble lectins, which are typically toxins. These toxins classically bind to oligosaccharide receptors on host cell surfaces and many of them show high specificity toward sialic acid residues, generally located on gangliosides (Karlsson, 1995). Many belong to the AB_5 family of toxins with an A-subunit carrying the catalytic domain of the toxin, while the B-subunit is responsible for binding the holotoxin. One of the best examples of a sialic acid-binding soluble lectin belonging to the AB_5 family is cholera toxin, produced by *V. cholerae*. The B-subunit exhibits specific binding to ganglioside GM_1, delivering the A-subunit to the cytosol. This results in the over-activation of an intracellular signalling pathway in gastrointestinal epithelial cells, causing severe diarrhoea.

TABLE 29.2 Bacterial sialic acid-transporters

Species	Sia-binding protein	Transporter system	Reference
Escherichia coli	NanT	Major facilitator superfamily (MFS)	Martinez *et al.*, 1995
	NanC	Porin (outer membrane channel protein)	Condemine *et al.*, 2005
Haemophilus ducreyi	SatA	ATP-binding cassette (ABC)	Post *et al.*, 2005
Haemophilus influenzae	SiaP	Tripartite ATP-independent periplasmic (TRAP)	Allen *et al.*, 2005; Severi *et al.*, 2005
			Muller *et al.*, 2006; Johnston *et al.*, 2008
Pasturella multocida	NanP	Tripartite ATP-independent periplasmic (TRAP)	Steenbergen *et al.*, 2005

Other notable examples of sialic acid-dependent toxins are those from *Clostridium botulinum* and *Clostridium tetani*, the causative agents of botulism and tetanus, respectively, which both recognize glycosphingolipids of the ganglioside G_{1b} series, particularly G_{T1b} or G_{Q1b} (Kitamura *et al.*, 2005) and the recently described SSL5 and SSL11 lectins from *Staphylococcus aureus* that bind sialyl-Lewis^x structures (Baker *et al.*, 2007; Chung *et al.*, 2007).

Not only do microbes express sialic acid-specific lectins to aid in host cell attachment, but they also decorate their cell surfaces with sialic acid residues. This allows, for example pathogenic bacteria to evade the host immune response by mimicking host cells (see Chapters 42 and 43) and provides bacteria with the ability to interact with different host cells through specific interactions with host cell sialic acid-lectins (Vimr *et al.*, 2004; Severi *et al.*, 2007).

Bacteria obtain sialic acids by one of two mechanisms, *de novo* biosynthesis or acquisition from the environment (Vimr *et al.*, 2004; Severi *et al.*, 2007). Gram-negative bacteria acquire sialic acids from the environment by utilizing transport systems that specifically recognize and bind these compounds. Following transport into the cell, the sialic acid can then be channelled into the LPS/LOS sialylation pathway. Thus far, four separate systems have been identified that mediate the transport of sialic acid (Table 29.2):

(i) the major facilitator superfamily (MFS) type;
(ii) the general outer membrane porin;
(iii) the adenosine triphosphate-(ATP-)binding cassette (ABC) transporter system; and
(iv) the tripartite ATP-independent periplasmic (TRAP) transporter system.

Although differing in the mechanism used to capture sialic acids from the environment, all these systems share a lectin-like sialic acid-binding domain. The nature of the sialic acid-binding domain associated with bacterial sialic acid transport systems will be discussed later in this chapter.

2.3. Protozoa

Only a few sialic acid-specific lectins expressed by protozoa, particularly protozoal pathogens, have been reported. Thus far, protozoan sialic acid-binding proteins have been described in members of the kinetoplasid pathogens; *Trypanosoma*, *Leishmania*, *Tritichomonas*, *Babesia* and *Plasmodium* spp., with the latter genus being the most extensively studied (see Lehmann *et al.* (2006) and references therein). The erythrocyte binding antigen (EBA)-175 (Klotz *et al.*, 1992; Orlandi *et al.*, 1992) and its paralogue, EBA-140 (Mayer *et al.*, 2001; Lobo *et al.*, 2003) and EBA-181 (Gilberger *et al.*, 2003), are erythrocyte-binding proteins of *Plasmodium falciparum* that belong to the Duffy-binding-like protein family and require sialic acid on host receptors for binding and invasion (see von Itzstein *et al.* (2008) and references therein). The glycophorins (A, B and C), i.e. sialoglycoproteins present on the erythrocyte surface, serve as the major receptors for sialic acid-dependent invasion of erythrocytes (Hadley and Miller, 1988).

Whereas, some Gram-negative bacteria utilize sialic acid transport systems to acquire sialic acids from the environment, trypansomes such as *Trypanosoma cruzi* (the aetiological agent of Chagas' disease) express a surface-bound protein, called trans-sialidase (TS), that enables the parasite to acquire sialic acid from mammalian host glycoconjugates (Schenkman *et al.*, 1993). In *T. cruzi*, the TS family is encoded by approximately 140 genes (Cross and Takle, 1993), many of which encode inactive enzyme. An enzymatically inactive recombinant TS, which was able to agglutinate desialylated erythrocytes, was identified as a sialic acid-recognizing lectin capable of stimulating $CD4^+$ T-cell activation *in vitro* and *in vivo*. The sialomucin CD43 was identified as a counter-receptor for TS on $CD4^+$ T-cells, with the inactive TS displaying a similar specificity to that described for active TS (i.e. α-(2→3)-linked sialic acid) (Todeschini *et al.*, 2002). Interestingly, blood stream forms of *Trypanosoma congolense*, an African trypanosome, have been reported to possess a lectin-like domain, localized at distinct sites on their flagellar surface, that appear to interact with sialic acid-containing receptors on the endothelial cell plasma membrane (Hemphill *et al.*, 1994). Different *T. congolense* TS forms have been identified (Tiralongo *et al.*, 2003a,b), raising the possibility that inactive members in the blood stream form may function as sialic acid-binding lectins.

2.4. Micro-fungi

The first human pathogenic fungi species thought to process a sialic acid-specific lectin were *Chrysosporium keratinophilum* and *Anixiopsis stercoraria* (synonym of *Aphanoascus fulvescens*) (Chabasse and Robert, 1986), which cause skin infections and onychomycosis in humans. Later, the same group identified sialic acid-specific binding of 13 different species of dermatophytes, including those of the genera *Trichophyton* and *Microsporum*, to erythrocytes (Bouchara *et al.*, 1987). However, it was not until a decade later that the first reports of sialic acid-dependent binding of human pathogenic fungal conidia and yeast to extracellular matrix (ECM) glycoproteins began to emerge. The involvement of sialic acids in micro-fungal pathogenesis has been most extensively studied in *Aspergillus fumigatus*, with several groups having investigated the sialic acid-dependent adhesion of *A. fumigatus* conidia to purified ECM proteins (Tronchin *et al.*, 1997; Wasylnka and Moore, 2000; Warwas *et al.*, 2007; J. Tiralongo *et al.*, unpublished). A putative sialic acid-specific lectin from *A. fumigatus* has now been purified (Tronchin *et al.*, 2002), thus providing an opportunity for the identification of similar lectins from other species.

3. CONSERVED BINDING SITE OF SIALIC ACID-SPECIFIC MICROBIAL LECTINS

The vast majority of the available crystal structures of microbial sialic acid-specific lectins are those of viral origin (Lehmann *et al.*, 2006) (see Table 29.1). Of the other microbial sialic acid-specific lectin structures, the bulk are bacterial toxins (see Table 29.1 and references therein) and the remainder comprise the *N. meningitidis* OpcA (Prince *et al.*, 2002; Cherezov *et al.*, 2008), the sialic acid-specific lectin domain of the *V. cholerae* neuraminidase (VCNA) (Moustafa *et al.*, 2004) and the *P. falciparum* EBA-175 (Tolia *et al.*, 2005). The overall structures of these proteins often differ quite substantially with diverse architectures and folds. However, the strong conservation of certain structural elements within the sialic acid-binding site of the *S. aureus* toxins SSL5 (Baker *et al.*, 2007) and SSL11 (Chung *et al.*, 2007), the rotavirus VP8* (Dormitzer *et al.*, 2002; Blanchard *et al.*, 2007) and *Bordetella pertussis* toxin (Stein *et al.*, 1994), all of which are structurally unrelated proteins, illustrates that commonalities in the sialic acid-binding site architecture can be extrapolated.

In order to identify common sialic acid-binding site structural elements, the crystal structures of the microbial sialic acid-specific lectins listed in Table 29.1 were examined and the critical interactions have been extracted and summarized in Figure 29.2. Residues and their relative prominence involved in hydrogen

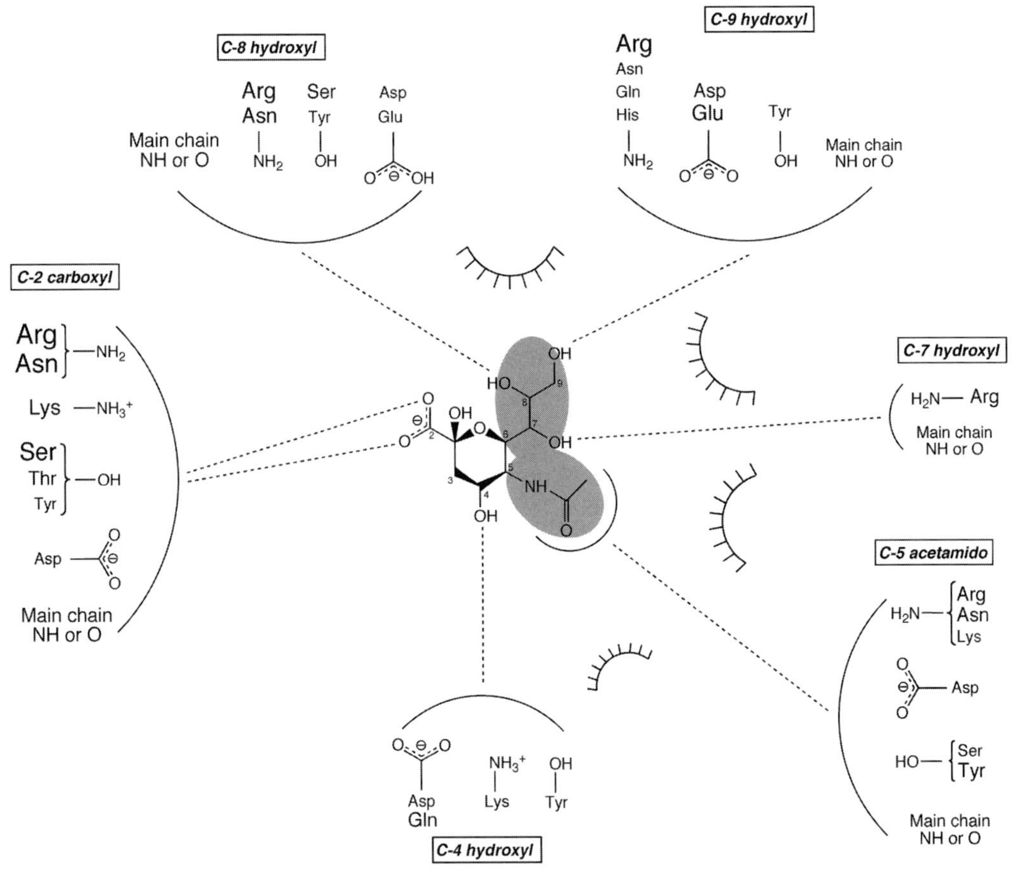

FIGURE 29.2 Common sialic acid-binding site structural elements based on the crystal structures of the microbial sialic acid-specific lectins listed in Table 29.1. Residues and their relative prominence involved in hydrogen bonding with sialic acid (dashed line), regions involved in van der Waals (grey) and hydrophobic contacts are indicated. Abbreviations: Arg, arginine; Asn, asparagine; Asp, aspartate; Gln, glutamine; Glu, glutamate; Lys, lysine; Ser, serine; Thr, threonine; Tyr, tyrosine.

bonding (dashed line) with sialic acids are indicated, as well as regions involved in van der Waals (grey) and hydrophobic contacts. Clearly, the interaction with the carboxyl group at C-2 of sialic acids is critical, with hydrogen bonding either directly or through water to side chains or main chain residues in all cases. Coordination of the C-2 carboxyl is achieved predominantly through interactions with either Arg (which also contributes a positive charge) or polar residues like asparagine (Asn) and serine (Ser). In the case of the *N. meningitidis* OpcA (Cherezov *et al.*, 2008), the sialic acid-specific lectin domain of VCNA (Moustafa *et al.*, 2004) and the *P. falciparum* EBA-175 (Tolia *et al.*, 2005), full coordination is achieved via multiple interactions involving charged and polar amino acids. Interactions with the C-2 carboxyl of sialic acids are the common thread linking all known prokaryotic and eukaryotic sialic acid-specific lectins. Nevertheless, unlike the eukaryotic sialic acid-binding immunoglobulin- (Ig-) like lectin (Siglec) family, where salt bridge formation between an Arg residue located within the V-set Ig-domain and the C-2 carboxyl of sialic acid is completely conserved (Crocker *et al.*, 2007; Zhuravleva *et al.*, 2008), this interaction in microbial sialic acid-specific lectins can eventuate through diverse residues (see Figure 29.2).

As is also the case for Siglecs (Zhuravleva *et al.*, 2008), Figure 29.2 reveals significant interactions of microbial sialic acid-specific lectins with the glycerol side chain of sialic acids, particularly with the C-8 and C-9 hydroxyl groups. Additional hydrogen bonding with the C-5 acetamido occurs mainly through main chain amide groups, i.e. -C(=O)NH-, whereas interactions with the C-7 hydroxyl, as well as the C-4 hydroxyl, are infrequent. van der Waals and hydrophobic contacts with the glycerol side chain and C-5 acetamido group of sialic acids are also of significance. van der Waals contacts between sialic acids and sialic acid-specific lectins often include packing interactions with aromatic amino acid side chains, as seen in the well-described human influenza virus A HA where a tryptophan (Trp) is in van der Waals contact with the methyl group of the C-5 acetamido (Skehel and Wiley, 2000).

4. SIALIC ACID-BINDING DOMAIN ASSOCIATED WITH BACTERIAL SIALIC ACID TRANSPORT SYSTEMS

The *E. coli* sialic acid transporter, NanT, is a periplasmic transporter of the MFS type that is essential for sialic acid uptake (Martinez *et al.*, 1995; Vimr *et al.*, 2004). The NanT transporter comprises 14 transmembrane domains, 2 domains more than that usually observed for MFS transporters. These 2 additional membrane-spanning domains (centrally located domains 7 and 8) are thought to be important for sialic acid specificity (Vimr *et al.*, 2004). A sialic acid-inducible porin (outer membrane channel protein), NanC, has also been characterized, with its expression essential for growth when the general porins OmpF and OmpC are absent (Condemine *et al.*, 2005).

The sialic acid transporter from *Haemophilus ducreyi* comprises a complex of four periplasmic proteins (SatABCD) and belongs to the family of high affinity ABC transporters. The periplasmic binding protein SatA represents the sialic acid-binding component, SatB the integral-membrane permease, SatD the ATPase and SatC contains a permease and ATPase domain. The SatA binding of sialic acid is believed to increase the affinity of SatA for the SatB/SatC periplasmic localized complex. This interaction with SatB and SatC initiates sialic acid transport following ATP hydrolysis (Post *et al.*, 2005).

The TRAP transport system represents the fourth proposed mechanism for sialic acid-uptake by bacteria. Two members of the *Pasteurellaceae* family, *Haemophilus influenzae* and *Pasturella multocida*, have now been shown to transport sialic acids via this mechanism (Allen *et al.*, 2005; Severi *et al.*, 2005; Steenbergen *et al.*, 2005). Like SatABCD, the TRAP sialic acid transport system comprises multiple periplasmic proteins (SiaPQM), however, unlike SatABCD, SiaPQM is ATP independent. The SiaPQM transport system utilizes an extracytoplasmic solute receptor (ESR) protein that binds sialic acid with high affinity. The crystal structure of the *H. influenzae* ESR, SiaP, has recently been resolved (Muller *et al.*, 2006; Johnston *et al.*, 2008).

The structure of SiaP identifies it as a member of the periplasmic binding protein type II superfamily of α/β proteins that includes several other sugar transport proteins and closely resembles the periplasmic binding protein of ABC transporters (Johnston *et al.*, 2008). The SiaP protein consists of two domains (α/β), separated by a large cleft where sialic acid is bound (Muller *et al.*, 2006). The high-affinity binding of sialic acid to SiaP is initiated by the interaction of the C-2 carboxyl group of Sia with Arg147 (Figure 29.3) in domain II of SiaP, with full coordination of the C-2 carboxyl achieved through interactions with Arg127 and Asn187 (see Figure 29.3) also located in domain II. It is believed that these interactions trigger hinge bending allowing interaction of Glu67 and Asn10 (located in domain I) with the glycerol side chain and the C-5 acetamido of sialic acids, respectively, resulting in tight binding, i.e. 0.12 μM (Muller *et al.*, 2006). As can be seen in Figure 29.3, these structural elements are comparable with those depicted

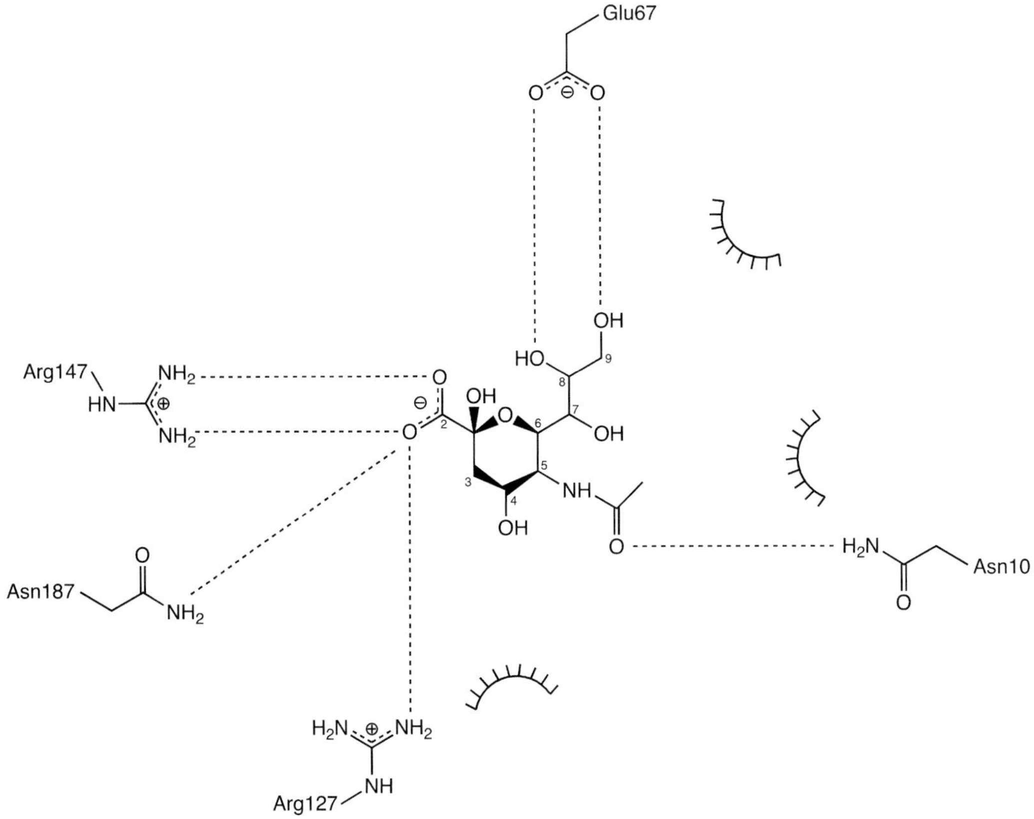

FIGURE 29.3 The interaction of sialic acid with the binding site of SiaP, the sialic acid-binding protein of the SiaPQM TRAP transporter system from *H. influenzae*. Residues involved in hydrogen bonding with sialic acid (dashed line) and hydrophobic contacts are indicated. Abbreviations: Arg, arginine; Asn, asparagine; Glu, glutamate.

in Figure 29.2. Multisequence alignment of SiaP with other TRAP ESR proteins has identified putative sialic acid-specific TRAP transporters in seven different bacterial species. Significantly, these putative sialic acid-specific TRAP ESRs have completely conserved all the residues involved in coordinating the C-2 carboxyl of sialic acid (Muller *et al.*, 2006; Johnston *et al.*, 2008).

5. CONCLUSIONS

The immense structural diversity and wide distribution of sialic acids suggest that sialobiology has only scratched the surface regarding the identification of sialic acid-specific lectins in Nature. This is particularly the case in the microbial world, where it seems probable that a vast array of sialic acid-specific lectins with unique specificities and functions exist that may represent novel targets for drug discovery. The biological roles of many of the microbial sialic acid-specific lectins described in the literature still remain unknown. Therefore, detailed investigations are necessary further to analyse the interaction of sialic acid-binding lectins with their counter-receptors, as well as the further elucidation of structural elements required for sialic acid-recognition. This will not only broaden our understanding of the role of sialic acids in biological systems but also their relevance in biomedical research (Research Focus Box).

RESEARCH FOCUS BOX

- Continued identification of novel microbial sialic acid-specific lectins and sialic acid-specific bacterial transport systems.
- Evaluation and characterization of the specificity of sialic acid-specific lectins, particularly aided by glycan array technology.
- Elucidation of the structural elements required for sialic acid-recognition.
- What is the role(s) of microbial sialic acid-specific lectins in pathogenesis?
- Can these interactions represent a target for drug discovery?

References

Allen, S., Zaleski, A., Johnston, J.W., Gibson, B.W., Apicella, M.A., 2005. Novel sialic acid transporter of *Haemophilus influenzae*. Infect. Immun. 73, 5291–5300.

Angata, T., Varki, A., 2002. Chemical diversity in the sialic acids and related α-keto acids: an evolutionary perspective. Chem. Rev. 102, 439–469.

Arias, C.F., Isa, P., Guerrero, C.A., et al., 2002. Molecular biology of rotavirus cell entry. Arch. Med. Res. 33, 356–361.

Aspholm, M., Olfat, F.O., Norden, J., et al., 2006. SabA is the *H. pylori* hemagglutinin and is polymorphic in binding to sialylated glycans. PLoS Pathog. 2, e110.

Baker, H.M., Basu, I., Chung, M.C., Caradoc-Davies, T., Fraser, J.D., Baker, E.N., 2007. Crystal structures of the staphylococcal toxin SSL5 in complex with sialyl Lewis X reveal a conserved binding site that shares common features with viral and bacterial sialic acid binding proteins. J. Mol. Biol. 374, 1298–1308.

Bennett, H.J., Roberts, I.S., 2005. Identification of a new sialic acid-binding protein in *Helicobacter pylori*. FEMS Immunol. Med. Microbiol. 44, 163–169.

Blanchard, H., Yu, X., Coulson, B.S., von Itzstein, M., 2007. Insight into host cell carbohydrate-recognition by human and porcine rotavirus from crystal structures of the virion spike associated carbohydrate-binding domain (VP8*). J. Mol. Biol. 367, 1215–1226.

Bouchara, J.P., Robert, R., Chabasse, D., Senet, J.M., 1987. Evidence for the lectin nature of some dermatophyte haemagglutinins. Ann. Inst. Pasteur Microbiol. 138, 729–736.

Burmeister, W.P., Guilligay, D., Cusack, S., Wadell, G., Arnberg, N., 2004. Crystal structure of species D adenovirus fiber knobs and their sialic acid binding sites. J. Virol. 78, 7727–7736.

Chabasse, D., Robert, R., 1986. Detection of lectin from *Chrysosporium keratinophilum* (Frey) Carmichael and *Anixiopsis stercoraria* (Hansen) Hansen by inhibition of haemagglutination. Ann. Inst. Pasteur Microbiol. 137B, 187–193.

Cherezov, V., Liu, W., Derrick, J.P., et al., 2008. *In meso* crystal structure and docking simulations suggest an alternative proteoglycan binding site in the OpcA outer membrane adhesin. Proteins 71, 24–34.

Chung, M.C., Wines, B.D., Baker, H., Langley, R.J., Baker, E.N., Fraser, J.D., 2007. The crystal structure of staphylococcal superantigen-like protein 11 in complex with sialyl Lewis X reveals the mechanism for cell binding and immune inhibition. Mol. Microbiol. 66, 1342–1355.

Ciarlet, M., Estes, M.K., 1999. Human and most animal rotavirus strains do not require the presence of sialic acid on the cell surface for efficient infectivity. J. Gen. Virol. 80, 943–948.

Condemine, G., Berrier, C., Plumbridge, J., Ghazi, A., 2005. Function and expression of an *N*-acetylneuraminic acid-inducible outer membrane channel in *Escherichia coli*. J. Bacteriol. 187, 1959–1965.

Crocker, P.R., Paulson, J.C., Varki, A., 2007. Siglecs and their roles in the immune system. Nat. Rev. Immunol. 7, 255–266.

Cross, G.A., Takle, G.B., 1993. The surface trans-sialidase family of *Trypanosoma cruzi*. Annu. Rev. Microbiol. 47, 385–411.

Delorme, C., Brussow, H., Sidoti, J., et al., 2001. Glycosphingolipid binding specificities of rotavirus: identification of a sialic acid-binding epitope. J. Virol. 75, 2276–2287.

Dormitzer, P.R., Sun, Z.Y., Wagner, G., Harrison, S.C., 2002. The rhesus rotavirus VP4 sialic acid binding domain has a galectin fold with a novel carbohydrate binding site. EMBO J. 21, 885–897.

Evans, D.G., Silver, R.P., Evans, D.J., Chase, D.G., Gorbach, S.L., 1975. Plasmid-controlled colonization factor associated with virulence in *Escherichia coli* enterotoxigenic for humans. Infect. Immun. 12, 656–667.

Evans, D.G., Evans, D.J., Clegg, S., Pauley, J.A., 1979. Purification and characterization of the CFA/I antigen of enterotoxigenic *Escherichia coli*. Infect. Immun. 25, 738–748.

Fotinou, C., Emsley, P., Black, I., et al., 2001. The crystal structure of tetanus toxin Hc fragment complexed with a synthetic GT1b analogue suggests cross-linking between ganglioside receptors and the toxin. J. Biol. Chem. 276, 32274–32281.

Gilberger, T.W., Thompson, J.K., Triglia, T., Good, R.T., Duraisingh, M.T., Cowman, A.F., 2003. A novel erythrocyte binding antigen-175 paralogue from *Plasmodium falciparum* defines a new trypsin-resistant receptor on human erythrocytes. J. Biol. Chem. 278, 14480–14486.

Ha, Y., Stevens, D.J., Skehel, J.J., Wiley, D.C., 2001. X-ray structures of H5 avian and H9 swine influenza virus hemagglutinins bound to avian and human receptor analogs. Proc. Natl. Acad. Sci. USA 98, 11181–11186.

Hacker, J., Kestler, H., Hoschutzky, H., Jann, K., Lottspeich, F., Korhonen, T.K., 1993. Cloning and characterization of the S fimbrial adhesin II complex of an *Escherichia coli* O18:K1 meningitis isolate. Infect. Immun. 61, 544–550.

Hadley, T.J., Miller, L.H., 1988. Invasion of erythrocytes by malaria parasites: erythrocyte ligands and parasite receptors. Prog. Allergy 41, 49–71.

Hanisch, F.G., Hacker, J., Schroten, H., 1993. Specificity of S fimbriae on recombinant *Escherichia coli*: preferential binding to gangliosides expressing NeuGcα(2-3)Gal and NeuAcα(2-8)NeuAc. Infect. Immun. 61, 2108–2115.

Hemphill, A., Frame, I., Ross, C.A., 1994. The interaction of *Trypanosoma congolense* with endothelial cells. Parasitology 109, 631–641.

Johnston, J.W., Coussens, N.P., Allen, S., et al., 2008. Characterization of the *N*-acetyl-5-neuraminic acid-binding site of the extracytoplasmic solute receptor (SiaP) of nontypeable *Haemophilus influenzae* strain 2019. J. Biol. Chem. 283, 855–865.

Karlsson, K.A., 1995. Microbial recognition of target-cell glycoconjugates. Curr. Opin. Struct. Biol. 5, 622–635.

Kitamura, M., Igimi, S., Furukawa, K., 2005. Different response of the knockout mice lacking b-series gangliosides against botulinum and tetanus toxins. Biochim. Biophys. Acta 1741, 1–3.

Klotz, F.W., Orlandi, P.A., Reuter, G., et al., 1992. Binding of *Plasmodium falciparum* 175-kilodalton erythrocyte binding antigen and invasion of murine erythrocytes requires *N*-acetylneuraminic acid but not its *O*-acetylated form. Mol. Biochem. Parasitol. 51, 49–54.

Lanctot, P.M., Gage, F.H., Varki, A.P., 2007. The glycans of stem cells. Curr. Opin. Chem. Biol. 11, 373–380.

Lehmann, F., Tiralongo, E., Tiralongo, J., 2006. Sialic acid-specific lectins: occurrence, specificity and function. Cell. Mol. Life Sci. 63, 1331–1354.

Levine, M.M., Ristaino, P., Marley, G., et al., 1984. Coli surface antigens 1 and 3 of colonization factor antigen II-positive enterotoxigenic *Escherichia coli*: morphology, purification, and immune responses in humans. Infect. Immun. 44, 409–420.

Lindahl, M., Faris, A., Wadstrom, T., 1982. Colonization factor antigen on enterotoxigenic *Escherichia coli* is a sialic-specific lectin. Lancet 2, 280.

Lindahl, M., Brossmer, R., Wadstrom, T., 1988. Sialic acid and *N*-acetylgalactosamine specific bacterial lectins of enterotoxigenic *Escherichia coli* (ETEC). Adv. Exp. Med. Biol. 228, 123–152.

Lobo, C.A., Rodriguez, M., Reid, M., Lustigman, S., 2003. Glycophorin C is the receptor for the *Plasmodium falciparum* erythrocyte binding ligand PfEBP-2. Blood 101, 4628–4631.

Lopez, S., Arias, C.F., 2004. Multistep entry of rotavirus into cells: a Versaillesque dance. Trends Microbiol. 12, 271–278.

Martinez, J., Steenbergen, S., Vimr, E., 1995. Derived structure of the putative sialic acid transporter from *Escherichia coli* predicts a novel sugar permease domain. J. Bacteriol. 177, 6005–6010.

Mayer, D.C., Kaneko, O., Hudson-Taylor, D.E., Reid, M.E., Miller, L.H., 2001. Characterization of a *Plasmodium falciparum* erythrocyte-binding protein paralogous to EBA-175. Proc. Natl. Acad. Sci. USA 98, 5222–5227.

Merritt, E.A., Kuhn, P., Sarfaty, S., Erbe, J.L., Holmes, R.K., Hol, W.G., 1998. The 1.25 Å resolution refinement of the cholera toxin B-pentamer: evidence of peptide backbone strain at the receptor-binding site. J. Mol. Biol. 282, 1043–1059.

Monnier, N., Higo-Moriguchi, K., Sun, Z.Y., Prasad, B.V., Taniguchi, K., Dormitzer, P.R., 2006. High-resolution molecular and antigen structure of the VP8* core of a sialic acid-independent human rotavirus strain. J. Virol. 80, 1513–1523.

Moore, J., Bailey, S.E., Benmechernene, Z., et al., 2005. Recognition of saccharides by the OpcA, OpaD, and OpaB outer membrane proteins from *Neisseria meningitidis*. J. Biol. Chem. 280, 31489–31497.

Morschhauser, J., Hoschutzky, H., Jann, K., Hacker, J., 1990. Functional analysis of the sialic acid-binding adhesin SfaS of pathogenic *Escherichia coli* by site-specific mutagenesis. Infect. Immun. 58, 2133–2138.

Moustafa, I., Connaris, H., Taylor, M., et al., 2004. Sialic acid recognition by *Vibrio cholerae* neuraminidase. J. Biol. Chem. 279, 40819–40826.

Muller, A., Severi, E., Mulligan, C., et al., 2006. Conservation of structure and mechanism in primary and secondary transporters exemplified by SiaP, a sialic acid binding virulence factor from *Haemophilus influenzae*. J. Biol. Chem. 281, 22212–22222.

Nassif, X., Pujol, C., Morand, P., Eugene, E., 1999. Interactions of pathogenic *Neisseria* with host cells. Is it possible to assemble the puzzle? Mol. Microbiol. 32, 1124–1132.

Ono, E., Abe, K., Nakazawa, M., Naiki, M., 1989. Ganglioside epitope recognized by K99 fimbriae from enterotoxigenic *Escherichia coli*. Infect. Immun. 57, 907–911.

Orlandi, P.A., Klotz, F.W., Haynes, J.D., 1992. A malaria invasion receptor, the 175-kilodalton erythrocyte binding antigen of *Plasmodium falciparum* recognizes the terminal Neu5Ac(α2-3)Gal-sequences of glycophorin A. J. Cell Biol. 116, 901–909.

Post, D.M., Mungur, R., Gibson, B.W., Munson, R.S., 2005. Identification of a novel sialic acid transporter in *Haemophilus ducreyi*. Infect. Immun. 73, 6727–6735.

Prince, S.M., Achtman, M., Derrick, J.P., 2002. Crystal structure of the OpcA integral membrane adhesin from *Neisseria meningitidis*. Proc. Natl. Acad. Sci. USA 99, 3417–3421.

Schauer, R., Kamerling, J.P., 1997. Chemistry, biochemistry and biology of sialic acids. In: Montreuil, J., Vliegenthart, J.F.G., Schachter, H. (Eds.), Glycoproteins II. Elsevier B.V., Amsterdam, pp. 243–402.

Schenkman, S., Ferguson, M.A., Heise, N., de Almeida, M.L., Mortara, R.A., Yoshida, N., 1993. Mucin-like glycoproteins linked to the membrane by glycosylphosphatidylinositol anchor are the major acceptors of sialic acid in a reaction catalyzed by trans-sialidase in metacyclic forms of *Trypanosoma cruzi*. Mol. Biochem. Parasitol. 59, 293–303.

Severi, E., Randle, G., Kivlin, P., et al., 2005. Sialic acid transport in *Haemophilus influenzae* is essential for lipopolysaccharide sialylation and serum resistance and is dependent on a novel tripartite ATP-independent periplasmic transporter. Mol. Microbiol. 58, 1173–1185.

Severi, E., Hood, D.W., Thomas, G.H., 2007. Sialic acid utilization by bacterial pathogens. Microbiology 153, 2817–2822.

Sharon, N., 2006. Carbohydrates as future anti-adhesion drugs for infectious diseases. Biochim. Biophys. Acta 1760, 527–537.

Sjoberg, P.O., Lindahl, M., Porath, J., Wadstrom, T., 1988. Purification and characterization of CS2, a sialic acid-specific haemagglutinin of enterotoxigenic *Escherichia coli*. Biochem. J. 255, 105–111.

Skehel, J.J., Wiley, D.C., 2000. Receptor binding and membrane fusion in virus entry: the influenza hemagglutinin. Annu. Rev. Biochem. 69, 531–569.

Steenbergen, S.M., Lichtensteiger, C.A., Caughlan, R., Garfinkle, J., Fuller, T.E., Vimr, E.R., 2005. Sialic acid metabolism and systemic pasteurellosis. Infect. Immun. 73, 1284–1294.

Stehle, T., Harrison, S.C., 1997. High-resolution structure of a polyomavirus VP1-oligosaccharide complex: implications for assembly and receptor binding. EMBO J. 16, 5139–5148.

Stein, P.E., Boodhoo, A., Armstrong, G.D., et al., 1994. Structure of a pertussis toxin-sugar complex as a model for receptor binding. Nat. Struct. Biol. 1, 591–596.

Stevens, J., Blixt, O., Paulson, J.C., Wilson, I.A., 2006. Glycan microarray technologies: tools to survey host specificity of influenza viruses. Nat. Rev. Microbiol. 4, 857–864.

Suzuki, Y., 2005. Sialobiology of influenza: molecular mechanism of host range variation of influenza viruses. Biol. Pharm. Bull. 28, 399–408.

Swaminathan, S., Eswaramoorthy, S., 2000. Structural analysis of the catalytic and binding sites of *Clostridium botulinum* neurotoxin B. Nat. Struct. Biol. 7, 693–699.

Teneberg, S., Willemsen, P.T., de Graaf, F.K., Karlsson, K.A., 1993. Calf small intestine receptors for K99 fimbriated enterotoxigenic *Escherichia coli*. FEMS Microbiol. Lett. 109, 107–112.

Teneberg, S., Miller-Podraza, H., Lampert, H.C., et al., 1997. Carbohydrate binding specificity of the neutrophil-activating protein of *Helicobacter pylori*. J. Biol. Chem. 272, 19067–19071.

Tiralongo, E., Martensen, I., Grotzinger, J., Tiralongo, J., Schauer, R., 2003a. Trans-sialidase-like sequences from *Trypanosoma congolense* conserve most of the critical active site residues found in other trans-sialidases. Biol. Chem. 384, 1203–1213.

Tiralongo, E., Schrader, S., Lange, H., Lemke, H., Tiralongo, J., Schauer, R., 2003b. Two trans-sialidase forms with different sialic acid transfer and sialidase activities from *Trypanosoma congolense*. J. Biol. Chem. 278, 23301–23310.

Todeschini, A.R., Girard, M.F., Wieruszeski, J.M., et al., 2002. Trans-sialidase from *Trypanosoma cruzi* binds host T-lymphocytes in a lectin manner. J. Biol. Chem. 277, 45962–45968.

Tolia, N.H., Enemark, E.J., Sim, B.K., Joshua-Tor, L., 2005. Structural basis for the EBA-175 erythrocyte invasion pathway of the malaria parasite *Plasmodium falciparum*. Cell 122, 183–193.

Tronchin, G., Esnault, K., Renier, G., Filmon, R., Chabasse, D., Bouchara, J.P., 1997. Expression and identification of a laminin-binding protein in *Aspergillus fumigatus* conidia. Infect. Immun. 65, 9–15.

Tronchin, G., Esnault, K., Sanchez, M., Larcher, G., Marot-Leblond, A., Bouchara, J.P., 2002. Purification and partial characterization of a 32-kilodalton sialic acid-specific lectin from *Aspergillus fumigatus*. Infect. Immun. 70, 6891–6895.

Varki, N.M., Varki, A., 2007. Diversity in cell surface sialic acid presentations: implications for biology and disease. Lab. Invest. 87, 851–857.

Vimr, E.R., Kalivoda, K.A., Deszo, E.L., Steenbergen, S.M., 2004. Diversity of microbial sialic acid metabolism. Microbiol. Mol. Biol. Rev. 68, 132–153.

von Itzstein, M., Plebanski, M., Cooke, B.M., Coppel, R.L., 2008. Hot, sweet and sticky: the glycobiology of *Plasmodium falciparum*. Trends Parasitol. 24, 210–218.

Warwas, M.L., Watson, J.N., Bennet, A.J., Moore, M.M., 2007. Structure and role of sialic acids on the surface of *Aspergillus fumigatus* conidiospores. Glycobiology 17, 401–410.

Wasylnka, J.A., Moore, M.M., 2000. Adhesion of *Aspergillus* species to extracellular matrix proteins: evidence for involvement of negatively charged carbohydrates on the conidial surface. Infect. Immun. 68, 3377–3384.

Wenneras, C., Holmgren, J., Svennerholm, A.M., 1990. The binding of colonization factor antigens of enterotoxigenic *Escherichia coli* to intestinal cell membrane proteins. FEMS Microbiol. Lett. 54, 107–112.

Wilson, I.A., Skehel, J.J., Wiley, D.C., 1981. Structure of the haemagglutinin membrane glycoprotein of influenza virus at 3 Å resolution. Nature 289, 366–373.

Zhou, L., Luo, Y., Wu, Y., Tsao, J., Luo, M., 2000. Sialylation of the host receptor may modulate entry of demyelinating persistent Theiler's virus. J. Virol. 74, 1477–1485.

Zhuravleva, M.A., Trandem, K., Sun, P.D., 2008. Structural implications of siglec-5-mediated sialoglycan recognition. J. Mol. Biol. 375, 437–447.

CHAPTER

30

Bacterial toxins and their carbohydrate receptors at the host–pathogen interface

Clifford A. Lingwood and Radia Mahfoud

SUMMARY

The binding of the pentameric B-subunits of cholera toxin and verotoxin to GM_1 ganglioside and globotriaosyl ceramide (Gb_3), respectively, initiates the process of internalization and intracellular trafficking of the A-subunit to the endoplasmic recticulum for cytosolic translocation. The diseases that can subsequently occur, i.e. dysentery and haemolytic uraemic syndrome, respectively, depend on the organ distribution of these glycosphingolipids (GSLs) and the catalytic activity of the A-subunit, in other words, the ability of adenosine diphosphate ribosyl transferase to activate chloride transport/water efflux and protein synthesis inhibition via depurination of the 60S ribosomal subunit, respectively. Both toxins require the GSL receptor to be present in detergent-resistant domains for cytopathology. In each case, GSL receptor-binding initiates a transmembrane signalling cascade, despite the fact that GSLs are not transmembrane components. The cholera toxin B-subunit provides a powerful tool in modulation of the immune system and verotoxin 1 (VT1) provides insight into the relationship between cell drug resistance and GSL metabolism, Gb_3 expression and cancer and the link between Gb_3 and human immunodeficiency virus susceptibility. Generation of receptor mimics may provide a fruitful approach to counter toxicity but have yet to prove effective clinically.

Keywords: Verotoxin; Cholera toxin; Glycosphingolipids; Retrograde transport; Glycosphingolipid signalling; Haemolytic uraemic syndrome; Dysentery; Glycosphingolipid receptor mimics

1. INTRODUCTION

Many bacteria choose host cell carbohydrates as their initial means to interact with their eukaryotic target cells. While such cells are normally surrounded by a candy coat to provide an entrée for a broad prokaryotic diet, bacterial toxins can have an exquisitely glycosphingolipid (GSL)-selective palate. These are listed in Table 30.1. Despite the extensive study of bacterial subunit toxins, their benefits to the bacterium remain in question. The three most significant bacterial subunit toxins share this property of GSL receptor binding. These three A_1B_5-subunit toxins, cholera toxin (CT), Shiga toxin (Stx) and verotoxin (VT), will be considered in this review and this is largely reduced to two, since VT and Stx are essentially the same toxin.

TABLE 30.1 Glycosphingolipid (GSL)-binding bacterial toxins

Toxin	GSL binding
Clostridium botulinum	Ganglioside GT_{1b} (Kozaki *et al.*, 1998)
Clostridium difficile toxin A	α-D-Gal-(1→3)-β-D-Gal-(1→4)-GlcNAc (Krivan *et al.*, 1986)
Cholera toxin	GM_1 ganglioside (Critchley *et al.*, 1982)
E. coli heat labile toxin	Gangliosides (Fukuta *et al.*, 1988)
Vero(shiga) toxins	Globotriaosyl ceramide (Lingwood, 1993)
Tetanus toxin	Ganglioside GT_{1b}, GD_{1a} (Louch *et al.*, 2002)
Bacillus thuringiensis crystal toxin	Invertebrate neutral α-D-(1→3)-Gal terminal arthro series GSLs (Mucha *et al.*, 2004; Griffitts *et al.*, 2005)

1.1. Cholera toxin and dysentery

Cholera toxin is the major virulence factor of the waterborne Gram-negative *Vibrio cholerae* responsible for dysentery. The organism colonizes the small bowel and secretes a protease which activates CT. As with VT and Stx, the A-subunit of CT is non-covalently associated with the B-subunit pentamer via the disulfide-linked C-terminal A_2 fragment. This fragment is cleaved by the *Vibrio* protease but Stx/VT requires a host cell protease, i.e. furin, for cleavage during intracellular transport (Garred *et al.*, 1995). The A- and B-subunits then remain attached by the disulfide bridge which is cleaved in the endoplasmic reticulum (ER). Cholera toxin A-subunit is an adenosine diphosphate (ADP)-ribosyl transferase, which accesses the cell cytosol to ADP-ribosylate the regulatory G protein ($G_{S\alpha}$) (de Haan *et al.*, 1998) constitutively to activate adenylate cyclase, to generate cyclic adenosine monophosphate (cAMP) for protein kinase A-mediated activation of the cystic fibrosis transmembrane conductance regulator (CFTR) chloride channel and promote massive water and electrolyte loss (Thiagarajah and Verkman, 2003). This activity on the epithelial cells of the gastrointestinal tract provides the basis of the torrid watery diarrhoea of dysentery which, by careful rehydration and electrolyte replacement, is no longer necessarily fatal (Bhattacharya, 2003). Interestingly, cystic fibrosis can ameliorate CT sensitivity (Gabriel *et al.*, 1994). Cholera toxin is closely related to the *Eschericia coli*-derived heat-labile enterotoxin family (LT) responsible for traveller's diarrhoea (Middlebrook and Dorland, 1984). These toxins have the same mechanism of action as CT and LT has the same GM_1 receptor, though other members of the LT family preferentially bind other gangliosides (Fukuta *et al.*, 1988).

1.2. Verotoxin (Shiga toxin) and the haemolytic uraemic syndrome

Verotoxin was discovered in Canada in 1976 as an additional cytopathic effect on vero cells exhibited by extracts of certain *E. coli* strains (Konowalchuk *et al.*, 1977). The purified toxin was later shown to be equivalent to Shiga toxin from *Shigella dysenteriae* (O'Brien *et al.*, 1982) and has therefore been termed Shiga-like toxin or Shiga toxin (Stx) also. There is a single amino acid difference between Stx1 and VT1 (Strockbine *et al.*, 1988). Interest in VT was sparked by the discovery that gastrointestinal infection with VT-producing *E. coli* (VTEC) was associated with first, haemorrhagic colitis (HC) (Riley *et al.*, 1983) and then a previously idiopathic syndrome termed haemolytic uraemic syndrome (HUS) (Karmali *et al.*, 1985). Various aetiological bases had been ascribed to HUS, but it was the careful epidemiological study of Karmali and co-workers who established the linkage to VTEC, particularly the O157:H7 serotype.

It is clear that HC is the prelude to HUS (Tarr *et al.*, 1990). Although there is no receptor for VT within the human gastrointestinal tract, in terms of gastrointestinal mucosal epithelial cells, the attaching and effacing lesion by which the enterohaemorrhagic *E. coli* attach to and manipulate the surface of gastrointestinal epithelial cells (Louie *et al.*, 1993; Ismaili *et al.*, 1998) compromises the mucosal barrier such that the toxin can cross either by transcytosis (Acheson *et al.*, 1996; Philpott *et al.*, 1997) or pericellular transit (Philpott *et al.*, 1998) to target submucosal receptor-positive endothelial cells in the microvasculature. This results in local haemorrhage giving rise to HC and provides greater toxin access to the systemic circulation, although bacteraemia and fever are not features of HUS. Systemic verotoxaemia then targets renal endothelial cells because of their higher Gb_3 content. Endothelial cell death results in microvascular occlusion and the infarction of the glomeruli, resulting in renal failure and even fatal outcome. Importantly, HUS is more prevalent in the very young and the very old following gastrointestinal VTEC infection (Lieberman *et al.*, 1966). The lipopolysaccharide- (LPS-) induced response (Louise and Obrig, 1992; Siegler *et al.*, 2001) and other cytokines (Louise and Obrig, 1991) can upregulate globotriaosyl ceramide (Gb_3) synthesis of endothelial cells to amplify VT cytopathology. Although the O157:H7 *E. coli* strain serotype remains the most frequently associated with outbreaks of clinical disease, VT is produced by a variety of non-O157 serotypes (Griffin and Tauxe, 1991; Karch *et al.*, 1999) which can induce similar pathology. Serotypes producing both VT1 and VT2 are common and VT2-producing VTEC infections are more commonly associated with the most severe clinical sequelae (Friedrich *et al.*, 2002; Jelacic *et al.*, 2002).

Verotoxins are a family of *E. coli*-elaborated toxins comprising primarily of VT1 and VT2 but also some several subvariants of VT2, VT2e, associated primarily with oedema disease in pigs (Johnson *et al.*, 1990), VT2d, an elastase activated form of VT2 (Melton-Celsa *et al.*, 2002) and VT2c, a rare but disease associated variant of VT2 (Friedrich *et al.*, 2002), are the primary members. Interestingly, VT2c infections have recently become more prominent (Uhlich *et al.*, 2008). Verotoxins are subunit toxins which comprise a single A-subunit responsible for the depurination of the 28S RNA of the 60S ribosomal subunit to inhibit protein synthesis (Saxena *et al.*, 1989).

1.3. Shiga toxin and shigellosis

The syndrome HUS is a major complication of shigellosis resulting from infection with *Shigella dysenteriae* but not other *Shigella* species (Shears, 1996). As for VTEC infections, *S. dysenteriae* HUS is primarily associated with paediatric infection (Nathoo *et al.*, 1998) but, unlike for VTEC (Wong *et al.*, 2000), early antiobiotic therapy is effective to prevent HUS (Bennish *et al.*, 2006), probably because the toxin gene is on a phage in *E. coli* (Zhang *et al.*, 2000; Wagner *et al.*, 2002).

2. TOXIN RECEPTOR GSL-BINDING

2.1. Verotoxin receptor

The VT1/Stx receptor was identified as a neutral GSL. The B-subunit pentamer mediates the binding to Gb_3, α-Gal-(1→4)-β-Gal-(1→4)-Glc-ceramide (where Gal, D-galactose; Glc, D-glucose) (Jacewicz *et al.*, 1986; Lindberg *et al.*, 1987; Lingwood *et al.*, 1987) (Figure 30.1). All members of the VT family bind Gb_3 (Lingwood, 1993) but VT2e, associated with oedema disease in pigs, preferentially binds Gb_4 (DeGrandis *et al.*, 1989). Reconstitution of receptor-negative cells with Gb_3 results in induction of cell sensitivity to VT (Waddell *et al.*, 1990) and ablation of Gb_3 synthase in mice engenders complete resistance to both VT1 and VT2 (Okuda *et al.*, 2006). Also, Gb_3 has been defined as the human germinal centre B-cell antigen, CD77 (Mangeney *et al.*, 1991)

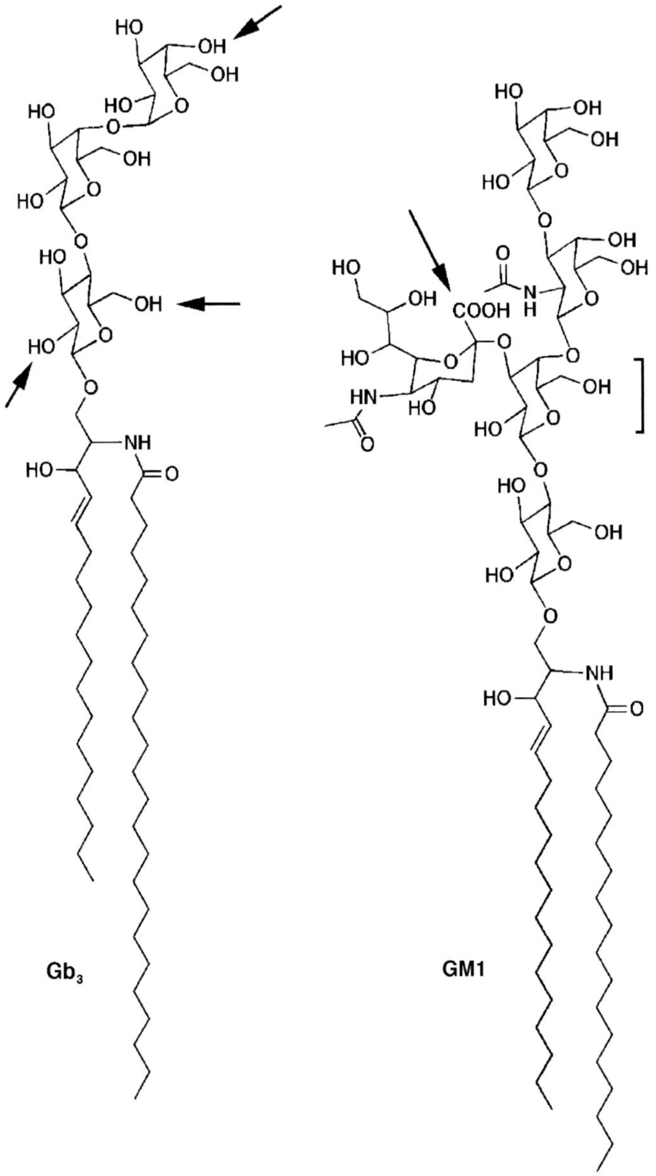

FIGURE 30.1 Structure of Gb_3 and GM_1 ganglioside. Structurally Gb_3, α-D-Gal-(1→4)-β-D-Gal-(1→4)-Glc ceramide, and GM_1, β-D-Gal-(1→4)[α-Neu5Ac-(2→3)-]-β-D-GalNac-(1→4)-β-D-Gal-(1→4)-Glc ceramide (where Gal, D-galactose; Glc, D-glucose; Neu5Ac, *N*-acetylneuraminic acid) are the respective receptors for VT and CT. Arrows in Gb_3 indicate hydroxyl groups which are important for VT2 but not VT1 binding (Chark *et al.*, 2004). The arrow in GM_1 indicates carboxyl group of sialic acid which can be replaced entirely by lactic acid. The bracket indicates β-Gal which can be replaced by cyclohexan-ediol (Bernardi *et al.*, 2001).

and is the Pk antigen in the P blood group system (Spitalnik and Spitalnik, 1995). Binding to the terminal disaccharide, α-Gal-(1→4)-Gal, moiety is the primary epitope since galabiosyl ceramide is also avidly bound by VT1 and VT2 (Chark *et al.*, 2004). However, this is a very rare GSL and *in vivo* pathology is entirely Gb_3-mediated. Although VT1 and VT2 bind to the same carbohydrate sequence, they do so in slightly different manners. The GSL receptor function within membranes is often dependent on the lipid moiety of the GSLs and the membrane environment (Lingwood, 1996). These aglycone components serve to constrain the conformation of the carbohydrate moiety and thereby modulate receptor function. This activity is mediated by the interface between the hydrophobic and hydrophilic domains of the membrane-bound GSL (Mylvaganam and Lingwood, 2003). This in turn will be strongly influenced by the lipid order of the bilayers, i.e. whether the GSLs are organized together with cholesterol in what are termed lipid "rafts", functionally monitored by resistance to detergent extraction, i.e. detergent-resistant membranes (DRMs) (London and Brown, 2000). These domains provide foci for cell signalling and host cell interaction with microbial pathogens and virulence factors (Heung *et al.*, 2006). These domains are considered more ordered and binding of CT to its GSL receptor GM_1 ganglioside is considered the gold standard marker of such domains within the plasma membranes of eukaryotic cells (Fujinaga *et al.*, 2003). While Gb_3 can also be found in such cholesterol-enriched domains, internalization of CT and VT via co-localized receptors on the cell surface occurs independently (Schapiro *et al.*, 1998). Verotoxin has an extremely high binding affinity for Gb_3 within the membrane of target cells which can be of the order of 10^{-13} or 10^{-14}M (Robinson *et al.*, 1995). Binding of VT1 to purified Gb_3 shows a binding affinity of 10^{-8}–10^{-9}M (Peter and Lingwood, 2000) indicating that membrane Gb_3 organization has a major affect on the affinity of this recognition system.

The Gb_3-binding site on the VT1 B-subunit pentameric array has still not completely been defined. In the original description of the B-subunit crystal structure, it was proposed that the binding site lay in the inter-subunit cleft (Stein *et al.*, 1992). Molecular modelling subsequently supported this location (site 1) (Nyholm *et al.*, 1995, 1996) together with a lower affinity site (site 2) in a shallow groove on the A-subunit distal surface of each B-subunit. Site 1 binding explained the finding that, although Gb_4, which is bound by the pig oedema disease toxin (DeGrandis *et al.*, 1989), is not bound by VT1, amino-Gb_4, generated by removal of the acetate of the terminal amino sugar, is strongly bound by all members of the VT family (Nyholm *et al.*, 1996) since this allowed the formation of a salt bridge with aspartic acid reside 17 (Asp17). The GSL-binding specificity of the VT family members has recently been further subdivided in that while VT1 prefers α-Gal-(1→4)-Gal, VT2 prefers α-GalNAc-(1→4)-β-Gal-(1→4)-Glc (where GalNAc is *N*-acetyl-D-galactosamine) and VT2c prefers α-GalNAc-(1→4)-GalNAc (Kale *et al.*, 2008). These latter sugar sequences are expressed on bacterial LPS rather than host cell GSLs, but this provides a clue as to the "evolutionary" origin of the Gb_3-binding specificity. The selective role for the glucose moiety in VT2-binding is consistent with deoxy-Gb_3 studies showing deoxyglucose Gb_3 analogues were bound by VT1 but not VT2 (Chark *et al.*, 2004) (see Figure 30.1). While the 3-OH of the terminal galactose is necessary for VT2 binding, this hydroxyl is substituted in amino-Gb_4 which binds all VTs, including VT2 (Nyholm *et al.*, 1996). Thus, substitution of important residues can be tolerated if compensatory ligand–receptor interactions are generated.

The lipid moiety of Gb_3 is important in the VT1-binding affinity since for the globotriaose oligosaccharide there is a 10^6 reduction in binding affinity (St Hilaire *et al.*, 1994). In addition, different lipid substitutions within the ceramide moiety can markedly affect Gb_3 binding

(Kiarash *et al.*, 1994; Binnington *et al.*, 2002). Nevertheless, the co-crystal structure of the VT1 B-subunit with the lipid-free Gb_3 oligosaccharide identified three potential Gb_3-binding sites (Ling *et al.*, 1998). The major site for bound Gb_3 oligosaccharide was a shallow groove within the face that opposes the plasma membrane when the toxin is bound (site 2), although the orientation of the carbohydrate was opposite from the model prediction. Site 1 was identified as the second most frequently occupied site by the Gb_3 oligosaccharide. A third site essentially comprising only tryptophan residue 34 (Trp 34) was identified. The structural solution from nuclear magnetic resonance spectroscopic analysis of the same B-subunit pentamer–Gb_3 oligosaccharide complex confirmed the importance of site 2 (Shimizu *et al.*, 1998) but had no evidence for site 3. Site-specific mutational studies showed that each site was required intact for full Gb_3-binding activity (Bast *et al.*, 1999). However, the site selectivity of the mutations was not absolute and interpretation is complex (Soltyk *et al.*, 2002). Detailed analysis of the binding of the wild-type and mutated toxins to Gb_3 oligosaccharide and Gb_3 glycosphingolipid have shown that mutations in site 1 could have a major effect on Gb_3 GSL-binding and holotoxin cytotoxicity, while site 2 Gb_3 oligosaccharide-binding was unaffected. From the crystal structure of VT2 holotoxin (Fraser *et al.*, 2004), site 2 is unavailable and site 3 is compromised by the position of the C-terminus of the A-subunit within the pentameric receptor-binding B-subunit central pore. In VT1, the A-subunit C-terminal extension through the B-subunit pentamer has been implicated in modulating the membrane Gb_3-binding of the holotoxin (Torgersen *et al.*, 2005).

Several Gb_3 sugar dendrimers have been constructed as inhibitors of VT1 and VT2 Gb_3-binding to protect cells (and animal models) against cytopathology (Nishikawa *et al.*, 2005; Kale *et al.*, 2008). Despite elegant chemistry to mimic the pentameric array presentation of the Gb_3-binding sites within the holotoxin (Kitov *et al.*, 2000) and the design of analogues simultaneously to bind in sites 1 and 2 (Kitov *et al.*, 2003), the random substitution of Gb_3 sugars within a glycopolymer has been shown to be more effective (Watanabe *et al.*, 2004), indicating that there is still a large unknown component to the receptor binding mechanism of these toxins.

Two eukaryotic proteins were shown to have sequence similarity to sequences within the VT1 B-subunit. These sequences were centred on amino acids within the inter-subunit cleft Gb_3-binding site (site 1) and it was proposed that these proteins could recognize Gb_3. These two proteins are the B-cell marker, CD19 (Maloney and Lingwood, 1994) and the α-2 interferon receptor, IFNR1 (Lingwood and Yiu, 1992). The Gb_3 deficiency was shown to compromise the signalling mediated by CD19 ligation and α_2 interferon. Both the growth inhibitory (Ghislain *et al.*, 1992, 1994) and antiviral activity of α-2 interferon (Khine and Lingwood, 2000) were compromised following Gb_3 deficiency. In Gb_3-deficient cells, the intracellular trafficking of CD19 following cell surface ligation was different and only in Gb_3-positive cells was nuclear envelope targeting observed (Khine *et al.*, 1998) when homotypic cell adhesion was induced (Maloney and Lingwood, 1994).

2.2. Cholera toxin receptor

The GSL receptor for CT was suggested by the inactivation of the toxin by gangliosides (Van Heyningen *et al.*, 1971). This ganglioside was later found to be resistant to sialidase (King and Van Heyningen, 1973), a property of the galactose-linked sialic acid of GM_1 (see Figure 30.1). Only the smaller choleragen subunit bound ganglioside GM_1 which was identified as the toxin receptor (Heyningen, 1974; Gascoyne and Van Heyningen, 1975). Ganglioside GM_1 added within the gastrointestinal tract can be used to block initial fluid loss but later fluid loss becomes resistant to lumenal receptor protection indicating the toxin is internalized within the epithelial cells (Stoll *et al.*, 1980). This process of transcytosis is also dependent on CT binding to GM_1 on the target cells, i.e. M cells (Blanco and DiRita, 2006). Ganglioside GM_1 is essentially present in all mammalian cells, so CT sensitivity is widespread.

Cholera toxin binding to the oligosaccharide of GM1 ganglioside is largely independent of the lipid moiety to which it is attached (Pacuszka and Fishman, 1990). Ganglioside GM_1 is a ceramide pentahexose-containing sialic acid and, as such, is one of the more soluble of the common GSLs. This relative solubility of the GSL and the independence of receptor function on the ceramide moiety would predict that the receptor binding of CT B-subunit should be more straightforward than that of VT1 B-subunit. While the co-crystal structure of the CT B-subunit and the intact GM_1 ganglioside has yet to be reported, studies with the GM_1 oligosaccharide have clearly identified the relevant receptor-binding site (Merritt *et al.*, 1994) and mutations within this site are supported by subsequent co-crystal structures (Merritt *et al.*, 1997). Each GM_1-binding site is found essentially within each B-subunit monomer in a shallow groove which opposes the target cell membrane. There are no interactions with the glucose moiety (which links to the ceramide in the intact GM_1 ganglioside) and only one interaction with the core galactose moiety (Merritt *et al.*, 1994), further supporting the relevance of the GM_1 oligosaccharide as a probe for the GM_1 ganglioside-binding site. Two domains for interaction of the sugar are seen, corresponding to the neutral and sialic acid-containing branches of the GM_1 oligosaccharide. The galactose-containing arm is most deeply buried in the binding pocket. The galactose stacks with trp88 to induce a blueshift on binding. Only one hydrogen bond is involved with the adjacent B-subunit, that of the glycine residue 33 (Gly33) whose hydrogen binds to both the galactose and sialic acid arms. This residue was subsequently mutated and the effect on binding found to be dependent on the substituting amino acid (Merritt *et al.*, 1997). Galactose analogues, initially *m*-nitrophenyl α-galactoside (Minke *et al.*, 1999) and later 3,5-substituted phenyl galactosides (Mitchell *et al.*, 2004) have been developed as receptor mimics to compete for GM_1 ganglioside-binding, but are significantly less active than GM_1 oligosaccharide. More effective mimics have been generated by reconstructing the core features of GM_1 sugar required for binding (Bernardi *et al.*, 2000, 2002; Arosio *et al.*, 2004). There are many interactions with the galactose arm of GM_1, but lactic acid can substitute for the sialic acid arm and a clever conformationally restricted cyclohexanediol can replace the core galactose (Bernardi *et al.*, 1999, 2001) (see Figure 30.1). Dimerizing this mimic on a cone-shaped scaffold gave a receptor mimetic with a higher CT B-subunit-binding affinity than GM_1 oligosaccharide (Arosio *et al.*, 2005).

3. INTRACELLULAR TRAFFICKING OF GSLs

Glycosphingolipids are transported to the cell surface from their site of synthesis in the Golgi apparatus via the vesicular secretory pathway. This is largely a bulk flow process and no preferential routing of newly synthesized GSLs has thus far been described. Differential retrograde GSL trafficking has only been recorded from the cell surface. This differential transport is essentially a function of the molecular organization of GSLs within lipid rafts, experimentally monitored by resistance to detergent extraction. Cell surface GSLs can be degraded *in situ* by a restricted number of cell surface glycohydrolases (Valaperta *et al.*, 2006). In this case, the ceramide generated can be re-utilized for GSL (mostly ganglioside) synthesis via shuttling to the trans-Golgi network. For the most part, cell surface GSLs are cycled via endosomes to the acidic lysosome for degradation.

Trafficking of GSLs from the cell surface has been monitored to a large extent, by observation of the bacterial toxins which bind them, i.e. CT and VT. These toxins first defined GSL retrograde trafficking pathway from the cell surface via endosomes to the Golgi apparatus and ER.

Defective GSL catabolism results in the GSL lysosomal storage diseases (LSDs), e.g. Tay-Sachs disease involving GM_2, Fabry disease involving Gb_3, Gaucher disease involving glucoceramide (GlcCer), metachromatic leukodystrophy disease (MLD) involving sulfatide, in which, as the name implies, the substrate for the missing hydrolase accumulates in the lysosome, generating multimembranous GSL vesicular inclusions which mediate the pathology of the various diseases. These diseases have been very instructive in relation to intracellular GSL trafficking. In the LSDs, GSL traffic abnormally. The retrograde transport pathway from the cell surface to the Golgi apparatus/ER is lost and traffic to the lysosome only is seen (Marks and Pagano, 2002). There is an intimate relationship between GSL and cholesterol traffic such that, in LSDs, cholesterol traffic is also aberrant (Puri *et al.*, 1999) and can be used diagnostically (Chen *et al.*, 1999). In Nieman–Pick disease, the primary lesion is in cholesterol metabolism (Ribeiro *et al.*, 2001) but GSL, as well as cholesterol, trafficking is abnormal (Pagano, 2003). Correction of the GSL trafficking simultaneously corrects the cholesterol accumulation (Choudhury *et al.*, 2002; Gondré-Lewis *et al.*, 2003; Lachmann *et al.*, 2004). This may relate to the intimate association of GSLs and cholesterol in membrane "rafts" (Sillence, 2007).

4. INTRACELLULAR TOXIN TRAFFIC

Both CT and VT hijack the retrograde transport pathways of GSLs to mediate A-subunit function (Figure 30.2). Pentavalent CT binding recruits GM_1 to lipid rafts (Panasiewicz *et al.*, 2003) and GM_1 containing unsaturated fatty acids is less prone to associate with DRMs. The CT binding, however, is largely unaffected by the ceramide fatty acid and the ceramide can be replaced by various lipid moieties (Pacuszka *et al.*, 1991) with little effect on binding. However, cell intoxication was consistent with the ability of the neo-GM_1 to partition into rafts (long aliphatic chain or coupling to cholesterol) (Pacuszka *et al.*, 1991).

For VT, the situation is more complex. As for CT, GSL receptor within DRMs is required for toxicity (Falguieres *et al.*, 2001; Hoey *et al.*, 2003). However, the Gb_3 fatty acid content does affect VT binding (Boyd *et al.*, 1994; Kiarash *et al.*, 1994; Binnington *et al.*, 2002) and intracellular trafficking (Arab and Lingwood, 1998).

In VT-sensitive cells, we have found that VT1 and VT2 bind to different raft formats of Gb_3 and *in vitro* DRMs constructs (Tam *et al.*, 2008). The Gb_3 DRM-binding of VT1 is greater than that of VT2 and, increased cholesterol can inhibit VT1 but not VT2 DRM-binding. In addition, sphingomyelin promotes VT2 DRM-binding but not that of VT1 (Tam *et al.*, 2008). Double-labelled cell surface binding studies showed that VT1 and VT2 bound to both distinct and overlapping punctate domains on the cell surface. Cell binding by VT1 and intracellular routing is more detergent-resistant than that of VT2; VT2 is found in transferrin containing intracellular endosomes, whereas VT1 is preferentially excluded from such vesicles following internalization (Tam *et al.*, 2008). Nevertheless, both VT1 and VT2 coalesce in retrograde transport to the Golgi apparatus but VT1 transits to the ER and nuclear envelope more rapidly than does the VT2 once it accumulates in the Golgi apparatus. In addition, we found a Gb_3-dependent, VT2-selective pathway which induces extensive acidic vacuolation in a subset of cultured vero cells (see Figure 30.2). Thus, differential Gb_3 DRM assemblies may

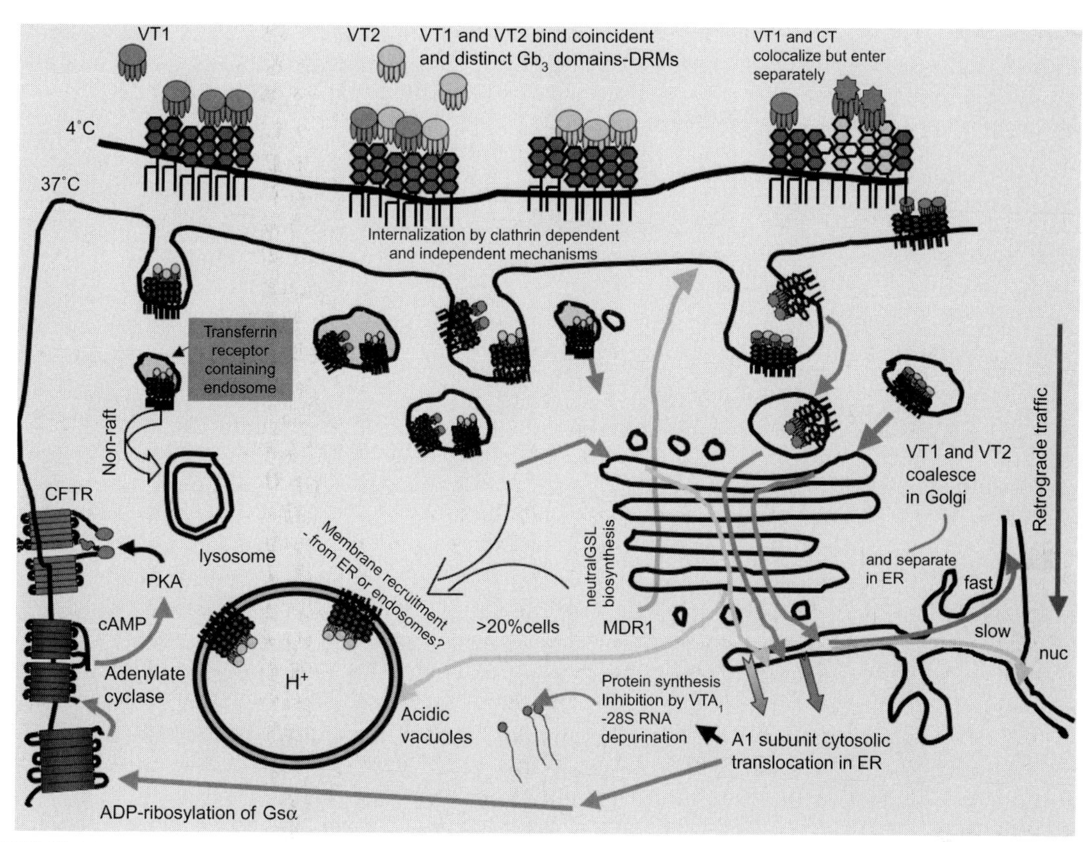

FIGURE 30.2 Retrograde trafficking scheme for the toxins VT and CT. The toxins VT1/2 bind punctate domains on the cell surface at 4°C. Pentameric CT binding recruits GM_1 to DRMs, but this has yet to be shown for VT. Both VT1/2 and CT can be internalized by clathrin-dependent and independent mechanisms (Torgersen *et al.*, 2001). The toxins VT1 and VT2 differentially retrograde traffic to the Golgi apparatus and are further transported to the ER and nuclear envelope (Arab and Lingwood, 1998; Tam *et al.*, 2008). The A1-subunits of CT and VT translocate to the cytosol from the ER but have distinct cytosolic targets (Gsα versus 28S RNA). (See colour plates section.)

mediate the distinct and shared surface-binding and intracellular retrograde trafficking of VT1 and VT2. This sets a precedent in intracellular GSL trafficking. In contrast to CT (Bastiaens *et al.*, 1996), only a small subfraction (≈2%) of the VT1 A- and B-subunits separate in the ER for A1-subunit translocation into the cytosol (Tam and Lingwood, 2007). Subunit separation of differentially labelled A/B-subunit VT1 holotoxin during retrograde transport from the Golgi apparatus to ER cannot be detected microscopically (Figure 30.3). As for CT (Nambiar *et al.*, 1993), brefeldin A- (BFA-) induced Golgi apparatus collapse protects against VT1 and VT2 (Donta *et al.*, 1995; Khine *et al.*, 2004; Tam *et al.*, 2008).

Cholera toxin binding to GM_1 within cell surface lipid rafts also mediates the retrograde transport through endosomes through the Golgi apparatus and ER. Cholera toxin A-subunit, unlike VT A-subunit, contains a KDEL sequence for ER retrieval and it was thought that the retrograde transport of CT was encoded within its protein sequence (Lencer *et al.*, 1995;

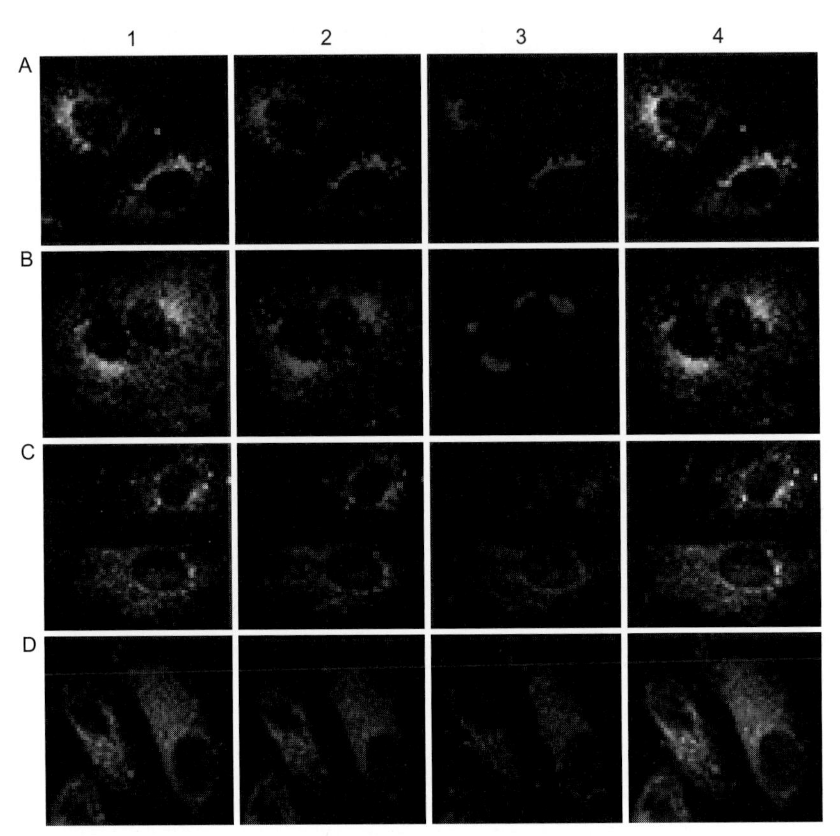

FIGURE 30.3 The VT1 A- and B-subunits remain coincident during retrograde transport. The VT1 holotoxin comprising fluorescein isothiocyanate-labelled A- and tetramethyl-rhodamine-labelled B-subunits (Tam and Lingwood, 2007) was added to Vero (A,C) or HeLa (B,D) cells for 1 h (A,B) or 6 h (C,D) at 37°C. Cells were fixed and stained with anti-Rab6 (a Golgi marker) (A3,B3) or anti-GRP78 (an ER marker) antibodies (C3,D3). Cells were imaged by confocal microscopy for each fluor separately and the images merged (n = 4) to define coincidence. The VT1 A and B subunits travel together to the ER via the Golgi apparatus. Reproduced, with permission, from Tam and Lingwood (2007) Membrane-cytosolic translocation of verotoxin A1-subunit in target cells. Microbiology 153, 2700–2710. (See colour plates section.)

Majoul *et al.*, 1998). However, this view has changed since the KDEL is evidently unnecessary and GM_1 ganglioside binding alone is sufficient for ER targeting by the CT B-subunit (Fujinaga *et al.*, 2003). For VT1, the retention of Gb_3 within a raft format is necessary for the retrograde trafficking pathway entirety through to the translocon within the ER (Smith *et al.*, 2006). Similarly, mutation of the GM_1-binding site of CT, such that a non-raft ganglioside (GD_{1a}) is bound (Wolf *et al.*, 1998), results in the loss of cell cytotoxicity, i.e. ADP-ribosylation. The increased association of VT2 infection with clinical disease (Miceli *et al.*, 1999) is somewhat counter intuitive since VT2 is less cytotoxic than VT1 *in vitro* (Tam *et al.*, 2008) and the receptor-binding affinity is somewhat less than that of VT1 (Nakajima *et al.*, 2001). However, this may provide an advantage for VT2 since it has been proposed that monocytes provide a shuttling mechanism to deliver toxin from the site of infection to the site of renal pathology (Te Loo *et al.*, 2001;

Geelen *et al.*, 2007) and the lower receptor-binding affinity of VT2 may allow this exchange between monocytes and renal glomerular endothelial Gb_3.

5. GLYCOSPHINGOLIPID RECEPTORS AND TOXIN-INDUCED PATHOLOGY

Globotriaosyl ceramide is the cellular receptor for VT. Incorporation of Gb_3 into Gb_3-negative cells induces VT sensitivity (Waddell *et al.*, 1990) and it is the only receptor mediating cytopathology *in vivo* since α-galactosyl transferase knockout mice are entirely resistant to both VT1 and VT2 (Okuda *et al.*, 2006). While Gb_3 is the VT receptor, all formats of Gb_3 do not mediate VT-induced cytopathology. In cell culture, cell membrane Gb_3 that is not within a lipid raft format (as defined by resistance to detergent extraction (London and Brown, 2000)) mediates internalization to lysosomes and bound toxin is degraded (Falguieres *et al.*, 2001). Within a detergent-resistant format in the cell plasma membrane, Gb_3 mediates the retrograde transport of the toxin intracellularly to endosomes, the trans-Golgi network, Golgi apparatus and ER for cytosolic translocation and cell death (Smith *et al.*, 2006). Thus, *in vivo* VT-induced pathology is not observed for every cell that expresses Gb_3. This is well illustrated by the fact that, in the bovine gastrointestinal tract, Gb_3 is expressed within the crypt cells, but this Gb_3 is in a non-DRM format and the bovine gastrointestinal tract is unaffected by VTEC colonization (Hoey *et al.*, 2003). Since, in humans, there is no mucosal cell Gb_3, the toxin gains systemic access by a transcellular route (Acheson *et al.*, 1996). In the bovine tract, binding to Gb_3 and intracellular targeting to lysosomes for degradation could be regarded as a protective mechanism. In humans, non-receptor-mediated transit of the mucosal barrier is followed by Gb_3-dependent binding to submucosal microvascular endothelial cells. The death of these cells results in haemorrhage and blood in the stools characteristic of HC. The HUS which may follow HC is characterized by renal glomerular microangiopathy, thrombocytopenia and haemolytic anaemia. Thrombocytopenia is likely mediated via Gb_3 expression on activated, but not quiescent platelets (Ghosh *et al.*, 2004) and may be mediated via monocyte transport of the toxin from the gastrointestinal tract to inflammatory sites within the kidney (Geelen *et al.*, 2007). These inflammatory sites are initially induced via VT targeting Gb_3 within the renal glomerular endothelial cells (Keepers *et al.*, 2007). Such microvasculature cells in culture have been shown to show high levels of Gb_3 and VT1 sensitivity (Obrig *et al.*, 1993), which can be increased via cytokine exposure (Warnier *et al.*, 2006).

Following gastrointestinal VTEC infection, HUS is essentially restricted to the young and the elderly (Lieberman *et al.*, 1966; Ochoa and Cleary, 2003). Our initial studies using fluorescein isothiocyanate- (FITC-) conjugated B-subunit showed that glomeruli within adult kidneys were essentially unreactive with VT1 and only glomeruli in the very young bound FITC-labelled VT1 B-subunit. In terms of renal Gb_3 content, Gb_3 increased as a function of renal age. Immunohistochemistry indicated that this is largely due to Gb_3 within the renal tubules which are not primarily affected in HUS (only at later more severe stages). This age-dependent expression of Gb_3 within the renal glomerulus was later questioned (Ergonul *et al.*, 2003) in studies in which binding was only detected at 4°C. In addition to the physiological relevance, low temperature affects the order of the membrane bilayer which can affect GSL receptor function (Tam *et al.*, 2008). These studies indicated that glomerular Gb_3 expression did not vary with age and that adult glomeruli expressed Gb_3. We, therefore, re-investigated our original observations and confirmed, when using FITC-conjugated VT1 subunit-B binding at room temperature, that only paediatric renal glomeruli were bound. However, using indirect immunostaining with the holotoxin at room temperature, we found a significant proportion of adult renal glomeruli samples were indeed positive, showing precisely coincident staining for VT1, VT2 and anti-Gb3 monoclonal antibody (Chark *et al.*, 2004). It has since been shown that the A-subunit contributes to the VT1 B-subunit membrane-associated Gb_3-binding (Torgersen *et al.*, 2005) and this may provide the basis for the differential glomerular binding of the FITC-conjugated VT1 B-subunit, as compared to the intact holotoxin. Many adult renal samples were, however, positive only using VT2 or anti-Gb_3 antibody as the ligand probe. Thus, a significant proportion of adult renal glomeruli are negative for VT1 holotoxin binding at room temperature (Chark *et al.*, 2004). Therefore, Gb_3 is present in the renal glomerulus of adults but can be presented in a format which is not bound by VT1. Extraction of such renal tissue cholesterol results in the induction of VT1 binding to such renal glomeruli, suggesting that the Gb_3/cholesterol complex in these glomeruli has too much cholesterol to allow VT1 binding (Chark *et al.*, 2004). In separate cell studies, we found that VT1 binding to artificial Gb_3/cholesterol DRMs could be inhibited by excess of cholesterol, whereas VT2 binding was unaffected (Tam *et al.*, 2008). Thus, the presentation of Gb_3 within this lipid raft format can affect the relative binding efficiency for VT1 and VT2 and may change as a function of age.

For CT, the role of the receptor in toxin-mediated pathology is more straightforward. Virtually all cells contain GM_1 ganglioside and are, therefore, CT-sensitive. The gastrointestinal tract is the primary site of *V. cholerae* colonization and therefore the gastrointestinal mucosal epithelial cells are on the front line of the host–parasite interaction since CT itself does not kill cells.

6. TOXIN-GSL-MEDIATED SIGNALLING

6.1. Verotoxin

The GSL, Gb_3, can occur in lipid rafts as defined by resistance to detergent extraction. Such DRMs can function as foci for signal transduction. Despite the fact that GSLs are not transmembrane species, carbohydrate-mediated signalling within such domains can occur and have been termed glyco-signalling domains (Hakomori *et al.*, 1998) or a glycosynapse (Hakomori, 2004). Verotoxin binding to plasma membrane Gb_3 within such domains has been shown to mediate signalling pathways in addition to those required for endocytosis of the toxin (Lauvrak *et al.*, 2005). Members of the Src family of the tyrosine kinases are associated with the cytosolic surface of such DRMs (Pawson, 2004). Binding of VT to "raft" Gb_3 has been shown to activate the Yes Src kinase (Katagiri *et al.*, 1999) in renal tubular epithelial cells to induce apoptosis (Taguchi *et al.*, 1998) and Lyn and Syk Src kinase (Mori *et al.*, 2000) to induce B-cell receptor-mediated apoptosis in human B-cells. Binding of VT1 has also been shown to activate Syk kinase which is involved in the mechanism of internalization (Lauvrak *et al.*, 2005). Both the activation of Yes kinase and Lyn kinase following the binding of VT1 was shown to release the src kinase from DRMs followed by the tyrosine phosphorylation of endogenous substrates (Katagiri *et al.*, 1999; Mori *et al.*, 2000). In the renal tubule cells, VT1 subunit B Gb_3-binding induced cytoskeletal re-arrangements involving src, PI-3 and Rho kinases (Takenouchi *et al.*, 2004). Induction of Verotoxin-induced apoptosis may be important in clinical pathology (Cherla *et al.*, 2003). The VT1 B-subunit can induce apoptosis in Burkitt lymphoma cells via caspase-independent pathways, that are yet to be clearly elucidated (Tetaud *et al.*, 2003), but which may vary from cell type to cell type. In HeLa cells, the VT1 apoptotic pathway has been shown to be unique in that the mitochondrial pathways are not involved [compare to lymphoma cells (Tetaud *et al.*, 2003)]. In HeLa cells, BFA protects and caspases 3, 6 and 8 are involved (Fujii *et al.*, 2003). Verotoxin has been shown to increase the renal production of chemokines, which are chemoattractants for neutrophils (Roche *et al.*, 2007). Verotoxins induce the ribotoxin stress response (Lee *et al.*, 2008) and have been shown to activate the stress kinases (Foster and Tesh, 2002). This has been shown to be mediated by the zipper sterile-alpha-motif kinase (ZAK) MAP3 kinase, both for VT and ricin (Jandhyala *et al.*, 2008).

Several cellular effects of VT are independent of gross protein synthesis inhibition and, therefore, are mediated by undefined signalling pathways. Endothelin-I is upregulated by VT1 and VT2 by mRNA stabilization (Bitzan *et al.*, 1998) and, although not detected in this bovine study, VT1 can reduce human endothelial and mesangial cell nitric oxide production without protein synthesis inhibition (Te Loo *et al.*, 2006), but stimulation of macrophage nitroc oxide synthesis is reported (Yuhas *et al.*, 1996). Proinflammatory cytokines are also induced by VT (Foster *et al.*, 2000; Harrison *et al.*, 2004) and these in turn, can upregulate cellular Gb_3 synthesis (Louise and Obrig, 1991) to provide a unique cyclic pathogenic amplification.

6.2. Cholera toxin B-subunit signalling

In addition to its A-subunit-mediated ADP ribosylation of Gsα, CT has a profound effect on the immune system and is an effective mucosal adjuvant and, together with LT, the *E. coli* derived homologue, is studied intensively for the development of mucosal vaccines (Cox *et al.*, 2006). The effect is mediated by the B-subunit binding to GM_1 ganglioside. In antigen presenting cells, CT B-subunit–GM_1 binding/internalization can induce the phosphorylation of several mitogen-activated protein (MAP) kinases including ERK1/2 and p38 (Schnitzler *et al.*, 2007). Phospholipase-C $\gamma 2$ is phosphorylated and B-cell differentiation markers CD86 and major histocompatibility complex (MHC) class II expression are upregulated. The phosphorylation of p90rsk, a substrate for ERK, was also induced, although activation of stress kinases was not seen. Transcription factors, NFκB and CERB, are upregulated and interleukin- (IL-) 6 was induced. Many of these pathways are shared with activators of the innate immune system (Liu *et al.*, 2007). Of note, MAPK1/2 inhibition prevents CT B-subunit induction of IL-6 secretion, indicating that CT B-subunit phosphorylation of MAPK1/2 drives the B-cells towards the T helper- (Th-) cell-2 response. Moreover, CT B-subunit can inhibit Th1 responses via inhibition of IL-12 and interferon-γ production (Coccia *et al.*, 2005). Ganglioside GM_1 within DRMs augments the effect of nerve growth factor (Mutoh *et al.*, 1995) but, surprisingly, CT B-subunit promotes NGF signalling (Mutoh *et al.*, 1993).

7. Gb_3 AND DRUG RESISTANCE

The correlation between increased Gb3 synthesis and resistance to cytosolic drugs in tumour cell lines was first observed with the ovarian carcinoma SKOV3 cell line (Farkas-Himsley *et al.*, 1995). These cells were sensitive to VT cytotoxicity but the spontaneous derived SKVLB drug-resistant tumour, which showed significant levels of the mutlidrug-resistant- (MDR-) 1 complex expression, was considerably more sensitive to VT cytotoxicity (by several logs). This association was more thoroughly investigated by looking at Madin-Darby canine kidney (MDCK) cells transfected with the MDR-1 gene (Lala *et al.*, 2000). These cells express high levels of MDR-1 protein and, unlike their parental cells, expressed high levels of globo series GSLs. These cells were 4–5 logs more sensitive to VT cytotoxicity than the parental MDCK cells. Moreover, the use of MDR-1 inhibitors such as cyclosporine A inhibited the drug resistance of these cells with a concomitant loss of globo series GSL biosynthesis. Glycosphingolipids are synthesized for the most part within the Golgi apparatus by membrane-bound glycosyltransferases located within the lumen. The exception is the glucosyl ceramide synthase which is a cytosolic enzyme and thus glucosyl ceramide is made on the cytosolic surface of the Golgi apparatus (Jeckel *et al.*, 1992; Lannert *et al.*, 1998). The question as to how the glucosyl ceramide then accesses the luminal Golgi compartment for the extension of the carbohydrate chain had not been addressed, but we postulated that the MDR-1 was capable of carrying out this function (De Rosa *et al.*, 2004). Immunoelectron microscopy confirmed that a significant proportion of MDR-1 was intracellular within Golgi-associated vesicles (De Rosa *et al.*, 2004). Thus, when cells become drug-resistant by upregulation of MDR-1, if they have the enzymes capable of synthesizing Gb_3, more Gb_3 will be made since the MDR-1 will flip more glucosyl ceramide into the Golgi apparatus to provide precursors for Gb_3. Thus, drug-resistant cells become hypersensitive to VT1 cytotoxicity. Also, Gb_3 was shown to be increased in ovarian carcinomas and, particularly in their drug-resistant variants, suggesting that this mechanism operates *in vivo* (Arab *et al.*, 1997). The MDR-1-mediated glycosyl ceramide flippase activity was shown to be important in glycolipid biosynthesis in many cell lines *in vitro* since use of the MDR-1 inhibitor of cyclosporine A was shown selectively to inhibit GSL biosynthesis (De Rosa *et al.*, 2004). In these studies, only neutral glycolipid biosynthesis was affected. Gangliosides remained relatively unaltered in the presence of cyclosporine suggesting that gangliosides could obtain their glucosyl ceramide and lactosyl ceramide from a source unaffected by MDR-1 inhibition. Metabolic labelling studies in microsomes showed that exogenous glucosyl ceramide could be converted into lactosyl ceramide and that this process was inhibited in the presence of cyclosporine. Thus, in microsomes, a liposomal format of glucosyl ceramide could be translocated into the microsomal membrane and flipped into the Golgi lumen to provide a substrate for lactosyl ceramide synthase via an MDR-1-dependent mechanism. Cell surface MDR-1 was found to be in part co-localized with cell surface Gb_3 (De Rosa *et al.*, 2004) and inhibition of GSL synthesis was shown to deplete MDR-1 from the cell surface. Functional cell surface MDR-1 requires expression within lipid rafts (Barakat *et al.*, 2005) and, potentially, inhibition of GSL synthesis could interfere with such a raft location to disturb the MDR-1 cell surface location. We showed that adamantylGb_3 functioned as a new physiologically based inhibitor of MDR-1 able to reverse drug resistance and deplete MDR-1 from the cell surface (De Rosa *et al.*, 2008). Interestingly, treatment of Gb_3-expressing MDR-1-containing cells with VT resulted in the temporary inhibition of MDR-1 pump activity, suggesting that the internalization of Gb_3, induced by VT, compromises MDR-1 function. Since Gb_3 was upregulated in several tumour cell types and cell lines derived from such tumours were VT1-sensitive *in vitro*, we suggested that VT1 is a potential novel anti-neoplastic (Farkas-Himsley *et al.*, 1995).

8. VEROTOXIN1 AS AN ANTI-NEOPLASTIC

Verotoxin1 was originally described as a novel vero cell cytotoxic activity derived from an *E. coli* strain O157:H7 (Konowalchuk *et al.*, 1977). This same *E. coli* strain was the original source of a bacteriocytin shown to have anti-cancer activity (Farkas-Himsley and Cheung, 1976). It was shown that this bacteriocytin activity was due to the presence of VT1 (Farkas-Himsley *et al.*, 1995). Anti-VT1 antibodies neutralize this effect and VT1 could mimic the effect of the bacteriocytin on tumour cell viability both in cell culture and in an animal model system. The efficacy of this approach against human tumours *in vivo* was first demonstrated using astrocytoma xenografts grown in nude mice (Arab *et al.*, 1999). Intra-tumoural injection of 2–4 μg/kg VT1 resulted in the rapid and complete elimination of subcutaneously growing glioblastoma xenografts without any apparent deleterious side-effect. Tumour cells and their neovasculature underwent massive apoptosis following VT1 injection. This approach was subsequently verified in cranial meningioma xenografts (Salhia *et al.*, 2002) and in subcutaneous ECV304 carcinoma xenografts (Heath-Engel and Lingwood, 2003). Verotoxin was shown to react with the neovasculature of many tumours irrespective of their Gb_3 status, suggesting that VT may have an anti-angiogenic as well as anti-neoplastic effect. These studies have been verified by other groups using colon carcinoma (Kovbasnjuk *et al.*, 2005) and renal carcinoma xenografts (Ishitoya *et al.*, 2004) and lymphomas (LaCasse *et al.*, 1996, 1999). Significantly, the studies on colon carcinomas showed that Gb_3 was necessary and sufficient for the induction of invasive metastases in an *in vitro* trans-epithelial migration model. VT1 was shown to target a spontaneous digestive system carcinoma in mice and the use of a modified B-subunit allowed the imaging of the tumour within the living animal (Janssen *et al.*, 2006).

9. VEROTOXIN OPENS A WINDOW FOR HUMAN IMMUNODEFICIENCY VIRUS (HIV) THERAPY

We developed the soluble Gb_3 anlogue, adamantylGb_3, specifically to bind VT to prevent the toxin from binding to Gb_3 expressed on target cell membranes (Mylvaganum and Lingwood, 2003). While this is highly effective *in vitro* against VT1 and VT2, efficacy *in vivo* was not observed. Indeed, the presence of this soluble analogue increased the toxicity of VT2 *in vivo* (Rutjes *et al.*, 2002). This is due, as we have subsequently determined, to the ability of adamantylGb_3 to partition into Gb_3-negative cell membranes and thus provide a functional receptor to mediate internalization on VT to kill cells. The reported selective role of Gb_3 in HIV-mediated host cell fusion (Puri *et al.*, 1998) prompted us, however, to investigate the interaction of adamantylGb_3 with HIV gp120. We showed that adamantylGb_3 was a highly effective receptor for this viral adhesin; about 1000-fold better than the native Gb_3 (Mahfoud *et al.*, 2002). Native Gb_3 interaction with gp120 was markedly enhanced by the presence of cholesterol, suggesting that the recognition by gp120 was of Gb_3 within a lipid raft format. Pretreatment of virus with adamantylGb_3 was found to protect sensitive cells against subsequent infection (Lund *et al.*, 2006). Inhibition was independent of the tropism of the virus, i.e. R5 (monocytropic) or T4 (T-cell tropic) viral infection was inhibited. Infection of cell lines and primary peripheral blood mononuclear cells was inhibited. In addition, virus strains which were resistant to various clinical antiviral agents (e.g. nucleosides, protease inhibitors, etc.) retained wild-type sensitivity to adamantylGb_3 inhibition. Using a model system in which the gp160 was expressed within 3T3 cells and CD4 and appropriate chemokine receptors expressed in HeLa cells, we showed that adamantylGb_3 was a fusion inhibitor preventing the mixing of different dyes contained within these two cell lines when mixed. Adamantyl derivates of other GSLs bound by gp120, galactosyl ceramide and sulfogalactosyl ceramide, were not fusion inhibitors. Lymphocytes from Fabry disease patients, in which Gb_3 accumulates, were resistant to HIV infection *in vitro* (Lund *et al.*, 2005). The protective effect of cellular Gb_3 levels against HIV infection (Ramkumar *et al.*, 2009) was further verified by examination of HIV susceptibility within the P blood group system and by transfecting CD4-expressing HeLa cells with the Gb_3 synthase gene (Lund *et al.*, 2009). Increase in the Gb_3 levels correlated with reduced susceptibility to HIV. Fusion of Gb_3-containing liposomes with Jurkat cells reduces susceptibility to infection, whereas Gb_4-containing liposomes had no effect. Pharmacological manipulation of cellular Gb_3 levels showed HIV susceptibility is an inverse function of Gb_3 expression. The efficacy of Gb_3 as a resistance factor against HIV is interesting to consider in the light of the role we have shown for Gb_3 to play in α2-interferon signalling, discussed above.

10. CONCLUSIONS

Although unrelated by sequence, CT and VT have adopted a similar structural format and selected GSL receptors that define cell targets. These receptors, though independent, share similar internalization and intracellular trafficking parameters. Apart from the diseases they induce, these toxins provide useful tools to investigate cellular GSL functions. The *E. coli* VT and *V. cholerae* CT-receptor GSLs, Gb_3 and GM_1 ganglioside, respectively, show a high degree of bioactivity as compared to other membrane GSLs, but the structural basis of these functions and the endogenous ligands for these GSLs have yet to be clearly defined. It is of interest to consider whether the high bioactivity of Gb_3 and GM_1 ganglioside is the reason these microbial pathogens elected to produce toxins of such specificity or whether the availability of these toxins provides probes of the function of these, but not other, GSLs. Key areas for further investigation are detailed in the Research Focus Box.

RESEARCH FOCUS BOX

- The importance of GSL lipid heterogeneity in membrane organization and toxin intracellular trafficking requires further investigation.
- The role of GSLs in vesicular sorting and membrane fusion/budding requires investigation.
- Linkage between cholesterol and GSL intracellular trafficking needs examination.

References

Acheson, D.W., Moore, R., De Breucker, S., et al., 1996. Translocation of Shiga toxin across polarized intestinal cells in tissue culture. Infect. Immun. 64, 3294–3300.

Arab, S., Lingwood, C., 1998. Intracellular targeting of the endoplasmic reticulum/nuclear envelope by retrograde transport may determine cell hypersensitivity to verotoxin: sodium butyrate or selection of drug resistance may induce nuclear toxin targeting via globotriaosyl ceramide fatty acid isoform traffic. J. Cell Physiol. 177, 646–660.

Arab, S., Russel, E., Chapman, W.B., Rosen, B., Lingwood, C.A., 1997. Expression of the verotoxin receptor glycolipid, globotriaosylceramide, in ovarian hyperplasias. Oncol. Res. 9, 553–563.

Arab, S., Rutka, J., Lingwood, C., 1999. Verotoxin induces apoptosis and the complete, rapid, long-term elimination of human astrocytoma xenografts in nude mice. Oncol. Res. 11, 33–39.

Arosio, D., Vrasidas, I., Valentini, P., Liskamp, R.M., Pieters, R.J., Bernardi, A., 2004. Synthesis and cholera toxin binding properties of multivalent GM_1 mimics. Org. Biomol. Chem. 2, 2113–2124.

Arosio, D., Fontanella, M., Baldini, L., et al., 2005. A synthetic divalent cholera toxin glycocalix[4]arene ligand having higher affinity than natural GM_1 oligosaccharide. J. Am. Chem. Soc. 127, 3660–3661.

Barakat, S., Gayet, L., Dayan, G., et al., 2005. Multidrug-resistant cancer cells contain two populations of P-glycoprotein with differently stimulated P-gp ATPase activities: evidence from atomic force microscopy and biochemical analysis. Biochem. J. 388, 563–571.

Bast, D.J., Banerjee, L., Clark, C., Read, R.J., Brunton, J.L., 1999. The identification of three biologically relevant globotriaosyl ceramide receptor binding sites on the verotoxin 1 B subunit. Mol. Microbiol. 32, 953–960.

Bastiaens, P.I., Majoul, I.V., Verveer, P.J., Söling, H.D., Jovin, T.M., 1996. Imaging the intracellular trafficking and state of the AB_5 quaternary structure of cholera toxin. EMBO J. 15, 4246–4253.

Bennish, M.L., Khan, W.A., Begum, M., et al., 2006. Low risk of hemolytic uremic syndrome after early effective antimicrobial therapy for *Shigella dysenteriae* type 1 infection in Bangladesh. Clin. Infect. Dis. 42, 356–362.

Bernardi, A., Checchia, A., Brocca, P., Sonnino, S., Zuccotto, F., 1999. Sugar mimics: an artificial receptor for cholera toxin. J. Am. Chem. Soc. 121, 2032–2036.

Bernardi, A., Carrettoni, L., Ciponte, A.G., Monti, D., Sonnino, S., 2000. Second generation mimics of ganglioside GM_1 as artificial receptors for cholera toxin: replacement of the sialic acid moiety. Bioorg. Med. Chem. Lett. 10, 2197–2200.

Bernardi, A., Arosio, D., Manzoni, L., Micheli, F., Pasquarello, A., Seneci, P., 2001. Stereoselective synthesis of conformationally constrained cyclohexanediols: a set of molecular scaffolds for the synthesis of glycomimetics. J. Org. Chem. 66, 6209–6216.

Bernardi, A., Arosio, D., Sonnino, S., 2002. Mimicking gangliosides by design: mimics of GM_1 headgroup. Neurochem. Res. 27, 539–545.

Bhattacharya, S.K., 2003. An evaluation of current cholera treatment. Expert Opin. Pharmacother. 4, 141–146.

Binnington, B., Lingwood, D., Nutikka, A., Lingwood, C.A., 2002. Effect of globotriaosyl ceramide fatty acid hydroxylation on the binding by verotoxin 1 and verotoxin 2. Neurochem. Res. 27, 807–813.

Bitzan, M.M., Wang, Y., Lin, J., Marsden, P.A., 1998. Verotoxin and ricin have novel effects on preproendothelin-1 expression but fail to modify nitric oxide synthase (ecNOS) expression and NO production in vascular endothelium. J. Clin. Invest. 101, 372–382.

Blanco, L.P., DiRita, V.J., 2006. Bacterial-associated cholera toxin and GM_1 binding are required for transcytosis of classical biotype *Vibrio cholerae* through an *in vitro* M cell model system. Cell Microbiol. 8, 982–998.

Boyd, B., Magnusson, G., Zhiuyan, Z., Lingwood, C.A., 1994. Lipid modulation of glycolipid receptor function: presentation of galactose α1-4 galactose disaccharide for verotoxin binding in natural and synthetic glycolipids. Eur. J. Biochem. 223, 873–878.

Chark, D., Nutikka, A., Trusevych, N., Kuzmina, J., Lingwood, C., 2004. Differential carbohydrate epitope recognition of globotriaosyl ceramide by verotoxins and monoclonal antibody: role in human renal glomerular binding. Eur. J. Biochem. 271, 1–13.

Chen, C., Patterson, M.C., Wheatley, C.L., O'Brien, J.F., Pagano, R.E., 1999. Broad screening test for sphingolipid-storage diseases. Lancet 354, 901–905.

Cherla, R.P., Lee, S.Y., Tesh, V.L., 2003. Shiga toxins and apoptosis. FEMS Microbiol. Lett. 228, 159–166.

Choudhury, A., Dominguez, M., Puri, V., et al., 2002. Rab proteins mediate Golgi transport of caveola-internalized glycosphingolipids and correct lipid trafficking in Niemann-Pick C cells. J. Clin. Invest. 109, 1541–1550.

Coccia, E.M., Remoli, M.E., Di Giacinto, C., et al., 2005. Cholera toxin subunit B inhibits IL-12 and IFN-γ production and signaling in experimental colitis and Crohn's disease. Gut 54, 1558–1564.

Cox, E., Verdonck, F., Vanrompay, D., Goddeeris, B., 2006. Adjuvants modulating mucosal immune responses or directing systemic responses towards the mucosa. Vet. Res. 37, 511–539.

Critchley, D.R., Streuli, C.H., Kellie, S., Ansell, S., Patel, B., 1982. Characterization of the cholera toxin receptor on Balb/c 3T3 cells as a ganglioside similar to, or identical with, ganglioside GM_1. No evidence for galactoproteins with receptor activity. Biochem. J. 204, 209–219.

DeGrandis, S., Law, H., Brunton, J., Gyles, C., Lingwood, C.A., 1989. Globotetraosyl ceramide is recognized by the pig edema disease toxin. J. Biol. Chem. 264, 12520–12525.

de Haan, L., Verweij, W., Agsteribbe, E., Wilschut, J., 1998. The role of ADP-ribosylation and GM_1-binding activity in the mucosal immunogenicity and adjuvanticity of the *Escherichia coli* heat-labile enterotoxin and *Vibrio cholerae* cholera toxin. Immunol. Cell Biol. 76, 270–279.

De Rosa, M.F., Sillence, D., Ackerley, C., Lingwood, C., 2004. Role of multiple drug resistance protein 1 in neutral but not acidic glycosphingolipid biosynthesis. J. Biol. Chem. 279, 7867–7876.

De Rosa, M.F., Ackerley, C., Wang, B., Ito, S., Clarke, D.M., Lingwood, C., 2008. Inhibition of multidrug resistance 1 (MDR1) by adamantylGb3, a globotriaosylceramide analog. J. Biol. Chem. 283, 4501–4511.

Donta, S., Tomicic, T.K., Donohue-Rolfe, A., 1995. Inhibition of Shiga-like toxins by brefeldin A. J. Infect. Dis. 171, 721–724.

Ergonul, Z., Clayton, F., Fogo, A.B., Kohan, D.E., 2003. Shigatoxin-1 binding and receptor expression in human kidneys do not change with age. Pediatr. Nephrol. 18, 246–253.

Falguieres, T., Mallard, F., Baron, C., et al., 2001. Targeting of Shiga toxin b-subunit to retrograde transport route in association with detergent-resistant membranes. Mol. Biol. Cell 12, 2453–2468.

Farkas-Himsley, H., Cheung, R., 1976. Bacterial proteinaceous products (bacteriocins) as cytotoxic agents of neoplasia. Cancer Res. 36, 3561–3567.

Farkas-Himsley, H., Hill, R., Rosen, B., Arab, S., Lingwood, C.A., 1995. Bacterial colicin active against tumour cells *in vitro* and *in vivo* is verotoxin 1. Proc. Natl. Acad. Sci. USA 92, 6996–7000.

Foster, G.H., Armstrong, C.S., Sakiri, R., Tesh, V.L., 2000. Shiga toxin-induced tumor necrosis factor alpha expression: requirement for toxin enzymatic activity and monocyte protein kinase C and protein tyrosine kinases. Infect. Immun. 68, 5183–5189.

Foster, G.H., Tesh, V.L., 2002. Shiga toxin 1-induced activation of c-Jun NH(2)-terminal kinase and p38 in the human monocytic cell line THP-1: possible involvement in the production of TNF-alpha. J. Leukoc. Biol. 71, 107–114.

Fraser, M.E., Fujinaga, M., Cherney, M.M., et al., 2004. Structure of Shiga toxin type 2 (Stx2) from *Escherichia coli* O157:H7. J. Biol. Chem. 279, 27511–27517.

Friedrich, A.W., Bielaszewska, M., Zhang, W.L., et al., 2002. *Escherichia coli* harboring Shiga toxin 2 gene variants: frequency and association with clinical symptoms. J. Infect. Dis. 185, 74–84.

Fujii, J., Matsui, T., Heatherly, D.P., et al., 2003. Rapid apoptosis induced by Shiga toxin in HeLa cells. Infect. Immun. 71, 2724–2735.

Fujinaga, Y., Wolf, A.A., Rodighiero, C., et al., 2003. Gangliosides that associate with lipid rafts mediate transport of cholera and related toxins from the plasma membrane to ER. Mol. Biol. Cell 14, 4783–4793.

Fukuta, S., Magnani, J.L., Twiddy, E.M., Holmes, R.K., Ginsburg, V., 1988. Comparison of the carbohydrate-binding specificities of cholera toxin and *Escherichia coli* heat-labile enterotoxins LTh-I, LT-IIa, and LT-IIb. Infect. Immun. 56, 1748–1753.

Gabriel, S.E., Brigman, K.N., Koller, B.H., Boucher, R.C., Stutts, M.J., 1994. Cystic fibrosis heterozygote resistance to cholera toxin in the cystic fibrosis mouse model. Science 266, 107–109.

Garred, Ø., van Deurs, B., Sandvig, K., 1995. Furin-induced cleavage and activation of Shiga toxin. J. Biol. Chem. 270, 10817–10821.

Gascoyne, N., Van Heyningen, W.E., 1975. Binding of cholera toxin by various tissues. Infect. Immun. 12, 466–469.

Geelen, J.M., van der Velden, T.J., van den Heuvel, L.P., Monnens, L.A., 2007. Interactions of Shiga-like toxin with human peripheral blood monocytes. Pediatr. Nephrol. 22, 1181–1187.

Ghislain, J., Lingwood, C., Maloney, M., Penn, L., Fish, E., 1992. Association between the ifn alpha receptor and the membrane glycolipid globotriaosyl ceramide. J. Interferon Res. 12, 114.

Ghislain, J., Lingwood, C.A., Fish, E.N., 1994. Evidence for glycosphingolipid modification of the type 1 IFN receptor. J. Immunol. 153, 3655–3663.

Ghosh, S.A., Polanowska-Grabowska, R.K., Fujii, J., Obrig, T., Gear, A.R., 2004. Shiga toxin binds to activated platelets. J. Thromb. Haemost. 2, 499–506.

Gondré-Lewis, M.C., McGlynn, R., Walkley, S.U., 2003. Cholesterol accumulation in NPC1-deficient neurons is ganglioside dependent. Curr. Biol. 13, 1324–1329.

Griffin, P.M., Tauxe, R.V., 1991. The epidemiology of infections caused by *Escherichia coli* O157:H7, other enterohemorrhagic *E. coli*, and the associated hemolytic uremic syndrome. Epidem. Rev. 13, 60–98.

Griffitts, J.S., Haslam, S.M., Yang, T., et al., 2005. Glycolipids as receptors for *Bacillus thuringiensis* crystal toxin. Science 307, 922–925.

Hakomori, S., 2004. Carbohydrate-to-carbohydrate interaction, through glycosynapse, as a basis of cell recognition and membrane organization. Glycoconj. J. 21, 125–137.

Hakomori, S., Handa, K., Iwabuchi, K., Yamamura, S., Prinetti, A., 1998. New insights in glycosphingolipid function: "glycosignaling domain", a cell surface assembly of glycosphingolipids with signal transducer molecules, involved in cell adhesion coupled with signaling. Glycobiology 8, xi–xix.

Harrison, L.M., van Haaften, W.C., Tesh, V.L., 2004. Regulation of proinflammatory cytokine expression by Shiga toxin 1 and/or lipopolysaccharides in the human monocytic cell line THP-1. Infect. Immun. 72, 2618–2627.

Heath-Engel, H.M., Lingwood, C.A., 2003. Verotoxin sensitivity of ECV304 cells *in vitro* and *in vivo* in a xenograft tumour model: VT1 as a tumour neovascular marker. Angiogenesis 6, 129–141.

Heung, L.J., Luberto, C., Del Poeta, M., 2006. Role of sphingolipids in microbial pathogenesis. Infect. Immun. 74, 28–39.

Heyningen, S.V., 1974. Cholera toxin: interaction of subunits with ganglioside GM_1. Science 183, 656–657.

Hoey, D.E., Sharp, L., Currie, C., Lingwood, C.A., Gally, D.L., Smith, D.G., 2003. Binding of verotoxin 1 to primary intestinal epithelial cells expressing Gb3 results in trafficking of toxin to lysosomal compartments. Cell Microbiol. 5, 85–97.

Ishitoya, S., Kurazono, H., Nishiyama, H., et al., 2004. Verotoxin induces rapid elimination of human renal tumor xenografts in SCID mice. J. Urol. 171, 1309–1313.

Ismaili, A., McWhirter, E., Handelsman, M.Y., Brunton, J.L., Sherman, P.M., 1998. Divergent signal transduction responses to infection with attaching and effacing *Escherichia coli*. Infect. Immun. 66, 1688–1696.

Jacewicz, M., Clausen, H., Nudelman, E., Donohue-Rolfe, A., Keusch, G.T., 1986. Pathogenesis of *Shigella diarrhea*. XI. Isolation of a shigella toxin-binding glycolipid from rabbit jejunum and HeLa cells and its identification as globotriaosylceramide. J. Exp. Med. 163, 1391–1404.

Jandhyala, D.M., Ahluwalia, A., Obrig, T., Thorpe, C.M., 2008. ZAK: a MAP3 Kinase that transduces Shiga toxin and ricin induced proinflammatory cytokine expression. Cell Microbiol. 10, 1468–1477.

Janssen, K.P., Vignjevic, D., Boisgard, R., et al., 2006. *In vivo* tumor targeting using a novel intestinal pathogen-based delivery approach. Cancer Res. 66, 7230–7236.

Jeckel, D., Karrenbauer, A., Burger, K.N., van Meer, G., Wieland, F., 1992. Glucosylceramide is synthesized at the cytosolic surface of various Golgi subfractions. J. Cell Biol. 117, 259–267.

Jelacic, S., Wobbe, C.L., Boster, D.R., et al., 2002. ABO and P1 blood group antigen expression and stx genotype and outcome of childhood *Escherichia coli* O157:H7 infections. J. Infect. Dis. 185, 214–219.

Johnson, W.M., Pollard, D.R., Lior, H., Tyler, S.D., Rozee, K.R., 1990. Differentiation of genes coding for *Escherichia coli* verotoxin 2 and the verotoxin associated with porcine edema disease (VTE) by the polymerase chain reaction. J. Clin. Microbiol. 28, 2351–2353.

Kale, R.R., McGannon, C.M., Fuller-Schaefer, C., et al., 2008. Differentiation between structurally homologous Shiga 1 and Shiga 2 toxins by using synthetic glycoconjugates. Angew. Chem. Int. Ed. Engl. 47, 1265–1268.

Karch, H., Bielaszewska, M., Bitzan, M., Schmidt, H., 1999. Epidemiology and diagnosis of Shiga toxin-producing *Escherichia coli* infections. Diag. Microbiol. Infect. Dis. 34, 229–243.

Karmali, M.A., Petric, M., Lim, C., Fleming, P.C., Arbus, G.S., Lior, H., 1985. The association between hemolytic uremic syndrome and infection by verotoxin-producing *Escherichia coli*. J. Infect. Dis. 151, 775–782.

Katagiri, Y., Mori, T., Nakajima, H., et al., 1999. Activation of Src family kinase induced by Shiga toxin binding to globotriaosyl ceramide (Gb3/CD77) in low density, detergent-insoluble microdomains. J. Biol. Chem. 274, 35278–35282.

Keepers, T.R., Gross, L.K., Obrig, T.G., 2007. Monocyte chemoattractant protein 1, macrophage inflammatory protein 1 alpha, and RANTES recruit macrophages to the kidney in a mouse model of hemolytic-uremic syndrome. Infect. Immun. 75, 1229–1236.

Khine, A.A., Lingwood, C.A., 2000. Functional significance of globotriaosylceramide in α2 interferon/type I interferon receptor mediated anti viral activity. J. Cell. Physiol. 182, 97–108.

Khine, A.A., Firtel, M., Lingwood, C.A., 1998. CD77-dependent retrograde transport of CD19 to the nuclear membrane: functional relationship between CD77 and CD19 during germinal center B-cell apoptosis. J. Cell Physiol. 176, 281–292.

Khine, A.A., Tam, P., Nutikka, A., Lingwood, C.A., 2004. Brefeldin A and filipin distinguish two globotriaosyl ceramide/verotoxin-1 intracellular trafficking pathways involved in Vero cell cytotoxicity. Glycobiology 14, 701–712.

Kiarash, A., Boyd, B., Lingwood, C.A., 1994. Glycosphingolipid receptor function is modified by fatty acid content: verotoxin 1 and verotoxin 2c preferentially recognize different globotriaosyl ceramide fatty acid homologues. J. Biol. Chem. 269, 11138–11146.

King, C.A., Van Heyningen, W.E., 1973. Deactivation of cholera toxin by a sialidase-resistant monosialosylganglioside. J. Infect. Dis. 127, 639–647.

Kitov, P.I., Sadowska, J.M., Mulvey, G., et al., 2000. Shiga-like toxins are neutralized by tailored multivalent carbohydrate ligands. Nature 403, 669–672.

Kitov, P.I., Shimizu, H., Homans, S.W., Bundle, D.R., 2003. Optimization of tether length in nonglycosidically linked bivalent ligands that target sites 2 and 1 of a Shiga-like toxin. J. Am. Chem. Soc. 125, 3284–3294.

Konowalchuk, J., Speirs, J.I., Stavric, S., 1977. Vero response to a cytotoxin of *Escherichia coli*. Infect. Immun. 18, 775–779.

Kovbasnjuk, O., Mourtazina, R., Baibakov, B., et al., 2005. The glycosphingolipid globotriaosylceramide in the metastatic transformation of colon cancer. Proc. Natl. Acad. Sci. USA, 102, 19087–19092.

Kozaki, S., Kamata, Y., Watarai, S., Nishiki, T., Mochida, S., 1998. Ganglioside GT_{1b} as a complementary receptor component for *Clostridium botulinum* neurotoxins. Microb. Pathog. 25, 91–99.

Krivan, H.C., Clark, G.F., Smith, D.F., Wilkins, T.D., 1986. Cell surface binding site for *Clostridium difficile* enterotoxin: evidence for a glycoconjugate containing the sequence Gal alpha 1-3Gal beta 1-4GlcNAc. Infect. Immun. 53, 573–581.

LaCasse, E.C., Saleh, M.T., Patterson, B., Minden, M.D., Gariépy, J., 1996. Shiga-like toxin purges human lymphoma from bone marrow of severe combined immunodeficient mice. Blood 88, 1561–1567.

LaCasse, E.C., Bray, M.R., Patterson, B., et al., 1999. Shiga-like toxin I receptor on human breast cancer, lymphoma, and myeloma and absence from CD34+ hematopoietic stem cells: implications for *ex vivo* tumor purging and autologous stem cell transplantation. Blood 94, 1–12.

Lachmann, R.H., te Vruchte, D., Lloyd-Evans, E., et al., 2004. Treatment with miglustat reverses the lipid-trafficking defect in Niemann-Pick disease type C. Neurobiol. Dis. 16, 654–658.

Lala, P., Ito, S., Lingwood, C.A., 2000. Retroviral transfection of Madin-Darby canine kidney cells with human *MDR1* results in a major increase in globotriaosylceramide and 10^5- to 10^6-fold increased cell sensitivity to verocytotoxin. Role of P-glycoprotein in glycolipid synthesis. J. Biol. Chem. 275, 6246–6251.

Lannert, H., Gorgas, K., Meissner, I., Wieland, F.T., Jeckel, D., 1998. Functional organization of the Golgi apparatus in glycosphingolipid biosynthesis. Lactosylceramide and subsequent glycosphingolipids are formed in the lumen of the late Golgi. J. Biol. Chem. 273, 2939–2946.

Lauvrak, S.U., Wälchli, S., Iversen, T.G., et al., 2005. Shiga toxin regulates its entry in a Syk-dependent manner. Mol. Biol. Cell 17, 1096–1109.

Lee, S.Y., Lee, M.S., Cherla, R.P., Tesh, V.L., 2008. Shiga toxin 1 induces apoptosis through the endoplasmic reticulum stress response in human monocytic cells. Cell Microbiol. 10, 770–780.

Lencer, W.I., Constable, C., Moe, S., et al., 1995. Targeting of cholera toxin and *E. coli* heat labile toxin in polarized epithelia: role of C-terminal KDEL. J. Cell Biol. 131, 951–962.

Lieberman, E., Heuser, E., Donnell, G.N., Landing, B.H., Hammond, G.D., 1966. Hemolytic-uremic syndrome. Clinical and pathological considerations. N. Engl. J. Med. 275, 227–236.

Lindberg, A.A., Brown, J.E., Strömberg, N., Westling-Ryd, M., Schultz, J.E., Karlsson, K.A., 1987. Identification of the carbohydrate receptor for Shiga toxin produced by *Shigella dysenteriae* type 1. J. Biol. Chem. 262, 1779–1785.

Ling, H., Boodhoo, A., Hazes, B., et al., 1998. Structure of the Shiga toxin B-pentamer complexed with an analogue of its receptor Gb_3. Biochemistry 37, 1777–1788.

Lingwood, C.A., 1993. Verotoxins and their glycolipid receptors. In: Bell, R., Hannun, Y.A., Merrill, A. (Eds.), Sphingolipids. Part A: Functions and Breakdown Products. Advances in Lipid Research, Vol. 25. Academic Press, San Diego, pp. 189–212.

Lingwood, C.A., 1996. Aglycone modulation of glycolipid receptor function. Glycoconj. J. 13, 495–503.

Lingwood, C.A., Yiu, S.C.K., 1992. Glycolipid modification of α2-interferon binding: sequence similarity between α2-interferon receptor and the verotoxin (Shiga-like toxin) B-subunit. Biochem. J. 283, 25–26.

Lingwood, C.A., Law, H., Richardson, S., et al., 1987. Glycolipid binding of purified and recombinant *Escherichia coli*-produced verotoxin *in vitro*. J. Biol. Chem. 262, 8834–8839.

Liu, Y., Shepherd, E.G., Nelin, L.D., 2007. MAPK phosphatases – regulating the immune response. Nat. Rev. Immunol. 7, 202–212.

London, E., Brown, D.A., 2000. Insolubility of lipids in triton X-100: physical origin and relationship to sphingolipid/cholesterol membrane domains (rafts). Biochim. Biophys. Acta 1508, 182–195.

Louch, H.A., Buczko, E.S., Woody, M.A., Venable, R.M., Vann, W.F., 2002. Identification of a binding site for ganglioside on the receptor binding domain of tetanus toxin. Biochemistry 41, 13644–13652.

Louie, M., de Azavedo, J.C., Handelsman, M.Y., et al., 1993. Expression and characterization of the *eae* gene product of *Escherichia coli* serotype O157:H7. Infect. Immun. 61, 4085–4092.

Louise, C.B., Obrig, T.G., 1991. Shiga toxin-associated hemolytic-uremic syndrome: combined cytotoxic effects of Shiga toxin, interleukin-1β, and tumor necrosis factor alpha on human vascular endothelial cells *in vitro*. Infect. Immun. 59, 4173–4179.

Louise, C.B., Obrig, T.G., 1992. Shiga toxin-associated hemolytic uremic syndrome: combined cytotoxic effects of Shiga toxin and lipopolysaccharide (endotoxin) on human vascular endothelial cells *in vitro*. Infect. Immun. 60, 1536–1543.

Lund, N., Branch, D.R., Sakac, D., et al., 2005. Lack of susceptibility of cells from patients with Fabry disease to infection with R5 human immunodeficiency virus. AIDS 19, 1543–1546.

Lund, N., Branch, D.R., Mylvaganam, M., et al., 2006. A novel soluble mimic of the glycolipid globotriaosylceramide inhibits HIV infection. AIDS 20, 1–11.

Lund, N., Ramkumar, S., Olsson, M.L., et al., 2009. The human P_k histo-blood group antigen provides protection against HIV infection. Blood 113, 4980–4991.

Mahfoud, R., Mylvaganam, M., Lingwood, C.A., Fantini, J., 2002. A novel soluble analog of the HIV-1 fusion cofactor, globotriaosylceramide (Gb_3), eliminates the cholesterol requirement for high affinity gp120/Gb_3 interaction. J. Lipid Res. 43, 1670–1679.

Majoul, I., Sohn, K., Wieland, F.T., et al., 1998. KEDL receptor (Erd2p)-mediated retrograde transport of the cholera toxin A subunit from Golgi involves COPI, p23, and the COOH terminus of Erd2p. J. Cell Biol. 143, 601–612.

Maloney, M.D., Lingwood, C.A., 1994. CD19 has a potential CD77 (globotriaosyl ceramide)-binding site with sequence similarity to verotoxin B-subunits: implications of molecular mimicry for B cell adhesion and enterohemorrhagic *Escherichia coli* pathogenesis. J. Exp. Med. 180, 191–201.

Mangeney, M., Richard, Y., Coulaud, D., Tursz, T., Wiels, J., 1991. CD77: an antigen of germinal center B cells entering apoptosis. Eur. J. Immunol. 21, 1131–1140.

Marks, D., Pagano, R., 2002. Endocytosis and sorting of glycosphingolipids in sphingolipid storage disease. Trends Cell Biol. 12, 605–613.

Melton-Celsa, A.R., Kokai-Kun, J.F., O'Brien, A.D., 2002. Activation of Shiga toxin type 2d (Stx2d) by elastase involves cleavage of the C-terminal two amino acids of the A2 peptide in the context of the appropriate B pentamer. Mol. Microbiol. 43, 207–215.

Merritt, E.A., Sarfaty, S., van den Akker, F., L'Hoir, C., Martial, J.A., Hol, W.G., 1994. Crystal structure of cholera toxin B-pentamer bound to receptor GM_1 pentasaccharide. Protein Sci. 3, 166–175.

Merritt, E.A., Sarfaty, S., Jobling, M.G., et al., 1997. Structural studies of receptor binding by cholera toxin mutants. Protein Sci. 6, 1516–1528.

Miceli, S., Jure, M.A., de Saab, O.A., et al., 1999. A clinical and bacteriological study of children suffering from haemolytic uraemic syndrome in Tucuman, Argentina. Jpn J. Infect. Dis. 52, 27–33.

Middlebrook, J.L., Dorland, R.B., 1984. Bacterial toxins: cellular mechanisms of action. Microbiol. Rev. 48, 199–221.

Minke, W.E., Roach, C., Hol, W.G., Verlinde, C.L., 1999. Structure-based exploration of the ganglioside GM_1 binding sites of *Escherichia coli* heat-labile enterotoxin and cholera toxin for the discovery of receptor antagonists. Biochemistry 38, 5684–5692.

Mitchell, D.D., Pickens, J.C., Korotkov, K., Fan, E., Hol, W.G., 2004. 3,5-Substituted phenyl galactosides as leads in designing effective cholera toxin antagonists; synthesis and crystallographic studies. Bioorg. Med. Chem. 12, 907–920.

Mori, T., Kiyokawa, N., Katagiri, Y.U., et al., 2000. Globotriaosyl ceramide (CD77/Gb3) in the glycolipid-enriched membrane domain participates in B-cell receptor-mediated apoptosis by regulating Lyn kinase activity in human B-cells. Exp. Hemat. 28, 1260–1268.

Mucha, J., Domlatil, J., Lochnit, G., et al., 2004. The *Drosophila melanogaster* homologue of the human histo-blood group Pk gene encodes a glycolipid-modifying α1,4-*N*-acetylgalactosaminyltransferase. Biochem. J. 382, 67–74.

Mutoh, T., Tokuda, A., Guroff, G., Fujiki, N., 1993. The effect of the B subunit of cholera toxin on the action of nerve growth factor on PC12 cells. J. Neurochem. 60, 1540–1547.

Mutoh, T., Tokuda, A., Miyadai, T., Hamaguchi, M., Fujiki, N., 1995. Ganglioside GM_1 binds to the Trk protein and regulates receptor function. Proc. Natl. Acad. Sci. USA 92, 5087–5091.

Mylvaganam, M., Lingwood, C.A., 2003. A preamble to aglycone reconstruction for membrane-presented glycolipids. In: Wong, C.-H. (Ed.), Carbohydrate-Based Drug Discovery, Vol. 2. Wiley-VCH Verlag, Weinheim, pp. 761–780.

Nakajima, H., Kiyokawa, N., Katagiri, Y.U., et al., 2001. Kinetic analysis of binding between Shiga toxin and receptor glycolipid Gb3Cer by surface plasmon resonance. J. Biol. Chem. 276, 42915–42922.

Nambiar, M.P., Oda, T., Chen, C., Kuwazuru, Y., Wu, H.C., 1993. Involvement of the Golgi region in the intracellular trafficking of cholera toxin. J. Cell Physiol. 154, 222–228.

Nathoo, K.J., Porteous, J.E., Siziya, S., Wellington, M., Mason, E., 1998. Predictors of mortality in children hospitalized with dysentery in Harare, Zimbabwe. Cent. Afr. J. Med. 44, 272–276.

Nishikawa, K., Matsuoka, K., Watanabe, M., et al., 2005. Identification of the optimal structure required for a Shiga toxin neutralizer with oriented carbohydrates to function in the circulation. J. Infect. Dis. 191, 2097–2105.

Nyholm, P.G., Brunton, J.L., Lingwood, C.A., 1995. Modelling of the interaction of verotoxin-1 (VT1) with its glycolipid receptor, globotriaosylceramide (Gb_3). Int. J. Biol. Macromol. 17, 199–205.

Nyholm, P.G., Magnusson, G., Zheng, Z., Norel, R., Binnington-Boyd, B., Lingwood, C.A., 1996. Two distinct binding sites for globotriaosyl ceramide on verotoxins: molecular modelling and confirmation by analogue studies and a new glycolipid receptor for all verotoxins. Chem. Biol. 3, 263–275.

O'Brien, A.D., LaVeck, G.D., Thompson, M.R., Formal, S.B., 1982. Production of *Shigella dysenteriae* type 1-like cytotoxin by *Escherichia coli*. J. Infect. Dis. 146, 763–769.

Obrig, T., Louise, C.B., Lingwood, C.A., Boyd, B., Barley-Maloney, L., Daniel, T.O., 1993. Endothelial heterogeneity in Shiga toxin receptors and responses. J. Biol. Chem. 268, 15484–15488.

Ochoa, T.J., Cleary, T.G., 2003. Epidemiology and spectrum of disease of *Escherichia coli* O157. Curr. Opin. Infect. Dis. 16, 259–263.

Okuda, T., Tokuda, N., Numata, S., et al., 2006. Targeted disruption of Gb3/CD77 synthase gene resulted in the complete deletion of globo-series glycosphingolipids and loss of sensitivity to verotoxins. J. Biol. Chem. 281, 10230–10235.

Pacuszka, T., Fishman, P.H., 1990. Generation of cell surface neogangliroproteins. J. Biol. Chem. 265, 7673–7678.

Pacuszka, T., Bradley, R.M., Fishman, P.H., 1991. Neoglycolipid analogues of ganglioside GM_1 as functional receptors of cholera toxin. Biochemistry 30, 2563–2570.

Pagano, R.E., 2003. Endocytic trafficking of glycosphingolipids in sphingolipid storage diseases. Philos. Trans. Roy. Soc. London B. Biol. Sci. 358, 885–891.

Panasiewicz, M., Domek, H., Hoser, G., Kawalec, M., Pacuszka, T., 2003. Structure of the ceramide moiety of GM_1 ganglioside determines its occurrence in different detergent-resistant membrane domains in HL-60 cells. Biochemistry 42, 6608–6619.

Pawson, T., 2004. Specificity in signal transduction: from phosphotyrosine-SH2 domain interactions to complex cellular systems. Cell 116, 191–203.

Peter, M., Lingwood, C., 2000. Apparent cooperativity in multivalent verotoxin globotriaosyl ceramide binding: kinetic and saturation binding experiments with radiolabelled verotoxin [^{125}I]-VT1. Biochim. Biophys. Acta 1501, 116–124.

Philpott, D.J., Ackerley, C.A., Kiliaan, A.J., Karmali, M.A., Perdue, M.H., Sherman, P.M., 1997. Translocation of verotoxin-1 across T84 monolayers: mechanism of bacterial toxin penetration of epithelium. Am. J. Physiol. 273, G1349–G1358.

Philpott, D., McKay, D.M., Mak, W., Perdue, M.H., Sherman, P.M., 1998. Signal transduction pathways involved in enterohemorrhagic *Escherichia coli*-induced alterations in T84 epithelial permeability. Infect. Immun. 66, 1680–1687.

Puri, A., Hug, P., Jernigan, K., et al., 1998. The neutral glycosphingolipid globotriaosylceramide promotes fusion mediated by a CD4-dependent CXCR4-utilizing HIV type 1 envelope glycoprotein. Proc. Natl. Acad. Sci. USA 95, 14435–14440.

Puri, V., Watanabe, R., Dominguez, M., et al., 1999. Cholesterol modulates membrane traffic along the endocytic pathway in sphingolipid-storage diseases. Nat. Cell Biol. 1, 386–388.

Ramkumar, S., Sakac, D., Binnington, B., Branch, D.R., Lingwood, C.A., 2009. Induction of HIV-1 resistance: cell susceptibility to infection is an inverse function of globotriaosyl ceramide levels. Glycobiology 19, 76–82.

Ribeiro, I., Marcão, A., Amaral, O., Sá Miranda, M.C., Vanier, M.T., Millat, G., 2001. Niemann-Pick type C disease: NPC1 mutations associated with severe and mild cellular cholesterol trafficking alterations. Hum. Genet. 109, 24–32.

Riley, L.W., Remis, R.S., Helgerson, S.D., et al., 1983. Hemorrhagic colitis associated with a rare *Escherichia coli* serotype. N. Engl. J. Med. 308, 681–685.

Robinson, L.A., Hurley, R.M., Lingwood, C., Matsell, D. G., 1995. *Escherichia coli* verotoxin binding to human paediatric glomerular mesangial cells. Pediatr. Nephrol. 9, 700–704.

Roche, J.K., Keepers, T.R., Gross, L.K., Seaner, R.M., Obrig, T.G., 2007. CXCL1/KC and CXCL2/MIP-2 are critical effectors and potential targets for therapy of *Escherichia coli* O157:H7-associated renal inflammation. Am. J. Pathol. 170, 526–537.

Rutjes, N., Binnington, B.A., Smith, C.R., Maloney, M.D., Lingwood, C.A., 2002. Differential tissue targeting and pathogenesis of verotoxins 1 and 2 in the mouse animal model. Kidney Intl. 62, 832–845.

Salhia, B., Rutka, J.T., Lingwood, C., Nutikka, A., Van Furth, W.R., 2002. The treatment of malignant meningioma with verotoxin. Neoplasia 4, 304–311.

Saxena, S.K., O'Brien, A.D., Ackerman, E.J., 1989. Shiga toxin, Shiga-like toxin II variant, and ricin are all single-site RNA *N*-glycosidases of 28 S RNA when microinjected into *Xenopus* oocytes. J. Biol. Chem. 264, 596–601.

Schapiro, F., Lingwood, C., Furuya, W., Grinstein, S., 1998. pH-independent targeting of glycolipids to the Golgi complex. Am. J. Physiol. 274, 319–332.

Schnitzler, A.C., Burke, J.M., Wetzler, L.M., 2007. Induction of cell signaling events by the cholera toxin B subunit in antigen-presenting cells. Infect. Immun. 75, 3150–3159.

Shears, P., 1996. Shigella infections. Ann. Trop. Med. Parasitol. 90, 105–114.

Shimizu, H., Field, R.A., Homans, S.W., Donohue-Rolfe, A., 1998. Solution structure of the complex between the B-subunit homopentamer of verotoxin VT-1 from *Escherichia coli* and the trisaccharide moiety of globotriaosylceramide. Biochemistry 37, 11078–11082.

Siegler, R.L., Pysher, T.J., Lou, R., Tesh, V.L., Taylor, F.B., 2001. Response to Shiga toxin-1, with and without lipopolysaccharide, in a primate model of hemolytic uremic syndrome. Am. J. Nephrol. 21, 420–425.

Sillence, D.J., 2007. New insights into glycosphingolipid functions – storage, lipid rafts, and translocators. Int. Rev. Cytol. 262, 151–189.

Smith, D.C., Sillence, D.J., Falguières, T., et al., 2006. The association of Shiga-like toxin with detergent-resistant membranes is modulated by glucosylceramide and is an essential requirement in the endoplasmic reticulum for a cytotoxic effect. Mol. Biol. Cell 17, 1375–1387.

Soltyk, A.M., MacKenzie, C.R., Wolski, V.M., et al., 2002. A mutational analysis of the globotriaosylceramide binding sites of verotoxin VT1. J. Biol. Chem. 277, 5351–5359.

Spitalnik, P., Spitalnik, S., 1995. The P blood group system: biochemical, serological and clinical aspects. Transfus. Med. Rev. 9, 110–122.

St Hilaire, P.M., Boyd, M.K., Toone, E.J., 1994. Interaction of the Shiga-like toxin type 1 B-subunit with its carbohydrate receptor. Biochemistry 33, 14452–14463.

Stein, P.E., Boodhoo, A., Tyrrell, G.J., Brunton, J.L., Read, R.J., 1992. Crystal structure of the cell-binding B oligomer of verotoxin-1 from *E. coli*. Nature 355, 748–750.

Stoll, B.J., Holmgren, J., Bardhan, P.K., et al., 1980. Binding of intraluminal toxin in cholera: trial of GM_1 ganglioside charcoal. Lancet 2, 888–891.

Strockbine, N.A., Jackson, M.P., Sung, L.M., Holmes, R.K., O'Brien, A.D., 1988. Cloning and sequencing of the genes for Shiga toxin from *Shigella dysenteriae* type 1. J. Bacteriol. 170, 1116–1122.

Taguchi, T., Uchida, H., Kiyokawa, N., et al., 1998. Verotoxins induce apoptosis in human renal tubular epithelium derived cells. Kidney Intl. 53, 1681–1688.

Takenouchi, H., Kiyokawa, N., Taguchi, T., et al., 2004. Shiga toxin binding to globotriaosyl ceramide induces intracellular signals that mediate cytoskeleton remodeling in human renal carcinoma-derived cells. J. Cell Sci. 117, 3911–3922.

Tam, P., Lingwood, C., 2007. Membrane-cytosolic translocation of verotoxin A1-subunit in target cells. Microbiology 153, 2700–2710.

Tam, P., Mahfoud, R., Nutikka, A., et al., 2008. Differential intracellular trafficking and binding of verotoxin 1 and verotoxin 2 to globotriaosylceramide-containing lipid assemblies. J. Cell Physiol. 216, 750–763.

Tarr, P.I., Neill, M.A., Clausen, C.R., Watkins, S.L., Christie, D.L., Hickman, R.O., 1990. *Escherichia coli* O157:H7 and the hemolytic uremic syndrome: importance of early cultures in establishing the etiology. J. Infect. Dis. 162, 553–556.

Te Loo, D.M., van Hinsbergh, V.W., van den Heuvel, L.P., Monnens, L.A., 2001. Detection of verocytotoxin bound to circulating polymorphonuclear leukocytes of patients with hemolytic uremic syndrome. J. Am. Soc. Nephrol. 12, 800–806.

Te Loo, D.M., Monnens, L., van der Velden, T., Karmali, M., van den Heuvel, L., van Hinsbergh, V., 2006. Shiga toxin-1 affects nitric oxide production by human glomerular endothelial and mesangial cells. Pediatr. Nephrol. 21, 1815–1823.

Tetaud, C., Falguiàres, T., Carlier, K., et al., 2003. Two distinct Gb3/CD77 signaling pathways leading to apoptosis are triggered by anti-Gb3/CD77 mAb and verotoxin-1. J. Biol. Chem. 278, 45200–45208.

Thiagarajah, J.R., Verkman, A.S., 2003. CFTR pharmacology and its role in intestinal fluid secretion. Curr. Opin. Pharmacol. 3, 594–599.

Torgersen, M., Skretting, G., van Deurs, B., Sandvig, K., 2001. Internalization of cholera toxin by different endocytic mechanisms. J. Cell Sci. 114, 3734–3737.

Torgersen, M.L., Lauvrak, S.U., Sandvig, K., 2005. The A-subunit of surface-bound Shiga toxin stimulates clathrin-dependent uptake of the toxin. FEBS J. 272, 4103–4113.

Uhlich, G.A., Sinclair, J.R., Warren, N.G., Chmielecki, W.A., Fratamico, P., 2008. Characterization of Shiga toxin-producing *Escherichia coli* isolates associated with two multi-state food-borne outbreaks that occurred in 2006. Appl. Environ. Microbiol. 74, 1268–1272.

Valaperta, R., Chigorno, V., Basso, L., et al., 2006. Plasma membrane production of ceramide from ganglioside GM3 in human fibroblasts. FASEB J. 20, 1227–1229.

Van Heyningen, W.E., Carpenter, C.C., Pierce, N.F., Greenough, W.B., 1971. Deactivation of cholera toxin by ganglioside. J. Infect. Dis. 124, 415–418.

Waddell, T., Cohen, A., Lingwood, C.A., 1990. Induction of verotoxin sensitivity in receptor deficient cell lines using the receptor glycolipid globotriosyl ceramide. Proc. Natl. Acad. Sci. USA 87, 7898–7901.

Wagner, P.L., Livny, J., Neely, M.N., Acheson, D.W., Friedman, D.I., Waldor, M.K., 2002. Bacteriophage control of Shiga toxin 1 production and release by *Escherichia coli*. Mol. Microbiol. 44, 957–970.

Warnier, M., Römer, W., Geelen, J., et al., 2006. Trafficking of Shiga toxin/Shiga-like toxin-1 in human glomerular microvascular endothelial cells and human mesangial cells. Kidney Int. 70, 2085–2091.

Watanabe, M., Matsuoka, K., Kita, E., et al., 2004. Oral therapeutic agents with highly clustered globotriose for treatment of Shiga toxigenic *Escherichia coli* infections. J. Infect. Dis. 189, 360–368.

Wolf, A.A., Jobling, M.G., Wimer-Mackin, S., et al., 1998. Ganglioside structure dictates signal transduction by cholera toxin and association with caveolae-like membrane domains in polarized epithelia. J. Cell Biol. 141, 917–927.

Wong, C.S., Jelacic, S., Habeeb, R.L., Watkins, S.L., Tarr, P.I., 2000. The risk of the hemolytic-uremic syndrome after antibiotic treatment of *Escherichia coli* O157:H7 infections. N. Engl. J. Med. 342, 1930–1936.

Yuhas, Y., Kaminsky, E., Mor, M., Ashkenazi, S., 1996. Induction of nitric oxide production in mouse macrophages by Shiga toxin. J. Med. Micobiol. 45, 97–102.

Zhang, X., McDaniel, A.D., Wolf, L.E., Keusch, G.T., Waldor, M.K., Acheson, D.W., 2000. Quinolone antibiotics induce Shiga toxin-encoding bacteriophages, toxin production, and death in mice. J. Infect. Dis. 181, 664–670.

CHAPTER

31

Toll-like receptor recognition of lipoglycans, glycolipids and lipopeptides

Holger Heine and Sabine Riekenberg

SUMMARY

The innate immune system as the first and, for many species, sole line of defence against pathogenic intruders must be capable of recognizing the vast diversity of harmful microbes to protect the host. Toll-like receptors (TLRs) are germ-line encoded and evolutionarily conserved receptors that are pivotal to this recognition. First found and characterized in the fruit fly, *Drosophila melanogaster*, they now have been studied in many different species across the animal kingdom. Over the last 10 years tremendous progress has been made to decipher their different ligands, the molecular mechanisms of ligand recognition, as well as the signal transduction pathways that are activated upon engagement of TLRs. This concise review describes our current knowledge about TLRs in innate immune defence, with a specific focus on TLR-2 and TLR-4, the two receptors essential for the recognition of lipopeptides and glycolipids.

Keywords: Toll-like receptor (TLR); Innate immunity; Adaptor molecule; Toll-like receptor 2 (TLR-2); Toll-like receptor 4 (TLR-4)

1. INTRODUCTION

The Toll protein has been identified as a crucial member of the innate immune response of the fly to combat and contain microbial infections. Toll is a type I transmembrane receptor (Hashimoto *et al.*, 1988) initially identified as an essential receptor controlling the dorsoventral polarization during the embryonic development of the fly larvae (Anderson and Nusslein-Volhard, 1984; Anderson *et al.*, 1985). In other laboratories, researchers had studied the induction of antimicrobial peptides in *Drosophila*, such as drosomycin, diptericin and cecropin (Lemaitre *et al.*, 1997), in response to infectious challenge. In the year 1996, Lemaitre *et al.* (1996) reported that the *spätzle/Toll/cactus* gene cassette was required for an antifungal response of *Drosophila*, demonstrating that the Toll receptor not only had a pivotal role during embryogenesis but also for the immune response of the fruit fly.

Today, the innate immune responses of many more non-vertebrates are being thoroughly investigated. Surprisingly, extreme variations in the expression of Toll-like receptors (TLRs) and their relevance in the innate immune responses of invertebrate species have been found: from presumably no expression in certain gastropods (Walker, 2006), over single TLR genes with apparent immune function as in the oyster or in *Caenorhabditis elegans* (Pradel *et al.*, 2007) to no immune function as in the horseshoe crab (Inamori *et al.*, 2004); or three TLR genes in *Ciona intestinalis* (Azumi *et al.*, 2003); multiple, immune-relevant TLR genes in mosquitos (Shin *et al.*, 2006); to finally, the sea urchin, where 222 TLRs have been identified (Rast *et al.*, 2006). However, in almost all cases, functional studies must be performed to analyse further the actual function of these TLRs in the immune response of their host.

This review gives a brief overview of TLRs and their role in innate immunity with particular focus on the recognition of lipoglycans, glycolipids and lipopeptides by TLR-2 and TLR-4 and their activation of different signalling pathways.

2. TOLL-LIKE RECEPTORS IN VERTEBRATES

In vertebrates, six major TLR families are found (Roach *et al.*, 2005): the TLR-1 group, including TLR-1, TLR-2, TLR-6 and TLR-10, for the detection of lipopeptides; TLR-4 for lipopolysaccharide; TLR-3 for double-stranded RNA; TLR-5 for flagellin; the TLR-7 family, including TLR-7, TLR-8 and TLR-9, for the detection of nucleic acids; and, finally, the TLR-11 family, including TLR-11–13 and TLR-21–23. In this family, TLR-11 has been shown to detect uropathogenic bacteria, as well as the protozoan ligand profilin (Zhang *et al.*, 2004; Yarovinsky *et al.*, 2005) (Figure 31.1). In general, all vertebrates express at least one member of these major families to cover the wide variety of different pathogen-associated molecular patterns (PAMPs) (Janeway, 1989). However, the TLR-11 family is only represented by a pseudogene in humans (Roach *et al.*, 2005).

2.1. Expression of TLRs in different cell types and regulation of their expression

The expression of TLRs varies among different cell types. The cells that have the most contact with potential pathogens and their PAMPs, such as dendritic cells (DCs) and other antigen-presenting cells, express the widest selection of TLRs. In particular, DCs show a very distinctive pattern of TLR expression and responses. Whereas conventional DCs express basically all TLRs and respond with a strong nuclear factor-kappaB (NF-κB)-driven innate immune response, plasmacytoid DCs preferentially express TLR-7 and TLR-9, which are essential for mounting the interferon- (IFN-) α response against viruses (Pichlmair and Sousa, 2007). In addition, the expression of TLRs is not restricted to immune cells, since they can also be found on cells such as fibroblasts and epithelial cells (Akira *et al.*, 2006).

2.2. Recognition of microbial components

How do TLRs actually recognize their ligands? All TLRs consist of repeating motifs of leucine rich repeats (LRR) in their ectodomain, leading to a horseshoe-shaped form with parallel β-strands forming the concave face and the convex surface composed of loops. Recently, a number of crystal structures of TLRs, with or without their ligands, have been obtained and provide us with more insight into the molecular basis of TLR-ligand interaction. Interestingly, TLR-ligand interaction can apparently occur on

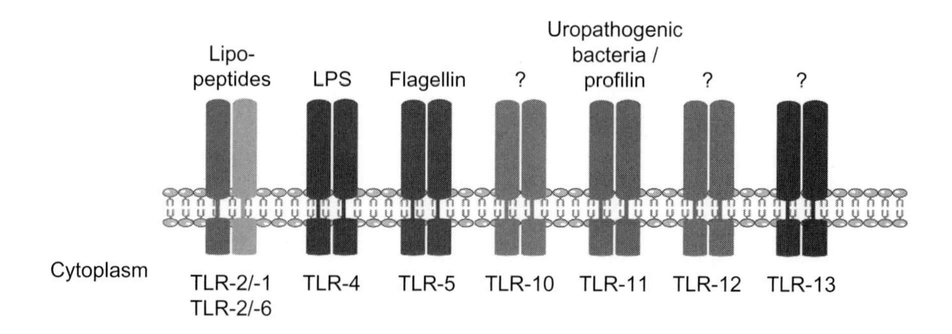

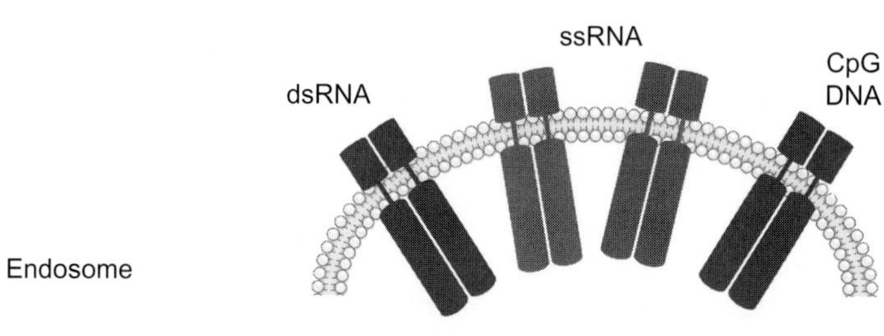

FIGURE 31.1 Toll-like receptors (TLRs) are transmembrane proteins that are located in the plasma membrane or in endosomal compartments. Theses receptors consist of a leucine-rich repeat (LRR) extracellular domain, a transmembrane region and an intracellular TIR domain. Variations in TLR ligands initiate specific immunological responses. For recognition of lipopeptides, TLR-2 forms heterodimers with TLR-1 or TLR-6, whereas LPSs are recognized by TLR-4. TLR-5 activation occurs following interaction with bacterial flagellin. Uropathogenic bacteria, as well as a protein from *Toxoplasma gondii* called profiling, activate cells through TLR-11, ligands for TLR-10, TLR-12 and TLR-13 have not been identified so far. The receptors TLR-1, TLR-2, TLR-4, TLR-5, TLR-6, TLR-10, TLR-11, TLR-12 and TLR-13 are all located on the cell surface; other TLR receptors like TLR-3, TLR-7, TLR-8, TLR-9 are located in the endosomal membrane. Activation of TLR-3 occurs by viral dsDNA, TLR-7 and TLR-8 by viral ssRNA and TLR-9 by bacterial CpG DNA, respectively. Abbreviations: CpG DNA, unmethylated cytosine-phosphatidyl-guanosine motif-containing deoxyribonucleic acid; dsRNA, double-stranded ribonucleic acid; LPS, lipopolysaccharide; ssRNA, single-stranded RNA; TLR, Toll-like receptor.

either side of the ectodomain (Jin *et al.*, 2007; Kim *et al.*, 2007; Liu *et al.*, 2008). Further details are given in the TLR-2 and TLR-4 sections below.

3. TLR-2

3.1. Overview of ligands and ligand binding

The receptor TLR-2 possesses a unique characteristic as it is apparently the only TLR that signals as a heterodimer. This receptor, in association with TLR-1 or TLR-6, is essential for the recognition of different and most diverse microbial components from a variety of microorganisms (Ozinsky *et al.*, 2000). Whether TLR-2 can also signal as a homodimer still needs to be investigated. *In vitro* experiments indicate that there is only a functional signalling together with TLR-1 or TLR-6 (Ozinsky *et al.*, 2000). The receptor is involved in detection of lipoproteins from pathogens such as Gram-positive and Gram-negative bacteria (Takeuchi *et al.*, 1999a; Adachi *et al.*, 2001), mycoplasma (e.g. *Mycoplasma fermentans*) and spirochetes (e.g. *Borrelia burgdorferi*, *Treponema pallidum*) (Lien *et al.*, 1999) and is also involved in recognition of glycolipids from the spirochetal *Treponema maltophilum* (Opitz *et al.*, 2001). Further ligands for TLR-2 are lipoarabinomannan (LAM) from mycobacteria (Means *et al.*, 1999) (see Chapter 9), glycosylphosphatidylinositol- (GPI-) anchored lipids from *Trypanosoma cruzi* (Campos *et al.*, 2001) (see Chapter 12), a phenol-soluble modulin from *Staphylococcus epidermis* (Hajjar *et al.*, 2001), lyso-phosphatidylserine from *Schistosoma mansoni* (van der Kleij *et al.*, 2002), porins from *Neisseria meningitides* (Massari *et al.*, 2002) and *Haemophilus influenzae* (Bieback *et al.*, 2002), lipophosphoglycan from *Leishmania major* (de Veer *et al.*, 2003; Tuon *et al.*, 2008) (see Chapter 12) and zymosan from fungi (Underhill *et al.*, 1999). Additional ligands are different viruses, such as herpes simplex (Kurt-Jones *et al.*, 2004), varicella-zoster (Wang *et al.*, 2005) and cytomegalovirus (Compton *et al.*, 2003), as well as whole bacteria, e.g. *Chlamydia pneumoniae* (Netea *et al.*, 2002). Also discussed as putative ligands (Takeda *et al.*, 2003) are peptidoglycan, a long linear sugar chain of alternating *N*-acetylglucosamine and *N*-acetylmuramic acid cross-linked by peptide chains (see Chapter 2) and lipoteichoic acid, an amphiphilic, negatively charged glycolipid (see Chapter 5). However, in many cases, these dependencies are due to lipopeptide contaminations (Hashimoto *et al.*, 2006). Travassos *et al.* (2004) showed that highly purified peptidoglycan is not able to signal via TLR-2 and also synthetic peptidoglycan cannot activate this receptor (Inamura *et al.*, 2006). Finally, some lipopolysaccharide (LPS) variants from Gram-negative bacteria like *Leptospira interrogans* (Werts *et al.*, 2001) and *Porphyromonas gingivalis* (Hirschfeld *et al.*, 2001) have also been reported to signal via TLR-2. These LPSs differ structurally in the number of acyl chains in the lipid A component in contrast to the typical LPS structures of enterobacteria like *Escherichia coli* and *Salmonella* spp. which are recognized by TLR-4. But more analyses are required to exclude lipoprotein contaminations in many LPS preparations and other putative TLR ligands.

Bacterial lipopeptides are the best characterized TLR-2 ligands and, by the use of synthetic lipopeptides free of possible contaminations, great progress has been made in understanding the ligand receptor dependency and interaction. Lipopeptides are composed of di-*O*-acylated-*S*-(2,3-dihydroxypropyl)-cysteinyl residues N-terminally coupled to different polypeptides through which they are anchored to the bacterial cell wall. Additionally, in Gram-negative bacteria like *E. coli*, the *S*-(2,3-dihydroxypropyl)-cysteine can be acylated with a third amide-bound fatty acid (Morr *et al.*, 2002; Buwitt-Beckmann *et al.*, 2006). As described before, TLRs are type I transmembrane receptors with an N-terminal, extracellular recognition domain of 19 to 25 leucine-rich repeats (LRRs), a single transmembrane domain and an intracellular Toll/interleukin- (IL-) 1 receptor homology (TIR) signalling domain. Proteins which contain an LRR domain form horseshoe-like structures with parallel β-strands forming the concave face and the convex site is built by α-helices and loops. Recently, the crystal structure of a TLR-2/-1 heterodimer with the triacylated synthetic lipopeptide $Pam_3C\text{-}SK_4$ has been obtained and demonstrated that the binding of the ligand occurs at the convex site of the extracellular LRR domain (Jin *et al.*, 2007). Accordingly, the ligand induces the heterodimerization of the ectodomains of TLR-2 and TLR-1, whereas antagonists have the possibility to obviate this dimerization. Two of the three lipid chains of $Pam_3C\text{-}SK_4$ interact with a pocket formed at the convex site of TLR-2 and the amide-bound fatty acid is inserted into a hydrophobic channel in TLR-1. By further hydrophobic, hydrogen-bonding and ionic interactions, TLR-1 and TLR-2 stabilize the heterodimer. The complex forms a so-called "m-" shaped heterodimer after ligand binding, whereas binding of diacylated $Pam_2C\text{-}SK_4$ did not (Jin *et al.*, 2007). Nevertheless, ligand binding leads to a dimerization of the extracellular domains followed by a connection of the two TIR domains and subsequently initiates signalling.

3.2. Differences between TLR-2/-1 and TLR-2/-6

To date, the TLR family consists of 13 mammalian members (Hoebe *et al.*, 2004), which can be divided into six subfamilies based on their primary sequence and genomic structure (Roach *et al.*, 2005). The TLR-1 subfamily contains the receptors TLR-1, TLR-2, TLR-6 and TLR-10. The identity of the amino acid sequence between TLR-1 and TLR-6 is about 69.3% and the TIR domains of both proteins are highly conserved with an identity of over 90% (Takeuchi *et al.*, 1999b). Due to their similar genomic structures, which consists of one exon, located in tandem in the same chromosome and their amino acid sequence identities, it is possible that they are products of evolutionary duplication. All biological active lipopeptides are absolutely TLR-2-dependent but differ in their requirement for TLR-1 and TLR-6, according to the number and length of their fatty acids and the amino acid sequence of their peptide tail. For example, mycoplasmal macrophage-activating lipopeptide-2kD (MALP-2) is a synthetic diacylated lipopeptide resembling the peptide tail from *M. fermentans* and FSL-1 (fibroblast stimulating lipopeptide-1), a synthetic lipopeptide from *Mycoplasma salivarium*. The lipopeptide MALP-2 signals through TLR-2/-6, whereas triacylated synthetic lipopeptide $Pam_3C\text{-}SK_4$ is recognized by the TLR-2/-1 heterodimers. These findings and additional results led to the hypothesis that triacylated lipoproteins of bacteria are recognized by TLR-2-/1, whereas diacylated lipoproteins of mycoplasma are recognized by TLR-2-/6. This assumption has recently been modified, because there are also diacylated lipopeptides like $Pam_2C\text{-}SK_4$ and MALP-2-SK_4, which show a TLR-6-independent signalling (Buwitt-Beckmann *et al.*, 2006) and triacylated lipopeptides like MALP-3 that can activate TLR-1-deficient cells.

Interestingly, sequence analysis of TLR-6 predicts no obtainable hydrophobic pocket in this receptor, because phenylalanines blocked the potential channel. Therefore, it is not possible that the amide-bound fatty acid of triacylated lipopeptides bind to TLR-6 in a TLR-1-manner (Jin *et al.*, 2007). Conceivably, TLR-6 interacts directly with the peptide tail of the di- and triacylated lipopeptides or with the glycerol-cysteine-backbone via ionic contacts. However, to understand fully the binding of TLR-6 to different ligands, a crystal structure of the ligand-TLR-6 complex is still needed. Nevertheless, the reason for interaction of TLR-2 with TLR-1 or TLR-6 seems to be an evolutionary mechanism to expand the ligand spectrum, since TLR-1 and TLR-6 recognize a variety of different lipopeptides but activate identical signalling pathways (Farhat *et al.*, 2008).

3.3. Signal transduction pathways of TLR-2

Activation of TLR-2 finally leads to an activation of NF-κB and mitogen-activated protein (MAP) kinases (see Figure 31.1). After ligand binding, TLR-2 dimerizes with TLR-1 or TLR-6 and undergoes conformational changes that are necessary for the recruitment of the myeloid differentiation primary-response protein 88 (MyD-88) and the MyD-88 adapter-like protein (MAL), also called TIR-associated protein (TIRAP). The binding between MyD-88 and the TLR occurs via TIR/TIR domain interaction. Additionally, MyD-88 can directly bind through a homophilic interaction to the death domain of the IL-1 receptor-associated kinase 4 (IRAK-4) (Akira *et al.*, 2006). The complex of MyD-88 and TLR triggers the activation of IRAK-4 which is associated to IRAK-1 and the

Toll-interacting protein (TOLLIP). The essential roles of MyD-88 and IRAK-4 in the signalling of TLR are demonstrated in MyD-88$^{-/-}$ and IRAK-4$^{-/-}$ mice. Macrophages from these mice do not produce any inflammatory cytokines in response to microbial components (Suzuki *et al.*, 2002; Takeda *et al.*, 2003). Another member of the IRAK-family, IRAK-M, is responsible for the negative regulation of this pathway, as deficient mice show an increased inflammatory response (Kobayashi *et al.*, 2002). Subsequently, after TLR stimulation, IRAK-1 is phosphorylated. This activation increases the kinase activity of IRAK-1 leading to an autophosphorylation of its N-terminus (Akira and Takeda, 2004). Hyper-phosphorylated IRAK-1 then associates with the tumour necrosis factor (TNF) receptor-associated factor 6 (TRAF-6) (Qian *et al.*, 2001). Activated TRAF-6 acts as a ubiquitin protein ligase (E3) together with a ubiquitination E2 enzyme complex and catalyses the additions of a ubiquitin chain on TRAF-6 itself. This ubiquitination activates another complex consisting of transforming growth factor-β (TGF-β)-activated kinase (TAK-1), TAK-1 binding protein-1 (TAB-1) and TAB-2/TAB-3 and results in a phosphorylation of NF-κB essential modulator (NEMO), also known as inhibitor of NF-κB–kinase γ (IKK-γ). Finally, activation of the TRAF-6 complex and NEMO leads to an activation of the I-κB kinase (IKK) complex and the MAP kinases (Akira and Takeda, 2004). The IKK complex consists of two catalytic subunits, IKK-α and IKK-β, and a regulatory subunit IKK-γ. After activation of the IKK complex, the inhibitor of NF-κB (I-κB) is phosphorylated, then ubiquitinylated and degraded by the ubiquitin-dependent proteasome. After proteolysis of the I-κB proteins, NF-κB dissociates from I-κB and is then able to translocate in the nucleus. In mammals, five family members of NF-κB have been identified: v-rel reticuloendotheliosis viral oncogene homologue B (RelB), c-Rel, RelA (p65), NF-κB-2 (p100/p52) and NF-κB-1 (p105/p50) (Ghosh *et al.*, 1998). The transcription of genes encoding inflammatory and immune mediators, including TNF-α, IL-1β and IL-6 is regulated by NF-κB (Kawai and Akira, 2005). Mice that lack several NF-κB subunits are more susceptible than wild type mice (Takeda *et al.*, 2003). In addition to the modulation of NF-κB, the MAP kinases are enabled by TAK-1 leading to an activation of p38, the c-Jun N-terminal kinase (JNK) and the extracellular signal-regulated kinases (ERK-1/-2) followed by an activation of AP-1 (activating protein-1) transcription factors and induction of genes expressing proinflammatory cytokines and chemokines (Takeda and Akira, 2005).

4. TLR-4

4.1. Overview of ligands and ligand binding

The first microbial ligand that was associated with TLR-signalling was the lipoglycan LPS and glycolipid lipo-oligosaccharide (LOS), the main components of the outer leaflet of the cell wall of Gram-negative bacteria. At first, LPS was erroneously believed to signal through TLR-2 (Yang *et al.*, 1998). However, this first definition of the signal transducing receptor for LPS already documented a serious problem which still hinders investigations in identifying true TLR ligands: the presence of hardly detectable or unknown contaminants in preparations that are believed to signal through TLRs. Later on, it was shown that TLR-2 was not required for LPS responses (Heine *et al.*, 1999; Takeuchi *et al.*, 1999a) and that it was not LPS itself which induces TLR-2-signalling, but a contamination within the commercial LPS preparation, namely bacterial lipoprotein (Lee *et al.*, 2002).

Mice strains C3H/HeJ and C57BL/10ScCr are unresponsive to LPS and, by defining the gene that is responsible for this genetic defect, the actual LPS-signalling receptor was identified as TLR-4 (Poltorak *et al.*, 1998). For LPS-induced signalling, which is greatly amplified by the

GPI-linked CD14 molecule (Wright *et al.*, 1990), TLR-4 requires the co-expression of an adaptor protein named myeloid differentiation-2 (MD-2) (Shimazu *et al.*, 1999). This protein consists of large disulfide-linked oligomers of dimeric subunits and is produced and secreted by monocytes and dendritic cells. MD-2 binds closely to TLR-4 and is essential for LPS responses (Akashi *et al.*, 2000; Visintin *et al.*, 2001).

Human and animal cells display remarkable species-specific responses to certain lipid A structures. For example, the synthetic lipid A precursor compound 406 possesses endotoxic activity in mice (Galanos *et al.*, 1984), but is inactive in human cells (Loppnow *et al.*, 1986). Moreover, *Rhodopseudomonas sphaeroides* lipid A is a potent antagonist in humans and mice, but acts as an LPS agonist in cells of hamster origin (Golenbock *et al.*, 1991). Today, at least two reasons for this differential activity have been identified:

(i) The species-specific activity of lipid A analogues depends on the species of TLR-4 expressed (Lien *et al.*, 2000); antagonistic lipid IVa binds to TLR-4/MD-2 complexes in a similar strength like lipid A but does not induce human TLR-4 oligomerization (Saitoh *et al.*, 2004);

(ii) It has also been found that the antagonistic activity of lipid IVa is defined by human MD-2 and not by human TLR-4 (Akashi *et al.*, 2001).

Genetic data clearly implicate physical contact of the lipid A portion of LPS with TLR-4 (Poltorak *et al.*, 2000; Lien *et al.*, 2000). Furthermore, using modified radiolabelled LPS molecules, physical contact between LPS, TLR-4 and MD-2 could be demonstrated, but only when CD14 is present (da Silva *et al.*, 2001). So far, results concerning the mode of action of MD-2 are conflicting: whereas Visintin *et al.* (2003) found that MD-2 must be bound to TLR-4 on the cell membrane before binding of LPS can occur, Kennedy *et al.* (2004) identified the activating ligand of TLR-4 as the MD-2/LPS complex. The functional region of TLR-4 required for the association with MD-2 has been identified as the amino-terminal subdomain with eight LRRs (Fujimoto *et al.*, 2004).

Recently, new data from crystal structures of TLR-4, in combination with LPS antagonists, have provided key information on how the interaction of TLR-4, LPS and MD-2 might actually occur: in contrast to Pam_3C-SK_4 and TLR-2, TLR-4 binds to MD-2 and to MD-2-eritoran (the LPS antagonist that has been used in this study) at the concave face of the LRR (Kim *et al.*, 2007). In addition, there is apparently no direct contact between TLR-4 and eritoran. Similar findings with the LPS precursor lipid IVa were obtained by Ohto *et al.* (2007). However, whether agonistic LPS molecules directly bind to TLR-4 and if this is the prerequisite for TLR-4 dimerization and signalling, remains to be seen.

Besides LPS, a number of other microbial and endogenous compounds have been reported to signal through TLR-4 (see reviews Tsan and Gao, 2004; Akira *et al.*, 2006). For many of them, convincing experimental evidence suggests that they indeed signal through TLR-4. Although, in most cases, possible contamination with LPS was carefully considered, recent publications cast doubts about the normally used procedures to rule out LPS-contamination as responsible for the detected TLR-4 activity (Tsan and Gao, 2004). Thus, one has to be extremely careful when investigating TLR-4 ligands other than LPS. Highly purified or synthetic compounds and sophisticated chemical analyses are required to identify further true TLR-4 ligands.

4.2. Signal transduction pathways of TLR-4

4.2.1. *MyD-88-dependent signalling*

The signal transduction of TLR-4 is unique among all TLRs, since it involves not only MyD-88 and Mal, but also a second set of adaptor proteins, namely TIR-domain-containing adaptor-inducing IFN-β (TRIF) and TRIF-related adaptor molecule (TRAM), that enable TLR-4, similar to TLR-3, to activate the IFN pathway as well (Yamamoto *et al.*, 2003; Akira and Takeda, 2004). The MyD-88-dependent signalling uses the same adaptors and molecules as TLR-2 (Figure 31.2), since both use MyD-88 and Mal.

4.2.2. MyD-88-independent signalling

When Akira and coworkers investigated the responses of MyD-88-deficient mice to LPS, it was evident that these animals were resistant to endotoxic shock because they could not produce any cytokines. Unexpectedly, activation of NF-κB and MAP kinases still occurred, albeit with delayed kinetics (Kawai *et al.*, 1999) and also activation of the IFN-regulatory factor 3 (IRF-3) was unimpaired (Kawai *et al.*, 2001). This so-called MyD-88-independent pathway was also employed by a TLR-3 ligand, polyinosine-polycytidylic acid (poly[I:C]), but no other TLR ligand, indicating that this pathway was specific for only some TLRs (Alexopoulou *et al.*, 2001).

The responsible adaptor protein for activating the TLR-3/TLR-4-specific induction of IFN-β was found to be the TIR domain-containing adaptor inducing IFN-β, TRIF (Yamamoto *et al.*, 2002). Interestingly, TRIF is able to activate the late phase NF-κB similar to MyD-88 and TIRAP/Mal but, in contrast, only TRIF can activate the IFN-β promoter. In order to recruit TRIF, TLR-4 (and only TLR-4) requires the bridging adaptor protein TRAM. Elimination of TRAM abolishes TLR-4 responses, but has no impact on TLR-3-mediated TRIF activation (Yamamoto *et al.*, 2003). But how can TRIF activate both pathways? The C-terminal region of TRIF contains a Rip homotypic interaction motif (RHIM) that mediates the interaction with RIP-1 (receptor-interacting protein 1). In the absence of RIP-1, TRIF-dependent NF-κB activation no longer occurs (Meylan *et al.*, 2004). The induction of Type I IFNs by LPS/TLR-4 requires the activation of IRF-3 and IRF-7 (Fitzgerald *et al.*, 2003). For this activation, the recruitment of TRAF-3 (Hacker *et al.*, 2006), which subsequently can associate with TANK (i.e. TRAF family member-associated NF-κB activator), TANK binding kinase 1 (TBK-1) and IKK-ε are required. Recently, another molecule critically involved in this pathway has been identified: SINTBAD (similar to NAP-1 TBK1 adaptor) (Ryzhakov and Randow, 2007), together with TANK and NAK-associated protein 1 (NAP1), links TBK-1 and IKK-ε to mediate downstream phosphorylation of IRF-3 and IRF-7 (Oganesyan *et al.*, 2006).

5. CONCLUSIONS

The TLRs are the prototypic innate immune receptors. By recognizing fundamental microbial structures they are able immediately to signal the host of any imminent danger. However, sepsis and septic shock arising from infectious diseases are still increasing in incidence and have a high mortality rate. In order to design better therapeutic approaches, a detailed knowledge of the molecular mechanisms of the recognition and activation pathways, of which TLRs are an important part, is desperately needed to fine-tune and antagonize an overwhelming immune response that is detrimental to the host.

Open issues and areas for future investigation are listed in the Research Focus Box.

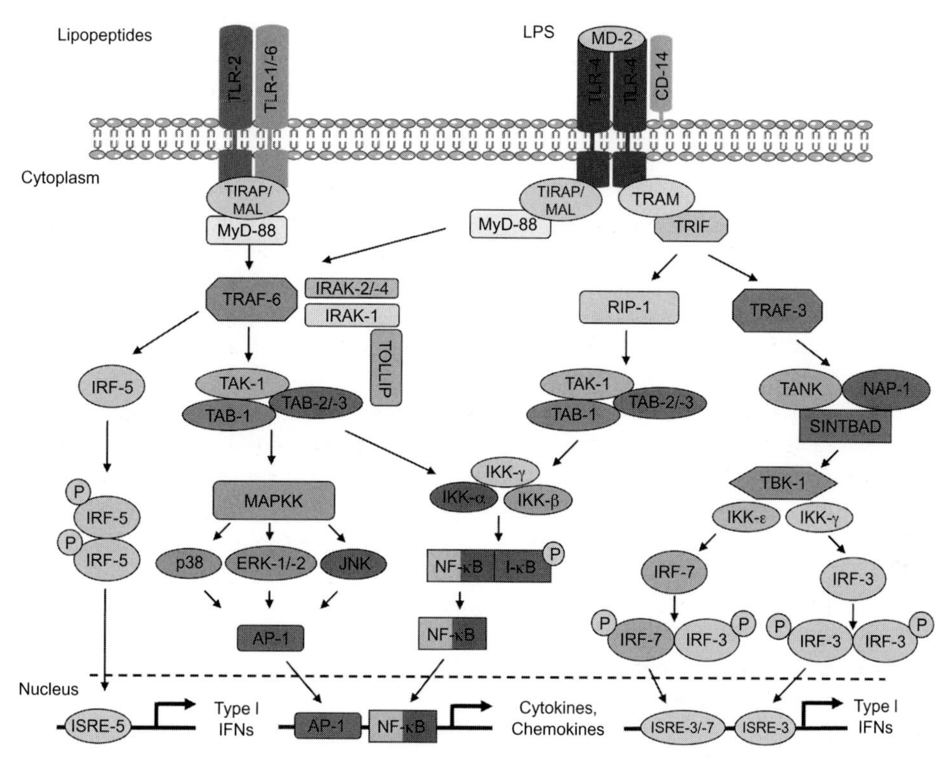

FIGURE 31.2 Overview of the TLR-2 and TLR-4 signal transduction pathways. The signalling of TLR-2 is initiated after binding of different lipopeptides. After activation TLR-2 dimerizes with TLR-1 or TLR-6 which is necessary for the recruitment of MyD-88 and Mal. The complex of MyD-88 and TLR triggers the activation of TRAF-6, IRAK-1, IRAK-2/-4 and TOLLIP. Activated TRAF-6 forms a complex with the kinases TAK-1, TAB-1 and TAB-2/-3, which results in a phosphorylation of IKK-γ. After activation of the IKK complex, I-κB is phosphorylated and subsequently ubiquitinylated which leads to degradation by the proteasome. After proteolysis of I-κB, NF-κB dissociates from I-κB and translocates in the nucleus. NF-κB regulates the transcription of different cytokines and chemokines. In addition, TAK-1 is able to activate MAPKK, which leads to phosphorylation of p38, ERK1/2 and JNK followed by an activation of AP-1 and the induction of several cytokines and chemokines. The production of Type I IFNs is due to the activation of TRAF-6, which also controls the phosphorylation of IRF-5. In the case of TLR-4, activation of the receptor leads to a recruitment of MyD-88 with Mal and, additionally, TRIF with TRAM. Subsequently, TRIF activates NF-κB through RIP-1, TAK-1 and the IKK complex. Moreover, TRIF also induces Type I IFNs by recruitment of TRAF-3, which then associates with TANK, NAP-1 and SINTBAD followed by an activation of TBK-1. Both TBK-1 and IKK-ε are required for the phosphorylation of IRF-7 and IRF-3 which is followed by Type I IFN induction. Abbreviations: AP-1, activator protein 1; CD-14, cluster differentiation marker 14; ERK-1/-2, extracellular signal-regulated kinases; Type I IFNs, Type I interferons α and β; I-κB, inhibitor of NF-κB; IKK, I-κB kinase; IRAK, interleukin-1 receptor-associated kinase; IRF, interferon regulatory factor; ISRE, interferon-stimulated response element; JNK, c-Jun N-terminal kinase; LPS, lipopolysaccharide; Mal, MyD-88 adapter-like protein; MAPKK, mitogen-activated protein kinase kinase; MD-2, myeloid differentiation-2; MyD-88, myeloid differentiation primary-response protein 88; NF-κB, nuclear factor-κB; NAP-1, NAK-associated protein 1; RIP-1, receptor-interacting protein 1; SINTBAD, similar to NAP-1 TBK-1 adaptor; TAB, TAK-1 binding protein; TAK1, TGF-β activated kinase; TANK, TRAF-family-member-associated NF-κB activator; TBK-1, TANK binding kinase; TLR, Toll-like receptor; TOLLIP, Toll-interacting protein; TIRAP, TIR-associated protein; TRAF, tumour necrosis factor receptor-associated factor; TRAM, TRIF-related adaptor molecule; TRIF, TIR-domain-containing adaptor-inducing interferon-β.

RESEARCH FOCUS BOX

- The receptor TLR-2, in association with TLR-1 and TLR-6, plays an essential role in the recognition of microbial lipopeptides, but it is still unclear whether TLR-2 can also signal as a homodimer.
- In addition to TLR-1 and TLR-6, the TLR-1 subfamily also contains TLR-10. It has been speculated that TLR-10 could act as a TLR-2 partner. However, up until now no specific ligand for TLR-10 has been identified.
- The ligands for TLR-2 are the most diverse group among all TLR ligands and there are still many controversies in the literature. Thus, careful analyses with highly purified ligands, e.g. to exclude lipoprotein contaminations in different ligand preparations, are required.
- In addition to LPS, other ligands such as streptococcal pneumolysin, different endogenous proteins like heat-shock protein 60 and the extra domain A of fibronectin have been discussed as TLR-4 ligands. Similar to TLR-2, it has to be carefully evaluated whether these ligands represent true TLR-4 agonists.
- Moreover, a comprehensive analysis of co-receptors in TLR-4 signalling is required. The adaptor protein MD-2 and CD14 are well established, but the exact molecular mechanisms of LPS binding and the activation of the LPS receptor complex are still unknown.

References

Adachi, K., Tsutsui, H., Kashiwamura, S., et al., 2001. *Plasmodium berghei* infection in mice induces liver injury by an IL-12- and Toll-like receptor/myeloid differentiation factor 88-dependent mechanism. J. Immunol. (167), 5928–5934.

Akashi, S., Shimazu, R., Ogata, H., et al., 2000. Cutting edge: cell surface expression and lipopolysaccharide signaling via the Toll-like receptor 4-MD-2 complex on mouse peritoneal macrophages. J. Immunol. 164, 3471–3475.

Akashi, S., Nagai, Y., Ogata, H., et al., 2001. Human MD-2 confers on mouse Toll-like receptor 4 species-specific lipopolysaccharide recognition. Int. Immunol. 13, 1595–1599.

Akira, S., Takeda, K., 2004. Toll-like receptor signalling. Nat. Rev. Immunol. 4, 499–511.

Akira, S., Uematsu, S., Takeuchi, O., 2006. Pathogen recognition and innate immunity. Cell 124, 783–801.

Alexopoulou, L., Holt, A.C., Medzhitov, R., Flavell, R.A., 2001. Recognition of double-stranded RNA and activation of NF-κB by Toll-like receptor 3. Nature 413, 732–738.

Anderson, K.V., Nusslein-Volhard, C., 1984. Information for the dorsal-ventral pattern of the *Drosophila* embryo is stored as maternal mRNA. Nature 311, 223–227.

Anderson, K.V., Bokla, L., Nusslein-Volhard, C., 1985. Establishment of dorsal-ventral polarity in the *Drosophila* embryo: the induction of polarity by the Toll gene product. Cell 42, 791–798.

Azumi, K., De Santis, R., De Tomaso, A., et al., 2003. Genomic analysis of immunity in a Urochordate and the emergence of the vertebrate immune system: "waiting for Godot". Immunogenetics 55, 570–581.

Bieback, K., Lien, E., Klagge, I.M., et al., 2002. Hemagglutinin protein of wild-type measles virus activates Toll-like receptor 2 signaling. J. Virol. 76, 8729–8736.

Buwitt-Beckmann, U., Heine, H., Wiesmuller, K.H., et al., 2006. TLR1- and TLR6-independent recognition of bacterial lipopeptides. J. Biol. Chem. 281, 9049–9057.

Campos, M.A., Almeida, I.C., Takeuchi, O., et al., 2001. Activation of Toll-like receptor-2 by glycosylphosphatidylinositol anchors from a protozoan parasite. J. Immunol. 167, 416–423.

Compton, T., Kurt-Jones, E.A., Boehme, K.W., et al., 2003. Human cytomegalovirus activates inflammatory cytokine responses via CD14 and Toll-like receptor 2. J. Virol. 77, 4588–4596.

da Silva, C.J., Soldau, K., Christen, U., Tobias, P.S., Ulevitch, R.J., 2001. Lipopolysaccharide is in close proximity to each of the proteins in its membrane receptor complex: transfer from CD14 to TLR4 and MD-2. J. Biol. Chem. 276, 21129–21135.

de Veer, M.J., Curtis, J.M., Baldwin, T.M., et al., 2003. MyD88 is essential for clearance of *Leishmania major*: possible role for lipophosphoglycan and Toll-like receptor 2 signaling. Eur. J. Immunol. 33, 2822–2831.

Farhat, K., Riekenberg, S., Heine, H., et al., 2008. Heterodimerization of TLR2 with TLR1 or TLR6 expands the ligand spectrum but does not lead to differential signaling. J. Leukoc. Biol. 83, 692–701.

Fitzgerald, K.A., Rowe, D.C., Barnes, B.J., et al., 2003. LPS-TLR4 signaling to IRF-3/7 and NF-κB involves the toll adapters TRAM and TRIF. J. Exp. Med. 198, 1043–1055.

Fujimoto, T., Yamazaki, S., Eto-Kimura, A., Takeshige, K., Muta, T., 2004. The amino-terminal region of Toll-like receptor 4 is essential for binding to MD-2 and receptor translocation to the cell surface. J. Biol. Chem. 279, 47431–47437.

Galanos, C., Lehmann, V., Luderitz, O., et al., 1984. Endotoxic properties of chemically synthesized lipid A part structures. Comparison of synthetic lipid A precursor and synthetic analogues with biosynthetic lipid A precursor and free lipid A. Eur. J. Biochem. 140, 221–227.

Ghosh, S., May, M.J., Kopp, E.B., 1998. NF-κB and Rel proteins: evolutionarily conserved mediators of immune responses. Annu. Rev. Immunol. 16, 225–260.

Golenbock, D.T., Hampton, R.Y., Qureshi, N., Takayama, K., Raetz, C.R., 1991. Lipid A-like molecules that antagonize the effects of endotoxins on human monocytes. J. Biol. Chem. 266, 19490–19498.

Hacker, H., Redecke, V., Blagoev, B., et al., 2006. Specificity in Toll-like receptor signalling through distinct effector functions of TRAF3 and TRAF6. Nature 439, 204–207.

Hajjar, A.M., O'Mahony, D.S., Ozinsky, A., et al., 2001. Cutting edge: functional interactions between Toll-like receptor (TLR) 2 and TLR1 or TLR6 in response to phenol-soluble modulin. J. Immunol. 166, 15–19.

Hashimoto, C., Hudson, K.L., Anderson, K.V., 1988. The Toll gene of *Drosophila*, required for dorsal-ventral embryonic polarity, appears to encode a transmembrane protein. Cell 52, 269–279.

Hashimoto, M., Tawaratsumida, K., Kariya, H., et al., 2006. Not lipoteichoic acid but lipoproteins appear to be the dominant immunobiologically active compounds in Staphylococcus aureus. J. Immunol. 177, 3162–3169.

Heine, H., Kirschning, C.J., Lien, E., Monks, B.G., Rothe, M., Golenbock, D.T., 1999. Cutting edge: cells that carry A null allele for Toll-like receptor 2 are capable of responding to endotoxin. J. Immunol. 162, 6971–6975.

Hirschfeld, M., Weis, J.J., Toshchakov, V., et al., 2001. Signaling by Toll-like receptor 2 and 4 agonists results in differential gene expression in murine macrophages. Infect. Immun. 69, 1477–1482.

Hoebe, K., Janssen, E., Beutler, B., 2004. The interface between innate and adaptive immunity. Nat. Immunol. 5, 971–974.

Inamori, K., Ariki, S., Kawabata, S., 2004. A Toll-like receptor in horseshoe crabs. Immunol. Rev. 198, 106–115.

Inamura, S., Fujimoto, Y., Kawasaki, A., et al., 2006. Synthesis of peptidoglycan fragments and evaluation of their biological activity. Org. Biomol. Chem. 4, 232–242.

Janeway, C.A., 1989. Approaching the asymptote? Evolution and revolution in immunology. Cold Spring Harb. Symp. Quant. Biol. 54, 1–13.

Jin, M.S., Kim, S.E., Heo, J.Y., et al., 2007. Crystal structure of the TLR1-TLR2 heterodimer induced by binding of a tri-acylated lipopeptide. Cell 130, 1071–1082.

Kawai, T., Akira, S., 2005. Pathogen recognition with Toll-like receptors. Curr. Opin. Immunol. 17, 338–344.

Kawai, T., Adachi, O., Ogawa, T., Takeda, K., Akira, S., 1999. Unresponsiveness of MyD88-deficient mice to endotoxin. Immunity 11, 115–122.

Kawai, T., Takeuchi, O., Fujita, T., et al., 2001. Lipopolysaccharide stimulates the MyD88-independent pathway and results in activation of IFN-regulatory factor 3 and the expression of a subset of lipopolysaccharide-inducible genes. J. Immunol. 167, 5887–5894.

Kennedy, M.N., Mullen, G.E., Leifer, C.A., et al., 2004. A complex of soluble MD-2 and lipopolysaccharide serves as an activating ligand for Toll-like receptor 4. J. Biol. Chem. 279, 34698–34704.

Kim, H.M., Park, B.S., Kim, J.I., et al., 2007. Crystal structure of the TLR4-MD-2 complex with bound endotoxin antagonist Eritoran. Cell 130, 906–917.

Kobayashi Jr., K., Hernandez, L.D., Galan, J.E., Janeway, C.A., Medzhitov, R., Flavell, R.A., 2002. IRAK-M is a negative regulator of Toll-like receptor signaling. Cell 110, 191–202.

Kurt-Jones, E.A., Chan, M., Zhou, S., et al., 2004. Herpes simplex virus 1 interaction with Toll-like receptor 2 contributes to lethal encephalitis. Proc. Natl. Acad. Sci. USA 101, 1315–1320.

Lee, H.K., Lee, J., Tobias, P.S., 2002. Two lipoproteins extracted from *Escherichia coli* K-12 LCD25 lipopolysaccharide are the major components responsible for Toll-like receptor 2-mediated signaling. J. Immunol. 168, 4012–4017.

Lemaitre, B., Nicolas, E., Michaut, L., Reichhart, J.M., Hoffmann, J.A., 1996. The dorsoventral regulatory gene cassette spatzle/Toll/cactus controls the potent antifungal response in *Drosophila* adults. Cell 86, 973–983.

Lemaitre, B., Reichhart, J.M., Hoffmann, J.A., 1997. *Drosophila* host defense: differential induction of antimicrobial peptide genes after infection by various classes of microorganisms. Proc. Natl. Acad. Sci. USA 94, 14614–14619.

Lien, E., Sellati, T.J., Yoshimura, A., et al., 1999. Toll-like receptor 2 functions as a pattern recognition receptor for diverse bacterial products. J. Biol. Chem. 274, 33419–33425.

Lien, E., Means, T.K., Heine, H., et al., 2000. Toll-like receptor 4 imparts ligand-specific recognition of bacterial lipopolysaccharide. J. Clin. Invest. 105, 497–504.

Liu, L., Botos, I., Wang, Y., et al., 2008. Structural basis of Toll-like receptor 3 signaling with double-stranded RNA. Science 320, 379–381.

Loppnow, H., Brade, L., Brade, H., et al., 1986. Induction of human interleukin 1 by bacterial and synthetic lipid A. Eur. J. Immunol. 16, 1263–1267.

Massari, P., Henneke, P., Ho, Y., Latz, E., Golenbock, D.T., Wetzler, L.M., 2002. Cutting edge: immune stimulation by neisserial porins is Toll-like receptor 2 and MyD88 dependent. J. Immunol. 168, 1533–1537.

Means, T.K., Wang, S., Lien, E., Yoshimura, A., Golenbock, D.T., Fenton, M.J., 1999. Human Toll-like receptors mediate cellular activation by *Mycobacterium tuberculosis*. J. Immunol. 163, 3920–3927.

Meylan, E., Burns, K., Hofmann, K., et al., 2004. RIP1 is an essential mediator of Toll-like receptor 3-induced NF-κB activation. Nat. Immunol. 5, 503–507.

Morr, M., Takeuchi, O., Akira, S., Simon, M.M., Muhlradt, P.F., 2002. Differential recognition of structural details of bacterial lipopeptides by Toll-like receptors. Eur. J. Immunol. 32, 3337–3347.

Netea, M.G., Kullberg, B.J., Galama, J.M., et al., 2002. Non-LPS components of Chlamydia pneumoniae stimulate cytokine production through Toll-like receptor 2-dependent pathways. Eur. J. Immunol. 32, 1188–1195.

Oganesyan, G., Saha, S.K., Guo, B.C., et al., 2006. Critical role of TRAF3 in the Toll-like receptor-dependent and -independent antiviral response. Nature 439, 208–211.

Ohto, U., Fukase, K., Miyake, K., Satow, Y., 2007. Crystal structures of human MD-2 and its complex with antiendotoxic lipid IVa. Science 316, 1632–1634.

Opitz, B., Schroder, N.W., Spreitzer, I., et al., 2001. Toll-like receptor-2 mediates Treponema glycolipid and lipoteichoic acid-induced NF-κB translocation. J. Biol. Chem. 276, 22041–22047.

Ozinsky, A., Underhill, D.M., Fontenot, J.D., et al., 2000. The repertoire for pattern recognition of pathogens by the innate immune system is defined by cooperation between Toll-like receptors. Proc. Natl. Acad. Sci. USA 97, 13766–13771.

Pichlmair, A., Sousa, C.R.E., 2007. Innate recognition of viruses. Immunity 27, 370–383.

Poltorak, A., He, X., Smirnova, I., et al., 1998. Defective LPS signaling in C3H/HeJ and C57BL/10ScCr mice: mutations in Tlr4 gene. Science 282, 2085–2088.

Poltorak, A., Ricciardi-Castagnoli, P., Citterio, S., Beutler, B., 2000. Physical contact between lipopolysaccharide and Toll-like receptor 4 revealed by genetic complementation. Proc. Natl. Acad. Sci. USA 97, 2163–2167.

Pradel, E., Zhang, Y., Pujol, N., Matsuyama, T., Bargmann, C.I., Ewbank, J.J., 2007. Detection and avoidance of a natural product from the pathogenic bacterium Serratia marcescens by Caenorhabditis elegans. Proc. Natl. Acad. Sci. USA 104, 2295–2300.

Qian, Y., Commane, M., Ninomiya-Tsuji, J., Matsumoto, K., Li, X., 2001. IRAK-mediated translocation of TRAF6 and TAB2 in the interleukin-1-induced activation of NFκB. J. Biol. Chem. 276, 41661–41667.

Rast, J.P., Smith, L.C., Loza-Coll, M., Hibino, T., Litman, G.W., 2006. Genomic insights into the immune system of the sea urchin. Science 314, 952–956.

Roach, J.C., Glusman, G., Rowen, L., et al., 2005. The evolution of vertebrate Toll-like receptors. Proc. Natl. Acad. Sci. USA 102, 9577–9582.

Ryzhakov, G., Randow, F., 2007. SINTBAD, a novel component of innate antiviral immunity, shares a TBK1-binding domain with NAP1 and TANK. EMBO J. 26, 3180–3190.

Saitoh, S., Akashi, S., Yamada, T., et al., 2004. Lipid A antagonist, lipid IVa, is distinct from lipid A in interaction with Toll-like receptor 4 (TLR4)-MD-2 and ligand-induced TLR4 oligomerization. Int. Immunol. 16, 961–969.

Shimazu, R., Akashi, S., Ogata, H., et al., 1999. MD-2, a molecule that confers lipopolysaccharide responsiveness on Toll-like receptor 4. J. Exp. Med. 189, 1777–1782.

Shin, S.W., Bian, G., Raikhel, A.S., 2006. A toll receptor and a cytokine, Toll5A and Spz1C, are involved in toll antifungal immune signaling in the mosquito Aedes aegypti. J. Biol. Chem. 281, 39388–39395.

Suzuki, N., Suzuki, S., Duncan, G.S., et al., 2002. Severe impairment of interleukin-1 and Toll-like receptor signalling in mice lacking IRAK-4. Nature 416, 750–756.

Takeda, K., Akira, S., 2005. Toll-like receptors in innate immunity. Int. Immunol. 17, 1–14.

Takeda, K., Kaisho, T., Akira, S., 2003. Toll-like receptors. Annu. Rev. Immunol. 21, 335–376.

Takeuchi, O., Hoshino, K., Kawai, T., et al., 1999a. Differential roles of TLR2 and TLR4 in recognition of gram-negative and gram-positive bacterial cell wall components. Immunity 11, 443–451.

Takeuchi, O., Kawai, T., Sanjo, H., et al., 1999b. TLR6: A novel member of an expanding Toll-like receptor family. Gene 231, 59–65.

Travassos, L.H., Girardin, S.E., Philpott, D.J., et al., 2004. Toll-like receptor 2-dependent bacterial sensing does not occur via peptidoglycan recognition. EMBO Rep. 5, 1000–1006.

Tsan, M.F., Gao, B., 2004. Endogenous ligands of Toll-like receptors. J. Leukoc. Biol. 76, 514–519.

Tuon, F.F., Amato, V.S., Bacha, H.A., Almusawi, T., Duarte, M.I., Amato, N.V., 2008. Toll-like receptors and leishmaniasis. Infect. Immun. 76, 866–872.

Underhill, D.M., Ozinsky, A., Hajjar, A.M., et al., 1999. The Toll-like receptor 2 is recruited to macrophage phagosomes and discriminates between pathogens. Nature 401, 811–815.

van der Kleij, D., Latz, E., Brouwers, J.F., et al., 2002. A novel host-parasite lipid cross-talk. Schistosomal lysophosphatidylserine activates Toll-like receptor 2 and affects immune polarization. J. Biol. Chem. 277, 48122–48129.

Visintin, A., Mazzoni, A., Spitzer, J.A., Segal, D.M., 2001. Secreted MD-2 is a large polymeric protein that efficiently confers lipopolysaccharide sensitivity to Toll-like receptor 4. Proc. Natl. Acad. Sci. USA 98, 12156–12161.

Visintin, A., Latz, E., Monks, B.G., Espevik, T., Golenbock, D.T., 2003. Lysines 128 and 132 enable lipopolysaccharide binding to MD-2, leading to Toll-like receptor-4 aggregation and signal transduction. J. Biol. Chem. 278, 48313–48320.

Walker, A.J., 2006. Do trematode parasites disrupt defence-cell signalling in their snail hosts? Trends Parasitol. 22, 154–159.

Wang, J.P., Kurt-Jones, E.A., Shin, O.S., Manchak, M.D., Levin, M.J., Finberg, R.W., 2005. Varicella-zoster virus activates inflammatory cytokines in human monocytes and macrophages via Toll-like receptor 2. J. Virol. 79, 12658–12666.

Werts, C., Tapping, R.I., Mathison, J.C., et al., 2001. Leptospiral lipopolysaccharide activates cells through a TLR2-dependent mechanism. Nat. Immunol. 2, 346–352.

Wright, S.D., Ramos, R.A., Tobias, P.S., Ulevitch, R.J., Mathison, J.C., 1990. CD14, a receptor for complexes of lipopolysaccharide (LPS) and LPS binding protein. Science 249, 1431–1433.

Yamamoto, M., Sato, S., Mori, K., et al., 2002. Cutting edge: a novel Toll/IL-1 receptor domain-containing adapter that preferentially activates the IFN-β promoter in the Toll-Like receptor signaling. J. Immunol. 169, 6668–6672.

Yamamoto, M., Sato, S., Hemmi, H., et al., 2003. TRAM is specifically involved in the Toll-like receptor 4-mediated MyD88-independent signaling pathway. Nat. Immunol. 4, 1144–1150.

Yang, R.B., Mark, M.R., Gray, A., et al., 1998. Toll-like receptor-2 mediates lipopolysaccharide-induced cellular signalling. Nature 395, 284–288.

Yarovinsky, F., Zhang, D.K., Andersen, J.F., et al., 2005. TLR11 activation of dendritic cells by a protozoan profilin-like protein. Science 308, 1626–1629.

Zhang, D.K., Zhang, G.L., Hayden, M.S., et al., 2004. A Toll-like receptor that prevents infection by uropathogenic bacteria. Science 303, 1522–1526.

CHAPTER

32

NOD receptor recognition of peptidoglycan

Ivo Gomperts Boneca

SUMMARY

Peptidoglycan has been known since the 1970s to have potent biological activity on higher eukaryotes. Most of the activities that were described could be assigned to degradation products released by bacteria during cell growth and division. The molecular basis of this potent activity was elucidated recently with the discovery of a new family of intracellular receptors of the innate immune system, the Nod-like receptors. In particular, Nod1 and Nod2 have been shown to sense different muropeptides structures. Thus, Nod1 senses specifically *meso*-diaminopimelic-type muropeptides originating from mostly Gram-negative bacteria, whereas Nod2 is a more general sensor of muropeptides. The discovery of the Nod-like receptors (NLRs) opened a new field of research on host–microbe interactions in which NLRs work in parallel and/or with Toll-like receptors to sense bacteria. Of note, NLRs have been shown to fulfil important roles in detecting pathogens such as *Helicobacter pylori*, *Listeria monocytogenes* or *Shigella flexneri*. In exchange, bacteria have evolved diverse strategies to modulate the host response by modifying their peptidoglycan or the amount of muropeptide shed.

Keywords: Peptidoglycan; NLRs; Nod1; Nod2; NLRP3; NLRP1; Muramyl dipeptide (MDP); Muramyl tripeptide; Tracheal cytotoxin; Muropeptides

1. BIOLOGICAL ACTIVITIES OF PEPTIDOGLYCAN: AN HISTORICAL PERSPECTIVE

As described in previous chapters, peptidoglycan (PG) is a major and almost ubiquitous component of the bacterial cell wall (see Chapters 1 and 2). Hence, it is an excellent marker of bacterial origin, in particular, due to its unique composition of alternative levo (L-) and dextrogiro (D-) amino acids. In fact, its minimal building block is conserved among bacteria composed very often of the disaccharide β-*N*-acetylglucosamine-(1→4)-*N*-acetylmuramic acid (MurNAc) [β-GlcNAc-(1→4)-MurNAc] linked to the pentapeptide L-alanyl-γ-D-glutamyl-diaminopimelyl-D-alanyl-D-alanine (L-Ala-γ-D-Glu-DAP-D-Ala-D-Ala). During evolution, PG has been selected by eukaryotes to detect the presence of bacteria and mount an appropriate response. Conversely, bacteria have evolved many variations to the minimal building block due to the collective action of PG synthetases, PG hydrolases, modifications of the amino sugars and the grafting of additional peptide cross-bridges (Schleifer and Kandler, 1972).

Historically, the PG was viewed as a rather inert exoskeleton, which served as armour. However, in the 1960–1970s, detailed studies on PG metabolism with different bacterial models revealed a more dynamic scenario with constant synthesis, degradation of the PG layer and, consequently, shedding of soluble PG fragments including muropeptides and stem peptides (Mauck and Glaser, 1970; Mauck *et al.*, 1971; Boothby *et al.*, 1973; Hebeler and Young, 1976; Goodell *et al.*, 1978). The 1970s–1980s were very productive in identifying a variety of potent biological activities for these muropeptides, including adjuvant activity, enhanced antibody production, macrophage priming, antimicrobial peptide, cytokine, nitric oxide and superoxide anion production, cytotoxicity, complement activation, induced autoimmunity, arthritis, slow-wave sleep and reduced appetite (Boneca, 2005). Such a diverse range of responses to similar molecules of bacterial origin led scientists to search actively for the host counterparts involved in PG detection.

The first report of a candidate eukaryotic receptor came from studies on CD14 (Dziarski *et al.*, 1998; Pugin *et al.*, 1994). However, CD14 seemed to function as a co-receptor and was not essential in mediating the biological activity of PG. With the discovery of the role of TOLL in *Drosophila* innate immunity against Gram-positive bacteria and fungi (Lemaitre *et al.*, 1996), the field of pattern-recognition receptors exploded in the 1990s. In mammals, Toll-like receptors (TLRs) were found to mediate the detection of a variety of bacterial products, such as lipopolysaccharide (LPS), lipoproteins, lipoteichoic acid (LTA), flagellin, bacterial DNA and viral RNA (see Chapter 31). Soon, TLR-2 was proposed to be the receptor candidate for PG acting synergistically with CD14 (Iwaki *et al.*, 2002; Schwandner *et al.*, 1999; Yoshimura *et al.*, 1999). Nevertheless, several additional reports indicated that, besides PG, TLR-2 detected a variety of structurally unrelated bacterial products, e.g. lipoproteins, LTA or lipoarabinomannan (LAM) (Aliprantis *et al.*, 1999; Means *et al.*, 1999; Morath *et al.*, 2002; Schwandner *et al.*, 1999). Since then, TLR-2 has been shown to sense preferentially lipophilic components of the bacterial cell wall that often contaminate PG preparations (Travassos *et al.*, 2004) (for a recent review, see Zähringer *et al.*, 2008). Some researchers still claim that TLR-2 retains a low affinity for soluble PG; however, the biological significance for this remains unclear and controversial (Dziarski and Gupta, 2005).

2. THE NOD-LIKE PROTEINS: RECEPTORS OF THE INNATE IMMUNE SYSTEM

In 1999, a new intracellular protein, nucleotide-binding oligomerization domain (Nod) 1, was identified as inducing the nuclear factor kappa-light chain enhancer of B-cells pathway (Inohara *et al.*, 1999). Soon, a close homologue, Nod2 was also characterized (Ogura *et al.*, 2001b). The first reports indicated that both Nod1 and Nod2 sensed LPS (Inohara *et al.*, 2001; Ogura *et al.*, 2001a), although these commercial LPS preparations were shown to be contaminated with PG fragments (Girardin *et al.*, 2003a). Importantly, Nod1 and Nod2 turned out to be the long sought PG receptors (Chamaillard *et al.*, 2003a; Inohara *et al.*, 2003; Girardin *et al.*, 2003a,b,c). Nod1 and Nod2 became the archetype of a new family of intracellular innate immune receptors termed Nod-like receptors (NLRs). These are modular proteins with, in general, an N-terminal effector domain, a central nucleotide-binding site (NBS or NATCH domain) and a ligand-sensing C-terminal domain of leucine-rich repeats (LRRs) similar to TLRs (for review, see Fritz *et al.*, 2006 and Chapter 31). Both proteins, Nod1 and Nod2, have caspase-recruiting domains (CARD) at the N-terminus (1 and 2, respectively), which are essential for signalling and recruiting the adaptor protein, receptor-interacting serine-threonine

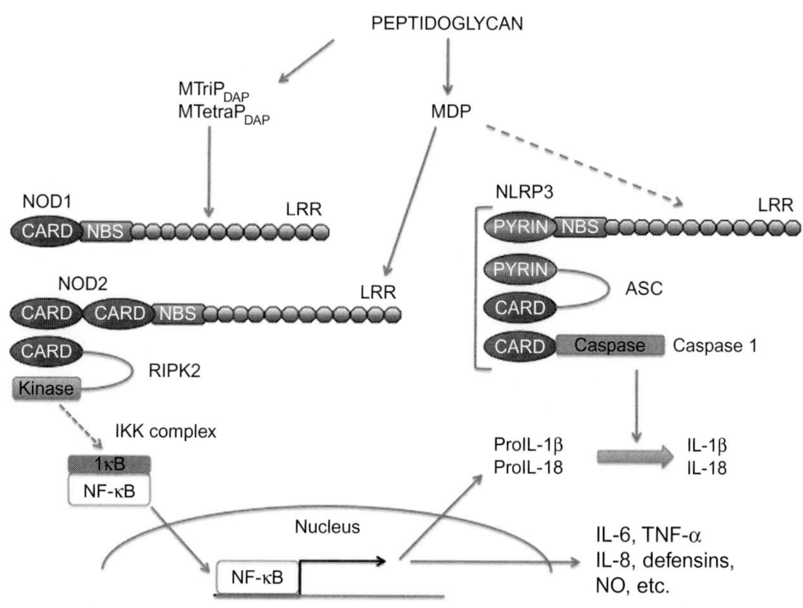

FIGURE 32.1 Peptidoglycan-sensing by the Nod proteins and NLRP3 (NLR family pyrin domain-containing 3, formerly NALP3). Both proteins, Nod1 and Nod2, have caspase-recruiting domains (CARD) at the N-terminus (1 and 2, respectively). The PG fragments such as the muropeptides, *meso*-diaminopimelic acid (*meso*-DAP)-containing muramyl tripeptide and tetrapeptide (MTriP$_{DAP}$, MTetraP$_{DAP}$) and muramyl dipeptide (MDP) are sensed presumably by the C-terminal leucine-rich repeats (LRRs) leading to activation and oligomerization of the NLR. Oligomerization requires adenosine triphosphate (ATP) allowing recruitment of the adaptor molecules, receptor-interacting serine-threonine kinase 2 (RIPK2) for Nod1 and Nod2 and apoptosis-associated speck-like protein (ASC) for NLRP3. Auto-phosphorylation of RIPK2 leads to the activation of the IkappaB kinase (IKK) complex, phosphorylation of IkB and release of nuclear factor-kappaB (NF-κB). Translocation of NF-κB into the nucleus induces the transcription of a variety of genes involved in the host response, including the *il1β* gene. Activation of NLRP3 by MDP remains poorly characterized (dashed arrow). Although it is assumed that the LRRs bind directly to the agonists, in the case of NLRP3, several unrelated bacterial products and cellular danger signals have been shown to activate NLRP3, raising the possibility that NLRP3 functions as a docking or scaffolding protein for a multiprotein complex. Hence, the NLRP3 LRRs may be involved in protein–protein interactions rather than directly with the agonists. Activated NLRP3 recruits the adaptor protein ASC, which activates caspase-1, leading to the proteolysis of pro-interleukin-(IL-) 1β and pro-IL18 into their active forms.

kinase 2 (RIPK2 also called cardiak, RIP2 or RICK kinase), through CARD–CARD interactions (Inohara *et al.*, 2000). The Nod1 (or 2)-RIPK2 complex then recruits the IkappaB kinase (IKK) complex leading to NF-κB activation (Figure 32.1). Also, Nod1 and Nod2 have been shown to activate the mitogen-activated protein (MAP) kinases and Jun N-terminal kinase (JNK), although the molecular bases remain unknown (Girardin *et al.*, 2001; Opitz *et al.*, 2006). Activation of p38 and JNK kinases by Nod2 would appear to require CARD9 associated with at least RIPK2 (Hsu *et al.*, 2007). Knowledge of the signalling cascades of Nod1 and Nod2 is still in its infancy, although some regulatory molecules have been described. Both pathways can be positively modulated by thyroid hormone receptor interactor 6 (TRIP6) and CARD6 (Dufner *et al.*, 2006; Li *et al.*, 2005). Besides, the ubiquitin ligase-associated protein, SGT1, is also required for Nod1 activation (da Silva Correia *et al.*, 2007), while the cell death-regulatory protein, GRIM19, caspase-12, the transforming growth factor β-activated kinase 1 (TAK1) and Erbin modulate Nod2 signalling

(Chen *et al.*, 2004; Barnich *et al.*, 2005; McDonald *et al.*, 2005; Kufer *et al.*, 2006; LeBlanc *et al.*, 2008).

More recently, another member of the NLR family has been described as sensing PG fragments, i.e. NLR family pyrin domain-containing 3 (NLRP3, formerly Nalp3) (Martinon *et al.*, 2004). Thus, NLRP3 is part of a multiprotein complex, a so-called inflammasome (Martinon and Tschopp, 2007), which activates caspase-1 and processing of pro-interleukin (IL-) 1β and pro-IL-18 into their active forms. However, the initial data on NLRP3 have been controversial. Previous data showed that PG-dependent induction and secretion of IL-1β required Nod2 (Netea *et al.*, 2005; van Heel *et al.*, 2005). Other work has elucidated the precise role of NLRP3 in PG-sensing. Detection by Nod2 of PG fragments leads to NF-κB activation and transcription of the *il1β* gene, whereas NALP3-sensing of the same PG fragments is involved exclusively in the activation of the inflammasome and secretion of IL-1β and IL-18 (Ferwerda *et al.*, 2008; Marina-García *et al.*, 2008). Nevertheless, since NLRP3 appears to participate in the sensing of a variety of different environmental cues, including different bacterial products as well as cellular danger signals (Kanneganti *et al.*, 2006, 2007; Mariathasan *et al.*, 2006; Martinon *et al.*, 2006), it is more likely that NLRP3 serves as a central adaptor rather than as a pattern-recognition receptor (Mariathasan, 2007). Hence, it is possible that PG-dependent activation of the inflammasome requires a multiprotein complex involving Nod2 and NLRP3 (see Figure 32.1). Interestingly, a recent work has shown that muramyl dipeptide (MDP) and *Bacillus anthracis* also induce NLRP1 (formerly Nalp1) and IL-1β secretion by a Nod2-NLRP1 (formerly Nalp1) complex (Hsu *et al.*, 2008).

The importance of these proteins in immunity has been extensively described in this last decade. NLRs, particularly in humans, have been associated with several immunological disorders (reviewed in Rosenstiel *et al.*, 2007). The first reports associated mutations of *card15* (Nod2) with Crohn's disease (Ogura *et al.*, 2001a; Hugot *et al.*, 2001; Hampe *et al.*, 2001), a chronic inflammatory condition of the bowel, and which were characterized by a loss of function of Nod2 and unable to sense PG fragments (Bonen *et al.*, 2003; Chamaillard *et al.*, 2003b). In contrast, gain-of-function mutations that constitutively activated NF-κB were associated with Blau syndrome and sarcoidosis (Miceli-Richard *et al.*, 2001; Kanazawa *et al.*, 2005). Loss-of-function alleles were also associated with increased risk of developing malignant disorders such as gastric mucosa-associated lymphoid tissue lymphoma (Kurzawski *et al.*, 2004; Rosenstiel *et al.*, 2006). Furthermore, mutations of *card4* (Nod1) have also been associated with inflammatory bowel disease (McGovern *et al.*, 2005), asthma (Hysi *et al.*, 2005) and atopic dermatitis (Kay, 2001). An increased serum immunoglobulin (Ig) E level characterizes allergic disorders. Interestingly, studies on *card4*-deficient mice showed Nod1 regulates IgE levels, indicating a role in linking innate and adaptive immunity (Fritz *et al.*, 2007). In addition, Nod2 has been shown to induce adaptive immunity mediated by PG fragments (Kobayashi *et al.*, 2005; Meshcheryakova *et al.*, 2007). These results are consistent with the adjuvant properties of PG fragments originally described in the 1970s (Adam *et al.*, 1974; Ellouz *et al.*, 1974; Kotani *et al.*, 1976) and show that Nod1 and Nod2 mainly mediate these effects.

While the association of gain-of-function mutations with chronic inflammation can be easily explained as leading to constitutive activation of NF-κB, the mechanistic basis for the onset of Crohn's disease due to loss-of-function mutations has remained elusive. These mutations result in a lack of sensing of PG fragments. It is now assumed that the aetiology of Crohn's disease involves inappropriate immune responses driven by the intestinal commensal floral due to an impaired epithelial barrier function. Reduced NF-κB activation impacts on the local production of antimicrobial products such as defensins (Kobayashi *et al.*, 2005; Boughan *et al.*, 2006; Peyrin-Biroulet *et al.*, 2006), presumably leading to an uncontrolled proliferation of the commensal flora. Nonetheless, Crohn's disease is characterized by increased NF-κB activation and a predominant T-helper cell (Th1) cytokine profile. Watanabe and colleagues provided an elegant explanation for this apparent paradox. They showed that, in the absence of functional Nod2, induction of TLR-2-pathway led to an enhanced production of IL-12 and a bias towards a Th1 profile (Watanabe *et al.*, 2004, 2006). Several groups have shown that there is extensive cross-talk between the TLR- and NLR-pathways with both synergistic and antagonistic effects (for a review, see Werts *et al.*, 2006).

In contrast to Nod1 and Nod2, for which both loss- and gain-of function mutations have been associated with several immunological disorders, mutations of the *cias1* gene (NLRP3) are autosomal dominant autoinflammatory diseases in the absence of any infection (Hoffman *et al.*, 2001). These have been associated with three diseases: familial cold autoinflammatory syndrome, Muckle-Wells syndrome and neonatal onset multisystem inflammatory disease. All these disease-associated mutations are located in the NBS (or NATCH) domain of NLRP3 and lead to high constitutive levels of IL-1β secretion in the absence of infection (Agostini *et al.*, 2004).

3. STRUCTURAL REQUIREMENTS OF PG FOR NOD PROTEINS DETECTION

Historically, Nod1 was first described as an intracellular receptor for LPS (Inohara *et al.*, 2001; Ogura *et al.*, 2001a). Subsequently, PG exclusively from Gram-negative bacteria was demonstrated to be the real agonist of Nod1 (Chamaillard *et al.*, 2003a; Girardin *et al.*, 2003a). Nod1 recognizes the naturally occurring muropeptide, GlcNAc-MurNAc-L-Ala-γ-D-Glu-*meso*-DAP ($GMTriP_{DAP}$) explaining Nod1 specificity for Gram-negative bacteria (Figure 32.2). The minimal structure capable of stimulating the Nod1 pathway is the dipeptide γ-D-Glu-*meso*-DAP (iE-DAP), albeit less efficiently. Although *meso*-diaminopimelic acid (*meso*-DAP) or *meso*-lanthionine (*meso*-Lan) alone have been shown to activate Nod1 as well, the biological relevance of this remains unclear as the amounts required to induce Nod1 exceed the levels available in a bacterial cell (Uehara *et al.*, 2006). The presence of saccharides is not essential for stimulating NF-κB activation via Nod1 (Girardin *et al.*, 2003c). Interestingly, the cytoplasmic PG precursor UDP-MurNAc-$TriP_{DAP}$ can also activate NF-κB via Nod1 (Girardin *et al.*, 2003c). While modifications on the saccharides do not influence PG recognition by the Nod1 pathway, amidation of *meso*-DAP or substitutions of *meso*-DAP by L-lysine (L-Lys) or L-ornithine (L-Orn) abolishes Nod1-dependent activation of NF-κB (Girardin *et al.*, 2003c).

Surprisingly, before Nod1 discovery, $GMTriP_{DAP}$ was not known to be biologically active on eukaryotic cells. Historically, it was the GlcNAc-anhMurNAc-tetrapeptide motif with *meso*-DAP ($GanhMTetraP_{DAP}$) that was considered active and which was identified in 1982 by Goldman as a *Bordetella pertussis* cytotoxin, called tracheal cytotoxin (TCT) (Goldman *et al.*, 1982). More recently, murine Nod1 (mNod1) rather than the human Nod1 (hNod1) was shown to sense TCT (Magalhaes *et al.*, 2005). In fact, while mNod1 senses TCT better, it also mediates sensing of $GMTriP_{DAP}$, although less efficiently than hNod1. Conversely, hNod1 also senses TCT but at much higher doses than those required for $GMTriP_{DAP}$-sensing (Magalhaes *et al.*, 2005). Despite the structural difference between the two Nod1 agonists, Nod1-sensing specificity for *meso*-DAP-type PG is conserved. Stimulation of mouse peritoneal macrophages with TCT, lactyl-tetrapeptide with *meso*-DAP or FK-156 led to cytokines production such as tumour-necrosis factor alpha (TNF-α), IL-6,

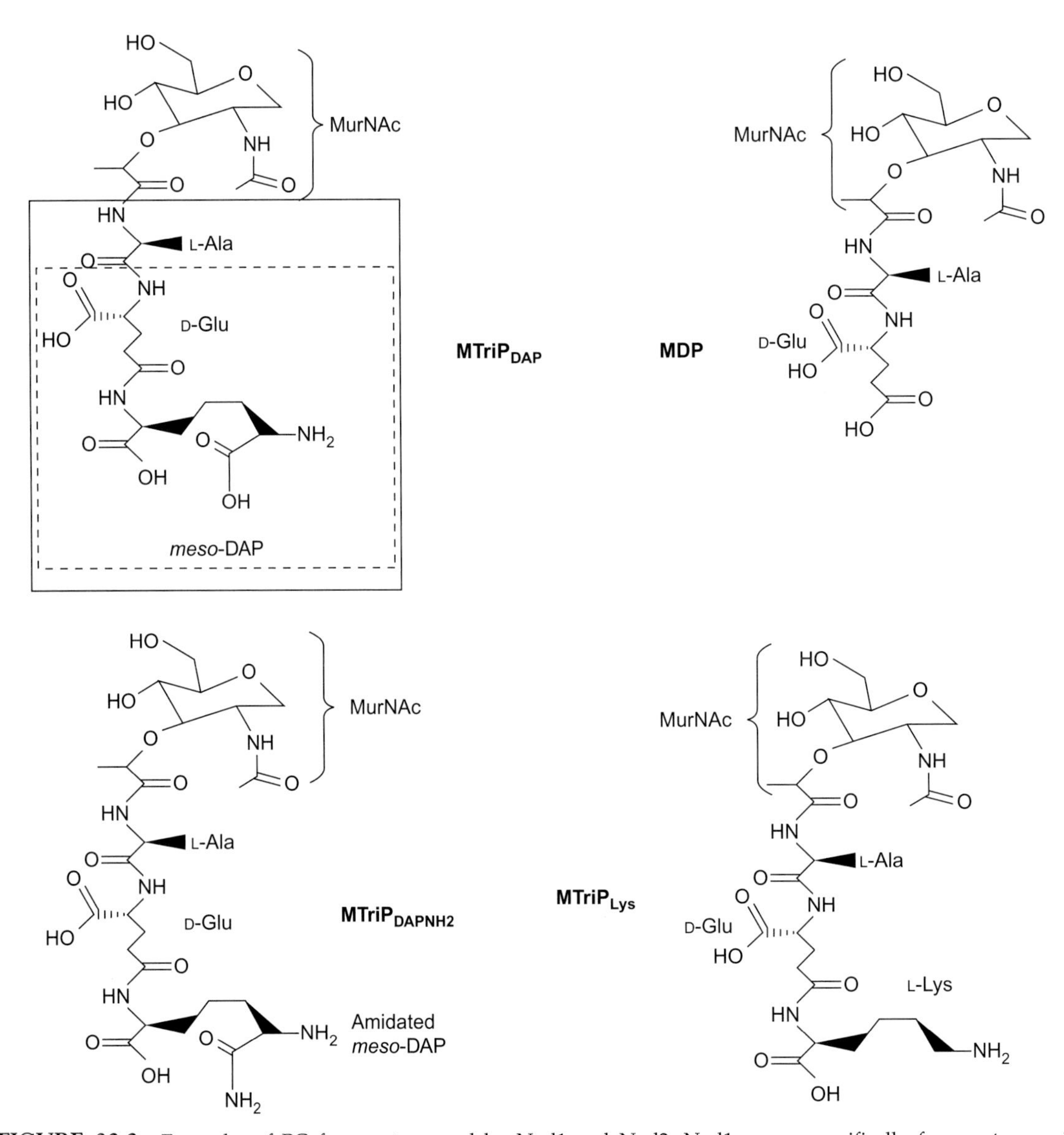

FIGURE 32.2 Examples of PG fragments sensed by Nod1 and Nod2. Nod1 senses specifically fragments carrying the *meso*-diaminopimelic acid (*meso*-DAP) at the third position of the stem peptide (see $MTriP_{DAP}$). The stem tripeptide (box) is the minimal structure that is as active as the naturally occurring GlcNAc-MurNAc-L-alanyl-γ-D-glutamyl-*meso*-diaminopimelic acid ($GMTriP_{DAP}$), indicating that the amino sugar backbone is not required for sensing. Hence, *N*-acetyl-muramoyl-L-alanyl amidases that cleave the bond between the *N*-acetylmuramic acid (MurNAc) and the L-alanine (L-Ala) residues do not impact on the biological activity of the released PG fragments towards Nod1. The dipeptide γ-D-glutamyl-*meso*-DAP (iE-DAP) (dashed box) is the minimally active fragment, although to a lesser extent than the tripeptide. However, Nod2 has a broader sensing specificity as it detects several distinct muropeptides, such as muramyl dipeptide (MDP), $MTriP_{DAPNH2}$ and $GMTriP_{Lys}$. The minimal structure sensed by Nod2 is MDP and, in contrast to Nod1, the amino sugar MurNAc is essential for biological activity. Thus, the stem peptides are biologically inactive towards Nod2.

IL-10 and keratinocyte chemoattractant (KC, the functional homologue to human IL-8) and also nitric oxide (NO). This stimulation was Nod1-dependent because macrophages from *card4*-deficient mice did not respond to PG fragments (Magalhaes *et al.*, 2005). The human and mouse Nod1 specificities may come from their distinct ecological niches and pathogens encountered during evolution. Recent studies have also addressed the PG fragment specificity of Nod1 proteins from additional species (Tohno *et al.*, 2008).

Interestingly, mutagenesis studies of hNod1 showed that Nod1 agonist specificities could be changed by two amino acid substitutions in the LRR domain (Girardin *et al.*, 2005). Although until now there is no direct evidence that hNod1 and, particularly, its LRR domain, binds directly to $GMTriP_{DAP}$, the mutagenesis studies favour the conclusion of a direct interaction. Interestingly, NLRs appear to have evolved from a very ancient ancestor protein since the yeast, *Candida albicans*, was shown to sense muramyl dipeptides by a related NLR protein, Cyr1p (Xu *et al.*, 2008). These authors showed that sensing was mediated by direct binding of muramyl dipeptides to the LRR domain of Cyr1p (Xu *et al.*, 2008).

Like Nod1, Nod2 senses PG fragments leading to NF-κB activation. However, unlike Nod1, Nod2 is able to sense PG from both Gram-negative and -positive bacteria. Moreover, the LRR domain is essential for PG recognition. Modification of one amino acid abolished Nod2-sensing or inversely increased NF-κB activation in an unregulated fashion (Tanabe *et al.*, 2004). After recognition of PG fragments via Nod2, NF-κB activation leads to production of cytokines, e.g. IL-1, IL-6 and TNF-α (Kobayashi *et al.*, 2005; Magalhaes *et al.*, 2005).

The minimal structure for recognition is the MDP motif (Inohara *et al.*, 2003; Girardin *et al.*, 2003b) known for 25 years for its adjuvant properties (Adam *et al.*, 1974; Ellouz *et al.*, 1974; Kotani *et al.*, 1976). Other PG fragments can be recognized via Nod2, i.e. MurNAc-tripeptide with L-Lys, with L-Orn or with amidated *meso*-DAP in third position and UDP-MurNAc–dipeptide (Girardin *et al.*, 2003c). Thus, Nod2 has a larger spectrum of PG recognition than Nod1, explaining its ability to sense PG from both Gram-negative and -positive bacteria. In contrast to Nod1, the MurNAc residue is essential for Nod2 stimulation because PG fragments hydrolysed by human serum amidase, chemically reduced or carrying anhMurNAc instead of MurNAc are not able to activate NF-κB via Nod2 (Girardin *et al.*, 2003c; Chaput *et al.*, 2006). Finally, increased polymerization of the sugar backbone to generate glycan chains with attached dipeptides drastically reduces Nod2-sensing of PG fragments, clearly demonstrating that the agonists are muropeptides instead of soluble polymeric fragments of PG (Inohara *et al.*, 2003). These observations have important implications as the initial and classical experiments demonstrating Nod1- and Nod2-sensing of PG were based on transfection assays into epithelial cells. Hence, stimulation of the Nod1 and Nod2 pathway by the insoluble PG requires processing of the PG into muropeptides by the host machinery. It can be speculated that this could involve host lysozyme, though it remains to be demonstrated.

4. THE ROLE OF NOD PROTEINS IN SENSING OF BACTERIAL INFECTIONS

The discovery of TLRs in 1996 (Lemaitre *et al.*, 1996), which are transmembrane proteins with their LRR domain exposed to the extracellular milieu, raised a paradox in understanding host–microbe interactions: how does a host distinguish the commensal bacteria from pathogenic bacteria? During the subsequent years it became apparent that most TLRs are absent, sequestered, downregulated or non-functional in most mucosal surfaces exposed to commensal flora (Abreu *et al.*, 2005). The presumed cytosolic localization of NLRs led to the hypothesis that NLRs' role might be to discriminate between harmless commensal bacteria and pathogens. Of note, Nod1 was detected in almost all tissues and cell types tested, in particular in epithelial cells, whereas Nod2 seemed to have a more restricted cell type distribution. Specifically, Nod2 was predominantly expressed in cells of the myeloid lineage (Ogura *et al.*, 2001b), although it can be upregulated in an NF-κB-dependent manner (Rosenstiel *et al.*, 2003). The first evidence for a role of NLRs in bacterial sensing came for experiments with *Shigella flexneri* (Girardin *et al.*, 2001). Accordingly, consistent with a role for NLRs in detecting only pathogens, the detection of invasive *S. flexneri* was found to be Nod1-mediated in epithelial cells (Girardin *et al.*, 2001). These observations were extended to other intracellular pathogens such as *Chlamydia* spp., enteroinvasive *Escherichia coli* and *Listeria monocytogenes* (Kim *et al.*, 2004; Opitz *et al.*, 2005, 2006; Welter-Stahl *et al.*, 2006). Consistent with an intracellular sensing pathway, Nod1 is localized at the membrane at the site of bacterial entry (Kufer *et al.*, 2008). Similarly, Nod2 was localized to the plasma membrane (Lécine *et al.*, 2007). The membrane localization of both Nod1 and Nod2 seems essential for signalling. Indeed, variants of Nod1 defective in signalling remain cytosolic (Kufer *et al.*, 2008). Despite a more restricted pattern of expression for Nod2, several groups have shown its role in sensing both Gram-positive and Gram-negative bacteria, including *Streptococcus pneumoniae*, *Mycobacterium tuberculosis*, *L. monocytogenes*, *Salmonella enterica* sv. Typhimurium and *S. flexneri* among others (Hisamatsu *et al.*, 2003; Opitz *et al.*, 2004; Kobayashi *et al.*, 2005; Ferwerda *et al.*, 2005; Kufer *et al.*, 2006).

As expected, most studies addressed NLRs sensing of intracellular pathogens, thereby favouring a role for NLRs to detect invasive pathogens complementary to TLRs' extracellular detection of bacteria. However, this model was challenged by two examples of exclusively extracellular pathogens, *Helicobacter pylori* and *Pseudomonas aeruginosa* (Viala *et al.*, 2004; Travassos *et al.*, 2005). In the *H. pylori* example, Nod1-sensing of the infection by epithelial cells was dependent on the presence of a type IV secretion system, the Cag apparatus, which facilitated delivery of PG fragments into the epithelial cytosol by a yet unknown mechanism. Additionally, *P. aeruginosa* induction of Nod1 remains so far unexplained, but not unique. Indeed, the different pathogens described historically as releasing TCT that cytotoxically targets ciliated epithelial cells, *B. pertussis* and *Neisseria gonorrhoeae* (Goldman *et al.*, 1982; Melly *et al.*, 1984), are exclusively extracellular pathogens. Hence, these observations indicate that non-professional phagocytes, such as epithelial cells, under certain conditions, could sense muropeptides present in the extracellular milieu. This hypothesis was validated at least for MDP, which can be transported into epithelial cells by the hPepT1 transporter (Vavricka *et al.*, 2004; Ismair *et al.*, 2006). In phagocytes, the hemi-channel pannexin-1 was shown to mediate cytosolic delivery of several distinct classes of pathogen-associated molecular patterns (PAMPs) including PG fragments (Kanneganti *et al.*, 2007; Marina-García *et al.*, 2008). Nevertheless, it still remains unknown how Nod1 and Nod2 ligands other than MDssP gain access to the cytosol of non-phagocytic cells. A partial answer may come from the pathogens themselves. For instance, certain extracellular pathogens produce pore-forming toxins that can mediate transient delivery of muropeptides into target cells (Ratner *et al.*, 2007).

Despite the increasing interest given to the role of NLRs in bacterial sensing and infections, most studies were done on cellular models and very few addressed their role *in vivo*. The first study linking an NLR in susceptibility to a bacterial infection came from *H. pylori*. Viala *et al.* (2004) showed that *card4*-deficient mice were unable to control the colonization of the gastric mucosa by strains carrying the Cag apparatus. *H. pylori* induced the production of cytokines and β-defensins in a Nod1-dependent manner that could participate actively in controlling the bacterial load (Wada *et al.*, 2001; Viala *et al.*, 2004; Boughan *et al.*, 2006). Interestingly, most studies that show a role for Nod2 in controlling infections are also restricted to gastrointestinal pathogens such as *L. monocytogenes*, *Yersinia pseudotuberculosis* and *S. enterica* sv. Typhimurium delivered by the oral route (Kobayashi *et al.*, 2005; Meinzer *et al.*, 2008). In these examples, increased susceptibility was also correlated with decreased expression of α-defensins (Kobayashi *et al.*, 2005; Meinzer *et al.*, 2008). In the few studies performed to date, neither Nod1 nor Nod2 appeared to play a major role in controlling systemic infections (Kobayashi *et al.*, 2005; Meinzer *et al.*, 2008). A possible explanation might come from the redundancy of Nod1 and Nod2, or of NLRs with TLRs during systemic infection. Indeed, mice deficient for Rip2, the Nod1 and Nod2 adaptor molecule, appeared more susceptible to *L. monocytogenes* delivered by the intravenous route (Chin *et al.*, 2002). More recently, Kim *et al.* (2008) addressed the issue differently. They showed that Nod1 and Nod2 play an important role in systemic infection by *L. monocytogenes* when mice are pre-exposed and become tolerant to TLR agonist. These observations highlight a complex interplay of NLRs and TLRs that depends on the route of infection, the host tissue and the cell type targeted by the pathogen. Furthermore, during an infection, the host may be faced with several distinct pathogens (a virus and bacterium or two bacteria). For example, *S. pneumoniae* has been shown to induce exclusively the Nod2-pathway (Opitz *et al.*, 2004) but, during co-infection with *Haemophilus influenzae*, a Gram-negative bacterium with a similar niche and disease spectrum as *S. pneumoniae*, is cleared faster by neutrophil-killing in a Nod1-dependent fashion, thus highlighting a distinct role for Nod1 compared to Nod2 in priming neutrophils (Lysenko *et al.*, 2007). One of the major challenges in the future will be to dissect precisely the role of NLRs and TLRs during infection.

5. PEPTIDOGLYCAN METABOLISM AND MODULATION OF NOD-DEPENDENT RESPONSES

Detection of the pathogen requires a sensing mechanism in the host. Conversely, it also requires for the pathogen to present its NLR or TLR agonist to the host-sensing machinery. In the case of the Nod proteins, during an infection, the PG metabolism is of seminal importance for the host to detect the presence of the bacterium. Hence, all strategies to modulate the release of PG fragments and manipulate the Nod-pathways may give the pathogen an advantage and access to new niches. Obviously, while some pathogens might benefit from a stealth mechanism by minimizing release of certain muropeptides or stem peptides, others can benefit from an excess of inflammation.

To date, few studies have addressed the role of PG metabolism in the context of host–microbe interactions, despite the fact that most studies on bioactive PG fragments such as TCT clearly indicated that PG degradation by hydrolases was central to this process (Rosenthal *et al.*, 1987). Furthermore, with the advent of genomics and high throughput screening of virulence factors, several genes involved in PG metabolism have been identified that are crucial for pathogenesis (reviewed in Boneca, 2005). Stimulation of the Nod receptors depends on the shedding of PG fragments and this is species-dependent (Hasegawa *et al.*, 2006). Most studies have shown that Gram-negative bacteria release essentially monomeric anhydromuropeptides (Rosenthal *et al.*, 1987). Therefore, it is to be expected that enzymes involved in their generation, the lytic transglycosylases, would be crucial for inducing the Nod1 pathway (anhydro-muropeptides are not Nod2 ligands, see above). Indeed, Viala *et al.* (2004) showed that the Nod1 stimulatory activity of *H. pylori* depended on the activity of its major lytic transglycosylase, Slt. Similarly, the lytic transglycosylase MltB was identified as important in a mutagenesis screen of *Neisseria meningitidis* mutants attenuated *in vivo*, although the molecular basis for influencing activity remains to be studied (Sun *et al.*, 2000). Lytic transglycosylases are responsible for the degradation of half of the PG layer during a growth cycle (Park and Uehara, 2008) and a transglycosylase-mutated *E. coli* strain had a drastically decreased turnover of its PG (Kraft *et al.*, 1999). But, despite such a high level of turnover, *E. coli* and other enterobacteria shed only 5% of the PG fragments into the environment (Park and Uehara, 2008) by recycling them very efficiently (Figure 32.3). Accordingly, Hasegawa *et al.* (2006) observed that enterobacteria released into the supernatant had less Nod1 stimulatory molecules than *Bacillus*, despite the fact that enterobacteria have exclusively *meso*-DAP-type peptidoglycan, whereas *Bacillus* spp. have most of their *meso*-DAP amidated (hence, a Nod2 ligand) (Girardin *et al.*, 2003c). The same study also confirmed the low Nod2 stimulatory activity of supernatants of Gram-negative bacteria in general (Hasegawa *et al.*, 2006). This observation is consistent with the inability of Nod2 to sense anhydro-muramyldipeptide (Chaput *et al.*, 2006). Proof that the recycling pathway was involved in modulating Nod1 stimulation came from a recent study of *S. flexneri* (Nigro *et al.*, 2008). Both *mppA* and *ampG* mutants, which have decreased ability to recycle PG degradation products, activated NF-κB more strongly and secreted more IL-8 in a Nod1-dependent manner in HEK293 epithelial cells. Similarly, even the *mppA* and *ampG S. flexneri* mutants had a very reduced ability to induce Nod2 (Nigro *et al.*, 2008). Hence, Gram-negative bacteria naturally produce less Nod2 ligands. In fact, *H. pylori* has refined this strategy into a unique stealth mechanism to avoid also Nod1-sensing. *H. pylori* is a spiral-rod shaped bacteria that undergoes a morphological transition into a coccoid shape. The shape transition has been correlated with the conversion of most of the Nod1 ligand, $GMTriP_{DAP}$, into the Nod2 ligand, GMDiP, resulting in a drastic decrease in the ability of coccoid-shaped bacteria to induce NF-κB and IL-8 secretion (Chaput *et al.*, 2006). Interestingly, the morphological transition required the activity of another PG hydrolase, the AmiA protein, which encodes a putative *N*-acetylmuramoyl-L-alanyl amidase. Several other studies have highlighted the role of PG hydrolases in virulence and modulation of innate immune response (Lenz *et al.*, 2003; Cabanes *et al.*, 2004; Humann *et al.*, 2007). However, very few have addressed the potential role for the Nods in this mechanism.

Although only a few examples have been studied, bacteria have developed remarkable weaponry to manipulate the host responses. Conversely, the host has also strategies to enhance pathogen detection and mount the best response. The host produces its own PG hydrolases (Fleming, 1922; Valinger *et al.*, 1982; Striker *et al.*, 1987) of which lysozyme is the best characterized. Lysozyme, as a lytic transglycosylase, hydrolyses the γ-(1→4) bond between the GlcNAc and the MurNAc, but generates instead fully hydrated muropeptides. Hence, lysozyme digestion products can induce both Nod1 and Nod2 activation in contrast to lytic transglycosylases. However, to date there is no direct evidence that lysozyme participated in the process of Nod1 and Nod2 signalling.

Since 1959, it has been known that bacteria modify their PG to inhibit the action of lysozyme on their cell walls (reviewed in Clarke and Dupont, 1992). Several studies have characterized the mechanisms of lysozyme resistance both in Gram-positive and Gram-negative bacteria (Vollmer and Tomasz, 2000; Weadge *et al.*, 2005; Raymond *et al.*, 2005; Bera *et al.*, 2005;

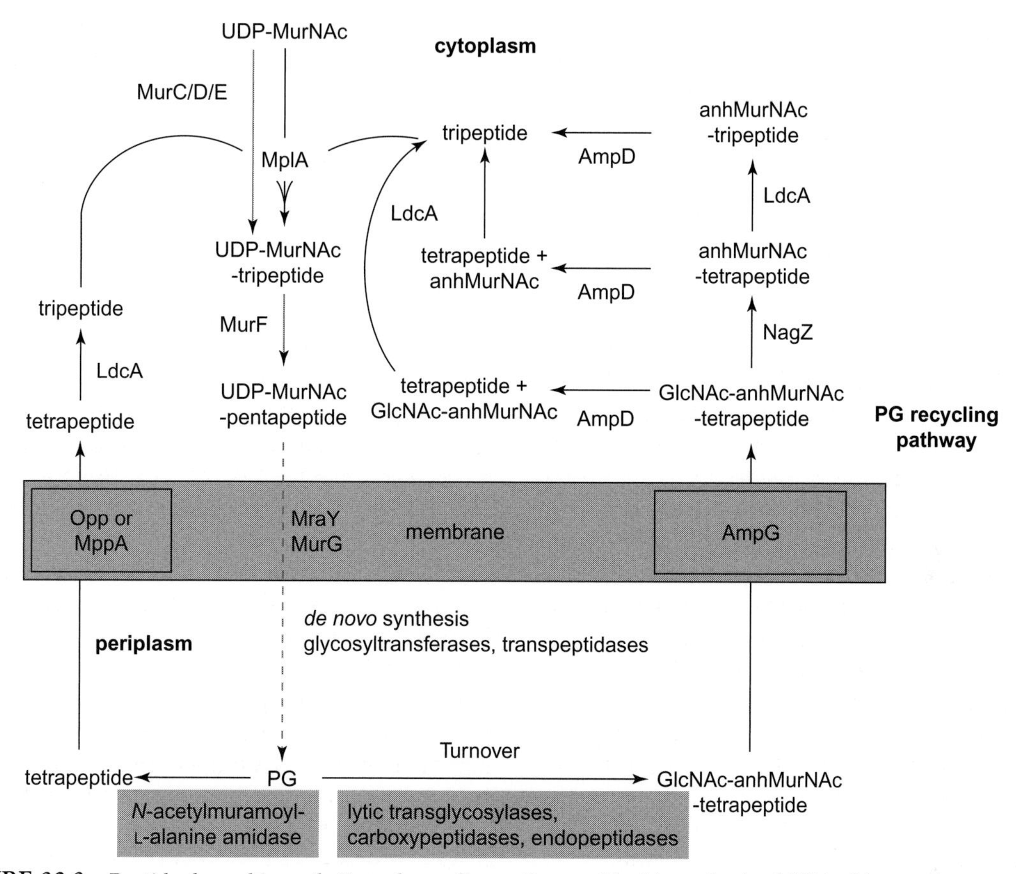

FIGURE 32.3 Peptidoglycan biosynthetic and recycling pathways. The biosynthesis of PG is able to re-incorporate PG degradation products that are recycled by a dedicated pathway. The PG layer is constantly being re-modelled during cell growth and division by PG hydrolases and results in the release of 50% of the pre-existing layer. These degradation products are recycled by dedicated transporters into the cytoplasm (AmpG, Opp system and MppA) and hydrolysed down to the tripeptide and the amino sugars. Furthermore, the tripeptide can be metabolized into amino acids or re-incorporated into the PG biosynthetic pathway by the MlpA protein. Abbreviations: anhMurNAc, *N*-acetylanhydromuramic acid; GlcNAc, *N*-acetylglucosamine; MurNAc, *N*-acetylmuramic acid; PG, peptidoglycan; UDP-MurNAc, uridine diphosphate-MurNAc.

Dillard and Hackett, 2005; Boneca *et al.*, 2007; Veiga *et al.*, 2007; Hébert *et al.*, 2007). Of particular note, bacteria that modify their PGs to resistant lysozyme, e.g. *S. pneumoniae* or *L. monocytogenes*, have a poor ability to induce Nod2 or Nod1 (Girardin *et al.*, 2003a; Travassos *et al.*, 2004; Hasegawa *et al.*, 2006), respectively. Boneca (Boneca *et al.*, 2007) addressed the role of PG modifications by *L. monocytogenes* as a mechanism to escape Nod-sensing. The PG deacetylase of *L. monocytogenes*, PgdA, deacetylated the PG of this bacterial species by removing the acetyl group from the GlcNAc saccharide. The authors showed that PgdA is essential for lysozyme resistance and Nod1-/Nod2-sensing *in vitro*, survival in macrophages and colonization of mice. Furthermore, in the absence of PgdA, *L. monocytogenes* induced a stronger cytokine production, e.g. IL-6 and TNF-α, due to a synergistic effect between Nod1 and TLR-2, whereas wild-type *L. monocytogenes* mainly induced responses in a TLR-2-dependent manner. Besides mechanisms of resistance to lysozyme, other modifications of the PG could also contribute to modulating Nod-sensing as has been suggested by Wolfert *et al.* (2007).

6. CONCLUSIONS AND PERSPECTIVES

The discovery of NLRs has opened a new avenue of research in the field of infectious diseases and in the investigation of innate and adaptive immunity that is just in its infancy. Future challenges (see Research Focus Box) lie in dissecting the signalling mechanisms of NLRs and their interactions with other pathways, such as those of the TLRs. Despite these inducing the same transcription factors, the downstream effects are often distinct. Furthermore, it will be interesting to attempt to understand the spatial and temporal activation of these different innate immune receptors during homeostasis, as opposed to during an infection, from the cellular level up to the entire host.

Conversely, bacteria have evolved with their hosts and developed ingenious strategies to manipulate the host response to their benefit. Such strategies have been extensively studied for other bacterial structures, such as LPS, or secondary polysaccharides, e.g. as in capsules. With the discovery of the Nods, the study of PG has been brought back into the spotlight and has already led to the unravelling of some novel aspects of infectious diseases. Nevertheless, much lies ahead and it can be speculated that we have only scratched the surface of the biological importance of PG metabolism in host–microbe interactions. Despite a focus on Nod-sensing in mammals, PG metabolism is also central to sensing of bacteria in other eukaryotic systems from yeast to insects to plants (Lemaitre and Hoffmann, 2007; Gust *et al.*, 2007; Xu *et al.*, 2008). Thus, the field will benefit from studies not only in mammals but also from studies on host–microbe interactions in alternate models.

RESEARCH FOCUS BOX

- Do Nod1 and Nod2 bind muropeptides directly?
- What are the signalling mechanisms of NLRs?
- How do NLRs interact with the TLRs?
- How do different PRRs such as NLRs and TLRs that use the same transcription factors often induce distinct responses?
- How are the PRRs regulated both spatially and temporally in a cell?
- What differentiates homeostasis from an inflammatory response?
- How do commensals and pathogens differentially engage the NLRs?
- How do pathogens regulate peptidoglycan metabolism to modulate the host response?
- Are there other peptidoglycan modifications involved in modulation of NLR sensing?
- What is the contribution of lysozyme and other host peptidoglycan hydrolases to initiate/terminate the host response?
- What are the common and distinct features of PG-sensing in different eukaryotic systems such as yeast, insects, plants or mammals?

References

Abreu, M.T., Fukata, M., Arditi, M., 2005. TLR signaling in the gut in health and disease. J. Immunol. 174, 4453–4460.

Adam, A., Ciorbaru, R., Ellouz, F., Petit, J.F., Lederer, E., 1974. Adjuvant activity of monomeric bacterial cell wall peptidoglycans. Biochem. Biophys. Res. Commun. 56, 561–567.

Agostini, L., Martinon, F., Burns, K., McDermott, M.F., Hawkins, P.N., Tschopp, J., 2004. NALP3 forms an IL-1β-processing inflammasome with increased activity in Muckle-Wells autoinflammatory disorder. Immunity 20, 319–325.

Aliprantis, A.O., Yang, R.B., Mark, M.R., et al., 1999. Cell activation and apoptosis by bacterial lipoproteins through toll-like receptor-2. Science 285, 736–739.

Barnich, N., Hisamatsu, T., Aguirre, J.E., Xavier, R., Reinecker, H.C., Podolsky, D.K., 2005. GRIM-19 interacts with nucleotide oligomerization domain 2 and serves as downstream effector of anti-bacterial function in intestinal epithelial cells. J. Biol. Chem. 280, 19021–19026.

Bera, A., Herbert, S., Jakob, A., Vollmer, W., Gotz, F., 2005. Why are pathogenic staphylococci so lysozyme resistant? The peptidoglycan *O*-acetyltransferase OatA is the major determinant for lysozyme resistance of *Staphylococcus aureus*. Mol. Microbiol. 55, 778–787.

Boneca, I.G., 2005. The role of peptidoglycan in pathogenesis. Curr. Opin. Microbiol. 8, 46–53.

Boneca, I.G., Dussurget, O., Cabanes, D., et al., 2007. A critical role for peptidoglycan *N*-deacetylation in *Listeria* evasion from the host innate immune system. Proc. Natl. Acad. Sci. USA 104, 997–1002.

Bonen, D.K., Ogura, Y., Nicolae, D.L., et al., 2003. Crohn's disease-associated NOD2 variants share a signaling defect in response to lipopolysaccharide and peptidoglycan. Gastroenterology 124, 140–146.

Boothby, D., Daneo-Moore, L., Higgins, M.L., Coyette, J., Shockman, G.D., 1973. Turnover of bacterial cell wall peptidoglycans. J. Biol. Chem. 248, 2161–2169.

Boughan, P.K., Argent, R.H., Body-Malapel, M., et al., 2006. Nucleotide-binding oligomerization domain-1 and epidermal growth factor receptor: critical regulators of β-defensins during *Helicobacter pylori* infection. J. Biol. Chem. 281, 11637–11648.

Cabanes, D., Dussurget, O., Dehoux, P., Cossart, P., 2004. Auto, a surface associated autolysin of *Listeria monocytogenes* required for entry into eukaryotic cells and virulence. Mol. Microbiol. 51, 1601–1614.

Chamaillard, M., Hashimoto, M., Horie, Y., et al., 2003a. An essential role for NOD1 in host recognition of bacterial peptidoglycan containing diaminopimelic acid. Nat. Immunol. 4, 702–707.

Chamaillard, M., Philpott, D., Girardin, S.E., et al., 2003b. Gene-environment interaction modulated by allelic heterogeneity in inflammatory diseases. Proc. Natl. Acad. Sci. USA 100, 3455–3460.

Chaput, C., Ecobichon, C., Cayet, N., et al., 2006. Role of AmiA in the morphological transition of *Helicobacter pylori* and in immune escape. PLoS Pathog. 2, e97.

Chen, C.M., Gong, Y., Zhang, M., Chen, J.J., 2004. Reciprocal cross-talk between Nod2 and TAK1 signaling pathways. J. Biol. Chem. 279, 25876–25882.

Chin, A.I., Dempsey, P.W., Bruhn, K., Miller, J.F., Xu, Y., Cheng, G., 2002. Involvement of receptor-interacting protein 2 in innate and adaptive immune responses. Nature 416, 190–194.

Clarke, A.J., Dupont, C., 1992. *O*-acetylated peptidoglycan: its occurrence, pathobiological significance, and biosynthesis. Can. J. Microbiol. 38, 85–91.

da Silva Correia, J., Miranda, Y., Leonard, N., Ulevitch, R., 2007. SGT1 is essential for Nod1 activation. Proc. Natl. Acad. Sci. USA 104, 6764–6769.

Dillard, J.P., Hackett, K.T., 2005. Mutations affecting peptidoglycan acetylation in *Neisseria gonorrhoeae* and *Neisseria meningitidis*. Infect. Immun. 73, 5697–5705.

Dufner, A., Pownall, S., Mak, T.W., 2006. Caspase recruitment domain protein 6 is a microtubule-interacting protein that positively modulates NF-κB activation. Proc. Natl. Acad. Sci. USA 103, 988–993.

Dziarski, R., Gupta, D., 2005. *Staphylococcus aureus* peptidoglycan is a toll-like receptor 2 activator: a reevaluation. Infect. Immun. 73, 5212–5216.

Dziarski, R., Tapping, R.I., Tobias, P.S., 1998. Binding of bacterial peptidoglycan to CD14. J. Biol. Chem. 273, 8680–8690.

Ellouz, F., Adam, A., Ciorbaru, R., Lederer, E., 1974. Minimal structural requirements for adjuvant activity of bacterial peptidoglycan derivatives. Biochem. Biophys. Res. Commun. 59, 1317–1325.

Ferwerda, G., Girardin, S.E., Kullberg, B.J., et al., 2005. NOD2 and toll-like receptors are nonredundant recognition systems of *Mycobacterium tuberculosis*. PLoS Pathog. 1, 279–285.

Ferwerda, G., Kramer, M., de Jong, D., et al., 2008. Engagement of NOD2 has a dual effect on proIL-1β mRNA transcription and secretion of bioactive IL-1β. Eur. J. Immunol. 38, 184–191.

Fleming, A., 1922. On a remarkable bacteriolytic element found in tissues and secretions. Philos. Trans. Roy. Soc. Lond., Sect. B, Biol. Sci. 93, 306–317.

Fritz, J.H., Ferrero, R.L., Philpott, D.J., Girardin, S.E., 2006. Nod-like proteins in immunity, inflammation and disease. Nat. Immunol. 7, 1250–1257.

Fritz, J.H., Le Bourhis, L., Sellge, G., et al., 2007. Nod1-mediated innate immune recognition of peptidoglycan contributes to the onset of adaptive immunity. Immunity 26, 445–459.

Girardin, S.E., Tournebize, R., Mavris, M., et al., 2001. CARD4/Nod1 mediates NF-κB and JNK activation by invasive *Shigella flexneri*. EMBO Rep. 2, 736–742.

Girardin, S.E., Boneca, I.G., Carneiro, L.A., et al., 2003a. Nod1 detects a unique muropeptide from gram-negative bacterial peptidoglycan. Science 300, 1584–1587.

Girardin, S.E., Boneca, I.G., Viala, J., et al., 2003b. Nod2 is a general sensor of peptidoglycan through muramyl dipeptide (MDP) detection. J. Biol. Chem. 278, 8869–8872.

Girardin, S.E., Travassos, L.H., Herve, M., et al., 2003c. Peptidoglycan molecular requirements allowing detection by Nod1 and Nod2. J. Biol. Chem. 278, 41702–41708.

Girardin, S.E., Jehanno, M., Mengin-Lecreulx, D., Sansonetti, P.J., Alzari, P.M., Philpott, D.J., 2005. Identification of the critical residues involved in peptidoglycan detection by Nod1. J. Biol. Chem. 280, 38648–38656.

Goldman, W.E., Klapper, D.G., Baseman, J.B., 1982. Detection, isolation, and analysis of a released *Bordetella pertussis* product toxic to cultured tracheal cells. Infect. Immun. 36, 782–794.

Goodell, E.W., Fazio, M., Tomasz, A., 1978. Effect of benzylpenicillin on the synthesis and structure of the cell envelope of *Neisseria gonorrhoeae*. Antimicrob. Agents Chemother. 13, 514–526.

Gust, A.A., Biswas, R., Lenz, H.D., et al., 2007. Bacteria-derived peptidoglycans constitute pathogen-associated molecular patterns triggering innate immunity in *Arabidopsis*. J. Biol. Chem. 282, 32338–32348.

Hampe, J., Cuthbert, A., Croucher, P.J., et al., 2001. Association between insertion mutation in NOD2 gene and Crohn's disease in German and British populations. Lancet 357, 1925–1928.

Hasegawa, M., Yang, K., Hashimoto, M., et al., 2006. Differential release and distribution of Nod1 and Nod2 immunostimulatory molecules among bacterial species and environments. J. Biol. Chem. 281, 29054–29063.

Hebeler, B.H., Young, F.E., 1976. Chemical composition and turnover of peptidoglycan in *Neisseria gonorrhoeae*. J. Bacteriol. 126, 1180–1185.

Hébert, L., Courtin, P., Torelli, R., et al., 2007. *Enterococcus faecalis* constitutes an unusual bacterial model in lysozyme resistance. Infect. Immun. 75, 5390–5398.

Hisamatsu, T., Suzuki, M., Reinecker, H.C., Nadeau, W.J., McCormick, B.A., Podolsky, D.K., 2003. CARD15/NOD2 functions as an antibacterial factor in human intestinal epithelial cells. Gastroenterology 124, 993–1000.

Hoffman, H.M., Mueller, J.L., Broide, D.H., Wanderer, A.A., Kolodner, R.D., 2001. Mutation of a new gene encoding a putative pyrin-like protein causes familial cold autoinflammatory syndrome and Muckle-Wells syndrome. Nat. Genet. 29, 301–305.

Hsu, Y.M., Zhang, Y., You, Y., et al., 2007. The adaptor protein CARD9 is required for innate immune responses to intracellular pathogens. Nat. Immunol. 8, 198–205.

Hsu, L.C., Ali, S.R., McGillivray, S., et al., 2008. A NOD2-NALP1 complex mediates caspase-1-dependent IL-1β secretion in response to *Bacillus anthracis* infection and muramyl dipeptide. Proc. Natl. Acad. Sci. USA 105, 7803–7808.

Hugot, J.P., Chamaillard, M., Zouali, H., et al., 2001. Association of NOD2 leucine-rich repeat variants with susceptibility to Crohn's disease. Nature 411, 599–603.

Humann, J., Bjordahl, R., Andreasen, K., Lenz, L.L., 2007. Expression of the p60 autolysin enhances NK cell activation and is required for *Listeria monocytogenes* expansion in IFN-γ-responsive mice. J. Immunol. 178, 2407–2414.

Hysi, P., Kabesch, M., Moffatt, M.F., et al., 2005. NOD1 variation, immunoglobulin E and asthma. Hum. Mol. Genet. 14, 935–941.

Inohara, N., Koseki, T., del Peso, L., et al., 1999. Nod1, an Apaf-1-like activator of caspase-9 and nuclear factor-B. J. Biol. Chem. 274, 14560–14567.

Inohara, N., Koseki, T., Lin, J., et al., 2000. An induced proximity model for NF-κB activation in the Nod1/RICK and RIP signaling pathways. J. Biol. Chem. 275, 27823–27831.

Inohara, N., Ogura, Y., Chen, F.F., Muto, A., Nunez, G., 2001. Human Nod1 confers responsiveness to bacterial lipopolysaccharides. J. Biol. Chem. 276, 2551–2554.

Inohara, N., Ogura, Y., Fontalba, A., et al., 2003. Host recognition of bacterial muramyl dipeptide mediated through NOD2. Implications for Crohn's disease. J. Biol. Chem. 278, 5509–5512.

Ismair, M.G., Vavricka, S.R., Kullak-Ublick, G.A., Fried, M., Mengin-Lecreulx, D., Girardin, S.E., 2006. hPepT1 selectively transports muramyl dipeptide but not Nod1-activating muramyl peptides. Can. J. Physiol. Pharmacol. 84, 1313–1319.

Iwaki, D., Mitsuzawa, H., Murakami, S., et al., 2002. The extracellular toll-like receptor 2 domain directly binds peptidoglycan derived from *Staphylococcus aureus*. J. Biol. Chem. 277, 24315–24320.

Kanazawa, N., Okafuji, I., Kambe, N., et al., 2005. Early-onset sarcoidosis and CARD15 mutations with constitutive nuclear factor-kappaB activation: common genetic etiology with Blau syndrome. Blood 105, 1195–1197.

Kanneganti, T.D., Ozoren, N., Body-Malapel, M., et al., 2006. Bacterial RNA and small antiviral compounds activate caspase-1 through cryopyrin/Nalp3. Nature 440, 233–236.

Kanneganti, T.D., Lamkanfi, M., Kim, Y.G., et al., 2007. Pannexin-1-mediated recognition of bacterial molecules activates the cryopyrin inflammasome independent of Toll-like receptor signaling. Immunity 26, 433–443.

Kay, A.B., 2001. Allergy and allergic diseases. First of two parts. N. Engl. J. Med. 344, 30–37.

Kim, J.G., Lee, S.J., Kagnoff, M.F., 2004. Nod1 is an essential signal transducer in intestinal epithelial cells infected with bacteria that avoid recognition by Toll-like receptors. Infect. Immun. 72, 1487–1495.

Kim, Y.G., Park, J.H., Shaw, M.H., Franchi, L., Inohara, N., Náñez, G., 2008. The cytosolic sensors Nod1 and Nod2 are critical for bacterial recognition and host defense after exposure to Toll-like receptor ligands. Immunity 28, 246–257.

Kobayashi, K.S., Chamaillard, M., Ogura, Y., et al., 2005. Nod2-dependent regulation of innate and adaptive immunity in the intestinal tract. Science 307, 731–734.

Kotani, S., Watanabe, Y., Shimono, T., Harada, K., Shiba, T., 1976. Correlation between the immunoadjuvant activities and pyrogenicities of synthetic *N*-acetylmuramyl-peptides or -amino acids. Biken J. 19, 9–13.

Kraft, A.R., Prabhu, J., Ursinus, A., Holtje, J.V., 1999. Interference with murein turnover has no effect on growth but reduces β-lactamase induction in *Escherichia coli*. J. Bacteriol. 181, 7192–7198.

Kufer, T.A., Kremmer, E., Banks, D.J., Philpott, D.J., 2006. Role for erbin in bacterial activation of Nod2. Infect. Immun. 74, 3115–3124.

Kufer, T.A., Kremmer, E., Adam, A.C., Philpott, D.J., Sansonetti, P.J., 2008. The pattern-recognition molecule Nod1 is localized at the plasma membrane at sites of bacterial interaction. Cell. Microbiol. 10, 477–486.

Kurzawski, G., Suchy, J., Kladny, J., et al., 2004. The *NOD2* 3020insC mutation and the risk of colorectal cancer. Cancer Res. 64, 1604–1606.

LeBlanc, P.M., Yeretssian, G., Rutherford, N., et al., 2008. Caspase-12 modulates NOD signaling and regulates antimicrobial peptide production and mucosal immunity. Cell Host Microbe 3, 146–157.

Lécine, P., Esmiol, S., Metais, J.Y., et al., 2007. The NOD2-RICK complex signals from the plasma membrane. J. Biol. Chem. 282, 15197–15207.

Lemaitre, B., Nicolas, E., Michaut, L., Reichhart, J.M., Hoffmann, J.A., 1996. The dorsoventral regulatory gene cassette spatzle/Toll/cactus controls the potent antifungal response in *Drosophila* adults. Cell 86, 973–983.

Lemaitre, B., Hoffmann, J., 2007. The host defense of *Drosophila melanogaster*. Annu. Rev. Immunol. 25, 697–743.

Lenz, L.L., Mohammadi, S., Geissler, A., Portnoy, D.A., 2003. SecA2-dependent secretion of autolytic enzymes promotes *Listeria monocytogenes* pathogenesis. Proc. Natl. Acad. Sci. USA 100, 12432–12437.

Li, L., Bin, L.H., Li, F., et al., 2005. TRIP6 is a RIP2-associated common signaling component of multiple NF-κB activation pathways. J. Cell. Sci. 118, 555–563.

Lysenko, E.S., Clarke, T.B., Shchepetov, M., et al., 2007. Nod1 signaling overcomes resistance of *S. pneumoniae* to opsonophagocytic killing. PLoS Pathog. 3, e118.

Magalhaes, J.G., Philpott, D.J., Nahori, M.A., et al., 2005. Murine Nod1 but not its human orthologue mediates innate immune detection of tracheal cytotoxin. EMBO Rep. 6, 1201–1207.

Mariathasan, S., 2007. ASC, Ipaf and Cryopyrin/Nalp3: bona fide intracellular adapters of the caspase-1 inflammasome. Microbes Infect. 9, 664–671.

Mariathasan, S., Weiss, D.S., Newton, K., et al., 2006. Cryopyrin activates the inflammasome in response to toxins and ATP. Nature 440, 228–232.

Marina-García, N., Franchi, L., Kim, Y.G., et al., 2008. Pannexin-1-mediated intracellular delivery of muramyl dipeptide induces caspase-1 activation via cryopyrin/NLRP3 independently of Nod2. J. Immunol. 180, 4050–4057.

Martinon, F., Tschopp, J., 2007. Inflammatory caspases and inflammasomes: master switches of inflammation. Cell Death Differ. 14, 10–22.

Martinon, F., Agostini, L., Meylan, E., Tschopp, J., 2004. Identification of bacterial muramyl dipeptide as activator of the NALP3/cryopyrin inflammasome. Curr. Biol. 14, 1929–1934.

Martinon, F., Petrilli, V., Mayor, A., Tardivel, A., Tschopp, J., 2006. Gout-associated uric acid crystals activate the NALP3 inflammasome. Nature 440, 237–241.

Mauck, J., Glaser, L., 1970. Turnover of the cell wall of Bacillus subtilis W-23 during logarithmic growth. Biochem. Biophys. Res. Commun. 39, 699–706.

Mauck, J., Chan, L., Glaser, L., 1971. Turnover of the cell wall of Gram-positive bacteria. J. Biol. Chem. 246, 1820–1827.

McDonald, C., Chen, F.F., Ollendorff, V., et al., 2005. A role for Erbin in the regulation of Nod2-dependent NF-κB signaling. J. Biol. Chem. 280, 40301–40309.

McGovern, D.P., Hysi, P., Ahmad, T., et al., 2005. Association between a complex insertion/deletion polymorphism in NOD1 (CARD4) and susceptibility to inflammatory bowel disease. Hum. Mol. Genet. 14, 1245–1250.

Means, T.K., Lien, E., Yoshimura, A., Wang, S., Golenbock, D.T., Fenton, M.J., 1999. The CD14 ligands lipoarabinomannan and lipopolysaccharide differ in their requirement for Toll-like receptors. J. Immunol. 163, 6748–6755.

Meinzer, U., Esmiol-Welterlin, S., Barreau, F., et al., 2008. Nod2 mediates susceptibility to *Yersinia pseudotuberculosis* in mice. PLoS ONE 3, e2769.

Melly, M.A., McGee, Z.A., Rosenthal, R.S., 1984. Ability of monomeric peptidoglycan fragments from *Neisseria gonorrhoeae* to damage human fallopian-tube mucosa. J. Infect. Dis. 149, 378–386.

Meshcheryakova, E., Makarov, E., Philpott, D., Andronova, T., Ivanov, V., 2007. Evidence for correlation between the intensities of adjuvant effects and NOD2 activation by monomeric, dimeric and lipophylic derivatives of *N*-acetylglucosaminyl-*N*-acetylmuramyl peptides. Vaccine 25, 4515–4520.

Miceli-Richard, C., Lesage, S., Rybojad, M., et al., 2001. CARD15 mutations in Blau syndrome. Nat. Genet. 29, 19–20.

Morath, S., Stadelmaier, A., Geyer, A., Schmidt, R.R., Hartung, T., 2002. Synthetic lipoteichoic acid from *Staphylococcus aureus* is a potent stimulus of cytokine release. J. Exp. Med. 195, 1635–1640.

Netea, M.G., Ferwerda, G., de Jong, D.J., Girardin, S.E., Kullberg, B.J., van der Meer, J.W., 2005. *NOD2* 3020insC mutation and the pathogenesis of Crohn's disease: impaired IL-1β production points to a loss-of-function phenotype. Neth. J. Med. 63, 305–308.

Nigro, G., Fazio, L.L., Martino, M.C., et al., 2008. Muramylpeptide shedding modulates cell sensing of *Shigella flexneri*. Cell. Microbiol. 10, 682–695.

Ogura, Y., Bonen, D.K., Inohara, N., et al., 2001a. A frameshift mutation in NOD2 associated with susceptibility to Crohn's disease. Nature 411, 603–606.

Ogura, Y., Inohara, N., Benito, A., Chen, F.F., Yamaoka, S., Nunez, G., 2001b. Nod2, a Nod1/Apaf-1 family member that is restricted to monocytes and activates NF-κB. J. Biol. Chem. 276, 4812–4818.

Opitz, B., Puschel, A., Schmeck, B., et al., 2004. Nucleotide-binding oligomerization domain proteins are innate immune receptors for internalized *Streptococcus pneumoniae*. J. Biol. Chem. 279, 36426–36432.

Opitz, B., Forster, S., Hocke, A.C., et al., 2005. Nod1-mediated endothelial cell activation by *Chlamydophila pneumoniae*. Circ. Res. 96, 319–326.

Opitz, B., Puschel, A., Beermann, W., et al., 2006. *Listeria monocytogenes* activated p38 MAPK and induced IL-8 secretion in a nucleotide-binding oligomerization domain 1-dependent manner in endothelial cells. J. Immunol. 176, 484–490.

Park, J.T., Uehara, T., 2008. How bacteria consume their own exoskeletons (turnover and recycling of cell wall peptidoglycan). Microbiol. Mol. Biol. Rev. 72, 211–227.

Peyrin-Biroulet, L., Vignal, C., Dessein, R., Simonet, M., Desreumaux, P., Chamaillard, M., 2006. NODs in defence: from vulnerable antimicrobial peptides to chronic inflammation. Trends Microbiol. 14, 432–438.

Pugin, J., Heumann, I.D., Tomasz, A., et al., 1994. CD14 is a pattern recognition receptor. Immunity 1, 509–516.

Ratner, A.J., Aguilar, J.L., Shchepetov, M., Lysenko, E.S., Weiser, J.N., 2007. Nod1 mediates cytoplasmic sensing of combinations of extracellular bacteria. Cell. Microbiol. 9, 1343–1351.

Raymond, J.B., Mahapatra, S., Crick, D.C., Pavelka, M.S., 2005. Identification of the *namH* gene, encoding the hydroxylase responsible for the *N*-glycolylation of the mycobacterial peptidoglycan. J. Biol. Chem. 280, 326–333.

Rosenstiel, P., Fantini, M., Brautigam, K., et al., 2003. TNF-α and IFN-γ regulate the expression of the NOD2 (CARD15) gene in human intestinal epithelial cells. Gastroenterology 124, 1001–1009.

Rosenstiel, P., Hellmig, S., Hampe, J., et al., 2006. Influence of polymorphisms in the NOD1/CARD4 and NOD2/CARD15 genes on the clinical outcome of *Helicobacter pylori* infection. Cell. Microbiol. 8, 1188–1198.

Rosenstiel, P., Till, A., Schreiber, S., 2007. NOD-like receptors and human diseases. Microbes Infect. 9, 648–657.

Rosenthal, R.S., Nogami, W., Cookson, B.T., Goldman, W.E., Folkening, W.J., 1987. Major fragment of soluble peptidoglycan released from growing *Bordetella pertussis* is tracheal cytotoxin. Infect. Immun. 55, 2117–2120.

Schleifer, K.H., Kandler, O., 1972. Peptidoglycan types of bacterial cell walls and their taxonomic implications. Bacteriol. Rev. 36, 407–477.

Schwandner, R., Dziarski, R., Wesche, H., Rothe, M., Kirschning, C.J., 1999. Peptidoglycan- and lipoteichoic acid-induced cell activation is mediated by toll-like receptor 2. J. Biol. Chem. 274, 17406–17409.

Striker, R., Kline, M.E., Haak, R.A., Rest, R.F., Rosenthal, R.S., 1987. Degradation of gonococcal peptidoglycan by granule extract from human neutrophils: demonstration of *N*-acetylglucosaminidase activity that utilizes peptidoglycan substrates. Infect. Immun. 55, 2579–2584.

Sun, Y.H., Bakshi, S., Chalmers, R., Tang, C.M., 2000. Functional genomics of *Neisseria meningitidis* pathogenesis. Nat. Med. 6, 1269–1273.

Tanabe, T., Chamaillard, M., Ogura, Y., et al., 2004. Regulatory regions and critical residues of NOD2 involved in muramyl dipeptide recognition. EMBO J. 23, 1587–1597.

Tohno, M., Shimazu, T., Aso, H., et al., 2008. Molecular cloning and functional characterization of porcine nucleotide-binding oligomerization domain-1 (NOD1) recognizing minimum agonists, meso-diaminopimelic acid and meso-lanthionine. Mol. Immunol. 45, 1807–1817.

Travassos, L.H., Girardin, S.E., Philpott, D.J., et al., 2004. Toll-like receptor 2-dependent bacterial sensing does not occur via peptidoglycan recognition. EMBO Rep. 5, 1000–1006.

Travassos, L.H., Carneiro, L.A., Girardin, S.E., et al., 2005. Nod1 participates in the innate immune response to *Pseudomonas aeruginosa*. J. Biol. Chem. 280, 36714–36718.

Uehara, A., Fujimoto, Y., Kawasaki, A., Kusumoto, S., Fukase, K., Takada, H., 2006. *Meso*-diaminopimelic acid and *meso*-lanthionine, amino acids specific to bacterial peptidoglycans, activate human epithelial cells through NOD1. J. Immunol. 177, 1796–1804.

Valinger, Z., Ladesic, B., Tomasic, J., 1982. Partial purification and characterization of *N*-acetylmuramyl-L-alanine amidase from human and mouse serum. Biochim. Biophys. Acta 701, 63–71.

van Heel, D.A., Ghosh, S., Butler, M., et al., 2005. Muramyl dipeptide and toll-like receptor sensitivity in NOD2-associated Crohn's disease. Lancet 365, 1794–1796.

Vavricka, S.R., Musch, M.W., Chang, J.E., et al., 2004. hPepT1 transports muramyl dipeptide, activating NF-κB and stimulating IL-8 secretion in human colonic Caco2/bbe cells. Gastroenterology 127, 1401–1409.

Veiga, P., Bulbarela-Sampieri, C., Furlan, S., et al., 2007. SpxB regulates *O*-acetylation-dependent resistance of *Lactococcus lactis* peptidoglycan to hydrolysis. J. Biol. Chem. 282, 19342–19354.

Viala, J., Chaput, C., Boneca, I.G., et al., 2004. Nod1 responds to peptidoglycan delivered by the *Helicobacter pylori cag* pathogenicity island. Nat. Immunol. 5, 1166–1174.

Vollmer, W., Tomasz, A., 2000. The *pgdA* gene encodes for a peptidoglycan *N*-acetylglucosamine deacetylase in *Streptococcus pneumoniae*. J. Biol. Chem. 275, 20496–20501.

Wada, A., Ogushi, K., Kimura, T., et al., 2001. *Helicobacter pylori*-mediated transcriptional regulation of the human beta-defensin 2 gene requires NF-κB. Cell. Microbiol. 3, 115–123.

Watanabe, T., Kitani, A., Murray, P.J., Strober, W., 2004. NOD2 is a negative regulator of Toll-like receptor 2-mediated T helper type 1 responses. Nat. Immunol. 5, 800–808.

Watanabe, T., Kitani, A., Murray, P.J., Wakatsuki, Y., Fuss, I.J., Strober, W., 2006. Nucleotide binding oligomerization domain 2 deficiency leads to dysregulated TLR2 signaling and induction of antigen-specific colitis. Immunity 25, 473–485.

Weadge, J.T., Pfeffer, J.M., Clarke, A.J., 2005. Identification of a new family of enzymes with potential *O*-acetylpeptidoglycan esterase activity in both Gram-positive and Gram-negative bacteria. BMC Microbiol. 5, 49.

Welter-Stahl, L., Ojcius, D.M., Viala, J., et al., 2006. Stimulation of the cytosolic receptor for peptidoglycan, Nod1, by infection with *Chlamydia trachomatis* or *Chlamydia muridarum*. Cell. Microbiol. 8, 1047–1057.

Werts, C., Girardin, S.E., Philpott, D.J., 2006. TIR, CARD and PYRIN: three domains for an antimicrobial triad. Cell. Death Differ 13, 798–815.

Wolfert, M.A., Roychowdhury, A., Boons, G.J., 2007. Modification of the structure of peptidoglycan is a strategy to avoid detection by nucleotide-binding oligomerization domain protein 1. Infect. Immun. 75, 706–713.

Xu, X.L., Lee, R.T., Fang, H.M., et al., 2008. Bacterial peptidoglycan triggers *Candida albicans* hyphal growth by directly activating the adenylyl cyclase Cyr1p. Cell Host Microbe 4, 28–39.

Yoshimura, A., Lien, E., Ingalls, R.R., Tuomanen, E., Dziarski, R., Golenbock, D., 1999. Cutting edge: recognition of Gram-positive bacterial cell wall components by the innate immune system occurs via Toll-like receptor 2. J. Immunol. 163, 1–5.

Zähringer, U., Lindner, B., Inamura, S., Heine, H., Alexander, C., 2008. TLR2 – promiscuous or specific? A critical re-evaluation of a receptor expressing apparent broad specificity. Immunobiology 213, 205–224.

CHAPTER

33

Microbial interaction with mucus and mucins

Stephen D. Carrington[1], Marguerite Clyne[1], Colm J. Reid[1], Eamonn FitzPatrick[1] and Anthony P. Corfield[1]

SUMMARY

A layer of mucus represents the "front-line" of an integrated mucosal barrier surmounting all mucosal surfaces. The molecular scaffolding of mucus comprises hydrated mucins, which are very high-molecular mass, glycosylated glycoproteins presenting arrays of *O*-linked glycans. The protective physicochemical properties of mucus are substantially attributable to their high carbohydrate content. Mucins are integral to microbial interactions with epithelial surfaces and have potential roles in microbial trophism, the presentation of ligands to block microbial binding or stabilize colonization and the provision of feedstocks for microbial metabolism. Microbes are capable of expressing specific lectins and mucin-degrading enzymes to engage with, utilize and penetrate mucus gels. Normally, the dynamic nature of such gels represents an optimal adaptation through host–microbial cross-talk, enabling the maintenance of barrier function through the co-regulation of mucus turnover and microbial ecology. Mucin expression and glycosylation are biologically responsive to changes in the sub- and supra-mucosal environment and are significantly influenced by inflammation and microbial colonization. Impairment of mucus gel turnover may lead to abnormal colonization, inflammatory pathology and biofilm formation. Novel antimicrobial strategies exploiting new knowledge of mucin–microbial interactions are a real prospect. However, these will require the development of new research tools and models.

Keywords: Mucin; Mucus; Microbial binding; Microbial signalling; Glycosylation; *O*-Linked glycans

1. INTRODUCTION

Most microbial pathogens of higher animals enter their hosts via the moist mucosal surfaces of the respiratory, gastrointestinal, reproductive or ocular mucosae. Some of these surfaces support diverse communities of commensal organisms, which are increasingly viewed as an integral component of the mucosal barrier (Hooper and Gordon, 2001; Corfield *et al.*, 2003; Sonnenburg *et al.*, 2005; Lievin-Le Moal and Servin, 2006). All mucosal surfaces have in common the presence of a surmounting layer of mucus. This is a semi-adherent defensive supra-mucosal gel that presents a highly hydrated, mechanical and lubricative barrier, as well as a

[1]All authors contributed equally to this work.

matrix for immune-related and surveillance proteins (Corfield *et al.*, 2001, 2003; Corfield, 2007; Hattrup and Gendler, 2008; Thornton *et al.*, 2008).

Mucus represents the front-line defensive barrier between the external environment and the tissues of the host (Corfield *et al.*, 2001). It is, therefore, surprising that its relevance to the colonization and penetration of mucosal epithelium by microbes has not historically been a high priority for mucosal- and microbiologists. The second line of defence is constituted by the epithelial glycocalyx, which is partially integrated with the overlying gel (Hattrup and Gendler, 2008). Bacterial interaction with supramucosal gels may lead to chronic colonization of the mucus, blockage of access to underlying cells, and the elimination of microbes or, specifically in the case of pathogens, the engagement of gel penetration strategies that allow cellular adhesion (Chen and LaRusso, 2000; de Repentigny *et al.*, 2000; Byrne *et al.*, 2007).

The polymeric molecular scaffolding of mucus is formed by secreted mucus glycoproteins (mucins), which are variously produced by glands, surface epithelial cells and epithelial goblet cells specialized for mucin storage and secretion. They have a very high molecular mass (5×10^5–30×10^6) and up to 80% of their mass is constituted by *O*-linked glycan chains, which are linked to serine (Ser) and threonine (Thr) residues in extended linear domains (Corfield *et al.*, 2001, Thornton *et al.*, 2008). These domains usually display tandem repeated sequence motifs, the number of which varies between different mucins and mucin alleles. Secreted mucins are most commonly polymeric and are, therefore, capable of forming viscoelastic gels. Membrane-bound mucins, which form part of the epithelial glycocalyx, are monomeric and may be shed into the overlying gel. They contribute to viscosity but not elasticity. Other poorly-glycosylated globular domains are present at the C and N termini of mucins and have a variety of different characteristics and proposed functions. Eighteen mucin (MUC) genes have been identified. A summary of their molecular characteristics and expression in human tissues is given in Tables 33.1 and 33.2. The molecular configuration of some typical examples is given in Figure 33.1. The nature of the mucus varies at different locations and constitutes a responsive system that can be adapted to local physiological requirements. This occurs through the expression of different mucins, variations in mucin glycosylation and the co-secretion of mucin-associated molecules.

The protective physicochemical properties of mucus are substantially attributable to the high carbohydrate content of mucins. This also presents considerable potential for interaction between microbial lectins and *O*-linked glycans. The glycosylation machinery of mucosal epithelia not only varies with the cell type but is also influenced by the sub- and supra-mucosal environment. For example, hormonal status, inflammation and microbial colonization have all been shown to influence the glycosylation of epithelial glycoproteins (Corfield *et al.*, 2001, 2003; Singh and Hollingsworth, 2006; Corfield, 2007). In addition, glycosidases and other enzyme activities secreted by microbes have the capacity to modify directly these glycans after the mucins have been secreted or exported to the cell membrane (Hooper and Gordon, 2001; Roberton *et al.*, 2005; Sonnenburg *et al.*, 2005). This strategy is significant as a mechanism for generating binding ligands and in the provision of a carbon source to fuel microbial metabolism.

Mucin glycans may be neutral, sialylated or sulfated (Brockhausen, 2007). They are composed of L-fucose (L-Fuc), *N*-acetylneuraminic acid (sialic acid, Neu5Ac), galactose (Gal), *N*-acetylgalactosamine (GalNAc) and *N*-acetylglucosamine (GlcNAc) in various linkages (Corfield *et al.*, 2001; Brockhausen, 2007). Their low content of mannose and uronic acid distinguishes them from the glycans of other glycoproteins and the proteoglycans. The size, charge and branching of mucin glycans varies considerably between different mucosal locations.

TABLE 33.1 Human mucin genes

Name	Chromosomal location	Size of tandem repeat (amino acid)	Domains present[a]
Membrane-associated mucins			
MUC1	1q21	20	SEA, β-Catenin
MUC3A	7q22	17	SEA, EGF
MUC3B	7q22	17	SEA, EGF
MUC4	3q29	16	EGFs, AMOP
MUC12	7q22	28	SEA, EGF
MUC13	3q21.2	27	SEA, EGF
MUC15	11p14.3	None	n.a.
MUC16	19p13.2	156	SEA
MUC17	7q22	59	SEA, EGF
MUC20	3q29	18	n.a.
MUC21	6p21?	15	SEA
Secreted mucins, gel-forming			
MUC2	11p15.5	23	VW
MUC5AC	11p15.5	8	VW
MUC5B	11p15.5	29	VW
MUC6	11p15.5	169	VW
MUC19	12q12	19	VW
Secreted mucins, non-gel forming			
MUC7	4q13-q21	23	HIS
MUC8	12q24.3	13/41	n.a.

[a]Domains not uniquely associated with mucins. Abbreviations: AMOP, adhesion-associated domain in MUC4 and other proteins; EGF, epidermal growth factor; HIS, histatin domain; n.a., not applicable; SEA, sea urchin sperm protein, enterokinase and agrin; VW, von Willebrand.

The *O*-glycosidic linkage is formed through GalNAc, which links the glycans to Ser and Thr residues in the VNTR (variable number tandem repeat) domains of the core apomucin. This linkage gives rise to a series of up to eight core glycan structures, of which β-Gal-(1→3)-GalNAc-Ser/Thr (core 1) and β-Gal-(1→3)[β-GlcNAc-(1→6)]-GalNAc-Ser/Thr (core 2) are the most widespread (Brockhausen, 2007). A diversity of glycan chains can be generated by extension of these cores through the sequential action of glycosyltransferases and sulfotransferases organized along defined pathways. *O*-Linked glycans on mucins are relatively short compared to *N*-linked glycans and frequently have Fuc or sialic acid as a terminal capping structure (Brockhausen, 2007). In some cases, major glycosylation variants are found. These are referred to as mucin "glycoforms". Many of the *O*-glycan structures carried by mucins are also common to other glycoproteins and glycolipids. Some typical structures of mucin glycans are shown in Tables 33.3 and 33.4 and Figure 33.2. A review of the "building

TABLE 33.2 Human mucin gene expression[a]

Organ system	Mucin																			
	1	**2**	**3A**	**3B**	**4**	**5AC**	**5B**	**6**	**7**	**8**	**9**	**11**	**12**	**13**	**15**	**16**	**17**	**19**	**20**	**21**
Ocular	+				+	+			+							+				
Salivary gland	+		+	+	+		+		+									+		
Airway	+	+	+	+	+	+	+		+	+		+		+					+	+
Oesophagus	+				+		+												+	
Stomach	+		+	+		+		+						+					+	
Small intestine	+	+	+	+				+									+		+	
Large intestine	+	+	+	+	+							+	+	+			+		+	+
Hepatobiliary	+	+	+	+		+	+	+												
Pancreas	+					+	+	+				+	+						+	
Female repro.	+				+	+	+	+		+	+		+		+	+				
Male repro.	+									+										

[a]+, Gene expressed.

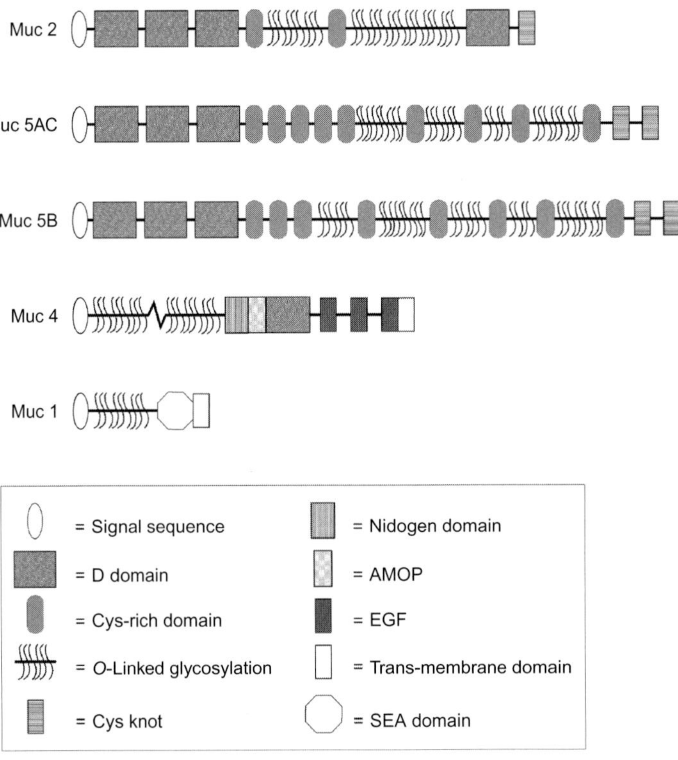

FIGURE 33.1 Typical examples of membrane-associated and secreted mucins.

TABLE 33.3 Examples of simple O-linked glycans

Name and sequence[a]	Structure
Alpha 2-3 sialyl-T α-Neu5Ac-(2→3)-β-Gal-(1→3)-GalNAc-Ser/Thr	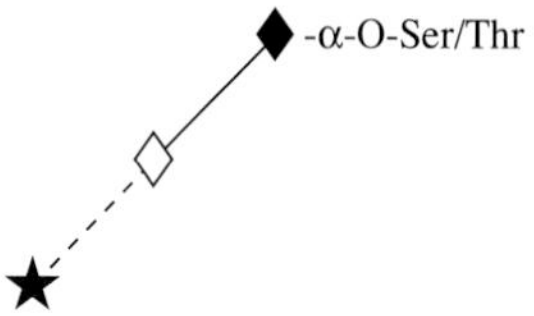 -α-O-Ser/Thr
Sialyl-Tn α-Neu5Ac-(2→6)-GalNAc-Ser/Thr	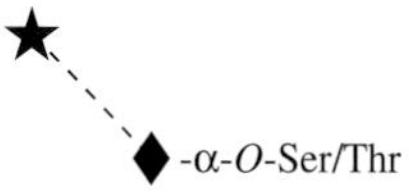 -α-*O*-Ser/Thr
Mono-sialyl 2-6 core 1 β-Gal-(1→3)-[α-Neu5Ac-(2→6)]-GalNAc-Ser/Thr	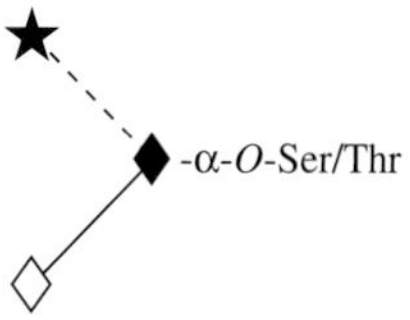 -α-*O*-Ser/Thr
Alpha 2-3 sialyl core 2 α-Neu5Ac-(2→3)-β-Gal-(1→3)-[β–GlcNAc-(1→6)]-GalNAc-Ser/Thr	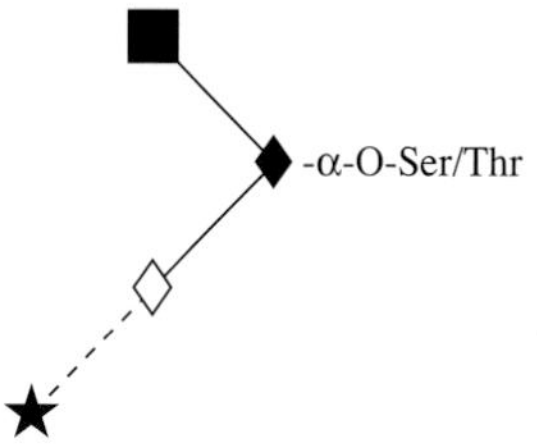 -α-O-Ser/Thr

[a]Abbreviations: Gal, galactose; GalNAc, *N*-acetylgalactosamine; GlcNAc, *N*-acetylglucosamine; Neu5Ac, *N*-acetylneuraminic acid (sialic acid); Ser, serine; Thr, threonine.

TABLE 33.4 Examples of complex O-linked glycans[a]

Characteristics and sequence[b]	Structure
Core =1 (C1) Backbone =Type 1 lactosamine (B) Peripheral =6′-Sulpho-sialyl-Lewis^{x} (P) α-Neu5Ac-(2→3)-β–Gal-(1→4)-[α–Fuc-(1→3)][6-SO^{3-}]-β-GlcNAc-(1→4)-β-Gal-(1→3)-β-GlcNAc-(1→3)-β-Gal-(1→3)-GalNAc-Ser/Thr	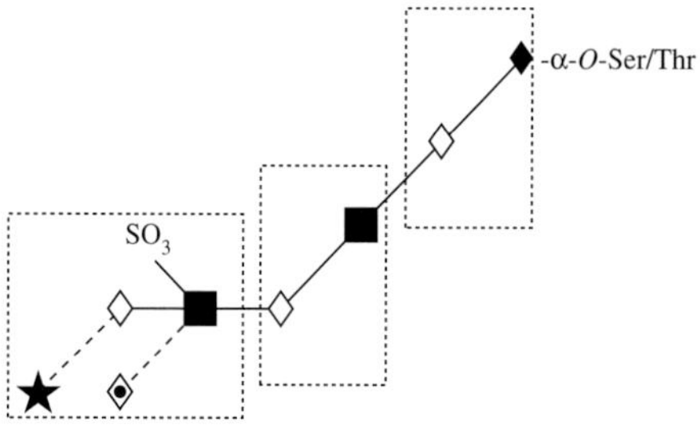 -α-*O*-Ser/Thr SO_3

(Continues)

TABLE 33.4 (Continued)

Characteristics and sequence[b]	Structure
Core = 2 (C2) Backbone = Type 1 lactosamine (B) Peripheral = Blood group H (P) α–Fuc-(1→2)-β-Gal-(1→3)-β-GlcNAc-(1→3)-β-Gal-(1→3)[β-GlcNAc-(1→6)]-GalNAc-Ser/Thr	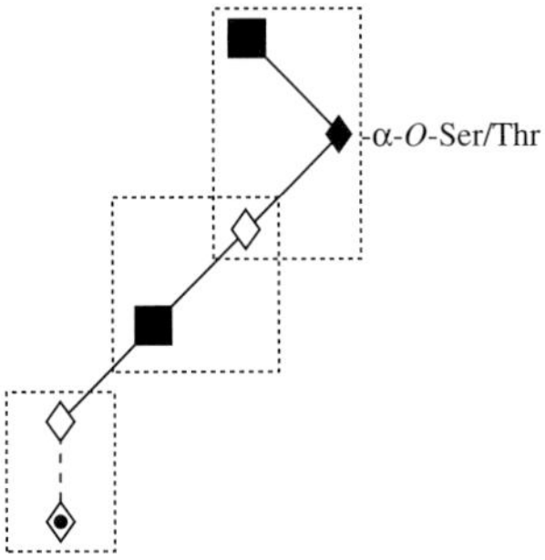 α-*O*-Ser/Thr
Core = 3 (C3) Backbone = Type 2 lactosamine (B) Peripheral = Lewis[a] (P) β–Gal-(1→3)[α–Fuc-(1→4)]-β-GlcNAc-(1→3)-β-Gal-(1→4)-β-GlcNAc-(1→3)-GalNAc-Ser/Thr	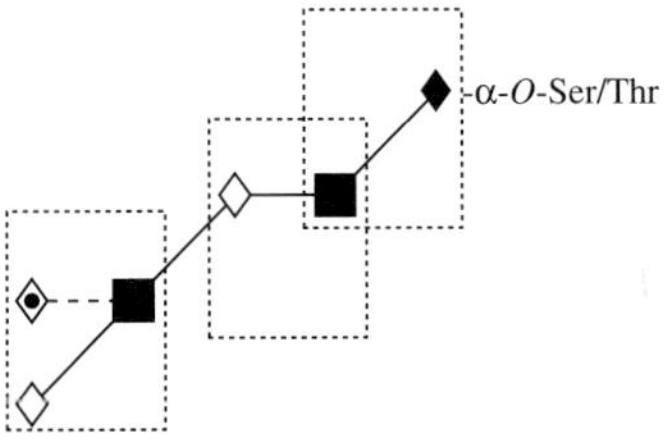 α-*O*-Ser/Thr
Core = 4 (C4) Backbone = Type 2 polylactosamine (B) Peripheral = Blood group A (P) α-GalNAc-(1→3)[α–Fuc-(1→2)]-β-Gal-(1→4)-β–GlcNAc-(1→3)-β-Gal-(1→4)-β-GlcNAc-(1→3)[β–Gal-(1→4)-β–GlcNAc-(1→3)-β–Gal-(1→4)-β–GlcNAc-(1→6)]-GalNAc-Ser/Thr	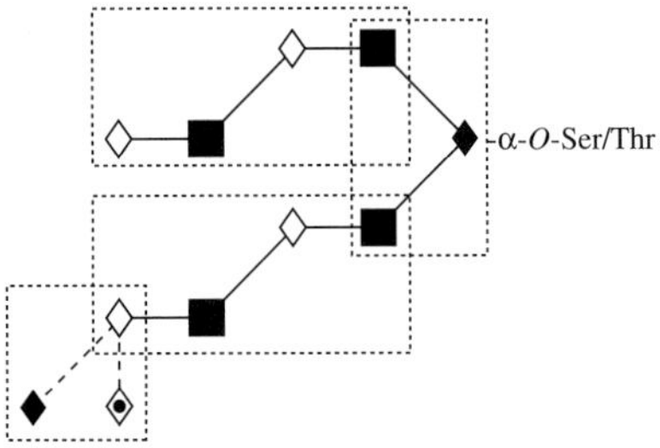 α-*O*-Ser/Thr

[a]The *O*-linked glycans of mucins are linked to serine (Ser) and threonine (Thr) residues in VNTR domains of the apomucin, through oxygen. The linkage sugar at the reducing terminal of the glycan is *N*-acetylgalactosamine (GalNAc). Truncated structures are found (e.g. Sialyl-TN). A range di- and tri-saccharide "cores" numbered from 1 to 8 have been identified; of which cores 1–4 are the most common. These cores may be extended by linear or branched *N*-acetyllactosamine-bearing backbones of type 1, β-(1→3)-linked, or type 2, β-(1→4)-linked. Finally, they may be terminated by peripheral structures including fucose (Fuc), generating a neutral glycan; or sialic acid, generating an acidic glycan; or a range of blood group antigens which may be neutral or acidic, depending on sialylation and/or sulfation. Diagrams were drawn with GlycanBuilder (see Ceroni *et al.*, 2007).

[b]Abbreviations: L-fucose (L-Fuc), Gal, galactose; GalNAc, *N*-acetylgalactosamine; GlcNAc, *N*-acetylglucosamine; Neu5Ac, *N*-acetylneuraminic acid (sialic acid).

blocks" of *O*-linked glycans can be found in Gabius (2008).

A comprehensive discussion of all defensive molecules integrated into mucus gels is beyond the scope of this review. However, Trefoil factor family (TFF) peptides are worthy of mention because their role in interactions with microbes has only recently emerged. These peptides are a group of low molecular mass cysteine- (Cys-) rich proteins expressed mainly by mucin-secreting cells of the gut (Wright *et al.*, 1997). They appear to have the capacity both to cross-link mucins and

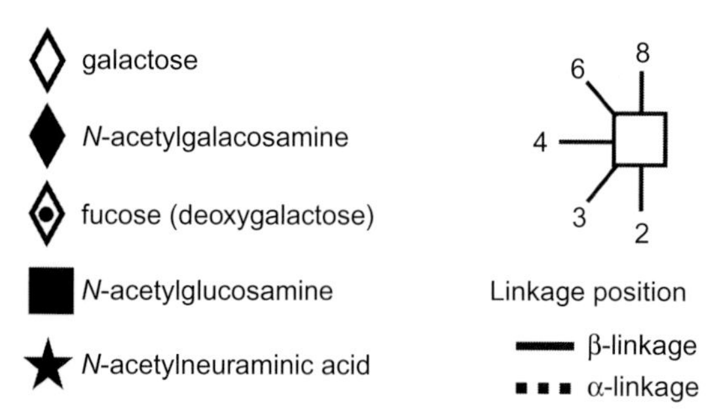

FIGURE 33.2 Key to glycan structural diagrams.

promote the integrity of the mucosal epithelium through a variety of different mechanisms (Thim and May, 2005; Hattrup and Gendler, 2008). There are three mammalian trefoil proteins, TFF1, TFF2 and TFF3, that co-localize with specific mucins. As to expression, TFF1 is co-expressed with MUC5AC, TFF2 is co-expressed with MUC6 and TFF3 and MUC2 are found throughout the large and small bowel mucosa. Also, TFF1 is present in three molecular forms in normal human gastric mucosa: TFF1 monomer, TFF1 dimer and a TFF1 heterodimer with a novel secreted protein TFIZ1 or gastrokine 2. The TFF1 monomer has been shown to interact directly with MUC5AC.

2. OVERALL EFFECT OF EPITHELIAL COLONIZATION ON MUCUS AND MUCINS

Many bacteria can regulate mucus synthesis and secretion from host goblet cells. Thus, the normal microflora may provide protection against infection by pathogens by altering mucus composition. The number of goblet cells in the gut of germ-free mice is reduced in comparison to conventionally raised mice and results in a mucus gel of reduced thickness. Probiotic bacteria have been shown to be capable of enhancing mucin secretion in colonic epithelial cells (Caballero-Franco *et al.*, 2007).

Evidence from intestinal diversion surgery in humans indicates that faecal contact modulates both goblet cell numbers and the glycosylation profile of colonic mucins (Corfield *et al.*, 2003). Since the colon is the most heavily colonized mucosal surface in the body, this effect may result from a change in the colonic microbiome (Hooper and Gordon, 2001; Roberton *et al.*, 2005; Sonnenburg *et al.*, 2005). However, further studies are required to confirm this.

Cholera toxin has been shown to exert a strong stimulatory effect on mucin secretion by colonocytes after incubations longer than 1 hour. Apical infection of HT29-MTX cells by *Listeria monocytogenes* also induces mucin exocytosis attributable to the listeriolysin O (LLO) toxin. In this system, both mucin secretion and the transcription of membrane-bound mucin genes were increased, resulting in an inhibition of cellular invasion (Lievin-Le Moal and Servin, 2006). Mucin over-production is a hallmark of chronic diseases of respiratory epithelia such as chronic obstructive pulmonary disease, asthma and otitis media. Both non-typeable *Haemophilus influenzae* and *Streptococcus pneumoniae* co-exist under these disease conditions and have been shown to act synergistically to induce mucin transcription in human epithelial cell lines (Shen *et al.*, 2008).

Total mucin synthesis was inhibited in a human gastric cell line exposed to *Helicobacter pylori* (Byrd *et al.*, 2000). Downregulation of MUC5AC and upregulation of MUC6 has been reported in gastric mucosa infected by *H. pylori* (Byrd *et al.*, 1997). Reduced MUC5AC synthesis may impair the gastric mucus gel layer since this mucin is a major component of it. However, the upregulation of MUC6 synthesis may be a host defence mechanism, since the *O*-glycans of MUC6 gastric mucin display antimicrobial activity against *H. pylori.* They achieve this through inhibiting its biosynthesis of cholesterol-α-D-glucopyranoside, a major cell wall component (Kawakubo *et al.*, 2004). This shows that changes in mucin expression can directly modulate microbial colonization.

Like the stomach, the lung normally expresses MUC5AC. However, in the lungs of cystic fibrosis (CF) patients, MUC2 (which is the principal intestinal mucin) is specifically upregulated after microbial colonization. The lipopolysaccharides (LPS) of *Pseudomonas aeruginosa* may mediate this effect. The lipoteichoic acid of the Gram-positive organisms *Staphylococcus aureus* and *Streptococcus pyogenes* have also been shown to be capable of stimulating MUC2 production (Lemjabbar and Basbaum, 2002).

3. THE ROLE OF MICROBES IN THE NORMAL TURNOVER OF MUCUS GELS

Microorganisms and other contaminants commonly adhere to mucus (Lievin-Le Moal and Servin, 2006; Miyake *et al.*, 2006). This dictates a physiological requirement for its continuous removal and renewal, to ensure the maintenance of an efficient barrier. In addition, the bulk secretion of mucus is responsive to factors in the sub- and supra-mucosal environment through the action of secretagogues (Hooper and Gordon, 2001; Sonnenburg *et al.*, 2005; Miyake *et al.*, 2006). At some mucosal surfaces, turnover involves the mechanical removal of mucus by local anatomical adaptations (e.g. peristalsis, mucociliary clearance). At other sites, it is likely that enzymic degradation plays a substantial role. In surfaces colonized by microbes, such as the gut and the lower female reproductive tract, the standing flora may contribute to a mixture of activities that collaborate in the degradation of mucins. These are collectively known as "mucinase" and their composition varies with the anatomical location of the epithelium concerned (Corfield *et al.*, 2001; Wiggins *et al.*, 2001). In particular, the lower gut presents stable populations of commensal organisms. The dynamic nature of the overlying supra-mucosal gel represents an optimal co-adaptation between host and bacterial populations, enabling the maintenance of barrier function through the regulation of mucus turnover.

The role of host microbial cross-talk in the provision of nutritional feedstocks is well established. Microbes have the capacity to undertake "glycan foraging" of host epithelial glycans to service their nutritional needs (Sonnenburg *et al.*, 2005). Mucins are an abundant source of carbohydrate for this process and a subset of epithelial-resident microflora is specialized for mucin degradation (Sonnenburg *et al.*, 2005; Derrien *et al.*, 2008). This process is particularly relevant at epithelia where mucins may represent the principal source of nutrition in the supra-mucosal environment. Indeed, such organisms could be particularly important in the gut during periods of starvation, where their capacity to degrade mucins may be important to the survival of other residents that normally derive their nutrition from dietary sources. Mucins may also be implicated in the generation of ligands for microbial binding that are important for niche stabilization.

4. MICROBIAL INTERACTIONS WITH MEMBRANE-ASSOCIATED MUCINS

The division of the mucin-bearing components of the epithelial barrier into supra-mucosal gel and glycocalyx has functional relevance. Microorganisms target the glycocalyx as a means to interact with epithelial cells through direct contact (Hooper and Gordon, 2001). Membrane-bound mucins are a potential component of this interaction and there is evidence that some can directly transduce signals to the intracellular environment (Singh and Hollingsworth, 2006).

One of the best-characterized membrane-bound mucins is MUC1. It consists of an extensively *O*-glycosylated extracellular region with tandem repeats, a transmembrane region and a cytoplasmic tail. This mucin is ubiquitously expressed on the surface of mucosal epithelial cells and has been reported to be a receptor for *P. aeruginosa*, through which downstream signalling involves the activation of MAP kinases and the phosphorylation of the cytoplasmic tail of MUC1 (Lillehoj *et al.*, 2004).

The role of membrane-bound mucins in enhancing the mucosal barrier to microbial adhesion and translocation is also significant. It is likely that the high molecular mass and extended linear configuration of these molecules cause them to extend beyond other components of the glycocalyx and thereby influence access to them.

There is a correlation between the absence of apical MUC1 tandem repeat epitopes and a greater frequency of MUC1 short alleles in patients with *H. pylori* gastritis, suggesting that short forms of MUC1 may enhance access of *H. pylori* to the gastric epithelium (Vinall *et al.*, 2002). Furthermore, the importance of this mucin in mucosal defence is also indicated by the enhanced susceptibility of Muc1 knockout mice to infection (McAuley *et al.*, 2007).

5. MICROBIAL BINDING AND HOMING TO SECRETED MUCINS

Secreted mucins also offer a range of possible interactions with microbes in the supra-mucosal environment. These include the provision of ligands for niche stabilization, the capacity to act as trophic factors to assist in microbial homing and the presentation of sacrificial ligands capable of blocking microbial binding.

Numerous bacterial species, including both pathogens and commensals, have been shown to interact with mucins and utilize supra-mucosal gels as a long-term niche (Figure 33.3). Nevertheless, the exact nature of the relevant receptor–adhesin interactions has not been well characterized in many cases.

Group A streptococci bind to mucin through α-(2→6)-linked sialic acid and this interaction is mediated through the fibrillar surface M-protein of this organism, which is a major virulence factor (Ryan *et al.*, 2001). Binding of *S. aureus* to human nasal mucin has been shown to occur via a specific adhesion–receptor interaction involving a bacterial protein and a carbohydrate moiety present on the mucin (Shuter *et al.*, 1996). Defined LPS and capsule-deficient mutants of *Vibrio cholerae* exhibited up to 6-fold reduction in adherence to a mucus secreting cell line compared with wild-type strains. This indicates that LPS and capsule play a role in mediating adherence of *V. cholerae* to mucus (Schild *et al.*, 2005). *H. pylori* binds to the gastric mucin MUC5AC through the bacterial outer membrane protein BabA which, in turn, binds to the fucosylated Lewis^{b} blood group antigen expressed on the mucin (Lindén *et al.*, 2002).

Probiotic bacteria used to prevent infection by other harmful organisms often do so by adhering to mucosal surfaces. The adherence of *Lactobacillus plantarum* strain LA 318 to human colonic mucin is mediated through glyceraldehyde-3-phosphate dehydrogenase (GAPDH) expressed on the surface of the bacteria (Kinoshita *et al.*, 2008). Some pathogens also express this adhesin indicating that *L. plantarum* may prevent infection by competing with pathogens for receptors in the intestine or its surmounting mucus gel. Mucin-associated molecules may also be important in facilitating bacterial colonization of epithelial surfaces. The TFF is a contemporary example, which is a current research focus because of the many physiological effects of trefoil proteins. Namely, TFF1 and TFF3 adhere to the surface of bacterial pathogens *S. aureus* and *P. aeruginosa* in infected CF airways and roles in host–bacterial interactions such as adhesion, phagocytosis and signalling to the immune system have been suggested (dos Santos Silva *et al.*, 2000).

The hypothesis that TFF1 can act as a receptor for *H. pylori* has been tested (Clyne *et al.*, 2004). All strains of *H. pylori* that were examined bound specifically to recombinant dimeric human TFF1, which is associated with MUC5AC in the stomach, but not to *E. coli* or *C. jejuni*. This finding presents a unique opportunity to study the effect of inhibiting bacterial receptor interactions within a

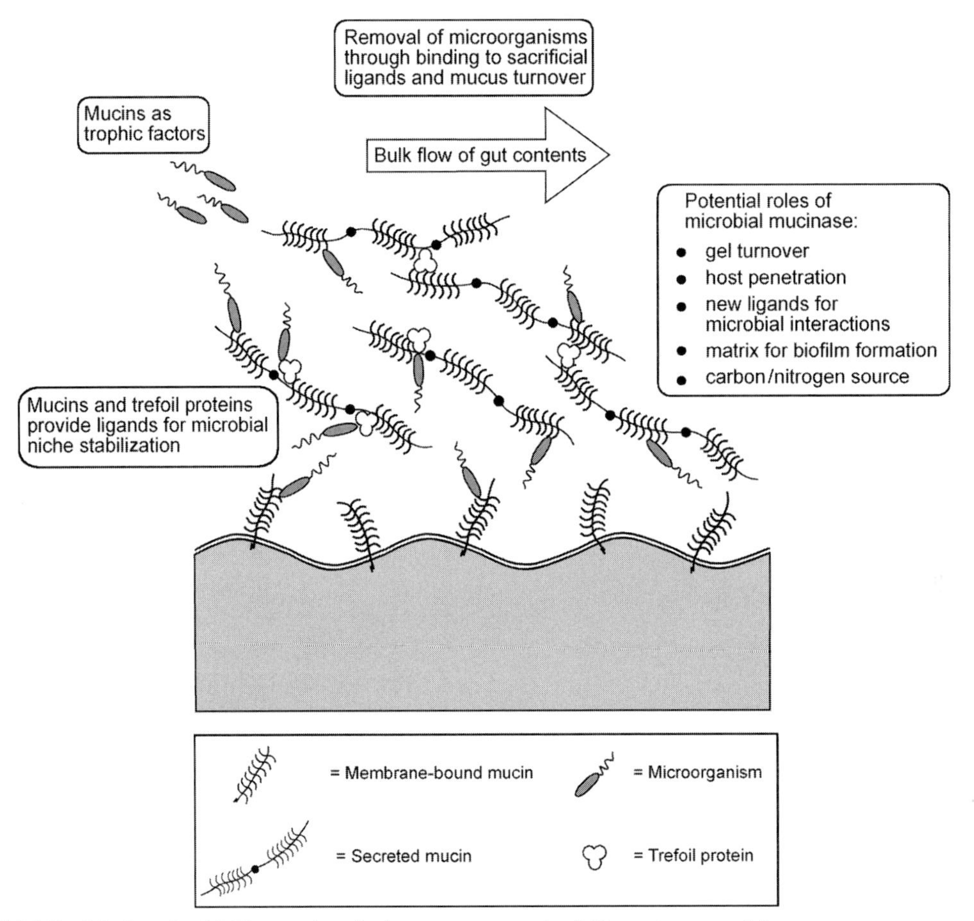

FIGURE 33.3 Mucin–microbial interactions in the supra-mucosal gel. The gut as a model.

clinically accessible mucus gel colonized by a persistent pathogen. The absence of a complex indigenous microflora in the stomach further enhances the potential value of the *H. pylori* model.

A striking feature of *H. pylori* is its strict host and species specificity and its gastric tropism. The organism is normally only found *in vivo* in association with gastric mucus secreting cells in the stomach or at sites of gastric metaplasia in the duodenum and oesophagus (Wyatt *et al.*, 1990). The localization of *H. pylori* within these tissues matches that of MUC5AC (Van den Brink *et al.*, 2000), implicating this mucin in both its tissue tropism and its sequestration within the gastric mucus layer.

It is logical to conclude that a prime function of mucus is to present decoy ligands that inhibit microbial invasion and penetration of epithelial tissues. Indeed, there are many studies showing that mucus or semi-pure mucins can block microbial binding to cell lines and tissues. This review can mention only a few examples.

Purified mucins from rabbit small intestinal mucus inhibit the attachment of the rabbit enteropathogen *E. coli* RDEC-1 to intestinal microvillus membranes (Drumm *et al.*, 1988). Chicken mucus, but not human mucus, inhibits epithelial cell invasion by *Campylobacter jejuni*, suggesting that its composition may contribute to the commensal status of this species (Byrne *et al.*, 2007). In another study, purified intestinal chicken mucin, but not bovine mucin, inhibited invasion of Madin-Darby bovine kidney cells by the coccidian parasite *Eimeria tenella* (Tierney *et al.*, 2007). This work demonstrated *E. tenella* interaction with native chicken intestinal mucin which, in turn, inhibited parasite invasion into epithelial cells *in vitro*.

Numerous previous studies have variously used chemical oxidation, exoglycosidase digestion and competition assays using lectins, saccharides, oligosaccharides or neo-glycoconjugates to indicate the relevance of protein–carbohydrate recognition in host–microbial interactions. Many examples of this can be found in the literature. Much of this work infers the involvement of the *O*-linked glycans of mucins in these effects. However, reports of specific microbial interactions with individual mucins or structurally-defined *O*-glycans are rare. Examples where some evidence of specific protein–carbohydrate interactions has been obtained include *Cryptosporidium parvum* (Gal-GalNAc) (Chen and LaRusso, 2000) and *H. pylori* (Lewis[b]) (Borén *et al.*, 1993). A recent comparative study of the glycans of ocular mucins in 3 species reviews their possible relevance to the binding of several ocular pathogens (Royle *et al.*, 2008).

The paucity of detailed structural information on specific mucin glycans and individual mucins and mucin domains involved in mucin–microbial interactions, results from the relatively limited repertoire of *O*-linked structures on currently available glycan arrays and the lack of commercially available sources of individual mucins of high purity.

6. MICROBIAL PENETRATION OF MUCUS GELS

In addition to the obvious relevance of microbial motility to gel penetration, some bacterial, fungal and protozoal pathogens can actively degrade mucins to gain access to the underlying mucosal epithelium. Mucosal pathogens displaying a capacity to degrade mucins may utilize secreted or cell-surface proteases and glycosidases for this purpose.

The ability of *C. albicans* to bind to purified small intestinal mucus is a key virulence factor. Adhesion triggers the production of secretory aspartyl proteinase, enabling the organism to move deeper into the mucus layer (de Repentigny *et al.*, 2000). *H. pylori* possesses the enzymatic ability to disrupt the oligomeric structure of mucin enabling the pathogen to move freely within gastric mucus. In bacterial vaginosis, the normal bacterial flora is replaced by pathogenic Gram-negative anaerobes, which produce glycosidases and proteases. These appear to play a role in degrading vaginal mucus, thereby aiding colonization (Roberton *et al.*, 2005). Isolates of *P. aeruginosa* involved in microbial keratitis produce proteases capable of degrading mucin and aiding corneal colonization (Aristoteli and Willcox, 2003). The Cys proteases of *Entamoeba histolytica* have been shown to play a key role in disrupting colonic mucus and aiding pathogenesis through cleavage of the C terminus of MUC 2 (Lidell *et al.*, 2006). A number of protozoan parasites also have the capacity to degrade mucins to facilitate colonization (Hicks *et al.*, 2000). The literature on microbial degradation of mucin glycans is extensive and readers are directed to Corfield *et al.* (2001) and Corfield (2007).

7. MICROBIALLY MEDIATED RE-PROGRAMMING OF HOST GLYCOSYLATION

Cross-talk between microbes in the supra-mucosal environment and host epithelia is capable of altering mucin glycosylation (Figure 33.4). The physiological importance of such responses is likely to be significant, but the mechanisms involved are poorly understood in all but a small number of model organisms and systems.

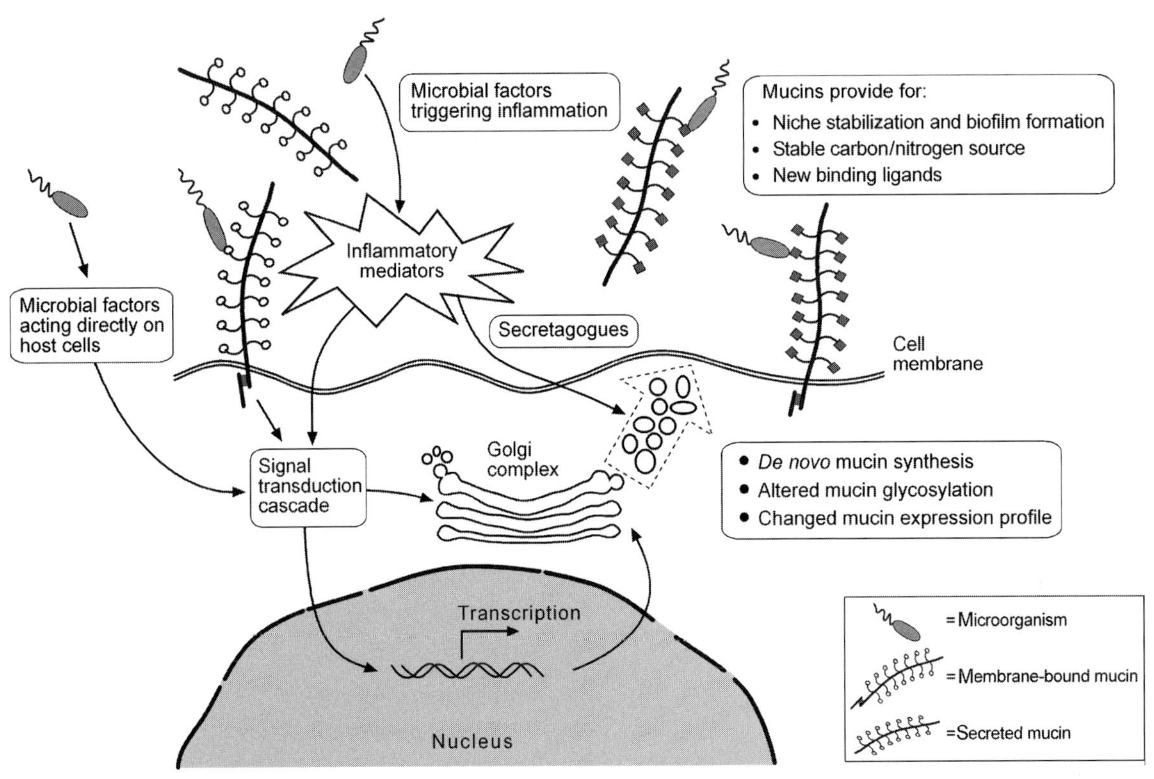

FIGURE 33.4 Microbial influences on mucin expression and biosynthesis.

An ongoing study of intestinal diversion surgery in humans shows that the temporary removal of colonic faecal flow results in reproducible changes in colonic mucin glycosylation. In particular, the pattern of sialoglycans is changed (Edwards and Corfield, 2000). This model presents a useful opportunity to correlate changes in the colonic microflora with glycosylation programming in the human colonic mucosa.

A longitudinal comparison of the glycosylation profile of the gut in germ-free and conventional mice has been undertaken. It showed that the resident microflora quantitatively and qualitatively modulate the epithelial glycosylation pattern by changing the cellular and subcellular distribution of glycans (Freitas *et al.*, 2002). *Bacteroides thetaiotaomicron* has been used as a model organism to study this effect in gnotobiotic mice, in which it specifically induces the expression of α-Fuc-(1→2)-β-Gal-glycans through the action of a diffusible factor or factors. This particular glycosylation change was specific for *B. thetaiotaomicron* and could not be induced by other Gram-negative organisms or by an isogenic strain carrying a transposon insertion that disrupted L-Fuc utilization. The ability of this organism to induce the fucosylation of mucosal glycoconjugates conferred a significant benefit by supplying Fuc as a carbon source (Hooper and Gordon, 2001).

A comparison of CF patients with and without *P. aeruginosa* colonization has demonstrated an upregulation of sialyltransferases and α-(1→3)-fucosyltransferases in the bronchial mucosa of colonized individuals. Mucins from such patients had a higher sialic acid content and the sialyl-Lewis^x determinant was also overexpressed in the mucins of severely infected patients (Davril *et al.*, 1999).

H. pylori infection modifies the glycosylation patterns of gastric mucins and promotes colonization (Mahdavi *et al.*, 2002). Experimental infection of Rhesus monkeys results in lower relative abundance of mucin oligosaccharides and a loss of mucin type 2 core structures compared to uninfected monkeys (Cooke *et al.*, 2007). Prolonged infection (10 months) decreases fucosylation and increases sialyation (Lindén *et al.*, 2008). These changes affect *H. pylori* adhesion targets and may play a role in modulating host–bacterial interactions.

8. PATHOLOGIES OF MUCUS TURNOVER, ABNORMAL COLONIZATION, AND BIOFILM FORMATION

Derangements of the normal turnover of supra-mucosal gels can lead to mucus accumulation, but may also cause mucus depletion and defective barrier function. These defects are often characterized by abnormal microbial colonization, which has the further capacity to influence epithelial phenotype through direct cross-talk or interaction with the innate and adaptive immune systems and inflammatory cascades. Several studies have shown the capacity of abnormal or altered colonization of mucosal surfaces to influence mucin expression, post-translational modification and turnover (Lievin-Le Moal and Servin, 2006; Miyake *et al.*, 2006). This effect may be part of a defensive response capable of modulating microbial binding. Paradoxically, such changes have the potential to promote chronic abnormal colonization where gel turnover is reduced, as in CF, or where barrier function is impaired, as in ulcerative colitis. There is evidence that some mucosal pathogens (e.g. *H. pylori*) may have the capacity to exploit ligands generated in epithelial responses to abnormal colonization (Mahdavi *et al.*, 2002).

It would be logical to consider the standing flora of colonized epithelia to constitute an organized biofilm (Macfarlane and Macfarlane, 2000). Nevertheless, biofilms are defined in part by their ability to secrete a polymeric matrix that maintains their topographical organization. Healthy epithelial surfaces utilize mucins to generate a host-derived polymeric matrix. This implies a functional requirement for host–microbial cross-talk in the establishment of supra-mucosal biofilms through the ability of commensal organisms to direct the makeup of the matrix in which they must maintain station. The contribution of microbially-derived glycopolymers to this process is as yet poorly defined, but clearly has a role in a range of mucosal pathologies (Macfarlane and Macfarlane, 2000).

Cystic fibrosis presents an example of a pathological phenotype in which gel turnover is severely impaired. Whereas normal turnover seems largely to preclude the formation of classical biofilms based on microbial glycopolymer production, pseudomonad biofilms based on alginates are a major feature of CF (Costerton, 2002). These utilize the static airway mucus as a basis for their formation. The resulting inflammation further modulates mucin biosynthesis and escalates mucosal pathology. *P. aeruginosa* binding to human respiratory mucin is neuraminidase-sensitive and sialyl-Lewis^x has been identified as a receptor for *P. aeruginosa* in respiratory mucin (Xia *et al.*, 2007).

9. CONCLUSIONS

Epithelial mucins are a fundamental and responsive component of the barrier at moist mucosal surfaces. They have the capacity to modulate microbial binding, penetration and colonization. Our understanding of the structural basis of mucin–microbial interactions and the cell biology underlying the biosynthesis of physiologically competent supra-mucosal gels is incomplete. However, the current body of literature provides a sound "proof of concept" that novel antimicrobial strategies, exploiting new knowledge of such interactions, could soon be within our grasp. The capacity of microbes to modulate glycosylation pathways in epithelial tissues also has important implications for future research, but new tools, approaches and models are required for further investigations (see Research Focus Box).

Progress in this area is hampered by the large size, complexity and natural heterogeneity of mucins. This makes the preparation of pure and homogeneous glycoprotein with consistent glycosylation difficult. It also hinders the characterization of mucin-related biological responses. In addition, reliable model systems are needed which allow controlled long-term experimental colonization in cell lines, tissues or whole animals. A number of different approaches could provide the toolkit for this work. *In vitro* models of the intestine are presented by gut diversion surgery in man (Edwards and Corfield, 2000). Similar models in pigs are also possible, particularly because of the widespread use of this species in gastrointestinal research. Likewise, the germ-free mouse is also a suitable model for studies of host–microbial cross-talk leading to changes in mucin biosynthesis. Most epithelial cell lines used previously to study host–microbial interactions lack a supra-mucosal gel and this has hampered understanding of mucin–microbial interactions. However, *in vitro* cell lines that mimic mucosal surfaces with an intact supra-mucosal gel are available. These include the HT-29–E12 cell line (Behrens *et al.*, 2001) which has proved a particularly useful model in our hands.

Recombinant versions of the secreted mucins, i.e. mucin mini-gene, are under development (C.J. Reid *et al.*, unpublished results). These contain all domains relevant to processing, but with reduced VNTR domains and hence reduced size. When expressed in relevant cell lines, including HT-29–E12 cells, these will serve as flexible model systems in which the proportion of mucin gene products and mucin glycosylation, can both be manipulated and characterized.

Many microorganisms present repetitive molecular patterns based on carbohydrate-containing structures that are recognized by the innate immune system. Most protein–carbohydrate interactions in Nature have low binding energies. High avidity is usually achieved through multimeric interaction. Therefore, it is significant that mucins present similar repetitively-arrayed structures on the host barrier. This infers a need for spatial concordance between the topographical distribution of microbial surface lectins and the density of the arrayed *O*-linked glycans on mucins, which may be a significant factor in host–microbial interactions. Existing experimental glycan arrays, such as those provided by the consortium for functional glycomics (see http://www.functionalglycomics.org/static/index.shtml), mainly comprise panels of isolated oligosaccharides where the ligand density may differ from that found in Nature. Newer versions of these microarrays, with glycans presented in their natural density and spatial configuration, are required to interrogate comprehensively the interactions between microbial surfaces and mucins. Such arrays could be populated by both natural and recombinant mucins presenting a range of glycosylation patterns, some of which could be induced by experimental microbial colonization *in vitro*. When coupled with high-level glycoanalysis of *O*-linked glycans (Royle *et al.*, 2008), a powerful combinatorial approach to the interrogation of mucin–microbial interactions would be provided, potentially identifying interventions that are informed by a detailed understanding of the relevant biology. Such interventions could include the production and use of glycan mimics and synthetic neo-glycoconjugates as decoys to block microbial colonization. The targeted therapeutic modulation of endogenous arrayed mucin glycans within supra-mucosal gels would also be possible, at both the biosynthetic and post-secretional levels. This could be achieved through a range of strategies including the use of microbial products, probiotics and nutraceuticals. Such approaches may also modulate the capacity of mucins within supra-mucosal gels to act as metabolic feedstocks for commensal organisms and provide a means to alter mucosal microbial ecology, thereby promoting beneficial colonization and blocking pathogen entry.

RESEARCH FOCUS BOX

- Investigations to be undertaken of the homeostatic mechanisms required for the maintenance of mucosal barrier function through the formation of competent mucus gels.
- Development of new screening tools to interrogate mucin–microbial interactions, including mucus gel-forming cell lines and generation of recombinant forms of mucin with defined glycosylation.
- Development of an enhanced capacity for *O*-glycan analysis to facilitate the interrogation of mucosal glycome responses through the analysis of mucins.
- Definition of the capacity of key commensals and pathogens to influence mucin biosynthesis and glycosylation and the mechanisms by which such changes occur.
- Development of ligand decoys for key pathogens based on this latter knowledge and testing of these in suitable *in vivo* models.

References

Aristoteli, L.P., Willcox, M.D., 2003. Mucin degradation mechanisms by distinct *Pseudomonas aeruginosa* isolates in vitro. Infect. Immun. 71, 5565–5575.

Behrens, I., Stenberg, P., Artursson, P., Kissel, T., 2001. Transport of lipophilic drug molecules in a new mucus-secreting cell culture model based on HT29-MTX cells. Pharm. Res. 18, 1138–1145.

Borén, T., Falk, P., Roth, K.A., Larson, G., Normark, S., 1993. Attachment of *Helicobacter pylori* to human gastric epithelium mediated by blood group antigens. Science 262, 1892–1895.

Brockhausen, I., 2007. Glycosyltransferases specific for the synthesis of mucin type O glycans. In: Samson, C. (Ed.), Glycobiology. Scion Publishing Limited, Bloxhan, pp. 217–234.

Byrd, J.C., Yan, P., Sternberg, L., Yunker, C.K., Scheiman, J.M., Bresalier, R.S., 1997. Aberrant expression of gland-type gastric mucin in the surface epithelium of *Helicobacter pylori*-infected patients. Gastroenterology 113, 455–464.

Byrd, J.C., Yunker, C.K., Xu, Q.S., Sternberg, L.R., Bresalier, R.S., 2000. Inhibition of gastric mucin synthesis by *Helicobacter pylori*. Gastroenterology 118, 1072–1079.

Byrne, C.M., Clyne, M., Bourke, B., 2007. *Campylobacter jejuni* adhere to and invade chicken intestinal epithelial cells *in vitro*. Microbiology 153, 561–569.

Caballero-Franco, C., Keller, K., De Simone, C., Chadee, K., 2007. The VSL#3 probiotic formula induces mucin gene expression and secretion in colonic epithelial cells. Am. J. Physiol. Gastrointest. Liver Physiol. 292, G315–G322.

Ceroni, A., Dell, A., Haslam, S.M., 2007. The GlycanBuilder: a fast, intuitive and flexible software tool for building and displaying glycan structures. Source Code Biol. Med. 2, 3. <http://www.scfbm.org/content/2/1/3>.

Chen, X.M., LaRusso, N.F., 2000. Mechanisms of attachment and internalization of *Cryptosporidium parvum* to biliary and intestinal epithelial cells. Gastroenterology 118, 368–379.

Clyne, M., Dillon, P., Daly, S., et al., 2004. *Helicobacter pylori* interacts with the human single-domain trefoil protein TFF1. Proc. Natl. Acad. Sci. USA 101, 7409–7414.

Cooke, C.L., An, H.J., Kim, J., Solnick, J.V., Lebrilla, C.B., 2007. Method for profiling mucin oligosaccharides from gastric biopsies of rhesus monkeys with and without *Helicobacter pylori* infection. Anal. Chem. 79, 8090–8097.

Corfield, A., 2007. Glycobiology of mucins in the human gastrointestinal tract. In: Samson, C. (Ed.), Glycobiology. Scion Publishing Limited, Bloxhan, pp. 248–260.

Corfield, A.P., Carroll, D., Myerscough, N., Probert, C.S., 2001. Mucins in the gastrointestinal tract in health and disease. Front. Biosci. 6, D1321–D1357.

Corfield, A.P., Wiggins, R., Edwards, C., et al., 2003. A sweet coating – how bacteria deal with sugars. Adv. Exp. Med. Biol. 535, 3–15.

Costerton, J.W., 2002. Anaerobic biofilm infections in cystic fibrosis. Mol. Cell 10, 699–700.

Davril, M., Degroote, S., Humbert, P., et al., 1999. The sialylation of bronchial mucins secreted by patients suffering from cystic fibrosis or from chronic bronchitis is related to the severity of airway infection. Glycobiology 9, 311–321.

de Repentigny, L., Aumont, F., Bernard, K., Belhumeur, P., 2000. Characterization of binding of *Candida albicans* to small intestinal mucin and its role in adherence to mucosal epithelial cells. Infect. Immun. 68, 3172–3179.

Derrien, M., Collado, M.C., Ben-Amor, K., Salminen, S., de Vos, W.M., 2008. The mucin degrader *Akkermansia muciniphila* is an abundant resident of the human intestinal tract. Appl. Environ. Microbiol. 74, 1646–1648.

dos Santos Silva, E., Ulrich, M., Doring, G., Botzenhart, K., Gott, P., 2000. Trefoil factor family domain peptides in the human respiratory tract. J. Pathol. 190, 133–412.

Drumm, B., Roberton, A.M., Sherman, P.M., 1988. Inhibition of attachment of *Escherichia coli* RDEC-1 to intestinal microvillus membranes by rabbit ileal mucus and mucin *in vitro*. Infect. Immun. 56, 2437–2442.

Edwards, C.M., Corfield, A.P., 2000. Mucin bound *O*-acetylated sialic acids and diversion colitis. Gut 46, A81.

Freitas, M., Axelsson, L.G., Cayuela, C., Midtvedt, T., Trugnan, G., 2002. Microbial–host interactions specifically control the glycosylation pattern in intestinal mouse mucosa. Histochem. Cell. Biol. 118, 149–161.

Gabius, H.-J., 2008. The Sugar Code. Fundamentals of Glycoscience. Wiley-VCH Verlag, Weinheim.

Hattrup, C.L., Gendler, S.J., 2008. Structure and function of the cell surface (tethered) mucins. Annu. Rev. Physiol 70, 431–457.

Hicks, S.J., Theodoropoulos, G., Carrington, S.D., Corfield, A.P., 2000. The role of mucins in host–parasite interactions. Part I protozoan parasites. Parasitol. Today 16, 476–481.

Hooper, L.V., Gordon, J.I., 2001. Glycans as legislators of host–microbial interactions: spanning the spectrum from symbiosis to pathogenicity. Glycobiology 11, 1R–10R.

Kawakubo, M., Ito, Y., Okimura, Y., et al., 2004. Natural antibiotic function of a human gastric mucin against *Helicobacter pylori* infection. Science 305, 1003–1006.

Kinoshita, H., Uchida, H., Kawai, Y., et al., 2008. Cell surface *Lactobacillus plantarum* LA 318 glyceraldehyde-3-phosphate dehydrogenase (GAPDH) adheres to human colonic mucin. J. Appl. Microbiol. 104, 1667–1674.

Lemjabbar, H., Basbaum, C., 2002. Platelet-activating factor receptor and ADAM10 mediate responses to *Staphylococcus aureus* in epithelial cells. Nat. Med. 8, 41–46.

Lidell, M.E., Moncada, D.M., Chadee, K., Hansson, G.C., 2006. *Entamoeba histolytica* cysteine proteases cleave the MUC2 mucin in its C-terminal domain and dissolve the protective colonic mucus gel. Proc. Natl. Acad. Sci. USA 103, 9298–9303.

Lievin-Le Moal, V., Servin, A.L., 2006. The front line of enteric host defense against unwelcome intrusion of harmful microorganisms: mucins, antimicrobial peptides, and microbiota. Clin. Microbiol. Rev. 19, 315–337.

Lillehoj, E.P., Kim, H., Chun, E.Y., Kim, K.C., 2004. *Pseudomonas aeruginosa* stimulates phosphorylation of the airway epithelial membrane glycoprotein Muc1 and activates MAP kinase. Am. J. Physiol. Lung Cell. Mol. Physiol. 287, L809–L815.

Lindén, S., Nordman, H., Hedenbro, J., Hurtig, M., Borén, T., Carlstedt, I., 2002. Strain- and blood group-dependent binding of *Helicobacter pylori* to human gastric MUC5AC glycoforms. Gastroenterology 123, 1923–1930.

Lindén, S., Mahdavi, J., Semino-Mora, C., et al., 2008. Role of ABO secretor status in mucosal innate immunity and *H. pylori* infection. PLoS Pathog. 4, e2.

Macfarlane, G.T., Macfarlane, S., 2000. Growth of mucin degrading bacteria in biofilms. Methods Mol. Biol. 125, 439–452.

Mahdavi, J., Sonden, B., Hurtig, M., et al., 2002. *Helicobacter pylori* SabA adhesin in persistent infection and chronic inflammation. Science 297, 538–573.

McAuley, J.L., Lindén, S.K., Png, C.W., et al., 2007. MUC1 cell surface mucin is a critical element of the mucosal barrier to infection. J. Clin. Invest. 117, 2313–2324.

Miyake, K., Tanaka, T., McNeil, P.L., 2006. Disruption-induced mucus secretion: repair and protection. PLoS Biol. 4, e276.

Roberton, A.M., Wiggins, R., Horner, P.J., et al., 2005. A novel bacterial mucinase, glycosulfatase, is associated with bacterial vaginosis. J. Clin. Microbiol. 43, 5504–5508.

Royle, L., Matthews, E., Corfield, A., et al., 2008. Glycan structures of ocular surface mucins in man, rabbit and dog display species differences. Glycoconj. J. 25, 763–773.

Ryan, P.A., Pancholi, V., Fischetti, V.A., 2001. Group A streptococci bind to mucin and human pharyngeal cells through sialic acid-containing receptors. Infect. Immun. 69, 7402–7412.

Schild, S., Lamprecht, A.K., Fourestier, C., Lauriano, C.M., Klose, K.E., Reidl, J., 2005. Characterizing lipopolysaccharide and core lipid A mutant O1 and O139 *Vibrio cholerae* strains for adherence properties on mucus-producing cell line HT29-Rev MTX and virulence in mice. Int. J. Med. Microbiol. 295, 243–251.

Shen, H., Yoshida, H., Yan, F., et al., 2008. Synergistic induction of MUC5AC mucin by nontypeable *Haemophilus influenzae* and *Streptococcus pneumoniae*. Biochem. Biophys. Res. Commun. 365, 795–800.

Shuter, J., Hatcher, V.B., Lowy, F.D., 1996. *Staphylococcus aureus* binding to human nasal mucin. Infect. Immun. 64, 310–318.

Singh, P.K., Hollingsworth, M.A., 2006. Cell surface-associated mucins in signal transduction. Trends Cell Biol. 16, 467–476.

Sonnenburg, J.L., Xu, J., Leip, D.D., et al., 2005. Glycan foraging in vivo by an intestine-adapted bacterial symbiont. Science 307, 1955–1959.

Thim, L., May, F.E., 2005. Structure of mammalian trefoil factors and functional insights. Cell. Mol. Life Sci. 62, 2956–2973.

Thornton, D.J., Rousseau, K., McGuckin, M.A., 2008. Structure and function of the polymeric mucins in airways mucus. Annu. Rev. Physiol. 70, 459–486.

Tierney, J.B., Matthews, E., Carrington, S.D., Mulcahy, G., 2007. Interaction of *Eimeria tenella* with intestinal mucin *in vitro*. J. Parasitol. 93, 634–638.

Van den Brink, G.R., Tytgat, K.M., Van der Hulst, R.W., et al., 2000. *H. pylori* colocalises with MUC5AC in the human stomach. Gut 46, 601–607.

Vinall, L.E., King, M., Novelli, M., et al., 2002. Altered expression and allelic association of the hypervariable membrane mucin MUC1 in *Helicobacter pylori* gastritis. Gastroenterology 123, 41–49.

Wiggins, R., Hicks, S.J., Soothill, P.W., Millar, MR., Corfield, A.P., 2001. Mucinases and sialidases: their role in the pathogenesis of sexually transmitted infections in the female genital tract. Sex. Transm. Infect. 77, 402–408.

Wright, N.A., Hoffmann, W., Otto, W.R., Rio, M.C., Thim, L., 1997. Rolling in the clover: trefoil factor family (TFF)-domain peptides, cell migration and cancer. FEBS Lett. 408, 121–123.

Wyatt, J.I., Rathbone, B.J., Sobala, G.M., et al., 1990. Gastric epithelium in the duodenum: its association with *Helicobacter pylori* and inflammation. J. Clin. Pathol. 43, 981–986.

Xia, B., Sachdev, G.P., Cummings, R.D., 2007. *Pseudomonas aeruginosa* mucoid strain 8830 binds glycans containing the sialyl-Lewis x epitope. Glycoconj. J. 24, 87–95.

CHAPTER

34

Mannose–fucose recognition by DC-SIGN

Jeroen Geurtsen, Nicole N. Driessen and Ben J. Appelmelk

SUMMARY

Dendritic cell-specific ICAM-3-grabbing non-integrin (DC-SIGN) is a C-type lectin receptor that recognizes *N*-linked high-mannose oligosaccharides and branched fucosylated structures. It is primarily expressed by dendritic cells and mediates the capture, destruction and presentation of microbial pathogens to induce successful immune responses. Furthermore, DC-SIGN is involved in the priming of T-cell responses and facilitates dendritic cell homeostasis by controlling extravasation into peripheral tissues. However, an increasing amount of evidence suggests that pathogens also exploit DC-SIGN to subvert host immune responses. Herein, we discuss the current state of knowledge of DC-SIGN-carbohydrate interactions and investigate how these interactions influence dendritic cell functioning. First, an overview of the structure of DC-SIGN is provided and its expression pattern among immune cells is discussed. After this, the molecular aspects that underlie the selectivity of DC-SIGN for mannose- and fucose-containing carbohydrates are detailed. Finally, the role of DC-SIGN in dendritic cell biology is discussed and how certain bacterial pathogens exploit DC-SIGN to escape immune surveillance.

Keywords: C-Type lectin; Dendritic cell-specific ICAM-3-grabbing non-integrin (DC-SIGN); High-mannose; Lewis antigens; Carbohydrate recognition; Dendritic cell; Immune suppression; Host-pathogen interaction; Pathogenesis; Virulence

1. INTRODUCTION

The capability of the innate immune system quickly to detect and respond to invading pathogens is essential for controlling infection. Pattern-recognition receptors (PRRs) have evolved to recognize a wide variety of pathogen-associated molecular patterns (PAMPs) and trigger antimicrobial responses in immune cells. Recognition of PAMPs by PRRs is an important determinant for the overall quality and effectiveness of immune responses by mediating not only direct effector functions, such as phagocytosis and degranulation, but also by transmitting signals that regulate the expression of genes important for both innate and adaptive immune responses. One of the best studied classes of PRRs are the Toll-like receptors (TLRs). This receptor family recognizes a diverse range of microbial compounds and, upon ligation, trigger robust innate immune responses by inducing proinflammatory cytokine and chemokine production and by upregulating co-stimulatory molecule expression (Akira and Takeda, 2004). Although TLRs are an important warning system that alerts host immune cells to the presence of invading microbes, it remains unclear whether they also facilitate antigen uptake and it seems that this process is mediated by other types of PRRs, such as scavenger receptors (Murphy *et al.*, 2005) and C-type lectin receptors (CLRs) (Cambi *et al.*, 2005; McGreal *et al.*, 2005). The C-type lectin receptors encompass a large family of proteins that are best known for their ability to detect and capture microbe-derived materials. Classical CLRs contain calcium- (Ca^{2+}-) dependent carbohydrate recognition domains (CRDs) which are involved in the recognition of carbohydrate structures on both self and non-self ligands. The C-type lectin CRDs are part of a large family of protein domains called C-type lectin domains (CTLDs) which are characterized by a common protein-fold consisting of two anti-parallel β-strands and two α-helices (Weis *et al.*, 1998; Drickamer, 1999). Although the name C-type lectin refers to a Ca^{2+}-dependent carbohydrate-binding protein, it is now clear that this name is a misnomer and that many C-type lectins are calcium-independent and do not bind carbohydrates but rather recognize proteins or lipids. Therefore, the term C-type lectin does not refer to a protein family which shares functional similarities but rather defines a class of proteins that all contain one or more CTLDs.

The C-type lectin receptors exhibit important functions during primary immune responses (Robinson *et al.*, 2006). Besides their role in pathogen detection and uptake via PAMPs, they also mediate cell–cell interactions by the recognition of endogenous ligands (Cambi and Figdor, 2003). Furthermore, they are involved in the induction of immune tolerance and play a role in endogenous glycoprotein homeostasis (Cambi and Figdor, 2003). In addition, C-type lectins have been shown to influence TLR-mediated responses (Mukhopadhyay *et al.*, 2004; Cambi *et al.*, 2005). Moreover, C-type lectins can be produced as transmembrane proteins or as soluble, secreted proteins. Examples of soluble C-type lectins are the members of the collectins family, such as lung surfactant proteins A and D (Pastva *et al.*, 2007), and the plasma-localized mannose-binding lectin (MBL) (Takahashi *et al.*, 2006) (see Chapter 35). Transmembrane C-type lectins can be divided into two groups based upon the orientation of their amino (N)-terminus. Myeloid cells predominantly express type-II C-type lectins. These lectins all contain a single CRD and have been shown to bind a wide variety of carbohydrate ligands. Examples of myeloid type-II C-type lectins are the dendritic cell-specific intercellular adhesion molecule-3-grabbing non-integrin (DC-SIGN) and its close relative DC-SIGNR (L-SIGN), the related murine receptor family termed SIGN-related (SIGNR) 1-4, Langerin, blood DC antigen 2 (BDCA-2), C-type lectin receptors 1 and 2 (CLEC-1 and CLEC-2, respectively), macrophage galactose-type lectin (MGL), DC immunoreceptor (DCIR) and Dectin-1 and -2 (Cambi *et al.*, 2005; McGreal *et al.*, 2005; Robinson *et al.*, 2006). Myeloid type-I C-type lectins contain multiple CRDs. Examples of these lectins are the macrophage mannose receptor (MR), the phosholipase A_2 receptor, DEC-205 and Endo-180 (Apostolopoulos and McKenzie, 2001; East and Isacke, 2002). Investigations on the carbohydrate-binding specificities of Ca^{2+}-dependent CRDs have shown that they fall within two broad categories: those recognizing D-mannose- (Man-) like structures and those recognizing D-galactose- (Gal-) like structures. This subdivision is supported at the molecular level by studies demonstrating that Man- and Gal-binding CRDs contain characteristic amino acid triplets: EPN (in one-letter amino-acid code) for Man-binding CRDs and QDP (in one-letter amino-acid code) for Gal-binding CRDs (Drickamer, 1992). In addition, some CLRs (e.g. MR) can recognize sulfated carbohydrates independent of their CRD. This interaction is mediated by a cysteine-rich domain which is present in these proteins (Fiete *et al.*, 1998).

Indeed, DC-SIGN is a dendritic cell (DC)-specific type-II C-type lectin which was originally identified as a receptor that specifically interacts with the intercellular adhesion molecule (ICAM)-3 on T-cells, thereby mediating DC–T-cell interactions (Geijtenbeek *et al.*, 2000b). Later on, DC-SIGN was also shown to be involved in DC migration by interacting with ICAM-2 on vascular endothelial cells (Geijtenbeek *et al.*, 2000a). Ever since its discovery, DC-SIGN received major scientific interest. This interest grew further when it became clear that DC-SIGN has a dual function in that it also recognizes a wide variety of pathogens (Table 34.1). Investigations on its carbohydrate-specificity using glyco-arrays have shown that DC-SIGN preferentially binds Man- and L-fucose- (Fuc-) containing structures. In this chapter, we summarize the current state of knowledge on DC-SIGN-carbohydrate interactions. First, the structure of DC-SIGN and its expression patterns among various immune cells is discussed. Then the molecular aspects that underlie the selectivity of DC-SIGN for Man- and Fuc-containing carbohydrates are detailed. Finally, the role of DC-SIGN in DC biology and how certain pathogens exploit DC-SIGN to escape immune surveillance are reviewed.

2. DC-SIGN STRUCTURE AND EXPRESSION

2.1. Structure of DC-SIGN

Human DC-SIGN (CD209) is a 404 amino acid protein encoded in seven exons on chromosome 19p13. Upstream, in the reverse orientation, and downstream of DC-SIGN, two DC-SIGN homologues are encoded: DC-SIGNR which shows 77% amino acid sequence identity to DC-SIGN and LSECtin which has 31% identity to DC-SIGN, respectively (Soilleux *et al.*, 2000; Liu *et al.*, 2004b). Further downstream lies the gene encoding the low affinity IgE receptor (CD23), a C-type lectin with a gene organization similar to DC-SIGN (Soilleux *et al.*, 2000). Structurally, DC-SIGN itself is a prototype type II transmembrane protein consisting of a carboxy (C)-terminal CRD followed by a neck domain fused to a transmembrane region and ending with a N-terminal cytoplasmic tail, which harbours internalization and recycling motifs, such as a di-leucine (LL) and a tri-acidic (EEE) motif, and an incomplete immunoreceptor tyrosine-based activation motif (ITAM) (van Kooyk and Geijtenbeek, 2003) (Figure 34.1A). Crystal structures of the CRDs of DC-SIGN and DC-SIGNR revealed that both CRDs exhibit a typical long-form C-type lectin fold (Feinberg *et al.*, 2001) (Figure 34.1B). The structure harbours three Ca^{2+} ions, of which one, the principal Ca^{2+} (see Figure 34.1B; Ca^{2+} no. 2), is common to all Ca^{2+}-dependent C-type lectins and dictates the recognition of specific carbohydrate structures. The principal Ca^{2+} interacts with four amino acids; Glu347, Asn349, Glu354 and Asn365 in DC-SIGN (where Glu, glutamic acid; Asn, asparagine) and Glu359, Asn361, Glu366 and Asn377 in DC-SIGNR and mutations in these amino acids lead to loss of ligand interaction (Geijtenbeek *et al.*, 2002b). The DC-SIGN neck domain is composed of one incomplete and seven complete 23-amino acid tandem repeats which are encoded in a single exon. The number of neck repeats is highly conserved, although polymorphisms have been reported (Liu *et al.*, 2004a; Barreiro *et al.*, 2007; Ben-Ali *et al.*, 2007; Rathore *et al.*, 2008). In contrast, the number of neck repeats in DC-SIGNR is much more variable and varies between four and nine (Bashirova *et al.*, 2001; Feinberg *et al.*, 2005). Cross-linking and analytical ultracentrifugation experiments have shown that purified recombinant DC-SIGN and DC-SIGNR CRDs

TABLE 34.1 Pathogens recognized by DC-SIGN

Pathogen	Ligand[b]	Carbohydrate epitope
Viruses[a]		
HIV-1	gp120	High-mannose
HIV-2	gp120	Unknown
SIV	gp120	Unknown
FIV	gp95	Unknown
Ebola virus	Glycoprotein	High-mannose
Marburg virus	Glycoprotein	Unknown
Cytomegalovirus	Glycoprotein B	Unknown
Hepatitis C virus	E1/E2 glycoproteins	Unknown
Dengue virus	Glycoprotein E	*N*-linked glycans
Alpha viruses	Unknown	Unknown
SARS corona virus	S protein	*N*-linked glycans
West Nile virus	Glycoprotein E	*N*-linked glycans
Human herpes virus 8	Unknown	Unknown
Measles virus	Glycoproteins F and H	Unknown
Bacteria		
Helicobacter pylori	LPS	Lewis antigens
Mycobacterium spp.	ManLAM, PIMs, LM, AM, 19- and 45-kDa antigens	Di- and tri-mannose
Lactobacillus spp.	S-layer protein A	Unknown
Escherichia coli	LPS	*N*-acetyl-D-glucosamine
Salmonella enterica	LPS	*N*-acetyl-D-glucosamine
Haemophilus ducreyi	LPS	*N*-acetyl-D-glucosamine
Neisseria spp.	*lgtB* LPS	*N*-acetyl-D-glucosamine
Streptococcus pneumoniae	Capsular polysaccharides and unknown	Unknown
Yersinia pestis	LPS	Unknown
Fungi		
Candida albicans	*N*-linked mannan	Mannose
Aspergillus fumigatus	Unknown	Unknown
Chrysosporium tropicum	Unknown	Unknown
Parasites		
Leishmania spp.	Unknown	Unknown
Schistosoma mansoni	SEAs and glycolipids	Lewis^{x} and peudo-Lewis^{y}
Toxocara canis	Excretory/secretory products	Unknown

[a]Viral abbreviations: HIV, human immunodeficiency virus; SIV, simian immunodeficiency virus; FIV, feline immunodeficiency virus; SARS, severe acute respiratory syndrome.
[b]Compound abbreviations: AM, arabinomannan; LM, lipomannan; LPS, lipopolysaccharide; ManLAM, mannose-capped lipoarabinomannan; PIMs, phosphatidylinositol-mannosides; SEAs, soluble egg antigens.

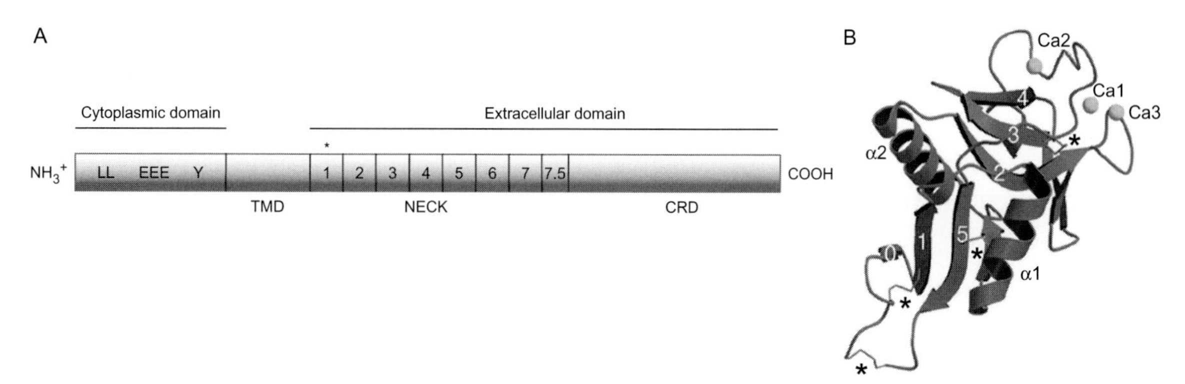

FIGURE 34.1 Structure of DC-SIGN. (A) Schematic representation of DC-SIGN. DC-SIGN is a prototype type II transmembrane protein consisting of a carboxy- (COOH-) terminal carbohydrate recognition domain (CRD), followed by a neck domain consisting of one incomplete and seven complete tandem repeats (numbered 1–7.5) fused to a transmembrane domain (TMD) and ending with an amino- (NH_3^+-) terminal cytoplasmic tail. The latter harbours internalization and recycling motifs, such as a di-leucine (LL) and a tri-acidic (EEE) motif, and an incomplete immunoreceptor tyrosine- (Y-) based activation motif (ITAM). An asterisk indicates the *N*-glycosylation site situated in neck repeat 1. (B) Ribbon diagram of the CRD of DC-SIGN. Large spheres represent the three Ca^{2+} ions (ions are numbered Ca1–Ca3). Disulfide bonds are indicated by an asterisk. The CRD of DC-SIGNR resembles that of DC-SIGN, except for the disulfide bond connecting the carboxy- and amino-termini, which cannot be seen in DC-SIGNR. Figure 34.1B was adapted, with permission, from Feinberg *et al.* (2001) Structural basis for selective recognition of oligosaccharides by DC-SIGN and DC-SIGNR. Science 294, 2163–2166. © AAAS.

exist as monomers, whereas the complete extracellular domains form tetramers (Mitchell *et al.*, 2001; Feinberg *et al.*, 2005). Analysis by circular dichroism spectroscopy indicated that the neck region has a high α-helical content and forms a tetrameric coiled-coil structure that projects the CRDs away from the cell surface (Mitchell *et al.*, 2001). Furthermore, it was demonstrated that tetramerization enhances the affinity for multivalent ligands (Mitchell *et al.*, 2001). A disadvantage of using recombinant proteins is that they may exhibit differential glycosylation and therefore behave differently from their natural equivalents. In a recent study, which investigated the multimerization and ligand-binding activities of various DC-SIGN splice forms, it was shown that *N*-glycosylation of Asn80 in repeat 1 negatively influences multimer formation (Serrano-Gomez *et al.*, 2008). Furthermore, it was demonstrated that the presence of the two most N-terminal repeats is sufficient for multimerization, that the specific order of the neck repeats is important for functionality and that the CRD contributes to tetramer stabilization via cysteine-mediated interactions (Serrano-Gomez *et al.*, 2008). Interestingly, the authors also found that, although the capacity to multimerize correlates with the ability to bind soluble carbohydrates, it does not correlate with the capability to interact with certain pathogens, such as *Candida albicans* and *Leishmania infantum* (Serrano-Gomez *et al.*, 2008). A possible explanation is that the large amount of DC-SIGN ligands on the surface of interacting pathogens compensates for the distinct affinities and multimerization abilities of DC-SIGN isoforms.

2.2. Expression of DC-SIGN

DC-SIGN is expressed by DCs both *in vitro* and *in vivo* (Bleijs *et al.*, 2001). Monocyte- and $CD34^+$-derived DCs abundantly express DC-SIGN (Geijtenbeek *et al.*, 2000b). It is not expressed by monocytes, monocytic cell lines, activated monocytes, granulocytes, $CD34^+$ bone marrow cells or T- and B-lymphocytes (Geijtenbeek *et al.*, 2000b). *In vivo*, DC-SIGN is expressed by immature DCs (iDCs) in lymphoid and fibrous connective tissues and in mucosae, but not by epidermal $CD1a^+$ Langerhans cells (Geijtenbeek *et al.*, 2000a,c; Soilleux and Coleman, 2001; Soilleux *et al.*, 2002). The percentage of DC-SIGN-positive blood cells is very low and only certain subsets of blood DCs seem to express DC-SIGN (Engering *et al.*, 2002b; Soilleux *et al.*, 2002). Although DC-SIGN was thought to be DC-specific, its expression has now been demonstrated on specific macrophage subsets, such as decidual and alveolar macrophages and on Hofbauer cells in the placenta (Geijtenbeek *et al.*, 2001; Lee *et al.*, 2001; Soilleux *et al.*, 2001, 2002). Interestingly, studies in fetal tissues have demonstrated that while fetal iDCs normally express DC-SIGN, alveolar macrophages do not (Soilleux *et al.*, 2002). This indicates that DC-SIGN expression in this type of macrophage may be dependent on antigenic stimulation. Consistent with this hypothesis is the finding that interleukin- (IL-) 13 treatment of monocyte-derived macrophages induces DC-SIGN expression (Soilleux *et al.*, 2002). This observation, together with the ability of IL-4 to induce monocytic DC-SIGN expression (Relloso *et al.*, 2002) and the genetic linkage between DC-SIGN and CD23, which is part of the T-helper-2 cell (Th2) axis of immunity, suggests that DC-SIGN plays a role in Th2-type immune responses (Soilleux *et al.*, 2000; Park *et al.*, 2001; Soilleux, 2003).

3. SELECTIVE RECOGNITION OF MAN- AND FUC-CONTAINING GLYCANS BY DC-SIGN

Glycan array screening using recombinant DC-SIGN fragments or DC-SIGN extracellular domains fused to an Fc domain (DC-SIGN–Fc) have demonstrated that DC-SIGN recognizes two classes of carbohydrates: *N*-linked high-Man oligosaccharides and branched fucosylated structures such as the Lewis (Le) blood group antigens, Le^a, Le^b, Le^x, and Le^y (Feinberg *et al.*, 2001, 2007; Appelmelk *et al.*, 2003; Guo *et al.*, 2004; van Liempt *et al.*, 2006). Also, DC-SIGNR, which is 77% homologous to DC-SIGN, displays a similar specificity, except for the Le^x epitope which is only recognized by DC-SIGN (Appelmelk *et al.*, 2003; van Kooyk *et al.*, 2003; van Die *et al.*, 2003; Guo *et al.*, 2004; van Liempt *et al.*, 2004). The hallmark of carbohydrate binding to C-type lectins is the primary interaction between the principal Ca^{2+} and the vicinal hydroxyl groups of a pyranose ring. Further substrate specificity comes from secondary interactions between the carbohydrate and CRD-specific amino acids.

3.1. DC-SIGN binding to high-Man oligosaccharides

Solid-phase competition and binding assays have shown that DC-SIGN and DC-SIGNR interact with *N*-linked high-Man structures (Feinberg *et al.*, 2001, 2007; Guo *et al.*, 2004; van Liempt *et al.*, 2006). Both proteins show the highest apparent affinity towards structures containing nine mannosyl residues, i.e. $Man_9GlcNAc_2$ (where GlcNAc, *N*-acetyl-D-glucosamine) (Figure 34.2(A). Co-crystallization experiments have shown that the interaction between DC-SIGN/DC-SIGNR and high-Man structures is mediated by vicinal, equatorial 3- and 4-OH groups of internal mannosyl residues (Feinberg *et al.*, 2001, 2007; Guo *et al.*, 2004) (Figure 34.3A). The preference for internal residues is unusual since most Man-binding lectins (e.g. MR and MBL) recognize terminal residues (Weis *et al.*, 1992; Hitchen *et al.*, 1998). In the first crystal structure that was published, DC-SIGN and DC-SIGNR were co-crystallized with the pentasaccharide

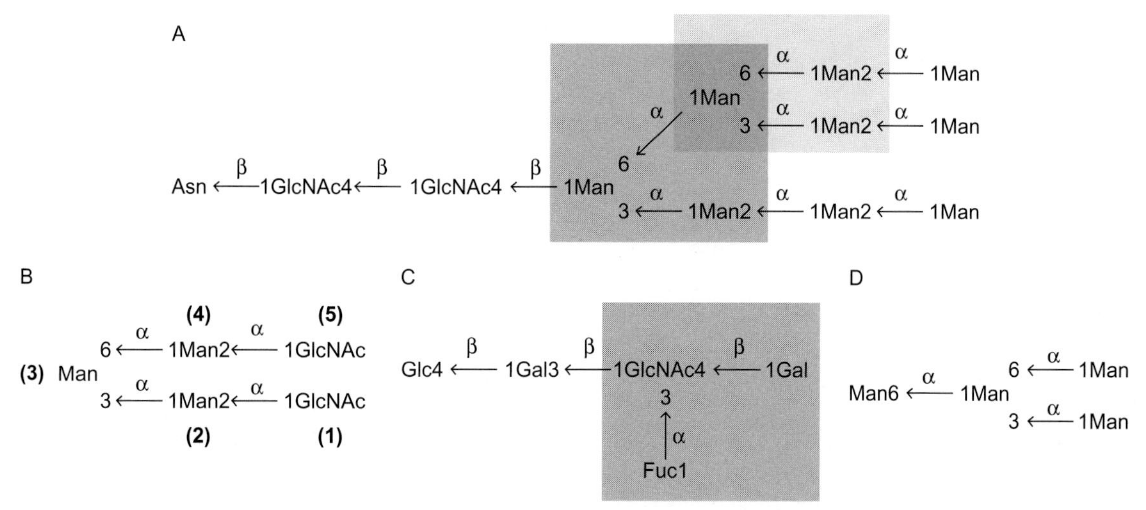

FIGURE 34.2 Schematic representations of mannose- (Man-) containing structures that interact with DC-SIGN and DC-SIGNR. (A) the *N*-linked high-mannose (Man) structure $Man_9GlcNAc_2$; the inner branched trimannose structure α-Man-(1→3)-[α-Man-(1→6)-]Man is indicated by the dark grey box, whereas the outer tri-Man structure is indicated by the light grey box; (B) the pentasaccharide $Man_3GlcNAc_2$ that has been co-crystallized with DC-SIGN and DC-SIGNR (residues are numbered as mentioned in the text); (C) the lacto-*N*-fucopentaose III pentasaccharide; and (D) the Man_4 oligosaccharide structures that were co-crystallized with DC-SIGN. For lacto-*N*-fucopentaose III, the Le^x structure is highlighted by the grey box. Abbreviations: Asn, asparagine; Fuc, L-fucose; Man, D-mannose; Gal, D-galactose; Glc, D-glucose; GlcNAc, *N*-acetyl-D-glucosamine. In part adapted, with permission, from Feinberg *et al.* (2001) Structural basis for selective recognition of oligosaccharides by DC-SIGN and DC-SIGNR. Science 294, 2163–2166. © AAAS.

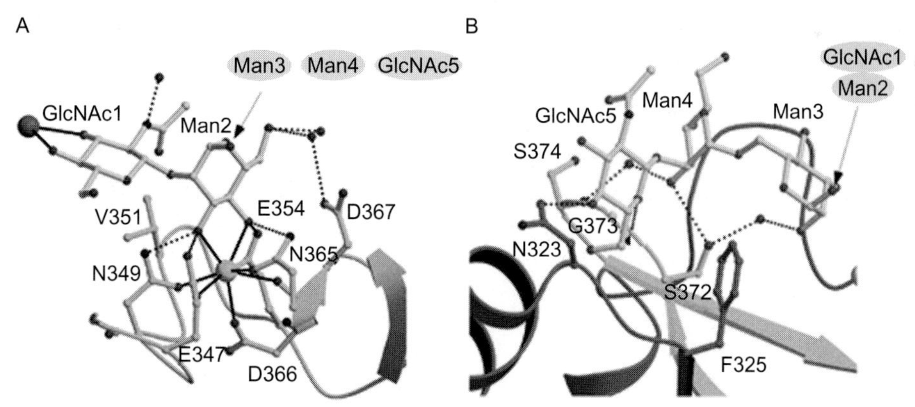

FIGURE 34.3 Co-crystallization of $Man_3GlcNAc_2$ with DC-SIGN and DC-SIGNR. (A) Interaction of the α-(1→3)-linked branch of $Man_3GlcNAc_2$ with the carbohydrate recognition domain (CRD) of DC-SIGN. For clarity, some of the sugar residues are shown schematically. Large spheres represent Ca^{2+} ions. The Ca^{2+} coordination bonds are shown as solid black lines; hydrogen bonds as dashed lines. Important residues are numbered. Of note, the terminal GlcNAc1 interacts with the principal Ca^{2+} ion of another DC-SIGN CRD (Note the second CRD is not shown). (B) Interaction of the α-(1→6)-linked branch of $Man_3GlcNAc_2$ with the CRD of DC-SIGNR. Hydrogen bonds are shown as dashed lines and important residues are numbered. Abbreviations: GlcNAc, *N*-acetyl-D-glucosamine; Man, D-mannose. Derived, with permission, from Feinberg *et al.* (2001) Structural basis for selective recognition of oligosaccharides by DC-SIGN and DC-SIGNR. Science 294, 2163–2166. © AAAS.

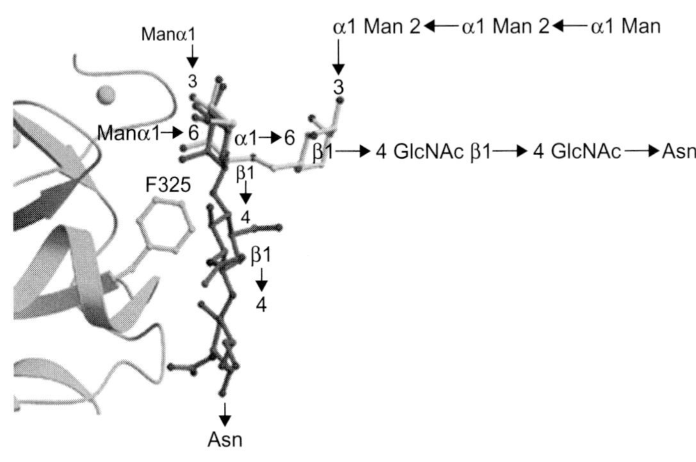

FIGURE 34.4 Ribbon diagram of the carbohydrate recognition domain (CRD) of DC-SIGNR with the Phe325 side chain in a ball-an-stick representation. The inner (in dark grey) and outer (in light grey) branched trimannose structures of $Man_9GlcNAc_2$ were superimposed on the central (reducing) mannose of the $Man_3GlcNAc_2$ structure that was co-crystallized with DC-SIGNR. The model was made by using average torsion angle values, with some small adjustments to overlay the structures precisely. Abbreviations: Asn, asparagine; GlcNAc, *N*-acetyl-D-glucosamine; Man, D-mannose. Derived, with permission, from Feinberg *et al.* (2001) Structural basis for selective recognition of oligosaccharides by DC-SIGN and DC-SIGNR. Science 294, 2163–2166. © AAAS.

$Man_3GlcNAc_2$; β-GlcNAc-(1→2)-α-Man-(1→3)-[β-GlcNAc-(1→2)-α-Man-(1→6)-]Man (Feinberg *et al.*, 2001) (see Figures 34.2B and 34.3A). The structures revealed that the interaction with the principal Ca^{2+} was mediated by the 3- and 4-OH groups of the α-(1→3)-linked mannosyl residue (Man2) (see Figure 34.3A; see Figure 34.2B for residue numbering). For DC-SIGNR, secondary interactions were mediated by phenylalanine residue 325 (Phe325) and serine residue 372 (Ser372) [Phe313 and Ser360 in DC-SIGN], in which Phe325 was located in the crevice between Man3 and Man4 and formed van der Waals contacts with Man3, and Ser372 participated in a hydrogen-bonding network involving both Man3 and Man4 (see Figure 34.3B). The structural information was used to obtain insight into the selectivity of DC-SIGN and DC-SIGNR for high-Man glycans. Superimposing the two distinct α-Man-(1→3)-[α-Man-(1→6)-]Man moieties present in $Man_9GlcNAc_2$ (see Figure 34.2A; light and dark grey boxes, respectively) on the equivalent portions of the oligosaccharide complexes in DC-SIGNR revealed that the β-(1→4)-linked GlcNAc of the inner tri-Man branch point (see Figure 34.2A; dark grey box) clashed with Phe325, whereas the outer tri-Man branch point (see Figure 34.2A; light grey box) did not (Feinberg *et al.*, 2001) (Figure 34.4). Therefore, although DC-SIGN and DC-SIGNR recognize the α-Man-(1→3)-[α-Man-(1→6)-]Man trisaccharide, they can only do so when the central Man is linked in an α-anomeric conformation, a feature only found in high-Man structures. Taken together, these data demonstrated that, besides the principal Ca^{2+}, DC-SIGN and DC-SIGNR harbour a secondary carbohydrate binding site (formed by Phe313 and Ser360 in DC-SIGN and Phe325 and Ser372 in DC-SIGNR) which makes additional contacts with the carbohydrate and explains the increased affinity towards higher order Man structures, as compared to Man alone.

Interestingly, it was recently shown that DC-SIGN is able to bind an α-Man-(1→2)-Man disaccharide (either by itself or as part of a Man_6 structure) independently of the α-Man-(1→3)-[α-Man-(1→6)-]Man trisaccharide (Feinberg *et al.*, 2007). This interaction encompassed not only the principal Ca^{2+} but also included a valine residue at position 351. This result suggests that the observed affinity enhancement towards oligosaccharides that besides a tri-Man core also contain terminal α-Man-(1→2)-Man groups (e.g. $Man_9GlcNAc_2$) is due to multiple binding modes to the CRD, which provide both additional contacts (mediated by Val351) and a statistical (entropic) enhancement of binding (Feinberg *et al.*, 2007).

3.2. DC-SIGN binding to branched Fuc structures

The first observations that DC-SIGN binds branched fucosylated structures were made in studies examining the interaction between DC-SIGN and surface glycans of human pathogens (Appelmelk *et al.*, 2003; van Die *et al.*, 2003). In one of these studies, it was observed that the interaction between *Schistosoma mansoni* soluble egg antigens (SEAs) and DC-SIGN could be inhibited by antibodies recognizing Le^x, β-Gal-(1→4)-[α-Fuc-(1→3)-]GlcNAc and *N*-acetyl-D-galactosamine- (GalNAc-) containing LDNF, β-GalNAc-(1→4)-[α-Fuc-(1→3)-]GlcNAc, epitopes (both present on SEAs), thereby suggesting that these structures bind to DC-SIGN (van Die *et al.*, 2003). This was supported by the observation that DC-SIGN–Fc specifically bound to polyvalent neo-glycoconjugates harbouring the Le^x epitope. In addition, it was shown that DC-SIGN binds to related Le epitopes, such as: Le^a, β-Gal-(1→3)-[Fuc-α-(1→4)-]GlcNAc; Le^b, α-Fuc-(1→2)-β-Gal-(1→3)-[Fuc-α-(1→4)-]GlcNAc; and Le^y, α-Fuc-(1→2)-β-Gal-(1→4)- [α-Fuc-(1→3)-]GlcNAc but not to sulfo- or sialyl-Le^x (Appelmelk *et al.*, 2003). Similar to DC-SIGN, DC-SIGNR can recognize Le^a, Le^b and Le^y epitopes (van Liempt *et al.*, 2004). However, DC-SIGNR does not bind to the Le^x epitope (Guo *et al.*, 2004; van Liempt *et al.*, 2004). Insight into the molecular mechanism underlying this differential recognition was provided when the CRD of DC-SIGN was co-crystallized with the Le^x-containing pentasaccharide lacto-*N*-fucopentaose III (Guo *et al.*, 2004) (see Figures 34.2C and 34.5A). The structure revealed that the interaction between Le^x and the principal Ca^{2+} was mediated by the α-(1→3)-linked Fuc residue. Because in Fuc the 3-OH group is in equatorial and the 4-OH group is in axial conformation, the Fuc ring was tipped compared to the mannosyl residue (for Man both the 3- and 4-OH groups are in equatorial conformation) in a structure of DC-SIGN co-crystallized with a Man_4 oligosaccharide (Guo *et al.*, 2004) (see Figures 34.2D and 34.5B). As a consequence of this orientation, the Fuc ring is close to Val351, which forms tight van der Waals contacts with its 2-OH group. In addition, the terminal Gal residue binds to a secondary binding site encompassing Glu358, Asp367, Lys368, Leu371 and Lys373 (where Glu, glutamic acid; Asp, aspartic acid; Lys, lysine; Leu, leucine) (Guo *et al.*, 2004).

Interestingly, the Val351 residue is substituted for a serine residue (Ser363) in DC-SIGNR, eliminating the van der Waals contact with the 2-OH group. This feature, together with a subtle difference in ligand orientation due to differences in amino acid sequence can explain the inability of DC-SIGNR to bind Le^x. This is supported by the observation that substituting Ser363 in DC-SIGNR for a valine residue enables it to bind to Le^x epitopes (Guo *et al.*, 2004; van Liempt *et al.*, 2004). The reason why DC-SIGNR does not bind Le^x but is able to interact with closely related structures such as Le^a and Le^y remains unclear and probably involves differences in stericity and/or bulkiness of the corresponding structures. Further co-crystallization experiments of DC-SIGN and DC-SIGNR with other Le antigens

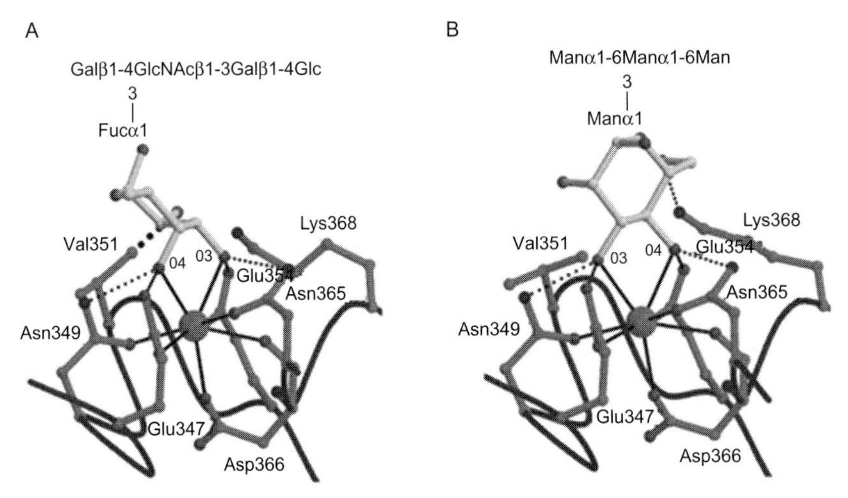

FIGURE 34.5 Co-crystallization of lacto-*N*-fucopentaose III and a Man_4 oligosaccharide with DC-SIGN. Close-up of (A) the α-(1→3)-linked fucose residue of lacto-*N*-fucopentaose III or (B) the terminal, non-reducing α-(1→3)-linked mannose residue of the Man_4 oligosaccharide in the primary binding site of DC-SIGN. The large spheres represent the principal Ca^{2+} ions. The Ca^{2+} coordination bonds are shown as solid black lines, hydrogen bonds as thin dashed lines and van der Waals contacts as thick dashed lines. Important residues are numbered. Abbreviations: Asn, asparagines; Asp, aspartic acid; Fuc, L-fucose; Lys, lysine; Man, D-mannose; Gal, D-galactose; Glc, D-glucose; GlcNAc, *N*-acetyl-D-glucosamine; Glu, glutamic acid; Val, valine. Reproduced, with permission, from Guo *et al.* (2004) Structural basis for distinct ligand-binding and targeting properties of the receptors DC-SIGN and DC-SIGNR. Nat. Struct. Mol. Biol. 11, 591–598.

are needed to obtain insight into the mechanism underlying this differential recognition.

4. *IN VIVO* FUNCTION AND ROLE IN DENDRITIC CELL BIOLOGY OF DC-SIGN

Dendritic cells are professional antigen presenting cells crucial for mounting successful immune responses against invading pathogens and for maintaining immune tolerance towards endogenous ligands. The iDCs originate from the bone marrow where they develop from both myeloid- and lymphoid-committed progenitors (Wu and Liu, 2007). After entering the blood stream, iDCs extravasate into peripheral tissues where they continuously sample their environment by capturing antigens and subsequently presenting them on major histocompatibility (MHC) complexes. Upon pathogen encounter, iDCs receive activation signals that trigger their maturation and stimulate migration into secondary lymphoid organs where they interact with naïve T-cells and initiate an adaptive immune response. Both DC maturation and migration are tightly controlled processes dictated by a variety of cytokines, chemokines and adhesion molecules. Adhesion molecules are not only important for facilitating cellular interactions involved in DC migration, i.e. DC–endothelial cell interactions, but also play a role in establishing DC–T-cell contacts that enable MHC scanning and T-cell receptor activation.

4.1. DC-SIGN facilitates DC migration from the blood into tissues

Migration of iDCs into surrounding tissues is an important process that is pivotal for the successful detection and eradication of invading pathogens. In recent years, the molecular mechanisms underlying DC migration have become increasingly clear and several molecules are now known to be involved, including various C-type lectins such as the MR, the selectins and DC-SIGN (Irjala *et al.*, 2001; Geijtenbeek *et al.*, 2002a, 2004). Indeed, DC-SIGN was shown to mediate the rolling and adhesion of iDCs on vascular endothelial cells by interacting with ICAM-2 (Geijtenbeek *et al.*, 2000a). Furthermore, the DC-SIGN–ICAM-2 interaction was shown to facilitate the transmigration of iDCs across both resting and activated endothelia (Geijtenbeek *et al.*, 2000a). Since ICAM-2 expression is constitutive and does not depend on endothelial activation (Nortamo *et al.*, 1991), DC-SIGN–ICAM-2 interactions probably play a central role in homeostatic extravasation of iDCs into surrounding tissues. Due to its heavy glycosylation, it has long been unclear what particular ICAM-2 structure is recognized by DC-SIGN. Although high-Man structures were initially suspected (Jimenez *et al.*, 2005), it was recently demonstrated that the interaction is primarily dependent on Le^y epitopes present on ICAM-2 (Garcia-Vallejo *et al.*, 2008).

4.2. DC-SIGN mediates DC–T-cell interactions

After receiving activation signals, iDCs become mature and migrate to secondary lymphoid organs where they interact with T-cells and stimulate an adaptive immune response. The initial DC–T-cell contact is transient and allows for the rapid scanning of MHC complexes by T-cells. Studies aimed at identifying molecules important in this process suggested the involvement of ICAM-3 (Hauss *et al.*, 1995; Starling *et al.*, 1995). Initially, it was thought that T-cell ICAM-3 interacted with leukocyte function antigen-1 (LFA-1) on DCs (Hauss *et al.*, 1995), however, later on, it was shown that the affinity of this interaction is low and that the actual DC counter-receptor is DC-SIGN (Geijtenbeek *et al.*, 2000b). It was demonstrated that DC-SIGN binds ICAM-3 with high affinity and forms the major ICAM-3 receptor on DCs. Monoclonal antibodies blocking DC-SIGN inhibited DC–T-cell clustering and prevented DC-dependent proliferation of resting T-cells. Furthermore, it was shown that the DC-SIGN-ICAM-3 interaction is transient, which allows for the rapid screening of MHC-peptide complexes (Geijtenbeek *et al.*, 2000b). Although these data seem strong and the interaction between DC-SIGN and ICAM-3 has been confirmed independently (Geijtenbeek *et al.*, 2002b; Su *et al.*, 2004; Jimenez *et al.*, 2005), a potential problem may be that these studies were performed using recombinant proteins. It is well known that protein glycosylation is not only dependent on the proteins themselves but is also influenced by the nature of the cell line in which they are expressed and the cell culture conditions that are used (Brooks, 2006; Devasahayam, 2007). The possibility that the utilization of recombinant proteins may have led to misinterpretation was raised in a recent paper in which the interactions between DC-SIGN–Fc and ICAM-3 molecules directly purified from blood leukocytes were investigated (Bogoevska *et al.*, 2007). Unexpectedly, this study found that DC-SIGN–Fc could only bind to ICAM-3 molecules that were derived from granulocytes and not to ICAM-3 that was isolated from monocytes, B-cells and, most importantly, T-cells. Moreover, it was found that the interaction was dependent on Le^x and could be abolished by pretreating ICAM-3 with fucosidase III (Bogoevska *et al.*, 2007). The presence of Le^x epitopes on ICAM-3 was confirmed using matrix-assisted laser desorption ionization time-of-flight (MALDI-TOF) mass spectrometry and an antibody direct against Le^x, which only recognized granulocyte ICAM-3 and not the ICAM-3 of other cell types. Surprisingly, these observations do not support the earlier finding that ICAM-3 plays an important role in DC–T-cell interactions (Geijtenbeek *et al.*, 2000b). Furthermore, they contradict an earlier study in which it was shown that native, neutrophil-derived ICAM-3 binds DC-SIGN–Fc with poor affinity (van Gisbergen *et al.*, 2005a). Possible explanations for these conflicting results may be:

(i) the genetic variability between different donors;
(ii) differences in ICAM-3 preparations due to alternative isolation procedures (immunoprecipitation versus affinity chromatography); and
(iii) structural differences between the DC-SIGN–Fc constructs that were used.

Whereas Bogoevska *et al.* (2007) used an N-terminal DC-SIGN–Fc fusion protein produced in human embryonic kidney 293 cells, van Gisbergen *et al.* (2005a) used a C-terminal DC-SIGN–Fc fusion protein prepared from Chinese hamster ovary K1 cells. That the origin of recombinant soluble lectin can indeed determine specificity was recently shown for the interaction of the NK-cell lectin NKp30 with heparan sulfate (Hershkovitz *et al.*, 2008). Furthermore, van Gisbergen *et al.* (2005a) did not include a control for the amount of ICAM-3 that was present on their blots. Nevertheless, the observation that the interaction of T-cells with iDCs can be blocked by monoclonal antibodies against DC-SIGN (Geijtenbeek *et al.*, 2000b) suggests that DC-SIGN plays an important role in early DC–T-cell contacts. The question whether these interactions are indeed dependent on ICAM-3 or, as has now been suggested, are mediated by different molecules remains open and awaits further investigation.

4.3. Role of DC-SIGN in DC-neutrophil interactions

Neutrophils are phagocytic cells that play an important role in innate immunity by engulfing and killing extracellular pathogens. Upon infection, neutrophils infiltrate inflamed tissues where they kill pathogens through phagocytosis and the release of antimicrobial compounds. Although primarily beneficial, neutrophil-derived oxidants, proteinases, cationic peptides and reactive oxygen species can damage surrounding tissues (Moraes *et al.*, 2006). For this reason, neutrophil turnover is a tightly controlled process which is dependent on a fine balance between pre- and anti-apoptotic signals (Walker *et al.*, 2005). In addition, neutrophils play a role in adaptive immune responses not only by recruiting additional immune cells to the site of infection but also by directly influencing their activity. More recently, it was shown that neutrophils can directly interact with iDCs (van Gisbergen *et al.*, 2005a). It was demonstrated that this interaction is dependent on DC-SIGN (van Gisbergen *et al.*, 2005a). Up until now, three potential neutrophil DC-SIGN-ligands have been identified. Besides ICAM-3, also Mac-1 (CD11b/CD18) and the carcinoembryonic antigen-related cellular adhesion molecule (CEACAM) 1 were shown to bind DC-SIGN (van Gisbergen *et al.*, 2005a,b; Bogoevska *et al.*, 2006, 2007). In all cases, binding of DC-SIGN to its neutrophil counter-receptor was dependent on Le^x epitopes that were expressed on these proteins. What effect DC-SIGN binding has on the neutrophil response is currently not well understood, however, it is known that ICAM-3, Mac-1 and CEACAM1 are involved in the regulation of neutrophil apoptosis (Yan *et al.*, 2004; Singer *et al.*, 2005; Kessel *et al.*, 2006). Therefore, DC-SIGN-dependent DC–neutrophil interactions may affect neutrophil survival (Ludwig *et al.*, 2006). This hypothesis is supported by the finding that the interaction of DCs with neutrophils prevents the downregulation of several markers, i.e. CD13, CD15, CD16 and Mac-1, on neutrophils (Megiovanni *et al.*, 2006). Downregulation of these markers is associated with increased neutrophil apoptosis (Dransfield and Rossi, 2004). Besides modulating neutrophil responses, DC-SIGN–neutrophil interactions also influence the DCs themselves. This was demonstrated by the observation that neutrophil-induced DC maturation provokes a strong T-helper-1 cell (Th1-) polarized DC response (van Gisbergen *et al.*, 2005a). Collectively, these findings illustrate the importance of DC-SIGN in DC–neutrophil interactions. Furthermore, they show that these interactions influence both the neutrophil and the DC.

4.4. DC-SIGN functions as an endocytic antigen receptor

The C-type lectins are well known for their capability to recognize antigens and mediate their uptake by antigen presenting cells. Besides the macrophage MR (Apostolopoulos and McKenzie, 2001), also DEC-205 (Mahnke *et al.*, 2000), BDCA-2 (Dzionek *et al.*, 2001), and DC-SIGN (Engering *et al.*, 2002a; Schjetne *et al.*, 2002) have been shown to perform such a function. By using anti-DC-SIGN monoclonal antibodies, it was demonstrated that DC-SIGN quickly internalizes upon antigen binding and subsequently traffics to late endosomal or lysosomal compartments where the complexes are degraded, loaded onto MHC complexes and presented to T-cells (Engering *et al.*, 2002a). In addition, DC-SIGN has been shown to mediate the recognition and uptake of various pathogens including human immunodeficiency virus (HIV)-1 and *Leishmania* amastigotes (Kwon *et al.*, 2002; Colmenares *et al.*, 2002). Notably, DC-SIGN endocytosis is dependent upon the di-leucine motif in its cytoplasmic tail (Engering *et al.*, 2002a) (see Figure 34.1A). Whereas wild-type DC-SIGN was efficiently taken up, a mutated form, in which the leucines were replaced by alanines, endocytosed much less efficiently (Engering *et al.*, 2002a). In addition, DC-SIGN harbours a tri-acidic cluster: EEE (in one-letter amino-acid code) (see Figure 34.1A). For DEC-205, it has been shown that a comparable tri-acidic motif (EDE instead of EEE) mediates targeting to late endosomal compartments (Mahnke *et al.*, 2000). Thus, in conclusion, DC-SIGN not only functions as an adhesion receptor but also mediates the uptake and presentation of pathogen-derived antigens by antigen presenting cells. While these functions suggest a primary role in host defence mechanisms, an increasing amount of evidence suggests that DC-SIGN–pathogen interactions may also be part of pathogenic strategies to increase the efficacy of infection and/or escape immune surveillance by modulating host immune responses.

5. PATHOGENS TARGET DC-SIGN TO SUBVERT HOST IMMUNE RESPONSES

As discussed above, DC-SIGN functions as an antigen receptor that mediates the recognition and uptake of a wide variety of pathogens (see Table 34.1). The first pathogen that was shown to interact with DC-SIGN is HIV-1 (Geijtenbeek *et al.*, 2000c). It was demonstrated that binding of HIV-1 is mediated by the interaction between DC-SIGN and the HIV-1 glycoprotein gp120. Furthermore, it was shown that HIV-1 binding by DCs facilitates infection of HIV-1 permissive cells *in trans* (Geijtenbeek *et al.*, 2000c). Later on, several other viral pathogens were also shown to interact with DC-SIGN, demonstrating its function as a broad viral receptor. The interaction of DC-SIGN with viral pathogens and its role in viral infectivity has been the subject of many studies. Several high-quality reviews on DC-SIGN–virus interactions have recently appeared, including ones discussing the interaction of DC-SIGN with HIV-1 (Lekkerkerker *et al.*, 2006; Wu and KewalRamani, 2006), hepatitis C virus (Cocquerel *et al.*, 2006), Ebola virus (Baribaud *et al.*, 2002) and severe acute respiratory syndrome (SARS) corona virus (Chen and Subbarao, 2007). For this reason, DC-SIGN-virus interactions will not be discussed in further detail. Rather we will focus on the interaction of DC-SIGN with another important class of pathogens, i.e. bacterial pathogens. In the next section, we will first discuss the ability of mycobacterial species to target DC-SIGN and thereby subvert DC functioning. After this, we continue with another pathogen, i.e. *Helicobacter pylori*, and see how this bacterium specifically modulates the Th1/Th2 balance through phase-variable interactions between its lipopolysaccharide (LPS) and DC-SIGN.

5.1. Mycobacteria target DC-SIGN to subvert DC functioning

Tuberculosis (TB), caused by the bacterium *Mycobacterium tuberculosis*, is a major cause of death worldwide. The bacterium is transmitted through aerosols spread by people suffering from active clinical disease. After inhalation, *M. tuberculosis* infects alveolar macrophages in which it is able to persist for extensive periods of time. Normally, the infected macrophages are contained within so-called granulomas, however, in a substantial number of cases (≈10%), the bacterium is released from its containment and causes active disease (Russell, 2007). To date, the only licensed TB vaccine is the so-called Bacille Calmette-Guérin (BCG) vaccine. Although effective against disseminated TB in children, it does not protect well against pulmonary TB later in life. Moreover, the prevalence of multiple- and extensive drug-resistant *M. tuberculosis* strains increases each year.

The hallmark of mycobacterial disease is the bacterium's ability to persist in host tissues for many years. This feature, which is also referred to as latency, is dependent on at least two different processes. The first one is the ability of the bacterium to inhibit phago–lysosome fusion. Normally, ingested bacteria are contained within phagosomes which later on fuse with lysosomes leading to bacterial destruction and subsequent presentation on MHC complexes. However, pathogenic mycobacteria interfere with this process by actively blocking phago–lysosome fusion thereby creating a unique niche in which they are able to survive for many years (Rohde *et al.*, 2007). For the past decades, the central dogma has been that pathogenic mycobacteria exclusively reside within this phagosomal compartment until they are released from the cell and cause active disease. However, recently, this view was challenged by the observation that *M. tuberculosis*-containing phagosomes rapidly fused with lysosomes in both monocyte-derived DCs and macrophages (van der Wel *et al.*, 2007). Yet, at day 2 post infection, *M. tuberculosis* was able to escape from the phago–lysosomes into the cytosol where they were able to replicate (van der Wel *et al.*, 2007). This same phenomenon was observed for *Mycobacterium leprae* but not for the vaccine strain *Mycobacterium bovis* BCG or for heat-killed mycobacteria, suggesting that the process is specific for pathogenic mycobacteria. Although these findings contradict earlier observations and await independent confirmation, they may have important implications for the current view on host–mycobacterial interactions.

A second feature that supports the ability of pathogenic mycobacteria to cause chronic infections is their capacity to suppress host immune responses. To this end, DCs form an interesting target due to their central role in the induction of adaptive immunity. Although early data already indicated that mycobacteria influence DC functioning, the underlying mechanisms remained poorly understood. Major progress into this field was made when it was found that the interaction between *M. tuberculosis* and DCs is almost exclusively dependent on the binding to DC-SIGN (Geijtenbeek *et al.*, 2003; Tailleux *et al.*, 2003) (see also Chapter 9). Although DCs express a wide variety of C-type lectins, such as the MR and dectin-1, only antibodies directed against DC-SIGN could block the *Mycobacterium*–DC interaction (Tailleux *et al.*, 2003). Comparable results were obtained with the *M. bovis* BCG vaccine strain (Tailleux *et al.*, 2003; Geijtenbeek and van Kooyk, 2003).

In an effort to identify the ligand responsible for binding to DC-SIGN, it was found that binding to *M. tuberculosis* could almost completely be inhibited by pre-incubating DC-SIGN with Man-capped lipoarabinomannan (ManLAM), a major glycolipid of the *M. tuberculosis* cell wall, suggesting that this component forms an important ligand for DC-SIGN (Tailleux *et al.*, 2003). Moreover, it was shown that the interaction of ManLAM with DC-SIGN was dependent on the terminal Man caps as AraLAM, an uncapped form of ManLAM, did not bind to DC-SIGN (Geijtenbeek *et al.*, 2003). Later on, these findings were confirmed by the demonstration that DC-SIGN specifically interacts with neo-glycoconjugates that resemble the Man-cap of ManLAM (Koppel *et al.*, 2004). Furthermore, it was shown that a reduction of the number of mannosyl residues in the cap-derived neo-glycoconjugates leads to a decreased affinity for DC-SIGN (Koppel *et al.*, 2004). Interestingly, ManLAM is exclusively found in pathogenic, slow-growing mycobacteria, whereas AraLAM is only found in avirulent, fast-growing species, suggesting that ManLAM-DC-SIGN interactions may be important for mycobacterial pathogenesis. These data, together with the finding that also PI-LAM, a phosphoinositide-capped form of LAM, poorly inhibits DC-SIGN–*M. tuberculosis* interactions, led to the hypothesis that DC-SIGN can discriminate between mycobacterial species through the recognition of the Man caps on LAM (Maeda *et al.*, 2003). However, later on, this view was found to be too simplistic as it was demonstrated that not all mycobacterial species that contain ManLAM are bound by DC-SIGN (Pitarque *et al.*, 2005). This was not due to intrinsic differences between the different ManLAMs, as it was shown that the purified ManLAMs, also the ones from the species that did not bind DC-SIGN, could all efficiently block DC-SIGN–*M. tuberculosis* interactions (Pitarque *et al.*, 2005). Furthermore, experiments addressing the cell-surface exposure of ManLAM showed that between species only minor differences exist (Pitarque *et al.*, 2005). These data strongly suggested that ManLAM was not the only DC-SIGN ligand present on mycobacteria and further investigations led to the discovery of at least four additional ligands: lipomannan, Man-capped arabinomannan and two mannosylated glycoproteins (19 and 45 kDa antigens) (Pitarque *et al.*, 2005). In a later study, also phosphatidylinositol-mannosides were shown to bind to DC-SIGN (Torrelles *et al.*, 2006).

The strongest evidence that the presence of Man caps on LAM is not essential for the binding of mycobacteria to DC-SIGN was provided in a study using *Mycobacterium marinum* and *M. bovis* BCG strains devoid of Man caps on LAM (Appelmelk *et al.*, 2008). It was demonstrated that the mutant strains bound as efficiently to DC-SIGN as the wild-type strains. Furthermore, the interaction could still be blocked with mannan. In addition, the mutants did not show any significant differences in *in vivo* survival and induced similar cytokine profiles (Appelmelk *et al.*, 2008). The only difference observed was that the phagocytosis of the capless *M. marinum* strain by macrophages was slightly reduced and that the mutant induced somewhat more phagosome–lysosome fusion, which is consistent with the earlier finding that ManLAM inhibits this process (Fratti *et al.*, 2003; Hmama *et al.*, 2004). Overall, this study demonstrated that the Man caps of LAM do not dominate the *Mycobacterium*–host interaction. This finding was unexpected since ManLAM was not only shown to bind to DC-SIGN but was also known to modulate DC cytokine secretion by suppressing the production of IL-12 and/or upregulating the production of IL-10, thereby inhibiting Th1-type immune responses (Nigou *et al.*, 2001; Geijtenbeek *et al.*, 2003). Interestingly, Geijtenbeek *et al.* (2003) observed that ManLAM, but not AraLAM, induces IL-10 secretion in LPS-primed DCs. By using monoclonal antibodies, they demonstrated that the phenomenon was dependent on DC-SIGN, suggesting that DC-SIGN ligation can modulate TLR responses (Geijtenbeek *et al.*, 2003).

Recently, the molecular signalling pathway underlying this modulatory activity was identified (Gringhuis *et al.*, 2007). It was demonstrated that pathogens target DC-SIGN to activate the serine/threonine kinase Raf-1, which subsequently leads to acetylation of the nuclear factor-kappaB (NF-κB) subunit p65, but only after NF-κB has been activated via TLRs. Acetylation of p65 both prolonged and increased *IL-10* transcription to enhance the DCs anti-inflammatory cytokine responses (Gringhuis *et al.*, 2007). These results demonstrate that the role of DC-SIGN in host–pathogen interactions can be interpreted in two ways. On the one hand, DC-SIGN forms an important pathogen receptor that mediates the uptake and destruction of a large panel of pathogenic microorganisms. Yet, on the other hand, pathogens can also exploit DC-SIGN to modulate DC responses and thereby subvert host immunity. The potential importance of DC-SIGN ligation during host infection is illustrated by the observation that microorganisms, such as mycobacteria, may express a large number of DC-SIGN ligands. Furthermore, pathogens may actively induce the expression of DC-SIGN. Interestingly, in the case of *M. tuberculosis*, it has been shown that in patients with TB, up to 70% of alveolar macrophages express DC-SIGN. By contrast, the lectin was hardly detected in alveolar macrophages from healthy individuals or in patients with unrelated lung pathologies (Tailleux *et al.*, 2005). Moreover, promoter polymorphisms that influence DC-SIGN expression have been associated with an altered susceptibility to mycobacterial infections (Barreiro *et al.*, 2006; Vannberg *et al.*, 2008).

5.2. *H. pylori* modulates Th1/Th2 polarization through interactions with DC-SIGN

H. pylori is a common human pathogen that has persistently colonized about 50% of the total human population. Colonization typically occurs early in life and may persist during the host's entire lifetime. In most cases, the infection is asymptomatic, however, in about 10% of the cases, the infection leads to disease, which can range from relatively mild gastritis and peptic ulcers to life-threatening diseases such as gastric cancer (Ernst and Gold, 2000; Makola *et al.*, 2007). *H. pylori* expresses a wide variety of virulence factors including LPS. In general, LPS is an amphiphilic molecule that is located in the outer membrane of Gram-negative bacteria. It consists of three distinct domains: lipid A, a core oligosaccharide and an O-antigen (Raetz and Whitfield, 2002) (see Chapters 3 and 4). Although LPSs are considered to have a strong immunostimulatory activity which is primarily dependent on the interaction between lipid A and TLR-4 (Palsson-McDermott and O'Neill, 2004), *H. pylori* lipid A is unusual in that it contains a variety of modifications that abrogate the interaction with TLR-4, thereby lowering the endotoxic activity of the LPS (Moran, 2007). The *H. pylori* core oligosaccharide consists of an inner core of a 3-deoxy-D-*manno*-oct-2-ulosonic acid (Kdo) moiety substituted with a linear heptose tetrasaccharide and an outer core encompassing Gal and glucose residues (Aspinall *et al.*, 1996; Aspinall and Monteiro, 1996). The O-antigen is linked to the heptose core and is composed of a poly-(*N*-acetyl-β-lactosamine) chain decorated at some positions with L-Fuc residues to produce internal Le^x determinants with terminal Le^x or Le^y moieties (Aspinall *et al.*, 1996; Aspinall and Monteiro, 1996). Furthermore, some strains may express Le^a, Le^b, Le^c and sialyl-Le^x units, as well as H-1 and blood groups A and B antigens, giving rise to a large variation in LPS-core composition (Moran, 2008). Importantly, Le^x and Le^y expression is a common property of *H. pylori* strains and it is found in 80–90% of all cases. Yet, expression of Le epitopes can vary within a single strain as a consequence of the phase variable expression of the responsible fucosyltransferases, i.e. FutA, FutB and FutC (Wang *et al.*, 2000; Bergman *et al.*, 2006; Sanabria-Valentin *et al.*, 2007). The frequency of "on–off" switching of Le antigen expression has been reported to be between 0.2 and 0.5% (Appelmelk *et al.*, 1998). Biologically, Le epitope expression is thought to mediate the evasion of host immune responses and influence bacterial colonization and adhesion (Moran, 2008). Furthermore, Le antigen expression has been shown to mediate the binding of *H. pylori* to DC-SIGN (Bergman *et al.*, 2004). Whereas strains harbouring Le^x and Le^y epitopes strongly bound to DC-SIGN, strains devoid of these structures did not (Bergman *et al.*, 2004). Interestingly, it was demonstrated that the Le-dependent interaction with DC-SIGN inhibited Th1 responses by inducing the production of IL-10 (Bergman *et al.*, 2004). Thus, since Le expression is phase variable, a typical *H. pylori* population will contain a mixture of DC-SIGN-binding and non-binding bacteria. Bacteria unable to bind DC-SIGN will primarily induce a Th1-type response. Yet, this Th1 polarization is counterbalanced by bacteria that target DC-SIGN, thereby leading to a mixed Th1/Th2 response. It is hypothesized that the induction of a mixed response promotes the establishment of chronic infections. This is supported by the observation that people suffering from chronic gastritis display a combined secretion of both Th1 and Th2 cytokines (D'Elios *et al.*, 2005). Furthermore, it is known that Th1 polarization of *H. pylori*-specific T-cell responses is associated with more severe disease (D'Elios *et al.*, 2005).

6. CONCLUSIONS

Ever since its discovery, DC-SIGN has received major scientific attention. It is now clear that the biological role of DC-SIGN is two-fold. On the one hand, DC-SIGN fulfils important functions necessary for the induction of successful immune responses that are essential for the clearance of microbial infections. Yet, on the other hand, pathogens may also exploit DC-SIGN to modulate DC functioning thereby skewing the immune response and promoting their own survival. Currently, a lot is known about the structure and carbohydrate specificity of DC-SIGN and crystallographic studies have provided us with detailed insights into the molecular mechanisms underlying DC-SIGN–carbohydrate interactions. However, much less is known about the role of DC-SIGN during *in vivo* infection. Obviously, there are good indications that suggest it plays an important role, however, these are mostly based upon *in vitro* experiments. Hence, one important goal is to obtain insight into the role of DC-SIGN during infection and see how it functions within the complexity of the immune system.

To enlighten this issue, the use of appropriate animal models seems essential. Nevertheless, one major problem is that, although DC-SIGN homologues can be found in many organisms, their carbohydrate specificity, expression pattern and immunological function may vary significantly. One example is the murine model. Like humans, mice express a set of DC-SIGN-like molecules, i.e. mDC-SIGN and mSIGNR1-4, of which some (mSIGNR1 and mSIGNR3) share a carbohydrate specificity similar to human DC-SIGN (Galustian *et al.*, 2004). Yet, unlike human DC-SIGN, mSIGR1 is not expressed by DCs and mSIGNR3 only at low levels (Koppel *et al.*, 2005), suggesting that these molecules exhibit distinct functions and, thus, that the murine model is unsuitable for studying the role of human DC-SIGN. One possibility is to use primates as these animals possess DC-SIGN homologues that resemble both the specificity and expression pattern of human DC-SIGN. However, due to both ethical and technical difficulties their use may not be feasible. A second option is to make use of humanized model systems. By reconstituting mice with human immune cells or by heterologously expressing human DC-SIGN in murine strains, it may be possible to study DC-SIGN in *in vivo* relevant situations. Both these approaches have recently been applied to study the function of DC-SIGN (Kretz-Rommel *et al.*, 2007; Schaefer *et al.*, 2008).

Another important goal is to dissect further the molecular signalling cascades downstream of DC-SIGN. It is clear that DC-SIGN can interfere with TLR signalling. It has been shown that DC-SIGN directly influences the acetylation of NF-κB p65 via Raf-1. Nonetheless, it remains unclear whether this route is the only way by which DC-SIGN influences TLR-signalling or whether alternative regulatory mechanisms may exist. Furthermore, it is unclear whether DC-SIGN can also influence unrelated signalling pathways. Moreover, it will be interesting to see whether distinct DC-SIGN ligands all function in a similar way.

The increasing knowledge on DC-SIGN has also led to some, potentially, interesting applications. One idea is that DC-SIGN can be used to target antigens specifically to DCs, thereby generating a more efficient immune response. The feasibility of such an approach was recently demonstrated when it was shown that glycan modification of the tumour antigen gp100 targets it to DC-SIGN and enhances DC-induced antigen presentation to T-cells (Aarnoudse *et al.*, 2008). Furthermore, it has been shown that administration of anti-DC-SIGN antibodies fused with either tetanus toxoid peptides or keyhole limpet haemocyanin raised efficient T-cell responses without additional adjuvant requirements (Kretz-Rommel *et al.*, 2007). Therefore, targeting of antigens to DC-SIGN may be a promising strategy to induce enhanced immune responses against both cancer and microbial antigens. Currently, the picture is emerging that pathogens target DC-SIGN specifically to suppress Th1 immunity, as exemplified by the observations that both mycobacterial ManLAM and *H. pylori* LPS suppress IL-12 secretion and/or induce IL-10 production by immune cells. In principle, this knowledge can be used to counteract the bacteria as the removal of DC-SIGN ligands may thus lead to improved Th1 responses and, thereby, enhance the efficacy of certain vaccines. One important vaccine to which this approach may apply is the BCG vaccine which is currently used to prevent *M. tuberculosis* infection. This vaccine is effective in preventing disseminated TB in children, but is poorly able to prevent pulmonary TB later in life. The general idea is that the BCG vaccine induces a sub-optimal immune response thereby abrogating protective immunity. As DC-SIGN seems to be involved in mycobacterial immunosuppression, it would be interesting to see whether the removal of mycobacterial DC-SIGN ligands can increase the potency of BCG vaccines.

Overall, the discovery of DC-SIGN has led to important insights into the interactions between pathogens and their hosts. Yet, important questions remain to be resolved (see Research Focus Box). Nevertheless, DC-SIGN forms an interesting target that may pave the way for the design of new therapeutic approaches against both microbial infections and cancer.

RESEARCH FOCUS BOX

- Why does DC-SIGNR bind to Le^a and Le^y but not to the closely related Le^x structure?
- How important is ICAM-3 in DC-SIGN-dependent DC and T-cell interactions?
- What is the functional role of DC-SIGN splice variants?
- How important is DC-SIGN ligation during *in vivo* infection?

RESEARCH FOCUS BOX (*cont'd*)

- Do all DC-SIGN ligands behave similarly in terms of induced downstream signalling cascades?
- Does DC-SIGN ligation modulate other host cell responses than IL-10/IL-12 cytokine production?
- How important are DC-SIGN-related polymorphisms for susceptibility to and outcome of microbial infections?
- How feasible is the approach to target DC-SIGN for the development of novel therapeutics against microbial infections, cancer and, maybe, autoimmune diseases?

References

Aarnoudse, C.A., Bax, M., Sanchez-Hernandez, M., Garcia-Vallejo, J.J., van Kooyk, Y., 2008. Glycan modification of the tumor antigen gp100 targets DC-SIGN to enhance dendritic cell induced antigen presentation to T cells. Int. J. Cancer 122, 839–846.

Akira, S., Takeda, K., 2004. Toll-like receptor signalling. Nat. Rev. Immunol. 4, 499–511.

Apostolopoulos, V., McKenzie, I.F., 2001. Role of the mannose receptor in the immune response. Curr. Mol. Med. 1, 469–474.

Appelmelk, B.J., Shiberu, B., Trinks, C., et al., 1998. Phase variation in *Helicobacter pylori* lipopolysaccharide. Infect. Immun. 66, 70–76.

Appelmelk, B.J., van Die, I., van Vliet, S.J., Vandenbroucke-Grauls, C.M., Geijtenbeek, T.B., van Kooyk, Y., 2003. Cutting edge: carbohydrate profiling identifies new pathogens that interact with dendritic cell-specific ICAM-3-grabbing nonintegrin on dendritic cells. J. Immunol. 170, 1635–1639.

Appelmelk, B.J., den Dunnen, J., Driessen, N.N., et al., 2008. The mannose cap of mycobacterial lipoarabinomannan does not dominate the *Mycobacterium*–host interaction. Cell Microbiol. 10, 930–944.

Aspinall, G.O., Monteiro, M.A., 1996. Lipopolysaccharides of *Helicobacter pylori* strains P466 and MO19: structures of the O antigen and core oligosaccharide regions. Biochemistry 35, 2498–2504.

Aspinall, G.O., Monteiro, M.A., Pang, H., Walsh, E.J., Moran, A.P., 1996. Lipopolysaccharide of the *Helicobacter pylori* type strain NCTC 11637 (ATCC 43504): structure of the O antigen chain and core oligosaccharide regions. Biochemistry 35, 2489–2497.

Baribaud, F., Doms, R.W., Pohlmann, S., 2002. The role of DC-SIGN and DC-SIGNR in HIV and Ebola virus infection: can potential therapeutics block virus transmission and dissemination? Expert Opin. Ther. Targets 6, 423–431.

Barreiro, L.B., Neyrolles, O., Babb, C.L., et al., 2006. Promoter variation in the DC-SIGN-encoding gene CD209 is associated with tuberculosis. PLoS. Med. 3, e20.

Barreiro, L.B., Neyrolles, O., Babb, C.L., et al., 2007. Length variation of DC-SIGN and L-SIGN neck-region has no impact on tuberculosis susceptibility. Hum. Immunol. 68, 106–112.

Bashirova, A.A., Geijtenbeek, T.B., van Duijnhoven, G.C., et al., 2001. A dendritic cell-specific intercellular adhesion molecule 3-grabbing nonintegrin (DC-SIGN)-related protein is highly expressed on human liver sinusoidal endothelial cells and promotes HIV-1 infection. J. Exp. Med. 193, 671–678.

Ben-Ali, M., Barreiro, L.B., Chabbou, A., et al., 2007. Promoter and neck region length variation of DC-SIGN is not associated with susceptibility to tuberculosis in Tunisian patients. Hum. Immunol. 68, 908–912.

Bergman, M.P., Engering, A., Smits, H.H., et al., 2004. *Helicobacter pylori* modulates the T helper cell 1/T helper cell 2 balance through phase-variable interaction between lipopolysaccharide and DC-SIGN. J. Exp. Med. 200, 979–990.

Bergman, M., Del, P.G., van Kooyk, Y., Appelmelk, B.J., 2006. *Helicobacter pylori* phase variation, immune modulation and gastric autoimmunity. Nat. Rev. Microbiol. 4, 151–159.

Bleijs, D.A., Geijtenbeek, T.B., Figdor, C.G., van Kooyk, Y., 2001. DC-SIGN and LFA-1: a battle for ligand. Trends Immunol. 22, 457–463.

Bogoevska, V., Horst, A., Klampe, B., Lucka, L., Wagener, C., Nollau, P., 2006. CEACAM1, an adhesion molecule of human granulocytes, is fucosylated by fucosyltransferase IX and interacts with DC-SIGN of dendritic cells via Lewis x residues. Glycobiology 16, 197–209.

Bogoevska, V., Nollau, P., Lucka, L., et al., 2007. DC-SIGN binds ICAM-3 isolated from peripheral human leukocytes through Lewis x residues. Glycobiology 17, 324–333.

Brooks, S.A., 2006. Protein glycosylation in diverse cell systems: implications for modification and analysis of recombinant proteins. Expert. R. Proteomics 3, 345–359.

Cambi, A., Figdor, C.G., 2003. Dual function of C-type lectin-like receptors in the immune system. Curr. Opin. Cell Biol. 15, 539–546.

Cambi, A., Koopman, M., Figdor, C.G., 2005. How C-type lectins detect pathogens. Cell Microbiol. 7, 481–488.

Chen, J., Subbarao, K., 2007. The immunobiology of SARS*. Annu. Rev. Immunol. 25, 443–472.

Cocquerel, L., Voisset, C., Dubuisson, J., 2006. Hepatitis C virus entry: potential receptors and their biological functions. J. Gen. Virol. 87, 1075–1084.

Colmenares, M., Puig-Kroger, A., Pello, O.M., Corbi, A.L., Rivas, L., 2002. Dendritic cell (DC)-specific intercellular adhesion molecule 3 (ICAM-3)-grabbing nonintegrin (DC-SIGN, CD209), a C-type surface lectin in human DCs, is a receptor for *Leishmania* amastigotes. J. Biol. Chem. 277, 36766–36769.

D'Elios, M.M., Amedei, A., Benagiano, M., Azzurri, A., Del, P.G., 2005. *Helicobacter pylori*, T cells and cytokines: the "dangerous liaisons". FEMS Immunol. Med. Microbiol. 44, 113–119.

Devasahayam, M., 2007. Factors affecting the expression of recombinant glycoproteins. Indian J. Med. Res. 126, 22–27.

Dransfield, I., Rossi, A.G., 2004. Granulocyte apoptosis: who would work with a "real" inflammatory cell? Biochem. Soc. Trans. 32, 447–451.

Drickamer, K., 1992. Engineering galactose-binding activity into a C-type mannose-binding protein. Nature 360, 183–186.

Drickamer, K., 1999. C-type lectin-like domains. Curr. Opin. Struct. Biol. 9, 585–590.

Dzionek, A., Sohma, Y., Nagafune, J., et al., 2001. BDCA-2, a novel plasmacytoid dendritic cell-specific type II C-type lectin, mediates antigen capture and is a potent inhibitor of interferon α/β induction. J. Exp. Med. 194, 1823–1834.

East, L., Isacke, C.M., 2002. The mannose receptor family. Biochim. Biophys. Acta 1572, 364–386.

Engering, A., Geijtenbeek, T.B., van Vliet, S.J., et al., 2002a. The dendritic cell-specific adhesion receptor DC-SIGN internalizes antigen for presentation to T cells. J. Immunol. 168, 2118–2126.

Engering, A., Van Vliet, S.J., Geijtenbeek, T.B., van Kooyk, Y., 2002b. Subset of DC-SIGN$^+$ dendritic cells in human blood transmits HIV-1 to T lymphocytes. Blood 100, 1780–1786.

Ernst, P.B., Gold, B.D., 2000. The disease spectrum of *Helicobacter pylori*: the immunopathogenesis of gastroduodenal ulcer and gastric cancer. Annu. Rev. Microbiol. 54, 615–640.

Feinberg, H., Mitchell, D.A., Drickamer, K., Weis, W.I., 2001. Structural basis for selective recognition of oligosaccharides by DC-SIGN and DC-SIGNR. Science 294, 2163–2166.

Feinberg, H., Guo, Y., Mitchell, D.A., Drickamer, K., Weis, W.I., 2005. Extended neck regions stabilize tetramers of the receptors DC-SIGN and DC-SIGNR. J. Biol. Chem. 280, 1327–1335.

Feinberg, H., Castelli, R., Drickamer, K., Seeberger, P.H., Weis, W.I., 2007. Multiple modes of binding enhance the affinity of DC-SIGN for high mannose *N*-linked glycans found on viral glycoproteins. J. Biol. Chem. 282, 4202–4209.

Fiete, D.J., Beranek, M.C., Baenziger, J.U., 1998. A cysteine-rich domain of the "mannose" receptor mediates GalNAc-4-SO_4 binding. Proc. Natl. Acad. Sci. USA 95, 2089–2093.

Fratti, R.A., Chua, J., Vergne, I., Deretic, V., 2003. *Mycobacterium tuberculosis* glycosylated phosphatidylinositol causes phagosome maturation arrest. Proc. Natl. Acad. Sci. USA, 100, 5437–5442.

Galustian, C., Park, C.G., Chai, W., et al., 2004. High and low affinity carbohydrate ligands revealed for murine SIGN-R1 by carbohydrate array and cell binding approaches, and differing specificities for SIGN-R3 and langerin. Int. Immunol. 16, 853–866.

Garcia-Vallejo, J.J., van Liempt, E., da Costa, M.P., et al., 2008. DC-SIGN mediates adhesion and rolling of dendritic cells on primary human umbilical vein endothelial cells through Lewis^Y antigen expressed on ICAM-2. Mol. Immunol. 45, 2359–2369.

Geijtenbeek, T.B., van Kooyk, Y., 2003. Pathogens target DC-SIGN to influence their fate DC-SIGN functions as a pathogen receptor with broad specificity. APMIS 111, 698–714.

Geijtenbeek, T.B., Krooshoop, D.J., Bleijs, D.A., et al., 2000a. DC-SIGN-ICAM-2 interaction mediates dendritic cell trafficking. Nat. Immunol. 1, 353–357.

Geijtenbeek, T.B., Torensma, R., van Vliet, S.J., et al., 2000b. Identification of DC-SIGN, a novel dendritic cell-specific ICAM-3 receptor that supports primary immune responses. Cell 100, 575–585.

Geijtenbeek, T.B., Kwon, D.S., Torensma, R., et al., 2000c. DC-SIGN, a dendritic cell-specific HIV-1-binding protein that enhances trans-infection of T cells. Cell 100, 587–597.

Geijtenbeek, T.B., van Vliet, S.J., van Duijnhoven, G.C., Figdor, C.G., van Kooyk, Y., 2001. DC-SIGN, a dentritic cell-specific HIV-1 receptor present in placenta that infects T cells in trans – a review. Placenta 22 (Suppl. A), S19–S23.

Geijtenbeek, T.B., Engering, A., van Kooyk, Y., 2002a. DC-SIGN, a C-type lectin on dendritic cells that unveils many aspects of dendritic cell biology. J. Leukoc. Biol. 71, 921–931.

Geijtenbeek, T.B., van Duijnhoven, G.C., van Vliet, S.J., et al., 2002b. Identification of different binding sites in the dendritic cell-specific receptor DC-SIGN for intercellular adhesion molecule 3 and HIV-1. J. Biol. Chem. 277, 11314–11320.

Geijtenbeek, T.B., van Vliet, S.J., Koppel, E.A., et al., 2003. Mycobacteria target DC-SIGN to suppress dendritic cell function. J. Exp. Med. 197, 7–17.

Geijtenbeek, T.B., van Vliet, S.J., Engering, A., 't Hart, B.A., van Kooyk, Y., 2004. Self- and nonself-recognition by C-type lectins on dendritic cells. Annu. Rev. Immunol. 22, 33–54.

Gringhuis, S.I., den Dunnen, J., Litjens, M., van het Hof, B., van Kooyk, Y., Geijtenbeek, T.B., 2007. C-type lectin DC-SIGN modulates Toll-like receptor signaling via Raf-1 kinase-dependent acetylation of transcription factor NF-κB. Immunity 26, 605–616.

Guo, Y., Feinberg, H., Conroy, E., et al., 2004. Structural basis for distinct ligand-binding and targeting properties of the receptors DC-SIGN and DC-SIGNR. Nat. Struct. Mol. Biol. 11, 591–598.

Hauss, P., Selz, F., Cavazzana-Calvo, M., Fischer, A., 1995. Characteristics of antigen-independent and antigen-dependent interaction of dendritic cells with $CD4^+$ T cells. Eur. J. Immunol. 25, 2285–2294.

Hershkovitz, O., Jarahian, M., Zilka, A., et al., 2008. Altered glycosylation of recombinant NKp30 hampers binding to heparan sulfate: a lesson for the use of recombinant immunoreceptors as an immunological tool. Glycobiology 18, 28–41.

Hitchen, P.G., Mullin, N.P., Taylor, M.E., 1998. Orientation of sugars bound to the principal C-type carbohydrate-recognition domain of the macrophage mannose receptor. Biochem. J. 333, 601–608.

Hmama, Z., Sendide, K., Talal, A., Garcia, R., Dobos, K., Reiner, N.E., 2004. Quantitative analysis of phagolysosome fusion in intact cells: inhibition by mycobacterial lipoarabinomannan and rescue by a 1α,25-dihydroxyvitamin D3-phosphoinositide 3-kinase pathway. J. Cell Sci. 117, 2131–2140.

Irjala, H., Johansson, E.L., Grenman, R., Alanen, K., Salmi, M., Jalkanen, S., 2001. Mannose receptor is a novel ligand for L-selectin and mediates lymphocyte binding to lymphatic endothelium. J. Exp. Med. 194, 1033–1042.

Jimenez, D., Roda-Navarro, P., Springer, T.A., Casasnovas, J.M., 2005. Contribution of *N*-linked glycans to the conformation and function of intercellular adhesion molecules (ICAMs). J. Biol. Chem. 280, 5854–5861.

Kessel, J.M., Sedgwick, J.B., Busse, W.W., 2006. Ligation of intercellular adhesion molecule 3 induces apoptosis of human blood eosinophils and neutrophils. J. Allergy Clin. Immunol. 118, 831–836.

Koppel, E.A., Ludwig, I.S., Hernandez, M.S., et al., 2004. Identification of the mycobacterial carbohydrate structure that binds the C-type lectins DC-SIGN, L-SIGN and SIGNR1. Immunobiology 209, 117–127.

Koppel, E.A., van Gisbergen, K.P., Geijtenbeek, T.B., van Kooyk, Y., 2005. Distinct functions of DC-SIGN and its homologues L-SIGN (DC-SIGNR) and mSIGNR1 in pathogen recognition and immune regulation. Cell Microbiol. 7, 157–165.

Kretz-Rommel, A., Qin, F., Dakappagari, N., et al., 2007. In vivo targeting of antigens to human dendritic cells through DC-SIGN elicits stimulatory immune responses and inhibits tumor growth in grafted mouse models. J. Immunother. 30, 715–726.

Kwon, D.S., Gregorio, G., Bitton, N., Hendrickson, W.A., Littman, D.R., 2002. DC-SIGN-mediated internalization of HIV is required for trans-enhancement of T cell infection. Immunity 16, 135–144.

Lee, B., Leslie, G., Soilleux, E., et al., 2001. *cis* Expression of DC-SIGN allows for more efficient entry of human and simian immunodeficiency viruses via CD4 and a coreceptor. J. Virol. 75, 12028–12038.

Lekkerkerker, A.N., van Kooyk, Y., Geijtenbeek, T.B., 2006. Viral piracy: HIV-1 targets dendritic cells for transmission. Curr. HIV Res. 4, 169–176.

Liu, H., Hwangbo, Y., Holte, S., et al., 2004a. Analysis of genetic polymorphisms in CCR5, CCR2, stromal cell-derived factor-1, RANTES, and dendritic cell-specific intercellular adhesion molecule-3-grabbing nonintegrin in seronegative individuals repeatedly exposed to HIV-1. J. Infect. Dis. 190, 1055–1058.

Liu, W., Tang, L., Zhang, G., et al., 2004b. Characterization of a novel C-type lectin-like gene, LSECtin: demonstration of carbohydrate binding and expression in sinusoidal endothelial cells of liver and lymph node. J. Biol. Chem. 279, 18748–18758.

Ludwig, I.S., Geijtenbeek, T.B., van Kooyk, Y., 2006. Two way communication between neutrophils and dendritic cells. Curr. Opin. Pharmacol. 6, 408–413.

Maeda, N., Nigou, J., Herrmann, J.L., et al., 2003. The cell surface receptor DC-SIGN discriminates between *Mycobacterium* species through selective recognition of the mannose caps on lipoarabinomannan. J. Biol. Chem. 278, 5513–5516.

Mahnke, K., Guo, M., Lee, S., et al., 2000. The dendritic cell receptor for endocytosis, DEC-205, can recycle and enhance antigen presentation via major histocompatibility complex class II-positive lysosomal compartments. J. Cell Biol. 151, 673–684.

Makola, D., Peura, D.A., Crowe, S.E., 2007. *Helicobacter pylori* infection and related gastrointestinal diseases. J. Clin. Gastroenterol. 41, 548–558.

McGreal, E.P., Miller, J.L., Gordon, S., 2005. Ligand recognition by antigen-presenting cell C-type lectin receptors. Curr. Opin. Immunol. 17, 18–24.

Megiovanni, A.M., Sanchez, F., Robledo-Sarmiento, M., Morel, C., Gluckman, J.C., Boudaly, S., 2006. Polymorphonuclear neutrophils deliver activation signals and antigenic molecules to dendritic cells: a new link between leukocytes upstream of T lymphocytes. J. Leukoc. Biol. 79, 977–988.

Mitchell, D.A., Fadden, A.J., Drickamer, K., 2001. A novel mechanism of carbohydrate recognition by the C-type lectins DC-SIGN and DC-SIGNR. Subunit organization and binding to multivalent ligands. J. Biol. Chem. 276, 28939–28945.

Moraes, T.J., Zurawska, J.H., Downey, G.P., 2006. Neutrophil granule contents in the pathogenesis of lung injury. Curr. Opin. Hematol. 13, 21–27.

Moran, A.P., 2007. Lipopolysaccharide in bacterial chronic infection: insights from *Helicobacter pylori* lipopolysaccharide and lipid A. Int. J. Med. Microbiol. 297, 307–319.

Moran, A.P., 2008. Relevance of fucosylation and Lewis antigen expression in the bacterial gastroduodenal pathogen *Helicobacter pylori*. Carbohydr. Res. 343, 1952–1965.

Mukhopadhyay, S., Herre, J., Brown, G.D., Gordon, S., 2004. The potential for Toll-like receptors to collaborate with other innate immune receptors. Immunology 112, 521–530.

Murphy, J.E., Tedbury, P.R., Homer-Vanniasinkam, S., Walker, J.H., Ponnambalam, S., 2005. Biochemistry and cell biology of mammalian scavenger receptors. Atherosclerosis 182, 1–15.

Nigou, J., Zelle-Rieser, C., Gilleron, M., Thurnher, M., Puzo, G., 2001. Mannosylated lipoarabinomannans inhibit IL-12 production by human dendritic cells: evidence for a negative signal delivered through the mannose receptor. J. Immunol. 166, 7477–7485.

Nortamo, P., Li, R., Renkonen, R., et al., 1991. The expression of human intercellular adhesion molecule-2 is refractory to inflammatory cytokines. Eur. J. Immunol. 21, 2629–2632.

Palsson-McDermott, E.M., O'Neill, L.A., 2004. Signal transduction by the lipopolysaccharide receptor, Toll-like receptor-4. Immunology 113, 153–162.

Park, C.G., Takahara, K., Umemoto, E., et al., 2001. Five mouse homologues of the human dendritic cell C-type lectin, DC-SIGN. Int. Immunol. 13, 1283–1290.

Pastva, A.M., Wright, J.R., Williams, K.L., 2007. Immunomodulatory roles of surfactant proteins A and D: implications in lung disease. Proc. Am. Thorac. Soc. 4, 252–257.

Pitarque, S., Herrmann, J.L., Duteyrat, J.L., et al., 2005. Deciphering the molecular bases of *Mycobacterium tuberculosis* binding to the lectin DC-SIGN reveals an underestimated complexity. Biochem. J. 392, 615–624.

Raetz, C.R., Whitfield, C., 2002. Lipopolysaccharide endotoxins. Annu. Rev. Biochem. 71, 635–700.

Rathore, A., Chatterjee, A., Sood, V., et al., 2008. Risk for HIV-1 infection is not associated with repeat-region polymorphism in the DC-SIGN neck domain and novel genetic DC-SIGN variants among North Indians. Clin. Chim. Acta. 391, 1–5.

Relloso, M., Puig-Kroger, A., Pello, O.M., et al., 2002. DC-SIGN (CD209) expression is IL-4 dependent and is negatively regulated by IFN, TGF-β, and anti-inflammatory agents. J. Immunol. 168, 2634–2643.

Robinson, M.J., Sancho, D., Slack, E.C., LeibundGut-Landmann, S., Reis e Sousa, E., 2006. Myeloid C-type lectins in innate immunity. Nat. Immunol. 7, 1258–1265.

Rohde, K., Yates, R.M., Purdy, G.E., Russell, D.G., 2007. *Mycobacterium tuberculosis* and the environment within the phagosome. Immunol. Rev. 219, 37–54.

Russell, D.G., 2007. Who puts the tubercle in tuberculosis? Nat. Rev. Microbiol. 5, 39–47.

Sanabria-Valentin, E., Colbert, M.T., Blaser, M.J., 2007. Role of *futC* slipped strand mispairing in *Helicobacter pylori* Lewis Y phase variation. Microbes Infect. 9, 1553–1560.

Schaefer, M., Reiling, N., Fessler, C., et al., 2008. Decreased pathology and prolonged survival of human DC-SIGN transgenic mice during mycobacterial infection. J. Immunol. 180, 6836–6845.

Schjetne, K.W., Thompson, K.M., Aarvak, T., Fleckenstein, B., Sollid, L.M., Bogen, B., 2002. A mouse C kappa-specific T cell clone indicates that DC-SIGN is an efficient target for antibody-mediated delivery of T cell epitopes for MHC class II presentation. Int. Immunol. 14, 1423–1430.

Serrano-Gomez, D., Sierra-Filardi, E., Martinez-Nunez, R.T., et al., 2008. Structural requirements for multimerization of the pathogen receptor DC-SIGN (CD209) on the cell surface. J. Biol. Chem. 283, 3889–3903.

Singer, B.B., Klaile, E., Scheffrahn, I., et al., 2005. CEACAM1 (CD66a) mediates delay of spontaneous and Fas ligand-induced apoptosis in granulocytes. Eur. J. Immunol. 35, 1949–1959.

Soilleux, E.J., 2003. DC-SIGN (dendritic cell-specific ICAM-grabbing non-integrin) and DC-SIGN-related (DC-SIGNR): friend or foe? Clin. Sci. (London), 104, 437–446.

Soilleux, E.J., Coleman, N., 2001. Langerhans cells and the cells of Langerhans cell histiocytosis do not express DC-SIGN. Blood 98, 1987–1988.

Soilleux, E.J., Barten, R., Trowsdale, J., 2000. DC-SIGN; a related gene, DC-SIGNR; and CD23 form a cluster on 19p13. J. Immunol. 165, 2937–2942.

Soilleux, E.J., Morris, L.S., Lee, B., et al., 2001. Placental expression of DC-SIGN may mediate intrauterine vertical transmission of HIV. J. Pathol. 195, 586–592.

Soilleux, E.J., Morris, L.S., Leslie, G., et al., 2002. Constitutive and induced expression of DC-SIGN on dendritic cell and macrophage subpopulations in situ and in vitro. J. Leukoc. Biol. 71, 445–457.

Starling, G.C., McLellan, A.D., Egner, W., et al., 1995. Intercellular adhesion molecule-3 is the predominant co-stimulatory ligand for leukocyte function antigen-1 on human blood dendritic cells. Eur. J. Immunol. 25, 2528–2532.

Su, S.V., Hong, P., Baik, S., Negrete, O.A., Gurney, K.B., Lee, B., 2004. DC-SIGN binds to HIV-1 glycoprotein 120 in a distinct but overlapping fashion compared with ICAM-2 and ICAM-3. J. Biol. Chem. 279, 19122–19132.

Tailleux, L., Schwartz, O., Herrmann, J.L., et al., 2003. DC-SIGN is the major *Mycobacterium tuberculosis* receptor on human dendritic cells. J. Exp. Med. 197, 121–127.

Tailleux, L., Pham-Thi, N., Bergeron-Lafaurie, A., et al., 2005. DC-SIGN induction in alveolar macrophages defines privileged target host cells for mycobacteria in patients with tuberculosis. PLoS Med. 2, e381.

Takahashi, K., Ip, W.E., Michelow, I.C., Ezekowitz, R.A., 2006. The mannose-binding lectin: a prototypic pattern recognition molecule. Curr. Opin. Immunol. 18, 16–23.

Torrelles, J.B., Azad, A.K., Schlesinger, L.S., 2006. Fine discrimination in the recognition of individual species of phosphatidyl-*myo*-inositol mannosides from *Mycobacterium tuberculosis* by C-type lectin pattern recognition receptors. J. Immunol. 177, 1805–1816.

van der Wel, N., Hava, D., Houben, D., et al., 2007. *M. tuberculosis* and *M. leprae* translocate from the phagolysosome to the cytosol in myeloid cells. Cell 129, 1287–1298.

van Die, I., van Vliet, S.J., Nyame, A.K., et al., 2003. The dendritic cell-specific C-type lectin DC-SIGN is a receptor for *Schistosoma mansoni* egg antigens and recognizes the glycan antigen Lewis x. Glycobiology 13, 471–478.

van Gisbergen, K.P., Sanchez-Hernandez, M., Geijtenbeek, T.B., van Kooyk, Y., 2005a. Neutrophils mediate immune modulation of dendritic cells through glycosylation-dependent interactions between Mac-1 and DC-SIGN. J. Exp. Med. 201, 1281–1292.

van Gisbergen, K.P., Ludwig, I.S., Geijtenbeek, T.B., van Kooyk, Y., 2005b. Interactions of DC-SIGN with Mac-1 and CEACAM1 regulate contact between dendritic cells and neutrophils. FEBS Lett. 579, 6159–6168.

van Kooyk, Y., Geijtenbeek, T.B., 2003. DC-SIGN: escape mechanism for pathogens. Nat. Rev. Immunol. 3, 697–709.

van Kooyk, Y., Appelmelk, B., Geijtenbeek, T.B., 2003. A fatal attraction: *Mycobacterium tuberculosis* and HIV-1 target DC-SIGN to escape immune surveillance. Trends Mol. Med. 9, 153–159.

van Liempt, E., Imberty, A., Bank, C.M., et al., 2004. Molecular basis of the differences in binding properties of the highly related C-type lectins DC-SIGN and L-SIGN to Lewis X trisaccharide and *Schistosoma mansoni* egg antigens. J. Biol. Chem. 279, 33161–33167.

van Liempt, E., Bank, C.M., Mehta, P., et al., 2006. Specificity of DC-SIGN for mannose- and fucose-containing glycans. FEBS Lett. 580, 6123–6131.

Vannberg, F.O., Chapman, S.J., Khor, C.C., et al., 2008. CD209 genetic polymorphism and tuberculosis disease. PLoS. ONE. 3, e1388.

Walker, A., Ward, C., Taylor, E.L., et al., 2005. Regulation of neutrophil apoptosis and removal of apoptotic cells. Curr. Drug Targets. Inflamm. Allergy 4, 447–454.

Wang, G., Ge, Z., Rasko, D.A., Taylor, D.E., 2000. Lewis antigens in *Helicobacter pylori*: biosynthesis and phase variation. Mol. Microbiol. 36, 1187–1196.

Weis, W.I., Drickamer, K., Hendrickson, W.A., 1992. Structure of a C-type mannose-binding protein complexed with an oligosaccharide. Nature 360, 127–134.

Weis, W.I., Taylor, M.E., Drickamer, K., 1998. The C-type lectin superfamily in the immune system. Immunol. Rev. 163, 19–34.

Wu, L., KewalRamani, V.N., 2006. Dendritic-cell interactions with HIV: infection and viral dissemination. Nat. Rev. Immunol. 6, 859–868.

Wu, L., Liu, Y.J., 2007. Development of dendritic-cell lineages. Immunity 26, 741–750.

Yan, S.R., Sapru, K., Issekutz, A.C., 2004. The CD11/CD18 (β_2) integrins modulate neutrophil caspase activation and survival following TNF-α or endotoxin induced transendothelial migration. Immunol. Cell Biol. 82, 435–446.

CHAPTER

35

Host surfactant proteins in microbial recognition

Mark R. Thursz and Wafa Khamri

SUMMARY

Host defence against microbial invasion involves both innate and adaptive immune responses. The cellular responses require pathogen-associated molecular pattern (PAMPs) recognition molecules. Surfactant proteins (SP) SP-A and SP-D, members of the collectin family, recognize PAMPs and act as important mediators of the immune system. Both SP-A and SP-D interact with various microorganism- and pathogen-derived components by binding to complex carbohydrate structures of bacterial, viral and fungal cells. The binding occurs via the surfactant carbohydrate recognition domain and leads to the agglutination and enhancement of pathogen clearance by specialized immune cells. In addition to their interaction with bacterial cells, SP-A and SP-D have a direct effect on immune cell function. This chapter reviews the structure, biosynthesis, ligands and functions of SP-A and SP-D. In addition, the possible clinical applications for both SP-A and SP-D are highlighted.

Keywords: Surfactant protein A (SP-A); Surfactant protein D (SP-D); Collectins; Immunity; Pathogens; Immune cells; Pathogenesis; Virulence

1. INTRODUCTION – GENERAL OVERVIEW ON SURFACTANT PROTEINS

Host defence against invading microbes is initiated through the recognition of conserved structures, i.e. lipids, proteins, nucleic acid or carbohydrates, expressed selectively by pathogens and known as pathogen-associated molecular pattern (PAMPs). Surfactant proteins belong to the family of innate immunity proteins that play a major role as PAMP recognition receptors (PRRs) capable of recognizing, binding and subsequently clearing infections agents. The surfactant proteins were originally described in the lungs as part of the pulmonary surfactant, which is a complex mixture of lipids (90%) and proteins (5–10%). The pulmonary surfactant plays two major roles. First, it reduces surface tension at the air–liquid interface within the lung in order to avoid lung collapse during respiration. Second, it plays a key role in regulating immune cell function. Four

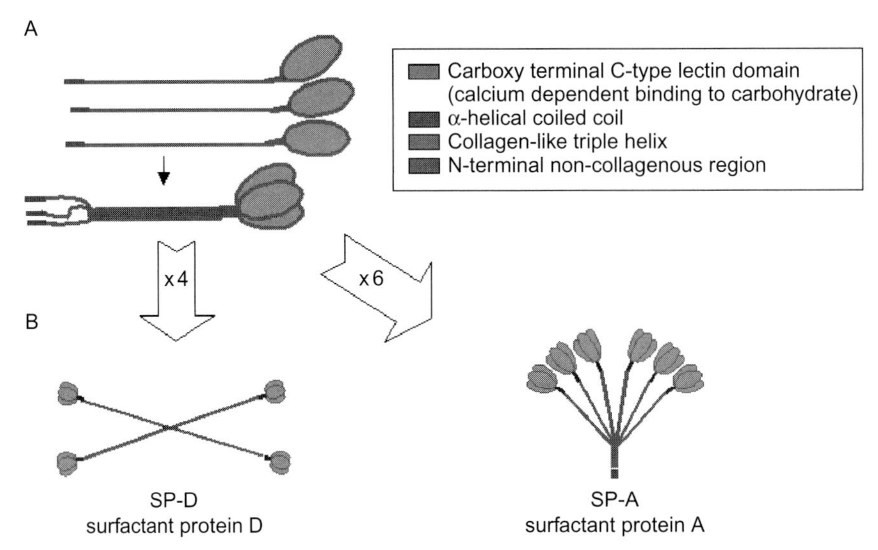

FIGURE 35.1 Domain organization and assembly of surfactant proteins molecules. (A) Schematic representation illustrating the trimeric subunit structure of the surfactant proteins which consist of a short amino terminal segment, a collagen-like region, a short α-helical neck region and the C-terminal C-type (calcium-dependent), carbohydrate recognition domain (CRD). (B) The SP-D molecule appears predominantly as a cruciform structure consisting of four trimers, whereas SP-A appears as a bouquet-like structure of six trimers.

surfactant proteins (SPs) have been described: SP-A, SP-B, SP-C and SP-D. Specifically, SP-B and SP-C are small hydrophobic proteins which, together with the lipid component, ensure the main function of stabilizing the surface tension at the air–liquid interface. In contrast, the surfactant immune function is mainly attributed to the large hydrophilic proteins, SP-A and SP-D. Surfactant proteins A and D are members of the collectins (*collagenous lectins*), a family of soluble oligomeric mammalian proteins. These are C-type lectins which bind to carbohydrate structures in a calcium-dependent manner. Collectins are defined by the presence of two discrete structural regions: a fibrillar collagen-like region and a globular carbohydrate-recognition domain. Structurally, these proteins assemble from multiple copies of a single polypeptide chain organized in four major domains (Figure 35.1). A short cystein-rich amino terminal region attached to a long collagen-like domain formed of Gly-Xaa-Yaa triplet repeats (where Gly, glycine; Xaa and Yaa, any amino acid). This domain is bound via an α-helical coiled neck to the calcium-dependent carbohydrate recognition domain (CRD) at the C-terminal end of each polypeptide chain.

Both SP-A and SP-D have been extensively studied and there is accumulating evidence showing that they are involved in antibody-independent pathogen recognition and clearance (Holmskov *et al.*, 2003). Their direct participation in innate immunity functions, such as lipopolysaccharide- (LPS-) binding molecules and opsonins, is now well recognized (Holmskov, 2000). Surfactants proteins interact with exogenous as well as endogenous ligands via the CRD and protein–protein interaction, respectively. Therefore, they have a dual function; one is to bind to the carbohydrate structures on the surface of a pathogen and the other is subsequently to recruit other cells and molecules to destroy the pathogen.

2. GENOMIC ORGANIZATION

The genomic organization of the known collectins closely reflects the domain structure of the encoded proteins. The signal peptide, the N-terminal region and the first part of the collagenous domain are encoded by the first translated exon. The remainder of the collagen region in SP-A is encoded by the second translated exon (Sastry *et al.*, 1989; Taylor *et al.*, 1989; Katyal *et al.*, 1992). The neck and the CRD are each encoded by a single exon in all the genes. The genes for SP-A and SP-D have been mapped to a cluster on the long arm of the human chromosome 10, spanning a region from 10q21 to 10q24 (Sastry *et al.*, 1989; Kolble and Reid, 1993) and are described by a similar intron–exon organization in all known collectins genes. In mouse, SP-D and SP-A are closely linked on chromosome 14 within a 55 kilobase (Kb) region (White *et al.*, 1994).

Two highly homologous human SP-A genes (*SP-A1* and *SP-A2* at 4.5 kb each), corresponding to two different copy DNA (cDNA) sequences, have been identified in humans (McCormick *et al.*, 1994). Moreover, an additional SP-A pseudogene (*SP-A*) has been characterized, which shows a high degree of homology to exons encoding the α-helical neck region and the CRD of the SP-A (Korfhagen *et al.*, 1991). The *SP-A1* and *SP-A2* gene products differ from each other at only 6–11 residues, mainly located in the collagenous region encoded by the first and second translated exons (Rishi *et al.*, 1992). It has been shown that both gene products are necessary for the formation of a correctly assembled SP-A *in vitro*. Macromolecular structural analyses of both gene products suggest that the SP-A subunit is a heterodimer of two *SP-A1* gene products and a single *SP-A2* gene product (Voss *et al.*, 1991). The SP-A gene is characterized by a high degree of polymorphism (Floros *et al.*, 1996). This makes SP-A a good marker for genetic studies, particularly in respiratory diseases.

Notably, SP-D is encoded by a single 11 kb gene containing eight exons and the gene has been localized to 10q22.2-q23.1. The signal peptide, amino-terminal domain, the CRD and the linking sequence between the collagenous region and the CRD are each encoded by a single exon, as for SP-A. However, a unique intron–exon structure for the collagen domain is encoded on five exons, including four tandem repeats of 117 base pair exons (Crouch *et al.*, 1993).

3. MOLECULAR STRUCTURE

Host SP-A and SP-D are characterized by their unusual primary structures and also by their unique three-dimensional structures. Each polypeptide (30 kDa–45 kDa, 222–355 amino acid residues) contains a short N-terminal cysteine-(Cys-) rich region of seven amino acids (in SP-A) and 25 amino acids (in SP-D), followed by a collagen-like sequence, the neck region and then by a globular CRD. The structural subunit consists of three identical or very similar polypeptides which bind together covalently by disulfide bonds in the N-terminal segments and non-covalently by interactions between the collagen and the neck regions, forming a triple helix and a triple coiled-coil of α-helices in the respective regions. The collectins consist of 3–6 homotrimeric structural units, except human SP-A which is heterotrimeric, these units are held together by further disulfide linkages at the N-terminus to organize "flower bouquet" structures (in SP-A), or form tetramers which have "X-like" structures (in SP-D) (see Figure 35.1).

3.1. C-Terminal lectin domain (CRD)

Crystallographic studies have determined the three-dimensional structure of human SP-D showing a highly conserved overall folding (Sheriff *et al.*, 1994; Hoppe and Reid, 1994;

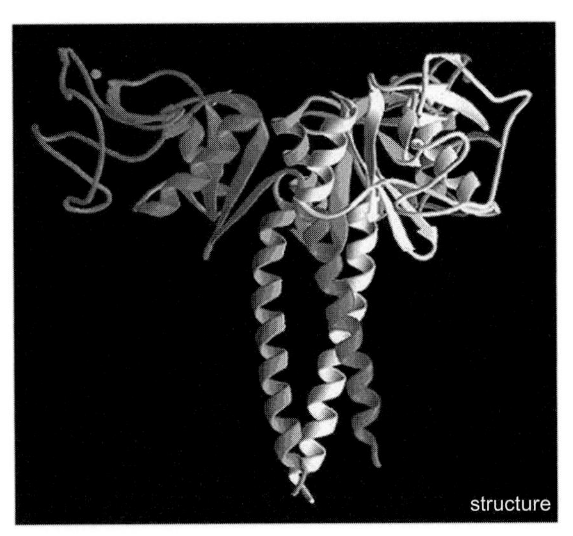

FIGURE 35.2 Three-dimensional structures of trimeric recombinant α-helical neck-CRD fragments of human SP-D. Each monomer is shown in a different colour and calcium ions are represented as green spheres. Reproduced, with permission, from Hakansson *et al.* (1999) Crystal structure of the trimeric α-helical coiled-coil and the three lectin domains of human lung surfactant protein D. Structure 7, 255–264. Elsevier © 1999. (See colour plates section.)

Hakansson *et al.*, 1999). This suggests that the spatial distribution of the CRDs within a trimeric subunit permits simultaneous and cooperative interactions with carbohydrate ligands arranged in closely spaced repeats, which are more commonly displayed on the surface of microorganisms than on mammalian cells. Therefore, this structure is important for distinguishing between self and non-self (Sheriff *et al.*, 1994).

Solid-phase binding studies have shown that monomeric CRDs have an approximately 10-fold lower binding affinity for multivalent ligands than the trimeric CRDs (Persson *et al.*, 1990). Consequently, a trimeric or multimeric cluster of CRDs within each surfactant protien is necessary for high-affinity binding to carbohydrate. The CRD structure is stabilized by two disulfide bridges and by binding calcium to two conserved sites which are located in the upper part of the domain (Weis *et al.*, 1992). The three-dimensional structure of the trimeric CRD and α-helical neck region of the human SP-D has been determined by X-ray crystallography (Figure 35.2). It has been shown that this trimeric structure presents a large positively charged surface between the three CRD sites, which is believed to facilitate binding to negatively charged structures, such as LPS (Hakansson *et al.*, 1999).

3.2. α-Helical neck domain

The three-dimensional studies of the human SP-D suggest that the spatial organization of the CRDs within a trimer is stabilized by interactions of the C-terminal sequence with the trimeric neck domain. This is a protease resistant fragment comprising a short stretch of amino acids linking the collagenous region to the "globular domain". The 34–41 amino acid neck region forms a triple-stranded coil of α-helices. Each helix interacts with the neighbouring CRD and, therefore, determines the spatial orientation of the CRDs. A recombinant peptide equivalent to this region in SP-D was found to form a trimer in solution. The structure is held together by strong non-covalent hydrophobic forces and is very stable against denaturation by heat (temperatures >55°C) or pH (pH 3.0–8.5). The tight association of the three chains in the neck region may serve as the nucleation point for the formation of the collagen triple helix. The formation of a collagen helix proceeds in a zipper-like fashion from a single nucleation point at the C-terminal end of the three chains (Hoppe and Reid, 1994). In addition to its role in protein folding, it is the neck domain that determines the spatial orientation of the CRD for selective ligand binding.

3.3. Collagen-like region

The collagen like-region is composed of three polypeptide chains. Each chain is composed primarily of repeating Gly-Xaa-Yaa triplets, a requirement of a collagen triple helix (Traub and Piez, 1971). There are significant sequence differences between collagen-like region of the different collectins seen in their lengths and in the number of Cys residues. It is not known whether the collagen-like region is involved in direct protein–protein interactions or whether it is primarily a scaffolding molecule that connects, orientates and amplifies the ligand binding activities of other domains of the molecule, such as the CRD.

Studies of the collagen region of the human SP-A showed that the Gly-Xaa-Yaa motifs are interrupted. The SP-A collagenous region is relatively short, containing 23 Gly-Xaa-Yaa repeats with proline (Pro) predicted at the Xaa position of 4 or 5 repeats and in the Yaa position of 15 or 16 repeats. All the Pro residues are hydroxylated, which acts as a stabilizing factor. The Gly-Xaa-Yaa repeat-sequence in SP-A is interrupted after the 13th repeat (in humans) by the motif Pro-Cys-Pro-Pro (Voss *et al.*, 1991). However, this is not the case in SP-D, where the collagen region is highly conserved and lacks interruption in the Gly- Xaa-Yaa sequence.

In contrast to SP-A, SP-D, which contains 59 Gly-Xaa-Yaa repeats, is Cys-free and contains hydroxyl-glycosides (Persson *et al.*, 1989; Crouch *et al.*, 1994). The first translated exon of SP-D contains a highly conserved and unusually hydrophilic Gly-Xaa-Yaa repeat that shows little homology with the rest of the collagen region. The functional significance of this region is unknown. However, it has been suggested that this might contribute to oligomer assembly or mediates interactions with cellular receptors (Crouch, 2000). The collagen-like domain is believed to play a role in determining the maximal spatial separation of trimeric, C-terminal lectin region within the SP-D molecules. Furthermore, a study has shown that the deletion in the total collagenous domain of the rat SP-D results in the secretion of trimers rather than dodecamers SP-D molecules. This evidence suggests that the collagen-like region may also be involved in the normal oligomeric assembly and secretion of the collectins (Ogasawara and Voelker, 1995).

3.4. N-Terminal sequence

The N-terminal region, 7–28 amino acids, non-collagenous sequences of the surfactants show only limited similarities to each other and no homologous regions have been found in other proteins. However, the conservation of Cys residues between is indicative of a similar role for these residues in the association to higher order oligomer (Hoppe and Reid, 1994). It has been shown that these residues participate in disulfide cross-links that stabilize the trimer, as well as the N-terminal association of four or more trimeric subunits (Brown-Augsburger *et al.*, 1996a).

3.5. Overall 3D structure

Information on the higher order oligomerization of the surfactant molecules has been determined by electron microscopy (Lu *et al.*, 1993). Although SP-D (43 kDa reduced) appear predominantly as cruciform structures (4×43 kDa trimers), electron microscopy studies showed that SP-D form multimers of up to 8 dodecamers, which are linked at the N-terminal region (Crouch *et al.*, 1994). It is specifically the presence of the two Cys residues at the N-terminal region that contributes to this formation of oligomers (Brown-Augsburger *et al.*, 1996b). These multimers, which can present up to 96 CRDs in the same molecule, show higher binding affinity to a variety of ligands (Brown-Augsburger *et al.*, 1996b). Therefore, this is considered as a potentially important feature in mediating microbial aggregation (Hartshorn *et al.*, 1996). The SP-A oligomer, which appears as a bouquet structure, consists of six subunits.

Human SP-A and SP-D are *N*-glycosylated in the CRD and the collagen regions, respectively. At present, the functional significance of these sugars is still unclear. Nevertheless, mutational analysis might suggest the involvement of the *N*-linked sugar in rat SP-D is not required for secretion, dodecamer formation or other interaction with other microorganisms (Brown-Augsburger *et al.*, 1996b; Hartshorn *et al.*, 1997).

4. BIOSYNTHESIS AND TISSUE DISTRIBUTION OF SP-A AND SP-D

In the lung, SP-D and SP-A are synthesized by alveolar type II cells and unciliated bronchial epithelial cells (Clara cells). In addition to its primary site of production in the lungs, SP-A has also been seen in the serous glands of the proximal trachea (Khoor *et al.*, 1993) and has been detected in the rat small intestine and the colon (Rubio *et al.*, 1995) and in the laminar bodies in synoviocytes and in lachrymal and salivary epithelia (Dobbie, 1996). Detection was seen in the human thymus, prostate and mesentery. There is also evidence of the presence of SP-A in the middle-ear effusion. Compared to SP-A, SP-D has been shown to be more widespread in extrapulmonary tissues of different species. Additional to its expression in the human lung and trachea, SP-D has been found to be strongly expressed in the kidney, brain, testis, pancreas, salivary gland, heart, prostate, small and large intestine, stomach and placenta (Madsen *et al.*, 2000; Murray *et al.*, 2002). Moreover, SP-D messenger RNA (mRNA) was also identified in amniotic fluid, sweat and mammary glands. In the rat, SP-D has been found in the gastric mucosa and mesentery cells. In the mouse, SP-D signal has been detected in the stomach, kidney and heart. Such a widespread expression at different mucosal surfaces and extra-pulmonary tissues is a strong indicator of the importance of SP-A and SP-D in participating in primary immune defence.

5. REGULATION OF GENE EXPRESSION

The expression of both SP-A and SP-D is increased by inflammation and infection. This is clearly illustrated by the 22-fold increase in serum SP-D levels during bacterial pneumonia (Leth-Larsen *et al.*, 2003). Systemic administration of LPS in rats leads to a marked increase in pulmonary SP-D expression 24–72 hours after the challenge (Shu *et al.*, 2007). The expression of SP-D in the uninfected, uninflamed stomach is difficult to detect but the expression increases markedly in mice stomachs with either acute or chronic *Helicobacter* infection (Murray *et al.*, 2002).

The mechanism through which infection or inflammation upregulate the expression of SP-A and SP-D is not clearly mapped. Cyclic adenosine monophosphate (cAMP) and interleukin- (IL-) 1 stimulate SP-A expression in human fetal lung cells mediated through the binding of thyroid transcription factor 1 (TTF-1) and nuclear factor-kappaB (NF-κB) binding to a TTF-1-binding element in the promoter region of the gene. This interaction modulates histone deacetylase expression and increases acetylation of histone H3 (Islam and Mendelson, 2008). Opposing effects are seen in response to glucocorticoids working through the endogenous glucocorticod receptor.

T-Helper-2 cell (Th2) cytokines, such as IL-4 and IL-13, appear to have divergent effects on the regulation of SP-A and SP-D expression. In cultured human sino-nasal epithelial cells, IL-4 and IL-13 suppressed SP-A expression which is postulated to explain the low levels of this collectin in the nasal mucosa of patients with allergic rhinitis (Ramanathan, *et al.*, 2008). In contrast, IL-4 and IL-13 increase SP-D expression in mice sensitized to *Aspergillus fumigatus* allergen (Haczku *et al.*, 2006). Interestingly, in this model, elimination of the allergen by SP-D provided a negative feedback loop by downregulating cytokine induction.

Serum levels of SP-D demonstrate wide inter-individual variation. In twin studies, this variation is accounted for by a combination of environmental and genetic factors (Sorensen *et al.*, 2006). Male gender, increasing age and smoking were associated with increased levels of SP-D. The Met-11-Thr variant (where Met, methionine; Thr, threonine) in the *SP-D* gene explained a significant proportion of the heritable component of the variation in SP-D levels. There are five allelic variants in the *SP-A1* gene and six allelic variants in the *SP-A2* gene. Results indicate that SP-A levels are also, to some degree, genetically controlled.

Sex hormones appear to influence the expression of SP-D in both the male and female reproductive tracts. In the male reproductive tract, SP-D expression is increased after castration suggesting a regulatory influence of androgens (Oberley *et al.*, 2007a), whereas in the female reproductive tract, levels of SP-D are highest during the oestrous phase and lowest in the dioestrous phase (Oberley *et al.*, 2007b). Signalling via the oestrogen-related receptor α working through a nuclear receptor element, NRE(SP-A), leads to increased SP-A expression *in vitro*.

6. PUTATIVE RECEPTORS FOR SP-A AND SP-D

In addition to binding to carbohydrate structures on the surface of pathogenic microorganisms, surfactants act as a cross-link molecule by binding to a variety of cells participating in the immune response, including alveolar macrophages, monocytes, neutrophils and type II pneumocytes. A number of surfactant-binding proteins, both soluble and surface-bound, have been identified. However, their specific contribution to the biological activity of the surfactants is not fully understood.

6.1. Putative receptors for SP-A

The C1q receptor (C1qR), first isolated as a receptor for the collagen part of the human complement component C1q (Ghebrehiwet *et al.*, 1984), has been shown to bind SP-A (Malhotra *et al.*, 1992). The binding is calcium-independent, which suggests the involvement of a protein–protein interaction rather than a lectin–carbohydrate binding. Subsequently, however, this putative receptor was found to be calreticulin, an intracellular endoplasmic reticulum (ER) protein, which helps to mediate the correct folding of glycoproteins prior to their removal from the ER. The C1q receptor has an almost complete amino acid sequence identity with calreticulin and, therefore, antibodies raised against calreticulin and C1qR demonstrated a high degree of immunological cross-reactivity. Subsequently, the role of calreticulin as a receptor has been re-evaluated (Eggleton *et al.*, 1998). Calreticulin is a cell-surface receptor that lacks a transmembrane domain. A number of studies have demonstrated that when bound to the collagen region of SP-A, calreticulin uses the endocytic receptor CD91, also known as low-density lipid receptor-related protein (LPR) or α_2 macroglobulinreceptor, as an adaptor molecule for signal transduction and phagocytosis. Another protein, termed C1qRp (C1q receptor that mediates phagocytosis; also designated CD93), was shown to be a membrane-spanning surface molecule capable of binding SP-A.

Two more SP-A putative receptors have been characterized. First, the specific 210 kDa SP-A binding protein (SP-R210), which was shown by ligand blot analyses to be present on both rat alveolar type II cells and alveolar macrophages. The SP-R210 protein was first isolated from U937 macrophage membranes by affinity chromatography (Chroneos *et al.*, 1996). The functional importance of this cell-surface receptor became clear when antibodies raised against SP-R210 blocked SP-A-mediated uptake of *Mycobacterium bovis* by macrophages and blocked the SP-A inhibitory effect on the phospholipids secretion by the alveolar type II cells. This evidence suggests that SP-R210 is involved in SP-A-mediated clearance of pathogen. Second, a protein was identified on cell membranes of rat type II pneumocytes and designated the binding molecule of reduced molecular mass 55 kDa (BP55) (Stevens *et al.*, 1995). The BP55 protein was identified using antibodies against a type-II pneumocyte component which were developed by an anti-idiotypic approach, using SP-A as the immunogen, i.e. antibodies directed against the surfactant protein binding region of anti-surfactant antibodies. Co-incubation of the anti-idiotype antibody with SP-A inhibited SP-A-mediated lipid uptake by alveolar type II cells. From the inhibition study, it seems as if BP55 is involved in SP-A-mediated lipid uptake by alveolar type II cells (Wissel *et al.*, 1996).

6.2. Putative receptors for SP-D

Various studies have demonstrated the binding of SP-D to alveolar macrophages. However, the identification of putative receptors has proved more difficult. The 340-kDa glycoprotein, gp-340, was shown to bind to SP-D in a calcium-dependent manner, not inhibited by maltose. This suggests the involvement of a specific protein–protein interaction between SP-D and gp-340 rather than a lectin–carbohydrate recognition.

It is generally accepted that the protein–protein interaction of the collectins and their receptors occurs on the binding sites in the collagenous region of the collectins. This is supported by evidence that both SP-D and the recombinant neck-CRD SP-D bind to gp-340. Furthermore, this interaction is independent of the calcium binding sites responsible for the lectin–carbohydrate interactions (Holmskov *et al.*, 1997). The gp-340 glycoprotein is a member of the scavenger receptor Cys-rich superfamily (SRCR). It was isolated from human broncho-alveolar lavage (BAL) and, more interestingly, it was identified as a molecule that binds SP-D because it co-purified with SP-D on a carbohydrate affinity column. The tissue distribution of the gp-340 and its high expression in the alveolar macrophages is compatible with the idea that gp-340 forms an opsonin receptor for SP-D.

6.3. Putative receptors common for both SP-A and SP-D

Both SP-A and SP-D have been shown to have some common receptors. In particular, SP-A and SP-D can bind to CD14, a known receptor to LPS, suggesting that both surfactants contribute to the LPS-mediated cell response (Sano *et al.*, 2000). Of note, SP-A specifically binds to the peptide portion containing the leucine-rich repeats of CD14, whereas SP-D binds to the carbohydrate moiety of CD14 (Sano *et al.*, 2000). In addition, SP-A has been shown to bind to Toll-like receptors (TLRs), a family of PRRs (see Chapter 31); SP-A binds to the extracellular domains of TLR-2 and TLR-4 and to myeloid differentiation-2 (MD-2) adaptor protein (Murakami *et al.*, 2002; Yamada *et al.*, 2006). Furthermore, reports have shown that SP-D also binds the extracellular domains of TLR-2 and TLR-4 through its CRD (Ohya *et al.*, 2006).

It has been shown that some of the receptors initially thought to be specific for either SP-A or SP-D, are in fact common to both surfactants. Notably, SP-A has been shown to bind gp-340 in a lectin-independent manner. On the other hand, SP-D can also bind to calreticulin and CD91.

TABLE 35.1 Interaction of SP-D and SP-A with microorganisms

Microorganism	Surfactant	Reference
Gram-negative bacteria		
Escherichia coli	SP-D	Kuan *et al.*, 1992
Pseudomonas aeruginosa	SP-D, SP-A	Mariencheck *et al.*, 1999
Klebsiella pneumoniae	SP-D, SP-A	Kabha *et al.*, 1997; Keisari *et al.*, 2001
Helicobacter pylori	SP-D	Khamri *et al.*, 2007; Murray *et al.*, 2002
Gram-positive bacteria		
Group B streptococci	SP-A	LeVine *et al.*, 1997; LeVine *et al.*, 1999b
Staphylococcus aureus	SP-A	Manz-Keinke *et al.*, 1992
Streptococcus pneumoniae	SP-A	Manz-Keinke *et al.*, 1992
Viruses		
Influenza type A	SP-D, SP-A	Hartshorn *et al.*, 1994; Benne *et al.*, 1997
Fungi		
Cryptococcus neoformans	SP-D	Schelenz *et al.*, 1995
Aspergillus fumigatus	SP-D	Madan *et al.*, 1997
Saccharomyces cerevisiae	SP-D	Allen *et al.*, 2001
Mycobacteria		
Mycobacterium tuberculosis	SP-D	Ferguson *et al.*, 1999

Nonetheless, SP-A and SP-D bind another receptor, termed signal inhibitory regulatory protein α (SIRPα). A major step forward in understanding the biological contribution of the receptors to the activity of SP-A and SP-D came from an interesting study by Gardai and colleagues. The study suggested a distinct mechanism for surfactant-mediated modulation of inflammation. The direct binding of the globular head (i.e. CRD) of SP-A and SP-D to SIRPα initiates a signalling pathway that blocks proinflammatory mediator production. In contrast, in the presence of microbes, which interact with the CRD, the collagenous tails stimulates proinflammatory mediator production through binding to calreticulin/CD91 (Gardai *et al.*, 2003). Therefore, SP-A and SP-D can act in a dual manner, to enhance or suppress inflammation depending on binding orientation.

7. DIVERSE FUNCTIONS OF SP-A AND SP-D

7.1. Direct microbial interaction

The differences observed in the glycosylation pattern between the invading organisms and the host allows the discrimination by SP-A and SP-D between self and non-self. In addition, the repetitive carbohydrate structures on microbial surfaces allow the multimers of the CRDs of SP-A and SP-D to increase the affinity to a level where effector mechanisms can be initiated. Indeed, SP-A and SP-D bind to a wide range of microorganisms including bacteria, fungi, viruses and parasites (Table 35.1).

Gram-negative and -positive bacterial cells express ligands for the host surfactants. Bacterial

LPSs were among the first specific ligands identified for both SP-A and SP-D and these interactions make a significant contribution to the host response to LPS. Both SP-A and SP-D modify the response to instilled LPS *in vivo* with decreased lung injury and inflammatory cell recruitment (Tino and Wright, 1999; Crouch, 2000). In particular, SP-A and SP-D bind the smooth form of LPS and can selectively modify the cellular responses to rough form LPS. In general, SP-A binds via the lipid A moiety of LPS, while SP-D binds through the core saccharide. Nevertheless, SP-D can bind to the gastroduodenal pathogen, *Helicobacter pylori*, through the LPS O-chain of the bacterial cell wall (Khamri *et al.*, 2005). Some viruses have also been shown to interact with SP-D. For example, SP-D binds the head region of the influenza A virus haemagglutinin (Reid, 1998). In the case of influenza A virus, interaction with SP-D may suppress inflammatory responses. Viral haemagglutination activity, which would normally increase neutrophil activation, is inhibited by SP-D (Hartshorn *et al.*, 1994).

Binding of surfactants to the bacterial cells leads to their agglutination; the surfactants act as opsonins to facilitate pathogen clearance through phagocytosis. *Staphylococcus aureus* uptake by monocytes is enhanced when SP-A acts as an opsonin and coats the microorganism (Hickling *et al.*, 1998). Both SP-D and SP-A levels are dramatically reduced in BAL fluid from children suffering from cystic fibrosis, which may play a major role in increasing risks of lung infections in these patients (Postle *et al.*, 1999). The roles of SP-A and SP-D in surfactant metabolism, and the host's response to inflammatory or infectious challenge, have been further elucidated by studying the phenotypes of mice deficient in these two proteins. Notably, *SP-A* gene knockout mice have been shown to be less effective at clearing *S. aureus* and *Pseudomonas aeruginosa* infections and more susceptible to lung inflammation (LeVine *et al.*, 1997). Further investigation, where the *SP-A* knockout mice were challenged by respiratory syncytial virus (RSV), showed increased pulmonary infiltration with neutrophils, increased tumour necrosis factor-alpha (TNF-α) and IL-6 production, together with impaired macrophage respiratory bursts (LeVine *et al.*, 1999a).

In addition to their role as opsonins, SP-A and SP-D have a direct effect on bacterial cell physiology, growth and motility. There is evidence that surfactants directly permeablize the bacterial cell wall in an LPS-dependent and rough-form LPS-specific manner. This antimicrobial activity results in direct inhibition of the proliferation of Gram-negative bacteria. Indeed, SP-D has an added effect that was observed following binding to the motile *H. pylori*. Motility is a major virulence factor that allows the pathogen to establish colonization of the stomach mucosa. Importantly, SP-D interaction with *H. pylori* induces aggregation of the bacterial cells and impairs their motility by approximately 50% (Murray *et al.*, 2002; Moran *et al.*, 2005).

7.2. Interaction with immune cells

Independent of microbial binding, SP-A and SP-D can directly affect the functions of a variety of immune cells including macrcophages, neutrophils, dendritic cells and lymphocytes. In fact, SP-D increases the internalization of microbes by directly modulating the function of phagocytic cells. Pre-incubation of neutrophils with SP-D enhances the uptake of *E. coli* by neutrophils. This has been attributed to the upregulation of the mannose receptor; SP-A and SP-D increase the cell surface expression of the mannose receptor on monocyte-derived macrophages (Beharka *et al.*, 2002; Kudo *et al.*, 2004). In addition to stimulating phagocytosis, the interaction of SP-A and SP-D with macrophages and neutrophils influences their chemotaxis, free radical and cytokine production. Notably, SP-D is a potent chemoattractant for both monocytes and neutrophils

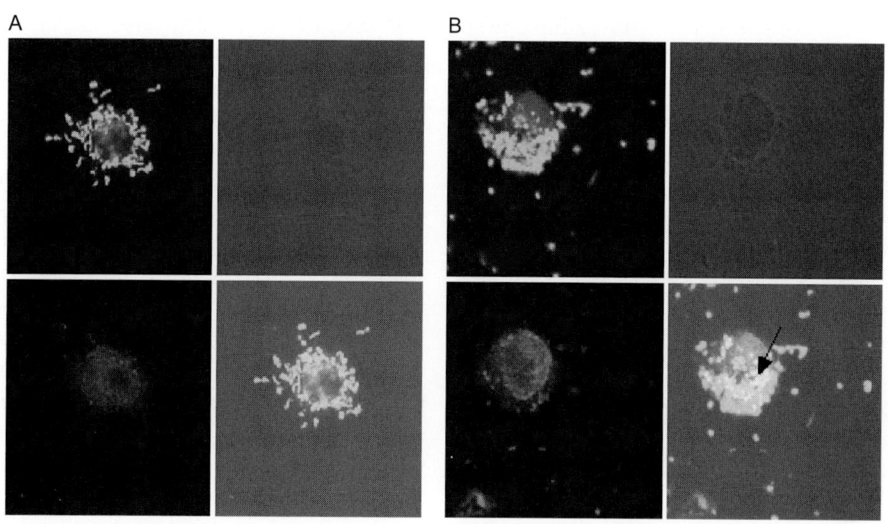

FIGURE 35.3 Enhanced *H. pylori* uptake by dendritic cells (DCs) in the presence of SP-D. Confocal microscopy of the uptake of *H. pylori* (green) by DCs (red) in the absence (A) and presence (B) of SP-D. Carboxyfluorescein diacetate succinimidyl ester (CFDA-SE) labelled *Helicobacter* cells (green) is co-localized with the early endosomal compartments in DCs (red) and thus appears yellow (arrowhead in Panel B). The uptake is more marked in the presence of SP-D and calcium and co-localization is seen in yellow (B). Reproduced, with permission, from Khamri *et al.* (2007) *Helicobacter* infection in the surfactant protein D-deficient mouse. Helicobacter 12, 112–123. (See colour plates section.)

(Crouch *et al.*, 1995; Madan *et al.*, 1997) which has been attributed to binding of CRD to carbohydrate structures on the phagocytic cell. *In vivo* models have shown that lack of SP-D expression in *H. pylori* infection is associated with reduced neutrophil recruitment, reflecting an important role for SP-D in neutrophil chemotaxis (Khamri *et al.*, 2007).

Surfactants affect reactive oxidant species production in phagocytic cells. Rat SP-D has been reported to stimulate directly the respiratory burst response of alveolar macrophages (Van Iwaarden *et al.*, 1992). However, SP-D deficiency results in a chronic inflammatory response characterized by increased levels of reactive oxygen species which may be due to higher numbers of the organisms present. In particular, SP-A enhances killing of *Mycoplasma pneumoniae* by increasing nitric oxide production (Hickman-Davis *et al.*, 1998).

Another important mechanism of host defence is that surfactants play a role in antigen presentation through interaction with antigen-presentation cells, predominantly dendritic cells (DCs). There is growing evidence that SP-D can directly bind to immature DCs in a dose, calcium- and carbohydrate-dependent manner. It has been shown that SP-D binding to Gram-negative bacteria, such as *E. coli* (Brinker *et al.*, 2001) and *H. pylori* (Khamri *et al.*, 2007), enhances their uptake by DCs (Figure 35.3). Subsequently, this upregulates surface expression of DCs' maturation markers and increases the proliferation of T lymphocytes and IL-2 production (Brinker *et al.*, 2001). However, this property is only attributed to SP-D and not SP-A, as SP-A has been reported to inhibit DCs' maturation (Brinker *et al.*, 2003).

7.3. Anti-allergic role of the surfactants

Knockout mice models have brought more insight into the role of surfactants in resistance to allergen challenge and pulmonary hypersensitivity. A number of studies have suggested that both SP-A and SP-D play a role in the anti-allergic response. Experiments carried out on purified allergen from *A. fumigatus*, which causes allergic symptoms such as hypersensitivity pneumonitis and allergic bronchopulmonary aspergillosis, showed that both SP-A and SP-D inhibited the allergen-induced release of histamine from basophils and mast cells *in vitro* (Madan *et al.*, 1997).

7.4. Removal of apoptotic cells

Importantly, SP-A and SP-D play a further physiological role by binding to DNA (Palaniyar *et al.*, 2003a). This binding controls inflammation for promoting the clearance of apoptotic cells (Palaniyar *et al.*, 2003b). Apoptotic cell death results in the fragmentation of DNA and its subsequent display. Inefficient clearance of these cells will lead to tissue inflammation. *In vitro*, SP-D enhances the uptake of DNA and DNA-containing beads by macrophages (Palaniyar *et al.*, 2005). *In vivo*, the absence of SP-D leads to increased anti-DNA autoantibody generation (Palaniyar *et al.*, 2005). In addition to binding DNA from apoptotic cells, it has been shown that bacterial cells, such as *Pseudomonas aerogenosa*, can actively secrete DNA into the extracellular matrix to form active biofilm, which is a virulence factor that confers the generation of chronic infection (Whitchurch *et al.*, 2002). Deficiency in SP-A and SP-D has been observed in patients suffering from cystic fibrosis where lung pathology reveals chronic microbial infection and accumulation of apoptotic and necrotic neutrophils (Postle *et al.*, 1999).

8. CLINICAL APPLICATIONS AND SIGNIFICANCE

Roles may be found for SP-A and SP-D in clinical practice as either biomarkers or as therapeutic interventions. Levels of SP-D in serum rise in response to acute lung injury (Holmskov *et al.*, 2003). However, SP-D and SP-A levels are diminished in the BAL fluid from patients with cystic fibrosis, emphysema and asthma, indicating a potential susceptibility to pneumonia.

In cystic fibrosis, the levels of SP-A and SP-D in BAL fluid are markedly reduced. In this group of patients, colonization and infection with *P. aeruginosa* are associated with a rapid decline in the patient's clinical condition. Colonization with *A. fumigatus* may induce allergic responses which further compromise respiratory function. Of note, SP-D binds to *P. aeruginosa* and promotes agglutination and phagocytosis of the bacterium. In addition, SP-D binds to *A. fumigatus* and inhibits cell growth. In mouse models of *Aspergillus*-induced allergic responses, administration of natural human or recombinant SP-D alleviates the inflammatory response (Strong *et al.*, 2002). Inflammation is also provoked in cystic fibrosis by the failure to clear apoptotic cell debris. Thus, SP-D promotes the elimination of apoptotic cells and may, therefore, help to reduce the inflammatory lung damage (Vandivier *et al.*, 2002). Hence, there is clear potential for the use of recombinant SP-D in the treatment of cystic fibrosis.

Both SP-A and SP-D bind to many of the important antigens responsible for allergic responses, e.g. the house dust mite antigen Derp, *A. fumigatus* allergens, etc. Mouse models that faithfully reproduce allergic response to Derp and *Aspergillus* lead to bronchial hyperactivity, eosinophilia and systemic immunoglobulin class E responses. Administration of recombinant human SP-D in these models ameliorates each of these effects providing a rationale for trials of SP-D in human asthmatics who are sensitized to these agents (Strong *et al.*, 2003; Singh *et al.*, 2003).

Of central importance, SP-D enhances the uptake of microorganisms into DCs and helps to promote the T-cell response against bacterial antigens (Brinker *et al.*, 2001; Khamri *et al.*, 2007). Peptide or polypeptide antigens on their own are relatively poor stimulators of T-cell- or antibody-mediated immune responses. However, targeting peptide antigens for SP-D binding by attaching carbohydrate moieties is an approach which may enhance the immunogenicity of peptide-based vaccines.

9. CONCLUSIONS

The collectins, SP-A and SP-D are an interesting and important component of the innate immune response. There has been enormous progress in our understanding of the interaction of these molecules with the pathogen-associated targets, but knowledge of the complex interactions of these collectins with other components of both innate and adaptive immunity is far from complete. Areas for further investigation within this field are listed in the Research Focus Box. Nevertheless, there is already potential for therapeutic interventions with these molecules, primarily in chronic lung disorders.

RESEARCH FOCUS BOX

- The definitive receptors for surfactant proteins A and D need to be identified.
- The regulation of surfactant protein expression and secretion requires further investigation.
- It remains to be determined how commonly microorganisms vary their carbohydrate structures in order to evade surfactant protein binding and whether this variation results in a replication disadvantage.
- While preliminary studies outlined in this chapter demonstrate that surfactant proteins may influence the function of dendritic cells and the adaptive immune response, detailed investigations on the impact of these molecules on the cellular immune response to target antigens are still required.

References

Allen, M.J., Voelker, D.R., Mason, R.J., 2001. Interactions of surfactant proteins A and D with *Saccharomyces cerevisiae* and *Aspergillus fumigatus*. Infect. Immun. 69, 2037–2044.

Beharka, A.A., Gaynor, C.D., Kang, B.K., Voelker, D.R., McCormack, F.X., Schlesinger, L.S., 2002. Pulmonary surfactant protein A up-regulates activity of the mannose receptor, a pattern recognition receptor expressed on human macrophages. J. Immunol. 169, 3565–3573.

Benne, C.A., Benaissa-Trouw, B., van Strijp, J.A., Kraaijeveld, C.A., van Iwaarden, J.F., 1997. Surfactant protein A, but not surfactant protein D, is an opsonin for influenza A virus phagocytosis by rat alveolar macrophages. Eur. J. Immunol. 27, 886–890.

Brinker, K.G., Martin, E., Borron, P., et al., 2001. Surfactant protein D enhances bacterial antigen presentation by bone marrow-derived dendritic cells. Am. J. Physiol. Lung Cell Mol. Physiol. 281, L1453–L1463.

Brinker, K.G., Garner, H., Wright, J.R., 2003. Surfactant protein A modulates the differentiation of murine bone marrow-derived dendritic cells. Am. J. Physiol. Lung Cell Mol. Physiol. 284, L232–L241.

Brown-Augsburger, P., Hartshorn, K., Chang, D., et al., 1996a. Site-directed mutagenesis of Cys-15 and Cys-20 of pulmonary surfactant protein D. Expression of a trimeric protein with altered anti-viral properties. J. Biol. Chem. 271, 13724–13730.

Brown-Augsburger, P., Chang, D., Rust, K., Crouch, E.C., 1996b. Biosynthesis of surfactant protein D. Contributions of conserved NH_2-terminal cysteine residues and collagen helix formation to assembly and secretion. J. Biol. Chem. 271, 18912–18919.

Chroneos, Z.C., Abdolrasulnia, R., Whitsett, J.A., Rice, W.R., Shepherd, V.L., 1996. Purification of a cell-surface receptor for surfactant protein A. J. Biol. Chem. 271, 16375–16383.

Crouch, E., Rust, K., Veile, R., Donis Keller, H., Grosso, L., 1993. Genomic organization of human surfactant protein D (SP-D). SP-D is encoded on chromosome 10q22.2-23.1. J. Biol. Chem. 268, 2976–2983.

Crouch, E., Persson, A., Chang, D., Heuser, J., 1994. Molecular structure of pulmonary surfactant protein D (SP-D). J. Biol. Chem. 269, 17311–17319.

Crouch, E.C., 2000. Surfactant protein-D and pulmonary host defense. Respir. Res. 1, 93–108.

Crouch, E.C., Persson, A., Griffin, G.L., Chang, D., Senior, R.M., 1995. Interactions of pulmonary surfactant protein D (SP-D) with human blood leukocytes. Am. J. Respir. Cell Mol. Biol. 12, 410–415.

Dobbie, J.W., 1996. Surfactant protein A and lamellar bodies: a homologous secretory function of peritoneum, synovium, and lung. Perit. Dial. Int. 16, 574–581.

Eggleton, P., Reid, K.B., Tenner, A.J., 1998. C1q – how many functions? How many receptors? Trends Cell Biol. 8, 428–431.

Ferguson, J.S., Voelker, D.R., McCormack, F.X., Schlesinger, L.S., 1999. Surfactant protein D binds to *Mycobacterium tuberculosis* bacilli and lipoarabinomannan via carbohydrate-lectin interactions resulting in reduced phagocytosis of the bacteria by macrophages. J. Immunol. 163, 312–321.

Floros, J., DiAngelo, S., Koptides, M., et al., 1996. Human SP-A locus: allele frequencies and linkage disequilibrium between the two surfactant protein A genes. Am. J. Respir. Cell Mol. Biol. 15, 489–498.

Gardai, S.J., Xiao, Y.Q., Dickinson, M., et al., 2003. By binding SIRPα or calreticulin/CD91, lung collectins act as dual function surveillance molecules to suppress or enhance inflammation. Cell 115, 13–23.

Ghebrehiwet, B., Silvestri, L., McDevitt, C., 1984. Identification of the Raji cell membrane-derived C1q inhibitor as a receptor for human C1q. Purification and immunochemical characterization. J. Exp. Med. 160, 1375–1389.

Haczku, A., Cao, Y., Vass, G., et al., 2006. IL-4 and IL-13 form a negative feedback circuit with surfactant protein-D in the allergic airway response. J. Immunol. 176, 3557–3565.

Hakansson, K., Lim, N.K., Hoppe, H.J., Reid, K.B., 1999. Crystal structure of the trimeric α-helical coiled-coil and the three lectin domains of human lung surfactant protein D. Structure 7, 255–264.

Hartshorn, K.L., Crouch, E.C., White, M.R., et al., 1994. Evidence for a protective role of pulmonary surfactant protein D (SP-D) against influenza A viruses. J. Clin. Invest. 94, 311–319.

Hartshorn, K.L., Reid, K.B., White, M.R., et al., 1996. Neutrophil deactivation by influenza A viruses: mechanisms of protection after viral opsonization with collectins and hemagglutination-inhibiting antibodies. Blood 87, 3450–3461.

Hartshorn, K.L., White, M.R., Shepherd, V., Reid, K., Jensenius, J.C., Crouch, E.C., 1997. Mechanisms of anti-influenza activity of surfactant proteins A and D: comparison with serum collectins. Am. J. Physiol. 273, L1156–L1166.

Hickling, T.P., Malhotra, R., Sim, R.B., 1998. Human lung surfactant protein A exists in several different oligomeric states: oligomer size distribution varies between patient groups. Mol. Med. 4, 266–275.

Hickman-Davis, J.M., Lindsey, J.R., Zhu, S., Matalon, S., 1998. Surfactant protein A mediates mycoplasmacidal activity of alveolar macrophages. Am. J. Physiol. 274, L270–L277.

Holmskov, U., Lawson, P., Teisner, B., et al., 1997. Isolation and characterization of a new member of the scavenger receptor superfamily, glycoprotein-340 (gp-340), as a lung surfactant protein-D binding molecule. J. Biol. Chem. 272, 13743–13749.

Holmskov, U., Thiel, S., Jensenius, J.C., 2003. Collections and ficolins: humoral lectins of the innate immune defense. Annu. Rev. Immunol. 21, 547–578.

Holmskov, U.L., 2000. Collectins and collectin receptors in innate immunity. APMIS 108 (Suppl. 100), 1–59.

Hoppe, H.J., Reid, K.B., 1994. Trimeric C-type lectin domains in host defence. Structure 2, 1129–1133.

Islam, K.N., Mendelson, C.R., 2008. Glucocorticoid/glucocorticoid receptor inhibition of surfactant protein-A (SP-A) gene expression in lung type II cells is mediated by repressive changes in histone modification at the SP-A promoter. Mol. Endocrinol. 22, 585–596.

Kabha, K., Schmegner, J., Keisari, Y., Parolis, H., Schlepper-Schaeffer, J., Ofek, I., 1997. SP-A enhances phagocytosis of *Klebsiella* by interaction with capsular polysaccharides and alveolar macrophages. Am. J. Physiol. 272, L344–L352.

Katyal, S.L., Singh, G., Locker, J., 1992. Characterization of a second human pulmonary surfactant-associated protein SP-A gene. Am. J. Respir. Cell Mol. Biol. 6, 446–452.

Keisari, Y., Wang, H., Mesika, A., et al., 2001. Surfactant protein D-coated *Klebsiella pneumoniae* stimulates cytokine production in mononuclear phagocytes. J. Leukoc. Biol. 70, 135–141.

Khamri, W., Moran, A.P., Worku, M.L., et al., 2005. Variations in *Helicobacter pylori* lipopolysaccharide to evade the innate immune component surfactant protein D. Infect. Immun. 73, 7677–7686.

Khamri, W., Worku, M.L., Anderson, A.E., et al., 2007. *Helicobacter* infection in the surfactant protein D-deficient mouse. Helicobacter 12, 112–123.

Khoor, A., Gray, M.E., Hull, W.M., Whitsett, J.A., Stahlman, M.T., 1993. Developmental expression of SP-A and SP-A mRNA in the proximal and distal respiratory epithelium in the human fetus and newborn. J. Histochem. Cytochem. 41, 1311–1319.

Kolble, K., Reid, K.B., 1993. The genomics of soluble proteins with collagenous domains: C1q, MBL, SP-A, SP-D, conglutinin, and CL-43. Behring Inst. Mitt. 93, 81–86.

Korfhagen, T.R., Glasser, S.W., Bruno, M.D., McMahan, M.J., Whitsett, J.A., 1991. A portion of the human surfactant protein A (SP-A) gene locus consists of a pseudogene. Am. J. Respir. Cell Mol. Biol. 4, 463–469.

Kuan, S.F., Rust, K., Crouch, E., 1992. Interactions of surfactant protein D with bacterial lipopolysaccharides. Surfactant protein D is an *Escherichia coli*-binding protein in bronchoalveolar lavage. J. Clin. Invest. 90, 97–106.

Kudo, K., Sano, H., Takahashi, H., et al., 2004. Pulmonary collectins enhance phagocytosis of *Mycobacterium avium* through increased activity of mannose receptor. J. Immunol. 172, 7592–7602.

Leth-Larsen, R., Nordenbaek, C., Tornoe, I., et al., 2003. Surfactant protein D (SP-D) serum levels in patients with community-acquired pneumonia small star, filled. Clin. Immunol. 108, 29–37.

LeVine, A.M., Bruno, M.D., Huelsman, K.M., Ross, G.F., Whitsett, J.A., Korfhagen, T.R., 1997. Surfactant protein A-deficient mice are susceptible to group B streptococcal infection. J. Immunol. 158, 4336–4340.

LeVine, A.M., Gwozdz, J., Stark, J., Bruno, M., Whitsett, J., Korfhagen, T., 1999a. Surfactant protein-A enhances respiratory syncytial virus clearance *in vivo*. J. Clin. Invest. 103, 1015–1021.

LeVine, A.M., Kurak, K.E., Wright, J.R., et al., 1999b. Surfactant protein-A binds group B streptococcus enhancing phagocytosis and clearance from lungs of surfactant protein-A-deficient mice Am. J. Respir. Cell Mol. Biol. 20, 279–286.

Lu, J., Wiedemann, H., Timpl, R., Reid, K.B., 1993. Similarity in structure between C1q and the collectins as judged by electron microscopy. Behring Inst. Mitt. 93, 6–16.

Madan, T., Eggleton, P., Kishore, U., et al., 1997. Binding of pulmonary surfactant proteins A and D to *Aspergillus fumigatus* conidia enhances phagocytosis and killing by human neutrophils and alveolar macrophages. Infect. Immun. 65, 3171–3179.

Madsen, J., Kliem, A., Tornoe, I., Skjodt, K., Koch, C., Holmskov, U., 2000. Localization of lung surfactant protein D on mucosal surfaces in human tissues. J. Immunol. 164, 5866–5870.

Malhotra, R., Haurum, J., Thiel, S., Sim, R.B., 1992. Interaction of C1q receptor with lung surfactant protein A. Eur. J. Immunol. 22, 1437–1445.

Manz-Keinke, H., Plattner, H., Schlepper-Schafer, J., 1992. Lung surfactant protein A (SP-A) enhances serum-independent phagocytosis of bacteria by alveolar macrophages. Eur. J. Cell Biol. 57, 95–100.

Mariencheck, W.I., Savov, J., Dong, Q., Tino, M.J., Wright, J.R., 1999. Surfactant protein A enhances alveolar macrophage phagocytosis of a live, mucoid strain of *P. aeruginosa*. Am. J. Physiol. 277, L777–L786.

McCormick, S.M., Boggaram, V., Mendelson, C.R., 1994. Characterization of mRNA transcripts and organization of human *SP-A1* and *SP-A2* genes. Am. J. Physiol. 266, L354–L366.

Moran, A.P., Khamri, W., Walker, M.M., Thursz, M.R., 2005. Role of surfactant protein D (SP-D) in innate immunity in the gastric mucosa: evidence of interaction with *Helicobacter pylori* lipopolysaccharide. J. Endotoxin Res. 11, 357–362.

Murakami, S., Iwaki, D., Mitsuzawa, H., et al., 2002. Surfactant protein A inhibits peptidoglycan-induced tumor necrosis factor-α secretion in U937 cells and alveolar macrophages by direct interaction with toll-like receptor 2. J. Biol. Chem. 277, 6830–6837.

Murray, E., Khamri, W., Walker, M.M., et al., 2002. Expression of surfactant protein D in the human gastric mucosa and during *Helicobacter pylori* infection. Infect. Immun. 70, 1481–1487.

Oberley, R.E., Goss, K.L., Quintar, A.A., Maldonado, C.A., Snyder, J.M., 2007a. Regulation of surfactant protein D in the rodent prostate. Reprod. Biol. Endocrinol. 5, 42.

Oberley, R.E., Goss, K.L., Hoffmann, D.S., et al., 2007b. Regulation of surfactant protein D in the mouse female reproductive tract *in vivo*. Mol. Hum. Reprod. 13, 863–868.

Ogasawara, Y., Voelker, D.R., 1995. The role of the amino-terminal domain and the collagenous region in the structure and the function of rat surfactant protein D. J. Biol. Chem. 270, 19052–19058.

Ohya, M., Nishitani, C., Sano, H., et al., 2006. Human pulmonary surfactant protein D binds the extracellular domains of Toll-like receptors 2 and 4 through the carbohydrate recognition domain by a mechanism different from its binding to phosphatidylinositol and lipopolysaccharide. Biochemistry 45, 8657–8664.

Palaniyar, N., Nadesalingam, J., Reid, K.B., 2003a. Innate immune collectins bind nucleic acids and enhance DNA clearance in vitro. Ann. NY Acad. Sci. 1010, 467–470.

Palaniyar, N., Clark, H., Nadesalingam, J., Hawgood, S., Reid, K.B., 2003b. Surfactant protein D binds genomic DNA and apoptotic cells, and enhances their clearance, in vivo. Ann. NY Acad. Sci. 1010, 471–475.

Palaniyar, N., Clark, H., Nadesalingam, J., Shih, M.J., Hawgood, S., Reid, K.B., 2005. Innate immune collectin surfactant protein D enhances the clearance of DNA by macrophages and minimizes anti-DNA antibody generation. J. Immunol. 174, 7352–7358.

Persson, A., Chang, D., Rust, K., Moxley, M., Longmore, W., Crouch, E., 1989. Purification and biochemical characterization of CP4 (SP-D), a collagenous surfactant-associated protein. Biochemistry 28, 6361–6367.

Persson, A., Chang, D., Crouch, E., 1990. Surfactant protein D is a divalent cation-dependent carbohydrate-binding protein. J. Biol. Chem. 265, 5755–5760.

Postle, A.D., Mander, A., Reid, K.B., et al., 1999. Deficient hydrophilic lung surfactant proteins A and D with normal surfactant phospholipid molecular species in cystic fibrosis. Am. J. Respir. Cell Mol. Biol. 20, 90–98.

Ramanathan Jr, M., Lee, W.K., Spannhake, E.W., Lane, A.P., 2008. Th2 cytokines associated with chronic rhinosinusitis with polyps down-regulate the antimicrobial immune function of human sinonasal epithelial cells. Am. J. Rhinol. 22, 115–121.

Reid, K.B., 1998. Interactions of surfactant protein D with pathogens, allergens and phagocytes. Biochim. Biophys. Acta 1408, 290–295.

Rishi, A., Hatzis, D., McAlmon, K., Floros, J., 1992. An allelic variant of the 6 A gene for human surfactant protein A. Am. J. Physiol. 262, L566–L573.

Rubio, S., Lacaze-Masmonteil, T., Chailley-Heu, B., Kahn, A., Bourbon, J.R., Ducroc, R., 1995. Pulmonary surfactant protein A (SP-A) is expressed by epithelial cells of small and large intestine. J. Biol. Chem. 270, 12162–12169.

Sano, H., Chiba, H., Iwaki, D., Sohma, H., Voelker, D.R., Kuroki, Y., 2000. Surfactant proteins A and D bind CD14 by different mechanisms. J. Biol. Chem. 275, 22442–22451.

Sastry, K., Herman, G.A., Day, L., et al., 1989. The human mannose-binding protein gene. Exon structure reveals its evolutionary relationship to a human pulmonary surfactant gene and localization to chromosome 10. J. Exp. Med. 170, 1175–1189.

Schelenz, S., Malhotra, R., Sim, R.B., Holmskov, U., Bancroft, G.J., 1995. Binding of host collectins to the pathogenic yeast *Cryptococcus neoformans*: human surfactant protein D acts as an agglutinin for acapsular yeast cells. Infect. Immun. 63, 3360–3366.

Sheriff, S., Chang, C.Y., Ezekowitz, R.A., 1994. Human mannose-binding protein carbohydrate recognition domain trimerizes through a triple α-helical coiled-coil. Nat. Struct. Biol. 1, 789–794.

Shu, L.H., Xue, X.D., Shu, L.H., et al., 2007. Effect of dexamethasone on the content of pulmonary surfactant protein D in young rats with acute lung injury induced by lipopolysaccharide. Zhongguo Dang Dai Er Ke Za Zhi 9, 155–158.

Singh, M., Madan, T., Waters, P., Parida, S.K., Sarma, P.U., Kishore, U., 2003. Protective effects of a recombinant fragment of human surfactant protein D in a murine model of pulmonary hypersensitivity induced by dust mite allergens. Immunol. Lett. 86, 299–307.

Sorensen, G.L., Hjelmborg, J.B., Kyvik, K.O., et al., 2006. Genetic and environmental influences of surfactant protein D serum levels. Am. J. Physiol. Lung Cell Mol. Physiol. 290, L1010–L1017.

Stevens, P.A., Wissel, H., Sieger, D., Meienreis-Sudau, V., Rustow, B., 1995. Identification of a new surfactant protein A binding protein at the cell membrane of rat type II pneumocytes. Biochem. J. 308, 77–81.

Strong, P., Reid, K.B., Clark, H., 2002. Intranasal delivery of a truncated recombinant human SP-D is effective at down-regulating allergic hypersensitivity in mice sensitized to allergens of Aspergillus fumigatus. Clin. Exp. Immunol. 130, 19–24.

Strong, P., Townsend, P., Mackay, R., Reid, K.B., Clark, H.W., 2003. A recombinant fragment of human SP-D reduces allergic responses in mice sensitized to house dust mite allergens. Clin. Exp. Immunol. 134, 181–187.

Taylor, M.E., Brickell, P.M., Craig, R.K., Summerfield, J.A., 1989. Structure and evolutionary origin of the gene encoding a human serum mannose-binding protein. Biochem. J 262, 763–771.

Tino, M.J., Wright, J.R., 1999. Surfactant proteins A and D specifically stimulate directed actin-based responses in alveolar macrophages. Am. J. Physiol. 276, L164–L174.

Traub, W., Piez, K.A., 1971. The chemistry and structure of collagen. Adv. Protein Chem. 25, 243–352.

Van Iwaarden, J.F., Shimizu, H., Van Golde, P.H., Voelker, D.R., van Golde, L.M., 1992. Rat surfactant protein D enhances the production of oxygen radicals by rat alveolar macrophages. Biochem. J. 286, 5–8.

Vandivier, R.W., Fadok, V.A., Ogden, C.A., et al., 2002. Impaired clearance of apoptotic cells from cystic fibrosis airways. Chest 121, 89S.

Voss, T., Melchers, K., Scheirle, G., Schafer, K.P., 1991. Structural comparison of recombinant pulmonary surfactant protein SP-A derived from two human coding sequences: implications for the chain composition of natural human SP-A. Am. J. Respir. Cell Mol. Biol. 4, 88–94.

Weis, W.I., Drickamer, K., Hendrickson, W.A., 1992. Structure of a C-type mannose-binding protein complexed with an oligosaccharide. Nature 360, 127–134.

Whitchurch, C.B., Tolker-Nielsen, T., Ragas, P.C., Mattick, J.S., 2002. Extracellular DNA required for bacterial biofilm formation. Science 295, 1487.

White, R.A., Dowler, L.L., Adkison, L.R., Ezekowitz, R.A., Sastry, K.N., 1994. The murine mannose-binding protein genes (*Mbl 1* and *Mbl 2*) localize to chromosomes 14 and 19. Mamm. Genome 5, 807–809.

Wissel, H., Looman, A.C., Fritzsche, I., Rustow, B., Stevens, P.A., 1996. SP-A-binding protein BP55 is involved in surfactant endocytosis by type II pneumocytes. Am. J. Physiol. 271, L432–L440.

Yamada, C., Sano, H., Shimizu, T., et al., 2006. Surfactant protein A directly interacts with TLR4 and MD-2 and regulates inflammatory cellular response. Importance of supratrimeric oligomerization. J. Biol. Chem. 281, 21771–21780.

CHAPTER

36

T-Cell recognition of microbial lipoglycans and glycolipids

Gennaro De Libero

SUMMARY

The T-cells recognize lipid antigens as complexes with CD1 antigen-presenting molecules. Responding T-cells use the T-cell receptor (TCR) $\alpha\beta$- or $\gamma\delta$-chains and exert proinflammatory and protective functions during infections. Most of the so far identified lipid antigens are produced by *Mycobacterium tuberculosis*, although lipids from *Nocardia*, *Corynebacteria*, *Sphingomonas*, *Ehrlichia* and *Borrelia* spp. may also stimulate CD1-restricted T-cells. Lipid immunogenicity is dictated by their structural features which control lipid diffusion in body fluids, in membranes, intra-cellular vesicular trafficking, CD1 binding and establishment of stimulatory interactions with T-cells. All lipid antigens share amphipatic properties, with hydrophobic parts responsible for insertion into CD1 antigen-binding pockets and hydrophilic moieties that interact with the TCR. Differences in the number, length and saturation of alkyl chains, as well as the chemistry and chirality of the lipid head group, dominantly influence lipid antigenicity by affecting the mode of CD1 binding and interaction with the TCR. The nature of lipid structures also influences their processing, mechanism of CD1 loading and cellular compartmentalization. Herein, how microbial lipid antigens stimulate specific T-cells is reviewed, and their possible use as subunit vaccines is discussed.

Keywords: Lipid antigens; T-cells; Vaccination; T-cell receptor; *Mycobacterium tuberculosis*; Lipoglycans

1. INTRODUCTION

T- and B-lymphocytes recognize antigens with two different modes. The B-cells express on the plasma membrane surface immunoglobulins that bind the hydrophilic parts of self and foreign molecules, irrespective of their chemical nature and provided they are exposed and readily accessible. On the other hand, T-cells utilize membrane-bound receptors, called T-cell receptors (TCRs), which interact with complexes formed by antigen-presenting molecules and antigens. The T-cell antigens can be proteic or lipidic in nature. Foreign protein antigens, e.g. those of microbial origin, are internalized by antigen-presenting cells (APCs) and digested within late endosomes (LE), lysosomes (Ly) or in the cytoplasm into 8–12 amino acid-long peptides, which then become associated with major histocompatibility complex (MHC) molecules (Rock and Goldberg, 1999). Peptide trimming occurs at defined sites, according to the sequences recognized by proteases involved in protein antigen processing. Peptide–MHC complexes are formed within LE/Ly or in the endoplasmic reticulum and then are transported to the cell surface (Trombetta and Mellman, 2005). During this intracellular traffic, peptides remain tightly bound to antigen-presenting molecules and are not readily displaced by other peptides. This is an important biological feature that provides the antigenic identity of APCs, i.e. the APCs that experience antigens generated during infection present microbial antigens until the complexes become degraded. After a series of intracellular recycling rounds, the MHC–peptide complexes are destroyed and the antigen is no longer available to the immune system if the generating infectious agent is also no longer present. This mechanism allows a progressive reduction of antigenic complexes and a concomitant dampening of the specific immune response. The number of complexes and their half-lives contribute to the immunogenicity of any particular antigen, which is important to induce T-cell clonal expansion and generation of immunological memory for infectious agents.

Similar rules control immunogenicity of lipid antigens, which are recognized by T-cells expressing the TCR α- and β-chains or the γ- and δ-chains. The biochemical nature of this class of antigenic molecules imposes important differences with T-cell recognition of protein antigens. For example, lipid antigens form complexes with dedicated antigen-presenting molecules which are members of the CD1 family. Individual CD1 molecules recycle through different cellular compartments that are enriched in different types of lipids. Thus, this mechanism selects the type of lipid antigens that form immunogenic complexes. Mycobacterial lipids generated by intracellular *Mycobacterium tuberculosis* residing within the phagosomal compartment become preferentially associated with CD1 molecules which recycle within the same organelle. However, mycobacterial lipids released by infected cells can be internalized by other APCs forming complexes, also with CD1 molecules not present within phagosomes (Schaible *et al.*, 2003).

An important issue is that the hydrophobic nature of lipids imposes association with lipid-binding proteins (LTPs) in body fluids and also within cells during lipid internalization and intracellular traffic. This makes LTPs very important in lipid uptake, intracellular sorting and loading on CD1 molecules.

The structure of glycolipid antigens plays a fundamental role in inducing specific immunity by different means. The lipid structure dictates the mode of intracellular trafficking, the capacity to bind to CD1 molecules and the requirement for partial degradation (processing) before forming complexes with CD1 (De Libero and Mori, 2005). Therefore, not all microbial glycolipids induce specific immunity.

2. STRUCTURE AND CELL BIOLOGY OF CD1 ANTIGEN-PRESENTING MOLECULES

In humans, five genes encode CD1 molecules that are all heterodimers composed of a heavy chain non-covalently associated with β_2-microglobulin, thus resembling MHC class I molecules. According to their homology, CD1 molecules are divided into three groups, with CD1a, CD1b and CD1c belonging to Group 1, CD1d to Group 2 and CD1e to Group 3. CD1 heavy chains are made of α1- and α2-domains that, similar to MHC molecules, form two anti-parallel α-helices located on the top of a β-pleated sheet and contain the antigen-binding region. The α1- and α2-domains are connected with the α3-domain that has an immunoglobulin-like structure, is attached to the membrane and terminates with a short cytoplasmic tail. The antigen-binding part of CD1 molecules is made of hydrophobic channels which are interconnected and present a common entrance permitting loading of lipid antigens. The size and structure of these hydrophobic pockets is different in each CD1 molecule and allow them to bind distinct lipids (Moody *et al.*, 2005). The CD1a molecule shows two pockets, called A′ and F′ according to the nomenclature of peptide-binding pockets in MHC molecules. The A′ pocket is closed at one end and is fused with the F′ pocket on the other end. The F′ pocket, in contrast, is partially open to the surface of the CD1a molecule, thus permitting protrusion of long lipid tails and also facilitating rapid loading and exchange of lipid antigens.

The CD1b molecule is made of four hydrophobic pockets, A′ C′, F′ and T′. While the A′, C′ and F′ pockets are fused in a narrow pocket open to the cell surface, the T′ pocket is deeply buried in between the α1- and α2-CD1b domains and connects the A′ and F′ pockets. This particular structure permits binding of lipid antigens with very long acyl chains (up to 60 carbon atoms), such as the mycobacterial lipids containing mycolic acid.

CD1d, like CD1a, is also made of two hydrophobic channels which are fused in a narrow entrance. Both pockets are closed at one end, which limits the lipid tails to lengths of 18–24 carbons. Although no crystal structure has been solved for CD1c and CD1e molecules, molecular modelling also predicts the presence of hydrophobic pockets in these two CD1 isoforms. In all CD1 molecules, amino acids in the α-helices also interact with the hydrophilic part of the antigen. These hydrogen bonds have the important function in positioning the sugar part of glycolipids and placing them in the right orientation for recognition by the TCR.

Recently, the first structure of the TCR–CD1d-lipid antigen ternary complex was solved (Borg *et al.*, 2007). As expected, the TCR makes cognate interaction with both the sugar part of the glycolipid and amino acids from the α1- and α2-helices of CD1d. In contrast to the foot print of TCR on the MHC-peptide complexes, the footprint of the TCR on the CD1d-α-galactosylceramide complex is parallel to the long axis of the CD1d α-helices. The TCR α-chain interacts laterally with the galactose (Gal) of the glycolipid antigen and also with CD1d, whereas the TCR β-chain only interacts with CD1d in a lateral position above the F′ pocket. This type of interaction, if conserved in other TCR–CD1-lipid complexes, may permit the presence of large sugar heads on glycolipid antigens, since they will not be positioned below but rather adjacent to the TCR.

An important difference between CD1 and MHC proteins is that allelic variation of CD1-encoding genes is extremely limited. Polymorphisms have been detected in all CD1 genes. However, while nucleotide substitutions in CD1B and CD1C genes are silent, substitutions in CD1A, CD1E and CD1D result in amino acid changes (Oteo *et al.*, 1999). With six alleles found so far, the gene encoding CD1E is the most polymorphic (Mirones *et al.*, 2000). None of these polymorphisms are located on the α-helices, which is an important difference with amino acid substitutions present in polymorphic MHC molecules that, in contrast, are most frequently localized around the antigen-binding groove. The polymorphisms of CD1 molecules may affect antigen presentation and can be associated with disease predisposition. The CD1E allele 4 with a proline in position-194 is very inefficient in exiting from the endoplasmic reticulum (ER) and in presenting the mycobacterial hexamannosylated phosphatidyl-*myo*-inositol (PIM_6) antigen (Tourne *et al.*, 2008). Genotype variants in exon 2 of the CD1A- and CD1E-encoding genes have been found associated with an increased susceptibility to Guillain-Barré syndrome (GBS) (Caporale *et al.*, 2006). However, this study was performed on a small number of donors and should be confirmed by investigating a larger cohort of GBS patients. Whether polymorphisms in CD1-encoding genes affect lipid-specific immune response is of main importance in the potential application of lipids as vaccines. Protection by a lipid antigen presented by non-polymorphic antigen-presenting molecules will probably be efficient in the entire population and will not show the caveats of protein-based vaccines, whose efficacy can be limited by MHC-endoing gene polymorphisms.

Other important aspects of CD1 biology are how CD1 proteins assemble, giving rise to mature forms, and how they recycle within the cell after maturation in the ER. Similar to MHC class I, CD1 molecules are translocated into the ER, where they are glycosylated, fold with the help of ER chaperones and associate with β_2-microglobulin. Studies performed with CD1d showed that it forms transient complexes with calnexin, calreticulin and the thiol oxidoreductase ERp57 (Sugita *et al.*, 1997; Huttinger *et al.*, 1999; Kang and Cresswell, 2002). In the ER, CD1d may associate with phosphatidylinositol (De Silva *et al.*, 2002) and CD1b with a 42–44 carbons long hydrophobic lipid, whose nature remains poorly understood (Garcia-Alles *et al.*, 2006). Binding of lipid spacers stabilizes nascent CD1 molecules which, however, have to be removed when microbial lipid antigens are loaded in LE/Ly. Upon maturation, CD1b and CD1d molecules rapidly appear on the cell membrane, probably along the secretory pathway (Jayawardena-Wolf *et al.*, 2001; Briken *et al.*, 2002) and then recycle within the cell where they finally are loaded with microbial lipids.

After surface appearance, CD1 molecules are internalized into different cellular compartments according to their cytoplasmic tail. The CD1b molecule is internalized through coated pits and reaches LE/Ly (Sugita *et al.*, 1996) where, in the presence of low pH, it is partially denatured and new lipid antigens can be loaded (Prigozy *et al.*, 1997). The cytoplasmic domain of CD1b contains a tyrosine- (Tyr-) based sorting motif that promotes binding of the adaptor proteins AP2 and AP3 (Briken *et al.*, 2002; Sugita *et al.*, 2002). In the absence of AP3, CD1b accumulates on the plasma membrane and recycles in early endosomes (EE) (Sugita *et al.*, 2002). In AP3-deficient cells, CD1b does not load or present mycobacterial lipids, further suggesting that recycling within LE/Ly is important for its immunological function.

In contrast with CD1b, CD1a is mostly found on the plasma membrane and recycles through EE (Sugita *et al.*, 1999). In human Langerhans cells, CD1a traffics with Rab 11 through recycling endosomes and in Birbeck granules (Salamero *et al.*, 2001). The CD1a molecule is internalized in a Rab22 and ARF6-dependent pathway and independently form clathrin and dynamin (Barral *et al.*, 2008) and, in immature dendritic cells (DCs), it co-precipitates with MHC class II, invariant chain (Ii), and CD9 molecules. The Ii only associates with CD1a in cells with lipid raft integrity and contributes to CD1a internalization and antigen presentation (Sloma *et al.*, 2008). Finally, surface expression of intact and functional CD1a is dependent on the presence of lipids in the extracellular microenvironment (Manolova *et al.*, 2006), probably as a result of the low stability of the CD1a/β_2-microglobulin heterodimer (Kefford *et al.*, 1984). It remains to be elucidated whether the Ii only associates with newly synthesized CD1a molecules and is important in protecting CD1a groove from binding lipid antigens in cellular compartments where foreign microbial lipids are not present.

The CD1c molecule recycles through both early recycling endosomal compartments and LE/Ly and thus broadly surveys intracellular compartments (Sugita *et al.*, 2000).

Human CD1d also recycles in LE/Ly, although it does not reach the lysosomal compartments with the same efficiency as CD1b (Sugita *et al.*, 2002). Apparently, this is due to the inability of the CD1d cytoplasmic tail to associate with the adaptor protein AP3 and this may explain why this CD1 molecule is not frequently used to present microbial lipid antigens. In mice, CD1d associates with AP3 and also with the Ii (Jayawardena-Wolf *et al.*, 2001). Association with endogenous lipids, stimulating the natural killer T-cells which are CD1d-restricted and use a semi-invariant TCR (termed iNKT-cells), occurs in the lysosomal compartment (Jayawardena-Wolf *et al.*,

2001) and requires the presence of two proteases, cathepsin S and cathepsin L (Riese *et al.*, 2001; Honey *et al.*, 2002), which probably are involved in maturation of resident LTPs, e.g. saposins (Zhou *et al.*, 2004a).

In conclusion, the intracellular trafficking of CD1 molecules is central to their ability to intersect and bind self and microbial antigens that have entered cells. How this affects priming and expansion of T-cells with protective functions during infections remains to be investigated.

3. STRUCTURE OF GLYCOLIPID ANTIGENS

The molecules CDla, CDlb, CDlc and CDld present foreign lipid antigens derived from mycobacteria, Gram-negative bacteria and also protozoan species to T-cells. These antigens are extremely diverse chemically and include naturally occurring lipopeptide, glycolipids and phospholipid structures that are distinct from mammalian lipids. The best studied and most frequent lipid antigens of microbial origin are derived from *M. tuberculosis*. These lipids are normal constituents of the bacterial cell envelope and exert important physiological functions for the microbe, such as formation of the cell membrane and of the protective cell wall and are also metabolic intermediates involved in iron-scavenging pathways. Some antigens are shed by bacteria, whereas others are instead attached to the microbial cell and require partial degradation by host cells in order to become soluble and immunogenic.

Mycolic acids were the first lipids stimulating T-cells to be identified (Beckman *et al.*, 1994). They are key components of the cell wall of mycobacteria, corynebacteria and *Nocardia* spp. Mycolic acids are high-molecular-mass, α-alkyl, β-hydroxy fatty acids, present as such or as bound esters of arabinogalactan and trehalose (Figure 36.1). The hydroxyl and carboxyl groups are important for recognition by T-cells (Beckman *et al.*, 1994). Mycobacterial mycolic acids are longer than those of other genera, have the largest side chains and the mero chain contains some decorations, such as one or two unsaturations, which can be double bonds of cyclopropane rings. Mycolic acids also contain oxygen functions additional to the β-hydroxy group and have methyl branches in the main backbone. These structural characteristics give

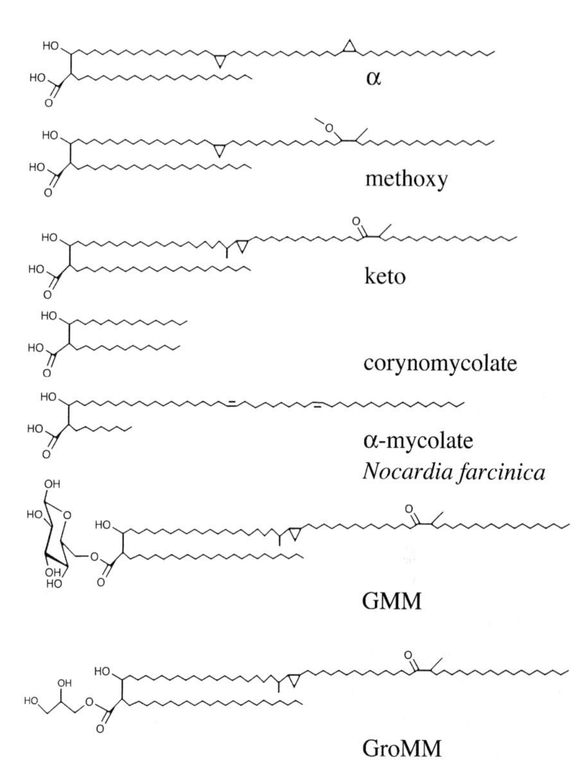

FIGURE 36.1 Structure of mycolic acids (MAs) and derived glycolipids. The MAs are produced by actinomycetes bacteria and are presented by CD1b. The antigens shown share the characteristic α-branched, β-hydroxy structure of mycolic acid, but differ in chain length and headgroup derivatization. Mycobacterial MAs have longer hydrocarbon chains than those produced by corynebacteria (corynomycolate) or *Nocardia* spp. (α-mycolate). Mycobacterial MAs are separated into three classes (α, methoxy, keto) according to the presence of different functional groups on the meromycolate chain. Glucose- (GMM) and glycerol-monomycolates (GroMM) are esters of MAs.

mycolic acids unique functions, such as the capacity to induce differentiation of macrophages into foamy cells and exhibiting increased proinflammatory functions (Korf *et al.*, 2005).

Glucose monomycolate (GMM) was the second immunogenic lipid to be discovered (Moody *et al.*, 1997). It is composed of a single glucose (Glc) residue that is attached to mycolic acid (see Figure 36.1). The structure of the Glc moiety is critical to GMM recognition, since the stereoisomers mannose- (Man-) or Gal-monomycolate are not or weakly recognized. Recognition is also specific for the natural glycosidic bond, which involves esterification of the 6-hydroxyl to mycolic acid, since glucose-3-monomycolate is not stimulatory (Moody *et al.*, 2000a).

Another antigen with modified mycolic acid is glycerol monomycolate (GroMM) (G. De Libero, unpublished data). This molecule is made of a glycerol moiety attached to a mycolic acid and is also generated by *Corynebacteria* and *Nocardia* spp. As with mycolic acids and with GroMM, the glycerol hydroxyl moiety and the fatty acid length influence T-cell response. Indeed, GroMM-specific T-cells cross-react with GMM, thus suggesting that the minimal glycerol structure substituting that of glucose is sufficient to establish productive interactions with the TCR. All these lipids stimulate CD1b-restricted T-cells, probably because of its unique structure and recycling properties. Therefore, mycolid acids constitute a scaffold that confers selective binding to this CD1 molecule.

Phosphatidylinositol-mannoside (PIM) and lipoarabinomannan (LAM) (see Chapters 9 and 21) stimulate CD1b-restricted T-cells (Sieling *et al.*, 1995). Both these molecules share a phosphatidyl-*myo*-inositol (PI) part that allows anchoring to the bacterial plasma membrane and to the mycoloyl layer and contain very different hydrophilic parts (Figure 36.2). The hexamannosylated form of PIM, PIM_6, is immunogenic after reduction to the di-mannosylated form (de la Salle *et al.*, 2005). Structurally, LAM is made by a core region, in which Man residues are linked to PI, and by a terminal part with many arabinoses that can be capped with terminal Man residues in both virulent and opportunistic mycobacterial strains (see Chapter 9). Lipomannan (LM), which is essentially similar to LAM but does not contain arabinose, can be also recognized by T-cells (Sieling *et al.*, 1995). Importantly, LAM is recognized by CD1b-restricted T-cells only after internalization and processing within LE (Ernst *et al.*, 1998), suggesting that the active antigen is smaller than the intact LAM. Also, T-cells may discriminate different types of lipoglycans. For example, some T-cells recognize LAM from *Mycobacterium leprae*, but not LAM from *M. tuberculosis*, which differ in the number of Man linkages and branching points (Sieling *et al.*, 1995).

Sulfoglycolipids (SGL) represent another class of CD1b-presented antigens (Gilleron *et al.*, 2004). SGL are present in the cell envelope and contain a trehalose-2′-sulfate core, acylated by two to four fatty acids (Figure 36.3). The abundance of SGL has been associated with strain virulence (Goren, 1982). This antigen was isolated from virulent mycobacteria strains and proved to a be a mixture of structurally related di-acylated SGL (2-palmitoyl-3-hydroxy-phthioceranoyl-2′-sulfate-α,α′-D-trehalose, Ac2SGL) containing hydroxyphtioceranoic and palmitic or stearic acid, both linked to the Glc without the sulfate in the trehalose. According to the structure of CD1b, it is probable that the long hydroxyphtioceranoic acid (C_{32}) is inserted into the A′T′F′ channel, whereas the shorter palmitic (C_{16}) or stearic (C_{18}) chains are inserted into the C′ pocket, with the sugar part protruding outside the binding groove. This model is supported by the finding that removal of the sulfate group completely abolishes T-cell activation, which is also indicative of a fine antigen specificity of the TCR for the strong negative charges of sulfate (Gilleron *et al.*, 2004).

A fourth group of mycobacterial lipids which stimulate specific T-cells comprises molecules closely related to mannosyl β-1-phosphomycoketides

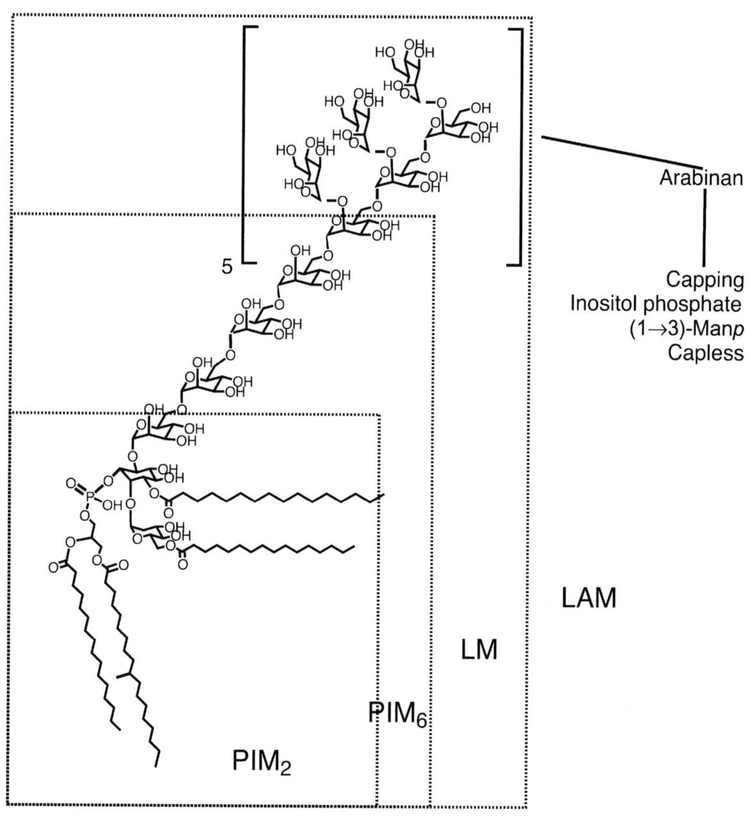

FIGURE 36.2 Structure of mycobacterial glycosylphosphatidylinositols presented by CD1b. Phosphatidylinositol-mannosides (PIM), lipomannan (LM) and lipoarabinomannan (LAM) all share a phosphatidylinositol core structure that anchors the lipids to the outer leaflet of mycobacterial plasma membrane. The PIMs may vary in the degree of headgroup mannosylation and number of attached acyl chains (from 1 to 4). The LM molecule is formed by addition of mannan repeats to a PIM structure, whereas LAM is made of arabinan linked to an LM structure, after which the headgroup can be capped with inositol phosphate, mannan or may remain unmodified.

(MPM) (Moody *et al.*, 2000b). These molecules contain a single fully saturated alkyl chain with methyl branches at every fourth carbon (Figure 36.4). These lipids are produced only by mycobacteria that infect human cells, e.g. *M. tuberculosis*, including the Bacille Calmette-Guérin (BCG) vaccine strain BCG and *Mycobacterium avium*, whereas rapidly growing saprophytes, such as *Mycobacterium phlei*, *Mycobacterium fallax* and *Mycobacterium smegmatis* do not produce it (Matsunaga *et al.*, 2004). Interestingly, although these lipids are similar to phosphodolichols that in mammalian cells function as carbohydrate donors in glycan synthesis, they are synthesized through an isoprenoid-independent pathway. Biosynthesis of MPM involves the enzyme Pks12 and sequential condensation of malonate and methylmalonate units. The role of these polyketides remains unknown and perhaps they are involved in bacterial intracellular growth and in Man transmembrane transport. Similarly to other glycolipid-specific T-cells, also MPM-specific T-cells discriminate changes in both the hydrophilic and hydrophobic

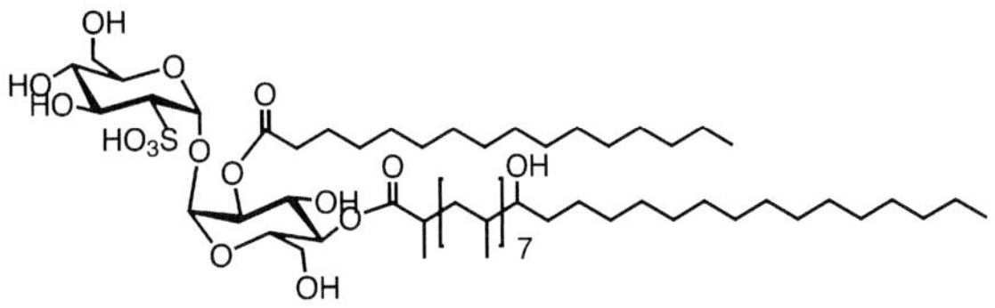

FIGURE 36.3 Structure of mycobacterial sulfoglycolipid presented by CD1b. The depicted antigen is 2-palmitoyl-3-hydroxy-phtioceranoyl-2′-sulfate-α,α′-D-trehalose (Ac2SGL) and is composed of a sulfate trehalose sugar esterified to two fatty acid chains. A stearic acid may substitute palmytic acid.

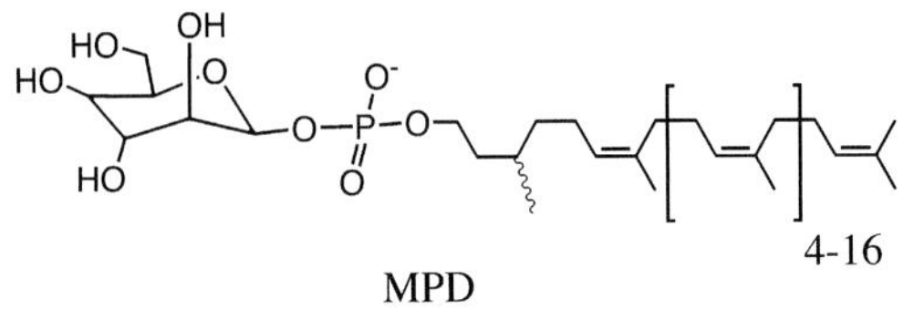

MPD

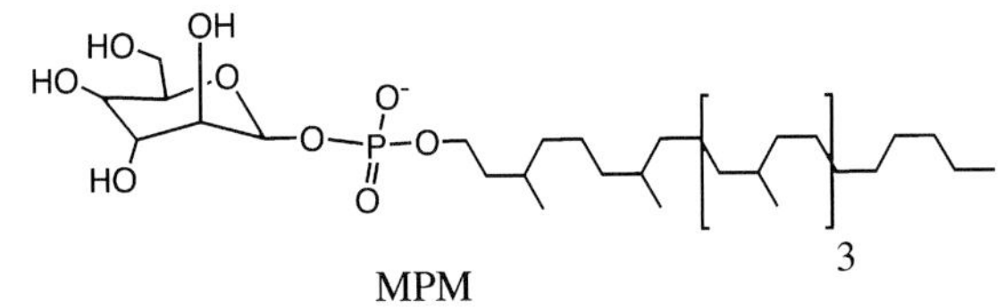

MPM

FIGURE 36.4 Structure of mycoketide antigens presented by CD1c. The upper structure is that of mannosyl-β-1-phosphodolichols (MPD), isoprenoid lipids present in many organisms including eukaryotic cells. Mammalian MPD are long (C95 lipid chain), whereas protozoan isoprenoiods are intermediate in length (C60). The lower structure is mannosyl β-1-phosphomycoketides (MPM) and is produced by *M. tuberculosis*. It has an identical headgroup to MPD but different saturated tail structures that are shorter than mammalian ones (C35). These lipids are synthesized by a polyketide mechanism not present in mammalian cells.

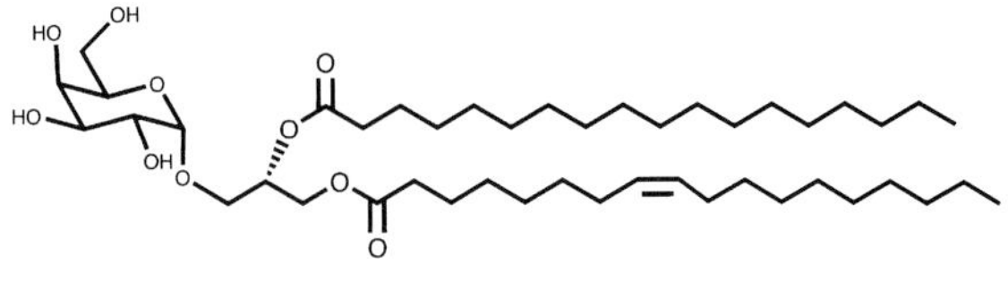

parts of the molecule. If Man is substituted with Glc, the glycolipid is no longer stimulatory. Synthetic analogues also showed that the saturated α-prenyl-like unit and the length of the prenyl chain are important for recognition (Moody *et al.*, 2000b; Matsunaga *et al.*, 2004). Moreover, variation in the number, length and saturation of alkyl chains, and the precise chemistry and chirality of the lipid head group,

α-Galactosyldiacylglycerol

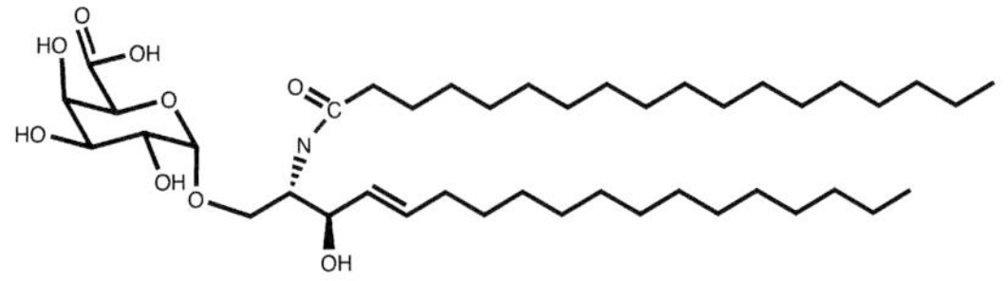

α-Galactouronosylceramide

FIGURE 36.5 Structure of glycosylated diacylglycerols and ceramides presented by CD1d. α-Galactosyldiacylglycerol (upper structure) is produced by *Borrelia burgdorferi* and has distinct fatty acids linked with ester bonds to a glycerol moiety. α-Glycosylated ceramides are present in *Sphingomonas* spp. and may present glucuronic or galacturonic sugars (in the lower structure) linked to ceramide. Both types of glycolipids are linked to the lipid part with an α-glycosidic bond, resembling the potent agonist α-galyctosylceramide found in the marine sponge *Agelas mauritianus*.

also exert dominant influences on antigenicity, probably by affecting the mode of CD1 binding and interaction with the TCR (de Jong *et al.*, 2007).

In addition to mycobacteria, *Sphingomonas* and *Borrelia* spp. also produce glycolipids that stimulate specific T-cells (Figure 36.5). Different strains of *Sphingomonas* produce α-galacturonosyl- and α-glucuronosyl-ceramide made by a ceramide with α-anomeric linkages to the carbohydrate moiety (Kinjo *et al.*, 2005; Mattner *et al.*, 2005; Sriram *et al.*, 2005). In comparison, *Borrelia* spp. produce diacylglcerols with linked Gal through an α-glycosidic bond (Kinjo *et al.*, 2006). These lipids have the capacity to stimulate CD1d-restricted T-cells that express markers of natural killer cells and a semi-invariant TCR, the so-called iNKT-cells (see above). The iNKT-cells represent an abundant population (0.1% of total T-cells) with important regulatory functions (Bendelac *et al.*, 2007). Both these bacteria

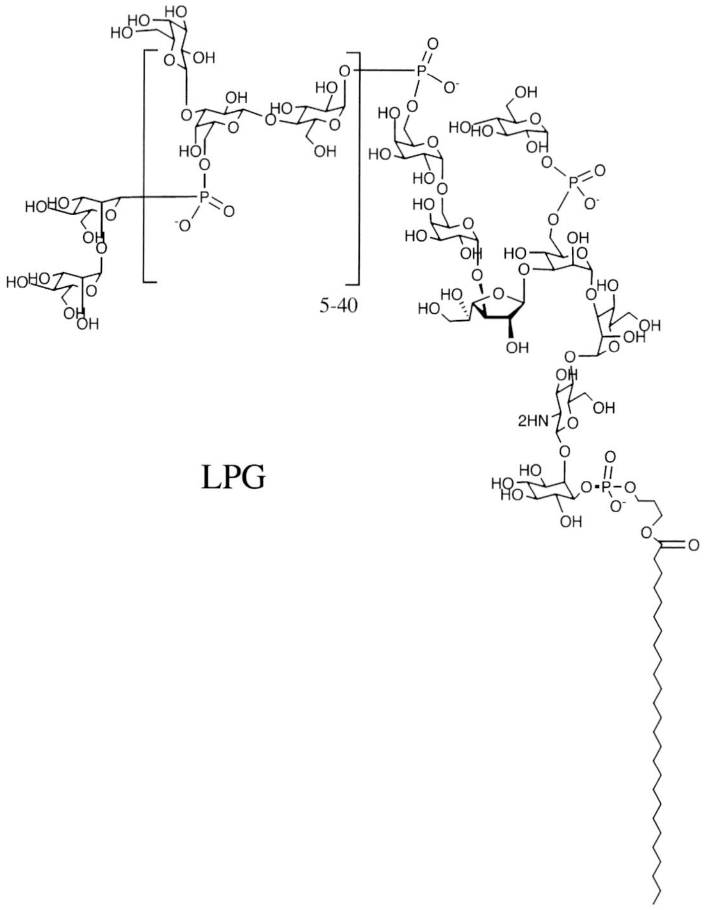

LPG

5-40

FIGURE 36.6 The structure of liposphosphoglycan (LPG) from *L. donovani* with two phosphate-galactose-mannose repeating units bracketed. This lipoglycan binds to CD1d and may stimulate iNKT-cells. It is unclear whether it is processed within APCs and forms immunogenic complexes that interact with the T-cell receptor (TCR) or instead it induces modifications in the lipid metabolism of antigen-representing cells (APCs) followed by synthesis of endogenous lipids that stimulate iNKT-cells.

do not synthesize lipopolysaccharide (LPS) and, thus, their unique glycolipids with α-anomeric bonds could represent an evolutionary selected mechanism to recognize these glycolipids as pathogen-associated molecular patterns similarly to Toll-like receptors.

A third type of bacterial lipids stimulating iNKT-cells is represented by lipophosphoglycan (LPG) (Figure 36.6). The LPG from the protozoan parasites *Leishmania donovani* (see Chapters 12 and 26) have been indicated as potential CD1d-restricted antigens (Amprey *et al.*, 2004). Although iNKT-deficient mice are more susceptible to *Leishmania* infection, suggesting that iNKT-cells have important functions in protection, it is still unknown how LPG stimulates these T-cells. LPG binds to CD1d, however, CD1d-LPG complexes do not stimulate, or stain, iNKT-cells. On the contrary, live DCs pulsed with LPG acquire the capacity to

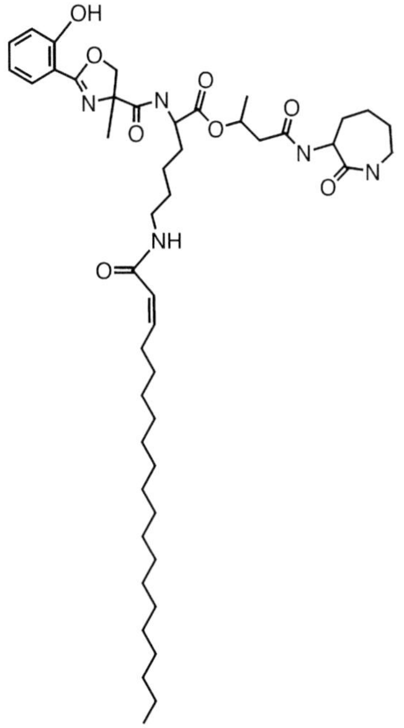

FIGURE 36.7 The structure of didehydroxymycobactin (DDM), a lipopeptide derived from *M. tuberculosis* and presented by CD1a. It consists of a peptide headgroup containing methylserine, lysine and cyclized lysine moieties connected by peptidic bonds. The DDM molecule is acylated by the central lysine residue to a C17 hydroxycarbon tail that permits binding to CD1a.

stimulate interferon-gamma (IFN-γ) release by iNKT-cells, leaving open the possibility that LPG stimulates the production of endogenous lipid ligands, which potently activate iNKT-cells. This is not a remote possibility, since bacterial infections may induce important changes in the lipid metabolism of infected cells. This, in turn, may induce accumulation of endogenous lipids that stimulate T-cells restricted by either Group 1 or Group 2 CD1 molecules (De Libero *et al.*, 2005; Paget *et al.*, 2007; Salio *et al.*, 2007).

The last class of lipid antigens is represented by lipopeptides. A unique lipopeptide, didehydroxymycobactin (DDM) (Figure 36.7) that is also produced by mycobacteria, stimulates CD1a-restricted T-cells (Moody *et al.*, 2004). The DDM molecule is the precursor of mycobactin, a siderophore with iron scavenging properties. The former is made of a complex head group synthesized by non-ribosomal mycobacterial enzymes, linked via acylation of a lysine to a single alkyl chain 20 carbons long. When a number of natural variants of DDM were tested, it was clear that the length and saturation of the alkyl chain are important, as well as the peptidic head group. The specificity of the CD1a-DDM interaction has been confirmed by the crystal structure of this complex (Zajonc *et al.*, 2005). Importantly, mycobacteria are induced to increase mycobactin synthesis during their intracellular growth. Therefore, recognition of this class of antigens may represent an efficient mechanism to identify cells harbouring living mycobacteria.

4. PRESENTATION OF LIPID ANTIGENS

Antigen presentation to T-cells can be dissected in a series of coordinated events leading to optimal formation of immunogenic complexes. This applies both to presentation of protein and lipid antigens. For clarity, four aspects are described here and their relevance in presentation of microbial lipids is discussed.

4.1. Antigen internalization and intracellular trafficking

The hydrophobic nature of lipid antigens makes it unlikely that they distribute within body fluids as free molecules. Lipids are associated with LTPs in serum such as low- and high-density lipoproteins, or are associated with individual proteins such as albumin, lipocalins, plasma phospholipids transfer protein or cholesteryl ester transfer protein. Lipids are also associated with cell debris or with intact cells, such as lipid constituents of bacterial cells. This has important consequences on lipid availability and their accessibility to APCs that express CD1 molecules. Moroever, LTPs bind a broad range of lipids without preferential binding for a unique lipid. However, despite lack of specificity they remain highly selective.

One example is that of ApoE, a lipoprotein that associates with high-density lipoproteins in serum. The ApoE lipoprotein may sequester glycolipid antigens and facilitates their transport and internalization in APCs expressing ApoE-specific receptors (van den Elzen *et al.*, 2005). This lipoprotein may also bind microbial lipids generated within infected cells and transport them in the extracellular milieu, thus facilitating their uptake by CD1-positive APCs. Lipid antigen transport between cells is instrumental for cross-priming of microbial lipid antigens (Schaible *et al.*, 2003). An important issue is that the lipoprotein-facilitated transport of lipids may result in both increased and decreased immunogenicity. This is the result of either lipoprotein selectivity for lipids or the type of cells expressing lipoprotein receptors. If lipoproteins are internalized by cells that do not express CD1 molecules, this becomes a cell sinking mechanism resulting in antigen sequestration (Schumann and De Libero, 2006).

When lipids are associated with microbial cells, the mechanism of their internalization is fully dependent on how microbes interact with host cells. A variety of receptors may be involved in microbial cell internalization and their intracellular trafficking may exert important effects on lipid immunogenicity (De Libero and Mori, 2005). If the receptor delivers microbial cells to LE/Ly, foreign lipids can be processed, become associated with resident LTPs, such as saposins (see paragraph below), and are loaded on CD1 molecules. If microbial lipids remain in other cellular compartments, such as cytoplasm and ER, the type of membranes with which they associate dictates their fate and immunogenicity.

Microbial lipids shed by bacterial cells may also become associated with the plasma membrane. In this case, they traffic to different organelles according to their structural characteristics. Lipid recycling is a highly regulated process and occurs through different routes. Sphingomyelin and sulfatide recycle through clathrin-coated vesicles, whereas lactosylceramide, globosides and gangliosides with clathrin-independent mechanisms. Although they are internalized by different routes, these glycolipids rapidly co-localize in an early sorting endosomal compartment (Sharma *et al.*, 2003).

The length and rigidity of lipids also control their capacity to diffuse into cellular organelles. Lipids with long and more rigid chains preferentially accumulate in LE, whereas those with short chains recycle through early recycling endosomes (Mukherjee *et al.*, 1999). Chain length, saturation and close packing of lipids increase membrane viscosity and reduce lipid movement. This prevents lipid distribution in recycling endosomes and facilitates lipid permanence into vesicles that mature into LE/Ly (Mukherjee and Maxfield, 2004). An example of how the structure of a microbial glycolipid affects its intracellular distribution and presentation by CD1b is that of mycobacterial glucosylmonomycolate (GMM) (Moody *et al.*, 2002). Synthetic GMM molecules with long acyl chains containing 80 carbons accumulate in LE, whereas a synthetic analogue with 32 carbons is three-fold less abundant in these compartments. This results in a different capacity of different GMM molecules to become loaded on CD1b molecules.

Another important aspect of lipid trafficking is where they localize in LE/Ly at the level of the multivesicular bodies (MVB) and whether they co-localize with CD1b and CD1d (De Libero and Mori, 2005). The co-localization of these CD1 molecules with microbial lipids in the MVB membranes might facilitate microbial lipoglycan processing and CD1 loading, similarly to peptide antigens that are preferentially loaded on MHC molecules with the same localization (Boes *et al.*, 2004).

4.2. Antigen processing

Large glycolipids cannot be accommodated between the TCR and CD1 molecules. This has been suggested by site-directed mutagenesis studies (Melian *et al.*, 2000) and recently confirmed by the first co-crystal structure of the CD1-glycolipid-TCR ternary complex (Borg *et al.*, 2007). Therefore, processing of glycolipids with large carbohydrate moieties is necessary to form immunogenic complexes. The importance of glycolipid processing has been shown in several studies. The hydrolase α-galactosidase generates a stimulatory glycolipid after digestion of the synthetic disaccharide α-Gal-(1→2)-Gal-ceramide (Prigozy *et al.*, 2001). Hexosaminidase B (HexB) (Zhou *et al.*, 2004b) and acidic mannosidase (de la Salle *et al.*, 2005), two other lysosomal hydrolases, also generate immunogenic glycolipids upon digestion of larger substrates.

Processing of lipid antigens might also involve partial degradation of the lipid moiety. Dendritic cells are able to cleave ester bonds present in the meromycolate chain of mycolic acid (Cheng *et al.*, 2006). However, this enzymatic activity has no functional consequences on T-cell recognition of this lipid antigen.

An important role in glycolipid antigen processing is exerted by CD1e (de la Salle *et al.*, 2005). The CD1e molecule is present in many different animal species and, like other CD1 molecules, is non-covalently associated with β_2-microglobulin. The CD1e-encoding gene was identified together with the other CD1-encoding genes (Calabi *et al.*, 1989), but the existence of the CD1e protein was confirmed much later (Angenieux *et al.*, 2000) and its function was recently identified (de la Salle *et al.*, 2005). After maturation in the ER, CD1e localizes in the Golgi compartment in immature DCs and, upon DC activation, rapidly traffics to LE where it is cleaved into a soluble form (Angenieux *et al.*, 2000, 2005).

Endosomal soluble CD1e facilitates processing of complex glycolipids, such as PIMs containing six Man residues (PIM_6), that are cleaved to a molecule containing only two Man residues (PIM_2) and stimulate specific T-cells. The CD1e molecule binds to PIM_6 and facilitates its cleavage by lysosomal mannosidase (de la Salle *et al.*, 2005). It is still unclear whether CD1e interacts with other CD1 molecules and also facilitates lipid loading.

An intriguing question is why the CD1 presentation system has evolved with one gene that never reaches the plasma membrane and participates in lipid presentation, notwithstanding the presence of other lysosomal LTPs. One hypothesis is that CD1e selects microbial lipids that, due to their intrinsic capacity to bind to CD1e, are more likely to become immunogenic and thus facilitates their processing and perhaps loading on other CD1 molecules. The CD1e molecule could also protect glycolipids from excessive degradation, thus participating in glycolipid editing.

4.3. Loading of CD1

The loading of CD1 with lipid antigens is facilitated by lysosomal resident LTPs, such as the saposin (SAP) family of compounds and GM_2-activator protein (GM2-A). Prosaposin, the precursor of SAPs, is first secreted and then internalized into LE/Ly, where it is cleaved into four <13 kDa mature glycoproteins, called SAP A, B, C, D, respectively, that form homodimers and have both overlapping and distinct functions (Sandhoff *et al.*, 2001). The GM2-A molecule is another ubiquitous LTP that directly traffics to lysosomes after synthesis. In general, LTPs are not redundant. For example, GM2-A is encoded by a SAP-unrelated gene, shares no primary

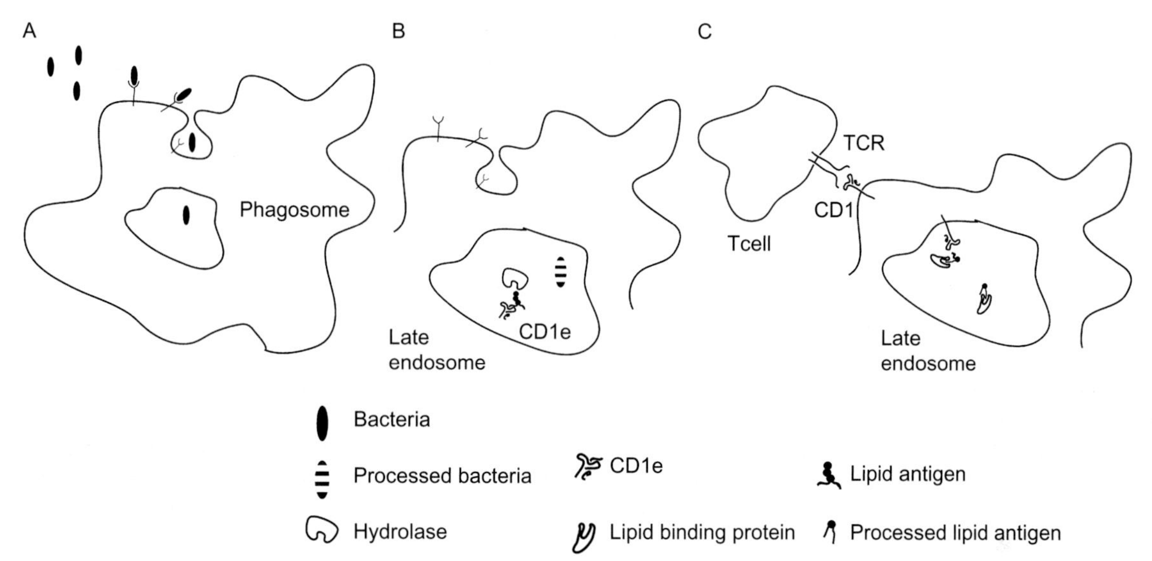

FIGURE 36.8 Mechanisms contributing to lipid immunogenicity. (A) Bacteria-associated lipid antigens are internalized during phagocytosis of bacterial cells. Cells exploit different surface receptors to internalize bacterial cells within phagosomes. (B) In acidified endosomes, complex bacterial lipids are processed, in some instances with the contribution of soluble CD1e, which offers glycolipid antigens to hydrolases. (C) Loading of CD1 molecules occurs within late endosomes and is facilitated by the presence of lipid-binding proteins (LTPs). Lipid-loaded CD1 molecules recycle to the plasma membrane and interact with the T-cell receptor (TCR).

structure homology with SAP A-D, is active as a monomer and its role cannot be replaced by SAP.

The LTPs extract glycolipids from the membrane, behaving as liftases, and offer them to hydrolases for degradation (Sandhoff and Kolter, 2003). Nevertheless, LTPs also participate in CD1 lipid loading as shown by *in vitro* CD1 loading assays (Zhou *et al.*, 2004a). Moreover, prosaposin-deficient mice lack CD1d-restricted T-cells and are deficient in lipid antigen presentation (Kang and Cresswell, 2004; Zhou *et al.*, 2004a). The SAP B molecule is the most important SAP facilitating loading of human CD1d (Yuan *et al.*, 2007), whereas SAP C is involved in loading of CD1b with mycobacterial lipids (Winau *et al.*, 2004).

Figure 36.8 illustrates the different steps of antigen internalization, processing, CD1 loading and presentation to T-cells, that contribute to lipid immunogenicity.

4.4. Fate and stability of CD1-glycolipid complexes

Once formed, the CD1-lipid complexes recycle on the plasma membrane where they may interact with the TCR of lipid-specific T-cells. Whether complexes formed within LE/Ly traffic through early recycling endosomes is not clear. It is important to understand whether these CD1 molecules have the possibility to intersect microbial lipids trafficking within other compartments. Once on the surface, CD1-lipid complexes may localize in discrete regions of the plasma membrane. For example, CD1a localizes in lipid rafts and disruption of rafts abolishes antigen presentation (Sloma *et al.*, 2008).

A prolonged half-life of complexes is instrumental to facilitate the interaction with specific TCR. In DCs, CD1a forms complexes surviving at least 3 days, whereas CD1b and CD1c form complexes with a half-life of 24 hours (Shamshiev *et al.*, 2002). These differences depend on the internalization of individual CD1 molecules and the short half-life is associated with recycling within lysosomes where low pH and proteases facilitate CD1 unloading and degradation.

5. PRIMING, EXPANSION AND GENERATION OF MEMORY GLYCOLIPID-SPECIFIC T-CELLS

Due to the lack of a suitable animal model expressing CD1a-1c-encoding genes, it is difficult to study the mechanisms that regulate priming, expansion and generation of lipid-specific memory T-cells. Priming probably occurs in peripheral lymphoid organs, like that of peptide-specific T-cells and DCs are the most likely candidate APCs since they express all CD1 molecules. However, it is not known whether lipid antigens are taken at peripheral sites, e.g. where infection occurs, and then are transported to draining lymph nodes.

T-cells that recognize microbial glycolipids are not detectable in donors who were never in contact with the specific pathogen. This was shown in donors that were not infected with *M. tuberculosis* and did not respond to the mycobacterial Ac2SGL antigen (Gilleron *et al.*, 2004). In guinea pigs (Hiromatsu *et al.*, 2002; Dascher *et al.*, 2003), lipid-specific T-cells are detected only after immunization with specific lipid. These findings confirm that lipid-specific T-cells undergo expansion after priming. However, it remains to be elucidated whether, like peptide-specific T-cells, lipid-specific T-cells show a contraction phase after expansion and generate long-term memory cells. The signals inducing memory cells remain to be investigated.

6. EFFECTOR FUNCTIONS OF GLYCOLIPID-SPECIFIC T-CELLS

Lipid-specific T-cells restricted by Group 1 CD1 molecules exert different functions, resembling peptide-specific T-cells. They release proinflammatory T-helper-1 cell (Th1) cytokines (Sieling *et al.*, 2000; Ulrichs *et al.*, 2003; Gilleron *et al.*, 2004) or T-helper-2 cell (Th2) cytokines (Bendelac *et al.*, 1996; Miyamoto *et al.*, 2001; Gumperz *et al.*, 2002; Akbari *et al.*, 2003; Fuss *et al.*, 2004). Moreover, they also efficiently kill infected cells and pathogens residing intracellularly (Gilleron *et al.*, 2004). Taken together, these findings strongly support the hypothesis that CD1-restricted T-cells may participate in protective immunity during infection. This is consistent with the finding that vaccination with mycobacterial lipids improves pathology in guinea pigs (Dascher *et al.*, 2003).

In accordance, CD1d-restricted T-cells recognizing microbial lipid antigens may also participate in protection. The protective role of iNKT-cells that recognize glycolipids produced by bacteria such as *Ehrlichia*, *Sphingomonas* and *Borrelia* spp. clearly support this conclusion. The iNKT-cells may also exert their effector function after recognition of self-antigens induced during infection and have a protective role through amplification of innate and adaptive responses. Nonetheless, due to the regulatory function of iNKT-cells, this cell population might also interfere with protection and have detrimental effects on infection (Kinjo and Kronenberg, 2005).

7. CONCLUSIONS

The T-cell immunity to microbial lipids is a key component of the antimicrobial response.

This seems to be very important for pathogens that have evolved intracellular parasitism and are shielded from humoral immunity. The response to *M. tuberculosis* lipids and the capacity of T-cells to distinguish between virulent and non-virulent mycobacterial strains may represent the rational basis for development of novel vaccine strategies. The use of lipids as subunit vaccines offers two important advantages. First, the complexity of lipid structures and their importance in microbial physiology prevent selection of microbial escape variants under immunological selective pressure. Second, the non-polymorphic nature of CD1 molecules may ensure efficacy of lipid-based vaccines in the entire population. To select the most appropriate strategies to translate these concepts into modern vaccinology, it will become important to improve our knowledge of lipid immunodominance, of kinetics of lipid-specific responses and of their correlation with protection during infection. Open questions in this field and where future studies are needed are detailed in the Research Focus Box.

RESEARCH FOCUS BOX

- Identification of non-mycobacterial lipid antigens remains an important issue. Most of the identified bacterial lipid antigens are produced by mycobacteria. Since other bacteria produce lipids which bind to CD1 molecules, it is possible that non-mycobacterial lipids are also immunogenic and stimulate specific T-cells. Future studies should clarify whether this is the case and whether lipid-specific T-cells also participate in the immune response during non-mycobacterial infections.
- A very important immunological question is whether lipid-specific memory T-cells are generated after priming with lipids. Indirect evidence suggests that priming with lipid antigens is necessary to expand specific T-cells. However, it remains unknown whether lipid-specific memory T-cells are generated and how long they persist in the body.
- Lipid-specific T-cells may exert a large variety of functions, thus resembling peptide-specific T-cells. It is important to investigate *in vivo* whether they also participate in protection during infections. These studies will be possible when *ad hoc* CD1 transgenic mice become available.
- The use of lipid antigens as vaccines is a new field which deserves proper experimental models. It remains to be investigated how lipids can be safely administered, how their solubility and availability can be preserved and which adjuvants facilitate priming of lipid-specific T-cells.
- The mechanisms of tolerance to self-lipids remain poorly characterized. Future studies are needed to identify the rules of thymic selection, which self-lipids induce negative selection and the mechanisms preventing autoreactivity in the periphery.

References

Akbari, O., Stock, P., Meyer, E., et al., 2003. Essential role of NKT cells producing IL-4 and IL-13 in the development of allergen-induced airway hyperreactivity. Nat. Med. 9, 582–588.

Amprey, J.L., Im, J.S., Turco, S.J., et al., 2004. A subset of liver NKT cells is activated during *Leishmania donovani* infection by CD1d-bound lipophosphoglycan. J. Exp. Med. 200, 895–904.

Angenieux, C., Salamero, J., Fricker, D., et al., 2000. Characterization of CD1e, a third type of CD1 molecule expressed in dendritic cells. J. Biol. Chem. 275, 37757–37764.

Angenieux, C., Fraisier, V., Maitre, B., et al., 2005. The cellular pathway of CD1e in immature and maturing dendritic cells. Traffic 6, 286–302.

Barral, D.C., Cavallari, M., McCormick, P.J., et al., 2008. CD1a, MHC class I and cargo internalized independently of clathrin follow a similar recycling pathway. Traffic 9, 1446–1457.

Beckman, E.M., Porcelli, S.A., Morita, C.T., Behar, S.M., Furlong, S.T., Brenner, M.B., 1994. Recognition of a lipid antigen by CD1-restricted $\alpha\beta^+$ T-cells. Nature 372, 691–694.

Bendelac, A., Hunziker, R.D., Lantz, O., 1996. Increased interleukin 4 and immunoglobulin E production in transgenic mice overexpressing NK1 T-cells. J. Exp. Med. 184, 1285–1293.

Bendelac, A., Savage, P.B., Teyton, L., 2007. The biology of NKT cells. Annu. Rev. Immunol. 25, 297–336.

Boes, M., Cuvillier, A., Ploegh, H., 2004. Membrane specializations and endosome maturation in dendritic cells and B cells. Trends Cell Biol. 14, 175–183.

Borg, N.A., Wun, K.S., Kjer-Nielsen, L., et al., 2007. CD1d-lipid-antigen recognition by the semi-invariant NKT T-cell receptor. Nature 448, 449.

Briken, V., Jackman, R.M., Dasgupta, S., Hoening, S., Porcelli, S.A., 2002. Intracellular trafficking pathway of newly synthesized CD1b molecules. EMBO J. 21, 825–834.

Calabi, F., Jarvis, J.M., Martin, L., Milstein, C., 1989. Two classes of CD1 genes. Eur. J. Immunol. 19, 285–292.

Caporale, C.M., Papola, F., Fioroni, M.A., et al., 2006. Susceptibility to Guillain-Barré syndrome is associated to polymorphisms of CD1 genes. J. Neuroimmunol. 177, 112–118.

Cheng, T.Y., Relloso, M., Van Rhijn, I., et al., 2006. Role of lipid trimming and CD1 groove size in cellular antigen presentation. EMBO J. 25, 2899–2989.

Dascher, C.C., Hiromatsu, K., Xiong, X., et al., 2003. Immunization with a mycobacterial lipid vaccine improves pulmonary pathology in the guinea pig model of tuberculosis. Int. Immunol. 15, 915–925.

de Jong, A., Arce, E.C., Cheng, T.Y., et al., 2007. CD1c presentation of synthetic glycolipid antigens with foreign alkyl branching motifs. Chem. Biol. 14, 1232–1242.

de la Salle, H., Mariotti, S., Angenieux, C., et al., 2005. Assistance of microbial glycolipid antigen processing by CD1e. Science 310, 1321–1324.

De Libero, G., Mori, L., 2005. Recognition of lipid antigens by T-cells. Nat. Rev. Immunol. 5, 485–496.

De Libero, G., Moran, A.P., Gober, H.-J., et al., 2005. Bacterial infections promote T-cell recognition of self-glycolipids. Immunity 22, 763–772.

De Silva, A.D., Park, J.J., Matsuki, N., et al., 2002. Lipid protein interactions: the assembly of CD1d1 with cellular phospholipids occurs in the endoplasmic reticulum. J. Immunol. 168, 723–733.

Ernst, W.A., Maher, J., Cho, S., et al., 1998. Molecular interaction of CD1b with lipoglycan antigens. Immunity 8, 331–340.

Fuss, I.J., Heller, F., Boirivant, M., et al., 2004. Nonclassical CD1d-restricted NKT cells that produce IL-13 characterize an atypical Th2 response in ulcerative colitis. J. Clin. Invest. 113, 1490–1497.

Garcia-Alles, L.F., Versluis, K., Maveyraud, L., et al., 2006. Endogenous phosphatidylcholine and a long spacer ligand stabilize the lipid-binding groove of CD1b. EMBO J. 25, 3684–3692.

Gilleron, M., Stenger, S., Mazorra, Z., et al., 2004. Diacylated sulfoglycolipids are novel mycobacterial antigens stimulating CD1-restricted T-cells during infection with *Mycobacterium tuberculosis*. J. Exp. Med. 199, 649–659.

Goren, M.B., 1982. Immunoreactive substances of mycobacteria. Am. Rev. Respir. Dis. 125, 50–69.

Gumperz, J.E., Miyake, S., Yamamura, T., Brenner, M.B., 2002. Functionally distinct subsets of CD1d-restricted natural killer T-cells revealed by CD1d tetramer staining. J. Exp. Med. 195, 625–636.

Hiromatsu, K., Dascher, C.C., LeClair, K.P., et al., 2002. Induction of CD1-restricted immune responses in guinea pigs by immunization with mycobacterial lipid antigens. J. Immunol. 169, 330–339.

Honey, K., Benlagha, K., Beers, C., et al., 2002. Thymocyte expression of cathepsin L is essential for NKT cell development. Nat. Immunol. 3, 1069–1074.

Huttinger, R., Staffler, G., Majdic, O., Stockinger, H., 1999. Analysis of the early biogenesis of CD1b: involvement of the chaperones calnexin and calreticulin, the proteasome and β_2-microglobulin. Int. Immunol. 11, 1615–1623.

Jayawardena-Wolf, J., Benlagha, K., Chiu, Y.H., Mehr, R., Bendelac, A., 2001. CD1d endosomal trafficking is independently regulated by an intrinsic CD1d-encoded tyrosine motif and by the invariant chain. Immunity 15, 897–908.

Kang, S.J., Cresswell, P., 2002. Calnexin, calreticulin, and ERp57 cooperate in disulfide bond formation in human CD1d heavy chain. J. Biol. Chem. 277, 44838–44844.

Kang, S.J., Cresswell, P., 2004. Saposins facilitate CD1d-restricted presentation of an exogenous lipid antigen to T-cells. Nat. Immunol. 5, 175–181.

Kefford, R.F., Calabi, F., Fearnley, I.M., Burrone, O.R., Milstein, C., 1984. Serum β2-microglobulin binds to a T-cell differentiation antigen and increases its expression. Nature 308, 641–642.

Kinjo, Y., Kronenberg, M., 2005. Vα14i NKT cells are innate lymphocytes that participate in the immune response to diverse microbes. J. Clin. Immunol. 25, 522–533.

Kinjo, Y., Wu, D., Kim, G., et al., 2005. Recognition of bacterial glycosphingolipids by natural killer T-cells. Nature 434, 520–525.

Kinjo, Y., Tupin, E., Wu, D., et al., 2006. Natural killer T-cells recognize diacylglycerol antigens from pathogenic bacteria. Nat. Immunol. 7, 978–986.

Korf, J., Stoltz, A., Verschoor, J., De Baetselier, P., Grooten, J., 2005. The *Mycobacterium tuberculosis* cell wall component mycolic acid elicits pathogen-associated host innate immune responses. Eur. J. Immunol. 35, 890–900.

Manolova, V., Kistowska, M., Paoletti, S., et al., 2006. Functional CD1a is stabilized by exogenous lipids. Eur. J. Immunol. 36, 1083–1092.

Matsunaga, I., Bhatt, A., Young, D.C., et al., 2004. *Mycobacterium tuberculosis* pks12 produces a novel polyketide presented by CD1c to T-cells. J. Exp. Med. 200, 1559–1569.

Mattner, J., DeBord, K.L., Ismail, N., et al., 2005. Both exogenous and endogenous glycolipid antigens activate NKT cells during microbial infections. Nature 434, 525–529.

Melian, A., Watts, G.F., Shamshiev, A., et al., 2000. Molecular recognition of human CD1b antigen complexes: evidence for a common pattern of interaction with αβ TCRs. J. Immunol. 165, 4494–4504.

Mirones, I., Oteo, M., Parra-Cuadrado, J.F., Martinez-Naves, E., 2000. Identification of two novel human CD1E alleles. Tissue Antigens 56, 159–161.

Miyamoto, K., Miyake, S., Yamamura, T., 2001. A synthetic glycolipid prevents autoimmune encephalomyelitis by inducing TH2 bias of natural killer T-cells. Nature 413, 531–534.

Moody, D.B., Reinhold, B.B., Guy, M.R, et al., 1997. Structural requirements for glycolipid antigen recognition by CD1b-restricted T-cells. Science 278, 283–286.

Moody, D.B., Guy, M.R., Grant, E., et al., 2000a. CD1b-mediated T-cell recognition of a glycolipid antigen generated from mycobacterial lipid and host carbohydrate during infection. J. Exp. Med. 192, 965–976.

Moody, D.B., Ulrichs, T., Muhlecker, W., et al., 2000b. CD1c-mediated T-cell recognition of isoprenoid glycolipids in *Mycobacterium tuberculosis* infection. Nature 404, 884–888.

Moody, D.B., Briken, V., Cheng, T.Y., et al., 2002. Lipid length controls antigen entry into endosomal and nonendosomal pathways for CD1b presentation. Nat. Immunol. 3, 435–442.

Moody, D.B., Young, D.C., Cheng, T.Y., et al., 2004. T cell activation by lipopeptide antigens. Science 303, 527–531.

Moody, D.B., Zajonc, D.M., Wilson, I.A., 2005. Anatomy of CD1-lipid-antigen complexes. Nat. Rev. Immunol. 5, 1–14.

Mukherjee, S., Maxfield, F.R., 2004. Lipid and cholesterol trafficking in NPC. Biochim. Biophys. Acta 1685, 28–37.

Mukherjee, S., Soe, T.T., Maxfield, F.R., 1999. Endocytic sorting of lipid analogues differing solely in the chemistry of their hydrophobic tails. J. Cell. Biol. 144, 1271–1284.

Oteo, M., Parra, J.F., Mirones, I., Gimenez, L.I., Setien, F., Martinez-Naves, E., 1999. Single strand conformational polymorphism analysis of human CD1 genes in different ethnic groups. Tissue Antigens 53, 545–550.

Paget, C., Mallevaey, T., Speak, A.O., et al., 2007. Activation of invariant NKT cells by toll-like receptor 9-stimulated dendritic cells requires type I interferon and charged glycosphingolipids. Immunity 27, 597–609.

Prigozy, T.I., Sieling, P.A., Clemens, D., et al., 1997. The mannose receptor delivers lipoglycan antigens to endosomes for presentation to T-cells by CD1b molecules. Immunity 6, 187–197.

Prigozy, T.I., Naidenko, O., Qasba, P., et al., 2001. Glycolipid antigen processing for presentation by CD1d molecules. Science 291, 664–667.

Riese, R.J., Shi, G.P., Villadangos, J., et al., 2001. Regulation of CD1 function and NK1.1$^+$ T-cell selection and maturation by cathepsin S. Immunity 15, 909–919.

Rock, K.L., Goldberg, AL., 1999. Degradation of cell proteins and the generation of MHC class I-presented peptides. Annu. Rev. Immunol. 17, 739–779.

Salamero, J., Bausinger, H., Mommaas, A.M., et al., 2001. CD1a molecules traffic through the early recycling endosomal pathway in human Langerhans cells. J. Invest. Dermatol. 116, 401–408.

Salio, M., Speak, A.O., Shepherd, D., et al., 2007. Modulation of human natural killer T-cell ligands on TLR-mediated antigen-presenting cell activation. Proc. Natl. Acad. Sci. USA 104, 20490–20495.

Sandhoff, K., Kolter, T., 2003. Biosynthesis and degradation of mammalian glycosphingolipids. Philos. Trans. R. Soc. Lond., Sect. B, Biol. Sci. 358, 847–861.

Sandhoff, K., Kolter, T., Harzar, K., 2001. Sphingolipid Activator Proteins. McGraw-Hill, New York.

Schaible, U.E., Winau, F., Sieling, P.A., et al., 2003. Apoptosis facilitates antigen presentation to T lymphocytes through MHC-I and CD1 in tuberculosis. Nat. Med. 9, 1039–1046.

Schumann, J., De Libero, G., 2006. Serum lipoproteins: Trojan horses of the immune response? Trends Immunol. 27, 57–59.

Shamshiev, A., Gober, H.J., Donda, A., Mazorra, Z., Mori, L., De Libero, G., 2002. Presentation of the same glycolipid by different CD1 molecules. J. Exp. Med. 195, 1013–1021.

Sharma, D.K., Choudhury, A., Singh, R.D., Wheatley, C.L., Marks, D.L., Pagano, R.E., 2003. Glycosphingolipids internalized via caveolar-related endocytosis rapidly merge with the clathrin pathway in early endosomes and form microdomains for recycling. J. Biol. Chem. 278, 7564–7572.

Sieling, P.A., Chatterjee, D., Porcelli, S.A., et al., 1995. CD1-restricted T-cell recognition of microbial lipoglycan antigens. Science 269, 227–230.

Sieling, P.A., Ochoa, M.T., Jullien, D., et al., 2000. Evidence for human CD4+ T-cells in the CD1-restricted repertoire: derivation of mycobacteria-reactive T-cells from leprosy lesions. J. Immunol. 164, 4790–4796.

Sloma, I., Zilber, M.T., Vasselon, T., et al., 2008. Regulation of CD1a surface expression and antigen presentation by invariant chain and lipid rafts. J. Immunol. 180, 980–987.

Sriram, V., Du, W., Gervay-Hague, J., Brutkiewicz, R.R., 2005. Cell wall glycosphingolipids of Sphingomonas paucimobilis are CD1d-specific ligands for NKT cells. Eur. J. Immunol. 35, 1692–1701.

Sugita, M., Jackman, R.M., van Donselaar, E., et al., 1996. Cytoplasmic tail-dependent localization of CD1b antigen-presenting molecules to MIICs. Science 273, 349–352.

Sugita, M., Porcelli, S.A., Brenner, M.B., 1997. Assembly and retention of CD1b heavy chains in the endoplasmic reticulum. J. Immunol. 159, 2358–2365.

Sugita, M., Grant, E.P., van Donselaar, E., et al., 1999. Separate pathways for antigen presentation by CD1 molecules. Immunity 11, 743–752.

Sugita, M., van Der Wel, N., Rogers, R.A., Peters, P.J., Brenner, M.B., 2000. CD1c molecules broadly survey the endocytic system. Proc. Natl. Acad. Sci. USA 97, 8445–8450.

Sugita, M., Cao, X., Watts, G.F., Rogers, R.A., Bonifacino, J.S., Brenner, M.B., 2002. Failure of trafficking and antigen presentation by CD1 in AP-3-deficient cells. Immunity 16, 697–706.

Tourne, S., Maitre, B., Collmann, A., et al., 2008. Cutting edge: a naturally occurring mutation in CD1e impairs lipid antigen presentation. J. Immunol. 180, 3642–3646.

Trombetta, E.S., Mellman, I., 2005. Cell biology of antigen processing *in vitro* and *in vivo*. Annu. Rev. Immunol. 23, 975–1028.

Ulrichs, T., Moody, D.B., Grant, E., Kaufmann, S.H., Porcelli, S.A., 2003. T-cell responses to CD1-presented lipid antigens in humans with *Mycobacterium tuberculosis* infection. Infect. Immun. 71, 3076–3087.

van den Elzen, P., Garg, S., Leon, L., et al., 2005. Apolipoprotein-mediated pathways of lipid antigen presentation. Nature 437, 906–910.

Winau, F., Schwierzeck, V., Hurwitz, R., et al., 2004. Saposin C is required for lipid presentation by human CD1b. Nat. Immunol. 5, 169–174.

Yuan, W., Qi, X., Tsang, P., et al., 2007. Saposin B is the dominant saposin that facilitates lipid binding to human CD1d molecules. Proc. Natl. Acad. Sci. USA 104, 5551–5556.

Zajonc, D.M., Crispin, M.D., Bowden, T.A., et al., 2005. Molecular mechanism of lipopeptide presentation by CD1a. Immunity 22, 209–219.

Zhou, D., Cantu, C., Sagiv, Y., et al., 2004a. Editing of CD1d-bound lipid antigens by endosomal lipid transfer proteins. Science 303, 523–527.

Zhou, D., Mattner, J., Cantu, C., et al., 2004b. Lysosomal glycosphingolipid recognition by NKT cells. Science 306, 1786–1789.

PART IV

BIOLOGICAL RELEVANCE OF MICROBIAL GLYCOSYLATED COMPONENTS

A. Environmental relevance

CHAPTER

37

Extracellular polymeric substances in microbial biofilms

Thomas R. Neu and John R. Lawrence

SUMMARY

In Nature, many, if not most, microorganisms grow as either suspended aggregates or films associated with interfaces. Both forms may be categorized as biofilm systems. In fact, the biofilm mode of growth represents the natural, microbial way of life for a vast majority of bacteria. Apart from cellular constituents, a second major component of biofilm systems is represented by the polymer matrix. The extracellular polymeric substances (EPS) of the biofilm matrix appear to have several key properties and functions. Consequently, in this chapter, the terms biofilm and EPS are briefly discussed. Subsequently, the nature of EPS, in general, and at interfaces is introduced. A survey of the main literature on the biofilm matrix is provided as well. The main portion of the chapter, however, is dedicated to the introduction of a novel concept on EPS functionality. The purpose of this concept is systematically to elaborate on EPS functions, to raise the discussion on biofilm EPS, to implement further studies using different biofilm systems and to trigger new approaches and hypotheses.

Keywords: Biofilms; Extracellular polymeric substances (EPS); Polysaccharides; Exopolysaccharides; EPS function; Biofilm organization; Pathogenesis

1. INTRODUCTION – BIOFILM SYSTEMS

In microbiology, during the last 30 years, a change in paradigms has been encountered. The two key issues were first, the pure culture approach and, secondly, the liquid culture approach. The change in paradigms was mainly triggered by microbial ecologists who realized that most environmental microbial communities are associated with interfaces. In addition, it is now accepted not only to work with pure cultures or defined mixed cultures, but also with multispecies cultures from the environment. In many research areas, microorganisms are now studied as complex three-dimensional communities, i.e. bioaggregates (mobile) and biofilms (non-mobile). At this point, some definitions of biofilm systems may be appropriate. Such a definition may be based on biofilm structure or on biofilm function. Structurally defined, biofilm systems are microorganisms and their extracellular polymeric substances associated with an interface (after T.R. Neu and J.R. Lawrence). In this definition, microorganisms are understood in the broadest sense including bacteria, Archaea, as well as eukaryotic microorganisms, i.e. protozoa, algae and fungi. The term "interfaces" does include any possible type: liquid–liquid, solid–liquid, gas–liquid or solid–gas. As a major component of biofilm systems, the extracellular polymeric substances produced by the members in a specific biofilm system are explicitly included in this definition. Functionally defined, biofilms are spatially structured communities of microbes whose function is dependent upon a complex web of symbiotic interactions (after S. Molin). Biofilms may be distinguished into those found in the natural environment, biofilms in technical systems which can be either very useful, e.g. in wastewater treatment or can be a nuisance and an economic or health problem, e.g. in case of biofouling and biodegradation. A third very important aspect are those biofilms directly related to the medical field, e.g. on animate surfaces or on invasive devices, implants and biomaterials. The now general acceptance that biofilm systems represent "the microbiological way of life" has led to an exponential growth in biofilm studies and biofilm-related publications.

TABLE 37.1 Use of the term EPS in microbiology

EPS translations/ suggestions	References
Exopolysaccharides	Sutherland, 1977a
Extracellular polymeric substances	Geesey, 1982
Extracellular polysaccharides	Sutherland, 1983
Exopolymers	No specific reference
Exopolymer secretions	Decho, 1990
Extracellular polymeric secretions	Kawaguchi and Decho, 2002
EPS as an acronym	Allison *et al.*, 2003
EPS versus xPS	Moran and Ljungh, 2003

2. EXTRACELLULAR POLYMERIC SUBSTANCES (EPS) IN BIOFILMS

The term EPS has been used in microbiology in a variety of ways. Possible translations and other acronyms are given in Table 37.1. It is obvious that the terms' meaning and use have changed over the years. Nevertheless, in the area of biofilm research, the term "EPS" is now used for extracellular polymeric substances representing an intrinsic constituent of biofilm systems. This may immediately require a further definition; EPS are organic polymers of microbiological origin which, in biofilm systems, are responsible for the interaction with interfaces, as well as with dissolved, colloidal and particulate compounds (after T.R. Neu and J.R. Lawrence). In this definition, organic polymers include polysaccharides (PSs), proteins, nucleic acids, amphiphilic polymers and bacterial refractory compounds. Microbiological origin should indicate that the polymers are, in fact, produced by the biofilm organisms. However, other secondary polymers of different origin, such as humic substances and colloids, may be embedded into the biofilm structure. It is important that possible functions of EPS are defined as generally as possible in order to include all known and not yet known roles of EPS in biofilm processes. Due to more recent research, these functions are becoming more and more diverse. Furthermore, new EPS compounds are being suggested which have to be integrated into the general biofilm–EPS concept (see section on EPS functionality).

At this point, some additional comments regarding other EPS abbreviations (see Table 37.1) have to be made. In order to keep the initial translation of EPS preserved for exopolysaccharides, the following acronyms were suggested for extracellular polymeric substances: EXPS (Allison *et al.*, 2003) or xPS (Moran and Ljungh,

2003). But, for example, XPS stands also for X-ray photoelectron spectroscopy, a technique which has been used, e.g. to analyse bacterial cell surfaces (Amory *et al.*, 1988). As a result, we are not in favour of these new acronyms as they will further complicate communication and may not allow progress in further understanding the role of EPS in biofilm systems. What we would suggest is to use EPS for "extracellular polymeric substances" if discussing specific aspects and to use the term "matrix" if the non-cellular structure of the biofilm is addressed.

3. NATURE AND APPEARANCE OF MICROBIAL EPS STRUCTURES

Microorganisms produce and express glycoconjugates with different appearances in suspended bioaggregates and in immobilized biofilms. Glycoconjugates or PSs represent a structural part of the cell surface, either as an integral part of the membrane, e.g. as lipopolysaccharide or lipoteichoic acid. However, PSs are also released from the membrane by cleavage of the lipid anchor. In this instance, the PSs build a more loosely associated layer around the cell which, eventually, will be excreted into the culture medium. With respect to interfaces, the microorganisms may produce specific adhesive PSs which can be excreted across the total cell area or at the cell pole. These PSs are produced as a response to the interface for the purpose of attachment only. This particular case is part of the life cycle of many microorganisms which are able to switch from planktonic to the adhered state and vice versa (Marshall, 1992). After artificially shearing microorganisms off a surface, these types of polymers are left on the surface as so-called microbial "footprints" (Neu and Marshall, 1991; Neu, 1992a). Other types of glycoconjugates can be produced by microorganisms, after they have attached to a surface, when they start growing and develop into microcolonies. These glycoconjugates may be produced in order to create a tailor-made microhabitat around the cells. Indeed, these microcolony-associated glycoconjugates have been found to be structured at different levels of complexity (Lawrence *et al.*, 2007). The various glycoconjugate regions can be detected by means of fluorescence lectin-binding analysis (FLBA). It allows the allocation of three different glycoconjugate signals, one to the bacterial cell surface, one to the space in between the microcolony cells and one to a shell around the microcolony. In mature biofilms, glycoconjugate clusters can be found which are not cell-associated. These cloud-like masses of glycoconjugates do not contain any bacterial cells. They may have their origin as released PSs, or they may represent the left-overs of microcolonies, which were partly abandoned by their producers in order to colonize new habitats. The latter has been shown by Webb *et al.* (2003) for *Pseudomonas* biofilms developed in flow cells. They reported that cell death and lysis occurred inside the microcolonies but, that at the same time, subpopulations of viable cells also appeared. This process facilitates dispersal of cells and, as a consequence, the bacteria leave empty cell-free colonies. It is obvious that this temporally and spatially controlled process must be linked to changes in the EPS of the microcolony. Unfortunately, this aspect was not addressed in this particular study.

Far less is known about the structural appearance of other EPS compounds. For example, extracellular proteins represent another major EPS class. Many of these proteins are likely enzymes. In contrast to our knowledge of the identity and activity of extracellular enzymes, hardly anything is known about their location within the biofilm or their behaviour at interfaces. Despite this lack of knowledge, the biofilm matrix may be regarded as a complex network consisting of PSs and associated extracellular enzymes (Hoffman and Decho, 1999). Recently, extracellular DNA (eDNA) was identified as a component of EPS, playing a structural role in developing biofilms. The potential roles of eDNA is an emerging research field which is discussed in detail below (see section on EPS functionality, informative EPS). Other EPS compounds are released in the form of lipid vesicles. Their presence opened up new issues which are still under investigation (see section on EPS functionality, constructive EPS). Finally, and surprisingly, amyloids were identified as common EPS compounds. Their presence in biofilm is of course important in relation to human health (see section on EPS functionality, constructive EPS).

4. THE BIOFILM MATRIX AND ITS LITERATURE RE-EXAMINED

Long before biofilms became a research topic or perhaps a research fashion, PSs were studied for a variety of different reasons. The previous research and literature published on microbial PSs may be divided into:

(i) PSs as structural bacterial cell constituents;
(ii) bacterial PS biosynthesis and chemistry;
(iii) bacterial PSs as immunogenic compounds in the medical field;
(iv) microbial PSs as products of potential industrial relevance; and
(v) microbiological and ecological roles of PSs.

In fact, the last point has driven the research on PSs towards the research on EPS. Historically, the publications on microbial PSs and polymers started with two comprehensive books (Sutherland, 1977b; Sandford and Laskin, 1977). The focus at that time was on the various PSs isolated from pure and liquid cultures, which were then characterized and analysed biochemically. After a long period of silence, the next overview was published in 1999 with focus on extracellular polymeric substances (Wingender *et al.*, 1999). This publication took the topic of EPS one step further and included PSs, as well as proteins, as EPS compounds. In addition, several book publications on microbial biofilms usually contain one chapter dedicated to EPS (e.g. Characklis and Marshall, 1990; Melo *et al.*, 1992; McLean and Decho, 2002; Lens *et al.*, 2003; Wuertz *et al.*, 2003; Ghannoum and O'Toole, 2004; Kjelleberg and Givskov, 2007). Apart from these books, and the chapters therein, several single outstanding reviews have tried to tackle the complex topic of microbial EPS. In Tables 37.2–37.4, an overview of the various chapters and reviews is supplied and sorted according to their focus.

In the following paragraphs, the main EPS reviews are briefly discussed to pave the ground for the subsequent discussion of a novel perspective on EPS functionality in biofilms. One of the first comprehensive reviews on microbial exopolymers was published by Geesey (1982). It was also the first paper using the acronym EPS for extracellular polymeric substances. Many, if not most, functions attributed to EPS go back to this manuscript. In summary, Geesey (1982) suggested the following roles for EPS:

(i) cell adhesion and anchoring;
(ii) stability of aggregates and films;
(iii) nutrient source;
(iv) conditioning of the microenvironment;
(v) matrix for extracellular enzymes;
(vi) adsorption of nutrients, particulates and detritus;
(vii) interaction with ions and trace metals;
(viii) protection against lytic agents (phages, *Bdellovibrio*);
(ix) re-distribution of nutrients;
(x) contribution to colloidal substances in aquatic habitats; and
(xi) interaction with pollutants (inorganic and organic).

Shortly after, biofilms and their EPS were discussed by Characklis and Cooksey (1983) with the focus on biofouling. They clearly indicated the need for further studies and listed several

TABLE 37.2 Reviews on extracellular polysaccharides in general and produced by specific microbial groups

Focus of article	References
Types of polysaccharides	Sandford, 1979
Components of polysaccharides	Kenne and Lindberg, 1983
Chemistry of bacterial polysaccharides	Lindberg, 1990
EPS of cyanobacteria	Bertocchi *et al.*, 1990; de Philippis and Vincenzini, 1998
Polysaccharides of Gram-negative bacteria	Whitfield and Valvano, 1993
Polysaccharides of lactic acid bacteria	Cerning, 1990; Vuyst and Degeest, 1999
Surface-active *Acinetobacter* polysaccharides	Rosenberg and Kaplan, 1987

TABLE 37.4 Reviews on the role of EPS in the aquatic environment

Focus of review	References
Polysaccharide function	Dudman, 1977
EPS in marine food webs and processes	Decho, 1990
EPS in stromatolites	Decho and Kawaguchi, 2003
Ecology of aquatic aggregates	Simon *et al.*, 2002
EPS in aquatic systems	Wotton, 2004
EPS in marine biogeochemical processes	Bhaskar and Bhosle, 2005
EPS of benthic diatoms	de Brouwer *et al.*, 2006
Dissolved organic matter and particulate organic matter in marine ecosystems	Azam and Malfatti, 2007

TABLE 37.3 Review articles on EPS in biofilms

Focus of review	References
First review on EPS	Geesey, 1982
EPS in biofilms and biofouling	Characklis and Cooksey, 1983
Role of EPS in biofilms	Christensen, 1989
EPS interactions at interfaces	Neu and Marshall, 1990
EPS as magic or mystic substance	Cooksey, 1992
Cohesiveness of EPS	Decho, 1994
Biofilm-specific EPS?	Sutherland, 1995
EPS properties	Sutherland, 1997
Function of EPS in biofilms	Wolfaardt *et al.*, 1999
EPS microdomains	Decho, 2000
EPS in films and flocs	Sutherland, 2001a
EPS dynamics	Sutherland, 2001b
Structure, ecology and function	Flemming and Wingender, 2001c
Structure, ecology and technical aspects	Flemming and Wingender, 2001a,b
Ecological and technical aspects	Flemming and Wingender, 2002
Role of EPS in biofilms	Flemming and Wingender, 2003
Physicochemistry of EPS	Moran and Ljungh, 2003
EPS matrix	Starkey *et al.*, 2004
EPS matrix in medical biofilms	Branda *et al.*, 2005
Biochemistry of EPS matrix	Pamp *et al.*, 2007

issues that are still valid today. For example, surface sensing by microbes, triggers and inhibition of EPS production, unequivocal analysis of EPS in biofilms, identification of physicochemical EPS properties, effects of biofouling through macrofouling, as well as interactions and cross-feeding in biofilms. They also challenged the focus of biochemists on *Escherichia coli* which, in the meantime has been extended to *P. aeruginosa*, as the standard biofilm bacterium. A critical review was published later by Cooksey (1992) discussing bacterial and microalgal EPS. He asked whether any EPS compound has been comprehensively and unequivocally analysed and related to biofilm processes. At the end of the chapter, he suggested that EPS may not only be a "magic" compound, as in Christensen and Characklis (1990), but also a "mystic" compound. An attempt to summarize the role of PSs as part of the EPS in biofilms was compiled and discussed with emphasis on biofilm development and detachment (Neu, 1992b). The aim of this review was an introduction of the research topic to the German-speaking scientific community. The molecular scale effects of EPS onto the macroscale were discussed by Decho (1994). He emphasized the significance of several parameters on the macroscale cohesiveness of EPS; specifically, the tertiary state, interactions between polymer strands, adsorption of metals and the effect of enzymes. In a later review (Decho, 2000), he introduced the concept of microdomains in the EPS matrix. Microdomains have different physicochemical properties in terms of the cohesiveness, the chemical structure, the degree of hydrophobicity and the enzymatic makeup. The postulation of microdomains was later confirmed by direct visualization of different glycoconjugate clusters within microcolonies (Lawrence *et al.*, 2007).

A special tribute should be given to Ian W. Sutherland. He started the discussion with the question: "Biofilm-specific polysaccharides – do they exist?" (Sutherland, 1995). In this short paper, he made clear that the polymers may be, in some cases, identical but different in their physical properties. In other cases, it was found that the polymers were not needed for the attachment to surfaces but they were produced during biofilm development. Depending on the system and species being investigated, different observations can be expected. In a subsequent paper, he stressed the solubility of PSs in biofilms (Sutherland, 1997). Usually, PSs are considered as being highly water soluble which is true for many. But there are several PSs that are virtually insoluble in water. They form rigid and highly ordered gels and, if present in biofilms, they will effectively stabilize the structure. Some of the insoluble polymers have a hydrophobic character and exclude water from certain areas within the biofilm. The solubility of PSs is also controlled by other dissolved compounds, e.g. chelators, ions and enzymes. He concluded that knowledge of the role(s) of EPS is still rather limited. An overview of the nature of PSs in biofilms and flocs did compare their composition, synthesis and physical properties (Sutherland, 2001a). In the conclusions, he stated that for PSs:

(i) they are always present;
(ii) they form a major component;
(iii) they vary widely in composition, structure and properties;
(iv) it is impossible to generalize about their contribution to the structure;
(v) that few polymers are obtained from biofilms;
(vi) that biofilm and floc PSs are similar to the planktonic ones;
(vii) they are sometimes produced due to stress; and
(viii) that polysaccharases extensively degrade EPS.

Another paper emphasized the complexity and dynamic of the biofilm matrix (Sutherland, 2001b). Sutherland employed several examples in order to discuss the microheterogeneity and the resulting microenvironments of the biofilm matrix. In the conclusions, he pointed out that: "the heterogeneous structure of the matrix will finally contribute to its functionality".

In the first book on EPS, its function was comprehensively discussed in one of the chapters (Wolfaardt *et al.*, 1999). The authors distinguished several major roles:

(i) in cellular associations;
(ii) in nutrition;
(iii) in the interaction of microorganisms with their biological, physical and chemical environment; and
(iv) a role directed towards the macroenvironment.

A series of reviews was published by H.-C. Flemming. He discussed the sorption properties of biofilms in three reviews and the general role of EPS in biofilms in several manuscripts. Here, we briefly discuss each of the most recent reviews. The sorption properties of biofilms were discussed in terms of sorption and re-mobilization of dissolved compounds (Flemming and Leis, 2002). Clearly, the EPS represents the first contact when a dissolved compound interacts with a microbial community. Therefore, the physicochemical characteristics of the EPS will determine its interaction with solutes. Other sorption sites may be present at the bacterial cell surface. Although discussing many examples of sorption in films and aggregates, the paper concluded that the binding capacity as well as the re-mobilization of dissolved compounds in biofilms remains unclear. The role of EPS in biofilms was mostly discussed with focus on ecological and technical aspects (Flemming and Wingender, 2003). In the later paper, the crucial role of EPS was emphasized in terms of the mechanical stability of films and flocs as well as in microbial aggregation. In addition, the mechanical stability and the resulting consequences for biofilms and bioaggregates were addressed. Interestingly, the functions of EPS as listed are essentially those discussed by Geesey more than 20 years ago (Geesey, 1982). In a review on the physicochemical properties of EPS, the focus is placed on extracellular PSs and proteins (Moran and Ljungh, 2003). With respect to the role of EPS, the discussion distinguished aggregate formation and biofilm functioning. An interesting aspect is the reaction-sink property of EPS for compounds from the environment, as well as for compounds produced by the biofilm organisms. A review discussing the different types of EPS and the different functions takes another look at the biofilm matrix (Starkey *et al.*, 2004). First of all, exopolysaccharides are discussed in general, then the PS intercellular adhesin (PIA) is addressed and alginate as well. The protein section critically looks at the concept of EPS as an enzymatically reactive matrix and the potential function of biofilm-specific proteins. For the first time in a review, Starkey *et al.* (2004) explicitly mentioned and discussed nucleic acids as an EPS component, although much earlier evidence has been published (Murakawa, 1973a,b). The functions of EPS were pooled in: attachment, architecture, protection and pathogenesis. A sort of revision of the article by Sutherland (2001b) is given in the form of a short review by Branda *et al.* (2005) with focus on carbohydrate-rich polymers and proteins. The authors have clearly a medically-related focus and, finally, discuss the meeting of the microbiological and mammalian matrices. This view is extended in terms of the cell surface diversity and advantage microbes may have in different habitats. In a recent comprehensive EPS review by Pamp *et al.* (2007), emphasis is put on three major EPS compounds, PSs, proteins and extracellular DNA. The PS section discusses:

(i) cellulose as a product of various bacteria;
(ii) PIA and poly-*N*-acetylglucosamine (PNAG) of *Staphylococcus*;
(iii) pellicle formation (PEL) and the PS synthesis locus (PSL) of *Pseudomonas*; and
(iv) *Vibrio* PS (VPS).

In the protein section, a distinction is made between multimeric cell appendages and surface proteins. These protein components are discussed in terms of their cell-to-surface and cell-to-cell adhesion. Nevertheless, no clear evidence for their extracellular matrix characteristic is given. The section on extracellular DNA has a focus on *Pseudomonas* and *Streptococcus*, as most evidence so far originates from these organisms (see details below).

5. ISSUES CONCERNING EPS FUNCTION IN BIOFILM SYSTEMS

Before discussing EPS functionality, several issues require further elaboration.

5.1. External microhabitat of microbial communities

The separation of the cell interior from the external environment is achieved by an energized membrane. Several structural constituents are known which are anchored inside the cell membrane, as in Gram-positive bacteria, or the outer membrane, as in Gram-negative bacteria. Other constituents are located completely outside of the membrane. Some of these, including EPS, may be shed into the space near the cell. As a consequence, starting from the cell membrane, there might be a sort of dilution of cellular constituents. The question is at what distance is the microbial cell able to control its microhabitat. Is it just around the cell, within the microcolony or even further?

5.2. Microbial carbohydrate metabolism

The biochemistry of carbohydrate synthesis and metabolism is well known for the intracellular space. Many of the cycles and pathways have been studied in detail and are part of microbial physiology textbooks. Recent investigations have indicated mechanisms of transport and excretion of polymers through the membrane and cell wall (Whitfield, 2006; Whitfield and Naismith, 2008). However, not much is known about the fate of released EPS, their lifetime, turnover and possible recycling. The question posed is: for how long (time) or how far (distance) can the microorganisms control the polymers they have produced?

5.3. Characterization of EPS

The chemical analysis of EPS is straightforward if a purified compound can be isolated, e.g. from a pure culture. Nonetheless, in environmental communities, it is very difficult, if not impossible, to approach the EPS chemically. One problem is possible cell lysis and contamination of the polymeric fraction by internal polymers. The second problem is the huge chemical variety of potential polymers. Consequently, the chemical data published are often a bulk analysis in the form of percent content of protein, PS, nucleic acid, etc. (Nielsen and Jahn, 1999). Therefore, a different approach by means of *in situ* imaging of EPS has been suggested. If pure cultures are targeted, the use of fluorescently-labelled antibodies against EPS will allow visualization of EPS distribution. Examples of this immunofluorescence approach are presented in the molecular microbial ecology manual (Dazzo, 2004). The immunogenic approach is, however, not possible in environmental samples due to the large variety of polymers present. In that case, FLBA may be the best strategy to characterize different types of glycoconjugates (Neu and Lawrence, 1999). In addition, by using other probes and advanced fluorescence techniques, a variety of different EPS types can be imaged and quantified.

5.4. Reactive groups in EPS

In EPS, charged and hydrophobic groups may be present. The negatively charged groups found include not only carboxyl, phosphate, sulfate, but also aspartic and glutamic acids. Positively charged groups may be found in amino sugars. Hydrophobic moieties can be present as acetyl groups, fatty acids, alanine, leucine, glycine and deoxy sugars. Due to these reactive groups in EPS, the polymers are affected by ions, chelators and surface-active compounds. This aspect has hardly been addressed in EPS studies (Keiding *et al.*, 2001).

5.5. Nature of the matrix

The EPS found in microbial communities are controlled by intrinsic and extrinsic factors (Wimpenny, 2000). The microorganisms present in a specific habitat have a genetic potential to synthesize polymers, but they may show a phenotypic expression of this ability. This, in turn, may be controlled by environmental factors. As a result, the EPS in a microbial community will change in time and space. In other words, there might be an evolution of the same polymers with different substituents or polymers with a completely different structure and very different properties.

5.6. Physicochemistry of EPS

One of the unsolved questions is the macromolecular conformation of the EPS matrix. The matrix could be composed of straight strands of polymers, or the polymers could be stochastically entangled depending on their production and interaction with other polymers, or they could be organized with specific cross-links forming a structured network. This aspect, also known as molecular crowding, has been raised as an underappreciated topic for intracellular environments (Ellis, 2001). However, molecular crowding of biological macromolecules may represent another issue in EPS matrices. An interesting aspect is the role of molecular chaperons which control the folding of proteins. One may speculate whether EPS-specific chaperons exist that are in charge of the physicochemistry of the biofilm matrix.

5.7. Common biofilm polymers?

Studies with pure cultures have led to speculations that there may be general features of biofilm EPS (Lasa, 2007). Nevertheless, this is only true for a very narrow spectrum of broadly related bacterial species. For example, PSs that are found in biofilm communities may include β-(1→4)-linked glucose (cellulose), β-(1→6)-linked *N*-acetylglucosamine (also known as PIA) found in staphylococci, or the alginate of *Pseudomonas* and *Azotobacter*. Similarly, biofilm-associated proteins (BAP) are discussed as common features of diverse species in biofilm development. The critical question of how truly general these findings are remains to be assessed.

5.8. Control of EPS production

Recently, cyclic-di-guanosine monophosphate (cyclic-di-GMP) was suggested as a key control element in bacteria switching from the planktonic to the adhered state (Jenal, 2004; D'Argenio and Miller, 2004; Römling *et al.*, 2005). Cyclic-di-GMP regulates a number of processes which seem to be crucial for biofilm development and detachment. Moreover, cyclic-di-GMP acts as a secondary messenger molecule, which antagonistically controls the motility of planktonic cells, as well as adhesion and persistence within microbial communities. A high concentration of cyclic-di-GMP is associated with attachment to surfaces, exopolysaccharide and EPS production, biofilm formation and communication. In contrast, it will also attenuate motility and virulence (Jenal and Malone, 2006; Dow *et al.*, 2007).

5.9. *Pseudomonas* and alginate?

It seems to be a never-ending story; *Pseudomonas* as the biofilm microorganism and alginate as the example for EPS. Alginate is actually an extremely unusual PS consisting of negatively-charged carbohydrates only. This PS is by no means a general example for PSs produced by environmental microbial communities. It rather seems to be the exception. A rough calculation of the possible number of different EPS molecules has revealed a potential for more than nine million types of extracellular polymers (Staudt *et al.*, 2003). This calculation takes the estimated number of bacteria, including uncultured ones, and assumes that each strain is able to produce at least one extracellular PS and one extracellular protein. This might be a gross under-estimation as algae, fungi and protists are neglected in the calculation. Yet, it is an aspect of EPS variety which is largely neglected when EPS molecules are discussed.

6. EPS FUNCTIONALITY – A NOVEL PERSPECTIVE

6.1. Constructive EPS

Polymers represent a structural component of microbial cell surfaces. As a result, a variety of polymeric compounds are exposed at the cell surface and often these compounds are shed from the cell surface into the extracellular space. Thereby, structural bacterial cell compounds may become enriched and form what is known as EPS or the biofilm matrix. Due to the complexity of the microbial cell surface, a range of chemically different compounds contributes to this EPS feature. Without any doubt, one major group of EPS compounds is certainly represented by PSs. For example, a study on biofilm development with focus on EPS glycoconjugates aimed at finding a lectin which bound to all glycoconjugates (Staudt *et al.*, 2003). For this purpose, a screening of all commercially available lectins (≈70) was performed on a variety of biofilms. By examination of the nucleic acid stained bacteria in relation to the glycoconjugate signals, it was found that the lectin from *Aleuria aurantia* stained nearly all of the polymers in a particular type of biofilm. After examination of the confocal laser scanning microscopy data, this became clearly visible as very few bacteria "floating" freely in three-dimensional (3-D) space were visible (Staudt *et al.*, 2003, 2004). An example of such a lectin-stained biofilm is presented in Figure 37.1A. The lectin approach (FLBA) was also employed in a study on phototrophic biofilms. The useful panel of lectins stained different clusters of glycoconjugates:

(i) occurring throughout the biofilm;
(ii) associated specifically with cyanobacteria; and
(iii) associated with non-phototrophic bacteria (B. Zippel and T.R. Neu, unpublished results).

In any case, every specific type of biofilm has to be screened in order to find suitable lectins which will bind to the feature of interest.

Another group of constructive EPS comprise cell surface-associated proteins, e.g. amyloids. These proteins are characterized by filamentous protein structures with a cross-β structure (Gebbink *et al.*, 2005). They are known to coat the cell surfaces of many bacteria and fungi. Very recently, they have been identified as an abundant compound in biofilms as well as in the biofilm matrix (Larsen *et al.*, 2007). This finding is of significance, not only for the EPS discussion, but also in relation to mammalian health, as amyloids have been associated with degenerative diseases such as Alzheimer's or Creutzfeld-Jakob. Examination of different habitats indicated that 5–40% of all 4′,6-diamino-2-phenylindole-(DAPI-) stained bacteria produced amyloids. A closer look at different bacterial groups confirmed the wide distribution across many phyla. Interestingly, the amyloids are

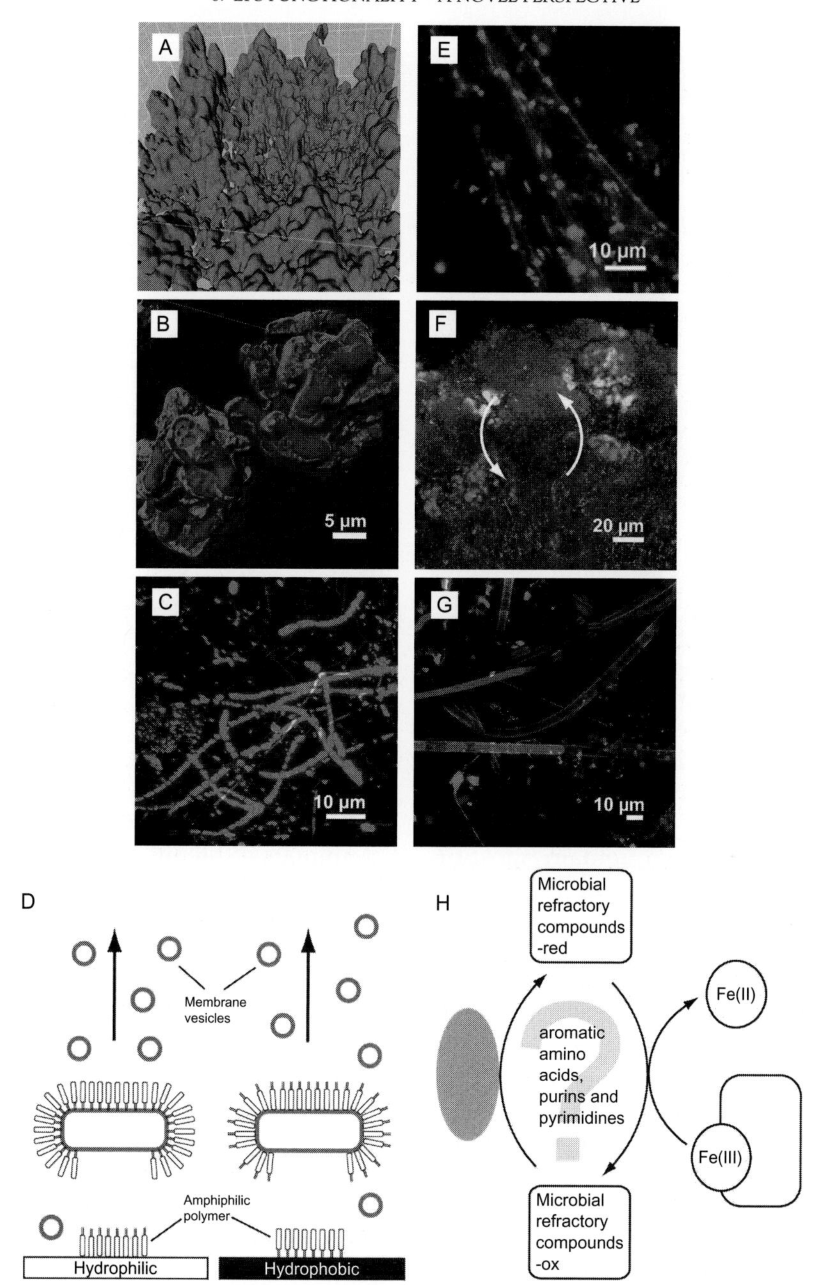

FIGURE 37.1 EPS functionality. (A) Constructive EPS; laser scanning microscopy (LSM) image of bulk EPS, with lectin-specific glycoconjugates, showing a wastewater biofilm. (B) Adsorptive EPS; LSM image of selective triple lectin-staining of a microcolony in a river biofilm. (C) Active EPS; LSM image of ELF enzyme activity (Paragas *et al.*, 2002) in a river biofilm. (D) Surface-active EPS; drawing of possible actions involving surface-active compounds. (E) Informative EPS; LSM image of F8 strain with eDNA filaments, whereby the sample was stained with a nucleic acid-specific stain (Syto9). (F) Nutritive EPS; LSM image of dense biofilm with nucleic acid stained cells (red) and glycoconjugates (green), the arrows indicate possible recycling. (G) Locomotive EPS; LSM image of gliding cyanobacteria with lectin-stained sheath. (H) Redox-active EPS; potential redox reactions of refractory microbial compounds. (See colour plates section.)

extremely stable and only specific proteases are able to degrade them. Nevertheless, the function of microbial amyloids in biofilms remains to be established (Larsen *et al.*, 2008).

6.2. Adsorptive EPS

EPS may carry charged or hydrophobic groups resulting in an adsorptive polymer. Examples for charged polymers are PSs containing uronic acids or carbohydrates with amino constituents. It was suggested that these types of PSs can act as an ion exchanger for dissolved compounds. There exists a substantial body of work indicating that biofilms and EPS bind various nutrients, metals and contaminants. In river biofilms, it was found that the EPS matrix served as a buffer against nutrient changes (Freeman and Lock, 1995). In the case of proteins, those containing a carboxyl or amino side group which are exposed to the exterior part of the protein may act as sorptive EPS. Hydrophobic constituents can be present in both types of EPS, PSs and proteins. In PSs, it is known that they often contain substituents such as acetyl or methyl groups (Lindberg, 1990). Some of the PSs even carry long chain fatty acids. These are either located at one end of the polymeric molecule, e.g. in the case of lipopolysaccharides and lipoteichoic acids or the fatty acids can be distributed across the entire polymer length, e.g. in emulsan (Belsky *et al.*, 1979). In addition, polysaccharides often contain one or more 6-deoxy-hexoses in their repeating unit, e.g. rhamnose or fucose, which possesses a hydrophobic group in the form of CH_3. This group may be minor, nevertheless, it will result in "hydrophobic" polysaccharides as it is accumulating across the entire polymer due to the repeating unit character of the polysaccharide. Proteins can also be involved in hydrophobic interactions as many amino acids carry a side group with hydrophobic character. All of these so-called adsorptive EPS can be involved in sorption, exchange and release processes of dissolved compounds. These environmental compounds can be organic, e.g. hydrocarbons, ionic, e.g. nutritive ions and metal ions, or reactive oxygen species. The concept of microdomains in the EPS of biofilms was originally suggested by Decho (2000). He discussed the physicochemical properties of these domains and addressed their hydrophobic and enzymatic character. An example of a structured microhabitat has been demonstrated recently using the FLBA approach for studying river biofilm microcolonies (Lawrence *et al.*, 2007). By using three different lectin specificities conjugated to three different fluorochromes, it could be shown that the various glycoconjugates produced are present at the bacterial cell surface, in between the cells, as well as around the microcolony (see Figure 37.1B).

6.3. Active EPS

Apart from PSs, enzymes are probably one of the major fractions of EPS. There is a substantial amount of literature on extracellular microbial enzymes, many of which were studied in microbial aggregates and biofilms. For an overview, see several chapters in Chrost (1991) and Wingender *et al.* (1999). Typically, extracellular enzymes are thought to be involved in degradation of organic polymers. It is out of the scope of this chapter to go into details of the various examples reported in the literature. In any case, this would be an advantage for the enzyme producers in order to have access to nutrients not readily available to non-enzyme producers. However, for bacteria, the expense of producing enzymes would require the anchoring of the extracellular protein in the nearby microhabitat in order to make sure that the substrate is available for the producing cell. To the best of our knowledge no such example has been reported. For some enzymes, their activity can be spatially located by using the ELF97 technique

(Paragas *et al.*, 2002). This substrate is ideal for laser scanning microscopy as it precipitates at the location of enzyme activity and, consequently, does not diffuse away. The technique was used to study enzyme activity in microbial aggregates and biofilms (Van Ommen Kloecke and Geesey, 1999; Baty *et al.*, 2001). An example of ELF97 alkaline phosphatase activity is presented in Figure 37.1C. Another function of extracellular enzymes may be in the control of other EPS constituents. It is largely unknown how much the active EPS components are involved in EPS turnover and recycling. Nevertheless, it is clear that EPS does accumulate to a certain level in the environment and remain "active" in a variety of biogeochemical processes in aquatic habitats (Wotton, 2004). Although not under the control of a specific organism these may act as abiontic enzymes and be an important resource for the biofilm community.

6.4. Surface-active EPS

Surface-active microbial compounds have been discussed as molecules which are able to condition interfaces (Neu, 1996). These low- or high-molecular mass compounds may condition either the microbial cell surface or the interface with which the bacteria interact. In addition, the production and release of such compounds may aid bacteria in detaching from interfaces. A schematic drawing of the possible roles is given (see Figure 37.1D). The already mentioned emulsan produced by *Acinetobacter* represents one type of these amphiphilic polymers (Rosenberg and Kaplan, 1987). Emulsan was found to be involved in the interaction of *Acinetobacter* with hydrocarbons. In the course of emulsan production, it first appears at the cell surface and later it is bound to the hydrocarbon droplets (Rosenberg *et al.*, 1989). Similar surface-active polymers are produced by a wide variety of microorganisms. It was suggested that these compounds may have a general role in desorption from hydrophobic interfaces (Rosenberg and Ron, 1997). The effect of low-molecular mass rhamnolipids on cell surface hydrophobicity and degradation of hydrocarbons has already been reported (Zhang and Miller, 1994). Initially, the production of surface-active compounds was linked to the degradation of hydrocarbons. More recently, rhamnolipids were detected in *Pseudomonas* biofilms developed without hydrocarbons (Davey *et al.*, 2003). The rhamnolipid produced by the bacteria within the biofilm was suggested to affect the architecture of the biofilm. By using biosurfactant mutants, the production of surface-active compounds was related to the maintenance of the channel structures in the biofilm. It was also found that rhamnolipids mediate the detachment of *Pseudomonas* from biofilms (Boles *et al.*, 2005). Similarly rhamnolipids modulate the swarming motility of *Pseudomonas* (Caiazza *et al.*, 2005). The conditioning of microbial cell surfaces by surface-active compounds has been described by Zhong *et al.* (2007) with respect to their cell surface hydrophobicity. These authors tested the adsorption and effect of rhamnolipids on cell surface hydrophobicity of different bacteria and found a dependency on the physiological status of the cultures. Multiple roles of biosurfactants were described in flow cell studies with colour-coded *Pseudomonas* strains (Pamp and Tolker-Nielsen, 2007). It was found that the biosurfactants were involved in microcolony formation in early biofilm development and that, in later stages, they facilitated the migration-dependent development of the biofilm structure.

The directed accumulation of surface-active compounds by bacteria may lead to another type of "surface-active" EPS compound. The result is lipid vesicles released from the cell membrane. Membrane vesicles were already known for some time to be produced by bacteria (Mayrand and Grenier, 1989; Beveridge, 1999). More recently, they were identified as an overlooked component of the biofilm matrix

(Schooling and Beveridge, 2006). This was not only demonstrated for *Pseudomonas* biofilms but also by examination of biofilms from different environmental habitats. The characterization of membrane vesicles derived from Gram-negative planktonic and biofilm bacteria revealed quantitative and qualitative differences. This indicates a functional role and it has been suggested that membrane vesicles serve as a transport vehicle for cellular components (inside) and for virulence factors (outside). This suggests the controlled production and release of EPS constituents via membrane vesicles. For example, it was shown that membrane vesicles contain enzymes which may result in an activity remote from the producing cell. In this context, membrane vesicles may be regarded as extensions of the bacterial cell and even of the microcolony. Thereby, the bacteria in a biofilm have the potential to alter their environment and control the geochemistry or biochemistry of their habitat, not at the microscopic, but at the macroscopic scale (Schooling and Beveridge, 2006).

6.5. Informative EPS

6.5.1. Nucleic acids

The first short report of extracellular nucleic acids as a structural compound in microbial communities caused a very controversial discussion (Whitchurch *et al.*, 2002). However, a few years later, several research groups provided evidence in one way or another that extracellular nucleic acids are present and may have a function in microbial communities. From different research perspectives, it is apparent that microorganisms can handle extracellular nucleic acids. For example, they are known to have a natural competence, meaning they can take up DNA, they also have a DNA release programme, they bind to DNA and they exchange genetic material. Consequently, eDNA must be present in microbial communities. So far it was just not looked at in terms of a structural constituent and as part of EPS. Evidence for eDNA originates from pure culture studies, from structural studies of biofilms and bioaggregates as well as from studies of eDNA in sediments. The latter aspect will be discussed together with nutritive EPS (see below). One important aspect of EPS as eDNA comprises natural transformation (Lorenz and Wackernagel, 1994). Indeed, transformation is facilitated if microbial communities are immobilized as bioaggregates and biofilms (Hausner and Wuertz, 1999; Wuertz *et al.*, 2001; Molin and Tolker-Nielsen, 2003). Several studies employed nucleic acid degrading enzymes in order to study the effect of DNA on biofilm formation or biofilm dispersal (Whitchurch *et al.*, 2002; Nemoto *et al.*, 2003; Petersen *et al.*, 2005; Moscoso *et al.*, 2006). For example, Whitchurch *et al.* (2002) demonstrated the inhibition of biofilm formation by deoxyribonuclease (DNase) I. Nemoto *et al.* (2003) removed established *P. aeruginosa* biofilms by varidase or DNase I. Petersen *et al.* (2005) studied the DNA uptake system of *Streptococcus mutans* wild-type and mutants using DNase I. Moscoso *et al.* (2006) could show that DNase I proteinase K impaired *Streptococcus pneumoniae* biofilm development and dramatically reduced the number of cells in established biofilms. Other investigations compared the composition of eDNA in comparison to cellular DNA (Steinberger and Holden, 2005; Allesen-Holm *et al.*, 2006; Böckelmann *et al.*, 2006). The examination of eDNA from different bacterial strains revealed differences when compared to cellular or total DNA of the same biofilm (Steinberger and Holden, 2005). For a deep branching *Gammaproteobacterium*, various degrees of homology between eDNA and cellular genomic DNA were reported (Böckelmann *et al.*, 2006). The analysis of eDNA and cellular DNA from *P. aeruginosa* biofilms showed similarities in composition and the dependency of eDNA generation from acylated homoserine lactone (AHL) as well as from flagella and type IV fimbriae (Allesen-Holm *et al.*, 2006). A subsequent study demonstrated that, for *P. aeruginosa*, the eDNA release is under iron regulation (Yang *et al.*, 2007). By using flow chambers, it was shown that high levels of iron decreased DNA release, structural biofilm development and the formation of subpopulations in combination with increased tolerance against antibiotics. In a recent review on eDNA as an EPS compound in biofilm formation, a focus was placed on pseudomonads and streptococci (Pamp *et al.*, 2007).

In a few reports, eDNA was linked to specific structural features of microbial communities. For *Pseudomonas* biofilms developed in flow cells, it was demonstrated that eDNA is located in distinctive zones inside the growing microcolonies (Allesen-Holm *et al.*, 2006). The stained eDNA was mainly found in the stalks of the "mushrooms" with the highest concentration in the outer areas of the stalks. In addition, eDNA production was found to be dependent on quorum sensing. An isolate (F8) from lotic aggregates was also found to release eDNA, thereby forming a three-dimensional network (Böckelmann *et al.*, 2006). The eDNA network was initially discovered by staining the F8 samples with the nucleic acid-specific fluorochrome, Syto 9. The F8 strain was identified as a deep branching *Gammaproteobacteria* with *Rheinheimeria baltica* as the closest relative. The formation of the network appeared to be initiated by the release of particles from the individual cells forming a matrix around each cell, through an unknown process, tendrils and filaments emerged from the cocoon of material around each cell extending into the surrounding environment and forming larger filaments, individual cells and cell masses appear to use the resulting matrix to form an aggregate. One example of the process described is shown in Figure 37.1E. Nevertheless, this process requires a more extensive and detailed examination for a full understanding. More recently, an *in vivo* study demonstrated the presence of both eDNA and pilin protein within the EPS of *Haemophilus influenzae* biofilms (Jurcisek and Bakaletz, 2007). The network of eDNA was found as thin and thick strands which could be stained by using DAPI. It was suggested that eDNA is important for the overall structural stability and the greater density in outer biofilm regions and it may be involved in protection from phagocytic cells. Of note, the eDNA structures look very similar to those of strain F8 which were stained by Syto 9 (Böckelmann *et al.*, 2006). It is obvious that eDNA may serve several functions, one of which is related to the structure and development of interfacial microbial communities. Up until now it can only be speculated what the advantages of eDNA are in this respect.

6.5.2. *Lectins*

Lectins are sugar-binding proteins with a very specific recognition function and, therefore, may be regarded as informative EPS compounds. Importantly, lectins may represent either cell surface- or EPS-associated proteins produced by the biofilm bacteria. In this case, they may have a function in the recognition of other glycoconjugates located on the cell surface of other bacteria and symbiotic organisms. This is well known for co-aggregation and co-adhesion of oral and freshwater bacteria (Rickard *et al.*, 2003). To make it unequivocally clear, in this case, lectins are not understood as fluorescently-labelled probes for biofilm analysis, rather they are produced by the microorganisms living inside the biofilm.

In addition to this established function, lectins may serve as a proteinaceous EPS compound binding to specific EPS glycoconjugates, thereby stabilizing the PS matrix in biofilms. This would result in a structured polymer matrix controlled by the microorganisms living in the biofilm. One may speculate as to how much the lectins really control the stability and possibly the degradation of the biofilm matrix. Lectins are not only known to bind to carbohydrate structures, but they are also known to have a second protein-specific binding site. This would allow the lectins first to bind to a specific location in the PS matrix and, secondly, bind another protein again at a specific location. This could be a possible way of building a structured EPS matrix in the form of a PS–protein network. Another function could be controlling the presence of enzyme activity near the cell or within the microcolony.

6.6. Nutritive EPS

Bacteria rely on nutrients such as carbon, nitrogen and phosphorous in order to keep their metabolism running. Extracellular polymeric substances as a biofilm constituent of variable biochemical composition represents an ideal source of nutrients. Carbon may originate from PSs, nitrogen may come from proteins and phosphorous may be accessed in extracellular nucleic acids. It has been shown that bacteria may use their EPS as a carbon source and cross-feeding between bacteria via polymers has also been demonstrated (Wolfaardt *et al.*, 1999). A variety of studies has also shown the important role of EPS as an accumulator of nutrients and carbon through sorptive processes and that these accumulated materials may be subsequently utilized (Decho and Lopez, 1993; Freeman *et al.*, 1995; Wolfaardt *et al.*, 1995). With respect to nucleic acids in environmental habitats, eDNA has a strong tendency to bind to sand and sediment (Lorenz and Wackernagel, 1987). The presence and fate of eDNA in sediments was also the subject of more recent studies which suggested new and improved isolation protocols (Dell'Anno *et al.*, 2002; Corinaldesi *et al.*, 2005). In subsequent investigations, the function of eDNA was discussed. Due to the high eDNA concentration, it was suggested that eDNA is an important nutrient source, especially in terms of P cycling (Dell'Anno and Corinaldesi, 2004; Dell'Anno and Danovaro, 2005). The high eDNA concentration in sediments raises issues with respect to molecular analysis techniques used as a measure to estimate microbial diversity (Luna *et al.*, 2006). The potential for eDNA to vary from the genomic DNA in specific bacteria also raises concerns. The distribution and possible recycling of EPS compounds in a biofilm is indicated in Figure 37.1F.

6.7. Locomotive EPS

Motility across interfaces without flagella is a characteristic feature of many bacteria that was reported some time ago (Burchard, 1981, 1984; Pate, 1988). However, several authors pointed out that gliding movement of bacteria is still a mystery. There are very likely different types of gliding mechanisms: type IV pilus-mediated gliding, cytoskeleton/surface protein-mediated gliding and gliding involving EPS constituents (McBride, 2001; Merz and Forest, 2002; Bardy *et al.*, 2003). The EPS compounds are excreted through specialized structures and may involve surface-active compounds and PSs. The translocation across an interface by surface-active compounds was reported for *Cytophaga* and *Myxococcus* (Keller *et al.*, 1983; Abbanat *et al.*, 1986). It was proposed that the surface tension gradient along the microbial cell is high enough to move a bacterium across an interface. Secretion of slime has been studied in detail in filamentous cyanobacteria and *Myxococcus xanthus*. *Phormidium* and *Anabaena* spp. possess a pore apparatus on both sides of the septum which is able to release slime fibrils (Hoiczyk and Baumeister, 1998; Hoiczyk, 2000). The release of the slime was found to be roughly equivalent to the speed of cellular movement. Most studies with a focus on bacterial gliding motility were done with *Myxococcus* and different mechanisms were suggested (Spormann, 1999). Later it was shown that *Myxococcus* has nozzle-like structures at both cell poles (Wolgemuth *et al.*, 2002). The functional principle is suggested to be based on the hydration of polymers resulting in a propulsive thrust. The released compounds will leave a trail behind which can be visualized by india ink or by electron microscopy. The EPS compounds involved in gliding motility are likely chemically different compared to other EPS compounds. An example of cyanobacteria gliding tracks is shown in Figure 37.1G.

6.8. Redox-active EPS

Although speculative, nevertheless, there is some evidence that there is a potential role for so-called redox-active EPS. The idea suggested is that refractory compounds originating from the remains of microbial cell structures may be used by certain bacteria for anaerobic respiration. The role of humic compounds was already suggested as they serve as electron donors or acceptors in anaerobic respiration (Lovley *et al.*, 1996). The question is what do microbial refractory compounds look like? Soil-derived humic compounds are characterized by multiple ring structures having many double bonds. Similarly, several microbial compounds may contain ring structures and double bonds, e.g. in certain amino acids or nucleotides (Ogawa *et al.*, 2001). A further aspect comes from so-called electroactive polymers, e.g. polyacetylene. In a recent paper, it has been discussed to what degree these man-made polymers can be replaced by natural PSs (Finkenstadt, 2005). Finally, the redox-related polymers could have an indirect function, e.g. as a support matrix for a variety of redox sensors (Green and Paget, 2004). Currently, it remains an open question as to how bacteria may take advantage of redox-active EPS, however, this type of EPS may be important in certain habitats where aerobic–anaerobic transitioning zones are present, e.g. in sediments or the deep biosphere. A speculative reaction scheme is shown in Figure 37.1H.

7. CONCLUSIONS

The functionality of EPS, as discussed above, has evolved over time and we have reviewed these functions in order. Nevertheless, it is evident that a strict separation of functionality in terms of specific polymer properties cannot be generalized. Contrary to the EPS types identified and the potential functions allocated, it is more likely that many, if not most, EPS compounds serve multiple functions. Because microbial EPS can be derived by its producers, the varying chemical constituents attached may result in polymers with very different physicochemical properties. This approach would allow the microorganism to use a single polymer type and fine tune its functionality. As a first result, a shortlist of EPS functionality has been provided (Figure 37.2). As a second result, a functionality network may be drafted which presents all potential cross-functionalities possible (Figure 37.3). Some of these cross-links have already been established, whereas others may be discovered in the future but, as always, some remain speculative.

During the course of preparing this review, another important consideration came up, i.e. system biology arose as a concern. This concept comprises a very broad approach to the investigation of complex biological systems. In other words the various "-omics" and their tools can be discussed with respect to biofilm systems. The genomic approach has already been applied to examine a simple acidophilic biofilm system dominated by *Leptospirillum* and *Ferroplasma* spp. (Tyson *et al.*, 2004). As a result of gene analysis, several pathways related to carbon and nitrogen fixation as well as energy generation were established. The proteomic approach was also used for the analysis and comparison of planktonic and biofilm populations (Oosthuizen *et al.*, 2002). A recent study even focused on the proteins in the EPS of *Haemophilus influenzae* biofilms (Gallaher *et al.*, 2006). The 265 proteins identified were analysed and annotated in a cluster of orthologous groups. The proteins included a range of over-represented proteins if compared to their genomic frequencies. This approach clearly indicates the way to proceed in future investigations. Indeed, it should be extended by the glycomic approach. However, to our knowledge, the glycomic

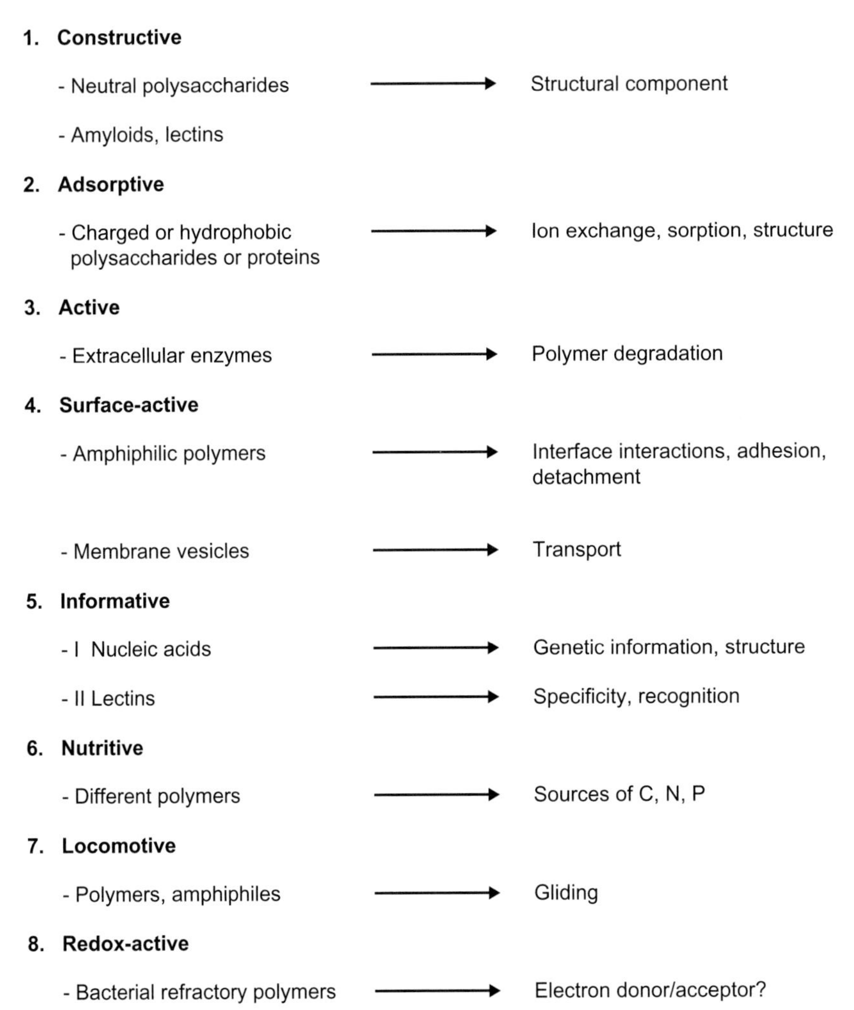

FIGURE 37.2 Summary of the novel concept on EPS functionality listed as discussed in the text. In the left column, the EPS type with examples is shown; in the right column, the main function is given.

approach has not been applied to biofilms, neither to the cellular nor the EPS constituents. This may be due to the complexity of glycomics which was outlined in an integrated system approach (Raman *et al.*, 2005). In a recent meeting report, the complexity of glycomics in relation to the genomic, transcriptomic and proteomic approach was illustrated. Of note, a log-scale was needed in order to graph out the relationship between the different types of "-omics" in terms of their information content (Turnbull and Field, 2007). It is likely necessary to adopt this point of view and, thus, consider the different "-omic" approaches for microbial EPS. This would facilitate the integration of different types of EPS, their currently accepted roles and yet to be discovered EPS functions. Future areas for investigation and important remaining issues in the study of EPS are detailed in the Research Focus Box.

RESEARCH FOCUS BOX

- Development of an improved protocol for isolating EPS is needed. The procedure should consider the complex biochemistry of the various types of EPS. In addition, it should be extended to environmental biofilm communities.
- Application of the *in situ* imaging approach would be beneficial. Advanced laser scanning microscopy of fully hydrated biofilms may allow localization and allocation of microbial cells and their different types of EPS. This should include probing of the microhabitat at the cell, microcolony and biofilm level. For this purpose, a correlative imaging approach may be required.
- Fluorescence lectin-binding analysis (FLBA), in combination with laser microscopy, has been demonstrated as a powerful tool for imaging of hydrated lectin-specific EPS glycoconjugates. Fine-tuning and control of the specificity is needed. The often asked for relationship of FLBA with chemical analysis remains a big challenge to achieve.
- Further studies are necessary in order to clarify the important issues of the dynamics of EPS production, modification of EPS by producers or neighbours, interactions of EPS compounds, lifetime and recycling of EPS, etc.
- Verification of the suggested functionalities of EPS will be a major challenge for many years to come, especially in defined mixed cultures and in microbial ecology.
- The complexity of EPS studies may require the application of the "-omic" approach, i.e. comprising genomics, proteomics and glycomics.

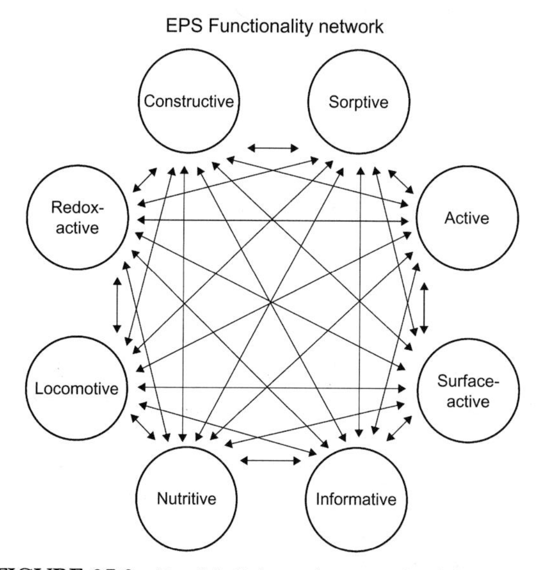

FIGURE 37.3 Possible linkages between the different EPS functions. The drawing seeks to indicate that an EPS compound may have multiple functions. Several of the functions have been established, whereas others are purely hypothetical.

ACKNOWLEDGEMENTS

The authors thank the following for the courtesy of providing images: C. Staudt (Figure 37.1A), U. Boeckelmann (Figure 37.1E), M. Boessman (Figure 37.1F), B. Zippel (Figure 37.1G).

References

Abbanat, D.R., Leadbetter, E.R., Godchaux, W., Escher, A., 1986. Sulphonolipids are molecular determinants of gliding motility. Nature 324, 367–369.

Allesen-Holm, M., Bundvik Barken, K., Yang, L., et al., 2006. A characterisation of DNA release in *Pseudomonas aeruginosa* cultures and biofilms. Mol. Microbiol. 59, 1114–1128.

Allison, D.G., Sutherland, I.W., Neu, T.R., 2003. EPS: What's in an acronym? In: McBain, A., Allison, D.G., Brading, M., Rickard, A., Verran, J., Walker, J. (Eds.), Biofilm Communities: Order from Chaos? Bioline, Cardiff, pp. 381–387.

Amory, D.F., Genet, M.J., Rouxhet, P.G., 1988. Application of XPS to the surface analysis of yeast cells. Surf. Interf. Anal. 11, 478–486.

Azam, F., Malfatti, F., 2007. Microbial structuring of marine ecosystems. Nat. Rev. Microbiol. 5, 782–791.

Bardy, S.L., Ng, S.Y.M., Jarrell, K.F., 2003. Procaryotic motility structures. Microbiology 149, 295–304.

Baty, A.M., Diwu, Z., Dunham, G., et al., 2001. Characterization of extracellular chitinolytic activity in biofilms. Methods Enzymol. 336, 279–301.

Belsky, J., Gutnick, D.L., Rosenberg, E., 1979. Emulsifier of *Arthrobacter* RAG-1: determination of emulsifier-bound fatty acids. FEBS Lett. 101, 175–178.

Bertocchi, C., Navarini, L., Cesaro, A., 1990. Polysaccharides from *Cyanobacteria*. Carbohydr. Polym. 12, 127–153.

Beveridge, T.J., 1999. Structures of Gram-negative cell walls and their derived membrane vesicles. J. Bacteriol. 181, 4725–4733.

Bhaskar, P.V., Bhosle, N.B., 2005. Microbial extracellular polymeric substances in marine biogeochemical processes. Curr. Sci. 88, 45–53.

Böckelmann, U., Janke, A., Kuhn, R., et al., 2006. Bacterial extracellular DNA forming a defined network like structure. FEMS Microbiol. Lett. 262, 31–38.

Boles, B.R., Thoendel, M., Singh, P.K., 2005. Rhamnolipids mediate detachment of *Pseudomonas aeruginosa* from biofilms. Mol. Microbiol. 57, 1210–1223.

Branda, S.S., Vik, A., Friedman, L., Kolter, R., 2005. Biofilms: the matrix revisited. Trends Microbiol. 13, 20–26.

Burchard, R.P., 1981. Gliding motility of procaryotes: ultrastructure, physiology, and genetics. Annu. Rev. Microbiol. 35, 497–529.

Burchard, R.P., 1984. Gliding motility and taxes. In: Rosenberg, E. (Ed.), Myxobacteria: Development and Cell Interactions. Springer Verlag, New York, pp. 139–161.

Caiazza, N.C., Shanks, R.M.Q., O'Toole, G.A., 2005. Rhamnolipids modulate swarming motility patterns of *Pseudomonas aeruginosa*. J. Bacteriol. 187, 7351–7361.

Cerning, J., 1990. Exocellular polysaccharides produced by lactic acid bacteria. FEMS Microbiol. Rev. 87, 113–130.

Characklis, W.G., Cooksey, K.E., 1983. Biofilms and microbial fouling. Adv. Appl. Microbiol. 29, 93–138.

Characklis, W.G., Marshall, K.C., 1990. Biofilms. John Wiley and Sons, New York.

Christensen, B.E., 1989. The role of extracellular polysaccharides in biofilms. J. Biotechnol. 10, 181–202.

Christensen, B.E., Characklis, W.G., 1990. Physical and chemical properties of biofilms. In: Characklis, W.G., Marshall, K.C. (Eds.), Biofilms. Wiley and Sons, New York, pp. 93–130.

Chrost, R.J., 1991. Microbial Enzymes in Aquatic Environments. Springer, New York.

Cooksey, K.E., 1992. Extracellular polymers in biofilms. In: Melo, L.F., Bott, T.R., Fletcher, M., Capdeville, B. (Eds.), Biofilms – Science and Technology. NATO ASI Series E: Applied Sciences, Vol. 23. Kluwer Academic Publishers, Dordrecht, pp. 137–147.

Corinaldesi, C., Danovaro, R., Dell'Anno, A., 2005. Simultaneous recovery of extracellular and intracellular DNA suitable for molecular studies from marine sediments. Appl. Environ. Microbiol. 71, 46–50.

D'Argenio, D.A., Miller, S.I., 2004. Cyclic di-GMP as a bacterial second messenger. Microbiology 150, 2497–2502.

Davey, M.E., Caiazza, N.C., O'Toole, G.A., 2003. Rhamnolipid surfactant production affects biofilm architecture in *Pseudomonas aeruginosa* PAO1. J. Bacteriol. 185, 1027–1036.

Dazzo, F.B., 2004. Production of anti-microbial antibodies and their utilisation in studies of autecology by immunofluorescene microscopy and in situ CMEIAS image analysis. In: Kowalchuk, G.A., de Bruijn, F.J., Head, I.M., Akkermans, A.D.L., van Elsas, J.D. (Eds.), Molecular Microbial Ecology Manual. Kluwer Academic Publishers, Dordrecht, pp. 911–931.

de Brouwer, J.F.C., Neu, T.R., Stal, L.J., 2006. On the function of secretion of extracellular polymeric substances by benthic diatoms and their role in intertidal mudflats: a review of recent insights and views. In: Kromkamp, J.C., de Brouwer, J.F.C., Blanchard, G.F., Forster, R.M., Creach, V. (Eds.), Functioning of Microphytobenthos in Estuaries. Royal Academy of Arts and Sciences, Amsterdam, pp. 45–61.

de Philippis, R., Vincenzini, M., 1998. Exocellular polysaccharides from cyanobacteria and their possible applications. FEMS Microbiol. Rev. 22, 151–175.

Decho, A., 1990. Microbial exopolymer secretions in ocean environments: their role(s) in food webs and marine processes. Oceanogr. Marine Biol. Annu. Rev. 28, 73–153.

Decho, A.W., 1994. Molecular-scale events influencing the macroscale cohesiveness of exopolymers. In: Krumbein, W. E., Paterson, D.M., Zavarzin, G.A. (Eds.), Biostabilisation of Sediments, a Natural History of Life on Earth. Kluwer Academic Publishers, Dordrecht, pp. 135–148.

Decho, A.W., 2000. Exopolymer microdomains as a structuring agent for heterogeneity within microbial biofilms. In: Riding, R.E., Awramik, S.M. (Eds.), Microbial Sediments. Springer, Heidelberg, pp. 9–15.

Decho, A.W., Kawaguchi, T., 2003. Extracellular polymers (EPS) and calcification within modern marine stromatolites. In: Krumbein, W.E., Paterson, D.M., Stal, L.J. (Eds.), Fossil and Recent Biofilms. BIS Verlag, Oldenburg, pp. 227–240.

Decho, A.W., Lopez, G.R., 1993. Exopolymer microenvironments of microbial flora: multiple and interactive effects on trophic relationships. Limnol. Oceanogr. 38, 1633–1645.

Dell'Anno, A., Corinaldesi, C., 2004. Degradation and turnover of extracellular DNA in marine sediments: ecological and methodological considerations. Appl. Environ. Microbiol. 70, 4384–4386.

Dell'Anno, A., Danovaro, R., 2005. Extracellular DNA plays a key role in deep-sea ecosystem functioning. Science 309, 2179.

Dell'Anno, A., Bompadre, S., Danovaro, R., 2002. Quantification, base composition, and fate of extracellular DNA in marine sediments. Limnol. Oceanogr. 47, 899–905.

Dow, J.M., Fouhy, Y., Lucey, J., Ryan, R.P., 2007. Cyclic di-GMP as an intracellular signal regulating bacterial biofilm formation. In: Kjelleberg, S., Givskov, M. (Eds.), The Biofilm Mode of Life. Horizon Bioscience, Wymondham, pp. 71–93.

Dudman, W.F., 1977. The role of polysaccharides in natural environments. In: Sutherland, I.W. (Ed.), Surface Carbohydrates of the Procaryotic Cell. Academic Press, New York, pp. 357–414.

Ellis, R.J., 2001. Macromolecular crowding: obvious but underappreciated. Trends Biochem. Sci. 26, 597–604.

Finkenstadt, V.L., 2005. Natural polysaccharides as electroactive polymers. Appl. Microbiol. Biotechnol. 67, 735–745.

Flemming, H.-C., Leis, A., 2002. Sorption properties of biofilms. In: Bitton, G. (Ed.), The Encyclopedia of Environmental Microbiology. John Wiley and Sons, New York, pp. 2958–2967.

Flemming, H.-C., Wingender, J., 2001a. Relevance of microbial extracellular polymeric substances (EPSs) – Part I: structural and ecological aspects. Water Sci. Technol. 43, 1–8.

Flemming, H.-C., Wingender, J., 2001b. Relevance of microbial extracellular polymeric substances (EPSs) – Part II: technical aspects. Water Sci. Technol. 43, 9–16.

Flemming, H.-C., Wingender, J., 2001c. Structural, ecological and functional aspects of EPS. In: Gilbert, P., Allison, D., Brading, M., Veran, J., Walker, J. (Eds.), Biofilm Community Interactions: Chance or Necessity? BioLine, Cardiff, pp. 175–189.

Flemming, H.-C., Wingender, J., 2002. Extracellular polymeric substances (EPS): structural, ecological and technical aspects. In: Bitton, G. (Ed.), The Encyclopedia of Environmental Microbiology. John Wiley and Sons, New York, pp. 1223–1231.

Flemming, H.-C., Wingender, J., 2003. The crucial role of extracellular polymeric substances in biofilms. In: Wuertz, S., Bishop, P.L., Wilderer, P.A. (Eds.), Biofilms in Wastewater Treatment. IWA Publishing, London, pp. 178–210.

Freeman, C., Lock, M.A., 1995. The biofilm polysaccharide matrix: a buffer against changing substrate supply. Limnol. Oceanogr. 40, 273–278.

Freeman, C., Chapman, P.J., Gilman, K., Lock, M.A., Reynolds, B., Wheater, H.S., 1995. Ion-exchange mechanisms and the entrapment of nutrients by river biofilms. Hydrobiologia 297, 61–65.

Gallaher, T.K., Wu, S., Webster, P., Aguilera, R., 2006. Identification of biofilm proteins in non-typeable *Haemophilus influenzae*. BMC Microbiol. 6, 1–9.

Gebbink, M.F.B.G., Claessen, D., Bouma, B., Dijkhuizen, L., Wösten, H.A.B., 2005. Amyloids – a functional coat for microorganisms. Nat. Rev. Microbiol. 3, 333–341.

Geesey, G.G., 1982. Microbial exopolymers: Ecological and economic considerations. ASM News 48, 9–14.

Ghannoum, M., O'Toole, G.A., 2004. Microbial Biofilms. ASM Press, Washington DC.

Green, J., Paget, M.S., 2004. Bacterial redox sensors. Nat. Rev. Microbiol. 2, 954–967.

Hausner, M., Wuertz, S., 1999. High rates of conjugation in bacterial biofilms as determined by quantitative in situ analysis. Appl. Environ. Microbiol. 65, 3710–3713.

Hoffman, M., Decho, A.W., 1999. Extracellular enzymes within microbial biofilms and the role of the extracellular polymer matrix. In: Wingender, J., Neu, T.R., Flemming, H.-C. (Eds.), Microbial Extracellular Polymeric Substances. Springer, Berlin, pp. 217–230.

Hoiczyk, E., 2000. Gliding motility in cyanobacteria: observations and possible explanations. Arch. Microbiol. 174, 11–17.

Hoiczyk, E., Baumeister, W., 1998. The junctional pore complex, a procaryotic secretion organelle, is the molecular motor underlying gliding motility in cyanobacteria. Curr. Biol. 8, 1161–1168.

Jenal, U., 2004. Cyclic di-guanosine-monophosphate comes of age: a novel secondary messenger involved in modulating cell surface structures in bacteria? Curr. Opin. Microbiol. 7, 185–191.

Jenal, U., Malone, J., 2006. Mechanisms of cyclic-di-GMP signaling in bacteria. Annu. Rev. Genetics 40, 385–407.

Jurcisek, J.A., Bakaletz, L.O., 2007. Biofilms formed by non-typeable *Haemophilus influenzae* in vivo contain both double-stranded DNA and type IV pilin protein. J. Bacteriol. 189, 3868–3875.

Kawaguchi, T., Decho, A.W., 2002. A laboratory investigation of cyanobacterial extracellular polymeric secretions (EPS) in influencing $CaCO_3$ polymorphism. J. Crystal Growth 240, 230–235.

Keiding, K., Wybrandt, L., Nielsen, P.H., 2001. Remember the water – a comment on EPS coligative properties. Water Sci. Technol. 43, 17–23.

Keller, K.H., Grady, M., Dworkin, M., 1983. Surface tension gradients: feasible model for gliding motility of *Myxococcus xanthus*. J. Bacteriol. 155, 1358–1366.

Kenne, L., Lindberg, B., 1983. Bacterial polysaccharides. In: Aspinall, G.O. (Ed.), The Polysaccharides, Vol. 2. Academic Press, New York, pp. 287–363.

Kjelleberg, S., Givskov, M., 2007. The Biofilm Mode of Life. Horizon Press, Wymondham.

Larsen, P., Nielsen, J.L., Dueholm, M.S., Wetzel, R., Otzen, D., Nielsen, P.H., 2007. Amyloid adhesins are abundant in natural biofilms. Environ. Microbiol. 9, 3077–3090.

Larsen, P.L., Nielsen, J.L., Otzen, D., Nielsen, P.H., 2008. Amyloid-like adhesins produced by floc-forming and filamentous bacteria in activated sludge. Appl. Environ. Microbiol. 74, 1517–1526.

Lasa, I., 2007. Towards the identification of the common features of bacterial biofilm development. Int. Microbiol. 9, 21–28.

Lawrence, J.R., Swerhone, G.D.W., Kuhlicke, U., Neu, T.R., 2007. In situ evidence for microdomains in the polymer matrix of bacterial microcolonies. Can. J. Microbiol. 53, 450–458.

Lens, P., Moran, A.P., Mahony, T., Stoodley, P., O'Flaherty, V., 2003. Biofilms in Medicine, Industry and Environmental Biotechnology. IWA Publishing, London.

Lindberg, B., 1990. Components of bacterial polysaccharides. Adv. Carbohydr. Chem. 48, 279–318.

Lorenz, M.G., Wackernagel, W., 1987. Adsorption of DNA to sand and variable degradation rates of absorbed DNA. Appl. Environ. Microbiol. 53, 2948–2952.

Lorenz, M.G., Wackernagel, W., 1994. Bacterial gene transfer by natural genetic transformation in the environment. Microbiol. Rev. 58, 563–602.

Lovley, D.R., Coates, J.D., Blunt-Harris, E.L., Phillips, J.P., Woodward, J.C., 1996. Humic substances as electron acceptors for microbial respiration. Nature 382, 445–448.

Luna, G.M., Dell'Anno, A., Danovaro, R., 2006. DNA extraction procedure: a critical issue for bacterial diversity assessment in marine sediments. Environ. Microbiol. 8, 308–320.

Marshall, K.C., 1992. Planktonic versus sessile life of procaryotes. In: Ballows, A., Trüper, H.G., Dworkin, M., Harder, W., Schleifer, K.-H. (Eds.), The Procaryotes. Springer Verlag, New York, pp. 262–275.

Mayrand, D., Grenier, D., 1989. Biological acitivities of outer membrane vesicles. Can. J. Microbiol. 35, 607–613.

McBride, M.J., 2001. Bacterial gliding motility: multiple mechanisms for cell movements over surfaces. Annu. Rev. Microbiol. 55, 49–75.

McLean, R.J.C., Decho, A.W., 2002. Molecular Ecology of Biofilms. Horizon Press, Wymondham.

Melo, L.F., Bott, T.R., Fletcher, M., Capdeville, B., 1992. Biofilms — Science and Technology. Kluwer Academic Publishers, Dordrecht.

Merz, A., Forest, K.T., 2002. Bacterial surface motility: slime trails, grappling hooks and nozzles. Curr. Biol. 12, R297–R303.

Molin, S., Tolker-Nielsen, T., 2003. Gene transfer occurs with enhanced efficiency in biofilms and induces enhanced stabilisation of the biofilm structure. Curr. Opin. Biotechnol. 14, 255–261.

Moran, A.P., Ljungh, Å., 2003. Physico-chemical properties of extracellular polymeric substances. In: Lens, P., Moran, A. P., Mahony, T., Stoodley, P., O'Flaherty, V. (Eds.), Biofilms in Medicine, Industry and Environmental Biotechnology. IWA Publishing, London, pp. 91–112.

Moscoso, M., Garcia, E., Lopez, R., 2006. Biofilm formation by *Streptococcus pneumoniae*: role of choline, extracellular DNA, and capsular polysaccharide in microbial accretion. J. Bacteriol. 188, 7785–7795.

Murakawa, T., 1973a. Slime production by *Pseudomonas aeruginosa* III. Purification of slime and its physicochemical properties. Jpn. J. Microbiol. 17, 273–281.

Murakawa, T., 1973b. Slime production by *Pseudomonas aeruginosa* IV. Chemical analysis of two varieties of slime produced by *Pseudomonas aeruginosa*. Jpn. J. Microbiol. 17, 513–520.

Nemoto, K., Hirota, K., Murakami, K., et al., 2003. Effect of varidase (streptodornase) on biofilm formed by *Pseudomonas aeruginosa*. Chemotherapy 49, 121–125.

Neu, T.R., 1992a. Microbial "footprints" and the general ability of microorganisms to label interfaces. Can. J. Microbiol. 38, 1005–1008.

Neu, T.R., 1992b. Polysaccharide in Biofilmen. In: Präve, P., Schlingmannn, M., Crueger, W., Esser, K., Thauer, R., Wagner, F. (Eds.), Jahrbuch Biotechnologie, Band 4. Carl Hanser Verlag, München, pp. 73–101.

Neu, T.R., 1996. Significance of bacterial surface-active compounds in interaction of bacteria with interfaces. Microbiol. Rev. 60, 151–166.

Neu, T.R., Lawrence, J.R., 1999. Lectin-binding-analysis in biofilm systems. Methods Enzymol. 310, 145–152.

Neu, T.R., Marshall, K.C., 1990. Bacterial polymers: physicochemical aspects of their interactions at interfaces. J. Biomater. Appl. 5, 107–133.

Neu, T.R., Marshall, K.C., 1991. Microbial "footprints" – a new approach to adhesive polymers. Biofouling 3, 101–112.

Nielsen, P.H., Jahn, A., 1999. Extraction of EPS. In: Wingender, J., Neu, T.R., Flemming, H.-C. (Eds.), Microbial Extracellular Polymeric Substances. Springer, Berlin, pp. 49–72.

Ogawa, H., Amagai, Y., Koike, I., Kaiser, K., Benner, R., 2001. Production of refractory organic matter by bacteria. Science 292, 917–920.

Oosthuizen, M.C., Steyn, B., Theron, J., et al., 2002. Proteomic analysis reveals differential protein expression by *Bacillus cereus* during biofilm formation. Appl. Environ. Microbiol. 68, 2770–2780.

Pamp, S.J., Tolker-Nielsen, T., 2007. Multiple roles of biosurfactants in structural biofilm development by *Pseudomonas aeruginosa*. J. Bacteriol. 189, 2531–2539.

Pamp, S.J., Gjermansen, M., Tolker-Nielsen, T., 2007. The biofilm matrix: a sticky framework. In: Kjelleberg, S., Givskov, M. (Eds.), The Biofilm Mode of Life. Horizon Press, Wymondham, pp. 37–69.

Paragas, V.B., Kramer, J.A., Fox, C., Hauglands, R.P., Singer, V.L., 2002. The ELF-97 phosphatase substrate provides a sensitive, photostable method for labelling cytological targets. J. Microsc. 206, 106–119.

Pate, J.L., 1988. Gliding motility in procaryotic cells. Can. J. Microbiol. 34, 459–465.

Petersen, F.C., Tao, L., Scheie, A.A., 2005. DNA binding-uptake system: a link between cel-to-cell communication and biofilm formation. J. Bacteriol. 187, 4392–4400.

Raman, R., Raguraman, S., Venkataraman, G., Paulson, J.C., Sasisekharan, R., 2005. Glycomics: an integrated system approach to structure-function relationships of glycans. Nat. Methods, 2, 817–824.

Rickard, A.H., Gilbert, P., High, N.J., Kolenbrander, P.E., Handley, P.S., 2003. Bacterial coaggregation: an integral process in the development of multi-species biofilms. Trends Microbiol. 11, 94–100.

Römling, U., Gomelsky, M., Galperin, M.Y., 2005. C-di-GMP: the dawning of a novel bacterial signalling system. Mol. Microbiol. 57, 629–639.

Rosenberg, E., Kaplan, N., 1987. Surface-active properties of *Acinetobacter* exopolysaccharides. In: Inouye, M. (Ed.), Bacterial Outer Membranes as Model Systems. John Wiley and Sons, New York, pp. 311–342.

Rosenberg, E., Ron, E.Z., 1997. Bioemulsans: microbial polymeric emulsifiers. Curr. Opin. Biotechnol. 8, 313–316.

Rosenberg, E., Rosenberg, M., Shoham, Y., Kaplan, N., Sar, N., 1989. Adhesion and desorption during growth of *Acinetobacter calcoaceticus* on hydrocarbons. In: Cohen, Y., Rosenberg, E. (Eds.), Microbial Mats. ASM, Washington DC, pp. 219–227.

Sandford, P.A., 1979. Exocellular, microbial polysaccharides. Adv. Carbohydr. Chem. Biochem. 36, 265–313.

Sandford, P.A., Laskin, A., 1977. Extracellular microbial polysaccharides. American Chemical Society, Washington DC.

Schooling, S.R., Beveridge, T.J., 2006. Membrane vesicles: an overlooked component of the matrices of biofilms. J. Bacteriol. 188, 5945–5957.

Simon, M., Grossart, H.-P., Schweitzer, B., Ploug, H., 2002. Microbial ecology of organic aggregates in aquatic ecosystems. Aquat. Microb. Ecol. 28, 175–211.

Spormann, A.M., 1999. Gliding motility in bacteria: insights from studies of *Myxococcus xanthus*. Microbiol. Mol. Biol. Rev. 63, 621–641.

Starkey, M., Gray, K.A., Chang, S.I., Parsek, M.R., 2004. A sticky business: the extracellular polymeric substance matrix of biofilms. In: Ghannoum, M., O'Toole, G.A. (Eds.), Microbial Biofilms. ASM Press, Washington DC, pp. 174–191.

Staudt, C., Horn, H., Hempel, D.C., Neu, T.R., 2003. Screening of lectins for staining lectin-specific glycoconjugates in the EPS of biofilms. In: Lens, P., Moran, A.P., Mahony, T., Stoodley, P., O'Flaherty, V. (Eds.), Biofilms in Medicine, Industry and Environmental Technology. IWA Publishing, pp. 308–327.

Staudt, C., Horn, H., Hempel, D.C., Neu, T.R., 2004. Volumetric measurements of bacterial cells and extracellular polymeric substance glycoconjugates in biofilms. Biotechnol. Bioeng. 88, 585–592.

Steinberger, R.E., Holden, P.A., 2005. Extracellular DNA in single- and multiple-species unsaturated biofilms. Appl. Environ. Microbiol. 71, 5404–5410.

Sutherland, I.W., 1977a. Bacterial exopolysaccharides their nature and production. In: Sutherland, I.W. (Ed.), Surface Carbohydrates of the Procaryotic Cell. Academic Press, New York, pp. 27–96.

Sutherland, I.W., 1977b. Surface Carbohydrates of the Procaryotic Cell. Academic Press, New York.

Sutherland, I.W., 1983. Extracellular polysaccharides. In: Rehm, H.-J., Reed, G. (Eds.), Biotechnology, Vol. 3. Verlag Chemie, Weinheim, pp. 531–574.

Sutherland, I.W., 1995. Biofilm-specific polysaccharides – do they exist? In: Wimpenny, J., Handley, P., Gilbert, P., Lappin-Scott, H. (Eds.), The Life and Death of Biofilm. BioLine, Cardiff, pp. 103–106.

Sutherland, I.W., 1997. Microbial biofilm exopolysaccharides – superglue or velcro? In: Wimpenny, J., Handley, P., Lappin-Scott, H.M. (Eds.), Biofilms: Community Interactions and Control. BioLine, Cardiff, pp. 33–39.

Sutherland, I.W., 2001a. Exopolysaccharides in biofilms, flocs and related structures. Water Sci. Technol. 43, 77–86.

Sutherland, I.W., 2001b. The biofilm matrix – an immobilized but dynamic microbial environment. Trends Microbiol. 9, 222–227.

Turnbull, J.E., Field, R.A., 2007. Emerging glycomics technologies. Nat. Chem. Biol. 3, 74–77.

Tyson, G.W., Chapman, J., Hugenholtz, P., et al., 2004. Community structure and metabolism through reconstruction of microbial genomes from the environment. Nature 428, 37–43.

Van Ommen Kloecke, F., Geesey, G.G., 1999. Localization and identification of populations of phosphatase – active bacterial cells associated with activated sludge flocs. Microb. Ecol. 38, 201–214.

Vuyst, L.D., Degeest, B., 1999. Heteropolysaccharides from lactic acid bacteria. FEMS Microbiol. Rev. 23, 153–177.

Webb, J.S., Thompson, L.S., James, S., et al., 2003. Cell death in *Pseudomonas aeruginosa* biofilms development. J. Bacteriol. 185, 4585–4592.

Whitchurch, C.B., Tolker-Nielsen, T., Ragas, P., Mattick, J.S., 2002. Extracellular DNA required for bacterial biofilm formation. Science 295, 1487.

Whitfield, C., 2006. Biosynthesis and assembly of capsular polysaccharides in *Escherichia coli*. Annu. Rev. Biochem. 75, 39–68.

Whitfield, C., Naismith, J.H., 2008. Periplasmic export machines for outer membrane assembly. Curr. Opin. Struct. Biol. 18, 466–474.

Whitfield, C., Valvano, M.A., 1993. Biosynthesis and expression of cell-surface polysaccharides in Gram-negative bacteria. Adv. Microb. Physiol. 35, 135–246.

Wimpenny, J., 2000. An overview of biofilms as functional communities. In: Allison, D.G., Gilbert, P., Lappin-Scott, H.M., Wilson, M. (Eds.), Community Structure and Co-Operation in Biofilms. Cambridge University Press, Cambridge, pp. 1–24.

Wingender, J., Neu, T.R., Flemming, H.-C., 1999. Microbial Extracellular Polymeric Substances. Springer, Heidelberg.

Wolfaardt, G.M., Lawrence, J.R., Robarts, R.D., Caldwell, D.E., 1995. Bioaccumulation of the herbicide diclofop in extracellular polymers and its utilisation by a biofilm community during starvation. Appl. Environ. Microbiol. 61, 152–158.

Wolfaardt, G.M., Lawrence, J.R., Korber, D.R., 1999. Function of EPS. In: Wingender, J., Neu, T.R., Flemming, H.-C. (Eds.), Microbial Extracellular Polymeric Substances. Springer, Berlin, pp. 171–200.

Wolgemuth, C., Hoiczyk, E., Kaiser, D., Oster, G., 2002. How myxobacteria glide. Curr. Biol. 12, 369–377.

Wotton, R.S., 2004. The ubiquity and many roles of exopolymers (EPS) in aquatic systems. Scientia Marina 68, 13–21.

Wuertz, S., Hendrickx, L., Kuehn, M., Rodenacker, K., Hausner, M., 2001. *In situ* quantification of gene transfer in biofilms. Methods Enzymol. 336, 129–143.

Wuertz, S., Bishop, P.L., Wilderer, P., 2003. Biofilms in wastewater treatment. IWA Publishing, London.

Yang, L., Barken, K.B., Skindersoe, M.E., Christensen, A.B., Givskov, M., Tolker-Nielsen, T., 2007. Effects of iron on DNA release and biofilm development by *Pseudomonas aeruginosa*. Microbiology 153, 1318–1328.

Zhang, Y., Miller, R.M., 1994. Effect of a *Pseudomonas* rhamnolipid biosurfactant on cell hydrophobicity and biodegradation of octadecane. Appl. Environ. Microbiol. 60, 2101–2106.

Zhong, H., Zeng, G.M., Yuan, X.Z., Fu, H.Y., Huang, G.H., Ren, F.Y., 2007. Adsorption of dirhamnolipid on four microorganisms and the effect on cell surface hydrophobicity. Appl. Microbiol. Biotechnol. 77, 447–455.

CHAPTER

38

Physicochemical properties of microbial glycopolymers

Klaus Brandenburg, Patrick Garidel and Thomas Gutsmann

SUMMARY

The functional role of glycolipids, such as lipopolysaccharides, polysaccharides or other glycoconjugates in membranes like the outer membrane of Gram-negative bacteria, arises from these glycopolymers acting as the first point of interaction of the bacterial cells with cells of the host's immune system. Therefore, their physical states, such as the fluidity of their acyl chains and the conformation of the backbones, are essential determinants of bacterial defence against the attack mounted by components of the human immune system, such as complement and antimicrobial proteins. The action of bacterial glycopolymers with molecules and cells of the human immune system, which eventually may lead to pathophysiological effects like septic shock, is exerted by isolated polymers after their release from the bacteria. Thus, the study of the direct, physical interaction of the pathogenicity factors with target structures, by applying biophysical means, is of high interest. The questions which arise in this context concern the biologically active unit of the glycopolymer (i.e. whether monomer, multimer or aggregate), the molecular conformation of the single molecules, the aggregate structures of the polymers, the strength of their binding to serum- and membrane-proteins, and/or unspecific binding to membrane phospholipids. For an elucidation of these questions, important physical-chemical parameters, such as the critical micellar concentration, the gel to liquid crystalline phase transition and type of aggregate structure, are essential parameters. Herein, recent progress is reviewed which has led to a deeper understanding of the bacterial interaction processes with target cell structures.

Keywords: Endotoxin; Lipopolysaccharide; Rhamnolipids; Mycobacterial glycolipids; Biophysical analysis

1. INTRODUCTION

Glycolipid polymers are compounds consisting of a sugar and a lipid moiety, in the main cases linked with ester or amide groups. Glycolipids are essential constituents of cellular membranes with a high number of functions. They may act as receptors, be important for cell aggregation and dissociation and may be responsible for specific cellular contact and for signal transduction (Curatolo, 1987). There is a large variety of glycolipids also in bacterial species and it can be assumed that only a small part of this variety is known up to now.

Glycolipids are amphiphilic molecules similar in their nature to other membrane constituents such as phospholipids. They form, above a critical concentration, known as the critical micellar concentration (CMC), aggregates in aqueous environments, depending on their hydrophobicity and, very importantly, on the conformation (i.e. shape) of the contributing molecules, which is again determined by their primary chemical structure and is influenced by ambient conditions like temperature, pH, water content and concentration of mono- and divalent cations (Israelachvili, 1991; Garidel *et al.*, 2004). These interconnections are schematically shown in Figure 38.1. According to this theory, the type of aggregate structure of amphiphilic molecules above the CMC can be assessed by a simple geometric model, which relates the resulting structure to the ratio of the effective cross-sectional areas, a_o, of the hydrophilic polar and, a_h, the hydrophobic non-polar regions, respectively. The dimensionless shape parameter, S, defined as $S = v/(a_o \cdot l_c) = a_h/a_o$ (where v, volume per molecule of the hydrophobic moiety; l_c, length of the fully extended hydrophobic portion), allows a prediction of the supramolecular aggregates. The value of S may, among others, be estimated from energy minimization calculations.

Furthermore, the molecular shape of a given amphiphilic molecule within a supramolecular aggregate is not a constant but will depend, e.g. on the fluidity (inversely correlated to the state of order) of its acyl chains which can assume two main phase states, the well-known gel phase (β) and liquid–crystalline phase (α). At a lipid-specific phase transition temperature (T_m), a reversible transition between these two phases takes place. This phase transition temperature, T_m, depends on the length and the degree of saturation of the acyl chains, as well as on the conformation and the charge density and its distribution within the headgroup region. From the variability in the acylation patterns and in the chemical structure and composition of the sugar moiety, a complex phase behaviour and structural polymorphism are to be expected for all glycolipid polymers.

This chapter focuses on the description of the physicochemical characteristics of essentially three groups of bacterial glycolipids: (i) lipopolysaccharides (LPSs), (ii) rhamnolipids and (iii) mycobacterial glycolipids.

2. LIPOPOLYSACCHARIDES

In 1971, it was found that chemical modifications of LPS structure did not affect endotoxic activities, which led to the conclusion that the lipid A represents the endotoxic principle (Rietschel *et al.*, 1971). From that time on, various research groups have studied a variety of different chemical structures of this macromolecule for

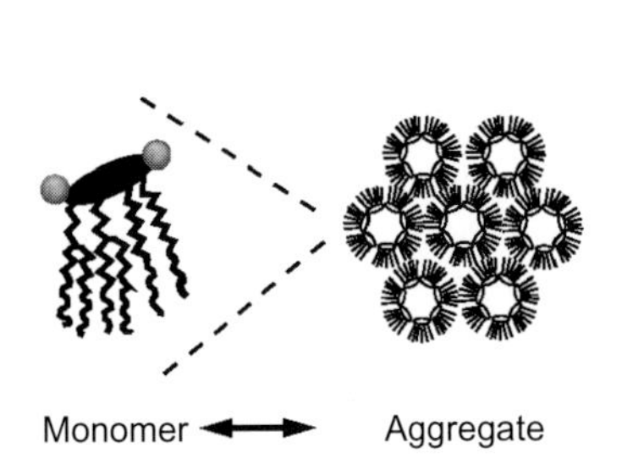

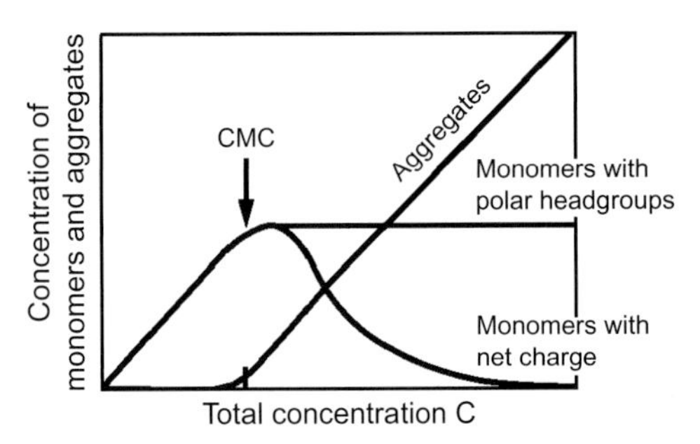

FIGURE 38.1 The interconnection between the total concentration of an amphiphilic molecule and the concentrations of monomers and aggregates.

correlating biological activity of LPS to specific chemical groups (Rietschel *et al.*, 1990; Loppnow *et al.*, 1993; Zähringer *et al.*, 1996). Roughly, from these investigations, it was deduced that the lipid A part is the "endotoxic principle" and that the sugar moiety may function merely as a solubilizing "carrier" for the essential biologically active part (Rietschel *et al.*, 1971) (see also Chapter 3). The relatively great variations in the core oligosaccharide and, even more, in that of the O-antigen structure between different bacterial strains and genera (Wilkinson, 1996; Holst *et al.*, 1998) (see Chapters 3 and 4), was assumed to be indicative of the main role of the sugar part as a defence factor of the bacterial cell, e.g. by contributing to the permeability barrier of the bacterial cell and functioning as the recognition structure for cells and components of the immune system. This is, however, not the only role of the sugar component of LPS. Although lipid A is the endotoxic principle, i.e. it exerts qualitatively all the characteristic biological activities observed in mammals, it was found that the addition of the sugar portion leads to an increase of bioactivity by at least 100-fold (Rietschel *et al.*, 1996). This is already valid for LPS with a very short sugar chain, i.e. Re-LPS, which contains, beside the lipid A moiety, only two monosaccharides of 3-deoxy-D-*manno*-oct-2-ulonic acid (Kdo). Because the added sugar residues by themselves are completely inactive, it can be assumed that it is not necessarily the single chemical groups within one molecule that are decisive, but that the physicochemical behaviour of whole aggregates of these amphiphiles is relevant, and which is discussed in detail in the following text.

In the first part of this chapter the basic physicochemical prerequisites for endotoxically active lipid A molecules are described, specifically, the supramolecular three-dimensional aggregate structure – which allows the determination of the molecular shape (i.e. conformation) – and the intramolecular conformation. Data are supplemented by the description of investigations into the morphology of endotoxins by laser light scattering and electron morphology. Importantly, it can be assumed that the sizes and the distribution of the endotoxin aggregates play a decisive role in the recognition process by immune cells.

2.1. The CMC, aggregate structure and phase states

As described, amphiphilic molecules, such as endotoxins, form in aqueous dispersions above the CMC three-dimensional supramolecular aggregates, sometimes also called micelles. Here, we use the term "micelles" for a particular aggregate structure (see below) and not as a synonym for aggregates. Reliable values for the CMC of lipid A and LPS have not been published so far. The data given by Aurell and Wistrom (1998) in the μM range reflect the sensitivity of the fluorescence and light scattering technique used rather than real CMC values. Measurements with partial structures indicate very low CMC values for lipid A, i.e. $<10^{-8}$M (Maurer *et al.*, 1991; Seydel *et al.*, 2000a) and the data from Takayama *et al.* (1990) indicate CMC values between 10^{-7} and 10^{-8}M for deep-rough mutant Re-LPS. This is in accordance with new data from Sasaki and White (2008) who found values in the nanomolar range. These investigators found a decrease of the "hydrodynamic radius" around 10 nM at 37°C for LPS Re (lipid A plus two Kdo monosaccharides). Unfortunately, they did not show the detection limit of the method by using, e.g. latex particles, so that the given value can be assumed as the upper limit of the actual CMC.

Due to the fact that LPS aggregates seem to be extremely stable, aggregates should predominate in the concentration range relevant for biological responses. However, as for every amphiphilic system, monomers are also present in a dynamic equilibrium depending on the actual concentration (Israelachvili, 1991) (see Figure 38.1). By applying synchrotron radiation X-ray diffraction, complete phase diagrams were established for enterobacterial lipid A (Brandenburg *et al.*, 1990,

1998). In all cases, at lower water content (<70%) and at a higher concentration of divalent cations such as Mg^{2+} ([lipid A]:[Mg^{2+}] < 3:1 molar ratio), only multilamellar structures are formed. At high water content and lower cation concentrations, non-lamellar structures are formed, in particular under near physiological conditions. This is exemplified in Figure 38.2 for lipid A at an 85% water content, showing the existence of a cubic inverted structure between 20° and 60°C and an inverted hexagonal phase between 70° and 80°C. The preferred cubic structures comprise the bicontinuous phases (Luzzati *et al.*, 1987) of symmetry Q^{224} (Pn3m), Q^{229} (Im3m) and Q^{230} (Ia3d). Such data, first published in 1990, were confirmed independently 18 years later (Reichelt *et al.*, 2008) without citing the former work. Also, hexa-acyl lipid A structures were investigated in which the phosphates were substituted by carboxymethyl groups. The aggregate structures were clearly identical to those of "normal" lipid A samples (Seydel *et al.*, 2005). The investigation of synthetic tria-acylated monosaccharide lipid A part structures, corresponding to the non-reducing moiety of enterobacterial lipid A (Matsuura *et al.*, 1999), showed a surprising behaviour in that the compound with an acyloxyacyl group linked to position-3 and an unbranched fatty acid to position-2 of the D-glucosamine (GlcN) exhibited a non-lamellar, probably cubic phase, whereas for the reversed linkage a multilamellar phase was observed (Brandenburg *et al.*, 2002). This could be correlated to different inclination angles of the backbone with respect to the direction of the acyl chains and, most interestingly, also to different biological activities (see later).

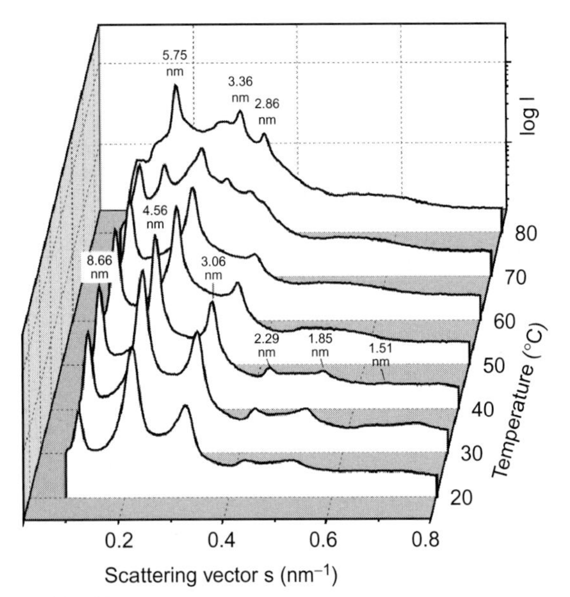

FIGURE 38.2 Synchrotron radiation small-angle X-ray scattering (SAXS) pattern of hexa-acyl lipid A from *S. enterica* sv. Minnesota. The logarithm of the scattering intensity log I is plotted versus the scattering vector ($s = 1/2d = \sin\theta/\lambda$, d = spacings, θ = scattering angle, λ = wavelength = 0.15 nm). The measurement was kindly performed by M.H.J. Koch at the European Molecular Biology Laboratory, Hamburg outstation, Germany.

Further studies showed that lipid A from other sources, such as from some phototropic strains, tend to form multilamellar structures (Brandenburg *et al.*, 1993; Schromm *et al.*, 1998, 2000). Also, enterobacterial lipid A with one or two cleaved acyl chains forms lamellar structures (Schromm *et al.*, 2000).

The structural polymorphism was also elucidated for deep-rough mutant Re-LPS (Brandenburg *et al.*, 1992) and other rough mutant LPS chemotypes, i.e. Rd through Ra, (Seydel *et al.*, 1993) from enterobacterial strains proving the existence of non-lamellar aggregate structures under near physiological conditions. Except at low water content and high divalent cation concentrations, no multilamellar structures were found.

It is important to note that there are lipid A variants which exhibit, despite possessing an identical backbone to that of enterobacterial lipid A, significantly reduced or completely absent biological activity which is connected with a changed acyl chain moiety. Such LPS may be able to act as antagonists to biologically active LPS or lipid A (Mayer *et al.*, 1990; Loppnow *et al.*, 1993).

2.2. Morphology and size distribution

A pioneer in investigating endotoxins by electron microscopy (EM) was the group of Shands (1971). They investigated smooth-form LPS and found long filaments and ribbon-like structures. These investigations, as well as those from Risco *et al.* (1993), were hampered by their use of wild-type LPS, which is a heterogeneous mixture of various compounds, and it is difficult to decide which fraction within this mixture corresponds to the biologically active fraction; an Ra- or Rb-LPS according to Jiao *et al.* (1989). Risco *et al.* (1993) have applied freeze-fracture in comparison to negative staining in EM and found that the former technique gave mainly spherical or elliptical particles of rather homogeneous size, whereas the latter technique showed a more heterogeneous population. These authors concluded that negative staining gives rise to artefactual forms due to the preparation technique. This interpretation is in full accordance to the small-angle X-ray scattering (SAXS) data cited above, in which, at low water content, only multilamellar structures were found. The negatively stained samples are dried before measurement in the electron microscope, thus changing the morphologies of the samples.

In recent investigations, freeze-fracture EM has been applied to analyse the morphologies of rough-LPS mutant forms. As an example of this, micrographs are presented for a deep-rough mutant Re-LPS from *Salmonella enterica* sv. Minnesota R595 (Figure 38.3) and a mutant Ra-LPS with a complete core oligosaccharide from *S. enterica* sv. Minnesota R60 (Figure 38.4) [See also the comparable micrographs of Andrä *et al.* (2007) and Chen *et al.* (2007).] It can be seen that the former sample exhibits spherical-like particles, but no closed liposomes, rather "open egg-shells" are formed. When increasing the oligosaccharide chain, the morphology completely changes, now giving rise to ribbon-like fibrillary structures (see Figure 38.4). The diameter of the fibrils is in the range of 10–12 nm, which would

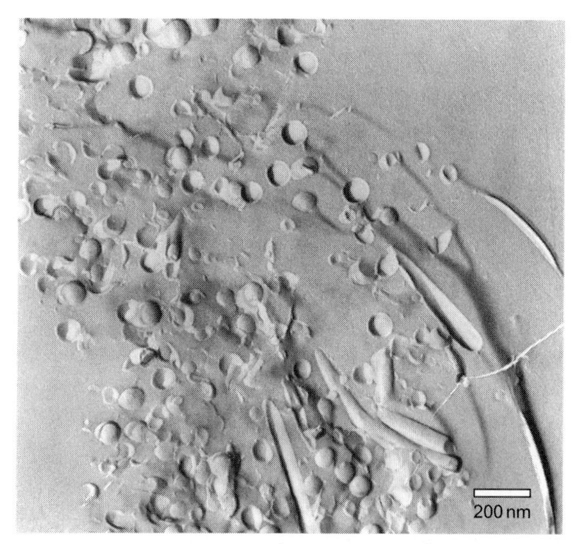

FIGURE 38.3 Freeze-fracture electron micrograph of LPS from *S. enterica* sv. Minnesota strain R595. The measurement was kindly performed by W. Richter, Friedrich-Schiller-Universität Jena, Germany.

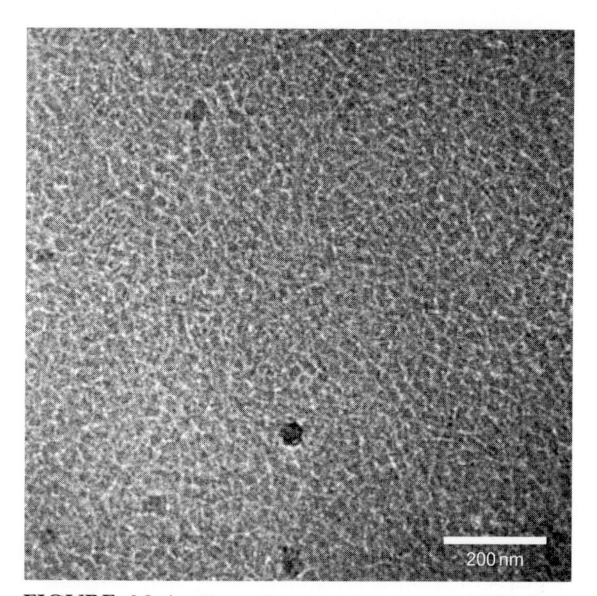

FIGURE 38.4 Cryo-electron micrograph of LPS from *S. enterica* sv. Minnesota strain R60. The measurement was kindly performed by W. Richter, Friedrich-Schiller-Universität Jena, Germany.

correspond to the bilayer repeat of Ra-LPS, which has been determined in the presence of Mg^{2+} (Seydel *et al.*, 1993).

The aggregate sizes of various LPS have been analysed by using laser light scattering techniques. Typical sizes for LPS rough mutants are 100–600nm (Howe *et al.*, 2007; M. Hammer, unpublished results). These size distributions, however, may not necessarily be very meaningful due to the – except for Re-LPS – very complex geometry of the aggregates deviating considerably from spherical forms.

2.3. Intramolecular conformation

Important aspects of the intra- and intermolecular conformation of lipid A, such as the inclination angle of the backbone, can be determined by applying Fourier-transform infra-red (FTIR) spectroscopy using an attenuated total reflectance unit with polarized infra-red light (Seydel *et al.*, 2000b). For enterobacterial lipid A, a strong inclination of the diGlcN backbone with respect to the direction of the hydrocarbon chains of 40°–55° was observed, while lesser acylated lipid A (penta- and tetraacyl), as well as some non-enterobacterial lipid A, exhibited a much lower angle (<20°). Thus, the backbones of lipid A with a conical shape are strongly inclined and those with a cylindrical shape have – if at all – only a slight inclination. The interconnections between the intramolecular conformations, molecular geometry and biological activity of lipid A-like molecules are presented in Figure 38.5. It has to be emphasized, however, that this molecular conformation only holds for the single molecules within an aggregate.

These findings apparently express the packing constraints of the hydrocarbon chains with respect to the available cross-sectional areas and determine, most importantly, also the ability of lipid A to act as agonistic or antagonistic, as will be discussed later in this chapter.

It is possible to calculate the potential conformations accessible to a given molecule, based on the primary chemical structure and by application of molecular modelling techniques. In an early paper, theoretical calculations on the conformation of the hexa- and hepta-acyl lipid A components of *S. enterica* sv. Minnesota suggested that an angle of 45° exists between the plane of the bisphosphoryl disaccharide backbone and the direction of the fatty acids chains and that the fatty acids occupy positions lying almost exactly on a hexagonal lattice, whereas the terminal methyl groups do not form a plane parallel to the membrane surface (Kastowsky *et al.*, 1991). A length of 2.6nm and a cross-section of 0.6–0.8nm for the smaller side and 1.2–1.6nm for the longer side, of the rectangular cross-section of the acyl chains was caculated. This model gives a hexagonal dense packing of the acyl chains and a tilt of the diGlcN axis with respect to the membrane surface of 53.7° and, with regard to the reducing side of the diGlcN, i.e. the 1-phosphate emerging

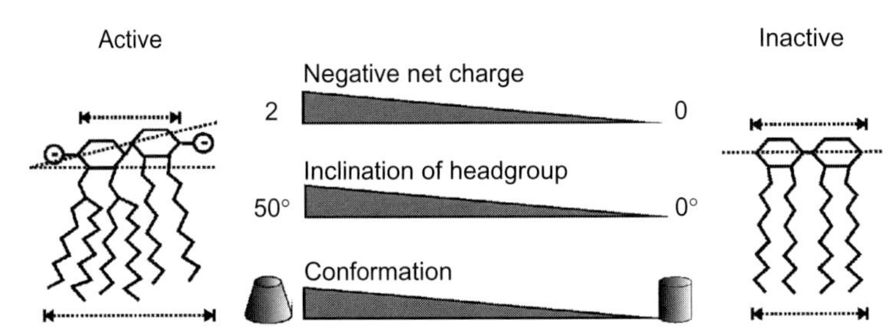

FIGURE 38.5 Connection between the chemical structure, net charge, inclination of head group, molecular conformation and biological activity of lipid A. Schematic representations of active and inactive lipid A molecules are shown on the left and right, respectively.

in the hydrophobic moiety of the neighbouring molecule, and the 4′-phosphate sticking out into the aqueous phase. In more recent papers, however, it has been suggested from infra-red spectroscopic experiments that the 1-phosphate is surrounded by water and the 4′-phosphate is buried in the backbone-close hydrophobic region, probably facing the 3-hydroxyl groups of the neighbouring molecules (Brandenburg *et al.*, 1997; Seydel *et al.*, 2000b). This was deduced from the observation that the 4′-monophosphoryl lipid A compound 504 showed nearly no phosphate band component corresponding to high hydration, in contrast to the 1-monophosphoryl compound 505. Despite the fact that this packing of the acyl chains is not optimal, this model could be verified by the following observations:

(i) Several lipid A analogues and partial structures do not have a hexagonal dense packing.
(ii) Lipid A and LPS have strongly reduced enthalpy changes (ΔH_c) of the gel-to-liquid crystalline phase transition as compared to saturated phospholipids with the same acyl chain length (Brandenburg and Seydel, 1984). A comparison on the basis of the number of methylene groups indicates a reduced packing density of LPS as compared to saturated phospholipids.

Frecer *et al.* (2000) studied, in a molecular dynamics investigation, a range of synthetic lipid A and partial structures, namely, synthetic bisphosphoryl compound 506 (LA-15-PP), monophosphoryl partial structures 504 and 505 (LA-15-PH and LA-15-HP) and dephosphorylated compound 503 (LA-15-HH), with respect to the conformation of the diGlcN headgroup, order and packing of the acyl chains, solvation of phosphate groups and other parameters. They also deduced from these results possible mechanisms of action with target cells. Frecer *et al.* (2000) did not find any significant geometrical differences in the molecular shape, mutual position and orientation of the GlcN rings, orientation of the phosphate groups or compact alignment of the fatty acid chains. The angle between the backbone axis of the diGlcN headgroup and the vector aligned parallel to the acyl chains was estimated to be 84°, 69°, 75° and 87° for compounds 503, 505, 504 and 506, respectively, which would correspond to an inclination angle as described above of 6°, 21°, 15° and 3°, respectively. These values are in contrast to those determined experimentally (Brandenburg *et al.*, 1997), at least for compounds 504 and 506 and, furthermore, the authors found an increase in the tendency of forming non-lamellar aggregates with decreasing number of phosphate groups, i.e. compound 503 has the highest tendency to adopt non-lamellar structures and compound 506 the lowest. This is the reversed sequence as found in experiments and can be readily explained by the non-consideration of the bridging function of divalent cations: The non-lamellar structure of compound 506 is essential due to the fact that cation bridging of neighbouring phosphate groups leads to a smaller backbone cross-section, which is less present in the monophosphoryl compound and absent for the dephospho-lipid A, thus explaining the preference for more lamellar structures.

2.4. Physicochemistry of agonism and antagonism

The biological activity of endotoxins has to be discussed in the context of the interaction of LPS molecules with the membranes of host immune cells, since the interaction of endotoxin molecules with membrane-associated proteins is an important step in the initiation of biological effects. For normal functioning of immune and other cells, the maintenance of a particular lipid composition of the membrane at given ambient conditions is a prerequisite (Shinitzky, 1984). An uptake/intercalation of exogeneous lipids, such as LPS, with differences in their chemical structures, such as acylation pattern, headgroup conformation and net electrical charge compared to those of the normal constituents of the cell matrix, can lead to alterations of membrane fluidity and/or permeability, phase separation and domain formation, disturbance of the lamellar membrane architecture and internalization of the extraneous lipids. If the cell is not able to alter the composition of the lipid matrix, so-called homoviscous adaptation (Cossins and Sinensky, 1984), these membrane alterations may cause severe dysfunctions of the cell, which may manifest themselves, e.g. in transient or permanent alterations in the functioning of transmembrane proteins which might be involved in signal transduction.

Our model of endotoxin activation consists of a first step of an LPS-binding protein (LBP)-induced intercalation of LPS into target immune cell membranes (Gutsmann *et al.*, 2000, 2007). Within the membrane, LPS forms domains due to the differences in chemical structures with respect to the membrane lipids. These domains may laterally diffuse within the target cell membrane and represent a strong disturbance of the membrane architecture because of the conical shape of the lipid A moiety of LPS. At the site of a signalling protein, this may lead to cell activation (Blunck *et al.*, 2001).

In a recent paper, Nomura *et al.* (2008) reported an interaction study of highly purified Re-LPS with phospholipid membranes by applying ^{31}P nuclear magnetic resonance (NMR) spectroscopy. These investigators found that LPS becomes part of the phospholipid membranes (phosphatidylcholine:phosphatidylglycerol, 9:1) in the form of aggregates, i.e. forming domains within the target cell membrane. This finding is a strong confirmation of the above described model of cell activation. Another aspect important for the biological activity of LPS is the mechanism of antagonism, i.e. the inhibitory effect of endotoxically non-active compounds that are able to inhibit cell activation by enterobacterial LPS, e.g. that by the tetra-acyl compound 406. When investigating the chemical composition of several natural lipid A preparations derived from *Escherichia coli*, it was found that besides the hexa-acylated lipid A, also significant amounts of penta- and tetra-acylated molecules were present in these isolates (Mueller *et al.*, 2004). Although the latter are known to antagonize the induction of cytokines by biologically active hexa-acylated endotoxins, their occurrence in these preparations did obviously not reduce the biological activity. In order to understand this phenomenon, separate aggregates of either synthetic compound 506 or 406 (tetra-acylated precursor IVa) were prepared by mixing these at different molar ratios, as well as preparing molecularly mixed aggregates containing both compounds in the same ratios. Surprisingly, the latter mixtures showed higher endotoxic activity than that of the pure compound 506, up to an admixture of 20% of compound 406 (Mueller *et al.*, 2004). These observations can only be understood by assuming that the active unit of endotoxins is the aggregate and may be interpreted as a decrease of the binding energy of 506 molecules within the aggregates induced by molecularly incorporated 406, thus allowing a better accessibility to binding proteins of the host (see also Müller *et al.*, 2002). This interpretation seems reasonable since it has been shown that the acyl chain order of compound 506 is relatively high, corresponding to a high gel to liquid crystalline phase transition temperature, T_c, of around 45°C (Brandenburg and Seydel, 1990). The addition of compound 406 with its $T_c < 20°C$ (Brandenburg *et al.*, 1997) leads to a decrease in acyl chain order and, thus, to a higher ability to interact with target structures. Besides compound 406, similar effects on the biological action of endotoxins were also reported for negatively charged phospholipids, such as cardiolipin, phosphatidylglycerol and phosphatidylinositol (Mueller *et al.*, 2005a,b).

Summarized, the presented concept of "endotoxic conformation", or better termed "endotoxic supramolecular conformation", is of general validity but does not necessarily need endotoxin chemical structures (see Figure 38.5). It has been shown also that non-LPS structures with six hydrocarbon chains and two phosphate groups, but with a serine- (Ser-) like backbone rather than the diGlcN part of lipid A, adopt cubic inverted structures and express full endotoxic activity (Seydel *et al.*, 2003; Brandenburg *et al.*, 2004). In a similar way, it was also found that bacterial lipopeptides may form non-lamellar cubic aggregate structures that are connected with considerable endotoxic activity (Schromm *et al.*, 2007). Thus, a necessary and sufficient condition for endotoxic activity is:

(i) The existence of a conical shape of the single monomers within the aggregates with a higher cross-section of the hydrophobic than the hydrophilic moiety.
(ii) The existence of charges, which maybe either negative (the normal case) or positive.
(iii) The transport of these aggregates into target cell membranes alone, i.e. without a carrier (high hydrophobicity) or by the action of binding proteins such as LBP.

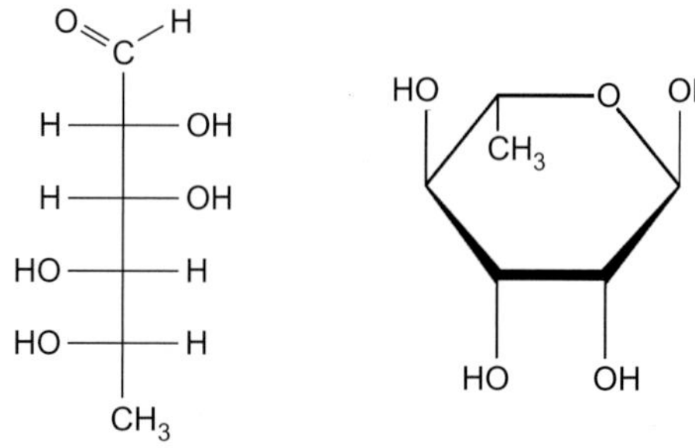

FIGURE 38.6 Chemical structures of L-rhamnose (left) and α-L-rhamnose (right).

3. RHAMNOLIPIDS

L-Rhamnose (L-Rha) (Figure 38.6), also denoted as 6-deoxymannose and isodulcit, is the best known 6-deoxyhexose. As a free sugar, it is present in wine (0.02–0.04%). This sugar also forms one unit of rutinose, a disaccharide, and is present in a number of naturally occurring glycosides, e.g. xanthorhamin, quercitrin, strophantin, naringin, hesperidin and diverse anthocyanes. Various *Pseudomonas* strains are able to synthesize surfactants containing this unusual sugar L-Rha. These surfactants, also denoted as biosurfactant, are glycolipids, composed of a polar part, the sugar moiety and a hydrophobic part, the hydrocarbon chains. Some prominent rhamnolipids are shown in Figure 38.7A and B. Rhamnolipids are anionic with a pK value of approx. 5–6. The hydrocarbon chains are often derived from 3-(3-hydroxyalkanoyloxy) alkanoic acids, as shown in Figure 38.7C (where m and n

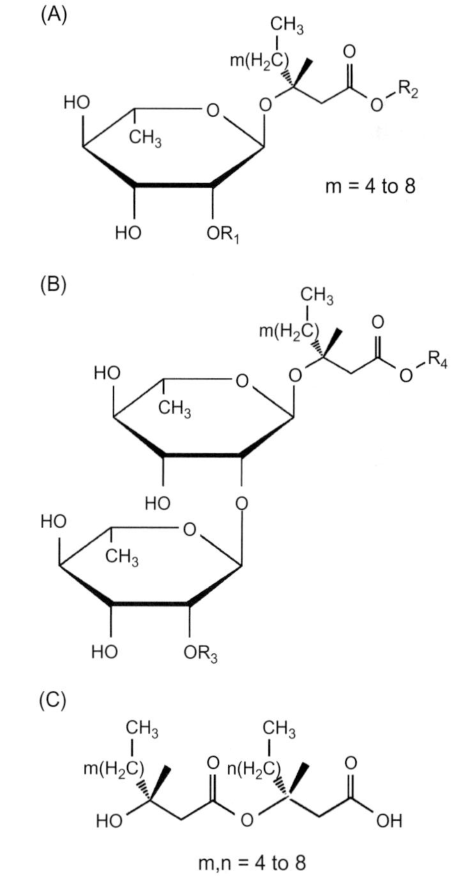

FIGURE 38.7 Chemical structures of the main rhamnolipid homologues RL-A, RL-B and RL-1 to RL-4. (A) Monorhamnolipid, (B) dirhamnolipid and (C) 3-(3-hydroxyalkanoyloxy)-alkanoic acids (HAAS). $R_1 - R_4$, Acyl chains of differing length (C_{12} to C_{16}).

vary between 4 and 8). Depending on the provenance of the rhamnolipid, saturated as well as unsaturated chains of different lengths have been isolated.

3.1. Isolation, purification and production

Rhamnolipids, as surface-active metabolites, are produced commercially from microorganisms (e.g. *Pseudomonas aeruginosa*) (Hauser and Karnovsky, 1959; Soberén-Chávez *et al.*, 2005), using biotechnological fermentation procedures (Desai and Banat, 1997; Perfumo *et al.*, 2006). The structure and amount of rhamnolipids produced by the microorganisms depend on the used strain, cell culture media substrate, etc. Rhamnolipids may vary in the number of Rha in the polar head group (see Figure 38.7), as well as on the composition of the acyl chains.

3.2. Glycolipid analytics

Glycolipid analytics is often quite complex (Rendell *et al.*, 1990). In the following, we present techniques that are used to separate, identify and characterize the Rha-glycolipids. This is a prerequisite for their physicochemical characterization.

3.2.1. *Surface tension*

Due to their amphiphilic nature, rhamnolipids are surface-active. The CMC has been determined using tensiometry (Guerra-Santos *et al.*, 1984) or an electro-analytical assay using mercury electrode (Emons *et al.*, 1989).

3.2.2. *Colorimetric assays*

Colorimetric assays are based on the binding of a specific dye to the rhamnolipid, e.g. cetyltrimethyl-ammonium bromide, or by a chemical reaction of the Rha moiety with a coloured chemical compound like 9,10-dihydro-9-oxoantracene, with photometry measured at 625 nm.

3.2.3. *Chromatography*

A number of chromatographical methods are available for the separation and identification of glycolipids. In particular, the coupling with mass spectrometry (MS) has been revealed to be a powerful method, e.g. for the identification of the chain length diversity observed for rhamnolipids.

In order to analyse rhamnolipids by gas chromatography (GC), the lipid sample has to be prepared, i.e. the rhamnolipids are hydrolysed using trifluoroacetic acid or a strong base like NaOH. Using these hydrolysis reagents, the protocol is such that the two fatty acid chains of the dirhamnolipid are not cleaved. For the detection, GC is coupled to a flame ionization detector or mass spectrometers.

High-performance liquid chromatography (HPLC) techniques allow an appropriate and complete separation of different rhamnolipid species as well as the identification and quantification of the lipids. This is done by coupling HPLC to specific detection devices like an ultraviolet spectrometer (after derivation to *p*-bromophenacyl-esters), mass spectrometer or evaporative light scattering detectors. For the elucidation of rhamnolipid isomers, HPLC has to be combined to tandem MS. The isomers are detected based on their distinct fragmentation pattern (Heyd *et al.*, 2008).

3.2.4. *Mass spectrometry*

The structural analysis as well as the composition of glycolipid mixtures is effectively analysed using tandem quadrupole mass spectrometers. For rhamnolipid ionization, electrospray ionization for direct analysis via HPLC-MS is routinely applied. Electrospray ionization is a soft ionization method, because the primary molecules show little fragmentation.

3.2.5. *The FTIR technique*

Application of FTIR is used for the identification of rhamnolipids (Andrä *et al.*, 2006), especially by analysing the infra-red fingerprint region. Particularly in the attenuated total reflectance mode, FTIR can be used as a process analytical technology. Also, the thermotropic properties of rhamnolipids are elegantly analysed by FTIR (Howe *et al.*, 2006).

3.2.6. *The NMR technique*

Accurate structural information is derived from NMR using solid-state and high resolution techniques. The NMR parameters of relevance for structural characterization of glycolipids are the chemical shifts of the absorption frequency, coupling constants and the integral height of the NMR signals. Additionally, different NMR techniques can be applied as correlation spectroscopy and heteronuclear multiple quantum coherence, or one- and two-dimensional ^1H and ^{13}C NMR spectroscopy. Choe *et al.* (1992) have reported ^1H chemical shifts of various rhamnolipids.

3.3. Physicochemical characterization

The major applications of biosurfactants, especially for rhamnolipids, are due to the efficiency and effectiveness of these surfactants to reduce surface and interfacial tension. Therefore, a number of studies are related to the surface-active properties of rhamnolipids (Özdemir *et al.*, 2004; Cohen and Exerowa, 2007; York and Firoozabadi, 2008). Table 38.1 summarizes some relevant data. The surface tension and the interfacial tension are important parameters for the evaluation of the surface-active properties of

TABLE 38.1 Surface tension, critical micelle concentration (CMC) and interfacial tension of various rhamnolipids and rhamnolipid fractions, respectively, obtained from different organisms as described in the literature

Pseudomonas aeruginosa strain number	Surface tension (mN/m)	CMC (mg/l)	Interfacial tension (mN/m)	References	Remarks
47T2	33	109	1.0[a]	Haba *et al.*, 2003a,b	RL[d] mixture
AT10	29	106	nd[b]	Abalos *et al.*, 2001	$Rha_2C_{10}C_{10}$
LB1	24	120	1.3[c]	Benincasa *et al.*, 2004	RL mixture
44T1	25	11	0.2[a]	Parra *et al.*, 1989	$RhaC_{10}C_{10}$
44T1	25	11	1.0[a]	Parra *et al.*, 1989	$Rha_2C_{10}C_{10}$
DSM2874	26	20	<1.0[c]	Syldatk *et al.*, 1985	$RhaC_{10}C_{10}$ + $Rha_2C_{10}C_{10}$
DSM2874	28	20	<1.0[c]	Syldatk *et al.*, 1985	RL mixture
DSM2874	26	20	4.0[c]	Syldatk *et al.*, 1985	$RhaC_{10}C_{10}$
DSM2874	27	10	<1.0[c]	Syldatk *et al.*, 1985	$Rha_2C_{10}C_{10}$
DSM2874	30	200	<1.0[c]	Syldatk *et al.*, 1985	Rha_2C_{10}
DSM2874	25	200	<1.0[c]	Syldatk *et al.*, 1985	$RhaC_{10}$
UG-2	31	53	nd	Mata-Sandoval *et al.*, 1999	$RhaC_{10}C_{10}$ + $Rha_2C_{10}C_{10}$

[a] Determined against kerosene;
[b] nd, not determined;
[c] determined against hexadecane;
[d] RL, rhamnolipid.

rhamnolipids, e.g. foam formation (Hung and Shreve, 2001; Cohen *et al.*, 2003; Chen, 2004). Studies on rhamnolipids film foam have shown the validity of the Derjaguin-Landau-Verwey-Overbek theory of the stability of lyophobic colloids (Cohen *et al.*, 2004). Of note, Cohen and Exerowa (2007) have described the formation of common, common black films and Newton black films for rhamnolipid films.

The CMC of rhamnolipids can be determined via tensiometry measurement, mercury electrodes or isothermal titration calorimetry (Emons *et al.*, 1989; Hildebrand *et al.*, 2004; Aranda *et al.*, 2007). The CMC is a crucial parameter for the evaluation of the membranolytic activity of surfactants (Sánchez *et al.*, 2007; Garidel *et al.*, 2007b) (see also below).

The CMC of rhamnolipids is quite low compared to other biosurfactants like bile salts (Garidel *et al.*, 2000). Due to their anionic properties, the CMC of rhamnolipids (pK $\approx$ 5–6) is strongly pH-dependent (Özdemir *et al.*, 2004). Sánchez *et al.* (2007) have determined the CMC at various pH via surface-tension measurements. At pH 7.4, the CMC of dirhamnolipids is 0.110mM, whereas at pH 4.0, the CMC decreases by one order of magnitude to 0.010mM. This indicates that the negatively charged dirhamnolipid has a much higher CMC value than the neutral chemical species. A decrease of the CMC is also observed by increasing the ionic strength of the medium.

The influence of the protonation degree with respect to the formed aggregate structure has been obtained by EM, fluorescence microscopy, dynamic light scattering, small-angle neutron scattering (SANS), as well as SAXS (Ishigami *et al.*, 1987, 1993; Ishigami, 1997; Champion *et al.*, 1995; Sánchez *et al.*, 2006, 2007; Andrä *et al.*, 2006; Howe *et al.*, 2006).

Above the CMC, large aggregates are formed. Using EM, Sánchez *et al.* (2007) have observed for a dirhamnolipid, isolated from the strain *P. aeruginosa* 47T2 ($Rha_2C_{10}C_{10}$), mainly the formation of multilamellar vesicles of heterogeneous size. The X-ray scattering gave a value of 3.2nm for the interlamellar repeat distance of these vesicles. Considering molecular modelling data for the Rha-Rha-C_{10}-C_{10} moiety with the structural results, Sánchez *et al.* (2007) concluded a possible intermolecular interaction responsible for the stabilization of dirhamnolipid aggregates. Recently, Dahrazma *et al.* (2008) published a SANS study for a rhamnolipid and dirhamnolipid isolated from *P. aeruginosa* obtained from a commercial source (JBR215) at pH 5 and 13. This study indicates the importance of the pH of the system for the formed aggregate morphology. Under alkaline conditions, large aggregates (>200nm) and micelles of approx. 1.7nm were formed, whereas under acidic conditions large polydispers vesicles with a radius from 55 to 60nm were detected. The morphology of the formed aggregates plays an important role for the application of rhamnolipids as a removal agent of heavy metals. Andrä *et al.* (2006) performed SAXS studies on various rhamnolipids. They were able to show that the investigated rhamnolipids show thermotropic as well as lyotropic properties. Lamellar as well as non-lamellar, cubic structures were detected. Using various synthetic rhamnolipids, these results were extended by Howe *et al.* (2006).

In-depth characterization of the physicochemical properties of rhamnolipids by, among others, infra-red spectroscopy by Howe *et al.* (2006) and Andrä *et al.* (2006) showed for the gel to liquid crystalline phases of different rhamnolipids a T_c at low temperatures, often below 0°C. Thus, under physiological temperature (37°C) most rhamnolipids are in the liquid crystalline phase. This may be of relevance to explaining the physiological effects of these lipids.

The interaction of rhamnolipids with phospholipid membranes showed that the partitioning is an entropically driven process. Aranda *et al.* (2007) showed that the presence of cholesterol in the membrane has an influence on the partitioning coefficient, which is decreasing with the increasing amount of cholesterol. They also described that phosphatidylethanolamine stimulates the binding of rhamnolipids, especially dirhamnolipids, whereas lysophosphatidyl-choline opposes binding. They proposed that the investigated dirhamnolipid behaves as an inverted cone-shaped molecule. The mixing properties of a dirhamnolipid with dielaidoylphosphatidyl-ethanolamine shows a solid-phase immiscibility (Sánchez *et al.*, 2006). The formation of domains in the presence of rhamnolipids with phospholipids has also been described.

3.4. Biological properties

It is known that *P. aeruginosa* is an opportunistic human pathogen. This bacterium causes chronic lung infection in patients having severe lung diseases, i.e. cystic fibrosis. It is considered that the reason for this effect lies in the excretion of rhamnolipids, i.e. exotoxins, by *P. aeruginosa*, which induces the solubilization of the lung surfactants and, as a consequence of the membranolytic activity, chronic infections and even lung insufficiency may occur (Zulianello *et al.*, 2006; Thanomsub *et al.*, 2007). Importantly, Zulianello *et al.* (2006) showed rhamnolipids from *P. aeruginosa* are the virulence factors promoting early infiltration of primary human airway epithelia.

Various studies have described the antimicrobial activity of rhamnolipids (Haba *et al.*, 2003b), their effects on cell aggregation and the influence of cell surface hydrophobicity (Herman *et al.*, 1997; Al-Tahhan *et al.*, 2000; Zhong *et al.*, 2007). Also, biophysical data have been correlated with biological data (Brandenburg *et al.*, 2003; Garidel *et al.*, 2007a) as a promising approach in order to gain more insights into the biological mechanisms. Haba *et al.* (2003 a,b) investigated the physicochemical characterization of rhamnolipids produced by *P. aeruginosa* 47T2 NCBIM 40044 and correlated these with their antimicrobial properties. They isolated via HPLC-MS up to 11 rhamnolipid homologues (Rha-Rha-C_8-C_{10}; Rha-C_{10}-C_8/Rha-C_8-C_{10}; Rha-Rha-C_8-$C_{12:1}$; Rha-Rha-C_{10}-C_{10}; Rha-Rha-C_{10}-$C_{12:1}$; Rha-C_{10}-C_{10}; Rha-Rha-C_{10}-C_{12}/Rha-Rha-C_{12}-C_{10}; Rha-C_{10}-$C_{12:1}$/Rha-$C_{12:1}$-C_{10}; Rha-Rha-$C_{12:1}$-C_{12}; Rha-Rha-C_{10}-$C_{14:1}$; Rha-C_{10}-C_{12}/Rha-C_{12}-C_{10}), which constituted the mixture denoted as RL47T2. The surface tension of RL47T2 decreased to 32.8 mN/m, compared to pure water and the interfacial tension against kerosene to 1 mN/m. The CMC for RL47T2 was 108.8 mg/ml.

Moreover, antimicrobial activity was evaluated according to the minimum inhibitory concentration (MIC), the lowest concentration of an antimicrobial agent that inhibits development of visible microbial growth. Comparatively, low MIC values were found for the bacteria *Serratia marcescens* (4 μg/ml), *Enterobacter aerogenes* (8 μg/ml), *Klebsiella pneumoniae* (0.5 μg/ml), *Staphylococcus aureus* and *Staphylococcus epidermidis* (32 μg/ml), *Bacillus subtilis* (16 μg/mL) and the phytopathogenic fungal species *Chaetonium globosum* (64 μg/ml), *Penicillium funiculosum* (16 μg/ml), *Gliocadium virens* (32 μg/ml) and *Fusarium solani* (75 μg/ml) (Haba *et al.*, 2003a,b).

Also, the permeabilization effect of a rhamnolipid on Gram-positive and Gram-negative bacterial strains was analysed by Sotirova *et al.* (2008). The study revealed that the rhamnolipids have a neutral or detrimental effect on the growth of Gram-positive strains and this was dependent on the surfactant concentration. The growth of Gram-negative strains was not influenced by the presence of the biosurfactants in the media. However, the detergent-like action of rhamnolipids is still controversially discussed.

Andrä *et al.* (2006) and Howe *et al.* (2006) have been the first to show that a special rhamnolipid exhibits strong stimulatory activity on human mononuclear cells to produce tumour necrosis factor-alpha (TNF-α). The over-production of TNF-α is known to cause sepsis and septic shock syndrome. Such a property has not been noted thus far for rhamnolipid exotoxins. Consequently, these investigators analysed the rhamnolipid, 2-*O*-α-L-rhamnopyranosyl-α-L-rhamnopyranosyl-α (*R*)-3-hydroxytetradecanoyl-(*R*)-3-hydroxytetradecanoate, with respect to its pathophysiological activities as a heat-stable extracellular toxin. Like LPS, the cell-stimulating activity of the rhamnolipid could be inhibited by incubation with polymyxin B. Nevertheless, immune cell activation by the rhamnolipid does not occur via receptors that are involved in LPS signalling, i.e. Toll-like receptor (TLR-) 4, or lipopeptide signalling, i.e. TLR-2 (see Chapter 31). Despite their completely different chemical structure, rhamnolipids exhibit a variety of endotoxin-related physicochemical characteristics, such as a cubic-inverted supramolecular structure. These data are in good agreement with the conformational concept of endotoxicity, i.e. intercalation of naturally occurring virulence factors into the immune cell membrane lead to strong mechanical stress on integral proteins, eventually causing cell activation (Andrä *et al.*, 2006). This has been confirmed by investigating synthetic, well-defined rhamnolipids (Howe *et al.*, 2006).

4. MYCOBACTERIAL GLYCOPOLYMERS

One of the most life-threatening diseases in the world is tuberculosis, with one-third of the world's population infected and more than 5000 deaths daily (Kaufmann, 2002). Due to an increase of multidrug resistant strains and an increase of people living with human immunodeficiency virus- (HIV-) acquired immunodeficiency syndrome (AIDS) and thus a strongly exacerbated risk of developing tuberculosis, mycobacterium–host cell interaction is an urgent research topic. Some mycobacteria, e.g. *Mycobacterium tuberculosis*, are intracellular pathogens which are able to persist inside macrophages. In contrast to Gram-negative bacteria, the cell wall of mycobacteria is more complex (see Chapters 1 and 9) and the knowledge of the individual cell-wall components being involved in the host immune response is limited. The results of a large number of studies suggest that, in contrast to LPS from Gram-negative bacteria, a single component alone is not responsible, but multiple components are required for the pathological mechanisms involved in infections with mycobacteria. The cell wall of mycobacteria adjacent to the phospholipid bilayer of the plasma membrane is comprised of a capsule, a thick, hydrophobic barrier and a thin peptidoglycan (murein) sacculus. This cell wall consists of genus or species-specific glycolipids and wax esters including long chain mycolic acids (MAs), derivatives thereof, trehalose dimycolate (TDM) and mycoloylarabinogalactan (MAG), as well as lipoarabinomannan (LAM) and phosphatidylinositol-mannosides (PIM) (see Chapters 9 and 21). The functions of these molecules, either in strengthening the mycobacterial outer barrier or in interacting with phagocyte membranes to facilitate intracellular survival, are not well studied.

4.1. Mycolylarabinogalactan–peptidoglycan (mAGP) complex

The basic scaffold of the mycobacterial cell wall is the mAGP complex. This complex is linked on one side to the thin peptidoglycan layer and on the other side to the mycolic acids. The galactan main chain is substituted by arabinan side chains and the hexa-arabinosyl-motif, which is substituted by five mycolic acids per arabinan chain (Besra *et al.*, 1995). The main and side chains are composed of β-D-Gal*f* (galactofuranose) and α-D-arabinofuranosyl (Ara*f*) residues, respectively. In the 23-sugar long galactan, 5- and 6-linked Gal*f* residues alternate (Brennan, 2003). The mAGP complex induces the release of the TNF-related apoptosis-inducing ligand (TRAIL) in neutrophils (Simons *et al.*, 2007).

4.2. Trehalose-dimycolate (TDM)

Trehalose 6,6′-dimycolate, also called cord factor (see Chapters 9 and 21), is a unique glycolipid in the cell wall of mycobacteria. The carbon chain length and the structure of the MAs have been found to be related to toxicity and bacterial intracellular survival (Gotoh *et al.*, 1991). The long MA chain in *M. tuberculosis* consists of 50 to 60 carbons and the α-branch is typically 24-carbons long. Among mycobacterial species and subspecies, the total number of carbons in the MA methyl ester of TDM can vary; numbers between 74 and 90 have been found (Fuijita *et al.*, 2005). It has been proposed that TDM may contribute to dehydration resistance of mycobacteria (Harland *et al.*, 2008).

The TDM glycolipid is able to induce a large number of cytokines and chemokines. It is able to stimulate innate, early adaptive and both humoral and cellular adaptive immunity (for review, see Ryll *et al.*, 2001). It has been shown that the *cis*-cyclopropanation of the α-MA is required for the full activation of the innate immune response during early infection (Rao *et al.*, 2005, 2006). In contrast, these same authors could show that the trans-cyclopropanation of the MAs on TDM leads to reduced virulence of *M. tuberculosis* and a suppression of the host immune response. Recently, it has been demonstrated that TDM-coated beads do not lead to an acidification of phagosomes, which is an essential step in killing of mycobacteria in macrophages (Axelrod *et al.*, 2008).

4.3. The PIMs, lipomannan (LM) and LAM

The PIMs and their multiglycosylated counterparts, LM and LAM, are major mycobacterial lipoglycans. The longest glycopolymer carries the LAM which is a complex of D-mannan and D-arabinan attached to a phosphatidyl-*myo*-inositol (PI) moiety that anchors the glycolipid in the mycobacterial cell wall (Hunter and Brennan, 1990). The D-mannan consist of a highly branched structure with an α-(1→6)-linked mannopyranose (Man*p*) backbone substituted at C-2 by single Man*p* units (Chatterjee *et al.*, 1992). The mannan size and the degree of branching can vary depending on the species. The arabinan consists of a linear α-(1→5)-linked Ara*f* backbone punctuated by branching produced by 3,5-*O*-linked α-D-Ara*f* residues. The lateral side chains are organized either as a linear tetra-arabinofuranoside or as a biantennary hexa-arabinofuranosides (Chatterjee *et al.*, 1991, 1993). *M. tuberculosis* and *Mycobacterium leprae* modify the termini with Man*p* residues (ManLAM), whereas fast-growing *Mycobacterium smegmatis* uses inositol phosphates (AraLAM) (Khoo *et al.*, 1995). These differences may be responsible for different biological functions.

The PIMs and LM have been reported to induce a strong proinflammatory response by inducing TNF-α, interleukin- (IL-) 8 and IL-12 secretion (Vignal *et al.*, 2003; Dao *et al.*, 2004). In contrast, ManLAM downregulates LPS-induced IL-12 production (Nigou *et al.*, 2001) and inhibits the phagosome/lysosome fusion (Chua *et al.*, 2004). As shown by Puissegur *et al.* (2007), all three compounds, PIM, LM and LAM, induce granuloma formation, however, only the proinflammatory PIMs and LM induce the fusion of granuloma into multinucleated giant cells. These same authors reported that LM induces large multinucleated giant cells resembling those found *in vivo* and that this process is mediated by TLR-2 and is dependent on the β1-integrin/ADAM9 cell fusion machinery.

Mycobacterium chelonae, a fast growing pathogenic mycobacterial species, expresses specific LM and LAM containing α-(1→3)-Man*p* side chains (Guérardel *et al.*, 2002). This motif reacts specifically with lectin from *Galanthus nivalis*. From cell activation experiments, Guérardel *et al.* (2002) suggested that the IP-capping in *M. smegmatis* may represent the major cytokine-inducing components of LAMs.

It has been shown that LM and LAM from *Mycobacterium kansasii* both bind CD14 and LBP immobilized on the surface of a surface plasmon resonance biosensor. The binding affinity of LM is higher to CD14 ($Kd = 2.89 \times 10^{-8}M$) than to LBP ($Kd = 0.84 \times 10^{-7}M$). These values are close to the ones published for the binding of LPS to

CD14 and LBP. The lipid anchor is responsible for the binding and not the carbohydrate domains (Elass *et al.*, 2007). Only the tri-acylated, but not the di-acylated, lipopeptides were shown to bind to CD14 (Nakata *et al.*, 2006). Moreover, LBP seems to enhance the stimulatory effect of LM and LAM on the matrix metalloproteinase-9 secretion in macrophages, through a TLR-1/TLR-2-dependent mechanism (Elass *et al.*, 2007).

4.4. Glycolipid 2,3-di-O-acyl-trehalose (DAT)

The cell-surface glycolipid DAT is found in all *M. tuberculosis* strains, but not in other species of the *M. tuberculosis* complex (Saavedra *et al.*, 2006). The DAT glycolipid contains multimethyl-branched mycocerosic fatty acids. It is able to reduce antigen-induced proliferation of human $CD4^+$ and $CD8^+$ T-cell subsets. This effect is associated with the down-modulation of IL-2, IL-12, TNF-α and IL-10 cytokines and with a decrease of cells expressing the T-cell surface activation markers CD25 and CD69. Thus, it is supposed that DAT is involved in the suppression of immune functions that are fundamental to host protection against *M. tuberculosis*.

4.5. Glycopeptidolipid (GPLs)

On the cell surface of bacteria of the *Mycobacterium avium* complex are antigenic GPLs (Aspinall *et al.*, 1995) (see Chapter 9). In the case of *M. smegmatis*, it has been shown that GPLs represent 85% of the surface-exposed lipids and play a role in the cell wall permeability (Etienne *et al.*, 2002). The C-type GPLs are composed of a common lipopeptidyl core consisting of a mixture of 3-hydroxy and 3-methoxy fatty acids with 26–34 carbon atoms amidated by a tripeptide, D-Phe-D-*allo*-Thr-D-Ala (where Ala, alanine; Phe, phenylalanine; Thr, threonine) and terminated by L-alaniol (Daffé *et al.*, 1983). This compound is glycosylated at the C terminal with 3,4-di-*O*-methyl-rhamnose and at the D-*allo*-Thr with a 6-deoxy-talose (6-dTal) (Aspinall *et al.*, 1995). Due to their species specificity, these compounds are responsible for distinguishing one serovar from another. It has been shown that modification of the GPL of *M. avium* can effectively neutralize ethambutol, which is known to inhibit arabinan synthesis (Khoo *et al.*, 1999). The modification is that instead of additional rhamnosylation on the 6-dTal, non-extended non-polar GPLs are expressed. The first observations showing their immunogenicity were made in 1982 (Barrow and Brennan, 1982). Importantly, GPLs participate in the receptor-dependent internalization of mycobacteria in human macrophages through complement receptor 3 and the mannose receptor (Villeneuve *et al.*, 2005). Kano *et al.* (2005) showed that when peripheral blood mononuclear cells were incubated with heat-killed *S. aureus* coated with GPL serovar 4, phagocytosis of monocytes increased and that after phagocytosis the phagosome/lysosome fusion in monocytes was inhibited.

5. FUTURE OUTLOOK

The open questions and areas for further research in this field are listed in the Research Focus Box.

ACKNOWLEDGEMENTS

This work was financially supported by the Deutsche Forschungsgemeinschaft (SFB470, project B5 and SFB 617, project A17). We thank Dr M.H.J. Koch at the European Molecular Biology Laboratory, Hamburg outstation for performing X-ray scattering and Dr W. Richter at the Universität Jena for performing electron microscopy.

RESEARCH FOCUS BOX

- There is still a severe lack of knowledge with respect to the physicochemical properties of glycopolymers from Gram-positive organisms, in particular from mycobacteria.
- Although the initial steps of the cell activation by Gram-negative glycopolymers like LPS and lipopeptides are well understood, the mechanisms of the final binding to cell surface receptors (e.g. TLR-2 and the TLR-4–MD2 complex) leading to intracellular signalling, have still to be elucidated.

References

Abalos, A., Pinazo, A., Infante, M.R., Casals, M., García, F., Manresa, A., 2001. Physicochemical and antimicrobial properties of new rhamnolipids produced by *Pseudomonas aeruginosa* AT10 from soybean oil refinery wastes. Langmuir 17, 1367–1371.

Al-Tahhan, R.A., Sandrin, T.R., Boour, A.A., Maier, R.M., 2000. Rhamnolipid-induced removal of lipopolysaccharide from *Pseudomonas aeruginosa*: effect on cell surface properties and interaction with hydrophobic substrates. Appl. Environ. Microbiol. 66, 3262–3268.

Andrä, J., Rademann, J., Koch, M.H.J., Heine, H., Zähringer, U., Brandenburg, K., 2006. Endotoxin-like properties of a rhamnolipid exotoxin from *Burkholderia* (*Pseudomonas*) *plantarii*: immune cell stimulation and biophysical characterization. Biol. Chem. 387, 301–310.

Andrä, J., Howe, J., Garidel, P., et al., 2007. Mechanism of interaction of optimized *Limulus*-derived cyclic peptides with endotoxins: thermodynamic, biophysical and microbiological analysis. Biochem. J. 406, 297–307.

Aranda, F.J., Espuny, M.J., Marqués, A., Teruel, J.A., Manresa, A., Ortiz, A., 2007. Thermodynamics of the interaction of a dirhamnolipid biosurfactant secreted by *Pseudomonas aeruginosa* with phospholipid membranes. Langmuir 23, 2700–2705.

Aspinall, G.O., Chatterjee, D., Brennan, P.J., 1995. The variable surface glycolipids of mycobacteria: structures, synthesis of epitopes, and biological properties. Adv. Carbohydr. Chem. Biochem. 51, 169–242.

Aurell, C.A., Wistrom, A.O., 1998. Critical aggregate concentration of Gram-negative bacterial lipopolysaccharides (LPS). Biochem. Biophys. Res. Commun. 253, 119–123.

Axelrod, S., Oschkinat, H., Enders, J., et al., 2008. Delay of phagosome maturation by a mycobacterial lipid is reversed by nitric oxide. Cell. Microbiol. 10, 1530–1545.

Barrow, W.W., Brennan, P.J., 1982. Immunogenicity of type-specific C-mycoside glycopeptidolipids of mycobacteria. Infect. Immun. 36, 678–684.

Benincasa, M., Abalos, A., Oliveira, I., Manresa, A., 2004. Chemical structure, surface properties and biological activities of the biosurfactant produced by *Pseudomonas aeruginosa* LBI from soapstock. Antonie van Leeuwenhoek 85, 1–8.

Besra, G.S., Khoo, K.H., McNeil, M.R., Dell, A., Morris, H.R., Brennan, P.J., 1995. A new interpretation of the structure of the mycolyl-arabinogalactan complex of *Mycobacterium tuberculosis* as revealed through characterization of oligoglycosylalditol fragments by fast-atom bombardment mass spectrometry and 1H nuclear magnetic resonance spectroscopy. Biochemistry 34, 4257–4266.

Blunck, R., Scheel, O., Müller, M., Brandenburg, K., Seitzer, U., Seydel, U., 2001. New insights into endotoxin-induced activation of macrophages: involvement of a K^+ channel in transmembrane signaling. J. Immunol. 166, 1009–1015.

Brandenburg, K., Seydel, U., 1984. Physical aspects of structure and function of membranes made from lipopolysaccharides and free lipid A. Biochim. Biophys. Acta 775, 225–238.

Brandenburg, K., Seydel, U., 1990. Investigation into the fluidity of lipopolysaccharide and free lipid A membrane systems by Fourier-transform infrared spectroscopy and differential scanninig calorimetry. Eur. J. Biochem. 191, 229–236.

Brandenburg, K., Koch, M.H.J., Seydel, U., 1990. Phase diagram of lipid A from *Salmonella minnesota* and *Escherichia coli* rough mutant lipopolysaccharide. J. Struct. Biol. 105, 11–21.

Brandenburg, K., Koch, M.H.J., Seydel, U., 1992. Phase diagram of deep rough mutant lipopolysaccharide from *Salmonella minnesota* R595. J. Struct. Biol. 108, 93–106.

Brandenburg, K., Mayer, H., Koch, M.H.J., Weckesser, J., Rietschel, E.Th., Seydel, U., 1993. Influence of the supramolecular structure of free lipid A on its biological activity. Eur. J. Biochem. 218, 555–563.

Brandenburg, K., Kusumoto, S., Seydel, U., 1997. Conformational studies of synthetic lipid A analogues and partial structures by infrared spectroscopy. Biochim. Biophys. Acta 1329, 193–201.

Brandenburg, K., Richter, W., Koch, M.H.J., Meyer, H.W., Seydel, U., 1998. Characterization of the nonlamellar cubic and H_{II} structures of lipid A from *Salmonella enterica* serovar Minnesota by X-ray diffraction and freeze-fracture electron microscopy. Chem. Phys. Lipids 91, 53–69.

Brandenburg, K., Matsuura, M., Heine, H., et al., 2002. Biophysical characterisation of triacyl monosaccharide lipid A partial structures in relation to bioactivity. Biophys. J. 83, 322–333.

Brandenburg, K., Andrä, J., Müller, M., Koch, M.H.J., Garidel, P., 2003. Physicochemical properties of bacterial glycopolymers in relation to bioactivity. Carbohydr. Res. 338, 2477–2489.

Brandenburg, K., Hawkins, L., Garidel, P., et al., 2004. Structural polymorphism and endotoxic activity of synthetic phospholipid-like amphiphiles. Biochemistry 43, 4039–4046.

Brennan, P.J., 2003. Structure, function, and biogenesis of the cell wall of *Mycobacterium tuberculosis*. Tuberculosis 83, 91–97.

Champion, J.T., Gilkey, J.C., Lamparski, H., Retterer, J., Niller, R.M., 1995. Electron microscopy of rhamnolipid (biosurfactant) morphology: effects of pH, cadmium, and octadecane. J. Colloid Interface Sci. 170, 569–574.

Chatterjee, D., Bozic, C.M., McNeil, M., Brennan, P.J., 1991. Structural features of the arabinan component of the lipoarabinomannan of *Mycobacterium tuberculosis*. J. Biol. Chem. 266, 652–660.

Chatterjee, D., Lowell, K., Rivoire, B., McNeil, M.R., Brennan, P.J., 1992. Lipidoarabinomannan of Mycobacterium tuberculosis. Capping with mannosyl residues in some strains. J. Biol. Chem. 267, 6234–6239.

Chatterjee, D., Khoo, K.H., McNeil, M.R., Dell, A., Morris, H.R., Brennan, P.J., 1993. Structural definition of the non-reducing termini of mannose-capped LAM from *Mycobacterium tuberculosis* through selective enzymatic degradation and fast atom bombardment-mass spectrometry. Glycobiology 3, 497–506.

Chen, G., 2004. Rhamnolipid biosurfactant behavior in solutions. J. Biomat. Sci. Polymer Ed. 15, 229–235.

Chen, X., Howe, J., Andra, J., et al., 2007. Biophysical analysis of the interaction of granulysin-derived peptides with enterobacterial endotoxins. Biochim. Biophys. Acta 1768, 2421–2431.

Choe, B.Y., Krishna, N.R., Pritchard, D.G., 1992. Proton NMR study on rhamnolipids produced by *Pseudomonas aeruginosa*. Magn. Reson. Chem. 30, 1025–1026.

Chua, J., Vernge, I., Master, S., Deretic, V., 2004. A tale of two lipids: *Mycobacterium tuberculosis* phagosome maturation arrest. Curr. Opin. Microbiol. 7, 71–77.

Cohen, R., Exerowa, D., 2007. Surface forces and properties of foam films from rhamnolipid biosurfactants. Adv. Colloid Interf. Sci. 134−135, 24–34.

Cohen, R., Ozdemir, G., Exerowa, D., 2003. Free thin liquid films (foam films) from rhamnolipids: type of the film and stability. Coll. Surf. B. 29, 197–204.

Cohen, R., Exerowa, D., Pigov, I., Heckmann, R., Lang, S., 2004. DLVO and non-DLVO forces in thin liquid fils from rhamnolipids. J. Adhesion 80, 875–894.

Cossins, A.R., Sinensky, M., 1984. Adaptation of membranes to temperature, pressure, and exogeneous lipids. In: Shinitzky, M. (Ed.), Physiology of Membrane Fluidity. CRC Press, Boca Raton, pp. 1–20.

Curatolo, W., 1987. Glycolipid function. Biochim. Biophys. Acta 906, 137–160.

Daffé, M., Laneelle, M.A., Puzo, G., 1983. Structural elucidation by field desorption and electron-impact mass spectrometry of the C-mycosides isolated from *Mycobacterium smegmatis*. Biochim. Biophys. Acta 751, 439–443.

Dahrazma, B., Mulligan, C.N., Nieh, M.-P., 2008. Effects of additives on the structure of rhamnolipid (biosurfactant): a small-angle neutron scattering (SANS) study. J. Colloid Interf. Sci. 319, 590–593.

Dao, D.N., Kremer, L., Guerardel, Y., et al., 2004. *Mycobacterium tuberculosis* lipomannan induces apoptosis and interleukin-12 production in macrophages. Infect. Immun. 72, 2067–2074.

Desai, J.D., Banat, I.M., 1997. Microbial production of surfactants and their commercial potential. Microb. Molec. Biol. Rev. 61, 47–64.

Elass, E., Codeville, B., Guerardel, Y., et al., 2007. Identification by surface plasmon resonance of the mycobacterial lipomannan and lipoarabinomannan domains involved in binding to CD14 and LPS-binding protein. FEBS Lett. 581, 1383–1390.

Emons, H., Werner, G., Haferburg, D., Kleber, H.P., 1989. Tensammetric determination of a rhamnolipid at mercury electrodes. Electroanalysis 1, 555–557.

Etienne, G., Villeneuve, C., Billman-Jacobe, H., Astarie-Dequeker, C., Dupont, M.A., Daffe, M., 2002. The impact of the absence of glycopeptidolipids on the ultrastructure, cell surface and cell wall properties, and phagocytosis of *Mycobacterium smegmatis*. Microbiology 148, 3089–3100.

Frecer, V., Ho, B., Ding, J.L., 2000. Molecular dynamics study on lipid A from *Escherichia coli*: insights into its mechanism of biological action. Biochim. Biophys. Acta 1466, 87–104.

Fuijita, Y., Naka, T., McNeil, M.R., Yano, I., 2005. Intact molecular characterization of cord factor (trehalose

6,6′-dimycolate) from nine species of mycobacteria by MALDI-TOF mass spectrometry. Microbiology 151, 3403–3416.

Garidel, P., Hildebrand, A., Neubert, R., Blume, A., 2000. Thermodynamic characterization of bile salt aggregation as a function of temperature and ionic strength using isothermal titration calorimetry. Langmuir 16, 5267–5275.

Garidel, P., Hildebrand, A., Blume, A., 2004. Understanding the self-organisation of colloidal systems. MicroCal Application Note 1–9.

Garidel, P., Andrä, J., Howe, J., Gutsmann, T., Brandenburg, K., 2007a. Novel anti-inflammatory and anti-infective agents. Anti-Infect. Agents Medic. Chem. 6, 185–200.

Garidel, P., Hildebrand, A., Knauf, K., Blume, A., 2007b. Membranolytic activity of bile salts: influence of biological membrane properties and composition. Molecules 12, 2292–2326.

Gotoh, K., Mitsuyama, M., Imaizumi, S., Kawamura, I., Yano, I., 1991. Mycolic acid-containing glycolipid as a possible virulence factor of *Rhodococcus equi* for mice. Microbiol. Immunol. 35, 175–185.

Guérardel, Y., Maes, E., Elass, E., et al., 2002. Structural study of lipomannan and lipoarabinomannan from *Mycobacterium chelonae*. Presence of unusual components with α1,3-mannopyranose side chains. J. Biol. Chem. 277, 30635–30648.

Guerra-Santos, L., Käppeli, O., Fiechter, A., 1984. *Pseudomonas aeruginosa* biosurfactant production in continuous culture with glucose as carbon source. Appl. Environm. Microbiol. 48, 301–305.

Gutsmann, T., Schromm, A.B., Koch, M.H.J., et al., 2000. Lipopolysaccharide-binding protein-mediated interaction of lipid A from different origin with phospholipid membranes. Phys. Chem. Chem. Phys. 2, 4521–4528.

Gutsmann, T., Schromm, A.B., Brandenburg, K., 2007. The physicochemistry of endotoxins in relation to bioactivity. Int. J. Med. Microbiol. 297, 341–352.

Haba, E., Abalos, A., Jauregui, O., Espuny, M.J., Manresa, A., 2003a. Use of liquid chromatography-mass spectroscopy for studying the composition and properties of rhamnolipids produced by different strains of *Pseudomonas aeruginosa*. J. Surfact. Deterg. 6, 155–161.

Haba, E., Pinazo, A., Jauregui, O., Espuny, M.J., Infante, M.R., Manresa, A., 2003b. Physicochemical characterisation and antimicrobial properties of rhamnolipids produced by *Pseudomonas aeruginosa* 47T2 NCBIM 40044. Biotechnol. Bioeng. 81, 316–322.

Harland, C.W., Rabuka, D., Bertozzi, C.R., Parthasarathy, R., 2008. The *M. tuberculosis* virulence factor trehalose dimycolate imparts desiccation resistance to model mycobacterial membranes. Biophys. J. 94, 4718–4724.

Hauser, G., Karnovsky, L., 1959. Studies on the biosynthesis of L-rhamnose. J. Biol. Chem. 233, 287–291.

Herman, D.C., Zhang, Y., Miller, R.M., 1997. Rhamnolipid (biosurfactant) effects on cell aggregation and biodegradation of residual hexadecane under saturated flow conditions. Appl. Environ. Microbiol. 63, 3622–3627.

Heyd, M., Kohnert, A., Tan, T.H., et al., 2008. Development and trends of biosurfactant analysis and purification using rhamnolipids as an example. Analyt. Bioanalyt. Chem. 391, 1579–1590.

Hildebrand, A., Garidel, P., Neubert, R., Blume, A., 2004. Thermodynamics of demicellisation of mixed micelles composed of sodium oleate and bile salts. Langmuir 20, 320–328.

Holst, O., Suesskind, M., Grimmecke, D., Brade, L., Brade, H., 1998. Core structures of enterobacterial lipopolysaccharides. Prog. Clin. Biol. Res. 397, 23–35.

Howe, J., Bauer, J., Andra, J., et al., 2006. Biophysical characterization of synthetic rhamnolipids. FEBS J. 273, 5101–5112.

Howe, J., Hammer, M., Alexander, C., et al., 2007. Biophysical characterization of the interaction of endotoxins with hemoglobins. Med. Chem. 3, 13–20.

Hung, H.C., Shreve, G.S., 2001. Effect of the hydrocarbon phase on interfacial and thermodynamic properties of two anionic glycolipid biosurfactants in hydrocarbon/water systems. J. Phys. Chem. B 105, 12596–12600.

Hunter, S.W., Brennan, P.J., 1990. Evidence for the presence of a phosphatidylinositol anchor on the lipoarabinomannan and lipomannan of *Mycobacterium tuberculosis*. J Biol. Chem. 265, 272–279.

Ishigami, Y., 1997. Characterisation of biosurfactants. In: Esumi, K., Ueno, M. (Eds.), Structure Performance Relationship in Surfactants. Dekker, New York, pp. 197–226.

Ishigami, Y., Gama, Y., Nahahora, H., Yamaguchi, M., Nakahara, H., Kamata, T., 1987. The pH-sensitive conversion of molecular aggregates of rhamnolipid biosurfactant. Chem. Lett. 7, 763–766.

Ishigami, Y., Gama, Y., Fumiyoshi, I., Choi, Y.K., 1993. Colloid chemical effect of polar head moieties of a rhamnolipid-type biosurfactant. Langmuir 9, 1634–1636.

Israelachvili, J.N., 1991. Intermolecular and Surface Forces. Academic Press, London.

Jiao, B., Freudenberg, M., Galanos, C., 1989. Characterization of the lipid A component of genuine smooth-form lipopolysaccharide. Eur. J. Biochem. 180, 515–518.

Kano, H., Doi, T., Fujita, Y., Takimoto, H., Yano, I., Kumazawa, Y., 2005. Serotype-specific modulation of human monocyte functions by glycopeptidolipid (GPL) isolated from *Mycobacterium avium* complex. Biol. Pharm. Bull. 28, 335–339.

Kastowsky, M., Sabisch, A., Gutberlet, T., Bradaczek, H., 1991. Molecular modelling of bacterial deep rough mutant lipopolysaccharide of *Escherichia coli*. Eur. J. Biochem. 197, 707–716.

Kaufmann, S.H., 2002. Protection against tuberculosis: cytokines, T cells, and macrophages. Annu. Rheum. Dis. 61 (Suppl. 2), 54–58.

Khoo, K.H., Dell, A., Morris, H.R., Brennan, P.J., Chatterjee, D., 1995. Inositol phosphate capping of the nonreducing termini of lipoarabinomannan from rapidly growing strains of *Mycobacterium*. J. Biol. Chem. 270, 12380–12389.

Khoo, K.H., Jarboe, E., Barker, A., Torrelles, J., Kuo, C.D.W., Chatterjee, D., 1999. Altered expression profile of the surface glycopeptidolipids in drug-resistant clinical isolates of *Mycobacterium avium* complex. J. Biol. Chem. 274, 9778–9785.

Loppnow, H., Flad, H.-D., Rietschel, E.Th., Brade, H., 1993. The active principle of bacterial lipopolysaccharides (endotoxins) for cytokine induction. In: Schlag, G., Redl, H. (Eds.), Pathophysiology of Shock, Sepsis, and Organ Failure. Springer-Verlag, Heidelberg, pp. 405–416.

Luzzati, V., Mariani, P., Gulik-Krzywicki, T., 1987. The cubic phases of liquid-containing systems: physical structure and biological implications. In: Mennier, J., Langevin, D., Boccara, N. (Eds.), Physics of Amphiphilic Layers, Vol. 21. Springer-Verlag, Berlin, pp. 131–137.

Mata-Sandoval, J.C., Karns, J., Torrents, A., 1999. HPLC method for the characterization of rhamnolipids mixtures produced by *Pseudomonas aeruginosa* UG2 on corn oil. J. Chromatogr. A. 864, 211–220.

Matsuura, M., Kiso, M., Hasegawa, A., 1999. Activity of monosaccharide lipid A analogues in human monocytic cells as agonists or antagonists of bacterial lipopolysaccharide. Infect. Immun. 67, 6286–6292.

Maurer, N., Glatter, O., Hofer, M., 1991. Determination of size and structure of lipid IV a vesicles by quasi-elastic light scattering and small-angle X-ray scattering. J. Appl. Cryst. 24, 832–835.

Mayer, H., Krauss, J.H., Yokota, A., Weckesser, J., 1990. Natural variants of lipid A. In: Friedman, H., Klein, T.W., Nakano, M., Nowotny, A. (Eds.), Endotoxin. Plenum Press, New York, pp. 45–70.

Mueller, M., Lindner, B., Kusumoto, S., Fukase, K., Schromm, A.B., Seydel, U., 2004. Aggregates are the biologically active units of endotoxin. J. Biol. Chem. 279, 26307–26313.

Mueller, M., Brandenburg, K., Dedrick, R., Schromm, A.B., Seydel, U., 2005a. Phospholipids inhibit lipopolysaccharide (LPS)-induced cell activation: a role for LPS-binding protein. J. Immunol. 174, 1091–1096.

Mueller, M., Lindner, B., Dedrick, R., Schromm, A.B., Seydel, U., 2005b. Endotoxin: physical requirements for cell activation. J. Endotoxin. Res. 11, 299–303.

Müller, M., Scheel, O., Lindner, B., Gutsmann, T., Seydel, U., 2002. The role of membrane-bound LBP, endotoxin aggregates, and the MaxiK channel in LPS-induced cell activation. J. Endotoxin Res. 9, 181–186.

Nakata, T., Yasuda, M., Fujita, M., et al., 2006. CD14 directly binds to tri-acylated lipopeptides and facilitates recognition of the lipopeptides by the receptor complex of Toll-like receptors 2 and 1 without binding to the complex. Cell Microbiol. 8, 1899–1909.

Nigou, J., Zelle-Rieser, C., Gilleron, M., Thurnher, M., Puzo, G., 2001. Mannosylated lipoarabinomannans inhibit IL-12 production by human dendritic cells: evidence for a negative signal delivered through the mannose receptor. J. Immunol. 166, 7477–7485.

Nomura, K., Inaba, T., Morigaki, K., Brandenburg, K., Seydel, U., Kusumoto, S., 2008. Interaction of lipopolysaccharide and phospholipid in mixed membranes: solid-state ^{31}P-NMR spectroscopic and microscopic investigations. Biophys. J. 95, 1226–1238.

Özdemir, G., Peker, S., Helvaci, S.S., 2004. Effect of pH on the surface and interfacial behaviour of rhamnolipids R1 and R2. Colloids Surf. A. 234, 135–143.

Parra, J.L., Guinea, J., Manresa, A., et al., 1989. Chemical characterization and physicochemical behavior of biosurfactants. J. Am. Oil Chem. Soc. 66, 141–145.

Perfumo, A., Banat, I.M., Canganella, F., Marchant, R., 2006. Rhamnolipid production by a novel thermophilic hydrocarbon-degrading *Pseudomonas aeruginosa* AP02-1. Appl. Microbiol. Biotech. 72, 132–138.

Puissegur, M.P., Gilleron, M., Botella, L., et al., 2007. Mycobacterial lipomannan induces granuloma macrophage fusion via a TLR2-dependent, ADAM9- and β1 integrin-mediated pathway. J. Immunol. 178, 3161–3169.

Rao, V., Fuijiwara, N., Porcelli, S.A., Glickman, M.S., 2005. *Mycobacterium tuberculosis* controls host innate immune activation through cyclopropane modification of a glycolipid effector molecule. J. Exp. Med. 201, 535–543.

Rao, V., Gao, F., Chen, B., Jacobs, W.R., Glickman, M.S., 2006. Trans-cyclopropanation of mycolic acids on trehalose dimycolate suppresses *Mycobacterium tuberculosis*-induced inflammation and virulence. J. Clin. Invest. 116, 1660–1667.

Reichelt, H., Faunce, C.A., Paradies, H.H., 2008. The phase diagram of charged colloidal lipid A-diphosphate dispersions. J. Phys. Chem. B 112, 3290–3293.

Rendell, N.B., Taylor, G.W., Somerville, M., Todd, H., Wilson, R., Coe, P.J., 1990. Characterisation of *Pseudomonas* rhamnolipids. Biochim. Biophys. Acta 1045, 189–193.

Rietschel, E.Th., Galanos, C., Tanaka, A., Ruschmann, E., Lüderitz, O., Westphal, O., 1971. Biological activities of chemically modified endotoxins. Eur. J. Biochem. 22, 218–224.

Rietschel, E.T., Brade, L., Holst, O., et al., 1990. Molecular structure of bacterial endotoxin in relation to bioactivity. In: Nowotny, A., Spitzer, J.J., Ziegler, E.J. (Eds.), Cellular and Molecular Aspects of Endotoxin Reactions. Elsevier Science Publishers, Amsterdam, pp. 15–32.

Rietschel, E.T., Brade, H., Holst, O., et al., 1996. Bacterial endotoxin: chemical constitution, biological recognition, host response, and immunological detoxification. Curr. Top. Microbiol. Immunol. 216, 39–81.

Risco, C., Carrascosa, J.L., Bosch, M.A., 1993. Visualization of lipopolysaccharide aggregates by freeze-fracture and negative staining. J. Electron. Microsc. 42, 202–204.

Ryll, R., Kumazawa, Y., Yano, I., 2001. Immunological properties of trehalose dimycolate (cord factor) and other mycolic acid-containing glycolipids. Microbiol. Immunol. 45, 801–811.

Saavedra, R., Segura, E., Tenorio, E.P., Lopez-Marin, L.M., 2006. Mycobacterial trehalose-containing glycolipid with immunomodulatory activity on human CD4+ and CD8+ T-cells. Microbes Infect. 8, 533–540.

Sánchez, M., Teruel, J.A., Espuny, M.J., et al., 2006. Modulation of the physical properties of dielaidoylphosphatidylethanolamine membranes by a dirhamnolipid biosurfactant produced by *Pseudomonas aeruginosa*. Chem. Phys. Lipids 142, 118–127.

Sánchez, M., Aranda, F.J., Espuny, M.J., et al., 2007. Aggregation behaviour of a dirhamnolipid biosurfactant secreted by *Pseudomonas aeruginosa* in aqueous media. J. Colloid Interf. Sci. 307, 246–253.

Sasaki, H., White, S.H., 2008. Aggregation behavior of an ultra-pure lipopolysaccharide (LPS) that stimulates TLR-4 receptors. Biophys. J. 95, 986–993.

Schromm, A.B., Brandenburg, K., Loppnow, H., et al., 1998. The charge of endotoxin molecules influences their conformation and interleukin-6 inducing capacity. J. Immunol. 161, 5464–5471.

Schromm, A.B., Brandenburg, K., Loppnow, H., et al., 2000. Biological activities of lipopolysaccharides are determined by the shape of their lipid A portion. Eur. J. Biochem. 267, 2008–2013.

Schromm, A.B., Howe, J., Ulmer, A.J., et al., 2007. Physicochemical and biological analysis of synthetic bacterial lipopeptides: validity of the concept of endotoxic conformation. J. Biol. Chem. 282, 11030–11037.

Seydel, U., Koch, M.H.J., Brandenburg, K., 1993. Structural polymorphisms of rough mutant lipopolysaccharides Rd to Ra from *Salmonella minnesota*. J. Struct. Biol. 110, 232–243.

Seydel, U., Ulmer, A.J., Uhlig, S., Rietschel, E.Th., 2000a. Lipopolysaccharide, a membrane-forming and inflammation-inducing bacterial macromolecule. In: Zimmer, G. (Ed.), Membrane Structure in Disease and Drug Therapy. Marcel Dekker Inc., New York, pp. 217–251.

Seydel, U., Oikawa, M., Fukase, K., Kusumoto, S., Brandenburg, K., 2000b. Intrinsic conformation of lipid A is responsible for agonistic and antagonistic activity. Eur. J. Biochem. 267, 3032–3039.

Seydel, U., Hawkins, L., Schromm, A.B., et al., 2003. The generalized endotoxic principle. Eur. J. Immunol. 33, 1586–1592.

Seydel, U., Schromm, A.B., Brade, L., et al., 2005. Physicochemical characterization of carboxymethyl lipid A derivatives in relation to biological activity. FEBS J. 272, 327–340.

Shands, J.W., 1971. The physical structure of bacterial lipopolysaccharides. In: Weinbaum, G., Kadis, S., Ajl, S.J. (Eds.), Microbial Toxins, Vol. IV: Bacterial Endotoxins. Academic Press, New York, pp. 127–144.

Shinitzky, M., 1984. Membrane fluidity and cellular functions. In: Shinitzky, M. (Ed.), Physiology of Membrane Fluidity. CRC Press, Boca Raton, pp. 1–51.

Simons, M.P., Moore, J.M., Kemp, T.J., Griffith, T.S., 2007. Identification of the mycobacterial subcomponents involved in the release of tumor necrosis factor-related apoptosis-inducing ligand from human neutrophils. Infect. Immun. 75, 1265–1271.

Soberón-Chávez, G., Lépine, F., Déziel, E., 2005. Production of rhamnolipids by *Pseudomonas aeruginosa*. Appl. Microbiol. Biotech. 68, 718–725.

Sotirova, A.V., Spasova, D.I., Galabova, D.N., Karpenko, E., Shulga, A., 2008. Rhamnolipid-biosurfactant permeabilizing effects on Gram-positive and Gram-negative bacterial strains. Curr. Microbiol. 56, 639–644.

Syldatk, C., Lang, S., Matulovic, U., Wagner, F., 1985. Production of four interfacial active rhamnolipids from *n*-alkanes or glycerol by resting cells of *Pseudomonas* species DSM 2874. Z. Naturforsch. 40c, 61–67.

Takayama, K., Din, Z.Z., Mukerjee, P., Cooke, P.H., Kirkland, T.N., 1990. Physicochemical properties of the lipopolysaccharide unit that activates B lymphocytes. J. Biol. Chem. 265, 14023–14029.

Thanomsub, B., Pumeechockchai, W., Limtrakul, A., et al., 2007. Chemical structures and biological activities of rhamnolipids produced by *Pseudomonas aeruginosa* B189 isolated from milk factory waste. Biores. Technol. 98, 1149–1153.

Vignal, C., Guérardel, Y., Kremer, L., et al., 2003. Lipomannans, but not lipoarabinomannans, purified from *Mycobacterium chelonae* and *Mycobacterium kansasii* induce TNF-α and IL-8 secretion by a CD14-toll-like receptor 2-dependent mechanism. J. Immunol. 171, 2014–2023.

Villeneuve, C., Gilleron, M., Marídonneau- Parini, I., Daffe, M., Astarie-Dequeker, C., Etienne, G., 2005. Mycobacteria use their surface-exposed glycolipids to infect human macrophages through a receptor-dependent process. J. Lipid Research. 46, 475–483.

Wilkinson, S.G., 1996. Bacterial lipopolysaccharides – themes and variations. Prog. Lipid Res. 35, 283–343.

York, J.D., Firoozabadi, A., 2008. Comparing effectiveness of rhamnolipid biosurfactant with a quaternary ammonium salt surfactant for hydrate anti-agglomeration. J. Phys. Chem. B. 112, 845–851.

Zähringer, U., Salvetzki, R., Ulmer, A.J., Rietschel, E.Th., 1996. Isolation, chemical analysis and biological investigations of natural lipid A and lipid A-antagonists. J. Endotoxin Res. 3 (Suppl. 1), 33.

Zhong, H., Zeng, G., Yuan, X.Z., Fu, H.Y., Huang, G.H., Ren, F.Y., 2007. Adsorption of dirhamnolipid on four microorganisms and the effect on cell surface hydrophobicity. Appl. Microbiol. Biotechnol. 77, 447–455.

Zulianello, L., Canard, C., Köhler, T., Caille, D., Lacroix, J.S., Meda, P., 2006. Rhamnolipids are virulence factors that promote early infiltration of primary human airway epithelia by *Pseudomonas aeruginosa*. Infect. Immunity 74, 3134–3147.

CHAPTER

39

Microbial biofilm-related polysaccharides in biofouling and corrosion

Heidi Annuk and Anthony P. Moran

SUMMARY

Understanding the mechanism of biofilm formation is the first step in determining its function and, thereby, its impact and role in the environment. The predominant components of biofilms are extracellular polysaccharides, which account for approximately 90% of the enveloping matrix polymers that have a significant impact on biofilm composition and characteristics. The unwanted accumulation of biofilms in the industrial setting is termed biofouling which includes biofilm formation in the domestic, industrial and health-related fields in various environments such as cooling towers, water treatment systems, paper mills and many surfaces within. In addition, the metabolic activities of microbial cells on a metal surface can cause biocorrosion. Practically, every industrial sector has accrued costs as a direct result of biofouling, such as decreased efficiency, increased running costs, system damage or breakdown, and deterioration in product quality. It is generally accepted that greater knowledge of the structure and functions of the extracellular polymeric substances, and their polysaccharide component in particular, would aid the development of methods to remove and/or prevent biofilm formation. Thus, the present overview includes a general discussion of the contributions of polysaccharides in biofouling and, subsequently, examples of environmental and industrial problems caused by polysaccharides in biofilms are described.

Keywords: Biofilms; Capsular polysaccharides; Exopolysaccharides; Slime; Extracellular polymeric substances; Biofouling; Biocorrosion; Microbially influenced corrosion

1. INTRODUCTION – BIOFILM SYSTEMS

The biofilm mode of growth encompasses surface-associated multispecies microbial communities that are embedded in a matrix of extracellular polymeric substances (EPS) (Donlan, 2002) and are associated with working materials, corrosion products, debris, soil, particles, etc. In natural and man-made environments, bacteria predominantly live in biofilms on biotic (e.g. tissue) or abiotic (e.g. plastic, metal, wood or mineral) surfaces. The composition of EPS is an important characteristic impacting on biofilm strength, absorption capacity of biofilms for pollutants, microbial disinfectant resistance and exo-enzyme degradation activity, which all affect the behaviour of biofilms in technical systems (Moran and Ljungh, 2003). For example, the characteristics of EPS are responsible for the erosion and sloughing from biofilms, frictional resistance in pipes, adsorption of solids from bulk liquid and clogging of filter membranes (Ceyhan and Ozdemir, 2008). In practice, biofilm formation itself is not a problem, but the effects it can have on a process, e.g. decreased efficiency, increased running costs, system damage or breakdown, and deterioration in product quality, can have major implications (Patching and Fleming, 2003).

A number of organisms, including bacteria, protozoa, fungi, barnacles and other marine organisms, have been shown to form biofilms (Qian *et al.*, 2007). Biofilms are innate to liquid–surface interfaces; all that is needed is a nutritious environment for microorganisms to grow and produce biofilms. Single-species biofilms are uncommon in the natural environment where they are usually composed of a mixture of microbial species (Sutherland, 2001a). Furthermore, as biofilms occur in a large and varied number of environments, the structural characteristics of any single biofilm formed, under a specific set of parameters, may be unique to that single environment and microflora (Sutherland, 2001b). Assuming this, within any biofilm, there will almost certainly be wide variation in composition, including that of the EPS produced (Moran and Ljungh, 2003). As the predominant components of EPS are polysaccharides (PSs), these will also vary in structure (see Chapter 37). Their hydrophilic and polymeric structure provides a permeable matrix suitable for cell growth in an aqueous environment, but also provides substantial mechanical properties (Moran and Ljungh, 2003). The function of EPS and exopolysaccharides in biofilm formation has been reviewed in detail elsewhere (Sutherland, 2001a,b; Flemming and Wingender, 2001; Branda *et al.*, 2005). In general, PS chains have chemical properties that make them polar and thus very "sticky", leading to surface adhesion and cell cohesion (Sutherland, 2001a). Moreover, this is one of the properties that makes biofilms difficult to remove from surfaces.

The unwanted accumulation of biofilms in an industrial setting is termed biofouling which includes biofilm formation in domestic, industrial and health-related fields in the varied environments of cooling water towers, heat exchangers, paper mills, food processing plants and implanted medical devices (e.g. catheters, implants, prosthetics, etc.). In particular, the latter has been the subject of extensive reviews and intensive investigations (Habash and Reid, 1999; Donlan, 2001, 2002; Götz, 2002; Hall-Stoodley *et al.*, 2004; Eiff *et al.*, 2005; Tamilvanan *et al.*, 2008), especially since this contamination can act to spread opportunistic infections, e.g. device-related infection with *Staphylococcus epidermidis* (Hall-Stoodley *et al.*, 2004). Thus, biofilm formation not only on medical devices but also in mucosae should be considered as a virulence factor (see review, Moran and Annuk, 2003). In addition, the metabolic activities of microbial cells on a metal surface can cause microbially influenced corrosion or so-called biocorrosion (Beech, 2003).

Practically, every industrial sector has accrued costs as a direct result of biofouling. It is generally accepted that a greater knowledge of the structure and functions of the EPS, and its PS component in particular, would aid the development of methods to remove and/or prevent biofilm formation. Thus, excluding the role of biofilms in implanted medical device contamination (which as described above has been reviewed extensively elsewhere), the present chapter includes a general discussion of the contributions of PSs in biofouling and, subsequently, examples of environmental and industrial problems caused by biofilms are described and, in particular, the participation of PSs in these.

2. BIOFILM-RELATED PSs IN BIOFOULING AND CORROSION

Extracellular polymeric substances (mostly PSs) surrounding the cells of microorganisms are a unique feature of biofilms. The PSs are present as thin strands between the cells and the surface, or as sheets of amorphous material on a surface, entrapping minerals or host-produced serum components (Donlan, 2001).

Extracellular PSs can be divided into capsular and exopolysaccharides according to separation by centrifugation of shaken liquid cultures. Capsular polysaccharides (CPSs) remain cell-associated, whereas cell-free supernatant contains exopolysaccharides. However, with biofilms this delineation is not so easy as many of the extracellular PSs produced in biofilms are insoluble, complicating the precise determination of their chemical structures (Branda *et al.*, 2005). Nevertheless, PSs generally are a main proportion of the dry matter of biofilms and play major roles in determining the physical properties and structures of microbial biofilms (Flemming and Wingender, 2001). Sutherland (2001b) speculated that the PS components of biofilms of microbial origin differ from those produced by planktonic bacteria due to minor structural differences, e.g. the degree of acetylation or their molecular mass, leading to their altered physical properties which can cause a change in viscosity or gel-forming capacity. In addition, considering the environmental differences between the niches which bacteria inhabit and the heterogeneous PSs produced, there can be a wide disparity in biofilm structure, hence all analysed samples have extensive variations in composition.

Biofilm chemical composition and exopolysaccharide production of surface-associated bacteria is dependent on their growth phase and nutrient status of the environment (Majumdar *et al.*, 1999). Furthermore, all bacterial species may excrete chemically and structurally different exopolysaccharides. For example, extracellular PSs have been shown to be distinct to a particular genus (e.g. pseudomonads), or can be even species-specific (e.g. *Escherichia coli*) (Lindberg *et al.*, 2001). Also, production of analogous extracellular PSs by different microorganisms have been observed, e.g. alginate by *Pseudomonas aeruginosa* and by *Azotobacter vinelandii* (Nivens *et al.*, 2001), however, the nature of these is different (Sutherland, 2001a). It is common for a given bacterial species to produce one or two different PSs, but rarely both at the same time (Sutherland, 2001b). Concerning adhesion, various microbial EPS-containing PSs, proteins and lipids have affinity for hydrophobic (low surface energy) substrates, whereas acidic and neutral PSs prefer attachment to hydrophilic (high surface energy) materials (Beech *et al.*, 2005).

Bacteria in biofilms produce exopolysaccharides that influence the corrosion behaviour of metals by providing a matrix with sites for aeration, ion-concentration and metal-binding (Majumdar *et al.*, 1999) and affect their capacity to concentrate metal ions from surfaces and bulk media. Binding of metals has importance in both passivation and activation reactions; a good example is biofilm production by the PS-producer *Delya marina* which has been shown to induce almost complete passivation of mild steel by 95% reduction of the corrosion rate (Ford *et al.*, 1988).

The capacity of EPS to bind metal ions is important in the development of corrosion (Kinzler *et al.*, 2003). These mechanisms are reviewed in detail in Beech and Sunner (2004). In summary, exopolysaccharide-associated functional groups (e.g. carboxyl, phosphate, sulfate, glycerate, pyruvate and succinate groups) have a high affinity towards certain metal ions (e.g. Ca^{2+}, Cu^{2+}, $Mg^{2+}$and Fe^{3+}) in different oxidation states, which can cause shifts in standard reduction potentials. Such EPS-bound metal ions can open up new redox reaction pathways, like direct electron transfer from the metal (e.g. iron) or a biomineral (e.g. FeS) which, with a suitable electron acceptor (e.g. oxygen or nitrate in aerobic or anaerobic systems, respectively), can cause cathodic depolarization and, consequently, increased corrosion. A good example of the process is a study of iron hydroxide-encrusted biofilms by Chan *et al.* (2004), which showed that bacterial exopolymers could act as a base for the accumulation of akaganeite (β-FeOOH) pseudo-single crystals through ferric iron-binding with carboxylic groups on the polymer. These authors proposed that the production of the exopolymer was for precipitating iron oxyhydroxide mineral outside the microbial cell in a process required for metabolic energy generation. In conclusion, these processes would assist the cathodic reaction with biominerals, thereby acting as electron-conducting fibres within the biofilm (Beech *et al.*, 2005).

Exopolysaccharides have been shown to contain uronic acid which endows the molecules with acidic properties and an overall negative surface charge; the latter is important in PS metal-complexing capacity. For example, alginates that are negatively charged and rich in uronic acids have a high metal-complexing capacity (Majumdar *et al.*, 1999). In addition, uronic acid has been shown to be the main component of *Klebsiella pneumoniae* CPS (Balestrino *et al.*, 2008). Moreover, a deficiency in uronic acid production by capsular mutants has been shown to influence *K. pneumoniae* initial adhesion to medical devices (Balestrino *et al.*, 2008). Additionally, exopolysaccharides with a significant amount of uronic acid show a high copper-binding capacity, which would be important to microorganisms for growing on surfaces coated with toxic and anti-fouling compounds, e.g. cuprous oxide (Majumdar *et al.*, 1999). In addition, in marine environments where biocorrosion occurs, bacterial and microalgal EPS may contain up to 20–25% uronic acids (Nichols and Nichols, 2008). Furthermore, the acidic PSs in the biofilms tend to ionize metals, thereby causing thinning of metal structures (Bhaskar and Bhosle, 2005).

The EPS in biofilms has several different functions, but protection and maintenance of matrix structure appears to be the most important (Moran and Annuk, 2003; Jain and Bhosle, 2008). Examples are colanic acid with β-(1→3)-, alginate with β-(1→4)- and PS intercellular adhesin (PIA) with β-(1→6)-linkages which are produced by *E. coli*, *P. aeruginosa* and *Staphylococcus aureus* and *S. epidermidis*, respectively (Danese *et al.*, 2000; Kaplan *et al.*, 2004; Chang *et al.*, 2007). These linkages between the subunits of the polymers provide a functional specificity to biofilm EPS and investigation of these particular structures would aid development of techniques to control biofouling in the marine environment. Furthermore, Jain and Bhosle (2008) studied cell attachment and formation of biofilms by a marine *Pseudomonas* sp. strain and showed the presence of calcofluor-binding and cellulase sensitive β-(1→4)-linked polymers, as well as α-D-glucose (Glc) and α-D-mannose (Man) in the biofilm matrix. A role for these polymers in the formation and maintenance of biofilms of *Pseudomonas fluorescens*, *E. coli*, *Salmonella enterica* sv. Enteritidis, *S. enterica* sv. Typhimurium and *Yersinia pestis* also has been reported (Zogaj *et al.*, 2001; Darby *et al.*, 2002; Spiers *et al.*, 2002; Solano *et al.*, 2002).

Hexoses and pentoses are the main building blocks of the biofilm PSs, but the exact monosaccharide content is dependent on environmental conditions (e.g. nutrients) and this variable monosaccharide composition influences also physicochemical properties of the PS (Bhaskar and Bhosle, 2005), e.g. arabinose (Ara) is important for cell aggregation (Bahat-Samet *et al.*, 2004) and deoxy sugars like fucose (Fuc) and rhamnose (Rha) from diatom EPS contribute to foaming and flocculation (Zhou *et al.*, 1998). Furthermore, EPS produced by microorganisms on submerged surfaces in aquatic environments are generally rich in monosaccharides, like Glc and Man, which may serve as cues for the settlement and colonization of surfaces by macrofoulers such as barnacles, oysters, mussels etc. (Bhaskar and Bhosle, 2005). These problems of biofouling are most common on ship hulls, navigational buoys, underwater equipment, seawater piping systems, beach well structures, industrial intakes, ocean thermal energy conversion plants, offshore platforms, moored oceanographic instruments and submarines (Azis *et al.*, 2003). Importantly, biofilm layers on a ship's hull can increase mass and hydrodynamic drag, increasing fuel costs by up to 15% (Molino and Wetherbee, 2008). Moreover, acetylated sugar moieties have been found of importance in biofilm initiation and cell adhesion (Tsang *et al.*, 2006). For example, poly-β-(1→6)-*N*-acetyl-D-glucosamine (GlcNAc) from staphylococci, a PS related to slime and biofilm production, mediates cell adherence to biomaterials (Sadovskaya *et al.*, 2005). As well, polymer-containing GlcNAc plays a critical role in the adhesive force of the holdfast of *Caulobacter crescentus*, the common stalked bacterium, which is among the first to colonize submerged surfaces in aquatic environments. The attachment between the PS component at the base of the stalk and a glass surface is one of the strongest biological adhesions measured to date (Tsang *et al.*, 2006). Moreover, Verhoef *et al.* (2005) showed the presence of *O*-acetyl groups in 50% of all EPSs in a large variety of paper mill slimes in addition to the 63% presence of uronic acid and pyruvate. The authors suggested that collectively these are responsible for the hydrophobic and strong polyanionic nature of particular EPSs, and are important factors for facilitating the involvement of EPSs in surface attachment and playing a central role for the structural unity of a biofilm.

Another PS complex, termed transparent exopolymer particles (TEPs), plays a role in biofilm formation and biofouling in the aqueous environment and is ubiquitous in large quantities in both marine and freshwaters (Passow, 2002). The TEPs, composed of acidic PSs, are one type of EPS produced by both diatoms and bacteria during their growth (Bhaskar and Boshle, 2005). The properties of these small (≈2–200 μm), negatively charged, mostly PS particles implies their potential to play a role in the initiation and growth of biofilms. By definition, TEPs are deformable, gel-like particles suspended in the water body and may be considered a "planktonic" subgroup of EPS (Wotton, 2004). Highly surface-active PSs are responsible for these particles forming metal ion bridges and hydrogen bonds (Mopper *et al.*, 1991). Consequently, TEPs tend to be very sticky, producing attachment upon collision, thus assisting biofilm development on the surfaces of filtration membranes used in water treatment plants. Additionally, TEPs may be the cause of biofouling of various infrastructure installations such as water delivery pipes, cooling towers, etc.

3. BIOFOULING CAUSING ENVIRONMENTAL AND INDUSTRIAL PROBLEMS

3.1. Contamination of cooling tower systems

Cooling towers are important for various industrial plants, such as petrochemical and wastewater treatment facilities (Bott, 1998). Biofouling of these systems causes severe economical loss by influencing the efficiency of the respective process in the following ways (Patching and Fleming, 2003):

(i) Hindrance of water flow by the biofilm because of occlusion of pipes and channels and high frictional resistance with a requirement for additional pumping and replacement costs (Michael, 2003).
(ii) Loss of heat transfer because of the biofilm having a low thermal conductivity (Flemming, 2002; Tanji *et al.*, 2007).
(iii) Biocorrosion caused by sulfate-reducing bacteria in an anaerobic biofilm layer and microbial biodeterioration of concrete and wood in cooling towers (Prince *et al.*, 2002).
(iv) Finally, biofilms can act as reservoirs for pathogens including *Legionella* spp. (Dondero *et al.*, 1980; Garcia-Fulgueiras *et al.*, 2003; Türetgen and Cotuk, 2007; Nygård *et al.*, 2008).

Because of the warm temperatures and continual flow of nutrients, cooling towers are a prime environment for the build-up of microorganisms. Ceyhan and Ozdemir (2008) studied biofilm samples from stainless steel cooling water towers for chemo-organotrophic bacteria and found that species isolated from cooling water towers were Gram-negative non-spore forming and catalase-positive bacteria (Table 39.1). They showed that all the isolated *Pasteurella* spp. had the highest production of PSs, with similar monosaccharide compositions of Glc, D-galactose (Gal), Man, Rha and ribose (Rib). Of note, four pseudomonads produced PSs with similar monosaccharide compositions and most extracellular PSs were enriched in Gal and Glc, but with lesser amounts of Man, Rha, fructose (Fru) and Ara; the EPS formed by *Burkholderia* spp. consisted mainly of Gal, Glc, Man, Ara, Rha and Rib (Ceyhan and Ozdemir, 2008). Consequently, as has been discussed, biofilm formation occurs in a nutrient-rich environment, especially production of EPSs with a high PS content and, accordingly, Zhang and Bishop (2003) have postulated the hypothesis that EPS is biodegradable and that mixed microbial cultures will degrade their own EPS material when they are in a starved state. This hypothesis has been supported by investigations (Ruijssenaars *et al.*, 1999; Rättö *et al.*, 2001). In addition, Ceyhan and Ozdemir (2008) observed that some bacterial species were able to utilize EPS from *Burkholderia cepacia*, *Pseudomonas* spp. and *Sphingomonas paucimobilis*, with the greatest viscosity reduction of *B. cepacia* induced by a *Pseudomonas* spp., thus showing that the bacteria in this particular system were able to degrade EPS from biofilms in cooling towers. Hence, some microorganisms could be potential agents for the biodegradation and bioconversion of wastes from agricultural, agro-industrial and food processing industries (Kumar and Takagi, 1999).

A range of materials can be used for tower manufacturing, but mainly galvanized steel, stainless steel and concrete are applied (Türetgen and Cotuk, 2007). Previous studies have shown that Legionnaires' disease outbreaks originate from cooling tower and air-conditioning contamination by *Legionella pneumophila* (Dondero *et al.*, 1980; Garcia-Fulgueiras *et al.*, 2003; Türetgen and Cotuk, 2007; Nygård *et al.*, 2008). Bacterial biofilm development is common on heat-transfer surfaces of cooling towers because of the higher temperatures present, which also favours the rapid growth of *L. pneumophila* (Harris, 2000). Rogers *et al.* (1994) studied the influence of the plumbing materials on biofilm formation and growth of *L. pneumophila* and found the lowest total flora on stainless steel. In contrast, a study by Türetgen and Cotuk (2007) reported that polyethylene and polyvinyl chloride supported significantly low heterotrophic bacterial numbers, whereas heterotrophic bacterial counts and those of *L. pneumophila* on galvanized steel were significantly higher than those on other materials. Nevertheless, Rogers *et al.* (1994) showed that *L. pneumophila* accounted for a low proportion of biofilm flora on polybutylene and chlorinated polyvinyl chloride, but no growth was observed on copper surfaces; hence, it was concluded that the copper material in water systems could limit the colonization of *L. pneumophila*. In addition, Assanta *et al.* (1998) showed a similar tendency of *Aeromonas hydrophilia* which preferentially colonized polybutylene, a hydrophobic surface, followed by stainless steel, but with only a few cells attached to copper surfaces. Furthermore, Türetgen and Cotuk (2007) also evaluated the quantity of EPS in biofilms by measuring carbohydrate content and showed that it was higher on galvanized steel and lowest on polyvinyl chloride surfaces. In summary, the corrosivity of galvanized steel in the aqueous environment results in more extensive corrosion products or surface area available for colonization. Thus, due to the rough corroded layer, galvanized steel may provide a suitable niche for microorganisms for rapid biofouling (Pasmore *et al.*, 2002).

TABLE 39.1 Genera of microorganisms found in different industrial environments

Environment	Genera of microorganisms	Reference
Cooling tower	*Burkholderia, Pseudomonas, Pasteurella, Alkaligenes, Aeromonas, Pantoea, Sphingomonas*	Ceyhan and Ozdemir, 2008
Reverse osmosis membrane	*Acinetobacter, Bacillus/Lactobacillus, Pseudomonas/Alcaligenes, Flavobacterium/Moraxella, Micrococcus, Serratia*	Ridgway *et al.*, 1983
Reverse osmosis membrane	*Bradyrhizobium, Rhodopseudomonas, Sphingomonas, Bacillus/ Clostridium*	Chen *et al.*, 2004
Reverse osmosis membrane	*Corynebacterium, Pseudomonas, Bacillus, Artrobacter, Flavobacterium, Penicillium, Trichoderma, Mucor, Fusarium, Aspergillus*; yeasts – occasionally	Baker and Dudley, 1998
Nanofiltration membrane	*Pseudomonas / Burkholderia, Ralstonia, Bacteroides, Sphingomonas*	Ivnitsky *et al.*, 2007
Microfiltration membrane	*Rhizobium, Agrobacterium, Zoogloea, Mesorhizobium, Caulobacter, Bradyrhizobium, Bosea, Aureobacterium, Brevibacterium, Gordonia, Bacillus, Staphylococcus, Bordetella, Stenotrophomonas, Flavobacterium*	Chen *et al.*, 2004
MBR	HDO-MBR: *Pandoraea, Bacillus, Sphingomonas, Citrobacter, Kluyvera*	Kim *et al.*, 2006
	LDO-MBR: *Azospirillum, Flavobacterium, Acinetobacter, Rhizobium, Stenotrophomonas, Haliscomenobacter*	
Concrete	Cyanobacteria – *Phormidium, Leptolyngbya, Calothrix*	Barberousse *et al.*, 2006c
	Algae – *Klebsormidium, Stichococcus, Chlorella, Bracteacoccus, Geminella*	
Concrete	*Pseudomonas, Bacillus, Micrococcus, Moraxella, Anthrobacter, Streptococcus, Staphylococcus, Acinetobacter*	Maruthamuthu *et al.*, 2008

MBR, Membrane bioreactor; HDO, high dissolved oxygen; LDO, low dissolved oxygen.

3.2. Biofouling in membrane-related processes

Reverse osmosis, ultrafiltration, nanofiltration (NF) and microfiltration are widely used to separate suspended or colloidal and dissolved (i.e. all other) materials from water solutions in numerous industrial, medical and drinking water supply applications. Reverse osmosis is a high-pressure, energy-efficient technique for dewatering process streams, concentrating low-molecular-mass substances in solution or purifying wastewater. Ultrafiltration is a pressurized membrane separation process used for removing high-molecular-mass substances, colloidal materials, and organic and inorganic polymeric molecules and can be applied for pretreatment of water for NF or reverse osmosis (Rosberg, 1997). Nanofiltration is a process selected when reverse osmosis and ultrafiltration are not the ideal choice for separation and is used for demineralization, desalination and also in drinking water purification process steps, such as water softening, decolouring and micropollutant removal (Schaefer *et al.*, 2004). Microfiltration is a low-pressure cross-flow membrane process for separating colloidal and suspended particles and is used for fermentation broth clarification and biomass clarification and recovery (Cheryan, 1998). Another system used for water purification is the membrane bioreactor (MBR). This system uses a combination of a membrane process like microfiltration or ultrafiltration with a suspended growth bioreactor, and is now widely used for municipal and industrial wastewater treatment (Judd, 2006).

In membrane systems, biofouling represents a major problem as organic and inorganic dissolved substances and particles can be removed by pretreatment, but various microorganisms present on the surface of the membrane (see Table 39.1) can multiply at the expense of biodegradable substances in the water (Flemming *et al.*, 1997). Once microorganisms attach to a membrane, formation of a biofilm starts within hours. The type and growth rate of bacteria forming the biofilm is dependent on feed-water composition, available food source and the presence or absence of a biocide. In addition, the extent of biofilm formation depends on the surface material, smoothness, nutrients, water properties (e.g. temperature and pH) and system design characteristics (e.g. flow and pressure) (Mulder, 1996; Chang *et al.*, 2002).

The problems caused by biofilm formation in membrane-related processes have been summarized by Patching and Fleming (2003):

(i) An increased hydraulic resistance can cause elevated pumping costs, lower permeate productivity, increase cleaning intervals and induce potential membrane damage (Chang *et al.*, 2002; Herzberg and Elimelech, 2007).
(ii) Biofouling of the feed channel spacers in desalination plants results in a pressure drop between the feed and brine lines (differential pressure, ΔP) leading to increased salt passage and, consequently, an impure product. The biofilm layer entraps salts and prevents the turbulent flow from thoroughly mixing the solutes in the feed stream, thus causing a "concentration polarization" effect (Paul and Abanmy, 1990).
(iii) Cellulose acetate membranes are disposed to biofilm-related degradation during periods of shutdown or storage, primarily caused by fungal species (Murphy *et al.*, 2001). Such contamination of the pure water on the permeate side will occur during no- or low-flow periods because of the growth of bacteria through the membrane.

Flux decline caused by the activity of microorganisms on membrane surfaces is one of the major problems in membrane facilities (Ridgway and Flemming, 1996; Vrouwenvelder *et al.*, 1998; Al-Ahmad *et al.*, 2000). Fonseca *et al.* (2007) studied EPS influence on the flux decline of model nanofiltration membrane systems. The authors showed that membrane flux decline was associated with an increase in surface extracellular PS mass since 30–80% of normalized flux decline was observed when membrane-associated extracellular PS content increased from 5 to $50\,\mu g/cm^2$. Above this PS level, higher surface-associated PSs did not result in a significant increase in flux decline, thus indicating a surface saturation effect concerning the hydrodynamics through the mature biofilm.

Furthermore, Herzberg and Elimelech (2007) studied the biofouling mechanisms of reverse osmosis membranes and the impact of biofouling on membrane performance with a model bacterial strain, *P. aeruginosa* PA01 under accelerated biofouling conditions. They showed that the drastic flux decline caused by the deposited bacterial cells on the membrane was lower than that caused by biofilm growth, despite the very high cell concentration added to the unit. It was shown that the additional decrease in permeate flux in the presence of the biofilm was caused by the hydraulic resistance produced by EPS. The authors suggested that the EPS matrix between the cells can give hydraulic resistance to permeate flow, comparable to fouling of reverse osmosis membranes by PSs, such as alginate (Ang *et al.*, 2006). With membrane biofilms, increased hydraulic resistance (ΔP) has often been attributed to the viscoelastic properties of biofilms (Baker and Dudley, 1998). The physical structure of biofilms found in membrane systems can be compact and "gel-like" or "slimy and adhesive" (Ye *et al.*, 2005), with some consisting of a high ratio of PS slime to viable microorganisms, i.e. 10^6–10^8 colony forming units of bacteria per cm^2 of membrane (Baker and Dudley, 1998).

In addition, Haberkamp *et al.* (2008) investigated the impact of EPS on ultrafiltration membrane fouling in tertiary sewage treatment by addition of an EPS mixture to different test solutions. The flux decline induced by bacterial EPS was influenced by the formation of a rather loose-bound, polarized concentration layer of large PSs, whereas the secondary effluent fouling layer consisted of the tighter adhesion of macromolecular substances with a variety of biopolymers present. However, ultrafiltration of the secondary effluent resulted in a higher flux decline than ultrafiltration of the EPS model solutions.

The influence of EPS on membrane fouling in the MBR system was studied by Rosenberger *et al.* (2006) who compared differences in membrane performance between two identical MBR pilots operated in parallel. They demonstrated that high PS concentrations in the sludge supernatant corresponded to high fouling rates. Also, the PS fraction was of microbiological origin, and the temperature and stress conditions of the microorganisms had an influence on the EPS produced, i.e. increased viscosity and PS concentration. Yun *et al.* (2006) also studied the effect of dissolved oxygen on membrane filterability by comparing two MBRs operated at aerobic and anoxic conditions for the treatment of synthetic dye wastewater. These investigators observed that the rate of membrane fouling for the anoxic MBR was greater than that for the aerobic MBR and production of more EPS in biofilms increased the membrane fouling in an aerobic MBR. It was concluded that the rate of membrane fouling in the anoxic phase was more rapid than that in the aerobic phase, regardless of EPS amount, and a greater hydraulic barrier could have been caused by more uniform distribution of PSs in the highly spread EPS in the anoxic biofilm. Similar tendencies were observed by Kim *et al.* (2006) who showed that the rate of membrane fouling for a low dissolved oxygen (LDO-) MBR was faster than that for a high dissolved oxygen (HDO-) MBR, although the thickness of the fouling layer was greater in the latter. In addition, the biofilm in the LDO-MBR contained a larger amount of EPS than that in the HDO-MBR, with the latter having a higher protein to PS ratio in the biofilm. The microbial communities have been shown to differ between the HDO- and LDO-MBRs, which likely explains the different structures and permeabilities of the respective biofilms.

3.3. Biofouling of concrete

Various microorganisms such as bacteria, fungi, algae and lichens are found to be important in concrete biodeterioration (see Table 39.1) (Gaylarde *et al.*, 2003). Green microalgae and cyanobacteria have been shown to be major constituents of biofilms on building façades (Barberousse *et al.*, 2006a,b) causing aesthetic damage, deterioration of the underlying material and aiding colonization by other microorganisms (Ortega-Calvo *et al.*, 1991). Also, biofilms growing on building walls have an influence on the heat uptake of buildings, increasing the energy cost required for additional air conditioning (Flemming, 2002). The biofouling contributes to "abiotic" deterioration processes due to the change of the material structure and their thermo-hygric characteristics (Sanchez-Silva and Rosowsky, 2008). Furthermore, mechanical pressure due to the shrinking and swelling of the biofilms may cause a further weakening of the surface layer (Warscheid and Braams, 2000), leading to the modification of the stone's pore size and causing changes of moisture circulation patterns and temperature response (Garty, 1991; Warscheid, 1996).

A greater understanding of the nature of the polymers produced by microorganisms would help development of anti-adhesive coatings for building façade protection. Barberousse *et al.* (2006c), who studied the released PSs (RPSs) and CPSs secreted by algal and cyanobacterial isolates from biofilms on building façades, found polymers with anionic and hydrophobic properties that varied between the different strains. In the algal isolates, the RPSs and CPSs were mainly composed of Man, Rha and Gal; in the cyanobacterial isolates constituents were Glc and Gal, with Ara, xylose (Xyl) and Fuc as minor residues (i.e. <15% of the molar composition). A significant correlation was reported between RPS and CPS compositions for some of the polymer extracts, thus validating the hypothesis that CPSs dissolve to form RPSs (Gloaguen *et al.*, 1995). In addition, some exopolymeric materials with multifunctional groups may participate in metal corrosion, e.g. be involved in adhesion of *Thiobacillus albertis* to elemental sulfur (Bryant *et al.*, 1984), but also aid in dissolving calcium from concrete, thereby degrading building material, especially in synergy with strains of the *Fusarium* fungus (Gu *et al.*, 1998). Overall loss of concrete strength, cracking and eventual failure can result.

In addition to biogenic transformation of the stone surface, which comprises agglutination of surface particles and re-mineralization, extracellular PSs protect microbial cellular and enzyme functioning during dry and hot periods. Kemmling *et al.* (2004) demonstrated that, in several polymers (i.e. alginate, dextran, and levan), a typical extracellular enzyme, the α-amylase, developed enhanced resistance against desiccation stress, hence the overall stabilizing effect of PS preserves the biological activity of the colonizing organisms against repeated drying and wetting.

3.4. Slimes in paper production

Paper manufacturing employs a warm and nutrient-rich environment advantageous for microbial growth, with the usage of a range of raw materials introducing a wide variety of microorganisms into the paper mill, which can grow on the surfaces of process equipment, to produce thick biofilm or "slime". The biofilm development is often noticed within paper-machines, particularly at interfaces between such a machine and water, and can account for 10–20% of all machine downtime (Ludensky, 2003). The slimy nature of paper mill biofilms is assignable to the presence of EPS, 90% of which is PSs (Allison, 1993). Slime deposits act as a continuous source of contamination of circulating waters and the final product (Ludensky, 2003). Importantly, it is essential to know the chemical composition of the slime deposits for optimal biofilm control by chemical and/or enzymatic methods.

Biofilm contamination is responsible for a wide range of operational problems, including runnability problems with wet end-breaks, paper sheet defects from slime sloughing onto the circulating wire, malodours that can be carried over onto the finished paper, etc. (Elsmore *et al.*, 1999). The main sources of microorganisms in the process are the water used and the recycled pulp, but also additives, fillers, dyes, pigments, starches and coatings can contribute, depending on the types of paper being manufactured and the mill in which they are produced (Ludensky, 2003). In addition, the paper industry is moving towards environmentally-friendly processes using a closed-loop procedure of water utilization and an increased consumption of recycled fibres as raw material (Blanco and Gaspar, 1998). This results in an increase in microbial activity and problems due to microbial-related deposits (Verhoef *et al.*, 2005).

A wide range of microorganisms is found in paper manufacturing mills (Blanco *et al.*, 1996; Desjardins and Beaulieu, 2003) (Table 39.2). According to Safade (1988), the organisms involved in slime production and colonization can be divided into two groups: primary organisms, e.g. *Bacillus* spp. and *Sphaerotilus natans*, that accumulate slime and aid secondary organisms to grow, e.g. *Klebsiella* spp., *Achromobacter* spp. and *Pseudomonas* spp. Filamentous bacteria, such as *Sphaerotilus* spp., are capable of accumulating

TABLE 39.2 Microbial genera found in the paper-mill environment

Genera of microorganisms	Reference
Bacillus, Brevundimonas, Cytophaga, Enterobacter, Klebsiella, Paenibacillus, Starkeya	Rättö *et al.*, 2005
Flavobacterium, Pseudomonas, Bacillus, Klebsiella, Sphaerotilus, Citrobacter, Burkholderia, Enterobacter	Verhoef *et al.*, 2005
Burkholderia, Alcaligenes, Bacillus	Lindberg *et al.*, 2001
Sinorhizobium, Rhizobium, Azorhizobium, α-, β-, γ-*Proteobacteria, Bacteroides*	Lahtinen *et al.*, 2006
Bacillus, Deinococcus, Ochrobactrum, Thermomonas	Kolari *et al.*, 2001
Flectobacillus, Cytophaga, Flavobacterium, Bacillus, Paenibacillus, Chryseobacterium, Norcardiopsis, Streptomyces	Oppong *et al.*, 2003
Acidovorax, Acinetobacter, Aureobacterium, Bacillus, Brevibacterium, Deinococcus, Enterococcus, Xanthobacter	Väisänen *et al.*, 1998

large amounts of slime that can clog pipes and filters, bind organic and inorganic deposits, and produce tenacious slimes (Pellegrin *et al.*, 1999). In particular, EPS produced by *S. natans* has been structurally characterized (Takeda *et al.*, 2002). The sheaths of that filamentous bacterium are covered with slime consisting of a gellan-like, acidic PS composed of pyranosidic Glc, Rha and D-glucuronic acid (GlcA) residues forming tetrasaccharide repeating units: →4)-α-D-Glc*p*-(1→2)-β-D-GlcA*p*-(1→2)-α-L-Rha*p*-(1→3)-β-L-Rha*p*-(1→. Takeda *et al.* (2002) suggested that the hygroscopic properties of the surface of the sheath of *S. natans*, due to PS, might also be the reason for bulking of the activated sludge in wastewater treatment systems. In addition, pink-pigmented filamentous bacteria have been found in paper mill deposits that contribute to slime production and are responsible for the pink stains in the finished paper (Oppong *et al.*, 2003). Despite representing only 10% of the total slime-producing population, *S. natans* usually contributes to formation of viscous and rubbery slime which, even in small amounts, causes a pink colour pigmentation in paper machines (Johnsrud, 1997).

Paper-machine slime is usually a mixture of microbial and chemical deposits containing wood fibres and other process water components, in addition to the biofilm material. It is well established that the type of EPS and PSs produced in biofilms depends on the surface-colonizing microorganisms, the environmental conditions and the available substrate (Allison, 1993). Thus, Lindberg *et al.* (2001) showed that *B. cepacia* and *Burkholderia pickettii* biofilm PSs consist mainly of Gal, Glc, Man and Rha sugar units, but that biofilms of different strains of *B. cepacia* have different proportions of Gal:Man:Rha. Also, it had been postulated that a microorganism produces only one or two different PSs and very seldom both at the same time (Schenker *et al.*, 1997). Importantly, Lindberg *et al.* (2001) observed that the change of the sugar composition of pure-culture biofilms depended on the composition of the medium. In addition, Rättö *et al.* (2005) observed minor amounts of additional sugars in almost all tested PS samples from paper-machine slime, which may reflect production of more than one PS by the isolates, but it cannot be ruled out that these were impurities liberated from contaminating bacterial cells or derived from the small amount of yeast extract in the medium.

According to Rättö *et al.* (2005), the EPS produced by slimy isolates contains heteropolysaccharides composed of varying amounts of the monosaccharides Fuc, Rha, Gal, Glc, Man and GlcA, which also have been reported by Väisänen *et al.* (1994) as the main constituents of bacterial PSs in paper-machine slimes.

Furthermore, Rättö *et al.* (2006) showed the presence of colanic acid produced by the enterobacteria isolated from the slime collected from the paper mills. The importance of colanic acid (Figure 39.1) in biofilm growth has been demonstrated for *E. coli* (Prigent-Combaret *et al.*, 1999). Colanic acid-deficient mutants of *E. coli* form a one- or two-layer thick biofilm, but not the usual three-dimensional complex structures (Prigent-Combaret *et al.*, 2000; Danese *et al.*, 2000). Thus, Danese *et al.* (2000) suggested that colanic acid is not so much needed for establishing biofilms, but for maintaining its architecture in the face of changing environmental conditions, such as alterations in medium bulk flow.

Of note, Kolari *et al.* (2001) found that *Bacillus* spp. are not primary biofilm formers in paper-machines but slime is responsible for their persistence in biofilms formed by primary colonizers, such as *Deinococcus geothermalis*. Consistent with this, Rättö *et al.* (2005) observed that *Bacillus cereus*, originally isolated from the slime of a paper-machine disc filter as a co-culture with *Brevundimonas vesicularis*, did not produce slime itself. Moroever, *B. vesicularis* was the most efficient PS producer and its PS was a linear polymer with a repeating unit containing Rha, Glc, GlcA and D-galacturonic acid (GalA).

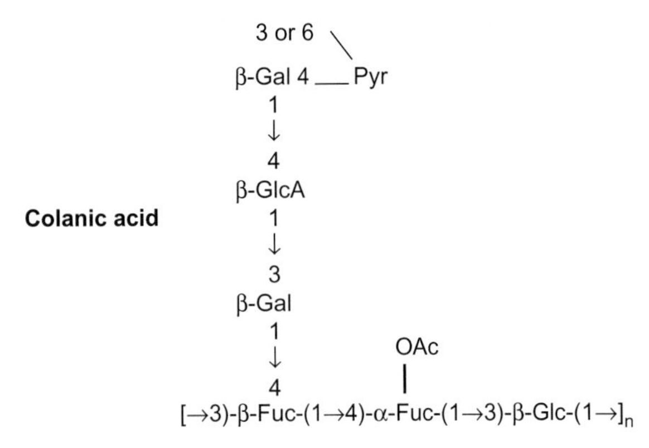

FIGURE 39.1 Structure of colanic acid monomer of *E. coli* K-12 (Danese *et al.*, 2000). Abbreviations: Fuc, L-fucose; Gal, D-galactose; GlcA, D-glucuronic acid; Glc, D-glucose; OAC, *O*-acetyl group; Pyr, pyruvate.

Furthermore, Verhoef *et al.* (2005) established the chemical structure of two exopolysaccharides that they considered were responsible for slime formation within pulp and paper manufacturing and were produced by *B. vesicularis* and *Methylobacterium* spp. The latter was found to be excreted as a highly pyruvated galactan with the repeating unit: →3)-[4,6-*O*-(1-carboxyethylidene)]-α-D-Gal*p*-(1→3)[4,6-*O*-(1-carboxyethylidene)]-α-D-Gal*p*-(1→3)-α-D-Gal*p*-(→ (Verhoef *et al.*, 2003). Furthermore, *B. vesicularis* produces a linear exopolysaccharide without non-sugar substituents containing a tetrasaccharide repeating unit: →4)-α-L-Glc*p*A-(1→4)-α-D-Gal*p*A-(1→4)-β-L-Rha*p*-(1→4)-β-D-Glc*p*-(1→ (Verhoef *et al.*, 2002).

In addition, Verhoef *et al.* (2005) studied the sugar composition of a large number of exopolysaccharide produced by bacterial strains isolated from different Finnish, French and Spanish paper mills using high performance size exclusion chromatography and principal component analysis. These investigators differentiated four groups of slime-producing species: *Bacillus* and related genera producing EPSs with high Man and/or Glc levels; *Klebsiella* with Gal and Rha as major sugar moieties of their EPSs; *Enterobacter* and related genera with EPS similar to colanic acid (containing the typical sugar moiety, fucose); and EPS-producing *Methylobacterium* with high Gal and pyruvate levels. However, in place of time-consuming sugar composition analysis Verhoef *et al.* (2005) also discussed the potential to use Fourier transform infrared spectroscopy (FT-IR), as a possible tool for pattern recognition and clustering of exopolysaccharide structures.

4. MICROBIALLY INFLUENCED CORROSION (MIC)

Corrosion is an electrochemical process that is described as chemical oxidation of metals and/or their mixtures (i.e. alloys), which involves the transfer of electrons in the presence of an electrolyte through a series of oxidation (anodic) and reduction (cathodic) reactions, with accumulation of the corrosion products on the surface (Beech and Gaylarde, 1999).

Biofilms influence the physicochemical interaction between a metal and its surroundings, causing a deterioration of the metal called biocorrosion or MIC (Beech, 2003). Importantly, Beech *et al.* (2006) suggested use of the term MIC for biofilm-influenced corrosion, as the term "biocorrosion" is used commonly to describe the corrosion of implant metal and alloys induced by host tissue.

Most metals are colonized by microorganisms but this does not mean they are susceptible to MIC. For example, titanium has excellent resistance against MIC, although it accumulates biofilms (Beech *et al.*, 2006). Overall, it has been proposed that MIC constitutes 10–50% of all metallic corrosion, depending on the industrial field (Booth, 1964; Flemming, 1996; Coetser and Cloete, 2005), and causing financial losses due to equipment biofouling and the cost requirement for prevention measures. Microorganisms within biofilms influence the kinetics of cathodic and/or anodic reactions (direct), assist in the establishment of the electrolyte in the cell (indirect) and modify the chemistry of the protective layers by acceleration or inhibition of corrosion (Beech and Sunner, 2004). The bacteria associated with corrosion of cast iron, and mild and stainless steel are sulfate-reducing bacteria (SRB), sulfur-oxidizing bacteria, iron-oxidizing/reducing bacteria, manganese-oxidizing bacteria and bacteria secreting organic acids and exopolymers or slime (Table 39.3). These organisms can coexist in biofilms and are able to influence electrochemical processes as a consequence of their metabolic activity (Beech and Gaylarde, 1999).

Corrosion of a technical material in the presence of biofilms, the role of EPS and, in particular, the role of PS in corrosion have been reviewed in detail elsewhere (Beech, 2004; Beech and Sunner, 2004). In summary, a large number of metals have been reported to cross-link microbial PSs (Geesey and Jang, 1989) with metal ions being essential for the growth of bacteria. Moreover, the interaction between PSs and/or EPS and metal ions, released at the metal–liquid interface, can be regulated by the amount of metal ion species. Hence, the colonization of a metal surface is dependent on the availability and the type of metal ions, i.e. the metal species and their reactivity as determined by the metal oxidation state (Beech, 2004). For example, the heterotrophic bacteria called "acid streamers", who produce slime consisting of neutral and acidic PSs and RNA (Johnson and Kelso, 1981), are able to participate in oxido-reduction of iron bound within their exopolymer matrix (Johnson *et al.*, 1993). Therefore, it has been hypothesized that carboxylic groups of the exopolysaccharides and phosphate groups from nucleic acid degradation could bind metals, hence increasing the overall metal-binding capacity of EPS (Beech, 2004). This would influence the electrochemical behaviour of a metal through the formation of metal concentration cells and galvanic coupling (Ford *et al.*, 1988).

Corrosion processes and microorganisms involved are extensively reviewed in Beech and Gaylarde (1999), Hamilton (2003) and Coetser and Cloete (2005) and here we concentrate on the MIC related to biofilms and EPS. Depending on the characteristics of the microorganisms, i.e. the aerobic and anaerobic groups (see Table 39.3), different types of corrosion reactions contribute to MIC:

(i) An aerobic biofilm causes corrosion by the formation of differential aeration or concentration cells (Ford and Mitchell, 1990), a process called tuberculation. The most common MIC processes take place in aerobic environments. Microorganisms from various classes such as bacteria, fungi, algae, etc., grow in a region where nutrients such as water and oxygen are available. The dissolution of metal (as

TABLE 39.3 Microbial genera involved in MIC in aerobic and anaerobic environments, as described by Beech and Gaylarde (1999) and Coetser and Cloete (2005)

Group	Genera of microorganisms
Aerobic MIC	
Sulfur oxidizers	*Thiobacillus*
Iron and manganese oxidizers	*Gallionella, Sphaerotilus, Crenothrix, Leptothrix, Clonothrix, Lieskeella, Siderocapsa, Metallogenium, Pseudomicrobium*
"Slime formers"	*Pseudomonas, Escherichia, Flavobacterium, Aerobacter, Clostridium, Bacillus, Desulfovibrio, Desulfotomaculum*
Anaerobic MIC	
SRB	*Desulfovibrio, Desulfobulbus, Desulfotomaculum, Desulfuromonas, Desulfobacter, Desulfosarcina, Desulfonema, Clostridium*
Iron-reducing bacteria	*Alteromonas, Shewanella*
Other hydrogenase-positive microorganisms	*Chlorophyta, Cyanophyta*

SRB: sulfate-reducing bacteria.

metallic cations) under aerobic conditions releases an excess of electrons that are captured by nearby cathodic sites, and the overall reaction leads to the formation and subsequent precipitation of insoluble corrosion products (Sanchez-Silva and Rosowsky, 2008). Aerobic corrosion takes place through the damage of an oxide film or rapid pitting which occurs under the biofilm when oxygen cannot reach the metal surface (Borenstein, 1994). Sulfur oxidizers may be involved, chiefly of the genus *Thiobacillus* that form sulfuric acid, which is a strongly corrosive agent, especially on concrete structures within which steel reinforcements are corroded and carbonates solubilized. Also involved are iron and manganese oxidizers causing corrosion through production of ferric chloride which is extremely aggressive and pits stainless steel (Coetser and Cloete, 2005). In particular, *Pedomicrobium manganicum* oxidizes manganese in drinking water systems and binds colloidal MnO_2 in the extracellular PS (Ridge *et al.*, 2007). The iron-oxidizing bacteria contribute to corrosion by excessive slime production creating differential oxygen concentrations and, thus, concentration cells (Jack, 2002). In particular, sheathed filamentous *S. natans* has been reported to be able to oxidize iron with the oxidation products deposited on the surface of the sheath (van Veen *et al.*, 1978). Nonetheless, Takeda *et al.* (2002) suggested that the gellan-like acidic PS of *S. natans* is responsible for this as the sheath itself does not contain acidic molecules.

Finally, other aerobic bacteria, called "slime formers", grow on the metal surface and exclude oxygen via respiration; the slime, e.g. excreted extracellular PSs, impedes oxygen diffusion, thereby creating an oxygen concentration cell and playing a role in the aggregation of bacteria on the surfaces, thus creating a unique milieu for further development of microbial communities and progressive MIC (Coetser and Cloete, 2005). Davidson *et al.* (1996) showed that the concentration of protein and carbohydrate produced by the bacterium *Acidovorax delafieldii* was correlated with the surface-associated copper found in the bulk aqueous phase (i.e. corrosion). In addition, *Pseudomonas* spp. have been reported to be involved in corrosion by colonizing metal surfaces, thus forming thin films with corrosion deposits on metal surfaces, thereby creating oxygen-free environments which harbour SRB (Iverson, 1987).

(ii) An anaerobic biofilm is where the mechanism of corrosion of iron and steel is by cathodic or anodic depolarization involving the removal of hydrogen by the hydrogenase system of the microorganism. From this microbial group, SRB have been recognized as the most significant contributor to MIC; the two main genera involved are *Desulfovibrio* and *Clostridium*. The MIC produced by SRB is closely linked to the presence of biofilms that are held together by microbial EPS (Coetser and Cloete, 2005). In the presence of SRB, steel and other iron alloys in an anaerobic aqueous environment corrode up to four times faster than during oxygen-promoted corrosion. Initially, microbial colonization starts with aerobic microorganisms which form biofilms with a strong reducing environment at their point of attachment where the SRB start to proliferate (Hamilton, 1985). The SRB perform dissimilatory reduction of sulfur compounds such as sulfate, sulfite, thiosulfate and sulfur itself to sulfide, with hydrogen sulfide also being important for the maintenance of anaerobiosis (Coetser and Cloete, 2005).

(iii) An aerobic–anaerobic biofilm can contribute to MIC, whereby oxygen is present but anaerobic microniches and/or an anaerobic layer exist. The SRB colonize these niches even when dissolved oxygen penetrates the entire biofilm at other locations (Lee *et al.*, 1995). Mixed population biofilms are heterogeneous (or "patchy") with the spatial distribution of the microbial species (Coetser and Cloete, 2005). Additionally, corrosion with different combinations of mixed bacterial cultures is notably higher than that in pure cultures, indicating the importance of microbial synergy (Pitonzo *et al.*, 2004). Variability in the bacterial species composition of a biofilm results in differences in metabolic activities and, thus, in similar systems under the same environmental conditions corrosion processes can vary significantly (Beech and Sunner, 2004).

5. CONCLUSIONS

Due to the variety of aqueous-surface interfaces being susceptible to biofouling processes, control of this fouling is a very multifarious problem and a proper understanding of the dynamics, mechanisms and factors controlling biofouling in various environments may help development of suitable strategies to solve the problem (see Research Focus Box). It is generally accepted that a greater knowledge of the structure and functions of the EPS, and its PS component in particular, would aid the development of methods to remove and/or prevent biofilm formation. Considering the diverse environments which bacteria inhabit and the heterogeneous PSs produced, there can be a wide disparity in biofilm structure, thus analysed samples may have extensive variations in composition. Hence, examining for similar structures between the subunits of the polymers that provide a functional specificity to biofilm EPS might aid development of enzymatic treatment protocols. For example, it has been shown that PS-degrading enzymes of microbial or phage origin can cause localized destruction and possible weakening of the biofilm matrix structure (Sutherland, 2001c). Also, the fast identification of EPS structures would be advantageous for effective control strategies in order to define the best anti-slime approach. Thus, in place of time-consuming sugar composition analysis, Verhoef *et al.* (2005) have investigated the potential for use of FT-IR as a possible tool for pattern recognition and clustering of EPS structures.

Another important aspect of biofouling is that the characteristics of the surface material have an effect on the density of biofilm formation. Even latex and other plastic surfaces have been shown to support microbial growth and biofilm formation (Rogers *et al.*, 1994). Furthermore, the corrosivity of galvanized steel in an aqueous environment results in more extensive corrosion products or surface area available for microbial colonization. Thus, due to the rough corroded layer, galvanized steel may provide a suitable niche for microorganisms participating in rapid biofouling (Pasmore *et al.*, 2002).

Of particular note, corrosion with different combinations of mixed bacterial cultures is notably higher than with pure cultures, hence indicating the importance of microbial synergy (Pitonzo *et al.*, 2004). Variability in bacterial species composition of a biofilm can result in differences in metabolic activities and, under the same environmental conditions, corrosion processes can vary significantly (Beech and Sunner, 2004). In addition, even if the microbial composition of the biofilm is identical, alterations in nutrient availability can influence the biofilm PS composition significantly. Thus, long-term biofilm community studies would aid knowledge about PS stability. In addition, good model systems are required to study specific biofouling environments which is a very time-consuming research field as there is no universal biofouling assemblage. Every industrial sector has to focus on particular microbial communities and the respective environmental conditions, e.g. oxygen availability, temperature, humidity, stress, etc., all of which affect biofilm structure, including biofilm-associated PSs.

As has been shown, a biofilm is a nutrient-rich environment, especially EPSs with a high PS content, although EPSs are potentially biodegradable (Zhang and Bishop, 2003) and may especially be utilized when the microbial community is in a starvation state (Ceyhan and Ozdemir, 2008). Therefore, the study of biofilm microbial communities could prove to be the source of possible anti-fouling agents. On the other hand, the introduction of beneficial microbial biofilms might protect against the damaging effects of biofouling, e.g. by passivation of a surface (Qian *et al.*, 2007).

RESEARCH FOCUS BOX

- There is a need for a methodological characterization of microbial communities in connection to their environments, i.e. nutrient and oxygen availability, in which they exist to provide a better understanding of bacterial involvement in PS production.
- Good laboratory models for biofilm formation, stability and longevity, as well as for studying the importance of PS production, are required.
- Chemical structure characterization and analysis of biofilm PSs from different biofouling environments would provide important insights.

ACKNOWLEDGEMENTS

The authors gratefully acknowledge the financial support of Griffith University, Science Foundation Ireland and the EU Marie Curie Programme (Grant No. MTKD-CT-2005-029774) for their studies.

References

Al-Ahmad, M., Aleem, F.A.A., Mutiri, A., Ubaisy, A., 2000. Biofouling in RO membrane systems. Part 1: Fundamentals and control. Desalination 132, 173–179.

Allison, D.G., 1993. Biofilm-associated exopolysaccharides. Microbiol. Eur. 1, 16–19.

Ang, W.S., Lee, S., Elimelech, M., 2006. Chemical and physical aspects of cleaning of organic-fouled reverse osmosis membranes. J. Membr. Sci. 272, 198–210.

Assanta, M.A., Roy, D., Monpetit, D., 1998. Adhesion of *Aeromonas hydrophilia* to water distribution system pipes after different contact times. J. Food Prot. 61, 1321–1329.

Azis, P.K.A., Al-Tisan, I., Al-Daili, M., et al., 2003. Marine macrofouling: a review of control technology in the context of an on-line experiment in the turbine condenser water box of Al-Jubail phase-I power/MSF plants. Desalination 154, 277–290.

Bahat-Samet, E., Castro-Sowinski, S., Okon, Y., 2004. Arabinose content of extracellular polysaccharide plays a role in cell aggregation of *Azospirillium brasilense*. FEMS Microbiol. Lett. 237, 195–203.

Baker, J.S., Dudley, L.Y., 1998. Biofouling in membrane systems – a review. Desalination 118, 81–89.

Balestrino, D., Ghigo, J.-M., Charbonnel, N., Haagensen, J.A.J., Forestier, C., 2008. The characterization of functions involved in the establishment and maturation of *Klebsiella pneumoniae in vitro* biofilm reveals dual roles for surface exopolysaccharides. Environ. Microbiol. 10, 685–701.

Barberousse, H., Tell, G., Yéprémian, C., Couté, A., 2006a. Diversity of algae and cyanobacteria growing on building facades in France. Algol. Stud. 120, 83–110.

Barberousse, H., Lombardo, R.J., Tell, G., Couté, A., 2006b. Factors involved in the colonisation of building facades by algae and cyanobacteria in France. Biofouling 22, 69–77.

Barberousse, H., Ruiz, G., Gloaguen, V., et al., 2006c. Capsular polysaccharides secreted by building facade colonisers: characterisation and adsorption to surfaces. Biofouling 22, 361–370.

Beech, I.B., 2003. Sulfate-reducing bacteria in biofilms on metallic materials and corrosion. Microbiol. Today 30, 115–117.

Beech, I.B., 2004. Corrosion of technical materials in the presence of biofilms – current understanding and state-of-the art methods of study. Int. Biodeterior. Biodegrad. 53, 177–183.

Beech, I.B., Gaylarde, C.C., 1999. Recent advances in the study of biocorrosion – an overview. Rev. Microbiol. 30, 177–190.

Beech, I.B., Sunner, J., 2004. Biocorrosion: towards understanding interactions between biofilms and metals. Curr. Opin. Biotechnol. 15, 181–186.

Beech, I.B., Sunner, J.A., Hiraoka, K., 2005. Microbe-surface interactions in biofouling and biocorrosion process. Int. Microbiol. 8, 157–168.

Beech, I.B., Sunner, J.A., Arciola, C.R., Cristiani, P., 2006. Microbially-influenced corrosion: damage to prostheses, delight for bacteria. Int. J. Artif. Organs. 29, 443–452.

Bhaskar, P.V., Bhosle, N.B., 2005. Microbial extracellular polymeric substances in marine biogeochemical processes. Curr. Sci. 88, 45–53.

Blanco, M.A., Gaspar, I., 1998. Microbial aspects in recycling paper industry. In: Blanco, M.A., Negro, C., Tijero, J. (Eds.), COST E1, Paper Recycling: An Introduction to Problems and Their Solutions. Office for Official Publications of the European Communities, Luxembourg, pp. 153–183.

Blanco, M.A., Negro, C., Gaspar, I., Tijero, J., 1996. Slime problems in the paper and board industry. Appl. Microbiol. Biotechnol. 46, 203–208.

Booth, G.H., 1964. Sulphur bacteria in relation to corrosion. J. Appl. Bacteriol. 27, 174–181.

Borenstein, S.W., 1994. Microbiologically Influenced Corrosion Handbook. Woodhead Publishing Ltd., Cambridge.

Bott, T.R., 1998. Techniques for reducing the amount of biocide necessary to counteract the effects of biofilm growth in cooling water systems. App. Therm. Eng. 18, 1059–1066.

Branda, S.S., Vik, A., Friedman, L., Kolter, R., 2005. Biofilms: the matrix revisited. Trends Microbiol. 13, 20–26.

Bryant, R.D., Costerton, J.W., Laishley, E.J., 1984. The role of *Thiobacillus aibertis* glycocalyx in the adhesion of cells to elemental sulfur. Can. J. Microbiol. 30, 81–90.

Ceyhan, N., Ozdemir, G., 2008. Extracellular polysaccharides produced by cooling water tower biofilm bacteria and their possible degradation. Biofouling 24, 129–135.

Chan, C.S., de Stasio, G., Welch, S.A., et al., 2004. Microbial polysaccharides template assembly of nanocrystal fibres. Science 303, 1656–1658.

Chang, I.-S., Le Clech, P., Jefferson, B., Judd, S., 2002. Membrane fouling in membrane bioreactors for wastewater treatment. J. Environ. Eng. 128, 1018–1029.

Chang, W.S., Mortel, M., Nielsen, L., Nino de Guzman, G., Li, X., Halverson, L.J., 2007. Alginate production by *Pseudomonas putida* creates a hydrated microenvironment and contributes to biofilm architecture and stress tolerance under water-limiting conditions. J. Bacteriol. 189, 8290–8299.

Chen, C.-L., Liu, W.-T., Chong, M.-L., et al., 2004. Community structure of microbial biofilms associated with membrane-based water purification processes as revealed using a polyphasic approach. Appl. Miocrobiol. Biotechnol. 63, 466–473.

Cheryan, M., 1998. Ultrafiltration and Microfiltration Handbook. CRC Press-Taylor & Francis, Boca Raton.

Coetser, S.E., Cloete, T.E., 2005. Biofouling and biocorrosion in industrial water systems. Crit. Rev. Microbiol. 31, 213–232.

Danese, P.N., Pratt, L.A., Kolter, R., 2000. Exopolysaccharide production is required for development of *Escherichia coli* K-12 biofilm architecture. J. Bacteriol. 182, 3593–3596.

Darby, C., Hsu, J.W., Ghori, N., Falkow, S., 2002. *Caenorhabditis elegans*: plague biofilm block food intake. Nature 417, 243–244.

Davidson, D., Beheshti, B., Mittelman, M.W., 1996. Effects of *Arthrobacter* sp., *Acidovorax delafieldii* and *Bacillus megatherium* colonization on copper solvency in a laboratory reactor. Biofouling 9, 279–292.

Desjardins, E., Beaulieu, C., 2003. Identification of bacteria contaminating pulp and a paper machine in a Canadian paper mill. J. Ind. Microbiol. Biotechnol. 30, 141–145.

Dondero, T.J., Rendtorff, R.C., Maluson, G.F., et al., 1980. An outbreak of legionnaires' disease associated with a contaminated air-conditioning cooling tower. N. Engl. J. Med. 302, 365–370.

Donlan, R.M., 2001. Biofilms and Device-Associated Infections. Emerg. Infect. Dis. 7, 277–281.

Donlan, R.M., 2002. Biofilms: microbial life on surfaces. Emerging Infect. Dis. 8, 881–890.

Eiff, C.V., Jansen, B., Kohnen, W., Becker, K., 2005. Infections associated with medical devices. Drugs 65, 179–214.

Elsmore, R., Gilbert, P., Ludensky, M., Coulburn, J., 1999. Biofilms in paper manufacturing. In: Wimpenny, G., Walker, B. (Eds.), Biofilms: The Good, the Bad and the Ugly. Bioline, Cardiff, pp. 81–89.

Flemming, H.-C., 1996. Biofouling and microbiologically influenced corrosion (MIC) – an economical and technical overview. In: Heitz, E., Sand, W., Flemming, H.-C. (Eds.), Microbial Deterioration of Materials. Springer-Verlag, Heidelberg, pp. 5–14.

Flemming, H.-C., 2002. Biofouling in water systems – cases, causes and countermeasures. Appl. Microbiol. Biotechnol. 59, 629–640.

Flemming, H.-C., Wingender, J., 2001. Relevance of microbial extracellular polymeric substances (EPSs) – part I: structural and ecological aspects. Water Sci. Technol. 43, 1–8.

Flemming, H.-C., Schaule, G., Griebe, T., Schmitt, J., Tamachkiarowa, A., 1997. Biofouling – the Achilles heel of membrane processes. Desalination 113, 215–225.

Fonseca, A.C., Summers, R.S., Greenberg, A.R., Hernandez, M.T., 2007. Extra-cellular polysaccharides, soluble microbial products, and natural organic matter impact on nanofiltration membranes flux decline. Environ. Sci. Technol. 41, 2491–2497.

Ford, T., Mitchell, R., 1990. The ecology of microbial corrosion. Adv. Microb. Ecol. 11, 231–262.

Ford, T.E., Maki, J.S., Mitchell, R., 1988. Involvement of bacterial exopolymers in biodeterioration of metals. In: Houghton, D.R., Smith, R.N., Eggins, H.O.W. (Eds.), Biodeterioration 7, Proceedings of the Seventh International Biodeterioration Symposium. Elsevier Applied Science, Barking, Essex, pp. 378–384.

Garcia-Fulgueiras, A., Navarro, C., Fenoll, D., Garcia, J., González-Diego, P., Jiménez-Buñuales, T., 2003. Legionnaires' disease outbreak in Murcia, Spain. Emerging Infect. Dis. 9, 915–921.

Garty, J., 1991. Influence of epilithic microorganisms on the surface temperature of building walls. Can. J. Bot. 68, 1349–1353.

Gaylarde, C., Ribas Silva, M., Warscheid, Th., 2003. Microbial impact on building materials: an overview. Mater. Struct. 36, 342–352.

Geesey, G.G., Jang, L., 1989. Extracellular polymers for metal binding. In: Beveridge, T.J., Doyle, R.J. (Eds.), Bacterial Interactions with Metallic Ions. Wiley, New York, pp. 223–247.

Gloaguen, V., Morvan, H., Hoffmann, L., 1995. Released and capsular polysaccharides of *Oscillatoriaceae* (*Cyanophyceae, Cyanobacteria*). Algol. Stud. 78, 53–69.

Götz, F., 2002. Staphylococcus and biofilms. Mol. Microbiol. 43, 1367–1378.

Gu, J.-D., Ford, T.E., Berke, N.S., Mitchell, R., 1998. Biodeterioration of concrete by the fungus *Fusarium*. Int. Biodeterior. Biodegrad. 41, 101–109.

Habash, M., Reid, G., 1999. Microbial biofilms: their development and significance for medical device-related infections. J. Clin. Pharmacol. 39, 887–898.

Haberkamp, J., Ernst, M., Böckelmann, U., Ulrich Szewzyk, U., Jekel, M., 2008. Complexity of ultrafiltration membrane fouling caused by macromolecular dissolved organic compounds in secondary effluents. Water Res. 42, 3153–3161.

Hall-Stoodley, L., Costerton, J.W., Stoodley, P., 2004. Bacterial biofilms: from the natural environment to infectious diseases. Nat. Rev. 2, 95–108.

Hamilton, W.A., 1985. Sulphate-reducing bacteria and anaerobic corrosion. Ann. Rev. Microbiol. 39, 195–217.

Hamilton, W.A., 2003. Microbially influenced corrosion as a model system for the study of metal microbe interactions: a unifying electron transfer hypothesis. Biofouling 19, 165–176.

Harris, A., 2000. Problems associated with biofilms in cooling tower systems. In: Keevil, C.V., Godfree, A., Holt, D., Dow, C. (Eds.), Biofilms in the Aquatic Environment. Springer, Berlin, pp. 139–144.

Herzberg, M., Elimelech, M., 2007. Biofouling of reverse osmosis membranes: role of biofilm-enchanced osmotic pressure. J. Membr. Sci. 295, 11–20.

Iverson, W.P., 1987. Microbial corrosion of metals. Adv. Appl. Microbiol. 32, 1–36.

Ivnitsky, H., Katz, I., Minz, D., et al., 2007. Bacterial community composition and structure of biofilms developing on nanofiltration membranes applied to wastewater treatment. Water Res. 41, 3924–3935.

Jack, T.R., 2002. Biological corrosion failures. ASM Handbook Volume 11: Failure Analysis and Prevention. ASM International, Materials Park, Ohio.

Jain, A., Bhosle, N.B., 2008. Role of β-(1→4)-linked polymers in the biofilm structure of marine *Pseudomonas* sp. CE-2 on 304 stainless steel coupons. Biofouling 24, 283–290.

Johnson, D.B., Kelso, W.I., 1981. Extracellular polymers of acid streamers from pyritic mines. Environ. Pollut. Ser. 24, 291–301.

Johnson, D.B., Ghauri, M.A., McGinness, S., 1993. Biogeochemical cycling of iron and sulphur in leaching environments. FEMS Microbiol. Rev. 11, 63–70.

Johnsrud, S.C., 1997. Biotechnology for solving slime problems in the pulp and paper industry. In: Scherper, P. (Ed.), Advances in Biochemical Engineering/Biotechnology. Springer-Verlag, Heidelberg, pp. 311–328.

Judd, S., 2006. The MBR Book: Principles and Applications of Membrane Bioreactors in Water and Wastewater Treatment. Elsevier, Oxford.

Kaplan, J.B., Velliyagounder, K., Ragunath, C., Rohde, H., Mack, D., Knobloch, J.K., 2004. Genes involved in the synthesis and degradation of matrix polysaccharide in *Actinobacillus actinomycetemcomitans* and *Actinobacillus pleuropneumoniae* biofilms. J. Bacteriol. 186, 8213–8220.

Kemmling, A., Kämper, M., Flies, C., Schieweck, O., Hoppert, M., 2004. Biofilms and extracellular matrices on geomaterials. Environ. Geol. 46, 429–435.

Kim, H.Y., Yeon, K.-M., Lee, Ch.-H., Lee, S., Swaminathan, T., 2006. Biofilm structure and extracellular polymeric substances in low and high dissolved oxygen membrane bioreactors. Sep. Sci. Technol. 41, 1213–1230.

Kinzler, K., Gehrke, T., Telegdi, J., Sand, W., 2003. Bioleaching – a result of interfacial processes caused by extracellular polymeric substances (EPS). Hydrometallurgy 71, 83–88.

Kolari, M., Nuutinen, J., Salkinoja-Salonen, M.S., 2001. Mechanisms of biofilm formation in paper machine by *Bacillus* species: the role of *Deinococcus geothermalis*. J. Ind. Microbiol. Biotechnol. 27, 343–351.

Kumar, C.G., Takagi, H., 1999. Microbial alkali proteases: from a bioindustrial viewpoint. Biotechnol. Adv. 17, 561–594.

Lahtinen, T., Kosonen, M., Tiirola, M., Vuento, M., Oker-Blom, C., 2006. Diversity of bacteria contaminanting paper machines. J. Ind. Microbiol. Biotechnol. 33, 734–740.

Lee, W., Lewandowski, Z., Nielsen, P.H., Hamilton, W.A., 1995. Role of sulfate-reducing bacteria in corrosion of mild steel: a review. Biofouling 8, 165–194.

Lindberg, L.E., Holmbom, B.R., Väisänen, O.M., Weber, A.M.-L., Salkinoja-Salonen, M.S., 2001. Sugar composition of biofilms produced by paper mill bacteria. Appl. Microbiol. Biotechnol. 55, 638–643.

Ludensky, M., 2003. Control and monitoring of biofilms in industrial applications. Int. Biodeterior. Biodegrad. 51, 255–263.

Majumdar, I., D'Souza, F., Bhosle, N.B., 1999. Microbial exopolysaccharides: effect on corrosion and partial chemical characterization. J. Indian Inst. Sci. 79, 539–550.

Maruthamuthu, S., Muthukumar, N., Natesan, M., Palaniswamy, N., 2008. Role of air microbes on atmospheric corrosion. Curr. Sci. 94, 359–363.

Michael, L., 2003. Control and monitoring of biofilms in industrial applications. Int. Biodeterior. Biodegrad. 51, 255–363.

Molino, P.J., Wetherbee, R., 2008. The biology of biofouling diatoms and their role in the development of microbial slimes. Biofouling 24, 365–379.

Mopper, K., Schultz, C., Chevolot, L., Germain, C., Revuelta, R., Dawson, R., 1991. Determination of sugars in unconcentrated seawater and other natural waters by liquid chromatography and pulsed amperometric detection. Environ. Sci. Technol. 26, 133–149.

Moran, A.P., Annuk, H., 2003. Recent advances in understanding biofilms of mucosae. Rev. Environ. Biotechnol. 2, 121–140.

Moran, A.P., Ljungh, Å., 2003. Physico-chemical properties of extracellular polymeric substances. In: Lens, P., Moran, A.P., O'Mahony, T., Stoodley, P., O'Flaherty, V. (Eds.), Biofilms in Medicine, Industry and Environmental Biotechnology – Characteristics, Analysis and Control. IWA Publishing, London, pp. 91–112.

Mulder, M., 1996. Basic Principles of Membrane Technology. Kluwer Academic Publishers, Dordrecht.

Murphy, A.P., Moody, C.D., Riley, R.L., Lin, S.W., Murugaverl, B., Rusin, P., 2001. Microbiological damage of cellulose acetate RO membranes. J. Membr. Sci. 193, 111–121.

Nichols, P.D., Nichols, C.A.M., 2008. Microbial signature lipid profiling and exopolysaccharides: experience initiated with professor David C White and transported to Tasmania. Australia. J. Microbiol. Methods. 74, 33–46.

Nivens, D.E., Ohman, D.E., Williams, J., Franklin, M.J., 2001. Role of alginate and its O acetylation in formation of *Pseudomonas aeruginosa* microcolonies and biofilms. J. Bacteriol. 183, 1047–1057.

Nygård, K., Werner-Johansen, Ø., Rønsen, S., et al., 2008. An outbreak of Legionnaires disease caused by long-distance spread from an industrial air scrubber in Sarpsborg, Norway. Clin. Infect. Dis. 46, 61–69.

Oppong, D., King, V.M., Bowen, J.A., 2003. Isolation and characterisation of filamentous bacteria from paper mill slimes. Int. Biodeterior. Biodegrad. 52, 53–62.

Ortega-Calvo, J.J., Hernandez-Marine, M., Saiz-Jimenez, C., 1991. Biodeterioration of building materials by cyanobacteria and algae. Int. Biodeterior. 28, 165–185.

Pasmore, M., Todd, P., Pfeifer, B., Rhodes, M., Bowman, C.N., 2002. Effect of polymer surface properties on the reversibility of attachment of *Pseudomonas aeruginosa* in the early stages of biofilm development. Biofouling 18, 65–71.

Passow, U., 2002. Transparent exopolymer particles (TEP) in aquatic environments. Prog. Oceanogr. 55, 287–333.

Patching, J.W., Fleming, G.T.A., 2003. Industrial biofilms: formation, problems and control. In: Lens, P., Moran, A.P., Mahony, T., Stoodley, P., O'Flaherty, V. (Eds.), Biofilms in Medicine, Industry and Environmental Biotechnology – Characteristics, Analysis and Control. IWA Publishing, London, pp. 568–590.

Paul, D., Abanmy, A.R.M., 1990. Reverse osmosis membrane fouling – the final frontier. Ultra Pure Water 7, 25–36.

Pellegrin, V., Juretschko, S., Wagner, M., Cottenceau, G., 1999. Morphological and biochemical properties of Sphaerotilus sp. isolated from paper mill slimes. Appl. Environ. Microbiol. 65, 156–162.

Pitonzo, B.J., Castro, P., Amy, P.S., Southam, G., Jones, D.A., Ringelberg, D., 2004. Microbiologically influenced corrosion capability of bacteria isolated from Yucca Mountain. Corrosion 60, 64–74.

Prigent-Combaret, C., Vidal, O., Dorel, C., Lejeune, P., 1999. Abiotic surface sensing and biofilm-dependent regulation of gene expression in *Escherichia coli*. J. Bacteriol. 181, 5993–6002.

Prigent-Combaret, C., Prensier, G., Le Thi, T.T., Vidal, O., Lejeune, P., Dorel, C., 2000. Developmental pathway for biofilm formation in curli-producing *Escherichia coli* strains: role of flagella, curli and colanic acid. Environ. Microbiol. 2, 450–464.

Prince, E.L., Muir, A.V.G., Thomas, W.M., Stollard, R.J., Sampson, M., Lewis, J.A., 2002. An evaluation of the efficacy of Aqualox cooling tower systems. J. Hosp. Infect. 52, 243–249.

Qian, P.-Y., Lau, S., Dahms, H.-U., Dobretsov, S., Harder, T., 2007. Marine biofilms as mediators of colonization by marine macroorganisms: implications for antifouling and aquaculture. Mar. Biotechnol. 9, 399–410.

Rättö, M., Mustranta, A., Siika-aho, M., 2001. Strains degrading polysaccharides produced by bacteria from paper machines. Appl. Microbiol. Biotechnol. 57, 182–185.

Rättö, M., Suihko, M.-L., Siika-aho, M., 2005. Polysaccharide-producing bacteria isolated from paper machine slime deposits. J. Ind. Microbiol. Biotechnol. 32, 109–114.

Rättö, M., Verhoef, R., Suihko, M.-L., et al., 2006. Colanic acid is an exopolysaccharide common to many enterobacteria isolated from paper machine slimes. J. Ind. Microbiol. Biotechnol. 33, 359–367.

Ridge, J.P., Lin, M., Larsen, E.I., Fegan, M., McEwan, A.G., Sly, L.I., 2007. A multicopper oxidase is essential for manganese oxidation and laccase-like activity in *Pedimicrobium* sp. ACM 3067. Environ. Microbiol. 9, 944–953.

Ridgway, H., Flemming, H.-C., 1996. Membrane biofouling. In: Mallevialle, J., Odendal, P.E., Wiesner, M.R. (Eds.), Water Treatment: Membrane Processes. McGraw-Hill, New York, pp. 6.1–6.61.

Ridgway, H.F., Kelly, A., Justice, C., Olson, B.H., 1983. Microbial fouling of reverse-osmosis membranes used in advanced wastewater treatment technology: chemical, bacteriological, and ultrastructural analyses. Appl. Environ. Microbiol. 45, 1066–1084.

Rogers, J., Dowsett, A.B., Dennis, P.J., Lee, J.V., Keevil, C. W., 1994. Influence of temperature and plumbing material on biofilm formation and growth of *Legionella pneumophila* in a model potable water system containing complex microbial flora. Appl. Environ. Microbiol. 60, 1585–1592.

Rosberg, R., 1997. Ultrafiltration (new technology), a viable cost-saving pretreatment for reverse osmosis and nanofiltration – a new approach to reduce cost. Desalination 110, 107–113.

Rosenberger, S., Laabs, C., Lesjean, B., et al., 2006. Impact of colloidal and soluble organic material on membrane performance in membrane bioreactors for municipal wastewater treatment. Water Res. 40, 710–720.

Ruijssenaars, H.J., De Bond, J.A.M., Hartmans, S., 1999. A pyruvated mannose-specific xanthan lyase involved in xanthan degradation by *Paenibacillus alginolyticus* XL-1. Appl. Environ. Microbiol. 65, 2446–2452.

Sadovskaya, I., Vinogradov, E., Flahaut, S., Kogan, G., Jabbouri, S., 2005. Extracellular carbohydrate-containing polymers of a model biofilm-producing strain, *Staphylococcus epidermidis* RP62A. Infect. Immun. 73, 3007–3017.

Safade, T.L., 1988. Tackling the slime problem in a paper-mill. Pap. Technol. Ind. September, 280–285.

Sanchez-Silva, M., Rosowsky, D.V., 2008. Biodeterioration of construction materials: state of the art and future challenges. J. Mater. Civ. Eng. 20, 352–365.

Schaefer, A., Fane, A.G., Waite, T.D. (Eds.), 2004. Nanofiltration: Principles and Applications. Elsevier Advanced Technology Books, Amsterdam.

Schenker, A., Popp, G., Papier, G., Schwalbach, E., 1997. Enzyme additions for biofilm control in PM circuits. Wochenbl. Papierfabr. 125, 702–709.

Solano, C.B., Garcia, J., Valle, C., Berasain, J.M., Ghigo, C., Gamazo Lasa, I., 2002. Genetic analysis of *Salmonella enteritidis* biofilm formation: critical role of cellulose. Mol. Microbiol. 43, 793–808.

Spiers, A.J., Kahn, S.G., Bohannon, J., Travisano, M., Rainey, P.B., 2002. Adaptive divergence in experimental populations of *Pseudomonas fluorescens*. I. Genetic and phenotypic bases of wrinkly spreader fitness. Genetics 161, 33–46.

Sutherland, I.W., 2001a. Biofilm exopolysaccharides: a strong and sticky framework. Microbiology 147, 3–9.

Sutherland, I.W., 2001b. Exopolysaccharides in biofilms, flocs and related structures. Water Sci. Technol. 43, 77–86.

Sutherland, I.W., 2001c. The biofilm matrix – an immobilized but dynamic microbial environment. Trends Microbiol. 9, 222–227.

Takeda, M., Nomoto, S., Koizumi, J., 2002. Structural analysis of the extracellular polysaccharide produced by *Sphaerotilus natans*. Biosci. Biotechnol. Biochem. 66, 1546–1551.

Tamilvanan, S., Venkateshan, N., Ludwig, A., 2008. The potential of lipid- and polymer-based drug delivery carriers for eradicating biofilm consortia on device-related nosocomial infections. J. Control. Release 128, 2–22.

Tanji, Y., Nishihara, T., Miyanaga, K., 2007. Monitoring of biofilm in cooling water system by measuring lactic acid consumption rate. Biochem. Eng. J. 35, 81–86.

Tsang, P.H., Li, G., Brun, Y.V., Freund, L.B., Tang, J.X., 2006. Adhesion of single bacterial cells in the micronewton range. Proc. Natl. Acad. Sci. 103, 5764–5768.

Türetgen, I., Cotuk, A., 2007. Monitoring of biofilm-associated *Legionella pneumophilia* on different substrata in model cooling tower system. Environ. Monit. Assess. 125, 271–279.

Väisänen, O.M., Nurmiaho-Lassila, E.-L., Marmo, S.A., Salkinoja-Salonen, M.S., 1994. Structure and composition of biological slimes on paper and board machines. Appl. Environ. Microbiol 60, 641–653.

Väisänen, O.M., Weber, A., Bennasar, A., Rainey, F.A., Busse, H.-J., Salkinoja-Salonen, M.S., 1998. Microbial communities of printing paper machines. J. Appl. Microbiol. 84, 1069–1084.

van Veen, W.L., Mulder, E.G., Deinema, M.H., 1978. The Sphaerotilus-Leptothrix group of bacteria. Microbiol. Rev. 42, 329–356.

Verhoef, R., Waard de, P., Schols, H.A., Rättö, M., Siika-aho, M., Voragen, A.G.J., 2002. Structural elucidation of the EPS of slime producing *Brevundimonas vesicularis* sp. isolated from a paper machine. Carbohydr. Res. 337, 1821–1831.

Verhoef, R., Waard de, P., Schols, H.A., Siika-aho, M., Voragen, A.G.J., 2003. *Methylobacterium* sp. isolated from a Finnish paper machine produces highly pyruvated galactan exopolysaccharide. Carbohydr. Res. 338, 1851–1859.

Verhoef, R., Schols, H.A., Blanco, A., et al., 2005. Sugar composition and FT-IR analysis of exopolysaccharides produced by microbial isolates from paper mill slime deposits. Biotechnol. Bioeng. 91, 91–105.

Vrouwenvelder, H.S., Paassen, J.A.M.v., Folmer, H.C., Nederlof, J.A.M.H., Kooij, D.v.d., 1998. Biofouling of membranes for drinking water production. Desalination. 118, 157–166.

Warscheid, Th., 1996. Biodeterioration of stones: analysis, quantication and evaluation. In: Proceedings of the 10th International Biodeterioration and Biodegradation Symposium, Dechema-Monograph No. 133. Dechema, Frankfurt, pp. 115–120.

Warscheid, Th., Braams, J., 2000. Biodeterioration of stone: a review. Int. Biodeterior. Biodegrad. 46, 343–368.

Wotton, R., 2004. The essential role of exopolymers (EPS) in aquatic systems. Oceanogr. Mar. Biol. Annu. Rev. 42, 57–94.

Ye, Y., Le Chlech, P., Chen, V., Fane, A.G., 2005. Evolution of fouling during crossflow filtration of model EPS solutions. J. Membr. Sci. 264, 190–199.

Yun, M.-A., Yeon, K.-M., Park, J.-S., Lee, Ch.-H., Chun, J., Lim, D.J., 2006. Characterization of biofilm structure and its effect on membrane permeability in MBR for dye wastewater treatment. Water Res. 40, 45–52.

Zhang, X., Bishop, P., 2003. Biodegradability of biofilm extracellular polymeric substances. Chemosphere 50, 63–69.

Zhou, J., Mopper, K., Passow, U., 1998. The role of surface-active carbohydrates in the formation of transparent exopolymer particles by bubble adsorption of seawater. Limnol. Oceanogr. 43, 1860–1871.

Zogaj, X., Nimtz, M., Rohde, M., Bokranj, W., Romling, U., 2001. The multicellular morphotypes of *Salmonella typhimurium* and *E. coli* produce cellulose as the second component of the extracellular matrix. Mol. Microbiol. 39, 1452–1463.

CHAPTER

40

Microbial glycosylated components in plant disease

Max Dow, Antonio Molinaro, Richard M. Cooper and Mari-Anne Newman

SUMMARY

Microbial glycosylated components can play important and sometimes opposing roles to influence the outcome of plant interactions with potential pathogens. Components such as lipopolysaccharides, peptidoglycan, chitin and glucan are considered to be microbe-associated molecular patterns, molecules that comprise (relatively) invariant structural moieties and that are indispensable to the pathogen, but which are recognized by plants to elicit defence responses. In order to be successful, a pathogen must be able to tolerate or suppress the induction of these responses. Pathogens have evolved multiple mechanisms to achieve this, some of which involve glycosylated components. In phytopathogenic bacteria, extracellular polysaccharides and extracellular cyclic glucan have been implicated in suppression of particular defence responses. Bacterial polysaccharides can also have additional roles in other aspects of pathogen behaviour that may include colonization of plant surfaces, biofilm formation and tolerance to the host environment. Although mechanisms for plant recognition of fungal chitin and oomycete glucan have been described, little is known about plant perception of bacterial lipopolysaccharides and peptidoglycan. Equally, the mechanistic basis for suppression of plant defences by cyclic glucan is unknown, although recent work has suggested that polyanionic bacterial extracellular polysaccharides may exert this action through sequestration of extracellular (apoplastic) calcium ions.

Keywords: Lipopolysaccharides; Extracellular polysaccharides; Peptidoglycan; Cyclic glucan; Fungal chitin; β-glucan; Microbe-associated molecular pattern; Plant defence induction; Defence response suppression

1. INTRODUCTION

The ability of microbes to cause disease in plants depends upon synthesis of a range of factors that act in different ways to promote virulence. To counter disease, plants have evolved both preformed and induced defence mechanisms that serve to eliminate or restrict the growth of potential pathogens (Hammond-Kosack and Jones, 1996; Lamb and Dixon, 1997; Greenberg, 1997). Sophisticated surveillance mechanisms in plants detect the presence of pathogen-derived molecules leading to triggering of induced defence responses following microbial attack. The co-evolution of plants and pathogens has led to the expression of a hierarchy of such responses, triggered by different pathogen molecules (Jones and

Dangl, 2006). In order to be successful, a pathogen must be able to tolerate the host environment, evade or overcome preformed defences and suppress or fail to elicit induced defences (Abramovitch *et al.*, 2006; He *et al.*, 2007). It is evident that microbial glycosylated components can play important, and sometimes opposing, roles in plant–microbe interactions. Roles such as elicitation of plant defences are of benefit to the host. Conversely, other roles such as promotion of colonization of plant surfaces, suppression of basal defence responses and increasing microbial tolerance to the host environment, are of benefit to the pathogen. Here, the current state of knowledge in this area is reviewed, beginning with a brief overview of induced defence responses in plants.

2. INDUCED DEFENCE RESPONSES IN PLANTS

2.1. Resistance responses in plants

Induced defence responses include the production of reactive oxygen species (i.e. the oxidative burst), production of reactive nitrogen species, e.g. nitric oxide (NO), alterations in the plant cell wall, induction of antimicrobial compounds (e.g. phytoalexins), the synthesis of pathogenesis-related (PR) proteins and the triggering of a programmed cell death response called the hypersensitive response (HR) (for reviews, see Hammond-Kosack and Jones, 1996; Lamb and Dixon, 1997; Greenberg, 1997). An early event following perception of certain pathogen-derived molecules is a rapid K^+/H^+ exchange at the plasma membrane resulting in a significant increase in apoplastic pH. Calcium influx from the cell exterior to the cytosol is also a prerequisite for most induced defence responses. Approximately concurrently, reactive oxygen species are generated, primarily by activation of plasma membrane-associated nictotinamide adenine dinucleotide phosphate (NADPH) oxidase. Reactive oxygen species and NO can have roles in plant signalling for defence and can have direct antimicrobial effects. Reactive oxygen species can drive oxidative cross-linking of polymers in the plant cell wall to strengthen it against degradation and to restrict pathogen spread. Other alterations in the plant wall, in addition to cross-linking of existing plant cell wall polymers, include the deposition of the β-(1→3)-linked glucan, callose and of the polyphenolic lignin. Phytoalexins are antimicrobial compounds of diverse chemical classes that are absent in healthy plants but induced upon infection. The PR proteins comprise a number of families that include enzymes, e.g. chitinase and β-(1→3)-glucanase, which can directly attack pathogen structures, antimicrobial peptides and small proteins, and PR1, which is of unknown function. Resistance of plants to attempted microbial attack is usually associated with more rapid and more intense induction of these responses which, nevertheless, also occur in susceptible plants at later stages of disease, hence the use of the term PR.

2.2. Microbe-associated molecular patterns and triggering of basal defences

An emerging view is that the oxidative burst, NO generation, cell wall alterations and PR protein induction are basal defences that are induced by pathogen surface-derived molecules (Gómez-Gómez and Boller, 2000; Parker, 2003; Nürnberger and Kemmerling, 2006). This is analogous to the innate immune system in vertebrate and invertebrate organisms, where the term pathogen-associated molecular pattern (PAMP) has been introduced to describe the molecules recognized (Nürnberger and Brunner, 2002). Since PAMPs are molecules found in both pathogenic and non-pathogenic microbes, the use of the alternative term microbe-associated molecular pattern (MAMP) has been suggested (Ausubel, 2005). The MAMPs are usually indispensable for microbial fitness and have relatively invariant molecular structures that are shared by many related microbes, but not with their host. In addition to proteins, such as bacterial flagellin (Gómez-Gómez and Boller, 2000) and Ef-Tu elongation factor (Zipfel *et al.*, 2006), glycosylated compounds, such as bacterial lipopolysaccharides (LPSs), bacterial peptidoglycan (PG) and fungal oligosaccharides derived from chitin and glucan, can act as MAMPs. These will be discussed in detail below.

Pathogens have evolved a number of methods to suppress the action of MAMPs in elicitation of basal defences (Abramovitch *et al.*, 2006; Jones and Dangl, 2006; He *et al.*, 2007). One key strategy is the delivery into the plant cell of effector proteins which, in bacterial pathogens, is mediated by type III-secretion systems. These effectors function to suppress basal defences and to alter host metabolism in order to promote disease. Alternative mechanisms of suppression have also been described. Of particular interest here are newly discovered roles for bacterial extracellular polysaccharides and bacterial cyclic glucan in suppression of basal resistance (see below). In some plants or plant varieties, certain type III-secreted effectors are specifically recognized by plant receptors, thus resulting in the triggering of the programmed cell death of the HR (Jones and Dangl, 2006). In this way, specific recognition, leading to HR-associated resistance, is superimposed on basal defence responses that are triggered by MAMPs. Subsequent variations in effector structure can allow the pathogen to evade detection, a situation that can be countered by evolution of plant receptors with appropriate altered specificities (Jones and Dangl, 2006).

3. THE DIVERSE ROLES OF BACTERIAL LPSs IN PLANT DISEASE

As ubiquitous and vital components of the cell surface of Gram-negative bacteria, LPSs have multiple roles in plant–microbe interactions (see also Chapter 3).

3.1. Lipopolysaccharide as a barrier to host factors

Importantly, LPS contributes to the reduced permeability of the Gram-negative outer membrane. This barrier function allows growth of bacteria in unfavourable conditions that may be encountered in niches within plant hosts. Notably, LPS-defective mutants show increased *in vitro* sensitivity to antibiotics and antimicrobial peptides and, upon introduction into susceptible plants, the numbers of viable bacteria often decline very rapidly (reviewed in Dow *et al.*, 2000; Newman *et al.*, 2007). This may in part be due to an impaired exclusion of preformed or induced antimicrobial substances of plant origin.

3.2. Lipopolysaccharide as an inducer of plant defences

A number of reports detail the effects of LPS on the induction of basal plant defences including the oxidative burst, NO production, calcium influx, the induction of *PR* gene expression and cell wall alterations that include deposition of callose and phenolics (Table 40.1). These observations have been made with LPS preparations from a range of bacteria used in cell suspension cultures or leaves from various plants, which may explain some variations in outcome. For example, LPS can induce the production of active oxygen species (Albus *et al.*, 2001; Meyer *et al.*, 2001; Gerber *et al.*, 2004; Braun *et al.*, 2005; Desaki *et al.*, 2006), but this is not always observed (Dow *et al.*, 2000; Desender *et al.*, 2006). The LPSs from a range of bacteria induced NO synthesis in suspension cultures and leaves of the crucifer *Arabidopsis thaliana* (Zeidler *et al.*, 2004). This common effect of LPSs from diverse bacteria suggested the involvement of a shared molecular determinant, the lipid A moiety and, indeed, isolated lipid A was also active (Zeidler *et al.*, 2004). This is consistent with the notion of LPS as a MAMP, since

TABLE 40.1 Plant responses induced by glycosylated microbe-associated molecular patterns (MAMPs)

MAMP and plant response	Experimental systems used		Reference
	Source(s) of MAMP	Plant system	
Bacterial peptidoglycan			
Induction of *PR1* gene transcription, callose deposition, influx of calcium, medium acidification	*Staphylococcus aureus* *Xanthomonas campestris* *Agrobacterium tumefaciens*	Cell suspension cultures and leaves of *Arabidopsis thaliana*	Gust *et al.*, 2007 Erbs *et al.*, 2008
Bacterial lipopolysaccharides			
Induction of the synthesis of reactive oxygen species (the oxidative burst)	*Burkholderia cepacia* *Xanthomonas campestris* *Sinorhizobium meliloti*	Cell suspension cultures of *Nicotiana tabacum*	Meyer *et al.*, 2001; Gerber *et al.*, 2004; Braun *et al.*, 2005; Albus *et al.*, 2001
Suppression of the oxidative burst	*Sinorhizobium meliloti*	Cell suspension cultures of *Medicago sativa* and *Medicago truncatula*	Albus *et al.*, 2001; Scheidle *et al.*, 2005
Induction of NO synthesis	*Escherichia coli, Pseudomonas* spp., *Erwinia* spp., *Xanthomonas campestris, Ralstonia solanacearum*	Cell suspension cultures of *Arabidopsis thaliana*	Zeidler *et al.*, 2004
Cell wall alterations, deposition of callose	*Xanthomonas campestris* pv. *vesicatoria*	Leaves of pepper (*Capsicum annuum*)	Keshavarzi *et al.*, 2004
Induction of defence-related β-(1→3)-glucanase	*Xanthomonas campestris* pv. *campestris*	Leaves of turnip (*Brassica campestris*)	Newman *et al.*, 1995
Induction of pathogenesis-related proteins	*Escherichia coli, Pseudomonas* spp., *Erwinia* spp., *Xanthomonas campestris, Ralstonia solanacearum,* Synthetic O-antigen oligosaccharides	Leaves of *Arabidopsis thaliana*	Zeidler *et al.*, 2004; Silipo *et al.*, 2005; Bedini *et al.*, 2005a
Potentiation of synthesis of hydroxycinnamoyl tyramines	*Xanthomonas campestris* pv. *campestris, Escherichia coli, Salmonella enterica* sv. Minnesota	Leaves of pepper (*Capsicum annuum*)	Newman *et al.*, 2002
Potentiation of PR gene induction	*Xanthomonas campestris* pv. *campestris, Escherichia coli, Salmonella enterica* sv. Minnesota	Leaves of pepper (*Capsicum annuum*)	Newman *et al.*, 2002
Calcium influx	*Xanthomonas campestris* pv. *campestris* *Burkholderia cepacia*	Cell suspension cultures of *Nicotiana tabacum*	Meyer *et al.*, 2001; Gerber *et al.*, 2004
Acidification of culture medium	*Pectobacterium atrosepticum* *Pseudomonas corrugata*	Cell suspension cultures of tobacco, tomato and potato	Desender *et al.*, 2006
Phosphorylation of MAP kinases and other proteins	*Burkholderia cepacia*	Cell suspension cultures of *Nicotiana tabacum*	Gerber *et al.*, 2004, 2006
		Leaves of *Nicotiana tabacum*	Piater *et al.*, 2004

(*Continues*)

TABLE 40.1 Continued

MAMP and plant response	Experimental systems used		Reference
	Source(s) of MAMP	Plant system	
Prevention of the hypersensitive response, a programmed cell death response	Wide number of sources including enteric bacteria, *Ralstonia solanacearum* and *Xanthomonas* spp.	Leaves of solanaceous (tobacco, pepper) and cruciferous (brassicas, *Arabidopsis*) plants	Reviewed in Dow *et al.*, 2000; Newman *et al.*, 1997; Silipo *et al.*, 2005
Induction of programmed cell death	Various sources including *Pseudomonas aeruginosa, Ralstonia solanacearum* and enteric bacteria	Cell suspension cultures of rice	Desaki *et al.*, 2006
Chitin oligosaccharides			
Oxidative burst, induction of PR gene expression and PAL transcription, MAP kinase activation	Synthetic $(GlcNAc)_8$	Seedlings of *Arabidopsis thaliana*	Miya *et al.*, 2007
Oxidative burst, induction of genes for PAL and peroxidase and endochitinase	Synthetic $(GlcNAc)_8$	Cell suspension cultures of rice	Kaku *et al.*, 2006
Lignin formation	Chitin oligomers	Wheat leaves	Pearce and Ride, 1982
Glucan oligosaccharides			
Accumulation of the phytoalexin glyceollin	Heptaglucosyl elicitor from β-glucan of *Phytophthora megasperma* f.sp. *glycinea*	Wounded soybean hypocotyls	Cheong *et al.*, 1991
Induction of genes encoding enzymes in phytoalexin synthesis; phytoalexin synthesis	Mixed oligosaccharides derived from β-glucan of *Phytophthora megasperma* f.sp. *glycinea*		Schmidt and Ebel, 1987 Umemoto *et al.*, 1997

most of the distinguishing structural features of lipid A are conserved between LPS from different bacteria. In these experiments, LPS activated also the accumulation of transcripts of defence-related genes, including *PR-1*, an effect mediated by NO (Zeidler *et al.*, 2004).

Lipo-oligosaccharides (LOSs) from the crucifer pathogen *Xanthomonas campestris* pv. *campestris* were also active in inducing expression of the defence-related genes *PR-1* and *PR-2* in *A. thaliana* (Silipo *et al.*, 2005). In some cases, effects on plant gene induction that are specific to a particular LPS/LOS have been observed. This may reflect the ability of particular plants to recognize structural features within LPS/LOS that are not widely conserved. For example, in turnip (*Brassica campestris*), LOS of *X. campestris* pv. *campestris* induced expression of a gene encoding a defence-related β-(1→3)-glucanase when applied to leaves at 1 μg per ml (Newman *et al.*, 1995). In contrast, LOSs (rough-form LPSs) of *Escherichia coli* and *Salmonella enterica* sv. Minnesota were ineffective at concentrations up to 50 μg per ml (Newman *et al.*, 1995). Nevertheless, LPSs from these enteric bacteria can elicit defence-related gene induction in different plants at 50 μg per ml (see Table 40.1).

The concentrations of LPS required to elicit most of the effects described above are in the 5–100 μg per ml range, which suggests that plants do not have the exquisite sensitivity to LPS shown by mammalian cells; the latter can respond at concentrations in the pg to ng per ml range. These considerations have led to suggestions that plants possess only low affinity systems to detect LPS (Zeidler *et al.*, 2004), although plants can detect other bacterial MAMPs, such as peptides derived from flagellin and Ef-Tu elongation factor at sub-nM levels. One complicating factor is the aggregation of LPS molecules within the purified preparations used, which may affect the ability of LPS to traverse the matrix of the plant cell wall and hence reach presumed membrane-associated receptors.

3.3. Lipopolysaccharide acts to potentiate the plant defence response

In a number of cases, LPS does not act in direct induction of plant defence responses, but increases the speed and/or degree of induction upon subsequent pathogen inoculation (Newman *et al.*, 2002), an effect also seen with a number of other biological agents and synthetic compounds (Prime-a-Plant group *et al.*, 2006). One example of potentiation or priming by LPS/LOS is the synthesis of the hydroxycinnamoyl-tyramine conjugates feruloyl tyramine (FT) and *p*-coumaroyl tyramine (CT), which are associated with the resistance of pepper to *X. campestris* pv. *campestris* (Newman *et al.*, 2001). The LPSs from *X. campestris* pv. *campestris* and *S. enterica* sv. Minnesota act to potentiate the accumulation of these compounds occurring in response to subsequent pathogen inoculation (Newman *et al.*, 2002). Similar potentiation effects have been observed for the expression of genes for acidic and basic β-(1→3)-glucanase and P6 (a PR1 homologue) in leaves of pepper (Newman *et al.*, 2002). The molecular basis of these effects is however unknown. Intriguingly, application of LPSs and other MAMPs to plants can lead to an increased expression of surveillance systems for bacterial type III-secreted effectors and viral proteins (Gerber *et al.*, 2004). This activation of additional surveillance mechanisms may be related to the potentiation phenomenon.

3.4. Lipopolysaccharide acts to suppress the HR in dicot plants

Perhaps the first effect of LPS on plants to be described was the ability to prevent the HR programmed cell death response in leaves (reviewed in Dow *et al.*, 2000; Newman *et al.*, 2007). This phenomenon was first described over three decades ago, although the molecular basis remains obscure. The effect is highly localized to the site of LPS inoculation and requires several hours to become established, which suggests that the protective mechanism depends on a plant response to LPS. In some cases, the refractory state is only present in a temporal window and, if the period between LPS and bacterial inoculation is extended, HR is again observed.

These findings present a puzzle. If basal resistance responses and HR both contribute to plant defence, why does LPS activate the former but block the latter? One possible answer is that the effects of LPS in prevention of HR and triggering basal defences may allow the plant to express resistance without catastrophic tissue collapse. This contention is supported by observations that the prevention of HR does not apparently lead to an increased susceptibility of the tissue. The onset of HR is generally associated with a decline in the number of bacteria that can be recovered from the leaf. In contrast, in leaf tissue pretreated with LPS, although bacterial survival was considerably promoted, there was no overall increase in population size.

In contrast to these effects in dicot plants, LPSs from plant pathogens and non-pathogens induce programmed cell death in cultured cells of monocot rice (Desaki *et al.*, 2006). However, LPSs used in these experiments were unable to trigger programmed cell death in cultured *Arabidopsis* cells. Further experimentation is required to establish whether induction, rather than prevention, of programmed cell death is a fundamental difference in the response to LPS by monocots and dicots.

3.5. Plants can recognize different structures within LPS to elicit defence responses

As in the animal innate immune system, recognition of the lipid A moiety may be at least partially responsible for many of the effects of LPS in plants. Nevertheless, there is evidence that both the core oligosaccharide and O-antigen can elicit plant responses. There is considerable O-antigen structural variability in plant-associated bacteria (Corsaro *et al.*, 2001; Lerouge and Vanderleyden, 2002), the significance of which remains unclear. The role of the O-antigen has been examined through the use of synthetic O-antigen oligosaccharides (Bedini *et al.*, 2005a). The O-antigen of the LPSs from many phytopathogenic bacteria is comprised of a rhamnan backbone with the trisaccharide repeating unit, α-L-Rha-(1→3)-α-L-Rha-(1→2)-α-L-Rha-(1→3). This rhamnose- (Rha-) based trisaccharide was synthesized and oligomerized to obtain hexa- and nona-saccharides. These rhamnans were effective in elicitation of transcription of the defence-related genes *PR1* and *PR2* and in suppression of the HR in *A. thaliana* (Bedini *et al.*, 2005a). Molecular dynamics calculations and conformational analysis of the oligorhamnans by nuclear magnetic resonance (NMR) spectroscopy revealed that a coiled structure develops with increasing chain length of the oligosaccharide. This is associated with increasing efficacy in HR suppression and *PR* gene induction (Bedini *et al.*, 2005a). The coil structure, therefore, may be a plant-recognizable PAMP. These findings are currently being extended through the examination of other, structurally distinct, synthetic O-antigens (Bedini *et al.*, 2005b). This work should provide insight into some of the potential roles of O-antigen variation in triggering responses in different plants.

The core region of LPS has an established influence on lipid A toxicity in animals, thus indicating that elucidation of the structure and of the biological activity of both lipid A and core region are of high importance for a better understanding of LPS action in plants (Schromm *et al.*, 2000). As with synthetic O-antigen, LOS of *X. campestris* pv. *campestris* is able to elicit defense-related genes *PR1* and *PR2* and to prevent the HR in *A. thaliana* (Silipo *et al.*, 2005). Both the lipid A and core oligosaccharides derived from the LOS have activity. Although LOS induced defence-related gene transcription in two temporal phases, the core oligosaccharide induced only the earlier phase and lipid A induced only the later phase. These findings suggest that plant cells can recognize lipid A and core oligosaccharide structures within LOSs to trigger defensive cellular responses and that this may occur via two distinct recognition events (Silipo *et al.*, 2005).

3.6. Variations in LPS structure that alter the biological activity in plants

Work on the effects of LPS from bacterial symbionts of plants indicates the occurrence of variations in LPS structure that influence the activity in elicitation of defences. Although LPS from *Sinorhizobium meliloti* elicited an oxidative burst in cell suspension cultures of the non-host plant, tobacco (*Nicotiana tabacum*), it had no apparent activity in cell suspension cultures of host plants, alfalfa (*Medicago sativa*) and the model legume, *Medicago truncatula* (Albus *et al.*, 2001; Scheidle *et al.*, 2005). Furthermore, *S. meliloti* LPS was able to suppress the production of reactive oxygen species in cultures of these host plants. Lipid A alone was sufficient for the suppression of the oxidative burst (Scheidle *et al.*, 2005).

Although a general conservation of lipid A structure is seen between many plant-associated bacteria, considerable divergence of structure occurs within the *Rhizobiaceae*. Lipid A from *S. meliloti* comprises two β-(1→6)-linked D-glucosamine (GlcN) residues, with two phosphate substituents and up to four acyl primary chains

(Gudlavalleti and Forsberg, 2003). A long chain fatty acid, 27-hydroxyoctacosanoate [28:0(27-OH)], is present and attached as secondary acyl substituent. In marked contrast, LPS from *Rhizobium leguminosarum* and *R. etli* contain an unusual lipid A in which a galacturonic acid (GalA) residue is present at O-4 of non-reducing GlcN, a 2-aminogluconate residue replaces the anomeric phosphate group, and a long chain secondary fatty acid is present on the non-reducing GlcN (Bhat *et al.*, 1994).

Furthermore, the core region of LPS in these *Rhizobium* spp. lacks heptose residues. To date, it is unreported whether LPS and lipid A from these bacteria have the same suppressive effects on the oxidative burst in legumes as LPS and lipid A of *S. meliloti*. More recent attention has focused on other *Rhizobiaceae* within the genus *Agrobacterium*, which are pathogens that attack a range of plants. Intriguingly, almost all agrobacterial LPSs contain the same lipid A as *S. meliloti* (Silipo *et al.*, 2004), which opens the possibility that LPS suppression of host defences may have a role in microbe–host interactions involving these more subtle pathogens.

Alterations in lipid A or other structures within LPS are known to occur during symbiotic interactions with plants (Kannenberg and Carlson, 2001) and in response to compounds in plant root exudates (Fischer *et al.*, 2003) and may occur during plant pathogenesis. These alterations may serve both to increase the resistance of the bacteria against host defences and to attenuate the activity of lipid A or LPS in triggering these defences. Regulated palmitoylation of lipid A has been shown to have such a dual role in the pathogenesis of bacterial pathogens of mammals (Bishop, 2005; Bishop *et al.*, 2005). Palmitoylation occurs by transfer of a palmitate chain, i.e. hexadecanoic acid (16:0), from a phospholipid to lipid A and is catalysed by an outer membrane enzyme, PagP. Homologues of *pagP* occur in the plant pathogens *Erwinia chrysanthemi* and *Erwinia carotovora*, although it is not yet known whether they have a role in virulence (Bishop *et al.*, 2005). Studies of the LOS of a mutant of *X. campestris* pv. *campestris* defective in core completion have shown that loss of the glycosyl residues of the outer core is accompanied by alterations in the lipid A, which becomes penta-acetylated (rather than hexa-acetylated) and is substituted with phosphoethanolamine moieties (Silipo *et al.*, 2008). This modified lipid A is not able to induce defence-related responses such as *PR* gene induction in *A. thaliana*. This suggests that *X. campestris* pv. *campestris* has the capacity to modify the structure of lipid A in order to reduce its activity as a MAMP. Whether alterations in the acylation pattern or phosphoethanamine substitution of *X. campestris* pv. *campestris* lipid A occur in the host, and whether they are triggered by specific plant environmental cues, are as yet unknown.

4. BACTERIAL PG AS A MAMP

Peptidoglycan is an essential and unique cell wall component of bacteria that imparts rigidity and structure to the bacterial envelope (see Chapter 2). The PG from *Staphylococcus aureus*, a Gram-positive human pathogen, and from the Gram-negative plant pathogens, *X. campestris* pv. *campestris* and *Agrobacterium tumefaciens*, have been shown to induce a range of defence responses in leaves or cell suspension cultures of *A. thaliana* (Gust *et al.*, 2007; Erbs *et al.*, 2008). These findings indicate that PG acts as a MAMP in plants, as it does in animals. Many of the plant responses observed are characteristic of those induced by other MAMPs and include *PR-1* induction, deposition of callose, acidification of the cell culture medium and induction of the oxidative burst. However, unlike the action of other MAMPs, no substantial calcium influx has been observed.

Differences are seen in the ability of PG or derived muropeptides from these contrasting pathogens to activate defence responses.

Muropeptides from *X. campestris* pv. *campestris* are more effective than native PG in triggering short-term responses, e.g. generation of reactive oxygen species, suggesting that they are generally recognized earlier by the plant than native PG (Erbs *et al.*, 2008). Muropeptides from *X. campestris* pv. *campestris* are also more efficient inducers of *PR1* gene transcription than PG; approximately equivalent effects have been seen with 0.1 μg ml of muropeptides and 50 μg ml of intact PG (Erbs *et al.*, 2008). Biologically active muropeptide fragments may be generated by the turnover of PG that occurs continuously in bacteria, or by enzymatic degradation by bifunctional chitinase/lysozyme enzymes from the host (Cloud-Hansen *et al.*, 2006). The greater activity of muropeptides than native PG for *X. campestris* pv. *campestris* contrasts markedly with perception by *A. thaliana* of the *S. aureus* PG, where the converse effect was seen (Gust *et al.*, 2007).

Fundamental chemical differences in PG from Gram-positive and Gram-negative organisms may explain the disparity in MAMP activities. Of note, PG consists of glycan chains of β-(1→4)-linked-*N*-acetylglucosamine (GlcNAc) and *N*-acetylmuramic acid (MurNAc). The presence of the lactyl group of the muramic acid allows the covalent attachment of a short peptide stem that typically contains alternating L- and D-amino acids. The structure of the carbohydrate backbone is generally conserved in all bacteria, with different degrees of acetylation the only identified chemical variation. In contrast, the peptide moiety displays considerable diversity. In general, the third position amino acid in Gram-positive bacteria is L-lysine (Lys) while in Gram-negative bacteria it is *meso*-2,6-diaminopimelic (DAP) acid. Furthermore, Gram-positive bacteria have peptide stems usually cross-linked through an inter-peptide bridge, generally glycine (Gly), whereas Gram-negative bacteria peptide stems are usually directly cross-linked.

The nature of the eliciting moiety within PG is unknown. The weaker activity of degradation products of PG from *S. aureus* compared to the native PG has led to speculation that the glycan backbone is the active elicitor moiety (Gust *et al.*, 2007). Alternative suggestions have come from structural comparisons of muropeptides from *A. tumefaciens*, which are weak elicitors, with those from *X. campestris* pv. *campestris*, which are more effective (Erbs *et al.*, 2008). These Gram-negative muropeptides differ by the presence of a Gly residue replacing alanine (Ala) in the case of *Agrobacterium* PG and by the lack of an acetyl group in *X. campestris* pv. *campestris* PG. There are only a few examples in which Gly is reported as a component of PG of Gram-negative bacteria and no correlation with the biological activity has been reported (Vollmer *et al.*, 2008). These variations might explain their different eliciting activities; chemical synthesis of these and related components should allow this question to be addressed.

The alteration of PG in *Agrobacterium* to reduce its effectiveness as a MAMP is redolent of the alteration of the lipid A moiety of LPSs in this genera, which may have a similar effect on the ability of the lipid A to induce defences responses (see above). Furthermore, although flagellin from many bacteria is a potent MAMP, that from *A. tumefaciens* is not recognized in *A. thaliana* (Zipfel *et al.*, 2006). This may reflect the lifestyle of *A. tumefaciens*, which is a subtle pathogen dependent on maintaining viability of host cells for transfer of T-DNA and consequent nutritional benefits from the transformed host cells.

5. THE MULTIPLE ROLES OF EXOPOLYSACCHARIDES (EPSs)

Many plant pathogenic bacteria produce high-molecular mass polysaccharides that are secreted into the bacterial environment (Table 40.2). The synthesis of these polymers is usually directed by operons or clusters of genes encoding appropriate glycosyl transferases and is subject to complex regulation. Mutational analyses have established that EPSs are required for the full virulence of a number of plant pathogenic bacteria, including *Ralstonia solanacearum*, *Erwinia amylovora*, *Erwinia stewartii*, *Pseudomonas syringae*, *X. campestris* and *Xanthomonas axonopodis* (Dolph *et al.*, 1988; Kao *et al.*, 1992; Geier and Geider, 1993; Newman *et al.*, 1994; Saile *et al.*, 1997; Yu *et al.*, 1999; Kemp *et al.*, 2004) (Figure 40.1). Although these bacteria produce EPSs of different composition and structure, the polymers share the common feature that they are all polyanionic. This property is conferred through the occurrence of uronic acid residues as part of the repeating oligosaccharide unit and, in some instances, through additional non-carbohydrate decorations such as pyruvyl and succinyl residues (see Table 40.2). As with many bacteria, a number of plant pathogens have the ability to synthesize cellulose in addition to the EPSs described above. In some

TABLE 40.2 Bacterial extracellular polysaccharides with proven or potential roles in plant disease

Polymer	Producing organism	Monosaccharide components[a]	Non-carbohydrate constituents	References
Heteropolymers				
Xanthan	*Xanthomonas campestris* pv. *campestris* (and related *Xanthomonas* spp.)	D-Glc, D-GlcA, D-Man	Pyruvate, acetate	Jansson *et al.*, 1975; Kemp *et al.*, 2004
Fastidian	*Xylella fastidiosa*	D-Glc, D-GlcA, D-Man	Acetate?	da Silva *et al.*, 2001
Alginate	*Pseudomonas syringae* pathovars	D-ManA, L-GulA	Acetate	Yu *et al.*, 1999
Amylovoran	*Erwinia amylovora*	D-Gal, D-GlcA	Pyruvate, acetate	Nimtz *et al.*, 1996a
Stewartan	*Erwinia stewartii*	D-Glc, D-Gal, D-GlcA		Nimtz *et al.*, 1996b
Cepacian	*Burkholderia cepacia* complex	D-Gal, D-Man, D-Glc, D-GlcA, D-Rha	Acetate	Cescutti *et al.*, 2000
Succinoglycan	*Agrobacterium* spp. Rhizobium spp.	D-Glc, D-Gal	Pyruvate, succinate	Matulová *et al.*, 1994; Fraysse *et al.*, 2003
Major EPS polymer	*Ralstonia* (formerly *Pseudomonas*) *solanacearum*	D-GlcNAc, 2-*N*-acetyl-2-deoxy-L-GalA, 2-*N*-acetyl-4-*N*-(3-hydroxybutanoyl)-2, 4,6-trideoxy-D-Glc	Acetate, 3-hydroxybutanoate	Orgambide *et al.*, 1991
Homopolymers				
Cellulose	Many plant pathogens	D-Glc	Acetate (in some cases)	Leigh and Coplin, 1992; Spiers *et al.*, 2003
Curdlan	*Agrobacterium* spp.	D-Glc		Stasinopoulos *et al.*, 1999
Levan	*Pseudomonas syringae*, *Erwinia amylovora* and other bacterial species	D-Fru, (D-Glc)		Gross *et al.*, 1992; Hettwer *et al.*, 1998

[a]Abbreviations: Fru, fructose; Gal, galactose; GalA, galacturonic acid; Glc, glucose; GlcA, glucuronic acid; GlcNAc, *N*-acetylglucosamine; GulA, guluronic acid; Man, mannose; ManA, mannuronic acid; Rha, rhamnose.

organisms, the cellulose polymer may be substituted with acetyl moieties.

5.1. Role of EPS in plant disease

Probably, EPS molecules have multiple roles during plant disease. Cellulose has been implicated

FIGURE 40.1 Symptoms caused by wild-type *Xanthomonas axonopodis* pv. *manihotis* and a xanthan-deficient mutant on cassava leaves. Upper panel: 28 days after petiole inoculation with 10^6 bacteria per ml, wild-type *X. axonopodis* pv. *manihotis* (left hand leaf) has caused wilting and desiccation, whereas the EPS-negative mutant (right hand leaf) has caused no visible symptoms. Lower panel: 28 days after leaf infiltration of 10^4 bacteria per ml, the wild-type (left) has caused extensive chlorosis and desiccation near the inoculation site, whereas the EPS-negative mutant (right) has only caused a small chlorotic halo to appear around the initial infection point. Reproduced, with permission, from Kemp *et al.* (2004) *Xanthomonas axonopodis gumD* gene is essential for EPS production and pathogenicity and enhances epiphytic survival on cassava (*Manihot esculenta*). Physiol. Molec. Plant Pathol. 64, 209–218. Elsevier © 2004.

in the attachment of bacteria to plant surfaces, which may be necessary for colonization of leaf surfaces and for intimate interactions, e.g. the transfer of DNA from *Agrobacterium* spp. Cellulose and other EPSs have been implicated in the formation of biofilms, which are matrix-enclosed bacterial populations in which the organisms are adherent to each other and/or to surfaces or interfaces (see Chapters 37 and 39). Bacteria in biofilms are generally more resistant to external stresses and to the action of host defence factors. For a detailed description of biofilm formation in plant pathogenic bacteria please consult Ramey *et al.* (2004) and Danhorn and Fuqua (2007). Importantly, EPSs can help bacteria to withstand desiccation, exposure to ultraviolet irradiation and the action of reactive oxygen species (Kemp *et al.*, 2004). Polyanionic EPSs have been implicated in the symptoms of disease, such as water stress induced by vascular (i.e. xylem-invading) pathogens and water soaking seen with some foliar pathogens. These polymers could also presumably act to concentrate trace metals from the plant environment prior to uptake by bacteria.

An intriguing problem is why plants have not evolved sensors that trigger defences upon detection of bacterial EPSs, which are apparently indispensable for optimal virulence. In part, this may be due to the inaccessibility of such large polymers to putative receptors located in the plasmalemma, coupled with the apparent inability of plants to degrade such polymers to readily diffusible components. A second explanation is afforded by the recently discovered role for EPSs in suppression of plant defence responses, which is the subject of the next section.

5.2. EPS and the suppression of plant defences

X. campestris pv. *campestris* synthesizes the EPS, xanthan, which is comprised of a cellulosic backbone with β-(1→4)-linked D-glucosyl residues and a trisaccharide side chain of β-D-Man-(1→4)-β-D-GlcA-(1→2)-α-D-Man-(1→3)- (where GlcA, glucuronic acid; Man, mannose) linked to every two glucosyl residues. Mutants that can no longer synthesize EPSs, or that synthesize a truncated polymer, lacking glucuronosyl, mannosyl residues and ketal pyruvate residues are unable to cause disease in *Nicotiana benthamiana* and *A. thaliana* plants (Yun *et al.*, 2006). No bacterial growth was seen in plants infected with the mutant incapable of xanthan synthesis, whereas the strain producing truncated xanthan showed reduced growth in plants compared to the wild-type. These effects were associated with an enhanced deposition of callose in the plant cell wall. Pre-inoculation or co-inoculation of either mutant with xanthan from the wild-type restored the wild-type level of growth and suppressed the synthesis of callose (Yun *et al.*, 2006). Truncated xanthan was not able to induce these effects, however. A number of lines of evidence suggest that suppressive action of xanthan was exerted through sequestration of extracellular Ca^{2+}. Callose synthase is activated by increased Ca^{2+} concentration within the plant cell cytoplasm. As outlined above, Ca^{2+} influx from the extracellular apoplastic pool occurs as an early local response to pathogen attack. Xanthan has the ability to chelate divalent metal ions, such as Ca^{2+}, through the negatively charged GlcA and ketal-pyruvate moieties. The truncated xanthan has reduced binding ability. Taken together, these considerations suggest a scenario in which sequestration of extracellular Ca^{2+} by xanthan prevents influx into the plant cell and activation of callose synthase (Yun *et al.*, 2006).

Concurrent with this work, the influence of xanthan and other polyanionic EPSs from plant pathogenic bacteria on Ca^{2+} influx has been established directly, using plant tissues in which the cells express the Ca^{2+}-sensitive luminescent protein, aqueorin. Highly purified preparations of xanthan, alginate from *P. syringae*, amylovoran from *E. amylovora* and EPSs from *S. meliloti* and *R. solanacearum* have been shown to bind Ca^{2+} and to prevent the Ca^{2+} influx into plant cells that is triggered by the MAMPs flg22, a peptide derived from bacterial flagellin, and elf18 derived from EF-Tu (Aslam *et al.*, 2008). The EPS polymers exert this effect without blocking binding of flg22 to its receptor (FLS2) on *Arabidopsis* cells, suggesting that they act via Ca^{2+} chelation (Aslam *et al.*, 2008). Pretreatment with EPS also suppresses induction of defence genes and other downstream defence responses. A potential suppressive effect of EPS is reflected in the more rapid and intense elicitation of plant defences by EPS-deficient mutants of *X. campestris* and *P. syringae* strains than by the cognate wild-type bacteria (Aslam *et al.*, 2008).

6. BACTERIAL CYCLIC GLUCAN AND SUPPRESSION OF HOST DEFENCES

Cyclic glucans are produced by a number of bacteria, including the plant pathogens *X. campestris* pv. *campestris*, *R. Solanacearum*, and *A. tumefaciens* (Breedveld and Miller, 1994; Talaga *et al.*, 1996; Vojnov *et al.*, 2001). These molecules have been shown to be important for a number of symbiotic and pathogenic plant–microbe interactions (Breedveld and Miller, 1994; Mithöfer, 2002). They are required for effective nodule invasion in symbiotic nitrogen-fixing *R. meliloti* and for crown gall tumour induction in *A. tumefaciens*. The precise roles of cyclic glucans during plant–bacterial interactions are, however, not clear. It has been suggested that a cyclic glucan from *B. japonicum* promotes symbiosis by suppression of defence responses, where it may act to antagonize the binding of microbial elicitors of plant defences (reviewed by Mithöfer, 2002).

X. campestris pv. *campestris* produces an extracellular cyclic glucan that contains sixteen glucosyl residues with fifteen β-(1→2)-linkages and one α-(1→6)-linkage (York, 1995). More recent work has indicated that this glucan has a role in suppression of plant defence responses, e.g. *PR-1* induction and callose deposition, and that it acts not only locally but also in a systemic fashion (Rigano *et al.*, 2007). Disruption of an *X. campestris* pv. *campestris* gene with homology to *ndvB* of *S. meliloti*, which encodes a glycosyltransferase required for cyclic glucan synthesis, generated a mutant that failed to synthesize extracellular cyclic β-(1→2)-glucan and which was compromised in virulence in *A. thaliana* and *N. benthamiana* (Rigano *et al.*, 2007). These effects in *N. benthamiana* were associated with enhanced callose deposition and earlier expression of *PR-1* than seen with the wild-type. Application of purified cyclic β-(1→2)-glucan prior to inoculation of the *ndvB* mutant suppressed callose accumulation and the expression of *PR-1* transcripts in *N. benthamiana* and restored virulence in both plants. These effects were seen when cyclic glucan and bacteria were applied either to the same or to different leaves. Cyclic β-(1→2)-glucan-induced systemic suppression was associated with transport of the molecule throughout the plant (Rigano *et al.*, 2007).

These results suggest that cyclic β-(1→2)-glucan, generated by *X. campestris* pv. *campestris* colonizing one leaf, can be translocated to other leaves to induce susceptibility, thus promoting bacterial spread through the plant. The ability of this cyclic β-(1→2)-glucan to act as a systemic effector of suppression of host defence responses distinguishes it from the action of almost all suppressors so far described and that have only been reported to act locally. The exception is the *P. syringae* phytotoxin coronatine, which induces systemic susceptibility in *A. thaliana*. It is still unknown if coronatine is itself systemically translocated or if it exerts its effect via local activation of the jasmonic acid pathway (Cui *et al.*, 2005). The mechanism by which the cyclic glucan exerts its action, either locally or systemically, is unknown.

7. FUNGAL CHITIN AND OOMYCETE GLUCAN AS MAMPs

It is now established that oligosaccharides derived from cell wall polymers of fungi and oomycetes can act as MAMPs and some mechanistic details of the perception of these signals by plants have emerged (Cheong and Hahn, 1991; Umemoto *et al.*, 1997; Kaku *et al.*, 2006; Miya *et al.*, 2007). Fungal chitin and its degradation products *N*-acetyl-chito-oligosaccharides, i.e. chitin oligomers, act as MAMPs to induce various defence responses in both monocot and dicot plants (Kaku *et al.*, 2006; Miya *et al.*, 2007). These responses include the induction of reactive oxygen species and the activation of expression of defence-related genes, such as *PR-2* and *PAL*. In contrast to many of the glycosylated MAMPs described above, some details of the mechanism of plant perception of chito-oligosaccharides have emerged. In rice, a plasma-membrane glycoprotein, chitin elicitor-binding protein (CEBiP), links the perception of chitin oligomers to the triggering of plant defences (Kaku *et al.*, 2006). Interestingly, CEBiP contains two extracellular Lysin Motif (LysM) domains and a transmembrane domain, but lacks cytoplasmic domains that would be expected to be required for signal transduction. This suggests that although CEBiP can bind chito-oligosaccharides, additional components may be required to effect signal transduction upon elicitor binding. In the dicot *A. thaliana*, the chitin elicitor receptor kinase 1 (CERK1)/LysM receptor-like kinase 1 (RLK1) is essential for chitin elicitor signalling and contributes to resistance to fungal but not to bacterial pathogens (Miya *et al.*, 2007; Wan *et al.*, 2008). The CERK1/LysM RLK1 is a plasma-membrane protein with three extracellular LysM motifs and has a cytoplasmic serine (Ser)/threonine (Thr) kinase domain, with a potential role in signal transduction. Although genetic evidence indicates that CERK1 is involved in elicitation of plant defences by chitin oligomers, it was not possible to demonstrate that the protein bound the oligosaccharide elicitor. Nevertheless, the presence of LysM motifs in the extracellular domain of the CEBiP protein from rice, which can bind chito-oligosaccharides, and in CERK1 from *Arabidopsis*, suggests that the latter protein has the same binding activity. One elegant suggestion is that both rice and *Arabidopsis* have CERK and CEBiP proteins, which act as a signalling complex, to sense the elicitor and initiate signal transduction. To date, there is no evidence to support this contention, as attempts to identify a CEBiP-like protein in *Arabidopsis* by methods successful for rice have thus far been unsuccessful. Interrogation of the proteome of *A. thaliana* with the CEBiP amino acid sequence (in BlastP searches) yields several related proteins of similar length and domain structure.

In the oomycetes, the cell walls are composed of β-glucans and cellulose, rather than chitin, as in the true fungi. Some of the earliest work on the role of glycosylated compounds in triggering plant defences has examined the effects of β-(1→3/1→6)-linked glucans from the cell walls of *Phytophthora megasperma* f. sp. *glycinea* on the induction of phytoalexin accumulation in soybean (reviewed in Cheong and Hahn, 1991). Early events in the host–pathogen interaction may involve the release of elicitor-active fragments from the *Phytophtora* cell wall by a β-(1→3)-glucanase from soybean. These oligosaccharides are perceived by the plant to trigger accumulation of pterocarpan phytoalexins, including glyceollin. Partial hydrolysis of the extracted β-glucan, followed by fractionation and structural analysis has allowed identification of the smallest elicitor-active fragment as a branched hepta β-glucoside, with a β-(1→6)-linked glucan backbone carrying two β-(1→3)-linked glucosyl residues. High-affinity binding sites for this heptaglucoside were detected in soybean membranes and the proteins involved were purified (Cheong and Hahn, 1991; Umemoto *et al.*, 1997). Reverse genetic methods then allowed the cloning of the cognate copy DNA (cDNA). Antibodies against the purified protein blocked elicitor induction of phytoalexin synthesis, thus indicating a role in signal transduction (Umemoto *et al.*, 1997). Northern and Western analyses using these probes have provided evidence for the occurrence of heptaglucoside-binding proteins in other species of plant within the *Fabaceae* that are thus related to soybean. Furthermore, genes encoding heptaglucoside-binding proteins can be found in many unrelated plant species, such as *A. thaliana* and rice, that do not respond to the elicitor, suggesting that the presence of the binding protein is essential but not sufficient for elicitation of host defences. The glucan-binding proteins encoded by these genes comprise two domains; one contains a β-glucan binding site and the other is related to fungal glucan endohydrolases and is enzymatically active (Fliegmann *et al.*, 2004). This had led to the suggestion that this second domain is responsible for release of elicitor-active glucan fragments from the *Phytophthora* cell wall during contact of the pathogen with the host.

8. CONCLUDING REMARKS

It is evident from the body of work described herein that microbial glycosylated components can have diverse effects on plants. Bacterial LPS and PG and chitin and glucan from fungi or oomycetes are MAMPs that activate plant disease resistance responses. Conversely, bacterial extracellular polysaccharides and cyclic glucan act to suppress basal plant resistance responses. These interactions constitute part of the molecular dialogue between microbes and plants that determines the outcome of plant–pathogen interactions. Such a dialogue can clearly involve other players such as type III-secreted effectors, which raises the important issue of the significance of MAMP-induced resistance responses and EPS-induced suppressive effects to overall

RESEARCH FOCUS BOX

- How do plants perceive MAMPs such as LPSs and PG and transduce the signal?
- What contribution does MAMP perception make to the outcome of plant–microbe interactions, including both local and systemic resistance?
- What are the targets of bacterial cyclic glucan that allow plant defence suppression?
- Which modifications in LPSs and PG from symbiotic bacteria account for their altered activities in defence induction?
- Do modifications of LPSs and other polymers occur when bacteria are in plants? Does LPS O-antigen variation have a role in plant disease specificity?

plant resistance and/or pathogen virulence. The reduced virulence and increased activation of plant defences by bacterial mutants unable to synthesize EPS or cyclic glucan indicates a role in suppression that contributes to disease, although the mechanistic basis of suppression by cyclic glucan is as yet unknown. The generally indispensable nature of many MAMPs precludes a similar genetic analysis using bacterial null mutants. In this case, one approach is through identification of genes encoding plant receptors for MAMPs, followed by assessment of the effects of mutagenesis or silencing of these genes on pathogen virulence. This has been done, for example, to assess the role of flagellin recognition by FLS2 in disease resistance in *A. thaliana*. Knowledge obtained from such experiments may also have a bearing on other aspects of plant–microbe interactions, e.g. the induction of systemic resistance by beneficial (non-pathogenic) bacteria such as *Pseudomonas fluorescens*, in which LPS is implicated, and signalling between bacterial symbionts such as rhizobial species and their plant hosts. The further understanding of the molecular basis of plant perception of glycoconjugates may have substantial practical impact for improvement of plant health through the creation of new plant varieties, either through transgenic technology or guidance of breeding programmes.

Important questions to be addressed in future research are listed in the Research Focus Box.

ACKNOWLEDGEMENTS

M.D. acknowledges funding by the Science Foundation of Ireland through Principal Investigator Awards 03/IN3/B373 and 07/IN.1/B955. The work at the Università di Napoli (A.M.) is supported by MIUR, Rome (Progetti di Ricerca di Interesse Nazionale). Work at the University of Bath (R.M.C) was funded by a grant from the Leverhulme Trust. Work at the University of Copenhagen (M.-A. N) is supported by grants from The Danish Council for Technology and Innovation (Copenhagen, Denmark). M.-A. N. also acknowledges financial support from Købmand Sven Hansen and Hustru Ina Hansens Fond, Sorø, Denmark and grants from the Danish Agricultural and Veterinary Research Council and by the Carlsbergfondet (Copenhagen, Denmark).

References

Abramovitch, R.B., Anderson, J.C., Martin, G.B., 2006. Bacterial elicitation and evasion of plant innate immunity. Nat. Rev. Mol. Cell Biol. 7, 601–611.

Albus, U., Baier, R., Holst, O., Pühler, A., Niehaus, K., 2001. Suppression of an elicitor-induced oxidative burst reaction in *Medicago sativa* cell cultures by *Sinorhizobium meliloti* lipopolysaccharides. New Phytol. 151, 597–606.

Aslam, S.N., Newman, M.-A., Erbs, G., et al., 2008. Bacterial extracellular polysaccharides suppress induced innate immunity by calcium chelation. Curr. Biol. 18, 1078–1083.

Ausubel, F., 2005. Are innate immune signalling pathways in plants and animals conserved? Nat. Immunol. 6, 973–979.

Bedini, E., De Castro, C., Erbs, G., et al., 2005a. Structure-dependent modulation of a pathogen response in plants by synthetic O-antigen polysaccharides. J. Am. Chem. Soc. 127, 2414–2416.

Bedini, E., Carabellese, A., Barone, G., Parrilli, M., 2005b. First synthesis of the β-D-rhamnosylated trisaccharide repeating unit of the O-antigen from *Xanthomonas campestris* pv. *campestris* 8004. J. Org. Chem. 70, 8064–8070.

Bhat, U.R., Forsberg, L.S., Carlson, R.W., 1994. Structure of lipid A component of *Rhizobium leguminosarum* pv. *phaseoli* lipopolysaccharide. Unique nonphosphorylated lipid A containing 2-amino-2-deoxygluconate, galacturonate, and glucosamine. J. Biol. Chem. 269, 14402–14410.

Bishop, R.E., 2005. The lipid A palmitoyltransferase PagP: molecular mechanisms and role in bacterial pathogenesis. Mol. Microbiol. 57, 900–912.

Bishop, R.E., Kim, S.H., El Zoeiby, A., 2005. Role of lipid A palmitoylation in bacterial pathogenesis. J. Endotoxin Res. 11, 174–180.

Braun, S.G., Meyer, A., Holst, O., Pühler, A., Niehaus, K., 2005. Characterization of the *Xanthomonas campestris* pv. *campestris* lipopolysaccharide substructures essential for elicitation of an oxidative burst in tobacco cells. Mol. Plant Microbe Interact. 18, 674–681.

Breedveld, M.W., Miller, K.J., 1994. Cyclic β-glucans of members of the family *Rhizobiaceae*. Microbiol. Rev. 58, 145–161.

Cescutti, P., Bosco, M., Picotti, F., et al., 2000. Structure of the exopolysaccharide produced by a clinical isolate of *Burkholderia cepacia*. Biochem. Biophys. Res. Comm. 273, 1088–1094.

Cheong, J.-J., Hahn, M.G., 1991. A specific, high-affinity binding site for the hepta-β-glucoside elicitor exists in soybean membranes. Plant Cell 3, 137–147.

Cheong, J.-J., Birberg, W., Fügedi, P., et al., 1991. Structure-activity relationships of oligo-β-glucoside elicitors of phytoalexin accumulation in soybean. Plant Cell 3, 17–136.

Cloud-Hansen, K.A., Brook Petersen, S., Stabb, E.V., Goldman, W.E., McFall-Ngai, M.J., Handelsman, J., 2006. Breaching the great wall: peptidoglycan and microbial interactions. Nat. Rev. Microbiol. 4, 710–716.

Conrath, U., Beckers, G.J., Flors, V.Prime-A-Plant Group, et al., 2006. Priming: getting ready for battle. Mol. Plant Microbe Interact. 19, 1062–1071.

Corsaro, M.M., De Castro, C., Molinaro, A., Parrilli, M., 2001. Structure of lipopolysaccharides from phytopathogenic bacteria. Recent Res. Develop. Phytochem. 5, 119–138.

Cui, J., Bahrami, A.K., Pringle, E.G., et al., 2005. *Pseudomonas syringae* manipulates systemic plant defenses against pathogens and herbivores. Proc. Natl. Acad. Sci. USA 102, 1791–1796.

Danhorn, T., Fuqua, C., 2007. Biofilm formation by plant-associated bacteria. Annu. Rev. Microbiol. 61, 401–422.

da Silva, F.R., Vettore, A.L., Kemper, E.L., Leite, A., Arruda, P., 2001. Fastidian gum: the *Xylella fastidiosa* exopolysaccharide possibly involved in bacterial pathogenicity. FEMS Microbiol. Lett. 203, 165–171.

Desaki, Y., Miya, A., Venkatesh, B., et al., 2006. Bacterial lipopolysaccharides induce defense responses associated with programmed cell death in rice cells. Plant Cell Physiol. 47, 1530–1540.

Desender, S., Klarzynski, O., Potin, P., Barzic, M.R., Andrivon, D., Val, F., 2006. Lipopolysaccharides of *Pectobacterium atrosepticum* and *Pseudomonas corrugata* induce different defence response patterns in tobacco, tomato, and potato. Plant Biol. (Stuttgart) 8, 636–645.

Dolph, P.J., Majerczak, D.R., Coplin, D.L., 1988. Characterization of a gene cluster for exopolysaccharide biosynthesis and virulence in *Erwinia stewartii*. J. Bacteriol. 170, 865–871.

Dow, M., Newman, M.A., von Roepenack, E., 2000. The induction and modulation of plant defense responses by bacterial lipopolysaccharides. Annu. Rev. Phytopathol. 38, 241–261.

Erbs, G., Silipo, A., Aslam, S., et al., 2008. Peptidoglycan and muropeptides from pathogens *Agrobacterium* and *Xanthomonas* elicit plant innate immunity: structure and activity. Chem. Biol. 15, 438–448.

Fischer, S.E., Miguel, M.J., Mori, G.B., 2003. Effect of root exudates on the exopolysaccharide composition and the lipopolysaccharide profile of *Azospirillum brasilense* Cd under saline stress. FEMS Microbiol. Lett. 219, 53–62.

Fliegmann, J., Mithöfer, A., Wanner, G., Ebel, J., 2004. An ancient enzyme domain hidden in the putative β-glucan elicitor receptor of soybean may play an active part in the perception of pathogen-associated molecular patterns during broad host resistance. J. Biol. Chem. 279, 1132–1140.

Fraysse, N., Couderc, F., Poinsot, V., 2003. Surface polysaccharide involvement in establishing the rhizobium-legume symbiosis. Eur. J. Biochem. 270, 1365–1380.

Geier, G., Geider, K., 1993. Characterization and influence on virulence of the levansucrase gene from the fireblight pathogen *Erwinia amylovora*. Physiol. Mol. Plant Pathol. 42, 387–404.

Gerber, I.B., Zeidler, D., Durner, J., Dubery, I.A., 2004. Early perception responses of *Nicotiana tabacum* cells in response to lipopolysaccharides from *Burkholderia cepacia*. Planta 218, 647–657.

Gerber, I.B., Laukens, K., Witters, E., Dubery, I.A., 2006. Lipopolysaccharide-responsive phosphoproteins in *Nicotiana tabacum* cells. Plant Physiol. Biochem. 44, 369–379.

Gómez-Gómez, L., Boller, T., 2000. FLS2: an LRR receptor-like kinase involved in the perception of the bacterial elicitor flagellin in *Arabidopsis*. Mol. Cell 5, 1003–1011.

Greenberg, J.T., 1997. Programmed cell death in plant-pathogen interactions. Annu. Rev. Plant Physiol. Plant Mol. Biol. 48, 525–545.

Gross, M., Geier, G., Rudolph, K., Geider, K., 1992. Levan and levansucrase synthesized by the fireblight pathogen *Erwinia amylovora*. Physiol. Mol. Plant Pathol. 40, 371–381.

Gudlavalleti, S.K., Forsberg, L.S., 2003. Structural characterization of the lipid A component of *Sinorhizobium* sp. NGR234 rough and smooth form lipopolysaccharide. Demonstration that the distal amide-linked acyloxyacyl residue containing the long chain fatty acid is conserved in *Rhizobium* and *Sinorhizobium* spp. J. Biol. Chem. 278, 3957–3968.

Gust, A., Biswas, R., Lenz, H.D., et al., 2007. Bacteria-derived peptidoglycans constitute pathogen-associated molecular patterns triggering innate immunity in *Arabidopsis*. J. Biol. Chem. 282, 32338–32348.

Hammond-Kosack, K.E., Jones, J.D., 1996. Resistance gene-dependent plant defense responses. Plant Cell 8, 1773–1791.

He, P., Shan, L., Sheen, J., 2007. Elicitation and suppression of microbe-associated molecular pattern-triggered immunity in plant-microbe interactions. Cell. Microbiol. 9, 1385–1396.

Hettwer, U., Jaeckel, F.R., Boch, J., Meyer, M., Rudolph, K., Ullrich, M.S., 1998. Cloning, nucleotide sequence, and expression in *Escherichia coli* of levansucrase genes from the plant pathogens *Pseudomonas syringae* pv. *glycinea* and *P. syringae* pv. *phaseolicola*. Appl. Environ. Microbiol. 64, 3180–3187.

Jansson, P.E., Kenne, L., Lindberg, B., 1975. Structure of extracellular polysaccharide from *Xanthomonas campestris*. Carbohydr. Res. 45, 274–282.

Jones, J.D.G., Dangl, J.L., 2006. The plant immune system. Nature 444, 323–329.

Kaku, H., Nishizawa, Y., Ishii-Minami, N., et al., 2006. Plant cells recognise chitin fragments for defense signaling through a plasma membrane receptor. Proc. Natl. Acad. Sci. USA 103, 11086–11091.

Kannenberg, E.L., Carlson, R.W., 2001. Lipid A and O-chain modifications cause *Rhizobium* lipopolysaccharides to become hydrophobic during bacteroid development. Mol. Microbiol. 39, 379–391.

Kao, C.C., Barlow, E., Sequeira, L., 1992. Extracellular polysaccharide is required for wild-type virulence of *Pseudomonas solanacearum*. J. Bacteriol. 174, 1068–1071.

Kemp, B.P., Horne, J., Bryant, A., Cooper, R.M., 2004. *Xanthomonas axonopodis gumD* gene is essential for EPS production and pathogenicity and enhances epiphytic survival on cassava (*Manihot esculenta*). Physiol. Molec. Plant Pathol. 64, 209–218.

Keshavarzi, M., Soylu, S., Brown, I., et al., 2004. Basal defenses induced in pepper by lipopolysaccharides are suppressed by *Xanthomonas campestris* pv. *vesicatoria*. Mol. Plant Microbe Interact. 17, 805–815.

Lamb, C., Dixon, R.A., 1997. The oxidative burst in plant disease resistance. Annu. Rev. Plant Physiol. Plant Mol. Biol. 48, 251–275.

Leigh, J.A., Coplin, D.L., 1992. Exopolysaccharides in plant-bacterial interactions. Annu. Rev. Microbiol. 46, 307–346.

Lerouge, I., Vanderleyden, J., 2002. O-antigen structural variation: mechanisms and possible roles in animal/plant-microbe interactions. FEMS Microbiol. Rev. 26, 17–47.

Matulová, M., Toffanin, R., Navarini, L., Gilli, R., Paoletti, S., Cesàro, A., 1994. NMR analysis of succinoglycans from different microbial sources: partial assignment of their 1H and 13C NMR spectra and location of the succinate and the acetate groups. Carbohydr. Res. 265, 167–179.

Meyer, A., Pühler, A., Niehaus, K., 2001. The lipopolysaccharides of the phytopathogen *Xanthomonas campestris* pv. *campestris* induce an oxidative burst reaction in cell cultures of *Nicotiana tabacum*. Planta 213, 214–222.

Mithöfer, A., 2002. Suppression of plant defence in rhizobia-legume symbiosis. Trends Plant Sci. 7, 440.

Miya, A., Albert, P., Shinya, T., et al., 2007. CERK1, a LysM receptor kinase, is essential for chitin elicitor signaling in *Arabidopsis*. Proc. Natl. Acad. Sci. USA 104, 19613–19618.

Newman, M.A., Conrads-Strauch, J., Scofield, G., Daniels, M.J., Dow, J.M., 1994. Defense-related gene induction in *Brassica campestris* in response to defined mutants of *Xanthomonas campestris* with altered pathogenicity. Mol. Plant Microbe Interact. 7, 553–563.

Newman, M.A., Daniels, M.J., Dow, J.M., 1995. Lipopolysaccharide from *Xanthomonas campestris* induces defense-related gene expression in *Brassica campestris*. Mol. Plant Microbe Interact. 8, 778–780.

Newman, M.A., Daniels, M.J., Dow, J.M., 1997. The activity of lipid A and core components of bacterial lipopolysaccharides in the prevention of the hypersensitive response in pepper. Mol. Plant Microbe Interact. 10, 926–928.

Newman, M.A., von Roepenack-Lahaye, E., Parr, A., Daniels, M.J., Dow, J.M., 2001. Induction of hydroxycinnamoyl-tyramine conjugates in pepper by *Xanthomonas campestris*, a plant defense response activated by *hrp* gene-dependent and *hrp* gene-independent mechanisms. Mol. Plant Microbe Interact. 14, 785–792.

Newman, M.A., von Roepenack-Lahaye, E., Parr, A., Daniels, M.J., Dow, J.M., 2002. Prior exposure to lipopolysaccharide potentiates expression of plant defenses in response to bacteria. Plant J. 29, 487–495.

Newman, M.-A., Dow, J.M., Molinaro, A., Parrilli, M., 2007. Priming, induction and modulation of plant defence responses by bacterial lipopolysaccharides. J. Endotoxin Res. 13, 68–79.

Nimtz, M., Mort, A., Domke, T., et al., 1996a. Structure of amylovoran, the capsular exopolysaccharide from the fire blight pathogen *Erwinia amylovora*. Carbohydr. Res. 287, 59–76.

Nimtz, M., Mort, A., Wray, V., et al., 1996b. Structure of stewartan, the capsular exopolysaccharide from the corn pathogen *Erwinia stewartii*. Carbohydr. Res. 288, 189–201.

Nürnberger, T., Brunner, F., 2002. Innate immunity in plants and animals: emerging parallels between the recognition of general elicitors and pathogen-associated molecular patterns. Curr. Opin. Plant Biol. 5, 318–324.

Nürnberger, T., Kemmerling, B., 2006. Receptor protein kinases – pattern recognition receptors in plant immunity. Trends Plant Sci. 11, 519–522.

Orgambide, G., Montrozier, H., Servin, P., Rousset, J., Trigalet-Demery, D., Trigalet, A., 1991. High heterogeneity of the exopolysaccharides of *Pseudomonas solanacearum* strain GMI 1000 and the complete structure of the major polysaccharide. J. Biol. Chem. 266, 8312–8321.

Parker, J.E., 2003. Plant recognition of microbial patterns. Trends Plant Sci. 8, 245–247.

Pearce, R.B., Ride, J.P., 1982. Chitin and related compounds as elicitors of the lignification response in wounded wheat leaves. Physiol. Plant Pathol. 20, 119–123.

Piater, L.A., Nürnberger, T., Dubery, I.A. 2004. Identification of a lipopolysaccharide responsive erk-like MAP kinase in tobacco leaf tissure. Mol. Plant Pathol. 5, 331–338.

Ramey, B.E., Koutsoudis, M., von Bodman, S., Fuqua, C., 2004. Biofilm formation in plant-microbe associations. Curr. Opin. Microbiol. 7, 602–609.

Rigano, L.A., Payette, C., Brouillard, G., et al., 2007. Bacterial cyclic β-(1, 2) glucan acts in systemic suppression of plant immune responses. Plant Cell 19, 2077–2089.

Saile, E., McGarvey, J.A., Schell, M.A., Denny, T.P., 1997. Role of extracellular polysaccharide and endoglucanase in root invasion and colonization of tomato plants by *Ralstonia solanacearum*. Phytopathology 87, 1264–1271.

Scheidle, H., Gross, A., Niehaus, K., 2005. The lipid A substructure of the *Sinorhizobium meliloti* lipopolysaccharides is sufficient to suppress the oxidative burst in host plants. New Phytol. 165, 559–565.

Schmidt, W.E., Ebel, J., 1987. Specific binding of a fungal glucan phytoalexin elicitor to membrane fractions from soybean *Glycine max*. Proc. Natl. Acad. Sci. USA 84, 4117–4121.

Schromm, A.B., Brandenburg, K., Loppnow, H., et al., 2000. Biological activities of lipopolysaccharides are determined by the shape of their lipid A portion. Eur. J. Biochem. 267, 2008–2013.

Silipo, A., De Castro, C., Lanzetta, R., Molinaro, A., Parrilli, M., 2004. Full structural characterization of the lipid A components from the *Agrobacterium tumefaciens* strain C58 lipopolysaccharide fraction. Glycobiology 14, 805–815.

Silipo, A., Molinaro, A., Sturiale, L., et al., 2005. The elicitation of plant innate immunity by lipooligosaccharide of *Xanthomonas campestris*. J. Biol. Chem. 280, 33660–33668.

Silipo, A., Sturiale, L., Erbs, G., et al., 2008. The acylation and phosphorylation pattern of lipid A from *Xanthomonas campestris* strongly influence its ability to trigger the innate immune response in *Arabidopsis*. ChemBioChem 9, 896–904.

Spiers, A.J., Bohannon, J., Gehrig, S.M., Rainey, P.B., 2003. Biofilm formation at the air-liquid interface by the *Pseudomonas fluorescens* SBW25 wrinkly spreader requires an acetylated from of cellulose. Mol. Microbiol. 50, 15–27.

Stasinopoulos, S.J., Fisher, P.R., Stone, B.A., Stanisich, V.A., 1999. Detection of two loci involved in (1,3)-β-glucan (curdlan) biosynthesis by *Agrobacterium* sp. ATCC31749, and comparative sequence analysis of the putative curdlan synthase gene. Glycobiology 9, 31–41.

Talaga, P., Stahl, B., Wieruszeski, J.M., et al., 1996. Cell-associated glucans of *Burkholderia solanacearum* and *Xanthomonas campestris* pv. *citri*: a new family of periplasmic glucans. J. Bacteriol. 178, 2263–2271.

Umemoto, N., Kakitani, M., Iwamatsu, A., Yoshikawa, M., Yamaoka, N., Ishida, I., 1997. The structure and function of a soybean β-glucan-elicitor-binding protein. Proc. Natl. Acad. Sci. USA 94, 1029–1034.

Vojnov, A.A., Slater, H., Newman, M.A., Daniels, M.J., Dow, J.M., 2001. Regulation of the synthesis of cyclic glucan in *Xanthomonas campestris* by a diffusible signal molecule. Arch. Microbiol. 176, 415–420.

Vollmer, W., Joris, B., Charlier, P., Foster, S., 2008. Bacterial peptidoglycan (murein) hydrolases. FEMS Microbiol. Rev. 32, 259–286.

Wan, J., Zhang, X.-C., Neece, D., et al., 2008. A LysM receptor-like kinase plays a critical role in chitin signaling and fungal resistance in *Arabidopsis*. Plant Cell 20, 471–481.

York, W.S., 1995. A conformational model for cyclic β-(1-2)-linked glucans based on NMR analysis of the β-glucans produced by *Xanthomonas campestris*. Carbohydr. Res. 278, 205–225.

Yu, J., Peñaloza-Vázquez, A., Chakrabarty, A.M., Bender, C.L., 1999. Involvement of the exopolysaccharide alginate in the virulence and epiphytic fitness of *Pseudomonas syringae* pv. *syringae*. Mol. Microbiol. 33, 712–720.

Yun, M.H., Torres, P.S., El Oirdi, M., et al., 2006. Xanthan induces plant susceptibility by suppressing callose deposition. Plant Physiol. 141, 178–187.

Zeidler, D., Zahringer, U., Gerber, I., et al., 2004. Innate immunity in *Arabidopsis thaliana*: lipopolysaccharides activate nitric oxide synthase (NOS) and induce defense genes. Proc. Natl. Acad. Sci. USA 101, 15811–15816.

Zipfel, C., Kunze, G., Chinchilla, D., et al., 2006. Perception of the bacterial PAMP EF-Tu by the receptor EFR restricts *Agrobacterium*-mediated transformation. Cell 125, 749–760.

PART IV

BIOLOGICAL RELEVANCE OF MICROBIAL GLYCOSYLATED COMPONENTS

B. Medical relevance

CHAPTER

41

Antigenic variation of microbial surface glycosylated molecules

Daniel C. Stein and Volker Briken

SUMMARY

Post-translational modifications of proteins represent an additional mechanism of generating surface variability in bacteria. Glycosylation of proteins, long thought to be unique to Eukarya, have now been demonstrated in both Bacteria and Archaea. However, while the eukaryotic glycosylation pathways are relatively well defined, little is known of the process in bacteria. Investigations are just beginning to unravel the biological implications of this modification. Bacteria are capable of forming both protein-attached *N*-glycans and *O*-glycans. These glycans are quite diverse. Variation in expression of genes that encode the sugar transferases allows the organism rapidly to adapt to changing environments, with the importance of these adaptive responses in human pathogens only now beginning to be understood.

Keywords: Pilin glycosylation; Flagellar glycosylation; Capsular variation; Glycoproteins; Lipopolysaccharide; Lipo-oligosaccharide

1. INTRODUCTION

Glycosylation of proteins can substantially influence and modulate protein structure and function and appears to be involved in the fine tuning of cell–cell recognition and signalling. Until the late 1970s, it was generally thought that protein glycosylation was restricted to eukaryotes. It has now been demonstrated that virtually all bacteria possess the ability to glycosylate proteins (see Chapters 7 and 8). Biological analysis of bacterial glycosylation has shown that most glycoproteins are associated with virulence factors of medically significant pathogens, suggesting that glycosylation serves specific functions in infection and pathogenesis. Since it has been demonstrated that virtually all surface components are variably glycosylated in one species or another, it implies that variable glycosylation plays a major role in bacterial pathogenesis. Antigenic variation of glycosylated molecules can occur at two levels;

the on–off switch where a molecule is either glycosylated or not, or the genetic substitution of the genes of one glycosylation pathway for another. In this chapter, we will review selected examples of bacterial glycosylation, with specific emphasis on the biological outcomes of the variable glycosylation.

2. PROTEIN-LINKED GLYCOSYLATION

2.1. Pilin glycosylation

Most bacterial pathogens have long filamentous structures, typically described as pili in Gram-negative organisms or fimbriae in Gram-positive organisms. These structures extend from their surface and are often involved in the initial adhesion of the bacterium to host tissues during colonization (Telford *et al.*, 2006). These proteins can be post-translationally modified and the presence or absence of the modification has significant biological consequences.

Post-translational modifications of pili have been most widely studied in *Neisseria* spp., however, other genera, such as *Pseudomonas*, have also been extensively studied (Castric, 1995). Glycosylation of gonococcal pili was first reported in 1977 by Robertson *et al.* Lambden and co-workers (1980) recognized very early on that the gonococcus was able to express two distinct pilus types and that these pilus types mediated differences in attachment to buccal epithelial cells. However, available technology precluded them from identifying the chemical basis for this difference. While early studies suggested that pilin glycosylation did not play a role in this pilus-mediated adhesion (Marceau *et al.*, 1998), we now know that this conclusion was due to the choice of cell lines employed (Virji *et al.*, 1993). This shows the fine level of specificity that glycosylation can impart on a protein. Data from a variety of laboratories have now clearly demonstrated that pili are essential for the adherence of pathogenic *Neisseria* to host cells and that pilins of *Neisseria meningitidis* and *Neisseria gonorrhoeae* can be *O*-glycosylated at serine residue 63 (Ser63) (Parge *et al.*, 1995; Stimson *et al.*, 1995). However, significant genetic evidence suggests that alternate glycosylation of pili also can occur (Kahler *et al.*, 2001; Power *et al.*, 2003). Because pilin glycosylation is phase-variable in the *Neisseria* (Banerjee *et al.*, 2002), the biological function of pilin glycosylation has not been resolved. The glycosylated region of pilin on the surface of the pilus structure in a region that is composed of conserved amino acid sequence suggests that glycosylation may shield the bacterium from the host immune response. Glycosylation of pili may also play a significant role in the resistance of meningococci to complement-mediated lysis by normal human serum as the binding of naturally occurring anti-galactose (Gal) antibodies to the terminal Gal may interfere in complement-mediated lysis (Hamadeh *et al.*, 1995).

The role of pilin glycosylation in adhesion can be mediated through several molecular mechanisms. Glycosylation may modulate the properties of pilin that are important determinants of the adhesive interaction of pilin with the host. Alternatively, glycosylation could influence the interaction of pilin with PilC, an essential pilus-associated adhesin, in such a way that it affects the overall adhesion properties of gonococci (Rudel *et al.*, 1992, 1995). While studies on the role of pili in infection have focused on the first step of disease, they may have other pathogenic implications. Strains that are able to vary antigenically their ability to glycosylate pili are associated with disseminated gonococcal disease (Banerjee *et al.*, 2002).

Pathogenic streptococci are decorated with long protruding pilus-like structures. A 200 kDa glycoprotein, Fap1, is the major subunit of streptococcal fimbriae and is involved in bacterial adhesion to saliva-coated hydroxyl apatite, an *in vitro* tooth adhesion model (Wu *et al.*, 1998). It is a glycoprotein essential for biofilm formation of *Streptococcus parasanguinis* (Stephenson *et al.*, 2002). In the absence of Fap1 glycosylation, the protein is poorly localized to the surface of the bacterium (Peng *et al.*, 2008). Since unglycosylated Fap1 is not surface localized, cells that cannot glycosylate Fap1 are defective in adhesion and diminished in biofilm formation.

Colonization leads to biofilm formation, a key intermediate in streptococcal pathogenesis. Akiyama and co-workers reported the presence of biofilm-like structures in skin sections from patients with impetigo and atopic dermatitis (Akiyama *et al.*, 2003); Takemura *et al.* (2004) identified the importance of biofilms in root canals, while Hidalgo-Grass and co-workers (2004) demonstrated the importance of biofilms in necrotizing fasciitis lesions. Manetti *et al.* (2007) concluded that one of the main roles of streptococcal pili is to facilitate the interaction with host cells and allow the bacteria to switch from planktonic to biofilm growth. It is thought that biofilm formation offers the pathogen a competitive advantage to counteract the defensive responses the human host mounts.

2.2. Flagellar glycosylation

The structural diversity in flagellar glycosylation systems, the consequences of the unique assembly processes, and the type of glycosidic linkages found on archaeal and bacterial flagellins has been reviewed by Logan (2006) (see also Chapter 8). The best-characterized polar flagellar glycosylation system is that of *Campylobacter jejuni*. This organism cannot assemble a flagellum in the absence of glycosylation (Goon *et al.*, 2003). Since motility is crucial to pathogenesis, glycosylation is directly linked to virulence (Morooka *et al.*, 1985). *Campylobacter* flagella are heavily glycosylated at multiple serine or threonine residues (Thibault *et al.*, 2001). Flagellin is also the immunodominant protein recognized during infection and has been suggested to be an immunoprotective antigen. This glycan variability among strains can confer serospecificity (Logan *et al.*, 2002).

Our understanding of the role of flagellar glycosylation in pathogen biology is evolving. *Pseudomonas aeruginosa* can be classified into two groups (a-type and b-type) on the basis of the amino acid sequence of flagellin (Spangenberg *et al.*, 1996). It was originally reported that the a-type flagellas were glycosylated, while the b-type flagella were not (Brimer and Montie, 1998). This conclusion was based on the fact that the b-type flagellin migrated as expected in electrophoretic (SDS-PAGE) gels, while a-type flagellin migrated much slower than expected. In addition, the glycosylation machinery of the a-type flagellin could not modify b-type flagellin (Arora *et al.*, 2001). This led to the hypothesis that because flagellin is not glycosylated in *P. aeruginosa* expressing the b-type flagellin, and that the bacteria are motile by standard motility assay, glycosylation is not required for motility. However, the strong association of glycosylation with certain pathogenic strains of *P. aeruginosa* led to further studies on the b-type flagella. More recent data indicate that these flagella can be glycosylated (Verma *et al.*, 2006).

In animals, flagellin is recognized by host cells through the Toll-like receptor (TLR-) 5 (Hayashi *et al.*, 2001). The impact of flagellin glycosylation on innate immunity is just beginning to be elucidated. *P. aeruginosa* strains that fail to glycosylate their flagella are significantly attenuated (Arora *et al.*, 2005). Flagellin glycosylation also plays an important role in the ability of flagellin to stimulate interleukin- (IL-) 8 release from human lung carcinoma cells (Verma *et al.*, 2005). This suggests that flagellin glycans might be responsible for the stimulation of inflammation.

Attachment by *C. jejuni* to a surface appears to be mediated by flagella. This attachment leads to the formation of microcolonies, which is often the first step in biofilm formation (Joshua *et al.*, 2006). Biofilm formation is crucial to the survival of the pathogen because it facilitates survival of this microaerophile. Changes in glycan composition affect autoagglutination and microcolony formation on intestinal epithelial cells; traits that are associated with disease in animals. Motility is required for the invasion of intestinal epithelial cells *in vitro* and adherence and invasion increase under viscous conditions (Szymanski *et al.*, 1995). Secretion of virulence proteins from *C. jejuni* is dependent on a functional flagellar export apparatus (Konkel *et al.*, 2004). Since organisms that fail to glycosylate their flagella cannot assemble the flagellum, this is an example of an indirect effect of protein glycosylation on virulence.

2.3. Cell surface adhesins

Non-typeable *Haemophilus influenzae* is a common human pathogen and initiates infection by colonizing the upper respiratory tract (Turk, 1984). Grass *et al.* (2003) demonstrated that attachment to human epithelial cells is mediated by an adhesin, HMW1, which is glycosylated. Glycosylation also protected HMW1 against premature degradation and appeared to influence HMW1 tethering to the bacterial surface, a prerequisite for HMW1-mediated adherence. Finally, shedding of the non-glycosylated adhesins decreases adherence to epithelial cells, demonstrating that glycosylation directly modulates cellular interactions and might also influence pathogenesis (Grass *et al.*, 2003).

A few glycoproteins have been identified in *Escherichia coli*. They include the TibA adhesion–invasion protein (Lindenthal and Elsinghorst, 1999), the autoaggregation factor antigen 43 (Ag43) (Sherlock *et al.*, 2006) and an adhesin involved in diffuse adherence (AIDA-I) (Benz and Schmidt, 2001).

E. coli can form multicellular aggregates when it expresses Ag43. The expression of Ag43 is co-regulated with type 1 fimbriae, where autoaggregating cells are non-fimbriated and non-aggregating cells are (Hasman *et al.*, 1999). Both systems are subject to phase variation and the phase variation frequencies are similar (Owen *et al.*, 1996). The multifunctional protein Ag43 promotes bacterial binding to some human cells as well as biofilm formation on various surfaces (Klemm *et al.*, 2004). *E. coli* cells expressing Ag43 are efficiently taken up by polymorphonuclear neutrophils (PMNs) in an opsonin-independent manner, the phagocytosed bacteria are not immediately killed but reside as tight aggregates within the PMNs, demonstrating that Ag43-mediated uptake allows the pathogen to subvert one of the primary defense mechanisms of the human body (Fexby *et al.*, 2007).

The TibA protein is a 104 kDa outer membrane autotransporter protein which forms a non-organelle adhesin in enterotoxigenic *E. coli*. It is produced as a precursor which is extensively modified by proteolytic cleavage and glycosylation (Lindenthal and Elsinghorst, 1999). Introduction of the *tib* locus into non-adherent *E. coli* K-12 hosts directs adhesion to epithelial cell line cells (Elsinghorst and Kopecko, 1992). Its protein product, TibA, appears to be a multipurpose protein. It is a potent adhesin with affinity for a number of different human cells. It is an efficient invasin that results in bacterial invasion of human cells. It is capable of mediating bacterial aggregation via intercellular self-recognition. Finally, it is a highly efficient initiator of biofilm formation (Sherlock *et al.*, 2005).

Unglycosylated AIDA-I is expressed in smaller amounts than its glycosylated counterpart and shows extensive signs of degradation upon heat extraction. It is also more sensitive to proteases and induces extracytoplasmic stress. Glycosylation is required for AIDA-I to mediate adhesion to cultured epithelial cells, but purified mature AIDA-I fused to glutathione *S*-transferase failed to bind *in vitro* to cells whether or not it was glycosylated (Charbonneau *et al.*, 2007). These results suggest that glycosylation is required to ensure a normal conformation of AIDA-I and may be only indirectly necessary for its cell-binding function. Given that biological functions of various proteins can be affected by glycosylation, such as maintenance of protein conformation, resistance against proteases and modulation of intermolecular interactions, it indicates that glycosylation provides a higher level of structure to bacterial proteins.

3. LIPID-LINKED GLYCOSYLATION

3.1. Lipopolysaccharide/ lipo-oligosaccharide

Lipopolysaccharide (LPS) is localized to the outer leaflet of the outer membrane of Gram-negative bacteria and is a major surface component of the bacterial cell envelope (see Chapter 1). It is composed of lipid A, a central core oligosaccharide and a variable length repeating polysaccharide portion referred to as the O-antigen. While an integral component of the outer membrane of Gram-negative bacteria, this molecule possesses significant structural heterogeneity, both in the lipid-containing region and in the carbohydrate region. Variations in both the lipid- and carbohydrate-containing regions have been shown to elicit differential responses in the host and to possess very different biological properties. Since LPS is a major surface molecule and a primary virulence determinant, it is not surprising that bacteria have evolved mechanisms to modify their LPS structure in order to promote their survival. Since LPS possesses multiple chemical domains, this can explain why LPS can initiate so many different signalling pathways and how it can alter host responsiveness and thus the outcome of bacterial–host interactions (Gioannini and Weiss, 2007).

These outer membrane lipoglycans are classified as LPS if the molecule contains a repeating polysaccharide that has been added as a unit onto a core structure. Strains that lack this O-antigen are defined as expressing the glycolipid, lipo-oligosaccharide (LOS). It is generally thought that mucosal pathogens, such as *N. gonorrhoeae*, *H. influenzae* and *Bordetella pertussis* (Schneider *et al.*, 1984; Inzana *et al.*, 1988; Caroff *et al.*, 1990), produce LOS, while enteric pathogens, such as *Salmonella enterica* sv. Typhimurium, or opportunistic pathogens like *P. aeruginosa* produce LPS (Heinrichs *et al.*, 1998; Rocchetta *et al.*, 1999). However, a third form of this molecule represents a structural intermediate between LPS and LOS. *Haemophilus ducreyi* can make an LPS whose oligosaccharide (OS) consists solely of a polylactosamine repeat, devoid of other core structures (Schilling *et al.*, 2002).

N. gonorrhoeae and *H. ducreyi* strains expressing polylactosamine are more infectious or induce more severe symptomatic infections (Gibson *et al.*, 1997; John *et al.*, 1999). The ability of this glycoform to induce an elevated immune response may occur through a specific innate immunity-stimulating signalling pathway. For example, the polysaccharide portion of the molecule could signal through the extracellular matrix protein mindin (He *et al.*, 2004). Mindin binds directly to oligosaccharide repeat portion of LPS and this binding initiates an inflammatory response. While the nature of the polysaccharide recognized by mindin is unclear, it represents a unique pattern-recognition molecule in the extracellular matrix for microbial pathogens and this pathway activation may explain how gonococci and *H. ducreyi* initiate disease. It is important to note that this signalling pathway operates in the absence of TLR-4 signalling.

The gonococcus variably expresses some of the genes encoding the glycosyl transferases needed for LOS biosynthesis and this results in rapid and reversible alterations in the oligosaccharide structure (Danaher *et al.*, 1995). This variation has biological significance at several levels. Modifications of the oligosaccharide alter gonococcal–host cell interactions. Song *et al.* (2000) demonstrated that, in the absence of Opa protein expression, strains expressing a specific

LOS (containing a terminal lacto-*N*-neotetraose) promoted increased gonococcal invasion of epithelial monolayers. This is probably due to the interaction between gonococci expressing lacto-*N*-neotetraose (LNT) LOS and the asialoglycoprotein (ASGP) receptor of host cells (Harvey *et al.*, 2001). Human challenge studies have demonstrated that, after inoculating volunteers with low-molecular-mass LOS variants, urethritis developed and primarily high-molecular-mass LOS (with terminal LNT) was purified from the exudates of symptomatic individuals (Schneider *et al.*, 1991). We demonstrated that alterations in the carbohydrate moiety of LOS do not impact the production of most cytokines by human monocytes (Patrone and Stein, 2007). Thus, while the carbohydrate structures of LOS also play an important function in disease pathogenesis, the lipid A portion of the LOS is responsible for immunological activation.

The LNT-bearing LOS, which is identical to a human blood group antigen, can be modified by the addition of sialic acid (Parsons *et al.*, 1989). The impact of sialylation of gonococcal LOS seems to differ from that seen in meningococcal LOS. Sialylation of gonococcal LOS has been shown to convert serum sensitive strains to a serum resistant phenotype (Nairn *et al.*, 1988). However, sialylation plays a role in other aspects of gonococcal pathogenesis by interfering with entry to epithelial cells (van Putten, 1993), with killing by neutrophils (Kim *et al.*, 1992) or with the bactericidal and opsonophagocytic action of antisera against gonococcal proteins (Demarco de Hormaeche *et al.*, 1991).

Sialylation of LOS affects the pathogenicity of meningococci but not as profoundly as for gonococci. Studies have demonstrated a correlation between the complement sensitivity of serogroup C isolates and their degree of LPS sialylation (Estabrook *et al.*, 1997), but these results differed when serogroup B mutants were examined (Kahler *et al.*, 1998). Vogel *et al.* (1997) found that mutants that lacked LOS sialylation were more serum-sensitive than the wild-type strains but only at high serum concentrations. Therefore, it is likely that LOS sialylation is only of minor importance for serum resistance in the meningococcus. However, truncation of LOS renders the meningococcus sensitive to complement. This can contribute to the increased susceptibility to meningococcal disease in those individuals deficient in complement (Bishof *et al.*, 1990) and has important implications for the overall pathogenesis of meningococcal disease.

Lipid A is the structure responsible for the activation of host cells through the TLR4 receptor (Poltorak *et al.*, 1998). Variation of the lipid A domain can occur on both the hydrophilic disaccharide region and the hydrophobic acyl chain region (see Chapter 3). Modifications to lipid A, e.g. addition of 2-hydroxy-dodecanoic acid (2-hydroxy-laurate), deacylation and addition of hexadecanoic acid (palmitate), are associated with an altered immune response (Ernst *et al.*, 2003). The PhoP/PhoQ two component regulatory system is responsible for the activation of several genes involved in *Salmonella* virulence which are involved in lipid A modification (Miller *et al.*, 1989). This remodelling includes enzymes that modify LPS. Modified LPS promotes bacterial survival by increasing resistance to cationic antimicrobial peptides and by altered host recognition of LPS (Ernst *et al.*, 2001). The lipid A of *P. aeruginosa* can be variably penta-, hexa- or hepta-acylated and these isoforms have differing potencies when activating host innate immunity (see also Chapter 17).

Clinical isolates of *Helicobacter pylori* produce high-molecular-mass LPS but, when repeatedly cultured on solid media, it may produce LPS that lacks an O-antigen (Moran *et al.*, 1997). Since the repeating units of the O-antigen of certain *H. pylori* strains mimic Lewis x and Lewis y blood group antigens in structure (Aspinall and Monteiro, 1996), this mimicry may camouflage the bacterium and hence aid colonization (Moran, 1996) or prolong infection (see Chapter 43). Variation in the composition of the O-antigen leads to differential binding of the innate immune component surfactant protein D (Khamri *et al.*, 2005). Since the expression of the O-antigen is modulated by a genetic mechanism (Appelmelk *et al.*, 1998), it demonstrates the importance of the oligosaccharide in disease.

In *P. aeruginosa*, LPS plays an important immunogenic and structural role, mediating the interaction between the bacterial cell surface and the external environment (Rocchetta *et al.*, 1999). *P. aeruginosa* produces two different forms of O-antigen: A-band (also known as common antigen), which is a homopolymer of D-rhamnose, and B-band, which is the heteropolymer responsible for serogroup specificity (Lam *et al.*, 1992). O-Antigen composition can change, depending on whether the organism is growing planktonically or if growing on biofilms (Beveridge *et al.*, 1997). The presence of particular O-antigens on the surface of *P. aeruginosa* affects the overall charge and physicochemistry of the bacterial cell and strains lacking the B-band O-antigen are more adherent to hydrophobic surfaces. It is widely believed that these rapid surface-associated changes have broad implications for pathogenesis. Such modifications appear to be associated with chronic infection, since they are commonly observed in cystic fibrosis patients with long-term *P. aeruginosa* infections and have not been detected in cases of acute infection.

P. aeruginosa undergoes differential expression of some of its virulence factors during infection; changes that have typically been observed in pulmonary infections in cystic fibrosis patients. The most dramatic change is the conversion of the organism from a non-mucoid to a mucoid phenotype due to the production of copious amounts of the exopolysaccharide alginate, reviewed in Govan and Deretic (1996). *P. aeruginosa* is one of the most important bacterial pathogens encountered by immunocompromised hosts and patients with cystic fibrosis and the LPS elaborated by this organism is a key factor in virulence as well as both innate and acquired host responses to infection (Pier, 2007). Its LPS has been extensively studied and plays a prominent role in bacterial virulence with the O-polysaccharide portion of the LPS serving as a target for protective antibodies.

Hadina *et al.* (2008) compared the inflammatory potency of hexa-acylated and penta-acylated LOS isolated from *N. meningitidis* and found that inhalation of hexa-acylated LPS resulted in strong inflammatory responses. Penta-acylated LOS was much less potent in inducing increases of neutrophils, tumour necrosis factor-alpha (TNF-α), macrophage inflammatory protein-1 alpha (MIP-1α), IL-6, granulocyte colony-stimulating factor (G-CSF) and IL-1β. Zhang *et al.* (2008), using synthetic lipid A derivatives, found that lipid A containing 3-deoxy-D-*manno*-oct-2-ulosonic acid (Kdo) was much more active than lipid A alone and just slightly less active than its parent LOS, indicating that one Kdo moiety is sufficient for full activity of TNF-α and interferon- (IFN-) β induction. They also found that the lipid A of *N. meningitidis* was a significantly more potent inducer of TNF-α and IFN-β than *E. coli* lipid A, due to a number of shorter fatty acids.

Mutagenesis of the genes involved in LOS biosynthesis has resulted in a series of mutants with stepwise truncations of the oligosaccharide chain. The isolation of the meningococcal *lpxD-fabZ-lpxA* locus involved in lipid A biosynthesis produced a strain that is completely devoid of LOS (Steeghs *et al.*, 1998). This strain has allowed researchers directly to demonstrate the role of LOS in various biological processes. While it has been known that LPS functions as an adjuvant and enhances the immune response to antigens, Steeghs and co-workers (1999) demonstrated that outer membrane complexes isolated from an LOS-deficient meningococcal mutant could be enhanced by the addition of external LPS. Since LOS was not able to enhance the immunogenicity of LOS-deficient heat-inactivated bacteria, they proposed that a direct physical interaction of LOS with the antigen is required to establish its function as an adjuvant. Albiger *et al.* (2003) showed that LOS deficiency was, without exception, accompanied by altered colony opacity and morphology and reduced levels of the iron-regulated proteins FetA and FbpA. They also showed that LOS is essential for pilus-associated adherence but dispensable for fibre formation and twitching motility. Finally, they demonstrated that LOS mutants do not invade host cells and have lost the natural competence for genetic transformation.

3.2. Other glycolipids

Lipoarabinomannan (LAM) is a complex lipoglycan making up a major component of the cell wall of mycobacteria (see Chapter 9). It is an example of interspecies variation in the glycosylation of lipids. The terminal arabinan component of LAM can either be left unmodified to form AraLAM in *Mycobacterium chelonae*, capped with phosphoinositol (PI) to form PILAM in the non-pathogenic *Mycobacterium smegmatis* or capped with mannose-residues to form ManLAM in virulent *Mycobacterium tuberculosis* (Briken *et al.*, 2004; Berg *et al.*, 2007). The mannose cap of the latter form has been implicated in interaction with the mannose receptor and dendritic cell-specific ICAM-3 grabbing non-integrin (DC-SIGN) to mediate the inhibition of proinflammatory responses such as IFN-γ and IL-12 cytokine secretion (Briken *et al.*, 2004; Appelmelk *et al.*, 2008). On the other hand, PILAM induces a strong proinflammatory response in macrophages, which might be mediated by the PI residue, since the uncapped AraLAM of *M. chelonae* does not induce cytokine secretion or apoptosis of macrophages (Guerardel *et al.*, 2002). All of the described results have been obtained via *in vitro* assays, by incubating cells with purified LAM fractions, however, the recent finding that a mutant of *M. tuberculosis* that lacks the mannose caps is not attenuated and does not induce an increase in proinflammatory response of macrophages highlights the limits of the biochemical approach to understand the function of bacterial cell wall components (Appelmelk *et al.*, 2008). Does this mean that the mannose caps are not important at all for the virulence of mycobacteria? This could be true but it is more likely that mycobacteria have developed redundant mechanisms to mediate the manipulation of the host cell.

An example for intraspecies variation in the glycosylation of an important component of the cell wall is the polyketide synthase-derived phenolic glycolipid (PGL). This glycolipid has been shown to be important for virulence of the aetiological agent of leprosy, *Mycobacterium leprae*, and it is implicated in the hypervirulence of some isolates of *M. tuberculosis* (Reed *et al.*, 2004). The precise mechanism of PGL action on the host remains to be determined but isolated PGL has anti-inflammatory capacity on primary murine macrophages, e.g. it reduces the secretion of the proinflammatory cytokine TNF-α (Reed *et al.*, 2004). Interestingly, not all isolates of *M. tuberculosis* express PGL which is due to a constitutive expression of the triacylglyceride synthetase (Rv3130) in members of Beijing family of *M. tuberculosis* (Reed *et al.*, 2007). The core structure of PGL is shared by most mycobacterial species, but the terminal sugar modification of the aromatic residues is distinct. Thus, the PGL in the cattle pathogen *M. bovis* is monoglycosylated, whereas the PGL of *M. tuberculosis* is triglycosylated (PGL-tb). The diversity of the PGL modification could be related to deletions and/or point mutations in three genes (Rv2958c, Rv1511, Rv1512) important for adding the two additional sugar residues of PGL-tb in the genomes of *Mycobacterium bovis* and other mycobacterial species. The deletion of Rv2958c in *M. tuberculosis* abolishes PGL-tb synthesis and results in a single sugar residue on the lipid core (Mycoside B) (Malaga *et al.*, 2008). It has yet to be determined if Mycoside B does not reproduce the anti-inflammatory activity of PGL-tb *in vitro* and what the impact of the loss of the two terminal sugar residues for the virulence of the bacteria *in vivo* is. Thus, the recently discovered mutants will be very valuable tools to address the impact of the differences in the terminal sugar modification of PGL on the virulence of the bacteria.

4. CAPSULES

Both pathogenic and non-pathogenic bacteria are capable of producing polysaccharide capsules (see Chapter 6). In general, capsules are highly antigenic and elicit strong antibody responses, which confer protection against subsequent infection. However, there is a subset of capsules that is poorly immunogenic. Many pathogens are classified into different serological groups based on the chemical composition of their capsules. *E. coli* can synthesize at least 80 different capsular polysaccharides, a diversity that results in vast intraspecies antigenic variability (Orskov *et al.*, 1977). Thus, infection with a bacterium of one capsular type does not elicit a protective immune response against infection with the same species of bacterium of a different capsular type.

Carbohydrate capsules serve a number of important biological functions (Taylor and Roberts, 2005). They are a physical barrier between the surroundings and the membrane of the organism. Capsule production is required for virulence in numerous animal models of infection, where the capsule contributes to virulence through several mechanisms. Capsules are anti-phagocytic and anti-bacteriolytic, allowing the organism to evade the adaptive and innate immune responses. They can inhibit phagocytosis by interfering with complement deposition. Some capsules even provide immune evasion through molecular mimicry to host glycans. Bacteria that lack a capsule are typically killed through the action of normal serum and polymorphonuclear leukocytes in the absence of specific antibodies. While antibody directed against capsule can opsonize an encapsulated organism, killing requires engulfment by polymorphonuclear leukocytes. The induction of specific antibodies through immunization, with consequent complement deposition on an organism, is a major factor in successful vaccination.

The genetic basis for intraspecies diversity of capsular polysaccharides has been studied in many different species of bacteria. The genes involved in capsule biosynthesis are usually clustered in operons allowing for frequent genetic exchange between related species. For a recent review of capsule biosynthesis, see Whitfield (2006). Exchange of capsule biosynthetic genes can result in the production of novel pathogens and has been documented in *Streptococcus pneumoniae* and *N. meningitidis* (Swartley *et al.*, 1997; Jiang *et al.*, 2001).

Bacteroides fragilis synthesize at least eight capsular polysaccharides whose expression is subject to phase variation, a reversible on–off phenotype. Phase variation is dictated by DNA inversions of the polysaccharide biosynthesis loci promoters (Krinos *et al.*, 2001). The amount of genetic material and energy dedicated to the biosynthesis of these structures in *Bacteroides* spp. strongly suggests that it confers a survival advantage during the life-long association of these organisms with the host and other members of the colonic microbiota. The production of at least one capsular polysaccharide is required for the viability of the microorganism; inhibiting the ability of *Bacteroides fragilis* to modulate its surface renders it incapable of competing with other bacteria for colonization of the gastrointestinal tract of animals (Liu *et al.*, 2008). The capsule of *B. fragilis* initiates a unique immune response in the host where injection of capsules alone was sufficient to induce abscess formation in the host (Coyne *et al.*, 2000). The *B. fragilis* capsule can also mediate resistance to complement-mediated killing and to phagocytic uptake and killing (Simon *et al.*, 1982).

In mammalian cells, hyaluronic acid is an important component of the extracellular matrix.

Group A streptococci produce a hyaluronic acid capsule (Wessels *et al.*, 1991), successfully mimicking host tissue. This allows the organism to evade the host immune system. The sialic acid capsule of *N. meningitidis* performs a similar function in that it lacks immunogenicity in humans due to its structural similarity to glycoproteins in human fetal brain tissues (Finne *et al.*, 1983).

The polysaccharide capsule constitutes the most important virulence factor of *S. pneumoniae* (Kim and Weiser, 1998). While the serotype of the capsule correlates with invasive potential, the *in vitro* growth rate differs in the length of their lag phase, with a long lag phase preferentially seen in serotypes with high invasive potential and serotypes associated more with colonization tended to have a shorter lag phase (Battig *et al.*, 2006). The zwitterionic polysaccharides of *S. pneumoniae* are an exception to the paradigm for how carbohydrates induce immune responses. While polysaccharides are usually classified as T-cell-independent antigens and fail to induce activation of helper T-cells, zwitterionic polysaccharides are processed and presented by the major histocompatibility complex (MHC) class II pathway (Cobb *et al.*, 2004). Hence, these capsules are highly immunogenic.

Cryptococcus neoformans is a soil-dwelling fungus that causes life-threatening illness in immunocompromised individuals (see review, Lin and Heitman, 2006). It is the only encapsulated human eukaryotic pathogen that has been described to date. The biological properties of its capsule are comparable to those of encapsulated bacteria. Although the capsule is a crucial virulence determinant, the primary habitats of this fungus are soils and trees. Hence, it is likely that the primary biological function of this structure must be related to environmental survival (McFadden *et al.*, 2007). However, the capsule has been shown to modulate host immune responses on many levels (Bose *et al.*, 2003). The presence of a capsule enhances adherence to endothelial cells, as well as the rate of transcytosis (Chen *et al.*, 2003). It alters the efficiency of phagocytosis and killing by immune effector cells, production of proinflammatory cytokines and antigen presentation to T-cells. It readily fixes complement components by the alternative pathway and can effectively deplete complement in infected hosts. Capsular material also alters leukocyte function, leading to reduced leukocyte entry into sites of inflammation. Finally, capsular polysaccharide produced in phagocytic cells accumulates in the cytosol and contributes to cell destruction.

5. FUTURE PERSPECTIVES

Glycosylation, the post-translational modification of proteins by carbohydrates, has long been recognized as a fundamental strategy used by eukaryotes to influence and modulate protein structure and function (Taylor and Drickamer, 2002). Prokaryotic glycoproteins contain significantly more structural diversity than eukaryotic glycoproteins. While we possess significant databases that allow us to perform comparative genomic studies and we have sophisticated computational tools for analysing protein structures, the identification of glycoproteins and, hence, their biological functions, is still in its infancy (see Research Focus Box).

The precise roles of glycosylation in most bacteria have yet to be resolved because the identification of the sugars and their chemical linkages is costly and must be performed on each glycoprotein. However, given that virtually all bacteria possess the ability to glycosylate at least one surface protein and the glycosylation of most of these proteins is variably expressed, it indicates that the functions of these modifications are poorly understood. The influence of bacterial glycosylation on the host immune response is an important question that needs to be addressed. We need to develop a systems biology approach in our effort to understand the influence of the bacterial glycome in the biological complexity of host–pathogen interactions. This will require the development of multidisciplinary research teams who combine genomics and proteomics with mathematical and computer modelling. The application of structural biology technologies will need to be integrated into our study of pathogens if we hope to develop a thorough understanding of host–pathogen interactions.

RESEARCH FOCUS BOX

- Does protein glycosylation represent a new target for therapeutics?
- What proteins are glycosylated in bacteria?
- How do various sugars get added onto proteins?
- What are the links between protein glycosylation and bacterial virulence?

References

Akiyama, H., Morizane, S., Yamasaki, O., Oono, T., Iwatsuki, K., 2003. Assessment of *Streptococcus pyogenes* microcolony formation in infected skin by confocal laser scanning microscopy. J. Dermatol. Sci. 32, 193–199.

Albiger, B., Johansson, L., Jonsson, A.B., 2003. Lipooligosaccharide-deficient *Neisseria meningitidis* shows altered pilus-associated characteristics. Infect. Immun. 71, 155–162.

Appelmelk, B.J., Shiberu, B., Trinks, C., et al., 1998. Phase variation in *Helicobacter pylori* lipopolysaccharide. Infect. Immun. 66, 70–76.

Appelmelk, B.J., den Dunnen, J., Driessen, N.N., et al., 2008. The mannose cap of mycobacterial lipoarabinomannan does not dominate the *Mycobacterium*-host interaction. Cell. Microbiol. 10, 930–944.

Arora, S., Bangera, M., Lory, S., Ramphal, R., 2001. A genomic island in *Pseudomonas aeruginosa* carries the determinants of flagellin glycosylation. Proc. Natl. Acad. Sci. USA 98, 9342–9347.

Arora, S.K., Neely, A.N., Blair, B., Lory, S., Ramphal, R., 2005. Role of motility and flagellin glycosylation in the pathogenesis of *Pseudomonas aeruginosa* burn wound infections. Infect. Immun. 73, 4395–4398.

Aspinall, G.O., Monteiro, M.A., 1996. Lipopolysaccharides of *Helicobacter pylori* strains P466 and MO19: structures of the O antigen and core oligosaccharide regions. Biochemistry 35, 2498–2504.

Banerjee, A., Wang, R., Supernavage, S.L., et al., 2002. Implications of phase variation of a gene (*pgtA*) encoding a pilin galactosyl transferase in gonococcal pathogenesis. J. Exp. Med. 196, 147–162.

Battig, P., Hathaway, L.J., Hofer, S., Muhlemann, K., 2006. Serotype-specific invasiveness and colonization prevalence in *Streptococcus pneumoniae* correlate with the lag phase during in vitro growth. Microbes Infect. 8, 2612–2617.

Benz, I., Schmidt, M.A., 2001. Glycosylation with heptose residues mediated by the *aah* gene product is essential for adherence of the AIDA-I adhesin. Mol. Microbiol. 40, 1403–1413.

Berg, S., Kaur, D., Jackson, M., Brennan, P.J., 2007. The glycosyltransferases of *Mycobacterium tuberculosis* – roles in the synthesis of arabinogalaetan, lipoarahinolrnlannan, and other glyeoeonjugates. Glycobiology 17, 35r–56r.

Beveridge, T.J., Makin, S.A., Kadurugamuwa, J.L., Li, Z., 1997. Interactions between biofilms and the environment. FEMS Microbiol. Rev. 20, 291–303.

Bishof, N.A., Welch, T.R., Beischel, L.S., 1990. C4B deficiency: a risk factor for bacteremia with encapsulated organisms. J. Infect. Dis. 162, 248–250.

Bose, I., Reese, A.J., Ory, J.J., Janbon, G., Doering, T., 2003. A yeast under cover: the capsule of *Cryptococcus neoformans*. Eukaryot. Cell 2, 655–663.

Briken, V., Porcelli, S.A., Besra, G.S., Kremer, L., 2004. Mycobacterial lipoarabinomannan and related lipoglycans: from biogenesis to modulation of the immune response. Mol. Microbiol. 53, 391–403.

Brimer, C.D., Montie, T.C., 1998. Cloning and comparison of *fliC* genes and identification of glycosylation in the flagellin of *Pseudomonas aeruginosa* a-type strains. J. Bacteriol. 180, 3209–3217.

Caroff, M., Chaby, R., Karibian, D., Perry, J., Deprun, C., Szabo, L., 1990. Variations in the carbohydrate regions of *Bordetella pertussis* lipopolysaccharides: electrophoretic, serological and structural features. J. Bacteriol. 172, 1121–1128.

Castric, P., 1995. *pilO*, a gene required for glycosylation of *Pseudomonas aeruginosa* 1244 pilin. Microbiology 141, 1247–1254.

Charbonneau, M.-E., Girard, V., Nikolakakis, A., et al., 2007. *O*-Linked glycosylation ensures the normal conformation of the autotransporter adhesin involved in diffuse adherence. J. Bacteriol. 189, 8880–8889.

Chen, S.H., Stins, M.F., Huang, S.H., et al., 2003. *Cryptococcus neoformans* induces alterations in the cytoskeleton of human brain microvascular endothelial cells. J. Med. Microbiol. 52, 961–970.

Cobb, B., Wang, Q., Tzianabos, A., Kasper, D.L., 2004. Polysaccharide processing and presentation by the MHCII pathway. Cell 117, 677–687.

Coyne, M.J., Kalka-Moll, W., Tzianabos, A.O., Kasper, D.L., Comstock, L.E., 2000. *Bacteroides fragilis* NCTC9343 produces at least three distinct capsular polysaccharides: cloning, characterization, and reassignment of polysaccharide B and C biosynthesis loci. Infect. Immun. 68, 6176–6181.

Danaher, R.J., Levin, J.C., Arking, D., Burch, C.L., Sandlin, R., Stein, D.C., 1995. Genetic basis of *Neisseria gonorrhoeae* lipooligosaccharide antigenic variation. J. Bacteriol. 177, 7275–7279.

Demarco de Hormaeche, R., van Crevel, R., Hormaeche, C.E., 1991. *Neisseria gonorrhoeae* LPS variation, serum resistance and its induction by cytidine 5′-monophospho-*N*-acetylneuraminic acid. Microb. Pathog. 10, 323–332.

Elsinghorst, E.A., Kopecko, D.J., 1992. Molecular cloning of epithelial cell invasion determinants from enterotoxigenic *Escherichia coli*. Infect. Immun. 60, 2409–2417.

Ernst, R.K., Guina, T., Miller, S.I., 2001. *Salmonella typhimurium* outer membrane remodeling: role in resistance to host innate immunity. Microbes Infect. 3, 1327–1334.

Ernst, R.K., Hajjar, A.M., Tsai, J.H., Moskowitz, S.M., Wilson, C.B., Miller, S.I., 2003. *Pseudomonas aeruginosa* lipid A diversity and its recognition by Toll-like receptor 4. J. Endotoxin Res. 9, 395–400.

Estabrook, M.M., Griffiss, J.M., Jarvis, G.A., 1997. Sialylation of *Neisseria meningitidis* lipooligosaccharide inhibits serum bactericidal activity by masking lacto-*N*-neotetraose. Infect. Immun. 65, 4436–4444.

Fexby, S., Bjarnsholt, T., Jensen, P.Ø., et al., 2007. Biological Trojan horse: Antigen 43 provides specific bacterial uptake and survival in human neutrophils. Infect. Immun. 75, 30–34.

Finne, J., Leinonen, M., Makela, P.H., 1983. Antigenic similarities between brain-components and bacteria causing meningitis. Implications for vaccine development and pathogenesis. Lancet 2, 355–357.

Gibson, B.W., Campagnari, A.A., Melaugh, W., et al., 1997. Characterization of a transposon Tn916-generated mutant of *Haemophilus ducreyi* 35000 defective in lipooligosaccharide biosynthesis. J. Bacteriol. 179, 5062–5071.

Gioannini, T.L., Weiss, J.P., 2007. Regulation of interactions of Gram-negative bacterial endotoxins with mammalian cells. Immunol. Res. 39, 249–260.

Goon, S., Kelly, J.F., Logan, S.M., Ewing, C.P., Guerry, P., 2003. Pseudaminic acid, the major modification on *Campylobacter* flagellin, is synthesized via the Cj1293 gene. Mol. Microbiol. 50, 659–671.

Govan, J.R.W., Deretic, V., 1996. Microbial pathogenesis in cystic fibrosis: mucoid *Pseudomonas aeruginosa* and *Burkholderia cepacia*. Microbiol. Rev. 60, 539–574.

Grass, S., Buscher, A.Z., Swords, W.E., et al., 2003. The *Haemophilus influenzae* HMW1 adhesin is glycosylated in a process that requires HMW1C and phosphoglucomutase, an enzyme involved in lipooligosaccharide biosynthesis. Mol. Microbiol. 48, 737–751.

Guerardel, Y., Maes, E., Elass, E., et al., 2002. Structural study of lipomannan and lipoarabinomannan from *Mycobacterium chelonae*. Presence of unusual components with α1,3-mannopyranose side chains. J. Biol. Chem. 277, 30635–30648.

Hadina, S., Weiss, J.P., McCray Jr., P.B., Kulhankova, K., Thorne, P.S., 2008. MD-2-dependent pulmonary immune responses to inhaled lipooligosaccharides: effect of acylation state. Am. J. Respir. Cell Mol. Biol. 38, 647–654.

Hamadeh, R.M., Estabrook, M.M., Zhou, P., Jarvis, G.A., Griffiss, J.M., 1995. Anti-Gal binds to pili of *Neisseria meningitidis*: the immunoglobulin A isotype blocks complement-mediated killing. Infect. Immun. 63, 4900–4906.

Harvey, H.A., Jennings, M.P., Campbell, C.A., Williams, R., Apicella, M.A., 2001. Receptor-mediated endocytosis of *Neisseria gonorrhoeae* into primary human urethral epithelial cells: the role of the asialoglycoprotein receptor. Mol. Microbiol. 42, 659–672.

Hasman, H., Chakraborty, T., Klemm, P., 1999. Antigen-43-mediated autoaggregation of *Escherichia coli* is blocked by fimbriation. J. Bacteriol. 181, 4834–4841.

Hayashi, F., Smith, K.D., Ozinsky, A., et al., 2001. The innate immune response to bacterial flagellin is mediated by Toll-like receptor 5. Nature 410, 1099–1103.

He, Y., Li, H., Zhang, J., et al., 2004. The extracellular matrix protein mindin is a pattern-recognition molecule for microbial pathogens. Nat. Immunol. 5, 88–97.

Heinrichs, D.E., Yethon, J.A., Whitfield, C., 1998. Molecular basis for structural diversity in the core regions of the lipopolysaccharides of *Escherichia coli* and *Salmonella enterica*. Mol. Microbiol. 30, 221–232.

Hidalgo-Grass, C., Dan-Goor, M., Maly, A., et al., 2004. Effect of a bacterial pheromone peptide on host chemokine degradation in group A streptococcal necrotising soft-tissue infections. Lancet 363, 696–703.

Inzana, T.I., Iritani, B., Gogolewski, R.P., Kania, S.A., Corbeil, L.B., 1988. Purification and characterization of lipopolysaccharide from four strains of *Haemophilus somnus*. Infect. Immun. 56, 2830–2837.

Jiang, S.M., Wang, L., Reeves, P.R., 2001. Molecular characterization of *Streptococcus pneumoniae* type 4, 6B, 8, and 18C capsular polysaccharide gene clusters. Infect. Immun. 69, 1244–1255.

John, C.M., Schneider, H., Griffiss, J.M., 1999. *Neisseria gonorrhoeae* that infect men have lipooligosaccharides with terminal *N*-acetyl-lactosamine repeats. J. Biol. Chem. 274, 1017–1025.

Joshua, G.W.P., Guthrie-Irons, C., Karlyshev, A.V., Wren, B.W., 2006. Biofilm formation in *Campylobacter jejuni*. Microbiology 152, 387–396.

Kahler, C.M., Martin, L.E., Shih, G.C., Rahman, M.M., Carlson, R.W., Stephens, D.S., 1998. The ($\alpha 2 \rightarrow 8$)-linked polysialic acid capsule and lipooligosaccharide structure both contribute to the ability of serogroup B *Neisseria meningitidis* to resist the bactericidal activity of normal human serum. Infect. Immun. 66, 5939–5947.

Kahler, C.M., Martin, L.E., Tzeng, Y.L., et al., 2001. Polymorphisms in pilin glycosylation locus of *Neisseria meningitidis* expressing class II pili. Infect. Immun. 69, 3597–3604.

Khamri, W., Moran, A.P., Worku, M.L., et al., 2005. Variations in *Helicobacter pylori* lipopolysaccharide to evade the innate immune component surfactant protein D. Infect. Immun. 73, 7677–7686.

Kim, J.J., Zhou, D., Mandrell, R.E., Griffiss, J.M., 1992. Effect of exogenous sialylation of the lipooligosaccharides of *Neisseria gonorrhoeae* on opsonophagocytosis. Infect. Immun. 60, 4439–4442.

Kim, J.O., Weiser, J.N., 1998. Association of intrastrain phase variation in quantity of capsular polysaccharide and teichoic acid with the virulence of *Streptococcus pneumoniae*. J. Infect. Dis. 177, 368–377.

Klemm, P., Hjerrild, L., Gjermansen, M., Schembri, M.A., 2004. Structure-function analysis of the self-recognizing antigen 43 autotransporter protein from *Escherichia coli*. Mol. Microbiol. 51, 283–296.

Konkel, M.E., Klena, J.D., Rivera-Amill, V., et al., 2004. Secretion of virulence proteins from *Campylobacter jejuni* is dependent on a functional flagellar export apparatus. J. Bacteriol. 186, 3296–3303.

Krinos, C.M., Weinacht, K.G., Tzianabos, A.O., Kasper, D. L., Comstock, L.E., 2001. Extensive surface diversity of a commensal microorganism by multiple DNA inversions. Nature 414, 555–558.

Lam, J.S., Graham, L.L., Lightfoot, J., Dasgupta, T., Beveridge, T.J., 1992. Ultrastructural examination of the lipopolysaccharides of *Pseudomonas aeruginosa* strains and their isogenic rough mutants by freeze-substitution. J. Bacteriol. 174, 7159–7167.

Lambden, P.R., Robertson, J.N., Watt, P.J., 1980. Biological properties of 2 distinct pilus types produced by isogenic variants of *Neisseria gonorrhoeae* P9. J. Bacteriol. 141, 393–396.

Lin, X., Heitman, J., 2006. The biology of the *Cryptococcus neoformans* species complex. Annu. Rev. Microbiol. 60, 69–105.

Lindenthal, C., Elsinghorst, E.A., 1999. Identification of a glycoprotein produced by enterotoxigenic *Escherichia coli*. Infect. Immun. 67, 4084–4091.

Liu, C.H., Lee, S.M., Vanlare, J.M., Kasper, D.L., Mazmanian, S.K., 2008. Regulation of surface architecture by symbiotic bacteria mediates host colonization. Proc. Natl. Acad. Sci. USA 105, 3951–3956.

Logan, S.M., 2006. Flagellar glycosylation – a new component of the motility repertoire? Microbiology 152, 1249–1262.

Logan, S.M., Kelly, J., Thibault, P., Ewing, C.P., Guerry, P., 2002. Structural heterogeneity of carbohydrate modifications affects serospecificity of *Campylobacter* flagellins. Mol. Microbiol. 46, 587–597.

Malaga, W., Constant, P., Euphrasie, D., et al., 2008. Deciphering the genetic bases of the structural diversity of phenolic glycolipids in strains of the *Mycobacterium tuberculosis* complex. J. Biol. Chem. 283, 15177–15184.

Manetti, A.G.O., Zingaretti, C., Falugi, F., et al., 2007. *Streptococcus pyogenes* pili promote pharyngeal cell adhesion and biofilm formation. Mol. Microbiol. 64, 968–983.

Marceau, M., Forest, K., Beretti, J.L., Tainer, J., Nassif, X., 1998. Consequences of the loss of *O*-linked glycosylation of meningococcal type IV pilin on piliation and pilus-mediated adhesion. Mol. Microbiol. 27, 705–715.

McFadden, D.C., Fries, B.C., Wang, F., Casadevall, A., 2007. Capsule structural heterogeneity and antigenic variation in *Cryptococcus neoformans*. Eukaryot. Cell 6, 1464–1473.

Miller, S.I., Kukral, A.M., Mekalanos, J.J., 1989. A 2-component regulatory system (*phoP-phoQ*) controls *Salmonella typhimurium* virulence. Proc. Natl. Acad. Sci. USA 86, 5054–5058.

Moran, A.P., 1996. The role of lipopolysaccharide in *Helicobacter pylori* pathogenesis. Aliment. Pharmacol. Ther. 10, 39–50.

Moran, A.P., Lindner, B., Walsh, E.J., 1997. Structural characterization of the lipid A component of *Helicobacter pylori* rough- and smooth-form lipopolysaccharides. J. Bacteriol. 179, 6453–6463.

Morooka, T., Umeda, A., Amako, K., 1985. Motility as an intestinal colonization factor for *Campylobacter jejuni*. J. Gen. Microbiol. 131, 1973–1980.

Nairn, C.A., Cole, J.A., Patel, P.V., Parsons, N.J., Fox, J.E., Smith, H., 1988. Cytidine 5′-monophospho-*N*-acetylneuraminic acid or a related compound is the low

Mr factor from human red blood cells which induces gonococcal resistance to killing by human serum. J. Gen. Microbiol. 134, 3295–3306.

Orskov, I., Orskov, F., Jann, B., Jann, K., 1977. Serology, chemistry, and genetics of O and K antigens of *Escherichia coli*. Bacteriol. Rev. 41, 667–710.

Owen, P., Meehan, M., de Loughry-Doherty, H., Henderson, I., 1996. Phase-variable outer membrane proteins in *Escherichia coli*. FEMS Immunol. Med. Microbiol. 16, 63–76.

Parge, H.E., Forest, K.T., Hickey, M.J., Christensen, D.A., Getzoff, E.D., Tainer, J.A., 1995. Structure of the fiber-forming protein pilin at 2.6-angstrom resolution. Nature 378, 32–38.

Parsons, N.J., Andrade, J.R.C., Patel, P.V., Cole, J.A., Smith, H., 1989. Sialylation of lipopolysaccharide and loss of absorption of bactericidal antibody during conversion of gonococci to serum resistance by cytidine 5′-monophosphate-*N*-acetyl neuraminic acid. Microb. Pathog. 7, 63–72.

Patrone, J.B., Stein, D.C., 2007. Effect of gonococcal lipooligosaccharide variation on human monocytic cytokine profile. BMC Microbiol. 7, 7.

Peng, Z., Wu, H., Ruiz, T., et al., 2008. Role of *gap3* in Fap1 glycosylation, stability, in vitro adhesion, and fimbrial and biofilm formation of *Streptococcus parasanguinis*. Oral Microbiol. Immunol. 23, 70–78.

Pier, G.B., 2007. *Pseudomonas aeruginosa* lipopolysaccharide: a major virulence factor, initiator of inflammation and target for effective immunity. Intl. J. Med. Microbiol. 297, 277–295.

Poltorak, A., He, X., Smirnova, I., et al., 1998. Defective LPS signaling in C3H/HeJ and C57BL/10ScCr mice: mutations in Tlr4 gene. Science 282, 2085–2088.

Power, P.M., Roddam, L.F., Rutter, K., Fitzpatrick, S.Z., Srikhanta, Y.N., Jennings, M.P., 2003. Genetic characterization of pilin glycosylation and phase variation in *Neisseria meningitidis*. Mol. Microbiol. 49, 833–847.

Reed, M.B., Domenech, P., Manca, C., et al., 2004. A glycolipid of hypervirulent tuberculosis strains that inhibits the innate immune response. Nature 431, 84–87.

Reed, M.B., Gagneux, S., DeRiemer, K., Small, P.M., Barry, C. E., 2007. The W-Beijing lineage of *Mycobacterium tuberculosis* overproduces triglycerides and has the DosR dormancy regulon constitutively upregulated. J. Bacteriol. 189, 2583–2589.

Robertson, J.N., Vincent, P., Ward, M.E., 1977. Preparation and properties of gonococcal pili. J. Gen. Microbiol. 102, 169–177.

Rocchetta, H.L., Burrows, L.L., Lam, J.S., 1999. Genetics of O-antigen biosynthesis in *Pseudomonas aeruginosa*. Microbiol. Mol. Biol. Rev. 63, 523–553.

Rudel, T., Scheurerpflug, I., Meyer, T.F., 1995. *Neisseria* PilC protein identified as type-4 pilus tip-located adhesin. Nature 373, 357–359.

Rudel, T., van Putten, J.P., Gibbs, C.P., Haas, R., Meyer, T. F., 1992. Interaction of two variable proteins (PilE and PilC) required for pilus-mediated adherence of *Neisseria gonorrhoeae* to human epithelial cells. Mol. Microbiol. 6, 3439–3450.

Schilling, B., Gibson, B.W., Filiatrault, M.J., Campagnari, A. A., 2002. Characterization of lipooligosaccharides from *Haemophilus ducreyi* containing polylactosamine repeats. J. Am. Soc. Mass Spectrom. 13, 724–734.

Schneider, H., Hale, T.L., Zollinger, W.D., Seid, R.C., Hammack, C.A., Griffiss, J.M., 1984. Heterogeneity of molecular size and antigenic expression within lipooligosaccharides of individual strains of *Neisseria gonorrhoeae* and *Neisseria meningitidis*. Infect. Immun. 45, 544–549.

Schneider, H., Griffiss, J.M., Boslego, J.W., Hitchcock, P. J., McJunkin, K.O., Apicella, M.A., 1991. Expression of paragloboside-like lipooligosaccharide may be a necessary component of gonococcal pathogenesis in men. J. Exp. Med. 174, 1601–1605.

Sherlock, O., Vejborg, R.M., Klemm, P., 2005. The TibA adhesin/invasin from enterotoxigenic *Escherichia coli* is self recognizing and induces bacterial aggregation and biofilm formation. Infect. Immun. 73, 1954–1963.

Sherlock, O., Dobrindt, U., Jensen, J.B., Vejborg, R.M., Klemm, P., 2006. Glycosylation of the self-recognizing *Escherichia coli* Ag43 autotransporter protein. J. Bacteriol. 188, 1798–1807.

Simon, G.L., Klempner, M.S., Kasper, D.L., Gorbach, S.L., 1982. Alterations in opsonophagocytic killing by neutrophils of *Bacteroides fragilis* associated with animal and laboratory passage: effect of capsular polysaccharide. J. Infect. Dis. 145, 72–77.

Song, W., Ma, L., Chen, W., Stein, D.C., 2000. Role of lipooligosaccharide in *opa*-independent invasion of *Neisseria gonorrhoeae* into human epithelial cells. J. Exp. Med. 191, 949–959.

Spangenberg, C., Heuer, T., Bürger, C., Tümmler, B., 1996. Genetic diversity of flagellins of *Pseudomonas aeruginosa*. FEBS Lett. 396, 213–217.

Steeghs, L., den Hartog, R., den Boer, A., Zomer, B., Roholl, P., van der Ley, P., 1998. Meningitis bacterium without endotoxin. Nature 392, 449–450.

Steeghs, L., Kuipers, B., Hamstra, H.J., Kersten, G., van Alphen, L., van der Ley, P., 1999. Immunogenicity of outer membrane proteins in a lipopolysaccharide-deficient mutant of *Neisseria meningitidis*: influence of adjuvants on the immune response. Infect. Immun. 67, 4988–4993.

Stephenson, A.E., Wu, H.J.N., Tomana, M., Mintz, K., Fives-Taylor, P., 2002. The Fap1 fimbrial adhesin is a glycoprotein: antibodies specific for the glycan moiety block the adhesion of *Streptococcus parasanguis* in an in vitro tooth model. Mol. Microbiol. 43, 147–157.

Stimson, E., Virji, M., Makepeace, K., et al., 1995. Meningococcal pilin: a glycoprotein substituted with digalactosyl 2,4-diacetamido-2,4,6-trideoxyhexose. Mol. Microbiol. 17, 1201–1214.

Swartley, J.S., Marfin, A.A., Stephens, D.S., 1997. Capsule switching of Neisseria meningitidis. Proc. Natl. Acad. Sci. USA 94, 271–276.

Szymanski, C.M., King, M., Haardt, M., Armstrong, G.D., 1995. *Campylobacter jejuni* motility and invasion of Caco-2 cells. Infect. Immun. 63, 4295–4300.

Takemura, N., Noiri, Y., Ehara, A., Kawahara, T., Noguchi, N., Ebisu, S., 2004. Single species biofilm-forming ability of root canal isolates on gutta-percha points. Eur. J. Oral Sci. 112, 523–529.

Taylor, C.M., Roberts, I.S., 2005. Capsular polysaccharides and their role in virulence. Contrib. Microbiol. Immunol. 12, 55–66.

Taylor, M.T., Drickamer, K., 2002. Introduction to Glycobiology. Oxford University Press, Oxford.

Telford, J.L., Barocchi, M.A., Margarit, I., Rappuoli, R., Grandi, G., 2006. Pili in gram-positive pathogens. Nat. Rev. Microbiol. 4, 509–519.

Thibault, P., Logan, S.M., Kelly, J.F., et al., 2001. Identification of the carbohydrate moieties and glycosylation motifs in *Campylobacter jejuni* flagellin. J. Biol. Chem. 276, 34862–34870.

Turk, D.C., 1984. The pathogenicity of Haemophilus influenzae. J. Med. Microbiol. 18, 1–16.

van Putten, J.P., 1993. Phase variation of lipopolysaccharide directs interconversion of invasive and immuno-resistant phenotypes of *Neisseria gonorrhoeae*. EMBO J. 12, 4043–4051.

Verma, A., Arora, S.K., Kuravi, S.K., Ramphal, R., 2005. Roles of specific amino acids in the N terminus of *Pseudomonas aeruginosa* flagellin and of flagellin glycosylation in the innate immune response. Infect. Immun. 73, 8237–8246.

Verma, A., Schirm, M., Arora, S.K., Thibault, P., Logan, S.M., Ramphal, R., 2006. Glycosylation of b-type flagellin of *Pseudomonas aeruginosa*: structural and genetic basis. J. Bacteriol. 188, 4395–4403.

Virji, M., Saunders, J.R., Sims, G., Makepeace, K., Maskell, D., Ferguson, D.J., 1993. Pilus-facilitated adherence of *Neisseria meningitidis* to human epithelial and endothelial cells: modulation of adherence phenotype occurs concurrently with changes in primary amino acid sequence and the glycosylation status of pilin. Mol. Microbiol. 10, 1013–1028.

Vogel, U., Claus, H., Heinze, G., Frosch, M., 1997. Functional characterization of an isogenic meningococcal α-2,3-sialyltransferase mutant: the role of lipooligosaccharide sialylation for serum resistance in serogroup B meningococci. Med. Microbiol. Immunol. (Berl), 186, 159–166.

Wessels, M.R., Moses, A.E., Goldberg, J.B., DiCesare, T.J., 1991. Hyaluronic acid capsule is a virulence factor for mucoid group A streptococci. Proc. Natl. Acad. Sci. USA 88, 8317–8321.

Whitfield, C., 2006. Biosynthesis and assembly of capsular polysaccharides in *Escherichia coli*. Annu. Rev. Biochem. 75, 39–68.

Wu, H., Mintz, K.P., Ladha, M., Fives-Taylor, P.M., 1998. Isolation and characterization of Fap1, a fimbriae-associated adhesin of *Streptococcus parasanguis* FW213. Mol. Microbiol. 28, 487–500.

Zhang, Y., Gaekwad, J., Wolfert, M.A., Boons, G.J., 2008. Innate immune responses of synthetic lipid A derivatives of *Neisseria meningitidis*. Chemistry 14, 558–569.

CHAPTER

42

Phase variation of bacterial surface glycosylated molecules in immune evasion

Michael A. Apicella and Michael P. Jennings

SUMMARY

A number of bacteria that exclusively colonize human mucosal surfaces express antigens on their surfaces which are molecular mimics of glycosphingolipids found on human cells. These glycolipid structures are important in the pathogenesis of *Neisseria* and *Haemophilus* spp. for both immune evasion and in the adherence and invasion of human cells. In several instances, this mimicry necessitates that these organisms parasitize specific sugars from the human in order to assemble the appropriate structure. Both species have developed mechanisms randomly to switch "on and off" expression of the enzymes responsible for assembly of the terminal sugars on these glycolipids by a process termed "phase variation". This allows members of the respective population to survive in a wide range of host environments assuring continuation of transmission, colonization and infection.

Keywords: Molecular mimicry; Phase variation; *Neisseria gonorrhoeae; Neisseria meningitidis; Haemophilus influenzae;* Lipo-oligosaccharide; Glycosphingolipid; Paragloboside; Phosphorylcholine; Asialoglycoprotein receptor; Platelet-activating factor receptor; Sialic acid

1. INTRODUCTION

Environmental mimicry as a means of survival exists within most living organisms. Some bacterial species which commonly infect humans have developed these strategies by mimicking the carbohydrate structures present on human cell surfaces. A number of the enzymes involved in the synthesis of these bacterial mimics can undergo random variation. This ability to modify surface carbohydrate expression in a random manner enhances selection for the most "fit" expression for any environment that the bacteria would encounter. This chapter will discuss this phenomenon in two human pathogens, *Neisseria gonorrhoeae* and *Haemophilus influenzae.*

Autoimmune disease caused by immunological reactivity to bacterial antigens has been recognized for many years. This has been of particular concern to investigators developing vaccines using bacterial products from human pathogens as immunogens. Bacterial proteins, capsular carbohydrates and glycolipid antigens have also been shown to cross-react with a variety of human antigens. The immunological cross-reactivity between the M protein of *Streptococcus pyogenes* and cardiac myosin has been well studied (Zabriskie *et al.*, 1970). Other examples of such host–pathogen relationships include the sharing of epitopes of *Klebsiella, Yersinia, Shigella* and *Salmonella* spp. with HLA-B27 resulting in reactive arthritis (van Bohemen *et al.*, 1984; Granfors and Toivanen, 1986; Moe *et al.*, 1999). The lipopolysaccharides (LPSs) of some strains of *Campylobacter jejuni* have been shown to cross-react with human GM_1 and this organism has been identified as the single most common pathogen associated with the development of Guillian-Barré syndrome (Moran and Prendergast, 2000; Neisser *et al.*, 2000) (see also Chapter 43). The capsular polysaccharide of *Neisseria meningitidis* serogroup B which is a homopolymer of α-(2$\rightarrow$8)-linked *N*-acetylneuraminic acid has been shown to elicit autoantibodies that cross-react with embryonic, but not the adult, form of neural cell adhesion molecules (NCAM) (Moe *et al.*, 1999). The risk of immunopathology has prevented use of the serogroup B capsule in conjugate vaccines that have been successfully implemented for other serogroups.

Endotoxin or LPS is usually the major glycolipid present in Gram-negative bacterial outer membranes. It is an amphopathic molecule composed of a carbohydrate portion and a lipid portion, designated the lipid A (Raetz and Whitfield, 2002) (see Chapter 3). In many species of bacteria, the carbohydrate portion is composed of a core structure which is substituted by multiple repeating unit of 4–6 sugars. These are known as O-antigens and there is a wide range of diverse antigenic properties on these structures which serve as the basis for many bacterial serotyping systems. Mutants whose synthesis or assembly of the O-antigen in bacteria, that would normally produce this structure, is defective produce a glycolipid that is referred to as a "rough" LPS (see Chapters 1, 4 and 18). A subset of bacteria lacks the appropriate enzymes to add these O-antigen subunits. The endotoxins of these bacteria are referred to as lipo-oligosaccharides (LOSs) by some investigators. The human mucosal pathogens, *Neisseria meningitidis, N. gonorrhoeae* and *Haemophilus influenzae* are such bacterial species (Swords *et al.*, 2003). Two characteristics of these LOSs make them different from "rough" type LPSs. The first is their ability to mimic, at the structural and antigenic level, human glycolipids (Mandrell *et al.*, 1992) and the second is the potential for these LOSs to be modified *in vivo* by host substances and secretions (Nairn *et al.*, 1988; Johnston *et al.*, 2008). The remainder of this chapter will focus on the aspects of the mimicry and phase variation of the LOS by *N. gonorrhoeae* and *H. influenzae*.

2. MOLECULAR MIMICRY OF HUMAN ANTIGENS BY THE LOSs OF PATHOGENIC *NEISSERIA* AND *HAEMOPHILUS*

Bacteria that inhabit human mucosal surfaces including members of the genera *Neisseria, Bordetella* and *Branhamella*, express shorter, more branched LPS structures than the LPSs of enteric bacteria. Also, these structures do not contain the repetitive O-antigen side chains of enteric bacteria. To highlight this distinction, these structures are termed LOSs, although this term is not universally used. An extensive review has appeared covering the LOSs of Gram-negative bacteria in general (Preston *et al.*, 1996). The LOSs are highly variable; a single strain simultaneously synthesizes multiple LOS structures (Gibson *et al.*, 1993). These are often present in unequal amounts such that a majority of LOS structures reported represent the major LOS structure of that strain. This diversity is generated by two distinct processes: first by incomplete synthesis of structures

TABLE 42.1 Bacterial carbohydrate antigens and human tissue mimics

LOS oligosaccharide branch structures[a]	Antigen mimics
β-Gal-(1→4)-β–Glc-(1→-	Lactosyl
β-Gal-(1→4)-β-Glc-(1→ (I), β-Gal-(1→4)-α-Glc-(1→	Lactosyl(α,β)
α-Gal-(1→4)-β-Gal-(1→4)-β-Glc-(1→	P^k-antigen
β-Gal-(1→4)-β–GlcNAc-(1→3)-β–Gal-(1→4)-β-Glc-(1→-	Lacto-*N*-neotetraose (paragloboside)
α-Neu5Ac-(2→3)-β-Gal-(1→4)-β-GlcNAc-(1→3)-β-Gal-(1→4)-β–Glc-(1→	Sialyllactosamine (i-antigen)
ChoP-Glc-	Platelet-activating factor (PAF)

[a]Abbreviations: Gal, galactose; Glc, glucose; GlcNAc, *N*-acetylglucosamine; Neu5Ac, 5-*N*-acetylneuraminic acid or sialic acid; ChoP, phosphorylcholine.

within individual cells in the population; or second, by synthesis of different structures made by cells within the population that are genetically different due to phase variation. Phase variation is the high frequency reversible on/off switching of bacterial surface antigens and, in the case of LOS structures, refers to the loss or gain of LOS structures, often described in terms of loss/gain of anti-LOS monoclonal antibody epitopes (Kimura and Hansen, 1986). Phase variation occurs at high frequency (often 10^{-2} per bacterium per generation) and is mediated by high frequency, reversible mutations in hypermutagenic DNA repeat sequences located in the genes encoding glycosyltransferases for LOS biosynthesis. The repertoire of LOS structures that can be synthesized by a particular strain is determined by the combination of the glycosyltransferase genes present and their state of expression (Jennings *et al.*, 1999). The role of phase variation in the biology of infection or colonization is unclear. It is possible that high frequency switching of exposed structures and generation of a repertoire of structures is a way of avoiding host defence mechanisms or an adaptation to a particular microenvironment within the host.

The LOSs of the exclusive human bacterial pathogens *N. meningitidis*, *N. gonorrhoeae* and *H. influenzae* express structures which are immunochemically identical to a group of different human glycosphingolipid and glycolipid antigens (Table 42.1). In spite of the high frequency of human exposure to these organisms, there is little evidence that they are involved in autoimmune disease in humans. *N. meningitidis* and *H. influenzae* are frequent colonizers of the upper respiratory tract (Murphy and Apicella, 1987; Apicella, 2003). These bacteria can cause a wide range of local infections, e.g. sinusitis and otitis media [caused by non-typeable *H. influenzae* (NTHi)] and disseminated infections, e.g. sepsis, pneumonia, acute bronchitis and meningitis (caused by *N. meningitidis* and *H. influenzae*) in humans. *Neisseria gonorrhoeae* is a strict human pathogen which causes the sexually transmitted disease, gonorrhoea (Workowski *et al.*, 2008). The presence of *N. gonorrhoeae* in a human is indicative of disease. Initial studies suggested that the molecular mimicry of human tissue antigens expressed on the LOS may serve the function of immune evasion. Recent studies, some of which are described below, indicate that this mimicry also allows these bacterial species to utilize human cell ligand–receptor interactions facilitating adherence to and entry into mucosal epithelial cells whose surface they infect.

3. THE LOS OF *N. GONORRHOEAE* MIMICS HUMAN PARAGLOBOSIDE AND SERVES AS A LIGAND TO THE ASIALOGLYCOPROTEIN RECEPTOR

Several studies have demonstrated that gonococcal LOS could act as a ligand for the human asialoglycoprotein receptor (ASGP-R) and this receptor could function to internalize gonococci into hepatocytes (Porat *et al.*, 1995a,b). Using the hepatoma cell line, HepG2, it has been shown that gonococcal LOS with the paragloboside structure could competitively inhibit binding of asialoorosomucoid (a purified ligand) which selectively binds ASGP-R (Porat *et al.*, 1995a,b).

A study of the molecular pathogenesis of urethral gonorrhoea in men indicated that the ASGP-R could be detected in infected urethral epithelial cells that were shed into the urethral exudate (Harvey *et al.*, 1997). This study also demonstrated that invasion of the male urethral cell by gonococci was a receptor-mediated process. In order to study the invasion of urethral epithelial cells by gonococci and to confirm the role of the gonococcal LOS–ASGP-R interaction, a model system of gonococcal infection was developed which used human primary urethral epithelial cells grown in culture (Harvey *et al.*, 1997, 2000, 2001). The results of these studies showed that the infection process mirrored the sequence of events that was seen during natural gonococcal infection in men (Apicella *et al.*, 1996). That is, the majority of bacteria were entering the urethral cell by a receptor-mediated process.

In order to determine if the gonococcal LOS–ASGP-R interaction was involved in this process, an ASGP-R specific monoclonal antibody was generated to the extracellular portion of the ASGP-R. This antibody was used to track the cycling of this receptor during infection of the urethral epithelial cell. Laser scanning confocal microscopy (LSCM) showed minimal amounts of receptor on the epithelial cell surface prior to infection. Trafficking of the ASGP-R to the apical surface of the urethral epithelial cells could be seen over the first 4 hours during the infection process (Figure 42.1). By 24 hours, almost all cells in the monolayer showed surface expression of the ASGP-R. Reverse transcriptase-polymerase chain reaction (RT-PCR) and Western blot analysis of the ASGP-R message and protein demonstrated that both were present at similar levels during all periods of infection. This strongly suggested that the ASGP-R was being recycled within the cell rather than increased amount of receptor production during infection. These strains also showed reduced association with the epithelial cells. Examination by LSCM of epithelial cell infections with gonococcal strains lacking the LOS paragloboside mimic showed minimal surface expression of the receptor over the infection period and no invasion of these cells. Asialofetuin (ASF), a purified glycoprotein that selectively binds ASGP-R, could inhibit the engagement of the wild-type gonococcal LOS to the receptor. Examination of urethral epithelial cells treated with asialofetuin prior to and during infection in LSCM showed minimal bacterial attachment but increased surface expression of the receptor. These results were confirmed by direct measurement of epithelial cell invasion by gonococci, which showed reduced invasion in the presence of ASF. Asialoglycoprotein receptor-mediated endocytosis is clathrin-dependent. Microscopy studies (LSCM) using fluorescent-labelled antibodies to clathrin and gonococci showed the majority of the organisms on the surface co-localized to clathrin. Clathrin was also demonstrated to be recruited to the sites of gonococcal entry into the epithelial cells.

Confirmation of the role the LOS paragloboside plays in the invasion of urethral epithelial cells was accomplished using the invasion assay devised by Falkow and co-workers (Shaw and Falkow, 1988). Studies of invasion with the parent gonococcal strain showed that between 2 and 3% of the initial inoculum entered the epithelial cells during the first 4 hours of infection.

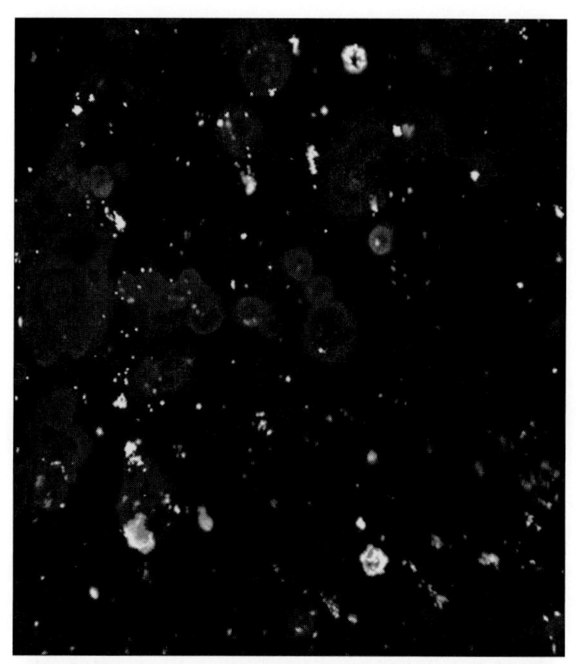

FIGURE 42.1 Co-localization of an antibody to a terminal lactosamine epitope (red) on the LOS of *N. gonorrhoeae* with an antibody to the asialoglycoprotein receptor (green) on primary male urethral epithelial cells. (See colour plates section.)

N. gonorrhoeae with mutations in glycosyltransferases responsible for assembling the LOS were used to determine the effect of progressive single carbohydrate deletions within the paragloboside structure on the ability of the gonococci to invade. The deletion of the terminal β-(1→4)-linked galactose (Gal) and the subsequent sugars reduced invasion to 0.1%.

Gonococci also adhere to human sperm but the mechanism by which this adherence occurs had never been elucidated. Mouse spermatids have been previously shown to express the ASGP-R receptor (Monroe and Huber, 1994). Using an ASGP-R specific monoclonal antibody, sperm samples from five men were examined and showed the presence of ASGP-R on the head and tail regions of the sperm cells (Harvey *et al.*, 2000). Examination of sperm cells infected with gonococci in LSCM showed co-localization of the gonococcus to the ASGP-R on sperm. Using nearest neighbour analysis with light-activated ligands, gonococcal LOS was demonstrated to bind to the ASGP-R (Harvey *et al.*, 2000). Purified gonococcal LOS that lacked the complete paragloboside structure had reduced adherence to sperm compared to LOS purified from the parent strain of gonococci. Also, inhibition studies confirmed the specificity of the gonococcus binding to the sperm cell via the ASGP-R. It is interesting to hypothesize that gonococcal adherence to sperm enhances the infectious ability of the organism as it "hitches" a ride on a motile cell form like sperm.

These studies demonstrate the role of gonococcal molecular mimicry of human paragloboside in the pathogenesis of gonococcal infection. The organism utilizes normal human cell ligand–receptor interactions to gain entry into urethral epithelial cells in men and to adhere to the surface of sperm. Thus, this molecular mimicry serves not only to assist the bacterium in immune evasion but also as a surrogate for the human ligand. A human experimental model may give clues to the importance of the gonococcal paragloboside mimicry in infectivity. The original US Army human experimental model of gonococcal infection in men utilized the MS11 strain of *N. gonorrhoeae*. Analysis of the organisms which were isolated from these men revealed a shift in the LOS phenotype from a lactosyl-type structure of the original inoculum to a paragloboside-like structure from the urethral isolate as the infection progressed (Griffiss *et al.*, 1991). This organism, designated MS11mKc, was used in a second series of human infection experiments (Schneider *et al.*, 1995, 1996; Ramsey *et al.*, 1995). In these experiments, as few as 250 organisms were capable of infecting the subjects and the LOS retained the paragloboside-like structure (Schneider *et al.*, 1995).

In addition to direct involvement in receptor–ligand interactions, the LOS structure expressed can influence the function of protein adhesions. Both *N. gonorrhoeae* and *N. meningitidis*

LOSs express structure with a terminal sialic acid. In *N. meningitides* adhesin function, strains expressing the adhesin Opc and the L3,7,9 immunotype structure (lacto-*N*-neotetraose, containing a terminal sialic acid), is far less invasive compared to bacteria expressing the L8 immunotype, non-sialylated, LOS structure (Dunn *et al.*, 1995). The L3,7,9 immunotype LOS structures are relatively serum-resistant compared to the L8 immunotype. In the mouse model of infection, it was revealed that the L8 immunotype predominates in the nasopharynx while the L3,7,9 immunotype is most commonly found in the blood of the infected mice (Jones *et al.*, 1992). Switching between the L3,7,9 and L8 immunotype is mediated by phase variation of the glycosyltransferase lgtA. These studies are strongly suggestive of key selective and counter selective pressures for these terminal LOS structures and for the observed phase variation of their expression.

4. THE LOS OF NTHi MIMICS THE HUMAN PLATELET-ACTIVATING FACTOR (PAF) AND CAN ACT AS A LIGAND FOR THE PLATELET-ACTIVATING FACTOR RECEPTOR (PAF-R) ON AIRWAY EPITHELIAL CELLS

In vitro infection of primary and immortalized human airway epithelial cells has been used to model interactions involved in NTHi colonization and pathogenesis (Ketterer *et al.*, 1999). Infections were performed using undifferentiated cells grown in submerged culture, as well as cells grown at an air–fluid interface to generate a polarized, differentiated monolayer. Infected monolayers were examined by LSCM, scanning electron microscopy and transmission electron microscopy. These results indicate that NTHi bacilli adhere preferentially to a subpopulation of non-ciliated bronchial epithelial cells. These cells undergo cytoskeletal changes, including the formation of lamellipodia and filopodia, which extend to engage adherent bacilli. Dextran-uptake experiments were performed to characterize the intracellular compartment containing NTHi bacilli and revealed that NTHi bacilli are found within macropinosomes (Ketterer *et al.*, 1999). This was confirmed by both transmission electron microscopy and LSCM.

The LOS is a key antigenic component of the NTHi cell surface and contains a variety of host structures. The role of the LOS in NTHi adherence and invasion was assessed in infection experiments using mutants with expression of altered LOS glycoforms. An NTHi 2019 *pgmB* null mutant expressing LOS truncated at the core region was significantly less adherent and invasive than the parental strain (Swords *et al.*, 2000). In order to determine the specific LOS sugar(s) responsible for adherence and invasion, polystyrene microbeads were coated with NTHi wild-type and mutated LOS structures and allowed to interact with airway epithelial cells in culture. Beads coated with wild-type NTHi 2019 LOS bound to airway cells in significantly greater numbers than did beads coated with LOS derived from NTHi 2019 *pgm*B (Swords *et al.*, 2000). The LOS of NTHi consists of a large number of glycoforms and a variety of molecular mimics of human antigens have been found among these glycoforms. These include i, paragloboside, p^k and sialoparagloboside (Mandrell *et al.*, 1992) (see Table 42.1). Recent studies have also shown that the host structure phosphorylcholine is found as a terminal structure within the oligosaccharide region of the NTHi LOS (Kim and Weiser, 1998). Using a panel of compounds containing structures found in *H. influenzae* LOS glycoforms, a series of studies were initiated to determine if any of these could inhibit the attachment of polystyrene coated with NTHi strain 2019 LOS to airway epithelial cells. These results indicated that phosphorylcholine

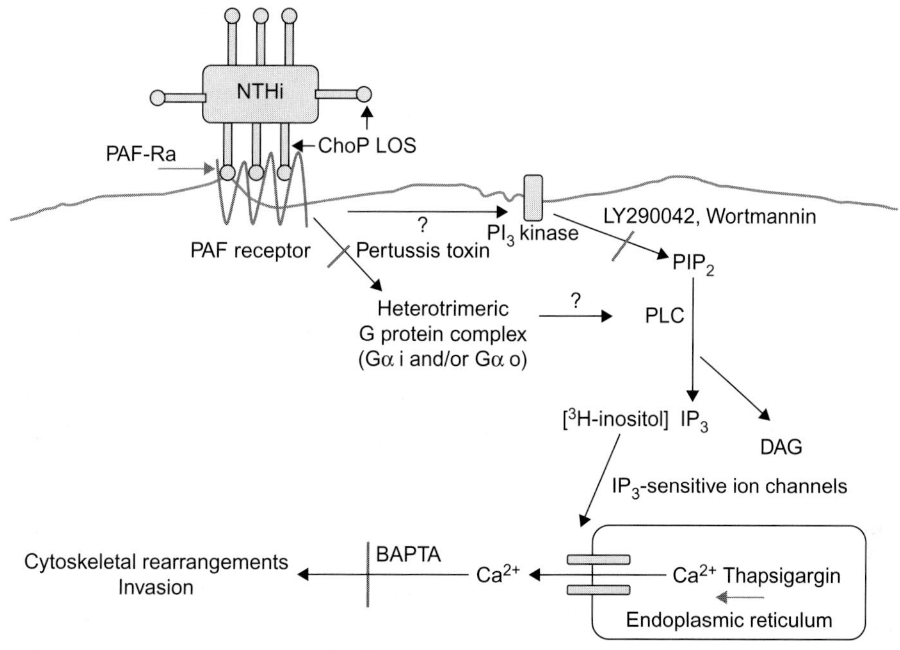

FIGURE 42.2 A model for non-typeable *H. influenzae* (NTHi) entry into primary human bronchial epithelial cells. The phosphorylcholine residues (ChoP) on the LOS engage the platelet-activating factor receptor (PAF-Ra) initiating a signalling response through heterotrimeric G protein-linked complexes and PI_3 kinase. The data suggest that this results in the generation of inositol 1,4,5-trisphosphate (IP_3) which interacts with IP_3- sensitive Ca^{2+} channels in the endoplasmic reticulum (ER). This results in Ca^{2+} flux from the ER into the cytoplasm resulting in actin network rearrangement and bacterial entry into the cell. The lines across the arrows mark the points in the NTHi signalling pathway that have been demonstrated can be blocked by the agents listed next to them.

contributes to the LOS-dependent adherence of NTHi. Figure 42.2 shows this model of NTHi entry into human airway cells in more detail.

Phosphorylcholine is also found within the pneumococcal C polysaccharide, where it contributes to the adherence and invasion of activated endothelial cells by pneumococci (Cundell *et al.*, 1996). These studies showed that phosphorylcholine bound to the PAF-R on the endothelial cells and that the interaction could be inhibited by chemical inhibitors of PAF-R signalling. Based on these observations, studies were instituted to determine if the PAF-R was involved in NTHi interaction with airway epithelial cells. Laser scanning confocal microscopy showed co-localization of surface NTHi, phosphorylcholine expressing NTHi and the PAF-R. Using a gentamycin-survival invasion assay, the effect of inhibitors of PAF-R on the invasion process was examined. These studies showed that invasion of the airway cells was significantly reduced in the presence of the PAF-R inhibitors. The inhibition was dose-dependent and ranged from 10nM to 100nM concentrations of inhibitor. Two different chemical inhibitors were studied and both inhibited invasion equally well (Swords *et al.*, 2000).

Analysis of the phosphorylcholine residue on the surface of a group of NTHi strains by colony immunoblots showed significant variability between strains, as has been reported previously (Rosenow *et al.*, 1997). Further, the reactivity of particular strains with a T-15 idiotype anti-phosphorylcholine monoclonal antibody

TABLE 42.2 Relationship between bacterial glyco-structures and environmental selection

Structure	Positive selective pressure	Negative selective pressure
ChoP[a], non-typeable *H. influenzae* LOS	Adherence to PAF	Target for CRP
Sialic acid, *N. meningitidis* LOS	Serum resistance	Reduced *opc*-mediated adherence
Sialic acid, *N. gonorrhoeae* LOS	Serum resistance	Reduced adherence to ASGP-R

[a]Abbreviations: ASGP-R, asialoglycoprotein receptor; ChoP, phosphorylcholine; CRP, C-reactive protein; PAF, platelet-activating factor.

on immunoblots appeared to be directly related to the capacity of the bacteria to adhere and invade. That is, strains that reacted strongly with T-15 invaded significantly better than those strains that reacted weakly or not at all. In addition, analysis of mutants of NTHi, which either expressed no LOS chain extensions from the triheptose core (*pgmB*) or lacked the phosphorylcholine transferase (*licD*) showed that they invaded airway epithelial cells significantly less than wild-type NTHi strains.

The PAF-R is a key mediator of the CD14-independent arm of the inflammatory response and initiates a multifactorial host cell signal cascade in response to agonists such as PAF. The capacity of NTHi to initiate PAF-R-mediated signal events was examined using a well-described radioassay to track the release of inositol phosphate signal mediators into the cytoplasm. The infection of bronchial cells with NTHi 2019 caused increases in cytoplasmic inositol levels comparable to those observed following exposure to PAF. Further, significantly less inositol was observed in cells infected with NTHi 2019 *pgmB* null mutants or with NTHi 2019 *licD* mutants, suggesting that phosphorylcholine expression is required for inositol production. The increased inositol levels occurred in a PAF-R-dependent manner, as pretreatment with a PAF-R antagonist significantly reduced inositol production following NTHi infection. Significant reduction in NTHi invasion was also observed following inhibition of heterotrimeric G proteins with pertussis toxin, suggesting that PAF-R coupling to a G protein is involved.

C-Reactive protein (CRP) binds to ChoP of *S. pneumoniae*, *H. influenzae* and commensal *Neisseria* (Kim and Weiser, 1998; Weiser *et al.*, 1998). In the nasopharynx, CRP binding can have an effect on the binding to the PAF-R resulting in decreased attachment (Gould and Weiser, 2002). In the blood, CRP binding also acts as an opsonin activating complement by the classical pathway resulting in a bactericidal effect (Kim and Weiser, 1998; Weiser *et al.*, 1997). Therefore, the phase variable expression of ChoP seems to be a critical mechanism in respiratory bacterial colonization and pathogenesis.

5. CONCLUSIONS

The ability randomly to regulate the mimicry of human glysolipid antigens has served the pathogenic *Neisseria* and *Haemophilus* spp. quite well. It allows them to survive in different host environments dictated by the relative protection afforded by having or not having the mimicry (Tables 42.1 and 42.2). It is important in their pathogenesis by ensuring human epithelial cell entry utilizing receptor–ligand systems normally found in these cells. In these very strict human pathogens, this is an excellent example of co-evolution between extremely distinct species. Areas for futher investigation and open questions are detailed in the Research Focus Box.

RESEARCH FOCUS BOX

- Do exclusive human pathogens such as *H. influenzae* and *N. meningitidis* play a role in the development of autoimmune diseases in subgroups of the human population?
- Can structures which are mimicks of human antigens be modified to serve as possible vaccines for prevention of infection by *N. gonorrhoeae*, *N. meningitidis* and *H. influenzae*?
- Can approaches be developed to prevent these pathogens from incorporating structures onto their surfaces that mimic human antigens.

References

Apicella, M.A., Ketterer, M., Lee, F.K.N., Zhou, D., Rice, P.A., Blake, M.S., 1996. The pathogenesis of gonococcal urethritis in men: confocal and immunoelectron microscopic analysis of urethral exudates from men infected with *Neisseria*. J. Infect. Dis. 173, 636–646.

Apicella, M.A., 2003. *Neisseria meningitidis*. In: Mandell, G., Bennett, J.E., Dolin, R. (Eds.), Principles and Practices of Infectious Diseases, Vol. 2. Elsevier, Philadelphia, p. 3662.

Cundell, D.R., Gerard, C., Idanpään-Heikkilä, I., Tuomanen, E.I., Gerard, N.P., 1996. PAF receptor anchors *Streptococcus pneumoniae* to activated human endothelial cells. Adv. Exp. Med. Biol. 416, 89–94.

Dunn, K.L., Virji, M., Moxon, E.R., 1995. Investigations into the molecular basis of meningococcal toxicity for human endothelial and epithelial cells: the synergistic effect of LPS and pili. Microb. Pathog. 18, 81–96.

Gibson, B.W., Melaugh, W., Phillips, N.J., Apicella, M.A., Campagnari, A.A., Griffiss, J.M., 1993. Investigation of the structural heterogeneity of lipooligosaccharides from pathogenic *Haemophilus* and *Neisseria* species and of R-type lipopolysaccharides from *Salmonella typhimurium* by electrospray mass spectrometry. J. Bacteriol. 175, 2702–2712.

Gould, J.M., Weiser, J.N., 2002. The inhibitory effect of C-reactive protein on bacterial phosphorylcholine platelet-activating factor receptor-mediated adherence is blocked by surfactant. J. Infect. Dis. 186, 361–371.

Granfors, K., Toivanen, A., 1986. IgA-anti-*Yersinia* antibodies in yersinia triggered reactive arthritis. Ann. Rheum. Dis. 45, 561–565.

Griffiss, J.M., Jarvis, G.A., O'Brien, J.P., Eads, M.M., Schneider, H., 1991. Lysis of *Neisseria gonorrhoeae* initiated by binding of normal human IgM to a hexosamine-containing lipooligosaccharide epitope(s) is augmented by strain-specific, properdin-binding-dependent alternative complement pathway activation. J. Immunol. 147, 298–305.

Harvey, H.A., Ketterer, M.R., Preston, A., Lubaroff, D., Williams, R., Apicella, M.A., 1997. Ultrastructural analysis of primary human urethral epithelial cell cultures infected with *Neisseria gonorrheae*. Infect. Immun. 65, 2420–2427.

Harvey, H.A., Porat, N., Campbell, C.A., et al., 2000. Gonococcal lipooligosaccharide is a ligand for the asialoglycoprotein receptor on human sperm. Mol. Microbiol. 36, 1059–1070.

Harvey, H.A., Jennings, M.P., Campbell, C.A., Williams, R., Apicella, M.A., 2001. Receptor-mediated endocytosis of *Neisseria gonorrhoeae* into primary human urethral epithelial cells: the role of the asialoglycoprotein receptor. Mol. Microbiol. 42, 659–672.

Jennings, M.P., Srikhanta, Y.N., Moxon, E.R., et al., 1999. The genetic basis of the phase variation repertoire of lipopolysaccharide immunotypes in *Neisseria meningitidis*. Microbiology 145, 3013–3021.

Johnston, J.W., Coussens, N.P., Allen, S., et al., 2008. Characterization of the *N*-acetyl-5-neuraminic acid-binding site of the extracytoplasmic solute receptor (SiaP) of nontypeable *Haemophilus influenzae* strain 2019. J. Biol. Chem. 283, 855–865.

Jones, D.M., Borrow, R., Fox, A.J., Gray, S., Cartwright, K.A., Poolman, J.T., 1992. The lipooligosaccharide immunotype as a virulence determinant in *Neisseria meningitidis*. Microb. Pathog. 13, 219–224.

Ketterer, M.R., Shao, J.Q., Hornick, D.B., Buscher, B., Bandi, V.K., Apicella, M.A., 1999. Infection of primary human bronchial epithelial cells by *Haemophilus influenzae*: macropinocytosis as a mechanism of airway epithelial cell entry. Infect. Immun. 67, 4161–4170.

Kim, J.O., Weiser, J.N., 1998. Association of intrastrain phase variation in quantity of capsular polysaccharide and teichoic acid with the virulence of *Streptococcus pneumoniae*. J. Infect. Dis. 177, 368–377.

Kimura, A., Hansen, E.J., 1986. Antigenic and phenotypic variations of *Haemophilus influenzae* type b lipopolysaccharide and their relationship to virulence. Infect. Immun. 51, 69–79.

Mandrell, R.E., McLaughlin, R., Aba Kwaik, Y., et al., 1992. Lipooligosaccharides (LOS) of some *Haemophilus* species mimic human glycosphingolipids, and some LOS are sialylated. Infect. Immun. 60, 1322–1328.

Moe, G.R., Tan, S., Granoff, D.M., 1999. Molecular mimetics of polysaccharide epitopes as vaccine candidates for prevention of *Neisseria meningitidis* serogroup B disease. FEMS Immunol. Med. Microbiol. 26, 209–226.

Monroe, R.S., Huber, B.E., 1994. Characterization of the "hepatic" asialoglycoprotein receptor in rat late-stage spermatids and epididymal sperm. Gene 148, 261–268.

Moran, A.P., Prendergast, M.M., 2000. Lipopolysaccharides in the development of the Guillain-Barré syndrome and Miller Fisher syndrome forms of acute inflammatory peripheral neuropathies. J. Infect. Dis. 178, 1549–1551; J. Endotoxin Res. 6, 341–359.

Murphy, T.F., Apicella, M.A., 1987. Nontypable *Haemophilus influenzae*: a review of clinical aspects, surface antigens and the human response to infection. Rev. Infect. Dis. 9, 1–15.

Nairn, C.A., Cole, J.A., Patel, P.V., Parsons, N.J., Fox, J.E., Smith, H., 1988. Cytidine 5′-monophospho-*N*-acetylneuraminic acid or a related compound is the low Mr factor from human red blood cells which induces gonococcal resistance to killing by human serum. J. Gen. Microbiol. 134, 3295–3306.

Neisser, A., Schwerer, B., Bernheimer, H., Moran, A.P., 2000. Ganglioside-induced antiganglioside antibodies from a neuropathy patient cross-react with lipopolysaccharides of *Campylobacter jejuni* associated with Guillain-Barré syndrome. J. Neuroimmunol. 102, 85–88.

Porat, N., Apicella, M.A., Blake, M.S., 1995a. A lipooligosaccharide binding site on HepG2 cells similar to the gonococcal opacity-associated surface protein Opa. Infect. Immun. 63, 2164–2172.

Porat, N., Apicella, M.A., Blake, M.S., 1995b. *Neisseria gonorrhoeae* utilizes and enhances the biosynthesis of the asialoglycoprotein receptor expressed on the surface of the hepatic HepG2 cell line. Infect. Immun. 63, 1498–1506.

Preston, A., Mandrell, R.E., Gibson, B.W., Apicella, M.A., 1996. The lipooligosaccharides of pathogenic gram-negative bacteria. Crit. Rev. Microbiol. 22, 139–180.

Raetz, C.R., Whitfield, C., 2002. Lipopolysaccharide endotoxins. Annu. Rev. Biochem. 71, 635–700.

Ramsey, K.H., Schneider, H., Cross, A.S., et al., 1995. Inflammatory cytokines produced in response to experimental human gonorrhea. J. Infect. Dis. 172, 186–191.

Rosenow, C., Ryan, P., Weiser, J.N., et al., 1997. Contribution of novel choline-binding proteins to adherence, colonization and immunogenicity of *Streptococcus pneumoniae*. Mol. Microbiol. 25, 819–829.

Schneider, H., Cross, A.S., Kuschner, R.A., et al., 1995. Experimental human gonococcal urethritis: 250 *Neisseria gonorrhoeae* MS11mkC are infective. J. Infect. Dis. 172, 180–185.

Schneider, H., Schmidt, K.A., Skillman, D.R., et al., 1996. Sialylation lessens the infectivity of *Neisseria gonorrhoeae* MS11mkC. J. Infect. Dis. 173, 1422–1427.

Shaw, J.H., Falkow, S., 1988. Model for invasion of human tissue culture cells by *Neisseria gonorrhoeae*. Infect. Immun. 56, 1625–1632.

Swords, W.E., Buscher, B.A., Ver Steeg Ii, K., et al., 2000. Non-typeable *Haemophilus influenzae* adhere to and invade human bronchial epithelial cells via an interaction of lipooligosaccharide with the PAF receptor. Mol. Microbiol. 37, 13–27.

Swords, W.E., Jones, P.A., Apicella, M.A., 2003. The lipo-oligosaccharides of *Haemophilus influenzae*: an interesting array of characters. J. Endotoxin. Res. 9, 131–144.

van Bohemen, C.G., Grumet, F.C., Zanen, H.C., 1984. Identification of HLA-B27M1 and -M2 cross-reactive antigens in *Klebsiella*, *Shigella* and *Yersinia*. Immunology 52, 607–610.

Weiser, J.N., Shchepetov, M., Chong, S.T., 1997. Decoration of lipopolysaccharide with phosphorylcholine: a phase-variable characteristic of *Haemophilus influenzae*. Infect. Immun. 65, 943–950.

Weiser, J.N., Goldberg, J.B., Pan, N., Wilson, L., Virji, M., 1998. The phosphorylcholine epitope undergoes phase variation on a 43-kilodalton protein in *Pseudomonas aeruginosa* and on pili of *Neisseria meningitidis* and *Neisseria gonorrhoeae*. Infect. Immun. 66, 4263–4267.

Workowski, K.A., Berman, S.M., Douglas, J.M., 2008. Emerging antimicrobial resistance in *Neisseria gonorrhoeae*: urgent need to strengthen prevention strategies. Ann. Intern. Med. 148, 606–613.

Zabriskie, J.B., Hsu, K.C., Seegal, B.C., 1970. Heart-reactive antibody associated with rheumatic fever: characterization and diagnostic significance. Clin. Exp. Immunol. 7, 147–159.

CHAPTER

43

Molecular mimicry of host glycosylated structures by bacteria

Anthony P. Moran

SUMMARY

Molecular mimicry by a range of viral, parasitic and bacterial pathogens has been described, but has largely been associated with the development and pathogesnesis of a variety of autoimmune diseases. Although the focus in autoimmune disease pathogenesis has been on peptides/proteins, molecular mimicry of glycosylated structures and glycomolecules can occur in microorganisms, especially bacteria, and has consequences other than those resulting in autoimmunity. Two well-established examples, the molecular mimicry of gangliosides by *Campylobacter jejuni* and of Lewis antigens by *Helicobacter pylori*, are reviewed in this chapter. Discussion of ganglioside mimicry by *C. jejuni* is used to illustrate important aspects and concepts of bacterial glycomimicry in autoimmune disease development, whereas Lewis antigen mimicry is utilized in demonstrating that molecular mimicry can have multiple biological effects in microbial pathogenesis.

Keywords: Helicobacter pylori; Campylobacter jejuni; Molecular mimicry; Gangliosides; Lewis antigens; Baterial pathogenesis; Autoimmunity; Immune response; Guillain-Barré syndrome; Bacterial adhesion; Bacterial colonization

1. INTRODUCTION

Molecular mimicry can be defined as the structural similarity between foreign (microbial) and self-molecules of the mammalian host. Much interest has focused on the immunological context of this phenomenon where molecular mimicry between microbial and host antigens can result in the generation of cross-reacting, autoreactive B- and T-cells leading to the development of autoimmune dieases or so-called "friendly fire" (Isenberg and Morrow, 1995). Indeed, molecular mimicry by a range of viral, parasitic and bacterial pathogens has been associated with the development of a variety of autoimmune diseases, including rheumatic heart diease, multiple sclerosis, myasthenia gravis, systemic lupus erythematosus and ankylosing spondylitis (Abu-Shakra *et al.*, 1999; Hughes *et al.*, 2001; Ebringer *et al.*, 2005, 2007b; Olson *et al.*, 2005; Zandman-Goodard and Shoenfeld, 2005; Faé *et al.*, 2006; Westall, 2006; Hogan *et al.*, 2007), although the strength of evidence for the different associations varies. On the other hand, molecular mimicry has been implicated in other diseases, e.g. Crohn's disease (Ebringer *et al.*,

2007b), in which an aberrant immune response plays an important role. Moreover, there has even been debate as to whether molecular mimcry by certain bacterial species plays a role in transmissible spongiform encephalopathies (Ebringer *et al.*, 2005, 2007a). However, the molecular mimcry that has been discussed in these examples concerns sequence and structural similarities between microbial and self-peptides/proteins.

Nevertheless, molecular mimicry of glycosylated structures and glycomolecules can occur in microorganisms, especially bacteria (Moran and Prendergast, 2001). Two well-established examples, the molecular mimicry of gangliosides by *Campylobacter jejuni* and of Lewis (Le) antigens by *Helicobacter pylori*, are reviewed in this chapter. Importantly, it should be noted that the impact of molecular mimicry is not solely limited to influencing the immune response of the host, but that it also impacts on microbial pathogenesis in a variety of ways (see also Chapters 42 and 44). Thus, in this chapter, ganglioside mimicry by *C. jejuni* is used to illustrate aspects and concepts concerning bacterial glycomimicry in autoimmune disease development, and Le antigen mimicry is utilized in demonstrating that mimicry potentially can have multiple biological effects in microbial pathogenesis.

2. EXPRESSION OF GANGLIOSIDE MIMICRY BY C. JEJUNI

C. jejuni lipo-oligosaccharides (LOSs) are glycolipids comprised of core oligosaccharide (OS) and lipid A components, but devoid of O-antigen, thus resembling those of *Haemophilus* and *Neisseria* spp. (see Chapter 42). The core OSs of *C. jejuni* are structurally diverse and, in part with serodominant capsular polysaccharide, account for heat-stable antigen serospecificity (Moran and Penner, 1999; Moran *et al.*, 2001; Karlyshev *et al.*, 2008). This is similar to *Campylobacter coli* and *Campylobacter lari* where extracellular polysaccharides associated with, but independent of, lipopolysaccharide (LPS) have been described (Moran *et al.*, 2000a). The discovery of sialic acid (*N*-acetyl-neuraminic acid, Neu5Ac) in the LOSs of some *C. jejuni* strains (Moran *et al.*, 1991) induced intensive investigations of the LOS core OSs. As reviewed elsewhere (Moran *et al.*, 2000a, 2002b; Prendergast and Moran, 2000; Moran and Prendergast, 2001; Yuki, 2001; Gilbert *et al.*, 2008), these analytical studies have revealed that the core OSs of certain serotypes mimic the carbohydrate moieties of human gangliosides. Mimicry of a variety of gangliosides has been documented, including GM_1, GD_{1a}, GD_{1b}, GD_{1c}, G_{T1a} and GM_2, GD_2, GM_2, GD_3, examples of which are shown in Figure 43.1.

Importantly, *C. jejuni* LOS can undergo phase variation and this impacts on the expression of ganglioside mimicry by the bacterium, both *in vitro* and *in vivo* (Guerry *et al.*, 2002; Prendergast *et al.*, 2004). The genetic determination of *C. jejuni* LOS and ganglioside expression is well established and has been reviewed in detail (Yuki, 2007; Gilbert *et al.*, 2008), and shown to influence neuropathy development (see Yuki, 2005), but will not be examined here.

However, not all *C. jejuni* LOSs are sialylated and some do not exhibit mimicry of gangliosides (Moran *et al.*, 1991; Aspinall *et al.*, 1995; Poly *et al.*, 2008) and such strains have not been found in association with neuropathy, e.g. *C. jejuni* serotype HS:3 strains. Thus, although ganglioside mimicry is a common feature shared by a number of *C. jejuni* serotypes, this mimicry is not observed in all strains or serotypes. On the other hand, strains that have been isolated from patients who developed enteritis alone, can also exhibit ganglioside mimicry (Moran and O'Malley, 1995).

3. C. *JEJUNI* GANGLIOSIDE MIMICRY AND PATHOGENIC ANTI-GANGLIOSIDE ANTIBODIES

Previous to the observation of ganglioside mimcry in *C. jejuni* LOS, infection with this bacterium, a leading cause of acute bacterial

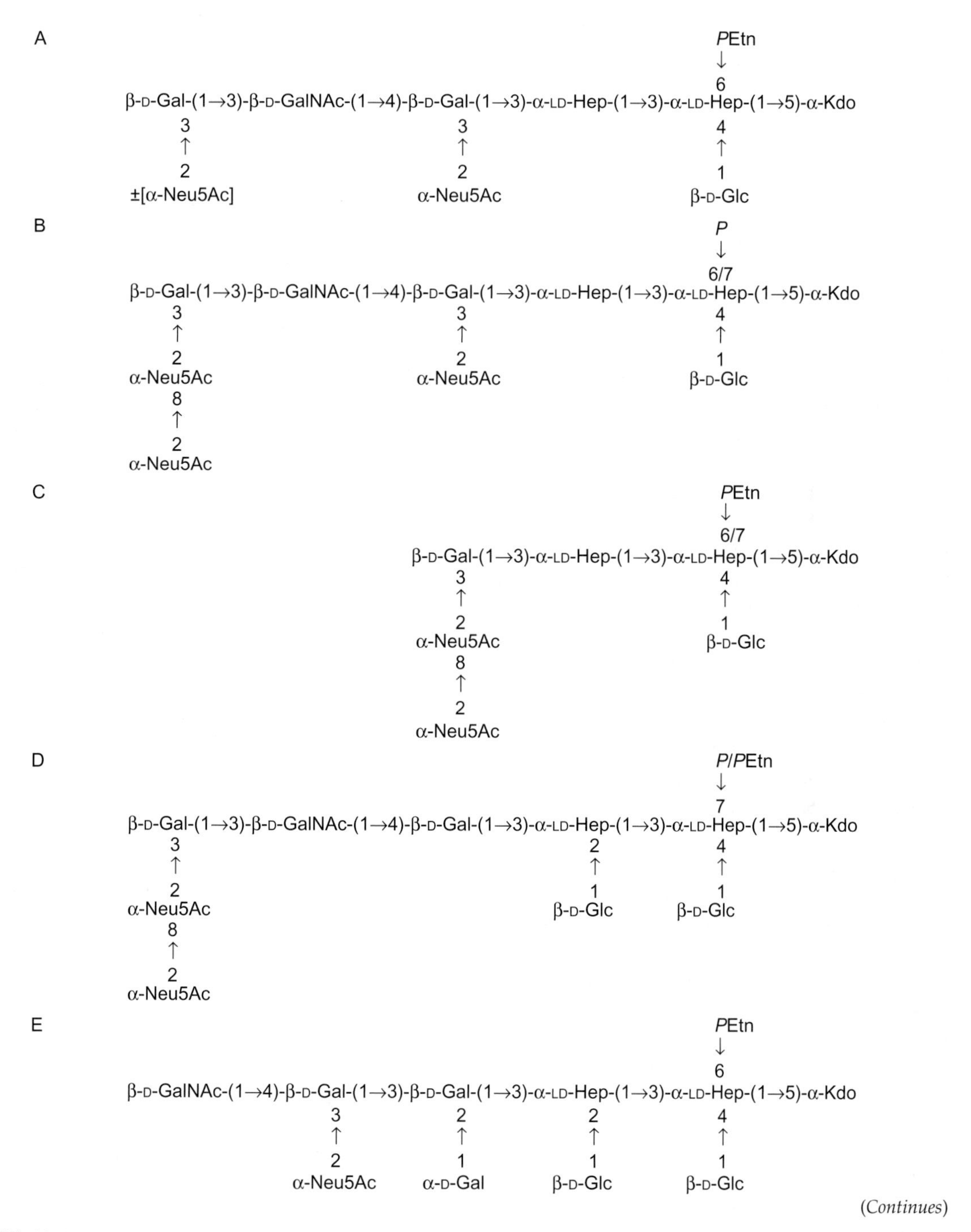

(Continues)

FIGURE 43.1 Structures of the core OSs of *C. jejuni* LOSs from different serostrains and isolates. (A) Serostrains HS:4 and HS:19, GM_1 and GD_{1a} mimics; (B) isolate OH4384, GT_{1a} mimic; (C) isolate 4382, a GD_3 mimic; (D) serostrain HS:10 and isolate PG836, GD_3 and GD_{1c} mimics; (E) serostrain HS:1, a GM_2 mimic.

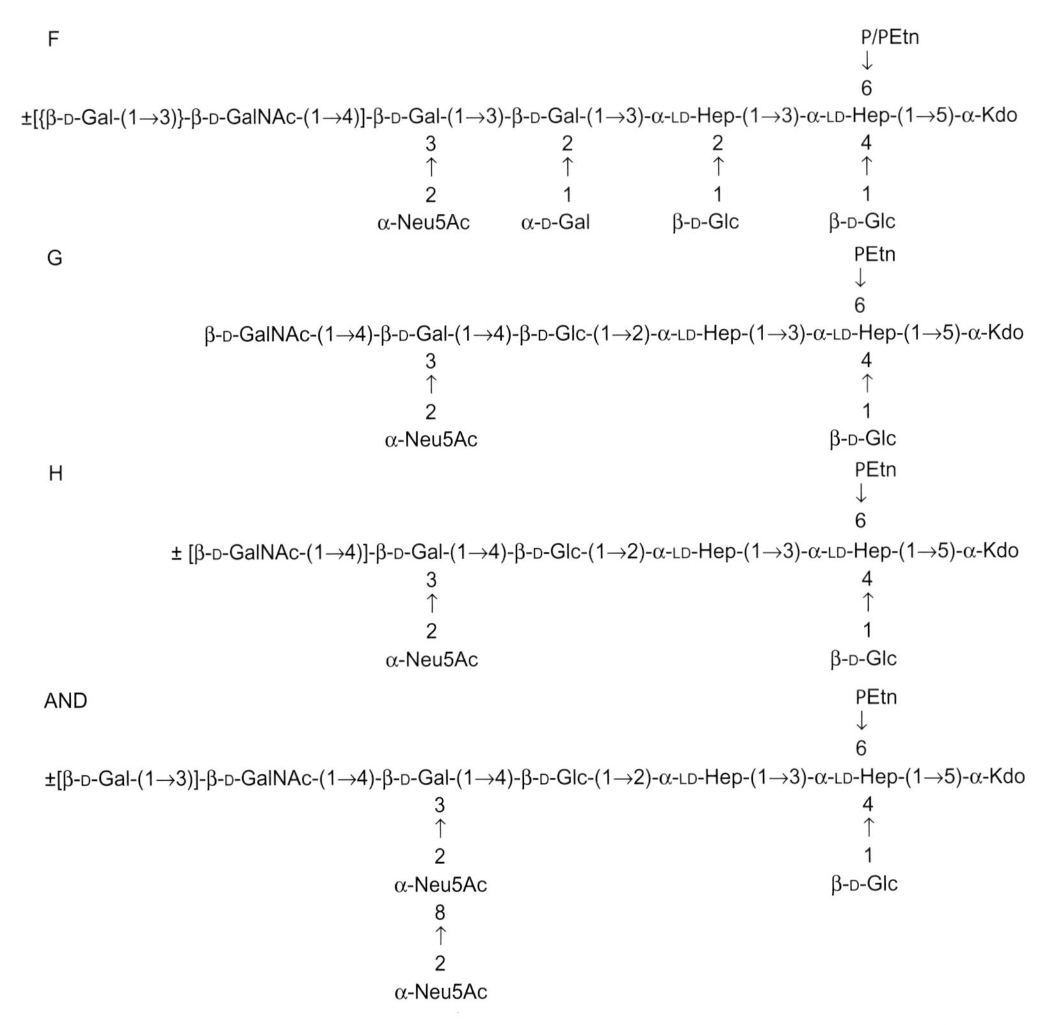

FIGURE 43.1 (F) serostrain HS:2 without the terminal disaccharide β-D-Gal-(1→3)-β-D-GalNAc, a GM_3 mimic, and NCTC 11637 (also serotype HS:2), GM_1 and GM_2 mimics; (G) serostrains HS:23 and HS:36, a GM_2 mimic; and (H) phase-variable strain 81–176, GM_2 and GM_3 mimics predominantly, but also GD_{1b} and GD_2 mimics. All sugars are in the pyranosidic form. Abbreviations: Gal, galactose; GalNAc, *N*-acetylgalactosamine; Glc, glucose; LD-Hep, L-*glycero*-D-*manno*-heptose; Kdo, 3-deoxy-D-*manno*-oct-2-ulosonic acid; Neu5Ac, *N*-acetylneuraminic acid; *P*, phosphate; *P*Etn, phosphoethanolamine. For the literature from which these structures are derived, see the reviews of Moran *et al.* (2000a), Prendergast and Moran (2000) and Gilbert *et al.* (2008).

gastroenteritis in humans, had been associated with the development of a neurological disorder of the peripheral nervous system, Guillain-Barré syndrome (GBS) (reviewed in Prendergast and Moran, 2000; Yuki, 2007). This was noteworthy since gangliosides are a family of Neu5Ac-containing glycosphingolipids present in the outer leaflet of the plasma membrane and are the major surface molecules in both the peripheral and central nervous systems.

Moreover, as reviewed elsewhere (Prendergast and Moran, 2000; Moran and Prendergast, 2001; Yuki, 2007), reports over many decades have observed anti-ganglioside antibodies with a variety of specificities in GBS patients and which are considered to play a role in the pathogenesis of this syndrome. This view is supported by the observation that plasma exchange facilitates the rate of recovery from GBS (Guillain-Barré syndrome Study Group, 1985; Prendergast and Moran, 2000).

Clinically, GBS is considered an acute, progressive and symmetrical motor weakness of the extremities with loss of tendon reflexes, which has replaced polio as the commonest cause of neuromuscular paralysis in most developed countries. At least four distinct subtypes of the disorder are recognized (Hughes and Cornblath, 2005). Of these, a severe, purely motor form known as acute motor axonal neuropathy (AMAN) and Miller Fisher syndrome (MFS), manifesting ophthalmoplegia, ataxia and areflexia, have received most attention in *C. jejuni*-associated disease. The wide range of anti-ganglioside antibody specificities reported in GBS contrasts with their limited range in chronic neuropathies, e.g. anti-GM_1 antibodies in multifocal motor neuropathy and a neuropathy induced by parenteral GM_1 therapy (Moran *et al.*, 2002b). In previous literature surveys (Prendergast and Moran, 2000; Moran and Prendergast, 2001; Moran *et al.*, 2002b; Yuki, 2007), reports included multiple findings of anti-ganglioside antibodies to GM_1, GM_{1b}, GQ_{1b}, LM_1, GalNAc-GD_{1a} and less frequent finding of antibodies to GM_2, GD_{1a}, GD_{1b}, GT_{1b} and asialo-GM_1. Furthermore, a correlation between antecedent infection with *C. jejuni* and the presence of anti-GM_1 antibodies has been demonstrated (see Moran and Prendergast, 2001 and Yuki, 2007). In general, half of GBS patients who are seropositive for *C. jejuni* also have anti-GM_1 antibodies. Also, AMAN is more strongly associated with *C. jejuni* infection than other forms of GBS and has been associated strongly with anti-GM_1, anti-GD_{1a} and anti-GalNAc-GD_{1a} antibodies (Prendergast and Moran, 2000; Yuki, 2007). Anti-GQ_{1b} antibodies are found in a high proportion of MFS patients, in up to 100% of cases (Willison and O'Hanlon, 2000) and, importantly for pathogenesis, GQ1b is enriched in the cranial nerves that innervate the extraocular muscles (Yuki, 2001). In contrast to the immunoglobulin (Ig) M isotype of anti-GM_1 antibodies found in chronic neuropathies, GBS sera contain IgG or less commonly IgA isotypes (Moran and Prendergast, 2001). The IgG subclasses are restricted to IgG1 and IgG3, subclasses which have been taken to be indicative of T-cell-dependent responses to proteins (Willison and Veitch, 1994) but which, nevertheless, can be induced by *C. jejuni* LOSs (Moran *et al.*, 2001; Matsumoto *et al.*, 2008). Also, anti-ganglioside antibodies are often present in disease-free individuals (Latov, 1990), but are not considered to be autopathogenic, since they differ from anti-ganglioside antibodies in GBS in their IgG subtype and they are not derived from antigen-driven mechanisms (Paterson *et al.*, 1995).

The observation that GBS-associated *C. jejuni* isolates exhibit ganglioside mimicry prompted a number of studies that showed the reactivity of serum antibodies in GBS patients against not only *C. jejuni* LOS, but particularly against the core OS mimicking gangliosides (Yuki *et al.*, 1993; Schwerer *et al.*, 1995; Neisser *et al.*, 1997, 2000; Prendergast *et al.*, 1999). Moreover, cross-reactive GM_1 and GD_{1a} antibodies were observed and the sialosyl-D-galactose (Gal) structure α-Neu5Ac-(2→3)-β-Gal-(1→ proposed as the minimal motif associated with GBS development (Moran *et al.*, 2002b). On the other hand, MFS is of great interest because of its specific correlation with anti-GQ_{1b} antibodies, although to date none of the chemically analysed LOSs from neuropathy-associated strains exhibits a complete GQ_{1b} ganglioside mimic. Of note, strains that have been isolated from MFS patients exhibit a GD_2/GD_3 mimic or GT_{1a}

and GD_{1c}-like mimicry (Moran *et al.*, 2000a; Prendergast and Moran, 2000; Yuki, 2005) and whose LOSs show cross-reaction with MFS sera. Since, to date, none of the chemically analysed LOSs from neuropathy-associated strains exhibit a complete GQ_{1b} ganglioside mimic, it has been suggested that the minimal required motif for MFS development is a disialosyl-Gal structure, α-Neu5Ac-(2→8)-α-Neu5Ac-(2→3)-β-Gal-(1→ (Moran *et al.*, 2002b; Yuki, 2005).

Overall, analyses of *C. jejuni* LPSs with anti-ganglioside antibodies has verified the presence of ganglioside-related epitopes in *C. jejuni* LPS (Schwerer *et al.*, 1995; Neisser *et al.*, 1997, 2000) and, thus, it has been inferred that infection with ganglioside-bearing *C. jejuni* strains may elicit high titres of anti-ganglioside autoantibodies in GBS patients. Although it is possible that anti-ganglioside antibodies may be formed secondarily by autoimmunization against gangliosides released from previously damaged nerves, such a process for the generation of anti-ganglioside antibodies is unlikely in GBS since the antibodies are present on the first day of neurological symptoms and high titres of these antibodies are not found in patients with peripheral nerve damage due to diseases other than GBS (Prendergast and Moran, 2000). Thus, the molecular mimicry of gangliosides by *C. jejuni* LOSs has been proposed as a pathogenic mechanism in GBS whereby the immune-mediated response, particularly the humoral response, elicits the production of cross-reactive anti-ganglioside antibodies that target and damage the neural gangliosides.

4. RELEVANCE OF MOLECULAR MICMICRY IN THE PATHOGENESIS OF GBS

It has been stated that many studies are consistent with the molecular mimicry hypothesis in autoimmune disease development but that none has convincingly demonstrated that mimicry is an important mechanism in disease in humans (Marrack *et al.*, 2001). However, others have contended that evidence concerning *C. jejuni* ganglioside mimicry fulfils such requirements (Ang *et al.*, 2004; Yuki, 2005). Nevertheless, in the case of infection-induced autoimmunity, the precepts that would need to be fulfilled include those of Koch's postulates for infectious disease and those of Witebsky for autoimmunity and, thus, a combination of these has been proposed in the so-called Galway postulates (Moran *et al.*, 2002b). Hence, criteria for determination of the role of molecular mimicry in autoimmune disease are:

(i) Evidence for antedent infection in the host is to be established.
(ii) Cross-recativity of the pothogen with a host target molecule has to be demonstrated.
(iii) Experimental disease reproduction in an animal by administration of the tissue target agent, a molecular mimic, or anti-target antibodies is required.
(iv) Disease reproduction in an experimental host by the pathogen must be demonstrated.

Although the first three precepts have been fulfilled, the last has proven more problematic for *C. jejuni*-related GBS. This is because an animal model allowing colonization by the infectious agent, plus the generation of a humoral response, like that seen in humans with GBS, and inducing a relevant pathology in animals, is required.

Experimental inoculation of humans with *C. jejuni* in a vaccine trial has been shown to induce anti-ganglioside antibodies, but no neuropathy developed, potentially due to the absence of a persistent antibody response, thereby reflecting the role of host factors in GBS development (Prendergast *et al.*, 2004). Nonetheless, experimental ataxic neuropathy (EAN) is considered the animal model for GBS because it shares a striking similarity to the classical form of GBS, acute inflammatory demyelinating polyradiculoneuropathy (AIDP), in clinical symptoms and pathological findings (Notterpek and Tolwani, 1999). However, it does not model the AMAN form of disease, nor is this T-cell-mediated model an animal model for *Campylobacter*-induced GBS. Despite this, GD_{1b} immunization has produced sensory ataxic neuropathy in rabbits (Kusunoki *et al.*, 1996) and the passive transfer of anti-GD_{1b} antibodies gives rise to axonal degeneration and macrophage infiltration. Intraneural injection of GBS serum or Ig into rat sciatic nerves gives variable results but, in most reports, significant conduction block and demyelination at levels higher than control human serum has been observed (Sumner *et al.*, 1982; Saida *et al.*, 1982; Feasby *et al.*, 1982; Harrison *et al.*, 1984). Importantly, Illa *et al.* (1995) demonstrated that purified anti-GM_1 antibodies from patients, who exhibited AMAN after immunization with a ganglioside preparation, recognized epitopes at the nodes of Ranvier and at the presynaptic nerve terminals of motor end plates from human nerve biopsies. Accumulation of these antibodies at the nodes of Ranvier can cause disruption of Na^+ and K^+ channels and, thus, interfere with nerve conduction. With regard to MFS-associated anti-GQ_{1b} antisera, there is strong supportive evidence that these antibodies can mediate pathological changes. In the mouse hemi-nerve diaphragm model, administration of anti-GQ_{1b}-positive sera decreased nerve excitability and neurotransmitter release and, subsequently, induced nerve terminal paralysis (Willison and O'Hanlon, 2000), although not in human limb muscle (Kuwabara *et al.*, 2007).

Active immunization of rabbits with GM_1 ganglioside-mimicking *C. jejuni* LOSs induced the production of high titres of IgG anti-LOS antibodies that were cross-reactive with GM_1 ganglioside (Ang *et al.*, 2000). None of the animals developed overt signs of limb weakness, but this may have been due to the short duration of the experiment. Other groups have induced anti-ganglioside antibodies with *C. jejuni* LOSs in animals (Ritter *et al.*, 1996; Wirguin *et al.*, 1997; Goodyear *et al.*, 1999), and rabbit antisera with anti-GM_1 specificity induced by a *C. jejuni* GM_1 mimic have been shown to bind at the nodes of Ranvier of human sciatic nerve *in vitro* (Moran *et al.*, 2005a). Moreover, this binding pattern was identical to that seen with anti-GM1 sera from a GBS patient. Immunization of mice with GT_{1a}-containing LOS can produce a serum with anti-GQ_{1b} antibodies which bind to ganglioside-rich sites causing release of acetylcholine and complement-mediated conduction block (Goodyear *et al.*, 1999).

An animal model in Japanese white rabbits has been established by sensitization with a bovine brain ganglioside mixture or isolated GM_1, in which rabbits developed high titre anti-GM_1 IgG antibodies and flaccid limb weakness (Yuki *et al.*, 2001). The disease characteristics in this animal model correspond to the pathological findings for human AMAN (Hughes and Cornblath, 2005). Building upon this model, a putative AMAN model has been reported after immunization of Japanese white rabbits with *C. jejuni* LOS bearing a GM_1-like structure (Yuki *et al.*, 2004). After sensitization with this GM_1-like LOS, rabbits developed high titre anti-GM_1 IgG antibodies and subsequent flaccid limb weakness. Additionally, axons of these nerve fibres showed various degrees of degeneration, but demyelination and remyelination were rare. The model has been proposed as a replica of *C. jeuni* LOS-induced GBS and even claimed as the first definitive replica of a human autoimmune disease produced by immunization with an infectious mimic (Yuki, 2005). Nevertheless, since this model requires multiple immunizations with LOS and haemocyanin as adjuvant (Yuki *et al.*, 2004), it cannot be excluded that because of the aggressive nature of haemocyanin, auto-presentation of host gangliosides may occur in this model. Also, this is not an infective animal model of *C. jejuni*-induced GBS and thus does not fulfil the Galway postulates

(Moran *et al.*, 2002b). On the other hand, a chicken model developed by Li *et al.* (1996), in which outbred chickens were fed an AMAN-related *C. jejuni* isolate that exhibits GM_1/GD_{1a} mimicry, showed a neuropathy in 30% of animals and pathology similar to human AMAN. Despite this potential advance, but disappointingly, extensive experiments by others to reproduce this model have not proven successful (see Prendergast and Moran, 2000).

In contrast to the IgM response generally seen against glycolipid antigens, long-lasting titres of IgG and IgA antibodies are present in GBS (Moran and Prendergast, 2001) which suggests that T-cell-dependent mechanisms are involved and that the mechanisms of glycolipid recognition in GBS do not follow the normal rules of T-independent carbohydrate antigen recognition. The T-cells bearing $\alpha\beta$ receptors are found in GBS (Cornblath *et al.*, 1990), but cells associated with recognition of glycolipid targets, $\gamma\delta$ T-cells, are also found (Khalili-Shirazi *et al.*, 1999). Also, the expression of major histocompatibility complex class II (MHC II) molecules is increased on circulating T-cells from GBS patients (Taylor and Hughes, 1989). Nonetheless, MHC molecules do not bind to glycolipids, suggesting that these compounds cannot directly activate T-cells in an MHC-restricted manner (Moran and Prendergast, 1998). Alternatively, as CD1 molecules are present on endoneurial macrophages and are upregulated in GBS, presentation to T-cells could occur in association with CDl molecules, as has been demonstrated for lipoglycans (Sieling *et al.*, 1995; Moran and Prendergast, 1998; Prendergast and Moran, 2000). Thus, $\gamma\delta$ T-cells and CDl molecules may play a role in the prominent anti-glycolipid antibody response in GBS patients. Interestingly, *C. jejuni* LOSs have been shown to increase endogenous glycosphingolipid synthesis and stimulate autoreactive, ganglioside-specific T-cells in a CD1-dependent manner (De Libero *et al.*, 2005). Despite these observations, the extent to which T-cells contribute to the pathophysiology of GBS and the mechanism by which they do so is unexplained. They could be involved in helping B-cells to secrete autoantibodies, but this response is CD1-independent (Matsumoto *et al.*, 2008), or the T-cells may produce harmful cytokines which injure the myelin sheath directly.

A simplistic pathogenesis model for *C. jejuni*-induced GBS proposes that strains bearing ganglioside-mimicking epitopes produce an aberrant immune response whereby the antibodies directed at *C. jejuni* LPS structures cross-react with ganglioside-rich neural tissues on the axolemma at the nodes of Ranvier or the terminal motor axons (Prendergast and Moran, 2000). The T-cells recruited by antibody-producing B-cells cross the blood–nerve barrier and the activated T-cells produce cytokines, e.g. tumour necrosis factor-alpha (TNF-α), which open the blood–nerve barrier allowing antibody access. Subsequently, opsonization of Schwann cells by anti-ganglioside antibodies occurs and accumulations of these antibodies at the nodes of Ranvier can cause disruption of Na^+ and K^+ channels and, thus, interfere with the propagation of nerve impulses. The final event in demyelination is invasion of the myelin sheath by macrophages, whereby macrophage attachment occurs via F_c receptors on opsonizing antibodies. Segmental demyelination and conduction block ensues when macrophage processes ramify between the lamellae of compact myelin, stripping the Schwann cells from the intact myelin sheath. Many of the detailed mechanisms of this model remain unresolved, however, in particular, the role played by host factors.

5. EXPRESSION OF LE ANTIGENS IN *H. PYLORI* LPS

H. pylori is a highly prevalent, bacterial gastroduodenal pathogen of humans (Kusters *et al.*, 2006) that can express fucosylated Le and related antigens in the O-antigens of its surface

LPS (Moran, 2001b, 2008). Infection once established can persist for life if left untreated (Kusters *et al.*, 2006) and is associated with an active inflammation in the gastric mucosa, termed gastritis. Although infection outcome is diverse, including the development and recurrence of gastritis, and gastric and duodenal ulcers, some individuals with long-term infection develop atrophic gastritis, which can have an autoimmune contributory background and is a precursor state in gastric cancer development (Ernst and Gold, 2000). Abetting the chronicity of this long-term infection, extensive studies of *H. pylori* LPS have revealed significantly lower endotoxic and immunological activities compared with those of enterobacterial LPS (Moran, 1999), attributable to the structural characteristics of the lipid A component (Moran *et al.*, 1997; Moran, 2007). Nevertheless, *H. pylori* has been implicated in the induction of autoreactive immune responses, including responses against LPS-expressed Le antigens (see below), that may contribute to atrophic gastritis development.

The Le antigens are biochemically related to the ABH blood group antigens of humans that are formed by the sequential addition of L-fucose (Fuc), Gal and *N*-acetyl-D-galactosamine (GalNAc) to the backbone carbohydrate chains of both lipids and proteins. The structural basis of type 1 and type 2 backbone chains, as well as Le and related antigens (e.g. H-1 and H-2), have been detailed elsewhere (Moran, 2001b, 2008). The mammalian genes controlling the expression and secretion of these molecules have been described (see Green, 1989). Of note, the terms secretor and non-secretor indicate the capacity of an individual to secrete such substances; secretors produce ABH, Le^b and Le^y, whereas non-secretors produce Le^a and Le^x in their saliva (Sakamoto *et al.*, 1984).

Expression of Le^x or Le^y antigens is a common property of *H. pylori* strains, as 80–90% of isolates from various geographical regions worldwide, that have been screened using anti-Le antibodies as probes, express these antigens (Moran, 2008). On the other hand, serological analysis can underestimate Le^x and Le^y expression in a population of strains (Hynes and Moran, 2000), as some strains that are non-typeable with anti-Le antibodies have been shown to express these antigens when examined in structural studies (Knirel *et al.*, 1999). The genetic determination of the Le antigens and their biosynthetic pathways in *H. pylori* have have been extensively reviewed elsewhere (Moran, 2001b, 2008; Ma *et al.*, 2006; Moran and Trent, 2008). Collectively, factors affecting Le antigen expression in *H. pylori* that can influence the biological impact of this molecular mimicry include regulation of fucosyltransferase (FucT) genes through slipped-strand mispairing, the activity and expression levels of the functional enzymes, the preferences of the expressed enzyme for distinctive acceptor molecules and the availability of activated sugar intermediates. With the accessibility of the crystal structure of *H. pylori* FucTs, further insights should be gained into the molecular recognition and functioning of these enzymes (Sun *et al.*, 2007).

Structurally, the O-chains of *H. pylori* clinical isolates have a poly-*N*-acetyllactosamine (LacNAc) chain decorated with multiple lateral α-L-Fuc residues forming internal Le^x determinants with terminal Le^x or Le^y determinants (Figure 43.2) or, in some strains, with additional glucose (Glc) or Gal residues (Moran 2001a,b, 2008; Monteiro, 2001). Moreover, Le^a, Le^b, sialyl-Le^x and H-1 antigens have been structurally described in other strains, as well as the related blood groups A and B, but these occur in association with Le^x and LacNAc chains. Overall, a mosaicism of Le antigen and blood group expression can occur in the same O-antigen and, thereby, along with some variability in the core oligosaccharide of LPS, gives rise to antigenic diversity that can be detected by antibody and lectin probing (Moran, 2008). On the other hand, there are *H. pylori* strains, isolated from asymptomatic individuals, that do

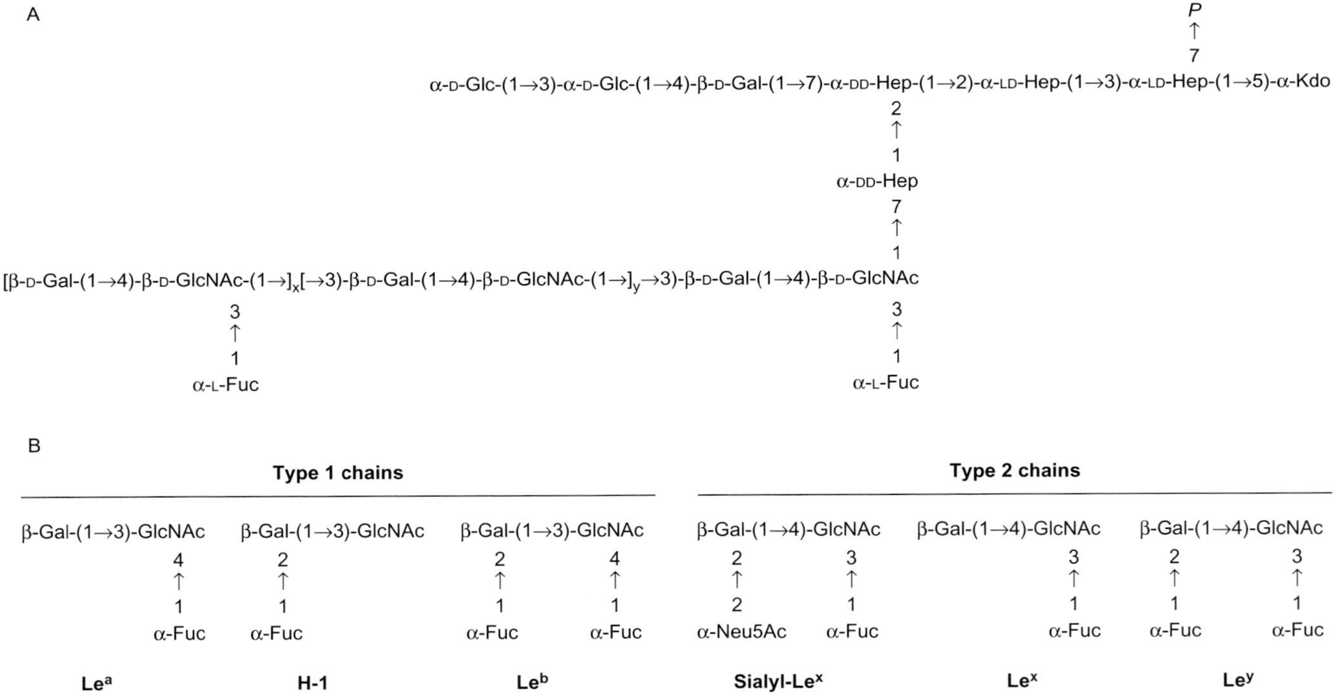

A

```
                                                                                    P
                                                                                    ↑
                                                                                    7
α-D-Glc-(1→3)-α-D-Glc-(1→4)-β-D-Gal-(1→7)-α-DD-Hep-(1→2)-α-LD-Hep-(1→3)-α-LD-Hep-(1→5)-α-Kdo
                                                  2
                                                  ↑
                                                  1
                                              α-DD-Hep
                                                  7
                                                  ↑
                                                  1
[β-D-Gal-(1→4)-β-D-GlcNAc-(1→]x[→3)-β-D-Gal-(1→4)-β-D-GlcNAc-(1→]y→3)-β-D-Gal-(1→4)-β-D-GlcNAc
                  3                                                               3
                  ↑                                                               ↑
                  1                                                               1
               α-L-Fuc                                                         α-L-Fuc
```

B

Type 1 chains

```
β-Gal-(1→3)-GlcNAc    β-Gal-(1→3)-GlcNAc    β-Gal-(1→3)-GlcNAc
              4        2                     2           4
              ↑        ↑                     ↑           ↑
              1        1                     1           1
           α-Fuc    α-Fuc                 α-Fuc       α-Fuc

      Le^a                 H-1                  Le^b
```

Type 2 chains

```
β-Gal-(1→4)-GlcNAc    β-Gal-(1→4)-GlcNAc    β-Gal-(1→4)-GlcNAc
   2          3                    3         2           3
   ↑          ↑                    ↑         ↑           ↑
   2          1                    1         1           1
α-Neu5Ac   α-Fuc                α-Fuc     α-Fuc       α-Fuc

   Sialyl-Le^x             Le^x                 Le^y
```

FIGURE 43.2 Structure of (A) the saccharide component of LPS of the *H. pylori* type strain NCTC 11637 (Aspinall *et al.*, 1996) and (B) structures of the Lewis antigens expressed in *H. pylori* strains (Moran, 2008). Abbreviations: Fuc, fucose; Gal, galactose; Glc, glucose; GlcNAc, *N*-acetylglucosamine; DD-Hep, D-*glycero*-D-*manno*-heptose; LD-Hep, L-*glycero*-D-*manno*-heptose; Kdo, 3-deoxy-D-*manno*-oct-2-ulosonic acid; *P*, phosphate.

A

→2)-α-D-Man3CMe-(1→3)-α-L-Rha-(1→3)-α-D-Rha-α-(1→

B

→2)-α-DD-Hep-(1→3)-α-DD-Hep-(1→3)-α-DD-Hep-(1→

C

→3)-α-DD-Hep-(1,

→2)-α-DD-Hep-(1→

→3)-α-DD-Hep-(1→2)-α-DD-Hep-(1→3)-α-DD-Hep-(1→

[→3)-α-DD-Hep-(1→]$_n$ 2)-α-DD-Hep-(1→ [2)-α-DD-Hep-(1→]$_n$

D

→2)-α-D-Glc-(1→3)-α-D-Glc-(1→

FIGURE 43.3 Structures of the non-Lewis mimicking repeating units of O-chains of *H. pylori* strains (A) D1, D3 and D6 (Kocharova *et al.*, 2000), (B) D2, D4 and D5 (Senchenkova *et al.*, 2001), (C) Hp1C2, Hp12C2, Hp62C, Hp7A, Hp77C and HpPJ1 (Monteiro *et al.*, 2001), (D) serotype O2 (Britton *et al.*, 2005). Abbreviations: Glc, glucose; DD-Hep, D-*glycero*-D-*manno*-heptose; Man3CMe, 3-C-methylmannose; Rha, rhamnose.

not express Le antigens (Kocharova *et al.*, 2000; Senchenkova *et al.*, 2001; Monteiro *et al.*, 2001; Britton *et al.*, 2005) (Figure 43.3). The absence of Le antigen mimicry in these strains and animal studies, in which a genetically modified *H. pylori* strain lacking Le antigen expression failed to induce gastritis compared to the parental strain (Eaton *et al.*, 2004), support the view that Le antigen-expressing LPS contributes directly to disease development.

6. ROLES OF LPS-EXPRESSED LE ANTIGENS IN *H. PYLORI* PATHOGENESIS

6.1. Gastric adaptation

In the human stomach, Le^a and Le^b are mainly expressed on the surface and foveolar epithelia, whereas Le^x and Le^y are predominantly expressed by the mucous, chief and parietal cells of the gastric glands (Kobayashi *et al.*, 1993), though the physical separation of Le antigens along the gastric pits is not always obvious (Murata *et al.*, 1992). More specifically, in non-secretors, the surface and foveolar epithelia express Le^a, whereas these epithelia express Le^b and some Le^y in secretors but, irrespective of secretor status of the individual, the glandular epithelium lacks type 1 antigens (Le^a and Le^b) and expresses type 2 antigens (Le^x and Le^y), although in non-secretors Le^y expression is weaker in the glands (Sakamoto *et al.*, 1989). Thus, a variety of Le antigens are expressed within the ecological niche of *H. pylori*. On the basis of this known Le antigen expression in the gastric mucosa, it has been proposed that bacterial molecular mimicry and its adaptation could provide an escape for *H. pylori* from the humoral response by preventing the formation of antibacterial antibodies because of cross-reaction with the host (Wirth *et al.*, 1997; Appelmelk *et al.*, 2000).

In particular, an initial study that examined the relationship between infecting *H. pylori* isolates and gastric Le expression in the human host, as determined by erythrocyte Le(a,b) phenotype (i.e. secretor status), reported that the relative expression of Le^x or Le^y by *H. pylori* strains corresponded to Le(a+ ,b−) and Le(a− ,b+) blood group phenotypes, respectively, of the hosts from whom the individual strains were isolated (Wirth *et al.*, 1997). Thus, it was inferred that because Le^a and Le^b are isoforms of Le^x and Le^y, and surface and foveolar epithelia express Le^a in non-secretors and Le^b in secretors, that the correlations observed were a form of adaptation or selection of *H. pylori* strains to the gastric mucosa of the individual host (Wirth *et al.*, 1997). A second study, performed in an identical manner, but in a different patient population, did not find a correlation between Le^x/Le^y on colonizing *H. pylori* isolates and host Le expression (Heneghan *et al.*, 2000). Likewise, another study, that used the different approach of directly examining Le expression on gastric epithelial cells of infected humans, did not find a correlation (Taylor *et al.*, 1998). Nevertheless, the discrepancies between these patient studies may reflect characteristics of the human populations under investigation and the study procedures employed (Moran, 1999; Broutet *et al.*, 2002).

Of considerable importance is that in the human study population where the correlation was observed, 26% of individuals were determined to be of the recessive Le phenotype, Le(a− ,b−), based on erythrocyte testing (Wirth *et al.*, 1997). This is higher than that usually observed for a Caucasian or European population from which the two other study populations were drawn (Taylor *et al.*, 1998; Heneghan *et al.*, 2000). Importantly, using salivary testing rather than erythrocyte typing would have allowed the distribution of this phenotypically recessive group of patients to their true secretor/non-secretor phenotype and, based on theoretical calculations, would have influenced the correlative results (Heneghan *et al.*, 1998a). Also, in the two other studies, no patients of a recessive phenotype were observed (Taylor *et al.*, 1998; Heneghan *et al.*, 2000). Finally, despite high anti-LPS antibody titres (Appelmelk *et al.*, 1996; Heneghan *et al.*, 2001), *H. pylori* eradication and protection is mediated by cellular not humoral immunity (O'Keeffe and Moran, 2008) and this challenges whether anti-Le and, hence, anti-LPS, antibodies can act as a selective pressure in humans (Appelmelk *et al.*, 2000).

On the other hand, in an animal model, using experimental infection of rhesus monkeys, an *H. pylori* strain isolated from infected monkeys that are of the secretor phenotype expressed more Le^y than Le^x; conversely, the same infective strain expressed more Le^x than Le^y when isolated following colonization of non-secretor monkeys (Moran *et al.*, 2002c; Wirth *et al.*, 2006). Therefore, in this experimental system, Le^x/Le^y expression by *H. pylori* depended on the host secretor status. This at first would appear to support the view that Le antigen-based mimicry could provide an escape for *H. pylori* from the host humoral response by a camouflage-type mechanism but, unlike in humans (Appelmelk *et al.*, 1996; Heneghan *et al.*, 2001), no data are available as to whether the *H. pylori*-infected monkeys formed antibodies against Le^x and Le^y that would have acted as a selective pressure (Appelmelk *et al.*, 2000).

Of note, it has been demonstrated that *H. pylori* strains expressing Le^x and those expressing Le^y can be isolated from the same host (Wirth *et al.*, 1999) and that extensive diversity in expression of Le^x and Le^y in *H. pylori* O-chains can occur over time and in different regions of the human stomach (Nilsson *et al.*, 2006). Moreover, Nilsson *et al.* (2008) detected genetic modifications in *H. pylori* FucT genes that could be attributable to recombination events within and between these genes that creates diversity and which, together with phase variation, contributes to divergent LPS expression within an *H. pylori* colonizing community. This would apparently contradict the hypothesis of bacterial Le antigen adaptation to that of gastric Le expression as determined simply by the secretor status of the host (Wirth *et al.*, 1997).

Notwithstanding this, the diversity of Le expression by *H. pylori* subclonal isolates in one host (Nilsson *et al.*, 2006) may reflect an ability, and potential, of the bacterium to adapt to differing microniches and environmental conditions within the human stomach (Moran *et al.*, 2002a; Keenan *et al.*, 2008). For example, pH may vary in the stomach and has been shown to influence relative Le^x/Le^y expression by *H. pylori* (Moran *et al.*, 2002a). The pH may vary between the differing regions of the stomach (i.e. the antrum versus the corpus where the stomach's acid-secreting cells are located), as well as varying across the microbial niches of the mucosa (i.e. pH 1–3 on the luminal surface of gastric mucus, about pH 4–5 within the mucus and pH 7 on the epithelial cell surface). Likewise, dietary iron can vary as can accessible iron for *H. pylori* acquisition (Keenan *et al.*,

2004) which, in turn, affects pathogenesis of *H. pylori* and can influence the quantitative expression of LPS and the nature of Le antigen mimicry occurring in *H. pylori* O-antigens (Keenan *et al.*, 2008).

6.2. Role in adhesion

Polymeric Le^x expression in the *H. pylori* O-chains has been shown to mediate, at least in part, *H. pylori* adhesion to the human antral gastric mucosa (Edwards *et al.*, 2000). The gastric receptor has been identified as the galactoside-binding lectin, galectin-3 (Fowler *et al.*, 2006). Emphasizing the role of Le^x in adhesion–recognition phenomena, fluorescently labelled latex beads bearing polymeric Le^x demonstrated the same tropic binding to human gastric tissue as fluorescently labelled *H. pylori* bacteria (Edwards *et al.*, 2000). It is important to note, however, that *H. pylori* adhesion is mediated by a number of adhesins (Moran, 1995, 1996; Moran and Wadström, 1998; Ilver *et al.*, 1998; Mahdavi *et al.*, 2002) with polymeric Le^x considered to play a distinct role (Mahdavi *et al.*, 2003).

Nevertheless, consistent with an adhesion role, strains with a high expression of Le^x cause a higher colonization density in patients than those with weaker expression (Heneghan *et al.*, 2000). By assisting bacterial adhesion and interaction with the gastric mucosa, like other adhesins, bacterial Le^x expression may enhance delivery of secreted products into the gastric mucosa (Rieder *et al.*, 1997; Moran, 1999). This would, in turn, promote chemotaxis and leukocyte infiltration, as has been observed in an adherence-dependent mouse model of *H. pylori*-induced disease (Guruge *et al.*, 1998). Noteworthy are the differences in pH within mucus and on the gastric cell surface, i.e. pH 2 on the luminal side of the gastric mucus layer to almost pH 7 on the cell surface, and that environmental pH influences Le^x expression by *H. pylori* (Moran *et al.*, 2002a). Optimal expression of Le^x occurs at neutral pH, as on the cell surface, but reduced expression occurs at lower pH, as in the mucus layer, thereby modulating expression of the Le^x-adhesin and allowing free-swimming *H. pylori* in the mucus layer. Potentially, these bacteria in the mucus layer can act as a reservoir for continued infection of the mucosa. Hence, the observed on–off switching of FucT activities and resulting phase variation in Le expression (Appelmelk *et al.*, 1999; Ma *et al.*, 2006; Moran and Trent, 2008) would be consistent with this mechanism of adhesin modulation and prolonged infection.

6.3. Influence on colonization

Construction of an isogenic *H. pylori galE* (HP0360/JHP1020) mutant, affecting Gal incorporation, resulted in truncation of LPS and loss of O-chain expression and led to the inability of this mutant to colonize a number of mouse strains compared with the Le^x-expressing parental strain (Moran *et al.*, 2000b). Mutation of a β-(1→4)-galactosyltransferase gene (HP0826/JHP0765) that affected synthesis of the O-antigen backbone resulted in less efficient colonization of the murine stomach by this engineered strain (Logan *et al.*, 2000). Inactivation of the *rfbM* gene (HP0043/JHP0037) that encodes a guanosine diphosphate- (GDP-) D-mannose pyrophosphorylase required for GDP-Fuc synthesis resulted in an i-antigen-expressing O-chain lacking Fuc, which exhibited reduced mouse colonization (Moran *et al.*, 2000b) and ablated interaction with the human gastric mucosa of biopsy specimens *in situ* (Edwards *et al.*, 2000). Although one study reported no significant changes in mouse colonization by *H. pylori* mutants with ablated Le^x/Le^y-expression (Takata *et al.*, 2002), this study used C3H/HeJ ($TLR4^{-/-}$) mice, so-called LPS non-responder mice, which may not be the most appropriate animal model of infection since *H. pylori* LPS induces Toll-like receptor (TLR) signalling

(Moran, 2007). In another study, using a more appropriate mouse model, however, a mutated *H. pylori* strain with a double knockout in both α-(1→3)-FucTs did not colonize, in contrast to the Le^x/Le^y-expressing parental strain (Appelmelk *et al.*, 2000). Taken together, these data indicate that not only O-chain occurrence, but specifically Le^x expression, is required for colonization. Whether the role of Le mimicry in adhesion is related to the influence of this mimicry in colonization remains an open question.

6.4. Influence on the innate and inflammatory response

Surfactant protein D (SP-D), a C-type lectin involved in antibody-independent pathogen recognition and clearance (Moran *et al.*, 2005b) (see also Chapter 35), binds *H. pylori* LPS, in particular by interaction with the O-antigen moiety, resulting in bacterial immobilization and aggregation (Murray *et al.*, 2002; Khamri *et al.*, 2005). Levels of expression of SP-D are significantly increased in *H. pylori*-associated antral gastritis compared to normal gastric mucosa (Murray *et al.*, 2002). Moreover, SP-D expression colocalizes with *H. pylori* organisms in the gastric mucosa (Murray *et al.*, 2002; Moran *et al.*, 2005b). Also, experiments in SP-D$^{-/-}$ mice have revealed that *Helicobacter* colonization is more common in the absence of SP-D (Moran *et al.*, 2005b; Khamri *et al.*, 2007) emphasizing an important influence of SP-D binding on the establishment of infection. To evade this important mechanism of innate immune recognition of *H. pylori*, escape variants can arise within the bacterial cell population with modifications in O-chain glycosylation decreasing their interaction with SP-D (Khamri *et al.*, 2005).

Contributing to the chronicity of *H. pylori* infection by modulation of the immune response, strain-to-strain variability in Le expression can influence the interaction of *H. pylori* LPS with the cellular innate immune receptor, dendritic cell-specific intercellular adhesion molecule 3-grabbing non-integrin (DC-SIGN), another C-type lectin, on gastric dendritic cells (Bergman *et al.*, 2004) (see also Chapter 34). This contributes to a changed T-cell polarization after innate immune activation (Bergman *et al.*, 2006). In the effector phase of an immune response, different T-cell subsets, called T-helper-1- (Th1-) and T-helper-2- (Th2-) cells, expand; Th1 cells promote proinflammatory cell-mediated immunity, whereas Th2-cells promote humoral immunity that induces B-cells to produce antibodies (O'Keeffe and Moran, 2008). Dendritic cells, in response to *H. pylori*, secrete a range of cytokines, but preferentially interleukin- (IL-) 12 that induces a Th1 response and also lesser amounts of IL-6 and IL-10 (Guiney *et al.*, 2003). Since Le antigen expression is subject to phase variation (Appelmelk *et al.*, 1999; Ma *et al.*, 2006; Moran, 2008), a significant proportion of Le-negative variants can occur within an isolate population of *H. pylori.* Importantly, Le-negative variants of *H. pylori* escape binding to dendritic cells and induce a strong Th1-cell response (Bergman *et al.*, 2004). On the other hand, *H. pylori* variants that express Le^x/Le^y can bind to DC-SIGN on dendritic cells and enhance the production of IL-10 which promotes a Th2-cell response and blocking of Th1-cell activation (Bergman *et al.*, 2006). Thus, a polarized Th1 effect can change to a mixed Th1/Th2-cell response through the extent of Le antigen-DC-SIGN interaction (Bergman *et al.*, 2004). This modulation of the host response allows a switch from an acute infection response to one that supports development of chronic infection since the humoral response and antibody production are not associated with *H. pylori* eradication or protection (O'Keeffe and Moran, 2008). A similar alteration in the T-cell response, although based on over-expression of IL-10 but induced by polymeric Le^x, has been observed with infection by eggs of the parasitic worm *Schistosoma mansoni* (Velupillai and Harn, 1994; Moran *et al.*, 1996).

Additionally, in patients with more aggressive inflammation and pathologies, the expression of both Le^x and Le^y by *H. pylori* may aid persistence of proinflammatory *cagA*-positive strains (Wirth *et al.*, 1996), i.e. strains carrying the *cag* pathogenicity island-encoding a type IV secretion system (Censini *et al.*, 1996). This hypothesis was based on observations in a sample human population that *H. pylori* isolates positive for Le^x/Le^y were predominantly *cagA*-positive and that a genetically engineered *cagA*-negative strain had diminished expression of Le^y (Wirth *et al.*, 1996). In contrast, two other studies drawn from a different human study population did not find an association between bacterial *cagA* status and Le^x/Le^y expression (Marshall *et al.*, 1999; Heneghan *et al.*, 2000). The discrepancy between these findings has been attributed to adaptation of *H. pylori* strains with differing properties to the different human populations studied (Moran, 1999), a conclusion which was later supported in a pan-European study examining *cagA* status and Le expression of isolates in differing ethnic populations (Broutet *et al.*, 2002). Therefore, Le^x/Le^y expression by *H. pylori* may aid persistence of more aggressive strains depending on the human host population.

6.5. Putative role in the development of gastric autoimmunity

Anti-Le^x and anti-Le^y antibodies have been detected in the sera of *H. pylori*-infected patients, particularly those with prolonged infection, using *H. pylori* LPS (Appelmelk *et al.*, 1996; Heneghan *et al.*, 2001; Hynes *et al.*, 2005). Importantly, the use of synthetic glycoconjugates in solution or on a solid phase does not optimally detect *H. pylori*-induced anti-Le antibodies (Heneghan *et al.*, 2001), thereby explaining the inability to detect these antibodies in some patient studies (Amano *et al.*, 1997). Although it has become apparent that bacterial colonization density and the ensuing inflammatory response can be influenced by host expression of ABO and Le^a blood group determinants (Heneghan *et al.*, 1998b), bacterial Le^x expression is associated with peptic ulcer disease (Marshall *et al.*, 1999) and is statistically related to neutrophil infiltration (Heneghan *et al.*, 2000). Interestingly, neutrophils are a potential target recognized by anti-Le^x antibodies (Appelmelk *et al.*, 1996). Neutrophils express CD15 (Le^x) on members of the adhesion-promoting glycoprotein family (CD11/CD18) and it has been speculated that cross-linking of CD15 by *H. pylori*-induced anti-Le^x autoantibodies could potentiate polymorph adhesiveness to the endothelium (Appelmelk *et al.*, 1996; Moran *et al.*, 1996). Supporting this, anti-Le^x monoclonal antibodies, including those induced by *H. pylori*, can activate and cause enhanced adherence of these cells, which may result in tissue damage and inflammation (Stöckl *et al.*, 1993; Appelmelk *et al.*, 1996; Moran *et al.*, 1996).

During chronic infection by *H. pylori*, expression of Le^y has been implicated in the pathogenesis of atrophic gastritis, a precursor pathological state before the development of gastric cancer, through the induction of autoreactive antibodies against the gastric mucosa (Negrini *et al.*, 1991, 1996; Appelmelk *et al.*, 1996; Faller *et al.*, 1997; Moran, 1999, 2001a). Such autoantibodies have been found in patients with atrophic gastritis and gastric cancer (Negrini *et al.*, 1996; Heneghan *et al.*, 2001; Hynes *et al.*, 2005). Specifically, the β-chain of the gastric proton pump, H^+,K^+-adenosine triphosphatase (H^+,K^+-ATPase), which is contained within parietal cell canaliculi, is glycosylated predominantly by Le^y and anti-Le^y antibodies have been implicated in the pathogenic autoimmune responses in atrophic gastritis (Appelmelk *et al.*, 1996; Moran *et al.*, 1996; Moran, 1999). Anti-Le^y *H. pylori*-induced antibodies have been shown to react with the purified human and murine β-chain (Appelmelk *et al.*, 1996) and, in an adoptive transfer experiment, growth in mice of an

H. pylori-induced anti-Le^y-secreting hybridoma resulted in gastric histopathological changes consistent with those of gastritis (Negrini *et al.*, 1991). Additionally, anti-Le and anti-parietal cell-related antibodies have been induced in a transgenic mouse model of *H. pylori* infection in which gastric pathology developed (Negrini *et al.*, 1991). Based on these data, it was inferred that Le^y mimicry plays a role in the induction of *H. pylori*-related atrophic gastritis (Moran *et al.*, 1996; Moran, 1999, 2001a,b) but, since these experiments were performed largely in mice, questions were raised as to whether a similar phenomenon occurs in humans (Appelmelk *et al.*, 1998).

Despite this, a correlation has been established between the presence of autoantibodies in *H. pylori*-infected human subjects (Faller *et al.*, 1997) and with the degree of gastric infiltration, numbers of inflammatory cells and glandular atrophy (Negrini *et al.*, 1996). In the case of patients suffering from severe gastric atrophy, *H. pylori* isolates were found more likely to express Le^x/Le^y antigens, whereas isolates from individuals with near-normal mucosa were less likely to express these antigens and to induce autoantibodies in experimental animals (Negrini *et al.*, 1996). Additionally, the anti-canalicular autoantibodies have been shown to increase significantly in patients with the duration of *H. pylori* gastritis and to correlate with gastric corpus atrophy (Vorobjova *et al.*, 2000). Nonetheless, based on the classical model of organ-specific autoimmunity in which there is a central role for increased autoantigen (i.e. Le) presentation in the gastric mucosa because of *H. pylori*-induced damage (Appelmelk *et al.*, 1998; Faller *et al.*, 1998), it has been suggested that these pathogenic antibodies are the consequence, rather than the causative factor, in atrophic gastritis. Such a view has been largely based on certain absorption experiments attempting to abolish the reactivity of autoantibodies from human sera using *H. pylori* preparations. In one such series of experiments using *H. pylori* lysates, a decrease in anti-gastric autoreactivity was observed (Negrini *et al.*, 1991), whereas in another, no significant reduction was found (Faller *et al.*, 1998), and anti-H^+,K^+-ATPase serum autoantibodies were not absorbed with *H. pylori* whole cells in another study (Ma *et al.*, 1994). Critically examining these studies, it was observed that many of these experiments did not use matched isolates with the sera of the patient from whom they were isolated (Moran, 2008). Although patient sera were shown to react with recombinant H^+,K^+-ATPase expressed in *Xenopus* oocytes, which was deduced to indicate that the reactivity observed was based on protein epitopes rather than Le glycosylation (Claeys *et al.*, 1998), nevertheless, the nature or lack of any glycosylation occurring in the oocytes was not confirmed (Moran, 1999, 2001a).

Importantly, it has been established that antigen presentation and the serological assay format that is used plays an important role in successful detection of Le antigen expression displayed in the O-chains of *H. pylori* (Appelmelk *et al.*, 1996; Hynes and Moran, 2000; Heneghan *et al.*, 2001; Moran *et al.*, 2004). Additionally, the use of synthetic glycoconjugates in solution or in solid phase does not optimally detect *H. pylori*-induced anti-Le antibodies (Heneghan *et al.*, 2001). For these reasons, it has been speculated that the form of antigenic presentation of Le^y could potentially influence the ability to absorb *H. pylori*-derived anti-Le antibodies from patient sera (Moran, 1999, 2001b). It is important to bear in mind that the outer membrane of *H. pylori* undergoes blebbing, producing vesicles, the membranes of which contain LPS with associated Le^x/Le^y expression (Hynes *et al.*, 2005). In absorption experiments, using these *H. pylori*-derived vesicles and sera from gastric cancer patients, it has been possible to remove 83–100% of the anti-Le^y antibodies from the sera (Figure 43.4) and cause significant reduction, though not complete ablation, of antibodies reactive with the canaliculi of parietal cells (Hynes *et al.*, 2005). This confirms that other protein-based epitopes of the H^+,K^+-ATPase are involved in the autoreactivity observed (Faller *et al.*, 1998).

Using available knowledge, a model can be proposed (Moran, 2008), whereby it can be deduced that the mechanisms underlying anti-gastric autoantibody production likely include, but are not exclusive to, molecular mimicry between *H. pylori*-expressed Le^y and the gastric mucosa, as well as facilitated exposition and presentation of the bacterial epitopes to recruited immune cells within the gastric mucosa. This, in turn, will be influenced by host immune-regulation and environmental factors. In this model, complex interactions can occur. A cross-reactive antibody, such as anti-Le^y, can initiate damage to the proton pump (Appelmelk *et al.*, 1996), with subsequent alteration in acid output. Changes in acid output would regulate LPS glycosylation (Moran *et al.*, 2002a), thereby influencing the LPS structure of the infecting *H. pylori* strain such that phase variation occurs in the expression of Le epitopes. Thus, a secondary reduction in autoreactive antibody production could occur, thereby confounding interpretation of the role of *H. pylori*-induced anti-Le antibodies. Furthermore, the initiation of the inflammatory cascade against *H. pylori* and the activities of the virulence factors produced by the bacterium (Moran, 1995, 1996; Moran and Wadström, 1998), and even anti-Le^x antibodies causing complement-mediated lysis of host cells (Appelmelk *et al.*, 1996; Moran *et al.*, 1996), would cause histological damage in the gastric mucosa. Subsequently, host structural epitopes would become exposed and presented to the immune system further to drive the autoreactive response, through humoral and T-cell responses to peptides (Faller *et al.*, 1998; Appelmelk *et al.*, 1998; D'Elios *et al.*, 2004). Moreover, molecular mimicry by *H. pylori* peptides also may contribute to the gastric autoimmunity observed (Amedei *et al.*, 2003). Therefore, it is unlikely that the presence of anti-Le^x/Le^y antibodies would be solely responsible for both the initiation and maintenance of the anti-gastric autoimmune response in *H. pylori*-positive patients.

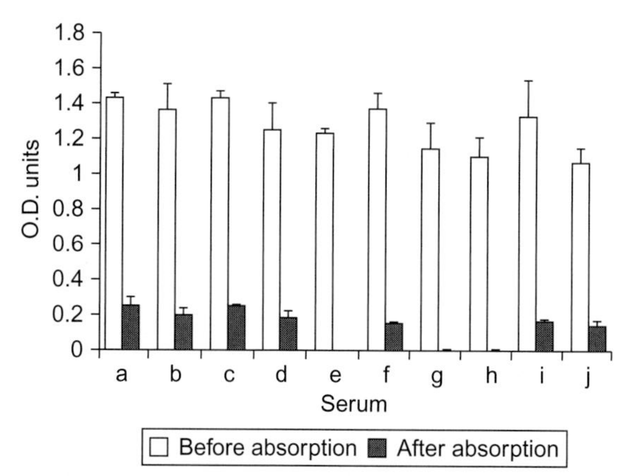

FIGURE 43.4 Effect of the absorption of sera from *H. pylori*-positive gastric cancer patients with outer membrane vesicles on the occurrence of anti-Le^y antibodies. Significant reductions (83–100%) in reactivity, as expressed in optical density (O.D.) units, were observed with the sera of different individuals.

7. CONCLUSIONS AND FUTURE OUTLOOK

Despite the evidence for molecular mimicry of gangliosides in the development of GBS, particularly the AMAN and MFS clinical forms, mimicry alone may not be sufficient to explain the induction of GBS by *C. jejuni*. Ganglioside-like structures (GM_1 and GD_{1a}) have been demonstrated in *C. jejuni* isolates from patients with enteritis who did not develop neuropathy (Moran and O'Malley, 1995; Prendergast *et al.*, 1998). Also, in a US survey of 275 enteritis isolates, one-quarter expressed the GM_1 epitope (Nachamkin *et al.*, 1999), though patients with enteritis rarely develop anti-ganglioside antibodies. Therefore, the presence of ganglioside-like epitopes on *C. jejuni* strains alone does not itself trigger GBS, but is a prerequisite for this autoimmune disease. Although studies have attempted to define host and immunogenetic factors in *C. jejuni*-related GBS (see Prendergast and Moran, 2000), these have not been convincing. Other bacterial factors, such as capsular polysaccharides, which are potentially virulence factors of the bacterium (Karlyshev *et al.*, 2008), are differentially expressed in GBS-related compared to enteritis-associated strains (Moran and O'Malley, 1995). In addition, attention, not surprisingly, has focused on the humoral response in GBS, but consideration should be made of T-cells in the pathogenic process. Notably, because subclasses IgG1 and IgG3 are characteristic of a T-cell-dependent antibody response, T-helper cells may be associated with B-cell production of anti-ganglioside antibodies. This and other unresolved issues in *C. jejuni*-related autoimmunity require investigation (see Research Focus Box).

H. pylori Le expression has been implicated in evading the immune response upon initial infection and in influencing bacterial colonization and adhesion (Moran 1999, 2001a,b) but, with the progress of chronic infection, has been suggested to contribute to gastric autoimmunity (Appelmelk *et al.*, 1996; Heneghan *et al.*, 2001; Broutet *et al.*, 2002; Hynes *et al.*, 2005) that leads to gastric atrophy, a precursor state of gastric cancer. Despite the latter being considered controversial, evidence supports a contributing role of Le mimicry-induced autoantibodies (Appelmelk *et al.*, 1998; Moran 1999, 2001a), as well as peptide-directed autoreactive T-cells (D'Elios *et al.*, 2004), in the development of gastric atrophy. Importantly, modulation of LPS structure could influence immune recognition within the gastric mucosa and, thus, influence colonization and its disease outcome. The interaction of *H. pylori* LPS with SP-D and the influence of phase variation in FucTs on *H. pylori*-SP-D interaction serve as an obvious paradigm (Khamri *et al.*, 2005). Likewise, the influence of Le antigen expression and fucosylation on T-cell polarization and the inflammatory response that develops after infection is another example (Bergman *et al.*, 2004, 2006). Hence, Le antigen expression and fucosylation in *H. pylori* should be viewed as producing multiple biological effects in pathogenesis and disease outcome, rather than affecting a single response or attribute. Needless to say, further studies are required to establish these roles (Research Focus Box).

The approaches and research methods used in investigations on *C. jejuni-* and *H. pylori*-associated autoimmunity should help in clarifying the molecular pathogenesis of other immune-mediated diseases. More specifically, important insights in bacterial pathogenesis have, and are likely to continue to be gained, by investigating the mimicry expressed by these important bacterial pathogens.

ACKNOWLEDGEMENTS

The author gratefully acknowledges Griffith University, Science Foundation Ireland and the EU Marie Curie Programme (Grant no. MTKD-CT-2005-029774) for financial support of his research.

RESEARCH FOCUS BOX

- An infection-based, *C. jejuni*-related GBS animal model is required to answer unresolved questions in the pathogenesis of GBS.
- The role of T-cells and the extent to which T-cells contribute to the pathophysiology of GBS, and the mechanism by which they do so, remains unexplained.
- What role, if any, do capsular polysaccharides play in the virulence of *C. jejuni* and hence in the pathogenesis of *C. jejuni*-related GBS?
- How tolerance is broken to carbohydrate-based self-structures and the role this plays in *C. jejuni*- and *H. pylori*-related diseases requires intensive investigation.
- The environmental conditions and factors that regulate the expression of FucTs and LPS/LOS genes require more detailed examination in *H. pylori* and *C. jejuni* in order to better understand the roles of mimicry in both microbial pathogenesis and autoimmune disease.

References

Abu-Shakra, M., Buskila, D., Shoenfeld, Y., 1999. Molecular mimicry between host and pathogen: examples from parasites and implication. Immunol. Lett. 67, 147–152.

Amano, K.-I., Hayashi, S., Kubota, T., Fujii, N., Yokota, S.-I., 1997. Reactivities of Lewis antigen monoclonal antibodies with the lipopolysaccharides of *Helicobacter pylori* strains isolated from patients with gastroduodenal diseases in Japan. Clin. Diagn. Lab. Immnunol. 4, 540–544.

Amedei, A., Bergman, M.P., Appelmelk, B.J., et al., 2003. Molecular mimicry between *Helicobacter pylori* antigens and H^+,K^+ -adenosine triphosphatase in human gastric autoimmunity. J. Exp. Med. 198, 1147–1156.

Ang, C.W., Endtz, H.P., Jacobs, B.C., et al., 2000. *Campylobacter jejuni* lipopolysaccharides from Guillain-Barré syndrome patients induce IgG anti-GM_1 antibodies in rabbits. J. Neuroimmunol. 104, 133–138.

Ang, C.W., Jacobs, B.C., Laman, J.D., 2004. The Guillain-Barré syndrome: a true case of molecular mimicry. Trends Immunol. 25, 61–66.

Appelmelk, B.J., Simoons-Smit, I., Negrini, R., et al., 1996. Potential role of molecular mimicry between *Helicobacter pylori* lipopolysaccharide and host Lewis blood group antigens in autoimmunity. Infect. Immun. 64, 2031–2040.

Appelmelk, B.J., Faller, G., Claeys, D., Kirchner, T., Vandenbroucke-Grauls, C.M.J.E., 1998. Bugs on trial: the case of *Helicobacter pylori* and autoimmunity. Immunol. Today 19, 296–299.

Appelmelk, B.J., Martin, S.L., Monteiro, M.A., et al., 1999. Phase variation in *Helicobacter pylori* lipopolysaccharide due to changes in the lengths of poly(C) tracts in α3-fucosyltransferase genes. Infect. Immun. 67, 5361–5366.

Appelmelk, B.J., Monteiro, M.A., Martin, S.L., Moran, A.P., Vandenbroucke-Grauls, C.M.J.E., 2000. Why *Helicobacter pylori* has Lewis antigens. Trends Microbiol. 8, 565–570.

Aspinall, G.O., Lynch, C.M., Pang, H., Shaver, R.T., Moran, A.P., 1995. Chemical structures of the core region of *Campylobacter jejuni* O:3 lipopolysaccharide and an associated polysaccharide. Eur. J. Biochem. 231, 570–578.

Aspinall, G.O., Monteiro, M.A., Pang, H., Walsh, E.J., Moran, A.P., 1996. Lipopolysaccharide of the *Helicobacter pylori* type strain NCTC 11637 (ATCC 43504): structure of the O antigen chain and core oligosaccharide regions. Biochemistry 35, 2489–2497.

Bergman, M.P., Engering, A., Smits, H.H., et al., 2004. *Helicobacter pylori* modulates the T helper cell 1/T helper cell 2 balance through phase-variable interaction between lipopolysaccharide and DC-SIGN. J. Exp. Med. 200, 979–990.

Bergman, M., Del Prete, G., van Kooyk, Y., Appelmelk, B., 2006. *Helicobacter pylori* phase variation, immune modulation and gastric autoimmunity. Nat. Rev. Microbiol. 4, 151–159.

Britton, S., Papp-Szabo, E., Simala-Grant, J., Morrison, L., Taylor, D.E., Monteiro, M.A., 2005. A novel *Helicobacter pylori* cell-surface polysaccharide. Carbohydr. Res. 340, 1605–1611.

Broutet, N., Moran, A., Hynes, S., Sakarovitch, C., Mégraud, F., 2002. Lewis antigen expression and other pathogenic factors in the presence of atrophic chronic gastritis in a European population. J. Infect. Dis. 185, 503–512.

Censini, S., Lange, C., Xiang, Z., et al., 1996. *cag*, a pathogenicity island of *Helicobacter pylori*, encodes type I-specific and disease-associated virulence factors. Proc. Natl. Acad Sci. USA 93, 14648–14653.

Claeys, D., Faller, G., Appelmelk, B.J., Negrini, R., Kirchner, T., 1998. The gastric H^+,K^+-ATPase is a major autoantigen in chronic *Helicobacter pylori* gastritis with body mucosa atrophy. Gastroenterology 115, 340–347.

Cornblath, D.R., Griffin, D.E., Welch, D., Griffin, J.W., McArthur, J.C., 1990. Quantitative analysis of endoneurial T-cells in human sural nerve biopsies. J. Neuroimmunol. 26, 113–118.

D'Elios, M.M., Appelmelk, B.J., Amedei, A., Bergman, M.P., del Prete, G., 2004. Gastric autoimmunity: the role of *Helicobacter pylori* and molecular mimicry. Trends Mol. Med. 10, 316–323.

De Libero, G.O., Moran, A.P., Gober, H.J., et al., 2005. Bacterial infections promote T cell recognition of self-glycolipids. Immunity 22, 763–772.

Eaton, K.A., Logan, S.M., Baker, P.E., Peterson, R.A., Monteiro, M.A., Altman, E., 2004. *Helicobacter pylori* with a truncated lipopolysaccharide O-chain fails to induce gastritis in SCID mice injected with splenocytes from wild-type C57BL/6J mice. Infect. Immun. 72, 3925–3931.

Ebringer, A., Rashid, T., Wilson, C., 2005. Bovine spongiform encephalopathy, multiple sclerosis, and Creutzfeldt-Jakob disease are probably autoimmune diseases evoked by *Acientobacter* bacteria. Ann. N.Y. Acad. Sci. 1050, 417–428.

Ebringer, A., Rashid, T., Jawad, N., Wilson, C., Thompson, E.J., Ettelaie, C., 2007a. From rabies to transmissible spongiform encephalopathies: an immune-mediated microbial trigger involving molecular mimicry could be the answer. Med. Hypotheses 68, 113–124.

Ebringer, A., Rashid, T., Tiwana, H., Wilson, C., 2007b. A possible link between Crohn's disease and ankylosing spondylitis via *Klebsiella* infections. Clin. Rheumatol. 26, 289–297.

Edwards, N.J., Monteiro, M.A., Faller, G., et al., 2000. Lewis X structures in the O antigen side-chain promote adhesion of *Helicobacter pylori* to the gastric epithelium. Mol. Microbiol. 35, 1530–1539.

Ernst, P.B., Gold, B.D., 2000. The disease spectrum of *H. pylori*: the immunopathogenesis of gastroduodenal ulcer and gastric cancer. Annu. Rev. Microbiol. 54, 615–640.

Faé, K.C., da Silva, D.D., Oshiro, S.E., et al., 2006. Mimicry in recognition of cardiac myosin peptides by heart-intralesional T cell clones from rheumatic heart disease. J. Immunol. 176, 5662–5670.

Faller, G., Steininger, H., Kränzlein, J., et al., 1997. Antigastric autoantibodies in *Helicobacter pylori* infection: implications of histological and clinical parameters of gastritis. Gut 41, 619–623.

Faller, G., Steininger, H., Appelmelk, B., Kirchner, T., 1998. Evidence of novel pathogenic pathways for the formation of antigastric autoantibodies in *Helicobacter pylori* gastritis. J. Clin. Pathol. 51, 244–245.

Feasby, T.E., Hahn, A.F., Gilbert, J.J., 1982. Passive transfer studies in Guillain-Barré polyneuropathy. Neurology 32, 1159–1167.

Fowler, M., Thomas, R.J., Atherton, J., Roberts, I.S., High, N.J., 2006. Galectin-3 binds to *Helicobacter pylori* O-antigen: it is upregulated and rapidly secreted by gastric epithelial cells in response to *H. pylori* adhesion. Cell. Microbiol. 8, 44–54.

Gilbert, M., Parker, C.T., Moran, A.P., 2008. *Campylobacter jejuni* lipooligosaccharides: structure and biosynthesis. In: Nachamkin, I., Szymanski, C.M., Blaser, M.J. (Eds.) *Campylobacter*, 3rd edn. ASM Press, Washington, DC, pp. 483–504.

Goodyear, C.S., O'Hanlon, G.M., et al., 1999. Monoclonal antibodies raised against Guillain-Barré syndrome-associated *Campylobacter jejuni* lipopolysaccharides react with neuronal gangliosides and paralyze muscle-nerve preparations. J. Clin. Invest. 104, 697–708.

Green, C., 1989. The ABO, Lewis and related blood group antigens: a review of structure and biosynthesis. FEMS Microbiol. Immunol. 47, 321–330.

Guerry, P., Szymanski, C.M., Prendergast, M.M., et al., 2002. Phase variation of *Campylobacter jejuni* 81–176 lipooligosaccharide affects ganglioside mimicry and invasiveness *in vitro*. Infect. Immun. 70, 787–793.

Guillain-Barré syndrome Study Group, Plasmapheresis and acute Guillain-Barré syndrome 1985. Neurology 35, 1096–1104.

Guiney, D.G., Hasegawa, P., Cole, S.P., 2003. *Helicobacter pylori* preferentially induces interleukin 12 (IL-12) rather than IL-6 or IL-10 in human dendritic cells. Infect. Immun. 71, 4136–4163.

Guruge, J.L., Falk, P.G., Lorenz, R.G., et al., 1998. Epithelial attachment alters the outcome of *Helicobacter pylori* infection. Proc. Natl. Acad. Sci. USA 95, 3925–3930.

Harrison, B.M., Hansen, L.A., Pollard, J.D., McLeod, J.G., 1984. Demyelination induced by serum from patients with Guillain-Barré syndrome. Ann. Neurol. 15, 163–170.

Heneghan, M.A., McCarthy, C.F., Moran, A.P., 1998a. Is colonisation of gastric mucosa by *Helicobacter pylori* strains expressing Lewis x and y epitopes dependent on host secretor status? In: Lastovica, A.J., Newell, D.G., Lastovica, E.E. (Eds.), *Campylobacter*, *Helicobacter* and Related Organisms. Institute of Child Health and University of Cape Town, Cape Town, pp. 473–477.

Heneghan, M.A., Moran, A.P., Feeley, K.M., et al., 1998b. Effect of host Lewis and ABO blood group antigen expression on *Helicobacter pylori* colonisation density and the consequent inflammatory response. FEMS Immunol. Med. Microbiol. 20, 257–266.

Heneghan, M.A., McCarthy, C.F., Moran, A.P., 2000. Relationship of blood group determinants on *Helicobacter pylori* lipopolysaccharide with host Lewis phenotype and inflammatory response. Infect. Immun. 68, 937–941.

Heneghan, M.A., McCarthy, C.F., Janulaityte, D., Moran, A.P., 2001. Relationship of anti-Lewis x and anti-Lewis y antibodies in serum samples from gastric cancer and chronic gastritis patients to *Helicobacter pylori*-mediated autoimmunity. Infect. Immun. 69, 4774–4781.

Hogan, E.L., Barry, T., Dasgupta, S., et al., 2007. Possible mycoplasma infection in MS: study of immunogenic glycolipids and search for mycoplasma DNA. In: Abstracts of the FEMS International Conference on *Chlamydia* and *Mycoplasma* Human Infections. University of Ferrara, Italy, p. 57.

Hughes, R.A.C., Cornblath, D.R., 2005. Guillain-Barré syndrome. Lancet 366, 1653–1666.

Hughes, L.E., Bonell, S., Natt, R.S., et al., 2001. Antibody responses to *Acinetobacter* spp. and *Pseudomonas aeruginosa* in multiple sclerosis: prospects for diagnosis using the myelin-acinetobacter-neurofilament antibody index. Clin. Diagn. Lab. Immunol. 8, 1181–1188.

Hynes, S.O., Moran, A.P., 2000. Comparison of three serological methods for detection of Lewis antigens on the surface of *Helicobacter pylori*. FEMS Microbiol. Lett. 190, 67–72.

Hynes, S.O., Keenan, J.I., Ferris, J.A., Annuk, H., Moran, A.P., 2005. Lewis epitopes on outer membrane vesicles of relevance to *Helicobacter pylori* pathogenesis. Helicobacter 10, 146–156.

Illa, I., Ortiz, N., Gallard, E., Juarez, C., Grau, J.M., Dalakas, M.C., 1995. Acute axonal Guillain-Barré syndrome with IgG antibody against motor axons following parenteral gangliosides. Ann. Neurol. 38, 218–224.

Ilver, D., Arnqvist, A., Ogren, J., et al., 1998. *Helicobacter pylori* adhesin binding fucosylated histo-blood group antigens revealed by retagging. Science 279, 373–377.

Isenberg, D., Morrow, J., 1995. Friendly Fire. Explaining Autoimmune Disease. Oxford University Press, Oxford.

Karlyshev, A.J., Wren, B.J., Moran, A.P., 2008. *Campylobacter jejuni* capsular polysaccharide. In: Nachamkin, I., Szymanski, C.M., Blaser, M.J. (Eds.), *Campylobacter*, 3rd edn. ASM Press, Washington, DC, pp. 505–521.

Keenan, J.I., Peterson, R.A., Fraser, R., et al., 2004. The effect of *Helicobacter pylori* infection and dietary iron deficiency on host iron homeostasis: a study in mice. Helicobacter 9, 643–650.

Keenan, J.I., Davis, K.A., Beaugie, C.R., McGovern, J.J., Moran, A.P., 2008. Alterations in *Helicobacter pylori* outer membrane and outer membrane vesicle-associated lipopolysaccharides under iron-limiting growth conditions. Innate Immun. 14, 279–290.

Khalili-Shirazi, A., Gregson, N., Gray, I.A., Rees, J.H., Winer, J.B., Hughes, R.A.C., 1999. Anti-ganglioside antibodies in Guillain-Barré syndrome following recent cytomegalovirus infection. J. Neurol. Neurosurg. Psychiatr. 66, 376–379.

Khamri, W., Moran, A.P., Worku, M.L., et al., 2005. Variations in *Helicobacter pylori* lipopolysaccharide to evade the innate immune component surfactant protein D. Infect. Immun. 73, 7677–7686.

Khamri, W., Worku, M.L., Anderson, A.E., et al., 2007. *Helicobacter* infection in the surfactant protein D-deficient mouse. Helicobacter 12, 112–123.

Knirel, Y.A., Kocharova, N.A., Hynes, S.O., et al., 1999. Structural studies on lipopolysaccharides of serologically non-typable strains of *Helicobacter pylori*, AF1 and 007, expressing Lewis antigenic determinants. Eur. J. Biochem. 266, 123–131.

Kobayashi, K., Sakamoto, J., Kito, T., et al., 1993. Lewis blood group-related antigen expression in normal gastric epithelium, intestinal metaplasia, gastric adenoma, and gastric carcinoma. Am. J. Gastroenterol. 88, 919–924.

Kocharova, N.A., Knirel, Y.A., Widmalm, G., Jansson, P.-E., Moran, A.P., 2000. Structure of an atypical O-antigen polysaccharide of *Helicobacter pylori* containing a novel monosaccharide 3-C-methyl-D-mannose. Biochemistry 39, 4755–4760.

Kusters, J.G., van Vliet, A.H.M., Kuipers, E.J., 2006. Pathogenesis of *Helicobacter pylori* infection. Clin. Microbiol. Rev. 19, 449–490.

Kusunoki, S., Iwamori, M., Chiba, A., Hitoshi, S., Arita, M., Kanazawa, I., 1996. GM_{1b} is a new member of antigen for serum antibody in Guillain-Barré syndrome. Neurology 47, 237–242.

Kuwabara, S., Misawa, S., Takahashi, H., et al., 2007. Anti-GQ1b antibody does not affect neuromuscular transmission in human limb muscle. J. Neuroimmunol. 189, 158–162.

Latov, N., 1990. Antibodies to glycoconjugates in neurological disease. Clin. Aspects Autoimmun. 4, 18–29.

Li, C.Y., Xue, P., Tian, W.Q., Liu, R.C., Yang, C., 1996. Experimental *Campylobacter jejuni* infection in the chicken: an animal model of axonal Guillain-Barré syndrome. J. Neurol. Neurosurg. Psychiatr. 61, 279–284.

Logan, S.M., Conlan, J.W., Monteiro, M.A., Wakarchuk, W. W., Altman, E., 2000. Functional genomics of *Helicobacter pylori*: identification of a β-1,4-galactosyltransferase and generation of mutants with altered lipopolysaccharide. Mol. Microbiol. 35, 1156–1167.

Ma, B., Simala-Grant, J.L., Taylor, D.E., 2006. Fucosylation in prokaryotes and eukaryotes. Glycobiology 16, 158R–184R.

Ma, J.Y., Borch, K., Sjöstrand, S.E., Janzon, L., Mårdh, S., 1994. Positive correlation between H,K-adenosine triphosphatase autoantibodies and *Helicobacter pylori* antibodies in patients with pernicious anemia. Scand. J. Gastroenterol. 29, 961–965.

Mahdavi, J., Sondén, B., Hurtig, M., et al., 2002. *Helicobacter pylori* SabA adhesin in persistent infection and chronic inflammation. Science 297, 573–578.

Mahdavi, J., Borén, T., Vandenbroucke-Grauls, C., Appelmelk, B.J., 2003. Limited role of lipopolysaccharide Lewis antigens in adherence of *Helicobacter pylori* to the human gastric epithelium. Infect. Immun. 71, 2876–2880.

Marrack, P., Kappler, J., Kotzin, B.L., 2001. Autoimmune disease: why and where it occurs. Nat. Med. 7, 899–905.

Marshall, D.G., Hynes, S.O., Coleman, D.C., O'Morain, C.A., Smyth, C.J., Moran, A.P., 1999. Lack of a relationship between Lewis antigen expression and *cag A*, CagA, *vacA* and VacA status of Irish *Helicobacter pylori* isolates. FEMS Immunol. Med. Microbiol. 24, 79–90.

Matsumoto, Y., Yuki, N., Van Kaer, L., Furukawa, K., Hirata, K., Sugita, M., 2008. Cutting edge: Guillain-Barré syndrome-associated IgG responses to gangliosides are generated independently of CD1 function in mice. J. Immunol. 180, 39–43.

Monteiro, M.A., 2001. *Helicobacter pylori*: a wolf in sheep's clothing. The glyconjugate families of *Helicobacter pylori* lipopolysaccharides expressing histo-blood groups: structure, biosynthesis, and role in pathogenesis. Adv. Carbohydr. Chem. Biochem. 57, 99–158.

Monteiro, M.A., St Michael, F., Rasko, D., et al., 2001. *Helicobacter pylori* from asymptomatic hosts expressing heptoglycan but lacking Lewis O-chains: Lewis blood-group O-chains may play a role in *Helicobacter pylori* induced pathology. Biochem. Cell Biol. 79, 449–459.

Moran, A.P., 1995. Cell surface characteristics of *Helicobacter pylori*. FEMS Immunol. Med. Microbiol. 10, 271–280.

Moran, A.P., 1996. Pathogenic properties of *Helicobacter pylori*. Scand. J. Gastroenterol. 31 (Suppl. 215), 22–31.

Moran, A.P., 1999. *Helicobacter pylori* lipopolysaccharide-mediated gastric and extragastric pathology. J. Physiol. Pharmacol. 50, 787–805.

Moran, A.P., 2001a. *Helicobacter pylori* lipopolysaccharides. In: Achtman, M., Suerbaum, S. (Eds.), *Helicobacter pylori*: Molecular and Cellular Biology. Horizon Scientific Press, Wymondham, pp. 207–226.

Moran, A.P., 2001b. Molecular structure, biosynthesis and pathogenic roles of lipopolysaccharides. In: Mobley, H.L.T., Mendz, G.L., Hazell, S.L. (Eds.), *Helicobacter pylori:* Physiology and Genetics. ASM Press, Washington, DC, pp. 81–95.

Moran, A.P., 2007. Lipopolysaccharide in bacterial chronic infection: insights from *Helicobacter pylori* lipopolysaccharide and lipid A. Int. J. Med. Microbiol. 297, 307–319.

Moran, A.P., 2008. Relevance of fucosylation and Lewis antigen expression in the bacterial gastroduodenal pathogen *Helicobacter pylori*. Carbohydr. Res. 343, 1952–1965.

Moran, A.P., O'Malley, D.T., 1995. Potential role of lipopolysaccharides of *Campylobacter jejuni* in the development of Guillain-Barré syndrome. J. Endotoxin Res. 2, 233–235.

Moran, A.P., Penner, J.L., 1999. Serotyping of *Campylobacter jejuni* based on heat-stable antigens: relevance, molecular basis and implications in pathogenesis. J. Appl. Microbiol. 86, 361–377.

Moran, A.P., Prendergast, M.M., 1998. Molecular mimicry in *Campylobacter jejuni* lipopolysaccharides and the development of Guillain-Barré syndrome. J. Infect. Dis. 178, 1549–1550.

Moran, A.P., Prendergast, M.M., 2001. Molecular mimicry in *Campylobacter jejuni* and *Helicobacter pylori* lipopolysaccharides: contribution of gastrointestinal infections to autoimmunity. J. Autoimmun. 16, 241–256.

Moran, A.P., Trent, M.S., 2008. Helicobacter pylori lipopolysaccharides and Lewis antigens. In: Yamaoka, Y. (Ed.), *Helicobacter pylori*: Molecular Genetics and Cellular Biology. Caister Academic Press, Norfolk, pp. 7–36.

Moran, A.P., Wadström, T., 1998. Pathogenesis of Helicobacter pylori. Curr. Opin. Gastroenterol. 14 (Suppl. 1), S9–S14.

Moran, A.P., Rietschel, E.T., Kosunen, T.U., Zähringer, U., 1991. Chemical characterization of *Campylobacter jejuni* lipopolysaccharides containing *N*-acetylneuraminic acid and 2,3-diamino-2,3-dideoxy-D-glucose. J. Bacteriol. 173, 618–626.

Moran, A.P., Appelmelk, B.J., Aspinall, G.O., 1996. Molecular mimicry of host structures by lipopolysaccharides of *Campylobacter* and *Helicobacter* spp.: implications in pathogenesis. J. Endotoxin Res. 3, 521–531.

Moran, A.P., Lindner, B., Walsh, E.J., 1997. Structural characterization of the lipid A component of *Helicobacter pylori* rough- and smooth-form lipopolysaccharides. J. Bacteriol. 179, 6453–6463.

Moran, A., Penner, J., Aspinall, G., 2000a. *Campylobacter* lipopolysaccharides. In: Nachamkin, I., Blaser, M.J. (Eds.), *Campylobacter*, 2nd edn. ASM Press, Washington, DC, pp. 241–257.

Moran, A.P., Sturegård, E., Sjunnesson, H., Wadström, T., Hynes, S.O., 2000b. The relationship between O-chain expression and colonisation ability of *Helicobacter pylori* in a mouse model. FEMS Immunol. Med. Microbiol. 29, 263–270.

Moran, A.P., Kosunen, T.U., Prendergast, M.M., 2001. Biochemical and serological analysis of the heat-stable antigens involved in serotyping *Campylobacter jejuni/coli*. Int. J. Med. Microbiol. 291 (Suppl. 31), 72.

Moran, A.P., Knirel, Y.A., Senchenkova, S.N., Widmalm, G., Hynes, S.O., Jansson, P.-E., 2002a. Phenotypic variation in molecular mimicry between *Helicobacter pylori* lipopolysaccharides and human gastric epithelial cell surface glycoforms. Acid-induced phase variation in Lewis^x and Lewis^y expression by *H. pylori* lipopolysaccharides. J. Biol. Chem. 277, 5785–5795.

Moran, A.P., Prendergast, M.M., Hogan, E.L., 2002b. Sialosyl-galactose: a common denominator of Guillain-Barré and related disorders? J. Neurol. Sci. 196, 1–7.

Moran, A.P., Ferris, J.A., Perepelov, A.V., et al., 2002c. Structural examination of Lewis expression and adaptation in lipopolysaccharides of *Helicobacter pylori* from experimentally infected rhesus monkeys. Gut 51 (Suppl. II), A15.

Moran, A.P., Perepelov, A.V., Knirel, Y.A., Amano, K.-i., McGovern, J.J., Jansson, P.-E., 2004. Lewis expression on LPS of Japanese *Helicobacter pylori*-associated cancer strains and importance of serological test format for detection. J. Endotoxin Res. 10, 315.

Moran, A.P., Annuk, H., Prendergast, M.M., 2005a. Antibodies induced by ganglioside-mimicking *Campylobacter jejuni* lipooligosaccharides recognise epitopes at the nodes of Ranvier. J. Neuroimmunol. 165, 179–185.

Moran, A.P., Kharmi, W., Walker, M.M., Thursz, M.R., 2005b. Role of surfactant protein D (SP-D) in innate immunity in the gastric mucosa: evidence of interaction with *Helicobacter pylori* lipopolysaccharide. J. Endotoxin Res. 11, 357–362.

Murata, K., Egami, H., Shibata, Y., Sakamoto, K., Misumi, A., Ogawa, M., 1992. Expression of blood group-related antigens, ABH, Lewis^a, Lewis^b, Lewis^x, Lewis^y, CA19-9, and CSLEX1 in early cancer, intestinal metaplasis, and uninvolved mucosa of the stomach. Am. J. Clin. Pathol. 98, 67–75.

Murray, E., Khamri, W., Walker, M.M., et al., 2002. Expression of surfactant protein D in human gastric mucosa and *Helicobacter pylori* infection. Infect. Immun. 70, 1481–1487.

Nachamkin, I., Ung, H., Moran, A.P., et al., 1999. Ganglioside GM1 mimicry in *Campylobacter* strains from sporadic infections in the United States. J. Infect. Dis. 179, 1183–1189.

Negrini, R., Lisato, L., Zanella, I., et al., 1991. *Helicobacter pylori* infection induces antibodies cross-reacting with human gastric mucosa. Gastroenterology 101, 437–445.

Negrini, R., Savio, A., Poiesi, C., et al., 1996. Antigenic mimicry between *Helicobacter pylori* and gastric mucosa in the pathogenesis of body atrophic gastritis. Gastroenterology 111, 655–665.

Neisser, A., Bernheimer, H., Berger, T., Moran, A.P., Schwerer, B., 1997. Serum antibodies against gangliosides and *Campylobacter jejuni* lipopolysaccharides in Miller Fisher syndrome. Infect. Immun. 65, 4038–4042.

Neisser, A., Schwerer, B., Bernheimer, H., Moran, A.P., 2000. Ganglioside-induced antiganglioside antibodies from a neuropathy patient cross-react with lipopolysaccharides of *Campylobacter jejuni* associated with Guillain-Barré syndrome. J. Neuroimmunol. 102, 85–88.

Nilsson, C., Skoglund, A., Moran, A.P., Annuk, A., Engstrand, L., Normark, S., 2006. An enzymatic ruler modulates Lewis antigen glycosylation of *Helicobacter pylori* LPS during persistent infection. Proc. Natl. Acad. Sci. USA 103, 2863–2868.

Nilsson, C., Skoglund, A., Moran, A.P., Annuk, H., Engstrand, L., Normark, S., 2008. Lipopolysaccharide diversity evolving in *Helicobacter pylori* communities through genetic modifications in fucosyltransferases. PLoS ONE 3, e3811.

Notterpek, L., Tolwani, R.J., 1999. Experimental models of peripheral neuropathies. Lab. Animal Sci. 49, 588–599.

O'Keeffe, J., Moran, A.P., 2008. Conventional, regulatory and unconventional T-cells in the immunological response to *Helicobacter pylori*. Helicobacter 13, 1–19.

Olson, J.K., Ercolini, A.M., Miller, S.D., 2005. A virus-induced molecular mimicry model of multiple sclerosis. Curr. Top. Microbiol. Immunol. 296, 39–53.

Paterson, G., Wilson, G., Kennedy, P.G.E., Willison, H.J., 1995. Analysis of anti-GM_1 antibodies cloned from motor neuropathy patients demonstrates diverse variable region gene usage with extensive somatic mutation. J. Immunol. 155, 3049–3059.

Poly, F., Read, T.D., Chen, Y.H., et al., 2008. Characterization of two *Campylobacter jejuni* strains for use in volunteer experimental-infection studies. Infect. Immun. 76, 5655–5667.

Prendergast, M.M., Moran, A.P., 2000. Lipopolysaccharides in the development of the Guillain-Barré syndrome and Miller Fisher syndome forms of acute inflammatory peripheral neuropathies. J. Endotoxin Res. 6, 341–359.

Prendergast, M.M., Lastovica, A.J., Moran, A.P., 1998. Lipopolysaccharides from *Campylobacter jejuni* O:41 strains associated with Guillain-Barré syndrome exhibit mimicry of GM_1 ganglioside. Infect. Immun. 66, 3649–3655.

Prendergast, M.M., Willison, H.J., Moran, A.P., 1999. Human monoclonal IgM antibodies to GM_1 ganglioside show diverse cross-reactivity with lipopolysaccharides of *Campylobacter jejuni* strains associated with Guillain-Barré syndrome. Infect. Immun. 67, 3698–3701.

Prendergast, M.M., Tribble, D.R., Baqar, S., et al., 2004. *In vivo* phase variation and serological response to lipooligosaccharide of *Campylobacter jejuni* in experimental human infection. Infect. Immun. 72, 916–922.

Rieder, G., Hatz, R.A., Moran, A.P., Walz, A., Stolte, M., Enders, G., 1997. Role of adherence in interleukin-8 induction in *Helicobacter pylori*-associated gastritis. Infect. Immun. 65, 3622–3630.

Ritter, G., Fortunato, S.R., Cohen, L., et al., 1996. Induction of antibodies reactive with GM_2 ganglioside after immunization with lipopolysaccharides from *Campylobacter jejuni*. Int. J. Cancer 66, 184–190.

Saida, T., Saida, K., Lisak, R.P., Brown, M.J., Silberberg, D.H., Asbury, A.K., 1982. *In vivo* demyelinating activity of sera from patients with Guillain-Barré syndrome. Ann. Neurol. 11, 69–75.

Sakamoto, J., Yin, B.T., Lloyd, K.O., 1984. Analysis of the expression of H, Lewis, X, Y and precursor blood group determinants in saliva and red cells using a panel of mouse monoclonal antibodies. Mol. Immunol. 21, 1093–1098.

Sakamoto, J., Watanabe, T., Tokumare, T., Takagi, H., Nakazato, H., Lloyd, K.O., 1989. Expression of Lewis[a], Lewis[b], Lewis[x], Lewis[y], silayl-Lewis[a] and silayl-Lewis[x] blood group antigens in human gastric carcinoma and in normal gastric tissue. Cancer Res. 49, 745–752.

Schwerer, B., Neisser, A., Polt, R.J., Bernheimer, H., Moran, A.P., 1995. Antibody crossreactivities between gangliosides and lipopolysaccharides of *Campylobacter jejuni* serotypes associated with Guillain-Barré syndrome. J. Endotoxin Res. 2, 395–403.

Senchenkova, S.N., Zatonsky, G.V., Hynes, S.O., et al., 2001. Structure of a D-*glycero*-D-*manno*-heptan from the lipopolysaccharide of *Helicobacter pylori*. Carbohydr. Res. 331, 219–224.

Sieling, P.A., Chatterjee, D., Porcelli, S.A., et al., 1995. CD1-restricted T-cell recognition of microbial lipoglycans. Science 269, 227–230.

Stöckl, J., Majdic, O., Rosenkranz, A., et al., 1993. Monoclonal antibodies to the carbohydrate structure Lewis[x] stimulate the adhesive activity of leukocyte integrin CD11b/CD18 (CR3, Mac-1, $\alpha_m\beta_2$) on human granulocytes. J. Leukocyte Biol. 53, 541–549.

Sumner, A., Said, G., Idy, I., Metral, S., 1982. Demyelinative conduction block produced by intraneural injection of human Guillain-Barré serum into rat sciatic nerve. Neurology 32, 106.

Sun, H.Y., Lin, S.W., Ko, T.P., et al., 2007. Structure and mechanism of *H. pylori* fucosyltransferase. A basis for lipopolysaccharide variation and inhibitor design. J. Biol. Chem. 282, 9973–9982.

Takata, T., El-Omar, E., Camorlinga, M., et al., 2002. *Helicobacter pylori* does not require Lewis X and Lewis Y expression to colonize C3H/HeJ mice. Infect. Immun. 70, 3073–3079.

Taylor, D.E., Rasko, D.A., Sherburne, R., Ho, C., Jewell, L.D., 1998. Lack of correlation between Lewis antigen expression by *Helicobacter pylori* and gastric epithelial cells in infected patients. Gastroenterology 115, 1113–1122.

Taylor, W.A., Hughes, R.A.C., 1989. T lymphocyte activation antigens in Guillain-Barré syndrome and chronic idiopathic demyelinating polyradiculoneuropathy. J. Neuroimmunol. 24, 33–39.

Velupillai, P., Harn, D.A., 1994. Oligosaccharide-specific induction of interleukin 10 production by B220$^+$ cells from schistosome-infected mice: a mechanism for regulation of CD4$^+$ T-cell subsets. Proc. Natl. Acad. Sci. USA 91, 18–22.

Vorobjova, T., Faller, G., Maaroos, H.-I., et al., 2000. Significant increase in antigastric autoantibodies in a long-term follow-up study of *H. pylori* gastritis. Virchows Arch. 437, 37–45.

Westall, F.C., 2006. Molecular mimicry revisited: gut bacteria and multiple sclerosis. J. Clin. Microbiol. 44, 2099–2104.

Willison, H.J., O'Hanlon, G.M., 2000. Anti-glycospingolipid antibodies and Guillain-Barré syndrome. In: Nachamkin, I., Blaser, M.J. (Eds.), *Campylobacter*, 2nd edn. ASM Press, Washington, DC, pp. 259–285.

Willison, H.J., Veitch, J., 1994. Immunoglobulin subclass distribution and binding characteristics of anti-GQ_{1b} antibodies in Miller Fisher syndrome. J. Neuroimmunol. 50, 159–165.

Wirguin, I., Briani, C., Suturkova-Milosevic, L., et al., 1997. Induction of anti-GM_1 antibodies by *Campylobacter jejuni* lipopolysaccharides. J. Neuroimmunol. 78, 138–142.

Wirth, H.P., Yang, M., Karita, M., Blaser, M.J., 1996. Expression of the human cell surface glycoconjugates Lewis X and Lewis Y by *Helicobacter pylori* isolates is related to *cagA* status. Infect. Immun. 64, 4598–4605.

Wirth, H.P., Yang, M., Peek, R.M., Tham, K.T., Blaser, M.J., 1997. *Helicobacter pylori* Lewis expression is related to the host Lewis phenotype. Gastroenterology 113, 1091–1098.

Wirth, H.P., Yang, M., Peek, R.M., Höök-Nikanne, J., Fried, M., Blaser, M.J., 1999. Phenotypic diversity in Lewis expression of *Helicobacter pylori* isolates from the same strain. J. Lab. Clin. Med. 133, 488–500.

Wirth, H.P., Yang, M., Sanabria-Valentin, E., Berg, D.E., Dubois, A., Blaser, M.J., 2006. Host Lewis phenotype-dependent *Helicobacter pylori* Lewis antigen expression in rhesus monkeys. FASEB J. 20, 1534–1536.

Yuki, N., 2001. Infectious origins of, and molecular mimicry in, Guillain-Barré and Fischer syndromes. Lancet Infect. Dis. 1, 29–37.

Yuki, N., 2005. Carbohydrate mimicry: a new paradigm of autoimmune diseases. Curr. Opin. Immunol. 17, 577–582.

Yuki, N., 2007. Ganglioside mimicry and peripheral nerve disease. Muscle Nerve 35, 691–711.

Yuki, N., Taki, T., Inagaki, F., et al., 1993. A bacterium lipopolysaccharide that elicits Guillain-Barré syndrome has a GM_1 ganglioside-like structure. J. Exp. Med. 178, 1771–1775.

Yuki, N., Yamada, M., Koga, M., et al., 2001. Aninal model of axonal Guillain-Barré syndrome induced by sensitization with GM1 ganglioside. Ann. Neurol. 49, 712–720.

Yuki, N., Susuki, K., Koga, M., et al., 2004. Carbohydrate mimicry between human ganglioside GM1 and *Campylobacter jejuni* lipooligosaccharide causes Guillain-Barré syndrome. Proc. Natl. Acad. Sci. USA 101, 11404–11409.

Zandman-Goodard, G., Shoenfeld, Y., 2005. Infections and SLE. Autoimmunity 38, 473–485.

CHAPTER

44

Role of microbial glycosylation in host cell invasion

Margaret I. Kanipes and Patricia Guerry

SUMMARY

Gram-negative bacteria possess a number of cell surface glycans that can play variable roles in pathogenesis. This chapter will discuss the role of surface carbohydrate structures in the ability of several invasive Gram-negative pathogens to interact with host cells. The role of microbial glycosylation systems, including lipopolysaccharides, lipo-oligosaccharides, capsular polysaccharides and glycoproteins, in host cell invasion of bacteria will be compared. Discussion of glycosylation systems that are covered in other chapters will be minimized and emphasis will be on selected, representative invasive organisms, particularly *Campylobacter jejuni*.

Keywords: Lipopolysaccharides; Lipo-oligosaccharides; Capsules; Capsular polysaccharide; Glycoproteins; *Campylobacter*; *Salmonella*; Invasion; Pathogenesis; Virulence

1. INTRODUCTION

Cell surface glycans have been shown to play an important role in the biosynthesis and regulation of the cell wall of pathogenic Gram-negative bacteria. These glycans include lipopolysaccharides (LPSs), lipo-oligosaccharides (LOSs), capsular polysaccharides (CPSs) and *N*- and *O*-glycoproteins (see Chapters 1, 3, 4, 5, 7 and 9). The analysis of the carbohydrate modifications in these systems with regards to biosynthesis, genetics and structure–function studies varies tremendously among microorganisms. However, this information has been instrumental for studies on understanding the pathogenesis of Gram-negative infection. In this chapter, we summarize our present understanding of the role of the various microbial glycosylation systems in cell invasion.

2. LPSs AND LOSs IN CELL INVASION

The major structural component of the surface of all Gram-negative bacteria is either LPS or LOS. The LPS molecule consists of three domains: a membrane anchor, lipid A, a non-repeating core oligosaccharide consisting of an inner and outer core region, and the O-antigen repeat (Raetz and Whitfield, 2002; Raetz *et al.*, 2007). Many Gram-negative mucosal pathogens lack the O-antigen repeat and have LOS structures composed of lipid A attached to an inner and outer core. In bacteria with LPSs, the inner and outer cores are less variable than the O-antigen repeating units which, in the case of the *Enterobacteriaceae*, form the basis for O-antigen serotyping. In contrast, for bacteria with LOS cores, the inner cores are often quite conserved and the outer cores are often hypervariable. Both LPSs and LOSs can function to facilitate intracellular survival, to avoid the host immune response and/or to serve as ligands to interact with host cell receptors to mediate invasion (Nikaido, 1996).

Bacterial invasion of the classic enteropathogens *Salmonella* and *Shigella* requires a type 3-secretion system (T3SS) that secretes proteins that mediate bacterial uptake into eukaryotic cells. In the case of *Shigella*, which causes bacillary dysentery, invasion is limited to the superficial epithelial layer of the intestine and results in an intense inflammatory response and formation of abscesses in the mucosa. The LPS structure is critical for invasion by *Shigella* species. The *Shigella* O-antigen is a tri-rhamnose-GlcNAc tetrasaccharide (where GlcNAc, *N*-acetyl-D-glucosamine), modified by the addition of glucose (Glc) and/or acetate, depending on the serotype (West *et al.*, 2005). Loss of glucosylation in *Shigella flexneri* serotype 5 resulted in decreases in epithelial cell invasion *in vitro* and reduced inflammation *in vivo* (West *et al.*, 2005). These defects could be restored by complementation with either the homologous glucosylation locus or the corresponding loci from *Shigella* strains of other serotypes. Glucosylation is predicted to induce a transition from a linear to a helical conformation of serotype 5a LPS, resulting in a more compact structure than the unglucosylated form (Clement *et al.*, 2003). West *et al.* (2005) showed that this compact structure optimized exposure of the T3SS needle, facilitating interaction with the eukaryotic cell. Thus, while mutants with truncated LPSs are more proficient at invasion of epithelial cells, presumably because the T3SS needle is more exposed, this advantage is offset by increased susceptibility of the rough bacterial cell to the host immune system. The structural changes that occur following glucosylation of the O-antigen repeat allow the needle to interact with the host cell while retaining the selective advantages resulting from expression of full length LPS.

Most serovars of *Salmonella* are invasive for epithelial cells *in vitro* but, in contrast to *Shigellae*, are capable of deeper invasion and replication within the reticulo-endothelial system. Of note, LPS contributes to *Salmonella* virulence by protecting the cell from the immune system and providing a barrier function to the outer membrane. An early study indicated that rough strains of *Salmonella enterica* sv. Typhimurium are not affected in their ability to invade epithelial cells *in vitro* (Kihlstrom and Edebo, 1976). However, it was recently shown that an *S. enterica* sv. Typhimurium *rfaE* mutant, which had an LPS composed of only Kdo_3-lipid A (Kdo, 3-deoxy-D-*manno*-oct-2-ulosonic acid), had a significantly reduced ability to invade human intestinal epithelium Henle-407 cells and human larynx epidermal carcinoma HEp-2 cells (Kim, 2003). The ability of the *rfaE* mutant to invade epithelial cell lines was restored to wild-type levels when the *rfaE* gene was introduced back into the mutant cell. Nevertheless, this deep-rough mutant also showed increased sensitivity to antibiotics, indicative of membrane perturbations (Kim, 2003). This suggests that the contribution to invasion relates to survival rather than direct interaction of LPS with epithelial cells. Rough mutants of *S. enterica* sv. Cholerasuis are also deficient in invasion (Finlay *et al.*, 1988). Additionally, other studies suggest that LPS contributes to the invasive phenotype indirectly by promoting intracellular survival of salmonellae, probably by interacting with antimicrobial peptides (Guo *et al.*, 1998; Ernst *et al.*, 1999).

Although an early study had indicated that intact LPS was required for invasion of *S. enterica* sv. Typhi (Mroczenski-Wildey *et al.*, 1989), a more recent study found no effect of O-antigen on adherence or invasion of HEp-2 cells (Hoare *et al.*,

2006). However, the same study showed that in serovar Typhi, the terminal Glc residue of the outer core of the LPS (GlcII) was critical for the adherence and entry of *S. enterica* sv. Typhi into epithelial cells. This suggests that a terminal Glc moiety on the outer core functions as a ligand for interaction with epithelial cells. Nonetheless, the presence of O-repeats would be expected to inhibit interaction of this terminal outer core sugar with epithelial cells. In this study, the authors presented data suggesting that not all core structures on an individual cell were capped with O-antigen and that some terminal Glc moieties on the outer core remain exposed for ligand interactions. These data are consistent with data in *Pseudomonas aeruginosa* in which a complete outer core with a terminal Glc was necessary for maximal adherence and invasion of *P. aeruginosa* into corneal epithelial cells (Zaidi *et al.*, 1996).

Bordetella bronchiseptica is a human pathogen that colonizes the respiratory tract of its host. A *B. bronchiseptica waaC*, or deep rough mutant, showed reduced levels of invasion compared to the wild-type strain in human pulmonary epithelial cells (Sisti *et al.*, 2002). In contrast, a *waaC* mutant of *Burkholderia cepacia*, a pathogen of cystic fibrosis patients, invaded human monocyte derived macrophages at levels comparable to the *B. cepacia* ATCC 25416 wild-type strain. These results suggest that the core oligosaccharide and LPS are not crucial for invasion or survival of *B. cepacia* into its host cells (Gronow *et al.*, 2003).

The food-borne pathogen *Campylobacter jejuni* is a common cause of human gastroenteritis worldwide (Friedman *et al.*, 2000; Oberhelman and Taylor, 2000). The organism is generally considered to be invasive, although there is a remarkable range in the ability of individual strains to invade intestinal epithelial cells *in vitro*. *C. jejuni* expresses LOS, rather than LPS, an unusual trait for an enteric pathogen. Based on structural studies, the *Campylobacter* inner core region of LOS has been shown to be highly conserved (Moran *et al.*, 2000; Gilbert *et al.*, 2002). This region consists of a single Kdo attached to lipid A and two L-*glycero*-D-*manno*-heptose (heptose) residues attached to the Kdo. The inner core region of *C. jejuni* strains may also contain additional sugar residues, phosphoryl-substituents and, in some strains, the second heptose is modified with *O*-linked glycine (Houliston *et al.*, 2006; Dzieciatkowska *et al.*, 2007). The outer core LOS region of *Campylobacter* strains is variable among strains and is also subject to phase variation by slip-strand mismatch repair. Most strains of *C. jejuni* contain outer core structures that mimic human gangliosides (reviewed in Moran *et al.*, 2000) (see also Chapter 43). These structures include sialic acid (Neu5Ac, *N*-acetylneuraminic acid) that is synthesized endogenously by the bacterium and which can be *O*-acetylated (Houliston *et al.*, 2006; Dzieciatkowska *et al.*, 2007). It is not unusual for a given strain to undergo phase variation that allows for synthesis of mixtures of ganglioside mimics. For example, the outer core of strain 81–176 is a mixture of GM_2 and GM_3 ganglioside mimics due to slip-strand repair at an *N*-acetyl-D-galactosamine (GalNAc) transferase gene (Guerry *et al.*, 2002). Similarly, the outer core of strain NCTC 11168 is a mixture of GM_2 and GM_1 mimics due to phase variation at a D-galactose (Gal) transferase gene (Linton *et al.*, 2000). Antibodies generated against these ganglioside mimics can result in an autoimmune response that can attack peripheral nerves resulting in Guillain Barré syndrome (GBS) (see Chapter 43). However, it is important to point out that not all strains of *C. jejuni* express sialylated cores that mimic gangliosides. There is one reported strain of *C. jejuni*, which was isolated from chicken skin, that contains an outer core that mimics human P blood group antigen (Mandrell *et al.*, 2008) and there are other clinical isolates that lack all mimicry (Aspinall *et al.*, 1995; Poly *et al.*, 2008) in their LOS cores. Thus, the role of ganglioside mimics in development of autoimmune responses such as GBS has been established (see Chapter 43), but the role of the mimics in pathogenesis of diarrhoeal disease is less clear.

There are limited studies on the role of LOS cores in invasion by *C. jejuni*. Mutation of the *galE* gene in strain 81116, thought to encode a uridine diphosphate- (UDP-) Glc-4-epimerase, resulted in a truncated LOS core. The mutant invaded INT407 epithelial cells at less than 1% of the level of the parent strain (Fry *et al.*, 2000). However, additional studies on this gene (which has been renamed *gne*) in another strain of *C. jejuni*, NCTC 11168, showed that the gene encodes a bifunctional enzyme that can convert UDP-Glc to UDP-Gal and UDP-GlcNAc to UDP-GalNAc (Bernatchez *et al.*, 2005). Mutation of this gene in NCTC 11168 affected the structure of LOS, the CPS and the *N*-linked protein glycosylation system (see below). Hence, the invasion defect reported by mutation of the corresponding gene in 81116 is likely due to previously described defects in invasion resulting from loss of capsule and the *N*-linked protein glycosylation system, as discussed below.

More extensive studies on the role of LOS have been performed in strain 81–176, which invades epithelial cells *in vitro* at higher levels than most other *C. jejuni* strains. Mutants that have been examined are summarized in Figure 44.1. The wild-type core of 81–176 is composed of approximately equal mixtures of GM_2 and GM_3 gangliosides (only the GM_3 mimic is shown for wild-type in Figure 44.1). This mixture of

CgtA GalT LgtF WaaF WaaC

Core structure	Strain	% Relative Invasion
β-GalNAc-(1->4)-β-Gal-(1->4)-β-Glc-(1->2)-L-α-D-Hep-(1->3)-L-α-D-Hep-(1->5)-Kdo; α-Neu5Ac-(2→3, Cst) on Gal; β-Glc-(1→4, LgtF) on Hep	WT	100
β-Gal-(1->4)-β-Glc-(1->2)-L-α-D-Hep-(1->3)-L-α-D-Hep-(1->5)-Kdo; α-Neu5Ac-(2→3) on Gal; β-Glc-(1→4) on Hep	*cgtA*	229
β-GalNAc-(1->4)-β-Gal-(1->4)-β-Glc-(1->2)-L-α-D-Hep-(1->3)-L-α-D-Hep-(1->5)-Kdo; β-Glc-(1→4) on Hep	*neuC*	100
β–Glc-(1->2)-L-α-D-Hep-(1->3)-L-α-D-Hep-(1->5)-Kdo; β-Glc-(1→4) on Hep	*galT*	100
L-α–D-Hep-(1->3)-L-α-D-Hep-(1->5)-Kdo	*lgtF*	100
L-α-D-Hep-(1->5)-Kdo	*waaF*	100
Kdo	*waaC*	<1

FIGURE 44.1 Role of *C. jejuni* 81–176 LOS in invasion of INT407 cells. The core of wild-type 81–176 and a series of mutants is shown. The enzymes involved in synthesis of the core are shown above the appropriate linkage in the wild-type (WT) core. Wild-type 81–176 core is a mixture of the structures shown and a core equivalent to that of the *cgtA* mutant due to slip-strand mismatch repair in the *cgtA* gene (Guerry *et al.*, 2001). The outer core of the wild-type as drawn is a GM_2 ganglioside mimic and that of the *cgtA* mutant is a GM_3 ganglioside mimic. The relative invasion levels of each mutant into INT407 cells are shown relative to a control of wild-type. The *cgtA* and *neuC* mutants are described in Guerry *et al.* (2001); the *galT* and *waaF* mutants are described in Kanipes *et al.* (2004, 2008) and the *waaC* mutant are described in Kanipes *et al.* (2004). Invasion assays are described in the same references. All cores were decorated with phosphoethanolamine (PEA) at HepI (not shown in figure). Abbreviations: Gal, galactose; GalNAc, *N*-acetyl-D-galactosamine; Glc, glucose; Hep, heptose; Kdo, 3-deoxy-D-*manno*-oct-2-ulosonic acid; Neu5Ac, *N*-acetylneuraminic acid (sialic acid).

LOS cores results from slip-strand mismatch repair of the *cgtA* gene, which encodes a GlcNAc transferase. Mutation of this gene results in a core that is locked into a GM_3 ganglioside mimic (see Figure 44.1). The *cgtA* mutant invaded INT407 epithelial cells at levels that were approximately twice that of the wild-type strain expressing the mixture of GM_2 and GM_3 mimics. Interestingly, although loss of Neu5Ac from the outer core by mutation of a sialic acid biosynthetic gene affected serum resistance in another strain of *C. jejuni* (Guerry *et al.*, 2000), there was no measurable effect on invasion of 81–176 into INT407 cells. This is consistent with the observation of clinical isolates of *C. jejuni* that lack all Neu5Ac in their LOS cores, as mentioned above (Aspinall *et al.*, 1995; Poly *et al.*, 2008). Similarly, mutation of *galT*, encoding a galactosyltransferase, or *lgtF*, encoding a putative bifunctional glucosyltransferase, had no effect on the ability of 81–176 to invade epithelial cells (Kanipes *et al.*, 2008). The *waaC* and *waaF* genes encode the heptosyltransferases I and II that catalyse the transfer of the first and second heptose residues to the LOS core oligosaccharide, respectively. Mutation of *waaF* had no effect on invasion (Kanipes *et al.*, 2004, 2008), but the heptose-less *waaC* mutant had a severely truncated LOS composed of Kdo and lipid A and was non-invasive *in vitro* (Kanipes *et al.*, 2006, 2008). Nevertheless, mutation of *waaC* also affected CPS structure such that this mutant also lacked a 3-*O*-methyl group from the 6-deoxy-*altro*-heptose residue in the capsule (see below). Both LOS core and capsule structure, as well as invasion, were restored when the mutant was complemented (Kanipes *et al.*, 2006). Thus, the invasion defect could be due to either the defect in the core or capsule. Collectively, these studies suggest that the LOS core of the *waaF* mutant defines a minimal structure for effective invasion and/or intracellular survival of *C. jejuni* (Kanipes *et al.*, 2008). Since *waaC* mutants also show enhanced sensitivity to antibiotics such as polymyxin, it is likely that a minimum core is required to protect against intracellular killing rather than adherence/invasion *per se*. However, the data also suggest that the GalNAc moiety may play a direct role in interactions with the eukaryotic cell, but additional studies are needed to confirm this. Similar results have been reported for *Helicobacter pylori*, which is related to *C. jejuni*, but which contains LPS. A *waaF* mutant of *H. pylori*, which had lost O-antigen and contained a core composed of only heptose, Kdo and lipid A, was also unaffected in invasion of human gastric adenocarcinoma cells *in vitro* (Chandan *et al.*, 2007).

Haemophilus influenzae and *Haemophilus ducreyi* are both important human pathogens that synthesize LOSs. A number of LOS mutants (Gibson *et al.*, 1997; Bauer *et al.*, 1998, 1999) have also been studied for their role in virulence in *H. ducreyi*, which is the causative agent of the sexually transmitted disease known as chancroid. A mutant in the *waaF* gene, encoding heptosyltransferase II, was shown to be less virulent than the wild-type strain in a rabbit model for dermal lesion production (Bauer *et al.*, 1999). The *H. ducreyi* 35000*gmhA* mutant produced a truncated LOS and also showed reduced virulence in the rabbit model for dermal lesion production (Bauer *et al.*, 1998). Similar studies found that the *H. influenzae* type b *gmhA* mutant showed reduced virulence in an infant rat model (Cope *et al.*, 1990). The *gmhA* gene product has been shown to catalyse the conversion of sedoheptulose-7-phosphate into D-glycero-D-*manno*-heptose-7-phosphate. This is an important step in the synthesis of heptose for the core oligosaccharide of LOS in Gram-negative bacteria such as *H. influenzae* and *H. ducreyi*. Inactivation of the *H. ducreyi rfaK* gene resulted in a reduced ability to adhere to and invade human keratinocytes (Gibson *et al.*, 1997). However, the same mutant did not affect the ability of *H. ducreyi* to produce dermal lesions in a rabbit model (Stevens *et al.*, 1997).

H. influenzae is a common pathogen of the human upper respiratory tract and can cause sinusitis, otitis media and bronchitis (Rao *et al.*, 1999) and adherence and invasion have been shown to be critical to pathogenesis. The core oligosaccharide of *H. influenzae* consists of variable sugar branches containing sugar residues such as Glc, Gal, GlcNAc, Neu5Ac and phosphorylcholine (ChoP) (Masoud *et al.*, 1997). Swords *et al.* (2000) examined the role of non-typeable *H. influenzae* (NTHi) 2019 LOS structure in adherence of human bronchial epithelial cells. The NTHi 2019 *pgmB* gene encodes a phosphomutase and deletion of this gene resulted in an LOS lacking the oligosaccharide chain distal to the core domain. The NTHi *pgmB* LOS was less adherent and invasive in a human bronchial epithelial cell line when compared to the wild-type strain (Swords *et al.*, 2000). Further experiments showed that a ChoP moiety was involved in adherence and that LOS ChoP may influence NTHi invasion of bronchial cells via an interaction with the platelet activating factor (PAF) receptor (Swords *et al.*, 2000) (see Chapter 42). The roles of the LOSs and LPSs in host cell invasion with regards to the aforementioned Gram-negative bacteria are highlighted in Table 44.1.

3. ROLE OF BACTERIAL CAPSULES IN INVASIVENESS

Capsules are considered virulence factors for all bacteria that express these surface structures, although the function can vary (Schrager *et al.*, 1996; Magee and Yother, 2001; O'Riordan and Lee, 2004; Locke *et al.*, 2007; Nelson *et al.*, 2007). The capsule of *Neisseria meningitidis* inhibits bacterial adhesion, but is important for intracellular survival because it confers resistance to cationic peptides (Spinosa *et al.*, 2007). Similarly, the capsules of *Klebsiella pneumoniae* and *Streptococcus iniae* also confer resistance to antimicrobial peptides (Campos *et al.*, 2004; Locke *et al.*, 2007). The Neu5Ac-containing capsule of *E. coli* K1, which causes meningitis, contributes to resistance to phagocytosis and serum resistance and maintains viability within brain microvascular endothelial cells (BMEC), but is not required for invasion of BMECs (Hoffman *et al.*, 1999). The Vi capsular polysaccharide from *S. enterica* sv. Typhi has been shown to down-regulate the host's inflammatory responses. Loss of the Vi-antigen of *S. enterica* sv. Typhi results in increased levels of Toll-like receptor- (TLR-) 4 and TLR-5-mediated interleukin- (IL-) 8 release from Caco2 cells, human macrophage-like cells (THP-1) and human intestinal explant tissue (Raffatellu *et al.*, 2005). Purified Vi-antigen binds to Caco2 cells via prohibitin in lipid rafts and is reported to perturb tight junctions of Caco-2 cells (Sharma and Qadri, 2004), an event that may be related to invasion.

The polysaccharide capsule of *C. jejuni* represents one of the few well-defined virulence factors in this invasive pathogen. The CPS is the major serodeterminant of the Penner serotyping scheme and is structurally very diverse. Moreover, the CPS of *C. jejuni* strain 81–176, and presumably other strains, undergoes a rapid on/off phase variation by slip-strand mismatch repair at unidentified genes (Bacon *et al.*, 2001). Additionally, there are phase variable genes that encode modifications to the sugars, including methyl groups and the highly unusual methyl phosphoramidate group (Szymanski *et al.*, 2003; Kanipes *et al.*, 2006; McNally *et al.*, 2007). In general, polysaccharide capsules are important for bacterial survival in extracellular fluids and it is unusual to find a capsule on an intestinal pathogen like *C. jejuni*, which does not normally penetrate beyond the mucosal layer of the intestine. There have been surprisingly few studies on the role of the capsule in virulence of *C. jejuni*. A *kpsM* mutant of strain 81–176 showed a modest decrease in adherence to INT407 cells after 2h incubation (54% the level of wild-type) and invasion (about 11% that of wild-type) (Bacon *et al.*, 2001). Complementation of the mutation on a plasmid *in trans* resulted in partial restoration of both adherence and invasion. However, Bachtiar *et al.* (2007) found that a *kpsE* mutant of another strain, 81116, showed

TABLE 44.1 Summary of the role of LPSs and LOSs in bacterial invasion

Bacterium	LOS or LPS	Observation	Host cell type	References
S. flexneri, M90T	LPS	Truncated LPS mutants are highly proficient at invasion	Epithelial cells	West *et al.*, 2005
S. enterica sv. Typhimurium	LPS	Deep rough mutant shows reduced invasion ability due to survival	Henle-407 human intestinal epithelial and human larynx epidermal carcinoma cells	Kim, 2003
S. enterica sv. Cholerasuis	LPS	Deep rough mutant shows reduced invasion ability	Epithelial cells	Finlay *et al.*, 1988
S. enterica sv. Typhi	LPS	Terminal Glc residue on outer core LOS serves as a ligand for interaction with epithelial cells	Hep-2	Hoare *et al.*, 2006
P. aeruginosa	LPS	Complete outer core with an exposed terminal Glc residue is necessary for maximal adherence and entry	Corneal epithelial cells	Zaidi *et al.*, 1996
B. bronchiseptica	LPS	Deep rough mutant shows reduced levels of invasion	Human pulmonary epithelial cells	Sisti *et al.*, 2002
B. cepacia	LPS	Deep rough mutant has no effect on invasion levels	Human monocyte macrophages	Gronow *et al.*, 2003
C. jejuni 81116	LOS	Reduced levels of invasion that possibly relate to loss of polysaccharide capsule and *N*-linked glycosylation system	INT407 epithelial cells	Fry *et al.*, 2000
C. jejuni 81-176	LOS	Reduced levels of invasion relate to survival and not LOS	INT407 epithelial cells	Kanipes *et al.*, 2006, 2008
H. pylori	LOS	Deep rough mutant unaffected in invasion	Human gastric adenocarinoma cells	Chandan *et al.*, 2007
H. influenzae	LOS	Deep rough mutant is less virulent than wild-type strain	Infant rat model	Cope *et al.*, 1990
H. influenzae	LOS	LOS lacking the side chain was less invasive	Human bronchial epithelial cells	Swords *et al.*, 2000
H. ducreyi 35000	LOS	Deep rough mutants show reduced ability to adhere and invade	Rabbit model for dermal lesion	Bauer *et al.*, 1998
H. ducreyi	LOS	Reduced ability to adhere and invade	Human keratinocytes	Gibson *et al.*, 1997

a greater decrease in adherence (5.8% the level of wild-type) and invaded at about 50% wild-type levels. This mutant, however, was not complemented. Although the statistical differences between wild-type and mutant were significant in both studies, the biological significance of these relatively minor differences in *in vitro* assays is uncertain. Moreover, phase variation of both CPS expression and structure further complicates analyses of these data. Nonetheless, the *kpsM* mutant of 81–176 was more sensitive to killing by normal human serum and was attenuated in the ferret model of diarrhoeal disease compared to the wild-type (Bacon *et al.*, 2001). However, the specific role that CPS plays in *C. jejuni* disease remains somewhat obscure. Zilbauer *et al.* (2005) reported that the capsule of *C. jejuni* strain NCTC 11168 did not contribute to resistance to human defensins 2 and 3 (hBD-2 and hBD-3). Thus, it remains to be determined if CPS plays a direct role in *C. jejuni* adherence and/or invasion or if the effect is indirect by increasing survival intracellularly.

4. ROLE OF PROTEIN GLYCOSYLATION IN INVASION

Until the late 1970s, it was believed that protein glycosylation was exclusively a eukaryotic phenomenon. However, there is an increased awareness that there are numerous prokaryotic glycoproteins. Although these prokaryotic glycoproteins, many of which are surface-exposed, may play critical roles in virulence of some bacterial pathogens (Benz and Schmidt, 2001), in most cases, very few details are currently available.

4.1. N-linked glycosylation systems

N-linked glycosylation of several proteins in *Chlamydia trachomatis* has been linked to infectivity of elementary bodies (Swanson and Kuo, 1990, 1991, 1994) and glycoproteins of unknown function have been described in *Borrelia burgdorferi* (Sambri *et al.*, 1992, 1993).

The best-characterized bacterial *N*-linked glycoproteins are those found in *C. jejuni*. This glycosylation system, which is highly conserved among strains of *C. jejuni*, is responsible for addition of an *N*-linked glycan to at least 45 distinct proteins (Szymanski *et al.*, 1999; Wacker *et al.*, 2002, 2006). The glycan is a heptasaccharide with the structure, α-GalNAc-(1→4)-α-GalNAc-(1→4)-[β-Glc-(1→3)-] α-GalNAc-(1→4)-α-GalNAc-(1→4)-α- GalNAc-(1→3)- β-Bac-(1→N-Asn-Xaa [where Bac is 2,4-diacetamido-2,4,6-trideoxyglucopyranose (bacillosamine); Asn, asparagine] (Young *et al.*, 2002). Interestingly, the heptasaccharide structure was shown to be linked to proteins at a site that is a subset of the eukaryotic sequon (Kowarik *et al.*, 2006).

Mutants in this *N*-linked glycosylation system in *C. jejuni* (or *pgl* for protein glycosylation) show pleiotrophic defects. Mutants in the *pgl* system have reduced levels of invasion of intestinal epithelial cells and reduced ability to colonize the intestinal tracts of mice and chickens (Szymanski *et al.*, 2002; Hendrixson and DiRita, 2004; Jones *et al.*, 2004). There are also major defects in natural transformation (Larsen *et al.*, 2004). The defects may not be due to loss of glycosylation of single proteins but may be due to multiple defects. Thus, Kakuda and DiRita (2006) constructed mutations in 22 genes predicted to encode glycoproteins and identified one that was defective in invasion, Cj1496c. Invasion was restored when this mutant was complemented with the wild-type gene *in trans* as well as a mutant lacking both sites for *N*-linked glycosylation. Thus, while Cj1496c was required for invasion, glycosylation of Cj1496c was not required.

4.2. O-Linked protein glycosylation

The best-characterized bacterial proteins with *O*-linked glycans are the pili of *Neisseria* species which are described in detail in other chapters of this book. *O*-Linked glycosylation has also been reported on the pili of *P. aeruginosa* (Castric *et al.*, 2001), numerous proteins of *Ehrlichia* spp. (McBride *et al.*, 2000, 2007) and *Mycobacterium tuberculosis* (Dobos *et al.*, 1995). However, the role of the *O*-linked glycans in virulence, if any, has not been established for most of these examples. Nevertheless, two autotransporters in an enterotoxigenic strain of *E. coli* and a diffusely adherent strain of *E. coli* contain heptose in an *O*-linkage and this glycan is required for adherence (Lindenthal and Elsinghorst, 2001; Benz and Schmidt, 2002).

Over the past few years it has been evident that the flagellins of most polar flagellated bacteria are glycosylated by *O*-linkages (Logan *et al.*, 2002) (see also Chapter 8). Mutants that are defective in glycosylation systems of most of these bacteria are fully motile. However, loss of these pathways in *C. jejuni* and *Helicobacter pylori*, both members of the epsilon division of *Proteobacteria*, results in loss of flagella. In *C. jejuni*, the modifications are based on 9-carbon sugars, pseudaminic acid and/or legionaminic acid, as well as variants of these two sugars (Thibault *et al.*, 2001; McNally *et al.*, 2006; see Chapter 8). At least some of these glycans, most of which map to the D2 and D3 domains of flagellin, are surface exposed in the filament and can confer serospecificity on the filament (Logan *et al.*, 2002). The genes encoding the enzymatic machinery to glycosylate *C. jejuni* flagellin map adjacent to the structural flagellin genes in one or the more hypervariable regions of the chromosome (Guerry *et al.*, 2006). Flagella are multifaceted virulence factors of *C. jejuni* (Guerry, 2007). Motility is required for intestinal colonization and invasion of epithelial cells *in vitro*. Additionally, in the absence of a specialized T3SS, flagella secrete a number of non-flagellar proteins, some of which contribute to virulence (Konkel *et al.*, 2004; Song *et al.*, 2004; Poly *et al.*, 2007, 2008). Since glycosylation is required for filament assembly in *C. jejuni*, the flagellar glycans are critical for virulence. Moreover, the surface exposed sugars on the flagellar filament appear to mediate interactions between bacterial cells that ultimately result in autoagglutination that leads to formation of microcolonies on epithelial cells (Guerry *et al.*, 2006). Mutants of strain 81–176 that express flagella that are decorated exclusively with pseudaminic acid and lack an acetamidino form of pseudaminic acid, are reduced in adherence and invasion of epithelial cells, an attenuation that is reflected in reduced virulence in an *in vivo* diarrhoeal disease model (Guerry, 2007). Thus, the surface-exposed glycans on flagella may interact directly with eukaryotic cells.

5. CONCLUSIONS

In this chapter, we have examined the biological role of microbial glycosylation in invasion, particularly the roles of LOSs/LPSs, CPSs and *O*- and *N*-linked glycoproteins have been reviewed. It is clear that, in some cases, the communication between the microbial pathogen and its host is due to recognition of these cell surface glycomolecules, thereby resulting in the ability of these organisms potentially to cause a multitude of infections and disease. In certain species, such as *Shigella* and *Pseudomonas* spp., the LPS core oligosaccharide is required for invasion (West *et al.*, 2005; Zaidi *et al.*, 1996). However, for micoorganisms such as *S. enterica* sv. Typhimurium and *C. jejuni*, the observed reduced invasion was observed not to be a result of a reduction in direct interaction with epithelial cells but was possibly a result of reduced intracellular survival (Kim, 2003; Kanipes *et al.*, 2006, 2008).

Studies on the role of bacterial CPSs and the *N*- and *O*-linked glycoproteins in invasion in pathogenic Gram-negative bacteria are limited. Open questions for future investigation are listed in the Research Focus Box. Nevertheless, research on the generation of mutants with defined deletions for the synthesis and assembly of these glycosylation pathways will facilitate additional studies on bacterial–host cell interactions in the future.

RESEARCH FOCUS BOX

- What role do the glycosylation systems in host cell invasion in bacteria play *in vivo*?
- What is the role of *O*- and *N*-linked glycoproteins in virulence of bacterial pathogens?

ACKNOWLEDGEMENTS

MIK was supported by 1R15AI070358 from NIAID and PG was supported by the Military Infectious Diseases Research Program (work unit 6000.RAD1.DA3.A0308).

References

Aspinall, G.O., Lynch, C.M., Pang, H., Shaver, R.T., Moran, A.P., 1995. Chemical structures of the core region of *Campylobacter jejuni* O:3 lipopolysaccharide and an associated polysaccharide. Eur. J. Biochem. 231, 570–578.

Bachtiar, B.M., Coloe, P.J., Fry, B.N., 2007. Knockout mutagenesis of the *kpsE* gene of *Campylobacter jejuni* 81116 and its involvement in bacterium-host interactions. FEMS Immunol. Med. Microbiol. 49, 149–154.

Bacon, D.J., Szymanski, C.M., Burr, D.H., Silver, R.P., Alm, R.A., Guerry, P., 2001. A phase-variable capsule is involved in virulence of *Campylobacter jejuni* 81–176. Mol. Microbiol. 40, 769–777.

Bauer, B.A., Stevens, M.K., Hansen, E.J., 1998. Involvement of the *Haemophilus ducreyi gmhA* gene product in lipooligosaccharide expression and virulence. Infect. Immun. 66, 4290–4298.

Bauer, B.A., Lumbley, S.R., Hansen, E.J., 1999. Characterization of a WaaF (RfaF) homolog expressed by *Haemophilus ducreyi*. Infect. Immun. 67, 899–907.

Benz, I., Schmidt, M.A., 2001. Glycosylation with heptose residues mediated by the *aah* gene product is essential for adherence of the AIDA-I adhesin. Mol. Microbiol. 40, 1403–1413.

Benz, I., Schmidt, M.A., 2002. Never say never again: protein glycosylation in pathogenic bacteria. Mol. Microbiol. 45, 267–276.

Bernatchez, S., Szymanski, C.M., Ishiyama, N., et al., 2005. A single bifunctional UDP-GlcNAc/Glc 4-epimerase supports the synthesis of three cell surface glycoconjugates in *Campylobacter jejuni*. J. Biol. Chem. 280, 4792–4802.

Campos, M.A., Vargas, M.A., Regueiro, V., Llompart, C.M., Alberti, S., Bengoechea, J.A., 2004. Capsule polysaccharide mediates bacterial resistance to antimicrobial peptides. Infect. Immun. 72, 7107–7114.

Castric, P., Cassels, F.J., Carlson, R.W., 2001. Structural characterization of the *Pseudomonas aeruginosa* 1244 pilin glycan. J. Biol. Chem. 276, 26479–26485.

Chandan, V., Logan, S.M., Harrison, B.A., et al., 2007. Characterization of a *waaF* mutant of *Helicobacter pylori* strain 26695 provides evidence that an extended polysaccharide structure has a limited role in the invasion of gastric cancer cells. Biochem. Cell. Biol. 85, 582–590.

Clement, M.-J., Imberty, A., Phalipon, A., et al., 2003. Conformational studies of the O-specific polysaccharide of *Shigella flexneri* 5a and of four related synthetic pentasaccharide fragments using NMR and molecular modeling. J. Biol. Chem. 278, 47928–47936.

Cope, L.D., Yogev, R., Mertsola, J., Argyle, J.C., McCracken, G. H., Hansen, E.J., 1990. Effect of mutations in lipooligosaccharide biosynthesis genes on virulence of *Haemophilus influenzae* type b. Infect. Immun. 58, 2343–2351.

Dobos, K.M., Swiderek, K., Khoo, K.H., Brennan, P.J., Belisle, J.T., 1995. Evidence for glycosylation sites on the 45-kilodalton glycoprotein of *Mycobacterium tuberculosis*. Infect. Immun. 63, 2846–2853.

Dzieciatkowska, M., Brochu, D., van Belkum, A., et al., 2007. Mass spectrometric analysis of intact lipooligosaccharide: direct evidence for *O*-acetylated sialic acids and discovery of *O*-linked glycine expressed by *Campylobacter jejuni*. Biochemistry 46, 14704–14714.

Ernst, R.K., Guina, T., Miller, S.I., 1999. How intracellular bacteria survive: surface modifications that promote resistance to host innate immune responses. J. Infect. Dis. 179 (Suppl. 2), S326–S330.

Finlay, B.B., Starnbach, M.N., Francis, C.L., et al., 1988. Identification and characterization of *TnphoA* mutants of *Salmonella* that are unable to pass through a polarized MDCK epithelial cell monolayer. Mol. Microbiol. 2, 757–766.

Friedman, C., Neiman, J., Wegener, H., Tauxe, R., 2000. Epidemiology of *Campylobacter jejuni* infections in the United States and other industrilaized nations. In: Nachamkin, I., Blaser, M. (Eds.), *Campylobacter*, 2nd edn. ASM Press, Washington, DC, pp. 121–138.

Fry, B.N., Feng, S., Chen, Y.Y., Newell, D.G., Coloe, P.J., Korolik, V., 2000. The *galE* gene of *Campylobacter jejuni* is involved in lipopolysaccharide synthesis and virulence. Infect. Immun. 68, 2594–2601.

Gibson, B.W., Campagnari, A.A., Melaugh, W., et al., 1997. Characterization of a transposon Tn916-generated mutant of *Haemophilus ducreyi* 35000 defective in lipooligosaccharide biosynthesis. J. Bacteriol. 179, 5062–5071.

Gilbert, M., Karwaski, M.F., Bernatchez, S., et al., 2002. The genetic bases for the variation in the lipo-oligosaccharide of the mucosal pathogen, *Campylobacter jejuni*. Biosynthesis of sialylated ganglioside mimics in the core oligosaccharide. J. Biol. Chem. 277, 327–337.

Gronow, S., Noah, C., Blumenthal, A., Lindner, B., Brade, H., 2003. Construction of a deep-rough mutant of *Burkholderia cepacia* ATCC 25416 and characterization of its chemical and biological properties. J. Biol. Chem. 278, 1647–1655.

Guerry, P., 2007. *Campylobacter* flagella: not just for motility. Trends Microbiol. 15, 456–461.

Guerry, P., Ewing, C.P., Hickey, T.E., Prendergast, M.M., Moran, A.P., 2000. Sialylation of lipooligosaccharide cores affects immunogenicity and serum resistance of *Campylobacter jejuni*. Infect. Immun. 68, 6656–6662.

Guerry, P., Szymanski, C.M., Prendergast, M.M., et al., 2002. Phase variation of *Campylobacter jejuni* 81–176 lipooligosaccharide affects ganglioside mimicry and invasiveness *in vitro*. Infect. Immun. 70, 787–793.

Guerry, P., Ewing, C.P., Schirm, M., et al., 2006. Changes in flagellin glycosylation affect *Campylobacter* autoagglutination and virulence. Mol. Microbiol. 60, 299–311.

Guo, L., Lim, K.B., Poduje, C.M., et al., 1998. Lipid A acylation and bacterial resistance against vertebrate antimicrobial peptides. Cell 95, 189–198.

Hendrixson, D.R., DiRita, V.J., 2004. Identification of *Campylobacter jejuni* genes involved in commensal colonization of the chick gastrointestinal tract. Mol. Microbiol. 52, 471–484.

Hoare, A., Bittner, M., Carter, J., et al., 2006. The outer core lipopolysaccharide of *Salmonella enterica* serovar *Typhi* is required for bacterial entry into epithelial cells. Infect. Immun. 74, 1555–1564.

Hoffman, J.A., Wass, C., Stins, M.F., Kim, K.S., 1999. The capsule supports survival but not traversal of *Escherichia coli K1* across the blood-brain barrier. Infect. Immun. 67, 3566–3570.

Houliston, R.S., Endtz, H.P., Yuki, N., et al., 2006. Identification of a sialate *O*-acetyltransferase from *Campylobacter jejuni*: demonstration of direct transfer to the C-9 position of terminal alpha-2, 8-linked sialic acid. J. Biol. Chem. 281, 11480–11486.

Jones, M.A., Marston, K.L., Woodall, C.A., et al., 2004. Adaptation of *Campylobacter jejuni* NCTC11168 to high-level colonization of the avian gastrointestinal tract. Infect. Immun. 72, 3769–3776.

Kakuda, T., DiRita, V.J., 2006. Cj1496c encodes a *Campylobacter jejuni* glycoprotein that influences invasion of human epithelial cells and colonization of the chick gastrointestinal tract. Infect. Immun. 74, 4715–4723.

Kanipes, M.I., Holder, L.C., Corcoran, A.T., Moran, A.P., Guerry, P., 2004. A deep-rough mutant of *Campylobacter jejuni* 81–176 is noninvasive for intestinal epithelial cells. Infect. Immun. 72, 2452–2455.

Kanipes, M.I., Papp-Szabo, E., Guerry, P., Monteiro, M.A., 2006. Mutation of *waaC*, encoding heptosyltransferase I in *Campylobacter jejuni* 81–176, affects the structure of both lipooligosaccharide and capsular carbohydrate. J. Bacteriol. 188, 3273–3279.

Kanipes, M.I., Tan, X., Akelaitis, A., et al., 2008. Genetic analysis of lipooligosaccharide core biosynthesis in *Campylobacter jejuni* 81–176. J. Bacteriol. 190, 1568–1574.

Kihlstrom, E., Edebo, L., 1976. Association of viable and inactivated *Salmonella typhimurium* 395 MS and MR 10 with HeLa cells. Infect. Immun. 14, 851–857.

Kim, C.-H., 2003. A *Salmonella typhimurium rfaE* mutant recovers invasiveness for human epithelial cells when complemented by wild type *rfaE* (controlling biosynthesis of ADP-L-glycero-D-manno-Heptose-containing lipopolysaccharide). Mol. Cells 15, 226–232.

Konkel, M.E., Klena, J.D., Rivera-Amill, V., et al., 2004. Secretion of virulence proteins from *Campylobacter jejuni* is dependent on a functional flagellar export apparatus. J. Bacteriol. 176, 3296–3303.

Kowarik, M., Young, N.M., Numao, S., et al., 2006. Definition of the bacterial *N*-glycosylation site consensus sequence. EMBO J. 25, 1957–1966.

Larsen, J.C., Szymanski, C., Guerry, P., 2004. *N*-linked protein glycosylation is required for full competence in *Campylobacter jejuni* 81–176. J. Bacteriol. 186, 6508–6514.

Lindenthal, C., Elsinghorst, E.A., 2001. Enterotoxigenic *Escherichia coli* TibA glycoprotein adheres to human intestine epithelial cells. Infect. Immun. 69, 52–57.

Linton, D., Gilbert, M., Hitchen, P.G., et al., 2000. Phase variation of a beta-1,3 galactosyltransferase involved in generation of the ganglioside GM_1-like lipo-oligosaccharide of *Campylobacter jejuni*. Mol. Microbiol. 37, 501–514.

Locke, J.B., Colvin, K.M., Datta, A.K., et al., 2007. *Streptococcus iniae* capsule impairs phagocytic clearance and contributes to virulence in fish. J. Bacteriol. 189, 1279–1287.

Logan, S.M., Kelly, J.F., Thibault, P., Ewing, C.P., Guerry, P., 2002. Structural heterogeneity of carbohydrate modifications affects serospecificity of *Campylobacter* flagellins. Mol. Microbiol. 46, 587–597.

Magee, A.D., Yother, J., 2001. Requirement for capsule in colonization by *Streptococcus pneumoniae*. Infect. Immun. 69, 3755–3761.

Mandrell, R.E., Parker, C.T., Bates, A.H., et al., 2008. *Campylobacter jejuni* strain RM1221 lipooligosaccharide mimics globo- and lactoneo-series P blood group antigens. In: Abstracts of the 107th General Meeting of the American Society for Microbiology (ASM), Boston, Massachusetts. ASM Press, Washington, DC, p. 53.

Masoud, H., Moxon, E.R., Martin, A., Krajcarski, D., Richards, J.C., 1997. Structure of the variable and conserved lipopolysaccharide oligosaccharide epitopes expressed by *Haemophilus influenzae* serotype b strain Eagan. Biochemistry 36, 2091–2103.

McBride, J.W., Yu, X.-J., Walker, D.H., 2000. Glycosylation of homologous immunodominant proteins of *Ehrlichia chaffeensis* and *Ehrlichia canis*. Infect. Immun. 68, 13–18.

McBride, J.W., Doyle, C.K., Zhang, X., et al., 2007. Identification of a glycosylated *Ehrlichia canis* 19-kilodalton major immunoreactive protein with a species-specific serine-rich glycopeptide epitope. Infect. Immun. 75, 74–82.

McNally, D.J., Hui, J.P.M., Aubry, A.J., et al., 2006. Functional characterization of the flagellar glycosylation locus in *Campylobacter jejuni* 81–176 using a focused metabolomics approach. J. Biol. Chem. 281, 18489–18498.

McNally, D.J., Lamoureux, M.P., Karlyshev, A.V., et al., 2007. Commonality and biosynthesis of the *O*-methyl phosphoramidate capsule modification in *Campylobacter jejuni*. J. Biol. Chem. 282, 28566–28576.

Moran, A., Penner, J., Aspinall, G., 2000. *Campylobacter* lipopolysaccharides. In: Nachamkin, I., Blaser, M. (Eds.), *Campylobacter*, 2nd edn. ASM Press, Washington, DC, pp. 241–257.

Mroczenski-Wildey, M.J., Di Fabio, J.L., Cabello, F.C., 1989. Invasion and lysis of HeLa cell monolayers by *Salmonella typhi*: the role of lipopolysaccharide. Microb. Path. 6, 143–152.

Nelson, A.L., Roche, A.M., Gould, J.M., Chim, K., Ratner, A.J., Weiser, J.N., 2007. Capsule enhances pneumococcal colonization by limiting mucus-mediated clearance. Infect. Immun. 75, 83–90.

Nikaido, H., et al., 1996. Outer membrane. In: Curtiss, R.C., Ingraham, J.L., Lin, E.C.C. (Eds.) *Escherichia coli* and *Salmonella*: Cellular and Molecular Biology. ASM Press, Washington, DC, pp. 29–47.

O'Riordan, K., Lee, J.C., 2004. *Staphylococcus aureus* capsular polysaccharides. Clin. Microbiol. Rev. 17, 218–234.

Oberhelman, R., Taylor, D., 2000. *Campylobacter* infections in developing countries. In: Nachamkin, I., Blaser, M. (Eds.), *Campylobacter*, 2nd edn. ASM Press, Washington, DC, pp. 139–153.

Poly, F., Ewing, C., Goon, S., et al., 2007. Heterogeneity of a *Campylobacter jejuni* protein that is secreted through the flagellar filament. Infect. Immun. 75, 3859–3867.

Poly, F., Read, T.D., Chen, Y.-H., et al., 2008. Characterization of two *Campylobacter jejuni* strains for use in experimental infection studies. Infect. Immun. 76, 5655–5667.

Raetz, C.R., Whitfield, C., 2002. Lipopolysaccharide endotoxins. Annu. Rev. Biochem. 71, 635–700.

Raetz, C.R., Reynolds, C.M., Trent, M.S., Bishop, R.E., 2007. Lipid A modification systems in gram-negative bacteria. Annu. Rev. Biochem. 76, 295–329.

Raffatellu, M., Chessa, D., Wilson, R.P., Dusold, R., Rubino, S., Baumler, A.J., 2005. The Vi capsular antigen of *Salmonella enterica* serotype Typhi reduces Toll-like receptor-dependent interleukin-8 expression in the intestinal mucosa. Infect. Immun. 73, 3367–3374.

Rao, V.K., Krasan, G.P., Hendrixson, D.R., Dawid, S., St Geme, J.W., 1999. Molecular determinants of the pathogenesis of disease due to non-typable *Haemophilus influenzae*. FEMS Microb. Rev. 23, 99–129.

Sambri, V., Stefanelli, C., Cevenini, R., 1992. Detection of glycoproteins in *Borrelia burgdorferi*. Arch. Microbiol. 157, 205–208.

Sambri, V., Massaria, F., Ardizzoni, M., Stefanelli, C., Cevenini, R., 1993. Glycoprotein patterns in *Borrelia* spp. Zentralbl. Bakteriol. 279, 330–335.

Schrager, H.M., Rheinwald, J.G., Wessels, M.R., 1996. Hyaluronic acid capsule and the role of streptococcal entry into keratinocytes in invasive skin infection. J. Clin. Invest. 98, 1954–1958.

Sharma, A., Qadri, A., 2004. Vi polysaccharide of *Salmonella typhi* targets the prohibitin family of molecules in intestinal epithelial cells and suppresses early inflammatory responses. Proc. Natl. Acad. Sci. USA 101, 17492–17497.

Sisti, F., Fernandez, J., Rodriguez, M.E., Lagares, A., Guiso, N., Hozbor, D.F., 2002. In vitro and in vivo characterization of a *Bordetella bronchiseptica* mutant strain with a deep rough lipopolysaccharide structure. Infect. Immun. 70, 1791–1798.

Song, Y.C., Jin, S., Louie, H., et al., 2004. FlaC, a protein of *Campylobacter jejuni* TGH9011 (ATCC43431) secreted through the flagellar apparatus, binds epithelial cells and influences cell invasion. Mol. Microbiol. 53, 541–553.

Spinosa, M.R., Progida, C., Tala, A., Cogli, L., Alifano, P., Bucci, C., 2007. The *Neisseria meningitidis* capsule is important for intracellular survival in human cells. Infect. Immun. 75, 3594–3603.

Stevens, M.K., Klesney-Tait, J., Lumbley, S., et al., 1997. Identification of tandem genes involved in lipooligosaccharide expression by *Haemophilus ducreyi*. Infect. Immun. 65, 651–660.

Swanson, A.F., Kuo, C.C., 1990. Identification of lectin-binding proteins in *Chlamydia* species. Infect. Immun. 58, 502–507.

Swanson, A.F., Kuo, C.C., 1991. Evidence that the outer membrane protein of *Chlamydia trachomatis* is glycosylated. Infect. Immun. 59, 2120–2125.

Swanson, A.F., Kuo, C.C., 1994. Binding of the glycan of the major outer membrane protein of *Chlamydia trachomatis* to HeLa cells. Infect. Immun. 62, 24–28.

Swords, W.E., Buscher, B.A., Ver Steeg Ii, K., et al., 2000. Non-typeable *Haemophilus influenzae* adhere to and invade human bronchial epithelial cells via an interaction of lipooligosaccharide with the PAF receptor. Mol. Microbiol. 37, 13–27.

Szymanski, C.M., Yao, R., Ewing, C.P., Trust, T.J., Guerry, P., 1999. Evidence for a system of general protein glycosylation in *Campylobacter jejuni*. Mol. Microbiol. 32, 1022–1030.

Szymanski, C.M., Burr, D., Guerry, P., 2002. *Campylobacter* protein glycosylation affects immunogenicity and host cell interactions. Infect. Immun. 70, 2242–2244.

Szymanski, C.M., Michael, F.S., Jarrell, H.C., et al., 2003. Detection of conserved *N*-linked glycans and phase-variable lipooligosaccharides and capsules from *Campylobacter* cells by mass spectrometry and high resolution magic angle spinning NMR spectroscopy. J. Biol. Chem. 278, 24509–24520.

Thibault, P., Logan, S.M., Kelly, J.F., et al., 2001. Identification of the carbohydrate moieties and glycosylation motifs in *Campylobacter jejuni* flagellin. J. Biol. Chem. 276, 34862–34870.

Wacker, M., Linton, D., Hitchen, P.G., et al., 2002. *N*-linked glycosylation in *Campylobacter jejuni* and its functional transfer into *E. coli*. Science 298, 1790–1793.

Wacker, M., Feldman, M.F., Callewaert, N., et al., 2006. Substrate specificity of bacterial oligosaccharyltransferase suggests a common transfer mechanism for the bacterial and eukaryotic systems. Proc. Natl. Acad. Sci. USA 103, 7088–7093.

West, N.P., Sansonetti, P.J., Mounier, R.M., et al., 2005. Optimization of virulence functions through glucosylation of *Shigella* LPS. Science 307, 1313–1317.

Young, N.M., Brisson, J.-R., Kelly, J., et al., 2002. Structure of the *N*-linked glycan present on multiple glycoproteins in the Gram-negative bacterium, *Campylobacter jejuni*. J. Biol. Chem. 277, 42530–42539.

Zaidi, T.S., Fleiszig, S.M., Preston, M.J., Goldberg, J.B., Pier, G.B., 1996. Lipopolysaccharide outer core is a ligand for corneal cell binding and ingestion of *Pseudomonas aeruginosa*. Invest. Ophthalmol. Vis. Sci. 37, 976–986.

Zilbauer, M., Dorrell, N., Boughan, P.K., et al., 2005. Intestinal innate immunity to *Campylobacter jejuni* results in induction of bactericidal human beta-defensins 2 and 3. Infect. Immun. 73, 7281–7289.

PART V

BIOTECHNOLOGICAL AND MEDICAL APPLICATIONS

CHAPTER

45

Exopolysaccharides produced by lactic acid bacteria in food and probiotic applications

Patricia Ruas-Madiedo, Nuria Salazar and Clara G. de los Reyes-Gavilán

SUMMARY

Exopolysaccharides (EPSs) produced by lactic acid bacteria (LAB) are a topic of research for scientists due to their technological application in the dairy industry where they improve the viscosity, texture and mouth-feel of yoghurt and fermented milks and act as fat substitutes in cheese-making. The physical characteristics, the molar mass and stiffness, determined by the chemical composition and linkages in the EPS molecule, are the key parameters for their viscosity-intensifying ability. A great variety of EPS-producing LAB strains are currently employed in milk fermentations because the production *in situ* of these natural biothickeners avoids employment of emulsifiers and gelling agents. However, the use of purified EPSs as food additives is still limited due to their low yield of production. In recent years, EPSs from LAB are receiving a renewed interest because it has been claimed that they have benefits for human health, such as a cholesterol-lowering ability, a prebiotic effect and modulation of the immune response system. Nonetheless, most studies have been performed *in vitro* and there is a lack of clinical evidence demonstrating these putative probiotic characteristics after oral administration of EPSs in functional foods. In addition, the mechanism(s) and EPS-parameter(s) involved in these biological effects still remain poorly understood.

Keywords: Exopolysaccharide; Lactic acid bacteria; Dairy product; Viscosity; Probiotic; Prebiotic; Immune response

1. INTRODUCTION

Microbial exopolysaccharides (EPSs) are polymers associated with the external cell surface, which can be covalently bound forming capsules, or are more loosely attached thereby developing a "slimy" layer (Ruas-Madiedo and de los Reyes-Gavilán, 2005) (see also Chapter 6). Lactic acid bacteria (LAB) and other Gram-positive bacteria employed in the manufacture of foods, e.g. bifidobacteria and propionibacteria, are also able to produce EPS. These biopolymers have received increasing attention in recent years because of their technological application in dairy products and their potentially beneficial properties for human health (Ruas-Madiedo *et al.*, 2002a). Most of the EPS-producing LAB have been isolated from different fermented foods such as sourdoughs, sausages, table olives, cheeses, yoghurt, kefir, other fermented dairy products and some traditional foods from non-industrialized countries (Ruas-Madiedo and de los Reyes-Gavilán, 2005; Mozzi *et al.*, 2006). Recently, other environments have been explored to isolate EPS-producing strains. This is the case of the gut microbiota of different animals and humans (Tieking *et al.*, 2005; Ruas-Madiedo *et al.*, 2007). It is obvious that LAB strains from different ecological niches are able to produce EPS, but the physiological role that these polymers play in producing bacteria still remains unclear. It seems that EPS are not used as an energy source reservoir, but it has been suggested that they have a role in the recognition of different ecosystems and also that they have a protective function against detrimental environmental factors (Ruas-Madiedo *et al.*, 2008). This overview summarizes some of the information about the types, as well as the genetic organization of EPS from LAB; it is mainly focused on the technological functionality of EPS in the dairy industry and their potential benefits for human health in probiotic applications.

2. THE EPSs FROM LAB

2.1. Chemical composition and structure of EPSs

Depending on the chemical composition, the EPSs from LAB are classified as homopolysaccharides (HoPSs) which contain a single type of monosaccharide, mainly glucose (Glc) or fructose (Fru), termed glucans and fructans, respectively, and hetero-polysaccharides (HePSs), which are constituted of repeating units of different monosaccharides, substituted monosaccharides and other organic and inorganic molecules (Monsan *et al.*, 2001; De Vuyst *et al.*, 2001; Ruas-Madiedo *et al.*, 2002a).

According to their primary structure, i.e. the sugar linkage type, glucans can be classified as α-D-glucans (i.e. dextrans, mutans, alternans and reuterans) and β-D-glucans (Table 45.1). The α-glucans from LAB present in foods are mostly produced by members of the genera *Leuconostoc*, *Lactobacillus* and *Weisella* (Korakli and Vogel, 2006). Recently, the isolation of an α-glucan-producing *Lactococcus lactis* strain from raw milk has been described for the first time (van der Meulen *et al.*, 2007). β-glucans are not often produced by LAB and only a few strains, which belong to the species *Lactobacillus diolivorans*, *Pediococcus damnosus*, *Pedicocccus parvulus* and *Oenococcus oeni*, have been reported to be producers (Walling *et al.*, 2005; Werning *et al.*, 2006). Two types of β-fructans have been described in LAB: levan and inulin-like, which contain β-(2→6)- and β-(2→1)-linkages, respectively (Monsan *et al.*, 2001). β-fructans in food LAB (see Table 45.1) are associated to genera *Leuconostoc*, *Lactobacillus* and *Weissella* (Monsan *et al.*, 2001; Korakli and Vogel, 2006). Finally, the strain *Lactococcus lactis* subsp. *lactis* H414 is able to produce a polygalactan containing only galactose (Gal) residues but with different linkages that form a pentasaccharide repeated unit (Gruter *et al.*, 1992). In general, HoPSs have a main monosaccharide backbone with variable degrees of branching, length of side chains and sugar linkage types, which differ among bacterial strains. They are usually molecules of high molecular mass (up to 10^6) (Monsan *et al.*, 2001; Korakli and Vogel, 2006).

The HePSs from LAB are built of repeating units that constitute the backbone of the polymer which can be branched or unbranched. So far, different repeating units containing between three and eight residues have been described (De Vuyst *et al.*, 2001; Ruas-Madiedo *et al.*, 2002a). The most common monosaccharides are

TABLE 45.1 Species of LAB, bifidobacteria and propionibacteria present in foods producing homo-polysaccharides (HoPSs) and hetero-polysaccharides (HePSs)

	Genus	Species
HoPSs		
α-glucan	*Lactococcus*	*Lc. lactis*
	Lactobacillus	*Lb. reuteri, Lb. sakei, Lb. fermentum, Lb. plantarum*
	Leuconostoc	*Leu. mesenteroides* subp. *dextranicum, Leu. citreum*
	Weissella	*W. cibaria*
β-glucan	*Lactobacillus*	*Lb. diolivorans*
	Oenococcus	*O. oeni*
	Pediococcus	*P. parvulus, P. damnosus*
Fructan	*Lactobacillus*	*Lb. acidophilus, Lb. panis, Lb. plantarum, Lb. reuteri, Lb. sanfranciscensis*
	Leuconostoc	*Leu. mesenteroides*
Polygalactan	*Lactococcus*	*Lc. lactis* subsp. *lactis*
HePSs		
	Lactococcus	*Lc. lactis* subsp. *cremoris*
	Lactobacillus	*Lb. acidophilus*
		Lb. delbrueckii subsp. *bulgaricus*
		Lb. curvatus
		Lb. helveticus
		Lb. paracasei
		Lb. rhamnosus
		Lb. sakei
		Lb. sanfranciscensis
	Streptococcus	*S. macedonius*
		S. thermophilus
	Bifidobacterium	*B. lactis*
		B. longum
	Propionibacterium	*Pr. acidipropionici*
		Pr. freudenreichii subsp. *shermanii*

D-Glc, D-Gal and L-rhamnose (L-Rha) which are present in different ratios and, remarkably, Gal seems to be the most universal one (Mozzi *et al.*, 2006). To a lesser extent, the sugars *N*-acetylglucosamine (GlcNAc), *N*-acetylgalactosamine (GalNAc), fucose (Fuc), ribose (Rib) and mannose (Man), and other substituents like glucuronic acid (GlcA) and acetyl, glycerol and phosphate groups are also components of the HePS repeating units. More than 50 HePS structures from LAB have been determined using nuclear magnetic resonance (NMR) spectroscopy, 31 of which represent unique structures (Ruas-Madiedo *et al.*, 2008). These can vary in charge, rigidity and spatial rearrangements and their molar masses range in size from 4×10^4 to 6×10^6 (Ruas-Madiedo *et al.*, 2002a; Mozzi *et al.*, 2006). The genera *Lactococcus*, *Lactobacillus* and *Streptococcus*, as well as *Bifidobacterium* and *Propionibacterium*, which are employed in food processing, have HePS-producing species (see Table 45.1).

2.2. Biosynthesis and genetic organization of EPS

The HoPSs are extracellularly synthesized by means of enzymes named glycansucrases that use sucrose as a specific substrate for glycosyl (Fru or Glc) donation (Monsan *et al.*, 2001). Over the years, a large number of glucansucrases (GSs) and fructansucrases (FSs) coding genes and enzymes have been cloned and identified (Korakli and Vogel, 2006; van Hijum *et al.*, 2006). The GS for α-glucan synthesis are limited to LAB, whereas FS are present in Gram-positive and Gram-negative bacteria. Generally, these genes share a common structure formed by a signal peptide and variable N-terminal domain, a catalytic domain and the C-terminal or binding domain. The GSs and FSs are able to catalyse three types of reactions depending on the acceptor molecule:

(i) sucrose + $H_2O \rightarrow$ Glc + Fru
(ii) sucrose + (glucan or fructan)$_n \rightarrow$ [(fructan)$_{n+1}$ + Glc] or [(glucan)$_{n+1}$ + Fru]
(iii) sucrose + acceptor (i.e. maltose) $\rightarrow$ [fructo-oligosaccharide + Glc] or [gluco-oligosaccharide + Fru].

Finally, β-glucans produced by LAB of food origin are synthesized by the activity of membrane-bound glycosyltransferases (GTFs), which resemble those involved in the synthesis of β-glucan by the pathogen *Streptococcus pneumoniae* (Werning *et al.*, 2006).

The genetic organization of HePS biosynthesis is more complex because this process combines intra- and extracellular steps and several enzymes are involved in the biosynthesis of the repeating units, secretion and polymerization (Boels *et al.*, 2001b; Jolly and Stingelle, 2001; Broadbent *et al.*, 2003). The genetic determinants of HePS biosynthesis can be located either on plasmid or chromosomal DNA and the genes are organized in *eps* clusters, which are highly conserved with respect to their structural organization. The HePS gene clusters are generally divided into four regions: the first one contains regulatory genes; the second codes for proteins involved in polymerization and chain length determination; the third region contains genes coding for GTFs specifically required for the assembling of the repeating unit; and the fourth region includes genes involved in transport and polymerization (Jolly and Stingelle, 2001; Broadbent *et al.*, 2003). The precursor repeating units of HePSs are formed intracellularly. The first enzyme in the process is the glycosyl-1-phosphate transferase, also called priming-GTF, that anchors the first sugar nucleotide to a phosphorylated carrier lipid (C_{55}) located in the cell membrane. This priming-GTF recognizes the lipid carrier but does not catalyse the glycosidic linkage. More monosaccharides are added to the repeating unit by means of specific GTFs that employ activated sugar nucleotides as precursors. The repeating units are then translocated across the membrane and they are polymerized extracellularly (Boels *et al.*, 2001b). This working model of EPS biosynthesis has not been fully demonstrated due to the complexity of the *eps* cluster organization and only the functionality of some *gtf* genes has been corroborated (Boels *et al.*, 2001a).

2.3. Yield of EPS produced by LAB

The total yield of EPS produced by LAB can be influenced by the composition of the culture medium (carbon and nitrogen sources and other nutrients), growth conditions (pH, temperature, oxygen concentration, etc.) and incubation time (Degeest *et al.*, 2001). The amount of EPS synthesized is also largely influenced by the producing strain and yield also varies according to the culture media and the EPS-isolation method employed (Ruas-Madiedo and de los Reyes-Gavilán, 2005). In general, HoPS-producing LAB render high amounts of polymers, some strains such as *Lactobacillus reuteri* LB121 being able to produce up to 10 g per litre (Figure 45.1).

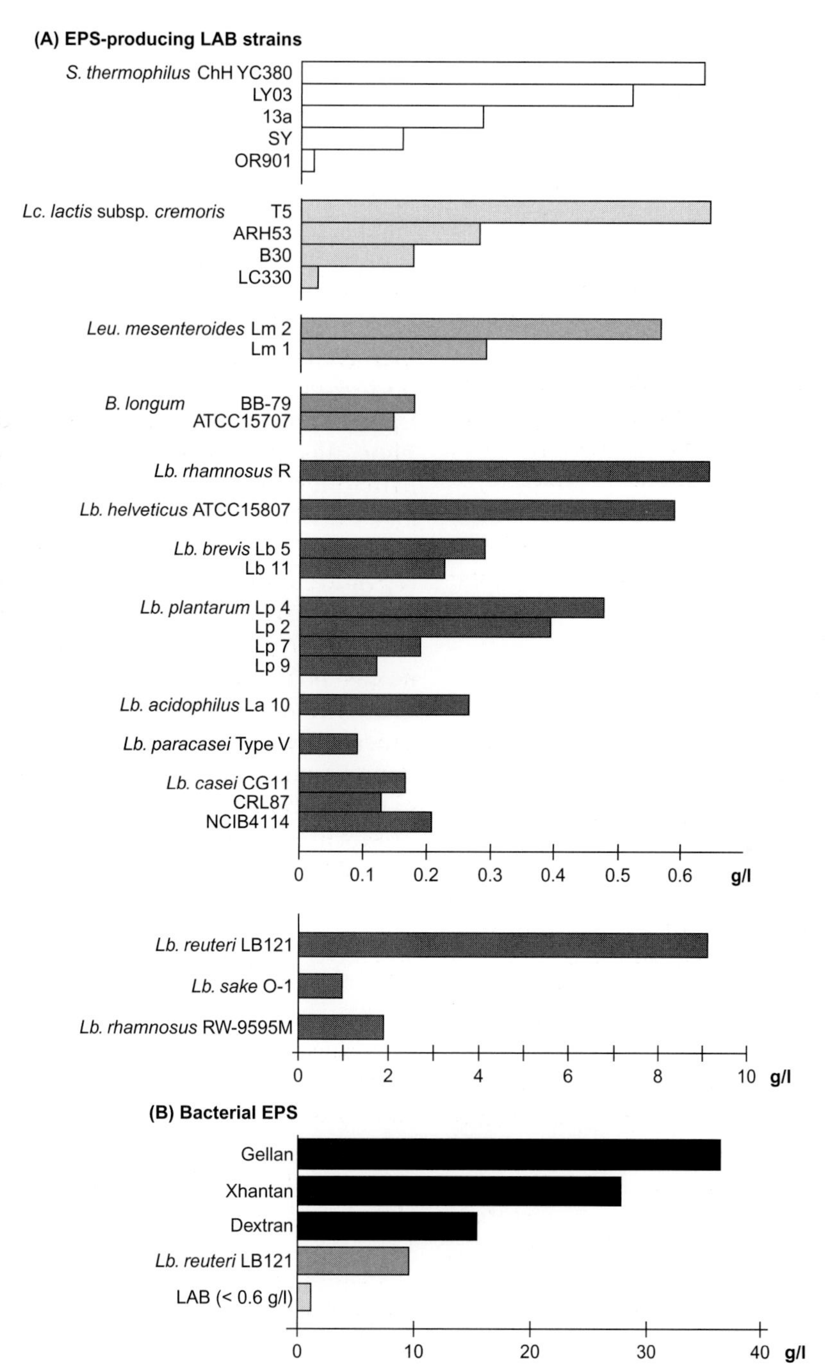

FIGURE 45.1 Comparison of the EPS yield (g/l) obtained from several LAB strains (A) and from other EPS-producing bacteria for food applications (B). Gellan is produced by *Sphingomonas paucimobilis* and xanthan by *Xanthomonas campestris*, both Gram-negative species, whereas dextran is produced by the Gram-positive LAB, *Leuconostoc mesenteroides*.

However, HePS from LAB and bifidobacteria are produced in considerably smaller amounts and range from 25 to 600 mg per litre, according to the strain (see Figure 45.1). Only a few exceptional strains are able to produce higher amounts of HePS under optimized culturing conditions. Such is the case of *Lactobacillus rhamnosus* RW-9595M which produces around 2 g per litre of polymer (Bergmaier *et al.*, 2005). The yield of EPS produced by LAB is very low as compared to other bacterial polymers, such as gellan, xhantan or dextran (see Figure 45.1), that are synthesized at industrial level for several applications (Sutherland, 1998).

3. TECHNOLOGICAL APPLICATIONS OF EPS FROM LAB IN DAIRY PRODUCTS

Several LAB species have Generally Recognized As Safe (GRAS) and Qualified Presumption of Safety (QPS) status, with a long history of safe use in food processing (Mogensen *et al.*, 2002). Some strains with scientifically demonstrated health benefits are being included in the manufacture of functional foods (Salminen and Gueimonde, 2004). Specifically, EPS-producing LAB have their most valuable application in the dairy industry to improve the rheology, texture and mouth-feel of fermented products (Welman and Maddox, 2003). However, other foods such as cider, beer and wine (Walling *et al.*, 2005; Werning *et al.*, 2006) or packaged meats (Korkeala and Björkroth, 1997) are spoiled by EPS-producing strains.

The EPSs can be considered as natural biothickeners because they are produced *in situ* by the LAB starters during milk fermentations. Thus, these polymers can be used as an alternative to thickeners, emulsifiers, gelling agents and stabilizers (Laws and Marshall, 2001; Ruas-Madiedo *et al.*, 2002a). Currently, a wide variety of EPS-producing LAB species are included in the formulation of dairy products (Duboc and Mollet, 2001) (Table 45.2). In yoghurt and fermented milks, EPS promote an increase of viscosity in the stirred products, improve the consistency and texture of the milk-coagulum and decrease the syneresis (or water release) in set yoghurts (Figure 45.2). Importantly, EPS-producing LAB can be used in the production of low-fat cheeses by replacing fat globules

TABLE 45.2 Some dairy products containing EPS-producing LAB, bifidobacteria and propionibacteria

Genus	Species	Dairy product
Lactococcus	*Lc. lactis* subp. *cremoris*	Scandinavian fermented milks (viili, långfil, etc.), cheeses
Lactobacillus	*Lb. delbrueckii* subsp. *bulgaricus*	Yoghurts, fermented milks
	Lb. acidophilus	Probiotic fermented milks
	Lb. kefiranofaciens	Kefir
Leuconostoc	*Leu. mesenteroides* subsp. *dextranicum*	Kefir
Streptococcus	*S. macedonicus*	Cheeses (Greek, Italian)
	S. thermophilus	Yoghurt, Mozzarella cheese
Bifidobacterium	*B. animalis* subsp. *lactis*	Probiotic fermented milks
	B. longum	Probiotic fermented milks
Propionibacterium	*Pr. freudenreichii* subsp. *shermanii*	Swiss cheeses

Modified from Duboc and Mollet, 2001.

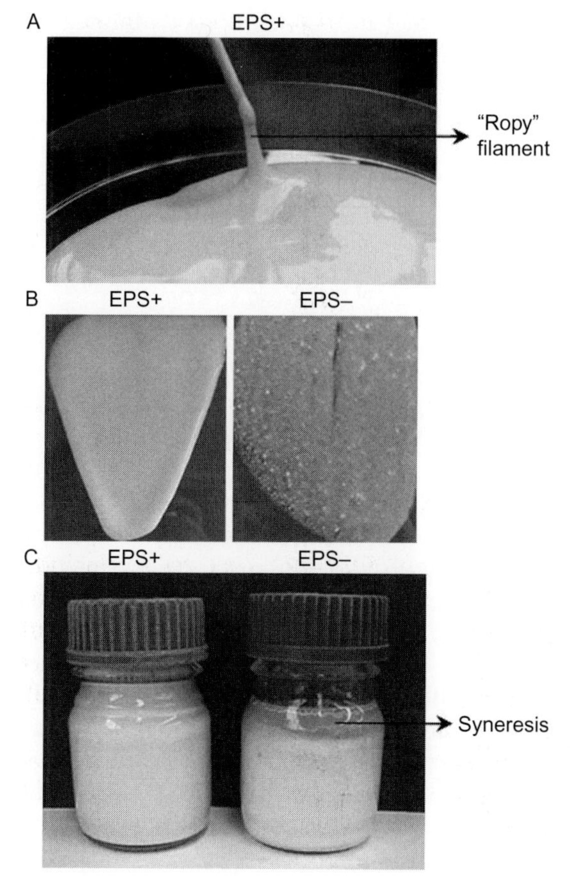

FIGURE 45.2 Dairy product fermented with an EPS-producing (EPS+) strain of *Lactococcus lactis* subsp. *cremoris*. (A) Increase of viscosity and formation of a "ropy" filament. (B) Smooth texture in the product containing the EPS+ strain as compared with roughness in the product fermented with a non-EPS producing strain (EPS−). (C) Syneresis (serum release) in the dairy product fermented with the EPS− strain as compared with absence of serum separation in the EPS+ product.

and, thereby, avoiding the rubbery consistency, improving the melting characteristics and also increasing moisture content due to the ability of EPS to retain water (Hassan *et al.*, 2007; Ruas-Madiedo *et al.*, 2008).

The effect of EPS on the physical properties of dairy products depends not only on the amount of EPS produced during milk fermentation, but also on the physicochemical characteristics of the polymers and on the interactions between EPS and the milk protein (Laws and Marshall, 2001). In general, an increase of EPS concentration leads to an increase in the viscosity of the stirred yoghurt, although the viscosifying effect of different polymers at similar concentrations is variable (Ruas-Madiedo *et al.*, 2002a). The intrinsic viscosity of EPS in watery solutions may predict the viscosity enhancement in the fermented product (Ruas-Madiedo *et al.*, 2002b). This intrinsic viscosity (η) can be calculated from the molar mass (M [kg mol^{-1}]) and the radius of gyration (Rg, [nm]) of the polymer ($\eta \approx 3.1 \times 10^{24} Rg^3 M^{-1}$ [m^{3}kg^{-1}]); parameters that are measured using light scattering techniques (Tuinier *et al.*, 1999a,b). For random coil EPS, Rg (≈size) depends on M (≈length) and also on the flexibility (or thickening efficiency, Ø) of the molecule ($Rg^2 \approx M\ Ø^{-1}$). These parameters are related to the chemical composition, sugar linkage type and presence of side chains of the repeating units that build the polymer. This means that in order to obtain high intrinsic viscosity, the M of EPS should be high, thereby Rg will increase and the repeating unit of EPS should be relatively stiff (Ruas-Madiedo *et al.*, 2002a,b). On the other hand, the structure of the protein network in the fermented milk gels, as well as the interactions of the EPS–bacteria–protein matrix, also play an important role in the texture and viscosity of yoghurt and fermented milks (Ruas-Madiedo *et al.*, 2008).

In spite of the high viscosifying capability at low concentrations of some EPS from LAB, these polymers have not been used until now as food additives. This is due to their low yield (see Figure 45.1), as compared to EPS produced by Gram-negative bacteria such as xanthan from *Xanthomonas campestris* and gellan from *Sphingomonas paucimobilis*. However, these two species are pathogens and only dextran, synthesized by the LAB *Leuconostoc mesenteroides*, is produced at an industrial level for food applications. On the other hand, kefiran, the EPS produced by *Lactobacillus kefiranofaciens* in high amounts in the dairy product kefir, could be an option for obtaining purified EPS with suitable physical properties for use as a food additive (Ruas-Madiedo *et al.*, 2008).

4. HEALTH BENEFITS OF EPS FROM LAB AND BIFIDOBACTERIA FOR PROBIOTIC APPLICATIONS

Lactobacillus and *Bifidobacterium* are the most common genera employed as probiotics for human consumption in functional foods. Probiotics have been defined by the World Health Organization (WHO)/Food and Agriculture Organization (FAO) (2006) as: "live microorganisms which when administered in adequate amounts confer a health benefit on the host". Importantly, we have shown that the human intestinal ecosystem is a good environment for the isolation of EPS-producing strains from these genera (Ruas-Madiedo *et al.*, 2007). In this niche, the EPSs probably play a role in the adhesion and transitory colonization of the intestinal mucosa by the producing strains (Ruas-Madiedo *et al.*, 2006). Besides, it seems that EPSs can be produced by *Bifidobacterium animalis* as a response to bile salts through a putative mechanism of protection (Ruas-Madiedo *et al.*, 2009). In addition to their physiological role and technological application, EPSs produced by LAB have been claimed to have health benefits, such as protection against gastric ulcers, a prebiotic effect, cholesterol-lowering activity and a capability to modulate the immune response and hence to exert anti-tumoural activity (Figure 45.3). Most studies have been done *in vitro* and, in a few cases, with animal models, but there is a lack of clinical trials demonstrating the health benefits of EPS after oral administration.

The ability to regulate cholesterol levels was first demonstrated in rats fed with a hypercholesterolaemic diet in combination with milk

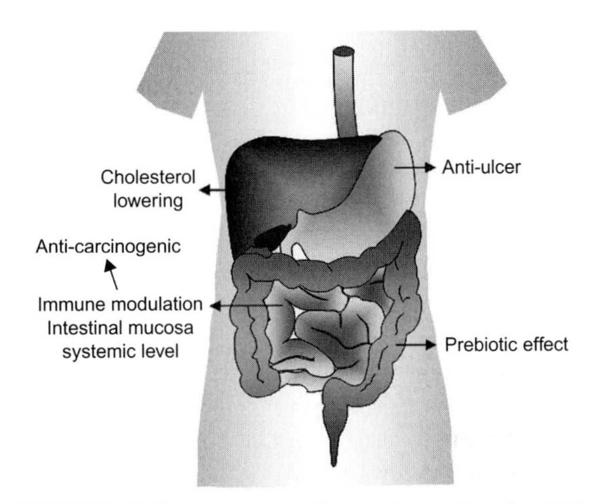

FIGURE 45.3 Putative health benefits attributed to EPSs produced by LAB.

fermented with the EPS-producing strain *Lc. lactis* subsp. *cremoris* SBT 0495 (Nakajima *et al.*, 1992). The high-density lipoproteins- (HDL-) cholesterol and the HDL-cholesterol:total-cholesterol ratio were higher in these rats compared to those fed with milk fermented with a non-EPS producing strain. A similar result was reported subsequently for the EPS-producing strain *Lb. kefiranofaciens* WT-2B, whereby a reduction in the lipid deposit in the aortic arch of rats fed with kefir containing this strain was shown (Maeda *et al.*, 2004). The mechanism(s) by which EPS could regulate the blood cholesterol levels still remain unknown (Ruas-Madiedo *et al.*, 2008). It has been shown that EPSs from several *Lactobacillus delbrueckii* subsp. *bulgaricus* and *Streptococcus thermophilus* strains are able to bind free bile acids, e.g. cholic acid, thus increasing their excretion after digestion (Pigeon *et al.*, 2002). This may reduce the pool of circulating cholesterol because more will be needed in the liver for the synthesis of new bile salts (Ruas-Madiedo *et al.*, 2008).

Prebiotics are: "non-digestible food ingredients that beneficially affect the host by selectively stimulating the growth and/or activity of a limited number of bacteria in the colon, and thus improve the host health" (Gibson and Roberfroid, 1995).

The most employed prebiotics are fructo-oligosaccharides (FOSs, including inulin), galacto-oligosaccharides (GOSs) and gluco-oligosaccharides such as β-glucans. The EPSs from LAB, mainly HoPSs, have similar chemical compositions to those of these prebiotic carbohydrates, but differ in the size of the molecules. The pool of microbial glycosidic enzymes present in the colon is one of the limiting factors for employing EPS as fermentable substrates that render short chain fatty acids, mainly acetate, propionate and butyrate. Another important factor is the accessibility of these enzymes to the monosaccharide linkages in the EPS structure. Until now, the studies about the prebiotic effect of EPSs from LAB are scarce and the most successful ones were those obtained with HoPSs. The levan-type HoPS produced by *Lactobacillus sanfranciscensis*, isolated from sourdoughs, promoted an enrichment of bifidobacteria in human faecal slurry cultures at similar levels to inulin (Dal Bello *et al.*, 2001). Furthermore, the bifidogenic effect of this fructan-like EPS was demonstrated in batch cultures on de Man-Rogosa-Sharpe (MRS) medium for several *Bifidobacterium* spp., i.e. *Bifidobacterium breve*, *Bifidobacterium bifidum*, *Bifidobacterium adolescentis* and *Bifidobacterium infantis* (Korakli *et al.*, 2002). However, this HoPS and other FOSs also produced an enhancement of *Eubacterium biforme* and *Clostridium perfringens*, respectively (Dal Bello *et al.*, 2001). First attempts to study HePSs as prebiotics showed that most of them are not employed by the human or animal faecal microbiota (Ruijssenaars *et al.*, 2000; Looijesteijn *et al.*, 2001; Cinquin *et al.*, 2006). In our group, we have found that 11 EPSs isolated from human *Bifidobacterium* strains were employed as fermentable substrates by the human intestinal microbiota (Salazar *et al.*, 2008). These EPSs also exerted a moderate bifidogenic effect comparable to that of inulin. In addition, changes in the banding patterns on polymerase chain reaction-denaturing gradient gel electrophoresis (PCR-DGGE) were obtained during incubation with EPS indicating microbial rearrangements other than bifidobacteria in the faecal microbiota (Salazar *et al.*, 2008). Metabolic cross-feeding or interactions among acetate or lactate-formers, poly- or oligosaccharide-degraders and butyrate or propionate-producers, or more complex relationships between different members of the colonic microbiota, have been suggested. Taking into account the complexity of the microbial interactions in the microbiota, the WHO/FAO have recommended a revision of the prebiotic concept which takes into consideration not only the effect on the bacteria traditionally considered beneficial, but also the ecological interactions among other members of the human microbiota (WHO/FAO, 2007).

Polysaccharides have long been considered as non-immune active molecules but, in recent years, certain microbial polymers have been shown to act as potent immunomodulating agents (Tzianabos, 2000; Cobb and Kasper, 2005). Specifically, several studies have reported the putative modulation of the immune response by the EPSs present on the extracellular surface of LAB (Table 45.3). The first indications of immune modulation by EPSs from LAB were indirectly shown by means of *in vivo* studies carried out in rats with Sarcoma-180 injected intraperitoneally with EPS isolated from *Lactobabillus helveticus* or *Lc. lactis* susbsp. *cremoris* KVS20 strains (Oda *et al.*, 1983; Kitazawa *et al.*, 1991). The survival of rats increased, but when Sarcoma-180 cells were cultivated *in vitro*, the EPS did not have a cytotoxic effect on the tumour cells. More recent *in vivo* studies in animal models with oral administration of EPSs from different LAB showed that these biopolymers are able to induce the immune response at both the intestinal mucosa and systemic levels (Makino *et al.*, 2006; Vinderola *et al.*, 2006). Indeed, a mild immune stimulation after oral administration would be a desirable characteristic for an EPS-producing (probiotic) strain to be included in a functional food. Vinderola and co-workers (2006) found that the EPS kefiran orally administrated to rats produced an increase in the number of immunoglobulin (Ig) A cells in the small intestine. Slightly higher values of interleukin (IL-) 10-, IL-6- and IL-12-positive cells were detected with respect to the control group, whereas interferon-gamma-positive (INF-γ^+) and tumour necrosis factor-alpha-positive (TNF-α^+) cells did not change. These effects were of greater magnitude in the large intestine than in the lamina propria of the small intestine. In the intestinal fluid, only cytokines IL-4 and IL-12 increased compared to the control and the cytokine blood serum pattern was similar to that found in the small intestine, indicating that cytokines produced there were released into the blood stream. This fact underlines the importance of blood serum as a tool for the study of the immunomodulation by orally administered antigens in animal models (Vinderola *et al.*, 2006). Most studies reporting the immunomodulating ability of EPS from LAB have been carried out *in vitro*. Models of studies employed cultures of spleen and cells from Peyer's patches, different lymphocyte populations and, in some cases, epithelial intestinal cellular lines (see Table 45.3). In addition, only a few different strains or EPSs have been tested and most immune effects reported are related to the induction of mitogenic activity of the eukaryotic cells and modulation of their cytokine production (see Table 45.3). These EPS-producing strains are aimed for inclusion in functional dairy or non-dairy foods. Thus, the first target site for immune response after oral intake is the intestinal mucosa. There, the gut-associated lymphoid tissue represents the highest density of immune system cells (more than 80% of the B-cells) in the body which means that the gut is the largest immune organ in humans (Brandtzaeg *et al.*, 1989). Thus, *in vitro* and *in vivo* models that simulate this environment are a suitable option to test new strains before undertaking the clinical studies of safety and probiotic efficiency.

The ability of EPS to modulate immune response is related to their chemical composition (Ruas-Madiedo *et al.*, 2008). In fact, not all EPS are able to induce an immune response, as is the case of the polymers synthesized by four *Bifidobacterium* strains isolated from infant faeces (Amrouche *et al.*, 2006). However, for the few strains characterized to date, it is worth noting that those EPS with immunomodulating capability have phosphate present in their composition. This is the case for strains *Lc. lactis* susbp. *cremoris* KVS20 and SBT0495 and *Lb. delbrueckii* subsp. *bulgaricus* OLL1073R-1 (see Table 45.3). In the latter bacteria, dephosphorylation by hydrofluoric acid degradation of its acidic EPS caused a reduction of the mitogenic activity in murine lymphocytes (Kitazawa *et al.*, 1998). In this way, it has been found that the mitogenic activity (B-cell specific) and cytokine expression (INF-γ and IL-10) in murine splenocytes were enhanced by the chemical phosphorylation of the dextran produced by *Leu. mesenteroides* (Sato *et al.*, 2004). Thereby, the phosphate substitution could act as a potent trigger for immune response induction. In addition, it is remarkable that the immune modulating-EPS produced by *Lb. rhamnosus* RW-9595M (Chabot *et al.*, 2001) presents a high Rha content (four molecules) in the heptasaccharide unit that builds this polymer (van Calsteren *et al.*, 2002). Previously, the anti-ulcer capability of several EPSs produced by *Bifidobacterium* (*B. breve* and *B. bifidum*) strains in rats with acid-induced gastric ulcer has been demonstrated. The EPSs favoured the reparation of the gastric tissues by increasing the levels of the epidermal and fibroblast growth factors. The most effective polymers were those containing more than 60% of Rha in their chemical composition (Nagaoka *et al.*, 1994). Therefore, it seems that EPSs with high content of Rha in their composition could also interact with the immune system.

TABLE 45.3 Studies reporting the putative immunomodulating effect of EPSs produced by LAB and bifidobacteria

	EPS-producing strain	Beneficial effect	Reference
***In vitro* studies**			
Stimulation of human peripheral blood lymphocytes	*Lc. lactis* subsp. *cremoris*[a]	Proliferation of T-cells	Forsén *et al.*, 1987
Cultures of murine spleen cells Stimulation of murine macrophages	*Lc. lactis* subsp. *cremoris* KVS20	B-cell mitogen Production of INF-γ and IL-1α	Kitazawa *et al.*, 1993 Kitazawa *et al.*, 1996
Murine splenocytes and Peyer's patches	*B. adolescentis* M101-4	Mitogenic activity	Hosono *et al.*, 1997
Cultures of murine splenocytes, Peyer's patches and lymphocytes	*Lb. delbureckii* subsp. *bulgaricus* OLL 1073R-1	Mitogenic response of spleen cells, Payer's patches and B-cells	Kitazawa *et al.*, 1998
Mouse splenocyte and human PBMC cultures	*Lb. rhamnosus* RW-9595M	Increase of TNF, IL-6 and IL12	Chabot *et al.*, 2001
Cultures of murine macrophage line	*Lb. delbrueckii* subsp. *bulgaricus* OLL1073R-1	Increase of IL-1α, IL-6, IL-10, IL-12-p40 and TNF-α. Decrease IL12-p35	Nishimura-Uemura *et al.*, 2003
Proliferation of the epithelial intestinal lines HT-29, HCT-116 and Caco-2	*B. bifidum* BGN4	Inhibit the proliferation of HT-29 and HCT-116	You *et al.*, 2004
Stimulation of mouse splenocytes	*Lb. delbrueckii* subsp. *bulgaricus* OLL1073R-1	Increase of INF-γ	Makino *et al.*, 2006
***In vivo* studies**			
Intraperitoneal injection of EPS in mice with Sarcoma S180	*Lb. helveticus* var. *jugurti* *Lc. lactis* subsp. *cremoris* KVS20	Increase of survival	Oda *et al.*, 1983 Kitazawa *et al.*, 1991
Intraperitoneal injection in mice of EPS+ 2,4-dinitrophenyl-ovoalbumin	*Lc. lactis* subsp. *cremoris* SBT0495	Production of antibodies (adjuvant effect)	Nakajima *et al.*, 1995
Intragastric administration of EPS on rats with induced colitis	*Lb. delbureckii* subsp. *bulgaricus* B3 *Lb. delbureckii* subsp. *bulgaricus* A13	Attenuation of the colonic inflammation	Sengül *et al.*, 2005
Oral administration of EPS to mice	*Lb. delbrueckii* subsp. *bulgaricus* OLL1073R-1	Increase of natural killer activity, increase of INF-γ and decrease IL-4	Makino *et al.*, 2006
Oral administration of EPS on rats	*Lb. kefiranofaciens* ATCC 43761	Induction of gut mucosa response (increase IgA) and systemic immunity (release of cytokines)	Vinderola *et al.*, 2006

[a]In this article, the strain is named *Streptococcus cremoris* and it was isolated from the Scandinavian fermented milk viili. Abbreviations: IgA: immunoglobulin A; IL: interleukin; INF-γ: interferon-γ; TNF: tumour necrosis factor.

5. CONCLUDING REMARKS AND FUTURE TRENDS

The key parameters that influence the viscosity-intensifying ability of EPS are well characterized. In general, EPSs which are able to enhance the viscosity of fermented dairy products are those which have high-molar mass and stiff structures. However, the EPS parameter(s) related to their biological functions deserves further investigation and its function has yet to be demonstrated. The sugar linkage types and spatial structure of the polymers, which influence the specificity and accessibility of glycolytic enzymes, could play a key role in the ability of EPSs to act as prebiotic substrates. Nonetheless, the monosaccharide composition, such as the presence of a high content of Rha, as well as the presence of other reactive substituents, e.g. phosphate, seems to be necessary for the induction of the immune response. More research on this topic is necessary in the near future in order to understand the relationship between the biological functions of EPSs and their physicochemical characteristics.

Significant progress in the understanding of EPS synthetic pathways, genetics, kinetics of production and structure–technological function relationships has been made in recent years. Taking into account the knowledge generated until now, some strategies may be undertaken in order to improve the technological and/or biological capabilities of EPS produced by LAB (Boels *et al.*, 2001b; Jolly *et al.*, 2002; Welman and Maddox, 2003). From an economical point of view, reasonable yields of EPSs to be used as food additives should range between 10 and 15 g per litre (Welman and Maddox, 2003) which is far from the reported yields of EPS synthesized by most LAB (see Figure 45.1). Optimizing the EPS production of a given strain by means of controlled culturing conditions will render only a slight increase in the amount of polymer produced. Another physiological approach to obtain new molecules with desirable biological functions is to restrict the length of the EPS chain synthesized. An example of this is the production of α-gluco-oligosaccharides from dextran synthesized by *Leu. mensenteroides* by adding maltose into the culture medium which limits the reaction of the dextransucrase that incorporates Glc into the HoPS in formation (Chung and Day, 2002). On the other hand, genetic engineering of EPS could be a valuable tool to increase the EPS production and also to modify the structure/composition of the polymers. However, the link between EPS synthesis and primary carbohydrate metabolism of the LAB on one hand and the broad range of structures and composition of EPSs (mainly HePSs) on the other hand, make it difficult to find a suitable way to undertake these tasks. Engineering of EPS production can be achieved by increasing the availability of the EPS precursors, i.e. sugar nucleotides, through the modification of the corresponding genes and/or by improving the efficiency of the EPS production machinery, through the increase of the copy number of *eps* genes. Heterologous expression of *eps* clusters has also been achieved (Jolly *et al.*, 2002; Welman and Maddox, 2003). The engineering of EPS structures could allow the production of "EPS à la carte" designed for specific functions. Currently, attempts to engineer the structure of EPS have been carried out in order to improve their viscosity-intensifying ability. Controlling the polymerization of the EPS repeating units will influence the molar mass of the biopolymer. Removing or introducing side chains and modifying the monosaccharide linkage types in the primary structure of EPS will influence the stiffness or flexibility of the molecule (Boels *et al.*, 2001b). Nevertheless, in spite of all efforts made in the metabolic engineering of EPS, few results have been obtained, mainly due to the lack of knowledge about the function and regulation of the genes forming the *eps* clusters. Finally, even if the genetic strategies to modify the EPS were successful, the resulting genetically modified microorganisms (GMO) must be subjected to strict regulatory control and public acceptance. For the moment, at least in Europe, the use of GMO in functional food applications is not well accepted.

Areas requiring further investigation in this field are listed in the Research Focus Box.

RESEARCH FOCUS BOX

- The EPS characteristics that influence the technological functionality of these biopolymers are well known but, in order to design EPSs for new applications, the mechanisms of biosynthesis, the full characterization of enzymes involved in this process and their link to the primary carbohydrate metabolism of LAB should be better understood.
- A future challenge for EPS research will be to gain insight into the role of these biopolymers for the producing strains. The production of EPS is in direct competition with primary carbohydrate metabolism and with the synthesis of the cell wall; thus, it seems that this phenotype must confer some competitive advantage in the ecological niche of these bacteria.
- The relationship between physicochemical EPS characteristics and their putative biological function still remains to be elucidated. Mechanisms of action to explain their putative prebiotic effect, immune modulation or cholesterol lowering ability should be studied.
- According to the WHO/FAO guidelines, to employ EPS-producing strains for functional food applications, their beneficial effect and safety must be proven in double-blind, randomized, placebo-controlled and well-designed clinical trials. Therefore, more intervention studies for specific strains and specific population targets must be performed.

ACKNOWLEDGEMENTS

The authors thank the financial support of the Spanish Ministry of Education and Science (MEC) for their research work on the projects AGL2006-03336 and AGL2007-62736/ALI. N. Salazar also acknowledges the Spanish MEC for her FPI pre-doctoral fellowship.

References

Amrouche, T., Boutin, Y., Prioult, G., Fliss, I., 2006. Effects of bifidobacterial cytoplasm, cell wall and exopolysaccharides on mouse lymphocyte proliferation and cytokine production. Int. Dairy J. 16, 70–80.

Bergmaier, D., Champagne, C.P., Lacroix, C., 2005. Growth and exopolysaccharide production during free and inmobilized cell chemostat culture of *Lactobacillus rhamnosus* RW-9595M. J. Appl. Microbiol. 98, 272–284.

Boels, I.C., Ramos, A., Kleerebezem, M., de Vos, W.M., 2001a. Functional analysis of the *Lactococcus lactis galU* and *galE* genes and their impact on sugar nucleotide and exopolysaccharide biosynthesis. Appl. Environ. Microbiol. 67, 3033–3040.

Boels, I.C., van Kranenburg, R., Hugenholtz, J., Kleerebezem, M., de Vos, W.M., 2001b. Sugar catabolism and its impact on the biosynthesis and engineering of exopolysaccharides production in lactic acid bacteria. Int. Dairy J. 11, 723–732.

Brandtzaeg, P., Halstensen, T.S., Kett, K., et al., 1989. Immunology and immunopathology of human gut mucosa. Humoral immunity and intraepithelial lymphocytes. Gastroenterology 97, 1562–1584.

Broadbent, J.R., McMahon, D.J., Welker, D.L., Oberg, C.J., Moineau, S., 2003. Biochemistry, genetics, and applications of exopolysaccharides production in *Streptococcus thermophilus*: a review. J. Dairy Sci. 86, 407–423.

Chabot, S., Yu, H.L., De Léséleuc, L., et al., 2001. Exopolysaccharides from *Lactobacillus rhamnosus* RW-9595M stimulate TNF, IL-6 and IL-12 in human and mouse culture immunocompetent cells, and INF-γ in mouse splenocytes. Lait 81, 683–697.

Chung, C.H., Day, D.F., 2002. Glucooligosaccharides from *Leuconostoc mesenteroides* B-742 (ATCC 13146): A potential prebiotic. J. Ind. Microbiol. Biotechnol. 29, 196–199.

Cinquin, C., Le Blay, G., Fliss, I., Lacroix, Ch., 2006. Comparative effects of exopolysaccharide from lactic acid bacteria and fructo-oligosaccharides on infant gut microbiota tested in an *in vitro* colonic model within mobilized cells. FEMS Microbiol. Ecol. 57, 226–238.

Cobb, B.A., Kasper, D.L., 2005. Coming of age: carbohydrates and immunity. Eur. J. Immunol. 35, 352–356.

Dal Bello, F., Walter, J., Hertel, C., Hammes, W.P., 2001. *In vitro* study of prebiotic properties of levan-type exopolysaccharides from lactobacilli and non digestible carbohydrates using denaturing gradient gel electrophoresis. System. Appl. Microbiol. 24, 232–237.

Degeest, B., Vaningelgem, F., De Vuyst, L., 2001. Microbial physiology, fermentation kinetics and process engineering of heteropolysaccharides production by lactic acid bacteria. Int. Dairy J. 11, 747–758.

De Vuyst, L., de Vin, F., Vaningelgem, F., Degeest, B., 2001. Recent developments in the biosynthesis and applications of heteropolysaccharides from lactic acid bacteria. Int. Dairy J. 11, 687–707.

Duboc, P., Mollet, B., 2001. Applications of exopolysaccharides in dairy industry. Int. Dairy J. 11, 759–768.

Forsén, R., Heiska, E., Herva, E., Arvilommi, H., 1987. Immunobiological effect of *Streptococcus cremoris* from cultured milk "viili"; application of human lymphocyte culture techniques. Int. J. Food Microbiol. 5, 41–47.

Gibson, G.R., Roberfroid, M.B., 1995. Dietary modulation of the human colonic microbiota: introducing the concept of prebiotics. J. Nutr. 125, 1401–1412.

Gruter, M., Leeflang, B.R., Kuiper, J., Kamerling, J.P., Vliegenthart, J.F.G., 1992. Structure of the exopolysaccharides produced by *Lactococcus lactis* subspecies *cremoris* H414 grown in a defined medium or skimmed-milk. Carbohydr. Res. 231, 273–291.

Hassan, AN., Awad, S., Mistry, V.V., 2007. Reduced fat process cheese made from young reduced fat Cheddar cheese manufactured with exopolysaccharide-producing cultures. J. Dairy Sci. 90, 3604–3612.

Hosono, A., Lee, J.W., Ametani, A., et al., 1997. Characterization of a water-soluble polysaccharide fraction with immunopotentiating activity from *Bifidobacterium adolescentis* M101-4. Biosci. Biotechnol. Biochem. 61, 312–316.

Jolly, L., Stingelle, F., 2001. Molecular organization and functionality of exopolysaccharide gene clusters in lactic acid bacteria. Int. Dairy J. 11, 733–745.

Jolly, L., Vincent, S.J.F., Duboc, P., Neeser, J.R., 2002. Exploiting exopolysaccharides from lactic acid bacteria. Antonie van Leeuwenhoek 82, 367–374.

Kitazawa, H., Toba, T., Itoh, T., Kumano, N., Adachi, S., Yamaguchi, T., 1991. Antitumoral activity of slime-forming, encapsulated *Lactococcus lactis* subsp. *cremoris* isolated from Scandinavian ropy sour milk, "viili". Anim. Sci. Technol. (Jpn) 62, 277–283.

Kitazawa, H., Yamaguchi, T., Miura, M., Saito, T., Itoh, T., 1993. B-cell mitogen produced by slime-forming, encapsulated *Lactococcus lactis* ssp. *cremoris* isolated from ropy sour milk. J. Dairy Sci. 76, 1514–1519.

Kitazawa, H., Itoh, T., Tomioka, Y., Mizugaki, M., Yamaguchi, T., 1996. Induction of IFN-γ and IL-1α production in macrophages stimulated with phosphopolysaccharide produced by *Lactococcus lactis* ssp. *cremoris*. Food Microbiol. 31, 99–106.

Kitazawa, H., Harata, T., Uemura, J., Siato, T., Kaneko, T., Itoh, T., 1998. Phosphate group requirement for mitogenic activation of lymphocytes by an extracellular phosphopolysaccharide from *Lactobacillus delbrueckii* ssp. *bulgaricus*. Int. J. Food Microbiol. 40, 169–175.

Korakli, M., Vogel, R.F., 2006. Structure/function relationship of hompolysaccharides producing glycansucrases and therapeutic potential of their synthesised glycans. Appl. Microbiol. Biotechnol. 71, 790–803.

Korakli, M., Gänzle, M.G., Vogel, R.F., 2002. Metabolism by bifidobacteia and lactic acid bacteria of polysaccharides from wheat and rye, and exopolysaccharides produced by *Lactobacillus sanfranciscensis*. J. Appl. Microbiol. 92, 958–965.

Korkeala, H.J., Björkroth, K.J., 1997. Microbial spoilage and contamination of vacuum-packaged cooked sausages. J. Food Prot. 60, 724–731.

Laws, A.P., Marshall, V.M., 2001. The relevance of exopolysaccharides to the rheological properties in milk fermented with ropy strains of lactic acid bacteria. Int. Dairy J. 11, 709–721.

Looijesteijn, P.J., Trapet, L., de Vries, E., Abee, T., Hugenholtz, J., 2001. Physiological function of exopolysaccharides produced by *Lactococcus lactis*. Int. J. Food Microbiol. 64, 71–80.

Maeda, H., Zhu, X., Suzuki, S., Suzuki, K., Kitamura, S., 2004. Structural characterization and biological activities of an exopolysaccharide kefiran produced by

Lactobacillus kefiranofaciens WT-2b^T. J. Agric. Food Chem. 52, 5533–5538.

Makino, S., Ikegami, S., Kano, H., et al., 2006. Immunomodulatory effects of polysaccharides produced by *Lactobacillus delbrueckii* ssp. *bulgaricus* OLL1073R-1. J. Dairy Sci. 89, 2873–2881.

Mogensen, G., Salminen, S., O'Brien, J., et al., 2002. Food microorganisms-health benefits, safety evaluation and strains with documented history of use in foods. IDF Bull. 377, 4–10.

Monsan, P.F., Bozonnet, S., Albenne, C., Joucla, G., Willemot, R.M., Remaud-Simeon, M., 2001. Homopolysaccharides from lactic acid bacteria. Int. Dairy J. 11, 675–685.

Mozzi, F., Vaningelgem, F., Hébert, E.M., et al., 2006. Diversity of heteropolysacharide-producing lactic acid bacterium strains and their biopolymers. Appl. Environ. Microbiol. 72, 4431–4435.

Nagaoka, M., Hashimoto, S., Watanabe, T., Yokokura, T., Mori, Y., 1994. Anti-ulcer effects of lactic acid bacteria and their cell wall polysaccharides. Biol. Pharm. Bull. 17, 1012–1017.

Nakajima, H., Suzuki, Y., Kaizu, H., Hirota, T., 1992. Cholesterol lowering activity of ropy fermented milk. J. Food Sci. 57, 1327–1329.

Nakajima, H., Toba, T., Toyoda, S., 1995. Enhancement of antigen-specific antibody production by extracellular slime products from slime-forming *Lactococcus lactis* subspecies *cremoris* SBT 0459 in mice. Int. J. Food Microbiol. 25, 153–158.

Nishimura-Uemura, J., Kitazawa, H., Kawai, Y., Itoh, T., Oda, M., Saito, T., 2003. Functional alteration of murine macrophages stimulated with extracellular polysaccharides from *Lactobacillus delbrueckii* ssp. *bulgaricus* OLL1073R-1. Food Microbiol. 20, 267–273.

Oda, M., Hasegawa, H., Komatsu, S., Kambe, M., Tsuchiya, F., 1983. Anti-tumor polysaccharide from *Lactobacillus* sp. Agric. Biol. Chem. 47, 1623–1625.

Pigeon, R.M., Cuesta, P., Gilliland, S.E., 2002. Binding of free bile acids by cells of yoghurt starter culture bacteria. J. Dairy Sci. 85, 2705–2710.

Ruas-Madiedo, P., de los Reyes-Gavilán, C.G., 2005. Methods for the screening, isolation and characterization of exopolysaccharides produced by lactic acid bacteria. J. Dairy Sci. 88, 843–856.

Ruas-Madiedo, P., Hugenholtz, J., Zoon, P., 2002a. An overview of the functionality of exopolysaccharides produced by lactic acid bacteria. Int. Dairy J. 12, 163–171.

Ruas-Madiedo, P., Tuinier, R., Kanning, M., Zoon, P., 2002b. Role of exopolysaccharide produced by *Lactococcus lactis* subp. *cremoris* on the viscosity of fermented milks. Int. Dairy J. 12, 689–695.

Ruas-Madiedo, P., Gueimonde, M., Margolles, A., de los Reyes-Gavilán, C.G., Salminen, S., 2006. Exopolysaccharides produced by probiotic strains modify the adhesion of probiotics and enteropathogens to human intestinal mucus. J. Food Prot. 69, 2011–2015.

Ruas-Madiedo, P., Moreno, J.A., Salazar, N., et al., 2007. Screening of exopolysaccharide-producing *Lactobacillus* and *Bifidobacterium* strains isolated from the human intestinal microbiota. Appl. Environ. Microbiol. 13, 4385–4388.

Ruas-Madiedo, P., Abraham, A., Mozzi, F., de los Reyes-Gavilán, C.G., 2008. Functionality of exopolysaccharides produced by lactic acid bacteria. In: Mayo, B., López, P., Pérez-Martínez, G. (Eds.), Molecular Aspects of Lactic Acid Bacteria for Traditional and New Applications. Research Signpost, Kerala, pp. 137–166.

Ruas-Madiedo, P., Gueimonde, M., Arigoni, F., de los Reyes-Gavilán, C.G., Margolles, A., 2009. Bile affects the synthesis of exopolysaccharides by *Bifidobacterium animalis*. Appl. Environ. Microbiol. 75, 1204–1207.

Ruijssenaars, H.J., Stingele, F., Hartmans, S., 2000. Biodegradability of food-associated extracellular polysaccharide. Curr. Microbiol. 40, 194–199.

Salazar, N., Gueimonde, M., Hernández-Barranco, A.M., Ruas-Madiedo, P., de los Reyes-Gavilán, C.G., 2008. Exopolysaccharides produced by intestinal *Bifidobacterium* strains act as fermentable substrates for human intestinal bacteria. Appl. Environ. Microbiol. 74, 4737–4745.

Salminen, S., Gueimonde, M., 2004. Human studies on probiotics: what is scientifically proven. J. Food Sci. 69, 137–140.

Sato, T., Nishimiura-Uemura, J., Shimosato, T., Hawai, Y., Kitazawa, H., Saito, T., 2004. Dextran from *Leuconostoc mesenteroides* augments immunostimulatory effects by the introduction of phosphate groups. J. Food Prot. 67, 1719–1724.

Sengül, N., Aslín, B., Uçar, G., et al., 2005. Effects of exopolysaccharides-producing probiotic strains on experimental colitis in rats. Dis. Colon Rectum. 49, 250–258.

Sutherland, I., 1998. Novel and established application of microbial polysaccharides. Trends Biotechnol. 16, 41–46.

Tieking, M., Kaditzky, S., Valcheva, R., Korakli, M., Vogel, R. F., Gänzle, M.G., 2005. Extracellular homopolysaccharides and oligosaccharides from intestinal lactobacilli. J. Appl. Microbiol. 99, 692–702.

Tuinier, R., Zoon, P., Olieman, C., Cohen-Stuart, M.A., Fleer, G., de Kruif, C.G., 1999a. Isolation and physical characterisation of an exocellular polysaccharide. Biopolymers 49, 1–9.

Tuinier, R., Zoon, P., Cohen-Stuart, M.A., Fleer, G., de Kruif, C.G., 1999b. Concentration and shear-rate dependence of an exocellular polysaccharide. Biopolymers 50, 641–646.

Tzianabos, A.O., 2000. Polysaccharide immunomodulators as therapeutic agents: structural aspects and biological functions. Clin. Microbiol. Rev. 13, 523–533.

van Calsteren, M.R., Pau-Roblot, C., Begin, A., Roy, D., 2002. Structure determination of the exopolysaccharides produced by *Lactobacillus rhamnosus* strains RW-9595M and R. Biochem. J. 363, 7–17.

van der Meulen, R., Grosu-Tudor, S., Mozzi, F., et al., 2007. Screening of lactic acid bacteria isolates from dairy and cereal products for exopolysaccharides production and genes involved. Int. J. Food Microbiol. 118, 250–258.

van Hijum, S.A.F.T., Kralj, S., Ozimek, L.K., Dijkhuizen, L., Geel-Schutten, I.G.H., 2006. Structure-function relationships of glucansucrase and fructansucrase enzymes from lactic acid bacteria. Microbiol. Mol. Biol. Rev. 70, 157–176.

Vinderola, G., Perdigón, G., Duarte, J., Farnworth, E., Matar, Ch., 2006. Effects of the oral administration of the exopolysaccharide produced by *Lactobacillus kefiranofaciens* on the gut mucosal immunity. Cytokine 36, 254–260.

Walling, E., Gindreau, E., Lonvaud-Funel, A., 2005. A putative glucan synthetase gene *dps* detected in exopolysaccharide-producing *Pediococcus damnosus* and *Oenococcus oeni* strains isolated from wine and cider. Int. J. Food. Microbiol. 98, 53–62.

Welman, A.D., Maddox, I.S., 2003. Exopolysaccharides from lactic acid bacteria: perspectives and challenges. Trends Biotechnol. 21, 269–274.

Werning, M.L., Ibarburu, I., Dueñas, M.Y., Irastorza, A., Navas, J., López, P., 2006. *Pediococcus parvulus gtf* gene encoding the GTF glycosyltransferase and its application for specific PCR detection of β-D-glucan-producing bacteria in foods and beverages. J. Food Prot. 69, 161–169.

WHO/FAO report, 2006. Probiotics in food. Health and nutritional properties and guidelines for evaluation. In: FAO Food and Nutrition Paper No. 85. <http://www.fao.org/icatalog/search/dett.asp?aries_id = 107387/>.

WHO/FAO report, 2007. Technical meeting on prebiotics. <http://www.fao.org/ag/agn/agns/files/Prebiotics_Tech_Meeting_Report.pdf/>.

You, H.J., Oh, D.K., Ji, G.E., 2004. Anticarcinogenic effect of a novel chiroinositol-containing polysaccharide from *Bifidobacterium bifidum* BGN4. FEMS Microbiol. Lett. 240, 131–136.

CHAPTER

46

Industrial exploitation by genetic engineering of bacterial glycosylation systems

Mario F. Feldman

SUMMARY

Several bacterial species synthesize glycoproteins using oligosaccharyltransferases (OTases) that transfer glycans preassembled onto a lipid donor to proteins. The PglB protein is the OTase from *Campylobacter jejuni* involved in the *N*-glycosylation of multiple proteins. Likewise, PilO from *Pseudomonas aeruginosa* and PglL from *Neisseria meningitidis* are OTases responsible for pilin *O*-glycosylation in these bacteria. These three OTases have been functionally expressed in *Escherichia coli* and exhibit relaxed glycan specificity, which opens the door to the possibility of engineering recombinant glycoproteins for biotechnological applications. Bacteria, in particular *E. coli* cells, constitute a perfect toolbox for glyco-engineering, as they tolerate the incorporation and manipulation of foreign bacterial glycosylation pathways. The most promising application for this type of glycoproteins is the design and synthesis of a new generation of conjugate vaccines and glycan-based therapeutics.

Keywords: Campylobacter; Glycoproteins; Glyco-engineering; *Neisseria*; PglB protein; Oligosaccharyl-transferase; Glycosyltransferase; Conjugate vaccine; Lipopolysaccharide

1. INTRODUCTION

Recombinant glycoproteins are a major fraction of the active compounds in the array of biotechnology drugs available today. Examples of therapeutic glycoproteins are recombinant human erythropoietin (rHuEPO), beta-interferon (IFN-β) and follicle stimulating hormone (FSH). While the biological function is typically determined by the protein component, carbohydrates can play a fundamental role in molecular stability, serum half-life, solubility, *in vivo* activity and immunogenicity. Until recently, glycoproteins were thought to be an exclusive feature of eukaryotic cells. However, advances in mass spectrometry (MS) techniques and the increase in the accessibility to MS equipment by the scientific community, have led to the discovery of bacterial glycoproteins containing both *N*- and *O*-linked glycans. To date, little is known about the biological role of the bacterial glycoproteins. Nevertheless, some of the bacterial protein glycosylation pathways have been characterized (Schmidt *et al.*, 2003; Szymanski and Wren, 2005). These systems can now be manipulated to generate novel glycoproteins, making the exploitation of these pathways a feasible approach for the production of glycan-based therapeutics and conjugate vaccines.

In *Helicobacter*, *Listeria*, *Campylobacter* spp., as well as several other species, a single monosaccharide is attached to serine (Ser) or threonine (Thr) residues in their flagellins (Logan, 2006) (see also Chapter 8). In some *E. coli* strains, certain adhesins, mainly autotransporters, are glycosylated with a single sugar. In these cases, it is accepted that the monosaccharides are transferred directly from the nucleotide-activated sugar donors to the protein in sequential manner by glycosyltransferases, most likely in the bacterial cytoplasm. In contrast, some oligosaccharides are transferred *en bloc* to the protein acceptors. In *Campylobacter jejuni*, the *N*-glycosylation machinery responsible for the glycosylation of more than 30 proteins has been identified (Wacker *et al.*, 2002; Young *et al.*, 2002; Linton *et al.*, 2005). The natural glycan moiety invariably consists of a heptasaccharide (Young *et al.*, 2002). Oligosaccharides are also linked to select Ser residues in the pilin proteins, the main components of the type IV pilus, in *Neisseria* spp. and certain *Pseudomonas aeruginosa* strains (Castric, 1995; Stimson *et al.*, 1995). In both *C. jejuni N*-glycosylation (Feldman *et al.*, 2005) and *Neisseria* pilin *O*-glycosylation (Faridmoayer *et al.*, 2007), as well as in S-layer biosynthesis (Messner *et al.*, 2008), the oligosaccharides are assembled on a lipid carrier, undecaprenyl-pyrophosphate (Und-PP) before being transferred *en bloc* by an oligosaccharyltransferase (OTase) to the proteins. Interestingly, the OTases exhibit relaxed glycan specificity (Feldman *et al.*, 2005; Faridmoayer *et al.*, 2007), making these glycosylation systems very attractive for engineering glycoproteins of biotechnological interest. The bacterial glycosylation systems that use OTases constitute the focus of this chapter.

2. N-GLYCOSYLATION PATHWAY IN *C. JEJUNI*

C. jejuni contains a locus, named *pgl*, which is responsible for the glycosylation of numerous periplasmic and membrane-bound proteins. Several of the proteins encoded in the *pgl* locus are homologous to enzymes involved in lipopolysaccharide (LPS) biosynthesis. Expression of this locus in *E. coli* resulted in modification of the LPS and, therefore, it was proposed that this locus was involved in LPS biosynthesis (Fry *et al.*, 1998). However, subsequent work demonstrated that the *pgl* locus encodes an *N*-linked protein glycosylation system. The key enzyme of this pathway is PglB, an OTase homologous to Stt3, the catalytic subunit of the eukaryotic OTase complex (Wacker *et al.*, 2002). The pathway for the synthesis of *N*-glycoproteins in *C. jejuni* has been studied in detail (Figure 46.1A). The heptasaccharide is assembled onto the undecaprenol Und-PP carrier, which is also the carrier used for the synthesis of O-antigen, exopolysaccharides, capsules and peptidoglycan in Gram-negative bacteria. The *pgl* locus also encodes sugar-modifying enzymes, glycosyltransferases and an adenosine triphosphate-(ATP-) binding cassette (ABC) transporter. The glycosylation process starts at the cytoplasmic side of the inner membrane, with the transfer of a sugar-phosphate to the Und-P carrier by PglC. The first sugar of the heptasaccharide is Bac2,4diNAc (2,4-diacetamido-2,4,6-trideoxyglucopyranose). Uridine diphosphate- (UDP-) Bac2,4diNAc is synthesized from UDP-*N*-acetylglucosamine (GlcNAc) by the concerted action of PglD, PglE and PglF. These three enzymes are also encoded in the *pgl* locus. By analysis of the glycosylation status in the respective mutants (Linton *et al.*, 2005) and subsequent experiments *in vitro* using chemically synthesized Und-PP-heptasaccharide (Glover *et al.*, 2005a), it was demonstrated that the glycosyltransferases PglA, PglJ, PglH and PglI

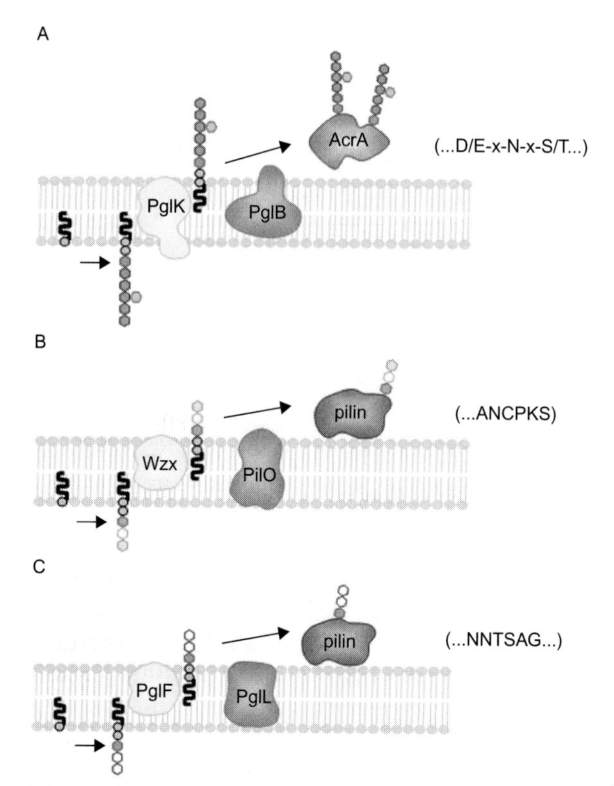

FIGURE 46.1 Proposed glycosylation pathways using oligosaccharyltransferases (OTases) in bacteria. (A) *N*-Linked protein glycosylation in *C. jejuni*. The flippase PglK, which contains an adenosine triphosphate- (ATP-) binding cassette, flips the undecaprenyl-pyrophosphate- (Und-PP-) heptasaccharide into the periplasm. The PglB protein, which is homologous to the eukaryotic OTase, transfers the glycan onto select Asn residues in more than 30 proteins. (B) *O*-Linked pilin glycosylation in *P. aeruginosa* 1244. The glycan attached to pilin is derived from the O-antigen biosynthesis. The PilO protein transfers an O-antigen subunit to the C-terminal Ser residue of pilin. Importantly, PilO can distinguish between short and fully polymerized O-antigen, transferring exclusively the former. (C) *O*-Linked pilin glycosylation in *Neisseria meningitidis*. The flippase PglF belongs to the Wzx family involved in O-antigen synthesis. The PglL OTase transfers a trisaccharide to an internal Ser residue of pilin, whereas PglL seems to possess highly relaxed glycan specificity.

sequentially incorporate the five *N*-acetylgalactosmine (GalNAc) residues and the glucose (Glc) needed to complete the heptasaccharide. Interestingly, PglH is responsible for the addition of the three terminal GalNAc residues (Glover *et al.*, 2005a). Once the heptasaccharide is completely assembled onto the UndPP-carrier, the flippase PglK translocates it into the periplasm. The ABC-type transporter PglK does not show any significant homology to the Wzx flippases involved in O-antigen synthesis. Nevertheless, although they likely use different mechanisms, PglK and Wzx are interchangeable (Alaimo *et al.*, 2006).

In the periplasm, PglB transfers the heptasaccharide to select asparagine (Asn) residues in about 40 proteins. Kowarik *et al.* have shown that PglB recognizes the consensus sequence D/E-Y-N-X-S/T in the protein substrates, where Y and X can be any amino acid except proline (Pro) (Kowarik *et al.*, 2006b). This sequence, which includes the eukaryotic *N*-glycosylation sequon N-X-S/T, is necessary but not sufficient for glycosylation. In *C. jejuni*, the glycosylation process is independent of the translocation of the protein into the periplasm. It has been shown that glycosylation can take place on folded proteins and it was suggested that the glycosylation sites are located in flexible loops within the target proteins (Kowarik *et al.*, 2006a).

The intact *N*-glycosylation machinery is required for efficient cell invasion and chicken gastrointestinal colonization by *C. jejuni*. Nonetheless, the exact role of the glycans attached to the proteins remains elusive. Orthologues of PglB have been found in all *Campylobacter* species sequenced so far, as well as in other Epsilon proteobacteria from the genera *Wolinella*, *Sulforovum* and *Nitratiruptors*, and in the Delta proteobacteria of the genera *Desulfovibrio*. However, the functionality and the pathways in these bacteria have not been yet investigated.

The *C. jejuni N*-glycosylation system has been functionally reconstituted in *E. coli*, the most generally used and genetically best-characterized bacterium in the world. This facilitated the study of the pathway and enzymes involved in the process, which opened up the way for their exploitation in glyco-engineering.

3. PILIN O-GLYCOSYLATION IN *P. AERUGINOSA*

Pilin glycosylation was initially described in *P. aeruginosa* strain 1244, by the pioneering work of Peter Castric (Castric, 1995) and, more recently, it has been found to occur in several clinical isolates (Kus *et al.*, 2004). An open-reading frame (ORF) named *pilO*, located adjacent to the gene encoding for the pilin protein (*pilA*), was identified in the 1244 strain. Mutagenesis of *pilO* resulted in loss of pilin glycosylation, leading to the proposal that PilO is the OTase responsible for the linkage of the glycan to pilin (Castric, 1995). Furthermore, pili and biofilm formation in a *P. aeruginosa* 1244 PilO mutant strain resembled the wild-type phenotype but exhibited reduced twitching motility (Castric, 1995). Competition index analysis using a mouse respiratory model comparing wild-type and PilO-deficient 1244 strains showed that the presence of the pilin glycan improves survival in the lung environment (Smedley *et al.*, 2005). These results collectively suggest that the pilin glycan is a significant virulence factor and may aid in the establishment of infection. Interestingly, strains carrying the PilO gene are more prevalent in cystic fibrosis patients than in the environment, suggesting that pilin glycosylation could be advantageous for the bacteria in colonization of these patients (Kus *et al.*, 2004). Despite these advances, the precise role for the pilin glycan remains unknown.

Pilin appears to be the only protein which becomes *O*-glycosylated by PilO. Unlike the *C. jejuni N*-glycosylation system, no consensus sequence for glycan attachment has been found. In the 1244 strain, pilin is glycosylated by the PilO OTase at its C-terminal Ser residue (Comer *et al.*, 2002). The glycan moiety consists of a trisaccharide, which is derived from the LPS biosynthetic pathway (Castric *et al.*, 2001) (see Figure 46.1B). Indeed, the glycan attached to pilin consists of a single O-antigen subunit. DiGiandomenico *et al.* (2002) expressed PilO heterologously in other *Pseudomonas* strains, where it was able to transfer a variety of O-antigen subunits to pilin. In mutant strains synthesizing truncated O-antigen subunits, PilO is able to transfer incomplete O-antigen subunits to pilin (Horzempa *et al.*, 2006a). These results demonstrated that PilO has relaxed glycan specificity.

More recently, the composition of the glycan attached to pilin in *P. aeruginosa* strain 5196 was determined (Voisin *et al.*, 2007). In this strain, the sugar moiety is not originated from the O-antigen biosynthetic pathway. Instead, the pilin glycan is composed of an unusual homo-oligomer of α-(1→5)-linked D-arabinofuranose (D-Araf). This sugar is uncommon in prokaryotes, but is one of the main components of the arabinogalactan and lipoarabinomannan (LAM) polymers found in mycobacteria cell walls. There are no obvious PilO homologues in the 5196 strain and, therefore, the OTase has not yet been identified.

4. PILIN O-GLYCOSYLATION IN *NEISSERIA* SPP.

Pilin glycosylation in *Neisseria meningitidis* and *Neisseria gonorrhoeae* was first described in 1993 (Virji *et al.*, 1993). In *Neisseria*, the glycan transferred to pilin is either a disaccharide or a trisaccharide and is not originated by the LPS biosynthetic machinery as in *P. aeruginosa* 1244. Instead, similar to the *N*-glycosylation system of *Campylobacter*, several enzymes are dedicated to the assembly of an Und-PP-glycan substrate that will be used exclusively in protein glycosylation (see Figure 46.1C). These enzymes are encoded in an operon also named *pgl*. A trisaccharide composed of β-Gal-(1→4)-α-Gal-(1→3)-Bac2,4diNAc was identified in *N. meningitidis* (Stimson *et al.*, 1995). Alternatively, a disaccharide missing the last galactose (Gal) residue can be found attached to pilin. Based on sequence homologies with PglD, PglE and PglF from

C. jejuni, it has been proposed that PglB, PglC and PglD are responsible for the synthesis of Bac2,4diNAc. Interestingly, PglB is a bifunctional enzyme, having the acetylation activity required for Bac2,4diNAc synthesis and the transferase activity necessary for transfer of Bac2,4diNAc-P to the Und-P carrier. In some strains, PglB has an insertion, resulting in the replacement of Bac2,4diNAc by glyceramido-acetamido-trideoxyhexose (GATDH), a sugar only identified in *Neisseria* (Chamot-Rooke *et al.*, 2007). Both PglE and PglA are the Gal transferases required to complete the trisaccharide and PglF is the hypothetical flippase involved in flipping the Und-PP-linked trisaccharide into the periplasm. The function of these enzymes in *N. gonorrhoeae* has recently been described (Aas *et al.*, 2007).

To date, there is only a single known *O*-glycosylation substrate in *Neisseria*. Prevention of pilin glycosylation by mutagenesis of the glycan attachment showed that the glycan increases the solubilization of pilin monomers and/or individual pilus fibres. However, this mutagenesis did not cause any significant phenotypic change. Thus, the exact role of pilin glycosylation remains unclear (Banerjee and Ghosh, 2003).

Power *et al.* (2006) showed that a mutation in a gene named *pglL* resulted in a shift in pilin electrophoretic mobility compatible with the loss of the pilin glycan. These investigators also found that PglL exhibits a homology to a domain present in several other enzymes: PilO OTase involved in *P. aeruginosa* pilin glycosylation, the O-antigen polymerases required to form the O-polysaccharide and the O-antigen ligases implicated in the transfer of O-antigen from its Und-PP carrier to the lipid A-core. Although these enzymes have different activities, they all use Und-PP-linked sugar as a substrate. Based on these observations, the authors hypothesized that PglL could be the OTase involved in pilin glycosylation in *Neisseria*. This hypothesis was supported by the work of Aas *et al.* (2007) who showed that a mutation in PglL in *N. gonorrhoeae*, renamed PglO, actually abolished pilin glycosylation. The final confirmation that PglL is an OTase was provided by the reconstitution of the *O*-pilin glycosylation process in *E. coli*, in which pilin PglL-dependent glycosylation was observed in presence of several Und-PP-linked glycan substrates. (Faridmoayer *et al.*, 2007).

5. RECOMBINANT PROTEIN N- AND O-GLYCOSYLATION IN E. COLI

The reconstitution of the *C. jejuni N*-linked protein glycosylation system in *E. coli* represented a breakthrough in the progress towards the exploitation of bacterial glycosylation systems for biotechnological purposes. In 2002, Wacker *et al.* expressed in *E. coli* the *C. jejuni* protein AcrA in the presence of the *pgl* locus, which contains PglB as well as the enzymes required for the synthesis of the lipid-linked heptasaccharide, resulting in incomplete glycosylation of AcrA. The panoply of mutants available in *E. coli* accelerated the study of the glycosylation machinery and the pathway of *N*-glycosylation. Using plasmids carrying mutations in all the genes of the *pgl* cluster in *E. coli*, Linton *et al.* (2005) proposed roles for most of the enzymes encoded in the *pgl* locus. Later work by the Szymanski group with *C. jejuni* and by the Imperiali group *in vitro* confirmed the roles assigned to the *pgl* genes in *E. coli* (Glover *et al.*, 2005a; Kelly *et al.*, 2006).

Another breakthrough in the bacterial glycoengineering field was the demonstration that PglB has relaxed glycan specificity. Feldman *et al.* (2005) exploited similarities between the LPS and glycosylation pathways to show that PglB can transfer a variety of different O-antigens to AcrA. Surprisingly, although PglB only transfers a heptasaccharide in *C. jejuni*, it was able to transfer fully polymerized O-antigen from different *E. coli* and *P. aeruginosa* serotypes in *E. coli*. The study of PglB glycan specificity was further examined by testing the capacity of PglB to link additional bacterial glycans to AcrA (Wacker *et al.*, 2006). The OTase PglB was able to transfer several polysaccharides as long as they contained an acetamido group at the position two of the sugar at the reducing end. Glover *et al.* (2005b) established an *in vitro* glycosylation which showed that PglB could transfer radiolabelled glycans to peptides. Using this assay, Chen *et al.* (2007) found that dolichyl-pyrophosphate-GlcNAc-GlcNAc disaccharide is a poor substrate for PglB. Although this experiment was not conclusive, the authors suggested that a β-(1→4)-linkage between the first two residues of the glycan substrate might not be accepted by PglB.

Kowarik *et al.* (2006a) showed that translocation of the protein into the periplasm and protein *N*-glycosylation are independent processes. *C. jejuni* AcrA could be glycosylated, translocated either via the Sec translocon or the Tat machinery. Furthermore, the authors inserted a fragment of AcrA carrying a glycosylation acceptor sequence into a loop of green fluorescent protein (GFP). Fluorescence of the hybrid protein was observed, indicating that it was folded. Efficient *in vitro* glycosylation of this protein was detected upon incubation with purified PglB and an extract containing Und-PP-oligosaccharide, indicating that fully folded proteins can be PglB substrates. Interestingly, ribonuclease (RNase) B, a eukaryotic protein, could be glycosylated by PglB *in vitro*. The glycosylation site of RNase B was modified by inserting an aspartic (Asp) acid residue at the minus two position. Glycosylation of folded RNase B was poor, but unfolding of RNAseB greatly favoured glycosylation. The authors proposed that in the *C. jejuni* acceptor proteins, glycosylation sites are located in flexible regions of folded proteins. Conversely, in eukaryotes, where glycosylation is co-translational, glycosylation evolved to a mechanism presenting the substrate in a flexible form before folding (Kowarik *et al.*, 2006a).

As already mentioned, the *N. meningitidis* and *P. aeruginosa* pilin *O*-glycosylation processes can also be reconstituted in *E. coli*. Faridmoayer *et al.* (2007) expressed both, PilO and PglL OTases, in the presence of their respective pilin substrates and two different lipid-linked glycan donors. Both OTases were able to transfer the glycans, indicating that the *O*-OTases are sufficient for glycosylation. Translocation of the lipid-linked glycan is a prerequisite for OTase activity, as glycosylation did not occur in an *E. coli* strain lacking all the flippases known to be involved in Und-PP-linked glycan translocation. By expressing PilO in different *P. aeruginosa* strains, DiGiandomenico *et al.* (2002) showed that PilO had relaxed glycan specificity. The same group confirmed this observation, showing that a single *N*-acetylfucosamine (FucNAc) residue can be transferred by PilO in a mutated *P. aeruginosa* strain (Horzempa *et al.*, 2006b). Faridmoayer *et al.* (2007) showed that PilO can also transfer the *C. jejuni* heptasaccharide found on *C. jejuni* *N*-glycoproteins. They also showed that PglL can transfer the *C. jejuni* heptasaccharide and fully polymerized *E. coli* O7-antigen. The *O*-OTase PglL has a more relaxed glycan specificity than *C. jejuni* *N*-OTase PglB, shown by its ability to transfer polysaccharides containing a hexose at the reducing end, which can be substituted at position 4 without affecting glycosylation (Faridmoayer *et al.*, 2008).

6. TOWARDS A NEW ERA IN GLYCO-ENGINEERING

Production of recombinant proteins for therapeutic applications constitutes a multibillion dollar industry. It is known that glycosylation affects the pharmacokinetic properties of recombinant proteins and, in some cases, specific glycans are essential for activity. However, different eukaryotic cell lines glycosylate their proteins with different glycans, limiting the use of

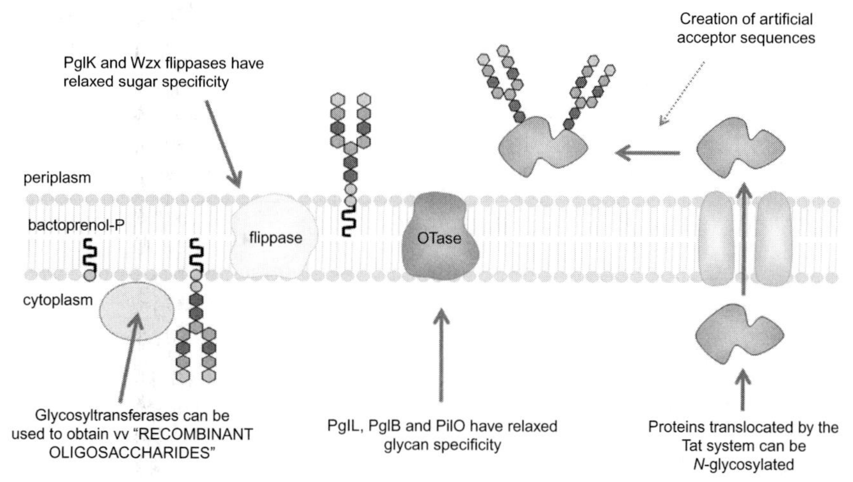

FIGURE 46.2 Glyco-engineering in bacteria. If "recombinant glycans" are synthesized onto the undecaprenyl-pyrophosphate (Und-PP carrier), they can be incorporated onto proteins in the periplasm of bacteria because the flippases and oligosaccharyltransferases (OTases) have relaxed glycan specificity. At least for *N*-glycosylation, translocation and glycosylation are independent processes, so proteins which are not translocated to the periplasm via the Sec system can be translocated via the Tat system without affecting glycosylation. Proteins can be engineered to contain glycosylation sites.

mammalian cells for production of recombinant glycoproteins for biotechnological purposes. Glycosylation is an essential process in mammalian cells and therefore manipulation of these pathways is extremely challenging. It has been shown that yeast can be modified to produce glycoproteins containing humanized *N*-glycans (Hamilton *et al.*, 2003). This is a big achievement that will likely revolutionize the production of recombinant glycoproteins. Although it has been shown that *N*-glycosylation pathways can be modified in yeast, it has not been demonstrated yet for the *O*-glycosylation pathway. Many glycans of high therapeutic potential are very different than the *N*-glycans found in eukaryotic glycoproteins and therefore cannot be engineered in yeast or mammalian cells.

Until recently, it would have been impossible to consider using *E. coli* to produce recombinant glycoproteins for biotechnological applications. Nonetheless, the discovery of the bacterial OTases with relaxed glycan specificity opens the door to this possibility. Bacteria, in particular *E. coli* cells, constitute a perfect toolbox for glyco-engineering. As glycosylation is not normally found in *E. coli*, it can tolerate the incorporation and manipulation of foreign bacterial glycosylation pathways. Figure 46.2 summarizes the current state of the bacterial protein glyco-engineering field.

6.1. Bacterial glycoproteins as conjugate vaccines

The most direct application of bacterial glycoproteins is in the area of the conjugate vaccines. Glycoconjugate vaccines are composed of a cell surface carbohydrate from a microorganism chemically attached to an appropriate carrier protein. These vaccines can provide an effective approach to generate a protective immune response against the glycan to prevent the establishment of a bacterial infection. The most successful example is the conjugate vaccine against *Haemophilus influenza* type b (Hib), which has drastically diminished the incidence of bacterial meningitis worldwide. Other conjugate vaccines have been licensed and several others are currently in clinical trials. Using

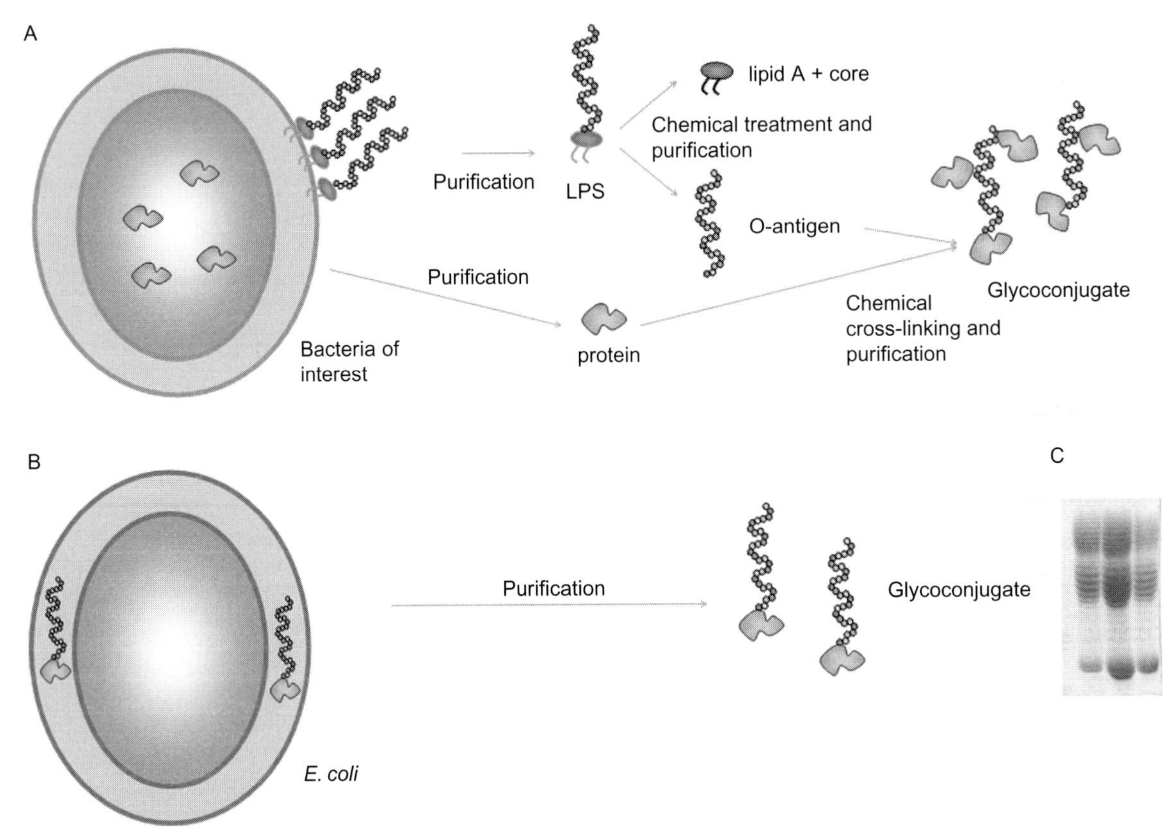

FIGURE 46.3 Synthesis of conjugate vaccines through glyco-engineering. Glycoconjugates can be obtained by conventional means (A) and by glyco-engineering *E. coli* cells (B). This approach has several advantages: minimal manipulation and simpler purification; no contamination with toxic components; defined glycoprotein structure; and no need to culture pathogenic bacteria. In (C), a conjugate composed of AcrA bound to *E. coli* O7 antigen (Feldman *et al.*, 2005) separated by electrophoresis (SDS-PAGE) and stained with Coommassie is shown.

Neisseria PglL or *Campylobacter* PglB, it is possible to attach polysaccharides to proteins, a step which is needed to elicit a T-cell memory that can provide immunological protection. Vaccine generation via the glyco-engineering method has considerable advantages over the traditional approach to glycoconjugate vaccines, in which heterogeneous mixtures are produced (Figure 46.3). Glyco-engineering in bacteria produces defined glycoconjugates because the length of the glycan chains and the sites of attachment of the sugar to the protein are well defined. Furthermore, engineering of glycoproteins does not require the culture of large amounts of pathogenic or slow-growing bacteria. Finally, certain vaccine candidates cannot be produced using the chemical coupling approach because some bacterial carbohydrates are labile to the acidic conditions employed.

The production of the vaccines by the glyco-engineering approach simply involves the expression of either OTase in presence of an appropriate acceptor protein and a plasmid containing the enzymes required for the synthesis of polysaccharide. Most often, these enzymes are encoded in single operons. Alternatively, the glycoconjugates can be produced directly in other bacteria with only the incorporation of the OTase and the acceptor protein. Of interest, PglL and PglB could be functionally expressed in *Yersinia entercolitica* cells, where they transferred the endogenous polysaccharides to their respective substrates (J. Iwashkiw and M. Feldman, unpublished results). The PglL protein is also active in *Salmonella* (Faridmoayer *et al.*, 2008). In *E. coli*, PglB has been shown to transfer diverse O polysaccharides of *Shigella flexneri*, *P. aeruginosa* and *E. coli*, among others. It is proposed that for efficient transfer, PglB requires an acetamido group in position-2 of the sugar at the reducing end. Futhermore, the linkage between this sugar and the following monosaccharide cannot be (1→4)-linked; PglL seems to have no specificity at all for the sugar substrate (Faridmoayer *et al.*, 2008). Several protein carriers are already used in conjugate vaccine production, including the tetanus toxoid, cholera toxin and exotoxin A, which could be modified to be substrates of *C. jejuni* PglB by simply adding the D/E-X-N-S/T sequon in flexible loops (Kowarik *et al.*, 2006a,b). Until now, the sequences or domains that are recognized by PglL remain undetermined. The use of *P. aeruginosa* PilO was thought to be limited due to the inability of this enzyme to transfer long-chain glycans. However, Horzempa *et al.* (2008) recently demonstrated that glycosylated *P. aeruginosa* 1244 pilin protects mice from a challenge with a pilus-null isogenic mutant, suggesting that the glycan-specific immune response was responsible for the protection. Although additional work is needed to determine the optimal glycan chain length, sugar-protein ratio and carriers, the use of these glycoproteins as a new generation of conjugate vaccine sounds feasible and exciting.

6.2. Bacterial glycoproteins as therapeutic agents

Human oligosaccharides chemically attached to proteins have been used for autoimmune diseases and cancer therapy. For example, Atochina and Harn showed that chemically obtained glyconjugates containing the Lewis X (Le^x) antigen can be efficient tools for the treatment of autoimmune diseases like psoriasis (Atochina and Harn, 2006). Also, the glycan moiety of human glycolipids attached to peptides or protein are promising instruments for cancer treatment (Bundle, 2007). The exploitation of bacterial protein glycosylation systems for the production of glycoproteins containing "human glycans" seems feasible in the near future. The selection and coordinated expression of different glycosyltransferases capable of assembling a human "recombinant" oligosaccharide onto the Und-PP-carrier constitute the most challenging step. Finding glycosyltransferases with high specificity will be crucial to obtain defined products with high yields. Oligosaccharides assembled onto the Und-PP carrier can be translocated into the periplasm by flippases, either from the Wzx family or from the ABC transporters family. Both families of flippases are known to be unspecific towards the glycan moiety. After translocation, the OTases will transfer the glycan onto a carrier protein.

The fact that some of the pathways for the synthesis of sugars common in eukaryotic glycans, like fucose and sialic acid, are unusual in *E. coli* constitutes an apparent difficulty. However, *E. coli* cells have been previously metabollically engineered to synthesize nucleotide-activated sugar donors like guanosine diphosphate- (GDP-) fucose or cytidene diphosphate- (CDP-) sialic acid, which are necessary for the synthesis of some human oligosaccharides (Antoine *et al.*, 2003; Dumon *et al.*, 2006). Therefore, although this may require additional genetic manipulation of the bacterial cells, it seems that this problem can be easily solved. Alternatively, it is possible to obtain a glycoprotein containing a basic glycan structure that can be further decorated with additional sugars by using glycosyltransferases *in vitro*.

TABLE 46.1 Characteristics of bacterial OTases

OTase	PglB	PglL	PilO
Organism	*Campylobacter*	*Neisseria*	*Pseudomonas*
Type of glycosylation	*N*	*O*	*O*
Compartment	Periplasm	Periplasm	Periplasm
Sequon recognized	D/E-X-N-X-S/T	Unknown, internal Ser residue	Unknown, C-terminal Ser residue
Endogenous glycan transferred	Heptasaccharide	Tri- or disaccharide	O-antigen subunit
Substrates identified	>30	Pilin	Pilin
Homologous to	STT3	WaaL, Wzy	WaaL, Wzy
Glycan requirement	Acetamido at position-2 of sugar at the reducing end	None	Not determined
Transfer of polysaccharides	Yes	Yes	No

Ser: serine.

7. CONCLUSIONS

The discovery of the bacterial OTases opened up the possibility of engineering glycoproteins in bacteria with applications in biotechnology. Conjugate vaccines and glycoproteins containing short human glycans constitute the most promising products that can be obtained using this bacterial glyco-engineering approach. To date, only three bacterial OTases (PglB, PilO and PglL) have been characterized (Table 46.1). A common denominator of the three OTases is that they have relaxed glycan specificity, being able to transfer diverse glycans to proteins. The PglB protein presents the advantage that it is possible to adapt a non-glycosylated protein to be a glycan acceptor by simply adding the acceptor consensus sequence in a loop of the protein. More work is needed to determine the structural features recognized by PglL and PilO in their substrates. Both PglL and PglB can transfer polysaccharides, whereas PilO activity is restricted to short glycan chains (Faridmoayer *et al.*, 2007). While PglB has certain limitations for the transfer of some important glycan targets, PglL has been able to transfer all the glycans tested. The versatility of PglL can be valuable for the synthesis of conjugates containing polysaccharides that have a hexose at the reducing end, which seem not to be transferred by PglB, but constitute a significant fraction of important targets for vaccines. For example, *Streptococcus* capsules contain a Gal residue at the reducing end and, as a result, PglB cannot be used for the generation of conjugates containing these capsular polysaccharides. Homologous proteins to both the *N*- and the *O*-OTases have been identified in the genomes of other bacteria. Therefore, it is expected to find new OTases with different or improved characteristics that could also be used for glyco-engineering. Considering that OTases have been just very recently identified and the notable advances made since then, it is reasonable to predict that the first glyco-engineered vaccines will be in clinical trials within a few years. Important questions for future research in this field are listed in the Research Focus Box.

RESEARCH FOCUS BOX

- What is the actual role of glycosylation during pathogenesis?
- How are proteins selected for *O*-glycosylation? Is there any target sequence for *O*-glycosylation?
- Is it possible to create customized glycan structures attached to proteins?
- Are glycoproteins containing bacterial polysaccharides protective against infections?
- Would it be possible to humanize bacterial glycoproteins?

ACKNOWLEDGEMENTS

I thank Adam Plumb, Parisa Shahrabadi and Jeremy Iwashkiw for the critical reading of this chapter and Dr Veronica Ielmini for assistance. I would like to thank all the members of my team for their constant input and enthusiasm in the study of bacterial glycoproteins. My laboratory is funded by grants from the Alberta Ingenuity Centre for Carbohydrate Science (AICCS), the Alberta Heritage Foundation for Medical Research (AHFMR), the Natural Sciences and Engineering Research Council of Canada (NSERC) and the Alberta Ingenuity Fund (AIF).

References

Aas, F.E., Vik, A., Vedde, J., Koomey, M., Egge-Jacobsen, W., 2007. *Neisseria gonorrhoeae O*-linked pilin glycosylation: functional analyses define both the biosynthetic pathway and glycan structure. Mol. Microbiol. 65, 607–624.

Alaimo, C., Catrein, I., Morf, L., et al., 2006. Two distinct but interchangeable mechanisms for flipping of lipid-linked oligosaccharides. EMBO J. 25, 967–976.

Antoine, T., Priem, B., Heyraud, A., et al., 2003. Large-scale in vivo synthesis of the carbohydrate moieties of gangliosides GM1 and GM2 by metabolically engineered *Escherichia coli*. Chembiochem 4, 406–412.

Atochina, O., Harn, D., 2006. Prevention of psoriasis-like lesions development in fsn/fsn mice by helminth glycans. Exp. Dermatol. 15, 461–468.

Banerjee, A., Ghosh, S.K., 2003. The role of pilin glycan in neisserial pathogenesis. Mol. Cell. Biochem. 253, 179–190.

Bundle, D.R., 2007. A carbohydrate vaccine exceeds the sum of its parts. Nat. Chem. Biol. 3, 605–606.

Castric, P., 1995. pilO, a gene required for glycosylation of *Pseudomonas aeruginosa* 1244 pilin. Microbiology 141, 1247–1254.

Castric, P., Cassels, F.J., Carlson, R.W., 2001. Structural characterization of the *Pseudomonas aeruginosa* 1244 pilin glycan. J. Biol. Chem. 276, 26479–26485.

Chamot-Rooke, J., Rousseau, B., Lanternier, F., et al., 2007. Alternative *Neisseria* spp. type IV pilin glycosylation with a glyceramido acetamido trideoxyhexose residue. Proc. Natl. Acad. Sci. USA 104, 14783–14788.

Chen, M.M., Glover, K.J., Imperiali, B., 2007. From peptide to protein: comparative analysis of the substrate specificity of *N*-linked glycosylation in *C. jejuni*. Biochemistry 46, 5579–5585.

Comer, J.E., Marshall, M.A., Blanch, V.J., Deal, C.D., Castric, P., 2002. Identification of the *Pseudomonas aeruginosa* 1244 pilin glycosylation site. Infect. Immun. 70, 2837–2845.

DiGiandomenico, A., Matewish, M.J., Bisaillon, A., Stehle, J.R., Lam, J.S., Castric, P., 2002. Glycosylation of *Pseudomonas aeruginosa* 1244 pilin: glycan substrate specificity. Mol. Microbiol. 46, 519–530.

Dumon, C., Bosso, C., Utille, J.P., Heyraud, A., Samain, E., 2006. Production of Lewis X tetrasaccharides by metabolically engineered *Escherichia coli*. Chembiochem 7, 359–365.

Faridmoayer, A., Fentabil, M.A., Mills, D.C., Klassen, J.S., Feldman, M.F., 2007. Functional characterization of bacterial oligosaccharyltransferases involved in *O*-linked protein glycosylation. J. Bacteriol. 189, 8088–8098.

Faridmoayer, A., Fentabil, M.A., Haurat, M.A., et al., 2008. Extreme substrate promiscuity of the *Neisseria* oligosaccharyltransferase involved in protein *O*-glycosylation. J. Biol. Chem. 283, 34596–34604.

Feldman, M.F., Wacker, M., Hernandez, M., et al., 2005. Engineering *N*-linked protein glycosylation with diverse O antigen lipopolysaccharide structures in *Escherichia coli*. Proc. Natl. Acad. Sci. USA 102, 3016–3021.

Fry, B.N., Korolik, V., ten Brinke, J.A., et al., 1998. The lipopolysaccharide biosynthesis locus of *Campylobacter jejuni* 81116. Microbiology 144, 2049–2061.

Glover, K.J., Weerapana, E., Imperiali, B., 2005a. *In vitro* assembly of the undecaprenylpyrophosphate-linked heptasaccharide for prokaryotic *N*-linked glycosylation. Proc. Natl. Acad. Sci. USA 102, 14255–14259.

Glover, K.J., Weerapana, E., Numao, S., Imperiali, B., 2005b. Chemoenzymatic synthesis of glycopeptides with PglB, a bacterial oligosaccharyl transferase from *Campylobacter jejuni*. Chem. Biol. 12, 1311–1315.

Hamilton, S.R., Bobrowicz, P., Bobrowicz, B., et al., 2003. Production of complex human glycoproteins in yeast. Science 301, 1244–1246.

Horzempa, J., Comer, J.E., Davis, S.A., Castric, P., 2006a. Glycosylation substrate specificity of *Pseudomonas aeruginosa* 1244 pilin. J. Biol. Chem. 281, 1128–1136.

Horzempa, J., Dean, C.R., Goldberg, J.B., Castric, P., 2006b. *Pseudomonas aeruginosa* 1244 pilin glycosylation: glycan substrate recognition. J. Bacteriol. 188, 4244–4252.

Horzempa, J., Held, T.K., Cross, A.S., et al., 2008. Immunization with a *Pseudomonas aeruginosa* 1244 pilin provides O-antigen-specific protection. Clin. Vaccine Immunol. 15, 590–597.

Kelly, J., Jarrell, H., Millar, L., Tessier, L., Fiori, L.M., Lau, P.C., Allan, B., and Szymanski, C.M., 2006. Biosynthesis of the *N*-linked glycan in *Campylobacter jejuni* and addition onto protein through block transfer. J. Bacteriol. 188, 2427–2434.

Kowarik, M., Numao, S., Feldman, M.F., et al., 2006a. *N*-linked glycosylation of folded proteins by the bacterial oligosaccharyltransferase. Science 314, 1148–1150.

Kowarik, M., Young, N.M., Numao, S., et al., 2006b. Definition of the bacterial *N*-glycosylation site consensus sequence. EMBO J. 25, 1957–1966.

Kus, J.V., Tullis, E., Cvitkovitch, D.G., Burrows, L.L., 2004. Significant differences in type IV pilin allele distribution among *Pseudomonas aeruginosa* isolates from cystic fibrosis (CF) versus non-CF patients. Microbiology 150, 1315–1326.

Linton, D., Dorrell, N., Hitchen, P.G., et al., 2005. Functional analysis of the *Campylobacter jejuni N*-linked protein glycosylation pathway. Mol. Microbiol. 55, 1695–1703.

Logan, S.M., 2006. Flagellar glycosylation – a new component of the motility repertoire? Microbiology 152, 1249–1262.

Messner, P., Steiner, K., Zarschler, K., Schaffer, C., 2008. S-layer nanoglycobiology of bacteria. Carbohydr. Res. 343 (12), 1934–1951.

Power, P.M., Seib, K.L., Jennings, M.P., 2006. Pilin glycosylation in *Neisseria meningitidis* occurs by a similar pathway to *wzy*-dependent O-antigen biosynthesis in *Escherichia coli*. Biochem. Biophys. Res. Commun. 347, 904–908.

Schmidt, M.A., Riley, L.W., Benz, I., 2003. Sweet new world: glycoproteins in bacterial pathogens. Trends Microbiol. 11, 554–561.

Smedley, J.G., Jewell, E., Roguskie, J., et al., 2005. Influence of pilin glycosylation on *Pseudomonas aeruginosa* 1244 pilus function. Infect. Immun. 73, 7922–7931.

Stimson, E., Virji, M., Makepeace, K., et al., 1995. Meningococcal pilin: a glycoprotein substituted with digalactosyl 2,4-diacetamido-2,4,6-trideoxyhexose. Mol. Microbiol. 17, 1201–1214.

Szymanski, C.M., Wren, B.W., 2005. Protein glycosylation in bacterial mucosal pathogens. Nat. Rev. Microbiol. 3, 225–237.

Virji, M., Saunders, J.R., Sims, G., Makepeace, K., Maskell, D., Ferguson, D.J., 1993. Pilus-facilitated adherence of *Neisseria meningitidis* to human epithelial and endothelial cells: modulation of adherence phenotype occurs concurrently with changes in primary amino acid sequence and the glycosylation status of pilin. Mol. Microbiol. 10, 1013–1028.

Voisin, S., Kus, J.V., Houliston, S., et al., 2007. Glycosylation of *Pseudomonas aeruginosa* strain Pa5196 type IV pilins with mycobacterium-like alpha-1,5-linked D-Araf oligosaccharides. J. Bacteriol. 189, 151–159.

Wacker, M., Linton, D., Hitchen, P.G., et al., 2002. *N*-linked glycosylation in *Campylobacter jejuni* and its functional transfer into *E. coli*. Science 298, 1790–1793.

Wacker, M., Feldman, M.F., Callewaert, N., et al., 2006. Substrate specificity of bacterial oligosaccharyltransferase suggests a common transfer mechanism for the bacterial and eukaryotic systems. Proc. Natl. Acad. Sci. USA 103, 7088–7093.

Young, N.M., Brisson, J.R., Kelly, J., et al., 2002. Structure of the *N*-linked glycan present on multiple glycoproteins in the Gram-negative bacterium, *Campylobacter jejuni*. J. Biol. Chem. 277, 42530–42539.

CHAPTER

47

Glycomimetics as inhibitors in anti-infection therapy

Milton J. Kiefel

SUMMARY

This chapter presents an overview of the general concepts that underpin the development of glycomimetics as pharmacologically useful molecules. Common structural changes to carbohydrate structures that are presented include replacement of the ring oxygen or substitution of the glycosidic oxygen, with nitrogen, sulfur or carbon. Such substitutions are usually made to impart a degree of metabolic or biological stability into the compounds being developed or to generate compounds that mimic transition-state structures from glycohydrolase reactions. Replacement of entire sialic acid moieties with simple charged functionality is also discussed, especially in regard to developing useful biological probes. A brief overview of those glycomimetics in current clinical use is also presented and emerging themes of potential future relevance to the field are described.

Keywords: Glycomimetics; Thioglycosides; Iminosugars; Carbasugars; Sialylmimetics

1. INTRODUCTION

The growing significance of the field of glycomimetics can perhaps best be gleaned from the fact that the journal *Carbohydrate Research* recently devoted an entire double issue to the topic of glycomimetics (2007, volume 342, issues 12 and 13). As our knowledge of the important and fundamental roles that complex carbohydrates play in a range of biological processes grows, so too does the need to develop compounds that can be used as tools to investigate such processes. The involvement of carbohydrates in many diseases has prompted considerable efforts into the development of carbohydrate-based molecules as biomedical investigation tools, as pharmaceutical agents or in the development of novel vaccines (Dwek, 1996; Butters *et al.*, 2000; McAuliffe and Hindsgaul, 2000).

Unfortunately, in many instances, the very thing that gives complex carbohydrates their unique ability to participate in biological processes is also their limitation with respect to developing pharmacologically relevant molecules based on carbohydrates. Carbohydrates are often chemically unstable compounds, being susceptible to degradation under acidic and basic environments, as well as being degraded by the many carbohydrate processing enzymes that are present in most living organisms. For this reason, glycomimetics have emerged as one potential solution to the inherent chemical instability of carbohydrates.

This chapter presents an overview of some of the more recent advances in the development of glycomimetics of biological and medical interest. Glycomimetics, at least for the purposes of this review, can be considered to be chemical entities that mimic the biological essence of carbohydrates. Glycomimetics presented in this article include carbohydrate-based molecules that have been chemically modified to reflect the electronics of enzyme/glycohydrolase transition states, such as replacement of the ring oxygen with non-oxygen atoms, or substitution of the glycosidic oxygen with a non-oxygen atom in order to impart more chemical stability. A second broad group of glycomimetics briefly discussed in this chapter are those compounds that are non-carbohydrate-based chemical entities that in some way mimic the relevant biological characteristics of particular carbohydrates. These non-carbohydrate glycomimetics are often designed to mimic the major structural feature of the carbohydrate, e.g. the acidic characteristic of sialic acid, such that the glycomimetic bears no structural resemblance to a carbohydrate and yet it is still able to interact with carbohydrate-binding proteins.

In light of the considerable volume of literature in this field, and in view of the space constraints within this chapter, certain types of glycomimetics have not been included. Their exclusion should in no way be taken as a reflection of their lack of importance to the field, but rather that they are outside the scope of this chapter. Many researchers have invested considerable time and effort in developing peptide-based molecules as mimics of carbohydrates. The use of peptides to mimic carbohydrates relies on the presence of functionalities within the peptide that participate in hydrogen bonding interactions; some of the achievements in this area with respect to mimics of sialyl-Lewis^{x} are particularly notable. A recent review (Murphy, 2007), on the use of peptide-based molecules as mimics of carbohydrates, is an excellent starting point for those interested. The next chapter also shows the importance of peptide-based glycomimetics in the development of new vaccines (see Chapter 48). Another important area of glycomimetics that will not be covered in this chapter relates to some very large dendrimer-type molecules that have been developed. Such compounds, often built on peptide or polyether templates, find relevance since they are designed to mimic the multivalent interactions that are usually a feature of biological interactions that occur with cell-surface carbohydrates. An enlightening article on the topic of glycan mimicry as the basis for the development of anti-infective agents is recommended for those interested (Mulvey *et al.*, 2001).

A final group of glycomimetics that is not covered in this chapter relates to the important carbohydrate heparin. Heparin, a highly sulfated oligosaccharide, has been widely used as an anticoagulant and thrombolytic since 1935. Difficulties associated with the isolation of heparin means that invariably it is obtained as a mixture of different polysaccharides, resulting in varied pharmacological activity and sometimes unwanted and dangerous side-effects due to variability in dose. For these reasons, specific chemical entities that retain the important and beneficial cardiovascular properties of heparin but without the variability of the natural material are extremely important molecules. Those interested are directed to some excellent review articles in this field (de Kort *et al.*, 2005; Hassan, 2007; Vlodavsky *et al.*, 2007).

2. REPLACEMENT OF THE RING OXYGEN

The majority of glycomimetics have been developed in the area of glycohydrolase inhibition, which is predominantly targeted to mammalian disease rather than microbial infection. Such examples are, however, relevant to the

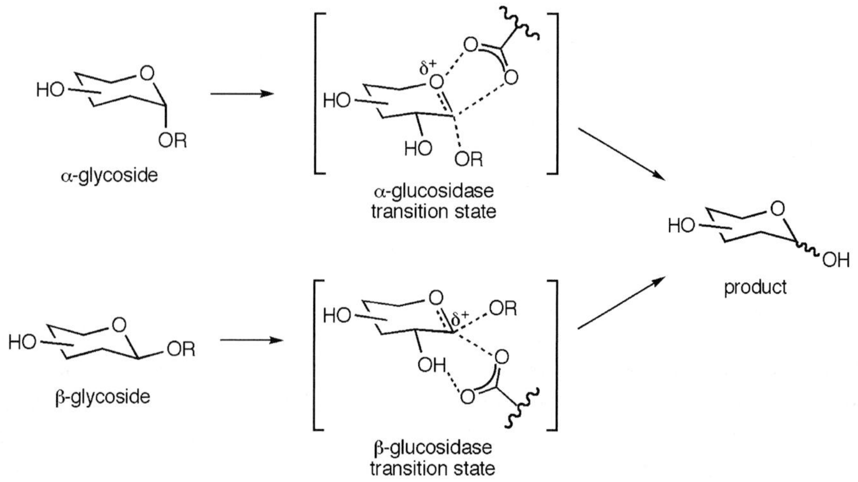

FIGURE 47.1 Representations of the transition states formed by the action of either α- or β-glycohydrolases.

discussion within this review since they show the types of chemical entities that have been developed. The role of glycohydrolases in the processing of carbohydrates is a fundamental process important to many intercellular recognition events and a range of metabolic pathways. Since glycohydrolases play a crucial role in a variety of diseases, including cancer (Paulsen and Brockhausen, 2001), diabetes (Kordik and Reitz, 1999), lysosomal storage diseases (Butters *et al.*, 2000) and viral and parasitic infections, there is substantial interest in developing inhibitors of these enzymes (Truscheit *et al.*, 1981; Lillelund *et al.*, 2002; Vasella *et al.*, 2002; Borges de Melo *et al.*, 2006).

Glycomimetics that have been developed as glycosidase inhibitors fall into two broad groups, based upon whether they mimic the transition state of either α- or β-glycohydrolases. The mechanisms associated with glycohydrolase function have been studied in great detail (Vasella *et al.*, 2002; Borges de Melo *et al.*, 2006) and it is generally accepted that α-glycosides are hydrolysed via a transition state that places a partial (or developing) positive charge on the ring oxygen (Figure 47.1). For β-glycohydrolases, the transition state is better represented (see Figure 47.1) by the development of a partial positive charge at the anomeric carbon.

Replacement of the ring oxygen is one of the strategies that has led to clinically relevant glycohydrolase inhibitors and an overview of the recent advances in this regard is presented below.

2.1. Iminosugars and azasugars (ring nitrogen)

Iminosugars like 1-deoxynojirimycin **1** (Figure 47.2) and related compounds derive their exquisite ability to inhibit glycohydrolases from the fact that the nitrogen in the ring can support a positive charge, thus enabling such compounds to mimic the transition state characteristic of the hydrolysis of α-glycosides, as shown in Figure 47.1. Compounds based upon 1-deoxynojirimycin, e.g. *N*-butyldeoxygalactonojirimycin, have been widely investigated for their ability to inhibit human immunodeficiency virus (HIV) replication (van den Broek, 1997). Since they are inhibitors of α-glucosidases, their inhibition of HIV stems from their ability to alter the glycosylation of the envelope glycoproteins gp41 and gp120, thus reducing

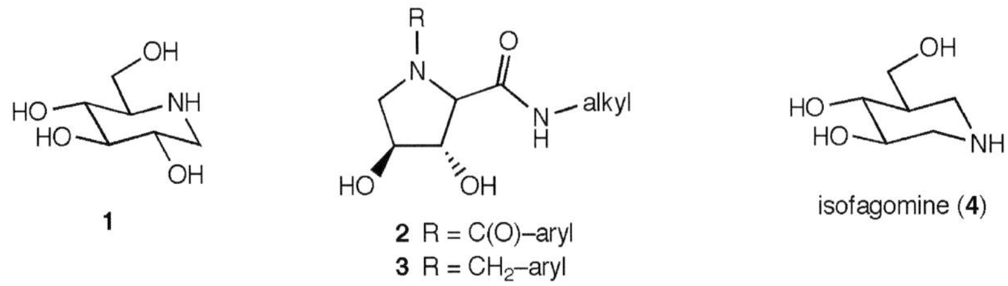

FIGURE 47.2 Examples of iminosugars that mimic the transistion state of α-glycohydrolases, and the azasugar isofagomine.

the infectivity of the virus. *N*-Alkyl analogues of deoxynojirimycin have been shown to possess antiviral activity against hepatitis B (Block *et al.*, 1994). Importantly, from a mode of action perspective, it appears that removing the basicity of the nitrogen in deoxynojirimycin analogues (by introducing *N*-acyl groups) is detrimental to their activity (Hughes and Rudge, 1994; Chapman *et al.*, 2005; Borges de Melo, *et al.*, 2006). This was particularly evident with the pyrrolidine glycomimetics represented by the general structure **2** (see Figure 47.2). While **2** exhibited no inhibition of glycosidase, even at 100 μM, reduction of the ring nitrogen, to give the amine **3**, resulted in good glycosidase inhibitory activity. Significantly, compounds like **2** with aromatic groups as part of the amide showed inhibition of bovine diarrhoea virus, a model for hepatitis C virus, comparable to 1-deoxynojirimycin (Chapman *et al.*, 2005).

An alternative group of ring nitrogen containing glycomimetics are the 1-azasugars, of which isofagomine **4** is a representative example (Bols, 1998). 1-Azasugars differ from the deoxynojirimycin-type iminosugars since the nitrogen is located at the position of the anomeric carbon and the ring oxygen is replaced by a carbon atom. The observation that azasugars like isofagomine **4** are potent inhibitors of β-glucosidases, but only moderate inhibitors of α-glucosidases, can be explained by the location of the nitrogen in the ring and the preference of β-glucosidases to form a stabilized transition state with the positive charge on the anomeric carbon (Sears and Wong, 1999; Borges de Melo *et al.*, 2006) (see Figure 47.1).

Some recent examples of advances in the area of iminosugars include the synthesis of some bicyclic analogues of nojirimycin (Cipolla *et al.*, 2007). Bicyclic iminosugars are believed to have potent glycosidase inhibitory activity because of the extra rigidity in the structures and the fact that the iminosugar ring often adopts a flattened-chair type conformation. The innovative feature of the compounds reported by Cipolla *et al.* (2007) (Figure 47.3) stems from the incorporation of pharmacophoric groups such as urea (e.g. **5**), guanidine (e.g. **6**) and cyclic carbamates (e.g. **7**). Of these bicyclic iminosugars, those with a cyclic carbamate group like **7** showed inhibition of α-glucosidase comparable to or better than 1-deoxynojirimycin. The structure–activity relationships of the glycosidase inhibitory activity of a series of azasugar glycomimetics based on the structure of the naturally occurring calystegines (e.g. **8**), have been described (Garcia-Moreno *et al.*, 2001, 2004a, 2007). Of the many structural analogues reported, it appears that replacing the sp^3 hybridized ring nitrogen with a pseudoamide-type nitrogen with more sp^2 character (e.g. **9**) gives excellent anomer specificity when inhibiting glycosidases. The presence of aromatic groups in this functionality results in potent inhibition of mammalian β-glucosidase/β-galactosidase (Garcia-Moreno *et al.*, 2001, 2004a,b).

With a view to gaining molecules that would be powerful probes for investigating glycosidases

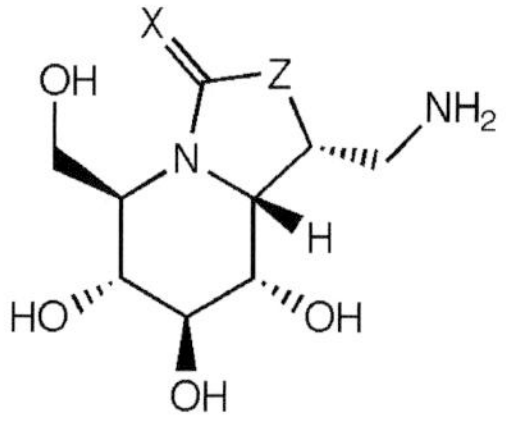

5 X = O, Z = NH
6 X = NH, Z = NH
7 X = O, Z = O

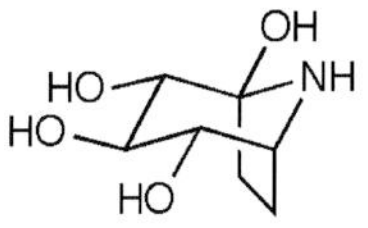

FIGURE 47.3 Bicyclic analogues of nojirimycin together with calystegines.

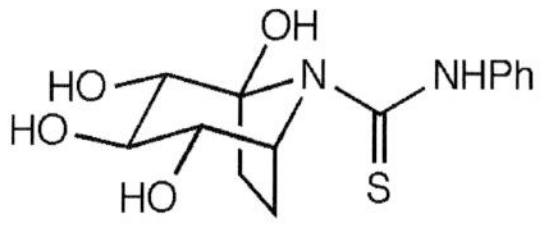

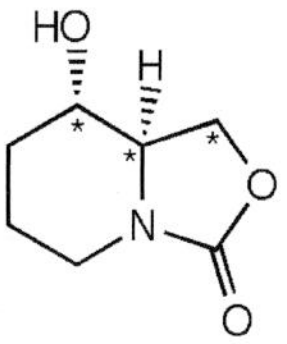

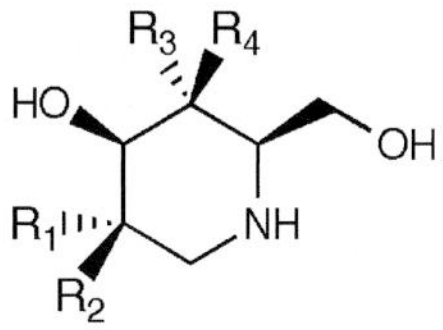

11 R_1 = H, R_2 = OH, R_3 = OH, R_4 = H
12 R_1 = OH, R_2 = H, R_3 = H, R_4 = OH

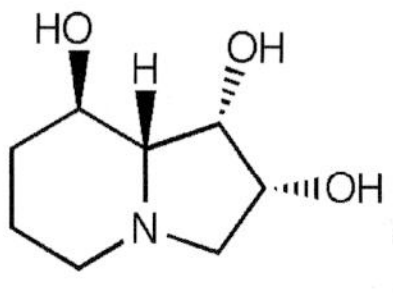

FIGURE 47.4 Examples of deoxymannojirimycin and swainsonine analogues.

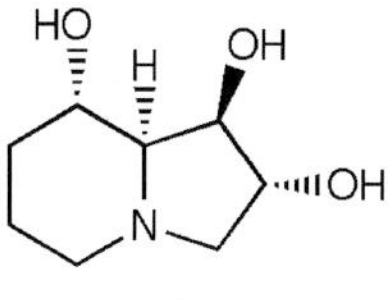

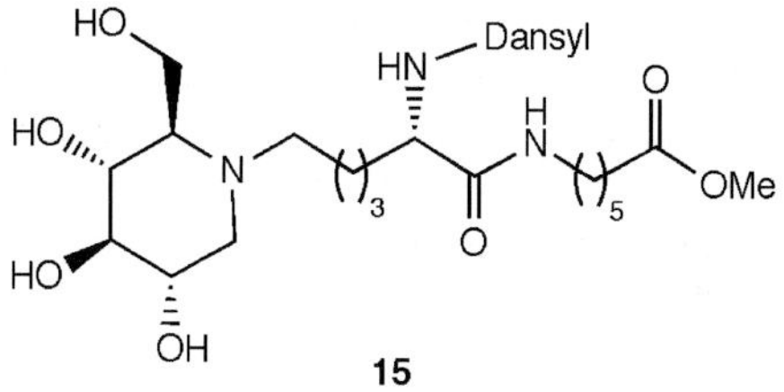

using nuclear magnetic resonance (NMR) spectroscopy, the specifically carbon-13 labelled piperidine **10** has been prepared using Sharpless asymmetric epoxidation (Murruzzu *et al.*, 2007). What makes this approach significant is that the single molecule **10** (Figure 47.4) can be readily transformed into a series of important and well-known glycosidase inhibitors, including 1-deoxymannojirimycin **11** (Ciufolini *et al.*, 1998), 1-deoxygalactostatin **12** (Asano *et al.*, 1999) and swainsonine **13** (Martin *et al.*, 2005). A series of stereochemical isomers of swainsonine have recently been described (Kumar *et al.*, 2008). These compounds, typified by the tri-*epi*-swainsonine derivative **14**, were all derived from D-glucose and studies into their inhibitory activity against glycosidases demonstrated that small structural changes in these rigid glycomimetics have a notable effect on potency and selectivity. The attachment of large fluorescent amino acid derived groups to the nitrogen of iminosugars resulted in compounds with comparable or improved inhibition of glycosidases (Steiner *et al.*, 2007). Interestingly, these types of compounds (e.g. **15**) contain a chemical functionality at the end of the chain that is suitable for immobilization of the compounds allowing use in micro-array screening of glycosidases.

Although outside the scope of this chapter, some iminosugars have found use as peptidomimetics, due to being able to attach pharmacophoric groups onto the ring nitrogen (Gouin and Murphy, 2005; Murphy, 2007).

2.2. Carbasugars (ring carbon)

Carbasugars, previously referred to as pseudosugars, are compounds that have had the ring oxygen replaced by a carbon. Such compounds have been widely investigated as inhibitors of glycosidases due to their relative chemical stability. The presence of a carbocyclic sugar ring within the antibacterial aminoglycosides (e.g. streptomycin) has been the impetus for many of the investigations into these compounds. In general terms, removal of the ring oxygen to generate carbasugars results in some changes to the chemical and biological characteristics of these compounds. Removal of the ring oxygen means that any important interactions this oxygen may have with a biological receptor, for example the stereoelectronic effects that relate to the anomeric centre, have been abolished. Although the loss of some potential electronic interactions may be detrimental, the gain in chemical stability mostly outweighs this. Importantly, in order to be valuable tools for biological investigations, carbasugars need to be readily available synthetically, which is perhaps one of the main challenges in this area of glycomimetics. Some outstanding recent reviews have been devoted to this topic (Ogawa, 2004; Arjona *et al.*, 2007), so the commentary below is to highlight recent additions to the field.

Carbasugars with *gluco*- (e.g. **16**), *galacto*- and L-*gulo*-configurations have all been prepared from D-mannose (Gómez *et al.*, 2004). The key feature of this approach is that carbasugar mimetics of rare sugars or those of the L-series can be obtained from readily available monosaccharides. In an elegant synthesis, the carbasugar analogue **17** of galactofuranose was prepared from glucose using a ring-closing metathesis approach (Frigell and Cumpstey, 2007). Significantly, the furanose form of galactose is unknown in mammals but very common in bacteria, especially in the cell wall of *Mycobacterium tuberculosis*. Therefore, compounds like **17** have great therapeutic interest, since they have the potential to inhibit the bacterial enzymes specific to galactofuranose biosynthesis and the incorporation of this sugar into bacterial cell wall structures.

An important class of carbasugar glycomimetics are the aminocyclitols (e.g. **18**), where the ring oxygen has been replaced by a carbon to give stability and the oxygen of the glycosidic bond has been replaced by a nitrogen, which can be protonated to give a positively charged compound that mimics the glycohydrolase transition state. The clinical relevance of the aminocyclitols relates to the natural product acarbose **19** (Figure 47.5), which is a potent glucosidase inhibitor used to treat postprandial hyperglycaemia in diabetic patients. Other aminocyclitols (e.g. voglibose **20**) are also used as antidiabetic agents. Galactose configured analogues of **18** with long alkyl chains appended to the nitrogen (e.g. **21**) show potential for the treatment of β-galactosidosis (Ogawa *et al.*, 2004).

Inositol **22** is a naturally occurring carbocycle substituted with hydroxyl groups. Phosphatidylinositol and phosphorylated inositol derivatives play crucial roles in cell signalling as part of the second messenger pathway involving phospholipases. The crucial role of inositol and its derivatives in cell function has prompted a number of investigations into inositol analogues, especially those where one or more of the hydroxyl groups are modified. Such compounds have great interest as potential analogues of the aminoglycoside antibiotics. Recently, a new synthesis of enantiomerically pure 1-amino-1-deoxy-*myo*-inositol **23** from achiral precursors has been described (Gonzalez-Bulnes *et al.*, 2007). The synthesis of **23** started with readily available *p*-benzoquinone and went via conduritol B **24**. Unsaturated

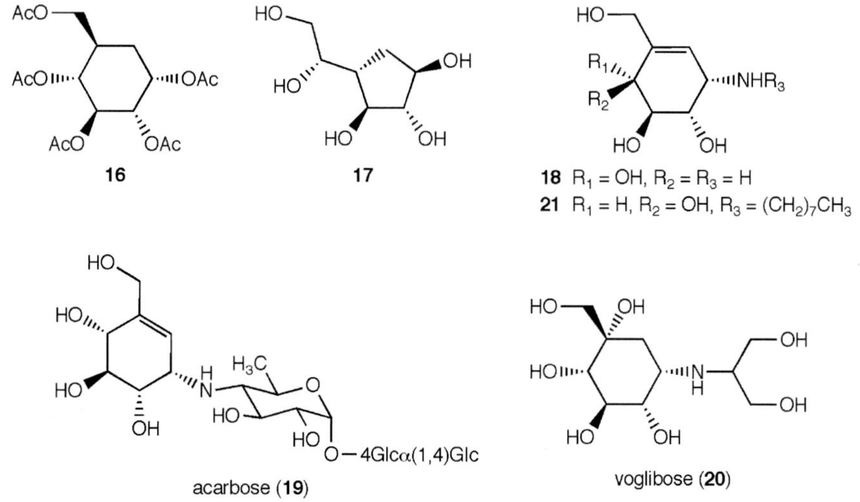

18 $R_1 = OH, R_2 = R_3 = H$
21 $R_1 = H, R_2 = OH, R_3 = (CH_2)_7CH_3$

acarbose (**19**) voglibose (**20**)

FIGURE 47.5 Examples of some simple carbasugars and some aminocyclitols.

cyclitols like conduritol B are part of a family of compounds often used as precursors to biologically interesting compounds (Figure 47.6). The conduritol family of compounds, including **24**, has shown potent hypoglycaemic activity, not only inhibiting intestinal glucose absorption but also preventing diabetic rats developing cataracts as a consequence of inhibition of lens aldose reductase (Borges de Melo *et al.*, 2006). The cyclitol epoxide **25** can be considered as a carbasugar analogue of D-glucose (Marco-Contelles, 2001). Compound **25** is one of the most potent and specific competitive inhibitors of almond and Molt-4 lysate β-glucosidases known (Borges de Melo *et al.*, 2006) and also shows no cytotoxicity in cultured cells, suggesting a possible role in the treatment of lysosomal storage disorders like Gaucher's disease (Butters *et al.*, 2000).

2.3. Others (e.g. S, Se)

Replacement of the ring oxygen with sulfur gives thiosugars (e.g. **26**). As with the iminosugars, the presence of the sulfur in the ring (as a sulfonium ion) means that a positive charge can be present, thus mimicking the charge of the oxocarbenium-ion transition state (see Figure 47.1) formed during glycoside hydrolysis by α-glycohydrolases. An excellent short review was recently published (Mohan and Pinto, 2007), so treatment of the topic here will be brief. The naturally occurring salacinol **27** (Figure 47.7) is a competitive inhibitor of intestinal α-glucosidases and was originally isolated from the roots and stems of *Salacia reticulata* (Yoshikawa *et al.*, 1997), a plant traditionally used in India and Sri Lanka for the treatment of diabetes. Several analogues of salacinol have been prepared, including nitrogen analogues (e.g. **28**), selenium analogues (e.g. **29**), sulfur analogues with different sugar stereochemistry, six-membered ring analogues (e.g. **30**) and chain-extended compounds (e.g. **31**) (Mohan and Pinto, 2007).

Several side chain modified salacinol analogues have been reported and it appears that the presence of side chain hydroxyls without a charged group (e.g. **32**) (Tanabe *et al.*, 2007) is more important than a charged group in a side chain devoid of hydroxyls (e.g. **33**) (Muraoka *et al.*, 2006). Replacement of the negatively charged sulfate group in salacinol with a carboxylate group gave a series of analogues that were evaluated for their inhibition of human inositol (**22**) **23** conduritol B (**24**) **25**

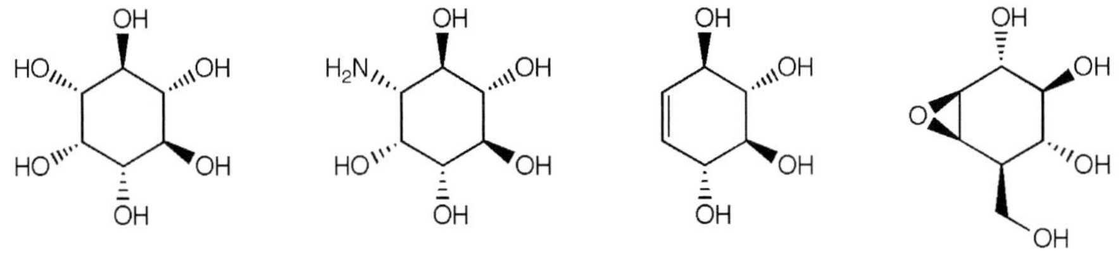

FIGURE 47.6 Some glycomimetics related to inositol **22** and conduritol B **24**.

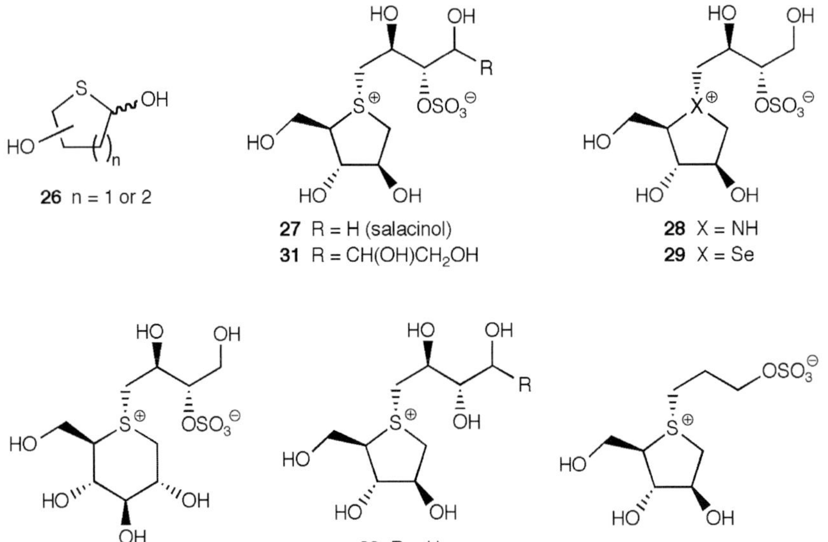

26 n = 1 or 2

27 R = H (salacinol)
31 R = $CH(OH)CH_2OH$

28 X = NH
29 X = Se

30

32 R = H
34 R = $CO_2^{\ominus}$

33

FIGURE 47.7 Representative sulfur and seleno-glycomimetics based on salacinol.

maltase glucoamylase, but only the one derived from D-arabinitol (e.g. **34**) showed inhibitory activity (Chen *et al.*, 2007).

The potent antiviral activity of some modified nucleosides, together with the recent report describing the incorporation of a thionucleotide into short RNA nucleotides as potential gene-silencing therapeutics (Dande *et al.*, 2006), is the driving force behind investigating both sulfur- (e.g. **35**) and seleno- (e.g. **36**) nucleosides (Figure 47.8). A series of selenonucleosides (e.g. **36**, with bases cytidine, thymidine, adenosine and uridine) were efficiently prepared from the key seleno-D-ribitol intermediate **37** (Jayakanthan *et al.*, 2008), although at this stage no biological data have been provided.

3. REPLACEMENT OF THE GLYCOSIDIC OXYGEN

Many of the characteristics of carbohydrates can be attributed to the presence of the glycosidic oxygen. From the perspective of the mechanism of glycohydrolases, the glycosidic oxygen is protonated during the hydrolysis reaction. Consequently, in addition to imparting

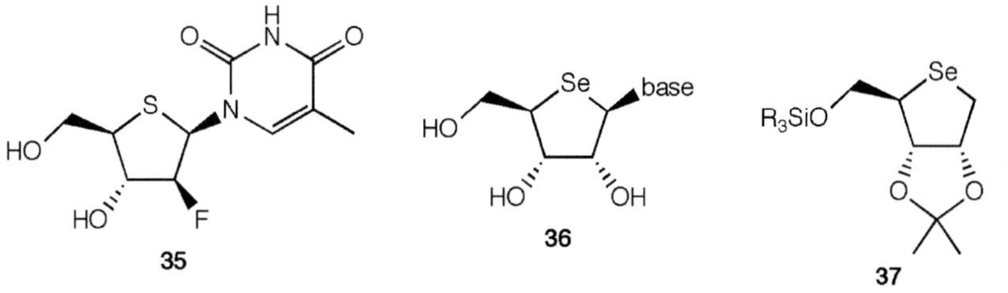

FIGURE 47.8 Sulfur and seleno-nucleosides with potential antiviral activity.

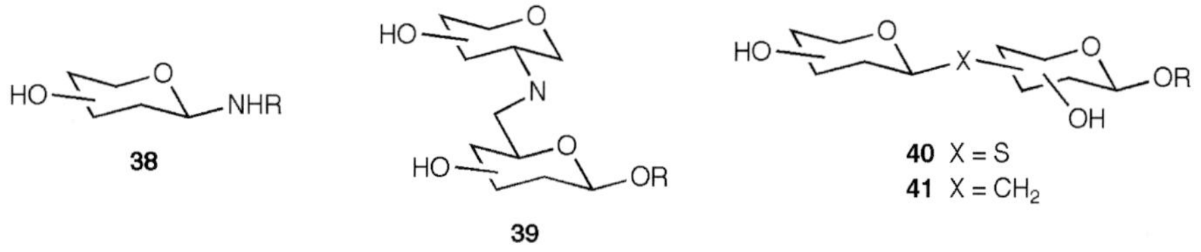

FIGURE 47.9 General examples of nitrogen, sulfur and carbon glycosides.

metabolic stability, replacement of the glycosidic oxygen can also serve the purpose of enabling compounds to better mimic the positively charged glycosidic oxygen that is present during glycoside hydrolysis. The most common form of glycosidic oxygen replacement to give glycohydrolase transition state analogues involves the use of amines (e.g. **38**), since the nitrogen can be protonated under physiological conditions (Figure 47.9). An excellent review (Lillelund *et al.*, 2002), describes several recent examples of compounds like **38**, including some glycohydrolase inhibition data. Some examples of these types of compounds are the aminocyclitols, which have been discussed above. An interesting new type of nitrogen containing glycomimetic (e.g. **39**) involves the connection of two sugars via non-anomeric positions, most notably using hexosamine sugars (e.g. *N*-acetylglucosamine) (Neumann *et al.*, 2007). The ability of these types of non-natural carbohydrate linkages to effectively mimic carbohydrate structures remains to be determined.

Aside from the use of nitrogen as a group that can bear a positive charge, the two most commonly encountered glycosidic oxygen replacements are sulfur (e.g. **40**) and carbon (e.g. **41**). The ability of compounds like **40** and **41** to impart metabolic stability into a molecule makes them extremely valuable as probes or potential inhibitors of carbohydrate recognizing proteins. In addition to their metabolic stability, thioglycosides also serve as useful and versatile glycosyl donors in the synthesis of *O*-linked saccharides.

3.1. Thioglycosides

Substitution of the glycosidic oxygen with sulfur is a common choice for a number of reasons, most notably because sulfur is in the same group on the periodic table as oxygen. Thioglycosides can therefore be expected to adopt similar conformations compared to *O*-glycosides, although there are some chemical differences due to the sulfur being less electronegative and more polarizable than oxygen. Due to their metabolic stability (Driguez, 2001), thioglycosides can be used as substrate analogues for carbohydrate-processing enzymes, as well as biological probes for carbohydrate-binding proteins. Several excellent reviews on the preparation of thioglycosides of biological interest have been published in recent years (Witczak and Culhane, 2005; Szilagyi and Varela, 2006; El Ashry *et al.*, 2006).

Thioglycosides of sialic acid have found some use as probes for sialic acid recognizing proteins (Kiefel and von Itzstein, 2002). In an attempt to elucidate the true nature of the cell-surface carbohydrate sequence bound by rotavirus, several thiosialosides (e.g. **42**) were evaluated for their inhibition of rotavirus from different sources (Kiefel *et al.*, 1996) (Figure 47.10). Thiogangliosides (Suzuki *et al.*, 1990) and, more recently, polymeric thiosialosides (e.g. **43**) (Matsuoka *et al.*, 2007) have been evaluated for their ability to inhibit influenza virus neuraminidase. Multivalent thiosialosides (Roy, 2002) have also been found to have potent inhibition of lectins including influenza virus haemagglutinin. The multivalent or polymeric presentation of sialosides enhances the inherently low binding affinity that monomeric sialosides have towards lectins.

Since mycobacteria contain galactofuranose as an essential component of their cell wall, the synthesis of compounds that mimic the structural characteristics of galactofuranosides are of interest as potential tuberculosis therapeutics. A series of galactofuranosyl sulfenamide (e.g. **44**) and sulfonamide (e.g. **45**) glycosides with varied chain lengths have been described, with the dioctyl series ($n = 7$) being potent inhibitors of mycobacterial growth (Owen *et al.*, 2007) (Figure 47.11). Elaboration of that original work produced the analogous thioglycoside (e.g. **46**) and sulfone (e.g. **47**) di-alkyl-galactofuranosides, which also showed inhibition of *Mycobacterium smegmatis* (Davis *et al.*, 2007). A detailed account describing the synthesis of glucosamine-based sulfonamide (e.g. **48**), sulfenamide (e.g. **49**) and sulfonate glycosides (e.g. **50**) has also been reported, although no biological data were presented (Knapp *et al.*, 2006). A series of thioglycosides of lactose (e.g. **51**) has recently been reported to have both antibacterial and antifungal activity (Mangte and Deshmukh, 2007), while the simple arylthioglycoside **52** of galactofuranose is an inhibitor of β-D-galactofuranosidase from *Penicillium fellutanum* and may have potential use in studies with organisms that contain galactofuranose (Mariño *et al.*, 2002).

3.2. C-Glycosides

Replacement of the glycosidic oxygen with carbon represents one of the largest classes of glycomimetics, with the synthesis of these molecules having been reviewed extensively over the last decade (Postema *et al.*, 2005; Zou, 2005; Yuan and Linhardt, 2005; Levy, 2006; Taylor *et al.*, 2006). This section will therefore focus on recent examples of pharmaceutical relevance. Importantly, introduction of the carbon in the linkage between two sugars changes the chemistry as well as the conformation of the disaccharides. The lack of the glycosidic oxygen means these compounds are metabolically more robust and the anomeric carbon has different electronic properties since it is no longer part of an acetal or ketal, but simply a cyclic ether. The conformation of *C*-glycosides has attracted considerable attention, with some recent articles focusing on *C*-glycoside mimics of blood group antigens (Vidal *et al.*, 2007; Garcia-Aparicio *et al.*, 2007) and

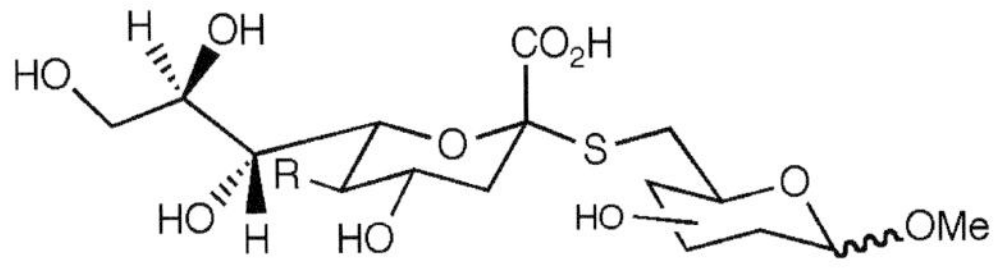

42 R = NHAc

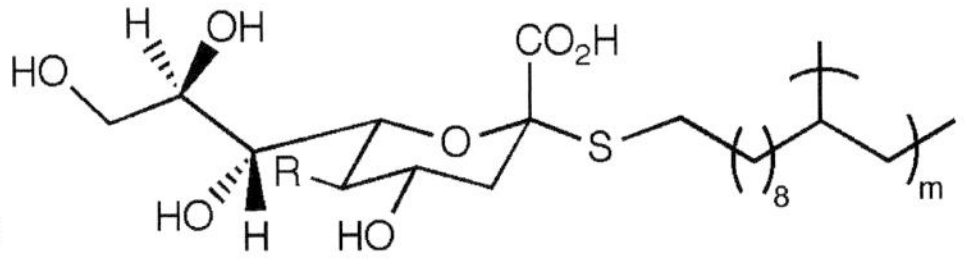

43 R = NHAc

FIGURE 47.10 Examples of thiosialosides with antiviral activity.

C-disaccharide mimics of sialyl-Lewis^{x} (Denton *et al.*, 2007; Pérez-Castells *et al.*, 2007). The conformation of *C*-galactosides has been used in a study aimed at developing potential small molecule inhibitors of cholera toxin (Podlipnik *et al.*, 2007), with simple cinnamic acid-based *C*-galactosides (e.g. **53**) showing IC_{50} values in the mM range (IC_{50} is defined as the amount of compound required to inhibit a given biological process by 50%). *C*-Cinnamoyl glycosides have also been prepared from the corresponding *C*-glycosyl ketones (e.g. **54**) via an aldol condensation (Bisht *et al.*, 2008) (Figure 47.12).

The use of indium chemistry has been employed to develop an extremely versatile and efficient synthesis of *C*-alkynyl-glycosides of both pyranose (e.g. **55**) and furanose sugars (Lubin-Germain *et al.*, 2008), while hetero-Diels-Alder chemistry was the key transformation in the efficient synthesis of Gal-(1→3)-*C*-GlcNAc disaccharides (e.g. **56**) (Wan *et al.*, 2008). The important biological activity of many naturally occurring naphthoquinones prompted the synthesis of 1,4-naphthoquinone glucosyl and galactosyl *C*-glycosides (e.g. **57**) using a Friedel-Crafts electrophilic substitution reaction (Lin *et al.*, 2008). These compounds were shown to have IC_{50} values in the low μM range against a human melanoma cell line, irrespective of the *gluco*- or *galacto*- configuration.

4. MIMICKING SPECIFIC FUNCTIONAL GROUPS WITHIN SUGARS – SIALYLMIMETICS

An important consideration in the development of carbohydrate-based pharmaceutical

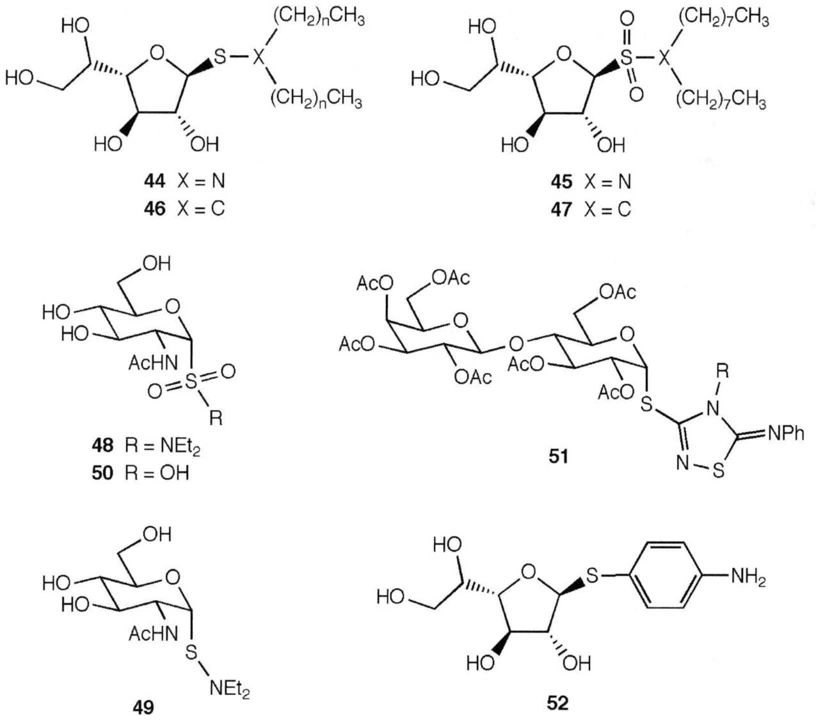

FIGURE 47.11 Examples of different sulfur-based glycomimetics with antibacterial and antifungal activity.

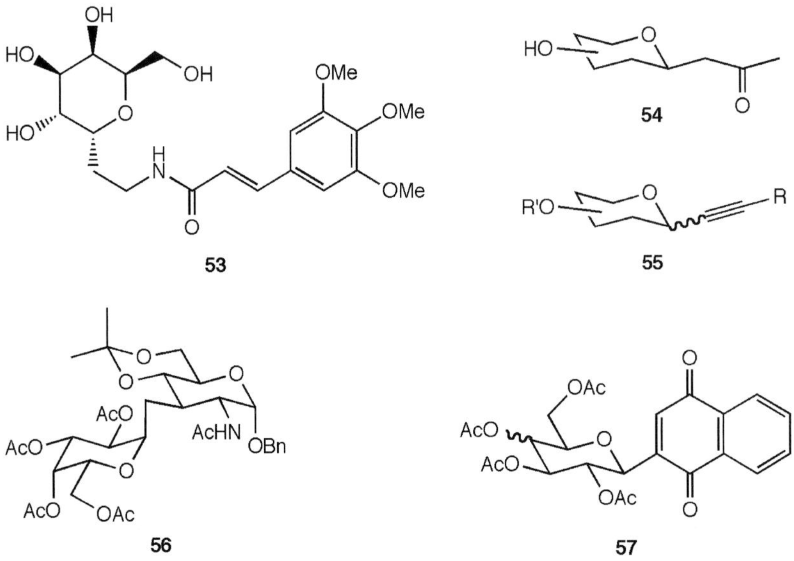

FIGURE 47.12 Representative examples of C-glycosides, including those with activity against cholera toxin (e.g. **53**) or with low μM activity against a human melanoma cell line (e.g. **57**).

agents is the fact that carbohydrates are highly oxygenated, water-soluble compounds. This can lead to problems, especially with respect to efficient delivery of the drug and the ability of the molecule to cross biological membranes. In some instances, glycomimetics have been prepared which bear no structural relationship to a carbohydrate, but rather they mimic a specific functional group within the sugar. This is particularly relevant to the acidic sugar sialic acid, wherein sialylmimetics have been prepared with the entire sialic acid unit mimicked by a simple charged functionality. Much of this work stems from studies into the role of sialyl-Lewis^{x} (**58**) in inflammation and the attempts to develop pharmacologically useful sialyl-Lewis^{x} mimics (Sears and Wong, 1999; Kiefel and von Itzstein, 2002; Wilson *et al.*, 2005). To highlight the potential of this concept of using a charged group to mimic sialic acid, the Gal-(1→1)-Man disaccharide **59** (Figure 47.13) and the C-disaccharide analogue **60**, both have comparable affinity with sialyl-Lewis^{x} towards E-selectin (Hiruma *et al.*, 1996; Cheng *et al.*, 2000). Analogous compounds to **59**, with a sulfate group as the sialic acid substitute have also been described (Kiefel and von Itzstein, 2002).

The use of a charged group to mimic sialic acid has also been employed in an attempt to inhibit cholera toxin, which binds to the ganglioside GM_1. The disaccharide **61**, with the sialic acid of GM_1 replaced by a carboxylate group and the carbocyclic ring acting as a mimic for galactose, exhibits μM affinity towards cholera toxin (Bernardi *et al.*, 2000). As part of a programme aimed at developing small molecule inhibitors of rotavirus, lactose-based sialylmimetics like **62** have been shown to inhibit human rotavirus (Fazli *et al.*, 2001). Other types of charged groups have also been investigated as mimics of sialic acid (Kiefel and von Itzstein, 2002). Molecules with a phosphonate group appended to a cyclohexene ring (e.g. **63**) have been shown to have promising sialylmimetic properties, showing inhibition against both parasitic and bacterial sialidases (Streicher and Busse, 2006).

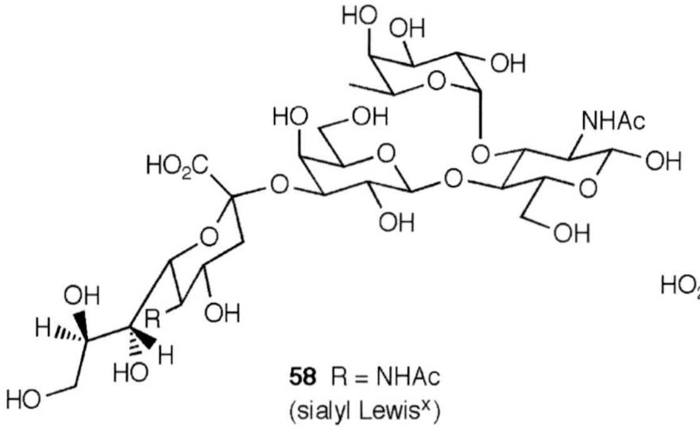

58 R = NHAc (sialyl Lewis^x)

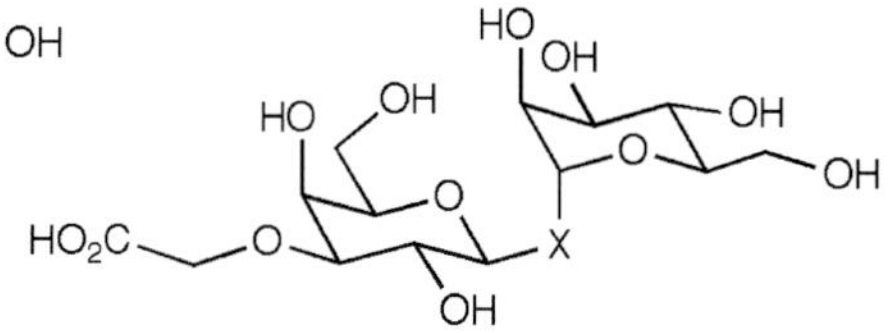

59 X = O
60 X = CH_2

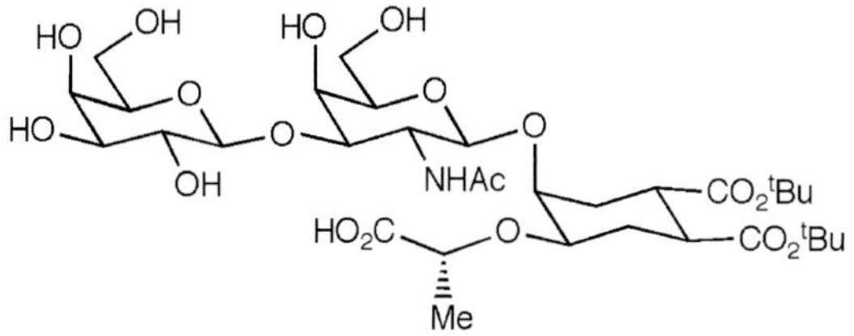

61

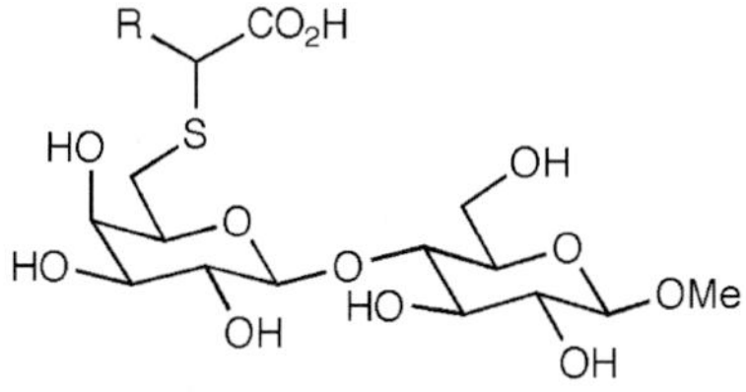

62 R = alkyl

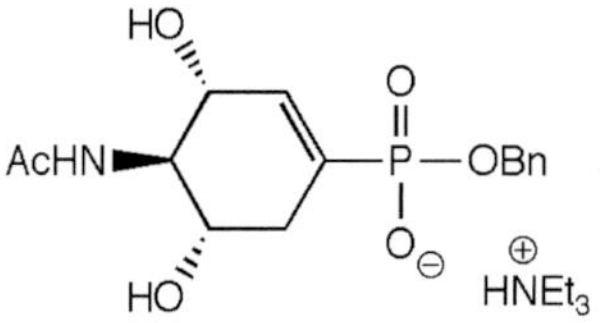

63

FIGURE 47.13 Examples of sialylmimetics where sialic acid is replaced by a charged group.

5. GLYCOMIMETICS IN CURRENT CLINICAL USE AS ANTI-INFECTIVES

The intimate involvement of carbohydrates in numerous disease processes has driven many of the investigations into developing glycomimetics as pharmaceutical agents (McAuliffe and Hindsgaul, 2000; Osborn *et al.*, 2004; Witczak, 2006). Aside from those carbohydrate analogues mentioned above that have clinical use in the treatment of diabetes (Section 2.1) or as components of the aminoglycoside antibacterial agents (Section 2.2), some simple glycomimetics have important clinical relevance. Perhaps one of the best examples is the anti-influenza drug Relenza® **64** (Figure 47.14), which was designed by detailed analysis of the catalytic site within influenza virus neuraminidase (von Itzstein *et al.*, 1993). Since its discovery in 1993, several groups have tried to develop analogues of Relenza® that mimic its exquisite ($IC_{50} \approx 10^{-11}$ M) inhibitory activity against influenza virus neuraminidase. The interested reader is directed to some excellent overviews of this topic (Thomson and von Itzstein, 2005; Islam and von Itzstein, 2007; von Itzstein, 2007). The one example of a Relenza® analogue that has progressed to clinical use is the carbocyclic anti-influenza drug Tamiflu® **65**. Both Relenza® and Tamiflu® contain a double bond within a

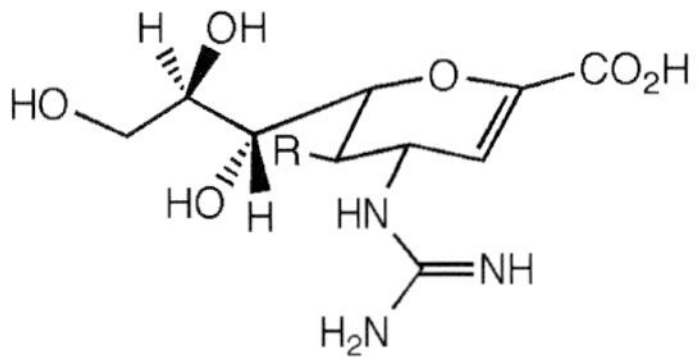

FIGURE 47.14 Glycomimetics used as anti-infective agents.

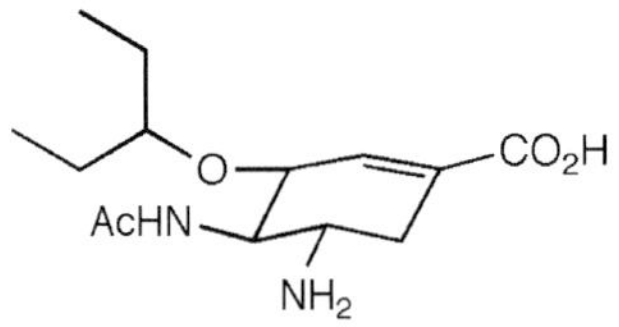

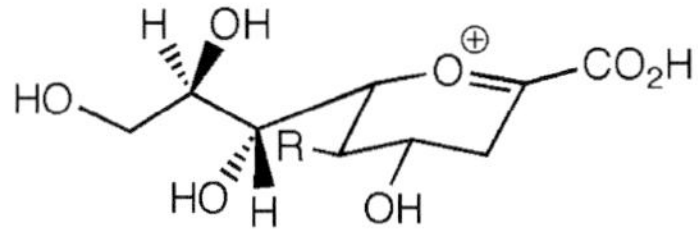

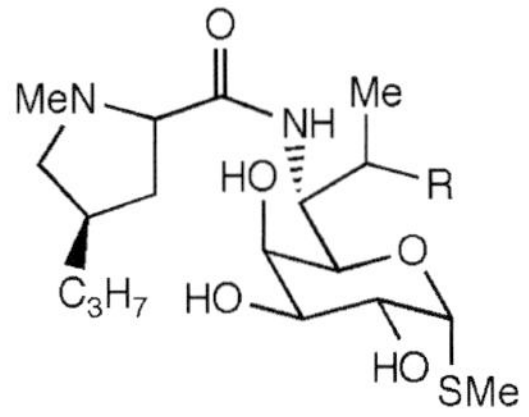

six-membered ring, which is considered to help these compounds better mimic the shape of the transition state sialosyl cation **66** that forms during sialoside cleavage (von Itzstein, 2007). Importantly, from a drug delivery perspective, Tamiflu® is much more lipophilic than Relenza®, allowing Tamiflu® to be delivered as its orally active ethyl ester prodrug, whereas Relenza® needs to be administered via inhalation directly into the respiratory tract.

The lincosamides are a family of unusual 8-carbon thio-galactoside antimicrobial agents that have been used clinically for more than 40 years (Guay, 2007; Rezanka *et al.*, 2007). Lincomycin **67** and clindamycin **68** are both used primarily for the treatment of Gram-positive bacterial infections, especially in patients who are allergic to penicillin.

Although outside the scope of this chapter, it is important to recognize that the carbohydrate component of many of the currently used antibiotics (e.g. macrolides, vancomycin) plays an important role in the efficacy of these drugs. Importantly, it is being increasingly recognized that modification of the carbohydrate component of drugs like vancomycin can both improve pharmacological activity and also circumvent bacterial resistance (Hubbard and Walsh, 2003).

6. CONCLUSIONS AND FUTURE DIRECTIONS

As we continue to learn more about the structure and function of carbohydrates in disease, so too will we look to glycomimetics as a new generation of pharmaceutical agents. The inherent chemical, metabolic and biochemical reactivity of carbohydrates, together with their high level of polarity, has driven much of the research into the development of molecules that have the biochemical properties of carbohydrates without some of the less advantageous chemical and physical properties. The glycomimetics presented in this chapter are, in many respects, simply a snapshot of a vast effort that has been made towards developing pharmacologically relevant carbohydrate-based molecules.

The exquisite selectivity and specificity of the anti-influenza drug Relenza® is a reflection of the potential of carbohydrate-based molecules as pharmaceutical agents. As we move towards the second decade of the 21st century, the future of glycomimetics looks promising. One area of great potential in this regard is the discovery of an efficient way to introduce complex functionality into carbohydrate-based molecules, via the use of "click chemistry". Developed by the Sharpless group (Kolb *et al.*, 2001; Rostovtsev *et al.*, 2002), "click chemistry" typically involves a 1,3-dipolar cyclo-addition reaction between an azide and an acetylene to give substituted triazoles. An overview of this approach for the synthesis of carbohydrate-based molecules of pharmacological interest has recently appeared (Wilkinson *et al.*, 2007). In the short time since its discovery, the "click chemistry" approach has been applied to the synthesis of analogues of Relenza® (Li *et al.*, 2006), multivalent glycoclusters as potential agents for targeting an adhesive lectin of *Pseudomonas aeruginosa* (Marotte *et al.*, 2007) and carbohydrate–β-lactam hybrids as lectin inhibitors (Palomo *et al.*, 2008).

While numerous challenges still remain, no doubt the coming years will build on our ever-expanding knowledge of carbohydrate chemistry and glycobiology, ultimately leading to an increasing number of glycomimetics entering clinical use. Future directions towards realizing the full potential of glycomimetic-based pharmaceuticals are outlined in the Research Focus Box.

RESEARCH FOCUS BOX

- The ability to incorporate improved pharmacokinetic properties into glycomimetics has the potential to open a new era of drug discovery.
- Better methods are required to allow the efficient incorporation of multiple functional groups onto small templates; "click chemistry" may prove quite valuable in this regard.
- Efficient ways to construct carbasugar ring systems require development.
- More in-depth investigations on the cross-talk and role of cell-surface glycans in disease would assist synthetic carbohydrate chemists in designing better glycomimetics.

References

Arjona, O., Gomez, A.M., Lopez, J.C., Plumet, J., 2007. Synthesis and conformational and biological aspects of carbasugars. Chem. Rev. 107, 1919–2036.

Asano, K., Hakogi, T., Iwama, S., Katsumara, S., 1999. New entry for asymmetric deoxyazasugar synthesis: syntheses of deoxymannojirimycin, deoxyaltrojirimycin and deoxygalactostatin. Chem. Commun., 41–42.

Bernardi, A., Carrettoni, L., Ciponte, A.G., Monti, D., Sonnino, S., 2000. Second generation mimics of ganglioside GM1 as artificial receptors for cholera toxin: replacement of the sialic acid moiety. Bioorg. Med. Chem. Lett. 10, 2197–2200.

Bisht, S.S., Pandey, J., Sharma, A., Tripathi, R.P., 2008. Aldol reaction of β-C-glycosylic ketones: synthesis of *C*-(*E*)-cinnamoyl glycosylic compounds as precursors for new biologically active *C*-glycosides. Carbohydr. Res. 343, 1399–1406.

Block, T.M., Lu, X., Platt, F.M., et al., 1994. Secretion of human hepatitis B virus is inhibited by the imino sugar *N*-butyldeoxynojirimycin. Proc. Natl. Acad. Sci. USA 91, 2235–2239.

Bols, M., 1998. 1-Aza sugars, apparent transition state analogues of equatorial glycoside formation/cleavage. Acc. Chem. Res. 31, 1–8.

Borges de Melo, E., da Silveira Gomes, A., Carvalho, I., 2006. α- and β-Glucosidase inhibitors: chemical structure and biological activity. Tetrahedron 62, 10277–10302.

Butters, T.D., Dwek, R.A., Platt, F.M., 2000. Inhibition of glycosphingolipid biosynthesis: application to lysosomal storage disorders. Chem. Rev. 100, 4683–4696.

Chapman, T.M., Davies, I.G., Gu, B., et al., 2005. Glyco- and peptidomimetics from three-component Joullié-Ugi coupling show selective antiviral activity. J. Am. Chem. Soc. 127, 506–507.

Chen, W., Sim, L., Rose, D.R., Pinto, B.M., 2007. Synthesis of analogues of salacinol containing carboxylate inner salt and their inhibitory activities against human maltase glucoamylase. Carbohydr. Res. 342, 1661–1667.

Cheng, X., Khan, N., Mootoo, D.R., 2000. Synthesis of the *C*-glycoside analogue of a novel sialyl Lewis x mimetic. J. Org. Chem. 65, 2544–2547.

Cipolla, L., Fernandes, M.R., Gregori, M., Airoldi, C., Nicotra, F., 2007. Synthesis and biological evaluation of a small library of nojirimycin-derived bicyclic iminosugars. Carbohydr. Res. 342, 1813–1830.

Ciufolini, M.A., Hermann, C.Y.W., Dong, Q., Shimizu, T., Swaminathan, S., Xi, N., 1998. Nitrogen heterocycles from furans: the Aza-Achmatowicz reaction. Synlett, 105–114.

Dande, P., Prakash, T.P., Sioufi, N., et al., 2006. Improving RNA interference in mammalian cells by 4′-thio-modified small interfering RNA (siRNA): Effect on siRNA activity and nuclease stability when used in combination with 2′-*O*-alkyl modifications. J. Med. Chem. 49, 1624–1634.

Davis, C.B., Hartnell, R.D., Madge, P.D., et al., 2007. Synthesis and biological evaluation of galactofuranosyl alkyl thioglycosides as inhibitors of mycobacteria. Carbohydr. Res. 342, 1773–1780.

de Kort, M., Buijsman, R.C., van Boeckel, C.A.A., 2005. Synthetic heparin derivatives as new anticoagulant drugs. Drug Discov. Today 10, 769–779.

Denton, R.W., Cheng, X., Tony, K.A., et al., 2007. *C*-disaccharides as probes for carbohydrate recognition – investigation of the conformational requirements for binding of disaccharide mimetics of sialyl Lewis x. Eur. J. Org. Chem., 645–654.

Driguez, H., 2001. Thiooligosaccharides as tools for structural biology. ChemBioChem 2, 311–318.

Dwek, R.A., 1996. Glycobiology: towards understanding the function of sugars. Chem. Rev. 96, 683–720.

El Ashry, E.S.H., Awad, L.F., Atta, A.I., 2006. Synthesis and role of glycosylthio heterocycles in carbohydrate chemistry. Tetrahedron 62, 2943–2998.

Fazli, A., Bradley, S.J., Kiefel, M.J., Jolly, C., Holmes, I.D., von Itzstein, M., 2001. Synthesis and biological evaluation of sialylmimetics as rotavirus inhibitors. J. Med. Chem. 44, 3292–3301.

Frigell, J., Cumpstey, I., 2007. First synthesis of 4a-carba-β-D-galactofuranose. Tetrahedron Lett. 48, 9073–9076.

Garcia-Aparicio, V., Malapelle, A., Abdallah, Z., et al., 2007. The solution conformation of *C*-glycosyl analogues of the sialyl-Tn antigen. Carbohydr. Res. 342, 1974–1982.

Garcia-Moreno, M.I., Benito, J.M., Mellet, C.O., Fernandez, J.M.G., 2001. Synthesis and evaluation of Calystegine B_2 analogues as glycosidase inhibitors. J. Org. Chem. 66, 7604–7614.

Garcia-Moreno, M.I., Mellet, C.O., Fernandez, J.M.G., 2004a. Synthesis of Calystegine B2, B3, and B4 analogues: mapping the structure-glycosidase inhibitory activity relationships in the 1-deoxy-6-oxacalystegine series. Eur. J. Org. Chem., 1803–1819.

Garcia-Moreno, M.I., Rodriguez-Lucena, D., Mellet, C.O., Fernandez, J.M.G., 2004b. Pseudoamide-type pyrrolidine and pyrrolizidine glycomimetics and their inhibitory activity against glycosidases. J. Org. Chem. 69, 3578–3581.

Garcia-Moreno, M.I., Mellet, C.O., Fernandez, J.M.G., 2007. Synthesis and biological evaluation of 6-oxa-nor-tropane glycomimetics as glycosidase inhibitors. Tetrahedron 63, 7879–7884.

Gómez, A.M., Moreno, E., Valverde, S., López, J.C., 2004. A stereodivergent approach to 5a-carba-α-D-*gluco*-, -α-D-*galacto* and -β-L-*gulo*pyranose pentaacetates from D-mannose, based on 6-*exo*-dig radical cyclisation and Barton-McCombie radical deoxygenation. Eur. J. Org. Chem., 1830–1840.

Gonzalez-Bulnes, P., Casas, J., Delgado, A., Llebaria, A., 2007. Practical synthesis of (-)-1-amino-1-deoxy-*myo*-inositol from achiral precursors. Carbohydr. Res. 342, 1947–1952.

Gouin, S.G., Murphy, P.V., 2005. Synthesis of somatostatin mimetics based on the 1-deoxymannojirimycin scaffold. J. Org. Chem. 70, 8527–8532.

Guay, D., 2007. Update on clindamycin in the management of bacterial, fungal, and protozoal infections. Expert Opin. Pharmacother. 8, 2401–2444.

Hassan, H.H.A.M., 2007. Chemistry and biology of heparin mimetics that bind to fibroblast growth factors. Mini-Rev. Med. Chem. 7, 1206–1235.

Hiruma, K., Kajimoto, T., Weitz-Schmidt, G., Ollmann, I., Wong, C.-H., 1996. Rational design and synthesis of a 1,1-linked disaccharide that is 5 times as active as sialyl Lewis x in binding to E-Selectin. J. Am. Chem. Soc. 118, 9265–9270.

Hubbard, B.K., Walsh, C.T., 2003. Vancomycin assembly: Nature's way. Angew Chem. Int. Ed. 42, 730–765.

Hughes, A.B., Rudge, A.J., 1994. Deoxynojirimycin: synthesis and biological activity. Nat. Prod. Rep. 11, 135–162.

Islam, T., von Itzstein, M., 2007. Anti-influenza drug discovery: are we ready for the next pandemic? Adv. Carbohydr. Chem. Biochem. 61, 293–352.

Jayakanthan, K., Johnston, B.D., Pinto, B.M., 2008. Stereoselective synthesis of 4′-selenonucleosides using the Pummerer glycosylation reaction. Carbohydr. Res. 343, 1790–1800.

Kiefel, M.J., Beisner, B., Bennett, S., Holmes, I.D., von Itzstein, M., 1996. Synthesis and biological evaluation of *N*-acetylneuraminic acid-based rotavirus inhibitors. J. Med. Chem. 39, 1314–1320.

Kiefel, M.J., von Itzstein, M., 2002. Recent advances in the synthesis of sialic acid derivatives and sialylmimetics as biological probes. Chem. Rev. 102, 471–490.

Knapp, S., Darout, E., Amorelli, B., 2006. New glycomimetics: anomeric sulfonates, sulfenamides, and sulfonamides. J. Org. Chem. 71, 1380–1389.

Kolb, H.C., Finn, M.G., Sharpless, K.B., 2001. Click chemistry: diverse chemical function from a few good reactions. Angew. Chem. Int. Ed. 40, 2004–2021.

Kordik, C.P., Reitz, A.B., 1999. Pharmacological treatment of obesity: therapeutic strategies. J. Med. Chem. 42, 181–201.

Kumar, K.S.A., Chaudhari, V.D., Dhavale, D.D., 2008. Efficient synthesis of (+)-1,8,8a-tri-*epi*-swainsonine, (+)-1,2-di-*epi*-lentiginosine, (+)-9a-*epi*-homocastospermine and (-)-9-deoxy-9a-*epi*-homocastanospermine from a D-glucose-derived aziridine carboxylate, and study of their glycosidase inhibitory activities. Org. Biomol. Chem. 6, 703–711.

Levy, D.E., 2006. Strategies towards *C*-glycosides. In: Levy, D.E., Fugedi, P. (Eds.), Organic Chemistry of Sugars. CRC Press-Taylor & Francis, Boca Raton, pp. 269–348.

Li, J., Zheng, M., Tang, W., He, P.-L., Zhu, W., Li, T., 2006. Synthesis of triazole-modified zanamivir analogues via click chemistry and anti-HIV activities. Bioorg. Med. Chem. Lett. 16, 5009–5013.

Lillelund, V.H., Jensen, H.H., Liang, X., Bols, M., 2002. Recent developments of transition-state analogue glycosidase inhibitors of non-natural product origin. Chem. Rev. 102, 515–553.

Lin, L., He, X.-P., Xu, Q., Chen, G.-R., Xie, J., 2008. Synthesis of β-*C*-glycopyranosyl-1,4-naphthoquinone derivatives and their cytotoxic activity. Carbohydr. Res. 343, 773–779.

Lubin-Germain, N., Baltaze, J.-P., Coste, A., et al., 2008. Direct *C*-glycosylation by indium-mediated alkynylation on sugar anomeric position. Org. Lett. 10, 725–728.

Mangte, D.V., Deshmukh, S.P., 2007. Synthesis and antimicrobial activities of 4-aryl-5-phenylimino-3-*S*-(hepta-*O*-acetyl lactosyl)-1,2,4-thiadiazolines. Heteroatom Chem. 18, 390–392.

Marco-Contelles, J., 2001. Cyclohexane epoxides – chemistry and biochemistry of (+)-cyclophellitol. Eur. J. Org. Chem., 1607–1618.

Mariño, K., Lima, C., Maldonado, S., Marino, C., de Lederkremer, R.M., 2002. Influence of exo β-D-galactofuranosidase inhibitors in cultures of *Penicillium fellutanum* and modifications in hyphal cell structure. Carbohydr. Res. 337, 891–897.

Marotte, K., Préville, C., Sabin, C., Moumé-Pymbock, M., Imberty, A., Roy, R., 2007. Synthesis and binding properties of divalent and trivalent clusters of the Lewis a disaccharide moiety to *Pseudomonas aeruginosa* lectin PA-IIL. Org. Biomol. Chem. 5, 2953–2961.

Martin, R., Murruzzu, C., Pericas, M.A., Riera, A., 2005. General approaches to glycosidase inihibitors. Enantioselective synthesis of deoxymannojirimycin and swainsonine. J. Org. Chem. 70, 2325–2328.

Matsuoka, K., Takita, C., Koyama, T., et al., 2007. Novel liner polymers bearing thiosialosides as pendant-type epitopes for influenza neuraminidase inhibitors. Bioorg. Med. Chem. Lett. 17, 3826–3830.

McAuliffe, J.C., Hindsgaul, O., 2000. Carbohydrates in medicine. In: Fukuda, M. (Ed.), Molecular and Cellular Glycobiology, 2nd edn. Oxford University Press, Oxford, pp. 249–285.

Mohan, S., Pinto, B.M., 2007. Zwitterionic glycosidase inhibitors: salacinol and related analogues. Carbohydr. Res. 342, 1551–1580.

Mulvey, G., Kitov, P.I., Marcato, P., Bundle, D.R., Armstrong, G.D., 2001. Glycan mimicry as the basis for novel anti-infective agents. Biochemie 83, 841–847.

Muraoka, O., Yoshikai, K., Takahashi, H., et al., 2006. Synthesis and biological evaluation of deoxy salacinols, the role of polar substituents in the side chain on the α-glucosidase inhibitory activity. Bioorg. Med. Chem. 14, 500–509.

Murphy, P.V., 2007. Peptidomimetics, glycomimetics and scaffolds from carbohydrate building blocks. Eur. J. Org. Chem., 4177–4187.

Murruzzu, C., Alonso, M., Canales, A., Jimenez-Barbero, J., Riera, A., 2007. Synthesis and NMR experiments of (4,5,6-13C)-deoxymannojirimycin. A new entry to 13C labelled glycosidase inhibitors. Carbohydr. Res. 342, 1805–1812.

Neumann, J., Weingarten, S., Theim, J., 2007. Synthesis of novel di- and trisaccharide mimetics with non-glycosidic amino bridges. Eur. J. Org. Chem., 1130–1144.

Ogawa, S., 2004. Design and synthesis of carba-sugars of biological interest. Trends Glycosci. Glycotechnol. 16, 33–53.

Ogawa, S., Sakata, Y., Ito, N., et al., 2004. Convenient synthesis and evaluation of glycosidase inhibitory activity of α- and β-galactose-type valienamines, and some *N*-alkyl derivatives. Bioorg. Med. Chem. 12, 995–1002.

Osborn, H.M.I., Evans, P.G., Gemmell, N., Osborne, S.D., 2004. Carbohydrate-based therapeutics. J. Pharm. Pharmacol. 56, 691–702.

Owen, D.J., Davis, C.B., Hartnell, R.D., et al., 2007. Synthesis and evaluation of galactofuranosyl *N,N*-dialkyl sulfenamides and sulfonamides as antimycobacterial agents. Bioorg. Med. Chem. Lett. 17, 2274–2277.

Palomo, C., Aizpurua, J.M., Balentová, E., et al., 2008. "Click" saccharide/β-lactam hybrids for lectin inhibition. Org. Lett. 10, 2227–2230.

Paulsen, H., Brockhausen, I., 2001. From iminosugars to cancer glycoproteins. Glycoconjugate J. 18, 867–870.

Pérez-Castells, J., Hernández-Gay, J., Denton, R.W., et al., 2007. The conformational behaviour and P-selectin inhibition of fluorine-containing sialyl LeX glycomimetics. Org. Biomol. Chem. 5, 1087–1092.

Podlipnik, C., Velter, I., La Ferla, B., et al., 2007. First round of a focused library of cholera toxin inhibitors. Carbohydr. Res. 342, 1651–1660.

Postema, M.H.D., Piper, J.L., Betts, R.L., 2005. Synthesis of stable carbohydrate mimetics as potential glycopharmaceuticals. Synlett, 1345–1358.

Rezanka, T., Spizek, J., Sigler, K., 2007. Medicinal use of lincosamides and microbial resistance to them. Anti-Infect. Agents Med. Chem. 6, 133–144.

Rostovtsev, V.V., Green, L.G., Fokin, V.V., Sharpless, K.B., 2002. A stepwise Huisgen cycloaddition process: copper(I) catalysed regioselective ligation of azides and terminal alkynes. Angew. Chem. Int. Ed. 41, 2596–2599.

Roy, R., 2002. Designing novel multivalent glycotools for biochemical investigations related to sialic acid. J. Carbohydr. Chem. 21, 769–798.

Sears, P., Wong, C.-H., 1999. Carbohydrate mimetics: a new strategy for tackling the problem of carbohydrate-mediated biological recognition. Angew. Chem. Int. Ed. 38, 2300–2324.

Steiner, A.J., Stutz, A.E., Tarling, C.A., Withers, S.G., Wrodnigg, T.M., 2007. Iminoalditol-amino acid hybrids: synthesis and evaluation as glycosidase inhibitors. Carbohydr. Res. 342, 1850–1858.

Streicher, H., Busse, H., 2006. Building a successful structural motif into sialylmimetics – cyclohexenephosphonate monoesters as pseudo-sialosides with promising inhibitory properties. Bioorg. Med. Chem. 14, 1047–1057.

Suzuki, Y., Sato, K., Kiso, M., Hasegawa, A., 1990. New ganglioside analogs that inhibit influenza virus sialidase. Glycoconj. J. 7, 349–356.

Szilagyi, S., Varela, O., 2006. Non-conventional glycosidic linkages: synthesis and structures of thiooligosaccharides and carbohydrates with three-bond glycosidic connections. Curr. Org. Chem. 10, 1745–1770.

Tanabe, G., Yoshikai, K., Hatanaka, T., et al., 2007. Biological evaluation of de-*O*-sulfonated analogs of salacinol, the role of sulfate anion in the side chain on the α-glucosidase inhibitory activity. Bioorg. Med. Chem. 15, 3926–3937.

Taylor, R.J.K., McAllister, G.D., Franck, R.W., 2006. The Ramberg-Baeckland reaction for the synthesis of *C*-glycosides, *C*-linked disaccharides and related compounds. Carbohydr. Res. 341, 1298–1311.

Thomson, R., von Itzstein, M., 2005. *N*-Acetylneuraminic acid derivatives and mimetics as anti-influenza drugs. In: Wong, C.-H. (Ed.), Carbohydrate-Based Drug Discovery. Wiley-VCH, Weinhelm, pp. 831–861.

Truscheit, E., Frommer, W., Junge, B., Müller, L., Schmidt, D.D., Wingender, W., 1981. Chemistry and biochemistry of microbial α-glycosidase inhibitors. Angew. Chem. Int. Ed. Engl. 20, 744–761.

van den Broek, L.A.G.M., 1997. Azasugars: chemistry and their biological activity as potential anti-HIV drugs. In: Witczak, Z.J., Nieforth, K.A. (Eds.), Carbohydrates in Drug Design. Marcel Dekker, New York, pp. 471–493.

Vasella, A., Davies, G.J., Böhm, M., 2002. Glycosidase mechanisms. Curr. Opin. Chem. Biol. 6, 619–629.

Vidal, P., Vauzeilles, B., Blériot, Y., et al., 2007. Conformational behaviour of glycomimetics: NMR and molecular modelling studies of the *C*-glycoside analogue of the disaccharide methyl β-D-galactopyranosyl-(1,3)-β-D-glucopyranoside. Carbohydr. Res. 343, 1910–1917.

Vlodavsky, I., Ilan, N., Naggi, A., Casu, B., 2007. Heparanase: structure, biological functions, and inhibition by heparin-derived mimetics of heparan sulfate. Curr. Pharm. Des. 13, 2057–2073.

von Itzstein, M., 2007. The war against influenza: discovery and development of sialidase inhibitors. Nat. Rev. Drug Discov. 6, 967–974.

von Itzstein, M., Wu, W.-Y., Kok, G.B., et al., 1993. Rational design of potent sialidase-based inhibitors of influenza virus replication. Nature 363, 418–423.

Wan, Q., Lubineau, A., Guillot, R., Scherrmann, M.-C., 2008. Synthesis of *C*-disaccharides via a hetero-Diels-Alder reaction and further stereocontrolled transformations. Carbohydr. Res. 343, 1754–1765.

Wilkinson, B.L., Bornaghi, L.F., Houston, T.A., Poulsen, S.-A., 2007. Click chemistry in carbohydrate based drug development and glycobiology. In: Kaplan, S.P. (Ed.), Drug Design Research Perspectives. Nova Science Publishers, New York, pp. 57–102.

Wilson, J.C., Kiefel, M.J., von Itzstein, M., 2005. Sialic acids and sialylmimetics: useful chemical probes of sialic acid-recognizing proteins. In: Yarema, K.J. (Ed.), Handbook of Carbohydrate Engineering. CRC Press-Taylor & Francis, Boca Raton, pp. 687–748.

Witczak, Z.J., 2006. Carbohydrate therapeutics: new developments and strategies. In: Klyosov, A.A., Witczak, Z.J., Platt, D. (Eds.), ACS Symposium Series 932 (Carbohydrate drug design). American Chemical Society, Washington, DC, pp. 25–46.

Witczak, Z.J., Culhane, J.M., 2005. Thiosugars: new perspectives regarding availability and potential biochemical and medicinal applications. Appl. Microbiol. Biotechnol. 69, 237–244.

Yoshikawa, M., Murakami, T., Shimada, H., et al., 1997. Salacinol, potent antidiabetic principle with unique thiosugar sulfonium sulfate structure from the Ayurvedic traditional medicine Salacia reticulata in Sri Lanka and India. Tetrahedron Lett. 38, 8367–8370.

Yuan, X., Linhardt, R.J., 2005. Recent advances in the synthesis of *C*-oligosaccharides. Curr. Top. Med. Chem. 5, 1393–1430.

Zou, W., 2005. C-glycosides and Aza-C-glycosides as potential glycosidase and glycosyltransferase inhibitors. Curr. Top. Med. Chem. 5, 1363–1391.

CHAPTER

48

Bacterial polysaccharide vaccines: glycoconjugates and peptide-mimetics

Harold J. Jennings and Robert A. Pon

SUMMARY

The use of bacterial capsular polysaccharides as vaccines in healthy adults has been established for several decades. However, the populations most at risk, children under the age of five years old and the immunocompromised, respond poorly to them and are thus inadequately protected against invasive bacterial disease. The simple act of conjugating capsular polysaccharides to protein carriers in large part overcomes many of these limitations and has realized the rapid development of conjugate vaccine technology. This review focuses on the current technology used in the preparation of glycoconjugate vaccines, their application and effectiveness towards bacterial pathogens of the highest importance and the state of the new generation of synthetic and peptide mimetic vaccines that are targeted to bacterial capsular polysaccharide antigens.

Keywords: Glycoconjugate; Conjugation; Capsular polysaccharide; Carbohydrate mimetic; Synthetic vaccine; T-dependent; Vaccine

1. INTRODUCTION

The early discovery that the capsular polysaccharides (CPS) of pneumococci were immunogenic in humans (Francis and Tillett, 1930) was an important development in vaccine technology, which eventually led to the licensing of polysaccharide vaccines against human bacterial diseases, such as *Neisseria meningitidis* (groups A, C, W-135, and Y), *Streptococcus pneumoniae* (23 serotypes), *Haemophilus influenzae* type b and *Salmonella enterica* sv. Typhi. However, despite the success of polysaccharide vaccines, most produce an inadequate immune response in infants (Gotschlich *et al.*, 1977), the cohort most vulnerable to bacterial infection, and some are even poorly immunogenic in adults (Baker and Edwards, 2003). These deficiencies have been largely overcome through the conjugation of CPS to protein carriers and the resultant conjugate vaccines have greatly extended the potential of polysaccharide vaccines.

The historical basis of polysaccharide conjugate vaccine technology can be found in research carried out nearly 80 years ago (Goebel and Avery, 1931), but the potential of glycoconjugates as human infant vaccines was only realized in the 1990s with the development and licensing of *H. influenzae* type b (Hib) polysaccharide–protein conjugates (Anderson *et al.*, 1989; Claesson *et al.*, 1990). The phenomenal success of these vaccines led to the rapid development of conjugate vaccines against other prevalent diseases, such as *N. meningitidis*, *S. pneumoniae*, group B *Streptococcus* and *S. enterica* sv. Typhi, and opened the door for the future development of many other commercial glycoconjugate vaccines. This review focuses on the synthesis and immunological performance of polysaccharide conjugate vaccines against bacterial pathogens having the highest priority and the application and potential of their associated synthetic saccharide- and peptide-mimetic–conjugate vaccines.

2. CAPSULAR POLYSACCHARIDE–PROTEIN CONJUGATES

2.1. *H. influenzae*

Type b *H. influenzae* (Hib) is the major cause of meningitis in children less than 5 years of age and has an extremely high mortality rate. Of equal importance is the high rate of severe and permanent neurological deficits within survivors, even those subjected to antibiotic therapy (Robbins, 1978). *H. influenzae* is classified into six different serotypes based on their CPS (a–f) but only the type b polyribosylribitol phosphate polysaccharide (PRP) (Table 48.1) causes meningitis, thus vaccines against this disease have the distinct advantage of being monovalent. The potential of PRP as a human vaccine was affirmed through the production of long-lived complement-mediated bactericidal antibodies in immunized adults (Anderson *et al.*, 1972) but, unfortunately in subsequent trials, the PRP vaccine failed to raise protective antibodies in the highly vulnerable infant cohort (Peltola *et al.*, 1977). However, because of the crippling nature of the disease, the polysaccharide vaccine was licensed in 1985 for use in children over the age of 18 months.

The inability of PRP to induce antibodies in infants is a phenomenon associated with nearly all CPS and is due mainly to the immaturity of their respective immune systems. Polysaccharides are categorized as T-cell-independent antigens and are only able to induce in infants low levels of non-boostable immunoglobulin (Ig) class M antibodies (Stein, 1992). In contrast, proteins are T-cell-dependent antigens, able to induce in infants high levels of isotype switched IgG antibodies and generate boostable memory responses. Enhancement of the immune response to PRP was accomplished through its covalent linkage to protein carriers forming glycoconjugate vaccines, the immune response of which, like that of the protein moiety, was T-cell-dependent (Insel and Anderson, 1986). Although not a new phenomenon, the inaugural PRP protein conjugates, due mainly to the development of more sophisticated and defined conjugate strategies (Jennings and Sood, 1994), have precipitated the era of modern conjugate vaccine technology.

Three different conjugation strategies were developed by Wyeth Lederle, Sanofi Pasteur and Merck which eventually led to the production of the following licensed Hib-conjugate vaccines; HbOC, PRP-T and PRP-OMP (Figure 48.1). The HbOC vaccine was produced by a method described by Anderson *et al.* (1986). This method was based on a single linkage aldehyde-reductive amination technology previously used in the preparation of meningococcal conjugates (Jennings and Lugowski, 1981), except due to the linearity of the PRP, extended periodate oxidation of the susceptible interchain ribitol also resulted in depolymerization of PRP with the spontaneous generation of two terminal aldehydes (see Figure 48.1A). Direct conjugation of the PRP oligomer to protein carrier CRM_{197}, a non-toxic variant of diphtheria toxin, was then achieved by reductive amination and, because the activated oligomer was divalent, the resultant conjugate was moderately cross-linked.

In contrast to terminal coupling, the remaining two Hib conjugates were produced by random activation of the depolymerized

TABLE 48.1 Structures of polysaccharides from *H. influenzae*, *N. meningitidis* and *S. enterica* sv. Typhi

Group/type	Structure[a,b]
H. influenzae [b]	→3)-β-D-Rib*f*-(1→1)-D-Ribitol-(5-OPO_3→
N. meningitidis	
A	→6)-α-D-Man*p*NAc-(3-*O*AC)(1-OPO_3→
B	→8)-α-D-Neu*p*5Ac-(2→
C	→9)-α-D-Neu*p*5Ac-(2→ 7/8 ↑ *O*Ac
W-135	→6)-α-D-Gal*p*-(1→4)-α-D-Neu*p*5Ac-(2→
Y	→6)-α-D-Glc*p*-(1→4)-α-D-Neu*p*5Ac-(2→(+ *O*Ac)
S. enterica sv. Typhi	
Vi	→4)-α-D-Gal*p*NAcA-(1→ 3 ↑ $OAc_{(0.6-0.9)}$
Di-*O*-acetylated pectin	→4)-α-D-Gal*p*A-(1→ 2,3 ↑ *O*Ac

[a]Individual references can be found in Jennings and Pon (1996).
[b]Abbreviations: Gal*p*, galactopyranose; Gal*p*A, pyranosidic galacturonic acid; Gal*p*NAcA, *N*-acetylgalacturonic acid (pyranose form); Man*p*NAc, *N*-acetyl-mannosamine (pyranose form); Neu*p*5Ac, *N*-acetylneuraminic (sialic) acid (pyranose form); *O*Ac, *O*-acetyl group; Rib*f*, ribosfuranose.

polysaccharide. Using this procedure, multiple coupling to a protein carrier occurs in the inner regions of a polysaccharide, thus having the potential to produce more extensively cross-linked conjugates. The Sanofi Pasteur procedure is based on a method first developed by the group of Robbins (Schneerson *et al.*, 1980), in which PRP depolymerized by thermal hydrolysis, was activated using cyanogen bromide and coupled to adipic dihydrazide (ADH)-derivatized tetanus toxoid (TT) through a covalent isourea linkage as depicted in Figure 48.1B. The Merck procedure employed a bigeneric spacer approach randomly to link carbonyldiimidazole activated PRP to *N. meningitidis* outer membrane protein (OMP) through a complex series of steps as outlined in Figure 48.1C (Marburg *et al.*, 1986).

Despite the structural diversity of the three conjugates, all have survived as commercial vaccines, with only minor differences in their efficacies being recorded (83%, 86%, 86% for PRP-T, HbOC and PRP-OMP respectively) (Chandran *et al.*, 2005). This indicates that there is a reasonable degree of flexibility in the design of glycoconjugate vaccines, although this is not always the case because one of the original Hib conjugates, PRP-D, which was also made following the procedure of Robbins (Schneerson *et al.*, 1980), is no longer in use because it induced low levels of antibodies in infants under 6 months of

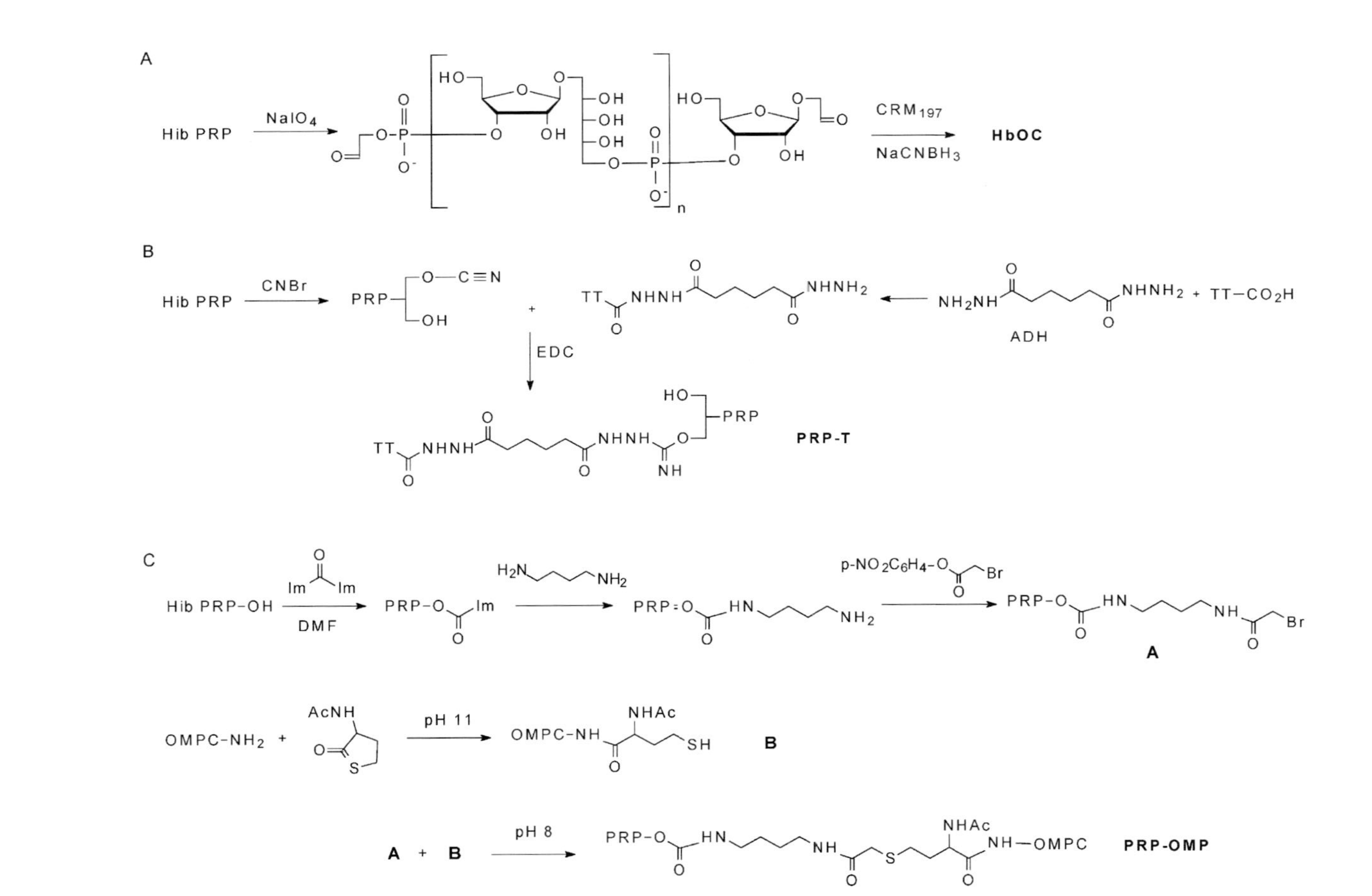

FIGURE 48.1 Key steps in the synthesis of the three licensed CPS-based *H. influenzae* type b conjugate vaccines, HbOC (A), PRP-T (B) and PRP-OMP (C). Abbreviations: PRP, type b capsular polysaccharide; CRM_{197}, non-toxic mutant diphtheria toxin; TT, tetanus toxoid; OMP, *N. meningitidis* outer membrane protein.

age (Anon, 1991) and was found to be only 32% effective in a population of youngsters with high endemic Hib disease rates (Ward *et al.*, 1990). Thus, this suggests that the sub-par performance of PRP-D, compared to the other Hib conjugate vaccines, can probably be attributed to a poorly executed conjugation procedure resulting in a poorly defined vaccine (Jennings and Pon, 1996). There is direct evidence that the immune performance of conjugate vaccines can be dependent on their structural parameters (Jennings and Sood, 1994). One example of this, carried out in infants using Hib-CRM_{197} conjugates constructed of different sized PRP fragments (Anderson *et al.*, 1987), demonstrated that the PRP fragment had to be at least 20 repeating units long in order optimally to prime infants for booster responses.

Ongoing surveillance of the three Hib conjugates has confirmed the importance of conjugate vaccines in vaccine technology. Immunized children are not only protected against invasive type b *H. influenzae* meningitis, but have also demonstrated a reduced nasopharyngeal carriage of the organism (Takala *et al.*, 1991), thereby reducing bacterial transmission. The latter unexpected effect indicates a direct link between vaccine-induced serum antibodies and mucosal immunity and it has been hypothesized that protection from any bacterial infection is mainly due to the ability of mucosally secreted IgG antibodies to neutralize bacterial colonization of the mucosa (Robbins *et al.*, 1996). This protective mechanism is the basis of herd immunity and could be beneficial in the eradication of diseases caused by bacteria for which humans are the only reservoir.

2.2. *S. pneumoniae*

Streptococcus pneumoniae (Pn) is the leading cause of bacteraemia and meningitis in young children, older adults and the immunocompromised, being responsible for both invasive disease with high levels of morbidity and mortality, as well as being the leading cause of otitis media (a bacterial infection of the middle ear) in children. The high rate of mortality prompted an early search for a preventative approach to its control, which led to the development of the first capsular polysaccharides used as human vaccines (Robbins *et al.*, 1983) and the eventual development of a licensed 23-valent pneumococcal polysaccharide vaccine. Because pneumococci are Gram-positive encapsulated organisms that can be divided into more than 90 different types based on the expression of their CPS, the final composition of the vaccine was a compromise whereby the 23-valent vaccine was formulated to cover >80% of Pn-related disease in the USA (Robbins *et al.*, 1983). Although used effectively in adults since 1981, the vaccine is poorly immunogenic in the elderly (Melegaro and Edmunds, 2004) and virtually ineffective in infants, the most susceptible cohort. Thus, a Pn–conjugate vaccine became a prime target for development.

Due to the plurality of serotypes, the development of a fully protective Pn–conjugate vaccine was infinitely more challenging than the development of the aforementioned monovalent Hib–conjugate. Therefore, it was decided to include in the vaccine the serotype polysaccharides most frequently associated with childhood disease. The first and currently only licensed Pn–conjugate was produced by Wyeth (Prevnar®) and consists of the CPS from serotypes 4, 6B, 9V, 14, 18C, 19F and 23F (Table 48.2), conjugated individually to CRM_{197} protein (PnC7–CRM_{197}). This formulation accounts for those serotypes causing 86% bacteraemia, 83% meningitis and 65% of otitis media in children <6 years of age in the USA (Farrell *et al.*, 2004). Prevnar® (PnC7–CRM_{197}) is produced by controlled periodate oxidation of the individual polysaccharides to generate aldehyde groups, followed by their linkage to CRM_{197} by reductive amination. The conjugation procedure is essentially the same as that used for the Hib–CRM_{197} conjugate (see Figure 48.1A), except

TABLE 48.2 Structures of the capsular polysaccharides of *Streptococcus pneumoniae* within the heptavalent conjugate vaccine

Type	Structure[a,b]
4	→4)-β-D-Man*p*NAc-(1→3)-α-L-Fuc*p*NAc-(1→3)-α-D-Gal*p*NAc-(1→4)-α-D-Gal*p*-(1→ 3 2 X H_3C CO_2H
6B	→2)-α-D-Gal*p*-(1→3)-α-D-Glc*p*-(1→3)-α-L-Rha*p*-(1→4)-Ribitol-(5-OPO_3→
9V	→4)-α-D-Glc*p*A-(1→3)-α-D-Gal*p*-(1→3)-β-D-Man*p*NAc-(1→4)-β-D-Glc*p*-(1→4)-α-D-Glc*p*-(1→ ↑ ↑ OAc OAc
14	→4)- β-D-Glc*p*-(1→6)-β-D-Glc*p*NAc-(1→3)-β-D-Gal*p*-(1→ 4 ↑ 1 β-D-Gal*p*
18C	αD-Glc*p* 1 ↓ 2 →4)-β-D-Glc*p*-(1→4)-β-D-Gal*p*-(1→4)-α-D-Glc*p*-(1→3)-α-L-Rha*p*-(1→ 3-O PO_3-1-Glycerol
19F	→4)-β-D-Man*p*NAc-(1→4)-α-D-Glc*p* (1→2)-α-L-Rha*p*-1-OPO_3→
23F	α-L-Rha*p* 1 ↓ 2 →4)-β-D-Glc*p*-(1→4)-β-D-Gal*p*-(1→4)-β-L-Rha*p*-(1→ 3-O PO_3-2-Glycerol

[a]Individual references can be found in Jennings and Pon (1996).
[b]Abbreviations: Gal*p*, glactopyranose; Glc*p*, glucopyranose; Rha*p*, rhamnopyranose; Fuc*p*NAc, *N*-acetylfucosamine (pyranose form); Gal*p*NAc, *N*-acetylgalactosamine (pyranose form); Glc*p*NAc, *N*-acetylglucosamine (pyranose form); Man*p*NAc, *N*-acetylmannosamine (pyranose form); Glc*p*A, glucuronic acid (pyranose form); OAc, *O*-acetyl group.

that, for some of the serotype polysaccharides, periodate-sensitive vicinal hydroxyls other than those associated with interchain polyols needed to be targeted.

The PnC7–CRM_{197} vaccine was shown to be highly effective against invasive pneumococcal disease in infants in a landmark efficacy trial (Black *et al.*, 2000), where over 90% of infants developed antibody to all 7 serotype polysaccharides, reducing the incidence of disease due to these serotypes by 97.4% and disease due to all pneumococci by 89%. Similar to that observed for other licensed conjugate vaccines (Hib and NmC), antibody titres dropped to slightly above baseline after the first year, but increased with memory kinetics following a booster immunization with the conjugate vaccine or polysaccharide alone (Choo *et al.*, 2000). Since the introduction of routine PnC7–CRM_{197} immunization in the USA, except for otitis media, vaccine serotype disease in children <5 years of age has declined 94% and for disease caused by all serotypes by 75% (Anon, 2005). The PnC7–CRM_{197} vaccine was also found to be effective against non-invasive otitis media (Eskola *et al.*, 2001), which is a major infant health problem in industrialized countries but, unfortunately, it was not as effective as against invasive disease. Prevention was estimated at 57–67% for the vaccine serotypes, 34% for all Pn serotypes and 6–9% for all cases of otitis media. However, despite its modest efficacy, PnC7–CRM_{197} has had a significant impact on otitis media by reducing the number of infant hospital visits (Fletcher and Fritzell, 2007) and spread of antibiotic resistance among pneumococci (Palmu *et al.*, 2004). The efficacy of the vaccine is due to reduction in serotype carriage of pneumococci at mucosal surfaces (Dagan *et al.*, 2003), presumably through IgG exudates (Bogaert *et al.*, 2005), that results in the generation of herd immunity (Hammitt *et al.*, 2006). Potential problems that still need to be addressed are the future possibility of non-vaccine pathogenic Pn strains replacing vaccine serotypes (Byington *et al.*, 2005) and the complexities arising from the addition of more serotype conjugates to provide additional coverage.

Attempts to include successfully more than seven serotype conjugates in a multivalent vaccine are being pursued. An 11-valent CPS conjugate vaccine (Sanofi Pasteur) employing mixed carrier proteins (types 1, 4, 5, 7F, 9V, 19F and 23F:TT; 3,6B, 14 and 18C:DT) has undergone clinical trials, however, the results were not promising due to carrier-induced epitope suppression (Dagan *et al.*, 2004). GlaxoSmithKline is currently testing a 10-valent conjugate vaccine (Streptorix®) which includes the above serotypes except for type 3, all conjugated to *H. influenzae* protein D (Prymula *et al.*, 2006). In this case, the carrier protein should also be functional against otitis media because its host organism is also responsible for a significant proportion of this non-invasive disease. Wyeth Lederle is also developing a 13-valent conjugate vaccine (PCV13) which includes types 1, 3, 5, 6A, 7F and 19A in addition to the components of their highly successful seven-valent vaccine (PnC7–CRM_{197}) (Scott *et al.*, 2007).

2.3. *N. meningitidis*

N. meningitidis (Nm) has at least 13 different serogroups of which groups A, B, C, W-135 and Y (see Table 48.1) are mainly associated with disease. Although the incidence of endemic meningococcal meningitis is relatively low (1–5/100 000) in developed countries, it remains a serious problem because of its association with both high mortality and morbidity rates with infants (6 months to 2 years) and adolescents. However, during epidemics, disease incidence in developing countries can reach 1000/100 000 due primarily to serogroup A Nm. Pioneering work by Gotschlich *et al.* (Goldschneider *et al.*, 1969) demonstrated that the capsular group A and C polysaccharides were immunogenic in humans and, subsequently, a vaccine composed of CPS from groups A, C, W-135 and Y has been used successfully to control outbreaks of disease caused by these serotypes. Similar to the Hib and Pn CPS vaccines, immunogenicity of the Nm tetravalent CPS vaccine, which does not include the important serogroup B component, was poor in infants and this was the driving force behind the development of the subsequent Nm conjugate vaccines. The first of the Nm conjugate vaccines developed was towards serogroup C disease due in large part to the disconcerting rise in NmC disease in the UK and Canada (Gray *et al.*, 2006).

Three group C conjugates were developed: Meningitec® (Wyeth Lederle), NeisVac-C® (Baxter International) and Menjugate® (Novartis).

The basic chemistry employing reductive amination used in the production of the first two vaccines shown in Figure 48.2A and 48.2B was pioneered by Jennings and Lugowski (1981). Terminal aldehyde groups were introduced into the NmC polysaccharides by controlled periodate oxidation with vaccine differences being the use of *O*-acetylated (OAc^+) NmC polysaccharide and CRM_{197} as a carrier (Meningitec®) versus the use of de-*O*-acetylated NmC and TT as a carrier protein (NeisVac-C®). Reductive amination resulting in a monovalent activation NmC was employed by Novartis in the synthesis of the third conjugate, in which the reducing end of acid depolymerized fragments of OAc^+ NmC polysaccharide were derivatized with ammonium chloride to yield primary amino groups (see Figure 48.2C). Using *N*-hydroxysuccinimide diesters of ADH as a linker, activated NmC fragments were then coupled to CRM_{197} via amide linkages (Costantino *et al.*, 1992).

To ensure the speedy introduction of the group C conjugates into routine vaccine programmes in the UK, the requirement for performing large-scale efficacy studies was waved and licensing was based on immune correlates for vaccine efficacy. Pre-licensure studies in adults, children and infants (Balmer *et al.*, 2002) demonstrated that all three conjugates were highly immunogenic and induced satisfactory serum bactericidal levels in more than 90% of all the above groups. Another study compared the performance of the three conjugates in UK toddlers (Richmond *et al.*, 2001a) and found that, in this cohort, the de-*O*-acetylated group C conjugate was more immunogenic than the others, capable of inducing higher bactericidal titres even against *O*-acetylated group C meningococci (Richmond *et al.*, 2001b). There are three possible explanations to account for these differences:

(i) NeisVac-C® used TT as opposed to CRM_{197} for Menjugate® and Meningitec® as carrier protein, although differences in immunogenicity attributable to these carriers were not observed previously with Hib conjugates (Granoff *et al.*, 1993).
(ii) The saccharide chain length used in NeisVac-C® was longer than that used in the other conjugates, a phenomenon which had been previously associated with superiority in Hib conjugates (Anderson *et al.*, 1987).
(iii) There is strong evidence to suggest that enhancement of the immunogenicity of NeisVac-C® is attributed to the removal of the *O*-acetyl groups from its saccharide component. In fact, even in adult antisera to a TT conjugate of the *O*-acetylated C saccharide, bactericidal activity against homologous meningococci could be inhibited more efficiently by the de-*O*-acetylated rather than the *O*-acetylated group C polysaccharide (Michon *et al.*, 2000). This infers that the most immunogenic epitopes are situated on their unacetylated residues, but whether these properties of NeisVac-C® will extrapolate to the future for more sustained antibody levels and improved memory effects cannot yet be determined.

Group A meningococci (NmA) are responsible for major meningitis epidemics in developing countries (Jodar *et al.*, 2002), but rarely cause disease in North America or Europe, a fact which has slowed the commercial development of group A conjugate vaccines. Although the NmA polysaccharide vaccine has had an impact in curbing major outbreaks (Robbins *et al.*, 2000) and is even uniquely immunogenic in infants (Mäkelä *et al.*, 1977), there is still a need to enhance its immune response in infants. Early studies focused on non-licensed bivalent A/C conjugate vaccines from Sanofi Pasteur and Novartis but have since been extended to tetravalent versions to include serogroups W-135 and Y CPS. The inclusion of conjugated W-135 and Y oligosaccharide components was to accommodate a situation unique to the USA,

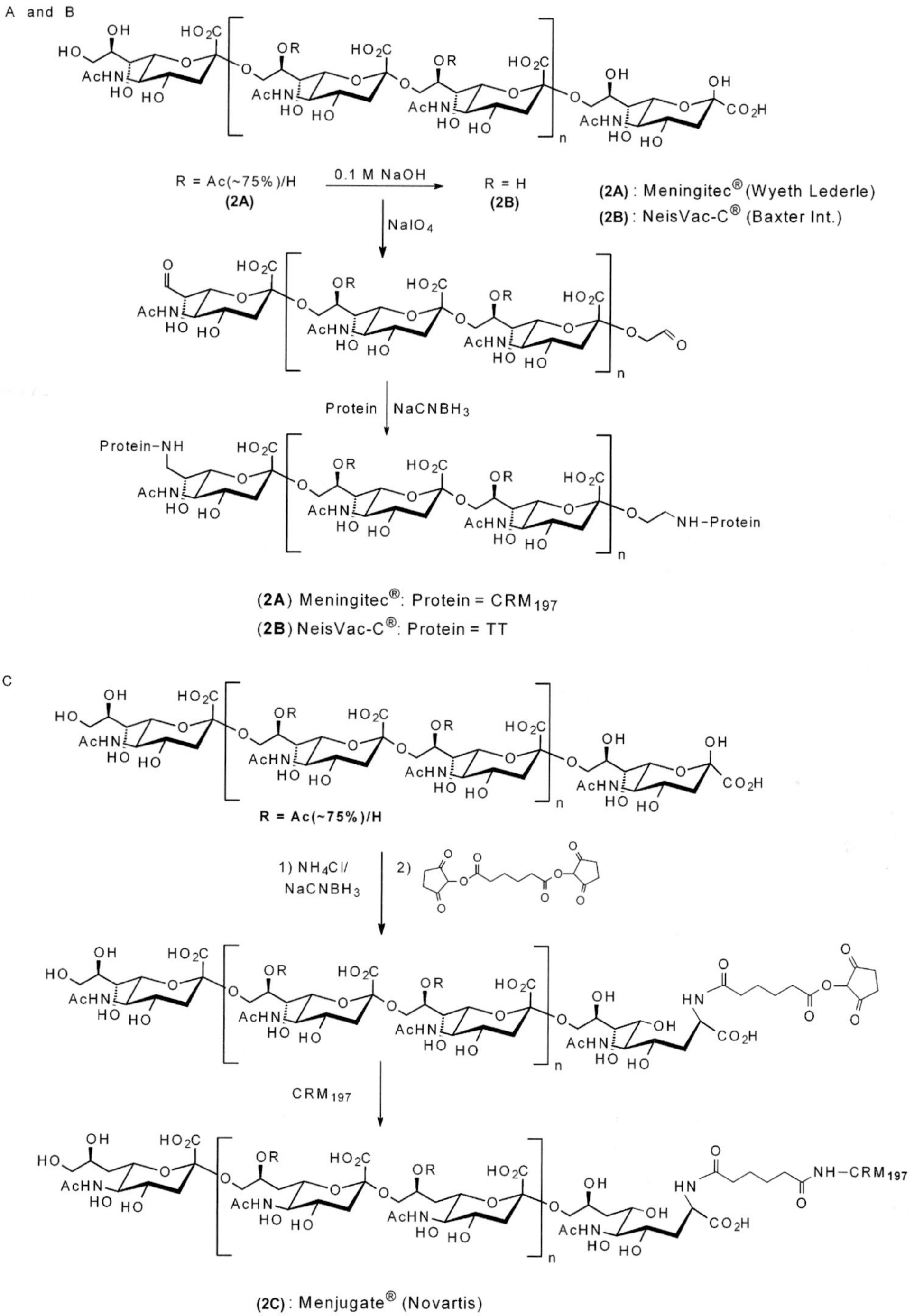

A and B

R = Ac(~75%)/H (2A) — 0.1 M NaOH → R = H (2B)

(2A) : Meningitec® (Wyeth Lederle)
(2B) : NeisVac-C® (Baxter Int.)

$NaIO_4$

Protein | $NaCNBH_3$

(2A) Meningitec®: Protein = CRM_{197}
(2B) NeisVac-C®: Protein = TT

C

R = Ac(~75%)/H

1) NH_4Cl/ $NaCNBH_3$ 2)

CRM_{197}

(2C) : Menjugate® (Novartis)

FIGURE 48.2 Synthetic schemes for licensed monovalent group C *N. meningitidis* conjugate vaccines, (A) Meningitec®, (B) NeisVac-C® and (C) Menjugate®.

where group Y disease now accounts for ≈33% of all meningococcal disease (Bilukha and Rosenstein, 2005). Menactra® (Sanofi Pasteur), in which each oligosaccharide component is linked individually to DT using technology developed for their Hib and Pn conjugate vaccines (see Figure 48.1B), was licensed in 2007 for use in individuals >2 years of age but showed no significant benefit in infants (Rennels *et al.*, 2004). Novartis, however, is currently evaluating a tetravalent upgrade of their original bivalent A/C vaccine (Men ACYW–CRM_{197}), which appears to be promising from the results of phase II trials in infants (Snape *et al.*, 2008) and is expected to be available for this age cohort in 2009 (Anon, 2008). Cost effectiveness is imperative in supplying vaccines to developing countries with epidemic NmA disease and, therefore, there is pressing need to develop a univalent group A conjugate. With the goal of supplying such a vaccine for 50 US cents a dose, the Serum Institute of India with funding from the Gates Foundation has used reductive amination (Jennings and Lugowski, 1981) and a modified hydrazide procedure to conjugate the NmA polysaccharide to TT (Jessouroun *et al.*, 2005). Phase II trials of the conjugate in toddlers are underway and it is expected that it will soon be introduced into mass vaccination programmes.

Meningitis and septicaemia due to group B *N. meningitidis* remain major global health problems and are responsible for 30–40% of cases, predominantly in infants, in North America and an even higher percentage in Europe (Girard *et al.*, 2006). As the only conserved antigenic structure on the surface of group B meningococci, the α-(2→8)-polysialic acid (PSA) capsule is the ideal molecule on which to base a vaccine, but even when conjugated to TT it is still poorly immunogenic (Jennings and Lugowski, 1981). This is attributed to the presence of similar structures in developing brain tissues (Finne *et al.*, 1983) which invoke a state of immune tolerance. The immunogenicity of PSA can be greatly improved by constructing *N*-propionylated (NPr) PSA conjugates (Jennings *et al.*, 1986), which were synthesized by treating PSA with base to remove the *N*-linked acetyl groups and replacing them with propionyl groups (Figure 48.3). The NPrPSA was then subjected to controlled periodate oxidation to introduce two terminal aldehydes into its structure, which were used to couple the saccharide to TT by reductive amination. This conjugate readily induced bactericidal antibodies in mice that were predominantly specific to NPrPSA but still retained the ability to bind and to kill group B meningococci, suggesting that NPrPSA mimics a unique protective bacterial surface epitope (Jennings *et al.*, 1987). A preclinical evaluation of an NPrPSA conjugate in baboons and rhesus monkeys, in which the carrier protein was a recombinant group B meningococcal porin (rPorB), was shown to elicit NPrPSA IgG antibodies which were bactericidal for group B meningococci with both monkey and human complement (Fusco *et al.*, 1997). These positive results suggest that NPrPSA conjugates would be excellent vaccine candidates, but commercial development has stalled because of the perceived risk of autoimmunity and the poor performance of an NPrPSA–TT conjugate vaccine in a limited human trial (Bruge *et al.*, 2004).

2.4. Group B *Streptococcus*

During the last decade, group B *Streptococcus* has become the leading cause of neonatal sepsis and meningitis, which is associated with significant morbidity and mortality (Law *et al.*, 2005) and it was envisioned that protection against invasive group B *Streptococcus* disease may be afforded by their CPS (Baker and Edwards, 2003). Group B streptococci are Gram-positive organisms which, based on their CPS, are classified into 9 different serotypes, of which types Ia, Ib, II, III and, more recently, type V (Table 48.3) (Jennings, 1998) account for 94–100% of North American disease isolates. The group B *Streptococcus* CPS are unique because, despite the fact that they

FIGURE 48.3 Preparation of *N*-propionyl-modified group B *N. meningitidis* conjugate vaccine.

all share extensive structural homologies to each other and to human tissue oligosaccharides (Jennings *et al.*, 1984), they remain immunospecific and immunogenic in adults (Baker and Kasper, 1985). With the exception of group V (Guttormsen *et al.*, 2008), this phenomenon is attributed to immune responses being directed to conformational epitopes. The unusual length dependency of these epitopes (Wessels *et al.*, 1987; Zou *et al.*, 1999), overcomes the immunosuppressive effect of the constituent human tissue oligosaccharides and they are hypothesized to be situated on intermittent extended helical segments within the random coil structure of the CPS. This is consistent with nuclear magnetic resonance (NMR) spectroscopic, computational and modelling studies on the type III CPS (Brisson *et al.*, 1997; Kadirvelraj *et al.*, 2006).

The strategy of using group B *Streptococcus* CPS to immunize pregnant women in order to protect their infants suffered a serious setback when it was discovered that a significant number of adults failed to give an adequate response to them (Baker and Kasper, 1985).

TABLE 48.3 Structures of the capsular polysaccharides of Group B *Streptococcus* for the putative pentavalent vaccine

Type	Structure[a,b]
Ia	→4)-β-D-Glc*p*-(1→4)-β-D-Gal*p* (1→ 3 ↑ 1 α-D-Neu*p*5Ac-(2→3)-β-D-Gal*p*-(1→4)-β-D-Glc*p*NAc
Ib	→4)-β-D-Glc*p*-(1→4)-β-D-Gal*p*-(1→ 3 ↑ 1 αD-Neu*p*5Ac-(2→3)-β-D-Gal*p*-(1→3)-β-D-Glc*p*NAc
II	→4)-β-D-Glc*p*NAc(1→3)-β-D-Gal*p*(1→4)-β-D-Glc*p*-(1→3)-β-D-Glc*p*-(1→2)-β-D-Gal*p*(1→ 6 3 ↑ ↑ 1 2 β-D-Gal*p* αD-Neu*p*5Ac
III	→4) βD-Glc*p* (1→6) βD-Glc*p*NAc (1→3)-β-D-Gal*p* (1→ 4 ↑ 1 α-D-Neu*p*5Ac-(2→3)-β-D-Gal*p*
V	→4)-α-D-Glc*p*-(1→4)-β-D-Gal*p*-(1→4)-β-D-Glc*p*-(1→ 6 6 ↑ ↑ 1 1 α-D-Neu*p*5Ac-(2→3)-β-D-Gal*p*-(1→4)-β-D-Glc*p*NAc β-D-Glc*p*

[a]Individual references can be found in Jennings (1998).
[b]Abbreviations: Gal*p*, galactopyranose; Glc*p*, glucopyranose; Glc*p*NAc, *N*-acetylglucosamine (pyranose form); Neu*p*5Ac, *N*-acetylneuraminic (sialic) acid (pyranose form).

Fortunately, however, group B *Streptococcus* CPS–protein conjugate vaccines were able to elicit more comprehensive responses in adults. The preferred method of conjugation, shown in Figure 48.4, is applicable to all group B *Streptococcus* CPS and involves the controlled periodate oxidation (Jennings and Lugowski, 1981) of the terminal sialic acid residues of their de-polymerized fragments (Wang *et al.*, 1998), resulting in the introduction of aldehydes into some of these residues. Although the sialic acid residues of some of the native group B *Streptococcus* CPS are reported to be variably *O*-acetylated (Lewis *et al.*, 2004), these groups are removed by base treatment of the native CPS prior to conjugation, a process which also removes the contaminating group antigen (Michon *et al.*, 1988). The activated fragments were then covalently linked to TT by reductive amination to produce cross-linked glyco-conjugates. Proof of principle of the immunization strategy was obtained when it was demonstrated that placentally transferred polysaccharide IgG antibodies of female mice, immunized with a mixture of types Ia, Ib, II and III polysaccharide–TT conjugates, were able to protect their newborn pups from lethal challenge with homologous group B streptococci (Paoletti *et al.*, 1994). The type Ia, Ib, II, III and V–TT conjugates have now undergone preliminary human trials in healthy adults (Baker and Edwards, 2003) and all generated >4-fold rise in type specific antibodies in 80–93% of the individuals and these antibodies were able uniformly to kill *O*-acetylated strains of their homologous type-specific organisms. The importance of *O*-acetylation patterns (Lewis *et al.*, 2004) on the native group B *Streptococcus* polysaccharides is not known at this time, however, this uniform *in vitro* killing of these streptococci suggests that antisera raised to the de-*O*-acetylated conjugates must cross-react with the *O*-acetylated polysaccharides on the surface of group B streptococci, as was observed with NmC conjugates (Richmond *et al.*, 2001b).

2.5. *S. enterica* sv. Typhi

Typhoid fever is a major cause of morbidity and mortality in developing countries, resulting in 200 000 deaths annually, and is a serious enteric disease in children and infants (Crump *et al.*, 2004). The aetiological agent, *S. enterica* sv. Typhi, colonizes the intestine and infection occurs when it invades the intestinal epithelium and is rapidly disseminated intracellularly. *S. enterica* sv. Typhi possesses a linear CPS (Vi-antigen) composed of a homopolymer of α-(1$\rightarrow$4)-D-galactosaminuronic acid, *N*-acetylated at C-2 and *O*-acetylated at C-3 (Heyns *et al.*, 1959) (see Table 48.1), of which these substituents have been shown to be immunodominant and vaccine required epitopes (Szu *et al.*, 1991). Clinical trials in which *O*-acetylation is preserved on the Vi polysaccharide vaccine resulted in a cumulative 3-year efficacy rate of 55% (Fraser *et al.*, 2007) and has been in use in persons >2 years of age since 1994. Surprisingly even though an enteric pathogen, protection correlates with the level of serum IgG antibodies and is consistent with the hypothesis that mucosally secreted IgG antibodies are preventing intestinal colonization (Robbins *et al.*, 1996) similar to that observed with Hib and NmC vaccines. As with other polysaccharides, routine vaccination with the Vi polysaccharide demonstrated that it was poorly immunogenic in infants <2 years of age and, therefore, the focus of research turned to the development of a glycoconjugate vaccine for *S. enterica* sv. Typhi.

Glycoconjugate vaccines were synthesized by conjugating the Vi polysaccharide to the recombinant exoprotein of *Pseudomonas aeruginosa* (rEPA) using two different basic synthetic strategies involving the use of either *N*-succinimidyl 3-(2-pyridyldithio)propionate (SPDP) or ADH linkers to cross-link the two molecules (Kossaczka *et al.*, 1999). The safety and immunogenicity of the Vi conjugates were then compared in humans and those synthesized with ADH linkers to rEPA were found to be the most immunogenic (Kossaczka *et al.*, 1999). The vaccine was prepared as shown in Figure 48.5, by ADH activation of rEPA prior to its coupling to the carboxylate groups of the Vi polysaccharide, because the alternate strategy involving the prior activation of the polysaccharide was not successful (Kossaczka *et al.*, 1997). The conjugate vaccine was successfully field tested in young children in Vietnam under elevated *S. enterica* sv. Typhi conditions (Lin *et al.*, 2001) and is now undergoing more extensive immunogenicity studies in infants (Anon, 2006a) with expectations of efficacy in this vulnerable population.

More recently, another interesting approach to developing an *S. enterica* sv. Typhi conjugate vaccine has been reported (Kossaczka *et al.*, 1997), in which 2,3-di-*O*-acetylated pectin (see Table 48.1) is used as a mimic of the Vi polysaccharide. This semi-synthetically derived polysaccharide is immunochemically indistinct with the Vi antigen and, when coupled to rEPA

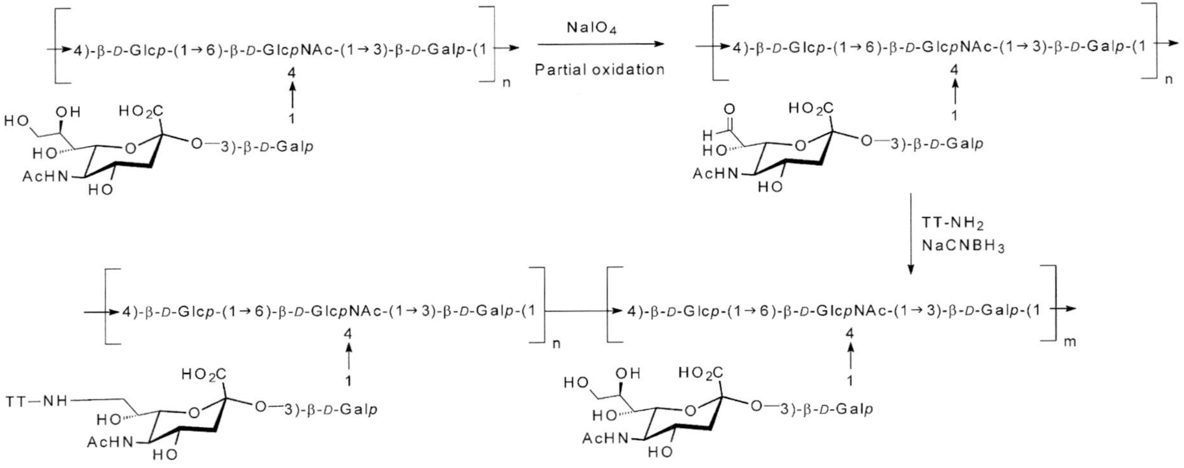

FIGURE 48.4 The use of controlled periodate oxidation for the conjugation of group B streptococcal polysaccharides to protein. An illustrative example using native type III group B streptococcal polysaccharide is depicted. Abbreviations: Gal*p*, galactopyranose; Glc*p*, glucopyranose; Glc*p*NAc, *N*-acetylglucosamine (pyranose form).

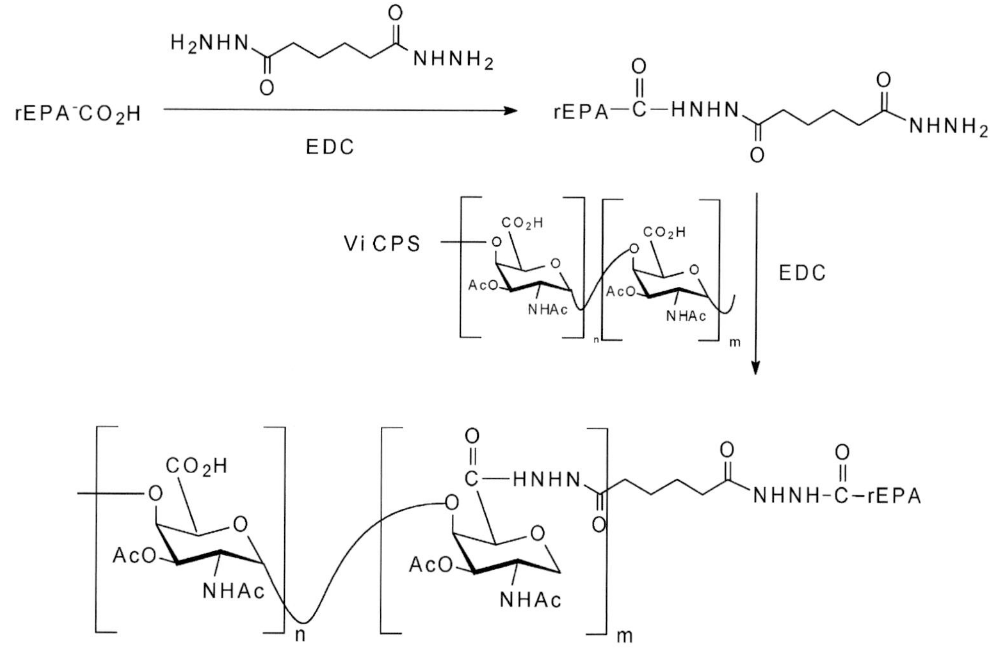

FIGURE 48.5 Synthesis of the *S. enterica* sv. Typhi conjugate vaccine currently being evaluated in clinical trials, using controlled carbodiimide chemistry. Abbreviations: rEPA, recombinant exoprotein of *P. aeruginosa*; Vi CPS, capsular polysaccharide of *S. enterica* sv. Typhi.

as above, induced an anamnestic antibody response to the native Vi antigen (Kossaczka *et al.*, 1997). Although it was found that the Vi conjugate was more immunogenic in animals than the pectin analogue (Szu *et al.*, 1994), there are still certain potential advantages in using pectin, such as its ready availability in the pure state, that merit further exploration. Therefore, conjugates made with the pectin derivative coupled to rEPA are currently undergoing safety evaluation in adults (Anon, 2006b).

3. SYNTHETIC GLYCOCONJUGATE VACCINES

3.1. *H. influenzae* type b

Due to recent advances in synthetic methodology, pure oligosaccharides can be obtained in improved yields and, therefore, the possibility of producing fully synthetic antibacterial glycoconjugate vaccines has been the focus of recent research (Pozsgay, 2001). Because of the relative simplicity of its structure and that it is an important pathogen, the most intensive research has focused on the development of synthetic glycoconjugate vaccines based on oligosaccharide fragments of type b *H. influenzae*. Two syntheses of up to three repeating units (RU) of ribosylribitol phosphate have been described (Wang and Just, 1988; Nilsson *et al.*, 1992), but further extrapolation to glycoconjugates has not been reported. Presumably, this is because of the necessity of utilizing multistep synthetic and purification procedures, in an iterative fashion for oligo repeating unit synthesis, which results in oligosaccharide yields being too low to contemplate for large scale production. Hoogerhout *et al.* (1987) were able to synthesize Hib oligosaccharides of up to 4RU, again using an iterative process, and to link these to protein. They went on to establish in animal studies that there is a length requirement for optimal immunogenicity of their synthetic conjugate vaccines because conjugates containing 4RU were more immunogenic than those containing 3RU (Peeters *et al.*, 1992). This is consistent with human studies using conjugates containing oligosaccharides obtained by depolymerization of the Hib polysaccharide (Anderson *et al.*, 1987). In these studies, the length requirement necessary to optimize conjugate performance in infants was established, where conjugates made from both 8RU and 20RU induced high antibody levels in adults, but only the latter was active in priming infants for a booster vaccination with the polysaccharide alone.

An improved synthesis involving a one-pot oligomerization process resulted in the synthesis of the first and only licensed (in Cuba) synthetic conjugate vaccine. Verez-Bencomo *et al.* (2004) synthesized a TT conjugate with PRP fragments (average size of 8RU), obtained in high yield, and amenable to large scale batch production, as illustrated in Figure 48.6, using H-phosphonate chemistry. Conjugation of maleimido functionalized oligomers was accomplished using thiolated TT. In preliminary trials in infants, the synthetic conjugate performed as well as a commercial non-synthetic Hib conjugate (Verez-Bencomo *et al.*, 2004) and has now become a commercial vaccine but so far has had limited licensure.

3.2. *S. pneumoniae*

The synthesis of oligosaccharide conjugate vaccines against pneumococcal infections has been another endeavour and includes those of types 3, 6B, 14, 17F and 23F. Type 3 di-, tri-, and tetrasaccharides, with 3-aminopropyl spacers, representing overlapping saccharides of 2RU of the type 3 polysaccharide, were synthesized and conjugated to CRM_{197}, keyhole limpet haemocyanin (KLH) and TT using squarate as a cross-linker (Lefeber *et al.*, 2001). Overlapping di- and trisaccharides and a 1RU tetrasaccharide of the type 6B polysaccharide were linked to KLH using a procedure which involved *S*-acetylation of the aminopropyl glycosidic spacer and bromoacetylation of the amino groups of KLH (Jansen *et al.*, 2001). Similar methods were also used in the synthesis of the type 17F (Alonso De Velasco *et al.*, 1993) and 23F (Alonso De Velasco *et al.*, 1994) conjugates where overlapping oligosaccharides representing less than 1RU of their respective polysaccharides were employed. In a similar study, aminopropyl glycosides of overlapping saccharide components of the full tetrasaccharide RU of the type 14 polysaccharide were conjugated to CRM_{197} using an *N*-hydroxysuccinimide-activated adipic acid diester (Mawas *et al.*, 2002).

While all the above conjugates, co-administered with strong adjuvant, were reported to induce polysaccharide-specific antibody in mice and/or rabbits, that was variably functional in opsonophagocytic assays and in mouse challenge experiments, these results are probably more of academic interest than of practical use. The oligosaccharides employed were extremely short and there is very strong evidence that a much longer length of saccharide would be required for the optimization of the efficacy of glycoconjugate vaccines. This has been demonstrated for the Hib oligosaccharide conjugate in monkeys, in adults and infants as described above and there is evidence in experiments in rabbits that this phenomenon also extends to the pneumococcal glycoconjugates (Laferriere *et al.*, 1998). Nevertheless, these studies do highlight the practical difficulties that would be encountered if the synthetic approach were extended to structurally more complicated polysaccharides.

4. PEPTIDE-MIMETICS OF POLYSACCHARIDE EPITOPES

Because peptide synthesis is more facile than the synthesis of large complex oligosaccharides and more adaptable to large-scale production, the use of peptide mimics of functionally protective bacterial polysaccharides as potential vaccines is worth exploration. Furthermore, creating a surrogate peptide image of the polysaccharide in the combining site of an anti-idiotype antibody also converts the thymus-independent polysaccharide into a thymus-dependent immunogen (Kieber-Emmons *et al.*, 1986). Westerink *et al.* (1988) developed a monoclonal anti-idiotypic antibody 6F9 (mAb6F9) by immunizing mice with an IgM monoclonal antibody specific

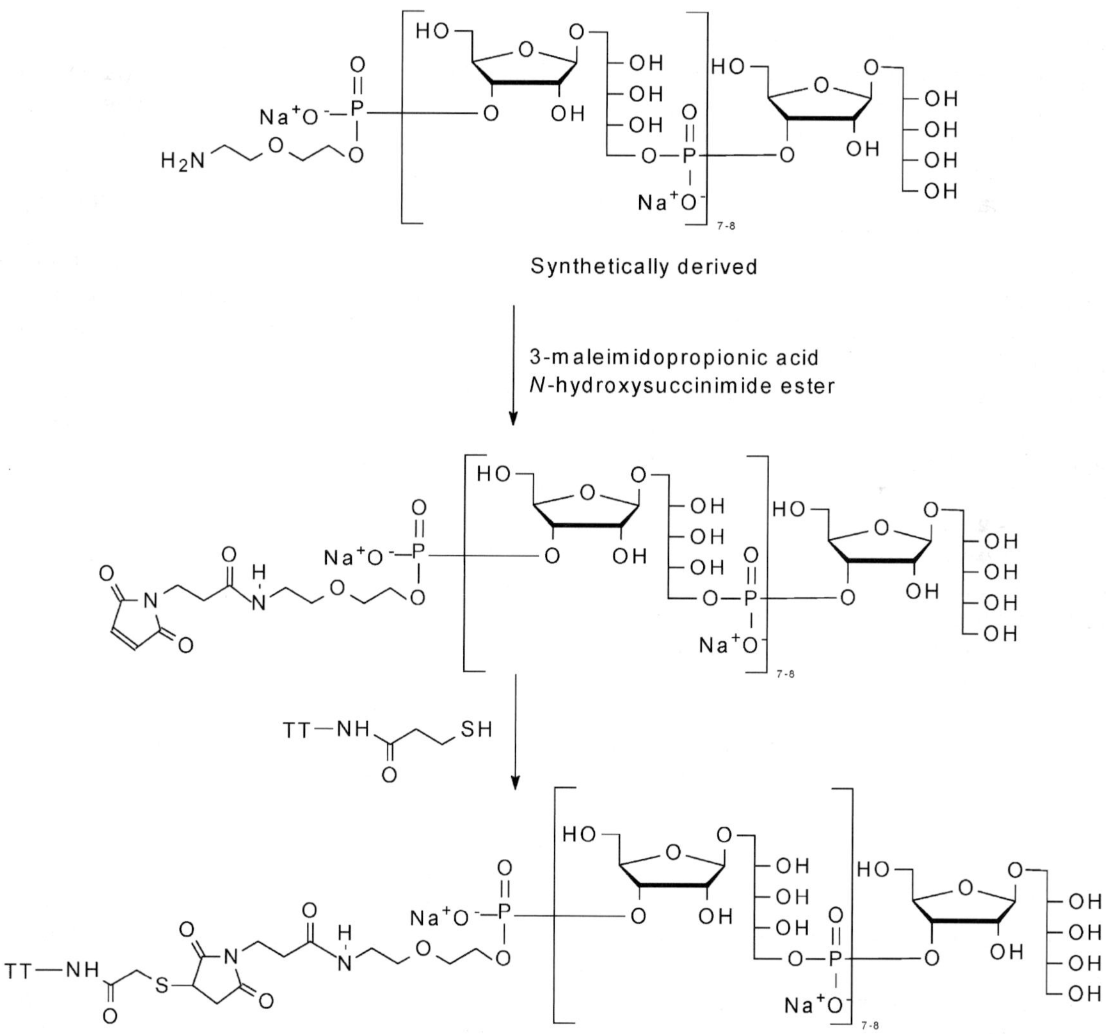

Synthetically derived

3-maleimidopropionic acid
N-hydroxysuccinimide ester

FIGURE 48.6 Key steps in the production of the first licensed synthetic antibacterial glycoconjugate vaccine to *H. influenzae* type b.

for the group C *N. meningitidis* polysaccharide (GCMP). As an immunogen, mAb6F9 was capable of inducing anti-anti-idiotypic antibodies (Ab3) in rabbits and mice, which were specific for the original GCMP and which were bactericidal to NmC in the presence of complement. In subsequent experiments, it was demonstrated that, when used as an immunogen, mAb6F9 resulted in a T-dependent anti-GCMP antibody response, which was reported to be more protective than the native GCMP polysaccharide in both mature and immature mice when challenged with live NmC (Westerink and Giardina, 1992).

In order to develop a peptide mimetic vaccine, sequence analysis of the heavy and light chain variable regions of mAb6F9 was performed and the immunogenic site responsible for inducing anti-GCMP antibodies was mapped by sequence and computer model analyses (Westerink *et al.*, 1995). A synthetic peptide spanning the CDR3 domain was synthesized and complexed to proteosomes (NmB outer membrane proteins) and the peptide mimetic vaccine induced GCMP-specific IgG antibodies in mice, which protected them from infection with a lethal dose of NmC. Comparing the copy DNA (cDNA) sequence of a hybridoma produced to GCMP with the cDNA sequence of a hybridoma raised to the peptide mimetic vaccine showed that both their binding sites (V_HCDR3 regions) were similar in length and sequence (Hutchins *et al.*, 1996). These results suggest that immunization with an anti-idiotype-derived peptide mimic of a polysaccharide epitope can induce the production of antibodies with combining sites structurally related to the combining site of antibodies raised to the carbohydrate themselves. The human immune response to the GCMP peptide mimetic was studied indirectly using an immunodeficient SCID mouse model (Hutchins *et al.*, 1999). The mice were reconstituted with human peripheral blood mononuclear cells (hu-PBMC-SCID) and following immunization with the above peptide mimetic, hydrophobically complexed to OMP, they were able to mount an IgG response that was polysaccharide-specific and bactericidal for NmC. All the above data support in principle the use of peptide mimics of polysaccharide epitopes as human vaccines but, so far, critical experiments comparing their immune performance with those of their equivalent CPS-conjugates have not yet been carried out.

An important alternate route to identify peptides that mimic polysaccharide epitopes is through screening phage display peptide libraries using polysaccharide-specific monoclonal antibodies. By screening the type III group B streptococcal polysaccharide (GBSP)-specific mAb S9 with such a library, Pincus *et al.* (1998) identified a dodecamer peptide which resembled the protective epitope of the type III GBSP both antigenically and immunogenically. Using a similar technique, Westerink *et al.* (Grothaus *et al.*, 2000), screened a group A meningococcal polysaccharide (GAMP)-specific mAb and identified a predominant peptide mimic which was capable of binding human hyperimmune GAMP-specific antisera and of inhibiting the binding of this antisera to its homologous polysaccharide. Immunization of mice with the peptide in the form of a proteosome complex induced an anamnestic GAMP-specific IgG response which was bactericidal for group A meningococci. Using phage display peptide libraries to screen selected group B meningococcal polysaccharide-specific (Shin *et al.*, 2001) and *N*-propionylated group B polysaccharide-specific mAbs (Moe *et al.*, 1999), a number of peptide mimics for each of the polysaccharides were identified. The weakness of the above screening technique is that, while they all generate peptide mimics, relatively few of them are functional. This situation can be improved by screening the peptide mimics with a human recombinant antibody (rAb) library as has been demonstrated (Weisser *et al.*, 2007) for the peptide mimics of the GBSP (see above). Finally, oligonucleotides encompassing peptide mimetics can be cloned into expression vectors and used as DNA vaccines. The feasibility of this approach was demonstrated when mice, immunized with a DNA construct encoding a peptide mimic of the type 4 pneumococcal polysaccharide (Lesinski *et al.*, 2001), generated antibodies to the homologous polysaccharide. In addition, a multiple epitope DNA vaccine encoding a T-helper-cell epitope together with a peptide mimic of the GCMP induced a significant anti-GCMP response that was bactericidal and subsequently protected immunized mice against a lethal challenge with NmC (Prinz *et al.*, 2003).

5. CONCLUSIONS

The development of bacterial polysaccharide conjugate vaccines, with their ability to enhance the immunogenicity of polysaccharides and thus extend prophylaxis into the vulnerable infant population, is one of the most significant advances in vaccine technology. One can predict with confidence that its use will be further applied to tackle other clinically important bacterial infections, both in the developed and underdeveloped world. Although there is adequate evidence to demonstrate a reasonable degree of flexibility in the design of these conjugates, some vaccine failures have indicated that there are indeed some limitations. The basic strategy for success is to utilize a simple coupling procedure, preferably carried out in aqueous solution, which is reproducible and mild enough to preserve the integrity of the component molecules of the conjugate.

Although of academic interest, due to the complexity of most polysaccharide structures, it is difficult to envisage, except in rare cases of structural simplicity, that the chemical synthesis of saccharide components of conjugate vaccines will be commercially viable. This is because most bacteria already produce adequate amounts of capsular polysaccharide for this purpose and which is readily available. Similarly, while the development of peptide conjugate vaccines solves the problem of synthesizing the more structurally complex saccharides, it has not yet been demonstrated that their performance is the equal of saccharide conjugate vaccines. Areas for further development and investigation are detailed in the Research Focus Box.

RESEARCH FOCUS BOX

- Research to develop the Group B component of a comprehensive combination vaccine against *N. meningitidis* is needed.
- With the widespread and effective use of conjugate vaccines, investigations will be required to study further the effects of vaccine pressure on microbial genetics.
- To develop novel carrier proteins and/or fragments that are in themselves beneficial, while relieving the potentially deleterious effects of over-using existing carrier proteins, especially in highly multivalent vaccine formulations.
- The emphasis in the development of synthetic vaccines to be placed on constructs where the native polysaccharides are poorly available or require chemical modification.
- Increased efforts to be placed on technology transfer and/or vaccine availability for the developing world.

References

Alonso De Velasco, E., Verheul, A.F., Veeneman, G.H., et al., 1993. Protein-conjugated synthetic di- and trisaccharides of pneumococcal type 17F exhibit a different immunogenicity and antigenicity than tetrasaccharide. Vaccine 11, 1429–1436.

Alonso De Velasco, E., Verheul, A.F., van Steijn, A.M.P., et al., 1994. Epitope specificity of rabbit immunoglobulin G (IgG) elicited by pneumococcal type 23F synthetic oligosaccharide- and native polysaccharide-protein conjugate vaccines: comparison with human anti-polysaccharide 23F IgG. Infect. Immun. 62, 799–808.

Anderson, P., Peter, G., Johnston, R.-B.J., Wetterlow, L.H., Smith, D.H., 1972. Immunization of humans with polyribophosphate, the capsular antigen of *Hemophilus influenzae*, type b. J. Clin. Invest. 51, 39–44.

Anderson, P.W., Pichichero, M.E., Insel, R.A., Betts, R., Eby, R., Smith, D.H., 1986. Vaccines consisting of periodate-cleaved oligosaccharides from the capsule of *Haemophilus influenzae* type b coupled to a protein carrier: structural and temporal requirements for priming in the human infant. J. Immunol. 137, 1181–1186.

Anderson, P., Pichichero, M., Edwards, K., Porch, C.R., Insel, R., 1987. Priming and induction of *Haemophilus influenzae* type b capsular antibodies in early infancy by Dpo20, an oligosaccharide-protein conjugate vaccine. J. Pediatr. 111, 644–650.

Anderson, P.W., Pichichero, M.E., Stein, E.C., et al., 1989. Effect of oligosaccharide chain length, exposed terminal group, and hapten loading on the antibody response of human adults and infants to vaccines consisting of *Haemophilus influenzae* type b capsular antigen unterminally coupled to the diphtheria protein CRM197. J. Immunol. 142, 2464–2468.

Anon, 1991. Haemophilus b conjugate vaccines for prevention of *Haemophilus influenzae* type b disease among infants and children two months of age and older. Recommendations of the immunization practices advisory committee (ACIP) 1991. MMWR Recomm. Rep. 40, 1–7.

Anon, 2005. Direct and indirect effects of routine vaccination of children with 7-valent pneumococcal conjugate vaccine on incidence of invasive pneumococcal disease – United States, 1998–20032005. MMWR Morb. Mortal. Wkly. Rep. 54, 893–897.

Anon, 2006a. Evaluation of the safety, immunogenicity, and compatibility with DTP of an investigational Vi-rEPA conjugate vaccine for typhoid fever when administered to infants in Vietnam concurrently with DTP. <http://clinicaltrials.gov/ct2/show/nct00342628/>.

Anon, 2006b. *Salmonella typhi* Vi *O*-acetyl pectin-rEPA conjugate vaccine. <http://clinicaltrials.gov/ct2/show/NCT00277147/>.

Anon, 2008. Novartis Menveo® vaccine shows superior immune response against four types of meningitis disease in pivotal phase III trial. <http://www.novartis.com/newsroom/media-releases/en/2008/1216115.shtml/>.

Baker, C.J., Edwards, M.S., 2003. Group B streptococcal conjugate vaccines. Arch. Dis. Child. 88, 375–378.

Baker, C.J., Kasper, D.L., 1985. Group B streptococcal vaccines. Rev. Infect. Dis. 7, 458–467.

Balmer, P., Borrow, R., Miller, E., 2002. Impact of meningococcal C conjugate vaccine in the UK. J. Med. Microbiol. 51, 717–722.

Bilukha, O.O., Rosenstein, N., 2005. Prevention and control of meningococcal disease. Recommendations of the advisory committee on immunization practices (ACIP). MMWR Recomm. Rep. 54, 1–21.

Black, S., Shinefield, H., Fireman, B., et al., 2000. Efficacy, safety and immunogenicity of heptavalent pneumococcal conjugate vaccine in children. Northern California Kaiser Permanente Vaccine Study Center Group. Pediatr. Infect. Dis. J. 19, 187–195.

Bogaert, D., Veenhoven, R.H., Ramdin, R., et al., 2005. Pneumococcal conjugate vaccination does not induce a persisting mucosal IgA response in children with recurrent acute otitis media. Vaccine 23, 2607–2613.

Brisson, J.R., Uhrinova, S., Woods, R.J., et al., 1997. NMR and molecular dynamics studies of the conformational epitope of the type III group B *Streptococcus* capsular polysaccharide and derivatives. Biochemistry 36, 3278–3292.

Bruge, J., Bouveret-Le-Cam, N., Danve, B., Rougon, G., Schulz, D., 2004. Clinical evaluation of a group B meningococcal *N*-propionylated polysaccharide conjugate vaccine in adult, male volunteers. Vaccine 22, 1087–1096.

Byington, C.L., Samore, M.H., Stoddard, G.J., et al., 2005. Temporal trends of invasive disease due to *Streptococcus pneumoniae* among children in the intermountain west: emergence of nonvaccine serogroups. Clin. Infect. Dis. 41, 21–29.

Chandran, A., Watt, J.P., Santosham, M., 2005. Prevention of *Haemophilus influenzae* type b disease: past success and future challenges. Expert Rev. Vaccines 4, 819–827.

Choo, S., Seymour, L., Morris, R., et al., 2000. Immunogenicity and reactogenicity of a pneumococcal conjugate vaccine administered combined with a *Haemophilus influenzae* type B conjugate vaccine in United Kingdom infants. Pediatr. Infect. Dis. J. 19, 854–862.

Claesson, B.A., Schneerson, R., Trollfors, B., Lagergard, T., Taranger, J., Robbins, J.B., 1990. Duration of serum antibodies elicited by *Haemophilus influenzae* type b capsular polysaccharide alone or conjugated to tetanus toxoid in 18- to 23-month-old children. J. Pediatr. 116, 929–931.

Costantino, P., Viti, S., Podda, A., Velmonte, M.A., Nencioni, L., Rappuoli, R., 1992. Development and phase 1 clinical testing of a conjugate vaccine against meningococcus A and C. Vaccine 10, 691–698.

Crump, J.A., Luby, S.P., Mintz, E.D., 2004. The global burden of typhoid fever. Bull. World Health Organ. 82, 346–353.

Dagan, R., Givon-Lavi, N., Zamir, O., Fraser, D., 2003. Effect of a nonavalent conjugate vaccine on carriage of antibiotic-resistant *Streptococcus pneumoniae* in day-care centers. Pediatr. Infect. Dis. J. 22, 532–540.

Dagan, R., Goldblatt, D., Maleckar, J.R., Yaich, M., Eskola, J., 2004. Reduction of antibody response to an 11-valent pneumococcal vaccine coadministered with a vaccine containing acellular pertussis components. Infect. Immun. 72, 5383–5391.

Eskola, J., Kilpi, T., Palmu, A., et al., 2001. Efficacy of a pneumococcal conjugate vaccine against acute otitis media. N. Engl. J. Med. 344, 403–409.

Farrell, D.J., Jenkins, S.G., Reinert, R.R., 2004. Global distribution of *Streptococcus pneumoniae* serotypes isolated from paediatric patients during 1999–2000 and the in vitro efficacy of telithromycin and comparators. J. Med. Microbiol. 53, 1109–1117.

Finne, J., Leinonen, M., Mäkelä, P.H., 1983. Antigenic similarities between brain components and bacteria causing meningitis. Implications for vaccine development and pathogenesis. Lancet ii, 355–357.

Fletcher, M.A., Fritzell, B., 2007. Brief review of the clinical effectiveness of PREVENAR against otitis media. Vaccine 25, 2507–2512.

Francis, T., Tillett, W.S., 1930. Cutaneous reactions in pneumonia. The development of antibodies following the intradermal injection of type-specific polysaccharide. J. Exp. Med. 52, 573–585.

Fraser, A., Paul, M., Goldberg, E., Acosta, C.J., Leibovici, L., 2007. Typhoid fever vaccines: systematic review and meta-analysis of randomised controlled trials. Vaccine 25, 7848–7857.

Fusco, P.C., Michon, F., Tai, J.Y., Blake, M.S., 1997. Preclinical evaluation of a novel group B meningococcal conjugate vaccine that elicits bactericidal activity in both mice and nonhuman primates. J. Infect. Dis. 175, 364–372.

Girard, M.P., Preziosi, M.P., Aguado, M.T., Kieny, M.P., 2006. A review of vaccine research and development: meningococcal disease. Vaccine 24, 4692–4700.

Goebel, W.F., Avery, O.T., 1931. Chemo-immunological studies on conjugated carbohydrate-proteins. IV. The synthesis of p-aminobenzyl ether of the soluble specific substance of type III *Pneumococcus* and its coupling with protein. J. Exp. Med. 54, 431–436.

Goldschneider, I., Gotschlich, E.C., Artenstein, M.S., 1969. Human immunity to the meningococcus. I. The role of humoral antibodies. J. Exp. Med. 129, 1307–1326.

Gotschlich, E.C., Goldschneider, I., Lepow, M.L., Gold, R., 1977. The immune responses to bacterial polysaccharides in man. In: Haber, E., Krause, R.M. (Eds.), Antibodies in Human Diagnosis and Therapy. Raven Press, New York, pp. 391–402.

Granoff, D.M., Holmes, S.J., Osterholm, M.T., et al., 1993. Induction of immunologic memory in infants primed with *Haemophilus influenzae* type b conjugate vaccines. J. Infect. Dis. 168, 663–671.

Gray, S.J., Trotter, C.L., Ramsay, M.E., et al., 2006. Epidemiology of meningococcal disease in England and Wales 1993/94 to 2003/04: contribution and experiences of the meningococcal reference unit. J. Med. Microbiol. 55, 887–896.

Grothaus, M.C., Srivastava, N., Smithson, S.L., et al., 2000. Selection of an immunogenic peptide mimic of the capsular polysaccharide of *Neisseria meningitidis* serogroup A using a peptide display library. Vaccine 18, 1253–1263.

Guttormsen, H.K., Paoletti, L.C., Mansfield, K.G., Jachymek, W., Jennings, H.J., Kasper, D.L., 2008. Rational chemical design of the carbohydrate in a glycoconjugate vaccine enhances IgM-to-IgG switching. Proc. Natl. Acad. Sci. USA 105, 5903–5908.

Hammitt, L.L., Bruden, D.L., Butler, J.C., et al., 2006. Indirect effect of conjugate vaccine on adult carriage of *Streptococcus pneumoniae*: an explanation of trends in invasive pneumococcal disease. J. Infect. Dis. 193, 1487–1494.

Heyns, K., Kiessling, G., Lindenberg, W., Paulsen, H., Webster, M.E., 1959. D-Galaktosaminuronsäure (2-Amino-2-desoxy-2-galakturonsäure) als Baustein des Vi-Antigens. Chem. Ber. 92, 2435–2438.

Hoogerhout, P., Evenberg, D., van Boeckel, C.A.A., et al., 1987. Synthesis of fragments of the capsular polysaccharide of *Haemophilus influenzae* type B, comprising two or three repeating units. Tetrahedron Lett. 28, 1553–1556.

Hutchins, W.A., Adkins, A.R., Kieber-Emmons, T., Westerink, M.A., 1996. Molecular characterization of a monoclonal antibody produced in response to a group C meningococcal polysaccharide peptide mimic. Mol. Immunol. 33, 503–510.

Hutchins, W.A., Kieber-Emmons, T., Carlone, G.M., Westerink, M.A., 1999. Human immune response to a peptide mimic of *Neisseria meningitidis* serogroup C in hu-PBMC-SCID mice. Hybridoma 18, 121–129.

Insel, R.A., Anderson, P.W., 1986. Oligosaccharide-protein conjugate vaccines induce and prime for oligoclonal IgG antibody responses to the *Haemophilus influenzae* b capsular polysaccharide in human infants. J. Exp. Med. 163, 262–269.

Jansen, W.T.M., Hogenboom, S., Thijssen, M.J., et al., 2001. Synthetic 6B di-, tri-, and tetrasaccharide-protein conjugates contain *Pneumococcal* type 6A and 6B common and 6B- specific epitopes that elicit protective antibodies in mice. Infect. Immun. 69, 787–793.

Jennings, H.J., 1998. Polysaccharide vaccines against disease caused by *Haemophilus influenzae*, group B *Streptococcus*, and *Salmonella typhi*. Carbohydr. Eur. 21, 17–23.

Jennings, H.J., Lugowski, C., 1981. Immunochemistry of groups A, B, and C meningococcal polysaccharide-tetanus toxoid conjugates. J. Immunol. 127, 1011–1018.

Jennings, H.J., Pon, R.A., 1996. Polysaccharides and glycoconjugates as human vaccines. In: Dimitriu, S. (Ed.), Polysaccharides in Medicinal Applications. Marcel Dekker, New York, pp. 473–479.

Jennings, H.J., Sood, R.K., 1994. Synthetic glycoconjugates as human vaccines. In: Lee, Y.C., Lee, R.T. (Eds.), Neoglycoconjugates. Preparation and Applications. Academic Press, London, pp. 325–371.

Jennings, H.J., Katzenellenbogen, E., Lugowski, C., Michon, F., Roy, R., Kasper, D.L., 1984. Structure, conformation, and immunology of sialic acid-containing polysaccharides of human pathogenic bacteria. Pure Appl. Chem. 56, 893–905.

Jennings, H.J., Roy, R., Gamian, A., 1986. Induction of meningococcal group B polysaccharide-specific IgG antibodies in mice by using an *N*-propionylated B polysaccharide-tetanus toxoid conjugate vaccine. J. Immunol. 137, 1708–1713.

Jennings, H.J., Gamian, A., Ashton, F.E., 1987. *N*-propionylated group B meningococcal polysaccharide mimics a unique epitope on group B *Neisseria meningitidis*. J. Exp. Med. 165, 1207–1211.

Jessouroun, E., Da Silveira, I.A.F., Bastos, R.C., Frasch, C.E., Lee, C.J., 2005. Process for preparing polysaccharide-protein conjugate vaccines. Patent PCT/US2004/026431 [WO/2005/037320], 1–14. 2005. 8-6-2004.

Jodar, L., Feavers, I.M., Salisbury, D., Granoff, D.M., 2002. Development of vaccines against meningococcal disease. Lancet 359, 1499–1508.

Kadirvelraj, R., Gonzalez, O.J., Foley, B.L., et al., 2006. Understanding the bacterial polysaccharide antigenicity of *Streptococcus agalactiae* versus *Streptococcus pneumoniae*. Proc. Natl. Acad. Sci. USA 103, 8149–8154.

Kieber-Emmons, T., Ward, R.E., Raychaudhuri, S., Rein, R., Kohler, H., 1986. Rational design and application of idiotope vaccines. Int. Rev. Immunol. 1, 1–26.

Kossaczka, Z., Bystricky, S., Bryla, D.A., Shiloach, J., Robbins, J.B., Szu, S.C., 1997. Synthesis and immunological properties of Vi and di-*O*-acetyl pectin protein conjugates with adipic acid dihydrazide as the linker. Infect. Immun. 65, 2088–2093.

Kossaczka, Z., Lin, F.Y., Ho, V.A., et al., 1999. Safety and immunogenicity of Vi conjugate vaccines for typhoid fever in adults, teenagers, and 2- to 4-year-old children in Vietnam. Infect. Immun. 67, 5806–5810.

Laferriere, C.A., Sood, R.K., deMuys, J.M., Michon, F., Jennings, H.J., 1998. *Streptococcus pneumoniae* type 14 polysaccharide-conjugate vaccines: length stabilization of opsonophagocytic conformational polysaccharide epitopes. Infect. Immun. 66, 2441–2446.

Law, M.R., Palomaki, G., Alfirevic, Z., et al., 2005. The prevention of neonatal group B streptococcal disease: a report by a working group of the Medical Screening Society. J. Med. Screen. 12, 60–68.

Lefeber, D.J., Kamerling, J.P., Vliegenthart, J.F.G., 2001. Synthesis of *Streptococcus pneumoniae* type 3 neoglycoproteins varying in oligosaccharide chain length, loading and carrier protein. Chem. Eur. J. 7, 4411–4421.

Lesinski, G.B., Smithson, S.L., Srivastava, N., Chen, D., Widera, G., Westerink, M.A., 2001. A DNA vaccine encoding a peptide mimic of *Streptococcus pneumoniae* serotype 4 capsular polysaccharide induces specific anti-carbohydrate antibodies in Balb/c mice. Vaccine 19, 1717–1726.

Lewis, A.L., Nizet, V., Varki, A., 2004. Discovery and characterization of sialic acid *O*-acetylation in group B *Streptococcus*. Proc. Natl. Acad. Sci. USA1 01, 11123–11128.

Lin, F.Y., Ho, V.A., Khiem, H.B., et al., 2001. The efficacy of a *Salmonella typhi* Vi conjugate vaccine in two-to-five-year-old children. N. Engl. J. Med. 344, 1263–1269.

Mäkelä, P.H., Peltola, H., Kayhty, H., Jousimies, K., Pettay, O., 1977. Polysaccharide vaccines of group A *Neisseria meningitidis* and *Haemophilus influenzae* type b: a field trial in Finland. J. Infect. Dis. 136 (Suppl.), 43–50.

Marburg, S., Jorn, D., Tolman, R.L., et al., 1986. Bimolecular chemistry of macromolecules: synthesis of bacterial polysaccharide conjugates with *Neisseria meningitidis* membrane protein. J. Am. Chem. Soc. 108, 5282–5287.

Mawas, F., Niggemann, J., Jones, C., Corbel, M.J., Kamerling, J.P., Vliegenthart, J.F.G., 2002. Immunogenicity in a mouse model of a conjugate vaccine made with a synthetic single repeating unit of type 14 pneumococcal polysaccharide coupled to CRM197. Infect. Immun. 70, 5107–5114.

Melegaro, A., Edmunds, W.J., 2004. The 23-valent pneumococcal polysaccharide vaccine. Part I. Efficacy of PPV in the elderly: a comparison of meta-analyses. Eur. J. Epidemiol. 19, 353–363.

Michon, F., Brisson, J.R., Dell, A., Kasper, D.L., Jennings, H. J., 1988. Multiantennary group-specific polysaccharide of group B *Streptococcus*. Biochemistry 27, 5341–5351.

Michon, F., Huang, C.H., Farley, E.K., Hronowski, L., Di, J., Fusco, P.C., 2000. Structure activity studies on group C meningococcal polysaccharide-protein conjugate vaccines: effect of *O*-acetylation on the nature of the protective epitope. Dev. Biol. (Basel) 103, 151–160.

Moe, G.R., Tan, S., Granoff, D.M., 1999. Molecular mimetics of polysaccharide epitopes as vaccine candidates for prevention of *Neisseria meningitidis* serogroup B disease. FEMS Immunol. Med. Microbiol. 26, 209–226.

Nilsson, S., Bengtsson, M., Norberg, T., 1992. Solid-phase synthesis of a fragment of the capsular polysaccharide of *Haemophilus Influenzae* Type B using *H*-phosphonate intermediates. J. Carbohydr. Chem. 11, 265–285.

Palmu, A.A., Verho, J., Jokinen, J., Karma, P., Kilpi, T.M., 2004. The seven-valent pneumococcal conjugate vaccine reduces tympanostomy tube placement in children. Pediatr. Infect. Dis. J. 23, 732–738.

Paoletti, L.C., Wessels, M.R., Rodewald, A.K., Shroff, A.A., Jennings, H.J., Kasper, D.L., 1994. Neonatal mouse protection against infection with multiple group B streptococcal (GBS) serotypes by maternal immunization with a tetravalent GBS polysaccharide-tetanus toxoid conjugate vaccine. Infect. Immun. 62, 3236–3243.

Peeters, C.C., Evenberg, D., Hoogerhout, P., et al., 1992. Synthetic trimer and tetramer of 3-beta-D-ribose-(1-1)-D-ribitol-5-phosphate conjugated to protein induce antibody responses to *Haemophilus influenzae* type b capsular polysaccharide in mice and monkeys. Infect. Immun. 60, 1826–1833.

Peltola, H., Häyhty, H., Sivonen, A., Mäkelä, P.H., 1977. *Haemophilus influenzae* type b capsular polysaccharide vaccine in children: a double blind field study of 100,000 children 3 months to 5 years of age in Finland. Pediatrics 60, 730–737.

Pincus, S.H., Smith, M.J., Jennings, H.J., Burritt, J.B., Glee, P.M., 1998. Peptides that mimic the group B streptococcal type III capsular polysaccharide antigen. J. Immunol. 160, 293–298.

Pozsgay, V., 2001. Oligosaccharide-protein conjugates as vaccine candidates against bacteria. Adv. Carbohydr. Chem. Biochem. 56, 153–199.

Prinz, D.M., Smithson, S.L., Kieber-Emmons, T., Westerink, M.A., 2003. Induction of a protective capsular polysaccharide antibody response to a multiepitope DNA vaccine encoding a peptide mimic of meningococcal serogroup C capsular polysaccharide. Immunology 110, 242–249.

Prymula, R., Peeters, P., Chrobok, V., et al., 2006. Pneumococcal capsular polysaccharides conjugated to protein D for prevention of acute otitis media caused by both *Streptococcus pneumoniae* and non-typable *Haemophilus influenzae*: a randomised double-blind efficacy study. Lancet 367, 740–748.

Rennels, M., King, J.J., Ryall, R., Papa, T., Froeschle, J., 2004. Dosage escalation, safety and immunogenicity study of four dosages of a tetravalent meninogococcal polysaccharide diphtheria toxoid conjugate vaccine in infants. Pediatr. Infect. Dis. J. 23, 429–435.

Richmond, P., Borrow, R., Goldblatt, D., et al., 2001a. Ability of 3 different meningococcal C conjugate vaccines to induce immunologic memory after a single dose in UK toddlers. J. Infect. Dis. 183, 160–163.

Richmond, P., Borrow, R., Findlow, J., et al., 2001b. Evaluation of de-*O*-acetylated meningococcal C polysaccharide-tetanus toxoid conjugate vaccine in infancy: reactogenicity, immunogenicity, immunologic priming, and bactericidal activity against *O*-acetylated and de-*O*-acetylated serogroup C strains. Infect. Immun. 69, 2378–2382.

Robbins, J.B., 1978. Vaccines for the prevention of encapsulated bacterial diseases: current status, problems and prospects for the future. Immunochemistry 15, 839–854.

Robbins, J.B., Austrian, R., Lee, C.J., et al., 1983. Considerations for formulating the second-generation pneumococcal capsular polysaccharide vaccine with emphasis on the cross-reactive types within groups. J. Infect. Dis. 148, 1136–1159.

Robbins, J.B., Schneerson, R., Szu, S.C., 1996. Hypothesis: how licensed vaccines confer protective immunity. In: Cohet, S., Shafferman, A. (Eds.). Novel Strategies in Design and Production of Vaccines. Plenum Press, New York, pp. 169–182.

Robbins, J.B., Schneerson, R., Gotschlich, E.C., 2000. A rebuttal: epidemic and endemic meningococcal meningitis in sub-Saharan Africa can be prevented now by routine immunization with group A meningococcal capsular polysaccharide vaccine. Pediatr. Infect. Dis. J. 19, 945–953.

Schneerson, R., Barrera, O., Sutton, A., Robbins, J.B., 1980. Preparation, characterization, and immunogenicity of *Haemophilus influenzae* type b polysaccharide-protein conjugates. J. Exp. Med. 152, 361–376.

Scott, D.A., Komjathy, S.F., Hu, B.T., et al., 2007. Phase 1 trial of a 13-valent pneumococcal conjugate vaccine in healthy adults. Vaccine 25, 6164–6166.

Shin, J.S., Lin, J.S., Anderson, P.W., Insel, R.A., Nahm, M.H., 2001. Monoclonal antibodies specific for *Neisseria meningitidis* group B polysaccharide and their peptide mimotopes. Infect. Immun. 69, 3335–3342.

Snape, M.D., Perrett, K.P., Ford, K.J., et al., 2008. Immunogenicity of a tetravalent meningococcal glycoconjugate vaccine in infants – a randomized controlled trial. J. Am. Med. Assoc. 299, 173–184.

Stein, K.E., 1992. Thymus-independent and thymus-dependent responses to polysaccharide antigens. J. Infect. Dis. 165 (Suppl. 1), S49–S52.

Szu, S.C., Li, X.R., Stone, A.L., Robbins, J.B., 1991. Relation between structure and immunologic properties of the Vi capsular polysaccharide. Infect. Immun. 59, 4555–4561.

Szu, S.C., Bystricky, S., Hinojosa-Ahumada, M., Egan, W., Robbins, J.B., 1994. Synthesis and some immunologic properties of an *O*-acetyl pectin [poly(1→4)-α-D-GalpA]-protein conjugate as a vaccine for typhoid fever. Infect. Immun. 62, 5545–5549.

Takala, A.K., Eskola, J., Leinonen, M., et al., 1991. Reduction of oropharyngeal carriage of *Haemophilus influenzae* type b (Hib) in children immunized with a Hib conjugate vaccine. J. Infect. Dis. 164, 982–986.

Verez-Bencomo, V., Fernandez-Santana, V., Hardy, E., et al., 2004. A synthetic conjugate polysaccharide vaccine against *Haemophilus influenzae* type b. Science 305, 522–525.

Wang, Y., Hollingsworth, R.I., Kasper, D.L., 1998. Ozonolysis for selectively depolymerizing polysaccharides containing β-aldosidic linkages. Proc. Natl. Acad. Sci. USA 95, 6584–6589.

Wang, Z.Y., Just, G., 1988. Synthesis of fragments of the capsular polysaccharide of Haemophilus influenzae type B. Tetrahedron Lett. 29, 1525–1528.

Ward, J., Brenneman, G., Letson, G.W., Heyward, W.L., 1990. Limited efficacy of a *Haemophilus influenzae* type b conjugate vaccine in Alaska Native infants. The Alaska H. influenzae Vaccine Study Group. N. Engl. J. Med. 323, 1393–1401.

Weisser, N.E., Almquist, K.C., Hall, J.C., 2007. A rAb screening method for improving the probability of identifying peptide mimotopes of carbohydrate antigens. Vaccine 25, 4611–4622.

Wessels, M.R., Munoz, A., Kasper, D.L., 1987. A model of high-affinity antibody binding to type III group B Streptococcus capsular polysaccharide. Proc. Natl. Acad. Sci. USA 84, 9170–9174.

Westerink, M.A.J., Giardina, P.C., 1992. Anti-idiotype induced protection against *Neisseria meningitidis* serogroup C bacteremia. Microb. Path. 12, 19–26.

Westerink, M.A., Campagnari, A.A., Wirth, M.A., Apicella, M.A., 1988. Development and characterization of an anti-idiotype antibody to the capsular polysaccharide of *Neisseria meningitidis* serogroup C. Infect. Immun. 56, 1120–1127.

Westerink, M.A.J., Giardina, P.C., Apicella, M.A., Kieber-Emmons, T., 1995. Peptide mimicry of the meningococcal group c capsular polysaccharide. Proc. Natl. Acad. Sci. USA 92, 4021–4025.

Zou, W., Mackenzie, R., Therien, L., et al., 1999. Conformational epitope of the type III group B *Streptococcus* capsular polysaccharide. J. Immunol. 163, 820–825.

CHAPTER

49

Immunomodulation by zwitterionic polysaccharides

Rachel M. McLoughlin and Dennis L. Kasper

SUMMARY

Traditionally, polysaccharides are regarded as T-cell-independent molecules. These molecules were considered incapable of activating T-cells directly and, consequently, have not proven very useful in the development of vaccines due to their inability to establish immunological memory. Seminal advances in the field have highlighted a group of capsular polysaccharide molecules that possess a unique zwitterionic charge motif in their repeating unit, i.e. they possess both a positive and a negative charge. The presence of this unique chemical structure confers upon these polysaccharides the ability to activate T-cells in a mechanism that was previously thought to be exclusive to protein antigens. In this chapter, the historical discovery of zwitterionic polysaccharides (ZPSs) as immunomodulators is discussed, the relatively exclusive number of organisms that currently are described expressing ZPS is explored and their unique chemical structure is characterized. Subsequently, the novel mechanisms by which ZPSs can modulate aspects of both the innate and adaptive arms of the host immune response to activate specific T-cell responses are reviewed. Moreover, further discussion involves the phenotypes of T-cells that are activated and consideration is made of the biological impact of this ZPS-mediated T-cell activation in maintaining a healthy immune system and protecting the host against the development of allergic and autoimmune diseases.

Keywords: Immunomodulation; Innate immunity; Adaptive immunity; T-cell; Antigen presenting cell; Toll-like receptor; Commensal bacteria; Allergic disease; Autoimmune disease; Immune development

1. INTRODUCTION

The immune response is the reaction of the vertebrate body to invasion by foreign materials (antigens) such as pathogens or tumour cells. The human body has evolved an intricate network of systems to detect the presence of these agents, distinguish them from its own healthy cells and mount an attack to neutralize them. Initial protection is provided by physical barriers, e.g. the skin, mucosal linings and tears, which prevent pathogens from entering the host. If these barriers are breached, the second line in host defence against the invading pathogen comes in the form of the innate immune system. If the innate immune response fails to clear completely the invading pathogen, vertebrates possess a third layer of defence, known as the adaptive immune response. Activation of the adaptive immune response is actually facilitated by the innate response and enables the host to recognize and remember specific pathogens, thereby generating immunological memory, which results in a more robust attack upon subsequent invasion by this pathogen or foreign antigen (Janeway *et al.*, 2005).

The innate immune response comprises the body's most basic defence against infection and injury. The innate immune system can detect and destroy pathogens very rapidly in a process that does not involve the expansion of antigen-specific lymphocytes and the development of immunological memory. This system recognizes features that are common to many pathogens, i.e. it is a non-specific form of defence (Janeway *et al.*, 2005). A key component of the innate response is the induction of effector molecules that impact directly upon the nature of the subsequent adaptive immune response, thereby facilitating the efficient clearance of the invading pathogen (Hoebe *et al.*, 2004). Macrophages, neutrophils and dendritic cells are the critical mediators of the innate response.

The adaptive immune system is composed of a variety of highly specialized cells and is activated and enhanced by signals from the innate immune system. The adaptive immune response facilitates recognition but also memory to specific pathogens. Lymphocytes make up the major cellular component of the adaptive immune system. The T-lymphocytes, also termed T-cells, are part of this group of white blood cells which play a critical role in the adaptive immune response (Janeway *et al.*, 2005). The T-cells can be distinguished from other lymphocyte subsets due to the expression of a specific T-cell receptor (TCR), composed of α- and β-chains, on their cell surface. Each TCR consists of two chains: a variable chain (V), which allows for the recognition of specific antigens; and a constant chain (C), which provides signalling functions. Upon maturation, each T-cell expresses a unique variant of the antigen receptor, i.e. TCR, that determines to which antigens the T-cell can bind. The entire population of T-cells within the host has an expansive repertoire of receptors that are highly diverse, conferring upon the T-cell the ability to recognize and mount a specific immune response against almost any foreign antigen (Kronenberg *et al.*, 1986).

2. IMMUNOMODULATION BY POLYSACCHARIDES

2.1. Immunomodulation

Compounds that can interact with various aspects of the immune response of the host, to either upregulate or downregulate specific events, are classified as immunomodulators. These compounds can be synthetically generated chemicals, e.g. immunomodulatory drugs, be naturally occurring, e.g. components of bacteria, or be engineered endogenous immune proteins, e.g. recombinant cytokines such as interferons (IFNs). The basic strategy underlying immunomodulatory therapy is to identify aspects of the host's response that can be enhanced or suppressed in such a way as to augment or complement a desired immune response. Immunomodulatory therapy, which is a rapidly growing area of biomedical research, can take many forms. Successful advances in this area include the development of vaccines, treatment of certain cancers with monoclonal antibodies (Plosker and Figgitt, 2003), treatment of various autoimmune diseases, e.g. treatment of multiple sclerosis with the recombinant cytokine IFN-β (McCormack and Scott, 2004). Immunotherapies which target the host's innate immune response in an attempt to enhance defence mechanisms against microbial invasion are currently being developed (Finlay and Hancock, 2004; Bowdish *et al.*, 2005; Carpenter and O'Neill, 2007; Scott *et al.*, 2007).

2.2. Traditional immune response elicited by polysaccharides

Traditionally, polysaccharides have been considered to be T-cell-independent antigens that cannot activate T-cell-mediated immune responses (Coutinho *et al.*, 1973; McGhee *et al.*, 1984; Barrett, 1985). Instead, they are considered simply to elicit a humoral immune response that results in immunoglobulin (Ig) M production, with very little IgG response detectable. This immune response to polysaccharides (Figure 49.1A) is relatively short-lived due to lack of T-cell activation and immunological memory generation. In addition, long-term antibody responses are absent due to the lack of antibody class switching from IgM to IgG. The inability of polysaccharides to activate T-cells was thought to result from the inability of antigen-presenting cells (APCs) to process and present these molecules for recognition by the T-cells. This is in contrast to protein antigens

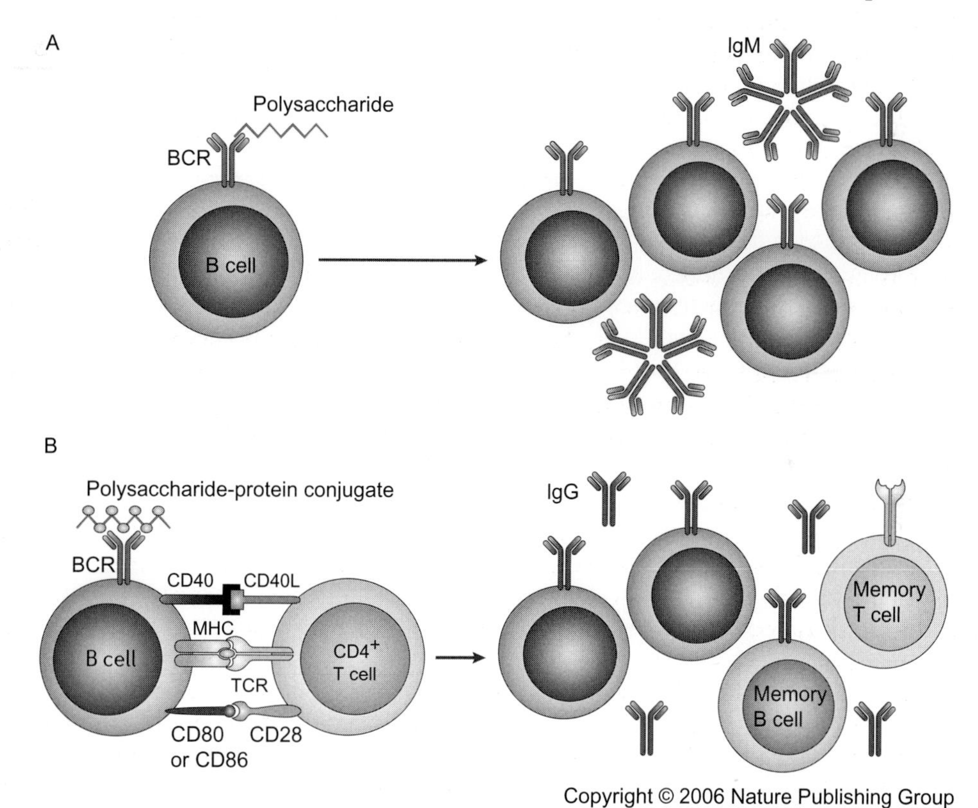

FIGURE 49.1 Immune cell activation by polysaccharides. (A) Bacterial polysaccharides are classically B-cell antigens. They are recognized by a B-cell receptor (BCR) that has the correct specificity and induces the signals that are required to stimulate clonal expansion of B-cells and antibody production, i.e. immunoglobulin M (IgM). However, this pathway by itself does not result in immunological memory. (B) Polysaccharide–protein conjugates interact with a BCR in a similar manner to pure polysaccharides, however, in addition, they elicit T-cell help, through antigen presentation of the protein component to the T-cell receptor (TCR) on $CD4^+$ T-cells, which provide the necessary co-stimulation to induce memory B-cells and memory T-cells. Antibody production to both the polysaccharide and the protein is achieved (IgG) in addition to immunological memory. Adapted, with permission, from Mazmanian and Kasper (2006) The love-hate relationship between bacterial polysaccharides and the host immune system. Nat. Rev. Immunol. 6, 849–858.

which are very efficiently taken up by APCs and presented to T-cells to induce help from T-cells in activating B-cells, which undergo proliferation and maturation, to become cells that are capable of making IgG antibody (Watts, 1997). Conjugation of polysaccharides to proteins is a method employed to generate carbohydrate-specific responses that result in T-cell activation (Sood and Fattom, 1998) and this technique has been used in the development of glycoconjugate vaccines (Kelly *et al.*, 2004) (see also Chapter 48). The mechanism of action of glycoconjugate vaccines is by "tricking" the T-cell into recognizing the anti-polysaccharide antibody-producing B-cell, so that the helper functions described above are initiated. It is hypothesized that the B-cells will specifically recognize the carbohydrate and take up the glycoconjugate, the B-cell then presents a peptide from the attached protein carrier to the T-cell, which recognizes the presented peptide. Stimulation of the B-cell, in conjunction with peptide-induced activation of the T-cell, promotes immunoglobulin class switching and B-cell memory (see Figure 49.1B).

3. ZWITTERIONIC POLYSACCHARIDES (ZPSs)

3.1. Structural attributes of capsular polysaccharides (CPSs)

The majority of extracellular bacteria express a thick slime-like layer of polysaccharide molecules on their surface, which is known as CPS. The major function of this CPS was traditionally thought to conceal antigenic proteins on the bacterial surface that would activate an unwanted immune response upon infection in the host. In addition, the CPS molecules protect extracellular bacteria from opsono-phagocytic killing by phagocytes (Silver *et al.*, 1988; Cobb and Kasper, 2005). However, recent advances in the area of CPS biology have attributed immunomodulatory functions to certain CPS molecules depending upon their chemical structure (Tzianabos *et al.*, 2003; Cobb and Kasper, 2005). Generally, CPS molecules consist of several hundred repeating units and these repeating units contain up to eight sugars usually linked by glycosidic bonds (see Chapters 6 and 20). Most CPSs found in Nature are neutral or contain a net negative charge, examples of which include most *Streptococcus pneumoniae* capsule types, the V1-antigen from *Salmonella enterica* sv. Typhi (Szu *et al.*, 1991) and the capsules of group B *Streptococcus* (Figure 49.2) (Wessels *et al.*, 1991). These molecules are incapable of activating a T-cell response (Tzianabos *et al.*, 2003; Cobb and Kasper, 2005). A novel category of CPS molecules has been identified that can activate T-cell responses and these effects are attributed to the unique chemical structure of these molecules (Tzianabos *et al.*, 1993, 2001; Cobb *et al.*, 2004). The ZPSs contain both positive (e.g. free amine) and negatively (e.g. phosphate or carboxylate) charged moieties within their repeating unit. The zwitterionic charge motif that characterizes these polysaccharide molecules is rare and confers upon them the unique ability to activate $CD4^+$ T-cells.

3.2. Historical discovery of ZPS

Seminal observations made about 30 years ago demonstrated that intra-abdominal abscesses that occurred in surgical patients were induced by bacteria that were present in the normal microbial environment of the gut and, in particular, by the Gram-negative obligate anaerobe *Bacteroides fragilis* (Onderdonk *et al.*, 1977). Extensive immunological investigations have revealed that abscess formation in response to this bacterium is T-cell-dependent and the ability of this organism to form abscesses was attributed to the expression of a high-molecular mass CPS complex on the surface of the bacterium (Lindberg *et al.*, 1982; Tzianabos and Kasper, 2002). Further studies then revealed that, in addition to its role in the development

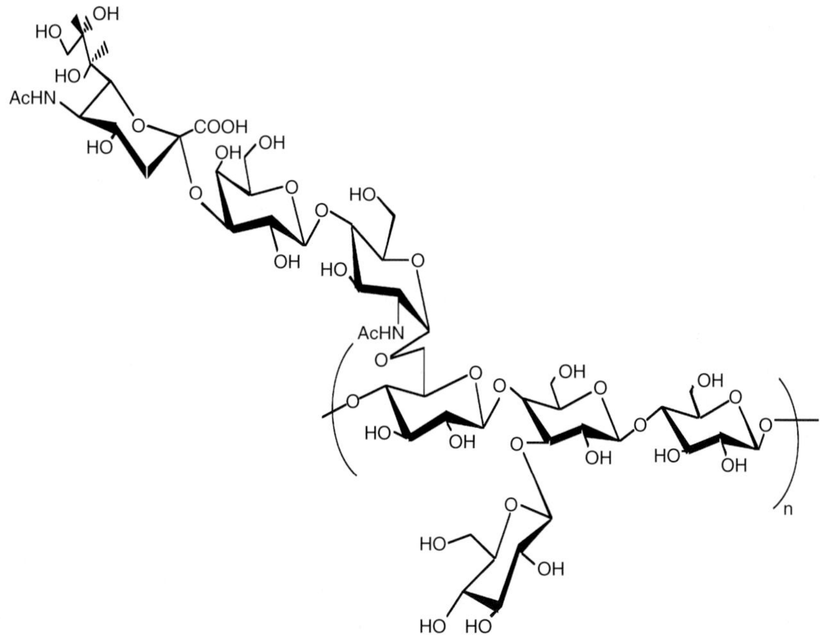

FIGURE 49.2 Non-zwitterionic polysaccharide. Capsular polysaccharide type V from group B *Streptococcus* contains a net negative charge.

of abscesses, this polysaccharide, when administered alone, could actually protect the animal against the development of abscesses in response to challenge with *B. fragilis* and other encapsulated bacteria (Bartlett, 1978; Kasper *et al.*, 1979). It turned out that this protection was T-cell-dependent and the unique zwitterionic structure of the polysaccharide was paramount in establishing the T-cell-mediated protection (Tzianabos *et al.*, 2000). Chemical neutralization of the positive charge in the polysaccharide repeating unit abrogated the ability of this polymer to protect against abscess formation (Tzianabos *et al.*, 1993, 1994). Thus, it was established that ZPS has the unique capacity to induce T-cell-dependent immunomodulation.

3.3. Occurrence of ZPSs in Nature

Expression of zwitterionic CPSs is relatively rare among bacterial species and, to date, the ZPSs from only three bacterial species have been characterized (Figure 49.3). Of these, the ZPSs of the commensal organism *B. fragilis* have been studied most extensively and, in particular, polysaccharide A. Genomic analysis has revealed that *B. fragilis* can express an unprecedented eight structurally distinct CPS species, i.e. PSA to PSH (Krinos *et al.*, 2001; Kuwahara *et al.*, 2004). High-resolution nuclear magnetic resonance (NMR) spectroscopic analysis has revealed that two of these CPSs, polysaccharide A (PSA) and B (PSB), are zwitterionic in nature (Baumann *et al.*, 1992).

The PSA molecule (see Figure 49.3A) consists of several hundred repeating units of a branched tetrasaccharide, with three monosaccharides in the polymer backbone and one residue on the side chain (Baumann *et al.*, 1992). It has a molecular mass of ≈110 000. The structure of PSA-2 (see Figure 49.3B), from an alternative *B. fragilis* strain (*B. fragilis* 638), has also previously been

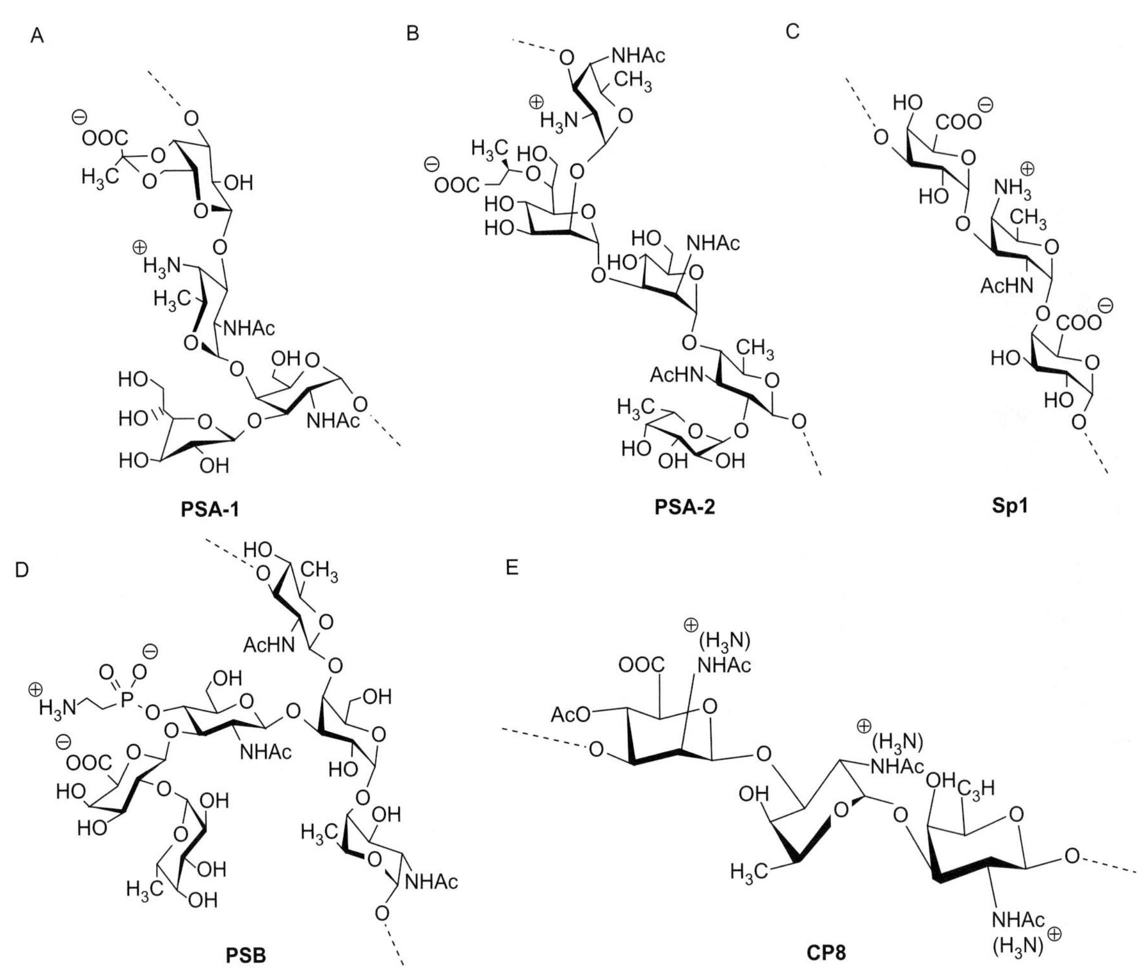

FIGURE 49.3 Polysaccharide repeating units. Zwitterionic capsular polysaccharides PSA-1, PSA-2 and PSB are from *B. fragilis*, Sp1 from *S. pneumoniae* and CP8 from *S. aureus*. Adapted, with permission, from Tzianabos et al. (2003) Biological chemistry of immunomodulation by zwitterionic polysaccharides. Carbohydr. Res. 338, 2531–2538.

investigated using total correlated spectroscopy and 2D nuclear Overhauser enhancement spectroscopy (NOESY) in NMR spectroscopic analysis, by which it was found that the average solution structure of PSA-2 is characterized by a right-handed helical conformation with two repeating units per turn and a pitch of 20 Å (Wang *et al.*, 2000; Choi *et al.*, 2002) (Figure 49.4A). The zwitterionic motif is formed with alternating anionic carboxylate lying in repeated grooves every 15 Å and cationic-free amines exposed on the outer surface of the carbohydrate (Wang *et al.*, 2000).

The PSB molecule (see Figure 49.3C) contains a hexasaccharide repeating unit, with three residues on the polymer backbone and three sugars forming the side chain. This molecule contains positively charged amines and negatively charged carboxylate and phosphonate groups.

The CPS from type 1 *S. pneumoniae* (Sp1) is a simpler polymer with a linear trisaccharide repeating unit (see Figure 48.3D). The overall

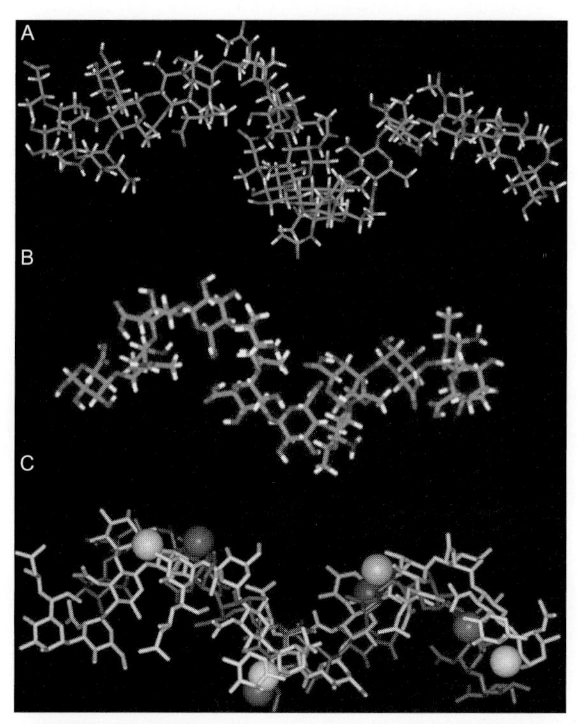

FIGURE 49.4 Stick models of the three-dimensional structures of PSA-2 (A) and Sp1 (B). Carbon atoms are coloured green, oxygen atoms are red, nitrogen atoms are coloured blue and hydrogen are white. (C) Conformations of Sp1 (red) and PSA-2 (yellow) superimposed on the basis of their amino groups. Red and yellow dots represent positive charges from Sp1 and PSA-2, respectively. Sp1 and PSA-2 display a similar zig-zag positive charge pattern with roughly equidistant charge separation of 15Å. Adapted, with permission, from Tzianabos *et al.* (2003) Biological chemistry of immunomodulation by zwitterionic polysaccharides. Carbohydr. Res. 338, 2531–2538. (See colour plates section.)

structure of Sp1 is an extended right-handed helix with eight residues per turn and a pitch of 20Å. The molecular surface is covered with a high density of both positive and negatively charged groups. Typically, within a repeating unit, the two carboxyl groups are in close proximity on one side of the helix, with the amino groups on the opposite side (Choi *et al.*, 2002) (see Figure 49.4B).

Capsular polysaccharide 8 (CP8) is expressed by the Gram-positive bacterium *Staphylococcus aureus* (see Figure 49.3E). *S. aureus* is an encapsulated bacterium that is highly associated with abscess formation and NMR spectroscopy has revealed that CP8 expressed by this organism is also zwitterionic in nature (Tzianabos *et al.*, 2001). Serotype 8 strains of *S. aureus*, that express CP8 belong to one of the commonest *S. aureus* serotypes isolated from human infections. The structure of CP8 consists of a linear trisaccharide repeating unit containing *N*-acetylmannosaminuronic acid (ManANAc) and two *N*-acetylfucosamine (FucNAc) residues. Importantly, CP8 is naturally polyanionic, but it is thought that, under certain environmental conditions, de-*N*-acetylation can occur to generate positively charged amines in the repeating unit, thereby creating a zwitterionic motif. Studies have revealed that purified CP8, like PSA and Sp1, can also modulate abscess formation and activate T-cells, hence justifying its classification as a zwitterionic polysaccharide (Tzianabos *et al.*, 2001; McLoughlin *et al.*, 2008). Interestingly, CP8 is distinct from the other previously mentioned ZPSs, which have regular repeating units, as partially de-*N*-acetylated CP8 carries its positive charges at random sites (Tzianabos *et al.*, 2001).

4. ACTIVATION OF T-CELLS BY ZPSs THROUGH INNATE AND ADAPTIVE IMMUNOMODULATION

4.1. The T-cell

4.1.1. Activation of T-cells by APCs

The main way in which T-cells become activated to carry out their designated function within the immune system occurs through interaction with, and receipt of signals from, accessory cells known as APCs. An APC is a cell that is capable of displaying, or presenting, a foreign antigen on its surface. This process is facilitated by a glycoprotein expressed on the surface of the APC known as the major histocompatibility complex (MHC). The APCs can express either MHC class I molecules, which result in antigen presentation and activation of $CD8^+$ T-cells, or class II molecules, which are responsible for activation of $CD4^+$ T-cells. Moreover, APCs can fall into two categories: professional APCs, e.g. macrophages, dendritic cells and B-cells, or non-professional APCs, which do not constitutively express MHC molecules on their surface but which can be upregulated to do so under certain inflammatory conditions; these cells include fibroblasts, vascular endothelial cells, epithelial cells and glial cells in the brain. Antigen-presenting cells are very efficient at engulfing foreign antigens, after which they present the antigen to naive T-cells and the process of T-cell activation ensues. Typically, this process of APC-induced T-cell activation involves three components (Figure 49.5). The first signal involves presentation of antigen on the cell surface by the MHC molecule, the T-cell then recognizes this antigen through its TCR. In order for antigen-specific T-cells to become activated and this population to expand in number, a second signal must be generated through the interaction of adhesion and co-stimulatory molecules present on the APC, e.g. B7.1 (CD80) and B7.2 (CD86), with CD28 on the surface of T-cells. The third signal involves production of cytokines by the APCs, which directs the differentiation of the activated T-cell into specific effector T-cell subsets. The generation of particular cytokines by the APC is critical for determining the appropriate type of T-cell needed to carry out the required immune response (Figure 49.6). Generally, all three phases of T-cell activation are required to mount a robust T-cell response to an invading antigen (Janeway *et al.*, 2005).

4.1.2. T-cell subsets

The T-cells develop in the thymus; they can be divided into two major classes depending on the expression of a specific cell surface marker, CD4 or CD8. These cells differ in the class of MHC molecule that they can recognize on the surface of the APC. The $CD4^+$ T-cells are activated through presentation of antigen by APCs in the context of MHC class II molecules.

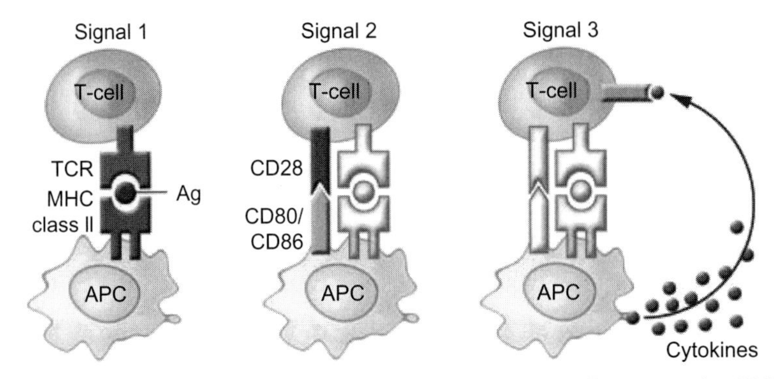

FIGURE 49.5 Activation of T-cells by antigen-presenting cells (APCs). The T-cell activation by APCs involves three components: Signal 1 involves the presentation of antigen, peptide or polysaccharide, in the context of MHC class II molecules, which is recognized by the antigen-specific T-cell receptor (TCR) on the T-cell. Signal 2 involves the stabilization of this interaction through adhesion molecules and the generation of signals via co-stimulatory molecules present on the surface of APCs and T-cells. The CD80/CD86 on APCs interact with their receptor, CD28, on T-cells. Signal 3 is produced by the secretion of cytokines by APCs, which signal via cytokine receptors on T-cells, in order to polarize them toward a specific T-cell phenotype. Adapted, with permission from Gutcher and Becher (2007) APC-derived cytokines and T-cell polarization in autoimmune inflammation. J. Clin. Invest. 117, 1119–1127.

Additionally, $CD4^+$ T-cells are known as T-helper cells; they function primarily to "help" fight against infection by differentiation into many different subtypes, which then contribute to activation of other cell types, e.g. macrophages and B-cells. Once the T-helper cell becomes activated, they divide rapidly and secrete small proteins called cytokines that regulate or "help" the immune response. Depending upon the signals received by these cells, they can differentiate into either effector or regulatory T-cells, which secrete different cytokines to enhance or control aspects of the immune response (see Figure 49.6). Different classes of T-cell can be identified based on their expression of unique intracellular signalling molecules that are responsible for driving the specific downstream effector functions of the T-cell subsets.

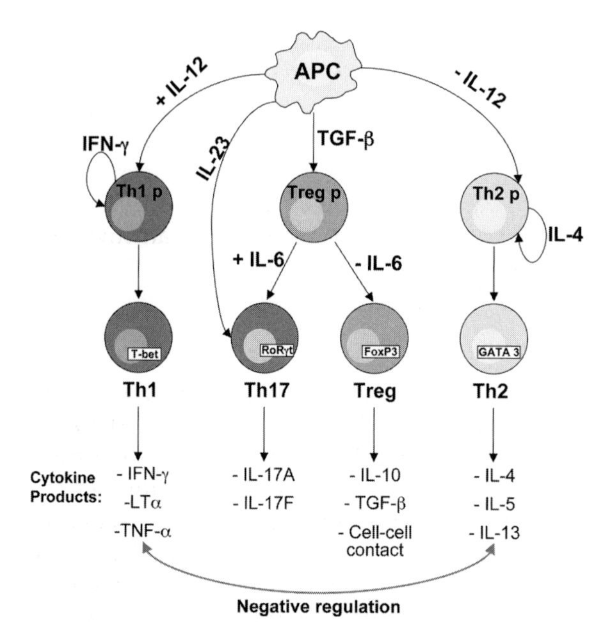

FIGURE 49.6 Differentiation of T-cell lineages. The antigen-presenting cell- (APC-) derived cytokines guide the differentiation of naive T-cells into an effector or regulatory T-cell subtype. Secretion of interleukin- (IL-) 12 by the APC acts directly on T-helper-1 (Th1-) precursor (Th1p) cells inducing interferon-gamma (IFN-γ) production, which then further expands a Th1-cell phenotype. The tissue growth factor-beta (TGF-β) secretion can polarize naive cells toward a regulatory phenotype (Treg) or a Th17-cell phenotype, depending on the cytokine environment. Secretion of TGF-β alone by APCs supports Treg formation from Treg precursor cells (Treg-p cells). However, the additional presence of IL-6 results in the production of Th17-cells. The pathogenic, APC-derived cytokine IL-23 is critical for the maintenance and survival of these Th17-cells. The interaction of APCs and T-helper-2 (Th2-) Th2-precursor (Th2p) cells in the absence of IL-12 induces the production of the Th2-cytokine, IL-4, by T-cells, which acts in an autocrine fashion to expand Th2-cells. All T-cell subsets express unique intracellular transcription factors (T-bet, RoRγT, FoxP3, GATA3) which, specifically, mediates their downstream effector functions and also facilitates the characterization of the individual T-cell phenotypes. Abbreviations: interleukin- (IL-); IFN-γ, interferon-gamma; TNF-α, tumour necrosis factor-alpha.

The $CD8^+$ T-cells are known as cytotoxic T-cells and they function primarily to target cells that have been infected with either viruses or intracellular bacteria. Also, $CD8^+$ T-cells are capable of recognizing tumour cells. Cytotoxic T-cells target infected cells and kill them, thereby eliminating the threat from the intracellular pathogen. The $CD8^+$ T-cells recognize their target infected cell through the presentation of foreign antigens in the context of the MHC class I molecule.

4.2. Antigen-mediated T-cell activation by ZPSs – adaptive immunomodulation

4.2.1. *ZPSs as T-cell antigens*

Activation of T-cells through the MHC II pathways was traditionally thought to be a method of T-cell activation that was exclusive to protein antigens. Advances in the field of CPS biology, however, have altered this "protein-only" paradigm to include zwitterionic polysaccharides. Seminal work on PSA expressed by *B. fragilis* identified a mechanism by which this carbohydrate molecule can be taken up by APCs, processed by these cells for eventual loading onto the MHC class II molecule on the surface of the cell and then recognition by the αβTCR-expressing T-cell (Cobb *et al.*, 2004). An interaction between ZPS and MHC class I molecules has not as yet been shown (Kalka-Moll *et al.*, 2002; Cobb *et al.*, 2004).

4.2.2. *Uptake of ZPSs by APC*

The first stage in the antigen-processing pathway requires uptake of foreign antigen into the endosome of an APC. To date, the mechanism by which ZPSs are taken up by APCs remains to be fully elucidated. The process of PSA uptake by APCs certainly involves cytoskeletal rearrangement and is not thought to be receptor-mediated, nevertheless, the specificity of this endocytosis is unclear (Cobb *et al.*, 2004). Given the relatively small size of PSA, it may be taken up by the cells in a process known as macropinocytosis (M. Kazmierczak and D. Kasper, unpublished data), otherwise known as high capacity fluid phase uptake, in a similar manner to the uptake of alternative non-zwitterionic carbohydrate molecules. However, the lack of ability of these other molecules to induce subsequent T-cell activation suggests that the ability to be processed and presented by MHC class II molecules is a feature exclusive to ZPSs.

4.2.3. *Endosomal processing of ZPSs*

Following endocytosis, the PSA must first be depolymerized, or processed into a much smaller molecular mass species, before it is capable of interaction with the protein components of the endosomal pathway and presentation on the surface on the APC by the MHC molecule. Upon uptake of foreign antigen, one of the earliest induced events within the APC is a process known as oxidative or respiratory burst. Oxidative burst is characterized by the generation of numerous oxidizing molecules stemming from two pathways, the superoxide and the nitric oxide pathways, which lead to the generation of reactive oxygen species (ROS) and reactive nitrogen species (RNS), respectively (Forman and Torres, 2001). The PSA molecule can induce nitric oxide (NO) production within the APC and studies have determined that RNS, including NO and nitroxyl, are the major reactive species capable of processing PSA through a deamination reaction (Lindberg *et al.*, 1980; Duan *et al.*, 2008).

Once ZPS antigens are processed, they tend to follow a vesicular path within the APC that is virtually indistinguishable from conventional protein antigens (Cobb *et al.*, 2004) (Figure 49.7). Endosomal vesicles, that carry processed ZPS, fuse with enzyme-loaded lysosomes to form the endo/lysosomal compartment. During the traditional processing of protein antigens, the delivery of these enzymes is crucial for processing of antigenic proteins into peptides (Watts, 1997), nonetheless, as discussed in the context of ZPSs, oxidative processing has already depolymerized the ZPS molecule. During this time the MHC class II proteins are being produced in the endoplasmic reticulum (ER). The MHC II proteins are associated with a homotrimer called the invariant chain (Li), which occupies the peptide-binding cleft on the MHC class II molecule, and directs the MHC II complex from the ER to the endo/lysosome containing the ZPS antigens. Upon fusion of these proteins with the endo/lysosome, the MHC class II compartment (MIIC) forms. Within this compartment a number of events take place before final presentation of the antigen by the MHC II molecule. First, the pH is reduced to create an acidic environment. This drop in pH triggers activation of the lysosomal enzymes that cleave il away from the MHC II binding cleft, however, an il-associated peptide, the class II-associated peptide (CLIP), remains in the binding cleft to act as a stabilizer. Next, a protein known as HLA-DM catalyses the exchange of CLIP with the foreign antigen present in the MIIC (Watts, 1997). Studies have shown a direct interaction between PSA and HLA-DM, suggesting that it is responsible for facilitating the exchange of processed ZPS into the MHC II binding cleft (Cobb *et al.*, 2004). The processed ZPS molecule is now tightly bound to the MHC II molecule and it traffics to the surface of the cell for presentation to the T-cell.

The MHC class II molecule is composed of an α- and a β-chain; in humans these chains are encoded by three pairs of genes called HLA-DR, -DP, -DQ. The polygenic nature of the MHC II

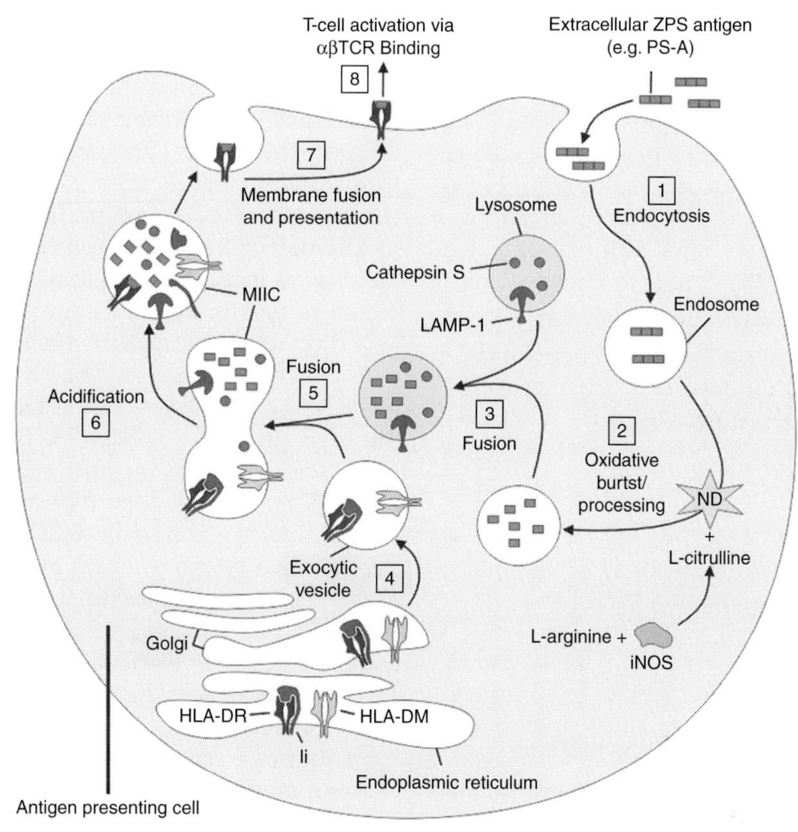

FIGURE 49.7 The major histocompatibility complex (MHC) II pathway model for ZPS antigens. [1] Extracellular antigens are endocytosed for presentation through a cytoskeleton-dependent mechanism. [2] Cells undergo an oxidative burst, including the production of nitric oxide (NO). The NO, and potentially other oxidants, process the antigen to low-molecular mass polysaccharides. [3] Endosomes fuse with resident lysosomes. [4] The MHC proteins HLA-DR, HLA-DM, and Ii are assembled in the endoplasmic recticulum, are transported through the Golgi apparatus and then budded into exocytic vesicles. [5] Antigen-rich endo/lysosomes fuse with exocytic vesicles, creating the MHC class II (MIIC) vesicle carrying HLA-DR, HLA-DM, LAMP-1 and antigen. [6] The MIIC acidification activates proteases responsible for processing protein antigens and cleavage of Ii, thus permitting the HLA-DM-mediated antigen exchange of the class II-associated peptide (CLIP). [7] Processed antigens are loaded onto MHC with the help of HLA-DM and then shuttled to the cell surface for recognition by βTCRs. [8] ZPS molecules mediate APC/T-cell engagement through αβTCR recognition. Adapted, with permission from Cobb *et al.* (2004) Polysaccharide processing and presentation by the MHC II pathway. Cell 117, 677–687. (See colour plates section.)

molecule means that every individual possesses a set of MHC II molecules with a different range of peptide-binding and polysaccharide-binding specificities. Studies involving ZPSs from *B. fragilis* (i.e. PSA) and *S. aureus* (i.e. CP8) have demonstrated that both ZPSs are presented in the context of the MHC II molecule HLA-DR, thus suggesting a specificity between HLA-DR and ZPSs (Kalka-Moll *et al.*, 2002; Duan *et al.*, 2008; McLoughlin *et al.*, 2008).

4.2.4. *Recognition of ZPSs by* αβ*TCR*$^+$ *T-cells*

Once the ZPS molecules are presented by MHC class II molecules, they are recognized by αβTCR-expressing T-cells and the T-cell becomes activated. To date, it is unclear what is the specificity of this interaction; it has been shown that T-cells that respond to ZPSs express a broad range of Vβ receptors and a T-cell activated by one ZPS can cross-react with an alternative ZPS, but not conventional protein antigens (Stingele *et al.*, 2004). Given the polyclonal nature of the T-cell response to ZPS, this argues in favour of recognition of a charge motif, as opposed to structural motifs that are specific to individual organisms.

4.2.5. The T-cell co-stimulation

Following presentation of antigen to the T-cell, a second component of APC-induced T-cell activation requires non-specific co-stimulatory signals through the interaction of co-stimulatory molecules on the surface of both the T-cell and the APC. The CD28 expressed on the T-cell interacts with its ligands, B7.1 (CD80) and B7.2 (CD86), on APCs to control the T-cell response to a variety of antigens. This co-stimulatory signal is maintained through a bi-directional activation signal transmitted through the interaction between CD40 ligand on the T-cell and CD40 receptor on the APC. Studies into the co-stimulatory mechanisms involved in ZPS-induced T-cell activation have revealed that inhibition of B7.2 and CD40L, but not B7.1, specifically block ZPS-mediated T-cell activation (Stephen *et al.*, 2005). In addition, surface expression of B7.2 and B7.1, but not of CD40, was enhanced by treatment of APCs with ZPS. Thus, the results demonstrate that the cellular immune response to ZPS requires co-stimulation via B7.2 and CD40 on activated APCs.

4.3. Cytokine-mediated T-cell activation by ZPSs – innate immunomodulation

4.3.1. Toll-like receptor signalling

Toll-like receptors (TLRs) play a critical role in innate immunity by sensing microbial pathogens (see Chapter 31) and initiating a response to clear them (Takeda *et al.*, 2004) (see Figure 49.7). These receptors are expressed on all the critical cellular components of the innate immune system and can recognize specific regions on the surface of invading pathogens (Figure 49.8). The surface of microorganisms typically expresses

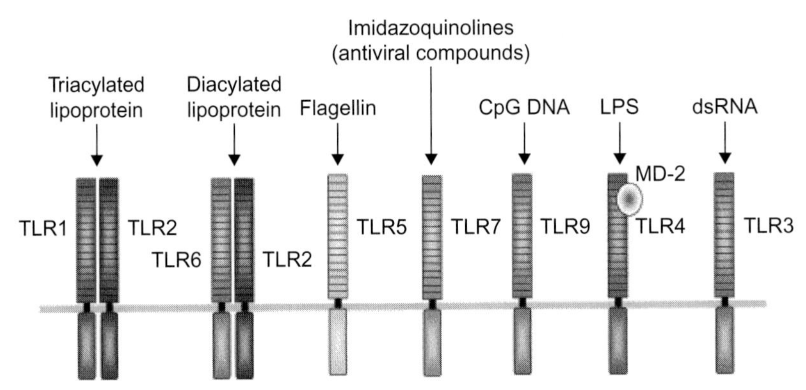

FIGURE 49.8 Toll-like receptors (TLRs) and their ligands. The TLR family of receptors is capable of distinguishing between specific patterns of microbial components: TLR-1–TLR-7 and TLR-9 have been characterized to recognize microbial components; TLR-2 is essential for the recognition of microbial lipopeptides; TLR-1 and TLR-6 associate with TLR-2 and discriminate subtle differences between triacyl- and diacyl-lipopeptides, respectively; TLR-4 recognizes lipopolysaccharide (LPS); TLR-9 is the CpG DNA receptor; TLR-3 is implicated in the recognition of viral double-stranded RNA (dsRNA); and TLR-5 is a receptor for flagellin. Adapted, with permission, from Takeda and Akira (2004) TLR signalling pathways. Semin. Immunol. 16, 3–9.

repeating patterns of molecular structure known as pathogen-associated molecular patterns (PAMPs) (Takeda *et al.*, 2004). Examples of these PAMPs are endotoxin or lipopolysaccharide (LPS), peptidoglycan and other bacterial cell-wall components including flagellin, bacterial DNA and viral double-stranded RNA. At least 11 TLRs and their distinct ligands have been identified so far in mammals (Pasare and Medzhitov, 2004). Among these receptors are TLR-4, which is activated by most bacterial LPSs (Miller *et al.*, 2005) and TLR-2, which is activated by bacterial products such as peptidoglycans, lipoproteins, mycobacterial lipoarabinomannan and what has been suggested as atypical LPSs, such as those found in *Porphyromonas* (Takeda *et al.*, 2004; Erridge *et al.*, 2004). Activation of TLR signalling can induce transcription of many immunologically important genes, such as those encoding MHC molecules and cytokines. It is through activation of these molecules that TLRs coordinate innate and adaptive immunity and prime naive T-cells ultimately to facilitate the elimination of pathogens (Iwasaki and Medzhitov, 2004; Gelman *et al.*, 2004; Pasare and Medzhitov, 2004).

4.3.2. *The TLR activation by ZPSs*

Many of the known TLR ligands contain carbohydrate moieties, however, it appears that the non-carbohydrate portion of these molecules (e.g. lipid A in LPS) is usually critical for TLR ligation and activation (Alexander and Rietschel, 2001). Until recently, the potential role of pure carbohydrates as ligands for TLRs was largely unknown. Having characterized the mechanisms by which ZPSs and, in particular PSA, could activate T-cell-driven adaptive immunity, the question remained as to whether this polysaccharide also induces an innate response directly to mediate or enhance antigen-mediated T-cell activation. One of the major mechanisms by which cells of the innate immune response can directly impact upon T-cell activation is through the production of specific cytokines. Importantly, cytokines are low-molecular mass proteins that act as signalling molecules within the immune system, transferring activation or inhibition signals from one cell type to another. Cytokines encompass a complex array of protein families all of which have specific and, in some cases, redundant functions within the immune system. In terms of T-cell activation and differentiation, cytokine signals can dictate the phenotypes of activated T-cells (see Figure 49.6).

Studies have demonstrated that PSA is capable of activating a TLR-2 signalling cascade in APCs and, hence, leading to several important biological events that impact directly upon T-cell activation. Activation of TLR-2 by PSA facilitates the antigen presentation process by increasing activation of NO production within the APC, increasing the expression of MHC class II molecules by these cells and enhancing expression of the co-stimulatory molecule CD86 (Wang *et al.*, 2006). In addition, TLR activation by PSA results in production of the cytokine interleukin-(IL-) 12 by the APC. Of note, IL-12 is known as a T-cell-stimulating factor and, specifically, it is involved in the differentiation of naive T-cells into a specific subset of T-cells known as T-helper-1 (Th1-) cells. Furthermore, IL-12 signals through the IL-12R receptor protein expressed on the surface of the T-cell. Experiments have shown that a critical stage in Th1-cell differentiation from naive $CD4^+$ T-cells involves the transcription factor, T-bet (i.e. T-box expressed in T-cells). The T-bet molecule induces the expression of the IL-12 receptor on the surface of the T-cell, thereby facilitating IL-12 binding, and subsequent Th1-cell development (Robinson and O'Garra, 2002). These Th1-cells, in turn, produce high levels of another cytokine known as IFN-γ, which has many functions within both the innate and adaptive immune response and is involved in mounting an efficient host response against microbial attack (Schroder *et al.*, 2004).

The identification of a critical role for TLR-2 signalling in PSA-mediated activation of APCs adds an essential piece of information to our understanding of how this polysaccharide antigen stimulates the host immune response. This effect appears to be partially charge-dependent because elimination of positively charged amine groups on PSA, via *N*-acetylation, resulted in a reduced ability of this polysaccharide to activate TLR-2–mediated pathways (Wang *et al.*, 2006). Interestingly, however, is that the TLR-2–mediated response of the APCs was not solely dependent on zwitterionic charge; Sp1, failed to activate TLR-2. Therefore, the zwitterionic charge motif of PSA is necessary, but not sufficient, to induce TLR-2-mediated APC activation and specific structure–function studies of PSA–TLR-2 interaction are needed.

4.4. Interplay between innate and adaptive immune responses to ZPSs

It is clear that PSA is capable of activating CD4$^+$ T-cells through an integrated mechanism

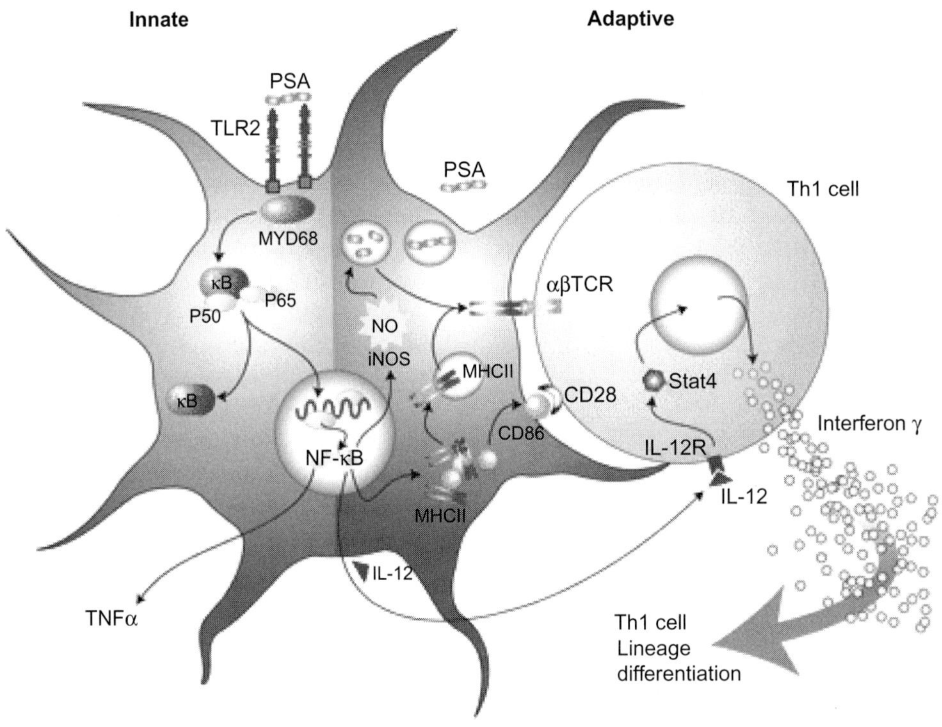

FIGURE 49.9 Interplay between innate and adaptive immune responses to PSA. The innate response begins with Toll-like receptor-2 (TLR2) recognition of PSA and the subsequent stimulation of the MyD88-mediated pathway inside the antigen-presenting cell (APC). Nuclear factor- (NF-) κB is translocated into the nucleus, stimulating the transcription of several important genes, including tumour necrosis factor-alpha (TNF-α) and interleukin- (IL-) 12. Meanwhile, NF-κB translocation also leads to enhanced inducible nitric oxide synthase (iNOS) expression, as well as upregulation of major histocompatibility complex class II (MHC II) molecules and CD86, thus facilitating PSA processing and presentation by MHC II proteins. Presentation of PSA on the cell surface by MHC class II molecules leads to adaptive CD4$^+$ T-cell activation and T-cell secretion of the cytokine interferon-gamma (IFN-γ). The IFN-γ production by the T-cell is optimized by TLR-2-mediated IL-12 production by the dendritic cell. Adapted, with permission, from Wang *et al.* (2006) A bacterial carbohydrate links innate and adaptive responses through Toll-like receptor 2. J. Exp. Med. 203, 2853–2863.

that involves considerable interplay of the innate and adaptive arms of the immune system (Figure 49.9). In the innate portion of this response, PSA stimulates APCs through a TLR-2–dependent mechanism. The TLR-2 activation then results in NO production, which is critical for PSA processing and presentation by MHC class II molecules and, thus, for initiation of the adaptive immune response. The PSA-mediated TLR-2 signalling also upregulates MHC class II molecule and co-stimulatory molecule expression on the surface of APCs, further linking the innate and adaptive responses to this antigen. In the adaptive arm of the immune response, antigen presentation by MHC class II molecules and recognition by TCR on the $CD4^+$ T-cell (along with co-stimulation), results in activation of T-cell proliferation and production of cytokines such as IFN-γ. This process is enhanced by the production of the cytokine IL-12 by DCs, which occurs as a consequence of TLR-2 stimulation and establishes the differentiation of a Th1 T-cell phenotype.

5. BIOLOGICAL IMPACT OF ZPS-INDUCED T-CELL ACTIVATION

5.1. Immune development

All mammals are born sterile and are subsequently colonized by a myriad of microorganisms. The lower gastrointestinal (GI) tract represents the major microbe-rich environment in the host. The GI tract harbours an astonishing number ($>10^{14}$) of microorganisms of $\approx$1000 species (Gill *et al.*, 2006). Moreover, it has been proposed that these commensal organisms are actually beneficial to their host. These microorganisms play important roles in the acquisition and absorption of nutrients from otherwise indigestible foods (Falk *et al.*, 1998) and, in addition, they have been shown to have potential immunomodulatory roles in the development of the gut-associated immune system (Hooper and Gordon, 2001). Commensal bacteria with known benefits to human health are known as probiotics and this is currently an area of very active research.

B. fragilis is a major component of the gut microflora and is found in the GI tract of almost all individuals. The previously documented ability of PSA expressed by *B. fragilis* to modulate T-cell activation and promote Th1-cytokine production, suggested the possibility that this organism may play a pivotal role in modulating the local gut-associated immune response. To investigate this concept, germ-free mice were employed as an immunological tool. Germ-free mice are animals that are born and raised in a sterile isolator, specifically, they are devoid of colonization by any micoorganism. Germ-free mice appear to have an underdeveloped immune system compared to normal mice. In addition, the proportions of $CD4^+$ T-cells is lower in germ-free mice (Dobber *et al.*, 1992; Mazmanian *et al.*, 2005) and the ultrastructural development of their lymphoid organs is altered, with smaller and more fragmented lymphoid follicles, compared to normal conventionally colonized mice (Mazmanian *et al.*, 2005). Notably, individual microbial species can be introduced into these mice to investigate the effects of a single organism on an aspect of the host immune response or development.

Therefore, to assess the contribution of a single commensal organism on the development of the host immune system, *B. fragilis* was introduced into germ-free mice and various immunological parameters assessed. Colonization of germ-free mice with *B. fragilis* was sufficient to correct the $CD4^+$ T-cell defect, as well as the defect in lymphoid follicle development. Importantly, colonization with a mutant strain of *B. fragilis* that did not produce PSA was unable to bring about these effects (Mazmanian *et al.*, 2005). When purified PSA was administered to the germ-free mice this polysaccharide alone was sufficient to correct the CD4 imbalance back to that seen in the normal conventionally colonized mice (Mazmanian *et al.*, 2005).

A further defect observed in germ-free mice is that their T-cells are skewed towards a T-helper-2 (Th2-) cell phenotype. The T-cells from these mice will naturally produce more of the Th2-cytokine IL-4, compared to T-cells from normal conventionally colonized mice, which have a normal balance of Th1/Th2 T-cells, as Th1- and Th2-cytokines can counter-regulate each others production (see Figure 49.6). If this balance is altered in any way, it can lead to pathogenesis, i.e. an over-production in Th2-cytokines is one of the underlying aetiologies of asthma and other allergic diseases (Umetsu and Dekruyff, 2006), whereas an over-production in Th1-cytokines can contribute to autoimmune disorders, such as type 1 diabetes (Yoon and Jun, 2005). Based on the previously generated data demonstrating the ability of PSA to induce IFN-γ production by T-cells, it was not surprising that colonization of germ-free mice with *B. fragilis* can promote a Th1 response and correct the Th1/Th2 imbalance seen in these mice (Mazmanian *et al.*, 2005). Once again, colonization of mice with a mutant strain of *B. fragilis* that lacked PSA expression was not capable of bringing about these effects (Mazmanian *et al.*, 2005). The PSA molecule expressed by *B. fragilis* during colonization stimulates IFN-γ production, which is critical in maintaining an appropriate Th1/Th2 balance in the host (Mazmanian and Kasper, 2006) (Figure 49.10).

5.2. ZPS protection against surgical fibrosis

Early work in the field of ZPS immunomodulation identified a protective role for ZPSs in the treatment of intra-abdominal abscess. Initially, investigations identified the mechanisms behind this protection to be T-cell-mediated (Tzianabos *et al.*, 1995, 2001; Tzianabos and Kasper, 2002). As the field evolved, this T-cell-mediated protection model became better understood, with the characterization of the ZPS antigen processing pathway, their interaction with MHC class II molecules and their TLR-2-mediated activation of T-cell cytokine production (Cobb *et al.*, 2004; Wang *et al.*, 2006).

To investigate further the protective properties of ZPSs, a model of post-surgical fibrosis was employed (Ruiz-Perez *et al.*, 2005). Adhesions are a common and severe fibrotic host response to trauma in the peritoneal or pelvic cavities that constitute a major clinical problem (Ellis, 1997). In response to intra-abdominal or pelvic injury, overwhelming inflammation and fibrin deposition combine to create dense fibrous bands that can lead to major organ failure, bowel obstruction and infertility in women (Chegini, 2002). The development of experimental adhesions, like abscess formation, is primarily a Th1, or perhaps Th17, T-cell-driven pathology (Chung *et al.*, 2002, 2003).

The ZPS molecule, Sp1, was capable of inhibiting adhesion formation in a mouse model of post-surgical adhesions (Ruiz-Perez *et al.*, 2005). The mechanism behind this effect was attributed to the ability of this molecule to activate a specific subset of $CD4^+$ regulatory T-cells (see Figure 49.5) known as $CD4^+CD45RB^{low}$ cells. When these cells were activated by Sp1 they produced the cytokine IL-10, which is an anti-inflammatory cytokine. In addition, the transfer of $CD4^+CD45RB^{low}$ T-cells that had been stimulated *in vitro* with Sp1, were capable of reducing adhesion formation in this model, thereby confirming a role for these regulatory T-cells in the protection against adhesion formation. It was, therefore, concluded that ZPSs are capable of activating not only Th1-cells, but can also activate a subset of regulatory T-cells and their production of the cytokine IL-10 mediates the anti-inflammatory protective effects of ZPSs.

5.3. Protection against colitis

The anti-inflammatory immunoprotective role of ZPS was further characterized under biologically

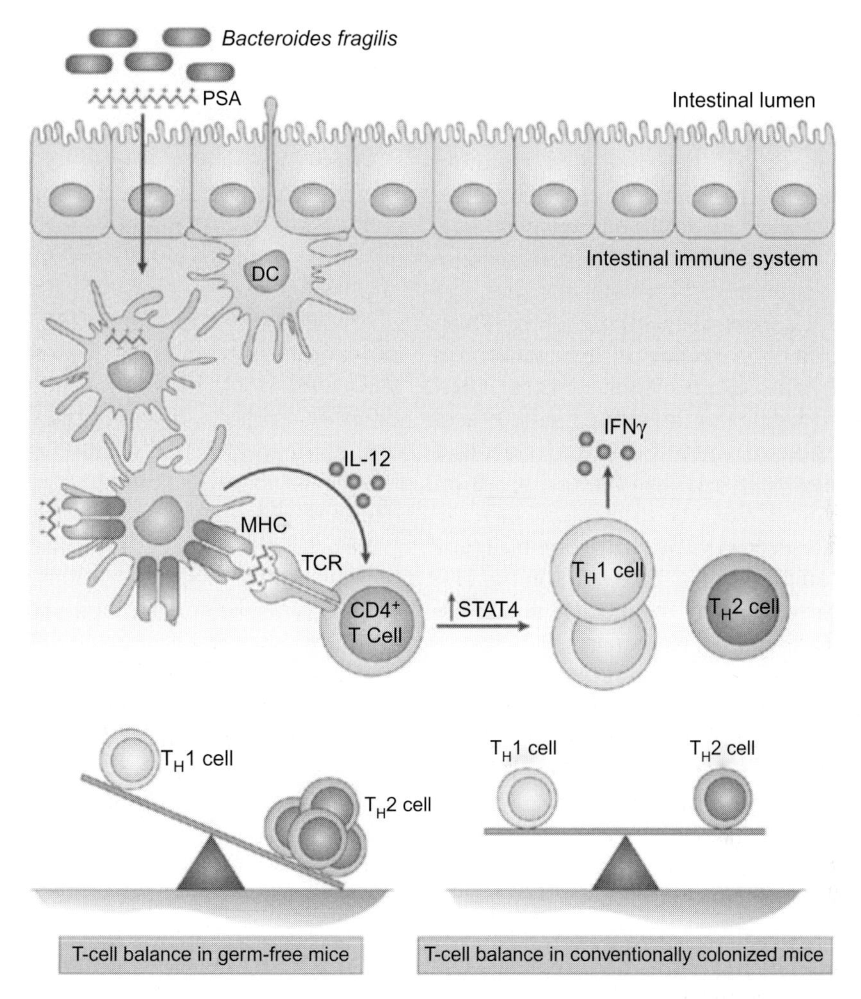

FIGURE 49.10 The PSA molecule from *B. fragilis* can maintain a Th1/Th2 balance in the host. *B. fragilis* is an abundant member of the colonic microflora of all mammals and produces the immunomodulatory polysaccharide, PSA. This molecule appears to be sampled by antigen-presenting cells, i.e. dendritic cells (DCs) of the gut-associated lymphoid tissue and these DCs then migrate to the mesenteric lymph nodes and initiate T-cell responses. The PSA is processed in the DCs and is presented to the T-cell receptor (TCR) of $CD4^+$ T-cells in the context of major histocompatibility complex (MHC) molecules. The immune system of germ-free mice is highly skewed towards a T-helper-2 (T_H2-) cell phenotype and colonization with PSA-producing *B. fragilis* corrects this immune defect. This correction is achieved through the PSA-induced secretion of interleukin- (IL-) 12 by DCs, leading to activation of T-helper-1 (T_H1-) cells and generation of interferon-gamma (IFN-γ). This process corrects the cytokine imbalance that is found in the absence of bacterial colonization. Adapted, with permission, from Mazmanian and Kasper (2006) The love-hate relationship between bacterial polysaccharides and the host immune system. Nat. Rev. Immunol. 6, 849–858.

relevant conditions by investigating a role for PSA from *B. fragilis* in protection against inflammatory bowel disease (Mazmanian *et al.*, 2008). In inflammatory bowel diseases, such as Crohn's disease and ulcerative colitis, commensal bacteria harboured within mammalian intestines are the targets of inflammatory responses (Poxton *et al.*, 1997; Sartor, 2006). Inflammatory responses are apparently directed towards specific subsets of commensal organisms that have pathogenic potential, but are not typically infectious pathogenic agents. These unwarranted inflammatory responses lead to extensive tissue damage in the gut due to the high levels of proinflammatory cytokines being produced (Bergenholt, 2000).

Using a murine model of experimental colitis, it was demonstrated that colonization with *B. fragilis*-expressing PSA was capable of decreasing the inflammatory response and associated pathologies (Mazmanian *et al.*, 2008). The same response was not seen when the mice were colonized with a PSA-negative mutant strain of *B. fragilis.* Furthermore, purified PSA administered orally to these mice also decreased disease pathogenesis (Mazmanian *et al.*, 2008). The mechanism by which PSA reduced disease in this model was attributed to its ability to decrease local inflammatory cytokine production and, in particular, the cytokine IL-17. Pathogenic Th17-producing T-cells can mediate colitis through the induction of further proinflammatory cytokine production (Bamias and Cominelli, 2007). In line with what was seen in the surgical adhesion model (Ruiz-Perez *et al.*, 2005), the immunoprotective role of PSA in the colitis model was also associated with its ability to induce IL-10. The anti-inflammatory cytokine IL-10 can be produced by many cell types, however, $CD4^+$ T-cells that express IL-10 have been shown to have immune-suppressive activities that inhibit inflammation during experimental colitis (Groux *et al.*, 1997). Similarly, in this model PSA protects against the development of colitis through a functional requirement for IL-10-producing $CD4^+$ T-cells (Mazmanian *et al.*, 2008).

5.4. The ZPSs' pro- versus anti-inflammatory effects

Together, these studies highlight the ability of ZPSs to modulate the immune system in different ways depending upon the challenges it is facing. The commensal nature of PSA in the gut drives a Th1-mediated response to maintain a healthy Th1/Th2 balance, thereby potentially protecting the host from the development of Th2-mediated allergic disease (Mazmanian *et al.*, 2005). Nevertheless, under pathological conditions, e.g. inflammatory bowel disease or fibrotic disease, it appears that ZPS can also facilitate regulatory cytokine production to dampen the host's immune response. In both cases, the ability of these molecules to activate T-cells is obviously paramount.

6. POTENTIAL FOR ZPSs AS THERAPEUTICS

6.1. The hygiene hypothesis

In the past two decades it has been observed that a reduction in exposure to infectious bacterial agents in early life, as a consequence of antibiotic use, vaccination and improved sanitation, results in increased incidences of asthma and allergic disease in later life. This phenomenon has become known as the "hygiene hypothesis" (Strachan, 1989; Wills-Karp *et al.*, 2001). Several epidemiological studies have shown that the composition of the gut microflora differs between atopic, i.e. predisposed to allergic disease, and non-atopic individuals (Bjorksten, 1999). In an example of such a study, analysis of intestinal bacteria showed that children with allergies had lower levels of *Bacteroides* spp. present in their gut flora than those children that did not develop allergies (Alm *et al.*, 2002).

The underlying immunological defect that occurs in asthmatic and allergic individuals is an over-production of Th2-cell responses (Umetsu and Dekruyff, 2006). As we now know, the expression of PSA by the commensal bacterium *B. fragilis* can correct the systemic Th2-cell bias that is found in the absence of bacterial colonization of germ-free mice, through induction of Th1-cytokines (Mazmanian *et al.*, 2005). Therefore, it is plausible that, in the developed world, widespread antibiotic use, together with improved sanitation, can decrease the levels of commensal bacteria, like *B. fragilis* in the gut, at an essential time in immune development. This would result in decreased exposure of the host to immunomodulatory molecules, such as PSA, which may be critical for development of an appropriate Th1/Th2 immune balance and protection against the development of allergic disease in later life.

In addition, PSA can mediate protection against the damaging proinflammatory effects occurring during experimental colitis by promoting local production of the anti-inflammatory cytokine IL-10 in the gut (Mazmanian *et al.*, 2008). Further studies are required to reconcile the role of PSA-induced IL-10 production and how this impacts upon the balance of Th1- and Th2-cytokines, however, one possibility is that PSA induces the differentiation of $CD4^+$ T-cells that produce both IFN and IL-10 (Stock *et al.*, 2004). It is clear that PSA is a potent immunomodulatory molecule that, if present in the gut, can contribute to the development of a healthy and balanced immune system. Further work is required to identify the precise molecular interactions that occur between ZPSs and the immune system during development. Notwithstanding, the idea that PSA, a normal component of the gut microbial environment, is capable of modulating the immune system of the host in such a way as to help protect it against future autoimmune and allergic disease, opens up exciting avenues for the development of new types of "probiotic therapy" involving ZPSs.

6.2. Generation of ZPS molecules

Polysaccharides are currently used in conjugate vaccines against infections caused by several organisms including *Haemophilus influenzae* type b, *S. pneumoniae* and *Neisseria meningitidis*. The polysaccharide from the pathogenic organism is conjugated to a carrier protein, which is required to elicit the desired T-cell activation and subsequent immunological memory (see Figure 49.1B). Developments in the field over the past two decades have now altered the way in which we think about polysaccharides and their interactions with the immune system. Characterization of ZPSs as potent immunomodulatory agents that can activate T-cells and potentially protect against various autoimmune and inflammatory responses, has been a seminal advancement of the field.

The vast majority of naturally occurring bacterial CPSs are either neutral or possess a net negative charge. To harness the beneficial immunomodulatory effects of ZPSs, it may be necessary to generate them either synthetically or, an alternative approach would be to target non-zwitterionic polysaccharides and chemically modify them to introduce a zwitterionic charge motif; the naturally anionic polysaccharides of group B *Streptococcus* were modulated in this way (Gallorini *et al.*, 2007). The CPS from group B *Streptococcus* serotype III was modified to become zwitterionic by means of a series of chemical modifications that either converted the *N*-acetyl groups to free amino groups (ZPS1) or through a periodate oxidation to generate an aldehyde group, which subsequently was converted to an amino group by a reductive amination reaction. All amino groups were then converted to tertiary amines that retain their positive charges (ZPS2). Thus, in each repeating unit, ZPS1 molecules contain two positive and one negative charge, whereas ZPS2 contains a balanced motif of one charge each per repeating unit (Figure 49.11).

Both of these ZPS molecules were shown to have similar biological activities to those described for the naturally occurring ZPS, PSA. They had the ability to activate TLR-2-mediated cytokine production by APCs, they increased the expression of MHC class II molecules by the APC and they could facilitate the APC-mediated activation of T-cell proliferation. The immunomodulatory effects of these molecules disappeared when the positive charge was chemically removed and the zwitterionic motif destroyed. These studies signal the start of an exciting new era in polysaccharide research. If naturally occurring polysaccharide molecules can be engineered to become zwitterionic immunomodulatory compounds, this will expand the repertoire of disease models in which we can study the effects of these molecules and, ultimately, advance their development as immunomodulatory therapies.

7. CONCLUSIONS

The ZPSs contain both a positive and a negative charge in their repeating subunit and it is

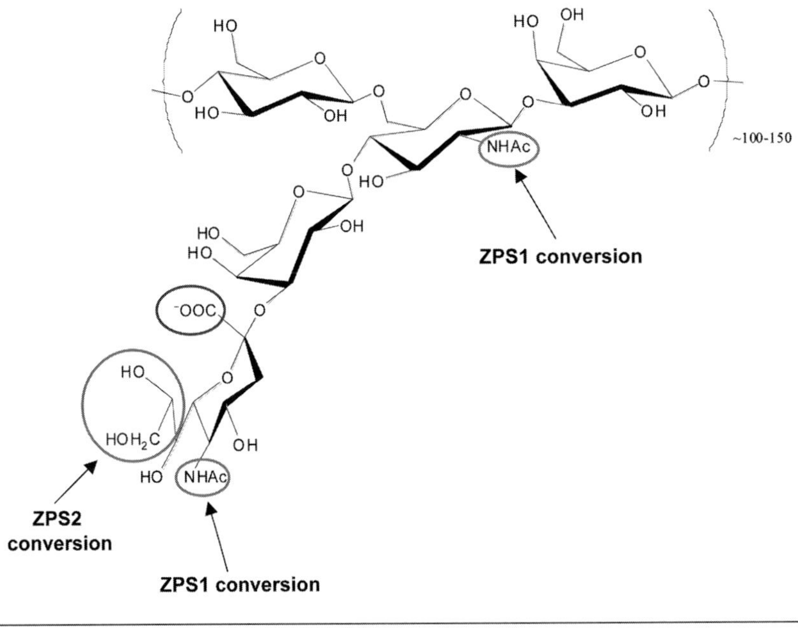

FIGURE 49.11 Chemical modifications of native group B *Streptococcus* serotype III capsular polysaccharide (CPS) to generate a ZPS motif. De-*N*-acetylation by basic treatment leads to ZPS1 and periodate oxidation and reductive amination leads to ZPS2 of one repeating unit of group B *Streptococcus* serotype III PS. The additional reaction with formaldehyde to obtain tertiary amines is also shown.

this zwitterionic charge motif that is essential for immunomodulation. These molecules have the ability to be taken up by APCs, to be depolymerized or processed by these cells, before being presented on the cell surface in the context of the MHC class II molecule, whereby they are recognized by the αβTCR on the surface of the T-cell which thereby results in T-cell activation. In addition, ZPS molecules can also activate TLR-2–mediated signalling in the APC, which enhances MHC II-mediated antigen activation of the T-cell. Activation of TLR by ZPS also induces IL-12 production by the APC which, in turn, interacts with the IL-12 receptor on the surface of the T-cell promoting production of the Th1-cytokine, IFN-γ by the T-cell.

The biological consequence of this ZPS-mediated T-cell activation appears to be quite profound. Expression of the ZPS molecule PSA by the gut commensal organism *B. fragilis* has been shown to promote $CD4^+$ T-cell development and also is critical in maintaining aTh1/Th2 T-cell balance within the host. Any alterations in the Th1/Th2 balance can predispose the host to the development of asthma and allergic diseases. As well as playing a role in maintaining immune homeostasis, the ZPS of *B. fragilis* can protect against the damaging inflammation occurring during inflammatory bowel disease and the development of post-surgical adhesions. Under these conditions, PSA induces the production of the anti-inflammatory cytokine IL-10.

Therefore, it is proposed that ZPS molecules expressed by commensal organisms may protect the host against the development of allergic and autoimmune diseases in later life. Thus, the potential development of ZPS molecules as immunomodulatory therapies presents an attractive option in the treatment of immunological and infectious diseases.

Open questions and areas for further investigation are detailed in the Research Focus Box.

RESEARCH FOCUS BOX

- Detailed structure–function analyses are required to define the nature of the binding interaction between ZPSs and their target molecules, TLR-2 and MHC class II.
- It remains to be established if ZPS molecules can activate other TLRs and/or other innate signalling pathways, e.g. the NOD pathway.
- The precise molecular interactions between ZPSs and specific aspects of the immune system that promote immune development need to be further investigated.
- It will be beneficial to explore the synthetic development of ZPSs and modulation of naturally occurring CPS to become zwitterionic.
- Ultimately, it is hoped that this research will lead to the development of ZPSs as immunomodulatory therapies in the treatment of allergic and autoimmune diseases.
- In addition, ZPSs could be used in vaccine development, i.e. ZPSs can directly activate a T-cell response and also ZPSs may have an adjuvant effect as a TLR agonist.

References

Alexander, C., Rietschel, E.T., 2001. Bacterial lipopolysaccharides and innate immunity. J. Endotoxin Res. 7, 167–202.

Alm, J.S., Swartz, J., Bjorksten, B., et al., 2002. An anthroposophic lifestyle and intestinal microflora in infancy. Pediatr. Allergy Immunol. 13, 402–411.

Bamias, G., Cominelli, F., 2007. Immunopathogenesis of inflammatory bowel disease: current concepts. Curr. Opin. Gastroenterol. 23, 365–369.

Barrett, D.J., 1985. Human immune responses to polysaccharide antigens: an analysis of bacterial polysaccharide vaccines in infants. Adv. Pediatr. 32, 139–158.

Bartlett, J.C., 1978. Antimicrobial therapy in experimental intra-abdominal sepsis. J. Antimicrob. Chemother. 4, 392–394.

Baumann, H., Tzianabos, A.O., Brisson, J.R., Kasper, D.L., Jennings, H.J., 1992. Structural elucidation of two capsular polysaccharides from one strain of *Bacteroides fragilis* using high-resolution NMR spectroscopy. Biochemistry 31, 4081–4089.

Bergenholt, S., 2000. Cells and cytokines in the pathogenesis of inflammatory bowel disease: new insights from mouse T-cell transfer models. Exp. Clin. Immunogenet. 17, 115–129.

Bjorksten, B., 1999. The environmental influence on childhood asthma. Allergy 54 (Suppl. 49), 17–23.

Bowdish, D.M., Davidson, D.J., Lau, Y.E., Lee, K., Scott, M.G., Hancock, R.E., 2005. Impact of LL-37 on anti-infective immunity. J. Leukoc. Biol. 77, 451–459.

Carpenter, S., O'Neill, L.A., 2007. How important are Toll-like receptors for antimicrobial responses? Cell Microbiol. 9, 1891–1901.

Chegini, N., 2002. Peritoneal molecular environment, adhesion formation and clinical implication. Front. Biosci. 7, e91–115.

Choi, Y.H., Roehrl, M.H., Kasper, D.L., Wang, J.Y., 2002. A unique structural pattern shared by T-cell-activating and abscess-regulating zwitterionic polysaccharides. Biochemistry 41, 15144–15151.

Chung, D.R., Chitnis, T., Panzo, R.J., Kasper, D.L., Sayegh, M.H., Tzianabos, A.O., 2002. $CD4^+$ T-cells regulate surgical and postinfectious adhesion formation. J. Exp. Med. 195, 1471–1478.

Chung, D.R., Kasper, D.L., Panzo, R.J., et al., 2003. $CD4^+$ T-cells mediate abscess formation in intra-abdominal sepsis by an IL-17-dependent mechanism. J. Immunol. 170, 1958–1963.

Cobb, B.A., Kasper, D.L., 2005. Zwitterionic capsular polysaccharides: the new MHC II-dependent antigens. Cell Microbiol. 7, 1398–1403.

Cobb, B.A., Wang, Q., Tzianabos, A.O., Kasper, D.L., 2004. Polysaccharide processing and presentation by the MHC II pathway. Cell 117, 677–687.

Coutinho, A., Moller, G., Anderson, J., Bullock, W.W., 1973. In vitro activation of mouse lymphocytes in serum-free medium: effect of T and B-cell mitogens on proliferation and antibody synthesis. Eur. J. Immunol. 3, 299–306.

Dobber, R., Hertogh-Huijbregts, A., Rozing, J., Bottomly, K., Nagelkerken, L., 1992. The involvement of the intestinal microflora in the expansion of $CD4^+$ T-cells with a naïve phenotype in the periphery. Dev. Immunol. 2, 141–150.

Duan, J., Avci, F.Y., Kasper, D.L., 2008. Microbial carbohydrate depolymerization by antigen-presenting cells: deamination prior to presentation by the MHCII pathway. Proc. Natl. Acad. Sci. USA 105, 5183–5188.

Ellis, H., 1997. The clinical significance of adhesions: focus on intestinal obstruction. Eur. J. Surg. Suppl., 5–9.

Erridge, C., Pridmore, A., Eley, A., Stewart, J., Poxton, I.R., 2004. Lipopolysaccharides of *Bacteroides fragilis*, *Chlamydia trachomatis* and *Pseudomonas aeruginosa* signal via toll-like receptor 2. J. Med. Microbiol. 53, 735–740.

Falk, PG., Hooper, L.V., Midtvedt, T., Gordon, J.I., 1998. Creating and maintaining the gastrointestinal ecosystem: what we know and need to know from gnotobiology. Microbiol. Mol. Biol. Rev. 62, 1157–1170.

Finlay, B.B., Hancock, R.E., 2004. Can innate immunity be enhanced to treat microbial infections? Nat. Rev. Microbiol. 2, 497–504.

Forman, H.J., Torres, M., 2001. Redox signaling in macrophages. Mol. Aspects Med. 22, 189–216.

Gallorini, S., Berti, F., Parente, P., et al., 2007. Introduction of zwitterionic motifs into bacterial polysaccharides generates TLR2 agonists able to activate APCs. J. Immunol. 179, 8208–8215.

Gelman, A.E., Zhang, J., Choi, Y., Turka, L.A., 2004. Toll-like receptor ligands directly promote activated $CD4^+$ T-cell survival. J. Immunol. 172, 6065–6073.

Gill, S.R., Pop, M., Deboy, R.T., et al., 2006. Metagenomic analysis of the human distal gut microbiome. Science 312, 1355–1359.

Groux, H., O'Garra, A., Bigler, M., et al., 1997. A $CD4^+$ T-cell subset inhibits antigen-specific T-cell responses and prevents colitis. Nature 389, 737–742.

Gutcher, I., Becher, B., 2007. APC-derived cytokines and T-cell polarization in autoimmune inflammation. J. Clin. Invest. 117, 1119–1127.

Hoebe, K., Janssen, E., Beutler, B., 2004. The interface between innate and adaptive immunity. Nat. Immunol. 5, 971–974.

Hooper, L.V., Gordon, J.I., 2001. Commensal host-bacterial relationships in the gut. Science 292, 1115–1118.

Iwasaki, A., Medzhitov, R., 2004. Toll-like receptor control of the adaptive immune responses. Nat. Immunol. 5, 987–995.

Janeway, C.A, Travers, P., Walport, M., Shlomchik, M., 2005. Immunobiology, 6th edn. Garland Science Publishing, London.

Kalka-Moll, W.M., Tzianabos, A.O., Bryant, P.W., Niemeyer, M., Ploegh, H.L., Kasper, D.L., 2002. Zwitterionic polysaccharides stimulate T-cells by MHC class II-dependent interactions. J. Immunol. 169, 6149–6153.

Kasper, D.L., Onderdonk, A.B., Crabb, J., Bartlett, J.G., 1979. Protective efficacy of immunization with capsular antigen against experimental infection with *Bacteroides fragilis*. J. Infect. Dis. 140, 724–731.

Kelly, D.F., Moxon, E.R., Pollard, A.J., 2004. *Haemophilus influenzae* type b conjugate vaccines. Immunology 113, 163–174.

Krinos, C.M., Coyne, M.J., Weinacht, K.G., Tzianabos, A.O., Kasper, D.L., Comstock, L.E., 2001. Extensive surface diversity of a commensal microorganism by multiple DNA inversions. Nature 414, 555–558.

Kronenberg, M., Siu, G., Hood, L.E., Shastri, N., 1986. The molecular genetics of the T-cell antigen receptor and T-cell antigen recognition. Annu. Rev. Immunol. 4, 529–591.

Kuwahara, T., Yamashita, A., Hirakawa, H., et al., 2004. Genomic analysis of *Bacteroides fragilis* reveals extensive DNA inversions regulating cell surface adaptation. Proc. Natl. Acad. Sci. USA 101, 14919–14924.

Lindberg, B., Lindqvist, B., Lonngren, J., Powell, D.A., 1980. Structural studies of the capsular polysaccharide from *Streptococcus pneumoniae* type 1. Carbohydr. Res. 78, 111–117.

Lindberg, A.A., Weintraub, A., Kasper, D.L., Lonngren, J., 1982. Virulence factors in infections with *Bacteroides fragilis*: isolation and characterization of capsular polysaccharide and lipopolysaccharide. Scand. J. Infect. Dis. Suppl. 35, 45–52.

Mazmanian, S.K., Kasper, D.L., 2006. The love-hate relationship between bacterial polysaccharides and the host immune system. Nat. Rev. Immunol. 6, 849–858.

Mazmanian, S.K., Liu, C.H., Tzianabos, A.O., Kasper, D.L., 2005. An immunomodulatory molecule of symbiotic bacteria directs maturation of the host immune system. Cell 122, 107–118.

Mazmanian, S.K., Round, J.L., Kasper, D.L., 2008. A microbial symbiosis factor prevents intestinal inflammatory disease. Nature 453, 602–604.

McCormack, P.L., Scott, L.J., 2004. Interferon-β-1b: a review of its use in relapsing-remitting and secondary progressive multiple sclerosis. CNS Drugs 18, 521–546.

McGhee, J.R., Michalek, S.M., Kiyono, H., et al., 1984. Mucosal immunoregulation: environmental lipopolysaccharide and GALT T lymphocytes regulate the IgA response. Microbiol. Immunol. 28, 261–280.

McLoughlin, R.M., Lee, J.C., Kasper, D.L., Tzianabos, A.O., 2008. IFN-γ regulated chemokine production determines the outcome of Staphylococcus aureus infection. J. Immunol. 181, 1323–1332.

Miller, S.I., Ernst, R.K., Bader, M.W., 2005. LPS, TLR4 and infectious disease diversity. Nat. Rev. Microbiol. 3, 36–46.

Onderdonk, A.B., Kasper, D.L., Cisneros, R.L., Bartlett, J.G., 1977. The capsular polysaccharide of *Bacteroides fragilis* as a virulence factor: comparison of the pathogenic potential of encapsulated and unencapsulated strains. J. Infect. Dis. 136, 82–89.

Pasare, C., Medzhitov, R., 2004. Toll-like receptors: linking innate and adaptive immunity. Microbes Infect. 6, 1382–1387.

Plosker, G.L., Figgitt, D.P., 2003. Rituximab: a review of its use in non-Hodgkin's lymphoma and chronic lymphocytic leukaemia. Drugs 63, 803–843.

Poxton, I.R., Brown, R., Sawyerr, A., Ferguson, A., 1997. Mucosa-associated bacterial flora of the human colon. J. Med. Microbiol. 46, 85–91.

Robinson, D.S., O'Garra, A., 2002. Further checkpoints in Th1 development. Immunity 16, 755–758.

Ruiz-Perez, B., Chung, D.R., Sharpe, A.H., et al., 2005. Modulation of surgical fibrosis by microbial zwitterionic polysaccharides. Proc. Natl. Acad. Sci. USA 102, 16753–16758.

Sartor, R.B., 2006. Mechanisms of disease: pathogenesis of Crohn's disease and ulcerative colitis. Nat. Clin. Pract. Gastroenterol. Hepatol. 3, 390–407.

Schroder, K., Hertzog, P.J., Ravasi, T., Hume, D.A., 2004. Interferon-gamma: an overview of signals, mechanisms and functions. J. Leukoc. Biol. 75, 163–189.

Scott, M.G., Dullaghan, E., Mookherjee, N., et al., 2007. An anti-infective peptide that selectively modulates the innate immune response. Nat. Biotechnol. 25, 465–472.

Silver, R.P., Aaronson, W., Vann, W.F., 1988. The K1 capsular polysaccharide of *Escherichia coli*. Rev. Infect. Dis. 10, S282–286.

Sood, R.K., Fattom, A., 1998. Capsular polysaccharide-protein conjugate vaccines and intravenous immunoglobulins. Expert Opin. Investig. Drugs 7, 333–347.

Stephen, T.L., Niemeyer, M., Tzianabos, A.O., Kroenke, M., Kasper, D.L., Kalka-Moll, W.M., 2005. Effect of B7-2 and CD40 signals from activated antigen-presenting cells on the ability of zwitterionic polysaccharides to induce T-Cell stimulation. Infect. Immun. 73, 2184–2189.

Stingele, F., Corthesy, B., Kusy, N., Porcelli, S.A., Kasper, D.L., Tzianabos, A.O., 2004. Zwitterionic polysaccharides stimulate T-cells with no preferential Vβ usage and promote anergy, resulting in protection against experimental abscess formation. J. Immunol. 172, 1483–1490.

Stock, P., Akbari, O., Berry, G., Freeman, G.J., Dekruyff, R.H., Umetsu, D.T., 2004. Induction of T helper type 1-like regulatory cells that express Foxp3 and protect against airway hyper-reactivity. Nat. Immunol. 5, 1149–1156.

Strachan, D.P., 1989. Hay fever, hygiene, and household size. Br. Med. J. 299, 1259–1260.

Szu, S.C., Li, X.R., Stone, A.L., Robbins, J.B., 1991. Relation between structure and immunologic properties of the Vi capsular polysaccharide. Infect. Immun. 59, 4555–4561.

Takeda, K., Akira, S., 2004. TLR signalling pathways. Semin. Immunol. 16, 3–9.

Tzianabos, A.O., Kasper, D.L., 2002. Role of T-cells in abscess formation. Curr. Opin. Microbiol. 5, 92–96.

Tzianabos, A.O., Onderdonk, A.B., Rosner, B., Cisneros, R.L., Kasper, D.L., 1993. Structural features of polysaccharides that induce intra-abdominal abscesses. Science 262, 416–419.

Tzianabos, A.O., Onderdonk, A.B., Smith, R.S., Kasper, D.L., 1994. Structure-function relationships for polysaccharide-induced intra-abdominal abscesses. Infect. Immun. 62, 3590–3593.

Tzianabos, A.O., Kasper, D.L., Cisneros, R.L., Smith, R.S., Onderdonk, A.B., 1995. Polysaccharide-mediated protection against abscess formation in experimental intra-abdominal sepsis. J. Clin. Invest. 96, 2727–3271.

Tzianabos, A.O., Finberg, R.W., Wang, Y., et al., 2000. T-cells activated by zwitterionic molecules prevent abscesses induced by pathogenic bacteria. J. Biol. Chem. 275, 6733–6740.

Tzianabos, A.O., Wang, J.Y., Lee, J.C., 2001. Structural rationale for the modulation of abscess formation by *Staphylococcus aureus* capsular polysaccharides. Proc. Natl. Acad. Sci. USA 98, 9365–9370.

Tzianabos, A., Wang, J.Y., Kasper, D.L., 2003. Biological chemistry of immunomodulation by zwitterionic polysaccharides. Carbohydr. Res. 338, 2531–2538.

Umetsu, D.T., Dekruyff, R.H., 2006. Immune dysregulation in asthma. Curr. Opin. Immunol. 18, 727–732.

Wang, Q., McLoughlin, R.M., Cobb, B.A., et al., 2006. A bacterial carbohydrate links innate and adaptive responses through Toll-like receptor 2. J. Exp. Med. 203, 2853–2863.

Wang, Y., Kalka-Moll, W.M., Roehrl, M.H., Kasper, D.L., 2000. Structural basis of the abscess-modulating polysaccharide A2 from *Bacteroides fragilis*. Proc. Natl. Acad. Sci. USA 97, 13478–13483.

Watts, C., 1997. Capture and processing of exogenous antigens for presentation on MHC molecules. Annu. Rev. Immunol. 15, 821–850.

Wessels, M.R., DiFabio, J.L., Benedi, V., et al., 1991. Structural determination and immunochemical characterization of the type V group B *Streptococcus* capsular polysaccharide. J. Biol. Chem. 266, 6714–6719.

Wills-Karp, M., Santeliz, J., Karp, C.L., 2001. The germless theory of allergic disease: revisiting the hygiene hypothesis. Nat. Rev. Immunol. 1, 69–75.

Yoon, J.W., Jun, H.S., 2005. Autoimmune destruction of pancreatic β cells. Am. J. Ther. 12, 580–591.

CHAPTER

50

Future potential of glycomics in microbiology and infectious diseases

Mark von Itzstein and Anthony P. Moran

SUMMARY

Glycomics, as the systematic study of carbohydrate polymers and oligosaccharides, seeks to interpret what can be called the third molecular language of cells. As carbohydrates and carbohydrate-recognizing proteins play pivotal roles in infection, pathogen survival and disease propagation, microbial glycomics is poised to revolutionize both antimicrobial vaccine and drug discovery. In a time when drug resistance of a number of clinically significant bacteria and fungi has occurred, and where only limited therapy is available for other pathogens such as viruses and parasites, new drug discovery and therapeutic strategies are required. The exploration of the role of carbohydrates and carbohydrate-recognizing proteins in infectious disease biology has been given a boost through significant technological advances, in analytical and synthetic chemistry, that are reviewed herein. Furthermore, the future outlook for the role of glycomics in the development of new therapeutic strategies is summarized.

Keywords: Carbohydrate technology; Nuclear magnetic resonance (NMR) spectroscopy; Mass spectrometry; Glycan array; Lectin array; Future directions; Drug discovery; Vaccines; Probiotics

1. INTRODUCTION

Carbohydrates and carbohydrate-recognizing proteins play pivotal roles in the microbial world (e.g. Schmidt *et al.*, 2003; Lehmann *et al.*, 2006; Moran, 2008; von Itzstein *et al.*, 2008a). These roles are broad ranging and, in the case of glycans, include the fundamental make-up of bacterial and parasite membranes as a protective barrier or involvement in host immune system evasion. Microbial surface-resident carbohydrate-recognizing proteins can be essential for infection and disease propagation (see Chapters 27, 28, 29, 34 and 35). Such proteins can act as homing devices that target the pathogen to specific host cells, e.g. influenza virus uses the sialic acid-recognizing protein haemagglutinin to target respiratory tract epithelial cells and rotavirus uses VP8* to target gastrointestinal tract epithelial cells.

Emerging and re-emerging infectious diseases present humanity with one of its greatest challenges. Many factors, such as climate change and drug resistance, can influence the rise and fall of human pathogens and the risk that they pose to the general population. For example, it is widely thought that with the current climate change cycle where the Earth is warming, an increase in tropical disease may occur. There is already some evidence for this, insofar as parts of the world have seen an increase in the number of reported cases of Dengue virus infection (WHO, 2008). Furthermore, the appearance of typical nosocomial pathogenic bacteria, e.g. vancomycin resistant enterococci (VRE) and methicillin resistant *Staphylococcus aureus* (MRSA) in the general community (Stevenson *et al.*, 2005; Munckhof *et al.*, 2009) have sharpened the focus of biomedical researchers to the task of developing novel strategies for next generation antimicrobial drug discovery. As an important part of this drug discovery strategy, the importance of the role of carbohydrates in the life-cycles of microbes, whether it be cell wall-related or related to immune system evasion or host cell invasion (see Chapters 42 and 44), has become of great significance.

To this end, significant technological developments, including those in nuclear magnetic resonance (NMR) spectroscopy, mass spectrometry (MS), microarray-based analysis and synthetic chemistry (Angulo *et al.*, 2006; Paulson *et al.*, 2006; Lee *et al.*, 2006; Seeberger and Werz, 2007), that facilitate a better understanding of the complexity of carbohydrates, have provided valuable insights into the structure–activity relationships of these molecules. A number of the preceding chapters provide many excellent examples that demonstrate the use of some of these important technological advances. These developments take the science of carbohydrates to a new level that will enable a better understanding of the carbohydrate language, the glycome. This language was once thought undecipherable and too complex. As a consequence, this belief very much isolated carbohydrate science from the well-accepted and more modern advancements in fields related to microbiology, such as genomics and proteomics. Nevertheless, glycomics as the systematic study of carbohydrate polymers and oligosaccharides, seeks to interpret what can be called "the third molecular language" of cells (Weiss and Iyer, 2007).

In this chapter, we provide a brief snapshot of some highlights from a number of the most recent technological developments that have made the study of carbohydrates and carbohydrate-recognizing proteins in infectious disease biology more accessible to the non-expert. A review by Turnbull and Field (2007) on emerging technologies in glycomics also provides valuable insights into the more recent technological advances in this field.

2. ADVANCES IN CARBOHYDRATE ANALYTICAL TECHNIQUES

2.1. Analysis by NMR spectroscopy

One of the most significant advances that has occurred in NMR spectroscopy as it relates to the understanding of the role of carbohydrates in infectious disease has been the development of the saturation transfer difference (STD-) NMR spectroscopic technique (Mayer and Meyer, 1999, 2001; Meyer and Peters, 2003; Angulo *et al.*, 2006). The technique of STD-NMR spectroscopy was initially used for screening a library of carbohydrate molecules for their affinity towards a lectin (Mayer and Meyer, 1999). Moreover, the technique allows the determination of the carbohydrate recognition region that interacts with the protein of interest.

The method has become a powerful tool for the solution-state study of protein–carbohydrate interactions at the atomic level without the need for the protein to be isotopically labelled. For example, a number of pathogen-associated carbohydrate-recognizing proteins, including those of bacterial (Haselhorst *et al.*, 2004), parasitic (Lamerz *et al.*, 2006) and viral (e.g. Benie *et al.*, 2003, Haselhorst *et al.*, 2007, 2008, 2009) origins have been investigated by this technique. In one of the most recent studies that employs this technique (Haselhorst *et al.*, 2009), the sialic acid-containing gangliosides GD_{1a} and GM_1 have been investigated for their specific structural interactions with the human and porcine rotavirus surface lectin, VP8*. This particular study has provided valuable insight into the key interactions of these complex gangliosides with this lectin. Additionally, the study has demonstrated the structural characteristics of both human and porcine rotavirus sialic acid-dependence in host cell infection.

Furthermore, technological advances in the hardware associated with NMR spectroscopy have also provided a more user-friendly scenario for non-NMR spectroscopy experts. One example of these advances has been the development of the cryoprobe in which high-level NMR spectroscopic investigations of protein–ligand interactions, at relatively low protein concentrations (typically μg per ml) can be achieved (for an example see Haselhorst *et al.*, 2009). Cryogenically-cooled probes have become invaluable in solution-state NMR studies of carbohydrate–protein interactions (Kovacs *et al.*, 2005). Importantly, one observes significant improvement in sensitivity, up to 3- to 6-fold, through detecting coil and pre-amplifier cooling, thus reducing thermal noise. Cryoprobes have now become a routine addition to the NMR structural biology laboratory and these probes are of great value in analysis when only low-level expression of recombinant microbial-associated proteins can be achieved.

2.2. Analysis by MS

One of the most valuable tools in defining glycan structure, i.e. glycan profiling, in the context of both carbohydrate content and linkage, has been MS. Over the years, significant technological advances in MS have been made (see Chapter 13) that now provides methodology for routine analysis of glycan composition (Haslam *et al.*, 2006; Hitchen and Dell, 2006; Geyer and Geyer, 2006; Wada *et al.*, 2007; Zaia, 2008; Dalpathado and Desaire, 2008; Estrella *et al.*, 2009), particularly as it relates to bacterial glycoproteomics (Hitchen and Dell, 2006).

These invaluable developments in MS have now provided the research scientist with the capacity to analyse more routinely complex glycan structures associated with cell walls of bacterial and other pathogens (Guerry *et al.*, 2002; Moran *et al.*, 2002; Wilson *et al.*, 2005; Peak *et al.*, 2007). Applications of these methods are elaborated in several foregoing chapters.

2.3. Microarray analysis

The fact that carbohydrates and their conjugates play significant roles in many biological processes, particularly processes related to infectious disease biology, the development of array technologies that enables the rapid analysis of interactions between carbohydrates and carbohydrate-recognizing proteins has been of great importance to the field of glycomics.

Glycan array technologies (Liang *et al.*, 2008; Horlacher and Seeberger, 2008) have revolutionized our capacity to interrogate carbohydrate–protein interactions and have become an invaluable research tool. Typically, these arrays consist of sugars either covalently or non-covalently associated to a surface in a well-defined manner; this technology has been extensively reviewed elsewhere (Paulson *et al.*, 2006; Liang *et al.*, 2008; Horlacher and Seeberger, 2008). As summarized in these reviews, the major benefit of these arrays is that the approach provides a highly-sensitive and rapid analysis of the carbohydrate specificity of isolated or recombinantly-expressed proteins and, indeed, whole infectious agents, including bacteria and viruses and, potentially, parasites. Moreover, the technology has potential applications in examining the extent and relevance of protein–carbohydrate interactions in immune recognition, as well as immune cell stimulation and proliferation.

In more recent years, lectin-based glycan profiling through the development and use of lectin arrays has also been an invaluable development in the field (Hirabayashi, 2008). While lectins, as a result of their glycan specificity, have been widely used to characterize and purify glycan structures, the concept of using lectins in a miniaturized format, such as on an array for rapid glycan detection and analysis, has been only a fairly recent development (e.g. Hirabayashi, 2004; Hsu and Mahal, 2006).

3. FUTURE OUTLOOK

The emerging field of microbial glycomics is poised to revolutionize both antimicrobial vaccine and drug discovery in the 21st century. At a time when drug resistance of a number of clinically-significant bacteria and fungi has occurred and where only limited therapy is available for other pathogens, such as viruses and parasites, new drug discovery and therapeutic strategies are required. In particular, the extent of pathogen engagement of carbohydrates, and the proteins that recognize them, in pathogen survival and infection provide outstanding opportunities and new direction for novel drug discovery. Such an approach has proven successful in the development of anti-influenza drugs (von Itzstein, 2007, 2008b) and other potential antiviral targets are under investigation (see Chapter 15). Additionally, with a greater understanding of the immunogenicity of microbial glycosylated structures and the cellular mechanisms of immune activation by carbohydrate-related molecules (see Chapters 36 and 48), a greater appreciation of the potential of carbohydrate moieties as vaccine candidates has been gained (Lucas *et al.*, 2005). Furthermore, additional therapeutic strategies involving engineered probiotics bearing toxin-binding carbohydrate-based surface structures promises a potentially new therapeutic approach for the treatment of bacterial enteric infections (Paton *et al.*, 2006).

As described in some of the preceding chapters, we have already seen the benefits of a better understanding of carbohydrate science as it relates to infectious disease biology. Importantly, the exploration of the role of carbohydrates and carbohydrate-recognizing proteins in infectious disease biology has been given a boost through significant technological advances that have been made. In a time where we see the rise of both emerging and re-emerging infectious diseases, all efforts to further crack the glycocode, i.e. glycome, which has been fuelled by many technological advances, must be strengthened. Given the extent of the use of carbohydrates by microbial pathogens in causing and propagating infection, there is no reason to doubt that many of the solutions to these diseases lie in targeting the carbohydrate-related mechanisms.

References

Angulo, J., Rademacher, C., Biet, T., et al., 2006. NMR analysis of carbohydrate-protein interactions. Methods Enzymol. 416, 12–30.

Benie, A.J., Moser, R., Bäuml, E., Blaas, D., Peters, T., 2003. Virus-ligand interactions: identification and characterization of ligand binding by NMR spectroscopy. J. Am. Chem. Soc. 125, 14–15.

Dalpathado, D.S., Desaire, H., 2008. Glycopeptide analysis by mass spectrometry. Analyst 133, 731–738.

Estrella, R.P., Whitelock, J.M., Roubin, R.H., Packer, N.H., Karlsson, N.G., 2009. Small-scale enzymatic digestion of glycoproteins and proteoglycans for analysis of oligosaccharides by LC-MS and FACE gel electrophoresis. Methods Mol. Biol. 534, 171–192.

Geyer, H., Geyer, R., 2006. Strategies for analysis of glycoprotein glycosylation. Biochim. Biophys. Acta 1764, 1853–1869.

Guerry, P., Szymanski, C.M., Prendergast, M.M., et al., 2002. Phase variation of *Campylobacter jejuni* 81-176 lipooligosaccharide affects ganglioside mimicry and invasiveness *in vitro*. Infect. Immun. 70, 787–793.

Haselhorst, T., Wilson, J.C., Thomson, R.J., et al., 2004. Saturation transfer difference (STD) ^{1}H experiments and *in silico* docking experiments to probe the binding of *N*-acetylneuraminic acid and derivatives to *Vibrio cholerae* sialidase. Proteins 56, 346–353.

Haselhorst, T., Blanchard, H., Frank, M., et al., 2007. STD NMR spectroscopy and molecular modeling investigation of the binding of *N*-acetylneuraminic acid derivatives to rhesus rotavirus VP8* core. Glycobiology 17, 68–81.

Haselhorst, T., Garcia, J.M., Islam, T., et al., 2008. Avian influenza H5-containing virus-like particles (VLPs): host-cell receptor specificity by STD NMR spectroscopy. Angew. Chem. Int. Ed. 47, 1910–1912.

Haselhorst, T., Fleming, F.E., Dyason, J.C., et al., 2009. Sialic acid dependence in rotavirus host cell invasion. Nat. Chem. Biol. 5, 91–93.

Haslam, S.M., North, S.J., Dell, A., 2006. Mass spectrometric analysis of *N*- and *O*-glycosylation of tissues and cells. Curr. Opin. Struct. Biol. 16, 584–591.

Hirabayashi, J., 2004. Lectin-based structural glycomics: glycoproteomics and glycan profiling. Glycoconj. J. 21, 35–40.

Hirabayashi, J., 2008. Concept, strategy and realization of lectin-based glycan profiling. J. Biochem. 144, 139–147.

Hitchen, P.G., Dell, A., 2006. Bacterial glycoproteomics. Microbiology 152, 1575–1580.

Horlacher, T., Seeberger, P.H., 2008. Carbohydrate arrays as tools for research and diagnostics. Chem. Soc. Rev. 37, 1414–1422.

Hsu, K.L., Mahal, K.L., 2006. A lectin microarray for the rapid analysis of bacterial glycans. Nat. Protoc. 1, 543–549.

Kovacs, H., Moskau, D., Spraul, M., 2005. Cryogenically cooled probes – a leap in NMR technology. Prog. NMR Spectr. 46, 131–155.

Lamerz, A-C., Haselhorst, T., Bergfeld, A.K., von Itzstein, M., Gerardy-Schahn, R., 2006. Molecular cloning of the *Leishmania major* UDP-glucose pyrophosphorylase, functional characterization and ligand binding analyses using NMR spectroscopy. J. Biol. Chem. 281, 16314–16322.

Lee, J.C., Wu, C.Y., Apon, J.V., Siuzdak, G., Wong, C.H., 2006. Reactivity-based one-pot synthesis of the tumor-associated antigen N3 minor octasaccharide for the development of a photocleavable DIOS-MS sugar array. Angew. Chem. Int. Ed. Engl. 45, 2753–2757.

Lehmann, F., Tiralongo, E., Tiralongo, J., 2006. Sialic acid-specific lectins: occurrence, specificity and function. Cell. Mol. Life Sci. 63, 1331–1354.

Liang, P.H., Wu, C.Y., Greenberg, W.A., Wong, C.H., 2008. Glycan arrays: biological and medical applications. Curr. Opin. Chem. Biol. 12, 86–92.

Lucas, A.H., Apicella, M.A., Taylor, C.E., 2005. Carbohydrate moieties as vaccine candidates. Clin. Infect. Dis. 41, 705–712.

Mayer, M., Meyer, B., 1999. Characterization of ligand binding by saturation transfer difference NMR spectroscopy. Angew. Chem. Int. Ed. 38, 1784–1788.

Mayer, M., Meyer, B., 2001. Group epitope mapping by saturation transfer difference NMR to identify segments of a ligand in direct contact with a protein receptor. J. Am. Chem. Soc. 123, 6108–6117.

Meyer, B., Peters, T., 2003. NMR spectroscopy techniques for screening and identifying ligand binding to protein receptors. Angew. Chem. Int. Ed. 42, 864–890.

Moran, A.P., 2008. Relevance of fucosylation and Lewis antigen expression in the bacterial gastroduodenal pathogen *Helicobacter pylori*. Carbohydr. Res. 343, 1952–1965.

Moran, A.P., Knirel, Y.A., Senchenkova, S.N., Widmalm, G., Hynes, S.O., Jansson, P.-E., 2002. Phenotypic variation in molecular mimicry between *Helicobacter pylori* lipopolysaccharides and human gastric epithelial cell surface glycoforms. Acid-induced phase variation in Lewis^x and Lewis^y expression by *H. pylori* lipopolysaccharides. J. Biol. Chem. 277, 5785–5795.

Munckhof, W.J., Nimmo, G.R., Schooneveldt, J.M., et al., 2009. Nasal carriage of *Staphylococcus aureus*, including community-associated methicillin-resistant strains, in Queensland adults. Clin. Microbiol. Infect. 15, 149–155.

Paton, A.W., Morona, R., Paton, J.C., 2006. Designer probiotics for prevention of enteric infections. Nat. Rev. Microbiol. 4, 193–200.

Paulson, J.C., Blixt, O., Collins, B.E., 2006. Sweet spots in functional glycomics. Nat. Chem. Biol. 2, 238–248.

Peak, I.R., Grice, I.D., Faglin, I., et al., 2007. Towards understanding the functional role of the glycosyltransferases involved in the biosynthesis of *Moraxella catarrhalis* lipooligosaccharide. FEBS J. 274, 2024–2037.

Schmidt, M.A., Riley, L.W., Benz, I., 2003. Sweet new world: glycoproteins in bacterial pathogens. Trends Microbiol. 11, 554–561.

Seeberger, P.H., Werz, D.B., 2007. Synthesis and medical applications of oligosaccharides. Nature 446, 1046–1051.

Stevenson, K.B., Searle, K., Stoddard, G.J., Samore, M., 2005. Methicillin-resistant *Staphylococcus aureus* and vancomycin-resistant *Enterococci* in rural communities, western United States. Emerg. Infect. Dis. 11, 895–903.

Turnbull, J.E., Field, RA., 2007. Emerging glycomics technologies. Nat. Chem. Biol. 3, 74–77.

von Itzstein, M., 2007. The war against influenza: discovery and development of sialidase inhibitors. Nat. Rev. Drug Discov. 6, 967–974.

von Itzstein, M., Plebanski, M., Cooke, B.M., Coppel, R.L., 2008a. Hot, sweet and sticky: the glycobiology of *Plasmodium falciparum*. Trends Parasitol. 24, 210–218.

von Itzstein, M., 2008b. Disease-associated carbohydrate-recognising proteins and structure-based inhibitor design. Curr. Opin. Struct. Biol. 18, 558–566.

Wada, Y., Azadi, P., Costello, C.E., et al., 2007. Comparison of the methods for profiling glycoprotein glycans – HUPO

Human Disease Glycomics/Proteome Initiative multi-institutional study. Glycobiology 17, 411–422.

Weiss, A.A., Iyer, S.S., 2007. Glycomics aims to interpret the third molecular language of cells. Some microbes exploit carbohydrate receptors to enter and disrupt particular cells. Microbe 2, 489–497.

WHO, 2008. World Health Organization – Dengue and dengue haemorrhagic fever. <http://www.who.int/mediacentre/factsheets/fs117/en/>.

Wilson, J.C., Hitchen, P.G., Frank, M., et al., 2005. Identification of a capsular polysaccharide from *Moraxella bovis*. Carbohydr. Res. 340, 765–769.

Zaia, J., 2008. Mass spectrometry and the emerging field of glycomics. Chem. Biol. 15, 881–892.

Index

H

N

R

S

T

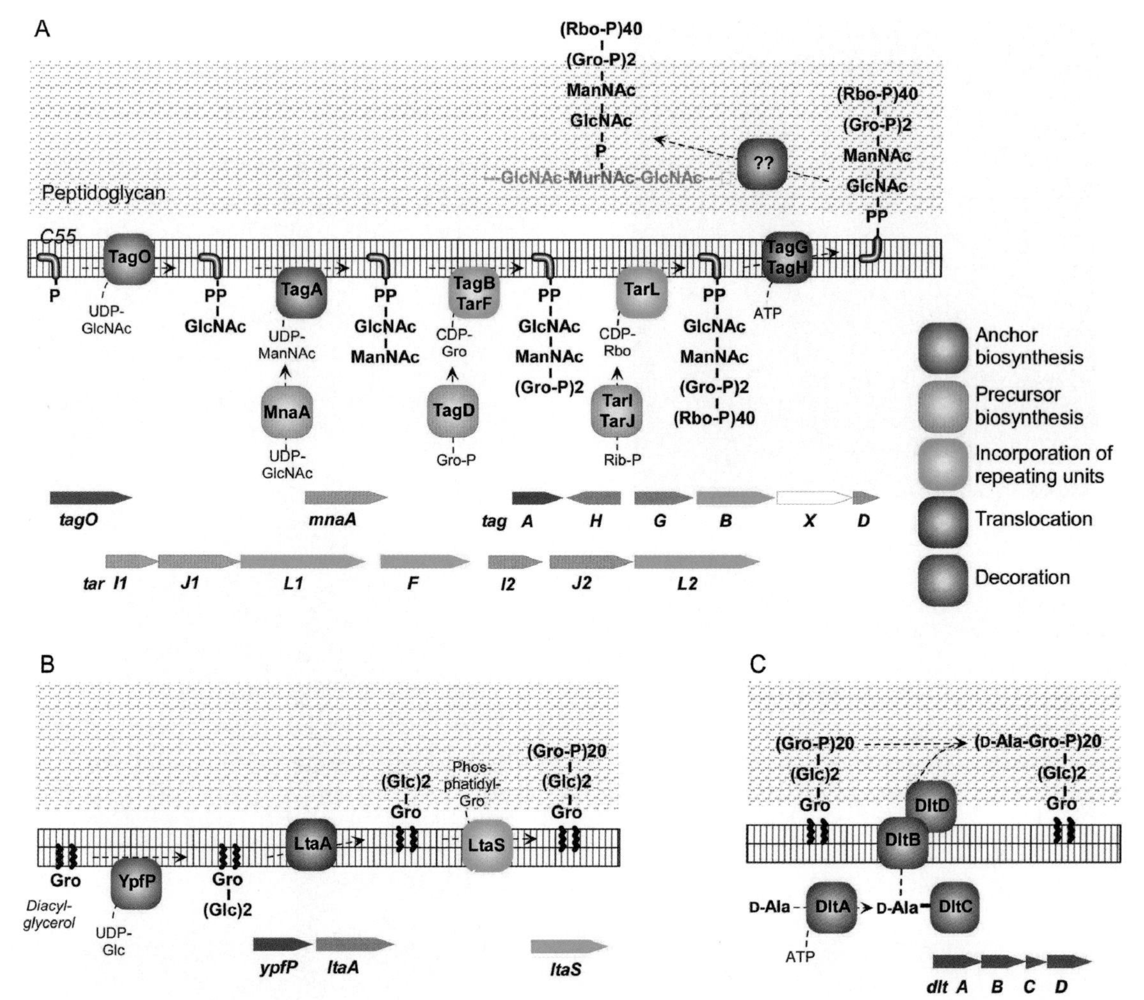

PLATE 5.3 Pathways of *S. aureus* wall teichoic acid (WTA) biosynthesis (A), lipoteichoic acid (LTA) biosynthesis (B), and D-alanine incorporation into LTA and WTA (C). CDP-Gro, cytidyldiphosphate-glycerol; CDP-Rbo, cytidyldiphosphate-ribitol; Glc, glucose; GlcNAc, *N*-acetylglucosamine; Gro, glycerol; Gro-P, glycerolphosphate; ManNAc, *N*-acetylmannosamine; MurNAc, *N*-acetylmuramic acid; Rbo-P, ribitol-phosphate; Rib-P, ribulose-5-phosphate; UDP-Glc, undecaprenylphosphate-glucose; UDP-GlcNAc, uridine-5´-diphosphate-glucose; UDP-GlcNAc, uridine-5´-*N*-acetylglucosamine; UDP-ManNAc, uridine-5´-*N*-acetylmannosamin.

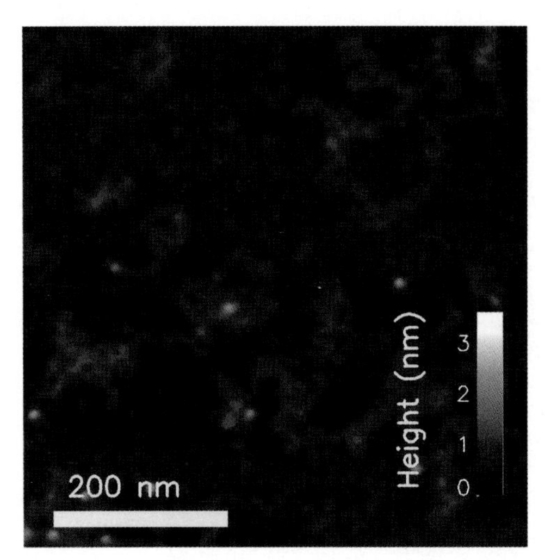

PLATE 14.2 Tapping mode AFM height topograph of mannuronan, the biological precursor PSs of alginate, as such molecules are being converted to alginate by the action of the C-5-mannuronan epimerase, AlgE4. The enzymes are identified as the higher globular domains in the image.

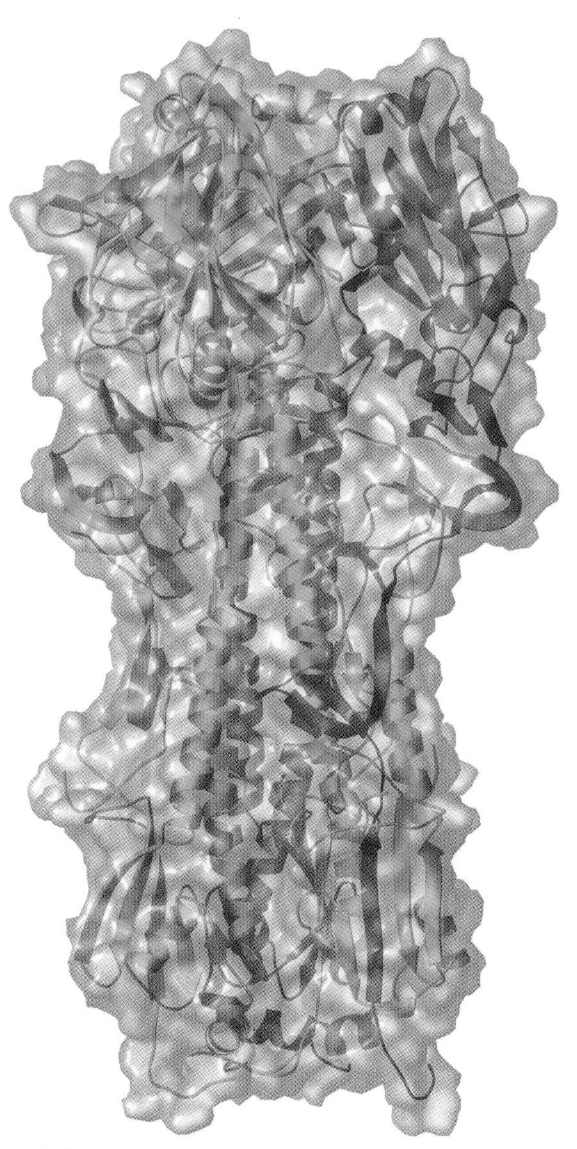

PLATE 15.1 The secondary structure of an influenza virus haemagglutinin trimer is shown with a translucent surface. The surface is coloured by electrostatic potential with red denoting areas of negative charge while blue denotes areas of positive charge.

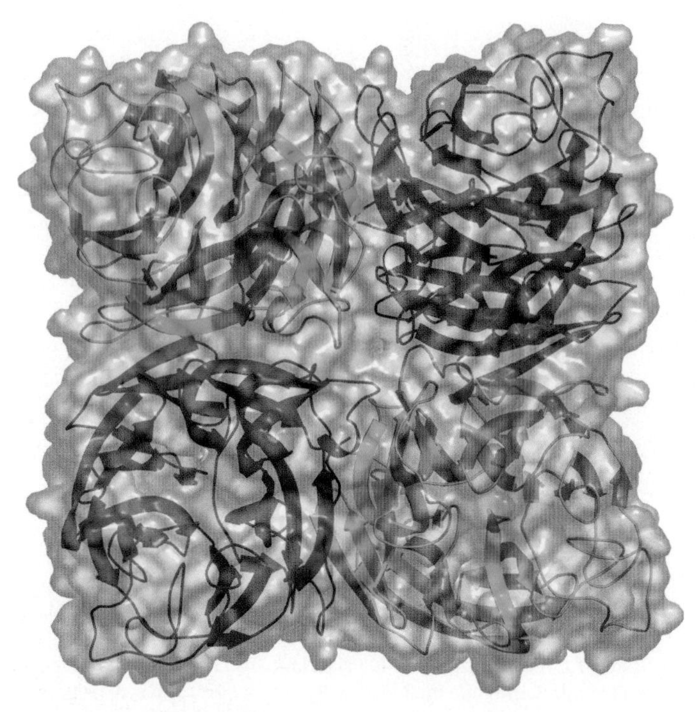

PLATE 15.2 The influenza virus sialidase tetramer is shown with a translucent surface, coloured by electrostatic potential overlaying the monomer secondary structure. Each monomer has six four-stranded anti-parallel β-sheets arranged as if on the blades of a propeller.

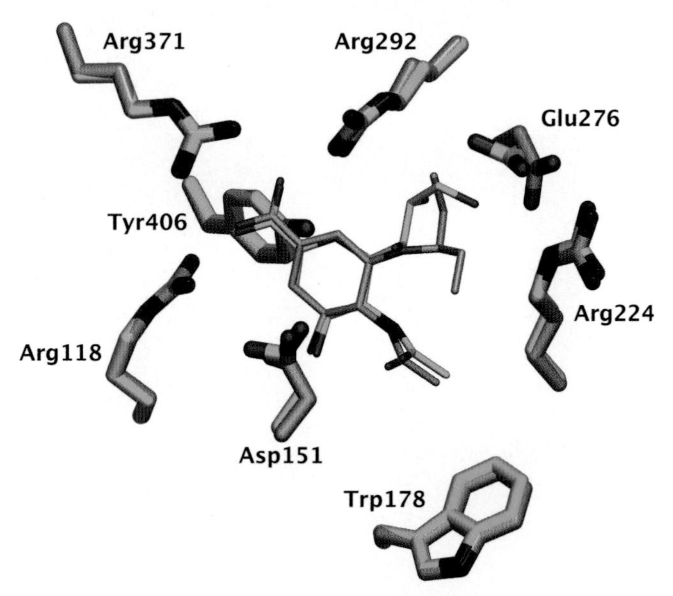

PLATE 15.3 Superimposition of two complexed N9 sialidase structures of influenza virus showing the induced fit of Glu276 when oseltamivir carboxylate (orange carbon atoms) is bound, compared with Neu5Ac2en (green carbon atoms).

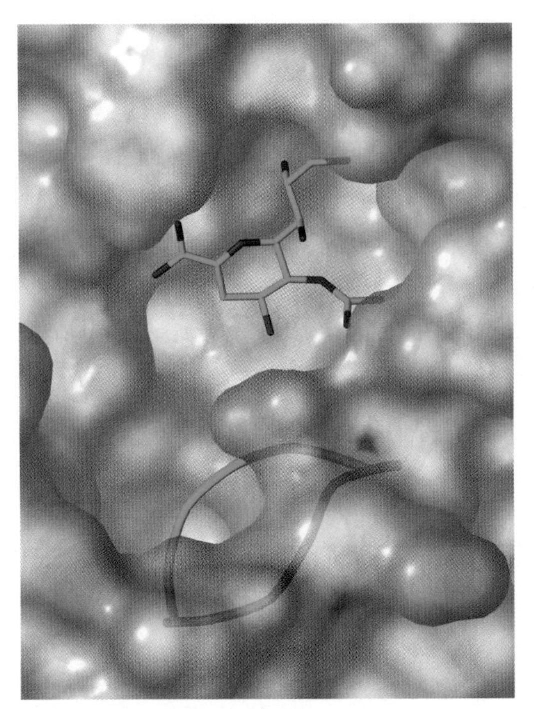

PLATE 15.4 Crystal structures of the influenza virus N8 sialidase. The surface and the magenta coloured 150 loop are from a structure of the uncomplexed protein while the Neu5Ac2en and green-coloured 150 loop are from a structure where Neu5Ac2en was soaked into the protein crystals. The surface is coloured by electrostatic potential.

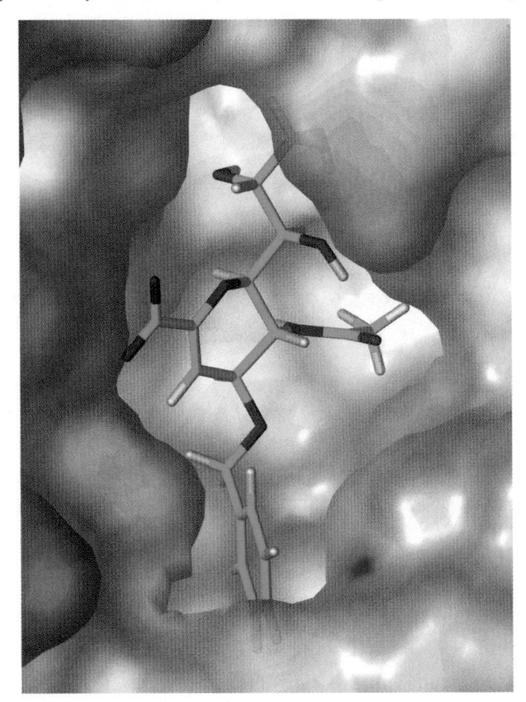

PLATE 15.5 Crystal structure of 4-*O*-benzyl-Neu5Ac2en soaked into the haemagglutinin-neuraminidase of Newcastle disease virus, showing the large pocket targeted by the benzyl group of the designed inhibitor.

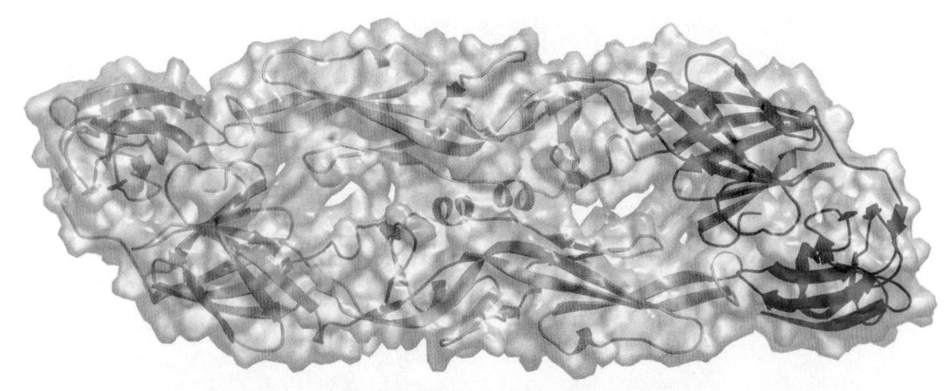

PLATE 15.6 Crystal structure of the E-glycoprotein dimer of dengue virus. The red-coloured ribbon is Domain 1, the yellow-coloured ribbon is Domain 2, while the blue-coloured ribbon is Domain 3. The surface is coloured by electrostatic potential.

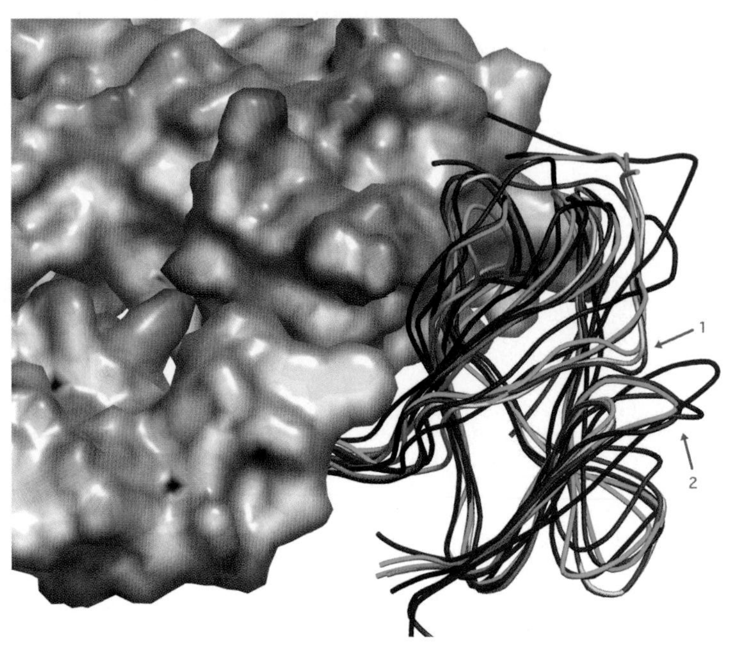

PLATE 15.7 Superimposition of the available flavivirus Domain 3 structures. Green, DENV2; red, DENV3; Blue, Japanese encephalitis virus; cyan, Langat virus; magenta, Omsk haemorrhagic fever virus; yellow, tick-borne encephalitis virus; and brown, West Nile virus. The surface represents the remaining portion of the EGP dimer and is coloured by electrostatic potential. A major difference can be seen in loop 1 for tick-borne encephalitis virus, Omsk haemorrhagic fever virus and Langat virus which bulges out because there is a 4 residue deletion in the loop 2 portions of these proteins.

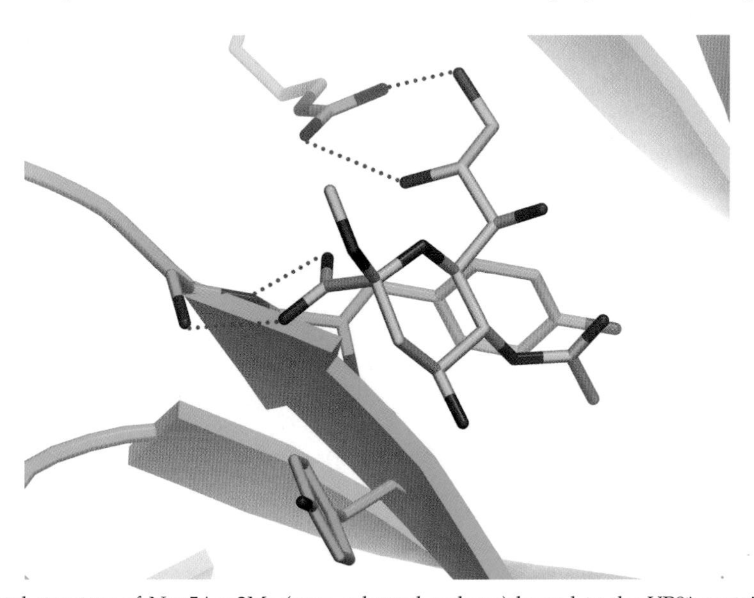

PLATE 15.8 Crystal structure of Neu5Acα2Me (cyan-coloured carbons) bound to the VP8* protein of rotavirus strain CRW-8. Hydrogen bonds to the serine (Ser) and arginine (Arg) are shown with magenta-coloured dotted lines while the two tyrosine (Tyr) residues which help position the sugar ring are also shown.

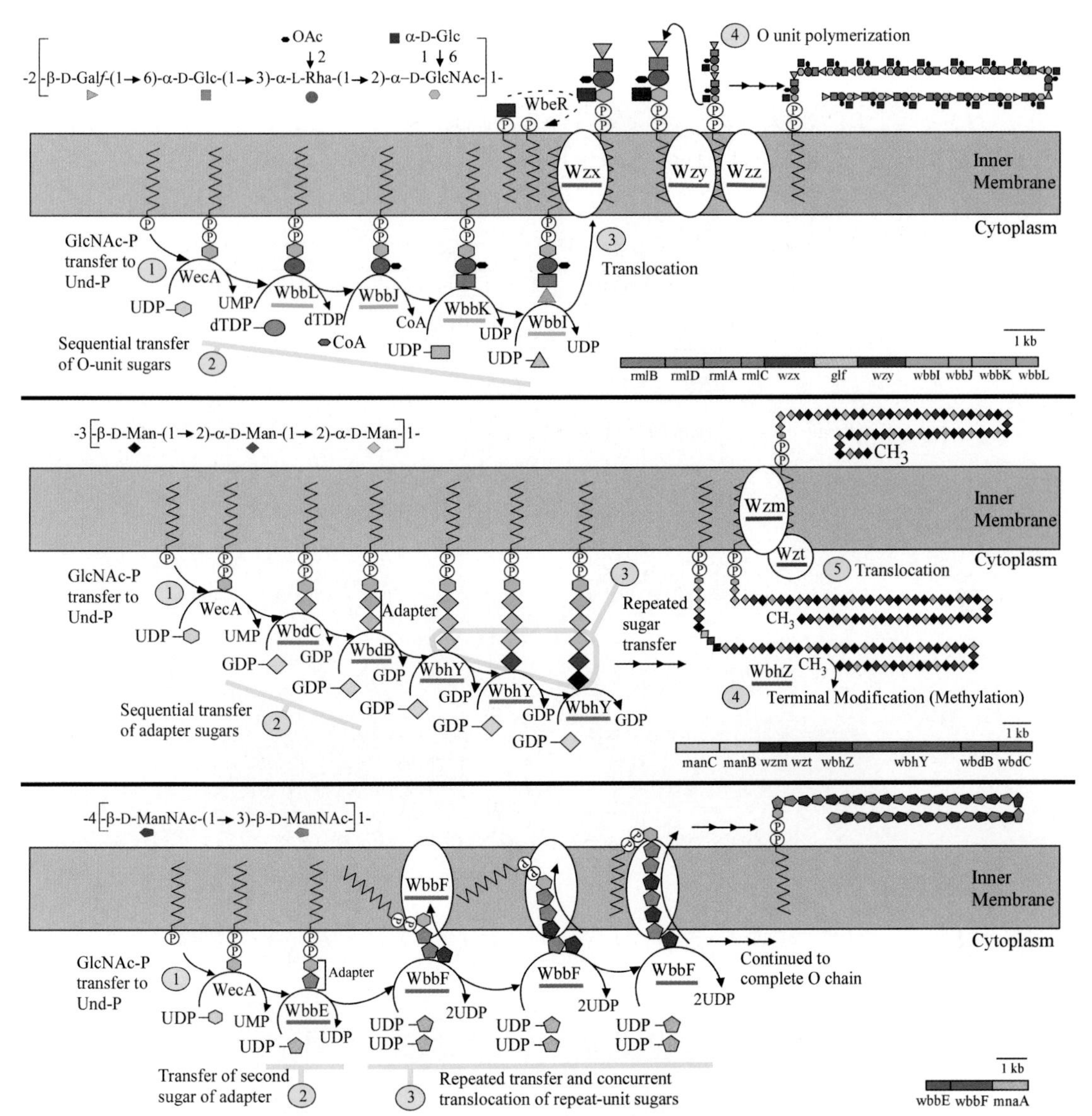

PLATE 18.1 O-Antigen biosynthesis pathways. Top panel: *E. coli* O16 (Wzx/Wzy), centre panel: *E. coli* O8 (ABC-transporter), and bottom panel: *S. enterica* O54 (Synthase), with each panel also indicating the gene cluster involved and the final repeating unit structure. Synthesis in all three pathways is initiated by transfer of GlcNAc-P onto UndP. Pathway steps are numbered sequentially. Each sugar residue is represented by a different symbol. Symbols for each nucleotide diphosphate-linked sugar, and the respective genes, have the same colouring. Transferases and processing proteins, and respective genes, are also colour coded. Residue colour varies in the repeating unit if more than one residue of that sugar is present, to indicate the different linkage formed. Each pathway has an independent colour scheme, except for UDP-GlcNAc, that is common to all three. O-Unit modification in *E. coli* O16 occurs after repeat unit synthesis and is indicated by a dashed line. The ligation of the O-antigen from Und-PP to the lipid-A-core by WaaL and the final export of the LPS molecule to the outer membrane, both common to all three pathways, are not shown (see Chapter 17). Not drawn to scale.

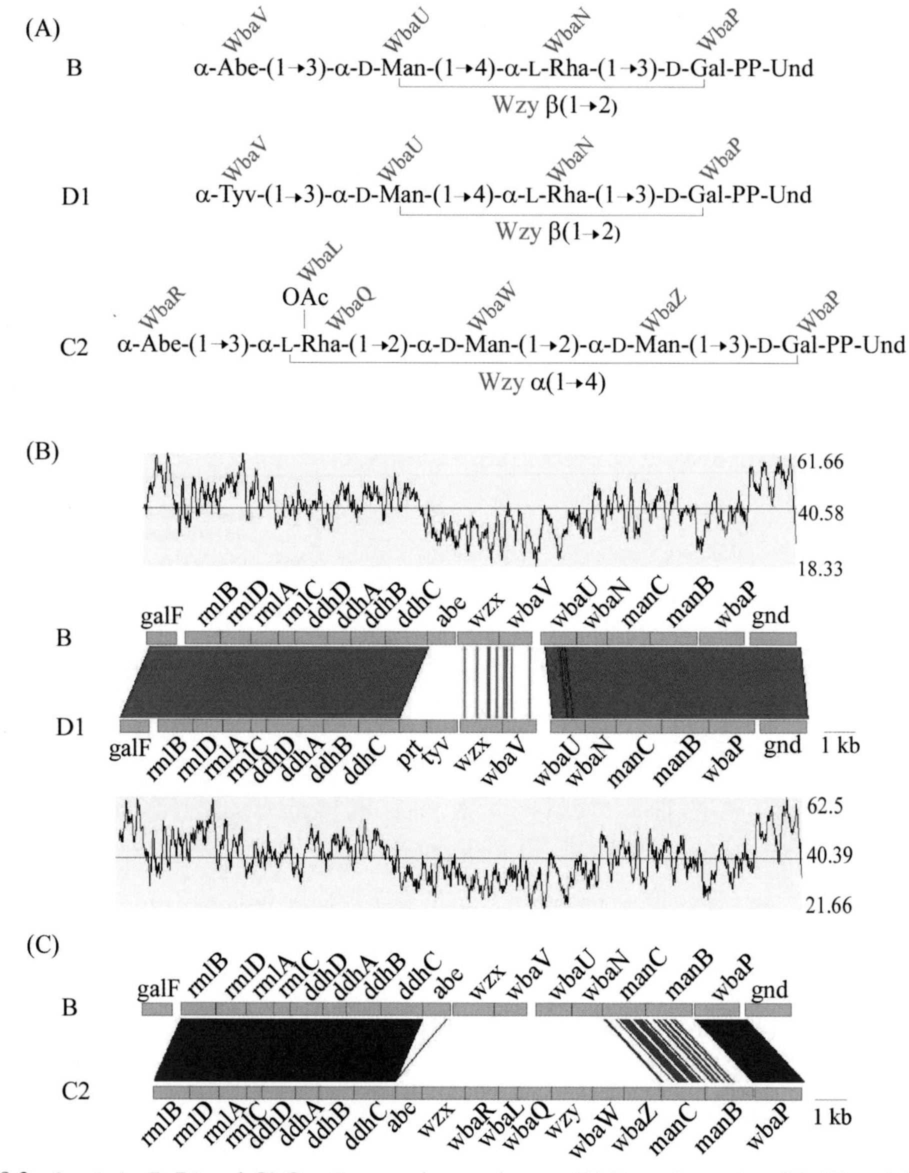

PLATE 18.2 *S. enterica* B, D1 and C2 O-antigens and gene clusters. (A) Repeating units of B, D1 and C2 O-antigens attached to UndPP, with the protein responsible for each linkage shown in red, and the polymerase linkage indicated below each structure. Only sugars and linkages in the basic structure encoded in the O-antigen gene clusters are shown, but most published structures include one or more additional Glc or *O*-acetyl side-branch residues. (B) Comparison of B and D1 gene clusters and flanking genes, generated by ARTEMIS, with G + C% distribution for each cluster. (C) Comparison of B and C2 gene clusters.

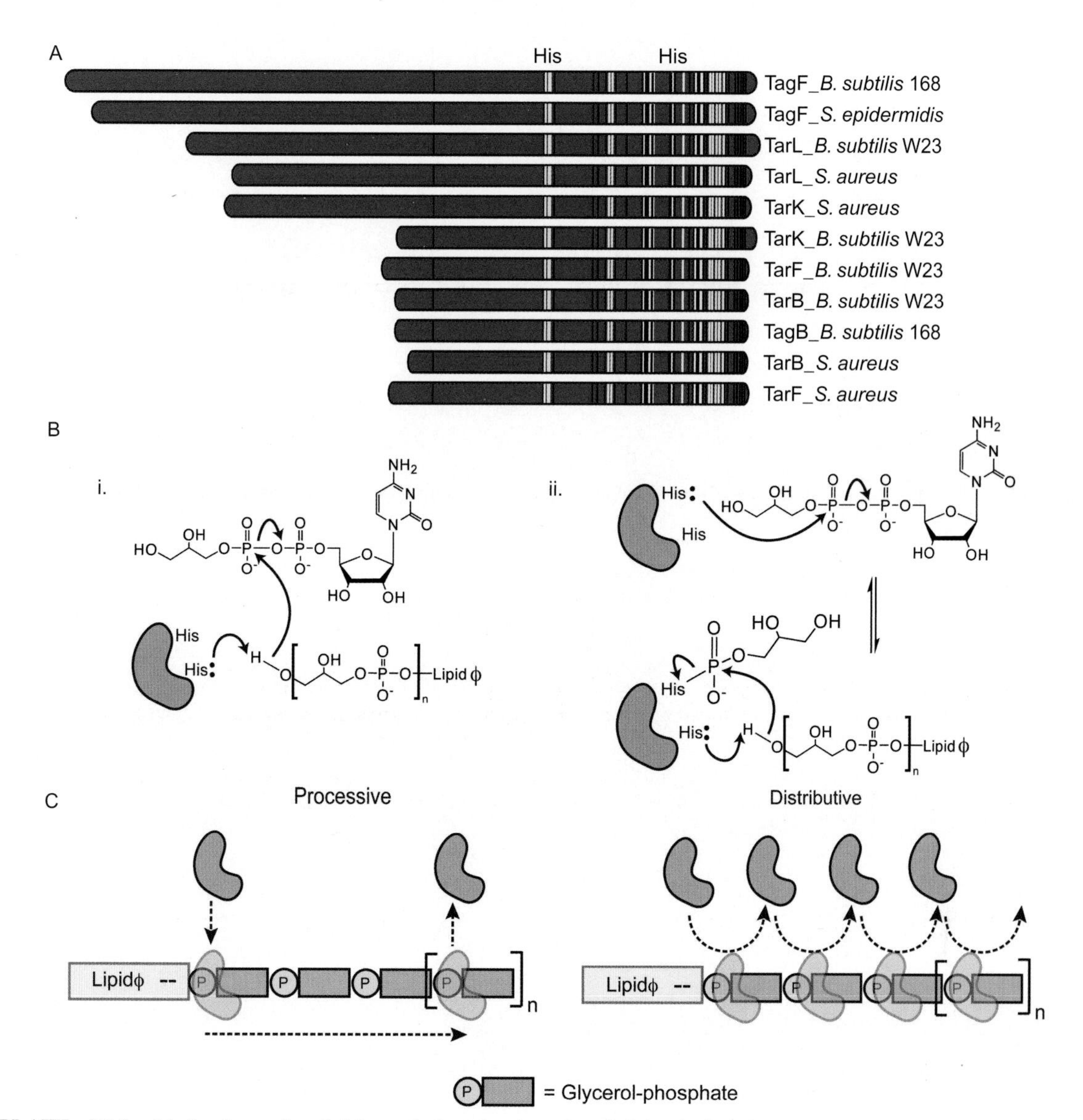

PLATE 19.3 Mechanism of poly(glycerol-phosphate) and poly(ribitol-phosphate) polymerization. (A) Alignment schematic of proteins involved in teichoic acid polyol-phosphate transfer. Highest levels of identity (yellow) and similarity (black) among the amino acid sequences are present in the C-terminal region. Highlighted are the conserved histidines Schertzer *et al.* (2005) implicated in TagF and TagB activity. One point of curiosity is the function of the extensive *N*-terminal region of TagF that is not homologous to any other sequenced protein. (B) Two potential catalytic mechanisms for glycerol-phosphate transfer are shown schematically. It is hypothesized that a histidine residue may function to deprotonate the terminal hydroxyl of the growing polymer for nucleophilic attack on the β-phosphate of cytidine diphosphate- (CDP-) glycerol. It is also conceivable that the histidine directly acts as a nucleophile forming a phosphor-histidine intermediate with glycerol-phosphate from CDP-glycerol to promote catalysis through a ping-pong mechanism. (C) The factors regulating polymer length are currently unknown. It is conceivable that the level of processivity of a polymerase may determine the length of polymer that is formed. A processive polymerase will remain associated with the growing polymer through the addition of many units in a single binding event, while a distributive polymerase dissociates and re-associates to the polymer repeatedly adding only as few as one unit in each binding event. Preliminary data studying the processivity of two teichoic acid polymerases provided differing models of elongation. The TarL enzyme from *S. aureus* was suggested to be processive (Brown *et al.*, 2008), while the TagF enzyme from *B. subtilis* did not appear to maintain association with the growing polymer during catalysis (Pereira *et al.*, 2008a).

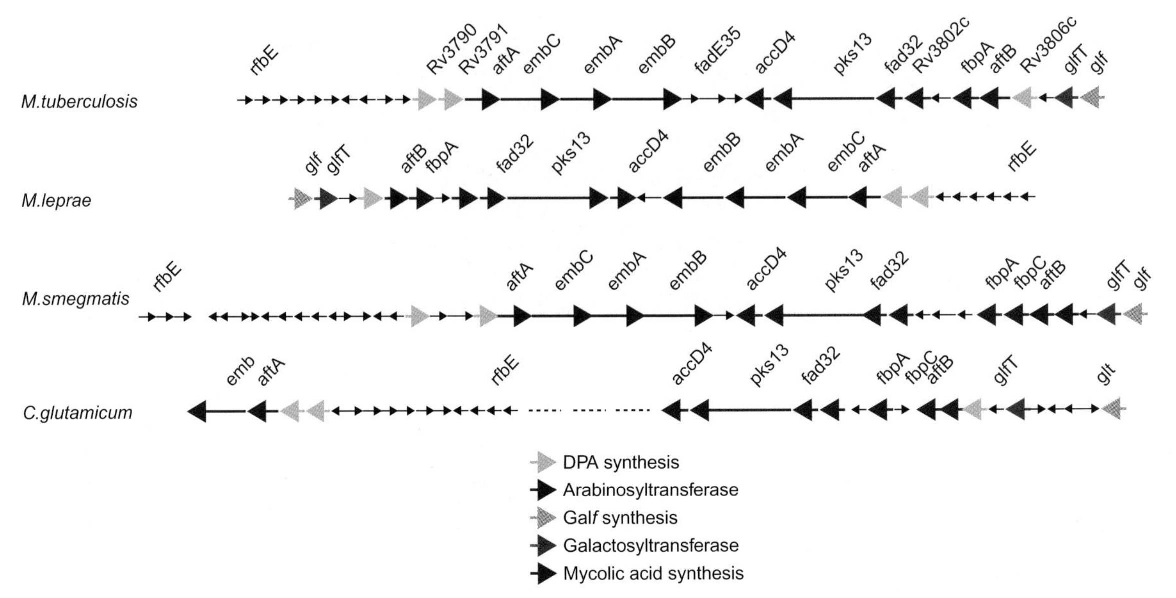

PLATE 21.2 Organization of the cell wall synthetic gene cluster of *M. tuberculosis*, *M. leprae*, *M. smegmatis* and *C. glutamicum* containing genes with confirmed or suspected roles in cell wall synthesis. See text for details of the functions of some of the annotated genes.

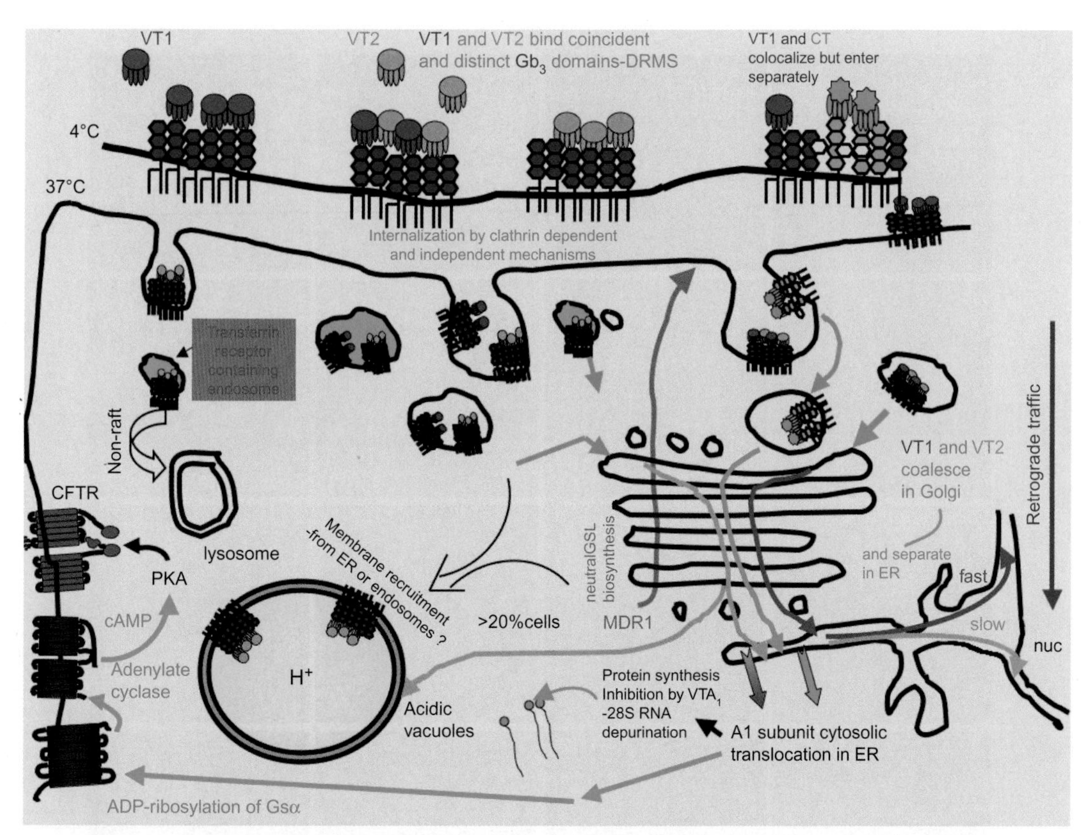

PLATE 30.2 Retrograde trafficking scheme for the toxins VT and CT. The toxins VT1/2 bind punctate domains on the cell surface at 4°C. Pentameric CT binding recruits GM_1 to DRMs, but this has yet to be shown for VT. Both VT1/2 and CT can be internalized by clathrin-dependent and independent mechanisms (Torgersen *et al.*, 2001). The toxins VT1 and VT2 differentially retrograde traffic to the Golgi apparatus and are further transported to the ER and nuclear envelope (Arab and Lingwood, 1998; Tam *et al.*, 2008). The A1-subunits of CT and VT translocate to the cytosol from the ER but have distinct cytosolic targets (Gsα versus 28S RNA).

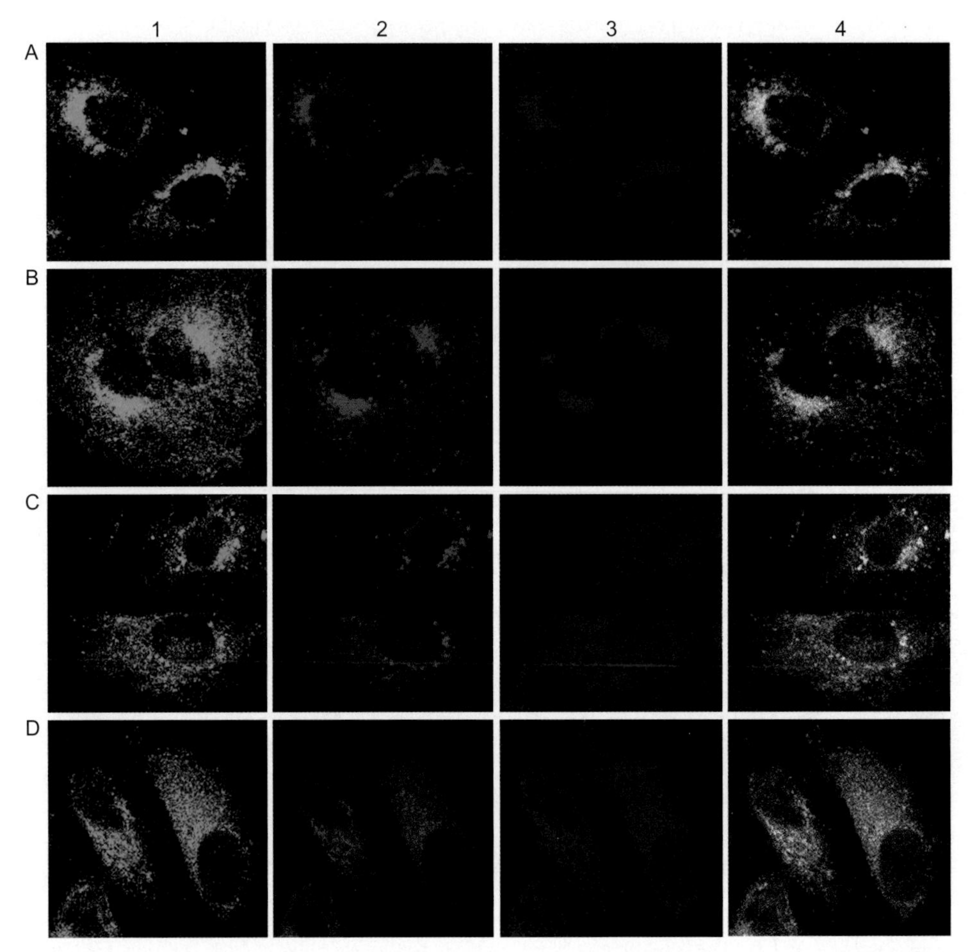

PLATE 30.3 The VT1 A- and B-subunits remain coincident during retrograde transport. The VT1 holotoxin comprising fluorescein isothiocyanate-labelled A- and tetramethyl-rhodamine-labelled B-subunits (Tam and Lingwood, 2007) was added to Vero (A,C) or HeLa (B,D) cells for 1 h (A,B) or 6 h (C,D) at 37°C. Cells were fixed and stained with anti-Rab6 (a Golgi marker) (A3,B3) or anti-GRP78 (an ER marker) antibodies (C3,D3). Cells were imaged by confocal microscopy for each fluor separately and the images merged (n = 4) to define coincidence. The VT1 A and B subunits travel together to the ER via the Golgi apparatus. Reproduced, with permission, from Tam and Lingwood (2007) Membrane-cytosolic translocation of verotoxin A1-subunit in target cells. Microbiology 153, 2700–2710.

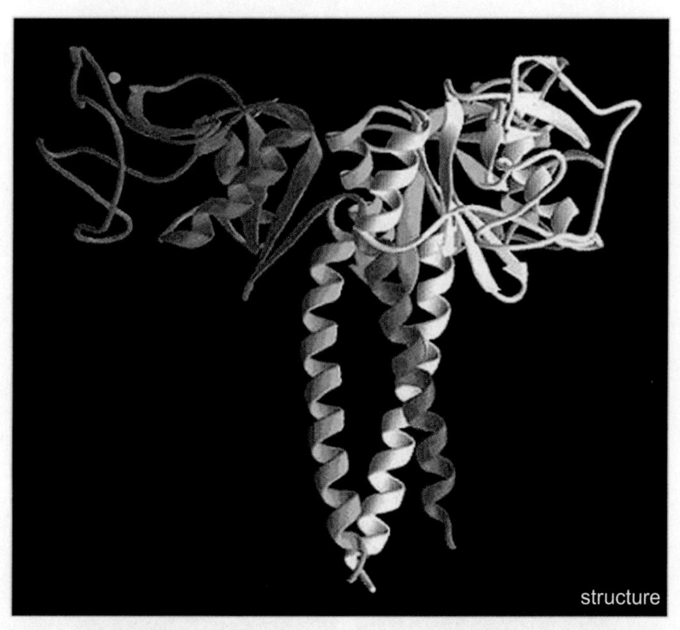

PLATE 35.2 Three-dimensional structures of trimeric recombinant α-helical neck-CRD fragments of human SP-D. Each monomer is shown in a different colour and calcium ions are represented as green spheres. Reproduced, with permission, from Hakansson *et al.* (1999) Crystal structure of the trimeric α-helical coiled-coil and the three lectin domains of human lung surfactant protein D. Structure 7, 255–264. Elsevier © 1999.

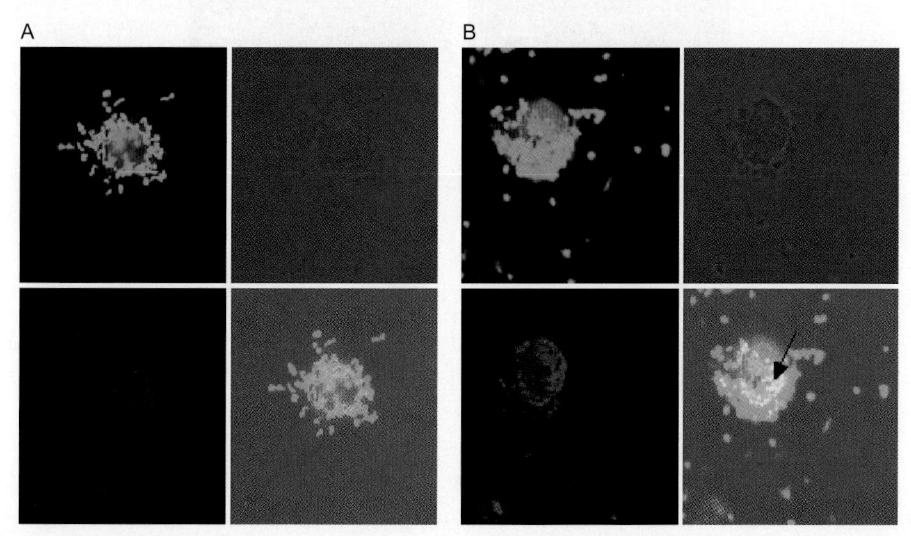

PLATE 35.3 Enhanced *H. pylori* uptake by dendritic cells (DCs) in the presence of SP-D. Confocal microscopy of the uptake of *H. pylori* (green) by DCs (red) in the absence (A) and presence (B) of SP-D. Carboxyfluorescein diacetate succinimidyl ester (CFDA-SE) labelled *Helicobacter* cells (green) is co-localized with the early endosomal compartments in DCs (red) and thus appears yellow (arrowhead in Panel B). The uptake is more marked in the presence of SP-D and calcium and co-localization is seen in yellow (B). Reproduced, with permission, from Khamri *et al.* (2007) *Helicobacter* infection in the surfactant protein D-deficient mouse. Helicobacter 12, 112–123.

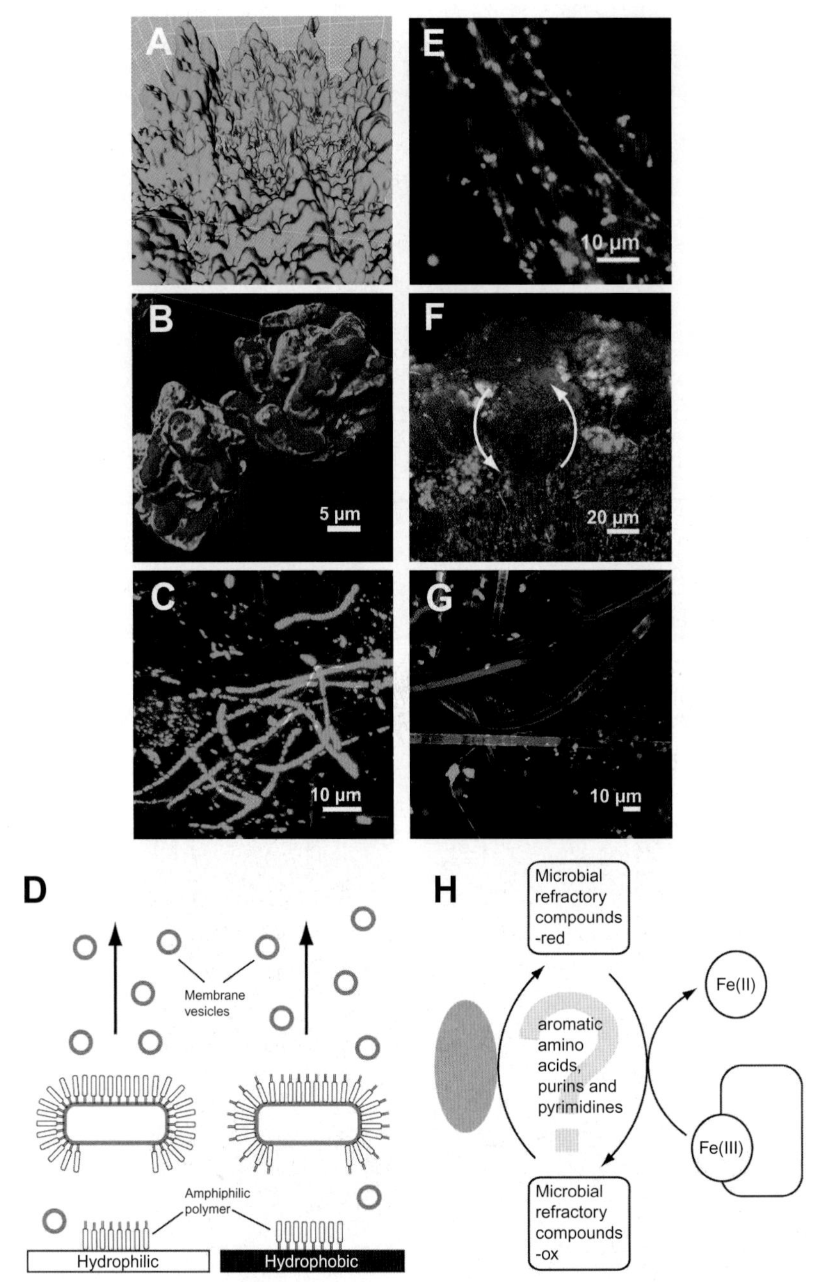

PLATE 37.1 EPS functionality. (A) Constructive EPS; laser scanning microscopy (LSM) image of bulk EPS, with lectin-specific glycoconjugates, showing a wastewater biofilm. (B) Adsorptive EPS; LSM image of selective triple lectin-staining of a microcolony in a river biofilm. (C) Active EPS; LSM image of ELF enzyme activity (Paragas *et al.*, 2002) in a river biofilm. (D) Surface-active EPS; drawing of possible actions involving surface-active compounds. (E) Informative EPS; LSM image of F8 strain with eDNA filaments, whereby the sample was stained with a nucleic acid-specific stain (Syto9). (F) Nutritive EPS; LSM image of dense biofilm with nucleic acid stained cells (red) and glycoconjugates (green), the arrows indicate possible recycling. (G) Locomotive EPS; LSM image of gliding cyanobacteria with lectin-stained sheath. (H) Redox-active EPS: potential redox reactions of refractory microbial compounds.

PLATE 42.1 Co-localization of an antibody to a terminal lactosamine epitope (red) on the LOS of *N. gonorrhoeae* with an antibody to the asialoglycoprotein receptor (green) on primary male urethral epithelial cells.

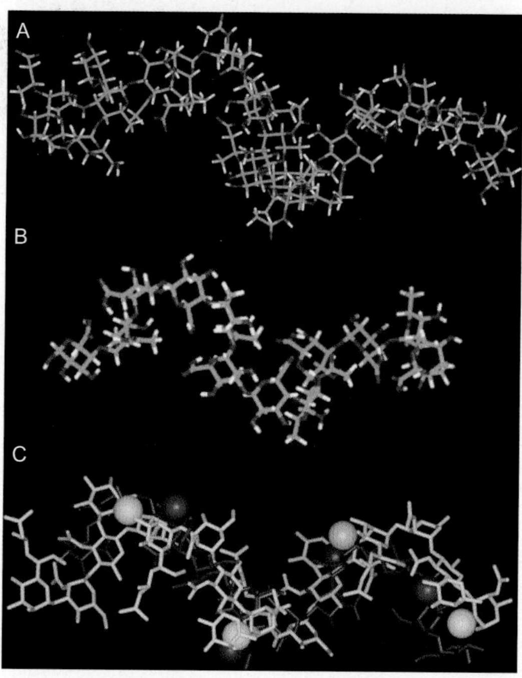

PLATE 49.4 Stick models of the three-dimensional structures of PSA-2 (A) and Sp1 (B). Carbon atoms are coloured green, oxygen atoms are red, nitrogen atoms are coloured blue and hydrogen are white. (C) Conformations of Sp1 (red) and PSA-2 (yellow) superimposed on the basis of their amino groups. Red and yellow dots represent positive charges from Sp1 and PSA-2, respectively. Sp1 and PSA-2 display a similar zig-zag positive charge pattern with roughly equidistant charge separation of 15 Å. Adapted, with permission, from Tzianabos *et al.* (2003) Biological chemistry of immunomodulation by zwitterionic polysaccharides. Carbohydr. Res. 338, 2531–2538.

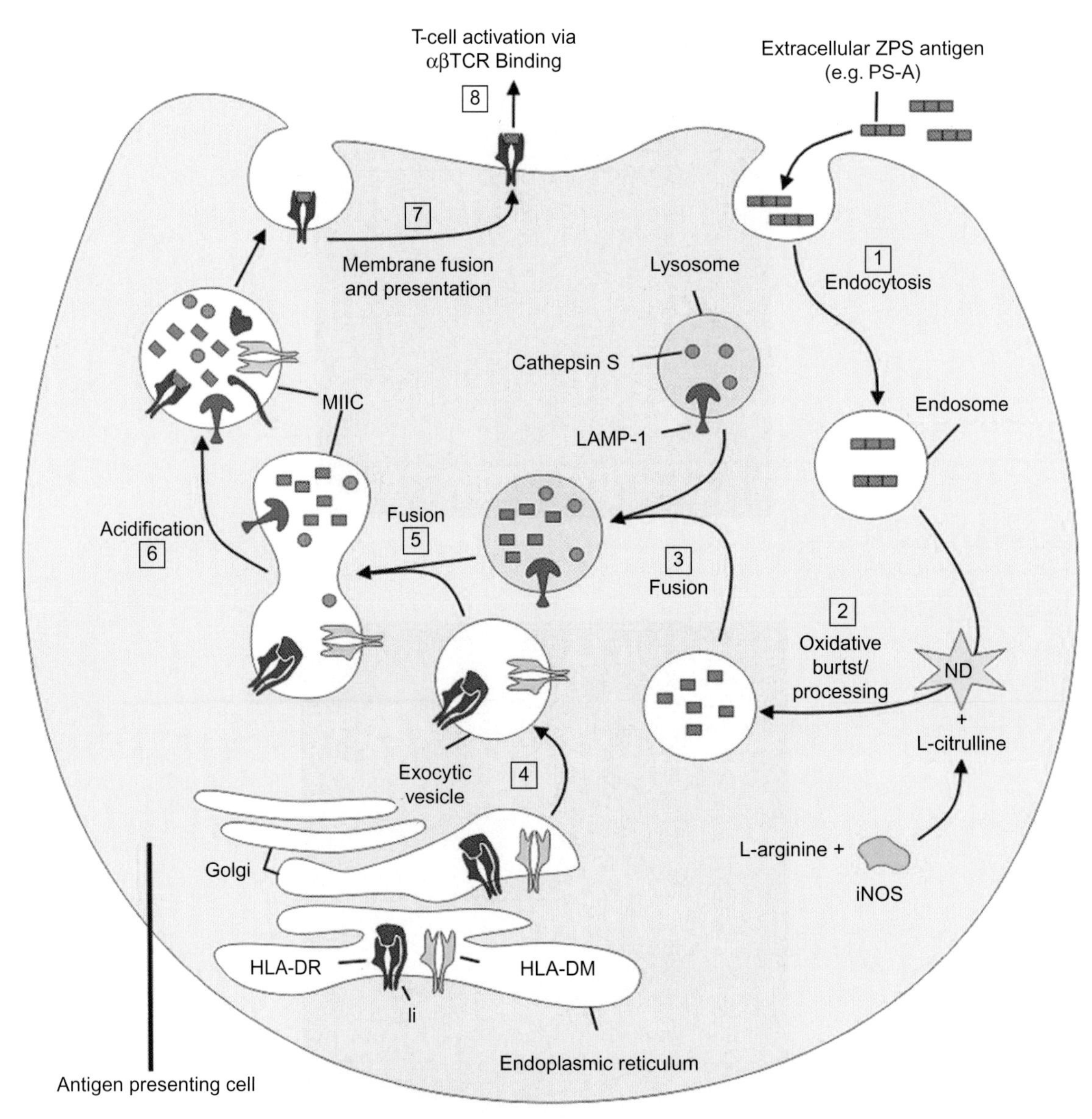

PLATE 49.7 The major histocompatibility complex (MHC) II pathway model for ZPS antigens. [1] Extracellular antigens are endocytosed for presentation through a cytoskeleton-dependent mechanism. [2] Cells undergo an oxidative burst, including the production of nitric oxide (NO). The NO, and potentially other oxidants, process the antigen to low-molecular mass polysaccharides. [3] Endosomes fuse with resident lysosomes. [4] The MHC proteins HLA-DR, HLA-DM, and Ii are assembled in the endoplasmic recticulum, are transported through the Golgi apparatus, and then budded into exocytic vesicles. [5] Antigen-rich endo/lysosomes fuse with exocytic vesicles, creating the MHC class II (MIIC) vesicle carrying HLA-DR, HLA-DM, LAMP-1, and antigen. [6] The MIIC acidification activates proteases responsible for processing protein antigens and cleavage of Ii, thus permitting the HLA-DM-mediated antigen exchange of the class II-associated peptide (CLIP). [7] Processed antigens are loaded onto MHC with the help of HLA-DM and then shuttled to the cell surface for recognition by βTCRs. [8] ZPS molecules mediate APC/T-cell engagement through αβTCR recognition. Adapted, with permission, from Cobb *et al*. (2004) Polysaccharide processing and presentation by the MHC II pathway. Cell 117, 677–687.